Chromatographic Analysis of the Environment

Third Edition

CHROMATOGRAPHIC SCIENCE SERIES

A Series of Textbooks and Reference Books

Editor: JACK CAZES

Chromatographic Analysis of the Environment

Third Edition

edited by
Leo M. L. Nollet
Hogeschool Gent
Ghent, Belgium

Taylor & Francis
Taylor & Francis Group
Boca Raton   London   New York

A CRC title, part of the Taylor & Francis imprint, a member of the
Taylor & Francis Group, the academic division of T&F Informa plc.

Published in 2006 by
CRC Press
Taylor & Francis Group
6000 Broken Sound Parkway NW, Suite 300
Boca Raton, FL 33487-2742

International Standard Book Number-10: 0-8247-2629-4 (Hardcover)
International Standard Book Number-13: 978-0-8247-2629-4 (Hardcover)
Library of Congress Card Number 2005048464

Library of Congress Cataloging-in-Publication Data

Chromatographic analysis of the environment / [edited by] Leo M.L. Nollet.--3rd ed.
    p. cm.
    Prev. ed. edited by Robert L. Grob.
  Includes bibliographical references and index.
  ISBN 0-8247-2629-4 (alk. paper)
    1. Chromatographic analysis. 2. Environmental chemistry. I. Nollet, Leo M.L., 1948- II. Grob, Robert Lee.

QD79.C4C48 2005
628.5'028'7--dc22   2005048464

Taylor & Francis Group
is the Academic Division of Informa plc.

Visit the Taylor & Francis Web site at
http://www.taylorandfrancis.com

and the CRC Press Web site at
http://www.crcpress.com

Preface

Since the publication of the second edition in 1983 tremendous changes have occurred in the field of environmental analytical chemistry. These changes have produced numerous approaches to sampling the air we breathe, the water we drink, and the soil in which we grow our fruits and vegetables. More dramatically, the changes have brought to the forefront the manner in which we regard waste problems. So writes Professor R.L. Grob, editor of the first and second edition.

Indeed since the 1980s tremendous changes have occurred; the most striking changes are miniaturization and automation of sample procedures and analytical techniques.

The philosophy of this third edition, totally different from the two former editions, is discussing most parameters of the different compartments of the environment (air, water, soil, waste) in a uniform structure: sample preparation techniques, separation methods, and detection modes. Most of the data are compiled in tables. Where necessary figures are added to elucidate the text.

The topic of sampling is discussed in two chapters. In chapter 1 attention is paid to specific aspects of sampling in the environment, whereas in chapter 2 the different sample preparation methods are explained. Chapters 3 and 4 complete the book with discussions on theoretical and practical aspects of chromatographic separations and detection methods. Finally, the importance of data processing is detailed in chapter 5.

In the second part, comprising chapters 6 through 32, the different, major and minor elements of the environment are dealt with. Special attention is given to volatile organic carbons (VOCs), peroxyacyl nitrates (PANs), and endocrine disrupting chemicals (EDCs). Of course, all other discussed environmental parameters are of equal importance.

All readers are aware that the project of preparing a text of this type is not possible without the assistance, support, and cooperation of many people. The third edition is certainly no exception. The finalization of such an undertaking can be frustrating in the least. The most important persons are the many authors, without whose hard work a task of this magnitude is not possible. I thank them very much.

For the understanding and patience, I wish to thank my wife and family.

I would like to dedicate this work to a fine friend of mine, José B., who died last year of cancer—a terrible disease. I hope people will be aware of the importance of good quality air, water, and soil.

Leo M.L. Nollet

The Editor

Leo M.L. Nollet is a professor of biotechnology at Hogeschool Gent, Ghent, Belgium. The author and co-author of numerous articles, abstracts, and presentations, Dr. Nollet is the editor of the three-volume *Handbook of Food Analysis, Second Edition*, *Handbook of Water Analysis* and *Food Analysis by HPLC, Second Edition* (all titles by Marcel Dekker Inc.).

His research interests include air and water pollution, liquid chromatography, and applications of different chromatographic techniques in food, water, and environmental parameters analysis.

He received M.S. (1973) and Ph.D. (1978) degrees in biology from the Katholieke Universiteit Leuven, Belgium.

Table of Contents

Contributors

Beatriz Almagro
Department of Analytical Chemistry,
  Nutrition and Bromatology
University of Alicante
Alicante, Spain

Willy Baeyens
Department of Analytical and Environmental
  Chemistry
Vrije Universiteit Brussel
Brussels, Belgium

Leon Barron
National Centre for Sensor Research
School of Chemical Sciences
Dublin City University
Dublin, Ireland

Kerstin Beiner
Centre for Environmental Research
  Leipzig-Halle
Department of Analytical Chemistry
Leipzig, Germany

Sara Bogialli
Dipartimento di Chimica
Università La Sapienza
Roma, Italy

Antonio Canals
Department of Analytical Chemistry,
  Nutrition and Bromatology
University of Alicante
Alicante, Spain

Alessio Ceccarini
Department of Chemistry and Industrial
  Chemistry
University of Pisa
Pisa, Italy

Haytham Chahin
High Institute for Environmental
  Research
Lattakia, Syria

Wei-Hsin Chen
Department of Marine Engineering
National Taiwan Ocean University
Keelung, Taiwan

Antonio Di Corcia
Dipartimento di Chimica
Università La Sapienza
Roma, Italy

Roberta Curini
Dipartimento di Chimica
Università La Sapienza
Roma, Italy

Thierry Dagnac
BRGM Laboratory of
  Enviromental Chemistry
Orleans, France

Filip D'hondt
Peakadilly Technologiepark
Gent, Belgium

Claudia E. Domini
Laboratorio FIA
Departamento de Química
Universidad Nacional del Sur.
Bahía Blanca, Argentina

Paul V. Doskey
Environmental Research Division
Argonne National Laboratory
Argonne, Illinois, USA

Merv Fingas
Emergencies Science and
  Technology Division
Environmental Technology Centre
Environment Canada, Ottawa
Ontario, Canada

Jeffrey S. Gaffney
Argonne National Laboratory
Argonne, Illinois, USA

Nacho Martín García
Universita de Valencia
Departamento de Ingenieria Química
Valencia, Spain

Carmen García-Jares
Departamento de Química Analítica,
  Nutrición y Bromatología
Facultad de Química
Instituto de Investigación y
  Análisis Alimentario
Universidad de Santiago de Compostela
Santiago de Compostela, Spain

Mohammad Ghafar
High Institute for Environmental
  Research
Lattakia, Syria

Stefania Giannarelli
Department of Chemistry and
  Industrial Chemistry
University of Pisa
Pisa, Italy

Marjan De Gieter
Department of Analytical and Environmental
  Chemistry (ANCH)
Vrige Universiteit Brussel
Brussels, Belgium

Tadeusz Górecki
University of Waterloo
Department of Chemistry
Waterloo, Canada

Dimitar Hristozov
Department of Analytical Chemistry
Plovdiv University
Plovdiv, Bulgaria

Tuulia Hyötyläinen
Department of Chemistry
Laboratory of Analytical Chemistry
University of Helsinki
Helsinki, Finland

S. Jayarama Reddy
Department of Chemistry
Sri Venkateswara University
Tirupati, India

Roger Jeannot
BRGM Laboratory of Enviornmental
  Chemistry
Orleans, France

Pedro Landín
Departamento de Química Analítica,
  Nutrición y Bromatologia
Facultad de Química, Instituto de Investigación
  y Análisis Alimentario
Universidad de Santiago de Compostela
Santiago de Compostela, Spain

Riccardo Leardi
Department of Pharmaceutical and
  Food Chemistry and Technology
University of Genova
Genova, Italy

Martine Leermakers
Department of Analytical and
  Environmental Chemistry (ANCH)
Vrije Universiteit Brussel
Brussels, Belgium

Congqiang Liu
State Key Laboratory of Environmental
  Geochemistry
Chinese Academy of Sciences
Guiyang, China

Maria Llompart
Departamento de Química Analítica,
  Nutrición y Bromatología
Facultad de Química, Instituto de Investigación
  y Análisis Alimentario
Universidad de Santiago de Compostela
Santiago de Compostela, Spain

Jau-Jang Lu
Department of Living Science
Tainan Woman's University
Tainan, Taiwan

Michela Maione
University of Urbino
Institute of Chemical Sciences
Urbino, Italy

Filippo Mangani
University of Urbino
Institute of Chemical Sciences
Urbino, Italy

Nancy A. Marley
Argonne National Laboratory
Argonne, Illinois, USA

Audrey E. McGowin
Wright State University
Dayton, Ohio, USA

Paal Molander
National Institute of Occupational Health
Oslo, Norway;
Department of Chemistry
University of Oslo
Oslo, Norway

Munro R. Mortimer
Queensland Environmental
  Protection Agency
Indooroopilly, Australia

Jochen F. Müller
National Research Centre for Environmental
  Toxicology
Coopers Plains, Australia

Manuela Nazzari
Dipartimento di Chimica
Università La Sapienza
Roma, Italy

Pavel Nesterenko
Department of Analytical Chemistry
Moscow State University
Moscow, Russian Federation

Leo M.L. Nollet
Hogeschool Gent
Department of Engineering Sciences
Ghent, Belgium

Declan Page
Power and Water Corporation
Darwin, Northern Territory, Australia

Brett Paull
National Centre for Sensor Research
School of Chemical Sciences
Dublin City University
Dublin, Ireland

Sigrid Peldszus
NSERC Chair in Water Treatment
Department of Civil Engineering
University of Waterloo
Waterloo, Ontario, Canada

Ana María Afonso Perera
Departamento De Química Analítica,
  Nutrición y Bromatología
Facultad de Química
Universidad de La Laguna
Avda. Astrofísico Francisco Sánchez
Tenerife, Spain

Anna Pielesz
Textile Institute
University of Bielsko-Biała
Bielsko-Biała, Poland

Peter Popp
Centre for Environmental Research
  Leipzig-Halle
Department of Analytical Chemistry
Leipzig, Germany

Soledad Prats
Department of Analytical Chemistry,
  Nutrition and Bromatology
University of Alicante
Alicante, Spain

Iván P. Román
Department of Analytical Chemistry,
  Nutrition and Bromatology
University of Alicante
Alicante, Spain

Antoine-Michel Siouffi
Université de Droit
Ave Escadrille Normandie
Normandy, France

Bjoern Thiele
Institute for Chemistry and Dynamics of
  the Geosphere
Institute III: Phytosphere
Research Centre Jülich
Jülich, Germany

Evaristo Ballesteros Tribaldo
University of Jaén
Jaén, Spain

Zhendi Wang
Emergencies Science and Technology Division
Environmental Technology Centre
Environment Canada
Ottawa, Ontario, Canada

Isabelle Windal
Department of Analytical and Environmental
  Chemistry (ANCH)
Vrije Universiteit Brussel
Brussels, Belgium

Fengchang Wu
State Key Laboratory of Environmental
  Geochemistry
Chinese Academy of Sciences
Guiyang, China

Guang-Guo Ying
CSIRO Land and Water
Adelaide Laboratory
Glen Osmond, Australia

1 The Sampling Process

Munro R. Mortimer and Jochen F. Müller

CONTENTS

I. INTRODUCTION

Any study is only as good as its weakest component. Accordingly, the quality of the output from an environmental sampling project is limited by whichever is the weakest component — sampling or analysis. The last 40 years have seen incredible improvements in analytical chemical techniques and precision.[1] However, the basic process of sampling the environment to acquire material for analysis has received comparatively little attention. Consequently, shortcomings in sampling aspects of environmental assessment often limit achievement of the data requirements. Moreover, how well the analyzed samples represent the target environment is unknown or at best uncertain.

In this chapter we provide an overview of essential steps in designing, organizing, and carrying out a successful program of environmental sampling, in which the requirements of adequate sampling are considered. We examine each of the basic considerations:

- the nature of the environment,
- the complexity of the sampling process including the range of media which can be sampled,
- the causes of variation in the material which can be sampled and how such variations can be addressed in sampling design,
- the spectrum of sampling methodologies applicable to sampling different media under a varied range of circumstances,
- important considerations related to "quality assurance (QA) and quality control (QC)" in the environmental sampling context.

It is not possible to cover each of the relevant topics in detail in a single chapter, but we have highlighted the issues to be addressed in planning and executing an environmental sampling project. For a more detailed coverage of individual topics, the reader should consult dedicated publications, for example, Ref. 2, and also take note of the relevant requirements of any environmental regulatory agencies with regard to data from the study being undertaken.

II. THE NATURE OF THE ENVIRONMENT AND THE COMPLEXITY OF THE SAMPLING TASK

Collecting truly representative samples from the natural environment is no simple task. The natural environment is both complex and heterogeneous, comprising a multitude of matrices (air, water, soil, sediment, biota), each with associated difficulties and potential sampling approaches. The concentrations of contaminants are usually nonuniform in space and over time and this adds to the complexities in sampling. In addition, the boundaries of the environment being sampled may not be sharply defined or indeed visible; the material sampled will rarely, if ever, be strictly uniform, and in many cases the properties of interest, for example, trace concentrations of contaminants, can be lost or at least altered in the sampling process through reactions with other components of the sample or with the materials used to collect and store the samples. All too often, conclusions based on laboratory results from the most careful analysis of the chemical and other properties of environmental samples are invalidated because the original collection of the environmental samples was inadequate or invalid.

III. THE COMPLEXITY OF THE SAMPLING PROCESS AND THE NEED FOR CLEAR OBJECTIVES

The planning of any study involving environmental sampling should begin with the determination of unambiguous sampling objectives defined by the data requirements of the study. These should be clearly stated at the outset.

At the core of any environmental sampling project are a series of questions relating to the purpose of sampling:

- What are the purposes or goals of the program or study being undertaken?
- What are the underlying questions to which answers are sought?
- What information is relevant, and over what spatial and temporal scales?
- How will the data be evaluated and presented?

Subsidiary questions include

- Where and when should the samples be collected?
- How many samples are needed?
- What matrices should be sampled and what equipment is appropriate?
- How will the samples be preserved and what containers should be used?
- What QC procedures and QA criteria and thresholds are required for the *sampling process* (as distinct from and additional to, QC/QA for laboratory analyses)?

Unless these questions are asked and answers are determined without ambiguity, before the collection of any environmental samples, there can be little confidence that the data derived from the analysis of the environmental samples, regardless of the precision and accuracy of the laboratory tasks, can provide useful or reliable insights as to the true state of the environment sampled.

IV. REPRESENTATIVENESS IN ENVIRONMENTAL SAMPLING

A environmental sample can be called *representative*, when it is collected and handled in a manner which preserves its original physical form and chemical composition. Accordingly, in the statistical sense, representative samples are an unbiased subset of the population measured.

To retain valid representativeness after collection, samples must be handled and preserved using methods adequate for preventing changes in the concentration of materials to be analyzed, by loss or by introduction of outside contamination. Failure to take account of each of the factors which can potentially reduce the representativeness of samples for the target environment is likely to result in analyzing samples which are not truly representative.

Analysis of samples which are not representative of the environment being assessed is inevitably a wasted effort and may lead to wrong and even expensive conclusions. The data from analyses may be precise and accurate in relation to contaminants of interest in the samples, but if the samples are not intrinsically representative of the environment, then the data has little relevance to the location or sites in question.

V. THE ROLE OF TIME IN ENVIRONMENTAL SAMPLING

The composition of the matrix to be sampled can often vary with time. If the rate of change is significant relative to the time needed to collect the sample, this alone makes meaningful interpretation difficult. The sampling of flowing water and gases frequently presents these kinds of challenges. Resolution of such sampling difficulties usually requires careful consideration of the use to which the data will be put. For example, if the measurements being made are for the purpose of assessing average exposures or loadings of contaminants, numerous samples can be taken over a long period of time and pooled prior to analysis to form a composite sample. However, if the release pattern of contaminants, and concentrations to which sensitive organisms are exposed is required, forming composites would not be appropriate. Instead the capture and analysis of many discrete samples, each taken over a short time interval may be more relevant. For some constituents of flowing material, continuous sampling using appropriate sensors and instrumentation may be appropriate.

VI. DESIGN AND IMPLEMENTATION OF A SAMPLING PROGRAM

To be successful, a sampling program must address and meet its objectives. There are a series of basic requirements in design and implementation to achieve that success. The American Chemical Society Committee on Environmental Improvement[3] suggests the following minimum requirements for an adequate sampling program:

- a proper statistical design which takes into account the goals of the study and its certainties and uncertainties;
- documentation of protocols for sample collection, labeling, preservation, and transport to the analytical facility. Such sampling protocols are "expressions of professional accountability" all too frequently lacking in environmental studies[4];
- adequate training of personnel in the sampling techniques and procedures specified.

However, even before these essential elements can be addressed, there are number of preliminary but equally important issues to be considered:

- the spatial boundaries of the sampling program should be defined;
- the scale of the sampling program should be defined;
- the timing and duration of the sampling program should be defined;
- potential sources of variability in data collected should be identified;

The flow diagram (Figure 1.1) shows the essential components of a well-designed environmental sampling project, each of which needs to be addressed from the preliminary planning stages through to the implementation of the sampling project.

A. SPATIAL BOUNDARIES FOR THE STUDY

The setting of boundaries should be based on the issues of concern giving rise to the need to conduct the sampling program, rather than on convenience or budgets. If boundaries are set inappropriately, significant and relevant sources of contamination and impact may be missed. For example, if the sampling is a component of a catchment study, the spatial boundaries for the sampling program should be those of the catchment of focus, or a series of sampling programs designed on a subcatchment by subcatchment basis.

B. SCALE OF THE STUDY

The scale of a study depends on the spatial and temporal ranges within which *in situ* measurements and samples for laboratory analysis will be taken. The appropriate scale is determined by consideration of the scale of the process or processes underlying the questions being addressed by sampling. For example, the rate of movement of contaminants in soils is slow relative to transport in flowing waters, so that the spatial scale appropriate for sampling downstream of a source of contamination will vary depending on the nature of the receiving environment.

An effect such as bioaccumulation in organisms exposed to contaminants is influenced by duration of exposure. This gives a temporal aspect to decisions concerning an appropriate scale of sampling.

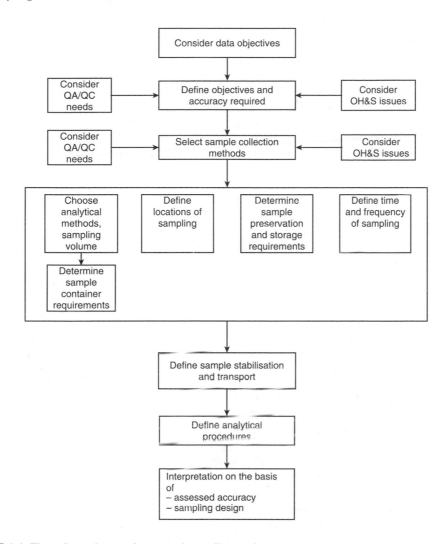

FIGURE 1.1 Flow chart of an environmental sampling project.

C. TIMING AND DURATION OF SAMPLING EFFORT

Timing and duration of sampling effort can be critical supplementary considerations in relation to the scale of sampling which is appropriate. This is particularly important when sampling involves collecting data from a system which is inherently variable in its capacity to transport contaminants. A flowing stream is a good example of a system with a variable transport capacity. For sampling in such a system, rainfall and stream flow patterns need to be taken into account, in terms of their potential effects both on the transport and on dilution of materials of interest for sampling. Sediment loads in streams, the base load and suspended material, are usually very different between storm run-off peaks and times of low flow. Contaminants washed out of a catchment by storm events often exhibit a "first-flush" concentration peak which may be of short duration relative to the overall period of high flow. Sampling of first-flush events requires a specific approach different from that of sampling a stream during normal flow conditions, but more importantly, data collected during such different flow conditions will, for most measurable variables, be quite different.

D. POTENTIAL SOURCES OF VARIABILITY IN DATA COLLECTED

Before a sampling program is commenced, all of the potential sources of variability of the data should be considered and minimized by prudent planning. For example, some potential sources of variability arise from sample handling and storage. These are discussed later in this chapter under the respective headings. However, actual environmental heterogeneity in respect of whatever variable is being measured, either *in situ* or in samples of actual materials, will be encountered, because that is the nature of the real environment. In addition, variability is potentially increased by seasonal and other time-related effects, disruptive processes such as soil disturbance, changes to drainage patterns, and patterns of chemical dispersion.

While a field sampling program is still at the planning stage, all potential sources of spatial and temporal variability should be assessed from sources such as published reports from similar sampling programs, and from consideration of the nature of the system to be sampled. For example, in planning sampling of soils, the type of soils to be encountered and attributes such as the particulate structure (which is likely to influence the distribution of metallic contaminants) and the presence of organic matter within the soil matrix, and often a surface layer of dead plant matter, should be considered. Organic matter usually accumulates organic chemical contaminants, so the question of whether a surface layer should be included with the soil sample, or scraped off before sampling the underlying soil, or taken as a discrete sample of material in addition to a companion sample of the underlying soil (to a prescribed depth), needs to be addressed at the planning stage. Similarly, in taking water samples from a waterbody which is unlikely to be well-mixed due to influences such as thermal stratification in a deep lake, the depth(s) from which samples are taken needs to be considered. Temporal contributions to variability can arise from diurnal and seasonal effects on water chemistry such as temperature and dissolved oxygen concentrations (particularly if algae are present in high numbers), as well as from the effects of flow variation and stormwater first-flush influences.

If all of the potential factors affecting variability are not considered and the sampling approach is not adapted to minimize each factor as much as practical, the resulting data set may prove to be so variable that impacts, disturbances and trends are totally obscured. If there is insufficient previous knowledge, the nature of the environment and media at the locations to be sampled, and, in particular, if the factors contributing to environmental heterogeneity are uncertain, a pilot study, or at least an on-the-ground inspection and assessment, should be done. Such assessment should involve testing of possible sampling approaches, and the taking of "typical samples" for assessment of factors which are likely to contribute to high variability is strongly recommended.

VII. THE NEED FOR A PROPER STATISTICAL DESIGN FOR SAMPLING

In practical terms, only small portions (samples) can be taken from an environment under assessment. Accordingly, the samples need to be representative of the media and the environment that is the focus of the study, within practical limitations determined by the resources available to those charged with the assessment task.

This requires a proper statistical design for sampling. The aim of the statistical design is to ensure the sampling effort maximizes every opportunity to be representative of the environment being sampled and minimizes errors. Statistical designs for sampling can be based on a number of different approaches, each based on the concept of randomness, i.e., every sample unit available in the population (the environment being sampled, as defined according to scale, spatial, and temporal boundaries defined in the sampling program) must have an equal probability of being included in the set of samples taken for assessment. There are three basic statistical designs available:

- simple random sampling;
- stratified random sampling; and
- systematic sampling.

A. SIMPLE RANDOM SAMPLING

The totality of available sampling units within the environment in terms of its spatial and temporal extent is sampled without conscious or unconscious selection or rejection of particular units. It should be noted that random sampling is not the same as "haphazard sampling." For example, random sampling may involve the use of a random number table or computer-generated set of random numbers to select sampling sites from within all possible coordinates defining a locality. However, procedures such as sticking a pin into a map while blindfolded are often subject to unconscious bias, resulting in a set of samples which provide a biased assessment of the environment being sampled.

Methods of assigning sampling locations without bias are discussed in standard statistical texts. Typically, all possible sampling locations within the spatial boundaries of the study location are defined using a grid system of coordinates superimposed on a large scale map or aerial photograph. Subsequently, actual sampling locations are chosen using a computer-generated series from all possible coordinates falling within the spatial boundaries, or by using a random number table. More often than not, box (square) grids are used to cover the area defined by the spatial boundaries for terrestrial sampling, but triangular or other shaped patterns are equally valid. Where contamination patterns relative to a point source are being assessed by sampling, it is often useful to construct a sampling grid as a series of rays projected outwards from the source. Sampling of water from lakes or embayments can be systematized by constructing a grid of transect lines (Figure 1 2). Similarly, the times for taking samples from a time-varying system can be randomized to avoid bias using a method which ensures that all possible sampling times have equal opportunity to be included.

Simple random sampling may not be the most cost-effective sampling design, for example, if there is prior knowledge of a particular spatial distribution pattern for the contaminants of interest. Similarly, if there is prior knowledge of temporal influences or media transport influences such as wind or stream flow on the likely concentration distribution for contaminants of interest, a simple random sampling at all possible times and all possible flow conditions may not be the most cost-effective strategy. In such circumstances, other approaches such as those described below can be a more efficient use of resources.

B. STRATIFIED RANDOM SAMPLING

In stratified random sampling, the system to be sampled is divided into a number of parts (strata) within each of which the contaminant concentration or other descriptor of interest is likely to be relatively consistent. Such strata need not be of equal size and the numbers of sampling units taken within each stratum can be set according to the anticipated degree of variability of whatever is being measured within that stratum. For example, in a stratum where variability encountered is expected to be relatively low, a small number of sampling units could prove adequate, whereas in another stratum with inherently high variability, a larger number of sampling units may be required.

Strata may be spatial, temporal, or determined by other relevant criteria. For example, in the spatial sense, a series of strata could comprise discrete areas associated with a study location, each with different geology, or different topography, or different history of contamination, or different soils (or aquatic sediments), or waters sampled at different depths within a lake to take account of stratification or, in an estuary, salinity gradient. See Ref. 5 for more detail of stratification issues in water sampling. In the temporal sense, different strata could comprise different seasons, or portions of the diurnal cycle, or time periods relative to a process such as an upstream effluent release.

Examples of strata determined by other criteria include sampling of biota for bioaccumulated contaminant content using species, or size (perhaps size class), or age (perhaps age class), or determinants such as sex, or reproductive status, or the assignment of specific organs or tissues as strata for sampling purposes.

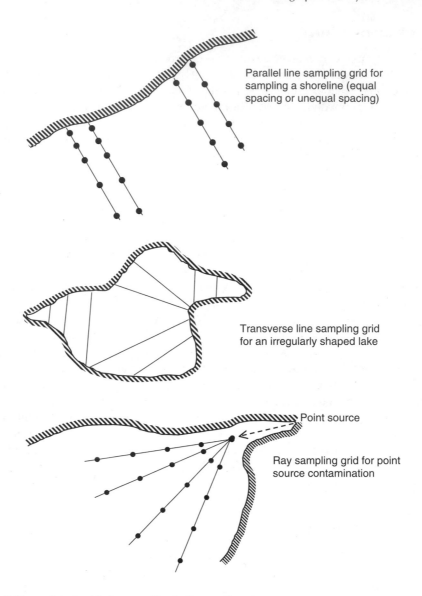

FIGURE 1.2 Examples of grids for sampling bottom sediments.

Methods of calculating statistical parameters from stratified sampling data are discussed in standard statistical texts.

C. SYSTEMATIC SAMPLING

In systematic sampling, sample units are collected at regular intervals in space or time. However, as is the case in simple random sampling, care is needed to ensure that bias is avoided. For example, temporal sampling conducted on a regular sampling schedule may coincide with periodicity within the site or system being assessed and thus risk collecting data which is not representative of conditions at other times. As an illustration, discharges of effluents are often periodic in nature in respect of constituent composition and relative concentrations, related to the process schedule which produces them. Accordingly, a systematic sampling taking samples according to a regular schedule involving collection at the same time every day would produce quite different data than a schedule involving collection every hour throughout the day. Similarly, in collecting water samples from tidal or estuarine waters, the values for variables of interest might depend on the state of the tide, so that samples taken on an ebb tide would not be representative of conditions during the flood tide or at slack water. See Ref. 5 for more details of temporal issues in water sampling.

VIII. SAMPLING SPECIFIC MATRICES

A. SAMPLING SOIL

Soils are a complex and usually heterogeneous matrix. Soil profile changes with depth; typically within the first meter there are marked changes in type, composition, texture, moisture, and organic matter. In addition, there are usually changes in factors such as porosity and the proportion of organic matter of plant origin (for example leaf litter) within the first centimeter.

A simple and direct method for taking near-surface soil samples is to use a spade and scoop. This approach is usually practical only to a depth of about half a meter, after which it becomes too labor intensive. It is important to use a precleaned spade, preferably with a stainless steel blade if metal analysis is to be conducted on samples. Similarly, a stainless steel (never chrome-plated) trowel or spoon should be used to manipulate sample material, for example to transfer it into storage containers.

The most complete method of sampling soils is to excavate and expose the profile, giving access to allow sampling of undisturbed material from the side of the excavation. Other common methods involve the use of augers to extract bore samples, or devices such as split-barrel samplers and thin-walled tubes which are inserted to take a core sample through the soil profile. However, adequate penetration of soils containing rock fragments may be difficult with such devices. Thin-walled tubes can be sealed with the undisturbed soil core retained inside and thus serve as sample containers for storage and transport for laboratory analysis. Soil samples taken from the sides of holes, auger spoil, and core materials removed on site from coring devices are commonly segregated according to depth and are sealed in appropriately prepared and labeled jars.

Prior to commencing sampling with a spade, drill, or coring device, the soil surface should be cleared of surface debris such as rocks, sticks, plant material, and litter including dry grass and leaf matter. In some circumstances such surface materials may also need to be taken as a sample for the purposes of the study. It is wise to clear this loose material for a sufficient distance from the point of excavation of insertion of a drill or coring device to ensure loose surface matter does not fall into the hole and contaminate subsurface samples.

Free air and gases held within the soil matrix can be sampled using metal tubes fitted with valves or seals at the upper end suitable for the extraction of a gas sample with a gas-tight syringe. Such tubes are driven into the soil to the required depth prior to taking of the gas sample.

B. SAMPLING WATERS

With respect to water sampling, common pitfalls in sampling strategy arise from making a false assumption that a waterbody is well-mixed and homogenous in time and space. This is rarely the situation even in vigorously flowing and shallow waters, and factors which may cause variability should be taken into account, along with the purpose for which sampling is being undertaken, when the sampling program is designed.

A false indication of apparent homogeneity within a waterbody can result from failure to take true replicates. For example, a series of aliquots taken from a single grab sample such as a bucket are not replicates from the waterbody being sampled. True replicates are a series of individually taken grab samples or buckets of water.

Major factors which can result in a lack of homogeneity within a waterbody are temperature and depth. In deep waters there is often an euphotic zone extending below the surface. Most of the biological productivity takes place within this relatively shallow layer and its chemistry is often quite different from that in the lower depths. In addition, thermal stratification is common in large waterbodies, sometimes forming and dissipating on a seasonal basis and in estuaries, salinity gradients and associated stratification are further compounding factors.[5] In lakes, the influences of wind and wave action, particularly near shores and in shallow areas, and the influence of tidal flow in estuaries can sometimes mix layers which on other occasions are distinctly different in physical and chemical properties. Rapidly flowing streams of considerable depth may exhibit chemical stratification even in the absence of accompanying thermal stratification.[6] All these factors add to the difficulties of taking truly representative samples.

Variations and cycles in the rate and volume of flow may be significant factors to take into account in designing a sampling program in a flowing stream, and if estuarine waters are involved, the tidal cycle is also likely to be significant. In slow moving or still water, diurnal cycles often have a significant effect on water chemistry. For example, the diurnal minimum concentration of dissolved oxygen usually occurs shortly before sunrise due to the relative roles of photosynthesis and respiration. Other related changes in water chemistry affected by the diurnal cycle are pH values, and dissolved carbon dioxide and carbonates.

It should also be recognized that there are a series of system processes operating in flowing waters which can have a profound effect on contaminant loads over a short period of time and these should be taken into account when considering the frequency and timing of sampling. These processes are run-off and storm events, remobilization of material previously deposited, as well as point source and nonpoint source influences on contaminant load. Rainfall events within a catchment can produce patterns in contaminant load which may be different from one location to another. Releases of contaminant, whether by design or circumstance, may be related to rainfall and flow conditions (or cycles in production in the case of anthropogenic releases) and under high flow may be diluted. During periods of high rainfall, bunds or other containment structures may be breached, leading to atypical contaminant releases. In the case of temporary or ephemeral watercourses or drainage lines, there may be a "first-flush" of water containing contaminant loads accumulated since the previous period of sustained flow.

Accordingly, if sampling is to provide data adequate to represent a waterbody, each of the above factors which potentially introduces variability into what is being sampled should be taken into account. For example, in flowing waters, consideration needs to be given to whether both base flow and storm event flow should be sampled and in the case of the latter, how first-flush and peak flow event data are captured. This may involve sampling the water and recording flow data. In designing the sampling of still waters, in many cases it is necessary to conduct a prior investigation of stratification as a prerequisite to designing a satisfactory sampling program.

A wide range of equipment types is available for collecting water samples. The appropriateness of each of these depends largely on the type of waterbody being sampled and the objectives of the sampling program.

For shallow (<1 m depth), well-mixed waterbodies, immersion of a sample bottle by hand to just below the surface (e.g., 0.3 to 0.5 m depth) may be satisfactory. However, to minimize the risk of contamination of the water sample, the bottle should be held in a plastic disposable glove, contributions to the sample from surface films or slicks avoided and if the sampler has to enter the waterbody to take the sample, the sampler should enter downstream of the sampling point and face upstream into the current while taking the sample. If a boat is used to access the sampling point, care is needed to avoid the risk of contamination from the vessel itself. This can include sampling from the bows while the vessel is held facing into the current, and/or the use of a sampling pole (preferably of acrylic material) typically 1 to 2 m in length and fitted with acrylic-coated jaws to hold the sampling bottle. A sampling pole is also appropriate to assist sampling from difficult locations (for example, from a bridge or jetty, or over steep banks). If a bucket and rope is used as an intermediary in collecting samples from such difficult locations, an acrylic rope is less likely to carry contamination from site to site and in any event, both rope and bucket should be cleaned between samples and transported in a plastic bag within a sealable plastic container to avoid contamination from vehicles used in transporting the equipment to the site.

The use of pumps to collect water samples demands particular care against the risks of contamination of the sample from tubing, or the adsorption of analytes of interest by tubing. The presence of tubing in a pumping apparatus provides a relatively high ratio of surface area of sampling equipment to volume of sample. Tubing may be of polyethylene, silicone, PTFE, or PVC and should be precleaned as well as conditioned by pumping through a large volume of water from the source to be sampled prior to taking the actual samples.

In deeper waters, sampling at the particular depths of interest can be accomplished using any of the range of purpose-built samplers available. Generally these operate by a system which allows a sample trapping vessel to be deployed to the required depth without collecting or retaining any sample, at which depth it can be triggered to close (often by releasing a weight from the surface, which travels down the deployment line to reach the sampling device). The closed device is then recovered by up hauling on the deployment line. Such devices, especially if they incorporate rubber components, have an associated risk of contaminating samples, which often increases with the age of the device. Accordingly, blank tests should be conducted with these samplers at regular intervals. Appropriate blank testing could involve filling the samplers with clean water for the longest likely time period that a sampling event could involve, then analyzing the water samples for analytes of interest.

A range of automated samplers is available for sampling water on a continuous or preprogrammed schedule related to flow or time-specific events. A typical system comprises a pump, a controller, and an array of sample bottles contained within a housing. The precautions applicable to the potential effects on samples through contamination from tubing or absorption of contamination by tubing and the recommended use of test blanks as noted above in respect of pump samplers and depth sampling devices, is also applicable to automatic samplers. In addition, for some analytes, the possibility that the integrity of samples may be compromised by delayed preservation needs to be considered. More sophisticated samplers which provide refrigeration of samples after collection and/or addition of appropriate preservatives are commercially available.

C. INTEGRATING SAMPLERS

The automated water samplers described above may be configured to act as "time-integrating samplers" whereby the sampler is programmed to collect a series of consecutive samples into a single collection vessel, effectively generating a composite sample covering the time period concerned.

Integrating samplers based on the passage of a known volume of water for a preprogrammed time period through an adsorbent–collector medium are available for a range of metals and organic compounds. Subsequently, the absorbent–collector medium is analyzed for the contaminants of interest. Such devices have the potential to extract and integrate contaminants for analysis from a much larger sample than would be practical to collect and transport using traditional grab-sampling techniques.

A range of passive sampling devices has also recently been developed. Such devices operate on the basis of diffusion of contaminants of interest across an appropriate membrane into a storage medium such as triolein or isooctane (in case of lipophilic contaminants). These latter devices are discussed more fully below.

D. SAMPLING GROUNDWATER

Of necessity, the sampling of groundwater usually involves disturbing the environment to gain access to the aquifer. Typically, this involves the construction of a test well (bore or access hole) and the use of pumps or similar equipment is often required to retrieve samples. Accordingly, a high degree of care is needed to avoid contamination of the groundwater as a consequence of the site development and the use of equipment involved in accessing the aquifer. As with sampling of surface waters using pumps and tubing, the risks of introducing contaminants or adsorbing contaminants of interest during sampling should be assessed and addressed.

Another important consideration in groundwater sampling is that standing water accumulated in the bottom of a test well may not be representative of the water contained within the adjacent aquifer. Accordingly, the standing water is usually removed (purged) to allow recharging of the test well from the aquifer prior to taking the samples. Sufficient water must be removed from the test well to ensure that the water present is newly derived from the aquifer. The means employed to purge test wells should also be carefully assessed for their potential to introduce contaminants to water which is subsequently sampled.

Sampling groundwater from test wells is also subject to variability induced by geological stratification, despite the standard precaution of sampling from recharge after purging. Recharge water chemistry can be influenced by substrata,[6] so that variation in the data can be introduced simply by sampling from different depths within the recharge. The purging process itself can have an influence on the water chemistry. Although the purpose of purging is to remove stagnant water from the well bore and adjacent sand packing, the extent of purging necessary varies with the hydraulic properties of the water-bearing unit in which the well is constructed. Too rapid a rate of purging may result in a steep local hydraulic gradient, resulting in the contamination of recharge by clay and silts, and/or aeration through turbulent flow. Each of these has the potential to alter the water chemistry of the sample.[7] Ideally, each sampling well should be tested to determine a protocol to ensure that before samples are taken, purging continues for a sufficient time to ensure that all water chemistry variables of interest achieve a steady state in the recharge water and that the rate of pumping does not exceed the rate at which excessive draw-down occurs.

E. SAMPLING PRECIPITATION (RAINFALL, ICE, AND SNOW)

While most water sampling involves collection of the liquid phase in flowing or standing waterbodies, the sampling of ice, snow, and rainfall have their own associated sources of variability which require consideration during the design of a sampling program. In the case of ice, airborne dust is a common source of entrapped contamination. This is usually an episodic source, which results in layers of contamination that vary with depth in the ice reflecting the history of deposition. Other sources of contamination in a floating ice layer are entrapped plankton and chemical contaminants in the water from which the ice was formed.[6] An appropriate sampling design needs to take account of the purpose of sampling, for example, if the study focus is the influence of melt water on a receiving waterbody, then bulk sampling over the full thickness may be appropriate. However, if an understanding of the sources of contamination is required, then a concentration versus depth focus would be more useful. Rainfall and snow are typically variable in contaminant concentration in relation to the duration of a precipitation event, and the temporal relationship between the commencement of the event and the taking of samples. The initial precipitation tends to be relatively concentrated in contaminants, but as the event progresses, the composition approaches that of distilled water.[6] Similarly, the pH of rainfall may be low initially, but becomes neutral as rainfall continues. Accordingly, the timing of sampling is highly relevant to the use to which the data will be put, and this should be taken into account in sampling design.

F. SAMPLING BOTTOM SEDIMENTS

Program design considerations for the sampling of bottom sediments from waterbodies such as lakes, estuaries, flowing streams, or offshore marine environments have much in common with soil sampling programs, but are confounded by greater difficulties of access. Bottom sediment may be variable in depth as well as in areal extent and if the thickness of sediment layer sampled is important, preservation of its surface to depth-of-sampling integrity is also important.

In natural waterbodies, sediment texture is often very variable, particularly in flowing waters. Depositional history is frequently an important potential confounding factor in designing sampling programs for bottom sediments. Near-bank deposits of sediments may be more recent than those in deeper areas.[8] The sampling design should take account of the likelihood that sediments near the banks will be different in texture as well as in depositional history and potential contaminant content from those in the deepest part of the channel. There are often differences in deposition and accumulation of fine sediments at different points of a stream channel. These differences are often determined by proximity to bends and obstructions. Generally, finer sediments accumulate in still or slack water flow conditions such as are found immediately downstream of obstructions which create back-eddies, and on the inside of bends.

The taking of representative sediment samples from a flowing stream is often a complex issue and is best addressed by forming composite samples taken along a cross-section or transect of the stream bed.

Near-bank and bottom sediments from small shallow streams may be sampled by wading into the water and scooping sediment into a suitable container, ensuring that the actual location to be sampled is approached against any current to avoid contamination from sediment resuspended by the operator's access. If currents are strong, the potential for finer particulate matter to be resuspended and lost during sample collection should be addressed. Hand-held coring devices (simple push corers or suction corers) may assist in reducing losses of fine particulate material. Suitable coring devices can be constructed from a variety of materials such as metal, PVC, perspex, and polycarbonate tubing (chosen with consideration of the potential for contamination of the sample), usually in the range 2.5 to 5 cm diameter with the leading edge beveled to ease penetration into the sediment. Such devices are capable of preserving the integrity of sediment continuity within the cores, which can be removed from the corer after retrieval and the depth of interest (for example, surface to 5 cm depth) sliced off and stored for analysis.

In deeper waters, a coring device such as a tube open at both ends and attached to a pole of sufficient length may be a practical means of acquiring samples of bottom sediments by pushing the coring device into the sediment layer. For some sediment types, to avoid loss of sample, it may be necessary to cap the upper end of the open tube with a one-way valve fitting which is arranged so that a vacuum is formed during the withdrawal of the coring device from the sediments. Hand-held corers can often be used by divers in a similar fashion to their use in shallow waters, although a strategy such as attaching end seals to sediment-filled coring tubes needs to be employed to preserve the contents of cores prior to them being brought to the surface for sample processing. In deep waters or when sediments have to be sampled to depths in the sediment layer beyond those which can be penetrated using hand-held devices, vibrocoring equipment may be employed. Such equipment often has provision for extrusion of the sample from the cores (e.g., by compressed gas) and uses plastic liners, which protect the samples from contamination.

A range of grab-sampling devices which can be deployed from boats are also available for use in waters too deep for sediment access by wading. There are a variety of designs for such devices and these are available in a range of sizes. Depending on the nature of the sampling task, some of these can be used effectively as a hand-held apparatus, but others may need the use of mechanical winches.

Whatever device is used to collect sediment, the device used and its material of construction should be such that contamination of samples is avoided. For example, if corers are used, they should be appropriately precleaned and sealed before use, and sediment grabs used to take a series of samples should be cleaned of any adhering sediment between each sample.

Ideally, the sampling equipment chosen should not significantly disturb the environment being sampled or alter the physical or chemical properties of the samples being taken. Grab samplers often do not enter the sediments perpendicularly and the sediment layers may be mixed during the mechanical closing of the sampling device. Most conventional mechanical grab samplers have jaws which close in a semicircular fashion and sediment layers below the level of initial penetration are not sampled on a fully quantitative basis. It is often necessary to know the depth and area which are sampled. Similarly, with coring devices, there is often a risk that easily resuspended surface materials are washed away during initial penetration or recovery and not included in the sample. Rotation of cores during penetration may mix layers and shorten core length, which distorts the sediment depth profile captured by the core.

If the chemical forms of contaminants such as metals in bottom sediments are to be preserved, the redox state of the sediments at the point of sampling needs to be considered. The potential that oxidizing or reducing conditions are introduced during sampling or storage of samples should be considered and addressed. Anoxic sediments can become oxygenated on contact with air, so that cores containing such sediments need to be capped immediately and stored under nitrogen, or at least kept frozen to minimize the rate of oxidation.

As with the sampling of soils, bottom sediments will in some circumstances have an associated vegetative layer on and below the surface. They may also contain animal life such as worms, molluscs, and crustaceans. Generally, if the objective is to sample sediment material without the confounding addition of plant and animal matter (the normal situation), then surface vegetation may have to be cleared prior to sampling, while roots and similar subsurface material can be removed from the sample at a later stage by sieving. Animal life can also be removed during postsampling sieving.

G. Sampling Suspended Particulate Matter

There are a variety of reasons for taking samples of suspended particulate matter, including the fact that many potential contaminants of interest may be associated with suspended particulate matter.

There are two different approaches available for taking samples of suspended matter from waterways. These are filtration or centrifugation of water samples and collecting settled particulate matter in a sediment trap. Filtration or centrifugation can be conducted on grab samples or on water which is pump-extracted from instream. It is necessary to measure volumes filtered or centrifuged to calculated concentrations or loads in water samples.

Due to clogging of fine filters (e.g., 0.45 μm is commonly regarded as the "soluble" threshold), it is often difficult in a realistic time frame to collect a volume of suspended matter sufficient for many chemical analyses of contaminants. The use of high flow commercial centrifugal separators (e.g., "Westfalia" and "Alfa Laval") can overcome this problem.[9,10]

In circumstances where particulate suspended matter occurs at very low concentrations (e.g., <1 mg l^{-1}), collection by filtration or centrifugation may require impractically large volumes of water to be processed in order to collect a sufficient quantity of sample for analysis. In such circumstances the deployment of sediment traps may be more practical.[11] Such traps typically comprise vertical tubes suspended in the water column on fixed lines between floating buoys and weighted anchors. The tops of the tubes are open to accept settling particulate matter and the bottoms of the traps are closed. Deployment may be brief or extended depending on the quantity of material required and the deployment period may also act as a means of time-integration of sediment characteristics (and contaminant content). However, deployment should not exceed 2 weeks, or possibly less in tropical waters, to avoid algal build up or decomposition of trapped organic matter.

The composition and nature of suspended particulates often varies with depth in the water column. For example, in a flowing stream the concentration and particle size distribution of suspended particulate matter usually increases with depth due to settling and resuspension dynamics, which are affected by flow rates. However, in the case of suspended particulate matter of biological origin, relatively high concentrations may be present near the surface, because this material is generally less dense than particulate matter of mineral origin. In addition, live particulate matter such as algal cells and zooplankton are often attracted to sunlight and, as a consequence, found at higher concentrations in surface layers during the day, but can disperse to greater depths at night.

Accordingly, the depth of collection for water samples filtered to extract particulate matter is a potential confounding factor in any sampling program involving suspended particulate matter and this should be considered at the design stage of the sampling program and incorporated into sampling protocols. Alternatively, depth-integrating samplers can be used, or composites formed from samples taken at a series of depths.

H. SAMPLING AIR

The atmosphere is an important media for the environmental distribution of chemicals. Air is a relatively dilute media and while this facilitates the sampling process (i.e., it allows high sampling rates and good detection limits) this poses a range of challenges for the sampling strategy. Sampling air contaminants can be required for a wide variety of reasons covering a wide range of pollution levels including emission evaluation, for example, from industrial stacks, workplace health and safety exposure evaluation, and the evaluation of background concentrations of pollutants in ambient air. In any given exposure situation, air pollutants can be differentiated into macro pollutants which may occur at relatively high concentrations, (i.e., in the mg m^{-3} or μg m^{-3} range) and can be quantified relatively easily, in contrast to trace or even ultra-trace pollutants which may need quantification in the subnanogram m^{-3} or subpicogram m^{-3} ranges.

For air sampling, site selection is integral to the success of a sampling program. Sampling site selection is often strongly influenced by the practical requirements to setup, maintain and secure the sampling equipment. However, it is essential that the site fulfils a set of criteria that assure the samples are representative and provide the required data for which the sampling program was initiated. Guidelines for selecting monitoring sites are provided in guidelines for monitoring air pollutants, for example, see Ref. 12.

Accurate determination of the sampling volume is integral to the air sampling process and the results of a study can never be more accurate than the accuracy of the determination of the sampling volume. Sampling volumes are typically measured either by measuring the gas flow rate or using volume meters.[13] Note that for accurate measurement of the sampling volume it is essential to determine and correct the measured volume for changes in pressure and temperature.

As with the sampling of other environmental phases, the air sampling strategy within the sampling design requires a good understanding of the analyte and contaminants to be sampled. In any given atmosphere, pollutants may occur either in the vapor phase or bound to atmospheric particles. Particulate matter as such is considered an air pollutant and the sampling of particulate matter is one of the most traditional approaches for evaluating air pollution. The basic problem for sampling particulate matter is efficiency in separating and collecting the particles from a given volume of air. Secondly, information on the type of particles and their size distribution may be relevant. A great variety of methods exist for sampling particulate matter in air samples. These include dust deposition gauges which evaluate the dust fall-out (i.e., particles >50 μm), electric and thermal precipitation techniques, various impingement techniques, filtration techniques, as well as laser particle samplers and nephelometers based on light scattering for the evaluation of particulate matter in air. The ideal system would allow artefact-free sampling of all particulate matter from a given air volume and allow further examination of the collected particulate matter using other techniques. Unfortunately, to the knowledge of the authors, no such system is available and the choice of the sampling system should be based on a good understanding of the limitations of the sampling systems available and the objectives of the study.

Sampling of vapor phase pollutants can be even more complex than sampling of particulate matters. Gases can be relatively reactive and a major requirement for any sampling method is that the concentration of the pollutant of interest must not be changed as it passes through the sampling unit. As is the case with the sampling of particulates, a wide variety of methods exist, ranging from continuous online monitoring techniques for gases that occur at relatively high concentrations in the atmosphere, such as nitrogen and carbon oxides, to high volume sampling systems equipped with large vapor-sorbent phases which are used for sorption of ultra-trace gases. In addition to these direct sampling techniques, diffusion-based sampling systems such as personal dosimeters have long been used to estimate air concentrations for vapor phase chemicals, and these techniques are discussed in more detail below.

Semivolatile air pollutants are those contaminants which may occur in ambient air in both the vapor and the particulate phases. Thus it is important to collect both phases during sampling. Semivolatile organic pollutants are particularly relevant as they include some of the most potent toxicants such as polyaromatic hydrocarbons, dioxin-like chemicals and various other persistent organic pollutants which are subject to a global treaty.[14] Typically, sampling of these pollutants is performed using filter–adsorbent type sampling systems (Figure 1.3). Separate analysis of the filter and the sorbent phases allows evaluation of the vapor/particle

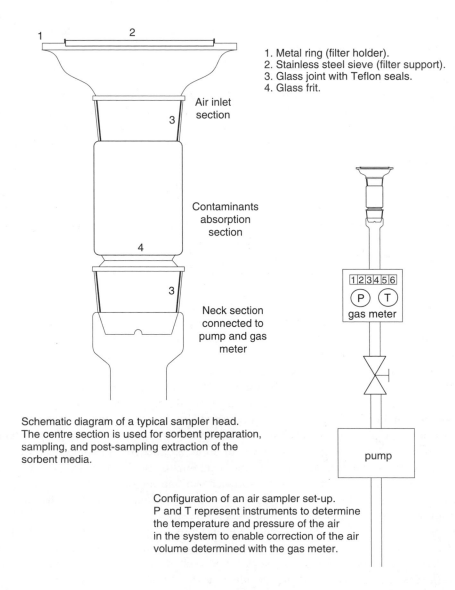

1. Metal ring (filter holder).
2. Stainless steel sieve (filter support).
3. Glass joint with Teflon seals.
4. Glass frit.

Air inlet section

Contaminants absorption section

Neck section connected to pump and gas meter

gas meter

pump

Schematic diagram of a typical sampler head.
The centre section is used for sorbent preparation, sampling, and post-sampling extraction of the sorbent media.

Configuration of an air sampler set-up.
P and T represent instruments to determine the temperature and pressure of the air in the system to enable correction of the air volume determined with the gas meter.

FIGURE 1.3 The air filter-absorption sampling system for semivolatile organic pollutants in ambient air.

distribution of chemicals where the vapor/particle distribution for a given group of chemicals can be related to the vapor pressure or the octanol/air partitioning coefficient of chemicals.[15]

I. SAMPLING BIOTA

Biota is an inherently highly variable media. Accordingly, nonstatistical strategies must be applied to biota sampling designs to eliminate as much as possible of this variability. The more that is known concerning the target biota and potential factors contributing to variability in measurable levels of the contaminants of interest in those biota, the more likely is success in reducing variability to manageable proportions.

Potential differences in contaminant accumulation due to seasonal factors such as availability of food, and life cycle factors such as fecundity, are common examples of complications in interpreting data from biota sampling and need to be considered and taken into account in the design of sampling programs. Biota sampling must be well focused on the species of interest, since for many contaminants there may be differences in contaminant accumulation between species. Similarly, the organs or tissues of choice must be a well-defined component of the sampling design. For example, if the contaminants of interest are known to accumulate in particular organs or tissues, or in individuals of a particular size or age range, then this knowledge may be turned to advantage by collecting samples of those organs, tissues, or individuals. Some metals such as Cd, Hg, and Pb accumulate preferentially in organs such as the liver, kidney, and hepatopancreas and persistent lipophilic compounds are usually preferentially sequestered in tissues with high lipid content. Accordingly, where relevant in terms of the contaminants of interest, these organs and tissues may be targeted in the sampling strategy. For example, investigations of the accumulation of lipophilic compounds in marine mammals often make the most of concentrations in the fat-rich tissues such as blubber.

Similar differentiation of sequestration in individual tissue types is commonly observed for accumulation of contaminants in plants. For example, *Typha* plants (bulrushes) are widely used as indicators of metal contamination in aquatic environments due to their capacity to tolerate high burdens of many potentially toxic metals, and to accumulate them from contaminated sediments and waters.[16,17] However, it is unlikely that useful data can be generated by sampling whole plants since these comprise different tissues (e.g., roots, rhizomes, stems, leaves, flowers) the relative masses of which vary between individual plants. An effective sampling design is likely to be one which involves the collection of one or more specific tissues, possibly with emphasis on the portion of the plant which is known to accumulate the highest concentrations of the contaminants of interest. In the case of *Typha*, this is most likely to involve sampling of root or rhizome material.

Many biota targeted for contaminant monitoring are immobile species, on the basis that their contaminant burden should reflect exposure and bioavailability at the location of collection, in contrast to the uncertainty of contaminant source when mobile species are collected. However, many investigations require that mobile species are sampled because the contaminant burdens in these species are the focus of investigation, for example, the assessment of the burden of contaminants of human health significance in commercially exploited species. Accordingly, methods employed for sampling such biota should be those which ensure samples are representative of the target population. For example, in targeting fish, if nets are used, the mesh size and method of deployment are likely to be significant factors in standardizing the size and range of species caught. Similarly, where biota are sampled by trapping, the bait used and placement of traps should be standardized.

The collection strategy, whether employing a random or stratified random collection approach, or a grid sampling approach, or some other design, will depend on the purpose of the sampling, in particular the use to which the data will be put and any hypotheses tested. For example, the collection strategy for a study investigating the level of contaminants of interest in marketable-size fish from a commercial fishery will be quite different to the collection strategy needed to assess contamination of fish populations exposed to a chemical effluent discharge point.

Sampling of benthic organisms may involve a range of equipment such as grabs, dredges and sieves depending on the nature and depth of the waterbody concerned.

A unique property of biota as a media for contaminant analysis is that individual organisms have a finite size. Accordingly, in contrast to sample quantity available from most inanimate media, the quantity of material available for analysis from an individual organism has an upper limit and to achieve levels of quantization, it may be necessary to form composite samples comprised of material from several individuals. Composite sampling is also useful as a means of reducing variability. However, depending on the underlying aims of the study, the use of composites may mask variability which is relevant.

J.   Use of Biota as Environmental Samplers

The capacity of many organisms to accumulate contaminants of interest from their environment (bioaccumulation) can be turned to advantage as a means of integrating exposure over time and indicating that the contaminants are present in a bioavailable form. For mobile organisms, integration within the space inhabited by the individuals or groups sampled may also be provided from the sampling of these species. Contaminants may be ingested along with or as a constituent of food. In the aquatic environment, absorption across gill membranes or other respiratory surfaces is often an important, if not the most important, route for many contaminants.[18] Aquatic filter feeders such as many bivalve molluscs often have limited ability to differentiate particulate material and since many potential environmental contaminants are attached or bound to particulate material, these organisms are often favored as "sentinel organisms" for indicating levels of contaminants present in host waterbodies. If there is a good correlation between the concentration of a contaminant in the tissues and the water in which the organism is exposed, measurements of contaminant concentrations in the tissues can be readily used to estimate concentrations of contaminants in the environment from which the organisms are collected. The estimation of exposure concentrations in the water column from tissue burdens of nonbiodegradable lipophilic contaminants is well developed since the relative contaminant concentrations between water and tissue, particularly the lipid fraction in tissues, usually form predictable bioconcentration relationships involving the octanol–water partition coefficient.[19] Bioaccumulation relationships are less predictable with contaminants which are readily biodegraded after assimilation, at rates that may be specific to the level of exposure (i.e., whether thresholds at which metabolism is activated have been exceeded).

The use of organisms as contaminant accumulators for environmental sampling purposes may be "passive" when the organisms are simply sampled from naturally occurring populations at the site of interest. Alternatively, in some circumstances suitable organisms can be "actively" exposed to potentially contaminated environments. In such cases it is common to deploy uncontaminated organisms from a "clean" reference site for a fixed period and then retrieve them for analysis. Such exposure, if animals are involved, may involve the use of cages constructed from inert material, or some equivalent means of retaining the sentinel organisms at the location of interest employed. In some cases, in addition to accumulation of contaminants of interest, exposed organisms may display histological changes characteristic of exposure to threshold concentrations of contaminant, thus providing additional information concerning contaminant exposure levels.

Criteria used to assess the suitability of organisms as contaminant samplers include:

- should accumulate the contaminants of interest at the levels present without suffering acute toxic effects;
- should accumulate contaminants of interest in a known relationship with the average contaminant exposure level. It is also useful if the organisms are easy to maintain under

laboratory conditions to enable the development of such relationships from controlled exposures;
- should be unambiguous to identify and easy to sample;
- organism behavior in relation to potential exposure should be sufficiently understood so that confounding effects such as season migration and feeding patterns can be taken into account. For this reason, immobile organisms are often preferred.

The popularity of bivalve molluscs such as oysters and mussels as sentinel organisms for passive and active sampling of contaminants in waterbodies relates to the good match of these organisms with the above criteria. These and other organisms which are immobile or unlikely to venture far from the point of sampling are the most commonly used organisms. A wide variety of plants and animals have been used in this manner to sample contaminants in the environment (for examples and techniques see Ref. 20–24).

The inherent variability in biota noted above is a complicating factor with the use of biota as environmental samplers, as are the many factors which influence the metabolism of organisms and potentially, the accumulation of contaminants. As noted above, age and associated factors such as size and weight of individuals used is often important. For example, some contaminants are accumulated throughout the life of the organisms, but others only during periods of growth. If accumulation does not keep pace with growth, dilution of contaminants previously accumulated within the tissues may occur. In addition, accumulated burdens of some contaminants are shed periodically in animals which molt, or with leaves shed from plants and some contaminants can be eliminated from the host animal or plant as a component of tissue lost or released during reproduction. Accordingly, knowledge of contaminant distribution in the organs and tissues, as well as factors such as physiological and nutritional status, may be important. Finally, it must be remembered that many potential contaminants are also essential elements, so that many organisms will have a consistent background concentration of some elements.

IX. TIME INTEGRATED MONITORING OF POLLUTANTS USING ABIOTIC PASSIVE SAMPLING DEVICES

Usually, sampling methods used for water or air quality monitoring are based on a "grab-sampling" regime which uses individual samples representative only of conditions during the relatively short time period in which the sample was collected. Furthermore, for many organic chemicals, the typical sample volumes normally collected by "grab-sampling" methods are often inadequate for obtaining analytical results in the concentration range necessary for assessment of whether the chemicals pose a risk to the environment or human health, since threshold concentrations of concern may be near or less than levels of quantization from a typical "grab-sample."

The problems associated with "grab-sampling" strategies and analytical method sensitivity for water and air samples have resulted in many attempts to develop alternative monitoring methods. These include sampling and analysis of sediment or biota. Unfortunately, prediction of contaminant concentrations in the water column based on concentrations measured in associated sediments and biota is complicated by a myriad of factors related to the complexity of the natural abiotic and biotic environments (i.e., sediment chemistry and particle distribution, biota mobility and metabolism and many others). In recognition of these problems, systematic attempts have been made to develop passive sampling systems which accumulate chemicals and from which reliable exposure concentrations can be calculated. The passive samplers used in such systems are usually designed either as "kinetic samplers" or as "equilibrium samplers."

Atmospheric kinetic samplers, sometimes referred to as "diffusion samplers", have a long history of use in roles such as personal monitors or dosimeters to evaluate personal exposure or work place exposure levels, typically to volatile air pollutants. However, in the last two decades there has been substantial progress towards using kinetics-based passive samplers for ultra-trace pollutants in water and air.[25-31] In principle, kinetic samplers rely on a large sampler capacity for the chemicals of interest, or with respect to water and air, a large sampler/water and sampler/air partition coefficient, respectively, for the contaminants to be sampled. This ensures that under sampling conditions the concentration of the chemicals within the sampler is not approaching an equilibrium state during sampler exposure. The models used to calculate the chemical concentration in the air or water phase (C_P) sampled, are based on the assumption that uptake is linearly related to the exposure concentration throughout exposure. The mean concentration in the air or water phase C_P can then be predicted from

$$C_P = \frac{C_S V_S}{R_S t}$$

where: C_S is the concentration of the analyte in the sampler, V_S is the sampler volume, R_S is the specific sampling rate and t is the exposure period. Note that the volume of a sampler with a given configuration is necessarily related to its sampling area for which R_S is determined in laboratory exposure experiments. Models such as the one above have been used to describe behavior of passive samplers and to predict the concentration of chemicals in water or air.

Time-integrated passive sampling techniques have become widely used in the last decade. In particular, the use of performance reference compounds introduced into the sampler to enable adjustment of field data from the samplers using kinetic data from the laboratory has increased user confidence in these sampling techniques. While some uncertainty factors associated with the use of these sampling techniques remain, such as their representativeness as mimics of biota uptake and measures of bioavailable fractions, as well as issues relating to biofouling and nonlinear uptake, the utility of these new sampling tools in part reflects the limitations of the other techniques available. Accordingly, these sampling techniques provide an additional set of tools often useful for modern monitoring programs.

Table 1.1 and Figure 1.4 and Figure 1.5 illustrate the range of passive sampling devices developed for application in sampling trace contaminants in the aquatic environment.

X. QA AND QC IN ENVIRONMENTAL SAMPLING

QA and QC procedures are needed to minimize sampling errors and to detect errors such as contamination or losses of analyte at all stages of sample handling. QA/QC include procedures to prevent, detect, and correct problems in the sampling process and to characterize errors through the use of QC samples which can assist in detecting changes in the samples prior to analysis. QA/QC also include documentation of protocols and training of sampling personnel, as well as checks as to how representative the samples taken are of the environment concerned.

A. THE NEED FOR DOCUMENTED ENVIRONMENTAL SAMPLING PROTOCOLS

It is as important to have a robust protocol in place to guide collection of the samples from the field, as it is for the receiving laboratory to follow an appropriate, documented and rigorously evaluated analytical protocol to determine the properties of interest in samples received.

Sampling protocols should be fully and unambiguously documented, including requirements for labeling samples and completing documentation such as field data sheets and chain-of-custody documents where appropriate. Locations from which sample collections are required should be specified in the protocols, along with the number and nature of samples required from each location. Typically, a protocol includes specifications of sample-taking and sample-handling equipment.

TABLE 1.1
Examples of Passive Samplers for Trace Aquatic Pollutants

Chemicals of Interest for Sampling Purposes	Sampler Types	Comments
Nonpolar organic pollutants, for example, PAHs, PCBs, organochlorines, and other nonpolar pesticides	SemiPermeable Membrane Device (SPMD). (1) PE only samplers. (2) Trimethylpentane filled PE sampler	A relatively large surface area of nonpolar membrane is required because these chemicals often occur at ultra-trace levels. SPMDs are commercially available and the most widely used type of passive samplers to date. Some advantages for SPMDs over other passive sampler types are the availability of the largest set of calibration data together with an extensive literature and the use of performance reference compounds for *in situ* calibration is routine. However, analysis of SPMDs is relatively complex compared to PE and trimethylpentane samplers. All samplers have biofouling issues
Polar chemicals	Polar Organic Chemical Integrated Sampler (POCIS); Portsmouth sampler using solid phase sorbents based samplers	Relatively new methods developed in parallel through the last decade. Limitations to the methods and availability of the devices at the present time should be solved in the near future. The optional use of different phases and membranes or even deployment without a membrane provides a high selectivity and flexibility in the analytes targeted. To date few calibration data are available and the polar samplers are not yet widely used for routine monitoring but show great potential
Metals, radionuclides and some nutrients	Diffusion Gradient in Thin films (DGT) and equilibrium techniques such as Peepers and DETs	Developed in the 1990s, these are now a widely used method for monitoring certain trace metals. To date primarily used in research and not for routine monitoring, although increasing numbers of studies show the potential of this technique for sampling metals. Discussion continuous concerning the fraction of metal collected with DGTs

In addition, it provides all necessary precautions and specifics for using sample collection and handling equipment, along with specifics for the study concerned, for example, depth at which samples are taken, cross-sectional area sampled and precautions necessary to avoid contamination of samples.

A length of sealed lay-flat PE tubing (the membrane) containing a small volume of trimethylpentane (the absorbant phase) is woven around a stainless steel frame.

The device is then inserted into a perforated stainless steel shroud for protection from mechanical damage during deployment.

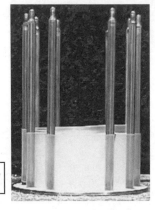

Photographs supplied by K.Manonmanii, EnTox, Brisbane, Australia.

FIGURE 1.4 An example of a SemiPermeable Sampling Device (SPMD).

If transects or grid patterns are involved in determining sampling points, these should be clearly specified in the protocols in terms of location, along with the method for their alignment. The protocols should be specific to each matrix and constituent, specifying the collection equipment, type of containment, and any preservation procedures. All essential details such as the process for cleaning sampling equipment between the consecutive samplings should be prescribed. Numbers and types of QC samples to be taken and their labeling should also be specified (which should be such that the QC samples are not distinguishable from other samples within the same batch) and contingency procedures should be in place to handle possible problems such as difficulties in obtaining the prescribed sample volume.

QA begins with the training of field staff in the correct operation of all items of sampling equipment and familiarity with the study protocols in such matters as what will be sampled, the methods of collecting samples, sample labeling, preservation, and storage. Staff should be able to demonstrate competency in the relevant field procedures according to the study protocols, including techniques needed to avoid contamination of samples, calibration of any instruments needed for taking measurements in the field and required labeling of samples and recording of field observations, before taking actual samples which are part of the study.

Every sampling method and related items of sampling equipment have recommended procedures associated with their use. These should be described in the protocols and followed on every sampling occasion. These may include basic "commonsense rules" such as avoiding any unnecessary disturbance to a site prior to collecting samples, for example, standing downstream and collecting upstream during sampling in flowing streams. In addition, items of equipment should be checked for cleanliness and their proper working order verified before they are taken into the field. Where automatic sampling devices are used, timing mechanisms should be checked to ensure that timed samples are able to be taken at the times intended.

Checks should be incorporated into the field sampling process to ensure that the sampling protocols are followed.

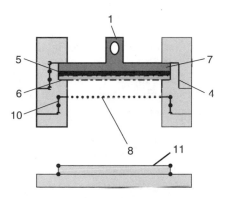
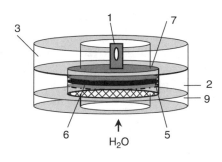

Sampler consists of 3 interlocking sections manufactured from PTFE (2, 3, 9) that screw together during deployment to form water-tight seals (4,10).

Integral to the device is a 50 mm rigid PTFE disk (7) designed to support both the chromatographic receiving phase (5) and the diffusion-limiting membrane (6).

On the reverse is a lug (1) for attaching the device during deployment.

The surface of the diffusion-limiting membrane is protected from mechanical damage during deployment by a mesh (8) of either stainless steel for organic analytes or nylon for inorganic analytes. This mesh is held in place during deployment by a removable PTFE ring (9).

> Figure and explanatory text supplied by Richard Greenwood, Portsmouth University, U.K.

FIGURE 1.5 An example of a passive sampler for organic or inorganic contaminants in the water column. Depending on the polarity range of the analytes of interest, a range of devices incorporating appropriate membranes and solid phase absorbant media have been developed. Illustrated is one of these devices based on the Empore™ disk. Other devices and their applications are described in Table 1.1. A PTFE screw cap (11) replaces the ring (9) during transport to and from the deployment site.

B. QA OF DATA

To assist effective recording of field measurements, field record sheets should be designed so that all essential information can be recorded at the point of sample collection, with the minimum of effort. As an example, preprinted recording sheets requiring minimal and simple entries are often used, because field sampling often involves working under adverse conditions. Unnecessarily complex record sheets can be counterproductive in wet and windy conditions.

It is important for data integrity that sampling personnel complete field record sheets during sample collection and field measurement taking before leaving a sampling location. The information recorded should include observations and information relating to conditions at the time of sampling or measurement and any other matters which might later be relevant in explaining unusual data. Basic information recorded on field record sheets for each sample includes:

- Time and date of collection;
- Exact location of the sampling point and related descriptive material (keyed to any grid reference, photograph taken, or relevant map);
- A unique sample number;
- Name and contact details of the persons taking the sample;
- Container type, sample storage, and preservation conditions;
- Where relevant, instrument calibration data.

Waterproof labels attached to sample containers with waterproof adhesive are essential, as is the use of permanent ink to enter particulars on the label. Generally it is expedient to attach and write labels before placing the sample in the container, since difficulties can arise in attempting to write on wet or soiled labels. Sample labels should clearly identify the sample and any treatment it has received (for example, filtration). Each label should, as a minimum, include the name of the sampler, date and time of sampling, the exact location of the sampling point and any treatment and preservatives added and, to avoid risk of ambiguity, a unique sample number.

C. Chain-of-Custody Record

To keep track of samples in storage or enroute to the analytical laboratory, it is good practice to record details of all samples in a chain-of-custody record. This is an essential document if data from field sampling is needed for legal proceedings.

Typical chain-of-custody records include:

sample identification particulars (as recorded on field record sheet and label);
transport and storage registration numbers with times and dates.

When samples are transferred from the custody of the sampling personnel to others (for example, for transport) or on arrival at the receiving laboratory, both parties should record this on the chain-of-custody record and the person relinquishing the samples should retain a copy.

D. QA of Sample Integrity

Potential sources of loss of sample integrity are:

- contamination from equipment;
- contamination from containers;
- contamination through diffusion;
- loss of analyte through adsorption to container;
- loss of analyte through volatilization;
- loss of analyte through biodegradation;
- loss of analyte through photo-degradation;
- loss of analyte through oxidation and precipitation.

Some environmental sampling may involve a series of different items of equipment for access to the medium to be sampled, collection of the sample, storage and preservation of sample material, and preparation prior to analysis. Accordingly, there must be accurate observation of documented protocols for the use of equipment and sample handling. These include, but are not limited to issues such as filtration, preservation, storage vessels, transportation, and storage conditions including permissible period of storage. Samples can be contaminated from equipment used during collection or after collection, inadequately or improperly cleaned glassware, filters and filtering apparatus, preservative chemicals, etc. As a consequence, thoroughness is needed in the cleaning of equipment before taking the first in a series of samples and also after each sample to avoid the risk of cross-contamination between samples. Gloves may be necessary, not only to protect sampling personnel, but to minimize the risks of contaminating the samples. Even when gloves are used, touching the inside surfaces of sample containers and lids should be avoided.

E. Use of Blanks

Periodic blank samples should be generated during field sampling (for example, one blank in every ten actual samples) to determine errors arising from contamination of samples during or subsequent to collection. Commonly for water sampling, sample storage containers identical with those used for actual samples, but prefilled with distilled water, are taken into the field during a sampling exercise to serve as field blanks. These are subjected in sequence in the field, along with a batch of actual samples, to all of the treatment steps such as filtration, addition of preservatives, and storage undergone by the actual samples. The blanks are then analyzed along with the batch of real samples collected on that sampling occasion.

F. QCs for Representativeness of Sampling

To check representativeness of sampling, a number of replicate samples should be taken to determine spatial and temporal variability. For example, temporal variability in a waterbody can be checked by taking a series of samples at the same point at different time intervals and spatial variability by simultaneously taking samples at different points of the waterbody. In a properly designed sampling program, potential temporal and spatial variability would be considered and investigated as a component of program design, including assessment by pilot sampling if warranted. For example, it might be necessary to check the magnitude of any diurnal or seasonal variations and the influence of flood events on the representativeness of samples from a waterbody. Regardless of the confidence placed in assumptions made, or conclusions reached and subsequent measures taken to remove or counter such variability at the sampling project design stage, the taking of occasional replicate samples can enable the veracity of such assumptions and assumptions to be checked.

G. Recording Locations of Sampling Sites

It is important to record the locations of sampling sites so that they can be revisited for resampling should additional information be required (for example, if data from analysis of the original samples proved difficult to interpret) or if further samples were needed at a future date to confirm the original findings or to monitor temporal changes. The exact spots where samples were collected can be recorded by reference to fixed features of the landscape, and by map grid references and use of global positioning system (GPS) equipment. In the case of GPS, it is important to be aware of the precision of the system being used (e.g., differential GPS equipment is needed for high accuracy) and to record the coordinate system used as well as the coordinates themselves. Video and photographic records of individual sites are highly desirable for illustration of site conditions and to assist accurate relocation for future reference.

H. Sample Containers

Sample containers must be of an appropriate capacity to hold the required volume of sample.

The choice of material chosen for sample containment is determined by the potential for particular container materials to irreversibly adsorb contaminants of interest from the sample, or conversely to leach material into the sample and change its composition. In some cases a sample may have to be split between two or more containers constructed from different materials and which have undergone different processes of decontamination. For example, if water, soil, or sediment samples are to be tested for metals as well as organic compounds, acid-washed glass, or plastic containers will be needed for sample material used for metals assessment as well as solvent-washed glass containers for sample material used for organic chemicals assessment. This requirement should be considered in determining the quantity of sample material collected. In addition, if sample analysis covers a range of metals of interest, samples taken for certain metals may need a different container material and different preservative. For example, the integrity of trace concentrations of most metals can be preserved for long periods in plastic, but if samples are to be analyzed for mercury content, plastics should be avoided.

The potential of contamination or adsorption problems due to the container closure (cap) must also be considered. The closure is often of a different material from the body of the container, but can be isolated from the sample by use of an appropriate liner. In some cases, subsamples may have to be stored in containers of more than one type to preserve all of the contaminants of interest free of the risk of contamination or adsorption.

In addition to the requirement that sample containers and closures are constructed from appropriate materials to avoid risks of contaminating the sample, sample containers and closures need to be precleaned to remove the risk of surface borne contaminants compromising the integrity of the samples. Metals in trace quantities can be present on the surfaces of glass and plastic containers and a range of organic compounds on the surfaces of plastic containers. Acid washing or solvent washing under laboratory "clean room" conditions is usually the minimum preparation required. For some analytes, other container preparation steps may be required. The adequacy of cleaning should be assessed by the use of blanks (for example, the analysis of clean water stored for the maximum expected storage time of real samples). After precleaning, the contaminant-free status of containers should be preserved by sealing them in appropriate containments such as plastic or aluminium foil bags before distribution from the cleaning facility. Sample containers should remain in such containment up to the point where the sample is to be placed into it. It may be necessary to "double-protect" sample containers during transport to a field sampling site by protecting the integrity of the sealing containment with a secondary containment such as a zip-lock bag to hold batches of sealed precleaned sample containers.

I. SAMPLE PRESERVATION/STABILIZATION AND STORAGE

In addition to ensuring that sample containers have sufficient capacity to store the minimum volume of sample needed for the range of tests to be applied, it is important to preserve the integrity of samples over the period between the taking of the samples and commencement of laboratory analysis. Essentially this involves preserving and/or stabilizing the samples against biological, chemical, and physical influences which may alter the condition or properties under investigation. Volatilization, adsorptive and diffusive gains or losses of analytes, as well as absorptive and diffusive additions, some of which have the potential to alter analyte concentrations or initiate chemical changes are each potential issues addressed by proper consideration of sample preservation/stabilization, containerization, and storage. In practice, complete preservation of samples is difficult to achieve. At best, preservation only retards chemical and other changes to sample material that are inevitable but there are a number of precautions which can be implemented to minimize undesirable alterations to the variables of interest in a sample. For this reason there are maximum limits of acceptable time after sampling applicable to most analytes and in general the shorter the time interval between collection and analysis, the better.

Basic precautions include the selection of an appropriate preservation technique and ensuring that the time lapse between sampling and analysis is kept within applicable limits.

Appropriate sealing of sample containers is important to avoid contamination from the atmosphere by absorption through the cap seal (which may, for example, result in oxidation of sulfides to sulfates) and to eliminate escape of potential analytes by volatilization and evaporation. Proprietary liners including tetrafluoroethylene (Teflon) are generally effective in minimizing contamination problems associated with diffusive processes. Filling sample containers completely to remove any head space reduces losses of analyte by volatilization.

Refrigeration and freezing of samples (including the use of water ice or dry ice in the field) is effective in extending the preservation period for most materials, however, if liquid samples are to be frozen, space to accommodate sample expansion on freezing is a necessity or containers are likely to burst. For this reason, providing zero headspace to minimize losses of analyte from volatilization is generally incompatible with freezing of samples.

It is common practice to add preservatives to many classes of samples, for example, nitric acid to water samples collected for metals assessment, since this minimizes adsorption to the container surface and the precipitation of insoluble oxides and hydroxides. Note however, that if water samples are to be preserved with acid, they must first be filtered, otherwise the acid is likely to solubilize metals from particulate matter included in the sample and give biased results on analysis. Nitric acid is generally preferred to hydrochloric or sulfuric acids for preservation of metal ion concentrations in water samples because these could enhance precipitation due to the low solubility of some chlorides and sulfates. Other preservatives are required for specific analytes, for example, sodium hydroxide solution to preserve cyanides.

If the analytes of interest are sensitive to photochemical changes, then amber rather than clear glass containers are likely to be more appropriate for containerization of samples. Alternatively, filled and sealed containers can be wrapped in aluminium foil to exclude light.

Recommendations regarding containers and preservation are published in Standard Methods and International Standards (ISO). For some analytes there may be a choice of method, for example alternative preservation methods may be applicable depending on whether it is practical to deliver samples to the analyzing laboratory within a few hours, or whether longer time periods cannot be avoided.

Overall, the assurance of adequate containerization and preservation of samples is a matter which needs to be built in to the preparatory planning of any environmental sampling exercise, in consultation with the receiving laboratory.

J. QC SAMPLES

To check for and quantify contamination which may occur during the field sampling process, blank samples are required

For field blanks, extra containers with appropriate contents (equivalent to the matrix being sampled in the field) are taken to the site and handled in a comparable way to the containers used for real samples. This includes uncapping the blank containers for the same time period and under the same conditions as apply to the real sample containers.

Trip blanks may be useful in addition to field blanks. These are similar to field blanks but are not uncapped in the field, thus being useful to delineate contamination which occurs from handling and storage from that which occurs from dust or other external sources during containment of real samples.

Container blanks are needed to check for and quantify contamination of samples from the containers. Such blanks are filled with deionized water (or water of the appropriate salinity where marine or brackish water is being sampled) and preserved in the same manner and for the maximum storage time as field samples.

Equipment blanks are required where items such as pumps, tubing, corers, water sampling bottles, or sediment grabs comes in contact with samples during collection, and in cases where water samples are filtered prior to containerization, filter blanks also. A filter blank is prepared in the field by filtering a sample volume of distilled water using the same equipment and handling as the field samples. An equipment blank consists of an unused sample of the water or other cleaning solvent used to rinse sampling equipment between samples.

Duplicate (or multiple) samples are useful for quantifying problems such as contamination or losses of analyte occurring between sampling and analysis. A duplicate or multiple sample is obtained by dividing one sample into two or more subsamples. This involves taking more than the usual quantity of sample material and subdividing it to form the required number of samples of the required size. Theoretically, in the absence of problems associated with containment, preservation, transport, and storage, the results from analysis of duplicate or multiple samples should be the same within the limits of analytical precision. Duplicate or multiple samples may also be used to test inter- and intra-laboratory accuracy and precision.

Replicate samples can be taken to check the reproducibility of the sampling process. Note that *replicate* samples are two or more separate samples collected from a single location. Normally, replicate samples are collected at the same time. However, they could be taken at different times to check the assumption that time is not a factor in determining the environmental property being studied.

Spiked field samples may be used to detect and quantify losses of analyte. These involve the addition of a known amount of an analyte to field blanks or subsamples of real samples as they are taken in the field.

In many situations it may be impossible to completely eliminate contamination of samples or changes to or losses of analyte, but a hierarchy of QC samples can be used to ensure that it is stable and quantified.

XI. HEALTH AND SAFETY OF SAMPLING PERSONNEL

All hazards and risks involved in field sampling should be identified and documented in the sampling protocols. In addition, sampling personnel should be briefed concerning the hazards and appropriate safety procedures. Where necessary, protective equipment should be available.

XII. SUMMARY AND CHECKLIST

The collection of samples from the environment is a multistage process which requires clear objectives and thorough planning if it is to succeed, and should incorporate appropriate QA/QC to ensure that the sampling process in the field is as rigorous as the laboratory analysis which follows. Without this approach to the sampling task, there is a high risk that the samples will prove to be the weakest component in the study.

A sampling operation includes the preparation prior to conducting sampling, the field collection of the samples and the post-sampling handling of samples and data. We conclude this overview of the environmental sampling process with the following checklist of essential components for a successful sampling operation:

1. Are the purpose and objectives of the sampling program clearly stated and understood?
2. Are data requirements stated and specific variables of interest determined?
3. Have appropriate measurement techniques been selected?
4. If *in situ* measurements are to be made, are calibration procedures in place?
5. Will the chosen sampling method and equipment collect a representative sample?
6. Will disturbance of the environment during sampling impact on the integrity of the sample?
7. Will the sample be altered physically or chemically by contact with sampling equipment?
8. If sampling equipment may be contaminated by a sample, how is the equipment to be cleaned between samples?
9. What steps are necessary to avoid contamination or alteration of the sample during collection?
10. What volume of sample is required and are the sample containers of an appropriate capacity?
11. Are the sample containers constructed of material which will not contaminate or affect the stability of the sample and are closures appropriate?

12. What preservation of samples is required and are the materials available?
13. Are protocols in place to identify and record sampling sites and to record *in situ* data?
14. Is a system in place to identify, measure and control sources of error (blanks, duplicates, and replicates)?
15. Have sampling protocols been documented?
16. How are sampling personnel to be trained?
17. How is the competence of sampling personnel to be assessed?
18. How are sampling equipment and personnel to be transported to the field and returned and how are samples to be transported from the field to the laboratory for analysis?
19. How are data from sample analysis to be recorded and stored?
20. Are hazards associated with sampling sites and equipment/materials identified and are appropriate measures in place to protect the health and safety of sampling personnel?

REFERENCES

1. Epstein, R. L., Residue Chemistry — past to present. Proceedings, 19CRC, *Residue Chemistry in the 21st Century, 19th Conference of Residue Chemists*, 16–18 September 2003, Brisbane, Australia.
2. Keith, L. H., Ed., *Principles of Environmental Sampling*, 2nd ed., American Chemical Society, Washington, DC, 1996.
3. American Chemical Society Committee on Environmental Improvement, *Anal. Chem.*, 52(14), 2242–2249, 1980.
4. Barcelona, M. J., Overview of the sampling process, In *Principles of Environmental Sampling*, 2nd ed., Keith, L. H., Ed., American Chemical Society, Washington, DC, 1996, chap. 2.
5. Liess, M. and Schulz, R., Sampling methods in surface waters, In *Handbook of Water Analysis*, Nollet, L. M. L., Ed., Marcel Dekker Inc., New York, 2000.
6. Cowgill, U. M., Sampling waters. The impact of sample variability on planning and confidence levels, In *Principles of Environmental Sampling*, 2nd ed., Keith, L. H., Ed., American Chemical Society, Washington, DC, 1996, chap. 18.
7. Kent, R. T. and Payne, K. E., Sampling groundwater monitoring wells. Special QA and QC considerations, In *Principles of Environmental Sampling*, 2nd ed., Keith, L. H., Ed., American Chemical Society, Washington, DC, 1996, chap. 21.
8. Thomas, R. and Maybeck, M., The use of particulate matter, In *Water Quality Assessments*, Chapman, D., Ed., Chapman & Hall, London, UK, 1992.
9. Burrus, D., Thomas, R. L., Dominik, J., and Vernet, J. P., Recovery and concentration of suspended solids in the Upper Rhone River by continuous flow centrifugation, *J. Hydrol. Process.*, 3, 65–74, 1988.
10. Horowitz, A. J., Elrick, K. A., and Hooper, R. C., A comparison of instrumental dewatering methods for the separation and concentration of suspended sediments for subsequent trace element analysis, *J. Hydrol. Process.*, 2, 163–184, 1989.
11. Bloesch, J. and Burns, N. M., A critical review of sedimentation trap technique, *Schwiz. Z. Hydrol.*, 42, 15–55, 1980.
12. Johnson, D. C., Selection of sample site, *Air Pollution Measurement Manual. A Practical Guide to Sampling and Analysis*, 5th ed., Vol. 1, The Clean Air Society of Australia and New Zealand, Victoria, Australia, 2000.
13. Clarke, P., Calibration of gas flow rate and volume meters, *Air Pollution Measurement Manual. A Practical Guide to Sampling and Analysis*, 5th ed., Vol. 1, The Clean Air Society of Australia and New Zealand, Victoria, Australia, 2000.
14. *Stockholm Convention on Persistent Organic Pollutants (POPs). Text and Annexes*, Interim Secretariat for the Stockholm Convention on Persistent Organic Pollutants, United Nations Environment Program (UNEP) Chemicals, Geneva, 2001.
15. Lohmann, R. and Jones, K. C., Dioxins and furans in air and deposition: a review of levels, behavior and processes, *Sci. Total Environ.*, 219, 53–81, 1998.

16. Maher, W. A., Norris, R. H., Curran, S., Gell, F., O'Connell, D., Taylor, K., Swanson, P., and Thurtell, L., Zinc in the sediments, water and biota of Lake Burley Griffin, Canberra, *Sci. Total Environ.*, 125, 235–252, 1992.

17. McNaughton, S. J., Folsom, T. C., Lee, T., Park, F., Price, C., Roeder, D., Schmitz, J., and Stockwell, C., Heavy metal tolerance in *Typha latifolia* without the evolution of tolerant races, *Ecology*, 55, 1163–1165, 1974.

18. Connell, D. W., Bioaccumulation behavior of persistent organic chemicals with organic chemicals, *Rev. Environ. Contam. Toxicol.*, 101, 117–154, 1988.

19. Connell, D. W. and Hawker, D. W., Use of polynomial expressions to describe the bioconcentration of hydrophobic chemicals by fish, *Ecotoxicol. Environ. Saf.*, 16, 242–257, 1988.

20. Phillips, D. J. H. and Rainbow, P. S., *Biomonitoring of Trace Aquatic Contaminants*, *Environmental Management Series*, Chapman & Hall, London, 1994.

21. Philips, D. J. H., *Quantitative Aquatic Biological Indicators*, Applied Science Publishers Ltd., London, 1980.

22. Burton, M. A. S., *Biological Monitoring of Environmental Contaminants (Plants)*, MARC Report No. 32, Monitoring and Assessment Research Centre, King's College, London, 1986.

23. Whitton, B. A., Use of plants to monitor heavy metals in rivers, In *Biomonitoring of Environmental Pollution*, Yasuno, M. and Whitton, B. A., Eds., Tokai University Press, Tokyo, 1988.

24. Samiullah, Y., *Biological Monitoring of Environmental Contaminants: (Animals)*, MARC Report No. 37, Monitoring and Assessment Research Centre, Kings's College, London, 1986.

25. Huckins, J. N., Petty, J. P., Prest, H. F., Clark, R. C., Alvarez, D. A., Orazio, D. C., Lebo, J. A., Cranor, W. L., and Johnson, B. T., *A Guide For The Use Of Semipermeable Membrane Devices (SPMDs) As Samplers Of Waterborne Hydrophobic Organic Contaminants*, American Petroleum Institute Publication, 2002, p. 4690.

26. Huckins, J., Manuweera, G., Petty, D., Mackay, D., and Lebo, J. C. N., Lipid-containing semipermeable membrane devices for monitoring organic contaminants in water, *Environ. Sci. Technol.*, 27, 2489–2496, 1993.

27. Kingston, J. K., Greenwood, R., Mills, G. A., Morrison, G. M., and Bjorklund Persson, L., Development of a novel passive sampling system for the time averaged measurement of a range of organic pollutants in aquatic environments, *J. Environ. Monit.*, 2, 487–495, 2000.

28. Alvarez, D. A., Huckins, J. N., Petty, J. D., and Manahan, S. E., Progress toward the development of a passive *in situ* SPMD-like sampler for hydrophilic organic contaminants in aquatic environments, *20th Annual Meeting of Society of Environmental Technology and Chemistry*, Philadelphia, November, 1999.

29. Davison, W. and Zhang, H., *In situ* speciation measurement of trace components in natural waters using thin-film gels, *Nature*, 367, 545, 1994.

30. Petty, J. D., Huckins, J. N., and Zajicek, J. L., Application of semipermeable membrane devices (SPMDs) as passive air samplers, *Chemosphere*, 27, 1609–1624, 1993.

31. Barthow, M. E., Booij, K., Kennedy, K. E., Müller, J. F., and Hawker, D. W., Passive air sampling theory for semivolatile organic compounds, *Chemosphere*, 60, 170–176, 2005.

2 Sample Preparation for Chromatographic Analysis of Environmental Samples

*Claudia E. Domini, Dimitar Hristozov, Beatriz Almagro,
Iván P. Román, Soledad Prats, and Antonio Canals*

CONTENTS

Abbreviations: APEO, alkylphenol ethoxylate; ASE, accelerated solvent extraction; BTEX, benzene, toluene, ethylbenzene and xylenes; BTX, benzene, toluene and xylenes; CDA, 2-(2-chlorovinyl)-1,3,2-dithiarsenoline; CEC-UV-Vis, Capillary electrochromatography ultraviolet-visible; CFLME, continuous flow membrane extraction; CFS, continuous flow system; CMP, chemical measurement process; CVAA, 2-chlorovinyl arsonous acid; CW (CWX), Carbowax; DAD, diode array detector; DBCP, 1,2-dibromo-3-chloropropane; DBT, dibutyltin; DCM, dichloromethane; DEDIA, desethyldesisopropyla-trazine; DIE-SPE, dynamic ion-exchange SPE; DIHA, desisopropyl-2-hydroxyatrazine; DMAE, dimethylaminoethanol or dynamic microwave assisted extraction; DPX-A, ethametsulfuron; DSASE, dynamic sonication assisted solvent extraction; DVB, divinylbenzene; DVB–PDMS, divinylbenzene–polydimethylsiloxane; ECD, electron capture detector; ELCD, electrolytic conductivity detector; EPA, environmental protection agency; FFF, field flow fractionation; FID, flame ionisation detector; FPD, flame photometric detector; FPD-P, flame photometric detection phosphorous mode; GC-AED, gas chromatography atomic emission detection; GCB, graphited carbon black or graphitized carbon blacks; GC–IT-MS, gas chromatography ion trap mass spectrometry; HF-MSME, hollow fiber membrane solvent microextraction; HPLC–FLD, HPLC coupled to fluorimetric detection; HRGC, high resolution gas chromatography; HRMS, high resolution mass spectroscopy; HS–SPME, headspace solid phase microextraction; ICP-MS, inductively coupled plasma mass spectroscopy; IEX, ionic exchangeable; IPA, iso-propyl alcohol; IP-LC–EI-MS, ion-pair liquid chromatography electrospray ionization mass spectrometry; IP-SPE, ion pair SPE; IP-SPME, ion pair solid phase microextraction; LC-ED, liquid chromatography electrochemical detection; LC–EI-MS, LC–ES-MS, LC–ESI-MS, liquid chromatography electrospray ionisation mass spectroscopy; LC-UV, liquid chromatography ultraviolet detection; LLE, liquid–liquid extraction; LLLME, liquid–liquid liquid microextraction; LOD, limit of detection; LOQ, limit of quantification; LPME, liquid phase microextraction; LPME/BE, liquid phase microextraction back extraction; LVI-GC–MS, large volume injection gas chromatography mass spectroscopy; MAE, microwave assisted extraction; MAE-HS–SPME, microwave-assisted extraction head space solid phase microextraction; MAP, microwave assisted process; MASD, microwave assisted steam distillation; MASE, microwave assisted solvent extraction; MBT, monobutyltin; MIP-SPE, molecularly imprinted polymer-SPE; MLLE, membrane liquid–liquid extraction; MMLLE, microporous membrane LLE; MSD, mass spectroscopy detector; MSM, metsulfuron methyl; MSPD, matrix solid phase dispersion; MTBE, methyl tert-butyl ether; NCI-MS, negative chemical ionisation mass spectrometry; NP, normal phase; NPD, nitrogen–phosphorus detector; OCPs, organochlorine pesticides; OPPs, organophosphorous pesticides; PA, polyacrylate; PAH, polycyclic aromatic hydrocarbon; PBTA-1, 2-[2-(acetylamino)-4-[bis(2-methoxyethyl)-amino]-5-amino-7-bromo-4-4-chloro-2H-benzotriazole; PBTA-2, 2-[2-(acetyl-amino)-4-[bis(2-cyanoethyl)-ethylamino]-5-amino-7-bromo-4-chloro-2H-benzotriazole; PCBs, polychlorinated biphenyls; PCDDs, polychlorinated dibenzo-P-dioxins; PCDFs, polychlorinated dibenzofurans; PDA, photodiode array; PDMS, polydimethyl siloxane; PF, polysilicone fullerene; PFE, pressurize fluid extraction; PHCs, petroleum hydrocarbons; PHWE, pressurized hot water extraction; PID, photoionization detector; PLE, pressure liquid extraction; PLRP, polymeric reversed phase (material); PTFE, polytetrafluoroethylene; PTV-GC, programmed temperature vaporizer gas chromatography; RP, reverse phase; RP-SPE, reversed phase-SPE; RSD, relative standard deviation; SAESC, sonication-assisted extraction in small columns; SAX, strong anion exchange; SBSE, stir-bar sorptive extraction; SCX, strong cation exchange; SD, standard deviation; SDB, styrene-divinyl-benzene copolymer; SDE, single drop extraction; SDSE, steam distillation-solvent extraction; SE, Soxhlet extraction; SF, supercritical fluid; SFC, supercritical fluid chromatography; SFE, supercritical fluid extraction; SFE–CZE, supercritical fluid extraction–capillary zone electrophoresis; SME, sorptive membrane extraction; SPE, solid-phase extraction; SPME, solid phase microextraction; SWE, superheated water extraction and supercritical water extraction; TBT, tributyltin; TCD, thermal conductivity detector; TCT, triclorinetin; TeBT, tetrabutyltin; TeET, tetraethyltin; TID, thermoionic detector; TMS, trimethyl silyl; TPhT, triphenyltin; TPR, templated resin; UAE, ultrasound assisted extraction; US-EPA, United States environmental protection agency; VOCs, volatile organic compounds; µTAS, miniaturized total analysis system.

I. INTRODUCTION

A. A Brief Description of the Analytical Approach

The main function of an analyst is to supply quality information which makes the correct decision about the studied problem possible. In order to provide this information, Analytical Chemistry, as all the other experimental sciences, follows a methodology called the "analytical approach." This analytical methodology could be described by the following goals:

- To correctly define the problem.
- To ensure that the available sample correctly represents the problem.
- To contact the client in order to obtain his/her knowledge about the problem and to define the available time and the accuracy required.
- To develop an analytical working plan where the sequence and the best methods to use are considered and evaluated.
- To perform the work with the highest level of quality and excellence, experience and knowledge of the analyst.
- To supply answers, not only raw data, including the precision and reliability of all numbers and specifying the limitations on the use of the data.
- To supply a clear, consistent and significant final report which interprets and explains the information and data given.

All these goals could be attained by means of a systematized set of steps that is referred to as the Total Analytical Process (TAP) or Chemical Measurement Process (CMP). However, these terms have different meanings. Hence, TAP is defined as "all steps that are between the general problem

FIGURE 2.1 Schematic definition of total analytical process (TAP).

(requested information) and the provided information." Figure 2.1 shows a scheme of this definition. On the other hand, CMP is defined as "all steps that are between the original sample (not sampled, not handled and not measured) and the obtained result given as the analytical problem was posed."[1] Hence, a chemical measurement process may be considered as divided into the three main parts as shown in Figure 2.2.

The overall number of steps in a TAP, and the limits of each one, is a subject of continuous change and it is still a matter of controversy.[1–3] Until the end of the 1940s the analyst paid special attention to the sample itself. The total analytical process started sampling, commonly upon arrival of the sample at the laboratory, and continued with the elimination of interfering species and the measurement of the analytical signal and finished with interpretation of the results. At that time it could be said that TAP was the same as CMP. After the Second World War, analysts paid special attention to the instrument. By then a new concept of solving problems suggested by the client (i.e., the analytical problem) was becoming important. The appearance of the concept of the analytical problem required the modification and expansion of old boundaries of analytical chemistry. The radical change from a passive attitude (to receive the samples and to supply results) to an active one has made analytical chemistry obtain the importance which it has today. In addition, as a consequence of this change in concept, the number of steps of TAP has increased.

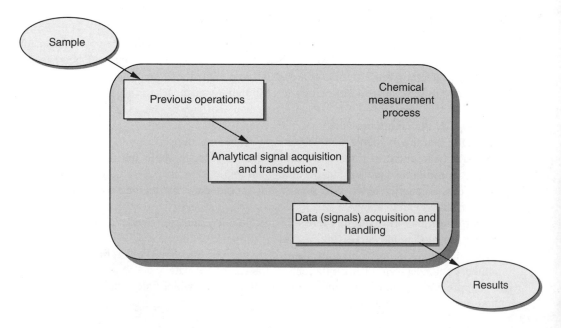

FIGURE 2.2 Scheme of the main steps of a chemical measurement process (CMP). (Reprinted from Valcárcel, M., *Principios de Química Analítica*, chaps. 4 and 7, Springer-Verlag Ibérica, Berlin, pp. 175–239 and 337–365, 1999. With permission.)

Many important authors are still supplying new points of view about TAP. As a compromise to all of them, we suggest that TAP consists of the following essentially steps:

1. general statement of the socio-economical or technical-scientific problem (client).
2. specific statement of the analytical problem and definition of objective(s) (client ↔ analyst).
3. selection of procedure (analyst).
4. sampling (client + analyst).
5. sample preparation (analyst).
6. measurement (analyst).
7. data handling and evaluation (analyst).
8. interpretation of results and suggestion of conclusions (analyst).
9. report (analyst ↔ client).

Steps four to eight could be considered as CMP.

This division of steps of the TAP is only a general approach. This general scheme has some exceptions, since in some cases the boundaries between steps are not clearly defined. For example, a chromatograph combines three of these steps, numbers five, six and seven, since it separates different analytes from interfering substances, has a detector and, nowadays, has a computer to acquire and handle signals and data and to control the instrument. In addition, in some cases, one or more steps could be removed as a function of the nature of problem studied or the information needed. However, it must be emphasized that all of the steps determine the quality of the analysis.

B. SAMPLE PREPARATION

Sample preparation is very important when a chromatographic analysis has to be done. This can be achieved by using a wide number of techniques or procedures. Many of these have not changed over the last 50 years and are as simple as filtration or centrifugation. Others have developed extensively in the last 10 years.

One concept of a sample preparation method is to convert a real matrix to a sample format which is suitable for analysis. According to this definition there are some steps, after sampling, which attempt to adapt the sample to the measurement step (i.e., second step of CMP). To this end, the physical state of the sample must be adequate for the requirements of the instrument used in testing. Figure 2.3 shows a scheme with some of the available possibilities.

Most samples have a complex matrix and, hence, analytes will coexist with other species. Interfering species must be removed using a suitable pretreatment. At this point, it is important to evaluate if the chromatographic techniques must be considered in sample preparation techniques or not. The authors' opinion is that, in the definition given above, chromatographic techniques should be considered in the sample preparation step of TAP since chromatographic techniques are, in themselves, separation techniques. On the other hand, sometimes the concentration of the species of interest is lower or higher than the limits of the range of concentrations required by the selected analytical technique and in these cases one additional step of preconcentration or dilution could be needed. In some cases, it may be necessary to perform a pretreatment in order to ensure that all the analyte is in the desired chemical forms. These transformation types are very common in titrimetric methods of analysis based on redox equilibria. In these methods, sometimes a pretreatment step of oxidation or reduction is needed in order to ensure that all the analyte will be in the same oxidation state before the analysis. In other cases, the chemical form of analyte cannot be detected with the selected analytical technique or its sensitivity is too low. Frequently, all these problems disappear if the analyte is transformed to a new product which can be detected (e.g., derivatization reactions).

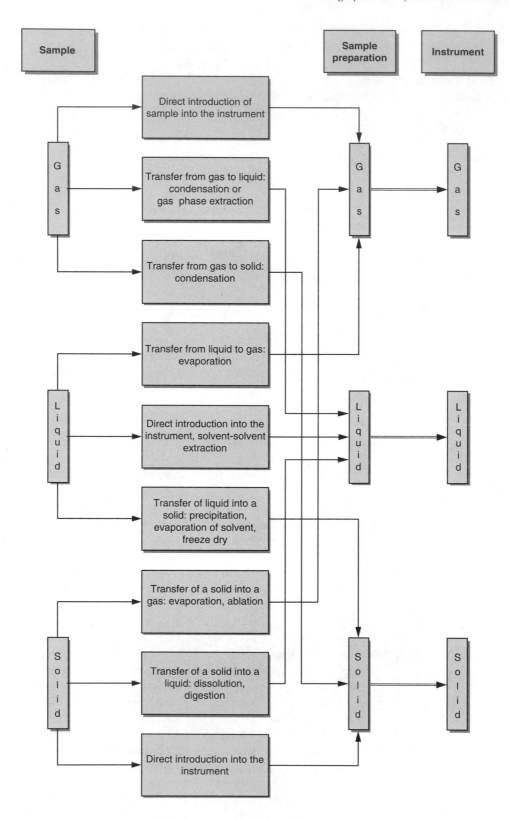

FIGURE 2.3 Pathways from sample to instrument.

Due to the great number of problems studied and the great number of matrices available it is impossible to make a detailed scheme of all the procedures for sample preparation. However, it is possible to summarize all the sample preparation methods with three principle objectives:

- *Sample conditioning:* Adapt the physical or chemical state to the requirements of the instrument (Figure 2.3).
- *Removal of interfering species:* Masking or separation techniques (e.g., adsorption, absorption, lixiviation, supercritical fluid (SF) extraction, dialysis, liquid–liquid extraction (LLE), solid phase extraction (SPE), precipitation, etc.).
- *Additional operations:* Dilution, preconcentration, chemical transformations and derivatization, etc.

Most environmental samples require some pretreatment before injection into a chromatograph. A search of works published between 1992 and 2002 was made using the three keywords "sample preparation," "chromatography" and "environmental analysis," with the result shown in Figure 2.4. The conclusion is that sample preparation is a topic of growing interest since the numbers of published papers have continuously increased during the last decade. Only two discontinuities on the growing trend are observed in the years 1997 and 1999, but the overall trend is increasing. Figure 2.5 shows the works published in this decade grouped by sample nature (i.e., soil, water, air and waste). From Figure 2.5a, it could be concluded that the greatest number of papers are devoted to water preparation, followed by soil, air and, finally, waste. Water sample preparation has significantly increased during recent years but less significant is the increase of papers published about soils and air pretreatment (Figure 2.5b). Finally, Figure 2.6 shows the number of papers published about one sample preparation topic during the studied decade. In this figure only sample preparation methods with a number of publications higher than 20 are shown. Derivatization, LLE, SPE, solid–liquid extraction, enrichment and solid phase microextraction (SPME) are the most popular sample preparation techniques in environmental applications of chromatography whereas field flow fractionation (FFF), biosorption, stir-bar extraction, grinding, ultrafiltration, wet digestion, pervaporation, fusion, internal standard addition and lyophilization are less used.

Nowadays, when separation methods can resolve complex mixtures of almost every matrix, from gases to biological macromolecules, and detection limits are down to femtograms or below, the whole advanced analytical process can still be wasted if an unsuitable nonchromatographic sample preparation method has been employed before the sample reaches the chromatograph.

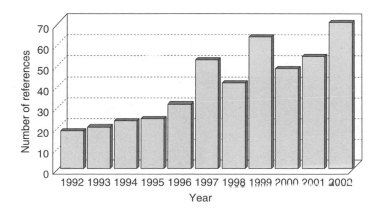

FIGURE 2.4 Number of references containing the three concepts stated in the text during the period 1992 to 2002. (*Source*: Scifinder; date 22/07/2003.)

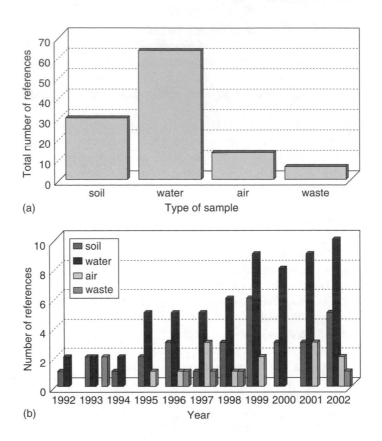

(a)

(b)

FIGURE 2.5 (a) Number of references containing the three concepts stated in the text during the period 1992 to 2002 according to sample type. (b) Number of references containing the three concepts stated in the text during the period 1992 to 2002 as sorted by sample type and year. (*Source*: Scifinder; date 22/07/2003.)

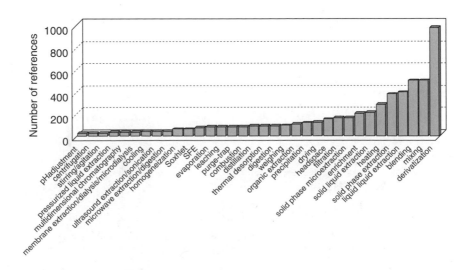

FIGURE 2.6 Number of papers published about one sample preparation method applied to chromatographic analysis of environmental samples during the period 1992 to 2002. (*Source*: Scifinder; date 22/07/2003.)

It must be taken into account that usually several sample pretreatment steps are necessary between sampling and placing the prepared sample into the chromatograph. However, the fewer the sample preparation techniques before injection, the better. A clear and optimized sample preparation strategy is necessary to minimize the number of steps because each step represents additional time and is a potential source of errors (contamination, loss of analytes and changes in actual composition).

C. FUTURE TRENDS

Sample preparation is under continuous evolution because it is a very time consuming and difficult stage in the whole analytical process and it is a source of errors. Hence, the recent trend is to try to reduce time and manual work to avoid introducing contaminants and losing analyte in the preparation process.

Samples must be of adequate form for the instrument used in the analysis. The liquid form is the most common way to introduce samples into instruments, such as chromatographs. Solid samples must be put into liquid form or the soluble components must be extracted. A recent trend is to decrease the volume of chemicals used in the sample preparation steps before injection as well as the injected volume itself. Microextraction on a drop[4,5] or small volume[6] is of increasing interest. Of course, the actual volume injected into a chromatograph depends upon the analyte concentration in the prepared samples. The current trends reveal that initial sample volumes for liquid samples are decreasing, less than 100 μl, or even smaller.[7] The use of smaller sample volumes is an indication of the increased sensitivity of analytical instrumentation.

Automation is a common trend in all branches of analytical chemistry. Increased sample loads should favor laboratory automation. Surprisingly, however, in a recent survey of trends in sample preparation for chromatography, a significant number of sample pretreatment users consider automation unnecessary mainly because, typically, the number of concurrent samples is small. Autosamplers are the most commonly used.[8] Strictly speaking, autosamplers only allow direct injection of liquid samples but do not perform automated sample preparation functions. Full laboratory robots which can automate many manual tasks are found in many high-throughput laboratories. However, it seems that the use of full laboratory robots is decreasing.[8] One area of automation which has gained acceptance is the use of dedicated sample preparation workstations.

Several sample preparation techniques have emerged during the last years. Among them, molecularly imprinted polymers (MIPs) and immunosorbents are new-selective stationary phases and sorbents for SPE.[9–13] High affinity and high selectivity are obtained because MIPs show three dimensional networks which have a "memory" of the shape and the functional group positioning of the template molecule and immunosorbents use natural antigen–antibody interactions.

In the search for more economical and environmentally friendly extraction solvents (green chemistry), sub- and super-critical water have been suggested for extraction and chromatographic separation.[14–18] In this way, both steps can be achieved without the need for organic solvents at any stage. When pressure and temperature are gradually increased water moves from subcritical to supercritical conditions and this change of conditions decreases the polarity of water. This property is very interesting for sequential extractions/elution of analytes of different polarities. In the last few years a lot of attention has been paid to solventless sample preparation techniques based upon sorptive extraction. Among these is the stir-bar sorptive extraction. Stir-bars are coated with a layer of polydimethylsiloxane and then used to stir aqueous samples, thereby extracting and enriching solutes into the polydimethylsiloxane coating. This sorptive extraction phase is the same as that used in SPME but the mass of the extraction phase available in the stir-bar extraction is between 50 to 250 times greater than in SPME which results in higher recoveries and sample capacity. In addition, stir-bar sorptive extraction (SBSE) provides much higher throughput with better sensitivity, reproducibility and accuracy.[19]

Recently, new separation principles have been introduced and although these are very promising, they have not been extensively used for environmental analysis. Among them are FFF,[20] pervaporation[21] and biosorption.[22-24] All of them are easy to handle and not very expensive. In addition, FFF has very simple fundamental principles while pervaporation is very prone to automation and miniaturization. Biosorption is especially interesting for metal concentration because biosorbents can accumulate up to 25% of their dry weight in heavy metals. Some of the biosorbents are waste by-products of large scale industrial fermentations or certain abundant seaweeds. Analytes are easily released from the biosorbent and the biosorbent is regenerated for subsequent reuse.[24]

Coupled-column separations or multidimensional chromatography can be considered as a sample preparation form, as one column is used to derive fractions for the second column.[25] It provides a two dimensional separation in which sample substances are distributed over a retention plane formed by the operation of two independent columns. This type of two dimensional based separation method is more powerful than a single dimensional based one. A retention plane has more peak capacity than a retention line and so can accommodate much more complex mixtures. Component identification is more reliable because each substance has two identifying retention measures rather than one.[26] These type of combinations offer high selectivity and high sensitivity, and could be used with less expensive and more robust detectors (e.g., flame ionization).[27]

Miniaturization has become a dominant trend in Analytical Chemistry during the last few years.[28-30] Miniaturization has some advantages for sample preparation (e.g., it is simple, rapid, inexpensive, minimizes sample handling steps and is more precise and sensitive). In addition, it minimizes sample quantities and use of expensive and/or toxic reagents. Developments in microscale sample preparation have been made in nonchromatographic (SPME, single-drop microextraction and small stir-bar sorptive extraction) and chromatographic (gas and liquid chromatography and capillary electroseparation) separation methods.[4,5,31-35] New polymeric filaments, small sections of packed tubes, miniaturized cartridge extraction and small pieces of polymeric membrane have been used in SPME. With column or packing material miniaturization more than 200,000 theoretical plates may be obtained.[36] A further step in this direction is the so-called lab-on-a-chip and more specifically on-chip sample preparation as a part of the concept of miniaturized total analysis systems (μTAS).[28-30]

II. SAMPLE PREPARATION METHODOLOGIES

Although chromatography is a separating technique it can not be expected to completely separate all components of a complex sample. Very often it is necessary to prepare the sample, filtrate, concentrate, clean up, etc. before chromatography. These steps vary depending on the nature of the sample. In this section a brief but recent view of these techniques is shown together with some specific applications to environmental analysis.

One important and general step in sample treatment, especially when liquid chromatography is to be used, is filtration of the solution. A first filtration may be necessary to separate large particulate matter from solvent since this may physically interfere with extractions in later stages. The final filtration before chromatography injection uses 0.45 μm to 0.20 μm or smaller disposable filters to prevent small particulate matter getting into the chromatography column and damaging it. Different types of filters can be used including filter membranes, centrifugal filters, and syringe filters. The nature of the filters depends on the type of solvent used. The commercially available materials are: cellulose acetate (aqueous solvents), polypropylene (aqueous samples), nylon (aqueous and most solvent based samples) and polytetrafluoroethylene (organic based, highly acidic or basic solvents) among others. When commercial syringe filters are used, different diameter sizes are offered depending on the volume of sample to filter, which is important when small samples have to be filtered.

An alternative to normal filtration is ultrafiltration. In this case, pressure is applied to a membrane and molecules smaller than the molecular weight cutoff can pass through while larger molecules are retained. Ultrafiltration can be used for sample concentration or to eliminate higher molecular weight compounds.

A. LIQUID SAMPLE PREPARATION

Depending on the nature of analyte(s) of interest, sample preparation processes will be quite different. In liquid sample preparation, as for water and wastewater analysis, the analytes from matrices can be separated in two different ways:

- by *extracting the analytes into a liquid phase* as in LLE, purge and trap technique, membrane extraction, and single drop extraction.
- or by *trapping the analytes in a solid phase* such as SPE, SPME, and stir-bar extraction.

1. LLE

The classical LLE method is still in use due to the simplicity of the instrumentation, just a separation funnel, and also because of its extensive implementation in official methods (U.S.-EPA methodology, EEC standard methods). As shown in Table 2.1 nearly all U.S.-EPA (United States Environmental Protection Agency) methods for nonvolatile and semivolatile analytes in environmental samples apply LLE even though there is a trend to change this.

When doing a LLE, a given volume of the sample, for example water, is shaken with a given volume of a suitable organic solvent so that the organic micropollutants migrate from the aqueous to the organic phase. Sometimes it is advisable to add a small quantity of sodium chloride to avoid foam formation and so obtain a better separation. Then the organic solvent with the analytes is separated and evaporated to concentrate the sample to a precise volume.

LLE can be done in different ways:

- *Discontinuous liquid extraction.* This is the most traditional extraction method which can be carried out in one or multiple steps.
- *Continuous LLE.* This is applied when the distribution constant is low or when the sample volume is large.
- *Countercurrent extraction.* This is advisable when complex samples with analytes of similar distribution are to be extracted.
- *Online LLE.* This is a dynamic process which allows the extraction of low volume samples and reduces the organic solvent consumption but it has the drawback of instrument complexity.

LLE has several disadvantages such as large volumes of generally toxic organic solvents. With some samples, the initial solvent extraction step results in the formation of an emulsion and hence prolongs the extraction process. A loss of sample frequently occurs during the concentration step and so reduces analyte recovery. To avoid these limitations a considerable interest in developing alternative sample preparation methods has been increasing in the last few years.

2. Liquid Membrane Extraction

Membrane LLE (MLLE) or sorptive membrane extraction (SME) is an alternative to LLE, and is an extension of the LLE principles. A membrane is used as a selective filter of the analytes, limiting diffusion between two solutions or as an active membrane in which its chemical

TABLE 2.1
U.S.-EPA Methods for Environmental Samples

Analyte Type	EPA Method Reference	Common Sample Preparation	Detector Types	Sample Matrix
Volatiles				
Trihalomethanes	501	Purge and trap, direct injection, headspace	ECD, ELCD	Drinking water
VOCs	502.2, 8021, CLP-volatiles	Purge and trap, direct injection, headspace	PID, ECD	Drinking water, waste water, solid wastes
Purgeable halogenated organics	601, 8010	Purge and trap, headspace for screening	PID, ECD	Waste water, solid wastes
Purgeable aromatic organics	503.1, 602, 8020	Purge and headspace for screening	PID	Drinking water, trap, waste water, solid wastes
VOCs using MSD	524.2, 624, 8240, 8260, CLP-VOCs	Purge and trap, direct injection, headspace	MSD	Drinking water, waste water, solid wastes
VOCs using 5973 MSD	524.2, 624, 8240, 8260, CLP-VOCs	Purge and trap, direct injection, headspace	MSD (5973)	Drinking water, waste water, solid wastes
EDB and DBCP	504.1, 8011	Microextraction with hexane	ECD	Drinking water, solid wastes
Acrylonitrile and acrolein	603, 8015, 8031	Purge and trap, liquid extraction, sonication	FID, NPD	Waste water, solid wastes
Semivolatiles				
Semivolatile organic compounds	525, 625, 8270	Liquid extraction, sonication, Soxhlet extraction, SPE	MSD	Drinking water, waste water, solid wastes
Phenols	528, 604, 8040, 8041	Liquid extraction, sonication, Soxhlet extraction, derivatization	ECD, FID	Waste water, solid wastes
Phthalate esters	506, 606, 8060, 8061	Liquid extraction, sonication, Soxhlet extraction, SPE	ECD, FID	Drinking water, waste water, solid wastes
Benzidines	605	Liquid extraction	ECD	Waste water
Nitrosamines	607, 8070	Liquid extraction, sonication, Soxhlet extraction, SPE	NPD	Waste water, solid wastes
Nitroaromatics and isophorone	609, 8090	Liquid extraction, sonication, Soxhlet extraction, SPE	ECD, FID	Waste water, solid wastes

Polynuclear aromatic hydrocarbons (PAHs)	610, 8100	Liquid extraction, sonication, Soxhlet extraction, SPE	FID	Waste water, solid wastes
Chlorinated hydrocarbons	612, 8120, 8121	Liquid extraction, sonication, Soxhlet extraction, SFE	ECD	Waste water, solid wastes
Chlorinated disinfection byproducts	551, 551.1A	Liquid extraction, derivatization	ECD	Drinking water
Halogenated acetic acids	552, 552.1, 552.2	Liquid extraction, derivatization	ECD	Drinking water
Pesticides, Herbicides and PCBs				
Organochlorine pesticides and PCBs	508.1, 608, 8081A, 8082, CLP-pesticides	Liquid extraction, sonication, Soxhlet extraction. SPE	ECD	Drinking water, waste water, solid wastes
Phenoxy acid herbicides	515, 615, 8150, 8151	Liquid extraction, sonication, Soxhlet extraction, SPE	ECD	Drinking water, waste water, solid wastes
N- and P-containing pesticides and herbicides	507, 614, 619, 622, 8140, 8141A	Liquid extraction, sonication, Soxhlet extraction, SPE	NPD, ELCD, FPD, MSD	Drinking water, waste water, solid wastes

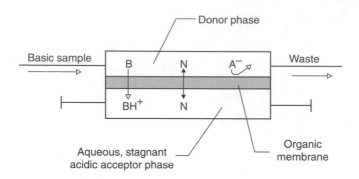

FIGURE 2.7 Schematic description of a membrane extraction process. (Reprinted from Jönsson, J. A. and Mathiasson, L., *Trends Anal. Chem.*, 18, 318–325, 1999. With permission from Elsevier.)

structure determines the selectivity of sample transference. It is especially suitable for nonpolar substances.

The methodology is based on a three-phase system with an organic phase sandwiched between two aqueous phases (a donor phase and an acceptor phase (Figure 2.7)).[37] The organic phase is immobilized in a porous hydrophobic membrane. Figure 2.7 shows, as an example, the extraction process of amines in which the sample is pumped through the donor channel and the uncharged amines (B) are extracted by the organic membrane phase. An amine molecule which has diffused through the membrane will be immediately protonated at the membrane acceptor, and thus prevented from reentering the membrane. The result is a completed transport of the amines from the donor to the acceptor phase and a concentration of the sample.

This type of extraction has a number of advantages:

high selectivity and very clean extracts,
high concentration enrichment, and it can be easily connected to analytical instruments (chromatographs, spectrographs) and
there is no need for organic eluents.

A variant of membrane extraction is called *microporous membrane LLE (MMLLE)*. Here the acceptor phase is an organic solvent, instead of an aqueous phase, which also fills the pores of the hydrophobic membrane. The system consists of two blocks, usually of polytetrafluoroethylene (PTFE) with identical grooves. The acceptor phase can be stagnant or flowing. If the acceptor is stagnant, the only driving force for the mass transfer is the attainment of a distribution equilibrium between the aqueous and organic phases. Microporous membrane extraction is complementary to the membrane extraction. It is applicable to hydrophobic compounds, preferably uncharged compounds. The analytes end up in an organic phase, not in an aqueous phase, so it is more easily interfaced to Gas Chromatography (GC) and normal phase HPLC, while membrane extraction is more compatible with reversed-HPLC.

Another recent and new variant of membrane liquid extraction was introduced by Cantwell,[38] and is known as liquid–liquid–liquid microextraction (LLLME). In this case three liquid phases are used — a_1 is the water sample where pH is adjusted to deionize the compounds, a_2 the acceptor aqueous phase with pH adjusted to ionize the compounds and an organic liquid phase (o), 40 μl or 80 μl of *n*-octane, which is layered over the donor phase. In this case no physical membrane is needed because the organic layer has this function. This modification is an appropriate application for preconcentration and purification for polar analytes in water samples such as amines.[39]

3. Single Drop Extraction

Single drop extraction (SDE) or liquid-phase microextraction (LPME) is a recently developed microscale extraction method. In this method a single liquid drop is used as a collection phase. Small volumes of organic solvent (from 0.5 μl to 2.5 μl) are used. The collection phase must have a sufficiently high surface tension to form a drop which can be exposed to the analyte solution.[40] When the extraction is finished, the single drop is injected into the GC. A scheme of such a device is shown in Figure 2.8.

SDE can be used in static and dynamic modes. When working in the static mode, steps in the extraction process are: (a) the magnetic stirrer is switched on to agitate the aqueous sample solution; (b) a specific volume of organic solvent is drawn into the syringe with the needle tip out of the solution and the plunger is depressed by 1 μl to 2 μl; (c) the needle is then inserted through the septum of the sample vial and immersed into the aqueous sample; (d) the plunger is depressed to expose the organic drop to the stirred aqueous solution for a period of time; (e) the drop is retracted into the microsyringe; and (f) finally, the organic solvent drop is transferred into a vial and subsequently injected in a chromatograph. In the dynamic mode all steps are done automatically. Static and dynamic modes were compared to extract polyaromatic hydrocarbons (PAHs) in water obtaining higher concentrations in the dynamic mode.[41]

SDE avoids the problems of solvent evaporation as in LLE. It is a fast, inexpensive, and simple method. However, the extraction is not exhaustive.

4. The Purge and Trap Technique

This method is recommended for extracting volatile organic components from water and other samples. Volatile components dissolved in water are expelled by flushing with an inert gas. These components are trapped in a cryogenic or a solid-phase trap at room temperature. Afterwards, the analytes are released by thermal desorption and transferred to the GC column. Another possibility is to elute these analytes from the solid-phase trap with an adequate liquid chromatographic (LC) solvent. This extraction method is highly sensitive and has the additional advantage of being a nonsolvent technique. The detection limit is in the range of ng/l. However, it has some drawbacks.

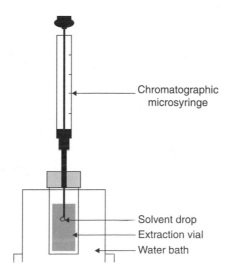

Chromatographic
microsyringe

Solvent drop
Extraction vial
Water bath

FIGURE 2.8 Schematic of a single-drop extraction apparatus. (Reprinted from Buszewski, B. and Ligor, T., *LC–GC Europe*, 15, 92–97, 2002. With permission from Advanstar Communications.)

It is quite time-consuming and labor intensive, particularly when many samples are involved. In addition it requires complex instrumentation.

An online sampling system connected through a purge-and-trap injector to a GC-FID for automatic sampling and analysis of volatile organic compounds (VOCs) in river water has recently been developed. A good review of this sample preparation technique applied to estuarine water has been written by Huybrechts et al.[42] In the review, advantages and disadvantages and novel developments of the technique are discussed.

5. SPE

Another sample preparation possibility is the method based on *trapping the analytes*. Using this methodology the analyte is trapped in an adsorbent material, and then washed off with a minimal volume of solvent. The most popular trapping method is SPE.

The history of SPE dates from more than fifty years ago, with granulated active carbons previously used in water treatment technologies. The pioneer work was conducted by the U.S. Public Health Service (Cincinnati, OH). After that, other approaches were investigated with petroleum pollutants, insecticides and VOCs. Disadvantages such as irreversible adsorption, analyte reactions on the activated carbon surface and low recoveries started research in new sorbent materials. See the historical review of SPE by I. Liska.[43]

SPE began to be extensively used in sample preparation processes in the 1970s but the great breakthrough in SPE occurred over the past seven or eight years with many improvements in formats, automation and introduction of new phases. In fact, SPE is currently accepted as an alternative extraction method to LLE for 22 of the official methods for the U.S. EPA.

Various kinds of commercial formats of adsorbent devices can be employed: syringe-barrels, cartridges and disks (Figure 2.9). The syringe barrel column is the most popular SPE configuration. The sorbent bed is held in place by porous polyethylene frits and the syringe barrel is typically manufactured from high purity materials. Cartridges have no reservoir capacity and are fitted with both male and female luer lock fittings. Disk membranes consist of sorbent particles enmeshed in a porous solid support. The selection of one of these formats will depend on the application and the quantity of sample available.

Three different strategies can be used to pass the analytes through the adsorbents, gravity, pressure or centrifugation. The first step in SPE is conditioning the adsorbent with an adequate eluent. Then the sample is passed through the column, where the analytes are retained. After that, some interfering species are eliminated by passing a well-selected solvent through the column (clean up step), but avoiding elution of the analytes. Finally, one or more (in the case of fractionating elution) eluent(s) carry away the analytes from the adsorbent obtaining an interference free, concentrated and isolated extract of a target pollutant group. The general steps involved in SPE are shown in Figure 2.10.

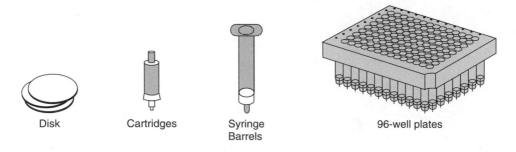

Disk   Cartridges   Syringe   96-well plates
                                                  Barrels

FIGURE 2.9 Typical SPE commercial product formats. (Reprinted with permission from Phenomenex.)

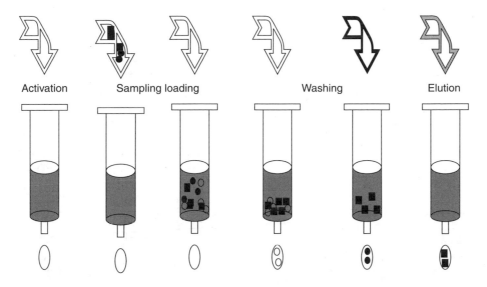

FIGURE 2.10 Steps involved in SPE. ■, analyte ○ and ●, interfering components. (Reprinted with permission from Phenomenex.)

The nature of the packing material differs depending on the analytes to be separated, so it is very important to have a good selection of materials and elution solvents. A general scheme of optimizing steps is shown in Figure 2.11. These steps depend on the nature of the solution.

To meet the varied needs of contemporary applications, there are an over-increasing range of sorbing phases commercially available (Table 2.2).[44,45] The nature of the sorbing phases are: Silica phases (normal phases and reversed phases) and ion-exchange phases.

- *Chemically bonded reversed-phase silicas*: C_{18} and to a lesser extent C_8 silica cartridges are packed with the same stationary phases as in LC columns, but with a larger particle size.[44] Although C_{18} sorbent with high carbon loading and some residual silanol are best suited, they are not able to extract polar compounds from large sample volumes. Therefore C_{18} silica is rarely used for multiresidue environmental analyses.
- *Carbon based sorbents*: The most widely used carbon based sorbents (CBS) are obtained from carbon blacks heated at high temperature (2700 to 3000°C) under inert atmosphere. This adsorbent is nonspecific and nonporous with a specific surface of about 100 cm^2/g. The adsorption is not only due to hydrophobic interactions, but also to specific electronic interactions with the analyte. The main disadvantage of this material is the almost irreversible adsorption of some analytes, which could be dissolved by carrying out reverse elution of the cartridges. Graphitized carbon blacks (GCBs) have higher efficiency than C_{18} for trapping polar pesticides.[44]
- *Polymeric adsorbents*: The more extensively used polymeric adsorbents are styrene-divinylbenzene co-polymers, with a hydrophobic surface. With these adsorbents the limitations of linked silicas are overcome, for example, more stability at higher pH range. Their efficiency depends on particle size, surface area, porous diameter, porous volume, reticulation degree and distribution of particle size.

Alternatively, in the last few years more analyte specific phases such as immunoadsorbents, and MIPs have been developed with the aim of obtaining better selectivity.

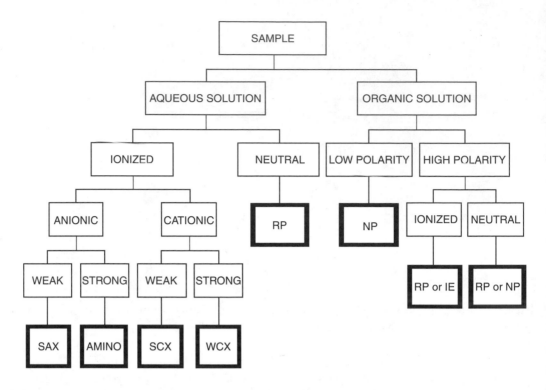

FIGURE 2.11 Method selection guide for the isolation of organic compounds from solution. SAX: strong anionic exchanger; SCX: strong cationic exchanger; WCX: weak cationic exchanger; RP: reversed phase sampling conditions; NP: normal phase sampling conditions; and IE: ion exchange sampling conditions. (Reprinted from Poole, C. F., *Trends Anal. Chem.*, 22, 362–373, 2003. With permission from Elsevier.)

TABLE 2.2
Typical SPE Sorbents

Support	Phase	Surface Modification	Surface Area Range (m²/g)	Particle Size Range (μ)	Pore Size Range (Å)	Retention Mechanism
Silica	C18	Octadecyl (polymeric)	450–550	50–60	65–75	RP
Silica	C18	Octadecyl (monomeric)	280–320	50–60	120–140	RP
Silica	C8	Octyl	450–550	50–60	60–75	RP
Silica	PH	Phenyl	450–550	50–60	60–75	RP
Silica	CN	Cyanobutyl	450–550	50–60	60–75	RP + NP
Silica	NH₂	Aminopropyl	450–550	50–60	60–75	NP + IEX
Silica	SCX	Phenylsulfonic acid	450–550	50–60	60–75	IEX
Silica	SAX	Me₂(propyl)ammonium Cl⁻	450–550	50–60	60–75	IEX
Silica	Silica	Acidic, neutral	250–600	50–60	60–75	NP
Alumina	Alumina	Acidic, neutral, basic	100–150	50–300	100–120	NP
Fluorisil	Fluorisil	None	300–600	50–200	60–80	NP
Polymer	SDB	None	500–1000	75–150	50–300	RP

- *Immunoadsorbents*. These are based on selective and reversible antigen-antibody interactions, in order to trap structurally related pollutants. These more selective sorbents extract trace levels of polar organic compounds from samples with higher amounts of interfering species. Although, up to now there are only a few immunosorbents available on the market. The main feature of these SPE materials is the selectivity. Then, immunoextraction exhibits a high degree of clean up, so the clean up step is faster and minimizes or avoids the use of an organic solvent.
- *Molecular imprinted polymers (MIPs)*. These consist of polymers with specific recognition sites for certain molecules. These polymers are synthesized using a template molecule and linking the monomers around it. Afterwards the template molecule is removed. The recognition sites are built so that the analyte is adapted to functional groups of the polymeric matrix. These sorbents show high affinity and selectivity as immunosorbents, and, moreover, have other advantages:
  - lower cost and preparation time,
  - higher preparation repeatability,
  - higher sample capacity and material stability.
- Other sorbents include mixed mode sorbents, normal-phase SPE sorbents or restricted access matrix sorbents.

SPE has many advantages over the more traditional sample preparation techniques. The main benefits are:

a reduction of organic solvent,
ease of automation,
a higher concentration of analytes, and
extracts are normally highly purified.

In the last few years, in order to reduce the cost and time of environmental monitoring there is a trend to automate the SPE process.[46] Now commercial workstations and extraction plates allowing numerous samples (up to 96) to be prepared simultaneously are available (Figure 2.9). Approaches to computer-aided method development are replacing tedious trial-and-error procedures with fast simulations based on suitable kinetic and retention models.

6. SPME

SPME was first reported by Pawliszyn and coworkers in 1990.[47,48] This method can be used for the extraction of organic compounds from aqueous samples or from a gas phase (headspace-SPME). SPME is by nature an equilibrium technique, based on the partitioning of the solutes between the silicone phase and the aqueous and/or gas matrix. The device employed for SPME is formed by a thin, fused silica fiber with a chemically bonded organic film on its surface (Figure 2.12).

SPME consists of two steps: extraction and desorption. In the extraction process the fiber is immersed into the sample with a syringe, vigorous stirring is applied, and the organic micropollutants are retained in the fiber depending on their distribution coefficients. Then, using the holder, the fiber is transferred to the analytical instrument for desorption, separation and quantification. The method has been automated and commercial systems are available which will extract, agitate, and inject the sample into a GC system. In HPLC the sample is extracted directly into the eluent stream rather than thermally desorbed.[49]

Various means of extraction exist. Direct extraction, headspace extraction and membrane extraction (Figure 2.13). In the direct process the fiber is introduced directly in the sample solution and so organic analytes are retained in the fiber. This extraction mode is specially suited for separating low volatile analytes.

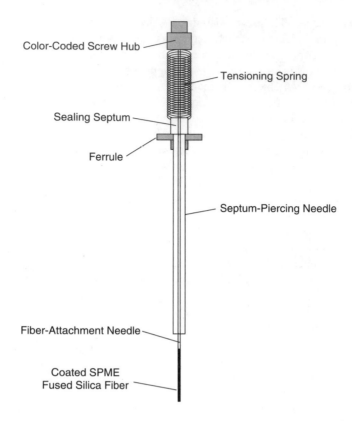

FIGURE 2.12 SPME fiber assembly details.

In the case of headspace extraction (HS–SPME) a fiber in the needle tip of a gas chromatographic microsyringe is exposed to the headspace above a sample. Volatile compounds are extracted and concentrated in the fiber. Next, the fiber is retracted into the microsyringe and injected directly into the gas chromatograph. This extraction technique has been successfully

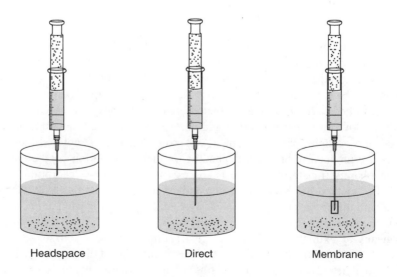

FIGURE 2.13 Different modes of extraction in SPME.

applied to determine volatile compounds as BTEX in water samples.[50] Because low volatile analytes are not absorbed in the fiber the extraction is more specific.

Another possibility is the use of a membrane to protect the fiber. This system is used when very complex and dirty samples have to be extracted and headspace SPME can not be applied.

The advantages of SPME over other classical extraction techniques are:

high sensitivity and low detection limit,
good precision,
wide selection of available solvents,
low cost,
simplicity and ease of use,
minimal solvent use,
short preconcentration time,
the possibility of automation and minimal sample preparation.

The drawbacks are that the fiber is fragile and can be damaged by a build-up of nonvolatile materials from the samples; the extraction process can be relatively slow because it relies on sufficient stirring or diffusion to bring the analytes into the fiber.

The development of new polymeric coatings will improve the selectivity and sensitivity of SPME for different classes of analytes. Today's commercially available fiber coatings are shown in Table 2.3 with some applications.

Many parameters can affect the absorption process and, hence, the amount of analyte extracted in the fiber, for example, the characteristics of the coating, the temperature and time of extraction, the addition of salt or an organic solvent to the sample, pH modification, agitation of the sample and the sample volume. To know more about this technique and how to optimize the sample preparation process see Refs. 48, 49, 51, and 52.

TABLE 2.3
Fiber Coatings Commercially Available for SPME Use, by Polarity

Fiber Coating	Film Thickness (μm)	Maximum Temperature (°C)	Applications
Non Polar Fibers			
Polydimethylsiloxane (PDMS)	100	280	Non polar compounds (VOCs, PAHs...)
	30	280	
	7	340	
Polar Fibers			
Polyacrylate (PA)	85	320	Polar compounds (pesticides and phenols)
CW–DVB	65	265	Polar organic compounds
Carbowax–templated resin (CW–TPR)	50	—	Anionic surfactants
Bi-Polar Fibers			
PDMS–DVB	65	270	Aromatic hydrocarbons, solvents...
	60	—	
Carboxen–polydimethylsiloxane (carboxen–PDMS)	75	320	Hydrocarbons and VOCs
Divinylbenzene–carboxen–PDMS	30	300	—
	55		

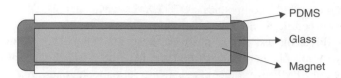

FIGURE 2.14 Scheme of a stir bar.

7. Stir-Bar Sorptive Extraction (SBSE)

A quite recent development in trap extraction, SBSE, attempts to avoid the drawbacks of other techniques. It was introduced by Baltussen et al.[53] The extraction device consists of a magnetic stir-bar coated with an adsorbent layer (Figure 2.14). The surface area of the stir-bar is greater than the fiber in the SPME (0.5 μl coating of polydimethylsiloxane, PDMS, in a fiber versus 50 μl to 300 μl in a stir-bar) allowing a higher extraction. This bar rotates in a sample (typically 10 ml to 25 ml) in a vial, for a certain time. After extraction, the stir-bar is put in a glass thermal desorption tube which is placed in a thermal desorption unit[54] and is then thermally desorbed (Figure 2.15). Alternatively, liquid desorption can be used. It is a very easy methodology, but the problem is that it is difficult to

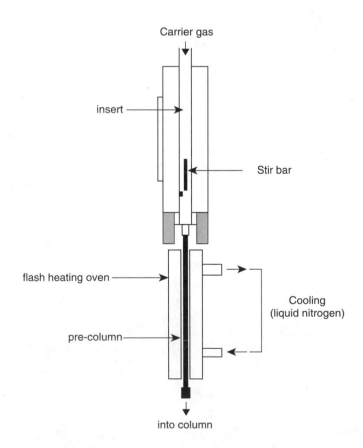

FIGURE 2.15 Schematic representation of the thermal desorption unit in SBSE. (Reprinted from Vercauteren, J., Peres, C., Devos, C., Sandra, P., Vanhaecke, F., and Moens, L., *Anal. Chem.*, 73, 1509–1514, 2001. With permission from American Chemical Society.)

automate the removal of the stir-bar from the sample matrix. PDMS coated stir-bars in different sizes are now commercially available, for example, from Twister, Gerstel GMBH.

Different applications of this methodology have been successfully carried out, for example, for PAHs,[55] pesticides, and PCBs[56] in water samples.

B. Solid Sample Preparation

Solid samples, such as soils and solid wastes which are quite insoluble in solvents, have to be submitted to an analyte extraction process. In some cases, such as in elemental analysis, it is adequate to digest the sample in a strong acid. However, in other cases, the analyte may be destroyed under strongly acidic conditions and alternative extraction methods have to be used. The most important and internationally used techniques are discussed in the following sections.

1. Soxhlet Extraction

The most widely used solid sample extraction method is *Soxhlet* extraction. This technique was introduced by Franz Von Soxhlet almost a century ago. Soxhlet extraction is described in U.S.-EPA method 3540[57] as "a procedure for extracting nonvolatile and semivolatile organic compounds from solids such as soils, sludges and wastes."

This is a classical extraction method in which samples are placed in cellulose thimbles covered with cotton-wool, as shown in Figure 2.16. A certain quantity of solvent is added into the system and is heated to an appropriate temperature which allows the solvent to boil. Then, the solvent vapor condenses in a condenser and drops of solvent fall down onto solid sample particles. Once the side-vessel is full, the solvent runs back to the bottom flask. This process is repeated for several hours until adequate recovery of the analytes is obtained.

In order to render a solid matrix permeable to the extracting solvent, some precautions have to be taken. If the sample matrix contains water, it is recommended to use a mixture of solvent miscible in water or it has to be air-dried or mixed with a drying agent (e.g., anhydrous sodium sulfate) before the extraction.

Nowadays there are various improved commercially available Soxhlet extraction systems, which allow faster extractions and the preparation of several samples at a time. These modern systems are more efficient than previous versions. They differ from the conventional Soxhlet method in that clean solvent is always in contact with the sample. This makes the extraction process more effective and shorter. Stages in the extraction process are as follows:

- The extraction thimble containing the sample is immersed in boiling solvent. The sample remains in contact with the solvent and the analytes are extracted.
- The thimble is elevated above the solvent and extracted in the normal mode to free the sample of entrained extract.
- The solvent is evaporated and separated from the sample and the extract in the same device.

Soxhlet extraction is very simple and easy to use but it is solvent consuming and requires laboratory facilities in which flammable and toxic solvents can be used safely. Cooling the condenser of the Soxhlet apparatus requires a constant supply of cooling water.

A more recent development applies microwave heating to the Soxhlet extraction apparatus (Figure 2.17). This modification reduces extraction time of PCBs from contaminated soils (70 min versus 24 h) and the organic solvent disposal, as 75% to 80% of the extractant is recycled.[58]

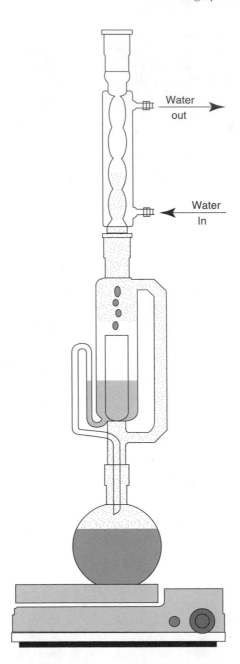

FIGURE 2.16 Scheme of a conventional Soxhlet extractor.

2. Pressurized Liquid Extraction (PLE)

PLE, pressurized fluid extraction (PFE), or accelerated solvent extraction (ASE) consists of a stainless-steel cell in which the sample is placed and kept at the selected temperature and pressure during the extraction, electronically controlled heaters and pumps for solvent delivery and a vial for the collection of the liquid extract. A schematic diagram of a PFE system is shown in Figure 2.18.

The commercial availability of ASE systems is limited nowadays. They are only available from the Dionex Corp., U.S.A. The ASE-200 model can reach temperatures up to 200°C and pressures up to 21 MPa in extraction cells of 1 ml, 5 ml, 11 ml, 22 ml or 33 ml.

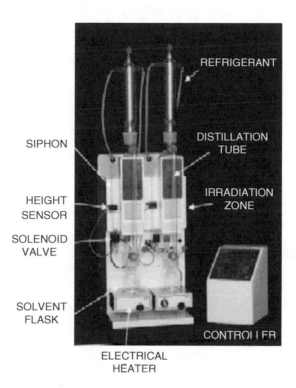

FIGURE 2.17 Automated focused microwave-assisted Soxhlet extractor. (Reprinted from Luque-García, J. L. and Luque de Castro, M. D., *J. Chromatogr. A*, 998, 21–29, 2003. With permission from Elsevier.)

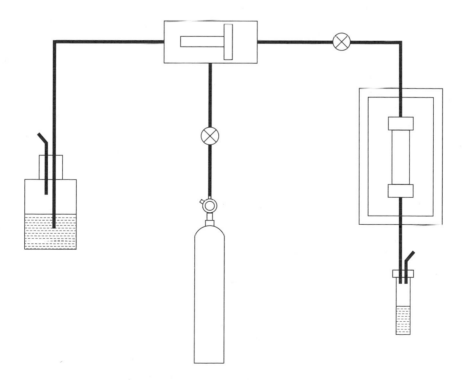

FIGURE 2.18 Scheme of an ASE system. (Reprinted with permission from Dionex.)

Employing a carousel, this extractor can operate with up to 24 extraction vessels. The speediness is a remarkable advantage over Soxhlet extraction.

Raising the temperature increases diffusion rates, solubility of the analytes and mass transfer, and decreases the viscosity and surface tension of the solvents. These changes improve contact of the analytes with the solvent and enhance the extraction efficiency, which can be achieved more rapidly and with less solvent consumption compared with classical methods. For example, ASE reduces solvent consumption by up to 95% compared to Soxhlet extraction. The only limitation is the thermal stability of the analyte of interest.

ASE can be carried out in static or dynamic mode. Most of the applications found in the literature are in the static mode and, until now, analyte recoveries obtained under dynamic mode are not quantitative.

The variables which affect the PLE process to great extent are the nature and temperature of the extraction solvent and the extraction time, while the pressure and flow rate of the extraction solvent in dynamic mode have little effect on the efficiency of the extraction.

This methodology has been accepted and introduced in the EPA methods for pesticides analysis in soils.[59]

3. Superheated Water Extraction or Subcritical Water Extraction (SWE)

Superheated water is water between 100°C to 374°C under sufficient pressure to keep it in the liquid state. Under these conditions the polarity of the water is lowered by the increased temperature. For that reason this solvent can act as a medium (at low temperatures) to a nonpolar solvent (at higher temperatures) for many analytes, like PAHs[60] or PCBs.[61] The equipment used in this application is laboratory-made because none has been commercially developed yet. Figure 2.19 shows the basic components used for continuous SWE. It consists of two pumps, one for deoxygenated water and one for the selected organic solvent, an oven containing a stainless-steel heating coil and the extraction cell, a stainless-steel cooling coil and a vial for collecting the extracts. The sample is introduced into the extraction cell, which is placed in the oven and the extraction starts by pumping both the water and the organic solvent at their selected flow rates until the selected pressure and

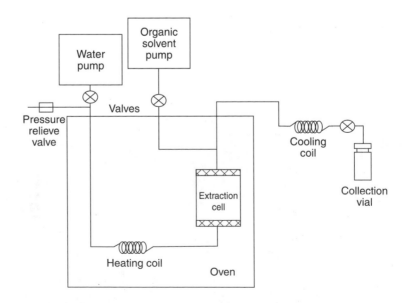

FIGURE 2.19 Schematic of the basic setup for SWE. (Reprinted from Ramos, L., Kristenson, E. M., and Brinkman, U. A. Th., *J. Chromatogr. A*, 975, 3–29, 2002. With permission from Elsevier.)

temperature are reached. Hot water containing the analytes is mixed with the organic solvent via a T junction placed in the oven at the outlet of the extraction cell. Before collection in a vial, the mixture is passed through a cooling coil where the temperature decreases rapidly, water becomes a more polar solvent and the less polar analytes previously dissolved in the SWE are partitioned to the less polar solvent preventing their adsorption to the tubing.

The SWE can be carried out in two modes, at a fixed temperature and pressure or at a variable temperature. In the latter mode, a more selective extraction can be obtained.

Although it can be very selective the main problem of this methodology is a lack of quantitative recovery, for example, recovery tends to be lower than that obtained with Soxhlet extraction.

4. Supercritical Fluid Extraction

a. Introduction

Supercritical fluid chromatography (SFC) was developed at an earlier stage (first demonstrated in 1962) than SF extraction (SFE), which emerged in the mid-1980s as a promising tool to overcome the difficulties of solid sample extractions.[62]

Most solid samples, such as soils, environmental solids, plant materials and polymers are largely insoluble and usually cannot be directly examined. In some cases, it is appropriate to digest the sample in strong acid but in most cases this will destroy the analytes. Most recent methods for extraction of solids aim to reduce the amount of solvent and sample, to decrease the time required for the analysis and to enhance selectivity of extraction. The latter is difficult to achieve as in any extraction process there is a balance between selective and complete extraction. There are two approaches, the use of conventional solvents in more efficient ways or the employment of alternative solvents, such as supercritical fluids, which have a higher diffusion rate. Apart from the higher diffusion rate, an additional reason for the introduction of SF as an extracting agent is the reduction of exposure of laboratory personnel to harmful organic solvents, supercritical fluids being environmentally friendly.

b. Supercritical Fluid Extraction Method

Supercritical fluid extraction is a method which uses a solvent in supercritical conditions as an extracting agent.

i. What is a SF?

A SF is defined as any substance which is above its critical temperature (T_C) and critical pressure (P_C).[63]

The critical region of a fluid begins at its critical point. A fluid has supercritical conditions when its pressure and temperature are above its critical point. A SF may be defined from a phase diagram, in which the regions corresponding to the solid, liquid and gaseous states are shown (Figure 2.20). From this figure the following parameters are defined:

- *Triple point (TP)*. This is the point at which all states (solid, liquid and gas) can coexist in equilibrium.
- *Critical temperature* (T_C). This is the highest temperature at which a gas can be converted to a liquid by an increase of pressure.
- *Critical pressure* (P_C). This is the highest pressure at which a liquid can be converted to a traditional gas by an increase of the liquid temperature.
- *Critical point (CP)*. This is defined by both critical temperature and pressure. Beyond this point, it is not possible to either liquefy or evaporate a liquid, by increasing temperature or pressure. The critical point is a characteristic of each substance.

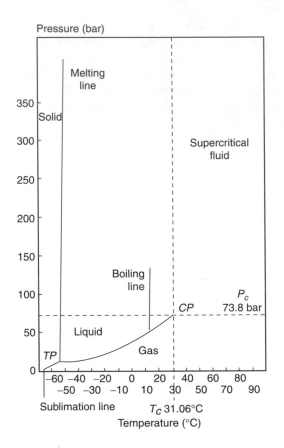

FIGURE 2.20 CO_2 phase diagram.

The region above the critical point, T_C and P_C, is the supercritical region. The fluid in this region is a supercritical fluid. In the supercritical region there is only one phase and it possesses the properties of both a gas and a liquid.

ii.   Supercritical Fluid Properties
In the supercritical region the properties of the fluids change markedly with temperature and pressure, reflecting the coexistence of properties of a gas and a liquid.[64]

Supercritical fluids exhibit physicochemical properties between those of liquids and gases. These properties favor their introduction into different matrices and also analyte solubility. In addition, SFs exhibit transport properties of gases (high diffusivity). Mass transfer is rapid with SFs.

Supercritical fluid density values are close to the characteristic liquid values, giving properties of a solvent showing strong fluid-analyte interaction. The SFE power is determined by the density of the fluid, and this parameter is influenced by the SF temperature and pressure. The solvent power of the SF increases with density at a given temperature and increases with temperature at a given density. The increase in density is not linear with pressure and the rate increases in the vicinity of the critical point. A large density increase results from a small pressure increase, under constant temperature. The fluid density is the easiest changeable property in the supercritical region, which makes this property the extraction key. It influences both fluid viscosity and dielectric constant, properties affecting the extraction power. Fluid density is of importance in this extraction method since it has an effect on the fluid solvent power and consequently on the yield and extraction time.

The dynamic viscosities of SFs are nearer to those found in a normal gaseous state. Low viscosities of SF enhance the extraction efficiency of an analyte in relation to the efficiency obtained with liquids, reducing the extraction time. Diffusion coefficient values of the SF are (in the vicinity of the critical point) more than ten times that of a liquid. The high diffusivity enhances analyte transport since the velocity and efficiency are also improved. As is the case for density, viscosity and diffusivity values are dependent on temperature and pressure. Diffusivity increases with an increase in temperature, whereas viscosity decreases with a temperature increase. In contrast, as pressure is increased the viscosity of the SF increases and the diffusivity decreases. Changes in viscosity and diffusivity are more pronounced in the region of the critical point. The properties of gaslike diffusivity and viscosity, and liquid like density combined with the pressure dependent solvating power are the main reasons for the growing interest in using SF technology in sample preparation.

It is possible to fine tune the solvating strength of the SF from an ideal gas to nearly that of a pure liquid. It is even possible, by adding small quantities of cosolvents (modifiers) to the SF, to custom design a SF for a specific application.

iii.   Solubility in a Supercritical Fluid

The prediction of analyte solubility in a SF is difficult; it depends on the SF density and dielectric constant and on the analyte vapor pressure. In addition, the polarities of the SF and the analyte should be as similar as possible in order to improve the solubility. Effects of some variables on analyte solubility are:

- *Pressure effect.* Under constant temperature, a pressure increase improves the extraction yield because the analyte solubility in the fluid is enhanced. Hence, a smaller volume of fluid will be necessary.
- *Temperature effect.* With a temperature increase there are two competitive effects: (i) at constant density, the analyte vapor pressure increases favoring analyte volatility and, hence, its solubility; (ii) at constant pressure, a density decrease is produced and therefore there is a solvent power decrease.
- *Analyte structure.* The effects described below may be controlled in order to optimize the extraction yield: (i) the higher the analyte molecular weight, the lower the solubility (low molecular weight PAHs can be extracted with pure CO_2, but high molecular weight PAHs can only be extracted efficiently using modifiers),[65,66] (ii) for the same molecular weight, linear compounds are less soluble than the ramified ones; (iii) analyte polarity (nonpolar compounds are the most soluble in supercritical CO_2); hence, the introduction of polar functional groups in the analyte structure decreases the solubility.

The ideal matrix for SFE is a finely powdered solid with good permeability, allowing a large surface area for fluid-solid interaction. Typical examples are soils, particulates, and powdered dried plant materials. Intermediate in suitability are semipermeable solids, such as polymers, which can be partially penetrated but giving no quantitative extractions. The worst types of samples are wet body tissues, such as fish, solid wood, rocks and liquid samples.[64]

iv.   Selection of a Supercritical Fluid

Supercritical carbon dioxide (supercritical CO_2) has been employed as an extracting agent in the majority of analytical SF extractions because of its advantageous characteristics: accessible critical properties (31.1°C, 72.8 bar), high purity, nontoxic, nonflammable, chemical inertness, nonpolluting and relatively inexpensive. It may be vaporized at atmospheric pressure allowing easier isolation of extracted analytes or it may be coupled with other analytical techniques, such as GC, high performance liquid chromatography (HPLC) or SF chromatography (SFC).

Supercritical CO_2 is an excellent extracting agent for lipophilic analytes (alkanes, terpenes) and suitable extracting agent for moderated polar analytes (PAHs, PCBs, organochlorides, pesticides, aldehydes and esters). The main problem is its strong apolar character, which reduces its use in polar analyte extractions. This problem may be solved, at least partially, by means of a modifier, or with small additions of polar organic solvents (i.e., methanol, ethanol…).

v.   Modifiers for SFE

When polar analytes and/or analytes strongly fixed to the matrix sample are extracted, it is necessary to increase the analyte solubility in the SF by means of a modifier. A modifier is commonly a polar organic solvent which is added in a small percentage to enhance the SF polarity. Hence, the polar organic analyte extraction efficiency is also enhanced. The most common modifiers employed are: methanol, propanol, tetrahydrofuran, acetonitrile, formic acid, acetone, ethyl acetate, toluene, methyl chloride, hexane and water.[63]

The modifier effect, "entrainer effect," is defined as the analyte solubility increase produced by adding a small amount of a second solvent to the primary one (supercritical fluid). This solubility increase is produced by analyte-modifier interactions in the supercritical phase through the intermolecular stresses (i.e., hydrogen bonds).[63,65]

The addition of a modifier can produce changes in the SF properties and in the sample matrix.[67] The fluid properties which may change are:

- *Density*. Commonly, the density of modified SF CO_2 increases in comparison with pure SF CO_2, under the same pressure and temperature conditions.
- *Polarity*. The polarity increases. Hence, it modifies the solubility features of the SF by increasing the solubility of polar analytes.

The modifier can operate in the sample matrix so that:

- it can fill the matrix sites avoiding analyte readsorptions,
- it can interact with the matrix-analyte system reducing the activation desorption energy, making the movement of analyte molecules from matrix active sites easier and,
- it can modify the matrix allowing the accessibility of the supercritical fluid, enhancing mass transfer and analyte diffusion.

The modifier can be added to supercritical CO_2 in two ways: (i) directly to supercritical CO_2 (dynamic addition) in the case of a weak matrix effect; or (ii) addition to the sample in the extraction cell (static addition) in the case of strong analyte–matrix interactions. A combination of static and dynamic addition is usually used.

vi.   Extraction Parameters

The selection of operating conditions is an important task in SFE. The main parameters affecting analyte extraction with supercritical fluids are fluid density, pressure, temperature, modifier nature and concentration, fluid flow rate, and extraction time. Fluid flow rate through the extraction cell has an important effect in the extraction efficiency. A low fluid flow rate gives a higher penetration depth in the sample, enhancing the extraction efficiency. The fluid flow rate can be changed by modifying the cell geometry. For cells with the same volume but a different diameter, a larger cell diameter results in higher extraction efficiencies.[63,68]

c. Instrumentation

The main components of a typical SF extractor are a gas supply, a high pressure pump, a controller used to pressurize the gas, a temperature controlled oven, an extraction vessel, a restrictor (backpressure regulator) and a collection device (Figure 2.21).

i.   Supply System, High Pressure Pump and Controller

The SF purity is a very important parameter in SFE (i.e., supercritical CO_2 purity is 99.9999%). The high purity is required for analytical extractions since if a universal detector is being used nonvolatile impurities in the CO_2 become trapped along with the separated analytes and these impurities may interfere with the analyte assay.

Usually, the gas supply system is simply a laboratory-sized cylinder. However, for applications where pure SF CO_2 is inadequate, mixtures of fluids may be used. Mixtures of fluids may be made in two different ways: (i) premixed fluids may be directly supplied in one unique cylinder or; (ii) both fluids are supplied in separate cylinders and mixed before the pump, where the mixture is pressurized to the desired value.

During the entire extraction process the fluid must be pumped at a constant pressure from the container to the extractor cell. Two major pump types are found in SFE instruments: syringe and piston. Nowadays, it appears that the majority of SFE instruments incorporate piston pumps. An auxiliary pump may be necessary when modifiers are used. Some manufacturers provide two pumps (usually as an option) with one delivering the primary fluid and the other delivering the desired level of cosolvent.

The majority of SFs are highly compressive, so pressure control must be adequate. Many SFE instruments control several gradients of CO_2 pressure. Also, fluid flows must be reproducible.

ii.   Extraction Vessel and Oven

From the pump the fluid travels to a heated zone, where it becomes supercritical, and then to an extraction vessel where the sample is contained. The extraction vessel and connecting tubes are housed in an oven, so the temperature is kept constant during extraction. Density determines the extraction efficiency and is dependent on pressure and temperature in the supercritical zone. It is essential to maintain a rigorous control of both parameters during the extraction.

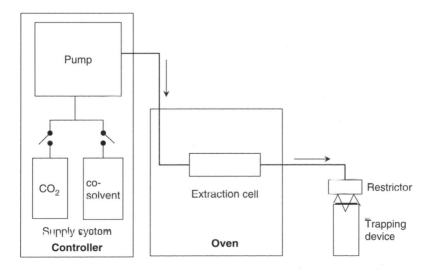

FIGURE 2.21 Basic instrumentation of SFE.

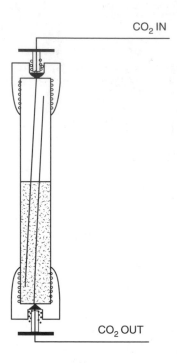

FIGURE 2.22 Liquid matrix extraction cell. (Reprinted from Heldrich, J. L. and Taylor, L. T., *Anal. Chem.*, 61, 1986–1988, 1989. With permission from American Chemical Society.)

The extraction vessel, usually made of stainless steel, is the compartment where the sample is placed in order to be extracted by the SF, and its temperature must remain greater than the critical temperature. The extraction cell must have the following characteristics:

- High pressure resistance (5,000 to 10,000 psi),
- Freedom from fluid losses,
- Suitable sample capacity (volumes from a fraction of a milliliter to more than 10 ml, although larger vessels are available, up to 50 ml),
- Chemically inert,
- Easy to assemble.

SFE is mainly used with solid samples, but recently extractions in liquid matrices have been performed. To this end special extraction cells have been designed (Figure 2.22). The fluid enters the cell through a first line where it is brought into intimate contact with the sample. Being less dense than water it rises in the cell and exits through a second line.[69]

iii.   Restrictor
The restrictor controls the backpressure and the SF flow rate which circulates through the vessel and, moreover, the SF is depressurized within the restrictor. When the SF passes through, the fluid decreases the pressure from the high pressure inside the extraction chamber to atmospheric pressure, passing the fluid from supercritical conditions in the extraction vessel to atmospheric conditions.
     Three main types of restrictors exist:

the fixed restrictor,
the linear restrictor and
high pressure electronically-controlled micrometering valves.

With a fixed restrictor the flow rate is dependent on the pressure. Hence, for a constant flow rate, the fixed restrictor should be replaced in order to change the SF density by means of pressure change. However, the diameter and length of the linear restrictors must be adjusted in order to maintain constant flow rates under different pressure conditions. Linear restrictors are more complex but these do not have to be changed during method development. High pressure electronically-controlled micrometering valves are becoming more popular in SF-based analytical techniques. These backpressure regulators, called variable restrictors, allow flow rates to be adjusted to constant levels at different densities (pressures).[63]

The narrow orifices of restrictors can be prone to plugging. Heating the restrictors helps to alleviate (although not eliminate) plugging problems.[70]

iv. Collection Device

The extracted analytes are collected after the depressurized step. The analytes travel through the restrictor, where the SF decompresses, and analyte deposits in some type of trapping device. Analyte trapping after the extraction step can be carried out with either a small amount of collection solvent (an appropriate solvent placed in a cooled vial) or in an adsorbent trap (solid surface cryogenically cooled by means of liquid nitrogen). The trapping system is selected depending on the nature of extracted analyte.

- *Adsorbent trap*. This method needs an additional desorption step with a small volume of eluent. For selecting suitable eluent one should consider:
  - analyte solubility,
  - restrictor temperature and trap temperature,
  - compatibility of the solvent with the trapping material,
  - compatibility of the solvent with the analytical method (GC, HPLC...).
- *Organic solvent*. To trap the extract in a liquid, the restrictor end is usually immersed in the solvent where the fluid bubbles. The decompressed CO_2 rises rapidly to the surface, leaving the extracted analytes in the collection solvent. In this method, the parameters to optimize are: analyte solubility, solvent temperature, solvent volume, contact time with the solvent and bubble size. This is the easier and more used method to trap analytes; moreover, it does not need the elution step. The main disadvantages are sample loss because of aerosol formation when high SF flow rates are used and solvent loss due to evaporation from the increasing temperature. Both effects make it impossible to calculate the concentration of the extracted analyte.

There are two common ways to operate SFE, in online mode or offline mode. In the online mode, the outlet of the SFE instrument is directly linked to an analytical instrument. Direct coupling with different chromatographic techniques allows simultaneous analyte preparation, separation and determination. The main drawback of the online extraction method is the limited sample size.[71] In the offline mode, the extracted analytes are trapped and later, the extracted analytes can be analyzed by means of different chromatographic techniques. The offline collection system is chosen according to the extracted analyte characteristics. The use of online methods reduces possible errors and sources of contamination. Studies on trapping methods have been performed to optimize the corresponding parameters.[72,73]

d. Coupling of SFE–Chromatographic Techniques

The direct coupling of SFE with different chromatographic techniques allows simultaneous preparation, separation and determination of analytes. In coupled methods, the analyte in the appropriate solvent is transferred directly from the extraction step to the chromatograph.

The preparation procedure is simplified and the analysis is more precise and sensitive. The analysis time is reduced and the sample contamination is minimized.

In order to obtain optimum performance of the tandem SFE–chromatograph, four fundamental conditions must be taken into account:

- The analytes must be efficiently extracted from the sample matrix,
- The analytes must be quantitatively transferred from the SFE system to the corresponding chromatographic system,
- The SF must be depressurized to a gas state and separated from the chromatographic system,
- The extracted analytes must be concentrated in a narrow band in order to obtain good sensitivity and resolution.

As in other tandem systems, the interface is a critical point which determines the quality of the SFE–chromatograph tandem system.

i.   SFE–GC
There are two possible means of analyte collection:

- *Analyte collection in the GC*. Analytes are directly injected in the chromatographic column which operates as a trapping system. In this way quantitative and reproducible recoveries are obtained.
- *Analytes are trapped in an external system to the GC*. Analytes remain in a trap while the fluid is eliminated. After that, there are two ways to drag the analytes: the carrier gas passes through the trap dragging the retained analytes, or the trap is heated carrying the analytes to the chromatographic column.

The quality of peaks depends on the choice of parameters in both systems.

ii.   SFE–High Performance Liquid Chromatography (SFE–HPLC)
Of all tandem methods, SFE–HPLC is the most difficult coupling system to use. The main reason lies in linking a separation step at high pressure and generating a gas in the interface with a chromatographic technique, which commonly works under lower pressure and uses a liquid as a mobile phase. For this reason, in most applications, the analytes are collected offline and later analyzed by HPLC. SFE–HPLC has been performed for only a few specific applications.[74,75]

iii.   SFE–Supercritical Phase Chromatography (SFE–SFC)
SFC is used for the separation of relatively nonpolar analytes, thermally unstable analytes and analytes with an elevated molecular weight. The key feature differentiating SFC from conventional techniques is the use of a significantly elevated pressure in the column. This allows the use of mobile phases which are either impossible or impractical to obtain under conventional LC and GC conditions. SFE is the ideal way to introduce a sample in SFC, since the same SF used as a solvent for the extraction acts as the mobile phase in the chromatograph.

The requirements for correct operation of SFE–SFC technique are: (i) the extraction chamber volume must be appropriate for the sample size in SFC; (ii) the SF pressure drop (mobile phase) during the extract transfer to the chromatograph must be minimum, and (iii) the chromatograph must be pressurized and equilibrated to the required pressure before the injection of extracts.

Another interesting tandem method is SFE–capillary electrophoresis (SFE–CE).[76,77]

e. Applications

Chemical derivatizations during SFE, for example, to convert carboxyl, sulfonic acid, and amino moieties to their alkyl, acyl, and silyl derivatives are described in Ref. 74. Such analyte derivatizations may be necessary to improve analyte solubility in the extraction fluid, to overcome analyte interactions with the sample matrix, or to facilitate subsequent analysis. This topic will be described in detail in Section II.D.

SFE extraction of liquid samples is done either by adsorbing the liquid onto a sorbent material or by direct extraction, the latter mode being less frequent.[74,78–81]

SFE has been used mainly in chromatographic analysis of solid samples: pesticides, hydrocarbons, phenolics, halogenated organics (particularly PCBs, PCDDs, and PCDFs),[82] organometallics,[83] metal chelates,[84] and metal ions. Selected applications of SFE on soil samples are summarized in Table 2.8.

5. Microwave Assisted Extraction (MAE)

a. Introduction

Very often quantitation is preceded by a sample preparation step. The demand for extraction techniques, amenable to automation, with shortened extraction times and reduced organic solvent consumption has resulted in a number of new techniques such as MAE,[6,85–87] SFE[88–90] and pressurized liquid extraction (PLE).[91] These techniques have the common advantage of working at elevated temperatures and pressures, which drastically improves the speed of the extraction process. Since MAE fulfills, to a great extent, all requirements for a modern sample preparation technique (i.e., it is fast, needs small amounts of solvent, allows simultaneous extractions, etc.) it is not surprising that the number of studies on MAE have increased in the last few years (Figure 2.23).

b. A Short History of Analytical-Scale MAE

The ability of microwaves to heat water very fast was discovered during the Second World War and the first application of microwaves in the field of sample preparation was described in 1975.[92,93] Microwave energy was used for the digestion of biological samples in Erlenmeyer flasks. The digestion time was decreased from 1–2 hours to 5 min to 15 min.

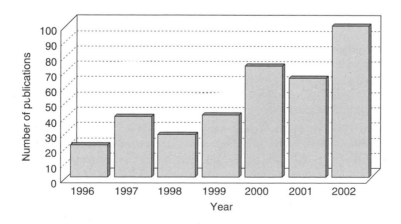

FIGURE 2.23 Number of studies utilizing the MAE technique during the last 7 years. (*Source*: CALPUS and MEDLINE databases.)

Surprisingly, it took 10 years to apply microwave energy to extraction. In 1986 Ganzler et al.[94] presented the extraction of crude fat and antinutrients from food and pesticides in soil using MAE. A patented variant of MAE has been developed (microwave-assisted process (MAP)) by Environment Canada.[95] MAP applications mainly cover extractions of substances from biological materials and extend from analytical-scale methods to industrial processes. The first application of MAP was reported in 1991 and dealt with the extraction of essential oils from plant products.[96] In the early 1990s various research groups started to investigate the potential of MAE. Because of the evaluation of new sample preparation techniques initiated by the US Environmental Protection Agency (EPA),[97] numerous laboratories have begun to study the analytical possibilities of MAE in environmental applications. Today MAE has matured and some standard methods have been published, mainly for organic compounds in solid matrices.[98–102]

c. Microwave Interaction with Matter

The microwave region of the spectrum is situated between the radiofrequency (RF) and infrared regions and corresponds to wavelengths between 1 cm and 1 m. So as not to interfere with telecommunication and RADAR systems, domestic and industrial microwave heaters operate at either 12.2 cm (2.45 GHz) or 33.3 cm (900 MHz).[92,102]

Both types of microwave instrument designs (with closed or with open vessels) use the direct absorption of microwave radiation by the reaction mixture through the walls of the vessels. The vessels are transparent to microwaves. Compared to conventional heating methods, the microwave systems with open vessels create more stable temperature conditions and are not limited by the heating mechanisms of convection or conduction. In addition, the open microwave systems permit the so-called "superheating" effect, which allows the heating of the reaction mixture to temperatures above the boiling point of the solvents. Microwave systems with closed vessels, on the other hand, are limited only by the temperature and pressure limits of the containment vessels and by the microwave-absorption characteristics of the solution. Another important difference with conventional heating is the internal reflux system, generated in microwave systems because of the significant difference between the energy absorption of liquid and gas phases inside the vessel.[92,102]

In discussing the interaction of microwave energy with matter it is very useful to suppose that a dielectric is exposed to an electromagnetic field with a wavelength comparable to the dimensions of the dielectric. In this simplistic explanation only dielectric effects associated with molecular movement will be needed. Dielectric materials can store electrical energy. This is accomplished by the displacement of positive and negative charges under the effect of an applied electric field and against the forces of atomic and molecular attraction. There are four main types of dielectric polarization (Figure 2.24)[103]:

- Electronic polarization, by realignment of electrons around specific nuclei.
- Atomic polarization, by the relative displacement of nuclei due to the unequal distribution of charges within the molecule.
- Orientation polarization results from the reorientation of permanent dipoles by the electric field.
- Space charge polarization occurs when the material contains free electrons whose displacement is restricted by grain boundaries. Hence, entire macroscopic regions of the material become either positive or negative. This mechanism is often called the Maxwell–Wagner effect and it takes place in low frequency fields.

In an alternating field the orientation of a polarization varies cyclically with the field. At low frequency, all types of polarization synchronize their orientation with the field, but as the frequency increases, the inertia of molecules causes certain modes of polarization to lag behind the field. In RF and microwave frequencies, electron and atomic polarization are much faster than the time

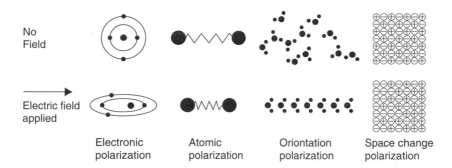

No
Field

Electric field
applied

Electronic
polarization

Atomic
polarization

Oriontation
polarization

Space change
polarization

FIGURE 2.24 Mechanisms of polarization. (Reprinted from Zlotorzynski, A., *Crit. Rev. Anal. Chem.*, 25, 43–76, 1995. With permission from Taylor and Francis Group.)

variation of the field and so these effects do not contribute to dielectric heating. Dipole and space charge polarization are on the same time-scale as the time variation of electromagnetic field, so there is an interaction producing energy transfer. The phase lag between polarization and the applied field leads to an absorption of energy and Joule heating.

A more simplistic mechanism to explain microwave heating, but very useful in the scope of this chapter, is to consider that microwaves transfer energy directly to the absorbing molecules by two general mechanisms: (i) ion conductivity and (ii) dipole rotation.

Ion conductivity is a process in which the ions in the solution move as a result of the applied electromagnetic field, thus causing friction and heating of the reaction mixture. This mechanism is less dependent on the frequency of the microwave field than the dipole rotation, which is the effect of the alignment of the molecule dipoles towards the applied electromagnetic field. The very fast movements of the molecules caused by the oscillations in the applied electromagnetic field cause the solution to heat.[92,102] In the microwave assisted digestion methods the most commonly used solvents are mineral acids, since these are polar and ionized in water solutions and, hence, absorb a high amount of microwave energy. The values of the absorption and the ability to predict them at a frequency of 2.45 GHz were derived in the very first publication dealing with closed Teflon vessels.[104]

d. Fundamental Relationships in Microwave Energy Absorption

Microwave energy absorption can be expressed in a fundamental thermodynamic relationship which relates energy absorption to the specific heat capacity, mass in the microwave field, temperature increase, and time of sample exposure as follows[92]:

$$P_{abs} = \frac{KC_p m \Delta T}{t} \tag{2.1}$$

where P_{abs} is the apparent power absorbed by the sample (in watts); K is a constant to convert calories/second to watts (4.184 J/cal); C_p is the heat capacity or specific heat of the microwave absorbing solvent (cal/g °C); m is the total mass of the sample (g); ΔT is the difference between the final temperature (T_f) and the initial temperature (T_i) (°C); and t is the time of microwave exposure (s). This equation can be rearranged to predict either the temperature (T_f) of the reagents at a certain moment in time (Equation 2.2) or the time required to reach a temperature (t) (Equation 2.3). These equations are valid for most vessels in the temperature range from 20°C to 250°C.[92]

$$T_f = T_i + \frac{P_{abs} t}{KC_p m} \tag{2.2}$$

$$t = \frac{KC_p m \Delta T}{P_{abs}} \tag{2.3}$$

In addition, a quartic model has been developed (Equation 2.4), which allows the prediction of the energy absorbed in closed vessels by a quantity of mineral acid used in decomposition.[92] This equation is valid within the mass range of 25 to 1000 g of reagent. Coefficients A through E are reported for specific concentrations of HNO_3, HCl, HF, H_2SO_4 and H_2O. Using this method the specific power absorption for other acid concentrations and mixtures can be derived as well.

$$\ln(P_{abs}) = A + B \ln(m) + C[\ln(m)]^2 + D[\ln(m)]^3 + E[\ln(m)]^4 \tag{2.4}$$

Equation 2.4 allows the calculation of the energy absorption with accuracy between 4% to 10% depending on the kind and concentration of the mineral acids used.

The ability of a solvent to absorb microwave energy and to convert it into heat is given by the dissipation factor ($\tan\delta$). The dissipation factor is given by the following equation[92,103,105]:

$$\tan \delta = \varepsilon''/\varepsilon' \tag{2.5}$$

where ε'' is the dielectric loss (a measure of the efficiency of converting microwave energy into heat) and ε' is the real permittivity or dielectric constant (a measure of the polarizibility of a molecule in an electric field).

Unlike mineral acids, which are 100% ionized in solution and have a permanent dipole moment which is affected by microwaves, some organic solvents such as hexane are nonpolar and therefore not heated when exposed to microwaves. Selected physical parameters, including dielectric constants and dissipation factors, are shown in Table 2.4 for solvents which are used in MAE applications.

A simple comparison between methanol and water shows that methanol has a lower dielectric constant but a higher dissipation factor, hence a higher dielectric loss (Equation 2.5) than water. This indicates that methanol, compared to water, has a lesser ability to obstruct the microwaves as they pass through, but a greater ability to dissipate the microwave energy into heat.

TABLE 2.4
Physical Constants and Dissipation Factors for Some Solvents Commonly Used in MAE

Solvent	Dielectric Constant,[a] ε'	Dipole Moment[b]	Dissipation Factor, $\tan \delta$ ($\times 10^{-4}$)	Boiling Point,[c] (°C)	Closed-Vessel Temperature,[d] (°C)
Acetone	20.7	2.69	—	56	164
Acetonitrile	37.5	3.44	—	82	194
Ethanol	24.3	1.69	2500	78	164
Hexane	1.88	<0.1	—	69	—[e]
Methanol	32.7	2.87	6400	65	151
2-Propanol	19.9	1.66	6700	82	145
Water	78.3	1.87	1570	100	—
Hexane—acetone (1:1)	—	—	—	52	156

Reprinted from Eskilsson, C. S. and Björklund, E., *J. Chromatogr. A*, 902, 227—250, 2000. With permission from Elsevier.

[a] Determined at 20°C.
[b] Determined at 25°C.
[c] Determined at 101.4 kPa.
[d] Determined at 1207 kPa.
[e] No microwave heating.

e. Instrumentation

Recently Luque-García and Luque de Castro have published an excellent review on modern microwave devices.[106] In the following paragraphs only a brief overview is given.

Microwave equipment used for sample pretreatment can be classified into two groups, according to how microwave energy is applied to the sample, namely:

- Multi-mode systems, in which the microwave radiation is allowed to disperse randomly in a cavity, so each zone in the cavity and the sample are evenly irradiated (Figure 2.25);
- Single-mode or focused systems, in which microwave radiation is focused on a restricted zone where the sample is subjected to a much stronger electric field than in the previous systems (Figure 2.26).

Usually, multi-mode systems use closed type vessels and focused systems use open vessels.[106]

i. Closed-Vessel Microwave Devices

Closed vessels have high upper pressure limits and normally are constructed in several layers of microwave transparent polymer. In well-insulated closed vessels, the temperature can be estimated by means of Equation 2.2 for temperatures higher than the boiling point of the reaction mixture. However, in most microwave systems the vessels are not completely insulated therefore a loss of heat occurs, introducing error in the calculation. A significant amount of heat is lost through the walls of the vessels to the cooling system of the microwave oven. Consequently, the pressure of gases inside the closed vessels is significantly lower than that predicted by the temperature of the liquid phase. In the microwave field the assumption that all components of the system

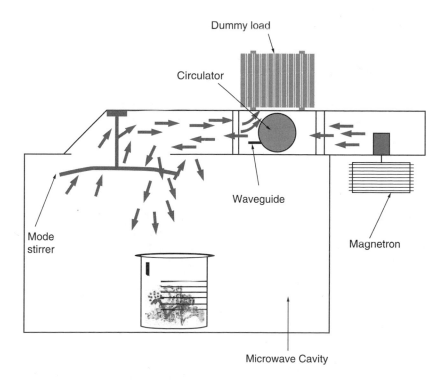

FIGURE 2.25 Schematic view of the microwave oven. (Reprinted from Zlotorzynski, A., *Crit. Rev. Anal. Chem.*, 25, 43–76, 1995. With permission from Taylor and Francis Group.)

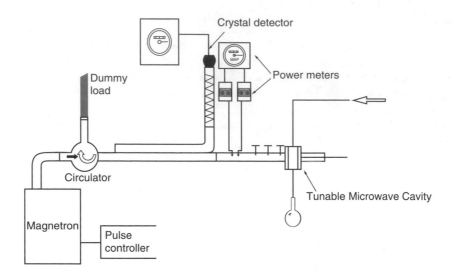

FIGURE 2.26 Single-mode resonant heating system. (Reprinted from Zlotorzynski, A., *Crit. Rev. Anal. Chem.*, 25, 43–76, 1995. With permission from Taylor and Francis Group.)

(liquid, gas and vessel) are in equilibrium is no longer valid because the gas phase is heated less effectively than the liquid phase. The ion conduction mechanism is not present in the gas phase because all free ions are left in solution thus leaving only the molecule rotation as a heating mechanism. In addition, the effectiveness of this mechanism is drastically decreased because of the statistically lower number of molecule–molecule collisions in the gas phase. For these reasons there is no thermal equilibrium reached between the liquid and gas phases. The lower temperature of the gas phase creates a vertical temperature gradient from the bottom of the vessel (the hottest part) to the top of it (the coolest part). This temperature gradient causes acid fumes to condense and creates an effective reflux system inside the vessel.[92]

The phenomenon of lower internal pressure at relatively high temperatures is one of the main advantages of the microwave-assisted sample preparation with closed vessels. The pressure inside a vessel may be additionally lowered by cooling the gas phase inside the vessel. There are some designs of closed microwave vessels which make use of heat loss to improve the safety and robustness of digestion procedures.[92]

ii.   Open-Vessel Microwave Devices

For microwave systems operating at atmospheric pressure, the temperature can be calculated by means of Equation 2.2 up to the boiling point of the solution, including any superheating effects. The boiling point limits the oxidation potential (i.e., the ability of reagents to destroy the matrix). This leads to different approaches for increasing the oxidation potential. In addition to using azeotropic mixtures and taking advantage of any superheating effect, acids with higher boiling points, such as H_2SO_4, are used to produce more rigorous digestion/leaching conditions. Another frequently used approach is the addition of H_2O_2, which may be safely used with open vessels. Hydrogen peroxide increases the oxidation potential and also improves the conversion of the microwave energy into heat due to its high dielectric constant value.

One useful aspect of microwave systems with open vessels is that they allow direct adaptation of already existing sample preparation methods. Of particular interest for chromatography is the focused microwave-assisted Soxhlet extractor (Figure 2.17 and Figure 2.27). It is based on the same principles as a conventional Soxhlet extractor but is modified to facilitate accommodation of the sample-cartridge compartment in the irradiation zone of a microwave oven. The modification

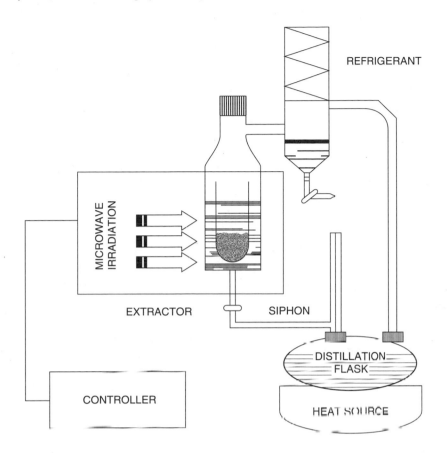

FIGURE 2.27 Focused microwave-assisted Soxhlet extractor. (Reprinted from Luque-García, J. L. and Luque de Castro, M. D., *Trends Anal. Chem.*, 22, 90–98, 2003. With permission from Elsevier.)

includes an orifice at bottom of the irradiation zone enabling connection of the cartridge compartment to the distillation flask through a siphon, as illustrated in Figure 2.27. The device, which enables FMAS extraction, retains the advantages of conventional Soxhlet extraction while overcoming restrictions. These restrictions are the difficulty of automation and the large volumes of organic solvents which are wasted, as well as the long extraction time and nonquantitative extraction of strongly retained analytes. The latter two drawbacks are alleviated because of the easier cleavage of analyte–matrix bonds by interactions with focused microwave energy.[106]

f. Applications

Eskilsson et al.[105] have written an excellent review covering the theory and applications of closed vessels MAE up till the year 2000. More recently, Nóbrega et al.[107] have compiled another review, this time dealing with focused-microwave-assisted techniques. In the following sections a brief overview of some recent work is given, divided into three fields (analysis of air, water and solids), and more detailed information is given in Table 2.8 and Table 2.9.

i. Air

Every year regulatory institutions decrease the amount of various toxic species allowed in air. In air analysis, air-borne particles are collected using various types of filters and MAE has been utilized successfully in the treatment of the collected particles. Table 2.9 summarizes some of the recently published studies concerning air analysis.

Ericsson and Colmsjo[108] have described an online method for the determination of organophosphate esters in air samples by large-volume injection gas chromatography. The extraction and cleanup step was performed by conducting dynamic microwave-assisted extraction (DMAE) coupled to SPE. The superiority of microwave-assisted extraction (MAE) compared to the other methods such as Soxhlet extraction, PLE and SFE was emphasized, with the main advantage being the achievement of higher extraction rates due to fast heating. The authors concluded that DMAE is a powerful tool for the analysis of organophosphate esters in indoor air. They achieved extraction efficiencies higher than 97%.

The superiority of microwave extraction compared to traditional extraction methods for the determination of polychlorinated biphenyl compounds in indoor air samples was also shown.[109] Again a decrease in the extraction time was highlighted; the microwave procedure needed only 10 min and, followed by GC-electron capture detection, was claimed as a valuable alternative to the Soxhlet method for the extraction of six noncoplanar PCBs associated with fly ashes.

Another interesting application of microwave energy was reported by Mueller et al.[110] The authors studied the amount of benzene and alkylated benzenes (BTX) in ambient and exhaled air by microwave desorption coupled to GC–MS. Microwave desorption was proved as an effective sample preparation technique for the analysis of BTX in air samples. A similar desorption technique was applied for the GC analysis of nicotine in indoor air as well.[111]

An interesting application of microwave energy for the preparation of standard gas of VOC was described by Xiong and Pawliszyn.[112] With a domestic microwave oven the authors developed a simple, powerful, rapid, accurate and safe procedure for preparation of VOC/semiVOC standard gas. Solid-phase micro extraction combined with GC was used for the gas analysis. Worth mentioning here is that because of the specific way in which microwaves interact with matter, an appropriate amount of water was introduced during the preparation of the gas mixtures since the molecules of the studied VOCs are weakly polar or nonpolar and therefore very poor absorbers of microwave energy.

ii.   Water
Liu et al.[113] have developed a new method for the determination of trace levels of bromate and perchlorate in drinking water by ion chromatography. A new evaporative preconcentration technique with the help of microwave energy was developed. The samples were subjected to microwave energy in the form of water solutions, containing the anions of interest. Eight anions were concentrated but their properties were unchanged. In addition, the effect of various microwave vessels (100 ml PTFE, quartz and glass beakers) was studied with best result obtained with the PTFE beakers. The authors came to the conclusion that with a household microwave oven, drinking water samples can be concentrated 20 fold in 15 min and excellent recoveries (95.3% to 96.7%) can be obtained for the studied species.

iii.   Soils and Sediments
In contrast with water analysis, where sample preparation is a relatively easy task, the analysis of solid matrices, such as soils and sediments, requires more steps before the measurements. One can save time and enhance the efficiency of extraction or the completeness of digestion. Table 2.8 summarizes some of the recent publications of MAE applied to soil and sediment analysis.

Yang et al.[114] studied MAE of polychlorinated biphenyls (PCBs) and polychlorinated dibenzodioxins (PCDDs) from fly ash and sea sediments. The effect of the addition of water to the extracting solvent (solvation) and sample matrices (wetting) on the variation of recoveries was studied. The results indicate that MAE, using 1% to 2% of solvation or 10% to 20% of wetting in 90:10 (v/v) of toluene-IPA mixtures, was the most effective treatment in isolating PCBs and PCDDs from the samples. Multi-layer column chromatography on neutral and acidic silica gel with *n*-hexane was used for cleaning up the extracts. Alumina column clean up was performed to remove interferences from sediment extracts. The addition of water to soil and sediment samples prior to the extraction procedure is not uncommon since water in the matrix may increase the extraction efficiency. This paper also examines one of the drawbacks of MAE (i.e., often a cleaning step is needed prior to the analysis).

Cleaning has been effectively avoided by Numata et al.[115] A novel sample extraction technique for PCBs and organochlorine pesticides (OCPs) analysis from marine sediments using a microwave-heating device was reported. MAE and steam distillation techniques were combined to create the microwave-assisted steam distillation (MASD) technique. Desorption of the analytes from solid matrices was accelerated with water vapor generated by microwave irradiation. A sample holder in a commercial microwave extraction cell kept the sample from direct contact with organic solvent used for analyte trapping during the treatment process. Therefore, relatively clean extracts were obtained with only a small amount of solvent. Without any cleanup steps, the obtained extracts could be analyzed with gas chromatograph–mass spectrometers (GC–MS). The proposed method was compared with other extraction methods such as exhaustive steam distillation, MAE, and traditional Soxhlet extraction. Low recoveries (30% to 60%) were found for highly chlorinated biphenyls (PCB 180, 194, 209) and relatively polar OCPs while reasonable recoveries (80% to 100%) were found for PCB 15, 28, 70, 101 and OCP 4,4'-DDE.

Although MAE is claimed as the best extraction technique in most published works, Soxhlet extraction was claimed as more precise for the determination of OCPs in sediments.[116] Additionally, higher recoveries were reported by using Soxhlet extraction.

Ghassempour et al.[117] published an interesting work comparing MAE and ultrasonic extraction (USE) for the extraction of diazinon from soil and the stems of rice plants. After optimizing the conditions, better results (98% recovery) were obtained by MAE with hexane as solvent. The USE method gave 91% recovery with the same solvent. The reduced extraction time, minimal amount of solvent, the fact that the soil and steam moisture did not influence MAE when a solvent such as acetone is used, and higher recoveries, made MAE the method of choice in this study.

On the contrary, Contat-Rodrigo et al.[118] proposed USE as a better method in terms of reproducibility and extraction efficiency. They studied the extraction of degradation products from degradable polyolefin blends aged in soil. Higher amounts of certain products (e.g., carboxylic acids) were extracted by USE than by MAE.

From the information in Table 2.8 it can be concluded that soils and sediments are quite often studied using MAE. In almost all the studies MAE is claimed to be better than Soxhlet extraction, with the exceptions of Refs. 116 and 118. However, some controversial results about the comparison MAE–USE were found.

g. Conclusions

MAE has found its place in the field of chromatographic analysis of environmental samples. From 1986 till now it has been accepted as an interesting alternative to the conventional extraction methods, as well as to some of the newly developed ones. Some of the main benefits and disadvantages of the MAE are summarized in Table 2.5, together with the Soxhlet and the ultrasound assisted extraction techniques.

6. Ultrasound Assisted Extraction (UAE)

a. Introduction

Sound is transmitted through a medium by inducing vibrational motion of the molecules through which it is traveling. Sound is a series of compression waves separated by rarefaction

TABLE 2.5
Comparison Between MAE, Soxhlet Extraction (SE) and Ultrasonic-Assisted Extraction (USE)

	MAE	SE	USE
Extraction time	3–30 min	3–48 h	10–60 min
Sample size (g)	1–10	1–30	1–30
Solvent usage (ml)	10–40	100–500	30–200
Investments	Moderate	Low	Low
Advantages	Fast and multiple extractions Low solvent volume Elevated temperature	No filtration required	Fast and multiple extractions Low temperature and pressure (safety) No waiting time for the vessels to cool down Thermolabile analytes could be extracted
Drawbacks	Extraction solvent must be able to absorb microwaves Clean up step needed Waiting time for the vessels to cool down High pressure and temperature are needed (no safety)	Long extraction time Large solvent volume Clean up step needed Thermolabile analytes are altered	Large solvent volume Repeated extractions may be required Clean up step needed Filtration and rinsing are needed Risk of losses and contamination of the extract during handling Ageing of the surface of the ultrasonic probe can alter extraction efficiency Particle size is a critical factor

(stretching) waves. Sound waves can be represented as a series of vertical lines or by shaded color with intensity related to the separation between lines or the color depth. Alternatively, it may be represented as a sine-wave where the intensity is related to amplitude (Figure 2.28). The wave moves but the water molecules which constitute the wave revert to their normal positions after the wave has passed and the sine-wave represents pressure variation with position at a fixed point in time.

At high sound frequencies the ear finds it difficult to respond and eventually the human hearing threshold is reached which is normally around 18 KHz to 20 KHz for adults. Sound beyond this limit is inaudible and is defined as ultrasound. Ultrasound comprises the region of frequencies between 20 KHz and 100 MHz, the upper limit not being sharply defined. This broad region can still be divided into two different regions: power and diagnostic ultrasounds, from 20 KHz to 100 KHz and from 1 MHz to 10 MHz, respectively. The former, which generates greater acoustic energy, induces cavitation in liquids and cavitation is the origin of mechanical and chemical effects of ultrasounds. Sonochemistry normally uses frequencies between 20 KHz and 40 KHz simply because this is the range employed in common laboratory equipment.

b. Cavitation — The Origin of Sonochemical Effects

The effects of ultrasound on chemical transformations are not the result of direct coupling of the sound field with the chemical species on a molecular level. Power ultrasound is able to produce

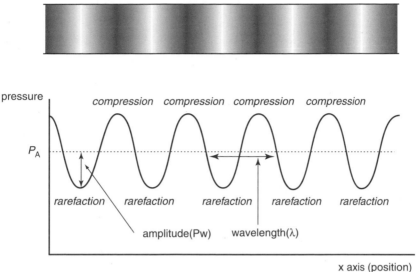

FIGURE 2.28 Representations of sound motion. P_A is the ambient pressure in the fluid. (Reprinted from Mason, T. J., *Sonochemistry*, Oxford University Press, Oxford, 1999. With permission.)

chemical effects through the phenomenon of cavitation. Cavitation is the production of microbubbles in a liquid when a large negative pressure is applied.[119]

Like any sound wave, ultrasound is propagated via waves which alternately compress and stretch the molecular spacing of the medium through which it passes (Figure 2.28). Thus the average distance between the molecules in a liquid will vary as the molecules oscillate around their mean position. When a large negative pressure (i.e., sufficiently below ambient pressure) is applied to the liquid the distance between molecules can overcome a critical molecular distance necessary to hold the liquid intact, below which the liquid breaks down so that cavitation bubbles form. Theoretical calculations predict a very high negative pressure to obtain cavitation bubbles. However, in practice cavitation can be produced at considerably lower acoustic pressures due to the presence of weak spots in the liquid which decrease its tensile strength.[120] Included among these spots are gas nuclei in the form of dissolved gas, gas-filled crevices in suspended particulate matter or transient microbubbles remaining from previous cavitation events.[119]

The formation of cavitation bubbles is initiated during the rarefaction cycle. These bubbles grow to an equilibrium size over a few cycles by taking in some vapor or gas from the medium to match the frequency of bubble resonance to that of the sound frequency (Figure 2.29). Some bubbles suffer sudden expansion to an unstable size and collapse violently. The whole process by which vapor bubbles form, grow and undergo implosive collapse takes place within about 400 μs.[120,121] It is generally accepted that the spectacular chemical and mechanical effects attributable to cavitation are entirely due to the collapse of transient cavities. In aqueous systems and under an ultrasonic frequency of 20 KHz each cavitation bubble collapse acts as a localized "hotspot" generating temperatures of about 4000 K (similar to the surface of the sun) and pressures in excess of 1000 atm (equivalent to that at the Marian Trench, the deepest point in the ocean). Hence, a collapsed cavitation bubble could be considered as a microreactor working under extreme conditions of temperature and pressure where chemical and mechanical effects are produced. Even at the extremely high temperature reached within a cavitation bubble, it is especially interesting for extraction, and, moreover, for speciation, that no significant change in the solution temperature is observed. Since the size of the bubbles is very small relative to the total liquid volume, the heat that they produce is rapidly dissipated.

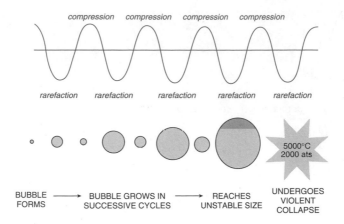

FIGURE 2.29 Development and collapse of cavitation bubbles. (Reprinted from Mason, T. J., *Sonochemistry*, Oxford University Press, Oxford, 1999. With permission.)

c. Parameters Affecting Cavitation

The most significant experimental variables are[119]:

- Frequency
- Solvent viscosity
- Solvent surface tension
- Solvent vapor pressure
- Bubbled gas
- External (applied) pressure
- Temperature, and
- Intensity

i. Frequency

The formation of cavitation bubbles decreases with increasing ultrasonic frequency. A simple qualitative explanation of this effect could be that, at very high frequency the rarefaction (and compression) cycle is extremely short and the formation of a cavity in the liquid requires a finite time to permit the molecules to pull apart. Hence, when the wavelength approaches or becomes shorter than this time, cavitation becomes more difficult to achieve. For this reason, and due to mechanical problems with transducers at high frequencies, the frequencies generally used for sonochemistry are between 20 KHz and 40 KHz.

ii. Solvent Viscosity

Since viscosity is a measure of resistance to shear it is more difficult to produce cavitation in a viscous liquid as a result of the increased negative pressure in the rarefaction region needed for disruption of the liquid.

iii. Solvent Surface Tension

As a general approximation, it might be expected that employing a solvent of low surface energy per unit area would lead to a reduction in the cavitation threshold.

iv. Solvent Vapor Pressure

During the expansion cycle, vapor from the surrounding liquid will permeate the interface. The higher the vapor pressure of the solvent, the more vapor in the bubbles. Hence, a more volatile solvent will support cavitation at lower acoustic energy and easily produce vapor filled bubbles, but their collapse is cushioned by the vapor in the bubble and therefore less energetic.

v.   Bubbled Gas

Dissolved gas or small gas bubbles in a fluid can act as nuclei for cavitation and in addition, ultrasound can be used to degas a liquid. At the beginning of the sonication of a liquid, any gas entrapped or dissolved in the liquid promotes cavitation and is removed. Hence, gas has been deliberately introduced into a reaction media in order to maintain uniform cavitation. From a theoretical point of view, the energy developed on collapse of gas-filled bubbles is greatest for gases with the largest ratio of specific heats (i.e., polytropic index). For this reason monoatomic gases (He, Ar, Ne) are recommended in preference to diatomic gases (N_2, air, O_2).

vi.   External Pressure

Increasing the external pressure will mean that a greater rarefaction pressure is required to initiate cavitation. However, the higher the external pressure, the higher the intensity of cavitational collapse, and consequently an enhanced sonochemical effect is obtained. It has been experimentally observed that at a specific frequency there is a particular external pressure which will provide an optimum sonochemical effect, and moreover the optimum pressure depends upon the frequency used.[122]

vii.   Temperature

Any increase in temperature will raise the vapor pressure of a medium and so lead to easier cavitation but less violent collapse (see above). This effect will be accompanied by a decrease in viscosity and surface tension. However, at temperatures approaching the solvent boiling point, a large number of cavitation bubbles are generated concurrently. These will act as a barrier to sound transmission and dampen the effective ultrasonic energy from the source which enters the liquid medium. The combination of all these effects shows a shape of maximum and the optimum temperature depends on the experimental conditions used and reaction studied.

viii.   Intensity

In general, an increase in sound intensity will provide an increase in sonochemical effects, but the intensity also shows an optimum value. When a large amount of ultrasound power enters a system, a great number of cavitation bubbles are generated in the solution and many of them coalesce forming larger, longer-lived bubbles. In addition, these bubbles decrease the energy transmission through the liquid. These effects, and the so-called decoupling effect, could explain the shape of the reaction product yield with respect to power, first increasing to reach a plateau, and then dropping dramatically above a power value where decoupling occurs.

Other significant variables influencing the solid–liquid extraction process are: sonication time, type and concentration of acid, particle size and solid concentration (i.e., sample mass and extraction volume).

d.   Mechanisms in Ultrasound Assisted Extraction

Sonochemistry is mainly concerned with reactions by inducing cavitation in a liquid component and this covers almost all possible chemical conditions. However, there are two heterogeneous reactions of interest for analytical extractions:

(i) those involving a solid and a liquid and;
(ii) those involving immiscible liquids.

In any heterogeneous system cavitation which occurs in the bulk liquid phase will be subject to the same conditions as have been described for homogeneous reactions.[119] However, there will be some differences when bubbles collapse at or near any interface and this will depend upon the type of interface involved, and, in the case of a solid–liquid interaction, it depends upon the morphology of solids (i.e., a large solid surface or a powder in suspension).

i.   Cavitation Involving a Solid–Liquid Interface

There exist two types of cavitational collapse which can affect the surface of a solid. One is cavitational collapse on the surface of the solid due to the presence of surface defects, entrapped gases, or impurities. The second is cavitational collapse close to a surface causing a microstreaming of solvent to impinge on the surface.[119,123] In the latter case the bubble collapse is asymmetrical. The large solid surface hinders liquid movement from one side and so the major liquid flow into the collapsing bubble will be from the opposite side of the bubble (Figure 2.30). As a result, a high-speed liquid jet will be formed which is targeted at the surface. Solid surfaces which have been subjected to ultrasonic irradiation reveal "pitting" that serves to expose new surface to the liquid reaction mixture and to increase the effective area available for reaction. These mechanical effects are particularly important for the solid–liquid extractions.

In addition, when the solid is a particulate, cavitation can produce a variety of effects depending on the size and type of material. Among them are mechanical deaggregation and dispersion of loosely held clusters, a local increase of temperature on the surfaces, and cleaning by desorption of reaction products onto surfaces.[119,123]

ii.   Cavitation Involving a Liquid–Liquid Interface

A problem with reactions involving immiscible liquids is that the reagents and analyte(s) are often dissolved in different phases. Hence, any reaction between these species can only occur in the interfacial region between the liquids and this is commonly a very slow process. Here sonication can be used to produce very fine emulsions from immiscible liquids. This is the result of cavitational collapse at or near the interface, which causes disruption and impels jets of one liquid into the other to form extremely fine emulsions (Figure 2.31). These emulsions increase the interfacial contact area between the liquids, dramatically increasing the reactivity between species dissolved in the separate liquids.

These processes, combined with a very high local effective temperature (which increases solubility and diffusivity) and pressures (which favor penetration and transport), along with the

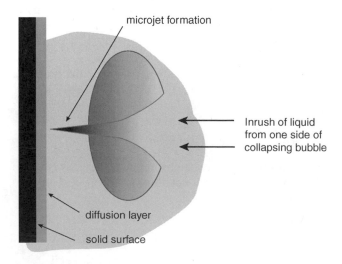

FIGURE 2.30 Cavitation bubble collapse near a solid surface. (Reprinted from Mason, T. J., *Sonochemistry*, Oxford University Press, Oxford, 1999. With permission.)

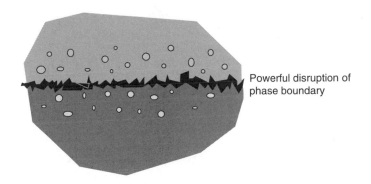

Powerful disruption of phase boundary

FIGURE 2.31 Cavitation bubble collapse in a biphasic medium. (Reprinted from Mason, T. J., *Sonochemistry*, Oxford University Press, Oxford, 1999. With permission.)

oxidative energy of radicals (hydroxyl and hydrogen peroxide for water) created during sonolysis may explain the strong extractive power of ultrasound.

e. Instrumentation

There are two common devices in ultrasound applications, bath and probe units. Both systems are based on an electromagnetic transducer (i.e., device capable of converting mechanical or electrical energy into high frequency sound) as a source of ultrasound power, commonly operating at a fixed frequency of 20 KHz. In bath systems the transducer is usually placed below a stainless steel tank, the base of which is the source of ultrasound (Figure 2.32). Some ultrasonic tanks are also provided with a thermostat. These systems are adequate for cleaning, degassing of solvents and extraction of adsorbed metals and organic pollutants from environmental samples, but less effective for extraction of analytes bound to the matrix. Although ultrasonic baths are more commonly used, these have two main drawbacks which adversely affect precision: (i) lack of uniformity in the distribution of ultrasound energy and; (ii) decline of power with time, so that the energy supplied to baths is wasted. Probe-type sonicators are able to deliver up to 100 fold greater power than the ultrasonic bath to the extraction medium (Figure 2.33). The main feature in the successful application of ultrasonic probes is that the ultrasonic energy is not transferred through the liquid medium to the extraction vessel but is focused on a localized sample zone, providing more efficient

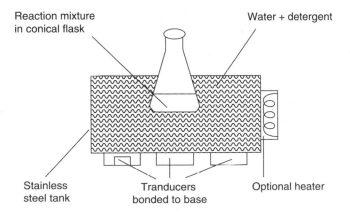

Reaction mixture in conical flask

Water + detergent

Stainless steel tank

Tranducers bonded to base

Optional heater

FIGURE 2.32 Schematic diagram of an ultrasonic bath. (Reprinted from Bendicho, C. and Lavilla, I., In *Encyclopedia of Separation Science*, Wilson, I. D., Adlard, T. R., Cooke, M., and Poole, C. F., Eds., Academic Press, New York, p. 1448–1454, 2000. With permission from Elsevier.)

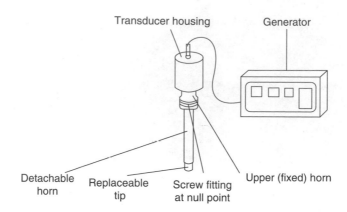

FIGURE 2.33 Schematic diagram of a probe-type sonicator. (Reprinted from Bendicho, C. and Lavilla, I., In *Encyclopedia of Separation Science*, Wilson, I. D., Adlard, T. R., Cooke, M., and Poole, C. F., Eds., Academic Press, New York, p. 1448–1454, 2000. With permission from Elsevier.)

cavitation. Although probe-type systems show improved performance for solid–liquid extractions compared with cleaning baths, they have some drawbacks. Volatile components can be lost due to the "degassing" effect of ultrasounds, and metallic probes may contaminate the samples due to erosion. In addition, although the temperature does not increase significantly a rigorous control of the sonication vessel is required.

Recently, dynamic ultrasonic systems have been suggested for sample preparation since they speed up the sample preparation process.[124] Other advantages of continuous ultrasound assisted extraction are the modest consumption of sample and reagents, the need for few or none of the chemicals required for dissolution in manual methods, and the possibility of developing fully automated methods. There are two designs for this approach: (i) the open system and; (ii) the closed system (Figure 2.34). In the former design fresh extractant flows continuously through the sample, whereas in the latter a preset volume of extractant is continuously circulated through the solid sample. Selection of the closed system has the advantage that the extract is less diluted.

Ultrasound irradiation has not been widely used for analytical extraction but it could be a powerful tool since, in some cases, ultrasound assisted extraction is an expeditious, inexpensive and efficient alternative to conventional extraction techniques such as SFE and MAE.[124] Operations with ultrasonic processors can be performed at ambient temperature and normal pressure, and under mild chemical conditions (Table 2.5).

f. Applications

It is well known that ultrasound has been applied to some organic substrates of environmental interest either to convert them to compounds which are less harmful than the original substrates or to extract species from particulate matter. Sonication is usually recommended for pretreatment of solid environmental samples for the extraction of nonvolatile and semivolatile organic compounds from solids such as soils, sludges and wastes since unsophisticated instrumentation may be used and separations can be performed in a short time using diluted reagents and low temperatures.

Ultrasonic extraction is an effective method for extracting a number of heavy metals from environmental and industrial hygiene samples.[125] In many cases, it provides quantitative recovery of metals and replaces drastic preparation procedures which would otherwise require the use of concentrated acids and the application of high temperatures and pressures (i.e., hot plate and/or microwave extraction). Quantitative extraction can be achieved for some analytes such as As, Cu, Pb, Cd, etc. from plant and animal tissues. Nevertheless, incomplete extraction has been observed from samples containing a typical inorganic matrix (e.g., sediment). A comparison of conventional

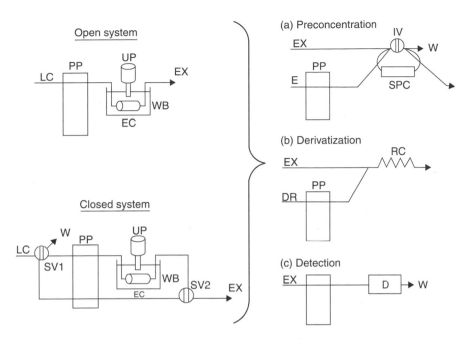

FIGURE 2.34 Experimental setup used for the two modes of continuous ultrasound-assisted leaching and the possibilities of their coupling to other steps of the analytical process. One, two or three steps can be used in a single method. Abbreviations used are: LC, leaching carrier; PP, peristaltic pump; UP, ultrasonic probe; EC, extraction chamber; WB, water bath; W, waste; SV, selection valve; EX, extract; E, eluent; IV, injection valve; SPC, solid-phase column; DR, derivatization reagent; RC, reaction coil; and D, detector. (Reprinted from Luque-García, J. L. and Luque de Castro, M. D., *Trends Anal. Chem.*, 22, 41–47, 2003. With permission from Elsevier.)

and ultrasound-accelerated sequential extraction schemes, in terms of extraction efficiency, precision, treatment time and partitioning patterns of metals revealed that the efficiency of the accelerated process depended strongly on the particular metal-matrix interaction.[123,125]

For organic species, ultrasound-assisted leaching is effective in extracting organic pollutants from various type of samples (e.g., separation prior to the determination of PAHs, PCBs,[126] nitrophenols[127] and pesticides[128]). An interesting review concerning ultrasound assisted extraction applied to inorganic and organic analytes has recently been published.[124]

Some recent examples of ultrasound assisted extractions are also given in Table 2.8 and Table 2.9.

C. GAS SAMPLE PREPARATION

Many gas samples do not require complex sample preparation processes since they can be directly injected in a gas chromatograph. However, analytes of interest are often at low concentration, near the limit of detection. It is interesting in such cases to concentrate the analytes in order to increase their sensitivity and transportability. Preconcentration of gas analytes is normally done by trapping them out in an absorbent.

The first attempts to trap analytes were done using a cold trap or a solvent trap (Figure 2.35b). However, misting, rather than condensation, can occur causing low yields and under-estimating real concentrations. To avoid these drawbacks new trapping methods have been developed in the last few years. An alternative is to pass the gas over a cold adsorption tube packed with a GC stationary phase material, such as Tenax, PDMS or polystyrene–divinyl benzene (Figure 2.35a). The trapped components are then usually desorbed thermally and passed directly into a gas chromatograph for separation.

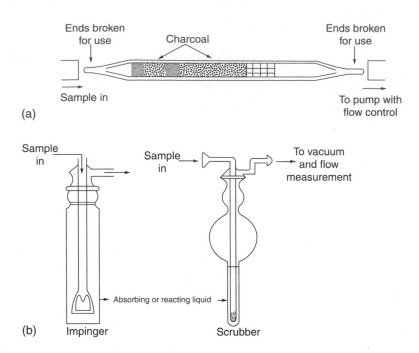

FIGURE 2.35 Scheme of some gas sample preparation systems. (a) absorber tube for gas sample preparation, (b) devices for concentration of gas samples in a liquid absorber.

Another recent alternative is the use of SPME for sampling and sample preparation. Currently eight SPME coatings are commercially available. The most popular for air sampling include polydimethylsiloxane (PDMS), PDMS/divinylbenzene (DVB), and carboxen/PDMS. Each coating has a different sensitivity and can be used to provide selective air sampling. A very promising methodology has been developed using SPME for field sampling combined with on-site analysis with a portable GC.[298] SPME offers some advantages for air analysis such as high sensitivity and precision, speed of extraction, a wide range of sampling times, applicability to a wide range of compounds, the possibility of automation and compatibility with conventional analytical equipment. Disadvantages are a lack of knowledge on the quantification processes, and the relative lack of comparisons with conventional methods because it is difficult to prepare standards for gas sampling in a range of concentrations for the calibration process.

D. DERIVATIZATION

Derivatization is the most studied sample preparation method for the chromatographic analysis of environmental samples (Figure 2.6). It is a sample preparation process in which analytes are chemically modified in order to optimize their possibility of separation and detection by chromatographic techniques. The aim of derivatization depends on the nature of analytes and it has different purposes as a function of chromatographic techniques used (GC or HPLC).

Derivatization in GC is used in order:

- to increase the volatility and decrease the polarity of compounds;
- to reduce thermal degradation of the samples by increasing their thermal stability;
- to increase detector response by incorporating functional groups into the derivative which produce a higher detector signal, for example CF_3 groups for electron capture detectors;

- to improve extraction efficiency from aqueous media (e.g., acylation of phenolic amines) and,
- to improve separation and reduce tailing.

Derivatization in HPLC is used in order:

- to increase the detection power by means of precolumn reagents which introduce chromophores or fluorophores to enhance detectability or reduce interaction problems on the column by decreasing the ability of analytes to ionize.
- to transform hydrophilic analytes into more hydrophobic derivatives.

For these reasons, derivatization is frequently employed in chromatographic analysis of polar compounds such as carboxylic acids, amino acids, and amines. It has a wide range of applications.

However, derivatization has a number of disadvantages. For example, the derivatizing agent may be difficult to remove, interfering in the analysis. Also the derivatization process increases the time and cost of analysis.

Various derivatization techniques are implemented. The principal ones can be classified into four general groups according to the reagents used and the reaction achieved: silylation, acylation, esterification and alkylation.

(a) *Silylation* is the most widely used derivatization technique. Nearly all functional groups which are problematic in GC (hydroxyl, carboxylic acids, phosphate, ...) can be derivatized by silylation reagents. The process involves the replacement of an acidic hydrogen on the compound with an alkylsilyl group, for example, $-SiMe_3-$. The most common reagents for silylation are the trimethylsilyl (TMS) reagents. The structures of the most widely used trimethylsilylating reagents are shown in Figure 2.36 and the reaction of derivatizing a phenol is shown in Figure 2.37.

(b) *Acylation* is a derivatization process in which the polarity of amino, hydroxyl or thiol groups are reduced (Figure 2.37), usually improving their chromatographic properties. It may also result in improved stability of compounds by protecting unstable functional groups or it can enhance the volatility of compounds such as carbohydrates or amino groups. The latter two have many polar functional groups, which can easily be decomposed by heating during GC operation. The most widely used acylating reagent is acetic anhydride, $(CH_3CO)_2O$. Some other acylating reagents are trifluoroacetic anhydride and N-fluoroacyl-imidazole.

(c) *Alkylation* is the replacement of an active hydrogen in an organic group like R-COOH, R-OH, R-SH, or R-NH$_2$ with an alkyl group or, sometimes, aryl group (Figure 2.37). The gas chromatographic properties of the compounds are enhanced because of the decreased polarity of the derivative as compared with the parent compound. One of the most important areas of chromatography where alkylation has been applied concerns carbohydrates. The most widely used reagents are diazomethane and pentafluorobenzyl bromide.

(d) *Esterification* is the first choice for derivatization of acids. It involves the condensation of the carboxyl group of an acid and the hydroxyl group of an alcohol, resulting in the elimination of water (Figure 2.37).

In order to know what is the most appropriate derivatization process for a particular analyte a summary is given in Table 2.6, but the reader is advised to consult the *"Handbook of Derivatives for Chromatography"*[129]; where all these aspects are treated in detail.

Since derivatization is a tedious process there is a trend to introduce it, when possible, in the extraction step. As an example, we note the derivatization in solid-phase microextraction. Derivatization may be used in SPME if very polar compounds must be extracted. It can be

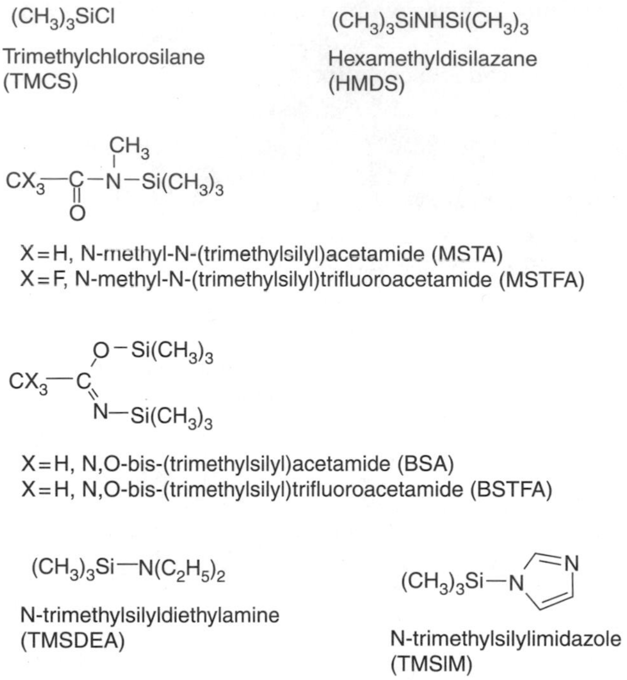

(CH$_3$)$_3$SiCl

Trimethylchlorosilane
(TMCS)

(CH$_3$)$_3$SiNHSi(CH$_3$)$_3$

Hexamethyldisilazane
(HMDS)

$$CX_3-\underset{\underset{O}{\|}}{C}-\underset{\overset{|}{CH_3}}{N}-Si(CH_3)_3$$

X = H, N-methyl-N-(trimethylsilyl)acetamide (MSTA)
X = F, N-methyl-N-(trimethylsilyl)trifluoroacetamide (MSTFA)

$$CX_3-C\overset{O-Si(CH_3)_3}{\underset{N-Si(CH_3)_3}{}}$$

X = H, N,O-bis-(trimethylsilyl)acetamide (BSA)
X = H, N,O-bis-(trimethylsilyl)trifluoroacetamide (BSTFA)

(CH$_3$)$_3$Si—N(C$_2$H$_5$)$_2$

N-trimethylsilyldiethylamine
(TMSDEA)

(CH$_3$)$_3$Si—N

N-trimethylsilylimidazole
(TMSIM)

FIGURE 2.36 Structures of the most commonly used trimethylsilylating reagents.

performed in three ways: (i) direct derivatization in the sample matrix; (ii) doping the fiber coating with the derivatization reagent, and; (iii) derivatization in the GC injection port. The most interesting and most useful is the simultaneous derivatization and extraction directly in the coating of the fiber (Figure 2.38), because it provides high efficiencies.

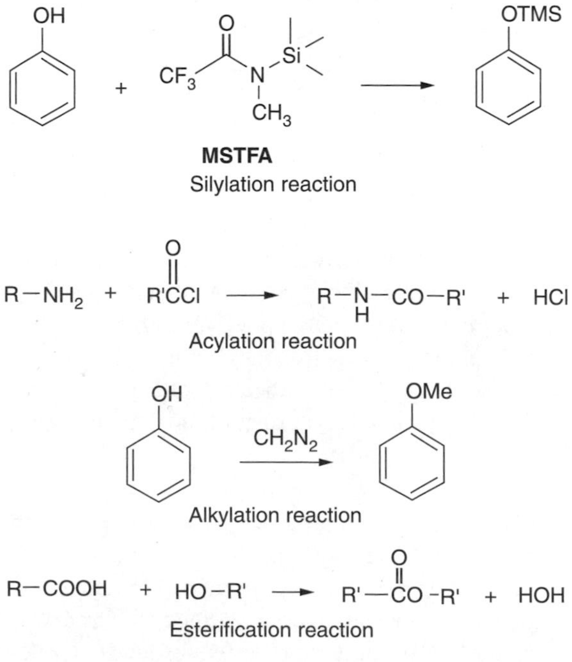

MSTFA
Silylation reaction

$$R-NH_2 \;+\; \underset{\underset{O}{\|}}{R'CCl} \longrightarrow R-\underset{\underset{H}{|}}{N}-CO-R' \;+\; HCl$$

Acylation reaction

Alkylation reaction

$$R-COOH \;+\; HO-R' \longrightarrow R'-\underset{\underset{O}{\|}}{CO}-R' \;+\; HOH$$

Esterification reaction

FIGURE 2.37 Common derivatization reactions.

TABLE 2.6
Selection of Some Derivatization Processes Depending on the Analyte to be Analyzed

Functional group	Derivatization
−OH (hydroxyl group) in primary, secondary and tertiary alcohols; phenols; carbohydrates	Silylation
	Acylation
	Alkylation
	Dansylation
	Ion pair formation, etc
R−COOH	Esterification
	Silylation
	Ion pair formation
R−C=O in aldehydes or ketones	Oxime formation
	Ketal formation
	Hydrazone formation
	Silylation
R−NH$_2$ in primary amines, amino acids and amino sugars	Acylation
	Benzoylation
	Silylation
	Thiourea formation
	Schiff's base formation
	Carbamate formation
	Alkylation
	Ion pair formation, etc
R′ NH−R in secondary amines, amino acids, substituted amino sugars	Acylation
	Silylation
	2,4-Dinitrophenylation
	Sulphonamide formation
	Ion pair formation
R−NH$_2$ and R−COOH in amino acids	Esterification + acylation
	Silylation

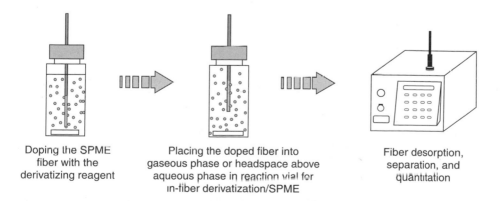

Doping the SPME fiber with the derivatizing reagent

Placing the doped fiber into gaseous phase or headspace above aqueous phase in reaction vial for in-fiber derivatization/SPME

Fiber desorption, separation, and quantitation

FIGURE 2.38 In-coating derivatization technique with fiber doping method. (Reprinted from Lord, H. and Pawilszyn, J., *J. Chromatogr. A*, 885, 153–193, 2000. With permission from Elsevier.)

TABLE 2.7
Different Sample Preparation Processes to Determine Pollutants in Water Samples

Analyte	Technique	Conditions	Comments	Ref.
Sulfur and VOCs	SPME	PDMS 100 μm, DVB–PDMS 65 μm and carboxen–PDMS 75 μm, all from Supelco	*Determination*: GC-pulsed flame photometric detection; good recoveries in all analytes	130
Naphthalene monosulphonates	IP-SPME online	PLRP-s sorbent	*Determination*: IP-LC with fast-scanning fluorescence detection, LOQ: 0.01 to 3 μg/l; *RSD*: between 0.5 and 4%	131
Alkyl sulfides	HS–SPME	Extraction parameters optimized for PDMS–carboxen fiber	*Determination*: GC–MS; *LOD*: 4 ng/l for dimethyl sulfide, 0.7 ng/l for ethylmethyl sulfide, 5 ng/l for diethyl sulfide and 1 ng/l for dimethyl disulfide; *RSD*: between 4 and 6%	132
Aromatic amines	LLLME	Donor solution: 0.1 M sodium hydroxide solution with 20% sodium chloride and 2% acetone; organic phase: di-n-hexyl ether; acceptor solution: 0.5 M hydrochloric acid and 500 mM 18-crown-6 ether; extraction time of 30 min; stirring at 1000 rpm	*Determination*: HPLC; *LOD*: <0; 10 μg/l; *recoveries*: >85%; RSDs of the four anilines were lower than 4.83% for inter-day experiments and 7.26% for intraday experiments	133
Amines	SPME	Three fibers prepared by sol–gel method, containing hydroxydibenzo-14-crown-4 (OH-DB14C4), dihydroxy-substituted saturated urushiol crown ether (DHSU14C4) and 3,5-dibutyl-unsymmetry-dibenzo-14-crown-4-dihydroxy crown ether (DBUD14C4)	*Determination*: GC, *LODs*: varied from 0.17 to 0.98 ng/ml; *RSD*: 3.23 to 6.20%	134
Amines	LLLME	Optimal conditions of the extraction were donor phase (a): 2 ml of water sample adjusted to pH 13 with NaOH–NaCl; organic phase (o); 150 μl ethyl acetate; and 1 receiving phase (a) of 2 μl aqueous solution at pH 2.1	*Determination*: HPLC; *LODs*: from 0.85 ng/ml to 1.80 ng/ml Enrichment factors ranged from 218 (for 4-nitroaniline) to 378 (for 4-chloro-2-aniline)	135
Amines	SPE	Empore disk C$_{18}$ (Sumitomo 3M, Tokyo, Japan)	*Determination*: LC–MS; LODs of PBTA-1 and PBTA-2: 1 ng/l and 2 ng/l; *recoveries*: 87 and 106% for PBTA-1 and PBTA-2, respectively; RSDs below 4%	136
Free volatile fatty acids	HS–SPME	PDMS–carboxen fiber	*Determination*: GC–CI-MS; LODs in the range of 150 μg/l for acetic acid and from 2 to 6 μg/l C$_3$–C$_7$ FA	137

			Determination	
Free volatile fatty acids	HS–SPME	Five different coatings and PDMS–carboxen; extraction parameters optimized	*Determination*: GC–FID/NCI-MS; *LODs*: μg/l levels and *RSDs*: between 5.6 and 13.3%; *linear dynamic range*: over 2 to 4 orders of magnitude	138
VOCs	HS–SPME	Fibers (100-μm PDMS)	*Determination*: GC–FID, *RSD*: <5%; concentration level 42.5 μg/l	139
VOCs	SPME	Carboxen–PDMS fiber	*Determination*: GC–FID; n.d	140
VOCs	HS–SPME	65-μm PDMS–DVB	*Determination*: GC–FID; *LOD*: 0.45 μg/l; RSD for 250 μg/l MTBE (*n* = 7) : 6.3%; *calibration linear range*: 5 to 500 μg/l	141
SemiVOCs	SPME	Polysilicone fullerene (PF) coating	*Determination*: GC–FID; *LODs*: 10 ng/l to 1 μg/l level; *RSD*: 7%	142
VOCs	SBSE	Optimized conditions	*Determination*: thermal desorption GC–MS; *RSD*: 10 to 15%	143
PAHs	LPME/SDE	—	*Determination*: HPLC; high enrichment (60- to 180-fold); RSD range of 4.7 to 9.0%; *LOD*: 0.35 to 0.60 μg/l; *recoveries*: >90% except for benzo[*b*]fluoranthene (88.7%) in tap water	144
PAHs	SPE	—	Review	145
PAHs and PCBs	SPE	500 mg of activated silica gel	*Determination*: GC–MS; recovery PCB (92 to 119%)	146
PAHs	RP-SPE; DIE-SPE	Ion exchange cartridges	Negative effect of humic acid	147
PAHs	SBSE	The optimal extraction time was found to be between 3 and 4 h	*Determination*: GC–MS; method *reproducibility*: (*n* = 9) : 5 and 15% at 10 ng/l and between 3 and 9% at 50 ng/l; detection limits are between 0.1 and 2 ng/l	148
PAHs	HS–SPME	85-μm polyacrylate (PA) and 100-μm PDMS fibers	*Determination*: GC–FID/GC–MS; the precisions of PA and PDMS fibers were from 3 to 24% and from 3 to 14%; respectively	149
PAHs	Sonication	Freeze-dried samples by a dichloromethane–methanol (2:1) mixture in a sonication bath; the sludge extracts were cleaned-up by an alumina column	*Determination*: GC–MS; recovery from 60 to 98%	150
Halogenated compounds	SBSE	The stir-bars (10 mm long × 3.2 mm O.D.), coated with an extracting phase of PDMS (63 μl)	*Determination*: GC–MS; *LODs*: 0.01 to 0.24 μg/l. The repeatability and reproducibility of the method (n = 5): below 13 and 23%; *recoveries*: between 42 and 96%	151
Halogenated compounds	SPE	Polystyrene–DVB sorbents LiChrolut EN, ER-P; Isolute ENV1, and Oasis HLB	*Determination*: IP-LC–EI-MS; *LOD*: 0.1 to 1.6 μg/l; LOQ: 0.1 to 2.4; highest recoveries were obtained with LiChrolut EN	152

Continued

TABLE 2.7
Continued

Analyte	Technique	Conditions	Comments	Ref.
Halogenated compounds	MIP-SPE	Molecularly imprinted polymer (MIP) was synthetized using the herbicide 2,4,5-trichlorophenoxyacetic acid as a template, 4-vinylpyridine as an interacting monomer, ethylendimethacrylate as a cross-linker and a methanol–water mixture as a porogen	*Determination*: reversed-phase liquid chromatography and capillary zone electrophoresis; concentration: 500-fold; quantitative recovery	153
Halogenated compounds	SPE	Sorbents: EnviCrom-P, Porapak, Oasis HLB, 30 mg, Oasis HLB, 60 mg, Oasis HLB, 200 mg	*Determination*: LC with electro-chemical and MS detection; *LODs*: 20 to 40 pg/μl; all types of sorbent gave recoveries around 100% for most of the studied compounds; method repeatability: from 2 to 8% in all cases	154
Halogenated compounds	SPME	Five commercially available fibers, PDMS, 100 μm; PA, 85 μm; CAR–PDMS, 75 μm; PDMS–DVB, 65 μm; and PDMS, 7 μm were purchased from Supelco	*Determination*: GC-ECD; *LOD*: 0.02 μg/l for Milli-Q water and 0.3 μg/l for tap water and river water; *RSD*: between 12 and 14%	155
Halogenated compounds	HS–SPME	PDMS fiber, sampling temperature of 25°C, an absorption time of 10 min, the addition of 0.1 g of anhydrous sodium sulfate and a desorption time of 2 min	*Determination*: GC–IT-MS; *LOD*: ranged from 10 to 200 ng/l; *RSD*: <10%	156
Halogenated compounds	HS–SPME	SPME fiber coated with 100-μm thick PDMS	*Determination*: GC–MS; LOD below 0.006 μg/l; linearity range was 0.02 to 20 μg/l; *RSD*: 1.19 to 8.19%; recovery: >90%	157
Halogenated compounds	SPME	Four types of fibers: 7-μm (7-PDMS), 100-μm (100-PDMS), 75-μm (CAR–PDMS) and 65-μm (CWX–DVB)	*Determination*: GC-ECD; linear range: from 0.1 μg/l to 20 μg/l; *LOD*: 2 and 3 ng/l; RSD for repeatability ranged from 2 to 7%; RSD for reproducibility ranged from 4 to 7%	158
PCBs PAH phthalate esters	SPME	30-μm PDMS fiber; multisimplex optimization of extraction parameters	*Determination*: GC–MS; *LOD*: 0.03 to 0.12 μg/l for PAHs, 0.03 to 0.11 μg/l for PCBs, 0.07 to 0.84 μg/l for phthalate esters (except DEHP 3.15 μg/l); *RSD*: 3 to 15 for PAHs, 8 to 13% for PCBs, 2 to 9 for phthalate esters; except nap (*RSDs*: 27%)	159

PCBs	MAE–HS–SPME	Irradiation of extraction solution (20 ml aq. sample in 40 ml HS vial with no additions of salt and MeOH) under 30 W microwave power for 15 cycles (1 min power on and 3 min power off of each cycle); desorption at 270°C for 3 min	*Determination*: GC-ECD; *LOD*: between 0.27 and 1.34 ng/l; linear dynamic range from 1 to 80 ng/l	160
Organometallic substances			Review	161
Organometallic substances	SPME	Sorption and thermal desorption optimized using experimental designs; two fibers investigated: 100-μm thickness PDMS-coated fiber and the 75-μm thickness (CAR–PDMS) fiber	*Determination*: GC coupled to pulsed flame photometric detection; LODs obtained for the 14 studied organotins compounds are widely sub-ng(Sn)/l; RSD ranges between 9 and 25% from five determinations of the analytes at 0.25 to 125 ng/l concentrations	162
Organotin compounds	HS–SPME	Manual SPME fiber holder and 100 μm film thickness PDMS fibers. Thermally desorbed in the split-split less injector of GC	*Determination*: multicapillary GC with atomic emission detection. *LOD*: 1 to 5 ng/l; *RSD*: 6 to 10% at concentrations of 20 ng/l	163
Organotin compounds	HS–SPME	HS extraction was performed for 10 to 20 min at 75 or 85°C with a 100 μm PDMS fiber	*Determination*: CGC-ICP-MS; *LODs*: 2 pg/l (instrumental) and 125 pg/l (procedure); repeatability of 8% RSD ($n = 10$); for triphenyltin (TPhT) and triclorinetin (TCT), average recoveries of 25 and 50%, respectively	164
Organotin compounds	HS–SPME	SPME holder and the fibers coated with 100 μm thickness PDMS	*Determination*: GC-FID; *LODs*: 28 ng/l (as Sn) for TeET and 20 ng/l (as Sn) for TeBT; DBT and TBT presented linear range from 0.5 to 10 mg/l, for MBT the linear large was check up to 50 mg/l; *RSD*: <10%	165
Phenols	SPE	Different SPE cartridges: EnviCarb, 250 mg graphitized carbon (Supelco, Bellefonte, PA, USA); Isolute ENV1, 200 mg styrene-DVB copclymer (IST, Mid Glamorgan, UK); Isolute C_{18}/ENV1 100 mg C_{18} silica and 100 mg styrene-DVB (IST); Oasis HLB, 200 mg N-vinylpyrrolidone-LVB copolymer (Waters, Milford, MA, USA)	*Determination*: GC–MS; *LODs*: 0.1 μg/l for most phenols (25), and for 12 phenols 0.01 μg/l; Isolute ENV + cartridges exhibited the best retention of phenols; *recoveries*: range 79 to 104% (SD 1 to 12%), except for phenol 26.6% (SD 1.3%) and 2-methoxyphenol 62.64 (SD 2%)	166
Phenols	SPE–Soxhlet	Automated system ASPEC XL (Gilson, Villiers-le-Bel, France): Oasis 60 mg cartridges were rinsed with 3 ml of dichloromethane; six-port processing system (Baker) was used; speed disks were of styrene-DVB and contained a mesh on top of the sorbent which acted as a filter	*Determination*: LC-ED; *LOD*: from 2 to 10 ng/ml for phenol compounds and dichlorophenols and from 20 to 60 ng/ml, for the other compounds; *recoveries*: between 60 and 120%; *RSD*: <12%	167

Continued

TABLE 2.7
Continued

Analyte	Technique	Conditions	Comments	Ref.
Phenols	LPME/BE	Extracted analytes from 2-ml aq. sample adjusted to pH 1 (donor solution) through a microliter-size organic solvent phase (400-μl n-hexane), confined inside a small PTFE ring, and finally into a 1-μl basic aq	*Determination*: HPLC; *LODs*: 0.5 to 2.5 μg/l; *RSD*: 5.4 to 11.5; *recovery*: >85%; large enrichment factor (more than 100-fold)	168
Phenols	SPE	Sorbents, mechanism, derivatization, matrix effects and storage reviewed	Review	169
Phenols	SPME	50 μm Carbowax-templated resin (CW-TPR) and 60 μm PDMS–DVB	*Determination*: HPLC; *LOD*: 1 to 10 μg/l; only about 1 to 16% of the phenols were extracted under optimum conditions; *RSD*: <5%	170
Phenols	Online SPE	Hysphere GP from Spark (polyDVB, 5 to 15 μm particle size, spherical shape) or Oasis HLB material (macroporous polyDVB-N-vinylpyrrolidone copolymer)	*Determination*: HPLC-atmospheric pressure CI-MS, *LODs*: from 40 to 280 ng/l; *RSDs*: <8%; the method was used to detect 2-nitrophenol and 4-nitrophenol in river water samples in the lower ng/l range; recovery 90 to 105%	171
Phenols	SPME	85 μm polyacrylate fibers; dynamic and static modes of desorption in both HPLC designs	*Determination*: HPLC; *LODs*: UV: 0.4 and 18 μg/l; ED: 0.01 and 12 μg/l	172
Phenols	HS–SPME direct-SPME	Three polar fibers: PA, 85 μm; CWX–DVB, 65 μm partially crosslinked, and StableFlex DVB–CAR–PDMS, 50 and 30 μm	*Determination*: GC-FID, GC–MS, *LODs*: from 30 to 150 ng/l; *linear dynamic range*: 0.11 to 2.5 μg/l; *RSD*: 4 to 15%	173
Phenols	HS–SPME	75 μm CAR–PDMS and 100 μm PDMS; optimization of temperature, type of microextraction fiber and volume of sample by means of a mixed-level categorical experimental design	*Determination*: GC–MS; *LOD*: <0.1 ng/ml for all phenols with both fibers; *linear dynamic range*: 0.1 to 10 ng/ml; RSD from 0.3 to 12% for PDMS and from 0.7 to 12% for CAR–PDMS at 10 ng/ml level of each phenol	174
Nitro-phenols	LLLME	The acceptor phase: NaOH solution used at various concentrations; organic solvent was immobilized into the pores of the hollow fiber	*Determination*: capillary liquid chromatography; *LODs*: 0.5 to 1.0 ng/ml; up to 380-fold enrichment. *RSDs*: <6.2%; *linear dynamic range*: from 1 to 200 μg/ml; *recoveries*: >90%	175
Nitro-phenols	Online SPE (MIP)	Molecularly imprinted polymer (MIP)	*Determination*: HPLC, *RSDs*: <12% in all cases; *recoveries*: 38 to 78%	176

Chloro-phenols	SPE	Sorbents: C$_{18}$ and activated carbon (AC) membranes by dynamic or static sorption	*Determination:* GC–MS; C$_{18}$ membranes *recoveries:* 70 to 102% for all compounds (except phenol) dynamic desorption with acetone as the eluting solvent; static desorption using C$_{18}$ membranes 65 to 80% for chlorophenols; static desorption using AC membranes resulted in low recoveries (<60%) for all compounds except phenol; dynamic desorption of both membranes in the reverse direction (AC on top of C$_{18}$) using 10 ml of acetone or toluene gave quantitative results for all compounds	177
Chloro-phenols	SPE	Three polymeric sorbents: poly-N-methylaniline (PNMA), polyaniline (PANI) and polydiphenylamine (PDPA)	*Determination:* GC-ECD, GC-FID; *LODs:* 1 and 40 ng/l; phenol recovery: 32% with PNMA, not concentrated on PANI and PDPA; *recoveries:* 55 and 72% for 2CP, 4CP and 24DMP and more than 86% for 246TCP and PCP	178
Chloro-phenols	SPME	Three fiber: 50 μm Carbowax-templated resin (CW-TPR), 60 μm polydimethylsiloxane–DVB (PDMS–DVB) and 85 μm polyacrylate (PA); the most suitable: CW-TPR	*Determination:* LC-ECD; *LODs:* 3 to 8 ng/l; *RSDs:* 4 and 11%; linear over three to four orders of magnitude	179
Insecticides (Chlorinated compounds)	LPME	—	*Determination:* HPLC, Concentration factors of 50-fold. *RSD:* range of 3.2% (lindane) to 10.7% (methoxychlor). *recoveries:* >90%; *LODs:* <0.05 μg/l	180
Insecticides (organophosphorus compounds)	HS–SPME	Fibers of PA 85 μm, and PDMS 100 μm	*Determination:* GC-FID, GC–MS; *LODs:* 0.01 to 0.04 μg/l; *RSD:* <17%; linear range of 0.05 to 1 μg/l; *recoveries:* from 80 to 120%	181
Insecticides (carbofurane)	SPE–SPME	SPE: 360 mg C$_{18}$ Sep-Pak cartridge; SPME: PDMS–DVB fiber	*Determination:* HPLC-PDA; *LOD:* 0.06 μg/l; *LOQ:* 0.08 μg/l; SPE: absolute recovery 90 ± 3.2% (at 1 μg/l); *reproducibility: RSD:* 7.0%, linearity range 0.1 to 50 μg/l; *LOD:* 8.9 μg/l; *LOQ:* 10.0 μg/l; SPME: relative recovery 100 ± 7.7% (at 10 μg/l); *reproducibility: RSD:* 5.1%, linearity range 10 to 50 μg/l	182
Insecticides (urea)	SPE–LLE, vs. SPE–SPME	SPE: Carbopack cartridges (Envi-Carb (Carbopack B), 250 mg, Supelco, Bellafonte, CA, USA). SPME: polyacrylate coated fiber (85 μm film thickness) from Supelco (Bellafonte, PA, USA)	*Determination:* GC–MS, *LODs:* 0.3 to 1.0 ng/l; *RSD:* <10%; SPE recoveries around 100%	183
Herbicides	SPME		Review	184
Herbicides	MLLE	Porous poly(tetrafluoroethylene) membrane, which is immobilized with a water-immiscible organic solvent, and are trapped in a stagnant acidic acceptor phase since these become protonated	*Determination:* HPLC; extraction efficiencies of 60% or better; about 0.03 μg/l were obtained by extraction of 1.0 μg/l	185

Continued

TABLE 2.7
Continued

Analyte	Technique	Conditions	Comments	Ref.
Herbicides	LLE	—	*Determination*: GC–MS; *LODs*: 10 to 60 ng/l; *RSDs*: 8 to 15%; quantitative recovery	186
Herbicides	SPE	Sep-Pak Plus C_{18} bonded phase	*Determination*: HPLC with photoinduced-fluorimetric detection; *LODs*: 0.07 to 0.35 μg/ml; *recoveries*: 70 to 130%, with a maximum RSD of 30%; calibration *RSD*: <2%	187
Herbicides (urea)	SPME	Sorbents: polydimethylsiloxanes and a polyacrylate fiber (PA)	*Determination*: GC–NPD; *LODs*: 0.04 μg/l for linuron to 0.1 μg/l for fluometuron and monuron; RSDs at the 1 μg/l level are between 15% and 9%	188
Herbicides (urea)	LLE–SPE	—	Review	189
Herbicides (urea)	SPE	Carbograph 4 cartridge	*Determination*: LC–MS, LODs for drinking water, groundwater and surface water were between 3 and 11 ng/l, 6 and 21 ng/l, 36 and 75 ng/l, respectively; *recoveries*: >85%; *RSDs*: 4.6 to 20% for drinking water, 4.3 to 15% for ground water, 5.9 to 13% for river water	190
Herbicides (urea)	MLLE	Online coupling continuous-flow liquid membrane extraction with HPLC	*Determination*: HPLC; *LODs*: 100 and 0.05 μg/l for MSM, and 96 and 0.1 μg/l for DPX-A 7881, respectively; linear range and precision (RSD): 0.1 to 50 μg/l and 7.0% for MSM, and 0.2 to 50 μg/l and 9.2% for DPX-A 7881, respectively; *recoveries*: 83 to 95% for MSM and 88 to 100% for DPX-A 7881, respectively	191
Herbicides	SPME	—	Review	192
Herbicides	SPE	Sorbent: C_{18} bonded silica	*Determination*: GC–MS; recovery over 80% except for Chloramben (50%), fenoprop (73%), MCPB (67%), and 2,4-DB (70%); *LODs*: 5 to 20 μg/l with 2 ml of sample injection	193
Herbicides	SPE	Sorbents: C_{18} on silica and polymeric sorbents (Oasis and LiChrolut EN), the best results being obtained with the 18 styrene-DVB cartridge and when the elution was performed with methanol and ethyl acetate	*Determination*: HPLC-PDA; *LODs*: 0.1 μg/l for DIHA and DEDIA and 0.02 μg/l for the other analytes; *RSDs*: <17.7%	194
Herbicides	SPE	Sorbent: polymeric cartridges (PLRP-s)	*Determination*: LC–MS; *LODs*: 0.005 μg/l; *recoveries*: ranging from 96 to 111%	195

Analyte	Method	Sorbent / Conditions	Determination	Ref.
Herbicides	SPE	*Sorbent:* 0.5 g GCB extraction cartridge	*Determination:* HPLC-ionspray interface-MS; *LODs:* <10 ng/l in drinking water samples; *recovery:* exceeded 83% for all the analytes	196
Herbicides (quaternary ammonia)	SPE	—	Review	197
Herbicides (quaternary ammonia)	Online IP-SPE	C$_8$ extraction disks	*Determination:* LC–MS; *recoveries:* higher than 70%; *LODs:* from 6 to 85 ng/l; RSDs in the range of 7 to 13%; *RSD:* between 9% and 14% (days-to-day); *accuracy:* 1 to 3% (run-to-run) and 2 to 9% (day-to-day)	198
Herbicides (quaternary ammonia)	SPE	Two different porous graphitic carbon cartridges	*Determination:* capillary electrophoresis; *LODs:* <0.3 μg/l for Milli-Q water and <2.2 μg/l for drinking water; *recoveries* > 80% for difenzoquat and around 40% for paraquat and diquat; *RSDs:* 6.3 to 7.6% (run-to-run), <12.3%	199
Herbicides	SPME	SPME fiber: a 60-μm crosslinked PDMS–DVB	*Determination:* HPLC; *LOD:* 0.27 ng/g; *LOQ:* 0.91 ng/g	200
Herbicides	SPME	Sorbent: CBX-DVB	*Determination:* GC-ECD, valid at concentration levels from 5 to 20 ng/l; *recoveries:* >98% with a *RSD:* <0.3%. *LODs:* 10 to 30 ng/l	201
Herbicides	SPME	Sorbent: LiChrolut EN (200 mg) styrene-DVB polymeric sorbents	*Determination:* micellar electrokinetic chromatography; *LODs:* between 0.13 and 2.73 μg/l	202
Pesticides	Online SPE	Commercially available C$_{18}$ columns (100 mg) were used for preconcentration and put directly into the CFS manifold after preconditioning with methanol, water and elution solvent	*Determination:* capillary electrophoresis; *recoveries:* 90 to 114% for most of the pesticides	203
Pesticides (organophosphorus)	SPME	Sorbent: 65-μm thickness PDMS–DVB solid-phase microextraction fiber	*Determination:* GC with flame photometric detector; GC–MS; *LODs:* 2 to 8 μg/l	204
Pesticides	SPE	Sorbent: C$_{18}$ cartridge	*Determination:* LC-ES-MS; *LODs:* 0.5 to 60 ng/l; *recoveries:* 70 to 120%; *precision:* RSD: <15%	205
Pesticides	SPME	CWX–DVB SPME fiber	*Determination:* GC–MS; *LODs:* <ng/l range; *Mean recoveries:* 81% for propanil (*RSD:* 20%), 99% for acetochlor (*RSD:* 22%), 97% for myclobutanil (*RSD:* 12%) and 110% for fenoxycarb (*RSD:* 14%)	206
Pesticides (phenylurea and carbamate)	Online SPME	Sorbents: a PPY-coated capillary, a PMPY-coated capillary, Omegawax 250 (0.25 μm film thickness, 0.25 mm I.D.), Supel-Q PLOT (thickness unknown, 0.32 mm I.D.), Supelcowax (0.1 μm film thickness, 0.25 mm I.D.), SPB-1 (0.25 μm film thickness, 0.25 mm I.D.), SPB-5 (0.25 μm film thickness, 0.25 mm I.D.	*Determination:* HPLC–MS; *LODs:* 0.01 to 0.32 ng/ml for PPY-coated capillary in-tube SPME-HPLC–ESI-MS and 2.1 to 4.5 ng/ml for PPY-coated capillary in-tube SPME-HPLC-UV	207

Continued

TABLE 2.7
Continued

Analyte	Technique	Conditions	Comments	Ref.
Pesticides	MIP-SPE	Preconcentration on a C_{18} disk coupled to selective clean up on the MIP	*Determination*: HPLC; The recovery for terbutylazine 70% ($n = 4$, SD = 16%), for atrazine 76% ($n = 4$, SD = 6%), for amertyn 45% ($n = 4$, SD = 14%) and for promeryn 38% ($n = 4$, SD = 25%)	208
Pesticides	SPE	—	Review	209
Pesticides	SPE	Sorbent: C_{18} cartridges	*Determination*: GC-ECD, GC-MS and GC-MS-MS; *recoveries*: 70 to 133% in water samples spiked at 100 ng/l and the relative standard deviations were in the range 5.3 to 17.4%	210
Pesticides	SPE	Extraction columns: 6 ml disposable inter extraction cartridges packed with 200 mg styrene DVB copolymer (SDB)	*Determination*: GC-MS; *LODs*: <0.025 µg/l, pendimethalin and metribuzin 0.062 and 0.035 mg/l, respectively; *linear dynamic range*: most of the pesticides from 0.015 to 0.04 ng/ml, pendimethalin and metribuzin from 0.15 to 0.5; *recoveries*: 75 to 125%	211
Pesticides	SPME	Six SPME fibers (7, 30 and 100 µm) PDMS, 85 µm PA, 60 µm PDMS-DVB and 65 µm CWX-DVB and the 60 µm PDMS-DVB selected for the simultaneous extraction of 34 compounds	*Determination*: GC-ECD and GC-thermoionic specific detector; *LODs*: in the range of 1 to 10 ng/l for OCPs, 1 to 30 ng/l for OPPs, 20 to 30 ng/l for pyrethroids and to 50 ng/l for triazines; *precision*: RSD <16%; *recovery*: 93.8 to 104.5%, four analytes shows matrix effect	212
Pesticides (triazines)	MLLE	A polymer membrane (nonporous polypropylene) to separate an aqueous sample from an organic extractant	*Determination*: GC-MS; *LODs*: 1 to 10 ng/l; *recoveries*: 60 to 90%	213
Pesticides (triazines)	SPME	Five different SPME fibers: PDMS 100 µm, PA 80 µm, PDMS-DVB 65 µm, CWX-DVB 65 µm, and CAR-PDMS 75 µm; PDMS-DVB was selected	*Determination*: GC-MS; *LODs*: 2 to 17 ng/l, and precision (*RSD*: <8%) for the selected PDMS-DVB fiber	214
Pesticides (triazines)	SPE	C_{18} SPE cartridges	*Determination*: HPLC; *LODs*: 4.9 to 16 µg/l and *LOQs*: 14.8 to 50.1 µg/l; *LODs*: 0.0098 to 0.034 µg/l; *LOQs*: 0.024 to 0.10 µg/l after 500-fold preconcentration; *recoveries*: 76 to 97%	215

Analyte	Technique	Sorbent / Details	Determination	Ref.
Pesticides (organochlorine)	SPME	30-μm thickness PDMS SPME fiber	*Determination*: dual column GC-ECD; *LODs*: 10 to 40 ng/l	216
Pesticides (organochlorine)	SPME	Sorbent: polyacrylate-coated fiber; optimized by using a response surface generated with a Doehlert design	*Determination*: GC-ECD; *LODs*: 0.15 to 0.35 ng/l; the linear range: 0.001 to 2.5 μg/l; repeatability (RSD): 5.7 to 25.6% and reproducibility: 7.6 to 26.5% (RSD) between days of the method at a level of 1 ng/l	217
Pesticides (carbamate)	SPE	Zorbax (Agilent Technologies — C$_{18}$; 3 ml, 500 mg), LC (Supelco—C$_{18}$; 3 ml, 250 mg, Envicarb (Supelco; 3 ml, 250 mg), Oasis HLB (Waters; 6 ml, 200 mg), Envi (IST; 3 ml, 100 mg) and Bond ElutEnvi (Varian; 3 ml, 500 mg)	*Determination*: LC–ESI-MS; *LOD*: 0.10 μg/l, except methomyl (LOD 0.50 μg/l); linear dynamic range: 1 to 50 μg/l and precision: RSD <7.8%; recoveries: 73.7 to 92.6%	218
Fungicides	SPME	Four types of fiber have been assayed and compared: PA (85 μm), PDMS (100 and 30 μm), CWX–DVB (65 μm) and PDMS–DVB (65 μm)	*Determination*: GC-ECD and GC–MS; LODs in the range of 1 to 60 ng/l, by using electron-capture and mass spectrometric detectors; *concentration range*: 0.1 to 10 μg/l; *recoveries of all fungicides*: 70.0 to 124.4%	219
Gasoline	SPE	Polystyrene–DVB sorbent	*Determination*: GC–MS; recovery from spiked 11 groundwater samples was 88 to 100%; the precision of the method, indicated by the RSD, was 4% and the detection limit was 0.2 μg/l	220
Phthalate esters	LPME	—	*Determination*: GC–MS; linear calibration curve from 0.02 to 10 μg/l; LOD 0.005 to 0.1 μg/l; *repeatability of the method*: 4 to 11%	221
Phthalate esters	SPME	Sorbents: 7 μm PDM; 100 μm PDMS: 85 μm PA (Polar); 65 μm PDMS–DVB (Bipolar); 50 to 30 μm DVB–carboxen–PDMS (Bipolar); 70 μm CW–DVB (Polar)	*Determination*: GC–MS; *LODs*: 0.015 to 0.06 μg/ml; precision: (RSD): 5.6 to 7%	222
Phthalate Esters	SPME	85 μm polyacrylate fiber	*Determination*: GC–MS; *LODs of the method*: 0.006 to 0.17 μg/l; *linear range*: 0.02 to 10 μg/l for most compounds	223
Surfactants	MMLLE	—	*Determination*: LC; enrichment of over 250 times; *LODs*: 0.7 to 5 μg/l	224
Surfactants	SPME	PDMS, 100 and 7 μm fibers and SPEC 3 ml strong anion exchange (SAX) columns	*Determination*: GC–MS; *recoveries*: 71 to 94%	225

TABLE 2.8
Different Sample Preparation Processes to Determine Pollutants in Soils and Sediments

Analyte	Technique	Conditions	Comments	Ref.
PAHs	PHWE HF-MMLLE	—	*Determination*: GC-FID, GC–MS; *LOD*: 50 to 890 pg; 0.11 to 1.22 μg/g; no extra clean up steps	226
PAHs	DMAE	—	*Determination*: GC-PID; extraction efficiency similar to Soxhlet, but lower sample preparation time; temperature and extraction time are found as important parameters by factorial design	227
Nitro-PAHs	DMAE & Soxhlet	—	*Determination*: GC-ECD; *optimized parameter by experimental design*: irradiation power, irradiation time, number of cycles and extractant volume	228
PAHs	PLE	Toluene was selected as extraction solvent and a total solvent volume of 100 ml was used for the 10 min static-dynamic PLE of 50-mg samples	*Determination*: GC–MS; *variable study by experimental design*: sample load, solvents used, solvent ratios, pressure, temperature, extraction time, and rinse volume	229
PAHs	PLE		*Determination*: LVI-GC–MS; *LOD*: below 9 ng/g soil for the 13 PAHs	230
PAHs	HF-MSME	Eight micro liter of octane extraction solvent was placed inside a porous, polypropylene fiber; following an 8 min analyte preconcentration step, 4 μl of extract was injected into a gas chromatograph	GC-FID; separation was achieved in less than 10 min with a detection limit of 0.13 mg/kg metylnaphthalene	231
PAHs	SWE, SPE	Static subcritical water extraction was coupled with styrene-DVB (SDB-XC) extraction discs; soil, water, and the SDB-XC disc are placed in a sealed extraction cell, heated to 250°C for 15 to 60 min, cooled, and the PAHs recovered from the disc with acetone/methylene chloride	PAHs with molecular weights from 128 to 276 are quantitatively (>90%) extracted and collected on the sorbent disc	232
PAHs	Soxhlet, PLE, SFE & SWE	With a Soxhlet apparatus (18 h), by PLE (50 min at 100°C), SFE (1 h at 150°C with pure CO_2), and subcritical water (1 h at 250°C, or 30 min at 300°C)	*Determination*: GC-FID, GC–MS; the organic solvent extracts (Soxhlet and PLE) were much darker, while the extracts from SWE (collected in toluene) were orange, and the extracts from SFE (collected in CH_2Cl_2) were light yellow	233
PAHs	DMAE	The optimal conditions: 30% of water, 30 ml of dichloromethane, 30 W, 10 min	*Determination*: GC–MS; Recovery of 90%	234

PAHs	SPE DMAE	Variables optimized by experimental design	Determination: GC–PID, GC–MS	235
PAHs	PLE	Dichloromethane–ethanol solvent mixture was found to be the most efficient solvent	Determination: GC–MS	236
PAHs	SFE	SFE conditions: 0.3 g urban air particulate matter, 0.5 g river sediment; modifier added: different modifiers (methanol, CH_2Cl_2, toluene, hexane, acetonitrile, aniline, diethylamine, acetic acid and its mixture) and concentrations (1 to 10%), 400 atm, 80°C, 1 ml/min, 5 min static, 10 min dynamic	Determination: PCBs: offline GC-ECD; PAHs: offline GC–MS; the modifier identity is more important than modifier concentration for enhancing extraction efficiencies of PAHs and PCBs; however, the concentration influences over high molecular weight PAHs recoveries; *PCBs best modifiers*: acidic/basic modifiers including methanol, acetic acid and aniline; *low molecular weight PAHs best modifiers*: aniline, acetic acid, acetonitrile, methanol/toluene, hexane and diethylamine; *high molecular weight PAHs best modifiers*: toluene, diethylamine and methylene chloride	65
PAHs	SFE	0.1 g soil: *Two-step extraction*: (1) 60 g CO_2, 450 atm, 80°C, 1 ml/min, T-trap − 10°C for collection and 10°C for desorption; (2) 100 g CO_2 + 10% aceton, 2 ml/min, T-trap 60°C desorption, other conditions were same as (1); *collection*: C-18 trap	Determination: Online SFE–LC–UV; extraction recovery: ≥95% PAHs; compared to EPA method (Soxhlet extraction following by GC–MS), online SFE–LC gives precise results in a much shorter time	237
PAHs	SFE	Pretreatment: 5 g of dried soil mixed with 1 g of copper powder and 10% (0.5 g) of water to increase the extraction efficiency; *SFE conditions*: *three procedures*: (1) pure CO_2, 121 atm, 80°C, 2 ml/min, 10 min static, 10 min dynamic; (2) CO_2 + 1% MeOH + 4% DCM, 339 bar, 80°C, 4 ml/min, 10 min static, 30 min dynamic; (3) pure CO_2, 334 bar, 100°C, 4 ml/min, 5 min static, 10 min dynamic; *collection*: (1) Octadecylsilane (ODS), T-nozzle 60°C Ttrap 5°C, elution with 0.5 ml of toluene; (2) T-nozzle and trap 80°C, no elution from the trap; (3) T-nozzle and trap 80°C, elution with 0.8 ml of toluene into the same vial (1.3 ml total volume); next extraction with 3.5 ml of toluene	Determination: offline GC–MS; (1) extraction of the more volatile PAHs, low density and temperature; (2) extraction of the less volatile PAHs, higher density and temperature and addition of modifiers; (3) purge the system for removing traces of modifier; studies comparing SFE to other extraction techniques (Soxhlet, sonication, KOH, ASE) indicate that SFE is capable of efficiently removing PAHs from solid matrices	66

Continued

TABLE 2.8
Continued

Analyte	Technique	Conditions	Comments	Ref.
PAHs	SFE	*Pretreatment:* sieved (<6 mm); *SFE conditions:* 2 g sample, pure CO_2, 400 bar, 150°C, 60 min dynamic mode, 1 ml/min; *collection solvent:* 15 ml CH_2Cl_2 (+60 ml CO_2 for collection)	*Determination:* offline GC-FID or MS; *recoveries:* low-molecular-mass PAH better than high; *comparison between different extraction methods:* SE, SFE, PLE and subcritical water; quantitative agreement of some PAHs between all of the methods; SFE has better selectivity for PAHs vs. bulk soil organic matter	238
PAHs	SFE	*Pretreatment:* Dried at 25°C during 48 h; *SFE conditions:* 2.3 g sample, CO_{2+} + methanol/dichloromethane (5:1), 45 MPa, 95°C, 15 min static, 60 min dynamic, 3 ml/min	*Determination:* offline HPLC-PDA/FD; *recoveries PAHs:* above 90%; *recoveries higher molecular weight:* around 70 to 90%; *comparison between SFE, sonication and Soxhlet:* SFE gives better or equal results, respectively, with significantly faster and easier procedure	239
PAHs	SFE	*Pretreatment:* railroad bed soil: air-dried and sieved (<2 mm); *SFE conditions:* 0.8 g railroad bed soil, 0.32 g urban air particulates, 0.2 g petroleum waste sludge (glass beads); 0.5 to 0.7 ml/min CO_2; *extraction sequence:* (1) pure CO_2, 30 min, 400 atm, 60°C, (2) CO_2 + 10% MeOH, 30 min, 400 atm, 60°C; *sample residues:* sonication in 6 ml pesticide-methylene chloride, 14 h; *SFE extracts:* 2.5 ml pesticide-methylene chloride	*Determination:* Petroleum waste sludge and railroad bed soil: offline GC-FID; *urban air particulate:* offline GC–MS; the use of different extractions approaches (e.g., sequential extraction) is more valid than the spike recovery studies for determining quantitative extraction conditions for native pollutants from complex matrix	240
PAHs	MAE	*Sample mass:* 1 g; *solvent volume:* 20 ml; *irradiation time:* 10 min; *MW power:* 500 W	A comparison between MAE and a 16-h SE method was made; both techniques gave comparable results with certified values. MAE has advantages over the SE technique since it is faster and uses lower quantity of solvent	241
PAHs	MAE	*Sample mass:* 1.5 g; *solvent:* acetone:hexane 1:1; *solvent volume:* 15 ml, *irradiation time:* 8 min (1 sample); 18 min (8 samples); *MW power:* 500 W	*Determination:* HPLC, GC; Comparative tables with concentrations for real samples and recoveries for reference materials are provided as well as such with comparison between different MAE conditions; optimization of MAE and comparison with the SE	242

| PAHs | Sonication | (a) *Sediment extraction with alumina cleanup:* sediment and sludge samples (previously lyophilized) were extracted by sonication for 20 min with dichloromethane/methanol (2:1) three times. centrifugation step, evaporated to dryness and redissolved in a mixture of hexane/dichloromethane (19:1); extracts purified following a cleanup procedure with an alumina column; (b) *Sediments extraction with immunosorbent (antifluorene) cleanup:* sediment and sludge samples (previously lyophilized) were extracted by sonication for 1 h with a mixture of dichloromethane/methanol (2:1) and preconcentrated in a rotary evaporator to 2 ml | *Determination:* LC-PDA; *recoveries:* 7 to 56%; *preconcentration:* cartridges prepacked with 0.5 g of silica and 10 mg of antifluorene antibodies and OSP-2 cartridges prepacked with 80 mg of silica and 2 mg of antifluorene antibodies; *other preconcentration:* precolumns prepacked with C18 silica; higher selectivity of the antifluorene immunosorbent compared to conventional cleanup (immunosorbents more reliable technique than use of alumina for cleanup of complex environmental samples); methodology using immunosorbents can be automated; *LOD:* 0.9 to 37 $\mu g/l$ | 243 |
| PAHs | Sonication | (a) *Sediment extraction with alumina cleanup:* sediment and sludge samples (previously lyophilized) were extracted by sonication for 20 min with dichloromethane/methanol (2:1) three times; centrifugation step, evaporated to dryness and redissolved in a mixture of hexane/dichloromethane (19:1); extracts purified following a cleanup procedure with an alumina column; (b) *Sediments extraction with IS (antifluorene and antipyrene) cleanup:* Sediment and sludge samples (previously lyophilized) were extracted by sonication for 1 h with a mixture of dichloromethane/methanol (2:1) and preconcentrated in a rotary evaporator to 2 ml | *Determination:* LC-PDA (validation using CG–MS); *LOD:* 1,6 to 19.3 $\mu g/l$ | 244 |

Continued

TABLE 2.8
Continued

Analyte	Technique	Conditions	Comments	Ref.
Nitro-PAHs	Ultrasound-assisted extraction	*Multivariate optimization*: Branson 450 sonicator (20 KHz, 100 W) equipped with a cylindrical titanium alloy probe (1.5 cm diameter); *temperature of water-bath*: 20°C; *extractant*: dichloromethane (8 ml), 10 min; *duty cycle*: 0.6 s, output amplitude 30% of the nominal amplitude; *applied power*: 100 W; during extraction, the direction of the extractant (at 2 ml/min) was changed each 90 s (minimizing dilution of the extractant and increased compactness of the sample in the extraction chamber)	*Determination*: GC–MS–MS; *extraction time*: 10 min; *extractant volume*: <10 ml, *detection limits*: low pg, *repeatability*: 4.21 to 5.70%; *reproducibility*: 5.20 to 7.23%	245
PCBs, PAHs & lipids	SPE/clean up	Different SPE phases were assayed	*Determination*: GC–MS; good recoveries when silica phases were used	246
TPH-PCBs & PAHs-cresol	SFE	—	*Determination*: GC-ECD, GC–MS; the recoveries were significantly higher than those achieved with solvent extraction	247
PCBs & PAHs	MASE	Two-level factorial designs have been used to optimize the microwave extraction process	*Determination*: HPLC; average recovery between 85 to 70 and 100 to 73%, with a relative standard deviation of 1.77 to 7.0%	248
PCBs	MASE	Optimized using experimental design methodology	*Determination*: GC–MS, Accuracy, precision, linear dynamic range and instrumental and method detection limits were evaluated for the analytical approach developed	249
PCBs	SFE	*SFE conditions*: *method 1 (Isco)*: CO$_2$, 400 bar, 150°C, 60 min dynamic; 1.5 ml/min; *method 2 (HP)*: CO$_2$, 0.75 g/ml, 305 bar, 80°C, 10 min static, 40 min dynamic, 1 ml/min; *collection*: *method 1 (Isco)*: 10 ml acetone, −10°C; *method 2 (HP)*: T-nozzle 45°C, T-trap 20°C, Florisil; *elution*: 2 × 1.5 ml n-heptane, 1.5 ml dichloromethane, 2 × 1.5 ml n-heptane	*Determination*: offline GC; the high temperature Isco method in some cases yields a more exhaustive extraction, but it presents co-extraction of unwanted matrix components	250

Analyte	Technique	Conditions	Determination / Notes	Ref.
PCBs	MAE	*Sample mass*: 0.5 g, *solvent*: toluene–IPA, *varying ratio*, *irradiation time*: 25.3 min; *temperature*: 100°C	*Determination*: HPLC; comparative tables are given; MAE was claimed as "a most effective method for extraction" for the selected analytes	114
PCBs	MAE	*For nonpolar solvent*: *sample mass*: 2.5 or 0.8 g; *solvent*: octane + water; *solvent volume*: 20 + 3 ml; *irradiation time*: 10 min; *temperature*: 150°C, *For polar solvent*: *sample mass*: 2.5 or 0.8 g; *solvent*: hexane/acetone (1:1); *solvent volume*: 20 ml, *irradiation time*: 10 min; *temperature*: 130°C	*Determination*: GC–MS; comparative tables with recoveries for the different methods are given; MAE and steam distillation techniques were combined; the use of isotope labeled internal standards for the MASD technique gave comparable results with the values obtained by other extraction methods and the certified values in the samples; developed MASD technique was claimed better than simple MAE, conventional steam distillation, and Soxhlet extraction	115
PCBs	MAE	*Sample mass*: 1 g (waste) or 2 g (soil): *solvent*: n-heptane; *solvent volume*: 15 ml; *irradiation time*: 15 min; *temperature*: 150°C	*Determination*: GC–MS; MAE and Soxhlet are compared; table with found PCB concentrations by both methods is given; recoveries ranged from 98 to 123% with MAE and from 86 to 111% with Soxhlet; comparison with other extraction techniques confirmed the efficiency of MAE	251
PCBs	MAE	*Sample mass*: 5 to 15 g; *solvent*: acetone/n-hexane (1:1, v/v); methanol and methanolic 1 M KOH; *solvent volume*: 30 ml; *irradiation time*: 6 min; *MW power*: 600 W	*Determination*: GC-ECD; *Three microwave-assisted techniques*: MAE, microwave- assisted saponification (MAS) and microwave-assisted decomposition (MAD) were studied and combined with success; tables with recoveries from studied methods are given	252
PCDDs, PCDFs, PCBs	SFE	*Pretreatment*: *sample*: 1 g sediment + 0.1 ml standard solution + 0.5 g activated copper powder − 2 g Al$_2$O$_3$ + 2 g Na$_2$SO$_4$; *SFE conditions*: CO$_2$, 40 J atm, 100°C, 10 min static, 60 min dynamic, 3 ml/min; *collection*: T-nozzle 45°C; T-trap 40°C; carbon (adsorbent) and Celite 545 (support material); *trapping method PCBs*: carbon:Celite 3:20 (w/w), 13% carbon, 9 ml hexane to eluted; cleaned with Al$_2$O$_3$ and eluted with 2% dichloromethane in hexane (17 ml); *trapping method PCDD/PCDF*: carbon:Celite 1:5 (w/w), 18% carbon, 4 ml hexane to elute impurities, 10 ml toluene to collect; addition of 0.05 ml toluene and concentrated to about 0.03 ml	*Determination*: offline HRGC–HRMS; *recoveries*: 60 to 95% PCBs; 60 to 90% PCDD/PCDF, SFE much faster than Soxhlet method; SFE gives results and precision comparable to Soxhlet	253

Continued

TABLE 2.8
Continued

Analyte	Technique	Conditions	Comments	Ref.
PCDDs, PCDFs	SFE	Dried at 40°C, grinding after removing large particles (<2 mm); *SFE conditions:* CO_2, 400 atm, 100°C, 10 min static, 60 min dynamic, 3 ml/min, T-nozzle 45°C; *collection:* solid phase trapping: T-trap 40°C, activated carbon–Celite (w/w) (A) 1:25, (B) 1:10, (C) 1:5; *elution:* 4 ml hexane to removal interferences, 10 ml toluene to elution; concentration to 0.03 ml using nitrogen flow	*Determination:* offline HRGC–MS; SFE values are between 65 to 126% of the value achieved by Soxhlet; addition of a modifier is not necessary to CO_2 during SFE for the quantitative extraction of PCDDs/PCDFs from soil contaminated with chlorophenols; the extraction efficiency of PCDDs/PCDFs might be affected by other substances which can act as a modifier	254
Organotin compounds	SBSE	—	*Determination:* thermal desorption-CGC-ICP-MS, *LODs:* 0.1 pg/l (procedure) and 10 fg/l (instrumental) and a repeatability of 12% RSD	255
Organotin compounds	SFE	*Pretreatment:* dried and sieved (<1 mm); *complexation:* diethylammonium diethyldithiocarbamate; *two systems:* (i) *Isco:* CO_2 + 5% MeOH, 0.92 g/ml, 450 atm, 60°C, 20 min static, 30 min dynamic, 1.5 ml/min; *direction of fluid flow:* down; *extraction vessel orientation:* vertical; (ii) *HP:* CO_2 + 5% MeOH, 0.92 g/ml, 365 atm, 45°C, 20 min static, 30 min dynamic, 2 ml/min; *Direction of fluid flow:* up; *extraction vessel orientation:* vertical; *SFE extracts collected in:* (i) *Isco:* 15 ml methylene chloride, (ii) *HP:* octadecyl-bonded silica trap and rinsed off the trap with 3 × 1.8 ml portions of methylene chloride; extracts transferred to a 60 ml reparatory funnel and concentrated by nitrogen to a final volume of approximately 0.5 ml; *derivatization step:* extract treated with pentylmagnesium bromide (PMB) to convert the ionic organotin compounds into their neutral derivatives	*Determination:* offline GC-AED; recoveries of >75% on the SFE/solvent system for all tri- and tetra substituted organotin compounds and for almost half of the disubstituted organotin compounds; some of the recoveries were lower on the SFE/sorbent system	256
Phenols	SPME	Extraction parameters studied		257

Analyte	Technique	Conditions	Determination	Ref.
Phenols	SFE	5 g sample (glass beads, 4 mm), CO_2 + 10% MeOH, 1200 psi, 50°C, 45 min static, 20 min dynamic, 0.2 ml/min (0.02 ml/min modifier + 018 ml/min CO_2), 2 ml methanol; volume of SFE extracts reduced to 0.01 ml by bubbling nitrogen; *collection*: 2 ml methanol; volume of SFE extracts reduced to 0.01 ml by bubbling nitrogen	*Determination*: offline CEC–UV–visible; *working ranges*: 0.019 to 2.72 mg analyte/kg soil; *detection limits ranges*: 0.0032 to 0.014 mg/kg soil for the alkyl-substituted phenols; rapid method for the direct determination of phenol and alkyl-substituted phenol in soils, with capability of interfacing with MS for confirmation of unknown peaks	258
Phenols	SPE	Sorbents, mechanism, derivatization, matrix effects and storage reviewed	Review	259
Phenols	SPE	Concentration using a C_{18} SPE; Continuous-flow (methanol), high-temperature (65°C), sonicated extraction system to isolate APEO metabolites from sediment samples (low-power ultrasonic energy); sediment extraction was complete after 7 min with a total solvent consumption of 3.5 ml/sample; *two-step cleanup*: normal-phase SPE, reversed-phase	*Determination*: mixed-mode HPLC–ES-MS; recovery for sediment and water; 78 to 94%; *LODs*: 1 to 20 pg injected on column, *RSD*: 3 to 5%; *run time*: 13 min	260
Acidic herbicides	SFE, SPE SWE, MAE, organic solvent extraction and aq. basic solutions; derivatization	—	Review	261
Acidic herbicides	SPE	—	Review	262
Herbicide (atrazine)	SFE	*Pretreatment*: sediments stirred during 2 h; dried and homogenized; *SFE conditions*: CO_2 + variable %methanol + %H_2O (10 to 2%, 15 to 2%, 15 to 10%) and 10% methanol:water:Et_3N (75:2:20), 300 bar, 65°C, 45 min, 1 ml/min	*Determination*: offline HPLC–MS, Modifier: methanol:water:Et_3N	263
Pesticides	MASE-SPE	—	*Determination*: online-HPLC–DAD recovery above 80% and limits of identification of 20 to 40 ng/g for 1.5% organic matter; for 3.5% organic matter identification limits of 30 to 50 ng/g	264

Continued

TABLE 2.8
Continued

Analyte	Technique	Conditions	Comments	Ref.
Insecticides	SWE	Water in the continuous mode at a flow-rate of 1 ml/min and 85°C was sufficient for quantitative extraction; minicolumn containing C_{18}-hydra	*Determination:* HPLC-DAD; recoveries of the target analytes ranged between 94.2 and 113.1%, and repeatabilities, expressed as relative standard deviations, were between 0.61 and 6.83%	265
Insecticides	SPME	Polydimethylsiloxane-coated fiber and high temperature alkaline hydrolysis	*Determination:* GC–MS, The method allows determination with 1.0 μg of VX spiked per g of agricultural soil	266
Insecticides	Sonication	The main factors optimized by means of a central composite design	*Determination:* continuous ultrasound-assisted extraction coupled to on line filtration–SPE–column liquid chromatography–post column derivatisation–fluorescence detection; *LOD:* 12 ng/g and *LOQ:* 40 ng/g; carbamates at 1 μg/g spiked level; recoveries similar to those provided by the EPA 8318 method; repeatability: 3.1%; reproducibility: 7.5%	267
Insecticides	LE-SPE	—	*Determination:* LC-UV detection, LC–MS, *LOD:* 10 and 50 μg/kg	268
Pesticides	MASE-SPME	*Extractant:* 1% methanol in water (LE); aqueous extractant at 105°C for 3 min, with 80% output of maximum power (1200 W)	*Determination:* GC–MS; *LOD:* 2 to 4 μg/kg; *precision:* <7% and *recoveries:* 76.1 to 87.2%	269
Pesticides	MASE	*Extractant:* 10 mM phosphate buffer, pH 7	*Determination:* HPLC-DAD; *LOD:* 5 μg/kg and *LOQ:* 10 μg/kg; *in soils with 1.5% organic matter content:* recoveries >80%; *in soils with organic matter content 3.5%:* recoveries < 70% and the respective LOD 10 μg/kg and LOQ 50 μg/kg	270
Pesticides	Ultrasounds SPME	Experimental design of the variable	*Determination:* GC-ECD, GC–MS; *recoveries:* 72 to 123%, SD < 16%	271
Pesticides	HS–SPME	100-μm polydimethylsiloxane (PDMS) and 65-μm PDMS–DVB	*Determination:* GC-ECD; *recoveries:* 68 to 127%. *RSD:* <25%	272

Analyte	Technique	Conditions	Determination/Results	Ref
Pesticides	SPME	Silica fibers coated with an 85-μm thick poly-acrylate (PA) or a 100-μm thick polydimethylsiloxane (PDMS) film and a manual SPME device (Supelco)	*Determination:* GC–MS, *LOD:* 1 ng/l	273
Pesticides	SWE	Quantitative extraction in 90 min at temperatures near to 300°C	*Determination:* GC–MS; *LOD:* between 3.2 and 137.1 μg/kg, *RSDs:* 2 to 34%	274
Pesticides	MSPD	—	*Recoveries:* 75 to 94%; *RSDs:* 1.5 to 6.5%	275
Pesticides	Sonication-assisted extraction in small columns	Ultrasonic water bath (*output:* 150 W; *frequency:* 33 KHz); Soil samples were fortified with 0.5 ml of a mixture of the different carbamates in methanol and left for 20 min at room temperature for solvent evaporation; Extraction with 5 ml of methanol for 15 min in an ultrasonic water bath at room temperature and another extraction with 4 ml of methanol for 15 min and 1 ml wash	*Determination:* reversed-phase HPLC with fluorescence determination after post-column derivatization; *recovery:* 82 to 99% (*RSD:* 0.4 to 10%); *linear range:* 0.1 to 1 μg/ml; *LODs:* 1.6 to 3.7 μg/kg; *LOQs:* 10 μg/kg; *advantages:* rapid extraction procedure with a small volume consumption of organic solvent	276
Pesticides	MAE & Soxhlet	—	*Determination:* HPLC–post-column fluorescence derivatization–detection; recoveries similar than EPA method 8318: 39.6 and 91.7%	277
Pesticides	MAE	—	*Determination:* LC-UV, *LODs:* between 5 and 50 μg/kg; *recoveries:* between 60 and 90%; *RSD:* between 5 and 25%	278
Organochlorine pesticides	MAE	Solvent: DCM, light petroleum, or hexane, Solvent volume: 10 ml, Irradiation time: 2 min, MW power: 10%	*Determination:* GC-EC, Comparison between LLE, SE and MAE. Authors concluded that SE could be a more accurate alternative to MAE	116
Triazine and Chloroacet-anilide Herbicides	MAE	*Sample mass:* 10 g; *solvent:* acetonitrile; *solvent volume:* 20 ml; *irradiation time:* 5 min; *MW power:* 900 W; *temperature:* 80°C; *pressure:* 100 psi	*Determination:* GC-NPD or GC–MS simple and rapid MAE was developed; *mean recovery:* >80%; *RSD:* <20%; *LOQ:* 10 μg/kg; *LOD:* 1 to 5 μg/kg	279
Diazinon	MAE	Sample mass: 3.7 g, Solvent: hexane/acetone (8:2), Solvent volume: 20 ml, Irradiation time: 2.5 min, MW power: 700 W	*Determination:* GC–MS comparison between MAE and USE; recovery at these conditions: 98%	117

Continued

TABLE 2.8
Continued

Analyte	Technique	Conditions	Comments	Ref.
Carbamates	MAE	*Sample mass:* 2.0 g; *solvent:* hexane/acetone (varying ratio); *solvent volume:* 30 ml; *irradiation time:* 4 or 6 min; *MW power:* 960 W; *temperature:* 115°C	*Determination:* HPLC; comparative tables with recoveries for the different solvents and matrixes are given. MAE was applied to study the thermal degradation of five carbamates; some nonpolar and polar pollutants spiked in soil, such as PAHs, PCBs, triazines (atrazine, simazine), and carbamates (propoxur, methiocarb, chlorpropham) subjected to MAE were also studied; the recoveries ranged between 70 and 99% with excellent reproducibility, except for carbamates	280
Pesticides	PLE	Soil samples were extracted by PLE using methanol–water (75:25) at 60°C	*Determination:* LC–MS–MS; *recoveries* 75% for metribuzin, DA and DADK and *LOD:* 1.25 μg/kg; DK *LOD:* 12.5 μg/kg and *recovery:* 50%	281
Fungicides	SAESC	Ultrasonic bath (*output:* 150 W; *frequency:* 35 KHz); samples sieved (2 mm) and stored at room temperature until fortified; soil samples were extracted with 4 ml of ethyl acetate for 15 min in an ultrasonic water bath at room temperature; the water level in the bath was adjusted to equal the extraction solvent level inside the columns; soil samples were extracted again with another 4 ml of ethyl acetate (15 min); *total extract:* 2 to 5 ml (stored at 4°C); *solvent:* methanol or ethyl acetate	*Determination:* GC–EC and GC-NPD; (confirmed by GC–MS); *recovery:* 80 to 104% (*RSD:* 1 to 8%); *linear range:* 0.05 to 0.5 μg/g; *LODs:* 2 to 10 μg/kg	282
Hydrocarbon	PLE	175°C with dichloromethane-acetone (1:1, v/v) with 8 min heat-up time and 5 min static time	*Determination:* GC-FID; *recoveries:* 115%	283

Analyte	Technique	Pretreatment / conditions	Determination	Ref.
PHCs	SFE	CO_2 + 5 or 10% acetone, 0.65 g/ml, 227 atm, 80°C, 15 min static CO_2 + modifier, 15 min dynamic CO_2 only, 1 ml/min; *collection*: solid phase: Octadecylsilane (ODS), Ttrap 30°C; extracted collected with 1.5 ml of *n*-hexane; *after each extraction*: solid trap washed with 1.5 ml acetone + 1.5 ml *n*-hexane which were checked for residual hydrocarbons	*Determination*: offline GC-FID; *recoveries*: 70 to 100% PHCs; the presence of the polar modifier improved the extraction of the aromatic fraction; higher concentration of modifier decreases the extraction efficiency; effect of the low and high MW-PAHs	284
OCPs, HCB	SFE	*Pretreatment*: dried at 35°C to 2% humidity; ground and sieved (<2 mm); *SFE conditions*: 5 g ground-dried soil + 2.5 g anhydrous Na_2SO_4 + 0.35 ml MeOH (modifier) CO_2, 0.95 g/ml, 380 bar, 40°C, 15 min static, 5 min dynamic; *collection*: trapped on a reversed-phase minicolumn, eluted with 1 ml petrolbenzine	*Determination*: offline GC-ECD; *comparison*: steam distillation–solvent extraction (SDSE), SFE and Soxhlet extraction (SE); *HCB recovery*: 100% SDSE, 98% SFE, 55% SE	285
Methylmercury	SFE	*Pretreatment*: 0.8 g sediment + 0.4 ml H_2O + 1 ml HCl + Celite mixture; *SFE conditions*: CO_2, 0.85 g/ml, 40°C, 5 min static, 45 min dynamic, 0.5 ml/min; *collection*: T-nozzle 60°C, T-trap 15°C, 2 × 1 ml toluene	*Determination*: offline GC-ECD; *recoveries*: 81 to 84%	286
Lipidic compounds	SFE	*Pretreatment*: air-dried; ground and sieved (<2 mm), homogenized and lyophilized; *SFE conditions*: 4 g sample, CO_2, 8 to 30 MPa, 40 to 120°C, 0.5 to 2.5 ml/min depending of the pressure, 10 min static, 20 min dynamic; *collection*: 2 ml methanol	*Determination*: offline GC–MS; the results show that SFE is an appropriate enrichment procedure for soil lipids; SFE and Soxhlet extraction give results which are in good agreement	287
29 polar (aromatic acids, phenols), slightly polar (herbicides, pesticides), nonpolar (PAHs) compounds	SFE	*Pretreatment*: air-dried; ground and sieved (<0.84 mm); pretreatment with 15% water, 5% (ethylenedinitrilo)-tetraacetic acid tetrasodium salt (Na_4EDTA) and 50% methanol; *SFE conditions*: 2 g sample (air-dried) + 1 ml MeOH, 5 min static–dynamic extraction, 30 ml CO_2, 34.5 MPa, 60°C; additional MeOH (1 ml) and reextraction (same conditions above); *collection*: methanol	*Determination*: polar analytes offline SFE–CZE; slightly polar analytes offline GC-NPD or ECD; nonpolar analytes offline GC-FID; 29 polar, slightly polar and nonpolar compounds recoveries ranging from 86 to 106%; the method is also available for risk assessment of parent pollutants and transformed products, particularly oxygen-borne metabolites in the environment	288

Continued

TABLE 2.8
Continued

Analyte	Technique	Conditions	Comments	Ref.
Hydrocarbons, dialkyl alkyl-phosphonates, alkyl alkyl-phosphonic acids	SFE	*SFE conditions: three-step extraction:* (1) pure CO_2, 0.5 g/ml, 129 atm, 60°C, 0.2 min static, 20 min dynamic, 1 ml/min; (2) CO_2 + MeOH, 0.6 g/ml, 202 atm, 80°C, 10 to 40 min static, 20 min dynamic, 1 ml/min; (3) pressurized liquid extraction (PLE); *collection:* trapping material: ENV + (styrene–DVB copolymer); elution with 1.8 ml of ethyl acetate; T-trap 25°C, T-nozzle 45°C; *derivatization:* methylated derivatives (diazomethane derivatization)	*Determination:* offline GC-FID or FPD-P; recoveries: (1) ≥95% of hydrocarbons; (2) quantitative extraction of phosphonates; (3) phosphonic acids; global recoveries are close to 80%, a loss of about 20% occurring during the derivatization process; successive implementation of three extraction steps giving a selective extraction method	289
2-Chlorovinyl-arsonous acid	SFE	*Pretreatment:* dried at 40°C for a week and sieved (<2 mm); *SFE conditions:* CVAA derivatization occurs during a 10 min static period; after that, CDA (product of derivatization) is extracted under dynamic conditions, 20 min, CO_2, 1 ml/min; extraction parameters are changed for evaluation; *collection:* Trapping material: Isolut ENV +; elution with 1.5 ml ethyl acetate	*Determination:* offline GC-FID; samples are allowed to age (up to 42 days) and periodically extracted; samples ageing leads to a recovery decrease due to a development as strong interactions between CVAA and matrix active sites, as time elapses; *comparison between three extraction methods:* SFE, ASE (accelerating solvent extraction) and USE (ultrasound extraction); SFE is the one which leads to the highest recoveries	290

			Determination	
2,4-Dichloro-phenoxyacetic acid and its major transformation products	Sonication	Soil samples were treated with 1 ml of acidified water at pH 1.0 and 15 ml of dichloromethane, extracted for 1 h under sonication	*Determination:* LC-UV, with diode array detector; *recovery:* 96% for 2,4-DCP and 99% for 2,4-D (calibration curves made in an extract of unfortified samples); the samples were analyzed five times from extraction to determination at each level; three injections were made for each extraction; 2,4-D and 2,4-DCP were detected up to the 15th day after the formulate application *LOD for 2,4-D and 2,4-DCP:* 0.02 mg/kg *LOQ for 2,4-D and 2,4-DCP:* 0.10 mg/kg	291
Total carbohydrate	Dialysis and lyophilization ultrasonic and hydrolysis	50 ml of CH_3COOH (1 or 2 *M*) or 1 *M* HCOOH; sonication time = 2 h; ultrasound cleaning bath works at 35 KHz	*Determination:* colorimetric analysis of total carbohydrates; determination of the mono-saccharide composition by GC–FID; *recovery* of glucose: 88, 1% (HCOOH 1 *M*); 99 and 96% (CH_3COOH 1 and 2 *M*, respectively)	292
Products from degradable polyolefin	MAE	*Sample mass:* 0.5 g; *solvent:* 98/2 (wt%) chloroform/2-propanol mixture; *solvent volume:* 10 ml; *irradiation time:* 30 min; *temperature:* 80°C	*Determination:* GC–MS; comparison between MAE and USE; amount of products obtained by MAE and USE are given; the USE was claimed better because of better reproducibility and better extraction efficiency	118

TABLE 2.9
Different Sample Preparation Processes to Determine Pollutants in Air Samples

Analyte	Technique	Conditions	Comments	Ref.
Acid gases (carboxylic acid and NO$_x$)	Derivatization	Cyclohexyl derivatives	*Determination*: GC–MS, Quantitative recoveries, *LOD*: 0.3 mg/m³	293
Volatile sulfur compounds	SPME	75 μm carboxen–polymethylsiloxane fiber at 22°C for 20 min	*Determination*: GC–MS, nine volatile sulfur compounds in complex gaseous samples. *LOD*: 1 to 450 ppt	294
Volatile and semivolatile airborne organic compounds	SPME	PDMS and PDMS–DVB fiber	*Determination*: GC-FID; quantification of target analytes in air using this method can be carried out without external calibration; no pumps and no polluting solvents	295
VOCs	Sorbent trapping/ SPME	Review of different sorbents according to the mechanism used to recover the trapped compound	Review	296
Odorants emissions	SPME	SPME on a three-phase fiber, DVB/carboxen/PDMS	*Determination*: GC–MS; odorants emissions form landfills (100 volatile organic compounds); the average removal efficiency was not very high (about 23.5%) due to scarce ability in removing low polarity compounds	297
VOCs	SPME	PDMS fiber	*Determination*: GC-FID; static and dynamic sampling were compared, in the dynamic mode narrower linear range were obtained	298
Benzene and alkylated benzene	MAE	The microwave desorber was operated under the following conditions: backflush (time of purging the sampling tube before desorption) 10 s, desorption power (applied microwave energy) 400 Ah 10-6, bypass delay (time of gas flow through the sampling tube after desorption) 30 s, desorption time (duration of the microwave pulse) 15 s, split ratio (ratio of gas flow through the GC column and total gas flow through the sampling tube) 10	*Determination*: GC-FID-MS; concentration of different compounds found in expired and inspired air was given; microwave desorption was utilized and claimed more useful than other techniques	110

Analyte	Technique	Conditions	Determination/Notes	Ref.
PAHs	SPME	PDMS fiber	*Determination:* GC–MS, *LOD:* 5 to 20 pg	299
PAHs	MAE	*Sample mass:* 2.6 g; *solvent:* hexane/acetone (1:1); *solvent volume:* 15 ml; *irradiation time:* 20 min; *MW power:* 400 W	*Determination:* HPLC-UV/FD, MAE with hexane/acetone (1:1) from real atmospheric particulate samples was investigated and the effect of microwave energy and irradiation time studied	300
Oxy-PAHs, nitro-PAHs (urban aerosol.)	SFE	*SFE conditions:* 2.3 g sample, 5 min static, 30 min dynamic, ml/min; *two extraction steps:* (1) CO_2 150 atm, 45°C, (2) CO_2 + toluene (5:95, v/v). 350 atm, 90°C; *collection system:* Supelclean LC-18 C_{18}; *elution:* 2 ml dichloromethane	*Determination:* offline GC-ECD/MS; the analytes studied are identified at concentration ranging between 10 and 364 pg/m^3; *SFE method compared with sonication:* good agreement; alternative method to conventional extraction techniques such as Soxhlet or sonication	301
Volatile halocarbons	Cold-SPME	Subambient temperature is used in order to enhance the retention capability of the fiber coating	*Determination:* GC–MS; results obtained showed that trace atmospheric halocarbons are detectable even when enriching very small air sample volumes	302
Organophosphate esters	Dynamic MAE (DMAE)	DMAE was coupled to SPE	*Determination:* GC-NPD; *recovery:* >97%; *repeatability:* 4.2 to 8.0 (*RSD:* 0.03%); total sampling and analysis time < 1.5 h; *LOD for studied phosphate esters, pg/filter:* tri-n-butyl phosphate 4,6; tris(2-chloroethyl) phosphate 8,2; tris(2-chloropropyl) phosphate 5,4; tris(2-butoxyethyl) phosphate 9,2; triphenyl phosphate 8,6; tris(2-ethylhexyl) phosphate 14; the superiority of MAE compared to the other methods such as Soxhlet extraction, PLE and SFE was underlined	108

Continued

TABLE 2.9
Continued

Analyte	Technique	Conditions	Comments	Ref.
Organophosphate esters	DSASE	Bransonic 52 ultrasonic bath (output power of 120 W; frequency of 35 KHz); *temperature of the ultrasonic bath:* 70°C; *duration of extraction:* 3 min, *flow-rate of the extraction solvent:* 200 μl/min; *pressure over the extraction cell during extraction:* 30 bar	*Determination:* DSASE online with large-volume injection GC-nitrogen–phosphorous detection, using a programmed-temperature vaporizer; PTV-GC with a nitrogen–phosphorous determination; *recovery:* >95% (86% for tri(*n*-butyl) phosphate); *extracted fraction:* 800 μl (hexane–methyl-tert-butyl ether, 7:3, v/v); *extraction and analysis:* 15 min; system cleaning performed during GC analysis step; *LODs:* 5 to 32 pg/filter; *peak area repeatability RSD:* 3.2 to 11.7% (*n* = 5); *peak area reproducibility-RSD:* 5.9 to 14;3% (*n* = 4)	303
Organophosphate esters	DSASE	Factorial design (center point); Bransonic 52 ultrasonic bath (output power of 120 W; frequency of 35 KHz); Tempunit TU-16A heater, outlet restrictor; *solvent mixture:* hexane–MTBE; *temperature of the ultrasonic bath:* 70°C; *duration of extraction:* 3 min, *flow-rate of the extraction solvent:* 200 μl/min; *extraction volume:* 600 μl	*Determination:* GC analysis (8000 Top gas chromatograph); *column:* DB-5MS (30 m × 0.32 mm i.d., 0.10 μm film thickness); nitrogen–phosphorus; *Detection* (thermionic determination, TID); *recovery:* > 95% (hexane-MTBE 7:3); highest extraction efficiency with smallest internal volume (0.25 ml); low time and solvent consumption	304
PCBs	MAE	*Sample mass:* 0.5 g; *solvent:* toluene-IPA, varying ratio; *irradiation time:* 25.3 min; *temperature:* 100°C	*Determination:* HPLC; comparative tables are given; MAE was claimed as "a most effective method for extraction" for the selected analytes	114

PCBs	MAE	Sample mass: 0.5 g; solvent: toluene/IPA (90:10, v/v); solvent volume: 5 ml; irradiation time: 35.3 min; temperature: 100°C	Determination: HPLC; the effects of extraction parameters on the MAE of PCBs and PCDDs from fly ash were compared	305
PCBs	MAE	Sample mass: 1 to 15 g; solvent: toluene; solvent volume: 30 ml; irradiation time: 10 min; temperature: 110°C	Determination: GC-ECD and GC–MS; recoveries: >80%; an alternative method for the extraction of PCBs in ash samples, which is less time and solvent consuming than SE, was presented; this method was optimized by experimental design	306
PCBs	MAE	Solvent: hexane/acetone (1:1); solvent volume: 15 ml; irradiation time: 10 min; temperature: 115°C	Determination: GC-ECD; recoveries (%) with SD in brackets: Analyte Soxhlet MAE PCB10 90.2 (8.2) 75.6 (6.7) PCB28 87.8 (3.3) 84.8 (8.7) PCB52 103.3 (7.3) 87.3 (9.2) PCB153 100.4 (6.9) 91.8 (8.2) PCB138 97.7 (6.7) 93.0 (8.7) PCB180 92.7 (7.4) 97.3 (9.5) The speed of MAE was underlined	109
Nicotine	MAE	GCB is used as a solid sorbent in quartz tubes; microwave thermal desorption is used after active sampling	Determination: GC	111

TABLE 2.10
Different Sample Preparation Processes to Determine Pollutants in Waste Samples

Analyte	Technique	Conditions	Comments	Ref.
PAHs	MAE	*Solvent:* hexane–acetone 1:1 v/v; *solvent volume:* 30 ml; *irradiation time:* 10 min; power 30 W	*Determination:* HPLC–FLD, in sewage sludges; *recoveries:* between 56 and 75%; this technique was compare to Soxhlet and sonification extraction and similar results were obtained	307
PAHs	SFE	*Pretreatment:* dried at 80°C during 24 h; milled, ground and sieved; *SFE conditions:* 1 g sample (glass beads) CO_2, 10 min static, CO_2 + 5% toluene, 30 min dynamic, 500 atm, 150°C, 1 ml/min; *collection:* 10 ml toluene	*Determination:* offline HPLC fluorescence and UV determinations; *recoveries:* 35 to 95%; comparison between SFE, PLE, focused microwave extraction in open vessels, Soxhlet and ultrasonic extraction; when optimized, the five extraction techniques are as much efficient with similar RSD; whatever the extraction techniques used, the whole analysis protocol permits to quantify PAHs in the range from 0.09 to 0.9 mg/kg of dried sludge	308
Pesticides, PCBs	Soxhlet/sound amplification and shaking	Different organic solvents were assayed	*Determination:* GC–MS, in all the cases the use of hexane increases the concentration, selectivity and recovery	309
PAHs, PCBs, OCPs	SFE	*Pretreatment:* precleaned with hexane and acetone; dried; grind; PAHs + desionized water to increase the extraction efficiency; *SFE conditions:* PAHs; *extraction step 1:* CO_2, 0.3 g/ml, 121 bar, 80°C, 10 min static, 10 min dynamic, 2 ml/min; *extraction step 2:* CO_2 + 1%methanol + 4% DCM, 0.63 g/ml, 335 bar, 120°C, 5 min static, 10 min static, 30 min dynamic, 4 ml/min; *extraction step 3:* CO_2, 0.63 g/ml, 335 bar, 120°C, 5 min static, 10 min dynamic, 4 ml/min; *PCBs:* CO_2, 0.75 g/ml, 305 bar, 80°C, 10 min static, 40 min dynamic, 2.5 ml/min; *OCPs:* CO_2, 0.87 g/ml, 299 bar, 50°C, 20 min static, 30 min dynamic, 1 ml/min; *collection: PAHs:* (1) T-nozzle 60°C, T-trap 5°C, ODS, 0.5 ml toluene (analysis) or isooctane (clean up); (2) T-nozzle and T-trap 80°C, no elution; *extraction:* 3 ml DCM–acetone (1:1) + 3.5 ml toluene or isooctane, respectively; (3) T-nozzle and T-trap 80°C, 0.7 ml toluene or isooctane, respectively; *PCBs:* T-nozzle 50°C, T-trap 15°C, Florising, 1.5 ml heptane; *OCPs:* T-nozzle 50°C, T-trap 20°C, ODS, 1.3 ml and 1.2 ml heptane (two vials)	*Determination:* offline HRGC–MS clean up step prior to HRGC–MS analysis; *recoveries PAHs:* 104 to 125% *recoveries PCBs:* 56 to 121%; *recoveries OCPs:* 75 to 106% *recoveries PAHs:* PAH extract could be measured by GC–MS after a short clean up step, PCBs could be determined directly after SFE, OCPs direct analysis is not possible, matrix compounds must be removed by a multi-step clean up	310

Analyte	Technique	Sample / Description	Comments	Ref.
PCBs	SFE	Syringe pump and liquid trapping (high temperature) vs. reciprocating pump and solid-phase trapping (medium temperature)	*Determination*: GC; two different SFE systems are compared; high temperature method in some cases yields a more exhaustive extraction, but also less clean extracts whereas medium temperature may sometimes cause problems with quantitative recoveries, but it yields very clean extracts	311
Phenols	SPE	Review	The use, advantages and disadvantages of silica sorbents, polymeric, functionalized, carbon-based and mixed available sorbents are discussed	312
Alkylphenolic compounds	PLE	*Sample*: river sediments	*Determination*: LC–MS; *best conditions*: *extraction solvent*: methanol–acetone (1:1, v/v); *temperature*: 50°C; *pressure*: 1500 psi; two static cycles; *recoveries*: >70%. *LOD*: between 1 to 5 μg/kg; loss of volatile molecules is produce at elevated temperatures	313
4-nonyl-phenols	PLE	*Sample*: sediments	*Determination*: GC–MS; *Best conditions*: *extractant*: methanol; *temperature*: 100°C; *pressure*: 100 atm combined with 15 min static and then 10 min dynamic; *recovery*: 111% (*RSD*: 4%) and 106% (*RSD*: 5%); extraction efficiency of the PLE was compared with conventional Soxhlet and bath ultrasonication	314
Nonylphenol bisphenol A and 17 α-ethinyl-estradiol	SPME	*Sample*: wastewater, different fibers were assayed, with polyacrylate proving most suitable; an automated SPME	GC–MS; an extraction time of 1 h was employed. *LOD*: 0.04 to 1 μg/l; *RSD*: 8%; linearity of calibration curves ranges over three orders of magnitude	315

E. SAMPLE PREPARATION FOR VARIOUS GROUPS OF COMPOUNDS

Various applications of environmental sample preparation for chromatographic analysis are shown in Table 2.7 to Table 2.10. In these tables recent applications of the new sample preparation techniques are shown classified in order of the different analytes. This compilation of applications is a representative overview of recent applications and new trends of sample preparation in the field of chromatographic techniques applied to environmental analysis. This compilation is not exhaustive; it is only a brief selection of the more significant works recently published.

ACKNOWLEDGMENTS

The authors wish to thank the Spanish Ministry of Science and Technology (MCYT, Spain) (projects: PTR1995-0581-OP-02-01 and DPI2002-04305-C02-01) for providing financial support.

REFERENCES

1. Valcárcel, M., *Principios de Química Analítica*, chaps. 4 and 7, Springer-Verlag Ibérica, Berlin, pp. 175–239 and 337–365, 1999.
2. Grasselli, J. G., Undergraduate education in analytical chemistry level in the US-industrial needs 200, In *Euroanalysis VII. Reviews on Analytical Chemistry*, Malissa, H., Ed., Springer Verlag, Berlin, 1991.
3. Grasselli, J. G., *The Analytical Approach*, American Chemical Society, Washington, 1983.
4. Liu, H. and Dasgupta, P. K., Analytical chemistry in a drop. Solvent extraction in a microdrop, *Anal. Chem.*, 68, 1817–1821, 1996.
5. Psillakis, E. and Kalogerakis, N., Developments in single-drop microextraction, *Trends Anal. Chem.*, 21, 53–63, 2002.
6. Jassie, L., Revesz, R., Kierstead, T., Hasty, E., and Matz, S., Microwave-assisted solvent extraction, In *Microwave Enhanced Chemistry. Fundamental, Sample Preparation, and Applications*, Kingston, H. M. and Haswell, S. J., Eds., American Chemical Society, Washington, p. 569–609, 1997.
7. Chiu, D. T., Hsiao, A., Gaggar, A., Garza-López, R. A., Orwar, O., and Zare, R. N., Injection of ultra small samples and single molecules into tapered capillaries, *Anal. Chem.*, 69, 1801–1807, 1997.
8. Majors, R. E., Trends in sample preparation, *LC–GC Europe*, 16, 71–81, 2003.
9. Remcho, V. T. and Tan, Z. J., MIPs as chromatographic stationary phases for molecular recognition, *Anal. Chem.*, 71, 248A–255A, 1999.
10. Chapuis, F., Pichon, V., Lanza, F., Sellergren, S., and Hennion, M. C., Optimization of the class-selective extraction of triazines from aqueous samples using a molecularly imprinted polymer by a comprehensive approach of the retention mechanism, *J. Chromatogr. A*, 999, 23–33, 2003.
11. Zhou, D., Zou, H., Ni, J., Yang, L., Jia, L., Zhang, Q., and Zhang, Y., Membrane supports as the stationary phase in high-performance immunoaffinity chromatography, *Anal. Chem.*, 71, 115–118, 1999.
12. Pichon, V., Bouzige, M., Miège, C., and Hennion, M. C., Immunosorbents: natural molecular recognition materials for sample preparation of complex environmental matrices, *Trends Anal. Chem.*, 18, 219–235, 1999.
13. Hennion, M. C. and Pichon, V., Immuno-based sample preparation for trace analysis, *J. Chromatogr. A*, 1000, 29–52, 2003.
14. Yang, Y. and Li, B., Subcritical water extraction coupled to high-performance liquid chromatography, *Anal. Chem.*, 71, 1491–1495, 1999.
15. Miller, D. J. and Hawthorne, S. B., Subcritical water chromatography with flame ionization detection, *Anal. Chem.*, 69, 623–627, 1997.
16. Hu, W. and Haddad, P. R., Electrostatic ion chromatography (EIC), *Trends Anal. Chem.*, 17, 73–79, 1998.
17. Tajuddin, R. and Smith, R. M., Online coupled superheated water extraction (SWE) and superheated water chromatography (SWC), *Analyst*, 127, 883–885, 2002.
18. Smith, R. M., Extractions with superheated water, *J. Chromatogr. A*, 975, 31–46, 2002.

19. David, F., Tienpont, B., and Sandra, P., Stir-bar sorptive extraction of trace organic compounds from aqueous matrices, *LC–GC Europe*, 16, 410–417, 2003.
20. Giddings, J. C., Factors influencing accuracy of colloidal and macromolecular properties measured by FFF, *Anal. Chem.*, 69, 552–557, 1997.
21. Luque de Castro, M. D. and Papaefstathiou, I., Analytical pervaporation: a new separation technique, *Trends Anal. Chem.*, 17, 41–49, 1998.
22. Madrid, Y. and Cámara, C., Biological substrates for metal preconcentration and speciation, *Trends Anal. Chem.*, 16, 36–44, 1997.
23. Madrid, Y., Barrio-Córdoba, M. E., and Cámara, C., Biosorption of antimony and chromium species by spirulina platensis and phaseolus. Applications to bioextract antimony and chromium for natural and industrial waters, *Analyst*, 123, 1593–1598, 1998.
24. http://ww2.mcgill.ca/biosorptio/publication/whatis.htm.
25. Smith, R. M., Before the injection — modern methods of sample preparation for separation techniques, *J. Chromatogr. A*, 1000, 3–27, 2003.
26. Phillips, J. B. and Beens, J., Comprehensive two dimensional gas chromatography: a hyphenated method with strong coupling between the two dimensions, *J. Chromatogr. A*, 856, 331–347, 1999.
27. Ljungkuist, G., Lärstad, M., and Mathiasson, L., Determination of low concentrations of benzene in urine using multi-dimensional gas chromatography, *Analyst*, 126, 41–45, 2001.
28. Reyes, D. R., Iossifidis, D., Auroux, P. A., and Manz, A., Micro total analysis systems. 1. Introduction, theory, and technology, *Anal. Chem.*, 74, 2623–2636, 2002.
29. Reyes, D. R., Iossifidis, D., Auroux, P. A., and Manz, A., Micro total analysis systems. 2. Analytical standard operations and applications, *Anal. Chem.*, 74, 2637–2652, 2002.
30. http://www.elsevier.com/vj/microTAS/us_index.html.
31. Saito, Y. and Jinno, K., Miniaturized sample preparation combined with liquid phase separations, *J. Chromatogr. A*, 1000, 53–67, 2003.
32. Psillakis, E. and Kalogerakis, N., Solid-phase microextraction versus single drop microextraction for the analysis of nitroaromatic explosives in water samples, *J. Chromatogr. A*, 938, 113–120, 2001.
33. Petersson, M., Wahlund, K. G., and Nilsson, S., Miniaturised online SPE for enhancement of concentration sensitivity in capillary electrophoresis, *J. Chromatogr. A*, 841, 249–261, 1999.
34. He, B., Tait, N., and Regnier, F., Fabrication of nanocolumns for liquid chromatography, *Anal. Chem.*, 70, 3790–3797, 1998.
35. Rohlícek, V. and Deyl, Z., Versatile tool for manipulation of electrophoresis chips, *J. Chromatogr. B*, 770, 19–23, 2002.
36. MacNair, J. E., Patel, K. D., and Jorgenson, J. W., Ultrahigh pressure reversed-phase capillary liquid chromatography: isocratic and gradient elution using columns packed with 1.0-μm particles, *Anal. Chem.*, 71, 700–708, 1999.
37. Jönsson, J. A. and Mathiasson, L., Liquid membrane extraction in analytical sample preparation, *Trends Anal. Chem.*, 18, 318–325, 1999.
38. Ma, M. and Cantwell, F. F., Solvent microextraction with simultaneous back-extraction for sample cleanup and preconcentration: preconcentration into a single microdrop, *Anal. Chem.*, 71, 388–393, 1999.
39. Zhu, L., Tay, C. B., and Lee, H. K., LLLME of aromatic amines from water samples combined with high-performance liquid chromatography, *J. Chromatogr. A*, 963, 231–237, 2002.
40. Buszewski, B. and Ligor, T., Single-drop extraction versus solid-phase microextraction for the analysis of VOCs in water, *LC–GC Europe*, 15, 92–97, 2002.
41. Hou, L. and Lee, H. K., Application of static and dynamic LPME in the determination of polycyclic aromatic hydrocarbons, *J. Chromatogr.*, 976, 377–385, 2002.
42. Huybrechts, T., Dewulf, J., and Langenhove, H. V., State-of-the-art of gas chromatography-based methods for analysis of anthropogenic volatile organic compounds in estuarine waters, illustrated with the river Scheldt as an example, *J. Chromatogr. A*, 1000, 283–297, 2003.
43. Liska, I., Fifty years of SPE in water analysis-historical development and overview, *J. Chromatogr. A*, 885, 3–16, 2000.
44. Hennion, M. C., SPE: method development, sorbents and coupling with liquid chromatography, *J. Chromatogr. A*, 856, 3–54, 1999.
45. Poole, C. F., New trends in SPE, *Trends Anal. Chem.*, 22, 362–373, 2003.

46. Rossi, D. T. and Zhang, N., Automating SPE: current aspects and future prospects, *J. Chromatogr. A*, 885, 97–113, 2000.
47. Arthur, C. L. and Pawliszyn, J., SPME with thermal desorption using fused silica optical fibers, *Anal. Chem.*, 62, 2145–2148, 1990.
48. Mindrup, R. and Shirey, R. E., Improved performance of SPME fibers and applications, SUPELCO, Supelco Park, Bellefonte, PA, 16823 USA, 2001, http://info.sial.com/Graphics/Supelco/objects/11000/10942.pdf.
49. Pawilszyn, J., *Solid-Phase Microextraction. Theory and Practice*, Wiley-VCH, New York, 1997.
50. Tankeviciute, A., Kazlauskas, R., and Vickackaite, V., Headspace extraction of alcohols into a single drop, *Analyst*, 126, 1674–1677, 2001.
51. Lord, H. and Pawilszyn, J., Evolution of solid-phase microextraction technology, *J. Chromatogr. A*, 885, 153–193, 2000.
52. Peñalver, A., Pocurull, E., Borrull, F., and Marcé, R. M., Trends in solid-phase microextraction for determining organic pollutants in environmental samples, *Trends Anal. Chem.*, 18, 557–568, 1999.
53. Baltussen, E., Sandra, P., David, F., and Cramers, C., Stir-bar sorptive extraction (SBSE), a novel technique for aqueous samples: theory and principles, *J. Microcol.*, 11, 737–747, 1999.
54. Vercauteren, J., Peres, C., Devos, C., Sandra, P., Vanhaecke, F., and Moens, L., Stir-bar sorptive extraction for the determination of ppq-level traces of organotin compounds in environmental samples with thermal desorption-capillary gas chromatography-ICP mass spectrometry, *Anal. Chem.*, 73, 1509–1514, 2001.
55. Kolahgar, B., Hoffmann, A., and Heiden, A. C., Application of SBSE to the determination of polycyclic aromatic hydrocarbons in aqueous samples, *J. Chromatogr.*, 963, 225–230, 2002.
56. Peñalver, A., García, V., Pocurull, E., Borrull, F., and Marce, R. M., Stir-bar sorptive extraction and large volume injection GC to determine a group of endocrine disrupters in water samples, *J. Chromatogr. A*, 1007, 1–9, 2003.
57. United States Environmental Protection Agency. *Method 3540 C: Organic extraction and sample preparation, revision 2, Methods for Evaluation of Solid Waste, Laboratory Manual, Physical/Chemical Methods*, 3rd ed., USEPA, Office of Solid Waste and Emergency Response, Washington, DC, 1986, (updated 1995).
58. Luque-García, J. L. and Luque de Castro, M. D., Extraction of PCBs from soils by automated FMAS extraction, *J. Chromatogr. A*, 998, 21–29, 2003.
59. Pressurised Fluid Extraction, EPA method 3545, EPA, Washington, DC, 1996.
60. Hawthorne, S. B., Yang, Y., and Miller, D. J., Extraction of organic pollutants form environmental solids with sub-and supercritical water, *Anal. Chem.*, 66, 2912–2920, 1994.
61. Yang, Y., Boward, S., Hawthorne, S. B., and Miller, D. J., Subcritical water extraction of polychlorinated biphenyls from soil and sediments, *Anal. Chem.*, 67, 4571–4576, 1995.
62. Luque de Castro, M. D. and Jiménez-Carmona, M. M., Where is SFE going?, *Trends Anal. Chem.*, 19, 223–228, 2000.
63. Taylor, L., Supercritical fluid chromatography and extraction, In *Handbook of Instrumental Techniques for Analytical Chemistry*, Settle, F. A., Ed., Prentice-Hall Inc., Englewood Cliffs, NJ, p. 183–197, 1997.
64. Smith, R. M., Supercritical fluids in separation science — the dreams, the reality and the future, *J. Chromatogr. A*, 856, 83–115, 1999.
65. Langenfeld, J. J., Hawthorne, S. B., Miller, D. J., and Pawliszyn, J., Role of modifiers for analytical scale SFE of environmental samples, *Anal. Chem.*, 66, 909–916, 1994.
66. Berset, J. D., Ejem, M., Holzer, R., and Lischer, P., Comparison of different drying, extraction techniques for the determination of priority polycyclic aromatic hydrocarbons in background contaminated soil samples, *Anal. Chim. Acta*, 383, 263–275, 1999.
67. Jeong, M. L. and Chesney, D. J., Investigation of modifiers effects in supercritical CO_2 extraction from various solid matrices, *J. Supercrit. Fluids*, 16, 33–42, 1999.
68. Smith, R. M., Ed., *Supercritical Fluid Chromatography*, The Royal Society of Chemistry, Cambridge, 1988.
69. Heldrich, J. L. and Taylor, L. T., Quantitative SFE/SFC of a phosphonate from aqueous media, *Anal. Chem.*, 61, 1986–1988, 1989.

70. Page, S. H., Benner, B. A., Small, J. A., and Choquete, S. J., Restrictor plugging in offline SFE of environmental samples. Microscopic, chemical, and spectroscopic evaluations, *J. Supercrit. Fluids*, 14, 257–270, 1999.
71. Motohashi, N., Nagashima, H., and Parkanyi, C., Supercritical fluid extraction for the analysis of pesticide residues in miscellaneous samples, *J. Biochem. Biophys. Methods*, 43, 313–328, 2000.
72. Turner, C., Eskilsson, C. S., and Björklund, E., Collection in analytical-scale SF extraction, *J. Chromatogr. A*, 947, 1–22, 2002.
73. van Bavel, B., Jaremo, M., Karlsson, L., and Lindstrom, G., Development of a solid phase carbon trap for simultaneous determination of PCDDs, PCDFs, PCBs, and pesticides in environmental samples using SFE–LC, *Anal. Chem.*, 68, 1279–1283, 1996.
74. Chester, T. L., Pinkston, J. D., and Raynie, D. E., SFC and extraction, *Anal. Chem.*, 70, 301R–319R, 1998.
75. Stone, M. A. and Taylor, L. T., Quantitative coupling of SFE and HPLC by means of a coated open-tubular interface, *J. Chromatogr. A*, 931, 53–65, 2001.
76. Mardones, C., Ríos, A., and Valcárcel, M., Automatic online coupling of SFE and capillary electrophoresis, *Anal. Chem.*, 72, 5736–5739, 2000.
77. Valcárcel, M., Arce, L., and Ríos, A., Coupling continuous separation techniques to capillary electrophoresis, *J. Chromatogr. A*, 924, 3–30, 2001.
78. Tai, C. Y., You, G. S., and Chen, S. L., Kinetics study on SFE of Zinc (II) ion from aqueous solutions, *J. Supercrit. Fluids*, 18, 201–212, 2000.
79. Jayasinghe, L. Y., Marriott, P. J., Carpener, P. D., and Nichols, P. D., SFE and GC electron capture detection method for sterol analysis of environmental water samples, *Anal. Commun.*, 35, 265–268, 1998.
80. Laitinen, A. and Kaunisto, J., Supercritical fluid extraction of 1-butanol from aqueous solutions, *J. Supercrit. Fluids*, 15, 245–252, 1999.
81. Erkey, C., Supercritical carbon dioxide extraction of metals from aqueous solutions: a review, *J. Supercrit. Fluids*, 17, 259–287, 2000.
82. Bjorklund, E., Nilsson, T., Bowadt, S., Pilorz, K., Mathiasson, L., and Hawthorne, S. B., Introducing selective SFE as a new tool for determining sorption/desorption behavior and bioavailability of persistent organic pollutants in sediment, *J. Biochem. Biophys. Methods*, 43, 295–311, 2000.
83. Gómez-Ariza, J. L., Morales, E., Giradles, I., Sánchez-Rodas, D., and Velasco, A., Simple treatment in chromatography-based speciation of organometallic pollutants, *J. Chromatogr. A*, 938, 211–224, 2001.
84. Khorassani, M. A. and Taylor, L. T., Supercritical fluid extraction of mercury (II) ion via *in situ* chelation and preformed mercury complexes from different matrices, *Anal. Chim. Acta*, 379, 1–9, 1999.
85. Lopez-Avila, V., Sample preparation for environmental analysis, *Crit. Rev. Anal. Chem.*, 29, 195–230, 1999.
86. Letellier, M. and Budzinski, H., Microwave assisted extraction of organic compounds, *Analysis*, 27, 259–271, 1999.
87. Camel, V., Microwave-assisted solvent extraction of environmental samples, *Trends Anal. Chem.*, 19, 229–248, 2000.
88. Hawthorne, S. B., Analytical-scale supercritical fluid extraction, *Anal. Chem.*, 62, 633A–642A, 1990.
89. Bøwadt, S. and Hawthorne, S. B., Supercritical fluid extraction in environmental analysis, *J. Chromatogr. A*, 703, 549–571, 1995.
90. Smith, R. M., Supercritical fluids in separation science — the dreams, the reality and the future, *J. Chromatogr. A*, 856, 83–115, 1999.
91. Bjorklund, E., Bøwadt, S., and Nilsson, T., Pressurised liquid extraction of persistent organic pollutants in environmental analysis, *Trends Anal. Chem.*, 19, 434–445, 2000.
92. Kingston, H. M., Walter, P. J., Chalk, S., Lorentzen, E., and Link, D., Environmental microwave sample preparation: fundamentals, methods, and applications, In *Microwave-Enhanced Chemistry*, Kingston, H. M. and Haswell, S. J., Eds., American Chemical Society, Washington, DC, p. 223–349, 1997.
93. Abu-Samra, A., Morris, J. S., and Koirtyohann, S. R., Wet ashing of some biological samples in a microwave oven, *Anal. Chem.*, 47, 1475–1478, 1975.

94. Ganzler, K., Salgo, A., and Valko, K., Microwave extraction: a novel sample preparation method for chromatography, *J. Chromatogr. A*, 371, 299–306, 1986.

95. Pare, J. R. J., Belanger, J. M. R., and Stafford, S. S., Microwave-assisted process (MAP™): a new tool for the analytical laboratory, *Trends Anal. Chem.*, 13, 176–184, 1994.

96. Pare, J. R. J, U.S. Patent, 5 002 784, 1991.

97. Lopez-Avila, V., Young, R., and Beckert, W. F., Microwave-assisted extraction of organic compounds from standard reference soils and sediments, *Anal. Chem.*, 66, 1097–1106, 1994.

98. ASTM D5258. *Standard Test Method For Acid Extraction of Elements from Sediments using Closed Vessel Microwave Heating*, Amcrican Society for Testing and Materials, Philadelphia, PA, 1996.

99. ASTM D5765. *Standard Practice for Solvent Extraction of Total Petroleum Hydrocarbons from Soils and Sediments using Closed Vessel Microwave Heating*, American Society for Testing and Materials, Philadelphia, PA, 1996.

100. ASTM D6010. *Standard Practice for Closed Vessel Micro-wave Solvent Extraction of Organic Compounds from Solid Matrices*, American Society for Testing and Materials, Philadelphia, PA, 1996.

101. Microwave Extraction of VOC's and SVOC's (organophosphorus pesticides, organochlorine pesticides, chlorinated herbicides, phenoxy acid herbicides, PCBs, etc.), EPA Method 3546, EPA SW-846 update III, US Environmental Protection Agency, 1999.

102. http://www.sampleprep.duq.edu/index.html.

103. Zlotorzynski, A., The application of microwave radiation to analytical and environmental chemistry, *Crit. Rev. Anal. Chem.*, 25, 43–76, 1995.

104. Kingston, H. M. and Jassie, L. B., Microwave energy for acid decomposition at elevated temperatures and pressures using biological and botanical samples, *Anal. Chem.*, 58, 2534–2541, 1986.

105. Eskilsson, C. S. and Björklund, E., Analytical-scale microwave-assisted extraction, *J. Chromatogr. A*, 902, 227–250, 2000.

106. Luque-García, J. L. and Luque de Castro, M. D., Where is microwave-based analytical equipment for solid sample pretreatment going?, *Trends Anal. Chem.*, 22, 90–98, 2003.

107. Nóbrega, J. A., Trevizan, L. C., Araújo, G. C. L., and Nogueira, A. R. A., Focused-microwave-assisted strategies for sample preparation, *Spectrochim. Acta*, 57B, 1855–1876, 2002.

108. Ericsson, M. and Colmsjo, A., Dynamic microwave-assisted extraction coupled online with SPE and large-volume injection gas chromatography: determination of organophosphate esters in air samples, *Anal. Chem.*, 75, 1713–1719, 2003.

109. Ramil Criado, M., Rodriguez Pereiro, I., and Cela Torrijos, R., Determination of polychlorinated biphenyl compounds in indoor air samples, *J. Chromatogr. A.*, 963, 65–71, 2002.

110. Mueller, W., Schubert, J., Benzing, A., and Geiger, K., Method for analysis of exhaled air by microwave energy desorption coupled with gas chromatography-flame ionization detection-mass spectrometry, *J. Chromatogr. B*, 716, 27–38, 1998.

111. Trinh, V. D. and Khanh, H. C., GC Bin quartz tubes for the sampling of indoor air nicotine and analysis by microwave thermal desorption-capillary gas chromatography, *J. Chromatogr. Sci.*, 29, 179–183, 1991.

112. Xiong, G. and Pawliszyn, J., Microwave-assisted generation of standard gas mixtures, *Anal. Chem.*, 74, 2446–2449, 2002.

113. Liu, Y., Mou, S., and Heberling, S., Determination of trace level bromate and perchlorate in drinking water by ion chromatography with an evaporative preconcentration technique, *J. Chromatogr. A*, 956, 85–91, 2002.

114. Yang, J. S., Lee, D. W., and Lim, H., Microwave-assisted extraction (MAE) of polychlorinated biphenyls and polychlorinated dibenzo-*p*-dioxins from fly ash and sea sediments: effect of water and removal of interferences, *J. Liq. Chromatogr. Relat. Technol.*, 26, 803–818, 2003.

115. Numata, M., Yarita, T., Aoyagi, Y., and Takatsu, A., Microwave-assisted steam distillation for simple determination of polychlorinated biphenyls and organochlorine pesticides in sediments, *Anal. Chem.*, 75, 1450–1457, 2003.

116. Fatoki, O. S. and Awofolu, R. O., Methods for selective determination of persistent organochlorine pesticide residues in water and sediments by capillary GC and electron-capture detection, *J. Chromatogr. A*, 983, 225–236, 2003.

117. Ghassempour, A., Mohammadkhah, A., Najafi, F., and Rajabzadeh, M., Monitoring of the pesticide diazinon in soil, stem and surface water of rice fields, *Anal. Sci.*, 18, 779–783, 2002.
118. Contat-Rodrigo, L., Haider, N., Ribes-Greus, A., and Karlsson, S., Ultrasonication and microwave-assisted extraction of degradation products from degradable polyolefin blends aged in soil, *J. Appl. Polym. Sci.*, 79, 1101–1112, 2000.
119. Mason, T. J., *Sonochemistry*, Oxford University Press, Oxford, 1999.
120. Suslick, K. S., Hammerton, D. A., and Cline, R. E., The sonochemical hot spot, *J. Am. Chem. Soc.*, 108, 5641–5642, 1986.
121. Suslick, K. S., The chemical effects of ultrasound, *Sc. Am.*, February, 80–86, 1989.
122. Cum, G., Gallo, R., Spadaro, A., and Galli, G., Effect of static pressure on the ultrasonic activation of chemical reactions. Selective oxidation at benzylic carbon in the liquid phase, *J. Chem. Soc. Perkin Trans.*, 2, 375–383, 1988.
123. Bendicho, C. and Lavilla, I., Ultrasound extractions, In *Encyclopedia of Separation Science*, Wilson, I. D., Adlard, T. R., Cooke, M., and Poole, C. F., Eds., Academic Press, New York, p. 1448–1454, 2000.
124. Luque-García, J. L. and Luque de Castro, M. D., Ultrasound: a powerful tool for leaching, *Trends Anal. Chem.*, 22, 41–47, 2003.
125. Bendicho, C. and Lavilla, I., Ultrasound-assisted metal extractions, In *Encyclopedia of Separation Science*, Wilson, I. D., Adlard, T. R., Cooke, M., and Poole, C. F., Eds., Academic Press, New York, pp. 4421–4426, 2000.
126. Beard, A., Naikwadi, K., and Karasek, F. W., Comparison of extraction methods for polychlorinated dibenzo-*p*-dioxins and dibenzofurans in fly ash using gas chromatography–mass spectrometry, *J. Chromatogr. A*, 589, 265–270, 1992.
127. Voznakova, Z., Podehradska, J., and Kohlickova, M., Determination of nitrophenols in soil, *Chemosphere*, 33, 285–291, 1996.
128. Babic, S., Petrovic, M., and Kastelan-Macan, M., Ultrasonic solvent extraction of pesticides from soil, *J. Chromatogr. A*, 823, 3–9, 1998.
129. Blau, K. and Halket, J. M., *Handbook of Derivatives for Chromatography*, 2nd ed., Wiley, Chichester, pp. 4–8, 1993.
130. Lestremau, F., Desauziers, V., Roux, J.-C., and Fanlo, J.-L., Development of a quantification method for the analysis of malodorous sulphur compounds in gaseous industrial effluents by solid-phase microextraction and gas chromatography-pulsed flame photometric detection, *J. Chromatogr. A*, 999, 71–80, 2003.
131. Gimeno, R. A., Beltrán, J. L., Marcé, R. M., and Borrull, F., Determination of naphthalenesulfonates in water by online ion-pair SPE and ion-pair liquid chromatography with fast-scanning fluorescence detection, *J. Chromatogr. A*, 890, 289–294, 2000.
132. Ábalos, M., Prieto, X., and Bayona, J. M., Determination of volatile alkyl sulfides in wastewater by headspace solid-phase microextraction followed by gas chromatography-mass spectrometry, *J. Chromatogr. A*, 963, 249–257, 2002.
133. Zhao, L., Zhu, L., and Lee, H. K., Analysis of aromatic amines in water samples by liquid–liquid–liquid microextraction with hollow fibers and high-performance liquid chromatography, *J. Chromatogr. A*, 963, 239–248, 2002.
134. Zeng, Z., Qiu, W., Yang, M., Wei, X., Huang, Z., and Li, F., Solid-phase microextraction of monocyclic aromatic amines using novel fibers coated with crown ether, *J. Chromatogr. A*, 934, 51–57, 2001.
135. Zhu, L., Tay, C. B., and Lee, H. K., Liquid–liquid–liquid microextraction of aromatic amines from water samples combined with high-performance liquid chromatography, *J. Chromatogr. A*, 963, 231–237, 2002.
136. Moriwaki, H., Harino, H., Hashimoto, H., Arakawa, R., Ohe, T., and Yoshikura, T., Determination of aromatic amine mutagens, PBTA-1 and PBTA-2, in river water by SPE followed by liquid chromatography-tandem mass spectrometry, *J. Chromatogr. A*, 995, 239–243, 2003.
137. Ábalos, M. and Bayona, J. M., Application of GC coupled to chemical ionisation mass spectrometry following headspace solid-phase microextraction for the determination of free volatile fatty acids in aqueous samples, *J. Chromatogr. A*, 891, 287–294, 2000.

138. Ábalos, M., Bayona, J. M., and Pawliszyn, J., Development of a headspace solid-phase microextraction procedure for the determination of free volatile fatty acids in waste waters, *J. Chromatogr. A*, 873, 107–115, 2000.

139. Matisova, E., Medved'ova, M., Vraniakov, J., and Simon, P., Optimisation of solid-phase microextraction of volatiles, *J. Chromatogr. A*, 960, 159–164, 2002.

140. Cho, H. J., Baek, K., Lee, H. H., Lee, S. H., and Yang, J. W., Competitive extraction of multi-component contaminants in water by carboxen-polydimethylsiloxane fiber during solid-phase microextraction, *J. Chromatogr. A*, 988, 177–184, 2003.

141. Dron, J., Garcia, R., and Millán, E., Optimization of headspace solid-phase microextraction by means of an experimental design for the determination of methyl tert-butyl ether in water by gas chromatography-flame ionization detection, *J. Chromatogr. A*, 963, 259–264, 2002.

142. Xiao, C., Han, S., Wang, Z., Xing, J., and Wu, C., Application of the polysilicone fullerene coating for solid-phase microextraction in the determination of semivolatile compounds, *J. Chromatogr. A*, 927, 121–130, 2001.

143. Leon, V. M., Alvarez, B., Cobollo, M. A., Munoz, S., and Valor, I., Analysis of 35 priority semivolatile compounds in water by stir-bar sorptive extraction-thermal desorption-gas chromatography-mass spectrometry I. Method optimisation, *J. Chromatogr. A*, 999, 91–101, 2003.

144. Hou, L. and Lee, H. K., Application of static and dynamic LPME in the determination of polycyclic aromatic hydrocarbons, *J. Chromatogr. A*, 976, 377–385, 2002.

145. Marcé, R. M. and Borrull, F., SPE of polycyclic aromatic compounds, *J. Chromatogr. A*, 885, 273–290, 2000.

146. Wolska, L., Miniaturised analytical procedure of determining polycyclic aromatic hydrocarbons and PCBs in bottom sediments, *J. Chromatogr. A*, 959, 173–180, 2002.

147. Li, N. and Lee, H. K., SPE of polycyclic aromatic hydrocarbons in surface water negative effect of humic acid, *J. Chromatogr. A*, 921, 255–263, 2001.

148. Kolahgar, B., Hoffmann, A., and Heiden, A. C., Application of SBSE to the determination of polycyclic aromatic hydrocarbons in aqueous samples, *J. Chromatogr. A*, 963, 225–230, 2002.

149. Doong, R., Chang, S., and Sun, Y., Solid-phase microextraction for determining the distribution of sixteen US Environmental Protection Agency polycyclic aromatic hydrocarbons in water samples, *J. Chromatogr. A*, 879, 177–188, 2000.

150. Pérez, S., Guillamón, M., and Barcelò, D., Quantitative analysis of polycyclic aromatic hydrocarbons in sewage sludge from wastewater treatment plants, *J. Chromatogr. A*, 938, 57–65, 2001.

151. Peñalver, A., García, V., Pocurull, E., Borrull, F., and Marcé, R. M., Stir-bar sorptive extraction and large volume injection GC to determine a group of endocrine disrupters in water samples, *J. Chromatogr. A*, 1007, 1–9, 2003.

152. Loos, R. and Barcelò, D., Determination of haloacetic acids in aqueous environments by SPE followed by ion-pair liquid chromatography-electrospray ionization mass spectrometric detection, *J. Chromatogr. A*, 938, 45–55, 2001.

153. Baggiani, C., Giovannoli, C., Anfossi, L., and Tozzi, C., Molecularly imprinted SPE sorbent for the clean up of chlorinated phenoxyacids from aqueous samples, *J. Chromatogr. A*, 938, 35–44, 2001.

154. Lacorte, S., Perrot, M. C., Fraisse, D., and Barceló, D., Determination of chlorobenzidines in industrial effluent by solid-phase extraction and liquid chromatography with electrochemical and mass spectrometric detection, *J. Chromatogr. A*, 833, 181–194, 1999.

155. Castells, P., Santos, F. J., and Galceran, M. T., Solid-phase microextraction for the analysis of short-chain chlorinated paraffins in water samples, *J. Chromatogr. A*, 984, 1–8, 2003.

156. Sarrión, M. N., Santos, F. J., and Galceran, M. T., Solid-phase microextraction coupled with gas chromatography-ion trap mass spectrometry for the analysis of haloacetic acids in water, *J. Chromatogr. A*, 859, 159–171, 1999.

157. He, Y., Wang, Y., and Lee, H. K., Trace analysis of ten chlorinated benzenes in water by headspace solid-phase microextraction, *J. Chromatogr. A*, 874, 149–154, 2000.

158. Cancho, B., Ventura, F., and Galceran, M. T., Solid-phase microextraction for the determination of iodinated trihalomethanes in drinking water, *J. Chromatogr. A*, 841, 197–206, 1999.

159. Cortazar, E., Zuloaga, O., Sanz, J., Raposo, J. C., Etxebarria, N., and Fernández, L. A., Multisimplex optimisation of the solid-phase microextraction-gas chromatographic-mass spectrometric

determination of polycyclic aromatic hydrocarbons, PCBs and phthalates from water samples, *J. Chromatogr. A*, 978, 165–175, 2002.

160. Shu, Y. Y., Wang, S. S., Tardif, M., and Huang, Y., Analysis of PCBs in aqueous samples by microwave-assisted headspace solid-phase microextraction, *J. Chromatogr. A*, 1008, 1–12, 2003.

161. Gómez-Ariza, J. L., Morales, E., Giráldez, I., Sánchez-Rodas, D., and Velasco, A., Sample treatment in chromatography-based speciation of organometallic pollutants, *J. Chromatogr. A*, 938, 211–224, 2001.

162. Gac, M. L., Lespes, G., and Potin-Gautier, M., Rapid determination of organotin compounds by headspace solid-phase microextraction, *J. Chromatogr. A*, 999, 123–134, 2003.

163. Carpinterio Botana, J., Rodríguez Pereiro, I., and Cela Torrijos, R., Rapid determination of butyltin species in water samples by multicapillary GC with atomic emission detection following headspace solid-phase microextraction, *J. Chromatogr. A*, 963, 195–203, 2002.

164. Vercauteren, J., De Meester, A., De Smaele, T., Vanhaecke, F., Moens, L., Damsa, R., and Sandrab, P., Headspace solid-phase microextraction-capillary gas chromatography-ICP mass spectrometry for the determination of the organotin pesticide fentin in environmental samples, *J. Anal. At. Spectrom.*, 15, 651–656, 2000.

165. Millan, E. and Pawliszyn, J., Determination of butyltin species in water and sediment by solid-phase microextraction–gas chromatography–flame ionisation detection, *J. Chromatogr. A*, 873, 63–71, 2000.

166. Reitzel, L. A. and Ledin, A., Determination of phenols in landfill leachate-contaminated groundwaters by SPE, *J. Chromatogr. A*, 972, 175–182, 2002.

167. Lacorte, S., Fraisse, Dl., and Barceló, D., Efficient SPE procedures for trace enrichment of priority phenols from industrial effluents with high total organic carbon content, *J. Chromatogr. A*, 857, 97–106, 1999.

168. Zhao, L. and Lee, H. K., Determination of phenols in water using liquid phase microextraction with back extraction combined with high-performance liquid chromatography, *J. Chromatogr. A*, 931, 95–105, 2001.

169. Rodríguez, I., Llompart, M. P., and Cela, R., SPE of phenols, *J. Chromatogr. A*, 885, 291–304, 2000.

170. González-Toledo, E., Prat, M. D., and Alpendurada, M. F., Solid-phase microextraction coupled to liquid chromatography for the analysis of phenolic compounds in water, *J. Chromatogr. A*, 923, 45–52, 2001.

171. Wissiack, R. and Rosenberg, E., Universal screening method for the determination of US environmental protection agency phenols at the lower ng l^{-1} level in water samples by online SPE–HPLC–atmospheric pressure chemical ionization mass spectrometry within a single run, *J. Chromatogr. A*, 963, 149–157, 2002.

172. Peñalver, A., Pocurull, E., Borrull, F., and Marcé, R. M., Solid-phase microextraction coupled to high-performance liquid chromatography to determine phenolic compounds in water samples, *J. Chromatogr. A*, 953, 79–87, 2002.

173. Díaz, A., Ventura, F., and Galceran, M. T., Development of a solid-phase microextraction method for the determination of short-ethoxy-chain nonylphenols and their brominated analogs in raw and treated water, *J. Chromatogr. A*, 963, 159–167, 2002.

174. Llompart, M., Lourido, M., Landín, P., García-Jares, C., and Cela, R., Optimization of a derivatization — solid-phase microextraction method for the analysis of thirty phenolic pollutants in water samples, *J. Chromatogr. A*, 963, 137–148, 2002.

175. Zhu, L., Zhu, L., and Lee, H. K., Liquid–liquid–liquid microextraction of nitrophenols with a hollowfiber membrane prior to capillary liquid chromatography, *J. Chromatogr. A*, 924, 407–414, 2001.

176. Caro, E., Masqué, N., Marcé, R. M., Borrull, F., Cormack, P. A. G., and Sherrington, D. C., Noncovalent and semicovalent MIPs for selective online SPE of 4-nitrophenol from water samples, *J. Chromatogr. A*, 963, 169–178, 2002.

177. Sojo, L. E. and Djauhari, J., Determination of chlorophenolics in waters by membrane solid-phase extraction: comparison between C$_{18}$ and activated carbon membranes and between modes of extraction and elution, *J. Chromatogr. A*, 840, 21–30, 1999.

178. Bagheri, H. and Saraji, M., Conductive polymers as new media for SPE: isolation of chlorophenols from water sample, *J. Chromatogr. A*, 986, 111–119, 2003.

179. Sarrión, M. N., Santos, F. J., and Galceran, M. T., Determination of chlorophenols by solid-phase microextraction and liquid chromatography with electrochemical detection, *J. Chromatogr. A*, 947, 155–165, 2002.

180. Zhao, L. and Lee, H. K., Liquid–liquid–liquid microextraction of aromatic amines from water samples combined with high-performance liquid chromatography, *J. Chromatogr. A*, 919, 381–388, 2001.

181. Lambropoulou, D. A. and Albanis, T. A., Optimization of headspace solid-phase microextraction conditions for the determination of organophosphorus insecticides in natural waters, *J. Chromatogr. A*, 922, 243–255, 2001.

182. Lópe-Blanco, M. C., Cancho-Grande, B., and Simal-Gandara, J., Comparison of SPE and solid-phase microextraction for carbofuran in water analyzed by high performance liquid chromatography — photodiode-array detection, *J. Chromatogr. A*, 963, 117–123, 2002.

183. Gerecke, A. C., Tixier, C., Bartels, T., Schwarzenbach, R. P., and Müller, S. R., Determination of phenylurea herbicides in natural waters at 21 concentrations below 1 ng/l using SPE, derivatization, and solid-phase microextraction– gas chromatography–mass spectrometry, *J. Chromatogr. A*, 930, 9–19, 2001.

184. Krutz, L. J., Senseman, S. A., and Sciumbato, A. S., Solid-phase microextraction for herbicide determination in environmental samples, *J. Chromatogr. A*, 999, 103–121, 2003.

185. Megersaab, N. and Jönsson, J. Á., Trace enrichment and sample preparation of alkylthio-s-triazine herbicides in environmental waters using a supported liquid membrane technique in combination with high-performance liquid chromatography, *The Analyst*, 123, 225–231, 1998.

186. Catalina, M. I., Dallüge, J., Vreuls, R. J. J., and Brinkman, U. A. Th., Determination of chlorophenoxy acid herbicides in water by *in situ* esterification followed by in-vial liquid–liquid extraction combined with large-volume on-column injection and gas chromatography–mass spectrometry, *J. Chromatogr. A*, 877, 153–166, 2000.

187. Muñoz de la Peña, A., Mahedero, M. C., and Bautista-Sánchez, A., Monitoring of phenylurea and propanyl herbicides in river water by solid-phase-extraction HPLC with photoinduced-fluorimetric detection, *Talanta*, 60, 279–285, 2003.

188. Berrada, H., Font, G., and Moltó, J. C., Indirect analysis of urea herbicides from environmental water using solid-phase microextraction, *J. Chromatogr. A*, 890, 303–312, 2000.

189. Berrada, H., Font, G., and Moltó, J. C., Determination of urea pesticide residues in vegetable, soil, and water samples, *Crit. Rev. Anal. Chem.*, 33, 19–41, 2003.

190. Corcia, A. D., Costantino, A., Crescenzi, C., and Samperi, R., Quantification of phenylurea herbicides and their free and humic acid-associated metabolites in natural waters, *J. Chromatogr. A*, 852, 465–474, 1999.

191. Chao, J., Liu, J., Wen, M., Liu, J., Cai, Y., and Jiang, G., Determination of sulfonylurea herbicides by continuous-flow liquid membrane extraction online coupled with high-performance liquid chromatography, *J. Chromatogr. A*, 955, 183–189, 2002.

192. Krutz, L. J., Senseman, S. A., and Sciumbato, A. S., Solid-phase microextraction for herbicide determination in environmental samples, *J. Chromatogr. A*, 999, 103–121, 2003.

193. Li, N. and Lee, H. K., Sample preparation based on dynamic ion-exchange SPE for GC/MS analysis of acidic herbicides in environmental waters, *Anal. Chem.*, 72, 3077–3084, 2000.

194. Carabias-Martínez, R., Rodríguez-Gonzalo, E., Herrero-Hernández, E., Javier Sánchez-San Román, F., and Prado-Flores, M. G., Determination of herbicides and metabolites by SPE and liquid chromatography. Evaluation of pollution due to herbicides in surface and groundwaters, *J. Chromatogr. A*, 950, 157–166, 2002.

195. Ferrer, I. and Barceló, D., Simultaneous determination of antifouling herbicides in marina water samples by online SPE followed by liquid chromatography–mass spectrometry, *J. Chromatogr. A*, 854, 197–206, 1999.

196. Curini, R., Gentili, A., Marchese, S., Marino, A., and Perret, D., SPE followed by high-performance liquid chromatography–ionspray interface–mass spectrometry for monitoring of herbicides in environmental water, *J. Chromatogr. A*, 874, 187–198, 2000.

197. Picó, Y., Font, G., Moltó, J. C., and Mañes, J., SPE of quaternary ammonium herbicides, *J. Chromatogr. A*, 885, 251–271, 2000.

198. Castro, R., Moyano, E., and Galceran, M. T., Online ion-pair SPE–liquid chromatography–mass spectrometry for the analysis of quaternary ammonium herbicides, *J. Chromatogr. A*, 869, 441–449, 2000.

199. Núñez, O., Moyano, E., and Galceran, M. T., SPE and sample stacking — capillary electrophoresis for the determination of quaternary ammonium herbicides in drinking water, *J. Chromatogr. A*, 946, 275–282, 2002.

200. González-Barreiro, C., Lores, M., Casais, M. C., and Cela, R., Optimisation of alachlor solid-phase microextraction from water samples using experimental design, *J. Chromatogr. A*, 896, 373–379, 2000.

201. Ramesh, A. and Elumalai Ravi, P., Applications of solid-phase microextraction (SPME) in the determination of residues of certain herbicides at trace levels in environmental samples, *J. Environ. Monit.*, 3, 505–508, 2001.

202. Carabias-Martínez, R., Rodríguez-Gonzalo, E., Revilla-Ruiz, P., and Domínguez-Alvarez, J., SPE and sample stacking — micellar electrokinetic capillary chromatography for the determination of multiresidues of herbicides and metabolites, *J. Chromatogr. A*, 990, 291–302, 2003.

203. Hinsmann, P., Arce, L., Ríos, A., and Valcárcel, M., Determination of pesticides in waters by automatic online SPE–capillary electrophoresis, *J. Chromatogr. A*, 866, 137–146, 2000.

204. Tomkins, B. A. and Ilgner, R. H., Determination of atrazine and four organophosphorus pesticides in ground water using SPME followed by GC with selected-ion monitoring, *J. Chromatogr. A*, 972, 183–194, 2002.

205. Hernández, F., Sancho, J. V., Pozo, O., Lara, A., and Pitarch, E., Rapid direct determination of pesticides and metabolites in environmental water samples at sub-mg/l level by online SPE-liquid chromatography–electrospray tandem mass spectrometry, *J. Chromatogr. A*, 939, 1–11, 2001.

206. Natangelo, M., Tavazzi, S., Fanelli, R., and Benfenati, E., Analysis of some pesticides in water samples using solid-phase microextraction — GC with different mass spectrometric techniques, *J. Chromatogr. A*, 859, 193–201, 1999.

207. Wu, J., Tragas, C., Lord, H., and Pawliszyn, J., Analysis of polar pesticides in water and wine samples by automated in-tube solid-phase microextraction coupled with high-performance liquid chromatography–mass spectrometry, *J. Chromatogr. A*, 976, 357–367, 2002.

208. Pap, T., Horváth, V., Tolokán, A., Horvai, G., and Sellergren, B., Effect of solvents on the selectivity of terbutylazine imprinted polymer sorbents used in SPE, *J. Chromatogr. A*, 973, 1–12, 2002.

209. Sabik, H., Jeannot, R., and Rondeau, B., Multiresidue methods using SPE techniques for monitoring priority pesticides, including triazines and degradation products, in ground and surface waters, *J. Chromatogr. A*, 885, 217–236, 2000.

210. Martínez Vidal, J. L., Pablos Espada, M. C., Garrido Frenich, A., and Arrebola, F. J., Pesticide trace analysis using SPE and GC with electron-capture and tandem mass spectrometric detection in water samples, *J. Chromatogr. A*, 867, 235–245, 2000.

211. Quintana, J., Martí, I., and Ventura, F., Monitoring of pesticides in drinking and related waters in NE Spain with a multiresidue SPE-GC–MS method including an estimation of the uncertainty of the analytical results, *J. Chromatogr. A*, 938, 3–13, 2001.

212. Gonçalves, C. and Alpendurada, M. F., Multiresidue method for the simultaneous determination of four groups of pesticides in ground and drinking waters, using solid-phase microextraction-GC with electron-capture and thermionic specific detection, *J. Chromatogr. A*, 968, 177–190, 2002.

213. Hauser, B., Popp, P., and Kleine-Bennc, E., Membrane-assisted solvent extraction of triazines and other semivolatile contaminants directly coupled to large-volume injection–gas chromatography–mass spectrometric detection, *J. Chromatogr. A*, 963, 27–36, 2002.

214. Frías, S., Rodríguez, M. A., Conde, J. E., and Pérez-Trujillo, J. P., Optimisation of a solid-phase microextraction procedure for the determination of triazines in water with gas chromatography–mass spectrometry detection, *J. Chromatogr. A*, 1007, 127–135, 2003.

215. Pinto, G. M. F. and Jardim, I. C. S. F., Use of SPE and high-performance liquid chromatography for the determination of triazine residues in water: validation of the method, *J. Chromatogr. A*, 869, 463–469, 2000.

216. Tomkins, B. A. and Barnard, A. R., Determination of OCPs in ground water using solid-phase microextraction followed by dual-column GC with electron-capture detection, *J. Chromatogr. A*, 964, 21–33, 2002.

217. Aguilar, C., Peñalver, A., Pocurull, E., Ferré, J., Borrull, F., and Marcé, R. M., Optimization of solid-phase microextraction conditions using a response surface methodology to determine OCPs in water by GC and electron-capture detection, *J. Chromatogr. A*, 844, 425–432, 1999.

218. Nogueira, J. M. F., Sandra, T., and Sandra, P., Considerations on ultra trace analysis of carbamates in water samples, *J. Chromatogr. A*, 996, 133–140, 2002.

219. Lambropoulou, D. A., Konstantinou, I. K., and Albanis, T. A., Determination of fungicides in natural waters using solid-phase microextraction and GC coupled with electron-capture and mass spectrometric detection, *J. Chromatogr. A*, 893, 143–156, 2000.

220. Reusser, D. E. and Field, J. A., Determination of benzylsuccinic acid in gasoline-contaminated groundwater by SPE coupled with gas chromatography–mass spectrometry, *J. Chromatogr. A*, 953, 215–225, 2002.

221. Psillakis, E. and Kalogerakis, N., Hollow-fibre LPME of phthalate esters from water, *J. Chromatogr. A*, 999, 145–153, 2003.

222. Luks-Betlej, K., Popp, P., Janoszka, B., and Paschke, H., Solid-phase microextraction of phthalates from water, *J. Chromatogr. A*, 938, 93–101, 2001.

223. Peñalver, A., Pocurull, E., Borrull, F., and Marcé, R. M., Determination of phthalate esters in water samples by solid-phase microextraction and GC with mass spectrometric detection, *J. Chromatogr. A*, 872, 191–201, 2000.

224. Norberg, J., Thordarson, E., Mathiasson, L., and Jonsson, J. A., Microporous membrane liquid–liquid extraction coupled online with normal-phase liquid chromatography for the determination of cationic surfactants in river and waste water, *J. Chromatogr. A*, 869, 523–529, 2000.

225. Alzaga, R., Pena, A., Ortiz, L., and Bayona, J. M., Determination of linear alkylbenzensulfonates in aqueous matrices by ion-pair solid-phase microextraction–in-port derivatization–gas chromato-graphy–mass spectrometry, *J. Chromatogr. A*, 999, 51–60, 2003.

226. Kuosmanen, K., Hyötyläinen, T., Hartonen, K., and Riekkola, M-L., Analysis of polycyclic aromatic hydrocarbons in soil and sediment with online coupled pressurised hot water extraction, hollow fibre microporous membrane liquid–liquid extraction and gas chromatography, *The Analyst*, 128, 434–439, 2003.

227. Ericsson, M. and Colmsjö, A., Dynamic microwave-assisted extraction, *J. Chromatogr. A*, 877, 141–151, 2000.

228. Priego-Capote, F., Luque-Garcia, J. L., and Luque de Castro, M. D., Automated fast extraction of nitrated polycyclic aromatic hydrocarbons from soil by FMAS extraction prior to gas chromatography–electron-capture detection, *J. Chromatogr. A*, 994, 159–167, 2003.

229. Lundstedt, S., van Bavel, B., Haglund, P., Tysklind, M., and Öberg, L., Pressurised liquid extraction of polycyclic aromatic hydrocarbons from contaminated soils, *J. Chromatogr. A*, 893, 151–162, 2000.

230. Ramos, L., Vreuls, J. J., and Brinkman, U. A. Th., Miniaturised pressurised liquid extraction of polycyclic aromatic hydrocarbons from soil and sediment with subsequent large-volume injection–gas chromatography, *J. Chromatogr. A*, 891, 275–286, 2000.

231. King, S., Meyer, J. S., and Andrews, A. R. J., Screening method for polycyclic aromatic hydrocarbons in soil using hollow fiber membrane solvent microextraction, *J. Chromatogr. A*, 982, 201–208, 2002.

232. Hawthorne, S. B., Trembley, S., Moniot, C. L., Grabanski, C. B., and Miller, D. J., Static subcritical water extraction with simultaneous SPE for determining polycyclic aromatic hydrocarbons on environmental solids, *J. Chromatogr. A*, 886, 237–244, 2000.

233. Hawthorne, S. B., Grabanski, C. B., Martin, E., and Miller, D. J., Comparisons of Soxhlet extraction, pressurized liquid extraction, SFE and subcritical water extraction for environmental solids: recovery, selectivity and effects on sample matrix, *J. Chromatogr. A*, 892, 421–433, 2000.

234. Budzinski, H., Letellier, M., Garrigues, P., and Menach, K. L., Optimisation of the MAE in open cell of polycyclic aromatic hydrocarbons from soils and sediments study of moisture effect, *J. Chromatogr. A*, 837, 187–200, 1999.

235. Ericsson, M. and Colmsjö, A., Dynamic MAE coupled online with SPE: determination of polycyclic aromatic hydrocarbons in sediment and soil, *J. Chromatogr. A*, 964, 11–20, 2002.

236. Zdráhal, Z., Karásek, P., Lojková, L., Bucková, M., Vecera, Z., and Vejrosta, J., Pressurised liquid extraction of ketones of polycyclic aromatic hydrocarbons from soil, *J. Chromatogr. A*, 893, 201–206, 2000.

237. Wang, Z., Ashraf-Korassani, M., and Taylor, L. T., Design for online coupling of SFC: quantitative analysis of polynuclear aromatic hydrocarbons in a solid matrix, *Anal. Chem.*, 75, 3979–3985, 2003.
238. Hawthorne, S. B., Grabanski, C. B., Martin, E., and Miller, D. J., Comparisons of Soxhlet extraction, pressurized liquid extraction, SFE and subcritical water extraction for environmental solids: recovery, selectivity and effects on sample matrix, *J. Chromatogr. A*, 892, 421–433, 2000.
239. Gonçalves, C., de Rezende Pinto, M., and Alpendurada, M. F., Benefits of a binary modifier with balanced polarity for an efficient SFE of PAHs from solid samples, followed by HPLC, *J. Liq. Chromatogr. Relat. Technol.*, 24, 2943–2959, 2001.
240. Buford, M. D., Hawthorne, S. B., and Miller, D. J., Extraction rates of spiked versus native PAHs from heterogeneous environmental samples using SFE and sonication in methylene chloride, *Anal. Chem.*, 65, 1497–1505, 1993.
241. Shu, Y. Y., Lao, R. C., Chiu, C. H., and Turle, R., Analysis of polycyclic aromatic hydrocarbons in sediment reference materials by microwave-assisted extraction, *Chemosphere*, 41, 1709–1716, 2000.
242. Vazquez Blanco, E., Lopez Mahia, P., Lorenzo, S. M., Rodriguez, D. P., and Fernandez Fernandez, E., Optimization of MAE of hydrocarbons in marine sediments: comparison with the Soxhlet extraction method, *Fresenius' J. Anal. Chem.*, 366, 283–288, 2000.
243. Pérez, S., Ferrer, I., Hennion, M. C., and Barceló, D., Isolation of priority polycyclic aromatic hydrocarbons from natural sediments and sludge reference materials by an antifluorene immunosorbent followed by liquid chromatography and diode array detection, *Anal. Chem.*, 70, 4996–5001, 1998.
244. Pérez, S. and Barceló, D., Evaluation of antipyrene and antifluorene immunosorbent clean up for PAHs from sludge and sediment reference materials followed by liquid chromatography and diode array detection, *Analyst*, 125, 1273–1279, 2000.
245. Priego López, E. and Luque de Castro, M. D., Ultrasound-assisted extraction of nitropolycyclic aromatic hydrocarbons from soil prior to gas chromatography-mass detection, *J. Chromatogr. A*, 1018, 1–6, 2003.
246. Dabrowska, H., Browski, L. D., Biziuk, M., Gaca, J., and Namiésnik, J., SPE cleanup of soil and sediment extracts for the determination of various types of pollutants in a single run, *J. Chromatogr. A*, 1003, 29–42, 2003.
247. Hartonen, K., Bøwadt, S., Dybdahl, H. P., Nylund, K., Sporring, S., Lund, H., and Oreld, F., Nordic laboratory intercomparison of SFE for the determination of total petroleum hydrocarbon, PCBs and polycyclic aromatic hydrocarbons in soil, *J. Chromatogr. A*, 958, 239–248, 2002.
248. Pino, V., Ayala, J. H., Afonso, A. M., and González, V., Determination of polycyclic aromatic hydrocarbons in marine sediments by high-performance liquid chromatography after MAE with micellar media, *J. Chromatogr. A*, 869, 515–522, 2000.
249. Luque-García, J. L., and Luque de Castro, M. D., Extraction of PCBs from soils by automated FMAS extraction, *J. Chromatogr. A*, 998, 21–29, 2003.
250. Nilson, T., Björklund, E., and Bowadt, S., Comparison of two extraction methods independently developed on two conceptually different automated SFE systems for the determination of PCBs in sediments, *J. Chromatogr. A*, 891, 195–199, 2000.
251. During, R. A. and Gath, S., Microwave assisted methodology for the determination of organic pollutants in organic municipal wastes and soils: extraction of PCBs using heat transformer disks, *Fresenius' J. Anal. Chem.*, 368, 684–688, 2000.
252. Xiong, G., He, X., and Zhang, Z., Microwave-assisted extraction or saponification combined with microwave-assisted decomposition applied in pretreatment of soil or mussel samples for the determination of polychlorinated biphenyls, *Anal. Chim. Acta*, 413, 49–56, 2000.
253. Mannila, M., Koistinen, J., and Vartiainen, T., Comparison of SFE with Soxhlet in the analyses of PCDD/PCDFs and PCBs in sediments, *J. Environ. Monit.*, 4, 1047–1053, 2002.
254. Mannila, M., Koistinen, J., and Vartiainen, T., Development of SFE with a solid-phase trapping for fast estimation of toxic load of polychlorinated dibenzo-p-dioxins-dibenzofurans in sawmill soil, *J. Chromatogr. A*, 975, 189–198, 2002.

255. Vercauteren, J., Peres, C., Devos, C., Sandra, P., Vanhaecke, F., and Monees, L., Stir-bar sorptive extraction for the determination of ppq-level traces of organotin compounds in environmental samples with thermal desorption-capillary gas chromatography-ICP mass spectrometry, *Anal. Chem.*, 73, 1509–1514, 2001.

256. Liu, Y., Lopez-Avila, V., Alcaraz, M., and Beckert, W. F., Offline complexation/SFE and GC with atomic emission detection for the determination and speciation of organotin compounds in soils and sediments, *Anal. Chem.*, 66, 3788–3796, 1994.

257. Baciocchi, R., Attinà, M., Lombardi, G., and Rosaria Boni, M., Fast determination of phenols in contaminated soils, *J. Chromatogr. A*, 911, 135–141, 2001.

258. Fung, Y. S. and Long, Y. H., Determination of phenols in soil by SF extraction-capillary electrochromatography, *J. Chromatogr. A*, 907, 301–311, 2001.

259. Rodríguez, I., Llompart, M. P., and Cela, R., SPE of phenols, *J. Chromatogr. A*, 885, 291–304, 2000.

260. Ferguson, P. L., Iden, C. R., and Brownawell, B., Analysis of alkylphenol ethoxylate metabolites in the aquatic environment using liquid chromatography-electrospray mass spectrometry, *Anal. Chem.*, 72, 4322–4330, 2000.

261. Macutkiewicz, E., Rompa, M., and Zygmunt, B., Sample preparation and chromatographic analysis of acidic herbicides in soils and sediments, *Crit. Rev. Anal. Chem.*, 33, 1–17, 2003.

262. Wells, M. J. M. and Yu, L. Z., SPE of acidic herbicides, *J. Chromatogr. A*, 885, 237–250, 2000.

263. Papilloud, S., Haerdi, W., Chiron, S., and Barcelo, D., Supercritical fluid extraction of atrazine and polar metabolites from sediments followed by confirmation with LC–MS, *Environ. Sci. Technol.*, 30, 1822–1826, 1996.

264. Patsias, J., Papadakis, E. N., and Papadopoulou-Mourkidou, E., Analysis of phenoxyalkanoic acid herbicides and their phenolic conversion products in soil by microwave assisted solvent extraction and subsequent analysis of extracts by online SPE-liquid chromatography, *J. Chromatogr. A*, 959, 153–161, 2002.

265. Luque-García, J. L. and Luque de Castro, M. D., Coupling continuous subcritical water extraction, filtration, preconcentration, chromatographic separation and UV detection for the determination of chlorophenoxy acid herbicides in soils, *J. Chromatogr. A*, 959, 25–35, 2002.

266. Hook, G. L., Kimm, G., Koch, D., Savage, P. B., Ding, B., and Smith, P. A., Detection of VX contamination in soil through solid-phase microextraction sampling and gas chromatography/mass spectrometry of the VX degradation product bis(diisopropylaminoethyl)disulfide, *J. Chromatogr. A*, 992, 1–9, 2003.

267. Caballo-Lopez, A. and Luque de Castro, M. D., Continuous ultrasound-assisted extraction coupled to on line filtration-SPE-column liquid chromatography-post column derivatisation-fluorescence detection for the determination of N-methylcarbamates in soil and food, *J. Chromatogr. A*, 998, 51–59, 2003.

268. Bossi, R., Vejrup, K., and Jacobsen, C. S., Determination of sulfonylurea degradation products in soil by liquid chromatography-ultraviolet detection followed by confirmatory liquid chromatography-tandem mass spectrometry, *J. Chromatogr. A*, 855, 575–582, 1999.

269. Shen, G. and Lee, H. K., Determination of triazines in soil by MAE followed by solid-phase microextraction and gas chromatography-mass spectrometry, *J. Chromatogr. A*, 985, 167–174, 2003.

270. Papadakis, E. N. and Papadopoulou-Mourkidou, E., Determination of metribuzin and major conversion products in soils by microwave-assisted water extraction followed by liquid chromatographic analysis of extracts, *J. Chromatogr. A*, 962, 9–20, 2002.

271. Bouaid, A., Ramos, L., Gonzalez, M. J., Fernandez, P., and Camara, C., Solid-phase microextraction method for the determination of atrazine and four organophosphorus pesticides in soil samples by gas chromatography, *J. Chromatogr. A*, 939, 13–21, 2001.

272. Doong, R. A. and Liao, P. L., Determination of OCPs and their metabolites in soil samples using headspace solid-phase microextraction, *J. Chromatogr. A*, 918, 177–188, 2001.

273. Zambonin, C. G. and Palmisano, F., Determination of triazines in soil leachates by solid-phase microextraction coupled to gas chromatography-mass spectrometry, *J. Chromatogr. A*, 874, 247–255, 2000.

274. Richter, P., Sepulveda, B., Oliva, R., Calderon, K., and Seguel, R., Screening and determination of pesticides in soil using continuous subcritical water extraction and gas chromatography-mass spectrometry, *J. Chromatogr. A*, 994, 169–177, 2003.

275. Li, Z. Y., Zhang, Z. C., Zhou, Q. L., Gao, R. Y., and Wang, Q. S., Fast and precise determination of phenthoate and its enantiomeric ratio in soil by the matrix solid-phase dispersion method and liquid chromatography, *J. Chromatogr. A*, 977, 17–25, 2002.

276. Sánchez-Brunete, C., Rodríguez, A., and Tadeo, J. L., Multiresidue analysis of carbamate pesticides in soil by sonication assisted extraction in small columns and liquid chromatography, *J. Chromatogr. A*, 1007, 85–91, 2003.

277. Prados-Rosales, R. C., Herrera, M. C., Luque-Garcia, J. L., and Luque de Castro, M. D., Study of the feasibility of FMAS extraction of *N*-methylcarbamates from soil, *J. Chromatogr. A*, 953, 133–140, 2002.

278. Hogendoorn, E. A., Huls, R., Dijkman, E., and Hoogerbrugge, R., Microwave assisted solvent extraction and coupled-column reversed-phase liquid chromatography with UV detection. Use of an analytical restricted-access-medium column for the efficient multi-residue analysis of acidic pesticides in soils, *J. Chromatogr. A*, 938, 23–33, 2001.

279. Vryzas, Z. and Papadopoulou-Mourkidou, E., Determination of triazine and chloroacetanilide herbicides in soils by microwave-assisted extraction (MAE) coupled to gas chromatographic analysis with either GC–NPD or GC–MS, *J. Agric. Food Chem.*, 50, 5026–5033, 2002.

280. Sun, L. and Lee, H. K., Microwave-assisted extraction behavior of nonpolar and polar pollutants in soil with analysis by high-performance liquid chromatography, *J. Sep. Sci.*, 25, 67–76, 2002.

281. Henriksen, T., Svensmark, B., and Juhler, R. K., Analysis of metribuzin and transformation products in soil by PLE and liquid chromatographic–tandem mass spectrometry, *J. Chromatogr. A*, 957, 79–87, 2002.

282. Sánchez-Brunete, C., Miguel, E., and Tadeo, J. L., Multiresidue analysis of fungicides in soil by sonication-assisted extraction in small columns and gas chromatography, *J. Chromatogr. A*, 976, 319–327, 2002.

283. Richter, B. E., Extraction of hydrocarbon contamination from soils using accelerated solvent extraction, *J. Chromatogr. A*, 874, 217–224, 2000.

284. Morselli, L., Setti, L., Iannuccilli, A., Maly, S., Dinelli, G., and Quattroni, G., Supercritical fluid extraction for the determination of petroleum hydrocarbons in soil, *J. Chromatogr. A*, 845, 357–363, 1999.

285. Seidel, V. and Lindner, W., Universal sample enrichment technique for OCPs in environmental and biological samples using a redesigned simultaneous steam distillation-solvent extraction apparatus, *Anal. Chem.*, 65, 3677–3683, 1993.

286. Lorenzo, R. A., Vazquez, M. J., Carro, A. M., and Cela, R., Methylmercury extraction from aquatic sediments. A comparison between manual, supercritical and microwave-assisted techniques, *Trends Anal. Chem.*, 18, 410–416, 1999.

287. Bautista, J. M., González-Vila, J. F., Martin, F., del Rio, J. C., Gutierrez, A., Verdejo, T., and Gustavo González, A., Supercritical-carbon-dioxide extraction of lipids from a contaminated soil, *J. Chromatogr. A*, 845, 365–371, 1999.

288. Guo, F., Li, Q. X., and Alcantara-Licudine, J. P., Na$_4$EDTA-Assisted sub/SFE procedure for quantitative recovery of polar analytes in soil, *Anal. Chem.*, 71, 1309–1315, 1999.

289. Chaudot, X., Tambute, A., and Caude, M., Selective extraction of hydrocarbons, phosphonates and phosphonic acids from soils by successive SF and pressurized liquid extractions, *J. Chromatogr. A*, 866, 231–240, 2000.

290. Chaudot, X., Tambute, A., and Caude, M., Simultaneous extraction and derivatization of 2-chlorovinylarsonous acid from soils using supercritical and pressurized fluids, *J. Chromatogr. A*, 888, 327–333, 2000.

291. de Amarante, O. P. Jr., Brito, N. M., dos Santos, T. C. R., Nunes, G. S., and Ribeiro, M. L., Determination of 2,4-dichlorophenoxyacetic acid and its major transformation product in soil samples by liquid chromatographic analysis, *Talanta*, 60, 115–121, 2003.

292. Mecozzi, M., Acquistucci, R., Amici, M., and Cardarilli, D., Improvement of an ultrasound assisted method for the analysis of total carbohydrate in environmental and food samples, *Ultrason. Sonochem.*, 9, 219–223, 2002.

293. Muir, B., Wilson, M., Rowley, L., Smith, F. J., and Hursthouse, A., Potential of electrophilic epoxide reactions for the monitoring of acid gases in the environment, *J. Chromatogr. A*, 977, 251–256, 2002.

294. Nielsen, A. T. and Jonson, S., Quantification of volatile sulfur compounds in complex gaseous matrices by solid-phase microextraction, *J. Chromatogr. A*, 963, 57–64, 2002.

295. Khaled, A. and Pawliszyn, J., Time-weighted average sampling of volatile and semivolatile airborne organic compounds by the SPME device, *J. Chromatogr. A*, 892, 455–467, 2000.

296. Harper, M., Sorbent trapping of volatile organic compounds from air, *J. Chromatogr. A*, 885, 129–151, 2000.

297. Davoli, E., Gangai, M. L., Morselli, L., and Tonelli, D., Characterisation of odorants emissions from landfills by SPME and GC/MS, *Chemosphere*, 51, 357–368, 2003.

298. Tuduri, L., Desauziers, V., and Fanlo, J. L., Dynamic versus static sampling for the quantitative analysis of volatile organic compounds in air with polydimethylsiloxane-carboxen solid-phase microextraction fibers, *J. Chromatogr. A*, 963, 49–56, 2002.

299. Moreira Vaz, J., Screening direct analysis of PAHS in atmospheric particulate matter with SPME, *Talanta*, 60, 687–693, 2003.

300. Piñeiro-Iglesias, M., Lopez-Mahia, P., Vazquez-Blanco, E., Muniategui-Lorenzo, S., Prada-Rodriguez, D., and Fernandez-Fernandez, E., Microwave assisted extraction of polycyclic aromatic hydrocarbons from atmospheric particulate samples, *Fresenius' J Anal. Chem.*, 367, 29–34, 2000.

301. Castells, P., Santos, F. J., and Galceran, M. T., Development of a sequential SFE method for the analysis of nitrated and oxygenated derivatives of polycyclic aromatic hydrocarbons in urban aerosols, *J. Chromatogr. A*, 1010, 141–151, 2003.

302. Mangani, G., Berloni, A., and Maione, M., "Cold" solid-phase microextraction method for the determination of volatile halocarbons present in the atmosphere at ultra-trace levels, *J. Chromatogr. A*, 988, 167–175, 2003.

303. Sanchez, C., Ericsson, M., Carlsson, H., and Colmsjö, A., Determination of organophosphate esters in air samples by dynamic sonication-assisted solvent extraction coupled online with large-volume injection GC utilizing a programmed-temperature vaporizer, *J. Chromatogr. A*, 993, 103–110, 2003.

304. Sanchez, C., Ericsson, M., Carlsson, H., Colmsjö, A., and Dyremark, E., Dynamic sonication-assisted solvent extraction of organophosphate esters in air samples, *J. Chromatogr. A*, 957, 227–234, 2002.

305. Yang, J. S., Lee, D. W., and Lee, S., Microwave-assisted extraction of polychlorinated biphenyls and polychlorinated dibenzodioxins from fly ash, *J. Liq. Chromatogr. Relat. Technol.*, 25, 899–911, 2002.

306. Ramil Criado, M., Rodriguez Pereiro, I., and Cela Torrijos, R., Optimization of a MAE method for the analysis of PCBs in ash samples, *J. Chromatogr. A*, 985, 137–145, 2003.

307. Flotron, V., Houessou, J., Bosio, A., Delteil, C., Bermond, A., and Camel, V., Rapid determination of polycyclic aromatic hydrocarbons in sewage sludges using microwave-assisted solvent extraction. Comparation with other extraction methods, *J. Chromatogr. A*, 999, 175–184, 2003.

308. Miege, C., Dugay, J., and Hennion, M. C., Optimization, validation and comparison of various extraction techniques for the trace determination of polycyclic aromatic hydrocarbons in sewage sludges by liquid chromatography coupled to diode-array and fluorescence detection, *J. Chromatogr. A*, 995, 87–97, 2003.

309. Sulkowski, W. and Rosinska, A., Comparison of the efficiency of extraction methods for PCBs from environmental wastes, *J. Chromatogr. A*, 845, 349–355, 1999.

310. Berset, J. D. and Holzer, R., Quantitative determination of polycyclic aromatic hydrocarbons, PCBs and OCPs in sewage sludges using SFE and mass spectrometric detection, *J. Chromatogr. A*, 852, 545–558, 1999.

311. Nilsson, T., Bjorklund, E., and Bowadt, S., Comparison of two extraction methods independently developed on two conceptually different automated SFE systems for the determination of PCBs in sediments, *J. Chromatogr. A*, 891, 195–199, 2000.

312. Rodríguez, I., Llompart, M. P., and Cela, R., SPE of phenols, *J. Chromatogr. A*, 885, 291–304, 2000.

313. Petrovic, M., Lacorte, S., Viana, P., and Barceló, D., Pressurized liquid extraction followed by liquid chromatography-mass spectrometry for the determination of alkylphenolic compounds in river sediment, *J. Chromatogr. A*, 959, 15–23, 2002.

314. Ding, W. H. and Fann, J. C. H., Application of PLE followed by gas chromatography-mass spectrometry to determine 4-nonylphenols in sediments, *J. Chromatogr. A*, 866, 79–85, 2000.

315. Braun, P., Moeder, M., Schrader, St., Popp, P., Kuschk, P., and Engewald, W., Trace analysis of technical nonylphenol, bisphenol A and 17 α-ethinylestradiol in wastewater using solid-phase microextraction and gas chromatography-mass spectrometry, *J. Chromatogr. A*, 988, 41–51, 2003.

3 Chromatography

Tadeusz Górecki

CONTENTS

I. INTRODUCTION

The year 2003 marks exactly 100 years since chromatography was introduced to the world by Mikhail Semyonovich Tswett, a Russian botanist working at the University of Warsaw, Poland. Tswett managed to separate plant pigments extracted from leaves by depositing them on top of a column packed with solid adsorbents and percolating various solvents through the column. In the process, the initially uniform band of pigment was separated into several bands of different colors. This activity of separating colors formed the basis of the name "chromatography." Chromatography means "color writing" in Greek (although some suspect that Tswett might have played on his name, which means "color" in Russian). Tswett presented the results of his research for the first time in a lecture to the Biological Section of the Warsaw Society of Natural Scientists.[1] They appeared in print in 1906, in two papers published in *Berichte der Deutschen Botanischen Gesellschaft*.[2,3] The technique as described by Tswett was largely ignored for nearly 30 years and it was rediscovered in 1931 by Edgar Lederer. The meteoric rise of chromatography from a relatively unknown, obscure technique to one of the most important methods of chemical analysis, started in the late thirties and early forties with the introduction of liquid–liquid chromatography by Martin and Synge,[4] for which the researchers were awarded the Nobel Prize in 1952. In the same year, together with A.T. James, Martin introduced the technique of gas liquid chromatography.[5] Today, chromatographic techniques are some of the most important tools in many areas, including environmental analysis.

Chromatography is a physical method of separation, in which the components to be separated are carried by a mobile phase along or through a layer of a stationary phase. In the process, the components move continuously between the two phases. Components momentarily associated with the stationary phase do not migrate, while all of the components momentarily present in the mobile phase migrate at the same speed. Thus, the separation does not occur in either of the two phases — rather, it is the result of the continuous movement of molecules between them. The components become separated when one of them spends a different length of time in the stationary phase than the rest. The time spent in the stationary phase by any of the components depends on its affinity to the stationary phase under given conditions. Figure 3.1 illustrates the principle of chromatographic separation.

Chromatographic methods can be classified using various criteria. Probably the most important one is the type of the mobile phase used in the separation process. In essence, any type of fluid can be used as the mobile phase. Historically, liquids were the first to be used, giving rise to a broad

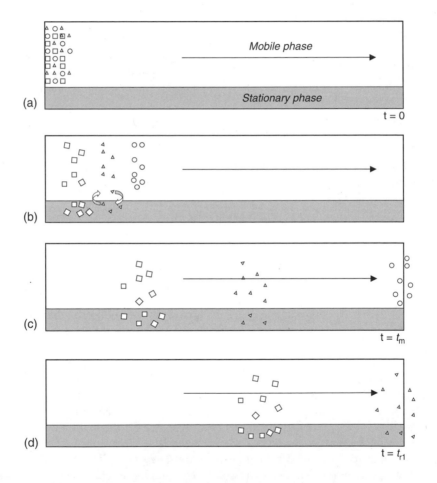

FIGURE 3.1 The principle of a chromatographic separation. (a) A mixture of solutes is introduced to the chromatographic system in the form of a sharp band and is carried through the system by the mobile phase. (b) Molecules of an unretained solute (marked by circles) do not interact with the stationary phase; other molecules partition into the stationary phase according to their partition coefficient. Partitioning is a dynamic process, with equal number of molecules going into and out of the stationary phase at any given time when the system is at equilibrium. (c) Unretained solute molecules travel at the same speed as the mobile phase and elute from the system in the time t_m. (d) Molecules of the first retained solute reach the outlet of the system in time t_{r1}. At any given time during the separation, the concentration of the solute marked by squares in the stationary phase is greater than the concentration of the solute marked by triangles. Thus, the solute marked by squares spends on average more time in the stationary phase and elutes from the system last.

range of methods classified as *liquid chromatography (LC)*. Gases can be used as the mobile phase whenever the components to be separated have appreciable vapor pressures. Methods based on this principle are classified as *gas chromatography (GC)*. Highly compressed, dense gases (fluids) kept above their critical temperatures are used in *supercritical fluid chromatography (SFC)*. This last type of fluid has very peculiar properties — its density and solvating power are close to those of a liquid, while its viscosity is only somewhat greater than that of a gas. Consequently, SFC bridges the gap between LC and GC. In spite of its potential, SFC did not find any significant applications in environmental analysis and hence it will not be discussed in more detail.

Another classification of chromatographic methods is based on the physical form of the stationary phase used. In the vast majority of chromatographic separations the stationary phase is confined within a tube through which the mobile phase is fed. The tube is called a chromatographic column, and all such methods are classified as column chromatography. The stationary phase in column chromatography can have the form of a compact bed of small, usually porous particles packed inside the column, or can be spread on the walls of the column. Columns of the first type are called *packed columns*, while columns of the second type are called *open tubular columns*. Alternatively, the stationary phase can be spread as a thin, homogenous layer on a flat, inert support. Methods utilizing this approach are termed thin-layer chromatography (TLC). In fact, TLC is a representative of a broader group of methods in which the stationary phase has a planar form, so called planar chromatography. Another representative of this group is paper chromatography, in which the support material itself (paper) constitutes the stationary phase.

LC uses mostly packed columns, as the use of open tubular columns in this method is not practical because of the extremely small column diameters required for good separation. In gas chromatography, both packed and open tubular columns can be used, but the latter are far more popular because of their vastly superior properties. The mobile phase is usually forced through the stationary phase at elevated pressure, although other approaches are also possible (e.g., electrically driven flow in electrochromatography (EC), gravity driven flow in classical LC or flow driven by capillary forces in TLC).

Chromatographic methods can be classified according to the type of interaction between the solute and the stationary phase. Of the several possibilities, sorption is by far the most common. The term applies to a class of processes in which one material (in this case the solute) is taken up by another (the stationary phase). Whenever the solute is confined to the surface of the stationary phase, we call the process adsorption, and the method is called adsorption chromatography. Whenever the solute penetrates the stationary phase and enters the bulk of it absorption occurs; however, chromatographic methods based on this principle are usually called partition chromatography. Ion exchange makes it possible to separate ions by liquid chromatography. Methods based on this principle are referred to as ion chromatography. Polymers and other high molecular weight compounds can be separated according to their size by using materials whose pores cover a specific range of sizes. Such methods are called size exclusion chromatography. Specific interactions between stationary phase and one particular type of solute form the basis of affinity chromatography. This last method differs from all the others in that it is typically used to isolate a single solute from a complex mixture rather than to separate the components of the mixture from each other.

II. FUNDAMENTAL RELATIONSHIPS

The result of chromatographic separation is usually presented in the form of a chromatogram, i.e., a plot of the concentration or mass of the sample components recorded as a function of the amount of mobile phase passed through the system. In column chromatography, instead of measuring the amount of mobile phase, one usually measures the time from the moment the sample was introduced to the column, and plots the chromatogram as a function of time. Figure 3.2 presents an example of a chromatogram together with some of the parameters which are used to characterize

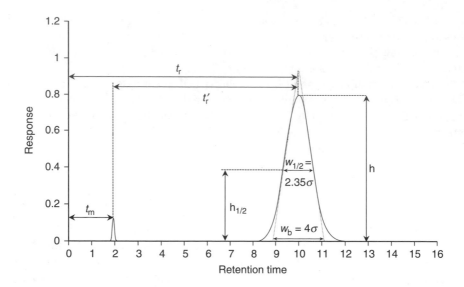

FIGURE 3.2 Example of a chromatogram including some important chromatographic parameters. The first peak eluting at time t_m represents a nonretained solute; t_r is the retention time of the main component, while t_r' is its adjusted retention time. The symbols h and $h_{1/2}$ represent peak height and one half of peak height, respectively; $w_{1/2}$ and w_b represent peak width at half height and at the base, respectively; σ is the standard deviation of the peak.

the separation process. Solute bands elute from the column in the form of peaks, whose profiles ideally are Gaussian. An unretained solute elutes from the column in time t_m, usually called dead time or hold-up time. A solute which interacts with the stationary phase elutes from the column in time t_r, called retention time. It is important to understand that whenever a retained solute molecule enters the mobile phase, it travels with it at the same speed as the unretained solute. Consequently, each solute spends exactly the same amount of time in the mobile phase before it reaches the detector, and this time is represented by t_m. So-called adjusted retention time is given by the difference between t_m and t_r:

$$t_r' = t_r - t_m \tag{3.1}$$

The adjusted retention time t_r' represents the additional time required for the solute to travel the length of the column. It therefore corresponds to the time spent by the solute in the stationary phase. The ratio of the time spent by the solute in the stationary phase to that spent in the mobile phase is called the solute capacity factor k:

$$k = \frac{t_r'}{t_m} = \frac{t_r - t_m}{t_m} \tag{3.2}$$

The capacity factor plays an important role in optimization of chromatographic separations. It is related to the partition coefficient of the solute between the mobile phase and stationary phase K in the following way:

$$k \left(= \frac{t_r'}{t_m} \right) = K \frac{V_s}{V_m} \tag{3.3}$$

where: $K = C_s/C_m$, C_s is the concentration of the solute in the stationary phase, C_m is its concentration in the mobile phase, V_s is the volume of the stationary phase and V_m is the volume of the mobile phase. Equation 3.3 relates the retention time of a solute to the partition coefficient and the volumes of the stationary and mobile phases.

Separation of two chromatographic peaks can be described in the first approximation by their relative retention α:

$$\alpha = \frac{t'_{r2}}{t'_{r1}} = \frac{k_2}{k_1} = \frac{K_2}{K_1} \tag{3.4}$$

By definition, $t'_{r2} > t'_{r1}$, therefore $\alpha > 1$. Relative retention is constant for a given set of analytical conditions (stationary phase, temperature, etc.) and is independent of the column dimensions. However, it is not the best measure of peak separation, because it takes into account only the separation of peak maxima while ignoring the fact that bands traveling along the column (or the stationary phase in general) become progressively broader. Consequently, two bands characterized by the same relative retention may either be completely separated if they are narrow, or may be effectively not separated at all if they are very broad. To account for this phenomenon, the degree of separation of two chromatographic peaks is usually described by their resolution R_s:

$$R_s = \frac{t_{r2} - t_{r1}}{\dfrac{(w_2 + w_1)}{2}} \tag{3.5}$$

where: t_{r2} and t_{r1} are the retention times of the two peaks ($t_{r2} > t_{r1}$), and w_1 and w_2 are their widths at the base (see Figure 3.2). The denominator of Equation 3.5 is therefore the average base width of the two peaks.

Figure 3.3 illustrates the concept of peak resolution. Peaks in Figure 3.3a are poorly resolved. Peaks in Figure.3.3b are characterized by the same relative retention, but are well resolved because of their significantly smaller widths. Peaks in Figure 3.3c and d have the same widths; their resolution is better in Figure 3.3d, because of greater relative retention. In general, resolution $R_s = 1$, corresponding to $\sim 94\%$ peak separation, is considered adequate and baseline resolution is achieved when $R_s = 1.5$ or more. Figure 3.3 illustrates two possible strategies which can be employed when optimizing the separation process: one can either try to produce narrower peaks, or to increase their relative retention. In practice, the best results are obtained when both approaches are used at the same time.

The width of the band eluting from the chromatographic system depends on the distance it has traveled. The proportionality factor between the two, termed "height equivalent to theoretical plate" (HETP), or plate height in short, is defined in the following way:

$$H = \sigma^2/x \tag{3.6}$$

where: H is the plate height, σ^2 is the variance of the band and x is the distance traveled by the band. The smaller the plate height, the narrower the band eluting from the column.

H depends on a number of parameters, the most prominent of which is the linear flow rate of the mobile phase u. The relationship between H and u is described by the van Deemter equation[6]:

$$H = A + \frac{B}{u} + (C_s + C_m) \cdot u \tag{3.7}$$

where: A is a term representing the contribution from eddy diffusion, B is the contribution from longitudinal diffusion, and C_s, C_m represent contributions from the mass transfer in the stationary and the mobile phases, respectively, to the total column plate height. Eddy diffusion occurs only in packed columns due to multiple paths which exist between the packing particles. Thus, the value of term A in open tubular columns is zero. The contribution of longitudinal diffusion to plate height is

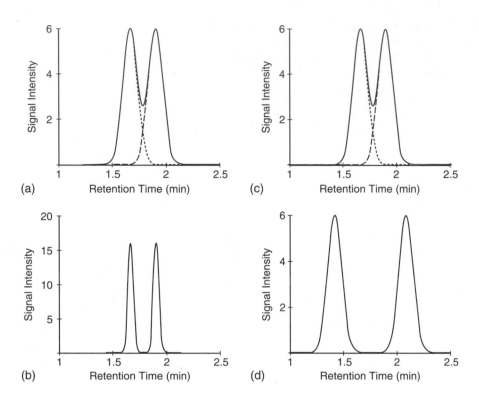

FIGURE 3.3 The concept of peak resolution. (a) A poorly resolved pair of peaks; (b) better resolution achieved for the same relative retention of the two peaks by making them narrower. Note that the peak height is significantly larger, which results in improved sensitivity; (c) A pair of peaks same as in (a); (d) better resolution achieved for the same peak width by increasing the relative retention of the peaks.

important in gas chromatography, but often negligibly small in LC due to small molecular diffusion coefficients of the solutes dissolved in the liquid phase. The contribution from mass transfer in the stationary phase is important in both gas and liquid chromatography. On the other hand, the contribution from mass transfer in the mobile phase is much more important in LC than in gas chromatography.

The plot of the van Deemter equation has the form presented in Figure 3.4. Since the B/u term decreases as the linear flow rate of the mobile phase increases, while the term $C \cdot u$ increases, there must be a minimum in the plate height. The linear flow rate at which the minimum occurs is considered optimal in chromatographic separations, as it corresponds to the least band broadening during separation. However, when optimizing the separation, one should also take into account the separation time. Minimum plate height often occurs at relatively low mobile phase flow rates, which translates into long separation times. In many cases, adequate resolution can be achieved at flow rates higher than optimal, resulting in faster separations.

For a column of length L, the number of theoretical plates can be calculated by dividing the length of the column by the plate height:

$$N = L/H = L^2/\sigma^2 \tag{3.8}$$

Peak variance for a Gaussian peak can be found easily from the peak width at the base ($w = 4\sigma$, see Figure 3.2), or at half-height ($w_{1/2} = 2.35\sigma$). The latter can usually be determined with better accuracy. If both the length of the column and peak variance are expressed in units of time,

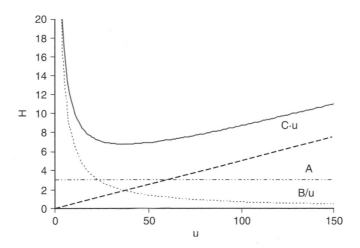

FIGURE 3.4 A plot of the van Deemter equation.

the number of theoretical plates can be easily determined from a chromatogram using the following relationships:

$$N = \frac{t_r^2}{\sigma^2} = \frac{16t_r^2}{w^2} = \frac{5.55t_r^2}{w_{1/2}^2} \tag{3.9}$$

For two peaks eluting close together on a reasonably efficient column, N is related to resolution in the following way[6]:

$$R_s = \frac{\sqrt{N}}{4}\left(\frac{\alpha - 1}{\alpha}\right)\left(\frac{k_2}{1 + k_2}\right) \tag{3.10}$$

where k_2 is the capacity factor of the later eluting peak. Equation 3.10 indicates that to double the resolution between two peaks, the number of theoretical plates has to be quadrupled. This could be accomplished for example by using a four times longer column; however, that would result in a concomitant increase in the separation time and pressure required to drive the mobile phase. Thus, increasing the length of the column is not a very efficient way of improving the resolution. Much better results can be achieved by changing the relative retention of the two peaks α. In liquid chromatography, this can often be accomplished by adjusting the composition of the mobile phase; in gas chromatography, it is usually necessary to change the stationary phase to achieve different selectivity.

III. GAS CHROMATOGRAPHY

GC is arguably the most widely used separation tool in environmental analysis. It owes its popularity to great separation power, relative simplicity of instrumentation and method development, and very good sensitivity. For a sample to be suitable for gas chromatographic separation, its components should be thermally stable and have appreciable vapor pressures in the temperature range typical for GC (typically up to ~ 320°C, although some columns can withstand temperature as high as 400°C). A molecular weight of about 1000 is considered the practical limit of gas chromatography. According to various estimates, this means only about 10% of all organic compounds known are amenable to gas chromatography. However, considering that a large number of environmentally relevant chemicals fall within this range, this is not a significant limitation.

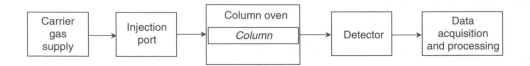

FIGURE 3.5 Block diagram of a gas chromatograph.

Figure 3.5 presents a block diagram of a gas chromatograph. The main components of this instrument include injector, chromatographic column, column oven, detector and data acquisition system. Separations in GC are achieved through analyte distribution between the gaseous mobile phase and the stationary phase. Today, the vast majority of all GC separations are carried out using liquid stationary phases, into which the sample components can partition *via* absorption. Methods based on this principle are referred to as gas-liquid chromatography (GLC). In certain applications (e.g., analysis of light hydrocarbons or permanent gases) better results can be achieved with solid stationary phases, which interact with the analytes *via* adsorption. Such methods are referred to as gas-solid chromatography (GSC).

A. COLUMNS

Historically, GC was first performed using packed columns. Such columns were usually made of coiled metal or glass tubes, with length of 1 to 5 m and internal diameter of a few millimeters. These were packed with small particles of solid support coated with nonvolatile liquid stationary phases. The solid particles themselves played the role of the stationary phase in GSC. Packed columns were characterized by poor efficiencies related to multiple flow paths among the packing particles (the A term in the van Deemter equation), as well as uneven distribution of the liquid phase within the particles and at the contact points between the particles.[6] The number of theoretical plates in packed columns was several thousand at the most, therefore improvements in resolution were usually achieved by the use of more selective stationary phases. Consequently, hundreds of different stationary phases were available at the peak of packed column development.

Today, packed columns are virtually no longer used in environmental analysis. They were replaced by open tubular columns containing no packing particles, usually called capillary columns. Open tubular columns are characterized by vastly superior efficiencies compared to packed columns as a result of elimination of the heterogeneities related to the use of packing particles. Their efficiencies are governed solely by longitudinal diffusion and resistance to mass transfer (the B and C terms in the van Deemter equation). The number of theoretical plates depends on the column diameter and the stationary phase film thickness, but it is typically several thousand plates per meter. Thus, a capillary column can easily have in excess of 100,000 theoretical plates, which is an improvement of nearly two orders of magnitude over packed columns. Consequently, satisfactory resolution can often be achieved even when the selectivity of the stationary phase towards the components undergoing separation is relatively poor.

The advantages of open tubular columns over packed columns were predicted theoretically by Marcel Golay in early 1957.[7] However, their acceptance was initially slow due to limited availability, lack of dedicated instrumentation and difficulties with handling fragile glass capillaries. The breakthrough proved to be the introduction of fused silica columns, first described in 1979.[8] Such columns are made of high purity silica coated with a temperature-resistant layer of polymeric coating (polyimide). Fused silica tubing is flexible, mechanically stable and easy to handle. Most capillary columns sold today are made of fused silica. Several manufacturers offer metal capillary columns made of stainless steel or nickel. Their introduction was made possible by the advances in surface deactivation techniques. Examples of metal columns include Silcosteel® tubing from Restek (Bellefonte, PA), which is stainless steel tubing deactivated inside with a very thin, flexible layer of fused silica, or Ultimetal® tubing from Varian (Middelburg, Holland).

In the most popular type of open tubular column in use today, the stationary phase has the form of a thin film of a viscous liquid coated on the walls of the capillary tube. Such columns are called wall-coated open tubular, or WCOT, and the separation is achieved through gas–liquid chromatography. The liquid stationary phase is usually immobilized in the tubing by chemical bonding to the tube wall. It is further stabilized by cross-linking of individual polymeric chains. Stationary phases prepared in this way are referred to as bonded and cross-linked. These are very stable and insoluble in solvents, which significantly increase the lifetime of a column. Because of the superior efficiency of capillary columns, the selection of stationary phases does not have to be as extensive as for packed columns. In fact, most separations can be accomplished using just a handful of stationary phases. Examples of the common stationary phases used in capillary GLC are listed in Table 3.1. In an apparent effort to broaden their markets, many manufacturers today offer a wide selection of stationary phases tailored to specific official methods. This approach helps the users meet the targets of regulatory compliance, but it also increases the expense incurred by the laboratories and sometimes the plethora of choices may create confusion.

GSC is carried out using a different type of open tubular column. The stationary phase in such columns has the form of a layer of porous material coated on the walls of the tube. Such columns are

TABLE 3.1
Examples of the Most Common Stationary Phases used in Capillary GLC

Structure	Polarity	Temperature Range (°C)	Trade Names (Examples)
	$x = 0$ Nonpolar	−60 to 360	DB-1, Rtx-1, SPB-1
	$x = 0.05$ Nonpolar	−60 to 360	DB-5, Rtx-5, SPB-5
	$x = 0.35$ Intermediate	−20 to 340	DB-35, Rtx-35, SPB-35
	$x = 0.50$ Intermediate	0 to 340	DB-17, Rtx-50, SPB-50
	$x = 0.65$ Intermediate	50 to 340	Rtx-65
	$x = 0.06$ Intermediate	−20 to 280	DB-1301, Rtx-1301, SPB-1301, DB-624, Rtx-624
	$x = 0.14$ Intermediate	−20 to 280	DB-1701, Rtx-1701, SPB-1701
	$x = 0.1$ Polar	0 to 275	Rtx-2330, SP-2330, SP-2560, BPX-70
	$x = 0$ Polar	0 to 275	Rt-2340, SP-2340
	Polar	40 to 260	Carbowax, Stabilwax, Supelcowax, DB-Wax

termed porous layer open tubular, or PLOT. The solid can be an inorganic adsorbent (e.g., alumina or silica), or a porous polymer (e.g., styrene-divinylbenzene copolymer). GSC is more trouble-prone than GLC because of the nonlinearity of the adsorption isotherms, high sorption energies and/or surface areas of the adsorbents, and surface inhomogeneities. This results in retention volumes depending on the sample size, asymmetric peaks, poor recovery of some solutes and excessively long retention times. As a result, applications of GSC are typically limited to solutes with boiling points lower than about 200°C. The main reasons why GSC is still in use is due to its ability to separate very light components (permanent gases, low molecular weight hydrocarbons) and unique selectivity offered in some applications.

Capillary columns usually have diameters ranging from 0.1 to 0.53 mm and length between 10 and 60 m. In general, columns with smaller diameters are characterized by better efficiencies (larger number of theoretical plates). However, very narrow bore columns (0.1 mm and below) require very high pressures to operate, and the amount of a sample which can be introduced to such columns without the risk of stationary phase overloading (so called sample capacity) is very limited. Columns with diameters greater than 0.53 mm are characterized by efficiencies similar to those of packed columns, therefore they offer no real advantages and are seldom used.

B. Ovens

Column ovens in typical laboratory gas chromatographs are forced circulation air thermostats, capable of maintaining constant temperature (within $\pm 0.1°C$). The geometry of column ovens is such that the temperature distribution inside the oven is as uniform as possible. The size of the oven has to be sufficient to make installation of GC columns easy. Since most analyses in GC are performed under temperature-programmed conditions, the ovens must have the capabilities to raise the temperature at a controlled rate. The minimum oven temperature is determined by ambient temperature and the amount of heat generated inside the oven by the heated zones (injector, detector, etc.). In moderate climates it is usually possible to cool the oven down to 35°C. The oven can be operated at subambient temperatures only if it is equipped with cryogenic cooling. Liquid CO_2 or N_2 are typically used as the coolants. The maximum oven temperature depends on the power of the heating element and the quality of the insulation around the oven. Typical GC ovens can be operated at temperatures between 350 and 400°C. Higher temperatures are not usually required as very few columns could withstand them. The heating rates of typical GC ovens can range from a fraction of a degree per minute to as much as 50°C/min with high-power heating elements. Special inserts reducing the volume of the oven are sometimes required to achieve high heating rates at high oven temperatures.

Recently, an alternative to conventional GC ovens has appeared in the form of compact column modules with integrated heaters. Such modules are offered by, among others, Thermo Orion[9] (EZFlash) and RVM Scientific[10] (LTM modules). In these modules, a temperature sensing element is combined with the GC column. Separately, an insulated heating wire element is paired with the GC column-temperature sensor combination. The entire assembly is wrapped with a conductive foil to assure uniform temperature distribution, coiled and placed in a small module which can be retrofitted to conventional GCs. Since the column is heated directly rather than through forced hot air convection and the thermal mass of the entire assembly is very low, it is possible to achieve much higher heating and cooling rates than with conventional GC ovens. The maximum heating rates for the two products mentioned above are 20 and 30°C/s, respectively. The very high heating rates make it possible to speed up the chromatographic separation quite dramatically. However, it should be kept in mind that under such conditions the short analysis time comes at a price of significantly reduced resolution, which has a negative impact on the usefulness of this technology in areas like environmental analysis, where sample matrices are usually very complex.

Field portable GC instruments most often do not have the capability of temperature programming because of the limited amount of power available. In such instruments the separation is typically carried out under isothermal conditions, which vastly simplifies the design of the instrument. Column assemblies in such systems are usually attached directly to heating blocks, the thermal mass and conductivity of which assure uniform temperature distribution in the column.

C. INJECTORS

With a few notable exceptions (e.g., headspace analysis), samples analyzed by GC are liquid. Thus, a GC inlet (injector) must be able to volatilize the sample components and mix them with the carrier gas before the GC analysis can commence. The task is relatively easy to accomplish when using packed columns because of the high carrier gas flow rates used in this technique. A direct injector for packed column GC has the form of a heated tube equipped with a septum on one end, and connected to the column on the other end. A glass tube (liner) is usually inserted into the injector to reduce its dead volume and improve its inertness. Carrier gas is supplied to the injector through a side port. The sample is introduced to the injector with a microsyringe. The septum is pierced with the syringe needle, and the sample is injected into the liner. The solvent and the sample components quickly evaporate at the elevated temperature of the injector and are swept rapidly into the column by the carrier gas flowing at a high volume flow rate.

Sample injection into an open tubular column cannot be performed in the same way because of the much lower carrier gas flow rates and limited sample capacity of such columns. The volume of the insert in the vaporizing chamber of the injector must be large enough to accommodate the vapors of the solvent and the sample components. On the other hand, because of the exponential dilution occurring in the insert when it is flushed with the carrier gas, at least three insert volumes of the carrier gas are required to transfer ~90% of the vapors from the injector to the column. At the low flow rates used in capillary gas chromatography, this would translate into unacceptably broad injection bands which would effectively eliminate all the advantages of open tubular columns. Historically, the first solution to this problem was the application of split injection. A schematic diagram of a split/splitless injector used for this purpose is presented in Figure 3.6. In this injector, the carrier gas is split at a controlled ratio between the column and the split vent. The flow rate through the liner is high enough to quickly transfer sample vapors from the injector to the column. However, only a small fraction of the sample is introduced to the column, which adversely affects the sensitivity of the method. Consequently, split injection is rarely used in environmental analysis, where ultimate sensitivity is often required. The goal can be achieved by using the same injector in the splitless mode. In this case, the injection is carried out with the split vent closed. The sample

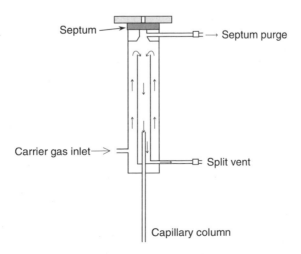

FIGURE 3.6 Schematic diagram of a split/splitless injector.

evaporates, and the vapors are transferred slowly to the column by the carrier gas. Injection band broadening is prevented by applying a so-called solvent effect. To produce this effect, during sample injection the column must be kept at a temperature lower by approximately 30°C than the boiling point of the solvent. Under such conditions, the solvent vapors condense at the beginning of the column and trap the analyte vapors. The solvent band gradually evaporates from the tail end to the front, focusing the analytes into a narrow band which starts to migrate only after the solvent evaporates completely. In this way, sharp analyte injection bands can be produced even though analyte transfer from the injector to the column is very slow. Solvent effect is efficient only for analytes whose boiling points are higher than that of the solvent. In addition, the sample components must be compatible with the solvent, and the solvent must be compatible with the stationary phase of the column. Analytes with boiling points higher by ~150°C than the oven temperature during the injection will produce sharp injection bands independent of the solvent effect *via* so-called thermal focusing. For such analytes, the partition coefficients between stationary phase and carrier gas at the initial oven temperature are high enough to trap the analyte at the head of the column in the form of a narrow band.

Vaporizing injectors (i.e., injectors in which the sample is vaporized before entering the column) have significant disadvantages. Thermally labile sample components might easily undergo degradation when subjected to the high temperature in the injector for a relatively long time (as is the case in splitless injection). Also, the efficiency of analyte transfer from the syringe to the injector depends strongly on the vapor pressure of the analyte. During injection, the sample is expelled from the syringe barrel by the movement of the plunger, but a certain volume of the sample is left in the syringe needle. In the hot injector, the solvent left inside the needle evaporates quickly, leaving behind sample components. The ones with higher vapor pressures also evaporate from the needle and contribute to the total amount of material introduced to the column. Compounds with lower vapor pressures evaporate only partially or not at all, therefore their relative abundance in the injection is lower. This phenomenon is called needle discrimination. Its severity can be reduced by applying special injection techniques, but it cannot be eliminated entirely unless very fast, automated injection is performed.

The drawbacks of vaporizing injectors can be eliminated by cold on-column injection. In this technique, a liquid sample is introduced directly into the GC column by means of a special syringe with a needle of a small enough external diameter to fit inside the column. The column is kept slightly below the boiling point of the solvent during the injection. Thus, the solvent vaporization–condensation cycle is avoided, and the sample components are never exposed to excessively high temperatures. Further, since the injection is carried out at a temperature below the boiling point of the solvent, the latter does not evaporate from the needle, and discrimination is eliminated. Analyte band focusing is achieved through a mechanism similar to that operating in splitless injection.

Even though the stationary phases in modern columns are usually insoluble in any solvents, when exposed to large amount of the sample these may swell and block the passage of the carrier gas. To avoid this problem, a segment of uncoated fused silica tubing (a precolumn) is usually attached upstream of the column when using on-column injection. The precolumn acts also as a so-called retention gap, helping focus the analyte bands. The retention gap effect is explained in Figure 3.7. When solute molecules encounter the stationary phase of the column, their migration rate decreases significantly. This allows solute molecules from the tail end of the band to catch up with its front, narrowing the band considerably. Another advantage of the precolumn is the fact that it can be easily replaced without affecting the performance of the column. Nonvolatile components of the sample gradually accumulate in the precolumn, which leads to deterioration of the GC separation. The problem can be easily cured by removing a segment of the precolumn or replacing it entirely.

Another alternative to splitless injectors is the programmed-temperature vaporization injector (PTV). The design of this injector resembles that of a split/splitless injector, except that PTV has a lower internal volume and lower thermal mass, and as a result can be heated very rapidly. One of the

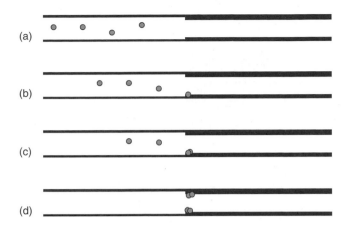

FIGURE 3.7 Retention gap effect. (a) nonretained molecules in the retention gap travel as a broad band; (b) the front of the band encounters the stationary phase in the column and partitions into it; (c) molecules from the tail end of the band catch up with the molecules in the stationary phase; (d) the initially broad band is focused into a narrow band.

main advantages of the PTV injector is its ability to introduce large sample volumes into the column. Large volume injection significantly improves detection limits, which is particularly important in trace environmental analysis. To perform large volume injection using a PTV, the sample is injected into a cold injector liner (usually packed with a suitable material to hold the liquid) with the split line open. The solvent gradually evaporates and its vapors are removed by the carrier gas through the split vent. Once the solvent is nearly gone, the split vent is closed and the injector is heated rapidly to vaporize the analytes and introduce them to the GC column. The technique works correctly only for semivolatile analytes — volatile compounds are partially or completely lost together with the solvent. This is not a major drawback, however, as the majority of environmentally relevant chemicals are semivolatile. Since the injection is carried out in this mode at a low temperature, needle discrimination is eliminated.

It should be pointed out that PTV is not the only way in which large volume injection can be accomplished. Recently a method has been proposed which allows the use of conventional split/splitless injectors for this purpose by a simple manipulation of the chromatographic parameters.[11] This is a promising method in environmental analysis.

D. DETECTORS

Historically, one of the first detectors used in GC was the thermal conductivity detector (TCD). A schematic diagram of this detector is shown in Figure 3.8. TCD measures the difference in thermal conductivities between pure carrier gas and carrier gas containing sample components, therefore its response is universal and nonselective. A typical TCD contains two heated filaments placed in a thermostated cavity. One filament is swept by the carrier gas from the chromatographic column, while the other is swept by pure carrier gas delivered under identical conditions through a reference column. Components of the sample eluting from the GC column change the thermal conductivity of the gas in the sensing arm, which causes a change in the temperature of the sensing filament. This, in turn, alters the resistance of the sensing filament compared to the reference filament, and this change is recorded as the detector signal. The nonselectivity of the TCD and its poor sensitivity limit its applications to permanent gases, light hydrocarbons and other compounds which respond poorly to other detectors. Sensitivity of TCD is much better in microfabricated devices, owing to much smaller dead volume and low thermal mass of the miniature detector elements. Such microfabricated thermal conductivity detectors found their niche in field portable gas

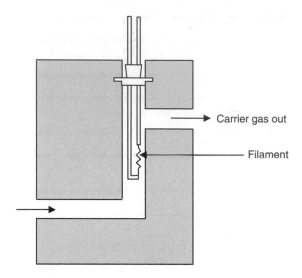

FIGURE 3.8 Schematic diagram of a thermal conductivity detector (TCD). Only one arm of the detector is shown; the reference arm looks identical.

chromatographs manufactured by Agilent Technologies and Varian, Inc. Their main advantage in this application is that they do not require additional gases to operate.

Today, the most popular detector in GC is the flame ionization detector (FID), presented schematically in Figure 3.9. The main advantages of this detector include a nearly universal response to organic compounds, long-term stability, reasonable sensitivity and very broad linear dynamic range spanning up to seven orders of magnitude. This last feature means that the FID can

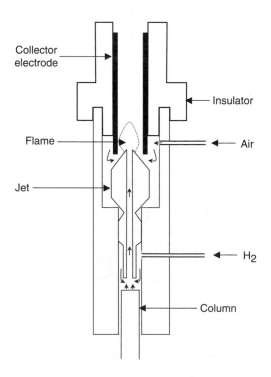

FIGURE 3.9 Schematic diagram of a flame ionization detector (FID).

be used with equal success to detect both major and minor components of a sample without the need to switch sensitivity ranges. In the FID, the effluent from the column is introduced to a hydrogen–air flame, where the organic analytes undergo combustion and generate ions. A collector electrode is placed above the flame jet, and a potential difference of several hundred volts is applied between them. The ions and electrons generated during the combustion process give rise to a small electric current between the collector electrode and the jet. On the other hand, very few ions are produced in a pure hydrogen flame, therefore in the absence of organic analytes a much smaller residual current is recorded. The response of the FID is generally proportional to the number of carbon atoms delivered to the detector in unit time; therefore the response factors for many different compounds are similar when using this detector. The universal response of the FID to organic compounds limits its usefulness in environmental analysis, especially when trace analytes have to be detected in complex matrices. Analyte identification based only on its retention time is in most cases not reliable enough, and the FID cannot supply any other information which would help with positive identification of the compound.

A relatively simple modification of the design of the FID changes it into a thermionic ionization detector (TID), known also as an alkali flame detector or nitrogen–phosphorus detector (NPD). In TID, an electrically heated glass or ceramic bead containing rubidium salts is placed a few millimeters above the flame jet. Hydrogen and air supplied to the detector create plasma next to the bead's surface. The bead is kept at a negative potential to suppress the FID signal. Under such conditions, nitrogen- and phosphorus-containing compounds undergo selective ionization in the bead region. The electrons generated in the process are collected by the collector electrode, which gives rise to the detector signal. The selectivity of the TID towards nitrogen or phosphorus can be changed by altering the temperature of the bead and the flow rate of hydrogen. TID is very sensitive and has a relatively broad linear range of four- to five- orders of magnitude. It is commonly used in environmental analysis for the determination of N- and P-containing pesticides.

Photoionization detector (PID) is another example of ionization detectors used in gas chromatography. In this detector, analytes are ionized after absorbing a photon of light of high energy. The ionization energy for organic compounds usually ranges from 5 to 20 eV. Light of this energy falls within the ultraviolet range. In PID, it comes from a discharge lamp containing an inert gas or gas mixture at low pressure. Such lamps emit monochromatic light, the wavelength of which depends on the choice of fill gases and the window material. The most popular choice is a 10.2 eV lamp, which strikes a reasonable compromise between selectivity and longevity. Lamps of higher energy can ionize more compounds, but their lifetime is usually much shorter. Ions and electrons generated in the ionization process are collected by a pair of electrodes placed inside a thermostated detector chamber. The resultant current is the signal of the detector. PID is nondestructive, which means that the sample leaving the detector has essentially the same composition as the sample entering the detector. Consequently, PID can be used as the first detector in a two-detector configuration. Its sensitivity strongly depends on the particular compound, but it is generally somewhat better than the sensitivity of the FID, especially for aromatic compounds. The linear range of this detector is similar to that of the FID. In environmental analysis, PID is used most often when selectivity towards aromatic compounds is required. It is also often found in field portable gas chromatographs and detectors, as it does not require any additional gases for operation.

The electron capture detector (ECD), presented schematically in Figure 3.10a, is also classified as an ionization detector, but the principle of its operation is different. In the ECD, a radioactive source (usually ^{63}Ni) emits β radiation. The high energy β electrons collide with molecules of a make-up gas (high purity nitrogen or argon-5% methane mixture), which leads to the release of thermal electrons. Each β electron may generate dozens or even hundreds of thermal electrons, which are collected by the collector electrode, giving rise to standing current. The collector electrode is usually polarized intermittently, so that the thermal electrons are collected in pulses. As long as the detector cell does not contain any species capable of capturing the thermal electrons, the magnitude of the standing current remains constant. Electron-capturing species reduce the concentration of

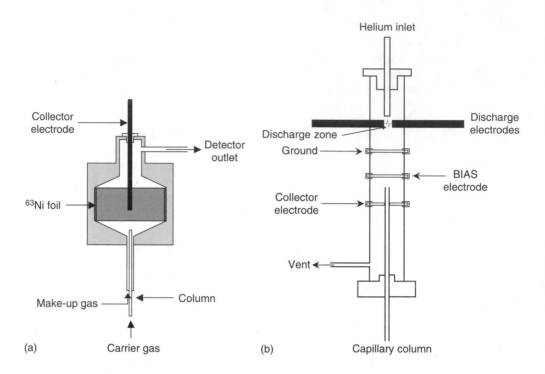

FIGURE 3.10 Schematic diagrams of: (a) conventional electron capture detector (ECD); (b) pulse discharge ECD.

free electrons in the detector, which in turn reduces the standing current. This decrease in the magnitude of detector current is recorded as the detector signal. ECD responds selectively to compounds with high electron affinity, with the strongest response to halogenated compounds and compounds containing nitro-groups. The magnitude of the response depends very strongly on the number and identity of the electron-capturing groups in the molecule, therefore ECD has to be calibrated individually for each and every analyte. It is one of the most sensitive detectors available in gas chromatography and it can detect as little as a few fg/s of some compounds. The linear dynamic range of the ECD is limited to three to four orders of magnitude. ECD remains a very popular choice in environmental analysis because it responds with very high sensitivity and selectivity towards many environmentally relevant chemicals, e.g., chlorinated solvents, organochlorine pesticides, polychlorinated biphenyls, etc.

While the use of a radioactive source to generate the electrons may be objectionable under certain circumstances, it is not always required. Recently, an alternative has been proposed in the form of pulse discharge electron capture detector,[12] in which the electrons are produced by a stable, low power DC discharge in helium atmosphere (Figure 3.10b). The performance of the pulse discharge ECD is the same as the performance of a conventional ECD. In addition, the pulse discharge detector can be used in a helium photoionization mode, in which it becomes a universal detector. All ECD detectors are nondestructive, therefore they can be used in multi-detector configurations (although the sample leaving the detector is significantly diluted by the make-up gas).

Flame photometric detector (FPD) is a representative of optical detectors for gas chromatography. In FPD, column effluent is introduced to a hydrogen flame, which breaks analyte molecules into atoms. The temperature of the flame is sufficient to excite some atoms, especially sulphur and phosphorus. These excited atoms emit characteristic lights on return to the ground state. The light emitted by the element of interest is selected by a suitable bandpass filter and measured by a photomultiplier tube. The response of the detector is linear for phosphorus and quadratic for sulphur, with the detection limits around single pg/s, and a dynamic range of three to five orders of magnitude. FPD is relatively popular in environmental analysis because many industrial or pest control products contain sulphur or phosphorus. Recently, an alternative to conventional FPD has been introduced in the form of a pulsed FPD (PFPD).[13] In this solution, the combustible mixture of gases inside the detector is ignited periodically rather than burning continuously as done in conventional FPD, and the flame is quickly terminated. This allows for time-resolved luminescence measurements, which increases both the sensitivity of the detector and its selectivity.

Chemiluminescence can also be used as the basis for selective detection in gas chromatography. In a chemiluminescence detector, analytes of interest are first converted to a form which is easy to detect. This can be accomplished, for example, by combusting the analytes in a hydrogen flame. The detected species next reacts with ozone to form an electronically excited product. This product then emits light on return to the ground state. Chemiluminescence detectors are used most often for detection of nitrogen- and sulphur-containing compounds. In the former case, nitric oxide is oxidized to excited nitrogen dioxide. In the latter case, sulphur monoxide is reacted with ozone to form excited sulphur dioxide. Chemiluminescence detectors are characterized by good sensitivity (fraction of pg/s) and a relatively wide linear dynamic range, but they are not as popular as some of the other selective detectors described above because of their relative complexities and high costs.

E. Hyphenated Techniques

Analyte identification based only on the retention time of a component is suspect even for simple mixtures. Selective detectors reduce the uncertainty to some extent by detecting only the components sharing a certain characteristic, but these do not eliminate the chance of false identification. On the other hand, spectroscopic techniques provide qualitative information about the analyte which is often specific enough to make the identification of a component certain. It is therefore not surprising that attempts to couple GC with various spectroscopic techniques were undertaken from the early days of GC. Today, combined instruments, often referred to as "hyphenated" systems, are used routinely, and gradually replace conventional gas chromatographs in many areas.

By far, the most important hyphenated technique is combination of GC with mass spectrometry (GC/MS). Mass spectrometry is a technique which allows determination of the masses of molecules or molecule fragments. Molecules which enter the mass spectrometer are ionized (and most often fragmented), and the ions are separated according to their mass to charge ratio (m/z). The plot of detector response vs. the m/z ratio is called a mass spectrum. Under given conditions, the mass spectrum of a given compound is characteristic for it and can be used to confirm its identity. Fragments produced during ionization yield information characteristic of the position and bonding order of the molecular substructures. This information helps to identify unknown compounds. Mass spectrometers coupled to gas chromatographs are usually delivered with computerized libraries, which typically contain mass spectra for over a hundred thousand compounds. Unknown analytes can be identified by comparing their spectra to the database spectra in the library for the best matches. This approach obviously fails when the mass spectrum of an unknown analyte is not found in the library. Moreover, correct identification may also not be possible when the unknown compound is not well separated from the matrix components.

Ionization of the molecules leaving the chromatographic column is carried out in the ion source of mass spectrometer. The vast majority of all GC/MS analyses are carried out using electron impact ionization (EI). In this technique, electrons emitted from a hot filament are accelerated in an electric field of 70 V, so that these achieve the energy of 70 eV. The accelerated electrons collide with molecules, creating excited molecular ions. These ions may then break apart into smaller fragments, each carrying an electric charge. The resultant mass spectrum will therefore usually contain signals from the molecular ions and all the fragments formed during the fragmentation process. If the molecular ion is not stable, its abundance in the mass spectrum may be very low, or it may even be completely absent. The abundance of the molecular ion can sometimes be increased by lowering the ionization energy. However, the mass spectra produced at lower energies are less reproducible. Moreover, the spectra in the library were all recorded at standard ionization energy of 70 eV and therefore library searches can be carried out efficiently only if the spectra of the unknown analytes were obtained at this matching energy level.

When the knowledge of the mass of the molecular ion is required, much better results can be achieved by using chemical ionization (CI). This soft ionization technique yields much less fragmentation than EI. To achieve chemical ionization, an enclosed ion source must be filled with a reagent gas such as methane, isobutene or ammonia, at a pressure of ~ 1 mbar. While a number of different reactions may contribute to the formation of ions in CI, the most common one is protonation, which yields a quasimolecular ion MH^+ according to the following two-step mechanism[14]:

$$\text{Primary reaction:} \qquad R \xrightarrow{70 \text{ eV}} RH^+ \text{ (Reagent gas cluster)}$$

$$\text{Secondary reaction:} \qquad RH^+ + M \xrightarrow{t} MH^+ + R \text{ (Protonation)}$$

Typical reagent gases for protonation include methane, water, methanol, isobutene or ammonia. Methane and water are hard ionisation reagents, which may cause extensive fragmentation of the molecules. Methanol is intermediate, while isobutene and ammonia are soft ionisation reagents which produce almost exclusively the quasimolecular ion. Other reactions which may take place during CI include hydride abstraction, charge exchange and adduct formation.

The ions created in the ion source are focused into a beam of narrow energy dispersion. The beam is accelerated in an electric field before entering the mass analyzer. A number of different analyzers can be used, but certain types are used more often than others in combination with gas chromatography. Typically, mass spectrometers used in GC/MS are low resolution instruments. Resolution A is a measure of the resolving power, i.e., the capacity to separate signals from ions which have similar masses. It is usually defined in the following way[14]:

$$A = \frac{m}{\Delta m}$$

where m is the mass of an ion and Δm is the closest mass which can be differentiated from m with an overlap of the two mass peaks not exceeding 10%. The vast majority of the GC/MS systems in use today are characterized by unit resolution, i.e., these can only deliver nominal rather than exact masses over the entire mass range. For such instruments, the definition of A *(resolution)* given above does not apply.

By far the most common mass analyzer in use today is a transmission quadrupole mass filter (Figure 3.11). The quadrupole consists of four parallel electrodes, usually in the form of metal rods with hyperbolic cross sections. The electrodes are arranged in a square array. Diagonally opposed rod pairs are connected to a DC power supply, so that one pair has a positive potential and the other has a negative potential. In addition, radiofrequency (RF) voltage is superimposed on the DC voltage. The RF voltage on the negative pair of electrodes is 180° out of phase with the RF voltage on the positive pair. Ions in a quadrupole filter travel along complex trajectories. For a given frequency and DC voltage ratio, only ions of a specific m/z value can pass through the filter. All other ions are deflected from the axis of the filter and are neutralized after striking the rods. The m/z value of transmitted ions can be changed by changing the magnitude of the alternating current and the DC voltage. To collect a mass spectrum, the voltages applied to the rods are continuously changed (scanned), so that at any given time, ions with only one particular m/z value reach the detector. Once the entire mass range has been covered, i.e., the mass scan is completed, the voltages are reset to their

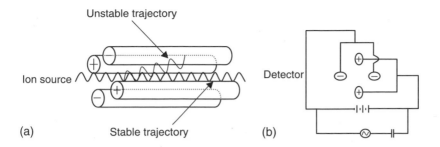

FIGURE 3.11 Schematic diagram of a transmission quadrupole mass filter (a) and the way voltages are applied to it (b).

initial values and the process is repeated. A single scan can often be completed in less than one second, although more reproducible results are obtained when several scans are averaged.

Quadrupole mass spectrometers were the first affordable benchtop instrument. They are responsible to a large extent for today's widespread acceptance of mass spectrometers as chromatographic detectors. Despite their great popularity, quadrupole mass spectrometers have disadvantages too. The biggest disadvantage is their relatively poor sensitivity in the full scan mode, which is inherently related to the principle of operation. Of all the ions generated in the ion source, only about 1% reach the detector at any given time, and the rest are lost on collision with the rods. In target compound analysis, this drawback can be overcome by operating the instrument in the selective ion monitoring (SIM) mode. In this mode, instead of scanning through the entire mass range, the voltages on the electrodes are kept constant for longer periods of time, allowing many more ions of a given m/z value to reach the detector. In order to be able to confirm the identity of the analyte, usually more than one mass is monitored for any given compound. This mode of operation often increases the sensitivity of the instrument by as much as two orders of magnitude. However, for obvious reasons it cannot be used when identifying unknown analytes.

Another drawback of quadrupole mass spectrometers is so-called spectral skewing, which is also related to the way in which the instrument operates. The phenomenon is illustrated in Figure 3.12. Ions in the ion source of quadrupole mass spectrometers are generated continuously, but at any given time only ions of one particular m/z value can reach the detector. With the narrow peaks generated by capillary columns and the relatively slow scan rate of the instrument, a different portion of the chromatographic peak is sampled at the beginning of a given scan compared to the end of the scan. As a result, the relative abundances of different masses in the mass spectrum of a compound may not reflect the true mass spectrum of this compound, which makes identification through library search difficult.

Ion trap mass spectrometers (Figure 3.13) function on the same mathematical basis, but their geometry and operating principle are completely different. An ion trap consists of a ring electrode and two end caps. Substances emerging from the GC column enter the ion trap through a heated transfer line. These are periodically ionized by electrons from a heated filament placed above the top end cap. The electrons are admitted to the trap only when the gate electrode is open, creating a packet of ions representative of the composition of the column effluent at the time of the ionization. The duration of the ionization pulse must be strictly controlled to maximize the number of ions created in the process while avoiding a so-called space charge effect. Space charge is created when too many ions are stored in the trap. It causes a distortion of the electrical fields leading to an overall reduction in performance of the instrument.

A constant RF voltage applied to the ring electrode causes the ions to circulate in stable trajectories inside the trap. Mass scanning is accomplished by increasing the amplitude of RF voltage applied to the ring electrode. Under such conditions, the trajectories of ions of gradually increasing m/z values become unstable and these are ejected from the trap through holes in the

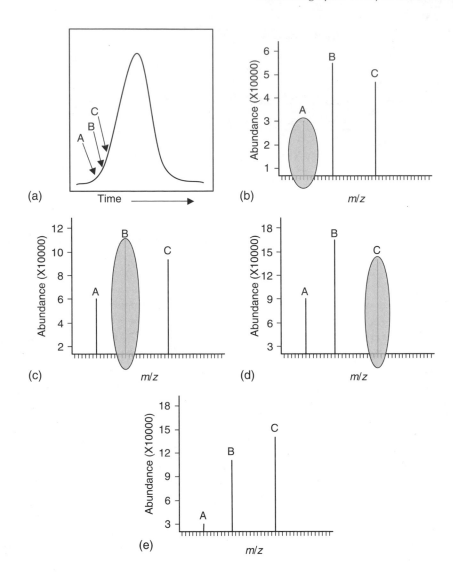

FIGURE 3.12 Spectral skewing and its causes. (a) A chromatographic peak with the times when ions A, B, and C are sampled during a single scan. (b) A is sampled at time A; its abundance is 30,000; (c) B is sampled at time B, with abundance of 110,000; (d) C is sampled at time C, with abundance of 140,000; (e) The mass spectrum recorded for the scan from time A to time C.

end caps. Since the direction of the ejection (up or down) is random, 50% of all the ions will leave the trap through each of the end caps. Consequently, half of all the ions of a given m/z value stored in the trap will reach the detector placed underneath the bottom end cap, which causes ion trap mass spectrometers to be inherently more sensitive in full scan mode than transmission quadrupole instruments. Spectral skewing does not occur in ion traps because of the discrete nature of the ionization process. However, the spectra can sometimes become heavily distorted at high analyte concentrations due to space charge effects, if the parameters of the method are not carefully optimized. Another problem is related to the fact that ions spend a considerable amount of time circulating inside the trap, which increases the probability of secondary reactions occurring in the trap. These reactions may significantly change the spectrum of some analytes, especially if their molecules are unstable.

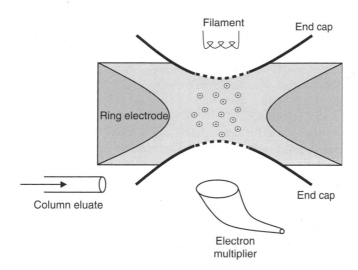

FIGURE 3.13 Schematic diagram of a quadrupole ion trap mass spectrometer.

Time-of-flight (TOF) mass spectrometers are experiencing a renaissance recently because of the unparalleled speed with which they can acquire mass spectra. This factor is becoming increasingly important as fast GC makes inroads into routine chromatographic analysis, especially as a part of comprehensive two-dimensional GC (see later). In the TOF-MS (Figure 3.14), ions generated in the ion source are periodically accelerated by an electric field of several thousand volts and injected into the drift region, where no electric or magnetic fields are applied. Ideally, all ions

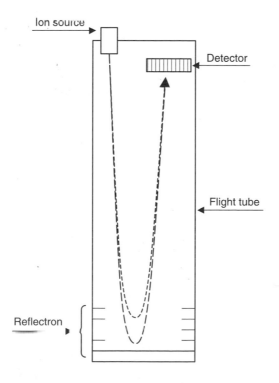

FIGURE 3.14 Schematic diagram of a time-of-flight mass spectrometer.

have the same kinetic energy $\frac{1}{2}mv^2$, where: m is the mass of the ion and v is its velocity. Consequently, ions of different masses migrate in the drift region at different speeds. The lightest ions travel the fastest and are the first to reach the detector at the end of the drift region, while the heaviest ions travel the slowest. Thus, a complete mass spectrum can be easily recorded by simply measuring the drift times. In practice, the kinetic energy of the ions emerging from the source is not perfectly uniform because of spatial distribution of ions in the source, which adversely affects the resolving power of the instrument. To circumvent this problem, a so-called reflectron or electrostatic mirror is often placed at the end of the drift tube. The reflectron is a series of hollow rings held at increasingly positive potentials and terminated with a grid held at a potential more positive than the accelerating potential. Ions reaching the reflectron are slowed down and ultimately reflected in the opposite direction towards the detector. Ions with higher kinetic energy penetrate deeper into the reflectron, therefore their flight paths become longer and these reach the detector at the same time as the ions with the same mass with lower kinetic energy.

TOF mass spectrometers operate at constant resolving power. The great speed of mass spectral acquisition is the biggest advantage of TOF mass spectrometers in GC applications. A complete mass spectrum can be acquired in as little as 20 μs. In practice, the highest frequency at which spectra can be acquired and recorded is about 100 Hz due to the limitations in the bandwidth and storage speed of the computers used for data acquisition. Speeds like these are necessary in fast gas chromatography, where peaks as narrow as 100 ms at the base may often occur. To maintain a good representation of the peak shape, it has to be sampled at least ten times across its profile. Thus, a 100 ms peak has to be sampled at least every 10 ms, which is currently the limit of TOF-MS.

When unequivocal analyte identification and ultimate sensitivity are required, the best solution is to use high resolution mass spectrometry. Such instruments yield exact masses of ions with an accuracy of four decimal places. This allows unequivocal determination of the elemental composition of the ions, as only one particular combination of different atoms can yield a given exact mass. The most popular high resolution mass spectrometer is the double focusing instrument and its principle is illustrated in Figure 3.15. Ions produced in the ion source are accelerated by an electric field and enter the electrostatic sector, which focuses the ion beam and diminishes the energy dispersion of the ions. The beam then enters the magnetic sector, which provides the dispersion of ion beam according to the m/z ratio of the ions. To record a complete mass spectrum, scanning of different masses is necessary. It is usually accomplished by varying the strength of magnetic field at constant accelerating voltage, although the opposite is also possible. High resolution mass spectrometers are extremely useful analytical tools, but a widespread adoption is stymied by the very high prices compared to the benchtop systems described above.

Tandem mass spectrometry (MS/MS) might be an alternative to high resolution MS under some circumstances (e.g., when detection with very high sensitivity is required in a complex matrix). In MS/MS, an ion characteristic for the analyte molecule is selected (so-called parent ion)

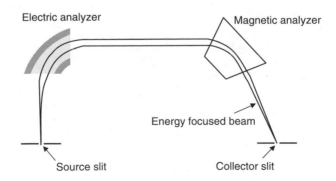

FIGURE 3.15 Schematic diagram of a double focusing mass spectrometer.

and subjected to further fragmentation *via* so-called collision induced dissociation (CID). The new ions of lower m/z formed in the process (daughter or product ions) are substance-specific, therefore they are very characteristic for the analyte. These ions are then detected in the second MS stage, which can operate either in scan mode (in which case a full spectrum of the daughter ions is obtained) or in the SIM mode (in which case only a single daughter ion is monitored). In the latter case, the single ion can be subjected to another CID event, and the product ions can be monitored. This method is called MS/MS/MS, or $(MS)^3$.

Depending on the type of the instrument, tandem MS can be carried out in space or in time. A typical representative of the first group is a triple quadrupole instrument (Figure 3.16). In this instrument, the first quadrupole filter is used to select the parent ion from the mixture of ions produced in the ion source. The parent ion is then transferred to the second quadrupole (or octapole), which serves as a transmission filter and a collision chamber. An inert gas (e.g., helium, nitrogen, argon, etc.) is added to the second quadrupole at a low pressure. Parent ions collide with the molecules of the gas and undergo fragmentation. The fragment ions then enter the third quadrupole, which either scans all the ions or selects a single ion before it reaches the detector.

MS/MS in time can be carried out in ion trap mass spectrometers. In the first stage of the analysis, a special isolation waveform is applied to the trap and causes all ions except the parent ion to be ejected from the trap. In the next stage, a resonant frequency is applied to the end caps of the trap, which imparts additional kinetic energy to the parent ions stored in the trap. The ions collide with atoms of helium present in the trap and undergo CID. The mixture of the daughter ions and the remnants of the parent ion are then analyzed in the usual way. Ion traps make it relatively easy to perform $(MS)^n$. Selected daughter ions can be stored in the trap while all other ions are ejected, and can be subjected to another CID event. Theoretically, the process can be continued until the smallest stable ion is produced. In the case of instruments in which multiple MS stages are carried out in space rather than in time, addition of another MS stage requires the physical addition of another set of quadrupoles, which is not very practical.

While GC/MS is the most popular hyphenated technique, it is not the only one. Atomic emission spectroscopy can also be easily coupled to gas chromatography. This technique allows selective monitoring of individual elements in the effluent from the column. Atomic emission detectors used in GC rely on microwave-induced plasma sources, which are capable of exciting nearly every element, including C, H, O, N, S, P and the halogens. Thus, they are particularly suitable for the analysis of organic compounds. The plasma is an ionized gas whose temperature ranges from 4000 to 10,000K. Molecules entering the plasma are broken down into atoms, which are then thermally excited and emit characteristic radiation. The response measured for each element depends only on the number of atoms in the plasma and is independent of the structure of the parent compounds. Consequently, it is possible to determine the empirical formula of a compound based on the emission intensities of its component atoms.

The first commercially available atomic emission detector (AED) was manufactured by Hewlett Packard. It allowed the detection of up to 15 elements automatically. In this instrument,

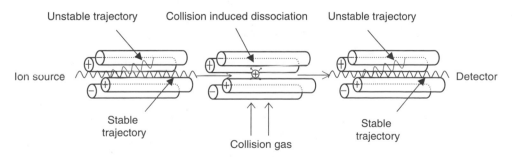

FIGURE 3.16 Schematic diagram of a triple quadrupole tandem mass spectrometer.

the effluent from the column was directed to a microwave cavity, where atomization and excitation took place. Light from the microwave source was directed onto a polychromator, and emission from the element of interest was detected with a movable photodiode array detector. Simultaneous detection of multiple elements was possible as long as their emission lines fell within the wavelength range spanned by the photodiode array. Agilent Technologies, the successor of Hewlett Packard, recently stopped manufacturing the AED and licensed the technology to JAS GmbH. Atomic emission detectors never captured a big market share because of the high prices, comparable to or even higher than that of most desktop GC/MS systems. Nevertheless, they can provide unique information and are the best tools for some applications.

Another spectroscopic technique which can be coupled to GC is infrared spectroscopy. It is particularly valuable when isomer identification is required. On its own, however, infra red (IR) spectroscopy is not very useful because it cannot differentiate homologues without additional molecular weight information. Besides, the gas phase spectra differ from the widely available condensed phase spectra, which means that new libraries would have to be built to enable analyte identification based on library searches. Thus, IR spectroscopy found its niche application in GC in combination with mass spectrometry.

Dispersive IR instruments are not suitable as detectors for GC because of the very low speed with which they acquire spectra. The only practical solution is the Fourier transform infrared (FTIR) spectrometer. In FTIR, spectra are recorded by a Michelson interferometer containing two mirrors, one stationary and one movable, at right angles to one another. The beam of light from the source is split by a semitransparent mirror. The two beams formed are reflected by the two mirrors (stationary and movable) and recombine at the beam splitter. Depending on the relative position of the movable mirror with respect to the stationary mirror, constructive or destructive interference occurs. The combined beam is then passed through the sample compartment to the liquid-nitrogen cooled detector. The interferogram recorded is converted into the IR spectrum using fast Fourier transformation. In GC/FTIR, the sample compartment has the form of a flow-through cell (lightpipe) through which the effluent from the GC column flows. The technique never found widespread application due to poor sensitivity, dead volume limitations, high price and limited usefulness of the information provided. Today, no major manufacturer of GC equipment offers GC/FTIR instruments.

IV. LIQUID CHROMATOGRAPHY

The term "liquid chromatography" (LC) is used to describe a number of different chromatographic methods whose common characteristic is the use of a liquid mobile phase. The technique is important in environmental analysis because many environmentally relevant compounds are not volatile enough to be analyzed by GC or are not thermally stable. LC differs fundamentally from GC in the way in which selectivity can be manipulated. In GC, at a given temperature, the relative retention of two peaks is determined only by the nature of the stationary phase and the properties of separated compounds. In order to change the selectivity, it is usually necessary to change the stationary phase of the column.* The nature of the mobile phase affects the speed of the separation and its efficiency, but not the selectivity. On the other hand, in LC, selectivity is determined to a large extent by the composition of the mobile phase. Since the selectivity of an LC system can be changed by changing the nature of the stationary phase or the composition of the mobile phase, LC is generally more flexible than GC. On the other hand, this additional functionality increases the number of parameters which can be manipulated during method optimization, making the process more complicated.

---

* There are exceptions to this rule. The relative retention of two compounds may change in GC when the temperature changes, sometimes leading to reversed elution order. This is often the case with polar stationary phases.

Another important difference between GC and LC is related to the fact that molecular diffusion is much slower in liquids than it is in gases (the difference between the molecular diffusion coefficients of a given compound in gases and in liquids may reach as much as four orders of magnitude). At the same time, viscosity of liquids is much higher than viscosity of gases. Because of the slow diffusion in liquids, the use of capillary columns of diameters similar to those used in GC is impractical. When used in LC, such columns would have very low efficiencies because equilibration between the mobile phase and the stationary phase would be very slow. To be useful for LC, open tubular columns would have to have much smaller diameters, but then pressures required to drive the mobile phase through the column would become impractical, and the sample capacity of such columns would be very low. Consequently, packed columns with a much smaller number of theoretical plates than open tubular columns used in GC are used almost exclusively in liquid chromatography. Thus, while most separations in GC are carried out under the conditions of moderate selectivity and high efficiency, typical LC separations are performed under the conditions of moderate efficiency and high selectivity.

A. Retention Mechanisms

Historically, the first liquid chromatographic separations were performed using unmodified solid particles as the stationary phases. In this scenario, the solute molecules interact with the particles *via* adsorption mechanisms. Typical stationary phases used in this technique include silica, alumina, carbon, as well as chemically bonded stationary phases with polar functional groups. The mobile phases in these types of separations are mixtures of nonaqueous polar solvents diluted to the desired strength with a nonpolar solvent, e.g., hexane. Because of once-widespread use of this technique, it is often referred to as normal phase chromatography. Other commonly used names include liquid solid chromatography or adsorption chromatography. During chromatographic separation by this method, solute molecules continuously become adsorbed to the surface of the stationary phase and then replaced by the solvent molecules, which compete for the active sites on the surface. The relative ability of the solvents to displace solutes from a given adsorbent is described by the solvent strength parameter ε^0 (also called eluent strength), which is defined as the free energy of adsorption of the solvent per unit surface area. By definition, ε^0 is set to zero for adsorption of pentane on unmodified silica. It is clear that while the numerical value of ε^0 depends on the type of the adsorbent, the general trends should be similar for different adsorbents. This is illustrated in Table 3.2,[6] which presents an example of a so-called eluotropic series (solvents ranked according to their solvent strength). In general, the greater the eluent strength, the more rapidly the solutes will be eluted from the column.

Normal phase chromatography is generally considered suitable for the separation of nonionic organic compounds soluble in organic solvents. However, the method is not as popular today as it was in the past because of a number of problems associated with the use of adsorbents as stationary phases. Adsorption isotherms are nonlinear, which leads to nonGaussian peaks at high solute concentrations. Retention of polar compounds may be irreproducible, e.g., as a result of irreversible adsorption. Traces of water in the mobile phase may deactivate the adsorbent, leading to irreproducible separations. Some of these problems can be eliminated by using polar, chemically bonded phases. Still, very often better results can be obtained by using other retention mechanisms. Today, normal phase LC remains the method of choice for separation of geometric isomers and class separations.

An alternative to normal phase chromatography is reversed phase LC (RPLC). This method is the most popular today owing to its unmatched simplicity, versatility and scope. In RPLC, the stationary phase is nonpolar, while the mobile phase is polar and usually contains water. The strength of the eluent increases as the polarity of the mobile phase decreases. This reversal of the properties of the stationary and the mobile phases compared to normal-phase chromatography

TABLE 3.2
Eluotropic Series for Alumina, Silica and Carbon[6]

Solvent	Solvent Strength Parameter (ε^0)		
	Alumina	Silica	Carbon
Pentane	0.00	0.00	—
Hexane	0.01	0.01	0.13–0.17
Carbon tetrachloride	0.17	0.11	—
1-Chlorobutane	0.26	0.20	0.09–0.14
Benzene	0.32	0.25	0.20–0.22
Methyl *tert*-butyl ether	0.48	—	—
Chloroform	0.36	0.26	0.12–0.20
Dichloromethane	0.40	0.30	0.14–0.17
Acetone	0.58	0.53	—
Tetrahydrofuran	0.51	0.53	0.09–0.14
Dioxane	0.61	0.51	0.14–0.17
Ethyl acetate	0.60	0.48	0.04–0.09
Acetonitrile	0.55	0.52	0.01–0.04
Pyridine	0.70	—	—
Methanol	0.95	0.70	0.00

led to the term "reversed phase" chromatography (even though today RPLC is so widespread that in fact it should be considered normal!). Retention in RPLC is due to hydrophobic interactions of the solute with the stationary phase. Since nearly all organic molecules have hydrophobic regions in their structure, this retention mechanism is nearly universal. Both neutral and ionic solutes can be separated by this technique.

Stationary phases in RPLC are usually solid particles with surfaces chemically modified by attachment of organic moieties. Silica is typically used as the support material, although recently zirconia is gaining ground owing to its better chemical stability, especially at pH extremes. Other materials used in RPLC include alumina, carbon and various polymers. The nature of the organic moiety determines the polarity of the stationary phases. Most separations are carried out using nonpolar stationary phases, including C-8 (octyl) and C-18 (octadecyl). Table 3.3 presents examples of LC stationary phases.

In RPLC, the solute continuously partitions between the stationary phase and the mobile phase. The nature of the partitioning between the two phases is very similar to partitioning between two immiscible liquids. For example, the process is noncompetitive and the sorption isotherms are linear. As a result, peaks are usually symmetrical, and the separations are very reproducible. RPLC is by far the most popular liquid chromatographic technique currently in use. Other separation modes are usually considered only after RPLC fails to deliver desirable results.

Ions and easily ionizable substances can be conveniently separated using ion-exchange chromatography. In this method, retention is based on electrostatic attraction between mobile phase ions and charged sites bound to the stationary phase. The sample ions are separated according to their relative affinity to the stationary phase compared to the mobile phase counter ions. In general, ion-exchangers tend to bind ions with multiple charges, small hydrated radius or large polarizability more strongly. Ion exchange finds application in nearly all areas of chemistry. In environmental analysis, it is most often used for the separation of inorganic and organic ions (both cations and anions). In this implementation, the technique is known simply as ion chromatography. Since its introduction, ion chromatography has revolutionized the analysis of ions and replaced many tedious wet chemical procedures.

TABLE 3.3
Examples of LC Stationary Phases

R =	Type of Phase	Trade Names (Examples)
$-(CH_2)_{17}CH_3$	Octadecyl (RP)	Supelcosil - LC18, - LC8
$-(CH_2)_7CH_3$	Octyl (RP)	Ultra C18, C8, Phenyl
$-(CH_2)_3C_6H_5$	Phenyl (RP)	Zorbax SB C18, SBC C8, SB Phenyl
$-(CH_2)_3NH_2$	Amino (NP)	Zorbax NH2, Hypersil APS
$-(CH_2)_3CN$	Cyano (NP)	Zorbax CN, Hypersil CPS
$-(CH_2)_2OCH_2CH(OH)CH_2OH$	Diol (NP)	Supelcosil LC Diol
−Phenylcarbamated beta cyclodextrin	Chiral	Shiseido CD-Ph
−None	Silica (NP)	Ultra silica, Zorbax Sil, Kromasil 60-A
−Modified amino acids/peptides	Chiral	Whelk O1, Chirex 3014

The stationary phases in ion-exchange chromatography are ion-exchange resins. These are usually made by copolymerization of styrene and divinylbenzene. The amount of divinylbenzene determines the extent of cross-linking of the resin. More highly cross-linked resins are more rigid, but less porous. Lightly cross-linked resins are more porous, which allows rapid equilibration of the solute between the inside and the outside of the particle, but these are less rigid and tend to swell in water. The phenyl rings of the styrene-divinylbenzene copolymer in ion-exchange resins are modified to provide acidic or basic functionality. Typical cation-exchange resins are modified with sulfonate groups (strongly acidic) or carboxylic acid groups (weakly acidic). Anion-exchangers usually contain quaternary ammonium groups (strongly basic) or polyalkylamine groups (weakly basic). In analytical ion-exchange chromatography, analyte resolution on a given column can be modified by adjusting the ionic strength of the mobile phase, its pH, temperature, flow rate and/or concentration of the buffer or organic modifier.

Size-exclusion (or molecular exclusion) chromatography finds limited applications in environmental analysis. Its uses are typically limited to the determination of classes of chemical compounds. In size-exclusion chromatography, solutes are separated according to the size of the molecule. The separation is carried out using stationary phases with well defined pore size distributions. Molecules which are too big to fit inside the pores are excluded from the stationary phase and migrate through the column in the dead time. Molecules small enough to penetrate the pores will spend some time in the stagnant portion of the mobile phase retained inside the pores and consequently will elute later. This is a unique feature of size-exclusion chromatography — in practically all other implementations of liquid chromatography, larger molecules tend to elute later from the chromatographic column. Another unique characteristic of size-exclusion chromatography is the lack of specific (enthalpic) interactions between the solute molecules and the stationary phase. The separation is driven mainly by the entropic contributions. Selectivity in size-exclusion chromatography can be changed only by changing the pore size distribution of the stationary phase, which is usually made of microporous polystyrene or silica. In the latter case, the stationary phase is usually modified with a hydrophilic coating to minimize solute adsorption. Pore size distribution is a critical factor — no separation will occur if all the solute molecules are excluded from the pores, or if all of them can fully penetrate the pores.

B. INSTRUMENTATION

Historically, liquid chromatographic separations were carried out using glass columns packed with relatively large particles of the stationary phase, through which the mobile phase flowed driven by gravity. Such separations took a very long time, and the efficiency of the columns was very poor. Today, all analytical liquid chromatographic separations are carried out using columns packed with much smaller particles (3 to 10 μm). Such columns constitute tremendous hydraulic resistance, therefore the mobile phase has to be forced through them under high pressure. On the other hand, these offer much higher efficiencies, hence all liquid chromatographic methods utilizing such columns are collectively termed High-Performance Liquid Chromatography (HPLC). A block diagram of an instrument for HPLC is shown in Figure 3.17. The basic components of the system include the pumping system, injector, column, detector and data acquisition system. A single pump is sufficient when all separations are carried out using the same mobile phase throughout the entire chromatographic run. This type of elution is called isocratic elution. In practice, it is usually difficult to obtain good separation of solutes whose properties vary widely using isocratic elution. If the mobile phase is optimized to provide good resolution of early eluting components (i.e., a weak solvent is used), the late eluting components might require inordinate amounts of time to elute, and may be completely lost in the baseline due to severe peak broadening. If, on the other hand, a stronger solvent is used as the mobile phase to elute the late components faster, the separation between the early eluting components is usually lost. A solution to this problem is gradient elution, in which the composition of the mobile phase is gradually changed during the run from low elution strength to high elution strength. Gradient elution can be accomplished with a single pump and a system of proportioning valves. However, much more reproducible results are usually obtained when the gradient is generated on the high pressure side of the pump, which requires the use of two (or more) pumps operating in parallel.

Pumps used in HPLC must be capable of delivering the mobile phases at flow rates of the order of a few ml/min at pressures as high as 400 bar. A number of different pump designs were used in the past, but today practically all HPLC instruments use single- or multi-head reciprocating piston pumps. During operation, each stroke of the piston displaces a small volume of the mobile phase from the pump cylinder to the column. The cylinder is refilled during the intake stroke. The direction of the mobile phase flow is controlled by check valves. The piston and elements of the check valves in the pump head are usually made of sapphire or ruby due to the high wear resistance of these materials. The high-pressure piston seals are typically made of fluoropolymers. A reciprocating pump delivers pulsating flow because of the periodic operation. The pulses can be almost entirely eliminated by using dual-head pumps, in which two pistons are operated 180° out of phase. Further reduction in the magnitude of pulses can be achieved using pulse dampeners.

In contrast to GC where several different types of injectors are used, modern HPLC instruments all use injection valves for sample introduction (Figure 3.18). The valves are equipped with six ports. One port is connected to the pump, one to the column, two to a sample loop and one to waste. The sixth port is modified in such a way that it can accommodate the sample syringe. In the load

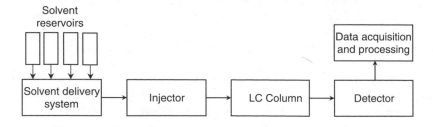

FIGURE 3.17 Block diagram of a liquid chromatograph.

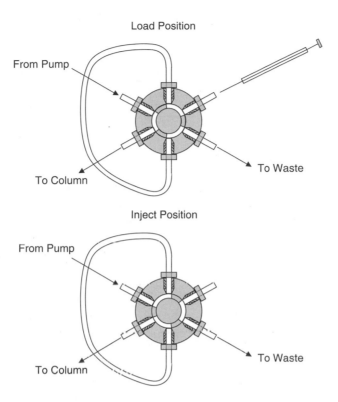

Load Position

From Pump

To Column

To Waste

Inject Position

From Pump

To Column

To Waste

FIGURE 3.18 Schematic diagram of an HPLC injector.

position, the mobile phase from the pump flows directly to the column, while the syringe port is connected through the sampling loop to waste (under atmospheric pressure). Thus, the sampling loop can be easily washed and filled with the sample. When the valve is switched to inject position, the mobile phase from the pump is directed through the sampling loop to the column, carrying the sample with it.

Only packed columns are routinely used in HPLC. Their length is typically 5 to 30 cm, and their diameter ranges from 1 to 5 mm. In the past, 4.6 mm I.D. columns were the most common. Today, narrower bore columns are becoming increasingly popular due primarily to the lower consumption of the expensive HPLC-grade solvents. The columns can be made of stainless steel or plastic. The packing particles usually have diameters ranging from 10 to 5 μm, although smaller packing particles (3 to 1 μm) are also available. Pressures required to force the mobile phase through columns packed with the smallest particles are very high and usually cannot be achieved with conventional HPLC pumps. In addition, a considerable amount of heat is generated in the column, due to frictional heating of the mobile phase, leading to temperature gradients in the column. Such temperature gradients contribute to band broadening. On the other hand, columns packed with the smallest particles have the highest efficiencies, and the van Deemter curves for such columns are very flat, which means that these can be operated at above-optimal flow rates with nearly no loss in efficiency. Solvent consumption can be further reduced when using micropacked capillary columns. Such columns have below 1 mm diameters, and lengths from a few dozen centimeters to a few meters. These are made of fused silica, which creates new possibilities in detection (on-column detection). However, there are significant obstacles which need to be overcome before micropacked columns can become more popular. For example, it is very difficult to create frits in the micropacked column which would hold the packing material in place under the high operating pressures, while not contributing to extracolumn band broadening. Also, void volumes in typical components of HPLC equipment may be too large considering the very low volumetric flow rate of the mobile phase.

Monolithic columns are an interesting recent alternative to conventional packed columns. Such columns are created by *in situ* polymerization from liquid precursors, usually organic polymer- or silica-based. When prepared, monolithic columns have the form of cylindrical rods. They are much more porous than typical packed particle beds, therefore they present significantly lower resistance to mobile phase flow. Consequently, these can be operated at much higher flow rates than conventional columns. The main application of monolithic columns is in high-throughput analysis.

Detectors used in HPLC should have low internal volumes to minimize extracolumn band broadening; in addition, they should be sensitive and should respond quickly to concentration changes. Few detectors fulfill all of these requirements. One of the oldest detectors used in HPLC is the refractive index detector, which detects subtle differences between the refractive index of the pure mobile phase and a mobile phase containing the solute. This detector is universal, i.e., it can detect any solute whose refractive index differs from that of the pure solvent. However, its sensitivity is poor, which practically precludes its use in trace analysis. Besides, refractive index detectors are very sensitive to changes in the composition of the mobile phase and to temperature fluctuations. The former makes their use in gradient elution impractical; the latter requires that the detector is thermostated to at least $\pm 0.01°C$.

The most common HPLC detector is the UV absorption detector. It is useful for any analyte which absorbs light at the wavelength(s) used. The detector cell should provide a long optical path length while minimizing the internal volume to avoid extracolumn band broadening. This can be accomplished for example by using the z-shaped configuration shown in Figure 3.19. Most organic compounds absorb most strongly in the wavelength range of 180 to 210 nm. However, this is also the range in which most solvents used in HPLC absorb light, making detection in this region difficult or impossible. Thus, longer wavelengths are typically used in practice. Simple detectors utilize the intense emission of a mercury lamp at 254 nm. More versatile detectors utilize wideband radiation sources (e.g., deuterium lamp) and monochromators to select the analytical wavelength. The photodiode array detector (Figure 3.20) measures the absorbance across the entire range of UV radiation, providing spectra of the peaks. Light from the source is directed through the flow cell onto a polychromator, which splits the beam according to the wavelengths. A photodiode detector is placed in the focal plane of the polychromator and collects information about the intensity of light at all wavelengths simultaneously. UV spectra are relatively simple and in most cases cannot be used for identification of unknown analytes. However, they can help confirm the identity of target analytes.

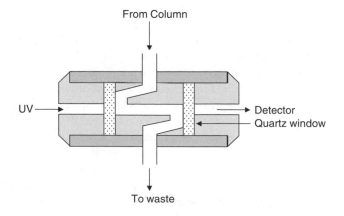

From Column

UV →

→ Detector
→ Quartz window

To waste

FIGURE 3.19 Schematic diagram of a flow cell for HPLC UV detector.

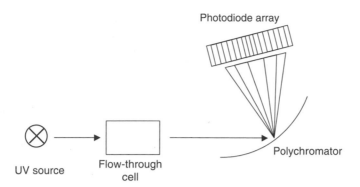

FIGURE 3.20 Schematic diagram of a photodiode array detector.

Fluorescence detectors are inherently more sensitive than absorption detectors, but their applicability is limited to naturally fluorescent compounds and those which can be easily converted to fluorescent derivatives. In the detector, fluorescent molecules are excited using intense light (usually from a laser), and emit light at a longer wavelength. This light is observed at a right angle to the exciting beam to minimize the effect of stray light from the latter. The great sensitivity of fluorescence detection is related to the fact that the faint fluorescence radiation is observed against a dark background; therefore it is much easier to measure than the small changes in the power of otherwise intense radiation observed in absorption detectors.

Several different electrochemical detectors have been developed for HPLC, but the most popular choice is the amperometric detector. It responds to analytes which can be oxidized or reduced electrochemically, such as phenols, aromatic amines, peroxides, mercaptans, ketones, aldehydes, conjugated nitriles, etc. An amperometric detector consists of a flow cell of a very small volume (a few µL is common) fitted with three electrodes: working, auxiliary and reference. The working electrode can be made of a variety of materials, but the most popular choice is glassy carbon. The choice of the material for the auxiliary electrode is less critical. Platinum or stainless steel are typically used. A potentiostat is used to keep the potential of the working electrode constant. When electroactive species enter the detector, these become oxidized or reduced (depending on the electrode potential) and the resulting current is recorded as the detector signal. The current is proportional to the solute concentration over up to six orders of magnitude. Amperometric detectors require the use of conductive mobile phases containing electrolytes, thus they are compatible with reversed phase and ion-exchange chromatography. Among other things, they are sensitive to flow rate changes, mobile phase pH, ionic strength, temperature, etc. Typically, only a small percentage of the analyte becomes oxidized or reduced in the cell. The efficiency of the process can be improved by increasing the surface area of the working electrode. Once it reaches 100%, the amperometric detector turns into a coulometric detector. Such detectors are insensitive to flow changes and do not require calibration, but they are more difficult to design and operate than regular amperometric detectors.

Another type of electrochemical detector is the conductivity detector. It consists of a chamber made of a nonconductive material and fitted with two electrodes. Alternating potential is applied to the electrodes to measure the electrolytic conductivity of the solution in the cell. Conductivity detectors found their main application in ion chromatography. In this method, the mobile phase by definition must contain an electrolyte at a relatively high concentration, therefore it is highly conductive. This would make the detection of small amounts of analyte ions in the column effluent impossible, as the background conductivity would be too high. To eliminate this problem, ion suppression is applied. The technique is based on the removal of highly conductive eluent ions. For example, in anion chromatography, KOH may be used as the mobile phase. In the suppressor, potassium ions are replaced with hydronium ions, which then react with hydroxyl ions in the mobile phase to form water. At the same time, analyte anions, in the form of their potassium salts, are converted to the respective acids with high specific conductivity. In cation chromatography, the acid in the mobile phase is converted to water, while the cations of interest are converted to their hydroxides. Ion suppression can be carried out using suppressor columns filled with an ion-exchanger or using suppressors based on ion-exchange membranes (flat sheet or hollow fiber). The most popular type today are micromembrane suppressors, in which the ion-exchange membrane is continuously regenerated with a constant flow of an electrolyte. The latest development in the area of ion suppression is so-called "Reagent-Free Ion Chromatography" introduced recently by Dionex.[15] In this technique, the eluents are prepared automatically from deionized water and chemicals supplied in special cartridges. Potassium hydroxide is generated for anion chromatography, and methanesulfonic acid is generated for cation-exchange applications. At the same time, the ions required for eluent suppression are generated by the continuous electrolysis of water. Thus, once equipped with the generator cartridges, the system requires only deionized water to operate, which vastly simplifies its operation and improves the reproducibility of the results by eliminating manual operations.

C. HYPHENATED TECHNIQUES

LC is easily coupled to molecular spectroscopy. The photodiode array detector described in the previous section is an example of such coupling. While this detector provides additional spectroscopic information which is independent of the chromatographic process, this information is not specific enough to form the basis for unequivocal analyte identification due to the simple nature of the UV spectra of most molecules. Consequently, the coupling of HPLC with a photodiode array detector is normally not considered to be a hyphenated technique.

Attempts to couple LC to mass spectrometry date back to the early 1970s. The problem is much more difficult than GC/MS coupling. Mass spectrometry requires vacuum to operate, while HPLC is by definition a high-pressure technique. In addition, the volume of the gas formed during evaporation of the liquid mobile phase eluting from the column is many orders of magnitude larger than the pumping capacity of the vacuum pumps used in mass spectrometers. This gas must be removed prior to ion separation in the MS. Over time, many different solutions were proposed for the coupling of LC with mass spectrometry. Some of them were quite exotic, like the mechanical moving belt interface.[16] Today, two techniques are predominantly used in LC/MS: pneumatically assisted electrospray (often called electrospray ionisation, or ESI) and atmospheric pressure CI (APCI). The principle of pneumatically assisted electrospray operation is illustrated in Figure 3.21. Liquid from the LC column enters a metal nebulizer capillary along with a coaxial flow of nitrogen. The nebulizer is held at a few thousand volts with respect to the spray chamber housing. The liquid passing through the nebulizer becomes charged to a high potential (1 in Figure 3.21). As it is forced to hold more and more charge, it becomes unstable and breaks into highly charged, small droplets (2). The flow of nitrogen helps the solvent to evaporate from the droplets. This causes the distances between the electrical charges in the droplets to decrease, which results in large repulsive forces. Once the critical limit is reached, the droplets violently explode, forming even smaller droplets which ultimately evaporate (3). The electrospray has a very characteristic shape, beginning as a cone, then changing into a fine filament of liquid and finally into a plume of fine spray. The conical portion is called Taylor cone, in honor of G.I. Taylor, who first described the phenomenon. The liquid at the outlet of the capillary assumes the characteristic cylindrical shape because this shape can hold more charge than a sphere. The plume of fine spray is formed when the filament of liquid becomes unstable due to high concentration of charges.

The most important characteristic of electrospray is that in fact it does not generate any new ions. The ions which reach the mass spectrometer were already in solution in one form or another in the chromatographic column. Electrospray helps transfer these solution ions into the gas phase, so that they can be analyzed by mass spectrometry. Examples of ions which can be analyzed by

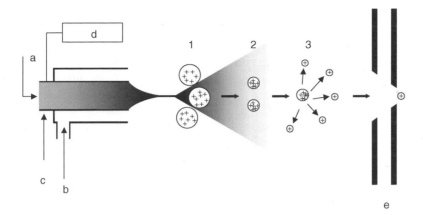

FIGURE 3.21 Schematic diagram of a pneumatically assisted electrospray; (a) effluent from the column; (b) nebulizing gas; (c) electrospray capillary; (d) high-voltage power supply; (e) skimmer cones. For a description of operation see text.

electrospray LC/MS include ionized acids and protonated bases, as well as adduct ions consisting of solute molecules and stable ions from the solution (e.g., H^+, Na^+, NH_4^+, HCO_2^-, etc.). The gas-phase ions enter the high vacuum region of the mass spectrometer through a series of differentially pumped skimmer cones, which remove excess vapor of the solvent. Because of the evaporation of the mobile phase, nonvolatile buffers are usually not recommended when using electrospray.

ESI is a soft ionization technique, and the mass spectra are usually simple. If required, fragmentation can be increased by CID in the skimmer cone region. CID also helps break some adduct ions. A very characteristic feature of electrospray is the formation of multiply charged ions, whose m/z ratio is lower than that of a singly charged species. This makes it possible to analyze very large molecules (molecular weight can be as high as 200,000[17]) with relatively simple mass spectrometers like quadrupole MS.

APCI is a technique which can be considered complementary to ESI. In APCI (Figure 3.22), the effluent from the column is nebulized by coaxial flow of nebulizing gas (nitrogen). The vaporizer tube is heated to 350 to 500°C, which causes the solvent in the fine aerosol formed to evaporate rapidly. On its way through the source, the gas–vapor mixture contacts a corona discharge needle, which is held at a voltage of 2.5 to 3 kV. The electric corona which forms around the needle is a plasma containing charged particles and free electrons. Any sample molecule which passes through

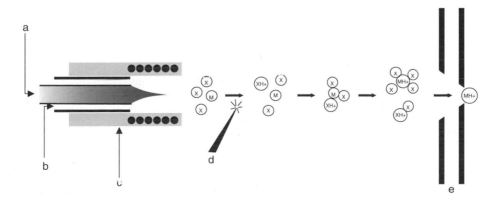

FIGURE 3.22 Schematic diagram of an atmospheric pressure chemical ionization source; (a) effluent from the column; (b) nebulizing gas; (c) heater; (d) corona discharge needle; (e) skimmer cones.

the corona region can be ionized by the transfer of a proton to produce $(M + H)^+$ or $(M - H)^-$ ions. Thus, APCI acts in a manner similar to CI in the ion source of a mass spectrometer. Most importantly, it does not require that ions are already present in the solution before the ionization takes places, as in ESI. Ions formed in APCI carry a single charge; therefore the technique is generally not suitable for the analysis of high molecular weight compounds. Typical classes of chemical compounds analyzed routinely by APCI include pesticides, drugs and their metabolites, surfactants, polycyclic aromatic hydrocarbons, and many other organic compounds.

In essence, any of the mass spectrometer types described in the GC/MS section can be used in combination with LC/MS. Double focusing mass spectrometers provide high resolution, but are very expensive. The most popular choices in LC/MS are quadrupole and ion trap mass spectrometers. TOF mass spectrometers are quickly gaining popularity in LC/MS. Recent advances in the technology make it possible to measure exact masses of the ions, which allows the unequivocal determination of their elemental composition. Thus, results comparable to those produced by high-resolution double focusing instruments can be obtained at a fraction of the cost.

The ESI and APCI are both soft ionization techniques, hence the spectra obtained using these methods are usually very simple and consist predominantly of the pseudomolecular ion. When more information is required, tandem mass spectrometry is a popular option. Several different types of tandem LC/MS systems can be found, with triple quadrupole and ion trap being the most popular choices. Hybrid systems including magnetic sector-quadrupole, magnetic sector-TOF, quadrupole-TOF (Q-TOF) and ion trap-TOF have been described.[18] The most successful among these hybrids is the Q-TOF instrument.[19]

Nuclear magnetic resonance (NMR) spectroscopy is probably the most widely used method for structural elucidation in organic chemistry. It is not surprising therefore to find that attempts have been undertaken to couple LC to NMR spectroscopy. The main limitation of this technique is its poor sensitivity. While MS analysis can be routinely carried out in sub-picogram range, high-field NMR spectrometers require hundreds of nanograms of the analyte at the least for real world samples. In addition, conventional solvents cannot be used in LC-NMR as these would obscure the signals from the sample components. Thus, only very expensive deuterated solvents can be used.

Coupling of LC to NMR is relatively simple. The effluent from the column is delivered through a polyether-ether ketone (PEEK) transfer line to the NMR flow cell, which typically has a volume of 60 μl. The measurement can be carried out in one of four modes: on-flow, stop-flow, time-sliced and loop collection.[20] In the on-flow mode, the effluent from the column flows continuously through the NMR flow cell. Because of the very short time available for the measurement when peaks elute in real time, this approach is limited to major components of a mixture. In the stop flow mode, peaks detected with a UV detector are transferred to the NMR flow cell, and the run is automatically stopped. The NMR spectra can then be acquired over a period of several minutes, hours or even days. In the time-sliced mode, the elution is stopped several times during the elution of the peak of interest. This mode is usually used when two analytes are poorly resolved. In the loop collection mode, the chromatographic peaks are stored in loops for offline NMR study. This approach is therefore not a real online hyphenated technique.

Mass spectrometric information is complimentary to NMR information in structure elucidation, therefore attempts have been undertaken to simultaneously couple both techniques to liquid chromatography. LC-MS-NMR is usually accomplished by splitting the flow coming from the UV detector into two streams, one going to the NMR and one going to the MS.[20] Because of the sensitivity difference, most of the effluent is directed to the former, with only a small fraction ($\sim 1\%$) going to the latter. Electrospray has to be used for the LC-MS part because it is the only source of ionization which can work with the very low resultant flow rate of the effluent. LC-MS-NMR is a powerful structure elucidation technique, but it has significant practical limitations including poor sensitivity and long time required to perform the measurements (especially in the stop-flow mode). In general, the limitations of the LC-NMR coupling cause this technique to find only limited applications in environmental analysis.

D. ELECTRICALLY DRIVEN CHROMATOGRAPHY

A number of different electrically driven separation methods exist. Some examples include capillary zone electrophoresis, isotachophoresis, isoelectric focusing, etc. However, most of these methods do not utilize chromatographic mechanisms to accomplish analyte separation, therefore these lie outside the scope of this chapter. The two exceptions to this rule include micellar electrokinetic chromatography (MEKC) and electrochromatography (EC). Both methods rely on electro-osmotic flow to accomplish the separation. The principle of electro-osmotic flow creation is illustrated in Figure 3.23. The surface of fused silica is covered with silanol groups, which are partly or completely ionized at pH > 2. Thus, the surface has a negative charge which attracts cations from the solution. Some of this negative charge is neutralized by tightly adsorbed cations, which are immobile. The balance of the negative charge is neutralized by cations which are attracted to the surface, but do not become adsorbed to it. These cations are mobile. The thin layer comprising the negatively charged surface and the region of the solution enriched in the cations is called the electrical double layer. Its thickness varies from ~ 10 nm in solutions of low ionic strength to less than 1 nm in solutions of high ionic strength. An electrical double layer is formed whenever an electrical insulator is immersed in a solution of an electrolyte, either by dissociation of surface functional groups (as is the case with fused silica), or due to the adsorption of ions from the solution.

The excess free cations in the diffuse part of the electrical double layer form a charged "sheath" which encloses the electrolyte. When a potential difference is applied across the length of tube filled with the electrolyte, these cations are drawn towards the cathode. This causes the "sheath" to move together with the electrolyte it surrounds. As long as the tube diameter is much larger than the thickness of the electrical double layer (which is usually the case), the profile of the flow is flat, or plug-like. This remains in stark contrast to hydrodynamic or pressure driven flow, which has a parabolic profile with zero flow velocity at the wall and twice the average velocity at the center. The differences in flow profiles have significant consequences in chromatography. The different flow velocities across the column diameter in pressure driven separations significantly contribute to band broadening. This effect is eliminated in electro-osmotic flow, which results in much less band broadening.

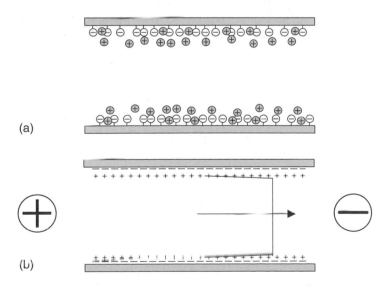

FIGURE 3.23 Principle of electro-osmotic flow generation. (a) formation of electrical double layer; (b) electro-osmotic flow profile.

MEKC is based on the use of a pseudostationary phase consisting of micelles. Micelles are aggregates of colloidal dimensions, which exist in equilibrium with the molecules or ions from which they are formed. In MEKC, they are created by adding a surfactant to the solution. At a high enough concentration, the surfactant molecules tend to aggregate, with the hydrophobic tails sticking to each other and kept inside the aggregate, and the hydrophilic ionic heads facing the solution. In effect, micelles act as charged microscopic oil droplets shielded from the solution by ionic groups. When an electric field is applied across the length of the tubing filled with a mobile phase for MEKC, the solution starts to migrate at a rate determined by the electroosmotic flow. The charged micelles migrate at a different mean velocity because they are attracted to one of the electrodes. For example, when sodium dodecyl sulphate is used as the surfactant, it is negatively charged in the solution and therefore the micelles are attracted to the anode. Since the electroosmotic flow is towards the cathode, the micelles tend to migrate more slowly than the electrolyte solution. The micelles form a pseudostationary phase because they are neither immobilized, nor consist of a distinct phase — they are distributed uniformly throughout the solution. Solutes added to the solution partition between the micelles and the aqueous phase. They can be both neutral and ionic. Their speed of migration is different in each of the two phases. The solutes which spend more time in the phase whose migration speed is lower, elute from the column later. The rate of migration of charged solutes depends also on their electrophoretic mobility. Thus, two separation mechanisms, electrophoresis and partition chromatography, are combined in MEKC. Band broadening in this method is low compared to HPLC because of the plug flow profile of the electrolyte. Resistance to mass transfer (the C term in the van Deemter equation) does occur, but its contribution is modest because mass transfer into and out of the micelles is quite fast.

EC differs from MEKC in that it uses a real stationary phase. In essence, it is conventional LC in which the mobile phase is driven by electro-osmotic flow rather than pressure. However, as in MEKC, the mechanism of separation is mixed, with electrophoretic mobility affecting the retention of charged solutes. The stationary phase particles can be very small, as there is no pressure drop in the column. Typically, 1.5 μm particles of C_{18}-modified silica are used. Promising results were also obtained with monolithic columns for EC.[21] Capillary EC provides about twice as many plates as HPLC for the same particle size and column length.

In spite of their great potential, neither of the two electrodriven chromatographic methods described in this section is very popular or widely used. The probable main reason for this is the great number of variables which affect the separation, rendering it difficult to obtain reproducible results routinely. Because of its reliance on electrically generated electroosmotic flow rather than pressure, EC bears great promise in microseparation systems ("lab-on-a-chip") and quite likely will become much more popular in the near future.

V. MULTIDIMENSIONAL TECHNIQUES

When dealing with complex mixtures of compounds, it is usually impossible to separate all of them using any single chromatographic method. This is related to the limited resolving power of chromatography. Because peaks eluting from chromatographic columns have finite, nonzero widths, the number of peaks which can be fully resolved at the outlet of the system (called peak capacity) is limited. For example, in GC using a 50 m column with 200,000 theoretical plates, the theoretical peak capacity is 260; to achieve the same peak capacity in HPLC it is necessary to use a 50 cm column packed with 1.5 μm particles.[22] In reality, the peak capacity is even lower because components of a mixture at the outlet of the column are distributed randomly rather than evenly. Thus, to fully resolve 90 peaks in a 100 component mixture, the theoretical peak capacity should be 1910, and the corresponding number of theoretical plates should be nearly 11 million.[22] In practice, it is impossible to achieve this number of plates, as it would require the use of GC columns measured in kilometers, or LC columns measured in tens of meters. Even worse, a mixture of only

100 components would often be considered simple in environmental analysis. Real samples may contain thousands of components (especially when these contain petroleum fractions).

The problem of limited peak capacity can often be solved by using multiple dimensions to perform separations of such samples. The dimensionality of a separation method can be viewed as the number of different separation mechanisms to which the sample is subjected. This idea of subjecting a sample to multiple types of separations to get improved resolution and separation power was discussed at length by Giddings.[23] For discrete separations, where the sample is separated first by one dimension and subsequently separated by the second dimension, the best results are obtained when the two separation mechanisms are independent or orthogonal.

The concept of multidimensional separation is the easiest to understand using TLC as an example. A multicomponent sample is spotted at one corner of a TLC plate, and developed with a solvent (Figure 3.24). The components of the sample undergo partial separation, which results in a number of spots along the edge of the plate. Some of these spots may be single compounds, while others may represent coelutions. The plate is then dried, rotated 90° and developed again in the second direction using a different solvent. Owing to the different selectivity of the solvent used for the second plate elution (second separation dimension), many of the coelutions present after the first elution may be resolved in the second dimension. The peak, or spot capacity, is in this case theoretically equal to the product of the individual peak capacities in each of the two dimensions. Thus, two-dimensional separation achieves in this case a much higher resolving power than even the best one-dimensional separation.

Practical implementation of two-dimensional separation is very easy in TLC. By definition, all the components of the sample are subjected to both separation dimensions, and components separated in the first dimension remain resolved in the second dimension. When a two-dimensional separation fulfills these requirements, it is considered comprehensive. From the practical point of view, multidimensional separations are much more difficult to implement in column chromatography. To perform the 2D separation "in space," i.e., in a manner analogous to

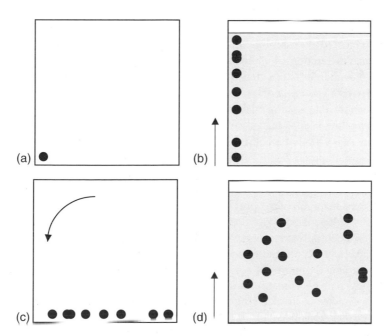

FIGURE 3.24 Two dimensional thin layer chromatography (2D-TLC). (a) A sample is spotted at one corner of the plate; (b) the plate is developed in one direction; (c) the plate is rotated 90°; (d) the plate is developed in the second direction with a different solvent.

2D-TLC, one would have to stop the separation before the first component elutes from the column, then cut the column lengthwise and somehow direct the components to a large number of parallel columns placed at a right angle to the first column. This is of course not practical, therefore multidimensional separations in column chromatography are performed "in time." In the simplest version, only a fraction of the sample is subjected to two-dimensional separation. The sample is injected to the column, and the separation initially proceeds as usual. When the fraction containing the analyte(s) of interest starts to emerge from the column, it is directed to a second column characterized by different selectivity rather than to the detector (see Figure 3.25a). The rest of the sample is again directed to the detector connected to the first column. This mode of operation is often referred to as "heart-cut" analysis. It is possible to analyze more than one heart-cut in a single run; however, if a single second dimension column is used, the time available for the separation in the second dimension must be shorter, so that separation of one fraction is completed before the components of the next fraction can reach the detector (Figure 3.25b). In the limiting case when the number of heart-cuts gets high enough and the time for the heart-cut separation gets short enough, a comprehensive multidimensional separation is accomplished (Figure 3.25c). Material exiting the first dimension is sampled periodically, frequently enough for the separation in the first dimension to be preserved, and all of the components of the sample are subjected to both separation dimensions. Thus, in column chromatography, comprehensive two-dimensional separation is in effect a separation where very many sequential heart-cuts are taken.

Multidimensional chromatographic separation can be accomplished by coupling two separation dimensions based on the same chromatographic method (e.g., GC–GC or HPLC–HPLC), or by coupling two different methods (e.g., HPLC–GC or HPLC–SFC). Following is a brief overview of the most popular two-dimensional separation techniques.

A. GC–GC AND GC × GC

Two dimensional heart-cut GC separations (GC–GC) were used in the analysis of crude oil and refinery products as early as the late sixties.[24] In the simplest implementation of heart-cut GC, a six-port valve is placed between the outlet of the first column and the first detector. The remaining ports of the valve are connected to a sampling loop, the second dimension column and an auxiliary carrier gas supply. When the fraction of interest starts to emerge from the first column, the valve is switched so that the fraction is directed to the sampling loop. The valve is then switched again, restoring the flow from the first column to the first detector and injecting the heart-cut into the second column connected to a second detector. This simple system allows only a few fractions to be analyzed in the two dimensions because the separation time in the second dimension is usually long. While definitely a limitation, this does not discount heart-cut GC as a useful analytical technique. This simple approach can be modified in many ways to improve its performance. For example, instead of using a single second dimension column and detector, one can use several column-detector assemblies in parallel. This makes it possible to inject a new fraction to the second dimension while the separation of the previous fraction still proceeds in the other column. Multiple traps between the two dimensions (e.g., sorbent traps or cryotraps) to collect fractions of the sample and analyze them sequentially on a single second dimension column can be used. As an additional advantage, the traps help refocus the fractions into narrow bands prior to reinjection to the second column, which significantly improves peak capacity in this column. The valves can be eliminated from the sample flow path by using pressure-based Deans switching. In this technique, effluent from the first column is directed to the detector or the second column by carefully adjusting pressures at strategically placed junctions between the two columns.[22]

Comprehensive two-dimensional GC (GC × GC) was introduced in the early nineties by the late John Phillips.[25] This technique differs from GC–GC in that the entire sample injected to the column is subjected to separation in both dimensions. In GC × GC, the sample injected into the system is first subjected to chromatographic separation in the first column (primary dimension), as

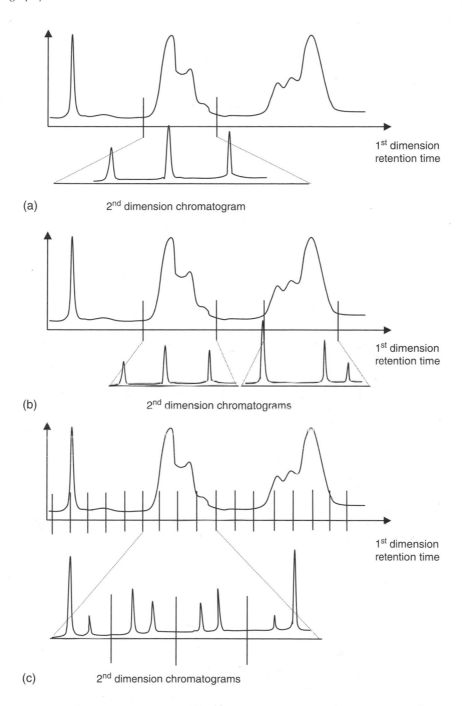

FIGURE 3.25 The concept of multidimensional GC.[26] (a) Single heart-cut GC analysis, in which a portion of the effluent from the primary column containing analytes of interest is diverted to the second dimension column and subjected to additional separation over an extended period of time. (b) Dual heart-cut GC analysis, in which two regions with coelutions are diverted to the second dimension column, with less time to perform each separation. (c) Comprehensive two dimensional GC analysis, in which the sizes of the sequential heart-cut fractions are very small, and the time to develop each sequential second dimension chromatogram is very short.

in one-dimensional GC. However, rather than reach a detector, the effluent from the primary column enters a special interface (modulator) placed between the first and second column. This modulator collects the material from the first column for a short period of time, and then injects the entire fraction which it has collected into the second dimension column as a short chromatographic pulse. It then collects another fraction of the effluent from the first column while the previous fraction is being separated on the second dimension column. This process of effluent collection and injection is repeated frequently throughout the entire analysis.[26] The second dimension column is short, so that the separation in this column can be completed before first components of subsequent fraction reach the detector (a few seconds). The stationary phase in the second column must have different selectivity than the first column to fulfill the condition of orthogonality. The material exiting the second dimension column is passed to the detector, so that a series of sequential short second dimension chromatograms is obtained. In order to preserve the separation achieved in the first dimension, each peak eluting from the first dimension should be sampled at least three times.[27] For example, if the peaks eluting from the first dimension have a width of 18s, the modulation period must be no longer than 6s.

The multiple second dimension chromatograms are recorded by the system as a single linear chromatogram. In this form, it is exceedingly difficult to interpret. For this reason, the data are usually converted into a three-dimensional plot with primary retention plotted along the X axis, secondary retention plotted along the Y axis and peak intensity plotted along the Z axis. This 3D plot is usually displayed as a top–down view in the form of a contour plot. The construction of such a plot is outlined in Figure 3.26. An appropriate software package uses the modulation period of the interface and the times at which the pulses to the second dimension column occur (t_1, t_2 and t_3 in Figure 3.26a) to slice the original chromatographic signal into its component second dimension chromatograms (Figure 3.26b). These chromatograms are then aligned side-by-side to form GC × GC retention plane (Figure 3.26c), which is then plotted top–down as in Figure 3.26d. The time at which a modulation pulse occurs provides the primary retention time for all of the peaks which elute between that pulse and the following pulse. The second dimension retention time of a peak is then its original (1D) retention time minus the primary dimension retention time.

The heart of any GC × GC system is the modulator. There are two basic types of modulators currently in use: thermal modulators and valve-based modulators. Thermal modulators are more popular; in fact, the commercial GC × GC systems are all based on thermal modulation. Early thermal modulators required moving parts, which made them not always reliable. Today, most thermal modulators are based on cryocooling, with no moving parts inside the oven. These are reliable enough to be used in routine applications. An example of a modern cryogenic modulator utilizing liquid CO_2 as the cryocoolant[28] is presented in Figure 3.27. When the downstream CO_2 jet (D) is on and the upstream jet (U) is off, material is focused into a narrow band within a cooled segment of the second dimension column. The upstream jet is then turned on so that it can trap the material eluting from the primary column while the downstream jet turns off to launch the focused band into the second dimension column. The downstream jet turns back on, before the upstream jet turns off, so that the material released from the upstream cold spot is retrapped in the downstream cold spot prior to injection into the second column. This two-stage mode of operation prevents breakthrough of the analytes through the trap while any of the jets is off.

GC × GC offers unparalleled resolving power. It can separate components of very complex mixtures, for example all 209 PCB congeners,[29] which is impossible using 1D GC. It can also potentially simplify sample preparation before chromatographic analysis by eliminating the need for extensive sample clean-up when the analytes of interest can be chromatographically separated from the matrix components. Consequently, GC × GC has tremendous potential in environmental analysis, especially in combination with TOF mass spectrometry.

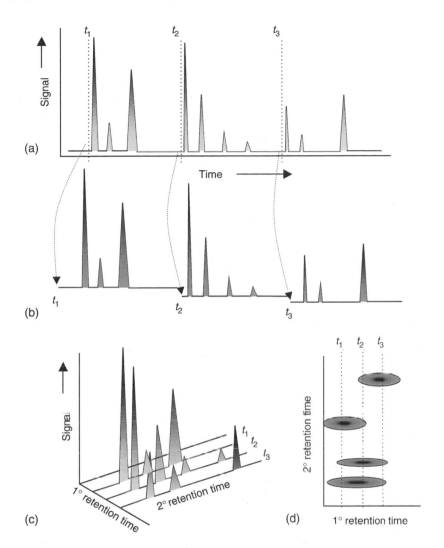

FIGURE 3.26 The interpretation of GC × GC data and generation of contour plots.[26] (a) Raw GC × GC chromatogram consisting of a series of short second dimension chromatograms; t_1, t_2, and t_3 indicate the times when injections to the second dimension column occurred. (b) The computer uses these injection times to slice the original signal into the individual second dimension chromatograms. (c) The second dimension chromatograms are aligned on a three dimensional plane with primary retention time and secondary retention time as the X and Y axes, respectively, and signal intensity as the Z axis. (d) When viewed from above, the peaks appear as rings of contour lines or color-coded spots.

B. COUPLED COLUMN LIQUID CHROMATOGRAPHY

In liquid chromatography, two dimensional separations in the vast majority of cases are not comprehensive. While comprehensive 2D-LC separations (LC × LC) can be accomplished and have been demonstrated (e.g., Refs. 30 and 31), the technique is not very popular. Probably one of the main reasons for this is the inability to perform very fast separations in liquid chromatography. In GC × GC, a typical second dimension separation can be completed in a few seconds. In LC, the separation time required is much longer. The problem can be overcome by stopping the flow in the first dimension column while the second dimension separation proceeds, but this causes the overall analysis times to be very long.

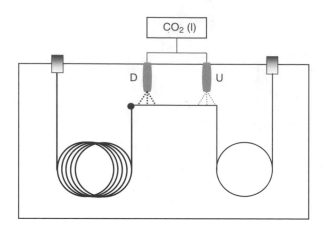

FIGURE 3.27 Schematic diagram of the dual cryojet interface.[26] When the downstream jet (D) is on and the upstream jet (U) is off, material from the primary column is trapped as a narrow band within the second dimension column. It is then released by turning the downstream jet off, and retrapped by the upstream jet. The downstream jet is turned back on before desorption from the second stage is effected to prevent breakthrough.

In LC–LC, two columns are linked *via* a switching valve so that any component flowing through the first column can be directed to the detector or to the second column. Two types of arrangements are used: the two columns may have the same stationary phase, but a different length, or they may have similar length, but different stationary phase. For reasons explained at the beginning of this section, only the second approach can be classified as two-dimensional separation. The most important applications of LC–LC include trace enrichment and sample clean-up. Both of them are important from the point of view of environmental analysis. In many cases, both sample clean-up and trace enrichment are employed in the same LC–LC scheme. Trace enrichment is based on the fact that the analytes may be retained as a narrow zone at the head of the first (preconcentrating) column while a large sample volume is pumped through this column. For example, nonpolar or weakly-polar analytes can be preconcentrated from aqueous solutions on a reversed phase column because water is a weak eluent in this scenario. This will also result in partial sample clean-up, as the polar sample components will not be retained. The preconcentrated sample can then be eluted with a stronger eluent into the second (analytical) column, where the proper separation takes place.

Apart from analyte preconcentration and sample clean-up, LC–LC can also be used to improve the separation of critical sample components. This is done by using heart-cut techniques similar in principle to those used in GC–GC. A high-resolving LC–LC system can be implemented by using columns packed with stationary phases offering different separation mechanisms. Examples of the possible combinations include size exclusion–ion-exchange; size exclusion–reversed phase; ion-exchange–reversed phase; reversed phase (alkyl ligand)–reversed phase (ion-pairing eluent); reversed phase–affinity, etc.[32] The resolving power of the system can be enhanced even further by coupling the LC–LC system with mass spectrometry.[33,34] It is also possible to couple the LC–LC system to other separation techniques like capillary zone electrophoresis, which creates a three-dimensional separation system.[35] However, such couplings are outside the scope of this chapter.

C. HPLC–GC

The combination of HPLC and GC offers two separation mechanisms which can be made entirely orthogonal. The main difficulty in this technique is the fact that the mobile phase in the two systems is in two different physical states (liquid and gas). The problem is exacerbated by the fact that the volume of the vapor is many times greater than the volume of the liquid from which it is formed.

Thus, special techniques must be used when coupling the two techniques online. Their main goal is to eliminate the large volumes of solvent vapors before introducing the analytes in the form of a sharp band to the GC column. A number of different approaches, both direct and indirect, can be used to accomplish this goal.[22] In the retention gap technique, the liquid fraction from the HPLC column is introduced into a retention gap, usually in the form of a long segment of deactivated tubing. The retention gap is kept below the boiling point of the LC eluent, so that the latter remains in the liquid form. As a result, all analytes are focused through the solvent effect. As the solvent evaporates from the back of the liquid film to the front, the analytes are focused into a narrow band. Additional focusing for analytes with higher boiling points can be accomplished through the retention gap effect and/or through thermal focusing in the analytical column. The solvent vapors are eliminated from the systems through an early vapor exit, a side line connected between the retention gap and the column. The early vapor exit is open at the beginning of the separation to prevent the solvent vapors from entering the column. Once the majority of the solvent is gone, the early vapor exit is closed, and the GC separation begins. The retention gap technique allows transfer of LC fractions as large as several hundred microliters. It can be simplified by employing partially concurrent solvent evaporation. In this technique, a large fraction of the solvent is evaporated during the transfer of the LC fraction to the GC column. This allows the use of shorter retention gaps or larger transfer volumes.

In loop-type interfaces, the fraction of interest is collected in a sampling loop attached to a six-port valve. When the valve is switched, carrier gas pushes the fraction from the loop to an uncoated GC inlet kept at or above the boiling point of the LC eluent. Solvent vapors are removed through early vapor exit, as in the retention gap technique. The main difference between this technique and the retention gap technique is that the solvent evaporates from the front and the precolumn is not flooded. Thus, focusing through solvent effect cannot be used. This technique leads to losses of volatile analytes, which leave the system through the early vapor exit together with the solvent. On the other hand, it allows the introduction of much larger volumes of the liquid than the retention gap technique.

LC–GC coupling can also be accomplished through the use of a programmed temperature vaporizing injector (PTV). In this case, the liquid is introduced to a PTV injector containing a packing material, which can be inert (e.g., glass spheres) or can act as a sorbent. The split vent of the injector is initially open. The evaporating solvent escapes from the system through the split vent, while the analytes are retained on the packing bed. Once the solvent is removed, the split vent is closed and the injector is heated to mobilize the analytes and initiate gas chromatographic separation. This technique allows the introduction of large LC fractions to GC, but it usually leads to losses of volatile compounds and may create problems when the analytes are thermally labile, as the injector must be heated above the column temperature to mobilize the analytes trapped by the packing material.

Water creates problems in LC–GC coupling. For example, the retention gap technique cannot be used due to unavailability of tubing which would be sufficiently inert while being wettable by water. This problem can be overcome by using indirect methods. For example, the aqueous sample can be introduced to an LC precolumn or a solid-phase extraction (SPE) cartridge, which retains the analytes. The precolumn is then flushed with nitrogen to remove water, and the analytes are desorbed into the gas chromatograph with an organic solvent. The procedure does not differ in principle from a regular SPE procedure. The main difference is that the entire process is in this case automated.

LC–GC is usually limited to sample preconcentration and clean-up or to heart-cut analysis. Comprehensive LC–GC is difficult to accomplish because the solvent evaporation process is slow. Besides, the analytes contained in a single LC fraction can differ vastly with respect to their volatilities, which often requires the GC part of the separation to be carried out under temperature programmed conditions. Consequently, the time required for a single second dimension GC separation is measured in the best case in minutes rather than seconds, as in GC × GC. Thus, the easiest way to accomplish LC × GC is to perform the second dimension separation offline.[36] This of course results in very long total analysis times, which would be impractical in routine applications. A system for comprehensive LC × GC has been described in the literature,[37] but it was only suitable for volatile analytes.

REFERENCES

1. Ettre, L. S., *LCGC North Am.*, 21(5), 458, 2003.
2. Tswett, M., *Ber. Dtsch. Botan. Ges.*, 24, 316, 1906.
3. Tswett, M., *Ber. Dtsch. Botan. Ges.*, 24, 384, 1906.
4. Martin, A. J. P. and Synge, R. L. M., *Biochem. J.*, 35, 1358, 1941.
5. James, A. T. and Martin, A. J. P., *Biochem. J.*, 50, 679, 1952.
6. Poole, C. F. and Poole, S. K., *Chromatography Today*, Elsevier, Amsterdam, 1991.
7. Golay, M., *Anal. Chem.*, 29, 928, 1957.
8. Dandeneau, R. D. and Zerenner, E. H., *J. High Res. Chromatogr. Chromatogr. Commun.*, 2, 351, 1979.
9. http://www.ezflash.com/
10. http://www.rvmscientific.com/index.htm
11. Cavagnino, D., Magni, P., Zilioli, G., and Trestianu, S., *J. Chromatogr. A*, 1019, 211, 2003.
12. http://www.vici.com/instr/pdd.htm
13. http://www.oico.com/descpfpd.htm
14. Huebschmann, H.-J., *Handbook of GC/MS. Fundamentals and Applications*, Wiley-VCH verlag GmbH, Weinheim, Germany, 2001.
15. http://www.dionex.com/app/tree.taf?asset_id = 283786
16. Ten Noever De Brauw, M. C., *J. Chromatogr.*, 165, 2, 1979.
17. Ashcroft, A. E., *Ionization Methods in Organic Mass Spectrometry*, The Royal Society of Chemistry, Cambridge, 1997.
18. Niessen, W. M. A., *J. Chromatogr. A*, 1000, 413, 2003.
19. Morris, H. R., Paxton, T., Dell, A., Langhorne, J., Berg, M., Bordoli, R. S., Hoyes, J., and Bateman, R. H., *Rapid Commun. Mass Spectrom.*, 11, 889, 1996.
20. Elipe, M. V. S., *Anal. Chim. Acta*, 497, 1, 2003.
21. Hayes, J. D. and Malik, A., *Anal. Chem.*, 72, 4090, 2000.
22. Mondello, L., Lewis, A. C., and Bartle, K., Eds., *Multidimensional Chromatography*, Wiley, Chichester, UK, 2001.
23. Giddings, J. C., *Anal. Chem.*, 56, 1258A, 1984.
24. Luke, L. A. and Brunnock, J. V., *Ger. Offen.*, 1(908), 418, 1968.
25. Phillips, J. B. and Liu, Z., *J. Chromatogr. Sci.*, 29, 227, 1991.
26. Górecki, T., Harynuk, J., and Panić, O., The evolution of comprehensive two dimensional gas chromatography (GC × GC), *J. Sep. Sci.*, 27, 431, 2004.
27. Murphy, R., Schure, M., and Foley, J., *Anal. Chem.*, 70, 1585, 1998.
28. Beens, J., Adahchour, M., Vreuls, R. J. J., van Altena, K., and Brinkman, U. A. Th., *J. Chromatogr. A*, 919, 127, 2001.
29. Harju, M. and Haglund, P., *J. Microcolumn Sep.*, 13, 300, 2001.
30. Holland, L. A. and Jorgenson, J. W., *Anal. Chem.*, 67, 3275, 1995.
31. Opiteck, G. J., Ramirez, S. M., Jorgenson, J. W., and Moseley, M. A. III, *Anal. Biochem.*, 258, 349, 1998.
32. Cortes, H. J., Ed., *Multidimensional Chromatography. Techniques and Applications*, Marcel Dekker, New York, 1990.
33. Opiteck, G. J., Lewis, K. C., Jorgenson, J. W., and Anderegg, R. J., *Anal. Chem.*, 69, 1518, 1997.
34. Creaser, C. S., Feely, S. J., Houghton, E., and Seymour, M., *J. Chromatogr.*, 794, 37, 1998.
35. Moore, A. W. Jr. and Jorgenson, J. W., *Anal. Chem.*, 67, 3456, 1995.
36. Janssen, H.-G., Boers, W., Steenbergen, H., Horsten, R., and Flöter, E., *J. Chromatogr. A*, 988, 117, 2003.
37. Quigley, W. W. C., Fraga, C. G. and Synovec, R. E., *J. Microcolumn Sep.*, 12, 160, 2003.

4 Detection in Chromatography

Antoine-Michel Siouffi

CONTENTS

I. INTRODUCTION

Analysts must fulfill many requirements, they must deliver reliable results in a minimum time with more and more miniaturized and sophisticated instruments. The race to trace levels is never ending and we are currently challenging the attomole.

Chromatography and capillary electrophoresis are the best performing separation techniques.

Gas Chromatography (GC) is well suited for solutes that can readily be volatilized, whereas liquid chromatography (LC) is well suited for thermally labile solutes. However, the separation power should be transferred to a detector that would not hamper the separation and to a fast data acquisition system. Analysts have to demonstrate that the results obtained under particular application conditions are reliable and fit for the purpose. The performances of any chromatographic system are changing with time. Two desirable features of chromatographic detectors are high sensitivity and high selectivity. Typically the plate number of a column is calculated from the standard deviation of a recorded peak. However, both the column and the instrument contribute to band broadening:

$$\sigma_{\text{tot}}^2 = \sigma_{\text{col}}^2 + \sigma_{\text{ext.col}}^2$$

σ^2_{col} for the chromatographic column variance, $\sigma^2_{ext.col}$ for extra column effects contribution to the variance of the peak. It is generally agreed that the loss in column plate count based on extra column effects should not exceed 10%.[1]

As an example for a specific LC system the maximum acceptable variance is

$$\sigma^2_{acc} = 0.10\sigma^2 \leq 0.10\pi^2 l^2 r^4 \varepsilon^2 (1+k)^2 / N$$

where N, l, and ε are the plate number, the length, and the porosity of the column, respectively, and k is the retention factor. Most of the real chromatographic peaks are not symmetrical. The significant deviation of the peak shape from the symmetrical peak makes difficult the acquisition of chromatographic signal information such as the retention time, the peak area, and the peak width at half height. The asymmetry factor b/a is usually measured at 0.1 h peak height. A chromatographic column looses its performances with continuous use. To overcome this drawback and check the time of changing the column we encourage the reader to use the deferred standard.[2]

We can roughly distinguish concentration sensitive detectors in which the signal is proportional to the concentration of the analyte in the mobile phase inside the detector cell and mass flow detectors where the signal is proportional to the mass flow rate of analyte swept by the mobile phase into the detector cell.

We can also distinguish universal detectors and selective detectors.

II. VALIDATION

No analytical method can be routinely used if it has not been fully validated.

Detection and quantification capabilities represent fundamental performance characteristics of measurement processes. New, coordinated documents prepared for the International Union of Pure and Applied Chemistry (IUPAC)[3] and the International Organization for Standardization (ISO)[4] provide, for the first time, a harmonized position on standards and recommendations for adoption by the international scientific community.

The reader is referred to ISO 17025[5] or ICH guidelines.[6] A validation procedure must check the fulfillment of certain requirements.

A. DETECTION

Dynamic range is the range of concentrations of the test substance over which a change in concentration produces a change in detector signal.

The lower limit of the dynamic range is defined as the concentration producing a detector output signal equal to a specified multiple of the detector short term noise level.

The upper limit of the dynamic range is the concentration at the point where the slope of the curve obtained by plotting detector response as a function of concentration becomes zero.

Linear range is the range of concentrations over which the sensitivity (S) is constant to within a defined tolerance (Figure 4.1).

Limit of detection (LOD) is defined as the concentration below which the analytical method cannot reliably detect a response.

A widely used detection limit technique is the 3σ approach that is mandated for EPA testing. The standard deviation in concentration units is calculated by computing the standard deviation of blank replicates (≥ 7) and dividing by the slope of the calibration curve. The number is multiplied by the appropriate value of the Student's t with $n-1$ degrees of freedom for the chosen α.

Limit of quantification (LOQ) is the smallest quantity of a compound to be determined in given experimental conditions with defined reliability and accuracy. A signal to noise ratio of ten is adequate.

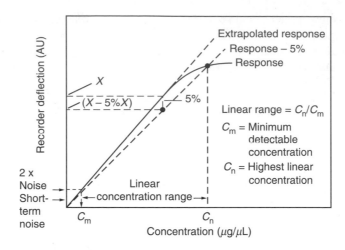

FIGURE 4.1 A typical calibration curve (peak area in absorbance units vs. concentration).

Detector noise is the short-term noise in the maximum amplitude of response for all random variations of the detector signal of a frequency greater than one cycle per minute. Long term noise is similar to short term except that the frequency range is between 6 and 60 cycles per hour. Drift is the measure of the amplitude of the deviation of detector response within 1 h.

Precision: The USP[7] defines precision as "the degree of agreement among individual test results when the method is applied repeatedly to multiple samplings of a homogeneous sample." Precision may be measured as repeatability, reproducibility, and intermediate precision.

Intermediate precision: the degree of agreement of test results obtained by the analysis of the same sample under various conditions (for example, different equipment).

Repeatability (ISO 3534)[8]

Qualitative: the closeness of agreement between the results obtained by the same method on identical test material under the same conditions (same operator, same laboratory, same apparatus, and short interval of time).

Quantitative: the value below which the absolute difference between two single test results obtained under above conditions may be accepted to lie with a specified probability (usually 95%).

The ICH documents[9] recommend that repeatability should be assessed using a minimum of nine determinations covering the specific range (i.e., three concentrations and three replicates for each concentration) or a minimum of six determinations of 100% of the test concentration.

Reproducibility (ISO 3531)[10]

Qualitative: the closeness of agreement between individual results obtained with the same method on identical test material but under different conditions (different operators, different apparatus, different times …).

Quantitative: the value below which the absolute difference between two single test results on individual material obtained by operators in different laboratories using the standardized test method may be expected to lie with a specific probability (usually 95%).

Selectivity: a measure of the extent to which the method is able to determine a particular compound in the matrices without interference from matrix components.

Other authorities (e.g., ICH) prefer the term specificity.

Linearity of analytical procedure: the capacity within a given interval to produce results that are directly proportional to concentration (or mass) of the compound to be determined in the sample. For the assessment of linearity, a minimum of five different concentrations should be used.

The analyst must determine:

- linear slope
- y intercept
- correlation coefficient
- relative standard deviation
- normalized intercept/slope.

Accuracy: measures the difference between the true value and the mean value obtained from repeated analysis. Accuracy is often calculated as percent recovery by the assay of known, added amounts of analyte to the sample. The ICH documents[5] recommend that accuracy should be assessed using a minimum of nine determinations over a minimum of three concentration levels, covering a specific range.

Reliability: the probability that the results lie in the interval defined by two selected limits. It gives a rigorous method for evaluating the correctness of a method of analysis in relation to two limits for error.

Ruggedness: the capacity of an analytical method to remain unaffected and produce accurate data in spite of small but deliberately introduced changes in experimental conditions. Fractional factorial or Plackett Burman designs are frequently used to screen the impact of those changes.

Internal Standard (IS): mostly used in Chromatography and Capillary Electrophoresis, it monitors the behavior of sample solutes to be analyzed and quantitatively determined.

IS must fulfill certain requirements:

- It must exhibit similar retention behavior as compared to the solutes.
- It must exhibit similar chemical functionalities and structure as the solute does.
- If a derivatization step is involved in the method, the same reaction must be applied to the internal standard.
- If a sample pretreatment is required it is better to submit the internal standard to the sample pretreatment and check recovery.

IS Analysis Procedure: a standard solution contains the sample and the internal standard at concentrations C_T and C_E, respectively.

$$m_T = C_T V_T = K_T A_T$$

$$m_E = C_E V_E = K_E A_E$$

V_T, V_E are the injected volume of the sample and the internal standard, respectively, K_T is the response coefficient of the sample and A_T the peak area, K_E is the response coefficient and A_E the peak area of the internal standard

$$\frac{C_T V_T}{C_E V_E} = \frac{K_T A_T}{K_E A_E}$$

Usually $V_T = V_E$.

The concentration ratio is kept constant whatever the injection volume. A sample solution contains the substance to be analyzed at concentration C_X. Internal standard is added at the same C_E concentration as in the previous standard solution.

We thus can write

$$m_X = C_X V_{\text{inj}} = K_X A_X$$

$$m_E = C_E V_{\text{inj}} = K_E A_E'$$

$A_E' \neq A_E$ since two injections are performed and V_{inj} is constant.

$$\frac{C_X}{C_E} = \frac{K_X}{K_E} = \frac{A_X}{A_E'}$$

$$\frac{K_X}{K_E} = \frac{K_T}{K_E}$$

then

$$\frac{C_X}{C_E} = \frac{C_T A_E}{C_E A_T} = \frac{K_X}{K_E}$$

$$\frac{C_X}{C_E} = \frac{C_T}{C_E} \frac{A_X}{A_T} = \frac{K_X}{K_E}$$

$$C_X = C_T \frac{A_E}{A_T} \frac{A_X}{A_E'}$$

if C_E is kept constant.

It is necessary to check the detector response.

Standard solutions are prepared.

$$C_{1T} + C_E \rightarrow A_{1T} + A_E'$$

$$C_{2T} + C_E \rightarrow A_{2T} + A_E''$$

$$C_{3T} + C_E \rightarrow A_{3T} + A_E'''$$

Sample solution

$$C_X + C_E \rightarrow A_E + A_E$$

A plot of A_T/A_E vs. C_T/C_E yields a regression line whose K_T/K_E is the slope.

The analyst records the peak area given by $\int_{t_1}^{t_2} h(t)\, dt$ where t_1 is the time of solute input and t_2 the solute output from the detector. If the peak is gaussian the peak area A and the peak height h_p are related through $h_p = A/\sigma\sqrt{2\pi}$.

If C_i is the solute concentration in the injection volume V_i then $A = C_i V_i$

$$C_{\text{max}} = C_i V_i // \sigma \sqrt{2\pi} = [C_i V_i // \sqrt{2\pi}][\sqrt{N}/V_r]$$

where V_r is the retention volume.

$$C_{\text{max}} = [C_i/\sigma\sqrt{2\pi}][V_i/V_0][\sqrt{N}/(1+k)]$$

where V_0 is the retention volume of the unretained solute.

The detected concentration depends on the column plate count. We can increase the sensitivity by increasing the efficiency, which means that the analyst must select the flow rate according to the optimum.

We shall successively examine GC and LC detectors. Since mass spectrometry is now widely used as the detector in both modes, it deserves a special section.

III. GC DETECTORS

A. OLFACTOMETRY

Olfactometry is surprisingly effective with some solutes that exhibit intense odor. 0.2 ppm can be detected. To carry out the sniffings a panel of judges is trained prior to the first run; a scale of odor intensity evaluation is established.

B. THERMAL CONDUCTIVITY DETECTORS (TCD)

TCD is a universal detector but suffers from lack of sensitivity compared to other detectors.

Principle of operation: a resistor is heated by a current and cooled by the gas stream from the carrier gas. The equilibrium temperature depends on the composition of the gas. The resistance of the resistor, in turn, depends on its temperature. In the detector device, resistors are connected to a Wheatstone bridge. Cells in one diagonal are swept by pure carrier gas, cells of the other diagonal by column effluent. When solutes are eluted the bridge experiences a desequilibrium which is amplified and recorded. Its highest sensitivity is obtained with carrier gases that exhibit a high thermal conductivity, e.g., hydrogen and helium. The rhenium–tungsten filaments having a resistance of 100 Ω each provide excellent stability and high reliability.

The linear dynamic range is five orders of magnitude.

The lowest detection limit is 10^{-8} g/ml of n-C12 hydrocarbon.

C. IONIZATION DETECTORS

All ionization detectors have the same base body. They are all miniaturized. They are not universal with the exception of the helium ionization detector.

1. Flame Ionization (FID) (Figure 4.2)

FID detects C and H. However, a response is observed for some other elements.

The measuring effect is based on ionization of carbon/hydrogen organic substances burned in an oxyhydrogen flame.

Principle of operation: a small hydrogen-air flame burns at a capillary jet. In the hottest part of a flame at high temperature, a certain amount of radicals are created (a few ions per million molecules). It generates a current between two electrodes. A collector electrode is located a few millimeters above the flame and the ion current is measured by establishing a potential between the jet tip and the collector electrode. When a carbon compound is eluted from the GC column into the hydrogen flame of the detector, current will pass between electrodes placed near the flame.

$$H_2 + 2O_2 \rightarrow 2O^{\cdot} + 2OH^{\cdot}$$

$$H_2 + \frac{1}{2}O_2 \rightarrow H^{\cdot} + OH^{\cdot}$$

$$H_2 + OH^{\cdot} \rightarrow H_2O + H^{\cdot}$$

$$CH^{\cdot} + O^{*} \rightarrow CHO^{+} + e^{-}$$

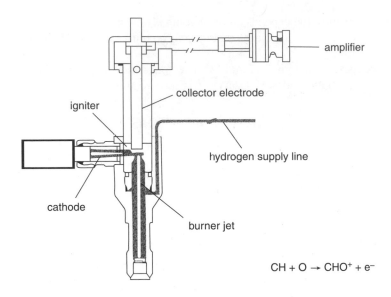

amplifier

collector electrode

igniter

hydrogen supply line

cathode

burner jet

$$CH + O \rightarrow CHO^+ + e^-$$

FIGURE 4.2 A scheme of flame ionization detector (by courtesy of Varian).

Empirical rules give contributions to effective carbon number:

C (aliphatic)   1.0
C (aromatic)   1.0
C (olefinic)   0.95
C (acetylenic) 1.3
C (nitrile)   0.3

The required ionization energy to form carbon ions in an oxyhydrogen flame mostly results from the high carbon oxidation energy released during the combustion reaction of carbon to carbon monoxide and carbon dioxide. The flame temperature itself is insufficient for a direct atom or molecule ionization.

The combustion of hydrogen produces some radicals and ions such as H_3O^+ and OH^- at such low concentrations that the effect on the ionization current is negligible. FID detectors exhibit a linearity over seven orders of magnitude.

FID detectors are more and more miniaturized and a micro FID where both oxygen and hydrogen are generated by a miniaturized electrolysis cell has been devised.[11]

0.1 pg hydrocarbon can be detected.

Modern FID detectors are miniaturized and work at 100 Hz.

Flame laser enhanced ionization and flame laser induced atomic fluorescence can be used as sensitive detectors for organo tin compounds.

2. Electron Capture Detector (ECD) (Figure 4.3)

ECD is very sensitive to any electrophilic compounds and particularly well suited for organochlorine species. It is very widely used to detect chlorinated compounds.

Principle of operation: a ^{63}Ni source emits a β electron beam. A current between two electrodes is generated. When electrophilic species enter the detector, a decrease in the detector background current is observed due to the capture of the electrons by the electrophilic species.

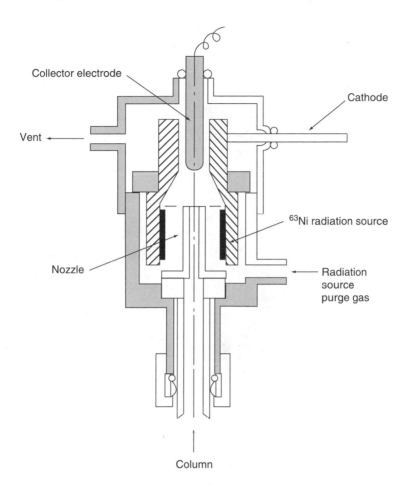

FIGURE 4.3 A scheme of electron capture detector (by courtesy of Shimadzu).

Constant current ECD is the most common mode of operation but fixed frequency is gaining acceptance. The thermal electron concentration in the detector cell is measured discontinuously by a pulsed voltage. A pulsed discharge ECD a nonradioactive source is more sensitive than the radioactive source for most compounds, covers a wide dynamic response range similar to the radioactive source. Cell volumes are typically 480 μl to 1.5 ml with 150 μl for the micro ECD at a data acquisition frequency of 50 Hz.

Detection limit: 10 to 15 pg Cl/sec, e.g., 8 fg lindane/sec. The common ECD can detect one sulfur compound (SF_6) at 0.2 fmole. Addition of ammonia to the nitrogen makeup gas may increase the response for various chlorinated compounds. The major drawback of ECD is its poor linearity. Multilevel calibration method in which five or six different dilutions were randomly spread over a series of samples injected is generally well accepted.

3. Pulsed Discharge Helium Ionization (PDHID)

PDHID is one of the most sensitive detectors available for GC.

Principle of operation: photo emission in pure helium arises from excited states of He_2 and consists of a continuum extending from 11.6 to 21.7 eV (Figure 4.4).

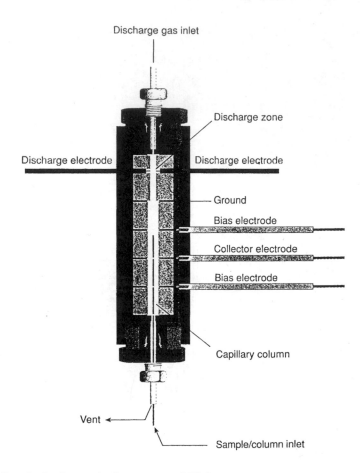

FIGURE 4.4 Helium ionization mode (by courtesy of Vici).

Since these energies are greater than the ionization potentials of all atoms and molecules, the photoionization detector is a universal detector. The photo emission distribution can be characterized by energies at half maximum (14.1 to 16.7 eV). This range contains 66% of the photon emission. The molar responses are correlated to the number of ionizable electrons in a molecule. See details in Ref. 12. PDHID is a truly universal detector capable of detecting H_2, O_2, CO, CO_2, H_2O as well as organic compounds. The PDHID has several variable operating parameters that affect its sensitivity and the linearity of response with concentration: pulse interval and power, the potential applied to repel the electrons to the collecting electrode, and the helium flow rate through the discharge region.

Detection limit is 1 to 20 pg (0.3 ppm for hydrogen). A fiber optic multiphoton ionization detection is able to detect 0.12 ng PAH.

4. Chlorine Sensitive Pulsed Discharge Emission Detector

It is based upon molecular emission from $KrCl^*$. Low concentrations of krypton in helium (0.1 to 0.4%) are sufficient to react with chlorinated compounds in the pulsed discharge emission detector to produce an excited state of $KrCl^*$ that emits in a fairly narrow molecular band at 221 to 222 nm. Spectrometers with conventional quartz optics are sufficient to detect this emission.

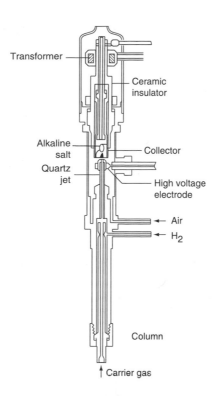

Transformer

Ceramic insulator

Alkaline salt

Quartz jet

Collector

High voltage electrode

Air

H₂

Column

↑ Carrier gas

FIGURE 4.5 A scheme of the thermionic detector (by courtesy of Shimadzu).

5. Thermionic Detector (Figure 4.5)

Principle of operation: adding an alkali metal salt to a flame enhances the response to compounds containing N_2, P, S. The alkali source is an electrically heated ceramic bead of a sintered complex of an alkaline salt and silicate. The usual salt is rubidium silicate. The mechanism is not fully understood. Gas phase reactions involve free alkali metal atoms in the flame that are ionized by collision with carrier gas molecules.

$$A + M \rightarrow A^+ + e^- + M$$

Free radicals resulting from the pyrolysis of organic compounds containing P or N react with alkali metal atoms. Frequent replacement of the alkali source is still necessary. To overcome this drawback the alkali salt may be dissolved in water and introduced in the detector sensing volume as an aerosol or by means of a syringe pump.

A halogen specific detection method is based on halogen induced thermal electron emission.

Detection limit 10^{-13} g of N/sec; 5.10^{-14} g of P/sec. The high sensitivity to nitrogen and phosphorous compounds makes this detector suited for pesticide residues and pharmaceuticals.

6. Surface Ionization Detector

The organic molecules from the GC are seeded in a hydrogen or helium supersonic beam and enter the vacuum chamber through a ceramic nozzle. The distance from the top of the nozzle to the surface is roughly 5 mm. In the vacuum chamber, the beam collides with ReO_2 or Pt surface for efficient positive ion production. The surface is always at a positive potential of 200 V against the collector electrode. The kinetic energy of the sample molecule, which is proportional to the nozzle temperature, and the surface temperature are the most relevant parameters. ReO_2 gives a 20 times higher sensitivity as positive ion-emitting surface.

The sensitivity is expressed as Coulomb per g of sample.

Limit of detection is in the nanogram range (e.g., 10^{-13} g/sec for pyrene) (linear dynamic range 10^6).

7. Ion Mobility

The first successful use of ion mobility spectrometer (IMS) as detector in GC was in 1982.

Ion mobility spectrometry provides a rapid response to trace gases by converting sample molecules to ions at atmospheric pressure and by characterizing these ions with the help of their gas phase mobilities in weak electric fields.

Next to radioactive isotopes such as, ^{63}Ni, ^{3}H, and ^{241}Am, which are still the ionization sources most commonly employed in IMS, other sources like photo-ionization, corona, or partial discharges, electrospray ionization, and flames have become increasingly popular. However, despite the rising number of regulatory requirements going along with the use of radioactive material, no nonradioactive ionization source unsurpassed the others because of their unique combination of simplicity, long-term stability, and robustness. Recently, manufacturers managed to phase out ^{63}Ni sources. Radio frequency IMS analyzer can be used as a small detector in GC separations of volatile organic compounds since it provides a second dimension of chemical identity.

Ion mobility spectrometers consist of three parts, namely an ionization region, a drift region separated from the ionization region by an ion gate (shutter grid), and a detector. Gaseous samples are transported by a carrier gas into the ionization region where, in the case of a radioactive source, carrier gas molecules are ionized by radiation. So-called reactant ions are created, which undergo a series of reactions with molecules of the analyte to generate product ions that are directed by an electric field E.

D. Photometric Detection

Photometric detectors can be divided into three classifications: emission, absorption, and scattering.

1. GC/AED (Atomic Emission Detector)

An AED detector is a multielement detector capable of detecting elements with atomic emission lines in the vacuum UV, UV–VIS, and near IR portions of the electromagnetic spectrum.

AED allows multielement measurement.

Plasma sources are capable of producing intense emission from the elements. Types of plasma used in chromatographic detection are microwave induced plasmas (MIP) and inductively coupled plasma (ICP). An argon plasma is sustained in a microwave cavity which focuses into a capillary discharge cell. The most widely used cavities are cylindrical resonance cavities and "surfatron" that operates by surface microwave propagation along a plasma column. Atmospheric pressure cavities are very simple to interface with capillary GC columns.

Other plasmas are glow discharge plasmas, and direct current plasmas with a continuous Direct Current arc. A typical AED uses a 50 W microwave generator and a reentrant cavity to focus the energy into a 1 mm i.d. fused silica tube in which a plasma is sustained by a steady flow of helium makeup gas. A spectrometer employing a diffraction grating and a movable photodiode array (PDA) views the plasma axially and can detect the emitted radiation in the 160 to 800 nm region with a 0.1 nm resolution at 400 nm. All major hetero atoms, the halogens, and most metals (e.g., Pb, As, Sn, Hg) can be detected with high sensitivity (LOD 0.1 to 30 pg/sec). In the Pulsed Discharge Emission Detector (PDED) the GC effluent is passed directly into the discharge and the resulting emission spectra are observed. Coupling with a vacuum UV monochromator allows observations of atomic emissions, e.g., Cl, Br, I, and S.

2. Flame Photometric Detector (FPD) (Figure 4.6)

Principle of operation: a flame breaks down large molecules. The high temperature of the flame stimulates atoms and species that are brought to an excited state (S_2^* or PO^*) and relax with emission of a light of characteristic wavelength. In a common burner design, the flame burns on a set of concentric tubes that deliver the reagent gases.

This detector is well adapted for sulfur, phosphorus, or tin determination. Two flames are often used to separate the region of sample decomposition to sample emission. Response is dependent on the environment of the sulfur atoms (thiols, sulfides, disulfides, thiophenes). The FPD can also detect iron.

Limit of detection is around 10^{-12} g P/sec or 10^{-10} g S/sec.

3. GC/FTIR (Fourier Transform Infra Red)

MS cannot distinguish closely related structural isomers because they exhibit very similar mass spectra. Infrared (IR) spectroscopy provides information on the intact molecule. There are three basic types of GC–FTIR instruments: (a) light pipe, (b) matrix isolation, and (c) subambient trapping. A light pipe is a narrow bore (100 to 200 μm i.d.) borosilicate capillary with a smooth thin layer of gold coated on the inside surface. Reflection occurs with gold coating thus increasing path length of the cell by a factor of ten or more according to Beer's law. A schematic of GC/FTIR instrumentation is displayed in Figure 4.7.

In the Michelson interferometer a collimated light beam is divided at a beam splitter into two coherent beams of equal amplitude that are incident normally on two plane mirrors. The reflected beams recombine coherently at the beam splitter to give circular interference fringes at infinity focused by a lens at the plane of the detector (see figure on GC–FTIR).

For monochromatic light of wavelength λ_0 and intensity $B(\lambda_0)$ the intensity at the center of the fringe pattern as a function of the optical path difference x between the two beams is given by

$$I_0 = B(\lambda_0)\left[1 + \cos\frac{2\pi x}{\lambda_0}\right] = B(\sigma_0)[1 + \cos 2\pi\sigma_0 x]$$

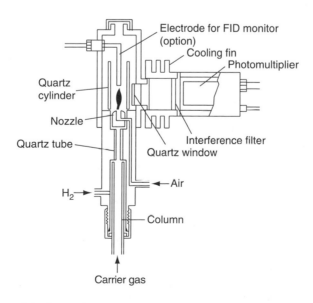

FIGURE 4.6 A scheme of the flame photometric detector (by courtesy of Shimadzu).

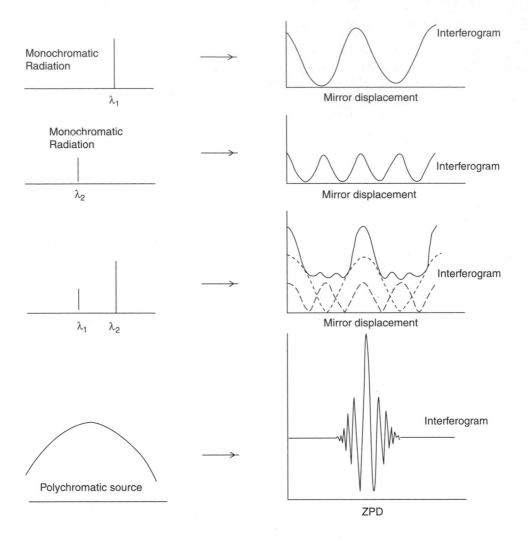

FIGURE 4.7 The Fourier transform signal.

where $\sigma = \nu/c$; σ is the wavenumber, ν is the frequency of the light in $\sec^{-1}$, c is the speed of light in cm $\sec^{-1}$.

If x is changed by scanning one of the mirrors, the recorded intensity (the interferogram) is a cosine of spatial frequency σ_0. Its temporal frequency is given by $f_0 = \nu \cdot \sigma_0$ where ν is the rate of change of optical path. If the source contains more than one frequency, the detector sees a superposition of such cosines (Figure 4.8).

$$I_0(x) = \int_0^\infty B(\sigma)(1 + \cos 2\pi\sigma x)\, d\sigma$$

Subtracting the constant intensity $\int_0^\infty B(\sigma)\, d\sigma$ corresponding to the mean value of the interferogram $\langle I(x) \rangle$ yields:

$$Ix = I_0(x) - \langle I(x) \rangle = \int_0^\infty B(\sigma)\cos (2\pi\sigma x)\, d\sigma$$

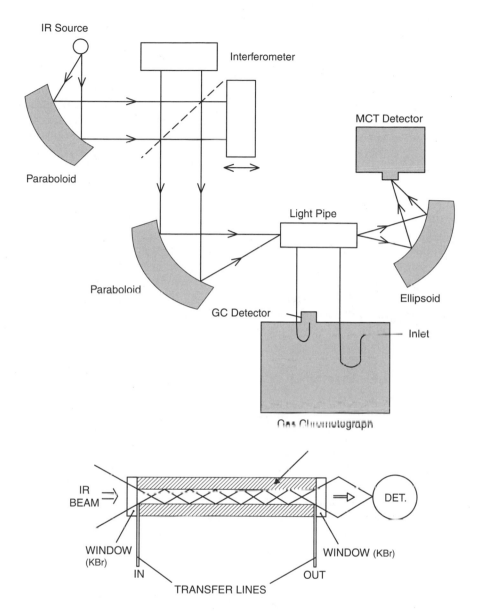

FIGURE 4.8 A scheme of typical GC–FTIR with light pipe.

The right hand side contains all the spectral information in the source and is the cosine Fourier transformed of the source distribution $B(\sigma)$.

The latter can be recovered by the inverse Fourier transform

$$B(\sigma) = \int_0^\infty I(x) \cos(2\pi\sigma x)\, dx$$

Actually the Fourier transform reproduces $B(\sigma)$ and adds a mirror image $B(-\sigma)$ at negative frequencies.

In fact, the interferogram is never totally symmetric about $x = 0$ and to recover the full spectral information, it is necessary to take the complex rather than the cosine Fourier transform. The interferogram is recorded to a finite path difference L rather than infinity. It is actually recorded by sampling it at discrete intervals Δt.

The matrix isolation technique involves mixing a gaseous sample with an inert gas, usually argon, and cryogenically freezing it onto a rotating gold disk maintained at liquid He temperature to form a solid matrix trace approximately 300 μm wide. Reflection–absorption spectra are obtained and very sharp peaks are produced. The detection limits of matrix isolation are in the tens of picogram range. In the subambient trapping method, the effluent is frozen onto a moving IR transparent window, usually made out of zinc selenide. One advantage of subambient trapping is that the IR spectra obtained can be searched against standard KBr spectra.

A sensitive technique used for real time reconstruction of chromatograms from the interferogram is the Gram Schmidt vector orthogonalization method. The Gram Schmidt method relies on the fact that the interferogram contains information on absorbing samples at all optical retardations less than the reciprocal of the width of each band in the spectrum.

In off-line systems, analytes eluting from the GC column are frozen as pure substances onto the surface of an IR transparent Zn/Se window. Immediately after deposition, peaks are passed under a transmittance IR microscope and scanned. This offline procedure permits rescanning.

4. Chemiluminescence Detectors

Chemiluminescence is the production of electromagnetic radiation (UV, VIS, or IR) by a chemical reaction between at least two reagents, A and B, in which an electronically excited intermediate or product C^* is obtained and subsequently relaxes to the ground state with emission of a photon or by donating its energy to another molecule that then luminesces. The intensity of light emission depends on the rate of the chemical reaction, the yield of excited state and the efficiency of light emission from the excited states.

A simple oxidative combustion does not generate a sulfur chemiluminescent species since SO_2 does not chemiluminesce with ozone. Sulfur chemiluminescence detection has the advantage that SO is produced during FID operation. When SO reacts with ozone, a strong blue chemiluminescence signal is emitted by the resulting SO_2^*. The signal is isolated from other radiations and detected by a photomultiplier tube.

Detection limit is around 10 pg of sulfur.

In CLND, (pyrochemiluminescent nitrogen detection), components eluting from the column undergo high temperature (1000°C) oxidation. All nitrogen containing compounds are converted into nitric oxide NO. The resulting gases are dried and mixed with ozone in a reaction chamber. This results in the formation of nitrogen dioxide NO_2^* in the excited state. Light is emitted by the chemical reaction and detected by a photomultiplier tube. Under optimized conditions the released radiation energy is proportional to the NO concentration.

$$NO + O_3 \rightarrow NO_2^* + h\nu(\text{NIR})$$

Ions travel towards the ion gate which periodically opens to permit a swarm of ions to enter the drift tube. While colliding with a counterflow of uncharged drift gas molecules, the ion swarm is separated into small clouds of ions according to their individual mobilities.

By measuring the drift time t_d needed by ions to overcome the distance l_d between the shutter grid and the detector (a Faraday plate), mobilities K are determined.

E. ELECTROCHEMICAL DETECTORS

Electrolytic conductivity detector (Hall detector) relies on the absorption of ionizable gases into liquid for conductivity measurement. These detectors are rarely advocated in EPA methods probably because the electrolyte must be kept extremely clean.

Limit of detection with sulfur is 1 pg of sulfur.

F. GC–ICP–MS (INDUCTIVELY COUPLED PLASMA–MASS SPECTROMETRY)

1. Principle of Operation

A plasma is a very hot gas in which a significant fraction of the atoms or molecules are ionized. A plasma is able to react by its ionic nature when it is submitted to electromagnetic beams. A plasma surrounded by a time varying magnetic field is inductively coupled, i.e., current flows are induced in the ionized medium. These current flows cause ohmic (resistive) heating of the plasma gas enabling the plasma to be self sustaining. At room temperature argon gas does not contain any ions to initiate the plasma formation. To create a small number of ions a high voltage discharge is required. When species are entering the plasma, nearly complete atomization occurs. In optical spectroscopy the plasma serves as excitation source and is combined with an optical spectrometer for selective elemental detection by observation of characteristic atomic emission line spectra. In the plasma positively charged gas ions Ar^+ and electrons e^- are produced. Singly charged ions are generated with a degree of ionization of over 80% for the majority of elements.

2. Instruments

In practice the source arrangement commonly used consists of a quartz tube surrounded by a multiturn copper induction coil connected to a radio frequency (RF) generator. The generator operates in the 20 to 50 MHz range with a variable output power of up to 2 kW. For hot plasmas RF power is 1200 W for cool plasmas 600 to 800 W. A Tesla coil initiates the operation. The plasma is prevented from touching the walls of the quartz tube by a thin screen of cool gas. The RF is tuned in such a way that a toroidal shape of plasma is obtained. In this way the axial zone in the center is relativity cool in comparison to the periphery. A gas stream containing sample aerosol is injected in the center of the toroid without disturbing plasma stability. A pneumatic nebulizer is utilized. Microconcentric pneumatic nebulizer operates at 30 μl/min. It is of primary importance to yield ideal electrostatic conditions for ion production while keeping a stable plasma. Automatic positioning of the torch and knitted induction coil increase reproducibility of ion production. Physical interferences are often caused by samples that contain high levels of dissolved solids such as sea water. Chemical interferences result from charges in vaporization or ionization (due, for example, to high amounts of sodium). Spectral interferences in atomic emission are caused by a continuum emission or overlapping emission light. Sequential and simultaneous multichannel instruments are available.

ICP–MS uses ICP as an atmospheric pressure ionization source. The use of ICP as an ionization and excitation source is largely determined by the shape of the plasma. The ions are transported through successive pumping stages into the mass spectrometer at low pressure. The plasma mass spectrometer interface is an ion lens system. The mass spectrometer may be set to monitor the isotopic signal of an element and the resulting chromatogram shows peaks which must contain the isotope of intent. In the HP 4500 (from Agilent), the lens system bends the ion beam into off axis quadrupole rather than defocusing the ion around a conventional photon stopper. High ion transmission is the requirement. The charge on the element's ions from the plasma source is usually $+1$. Given a mass resolution of 1 amu, one might expect minimal spectral overlap or complexity of the mass spectra.

GC–ICP–MS has the potential to facilitate simultaneous multielemental speciation analysis because species of Se, Pb, Hg, and Sn have volatile forms and can be analyzed in a single run. The use of ICP–MS as a detector enables calibration by isotope dilution mass spectrometry as well as providing very low limits of detection (pg–ng range).

IV. LC DETECTORS

A. OPTICAL DETECTORS

1. UV–VIS

UV–VIS is the most popular, and relies on Beer's law.

The total interaction index describing the interaction of light with matter has two parts. One is concerned with the change in intensity of an incident beam as it passes through an absorbance medium, the other derives from the associated change in the speed of light. The former is measured as an absorbance, the latter as the refractive index of a solute in a solvent.

UV–VIS: UV–VIS spectra arise from electronic transitions within molecules. Broad absorption bands are usually observed due to the contribution of vibrational and rotational energy levels. The principal characteristics of an absorption band are its position and intensity. The intensity of an absorption band is expressed by the transmittance $T = I_0/I$ where I_0 is the intensity of the radiant energy and I is the intensity of the radiation emerging from the sample. The Beer–Lambert law is expressed as

$$\log_{10}T = A = \varepsilon l c$$

where ε is the molar absorptivity of solute, l is the path length through sample, c the concentration of solute, and A is the absorbance.

The most popular LC detector is the UV absorbance detector because it is easy to use and has broad application due to the fact that many organic solutes exhibit a UV absorbance and to the possible selectivity of derivatization. These detectors measure changes in absorbance of light in the 190 to 350 nm region or 350 to 700 nm region.

Basic instrumentation includes a mercury lamp with strong emission lines at 254, 313, and 365 nm, cadmium at 229 and 326 nm, and zinc at 308 nm. Deuterium and xenon lamps exhibit a continuum in the 190 to 360 nm region requiring the use of a monochromator. A filter or grating is used to select a specific wavelength for measurement. Cutoff filters pass all wavelengths of light above or below a given wavelength. Band pass filters pass light in a narrow range (e.g., 5 nm). In the single beam mode, the energy from the source lamp passes through the sample flow cell to a photocell via some wavelength selection device. The double beam system is preferred. The optoacoustic filter has no moving parts, wider spectral tuning range, higher throughput, and higher resolution as compared to conventional grating monochromators.

> Flow cell is typically 8 μl with a 10 mm path length.
> Photodiode array (see below) is now the best sensor.

According to Beer's law, the higher the path length the higher the transmitted light. Most cells are Z shaped.

With capillaries such as LC capillaries or CE capillaries, there is only limited path length. A free portion of capillary is brought into the light path of a UV absorbance detector. When the aperture of the source is adjusted to the inside diameter of the capillary the effective light path is $I_{\mathrm{eff}} = 1/2\pi r$ where r is the radius of the capillary.

A U cell design provides a longer longitudinal light path and a substantial increase in signal to noise ratio.

UV detectors can be subject to baseline shifts due to changes in the refractive index of the carrier solvent. This effect can be bothersome when a gradient is carried out.

Limits of detection are highly dependent on the molar absorptivity of the solutes (see Beer's law). UV–VIS detectors must be checked for wavelength accuracy, absorption accuracy, scattered light, and spectral resolution. UV detectors are more accurate and precise than MS detection but they afford only minimum information regarding the identity of the analytes.

2. Photodiode Array (Figure 4.9)

The detection of structurally similar impurities eluting simultaneously with the analytes of interest is a problem. The analyst must detect the existence of peaks of interest, determine the extent of their purity and confirm their identity.

Photodiode operation relies on the photovoltaic effect. In the typical photodiode there are two components of semiconductor, called P and N. P is a very pure silicon with low levels of three valent impurities such as boron or gallium. Each impurity atom can accept an electron from the valence bonds giving rise to a hole that can take part in the electrical conduction process and an immobile negatively charged impurity ion. Since the hole is positively charged such a material is a P (positive) silicon crystal. If the impurity added is a pentavalent atom (As), the atoms behave as donors of electrons that can move through the entire silicon crystal. It is thus a N (negative) type. A photon of wavelength less than 1.1 nm is able to break a covalent bond between the silicon atoms. The free electron formed is free to move with the missing electron in the broken valence bond inducing electrical conduction by repeated replacement. The PDA detector passes the total light through the flow cell and disperses it with a diffraction grating. The dispersed light is measured by an array of photosensitive diodes. Diode arrays having number of elements ranging from 128 to 1024 and even up to 4096 are available. Adjacent diodes are usually 25.6 mm long and spaced 25 mm on centers. The array of photodiodes is scanned by the microprocessor (16 times a second is usual). The readings for each diode are summed and averaged. PDA detector can simultaneously measure the absorbance of all wavelength vs. time. The amount of data storage is a key feature in PDA. In a PDA instrument having a 1000 photodiode array, 1000 data points can be measured in 1 sec and it would take 1/1000 sec to achieve the same result obtainable in 1 sec in a conventional UV instrument. A run can easily take several megabytes of data storage. Dynamic range is usually 0.5 mAU to 2.0 AU.

Peak purity is based on the proprietary spectral contrast algorithm that converts spectral data into vectors used to compare spectra mathematically. This comparison is expressed as a purity angle. The purity angle is derived from the combined spectral contrast angles between the peak apex spectrum

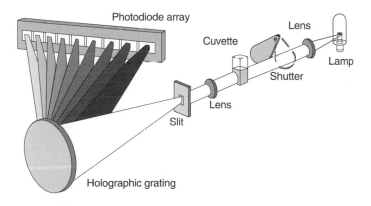

FIGURE 4.9 A photodiode array detector.

and all other spectra within that peak. To determine peak purity, the purity angle is compared to the purity threshold. For a pure peak the purity angle will be less than the purity threshold.

Spectral deconvolution techniques are used when two peaks coelute.

Identification of peaks is performed by comparison with spectra contained in a library of standards.

3. Fluorescence Detection

Fluorescence is a three-stage process.

The first step is excitation via the absorption of radiation when a photon of energy $h\nu$ is supplied by a source (lamp or laser). This process distinguishes fluorescence from chemiluminescence where the excited state is created by a chemical reaction.

Excitation occurs when the energy of the incident radiation corresponds to the energy spacing between the ground and one of the excited singlet states S_1, S_2, S_n and T_1, T_2, T_n for the excited triplet states, which by Hund's rule are lower in energy than the corresponding singlet states. A singlet state is one where all the electrons in a molecule have a paired electron with opposite sign. A triplet state exists when two unpaired electrons have the same spin.

The excited state exists for a very short time (typically 10^{-10} to 10^{-9} sec). This is the second step.

The third step is emission.

Emission is seldom observed from the higher singlet state because a radiationless process known as internal conversion results in a S_n to S_1 transition. From the excited singlet stage, S_1, a variety of transitions may occur. The most important are:

1. Radiationless internal conversion to S_0.
2. Radiationless intersystem crossing to the triplet state T_1.
3. Radiative transition to S_0.

These three transitions are all competing processes and only the third one leads to fluorescence (Figure 4.10). Process 2 under specialized conditions can lead to phosphorescence.

The quantum yield ϕ is a fundamental molecular property that describes the ratio of a number of photons emitted to the number of photons absorbed.

$$\phi = \frac{k_f}{k_f + \Sigma k_d}$$

where k_f is the rate constant for fluorescence emission, Σk_d is the sum of the rate constants for all the nonradiative processes that can depopulate S_1.

The rate constant for radiative transition should be large relative to those for nonradiative transitions.

Numerous factors can affect molecular fluorescence, e.g., the type of solvent, the pH etc.

A molecule exhibiting high fluorescence does not contain functional groups that enhance the rates of radiationless transitions. In addition such a molecule should possess a high molar absorptivity (ε).

The signal intensity I_f is given by Beer's law

$$I_f = I_0(1 - e^{\varepsilon lC})\phi k$$

When sample absorbance is small this expression is reduced to

$$I_f = I_0 \cdot 2.3 \cdot \varepsilon lC \cdot \phi k$$

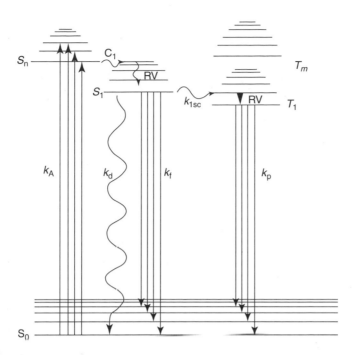

FIGURE 4.10 Jablonski diagram: k_A, Excitation rate constant; k_1, de-excitation rate constant (pathway No. 1); k_p, phosphorescence rate constant; S_0 is for ground state, S_n, S_1 are for singlet state, T is for triplet state.

where k is the instrumental efficiency for collecting the fluorescence emission and I_0 is the intensity of the incident radiation.

a. Excitation Sources

Gas discharge lamps containing deuterium, mercury, zinc, cadmium, or xenon are the most common sources. They exhibit high spectral radiance and good stability.

> Xenon lamp provides continuous emission in the 250 to 300 nm range; deuterium in the 200 to 300 nm range.
> Mercury exhibits lines at 254, 365, and 405 nm.
> Zinc exhibits lines at 214, 308, and 335 nm.

Laser radiation is monochromatic and the output beams of lasers are highly collimated. With a laser as the excitation source larger fluorescence signal levels are observed and nonlinear excitation is possible.

b. Excitation Wavelength Selection

Filters or monochromators are generally used. Filters are less expensive but monochromators provide greater versatility and selectivity for excitation. Grating monochromators have a constant band pass regardless of wavelength selection. Gratings are either ruled or holographics.

c. Emission Wavelength Selection

There are several sources of radiation that must be selectively prevented from reaching the photomultiplicator. These included Rayleigh scattering, Raman scattering, second order radiation, and solvent impurity emission. Appropriate emission wavelengths can be selected with monochromators or filters. Filters generally offer greater sensitivity than monochromators.

Fluorescence emission provides more selectivity and increased sensitivity compared to UV absorption. Xenon lamps are far superior to Hg lamps or D_2 lamps as light source. Selection of excitation or emission wavelength is done by a monochromator.

Laser induced fluorescence is of current use. Various lasers are utilized (e.g., He–Ne, diode, Argon ion). The diode laser seems the best choice. Due to the highly collimating nature of lasers, most scattering sources are eliminated. Increase of detection is carried out with pre or post column derivatization (see, for example, derivatization of amines with 9-fluorenyl chloroformate).

Attomole detection is possible.

One unique feature of fluorescence detection is that emission spectra can be utilized to reveal structural information of unknown compounds.

4. Infra Red

When coupling HPLC to IR detection the absorption of infrared radiation by the mobile phase results in strong spectral interferences, especially in the case of aqueous eluents. Using FTIR the optical path length must be in the low micrometer range (typically between 10 and 50 μm) to be able to perform measurements in the water window of the fingerprint region between 1600 and 950 cm^{-1}. The consequence of the short optical path length is a limitation in sensitivity. To overcome this shortcoming quantum cascade lasers are used.

5. Raman

Raman spectroscopy (RS) is making a comeback as a detection technique. Raman signals are obtained by irradiating a sample with monochromatic radiation and measuring the small portion of scattered radiation that is inelastic, i.e., has shifted in wavelength. The process is not efficient and the signal intensities are proportional to λ^{-4}. To enhance the sensitivity, special modes of Raman spectroscopy are designed: Resonance RS, Surface enhanced RS (SERS). The choice of the laser source is critical: red or near infra red (785 nm), Nd, or YAG laser are utilized. SERS can be coupled to LC via a TLC plate. A low percentage (3%) of the effluent from the LC column is immobilized on a moving TLC plate using a spray jet solvent elimination interface. Next, colloidal Ag is applied to the analyte spots and *in situ* SERS spectra are recorded with a multichannel micro Raman spectrometer.

6. Light Scattering Detector (LSD) (Figure 4.11)

Principle of operation: It is a three step process. The effluent of the LC column is vaporized in a nebulizer by means of a gas. The droplets pass through a heated drift tube at a temperature of 40 to 250°C, and the only particles left are the analyte and the solvent impurities. A laser (typically 1 mV He/Ne) irradiates the particles, and the scattered light is collected by a glass rod and transmitted to a photomultiplier tube at a fixed angle from the incident light. The light measured is proportional to the amount of sample in the light scattering chamber.

Parameters affecting the response are particle size, degree of nebulization (most critical), and nature of the solvent. When an LSD detector is used to detect thermally labile compounds, the temperature used to evaporate the mobile phase is critical. If the temperature is too high, the compounds of interest can be thermally decomposed and reduce the sensitivity of the assay. Temperature range of the nebulizer is 40 to 220°C. The design of the evaporation tube is critical. The amount of scattered light depends strongly on the molar absorptivity of the solute. The light-scattering detector is a universal detector but not a mass detector. Its response is nonlinear. The observed peak area (A) is related to the quantity of analyte oncolumn (m) through the relationship $A = am^x$ where x is the slope of the response line and a is the response factor. The calibration curve

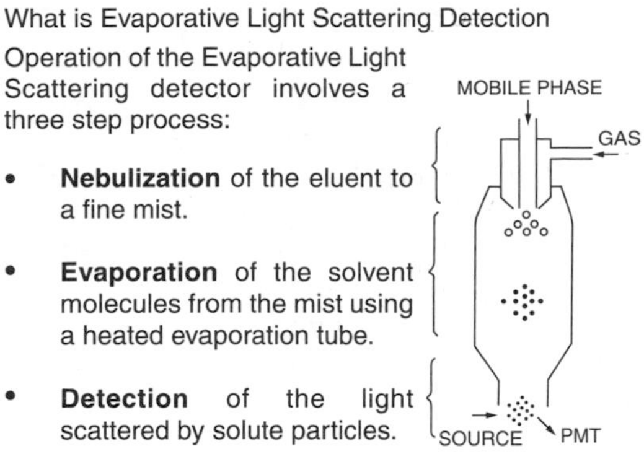

What is Evaporative Light Scattering Detection

Operation of the Evaporative Light Scattering detector involves a three step process:

MOBILE PHASE

GAS

- **Nebulization** of the eluent to a fine mist.

- **Evaporation** of the solvent molecules from the mist using a heated evaporation tube.

- **Detection** of the light scattered by solute particles.

SOURCE   PMT

FIGURE 4.11 The light scattering detection.

is log–log. It can be easily used with a gradient. A mobile phase suitable for MS detection is also suitable for LSD. The detector presents negligible back pressure and is well suited to analytes that lack chromophores. There is no need for derivatization but it is a destructive detector. The detector is suited for lipids and sugars. The LOD is about 20 to 100 ng.[13]

7.   Refractive Index Detector (RI)

RI is one of the very few universal detectors available. RI detector monitors both the eluent and the analyte. The output reflects the difference in refractive index between a sample flow cell and a reference flow cell. The refractive index of a mixture is given by

$$\eta - \eta_s + C(\eta_s^2 + 2)^2/6\eta_s \cdot M_s/\rho_s[(\eta_a^2 - 1)/((\eta_a^2 + 1) - (\eta_s^2 - 1))/(\eta_s^2 + 2)]$$

where η_s, M_s, ρ_s are the refractive index, molecular mass and density of the solvent, respectively, while C is the analyte concentration, and η_a is the refractive index of the analyte.

Obviously only analytes with an RI different from that of the solvent can be detected. The signal can be positive when $\eta_a > \eta_s$ or negative when $\eta_a < \eta_s$.

The measured RI response is determined by the volume fraction of the analyte in the flow cell (x) and the volume fraction of the eluent in the other flow cell ($1 - x$).

$$\eta - \eta_2 = v_1(\eta_1 - \eta_2)$$

where v_1 is the volume fraction of the analyte, η_1 refractive index of pure analyte, η_2 is the refractive index of pure solvent (contained in a reference cell) and refractive index of solution in sample cell.

There are four types of RI detectors:

- Deflection type which is by far the most popular. It relies on Snell's law governing the angles of incidence and refraction at an interface

$$\eta_1 \sin \theta_1 = \eta_2 \sin \theta_2$$

where θ_1 is the angle of the beam with respect to the normal of the interface in the medium with RI of η_1.
- Reflection type according to Fresnel's law of reflection. Fresnel's law describes the reflectivity and transmittance of light at an interface for the two types of linearly polarized light. Measurement of $\Delta\eta$ is a measure of change in reflectivity.

- Interference type (utilized in capillary LC).
- Christiansen effect type.

Refractive index back scattering (RIBS) is based on interferometry. The difference in phase of two coherent light beams is monitored. RIBS detection is based on the fact that the position of the fringes will shift if analyte molecules pass through the irradiated volume. He–Ne lasers are usually the light source, detection is generally performed by using the brightest interference fringe. A RI change induces a displacement of the fringe, which is recorded by a position-sensitive device.

Refractive index is very sensitive to temperature and pressure. For that reason the reference and the measurement cell are close since the difference in temperature between them is critical. Specifications of a RI detector are: refractive index range, linearity range, cell volume, maximum pressure in cell, temperature control.[14]

$$10^4 d\eta/dt = 0.67 \text{ for water} = 6.84 \text{ for dichloromethane}$$

$$10^5 d\eta/dP = 1.53 \text{ for water} = 5.56 \text{ for dichloromethane}$$

With usual RI detectors the LODs are about 10^{-6} RI units.

8. Optical Activity

The classical method of determining enantiomeric purity is by comparison of the optical rotation of an enantiomerically enriched sample with the value of the enantiomerically pure antipode. Chiral compounds rotate a plane of polarized light. When plane polarized light is passed through a solution containing an optically active compound, it will be rotated in a clockwise or counter-clockwise direction. According to Biot's formula, $^{\alpha}D = \alpha/l{\cdot}c$ where $^{\alpha}D$ is the specific rotation of the compound, α is the observed rotation in degrees, l is the light pathlength in dm, and c is the concentration of the compound in g/ml under conditions of temperature and wavelength. The molecular rotation is

$$\Phi_\lambda = {}^{\alpha}D{\cdot}M_w/100$$

Molecular circular dichroism is only observed in regions of light absorption and is quantified by a differential circular polarization extinction coefficient given by Beer's law

$$\Delta\varepsilon = (A_L - A_R)/cl = \Delta A/c$$

where A_L, A_R are the absorbances of left and right circularly polarized light, respectively, c is the concentration of the optically active species, and l is the pathlength of the cell. The ratio of this differential coefficient and the ordinary molar extinction coefficient is known as the Kuhn dissymmetry number. Circular dichroism detection is usually best achieved at the wavelength of maximum ordinary absorption about which optical rotation averages to zero.

The classical method of determining enantiomeric purity is by comparison of the optical rotation of an enantiomerically enriched sample with the value of the enantiomerically pure antipode. An optical rotation detector measures an angle of rotation when linearly polarized light passes through a flow cell containing optically active compound. This is due to the difference in refractive indices between right and left circularly polarized lights. The optical rotation of a chiral compound is greatly changed at its absorption band. Modulated polarimeters are very effective in measuring small optical rotations when they are used with low volume flow cells.

A circular dichroism detector (CD) can differentiate between enantiomers by measuring the difference in absorbance of right and left-handed circularly polarized light. Unlike single beam measurements made by optical rotation based detectors, the CD measurements of differential absorption is performed within 20 μsec, resulting in virtual dual beam detection. A multibeam circular dichroism detector utilizes the optical system of a conventional PDA.[15]

A CD detector instrument includes a Hg–Xe lamp providing a wavelength range of 220 to 420 nm with a 20 nm spectral bandwidth or an Ar laser. Sensitivity (given by Jasco Instrument) is 0.1 μg of camphor sulfonic acid at 290 nm and 0.01 μg for pantoyl lactone at 240 nm.

9. Chemiluminescence

It is a mass sensitive detector that can be used in ion pairing LC mode.

Nitric oxide when in contact with ozone, produces a metastable nitrogen dioxide molecule which relaxes to a stable state by emitting at a wavelength of 700 to 900 nm.

$$NO + O_3 \rightarrow NO_2^* + O_2$$

$$NO_2^* \rightarrow NO_2 + h\nu$$

To convert chemically bound nitrogen to nitric oxide, the sample is submitted to oxidative pyrolysis. As the light emission occurs, light intensity is measured by a photomultiplier tube through a band pass filter. *Application*: thermal energy analyzer (TEA) for nitrosamines.

Most chemiluminescence systems are those using:

- *peroxyoxalate*: reaction of hydrogen peroxide with an aryloxalate ester produces a high energy intermediate (1,2-dioxetane – 3,4-dione). In the presence of a fluorophore the intermediate forms a charge transfer complex that dissociates to yield an excited state fluorophore which then emits a photon. *Applications*: determination of hydrogen peroxide, polycyclic aromatic hydrocarbons, dansyl derivatives and nonfluorescers (sulfite, nitrite) that quench the emission.
- *acridinium esters*: oxidation of an acidinium ester by hydrogen peroxide in alkaline medium.
- *luminol*: luminol (5-amino-2,3 dihydro-1,4 phthalazinedione) reacts with an oxidant (in the presence of a catalyst) to produce 3-aminophthalate which emits at 425 to 435 nm in alkaline medium. All chemiluminescence reactions of luminol or isoluminol are oxidation reactions carried out in either aprotic solvents (DMSO, DMF) or protic solvents (water, lower alcohols). The chemiluminescent quantum yield of luminol is about 5% in DMSO and about 1 to 2% in water. To obtain chemiluminescence from luminol in an aqueous solution, an oxidizing reagent, e.g., hydrogen peroxide is needed. Isoluminol shows 10 to 100 times weaker luminescence than that of luminol.
- *firefly luciferase*: luciferin reacts with adenine triphosphate (ATP) to form adenylluciferine which oxidizes to form oxyluciferin, adenine monophosphate (AMP), CO_2, and light.

B. ELECTROCHEMICAL DETECTION

The three basic detection modes of electrochemical detection are conductivity, amperometric, and potentiometric detection.

1. Conductivity

Solution conductivity is due to ion mobility. The conductivity depends on the number of ions present. If the concentration is C (in moles per unit volume) the molar conductivity is $\Lambda_m = \kappa/C$. Since the resistance is measured in Ohm (Ω) the units are Ω^{-1} cm^{-1}. Conductivity is expressed in Siemens/cm or S/cm.

The detection method is based on the application of an alternative voltage E to the cell electrodes. The cell current i is directly proportional to the conductance G of the solution between the electrodes by Ohm's law.

$$G = i/E$$

The measured conductivity is the sum of individual contribution to the total conductivity of all the ions in solution. Kohlrausch's law states that

$$k = \sum_i \lambda_i^0 C_i$$

where C_i is the concentration of each ion i, λ_i^0 is the limiting equivalent conductivity which is the contribution of an ion to the total conductivity divided by its concentration extrapolated to infinite dilution.

Kohlrausch's law is only valid in dilute solutions (chromatography or electrophoresis). The magnitude of the signal is greatest for small high mobility ions with multiple charge such as sulfate.

Early ion chromatography systems detected ions eluted by strong eluents from high capacity ion exchange columns by measuring changes in conductivity. To achieve reasonable sensitivity, it was necessary to suppress the conductivity of eluent prior to detection in order to enhance the overall conductance of the analyte and lower the background conductance of the eluent. This was achieved by a "suppressor" column where counter ion were exchanged with H^+ or OH^-. Due to excessive band broadening column suppressors are no longer in use. Membrane based devices are utilized. The membrane suppressor incorporates two semi permeable ion exchange membranes sandwiched between sets of screens. The eluent passes through a central chamber. Regenerant flows in a counter current direction over the outer surfaces of the membranes providing constant regeneration. Electrolysis of water produces hydrogen or hydroxide ions required for regeneration. There is no contamination with carbonate (Figure 4.12).

In conductivity detectors the change in conductivity Δk depends on the concentration of the injected ion (A) and its equivalent ionic conductivity λ_A compared with that of the eluent ion λ_E.

$$\Delta k = (A)(\lambda_A - \lambda_E)$$

Conductivity detectors used previously were range dependent, which is a disadvantage when analyzing environmental samples where small amounts of one analyte are present together with a large amount of others. A single range digital conductivity detector eliminates the need for dilution.

2. Amperometric

Electrochemical detection is a concentration sensitive technique. In amperometric mode compounds undergo oxidation or reduction reaction through the loss or gain, respectively, of electrons at the working electrode surface. The working electrode is kept at constant potential against a reference electrode. Electrical current from the electrons passed to or from the electrode is recorded and is proportional to the concentration of the analyte present.

A thin layer cell is displayed in Figure 4.13.

A thin gasket with a slot cut in the middle is sandwiched between two blocks: one contains the working electrode, the other contains the counter electrode. The slot in the gasket forms the thin layer channel. The reference electrode is placed down stream from the working electrode. The thin layer design produces high mobile phase linear velocity which in turn produces high signal magnitude. The intensity of the current is

$$i = \phi n F u^{1/2} C D^{2/3} A$$

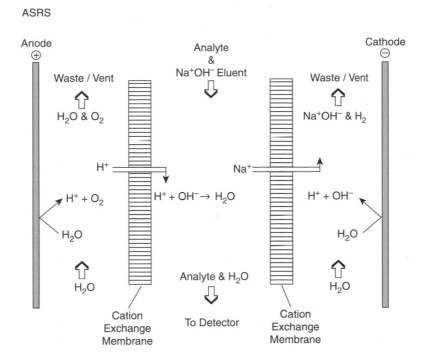

FIGURE 4.12 Anion self regenerating suppressor (ASRS) for detection in ion chromatography (by courtesy of Dionex).

where n is the number of electrons, F the Faraday constant, u the linear velocity of the mobile phase, C is the analyte concentration, D the diffusion coefficient of the solute, A the electrode surface, and ψ the geometrical constant of the cell.

The quality of the sample clean up procedure often determines the detection limits. The instability of the reference electrode is the source of voltage noise.

Parallel dual electrode may be used for a number of reasons:

- With one electrode at a positive and one electrode at a negative potential, oxidizable and reducible compounds can be detected in one single chromatographic run.
- When two solutes with different redox potentials coelute from the column, the potential of one electrode can be selected such that only the most easily oxidized (or reduced) compound is detected while on other electrode both compounds are converted. The concentration of the second compound is evaluated by substraction of the signal.

Series dual electrodes are set up such that one electrode is in oxidative and the other is in reductive mode. The downstream electrode measures the products of the upstream electrode. The second electrode only responds to compounds which are converted reversibly. The redox product is more selectively detected.

Voltammetric analysis is performed by scanning the potential or by applying triangular potential wave form to the electrode. Coeluting peaks are distinguished if their voltagrams are significantly different.

Coulometry permits determination of chemical substances by measuring the quantity of electricity required for their conversion to a different oxidation state. The quantity of electricity or charge is measured in Coulombs. (The coulomb is the quantity of charge that is transported in 1 sec

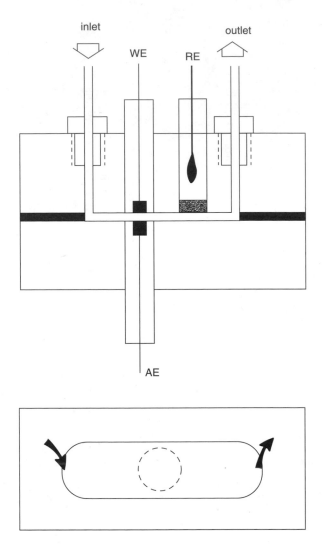

FIGURE 4.13 A thin layer cell for electrochemical detection: WE, working electrode; RE, reference electrode; AE, Auxiliary electrode.

by a constant current of 1 A). For a constant current of I amperes, the number of coulombs Q is $Q = It$. Faraday's law states that $Q = Fnz$, where F is the Faraday constant, Q is the number of coulombs required to convert n moles of reactant to product by a reaction involving z electrons per ion or molecule of reactant.

C. HYPHENATION WITH NMR

Online coupling of HPLC with nuclear magnetic resonance spectroscopy (NMR) has proved useful for a wide range of applications. The shortcoming of suppression of eluent signals can be circumvented by use of capillary separation technique. In this mode detection cells with internal volumes in the nanoliter scale and miniaturized probe heads have been developed by Albert et al. in Tuebingen.[16] The system can be used in either HPLC, CE, or CEC, and consists of a capillary inserted into a 2.5 or 2.0 mm NMR microprobe equipped with a Helmholtz coil. In experiments, a capillary tube of 315 μm can create a detection volume of 900 nl. The flow rate of the capillary

HPLC–NMR can be adjusted to 3 μl/min with the help of a T piece inserted between the HPLC pump and the capillary device. The polyimide coating is removed over the length of the NMR r.f. coil directly after the outlet frit of the capillary packing. In another design the packed capillary LC column is placed directly below the cryomagnet. With the help of a transfer capillary (400 × 50 μm^2) the eluate is transferred to the detection capillary with an internal diameter of 180 μm. The NMR detection volume is thus 200 nl. Assignment of vitamin E structures is possible.

D. Gas Chromatography Detectors in LC

Thermionic, flame photometric, and electron capture detectors can be connected to a LC column. The LC eluent can either be transported into the GC detector or be directly introduced. Suppressor must be incorporated.

V. DETECTION IN TLC

A. Densitometry

Densitometry is the mode for quantitative detection; optical measurement of a layer is difficult. Three types of measurements are in common use: transmission, reflection, and simultaneous transmission/reflection. The reflection mode is most popular.

A typical detecting device is shown in Figure 4.14.

Lamps are continuous-spectrum, halogen, or tungsten for visible spectrum, and deuterium or xenon for UV.

B. Videodensitometry

Unlike scanning densitometers, videodensitometers have no moving parts.

A video camera permits illumination by UV light at selected wavelength (254, 356 nm). The camera focuses on the media to be scanned and a video signal is sent to the digitizer board in the computer. Additionally the signal is also sent to a black and white video monitor displaying a real time image of the media. This helps the user to position a cursor to establish the boundaries. Parameters are set and the computer scans all lanes automatically. A chromatogram is produced for

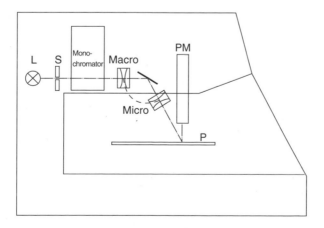

FIGURE 4.14 Scheme of a TLC Scanner (by courtesy of Camag) L, light source; S, slit; macro and micro, lens; PM, photomultiplicator; P, TLC plate.

each lane scanned. The charge coupling device (CCD) as an imaging detector has a number of features including high sensitivity, with spectral range and two-dimensional imaging ability.

C. Blotting

Blotting is a method in which compounds are blotted from a TLC plate to a membrane (polyvinylidene difluoride). Immunostaining on TLC is less sensitive than ELISA.

VI. DERIVATIZATION

Two general goals are achieved by chemical derivatization reactions: one is to increase detection sensitivity, the other is to increase selectivity.

There are two ways to perform chemical reaction detection: online following the elution of solutes of interest from the chromatographic column, or offline by carrying out the derivatization prior to chromatography. The former is mainly used in LC, the latter in GC.

GC: Derivatization is usually carried out to increase the volatility of substances that exhibit high boiling points or molecules that are thermally unstable and may decompose in the inlet port. Derivatization frequently makes it possible to resolve compounds. A derivatization reaction is performed prior to injection to make the solute more volatilizable. Each class of reactions replaces the active hydrogen of OH, NH, or SH. Alkylation, silylation, or acylation are very easily performed. Reagents are sold in vials to carry out the reaction. According to the type of solute different reagents are available.

Pierce company provides a useful directory to help the analyst in the selection.

Methyl esters are the most commonly used derivatives of the carboxylic groups. There are several methods for the preparation of methyl esters: diazomethane, methanolic solution of boron trifluoride, pyrolysis of tetramethylammonium salts, methylation with 2,2-dimethoxypropane, etc.

Silyl derivatives have received much attention, they can be utilized to block diverse polar groups (OH, COOH, SH, NH_2, =NH). The amino group is not very reactive in silylation reactions. Silyl derivatives are either trimethyl chlorosilanes (TMS), hexamethyldisilazane (HMDS) in conjunction with TMS, silylamines such as trimethylsilyl diethylamine and trimethylsilyl imidazole, silyl amides such as *N,O*-bistrimethylsilylfluoroacetamide (BSTTFA). Example is given by sterols. Isomers with a hydroxy group in the α position are not separated from the β isomers but when the OH group is converted into a suitable derivative (e.g., TMS) both isomers are well resolved.

Screening for estrogenic substances in fish is carried out by GC with silyl derivatives and ion trap detection. Use of silyl derivatives with FID produces silica. Nevertheless silyl derivatives are easily prepared and volatile.

$$R-OH + Cl-Si-R_3 \rightarrow R-O-Si-R_3$$

The low volatility of amino acids renders the GC of free acids impossible. *N*-trifluoroacetyl-*n*-butylesters proved to be most suitable and preparation is quite easy. Aldehydes and ketones can be converted to oximes, enamines, etc. New derivatization reactions are published every year

LC: Many solutes do not exhibit UV absorption or fluorescence, they can be converted in UV absorbing or fluorescent derivatives by pre or post column derivatization. The procedure has a wider range than in GC since the reaction can be performed following separation. When precolumn derivatization is carried out the chromatographic system is obviously different to the one selected for the nonderivatized solutes.

There is a large volume of literature dealing with post column reactions. It can be carried out in coils, in packed bed reactors or by photolysis. The main requirement is not the completion of the reaction but the reproducibility. Reaction vessels should not produce excessive band broadening. The reagent should be delivered continuously, pulselessly, and at a constant flow rate.

HPLC pumps are suitable for that purpose. The column effluent and the reagent should be mixed completely and quickly. In the simplest cases, either a T or Y union can be used. The cyclon type reduces the total volume variance. Table 4.1 displays some derivatizing reagents for the fluorescence labeling of functional groups.

TABLE 4.1
Derivatizing Reagents for the Fluorescence Labeling of Functional Groups

Reagent	Abbreviation	Functional Group
Aminoethyl-4-dimethylaminonaphthalene	DANE	Carboxyl
4-(Aminosulfonyl)-7-fluoro-2,1,3-benzoxadiazole	ABD-F	Thiol
Ammonium-7-fluorobenzo-2-oxa-1,3-diazole-4-sulfonate	SBD-F	Thiol
Anthracene isocyanate	AIC	(Amine), hydroxyl
9-Anthryldiazomethane	ADAM	Carboxyl (and other acidic groups)
Bimane, monobromo-	mBBr	Thiol
Bimane, dibromo-	bBBr	Thiol
Bimane, monobromotrimethylammonio-	qBBr	Thiol
4-Bromo-methyl-7-acetoxycoumarin	Br-Mac	See Br-Mmc
4-Bromo-methyl-7-methoxycoumarin	Br-Mmc	Carboxyl, imide, phenol, thiol
N-Chloro-5-dimethylaminonaphthalene-1-sulfonamide	NCDA	Amine (prim., sec.), thiol
9-(Chloromethyl)anthracene	9-CIMA	See Br-Mmc
7-Chloro-4-nitrobenzo-2-oxa-1,3-diazole	NBD Cl	Amine (prim., sec.), phenol
2-p-Chlorosulfophenyl-3-phenylindone	DIS-Cl	Amino acids, amino sugars
9,10-Diaminophenanthrene	9,10-DAP	Carboxyl
2,6-Diaminopyridine-Cu^{2+}	2,6-DAP-Cu	Amines (prim. aromatic)
4-Diazomethyl-7-methoxycoumarin	DMC	See ADAM
5-Di-n-butylaminonaphthalene-1-sulfonyl chloride	Bns-Cl	See Dns-Cl
Dicyclohexylcarbodiimide	DCC	Carboxyl
N,N'-Dicyclohexyl-O-(7-methoxycoumarin-4-yl)methylisourea	DCCl	Carboxyl
N,N'-Diisopropyl-O-(7-methoxycoumarin-4-yl)methylisourea	DICl	Carboxyl
4-Dimethylaminoazobenzene-4'-sulfonylchloride	Dbs-Cl	See Dns-Cl
N-(7-Dimethyl)amino-4-methyl-3-coumarinylmaleimide	DACM	Thiol
5-Dimethylaminonaphthalene-1-sulfonyl-aziridine	Dns-aziridine	Thiol
5-Dimethylaminonaphthalene-1-sulfonylchloride	Dns-Cl	Amine (prim., sec., tert.), (hydroxyl), imidazole, phenol, thiol
5-Dimethylaminonaphthalene-1-sulfonyl-hydrazine	Dns-hydrazine	Carbonyl
4-Dimethylamino-1-naphthoylnitrile	DMA-NN	Hydroxyl
9,10-Dimethoxyanthracene-2-sulfonate	DAS	Amine (sec., tert.)
2,2'-Dithiobis (1-aminonaphthalene)	DTAN	Aromatic aldehydes
1-Ethoxy-4-(dichloro-s-triazinyl)naphthalene	EDTN	Amine, hydroxyl (prim.)
9-Fluorenyl-methylchloroformate	FMOCCl	Amine (prim., sec.)
7-Fluoro-4-nitrobenzo-2-oxa-1,3-diazole	NBD-F	Amine (prim., sec.), phenol, thiol
4'-Hydrazino-2-stilbazole	—	α-Oxo acids
4-Hydroxymethyl-7-methoxycoumarin	Hy-Mmc	Carboxyl
4-(6-Methylbenzothiazol-2-yl)-phenyl-isocyanate	Mbp	Amine (prim., sec.), hydroxyl
N-Methyl-1-naphthalenemethylamine	—	Isocyanates (aliphatic, aromatic)
1,2-Naphthoylenebenzimidazole-6-sulfonyl chloride	NBI-SO$_2$Cl	See Dns-Cl
2-Naphthylchloroformate	NCF	Amine (tert.)
Naphthyl isocyanate	NIC	(Amine), hydroxyl
Ninhydrin	—	Amine (prim.)
4-Phenylspiro(furan-2(3H), 1'-phthalan)-3,3'-dione (fluorescamine)	Flur	Amine (prim., sec.), hydroxyl, (thiol)
o-Phthaldialdehyde (o-phthalaldehyde)	OPA	Amine (prim., sec.), thiol
N-(1-Pyrene)maleimide	PM	Thiol

VII. MASS SPECTROMETRY

The mass spectrometer is a mass flow sensitive detector. The peak area is independent of the mobile phase flow rate although the ionization efficiency of a LC–MS interface may be affected by the flow rate. Mass spectrometers are more and more miniaturized. They are routinely hyphenated to separation instruments, (GC–MS, LC–MS, SFC–MS, TLC–MS, CE–MS).

In a mass spectrometry experiment, the material in the gas phase is introduced to the high vacuum region of the ion source of the instrument. Here the molecules are ionized, usually by allowing them to interact with a beam of electrons typically in the energy region of 70 eV, but there are many other ionization sources.

From the ion source a mixture of molecular ions is produced, which gives molecular weight information and fragment ions containing the structural information.

Ions are separated according to their mass to charge ratio prior to detection. This is achieved by means of an external electric or magnetic field on the ion beam. It is the mass analyzer that yields a spectrum of abundance of ions (ion current) vs. mass to charge (m/z) ratio.

A mass spectrometer consists of an introduction device, an ion source, and an analyzer.

The resolving power of a mass spectrometer is a measure of its ability to distinguish between two neighboring masses. Resolution is Δamu/amu (atomic mass unit) or $M/\Delta M$ where M is the mass of the ion and ΔM is the width of the peak at half height of the Gaussian peak. Spectrometers easily perform resolutions of 50,000 (i.e., distinguish Δamu = 0.01 when $M = 500$). Resolution is often written in ppm (Δamu $\times 10^6/M$).

High resolution MS is of the double focusing type since a primary electrostatic analyzer lowers the dispersion of the ion beam, then a magnetic analyzer provides dispersion of the ion beam according to the mass to charge ratio.

A. ELECTRON IMPACT (EI)

In EI mode, relatively high energy electrons collide with analyte molecules producing positive ions and other species.

Sources of EI consist of a heated, evacuated chamber where a beam of electrons is generated from a heated metal filament. The energy of ionizing electrons is controlled by the voltage established between the cathode and the electron source filament. The standard practice is 70 eV which is large enough to cause ionization and fragmentation of organic moieties.

Bombardment of a neutral molecule with electron beam provides molecular ion (or parent ion) but in many cases this ion is too unstable to be present in the spectrum. The fragmentation process executed under constant conditions is well understood. The analyte fragmentations in EI mass spectra may provide information to determine molecular mass, elemental formula, and substitution patterns. In some cases EI does not provide sensitivity sufficient for the analysis of very small amounts of solute.

B. CHEMICAL IONIZATION (CI)

CI is an indirect process involving an intermediate chemical reagent.

In CI, mass spectra are produced by reaction between neutral organic molecules and reagent gas ion plasma. Concentration of reagent exceeds that of the sample by several orders of magnitude.

CI sources are operated at high energies (200 to 500 eV) which favor production of thermal electrons. CI produces stable molecular ions with little fragmentation.

Several gases are used in CI ionization: methane, propane, isobutane, hydrogen, ammonia, water, tetramethyl silane, or dimethyl amine. The CI ion source is similar to the EI source but is designed to have an ionization chamber. In positive chemical ionization, the ion source is filled with a reagent gas which is ionized to create a species of the proton donator type that can form a protonated molecule with an analyte; for example, with methane

$$CH_4 + e^- \rightarrow CH_4^+, CH_3^+, CH_2^+$$

$$CH_4^+ + CH_4 \rightarrow CH_5^+ + CH_3^{\cdot}$$

$$CH_2^+ + CH_4 \rightarrow C_2H_4^+ + H_2$$

$$CH_3^+ + CH_4 \rightarrow C_2H_3^+ + H_2$$

$$C_2H_3^+ + CH_4 \rightarrow C_3H_5^+ + H_2, \text{etc.}$$

All the processes proceed simultaneously.

Methane forms characteristic molecular adducts

$$CH_5^+ + M \rightarrow [MH]^+ + CH_4$$

$$C_2H_5^+ + M \rightarrow [M, C_2H_5]^+, \text{etc.}$$

Such processes are called proton affinity. In order to generate a protonated molecule the proton affinity of an analyte must be greater than that of the reagent gas ion. Ammonia has a high proton affinity value and provides in many cases a better differentiation between the sample matrix and an analyte.

In electron capture negative ionization, thermal electrons are generated by collision of electrons emitted from the filament with buffer gas molecules located at high pressure in the ionization chamber of the source.

$$e^*(70\ eV) + CH_4(\text{buffer gas}) \rightarrow e^*(\text{thermal electrons})(? \ eV)$$

$$e^* + M \rightarrow M^-$$

Suitable analytes exhibit high electron capture capacity or high electron affinity. CI is the technique of choice for the analysis of isomers in environmental samples. In APCI (Atmospheric Pressure Chemical Ionization) solvent evaporation and analyte ionization are two separate processes.

C. ATMOSPHERIC PRESSURE PHOTOIONIZATION (APPI)

The technique was introduced by Bruins et al.[17] In the APPI interface the corona discharge of the APCI source is replaced by a gas discharge lamp emitting photons in the vacuum UV region of the electromagnetic domain. When the energy of the photons is higher than the first ionization potential of a species in solution, then absorbance leads to single photon ionization.

D. ELECTROSPRAY (ESI) (FIGURE 4.15)

The electrospray process is initiated by applying an electrical potential of several kV to a liquid in a narrow bore capillary or electrospray needle.

There are three major processes in ES–MS:

(i) *Production of charged droplets at the ES capillary tip.* A voltage of 2 to 3 kV is applied to the metal capillary. When the capillary is the positive electrode, source positive ions in the liquid will drift toward the liquid surface and some negative ions drift away from it until the imposed field inside the liquid is essentially removed by this

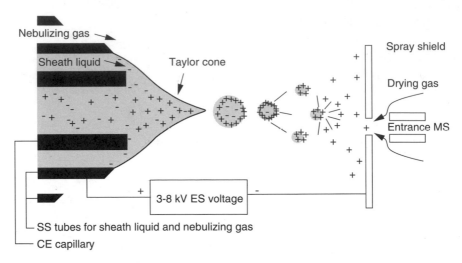

FIGURE 4.15 Principle of the electrospray ionization.

charge redistribution. However, the accumulated positive charge at the surface leads to destabilization of the surface because the positive ions are drawn downfield but cannot escape from the liquid. A liquid cone is produced. At a sufficiently high field E the liquid cone vanishes and a fine mist of small droplets is generated. The droplet surfaces are enriched with positive ions for which there are no negative counter ions.

(ii) *Shrinkage of charged ES droplets*. Charge and size of droplets are depending on spray conditions. When good conditions are provided the droplets are small and exhibit a narrow distribution of sizes. The droplets shrink by evaporation of solvent molecules until they come close to the Rayleigh limit that gives the condition in which the charges become sufficient to overcome the surface tension (that holds the droplet together). They undergo fission into smaller droplets.

(iii) *Highly charged droplets are capable of producing gas phase ions*. The droplet evaporation is stimulated by the use of a current heated gas or heated sampling capillary. Two different mechanisms have been proposed to account for the formation of gas phase ions from the small charged droplets. Extremely small droplets containing a single ion will give rise to a gas phase ion. The other mechanism assumes emission or ion evaporation. Under some conditions the droplets do not undergo fission but emit gas phase ions.

(iv) Gas phase ions are modified in the atmospheric and the ion sampling regions of the spectrometer. Analyte ion intensity depends on the analyte concentration and the pressure of the other electrolytes.

Pneumatically assisted electrospray is also called ion spray.

ESI is very popular due to the absence of critical temperature. It is a soft technique since very labile structures can be carried as ions.

E. FAST ATOM BOMBARDMENT (FAB)

If a solid is bombarded by high velocity particles, e.g., rare gas ions of about 8 keV energy, the material will be removed into the gas phase. Some of the sputtered material will be in the form of positively or negatively charged ions. A FAB source consists of:

(a) Atom gun
(b) Atom beam

(c) Sample holder

(d) Lens system leading to the mass analyzer.

In the mass spectrum one can find even electron molecular ions.

F. MALDI

Matrix Assisted Laser Desorption Ionization is well suited for macromolecules such as peptides, proteins, oligosaccharides, and oligonucleotides. In a MALDI experiment a proper organic matrix (e.g., glycerol) is required for mixing with the analyte in a ratio typically 500/1. The mixture is dried and inserted in a MS. A laser beam will desorb and ionize the matrix species, thus ionizing the analyte. Most MALDI–MS systems are based on Time of Flight (TOF) mass analyzers. Ions produced by the laser beam are extracted from the source and expelled to the flight tube (Figure 4.16).

G. MASS ANALYZERS

1. Quadrupole

The quadrupole mass filter consists of four parallel hyperbolic rods in a square array (Figure 4.17). Cylindrical rods are in current use. The inside radius (field radius) is equal to the smallest radius curvature of the hyperbola. Diagonally opposite rods are electrically connected to radio frequency/ direct current voltages, which create a hyperbolic field within the rods. For a given radio frequency/ current voltage ratio, only ions of a dedicated m/z value are transmitted to the filter and reach the detector. Ions with a different m/z ratio are deflected away from the principal axis and strike the rods. To scan the mass spectrum, the frequency of the radio frequency voltage and the ratio of

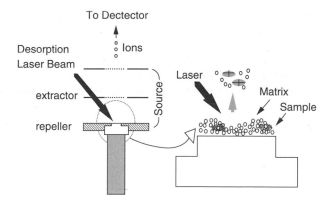

FIGURE 4.16 Principle of matrix-assisted laser desorption ionization.

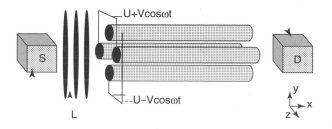

FIGURE 4.17 A quadrupole analyzer with S, source; L, lens; D, detector.

the ac/dc voltages are held constant while the magnitude of ac and dc voltages is varied. The transmitted ions of m/z are then linearly dependent on the voltage applied to the quadrupole producing m/z scale, which is linear with time. A triple quadrupole instrument uses two quadrupole MS analyzers for the actual MS experiments and a third quadrupole in RF mode that transmits all incoming ions from MS_1 to MS_2.

These analyzers are low cost instrument with rather low resolution.

2. Ion Trap

Ionization steps and ion separation are space separated in a quadrupole system whereas they are time spaced in ion traps. An ion trap (Figure 4.18) is a linear quadrupole bent to a close loop. Typically three electrodes are a common design of an ion trap: a ring electrode and two end cap electrodes. The outer rods form a ring and the inner rod is reduced to a mathematical point in the center of the trap. End electrodes have a hole in their center to allow for introduction of ions and ejection of these ions towards the detecting device. A radio frequency is applied to the ring electrode and by consequence a quadrupole field which traps ions is produced. Each ion is submitted to an oscillating motion, the amplitude of which depends on the RF and the m/z ratio. Ions of different masses are stored together in the trap and released one at a time by scanning the applied voltages. They can be ejected through the end caps and detected by applying an RF voltage with a frequency corresponding to the characteristic frequency of the ion moving through the ion trap or by scanning the amplitude of the applied RF voltage. With a reduced pressure of gas (He) (10^{-3} Torr), the motion of ions in the trap is dampened and the ions move closely around the center of the trap. The damping gas reduces the motion of ions and makes the ion trap ideally suited for hyphenation with GC. Ion traps are powerful tools in elucidating fragmentation mechanisms since they allow stepwise and controlled fragmentation in multistage MS. Ion traps are small benchtop instruments.

The full scan mode consists of acquisition of mass spectra with a wide range of m/z in order to detect all types of ions from the source. With chromatographic hyphenation it means one spectrum/sec. The full scan mode is identification of eluted analytes. Single Ion Monitoring (SIM) is the procedure where one single (or several) ions are solely detected. SIM is operated with quadrupole whereas SIS (single ion storage) is the acronym for the same procedure with ion traps. For quadrupole mass spectrometers, selected ion monitoring (SIM) yields significantly enhanced detection limits compared with scanning MS operation because of the greater dwell time for signal acquisition at each selected m/z value.

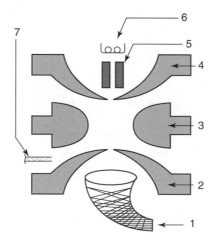

FIGURE 4.18 Ion trap detection in GC. 1, channeltron detection; 2, cap electrode; 3, ring electrode; 4, cap electrode; 5, electron gate; 6, electron emitting device; 7, capillary column.

3. Fourier Transform MS (FTMS)

In an FTMS instrument, detection of ions of interest is performed by applying a very fast frequency sweep voltage to the transmitter plates following ionization process. The frequency of this cyclotron motion is mass dependent. The coherent motion of the excited ions induces image currents in the receiver circuit. Positive ions approaching one receiver plate attract electrons. As they continue to move in their orbits they approach the opposite receiver plate and attract electrons on this surface. When the receiver plates are connected in a circuit the induced image current of ions can be detected in the form of a time domain signal resulting from the superposition of a number of individual frequencies produced by the different ion species coherently orbiting at the same time. A mass spectrum is obtained by amplification, digitization, and conversion of this time domain signal to a frequency domain spectrum using Fourier transformation.

4. Time of Flight

A scheme is displayed below.

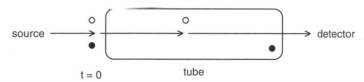

A small number of ions is extracted from the source in a few μsec, accelerated with a few kV, and they are directed to a field free light tube. The process can be repeated 100,000 times per second. Kinetic energy is similar for every ion. Ions with higher velocities (light ions) will reach the end of the tube before heavy ions. Instruments have two tubes with a mirror in the middle and resolution may reach 5000.

In a tube of length L the time of flight t is connected to the velocity v

$$t = L/v$$

Ions get a kinetic energy $E_k = 1/2 \, mv^2 = zV$ where V is the voltage. In a tube of length L the time of travel is $t = L/v$ thus $t = L(m/2Vz)^{1/2}$. The time of flight is proportional to the square root of the ion mass, which allows discrimination according to the m/z ratio.

As an example, if 3000 V are used to accelerate the ions in the flight tube of 1 m, an ion with $m = 200$ amu, and $z = 1$ the travel time is 18.597 msec, and with $m = 201$ amu, $t = 18.643$ msec. The difference is 48 nsec. It is thus necessary to use fast electronics. To cope with the small differences of velocities encountered by ions of the same m/z, a reflectron is used, which acts as a retarding electric field. It is a series of lenses with linearly increasing voltages. From ions of the same m/z value, those with the greater velocity will penetrate the reflectron further and take a longer time to turn around and leave the reflectron towards the detector. The ions of lower velocity will catch up with those of higher velocity and reach the detector at the same time. Orthogonal-acceleration reflectron TOF instruments combine the ability to perform accurate mass determination with an excellent fullscan sensitivity.

The hybrid quadrupole time of flight mass spectrometer (QTOF) was introduced as a mass spectrometer capable of tandem MS with particular emphasis on its applicability for protein and peptide analysis. It combines the simplicity of a quadrupole MS with the high efficiency of a

TOF analyzer. Key components of the instrument are the quadrupole, hexapole collision cell, and the reflectron–TOF analyzer. The sample is introduced through the interface and ions are focused using the hexapole ion bridge into the quadrupole. Here, the precursor ion is selected for later fragmentation and analysis with a mass window of approximately three mass units, which is a typical window to preserve the isotope envelopes into the product ion spectra. The ions are ejected into the hexapole collision cell, where argon is used for fragmentation. From this point, the ions are collected into the TOF region of the MS–MS. In an orthogonal TOF (oa-TOF) the flight path of the ions changes 90°. The ions are then accelerated by the pusher and travel about 1 m down the flight tube to the reflectron. Thus, the TOF side of the Q–TOF–MS achieves simultaneous detection of ions across the full mass range at all times. This continuous fullscan mass spectrum is in contrast to the tandem quadrupoles that must scan over one mass at a time. The Q–TOF–MS–MS is capable of 10,000 resolving power expressed at full width half maximum.

The electronics of the detector must record the complete mass spectrum within the flight time of the ions (1 to 100 μsec range) with peak widths in the ns range. This is possible since a high scan rate (up to 20,000 scans/sec) allows for the detection of narrow chromatographic peaks. There is virtually no limit on mass range and no ion loss. In TOF–MS there are instruments that can provide high resolution at a moderate scan speed, and instruments that can store 100 to 500 spectra/sec with unit mass resolution.

H. SECONDARY ION MASS SPECTROMETRY (SIMS)

High energy ions are fired at the surface of the solid material to be investigated. These "primary ions" penetrate the near surface atomic layers and set up chains of collisions between the surface atoms. The resulting disruption ejects some atoms and molecules from the surface, and these "secondary" ions are analyzed in a mass spectrometer. The technique is extremely sensitive to halides but very expensive.

I. LC–MS

Interfaces have been developed to solve the problem of handling high LC flow rates (1 ml/min) and the high vacuum required by mass spectrometer. LC is not nearly as compatible with MS as in GC.

Hyphenating LC and MS requires overcoming major difficulties.

- conventional packed column are operated at 1 ml/min
- LC separations make use of nonvolatile mobile phase and very often buffer solutions.

Ionization of nonvolatile or thermally labile solutes is difficult. The ionization and thus the response in LC–MS analysis is limited by the ability to protonate or deprotonate the analytes of interest. However, difficulties have been overcome to make LC–MS a robust and routinely applicable tool in environmental laboratories.

LC–MS interfacing has been achieved in a number of ways. The first successful commercially available LC–MS interface was the transport or moving belt system. Direct liquid introduction (DLI) was used in the 1980s and has disappeared. LC–Thermospray MS and LC–FAB have largely been replaced by electrospray (ESI) and atmospheric pressure chemical ionization (APCI) that rely on the formation of a continuous spray from the chromatography column effluent. According to Voyksner,[18] API techniques can handle liquid flow rates that are typically used in LC. API techniques are suitable for the analysis of nonvolatile, polar, and thermally unstable molecules. API–MS is sensitive, easy to use, and robust. An API interface/source consists of five parts: the liquid introduction device or spray probe; the actual atmospheric pressure ion source region where the ions are generated (ESI, APCI); an ion sampling aperture; an atmospheric pressure to vacuum interface; an ion optical system where the ions are subsequently transported to the mass analyzer.

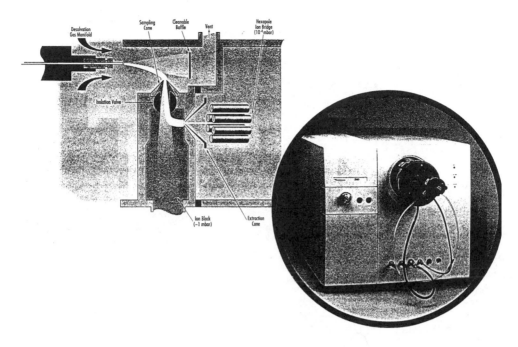

FIGURE 4.19 A LC–ESI/MS interface (by courtesy of Waters).

(i) *Electrospray Ionization*. A typical LC–MS with ESI interface is displayed in Figure 4.19. Orthogonal spray orientation reduces ion source contamination and keeps the capillary and ion optics cleaner. It also eliminates the need to adjust the position of the sprayer even when the flow rate or solvent composition changes. High flow rates (1 ml/min) can be handled. A great deal of effort is focused on miniaturization of the system. ESI is more likely to preserve the integrity of the particular species than APCI. ESI–MS techniques are recognized as having high potential for mass determination of food proteins. A multichannel device allows analysis of a series of samples in a very short time (96 peptides in 480 sec) (Figure 4.20).

(ii) In atmospheric pressure chemical ionization, the liquid flow from the LC is sprayed and rapidly evaporated by a coaxial nitrogen stream and heating the nebulizer to high temperature (350 to 500°). Additional ionization is achieved by means of a corona discharge (3 to 6 keV). The interface consists of a concentric pneumatic nebulizer and a large diameter heated quartz tube. Both analyte and solvent molecules are ionized, the solvent ions can react with the analytes in the gas phase in the same way as samples are ionized in CI mode. Solvent evaporation and ion formation processes are separated in APCI which allows the use of some unfavorable solvents. The major limitation is the strong dependency of the response on the nature of the analyte plus the mobile phase. APCI can stand high flow rates but miniaturization is more difficult than with ESI.

Nonpolar solutes that are not prone to undergo acid–base reactions are difficult to analyze with either ESI or APCI. For quantitative evaluations using APCI–MS detection mode it is mandatory to perform calibration runs with samples of known relative concentrations under the same conditions to be utilized in the actual analysis.

APPI interface is promising since the common LC solvents are characterized by high first ionization potentials with the consequence that selective ionization of the analytes may occur. Addition of a dopant to the mobile phase such as acetone or toluene offers increased selectivity.

In coordination ion-spray MS, various ionic reagents induce charges on the solutes eluting from the chromatography column. Ag(I) adducts with olefins or unsaturated triglycerides or fatty acids permit the sensitive detection of $(M^+Ag)^+$ species.

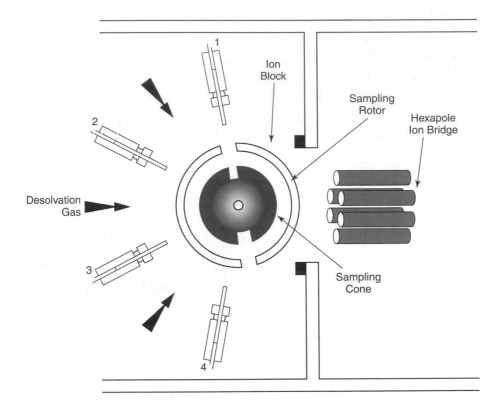

FIGURE 4.20 The multichannel (MUX) technology (by courtesy of Micromass).

J. GC–MS

Mass spectrometers can readily be interfaced to GC (Figure 4.21). As the compounds leave the chromatograph, they are introduced into the mass spectrometer operating in a vacuum. The great majority of GC–MS are benchtop instruments with linear quadrupoles and EI. Compound identification is performed by comparison of the spectrum with a data base for precise identification and confirmation. Huge EI mass spectral libraries such as NIST Library or Wiley Library contain

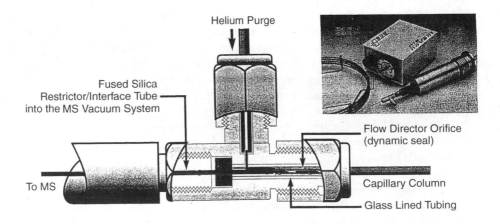

FIGURE 4.21 The GC–MS interface.

more than 230,000 spectra. Quantitation based upon a unique ion fragment also allows accurate determination of analyte concentration even when the GC separation of analytes is incomplete.

GC/MS is now the most widely used detection technique.

1. GC–Quadrupole MS with EI (Electron Impact)

The most common design is a single capillary column directly coupled to an EI quadrupole mass spectrometer. The sensitivity is very high allowing detection of extremely small quantities of solutes.

Electron Impact mode is 70 eV.

Typical scan range is 65 to 400 amu, scan time 0.5 sec with a delay time of 0.2 sec between individual scans.

Library spectra are capable of producing more than 130,000 spectra.

In GC–MS, as the ionization energy is rather constant, the 70 eV EI mass spectra are rather uniform and unique for a molecule whereas in LC–MS the fragmentation largely depends on the configuration of the LC–MS interface. By consequence mass spectral libraries can be used for identification of unknowns. Due to extensive fragmentation EI is not as sensitive as CI.

2. GC–Quadrupole MS with CI (Chemical Ionization)

Chemical Ionization (CI) is a soft ionization technique that produces molecular ions (M^+ or M^-), adduct ions (M^+CI reagent), and fragment ions. Instrumentation is more expensive but CI permits isomer differentiation. CI reagents are usually methane, isobutane, or ammonia. The degree of fragmentation is less than in EI.

Linear quadrupole instruments are widely spread. Recent developments in the technology now make it possible to work simultaneously with fullscan and selected ion monitoring (SIM) modes in a single run. In GC tandem MS, a first quadrupole acts as a mass selective filter and the second quadrupole is used as the collision cell with addition of a collision gas such as helium or argon. In a third quadrupole the full mode is performed to obtain the full mass spectrum of the product ions.

The mass range is 2 to 1000 Da. Triple quadrupole can solely perform MS–MS and the ratio performance/cost is low.

3. GC–Ion Trap MS

Ion trap operates in a pulsed mode so that ions are accumulated mass selectively over time. Collision-induced dissociation in the ion trap is produced by several hundred collisions of a mass-selected ion with helium buffer gas atoms. An advantage of ion-trap instruments is the ability to perform MS_n.

In this technique an ion of interest is selected by ejecting all unwanted ions and the selected ion is subsequently fragmented by collision with a neutral gas. The resulting mass spectrum is called a daughter or product ion spectrum and it is a characteristic of the secondary fragmentation process.

The mass range is 10 to 10,000 Da with a mass accuracy of $\pm 0.2 \; m/z$.

4. GC–TOF–MS

In most GC–TOF–MS instruments, an appropriate voltage pulse is applied to accelerate the ions in the direction orthogonal to their initial flight direction. In such oa-TOF–MS a nearly parallel ion beam ideally has no velocity spread, and the finite spatial spread is corrected with a linear or reflecting instrument geometry. Noise-free mass spectra are produced within a very short time (a few milliseconds). Only TOF–MS instruments have the capability required to detect peaks in GC * GC since the half widths of the peaks eluting from the second column are of the order of 200 msec. Selecting the proper scan rate is essential since an increase in the acquisition speed decreases the sensitivity expressed in terms of signal to noise. GC–TOF–MS is the detector of two-dimensional GC. In this mode peaks eluting from the second dimension exhibit a width of 60 to 200 msec. High acquisition rates are therefore needed.

The overall separation is displayed as a contour plot with two time axes. Peaks are presented as dots in this plane and the signal intensity by differences in color. TOF–MS provides an excellent means to use a chemometric approach.

Detection of a target analyte may be tedious with GC–SIM because signal interference may conceal the signal. A method for mass spectrum extraction of GC–MS data has been incorporated into an automated software program developed by the National Institute of Standards, called Automated Mass Spectral Deconvolution and Identification system (AMDIS). The LECO corporation has also developed a software program used to resolve overlapped signals from GC–TOF–MS.

The mass range is 5 to 1500 Da with very high mass accuracy and high ratio performance/cost.

K. TLC–MS

Considerable efforts have been made over the past few decades to combine TLC with (i) FAB, (ii) MALDI or SALDI, (iii) ESI, (iv) Laser two step MS, but TLC–MS is still only used by research groups. TLC–MS methods are mostly based on surface desorption and ionization techniques (FAB and MALDI). To overcome the bottleneck of the fact that ions would be removed from the surface only, the plate (after chromatography) is impregnated with a viscous liquid or low melting point solid to improve the sensitivity. Glycerol is often advocated but a variety of other organic compounds can be used. The combination of TLC and MALDI offers the potential advantage of minimal analyte spreading compared with TLC–FAB or TLC–LSI (liquid secondary ion). An interesting preparation of a TLC plate for MALDI has been described.[19]

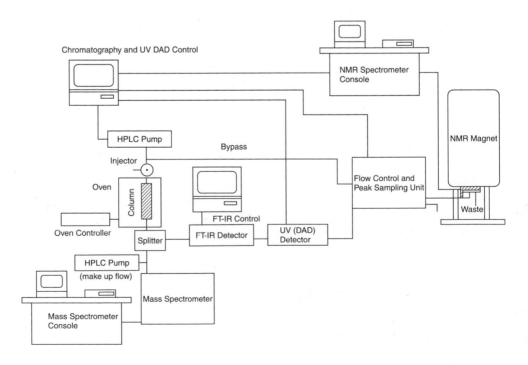

FIGURE 4.22 An example of hypernation. (Reproduced from the Journal of Chromatography with permission from Elsevier.)

The technique involves the preparation of a matrix (4-cyano-hydroxycinnamic acid) and L-fucose on a stainless steel plate, which is transferred onto the plate that has been prewetted with an extraction solvent. Busch et al.[20] have described a micro extraction technique that transfers the analyte to ESI.

VIII. HYPHENATION

Hyphenation refers to the online combination of a separation technique (mainly GC or LC) and a spectroscopic detection method that provides structural information for the analytes concerned. LC–MS and GC–MS are the most popular hyphenated techniques in use today, thanks to the incredible shrinking of mass spectrometers. However, MS does not provide the same information as IR or NMR. Hypernation (one higher than hyphenation) is the multiple hyphenation which may combine, for example, UV detection with PDA, NMR, FTIR, and MS as displayed in Figure 4.22, the last one is a destructive detector. At the present time most analysts are using a single detector but it can be guessed that hypernation will be used more and more as size and cost of the instruments decrease.

REFERENCES

1. Vissers, J. P. C., Claessens, H., and Cramers, C. A., *J. Chromatogr. A*, 779, 1, 1997.
2. Guiochon, G. and Guillemin, C. L., Quantitative gas chromatography, *J. Chromatogr. Library*, Vol. 42, Elsevier, Amsterdam, pp. 705–710, 1988.
3. Currie, L. A., IUPAC Commission on Analytical Nomenclature, Recommendations in Evaluation of Analytical Methods Including Detection and Quantification Capabilities, *Pure Appl. Chem.*, 67, 1699–1723, 1995.
4. ISO/11843-1, *Capability of Detection (Part 1): Terms and Definitions*, International Organization for Standardization, ISO/TC69/SC6, 1997.
5. ISO 17025, *General Rules on Laboratories*, Geneva, 2000.
6. ICH-Topic Q2B: Validation of Analytical Procedures: Methodology, *International Conference on Harmonization of Technical Requirements for Registration of Pharmaceuticals for Human Use*, Geneva, 1997. http://www.ich.org/pdfICH/Q2B.pdf (30 August 2002).
7. United States Pharmacopoeia 26, General information ⟨1225⟩, *Validation of Compendial Methods*, The United States Pharmacopoeia Inc., Rockville, MA, pp. 2439–2442, 2003.
8. ISO 3534-1, *Statistics — Vocabulary and Symbols. Part 1: Probability and Statistics*, 1993.
9. ICH-Topic Q2A: Validation of Analytical Procedures, *International Conference on Harmonization of Technical Requirements for Registration of Pharmaceuticals for Human Use*, Geneva, 1995. http://www.ich.org/pdfICH/Q2A.pdf (30 August 2002).
10. ISO 3531, *Applications of Statistics Accuracy*, Geneva, 1994.
11. Zimmermann, S., Wischhusen, S., and Mueller, J., *Sens. Actuators B: Chemical*, 63, 159, 2000.
12. Mendonca, S., Wentworth, W. E., Chen, E. C. M., and Streans, S. D., *J. Chromatogr.*, 749, 131, 1996.
13. Young, C. S. and Dolan, J. W., *LC–GC Eur.*, 17, 192, 2004.
14. Bornhop, D. J., Nolan, T. G., and Dovichi, N. J., *J. Chromatogr.*, 384, 181, 1987.
15. Yamamoto, A., Kodama, S., Matsunaga, A., Hayakawa, K., and Kitaoka, M., *Analyst*, 124, 483, 1999.
16. Andrade, F. D. P., Santos, L. C., Datchler, M., Albert, K., and Vilegas, W., *J. Chromatogr.*, 953, 287, 2002.
17. Robb, D. B., Covey, T. R., and Bruins, A. P., *Anal. Chem.*, 72, 36532, 2000.
18. Voyksner, R. D., Combining liquid chromatography with electrospray mass spectrometry, In *Electrospray Ionization Mass Spectrometry, Fundamentals, Instrumentation and Applications*, Cole, R. B., Ed., Wiley, New York, 1997.
19. Mehl, J. T., Gusev, A. I., and Hercules, D. M., *Chromatographia*, 46, 358, 1997.
20. Henderson, R. M. and Busch, K. L., *J. Planar Chromatogr.*, 11, 336, 1998.

FURTHER READING

A. GC–MS
Santos, F. J. and Galceran, M. T., *J. Chromatogr. A*, 1000, 125, 2003.

B. LC–MS.
Niessen, W. M. A. and Voyksner, R. D., Eds., *Current Practice of Liquid Chromatography–Mass Spectrometry*, Elsevier Science, New York, 1998.
Niessen, W. M. A., *J. Chromatogr. A*, 1000, 413, 2003.

C. Ion Chromatography Detection.
Buchberger, W. W., *J. Chromatogr. A*, 884, 3, 2000.

D. Liquid Chromatography.
Katz, E., Eksteen, R., Schoenmakers, P., and Miller, N., Eds., *Handbook of HPLC*, Marcel Dekker, New York, 1998.
Snyder, L. R., Kirkland, J. J., and Glajch, J. L., Eds., *Practical HPLC Development*, 2nd ed., Wiley, New York, 1997.

E. LC–NMR.
Albert, K., *J. Chromatogr. A*, 856, 199–211, 1999.

F. TLC.
Geiss, F., *Fundamentals of Thin Layer Chromatography*, Huethig, Heidelberg, 1987.

5 Chemometrics in Data Analysis

Riccardo Leardi

CONTENTS

I. INTRODUCTION

In this chapter the fundamentals of chemometrics will be presented by means of a quick overview of the most relevant techniques for data display, classification, modeling, and calibration. The goal of the chapter is to make people aware of the great superiority of multivariate analysis over the commonly used univariate approach. Mathematical and algorithmical details will not be presented, since the chapter is mainly focused on the general problems to which chemometrics can be successfully applied in the field of environmental chemistry.

As a matter of fact, many of the readers of this book may not be familiar with chemometrics, and a significant percentage of them may have never even heard of this "new" science (quite strange that it is still considered a "new" science, when the Chemometrics Society was founded 30 years ago and the most basic algorithms date back to the beginning of the 20th century). Furthermore, some of them could be quite put off by anything involving mathematical computations higher than a square root or statistical tests more complex than a *t*-test.

Therefore, the goal of this chapter is simply that of being read and understood by the majority of the readers of this book. This goal will be completely achieved if some of them, after having read it, could say: "Chemometrics is easy and powerful indeed, and from now on I will always think in a multivariate way."

Of course, to accomplish this goal in the limited space of a chapter the attractive sides of chemometrics must be highlighted. Therefore, the intuitive aspects of each technique will be shown, without giving too much relevance to the algorithms.

First of all, what is Chemometrics? According to the definition of the Chemometrics Society, it is "the chemical discipline that uses mathematical and statistical methods to design or select optimal procedures and experiments, and to provide maximum chemical information by analyzing chemical data."

One of the major mistakes people make about chemometrics is thinking that to use it one has to be a very good mathematician and to know the mathematical details of the algorithms being used.

From the definition itself, it is clear instead that a chemometrician is a *chemist* who can *use* mathematical and statistical methods.

If we want to draw a parallel with everyday life, how many of us really know in detail how a TV set, a telephone, a car, or a washing machine works? But everybody watches TV programs, makes phone calls, drives a car, and starts a washing machine. Of course, what is important is that people know what each instrument is made for and that nobody tries to watch inside a telephone, or to drive a TV set, or to speak inside a washing machine, or to do the laundry in a car.

Though chemometrics makes available a very wide range of techniques, some of them being very difficult to fully understand and use correctly, the great majority of real problems can be solved by applying one of the basic techniques, whose understanding, at least from an intuitive point of view, is relatively easy and does not require high-level mathematical skills.

II. DATA COLLECTION

Chemometrics works on data matrices. This means that on each sample a certain number of variables have been measured (in the "chemometrical jargon" we say that each object is described by v variables). Although some techniques can work with a limited number of missing values, a chemometrical data set must be thought of as a spreadsheet where all the cells are full.

Sometimes, instead, if data are gathered without having any specific project, it happens that the result is a "sparse" matrix containing some blank cells. In such cases, if the percentage of missing data is quite high, the whole data set is not suitable for a multivariate analysis; as a consequence, the variables and/or the objects with the lowest number of data must be removed, and therefore a huge amount of experimental effort can be lost.

All the chemometrical software allows the import of data from ASCII files or from spreadsheets. It is therefore suggested to organize the data in matrix form from the start, as shown in Figure 5.1, in such a way that the import can be performed in a single step. If, on the contrary, the data are spread in several files or sheets (e.g., one file for each sample or for each variable), then the import procedure would be much longer and more cumbersome.

III. DATA DISPLAY

The human mind can digest much more information when looking at plots rather than numbers. This is easily demonstrated by looking first at the sequence of numbers reported in Table 5.1, and then the plot in Figure 5.2. It is very clear that, even in a very simple data set like this one (just ten samples, and only one variable), the information obtained by looking at the plot is superior and much more easily available than the information one can get by analyzing the raw numbers. From the plot, it becomes evident that the samples are clustered into two groups of the same size, the one at higher values being much tighter than the one at low values. Much more time and effort is required when we want to get the same information from the table.

	var. 1	var. 2	var. 3	var. 4	var. 5	var. 6	var. 7	...	var. v
obj. 1									
obj. 2									
obj. 3									
obj. 4									
obj. 5									
obj. 6									
......									
obj. n									

FIGURE 5.1 The structure of a chemometrical data set.

TABLE 5.1
Ten Samples Described by One Variable

Sample	1	2	3	4	5	6	7	8	9	10
Value	25.3	22.1	25.5	25.6	19.4	25.7	20.2	21.3	25.9	21.8

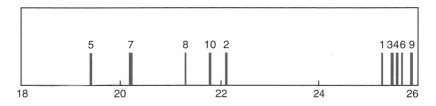

FIGURE 5.2 Scatter plot of the data in Table 5.1.

Let us now take into account a more complex data set, i.e., the one reported in Table 5.2, where each object is described by two variables. The same data are plotted in Figure 5.3. This bivariate data set, beyond showing once more that a plot is much more easily handled by the human brain than a data table, demonstrates that when dealing with more than one variable the analysis of just one variable at a time can lead to wrong results. In this data set we have 20 samples, supposed to belong to the same population. When looking at the plot, we realize that we are in a situation very similar to what we found with the univariate data set. The samples are split into two clusters of the same size, with the objects of the first one more tightly grouped than the objects of the second one. This conclusion cannot be reached when looking at one variable at a time, since neither of the two variables is able to discriminate between the two groups.

If we had a data set with three variables it would still be possible to visualize the whole information by a three-dimensional scatter plot, in which the coordinates of each object are the values of the variables. But what to do if there are more than three variables? What we need therefore is a technique permitting the visualization by simple bi- or tri-dimensional scatter plots of the majority of the information contained in a highly dimensional data set. This technique is Principal Component Analysis (PCA), one of the simplest and most used methods of multivariate analysis. PCA is very important especially in the preliminary steps of an elaboration, when one wants to perform an exploratory analysis in order to have an overview of the data.

It is quite common to have to deal with large data tables with, for instance, a series of samples described by a number (v) of chemico-physical parameters. Examples of such data sets can be samples of olive oils from different origins described by their content in fatty acids and sterols, or samples of wines described by Fourier-Transformed Infra-Red (FT-IR) spectra. It is easy to realize how, especially in spectral data sets, v can be very high (>1000). In such cases it would be impossible to obtain valuable information without the help of multivariate techniques.

From a geometrical point of view, we can consider a v-dimensional space, in which each dimension is associated to one of the variables. In this space each sample (object) has coordinates corresponding to the values of the variables describing it.

Since it is impossible to visualize all the information at once, one should be content with the analysis of several bi- or tri-dimensional plots, each of them showing a different part of the global information.

It is also evident that not all possible combinations of two or three variables will give the same quality of information. For instance, if some variables are very highly correlated, then the information brought by each of them would be almost the same. If two variables are perfectly correlated, then one of them can be discarded, losing no information at all. In this way, the

TABLE 5.2
Twenty Samples Described by Two Variables

Sample	Variable 1	Variable 2
1	21.2	32.5
2	16.2	21.0
3	13.1	21.7
4	11.6	21.3
5	20.8	29.9
6	10.4	20.6
7	19.5	26.8
8	9.8	25.2
9	15.2	31.2
10	12.0	26.0
11	17.6	28.5
12	24.0	30.0
13	17.8	33.1
14	15.0	24.0
15	11.0	24.2
16	24.8	25.3
17	12.8	23.3
18	26.5	30.6
19	22.9	27.5
20	9.7	22.8

dimensionality of our space will be reduced from v to $v - 1$. If two variables are very highly correlated, then the elimination of one of them would produce only a slight loss of information, while the dimensionality of the space would be reduced to $v - 1$. So, one can deduce that the information contained in the "lost" vth dimension was well below the average of the information contained in the other dimensions.

It is quite apparent now that not all the dimensions have the same importance, and that, owing to the correlations among the variables, the "real" dimensionality of our data matrix is somehow lower

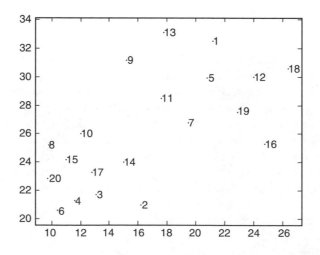

FIGURE 5.3 Scatter plot of the data in Table 5.2.

than v. Therefore, it would be very valuable to have a technique capable of concentrating in a few variables, and therefore in a few dimensions, the bulk of our information. This is exactly what is performed by PCA: it reduces the dimensionality of the data and extracts the most relevant part of the information, placing into the last dimensions the nonstructured information, i.e., the noise. According to these two characteristics, the information contained in very complex data matrices can be visualized in just one or a few plots.

From the mathematical point of view, the goal of PCA is to obtain, from v variables $(X_1, X_2, ..., X_v)$, v linear combinations having two important features: to be uncorrelated and to be ordered according to the explained variance (i.e., to the information they contain). The lack of correlation among the linear combinations is very important, since it means that each of them describes different "aspects" of the original data. As a consequence, the examination of a limited number of linear combinations (generally the first two or three) allows us to obtain a good representation of the studied data set.

From a geometrical point of view, what is performed by PCA corresponds to finding the direction which, in the v-dimensional space of the original variables, brings the greatest possible amount of information (i.e., explains the greatest variance). Once the first direction is identified, the second one is looked for: it will be the direction explaining the greatest part of the residual variance, under the constraint of being orthogonal to the first one. This process goes on until the vth direction has been found.

These new directions can be considered as the axes of a new orthogonal system, obtained after a simple rotation of the original axes. While in the original system each direction (i.e., each variable) brings with it, at least in theory, $1/v$ of total information, in the new system the information is concentrated in the first few directions and decreases progressively so that in the last ones no information can be found except noise.

The global dimensionality of the system is always that of the original data (v), but, since the last dimensions explain only a very small part of the information, they can be neglected and one can take into account only the first dimensions (the "significant components"). The projection of the objects in this space of reduced dimensionality retains almost all the information that can now also be analyzed in a visual way, by bi- or tri-dimensional plots. These new directions, linear combinations of the original ones, are the Principal Components (PC) or Eigenvectors.

With a mathematical notation, we can write:

$$\text{var}(Z_1) > \text{var}(Z_2) > ... > \text{var}(Z_v)$$

where $\text{var}(Z_i)$ is the variance explained by component i. Furthermore, since a simple rotation has been performed, the total variance is the same in the two systems of axes:

$$\sum \text{var}(X_i) = \sum \text{var}(Z_i)$$

The first PC is formed by the linear combination

$$Z_1 = a_{11}X_1 + a_{12}X_2 + \cdots + a_{1v}X_v$$

explaining the greatest variance, under the condition

$$\sum a_{1i}^2 = 1$$

This last condition notwithstanding, the variance of Z_1 could be made greater simply by increasing one of the values of a.

The second PC

$$Z_2 = a_{21}X_1 + a_{22}X_2 + \cdots + a_{2v}X_v$$

is the one having var(Z_2) as large as possible, under the conditions that

$$\sum a_{2i}^2 = 1$$

and that

$$\sum a_{1i}a_{2i} = 0$$

(this last condition assures the orthogonality of components one and two).

The lower order components are computed in the same way, always under the two conditions previously reported.

From a mathematical point of view, PCA is solved by finding the eigenvalues of the variance–covariance matrix; they correspond to the variance explained by the corresponding principal component. Since the sum of the eigenvalues is equal to the sum of the diagonal elements (trace) of the variance–covariance matrix, and since the trace of the variance–covariance matrix corresponds to the total variance, one has the confirmation that the variance explained by the principal components is the same as explained by the original data.

It is now interesting to locate each object in this new reference space. The coordinate on the first PC is computed simply by substituting into equation $Z_1 = a_{11}X_1 + a_{12}X_2 + \cdots + a_{1v}X_v$ the terms X_i with the values of the corresponding original variables. The coordinates on the other principal components are then computed in the same way.

These coordinates are named scores, while the constants a_{ij} are named loadings.

By taking into account the loadings of the variables on the different principal components, it is very easy to understand the importance of each single variable in constituting each PC. A high absolute value means that the variable under examination plays an important role for the component, while a low absolute value means that it has a very limited importance.

If a loading has a positive sign, it means that the objects with a high value of the corresponding variable have high scores on that component. If the sign is negative, then the objects with low values of that variable will have high scores. As already mentioned, after a PCA the information is mainly concentrated on the first components. As a consequence, a plot of the scores of the objects on the first components allows the direct visualization of the global information in a very efficient way. It is now very easy to detect similarity between objects (similar objects have a very similar position in the space), the presence of outliers (they are very far from all other objects), or the existence of clusters. Taking into account at the same time scores and loadings it is also possible to interpret very easily the differences among objects or groups of objects, since it is immediately understandable which are the variables giving the greatest contribution to the phenomenon under study.

Mathematically speaking, we can say that the original data matrix $\mathbf{X}_{o,v}$ (having as many rows as objects and as many columns as variables) has been decomposed into a matrix of scores $\mathbf{S}_{o,c}$ (having as many rows as objects and as many columns as retained components, with c usually $\ll v$) and a matrix of loadings $\mathbf{L}_{c,v}$ (having as many rows as retained components and as many columns as variables). If, as usual, $c < v$, a matrix of the residuals $\mathbf{E}_{o,v}$, having the same size as the original data set, contains the differences between the original data and the data reconstructed by the PCA model (the smaller the values of this matrix, the higher the variance explained by the model).

We can therefore write the following relationship:

$$\mathbf{X}_{o,v} = \mathbf{S}_{o,c} \times \mathbf{L}_{c,v} + \mathbf{E}_{o,v}$$

Now, let us see the application of PCA to a real data set.[1] Seven variables describing protein composition have been measured on 23 samples of peas, of different cultivars. Fifteen samples were from smooth pea cultivars, while eight samples were from wrinkled pea cultivars. The data are reported in Table 5.3. It could be interesting to check whether the protein composition of the smooth peas is different from that of the wrinkled peas. When looking separately at each of the seven variables, it can be seen that none of them completely separates the two categories. Therefore, one could say that, though some variables are on average higher in one category (e.g., the vicilin/legumin ratios are higher in the wrinkled peas), it is not possible to discriminate between smooth and wrinkled peas. As a consequence, one could look for different (and possible more expensive to determine) variables.

After a PCA (Figure 5.4), it is instead evident that the information present in the seven variables is sufficient to clearly discriminate the two categories. Once more, it has to be pointed out that taking into account all the variables at the same time gives much more information than just looking at one variable at a time.

Now, let us go one step back and try to understand how this result has been obtained. First, since the variables have different magnitudes and variances, a normalization has to be performed, in such a way that each variable will have the same importance. Autoscaling is the most frequently used normalization, which is done by subtracting from each variable its mean value and then dividing the result by its standard deviation. After that, each normalized variable will have mean = 0 and variance = 1. Table 5.4 shows the data after autoscaling.

TABLE 5.3
Protein Composition of Peas[1] (Reduced Data Set): (a) 1 = Smooth Pea Cultivars; 2 = Wrinkled Pea Cultivars; (b) Laurell's Technique; (c) Ultracentrifugation

Object	Category (a)	Protein	Nonprot. Material	Albumin	Globulin	Insoluble Prot. Fract.	Vicilin/ Legumin (b)	Vicilin/ Legumin (c)
1	1	219	20.7	24.3	55.7	20.0	2.2	2.0
2	1	273	30.2	12.3	61.0	26.6	1.3	1.5
3	1	255	17.8	19.3	53.8	26.9	1.5	2.0
4	1	262	30.2	13.1	63.2	23.5	1.6	2.3
5	1	242	20.8	20.8	52.6	26.5	0.8	1.3
6	1	235	16.1	23.2	60.8	16.0	0.8	1.4
7	1	272	14.9	17.9	62.1	19.9	0.8	1.3
8	1	235	24.5	25.1	59.6	14.9	0.8	1.4
9	1	225	22.0	25.0	58.8	16.1	1.9	1.8
10	1	195	20.0	15.1	58.6	26.2	2.1	2.1
11	1	181	18.7	16.1	65.4	18.4	2.7	3.2
12	1	236	16.6	20.0	57.0	23.0	1.2	1.6
13	1	261	22.1	19.2	63.7	17.0	1.3	1.6
14	1	244	21.9	19.6	65.0	22.2	1.8	1.9
15	1	239	32.1	27.9	58.0	14.1	1.6	1.6
16	2	263	19.8	21.9	59.4	18.6	2.5	2.5
17	2	263	20.3	22.8	60.3	16.8	2.9	2.8
18	2	309	18.5	24.6	58.5	16.8	2.2	2.5
19	2	241	16.7	24.0	58.6	17.3	2.5	3.7
20	2	241	19.3	24.6	55.6	19.7	3.2	3.2
21	2	292	21.3	20.0	54.6	25.3	2.0	3.0
22	2	287	21.2	21.5	54.7	23.7	4.3	3.3
23	2	278	20.0	23.1	55.6	21.3	2.5	4.7

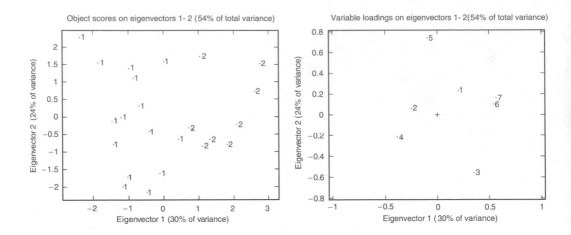

FIGURE 5.4 PCA of the data of Table 5.3. On the left, the score plot of the objects (coded according to the category number), on the right the loading plot of the variables (coded according to the order in Table 5.3).

The results of PCA are such that PC1 explains 30.3% of the total variance and PC2 23.6%. This means that the PC1–PC2 plots shown in Figure 5.4 explain 53.9% of total variance.

Table 5.5 shows the loadings of the variables on PC1 and PC2. From it, the loading plot in Figure 5.4 is obtained.

TABLE 5.4
Autoscaled Data

Protein	Nonprot. Material	Albumin	Globulin	Insoluble Prot. Fract.	Vicilin/Legumin (b)	Vicilin/Legumin (c)
−1.040	−0.094	0.837	−0.871	−0.115	0.304	−0.326
0.777	2.042	−2.144	0.614	1.495	−0.727	−0.887
0.171	−0.746	−0.405	−1.403	1.569	−0.498	−0.326
0.407	2.042	−1.946	1.230	0.739	−0.383	0.010
−0.266	−0.071	−0.032	−1.739	1.471	−1.300	−1.111
−0.502	−1.128	0.564	0.558	−1.090	−1.300	−0.999
0.743	−1.398	−0.753	0.922	−0.139	−1.300	−1.111
−0.502	0.760	1.036	0.222	−1.359	−1.300	−0.999
−0.838	0.198	1.011	−0.002	−1.066	−0.040	−0.551
−1.847	−0.251	−1.449	−0.058	1.398	0.189	−0.214
−2.318	−0.543	−1.200	1.846	−0.505	0.876	1.018
−0.468	−1.015	−0.231	−0.507	0.617	−0.842	−0.775
0.373	0.221	−0.430	1.370	−0.846	−0.727	−0.775
−0.199	0.176	−0.331	1.734	0.422	−0.154	−0.439
−0.367	2.469	1.732	−0.227	−1.554	−0.383	−0.775
0.440	−0.296	0.241	0.166	−0.456	0.647	0.234
0.440	−0.184	0.465	0.418	−0.895	1.105	0.570
1.988	−0.588	0.912	−0.086	−0.895	0.304	0.234
−0.300	−0.993	0.763	−0.058	−0.773	0.647	1.579
−0.300	−0.409	0.912	−0.899	−0.188	1.449	1.018
1.416	0.041	−0.231	−1.179	1.178	0.075	0.794
1.248	0.019	0.142	−1.151	0.788	2.709	1.130
0.945	−0.251	0.539	−0.899	0.203	0.647	2.699

TABLE 5.5
Loadings of the Variables on PC1 and PC2

	Protein	Nonprot. Material	Albumin	Globulin	Insoluble Prot. Fract.	Vicilin/Legumin (b)	Vicilin/Legumin (c)
PC1	0.214	−0.239	0.370	−0.372	−0.080	0.546	0.563
PC2	0.237	0.066	−0.557	−0.219	0.739	0.115	0.151

From the score plot in Figure 5.4 it can be seen that PC1 perfectly separates the two categories. By looking at the loading plot and at Table 5.5 it is possible to know which are the variables mainly contributing to PC1 (and therefore to the separation). Variables six and seven (the two vicilin/legumin ratios) have the loadings with the highest absolute values, both being positive. This means that these ratios are higher in the wrinkled peas (the objects of category two, being on the right side of the score plot, have higher scores on PC1) than in the smooth peas. Also albumin and globulin have high absolute values of their loadings on PC1, though having opposite signs (positive for albumin, negative for globulin). This means that wrinkled peas have a higher content of albumin and a lower content of globulin. Table 5.6 reports the scores of the objects on PC1 and PC2.

As previously shown, the scores of an object are computed by multiplying the loadings of each variable by the value of the variable. As an example, let us compute the score of sample one on PC1

TABLE 5.6
Scores of the Objects on PC1 and PC2

Object	Category	Score on PC1	Score on PC2
1	1	0.425	−0.627
2	1	−2.358	2.264
3	1	0.006	1.576
4	1	−1.841	1.548
5	1	−0.858	1.100
6	1	−1.023	−1.735
7	1	−1.453	−0.119
8	1	−1.153	−1.998
9	1	−0.099	−1.623
10	1	−0.978	1.386
11	1	−0.404	−0.439
12	1	−0.700	0.304
13	1	−1.408	−0.784
14	1	−1.217	−0.004
15	1	−0.466	−2.148
16	2	0.714	−0.312
17	2	1.151	−0.705
18	2	1.304	−0.647
19	2	1.781	−0.806
20	2	2.085	−0.226
21	2	1.040	1.724
22	2	2.797	1.535
23	2	2.653	0.737

(since the autoscaled data have been used, these are the values to be taken into account):

$$0.214(-1.040) + (-0.239)(-0.094) + 0.370 \times 0.837 + (-0.372)(-0.871)$$
$$+ (-0.080)(-0.115) + 0.546 \times 0.304 + 0.563(-0.326) = 0.425$$

IV. PROCESS MONITORING AND QUALITY CONTROL

When running a process it is very important to know whether it is under control (i.e., inside its natural variability) or out of control (i.e., in a condition which is not typical and therefore can lead to an accident).

Analogously, when producing a product it is very important to know whether each single piece is within specifications (i.e., close to the "ideal" product, inside its natural variability) or out of specifications (i.e., significantly different from the "standard" product and therefore in a condition possibly leading to a complaint by the final client).

PCA is the basis for a multivariate process monitoring and a multivariate quality control, which are much more effective than the usually applied univariate approaches.[2]

After having collected a relevant number of observations describing the "normally operating" process (or the "inside specification" products), encompassing all the sources of normal variability, it will be possible to build a PCA model defining the limits inside which the process (or the product) should stay.

Any new set of measurements (a vector $\mathbf{x}_{1,v}$) describing the process in a given moment (or a new product) will be projected onto the previously defined model by using the following equation: $\mathbf{s}_{1,c} = \mathbf{x}_{1,v} \times \mathbf{L}'_{c,v}$. From the computed scores, it can be estimated how far from the barycenter of the model, i.e., from the "ideal" process (or product) it is.

Its residuals can also be easily computed: $\mathbf{e}_{1,v} = \mathbf{x}_{1,v} - \mathbf{s}_{1,c} \times \mathbf{L}_{c,v}$ ($\mathbf{e}_{1,v}$ is the vector of the residuals, and each of its v elements corresponds to the difference between the measured and the reconstructed value of each variable). From them, it can be understood how well the sample is reconstructed by the PCA model, i.e., how far from the model space (a plane, in case $c2$) it lies.

Statistical tests make possible the automatic detection of an outlier in both cases (they are defined as T^2 outliers in the first case and Q outliers in the second case). With these simple tests it will be possible to detect a fault in a process or to reject a bad product by checking just two plots, instead of as many plots as variables as in the case of the Sheward charts commonly used when the univariate approach is applied. Furthermore, the multivariate approach is much more robust, since it will lead to a lower number of false negatives and false positives, and much more sensitive, since it allows the detection of faults at an earlier stage. Finally, the contribution plots will easily outline which variables are responsible for the sample being an outlier.

V. THREE-WAY PRINCIPAL COMPONENT ANALYSIS

It can happen that the structure of a data set is such that a standard two-way table (objects versus variables) is not enough to describe it. Let us suppose that the same analyses have been performed at different sampling sites on different days. A third way needs to be added to adequately represent the data set, which can be imagined as a parallelepiped of size $I \times J \times K$, where I is the number of sampling sites (objects), J is the number of variables, and K is the number of sampling times (conditions).[3,4]

To apply standard PCA, these three-way data arrays $\underline{\mathbf{X}}$ have to be matricized to obtain a two-way data table. This can be done in different ways, according to what one is interested in focusing on.

If we are interested in studying each "sampling", a matrix $\mathbf{X}'_b$ is obtained having $I \times K$ rows and J columns. This approach is very straightforward in terms of computation, but since $I \times K$ is usually a rather large number, the interpretation of the resulting score plot can give some problems.

To focus on the sampling sites, the data array $\underline{\mathbf{X}}$ can be matricized to $\mathbf{X}'_a$ (I rows, $J \times K$ columns). The interpretability of the score plot is usually very high, but since $J \times K$ is usually a rather large number, the interpretation of the loading plot is very difficult.

The same considerations can be made when focusing on the sampling times: in this case, $\mathbf{X}'_c$ is obtained (K rows, $I \times J$ columns).

Three-way PCA allows a much easier interpretation of the information contained in the data set, since it directly takes into account its three-way structure. If the Tucker3 model is applied, the final result is given by three sets of loadings together with a core array describing the relationship among them. If the number of components is the same for each way, the core array is a cube. Each of the three sets of loadings can be displayed and interpreted in the same way as a score plot of standard PCA.

In the case of a cubic core array a series of orthogonal rotations can be performed on the three spaces of the objects, variables, and conditions, looking for the common orientation for which the core array is as much body-diagonal as possible.

If this condition is sufficiently achieved, then the rotated sets of loadings can also be interpreted jointly by overlapping them.

An example of application of three-way PCA is a data set from the Venice lagoon.[5] In it, 11 chemical variables (chlorophyll-α, total suspended matter, water transparency, fluorescence, turbidity, suspended solids, NH_4^+, NO_3^-, P, COD, and BOD_5) have been measured monthly in 13 sampling sites (see Figure 5.5) during the period May 1987 to December 1990, for a total of 44 months.

The resulting loading plots (Figure 5.6) clearly show the effect of the sampling sites, with the pollution regularly decreasing from the industrial region to the open sea. The time effect can be split into a seasonal effect and a general trend, with an increase of eutrofication.

Table 5.7 shows some types of data sets on which three-way PCA can be successfully applied.

VI. CLASSIFICATION

In Section III we could verify that the smooth and the wrinkled peas are indeed well separated in the multivariate space of the variables. Therefore, we can say that we have two really different classes. Let us suppose we now get some smashed peas (so that we can not see if they are smooth or wrinkled) and we want to know what their class is. After having performed the chemical analyses, we can add these data to the previous data set, run a PCA and see where the new samples are placed. This will be fine if the new samples fall inside one of the clouds of points corresponding to a category, but what if they fall in a somehow intermediate position? How can we say with "reasonable certainty" that the new samples are from smooth or from wrinkled peas? We know that PCA is a very powerful technique for data display, but we realize that we need something different if we want to classify new samples. What we want is a technique producing some "decision rules" discriminating among the possible categories.

While PCA is an "unsupervised" technique, the classification methods are "supervised" techniques. In these techniques the category of each of the objects on which the model is built must be specified in advance.

The most commonly used classification techniques are Linear Discriminant Analysis (LDA) and Quadratic Discriminant Analysis (QDA). They define a set of delimiters (according to the number of categories under study) in such a way that the multivariate space of the objects is divided into as many subspaces as the number of categories, and that each point of the space belongs to one

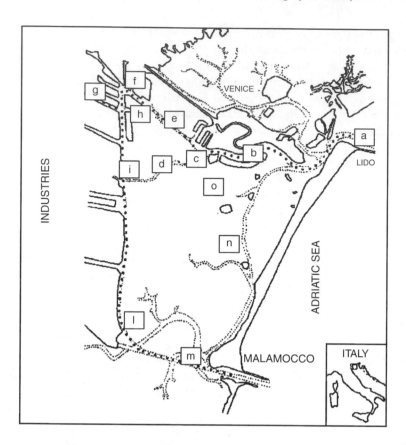

FIGURE 5.5 The location of the 13 sampling sites.

and only one subspace. Rather than describing in detail the algorithms behind these techniques, special attention will be given to the critical points of a classification.

As previously stated, the classification techniques use objects belonging to the different categories to define boundaries delimiting regions of the space. The final goal is to apply these classification rules to new objects for their classification into one of the existing categories. The performance of the technique can be expressed as classification ability and prediction ability. The difference between "classification" and "prediction", though quite subtle at first glance, is actually very important and its underestimation can lead to very bitter deceptions.

TABLE 5.7
Data Sets on which Three-Way PCA Can Be Applied

Field of Application	Objects	Variables	Conditions
Environmental analysis	Air or water samples	Chemico-physical analyses	Time
Environmental analysis	Water samples (different locations)	Chemico-physical analyses	Depth
Panel tests	Food products (oils, wines)	Attributes	Assessors
Food chemistry	Foods (cheeses, spirits,…)	Chemical composition	Ageing
Food chemistry	Foods (oils, wines,…)	Chemical composition	Crops
Sport medicine	Athletes	Blood analyses	Time after effort
Process monitoring	Batches	Chemical analyses	Time

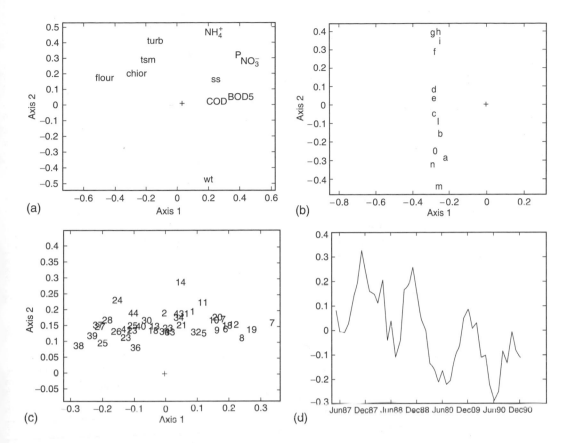

FIGURE 5.6 The results of the three-way PCA applied to the Venice data set. (a) Plot of the loadings of the variables; (b) plot of the loadings of the sampling sites; (c) plot of the loadings of the months; (d) plot of the loadings of the month on axis 1 versus time.

Classification ability is the capability of assigning to the correct category the same objects used to build the classification rules, while prediction ability is the capability of assigning to the correct category objects that have not been used to build the classification rules. Since the final goal is the classification of new samples, it has to be clear that the predictive ability is by far the most important figure-of-merit to be looked at.

The results of a classification method can be expressed in several ways. The most synthetic one is the percentage of correct classifications (or predictions). Note that in the following, only the term "classification" will be used, but it has to be understood as "classification or prediction." This can be obtained as the number of correct classifications (independently of the category) divided by the total number of objects, or as the average of the performance of the model over all the categories. The two results are very similar when the sizes of all the categories are very similar, but can be very different if the sizes are quite different. Let us consider the case shown in Table 5.8. The very poor performance of category three, by far the smallest one, almost does not affect the classification rate computed on the global number of classifications, while it produces a much lower result if the classification rate is computed as the average of the three categories.

A more complete and detailed overview of the performance of the method can be obtained by using the classification matrix that also allows to know the categories to which the wrongly classified objects are assigned (in many cases the cost of an error can be quite different according to the category the sample is assigned to). In it, each row corresponds to the true category and each column to the category to which the sample has been assigned. Continuing with the previous

TABLE 5.8
Example of the Performance of a Classification Technique

Category #	Objects	Correct Class	Correct Class (%)
1	112	105	93.8
2	87	86	98.9
3	21	10	47.6
Total	220	201	91.4/80.1

example, a possible classification matrix is the one shown in Table 5.9. From it, it can be seen that the 112 objects of category one were classified in the following way: 105 correctly to category one, none to category two, and seven to category three. In the same way, it can be deduced that all the objects of category three which were not correctly classified have been assigned to category one. Therefore, it is easy to conclude that category two is well defined and that the classification of its objects gives no problems at all, while categories one and three are quite overlapping. As a consequence, to have a perfect classification more effort must be put into better separating categories one and three. All this information cannot be obtained from just the percentage of correct classifications.

If overfitting occurs, then the prediction ability will be much worse than the classification ability. To avoid it, it is very important that the sample size is adequate to the problem and to the technique. A general rule is that the number of objects should be more than five times (at least, no less than three times) the number of parameters to be estimated. LDA works on a pooled variance–covariance matrix: this means that the total number of objects should be at least five times the number of variables. QDA computes a variance–covariance matrix for each category, which makes it a more powerful method than LDA, but this also means that each category should have a number of objects at least five times higher than the number of variables. This is a good example of how the more complex, and therefore "better" methods, sometimes cannot be used in a safe way because their requirements do not correspond to the characteristics of the data set.

VII. MODELING

In classification, the space is divided into as many subspaces as categories, and each point belongs to one and only one category. This means that the samples that will be predicted by such methods must belong to one of the categories used to build the models; if not, they will anyway be assigned to one of them. To make this concept clearer, let us suppose the use of a classification technique to discriminate between water and wine. Of course, this discrimination is very easy. Each sample of water will be correctly assigned to the category "water" and each sample of wine will be correctly assigned to the category "wine." But what happens with a sample of orange squash? It will be

TABLE 5.9
Example of a Classification Matrix

Category	1	2	3
1	105	0	7
2	1	86	0
3	11	0	10

assigned either to the category "water" (if variables such as alcohol are taken into account) or to the category "wine" (if variables such as color are considered). The classification techniques are therefore not able to define a new sample as being "something different" from all the categories of the training set. This is instead the main feature of the modeling techniques.

Though several techniques are used for modeling purposes, UNEQ (one of the modeling versions of QDA) and Soft Independent Model of Class Analogy (SIMCA) are the most used. While in classification every point of the space belongs to one and only one category, with these techniques the models (one for each category) can overlap and leave some regions of the space unassigned. This means that every point of the space can belong to one category (the sample has been recognized as a sample of that class), to more than one category (the sample has such characteristics that it could be a sample of more than one class), or to none of the categories (the sample has been considered as being different from all the classes).

Of course, the "ideal" performance of such a method would be not only to correctly classify all the samples in their categories (as in the case of a classification technique), but also be such that the models of each category could be able to accept all the samples of that category and to reject all the samples of the other categories. The results of a modeling technique are expressed the same way as in classification, plus two very important parameters: specificity and sensitivity. For category c, its specificity (how much the model rejects the objects of different categories) is the percentage of the objects of categories different from c rejected by the model, while its sensitivity (how much the model accepts the objects of the same category) is the percentage of the objects of category c accepted by the model.

While the classification techniques need at least two categories, the modeling techniques can also be applied when only one category is present. In this case the technique detects if the new sample can be considered as a typical sample of that category or not. This can be very useful in the case of Protected Denomination of Origin products, to verify whether a sample, declared as having been produced in a well-defined region, has indeed the characteristics typical of the samples produced in that region.

The application of a multivariate analysis will greatly reduce the possibility of frauds. While an "expert" can adulterate a product in such a way that all the variables, independently considered, still stay in the accepted range, it is almost impossible to adulterate a product in such a way that its multivariate "pattern" is still accepted by the model of the original product, unless the amount of the adulterant is so small that it becomes unprofitable from the economic point of view.

VIII. CALIBRATION

Let us imagine we have a set of wine samples and that on each of them the FT-IR spectrum is measured, together with some variables such as alcohol content, pH, or total acidity. Of course, chemical analyses will require much more time than a simple spectral measurement. It would therefore be very useful to find a relationship between each of the chemical variables and the spectrum. This relationship, after having been established and validated, will be used to predict the content of the chemical variables. It is easy to understand how much time (and money) this will save, since in a few minutes it will be possible to have the same results as previously obtained by a whole set of chemical analyses.

Generally speaking, we can say that multivariate calibration finds relationships between one or more response variables y and a vector of predictor variables $\mathbf{x}$. As the previous example should have shown, the final goal of multivariate calibration is not just to "describe" the relationship between the $\mathbf{x}$ and the y variables in the set of samples on which the relationship has been computed, but to find a real practical application for samples that in a following time will have the $\mathbf{x}$ variables measured.

The model is a linear polynomial ($y = b_0 + b_1 x_1 + b_2 x_2 + \cdots + b_K x_K + f$), where b_0 is an offset, the b_k ($k = 1, \ldots, K$) are regression coefficients, and f is a residual. The "traditional" method of calculating **b**, the vector of regression coefficients, is Ordinary Least Squares (OLS). However, this method has two major limitations that make it inapplicable to many data sets:

- It cannot handle more variables than objects.
- It is sensitive to collinear variables.

It can be easily seen that both these limitations do not allow the application of OLS to spectral data sets, where the samples are described by a very high number of highly collinear variables. If one wants to use OLS for such data anyway, the only way to do it is to reduce the number of variables and their collinearity through a suitable feature selection (see later).

When describing the PCA, it has been noticed that the components are orthogonal (i.e., uncorrelated) and that the dimensionality of the resulting space (i.e., the number of significant components) is much lower than the dimensionality of the original space. Therefore, it can be seen that both the aforementioned limitations have been overcome. As a consequence, it is possible to apply OLS to the scores originated by PCA. This technique is Principal Component Regression (PCR).

It has to be considered that Principal Components are computed by taking into account only the **x** variables, without considering at all the *y* variables, and are ranked according to the explained variance of the "*x* world." This means it can happen that the first PC has little or no relevance in explaining the response we are interested in. This can be easily understood by considering that, even when we have several responses, the PCs to which the responses have to be regressed will be the same.

Nowadays, the most favored regression technique is Partial Least Squares Regression (PLS or PLSR). As happens with PCR, PLS is based on components (or "latent variables"). The PLS components are computed by taking into account both the **x** and the **y** variables, and therefore they are slightly rotated versions of the Principal Components. As a consequence, their ranking order corresponds to the importance in the modeling of the response. A further difference with OLS and PCR is that, while the former must work on each response variable separately, PLS can be applied to multiple responses at the same time.

Because both PCR and PLS are based on latent variables, a very critical point is the number of components to be retained. Though we know that information is "concentrated" in the first few components and that the last components explain just noise, it is not always an easy task to detect the correct number of components (i.e., when information finishes and noise begins). Selecting a lower number of components would mean removing some useful information (underfitting), while selecting a higher number of components would mean incorporating some noise (overfitting).

Before applying the results of a calibration, it is very important to look for the presence of outliers. Three major types of outliers can be detected: outliers in the *x*-space (samples for which the *x*-variables are very different from that of the rest of the samples; they can be found by looking at a PCA of the *x*-variables), outliers in the *y*-space (samples with the *y*-variable very different from that of the rest of the samples; they can be found by looking at a histogram of the *y*-variable), and samples for which the calibration model is not valid (they can be found by looking at a histogram of the residuals).

The goodness of a calibration can be summarized by two values, the percentage of variance explained by the model and the Root Mean Square Error in Calibration (RMSEC). The former, being a "normalized" value, gives an initial idea about how much of the variance of the data set is "captured" by the model; the latter, being an absolute value to be interpreted in the same way as a standard deviation, gives information about the magnitude of the error.

As already described in the classification section and as pointed out at the beginning of this section, the goal of a calibration is essentially not to describe the relationship between the response and the *x*-variables of the samples on which the calibration is computed (training, or calibration, set), but to apply it to future samples where only the cheaper *x*-variables will be measured. In this case too, the model must be validated by using a set of samples different from those used to compute the model (validation, or test, set). The responses of the objects of the test set will be computed by applying the model obtained by the training set and then compared with their "true" response. From these values the percentage of variance explained in prediction and the Root Mean Square Error in Prediction (RMSEP) can be computed. Provided that the objects forming the two sets have been selected flawlessly, these values give the real performance of the model on new samples.

IX. VARIABLE SELECTION

Usually, not all the variables of a data set bring useful and nonredundant information. Therefore, a variable (or feature) selection can be highly beneficial, since from it the following results are obtained:

- Removal of noise and improvement of the performance
- Reduction of the number of variables to be measured and simplification of the model

The removal of noisy variables should always be looked for. Though some methods can give good results even with a moderate amount of noise disturbing the information, it is clear that their performance will increase when this noise is removed. So, feature selection is now widely applied also for those techniques (PLS and PCR) that in the beginning were considered to be almost insensitive to noise.

While noise reduction is a common goal for any data set, the relevance of the reduction of the number of variables in the final model depends very much on the kind of data constituting the data set, and a very wide range of situations are possible. Let us consider the extreme conditions:

- Each variable requires a separate analysis.
- All the variables are obtained by the same analysis (e.g., chromatographic and spectroscopic data).

In the first case, each variable not selected means a reduction in terms of cost and/or analysis time. The variable selection should therefore always be made on a cost/benefit basis, looking for the subset of variables leading to the best compromise between performance of the model and cost of the analyses. This means that, in the presence of groups of useful but highly correlated (and therefore redundant) variables, only one variable per group should be retained. With such data sets, it is also possible that a subset of variables giving a slightly worse result is preferred, if the reduction in performance is widely compensated by a reduction in cost or time.

In the second case, the number of retained variables has no effect on the analysis cost, while the presence of useful and correlated variables improves the stability of the model.

Intermediate cases can happen, when "blocks" of variables are present. As an example, take the case of olive oil samples, on each of which the following analyses have been run: a titration for acidity, the analysis of peroxides, a UV spectroscopy for ΔK, a GC for sterols, and another GC for fatty acids. In such a situation what counts is not the final number of variables, but the number of analyses one can save.

The only possible way to be sure that "the best" set of variables has been picked up is the "all-models" techniques testing all the possible combinations. Since, with k variables, the number of possible combinations is $2^k - 1$, it is easy to understand that this approach cannot be used unless the number of variables is really very low (e.g., with 30 variables more than 10^9 combinations should be tested).

The simplest (but least effective) way of performing a feature selection is to operate on a "univariate" basis, by retaining those variables having the greatest discriminating power (in case of a classification) or the greatest correlation with the response (in case of a calibration). By doing that each variable is taken into account by itself without considering how its information "integrates" with the information brought by the other (selected or unselected) variables. As a result, if several highly correlated variables are "good", they are all selected, without taking into account that, owing to their correlation, the information is highly redundant and therefore at least some of them can be removed without any decrease in the performance. On the other hand, those variables that, though not giving by themselves significant information, become very important when their information is integrated with that of other variables, are not taken into account.

An improvement is brought by the "sequential" approaches. They select the best variable first, then the best pair formed by the first and second, and so on in a forward or backward progression. A more sophisticated approach applies a look back from the progression to reassess previous selections. The problem with these approaches is that only a very small part of the experimental domain is explored and that the number of models to be tested becomes very high in case of highly dimensional data sets, such as spectral data sets. For instance, with 1000 wavelengths, 1000 models are needed for the first cycle (selection or removal of the first variable), 999 for the second cycle, 998 for the third cycle, and so on.

More "multivariate" methods of variable selection, especially suited for PLS applied to spectral data, are currently available. Among them, we can cite Interactive Variable Selection,[6] Uninformative Variable Elimination,[7] Iterative Predictor Weighting PLS,[8] and Interval PLS.[9]

X. FUTURE TRENDS

In future, multivariate analysis should be used more and more in everyday (scientific) life. Until recently, experimental work resulted in a very limited amount of data, the analysis of which was quite easy and straightforward. Nowadays, it is common to have instrumentation producing an almost continuous flow of data. One example is process monitoring performed by measuring the values of several process variables, at a rate of one measurement every few minutes (or even seconds). Another example is quality control of a final product of a continuous process on which an FT-IR spectrum is taken every few minutes (or seconds).

In Section VIII the case of wine FT-IR spectra was cited, from which the main characteristics of the product can be directly predicted. It is therefore clear that the main problem has shifted from obtaining a few data to the treatment of a huge amount of data. It is also clear that standard statistical treatment is not sufficient to extract all the information buried in them.

Many instruments already have some chemometrics routines built into their software in such a way that their use is totally transparent to the final user (and sometimes the word "chemometrics" is not even mentioned, to avoid possible aversion). Of course, they are "closed" routines, and therefore the user cannot modify them. It is quite obvious that it would be much better if chemometric knowledge were much more widespread, in order that the user could better understand what kind of treatment the data have undergone and eventually modify the routines in order to make them more suitable to the user's requirements. As computers become faster and faster, it is nowadays possible to routinely apply some approaches requiring very high computing power. Two of them are Genetic Algorithms (GA) and Artificial Neural Networks (ANN).

GA are a general optimization technique with good applicability in many fields, especially when the problem is so complex that it cannot be tackled with "standard" techniques. In chemometrics it has been applied especially in feature selection.[10] GA try to simulate the evolution of a species according to the Darwinian theory. Each experimental condition (in this case, each model) is treated as an individual, whose "performance" (in the case of a feature selection for a calibration problem, it can be the explained variance) is treated as its "fitness." Through operators simulating the fights among individuals (the best ones have a greatest probability of mating and thus spreading their genome), the mating among individuals (with the consequent "birth" of "offspring" having a genome that is derived by both the parents), and the occurrence of mutations, the GA result in a search pattern that, by mixing "logical" and "random" features, allows a much more complete search of complex experimental domains. ANN try to mimic the behavior of the nervous system to solve practical computational problems. As in life, the structural unit of ANN is the neuron. The input signals are passed to the neuron body, where they are weighted and summed, and then they are transformed, by passing through the transfer function into the output of the neuron. The propagation of the signal is determined by the connections between the neurons and by their associated weights. The appropriate setting of the weights is essential for the proper functioning of the network. Finding the proper weight setting is achieved in the training phase. The neurons are usually organized into three different layers: the input layer contains as many neurons as input variables, the hidden layer contains a variable number of neurons, and the output layer contains as many neurons as output variables. All units from one layer are connected to all units of the following layer. The network receives the input signals through the input layer. Information is passed to the hidden layer and finally to the output layer that produces the response.

These techniques are very powerful, but very often they are not applied in a correct way. In such cases, despite a very good performance on the training set (due to overfitting), they will show very poor results when applied to external data sets.

XI. CONCLUSION: THE ADVANTAGES AND DISADVANTAGES OF CHEMOMETRICS

In one of his papers, J. Workman Jr.[11] very efficiently depicts the advantages and disadvantages of multivariate thinking for scientists in industry.

From the eight advantages of chemometrics he clearly outlines, special relevance should be given to the following ones:

1. Chemometrics provides speed in obtaining real-time information from data.
2. It allows high quality information to be extracted from less resolved data.
3. It promises to improve measurements.
4. It improves knowledge of existing processes.
5. It has very low capital requirements, i.e., it is cheap.

The last point especially should convince people to give chemometrics a try. No extra equipment is required: just an ordinary computer and some chemometrical knowledge (or a chemometrical consultancy). It is certain that in the very worst cases the same information as found from a classical analysis will be obtained in a much shorter time and with much more evidence. In the great majority of cases, instead, even a simple PCA can provide much more information than what was previously collected. So, why are people so shy of applying chemometrics? In the same paper previously cited, Workman gives some very common reasons:

1. The perceived disadvantage of chemometrics is that there is widespread ignorance about what it is and what it can realistically accomplish.
2. This science is considered too complex for the average technician and analyst.
3. Chemometrics requires a change in one's approach to problem solving from univariate to multivariate thinking.

So, while chemometrics leads to several real advantages, its "disadvantages" lie only in the general reluctance to use it and accepting the idea that the approach followed over many years can turn out not to be the best one.

REFERENCES

1. Gueguen, J. and Barbot, J., Quantitative and qualitative variability of pea (*Pisum sativum L.*) protein composition, *J. Sci. Food Agric.*, 42, 209–224, 1988.
2. Kourti, T. and MacGregor, J. F., Process analysis, monitoring and diagnosis, using multivariate projection methods, *Chemom. Intell. Lab. Syst.*, 28, 3–21, 1995.
3. Geladi, P., Analysis of multi-way (multi-mode) data, *Chemom. Intell. Lab. Syst.*, 7, 11–30, 1989.
4. Smilde, A. K., Three-way analyses. Problems and prospects, *Chemom. Intell. Lab. Syst.*, 15, 143–157, 1992.
5. Leardi, R., Armanino, C., Lanteri, S., and Alberotanza, L., Three-mode principal component analysis of monitoring data from Venice lagoon, *J. Chemom.*, 14, 187–195, 2000.
6. Lindgren, F., Geladi, P., Rännar, S., and Wold, S., Interactive variable selection (IVS) for PLS. 1. Theory and algorithms, *J. Chemom.*, 8, 349–363, 1994.
7. Centner, V., Massart, D. L., de Noord, O. E., de Jong, S., Vandeginste, B. M., and Sterna, C., Elimination of uninformative variables for multivariate calibration, *Anal. Chem.*, 68, 3851–3858, 1996.
8. Forina, M., Casolino, C., and Pizarro Millán, C., Iterative predictor weighting (IPW) PLS: a technique for the elimination of useless predictors in regression problems, *J. Chemom.*, 13, 165–184, 1999.
9. Nørgaard, L., Saudland, A., Wagner, J., Nielsen, J. P., Munck, L., and Engelsen, S. B., Interval partial least-squares regression (iPLS): a comparative chemometric study with an example from near-infrared spectroscopy, *Appl. Spectrosc.*, 54, 413–419, 2000.
10. Leardi, R., Application of genetic algorithm-PLS for feature selection in spectral data sets, *J. Chemom.*, 14, 643–655, 2000.
11. Workman, J. Jr., The state of multivariate thinking for science in industry: 1980–2000, *Chemom. Intell. Lab. Syst.*, 60, 13–23, 2002.

FURTHER READING

A. Books

Beebe, K. R., Pell, R. J., and Seasholtz, M. B., *Chemometrics: A Practical Guide*, Wiley, New York, 1998.
Brereton, R. G., *Chemometrics — Data Analysis for the Laboratory and Chemical Plant*, Wiley, Chichester, 2003.
Leardi, R., Ed., *Nature-Inspired Methods in Chemometrics: Genetic Algorithms and Artificial Neural Networks*, Data Handling in Science and Technology Series, Vol. 23, Elsevier, Amsterdam, 2003.
Manly, B. F. J., *Multivariate Statistical Methods. A Primer*, Chapman & Hall, London, 1986.
Martens, H. and Naes, T., *Multivariate Calibration*, Wiley, New York, 1991.
Massart, D. L., Vandeginste, B. G. M., Deming, S. N., Michotte, Y., and Kaufman, L., *Chemometrics: A Textbook*, Data Handling in Science and Technology Series, Vol. 2, Elsevier, Amsterdam, 1990.
Massart, D. L., Vandeginste, B. G. M., Buydens, L. M. C., de Jong, S., Lewi, P. J., and Smeyers-Verbeke, J., *Handbook of Chemometrics and Qualimetrics. Part A*, Data Handling in Science and Technology Series, Vol. 20A, Elsevier, Amsterdam, 1997.
Massart, D. L., Vandeginste, B. G. M., Buydens, L. M. C., de Jong, S., Lewi, P. J., and Smeyers-Verbeke, J., *Handbook of Chemometrics and Qualimetrics. Part B*, Data Handling in Science and Technology Series, Vol. 20B, Elsevier, Amsterdam, 1998.
Meloun, M., Militky, J. and Forina, M., *Chemometrics for Analytical Chemistry*, PC-Aided Statistical Data Analysis, Vol. 1, Ellis Horwood, Chichester, 1992.
Meloun, M., Militky, J. and Forina, M., *Chemometrics for Analytical Chemistry*, PC-Aided Regression and Related Methods, Vol. 2, Ellis Horwood, Hemel Hempstead, 1994.
Sharaf, M. A., Illman, D. L. and Kowalski, B. R., In *Chemometrics*, Elving, P. J. and Winefordner, J. D., Eds., Chemical Analysis, A Series of Monographs on Analytical Chemistry and its Applications Series, Vol. 82, Wiley, New York, 1986.

B. Web Sites

http://ull.chemistry.uakron.edu/chemometrics/.
http://www.acc.umu.se/~tnkjtg/chemometrics/index.html.
http://www.statsoft.com/textbook/stathome.html.

6 Major Air Components

Wei-Hsin Chen and Jau-Jang Lu

CONTENTS

I. PHYSICAL PROPERTIES

Many different gases exist in our environment. Measurements of their concentrations are the prerequisites for understanding the physico-chemical processes in the environment. The observation of trends of greenhouse gases like CO_2 and CH_4 is an example. The intention of this chapter is to discuss the chromatographic analysis of the most abundant gases in the atmosphere, water, and soil. The analysis of gaseous forms of other chemicals, such as volatile organic compounds (VOCs), polyaromatic hydrocarbons (PAHs), etc., is beyond the scope of this chapter and will be presented elsewhere in this book.

A. ATMOSPHERE

Planet Earth is enveloped by a thin layer of gaseous medium in which we exist — the atmosphere. It is truly a thin layer compared with the radius of the earth. Almost 99% of the atmosphere's mass lies in the first 30 km above sea level.

The two most abundant gases in the atmosphere are nitrogen and oxygen, which account for 78.09 and 20.94% of the troposphere by volume, respectively. The rest of the ~1% is composed of other gases, such as argon, carbon dioxide, methane, hydrogen, carbon monoxide, and so on. These gases, along with nitrogen and oxygen, are considered to be the major air components.

Among these major air components, oxygen is primarily involved in the biosphere in the processes of photosynthesis and respiration, and has also long been recognized as an important component of biogeochemical processes on land, in the air, and in water. Changes in the oxygen concentration have been routinely used in biogeochemical studies.

Carbon dioxide is an important constituent of interest in the field of air pollution, from both local and global perspectives. Its anthropogenic sources include combustion of fossil fuels and depletion of rain forests. The background concentration of carbon dioxide in the northern hemisphere has risen from approximately 310 ppm in the middle of the 20th century to 369 ppm in 2000.[1] This situation is connected to the greenhouse effect and has become an important issue beyond national boundaries. Moreover, enormous amounts of CO are generated by incomplete combustion of carbonaceous fuels such as wood, coal, gasoline, and natural gas. From the human health point of view, the current U.S. Occupational Safety and Health Administration (OSHA) permissible exposure limit (PEL) for CO is 50 ppm.[2] The exposure standard recommended by the U.S. National Institute for Occupational Safety and Health (NIOSH) is 35 ppm with a ceiling value of 200 ppm. The U.S. Environmental Protection Agency (EPA) has set an ambient air quality standard of 9 ppm, averaged over an 8-h period, and 35 ppm for 1 h, not to be exceeded more than once a year. The American Conference of Governmental and Industrial Hygienists (ACGIH) has set a threshold limit value for CO of 25 ppm.[3]

Methane is the simplest of the hydrocarbon molecules. While it is colorless, odorless, nontoxic, and rather inert, it is one of the greenhouse gases and is estimated to contribute 15% of the greenhouse effect.[4,5] Its atmospheric concentration is increasing, even though its rate of increase has declined.[6] Naturally occurring methane can be classified as either a thermogenic or biogenic process. The former process refers to the thermal degradation of organic matter at depth within sedimentary basins and is commonly associated with coal and accumulations of petroleum and natural gas. Biogenic methane is produced under anaerobic, near-surface conditions by microbial degradation of organic matter, such as by ruminant animals[7] and in rice paddies.[8] Terrestrial wetlands are another important source of methane. Approximately 20 to 40% (120 to 200 Tg/year) of total methane emission comes from these wetland areas.[9] On the other hand, anthropogenic methane can be derived from natural gas pipeline leakages, oil and gas wells, sewer pipes and septic systems, burial compost, landfill sites, spilled petroleum, and so on.

B. WATER

Dissolved gases, in fresh and sea water, play an important role in marine and aquatic science. Deviations from equilibrium concentrations are common and reflect biological activity and physical processes acting on the system. Of the most concentrated gases in aerobic water, N_2 and O_2 are affected by biological and physical processes, whereas Ar is affected strictly by physical processes. This distinction has enabled limnologists and oceanographers to separate physically driven processes (such as bubble release or injection) from biological processes (such as net biological oxygen exchange). When the normal atmosphere contacts the water surface, gases can dissolve in water according to their pressures. This process plays an important role with regard to dissolved gases in water. The pressures of these different gases also determine their solubility in water. Henry's Law concludes the above description using the following equation:

$$C(mg/l) = K(mg/l\ atm) \times P(atm)$$

where C represents the gas solubility, P the partial pressure of the specific gas in the atmosphere, and K the Henry's Law constant of the specific gas. Henry's Law finds its application in chromatographic analysis of dissolved gases following headspace sampling methods.[10–12]

Most major air components are nonpolar molecules, such as nitrogen, oxygen, hydrogen, and carbon dioxide. Argon is a single-atom novel gas. Methane is also nonpolar according to its symmetric and tetrahedron structure. Nitrogen and oxygen have very low solubility in water because of their low-polarity structures. Solubility decreases from CH_4 (30 ml/l) to He (8.7 ml/l) according to the following sequence: $CH_4 > O_2 > CO > H_2 > N_2 > Ar > Ne > He$.

C. SOIL

Smith[13] has defined the soil atmosphere as the gaseous phase of heterogeneous porous material on the surface of the earth and the air that fills the pores between the solid particles which are not occupied by water. The air above the soil may significantly modify the soil atmosphere. Many gaseous compounds can be found in subsurface soil atmosphere, such as atmospheric gases, carbon- and nitrogen-containing gases, rare gases, hydrogen, and so on. Complex processes, including mantle degassing, crustal radiogenic production, rock alternation, biogenic activity and atmospheric dilution, at various depths, are involved in the generation of these gases prior to their reaching the ground surface. Furthermore, because several situations such as mixing, contamination, chemical reaction, and differential solubility in groundwater may change the original gas concentrations, a wide range of concentrations for a single gas is expected, even within a very small area. The study of origins and behaviors of gases, including their concentrations in soil atmospheres helps to understand a fracture network allowing degassing.[14]

Oxygen in soil atmospheres, for example, can act as a reactant in the aerobic decomposition of organic pollutants into smaller and perhaps less harmful products in soil. These reactions are possible because of the high concentration of oxygen in soil atmospheres (21% in most situations) compared with that in water (9.2 mg/l at 25°C, saturated). Table 6.1 lists selected native soil gases and their origins. Among these soil gases, oxygen, nitrogen, and methane will be discussed regarding their analyses by chromatographic methods.

II. SAMPLE PREPARATION

Although chromatographic systems are highly sensitive, it is also true that we must be concerned with concentrations of target materials and the detectability limit of the measurement system. Because of the relatively high concentrations of major air components in most of the studies, usually no further preconcentration of the analytes is necessary prior to chromatographic analysis. In some cases, sample volumes can even be reduced down to the level of μl. Carbon dioxide and carbon monoxide can first be reduced to methane if a flame ionization detector (FID) coupled with gas chromatography (GC) is used. However, moisture is considered a major interference and shall

---

TABLE 6.1
Native Soil Gases and their Origins[15]

Gases	Origins
Oxygen, nitrogen, and argon	Earth's atmosphere
Volatile fatty acids, aldehydes, alcohols, and ketones	Microbial decomposition of plant residues
Ammonia, nitrous oxide, nitric oxide, nitrogen dioxide, and nitrogen	Microbiological and chemical processes in the soil
Hydrogen	Microbial decomposition of organic matter under strong reducing conditions
Organosulfur compounds, carbon disulfide, carbonyl sulfide, and hydrogen sulfide	Microbial decomposition of sulfuramino acids and other sulfur-containing compounds
Carbon monoxide	Chemical decomposition of soil organic matter
Methane, ethylene, and other C_2–C_4 hydrocarbons	Microbial activity in the soil; diffusion from underlying oil- or gas-bearing strata

be removed before the chromatographic analysis. All samples except as mentioned otherwise should be analyzed as soon as possible after collection, either on-site or in the laboratory.

A. ATMOSPHERE

In a study to measure carbon monoxide levels in the ambient air,[16] a specialized sorbent designed exclusively for carbon monoxide is packed into a 6 mm outside diameter (4 mm inside diameter) glass tube and held in place by glass wool plugs. Carbon monoxide is then collected by drawing a fixed volume of air sample through the sorbent tube using a SKC pump at a flow rate of 100 ml/min. The sorbent (in the tube) is stored for chromatographic analysis.

Air samples from various shipboards and 36 fixed sites, which are part of a global network to measure atmospheric methane, were collected and analyzed for the period from 1983 to 1992.[6] The sites were chosen on the basis of their remoteness from known or suspected methane source regions. This results in air samples that are representative of large and well-mixed volumes of the atmosphere. Until 1989, air samples are collected in 0.5 l, cylindrical, Pyrex flasks fitted with a solid plug, greased, with a ground-glass stopcock on each end. Because of their better performance after long storage periods, cylindrical Pyrex flasks (2.5 l) with two glass piston stopcocks sealed with Teflon O-rings were introduced. All the flasks are designed to ensure vacuum-tight connections with the sampling and analytical system. To pump air samples into the flasks, a portable Martin and Kitzis Sampler (MAKS) was employed. Some features of the MAKS include a light shield to avoid exposure of the flask to sunlight, which is a possible source of CO and H_2 contamination, and a back-pressure regulator to ensure that the final pressure in the flasks is held constant after shutting off the pump, allowing the operator to check for gas leakages. All samples are sent to a designated laboratory for analysis. Samples that are identified to have sampling or analytical errors, or which appeared to be contaminated by CH_4 sources, were excluded from the data set.

B. WATER

When efforts are made to effectively collect water samples for the analysis of dissolved oxygen (DO) and $\delta^{18}O-O_2$ values, preevacuated 200 ml glass vessels fitted with a high-vacuum glass stopcock and 3/8 in. outside diameter, 90° arm inlet at one end are used.[17] Saturated mercuric chloride (1 ml) is added to each vessel and dried before evacuation to impede biological activity. On the sampling site, these vessels are flushed thoroughly with CO_2 gas. Water is collected using either a 12LPVC "Niskin-type" water sampler or a 3.2LBeta Bottle horizontal sampler. Flow is initiated from the sampler through a 3/16 in. outside diameter plastic tube inserted into a 20 cm Tygon tube on the glass vessel. The CO_2 line is then removed. Water is allowed to overflow the Tygon tube until the vessel is half full and any trapped CO_2 bubbles are removed. For storage, the neck of the vessel is flushed with CO_2 gas and maintained in place by a rubber serum septum thereafter. Gases within the headspace of these vessels are equilibrated in a 28°C water bath for 8 h with continuous agitation after returning to the laboratory.

In an 8-year field study to determine the concentrations of dissolved methane in groundwater,[10] water samples from field monitoring wells are collected into 60 ml serum bottles. Water is gently added down the side of the bottle until the bottle is completely filled so as not to agitate or create bubbles, which could strip gases dissolved in water. Several drops of 1:1 sulfuric acid are then added as a preservative. The bottle is capped, sealed, and kept cold in an ice chest in transit to the laboratory. Samples are kept at 4°C and analyzed within 14 days of collection. This sampling method, along with an analytical procedure using GC-FID, are adopted by U.S. EPA as a technical guide.[18]

The sampling apparatus used in another study of dissolved gases in thermal groundwater near an active volcano consists of a glass flask of known volume (122 cm^3), which can be sealed by gas-tight rubber/Teflon plugs.[19] The flask is totally filled with the water sample to be analyzed, purging out the gas bubbles, sealed, and kept for further analysis in the laboratory.

Three different sampling methods for measuring hydrogen concentrations in groundwater have been compared based upon the sampling times needed, the pumping equipment, effects of well casing materials, and ease of data computation.[20] All three methods are designed to create a headspace or gas bubble inside the samplers, and allow hydrogen gas to reach an equilibrium state with dissolved hydrogen in the aqueous phase. After the analysis of hydrogen gas, by gas chromatography, the concentrations of hydrogen in groundwater samples can be obtained by straightforward computations. Of these methods, the so-called gas-stripping method is best suited to field conditions in this study. Moreover, the well casing material is critical for measuring hydrogen concentrations in groundwater because, for example, iron or steel well casings can produce extra hydrogen and lead to a masking effect.

If water samples containing dissolved gases of interest are to be analyzed, caution must be taken to keep researchers away from potential hazards during sampling. A passive *in situ* headspace sampler has been developed by Sanford and co-workers[21] for this purpose, on a contaminated site, by collecting dissolved gases in groundwater without collecting the contaminated water. The device is a semipermeable membrane attached to a gas-filled reservoir that is immersed in a solution containing dissolved gases. The sampler is placed down, for example, a well into the groundwater and the gases in the sampler system eventually reach equilibrium with the dissolved gases in the groundwater. Any gas, such as H_2, He, Ar, O_2, N_2, CH_4, etc., which can diffuse through the membrane can be collected by this device. The collected gases are then analyzed by GC with a connection tool to extract the gases from the sampler.

C. SOIL

One noteworthy strategy for soil gas surveys is that sampling is accomplished within as few days as possible. Hinkle[22] has pointed out that short-term sampling performed in stable meteorological and soil moisture conditions (e.g., during the dry season) provides minimal soil gas variations. The homogeneity and stability of the soil samples should be tested in order to provide quality assurance of the analytical data.[23,24] Soil gas samples can be collected by means of a hollow steel probe or Teflon tube (0.5 m × 3 mm outside diameter × 1 mm inside diameter, internal volume < 20 cm^3) driven into the ground to a depth between 50 and 60 cm. Soil gas is withdrawn from the soil by a 60 cm^3 syringe after purging the system in order to avoid atmospheric contamination. Overpressurized gas of 50 cm^3 volume is stored in preevacuated stainless steel cylinders (30 cm^3) for laboratory analyses.[25–29] Because, for example, methane is a major component both in soil gas and automobile exhaust, caution should be taken to avoid contamination during transportation.[30,31]

While the most widely used method for measuring fluxes of trace gases such as carbon dioxide and methane between soil and the atmosphere to date is based on the closed chamber, underestimates of the real fluxes could happen because of a lack of proper assumptions. In a study[32] to obtain more reasonable results, a soil core (24 cm diameter, 13 cm height) with associated plants, encased in a PVC cylinder, is taken from a grassland site to the laboratory. Water lost by evapotranspiration is not added during the study period. The plastic cylinder enclosing the core is closed at the base with a cap of the same material, which has been fitted with a soil–air sampler consisting of a perforated plastic tube (0.45 m × 5 mm inside diameter), laid out in a circle and embedded in a 1 cm high layer of medium sand. The sides of the cores are sealed to the cylinder with wax to prevent gas movement. The 7 cm of the plastic cylinder protruding above the top of the core acted as a flux chamber that could be closed with a gas-tight plastic lid. To measure trace gas concentration, gas samples from the spaces below and above the core are taken every 20 min and analyzed by gas chromatography. In this case, a constant gas production source in the chamber soil is assumed and possible changes in gas concentrations at the source during chamber deployment are taken into account.

Baubron et al.[14] used a different approach by drilling holes 1 m deep and 3 cm in diameter in the ground along traverses with a mean interval of 10 m to measure soil gas concentration *in situ*. A 2.8 cm diameter and 80 cm long cardboard tube was then inserted in the hole to ensure an air-proof isolated small cavity at the base. Experiments assume the system to be at equilibrium with soil gases within 1 day after the setting up of the cardboard tube. This procedure allows both gas sampling to be performed at precisely the same depth, and the possibility to replicate sampling with identical conditions for some months.

III. ANALYSIS METHODS

The chromatographic separations of major air components are mainly accomplished by gas chromatography. Helium is used as the carrier gas in many applications. Any possible interference must be eliminated prior to chromatographic analysis. For example, if necessary, an ascarite trap is used to remove CO_2 in air samples and a LiOH trap can adsorb water efficiently. Residual water or CO_2, if any, entering the gas sampling loop can be trapped onto a molecular sieve column, and removed later by heating the column. The analyses are typically finished within 10 min; however, the separation of all constituents of major air components on a single column is difficult because of the wide range of their polarities. Oxygen and nitrogen are separable on 5A or 13X molecular sieves at ambient temperatures. Frequently, split columns and dual detectors[23,33] are used in order to determine all the major constituents at the same time.

A. Atmosphere

Miller and co-workers show an on-stream sampler and a portable GC system to analyze stack gases as part of the requirements of determination of particulate matter from incinerator stacks.[34] The concentrations of carbon dioxide, oxygen, and carbon monoxide are measured and that of nitrogen is calculated by difference. This method is of higher precision and can save time and labor when compared with the Orsat analyzer method. Further, this GC system is portable and all the components are mounted on a 2 ft by 3 ft laboratory cart, which makes mobile measurements possible. A summary of some of the techniques used is presented in Table 6.2.

A method verified by the Intersociety Committee in a manual on Methods of Air Sampling and Analysis has been used to separate and determine O_2, N_2, CO, CO_2, and CH_4 in gas samples by GC.[35] A dual column/dual thermal conductivity detector system is employed. The first column contains a very polar stationary liquid phase and retains CO_2 only, while the second column is packed with molecular sieve 13X and separates the rest of the components. A tube filled with 10/20 mesh Indicating Drierite is installed between the sample introduction system and the first column to retain water in the sample. The analysis is operated slightly above ambient temperature to obtain the best precision results. The detection limits for CO_2 and O_2 are 250 and 300 ppm, respectively. The separation can be completed within 8.5 min.

A customized Cu(I)Y zeolite is employed as a sorbent to actively collect CO in air samples to measure the concentration of CO in ambient air.[16] The interaction is selective to CO only, but not to N_2, O_2, and CO_2. The sorption process is facilitated by formation of Cu(I)–CO complexes, while CO can be desorbed at 300°C under helium flow for 2 min. Before the gas chromatographic analysis, a methanizer is used to reduce CO to CH_4, which can then be quantified by FID. Detection limit of methane by this method is approximately 0.2 ppm. The laboratory data shows the capacity of the Cu(I)Y zeolite sorbent as 2.74 mg CO/g of sorbent. For a typical sorbent tube containing 0.5 g of treated zeolite sampling at the PEL of 50 ppm with a nominal flow rate of 100 ml/min, sampling can last as long as approximately 4 h before a breakthrough point is reached. Furthermore,

TABLE 6.2
Methods of Chromatographic Analysis of Major Air Components

Sample Type	Column Conditions	Detectors	Detection Limit	References
Air	Porapak inner column (6 ft × 1/8 in.); mol. sieve 5 Å outer column (6 ft × 1/4 in.); carrier gas, helium, 50 ml/min; oven and injection port temp., ambient; TCD temp., 100°C	TCD	1000 ppm for N_2, CO, O_2, CO_2	34
	30% (w/w) hexamethylphosphoramide (HMPA) on 60/80 mesh Chromosorb P column (6 ft × 1/4 in.)	TCD	250 ppm for CO_2	35
	30% (w/w) di-2-ethyl-hexyl sebacate (DEHS) on 60/80 mesh Chromosorb P column (6 ft × 1/4 in.)	TCD	250 ppm for CO_2	35
	40/60 mesh mol. sieve 13X column (6.5 ft × 3/16 in.)	TCD	500 ppm for CO, 300 ppm for O_2, N_2, CH_4	35
	Carle 400 GC; mol. sieve 5 Å stainless steel column (1.1 m × 0.32 cm outside diameter); carrier gas, helium	FID for CH_4	N/A	6
	60/80 mol. sieve 5 Å column; carrier gas, argon	TCD for H_2 and CH_4	N/A	44
	PoraPack Q 50/80 column; carrier gas, helium	FID for CH_4	N/A	44
	PE Sigm 3B GC; Carbosieve SII steel column (1 m)	FID	8.8×10^{-11}/mol/min for CH_4	45
	Tracor 540 GC; 80/100 mesh, washed mol. sieve 5 Å stainless steel column (6 ft × 1/8 in. outside diameter); oven temp., 35°C; carrier gas, helium, 20 ml/min; methanizer temp., 250°C, with H_2 supply at 22.5 ml/min; FID temp., 325°C	FID	0.2 ppm for CO (converted to CH_4)	16
	Porapack Q column (for separation of CO_2); HayeSep D column (for separation of C_2 and C_3); mol. sieve 13 Å column (for separation of CH_4 and CO); oven temp., 40°C	Finnigan MAT 252 IRMS	N/A	46
	60/80 mesh mol. sieve 5 Å stainless steel column (1.8 m × 1/8 in. outside diameter)	FID	Approximately 2 ppm for CH_4	36
	HP 6890 GC; Porapack Q column (2.0 m × 1/8 in.) + HayeSep D column (0.7 m × 1/8 in.) + mol. sieve 13 Å column (2.0 m × 1/8 in.); oven temp., 55°C	Finnigan MAT 252 MS	4.65 ng carbon equivalent	38
	PoraPLOT Q column (10 m); oven temp., 40°C; carrier gas, helium	TCD for CO_2	N/A	47
	High resolution mol. sieve 5 Å column (20 m); oven temp., 60°C; carrier gas, argon	TCD for Ne	N/A	47

Continued

TABLE 6.2
Continued

Sample Type	Column Conditions	Detectors	Detection Limit	References
	Shimadzu GCMS-QP5050; Supelco SPB-5 column	MS for CH_4	N/A	48
	HP 5890 II GC; Chromopack PoroPLOT Q column (27.5 m long with a 2.5 m particle trap to separate N_2O from CO_2); oven temp., 25°C	Finnigan Delta S IRMS	10 ppm for CO_2	37
	Varian GC 3400; Porapack Q column	TCD for CO and CO_2	N/A	49
	CE Instruments 8000 GC; Chromosorb 101 stainless steel tube (5 m × 6 mm outside diameter × 5 mm inside diameter); carrier gas, helium, 45 ml/min; injector temp., 100°C; oven temp., 26°C; HWD temp., 160°C	Finnigan MAT Delta S MS for CO_2	N/A	39
	MTI P200H micro GC; mol. sieve 5Å + PoraPLOT U columns (for CO_2 and CO)	TCD	sdev for produced CO_2 0.057 g/kg coal, CO 0.013 g/kg coal	50
	PoraPLOT Q column (25 m × 10 μm inside diameter × 0.32 mm outside diameter); oven temp., 40°C; carrier gas, helium, 1.5 ml/min	Finnigan Delta S MS	sdev for CO_2 0.71 ppm	40
Water	Carbosieve S-II stainless steel column (0.5 m × 3.2 mm inside diameter)	Reduction gas detector for H_2	N/A	51
	SRI Instruments 8610B GC; mol. sieve column (3 ft × 1/8 in.); oven temp., 100°C; carrier gas, helium, 10 ml/min	TCD	100 ppm O_2 (for 100 μl air sample)	52
	HP 5890 GC; 80/100 Porapak Q column (6 ft × 1/8 in.); carrier gas, helium, 20 ml/min; injector temp., 200°C	FID	0.001 mg/l for CH_4 (dissolved in water)	10
	PE 8500 GC; carbosieve II column (4 m)	HWD (nondestructive TCD)	up to 2 ppm (v/v) for He, H_2, O_2, N_2	11
	PE 8500 GC; carbosieve II column (4 m)	FID-methanizer + FID	fraction of ppm (v/v) for CO, CO_2, CH_4	11
	HP 5890 GC; mol. sieve 5 Å column (5 m × 1/8 in. outside diameter); oven temp., 50°C (for separation N_2 and O_2)	Micromass Prism IRMS	N/A	17
	Mol. sieve 5 Å column (4 m × 0.32 mm inside diameter)	Micro-TCD	2 ppm (v/v) for He, 100 ppm for O_2, N_2, and CO_2	19
	Ultimetal PLOT column (10 m), mol. sieve 5 Å (for O_2); oven temp., −15 to −20°C; carrier gas, helium, 2 ml/min	Finnigan MAT Delta S CF-IRMS	N/A	53

Sample	Method	Detector	Detection limit	Ref.
Soil water	Carla Erba 1108 GC; Porapak Q column (2 m × 6.4 mm outside diameter); oven temp., 60°C; carrier gas, helium, 80 ml/min; as little as 40 μl water sample	Finnigan MAT Delta S IRMS	N/A	54
Mineral oil	HP 6890 GC; GS-Q column (40 m × 0.32 mm) + mol. sieve 5 Å column (20 m × 0.32 mm), both with a film thickness of 30 μm; carrier gas, helium, 8.2 ml/min	PDHID for H_2, O_2, N_2, CH_4, CO, CO_2	Low ppm level	55
	HP 6890 GC; stabilwax column (60 m × 0.25 mm) with a film thickness 0.25 μm; carrier gas, helium, 3.9 ml/min	PDHID for H_2O	Low ppm level	55
Soil atmosphere	80/100 mesh Porapak N column (2 m × 3.2 mm inside diameter)	FID for CH_4	N/A	43
	Varian 3400 GC; Porapak T column	FID for CH_4	N/A	27
	Shimadzu GC-8A; Porapak Q column; oven temp., 45°C; injector and detector temp., 250°C	ECD for CO_2	N/A	33
	Shimadzu GC-8A; 80/100 mesh HayeSep Q stainless steel column (91 cm × 3 mm inside diameter); oven temp., 100°C; injector and detector temp., 130°C	FID for CH_4	N/A	33
	Shimadzu GC-14A; Porapak N stainless steel column (2 m × 3.2 mm outside diameter)	FID for CH_4, TCD for CO_2	N/A	23
	PE 8500 GC; Carbosieve S II column (4 m); carrier gas, argon	TCD + FID	2 ppm (v/v) for H_2 and He, 500 ppm for O_2 and N_2, 1 ppm for CO, CH_4, and CO_2	28
	Shimadzu Mini-2 GC; Porapak Q column; oven temp., 50°C; injector and detector temp., 275°C	ECD for CO_2 and N_2O	N/A	31
	Shimadzu GC-8; Porapak Q column; injector and detector temp., 130°C; oven temp., 100°C	FID for CH_4	N/A	31
	Hitachi 263-50 GC; Unibeads C column (2 m × 1.5 mm); oven temp., 95°C; carrier gas, helium, 80 ml/min	FID for CH_4	N/A	29
	Varian 3400 GC; Carboxen 1000 stainless steel column (1/8 in.)	TCD for O_2 and CO_2	N/A	56

Abbreviations: GC, Gas Chromatography; IRMS, Isotope Ratio Mass Spectrometry; ECD, Electron Capture Detector; FID, Flame Ionization Detector; TCD, Thermal Conductivity Detector; MS, Mass Spectrometry; PDHID, Pulsed-Discharge Helium Ionization Detector; mol. sieve, Molecular Sieve; HP, Hewlett-Packard; PE, Perkin-Elmer.

the collected samples are stable at room temperature for 1 day and up to 7 days at 4°C if flushed with helium before storage.

To resolve the problem of long run time and also to make a fully automated gas chromatographic system for monitoring ambient methane concentration, a process controller with power supply heats up only the stainless steel column without heating the whole oven like a conventional GC.[36] With the column being the only thermal mass, the cooling is also fast. This system is tested for on-site automatic measurement of ambient methane with an uninterrupted time resolution of 30 min for 27 days. By testing 388 samples, this system shows good accuracy and reproducibility and reveals no noticeable systematic problems.

One technique to conquer the collection job of large sample volume (2.5 l) in glass containers in traditional measurements of the stable isotope of CO_2 is by using a gas chromatograph isotope ratio mass spectrometer (GC/IRMS) to analyze ^{13}C and ^{18}O of CO_2 at ambient CO_2 concentration (of approximately 369 ppm).[37] The sample volumes are reduced down to the level of ml and collection of large sample numbers in the field is allowable. Detection limit of this technique for CO_2 is 10 ppm and meets the requirement for global monitoring of atmospheric CO_2.

In a study to characterize and classify the composition of gases generated for pyrolysis of coals, a three-column GC system coupled with an IRMS has been used.[38] Control valves are operated electrically or electro-pneumatically to obtain essential accurate switching of the GC gas streams for the long-term stability and reliability of the whole unit. The first column is filled with Porapack Q (2.0 m × 1/8 in.) to retard higher hydrocarbons (>C_3) and separate carbon dioxide from the remaining gas components (e.g., CH_4, CO, and N_2). The second column is packed with HayeSep D (0.7 m × 1/8 in.) to separate the remaining hydrocarbons. CH_4, CO and N_2 first elute and are then separated in a third column (molecular sieve 13 Å, 2.0 m × 1/8 in.). Stable isotope analysis is performed on a Finnigan MAT 252 instrument. Each run takes about 12 min and the sensitivity is approximately 4.65 ng carbon equivalent/compound. One important parameter regarding the molecular sieve column is that it has a separate carrier gas supply to by-pass CO_2 (to avoid stagnation). This also speeds up the chromatographic separation of CH_4, CO, and N_2. An additional switching step transfers these three separated gas components to the IRMS interface after the analysis of the by-passing gases has been completed.

Demény and Haszpra[39] have demonstrated a procedure to effectively measure CO_2 concentration plus its stable isotope compositions. Air samples are pumped through silica gel and $Mg(ClO_4)_2$ traps and 3 l glass bottles equipped with two Teflon stopcocks. Moisture can thus be removed from the collected air. In spite of its small amounts in the air samples, N_2O must be separated from CO_2 because they have very similar m/z values. This effect is important especially when an IRMS is employed as the detector. A Chromosorb 101 column is used and shows acceptable separation characteristics. Another alternative to separate CO_2 from N_2O is to pump the sample by helium flow through a liquid nitrogen trap. Under the experimental conditions, CO_2 can be detected at about 4.5 min and N_2O at 6 min. The actual volume of CO_2 in the 3 l air sample at 1 atmospheric pressure is around 2 to 3 ml.

To automatically and simultaneously analyze CO_2 concentrations as well as $\delta^{13}C$ and $\delta^{18}O$ values in air samples, a so-called continuous-flow isotope ratio mass spectrometry (CF-IRMS) coupled with GC has been developed.[40] A relatively small sample size (500 μl of air) is required even where a highly variable CO_2 concentration range (250 to 1200 ppm) exists. A fixed quantity of air is first dried by passing it through a dry ice/ethanol trap, and CO_2 is further condensed on a liquid nitrogen trap. After getting vaporized into flowing helium, CO_2 is then passed through a capillary GC column to separate it from N_2O before entering CF-IRMS. Since repeated cycles of sample and reference measurements enhance the precision of this method, CO_2 from the sample and reference air are alternatively directed to the mass spectrometer. In the case of four replicate cycles, the standard error of the mean for a measurement is 0.7 ppm for CO_2, while this figure is 0.4 ppm with nine replicate cycles.

B. WATER

Gas chromatography offers a precision of measurement in the range of 0.3 to 1% for N_2, O_2, and Ar, extracted from water samples.[41] Techniques used in the sampling of dissolved gases include water vapor that must be removed from the sample prior to gas chromatography. This is accomplished by passing the sample in the vapor phase through a column of water-retaining material such as drierite, scarite, or $Mg(ClO_4)_2$. The GC is usually conducted on columns of molecular sieve packing.

Capasso and Inguaggiato[11,19] used a simple sampling apparatus consisting of a glass flask of known volume (122 ml of water sample), which can be sealed by gas-tight rubber/Teflon plugs. Argon, for example, is introduced into the water sample as a host gas. Gases dissolved in the water sample will leave and remain in equilibrium state between a liquid and a gaseous phase. The gaseous phase is analyzed by gas chromatography, and by means of the partitioning coefficients of different gas species, the initial concentrations of the dissolved gases can be derived. For example, oxygen extracted from a 10 ml water sample in Berenati Well on Vulcano Island, Italy, is reported to be 6.2% (v/v), and calculations show that DO in that sample is 7.2 ml/l at standard temperature and pressure (STP). Using Henry's Law, partial pressure of oxygen in its source, i.e., volcanic gases, is 0.34 atm in this case. Analyzed partial pressures of He, CO_2, and sometimes H_2, CO, and CH_4 are appreciably higher than those in water in equilibrium with the atmosphere, showing interaction between volcanic gases and groundwater. After some modifications, this method is adequate for routine analysis of dissolved gases in water.

C. SOIL

Air samples from soil are usually withdrawn directly with syringes having airtight stopcocks through stainless steel or nylon tubes installed into the soils or headspace inside the closed chambers, as described in the previous section. Methane, one of the target gases, can be analyzed within 6 h of collection by GC-FID. The column and detector parameters are very similar to those set for the analyses of major air components in the atmosphere. Experiments have been conducted by individual teams at different geological locations[42,43] to measure how soil consumes atmospheric methane, and how soil moisture and temperature control the consumption and production of CH_4 and CO_2 in it.[23,33]

As the analysis time is an important issue, especially when multiple analyses or continuous monitoring are to be done, the design of a small field-portable chromatograph is desirable. On the other hand, to meet the needs of on-site and real-time measurements without compromising the ability to provide consistent, defensible, and reliable results, considerable developments have been focused on field-portable chromatographic systems. For example, integration of flexible and expandable sample introduction techniques with software-configurable hardware platforms provide for easier control and use of multipurpose field-portable GC/MS systems. Advances in the design, miniaturization and automation of components will lead to a further reduction in power consumption, size, and weight of newer systems, which make field research even easier. Furthermore, there are modern chemistry models, which cannot be tested because of nonavailability of techniques to observe the two- and three-dimensional distribution of gases in the atmosphere on regional or even global scales at high spatial resolution. The future development of atmospheric research will, to a large extent, depend upon the progress in instrumentation development.

IV. DETECTION METHODS

More than 30 different GC methods have been described in the previous section. Detector selection for the GC methods depends on the target compound and concentration range. Among these detectors, TCD, FID, MS, and ECD are the most popular ones. While traditionally TCD has been

TABLE 6.3
Detectors in Gas Chromatographic Methods and the Compounds Analyzed by these Detectors

Detectors	Compounds Analyzed	References
TCD	CO_2	35
	CO_2, O_2	35,56
	H_2, CH_4	44
	CO, CO_2	49,50
	O_2	52
	He, H_2, O_2, N_2	11
	He, O_2, N_2, CO_2	19
	He, O_2, N_2, CO, CO_2	28
FID	CH_4	6,16,28,29,36,44,45,57,59
	CH_4, CO, CO_2	11
MS	CO_2	37,39,40
	CO_2, CH_4	38
	CH_4	48

the choice for major air components, improved overall sensitivity and ease of operation has lead to a gradual increase of MS detection in this category. In some cases, CO is first converted to CH_4 by a methanizer prior to FID because FID is not sensitive to CO.[16,57,58] Table 6.3 is organized by the detectors used coupled with gas chromatographic methods and the compounds analyzed by these detectors.

V. APPLICATIONS IN THE ANALYSIS OF THE ENVIRONMENT

Chromatographic analysis of major air components has been demonstrated for providing a powerful tool to be reliable, and even on-site and real-time data. Most of the time, the data is not meant only for the purpose of scientific researches but rather as a means to interpret targeted activities in the environment. Policy makers can thus reach their conclusions based on the interpretations and ask the general public or the manufacturers to comply with. In many cases, abrupt changes of major air components, for example, in groundwater, could be an early alert for natural disasters like volcanic eruptions. To save lives in advance has been an ultimate goal of such studies. More details about these applications are described below and listed in Table 6.4.

A. ATMOSPHERE

Carbon monoxide is one of the most common toxic gases and its concentration is regulated by OSHA and EPA, and recommended by many other institutions. Routine and accurate measurement of its level has become very important. By using a relatively nontoxic, convenient to prepare and use sorbent sampler, coupled to a GC and a FID, the concentration of carbon monoxide can be monitored in a working area or on workers.[16,57]

The threat of global warming has become one of the most widely accepted, important scientific and social issues in recent times.[60] Among the major factors, the increasing amounts of atmospheric carbon dioxide, which can be related to fossil fuel and biomass burning, land use changes, industrial processes, etc., receive most attention. The determination of global sinks and sources of atmospheric carbon dioxide requires maintenance of regional and global monitoring station networks. In the meanwhile, knowledge of local/regional reactions and processes is

TABLE 6.4
Some Application Sectors for Chromatographic Analysis of Major Air Components

Atmosphere	Continuous emission monitoring
	Worker safety/industrial hygiene
	Indoor air monitoring
	Global monitoring network for greenhouse effect
	Biological processes
Water	Wastewater effluent monitoring
	Earthquake prediction
	Volcanic eruption prediction
	Ecosystem monitoring
Soil	Energy exploration
	Geological evaluation
	Earthquake prediction
	Pollution remedial process feedback
	Site investigation
	Volcanic eruption prediction
	Soil biogenic emission measurement
	Environmental forensics

necessary for development of reliable global carbon budget models.[61] Certain chromatographic methods have been developed for this purpose that allow large sample numbers to be collected in the field and analyzed rapidly in the laboratory. Stable isotope geochemistry is one of the most powerful methods of CO_2 source investigation and can be facilitated by GC–IRMS.[39] For example, one study conducted in the East European Russian tundra zone region shows that over the critical value of 14°C, an increase of mean diurnal air temperature in these ecosystems lead to a change in the carbon net flux from sink to source.[62] The result is reversed when the temperature is under that critical value. This effect is primarily caused by the increase of gross ecosystem respiration at higher temperatures. This study provides evidence of possible positive feedback between climate warming and carbon emission to the atmosphere on regional and short-term scales.

Another interesting result indicates that anthropogenic and biogenic CO_2 have similar isotopic characteristics and polluted and unpolluted sites cannot be distinguished by the $\delta^{13}C–\delta^{18}O$ distribution. Further comparison of data from different sampling networks is needed.[39]

Like carbon dioxide, atmospheric methane has long been considered a major contributor to the greenhouse effect. Measurement of methane concentrations in the atmosphere by GC-FID can help to understand its distribution between global sources and sinks and its growth rate for a period of time. Dlugokencky et al.[6] have shown that a strong north–south gradient in methane with an annual mean difference of about 140 ppb between the northernmost and southernmost sampling sites exists. Methane time-series concentrations from the high southern latitude sites have a relatively simple seasonal cycle with a minimum during late summer–early fall, which can be explained by its photochemical destruction during that period. Typical seasonal cycle amplitudes there are about 30 ppb. This figure is almost twice in the high north region, possibly because of a more complex interaction between methane sources and sinks and atmospheric transport. Moreover, from a global perspective, the growth rate for methane has decreased from about 13.5 ppb/year in 1983 to 9.3 ppb/year in 1991, while the global burden of atmospheric methane increased at an average rate of 11.1 ± 0.2 ppb/yr. In the northern hemisphere alone, the growth rate of methane was near zero in 1992. The most acceptable reason is that a change in a methane source is influenced directly by human activities, such as fossil fuel production. The atmospheric measurements provide an integrated picture, which is useful in gaining a better understanding of the global methane budget especially when combined with a model of atmospheric transport and chemistry. The information from monitoring programs can be applied in decision-making for policies and regulations.

The fact that natural gas is generated by thermal degradation of sedimentary organic matter has been widely accepted. Because of the complex structure of kerogen, it is not well understood which reactions proceed during thermal gas generation. Gases generated from dry coal-pyrolysis reactions in an open system, such as methane, carbon dioxide, carbon monoxide, propane, etc., can be analyzed by GC coupled with IRMS.[46] Experimental results can be used to study the reaction mechanisms, kinetic order of the reactions, signature of the hydrocarbon precursors within the kerogen, and the burial and accumulation history of the kerogen as well. The results also help to provide the basis for an extension of compositional kinetic approaches in natural gas geochemistry to isotope level (isotope-specific kinetics). In light of this, the feasibility of a reaction–kinetic approach for prediction of isotope compositions of natural gas as a function of time–temperature history of the source material becomes achievable. This approach can ultimately be employed in geological petroleum and natural gas exploration.

In order to understand how H_2O and CO_2 are exchanged between leaf tissues and the air, measurement of the photosynthetic isotope fractionation by open flow gas analysis is conducted in a gas exchange chamber under controlled conditions of temperature, light intensity and humidity, and CF-IRMS and GC are employed.[40] When the same method is applied to ecosystem isotope fractionation in a field experiment, the variation in atmospheric CO_2 concentrations and $\delta^{13}C$ values primarily reflect variations in net photosynthesis (daytime) and respiration. The $\delta^{18}O$ and CO_2 data also reflect these processes, but with added complications because of oxygen isotope exchanges with soil and leaf water pools.

B. WATER

Gas concentrations in groundwater can provide information on large regional variations reflecting chemical and structural properties of aquifer systems as well as the temperature and pressure at which the groundwater has interacted with its adjacent rocks. Moreover, changes in aquifer properties caused by human activities, rainfall and tectonic activities may also result in temporal changes in gas concentrations in groundwater. Continuous monitoring of dissolved gas concentrations will provide important information regarding the aquifer system related to various events. For example, for research in geochemical surveying of active volcanism, the investigation of gaseous phases such as fumarolic emissions,[19] diffused exhalations from the soil,[30,63] and dissolved gases in groundwater,[11,19,64] is important because the high mobility of gases makes them excellent tracers. Many studies show the importance of some gaseous components as precursors to possible resumption of volcanic and seismic activity. For example, He and CO_2 are good tracers of degassing processes originating from deep-seated sources through faults and fissures. Significant variations in their natural water and soil content may represent reasonable predictions of such activities.[28]

The concentration of DO in water plays an important role in many industries and also in the aquatic environment. To measure the concentration of biodegradable organic matter in wastewater effluents to meet EPA discharge requirements, virtually all wastewater treatment plants conduct the conventional 5-day biochemical oxygen demand (BOD$_5$) test. However, this test is time-consuming and labor-intensive. A new method to measure BOD based on calculating oxygen demand from the decrease in oxygen concentrations in the gas phase, or headspace, of the sealed tube that contains the liquid sample has been developed.[52] This Headspace BOD (HBOD) method carries out gas phase oxygen concentration measurements on a GC-TCD system. The percentage of oxygen in the headspace is a function of the DO concentration and the percentage of the volume of liquid in the sealed tube. Headspace volumes should thus be selected to keep the final liquid DO at >2 mg/l and to obtain a DO depletion of >1 mg/l. The DO is calculated from gas phase measurements by assuming that the wastewater and gas phase are in equilibrium. This method provides a reliable estimate of BOD_5 by a 3-day HBOD test, which saves labor and time and possesses the potential of easy automation. The saving in analytical time (3 vs. 5 days) could provide an earlier warning of violations in discharge BOD_5, which in turn leads to reduced fines.

A study conducted in a perennially ice-covered lake is focused on how research diving can have an impact in water quality parameters such as DO concentrations.[65] Because ice covers on the lake minimize, at least to a limited extent, the exchange of gases between the water column and the atmosphere, gases exhaled from divers are suspected to be a potentially disruptive activity and requires objective evaluation. The results indicate that the impact of diver-exhaled gases on DO concentrations in the lake water column are negligible but, on the other hand, environmental drivers (e.g., stream inputs and wind energy) have even greater effects than diving. It is most likely that the majority of diver-generated gases escape to the atmosphere within minutes or even seconds of release or are frozen into the ice cover.

Soil water samples taken from a forest are analyzed for their $\delta^{18}O$ values by GC-CF-IRMS to compare with those obtained from tree water on the same site.[54] The tree water $\delta^{18}O$ values differ from that of soil water near the surface, which implies that tall trees (60 m) are likely to extract water from below the surface soil layers. In contrast, the 2 m tall shrubs have stem water $\delta^{18}O$ values that indicate the water source to be closer to the surface soil layers.

One modified chromatographic method for measuring dissolved gases in mineral insulating oils, although not in water, finds an important application in the power industry.[55] Several gases, including H_2, CH_4, CO, CO_2, etc., are produced with time in mineral insulating oils in high-voltage power transformers, which in turn cause chemical decomposition of the oil and eventually power malfunctions. It is thus crucial to detect the presence of these gases at an early stage of development and take preventive action before such failures do occur. As a result, routine analyses of dissolved gases in mineral oils become one of the most important assays for electric utilities worldwide.

C. SOIL

In situ analysis of soil atmospheres can reveal patterns of reactions in soil according to their products. Carbon dioxide and methane, for example, represent aerobic and anaerobic reactions, respectively. An investigation of soil gas composition on a petroleum hydrocarbon contamination site suggests that aerobic biodegradation of petroleum in soil accounts for the majority of CO_2 produced.[56] Also, methane produced in the anoxic region of wetlands is transported to the atmosphere via diffusion, ebullition, or through vascular plants.[66-68] A significant proportion (50 to 90%) of the methane produced is oxidized to carbon dioxide by methanotropic bacteria before reaching the soil surface. Depth profiles of the gas concentrations in the peatlands, with measurements from chromatographic methods, will help reveal the kinetics and mechanisms of how methane is produced and reacts.[9] This is important because our understanding of the global budget of methane and carbon dioxide is still far from complete. Unknown terrestrial sinks of 1.3 Tg C/year (equivalent to 30% of the contribution from fossil fuel combustion and cement production) are required to balance the CO_2 budget, and considerable uncertainties still remain regarding many components of the CH_4 budget.[27,31,42,43,69,70] Understanding the links of soil and atmospheric CH_4 and CO_2 will help resolve the puzzles of their global budgets as well as global environmental changes.

Gas, such as carbon dioxide, discharges over seismically active faults often links to a long-term, permanent phenomenon which indicates that active faults are characterized by a high permeability and act as preferential conduits in the crust. This permeability in fault gouges and intensely sheared zones can generate complex geochemical patterns in soil atmosphere. This characteristic is employed to search for active faults as well as for monitoring, in seismic and volcanic areas, distribution in soils, using carbon dioxide, along with other gases.[14,71] Two regional faults[72] and an underground hydrothermal aquifer[28] in Italy are, respectively, located by searching for abnormal CO_2 flux of soil origin. A soil gas emission monitoring program at Furnas volcano, Azores, shows that a potential health hazard of high concentrations of CO_2 from the hydrothermal system beneath that area could occur without warning and reach a disastrous level.[73] Determination made directly in the field becomes particularly important during periods of increased volcanic activity and movement of magma beneath the volcanic edifice. Chromatographic analyses of these gases thus become a useful tool.[19]

From the standpoint of environmental hazard and liability, identifying the source and origin of soil CH_4 accumulations, for example, near petroleum spill sites, has become very important. On-site GC measurement of gas composition in the soil and carbon isotopic composition analysis in the laboratory, along with geological, geochemical, and land use data, help distinguish biogenic and anthropogenic sources of CH_4. Appropriate methods for site remediation can also be chosen according to these results.[56,74] Similar evaluations have been conducted by using concentrations of dissolved H_2 to determine the distribution of redox processes in anoxic groundwater, which are heavily contaminated with organic compounds such as petroleum products or landfill leachate.[51]

Also, there is a need to identify soil emission hotspots such as accurate locations of organic soil pollution, and the effects of local soil composition and fertilizer distribution. Repeated sampling and analysis over long periods can be used to integrate cumulative emission rates of soil gases, such as methane and carbon dioxide. For example, Rahn et al.[59] conducted a series of measurements of the CO_2 emission that is concluded to be responsible for tree mortality at the site on the flanks of Mammoth Mountain. Dai and co-workers[75] found a strong correlation between cumulative mass of CO_2 from five incubated Arctic soils and the relative percentages of polysaccharides in these soil samples. Chromatographic methods are appropriate in terms of high spatial resolution, short analysis time, small sample volume, and little sample manipulation. The demand to improve the ability of equipment mobility has also been risen to make real-time analysis and monitoring in the field a routine job.

REFERENCES

1. Nemani, R. R., Keeling, C. D., Hashimoto, H., Jolly, W. M., Piper, S. C., Tucker, C. J., Myneni, R. B., and Running, S. W., Climate-driven increases in global terrestrial net primary production from 1982 to 1999, *Science*, 300, 1560–1563, 2003.
2. Sitting, M., *Handbook of Toxic and Hazardous Chemicals*, Noyes Publications, Park Ridge, NJ, pp. 137–138, 1981.
3. American Conference of Governmental Industrial Hygienists (ACGIH). *Threshold Limit Values for Chemical Substances and Physical Agents and Biological Exposure Indices for 1995–1996*, American Conference of Governmental Industrial Hygienists, Cincinnati, OH, 1995.
4. Khalil, M. A. K. and Rasmussen, R. A., Atmospheric methane: recent global trends, *Environ. Sci. Technol.*, 24, 549–553, 1990.
5. Rodhe, H., A comparison of the contribution of various gases to the greenhouse effect, *Science*, 248, 1217–1219, 1990.
6. Dlugokencky, E. J., Steele, L. P., Lang, P. M., and Masarie, K. A., The growth rate and distribution of atmospheric methane, *J. Geophys. Res.*, 99(D8), 17021–17043, 1994.
7. Wahlen, M., Tanaka, N., Henry, R., Deck, B., Zeglen, J., Vogel, J. S., Southon, J., Shemesh, A., Fairbanks, R., and Broecker, W., Carbon-14 in methane sources and in atmospheric methane: the contribution from fossil carbon, *Science*, 245, 236–245, 1989.
8. Lal, S., Venkataramani, S., and Subbaraya, B. H., Methane flux measurements from paddy fields in the tropical Indian region, *Atmos. Environ.*, 27A(11), 1691–1694, 1993.
9. Thomas, K. L., Price, D., and Lloyd, D., A comparison of different methods for the measurement of dissolved gas gradients in waterlogged peat cores, *J. Microbiol. Methods*, 24, 191–198, 1995.

10. Kampbell, D. H. and Vandegrift, S. A., Analysis of dissolved methane, ethane, and ethylene in ground water by a standard gas chromatographic technique, *J. Chromatogr. Sci.*, 36, 253–256, 1998.

11. Capssso, G. and Inguaggiato, S., A simple method for the determination of dissolved gases in natural water. An application to thermal water from Vulcano Island, *Appl. Geochem.*, 13(5), 631–642, 1998.

12. Emerson, S., Stump, C., Wilbur, D., and Quay, P., Accurate measurement of O_2, N_2, and Ar gases in water and the solubility of N_2, *Mar. Chem.*, 64, 337–347, 1999.

13. Smith, K. A., Gas-chromatographic analysis of the soil atmosphere, In *Advances in Chromatography*, Vol. 15, Giddings, J. C., Grushka, E., Cazes, J., and Brown, P. R., Eds., Marcel Dekker, New York, pp. 197–231, 1977.

14. Baubron, J.-C., Rigo, A., and Toutain, J.-P., Soil gas profiles as a tool to characterize active tectonic areas: the Jaut Pass example (Pyrenees, France), *Earth Planet. Sci. Lett.*, 196, 69–81, 2002.

15. Grob, R. L. and Kanatharana, P., Gas chromatographic analysis in soil chemistry, In *Chromatographic Analysis of the Environment*, 2nd ed., Grob, R. L., Ed., Marcel Dekker, New York, p. 396, 1985.

16. Juntarawijit, C., Poovey, H. G., and Rando, R. J., Determination of carbon monoxide with a modified zeolite sorbent and methanization–gas chromatography, *Am. Ind. Hyg. Assoc. J.*, 61(3), 410–414, 2000.

17. Roberts, B. J., Russ, M. E., and Ostrom, N. E., Rapid and precise determination of the $\delta^{18}O$ of dissolved and gaseous dioxygen *via* gas chromatography-isotope ratio mass spectrometry, *Environ. Sci. Technol.*, 34, 2337–2341, 2000.

18. U.S. EPA-Region 1. *Technical guidance for the natural attenuation indicators: methane, ethane, and ethene. Methane, Ethane, Ethene Analysis Guidance, Revision 1*, 2002.

19. Capasso, G., Favara, R. and Inguaggiato, S., Interaction between fumarolic gases and thermal groundwater at Vulcano Island (Italy): evidences from chemical composition of dissolved gases in water, *J. Volcanol. Geotherm. Res.*, 102, 309–318, 2000.

20. Chapelle, F. H., Vroblesky, D. A., Woodward, J. C., and Lovley, D. R., Practical considerations for measuring hydrogen concentrations in groundwater, *Environ. Sci. Technol.*, 31, 2873–2877, 1997.

21. Sanford, W. E. and Solomon, D. K., Site characterization and containment assessment with dissolved gases, *J. Environ. Eng.*, 124(6), 572–574, 1998.

22. Hinkle, M., Environmental conditions affecting concentrations of He, CO_2, O_2 and N_2 in soil gases, *Appl. Geochem.*, 9, 53–63, 1994.

23. Gulledge, J. and Schimel, J. P., Moisture control over atmospheric CH_4 consumption and CO_2 production in diverse Alaskan soils, *Soil Biol. Biochem.*, 30(8/9), 1127–1132, 1998.

24. Mäkinen, I., Suortti, A.-M., Pőnni, S., and Huhtala, S., Proficiency test on the determination of mineral oil from polluted soils, *Accredit. Quality Assurance*, 7, 209–213, 2002.

25. Grob, R. L. and Kanatharana, P., Gas chromatographic analysis in soil chemistry, In *Chromatographic Analysis of the Environment*, 2nd ed., Grob, R. L., Ed., Marcel Dekker, New York, pp. 352–353, 1985.

26. Klusman, R. W., *Soil Gas and Related Methods for Natural Resource Exploration*, Wiley, Chichester, 1993.

27. Dobbie, K. E., Smith, K. A., Priemé, A., Christensen, S., Degorska, A., and Orlanski, P., Effect of land use on the rate of methane uptake by surface soils in northern Europe, *Atmos. Environ.*, 30(7), 1005–1011, 1996.

28. Giammanco, S., Inguaggiato, S., and Valenza, M., Soil and fumarole gases of Mount Etna: geochemistry and relations with volcanic activity, *J. Volcanol. Geotherm. Res.*, 81, 297–310, 1998.

29. Ishizuka, S., Sakata, T., and Ishizuka, K., Methane oxidation in Japanese forest soils, *Soil Biol. Biochem.*, 32, 769–777, 2000.

30. Ciotoli, G., Guerra, M., Lombardi, S., and Vittori, E., Soil gas survey for tracing seismogenic faults: a case study in the Fucino basin, Central Italy, *J. Geophys. Res.*, 103(B10), 23781–23794, 1998.

31. Bowden, R. D., Rullo, G., Stevens, G. R., and Steudler, P. A., Soil fluxes of carbon dioxide, nitrous oxide, and methane at a productive temperate deciduous forest, *J. Environ. Quality*, 29, 268–276, 2000.

32. Conen, F. and Smith, K. A., An explanation of linear increases in gas concentration under closed chambers used to measure gas exchange between soil and the atmosphere, *Eur. J. Soil Sci.*, 51, 111–117, 2000.

33. Bowden, R. D., Newkirk, K. M., and Rullo, G. M., Carbon dioxide and methane fluxes by a forest soil under laboratory-controlled moisture and temperature conditions, *Soil Biol. Biochem.*, 30(12), 1591–1597, 1998.

34. Miller, D. L., Woods, J. S., Grubaugh, K. W., and Jordon, L. M., Analysis of stack gases using a portable gas chromatographic sampling and analyzing system, *Environ. Sci. Technol.*, 14(1), 97–100, 1980.

35. Lodge, J. P. Jr., *Methods of Air Sampling and Analysis*, 3rd ed., CRC Press LLC, Boca Raton, FL, pp. 303–306, 1988.

36. Wang, J.-L., Kuo, S.-R., Ma, S.-S., and Chen, T.-Y., Construction of a low-cost automated chromatographic system for the measurement of ambient methane, *Anal. Chim. Acta*, 448, 187–193, 2001.

37. Mortazavi, B. and Chanton, J. P., A rapid and precise technique for measuring $\delta^{13}C$–CO_2 and $\delta^{18}O$–CO_2 ratios at ambient CO_2 concentrations for biological applications and the influence of container type and storage time on the sample isotope ratios, *Rapid Commun. Mass Spectrom.*, 16, 1398–1403, 2002.

38. Gaschnitz, R., Krooss, B. M., Gerling, P., Faber, E., and Littke, R., On-line pyrolysis–GC–IRMS: isotope fractionation of thermally generated gases from coals, *Fuel*, 80, 2139–2153, 2001.

39. Demény, A. and Haszpra, L., Stable isotope compositions of CO_2 in background air and at polluted sites in Hungary, *Rapid Commun. Mass Spectrom.*, 16, 797–804, 2002.

40. Ribas-Carbo, M., Still, C., and Berry, J., Automated system for simultaneous analysis of $\delta^{13}C$, $\delta^{18}O$ and CO_2 concentrations in small air samples, *Rapid Commun. Mass Spectrom.*, 16, 339–345, 2002.

41. Craig, H. and Hayward, T., Oxygen supersaturation in the ocean: biological vs. physical contributions, *Science*, 235, 199–202, 1987.

42. Whalen, S. C. and Reeburgh, W. S., Consumption of atmospheric methane by tundra soils, *Nature*, 346, 160–162, 1990.

43. Striegl, R. G., McConnaughey, T. A., Thorstenson, D. C., Weeks, E. P., and Woodward, J. C., Consumption of atmospheric methane by desert soils, *Nature*, 357, 145–147, 1992.

44. Narisawa, M., Yoshida, T., Iseki, T., Katase, Y., and Okamura, K., γ-ray curing of polymethylsilane and polymethylsilane–demethylsilane for improved ceramic yields, *Chem. Mater.*, 12, 2686–2692, 2000.

45. Baumgarten, E. and Maschke, L., Hydrogen spillover through the gas phase reaction with graphite and activated carbon, *Appl. Catal. A General*, 202, 171–177, 2000.

46. Cramer, B., Faber, E., Gerling, P., and Krooss, B. M., Reaction kinetics of stable carbon isotopes in natural gas-insights from dry, open system pyrolysis experiments, *Energy Fuels*, 15, 517–532, 2001.

47. Mori, T., Hernández, P. A., Salazar, J. M. L., Pérez, N. M., and Notsu, K., An *in situ* method for measuring CO_2 flux from volcanic-hydrothermal fumaroles, *Chem. Geol.*, 177, 85–99, 2001.

48. Schrebler, R., Cury, P., Suárez, C., Muñoz, E., Gómez, H., and Córdova, R., Study of the electrochemical reduction of CO_2 on a polypyrrole electrode modified by rhenium and copper–rhenium microalloy in methanol media, *J. Electroanal. Chem.*, 533, 167–175, 2002.

49. Stevens, G. B., Reda, T., and Raguse, B., Energy storage by the electrochemical reduction of CO_2 to CO at a porous Au film, *J. Electroanal. Chem.*, 526, 125–133, 2002.

50. Wang, H., Dlugogorski, B. Z., and Kennedy, E. M., Examination of CO_2, CO, and H_2O formation during low-temperature oxidation of a bituminous coal, *Energy Fuels*, 16, 586–592, 2002.

51. Lovley, D. R., Chapelle, F. H., and Woodward, J. C., Use of dissolved H_2 concentrations to determine distribution of microbially catalyzed redox reactions in anoxic groundwater, *Environ. Sci. Technol.*, 28(7), 1205–1210, 1994.

52. Logan, B. E. and Patnaik, R., A gas chromatographic-based headspace biochemical oxygen demand test, *Water Environ. Res.*, 69(2), 206–214, 1997.

53. Baker, L., Franchi, I. A., Maynard, J., Wright, I. P., and Pillinger, C. T., A technique for the determination of $^{18}O/^{16}O$ and $^{17}O/^{16}O$ isotopic ratios in water from small liquid and solid samples, *Anal. Chem.*, 74, 1665–1673, 2002.

54. Fessenden, J. E., Cook, C. S., Lott, M. J., and Ehleringer, J. R., Rapid ^{18}O analysis of small water and CO_2 samples using a continuous-flow isotope ratio mass spectrometer, *Rapid Commun. Mass Spectrom.*, 16, 1257–1260, 2002.

55. Jalbert, J., Gilbert, R., and Tétreault, P., Simultaneous determination of dissolved gases and moisture in mineral insulating oils by static headspace GC with helium photoionization pulsed discharge detection, *Anal. Chem.*, 73, 3382–3391, 2001.
56. Aelion, C. M. and Kirtland, B. C., Physical vs. biological hydrocarbon removal during air sparging and soil vapor extraction, *Environ. Sci. Technol.*, 34, 3167–3173, 2000.
57. Lee, K., Yanagisawa, Y., Hishinuma, M., Spengler, J. D., and Billick, I. H., A passive sampler for measurement of carbon monoxide using a solid adsorbent, *Environ. Sci. Technol.*, 26, 697–702, 1992.
58. Helmig, D., Air analysis by gas chromatography, *J. Chromatogr. A*, 843, 129–146, 1999.
59. Rahn, T. A., Fessenden, J. E., and Wahlen, M., Flux chamber measurements of anomalous CO_2 emission from the flanks of Mammoth Mountain, California, *Geophys. Res. Lett.*, 23(14), 1861–1864, 1996.
60. Hansen, J., Ruedy, R., Sato, M., and Lo, K., Global warming continues, *Science*, 295, 275, 2002.
61. Chen, W.-H. and Lu, J.-J., Microphysics of atmospheric carbon dioxide uptake by a cloud droplet containing a solid nucleus, *J. Geophys. Res.*, 108(D15), 4470–4479, 2003.
62. Zamolodchikov, D., Karelin, D., and Ivaschenko, A., Sensivity of tundra carbon balance to ambient temperature, *Water, Air Soil Pollut.*, 119, 157–169, 2000.
63. Guerra, M. and Lombardi, S., Soil–gas method for tracing neotectonic faults in clay basins: the Pisticci field (Southern Italy), *Tectonophysics*, 339, 511–522, 2001.
64. Allard, P., Jean-Baptiste, P., Alessandro, W. D., Parello, F., Parisi, B., and Flehoc, C., Mantle-derived helium and carbon in groundwater and gases of Mount Etna, Italy, *Earth Planet. Sci. Lett.*, 148, 501–516, 1997.
65. Kepner, R. Jr., Kortyna, A., Wharton, R. Jr., Doran, P., Andersen, D., and Roberts, E., Effects of research diving on a stratified Antarctic lake, *Water Res.*, 34(1), 71–84, 2000.
66. Dörr, H., Katruff, L., and Levin, I., Soil texture parameterization of the methane uptake in aerated soils, *Chemosphere*, 26, 697–713, 1993.
67. Wang, Z., Zeng, D., and Patrick, W. H. Jr., Methane emissions from natural wetlands, *Environ. Monitor. Assess.*, 42, 143–161, 1996.
68. Segers, R. and Kengen, S. W. M., Methane production as a function of anaerobic carbon mineralization: a process model, *Soil Biol. Biochem.*, 30(8/9), 1107–1117, 1998.
69. Steudler, P. A., Bowden, R. D., Melillo, J. M., and Aber, J. D., Influence of nitrogen fertilization on methane uptake in temperate forest soils, *Nature*, 341, 314–316, 1989.
70. King, G. M. and Schnell, S., Effect of increasing atmospheric methane concentration on ammonium inhibition of soil methane consumption, *Nature*, 370, 282–284, 1994.
71. Baubron, J.-C., Allard, P., and Toutain, J.-P., Diffusive volcanic emissions of carbon dioxide from Vulcano Island, Italy, *Nature*, 344, 51–53, 1990.
72. Finizola, A., Sortino, F., Lénat, J.-F., and Valenza, M., Fluid circulation at Stromboli volcano (Aeolian Islands Italy) from self-potential and CO_2 surveys, *J. Volcanol. Geotherm. Res.*, 116, 1–18, 2002.
73. Baxter, P. J., Baubron, J.-C., and Coutinho, R., Health hazards and disaster potential of ground gas emissions at Furnas volcano São Miguel, Azores, *J. Volcanol. Geotherm. Res.*, 92, 95–106, 1999.
74. Lundegard, P. D., Sweeney, R. E., and Ririe, G. T., Soil gas methane at petroleum contaminated sites: forensic determination of origin and source, *Environ. Forensics*, 1, 3–10, 2000.
75. Dai, X. Y., White, D., and Ping, C. L., Comparing bioavailability in five Arctic soils by pyrolysis–gas chromatography/mass spectrometry, *J. Anal. Appl. Pyrolysis*, 62, 249–258, 2002.

7 The Determination of Phosphates in Environmental Samples by Ion Chromatography

Brett Paull, Leon Barron, and Pavel Nesterenko

CONTENTS

I. PHOSPHATES IN THE ENVIRONMENT

A. PHYSICAL AND CHEMICAL PROPERTIES OF PHOSPHOROUS

According to popular belief, elemental phosphorous (P) was discovered in 1669 by the German chemist Henning Brand, while trying to convert silver into gold. The name phosphorous derives form the Greek words *phôs* (light) and *phoros* (bearer). Phosphorous has the atomic number 15 and a mass of 30.97376 atomic mass units (a.m.u.) with a melting point of 44.1°C and a boiling point of 280°C. Phosphorous is commonly seen as a waxy white solid, although when pure it is colorless and transparent. Pure phosphorous is also insoluble in water.

Phosphorous is an essential element and a vital nutrient for all living organisms (human, animal, and plant life). For example, in man, phosphorous containing compounds are found in our genetic material (DNA) and in our proteins. Phosphorous is also essential for healthy teeth and bones, and is involved with transfer of energy within cells as adenosine triphosphate (ATP), thus being either directly or indirectly fundamental to all living processes. Humans and animals take in phosphorous in the form of phosphates, which is naturally present in a great many foodstuffs, including cheese, milk, meat, and cereals. Values for the minimum dietary intake for adult humans vary from country to country but generally lie between 500 to 800 mg/day, although the actual daily intake for most humans from food is estimated to be closer to 1200 to 2000 mg/day.

1. Oxides and Oxoacids of Phosphorous

The two most important and environmentally significant oxides of phosphorous include phosphorous in either $+3$ or $+5$ oxidation states. The basic structural units of elemental phosphorous are P_4 molecules, therefore, the oxide of phosphorous in the oxidation state $+3$ is P_4O_6, and the oxide of phosphorous in the $+5$ oxidation state is P_4O_{10}. Both of these oxides react with water to form subsequent oxoacids, as shown below:

$$P_4O_6 + 6H_2O \rightarrow 4H_3PO_3 \qquad \textit{ortho-phosphorous acid}$$

$$P_4O_{10} + 6H_2O \rightarrow 4H_3PO_4 \qquad \textit{ortho-phosphoric acid}$$

Other oxoacids of phosphorous include *pyro*-phosphoric acid ($H_4P_2O_7$), formed from the heat treatment ($>215°C$) of *ortho*-phosphoric acid, and *meta*-phosphoric acid (general formula $[HPO_3]_n$, where $n = 2, 3, 4, 6$), which has a ring-like structure and is formed from the heat treatment ($>300°C$) of either *ortho*-phosphoric or *pyro*-phosphoric acid. Table 7.1 shows the common oxoacids of phosphorous, with their respective pK_as and their chemical and structural formulas.

2. Polyphosphoric Acids and Polyphosphates

The dehydration reaction (elimination of H_2O) between phosphoric acid (H_3PO_4) and diphosphoric acid ($H_4P_2O_7$) results in the molecule $H_5P_3O_{10}$ or tripolyphosphoric acid. A further dehydration reaction with phosphoric acid results in the formation of $H_6P_4O_{13}$ or tetrapolyphosphoric acid.

TABLE 7.1
Oxoacids of Phosphorous

Oxidation State	Name	Chemical Formula	Structural Formula	pK_a
+3	*ortho*-Phosphorous acid (*Ortho*-Hydrogen phosphite)	H_3PO_3		1.5, 6.79
+5	*ortho*-Phosphoric acid (*Ortho*-Hydrogen phosphate)	H_3PO_4		2.15, 7.20, 12.15
+5	*pyro*-Phosphoric acid (diphosphoric acid)	$H_4P_2O_7$		
+5	*meta*-Phosphoric acid (*n* = 3, *cyclo*-tri-metaphosphoric acid, *n* = 4, *cyclo*-tetra-metaphosphoric acid, etc.)	$(HPO_3)_n$		

The structural formula shown for $(HPO_3)_n$ is *cyclo*-trimetaphosphoric acid.

This process can continue until the percentage of P in the polyacid reaches $\sim 39\%$. The chemical and structural formula of polyphosphoric acids is given below.

$$H_2PO_3(HPO_3)nPO_4H_2$$

The salts of the above acids are known as polyphosphates and are used extensively in industry (including water treatment), predominantly due to their strong metal ion complexing abilities (sequestering agents). A common example is sodium tripolyphosphate (STPP) or $Na_5P_3O_{10}$, which is used extensively as a builder for household detergents.

B. SOURCES AND OCCURRENCE OF PHOSPHATES IN THE ENVIRONMENT

Phosphorous constitutes one of the commonest substances in our environment, eleventh most abundant element in the earth's crust. However, free phosphorous is never found in nature, instead it is widely distributed in combination with many minerals. Phosphate rock (phosphorite), which has approximately the same composition as the mineral fluorapatite $[3Ca_3(PO_4)_2CaF_2]$, is a major source of the element. Large deposits of phosphorite can be found in the U.S.A. (principally Florida, and lesser deposits in Utah, Idaho and Tennessee), Morocco, and the Russian Federation.

Phosphorous containing species, including the oxoacids of phosphorous described in Section I are released into the environment through a number of natural and anthropogenic processes. The continuous weathering of phosphate containing rocks and the slow decomposition of flora and fauna lead to a fairly consistent background level of phosphates in soils and surface and ground waters (natural fertilization). These background levels of phosphates within nonpolluted surface waters are fairly low, at approximately 0.02 ppm, and are in fact a limiting nutrient for plant growth.

Anthropogenic activities leading to the release of phosphates into the environment are predominantly centered upon the release of industrial and domestic wastewaters and agricultural practices. The fertilizer industry, which uses large quantities of the acid to produce super-phosphate fertilizers, is a main user of phosphoric acid, and hence prime releaser of phosphates. Demand for these high-phosphate fertilizers by the agricultural industry has recently seen record worldwide demand for elemental phosphorous and concentrated phosphoric acid, and led to general rises in phosphate levels in natural surface and ground waters due to surface runoff and leaching of fertilized soils. Other agricultural practices which can lead to phosphate release on a more local level include leakage into ground and surface waters of animal sewage from slurry pits and holding tanks.

Domestic wastewater is also a major source of phosphate pollution, being relatively rich in phosphorous containing chemicals. This would include inorganic, and to a lesser extent organic, phosphates, originating from human waste and high levels of mono- and polyphosphates from use of household detergents (although levels in many detergent products have been significantly reduced in recent years and in some states of the U.S.A. phosphates in household detergents are now completely banned). Many sources of industrial wastewater also contain high levels of inorganic phosphates. Many mono- and polyphosphate species are used as inhibitor products, designed to control corrosion and scale formation in all kinds of industrial processes, including power generation. The water industry itself also uses a number of phosphate products in potable water treatment, for water softening as well as for elimination of red and black waters.

Removal of phosphates from the above wastewaters at the primary and secondary treatment stages prior to release into the water system, accounts for approximately 10 and 30% reductions in phosphate content, respectively, while the remaining 60% is discharged in receiving waters. Estimates compiled in the mid 1990s concluded that the sources of phosphates entering European surface waters were 49% from agriculture, 23% from human waste, 11% from detergent use, 7% from industry and 10% from natural bed rock erosion (Morse, Imperial College London, 1993).

C. THE NEED FOR MONITORING

The main problem associated with phosphates is over-fertilization of surface water, more commonly termed "eutrophication", referring to the "eutrophic" or enriched state of the water system. As a natural limiting nutrient, phosphate levels govern growth rates for vegetation, generally microscopic floating plants and algae (in particular blue-green algae which subsequently produces algal toxins). In water where other limiting nutrients or factors are low (such as nitrogen, silicate, temperature, and light), increase in phosphate levels may have little effect. However, where these other factors are sufficiently high, increasing phosphate levels will see immediate increases in biological growth and subsequent biological oxygen demand and also affect the pH balance of the water. This increased oxygen demand disturbs the delicate water ecosystem leading directly to fish kills and the growth of anaerobic bacteria.

Monitoring of phosphates levels, as released by industry, agriculture, and domestic sources, is essential for enforcement of regulations and imposition of limits. It also allows more accurate development of environmental models to determine the sources, fates, and risks associated with phosphates. Such monitoring occurs at point sources at the input and various treatment stages of water treatment plants, and throughout river and estuarine waterways. At a local level, individual catchment areas or river systems require regular monitoring to allow environmental scientists to determine past and predict future trends in phosphate levels. Regulatory agencies also depend upon accurate and regular data to identify isolated pollution incidents, particularly where legal action may follow.

Monitoring of phosphates also takes place within industry as part of environmental management and within processes that employ phosphate containing chemicals. Self-monitoring of nutrients in soils and run-off waters is also increasing within the agricultural sector.

D. Regulations

U.S.A. and European regulations regarding levels of phosphates being released into the environment are based upon the role of phosphorus as a nutrient chemical. The EU Directive 2000/60/EC for "establishing a framework for community action in the field of water policy" sets strict guidelines for monitoring, control, and limiting release of nutrient species into groundwater, rivers and streams, estuarine, and coastal waters.[1] The deadline for EU member states to implement this directive was set at 22nd December 2003. The Directive aims to regulate and contribute to a progressive reduction of emissions of all hazardous substances, including excessive nutrients into water bodies and to promote common principles among member states to coordinate efforts to improve the quality and quantity of water bodies for sustainable future usage. The Directive is also in place to control future transboundary water problems and stop further reduction in quality of currently polluted water bodies. The Directive calls for monitoring of water bodies under a number of headings, including biological elements, hydromorphological elements supporting the biological elements, chemical and physico-chemical elements supporting the biological elements, and specific pollutants. Nutrient conditions (and therefore phosphate levels) are included with the chemical and physico-chemical category. In river systems the Directive classifies as "high status" those waters where nutrient concentrations remain within the range normally associated with undisturbed conditions. Rivers classified as having "good status" have nutrient levels which do not affect the functioning of the ecosystem and the achievement of values specified under biological quality requirements.

II. COLLECTING AND PREPARING SAMPLES FOR PHOSPHATE DETERMINATIONS BY ION CHROMATOGRAPHY (IC)

Standard methods for phosphates, polyphosphates, and organic phosphates in environmental samples are predominantly nonchromatographic methods, which are based upon the molybdenum blue method. Within this colorimetric method ammonium molybdate and antimony potassium tartrate react under acidic conditions with dilute solution of phosphorous to form an antimony–phospho–molybdate complex which is then reduced to an intensely blue-colored complex by ascorbic acid. U.S. EPA Methods 365.1 to 365.4 are based upon this chemistry.

However, throughout the 1980s and 1990s IC emerged as the leading analytical method for the simultaneous determination of common inorganic anions (predominantly in nonsaline water samples and aqueous sample extracts), and several reviews have been published which detail the application of IC to environmental matrices.[2–4] Included within the modern definition of IC are all high-performance liquid chromatographic techniques, which are applied to the separation of ionic species, including inorganic and organic anions, cations, and charged complexes. Various modes of IC include:

ion exchange chromatography (inorganic anion and cations),
ion-exclusion chromatography (organic acids),
ion-interaction chromatography (also known as ion-pair chromatography – inorganic anions and cations)
chelation ion chromatography (inorganic cations).

However, for each of the above techniques, the sample is required in a fully dissolved aqueous format, and so sample clean-up and digestion methods require careful consideration.

A. Sample Pretreatment

1. Natural Waters

The concentration of phosphate in many nonsaline natural waters is often sufficiently high for direct determination of phosphate by IC techniques, in which case filtration or dilution is often required prior to analysis. In other cases, sample pretreatment may be required and include the complete or partial removal of excess interfering anions. For example, the high concentration of chloride in seawater overloads the ion chromatographic column and for direct analysis samples have to be diluted five to tenfold. This makes the determination of phosphate at low μM levels difficult. Therefore, the direct IC determination of phosphate only becomes possible after preelution of a substantial amount of interfering chloride through column switching techniques, such as those applied by Dahllof et al. using a Dionex AG4A guard column and a switching valve, positioned before the main Dionex AS4A column.[5]

An alternative interference removal step was developed by Ledo de Medina et al. who developed an IC method for phosphate in natural waters in the presence of high concentration of sulphates.[6] This interference was avoided by first precipitating sulphate as lead sulphate prior to IC analysis. Samples with high iron content were investigated by Simon.[7] Interferences caused by the precipitation of iron hydroxides from air oxidation of ferrous iron in anoxic water samples and from the alkaline eluent used in IC, were found to affect the determination of phosphate and other inorganic anions in riverine sediment interstitial water samples with high concentrations of dissolved iron (0.5 to 2.0 mmol/l). To eliminate this interference the complexation of iron with cyanide was used prior to IC analysis.[7]

For determination of low concentrations of phosphate in natural waters, the selective preconcentration of phosphate may be required. An example of this was recently shown by Yuchi et al. who used a chelating column loaded with Zr for this purpose (TSK gel AF-chelate Toyopearl 650, 40–90 μm, containing 30 μmol/g iminodiacetic acid group).[8]

2. Wastewaters

In the analysis of wastewaters to prolong the lifetime of the chromatographic column the removal of hydrophobic organic substances from the wastewater samples is strongly recommended.[9] A prescribed procedure for this uses a C_{18} cartridge (preconditioned with 5 ml methanol, followed by 5 ml of deionized water), through which the sample (approx. 5 ml) is passed, with the first 1 ml of the eluate being discarded. Using this method, the recovery of phosphate from domestic and industrial wastewaters is usually higher than 95%.[9–10] Buldini et al. recently reported an on line micro-dialysis-IC method for determination of inorganic anions, including phosphate, in olive oil mill wastewater.[11] The system removed the majority of organic load while maintaining recoveries for inorganic anions in wastewater and standard oil emulsions of between 96 and 104%.

B. Decomposition and Extraction

1. Plant Material

The extraction of phosphates from freshly cut plants containing different amounts of water at the surface of leaves or within other material could be a source of uncertainty in the analysis step. With this in mind, one solution is to freeze-dry the fresh plant material.[12] The dried material can then be milled to a fine powder, and the content of water is determined by the Karl–Fisher method, as modified by Moibroek and Shahwecker.[13] Finally, analyte anions can be extracted from the plant material in accordance with the AOAC method for dried vegetables and flours.[14]

The procedure developed by Bradfield and Cooke is an example of an extraction procedure used for the IC determination of anions in plant materials.[15] Here, the extraction of anions from 100 mg of fresh plant material (nondried rape) was carried out with 5 ml of deionized water at 70°C for 30 min, followed by filtration through Whatman No.541 filter paper, prior to analysis using ion-pair RP-HPLC. Slightly modified procedures followed by micro-filtration have been used for the determination of inorganic anions, including phosphate, in dried potatoes,[16] coffee beans and tea,[17] tomato, pumpkin, turnip, and lettuce leaves,[18] and spinach.[19] A more unusual extraction solution of 2.8 mM $NaHCO_3$ and 2.2 mM Na_2CO_3 in D_2O was used by Raiser for the extraction of inorganic anions, including phosphate, from ground tobacco.[20] In this case the recovery of phosphate spiked within tobacco was between 92 and 98%.

Work by Ruiz et al. in the analysis of water soluble inorganic phosphates in vegetables, such as tomato, lettuce, marrow, mushroom, celery, cauliflower, chard, onion, carrot, used SPE with Sep-Pak C_{18} cartridges for removal of organic compounds prior to IC analysis.[21] Spiro and co-workers studied the kinetics of extraction of inorganic anions, including phosphate, from different types of tea into water.[22–23]

Improved methods for the digestion of various plant materials (leaves, juices and extracts of vegetables, fruits, plants) were investigated by Buldini et al.[24] In order to dissolve samples prior to IC analysis, oxidative UV photolysis was used. This has the advantage over other sample decomposition methods of being a simple procedure with minimal reagent addition requirements, resulting in minimal contamination. Here the homogenized sample was mixed with a small amount of hydrogen peroxide and nitric acid and was subjected to UV-degradation in a digester equipped with a high-pressure mercury lamp (500 W) at a temperature of 85°C maintained by air–water cooling system. While organic constituents were degraded, metals and nonmetals remained unaffected by the UV radiation, except nitrate, iodide, and manganese. With this method phosphate could be determined quantitatively as total phosphate only, without any speciation.

2. Soils

Due to the nature of the sample, a more complex procedure is often required in the analysis of soils by IC. Soil samples should be air-dried, crushed, and sieved (<2 mm) before digestion/extraction. Then extraction of phosphate may be performed either with water or more complex solutions. Murcia et al.[25] compared the extraction of inorganic anions from soils with water, 10 mM LiCl, 10 mM KCl and a mixture of 30 mM $NaHCO_3$ and 18 mM Na_2CO_3. The work found little significant difference between each of the extractant solutions, and so deionized water is generally used in agreement with an early investigation by Dick and Tabatabai.[26] The general procedure would involve a sample of soil (~3 g) being mixed with a tenfold volume of extractant and sonicated in an ultrasonic bath for ~30 min. The resultant suspension would then be filtered through a 0.45 μm membrane filter or centrifuged, followed by solid phase extraction of organic substances on Sep-pak C_{18} cartridges (as described above for wastewaters[9,10]). Such a procedure results in 84 to 112% recovery of phosphate from spiked soil samples. Losses in recovery are generally due to the formation of insoluble salts (mostly with magnesium and calcium) or complexes with other metals, or with the possible hydrolysis of organic phosphates.[9,10] To suppress the formation of metal–phosphate species and increase recovery, EDTA[27] or sulphonated cation-exchange resins[15] may be added to the extraction mixture.

3. Sewage Sludge, Coal, Oil

Phosphorus content is an important component requiring quantification of any fuel or material for combustion, including coal, oil and sewage sludge. The digestion of coal, oil, and sewage sludge prior to IC analysis has been performed by combustion in a high-pressure bomb containing oxygen.[28] A less extreme approach which has been used involved digestion using *aqua-regia* followed by 1000-fold dilution of the resultant digest.[29]

4. Other Environmental Matrices

An IC based procedure for the determination of chlorine, sulphur, and phosphorous in general organic materials was proposed by Novic et al.[30] It includes the oxidation of nitrogen, phosphorus, and sulphur to nitrate, phosphate, and sulphate anions using a two stage closed-vessel microwave assisted digestion. The first step involves the microwave digestion of 5 g of sample in 10 ml of 22% (v/v) hydrogen peroxide with the addition of 50 μl formic acid. After cooling, an additional 10 ml portion of hydrogen peroxide is added to the sample and the microwave treatment is repeated, followed by cooling and dilution with deionized water to 100 ml prior to IC analysis. The method was successfully validated with the analysis of NIST SRM 1566a oyster tissue and NIST SRM 2704 Buffalo river sediment samples.

The oxidative decomposition of biological samples (0.15 to 0.30 g) by sodium peroxide (7.0 g) fusion in a Parr bomb, followed by dissolution in 250 ml water of deionized water and subsequent electrodialysis, was used prior to the IC determination of chlorine, sulphur, and phosphorus in SRM 1577b bovine liver standard sample by Nguyen and Rossbach.[31]

C. DERIVATIZATION

The most popular derivatization method for phosphate analysis involves the complexation of phosphate with the molybdate ion under acidic conditions, according to the following equation[32]:

$$PO_4^{3-} + 12MoO_4^{2-} + 27H^+ \rightarrow H_3PO_4(MoO_3)_{12} + 12H_2O$$

The resultant complex is reduced to a blue complex (phosphomolybdenum blue) in the presence of reducing agents such as ascorbic acid or stannous chloride, and in the presence of a catalyst, e.g., antimony potassium tartrate. This method may overestimate inorganic phosphate by including hydrolysable organic or condensed phosphorus compounds and, therefore, it is referred to as a measurement of soluble reactive phosphate (SRP). The intense blue color of this complex allows direct spectrophotometric determination of phosphate at 700 nm with detection limits of around 10 nM.[33] The molybdenum blue may be preconcentrated by SPE techniques on cartridges packed with either hydrophobic C_{18} silica or hydrophilic diolsilica particles and then analyzed using RP-HPLC. This approach has been carried out for phosphates in natural waters,[34] and hair samples after digestion with nitric acid.[27]

D. STORAGE

Water-based extracts containing phosphate should be kept in polyethylene bottles in a refrigerator at temperatures between 2 to 4°C.[35] Storage time should be limited to 2 days maximum.

III. ION CHROMATOGRAPHIC METHODS FOR PHOSPHATE DETERMINATIONS IN ENVIRONMENTAL MATRICES

The majority of published and standard IC methods for phosphate determinations are based upon anion exchange chromatography, coupled with various modes of detection. Standard IC methods for inorganic ion analysis are also now available and widely used by regulatory environmental monitoring agencies. Table 7.2 lists a selection of standard IC methods approved in the U.S.A. for environmental water and wastewater analysis.

A. SUPPRESSED IC METHODS

1. U.S. EPA Method 300.0 (Part A)

U.S. EPA Method 300.0 (Part A) recommends the use of IC for determination of *ortho*-phosphate in drinking water, surface water, mixed domestic and industrial wastewaters, groundwater, reagent waters, solids (after extraction), and leachates.[37] Method 300.0 prescribes the use of a Dionex AS4A anion exchange column (or alternative column producing similar resolution) with suppressed conductivity detection using a Dionex micro-membrane suppressor system (or alternative producing similar baseline conductivity and stability). Using the column and suppressor combination, together with a 1.7 mM NaHCO$_3$/1.8 mM Na$_2$CO$_3$ eluent (delivered at 2 ml/min), phosphate elutes after nitrate and before sulphate, at a retention time of ~5.4 min. The method proposes a detection limit for phosphate of 0.003 mg/l in reagent water.

2. Commercial Methods

A considerable number of application notes produced by IC manufacturers exist describing developed methods for phosphate determination in a wide variety of environmental matrices. For example, Dionex Corporation have produced a large number of detailed suppressed IC application notes for inorganic anions (including phosphate and polyphosphates) determination in natural waters, drinking water, wastewater, and high-purity waters. Metrohm also have an impressive range of applications based on the use of their own suppressed IC system. A number of these developed application methods, together with the recommended chromatographic conditions, are included in Table 7.3.

TABLE 7.2
Standard IC Methods Approved in the U.S.A. for Environmental Water and Wastewater Analysis

Method	Analytes	Matrices
U.S. EPA Method 300.0 (A)	F$^-$, Cl$^-$, NO$_2^-$, Br$^-$, NO$_3^-$, PO$_4^{3-}$, SO$_4^{2-}$	rw, dw, sw, ww, gw, se
U.S. EPA Method 300.1 (A)	F$^-$, Cl$^-$, NO$_2^-$, Br$^-$, NO$_3^-$, PO$_4^{3-}$, SO$_4^{2-}$	rw, dw, sw, gw
U.S. EPA SW-846 9056	F$^-$, Cl$^-$, Br$^-$, NO$_3^-$, PO$_4^{3-}$, SO$_4^{2-}$	Combustion extracts, all waters
ASTM D 4327-97	F$^-$, Cl$^-$, NO$_2^-$, Br$^-$, NO$_3^-$, PO$_4^{3-}$, SO$_4^{2-}$	dw, ww
Standard Methods 4110[36]	Cl$^-$, NO$_2^-$, Br$^-$, NO$_3^-$, PO$_4^{3-}$, SO$_4^{2-}$	rw, dw, ww
U.S. EPA Method 300.6	Cl$^-$, NO$_3^-$, PO$_4^{3-}$, SO$_4^{2-}$	Wet deposition, rain, snow, dew, sleet hail

rw, reagent water; dw, drinking water; sw, surface water; ww, wastewater (domestic and industrial); gw, groundwater; se, solid extracts.

Adapted from Jackson, P. E., Ion chromatography in environmental analysis, In *Encyclopedia of Analytical Chemistry*, Meyers, R. A., Ed., Wiley, Chichester, 2000.

TABLE 7.3
Environmental Water Samples (Marine, River, Ground, Surface, Drinking)

Sample	Ion	Column(s)	Eluent and Flow Rate	Inj. Vol. (μl)	Detection	Comments	References
Ultra pure water	PO_4^{3-}, F^-, Cl^-, NO_2^-, Br^-, NO_3^-, SO_4^{2-}	Phenomenex, Star-Ion-A300 (100×4.6 mm²)	3.6 mM NaHCO$_3$/3.75 mM Na$_2$CO$_3$ at 0.50 ml/min	20	Supp. Cond.	Preconcentration carried out with Metrosep A PCC 1 HC column	48
Marine and freshwater samples	PO_4^{3-} as SRP	LC-8 or C8-5	Complex mixed gradient	100	UV/VIS @ 700 nm	Complexation of SRP with Molybdenum Blue reagent Phosphomolybdenum complex: $H_3PO_4(MoO_3)_{12}$	34
Seawater	NO_2^- and PO_4^{3-}	2 × Dionex AG4A guard, 1 × Dionex AS4A analytical	NaHCO$_3$–Na$_2$CO$_3$ gradient	20	Supp. Cond.	Use of 2 precolumns and 2 injector valves to elute excess chloride to waste prior to redirection to analytical column	5
Estuary and lagoon samples	PO_4^{3-} and SO_3^{2-}	Dionex AS4A	1.7 mM NaHCO$_3$/1.8 mM Na$_2$CO$_3$ at 2ml/min	100	Supp. Cond.	Potentiometric titration of sulphate with lead perchlorate to remove sulphate and boost phosphate sensitivity	6
Wastewater	PO_4^{3-}, F^-, Cl^-, NO_3^-, SO_4^{2-}	Quatern. ammonium anion ex. (3×500 mm²)	3 mM NaHCO$_3$/2.4 mM Na$_2$CO$_3$ at 92 ml/h	100	Supp. Cond.	Simultaneous separation of anions and cations	49
Rain water	PO_4^{3-}, F^-, Cl^-, NO_3^-, SO_4^{2-}	Dionex anion ex. (3×500 mm²)	3 mM NaHCO$_3$/2.4 mM Na$_2$CO$_3$ at 138 ml/h	100	Supp. Cond.		50
Various natural and industrial waters	PO_4^{3-}, F^-, Cl^-, NO_2^-, Br^-, NO_3^-, SO_4^{2-}	(A) Dionex AS4A (B) Dionex AS9	(A) 1.7mM NaHCO$_3$/1.8 mM Na$_2$CO$_3$ at 2 ml/min (B) 1.7mM NaHCO$_3$/1.8 mM Na$_2$CO$_3$ at 1 ml/min	100–1000 100–1000	Supp. Cond.		37
Drinking water	PO_4^{3-}, F^-, Cl^-, NO_2^-, Br^-, NO_3^-, SO_4^{2-}	Dionex (6×250 mm²) anion ex.	4.5 mM NaHCO$_3$/2.4 mM Na$_2$CO$_3$ at 115 ml/h	100	Supp. Cond.	Comparison to potentiometric methods	51
Wastewater (Effluent/Influent)	PO_4^{3-}, Cl^-, NO_3^-	QS-A5 (150×2.0 mm²)	Eluent Stream: 10 mM NaHCO$_3$/1 mM NaHCO$_3$/1 mM Na$_2$CO$_3$ at 0.60 ml/min	25	Supp. Cond. (using cartridge)	Suppressor cartridge off line during injection and once ammonia elutes, the cartridge is brought on line	39

Sample	Analytes	Column		Eluent	Detection	Notes	Ref
Waters with high nitrate levels	PO_4^{3-}, F^-, Cl^-, Br^-, NO_3^-, SO_4^{2-}, CO_3^{2-}	Dionex AS15 (250 × 2 mm²)	1	KOH gradient	Supp. Cond.	Paper also includes anion determinations in molten glass	52
Wastewater	PO_4^{3-}, F^-, Cl^-, NO_2^-, Br^-, NO_3^-, SO_4^{2-}	Dionex 30827 (4 × 250 mm²)	250	3.0 mM NaHCO$_3$/2.4 mM Na$_2$CO$_3$ at 2.6–3.1 ml/min	Supp. Cond.		53
Wastewater	PO_4^{3-}, F^-, Cl^-, NO_3^-, SO_4^{2-}	Dionex 30170 (50 × 3 mm²)	60	2.1 mM NaHCO$_3$/17 mM Na$_2$CO$_3$ at 2 ml/min	Supp. Cond.	Phosphate LOD 5 µg/l Sodium carbonate fusion used to decompose sample	54
Rain Water	PO_4^{3-}, F^-, Cl^-, NO_2^-, Br^-, NO_3^-, SO_4^{2-}	Dionex (250 × 3 or 500 mm) anion ex.	300	3.0 mM NaHCO$_3$/2.4 mM Na$_2$CO$_3$ at 2.75 and 3 min	Supp. Cond.	Semi-automated IC (60 samples per day) Not sensitive to phosphate at concentrations lower than 0.1 mg/l	55
Rain and lake water	(A) PO_4^{3-}, Cl^-, F^-, NO_3^-, SO_4^{2-} (B) Li$^+$, Na$^+$, K$^+$, NH$_4^+$, Ca^{2+}, Mg^{2+}	(A) Dionex AS4 (250 × 4 mm²) (B) Dionex CS3 (250 × 4 mm²)	50	(A) 2.2 mM NaHCO$_3$/1.8 mM Na$_2$CO$_3$ at 2.75 and 1 ml/min (B) 4.8 mM HCl-4.0 mM DAP. HCl-4.0 mM histidine at 1.1 and 1.2 ml/min	Supp. Cond.	Simultaneous anion and mono/divalent cation analysis in 14 min. Phosphate LOD 0.05 mg/l	56
Waste water	PO_4^{3-}, Cl^-, NO_3^-, SO_4^{2-}	Metrosep Anion Dual two	20	2.0 mM NaHCO$_3$/1.2 mM Na$_2$CO$_3$ at 0.8 ml/min	Supp. Cond.	Phosphate determined at 173.0 mg/l in sample	57
Interstitial water	PO_4^{3-}, Cl^-, NO_3^-, SO_4^{2-}	Dionex AS1 (250 × 3 mm²)	100	36 mM NaHCO$_3$/24 mM Na$_2$CO$_3$ at 2.75 and 2.5 ml/min	Supp. Cond.	Samples taken had high iron content. Cyanide used to complex iron and remove interference. Phosphate LOD: 1.6 µM	7
Drinking water	PO_4^{3-}, Cl^-, Br^-, SO_4^{2-}, BrO_3^-, ClO_2^-, ClO_3^-, SeO_3^{2-}, SeO_4^{2-}, AsO_4^{3-}	Dionex AS12A (250 × 4 mm²)	50	11 mM ammonium carbonate (pH 11.2) at 2 ml/min	ICP–MS	Runtime 4 min Phosphate LOD: 36 µg/l	47
Olive oil mill wastewater	PO_4^{3-}, Cl^-, NO_2^-, NO_3^-, SO_4^{2-}	Metrosep Anion Dual two	20	2 mM NaHCO$_3$/1.3 mM Na$_2$CO$_3$ at 0.8 ml/min	Supp. Cond.	Method involves online micro-dialysis of anions Phosphate LOD: 10 µg/l	11
River water	PO_4^{3-}, F^-, Cl^-, NO_2^-, Br^-, NO_3^-, SO_4^{2-}, CO_3^{2-}, BrO_3^-	Phenomenex Kingsorb (30 × 4.6 mm²) 3 µm particle size, coated with DDAB	50	5 mM pthalate, pH 7.5 at 2 ml/min and column temperature of 45°C	Indirect UV @ 279 nm	Separation of 9 anions in 160 sec	46

Continued

TABLE 7.3
Continued

Sample	Ion	Column(s)	Eluent and Flow Rate	Inj. Vol. (µl)	Detection	Comments	References
Power plant water	PO_4^{3-}, F^-, Cl^-, NO_2^-, Br^-, NO_3^-, SO_4^{2-}, CO_3^{2-}, BrO_3^- organic acids	Dionex AS15 (150 × 3 mm²)	Electrolytically generated KOH gradient at a flow of 0.50 ml/min	1000	Supp. Cond.	Very low LODs between 9.9 and 79 ng/l	58
Seawater	PO_4^{3-}, Cl^-, NO_2^-, NO_3^-, SO_4^{2-}	2 × Dionex AG9-HC guard and 1 × AS9-HC (250 × 4 mm²) analytical	(A) Gradient of 3 mM NaHCO$_3$/14 mM Na$_2$CO$_3$ and 9 mM carbonate at 1 ml/min	25	Supp. Cond. and UV absorbance @ 225 nm	Uses a column switching method to remove excess chloride to waste prior to separation of other nutrient anions	38
River water	PO_4^{3-}, Cl^-, NO_3^-, SO_4^{2-}, oxalate	Dionex AS4A	1.5 mM Na$_2$CO$_3$/0.65 mM NaHCO$_3$ (no flow rate specified)	—	Supp. Cond.	Also determines anions in spinach and soil	25
Ultra pure water	PO_4^{3-}, F^-, Cl^-, NO_2^-, Br^-, NO_3^-, SO_4^{2-}, CO_3^{2-}, organic acids	Dionex AS15 (150 × 3 mm²)	KOH gradient at 0.7 ml/min	1000	Supp. Cond.	0.18 µg/l phosphate MDL	59
Drinking water	PO_4^{3-}, F^-, Cl^-, NO_2^-, Br^-, NO_3^-, SO_4^{2-}	Dionex AS14A (150 × 3 mm²)	1.0 mM NaHCO$_3$/8.0 mM Na$_2$CO$_3$ at 0.8 ml/min	25	Supp. Cond.	Compares Dionex AEES Atlas and ASRS Ultra suppressors	60
Wastewater	PO_4^{3-}, F^-, Cl^-, NO_2^-, Br^-, NO_3^-, SO_4^{2-}	(A) Dionex AS4A-SC (250 × 4 mm²)	(A) 1.7 mM NaHCO$_3$/1.8 mM Na$_2$CO$_3$ at 2 ml/min	50	Supp. Cond.	AS4A column method more sensitive to phosphate (17.8 Ppb as opposed to 20.2 Ppb with AS14 method)	10
		(B) Dionex AS4A-SC (250 × 4 mm²)	(B) 1.0 mM NaHCO$_3$/3.5 mM Na$_2$CO$_3$ at 1.2 ml/min	50			
Drinking water	PO_4^{3-}, condensed phosphates	Dionex AS11 (250 × 2 mm²)	Various NaOH gradients	10	Supp. Cond.	MDL of all phosphate species in 5–30 µg/l range	61
Drinking water	PO_4^{3-}, F^-, Cl^-, NO_2^-, Br^-, NO_3^-, SO_4^{2-}	(A) Dionex AS4A-SC (250 × 4 mm²)	(A) 1.7 mM NaHCO$_3$/1.8 mM Na$_2$CO$_3$ at 2 ml/min	50	Supp. Cond.	AS4A column method more sensitive to phosphate (17.8 Ppb as opposed to 20.2 Ppb with AS14 method)	9
		(B) Dionex AS4A-SC (250 × 4 mm²)	(B) 1.0 mM NaHCO$_3$/3.5 mM Na$_2$CO$_3$ at 1.2 ml/min	50			

3. Natural Waters

Much of the work in the past decade on IC development has been in the areas of selectivity improvement to allow more complex application (particularly relevant to environmental analysis), and improved detection methods for enhanced sensitivity and selectivity.

Ledo de Medina et al. developed an IC method for phosphate determination in natural water samples with high sulphate concentrations.[6] The developed method used a Dionex AS4A anion exchange column with a 1.7 mM NaHCO$_3$/1.8 mM Na$_2$CO$_3$ eluent and a Dionex AMMSII membrane suppressor. In this method much of the excess sulphate was precipitated using lead perchlorate prior to IC analysis, however, the sample still contained excess sulphate relative to the trace levels of phosphate present. The method was suitable for the determination of low μg/l phosphate in saline water samples, with the large chloride peak eluting close to the eluent dip.

Dahllof et al. also used suppressed IC to determine phosphate in saline water samples.[5] In this work, resolution of the analytes and the excess matrix anions was optimized using a factorial experimental design. The method used two precolumns prior to the separation column (Dionex AS4A) and the majority of excess chloride could be selectively eluted to waste by the use of a switching valve positioned between the precolumns and analytical column. The experimental design optimized eluent conditions and timing of the chloride removal valve. The optimized procedure resulted in a detection limit of 1 μM for phosphate in seawater, which was well resolved from the large chloride and sulphate peaks present. Column switching techniques were also employed by Bruno et al. to determine nutrient concentrations in seawater.[38] The instrumental set-up consisted of two in-line Dionex AG9 precolumns and an AG9-HC separator column, connected by way of a four way switching valve. The method allowed the direct injection of seawater without any pretreatment, and the elution to waste of excess chloride. However, detection limits for phosphate were rather poor, at 1000 μg/l.

4. Wastewaters

Recently, Karmarkar reported an impressive dual IC-flow injection analysis (FIA) method for the sequential determination of anionic (nitrate and phosphate) and cationic (ammonium) nutrients in wastewater samples.[39] The dual system was based upon the use of an anion exchange column (Lachat QS-A5) and two detectors, one suppressed conductivity detector using a Lachat Instruments QE-A1 small suppressor cartridge, which is regenerated between samples, and a second visible absorbance detector. Upon injection of the sample the conductivity detector was switched off line and the nonretained ammonium was passed through the analytical column and detected by the visible absorbance detector, following an on line colorimetric reaction. The conductivity detector was then immediately switched on line to detect the retained nutrient anions. The method reported detection limits for phosphate of 0.006 mg/l phosphate.

5. Solids, Soils, and Sediments

Where sample digestion is required, for example in the analysis of solids, soils, and sediments, the sample extracts will contain high levels of acid anions used to digest the sample, such as sulphate from sulphuric acid, chloride from hydrochloric acid, and nitrate from nitric acid. Shotyk et al. developed a suppressed IC method for the nitrate, phosphate, and organically bound phosphorous in digests of coral skeletons.[40] The samples were digested using concentrated hydrochloric acid and in this case the subsequent excess chloride within the digests was reduced using a Dionex On-Guard Ag cartridge for chloride reduction. The IC method used a Dionex AS4A anion exchange column with a 1.7 mM NaHCO$_3$/1.8 mM Na$_2$CO$_3$ eluent and a Dionex AMMSII membrane suppressor. Large matrix chloride and sulphate peaks were well resolved from the phosphate peak of interest. Colina et al.[41] were able to determine total phosphorous in marine sediments, which had been previously digested using a mixture of potassium persulphate and NaOH. The developed method used the same Dionex AS4A column, suppressor and eluent as Shotyk et al. In this case the phosphate eluted at ~4 min just prior to a large matrix sulphate peak. The method reported a linear range of 0.01 to 1.0 mg/l for phosphate and a detection limit of 0.006 mg P/g of sediment.

6. Plant Material

The use of suppressed IC in the analysis of plant material digests for nutrient anions has also been well documented. For example, Masson et al. used suppressed IC for the simultaneous determination of inorganic and organic anions (including phosphate) in plant sap.[42] The final method employed a Dionex AS11 column with gradient elution using NaOH and only a 5 min total run time. More recently, Casey et al. determined anionic nutrients in seed exudates by IC, using a Dionex AS4A column with the standard 1.7 mM NaHCO$_3$/1.8 mM Na$_2$CO$_3$ eluent. Phosphate eluted just after 9 min and was well resolved from other anions present.[43]

7. Determination of Polyphosphates (Condensed Phosphates)

A number of suppressed IC methods have emerged recently for the analysis of polyphosphates. For example, Baluyot and Hartford used a microbore Dionex AS11 column with NaOH gradient elution and obtained excellent resolution of polyphosphate species up to a chain length (n) of 45.[44] Although the method was used for the analysis of commercial polyphosphate mixtures, it was clearly also applicable to wastewater samples. Sekiguchi et al. recently used a similar IC method for the analysis of polyphosphates in complex food extracts.[45] Again the microbore Dionex AS11 column was used with an electrochemically generated NaOH eluent gradient. The method was applied to extracts from meat, fish paste, and cheese, with polyphosphates up to $n = 7$ identified latter.

B. Nonsuppressed IC Methods

Although the vast majority of published IC methods which include phosphate as an analyte use suppressed conductivity detection, a number of interesting alternative detection methods have been exploited.

1. Direct Conductivity

Although direct conductivity detection offers lower sensitivity for most inorganic species than the suppressed mode, for certain applications it offers a simpler and more robust alternative. Ruiz et al. developed a nonsuppressed IC method for the determination of water soluble extracts of inorganic phosphates in plant materials. The developed method employed a Waters IC-Pak anion HR column in combination with a low conductivity borate/gluconate eluent.[21] The method resulted in the elution of phosphate at approximately 11 min, between nitrate and sulphate and well resolved from excess nitrate in sample extracts. More recently, Alcazar et al. developed a similar direct conductivity IC method for the determination of water soluble organic acids, chlorides, and phosphates in extracts from coffee and tea.[17] This method employed a low capacity Hamilton PRP-X110 anion exchange column with a dilute potassium hydrogen phthalate eluent (0.6 mM, pH 4.0). The method resulted in good resolution and detection of phosphate in the sample extracts, with phosphate eluting between peaks for succinic acid and chloride. However, as with many direct conductivity methods, limits of detection were quite poor, stated as 4.1 mg/l for phosphate.

2. Spectrophotometric Detection

As phosphate does not absorb in either the usable UV region or within the visible spectrum, options for spectrophotometric detection are limited to either indirect methods or some form of precolumn or postcolumn derivitization. With indirect UV absorbance detection, the eluent used must be based upon a strongly UV absorbing eluent anion, which is displaced on a charge for charge basis by the analyte species. An example of this type of detection was recently shown by Connolly and Paull who used a 5 mM phthalate eluent with a short dynamically modified anion exchange column for the rapid separation and detection of inorganic anions, including phosphate.[46] Indirect detection was monitored at 279 nm and the method was applied to the determination of phosphate in a polluted river water sample. However, as the method was based upon indirect and not direct absorbance, background absorbance was high and this reduced the method sensitivity, giving a method quantification limit for phosphate of 0.5 mg/l.

A more complex but sensitive IC method was developed by Antony et al. based upon separation using a Dionex AS4A-SC column and postcolumn derivitization of phosphate using a solution containing 0.5% w/v ammonium molybdate and 0.5% w/v bismuth nitrate, in 1.75 M H_2SO_4 and 0.75% ascorbic acid.[27] The resultant reduced ion association complex absorbed strongly at 700 nm and a detection limit for phosphate (P) of an impressive 0.8 μg/l. The above chemistry has also been exploited in a recent publication by Haberer and Brandes[34] who carried out precolumn derivitization of phosphate within freshwater and saltwater samples and then solvent extracted the resultant molybdenum blue complex prior to separation and detection (at 700 nm) using reversed-phase HPLC.

3. Mass Spectrometric Detection

There is a growing interest in coupling IC with more powerful detection methods, particularly in the field of environmental analysis. Mass spectrometry is one such method and the coupling of suppressed IC with electrospray ionization mass spectrometry (ESI–MS) and inductively coupled plasma mass spectrometry (ICP–MS) can provide the solution to obtaining low limits of detection within complex environmental matrices. A recent example is given by Divjak et al. who used ICP–MS as an element-specific detector for a range of oxyanions including phosphate.[47] The method used a Dionex AS12A anion exchange column with a 11 mM $(NH_4)_2CO_3$ eluent, which was fed postcolumn directly into the ultrasonic nebulizer of the ICP–MS system. The method reported a detection limit of 36 μg/l for phosphate based upon a 50 μl sample injection, with single ion monitoring at a m/z value of 31.

IV. APPLICATIONS OF ION CHROMATOGRAPHY TO THE DETERMINATION OF PHOSPHATES IN ENVIRONMENTAL MATRICES

A. ENVIRONMENTAL WATER SAMPLES (MARINE, RIVER, GROUND, SURFACE, AND DRINKING WATERS)

Table 7.3 shows details from the IC methods developed for application for phosphate determination (together with other common anions) in environmental water samples.

B. SOILS, SEDIMENTS, AND SLUDGE SAMPLES (AND OTHER ORGANIC SOLIDS)

Table 7.4 shows details from the IC methods developed for application for phosphate determination (together with other common anions) in extracts from soil, sediment, and sludge samples.

C. PHOSPHATES IN BIOLOGICAL MATERIALS (NONPROCESSED FOODS, FAUNA, AND FLORA)

Table 7.5 shows details from the IC methods developed for application for phosphate determination (together with other common anions) in extracts from biological materials.

TABLE 7.4
Phosphates in Soils, Sediments and Sludge Samples and Other Organic Solids

Sample	Ion	Column(s)	Eluent and Flow Rate	Inj. Vol. (μl)	Detection	Comments	References
Soil and Plants	PO_4^{3-}, Cl^-, NO_3^-, SO_4^{2-}	Apex ODS (5 μm, 250 × 4.6 mm²) column	0.5 mM TBAOH in 5% MeOH at pH 7–7.1, adjusted with 0.1 M potassium hydrogen phthalate. Flow rate: 1.5 ml/min	20	UV @ 255 nm	Recovery 84–108%; runtime: 20 min; interference caused by aspartic acid	15
Soil	SeO_3^{2-}, PO_4^{3-}, Cl^-, NO_3^-, SO_4^{2-}	Vydac 302 IC (4.6 × 250 mm²) anion ex.	1.5 mM phthalic acid (pH 2.7)	100–7000	Nonsupp. Cond.	LOD of 0.3 μg/l PO_4^{3-} with 2000 μl loop, RSD 3.3% for PO_4^{3-} ($n = 10$)	62
Soil	MoO_4^{2-}, PO_4^{3-}, Cl^-, NO_3^-, SO_4^{2-}	Waters TSK Gel IC Pak IC-Pak Anion column (26770)	5 mM p-hydroxybenzoic acid (pH 8.25) at 1.8 ml/min	100–2000	Nonsupp. Cond.	Comparison with ICP–OES. Molybdate LOD was 45 μg/l with 2 ml injection loop	63
River water sediment	PO_4^{3-}, SO_4^{2-}	Dionex AG4A and AS4A	1.7 mM NaHCO₃/1.8 mM Na₂CO₃ at 2 ml/min	100	Supp. Cond. (column)	Uses high pressure bomb for persulphate digestion and converts all phosphorus to orthophosphate	31
Sewage sludge	PO_4^{3-}	Metrosep Anion Dual one	1.0 mM Na₂CO₃/4.0 mM NaHCO₃ at 0.5 ml/min	20	Supp. Cond.	Determination of phosphate at 1.314 g/l after *aqua-regia* digestion	64
Fertilizer	PO_4^{3-}, O_3^-, SO_4^{2-}	Metrosep Anion Dual one (70 mm)	2.4 mM Na₂CO₃/2.0 mM NaHCO₃ at 0.5 ml/min	20	Supp. Cond.	Determination of anions after digestion with HCl	65
Soil	PO_4^{3-}	Wescan 269-003 (4.6 × 40 mm²) guard and Vydac 302IC4.6 (4.6 × 250 mm²) analytical	1.5 mM phthalic acid at pH 2.7	100–2000	Nonsupp. Cond.	LOD for phosphate 0.3 μg/l	35
Soil	PO_4^{3-}, Cl^-, NO_3^-, SO_4^{2-}, oxalate	Dionex AG4A guard and AS4 analytical	1.5 mM Na₂CO₃/0.65 mM NaHCO₃ (no flow rate specified)	—	Supp. Cond.	Also determines anions in river water and spinach	25

Sample	Species	Column	Eluent	Conc.	Detection	Comments	Ref.
Soil	PO_4^{3-}, AsO_4^{3-}, Cl^-, NO_3^-, SO_4^{2-}	Wescan 269-003 guard and 29-029 (4.1 × 250 mm²) analytical	6 mM p-hydroxybenzoate, pH 6 at 2 ml/min	100–2000	Nonsupp. cond.	Comparison with ICP and AAS methods	12
Clay Extract	PO_4^{3-}, silicate, SO_4^{2-}	Phenomenex Starlon A300	Step gradient of 1.0 mM NaOH/0.1 mM Na_2CO_3 and 1.0 mM NaOH/1.0 Na_2CO_3 at 2 ml/min	100	Supp. Cond.	Involves valve switching	66
Incineration dusts and filter residues	PO_4^{3-}, F^-, Cl^-, NO_2^-, Br^-, NO_3^-, SO_4^{2-}, SO_3^{2-}, $S_2O_3^{2-}$	Dionex AS3	2.4 mM Na_2CO_3/3 mM $NaHCO_3$ at 253 ml/h	200	Supp. Cond.	No detected phosphate	67

TABLE 7.5
Phosphates in Biological Materials (Nonprocessed Foods, Fauna and Flora)

Sample	Ion	Column(s)	Eluent and Flow Rate	Inj. Vol. (μl)	Detection	Comments	References
Ham, fish paste and cheese	Phosphate and condensed phosphates	Dionex AS11 (2×250 mm^2)	KOH gradient from 30–200 mM over 25 min @ 0.5 ml/min	10	Supp. Cond.	Condensed phosphates extracted with trichloro-acetic acid from food samples	45
Coral skeletons	NO^{2-}, SO_4^{2-}, PO_4^{3-}	Dionex AG4A and AS4A	1.8 mM Na$_2$CO$_3$/1.7 mM NaHCO$_3$	250	Supp. Cond.	Samples dissolved in HCl and passed through chloride removal cartridges	40
Water and hair	PO_4^{3-} and AsO_4^{3-}	Dionex AG4A-SC and AS4A-SC	3.5 mM NaHCO$_3$ and 10 mM NaOH at 1.65 ml/min	200	UV/VIS @ 700 nm	IC separation followed postcolumn reaction	27
Plant sap	PO_4^{3-}, Cl^-, NO_3^-, SO_4^{2-}	(A) Dionex AG9-SC (50×4 mm^2) guard and AS9-SC (250×4 mm^2) analytical (B) Waters IC-PACK Anion HR and IC-PACK Anion Guard	(A) 2 mM Na$_2$CO$_3$/0.75 mM NaHCO$_3$ at 2 ml/min for isocratic runs. 17.75–31.5 mM NaOH for gradient elutions (B) Borate–gluconate, pH 8.5 and borate–tartrate, pH 4.0	—	(A) Supp. Cond. (B) Nonsupp. Cond.	Two methods compared. Isocratic suppressed conductivity method most sensitive to phosphate: LOD = 18 μg/l	42
Dried potatoes	PO_4^{3-}, Cl^-, NO_3^-, SO_4^{2-}, oxalate	Hamilton PRP X-100	3 mM p-hydroxybenzoic acid, 0.75 mM hydroxy-benzonitrile, 2.5% aceto-nitrile, pH 8.5 at 1.5 ml/min	100	Nonsupp. Cond.	Phosphate determined at 2.01 mg/g in sample	16
Seed exudate	PO_4^{3-}, F^-, Cl^-, NO_3^-, SO_4^{2-}, carboxylic acids, carbohydrates	(A) Aminex HPX-87H cation ex. (30×7.8 mm^2) (B) Dionex AG4A-SC (50×4 mm^2) and AS4A-SC (250×4 mm^2)	(A) 5 mM H$_2$SO$_4$ at 0.6 ml/min, temp 60°C (B) 1.7 mM Na$_2$CO$_3$/1.8 mM NaHCO$_3$ at 1 ml/min	(A) 20 (B) —	(A) Refractive index detection (B) Supp. Cond.	Neutral monosaccharides separated on Dionex CarboPac PA1 (250×4 mm^2) at 1 ml/min	43
Sugar cane	PO_4^{3-}, Cl^-, NO_3^-, SO_4^{2-}	Metrosep Anion Dual one (150×3 mm^2)	(A) 2.5 mM Na$_2$CO$_3$/2.4 mM NaHCO$_3$ at 0.5 ml/min (B) 8.0 mM phthalic acid — 2% acetonitrile, pH 4.25 with Tris at a flow of 0.5 ml/min	20 / 20	(A) Supp. Cond. (B) Nonsupp. Cond. UV @ 210 nm and RI detection	Comparison of suppressed IC method to ICP–AES method	68

Sample	Species	Column	Eluent		Detection	Comments	Ref.
Coffee and Tea	PO_4^{3-}, Cl^-, organic acids	Hamilton PRP-X110 (150×4.1 mm^2) anion ex. analytical and guard at 40°C	0.6 mM aqueous potassium hydrogenphthalate (pH 4.0) with 4% acetonitrile at a flow rate of 1 ml/min	20	Nonsupp. Cond.	Phosphate determined in 4 different coffee brands and 3 tea samples	17
Plants (*Arabidopsis thaliana*)	Organo-phosphates	Dionex CarboPac PA1 (250×4 mm^2) analytical and (50×4 mm^2) guard	75 mM NaOH to 75 mM NaOH and 500 mM sodium acetate gradient	25	Pulsed amperometric detection	Uses Titansphere TiO column to selectively retain phosphate compounds	69
Plant leaf extracts	PO_4^{3-}, Cl^-, NO_3^-, SO_3^{2-}	SUPER-SEP Anion Column	1.5 mM p-hydroxybenzoic acid, 5% acetonitrile pH 8.0 (Triethanolamine) at 1.5 ml/min	100	Nonsupp. Cond.	Phosphate determined at 17.1 μg/g in sample	70
Bovine liver	PO_4^{3-}, F^-, Cl^-, NO_2^-, Br^-, NO_3^-, SO_4^{2-}	Dionex HPIC AG4A (4 mm) guard and AS4A SC (4 mm) analytical	180 mM Na_2CO_3/1.7 mM $NaHCO_3$ at 2 ml/min	50	Supp. Cond.	Method uses electrodialysis sample pretreatment. Comparison with titrimetric method	30
Foods	Condensed phosphates	Dionex AG11 (2×50 mm^2) and AS11(2×250 mm^2)	NaOH gradient from 20–140 mM from 0.1–47 min at 0.3 ml/min	20	Supp. Cond.	Comparison with titration methods	44
Fruit juices	PO_4^{3-}, Cl^-, NO_3^-, SO_4^{2-}, organic acids	Dionex OmniPac Pax 500 (50×4 mm^2) guard and Pax 500 (250×4 mm^2) analytical	NaOH–ethanol–methanol gradient at 1 ml/min	20	Supp. Cond.	More than 500 fruit juice samples analyzed including grape, apple, cherry and blackcurrant juices of different origin	71
Tea	PO_4^{3-}, Cl^-, NO_2^-, Br^-, NO_3^-, SO_4^{2-} organic acids	Shimadzu Shim-Pack AC-A1 (100×4.6 mm^2) at 40°C	Potassium hydrogenphthalate/phthalic acid at 1 ml/min	20	Nonsupp. Cond.	LOD range for inorganic anions: 0.044–0.19 mg/l. LOD for organic acids: 0.48–1.34 mg/l	72
Rice flour and tea leaves	PO_4^{3-}, Cl^-, Br^-, SO_4^{2-}	Dionex AG12A and AS12A guard and analytical	2.7 mM Na_2CO_3/0.3 mM $NaHCO_3$ at 1.5 ml/min	50	Supp. Cond.	LOD for phosphate: 0.025 mg/l	24
Spinach	PO_4^{3-}, Cl^-, NO_3^-, SO_4^{2-}, oxalate	Dionex AG4A guard and AS4 analytical	1.5 mM Na_2CO_3/0.55 mM $NaHCO_3$	—	Supp. Cond.	Also determines anions in river water and soil	25
Tea	PO_4^{3-}, F^-, Cl^-, SO_4^{2-}, oxalate	Dionex IonPac AG4A guard and AS4 analytical	2.2 mM Na_2CO_3/0.75 mM $NaHCO_3$ at 2 ml/min	—	Supp. Cond.	Also determines cations in two tea samples	23

Continued

TABLE 7.5
Continued

Sample	Ion	Column(s)	Eluent and Flow Rate	Inj. Vol. (μl)	Detection	Comments	References
Spinach extract	PO_4^{3-}, Cl^-, NO_3^-, SO_4^{2-}, malate	Dionex HPIC AG4A (50 × 3 mm²) guard and AS4A SC (250 × 3 mm²) analytical	1.7 mM $NaHCO_3$/1.8 mM Na_2CO_3 at 2 ml/min	50	Supp. Cond.	Phosphate concentration not determined, but good resolution from other anions	19
Tea	PO_4^{3-}, Cl^-, SO_4^{2-} formate, acetate	Dionex IonPac AG4A guard and AS4 analytical	Not specified	—	Supp. Cond.	Also determines cations in two tea samples	22
Vegetable leaves	PO_4^{3-}, Cl^-, NO_3^-, SO_4^{2-}	Dionex HPIC AG3 and AS3	2.2 mM Na_2CO_3/2.8 mM $NaHCO_3$ at 2 ml/min	100	Supp. Cond.	Anion determinations in lettuce, turnip, tomato and pumpkin leaves	18
Carrot extract	PO_4^{3-}, Cl^-, Br^-, NO_3^-	Hamilton PRP-X (150 × 4.1 mm²)	0.5 mM pyromellitate buffer at pH 3 at 1 ml/min	20	Indirect UV @ 295 nm	Investigates pyromellitate as eluent with regard to molar absorptivity and convenient eluting power	73
Pea, tomato and cherry	Metabolic phosphates	Laboratory packed with BioRad AG MP-1 (3 × 150 mm²)	Borate–NH_4Cl pH 8.5–9.0 (no flow rate specified)	—	UV/VIS @ 820 nm	Uses molybdenum blue method for detection	74
Vegetables	PO_4^{3-}, Cl^-, NO_2^-, NO_3^-, SO_4^{2-}	Waters IC-Pak A HR analytical and Waters Guard Pak precolumn	Sodium borate/gluconate pH 8.5 at 0.9 ml/min	100	Nonsupp. Cond.	Phosphate determined for 9 commercial samples	21
Tobacco	PO_4^{3-}, Cl^-, NO_3^-, SO_4^{2-}, malate, oxalate	Dionex HPIC AG4 (50 × 4 mm²) guard and AS4 (250 × 4 mm²) analytical	65% 2.2 mM Na_2CO_3/2.8 mM $NaHCO_3$ at 2 ml/min and 35% D^2H_2O	100	Supp. Cond.	Phosphate determined in three types of commercial tobacco	20
Meat extract	PO_4^{3-}, Cl^-, NO_2^-, NO_3^-, SO_4^{2-}	Metrosep A SUPP 4-250	1.0 mM Na_2CO_3/4.0 mM $NaHCO_3$ at 1 ml/min	20	Supp. Cond.	Phosphate determined at 41 mg/l in sample	75
Grapefruit lemonade	PO_4^{3-}, Cl^-, NO_3^-, SO_4^{2-}, citrate	Metrosep A SUPP 5-100	Gradient of ultra pure water between 7–10 min to 17.5 mM Na_2CO_3/3.5 mM $NaHCO_3$ at 0.7 ml/min	20	Supp. Cond.	Phosphate determined at 42.4 mg/l in sample	76
Standard NIST oyster tissue and River sediment	Total N and P as nitrate and phosphate	Dionex AG4A and AS4A guard and analytical	1.7 mM $NaHCO_3$/1.8 mM Na_2CO_3 at 1 ml/min	100	Supp. Cond.	Oxidation of N and P using 22% (v/v) hydrogen per oxide and closed-vessel microwave assisted digestion in two stages % recoveries above 94.7% for phosphate	29

REFERENCES

1. EU Directive 2000/60/EC of the European Parliament and the Council of 23 October 2000 establishing a framework for community action in the field of water quality, *Official J. Eur. Communities*, 2000.
2. Frankenberger, W. T., Mehra, H. C., and Gjerde, D. T., Environmental applications of ion chromatography, *J. Chromatogr.*, 504, 211–245, 1990.
3. Woods, C. and Rowland, A. P., Applications of anion chromatography in terrestrial environmental research, *J Chromatogr. A*, 789, 287–299, 1997.
4. Jackson, P. E., Ion chromatography in environmental analysis, In *Encyclopedia of Analytical Chemistry*, Meyers, R. A., Ed., Wiley, Chichester, 2000.
5. Dahllof, I., Svensson, O., and Torstensson, C., Optimising the determination of nitrate and phosphate in seawater with ion chromatography using experimental design, *J. Chromatogr. A*, 771, 163–168, 1997.
6. Ledo de Medina, H., Gutierrez, E., Colina de Vargas, M., Gonzalez, G., Marin, J., and Andueza, E., Determination of phosphate and sulphite in natural waters in the presence of high sulphate concentrations by ion chromatography under isocratic conditions, *J. Chromatogr. A*, 739, 207–215, 1996.
7. Simon, N. S., The rapid determination of orthophosphate, sulfate, and chloride in natural water samples with high iron concentrations using ion chromatography, *Anal. Lett.*, 21, 319–330, 1988.
8. Yuchi, A., Ogiso, A., Muranaka, S., and Niwa, T., Preconcentration of phosphate and arsenate at sub-ng/ml level with a chelating polymer-gel loaded with zirconium(IV), *Anal. Chim. Acta*, 494, 81–86, 2003.
9. Determination of inorganic anions in drinking water by ion chromatography, *Dionex Application note 133*, Dionex Corporation, Sunnyvale, USA.
10. Determination of inorganic anions in wastewater by ion chromatography, *Dionex Application note 135*, Dionex Corporation, Sunnyvale, USA.
11. Buldini, P. L., Mevoli, A., and Quirini, A., On-line microdialysis-ion chromatographic determination of inorganic anions in olive oil mill wastewater, *J. Chromatogr. A*, 882, 321–328, 2000.
12. Mehra, H. C. and Frankenberger, W. T., Single-column ion chromatography: IV. Determination of arsenate in soils, *Am. J. Soil Sci. Soc.*, 52, 1603–1606, 1988.
13. Moibroek, K. and Shahwecker, P., *Moisture Measurements in Food and Related Products*, Mettler Instruments, Greifensee, Switzerland, 1984.
14. AOAC. *Official Methods of Analysis*, Association of Official Analytical Chemists, Washington, DC, 1980.
15. Bradfield, E. G. and Cooke, D. T., Determination of inorganic anions in water extracts of plants and soils by ion chromatography, *Analyst*, 110, 1409–1410, 1985.
16. Five anions in dried potatoes, *Metrohm IC Application Note No. 6*, Metrohm Ltd. CH-9101 Herisau, Switzerland.
17. Alcazar, A., Fernandez-Caceres, P. L., Martin, M. J., Pablos, F., and Gonzalez, A. G., Ion chromatographic determination of some organic acids, chloride and phosphate in coffee and tea, *Talanta*, 61, 95–101, 2003.
18. Grunau, J. A. and Swiader, J. M., Application of ion chromatography to anion analysis in vegetable leaf extracts, *Commun. Soil Sci. Plant Anal.*, 17, 321–335, 1986.
19. Smolders, E., Van Dael, M., and Merck, R., Simultaneous determination of extractable sulphate and malate in plant extracts using ion chromatography, *J. Chromatogr.*, 514, 371–376, 1990.
20. Risner, C. H., Quantitation of some tobacco anions by eluent suppressed anion exchange chromatography using conventional liquid chromatographic equipment, *Tobacco Sci.*, 1986, 85–90, 2004.
21. Ruiz, E., Santillana, M. I., Nieto, M. T., and Sastre, I., Determination of water soluble inorganic phosphates in fresh vegetables by ion chromatography, *J. Liq. Chromatogr.*, 18, 989–1000, 1995.
22. Chen, S. S. and Spiro, M., Rose-hip tea: equilibrium and kinetic study of mineral ion extraction, *Food Chem.*, 48, 47–50, 1993.
23. Spiro, M. and Lam, P. L. L., Kinetics and equilibria of tea infusion — Part 12. Equilibrium and kinetic study of mineral ion extraction from black Assam Bukial and green Chun Mee teas, *Food Chem.*, 54, 393–396, 1995.

24. Buldini, P. L., Cavalli, S., and Mevoli, A., Sample pretreatment by UV photolysis for the ion chromatographic analysis of plant material, *J. Chromatogr. A*, 739, 167–173, 1996.
25. Murcia, M. A., Vera, A., Ortiz, R., and Garcia-Garmona, F., Measurement of ion levels of spinach grown in different fertilizer regimes using ion chromatography, *Food Chem.*, 52, 161–166, 1995.
26. Dick, W. A. and Tabatabai, M. A., Ion chromatographic determination of sulfate and nitrate in soils, *Am. J. Soil Sci. Soc.*, 43, 899–904, 1979.
27. Antony, P. J., Karthikeyan, S., and Iyer, C. S. P., Ion chromatographic separation and determination of phosphate and arsenate in water and hair, *J. Chromatogr. B*, 767, 363–368, 2002.
28. Butler, F. E., Toth, F. J., Driscoll, D. J., Hein, J. N., and Jungers, R. H., *Analysis of fuels by ion chromatography: comparison with ASTM methods, Ion Chromatographic Analysis of Environmental Pollutants*, 1979.
29. Colina, M. and Gardiner, P. H. E., Simultaneous determination of total nitrogen, phosphorus and sulphur by means of microwave digestion and ion chromatography, *J. Chromatogr. A*, 847, 285–290, 1999.
30. Novic, M., Dovzan, A., Pihlar, B., and Hudnik, V., Determination of chlorine, sulphur and phosphorus in organic materials by ion chromatography using electrodialysis sample pretreatment, *J. Chromatogr. A*, 704, 530–534, 1995.
31. Nguyen, V. D. and Rossbach, M., Ion chromatographic investigation of brown algae (fucus-vesiculosus) of the German-environmental-specimen-bank, *J. Chromatogr.*, 643, 427–433, 1993.
32. McKelvie, I. D., Peat, D. M. W., and Worsfold, P. J., Techniques for the quantification and speciation of phosphorus in natural waters, *Anal. Proc.*, 32, 437–445, 1995.
33. Grasshoff, K., Kremling, K., and Ehrhardt, M., *Methods of Seawater Analysis*, Wiley-VCH, Weinheim, 1999.
34. Haberer, J. L. and Brandes, J. A., A high sensitivity, low volume HPLC method to determine soluble reactive phosphate in freshwater and saltwater, *Mar. Chem.*, 82, 185–196, 2003.
35. Karlson, U. and Frankenberger, W. T., Single column ion chromatography: III. Determination of orthophosphate in soils, *Am. J. Soil Sci. Soc.*, 51, 721987.
36. Greenberg, A. E., Clesceri, L. S., Eaton, A. D., Eds., *Standard Methods for the Examination of Water and Wastewater*, 18th ed., American Public Health Association, Washington, DC, 1992.
37. Pfaff, J. D., *Method 300.0, Revision 2.2: Determination of Inorganic Anions by Ion Chromatography*, United States EPA, Cincinnati, Ohio, USA, 1999.
38. Bruno, P., Caseli, M., de Grennaro, G., De Tommaso, B., Lastella, G., and Mastrolitti, S., Determination of nutrients in the presence of high chloride concentrations by column-switching ion chromatography, *J. Chromatogr. A*, 1003, 133–141, 2003.
39. Karmarkar, S. V., Analysis of wastewater for anionic and cationic nutrients by ion chromatography in a single run with sequential flow injection analysis, *J. Chromatogr. A*, 850, 303–309, 1999.
40. Shotyk, W., Immenhauser-Potthast, I., and Vogel, H. A., Determination of nitrate, phosphate and organically bound phosphorous in coral skeletons by ion chromatography, *J. Chromatogr. A*, 706, 209–213, 1995.
41. Colina, M., Ledo, H., Gutierrez, E., Vilalobos, E., and Marin, J., Determination of total phosphorous in sediments by means of high-pressure bombs and ion chromatography, *J. Chromatogr. A*, 739, 223–227, 1996.
42. Masson, P., Hilbert, G., and Plenet, D., Ion chromatography methods for the simultaneous determination of mineral anions in plant sap, *J. Chromatogr. A*, 752, 298–303, 1996.
43. Casey, C. E., O'Sullivan, O. B., O'Gara, F., and Glennon, J. D., Ion chromatographic analysis of nutrients in seed exudates for microbial colonisation, *J. Chromatogr. A*, 804, 311–318, 1998.
44. Baluyot, E. S. and Hartford, C. G., Comparison of polyphosphate analysis by ion chromatography and by modified end-group titration, *J. Chromatogr. A*, 739, 217–222, 1996.
45. Sekiguchi, Y., Matsunaga, A., Yamamoto, A., and Inoue, Y., Analysis of condensed phosphates in food products by ion chromatography with an on-line hydroxide eluent generator, *J. Chromatogr. A*, 881, 633–644, 2000.
46. Connolly, D. and Paull, B., Fast ion chromatography of common inorganic anions on a short ODS column coated with didodecyldimethylammonium bromide, *J. Chromatogr. A*, 953, 299–303, 2002.

47. Divjak, B., Novič, M., and Goessler, W., Determination of bromide, bromate and other anions with ion chromatography and an inductively coupled plasma mass spectrometer as element-specific detector, *J. Chromatogr. A*, 862, 39–47, 1999.

48. Kapinus, E. N., Revelsky, I. A., Ulogov, V. O., and Lyalikov, Yu. A., Simultaneous determination of fluoride, chloride, nitrite, bromide, nitrate, phosphate and sulphate in aqueous solutions at $10^{-9}\%$ to $10^{-8}\%$ level by ion chromatography, *J. Chromatogr. B*, 800, 321–323, 2004.

49. Rich, W. E. and Wetzel, R. A., La chromatographie ionique: une nouvelle technique pour l'analyse des ions, *L'actualité Chimique*, 6, 51–57, 1980.

50. Lindgren, M., Jonkromatografisk vattenanalys, *Vatten*, 36, 249–264, 1980.

51. Darimont, T., Schulze, G., and Sonneborn, M., Determination of nitrate in drinking water by means of ion chromatography, *Fres. Z Anal. Chem.*, 314, 383–385, 1983.

52. Kaiser, E., Rohrer, J. S., and Jensen, D., Determination of trace anions in high-nitrate matrices by ion chromatography, *J. Chromatogr. A*, 920, 127–133, 2001.

53. Mosko, J. A., Automated determination of inorganic anions in water by ion chromatography, *Anal. Chem.*, 56, 629–633, 1984.

54. Green, L. W. and Woods, J. R., Ion chromatographic determination of anions in wastewater precipitate, *Anal. Chem.*, 53, 2187–2189, 1981.

55. Crowther, J. and McBride, J., Determination of anions in atmospheric precipitation by ion chromatography, *Analyst*, 106, 702–709, 1981.

56. Cheam, V. and Chau, S. Y., Automated simultaneous analysis of anions and monovalent and divalent cations, *Analyst*, 112, 993–997, 1987.

57. Chloride, nitrate, phosphate and sulphate in wastewater, *IC Application Note No. S-7*, Metrohm Ltd. CH-9101 Herisau, Switzerland.

58. Lu, Z., Liu, Y., Barreto, V., Pohl, C., Avdalovic, N., Joyce, R., and Newton, B., Determination of anions at trace levels in power plant water samples by ion chromatography with electrolytic eluent generation and suppression, *J. Chromatogr. A*, 956, 129–138, 2002.

59. Improved determination of trace anions in high purity waters by high-volume direct injection with the EG40, *Application Update 142*, Dionex, Sunnyvale, CA, USA, 2001.

60. Fast analysis of anions in drinking water by ion chromatography, *Application Note 140*, Dionex, Sunnyvale, CA, USA, 2001.

61. Determination of polyphosphates using ion chromatography with suppressed conductivity detection, *Application Note 71*, Dionex, Sunnyvale, CA, USA, 2002.

62. Karlson, U. and Frankenberger, W. T. Jr., Single column ion chromatography of selenite in soil extracts, *Anal. Chem.*, 58, 2704–2708, 1986.

63. Mehra, H. and Frankenberger, W. T. Jr., Determination of trace amounts of molybdate in soil by ion chromatography, *Analyst*, 114, 707–710, 1989.

64. Phosphate in sewage sludge after aqua regia digestion, *IC Application Note No. S-97*, Metrohm Ltd. CH-9101 Herisau, Switzerland.

65. Nitrate, phosphate and sulphate in fertilizer after digestion with HCl, *IC Application Note No. S-49*, Metrohm Ltd. CH-9101 Herisau, Switzerland.

66. Silicate, sulphate and phosphate in a clay extract, *IC Application Note No. S-142*, Metrohm Ltd. CH-9101 Herisau, Switzerland.

67. Coerdt, W. and Mainka, E., Determination of inorganic anions in dusts and filter residues of waste incineration plants, *Frez. Z Anal. Chem.*, 320, 503–506, 1985.

68. Walford, S., Applications of ion chromatography in cane sugar research and process problems, *J. Chromatogr. A*, 956, 187–199, 2002.

69. Sekiguchi, Y., Mitsuhashi, N., Inoue, Y., Yagisawa, H., and Mimura, T., *J. Chromatogr. A*, 1039, 71–76, 2004.

70. Chloride, nitrate, phosphate and sulphate in plant leaf extracts, *IC Application Note No. N10*, Metrohm Ltd. CH-9101 Herisau, Switzerland.

71. Saccani, G., Gherardi, S., Trifiro, A., Soresi Bordini, C., Calza, M., and Freddi, C., Use of ion chromatography for the measurements of organic acids in fruit juices, *J. Chromatogr. A*, 706, 395–403, 1995.

72. Ding, M.-Y., Chen, P.-R., and Luo, G.-A., Simultaneous determination of organic acids and inorganic anions in tea by ion chromatography, *J. Chromatogr. A*, 764, 341–345, 1997.

73. Jardy, A., Caude, M., Diop, A., Curvale, C., and Rosset, R., Single-column anion chromatography with indirect UV detection using pyromellitate buffers as eluents, *J. Chromatogr.*, 439, 137–149, 1988.

74. Sayler, D. and Geiger, P., Extraction and analysis of metabolic phosphates in plants, *J. Food Sci.*, 58, 890–892, 1993.

75. Five anions in a meat extract, *IC Application Note No. S-123*, Metrohm Ltd. CH-9101 Herisau, Switzerland.

76. Chloride, nitrate, phosphate, sulphate and citrate in beverages, *IC Application Note No. S-141*, Metrohm Ltd. CH-9101 Herisau, Switzerland.

8 Characterization of Organic Matter from Air, Water, Soils, and Waste Material by Analytical Pyrolysis

Declan Page

CONTENTS

I. PHYSICAL AND CHEMICAL PROPERTIES OF ORGANIC MATTER IN WATER, SOILS, AIR, AND WASTE

The organic matter present in soils and sediments has been studied for the last few centuries; yet it is only relatively recently that similar studies have been performed on organic matter present in water, air, and waste materials.[1] In this chapter, the term organic matter is used to designate all organic matter found in soil, water, air or waste material other than living organisms and specific compounds of anthropogenic origin.

Organic matter is present in water, soil, air particulates, and waste materials as a conglomeration of carbonaceous chemical by-products of living and decaying plant and animal matter. Several factors control the range and types of chemical and structural characteristics of the

287

moieties within a sample of organic matter. These include the nature of origin of the material, the degradation processes to which it has been subjected and the physical and chemical environments with which it has come into contact.

For example, the organic matter present in water contains a mixture of components from various water sources which have been transported via aboveground or underground paths.[2,3] Hence, the character of organic matter can be considered unique for each sample and moreover is constantly changing. These changes can occur during transport and are associated with four main types of processes:

1. microbial
2. chemical
3. physical
4. photochemical[4-6]

Categorizing the many processes concerning organic matter in water, soils, air, and waste materials into general types greatly simplifies the overall view of the system. Hence, with choice of appropriate and sufficient parameters for each of these processes to account for the significant interactions within the environment, it may be possible to improve the understanding of the character of organic matter.

The organic matter found in soils, water, air, and waste materials is heterogeneous and difficult to characterize using traditional analytical methods. The organic matter is composed of an extremely complex mixture of compounds, most of which have not yet been structurally identified.[6,7] This complex mixture is formed by the natural and continuous processes of synthesis and degradation of organic matter and as such there is little hope of separating and characterizing all individual compounds. Consequently, studies of organic matter have nearly always been concerned not with pure compounds but rather with groups of compounds separated from the initial mixture by means of rather arbitrarily chosen techniques, for example, humic acids and related substances.[8] Since the properties of the compounds separated are largely unknown, it follows that this approach often yields results which depend on the origin of the samples or on the separation procedures, thereby rendering comparison of samples extremely difficult. Nevertheless, this approach has permitted gathering of substantial information on the nature and role of organic matter in soil, water, air, and waste materials.[9-12] Organic matter consists of a truly dynamic structure and hence organic matter belonging to different environments possesses its own particular properties, which in the last 20 years have become sufficiently well known to serve as a basis for theoretical classification and experimental fractionation.[13,14]

II. PRINCIPLES OF ANALYTICAL PYROLYSIS

The chemical properties of organic matter derived from soil, water, air, or waste materials has traditionally been analyzed by chemical degradation techniques such as cupric oxide oxidation.[15-19] These are often lengthy, tedious procedures, and involve derivatization of reaction mixtures before analysis. Furthermore the hydrolysis and degradation reactions applied are far from complete, resulting in a less reliable estimation of the character of organic matter.[16,17]

Analytical pyrolysis techniques, in which organic matter samples are thermally degraded in the absence of oxygen to form smaller more recognizable compounds, were developed during the 1960s and 1970s.[20] During pyrolysis, the organic matter sample absorbs thermal energy and this energy becomes distributed throughout the molecular structure. This gives rise to the excitation of chemical bond vibrational modes. Relaxation of these same bond vibrational modes causes cleavage, both heterolytic and homolytic, of weaker bonds. In this way, bond scission leads to high yields of pyrolysis products. In most pyrolysis reactions the bonds of strongly electron-withdrawing groups (for example oxygen) break preferentially giving rise to double bond structures (a conjugated chain of double bonds undergoes cyclic rearrangement to give benzene and other aromatic hydrocarbons, a process known as aromatization). By these pyrolysis mechanisms, heteroatom bridges break preferentially, and carboxylates decarboxylate readily.

A common error is to consider all compounds evolved from pyrolysis as pyrolysis products. There are three possible origins which can be traced for the majority of the compounds identified in pyrolysis:

1. *Pyrolysis*. Structural units of the organic matter split off under fast heating in an inert atmosphere, and lower molecular weight fragments evolve (termed the pyrolysis products).
2. *Evaporation*. Free compounds associated with the organic matter evaporate quickly under pyrolysis (e.g., fatty acids).
3. *Combustion*. Structural units of the organic matter split off through burning in the presence of oxygen.

The pyrolysis temperatures chosen are normally in the range of 500 to 800°C. This range provides sufficient energy input to the bond vibrational modes to cause scission and to give an acceptable yield.[20,21] Higher temperatures give fragments too small to be of much structural significance. Low temperature pyrolysis at 300 to 400°C can be used in special cases to activate low energy reactions, such as decarboxylations, dehydrations or the decomposition of quaternary amines.[22]

III. ANALYTICAL PYROLYSIS — INSTRUMENTATION

Analytical pyrolysis should aim to avoid combustion and secondary thermal reactions such that the larger molecular weight pyrolysis products of organic matter can be detected. Pyrolysis products therefore need to be removed rapidly from the reaction zone and sample size is typically in the microgram range. Attention should be paid to ensure a low dead volume and that pyrolysis products are readily removed by the carrier gas stream, with a linear gas flow of at least 150 mm/sec. The total heating time and the final temperature are decisive in determining the nature of pyrolysis products evolved.

To address this, analytical pyrolysis units usually incorporate design features such as:

1. pyrolysis in a vacuum, or in a rapid stream of inert gas such as helium
2. provision for operating the pyrolysis unit at 200°C and preheating of the carrier gas
3. the importance of adequate carrier gas velocity, plus small dead volume in order to avoid secondary reactions
4. the presence of a replaceable liner (usually quartz) tube surrounding the pyrolysis wire, to avoid cross contamination between samples
5. small sample sizes of less than 100 μg
6. temperature rise time of less than 200 msec in total

A diagram of a commonly employed pyrolysis unit incorporating these design features (Pyrojector, Scientific Glass Instruments, USA) is shown in Figure 8.1.

The true value of analytical pyrolysis for the characterization of organic matter is realized when combined with analytical methods such as mass spectrometry (Pyrolysis–Mass Spectrometry, Py–MS).[23] High sensitivity, specific and fast analysis are widely recognized characteristics of mass spectrometry (MS), which have earned this technique its reputation as one of the most powerful analytical tools for organic materials available today. With the total number of library mass spectra

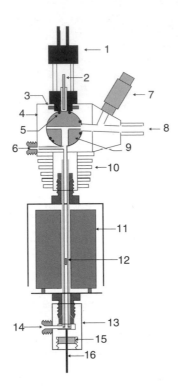

FIGURE 8.1 Pyrolysis unit, Scientific Glass Engineering (SGE) Pyrojector. 1. Sample loading area, 2. Quartz sample tube (reaction zone), 3. Sample loading cover interlock, 4. Valve body, 5. O-ring type seal, 6. Carrier gas inlet, 7. Check valve, 8. Carrier gas vent, 9. Valve spindle, 10. Heat deflectors, 11. Micro furnace, 12. Quartz replaceable liner, 13. GC interface nut, 14. Carrier gas inlet, 15. GC injection port septum, 16. Needle unit.

exceeding 100,000, it is tempting to credit MS with near universal applicability.[23] However, unfortunately most organic matter samples consist of molecular assemblies of a complexity and size far beyond the capabilities of even direct MS techniques. Hence, the use of MS does not provide positive identification of products. It also does not reliably detect products which have high ionization potentials distinguish between compounds of the same molecular weight. This can only be achieved when the products are first separated based on chemical or physical properties. Gas Chromatography/MS (GC/MS) provides a convenient separation and identification procedure and it can be readily interfaced with most pyrolysis systems. Indeed, commercial pyrolysis units are designed to be directly interfaced with the GC injection port. The GC column effluent can flow through the appropriate GC/MS interface, now a standard installation on all mass spectrometers, and positive identification of pyrolysis products may be made using the mass spectral library and associated software. Optional chemical ionization (CI) can aid in the identification by giving a molecular ion for compounds which do not yield a molecular ion in the electron ionization (EI) mode.[22]

To avoid dead space and secondary pyrolysis reactions due to low flow rate, a split stream should also be used. Most commercial gas chromatographs have gas splitters, to permit sufficient flow rate (20 to 30 cm^3/min) through the pyrolysis unit and the optimum flow rate through the column (1 to 5 cm^3/min). An alternative procedure is to cold trap (cyrofocus) the pyrolysis products on the head of the column, then optimize the column flow rate, and release the products by rapid warming to the starting temperature of the GC program (usually below 100°C).

Generally, one of three heating techniques is employed:

1. *Curie point pyrolysis* uses a ferromagnetic probe that is inductively heated. The sample is pyrolysed by a high-frequency field that causes inductive heating of the ferromagnetic wire. Depending on the strength of the field, the wire may heat up to the Curie point temperature of the ferromagnetic alloy in a time ranging 0.1 to 5 sec. Under appropriately selected frequency as well as the dimension and alloys of the wires, the temperature will automatically stabilize within a few degrees of the Curie-point, (358°C [Ni], 770°C [Fe], 1128°C [Co] and intermediate values for various ferromagnetic alloys).
2. *Microfurnaces* provide a constantly heated, isothermal pyrolysis zone into which solid samples are introduced.
3. *Flash pyrolysis* involves the exposure of the sample to high temperatures for very short periods of time, e.g., 0.001 to 0.1 sec.

With these points in consideration, a wide variety of commercial analytical pyrolysis units are now available. The great advantage of Py–GC/MS over chemical degradation methods is the small sample size required and that no sample pretreatment is needed after the initial isolation of organic matter. However, Py–GC/MS is limited by the lack of final identification of the pyrolysis products, since their mass determinations can only give the molecular formula. Therefore Py–GC/MS data should ideally be combined with other spectroscopic methods and chemical analyses.[18,19,24,25] Analytical pyrolysis is commonly used for the analysis of organic matter and the composition of the original sample is inferred from the pyrolysis products. Hence, Py–GC/MS can be used to develop an understanding of the types of molecular structural units present in the organic matter samples. The obtained pyro-chromatogram or "pyrogram" constitutes a fingerprint of the organic matter and gives information on the relative amounts of structural units present in organic matter. However, a drawback is that conventional pyrolysis has been demonstrated to be biased by the thermal degradation of carboxyl groups in building blocks and the adsorption of polar pyrolysis products.[26–32]

IV. DERIVATIZATION FOR ANALYTICAL PYROLYSIS

In the last decade, pyrolysis derivatization techniques have been increasingly used by many researchers in the characterization of organic matter.[14,32–40] Derivatization has the potential to provide additional information not readily obtainable by conventional pyrolysis techniques. Conventional pyrolysis of organic matter releases many compounds that are not volatile enough to pass through a gas chromatographic column.[27–29,33] Less polar pyrolysis products such as alkanes and alkylbenzenes are thus detected and analyzed, but the more polar pyrolysis products of similar mass such as alcohols and carboxylic acids are not detected.

Derivatization of the sample renders many of these polar pyrolysis products sufficiently volatile for gas chromatographic separation. Thus it is possible to separate and detect many more structurally significant products than observed by conventional pyrolysis techniques.[29,33] The most common of the derivatization processes is a methylation reaction where organic matter is mixed with tetramethylammonium hydroxide (TMAH) prior to pyrolysis. Throughout the literature, several different terms have been employed to describe derivatization reaction and in this chapter the term thermochemolysis will be used.

Thermochemolysis produces mostly long chain methyl esters and dimethyl esters via a mechanism which could first involve a pyrolysis to release an acid anion which is rapidly methylated in the gas phase or the TMAH may be saponifying/transesterifying long chain fatty acids esterified to a macromolecular network. This provides relatively good preservation of original carboxyl and hydroxyl structures in organic matter, due to formation of methyl-esters and methoxyls. Two processes, regarding functional groups, are clearly involved in the thermochemolysis of organic matter:

1. Partial decarboxylation of phenolic acids
2. Quantitative methylation of remaining carboxyls and partial methylation of hydroxyls

Presumably, as the pyrolysis cleaves polar fragments from organic matter, the TMAH methylates them in the chromatographic inlet, whereupon they undergo chromatographic separation. It is possible that the TMAH induces methylation of organic matter and this leads to higher product yields in the MS simply because charring or condensation reactions are minimized.

In practice, thermochemolysis of organic matter mainly produces esters of aliphatic and aromatic acids, methyl esters of aliphatic alcohols and phenols, and a variety of other methylated derivatization products.[14] Thermochemolysis is particularly suitable for the analysis of fatty acids associated with organic matter matrix which are evolved as methyl esters when TMAH is employed.

Thermochemolysis enhances yields of pyrolysis products which are probably more representative than conventional pyrolysis of the structures of organic matter.[27–29,33]

Off line thermochemolysis is also possible at much lower temperatures (250°C).[14] Thermochemolysis reactions with TMAH can be conducted in a sealed glass ampoule thus allowing for more controlled conditions, where internal standards can be added to provide quantitative measurement. Off line thermochemolysis gives similar results as the more time consuming chemical degradation methods used in the analysis of organic matter.[6,14,37,38]

V. ANALYTICAL PYROLYSIS OF ORGANIC MATTER AND RELATED BIOMATERIALS

Knowledge of pyrolysis mechanisms in organic matter and related biomacromolecules is most advanced for some classes of relatively simple compounds, such as amino acids. However, relatively little is known about these mechanisms for most biomacromolecules. Whereas some synthetic polymers such as polystyrene or polytetrafluoroethylene produce extremely simple pyrograms provided that soft ionization methods are used, the pyrolysis mass spectra of most biomacromolecules are usually more complex.[6,20,41–44] To some extent this is caused by the complexity of the structure or by general differences in pyrolysis mechanisms. For example, in polystyrene and polytetrafluoroethylene the major degradation reaction is a simple depolymerization by β bond scission (initiation followed by unzipping). This yields styrene and tetrafluoroethylene, respectively. Pyrolysis of biomacromolecules is governed by the same general principles as for these simpler synthetic polymers. However, none of the common biomacromolecules possesses a structure wherein each of the monomeric building blocks contributes a two-carbon segment to the polymer backbone which could lead to simple depolymerization by β scission. Natural rubber, a polymeric terpene, is perhaps the only biomacromolecule to yield marked amounts of monomeric units under analytical pyrolysis conditions.[20] Instead most biomacromolecules decompose by a variety of mechanisms often characterized by the elimination of an electron withdrawing group in the molecular structure, such as H_2O, HCN, CH_2O, CH_3OH, H_2S, CO, CO_2, C_2H_4, and H_2 accompanied by the break-up of the biomacromolecule into smaller fragments. Such reactions are most likely free radical in nature, but it is worth noting that the pyrolysis products evolved still bear the expected relationships to parent biomacromolecule structures.[6,22]

Analytical pyrolysis yields from such biomacromolecules are never 100%, and are frequently 50% or less, owing to competing char-forming reactions. It should be noted that since the initial pyrolytic reactions are heterolytic, the course of the thermal degradation and relative yields of

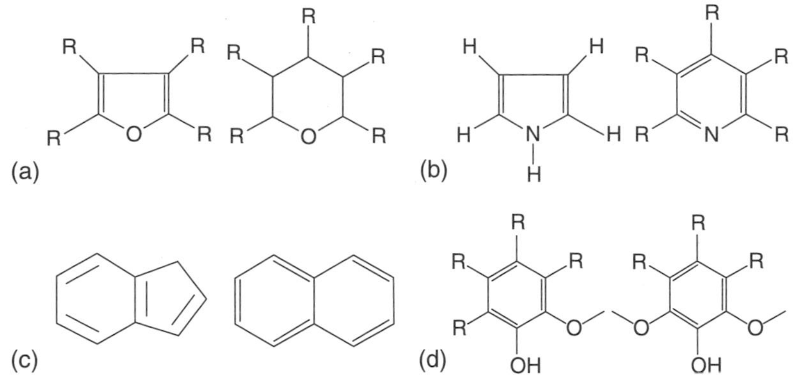

FIGURE 8.2 (a) *Polysaccharides*: Predominant pyrolysis fragments from polysaccharides include furan and pyran derived structures. Other common pyrolysis products also include pyranosides and cyclopentenes and alkyl derivatives. (b) *Proteins and nucleic acids*: Predominant pyrolysis fragments characteristic of proteins and nucleic acids include the nitrogen containing aromatic products such as pyrroles and pyridines and their alkyl derivatives. (c) *Lipids*: Lipids undergo thermally catalyzed cyclization reactions and decarboxylation reactions at subpyrolysis temperatures. Dominant products include benzene, indole, naphthalene and their alkyl derivatives. (d) *Lignins*: Lignins produce the most structurally similar monomers upon analytical pyrolysis. Common pyrolysis products include methoxy and dimethoxy phenols as well as their various substituted derivatives.

the various pyrolysis products can be significantly influenced by the presence of acidic or alkaline catalysts or salts.[44,45]

An improved understanding of the quantity and quality of more complex pyrolysis products of organic matter can be obtained by first examining the relatively simpler pyrolysis characteristics of the major types of biomacromolecules: polysaccharides, lignin, nucleic acids, proteins, and lipids (Figure 8.2).

A. ANALYTICAL PYROLYSIS OF POLYSACCHARIDES

The general pyrolysis mechanisms of polysaccharides have been determined from model studies on cellulose and involve the splitting of the polysaccharide structure by three basic chemical reaction mechanisms: dehydration, retroaldolization, and decarboxylation.[43] Using these basic pyrolysis mechanisms, it is possible to explain the pyrolysis of polysaccharides and evolved pyrolysis products. The hexose degradation pathway for cellulose results in formation of furan- and pyran-type fragments and smaller acyclic aldehyde and ketone fragments.[43,44,46]

However, relatively little work has been carried out to determine pyrolysis mechanisms in nonhexose polysaccharides, such as *N*-acetyl amino sugars. However, the similarity of pyrograms for cellulose and its *N*-acetylglucoamine analogue, chitin, indicates a marked degree of correspondence in basic pyrolysis mechanisms.[6,21] As this similarity also holds for other types of carbohydrates, the hexose model studies have considerably aided in the qualitative interpretation of nonhexose polysaccharides.

Even though there are apparent basic similarities in pyrolytic pathways, different types of sugar moieties, for example, pentoses, amino sugars, hexuronic acids, and deoxy- and anhydrosugars often contribute characteristic pyrolysis products by virtue of their different structures. Compounds detected from pyrolysis of cellulose at 650°C (Figure 8.3) and indicative of polysaccharides are chosen on the basis of being five- or six-member oxygen containing heterocyclic compounds fitting the basic pyrolysis mechanisms already known.[6] Levoglucosan is commonly detected as the major

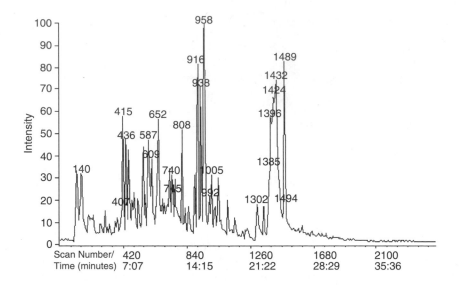

FIGURE 8.3 Pyrogram of cellulose.

pyrolysis product of cellulose (scan range 1350 to 1430 in Figure 8.3) as are anhydrohexoses in relatively small amounts (peaks in the scan range 400 to 1000 in Figure 8.3). Furan, pyran, cyclopentene and cyclopentenone derivatives are identified, together with less specific C_3–C_5 ketones, aldehydes, and alcohols especially at higher pyrolysis temperatures.

Similarly other Py–GC/MS analyses of polysaccharides[20] have shown that the simple units give rise to characteristic peak patterns, thus providing a basis for the distinguishing structural contributions of hexoses, pentoses, and hexuronic acids.

B. ANALYTICAL PYROLYSIS OF LIGNINS

Pyrolysis of lignin and wood at 650°C yields as pyrolysis products phenols, aldehydes, ketones, acids, and alcohols, generally with the retention of original substituents (OH, OCH_3) on the phenyl ring (Figure 8.4).[6]

Pyrolysis of lignin derived from softwoods yields mainly vanillin, vanillic acid and various amounts of other vanillyl-type pyrolysis products. However, minor amounts of syringaldehyde and syringic acid are also detectable. Pyrolysis of lignin derived from hard wood yields mainly syringylaldehyde, syringic acid, various amounts of sinapyl-type pyrolysis products, and lesser amounts of vanillyl-type products. Pyrolysis of lignin derived from grasses (macrophytes) contains primarily p-anisaldehyde, p-anisic acid and minor amounts of p-coumaryl-type pyrolysis products. While the presence of single vanillyl-, syringyl-, or coumaryl-type compounds in a pyrogram does not constitute a unique tracer for the original source of the lignin, the relative proportions of such compounds are often used to identify the nature of the source of lignin in organic matter, in soils, water, air, and waste materials.[6,47–49]

An additional group of compounds present in pyrograms from lignin, associated with organic matter, are the secondary dimer compounds derived by radical recombination of methoxyphenol pyrolysis products.[47] Lignin derived from softwoods contains divanillyl, divanillylmethane and divanillylethane, and minor amounts of diveratryl. In addition, lignin derived from hard woods yields the sinapyl-type compounds: bis-guaiacyl-syringyl and disyringyl. Pyrolysis of lignin derived from grasses has only dianisyl derived from a recombination of p-coumaryl-type pyrolysis products.[47–49] The lignins and secondary dimers have the same substitution pattern (OH, OCH_3) on the aromatic rings as the precursor aromatic alcohols from which they were derived. The relative

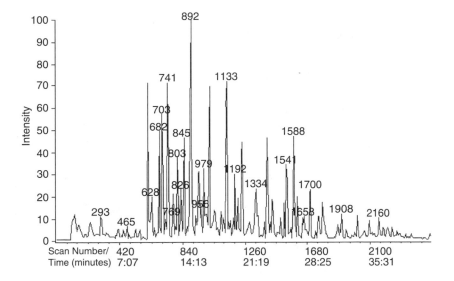

FIGURE 8.4 Pyrogram of lignin.

proportions of these precursors in the parent lignin materials are reflective of the proportions of the substituents found on the pyrolysis products. This in turn can be used to help identify the lignin class and hence, in organic matter characterization, the likely sources of lignin.

C. ANALYTICAL PYROLYSIS OF NUCLEIC ACIDS

The main pyrolysis mechanism of nucleic acids (occurring at temperatures as low as 180°C) is the expulsion of the polysaccharide moiety with the simultaneous formation of base-phosphate condensates.[20] The base phosphate complex is further pyrolysed, yielding the base fragments. Compared with polysaccharides and lignin, the application of Py–GC/MS techniques to the analysis of nucleic acids is still in its infancy.

D. ANALYTICAL PYROLYSIS OF PROTEINS

In the pyrolysis of proteins, the dominant mechanism is the splitting-off of the amino acids rather than break-up of the backbone into smaller fragments which are characteristic of the original amino acids.[20] Usually highly characteristic signals are found for the aromatic and sulphur containing amino acids, e.g., hydrogen sulphide from cystine and in combination with methionine; pyrrole; pyrrolidine and methylpyrrole for proline, phenyl, and cresol from tyrosine, toluene, styrene and phenylacetonitrile for phenylalanine; indole, and methylindol from tryptophan.[6,20,50]

E. ANALYTICAL PYROLYSIS OF LIPIDS

Lipids constitute another important group of biomacromolecules which are often insufficiently volatile to be analyzed by direct chromatographic techniques. With the exception of terpenes, which could be regarded as polymers of isoprene, lipids do not usually possess the repeating subunits which are hallmarks of the other biomacromolecules as above.

Pyrolysis has had only a limited success in the study of lipids. The rapid heating and short residence times of the products in the reaction zone (Figure 8.1) often result in escape of intact or at most minimally fragmented lipids, e.g., complete fatty acids which are subsequently lost by condensation on the walls of the reaction chamber. Thus, lipid moieties tend to be strongly under-represented in the pyrograms of lipid-containing samples.[27–29]

For many lipid marker compounds, the structural and stereochemical specificity enables them to serve as indicators for the sources of organic matter. These include sterols, which are ubiquitous in microorganisms and vascular plants but show structural variety allowing for different sources to be distinguished. Terpenoid-derived pyrolysis products can be detected in pyrograms of organic matter.[14] Terpenoids occur very frequently in plants and their distribution patterns can be of great diagnostic value in determining likely precursor materials. A series of sterols is commonly found, based on 24-ethylcholestane, which is most likely associated with vegetation-derived precursor materials. Resin acids, such as abietic or pimaric acids, are also present in the pyrograms of some organic matter, derivatives of which, at various stages of diagenic alteration, have been identified in pyrograms organic matter from soil and water.[6] Dehydroabietic and 7-oxo-dehydroabietic acids can be regarded as partially altered products from resin acids, and the other terpene-derived pyrolysis products are directly volatilized unaltered compounds. Retene is a higher temperature pyrolysis product of plant-derived lipids and has been identified in organic matter extracts at trace levels.[6,14]

F. ANALYTICAL PYROLYSIS OF ORGANIC MATTER

An understanding of the pyrolysis mechanisms of common biomacromolecules facilitates interpretation of analytical pyrolysis of organic matter present in soil, water, air, or waste materials. The structure of the organic matter often exhibits some resemblance to the structure of its original biomaterial precursor. For example, a pyrogram of sediment-derived organic matter is presented in Figure 8.5.

The analytical pyrolysis of the sedimentary organic matter presented in Figure 8.5 was performed using an (SGE) Pyrojector (Figure 8.1) interfaced to a Hewlett Packard 5890 gas chromatograph and a Hewlett Packard 5971 mass spectrometer. Approximately 5 mg of dried ground whole sediment was pyrolysed at 580°C. The MS was performed with electron impact ionization at 70 eV, over a mass range (m/z) 50 to 650. Pyrolysis products detected following application of Py–GC/MS on sedimentary organic matter were assigned to the same potential precursor biomacromolecules: polysaccharides, lignin, lipids, and proteins as described above (Table 8.1). Identification of

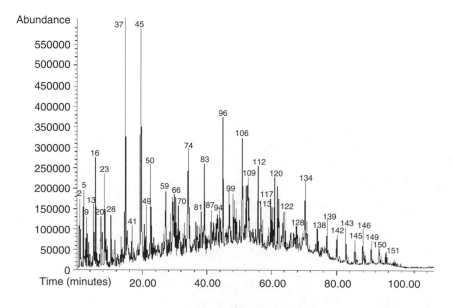

FIGURE 8.5 Representative pyrogram of whole sediment from Corin Reservoir. (Reprinted from Page, D. W., *J. Anal. Appl. Pyrolysis*, 70, 169–183, 2003. Copyright 2003, with permission from Elsevier Sciences.)

TABLE 8.1
Pyrolytic Products Obtained by Pyrolysis of Sedimentary Organic Matter at 500°C and Tentatively Identified by Library Matching

Retention Time	Peak Number	Compound	CAS Number	Possible Origin
0.76	1	1,3-Butadiyne	62283 000460-12-8	—
0.93	2	1,3-Butadiene	62293 000106-99-0	—
1.14	3	2-Butene	77 000590-18-1	—
1.93	4	Propenal	72 000107-02-8	—
2.12	5	Oropylene oxide	62320 000075-56-9	—
2.29	6	1,3-Pentadiene	178 000504-60-9	—
2.6	7	2-Methylbutene	225 000563-46-2	—
2.79	8	2-Methyl-1,3-butadiene	62407 000078-79-5	—
3.04	9	1,3-Cyclopentadiene	158 000542-92-7	Polysaccharides
3.14	10	2-Methyl-1,3-butadiene	184 000078-79-5	—
3.49	11	Cyclopentene	185 000142-29-0	Polysaccharides
3.93	12	2-Methylpropanal,	279 000078-84-2	—
5.04	13	1-Hexene	586 000592-41-6	Polysaccharides
5.18	14	2-Methyl-2-butene	62461 000513-35-9	—
5.41	15	Hexane	62873 000110-54-3	Polysaccharides
5.73	16	2-Methylfuran	463 000534-22-5	Polysaccharides
5.81	17	2-Hexyne	492 000764-35-2	Polysaccharides
6.16	18	Methylfuran	62666 000534-22-5	Polysaccharides
6.87	19	2,4-Hexadiene	474 000592-46-1	Polysaccharides
7.35	20	1,3-Cyclohexadiene	435 000592-57-4	Polysaccharides
7.64	21	5-Methyl-1,3-cyclopentadiene	429 000096-38-8	Polysaccharides
7.94	22	Methylcyclopentene	62682 000693-89-0	Polysaccharides
8.35	23	Benzene	401 000071-43-2	—
8.5	24	3-Methylbutanal	62812 000590-86-3	—
8.75	25	1,3-Cyclohexadiene	62650 000592-57-4	Polysaccharides
9.02	26	2-Methylbutanal	62844 000096-17-3	—
9.31	27	Cyclohexene	62694 000110-83-8	Polysaccharides
10.48	28	Heptene	63242 000592-76-7	Aliphatic
10.98	29	Heptane	1600 000142-82-5	Aliphatic
11.56	30	2,5-Dimethylfuran	63114 000625-86-5	Polysaccharides
13.38	31	1,3,5-Heptatriene	1018 017679-93-5	Aliphatic
13.59	32	Methyl-1H-pyrrole	62655 000096-54-8	Protein
13.94	33	3-Methyl-1,3,5-hexatriene	1012 024587-27-7	Polysaccharides
14.17	34	2-Ethylcyclohexanone	64898 004423-94-3	Polysaccharides
14.27	35	Methyl-1,4-cyclohexadiene	1030 004313-57-9	Polysaccharides
14.44	36	3-Hepten-1-yne	1026 000764-57-8	Aliphatic
14.77	37	Toluene	63028 000108-88-3	—
14.92	38	5-Hexenoic acid	2873 001577-22-6	Polysaccharides
15.25	39	Pyrrole	62389 000109-97-7	Protein
16.21	40	Cyclohexene	62694 000110-83-8	Polysaccharides
16.73	41	Octene	63996 000111-66-0	Aliphatic
17.26	42	Octane	64208 000111-65-9	Aliphatic
18.38	43	2-Furancarboxaldehyde	63104 000098-01-1	Polysaccharides
18.63	44	1,2-Nonadiene	4223 022433-33-6	—
19.78	45	2-Methylfuran	62666 000534-22-5	Polysaccharides
20.01	46	2-Cyclopenten-1-one	465 000930-30-3	Polysaccharides
20.65	47	1,3-Dimethylbenzene	63696 000108-38-3	—

Continued

TABLE 8.1
Continued

Retention Time	Peak Number	Compound	CAS Number	Possible Origin
21.13	48	1,2-Dimethylbenzene	63695 000108-38-3	—
21.42	49	5-Methyl-2-(5H)-furanone	63165 000591-11-7	—
22.43	50	Styrene	63646 000100-42-5	—
22.86	51	Nonene	64957 000124-11-8	Aliphatic
23.36	52	Nonane	65142 000111-84-2	Aliphatic
23.76	53	3-Methyl-2-cyclopenten-1-one	1084 002758-18-1	Polysaccharides
24.7	54	Hexene derivative	62750 013269-52-8	Polysaccharides
24.86	55	1,3-Nonadiene	4241 056700-77-7	Aliphatic
25.99	56	5-Methyl-2-(5H)-furanone	63165 000591-11-7	—
26.14	57	Propylbenzene	64584 000103-65-1	—
26.55	58	Benzaldehyde	63687 000100-52-7	—
27.1	59	5-Methyl-2-furancarboxaldehyde	63847 000620-02-0	Polysaccharides
27.6	60	1-Ethyl-2-methylbenzene	64560 000611-14-3	—
27.82	61	2,4-Dimethyl-1,3-pentadiene	1139 001000-86-8	Polysaccharide
27.99	62	Benzonitrile	63596 000100-47-0	Protein
28.37	63	1,2,3-Trimethylbenzene	64576 000526-73-8	—
28.47	64	Benzofuran	3591 000271-89-6	—
28.66	65	Decene	66063 000872-05-9	Aliphatic
29.37	66	Phenol	63072 000108-95-2	—
30.2	67	Methyl-2-(1-methylethyl)-benzene	65582 000527-84	—
30.37	68	Limonene	65775 000138-86-3	Terpene
30.51	69	Eucalyptol	67090 000470-82-6	Terpene
31.01	70	Indene	64386 000095-13-6	—
31.29	71	3-Methyl-1,2-cyclopentanedione	63937 000765-70-8	Polysaccharides
32.5	72	Acetophenone	64533 000098-86-2	—
32.87	73	2-Methylphenol	2122 000095-48-7	—
34.08	74	4-Methylphenol	63786 000106-44-5	—
34.54	75	2-Methylbenzofuran	65399 004265-25-2	—
35.81	76	10-Undecenol	68229 000112-43-6	Aliphatic
35.94	77	4-Cyclopentene-1,3-diol	1478 000694-47-3	Polysaccharides
36.48	78	2-Methylbenzonitrile	64399 000529-19-1	Protein
36.85	79	Benzeneacetonitrile	64396 000140-29-4	Protein
37.5	80	2,4-Dimethylphenol	3958 000105-67-9	—
38.17	81	Naphthalene	65150 000091-20-3	—
38.6	82	3-Ethylphenol	64691 000620-17-7	—
39.19	83	2-Methoxy-4-methylphenol	6909 000093-51-6	Lignin
39.56	84	Dodecane	68253 000112-40-3	Aliphatic
40.23	85	Octadecene	71502 000112-88-9	Aliphatic
40.56	86	Octanoic acid	66325 000124-07-2	Aliphatic
41.29	87	2,3-Dihydrobenzofuran	64532 000496-16-2	—
41.73	88	1,1-Dimethyl-1H-indene	8563 018636-55-0	—
41.86	89	1,2-Benzenediol	63845 000120-80-9	—
42.21	90	Benzylcyclopentane	12565 004410-78-0	—
43.13	91	2,3-Dihydro-1H-inden-1-one	65405 000083-33-0	—
43.27	92	4-Ethyl-2-methoxyphenol	10240 002785-89-9	Lignin
43.5	93	Methylnaphthalene	66231 000090-12-0	—
43.92	94	Octadecene derivative	34422 007206-25-9	Aliphatic
44.05	95	Indole	64403 000120-72-9	—

Continued

TABLE 8.1
Continued

Retention Time	Peak Number	Compound	CAS Number	Possible Origin
45	96	1-(3-Methoxyphenyl)-ethanone	66743 000586-37-8	Lignin
45.78	97	(3-Methyl-1-methylenebutyl)-benzene	12541 038212-1	—
46.55	98	Benzeneacetaldehyde	64543 000122-78-1	—
46.8	99	2,6-Dimethoxyphenol	67055 000091-10-1	Lignin
46.9	100	Eugenol	13470 000097-53-0	Lignin
47.26	101	2-Methylbenzofuran	5859 004265-25-2	—
47.46	102	Biphenyl	11094 000092-52-4	—
48.42	103	9-Octadecene	34422 007206-25-9	Aliphatic
49.24	104	2,6-Dimethylnaphthalene	11624 000581-42-0	—
49.82	105	1,5-Decadiyne	6199 053963-03-4	Aliphatic
50.93	106	2-Methoxy-4-(1-propenyl)-phenol	67804 000097-54-1	Lignin
51.15	107	3-Octadecene derivative	34414 007206-19-1	Aliphatic
51.91	108	Hexamethylbenzene	67705 000087-85-4	—
52.68	109	Octadecene	34422 007206-25-9	Aliphatic
52.99	110	Pentadecane	26001 000629-62-9	Aliphatic
55.03	111	1,6,7-Trimethylnaphthalene	68264 002245-38-7	—
55.78	112	3,5-Dimethoxyacetophenone	17729 039151-19-4	Lignin
56.7	113	Hexadecene	28779 000629-73-2	Aliphatic
56.99	114	Hexadecane	70785 000544-76-3	Aliphatic
57.22	115	2,6-Dimethoxy-4-(2-propenyl)-phenol	21233 006627-8	Lignin
59.1	116	2,5-Dimethoxy-4-ethylbenzaldehyde	21218 000000-00-0	Lignin
59.85	117	5,6-Dimethoxy-1-indanone	20685 002107-69-9	Lignin
60.52	118	Heptadecene	31657 006765-39-5	Aliphatic
60.79	119	Heptadecane	71193 000629-78-7	Aliphatic
60.95	120	2,6-Dimethoxy-4-(2-propenyl)-phenol	69395 006627-8	Lignin
63.04	121	Anthracene	68645 000120-12-7	—
63.64	122	Tetradecanoic acid	70842 000544-63-8	Aliphatic
64.14	123	Octadecene derivative	71502 000112-88-9	Aliphatic
64.39	124	Octadecane	71561 000593-45-3	Aliphatic
65.98	125	Cyclotetradecane	69525 000295-17-0	Aliphatic
66.96	126	Pentadecanoic acid	71238 001002-84-2	Aliphatic
67.6	127	Nonadecene	71890 018435-45-5	Aliphatic
67.83	128	Nonadecane	71949 000629-92-5	Aliphatic
68.48	129	Cyclododecene	14166 001486-75-5	Aliphatic
68.65	130	Heptadecyne	31178 026186-00-5	Aliphatic
68.73	131	Hexadecanoate	37803 000112-39-0	Aliphatic
69.69	132	Hexadecene	70724 000629-73-2	Aliphatic
69.98	133	1,13-Tetradecadiene	21425 021964-49-8	Aliphatic
70.46	134	Hexadecanoic acid	71606 000057-10-3	Aliphatic
70.92	135	Octadecene derivative	34426 000112-88-9	Aliphatic
71.13	136	Nonadecane	71950 000629-92-5	Aliphatic
74.07	137	10-Heneicosene	41868 000000-00-0	Aliphatic
74.28	138	Heptadecane	71193 000629-78-7	Aliphatic
77.11	139	Docosene	72943 001599-67-3	Aliphatic
77.28	140	Docosane	44318 000629-97-0	Aliphatic
80.01	141	Nonadecene	37066 018435-45-5	Aliphatic
80.18	142	Nonadecane	71950 000629-92-5	Aliphatic
82.8	143	Cyclotetracosane	47764 000297-03-0	Aliphatic

Continued

TABLE 8.1

Continued

Retention Time	Peak Number	Compound	CAS Number	Possible Origin
82.95	144	Tetracosane	73541 000646-31-1	Aliphatic
85.49	145	Eicosanol	72722 000629-96-9	Aliphatic
88.08	146	2-Nonadecanol	40239 026533-36-8	Aliphatic
88.18	147	Hexacosane	51010 000630-01-3	Aliphatic
88.31	148	Hexacosene	50820 018835-33-1	Aliphatic
90.68	149	2-Methyleicosane	72682 001560-84-5	Aliphatic
92.98	150	17-Pentatriacontene	58705 006971-40-0	Aliphatic
95.39	151	Pentatriacontane	58743 000630-07-9	Aliphatic

Reprinted from Page, D. W., *J. Anal. Appl. Pyrolysis*, 70, 169–183, 2003. Copyright 2003, With Permission from Elsevier Sciences.

compounds was based on comparison with those of a mass spectral library and by comparison with published literature.[51] The maintenance of functionality throughout the degradation procedure and the tracing of structures specific to particular biomacromolecules were features used for assigning the pyrolysis products as marker compounds.[52]

Most of the pyrolysis products detected during the pyrolysis of organic matter can usually be assigned to polysaccharides, lignin, lipids, or proteins and nucleic acids. This has revealed that earlier models of organic matter[53] had over emphasized the role of polyhydroxyaromatics and that Py–GC/MS studies determined the ubiquitous presence of lipids, polysaccharides, lignin and proteinaceous material. The term organic matrix was hence introduced to designate this mixture of biomacromolecules; such a term alleviates the need for ultimate knowledge of the structures present and their possible chemical bonds.[2,3]

VI. INTERPRETATION OF PYROGRAMS

The common applications of analytical pyrolysis techniques are sample-based classification and identification of pyrolysis products, using a library of reference compounds.[41,46] Computer evaluation of pyrograms has greatly increased the speed of analysis afforded by modern Py–GC/MS techniques and has opened up important new areas of application. Py–GC/MS techniques are increasingly employed to directly address problems concerning biochemical nature, composition, and structure of a sample.[54–56] The success of attempted biochemical interpretation of interesting features in pyrograms, whether done visually or with the aid of numerical computer techniques, depends critically on the following factors:

1. complexity of sample
2. availability of reference mass spectra from libraries or standard materials
3. knowledge of relevant pyrolysis mechanisms
4. availability of analytical methods

Any sample of organic matter generally provides a spectrum more difficult to interpret than that of a sample consisting of a single, pure component. However, often for such samples just one or two components suffice to adequately describe the analytical problem.[56] Also if suitable control samples are available, subtraction of patterns may yield a much simpler pattern, mainly representative of the components of interest.

An alternative to the sample-based methods of classification is provided by variable based techniques of multivariate analysis, such as, factor, principal component and various methods of discriminant analysis. The data used from each analysis can be used as a *fingerprint* of that organic matter sample for the particular pyrolysis conditions used. Because of the large amount of data collected using these analytical techniques, multivariate analysis with the aid of computers is used to collate the results. In each pyrogram the n peaks can be considered by representing a sample in n- dimensional space, of coordinates given by n intensities.[56] When samples are compared this way, similar spectra tend to cluster together in such space; this clustering may be examined mathematically by a number of statistical methods, based on the distance between the points. Alternatively the sample distribution may be examined visually by projection on a 2D plane, using methods of nonlinear mapping which preserves these distance relationships between the points.[22] Py–GC/MS techniques can be used to differentiate organic matter with respect to the nature and concentration of component moieties.[20] The relationships between the n peak intensities are examined and this usually leads to a marked reduction of dimensionality which facilitates comparison of sample.[14,20] For example in the plot in Figure 8.6, the 150 or similar number of peaks present for each pyogram may be reduced to two or three principal components which represent the original variance pattern of the pyrograms, and consist of linear functions of the original variables (peak intensities). A plot of these principal components reveals not only clustering or separation of samples, but also something of the underlying chemical tendencies responsible for the groupings. Ultimately, principal component analysis results in a distance matrix graphically represented in a nonlinear plot, in which each sample is given by a point. The relative distance between points reflects the dissimilarity of the pyrograms. Figure 8.6 shows a nonlinear map of the pyrograms of organic matter from a surface source water and alum (aluminum sulphate) treated drinking water.

In Figure 8.6, principal component analysis of the data extracted four principal components which represented 63.7% of the total variance between the pyrograms.[56] Principal components one and two are primarily clustering pyrograms on the basis of the miscellaneous pyrolysates, including alkylbenzene, alkylphenol, indole, and naphthalene derivatives. Polysaccharide-derived marker compounds also influence principal component two. Alum treated reservoir organic matter

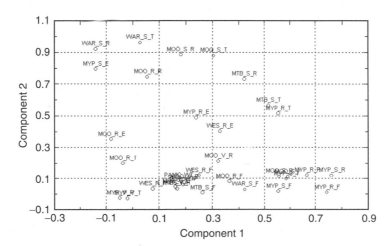

FIGURE 8.6 Plot of PC 1 and PC 2 from the principal component analysis of pyrograms. Samples are labeled as follows: Location, MTB, Mt. Bold; MYP, Myponga; MOO, Moorabool; WAR Warren; WES, West Gellibrand, PAN, Pankalak. DOM source: R, reservoir; S, soil; V, vegetation and sample type: R, raw, F, coagulated (flocculated); T, noncoagulated (treated); E, treated by enhanced coagulation. (Reprinted from Page, D. W., van Leeuwen, J. A., Spark, K. M., and Mulcahy, D. E., *J. Anal. Appl. Pyrolysis*, 67, 247–262, 2002. Copyright 2003, with permission from Elsevier Sciences.)

(denoted by an R_T), and raw and treated soil, and vegetation-derived organic matter (denoted by an S_R, S_T and V_R, V_T, respectively) tend to score higher than all the alum flocculated material and raw reservoir-derived organic matter. Differences in the relative ratios of nondiagnostic pyrolysis products (such as alkylbenzene and alkylphenol derivatives) account for the bulk of the variance described by this method. Most of the samples scored about 0.1 for principal component two. Soil-derived organic matter samples and most alum-flocculated materials score highly in components one and two because of the relatively high proportion of aromatic material, much of which is probably derived from secondary pyrolysis reactions.

The described Py–GC/MS analysis techniques have some limitations for the analysis of whole samples. The main limitation is the formation of isobaric and isomeric ions with a given integer mass to charge (m/z) ratio, which presents the assignment of an ion structure to a particular m/z ratio in a complex mass spectrum. The observation of a peak in a spectrum is thus not conclusive evidence for the presence of a particular pyrolysis product. For this reason it is important to utilize standard materials. Pattern recognition and multivariate analysis procedures have been also developed to specifically deal with this problem. These methods involve computer programs for "cluster analysis" which identifies groups of ions of common origin. Observation of these groups of ions then allows identification of a compound or type of compound with a greater degree of confidence.

VII. ANALYSIS OF ORGANIC MATTER ANALYSIS IN ENVIRONMENTAL EXAMPLES

The range of organic matter which can be analyzed by Py–GC/MS is extensive and due to the small samples required and minimal preparation, analytical pyrolysis is specially well suited to analyze environmental samples (Table 8.2).

Examples of analytical pyrolysis of organic matter from water, soils, air, and waste are further discussed below.

A. Water

The cost associated with the removal of organic matter is one of the major cost elements associated with treating water for drinking purposes. The flocculation and filtration processes easily remove

---

TABLE 8.2
Applications of Py–GC/MS Used to Study Organic Matter from Environmental Samples

	Organic Matter Source	Reference
Water	Dissolved organic matter in water	3–7,11,14,32,53,56,57,68,70
	Aquatic humic substances	1,2,24,25,61,71,74
Air	Air borne particles	9,47–49,69
Soil	Agricultural soils	29
	Soil particles	12
	Sub alpine soils	78
	Terrestrial humic substances	21–23,27,28,33,40,46,72,75,77
	Soil humic acid	1, 8
	Soil organic matter	13,18,19,26,54,62–64
	Humified sphagnum	76
	Sediments	2,52,55,73
Waste	Waste deposit leachates	10
	Chlorinated lignins	59,60,66
	Paper plant waste water	67

the particulate fraction of organic matter.[6] However, the dissolved fraction (<0.45 μm) of the organic matter is often more recalcitrant to conventional water treatment. To minimize water treatment costs, it is important to improve the understanding of the significant factors controlling the character of organic matter which impact directly on the treatability of water.

Many authors have previously studied dissolved organic matter in water (Table 8.2). In most of these studies, absolute quantification in analytical pyrolysis was not used. Hence a practical, relatively rapid method for quantification of pyrolysates of organic matter is needed for comparing the samples. Lack of even semiquantification of Py–GC/MS data prevents them from being readily compared to data obtained by other techniques such as spectroscopy or general chemical parameters of a sample. Previous studies[57] have used 1,3,5-tri-*tert*-butylbenzene and 1,3,5-trimethoxybenzene as internal standards in the pyrolysis of lignocellulosic materials. A similar approach using internal standards to semiquantify pyrolysis products from dissolved organic matter during the conventional water treatment process has been developed.[6] However, due to the possibility of decomposition of the 1,3,5-tri-*tert*-butylbenzene and production of 1,3,5-trimethoxybenzene by pyrolysis of organic matter, an anthropogenic chlorine-substituted compound with a highly characteristic mass spectrum, tetrachloroveratrole was used. A pyrogram of tetrachloroveratrole alone, with pyrolysis at 650°C, is shown in Figure 8.7.

The lack of any decomposition products in the above pyrogram suggests that tetrachloroveratrole is a suitable surrogate internal standard for use with organic matter. A pyrogram of organic matter from a freshwater reservoir, mixed with the tetrachloroveratrole internal standard, obtained using Pyroprobe 2000 (CDS Analytical Instruments, U.S.A.) at 650°C interfaced to a HP 5890 gas chromatograph and a VG-Tritech TS-250 mass spectrometer is shown in Figure 8.8.

Pyrolysis products similar to those listed in Table 8.1 were obtained. However, the important feature of the pyrogram is the clear internal standard peak at 1514 scans (the shift in the retention time of the internal standard [from 1020 in Figure 8.7 to 1540 in Figure 8.8] is due to changes made to the chromatographic procedure and a decrease in the flow rate). This internal standard peak can be used to semiquantify the pyrolysis products and facilitates comparison between samples. This approach has been used[2,3,58] to compare samples with respect to the different biomacromolecule classes in organic matter and relate this to its origin and treatability, using alum coagulation. Other studies have also used Py–GC–MS to study the character of organic

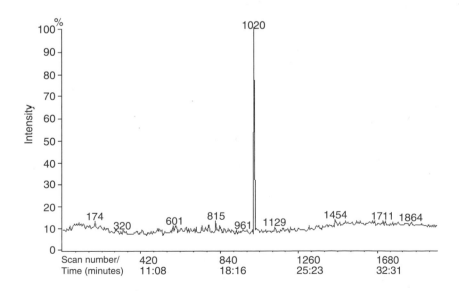

FIGURE 8.7 Pyrogram of internal standard tetrachloroveratrole.

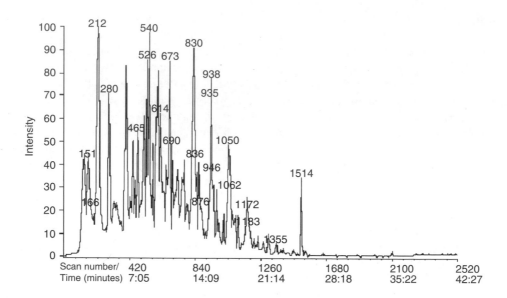

FIGURE 8.8 Pyrogram of organic matter derived from Moorabool Reservoir.

matter involved during the disinfection with chlorine and subsequent formation of chlorinated disinfection by-products.[59,60]

B. Soil

Many studies have been performed using Py–GC/MS to study the structure of organic matter from soils,[61–64] humic and fulvic acids, dissolved organic matter and whole soils and sediments (Table 8.2). Considerable work is also being done on soil organic matter, including humic acids, by both Py–GC and Py–MS. In many of these studies the residues of plants and biomacromolecules were important sources of organic matter in soils and sediments. The studies evaluated these residues, such as lignin and polysaccharides, as they were transformed by a combination of biological, physical, chemical, and photochemical processes. The information obtained from the analysis of organic matter can support interpretations on the character and origin of the soils or sediments.

From an agricultural research point of view, work has been done in two general areas; agricultural forage materials[51,65] and relationships to agriculture[30] and others in Table 8.2. Similarly to Figure 8.8, soil-derived organic matter has also been studied semiquantitatively by Py–GC–MS (Figure 8.9).[6]

Here the internal standard tetrachloroveratrole can be seen at 1542 scans (Figure 8.9). This organic matter sample yielded a relatively higher intensity of pyrolysis products, due to the higher carbon content per gram of sample. The pyrograms of Figure 8.8, (Moo_R_R) and Figure 8.9, (Moo_S_R) were compared using principal component analysis and the extracted components were plotted in Figure 8.7 and used to characterize organic matter before and after water treatment.

The characterization of organic matter in whole soils and sediments has also been studied using thermochemolysis techniques. The origin of sediment-derived organic matter was studied[55] using TMAH derivatization which enabled detection of C_2–C_{20} dicarboxylic acid methyl esters. Methoxybenzenes from phenols, benzenediols, benzenetriols, furancarboxylic acid methyl esters from carbohydrates were also identified. All of these pyrolysis products are not usually detected in pyrolysis of organic matter due to their polar nature and poor amenability to chromatographic analysis (e.g., Figure 8.6 and Table 8.1) and hence the results of the thermochemolysis have been complementary to the information obtained by traditional pyrolysis.

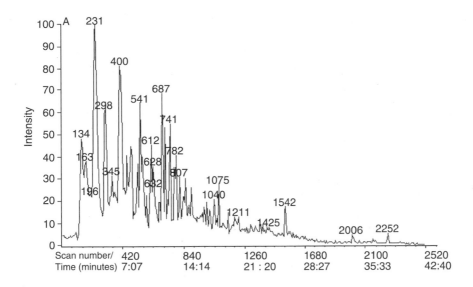

FIGURE 8.9 Pyrogram of organic matter derived from soil (Moorabool).

C. Air

Pyrolysis of organic matter derived from aerosols and emissions from the burning of woods and biomaterials has been studied for over 10 years.[47–49] Biomass combustion is an important source of particulate organic matter in the atmosphere. Although the molecular compositions of organic matter in smoke particles are highly varied the molecular tracer compounds are generally still source-specific. Degradation products from biomacromolecules (e.g., levoglucosan from cellulose, methoxyphenols from lignin) are also excellent tracers. The major marker compounds include dehydroabietic acid, the resin acids (pimaric, iso-pimaric, and abietic acids), retene, and polysaccharide-derived levoglucosan and lignin-derived methoxyphenols. The major sources of organic matter in the atmosphere include the products from burning of biomaterials (e.g., wood, vegetation, and similar materials) and particle emissions from anthropogenic materials (e.g., oils, coal, synthetic plastic compounds). Factors other than the nature of the starting materials also influence the character of particulate organic matter. For example, emissions of organic material in coal smoke particulate matter depend on pyrolysis temperature, ventilation, burn time, and coal geologic maturity. The pyrolysis of organic matter from peat and brown coal and to a lesser degree semibituminous coal, produces organic particulates composed mainly of hydrocarbons, lignin-derived methoxyphenols, and aromatic components, quite similar to burning of wood.

D. Waste Materials

Analytical pyrolysis of organic matter waste materials has seen similar applications to of organic matter present analyses in water, soil, and air. Similar pyrolysis products evolve from the pyrolysis of waste materials. Pyrolysis studies have also been used to investigate the organic matter present in wastewater from sewerage treatment plants (Page, unpublished results) and wastewater derived from landfill seepage[10] and chlorinated sediments.[66]

Py–GC/MS analyses have been used to study waste materials from pulp and paper mills and found some fractions to be constituted mainly of lignin-derived compounds. Py–GC/MS also showed alterations in the lignin units.[67]

VIII. CONCLUSIONS

Instrumental techniques of analytical pyrolysis have been used as important tools in the characterization of organic matter from water, soils, air, and waste materials. The advances in these techniques were based on improved instrumentation and increased understanding of the classes of compounds likely to evolve in analytical pyrolysis. Unraveling the origins of organic matter can be a formidable challenge due to its complex and somewhat recalcitrant nature. The use of pyrolysis or thermochemolysis in combination with GC/MS is one way of providing useful structural information on marker compounds of organic matter. However the potential may not be fully realized because the carbon skeleton structures are not always sufficiently diagnostic to resolve individual sources. The culmination of published data in the last 20 years gives numerous structural pieces which together help to understand of the complex character of organic matter in the environment.

REFERENCES

1. Wilson, M. A., Philip, R. P., Gillam, A. H., Gilbert, T. D., and Tate, K. R., Comparison of structures of humic substances from aquatic and terrestrial sources by pyrolysis–gas chromatography–mass spectroscopy, *Geochim. Cosmochim. Acta*, 47, 497–502, 1983.
2. Gadel, F. and Bruchet, A., Application of pyrolysis–gas chromatography–mass spectroscopy to the characterization of humic substances resulting from decay of aquatic plants in sediments and waters, *Water Res.*, 21, 1195–1206, 1987.
3. Bruchet, A., Rousseau, C., and Mallevialle, J., Pyrolysis–GC/MS for investigating high-molecular weight THM precursors and other refractory organics, *J. Am. Water Works Assoc.*, 82, 66–74, 1990.
4. Biber, M. V., Gulacar, F. O., and Buffle, J., Seasonal variations in principal groups of organic matter in a eutrophic lake using pyrolysis–GC/MS, *Environ. Sci. Technol.*, 30, 3501–3507, 1996.
5. Opsahl, S. and Benner, R., Photochemical reactivity of dissolved lignin in river and ocean waters, *Limnol. Oceanogr.*, 43, 1297–1304, 1998.
6. Page, D. W., *Characterization of terrestrially derived dissolved organic matter before and after water treatment*, Ph.D. Thesis, School of Chemical Technology, University of South Australia, 2000.
7. Heitz, A., Joll, C., Alexander, R., and Kagi, R. I., Characterization of aquatic natural organic matter in some Western Australian drinking water sources, In *Understanding and Managing Organic Matter in Soils, Sediments and Waters*, Swift, R. S. and Spark, K. M., Eds., International Humic Substances Society, Adelaide, Australia, 2001.
8. Schulten, H.-R. and Schnitzer, M., Structural studies on soil humic acids by Curie-point pyrolysis–gas chromatography–MS, *Soil Sci.*, 153, 205–224, 1992.
9. Simoneit, B. R. T., Rogge, W. F., Mazurek, M. A., Standley, l. J., Hildemann, L. M., and Cass, G. R., Lignin pyrolysis products, lignans and resin acids as specific tracers of plant classes in emissions from biomass combustion, *Environ. Sci. Technol.*, 27, 2533–2541, 1993.
10. Göbbels, F. J. and Püttmann, W., Structural investigations of isolated aquatic fulvic and humic acids in seepage water of waste deposits by pyrolysis–gas chromatography–MS, *Water Res.*, 31, 1609–1618, 1997.
11. Chow, C. W. K., van Leeuwen, J. A., Drikas, M., Fabris, R., Spark, K. M., and Page, D. W., The impact of natural organic matter in conventional treatment with alum, *Water Sci. Technol.*, 40, 97–104, 1999.
12. Schulten, H.-R., Leinweber, P., and Schnitzer, M., Analytical pyrolysis and computer modeling of humic soil particles, In *Structure and Surface Reactions of Soil Particles*, Huang, P. M., Senesi, N. and Buffle, J., Eds., Wiley, New York, 1998.
13. Haider, K. and Schulten, H.-R., Pyrolysis field ionization MS of lignins, soil humic compounds and whole soil, *J. Anal. Appl. Pyrolysis*, 8, 317–331, 1985.
14. Page, D. W., van Leeuwen, J. A., Spark, K. M., and Mulcahy, D. E., Tracing terrestrial compounds leaching from two reservoir catchments as input to dissolved organic matter (DOM), *J. Marine Freshwater Res.*, 52(2), 223–233, 2000.

15. Hedges, J. I. and Ertel, J. R., Characterization of lignin by gas capillary chromatography of cupric oxide oxidation products, *Anal. Chem.*, 54, 174–178, 1982.

16. Ertel, J. R. and Hedges, J. I., The lignin component of humic substances: distribution among soil and sedimentary humic, fulvic, and base-insoluble fractions, *Geochim. Cosmochim. Acta*, 48, 2065–2074, 1984.

17. Ertel, J. R. and Hedges, J. I., Sources of sedimentary humic substances: vascular plant debris, *Geochim. Cosmochim. Acta*, 49, 2097–2107, 1985.

18. Beyer, L., Sorge, C., and Schulten, H.-R., Studies on the composition and formation of soil organic matter in terrestrial soils by wet-chemical methods and pyrolysis–field ionization MS, *J. Anal. Appl. Pyrolysis*, 27, 169–185, 1993.

19. Beyer, L., Schulten, H.-R., Freund, R., and Irmler, U., Formation and properties of organic matter in a forest soil, as revealed by its biological activity, wet chemical analysis, CPMAS 13C-NMR spectroscopy and pyrolysis field ionization MS, *Soil Biol. Biochem.*, 25, 587–596, 1993.

20. Meuzelaar, H. L. C., Haverkamp, J., and Hileman, F. D., *Pyrolysis Mass Spectroscopy of Recent and Fossil Biomaterials, Compendium and Atlas*, 3rd ed., Elsevier Scientific Publishing Company, Amsterdam, 1991.

21. Marbot, R., The selection of pyrolysis temperatures for the analysis of humic substances and related materials 1. Cellulose and chitin, *J. Anal. Appl. Pyrolysis*, 39, 97–104, 1997.

22. Bracewell, J. M., Haider, K., Larter, S. R., and Schulten, H.-R., Thermal degradation relevant to structural studies of humic substances, In *Humic Substances 2*, Hayes, M. H. B., MacCarthy, P., Malcolm, R. L. and Swift, R. S., Eds., Wiley, New York, pp. 181–222, 1989.

23. Schulten, H.-R. and Gleixner, G., Analytical pyrolysis of humic substances and dissolved organic matter in aquatic systems: structure and origin, *Water Res.*, 33, 2489–2498, 1999.

24. Gonzáles-Vila, F. J., Lankes, U., and Ludemann, H.-D., Comparison of the information gained by pyrolytic techniques and NMR spectroscopy of the structural features of aquatic humic substances, *J. Anal. Appl. Pyrolysis*, 58–59, 349–359, 2001.

25. Sihombing, R., Greenwood, P. F., Wilson, M. A., and Hanna, J. V., Composition of size exclusion fractions of swamp water humic and fulvic acids as measured by solid state NMR and pyrolysis–gas chromatography–MS, *Org. Geochem.*, 24, 859–873, 1996.

26. Saiz-Jimenez, C. and de Leeuw, J. W., Chemical character of soil organic matter fractions by analytical pyrolysis–gas chromatography–MS, *J. Anal. Appl. Pyrolysis*, 9, 99–119, 1986.

27. Saiz-Jimenez, C., Production of alkylbenzenes and alkylnaphthalenes upon pyrolysis of unsaturated fatty acids: a model reaction to understand the origin of some pyrolysis products from humic substances, *Naturwissenschaften*, 81, 451–453, 1994.

28. Saiz-Jimenez, C., Conventional pyrolysis: a biased technique for providing structural information on humic substance?, *Naturwissenschaften*, 81, 28–29, 1994.

29. Saiz-Jimenez, C., Analytical pyrolysis of humic substances: pitfalls limitations and possible solutions, *Environ. Sci. Technol.*, 28, 1773–1779, 1994.

30. Saiz-Jimenez, C., Hermosin, B., Guggenberger, G. and Zech, W., Land use effects on the composition of organic matter in soil particle size separates. 3. Analytical pyrolysis, *Eur. J. Soil Sci.*, 47, 61–69, 1996.

31. Saiz-Jimenez, C., Reactivity of the aliphatic humic moiety in analytical pyrolysis, *Org. Geochem.*, 23(10), 955–961, 1996.

32. Saiz-Jimenez, C. and Hermosin, B., Thermally assisted hydrolysis and methylation of dissolved organic matter in dripping water from Altamira cave, *J. Anal. Appl. Pyrolysis*, 49, 337–347, 1999.

33. Hatcher, P. G. and Clifford, D. J., Flash pyrolysis and *in situ* methylation of humic acids from soil, *Org. Geochem.*, 21, 1081–1092, 1994.

34. Hatcher, P. G. and Minnard, R. D., Comparison of dehydrogenase polymer (DHP) lignin with native lignin from gymnosperm wood by thermochemolysis using TMAH, *Org. Geochem.*, 24, 593–600, 1996.

35. Hatcher, P. G. and Minnard, R. D., Comment on the origin of benzenecarboxylic acids in pyrolysis methylation studies, *Org. Geochem.*, 23, 991–994, 1996.

36. McKinney, D. E. and Hatcher, P. G., Characterization of peatified and coalified wood by TMAH thermochemolysis, *Int. J. Coal Geol.*, 32, 217–228, 1996.

37. Del Rio, J. C., McKinney, D. E., Knicker, H., Nanny, M. A., Minard, R. D., and Hatcher, P. G., Structural characterization of bio- and geo-macromolecules by off line thermochemolysis with tetramethylammonium hydroxide, *J. Chromatogr. A*, 823, 433–448, 1998.
38. Del Rio, J. C. and Hatcher, P. G., Analysis of aliphatic biopolymers using thermochemolysis with TMAH and gas chromatography–MS, *Org. Geochem.*, 29, 1441–1451, 1998.
39. Challinor, J. M., The development and application of thermally assisted hydrolysis and methylation reactions. Review, *J. Anal. Appl. Pyrolysis*, 61, 3–34, 2001.
40. Fabbri, D., Chiavari, G., and Galletti, G. C., Characterization of soil humin by pyrolysis(/methylation)–gas chromatography–mass spectroscopy: structural relationships with humic acid, *J. Anal. Appl. Pyrolysis*, 37, 161–172, 1996.
41. Almendros, G., Dorado, J., Gonzáles-Vila, F. J., and Martin, F., Pyrolysis of carbohydrate-derived macromolecules: It's potential in monitoring the carbohydrate structure of geopolymers, *J. Anal. Appl. Pyrolysis*, 40–41, 599–610, 1997.
42. van der Kaaden, A., Haverkamp, J., Boon, J. J., and de Leeuw, J. W., Analytical pyrolysis of carbohydrates. 1. Interpretation of matrix influences on pyrolysis-mass spectra of amylose using pyrolysis–gas chromatography–MS, *J. Anal. Appl. Pyrolysis*, 5, 199–220, 1983.
43. van der Kaaden, A., Boon, J. J., de Leeuw, J. W., de Lange, F., Schuyl, P. J. W., Schulten, H.-R., and Bahr, U., Comparison of analytical pyrolysis techniques in the characterization of chitin, *Anal. Chem.*, 56, 2160–2165, 1984.
44. van der Kaaden, A., Boon, J. J., and Haverkamp, J., The analytical pyrolysis of carbohydrates, *Biomed. Mass Spectrom.*, 11, 486–492, 1984.
45. Tromp, P. J. J., Moulijn, J. A., and Boon, J. J., Probing the influence of $K_2CO_3^-$ and $Na_2CO_3^-$ addition on the flash pyrolysis of a lignite and bituminous coal with Curie-point pyrolysis techniques, *Fuel*, 65, 960–967, 1986.
46. Almendros, G., Martin, F., Gonzáles-Vila, F. J., and del Rio, J. C., The effect of various chemical treatments on the pyrolysis of peat humic acid, *J. Anal. Appl. Pyrolysis*, 25, 137–147, 1993.
47. Simoneit, B. R. T., Biomass burning — a review of organic tracers for smoke from incomplete combustion, *Appl. Geochem.*, 17, 129–162, 2002.
48. Simoneit, B. R. T. and Elias, V. O., Detecting organic tracers from biomass burning in the atmosphere, *Marine Pollut. Bull.*, 42, 805–810, 2001.
49. Simoneit, B. R. T., Rogge, W. F., Lang, Q., and Jaffé, R., Molecular characterization of smoke from campfire burning of pine wood (*Pinus elliottii*), *Chemosphere: Global Change Sci.*, 2, 107–122, 2000.
50. Boon, J. J. and de Leeuw, J. W., Amino acid sequence information in proteins and complex proteinaceous material revealed by pyrolysis–capillary gas chromatography–low and high resolution MS, *J. Anal. Appl. Pyrolysis*, 11, 313–327, 1987.
51. Ralph, J. and Hatfield, R. D., Pyrolysis–GC/MS characterization of forage materials, *J. Agric. Food Chem.*, 39, 1426–1437, 1991.
52. van Heemst, J. D. H., van Bergen, P. F., Stankrewicz, B. A., and de Leeuw, J. W., Multiple sources of alkyl phenols produced upon pyrolysis of DOM, POM and recent sediments, *J. Anal. Appl. Pyrolysis*, 52, 239–256, 1999.
53. Stieglitz, L. and Roth, W. Characterization of non-volatile organics from water by pyrolysis–gas chromatography–MS. *Organic Micropollutants in Water: Proceedings of the Third European Symposium*, Oslo, Norway. 141–146, 1983.
54. Poerschmann, J., Kopinke, F.-D., Balcke, G., and Mothes, S., Pyrolysis pattern of anthropogenic and natural humic organic matter, *J. Microcolumn Separations*, 10, 401–411, 1998.
55. Pulchan, J., Teofilo, A., Abrajano, A., and Helleur, R., Characterization of TMAH thermochemolysis products of near-shore marine sediments using GC/MS and GC/combustion/isotope ratio MS, *J. Anal. Appl. Pyrolysis*, 42, 135–150, 1997.
56. Page, D. W., van Leeuwen, J. A., Spark, K. M., and Mulcahy, D. E., Application of pyrolysis–gas chromatography–MS for characterization of dissolved organic matter before and after water treatment, *J. Anal. Appl. Pyrolysis*, 67, 247–262, 2002.
57. Bocchini, P., Galletti, G. C., Camarero, S., and Martinez, A. T., Absolute quantification of lignin pyrolysis products using an internal standard, *J. Chromatogr. A*, 773, 227–232, 1997.

58. Page, D. W., van Leeuwen, J. A., Spark, K. M., and Mulcahy, D. E., Pyrolysis characterization of soil, litter and vegetation extracts from Australian catchments, *J. Anal. Appl. Pyrolysis*, 65(2), 249–265, 2002.
59. Flodin, C., Ekelund, M., Boren, H., and Grimvall, A., Pyrolysis–GC/AED and pyrolysis–GC/MS analysis of chlorinated structures in aquatic fulvic acids and chlorolignins, *Chemosphere*, 34, 2319–2328, 1997.
60. Frimmel, F. H. and Schmiedel, U., Pyrolysis GC/MS identification of halogenated furandiones and other products from chlorinated aqueous humic solutions, *Fresenius J. Anal. Chem.*, 346, 707–710, 1993.
61. Abbt-Braun, G., Frimmel, F. H., and Schulten, H.-R., Structural investigations of aquatic humic substances by pyrolysis–field ionization MS and pyrolysis–gas chromatography/mass spectroscopy, *Water Res.*, 23, 1579–1591, 1989.
62. Hempfling, R. and Schulten, H.-R., Chemical characterization of the organic matter in forest soils by Curie point pyrolysis–GC/MS and pyrolysis–field ionization MS, *Org. Geochem.*, 15, 131–145, 1990.
63. Kögel-Knaber, I., Hempfling, R., and Schulten, H.-R., Decomposition in forest humus layers studied by CPMAS ^{13}C NMR, pyrolysis–field ionization–MS and CuO oxidation, *Sci. Total Environ.*, 62, 111–113, 1987.
64. Leinweber, P. and Schulten, H.-R., Advances in analytical pyrolysis of soil organic matter, *J. Anal. Appl. Pyrolysis*, 49, 359–383, 1999.
65. Reeves, J. B. and Francis, B. A., Pyrolysis–gas chromatography–MS for analysis of forages and by-products, *J. Anal. Appl. Pyrolysis*, 40–41, 243–265, 1997.
66. Mansuy, L., Bourezgui, Y., Zarli, E. G., Jarde, E., and Reveille, V., Characterization of humic substances in highly polluted river sediments by pyrolysis–methylation–gas chromatography–MS, *Org. Geochem.*, 32, 223–231, 2001.
67. Calvo, A. N., Galletti, G. C., and González, A. F., Paper waste-water analyses by pyrolysis–gas chromatography/MS during biological decolorization with the fungi *Coriolopsis gallica* and *Paelcilomcyes variotti*, *J. Anal. Appl. Pyrolysis*, 33, 39–50, 1995.
68. Christy, A. A., Bruchet, A., and Rybacki, D., Characterization of natural organic matter by pyrolysis/GC/MS, *Environ. Int.*, 25, 181–189, 1999.
69. Didyk, B. M., Simoncit, B. R. T., Pezoa, L. A., Riveros, M. L., and Flores, A. A., Urban aerosol particles of Santiago, Chile: organic content and molecular characterization, *Atmos. Environ.*, 34, 1167–1179, 2000.
70. Harrington, G. W., Bruchet, A., Rybacki, D., and Singer, P. C., Characterization of natural organic matter and its reactivity with chlorine, In *Water Disinfection and Natural Organic Matter*, Minear, R. A. and Amy, G. L., Eds., American Chemical Society, USA, pp. 138–158, 1996.
71. Lehtonen, T., Peuravuori, J., and Pihlaja, K., Characterization of aquatic humic matter isolated from two different sorbing solid techniques: tetramethyl ammonium hydroxide treatment and pyrolysis–gas chromatography/mass spectrometry, *Anal. Chim. Acta*, 424, 91–103, 2000.
72. Martin, F., del Rio, J. C., Gonzáles-Vila, F. J., and Verdejo, T., Pyrolysis derivatization of humic substances 2. Pyrolysis of soil humic substances in the presence of tetramethyl ammonium hydroxide, *J. Anal. Appl. Pyrolysis*, 31, 75–83, 1995.
73. Page, D. W., Characterization of organic matter from Corin Reservoir, Australia, *J. Anal. Appl. Pyrolysis*, 70, 169–183, 2003.
74. Peuravuori, J., Lehtonen, T., and Pihlaja, K., Sorption of aquatic humic matter by DAX-8 and XAD-8 resins: comparative study using pyrolysis–gas chromatography, *Anal. Acta*, 471, 219–226, 2002.
75. Schnitzer, M., Kodama, H., and Schulten, H.-R., Mineral effects on the pyrolysis–field ionization MS of fulvic acid, *Soil Sci. Soc. Am. J.*, 58, 1100–1107, 1994.
76. van Smeerdijk, D. G. and Boon, J. J., Characterization of subfossil *Sphagnum* leaves, rootlets of *Ericaceae* and their peat by pyrolysis–high-resolution gas chromatography–MS, *J. Anal. Appl. Pyrolysis*, 11, 377–402, 1987.
77. Zhang, J., Zhai, J., Zhao, F., and Tao, Z., Study of soil humic substances by cross-polarization magic angle spinning ^{13}C nuclear magnetic resonance and pyrolysis–capillary gas chromatography, *Anal. Chim. Acta*, 378, 177–182, 1999.
78. Zech, W., Hempfling, R., Haumaier, L., Schulten, H.-R., and Haider, K., Humification in subalpine rendzinas: chemical analyses, IR and ^{13}C NMR spectroscopy and pyrolysis–field ionization MS, *Geoderma*, 47, 123–138, 1990.

9 Monitoring of Nitrogen Compounds in the Environment, Biota, and Food

S. Jayarama Reddy

CONTENTS

I. INTRODUCTION

The compounds of nitrogen are of great importance in water resources, in the atmosphere and in the life processes of all plants and animals. The chemistry of nitrogen is complex because of the several oxidation states exhibited by nitrogen and due to the fact that changes in oxidation state can be brought about by the living organisms. Environmental issues and global environmental changes are generating an increasing amount of attention worldwide. The occurrence and determination of nitrogen compounds in all oxidation states have received a great deal of attention in recent years. These compounds occur in a number of ambient environments, such as, air, water, soil and foods, and are sources of serious social and hygienic products. Their monitoring in various environments is important to preserve human health because these compounds have toxic effects.

Aliphatic and aromatic mono-, di-, and poly-amines are naturally occurring compounds formed as metabolic products in microorganisms, plants, and animals, in which the principal routes of amine formation include the decarboxylation of amino acids, amination of carbonyl compounds and degradation of nitrogen containing compounds. They are discharged into the atmosphere and water from anthropogenic sources such as cattle feeds and livestock buildings, waste incineration, sewage treatment, automobile exhausts, cigarette smoke, and various industries. Presumably, a natural background level of amines also exists originating from animal waste and microbiological activities. Some amines are also suspected to be allergenic, mutagenic, or carcinogenic substances owing to their tendency for adsorption in tissues. Amines are not only toxic themselves but can also become toxic nitrosamines through chemical reactions with nitrosating agents such as, nitrite or nitrate. Recent developments in environmental carcinogenesis have demonstrated that *N*-nitrosamines are potentially carcinogenic substances leading to a wide variety of tumors in many animals. Biogenic amines including catecholamines and indoleamines play a number of important functions in the peripheral and central nervous system. Levels of biogenic amines and their metabolites in tissues or biological fluids have been widely investigated for a variety of physiological and disease states in neurology and physiology of mental illness and neurological disorders. Therefore for clinical diagnosis or prognosis, and for basic research purposes, it is necessary to determine concomitant levels of biogenic amines such as serotonin, catecholamines and their metabolites in tissues and biological fluids. Many analytical techniques such as

311

radiochemistry, gas chromatography (GC), spectroscopic and HPLC techniques have been developed for the determination of amines, biogenic amines, their precursors, and metabolites.

The degree of environmental pollution is increasing with rapid industrial and economic growth. Anthropogenic and naturally supplied nitrogen is a key factor controlling primary production in N-limited environmental waters. Analysis of gas mixtures containing nitrogen, oxides of nitrogen ammonia has been undertaken by many chromatographic techniques providing the complete data with high sensitivity and reliability.

II. AMINES AND NITROSAMINES

Extensive research has been done on the determination of amines including aromatic, aliphatic, alkyl, poly, glyco, biogenic amines and nitrosamines by various chromatographic methods in which many applications are based on the derivatization procedures adopted through different detection systems. The determination of aromatic amines has been carried out by capillary zone electrophoresis (CZE), High-performance Liquid chromatography (HPLC), GC, or GC–mass spectrometry with selected ion monitoring (GC–MS–SIM). However, CZE with UV detection lacks sensitivity and requires preliminary cleanup of the samples by solid-phase extraction (SPE). HPLC analysis of aromatic amines by using UV and fluorescence detection requires tedious and time-consuming procedures for the preparation of the sample. Many of these aromatic amines occur in a number of ambient environments such as air, water and soil and are a source of serial social and hygiene problems as important occupational and environmental pollutants. They are known to be highly mutagenic and carcinogenic and to form adducts with proteins and DNA. Literature reports suggest that GC methods with electron capture and nitrogen phosphorous thermioinic detection are specific and sensitive, but these methods also require anhydrous conditions for derivatizations.

Trace amounts of aromatic amines have been measured in seawater by using HPLC with electrochemical detection.[1] LC–ED has been shown to be a satisfactory technique for the determination of aromatic amines such as aniline, methylaniline, 1-naphthylamine and diphenylamine at trace levels with detection limits of 15 and 1.5 mM using coulometric and amperometric electrochemical cells, respectively. The GC determinations of aromatic amines were extensively carried out in various environmental samples such as air, cigarette smoke, waste water, etc.,[2–4] in which the detection mode was based on the derivatization of amines with an acid anhydride. A capillary gas chromatographic method using nitrogen-selective detection was developed by Sparking et al. for the analysis of complex mixtures of aniline and related aromatic amines in work-place air.[2] Air samples were collected in alkaline ethanol solutions where the isocyanate group was converted directly in the air-sampling step into urethane by reaction with ethanol using KOH as a catalyst and the amine group was converted into corresponding pentafluoro propionic amides by an extractive derivatization procedure. The detection limit of this method is approximately 40 to 80 mol. Pieraccini and coworkers[3] reported the determination of 17 primary aromatic amines as their pentafluoropropinyl (PEP) amides in cigarette smoke and indoor air by GC–MS–SIM. A cigarette is smoked in a laboratory-made smoking machine and the amines in the main and sidestream smoke are trapped in dilute hydrochloric acid. It was confirmed that sidestream smoke contains total levels of aromatic amines approximately 50 to 60 times higher than those of mainstream smoke, and some aromatic amines in ambient air such as, offices and houses may be derived from considerable contamination of aromatic amines in sidestream smoke.

Aromatic amines were converted into their N-dimethylthiophosphoryl derivatives and measured by GC with flame photometric detection (FPD) using two connected fused-silica capillary column containing DB-1 and DB-17, respectively, (see Figure 9.1).[5] The N-dimethylthiophosphorylation of aromatic amines included a simple derivatization method using dimethylchlorothiophosphate (DMCTP) as a phosphorous-containing reagent and quantitative extraction of the derivatives in to n-hexane. The derivatives were sufficiently volatile and stable

FIGURE 9.1 Derivatization scheme of aromatic amine with DMCTP to form the corresponding *N*-dimethylthiophosphoryl derivative.

to give single symmetrical peaks. This method is selective and sensitive, and combustion smoke samples can be directly analyzed without prior cleanup and any interference from other substances.

Similar kinds of studies were also carried out by Longo and coworkers for the determination of aromatic amines at trace levels but by derivatization with heptafluorobutyric anhydride and gas chromatography electron capture negative-ion chemical ionization mass spectrometry.[6] Almost 73 primary and secondary aromatic amines (including alkyl-, chloro- and nitro-substituted anilines, benzidines, aminonaphthalenes, and aminobiphenlys) were simultaneously determined. These amines were derivatized by reaction with heptafluorobutyric anhydride to form the corresponding heptafluorobutyramides. With respect to sensitivity, the GC–EC–NICI–MS technique allowed to obtain LOD even lower than those achieved by means of electron-capture detection (ECD). The application of real samples showed that extraneous peaks did not interfere. Detection limits were in the range of 0.3 to 66.3 pg injected on the full scan acquisition mode and 0.01 to 0.57 pg injected in the selected ion monitoring acquisition mode. As a matter of fact, a GC–EC–NICI–MS method was proposed in 1989 for the determination of 2-amino-biphenyl and 4,4′-diaminodiphenylmethane in biological material after derivatization with pentafluoropropionic anhydride,[7] and the reported detection limits for these compounds were approximately 50 and 100 pg/ml, respectively.

Muller and coworkers[8] introduced a rapid and sensitive quantitative method for the detection of some aniline derivatives (*o*-toluidine, *p*-chloroaniline, 2,4-dichloroaniline, 2,5-dichloroanliline, 3,4-dichloroaniline, and 3,5-dichloroaniline) in environmental water samples using solid-phase micro extraction (SPME) in gas chromatography–mass spectrometry (GC–MS). It detects these polar compounds even in trace amounts, without derivatization in water samples. SPME is a fast, simple, solvent-free extraction technique that can be easily automated. It reduces analyte loss during extraction and requires only small water samples.[9,10] This extraction technique comprises of two simple steps. First, the fiber is exposed to the aqueous sample for extraction of the analytes by the stationary phase. The fiber is then removed from the solution and introduced into the GC injector where the analytes are thermally desorbed, separated on the column, and identified by a detector. The extraction can be carried out by direct immersion of the fiber in the aqueous sample or by exposure to the headspace of the water solution. Polyacrylate 85 μm, polydimethylsiloxane 100 μm, polydimethylsiloxane-divinylbenzene 65 μm, carbowax divinylbenzene 65 μm, and an experimental carbon fiber 80 μm can be used as many coating fibers. Total ion chromatogram of a spiked water solution (10 μg/l) extracted for 30 min with direct immersion of a carbowax dimethybenzene 65 μm fiber of some aniline derivatives is shown in Figure 9.2.

Heterocyclic amines (HCAs) formed during heating of aminoacids, proteins, creatinine, and sugars are potent mutagens in the Ames/Salmonella assay. Up to the present, 23 HCAs have been isolated as mutagens, and the structures of 19 of them were determined.[11] Many of these HCAs have been isolated from various proteinaceous foods including cooked meats and fish, and some of them have also been detected in environmental components such as airborne particles, indoor air, cigarette smoke, diesel exhaust particles, cooking fumes, and rainwater. Humans are continually exposed to HCAs in a number of ambient environments, some of the mutagenic HCAs have been verified to be carcinogenic to rodents and to be implicated in human carcinogenesis. They also possess cardiotoxic effects and various pharmaco-toxicological activities such as convulsant and potent inhibitory effects on platelet function and dopamine metabolism. The determination of HCAs has been carried out by capillary zone electrophoresis,[12] HPLC with UV,[13,14]

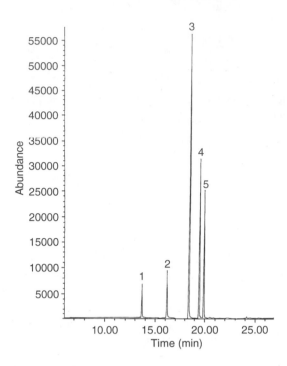

FIGURE 9.2 Total ion chromatogram of a spiked water solution (10 μg/l) extracted for 30 min with direct immersion of a carbowax divinylbenzene 65 μm fiber (1) toluidine (2) p-chloroanniline (3) 2,4 and 2,5-dichloroanniline (4) 3,5-dichloroanniline and (5) 3,4-dichloroanniline. (Reproduced with permission from Muller, L., Fattore, E., and Benfenati, E. *J. Chromatogr. A*, 791, 221–230, 1997. Copyright 1997, Elsevier Science.)

electrochemical[14-16] and fluorescence detection,[17-19] liquid chromatography–mass spectrometry,[20-23] and gas chromatography–mass spectrometry with selected ion monitoring.[24-27] GC has been widely utilized for amine analysis because of its inherent advantages of simplicity, high resolving power, high sensitivity, and low cost. Kataoka and coworkers reported the simultaneous determination of HCAs by GC with nitrogen-phosphorus selective detection using two connected fused-silica capillary columns containing DB-1 and DB-17 ht after simple derivatization with dimethylformamide dimethyl acetal (DMF-DMA) to corresponding N-dimethylaminomethylene derivatives.[28] The derivatives of ten HCAs were sufficiently volatile and stable to give single symmetrical peaks. Figure 9.3 shows the typical gas chromatogram obtained from standard heterocyclic amines. This method is simple, rapid, selective, and sensitive. HCAs can be simultaneously and quantitatively analyzed in 30 min. This method provides a useful tool for environmental analysis. Other derivatization agents tested for the determination of various HCAs include acetic, trifluoromethyl and heptafluorobutyric anhydrides, pentafluorobenzyl bromide (PFB-Br), 3,5-bis-trifluoromethylbenzyl bromide (bis-TFMBZ-Br), and bis-trifluoromethylbenzoyl bromide (bis-TFMBO-Cl). N,N-Dimethylformamide dimethyl acetal has been used not only for methyl esterification of carboxylic acid but also for one-step derivatization of amino acids into N,N-dimethylaminomethylene methyl esters. The reaction with the amino group is based on the Schiff base condensation of primary amines.

An accurate and very sensitive method was developed for the simultaneous determination of the diamines (1,3-diaminopropane (DAP), putrescine (Pu), and tryamine), polyamines (spermidine and spermine), and of the aromatic amines (β-phenylethylamine and tyramine) found in pool wines and grape juices.[29] This method combines a simple ion-pair extraction

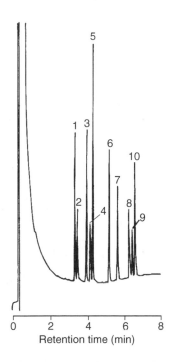

FIGURE 9.3 Total gas chromatogram obtained from standard heterocyclic amines (containing 5 ng of each amine). GC conditions–peaks: 1. 2-amino-9H-pyrido[2,3-b]indole (AαC) + 2-aminodipyrido [1,2-a:3'2'-d]imidazole (Glu-p-2); 2. 2-amino-6 methyldipyrido[1,2-a:3'2'-d]imidazole (Glu-P-1); 3. 3-amino-1,4-dimethyl 5H-pyrido[3,4-b]indole (Trp-P-1); 4. 3-amino-1-methyl-5H-pyrido[3,4-b]indole (Trp-P-2); 5. 2-amino 3-methyl-imidazo[4,5-f]quinoline (IQ); 6. 2-amino-3,4-dimethylimidazo[4,5-f]-quinoline (McIQ); 7. 2-amino 3,8-dimethylimidazo[4,5-f]quinoxaline (MeIQx); 8. 2-amino3,4,8-trimethyl-imidazo[4,5-f]quinoxaline (4,8-DiMeIQx); 9. 2-amino-1-methyl 6-phenylimidazo[4,5-b]pyridine (PhIP); 10. 2-amino-,4,7,8-trimethylimidazo[4,5-f]quinoxaline (4,7,8-TriMeIQx). (Reproduced with permission from Kataoka, H. and Kijima, K. *J. Chromatogr. A*, 767, 187–194, 1997. Copyright 1997, Elsevier Science.)

procedure using BEHPA with the GC–MS analysis of the HFB derivatives of the amines. Sample cleanup consisted of the extraction of the amines with the ion-pairing reagent bis-2-ethylhexyl phosphate (BEHPA) dissolved in chloroform followed by a back-extraction with 0.1 *M* HCl. The hydrochloric extract obtained was dried, the amines were further derivatized with heptafluoro-butyric anhydride, and analyzed by GC–MS in the selected ion-monitoring mode, with a total run time of 18 min. BEHPA has proved to be a powerful and very efficient agent for the extraction of different organic bases when used as an initial separation step of a chromatographic method. It was successfully applied in the extraction of 4-methylimidazole from caramel colors, of histamine (HA) from blood and plasma, preceding the HPLC determination of the analytes, in the extraction of catecholamines from rat brain, 3-*O*-methylated catecholamines from urine, and several sympathomimetics from biological materials preceding their GC analysis. Many authors describe the use of the acylating reagent HFBA for the conversion of the diamines and polyamines into volatile derivatives with good chromatographic properties.[30] It is known that the reagent also reacts under mild conditions with phenol groups originating the correspondents *O*-heptafluorobutyryl derivatives. The high levels of sensitivity, accuracy, and reproducibility achieved recommend the use of the proposed method for the quantification of biogenic amines in these types of matrices. Figure 9.4 shows the total ion chromatogram obtained under the conditions used for quantification (selected ion-monitoring of three different groups of ions) from a grape juice sample

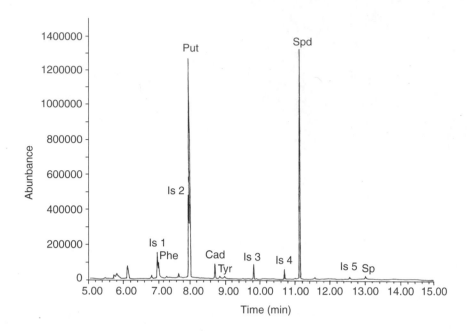

FIGURE 9.4 Total ion chromatogram obtained under the conditions used for quantification (selected ion monitoring of three different groups of ions) from a grape juice sample with β-phenylethylamine (0.271 mg/l); putrescine (3.096 mg/l); cadaverine (0.232 mg/l); tyramine (0.013 mg/l); spermidine (3.036 mg/l) and spermine (0.232 mg/l). (Reproduced with permission from Fernandes, J. O. and Ferreira, M. A. *J. Chromatogr. A*, 886, 183–195, 2000. Copyright 2000, Elsevier Science.)

with β-phenylethylamine (0.271 mg/l), putrescine (3.036 mg/l), cadaverine (0.232 mg/l), tryamine (0.013 mg/l), spermidine (3.036 mg/l), and spermine (0.232 mg/l).

The applicability of chromatographic techniques has been extended to the possible detection of related amines in vaginal fluid. Bacterial vaginosis (BV) commonly occurs in women at child-bearing age. Amsel et al.[31] proposed a set of practical dangerous criteria for the clinical diagnosis of BV, which is now accepted as the "gold standard". The syndrome of BV as defined by Amsel's clinical bacteria seems to be a well-defined polythetic concept of major importance to women's health. According to paper by Wolrath et al.[32] the amine content in the vaginal fluid and BV as scored according to the Nugent method are quantitatively related to BV. A sensitive and specific method for analysis of the amines isobutylamine, phenethylamine, putrescine, cadaverine, and tryamine with GC–MS was developed by them. A proper diagnosis was obtained using Gram-stained smears of the vaginal fluid that were Nugent scored according to the method of Nugent et al.[33] Putrescine, cadaverine, and tryamine were found only in low concentrations in vaginal fluid from women with Nugent scores of 0 to 3. There is a strong correlation between bacterial diagnosis and the presence of putrescine, cadaverine, and tryramine in high concentrations in vaginal fluid.

A novel method was described for the determination of two kinds of aromatic amine mutagens 2-[2-Acetylamino-4-bis(2-methoxyethyl)-amino]-5-amino-7-bromo-4-chloro-2H-benzotriazole (PBTA-1), and 2-(2-acetylamino)-4-(bis(2-cyanoethyl)-ethylamine)-5-amino-7-bromo-4-chloro-2H-benzotriazole (PBTA-2) in river water based on liquid chromatrography–electrospray ionization tandem mass spectrometry (LC–ESI–MS–MS) by Moriwaki et al.[34] These two aromatic amines are potent mutagens in salmonella and are presumed to be formed from desperse azo dyes used as industrial materials for dyeing through the reducing process in dyeing factories and chlorination in sewage plants for disinfection. The analytical method for the determination of

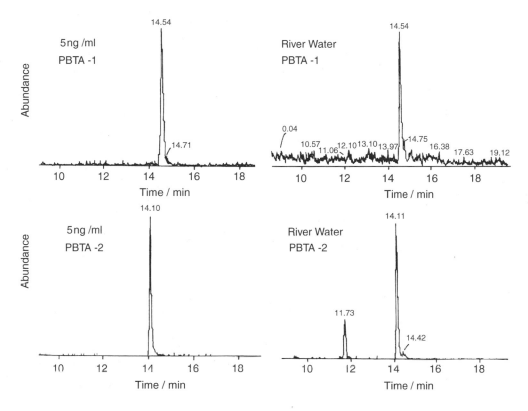

FIGURE 9.5 LC–MS–MS chromatograms for (a) standard solution (5 ng/ml) and (b) for the river water. (Reproduced with permission from Moriwaki, H., Harino, H., Hashimoto, H., AraKawa, R., Ohe, T., and Yoshikura, T. *J. Chromatogr. A*, 995, 239–243, 2003. Copyright 2003, Elsevier Science.)

PBTA-1 and PBTA-2 in river water requires troublesome pretreatment because the selection of the LC–ED system is somewhat poor, and this method is too tedious and time-consuming to be used for a detailed environmental survey of PBTA-1 and PBTA-2 in rivers. Furthermore, by using LC–MS–MS, there is the potential of simplifying the cleanup procedures for the analysis of the samples containing various environmental contaminants because LC–MS–MS generally has a higher selectivity than other detection methods coupled with LC. According to the paper by Moriwaki et al.[34] PBTA-1 and PBTA-2 could be detected in river water based on LC–MS–MS following a SPE. Due to this, the cleanup treatment was simplified and faster than the conventional

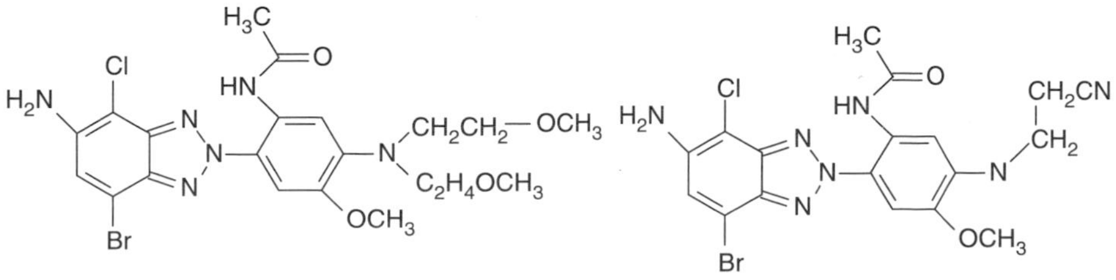

FIGURE 9.6 Proposed ion structures for the fragment ions of the protonated PBTA-1 and PBTA-2.

method. For the mass spectral investigation using ESI–MS by infusing a standard solution of the analytes, main peaks could be detected m/z 543 and 508 ions assigned as the $(M+H)^+$ ions. For PBTA-1 and PBTA-2, the major product ions were m/z 511 and 467, respectively. The productions would form through bond cleavages as the amino substitutes by collision-induced dissociation as shown below. The recoveries of PBTA ions after SPE were determined by the peak areas based on MRM chromatograms. The above procedure facilitates the monitoring of PBTAs in river water of the ultra trace level. Figure 9.5 shows LC–MS–MS chromatograms of PBTA-1 and PBTA-2 in the standard and river water by MRM. In Figure 9.6 the ion structures of PBTA-1 and PTBA-2 are shown.

Carcinogenic aromatic amines were analyzed from harmful azo colorants by Streptomyces SP.SS07 by Bhaskar et al.[35] Reduction of azo dyes to corresponding aromatic amines by extracellular fluid protein isolated from *streptomyces* species and a comparison with the dithionite reduction method has been made. Although both chemical and enzymatic reductions release similar amines, enzymatic reduction yielded higher percentage of major and minor amines. Several aromatic amines including benzidine, 4-aminobiphenyl, and 2-napthylamine have been classified by the International Agency for Research on Cancer (IARC) as human carcinogens. These aromatic amines pose health hazards to human beings in two ways, viz., direct contact and through environment.

Substituted aromatic amines have been widely used in the chemical industry as intermediates in the production of dyes, pesticides, pharmaceuticals, paints, etc. They may be released into the environment directly as a result of industrial discharge form factories or indirectly as a result of degradation of phenylcarbamates, phenylurea and anilide herbicides, and azodyes. Chromatographic separation and quantification of primary amines is hampered by their polarity that can cause tailing and irreversible adsorption. Aromatic primary amines were derivatized to *N*-allyl-*N'*-arylthiourea by

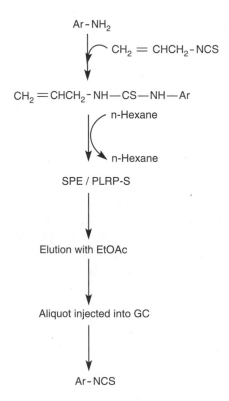

FIGURE 9.7 Reaction Scheme and analytical steps involved in the conversion of aromatic primary amine to their isocyanates.

reaction with allyl isothiocyanates.[36] Pyrolysis of the derivative always produced the corresponding arylisothiocyanate, and the total process can be regarded as a straightforward conversion of the aromatic primary amines into their isothiocyanates. The proposed method using precolumn derivatization to N-allyl-N'-arylthioureas and their thermolytic conversion into aryl isothiocyanates in the GC injector has been found to be convenient and selective to determine aromatic amines in their complex mixtures in aqueous samples (Figure 9.7). Styrene-divinylbenzene copolymer (PLRP-s) was a suitable sorbent owing to its higher retention efficiency and shorter drying time after sample loading. The specific mass fragmentation pattern of aryl isothiocyanates can be used for positive identification of aromatic amines. It is also possible to use sulfur-sensitive detection and obtain still clean chromatograms. The derivatization/pyrolysis can be studied further for their synthetic value in the preparation of aryl isothiocyanates.

A highly sensitive and specific GC–MS assay for the determination of 4,4'-methylene bis (2-chloroaniline) (MBOCA) in urine is reported.[37] It is based on the solvent extraction of the hydrolyzed MBOCA conjugates, together with deuterium-labeled benzidine-∂_8 added as an internal standard and a two-phase derivatization procedure involving use of pentafluoro propionic anhydride in the presence of ammonia as the phase transfer catalyst. In general, derivatization in a two-phase system is faster than in a single-phase system. In addition, the extractive derivatization combines extraction, derivatization, and removal of excess reagent in one step. For the determination of MBOCA in urine, acylation with PFPA to the corresponding amide is determined by the use of capillary column GC–MS with selected ion monitoring in the negative-ion chemical ionization mode. The use of the extractive derivatization technique improved the reaction yield and reduced the analysis time. The method eliminated interference from other urinary constituents, and provided a reliable and fairly precise tool for biological monitoring of workers exposed to MBOCA. Several analytical techniques have been supported for the determination of MBOCA in urine, including thin-layer chromatography followed by gas chromatography with flame ionization detection,[38] gas chromatography with electron-capture detection (GC-ECD),[39] HPLC with ultraviolet detection,[40] and electrochemical detection.[41,42] MBOCA is classified as a potential human carcinogen by the International Agency for Research on Cancer, and by the National Institute for Occupational Safety and Health (NIOSH).

Terashi et al.[43] reported the determination of primary and secondary aliphatic amines in the environment by gas chromatography–mass spectrometry. Many derivatization

$$R_1R_2NH + C_6H_5SO_2Cl \rightarrow C_6H_5SO_2NR_1R_2 + HCl$$

reagents for GC analysis of amines by using ECD,[44,45] flame thermionic detection (FTD),[46] FPD,[47] or GC–mass spectrometry with selected ion monitoring[48] (GC–MS–SIM) have also been reported. Benzene sulfonyl chloride (BSC)[46,47] and 2,4 dinitrofluoro benzene (DNFB)[49] have proved to be very useful for low molecular weight aliphatic amines, because they can convert the amines into hydrophobic and nonvolatile derivatives in water. Terashi et al. used BSC as a derivatization reagent, which can derive amines to form the corresponding sulfonamides. A standard solution of amines was added to river water, seawater, and sea sediment and distilled under alkaline conditions. The distillate was reacted with BSC to form corresponding sulfonamides. After extracting the derivatives into dichloromethane, the organic layer was concentrated to adequate volume. The determination was carried out by GC–MS with selected-ion monitoring. The detection limits of amines in water and sediment were 0.02 to 2 μg/l and 0.5 to 50 μg/kg, respectively.

A novel gas chromatographic–mass fragmentographic method was described by Rosenberg et al.[50] for the determination of exogenous aliphatic diamines in urine. Extracts of acid hydrolyzed samples were purified on Sep-Pak silica gel cartridges. The isolated diamine was converted into its perfluoro-substituted amide and determined by capillary gas chromatography–mass fragmentography. The peak hexane-1,6 diamine concentration in urine occurred 30 min after the end of the exposure. The determination of diisocyanate-derived diamines in urine offers a selective and

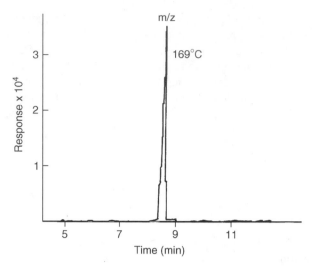

FIGURE 9.8 Total ion chromatogram of HFBA derivative of hexane-1,6,diamine (115 ng/ml). (Reproduced with permission from Rosenberg, C. and Savolainen, H. *Analyst*, 3, 1069–1071, 1986. Copyright 1986, Royal Chemical Society.)

sensitive means of biological monitoring of occupational isocyanate exposure. The recovery of hexane-1,6 diamine depends strongly on the alkalinity of the extraction mixture. Figure 9.8 shows the typical ion chromatogram of the heptafluoroamide of hexane-1,6 diamine in which the acylated derivative was synthesized using a standard procedure for symmetric diamides. The GC technique coupled with mass fragmentography allowed the detection of the isocyanate-derived amine in the 0.2 pmol/μl range. The total ion chromatogram of HFBA derivative of hexane-1,6 diamine is illustrated in Figure 9.8.

Schwarzenbach et al.[51] suggested for the determination of *N*-nitrosodiethanolamine (NDELA) in cosmetic products considering the alternatives of HPLC with UV detection and gas chromatographic–mass spectrometric method (Figure 9.9). The mass spectrometer, operated in the negative-ionization mode, has been shown to be more sensitive than the thermal energy analyzer detector. UV detection at the HPLC separation proved to be an easy method for routine determinations of NDELA in known products. Many analysts have successfully used direct UV detection as an alternative to chemiluminescence detection. The HPLC separation is achieved on a reversed-phase column. Only Lichrosorb RP-18 gives a reasonable retention for NDELA with water as the mobile phase. HPLC with organic mobile phases on silica gel or polar-bonded phases is also possible. However, the retention time of NDELA may be influenced by many factors and the possibility of coelution with other compounds is high, but for GC determinations, derivatization of NDELA is essential owing to the chromatographic behavior and lack of thermal stability of the compound. NDELA can be easily silylated by reaction with MSHFBA, MSTFA, or BSA to term NDELA bis-TMS ether.

A capillary GC method was developed for the determination of 1,6-hexamethylenediamine (HAD) in hydrolyzed human urine.[52] This method was based on a derivatization procedure using heptafluorobutyric anhydride. Capillary GC with thermionic specific detection (GC–TSD) made it possible to determine low concentrations (10 to 1000 μg/l) of HAD in urine after oral

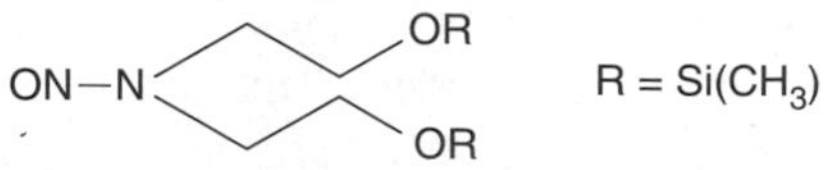

ON—N〈 ~~ OR / ~~ OR   R = Si(CH$_3$)

FIGURE 9.9 Structure of *N*-nitrosodiethanolamine.

administration of the compound. The amides formed were determined using capillary GC with selected ion monitoring in the chemical ionization mode with ammonia as reagent gas. Deuterium-labeled HAD [$H_2NC_2H_2$ $(CH_2)_4$ $C_2H_2NH_2$] was used as the internal standard. The chromatographic behavior of the amide derivatives was excellent. The use of a column with a polar stationary phase with relatively low film thickness was preferred owing to the lower temperature and column bleeding. Using ammonia as reagent gas, it showed ~ 10 times higher sensitivity than that with isobutane. The contamination of the ion source was much lower when ammonia was used. The use of ammonia as reagent gas was therefore concluded to be the best choice for the determination of HAD in hydrolyzed urine.

Direct determination of few alkylamines in aqueous solution using the indirect photometric chromatography technique was reported.[53] This method uses 3-cm columns packed with high capacity resins and copper sulfate solution as the eluent. The small particle size Aminex A-8 gave adequate resolution using a 3.0×0.41-cm^2 column. A flow rate of 1.0 ml/min gave a backpressure of 1500 lb/in.[2], and the column and flow-rate combination gave about 27,600 plates/m. The detection limits ranged from 0.5 to 2 μg/ml using a 1-ml sample loop. Various amines detected by indirect photometric chromatography technique are listed in Table 9.1.

A gas chromatographic–mass spectrometric screening procedure is described for the identification and differentiation of alkylamine, antihistamines, and their metabolites in urine such as azatadine, benzquinamide, brompheniramine, chlorphenamine, cyproheptadine, dimetindene, ketotifen, phenindamine, pheniramine, pyrrobutamine, tertenadine, and tolpropamine.[54] The antihistamines are one of the largest groups of drugs, usually classified into alkanolamine, alkylamine, ethylenediamine, piperazine, and phenothiazine derivatives, which can be found in higher concentration in urine than in plasma. Some of the alkylamine antihistamines are excreted in urine in a completely metabolized and conjugated form in the later phase of excretion. The conjugates are cleaved by acid hydrolysis, which can be completed more quickly than enzymatic hydrolysis. The above class of drugs have been determined using a Hewlett-Packard (HP) series 5890 gas chromatograph combined with an HPMSD series 5970 mass spectrometer and an HP series 59,970C workstation retaining GC conditions: splitless injection mode; column, HP capillary, crosslinked methylsilicone; column temperature programmed 100°C at 310°C at

---

TABLE 9.1
Various Amines Showing Retention Times and their Detection Limits by Indirect Photometric Chromatography

Amines	Retention (Time/min)	Detection (Limit/μg/ml)
Ethylamine	15.2	1
Diethylamine	17.6	1
Triethylamine	23.2	2
Ethanolamine	11.2	1
Diethanolamine	11.2	1
Triethanolamine	11.2	1
Methylamine	12.4	1
Dimethylamine	14.2	0.5
Trimethylamine	16.4	1
Sec-Butylamine	19.0	1
Dibutylamine	Not detected	—
Di-isopropylamine	15	Very broad peak

Reproduced with permission from Sithole, B. B. and Guy, R. D. *Analyst*, 3, 395–397, 1986. Copyright 1986, Royal Chemical Society.

30°C per minute; injection port temperature, 270°C; carrier gas, helium; flow rate, 1 ml/min and MS conditions-scan mode, ionization energy, 70 eV; ion source temperature 220°C; capillary direct interface heated at 260°C. After acid hydrolysis of the conjugates, extraction, and acetylation, the urine samples were analyzed by computerized gas chromatography–mass spectrometry. The presence of alkylamine antihistamines and their metabolites were indicated with the selected ions m/z 58, 169, 203, 205, 230, 233, 262, and 337.

Capillary gas chromatography–mass spectrometric detection method was developed for measuring dimethylamine (DMA), trimethylamine (TMA), and trimethylamine N-oxide (TMAO) in biological samples.[55] Various procedures exist for the analysis of these methylamines including thin-layer chromatography, ion chromatography, colorimetric assays, gas chromatography, and HPLC. GCIMS assay is preferable to existing methods because the derivatized amines are stable during storage. There is no ghosting or tailing, and internal standards labeled with stable isotopes can be used to correct for variations in recovery and for metabolic tracer studies. DMA, TMA, and TMAO were extracted from biological samples into acid after internal standards (labeled with stable isotopes) were added. p-Toluenesulfonyl chloride was used to form the tosylamide derivative of DMA. 2,2,2-Trichloroethyl chloroformate was used to form the carbamate derivative of TMA. TMAO was reduced with titanium (III) chloride to form TMA which was then analyzed (Figure 9.10). The derivatives were chromatographed using capillary gas chromatography. They were detected and quantified using electron ionization mass spectrometry (GC/MS). The derivatized amines are much more stable and less likely to be lost as gases when samples are stored. This assay is suitable for metabolic studies in humans using isotopically labeled methylamines.

Urinary polyamines are also extensively studied as biochemical markers of cancer. Measurement of polyamines in erythrocytes also appears to be the most promising area for assessing their levels in circulation. The majority of the circulating polyamines are associated with erythrocytes leading to the hypothesis that they may serve as passive "carriers" from sites of conjugation, catabolism, excretion, and reuptake.

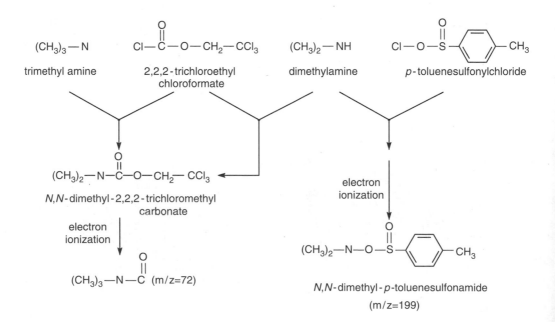

FIGURE 9.10 Derivatization of TMA and DMA. DMA and TMA were derivatized to form N,N-dimethyl-p-toluene sulfonamide and N,N-dimethyl-2,2,2-trichloroethyl carbamate. These were isolated using gas chromatography and fragmented using electron impact ionization.

The most commonly applied methods for the analysis of polyamines in erythrocytes make use of amino acid analyzers and HPLC techniques. A capillary gas chromatographic method with nitrogen-phosphorous detection was applied to the simultaneous determination of 1,3-diaminopropane, putrescine, cadaverine (Cad), spermidine (Sd), and spermine (Sp) in human erythrocytes.[56] Blood samples, collected by venipuncture into EDTA containing Venoject tubes, were subjected to the removal of plasma by centrifugation and erythrocytes were washed three times with two volumes of 0.9% NaCl. The stability of polyamines in erythrocyte suspensions was also investigated. Quantification of polyamines was done by comparing the peak-area ratio of each analyte and its internal standard with that of the standard. The polyamine samples were eluted with 0.1 M hydrochloric acid solutions. The eluate was evaporated to dryness at 120°C under a stream of air and 200 μl each of acetonitrile and heptafluorobutyric anhydride were added. The isolation of derivatives from the derivative-containing solution was performed by extraction with dichloromethane. The derivatives were dissolved in ethylacetate solution containing 0.2% (w/v) carbowax-1000 M.

Polyamines also find importance in numerous metabolic processes in plants and hence they have to be quantified accurately. Wehr[57] reported the detection of polyamines by HPLC analysis after purification of polyamines from plant materials. Plant material contains carbohydrates and phenolic compounds that interfere with derivatization. Using anion-exchange columns cleanup, plant extracts can be successfully purified prior to HPLC analysis. Polyamines can be eluted from the anion exchange resin without the need to concentrate the eluate, resulting in tremendous timesaving. Therefore, the use of anion exchange resins is a better option and many more samples can be processed per day. In comparison to cation-exchange cleanup, the recoveries of polyamines are similar and the quality of purification is comparable, if not superior. Furthermore, if the anion exchange resin is used in the OH-form, no complicating anions are present in the eluate allowing for sample and rapid derivatization by benzoylation. Phenolic substances, organic acid, and carbohydrates are optimally reformed on anion exchange resins in alkaline medium. Since polyamines do not carry a net charge in alkaline medium, they are unretained and eluted in the void fraction, and derivatization can be performed directly on the eluate resulting in enormous time savings. Figure 9.11 shows the HPLC-chromatogram of plant leaf extract purified with Dowex 50W-X8 cation exchange resins.

Ultra high sensitivity determination of primary amines by microcolumn liquid chromatography with laser-induced fluorescence detection has been reported using 3-benzoyl-2-quinolinecarboxaldehyde (BQCA) (Figure 9.12) as a precolumn fluorogenic reagent.[58] Synthesis of fluorogenic reagents containing a variety of functional moieties can be done so that the absorption maxima of their derivatives may be tuned to fit in with the desired detection scheme. BQCA, a fluorogenic reagent can yield derivatives with absorption maxima compatible with the 442 nm line of the He-Cd laser.

The reagent (BQCA), in the presence of primary amine and an appropriate nucleophile (cyanide), forms an intensely fluorescent isoindole with an absorption maximum close to the 442 nm line of the He-Cd laser. The reaction proceeds reproducibly under mild conditions (pH-8, room temperature), and the products are stable for several hours in solution. Through the use of micro-LC/LIF, detection limits to the femtomolar range were achieved avoiding interferences. Micro-LC/LIF is ideally suited for this type of sample-limited trace analysis. The inherently small volume and hence greater mass sensitivity featured in micro-LC make it the separation mode of choice, and LIF with proper derivitization can further enhance the sensitivity. Thus, micro-LC/LIF may be a very suitable tool for use by chemists or biochemists interested in extremely small biological objects and their trace analysis. A number of fluorogenic reagents specific for the amino-group, such as fluorescamine 7-chloro-4-nitrobenzene-2-oxa-1,3-diazole (NBD-chloride), o-phthaldehyde (OPA), ninhydrin, and naphthalene-2,3-dicarboxaldehyde (NDA) have been developed for this purpose and will give the detection of primary amines of picomolar concentrations. OPA also has become the most popular fluorogenic reagent for pre- and post-column derivatization in chromatographic determination of aminoacids. OPA, itself

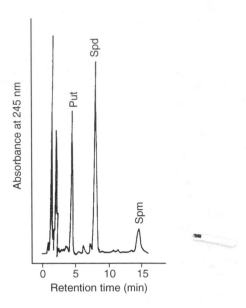

FIGURE 9.11 HPLC-chromatogram of plum leaf extract purified with Dowex 50W-X8 cation exchange resin. Concentration of polyamines: putrescine-0.9 nmol, spermidine-1.6 nmol and spermine-0.2 nmol. (Reproduced with permission from Wehr, J. B. *J. Chromatogr. A*, 709, 241–247, 1995. Copyright 1995, Elsevier Science.)

nonfluorescent, reacts with primary amines to form an intensely fluorescent isoindole with an absorption maximum at 430 nm, but its use, particularly in a precolumn, is somewhat restricted by its sensitivity to auto-oxidation and attack by light and acids. NDA also forms a more stable isoindole derivative than OPA product from a primary amine. The improvement in stability results, in part, from the use of cyanide instead of a thiol as a nucleophilic reagent, resulting in substitution of a nitrile group instead of a thiol group on the isoindole ring.

The determination of primary and secondary amines in foodstuffs was carried out by Pfundstein et al.[59] by gas chromatography with a modified thermal energy analyzer, operated in nitrogen mode. A few samples were subjected to mineral oil vacuum distillation, then the isolated amines were derivatized with BSC to form the corresponding sulfonamides which were fractionated to yield primary and secondary derivatives using a modified Husberg procedure. The choice of BSC derivatives for amine analyses by GC has two major advantages over other commonly used derivatization techniques. The identification of unknown compounds is simplified, as fractionation using a modified Husberg method immediately allows the distinction between primary and secondary amines. Further, benzene sulfonamides have very characteristic mass spectral fragmentation patterns of ion masses (if m/c 77 and 141) relating to the structure of the derivatization reagent, which can be easily located using GC–MS–SIM for gaining structural information on the suspected

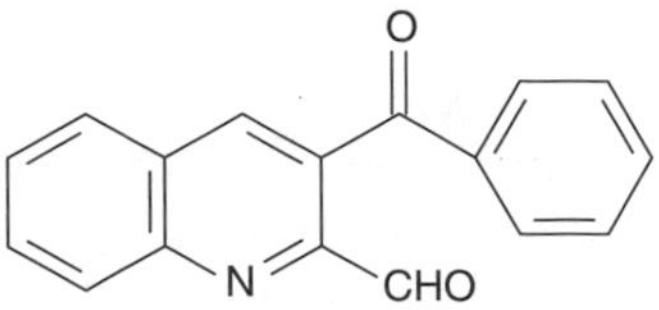

FIGURE 9.12 Structure of 3-benzoyl-2-quinolinecarboxaldehyde (BQCA).

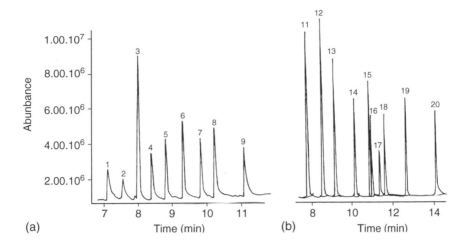

FIGURE 9.13 Gas chromatograms of (a) primary amines and (b) secondary amines, BSC derivatives on a 12 m × 0.2 mm i.d. U-1 fused silica column, carrier gas helium 2 ml/min, oven programmed from 70 to 300°C at 10°C with MS detection. Peaks: 1. methylamine; 2. ethylamine; 3. *tert*-butylamine; 4. propylamine; 5. isobutylamine; 6. *n*-butylamine; 7. isopentylamine; 8. pentylamine; 9. hexylamine; 10. phenethylamine; 11. dimethylamine; 12. methylethylamine; 13. diethylamine; 14. ethylpropylamine; 15. dipropylamine; 16. pyrrolidine; 17. morpholine; 18. piperidine; 19. dibutylamine; 20. methylbenzylamine. (Reproduced with permission from Pfundstein, B., Tricker, A. R., and Preussmann, R. *J. Chromatogr.*, 539, 141–148, 1991. Copyright 1991, Elsevier Science.)

amine parent molecule. Gas chromatograms of primary and secondary amines as BSC derivatives are shown in Figure 9.13.

In 1989, Glinsky[60] reported a method for estimation of total glycoamines in serum based on the earlier work of his research group. Glycoamines were purified from blood serum by ultrafiltration, followed by lyophilization, conventional chromatography with sephadex G-25, and size-exclusion HPLC with protein-pack I-60, then final separation of gradient reversed-phase HPLC with ultraviolet absorbance detection. This measurement was used for analysis of blood samples from more than 200 cancer patients and total serum glycoamines measurement of less than 10 kDa could detect 56 to 90% of certain types of human cancer in the early stages. These results indicated that the quantification of glycoamines in human serum has potential as a new biological marker for cancer. Kuo et al.[61] denoted their studies to improve the HPLC-UV method of glycoamine quantification in serum with emphasis on an assay that is sufficiently specific to detect the identical and structurally related glycoamine molecules in different samples with a high degree of reproducibility. The glycoamines, a newly recognized class of endogenous, low molecular mass biopolymers, are conjugates of amino acids and sugar units, containing 5 to 29 amino acid and 1 to 17 sugar units. After ultrafiltration of serum samples, reversed-phase HPLC separation with diode-array detection was used to obtain standard profiles of serum ultrafiltrates below M_r 10,000 in healthy subjects. These highly reproducible profiles utilized two-dimensional peak identification and were used to develop a statistical profile of the major glycoamine peaks in normal serum.

Biogenic amines, including catecholamines and indoleamines, play a number of important functions in the peripheral and central nervous systems. Levels of biogenic amines and their metabolites in tissues or biological fluids have been widely investigated for a variety of physiological and disease states, which include parent neurotransmitters, norepinephrine (NE), epinephrine (E), dopamine (DA), serotonin (5-HT), their precursors -3,4-dihydroxy phenylalanine, (DOPA), 5-hydroxy-tryptophan (5-HTP), their major metabolites 3-methoxy-4-hydroxyphenylethylene glycol (MHP$_4$), vanillymandelic acid (VMA), 3,4-dihydroxyphenylacetic acid (DOPAC), 3-methoxy tyramine (3-MT), homovanillic and (HVA), and 5-hydroxyindoleacetic acid (5-HIAA).

These biogenic amines and their related compounds have been implicated in the neurochemistry and physiology of mental illness and neurological disorders. The complexity of biological matrices and the diverse levels of biogenic amines, their precursors and metabolites require the use of an efficient separation technique and sensitive detection systems.

Many analytical techniques such as, radiochemistry, gas chromatography, and liquid chromatography have been developed for the determination of biogenic amines, their precursors, and metabolites.[62–64] HPLC with electrochemical detection is considered to be one of the most popular methods for determining biogenic amines, owing to its simplicity, versatility, sensitivity, and specificity.

N-acetylated metabolites of *p*-tyramine, *p*-octapamine, and dopamine were identified unambiguously, and quantitatively determined in a single ventral thoracic nerve cord of the locust, *Schistocerca gregaria*, by gas chromatography–negative-ion chemical ionization–mass spectrometry (GC–NICI–MS).[65] Deuterium-labeled analogues of each compound were added to a single ventral thoracic nerve cord in acetonitrile. The tissue was homogenized and the suspension centrifuged. The solvent was removed from the supernatent and the resultant residue was derivatized with trifluoroacetic anhydride. Under negative-ion chemical ionization conditions, the trifluoroacetyl derivatives produced ions that were sufficiently abundant to be suitable for selected ion monitoring. This method is highly specific and gave a limit of detection GC–NICIMC technique.

Shimzu et al.[66] investigated biogenic amines in the corpus cardiacum of *Periplaneta americana* using HPLC with a Neurochem neurochemical analyzer. Dopamine, tyramine, vanilic aid, and octopamine (OA) were detected in the corpus cardiacum. As amino acids, high levels of tyrosine (Tyr-4) and tryptophan were also detected at high levels. Octopamine levels in the corpus cardiacum were increased on injection of an acetone solution.

Various mechanisms have been studied in relation to the inactivation of monoamines in insects, which include *N*-acetylation,[67,68] oxidative deamination,[69,70] *O*-sulfate or β-alanyl conjugation,[71,72] and sodium-sensitive and sodium-insensitive uptake mechanisms.[73] GC and MS properties were determined for a variety of biogenic amines as their DTFMB-TMS and DTFMB-TBDMS derivatives. Since the first application of HPLC-ED, it has become an increasingly important analytical tool on neuroscience. Recently, microbore HPLC-ED has become the method of choice for the determination of trace biogenic amines and their metabolites.

The ease of sample preparation, versatility of applications, sensitivity, and ease of equipment maintenance make microbore HPLC-ED popular on neurochemical research. In addition, several factors, such as age, gender, diet, and drug interactions are known to affect the concentrations of biogenic amines and their metabolites in the body. Although very significant progress has been made recently, there is still room for improvement in microbore HPLC-ED. Simplified sample workup procedures and more reliable microbase columns with sufficient separation power are the major areas where further work is needed. Microdialysis technique and ultrafiltration techniques coupled with the microbase HPLC-ED system may become essential for overcoming analytical limitations in the determination of drug concentrations of biogenic amines and related compounds. For the determination of biogenic amines in tissues and biological fluids, a number of pretreatment protocols have been suggested. In order to achieve the best results, one has to consider the chemical nature of the analytes, the biological matrix, and the possible preference of interfering compounds.

Gas-chromatography-negative-ion chemical ionization mass spectrometry was described for the unambiguous identification of biogenic amines and their putative amino acid precursors and metabolites in single ventral thoracic nerve cords of the locust, *Schistocerca gregaria* with selective ion monitoring.[74] The configuration of the enantiomers of *p*-octopamine present in the thoracic nervous system of the locust was established as R using the chiral derivatization reagent, (-)heptafluorobutyryl phenylalanyl chloride.

A combined extraction system for the selective and quantitative isolation of the monoamines norepinephrine, epinephrine, dopamine, serotonin and their metabolites

3-methoxy-4 hydroxyphenylethylene glycol, 3,4-dihydroxy phenylacetic acid, 3-hydroxy indolecetic acid, homovanillic acid, and 3-methoxy tyramine from one single brain tissue sample is discussed by Herregodts et al.[75] The extraction system is a combination of an ethyl acetate extraction for 3-methoxy-4-hydroxy phenyl ethylene glycol, 3,4-dihydroxy phenyl acetic acid, 5-hydroxyl indolecetic acid, homovanillic acid and two successive ion-pair extractions. In the first step, the catecholamines are quantitatively isolated by extracting with heptane-octanol (99:1) containing 0.25% tetraoctylammonium bromide as an ion-pairing agent in the presence of 0.2% diphenyl borate. In a second step, 3-methoxytyramine and 5-hydroxy-tryptamine were isolated from aqueous phase with di(2-ethylhexyl) phosphoric acid as counter ion in chloroform.

Detection of DOPA metabolites in cells derived from the neutral crest cultured *in vitro* and in culture media might be useful for monitoring *in vitro* effects of agents such as, analogs of nucleotides on differentiation of pheochromocytoma or neuroblastomer cells. A highly sensitive method appears to the best chromatographic method according to work by Slingerland et al.[76] for the automatic quantitative detection of DOPA metabolites in low concentrations in cells derived from natural crest using reversed-phase HPLC in combination with fluorescence and electrochemical detection. The HPLC system was combined with online dialysis and online trace enrichment for the detection of small quantities of DOPA metabolites in culture media.

Mita and coworkers[77] have developed an assay for histamine, in which the amine is extracted from basified aqueous solution, derivatized with HFBA, and then with ethyl formate. The detection limit on GC-FID is claimed to be 20 ng. This technique can be applied to the analysis of HA and 1-methyl-HA in urine, whole blood, and leucocytes by GC–MS.

Edwards and colleagues have reported derivatization with 2,4-dinitrobenzene sulfonic acid, a procedure useful for quantification of a number of phenylethylamines,[78] to be unsuitable for analysis of HA. However, Doshi and Edwards[79] discovered that 2,6-dinitro-4-trifluoromethyl benzenesulfonic acid (DNTS) reacts readily with catecholamines, histamines, and related biogenic amines with very suitable properties for GC-ED. Formation of the trifluoroacetyl, trimethylsilyl, and heptafluoro butyryl derivatives of HA for GC have been reported,[80,81] but these compounds have been found to be unsuitable for quantification of HA because of excess tailoring.

Gas chromatography with NPD has been applied to the simultaneous analysis of HA and its basic metabolites in biological samples. HA and metabolites are first extracted from the biological material with an ion exchange resin. After elution and lyophilization, the sample is reacted with HFBA, and the derivatives purified on a silicone acid column. Acetylation of the ring NH group, if present, precedes GC analysis. Sensitivity was reported in the picomolar range.

Nitrosamines are potent mutagenic and carcinogenic compounds, which can be produced both *in vivo* and *in vitro* by bacteria. They are found in foodstuffs, drinking water, and various environmental sources such as rubber products, drug formulations, herbicide formulations, tobacco and tobacco smoke, and indoor and outdoor environments. The occurrence of *N*-nitrosamines in baby bottle rubber nipples and pacifiers is of special concern because traces of these amines may migrate into infant saliva during sucking and then be ingested. Detection of these compounds by using a selective and extremely sensitive method such as electron capture gas liquid chromatography (EC-GLC) is highly desirable. Brooks et al.[82] reported the reaction of nitrosamines with HFBA and pyridine (PY) to form derivative having a high affinity for free electrons. The pyridine-catalyzed reaction with HFBA and nitrosamines has been confirmed,[83] and some end products of the reaction were reported. However, neither the reaction mechanism nor all the end products have been clearly defined. Glass columns with good resolution are essential for detecting the most volatile HFBA-PY derivatives of *N*-nitrosodimethylamine. Nitrosamines are best obtained in the basic extraction. Chloroform has been found to be a good solvent for extracting many nitrosamines, and diethyl ether is a good solvent for the HFBA-PY derivatives of nitrosamines.

GC determinations of *N*-nitrosamines in environmental samples have been carried out in rubber nipples, pacifiers, and cigarette smoke samples. In most of them *N*-nitrosamines are directly determined as the free forms by GC-TEA, based on the detection of chemiluminescence emitted from a reaction between released NO radicals and ozone after thermal cleavage of the N–NO bond in N-nitroso compounds. Although GC-TEA is sensitive and specific for N-nitroso compounds, it is very expensive. Kataoka et al.[84] reported the determination of seven N-nitrosoamines by GC-FPD. The method is based on denitrosation with hydrobromic acid to produce the corresponding secondary amines and subsequent diethylthiophosphorylation of secondary amines. By using this method, it was confirmed that N-nitrodimethylamine, N-nitrosopyrrolidine, and N-nitrosopiperidine occur in main and sidestream smoke of cigarettes.

III. AMMONIA

Ammonia is principally manufactured by direct synthesis from nitrogen and hydrogen using promoted iron catalysts, although it is also obtained commercially cheaply as a byproduct in the manufacture of coke and gas from coal. Ammonia is largely used in agriculture as fertilizer (e.g., ammonia and solutions, ammonium nitrate, ammonium sulfate, ammonium phosphates, urea), in the production of nitric acid (by oxidation of anhydrous NH_3), in the lead chamber process for manufacturing sulfuric acid, in the purification and dehydration of NaOH, in cleaning agents such as household ammonia, soaps, in explosives via initial conversion to nitric acid and hence to basic ingredients such as nitrocellulose, nitroglycerin, TNT, ammonium nitrate, sodium nitrate, etc. Ammonia is also extensively used in the food and beverage industries as a fumigant and refrigerant, in metallurgy, in petroleum refining, as an intermediate in pharmaceutical manufacture, in paper and pulp industry as a substrate for calcium in the bisulfate pulping of wood, in textiles for the production of synthetic fibers such as Cuprammonium rayon, nylon, in the manufacture of Caprolactam, and in water purification in combination with chlorine. Ammonia occurs free in the form of salts or as traces in air, the level depending on the vicinity of natural or artificial decomposition processes. The mode of action of ammonia is via irritation to mucous membranes of the mouth and nose, as well as the upper respiratory tract. The most frequent cause of death in man from exposure to ammonia is pulmonary edema.

The determination of ammonia content in air by colorimetic techniques utilizing the Nesler's reagent,[85,86] the indophenol reaction[87] by direct UV Spectrophotometry[88,89] at 204.3 nm has been described. The trailing of chromatographic peaks noted with ammonia has been studied extensively by Sze et al.[90] with reference to the deactivation of the support material as a means of correction. Two percent of KOH on Chromosorb W has eliminated this trailing when used with 15% carbowax 400 or carbowax 1540 as the liquid phase. Also, 5% tetraethylene pentamine was effective when used with carbowax 400 or with 15% diglycerol. But for the separation of ammonia and the mono-, di-, and tri-methylamines, ethylamines and n-propylamines, a 15-ft length of column packed with Chromosorb W containing 5% tetrahydroxy ethylene-diamine and 15% tetraethylene pentamine was used. This column was operated at a temperature of 58°C.

The separation of ammonia, methylamine, dimethylamine, and ethylamine was also reported by Amell et al.[91] using a 6-ft length of column packed with 30% O-toludine on firebrick. A thermal conductivity cell was used as the chromatographic detector.

One of the earliest separations in gas liquid chromatography was that of James et al.[92] who used a mixture of hendecanol and liquid paraffin on celite using ammonia and the methyl amines as eluents in the order of their melting points. Other stationary phases used for this and for other similar separations include triethanolamine, a mixture of n-octadecane and n-hendecanol, and polyethylene oxide. "Titration cell," the first detector designed specifically for gas chromatography, was used in these early studies of the separation of ammonia and ethylamines. More recently thermal conductivity cells have been used for the detection of these compounds.

The determination of ammonia content in gas samples by vapor phase chromatographic analysis for nitrogen after catalytic decomposition was described by Diedrich et al.[93] A Dynatronics chrom-Analyzer model 100 gas chromatograph equipped with thermal conductivity detector was

TABLE 9.2
Sensitivity of some Gases Using Katharometer Detection

Gas	Column	Sensitivity (μg/0.01 mV Peak)	Retention Volume (ml)
Oxygen	48 in. molecular sieve 5 Å	0.87	96
Nitrogen	48 in. molecular sieve 5 Å	1.54	212
Nitric oxide	48 in. molecular sieve 5 Å	2.78	324
Nitrogen dioxide	48 in. molecular sieve 5 Å	8.34	324
Carbon dioxide	36 in. silica gel	13.3	340
Nitrous oxide	36 in. silica gel, 36 in. ascribe	6.1	342
Nitrous oxide	36 in. silica gel	6.4	306
Carbon dioxide	9 in. acid washed charcoal	13.0	456
Nitrous oxide	9 in. acid washed charcoal	9.7	312
Ammonia	600 in. NaOH washed but not rinsed back	19.6	298

Reproduced with permission from Chromatography of Environmental Hazards. Copyright 1973, Elsevier Science.

used with an 8 ft × 0.25- m diameter stainless steel tube fitted with 12.4 g of Linds 5 Å molecular sieve. The carrier gas was hydrogen at a flow rate of 40 ml/min, column temperature 30°C, and the thermal conductivity detector current 250 mA. The ammonia decomposition vessel was made of quartz tube and had a capacity of 25 ml.

Smith and Clark[94] compared the sensitivities of a number of gases using katharometer detection, and demonstrated the rather poor sensitivity of ammonia for this type of detector. Sensitivity of some gases using katharometer detection has been listed in Table 9.2.

Liquid anhydrous ammonia is used extensively as a coolant in heat exchange systems because of its chemical stability, low corrosiveness, and high latent heat of vaporization. However, anhydrous ammonia is readily contaminated during handling and storage. The gas chromatographic analysis of trace contaminants (O_2, N_2, CO, CH_4, CO_2, and water) in liquid ammonia was described by Mindrup and Taylor.[95] An F&M Model 5750 equipped with a Carle Microcavity Thermistor detector was used with dual columns for analysis. Table 9.3 lists the experimental parameters for both columns, which were conditioned for a minimum of 12 h at a temperature of 180°C and a

TABLE 9.3
Experimental Parameters for GC Analysis of Trace Contaminants in Liquid Ammonia

	Analysis of O_2, N_2, CO, CH_4, and CO	Analysis of H_2O	Analysis of CO_2
Column	6 ft × $\frac{1}{4}$ in. o.d., 0.028 in. gage stainless steel: molecular sieve 5 Å	18 ft × 1/8 in., 0.028 in. gage stainless steel 8 ft. porapak R, 80–100 mesh and 10 ft. porapak R, 80–100 mesh, with 10% PEI	
Detector	Thermal conductivity: thermistor at 15 mA	Thermal conductivity: thermistor at 15 mA	Thermal conductivity: thermistor at 15 mA
Temperature *Column*	600°C	800°C	600°C
Detector	Ambient	Ambient	Ambient
Injector	600°C	800°C	800°C
Flow rate	50 ml/min	70 ml/min	50 ml/min
Recorder	0.5 in./min	0.5 in./min	0.5 in./min

Reproduced with permission from Chromatography of Environmental Hazards. Copyright 1973, Elsevier Science.

30 ml/min helium flow rate. Detection limits of better than 3 ppm were attained for the above contaminants. Experimental parameters for GC analysis of trace contaminants in liquid NH_3 is given in Table 9.3.

The separation of ammonia from interfering compounds was also based on gaseous diffusion of ammonia from an alkaline medium and absorption by an acidic medium. Walker and Shipman[96] described the isolation of ammonia by the use of a zirconium phosphate cation exchanger. The adsorbed ammonia was displaced from the column by 1.24 M cesium chloride, then oxidized by hypochlorite, reacted with phenol to form a phenol-indophenol complex which was measured at 395 or 625 nm, depending on the concentration range.

A convenient and sensitive method for the determination of $^{15}NH_3$ has been developed by Fujihara et al.[97] The use of ^{15}N appears to have special advantages for the investigation of these reactions, particularly for studies tracing the metabolic fate of nitrogenous compounds in the lung system. In the determination of $^{15}NH_3$ two methods, i.e., isotope mass spectrophotometric and optical emission spectrophotometric methods, are available at present. Though these methods have been extensively employed in agricultural and biomedical fields, they suffer from lack of sensitivity and find limited application to the samples that contain relatively large amounts of ammonia. In these methods conversion of ammonia to nitrogen gas is a prerequisite prior to the ^{15}N determination. As contamination of nitrogen gas from air results in serious errors in the measurement of ^{15}N abundance ratios, a highly sophisticated technique is needed for accurate ^{15}N determination. A simple method of GC–MS analysis includes the purification of ammonia from sample solutions by a modified microdiffusion method, derivatized with pentaflurobenzoyl chloride to pentaflorobenzamide (PFBA) (Figure 9.14), and determined by GC–MS using multiple ion detector. PFBA was eluted from the gas chromatographic column within 2 min and resulted in a simple mass fragmentation pattern. The $^{15}N/^{14}N$ ratio was accurately determined with picomole amounts of PFBA by measuring the molecular ions of PFBA and (^{15}N)PFBA. This method was applied to the assay of putrescine oxidation by human plasma. Figure 9.15 shows the total ion chromatogram of mass spectrum of PFBA which is prepared from a standard solution of ammonium sulfate with a natural abundance of ^{15}N.

Another way of detection of ammonia has been reported by Kataoka et al.,[98] in which ammonia was converted into its benzenesulfonyl dimethylaminomethylene derivative by a convenient procedure involving benzenesulfonylation with benzenesulfonyl chloride and subsequent reaction with dimethyl formamide dimethyl acetal, and was determined by GC with FPD using a DB-I capillary column (Figures 9.16 and 9.17). The derivative was very stable on standing in ethylacetate, eluted as a single peak, and provided an excellent response in the FPD. Ammonia in environmental water samples could be measured without interference from coexisting substances.

The minimum detectable amount of ammonia required to give a signal three times as high as the noise under instrumental conditions was ~ 1.5 pmol injected. Mass spectrum obtained by GC–MS of the benzenesulfonyldimethylaminomethylene derivative of ammonia is shown in Figure 9.17.

A simultaneous way of determining ammonia including aliphatic amines, aromatic amines, and phenols in environmental samples by GC–MS method using a single derivatization reagent has been reported recently by Mishra et al.[99] The method consisted in precolumn formation of benzoate esters and benzamides under the conditions of the Schotten–Baumann procedure with benzoyl chloride and SPE of the derivatives. The limit of detection of ammonia was 20 $\mu g/l$ when 80 ml of sample was preconcentrated, after derivatization, on a styrene divinyl benzene copolymer sorbent.

$$C_6F_5COCl + NH_3 \longrightarrow C_6F_5CONH_2 + HCl$$

PFBC   PFBA

FIGURE 9.14 Conversion of pentaflurobenzoyl chloride to pentaflorobenazmide.

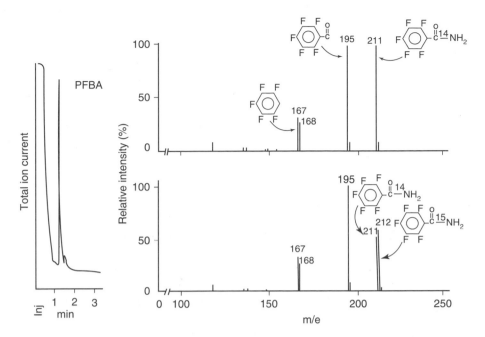

FIGURE 9.15 Total ion-current chromatogram of mass spectrum of PFBA. PFBA was prepared from a standard solution of ammonium sulfate with a natural abundance of ^{15}N (A) or with 50.8% of ^{15}N (B). (Reproduced with permission from Fujihara, S., Nakashima, T., and Kurogochi, Y. *J. Chromatogr.*, 383, 271–280, 1986. Copyright 1986, Elsevier Science).

The developed method was applied to spiked drinking water, ground water, and river water samples; and was used to detect halo-phenols in paper mill effluents. The described method is rapid and can be applied to control the water quality of environmental waters with respect to these important classes of organic pollutants and ammonia.

High performance liquid chromatography can also be applied towards the detection of ammonia in tobacco.[100] It can be detected in a cation exchange analytical column that uses a carboxylic acid/phosphoric acid functional group to achieve separation of ammonium and monovalent cations. In order to adequately resolve sodium from the ammonium cation for quantitation, sulfonic (3 mN) acid solution is used as the mobile phase. After the ammonium ion has eluted, a gradient using 0.2 N H_2SO_4 to 0.05 N H_2SO_4 concentration is used to remove any divalent cations and quaternary amines that may be present in the sample and may interfere with subsequent samples. Quantification is obtained from a five-point external standard calibration using the peak-height response of ammonium sulfate.

Ammonia was determined in the atmosphere by gas chromatographic system equipped with FTD by Yamamoto et al.[101] It is based on preconcentration with alkalized Forasil B and analysis

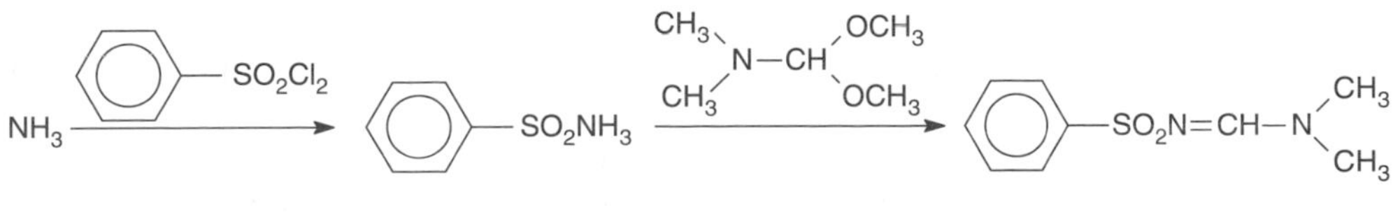

FIGURE 9.16 Ammonia derivatization process involving benzene sulfonylation with benzenesulfonyl chloride and subsequent reaction with dimethylformamide dimethyl acetal to form benzesulfonyldimethyl-aminomethylene derivative.

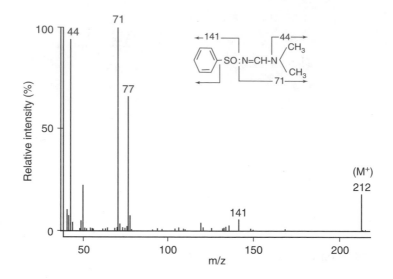

FIGURE 9.17 Mass Spectrum obtained by GC–MS of the benzenesulfonyldimethylaminomethylene derivative of ammonia. (Reproduced with permission from Kataoka, H., Ohrui, S., Kanemoto, A., and Makita, M. *J. Chromatogr.*, 633, 311–314, 1993. Copyright 1993, Elsevier Science.)

with GC-FTD. Collection and thermal release from the adsorbent are always quantified. This method is applicable to determine the atmospheric NH_3 up to ppb levels, with a sampling volume of 2 l/min.

IV. OXIDES OF NITROGEN

The oxides of nitrogen are normal constituents of air but generally present at low concentrations in the parts per billion ranges. Nitric oxide (NO) and nitrogen dioxide (NO_2) are the major oxides of nitrogen produced during combustion. The waste gas of gas turbines contains up to 2000 ppm of NO_2 while that from coal-processing thermal power plants contains between 200 and 1200 ppm of NO depending upon the type of heating. NO is formed by direct combustion of oxygen and nitrogen in air at elevated temperatures. NO is slowly oxidized to NO_2 at ordinary temperatures. Nitrogen oxides react in the atmosphere with many organic compounds, particularly hydrocarbons, to yield a spectrum of pollutants including formaldehyde, acrolein, peroxyacetyl nitrate and its analogues, as well as ketones, acids, ozonides, and ozonated olefins.

The determination of nitric oxide by gas chromatography is complicated by the equilibrium existing between nitric oxide and nitrogen dioxide in the presence of air or oxygen, and by the dimerization of the dioxide.

$$2NO + O_2 \rightarrow 2NO_2$$
$$2NO \rightarrow N_2O_4$$

The most frequently used stationary phase for the chromatographic separation of nitric oxide is probably still molecular sieve No. 5 Å. This particular separation has been found to provide an excellent example of the type of adsorption isotherm, in which the retention time varies with the amount of the component added. The oxide of nitrogen could be separated on a column 5 ft in length, packed with molecular sieve material activated in the usual way.

The retention times of oxygen and nitrogen measured using this column were found to be 0.8 and 1.2 min, respectively. The carrier gas used was argon and the detector a conventional argon ionization cell. This use of the argon ionization detector has enabled very much smaller amounts of nitric oxide to be determined, than had previously been noted using GC. The molecular sieve material packed into a glass column was operated at a temperature of 100°C and was conditioned by the addition of 170 μl of nitric oxide, added to the column before analysis. In the presence of oxygen some oxidation of the nitric oxide was observed to occur in the chromatographic column. The tailing of nitric oxide was eliminated by Dietz[102] with an elaborate pretreatment of the column packing. Molecular sieve 5 Å packed in the column was heated to 300°C in vacuum for 20 h to remove water and to activate the material, after which helium was passed through the column to minimize oxygen adsorption with a subsequent switch to a low flow rate of nitric oxide. After that, the temperature of the column was lowered to 20°C and the nitric oxide flow maintained for a further 0.5 h. The column was then flushed with helium to remove excess nitric oxide. Oxygen was introduced to convert the more tightly held nitric oxide to the dioxide. This oxygen flow was maintained at 25°C for 0.5 h and finally for 0.5 h at 100°C to complete the conversion to the dioxide. Helium was used as carrier gas.

The most striking advance in the analysis of the oxides of nitrogen, particularly for nitric oxide, has been the introduction of porous polymers such as Porapak; the column employed 12 ft of porapak Q at temperature 270°C with a helium flow rate of 50 ml/min, produced chromatograms with good separation of nitrous oxide from CO_2 and nitric oxide. Figure 9.18 shows the separation of oxides of nitrogen and CO_2 with the requirement of the above conditions. Columns packed with silica gel have also been used for the separation of nitric oxide from inorganic gases,[103] although Smith and Clark[94] reported failures in their attempts to use this material, and in similar attempts to use alumina and bentonite for this separation.

However, Szulczewski and Higuchi[104] were successful in using silica gel. They also had no difficulty in resolving nitrous oxide and CO_2 on a column 6 ft in length. Helium was used as a carrier gas and the column was operated at room temperature for this later separation, but at the

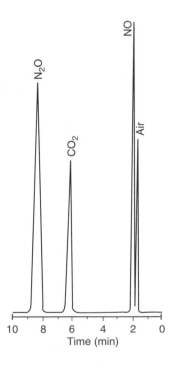

FIGURE 9.18 Separation of oxides of nitrogen and CO_2 in Porapak Q, Column 12 ft; temp. −27°C; carrier gas-helium at 50 ml/min. (Reproduced with permission from Gas Analysis by Gas Chromatography. Copyright 1972, Pergamon Press, Elsevier Science.)

TABLE 9.4
Experimental Conditions Required for NO Analysis

Column	2 ft × 0.25-in. Davidson silica gel grade 12, 28–200 mesh at 250°C	9 ft × 0.25-in. molecular sieves
Carrier	N_2 16 ml/min	N_2 50 ml/min
Scavenger	NO scavenger gas	NO scavenger gas
Pulse internal	150 μsec	150 μsec
Detector temperature	400°C	400°C
Sample size	5.0 ml	5.0 ml
Detection time	O_2, 0.9 min	O_2, 3.9 min
	NO, 5.0 min	NO, 5.9 min
	NO_2 irreversibly observed	NO_2 irreversibly observed

temperature of a solid CO_2-acetone mixture for the separation of oxygen (with argon), nitrogen, nitric oxide, and CO, which were eluted in that order.

Sakaida et al.[105] separated NO and N_2 on a 8-ft column packed with silica gel (48 to 60 mesh, Davison) at 28 to 31°C with helium carrier flows of 40 to 50 ml/min. Meckeev and Smirnova[106] separated a mixture of gases containing H_2, N_2, NO, and CO in NO_x molecular sieves (0.4 to 0.5 mm) at 136°C. The sorbent was activated for 3 h at 350°C in dry air and then for an additional hour at the same temperature in dry argon, free of oxygen. The mixtures were separated on a 3 m × 6 mm stainless steel column in 5 min with argon as carrier gas at 100 ml/min. Satisfactory separations were also obtained by using 3 m × 6 mm stainless steel columns filled with KSM-5 silica gel and an additional packing of NO_x molecular sieve (to ensure the separation of N_2 and O_2) as well as with CaA molecular sieves, but the analysis achieved with the latter columns was more time consuming.

The gas chromatographic determination of nitrogen dioxides in concentrations as low as 10 ppm has been accomplished by Lawson and McAdie[107] using electron capture detection and employing direct injection of the sample without previous trapping. A Hewlett-Packard model 5750 gas chromatography was used initially with two columns one for NO and one for NO_2 analysis. For NO analysis, either a 9 ft × 0.25-in. molecular sieve 5 Å or a 2 ft × 0.25-in. silica gel column was used. Operating conditions of NO and NO_2 are shown below for NO_2 analysis, a 20 ft × 0.125-in. column of 10% S F-96 on Fluoropak-80 was used. Conditioning of the column for NO analysis was achieved by prolonged injection of NO/air mixtures, which subsequently produce NO_2 on the

TABLE 9.5
Experimental Conditions Required for NO_2 or NO_x Analysis

Column	20 ft × 0.125-in. Fluropak-80, 10% SF-96 at 250
Carrier	N_2 4 ml/min
Scavenger	N_2 8 ml/min
Pulse internal	150 μsec
Detector temperature	400°C
Sample size	0.5 ml, 5.0 ml
Retention time	O_2, 1.6 min; NO, 1.6 min; NO_2, 2.3 min

column, or preferentially by direct injection of NO_2. Tables 9.4 and 9.5 give the experimental conditions required for NO and NO_x analysis.

Phillips and Coyne[108] separated NO and NO_2 in a variety of nitrogen containing organic compounds using a 6-ft column packed with 25% dinonylphthlate on Chromosorb B at 110°C with a hydrogen flow of 60 ml/min. The nitric oxide was quantitatively scrubbed out of the sample gas by acidified ferrous sulfate and determined by difference from samples.

Nitrous oxide (N_2O) usually occurs in low concentrations. It is a normal constituent of both unpolluted atmosphere as well as seawater. N_2O is formed upon decomposition of nitrogen containing inorganic and organic substances, and is also found in tobacco smoke. It is also used as an anesthetic in dental practice and in surgery. Commercially, heating pure ammonium nitrate to a temp of 245 to 270°C and allowing dissociating exothermically produces nitrous oxide.

$$NH_4NO_3 \rightarrow N_2O + 2H_2O + 106 \text{ kcal.}$$

Von Oettingen[109] and Parbrook[110] have cited the toxicity of nitrous oxide. Nitrous acid has been shown to be lethal to chick embryos, and teratogenic in rat and chick embryos. The effect of N_2O on RNA and DNA of rat bone marrow and thymus has also been described by Green.[111] A method was described by Buford[112] for quantitatively analyzing gaseous mixtures of N_2, N_2O, CO_2, A, and O_2 by gas chromatography using three columns of molecular sieve material at elevated ambient and subambient temperatures; with simple modifications, the analysis time of 5 min could be reduced, which is shown in Figure 9.19. The columns all of 3 mm i.d. were packed with Linde molecular sieves. The high temperature (HT) column contained molecular sieve 5 Å flour (<270 mesh) with non-acid-washed 60 to 80 mesh Chromosorb,[113] the medium temperature (MT) column molecular sieve 13 × (32 to 60 mesh). The HT, MT, and low temperature (LT) columns were 225, 38, and 75 cm long, respectively, and all packing were activated by drying in air at 105°C (16 h) and 350°C (40 h) after the columns had been packed. A Shimadzu GC-IC

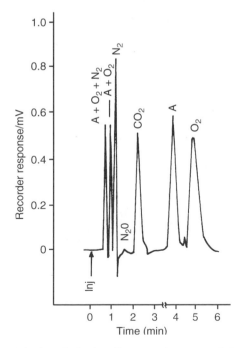

FIGURE 9.19 Quantitative analysis of 5 ml soil atmosphere containing gaseous mixtures by gas chromatography. (Reproduced with permission from Chromatography of Environmental Hazards. Copyright 1973, Elsevier Science.)

gas chromatograph was used with a thermal conductivity detector operated at 220°C, a bridge current of 100 mA and a recorder of 1 mV range. The carrier gas was helium with an inlet pressure of 3 kg/cm^2 and an outlet flow of 75 ml/min. Temperatures were controlled by using the column oven at 146°C (HT column), a water bath at 25°C (MT column), and freezing methanol bath at −98°C (LT column).

Gas absorption chromatography was used by Rozenberg et al.[114] to determine nitrous oxide in a mixture with nitrogen or nitric oxide. Silica gel KSK-2.5 (0.25 to 0.5 mm) heated preliminarily for 3 h at 350°C was used as the adsorbent. The analysis was performed using a column 163 × 3 mm^2, a valve that gave precise regulation of the gas flow, a monometer, flow meter, sampling apparatus, a katharometer, and automatic recorder. The flow rate of hydrogen gas was 30 ml/min.

Bock and Schutz[115] analyzed N_2O in air by an initial collection on molecular sieve 5 Å at room temperature, desorption at reduced pressure at 250 and 300°C, and finally determination by gas chromatography. An F&M Model 720 gas chromatograph was used with a thermal conductivity detector and a 1 m × 4 mm column containing molecular sieve 5 Å (Type 0.5 to 0.91, Perkin Elmer, Bodenserwerk) with helium carrier gas at 50 ml/min.

Bennett[116] described the use of two columns in series to affect complete separation of oxygen, nitrogen, methane, CO_2, and nitrous oxide. The first was packed with porous polymer beads and the second with molecular sieve 5 Å. A length of copper tubing between the columns enabled the gases that were separated on column I to be eluted before any emerge from column II. Column I was a 2 ft 3 in. length of 0.25 in. o.d. copper tubing filled with 50 to 80 mesh Porapak Q. The delay coil was a 7 ft × 0.25-in. o.d. copper tubing housed in the detector oven. Column II was a 6 ft × 0.25-in. o.d. copper tubing packed with 30 to 60 mesh molecular sieves 5 Å, which was activated prior to packing by heating at 2500 for 4 h under vacuum. A Gow-Mac type 9235 Thermal conductivity cell fitted with SS-W2 filaments was used in conjunction with a Gas Chromatography RY 100 bridge unit. The resolution of nitrogen oxides with other gas is shown in Figure 9.20. The resolution of mixtures of CO_2 and N_2O was effected by Degrazio[117] using a two-column system. An I&M Model

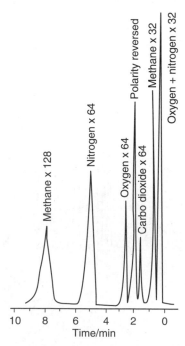

FIGURE 9.20 Gas chromatographic separation of methane, nitrogen, oxygen, nitrous oxide, and carbon dioxide obtained using Porapak Q and molecular sieve 5 Å columns in series. (Reproduced with permission from Chromatography of Environmental Hazards. Copyright 1973, Elsevier Science.)

720 gas chromatograph with a thermal conductivity detector was used with a small 4 in. precolumn insert of Linde Molecular sieve 13X connected with a 0.25-in. Swagelock union to a 6 ft × 0.25-in. o.d. stainless steel column packed with 30 to 60 mesh silica gel. The chromatographic conditions used for the resolution of CO_2 and N_2O were: column, detector, and injection port temperatures, respectively: helium carrier gas flow at 26 ml/min.

REFERENCES

1. Varshney, M. S. and Preston, M. R., Measurement of trace aromatic amines in seawater using high-performance liquid chromatography with electrochemical detection, *J. Chromatogr.*, 348, 265–274, 1985.

2. Sparking, G., Renman, L., Sango, C., Mathiasson, L., and Dalene, M., Capillary gas chromatographic method for the determination of complex mixture of isocyanates and amines, *J. Chromatogr.*, 346, 191–204, 1985.

3. Pieraccini, G., Luceri, F., and Moneti, G., New gas-chromatographic/mass-spectrometric method for the quantitative analysis of primary aromatic amines in main- and side-stream cigarette smoke, *Rapid Commun. Mass Spectrom.*, 6, 406–409, 1992.

4. Riggin, R. M., Cole, T. F., and Billets, S., Determination of aniline and substituted derivatives in wastewater by gas and liquid chromatography, *Anal. Chem.*, 55, 1862–1869, 1983.

5. Kijima, K., Kataoka, H., and Makita, M., Determination of aromatic amines as their *N*-dimethylthiophosphoryl derivatives by gas chromatography with flame photometric detection, *J. Chromatogr. A*, 738, 83–90, 1996.

6. Longo, M. and Cavallaro, A., Determination of aromatic amines at trace levels by derivatization with heptaflourobutyric anhydride and gas chromatography-electron capture negative-ion chemical ionization mass spectrometry, *J. Chromatogr. A*, 753, 91–100, 1996.

7. Avery, M. J., Determination of aromatic amines in urine and serum, *J. Chromatogr.*, 488, 470–475, 1989.

8. Muller, L., Fattore, E., and Benfenati, E., Determination of aromatic amines by solid-phase microextraction of gas chromatography–mass spectrometry in water samples, *J. Chromatogr. A*, 791, 221–230, 1997.

9. Fromberg, A., Nilsson, T., Larsen, B. R., Montanarella, L., Facchetti, S., and Madsen, J. Q., Analysis of chloro- and nitroanilines and -benzenes in soils by headspace solid-phase microextraction, *J. Chromatogr. A*, 746, 71–81, 1996.

10. Eisert, R. and Levsen, K., Solid-phase microextraction coupled to gas chromatography: a new method for the analysis of organics in water, *J. Chromatogr. A*, 733, 143–157, 1996.

11. Sugimura, T., Nagao, M., and Wakabayashi, K., Carcinogenicity of food mutagens, *Environ. Health Perspect.*, 104 (Suppl. 3), 429–433, 1996.

12. Wu, J., Wong, M. K., Li, S. F. Y., Lee, H. K., and Ong, C. N., Combination of orthogonal array design and overlapping resolution mapping for optimizing the separation of heterocyclic amines by capillary zone electrophoresis, *J. Chromatogr. A*, 709, 351–359, 1995.

13. Gross, G. A. and Gruter, A., Quantitation of mutagenic/carcinogenic heterocyclic aromatic amines in food products, *J. Chromatogr.*, 592, 271–278, 1992.

14. Schwarzenbach, R. and Gubler, D., Detection of heterocyclic aromatic amines in food flavors, *J. Chromatogr.*, 624, 491–495, 1992.

15. Galceran, M. T., Pais, P., and Puignas, L., High-performance liquid chromatographic determination of ten heterocyclic aromatic amines with electrochemical detection, *J. Chromatogr. A*, 655, 101–110, 1993.

16. Van Dyck, M. M. C., Rollmann, B., and De Meester, D., Quantitative estimation of heterocyclic aromatic amines by ion-exchange chromatography and electrochemical detection, *J. Chromatogr. A*, 697, 377–382, 1995.

17. Manabe, S., Wada, O., Morita, M., Izumikawa, S., Asakuno, K., and Suzuki, H., Occurrence of carcinogenic amino-α-carbolines in some environmental samples, *Environ. Pollut.*, 75, 301–305, 1992.

18. Manabe, S., Wada, O., and Kanai, Y., Simultaneous determination of amino-carbolines and amino-γ-carbolines in cigarette smoke condensate by high-performance liquid chromatography, *J. Chromatogr.*, 529, 125–133, 1990.

19. Manabe, S., Izumkawa, S., Asakuno, K., Wada, O., and Kanai, Y., Detection of carcinogenic amino-α-carbolines and amino-γ-carbolines in diesel-exhaust particles, *Environ. Pollut.*, 70, 255–265, 1991.

20. Edmonds, C. G., Sethi, S. K., Yamaizumi, Z., Kasai, H., Nishimura, S., and McCloskey, J. A., Analysis of mutagens from cooked foods by directly combined liquid chromatography–mass spectrometry, *Environ. Health Perspect.*, 67, 35–40, 1986.

21. Millon, H., Bur, H., and Turesly, R., Thermospray liquid chromatographic–mass spectrometric analysis of mutagenic substances present in tryptophan pyrolysates, *J. Chromatogr.*, 394, 201–208, 1987.

22. Galceran, M. T., Mayano, E., Puignas, L., and Pais, P., Determination of heterocyclic amines by pneumatically assisted electrospray liquid chromatography–mass spectrometry, *J. Chromatogr. A*, 730, 185–194, 1996.

23. Riching, E., Herderich, M., and Schreier, P., High performance liquid chromatography–electrospray tandem mass spectrometry (HPLC–ESI–MS–MS) for the analysis of heterocyclic aromatic amines (HAA), *Chromatographia*, 42, 7–11, 1996.

24. Vainiotalo, S., Matveinen, K., and Reunanen, A., GC/MS determination of the mutagenic heterocyclic amines MeIQX and DiMeIQX in cooking fumes, *Fresenius J. Anal. Chem.*, 345, 462–466, 1993.

25. Friesen, M. D., Garren, L., Bereziat, J.-C., Kadlubar, F., and Lin, D., Gas chromatography–mass spectrometry analysis of 2-amino-1-methyl-6-phenyl imidazo [4,5-6] pyridine in urine and feces, *Environ. Health Perspect.*, 99, 179–181, 1993.

26. Murray, S., Gooderham, N. J., Barnes, V. F., Boobis, A. R., and Davies, D. S., TrP-P-2 is not detectable in cooked fish, *Carcinogenesis*, 8, 937–940, 1987.

27. Tikkanen, L. M., Sauri, T. M., and Latva-kala, K. J., Screening of heat-processed Finnish foods for the mutagens 2-amino-3,8-dimethyl imidazole [4,5f] quinoxaline, 2-amino-3,4,8-trimethyl imidazole [4,5f] quinoxaline and 2-amino-1-methyl-6-phenyl imidazo [4,5b] pyridine, *Food Chem. Toxicol.*, 31, 717–721, 1993.

28. Kataoka, H. and Kijima, K., Analysis of heterocyclic amines as their *N*-dimethylaminomethylene derivatives by gas chromatography with nitrogen–phosphorous selective detection, *J. Chromatogr. A*, 767, 187–194, 1997.

29. Fernandes, J. O. and Ferreira, M. A., Combined ion-pair extraction and gas chromatography–mass spectrometry for the simultaneous determination of diamines, polyamines, and aromatic amines in port wine and grape juice, *J. Chromatogr. A*, 886, 183–195, 2000.

30. Niitsu, M., Samejima, K., Matsuzaki, S., and Hamana, K., Systematic analysis of naturally occurring linear and branched polyamines by gas chromatography and gas chromatography–mass spectrometry, *J. Chromatogr.*, 641, 115–123, 1993.

31. Amsel, R., Tottem, P. A., Spiegel, C. A., Chen, K. C., Eschenbach, D., and Holmes, K. K., Non specific vaginitis: diagnostic criteria and microbial and epidemiologic associations, *Am. J. Med.*, 74, 14–22, 1983.

32. Wolrath, H., Forsum, U., Larsson, P. G., and Boren, H., Analysis of bacterial vaginosis-related amines in vaginal fluid by gas chromatography and mass spectrometry, *J. Clin. Microbiol.*, 39(1), 4026–4031, 2001.

33. Nugent, R. P., Krohn, M. A., and Hillier, S. L., Reliability of diagnosing bacterial vaginosis is improved by a standardized method of gram stain interpretation, *J. Clin. Microbiol.*, 29, 297–301, 1991.

34. Moriwaki, H., Harino, H., Hashimoto, H., Ara Kawa, R., Ohe, T., and Yoshikura, T., Determination of aromatic amine mutagens, PBTA-1 and PBTA-2 in river water by solid-phase extraction followed by liquid chromatography–tandem mass spectrometry, *J. Chromatogr. A*, 995, 239–243, 2003.

35. Bhaskar, M., Gnanamani, A., Ganeshjeevan, R. J., Chandrasekar, R., Sadulla, S., and Radhakrishnan, G., Analysis of carcinogenic aromatic amines released from harmful azo colorance by *streptomyces* sp. SS07, *J. Chromatogr. A*, 1018, 117–123, 2003.

36. Singh, V., Gupta, M., Jain, A., and Verma, K. K., Determination of aromatic primary amines at μg/l level in environmental waters by gas chromatography–mass spectrometry involving *N*-allyl-*N'*-arylthiourea formation and their online pyrolysis to aryl isothiocyanates, *J. Chromatogr. A*, 101, 243–253, 2003.

37. Jedrzejczak, K. and Gaind, V. S., Determination of 4,4'-methylenebis(2-chlroaniline) in urine using capillary gas chromatography and negative-ion chemical ionization mass spectrometry, *Analyst*, 117, 1417–1420, 1992.

38. Van Roosmalen, P. B., Klein, A. L., and Drummond, I., An improved method for determination of 4,4'-methyline bis-(2-chloroaniline) (MBOCA) in urine, *Am. Ind. Hyg. Assoc. J.*, 40, 66–69, 1979.

39. Gristwood, W., Robertson, S. M. and Wilson, K. H., The determination of 4,4'-methyline bis-(2-chloroaniline) in urine by electron capture gas chromatography, *J. Anal. Toxicol.*, 8, 101–105, 1984.

40. Mckerell, P. J., Saunders, G. A., and Geyer, R., Determination of 4,4'-methylenebis(2-chloroaniline) in urine by high-performance liquid chromatography, *J. Chromatogr.*, 408, 399–401, 1987.

41. Okayama, A., Ichikawa, Y., Yoshida, M., Hare, I., and Morimoto, K., Determination of 4,4'-methyline bis-(2-chloroaniline) in urine by liquid chromatography with ion-paired solid phase extraction and electrochemical detection, *Clin. Chem.*, 34, 2122–2125, 1988.

42. Trippel-Schulte, P., Zeiske, J., and Kettrup, A., Trace analysis of selected benzine and diamino diphenyl methane derivatives in urine by means of liquid chromatography using precolumn sample preconcentration, UV and electrochemical detection, *Chromatographia*, 22, 138–146, 1986.

43. Terashi, A., Hanada, Y., Kida, A., and Shinohara, R., Determination of primary and secondary aliphatic amines in the environment as sulphonamide derivatives by gas chromatography–mass spectrometry, *J. Chromatogr.*, 503, 369–375, 1990.

44. Francis, A. J., Morgan, E. D., and Poole, C. F., Flophemesyl derivatives of alcohols, phenols, amines and carboxylic acids and their use in gas chromatography with electron-capture detection, *J. Chromatogr.*, 161, 111–117, 1978.

45. Scully, F. E. Jr., Howell, R. D., Penn, H. H., Mazina, E., and Johnson, J. D., Small molecular weight organic amino nitrogen compounds in treated municipal wastewater, *Environ. Sci. Technol.*, 22, 1186–1190, 1988.

46. Jacob, K., Falkner, C., and Vogt, W., Derivatization method for the high sensitive determination of amines and amino acids as dimethylthiophosphinic amides with the alkali flame-ionization detector, *J. Chromatogr.*, 167, 67–75, 1978.

47. Hamano, T., Mitsuhasi, Y., Hasegawa, A., Tanaka, K., and Matsuki, Y., Improved gas chromatographic method for the quantitative determination of secondary amines as sulphonamides formed by reaction with benzenesulphonyl chloride, *J. Chromatogr.*, 190, 462–465, 1980.

48. Jacob, K., Voty, W., Krauss, C., Schnabl, G., and Knedel, M., Selected ion monitoring determination of mono- and bi-functional amines by using phosphorus containing derivatives, *Biomed. Mass Spectrom.*, 10, 175–182, 1983.

49. Knapp, D. R., *Handbook of Analytical Derivatization Reactions*, Wiley, New York, 1979.

50. Rosenberg, C. and Savolainen, H., Determination in urine of diisocyanate-derived amines from occupational exposure by gas chromatography–mass fragmentography, *Analyst*, 3, 1069–1071, 1986.

51. Schwarzenbach, R. and Schmid, J. P., Determination of N-nitrosodiethanolamine in cosmetics. High performance liquid chromatography and gas chromatography–mass spectrometry as alternative methods to chemiluminescence detection, *J. Chromatogr.*, 472, 231–242, 1989.

52. Dalene, M., Skarping, G., and Brorson, T., Chromatographic determination of amines in biological fluids with special reference to the biological monitoring of isocyanates and amines, *J. Chromatogr.*, 516, 405–413, 1990.

53. Sithole, B. B. and Guy, R. D., Determination of alkylamines by indirect photometric chromatography, *Analyst*, 3, 395–397, 1986.

54. Maurer, H. and Pfleger, K., Identification and differentiation of alkyl amine antihistamines and their metabolites in urine by computerized gas chromatography–mass spectrometry, *J. Chromatogr.*, 430, 31–41, 1988.

55. Ann da costa, K., James Vibanac, J., and Zeisel, S. H., The measurement of dimethylamine, trimethylamine and trimethylamine N-oxide using capillary as gas chromatography–mass spectrometry, *Anal. Biochem.*, 187, 234–239, 1990.

56. Vanderberg, G. A., Kingma, A. W., and Muskiet, F. A. J., Determination of polyamines in human erythrocytes by capillary gas chromatography with nitrogen–phosphorous detection, *J. Chromatogr.*, 415, 27–34, 1987.

57. Wehr, J. B., Purification of plant polyamines with anion-exchange column cleanup prior to high-performance liquid chromatographic analysis, *J. Chromatogr. A*, 709, 241–247, 1995.
58. Beale, S. C., Hsich, Y.-Z., Savage, J. C., Wiesler, D., and Novotry, M., 3-Benzoyl-2-quinoline carboxaldehyde: a novel fluorogenic reagent for the high sensitivity chromatographic analysis of primary amines, *Talanta*, 36(1/2), 321–325, 1989.
59. Pfundstein, B., Tricker, A. R., and Preussmann, R., Determination of primary and secondary amines in foodstuffs using gas chromatography and chemiluminescence detection with a modified thermal energy analyzer, *J. Chromatogr.*, 539, 141–148, 1991.
60. Glinsky, G. V., *J. Tumor Marker Oncol.*, 4, 193, 1989.
61. Kuo, K. C., Gehrke, J. C., Allen, W. C., Holsbeke, M., Li, Z., Glinsky, G. V., Zumwalt, R. W., and Gehrke, C. W., High-performance liquid chromatographic analysis of glycoamines in serum, *J. Chromatrogr. B: Biomed. Sci. Appl. Sci.*, 656, 295–302, 1994.
62. Benjonothan, N. and Porter, J. C., A sensitive radioenzymatic assay for dopamine, norepinephrine and epinephrine in plasma and tissue, *Endocrinology*, 98, 1497–1507, 1976.
63. Davis, J. P., Schoumaker, H., Chen, A., and Yamamura, H. I., High performance liquid chromatography of pharmacologically active amines and peptides in biological materials, *Life Sci.*, 30, 971, 1982.
64. Anderson, G. M. and Young, J. G., Applications of liquid chromatographic–fluorometric systems in neurochemistry, *Life Sci.*, 28, 507, 1981.
65. Macfarlane, R. G., Midgley, J. M., and Watson, D. G., Identification and quantification of N-acetyl metabolites of biogenic amines in the thoracic nervous system of the locust, *Schistocerca gregaria*, by gas chromatography negative-ion chemical ionization mass spectrometry, *J. Chromatogr.*, 532, 13–25, 1990.
66. Schmizu, T. and Mihara, M., High-performance liquid chromatography of biogenic amines in the corpus cardiacum of the American cockroach, *Periplaneta americana*, *J. Chromatogr.*, 539, 193–197, 1991.
67. Evans, P. H., Soderlund, D. M., and Aldrich, J. R., *In vitro* N-acetylation of biogenic amines by tissues of the European corn borer, *Ostrinia nubilalis* Hübner, *Insect Biochem.*, 10, 375, 1980.
68. Evans, P. H. and Fox, P. M., Enzymatic N-acetylation of indolealkylamines by brain homogenates of the honeybee, *Apis mellifera*, *J. Insect Physiol.*, 21, 343, 1975.
69. Boadle, M. C. and Blaschko, H., Cockroach amine oxidase: classification and substrate specificity, *Comp. Biochem. Physiol.*, 25, 129–138, 1968.
70. Hark, E. J. and Beck, S. D., Monoamine oxidase in the brain of European corn borer larvae, *Ostrinia nubilalis* (Hübner), *Insect Biochem.*, 8, 231, 1978.
71. Bodnaryk, R. P., Brunet, P. C. J., and Koeppe, J. K., On the metabolism of N-acetyldopamine in *Periplaneta americana*, *J. Insect Physiol.*, 20, 911, 1974.
72. Hopkins, T. L., Morgan, T. D., Aso, Y., and Kramers, K. J., N-β-alanyldopamine: major role in insect cuticle tanning, *Science*, 217, 364–366, 1982.
73. Evans, P. D., Octopamine distribution in the insect nervous system, *J. Neurochem.*, 30, 1009–1013, 1978.
74. Macfarlane, R. G., Midgley, J. M., and Watson, D. G., Biogenic amines: their occurrence, biosynthesis of metabolism in the locust, *Schistocerca gregaria*, by gas chromatography-negative-ion chemical ionization mass spectrometry, *J. Chromatogr.*, 562, 585–598, 1991.
75. Herregodts, P., Michotte, Y., and Binger, G. E., Determination of the biogenic amines and their major metabolites in single human brain tissue samples using a combined extraction procedure and high-performance liquid chromatography with electrochemical detection, *J. Chromatogr. B: Biomed. Sci. Appl. Sci.*, 345, 33–42, 1985.
76. Slingerland, R. J., Van Kuilenburg, A. B. P., Bodlaender, J. M., Overmars, H., Voute, P. A., and Van Gennip, A. H., HPLC analysis of biogenic amines in cells and in culture media using online dialysis and trace enrichment, *J. Chromatogr. B*, 716, 65–75, 1998.
77. Mita, H., Yasueda, H., and Shida, T., Histamine derivative for quantitative determination by gas chromatography, *J. Chromatogr.*, 175, 339–342, 1979.
78. Edwards, D. J. and Blau, K., Phenethylamines in brain and liver of rats with experimentally induced phenylketonuria-like characteristics, *Biochem. J.*, 132, 95–100, 1973.
79. Doshi, P. S. and Edwards, D. J., Use of 2,6-dinitro-4-trifluoromethylbenzenesulfonic acid as a novel derivatizing reagent for the analysis of catecholamines, histamines and related amines by gas chromatography with electron-capture detection, *J. Chromatogr.*, 176, 359–366, 1979.

80. Mahy, N. and Gelpi, E., Gas chromatographic separation of histamine and its metabolites, *J. Chromatogr.*, 130, 237–242, 1977.

81. Navert, H., New approach to the separation and identification of some methylated histamine derivatives by gas chromatography, *J. Chromatogr.*, 106, 218–224, 1975.

82. Brooks, J. B., Alley, C. C., and Jones, R., Reaction of nitrosamine with fluorinated anhydrides and pyridine to form electron capturing derivatives, *Anal. Chem.*, 44, 1881–1884, 1972.

83. Gough, T. A., Sudgen, K., and Webb, K. S., Pyridene catalyzed reaction of volatile *N*-nitosamines with heptaflourobutyric anhydride, *Anal. Chem.*, 47, 509–512, 1975.

84. Kataoka, H., Shindoh, S., and Makita, M., Selective determination of volatile *N*-nitrosamines by derivatization with diethyl chlorothiophosphate and gas chromatography with flame photometric detection, *J. Chromatogr. A*, 723, 93–99, 1996.

85. Hanson, N. W., Reilly, D. A., and Stagg, H. E., *Determination of Toxic Substances in Air*, Heffer, Cambridge, 1965.

86. Elkins, H. B., *The Chemistry of Industrial Toxicology*, 2nd ed., Wiley, New York, 1959.

87. Leithe, W., *The Analysis of Air Pollutants*, Ann Arbor Science, Ann Arbor, MI, 1971.

88. Gunther, F. A., Barkley, J. H., Kolbezen, M. J., Blinn, R. C., and Staggs, E. A., Quantitative microdetermination of gaseous ammonia by its absorption at 204.3 nm, *Anal. Chem.*, 28, 1985–1989, 1956.

89. Kolbezen, M. J., Eckert, J. W., and Wilson, C. W., An apparatus for the automatic quantitative determination of ammonia concentrations in air, *Anal. Chem.*, 36, 593–596, 1964.

90. Sze, Y. L., Borke, M. L., and Otterstein, D. M., Separation of lower aliphatic amine by gas chromatography, *Anal. Chem.*, 35, 240–242, 1963.

91. Amell, A. R., Lamprey, D. S., and Schiek, R. C., Gas chromatographic separation of simple aliphatic amines, *Anal. Chem.*, 33, 1805–1806, 1961.

92. James, A. T., Martin, A. J. P., and Smith, G. H., Gas–liquid partition chromatography: the separation and microestimation of ammonia and methyl amines, *Biochem. J.*, 52, 238–242, 1952.

93. Diedrich, A. T., Bult, R. P., and Ramaredhya, J. M., Determination of ammonia in gas samples by vapour phase chromatographic anlaysis for nitrogen after catalytic decomposition, *J. Gas Chromatogr.*, 4, 241, 1966.

94. Smith, D. H. and Clark, F. E., Some useful techniques and accessories for adaptation of the gas chromatograph to soil nitrogen studies, *Soil Sci. Soc. Am. Proc.*, 24, 111–115, 1960.

95. Mindrup, R. F. Jr. and Taylor, J. H., Gas chromatographic analysis of trace contaminants in liquid ammonia, *J. Chromatogr. Sci.*, 8, 723–726, 1970.

96. Walker, R. I. and Shipman, W. H., Isolation of ammonia by use of zirconium phosphate cation exchanger, *J. Chromatogr.*, 50, 157, 1970.

97. Fujihara, S., Nakashima, T., and Kurogochi, Y., Determination of $^{15}NH_3$ by gas chromatography–mass spectrometry, application to the measurement of putrescine oxidation by human plasma, *J. Chromatogr.*, 383, 271–280, 1986.

98. Kataoka, H., Ohrui, S., Kanemoto, A., and Makita, M., Determination of ammonia as its benzenesulphonyldimethylaminomethylene derivative in environmental water samples by gas chromatography with flame photometric detection, *J. Chromatogr.*, 633, 311–314, 1993.

99. Mishra, S., Singh, V., Jain, A., and Verma, K. K., Simultaneous determination of ammonia, aliphatic amines, aromatic amines and phenols at $\mu g/l$ levels in environmental waters by solid-phase extraction of their benzoyl derivatives and gas chromatography–mass spectrometry, *Analyst*, 126, 1663–1668, 2001.

100. Risner, C. H. and Corner, J. M., A quantification of 4- to 6- ring polynuclear hydrocarbons in indoor air samples by high-performance liquid chromatography, *Environ. Toxicol. Chem.*, 10, 1417–1423, 1991.

101. Yamamoto, N., Nishiura, H., Honjo, T., and Inoue, H., Determination of ammonia in the atmosphere by gas chromatography with a flame thermionic detector, *Anal. Sci.*, 7, 1041–1044, 1991.

102. Dietz, R. N., Gas chromatographic determination of nitric oxide on treated molecular sieve, *Anal. Chem.*, 40, 1576–1578, 1968.

103. Marvillet, L. and Tranchant, J., *Proc. Third Symp. Gas Chromatogr.*, Edinburgh, Butterworths, London, p. 32, 1960.

104. Szulczewski, D. H. and Higuchi, T., Gas chromatographic determination some permanent gases on silica gel at reduced temperatures, *Anal. Chem.*, 29, 1541–1543, 1957.

105. Sakaida, R. R., Rinker, R. G., Cuffel, R. F., and Corcoran, W. H., Determination of nitric oxide in nitric oxide–nitrogen systems by gas chromatography, *Anal. Chem.*, 33, 32–34, 1961.

106. Mekeev, E. E. and Smirnora, G. D., Chromatographic analysis of gases containing hydrogen, oxygen, nitric oxide, nitrogen and carbon monoxide, *Izv Akad Nauk Kaz SSR Ser Khim*, 19(5), 87–89, 1969.

107. Lawson, A. and McAdie, H. G., Gas [gas chromatographic] determination of nitrogen oxides in air, *J. Chromatogr. Sci.*, 8, 731–734, 1970.

108. Phillips, L. V. and Coyne, D. M., A study of isobutylene-nitric oxide reaction products, *J. Org. Chem.*, 29, 1937–1942, 1964.

109. Von Oettingen, W.F., U.S. Public Health Report. No. 272, Federal Security Agency, Washington, 1941.

110. Parbrook, G.D., *Progress in Anesthesiology*, London, Excerpta Medica Foundation, Amsterdam, September 9–13, 1968, 1970.

111. Green, L. D., In *Toxicity of Anaesthetics*, Fink, B. R., Ed., Williams and Wilkins Co., Baltimore, p. 114, 1968.

112. Buford, J. R., *J. Chromatogr. Sci.*, 7, 760, 1969.

113. Bombaugh, K. J., Improved efficiency in gas chromatography by molecular sieve flour, *Nature*, 197, 1102–1103, 1963.

114. Rozenberg, G. I. and Kuznetsov-Fetsiov, L. I. Jr., Chromatographic analysis of a nitrous oxide–nitrogen system, *Tr Kazans Khim. Teknol. Inst.*, 36, 1620, 1967.

115. Bock, R. and Schutz, K. K., Gas chromatographic determination of a nitrous oxide traces in air, *Fresenius Z. Anal. Chem.*, 237, 321–330, 1968.

116. Bennett, J., Analysis of gas mixtures by gas chromatography, *J. Chromatogr.*, 26, 482, 1967.

117. Degrazio, R. P., *J. Gas Chromatogr.*, 26, 482, 1967.

10 Sulfur Compounds

Kerstin Beiner and Peter Popp

CONTENTS

I. PHYSICAL AND CHEMICAL PROPERTIES

With its sickly yellow color and rotten-egg odor, the ancients used to call it brimstone and mined it for religious purposes. In modern times, we call it sulfur and use it to make countless useful items.

Pure sulfur is actually tasteless and odorless. It combines with nearly all other elements and some of the resulting compounds are extremely reactive. This reactivity stems from the distribution of the electron density around the sulfur atoms in their various binding constellations. Because of their low electronegativity, free electron pairs of sulfur are available to electrophilic reaction sites. Consequently, sulfur compounds exhibit absorptive, adsorptive, and photooxidative behavior.

We use these properties in products such as matches, gunpowder, fungicides, bleaching agents, and medicine. Even the unpleasant smell of highly volatile compounds found application for safety purposes as additives for gases.

Yet sulfur compounds also cause problems for human civilization, acid rain being the best known. Furthermore, they poison catalysts involved in industrial chemical processes and impair the storage stability of petroleum products.

Furthermore, the high reactivity of sulfur compounds poses a host of difficulties during chemical analysis. Irreversible losses, elimination, and oxidation reactions catalyzed by heated metal surfaces easily take place during the sampling and transfer of sulfur compounds. Even oxidants in ambient air are known to oxidize analytes sampled cryogenically or on solid adsorbents. In addition, the dryers necessary for the cryogenic and adsorptive sampling of low molecular mass compounds can cause severe losses of sulfur compounds. Last but not least, the absorptive, adsorptive, and photooxidative behavior of sulfur compounds complicate their analysis.

A. OCCURRENCE IN THE ENVIRONMENT

Sulfur compounds are released into the environment by natural and anthropogenic sources (Table 10.1). The main natural sources are oceans, soil, vegetation, and volcanoes. Until 2000 years ago anthropogenic emissions were negligible and the sulfur cycle was solely determined by natural release. However, since the mid 19th century the global anthropogenic emission of sulfur compounds has approximately tripled owing to the population explosion and extensive industrialization. The sources of emissions are well known. The majority of emissions are accounted for by the combustion of fossil fuels and coal. The sulfur oxides formed cause acid rain which is responsible for the acidification of water, forest dieback, and corrosion of metal structures and historical buildings.

Additional sulfur-containing substances that penetrate the environment include intermediates, byproducts and waste products of the chemical, pulp, coal, and petrol industries, as well as compounds used in dye production such as thiols, thiophenes, sulfides, thiazoles, sulfoxides, sulfones, and sulfonic acids. Agriculture contributes to soil and water pollution through the widespread use of pesticides containing sulfur. Unfortunately, warfare agents are still produced,

---

TABLE 10.1
Overview of the Use of Sulfur Compounds and Their Deposition in Various Environmental Compartments

Compounds	Use and Formation	Occurrence in the Environment
Linear alkylsulfonates	Surfactants in detergents	Water
Aromatic sulfonates	Educts for azo and anthraquinone dyestuffs, intermediates in ion-exchange resin production, pesticides, pharmaceuticals, wetting agents, optical brighteners, tanning agents, plasticizers	Water
Sulfonamides	Pharmaceuticals, herbicides, enzymatic degradation product of corrosion inhibiting agents	Water
SO_2	Oxidation product	Air
PASH, sulfides, thioles	Matter of fossil origin as coal, mineral oil or derived products, released during combustion	Air, water, soil
Thiocarbamates	Herbicides	Air, water, soil
Sulfur mustard, thiodiglycols	Warfare agents and metabolites	Air, water, soil
H_2S, dimethylsulfon, dimethyl-sulfoxid, COS, dimethyl-sulfide, dimethyldi-, tri-, and tetra-sulfide, dimethyl sulfoniopropionate	Sulfur cycle in the different oxidation states	Air, water, soil

stored, and used all over the world. Moreover, sulfurous pharmaceutical products and their metabolites are also starting to appear in the environment owing to their high efficacy.

B. Why Do We Need to Detect Sulfur?

Sulfur is assimilated and used by several organisms in different ways. As a characteristic part of enzymes and structure proteins, it plays an important part in, for example, biological redox systems, blood coagulation, and the natural detoxification of many organisms; and is essential for their development. On the other hand, many organic sulfur compounds such as thiols, sulfides, and disulfides are a risk to human health and the environment.

Some thiols disrupt the central nervous system, while others have a more minor effect. Their ability to release primary irritations of the skin and mucous membrane varies. Toxicity decreases with increasing chain length and the recommended human handling levels depend on the substance concerned.

The physiological activity of sulfides increases with molecular weight and complexity. Diethyl sulfide has been found to cause gastroenteritis. More complex compounds like allyl sulfides display antiseptic properties. Polyvinyl sulfides have a bactericidal effect. Nitro-, chloro-, and hydroxy-substituted diaryl sulfides possess insecticidal properties.

Another important reason for the growing public interest in analytical developments regarding sulfur compounds is the improved awareness of global processes like the sulfur cycle. Consequently, increasing attention is being paid to:

The monitoring of the emission of organic sulfur compounds
Determining the potential formation of metabolites to predict possible risks
Product development, process monitoring, and quality control to minimize toxic, harmful,
   and dangerous products and byproducts.
Sulfur-containing polycyclic aromatic hydrocarbons (PAHs) (see Section V.B) feature in all three categories.

C. Regulations

Governments throughout the world have recognized the problems associated with emissions of sulfur gases and amend legislation to reduce them. Cutting "greenhouse gas" emissions (particularly carbon dioxide) is currently the subject of fierce debate. In recent decades, almost everyone has agreed that reducing sulfur emissions is a good idea. Differences of opinion now merely concern the level that should be allowed.

The easiest way to restrict the amount of sulfur dioxide emitted into the air is to limit the amount of sulfur in fuel. Another less practical way would be to fit stationary emitters with more efficient scrubbers and to install them on mobile emitters (such as trains, aircraft, and cars). Understandably, government agencies have opted for regulations governing the level of sulfur in fuel, especially petrol and diesel fuel. "Straight-run diesel" (taken directly from the crude distillation tower) can have sulfur levels ranging from <500 ppm to >5000 ppm, depending on the crude oil used and whether it is desulfurized in refineries. Towards the end of the 20th century, U.S. on-highway diesel fuel contained <500 ppm sulfur.

In December 2000, a new diesel program was launched in the U.S., under which, as of 2010, refineries will be required to produce diesel fuel for use in highway vehicles with a sulfur content of no more than 15 ppm. The new sulfur limits will be part of the system of advanced emission control. Related programs have also been approved in other countries.

Other topics of legislation include limiting sulfonates. The European Union is developing a new detergent directive containing rules for the crossborder movement of detergents.

In the United States the Environmental Protection Agency (EPA) issued a final rule in 2002 requiring companies to notify it before manufacturing or importing any of 13 chemicals, including polymers derived from perfluorooctane sulfonic acid (PFOSA). When incinerated, these chemicals may be converted into a category of compounds known as perfluorooctyl sulfonates (PFOs). A significant new use rule (SNUR) proposed by the EPA in March 2002 covers 75 perfluoroalkyl sulfonates (PFAS) having the potential to degrade into PFOSA, a substance which is highly persistent in the environment.

The analysis of chemical munitions, including their precursors and degradation products, is an important element of verification used to enforce the Chemical Weapons Convention (CWC). The CWC that entered into force in 1997 prohibits the development, production, stockpiling, and use of chemical weapons including sulfur compounds such as mustard gas.

II. SAMPLE PREPARATION

Sample preparation steps are important parts of analytical procedures designed to enrich low concentrated compounds. They are also used to separate undesirable matrix components and to avoid or reduce superimpositions of analyte and matrix in the chromatographic process. Therefore, the choice of suitable sample separation steps depends on the chemical and physical behavior, the concentration of analytes, and matrix components.

Organic sulfur compounds are usually found in the environment in low concentrations, necessitating both isolation and preconcentration prior to detection.

A. CLEANUP

The extraction techniques described in Section II.B provide a good basis of isolation steps to be used on environmental samples from different compartments. Other ways of eliminating interfering compounds are outlined below.

In air samples, the presence of atmospheric oxidants like ozone, nitrogen dioxide, and hydrogen peroxide, as well as a range of reactive radicals including the hydroxyl radical (OH) and peroxide radicals (HO_2, RO_2), can be a problem, complicating sampling for reduced gases in air. Such oxidants lead to variable and extensive sample loss unless removed prior to analyte trapping.

A variety of methods have been tested to remove oxidants from a sample. Sodium carbonate (Na_2CO_3) coated on a chromatographic support material was used by several authors.[1−7]

Other groups used carbonate scrubbers,[8] glass fiber filters impregnated with potassium hydroxide[9,10] and, prefilters impregnated with sodium hydroxide[11] to specifically remove sulfur dioxide from sample air streams. Other methods have included the use of ferrous sulfate ($FeSO_4$), potassium-iodide-coated filter paper,[12] and neutral aqueous potassium iodide solution.[13]

Another problem in air may be the presence of H_2S if carbonyl sulfide (COS) is to be quantified, two of the most common gases in the air. However, chromatographic separation may be a problem if the concentration of H_2S is much higher than that of COS.

Recommended are tubes packed with a gas-detecting reagent, such as $CuSO_4$ or $Pb(CH_3CO_2)_2$ (Figure 10.1).[14] The dry chemical reactions are reported as follows:

$$H_2S + CuSO_4 \rightarrow CuS + H_2SO_4 \tag{10.1}$$

$$H_2S + Pb(CH_3CO_2)_2 \rightarrow PbS + 2CH_3COOH \tag{10.2}$$

In terms of its chemical properties, COS cannot undergo any form of dry chemical reaction with $CuSO_4$ or $Pb(CH_3CO_2)_2$, although mercaptanes can. The color changes of detecting reagents have another positive effect: the extent (or length in a tube) of the discolored layer is proportional to the concentration of the gas or vapor in the sample when a fixed volume of sample is used. This enables

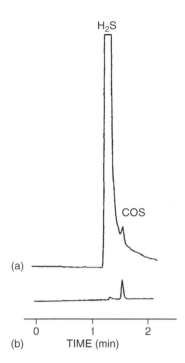

FIGURE 10.1 Chromatograms of the 0.1 Ppmv COS and 300 Ppmv H₂S test sample (a) without and (b) with Pb(CH₃CO₂)₂. (From Tang, H. M., Heaton, P., and Brassard, B., *Field Anal. Chem. Technol.*, 1, 171–174, 1997.)

the H₂S to be removed from gas samples and the discolored layer to be used to monitor its breakthrough.

The water content of the gas phase influences the adsorption dependency on numerous solids. In addition, the cotrapped water can cause certain problems in the desorption and chromatography steps unless it is removed, e.g., by using a calcium chloride tube before the actual trapping. The water content can also be reduced by membrane extraction[15] through silicone material. However, this procedure is not quantitative and should only be applied to minimize the water content in the air after purging the water.

Elemental sulfur is frequently formed by microbial activities in terrestrial and aquatic systems under anoxic conditions. It is often used in agriculture as a fungicide. As a result, sulfur levels of up to 2% are formed in sediment. Sulfur exists in the solid, liquid, and gas phases in several states. Above boiling point, S_8 molecules appear first of all, then decompose to lower aggregates ($S_6 \rightarrow S_4 \rightarrow S_2$) as the vapor temperature increases, observable as peaks in the gas chromatograms (Figure 10.2). Several methods for removing sulfur from extracts are recommended. Treatment with copper powder,[16] metallic mercury,[17] silver on silica gel,[18] and conversion with tetrabutylammonium sulfite to thiosulfate[19] or with polymeric triphenylphosphine to form the corresponding sulfide[20] have been advocated.

B. EXTRACTION

Enrichment techniques described in the literature for organic sulfur compounds include liquid/liquid extraction, static and dynamic gas extraction methods, trapping, solid-phase microextraction (SPME), and solid-phase extraction.

Sulfur compounds can be enriched directly, in a complex as sulfonic acids or sulfonates by ion-pair binding with tetrabutylammonium bromide, and as nonpolar oxidation products. Chromatography has also been used for extraction.

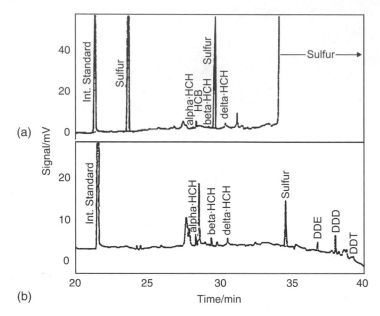

FIGURE 10.2 Gas chromatograms of the extract from a sediment sample showing (a) before and (b) after desulfurization with commercially available copper powder. (From Riis, V. and Babel, W., *Analyst*, 124, 1771–1773, 1999.)

1. Liquid/Liquid Extraction

Liquid/liquid extraction, based on the distribution balance of analyte between two immiscible phases, is mainly used for semivolatile and nonvolatile sulfur compounds, such as sulfur-containing pesticides, benzothiophenes, and their degradation products.[21–24]

The technique achieves good recoveries and reproducibility, especially for nonpolar and neutral compounds.

The method's selectivity and efficiency has been improved by adding complexing agent to the extract.[25,26]

2. Headspace Technique

The headspace technique, a static gas extraction method, is particularly suitable for the enrichment of volatile compounds. It enables the analysis of solid and liquid samples by direct sampling from the gas phase, and can be directly combined with gas chromatography. This principle is based on the distribution of analyte between the matrix and gas phase. It has been used successfully to determine volatile sulfur compounds from various matrices, such as wastewater,[27] body fluids, plants, and animal fatty tissue.

The dynamic headspace technique (purge and trap) comprises the continuous extraction of volatile compounds from the matrix into the gas phase followed by a concentration step. The transfer of volatile compounds into the gas phase is limited mainly by their volatility. Although this enables volatile substances to be distinguished from less volatile ones, there is no way of differentiating between substance classes of volatiles during the purge step.

Differentiation enables the enrichment step of the method, in which analytes are selectively adsorbed on suitable solids as described below.

Dynamic headspace has been especially applied for the enrichment of volatile sulfur compounds from complex samples like fuels, food, pharmaceuticals, water,[15,27–29] and polymers.

3. Adsorptive Sampling

Trapping is an enrichment step based on adsorption in solutions or on thermally desorbable sorbents, or on the cryogenic trapping of substances, and has been predominantly used in atmospheric chemistry.

Cryogenic trapping, an unselective enrichment technique, has been used in cryogenic traps and the heads of analytical columns for the purpose of focusing. By contrast, trapping by adsorption on solids can be very selective. A broad choice of sorbents with different enrichment behavior for organic compounds is commercially available. Selectivity is not limited to concentrated analytes, undesirable matrix compounds can also be removed as described in some exercises in the cleanup section. Alternatively, some mixtures of sorbents are available to cover substances with a broad substance spectrum. Sorbed substances can be mobilized by thermodesorption and elution.

Trapping is mainly applied to analyze volatile sulfur compounds in the air. It is also used in the same fields as the dynamic headspace technique. The key advantages of the dynamic headspace and trapping procedure are solvent-free handling and the elimination of time-consuming sample preparation. It also enables enrichment from large gas volumes. Furthermore, sampling can be automated with the right equipment. The drawback is that adsorption on numerous solids depends on the water content in the gas phase for several reasons, e.g., water competes with the analytes on the active adsorption sites or in condensation processes.

Comparative investigations have been carried out into different aspects of the behavior of organic sulfur compounds in connection with various sorbents,[30-36] e.g., breakthrough volumes[37,38] and temperature dependences.[34,37]

Using graphitized carbon proved successful for the enrichment of methyl mercaptane and CS_2.[39,40] In a comparison of 14 different sorbent materials, carbotrap achieved the best recovery rates after silica gel for enrichment from dry air, e.g., 99.7% for COS.[37]

Carbosive B was used to enrich CS_2.[41] The influence of temperature on the adsorption and breakthrough capacity of the sorbent was determined. When combined with GC–MS, detection limits in the ppt(v) range were achieved.

Carbotrap was used as a type of molecular sieve for the enrichment of CS_2 from seawater samples by purge and trap.[41,42]

Glass fiber materials have been used at extremely low temperatures for the sampling of volatile sulfur compounds from air.[32,37,43] The snag with the method is the simultaneous collection of water, which may impair the chromatographic process and decrease recovery rates. This usually makes the method impractical during field campaigns in view of storage and transport.

Leck and Bagander[44] extracted volatile sulfur compounds with purge and trap at $-196°C$ on a packing glass bed. After thermodesorption (90°C), the following detection limits were achieved: H_2S (1 ng/l), CS_2 (0.2 ng/l), dimethyl sulfide (DMS) (0.2 ng/l), methyl mercaptane (0.6 ng/l), dimethyl disulfide (DMDS) (0.4 ng/l).

The application of polymers is widespread due to their easy handling. Analytes can be desorbed more easily than with activated carbon. Because of their low affinity to water, their adsorption capacity is less dependent on the water content.

Tenax (2,6-diphenyl-p-phenylenoxid) has been used by several authors.[34,38,45-48] Tenax is thermally very stable (320°C), but has a relatively low enrichment capacity.[34] It is recommended for the analysis of compounds with more than six C atoms. It has also been proved suitable for compounds of lower molecular weights. Tangerman[34] enriched H_2S, COS, CS_2, thiols, sulfides, and disulfides from air at $-196°C$ on Tenax. After gas chromatographic separation, he detected the compounds by flame photometric detector (FPD) in the lower ng/l range.

Bandy, Tucher, and Moroulis[41] used Carbosive B. In combination with mass spectrometry, limits of detection at the ppt level were achieved.

Early investigations into the sorption of volatile sulfur compounds on metals partly resulted from efforts to explain poisoning on catalysts such as platinum.

Several authors[31,35,49,50] tested the enrichment of organic sulfur compounds on gold in atmospheric chemistry. Others[51,52] tried sorption on metal foils of Pd, Pt, Au, Ag, Rh, W, Mo, Sn, and Ni. Assays were carried out on chemically impregnated (Pb^{2+}, Hg^{2+}, Ag^+) filters.[53-55] Investigations were based on the fact that the nucleophilic sulfur atom facilitates adsorption on the electrophilic metals.

Braman et al.[31] achieved LODs of 0.1 ppt v/v for H_2S after enrichment on gold and detection with flame photometric detection. The method's potential was limited by the carrier gas helium, because reduction was detected for mercaptanes, CS_2, and COS.

By using GC/AED and helium as carrier gas, Swan et al.[35] succeeded in carrying out the quantitative desorption of DMS and CS_2 from gold wool. The wool was regenerated at 450°C under hydrogen.

Kagel et al.[51] tested several metal foils for the enrichment of H_2S, COS, CH_3SH, CS_2, DMS, DMDS, and SO_2. Desorption was performed by resistor heating. The best results were found for Pt, Pd, and Ag. The low adsorption capacities of the method allow measurements within a very small concentration range. The decomposition of analytes was assumed owing to maximum recovery rates of 45%.

The method has been enhanced by using palladium-coated platinum. Measurements of COS, DMS, SO_2, H_2S, and CS_2 in air were performed in the lower $\mu l/m^3$ range.[33,56]

Beiner et al.[15] tested the collecting capacity of several metal compounds for some volatile sulfur substances. They noted high enrichment rates and good selectivity for silver sulfide. It was used in combination with membrane extraction, thermodesorption, and GC–MS to analyze sulfides, thiols, and tetrahydrothiophene from water samples. Detection limits down to the lower ng/l range were achieved. The disadvantages of the method are the experimental equipment, the long analysis times, and displacement reactions between the matrix and analytes on the sorbent's surface.

4. Solid-Phase Microextraction

SPME has been used to analyze volatile, semivolatile, and several nonvolatile organic sulfur compounds. Only a few specific applications have been published for the extraction of sulfur compounds in environmental compartments, such as wastewater[57-59] and air.[60-63]

Determinations of semi and nonvolatile compounds have mainly focused on benzothiophene and other hetero aromatic substances. Popp et al.[58] favored polyacrylate for the enrichment of benzothiophenes. LODs between 0.4 and 5 ng/l were reached. Usable results were also found with the divinylbenzene/polydimethyl siloxane (PDMS) fiber recommended for compounds with a higher molecular weight. Johansen et al.[57] applied polyacrylate with limits of detection between 20 and 40 ng/l for thiophene and several benzothiophenes.

Sng and Ng[64] compared several fiber materials for the *in situ* derivatization of metabolites of chemical munitions such as thiodiglycerol and ethyl-2-hydroxyethyl sulfide.

According to the manufacturer's recommendation, Carboxen/PDMS is favored for applications involving low volatile sulfur compounds.

The use of PDMS and Carboxen/PDMS fibers followed by GC-AED was investigated for the analysis of volatile organic compounds (VOCs) in spiked air samples by Haberhauer-Troyer et al.[60] Detection limits down to the lower ppt range were achieved with Carboxen/PDMS.

Wardencki and Namiesnik[61] used PDMS fibers. For gaseous matrices they attained detection limits of 0.1 mg/m^{-3} with GC/FPD.

Kim et al.[62] compared Carboxen/PDMS and polyacrylate as well as static and dynamic conditions for enrichment from air.

Lestremau et al.[63] analyzed malodorous sulfur compounds in gaseous industrial effluents. Carboxen/PDMS fibers provided sufficient sensitivity for the $\mu g/m^3$ human perception levels of some VOCs.

Carboxen–PDMS fibers possess only limited numbers of adsorption sites due to the small volume of the Carboxen coating. The fiber thus becomes rapidly saturated, leading to competitive adsorption. Nevertheless, applications of Carboxen–PDMS fibers were found to be the most effective for the trace detection of sulfur compounds, and limits of detection in the ng/m^3 range were obtained using mass spectrometry[65] or atomic emission.[60]

However, several limitations have been observed with SPME concerning the decomposition or reaction of analytes in the GC injection port, such as the oxidation of DMS to dimethyl sulfoxide (DMSO) reported by Haberhauer-Troyer et al.[60] Lestremau et al.[63] observed the dimerization of methanthiol to DMDS.

5. Solid-Phase Extraction

The enrichment of organic sulfur compounds has been described on several solid-phase extraction materials, such as bounded silicates, polymers, ion-exchange materials, metal-loaded sorbents, activated carbon, and materials with several adsorption sites.

Extraction mechanisms on bounded silicates are based on polar and nonpolar interactions. Polar sorbents are recommended for organic sulfur compounds, because the sulfur atoms in organic substances often generate polar properties. However, these properties are also displayed by other hetero atoms, double bindings, and aromatic structures. Therefore, extraction on bounded silicates can be assessed as relatively unselective.

Bounded silicates have been used in combination with ion-pair extraction to develop an enrichment method for sulfonates.[66] Ionic substances such as organic sulfonates can be extracted from water in nonpolar solvents or on solid sorbents by adding ion-pair agents. Ion-pair agents are mainly organic substances which enrich on surfaces between polar and nonpolar phases. Long chain tetraalkylammonium salts are suitable ion-pair agents for organic sulfonates in water. Sulfonates are covered by the agents and can then be enriched with SPE. Reversed-phases such as C18 materials are used for adsorption, because long chain tetraalkylammonium salts tend to interact with nonpolar phases.

The method is easy to handle, readily adaptable to online SPE-HPLC systems, and gives good results for naphthalene sulfonates and benzene sulfonates with nitro, chloro, and alkyl groups. However, apart from the fact that a number of hydrophilic benzene sulfonates (e.g., 4-phenolsulfonate) are not enriched, ion-pair extraction is not very specific. Substances with a high molecular weight, such as humic matter and nonionic polar and nonpolar compounds are also enriched occurring in the HPLC samples.

Polymers have found a broad range of application for the extraction of organic sulfur compounds. Apart from nonpolar interactions, polymers can also interact over $\pi\pi$-bindings. Therefore, they are suitable for the extraction of semipolar to nonpolar analytes from polar matrices. Polymers enable the enrichment of substances over a broad polarity range. Other advantages over modified silica gels result from the high stability of polymers in acidic pH ranges.

Alkyl polymers have been tested for the determination of nonpolar compounds such as thiophenes from distillation residues.[24,67] Besides thiophenes, more polar substances, such as, thioles and disulfides have also been enriched on methacrylate (XAD-7) and polystyrene.[68,38]

Methacrylate[69] and LiChrolut EN (ethylvinylbenzene/divinylbenzene)[70] have also been used for the extraction of sulfur-containing phosphoro-acid esters and pesticides, such as malathion and endosulfan.

The most frequently used material is divinylbenzene/styrene. This polymer has been used for the enrichment of compounds like benzothiazol, malathion, atrazine, disulfoton, 2-(methylthio)-benzothiazol, and N-butylbenzene sulfonic acid from river water. In comparison with other polymers, Przyjazny[38] reached the best extraction rates for disulfides on divinylbenzene/styrene. The copolymer was found to be suitable for the extraction of aryl sulfonates after ion-pair formation. In addition, the polymer is also a suitable solid-phase extraction material for polar substances such as sulfotep, diazinon, prometryn, simetryn, dipropetryn, and dimethoate.

Both modified and unmodified ion-exchange materials have been used.

Attempts to enrich aromatic sulfonates from environmental samples on ion-exchange resins were not very successful, according to Zerbinati et al.[71] Although quantitative extraction was achieved, several compounds could not be desorbed further.

Beiner et al.[72] modified a divinylbenzene/N-vinyl pyrrolidone copolymer, containing sulfonic acid groups, with several metal ions. The modified cation-exchange material was used for the extraction of thiols, sulfides, and methyl thiophosphates from water samples, with LODs in the upper ng/l range after elution with a CS_2/toluene mixture and analysis by GC/MS.

Enrichment on metals and metal-loaded sorbents is based on the formation of coordinative bindings between sorbents and analytes. Metals and metal-loaded sorbents have been described as suitable for the selective extraction of organic sulfur compounds because of the enhanced electron density at the analytes' binding sites.

Such sorbents are mainly used for the extraction of organic sulfur compounds from organic solvents. Extensive determinations have been applied to extract and separate thiophenes, disulfides, mercaptanes, and polycyclic aromatic sulfur hydrocarbons (PASHs) from oils and fuels, as well as to separate PAHs and PASHs.

Back in 1978, Kaimai and Matsunaga[73] tried to separate PAHs and PASHs by thin layer chromatography on silica gel impregnated with several metal acetates. Mercury acetate showed the best results. Following that idea, separation was tested by Andersson[74] on a mercury-acetate-substituted phenyl kieselguhr. However, PASHs could not be separated selectively.

In 1983, separation on silica gel impregnated with 5% $PdCl_2$ were introduced. During fractionated elution, the sulfur compounds were displaced by diethylamine from the Pd complex. Problems resulted from the catalytic effects of $PdCl_2$, manifested by the partial desulfurization of compounds with terminal thiophene rings at elevated temperatures. Therefore Andersson[75] combined elution with a subsequent aminopropyl-silicate layer, in which the amino group competes with the PASHs and thus binds the Pd^{2+}.

Activated carbon has been used by several authors.[166,169-171] Problems mostly arose from the incomplete recovery of the sorbed substances.

Sorbents including several binding mechanisms were represented by, for example, Carbopack B, a graphitized carbon black with positively charged oxonium groups. The graphite structure and anion-exchange sites make it a very selective adsorbent for aromatic anions.

Carbopack B has been used for the enrichment of aromatic sulfonates by Di Corcia et al.[76,77] and Altenbach et al.,[78] and of aliphatic sulfonates and sulfates from aqueous samples by Benomar et al.[79] The fact that humic substances are almost completely absent in the final extracts is very promising with respect to surface water monitoring (Figure 10.3).

Membrane extraction was performed as a further separation method for sulfur compounds by Dercksen et al.[80] and Beiner et al.[15] VOSs were separated from water samples using silicon membranes.

C. DERIVATIZATION

Required derivatization steps have been published for various reasons (Table 10.2).

The derivatization of thiols resulted from the need to separate sulfides and thiols following the failure of SPE.

Initial derivatizations were performed using trifluoroacetic anhydride[81] and 4-fluorobenzoyl chloride.[82] Trifluoroacetylated thiols were found to coelute with sulfides from the same sample. In addition, during EI/MS spectral fragmentation, patterns of trifluoroacylarylthiols were less intense

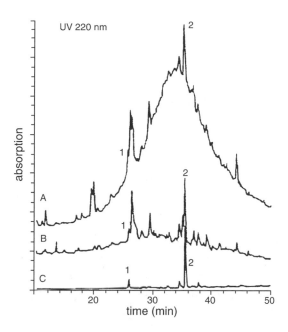

FIGURE 10.3 Chromatograms of extracts from 1 l river water spiked with 1 μg/l 3-nitrobenzene sulfonate (1) and naphthalene-2-sulfonate (2). The samples were extracted with (a) 1 g C18 adsorbent and 5 m*M* tributylammonim bromide, (b) 1 g Carbopack B, and (c) standard solution. (From Altenbach, B. and Giger, W., *Anal. Chem.*, 67, 2325–2333, 1995.)

than those of the unreacted compounds, 4-fluorobenzoylthiol peaks tailed badly on the GC-column. Thomson et al.[83] found an analogous fluorinated compound, pentafluorobenzoyl chloride (PFBC), producing thiol derivatives with symmetrical peak shapes. In addition, PFBC derivatives had the desired boiling point separation from sulfides and their identification was aided by intense 195 fragment ions in their EI/MS spectra.

Other derivatization possibilities of thiols include silylation (Figure 10.4) with *N-(tert-*butyldimethylsilyl)-*N*-methyltrifluoroacetamide (MSTFA) and acylation using *N*-methylbis (trifluoroacetamide) (MBTFA).

Silylation increases detection and raises the derivatives' thermal stability. The substitution of active hydrogen leads to the higher polarity of compounds and reduces hydrogen bridge bonds, so

TABLE 10.2
Suitable Derivatization Reagents for Sulfur Compounds

Derivatization Agents	Compounds	References
Trifluoroacetic anhydride	Thiols	81
4-Fluorobenzoyl chloride	Thiols	82
Pentafluorobenzoyl chloride (PFBC)	Thiols	83
*N-(tert-*Butyldimethylsilyl)-*N*-methyl-trifluoroacetamide (MSTFA)	Thiols, thiodiglycerol, ethyl-2-hydroxyethyl sulfide	64,86
N-Methylbis (trifluoroacetamide) (MBTFA).	Thiols, thiodiglycerol	86
Trimethylsulfonium hydroxide (TMSH)	Thiols, thiodiglycerol	86
Meta-chloroperbenzoic acid (MCPBA)	PASH	83
Hydrogen peroxide	PASH	83,84

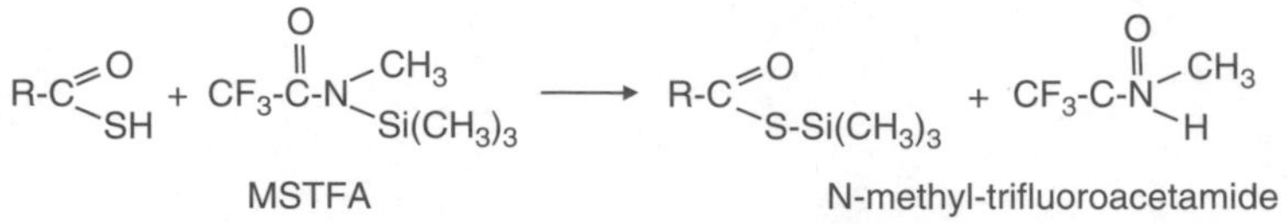

FIGURE 10.4 Silylation pattern of thiols with *N*-(*tert*-butyldimethylsilyl)-*N*-methyltrifluoroacetamide (MSTFA).

that in the end the silylated derivative becomes more volatile. Increased stability results from the reduced number of reactive centers that include active hydrogen.

Acylation converts compounds with active hydrogen, for example, in the −SH into thioesters by reaction with carboxyl acids or their derivatives. The presence of a carbonyl group next to the halogenated carbon increases the ECD response. For thiols, using *N*-methyl-bis(trifluoroacetamide) (MBTFA) is recommended.

Another substance group which has been derivatized is PASH. These compounds occur ubiquitously and as a matter of fact nearly always together with PAHs, which they resemble in many respects. Although attempts have been made to analyze PAHs for more than half a century, efforts have been hampered by several problems: PASHs nearly always occur as minor constituents together with related PAHs, their chromatographic properties are very similar, there is a larger number of isomeric parent structures among PASHs than among PAHs, and there is a larger number of alkylated isomers among PASHs than among PAHs. Despite the use of sulfur selective detectors extracts of sediments, air particles, polluted water, and other contaminated samples are too complex for the direct quantification of many compounds included.

Oxidation of sulfur's functionality to a sulfone is one way of separating PAHs and PASHs, because the polar sulfones differ from unreacted nonpolar PAHs in their chromatographic behavior. The favored oxidant is metachloroperbenzoic acid (MCPBA). One major drawback of the oxidation approach is that with the exception of benzothiophenes (BT) and some of their alkyl derivatives, PASHs containing terminal thiophene rings are largely lost, probably by oxidation of other molecular features. Thus, the naphthothiophenes cannot be analyzed after oxidation with MCPBA.

The application of hydrogen peroxide has been controversially discussed.[83,84] It was shown that oxidation in benzene/acetic acid not only built the desired sulfones of the PASHs but also that the aromatic rings of all types of polycyclic compounds were oxidized. Low or zero recovery of the analytes is often the result. Oxidation with hydrogen peroxide should therefore be avoided for any samples in which aromatic compounds are to be analyzed.

Attempts to convert the sulfones back into PASHs have been successful with a number of agents such as various metals (zinc, tin, magnesium, aluminum, iron, and nickel) in acetic acid, palladium on carbon with hydrazine, stannous chloride, lithium triethylborohydride, diphenylsilane, sodium borohydride, boron trifluoride, dicyclohexylcarbodiimide, triethyl phosphite, dimethyl dichlorosilane with lithium aluminum hydride, diphenylsilane, and triphenyl phosphine with iodine. However, none of them cleanly effect this conversion.

Several decomposition products of chemical munition ingredients, like sulfur mustard metabolites, also require derivatization processes (Figure 10.5). In environmental and biological matrices, sulfur mustard is predominantly hydrolyzed to the more polar and less volatile

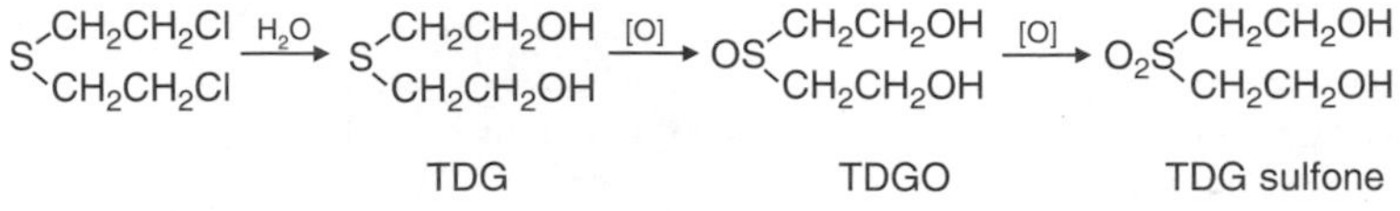

FIGURE 10.5 Hydrolytic and oxidative degradation of sulfur mustard.[86]

thiodiglycerol (TDG). Thiodiglycerol may be oxidized in soil to the sulfoxide (TDGO), further oxidation to the sulfone is less commonly observed in the environment. Thiodiglycolic acid has been found in certain soil types, possibly due to microbial assistance.

Although underivatized TDG can be analyzed by GC,[85] the peak shapes are not ideal, and derivatization is required for analysis at concentrations $< \sim 1$ ppm. Two types of derivative have been used for TDG. The ones most commonly used are silyl ethers, either trimethylsilyl, or *tert*-butyldimethylsilyl. Pentafluorobenzoyl and heptafluorobutyryl esters have been used for biomedical sample analysis. The conditions are extensively described by Black et al.[86]

The most convenient way to analyze TDGO is with LC–MS that achieves detection limits down to 10 ng/ml.[87] Required derivatization for analysis by GC was investigated with a number of agents. Derivatization is much more complex than with TDG because the sulfoxide oxygen forms an additional nucleophilic site for reaction. Three major types of derivative are formed depending on the reagent and conditions. These result from simple derivatization with the preservation of the sulfoxide function, the reduction of the corresponding TDG derivative, and Pummerer-type rearrangement to derivatives of 1-hydroxy-TDG which undergo elimination to olefinic products. Details of them and the β-Lyase metabolites have also been described by Black et al.[86]

Another analytical procedure for the extraction of degradation products of chemical munitions from water was developed by Sng and Ng.[64] The technique is based on *in situ* derivatization by SPME. They investigated, for example, the derivatization of thiodiglycerol and ethyl-2-hydroxyethyl sulfide. The derivatization reagent N-methyl-N-(*tert*-butyl-dimethylsilyl) trifluoroacetamide with 1% *tert*-butyldimethylsilyl chloride, which forms hydrolytically stable *tert*-butyldimethylsilyl derivatives, was used. Detection limits in the ppb range were realized with GC–MS in full scan mode.

III. ANALYSIS METHODS

A. GC

Sulfur gases pose a challenge in gas chromatographic analysis because they are both highly mobile and chemically very active molecules, which can lead to losses through adsorption and peak tailing. Gas chromatography is the method of choice for their separation due to their volatile nature.

With packed columns, losses through adsorption and peak tailing are greater (due to the larger surface area) than with capillary wall coated open tubular columns. The latter is preferred because it provides the required inertness, high resolution, and sharp signals.

Packed columns are used to separate H_2S, SO_2, COS, and other sulfur gases in the ppm and ppb concentrations. Chromosil 330, a modified silica gel for separating ppb concentrations of light sulfur gases, mercaptanes, and alkyl sulfides, has been used by Steudler et al.,[46] Bandy et al.,[41] and Leck et al.[44] for air analysis.

Devai et al.[39] used 40/60 CarbopackTMB HT 100, a material specialized in separating H_2S, SO_2, COS, and methanthiol at ppm and ppb levels, for the same purpose.

For the separation of sulfur compounds on capillary columns, nonpolar GC phases are recommended. Phases such as crosslinked PDMS and (5%)-diphenyl-(95%)dimethyl siloxane materials have been used for ethyl-2-hydroxyethyl sulfide and thiodiglycerol, PASHs, thiophene, benzothiophenes, benzothiophene sulfones, VOCs, and screening analysis.

Several column materials have been evaluated regarding separation between PAHs and PASHs because severe coelution is known to occur for many PASHs on such phases. Nonpolar GC phases such as DB5 materials are normally used for the separation of aromatics.

A biphenyl phase has been shown to provide a usable separation of the important C_1-dibenzothiophenes,[88] although it does not allow the complete separation of the four-ring parent compounds. On screening several columns, Andersson et al.[83] discovered that a very polar cyanopropyl phase was able to separate several three-ring PASHs. This polar stationary phase discriminates better between PAHs and PASHs than the nonpolar phases traditionally used. The introduction of a sulfur atom into the aromatic system of a terminal ring leads to stronger retention on the cyanopropyl phase compared to other phases. It was shown that the polar phase separates the analytes not only depending on the boiling points but electronic effects are also involved. The degree of methylation seems to be less important than for the nonpolar phase. One drawback is the fairly low upper thermal limit of 250°C of the cyanopropyl phase, which hampers the detection of four-ring aromatics.

In addition, the advantages of various injection techniques have been used. Oncolumn injection was used to avoid the decomposition of compounds within the injector.[89–92] A direct aqueous injection technique for the analysis of VOCs in water has been used in analysis for many years.[93] The advantages of the direct aqueous injection technique were summarized by Gurka et al.[94]

Ligand-exchange chromatography on cation-exchange materials has been used as a separation technique for PASHs, sulfides, polysulfides, and thiols in the petrol industry. The method is only partly suitable for thiols and disulfides, and the retention of compounds is only sufficient in individual cases for the clear separation of compounds. On the other hand, some substances showed small recoveries due to very strong binding to the stationary phases. Paper and thin-layer chromatography have been used as preexperimental qualitative tests in this field.

B. LC

Strongly acidic sulfur compounds such as sulfonates are very difficult to derivatize and cannot be suitably separated by classical GC methods. However, the rapid progress of HPLC technology over the past 10 to 15 years has proved very successful for the determination of organic anions. Apart from some methods based on anion-exchange chromatography, the most popular approach is ion-pair (IP) chromatography. When paired with a suitable ion, sulfonates behave in a substantially hydrophobic fashion. Due to the ion-pairing effect, even aromatic and aliphatic sulfonic acids, which exhibit pronounced hydrophilicity, can be sufficiently retained on standard RP-HPLC columns.

Many investigations have been conducted on alternatives to nonvolatile tetraalkylammonium salts such as ion-pair agents. Unfortunately, an electrospray ionization (ESI) unit, for which the volatility of the HPLC effluent is indispensable, is rapidly encrusted by nonvolatile ion-pairing agents. Likewise, an atmospheric pressure chemical ionization (APCI) unit is similarly contaminated, thus precluding the use of LC–MS as a standard analytical tool for this purpose. In this respect, ammonium acetate is well suited in many cases. The ammonium ion is the weakest ion-pairing agent. Despite not being volatile ion-pairing agents, satisfactory separation is often also achievable by sodium perchlorate[95] or sodium sulfate.

As a powerful alternative to ammonium acetate formate volatile ion-pairing agents, tri- and di-alkylamines[96] are becoming increasingly important in IP-RPLC. Of the series of volatile trialkylamine homologues, tributylamine proved to be the most efficient, as even trisulfonated compounds are satisfactorily retained on reversed phase materials and may offer the potential for future developments.

If ammonium acetate is used as a volatile ion-pairing agent for aromatic sulfonates, only monosulfonates are sufficiently retained.[97] The retention of these and other strongly acidic and polar analytes can be fine-tuned by selecting di- or tri-alkylamines with an appropriate number of aliphatic carbons (Figure 10.6). Triethylamine can be used for monosulfonated naphthalenes,[98] whereas the strong retention of even trisulfonated naphthalenes has been achieved with tributylamine (TrBA).[96] Recently, the dihexylammonium cation (with the same number of aliphatic carbons as TrBA) has been used for this purpose.[99]

While anion-exchange chromatography and capillary electrophoresis (CE) were used in very early studies,[100,101] RPLC with the addition of modifiers, such as ammonium acetate[102–104] or

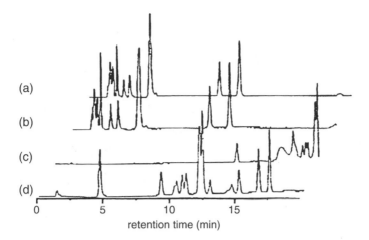

FIGURE 10.6 Retention of naphthalene sulfonates in RPLC with an eluent of 10% methanol and 2.5 mM of different ion-pairing agents. (a) triethylamine, (b) dimethylbutylamine, (c) tributylamine, and (d) tributylamine with 30% methanol. (From Holcapek, M., Jandera, P., and Zderadicka, P., *J. Chromatogr. A.*, 926, 175–186, 2001.)

acid,[105,106] has become the standard separation method for sulfonated azo dyes. Polysulfonated dyes may be too hydrophilic for conventional RPLC, and IP-RPLC may be required.[99]

Although the sulfonate group of linear alkylbenzene sulfonates (LAS) is quite acidic, the long hydrophobic alkyl chain provides sufficient retention in RPLC and so IP-RPLC is not needed. LAS mixtures are separated according to their alkyl chain length.[95,107] However, the more polar carboxylated degradation products, the sulfophenyl carboxylates (SPC) require additives such as triethylamine (TrEA)[95,108,109] or tetraethylammonium (TEA) acetate for sufficient retention (Figure 10.7). IP-RPLC with TrEA can also be used for the analysis of branched alkylbenzene sulfonates.[110] Alternatively, sulfophenyl carboxylates can be methylated to reduce their polarity prior to LC–MS analysis.[107] In all cases, the positional isomers of LAS coelute. Using TrEA[108,109] or TEA, SPC and LAS can be analyzed together, with the SPC eluting before the LAS (Figure 10.7).

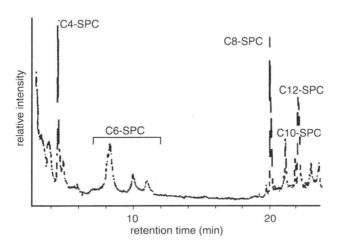

FIGURE 10.7 Separation of sulfophenyl carboxylates (SPC) by IP-RPLC and ESI-MS detection in the negative ion mode. The LAS elutes at 27.5 min (time window not shown). (From Eichhorn, P. and Knepper, T. P., *Environ. Toxicol. Chem.*, 21, 1–8, 2002.)

Aliphatic sulfonates and sulfates can be separated by RPLC with ammonium acetate and detected by their molecular anions in selected ion recording (SIR). Secondary alkane sulfonates may be analyzed under similar conditions to the LAS.[111] For alkylphenol ethoxysulfonates, mixed-mode RP/anion-exchange chromatography with ammonium acetate as buffer has been used, which provided separation according to the number of ethoxylate units.[79,112]

Another increasingly used rapidly developing separation technique for ionic analytes is capillary (zone) electrophoresis offering different selectivity from LC. CE can be coupled to MS and a number of publications on the CE/MS analysis of dyes[101] and aromatic sulfonates[108] have been published.

CE/MS was used by Loos et al.[13] for the analysis of polar hydrophilic aromatic sulfonates in wastewater treatment plants. One disadvantage of the method developed was that two and threefold negatively charged sulfonates could not be detected as these very polar negatively charged compounds migrate in the opposite direction from the electroosmotic flow (EOF). By comparison, LC/MS offered higher separation efficiency and sensitivity for LC/MS than the method developed.

Supercritical fluid chromatography (SFC) is not restricted by the volatility or thermal lability of the analyte, which is an advantage compared with GC. Much of the literature on SFC concerns the analysis of relatively nonpolar materials. However, applications of polar analytes are becoming increasingly prevalent with modified fluids. SFC of organic sulfur compounds has been limited to sulfonamides[114–118] and PASH.[119]

IV. DETECTION METHODS

Environmental samples usually contain a variety of organic compounds from which the sulfur compounds cannot be completely separated. A number of sensitive, universal, and selective detectors are available for this purpose.

The different detectors have been evaluated by various authors[37,120–122] in terms of sensitivity, selectivity, reproducibility, quenching effect, stability, and compound dependence of the sulfur response (Table 10.3).

Selective detectors are especially useful for the analysis of various contaminants in increasingly complex matrices because they largely avoid interference. They can shorten the analysis time by eliminating laborious, time-consuming sample preparation procedures that can also often cause the contamination or loss of analytes. For these reasons selective detectors have been extensively used in the determination of environmental sulfur compounds.

TABLE 10.3
Comparison of Commercially Available Detectors[37,120–122]

Detector	Limit of Detection (gS/s)	Linear Dynamic Range[a] (decades)	Selectivity[b]
FPD	10^{-11}	3	$10^3 - 10^6$
ECD	Variable up to 10^{-15}	4	Variable
TCD	10^{-11}	3	—
HECD	10^{-11}	3–5	$10^4 - 10^6$
PID	10^{-12}	6	Poor
SCD	10^{-13}	3–4	$10^6 - 10^7$
PFPD	10^{-12}	—	10^{5-6}
MSD	10^{-11}	5	Specific
AED	10^{-12}	3–4	10^4
FTIR	10^{-10}	4	Specific

[a] Linear section in the calibration graph on the log–log scale.
[b] Ratio of the response of sulfur relative to carbon.

Besides the analysis of individual substances, organic sulfur compounds are also analyzed as sum parameters. Examples include dissolved organic sulfur (DOS), adsorbable organic sulfur compounds (AOS), and ion-pair extractable sulfur compounds (IOS).

A. SELECTIVE DETECTION

In the early 1990s Amirav et al.[123] introduced a new strategy for the operation of FPD based on a pulsed flame instead of a continuous flame for the generation of flame chemiluminescence. This pulsed flame photometric detector (PFPD) is characterized by the additional dimension of a light emission time and the ability to separate in time the emission of sulfur species from those of carbon and phosphorus, resulting in considerable enhancement of detection selectivity. In addition, detection sensitivity is markedly improved, thanks to:

(a) Reduced flame background noise which is filtered in time
(b) Increased signal due to the higher brightness of the pulsed flame, stemming from a small combustion cell volume and low combustible gas flow rate
(c) The use of broad band color glass filters instead of interference filters.

The emission spectra detected by the PFPD are the product of the transmission of the emitted photons through the light pipe, filter, and the spectral sensitivity of the photomultiplier tube. Specific emission bands can be observed for species produced in the PFPD. With a conventional FPD, narrow band pass or interference filters are used to minimize interference from carbon in the continuous flame. In contrast, PFPD uses broad band pass filters and time-based selectivity to minimize interference and to increase optical throughput. The background emission associated with the H_2-rich flame in the combustor consists of CH^*, C_2^*, and OH^* species that exist in the flame. While this emission lasts less than 1 msec, the observed emission time of up to 4 msec is due to the dynamics of the flame propagating through the combustor. However, the sulfur emission reaches its maximum 5 to 6 msec after the background emission has nearly ended. Thus, sulfur has a chemiluminescence lifetime substantially longer than the background analyte-free emission. Detectivity and selectivity can be improved by using gated integration to reject the unwanted CH^*, C_2^*, and OH^* chemiluminescence response.[124]

Since the primary flow of gas is from the combustor towards the igniter at the top of the detector, there is a measurable time delay between the extinction of one flame pulse at the bottom of the combustor and the initiation of the next. To minimize this delay, a second flow of air and hydrogen is continuously directed around the outside of the combustor. The primary and secondary gas flows are known as combustor gas and wall gas, respectively (see Figure 10.8).

Pulsed flame propagation consists of a four-phase cycle: replenishment of combustible gases, ignition of gases, downward propagation of the flame, and extinction of the flame (see Figure 10.9). The cycle begins with the combustor gases (GC effluent, H_2, and air) flushing out the combustor. Simultaneously, the wall gases (hydrogen and air) sweep spent gases from the igniter region through a vent and fill this space with a combustible gas mixture. The flame is initiated when this gas mixture reaches the glowing igniter coil. The flame then propagates from the igniter region through a convoluted pathway (to prevent light from the igniter reaching the photomultiplier tube) and down into the combustor. If the gas composition within the combustor is set correctly, the flame continues to propagate towards the bottom of the combustor, where it terminates once all the combustor gas has been consumed. The method has been used in gaseous industrial effluents.[63]

An attractive alternative for selective detection is sulfur chemiluminescence (SCD). In 1989 Benner and Stedman[125] described a new ozone-based SCD. Owing to its high sensitivity, the great advantage of linear response vs. concentration of sulfur species, its wide linear dynamic range, high

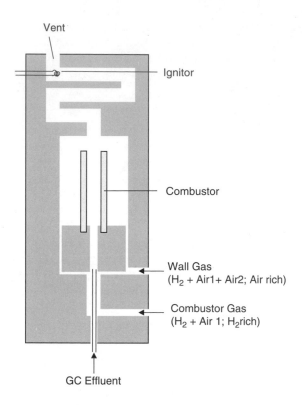

FIGURE 10.8 Combustor and wall gas pathways in PFPD. (From *Operator's Manual Model 5380 Pulsed Flame Photometric Detector*, OI Analytical, Texas, 1997.)

sensitivity, and absence of quenching effects, SCD has gained widespread acceptance for trace-level sulfur determination in various fields.

The following equations illustrate the principle of SCD[126]:

$$\text{Sulfur compound} + H_2 + O_2 \rightarrow SO + \text{other products}$$

$$SO + O_3 \rightarrow SO_2^* + O_2$$

$$SO_2^* \rightarrow SO_2 + h\nu$$

In the original design of SCD, the FID is used as an interface and the sulfur compounds are first combusted to produce SO. The SO and the other combustion products are collected with a ceramic sampling probe, and then transferred to a reduced-pressure reaction cell to react with ozone. The SO_2^* species relax by emission in the wavelength range of 280 to 420 nm.

Shearer et al.[127,128] published a new version of the combustion module for SCD and called it "flameless SCD." In flameless SCD, the SO is produced inside the combustion module, where the temperature is maintained at 780°C. The efficiency of the production and collection of SO is significantly increased. Unlike flame SCD, the ceramic probe in the combustion module of flameless SCD is less likely to be contaminated by ambient air. Flameless SCD consumes less fuel, thereby decreasing the pump loading to vent water vapor and extending the life of the pump. Furthermore, aligning the sampling probe is unnecessary. Flameless SCD is considered to be one of the most sensitive detectors for sulfur compounds.

SCD has been performed in gasoline streams[91] and wastewater.[92]

In the last few years, atomic emission detection (AED) has been found to feature a good combination of specificity and sensitivity for the analysis of organic sulfur compounds. AED can be used to confirm the elemental composition of a compound by its ability to monitor several atomic

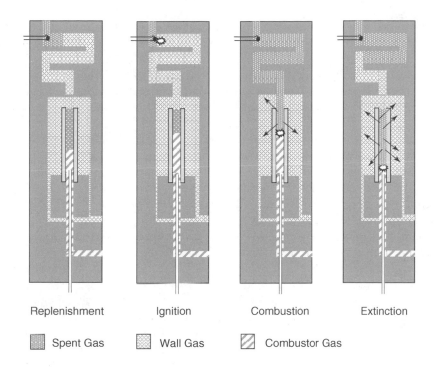

| | Replenishment | Ignition | Combustion | Extinction |

Spent Gas Wall Gas Combustor Gas

FIGURE 10.9 Diagram of four phase-cycle of propagating flame in PFPD. (From *Operator's Manual Model 5380 Pulsed Flame Photometric Detector*, OI Analytical, Texas, 1997.)

lines simultaneously, including (but not limited to) C, H, S, N, O, P, halogens, and many metals. The response of AED to sulfur at 180.7 nm is reported to have a linear range of 20,000, a sensitivity of 1.7 pgS/sec, and a selectivity over carbon of 15,000.[129] The use of AED for the quantitative analysis of certain sulfur species in various matrices has recently been described.[28,35,36,51,89,130–133]

The measurement of atmospheric sulfur compounds can be complicated by significant carbon interference on the sulfur channel (181 nm). This can be attributed to the quantity of coeluting carbon-containing material and the high intensity of the nearby 193.03 nm carbon emission wavelength. The AED software can be used to apply filtering algorithms to compensate for this unwanted response. For the best results, the unwanted response has to be suppressed and the optimum background compensation factor for this type of analysis should be frequently determined. The emission wavelength response obtained for a compound can be assessed by using the software's snapshot function. The sulfur emission wavelengths of 180.676, 181.978, and 182.568 nm, when all are present in the characteristic proportions, positively identify a sulfur-containing compound.

Combining chromatography with mass spectrometric detection (MSD) provides an extremely powerful tool for the analysis of unknown samples. MSD has a degree of selectivity which is unsurpassed by any of the other selective detection tools. Whereas selective detectors such as those described above only reveal the presence of a certain heteroatom, the mass spectrometer gives detailed information on various structural groups present in a molecule and enables unknown components to be identified. This makes it eminently suitable for screening analysis[119,132] and has enabled the identification of for instance aryl esters of alkylsulfonic acids[90] and pharmaceuticals[134] in water. Although the equipment is often too expensive for routine tests and process control, MSD is becoming increasingly popular for the analysis of environmental sulfur compounds.

Triska et al.[135] used the ion $CH_2=S^+H$ with m/z 47 to determine sulfur compounds (R–SH, R–S–R, R–SS–R with R = alkyl or aryl) in underground reservoirs of natural gas and town gas.

A GC/MS method for DMS and SO_2 determination in air in real time at the sub-ppt level involving a high pressure selected-ion chemical ionization flow reactor was developed by Ridgeway et al.[136] The use of isotope dilution GC–MS for DMS determination in seawater provided relatively good precision, better than 2%. Perdeuterated DMS (^2H_6DMS) was chosen as an internal standard to improve precision and to differentiate between aqueous and sampling-generated DMS. The relative signal obtained for the isotopomers was not affected by analyte losses occurring during sample collection and storage, or by fluctuation in detector sensitivity. In this case, instrumental drift as well as any losses in sampling ambient air were compensated for by using the ratio of the MS response at m/z 62 and m/z 68. Kelly and Kenny[137] demonstrated the highly sensitive and specific continuous measurement of DMS in air using triple quadrupole mass spectrometry with APCI.

Becker et al.[89] determined PASHs in airborne particulate by GC-AED and GC-MSD. For mass spectrometric ionization, electron impact (EI) and negative ion chemical ionization (NICI) were used, the latter proving to be the most sensitive (Figure 10.10).

Combining MS and liquid chromatography has enabled applications to be extended to nonvolatile and higher molecular compounds such as sulfonic acids and derivates. For example, LC–MS using atmospheric pressure ionization (LC–API–MS) has drastically changed the analytical methods used to detect polar pollutants in water.

In order to ensure the column effluents are compatible with a mass selective system, a micromembrane suppressor for trapping nonvolatile ion-pairing agents used in LC–MS with a moving-belt interface is described by Escott et al.[138] Both cationic and anionic membrane suppressors for LC–MS with an ion–spray interface were reported by Conboy et al.,[139] whereas Forngren et al.[140] removed nonvolatile mobile phase ingredients with an ion exchanger placed between the separation column and MS interface.

The parallel development of ion-pair agents led to the use of tributylamine. Although tributylamine seems to be suitable for negative ESI-MS (electrospray ionization), it exhibits marked carryover in the positive ion mode, despite the careful cleaning of the whole HPLC-MS apparatus, resulting in an intensive $[M + H]^+$ background ion at m/z 186.[140]

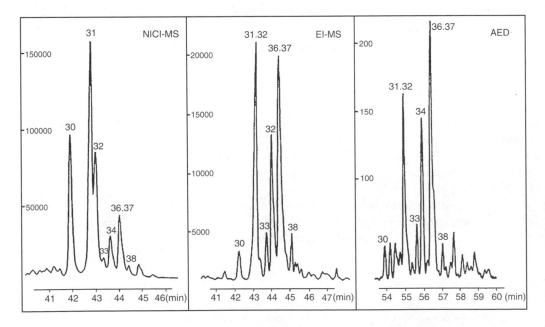

FIGURE 10.10 GC–NICI–MS and GC–EI–MS ion currents (SIM mode) for components with molecular mass of 258 and corresponding sulfur-selective GC-AED chromatogram. (From Becker, G., Nilsson, U., Colmsjö, A., and Östman, C., *J. Chromatogr. A*, 825, 57–66, 1998.)

The extraction and chromatography of polar aromatic sulfonates are usually based on ion-pair formation, traditionally with tetraalkylammonium cations,[141] because of their acidity. However, tetraalkylammonium cations are not suited for LC–MS coupling as they are virtually nonvolatile and tend to form adducts that complicate MS detection.[97]

IP-RPLC with TrBA is suitable for the LC–MS analysis of the degradation products of dyes.[142] However, the alkylamines used for ion-pairing need to be volatile, which limits the number of carbons they can bear. Contrary to the tetraalkylammonium counter-ions, di- and tri-alkylamines do not tend to form adducts in modern API interfaces. Instead, they can act as H donors in their ammonium form and may then influence the ionization process of sulfonates in ESI[143] by decreasing the sensitivity of detection.[144] In the same way, these amines diminish the risk of sodium adduct formation and of multiple charging. Due to their suppression effect, the concentration of alkylamines should be reduced to the lowest level acceptable for chromatographic retention.

The detection of strongly acidic sulfonates is performed by ESI-MS in the negative ion mode. In SIR the molecular anions, or in the case of polysulfonated compounds the dianions, are selected for detection, while multiple reaction monitoring (MRM) uses a loss of 64 amu [M-SO$_2$] or 80 amu [M-SO$_2$] from the parent anions.[96] It is still unclear which structural elements govern whether mainly SO$_2$ or SO$_3$ is eliminated. The selectivity of MS, especially MS–MS, allows ion-pair extraction to be used for the enrichment of aromatic sulfonates from aqueous samples without interference from humic material as in UV detection.[145] No reliable LC–MS method with volatile alkylamines has been developed yet, as problems occur in chromatographic and mass spectrometric separation.[146]

Sulfonated dyes were among the first compounds to illustrate the benefits of ESI-MS.[147] The analysis of dyes by LC–MS is a seasoned field (reviewed, for example, by Yinon et al.[148]), and LC–MS is frequently used to analyze dyes, especially sulfonated azodyes.[102,103,105,148,149] These anionic species can best be detected with ESI MS in the negative ion mode. A common feature of polysulfonated dyes is the formation of multiply charged molecular anions with variable numbers of sodium. Adding di- or tri-alkylammonium cations to the eluent helps suppress the formation of multiply charged alkali cations.[143,150] This improves the sensitivity, the clarity of the spectra, and the fragmentation behavior in collision-induced dissociation (CID), as the alkali cations of sulfonated dyes show only weak fragmentation. The addition of a volatile amine to the eluent in LC–MS, however, evokes an ion-pairing effect and increases retardation on the reversed-phase column.

Degradation products and byproducts of dyes with a lower molecular mass are more polar than the parent compounds. Thus, ion-pairing is mandatory to enhance the retention of these compounds. Dye metabolites formed from azo dyes[104,144] and sulfonated phthalocyanine dyes[99] have been analyzed by IP-RPLC.

Using a triple-quadrupole mass spectrometer, a parent ion scan of m/z 80 (SO$_3^-$) allows all the sulfonated dyes present in the mixture being analyzed to be detected,[147] although the intensity of this fragment may be low.[151] The cleavage of the azo bond can be induced, helping to confirm the dye structures.[105] A recent study reported that *ortho*-hydroxy azo dyes subjected to a hydrazo-azo tautomery have two fragmentation pathways for the azo bond.[151] The azo form is usually split at the C–N bond, while the homolytic cleavage of the azo bond occurs for the hydrazo form of these dyes (Figure 10.4). For those dyes also bearing a carboxylate group, decarboxylation is observed through insource fragmentation[102] as well as CID.[105]

Methods involving the use of LC–MS to analyze anionic surfactants such as household detergents and their metabolites have rapidly emerged. LC–MS provides access to the polar metabolites and biodegradation intermediates of surfactants, some of which escaped previous investigations based on GC–MS after derivatization.

Work has mainly been directed towards LAS, as these are still the most widely consumed group of anionic surfactants. Several studies on the detection of LAS in raw and treated sewage[152] have been published, together with their biodegradation intermediates, the SPC, and the byproducts, dialkyl tetraline sulfonates, in laboratory degradation experiments.[153] These compounds have also been detected in sewage treatment,[153] surface water,[95,108] and coastal water.[108] As with all sulfonates, the most suitable form of detection is ESI-MS in the negative ion mode. Using tetraethylammonium acetate for sufficient retention requires a suppressor that must be coupled between the LC and the MS to remove this nonvolatile cationic additive.

For quantification with a single MS, the molecular anions of SPC and LAS are used, and at higher cone voltages the styrene-4-sulfonate fragment (in m/z 183) can be detected to confirm the peak assignment.[108] It may thus be necessary to perform two analyses, for confirmation and quantification. Surprisingly, MRM detection has not yet been applied to this task.

The response factors for the molecular anions of the alkyl homologues can vary by a factor of six for LAS and three for SPC.[108] Thus, well described tenside mixtures and pure SPC alkyl homologues must be available for calibration prior to any quantitative analysis of LAS and SPC by LC–MS.

The potential of LC–ESI–MS to analyze alkyl ether sulfates was spotted early on[154] and it has been applied to raw and treated municipal wastewater as well as river water. The analytes were separated by RPLC with ammonium acetate and detected by their molecular anions in SIR. Secondary alkane sulfonates may be analyzed under similar conditions to LAS.[111]

For alkylphenol ethoxy sulfonates, mixed-mode RP/anion-exchange chromatography with ammonium acetate has been used. Surprisingly, ESI in the positive ion mode was used and no comparison was made with the negative ion mode.[79]

Ethane sulfonates may be formed from chloroacetanilide herbicides in soil if enzymatic activation proceeds via glutathione. As these compounds are more polar and, thus, more mobile in the soil–water system than the respective parent herbicide, ethane sulfonates are more frequently found in groundwater.[155] The use of ESI in the negative ion mode is self-evident from the acidity of the sulfonate group, but as these ethane sulfonates are less polar than the aromatic sulfonates considered above, ion-pairing is not required. Instead, conventional RPLC with an acidified eluent may be used.[156,157]

PASHs were determined by using LC combined with APCI-MS[158]

CE/MS is being rapidly developed and is catching on as it offers different selectivity from LC. Principally, the low flow rates used with CE, typically in the 100 nl/min range, are well suited for the introduction of the effluents into a mass spectrometer through an electrospray sheath flow interface. However, a sheath liquid has to be added to the CE electrolyte in order to establish the electrical contact of the capillary and to ensure sufficient flow for the electrospray.[113]

CE/MS was used by Loos et al.[113] for the analysis of polar hydrophilic aromatic sulfonates in wastewater treatment plants. Compounds were detected by negative ion electrospray ionization (NIESI) and selected by ion monitoring. In comparison with CE/UV, sensitivity was slightly better for CE/MS, but LC/MS proved more sensitive than CE/MS.

Since all sample manipulations are associated with the loss of analytes, internal standards are usually added to the sample before workup. In polycyclic aromatic carbon (PAC) work with mass spectrometric methods, deuterated aromatics have long been used for this purpose. Fluorinated PACs also offer several advantages. Andersson et al.[83] used 5-fluorobenzothiophene for benzothiophenes and 2-fluorodibenzothiophene for three-ring PAHs. These fluoroderivatives were chosen because they are well separated from other sample constituents typically occurring in complex oil samples and because they show suitable elution characteristics in the various separation steps of sample workup.

Although GC-methods are widely used, they are time-consuming. Rapid, sensitive online methods need to be developed, for which conventional GC techniques are not well suited. If species of interest with similar mass spectra do not need to be separated, membrane inlet mass spectrometry (MIMS) is perfectly adequate and fast.

The MIMS method has been used to analyze VOCs in environmental water and air samples as well as in the monitoring of chemical and biochemical processes.[159,160] MIMS is faster than GC-FID and GC-ELCD. In comparison, the MIMS method in single ion mode was the more sensitive and the linear dynamic ranges were similar to those of the GC-FID method, although not all compounds could be separated because of the similarity of their mass spectra.[159]

B. Universal Detection

Gas chromatography combined with FPD is now widely used for sulfur speciation. Although FPD has proven to be reliable and sensitive, it is also beset by a number of serious disadvantages such as the well-known quenching effect[161] and the inconvenient square dependence of the output signal on the concentrations of sulfur species.

The main snag with universal detectors such as FID or TCD is the lack of selectivity needed for sensitive trace analysis in complex matrices. If, however, the analytical system incorporates efficient sample pretreatment and the adequate separation of the compounds of interest from interfering compounds in the matrix, these universal detectors become very attractive because of their robustness, reliability, low cost, and ease of use.[159]

The use of electrolytic conductivity detectors (HECDs, also known as Hall detectors) for the analysis of organic sulfur compounds is limited, probably because they require high maintenance. The electrolyte must be kept extremely clean and its sulfur specificity is limited by high concentrations of cotrapped carbon dioxide. Despite these problems, HECD performed well in the sulfur detection mode.[159]

The electron capture detector (ECD), sensitive to electronegative elements, has been reported to be very sensitive for sulfur species.[121] One of the disadvantages of this detector, however, is the strong compound dependence of its response.

Until the 1990s, UV and fluorescence detection were the main detection methods used for a wide range of water pollutants such as sulfonates.[66,113,162]

One of the initial accounts of the use of RP-HPLC with fluorescence detection[163] centered on the determination of LAS in river water without any preconcentration. In this approach, a C18 column was used with a methanol–water eluent containing 0.1 M NaClO₄ and provided a partial separation of LAS isomers from clean aqueous samples with a concentration of 0.1% LAS. Kikuchi et al.[164] modified the method to make use of gradient elution with acetonitrile and water using UV detection. Mottaleb et al.[165] demonstrated the use of HPLC–UV–FTIR using the modified thermospray interface for the analysis of LAS.

V. APPLICATIONS IN ENVIRONMENTAL ANALYSIS

The determination of sulfur compounds in the real environment has mainly been concentrated on volatiles in the air, PASHs and thiophenes in the soil, and several compounds in water analyzed as sum parameters. Attention has also been devoted to sulfonamides, pharmaceuticals, munitions, individual sulfonates, and alkyl thiophosphates.

A. Volatile Sulfur Compounds in the Atmosphere

Analyzing atmospheric sulfur gases is important in order to understand the cycling of sulfur between the biosphere and the atmosphere. The origin and removal of ubiquitous sulfur-containing aerosols, which constitute the natural seeding agents postulated to regulate cloud formation and surface albedo, is of fundamental importance in the marine environment. The measurement and removal of reduced atmospheric gases have been conducted by many scientists.[35,49,50,52–56]

The volatile sulfur compound of most interest is DMS since it is believed to be the principal sulfur carrier in the global sulfur cycle.

DMS is derived primarily from the enzymatic hydrolysis of dimethylsulfoniopropionate (DMSP), an osmoregulatory compound produced by a wide variety of marine phytoplankton. Intercellular DMSP hydrolysis has been demonstrated in phytoplankton, microalgae, and also bacteria following the uptake of DMSP from seawater. Once released into seawater, the gas is transferred through the water–air interface as a result of its considerable concentration gradient. Indeed, relative to its DMS concentration in the air, ocean surface water is typically supersaturated by two orders of magnitude. The DMSP-transfer has been described in terms of the concentration gradient and a transfer velocity, which itself depends upon wind speed. This, along with a variety of methods based on ambient air concentration measurements, has allowed the global flux of DMS to be estimated. Reported seawater concentrations of dissolved DMS (<0.1 to 90 nM) and DMSP (1 to 1000 nM) vary with increasing depth, spatially from coastal areas to the open ocean, and also temporally from winter to summer.

On the other hand, DMS oxidation by the OH^*-radical is postulated to explain the occurrence of methane sulfonic acid (MSA), DMSO, and dimethyl sulfone ($DMSO_2$) in marine air. The MSA and SO_2 formed act as condensation nuclei during cloud formation.

Measuring the DMS oxidation products DMSO and $DMSO_2$ in rain water provides information about atmospheric concentrations of DMS, and their measurement in ice-core samples from polar regions can be helpful for assessing paleo-climatic conditions.

To fully understand the production of this sulfur particulate matter, the rates of emission and atmospheric concentrations of their precursor gases must be accurately measured.

B. POLYCYCLIC AROMATIC SULFUR HYDROCARBONS

PASHs are present in fossilized matter such as coal, mineral oil, and derived products, and are released into ambient air by combustion processes. Their chemical stability causes them to be further distributed to other environmental compartments. Many PASHs share the carcinogenic and mutagenic properties of PAHs, and are therefore a significant health issue. It is well known that subtle changes to the skeleton or alkylation pattern of aromatic compounds can drastically alter their carcinogenic potential, prompting demands for the improved identification and quantification of such compounds in the environment.

The chemical analysis of PASHs has been extensively investigated in order to characterize fossilized matter, to examine PASHs' degradation behavior, their microbial metabolism, their photoreactions in the aqueous phase after oil spills, and their occurrence in the aquatic environment, in effluents and exhausts from combustion processes, and to a lesser extent in ambient air and in the occupational environment, and the difficulty of desulfurizing them in the production of low sulfur fuels.[83,89,168]

The presence of sulfur compounds in fuels is also undesirable from the perspective of catalyst poisoning during refining, and therefore much research is being conducted into desulfurization processes. Particular attention needs to be focused on aromatics which many experiments have indicated are the hardest compounds to desulfurize.

The identification and quantification of organic sulfur compounds in these complex matrices is usually arduous and time-consuming, involving extensive fractionation and preconcentration steps to isolate a PASH fraction suitable for analysis. A variety of techniques have been employed to this end, including chemical transformations designed either to selectively concentrate these oxygen- and sulfur-containing compounds, or to effect selective chemical reduction, as described in previous chapters.

C. SUM PARAMETERS

Chromatographic methods are important tools for the classification of substances, and a huge number of organic substances have been identified and determined.

Chromatographic methods work well on most organic substances with a defined molecular structure. However, there is a lack of analytical methods that can be used on substances with a high molecular weight such as humic acids and lignin sulfonic acids, which may significantly influence water quality. High resolution chromatography, which can identify individual compounds, is often not the method of choice in this instance. The development of sum parameters has proved beneficial for water analysis. Sum parameters try to cover all the substances in a group by a suitable enrichment step combined with integrated quantification. So far two important sum parameters have proved successful: dissolved organic carbon (DOC) and adsorbable organic halogens (AOX).

Schnitzler[165] established a sum parameter for AOS compounds analogous to the AOX method. The technique is based on the adsorption of organic sulfur compounds on activated charcoal, coal combustion, and the detection of the sulfur dioxide formed.

Although no standard method has caught on, several labs use AOS to avoid time-consuming, elaborate individual substance analysis. The method has come in for some criticism due to the nonavailability of suitable charcoal, the unspecific adsorption behavior of sulfur compounds, the varying pH dependency of adsorption, and significant losses in the sulfur balance of combustion.[166]

Schullerer et al.[66] developed a method for the enrichment of IOS compounds for aromatic sulfonates from water samples. The sum parameter is based on the principle of ion-pair solid-phase extraction. Analogous to the ion-pair chromatography, a reversed-phase C_{18} material is used as adsorbent while tetrabutyl ammonium bromide is used as ion-pairing reagent, as already mentioned in previous chapters.

Another technique based on the principle of ion-pair extraction is the determination of DOS compounds.[167] In contrast to IOS, which analyzed sulfate separately by IC and subtracted it, DOS also gauges extracted sulfate.

REFERENCES

1. Andreae, M. O., Ferek, R. J., Bermond, F., Byrd, K. P., Engstrom, R. T., Hardin, S., Houmere, P. D., LeMarrec, F., and Raemdonck, H., Dimethylsulfide in the marine atmosphere, *J. Geophys. Res.*, 90, 12891–12900, 1985.
2. Andreae, M. O. and Raemdonck, H., Dimethylsulfide in the surface ocean and the marine atmosphere: a global view, *Science*, 221, 744–747, 1983.
3. Berresheim, H., Andreae, M. O., Ayers, G. P., Gillett, R. W., Merrill, J. T., Davis, V. J., and Chameides, W. L., Airborne measurements of dimethylsulfide, sulfur dioxide, and aerosol ions over the Southern Ocean south of Australia, *J. Atmos. Chem.*, 10, 341–370, 1990.
4. Berresheim, H., Andreae, M. O., Iverson, R. L., and Li, S. M., Seasonal variations of dimethylsulfide emissions and atmospheric sulfur and nitrogen species over the western North Atlantic Ocean, *Tellus*, 43b, 353–372, 1991.
5. Bürgermeister, S., and Georgii, H. W., Distribution of methanesulfonate, sulfate and dimethylsulfide over the Atlantic and the North Sea, *Atmos. Environ.*, 25A, 587–595, 1991.
6. Saltzman, E. S., and Cooper, D. J., Shipboard measurements of atmospheric dimethylsulfide and hydrogen sulfide in the Caribbean and the Gulf of Mexico, *J. Atmos. Chem.*, 7, 191–209, 1988.
7. Van Valin, C. C., Berresheim, H., Andreae, M. O., and Luria, M., Dimethylsulfide over the western Atlantic Ocean, *Geophys. Res. Lett.*, 14, 715–718, 1987.
8. Barnard, W. R., Andreae, M. O., Watkins, W. E., Bingemer, H., and Georgii, H. W., The flux of dimethylsulfide from the oceans to the atmosphere, *J. Geophys. Res.*, 87, 8787–8793, 1982.
9. Bates, T. S., Johnson, J. E., Quinn, P. K., Goldan, P. D., Kuster, W. C., Covert, D. C., and Hahn, C. J., The biogeochemical sulfur cycle in the marine boundary layer over the northeast Pacific Ocean, *J. Atmos. Chem.*, 10, 59–81, 1990.
10. Quinn, P. K., Bates, T. S., Johnson, J. E., Covert, D. S., and Charlson, R. J., Interactions between the sulfur and reduced nitrogen cycles over the central Pacific Ocean, *J. Geophys. Res.*, 95, 16405–16416, 1991.
11. Ayers, G. P., Ivey, J. P., and Gillett, R. W., Coherence between seasonal cycles of dimethylsulfide, methanesulfonate and sulfate in marine air, *Nature*, 349, 404–406, 1991.
12. Kittler, P., Swan, H., and Irvey, J., An indicating oxidant scrubber for the measurement of atmospheric dimethylsulfide, *Atmos. Environ.*, 26A, 2661–2664, 1992.
13. Cooper, D. J. and Saltzman, E. S., Measurement of atmospheric dimethylsulfide and carbon disulfide in the Western Atlantic boundary layer, *J. atmos. Chem.*, 12, 153–168, 1991.

14. Tang, H. M., Heaton, P., and Brassard, B., Analysis of trace COS in samples with high concentrations of H_2S by gas chromatography and a sulfur chemiluminescence detector, *Field Anal. Chem. Technol.*, 1, 171–174, 1997.

15. Beiner, K., Popp, P., Wennrich, R., and Salzer, R., Determination of volatile sulfur compounds in water samples by GC–MS with selective preconcentration, *Fresenius Environ. Bull.*, 10, 755–760, 2001.

16. Riis, V. and Babel, W., Removal of sulfur interfering in the analysis of organochlorines by GC-ECD, *Analyst*, 124, 1771–1773, 1999.

17. Gocrlitz, D. F. and Law, L. M., Note on removal of sulfur interferences from sediment extracts for pesticide analysis, *Bull. Environ. Contam. Toxicol.*, 6, 9–10, 1971.

18. Buchert, H., Bihler, S., and Ballschmiter, K., Untersuchungen zur globalen Grundbelastung mit Umweltchemikalien: VII. Hochauflösende Gas-Chromatographie persistenter Chlor-Kohlenwasserstoffe (CKW) und Polyaromaten (AKW) in limnischen Sedimenten unterschiedlicher Belastung, *Freseniu J. Anal. Chem.*, 313, 1–20, 1982.

19. Jensen, S., Renberg, L., and Reutergardh, L., Residue analysis of sediment and sewage sludge for organochlorines in the presence of elemental sulfur, *Anal. Chem.*, 49, 316–318, 1977.

20. Andersson, J. T. and Holwitt, U., An advantageous reagent for the removal of elemental sulfur from environmental samples, *Fresenius J. Anal. Chem.*, 350, 474–480, 1994.

21. Kropp, G. K., Goncalves, J. A., Andersson, J. T., and Fedorak, P. M., Bacterial transformations of benzothiophene and methylbenzothiophenes, *Environ. Sci. Technol.*, 28, 1348–1356, 1994.

22. Saftic, S., Fedorak, P. M., and Andersson, J. T., Diones, sulfoxides, and sulfones from the aerobic cometabolism of methylbenzothiophenes by Pseudomonas strain BT1, *Environ. Sci. Technol.*, 26, 1759–1764, 1992.

23. Saftic, S., Fedorak, P. M., and Andersson, J. T., Transformations of methyldibenzothiophenes by three Pseudomonas isolates, *Environ. Sci. Technol.*, 27, 2577–2584, 1993.

24. Johansen, S. S., Hansen, A. B., Mosbaek, H., and Arvin, E., Method development for trace analysis of heteroaromatic compounds in contaminated groundwater, *J. Chromatogr. A*, 738, 295–304, 1996.

25. Nguyen, B. C., Gaudry, A., Bonsang, B., and Lambert, G., Reevaluation of the role of dimethylsulfide in the sulfur budget, *Nature (London)*, 275, 637–639, 1978.

26. Gaudry, A., Bonsang, B., Nguyen, B. C., and Nadaud, Ph., Méthode de mesure des concentrations dans l'atmosphère et dans l'océan du dimethyl sulfure à un niveau voisin de la P.P.T./V, *Chemosphere*, 10, 731–744, 1981.

27. Hwang, Y., Matsuo, T., Hanaki, K., and Suzuki, N., Identification and quantification of sulfur and nitrogen containing odorous compounds in wastewater, *Water Res.*, 29, 711–718, 1995.

28. Gerbersmann, C., Lobinski, R., and Adams, F. C., Determination of volatile sulfur compounds in water samples, beer and coffee with purge and trap gas chromatogaphy-microwave-induced plasma atomic emission spectrometry, *Anal. Chim. Acta*, 316, 93–104, 1995.

29. Smith, G. C., Clark, T., Knutsen, L., and Barrett, E., Methodology for analyzing dimethylsulfide and dimethyl sulfoniopropionate in seawater using deuterated internal standards, *Anal. Chem.*, 71, 5563–5568, 1999.

30. de Souza, T. L. C. and Bhatia, S. P., Development of calibration systems for measuring total reduced sulfur and sulfur dioxide in ambient concentrations in the parts per billion range, *Anal. Chem.*, 48, 2234–2239, 1976.

31. Braman, R. S., Ammons, J. W., and Bricker, J. L., Preconcentration and determination of hydrogen sulfide in air by flame photometric detection, *Anal. Chem.*, 50, 992–996, 1978.

32. Farwell, S. O., Liebowitz, D. P., Kagel, R. A., and Adams, D. F., Determination of total biogenic sulfur gases by filter/flash vaporization/flame photometry, *Anal. Chem.*, 52, 2370–2375, 1980.

33. Farwell, S. O., Kagel, R. A., Barinaga, C. J., Goldan, P. D., Kuster, W. C., Fehsenfeld, F. C., and Albritton, D. L., Intercomparsion of two techniques for the preparation of gaseous sulfur calibration standards in the low to sub-ppb range, *Atmos. Environ.*, 21, 1983–1987, 1987.

34. Tangerman, A., Determination of volatile sulfur compounds in air at the parts per trillion level by Tenax trapping and gas chromatography, *J. Chromatogr.*, 366, 205–216, 1986.

35. Swan, H. B. and Ivey, J. P., Analysis of atmospheric sulfur gases by capillary gas chromatography with atomic emission detection, *J. High Resolut. Chromatogr.*, 17, 814–820, 1994.

36. Junyapoon, S., Bartle, K. D., Ross, A. B., and Cooke, M., Analysis of malodorous sulfur gases and volatile organometalloid compounds in landfill gas emissions using capillary gas chromatography

with programmed temperature vaporization injection and atomic emission detection, *Int. J. Environ. Anal. Chem.*, 82, 47–59, 2002.

37. Wardencki, W., Problems with the determination of environmental sulfur compounds by gas chromatography, *J. Chromatogr. A*, 793, 1–19, 1998.

38. Przyjazny, A., Evaluation of the suitability of selected porous polymers for preconcentration of organosulfur compounds from water, *J. Chromatogr.*, 346, 61–67, 1985.

39. Devai, I. and DeLaune, R. D., Trapping efficiency of various solid adsorbents for sampling and quantitative gas chromatographic analysis of carbonyl sulfide, *Anal. Lett.*, 30, 187–198, 1997.

40. Devai, I. and Delaune, R. D., Evaluation of various solid adsorbents for sampling trace levels of methanethiol, *Org. Geochim.*, 24, 941–944, 1996.

41. Bandy, A. R., Tucher, B. J., and Moroulis, P. J., Determination of part-per-trillion by volume levels of atmospheric carbon disulfide by gas chromatography/mass spectrometry, *Anal. Chem.*, 57, 1310–1314, 1985.

42. Kim, K. H. and Andreae, M. O., Carbon disulfide in seawater and the marine atmosphere over the North Atlantic, *J. Geophys. Res.*, 92, 14733–14738, 1987.

43. Sandalls, F. J. and Penkett, S. A., Measurements of carbonyl sulfide and carbon: disulfide in the atmosphere, *Atmos. Environ.*, 11, 197–199, 1977.

44. Leck, C. and Bagander, L. E., Determination of reduced sulfur compounds in aqueous solutions using gas chromatography flame photometric detection, *Anal. Chem.*, 60, 1680–1683, 1988.

45. Claus, D., Geypens, B., Ghoos, Y., Rutgeerts, P., Ghyselen, J., Hoshi, K., and Delanghe, G., Oral malodor, assessed by closed-loop, gas chromatography, and ion-trap technology, *J. High Resolut. Chromatogr.*, 20, 94–98, 1997.

46. Steudler, P. A. and Kijowski, W., Determination of reduced sulfur gases in air by solid adsorbent preconcentration and gas chromatography, *Anal. Chem.*, 56, 1432–1436, 1984.

47. Yasuhara, A. and Keeichiro, F., Hydrogen sulfide and dimethylsulfide in liquid swine manure, *Chemosphere*, 7, 833–838, 1978.

48. Yokouchi, Y., Bandow, H., and Akimoto, H., Development of automated gas chromatographic-mass spectrometric analysis for natural volatile organic compounds in the atmosphere, *J. Chromatogr.*, 642, 401–407, 1993.

49. Davision, B. M. and Allen, A. G., A method for sampling dimethylsulfide in polluted and remote marine atmospheres, *Atmos. Environ.*, 28, 1721–1729, 1994.

50. Davison, B., O'Dowd, C., Hewitt, C. N., Smith, M. H., Harrison, R. M., Peel, D. A., Wolf, E., Mulvaney, R., Schwikowsky, M., and Baltensperger, U., Dimethylsulfide and its oxidation products in the atmosphere of the atlantic and southern oceans, *Atmos. Environ.*, 30, 1895–1906, 1996.

51. Kagel, R. A. and Farwell, S. O., Evaluation of metallic foils for preconcentration of sulfur-containing gases with subsequent flash desorption /flame photometric detection, *Anal. Chem.*, 58, 1197–1202, 1986.

52. Ferek, R. J., Chatfield, R. B., and Andrae, M. O., Vertical distribution of dimethylsulfide in the marine atmosphere, *Nature*, 320, 514–516, 1986.

53. Natusch, D. F. S., Klonis, H. B., Axelrodt, H. D., Teck, R. J., and Lodge, J. P., Sensitive method for measurement of atmospheric hydrogen sulfide, *Anal. Chem.*, 44, 2067–2070, 1972.

54. Slatt, B. J., Natusch, D. F. S., Prospero, J. M., and Savoie, D. L., Hydrogen sulfide in the atmosphere of the northern equatorial Atlantic Ocean and its relation to the global sulfur cycle, *Atmos. Environ.*, 12, 981–991, 1978.

55. Jaeschke, W., New methods for the analysis of SO_2 and H_2S in remote areas and their application to the atmosphere, *Atmos. Environ.*, 12, 715–721, 1978.

56. Farwell, S. O., MacTaggart, L. D., and Chatham, W. H., Airborne measurements of total sulfur gases during NASA global tropospheric experiment/chemical instrumentation test and evaluation 3, *J. Geophys. Res.*, 100, 7223–7234, 1995.

57. Johansen, S. S. and Pawliszyn, J., Trace analysis of hetero aromatic compounds (NSO) in water and polluted groundwater by solid phase micro-extraction (SPME), *J. High Resolut. Chromatogr.*, 19, 627–632, 1996.

58. Popp, P., McCann, I., and Moeder, M., Determination of sulfur containing compounds in waste water, In *Applications of SPME*, Pawliszyn, J., Ed., Royal Society of Chemistry, Cambridge, 1999.

59. Abalos, M., Prieto, X., and Bayona, J. M., Determination of volatile alkyl sulfides in wastewater by headspace solid-phase microextraction followed by gas chromatography-mass spectrometry, *J. Chromatogr. A*, 963, 249–257, 2002.

60. Haberhauer-Troyer, C., Rosenberg, E., and Grasserbauer, M., Evaluation of solid-phase microextraction of sampling of volatile organic sulfur compounds in air for subsequent gas chromatographic analysis with atomic emission detection, *J. Chromatogr. A*, 848, 305–315, 1999.

61. Wardencki, W. and Namiesnik, J., Studies on the application of solid-phase microextraction for analysis of volatile organic sulfur compounds in gaseous and liquid samples, *Chem. Anal. (Warsaw)*, 44, 485–493, 1999.

62. Kim, H., Nochetto, C., and McConnell, L. L., Gas-phase analysis of trimethylamine, propionic and butyric acids, and sulfur compounds using solid-phase microextraction, *Anal. Chem.*, 74, 1054–1060, 2002.

63. Lestremau, R., Desauziers, V., Roux, J. C., and Fanlo, J. L., Development of a quantification method for the analysis of malodorous sulfur compounds in gaseous industrial effluents by solid-phase microextraction and gas chromatography-pulsed flame photometric detection, *J. Chromatogr. A*, 999, 71–80, 2003.

64. Sng, M. T. and Ng, W. F., *In situ* dramatization of degradation products of chemical warfare agents in water by solid-phase microextraction and gas chromatographc-mass spectrometric analysis, *J. Chromatogr. A*, 832, 173–182, 1999.

65. Nielsen, A. T. and Jonsson, S., Quantification of volatile sulfur compounds in complex gaseous matrices by solid-phase microextraction, *J. Chromatogr. A*, 963, 57–64, 2002.

66. Schullerer, S., Brauch, H. J., and Frimmel, F. H., Bestimmung organischer Sulfonsäuren in Wasser durch Ionenpaar-Chromatographie, *Vom. Wasser*, 75, 83–97, 1990.

67. Rostad, C. E., Pereira, W. E., and Ratcliff, S. M., Bonded-phase extraction column isolation of organic compounds in groundwater at a hazardous waste site, *Anal. Chem.*, 56, 2856–2860, 1984.

68. Binde, F. and Rüttiger, H. H., Isolation and determination of dissolved organic sulfur compounds-development of the organic group parameter DOS, *Fresenius J. Anal. Chem.*, 357, 411–415, 1997.

69. Tolosa, I., Readman, J. W., and Mee, L. D., Comparison of the performance of solid-phase extraction techniques in recovering organophosphorus and organochlorine compounds from water, *J. Chromatogr. A*, 725, 93–106, 1996.

70. Aguilar, C., Borull, F., and Marcé, R. M., Determination of pesticides in environmental waters by solid-phase extraction and gas chromatography with electron-capture and mass spectrometry detection, *J. Chromatogr. A*, 771, 221–231, 1997.

71. Zerbinati, O., Ostacoli, G., Gastaldi, D., and Zelano, V., Determination and identification by high-performance liquid chromatography and spectrofluorimetry of twenty-three aromatic sulfonates in natural waters, *J. Chromatogr.*, 640, 231–240, 1993.

72. Beiner, K., Popp, P., and Wennrich, R., Selective enrichment of sulfides, thiols and methylthiophosphates from water samples on metal-loaded cation-exchange materials for gas chromatographic analysis, *J. Chromatogr. A*, 968, 171–176, 2002.

73. Kaimai, T. and Matsunaga, A., Determination of sulfur compounds in high-boiling petroleum destillates by ligand-exchange thin-layer chromatography, *Anal. Chem.*, 50, 268–270, 1978.

74. Andersson, J. T., Separations on a mercuric-acetate-substituted phenylsilicia phase in normal-phase liquid chromatography, *Fresenius Z. Anal. Chem.*, 326, 425–433, 1987.

75. Andersson, J. T., Retention properties of a palladium chloride/silica sorbent for the liquid chromatographic separation of polycyclic aromatic sulfur heterocycles, *Anal. Chem.*, 59, 2207–2209, 1987.

76. Di Corcia, A. and Marchetti, M., Multiresidue method for pesticides in drinking water using a graphitized carbon black cartridge extraction and liquid chromatographic analysis, *Anal. Chem.*, 63, 580–585, 1991.

77. Di Corcia, A., Samperi, R., Marcomini, A., and Steiluto, S., Graphitized carbon black extraction cartridges for monitoring polar pesticides in water, *Anal. Chem.*, 65, 907–912, 1993.

78. Altenbach, B. and Giger, W., Determination of benzene- and naphthalenesulfonates in wastewater by solid-phase extraction with graphitized carbon black and ion-pair liquid chromatography with UV detection, *Anal. Chem.*, 67, 2325–2333, 1995.

79. Benomar, S. H., Clench, M. R., and Allen, D. W., The analysis of alkylphenol ethoxysulfonate surfactants by high-performance liquid chromatography, liquid chromatography-electrospray ionisation-mass spectrometry and matrix-assisted laser desorption ionisation-mass spectrometry, *Anal. Chim. Acta*, 445, 255–267, 2001.

80. Dercksen, A., Laurens, J., Torline, P., Axcell, B. C., and Rohwer, E., Quantitative analysis of volatile sulfur compounds in beer using a membrane extraction interface, *J. Am. Soc. Brew. Chem.*, 54, 228–233, 1996.

81. Green, J. B., Yu, S. K. T., and Vrana, R. P., GC–MS Analysis of phenolic compounds in fuels after conversion to trifluoroacetate esters, *J. High Resolut. Chromatogr.*, 17, 439–451, 1994.

82. Spratt, M. P. and Dorn, H. C., *P*-fluorobenzoyl chloride for characterization of active hydrogen functional groups by fluorine-19 nuclear magnetic resonance spectrometry, *Anal. Chem.*, 56, 2038–2043, 1984.

83. Andersson, J. T., Polycyclic aromatic sulfur heterocycles IV. Determination of polycyclic aromatic compounds in a shale oil with the atomic emission detector, *J. Chromatogr. A.*, 693, 325–338, 1995; Thomson, J. S., Green, J. B., McWilliams, T. B., and Sturm, G. P., *Analysis of Sulfur Compounds in Light Distillates*, Symposium on Petroleum Chemistry and Processing, Chicago IL, 1995, pp. 696–698.

84. Andersson, J. T., Polycyclic aromatic sulfur heterocycles. I. Use of hydrogen peroxide oxidation for the group separation of polycyclic aromatic hydrocarbons and their sulfur analogs, *Int. J. Environ. Anal. Chem.*, 48, 1–15, 1991.

85. Beck, N. V., Carrick, W. A., Cooper, D. B., and Muir, B., Extraction of thiodiglycol from soil using pressurized liquid extraction, *J. Chromatogr. A*, 907, 221–227, 2001.

86. Black, R. M. and Muir, B., Derivatisation reactions in the chromatographic analysis of chemical warfare agents and their degradation products, *J. Chromatogr. A*, 1000, 253–281, 2003.

87. Black, R. M. and Read, R. W., Application of liquid chromatography-atmospheric pressure chemical ionisation mass spectrometry, and tandem mass spectrometry, to the analysis and identification of degradation products of chemical warfare agents, *J. Chromatogr. A*, 759, 79–92, 1997.

88. Nishioka, M., Bradshaw, J. S., Lee, M. L., Tominaga, Y., Tedjamulia, M., and Castle, R. N., Capillary column gas chromatography of sulfur heterocycles in heavy oils and tars using a biphenyl polysiloxane stationary phase, *Anal. Chem.*, 57, 309–312, 1985.

89. Becker, G., Nilsson, U., Colmsjö, A., and Östman, C., Determination of polycyclic aromatic sulfur heterocyclic compounds in airborne particulate by gas chromatography with atomic emission and mass spectrometric detection, *J. Chromatogr. A*, 825, 57–66, 1998.

90. Franke, S., Schwarzbauer, J., and Francke, W., Arylesters of alkylsulfonic acids in sediments, Part III of organic compounds as contaminants of the river Elbe and its tributaries, *Fresenius J. Anal. Chem.*, 360, 580–588, 1998.

91. Di Sanzo, F. P., Bray, W., and Chawla, B., Determination of the sulfur components of gasoline streams by capillary column gas chromatography with sulfur chemiluminescence detection, *J. High Resolut. Chromatogr.*, 17, 255–258, 1994.

92. Tang, H. and Heaton, P., Analysis of volatile trace sulfur compounds in water with GC-SCD by direct aqueous injection, *Int. J. Environ. Anal. Chem.*, 62, 263–271, 1996.

93. Ellington, J. J. and Trusty, C. D., Quantitative analysis of alkyl phosphates using automated cool on-column aqueous injection, *J. High Resolut. Chromatogr.*, 12, 470–473, 1989.

94. Gurka, D. F., Pyle, S. M., and Titus, R., Environmental analysis by direct aqueous injection, *Anal. Chem.*, 64, 1749–1754, 1992.

95. González-Mazo, E., Honing, M., Barcelo, D., and Gomez-Parra, A., Monitoring long-chain intermediate products from the degradation of linear alkylbenzene sulfonates in the marine environment by solid-phase extraction followed by liquid chromatography/ionspray mass spectrometry, *Environ. Sci. Technol.*, 31, 504–510, 1997.

96. Storm, T., Reemtsma, T., and Jekel, M., Use of volatile amines as ion-pairing agents for the high-performance liquid chromatographic-tandem mass spectrometric determination of aromatic sulfonates in industrial wastewater, *J. Chromatogr. A*, 854, 175–185, 1999.

97. Suter, M. J.-F., Riediker, S., and Giger, W., Selective determination of aromatic sulfonates in landfill leachates and groundwater using microbore liquid chromatography coupled with mass spectrometry, *Anal. Chem.*, 71, 897–904, 1999.

98. Alonso, M. C., Castillo, M., and Barcelo, D., Solid-phase extraction procedure of polar benzene- and naphthalene-sulfonates in industrial effluents followed by unequivocal determination with ion-pair chromatography/electrospray-mass spectrometry, *Anal. Chem.*, 71, 2586–2596, 1999.

99. Holcapek, M., Jandera, P., and Zderadicka, P., High performance liquid chromatography-mass spectrometric analysis of sulfonated dyes and intermediates, *J. Chromatogr. A.*, 926, 175–186, 2001; Reemtsma, T., The use of liquid chromatography-atmospheric pressure ionization-mass spectrometry in water analysis — II Obstacles, *Trends Anal. Chem.*, 20, 533–542, 2001.

100. Kim, I. S., Sasinos, F. I., Stephens, R. D., and Brown, M. A., Anion-exchange chromatography particle beam mass spectrometry for the characterization of aromatic sulfonic acids as the major organic pollutants in leachates from Stringfellow, California, *Environ. Sci. Technol.*, 24, 1832–1836, 1990.

101. Lee, E. D., Mueck, W., Henion, J. D., and Covey, T. R., Capillary zone electrophoresis/tandem mass spectrometry for the determination of sulfonated azo dyes, *Biomed. Environ. Mass Spectrom.*, 18, 253–257, 1989.

102. Rafols, C. and Barcelo, D., Determination of mono- and di-sulfonated azo dyes by liquid chromatography-atmospheric pressure ionization mass spectrometry, *J. Chromatogr. A*, 777, 177–192, 1997.

103. Holcapek, M., Jandera, P., and Prikryl, J., Analysis of sulfonated dyes and intermediates by electrospray mass spectrometry, *Dyes Pigm.*, 43, 127–137, 1999.

104. McCallum, J. E. B., Madison, S. A., Alkan, S., Depinto, R. L., and Rojas Wahl, R. U., Analytical Studies on the oxidative degradation of the reactive textile dye uniblue, *Environ. Sci. Technol.*, 34, 5157–5164, 2000.

105. Edlund, P. O., Lee, E. D., Henion, J. D., and Budde, W. L., The determination of azo dyes in municipal waste water by ion spray LC/MS/MS, *Biomed. Environ. Mass Spectrom.*, 18, 233–240, 1989.

106. Smyth, W. F., McClean, S., O'Kane, E., Banat, I., and McMullan, G., Application of electrospray mass spectrometry in the detection and determination of Remazol textile dyes, *J. Chromatogr. A*, 854, 259–274, 1999.

107. DiCorcia, A., Casassa, F., Crescenzi, C., Marcomini, A., and Samperi, R., Investigation of the fate of linear alkyl benzenesulfonates and coproducts in a laboratory biodegradation test by using liquid chromatography/mass spectrometry, *Environ. Sci. Technol.*, 33, 4112–4118, 1999.

108. Riu, J., Gonzalez-Mazo, E., Gomez-Parra, A., and Barcelo, D., Determination of parts per trillion level of carboxylic degradation products of linear alkylbenzenesulfonates in coastal water by solid-phase extraction followed by liquid chromatography/ionspray/mass spectrometry using negative ion detection, *Chromatographia*, 50, 275–281, 1999.

109. Eichhorn, P. and Knepper, T. P., α,β-unsaturated sulfophenylcarboxylates as degradation intermediates of linear alkylbenzenesulfonates: evidence for Ω-oxygenation followed by β-oxidations by liquid chromatography-mass spectrometry, *Environ. Toxicol. Chem.*, 21, 1–8, 2002.

110. Eichhorn, P., Flavier, M. L., Paje, M. L., and Knepper, T. P., Occurrence and fate of linear and branched alkylbenzenesulfonates and their metabolites in surface waters in the Philippines, *Sci. Total Environ.*, 269, 75–85, 2001.

111. Castillo, M., Riu, J., Ventura, P., Boleda, R., Scheding, R., Schröder, H. F., Nistor, C., Emneus, J., Eichhorn, P., Knepper, T. P., Jonkers, C. C. A., de Voogt, P., Gonzalez-Majo, E., Leon, V. M., and Barcelo, D., Inter-laboratory comparison of liquid chromatographic techniques and enzyme-linked immunosorbent assay for the determination of surfactants in wastewaters, *J. Chromatogr. A*, 889, 195–209, 2000.

112. Reemtsma, T., Liquid chromatography-mass spectrometry and strategies for trace-level analysis of polar organic pollutants, *J. Chromatogr. A*, 1000, 477–501, 2003.

113. Loos, R., Riu, J., Alonso, M. C., and Barcelo, D., Analysis of polar hydrophilic aromatic sulfonates in waste water treatment plants by CE/MS and LC/MS, *J. Mass Spectrom.*, 35, 1197–1206, 2000.

114. Berry, A. J., Games, D. E., and Perkins, J. R., Supercritical fluid chromatographic and supercritical fluid chromatographic-mass spectrometric studies of some polar compounds, *J. Chromatogr.*, 363, 145–158, 1986.

115. Ramsey, E. D., Perkins, J. R., Games, D. E., and Startin, J. R., Analysis of drug residues in tissue by combined supercritical fluid extraction–supercritical fluid chromatography-mass spectrometry, *J. Chromatogr.*, 464, 353–364, 1989.

116. Perkins, J. R., Games, D. E., Startin, J. R., and Gilbert, J., Analysis of veterinary drugs using supercritical fluid chromatography-mass spectrometry, *J. Chromatogr.*, 540, 239–256, 1991.

117. Berger, T. A., Density of methanol–carbon dioxide mixtures at three temperatures: comparison with vapor–liquid equilibria measurements and results obtained from chromatography, *J. High Resolut. Chromatogr.*, 14, 312–316, 1991.

118. Combs, M. T., Ashraf-Khorassani, M., and Taylor, L. T., Method development for the separation of sulfonamides by supercritical fluid chromatography, *J. Chromatogr.*, 35, 176–180, 1997.

119. Nishioka, M., Aromatic sulfur compounds other than condensed thiophenes in fossil fuels: enrichment and identification, *Energy Fuels*, 2, 214–228, 1988.

120. Wardencki, W. and Zygmunt, B., Gas chromatographic sulfur-sensitive detectors in environmental analysis, *Anal. Chim. Acta*, 255, 1–13, 1991.

121. Tuan, H. P., Janssen, H. G. M., Cramers, C. A., Kuiper van Loo, E. M., and Vlap, H., Evaluation of the performance of various universal and selective detectors for sulfur determination in natural gas, *J. High Resolut. Chromatogr.*, 18, 333–342, 1995.

122. Eckert-Tilotta, S. E., Hawthorne, St. B., and Miller, D. J., Comparison of commercially available atomic emission and chemiluminescence detectors for sulfur-selective gas chromatographic detection, *J. Chromatogr.*, 591, 313–323, 1995.

123. Amirav, A. and Jing, H., Pulsed flame photometer detector for gas chromatography, *Anal. Chem.*, 67, 3305–3318, 1995.

124. *Operator's Manual Model 5380 Pulsed Flame Photometric Detector*, OI Analytical, Texas, 1997.

125. Benner, R. L. and Stedman, D. H., Universal sulfur detection by chemiluminescence, *Anal. Chem.*, 61, 1268–1271, 1989.

126. Chen, Y. C. and Lo, J. G., Gas chromatography with flame ionization and flameless sulfur chemiluminescence detectors in series for dual channel detection of sulfur compounds, *Chromatographia*, 43, 522–526, 1996.

127. Shearer, R. L., Development of flameless sulfur chemiluminescence detection: application to gas chromatography, *Anal. Chem.*, 64, 2192–2196, 1992.

128. Shearer, R. L., Poole, E. B., and Nowalk, J. B., Application of gas chromatography and flameless sulfur chemiluminescense detection to the analysis of petroleum products, *J. Chromatogr. Sci.*, 31, 82–87, 1993.

129. Quimby, B. D. and Sullivan, J. J., Evaluation of a microwave cavity, discharge tube, and gas flow system for combined gas chromatogrphy-atomic emission detection, *Anal. Chem.*, 62, 1027–1034, 1990.

130. Bjergaard-Pedersen, St, Asp, T. N., and Greibrokk, T., Determination of sulfur- and chlorine-containing compounds using capillary gas chromatography and atomic emission detection, *Anal. Chim. Acta*, 265, 87–92, 1992.

131. Pedersen-Bjergaard, St, Asp, T. N., Vedde, J., Carlberg, G. E., and Greibrokk, T., Identification of chlorinated sulfur compounds in pump mill effluents by GC–MS and GC-AED, *Chromatographia*, 35, 193–198, 1993.

132. Pedersen-Bjergaard, St, Semb, S. I., Vedde, J., Brevik, E. M., and Greibrokk, T., Environmental screening by capillary gas chromatography combined with mass spectrometry and atomic emission spectroscopy, *Chemosphere*, 32, 1103–1115, 1996.

133. Tanzer, D., and Heumann, K. G., Gas chromatographic determination of volatile organic sulfides and selenides and of methyl iodide at trace levels in atlantic surface water, *Int. J. Environ. Anal. Chem.*, 48, 17–31, 1992.

134. Reupert, R. and Brausen, G., Multiverfahren zur Bestimmung relevanter Arzneimittelwirkstoffe in Oberflächenwasser mittels HPLC-MS/MS, *GIT*, 9, 900–904, 2003.

135. Třiska, J., Kuraš, M., Zachař, P., and Vodička, L., Analysis of sulfur compounds in underground reservoirs of natural gas and town gas by gas chromatography and mass spectrometry, *Fresenius J. Anal. Chem.*, 338, 77–78, 1990.

136. Ridgeway, R. G., Bandy, A. R., and Thorton, D. C., Determination of aqueous dimethylsulfide using isotope dilution gas chromatography/mass spectrometry, *Mar. Chem.*, 33, 321–334, 1991.

137. Kelly, T. J. and Kenny, D. V., Continuous determination of dimethylsulfide at part-per-trillion concentrations in air by atmospheric pressure chemical ionization mass spectrometry, *Atmos. Environ. 25A*, 2155–2160, 1991.

138. Escott, R. E. A., McDowell, P. G., and Porter, N. P., Use of nonvolatile ion-pairing agents for liquid chromatographic-mass spectrometric analyses with a moving-belt interface, *J. Chromatogr.*, 554, 281–292, 1991.

139. Conboy, J. J., Henion, J. D., Martin, M. W., and Zweigenbaum, J. A., Ion chromatography/mass spectrometry for the determination of organic ammonium and sulfate compounds, *Anal. Chem.*, 62, 800–807, 1990.

140. Forngren, B. H., Samskog, J., Gustavson, S. A., Tyrefos, N., Markides, K. E., and Langström, B., Reversed-phase ion-pair chromatography coupled to electrospray ionisation mass spectrometry by online removal of the counter-ions, *J. Chromatogr. A.*, 854, 155–162, 1999; Socher, G., Nussbaum, R., Rissler, K., and Lankmayr, E., Analysis of sulfonated compounds by ion-exchange high-performance liquid chromatography-mass spectrometry, *J. Chromatogr. A.*, 912, 53–60, 2001.

141. Reemtsma, T., Methods of analysis of polar aromatic sulfonates from aquatic environments, *J. Chromatogr. A*, 733, 473–489, 1996.

142. Reemtsma, T., Analysis of sulfophthalimide and some of its derivatives by liquid chromatography-electrospray ionization tandem mass spectrometry, *J. Chromatogr. A*, 919, 289–297, 2001.

143. Ballantine, J. A., Games, D. E., and Slater, P. S., The use of amine base to enhance the sensitivity of electrospray mass spectrometry towards complex polysulfonated azo dyes, *Rapid Commun. Mass Spectrom.*, 9, 1403–1410, 1995.

144. Storm, T., Hartig, C., Reemtsma, T., and Jekel, M., Exact mass measurements online with high-performance liquid chromatography on a quadrupole mass spectrometer, *Anal. Chem.*, 73, 589–595, 2001.

145. Altenbach, B. and Giger, W., Determination of benzene- and naphthalene-sulfonates in wastewater by solid-phase extraction with graphitized carbon black and ion-pair liquid chromatography with UV-detection, *Anal. Chem.*, 67, 2325–2333, 1995.

146. Crescenzi, C., Di Corcia, A., Moarcomini, A., Pojana, G., and Samperi, R., Method development for trace determination of poly(naphthalenesulfonate)-type pollutants in water by liquid chromato-graphy-electrospray mass spectrometry, *J. Chromatogr. A*, 923, 97–105, 2001.

147. Bruins, A. P., Weidolf, L. O. G., Henion, J. D., and Budde, W. L., Determination of sulfonated azo dyes by liquid chromatography/atmospheric pressure ionization mass spectrometry, *Anal. Chem.*, 59, 2647–2652, 1987.

148. Yinon, J., Betowski, L. D., and Voyksner, R. D., Application of LC-MS. in environmental chemistry, In *Journal of Chromatography Library*, Barcelo, Ed., Elsevier, Amsterdam, pp. 187–235, 1996.

149. Preiss, A., Sänger, U., Karfich, N., and Levsen, K., Characterization of dyes and other pollutants in the effluent of a textile company by LC/NMR and LC/MS, *Anal. Chem.*, 72, 992–998, 2000.

150. Ballantine, J. A., Games, D. E., and Slater, P. S., The use of diethylamine to determine the number of sulfonate groups present within polysulfonated alkali metal salts by electrospray mass spectrometry, *Rapid Commun. Mass Spectrom.*, 11, 630–637, 1997.

151. Sullivan, A. G., Garner, R., and Gaskell, S. J., Structural analysis of sulfonated monoazo dyestuff intermediates by electrospray tandem mass spectrometry and matrix-assisted laser desorption/ionization postsource decay mass spectrometry, *Rapid Commun. Mass Spectrom.*, 12, 1207–1215, 1998.

152. Schröder, H. F., Identification of nonbiodegradable, hydrophilic, organic substances in industrial and municipal waste water treatment plant-effluents by liquid chromatography-tandem mass spec-trometry (LC/MS/MS), *Water Sci. Technol.*, 23, 339–347, 1991.

153. Di Corcia, A., Capuani, L., Casassa, F., Marcomini, A., and Sarnperi, R., Fate of linear alkyl benzenesulfonates, coproducts, and their metabolites in sewage treatment plants and in receiving river waters, *Environ. Sci. Technol.*, 33, 4119–4125, 1999.

154. Popenoe, D. D., Morris, S. J., Horn, P. S., and Norwood, K. T., Determination of alkyl sulfates and alkyl ethoxysulfates in wastewater treatment plant influents and effluents and in river water using liquid chromatography/ion spray mass spectrometry, *Anal. Chem.*, 66, 1620–1629, 1994.

155. Kolpin, D. W., Thurman, E. M., and Linhart, S. M., Finding minimal herbicide concentrations in ground water? Try looking for their degradates, *Sci. Total Environ.*, 248, 115–122, 2000.

156. Ferrer, I., Thurman, E. M., and Barcelo, D., Identification of ionic chloroacetanilide-herbicide metabolites in surface water and groundwater by HPLC/MS using negative ion spray, *Anal. Chem.*, 69, 4547–4553, 1997.

157. Yokley, R. A., Mayer, L. C., Huang, S. B., and Vargo, J. D., Analytical method for the determination of metolachlor, acetochlor, alachlor, dimethenamid, and their corresponding ethanesulfonic and oxanillic acid degradates in water using SPE and LC/ESI-MS/MS, *Anal. Chem.*, 74, 3754–3759, 2002.

158. Darren, T., Crain, S. M., Sim, P. G., and Benoit, F. M., Application of reversed phase liquid chromatography with atmospheric pressure chemical ionization tandem mass spectrometry to the determination of polycyclic aromatic sulfur heterocycles in environmental samples, *J. Mass Spectrom.*, 30, 1034–1040, 1995.

159. Ojala, M., Ketola, R., Mansikka, T., Kotiaho, T., and Kostiainen, R., Detection of volatile organic sulfur compounds in water by headspace gas chromatography and membrane inlet mass spectrometry, *High Resolut. Chromatogr.*, 20, 165–169, 1997.

160. Ketola, R. A., Mansikka, T., Ojala, M., Kotiaho, T., and Kostiainen, R., Analysis of volatile organic sulfur compounds in air by membrane inlet mass spectrometry, *Anal. Chem.*, 69, 4536–4539, 1997.

161. Kalontarov, L., Jing, H., Amirav, A., and Checkis, S., Mechanism of sulfur emission quenching in flame photometric detectors, *J. Chromatogr. A*, 696, 245–256, 1995.

162. Fichtner, S., Lange, Th. F., Schmidt, W., and Brauch, H. J., Determination of aromatic sulfonates in the river Elbe by online ion-pair chromatography, *Fresenius J. Anal. Chem.*, 353, 57–63, 1995.

163. Nakae, A., Tsuji, K., and Yamaunaka, M., Determination of alkyl chain distribution of alkylbenzenesulfonates by liquid chromatography, *Anal. Chem.*, 53, 1818–1821, 1981.

164. Kikuchi, M., Tokai, A., and Yoshida, T., Determination of trace levels of linear alkylbenzenesulfonates in the marine environment by high-performance liquid chromatography, *Water Res.*, 20, 643–650, 1986.

165. Mottaleb, M. A., Development of a HPLC method for analysis of linear alkylbenzene sulfonates and detection by UV and FTIR spectroscopy using thermospray interface, *Mikrochim Acta.*, 132, 31–39, 1999; Schnitzler, M. and Sontheimer, H., Eine Methode zur Bestimmung des gelösten organisch gebundenen Schwefels in Wässern, *Vom Wasser*, 59, 623–637, 1979.

166. Randt, C. H. and Altenbeck, R., Problems in the development of a method for the determination of the total amount of adsorbable organic sulfur compounds (AOS) in water/wastewater, *Vom Wasser*, 88, 217–225, 1997.

167. Rüttinger, H. H. and Binde, F., Determination of the organic group parameters DOS (dissolved organic sulfur) in rivers and in the process of drinking water production, *Vom Wasser*, 90, 1–8, 1998.

168. Knobloch, Th. and Engewald, W., Identification of some polar polycyclic compounds in emissions from brown-coal-fired residential stoves, *J. High Resolut. Chromatogr.*, 16, 239–242, 1993.

169. Branch H. J., Fleig, M., Kühn, W., and Lindner, K., Rhine Waterworks Registered Society Annual Report, 51, 1994.

170. Branch, H. J, and Jühlich, W., Rhine Report, International Work Pool of the Waterworks in the Rhine Catchment Area, 1993.

171. Kümmerer, K., Analysis of organic sulfur compunds dissolved in water, Theris, Tübinger, 1990.

11 Amines

Anna Pielesz

CONTENTS

I. INTRODUCTION

Amines occur, often at trace levels, in a number of ambient environments such as, air, water, soil, and waste, and pose serious hygienic problems, being highly toxic and reactive.[1,2] They are discharged into the atmosphere and water from anthropogenic sources such as cattle feeds and livestock buildings, waste incineration, sewage treatment, automobile exhausts, cigarette smoke, and various industries.[3] Naturally occurring aliphatic and aromatic mono-, di-, and polyamines are formed as metabolic products in microorganisms, plants, and animals.[3] These amines are widely used as industrial intermediates in the manufacture of carbamate and urethane pesticides, dyestuffs, cosmetics, and medicines. They are also employed in the rubber industry as antioxidants and antiozonants and as components in epoxy and polyurethane polymers.[4] High amounts of amines are released to sewage plants where they do not undergo complete degradation. Aromatic amines (AAs) are also generated in the environment via the degradation of pesticides, nitroaromatics, and azo dyes.

AAs are of considerable interest in environmental analytical chemistry on account of two factors. Firstly, they are very polar and water soluble, which make them easily transported to, and in, aquifers. Secondly, the amino group is capable of unique sorption processes with soil particles, either due to covalent binding or ion exchange processes. Because of their widespread distribution and mobility, a wide range of AAs have been found in environmental matrices; for example, in river water and ammunition wastewater.[5]

Amines are not only toxic themselves, but can change in toxic *N*-nitrosamines, potentially carcinogenic substances giving a wide variety of tumors in many animals,[3] or form adducts with proteins and DNA.[4] Most analyzed amines are allergens. Many others, including heterocyclic amines, have been proved to be mutagenic and carcinogenic. Many of these compounds have been isolated and identified not only from various proteinaceous foods, such as cooked meats and fish, but also from environmental components, such as outdoor air, indoor air, diesel-exhaust particles, cigarette smoke, cooking fumes, rain water, incineration ash, and soil.[6] Heterocyclic amines may be emitted into the atmosphere through combustion of various materials (e.g., food, wood, grass, garbage, petroleum) and discharged into the water through domestic and human waste, although their mechanism has not been determined.[7] To date the following compounds have been classified as carcinogens in humans: 4-aminobiphenyl, benzidine, 4-chloro-2-methylaniline, and 2-naphthylamine. The main exposure of the general population to AAs is through cigarette smoke or items containing products synthesized from AAs. These compounds may account for the positive correlation between cigarette smoking and the incidence of bladder cancer in humans.[8]

Because of their carcinogenicity, amines have to be monitored. For example, primary aromatic amines (PAAs) are substances that can be transferred from food packaging materials into foodstuffs. In the production of multilayer plastic materials it is common to use reactive adhesive mixtures containing aromatic isocyanate monomers. In cases of incomplete curing, residues of the aromatic isocyanates react with water to produce PAAs. Some of these amines, including 2,4-diaminotoluene and 4,4′-methylenedianiline, are classified as "possibly carcinogenic to humans" by the International Agency for Research on Cancer (IARC), and thus their appearance in foodstuffs should be prevented. According to European legislation, the total concentration of PAAs migrated into foodstuffs or food simulants should not be detectable using an analytical method with a detection limit of 20 μg/kg foodstuff.[9]

The carcinogenic nature of benzidine, 3,3′-dichlorobenzidine, and its congeners was recognized several years ago. Now they are included in all priority pollutant lists worldwide. 3,3′-Dichlorobenzidine is less toxic than benzidine, but studies have shown that in natural sediments it transforms to benzidine.[10] The U.S. Environmental Protection Agency (EPA) established water quality criteria for both compounds. In Europe, they are listed among the "very toxic substances for the environment" and required to be controlled in industrial effluents.[11,12] The textile industry is of major concern as are industries dealing with painting pigments, printing inks, and food coloring production. In this case, the source of the mutagenic activity is both the dyestuffs and the amines contained in their chemical structure. The dyestuffs themselves provide only a minor part of the mutagenic potential. Lists of dangerous azo dyes and amines can be found in literature.[13] The result is a systematic removal from the market of those dyes that are suspected to have carcinogenic or mutagenic influence on the human organism. The main hazard criterion for a dye is its ability to split AAs reductively in contact with sweat, saliva, or gastric juice. The process of the reduction of azo dyes with the cleavage of aromatic R-NH$_2$ amines is one way of degradation of those dyes. Other methods of splitting amines from dyestuffs are photodegradation and biodegradation by means of hydroxylation, oxidation, or hydrolysis. However, the biological reduction of an azo dye is responsible for the possible presence of toxic amines in the human organism.

AAs are specified in the groups III A1 and III A2 of the Maximale Arbeitsplatz Konzentration (MAK) list as well as in the IARC and Ecological and Toxicological Association of the Dyestuffs Manufacturing (ETAD) lists (e.g., benzidine, o-toluidine, 4-aminodiphenyl). Legally, the issue of toxic chemicals (including AAs that split from dyestuffs or dyed textiles) is regulated by a European Union directive.[13] However, according to EU law, each member state may define its own regulations to protect people's health. Thus, first Germany and then other EU countries approved bans on importing and marketing textiles containing dyes capable of reductively splitting carcinogenic amines. As for textile products, EU law not only provides a list of hazardous substances, but also specifies their maximum acceptable quantities and obligatory detection methods.[14] One of them is the German DIN 53316 method for determination of azo colorants in leather. Developed within a restricted time period in 1997 after the Germans had imposed a ban on certain azo colorants, it has some drawbacks which result in relatively low accuracy.[15]

Numerous legislative changes led to reduction in the levels of amine discharges, particularly to estuarine and coastal waters (e.g., the Dangerous Substances Directive 74/464/EEC and daughter directives, which set environmental quality standards for a number of substances; the Shellfish Waters Directive 79/923/EEC, which specified a range of further standards for waters supporting shellfish; and the Urban Waste Water Treatment Directive 91/271/EEC, which specified the degree of treatment to be given to various sewage discharges). More recently, the Environment Protection Act of 1990 and the EC Directive on Integrated Pollution, Prevention and Control (96/61/EEC) introduced the concept of Integrated Pollution Control in order to ensure that the effect of any release to the environment is minimized.[16]

II. SAMPLE PREPARATION

A. CLEANUP AND EXTRACTION

Identification and quantification of organic compounds in water or other matrices are both necessary for solving various environmental, biological, or clinical problems. Because of the high polarity and the corresponding high solubility in water, the extraction of AAs from water samples is difficult. In the past, the enrichment of aminoaromatic compounds was performed by liquid–liquid extraction (LLE). In recent years LLE has been increasingly replaced by solid-phase extraction (SPE).[17]

SPE with porous solid particles originated in the early 1970s. It was developed to replace many traditional LLE methods for the determination of organic analytes in aqueous samples. Traditional LLE procedures employ a serial extraction of an aqueous sample with an organic solvent, resulting in a relatively large volume of solvent that must be dried and concentrated prior to analysis. In SPE techniques, a solid sorbent material, usually an alkyl bonded silica, is packed into a cartridge or embedded in a disk and performs essentially the same function as the organic solvent in LLE. This allows samples to be processed quickly, eliminates some of the glassware necessary in LLE procedures, consumes much less solvent, isolates organic analytes from large volumes of water with minimal or no evaporation losses, reduces exposure of analysts to organic solvents relative to traditional methods, and can provide more reproducible results.[18]

The efficiency of SPE can be improved by the chemical introduction of acetyl or hydroxy-methyl groups into polymeric resins, which provide better surface contact with aqueous samples. Lightly sulfonated resins display excellent hydrophilicity and improved extraction efficiencies of polar organic compounds over underivatized resins. Sulfonated resins can also be used for group separation of neutral and basic organic compounds.[19] Particle-loaded membranes (known as Empore extraction disks), introduced in 1990, further improve extraction efficiency, reduce the use of solvent, and decrease plugging in SPE. SPE does have some limitations, however, such as low recovery caused by interaction between the sample matrix and analytes, and plugging of the cartridge or blocking of the pores in the sorbent by solid and oily components, which results in low breakthrough volume and low capacity. As SPE is a multistep approach involving concentration of the extract, it is limited to semivolatile compounds with boiling temperatures substantially above those of the solvents.[20]

Several steps are required in a typical reversed-phase SPE:

- Washing solid-phase sorbent with organic solvent or a mixture of solvents to remove potential interferents from SPE system
- Conditioning or activating the solid sorbent with organic solvents or mixture of organic solvents and reagent water to effectively extract organic analytes from the aqueous sample
- Preparation of sample, typically by addition of methanol, followed by extraction of sample by passing it through solid sorbent
- Drying solid sorbent by passing air or nitrogen through disk or cartridge
- Cleanup of sample extract to remove possible contaminants in sample trapped in sorbent

- Elution of organic analytes from solid sorbent with organic solvents or mixtures of solvents
- Drying sample eluate with sodium thiosulfate to remove any residual water
- Concentrating sample extract and solvent exchange if necessary

Examples of SPE for amines are shown in Table 11.1.

Volatiles are usually analyzed by the use of purge and trap (which is an EPA-approved technique[31]), stripping, and headspace (HS) analysis. These methods either require expensive instrumentation or are not sufficiently sensitive. Nonvolatiles are analyzed primarily by means of LLE (also an EPA-approved technique) and SPE. These methods are generally time consuming, difficult to automate, and use expensive high-purity toxic organic solvents.[32]

Both LLE and SPE require evaporation of the solvent to dryness and the reconstitution of the dry residue in a suitable solvent for high-performance liquid chromatograpgy (HPLC) or capillary electrophoresis (CE).[33] To eliminate both the solvent evaporation step and large sample volume consumption, one can perform solid-phase microextraction (SPME) onto chemically modified fused silica fibers with thermal desorption.[34] This solves the problems associated with SPE while retaining the advantages: solvents are completely eliminated, blanks are greatly reduced, and extraction time can be reduced to a few minutes.[35] SPME can be used with liquid, gaseous, or "dirty" samples. It consists of two processes: partitioning of analytes between the coating and the sample, and desorption of concentrated analytes into an analytical instrument. In the first process, the coated fiber is exposed to the sample and the target analytes are extracted from the sample matrix into the coating. The fiber with concentrated analytes is then transferred to an instrument for desorption, followed by separation and quantitation. SPME applications have focused on extracting organic compounds from various matrices, such as air, water, and soils, followed by directly transferring them into a gas chromatograph injector where they are thermally desorbed, separated on the column, and quantified by the detector.[20]

Examples of SPME for amines are shown in Table 11.2.

TABLE 11.1
Solid-Phase Extraction (SPE)

Amine	Environment	Solid Phase for Extraction	Relative Recovery (%)	R.S.D. (%)	Concentration Levels[a] or Detection Limit[b]	Reference
AA	W	PS/DVB	80–120	<5	[a]10–20 μg/l	17
AA	H	C_{18}	87–106	<4	[a]2–33 μg/l	21
PAA	H	PE	—	4–17	[a]5 μg/l	9
ABDAC	W	C_{18}	95	9	[a]<0.1 μg/l	22
T	A	COOH-resin	80–105	<2	[b]0.6–1.5 ppm	23
AA	W	PS/DVB	—	<7	[a]0.5–8 μg/l	24
AA	H	Hyspher-GP and -SH	98–116	<4	[b]0.05–0.5 μg/l	25
AA	H	PS/DVB	47–97	1–6	[b]0.06–1.8 μg/l	26
Al	W	Hypersyl ODS C_{18}	73–120	2–15	[b]1 μg/l	27
AA	A	MCX and HLB type	79–109	—	[b]0.02–1.4 ng/cigar	28
AME	W	Porapak Rdx type	28–96	1–39	[b]0.3–6 μg/l	29
AA	W	S/DVB	—	3–8	[b]0.1–1 μg/l	10
Al	H	PDMS	26–107	6–11	[a]10 μg/l	30

A, air; H, water; W, waste; Al, aliphatic amine; AA, aromatic amine; PAA, polyaromatic amine; ABDAC, alkylbenzyldimethylammonium chloride; AME, alkylamine ethoxylates; T, tertiary amine; PE, polyethylene; PS/DVB, polystyrene/divinylbenzene; PDMS, polydimethylsiloxane; PA, polyacrylate; PAB, polyacrilonitrilbutadiene.

TABLE 11.2
Solid-Phase Microextraction (SPME)

Amine	Environment	SPME Phase for Extraction	Concentration Levels[a] or Detection Limit[b] or Recovery[c]	R.S.D. (%)	Reference
AA	W	Carbon, PDMS, CW/DVB, PDMS/DVB, PA	[a]0.05–5 μg/l	~5	36
ClNA	S	PDMS, PA	[b]10–1000 ppb	8–20	37
ClAA	H	PAB	[b]<100 ppb	3–6	38
TMA	A	PDMS, PDMS/DVB	[c]100.4%	2	39
LMMA	A	CW/DVB, PDMS, PA	[b]0.01–0.1 ppm	—	40
AA	H	CW/TPR,CW/DWB, PDMS/DVB, PA	[b]0.33–2.4 ng/ml	2–8	41
HA	F	Supelcosil LC-CN	[b]0.2–3.1 ng/ml	0–10.7	42

AA, aromatic amine; TMA, trimethylamine; ClNA, chloronitroanilina; ClAA, chlorinated aromatic amine; LMMA, low molecular mass amine; HA, heterocyclic amine; PE, polyethylene; PS/DVB, polystyrene-divinylbenzene; PDMS, polydimethylsiloxane; PA, polyacrylate; PAB, polyacrilonitrilbutadiene; CW, carbowax; TPR, templated resin; F, food samples.

For gas chromatography (GC) and nonpolar compounds, liquid-phase microextraction (LPME)[43–46] is an alternative to SPME. Only one drop of organic solvent is used to extract compounds from water samples.

Finally, Ma and Cantwell[47,48] developed liquid liquid–liquid microextraction (LLLME) to achieve preconcentration and purification for polar analytes without the need for both solvent evaporation and analyte desorption. The compounds were extracted from aqueous samples (donor phase) into an organic phase, layered on the donor phase, then back extracted to the receiving phase, and suspended in the organic phase. After extraction, the microdrop was injected into the HPLC system directly for analysis[34] (Table 11.3).

For GC analysis, HS is a preconcentration technique particularly suitable for the sampling of volatile organic compounds in air, water, and solids. Few reports have been published on the use of static headspace in the analysis of free amines in aqueous samples because of the high polarity and solubility in water of these compounds.[49] In one experiment,[49] static headspace preconcentration was developed for the gas chromatographic analysis of aliphatic amines in aqueous samples. A liquid–gas ratio of 1, an incubation temperature of 80°C (15 min), a pH of 13.7, and a mixture of salts (NaCl and K_2SO_4) at saturation concentration gave a maximal headspace amine concentration (Table 11.4).

Microwave-assisted extraction (MAE) and supercritical fluid extraction (SFE) are well established techniques for the determination of different pollutants from solid samples, providing faster extractions and less usage of organic solvents than conventional solvent extraction.[51] SFE was proved to be useful in the selective removal of analytes in different types of samples.

TABLE 11.3
Liquid–Liquid–Liquid Microextraction (LLLME)

Amine	Conditions of the Extraction	Detection Limit	Reference
AA	aq (pH13) → org (ethyl acetate) → aq (pH 2.1)	0.85–1.80 ng/ml	34
AA	aq (pH13) → org (di-n-hexyl ether) → aq (acid)	0.05–0.10 μg/l	33

AA, aromatic amine.

TABLE 11.4
Static Headspace Analysis

Amine	Environment	Detection Limit	RSD (%)	Reference
Al	W	0.2–3000 μg/l	0.6–6.4	49
Al	W	120–1200 ng/l	12	50

Al, aliphatic amine; W, waste.

It also minimizes sample handling, provides fairly clean extracts, expedites sample preparation, and reduces the use of environmentally toxic solvents.[52] SFE has been applied to the extraction of carcinogenic AAs from soil and sand.[53–55] The paper[51] studies the possibilities of using MAE and SFE in determination of AAs by HPLC after reduction of the azo colorants. Two SFE pieces of equipment differing in the trapping step (solid-phase trap or solvent collection) were utilized for the extractions. The MAE experiments were then performed with a vessel system with temperature and pressure control.

B. DERIVATIZATION

GC is a technique for the separation of volatile compounds that are thermally stable during GC running. Unfortunately, there are many compounds of environmental interest, particularly those of high molecular weight or containing polar functional groups which cannot be readily analyzed by GC, either because they are not sufficiently volatile or because they tail badly or are too strongly attracted to the stationary phases or because they are thermally labile at the temperature required for GC running, and consequently are decomposed.

Derivatization for gas chromatographic separation is used to improve the thermal stability of those compounds. The main reasons for derivatization are:

- To increase the volatility and decrease the polarity of the compounds
- To reduce thermal degradation of the samples by increasing their thermal stability
- To increase detector response by incorporating into the derivative functional groups which produce a higher detector signal, for example CF_3 groups for electron capture detectors
- To improve separation and reduce tailing, and to improve extraction efficiency from aqueous media (e.g., acylation of phenolic amines).

In addition, there are also a number of disadvantages in derivatization:

- The derivatizing agent may be difficult to remove and may interfere in the analysis.
- The derivatization conditions may cause unintended chemical changes in a compound.
- The derivatization step increases the analysis time.

Derivatization reactions, often selective for amine type (primary, secondary, tertiary), have also been used to improve the detection and separation of these amines. Examples of derivatization reaction for GC determination of amines are shown in Table 11.5.

Acylation is one of the most popular derivatization reactions for primary and secondary amines (Figure 11.1). The reagents listed in Table 11.5 easily react with amino groups under mild reaction conditions. In the reactions of amines with acid anhydrides and acyl chlorides, it is usually necessary to remove excess reagent and byproduct acid because these compounds damage the GC column.

TABLE 11.5
Derivatization Reactions for Gas Chromatographic Determination of Amines

Reagent	Amine Type	Environment	Reference
Acylation			
Trifluoroacetic anhydride (TFA)	P, S, A	—	56
Trichloroacetyl chloride (TCA)	P, S	A	3
N-Methylbis(trifluoroacetamide) (MBTFA)	P, S	A	3
Trifluoroacetic anhydride and diethylether (1:1) (TFA)	AA	W	4,9
Heptafluorobutyric anhydride (HFBA)	AA, T	—	8,28,57
Pentafluoropropionic anhydride (PFPA)	T	—	8
Silylation			
N,O-Bis(trimethyl silyl)trifluoroacetamide (BSTFA)	P, S	A	3
N-Methyl-N-(tert-butyldimethylsilyl)acetamide (MTBSTFA)	P, S	A	3
Dinitrophenylation			
2,4-Dinitrofluorobenzene (DNFB)	P, S	A	58
2,4-Dinitrobenzenesulphonic acid (DNBS)	P, S, A	W	3
Permethylation			
Formamide–sodium borohydride	P, S	A	3
N-Dimethylaminimethylene (DMAM)	HA	W	59
Schiff base formation			
Furfural	P	A	3
Pentafluorobenzaldehyde (PFBA)	P, Al	W	40,49,60
Dimethylformamide dimethyl acetal (DMFDMA)	P	W	61
Carbamate formation			
Ethyl chloroformate (ECF)	P, S, T	W	62
Isobutyl chloroformate (IBCF)	P, S	W	63
Sulphonamide formation			
Benzenesulphonyl chloride (BSC)	P, S, A	W	55,58,61,64,65
p-Toluenesulphonyl chloride (TSC)	P, S	W	3
Phosphoamide formation			
Dimethylthiophosphoryl chloride (DMTPC)	P, S	W	66
Diethylthiophosphoryl chloride (DETPC)	P, S, N	W	67,68
N-Diethylthiophosphoryl (DETP)	N, AA	—	4,69
Diethyl chlorothiophosphate (DECTP)	N	—	67
Halogenation			
Iodine (I)	AA	W	5,24,70
Bromine (B)	AA	W	71,72
p-Nitrophenyl trifluoroacetate	LMMA	A	40

A, air; W, waste; P, primary amine; S, secondary amine; T, tertiary amine; A, ammonia; N, nitrosamine; AA, aromatic amine; HA, heterocyclic amine; Al, aliphatic amine; LMMA, low molecular mass amine.

MBTFA is very volatile, but methyltetrahydrofolic acid (MTFA) does not cause column damage. MBTFA can be useful for N-selective acylation after trimethylsilylation of hydroxyamino compounds.[3] A new GC–MS method has been developed for the determination of PAAs in water samples, using solid-phase analytical derivatization (SPAD) for the sample preparation.

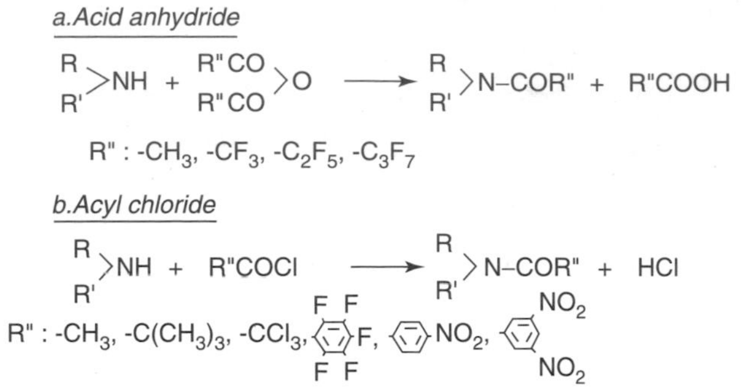

a.Acid anhydride

$$\underset{R'}{\overset{R}{\diagdown}}NH \; + \; \underset{R''CO}{\overset{R''CO}{\diagdown}}O \; \longrightarrow \; \underset{R'}{\overset{R}{\diagdown}}N\text{–}COR'' \; + \; R''COOH$$

R" : -CH$_3$, -CF$_3$, -C$_2$F$_5$, -C$_3$F$_7$

b.Acyl chloride

$$\underset{R'}{\overset{R}{\diagdown}}NH \; + \; R''COCl \; \longrightarrow \; \underset{R'}{\overset{R}{\diagdown}}N\text{–}COR'' \; + \; HCl$$

R" : -CH$_3$, -C(CH$_3$)$_3$, -CCl$_3$, ⟨F F / F F⟩F, ⟨⟩NO$_2$, ⟨NO$_2$ / NO$_2$⟩

FIGURE 11.1 Acylation of primary and secondary amines. R, alkyl or aryl; R', hydrogen, alkyl, or aryl. (From Kataoka, H., *J. Chromatogr.*, 733, 19–34, 1996.)

Extraction and derivatization supported on a neat polystyrene–divinylbenzene copolymer adsorbent material provided several advantages over conventional sample preparation techniques, including less solvent consumption, less time-consuming steps in the method, low detection limits, excellent repeatability, and no loss of volatiles.[9]

BSA, BSTFA, and MTBSTFA have been used as silylating reagents (Table 11.5, Figure 11.2). The amino group is not very reactive to silylating reagents, and its conversion into a silyl derivative is difficult. By using stronger silylating reagents and catalysts, however, the silyl derivatives of amines can be prepared. The addition of trimethylchlorosilane as a catalyst ensures the effective derivatization of amino groups.

Dinitrophenylation, which can be performed in aqueous media, is used for the derivatization of primary and secondary amines. These DNP derivatives are sensitive to electron-capture detection (ECD) and they are particularly suitable for molecular-mass amines that have inconveniently short retention times. DNBS has a greater specificity for the amino group, whereas DNFB reacts with thiol, imidazole hydroxyl groups along with amino groups. However, DNBS generally reacts more slowly than does DNFB, so that longer reaction times or more strongly alkaline conditions may be required to complete the reaction[3] (Figure 11.3).

Permethylation has been applied to the determination of polyamines. Permethyl derivatives eliminate the polar NH groups but retain the troublesome adsorptive properties of tertiary amines.

Schiff base-type reactions are employed to condense primary amines with a carbonyl compound (Figure 11.4). Furfural and PFBA have also been used for low-molecular-mass amines.

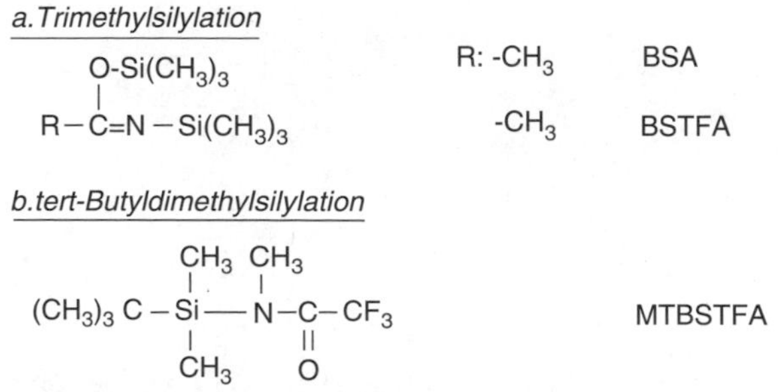

a.Trimethylsilylation

$$\underset{R-C=N-Si(CH_3)_3}{\overset{O\text{-}Si(CH_3)_3}{|}}$$

R: -CH$_3$ BSA

-CH$_3$ BSTFA

b.tert-Butyldimethylsilylation

$$(CH_3)_3\,C-\underset{\underset{CH_3}{|}}{\overset{\overset{CH_3}{|}}{Si}}-N\text{-}\underset{\overset{||}{O}}{C}\text{-}CF_3$$

MTBSTFA

FIGURE 11.2 Silylating reagents for primary and secondary amines. (From Kataoka, H., *J. Chromatogr.*, 733, 19–34, 1996.)

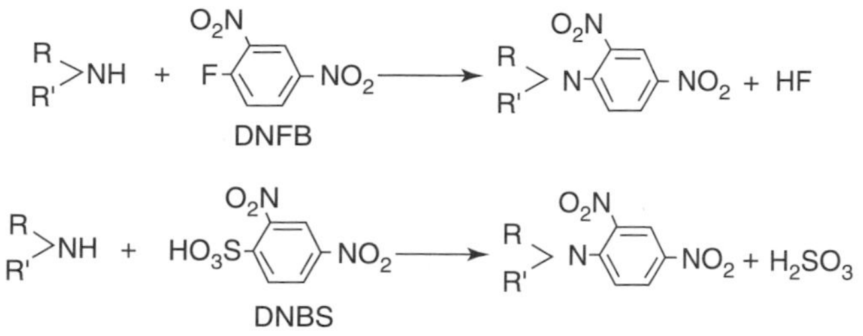

FIGURE 11.3 Dinitrophenylation of primary and secondary amines. R, alkyl or aryl; R′, hydrogen, alkyl, or aryl. (From Kataoka, H., *J. Chromatogr.*, 733, 19–34, 1996.)

The condensation reactions with these reagents proceed rapidly in aqueous medium at room temperature or on warming, and Schiff bases are obtainable in good yields. Dimethylformamide dialkyl acetal forms a Schiff base-type derivative with primary amines (Figure 11.4), but this reagent also reacts with carboxyl groups of fatty acids and amino acids to give the corresponding esters.[3]

The reaction of amines with alkyl chloroformates can be easily performed in aqueous alkaline media and the resulting carbamate derivatives have good GC properties (Figure 11.5). This reaction proceeds easily in aqueous alkaline media at room temperature.

BSC, DNFB, and *p*-toluenesulphonyl chloride have also been used for the selective determination of low-molecular-mass primary and secondary amines[3] (Figure 11.6). BSC and DNFB can convert the amines into hydrophobic and nonvolatile derivatives in water.[64]

By using dialkylthiophosphoryl chlorides in aqueous alkaline medium, a selective and sensitive method for the determination of aliphatic and aromatic amines by GC-FPD has been developed. In particular, secondary amines can be selectively converted into their *N*-diethylthiophosphoryl (DETP) derivatives with diethyl chlorothiophosphate (DECTP) after treatment with *o*-phthaldialdehyde (OPA), because OPA reacts only with primary amino groups. On the other hand, secondary amines are detected irrespective of pretreatment, because they do not react with OPA[3] (Figure 11.7 and Figure 11.8).

N-Nitrosamines can be accurately and precisely determined by GC-FPD as their *N*-diethylthiophosphoryl derivatives after denitrosation. This method is selective and sensitive, allowing cigarette smoke samples to be analyzed directly without pretreatment, except for separation from secondary amines by solvent extraction, and without any interference from other coexisting substances[67] (Figure 11.9).

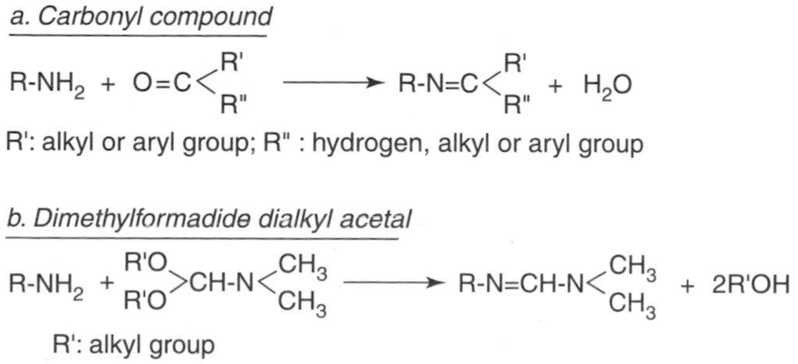

a. Carbonyl compound

R′: alkyl or aryl group; R″ : hydrogen, alkyl or aryl group

b. Dimethylformadide dialkyl acetal

R′: alkyl group

FIGURE 11.4 Schiff base-type condensation reactions of primary and secondary amines. (From Kataoka, H., *J. Chromatogr.*, 733, 19–34, 1996.)

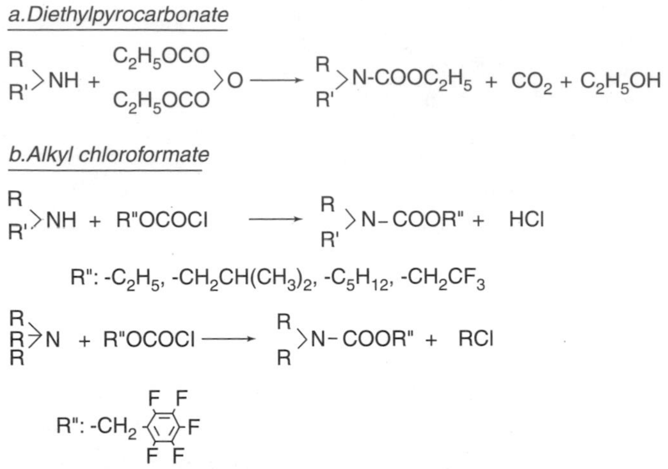

a. Diethylpyrocarbonate

$$\frac{R}{R'}{>}NH + \frac{C_2H_5OCO}{C_2H_5OCO}{>}O \longrightarrow \frac{R}{R'}{>}N\text{-}COOC_2H_5 + CO_2 + C_2H_5OH$$

b. Alkyl chloroformate

$$\frac{R}{R'}{>}NH + R''OCOCl \longrightarrow \frac{R}{R'}{>}N\text{-}COOR'' + HCl$$

R'': -C_2H_5, -CH_2CH(CH_3)_2, -C_5H_{12}, -CH_2CF_3

$$\frac{R}{R}{>}N + R''OCOCl \longrightarrow \frac{R}{R}{>}N\text{-}COOR'' + RCl$$

R'': -CH_2 ⟨F F F F F⟩ F

FIGURE 11.5 Carbamate formation from primary, secondary, and tertiary amines. R, alkyl or aryl; R', hydrogen, alkyl, or aryl. (From Kataoka, H., *J. Chromatogr.*, 733, 19–34, 1996.)

Two derivatization methods for AAs based on the halogenation of the aromatic ring are presented.[72] Bromination yields brominated anilines, in which all hydrogens in *ortho*- and *para*-positions are replaced by bromine via an electrophilic substitution. In contrast, iodination yields the corresponding iodobenzenes, in which all amino groups are substituted by iodine. Derivatization is

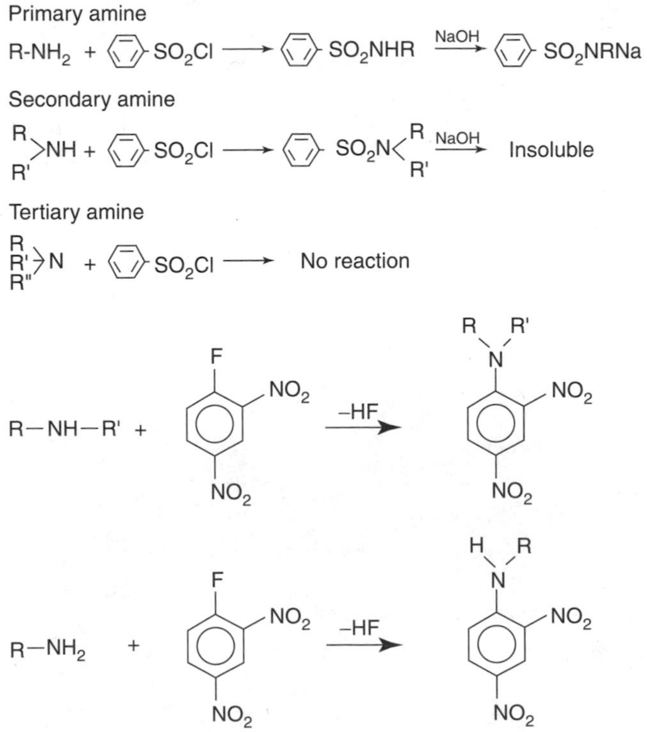

Primary amine

$$R\text{-}NH_2 + \langle\bigcirc\rangle\text{-}SO_2Cl \longrightarrow \langle\bigcirc\rangle\text{-}SO_2NHR \xrightarrow{NaOH} \langle\bigcirc\rangle\text{-}SO_2NRNa$$

Secondary amine

$$\frac{R}{R'}{>}NH + \langle\bigcirc\rangle\text{-}SO_2Cl \longrightarrow \langle\bigcirc\rangle\text{-}SO_2N{<}\frac{R}{R'} \xrightarrow{NaOH} \text{Insoluble}$$

Tertiary amine

$$\frac{R}{R'}{>}N + \langle\bigcirc\rangle\text{-}SO_2Cl \longrightarrow \text{No reaction}$$

$$R\text{—}NH\text{—}R' + \text{[F, NO}_2\text{, NO}_2\text{ benzene]} \xrightarrow{-HF} \text{[R, R', N, NO}_2\text{, NO}_2\text{ benzene]}$$

$$R\text{—}NH_2 + \text{[F, NO}_2\text{, NO}_2\text{ benzene]} \xrightarrow{-HF} \text{[H, R, N, NO}_2\text{, NO}_2\text{ benzene]}$$

FIGURE 11.6 Separation of primary, secondary, and tertiary amines. R, R', R'', alkyl or aryl. (From Kataoka, H., *J. Chromatogr.*, 733, 19–34, 1996.)

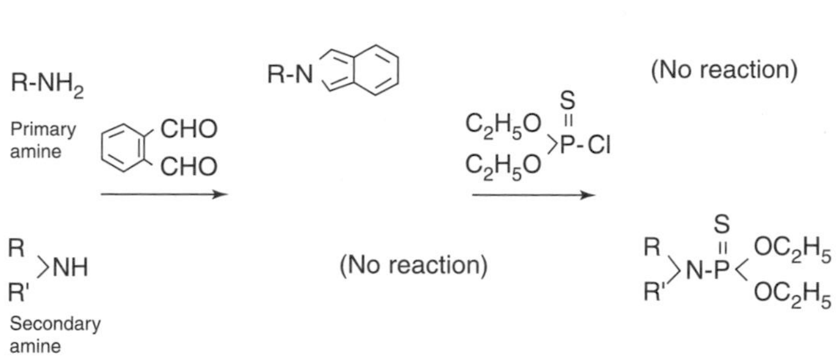

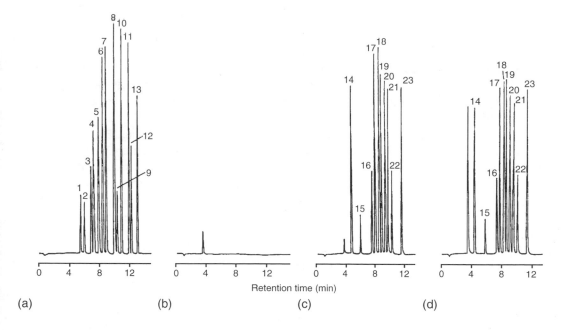

FIGURE 11.7 Selective derivatization of primary and secondary amines with OPA and DETPC. R, R', alkyl or aryl. (From Kataoka, H., *J. Chromatogr.*, 733, 19–34, 1996.)

usually done by perfluoracylation of the amines, and the derivatives can be detected by ECD or mass spectrometry (MS). However, many of the reagents in use for perfluoracylation need a strictly anhydrous medium. Thus it was necessary to develop an alternative derivatization method for AAs which would not be restricted to anhydrous media and would be comparable or even superior in terms of ECD sensitivity. One way to increase the sensitivity of ECD is to introduce heavier halogens instead of fluorine into the molecules because the detector response increases in the order F < Cl < Br < I.[73]

FIGURE 11.8 Gas chromatograms obtained from *N*-diethylthiophosphoryl derivatives of primary and secondary amines. (a) Primary amines; (b) primary amines pretreated with OPA; (c) secondary amines; (d) secondary amines pretreated with OPA. Each peak represents 20 pmol of amines. Peaks: 1, methylamine; 2, ethylamine; 3, propylamine; 4, isobutylamine; 5, *n*-butylamine; 6, isoamylamine; 7, *n*-amylamine; 8, hexylamine; 9, cyclohexylamine; 10, heptylamine; 11, octylamine; 12, benzylamine; 13, β-phenylethylamine; 14, dimethylamine; 15, diethylamine; 16, dipropylamine; 17, pyrrolidine; 18, piperidine; 19, morpholine; 20, dibutylamine; 21, hexamethyleneimine; 22, *N*-methylcyclohexylamine; 23, *N*-methylbenzylamine. (From Kataoka, H., *J. Chromatogr.*, 733, 19–34, 1996.)

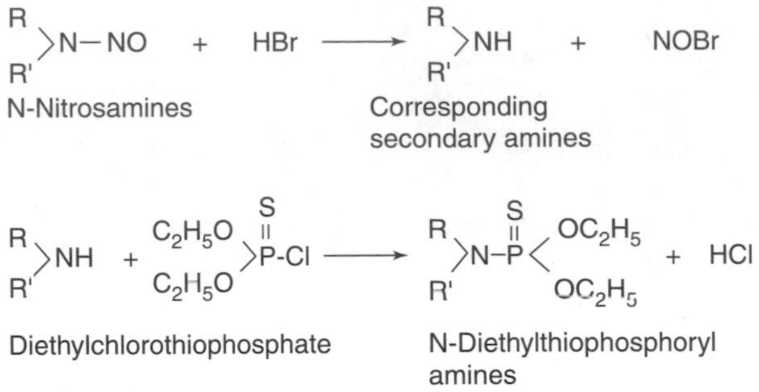

$$\begin{array}{c} R \\ \diagdown \\ R' \end{array} N-NO \quad + \quad HBr \quad \longrightarrow \quad \begin{array}{c} R \\ \diagdown \\ R' \end{array} NH \quad + \quad NOBr$$

N-Nitrosamines   Corresponding
                                        secondary amines

Diethylchlorothiophosphate   N-Diethylthiophosphoryl
                                       amines

FIGURE 11.9 Process for derivatization of *N*-nitrosamines. (From Kataoka, H., Shindoh, S., and Makita, M., *J. Chromatogr.*, 723, 93–99, 1996.)

The derivatization takes place in two steps following the reaction scheme given in Figure 11.10. The amino group is first diazotized at room temperature with nitrite in an acidic medium. In the second step, the diazo group is substituted by iodine at elevated temperatures.[17]

As well as improving the gas chromatographic properties of compounds, derivatives exhibit improved peak shape. Frequently, derivatization can also improve the detectability of compounds. For these reasons, derivatization is frequently employed in gas chromatographic analyses of polar compounds, such as carboxylic acids, amino acids, and amines, and has a wide range of applications.

Most primary and secondary amines exhibit poor chromatographic performance via direct HPLC approaches, making quantitative trace analysis difficult. Chemical derivatization in solution has long been accepted as an effective modification technique in HPLC, improving the overall specificity, chromatographic performance, and sensitivity for trace analysis.[74]

Most of these derivatization reactions proceed in organic media and require long reaction times. Extraction and preconcentration procedures are often needed before HPLC analysis.[75]

Examples of derivatization reaction for HPLC determination of amines are shown in Table 11.6.

Various derivatization reagents have been developed and used as labeling reagents for traces of primary and secondary aliphatic amines in HPLC, including OPA, 3,5-dinitrobenzoyl chloride, 8-quinolinesulfonylchloride, 1-naphthyl isocyanate, 9-fluorenylmethyl-chloroformate (FMOC-Cl), phthalimidylbenzoyl chloride (Phibyl-Cl), phenyl isothiocyanate, 5-dimethylaminonaphthalene 1-sulfonyl chloride (Dns-Cl), 3,4-dihydro-6,7-dimethoxy-4-methyl-3-oxoquinoxaline-2-carbonyl chloride (DMEQ-Cl), 6-methoxy-2-methyl-sulfonylquinoline-4-carbonyl chloride (MSQC-Cl), 7-chloro-4-nitrobenzo-2-oxa-1,3-diazole (NBDCl), 7-fluoro-4-nitrobenzo-2-oxa-1,3-diazole (NBD-F), *p*-nitrophenylacetamides, trinitrobenzene sulfonate, nitrophenyls, *p*-benzoquinone, 1,2-naphthoquinone-4-sulfonate, and some acridinium trifluromethanesulfonates (Figure 11.11).

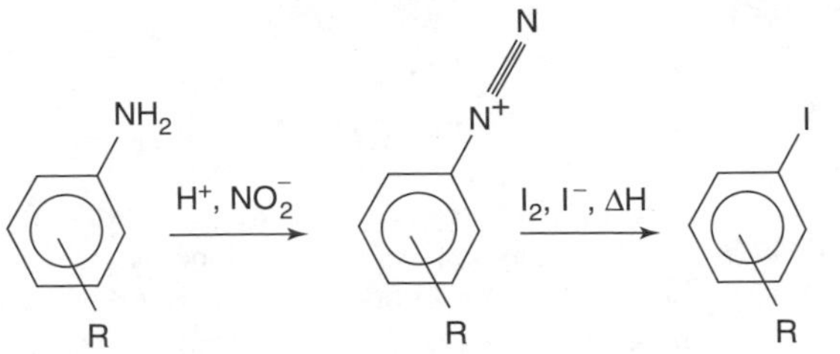

FIGURE 11.10 Reaction scheme of the iodination of aromatic amines.

TABLE 11.6
Derivatization Reactions for HPLC Determination of Amines

Reagent	Amine Type	Environment	Reference
Fluorescent reagent			
Fluorescamine	P, S	W	73,74
Fluorenylmethoxycarbonyl	—	A	76
N-Hydroxysuccinimidyl fluorescein-O-acetate (SIFA)	Al	W	1,60
9-Fluorenylmethyl chloroformate (FMOC)	P, S, Al	W	73,77,78
Chloride reagent			
Dansyl chloride	P, S	W	58,79,80
8-Quinolinesulfonyl chloride	P, S	—	81
2-Naphthyloxycarbonyl chloride	P, S	—	82
3,5-Dinitrobenzoyl chloride (DBN)	Al	H	49
Ester reagent			
Lumarin I and II	P, S	—	83,84
Other			
N-Hydroxysuccinimidyl 4,3,2′-naphthapyrone-4-acetate	P, S	H	75
(NPA-Osu)	B	—	85
2-Chloroethylnitrosourea (CENU)	PA	W	86–89
O-Phthalaldehyde (OPA)	P,S	W	90
1,2-Naphthoquinone-4-sulphonate	AA	—	91
Acridinium trifturomethanesulphonates	Al	A	92
1 Naphthyl isothiocyanate (NITC)	—	W	93
1-Naphthyl isothiocyanate (NITC)	Al	W	93
Bis(2-nitrophenyl)oxalate (2-NPO)	S	A	95
2,5-Dihydroxybenzaldehyde (2,5-DBA)	P	W	49
Acridine-9-acetyl-N-hydroxysuccinimide	Al	W	1
Phenylisothiocyanate	Al	—	96
4-(5′,6′-Dimethoxybenzothizolyl)phenylisothiocyanate	Al	H	97
Fluram	Al		

P, primary amine; S, secondary amine; T, tertiary amine; A, ammonia; N, nitrosamine; AA, aromatic amine; PA, polyamine; Al, aliphatic amine; A, air; H, water; S, soil; W, waste.

For example, 9-fluorenylmethylchloroformate (FMOC) (Figure 11.11) forms derivatives with both primary and secondary amines and may be used to protect hydroxy groups. The derivatives are fluorescent and absorb in the ultraviolet region, they are formed in a reaction time of less than 1 min in a buffered aqueous solution at room temperature, and yield stable derivatives. 9-FMOC is a derivatizing agent useful for reversed-phase liquid chromatographic (LC) determination of DMA in groundwater samples at low (μg/l) concentration levels. The reagent allows quick and selective amine quantitation with excellent recovery and linearity. The method is free from matrix effects and allows simultaneous detection of a number of other amines, polyamines, and amino acids.[73]

N-Hydroxysuccinimidyl fluorescein-O-acetate (SIFA) (Figure 11.11) is a new derivatizing reagent. SIFA reacts with amines to form derivatives in a pH 8.5 water–methanol solution modified with H_3BO_3–Na_2BO_7 buffer. The derivatives of SIFA with some amines were separated in a mobile phase of methanol–water (46/54, v/v) containing 10 mmol/l pH 5.40 citric acid–Na_2HPO_4 buffer within 18 min, with fluorescence detection at excitation and emission wavelengths of 488 and 516 nm, respectively[1] (Figure 11.12). The method using SIFA as the derivatizing reagents in

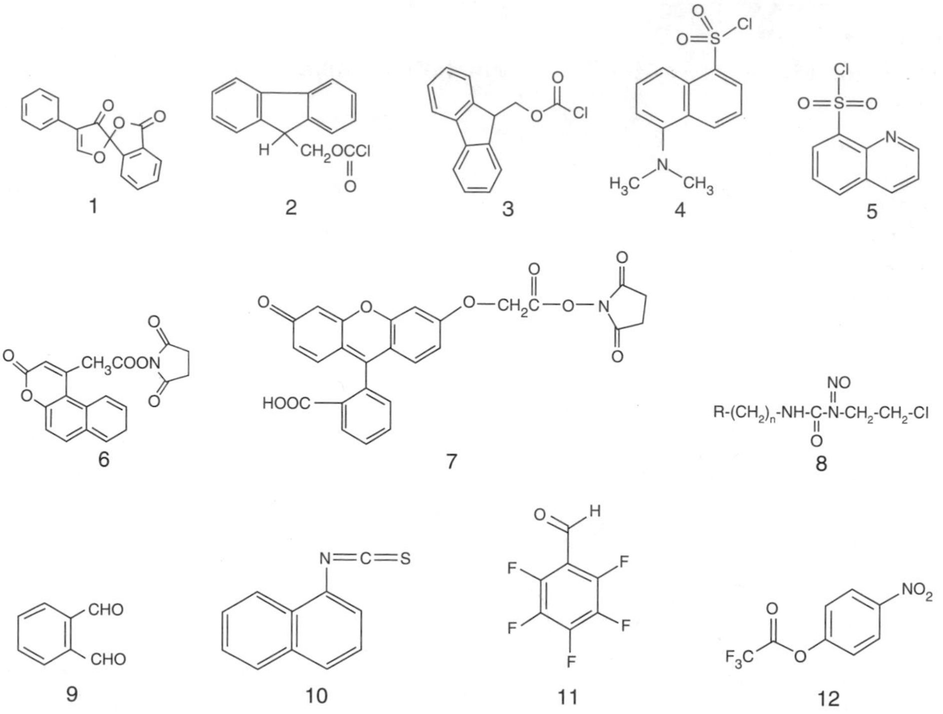

FIGURE 11.11 Derivatives used in HPLC: 1, fluorescamine; 2, fluorenylmethoxycarbonyl; 3, FMOC; 4, Dansyl chloride; 5, 8-quinolinesulfonyl chloride; 6, NPA-Osu; 7, SIFA; 8, CENU; 9, OPA; 10, NITC; 11, 2,3,4,5,6-pentafluorobenzylaldehyde (PFBAY); 12, p-nitrophenyl trifluoroacetate (NPTFA).

aliphatic amine analysis had the advantages of high sensitivity and selectivity in determination, and facility and convenience in handling.[1]

8-Quinolinesulfonyl chloride (Figure 11.13) is susceptible to nucleophilic attack by primary and secondary aliphatic amines, the strong UV-absorbing character of the sulfonamides formed. The optimum separation of the amine derivatives is achieved with the MeCN–acetate buffer–TEA ratio of 50:50:0.01.[81]

N-Hydroxysuccinimidyl ester reagents have been developed earlier for analyzing amines. N-Hydroxysuccinimidyl carbamate (AQC), N-hydroxysuccinimidyl-3-indoylacetate (SIIA), 9-flouorenylmethoxycarbonylsuccinimide (FMOC-Osu), 7-(diethylamino)coumarin-3-carboxylic acid succinimidyl ester (DCCS), N-hydoxysuccinimidyl fluorescein-O-acetate (SIFA), Lumarin I and II, etc., all readily react with primary and secondary amines under mild conditions to give the stable conjugates without forming byproducts.[75] In Liu's experiments, N-hydroxysuccinimidyl 4,3,2'-naphthapyrone-4-acetate (NPA-OSu) permits the sensitive and fairly selective derivatization of primary and secondary amines (Figure 11.11). NPA-OSu reacted with amines to form derivatives in a pH 8.50 aqueous buffer. Optimum separation for eight investigated amines was obtained using water–methanol (56:44, v/v). The detection limits were in the range 1 to 33 fmol for an injection volume of 20 μl. This method using NPA-OSu allowed the existence of salts and other organic substances, and was suitable for direct determination of aliphatic amines in a real environmental water sample.[75]

Examples of derivatization reactions of amines by other separation methods are shown in Table 11.7.

Fluorescein isothiocyanate isomer I (FITC), which provides good sensitivity for primary and secondary amines, was chosen as a reagent for derivatization of dimethylamine (DMA) and other

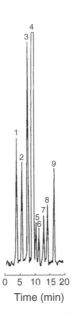

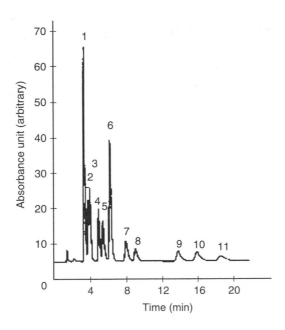

FIGURE 11.12 Typical chromatogram of SIFA derivatives with amines. (1) Methylamine, (2) ethylamine, (3) ethylene diamine, (4) FOAA (hydrolyzed product of SIFA), (5) isopropylamine, (6) isobutylamine, (7) n-butylamine, (8) 1,4-butylene diamine, (9) 1,5-amylene diamine. (From Wang, H., Li, J., Liu, X. and Zhang, H.-S., *Anal. Chim. Acta*, 423, 77–83, 2000.)

FIGURE 11.13 Liquid chromatogram of 8-quinolinesulfonamides on a RP-C₆ column. Sulfonamide of 1, methylamine; 2, dimethylamine; 3, ethylamine; 4, propylamine; 5, diethylamine; 6, butylamine; 7, pentylamine; 8, dipropylamine; 9, heptylamine; 10, dibutylamine; 11, octylamine. (From Saleh, M. I. and Pok, F. W., *J. Chromatogr.*, 763, 173–178, 1997.)

TABLE 11.7
Derivatization Reactions of Amines (Other Methods)

Reagent	Separation Method	Amine Type	Environment	Reference
Fluorescein isothiocyanate	CE	S, Al, AA	A, H, W	94,98,99
O-Phthalaldehyde (OPA)	IC	PA	W	100
Fluorescamine	CE, MEKC	AA	W	101
Catechol	Voltammetric	Aniline	H	102

MEKC, micellar electrokinetic chromatography; CE, capillary electrophoresis; IC, ion chromatography; S, secondary amine; A, ammonia; AA, aromatic amine; PA, polyamine; Al, aliphatic amine; A, air; H, water; S, soil; W, waste.

amines. To the standard solution containing DMA or mixture of amines, 0.2 M sodium bicarbonate (pH 8.8) and 1.1 mM FITC acetone solution were added, and the total volume was made up to 1 ml with deionized water. The screwed capped reaction vessel was allowed to stand overnight in darkness and at room temperature (21°C). Before CE analysis, the derivatization mixtures were diluted five times with a running electrolyte.

The conditions for the derivatization reaction were optimized using DMA. The general aim of these experiments was to achieve the best possible compromise between high fluorescence intensity of DMA derivative and low side reaction products. For optimization of derivatization conditions several parameters affecting the reaction were studied, including the chemical composition, concentration, and pH of the buffer used, the amount of FITC, addition of organic solvents, reaction time, and temperature. CE analysis of FITC-derivatized DMA was performed with a 20 mM borate buffer containing 10% acetone and with 25 kV voltage.[94]

III. ANALYSIS AND DETECTION METHODS

A. HPLC AND LC–MS

The most common techniques for the analysis of amines in the environment are GC and HPLC. HPLC analysis seems to be a good alternative to GC analysis since there are no derivatization requirements. A preconcentration step is necessary, however, in trace analysis owing to the relatively low sensitivity of HPLC detectors suitable for these compounds.[33]

The most widely used detection methods in HPLC analysis of anilines and phenols are UV (especially diode array) and electrochemical detection (ED). UV detectors provide very good signal stability and in the case of diode-array detectors they can be used for analyte tentative confirmation purposes using UV spectra libraries. Electrochemical detectors are more sensitive than the UV detectors; however, their performance is highly dependent on the type of samples analyzed. Components from dirty samples are deposited on the electrochemical cell and the detector sensitivity is rapidly decreased.[103–105] The heterocyclic amines can be measured with UV, electrochemical, and fluorescence detectors.

Examples of HPLC methods are shown in Table 11.8.

Liquid chromatography–mass spectrometry (LC–MS) ideally combines advantages of the gentle separation of HPLC with high sensitivity and selectivity of mass spectrometry. There are two methods of ionization. The electrospray LC–MS ionization process can effectively transform the heterocyclic amines from solution to protonated ions in the gas phase. The thermospray LC–MS ionization process produces abundant pseudo-molecular ions for this class of compounds.[7] As a result, in both ionization processes the base peaks in the mass spectra are

TABLE 11.8
HPLC Methods for Amines

Method	Detection	Amine	Environment	Reference
HPLC	UV	HA		106–108
HPLC	UV-FL	HA		109,110
HPLC	UV-ECD	HA		111
HPLC	ECD	HA		112–114
HPLC	FL	HA		115–119
HPLC	UV	AA	H, W	117–119,33
RP-HPLC	UV	PA	W	120
RP-HPLC	UV	PA	W	121,85
HPLC	FL	Al	H	75,1
RP-HPLC	UV	Al	—	81
RP-HPLC	UV	AA	H	122
HPLC	UV–ViS	Al	W	123,74
HPLC	ECD	Al	A	95
HPLC	ECD	AA	W	105
LC–MS	TSI-SIM	HA		110
LC–MS	TSI-SIM	HA		124
LC–MS	ESI-MS–MS	HA	H	21
LC–MS–MS	ESI	N	W	125

HA, heterocyclic amine; AA, aromatic amine; PA, polyamine; Al, aliphatic amine; N, nitrosamine; ECD, electrochemical; UV, ultraviolet; FL, fluorescence; TSI, thermospray ionization; ESI, electrospray ionization; MS, MS-tandem mass spectrometry; SIM, selected ion monitoring; A, air; H, water; W, waste.

detected at $[M + H]^+$ (Figure 11.14). The formulas of the amines presented are shown in Figure 11.15.

LC–MS using atmospheric pressure ionization (LC-API-MS) has dramatically changed the analytical methods used to detect polar pollutants in water. Most API mass spectrometers offer two interfaces: electrospray ionization (ESI) and atmospheric pressure chemical ionization (APCI), both of which can be operated in positive and negative ion mode. ESI transfers ions from solution into the gas phase, whereas APCI ionizes in the gas phase. Analytes occurring as ions in solution may be best analyzed by ESI, while nonionic analytes may be well suited for APCI. What must always be taken into consideration is the relations between analyte properties and the chosen method of chromatographic separation.[126]

Most amines could be separated using the columns and elution conditions presented in Table 11.9.

B. GC and GC–MS

GC has been widely used for amine analyses because of its simplicity, high resolving power, high sensitivity, and low cost.[7] Coupled with mass spectrometry (GC–MS), it is a technique most commonly employed for the analysis of volatile organic pollutants in environmental samples. In this combination, the GC separation usually provides isomer selectivity, while the MS shows compound class homologue specificity. The MS fragmentation pattern can provide unambiguous component identification by comparison with library spectra.

Several ionization techniques are used in GC–MS. One of them, popular because of spectra reproducibility, is electron ionization (EI). In EI, gas analyte molecules are bombarded by energetic electrons (typically 70 eV), which leads to the generation of a molecular radical ion (M^+) that can

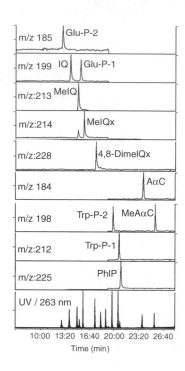

FIGURE 11.14 HPLC separation of 11 heterocyclic amine standards with thermospray MS detection. The lowest panel shows the chromatogram observed with UV detection, whereas the upper panels show the chromatograms observed in various mass selective detection channels. (From Kataoka, H., *J. Chromatogr.*, 774, 121–142, 1997.)

subsequently generate ionized fragments. In other cases, chemical ionization (CI) is applied. In CI, ion–molecule reactions take place between reagent gas ions and sample molecules. As a result, molecular ions, adduct ions, and fragment ions can be generated. Chemical ionization provides better sensitivity and selectivity than EI, but the number of applications is relatively low compared with EI.

The GC–MS instruments used range from simple linear quadrupoles to multisector analyzers with EI and positive/negative CI capabilities that allow for the achievement of low detection limits, because of increasing selectivity and detection limits parts per trillion (ppt). However, high-resolution mass spectrometry (HRMS) is recommended when an enhancement of the selectivity of MS detection is required, because this technique has the capacity of matrix-interfering compounds. In recent years, the use of MS instruments with quadrupole ion-trap (GC-ITMS-MS) or time-of-flight mass analyzers (TOF-MS) has come to play an important role in environmental analysis.

The new-generation sensitive detection instruments are ideal for combining with high-speed GC or comprehensive two-dimensional gas chromatography (GC × GC), the two most promising recent developments in GC, for the identification and quantification of complex environmental samples, which require an extremely fast acquisition rate.[128]

GC–MS bas been recognized as the method of choice in a wide series of environmental analyses because of its superiority in selectivity and sensitivity.[129] However, these methods have some inherent problems related to the difficulty in handling low molecular weight (LMW) amines because of their high water solubility and volatility.[61] Moreover, in GC analysis, amines are likely to be adsorbed and decompose in the column, and readily give tailed elution peaks.[64]

The direct GC determination of some underivatized AAs, particularly nitro-substituted anilines, requires optimal deactivation of the GC system and careful choice of experimental conditions to

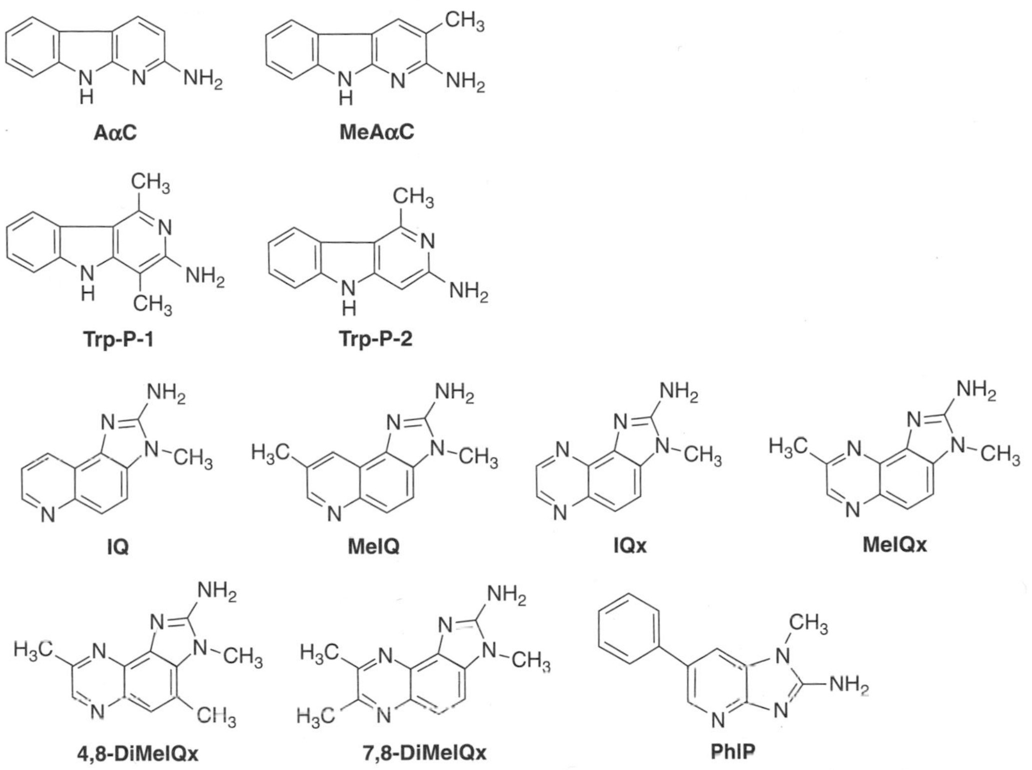

FIGURE 11.15 The following amines that are in Figure 11.14 and Figure 11.21.

obtain satisfactory peak shape and resolution. In this case, a set of repeated analyses of the same sample has to be carried out for the determination of a broad spectrum of compounds.[130] So, determination of AAs usually demands a derivatization step to lower the polarity of compounds, and to improve detection senility and selectivity.[24]

Many derivatization reagents for GC analyses of aliphatic primary amines by flame ionization detection (FID), electron capture detection, flame thermionic detection, GC–MS with selected ion monitoring (GC–MS-SIM), and electron-capture-negative-ion chemical ionization (GC-NICI-MS) detection with a modified thermal energy analyzer have been reported.[131]

Table 11.10 shows popular methods of amine detection used in GC analyses, while Table 11.11 shows examples of GC–MS analyses depending on the selection of column type, temperature, carrier gas, as well as the conditions of injection and detection.

C. CE AND IC

CE has emerged as a powerful alternative to HPLC in separation science. CE methods afford high-speed and high-efficiency separations, utilize relatively inexpensive and long-lasting capillary columns, and consume small volumes of sample and reagent.[99] CE is an extremely versatile separation method because selectivity can be changed essentially by addition of different modifiers to aqueous buffers or by changing buffer pH.[94] In CE, where the most widely applied detection method is UV absorption, the small injection volumes and short optical path lengths (25 to 75 μm) encountered in most systems make concentration sensitivity (using UV absorption) relatively low when compared with HPLC methods.[99]

TABLE 11.9
Analysis of Amines by HPLC

Column (C), Elution (E), Flow-rate (F)	Reference
C: 150 × 4.6 mm (Nucleosil 100-5 C_{18}, 5 μm) E: 5 mM phosphate buffer, pH 3-acetonitrile–water (90:10) F: 2 ml/min	25
C: 150 × 4.6 mm (Prodigy ODS RP-18, 5 μm) E: methanol–acetonitrile–buffer phosphate 0.01 M (30:30:40) F: 1 ml/min	95
C: 250 × 4.6 mm (Chrompack Inertsil ODS-2) E: methanol–water–buffer pH 3.40 (60:35:5) F: 0.2 ml/min	33
C: 100 × 4.6 mm (Partisil 5 μm C_{18}) E: methanol–water (45:55)	34
C: 150 × 3.9 mm (Nova-Pak 4 μm C_{18}) E: acetonitrile–acetate buffer pH 4.66 (40:60) F: 0.2 ml/min	119
C: 250 × 4.6 mm (I.D. ODS-2.5 μm) E: acetonitrile–water (30:70) F: 1 ml/min	93
C: 250 × 4.6 mm (10 μm Spherisorb Nitric and Hexyl) E: methanol–acetonitrile F: 1 ml/min	81
C: 250 × 4.6 mm (10 μm Lichrosorb RP-C_8) E: water–ammonium acetate buffer and acetonitrile–water (80:20) F: 0.8 ml/min, 40°C	85
C: 150 × 3.9 mm (Nova-Pak C_{18}) E: acetonitrile–acetate buffer pH 4.66 (40:60) F: 1 ml/min	117
C: 250 × 4.6 mm (ODS 80T, 5 μm) E: ammonium acetate pH 6.0–acetonitrile (90:10), ammonium acetate pH 6.0–acetonitrile (70:30)	112,113
C: 250 × 4.6 mm (I.D. 5 μm C_{18}) E: methanol–water (46:54) and buffer pH 5.40 F: 1 ml/min	1
C: 250 × 4 mm (Spherisorb ODS-2, 10 μm C_{18}) E: acetate buffer pH 4.0–acetonitrile (30:70) F: 1 ml/min	74
C: 250 × 4.6 mm (Nucleosil SA, 5 μm) E: various ratios of KNO_3, NaH_2PO_4, KH_2PO_4, H_2O, 85% H_3PO_4, acetonitrile F: 1 or 2 ml/min	127
C: 2.6 × 50 mm cation exchange resin E: 0.045 M sodium citrate, 0.061 M citric acid, 0.064 M NaCl and 0.20 M sodium citrate, 2 M NaCl, pH 7.0	89
C: 100 × 3 mm (Genesis C_{18}; I.D. 4 μm) E: methanol-10 mM ammonium formate, pH 4 in water–methanol (80:20) F: 0.3 ml/min	125

TABLE 11.10
GC Methods for Amines

Method	Detection	Amine	Environment	Reference
GC	NPD	HA		7
GC–MS	NICI-SIM	HA		7
GC–MS	EI-SIM	HA		7
GC–MS	NICI-SIM	HA		7
GC–MS	NICI-SIM	HA		7
GC–MS	SIM	AA	H	129
GC–MS	SIM	Al	H	64
GC	FID	S	W	69
GC	ECD	AA	H	71
GC	FPD	Al (P)	W	131
GC	FPD	NA	A	3
GC	EC-NICI-MS	AA	W	57
GC	FPD	AA	A	4
GC	NPD	HA	W	59
GC	ECD	AA	W	132
GC	ECD	AA	W	133
GC–MS	EI	ABDACs	H	22
GC–MS	SIM	AA	W	9
GC–MS	NICI-SIM	AA	A	28
GC	FID	Al	A	39
GC	FID	Al	A	40
GC–MS	EI	AA	W	134
GC	ECD, AED	AA	W	134
GC–MS	—	Al	W	58
GC	ECD	AA	W	135
HRGC, IT-MS/MS	NICI-SIM	MAM	W	136
GC–MS	—	AA	—	137
GC	—	AA	A	138
CGC	NPD	Al	H	139

HA, heterocyclic amine; AA, aromatic amine; PA, polyamine; Al, aliphatic amine; N, nitrosamine; MAM, Musk amino metabolites; ABDACs, alkylbenzyldimethylammonium chlorides; ECD, electron capture detection; AED, atomic emission detection; FID, flame ionization detection; FPD, flame photometric detection; GC–MS-SIM, GC–MS selected ion monitoring; NPD, nitrogen phosphorus detection; NICI, negative-ion chemical ionization; EI, electron ionization; CGC, capillary GC; A, air; H, water; W, waste.

Laser-induced fluorescence (LIF) detection offers a high sensitivity in CE; however, compared with ED, it is expensive and lacks universality. So, ED could become an important alternative detection mode for CE.[140]

Examples of CE and IC analyses are shown in Table 11.12 and CE conditions in Table 11.13.

Figure 11.16 shows an example of an electrophorogram of an amine standard solution by capillary zone electrophoresis.

CE appeared as a promising substitute for IC, mainly because of its higher speed of separation, but now IC remains the major analytical technique, not only for inorganic species. IC offers greater sensitivity and analytical ruggedness; for example, by column switching and enrichment trap columns, detection of ppb or ppt concentrations can easily be achieved. Classic applications of IC are strictly associated with suppressed conductometric detection.[147]

Examples of separation of amines by IC method are shown in Table 11.14.

TABLE 11.11
GC–MS Methods

Analytes (A), Column (m/mm/μm) (C), Temperature (T), Carrier Gas (Cg), Injection and Detector temperature (ID)	Reference
A: Al C: 35/0.25/0.25 fused-silica capillary DB5 T: 140:3 min; 3/min to 210, 10/min to 290, 290: 5min Cg: Helium, 1 ml/min ID: 290	58
A: AA C: 30/0.25/0/25 (5%-phenyl)-methlpolysiloxane DB5 T: 135: 21.5 min; 12.5/min to 235, and hold for 8.5 min Cg: Nitrogen ID: 250 and 300	17
A: AA C: 25/0.25/0.25 (5%-phenyl)-methlpolysiloxane ZB5 T: 40/5min, 10/min to 320 and hold for 15 min Cg: Helium ID: 250 and 280	9
A: Al C: 15/0.53/1.5 fused silica capillary DB-1 T: 150/5min; 5/min to 200, 20/min to 280 and hold for 2 min Cg: Nitrogen, 10 ml/min ID: 290	65
A: AA C: 50/0.20/0.5 fused silica capillary HP5 T: 50/0.5 min, 50/min to 110, 5/min to 225, 20/min to 280 and hold for 15 min Cg: Helium ID 250 and 280	130
A: Al C: 15/0.53/1.0 fused silica capillary DB-1701 T: 100/10min, 10/min to 260 Cg: Nitrogen, 10 ml/min ID: 280	69
A: AA C: 30/0.25/0.50 PTA-5 base-deactivated T: 60/5min, 10/min to 250 and hold at 250 for 3 min Cg: — ID: detector temperature 280	129
A: AA C: 30/0.25/0.25 (5%-phenyl)-methlpolysiloxane T: 170/18 min, 10/min to 230 and hold 10 min Cg: Nitrogen, 20 ml/min ID: 250 and 300	71
A: AA C: 30/0.25/0.25 (5%-phenyl)-methlpolysiloxane T: 135/20.5 min, 12.5/min to 235 and hold 8.5 min Cg: Helium ID: injection temperature 250	132

Continued

TABLE 11.11
Continued

Analytes (A), Column (m/mm/μm) (C), Temperature (T), Carrier Gas (Cg), Injection and Detector temperature (ID)	Reference
A: AA C: 30/0.25/0.25 (5%-phenyl)-methlpolysiloxane T: 170/18, 10/min to 230 and hold for 20 min Cg: Nitrogen, 20 ml/min ID: injection temperature 250	134
A: Al (determination by CGC) C: 15 m × 150 μm i.d. column coated with 2 μm film of CP SIL 5CB Cg: helium at a pressure of 145 kPa ID: thermal desorption and cryotrap reinjection	139

AA, aromatic amine; Al, aliphatic amine.

TABLE 11.12
Analysis of Amines by CE and IC

Method	Detection	Amine	Environment	Reference
CZE	UV	HA		1
CE	LIF	Al	A	94
CE (CZE)	ED	AA	H	141
CE (MEKC)	FL	AA	H	141
CE	FL	AA	H	99
CZE	UV	AA	H, W, S	26
CE	UV	Al	H	142
CE (MEKC)	—	AA	—	143
CE (MEKC)	UV	NA	—	144
CE	LIF	Al	W	98
CE (MEKC)	LIF	AA	—	145
IC	—	AL.	W	146
IC	ED	AA	W	147
IC	FL	PA	W	89
IC	UV	PA	W	100
IC	ED	Hydroxylamine	W	148
IC	CD, AD	Hydroxylamine	W	149
IC	PA	AlA	W	150
CE	ED	AA	—	151
CE	UV	AQA	—	152

HA, heterocyclic amine; AA, aromatic amine; PA, polyamine; Al, aliphatic amine; N, nitrosamine; CZE, capillary zone electrophoresis; MEKC, micellar electrokinetic capillary chromatography; LIF, laser induced fluorescence; ED, electrochemical detection; CD, conductivity detection; AD, amperometric detection; PD, potentiometric detection; AlA, alkylamines; AQA, alkyl and alkylbenzyl quaternary ammonium compound; A, air; H, water; S, soil; W, waste.

TABLE 11.13
CE Separation Conditions for Amines

Column (C), Elution (E), Separation Voltage (S)	Reference
C: 57 cm × 75 μm I.D. (effective length 50 cm) E: 40 mM phosphate buffer, pH 7.0 in 25% methanol + 0.1 M NaOH, H_2O and above buffer S: 20 or 25 kV	98
C: 37.5 cm × 75 μm fused silica E: 10 mM sodium acetate + 5 mM acetic acid + 1 mM NaCl, pH 5.0 S: 25 kV	141
C: 68 cm × 75 μm I.D. (effective length 53 cm) E: 5 mM sodium tetraborate + 4.5 mM boric acid + 20 mM SDS, pH 9 S: 30 kV	141
C: 27 cm × 75 μm I.D. (effective length 20 cm) E: 0.4 M NaOH + deionized water + phosphate buffer pH9 S: 16 kV, 93–104 μA	99
C: 47 cm × 75 μm I.D. (effective length 40 cm) E: 50 mM NaH_2PO_4 + 7mM 1.3-diaminopropane, pH 2.35 S: 30 kV, 55 μA	26
C: 82 cm × 50 μm I.D. (effective length 50 cm) E: 7 mM hydroxypropyl β-cyclodextrin + 13 mM sulfobutylether β-cyclodextrin + 50 mM phosphate buffer, pH 8 S: 20 kV	143
C:65 cm (effective length 56.5 cm) E: 75 mM SDS + 10 mM triethanolamine + 30 mM H_3PO_4, pH 1.9 S: 25 kV	145

IV. APPLICATIONS IN ENVIRONMENTAL ANALYSIS

A. AIR (A)

Examples of environmental analysis of amines in air are shown in Table 11.15.

Air pollution is a important problem for public health. For air analysis, the National Institute for Occupational Safety and Health specifies exposure limits[163] for amines in industrial air (10 to 30 ng/ml) and for amines in indoor air (10 to 300 pg/ml). The direct SPME could be used to monitor the amounts of the amines extracted from air in the following order: Carbowax divinylbenzene CAX(DAB) > poly(acrylate) PA > poly(dimethyl-siloxane) PDMS.[40]

An integral part of environmental analysis in air is the examination of the presence of amines (mainly heterocyclic) in an atmosphere polluted with cigarette smoke. Epidemiological studies have shown that cigarette smoking is associated with the development of human cancers, and cigarette smoke condensate has been known to be mutagenic in bacteria and carcinogenic to experimental animals. Cigarette smoke contains various pyrolysis products. Among them are known patent mutagens and carcinogens such as volatile N-nitrosamines, tobacco-specific nitrosamines, polycyclic aza-arenes, and mutagenic and carcinogenic heterocyclic amines.[7] Figure 11.17 shows an example of a chromatogram of AAs in Kentucky cigarettes, while Table 11.16 compares the results of a determination of AA amounts in ng/cigarette using the new SPE method, ISO method, and some other references.

It is also very important to investigate the contents of N-nitrosamines in cigarette smoke, and especially in mainstream and sidestream smokes. In Kataoka's experiment,[67] cigarette smoke was

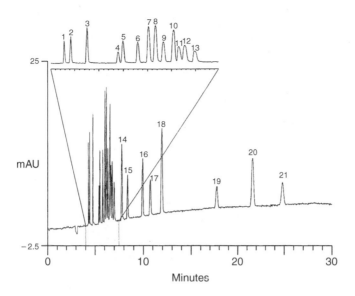

FIGURE 11.16 Electrophorogram for a standard solution of aromatic amines. Peaks: 1, pyridine; 2, *p*-phenylenediamine; 3, benzidine; 4, *o*-toluidine; 5, aniline; 6, *N,N*-dimethylaniline; 7, *p*-anisidine; 8, *p*-chloroaniline; 9, *m*-chloroaniline; 10, ethylaniline; 11, α-naphthylamine; 12, diethyl aniline; 13, *N*-(1-naphthyl)ethylenediamine; 14, 4-aminophenazone; 15, *o*-chloroaniline; 16, 3,4-dichloroaniline; 17, 3,3'-dichlorobenzidine; 18, 2-methyl-3-nitroaniline; 19, 2,4-dichloroaniline; 20, 2,3-dichloroaniline; 21, 2,5-dichloroaniline. (From Cavallaro, A., Piangerelli, V., Nerini, F., Cavalli, S., and Reschiotto, C., *J. Chromatogr.*, 709, 361–366, 1995.)

TABLE 11.14
IC Separation Conditions for Amines

Column (C), Elution (E), Flow-Rate (F)	Reference
C: 50 × 4 mm I.D. (Dionex IonPac CG 12A) and 250 × 4 mm I.D. (CS12A) E: 18 mM methanesulphonic acid F: 1 ml/min	94
C: IonPac CS 15 E: 9 mM methanesulphonic acid + 0.7% methyl ethyl ketone F: 1 ml/min	146
C: 50 × 4 mm (Dionex IonPac CG-5 and IonPac CG 10) E: water–methanol (90:10) + 1.5 mM sodium formate F: 1 ml/min	147
C: 250 × 4 mm (Dionex IonPac CS2, CS3 and CS 15) E: 14% CH_3CN + 18 mM HCl + 430 mM NaCl F: 1 ml/min	100
C: 250 × 4 mm (Dionex IonPac CS 14) and 50 × 4 mm (IonPac CG 14) E: 7.5–15 mM sulfuric acid F: 1 ml/min	148
C: 100 × 4.6 mm I.D. Alltech universal cation-exchange E: 5 mM HNO_3 + 5% acetonitrile F: 1 ml/min	150

TABLE 11.15
Contents of Amines in Air Environments

Environments	Method	Type of Amine	Reference
Cigarette smoke	GC-FPD	N	67,153
Cigarette smoke	GC	AA	4,28,154–155
Airborne	GC-FID	TMA	39
Air	GC-FID	Al	40
Air	—	AA	156–157
Waste gases	—	AA	158–162
Air	HPLC-ECD	Al	95
Outdoor air	HPLC-FL	HA	3
Indoor air	HPLC-FL	HA	3
Diesel exhaust particles	HPLC-FL	HA	67
Cigarette smoke (mainstream)	HPLC-FL	HA	115,116
Cigarette smoke (sidestream)	GC-NPD	HA	7
Incineration ash	HPLC-FL	HA	7
Air	GC	Al	163
Work place air	GC-NPD	AA	164,165
Cigarette smoke (indoor air)	GC–MS-SIM	AA	3
Cigarette smoke (mainstream)	GC–MS–MS	AA	166,167

HA, heterocyclic amine; AA, aromatic amine; Al, aliphatic amine; N, nitrosamine; TMA, trimethylamine.

collected with a laboratory-made smoking machine by bubbling in 5% hydrochloric acid. The mainstream and sidestream smoke samples could be separately collected by this apparatus. By extraction twice with diethyl ether containing 25% 2-propanol in acidic media, the *N*-nitrosamines were quantitatively transferred into the organic layer (*N*-nitrosamine fraction),

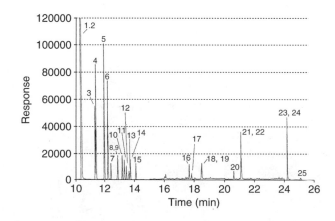

FIGURE 11.17 Reconstructed chromatogram of aromatic amine-HFBA derivatives from the particulate phase of 20 lR4F Kentucky reference cigarettes. Peaks: 1 and 2, aniline; 3 and 4, *o*-toluidine; 5, *m*-toluidine; 6, *p*-toluidine; 7, 2-ethylaniline; 8, 2,6-dimethylaniline; 9, 2,5-dimethylaniline; 10, 2,4-dimethylaniline; 11, 3-ethylaniline; 12, 3,5-dimethylaniline; 13, 2,3-dimethylaniline; 14, 4-ethylaniline; 15, 3,4-dimethylaniline; 16, 1-aminonaphthalene; 17, 2-aminobiphenyl; 18 and 19, 2-aminonaphthalene; 20, 21 and 22, 3-aminobiphenyl; 23 and 24, benzidine; 25, tolidine. (From Smith, C. J., Dooly, G. L., and Moldoveanu, S. C., *J. Chromatogr.*, 991, 99–107, 2003.)

TABLE 11.16
Comparison of AA in ng/cigarette

AA	Citation and Cigarette Type				
	Ref. 28 (Kentucky)	Ref. 155 (Kentucky)	Ref. 174 (Camel)	Ref. 168 (Kentucky)	Ref. 169 (Kentucky)
Aniline	331.4	212.4	220	—	251.6
1-Aminonaphtahalene	17.4	9.3	5.6	15.6	17
2-Aminonaphtahalene	9.5	9.8	3.8	10.4	8.6
3-Aminobiphenyl	3.4	6.3	0.5	3.2	2.95
4-Aminobiphenyl	2.1	5.4	0.3	1.9	1.6
Benzidine	0.1	2.2	—	—	0.1

and other amines containing secondary amines remained completely in the aqueous layer (amine fraction).

Cigarettes with a higher concentration of tar and nicotine tended to have higher concentrations of N-nitrosamines and secondary amines (Figure 11.18). The contents of secondary amines in mainstream smoke samples of filter-tipped cigarettes were very low or not detectable, but those of nonfiltered cigarettes were relatively high. These results indicate that a filter-tip is effective for trapping of secondary amines but is less effective for N-nitrosamines. These results also suggest that cancer risk due to exposure to cigarette smoke increases not only in smokers but in nonsmokers as well, because of the N-nitrosamine contents in smoke.[67]

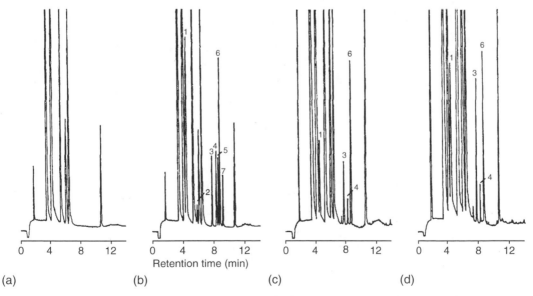

(a)   (b)   (c)   (d)

Retention time (min)

FIGURE 11.18 Typical gas chromatograms obtained from (a) reagent blank, (b) standard N-nitrosamine (containing 0.5 nmol of each N-nitrosamine), (c) mainstream smoke and (d) sidestream smoke. GC conditions are given under experimental. Peaks: 1, N-nitrosodimethylamine; 2, N-nitrosodiethylamine; 3, N-nitrosopyrrotidine; 4, N-nitrosopipendine; 5, N-nitrosomorpholine; 6, phenylphosphonic acid diethyl ester; 7, N-nitrosodibutylamine (From Kataoka, H., Shindoh, S. and Makita, M., *J. Chromatogr.*, 723, 93–99, 1996).

B. WATER (H)

Examples of environmental analysis of amines in water are shown in Table 11.17.

An example of analytical application was the determination of LMW aliphatic amines, a group of important compounds widely found in environmental samples, usually in aqueous solution. Detection of these amines at trace level is difficult due to their high basicity and strong adsorption on solid surfaces (Figure 11.19).

Moriwaki[21] describes a novel method for the determination of two kinds of aromatic amine mutagens, 2-[2-(acetylamino)-4-[bis(2-methoxyethyl)-amino]-5-amino-7-bromo-4-chloro-2H-benzotriazole (PBTA-l) and 2-[2-(acetylamino)-4-[bis(2-cyanoethyl)-ethylamino]-5-amino-7-bromo-4-chloro-2H-benzotriazole (PBTA-2), in river water based on liquid chromatography-electrospray ionization-tandem mass spectrometry (LC-ESI-MS–MS). By applying LC–MS–MS and solid-phase extraction to the analysis of both compounds, the cleanup treatment was simplified and faster than the conventional LC method.

Figure 11.20 shows the chromatograms of the standard solution and solution for the river water of the PBTA-2. Is the result of a Figure 11.21 satisfactory separation of the compound examined.

Another example described by Kataoka[59] is the identification of some mutagenic heterocyclic amines in river water from the Danube. Figure 11.21 shows a total ion chromatogram (TIC) (A and B) and selected ion monitoring (SIM) (C–E) of heterocyclic amines. The formulas of these compounds are shown in Figure 11.15. The method shown in Figure 11.22 is used in application analysis. SIM chromatograms reveal the separation of the following heterocyclic amines: AαC, Trp-P-1, and IQ.

TABLE 11.17
Contents of Amines in Water

Environment	Method	Amine	Reference
Water (river, sea)	GC–MS-SIM	Al	64,163
Water (river, lake, sea)	GC-FPD	Ammonia	69
Water (rain)	HPLC-FL	HA	7
Water (rain)	CZE-UV	HA	7
Water (river, sea)	GC–MS	Al	64
Water (river)	GC-NPD, GC–MS	HA	59
Water (river)	GC–MS	ABDACs	22
Water	CE	Aniline, 2,4-dimethlaniline	99
Water (surface)	CE	AA	101
Water (deionized)	CE-LIF	Al	98
Water (tap)	CZE	AA	26
Water	HS-GC-FID	Al	49
Water (lake)	HPLC	Al	75
Water (sea)	HPLC	AA	119
Water (tap, surface)	HPLC	AA	33
Water (river)	HPLC	Aniline	105
Water (tap, river, aquarium)	HPLC	Ammonia, Al	93
Water (river)	LC-ESI-MS–MS	AA	22
Water (tap)	CGC	Al	30
Water (lake)	HPLC	AA	170

HA, heterocyclic amine; AA, aromatic amine; Al, aliphatic amine; ABDACs, alkylbenzyldimethylammonium chlorides.

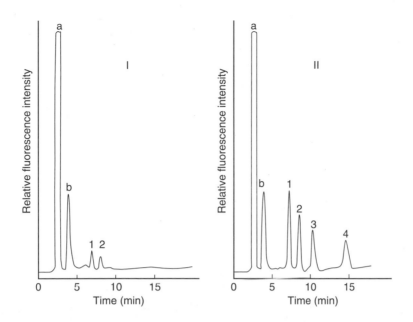

FIGURE 11.19 Chromatogram obtained from (I) lake water; (II) the same sample spiked with 600 nmol/l of methylamine, 500 nmol/l ethylamine, 1 μmol/l of n-propylamine and n-butylamine. (a) NPA-Osu; (b) unknown (1) methylamine, (2) ethylamine, (3) n-propylamine, (4) n-butylamine. (From Liu, X., Wang, H., Liang, S. C., and Zhang, H. S., *Anal. Chim. Acta* 441, 45–52, 2001.)

C. SOIL (S)

Examples of environmental analysis of amines in soil are shown in Table 11.18.

There are very few papers devoted to the determination of amines in soils. Finding an analytical method for determining amines in soil is possible but difficult because of the fact that extracting AAs involves both reversible and irreversible binding with humic acids in the soil working with spiked samples. The sensitivity of the method can be improved substantially by manipulation of the matrix, e.g., water addition, and by optimization of the extraction conditions, e.g., temperature, fiber coating material, mixing, and extraction time. However, matrix effects determined by the soil characteristics, especially the organic carbon content, are large.[37]

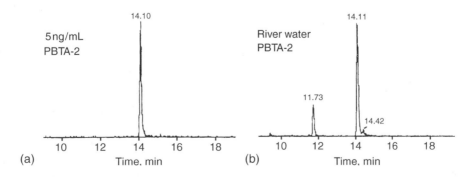

FIGURE 11.20 LC–MS–MS chromatograms for: (a) the standard solution (5 ng/ml) and (b) river water. (From Moriwaki, H., Harino, H., Hashimoto, H., Arakawa, R., Ohe, T., and Yoshikura, T., *J. Chromatogr.*, 995, 239–243, 2003.)

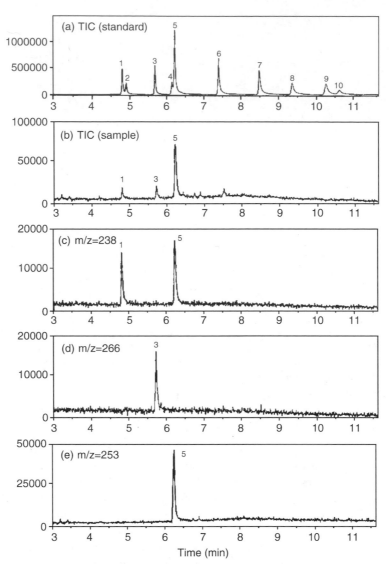

FIGURE 11.21 TIC and SIM chromatograms obtained from standard heterocyclic amine derivatives and the Danube water sample. (a) TIC obtained from standard heterocyclic amine derivatives, representing the sums of all selected ions; (b) TIC obtained from river water sample; (c) SIM chromatogram selected for m/z 238 (A α C); (d) SIM chromatogram selected for m/z 266 (Trp-P-1); (e) SIM chromatogram selected for m/z 253 (IQ). (b)–(e) were obtained from the river water sample. OC-MS conditions are given in the text. Peaks: 1, AαC; 2, Glu-P-1; 3, Trp-P-1; 4, Trp-P-2; 5, IQ; 6, MeIQ; 7, MeIQx; 8, 7,8-DiMeIQx; 9, 4,8-DiMeIQx; 10, PhIP. (From Kataoka, H., *J. Chromatogr.*, 774, 121–142, 1997.)

HS-SPME should be regarded as a rapid and very valuable screening technique in soil analysis. A rapid LC method with UV or fluorescence detection was developed for ppb levels of AAs in soils. 2,4-diaminotoluene, pyridine, aniline, 2-picoline, 2-toluidine, 5-nitro-2-toluidine, 2-methyl-6-ethylaniline, 4-aminobiphenyl, 4-nitroaniline, 1-naphthylamine, 2-methoxyaniline, and 2-naphthylamine were tested. The method involves extraction by sonication with 1% ammonium hydroxide-acetonitrile and analysis by LC using gradient elution with aqueous 0.01M ammonium acetate–0.0005% triethylamine and acetonitrile. Recoveries of 67 to 106% were obtained from sand and organic-containing soils spiked in the ppm range. Detection limits ranged tram 0.5 ppb for

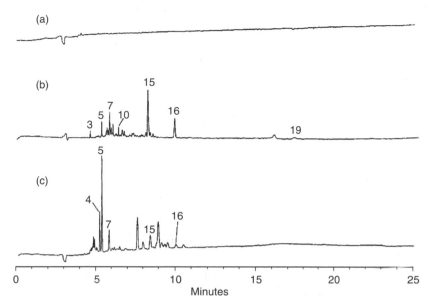

FIGURE 11.22 Electrophorograms of real samples after 1000-fold SPE preconcentration. (a) Tap water; (b) first layer groundwater. Peaks: 3, benzidine (2.7 μg/l); 5, aniline (1.8 μg/l); 7, *p*-anisidine (1.5 μg/l); 10, ethylaniline (0.5 μg/l); 15, *o*-chloroaniline (9.9 μg/l); 16, 3,4-dichloroaniline (2.9 μg/l); 19, 2,4-dichloroaniline (1.1 μg/l). (c) Soil sample from industrial plant. Peaks: 4, *o*-toluidine (600 μg/kg); 5, aniline (801 μg/kg); 7, *p*-anisidine (11.2 μg/kg); 15, *o*-chloroaniline (15.2 μg/kg); 16, 3,4-dichloroaniline (1.8 μg/kg). (From Cavallaro, A., Piangerelli, V., Nerini, F., Cavalli, S. and Reschiotto, C., *J. Chromatogr.*, 709, 361–366, 1995.)

highly fluoreseent 2-naphthylamine (by fluorescence detection) to 0.5 ppm for nonfluorescing pyridines (by UV detection).[178]

Examples of real CE analysis are presented in Figure 11.22. In Figure 11.22, a substantial enrichment of samples with the amines identified is noticeable. For example, 1.8 μg/l of aniline was found in a groundwater layer, while soil samples from a plant neighborhood contained as much as 801 μg/kg. Tap water samples have no amines.

TABLE 11.18
Contents of Amines in Soil

Environment	Method	Amine	Reference
Soil	CZE	AA	26
Soil	HPLC-FL	HA	7
Soil	GC	Al	64
Sediments	HPLC	FFA	171
Sea sediment	GC–MS	Al	64
Surface soil	HPLC-UV	AA	141,172,173
Soil	LC–MS	Nitramine	175
Soil	LC–MS	Aniline	176
Soil	GC-NPD	AA	177
Soil	GC-ECD	Chloro- and nitraniline	37
Soil	LC-UV or FL	AA	178

FFA, fluorfenicol amine; HA, heterocyclic amine; Al, aliphatic amine.

TABLE 11.19
Contents of Amines in Waste

Environment	Method	Amine	Reference
Waste water	HPLC-UV-FL	Al	27
Domestic surface water	HPLC-UV	Al	1
Waste water	HPLC-FL	Al	49
Water (waste, surface)	GC–MS	Al	58
Waste water	GC-FID	Al	40
Waste water	GC-NPD	Al	60
Water (from ammunition plants)	GC-ECD	AA	17,24,71,72
Groundwater	GC–MS	AA	36
Waste (urine sample)	GC-FPD	Al	61,131
Groundwater	GC-ECD	AA	130
Ground, leachate, wastewater	GC-ECD	AA	132
Water food	GC–MS	AA	9
Waste water (industrial sewage plant)	GC–MS and GC	AA	134
Waste (fish extracts)	CE-UV	Al	179
Groundwater	RP-LC	Al	74
Ground and surface water	LC–MS	AMEs	29

AA, aromatic amine; Al, aliphatic amine; ABDACs, alkylbenzyldimethylammonium chlorides.

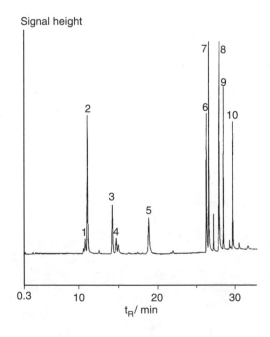

FIGURE 11.23 GC-ECD chromatogram of a ground water sample from a measuring point of a former ammunition plant in Mecklenburg-Vorpommern after enrichment of 100 ml water, followed by derivatization; peak identification: 1, 3NA (1.4); 2, 2,6-dinitrotoluene; 3, 2A6NT (I.S); 4, 4A2NT, (1.3); 5, 2A4NT (9.1); 6, 3,5DNA (12); 7, 4A2,6DNT (85); 8, 2A4,6DNT (71); 9, 2,4DANT (9.9); 10, 2,6 DANT (3.9). Concentrations of analytes in $\mu g/l$ given in parentheses. (From Less, M., Schmidt, T. C., Haas, R., von Löw, E. and Stork, G., *J. Chromatogr.*, 810, 173–182, 1998.)

D. Waste (W)

Examples of environmental analysis of amines in waste are shown in Table 11.19.

A good example of the samples discussed in Table 11.19 is the chromatogram of a water sample from a groundwater measuring point shown in Figure 11.23. The developed SPE procedure was applied to real samples from a disposal, a gas plant, and farmer ammunition plants in Stadtallendorf and Mecklenburg-Vorpommern. The eluates were derivatized and analyzed by GC-ECD. Satisfactory separations of AAs identified in groundwater of the ammunition plant were obtained.

REFERENCES

1. Wang, H., Li, J., Liu, X., and Zhang, H.-S., N-hydroxysuccinimidyl fluorescein-O-acetate as a highly fluorescent derivatizing reagent for aliphatic amines in liquid chromatography, *Anal. Chim. Acta*, 423, 77–83, 2000.
2. http://www.lakes-environmental.com/laketoxi.html.
3. Kataoka, H., Derivatization reactions for the determination of amines by gas chromatography and their applications in environmental analysis, *J. Chromatogr.*, 733, 19–34, 1996.
4. Kijima, K., Kataoka, H., and Makita, M., Determination of aromatic amines as their N-dimethylthiophosphoryl derivatives by gas chromatography with flame photometric detection, *J. Chromatogr.*, 738, 83–90, 1996.
5. Schmidt, T. C., Leß, M., Haas, R., Löw, E. V., and Steinbach, K., Determination of aromatic amines in ground and waste water by two new derivatization methods, *Intern. J. Environ. Anal. Chem.*, 74(1–4), 25–41, 1999.
6. EPA Method 8151, Chlorinated Herbicides By Gc Using Methylation Orpentafluorobenzylation Derivatization; Capillary Column Technique.
7. Kataoka, H., Methods for the determination of mutagenic hetrocyclic amines and their applications in environmental analysis, *J. Chromatogr.*, 774, 121–142, 1997.
8. Richter, E. and Branner, B., Biomonitoring of exposure to aromatic amines: haemoglobin adducts in humans, *J. Chromatogr.*, 778, 49–62, 2002.
9. Brede, C., Skjevrak, I., and Herikstad, H., Determination of primary aromatic amines in water food simulant using solid-phase analytical derivatization followed by gas chromatography coupled with mass spectrometry, *J. Chromatogr.*, 983, 35–42, 2003.
10. Bouzige, M., Legeay, P., Pichon, V., and Hennion, M. C., Selective on-line immunoextraction coupled to liquid chromatography for the trace determination of benzidine, congeners and related azo dyes in surface water and industrial effluents, *J. Chromatogr.*, 846, 317–329, 1999.
11. EU Directive 2001/405/EC.
12. http://www.cogentregs.com/cogentregs/SearchFRs.cfm.
13. EU Directive 2002/61/EC (list of aromatic amines).
14. Pielesz, A., Baranowska, I., Rybak, A., and Włochowicz, A., Detection and determination of aromatic amines as products of reductive splitting from selected azo dyes, *Ecotoxicol. Environ. Saf.*, 53, 42–47, 2002.
15. Eskilsson, C. S., Davidsson, R., and Mathiasson, L., Harmful azo colorants in leather. Determination based on their cleavage and extraction of corresponding carcinogenic aromatic amines using modem extraction techniques, *J. Chromatogr.*, 955, 215–227, 2002.
16. Matthiessen, P. and Law, R. J., Contaminants and their effects on estuarine and coastal organisms in the United Kingdom in the late twentieth century, *Environ. Pollut.*, 120, 739–757, 2002.
17. Less, M., Schmidt, T. C., Haas, R., von Löw, E., and Stork, G., Gas chromatographic determination of aromatic amines in water samples after solid-phase extraction and derivatization with iodine. II. Enrichment, *J. Chromatogr.*, 810, 173–182, 1998.
18. Ambrose, D. L., Fritz, J. S., Buchmeiser, M. R., Atzl, N., and Bonn, G. K., New, high-capacity carboxylic acid functionalized resins for solid phase extraction of a broad range of organic compounds, *J. Chromatogr.*, 786, 259–268, 1997.
19. Fritz, J. S., Dumont, P. J., and Schmidt, L. W., Methods and materials for solid-phase extraction, *J. Chromatogr.*, 691, 133–140, 1995.

20. Zhang, Z., Yang, M. J., and Pawliszyn, J., Solid-phase microextraction, *Anal. Chem.*, 66, 844–853, 1994.
21. Moriwaki, H., Harino, H., Hashimoto, H., Arakawa, R., Ohe, T., and Yoshikura, T., Determination of aromatic amine mutagens, PBTA-1 and PBTA-2, in river water by solid-phase extraction followed by liquid chromatography-tandem mass spectrometry, *J. Chromatogr.*, 995, 239–243, 2003.
22. Ding, W. H. and Liao, Y. H., Determination ot alkylbenzyldimethylammonium chlorides in river water and sewage effluent by solid-phase extraction and gas chromatography/mass spectrometry, *Anal. Chem.*, 73, 36–40, 2001.
23. Seeber, G., Buchmeiser, M. R., Bonn, G. K., and Bertsch, T., Determination of airborne, volatile amines fram polyurethane foams by sorptionan to a high-capacity cation-exchange resin based on poly(succinic acid), *J. Chromatogr.*, 809, 121–129, 1998.
24. Schmidt, T. C., Less, M., Haas, R., von Löw, E., Steinbach, K., and Stork, G., Gas chromatographic determination of aromatic amines in water samples after solid-phase extraction and derivatization with iodine. I. Derivatization, *J. Chromatogr.*, 810, 161–172, 1998.
25. Patsias, J. and Papadopoulou-Mourkidou, E., Development of an automated on-line solid-phase extraction-high performance liquid chromatographic method for the analysis of aniline, phenol, caffeine and various selected substituted aniline and phenol compounds in aqueous matrices, *J. Chromatogr.*, 904, 171–188, 2000.
26. Cavallaro, A., Piangerelli, V., Nerini, F., Cavalli, S., and Reschiotto, C., Selective determination of aromatic amines in water samples by capillary zone electrophoresis and solid-phase extraction, *J. Chromatogr.*, 709, 361–366, 1995.
27. Lloret, S. M., Legua, C. M., and Falco, P. C., Preconcentration and dansylation of aliphatic amines using C_{18} solid-phase packings. Application to the screening analysis in environmental water samples, *J. Chromatogr.*, 978, 59–69, 2002.
28. Smith, C. J., Dooly, G. L., and Moldoveanu, S. C., New technique using solid-phase extraction for the analysis of aromatic amines in mainstream cigarette smoke, *J. Chromatogr.*, 991, 99–107, 2003.
29. Krogh, K. A., Vejrup, K. V., Mogensen, B. B., and Sjjrensen, B. H., Liquid chromatography–mass spectrometry method to determine alcohol ethoxylates and alkylamine ethoxylates in soil interstitial, water, ground water and surface water samples, *J. Chromatogr.*, 957, 45–57, 2002.
30. Baltussen, E., David, F., Sandra, P., Janssen, H. G., and Cramers, C., Capillary GC determination of amines in aqueous samples using sorptive preconcentration on polimethylsiloxane and polyacrylate phases, *J. High Resolut. Chromatogr.*, 21, 645–648, 1998.
31. Clesceri, L. S., Greenberg, A. E., and Trussell, A. A., Eds., *Standard Methods for the Examination of Water and Wastewater*, 17th ed., American Public Health Aasociation, Washington, DC, 1989.
32. Louch, D., Motlagh, S., and Pawliszyn, J., Dynamics of organic compound extraetion from water using liquid-coated fused silica fibers, *Anal. Chem.*, 64, 1187–1199, 1992.
33. Zhao, L., Zhu, L., and Lee, H. K., Analysis of aromatic amines in water samples by liquid–liquid–liquid microextraction with hollow fibers and high-performance liquid chromatography, *J. Chromatogr.*, 936, 239–248, 2002.
34. Zhu, L., Tay, Ch. B., and Lee, H. K., Liquid–liquid–liquid microextraction of aromatic amines from water samples combined with high-performance liquid chromatography, *J. Chromatogr.*, 963, 231–237, 2002.
35. Arthur, C. L. and Pawliszyn, J., Solid phase microextraction with thermal desorption using fused silica optical fibers, *Anal. Chem.*, 62, 2145–2148, 1990.
36. Müller, L., Fattore, E., and Benfenati, E., Determination of aromatic amines by solid-phase microextraction and gas chromatography-mass spectrometry in water samples, *J. Chromatogr.*, 791, 221–230, 1997.
37. Fromberg, A., Nilsson, T., Larsen, B. R., Montanarella, L., Facchetti, S., and Madsen, J. O., Analysis of chloro- and nitroanilines and -benzenes in soils by headspace solid-phase microextraction, *J. Chromatogr.*, 71–81, 1996.
38. Yang, J. and Tsai, F. P., Development of a solid-phase microextraction/reflection-absorption infrared spectroscopic method for the detection of chlorinated aromatic amines in aqueous solutions, *Anal. Sci.*, 17, 751–756, 2001.

39. Chien, Y. C., Uang, S. N., Kuo, C. T., Shih, T. S., and Jen, J. F., Analytical method for monitoring airborne trimethylamine using solid phase micro-extraction and gas chromatography-flame ionization detection, *Anal. Chim. Acta*, 419, 73–79, 2000.

40. Pan, L., Chong, J. M. and Pawliszyn, J., Determination of amines in air and water using derivatization combined with solid-phase microextraction, *J. Chromatogr.*, 773, 249–260, 1997.

41. Wu, Y. Ch. and Huang, S. D., Solid-phase microextraction coupled with high-performance liquid chromatography for the determination of aromatic amines, *Anal. Chem.*, 71, 310–318, 1999.

42. Kataoka, H. and Pawliszyn, J., Development of in-tube solid-phase microextraction/liquid chromatography/electrospray ionization mass spectrometry for the analysis of mutagenic heterocyclic amines, *Chromatographia*, 50(9/10), 532–538, 1999.

43. Liu, H. and Dasgupta, P. K., Analytical chemistry in a drop. Solvent extraction in a microdrop, *Anal. Chem.*, 68, 1817–1821, 1996.

44. Jeannot, M. A. and Cantwell, F. F., Mass transfer characteristics of solvent extraction into a single drop at the tip of a syringe needle, *Anal. Chem.*, 69, 235–239, 1997.

45. He, Y. and Lee, H. K., Liquid-phase microextraction in a single drop of organic solvent by using a conventional microsyringe, *Anal. Chem.*, 69, 4634–4640, 1997.

46. Jeannot, M. A. and Cantwell, F. F., Solvent microextraction into a single drop, *Anal. Chem.*, 68, 2236–2240, 1996.

47. Ma, M. and Cantwell, F. F., Solvent microextraction with simultaneous back-extraction for sample cleanup and preconcentration: preconcentration into a single microdrop, *Anal. Chem.*, 71, 388–393, 1999.

48. Ma, M. and Cantwell, F. F., Solvent microextraction with simultaneous back-extraction for sample cleanup and preconcentration: quantitative extraction, *Anal. Chem.*, 70, 3912–3919, 1998.

49. Maris, C., Laplanche, A., Morvan, J., and Bloquel, M., Static headspace analysis of aliphatic amines in aqueous samples, *J. Chromatogr.*, 846, 331–339, 1999.

50. Tsukioka, T., Ozawa, H., and Murakami, T., Gas chromatographic-mass spectrometric determination of lower aliphatic tertiary amines in environmental samples, *J. Chromatogr.*, 642, 395–400, 1993.

51. Sparr Eskilsson, C., Davidsson, R., and Mathiasson, L., Harmfulazo colorants in leather. Determination based on their cleavage and extraction of corresponding carcinogenic aromatic amines using modern extraction techniques, *J. Chromatogr.*, 955, 215–227, 2002.

52. Reche, F., Garrigós, M. C., Marin, M. L., Cantó, A., and Jimenez, A., Optimization of parameters for the supercritical fluid extraction in the determination of *N*-nitrosamines in rubbers, *J. Chromatogr.*, 963, 419–426, 2002.

53. Oostdyk, T. S., Grob, R. L., Snyder, J. L., and McNally, M. E., Study of sonication and supercritical fluid extraction of primary aromatic amines, *Anal. Chem.*, 65, 596–600, 1993.

54. Ashraf-Khorassani, M., Taylor, L. T., and Zimmerman, P., Nitrous oxide versus carbon dioxide for supercritical-fluid extraction and chromatography of amines, *Anal. Chem.*, 62, 1177–1180, 1990.

55. Janda, V., Kriz, J., Vejrosta, J., and Bartle, K. D., Supercritical fluid extraction and chromatography of aromatic amines, *J. Chromatogr.*, 669, 241–245, 1994.

56. Miyamoto, Y., Kataoka, H., Ohrui, S., and Makita, M., Determination of primary amines as their *N*-benzenesulfonyl-*N*-trifluoroacetyl derivatives by GC with electron capture detection, *Bunseki Kagaku*, 43, 1113–1118, 1994.

57. Longo, M. and Cavallaro, A., Determination of aromatic amines at trace levels by derivatization with heptafluorobutyric anhydride and gas chromatography-electron-capture negative-ion chemical ionization mass spectrometry, *J. Chromatogr.*, 753, 91–100, 1996.

58. Sacher, F., Lenz, S. and Brauch, H. J., Analysis of primary and secondary aliphatic amines in waste water and surface water by gas chromatography-mass spectrometry after derivatization with 2,4-dinitrofluorobenzene or benzenesulfonyl chloride, *J. Chromatogr.*, 764, 85–93, 1997.

59. Kataoka, H., Hayatsu, T., Hietsch, G., Steinkellner, H., Nishioka, S., Narimatsu, S., Knasmüller, S., and Hayatsu, H., Identification of mutagenic heterocyclic amines (IQ, Trp-P-1 and AαC) in the water of the Danube River, *Mutat. Res.*, 466, 27–35, 2000.

60. Abalos, M., Bayona, J. M., and Ventura, F., Development of a SPE GC-NPD. Procedure for the determination of free volatile amines in wastewater and sewage-polluted water, *Anal Chem.*, 71, 3531–3537, 1999.

61. Kataoka, H., Ohrui, S., Kanemoto, A., and Makita, M., Determination of ammonia as its benzenesulphonyldimethylaminomethylene derivative in environmental water samples by gas chromatography with flame photometric detection, *J. Chromatogr.*, 633, 311–314, 1993.

62. Skarping, G., Dalene, M., Brorson, T., Sandström, J. F., Sangö, C., and Tiljander, A., Chromatographic determination of amines in biological fluids with special reference to the biological monitoring of isocyanates and amines: I. Determination of 1,6-hexamethylenediamine using glass capillary gas chromatography and thermionic specific detection, *J. Chromatogr.*, 479, 125–133, 1989.

63. Kataoka, H., Imamura, Y., Tanaka, H., and Makita, M., Determination of cysteamine and cystamine by gas chromatography with flame photometric detection, *J. Pharm. Biomed. Anal.*, 11, 963–969, 1993.

64. Terashi, A., Hanada, Y., Kido, A., and Shinohara, R., Determination of primary and secondary aliphatic amines in the environment as sulphonamide derivatives by gas chromatography-mass spectrometry, *J. Chromatogr.*, 503, 369–375, 1990.

65. Kataoka, H., Ohrui, S., Miyamoto, Y., and Makita, M., Determination of low molecular weight aliphatic primary amines in urine as their benzenesulphonyl derivatives by gas chromatography with flame photometric detection, *Biomed. Chromatogr.*, 6, 251–254, 1992.

66. Kataoka, H., Nagao, K., Nabeshima, N., Kiyama, M., and Makita, M., Selective determination of secondary amino acids as their *N*-dimethylthiophosphoryl methyl ester derivatives by gas chromatography with flame photometric detection, *J. Chromatogr.*, 626, 239–243, 1992.

67. Kataoka, H., Shindoh, S., and Makita, M., Selective determination of volatile *N*-nitrosamines by derivatization with diethylchlorothiophosphate and gas chromatography with flame photometric detection, *J. Chromatogr.*, 723, 93–99, 1996.

68. Kataoka, H., Shindoh, S., and Makita, M., Determination of secondary amines in various foods by gas chromatography with flame photometric detection, *J. Chromatogr.*, 695, 142–148, 1995.

69. Kataoka, H., Eda, M., and Makita, M., Selective determination of secondary amines as their *N*-diethylthiophosphoryl derivatives by gas chromatography with flame photometric detection, *Biomed. Chromatogr.*, 7, 129–133, 1993.

70. Haas, R., Schmidt, T. C., Steinbach, K., and von Löw, E., Derivatization of aromatic amines for analysis in ammunition wastewater II: Derivatization of methyl anilines by iodination with a Sandmeyer-like reaction. Frescenius, *J. Anal. Chem.*, 359, 497–501, 1997.

71. Schmidt, T. C., Haas, R., von Löw, E., and Steinbach, K., Derivatization of aromatic amines with bromine for improved gas chromatographic determination, *Chromatographia*, 48, 436–442, 1998.

72. Schmidt, T. C., Haas, R., Steinbach, K., and von Löw, E., Derivatization of aromatic amines in ammunition wastewater. I. Derivatization via bromination of the aromatic ring. Frescenius, *J. Anal. Chem.*, 357, 909–914, 1997.

73. Blau, K. and Halket, J. M., *Handbook of Derivatives for Chromatography*, Wiley, Chichester, 1993, pp. 175–213.

74. Lopez, M. R., Alvarez, M. J. G., Ordieres, A. J. M., and Blanco, P. T., Determination of dimethlamine in groundwater by LC and precolumn derivatization with 9-fluorenylmethlcholroformate, *J. Chromatogr.*, 721, 231–239, 1996.

75. Liu, X., Wang, H., Liang, S. C., and Zhang, H. S., Determination of primary and secondary aliphatic amines by *N*-hydroxysuccinimidyl 4,3,2′-naphthapyrone-4-acetate and reversed-phase high-performance liquid chromatography, *Anal. Chim. Acta*, 441, 45–52, 2001.

76. Gao, C. X., Krull, I. S. and Trainor, T. M., Determination of volatile amines in air by on-line solid-phase derivatization and high-performance liquid chromatography with ultraviolet and fluorescence detection, *J. Chromatogr.*, 463, 192–200, 1989.

77. Khalaf, H. and Steinert, J., Determination of secondary amines as highly fluorescent formamidines by high-performance liquid chromatography, *Anal. Chim. Acta*, 334, 45–50, 1996.

78. Andrés, J. V., Falcá, P. C., and Hernández, R. H., Liquid chromatographic determination of aliphatic amines in water using solid support assisted derivatization with 9-fluorenylmethyl chloroformate, *Chromatographia*, 55, 129–135, 2002.

79. Price, N. P. J., Firmin, J. L., and Gray, D. O., Screening for amines by dansylation and automated high-performance liqid chromatography, *J. Chromatogr.*, 598, 51–57, 1992.

80. Lloret, S. M., Legua, C. M., and Falco, P. C., Preconcentration and dansylation of aliphatic amines using C_{18} solid-phase packings. Application to the screening analysis in environmental water samples, *J. Chromatogr.*, 978, 59–69, 2002.

81. Saleh, M. I. and Pok, F. W., Separation of primary and secondary amines as their sulfonamide derivatives by reversed-phase high-performance liquid chromatography, *J. Chromatogr.*, 763, 173–178, 1997.

82. Kirschbaum, J., Busch, I., and Brückner, H., *Chromatographia*, 45, 2631997.

83. Tod, M., Legendre, J. Y., Chalom, J., Kouwatli, H., Poulou, M., Farinotti, R., and Mahuzier, G., Primary and secondary amine derivatization with luminarins 1 and 2: separation by liquid chromatography with peroxyoxalate chemiluminescence detection, *J. Chromatogr.*, 594, 386–391, 1992.

84. Kouwatli, H., Chalom, J., Tod, M., Farinotti, R., and Mahuzier, G., Precolumn derivatization of amines in liquid chromatography using luminescent probes: comparison of several reagents with luminarin 1, *Anal. Chim. Acta*, 266, 243–249, 1992.

85. Vandenabeele, O., Garrelly, L., Ghelfenstein, M., Commeyras, A., and Mion, L., Use of 2-chloroethylnitrosourea, a new type of pre-column derivatizing agent for the measurement of biogenic amines, by high-performance liquid chromatography with ultraviolet detection, *J. Chromatogr.*, 795, 239–250, 1998.

86. Molmlr-Perl, I., Quantitation of amino acids and amines in the same matrix by high-performance liquid chromatography, either simultaneously or separately, *J. Chromatogr.*, 987, 291–309, 2003.

87. Saito, K., Horie, M., Nose, N., Nakagomi, K., and Nakazawa, H., High-performance liquid chromatography of histamine and 1-methylhistamine with on-column fluroescence derivatization, *J. Chromatogr.*, 595, 163–168, 1992.

88. Turiák, G. and Volicer, L., Stability of *o*-phthalaldehyde—sulfite derivatives of amino acids and their methyl esters: electrochemical and chromatographic properties, *J. Chromatogr.*, 668, 323–329, 1994.

89. Nishibori, N., Nishii, A., and Takayama, H., Detection of free polyamine in coastal seawater using ion exchange chromatography, *J. Mar. Sci.*, 58, 1201–1207, 2001.

90. Smith, J. R. L., Smart, A. U., Hancock, F. E., and Twigg, M. V., High-performance liquid chromatographic determination of low levels of primary and secondary amines in aqueous media via derivatisation with 1,2-naphthoquinone-4-sulphonate, *J. Chromatogr.*, 483, 341–348, 1989.

91. Dunning, J. W. and Stewart, J. T., Some new acridinium trifluoromethanesulfonates as spectrophotometric derivatization reagents for aromatic and aliphatic primary amines, *Talanta*, 38, 631–635, 1991.

92. Lindahl, R., Levin, J. O., and Andersson, K., Determination of volatile amines in air by diffusive sampling, thiourea formation and high-performance liquid chromatography, *J. Chromatogr.*, 643, 35–41, 1993.

93. Sahasrabuddhey, B., Jain, A., and Verma, K. K., Determination of ammonia and aliphatic amines in environmental aqueous samples utilizing pre-column derivatization to their phenylthioureas and high performance liquid chromatography, *Analyst*, 124, 1017–1021, 1999.

94. Dabek-Zlotorzynska, E., and Maruszak, W., Determination of dimethylamine and other low-molecular-mass amines using capillary electrophoresis with laser-induced fluorescence detection, *J. Chromatogr.*, 714, 77–85, 1998.

95. Santagati, N. A., Bousquet, E., Spadaro, A., and Ronsisvalle, G., Analysis of aliphatic amines in air samples by HPLC with electrochemical detection, *J. Pharm. Biomed. Anal.*, 29, 1105–1111, 2002.

96. Andrés, J. V., Falcó, P. C., and Hernández, R. H., Determination of aliphatic amines in water by liquid chromatography using solid-phase extraction cartridges for preconcentration and derivatization, *Analyst*, 126, 1683–1689, 2001.

97. You, J., Lao, W., You, J., and Wang, G., Characterization and application of acridine-9-*N*-acetyl-*N*-hydroxysuccinimide as a pre-column derivatization agent for fluorimetric detection of amino acids in liquid chromatography, *Analyst*, 124, 1755–1761, 1999.

98. Brumley, W. and Kelliher, V., Determination of aliphatic amines in water using derivatization with fluorescein isothiocyanate and capillary electrophoresis/laser-induced fluorescence detection, *J. Liq. Chromatogr.*, 20, 2193–2205, 1997.

99. Leung, S. A. and de Mello, A. J., Electrophoretic analysis of amines using reversed-phase, reversed-polarity, head-column field-amplified sample stacking and laser-induced fluorescence detection, *J. Chromatogr.*, 979, 171–178, 2002.

100. Conca, R., Concetta Bruzzoniti, M., Mentasti, E., Sarzanini, C., and Hajos, P., Ion chromatographic separation of polyamines: putrescine, spermidine and spermine, *Anal Chim Acta*, 439, 107–114, 2001.

101. Asthana, A., Bose, D., Durgbanshi, A., Sanghi, S. K., and Kok, W. T., Determination of aromatic amines in water samples by capillary electrophoresis with electrochemical and fluorescence detection, *J. Chromatogr.*, 895, 197–203, 2000.

102. Seymour, E. H., Lawrence, N. S., Beckett, E. L., Davis, J., and Compton, R. G., Electrochemical detection of aniline: an electrochemically initiated reaction pathway, *Talanta*, 57, 233–242, 2002.

103. Lacorte, S., Fraisse, D., and Barcelo, D., Efficient solid-phase extraction procedures for trace enrichment of priority phenols from industrial effluents with high total organic carbon content, *J. Chromatogr.*, 857, 97–106, 1999.

104. Jáuregui, J. and Galceran, M. T., Determination of phenols in water by on-line solid-phase disk extraction and liquid chromatography with electrochemical detection, *Anal. Chim. Acta*, 340, 191–199, 1997.

105. Patsias, J. and Papadopoulou-Mourkidou, E., Development of an automated on-line solid-phase extraction-high-performance liquid chromatographic method for the analysis of aniline, phenol, caffeine and various selected substituted aniline and phenol compounds in aqueous matrices, *J. Chromatogr.*, 904, 171–188, 2000.

106. Gross, G. A., Simple methods for quantifying mutagenic heterocyclic aromatic amines in food products, *Carcinogenesis*, 11, 1597–1603, 1990.

107. Johansson, M. A. and Jagerstad, M., Occurrence of mutagenic/carcinogenic heterocyclic amines in meat and fish products, including pan residues, prepared under domestic conditions, *Carcinogenesis*, 15, 1511–1518, 1994.

108. Skog, K., Steineck, G., Augustsson, K., and Jagerstad, M., Effect of cooking temperature on the formation of heterocyclic amines in fried meat products and pan residues, *Carcinogenesis*, 16, 861–867, 1995.

109. Ushiyama, H., Wakabayashi, K., Hirose, M., Itoh, H., Sugimura, T., and Nagao, M., Presence of carcinogenic heterocyclic amines in urine of healthy volunteers eating normal diet, but not of inpatients receiving parenteral alimentation, *Carcinogenesis*, 12, 1417–1422, 1991.

110. Gross, G. A., Turesky, R. J., Fay, L. B., Stillwell, W. G., Skipper, P. L., and Tannenbaum, S. R., Heterocyclic aromatic amine formation in grilled bacon, beef and fish and in grill scrapings, *Carcinogenesis*, 14, 2313–2318, 1993.

111. Schwarzenbach, R. and Gubler, D., Detection of heterocyclic aromatic amines in food flavours, *J. Chromatogr.*, 624, 491–495, 1992.

112. Galceran, M. T., Pais, P., and Puignou, L., High-performance liquid chromatographic determination of ten heterocyclic aromatic amines with electrochemical detection, *J. Chromatogr.*, 655, 101–110, 1993.

113. Galceran, M. T., Pais, P., and Puignou, L., Isolation by solid-phase extraction and liquid chromatographic determination of mutagenic amines in beef extracts, *J. Chromatogr.*, 719, 203–212, 1996.

114. Van Dyck, M. M. C., Rollmann, B., and De Meester, C., Quantitative estimation of heterocyclic aromatic amines by ion-exchange chromatography and electrochemical detection, *J. Chromatogr.*, 697, 377–382, 1995.

115. Manabe, S., Tohyama, K., Wada, O., and Aramaki, T., Detection of a carcinogen, 2-amino-1-methyl-6-phenylimidazo[4,5-b]pyridine (PhIP), in cigarette smoke condensate, *Carcinogenesis*, 12, 1945–1947, 1991.

116. Kanai, Y., Wada, O., and Manabe, S., Detection of carcinogenic glutamic acid pyrolysis products in cigarette smoke condensate, *Carcinogenesis*, 11, 1001–1003, 1990.

117. Lu, C. S. and Huang, S. D., Trace determination of aromatic amines or phenolic compounds in dyestuffs by high-performance liquid chromatography with on-line preconcentration, *J. Chromatogr.*, 696, 201–208, 1995.

118. Wu, Y. C. and Huang, S. D., Solid-phase microextraction coupled with high-performance liquid chromatography for the determination of aromatic amines, *Anal. Chem.*, 71, 310–318, 1999.

119. Zhu, L., Tay, C. B., and Lee, H. K., Liquid–liquid–liquid microextraction of aromatic amines from water samples combined with high-performance liquid chromatography, *J. Chromatogr.*, 963, 231–237, 2002.

120. Taibi, G., Schiavo, M. R., Gueli, M. C., Rindina, P. C., Muratore, R., and Nicotra, C. M. A., Rapid and simultaneous high-performance liquid chromatography assay of polyamines and monoacetylpolyamines in biological specimens, *J. Chromatogr.*, 745, 431–437, 2000.

121. Shpigun, A., Shapovalova, E. N., Ananieva, I. A., and Pirogov, A. V., Separation and enantioseparation of derivatized amino acids and biogenic amines by high-performance liquid chromatography with reversed and chiral stationary phases, *J. Chromatogr.*, 979, 191–199, 2002.

122. Gennaro, M. C., Bertolo, P. L., and Marengo, E., Determination of aromatic amines at trace levels by ion interaction reagent reversed-phase high-performance liquid chromatography. Analysis of hair dyes and other water-soluble dyes, *J. Chromatogr.*, 518, 149–156, 1990.

123. Sahasrabuddhey, B., Jain, A., and Verma, K. K., Determination of ammonia and aliphatic amines in environmental aqueous samples utilizing pre-column derivatization to their phenylthioureas and high performance liquid chromatography, *Analyst*, 124, 1017–1021, 1999.

124. Galceran, M. T., Moyano, E., Puignou, L., and Pais, P., Determination of heterocyclic amines by pneumatically assisted electrospray liquid chromatography-mass spectrometry, *J. Chromatogr.*, 730, 185–194, 1996.

125. Jansson, Ch., Paccou, A , and Osterdahl, B.-G., Analysis of tobacco-specific *N*-nitrosamines in snuff by ethyl acetate extraction and liquid chromatography-tandem mass spectrometry, *J. Chromatogr.*, 1008, 135–143, 2003.

126. Reemtsma, T., Liquid chromatography–mass spectrometry and strategies for trace-level analysis of polar organic pollutants, *J. Chromatogr.*, 1000, 477–501, 2003.

127. Kamiński, M., Jastrzębski, D., Przyjazny, A., and Kartanowicz, R., Determination of the amount of wash amines and ammonium ion in desulfurization products of process gases and results of related studies, *J. Chromatogr.*, 947, 217–225, 2002.

128. Santos, F. J. and Galceran, M. T., Modern developments in gas chromatography–mass spectrometry based environmental analysis, *J. Chromatogr.*, 1000, 125–151, 2003.

129. Müller, L., Fattore, E., and Benfenati, E., Determination of aromatic amines by solid-phase microextraction and gas chromatography–mass spectrometry in water samples, *J. Chromatogr.*, 791, 221–230, 1997.

130. Longo, M. and Cavallaro, A., Determination of aromatic amines at trace levels by derivatization with heptafluorobutyric anhydride and gas chromatography-electron-capture negative-ion chemical ionization mass spectrometry, *J. Chromatogr.*, 753, 91–100, 1996.

131. Kataoka, H., Ohrui, S., Miyamoto, Y., and Makita, M., Determination of low molecular weight aliphatic primary amines in urine as their benzenesulphonyl derivatives by gas chromatography with flame photometric detection, *Biomed. Chromatogr.*, 6, 251–254, 1992.

132. Schmidt, T. C., Less, M., Haas, R., von Löw, E., Steinbach, K., and Stork, G., Gas chromatographic determination of aromatic amines in water samples after solid-phase extraction and derivatization with iodine. I. Derivatization, *J. Chromatogr.*, 810, 161–172, 1998.

133. Schmidt, T. C., Less, M., Haas, R., von Löw, E., Steinbach, K., and Stork, G., Gas chromatographic determination of aromatic amines in water samples after solid-phase extraction and derivatization with iodine. II. Enrichment, *J. Chromatogr.*, 810, 173–182, 1998.

134. Schmidt, T. C., Leß, M., Haas, R., Löw, E. V., and Steinbach, K., Determination of aromatic amines in ground and waste water by two new derivatization methods, *Intern. J. Environ. Anal. Chem.*, 74(1–4), 25–41, 1999.

135. Schmidt, T. C., Haas, R., Steinbach, K., and von Löw, E., Derivatization of aromatic amines in ammunition wastewater. I. Derivatization via bromination of the aromatic ring, *Frescenius J. Anal. Chem.*, 357, 909–914, 1997.

136. Herren, D. and Berset, J. D., Nitro musks, nitro musk amino metabolites and polycyclic musks in sewage sludges. Quantitative determination by HRGC-ion-trap-MS/MS and mass spectral characterization of the amino metabolites, *Chemoshere*, 40, 565–574, 2000.

137. Straub, R. F., Voyksner, R. D., and Keever, J. T., Determination of aromatic amines originating from azo dyes by hydrogen–palladium reduction combined with gas chromatography/mass spectrometry, *Anal. Chem.*, 65, 2131–2136, 1993.

138. Patil, S. F. and Lonkar, T., Thermal desorption—gas chromatography for the determination of benzene, aniline, nitrobenzene and chlorobenzene in workplace air, *J. Chromatogr.*, 600, 344–351, 1992.

139. Baltussen, E., David, F., Sandra, P., Janssen, H. G., and Cramers, C., Capillary GC determination of amines in aqueous samples using sorptive preconcentration on polimethylsiloxane and polyacrylate phases, *J. High Resolut. Chromatogr.*, 21, 645–648, 1998.

140. Durgbanshi, A. and Kok, W. Th., Capillary electrophoresis and electrochemical detection with a conventional detector cell, *J. Chromatogr.*, 798, 289–296, 1998.

141. Asthana, A., Bose, D., Durgbanshi, A., Sanghi, S. K., and Kok, W. Th., Determination of aromatic amines in water samples by capillary electrophoresis with electrochemical and fluorescence detection, *J. Chromatogr.*, 895, 197–203, 2000.

142. Matchett, W. H. and Brumley, W. C., Preconcentration of aliphatic amines from water determined by capillary electrophoresis with indirect UV detection, *J. Liq. Chromatogr. Relat. Technol.*, 20, 79–100, 1997.

143. Hilmi, A., Luong, J. H. T., and Nguyen, A. L., Capillary elecrophores applied to kinetic studies of photocatalitic oxidation of substituted anilines, *Chromosphere*, 36, 3137–3147, 1998.

144. McLaughlin, G. M., Weston, A., and Hauffe, K. D., Capillary electrophoresis methods development and sensitivity enhancement strategies for the separation of industrial and environmental chemicals, *J. Chromatogr.*, 744, 123–134, 1996.

145. Quirino, J. P. and Terabe, S., Sweeping of analyte zones in electrokinetic chromatography, *Anal. Chem.*, 71, 1638–1644, 1999.

146. Pohl, Ch., Rey, M., Jensen, D. and Kerth, J., Determination of sodium and ammonium ions in disproportionate concentration ratios by ion chromatography, *J. Chromatogr.*, 850, 239–245, 1999.

147. Sarzanini, C., Recent developments in ion chromatography, *J. Chromatogr.*, 956, 3–13, 2002.

148. Fernando, P. N., Egwu, I. N., and Hussain, M. S., Ion chromatographic determination of trace hydroxylamine in waste streams generated by a pharmaceutical reaction process, *J. Chromatogr.*, 956, 261–270, 2002.

149. Prokai, A. M. and Ravichandran, R. K., Simultaneous analysis of hydroxylamine, *N*-methylhydroxylamine and *N*,*N*-dimethylhydroxylamine by ion chromatography, *J. Chromatogr.*, 667, 298–303, 1994.

150. Poels, I. and Nagels, L. J., Potentiometric detection of amines in ion chromatography using macrocycle-based liquid membrane electrodes, *Anal. Chim. Acta*, 440, 89–98, 2001.

151. Huang, X., You, T., Li, T., Yang, X., and Wang, E., End-column electrochemical detection for aromatic amines with high performance capillary electrophoresis/end-column/electrochemical detection/aromatic amines/capillary electrophoresis, *Electroanalysis*, 11, 969–972, 1999.

152. Weiss, C. S., Hazlett, J. S., Datta, M. H., and Danzer, M. H., Determination of quaternary ammonium compounds by capillary electrophoresis using direct and indirect UV detection, *J. Chromatogr.*, 608, 325–332, 1992.

153. Tricker, A. R., Ditrich, C., and Preussmann, R., *N*-nitroso compounds in cigarette tobacco and their occurrence in mainstream tobacco smoke, *Carcinogenesis*, 12, 257–261, 1991.

154. Borgerding, M. F., Bodnar, J. A., Chung, H. L., Mangan, P. P., Morrison, C. C., Risner, C. H., Rogers, J. C., Simmons, D. F., Uhrig, M. S., Wendelboe, F. N., Wingate, D. E., and Winkler, L. S., Chemical and biological studies of a new cigarette that primarily heats tobacco: Part 1. Chemical composition of mainstream smoke, *Food Chem. Toxicol.*, 36, 169–182, 1997.

155. Forehand, J. B., Dooly, G. L., and Moldoveanu, S. C., Analysis of polycyclic aromatic hydrocarbons, phenols and aromatic amines in particulate phase cigarette smoke using simultaneous distillation and extraction as a sole sample clean-up step, *J. Chromatogr.*, 898, 111–124, 2000.

156. Fabrega, J. R., Jafvert, C. T., Li, H., and Lee, L. S., Modeling short-term soil-water distribution of aromatic amines, *Environ. Sci. Technol.*, 32, 2788–2794, 1998.

157. Querol, X., Alastuey, A., Lopez-Soler, A., Plana, F., Mesas, A., Ortiz, L., Alzaga, R., Bayona, J. M., and de la Rosa, J., Physico-chemical characterisation of atmospheric aerosols in a rural area affected by the aznalcollar toxic spill, south-west Spain during the soil reclamation activities, *Sci. Total Environ.*, 242, 89–104, 1999.

158. Busca, G. and Pistarino, C., Abatement of ammonia and amines from waste gases: a summary, *J. Loss Prevention Process Industries*, 16(2), 157–163, 2003.

159. Bonnin, C., Martin, G., and Gragnie, G., Biopurification of odourous gases in urban wastewater treartment plants, In *Proceedings of the Second International Symposium on Characterization and Control of Odours and VOC, in the Process Industry*, Elsevier, pp. 431–444, 1994.

160. Shendan, B. A., Curran, T. P., and Dodd, V. A., Assessment of the influence of media particle size on the biofiltration of odorous exhaust ventilation air from a piggery facility, *Bioresour. Technol.*, 84, 129–143, 2002.

161. Kapahi, R. and Gross, M., Biotiltration for VOC and ammonia emissions control, *BioCycle*, 36, 87–90, 1995.

162. Liang, Y., Quan, X., Chen, J., Chung, J. S., Sung, J. Y., Chen, S., Xue, D., and Zhao, Y., Long lenn results of ammonia removal and transformation by biofiltration, *J. Hazard. Mater.*, B80, 259–269, 2000.

163. Grönberg, L., Lövkvist, P., and Jönsson, J. Å., Measurement of aliphatic amines in ambient air and rainwater, *Chemosphere*, 33, 1533–1540, 1992.

164. Skarping, G., Renman, L., Sango, C., Mathiasson, L., and Dalene, M., Capillary gas chromatographic method for the determination of complex mixture of isocyanates and amines, *J. Chromatogr.*, 346, 191–204, 1985.

165. National Institute for Occupational Safety and Health Pocket Guide to Chemical Hazards, U.S. Department of Health and Human Services, Centres for Disease Control and Prevention. U.S. Government, Washington, DC, 1994.

166. Stabbert, R., Schäfer, K. H., Biefel, Ch., and Rustemeier, K., Analysis of aromatic amines in cigarette smoke, *Rapid Commun. Mass Spectrom.*, 17, 2125–2132, 2003.

167. Lucceri, F., Pieraccini, G., Moneti, G., and Dolara, P., Primary aromatic amines from side stream cigarette smoke are common contaminants of indoor air, *Toxicol. Ind. Health*, 9, 405–413, 1993.

168. Chen, P. X., Mainstream smoke chemical analyses for 2R4F Kentucky reference cigarette. *Beits Tabakforsch Int.*, 20, 448–458, 2003.

169. Cigarettes — Determination of Total and Nicotine Free Particulate Matter Using Routine Analytical Smoking Machine, International Organization for Standardization, Geneva, ISO 4387:1991 (E), 1991.

170. Chang, W. Y., Sung, Y. H., and Huang, S. D., Analysis of carcinogenic aromatic amines in water samples by solid-phase microextraction coupled with high-performance liquid chromatography, *Anal. Chim. Acta*, 495(1–2), 109–122, 2003.

171. Hormazabal, V., Steffenak, I., and Yndestad, M., Simultaneous extraction and determination of florfenicol and the metabolite florfenicol amine in sediment by high-performance liquid Chromatography, *J. Chromatogr.*, 724, 364–366, 1996.

172. Lee, L. S., Nyman, A. K., Li, H., Nyman, M. C., and Jafvert, Ch., Initial sorption of aromatic amines to surface soils, *Environ. Toxicol. Chem.*, 16, 1575–1582, 1997.

173. Li, H. and Lee, L. S., Sorption and abiotic transformation of aniline and α-naphtylamine by surface soils, *Environ. Sci. Technol.*, 33, 1964–1970, 1999.

174. Li, H., Lee, L. S., Fabrega, J. R., and Jafvert, Ch. T., Role of pH in partitioning and cation exchange of aromatic amines on water-saturated soil, *Chemosphere*, 44, 627–635, 2001.

175. Groom, C. A., Beaudet, S., Halasz, A., Paquet, L., and Hawari, J., Detection of cyclic nitramine explosives hexahydro-1,3,5-trinitro-1,3,5-triazine (RDX) and octahydro-1,3,5,7-tetranitro-1,3,5,7-tetrazine (HMX) and their degradation products in soil environments, *J. Chromatogr.*, 909, 53–60, 2001.

176. Pierpoint, A. C., Hapeman, C. J., and Torrents, A., Ozone treatment of soil contaminated with aniline and trifluralin, *Chemosphere*, 50, 1025–1034, 2003.

177. Alzaga, R., Mesas, A., Ortiz, L., and Bayona, J. M., Characterization of organic compounds in soil and water affected by pyrite tailing spillage, *Sci. Total Environ.*, 242, 167–178, 1999.
178. Pace, Ch. M., Donnelly, J. R., and Jeter, J. L., Determination of aromatic amines in soils, *J. AOAC Int.*, 79, 777–783, 1996.
179. Timm, M. and Jorgensen, B. M., Simultaneous determination of ammonia, dimethlamine, trimethlamine and trimethylamine-*N*-oxide in fish extracts by capillary electrophoresis with indirect UV-detection, *Food Chem.*, 76, 509–518, 2002.

12 N-Nitrosamines

Ana María Afonso Perera

CONTENTS

The N-nitrosamines represent the most relevant group among the N-nitroso compounds in relation to their carcinogenic features. Magee and Barnes[1] first indicated the carcinogenicity of N-nitrosamines in 1956. Since then, research on N-nitroso compounds, their formation from precursors, occurrence in the environment and food, analysis, and toxicological activity have been extensive. In fact, no other chemical compound has received so much attention in the etiology of human cancer. Around 300 N-nitroso compounds tested for carcinogenicity in laboratory animals have been positive.[2]

The N-nitrosamines are formed from different precursors and under a wide variety of conditions. They can be formed by the reaction of secondary amines with nitrosating agents, such as

419

nitrite or nitrate.[3] Nitrosamines and their precursors nitrite and amines are ubiquitous in the environment. Humans are exposed to N-nitroso compounds from a variety of sources including food, water, occupational environments, tobacco, cosmetics, and even formation within the human body.[4,5] The suspected carcinogenicity of N-nitrosamines, combined with their apparent ease of formation, makes analytical determination of these compounds important.

I. PROPERTIES AND OCCURRENCE IN THE ENVIRONMENT

A. PHYSICAL AND CHEMICAL PROPERTIES

Nitrosamines are derived from amines, and like all N-nitroso compounds, they have the N−N=O functional group. They can be classified as volatile and nonvolatile. Volatile nitrosamines, in contrast to nonvolatile nitrosamines, can be removed from the matrix by distillation techniques and can be analyzed by gas chromatography (GC) without prior chemical derivatization. Alkyl and cyclic nitrosamines, which are volatile, have been intensively studied. The most commonly encountered nonvolatile nitrosamines are aryl compounds, hydroxylated compounds, and N-nitrosated amino acids. The structures of several nitrosamines are shown in Figure 12.1.

N-Nitrosamines are extremely reactive. They are sensitive to prolonged thermal treatment as well as to photochemical irradiation. Most N-nitrosamines undergo reactions with inorganic acids, such as HCl, HBr, and HI. In fact, this process is the basis for the denitrosation of such compounds.

The reduction of the nitroso group to the amino group is one of the most characteristic reactions of N-nitroso compounds. This reaction can be used for checking the presence or absence of N-nitrosamines. Likewise, N-nitroso derivatives can be readily converted into their N-nitro analogs via various oxidizing agents. One of the most effective oxidizing agents is trifluoroperacetic acid. Several other organic reagents can convert N-nitrosamines into the corresponding N-nitramines. This oxidation reaction has been the basis of various analytical methods.

The relative ease of dissociation of the −N−NO bond is probably one of the most significant physical properties of the N-nitroso derivatives. The release of the nitric oxide group from the N-nitrosamines is accomplished with relatively low energy requirements. Hence, the exposure of gaseous N-nitroso compounds to high temperatures, between 400°C and 500°C, can be a selective method for the removal of nitric oxide without causing other major rearrangements or dissociations in the rest of the molecule. This physical property of the N-nitrosamines has allowed the development of the thermal energy analyzer (TEA), a highly selective detector for N-nitrosamines.

1. The Chemistry of N-Nitrosamines Formation

N-Nitrosamines are formed by a chemical reaction between a secondary or a tertiary amine and a nitrosating agent, such as chemicals derived from nitrites or nitrogen oxides. Primary amines react with nitrosating agents to form unstable N-nitroso derivatives which degrade to olefins and alcohols.

The nitrite ion and the nitrous acid are not capable of the N-nitrosation, but under moderately acid conditions, they form dinitrogen trioxide (N_2O_3) which is a strong nitrosating agent. Nitrous acid under strongly acidic conditions (pH < 2) can be converted into more powerful nitrosating agents such as nitrous acidium ion (H_2ONO^+) or nitrosinium ion (NO^+).[6]

Dinitrogen trioxide reacts with the unshared pair of electrons on unprotonated secondary amine by a nucleophilic substitution reaction to form nitrosamines. The rate of nitrosation of secondary amines in a weakly acidic aqueous solution is proportional to the concentration of the amines and to the square of the nitrite concentration.[6,7] The concentrations of these two precursors depend on the pH of the medium. While the concentration of unprotonated amines increases when pH increases, the concentration of nitrous acid increases when the pH decreases. Hence, the pH rate profile for the nitrosation of amines shows a maximum resulting from the interaction between these two opposite

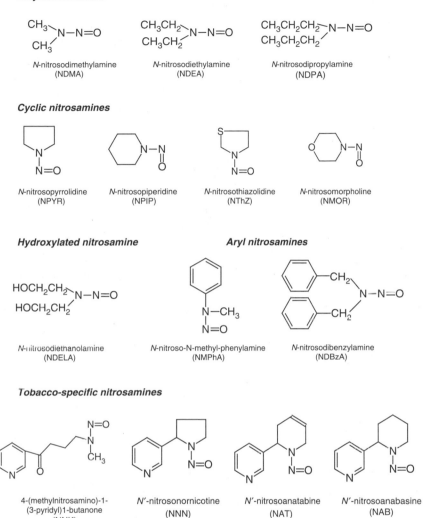

FIGURE 12.1 Structures of some nitrosamines.

reactions. The optimum pH for the nitrosation of most secondary amines is between 2.5 and 3.5, depending on the pK_a of the amine under consideration.[7]

Tertiary amines react with the nitrosating agents in an acidic aqueous solution to form *N*-nitrosamines but, in general, only after the complete formation of dialkylamines through dealkylation.

In the presence of a nucleophilic anion such as I^-, Br^-, Cl^-, SCN^-, acetate or phthalate, nitrous acid can be converted into more active nitrosating species. *N*-nitrosamines formation can be accelerated by certain microorganisms at acid pH values.[8] On the other hand, the nitrosation reactions can be inhibited by compounds such as ascorbic acid, sulfamic acid, tocopherol, and others.[9,10]

Dinitrogen trioxide (N_2O_3) and dinitrogen tetraoxide (N_2O_4) are effective reagents for nitrosation in both neutral and alkaline solutions.[11] Nitrosamines can be formed at basic pH because nitrogen oxides are direct nitrosating agents and do not require an acid medium to be transformed into reagents for nitrosation as in the case of the nitrite. The reaction is fast in the basic medium because the amines are unprotonated.

B. FORMATION AND OCCURRENCE IN THE ENVIRONMENT

Attending to the different pathways for the nitrosation of amines and amine derivatives, it is not unexpected for the N-nitrosamines to be found in many different areas of human environment. Nitrosamines and their precursors nitrite and amines, are ubiquitous in the environment. Nitrite can be formed by the nitrification of ammonia or by the denitrification of nitrate by microorganisms. The decomposition of organic materials from plants and animals, industrial (chemical) discharge, and pesticide preparations are primary sources of environmental amines.

Many N-nitroso-compounds are used as solvents in the fiber and plastic industries, antioxidants in fuels, insect repellants, insecticides, fungicides, and lubricating oil.[12-14] These compounds are discharged into the environment by fertilizers, industries, sewage output, and animal feedlots where nitrogen-related compounds are used. N-nitrosodimethylamine (NDMA) is very soluble and volatile, hence it can be spread out into our aquatic environment. NDMA has been detected in surface waters, seawaters, wastewater, and drinking water.[15,16] Therefore, NDMA is likely to be found especially where secondary amines and nitrite occur. Whereas food is generally the main source of nitrate intake in humans,[17] water with high nitrate levels can also be an important source of nitrate for regular consumers.[18-20] Contamination of groundwater by nitrate is a major concern in regions with intensive agricultural activities.[21,22]

Recently, NDMA was found in highly purified wastewaters intended for recycling, as well as in some treated drinking waters. The current investigations indicate that NDMA could be observed in treated waters, however, this was not previously suspected. This fact suggests that NDMA occurrence may be related to treatment and disinfection processes. It has been shown that NDMA is a disinfection by-product specifically produced by the reaction of monochloramine and dimethylamine in the absence of nitrite.[23-25] Monochloramine is purposely produced as a disinfectant, which may also be formed in chlorinated water in the presence of ammonia.[26] Other alkylamines or pesticides may also decompose to generate potential precursors of NDMA.

The occurrence of N-nitrodiethanolamine at various levels in all sorts of aqueous fluids in the mechanical industry results primarily from the reaction between alkanolamines and nitrite ions. The alkanolamines, such as di- or tri-ethanolamines (or their derivatives) are used as anticorrosives, lubricants or emulsifiers. The nitrite ions are frequently present in trace amounts or they are intentionally added to inhibit corrosion (metalworking fluids).[27,28]

Nitrosamines have been shown to be formed in soils in the presence of secondary amines and nitrite at acid pH values.[29,30] Many agricultural chemicals contain structures that can be degraded to secondary amines.[31] The excessive and widespread use of these pesticides may result in accumulation of secondary amines in localized environments, and may contribute to the formation of nitrosamines.

Epidemiological studies have suggested correlations between high nitrogen dioxide concentrations and cancer in urban populations, but since neither nitrogen dioxide nor nitrogen monoxide have yet been shown to be carcinogenic, it is possible that the cause may be the N-nitrosamines rather than the oxides of nitrogen. Certainly, oxides of nitrogen can be readily converted into nitrous acid under normal atmospheric conditions.[32]

The volatile N-nitrosamines present in environmental tobacco smoke are one of the classes of toxic air contaminants of particular concern because of their carcinogenicity. The most important four tobacco-specific nitrosamines (TSNA) are N'-nitrosonornicotine (NNN), 4-(methylnitros-amino)-1-(3-pyridyl)-1-butanone (NNK), N'-nitrosoanabasine (NAB), and N'-nitrosoanatabine (NAT). NNN and NNK are known to induce malignant tumors in mice, rats, and hamsters.[33] The other nitrosamines do not exhibit significant tumoral activity.

Tobacco alkaloids and nitrite are the major precursors of TSNA formation.[34] Green tobacco contains virtually no TSNA. These nitrosamines are generated during the postharvest treatment, i.e., curing (drying) and fermentation.[35] Available data suggest that nitrite and TSNA start to accumulate after approximately 2 to 3 weeks of air-curing. This is the time when the cells are disrupted due to moisture losses and then the nutrients are accessible to bacteria. Some of these bacteria reduce the nitrate with accumulation of nitrite. At the existing pH (5.5) the nitrite may form dinitrogen trioxide N_2O_3, which reacts with various tobacco constituents including the tobacco alkaloids.[36] TSNA present in the cured tobacco is partly transferred to tobacco smoke and a minor portion is generated during the smoking.[37] Among the TSNA, NDMA and N-nitrosopyrrolidine occur at the highest concentrations.[38]

It has been clearly reported that N-nitrosamines are also specific pollutants of the rubber industry. These are produced during the vulcanization steps as a result of nitrosation of aminated vulcanization accelerators by nitrogen oxides present in an industrial atmosphere, and/or by nitrites of uncertain origin.[39,40] For instance, widely used vulcanization agents such as tetramethylthiurame dusulphide, zinc diethyldithiocarbamate, and morpholinomercaptobenzothiazole are obvious precursors to the formation of NDMA, N-nitrosodiethylamine, and N-nitrosomorpholine, which are the most commonly found volatile nitrosamines in the atmospheres of vulcanization and posttreatment workshops, or in the storage of finished elastomer products.[41] Several reports have been published describing the presence of volatile N-nitrosamines in rubber products, especially baby bottle rubber nipples and pacifiers,[42-44] and likewise in nitrite-cured meats processed in elastic rubber netting.[45-47]

Convincing evidence also exists for endogenous formation of nitrosamines in human. Based on our knowledge of the nitrosation in acid aqueous media, it is not surprising that the nitrosamine formation has been demonstrated to occur in the stomach by interaction of secondary and tertiary amino compounds with nitrite or other nitrosating agents derived from the diet. This endogenous nitrosation has been extensively discussed for many years,[48] and has been demonstrated experimentally. It has also been studied epidemiologically in connection with human cancer.[49]

C. Toxicology

N-Nitroso compounds and specifically the N-nitrosamines exhibit mutagenic, carcinogenic, and teratogenic activities. Around 300 N-nitroso compounds have been tested to detect their carcinogenicity, with this activity found in most of them.[50-52] It is demonstrated that the nitrosamines develop a carcinogenic effect in a wide range of animal species like fishes, reptiles, birds, and mammals, including five species of primates.[52-54]

Studies in vivo with different N-nitrosamines and their effect on the glutathione levels in hepatocytes reveal that these compounds are inhibitors of the enzymatic mitochondrial activity.[54] The chemical structure of the N-nitrosamines plays an essential role in the alteration of these hepatic levels, confirming the hepatotoxic activity developing these compounds in the organism. On the other hand, teratogenic effects caused by the activity of N-nitrosamines have been detected, particularly at the level of the central nervous system.[55]

The N-nitrosamines develop carcinogenesis in different animal species. These compounds need a metabolic activation to become mutagens. The metabolic activation reaction of the N-nitrosamines is catalyzed for members of the enzymatic family P450. More precisely, the isoenzyme P450 2E1 (CYP2E1) is responsible for the metabolism of most N-nitrosamines.

The carcinogenicity of NOCs is indeed explained by their ability to form strong electrophilic alkylating agents, which can react with the nucleophilic sites of cellular macromolecules such as DNA, RNA, and proteins. The alkylation of bases can induce mutations and hence initiate carcinogenesis. Despite diversity of possible DNA adducts formed after exposure to NOCs, only the formation and persistence of O^6-alkylguanine and O^4-alkylthyamine in target organ DNA have been correlated with tumor development.[56] These alkylations cause alterations in the pairing of nucleic bases during the replication of DNA, leading respectively to the erroneous incorporation of thiamine instead of citosine[57] and of guanine instead of adenine.[58,59]

II. SAMPLE PREPARATION

A. Water and Wastewater Samples

There are basically two methods used for the analysis of N-nitrosamines in water samples. Those are liquid–liquid extraction using an organic solvent, and adsorption onto a sorbent material.

1. Liquid–Liquid Extraction

The most popular solvent is the dichloromethane (DCM) for two reasons: the volatile nitrosamines are highly soluble in this solvent, and the boiling point of the DCM is low, hence preventing the volatilization or the degradation of nitrosamines through a subsequent concentration step.

In this extraction method,[15,60–63] a known volume of water is filtered and adjusted to pH 12 with a 50% solution of NaOH in water. NaCl (14 g/l) can be added to break up any emulsion.[63] The sample is repeatedly extracted with (3 to 5 × 40) ml aliquots of DCM by shaking. The extracts are dried over anhydrous sodium sulfate or passed through a column containing anhydrous sodium sulfate to eliminate water. Then, they are concentrated on a Kuderna-Danish evaporative concentrator, which already contains a nitrosation inhibitor. Aliquots of the final concentrate are further analyzed by GC or high-performance liquid chromatography (HPLC). Separatory funnel extraction of wastewater samples was not practical due to the formation of emulsions during shaking. Therefore, wastewater samples are better extracted for 6 h[61] or overnight[62] with DCM by continuous liquid–liquid extraction.

Sen et al.[15] described a method for the determination of NDMA in drinking water practically free of artifactual formation. Such formation is minimized by extracting the samples in the presence of sulfamic acid, an excellent inhibitor of N-nitrosation. The water sample is acidified with the addition of H_2SO_4 and sulfamic acid and afterwards it is extracted with DCM. The extracts are washed with KOH solution.

2. Adsorption on Solid Supports

Several types of solid sorbents have been utilized for the removal of N-nitrosamines from water samples. They include active carbon, carbonaceous adsorbents, XAD resins, and others.

In the case of active carbon, the water samples are adjusted to pH 7 with hydrochloric acid and then the samples are passed through a mini activated carbon column by a sweep pump at a fixed flow rate. The carbon is then partly dried by flushing the column with purified nitrogen to remove most of the residual water before the column is eluted with acetone or chloroform.[64] The concentration of the organic extracts can be accomplished on a rotary evaporator under reduced pressure, at ambient conditions or slightly above ambient temperature. It is advisable to dry the organic extracts with anhydrous sodium sulfate prior to any final concentration step. This approach can be used only for the less volatile N-nitrosamines unless care is taken during the concentration step to avoid losses of NDMA or NDEA. Queiroz et al.[60] obtained better extraction efficiencies for the most volatile nitrosamines with two adsorbents coupled in series (C8 and active carbon) and using ethanol as eluent.

Carbonaceous adsorbents like Ambersorb 572 have been used to remove NDMA from water.[24] The sample is added with 200 mg of carbonaceous adsorbent and extracted by shaking the solution for 1 h at 200 rpm. Ambersorb beads are vacuum filtered onto a glass fiber filter, dried in air for 30 min and then soaked with methylene chloride for 20 min before the analysis.

Kimoto et al.[65] have used Ambersorb XE-340 to remove trace levels of volatile nitrosamines from tap water. In order to wet the adsorbent, it was first covered for 1 h with methanol and then washed with distilled water. Ambersorb XE-340 was packed in a 26×260 mm copper pipe equipped with a copper fitting on both ends. One fitting was connected to the faucet and the other to a valve which controls the flow rate. Water was sampled for 8.5 to 11.75 h. At the end of the water sampling, methanol was added to remove the water from the column. DCM (700 ml) was next passed through the column. The first 500 ml DCM contained more than 92% of the *N*-nitrosamines. The extracts were dried over anhydrous Na_2SO_4 and concentrated in a Kuderna-Danish evaporator.

The XAD-type resins have been used to remove *N*-nitrosamines from water samples and sewage effluent. The samples are filtered to remove suspended particles and then they are passed through a column at a flow rate between 11 and 15 ml/min. Acetone–DCM[66] or DCM–diethyl ether (75:25)[67] have been used to desorb the organic compounds retained by the XAD resin. After an appropriate concentration, the final solution can be analyzed directly by HPLC or GC. Alternatively, it may be necessary to perform a further clean-up of the extract prior to the analysis. This clean-up will depend largely on the nature of the water samples extracted and on the complexity of the organic materials present in the sample.

B. SOIL SAMPLES

Depending on the soil type, it may be possible to extract the *N*-nitrosamines using different solvents. The mineral oil distillation procedure can be used to extract volatile *N*-nitrosamines from soils which are not adequate to solvent extraction.

The mineral oil distillation procedure introduced by Fine et al.,[68] and modified by many authors,[69-74] is the most popular extraction method for volatile *N*-nitroso compounds and it has been applied to a wide variety of samples. This method consists of a distillation of the sample at reduced pressure from an alkaline medium containing mineral oil. The mineral oil ensures an effective and uniform heat distribution within the distillation mixture. It also improves the efficiency of distillation by reducing the distillation time and increasing the recoveries of the volatile nitrosamines with higher boiling points.[75] The first step is to homogenize the sample with a nitrosation inhibitor, such as ascorbic acid, sulfamic acid, or tocopherol. Afterwards, it is added to a round-bottom flask along with an equal volume of mineral oil. The mixture is then distilled under vacuum and the temperature is slowly increased up to 100°C over 40 to 50 min. The distillate is collected in a glass finger immersed in liquid nitrogen and quantitatively transferred to a separating funnel for further purification.

In determination of solid samples, Eisenbrand et al.[75] have pointed out the importance of adding water to the flask before distillation. The water increases recovery of the less volatile nitrosamines and avoids exposing the sample to excessive temperatures.

A modification of the mineral oil distillation procedure has been proposed using a gas purge system to trap nitrosamines on a ThermoSorb/N cartridge.[76] This procedure eliminates solvent extraction and evaporation steps, hence reducing the possibility of artifact formation. On the other hand, some methods have also been described for the vacuum distillation of alkaline suspensions without mineral oil.[77] The distillation is carried out from a slightly basified sample with carbonate potassium or from a highly basified sample with potassium hydroxide.

The direct extraction using different solvents is a more rapid alternative than the distillation for determination of *N*-nitrosamines in soils. The soil, altogether with suitable nitrosation inhibitors such as ascorbic acid and tocopherol, is homogenized with a suitable extraction solvent. The solvent is then filtered and purified prior to the analysis. You et al.[78] extracted 100 g of soil with a 50 ml portion of DCM during 20 min. The samples were then filtered to remove insoluble particulates. Afterwards, the filtrate was treated with anhydrous sodium sulfate and concentrated under a stream of dry nitrogen at 30°C. The nitrosamines were analyzed by HPLC with a precolumn fluorescence derivatization. A similar approach has been used by Ross et al.[79] obtaining a NDPA recovery of 80% at the 4 ppb level and of 45% at the 1 ppb level.

C. AIR SAMPLES

In order to determine the N-nitrosamines in the atmosphere, it is necessary to pass about 200 l of air through a suitable trap at a flowrate of roughly 2 l/min. The contents of the trap are then extracted into an organic solvent or desorbed by heating. Several trapping techniques have been employed and some of them will be described below. The N-nitrosamines precursors are generally present at levels from 100 to 1000 times greater than the N-nitrosamines. Therefore, great care must be taken to ensure that the concentration step does not also concentrate the precursors. The concentration of the precursors would lead to an important artifact formation of nitrosamines.

ThermoSorb/N cartridges are commercially available and have been specifically designed for the quantitative collection of N-nitrosamines in outdoor air.[80,81] The cartridge contains two sorbent zones. The first zone selectively traps and removes amines from the incoming air, hence preventing the subsequent formation of nitrosamines by airborne nitrogen oxides. The second zone contains a nitrosating inhibitor system which prevents the formation of N-nitrosamines followed by elution of the ThermoSorb/N cartridge. The cartridges have a relatively moderate sampling capacity (1500 ng/cartridge); however, they can be connected in series to increase the total capacity of the sampling system. N-Nitrosamines are removed from the cartridge by reverse elution with 0.5 ml of a special eluting solvent. An aliquot of the eluate is introduced into the GC or HPLC for further analysis. The concentration of the eluting solvent is not required, thereby eliminating the possibility of artifact formation during this step. Marano et al.[82] carried out determination of trace levels of nitrosamines in air using ThermoSorb/N cartridges. These cartridges were preeluted to remove interfering compounds prior to nitrosamine elution and selective ion monitoring MS detection. Enhanced sensitivity was achieved by using a commercial concentrator which allowed the introduction of 40 μl of eluent onto fused silica capillary and packed columns.

Sampling techniques using Tenax were developed by Pellizzari et al.[83-85] N-Nitrosamines and other organic vapors were collected from ambient air on a 1.5 × 6 bed of Tenax GC (35/60) in a glass cartridge. All cartridges were preconditioned by heating at 275°C prior to field sampling. They are desorbed by heating in a stream of helium and afterwards the contents are caught in a gold-lined trap held at −192°C. The gold trap is then flash heated, driving the contents directly into the capillary column of a GC−MS system. The main disadvantages of the method are: the Tenax may trap precursor amines, which could form N-nitrosamines during desorption and heating steps; and the Tenax has a relatively small breakthrough volume for NDMA, and this N-nitrosamine is often the one of maximum interest.

Rounbehler et al.[86] have examined several types of sorbents for their ability to collect and retain quantitatively a variety of volatile nitrosamines under simulated air-sampling conditions. Also, the artifactual formation of nitrosamines from trapped amines and air containing nitrogen oxides were studied. The dry solid sorbents included activated charcoal, activated alumina, silica gel, Florisil, Tenax, and ThermoSorb/N cartridges. It was found that a ThermoSorb/N cartridge was the only sorbent which was both free of artifact formation and capable of retaining 100% of the preloaded nitrosamines.

D. ENVIRONMENTAL TOBACCO SMOKE SAMPLES

The most highly developed and validated method for sampling volatile N-nitrosamines from tobacco smoke employs an aqueous buffered solution (pH 4.5 citrate−phosphate) with $2.10^{-2}M$ ascorbic acid contained in several impingers connected in series.[87-90] Ascorbic acid is added to the solution as a nitrosating inhibitor to prevent the formation of artifact N-nitrosamines from the amines and NO$_x$ in tobacco smoke.[89] An ascorbic acid-impregnated Cambridge filter is placed upstream of the impinger solution to remove particles. Buffer solutions and Cambridge filter are extracted with DCM. The extracts are concentrated and afterwards subjected to a clean-up on alumina. For elution 1% methanol in DCM is used.

For determining volatile nitrosamines in mainstream and sidestream, Kataoka et al.[91] employed 5% hydrochloric acid as the tripping solution and diethyl ether containing 25% 2-propanol as the extractant. A similar approach was used by Cárdenes et al.[92] using DCM as the extractant.

These methods are limited by the air-flow rate (especially for midget impingers), and also the aqueous solutions are not a very convenient sampling medium, particularly for field experiments. Sample recovery problems are often encountered in liquid–liquid extraction due to the aqueous buffers or to the emulsion formation. Labor-intensive sample preparation procedures and the relatively fast degradation of the aqueous ascorbic acid solution are the two additional drawbacks when using this sampling method.

The use of the ThermoSorb/N cartridges, specifically designed for the quantitative collection of *N*-nitrosamines in out air, eliminates the most common problems associated with the aqueous sampling medium. This approach has been evaluated by Mahanama et al.[93] for *N*-nitrosamines in a complex matrix like environmental tobacco smoke. Besides the TheromoSorb/N cartridges, ascorbic acid-impregnated Teflon filters were used to remove particles. The cartridges were back-flushed with 2 ml methanol. The extract was concentrated on a rotatory evaporator and loaded onto an alumina-B Sep-Pak, using 10% of chloroform in DCM to recover nitrosamines. This clean-up procedure eliminates polar interferences in the extracts, which could contaminate the capillary GC column and could also interfere in the later analysis. The procedure yields a 97.8 ± 2.8% recovery for the internal standard, NDEA, using nine cartridges loaded with environmental tobacco smoke samples.

Wu et al.[94] used an ascorbic acid-impregnated Cambridge filter pad to measure five TSNA in the particulate phase of mainstream tobacco smoke. Each pad with trapped smoke particulate was first spiked with an internal standard and then extracted with DCM, back extracted into an aqueous solution and further purified by solid-phase extraction (Water Oasis HLB 60 mg), using 100% ethanol as eluent.

Among other *N*-nitrosamines collection methods, the use of wet traps such as 1 *N* KOH, cold traps, and Tenax traps have been reported, each one with its own limitations.[95,96]

E. ADDITIONAL *N*-NITROSAMINES EXTRACTION METHODS

1. Solid-Phase Microextraction

The newly developed solid-phase microextraction (SPME) technique, first reported by Pawliszyn[97] in 1989, is increasingly used for the gas chromatographic determination of a wide variety of volatile and semivolatile organic compounds in water or aqueous extracts of different substrates. Basically, it involves the extraction of specific organic analytes directly from aqueous samples or from the headspace of these samples in closed vials. The extraction is achieved onto a fused-silica fiber coated with a polymeric liquid phase. After equilibration, the fiber containing the absorbed or adsorbed analyte is removed and thermally desorbed in the hot injector port of a gas chromatograph or in an appropriate interface of a liquid chromatograph.[97–102]

The technique is very simple, fast, and does not employ any organic solvents either for sample preparation or clean-up. This makes it highly desirable because, unlike other methods, it does not release environment-polluting organic solvents into atmosphere. Thus far, the technique has been successfully applied to the determination of a wide variety of organic compounds. However, the application of SPME to determine the nitrosamines could eliminate some problems like the widespread solvent use and the lengthy and time-consuming sample preparation steps (that are common in most of the published methods in this area).

Sen et al.[103] have reported a SPME analytical method for the determination of N-nitrosodibutylamina and N-nitrosodibenzylamina. This method is based on the isolation of the compounds by steam distillation, followed by the SPME in the distillate headspace using a polyacrylate coated silica fiber, and the determination by GC–TEA. Recoveries ranged between 41% and 112%. Nitrosamines in environmental matrixes (air, tobacco, and seawater) were preconcentrated on polydimethylsiloxane/divinylbenzene and the recovery of different nitrosamines from different matrixes varied between 95% and 98%.[104]

An automated in-tube solid-phase microextraction HPLC method for NNK and several metabolites have been developed by Mullett et al.[105] In-tube SPME is an on-line extraction technique where analytes are extracted and concentrated from the sample directly into a coated capillary by repeated draw–eject steps. A tailor-made polypyrrole-coated capillary was used to evaluate their extraction efficiencies for NNK and several metabolites in cell cultures. This automated extraction and analysis method simplified the determination of the tobacco-specific N-nitrosamines, requiring a total sample analysis time of only 30 min.

2. Supercritical Fluid Extraction

Because of U.S. Environmental Protection Agency regulations,[106] there is a strong incentive to reduce or replace organic solvents, particularly those containing halogens that are usually employed in residue analysis. These regulations are designed to reduce the use of solvents that are potentially harmful to the environment and to reduce the costs of solvent disposal. Supercritical fluid extraction (SFE) has the potential to achieve effectively the selective extraction in a single step and to concentrate the analyte as it is ready for instrumental analysis with a minimum amount of solvent.

SFE applications in environmental analysis, particularly for heterogeneous solid samples, are emerging as viable alternatives to more traditional methods employing polar liquids or mixtures as extractants. Most of these approaches utilize CO_2 as the fluid with or without methanol as modifier.[107]

The current emphasis on methods that use less solvent makes SFE an attractive alternative for the analysis of nitrosamines. However, only limited studies have been carried out on SFE with nitrosamines. Prokopczyk et al.[108,109] reported extraction efficiencies of 83% to 98% for the major nicotine-derived TSNAs in smokeless tobacco and snuff, with methanol-modified supercritical carbon dioxide. These compounds were extracted from cigarettes using SFE and purified by a sodium hydroxide wash of the ethyl acetate eluting solvent and solid-phase extraction.[110]

SFE has been demonstrated to be a good extraction technique for N-nitrosamines in rubber products. In addition, SFE allows fast analysis with a reduction in solvent waste, time, and manipulation. Although recoveries are not too good, especially for the smaller N-nitrosamines, SFE could be considered as a useful tool to determine these analytes, considering that through its selectivity it provides quite clean extracts in one step.[111] Reche et al.[112] determined N-nitrosamines in latex products by combining supercritical fluids and chemical derivatization. The addition of a denitrosation reagent into the extractor combined with an adequate liquid trap allows elucidation of the presence of N-nitrosamines as well as their potential precursors.

F. Derivatization Reactions for Gas Chromatographic Determination of N-Nitrosamines

Derivatization of N-nitrosamines can be employed not only to convert nonvolatile nitrosamines into volatile materials suitable for GC analysis, but also to improve selectivity and sensitivity, and to reduce the analysis time. The commonly used derivatization reactions for GC analysis of nitrosamines are stated below.

N-Nitroso compounds can be easily reduced to their hydrazines, which are further derivatized. The most satisfactory approach is the conversion of the hydrazine to its 3,5-dinitrobenzaldehyde hydrazone, which is then analyzed using GC with electron capture detection.[113]

In a similar way, the N-nitrosamines can be converted into secondary amines. The produced secondary amines can be converted into appropriate derivatives for their gas chromatographic analysis. One of the most common derivatives is the heptafluorobutyryl amide (HFB-amide), formed by the reaction of the HFB-acid chloride and the amine. Afterwards, this derivative can be analyzed by GC-electron capture or GC–MS.[114,115] Kataoka et al.[91] reported the determination of seven nitrosamines by GC-flame photometric detection. The method is based on denitrosation with hydrobromic acid and the subsequent diethylthiophosphorylation of secondary amines.

The N-nitroso compounds can be converted into their oxidation products, the nitramines, by several oxidation pathways. Thus, the peroxidation of nitrosamines with trifluoroperoxyacetic acid, prepared by the reaction of trifluoroacetic acid or anhydride with 30% to 90% of hydrogen peroxide, was used in the past for detecting nitrosamines. This is because the electron capture detector shows greater sensitivity and selectivity for nitramines than other detectors available for nitrosamines at that time. This peroxidation reaction has been applied for the detection of NDMA in ambient air and in cigarette smoke.[96] The peroxidation reaction with pentafluoroxybenzoic acid, a stable solid peroxyacid, has been used to confirm the NDMA and NPYR. This reaction has the advantage of minimizing the repeated use of concentrated hydrogen peroxide.[116] Cooper et al.[117] reported the conversion of a series of N-nitrosamines into their corresponding N-nitramines analogs by pertrifluoroacetic acid oxidation. The nitramines were detected by GC-electron capture.

The formation of ether derivatives of hydroxynitrosamines converts these nonvolatile N-nitroso derivatives into volatile compounds suitable for analysis by GC and their sensitive detection by TEA or MS.[118–121] In this sense, the trimethylsilylation using N-methyl trimethylsilyl trifluoroacetamide or tert-butyl dimethylchlorosilane/imidazole and trimethyldiclorosilane/hexamethyldisilazane produces trimethyl or butyldimethyl ethers of hydroxynitrosamines.[120,122,123]

III. ANALYSIS METHODS

Since the discovery of the carcinogenicity of N-nitrosamines and their occurrence in the environment, many separation and detection methods have been developed for their analysis. Polarographic, spectrophotometric, and thin-layer chromatographic methods have been the earliest techniques employed and have been reviewed by several authors.[124] However, the N-nitrosamines are now routinely analyzed by chromatographic procedures. Two major types of chromatography methods are commonly used, namely, GC and HPLC. Other techniques like capillary electrophoresis and capillary supercritical fluid chromatography have been introduced most recently.

A. GAS CHROMATOGRAPHIC SEPARATION

Different chromatographic columns, including packed and capillary columns, have been used to separate volatile nitrosamines or nonvolatile nitrosamines after derivatization.

The capillary columns are the most used columns for the separation of N-nitrosamines. These enable better separations and sensitivities than packed columns with comparatively lower temperatures, hence preventing the decomposition of thermally labile compounds. The most commonly used columns are made of siloxane polymers like DB-5 and of polyethylene glycol like DV-wax. Column lengths of 15 to 30 m and film thickness of 0.25 to 1.0 μm are used in many applications. The high capacity of 1.0 μm film allows the injection of large volumes of samples. This can be an important feature when TEA or MS detectors are used.

For packed columns, the most common stationary phase is Carbowax 20M generally coated on Chrosorb W with a mesh size of 80/100 and 100/120. There are some cases where the mesh size is 60/80. The columns have lengths between 1.5 and 4.5 m, and inner diameters between 2 and 4 mm. The common injection modes and the gas chromatographic conditions typically used for the separation of N-nitrosamines are summarized in Tables 12.1 and 12.2.

B. High-Performance Liquid Chromatographic Separation

GC coupled to TEA detection is the most suitable analytical method to determine volatile nitrosamines. However, a large number of N-nitrosamines are not generally adequate for direct analysis by GC, either because of low volatility or thermal instability. In these cases, HPLC seems to be the method of choice for the analysis of volatile and nonvolatile nitrosamines. The correct choice of stationary phase and type of composition of the mobile phase depends on the detection method used.

Reversed-phase chromatography of N-nitrosamines has been used in connection with UV, fluorimetric, amperometric, and MS detection. In many cases these methods are based on pre- or postcolumn denitrosation, and derivatization of the denitrosation products. Chemically bonded octadecylsilane, C_{18}, is the most often reported stationary HPLC phase. Isocratic or gradient elution is usually performed with mixtures of water with acetonitrile, methanol, ethanol, or propanol. In general, the mobile phase is acidified by the addition of acetic acid, ammonium acetate, or phosphate buffers. The HPLC conditions used for the separation and detection of N-nitrosamines are shown in Table 12.3.

HPLC techniques have been developed to allow interface with the TEA. The goal is to make use of the advantageous sensitivity of this detector. Some aspects concerning this subject will be discussed later. In some HPLC–TEA interfaces, the types of possible mobile phase are very restricted due to the impossibility of using aqueous solvent or inorganic buffers. Consequently, normal-phase HPLC is associated with TEA. Various silica supports have been used for adsorption chromatography like LiChrosorb and apolar phenyl-bonded silica, or polar cyano- or amino-bonded silica stationary phases in partitioning chromatography. Common organic eluents are mixtures of hexane with either acetone, chloroform, methanol, propanol, isopropanol, or DCM, and in the presence or absence of acetic or oxalic acid. Nevertheless, the development of the photolytic interface for HPLC–TEA[125] has allowed the use of reverse-phase chromatography.

C. Capillary Electrophoresis

Capillary electrophoresis (CE) has emerged as an efficient and rapid separation technique in recent years. Its high efficiency has been employed in many applications such as in the analysis of environmental pollutants.[138,139] Different approaches have been adopted to enhance selectivity for the analysis of different types of compounds. There are two approaches most commonly used to improve CE separations: the addition of modifiers into the electrophoretic medium and the modification of the column. Examples of the first approach include the addition of surfactants into the electrophoretic medium as in micellar electrokinetic chromatography (MEKC), and the use of organic solvents,[140] cyclodextrines,[141,142] or bile salts[143] as buffer modifiers. Examples of the second method include the use of gel-filled columns (capillary gel electrophoresis)[144] and the coating of the capillary wall surface.[145,146]

A considerable number of CE separation methods exist for a wide variety of analytes. However, nitrosamines separation and determination by CE requires additional development for its practical use.[147,148] For the separation of hydrophilic, low molecular weight, neutral, and polar compounds such as nitrosamines, it is necessary to develop CE techniques for enhancing the selectivity. The main reason is that these compounds do not interact strongly with the commonly used surfactants (e.g., sodium dodecyl sulfate, SDS) or other buffer modifiers such as cyclodextrins in electrokinetic chromatography. The separation depends on several factors which must be optimized to reach

TABLE 12.1
GC–TEA Methods

| Compound | Column (m/mm/μm)[a] | Temperature (°C) | | | Pyrolyzer Furnace | Oxygen Flow (ml/min) | Pressure Reaction Chamber | Carrier Gas | Ref. |
		Injector	Column	Interface					
NDMA, *N*-nitrosodiethylamine, *N*-nitrosopyrrolidine, *N*-nitrosomorpholine	DB-Wax, 30/0.32/0.25	—	50: 2 min, 8/min to 170, 170: 3 min	225	550	30	0.2 kPa at 50°C	Helium 103 kPa	163
NDMA	DB-225, 15/0.53/1	—	70: 5 min, 8/min to 110, 170: 2 min	200	—	—	—	—	164
Volatile and tobacco-specific *N*-nitrosamines	DB-5, 30/0.32/1.0	—	40: 1 min, 3/min to 90, 90: 5 min, 20/min to 200, 200: 15 min, 30/min to 300	b	575	25–35	0.3–0.5 Torr	Helium 12 psi	89
N-Nitrosodibenzylamine	SP-2401 DB, 100–200, 1.8/2.6	240	80: 5 min, 10/min to 220	275	475	—	1.0 mm	Helium 35 ml/min	46
NDMA	10% Carbowax1540/5%KOH, 100–200, 2.7/4	200	4/min: 100 to 180	—	450	—	—	Argon 40 ml/min	44
NDMA	10% Carbowax 20M/2%KOH, 80–100, 2.7/4	250	50: 1 min, 16/min to 190		500	10	0.9 Torr	Argon 40 ml/min	165
NDMA	10% Pennwalt 223/4%KOH, 80–100, 1.5/3	200	160	200	550	—	—	—	166
N-Nitrosodiethanolamine	10% Carbowax/0.5%KOH, 80–100, 1.8/2	170	130: 5 min, 4/min to 180, 180: 2 min	—	450	—	0.65 mmHg	Nitrogen 20 ml/min	117
NDMA, *N*-nitrosodiethylamine, *N*-nitrosodipropylamine, *N*-nitrosodibutylamine, *N*-nitrosopiperidine, *N*-nitrosopyrrolidine, *N*-nitrosomorpholine	Chromosorb W 10% Carbowax 20M/5%KOH, 100–120, 2.75/3.2	—	150: 2 min, 4/min to 180	—	475	—	1–1.8 Torr	Argon 30 ml/min	167

Continued

TABLE 12.1
Continued

Compound	Column (m/mm/μm)[a]	Injector	Temperature (°C) Column	Temperature (°C) Interface	Pyrolyzer Furnace	Oxygen Flow (ml/min)	Pressure Reaction Chamber	Carrier Gas	Ref.
NDMA, N-nitrosodiethylamine, N-nitrosomorpholine,	Gas-Chrom P 15% Carbowax/ 20M-TPA, 60–80, 2.8/3.2	200	4/min: 110 to 210	—	—	—	—	Helium 40 ml/min	65
NDMA	Chromosorb W 10% Carbowax 20M/5%KOH, 100–120, 3/2	200	140: 2 min, 6/min to 200	300	—	—	—	Argon 20–25 ml/min	15
NDMA	Rtx-200, 30/0.53/3.0	150	35: 5 min, 6/min to 175, 175: 6 min	280	900	80	—	Helium 4 ml/min	62
NDMA, N-nitrosodiethylamine, N-nitrosodipropylamine, N-nitrosodibutylamine, N-nitrosopiperidine, N-nitrosopyrrolidine, N-nitrosomorpholine	CP Wax 52 CB, 25/0.53/1	200	35: 0.5 min, 35/min to 100, 10/min to 160	250	400	—	—	Helium 7 ml/min	169
Tobacco-specific N-nitrosamines	Supelco SPB-1	—	60: 1 min, 40/min to 150, 2/min to 175, 40/min to 240	250	500	—	—	Helium	170
N-Nitrosodiethanolamine	AT-WAX, 10/0.53/1.2	220	210	250	450	10	15 mmHg	Helium 5 ml/min	168
N-Nitrosodibenzylamine	Supelcowax 10, 30/0.53/1	220	40: 1 min, 50/min to 160, 6/min to 220	375	800	—	—	Argon 8 ml/min	103
NDMA, N-nitrosodiethylamine, N-nitrosodibutylamine, N-nitrosopiperidine, N-nitrosomorpholine	Chromosorb W 10% Carbowax 20M, 60/80, 2.74/2	200	4/min: 110 to 150	300	—	—	—	—	134

Tobacco-specific *N*-nitrosamines	HP-5, 30/0.32/0.25	—	40: 1 min, 20/min to 160, 160: 1 min, 4/min to 200, 200: 1 min, 15/min to 260, 260: 1 min	200	500	—	—	Helium 40 cm/sg	110
NDMA, *N*-nitrosodiethylamine, *N*-nitrosodipropylamine, *N*-nitrosodibutylamine, *N*-nitrosopiperidine, *N*-nitrosopyrrolidine, *N*-Nitrosodibutylamine	15% Carbowax 20 M-TPA, 60–80, 1.8/6	190	120: 10 min, 4/min to 180, 180: 30 min. 15/min to 260, 260: 1 min	—	475	1.0 Torr	—	Helium 30 ml/min	171
N-nitrosodibenzylamine	15% Carbowax 20M-TPA, 60–80, 1.8/6	180	4/min: 120 to 200	—	475	0.4 mm	—	Helium 35 ml/min	45, 172–174

a Length/ID/film thickness.
b Pyrolyzer furnace connecting directly to the capillay column oulet.

TABLE 12.2
GC–MS Methods

Compound	Column (m/mm/μm)[a]	Injection Mode	Injector	Temperature (°C)			Carrier Gas	LOD	Ref.
				Column	Source	Interface			
NDMA	HP-5MS 30/0.25/0.50	Split 20:1	250	35: 3 min, 5/min to 100, 100: 5 min, 15/min to 280	150	230	Helium 30 cm/sg at 40°C	0.003 pg/μl	63
NDMA	DB-225 30/0.32/1	On column	—	70: 5 min, 8/min to 110, 170: 2 min	210	—	Helium 2–3 ml/min	—	164
NDMA	DB-Wax 30/0.22/0.25	—	60	60: 1.5 min, 10/min to 180, 180: 6.5 min		250	Helium 15 psi	1 pg/g	15
NDMA, N-nitrosomorpholine	Gas-Chrom P 15% Carbowax/20M-TPA 60–80 2.8/3.2	—	200	4/min: 90 to 140, 6/min: 140 to 180	150	180	15 ml/min	5 ng/μl	16
N-Nitrosodibenzylamine	DB-5-MS 30/0.32/1	Splitless	220	70: 5 min, 6/min to 220, 180: 6.5 min	190–200	300	Helium 1.2 ml/min	—	46
NDMA	SPB™-1701 30/0.25/0.25	Splitless	200	35: 1 min, 10/min to 70, 10/min to 220, 220: 2.4 min	150	260	—		61
N-Nitrosodibutylamine, N-nitrosodibenzylamine	DB-5 30/0.25/0.25	—	220–260	40: 2 min, 10/min to 260, 260: 4 min	—	—	—	—	103
N-Nitrosomorpholine	RSLSuerox FA 50/0.32/0.3	Splitless	120	90: 4 min, 5/min to 160, 160: 5 min	190	200	Helium 2 ml/min	10 pg	192
NDMA, N-nitrosodiethylamine, N-nitrosodibutylamine, N-nitrosopiperidine, N-nitrosomorpholine	CP-Wax 51 30/0.22	On column	—	120: 3 min, 10/min to 180, 160: 5 min	150	—	—	—	134

	Column[a]								
Tobacco-specific *N*-nitrosamines	J&W 30m DB 5/0.25/0.25	Splitless	250	40: 1 min, 20/min to 160, 160: 1 min, 4/min to 200, 200: 1 min, 15/min to 250, 260: 1 min	230	200	Helium 40 cm/sg	0.02–0.04 µg	110
NDMA	Chrompac 25/0.25/0.2	On column	—	35: 1 min, 70/min to 55, 55: 7 min, 3/min to 70, 20/min to 180	200	—	Helium 50 kPa	0.04 µg/kg	193

[a] Length/ID/film thickness.

TABLE 12.3
HPLC Analysis of N-Nitrosamines

Column[a]	Chromatographic Conditions	Detection (Detection Limit)	Compound	Ref.
Ultrasphere ODS 15/0.46/5, LiChrospher C_{18} 25/0.4/5, Nucleosil C_{18} 25/0.46/5	A/B (80:20, v/v), A: water (1% CH_3COOH and 5% acetonitrile), B: acetonitrile with 5% water containing 1% acetic acid	Postcolumn photohydrolysis Colorimetric detection by Griess reagent (20 pmoles)	19 Volatile dialkyl and seven nonvolatile N-nitrosamines	126
Brownlee Polymer RP (10 μm) 10/4.6/5, Sphersorb ODS (5 μm) 15/4.6/5 C_{18} (3 μm)	Acetonitrile/10 mM trifluoroacetic acid (gradient)	Postcolumn photolysis Chemiluminiscence detection (7–42 μg/l)	Nonvolatile N-nitroso compounds	125
	Actonitrile/water/ethanol (63.5/35.4/1.0 v/v), containing imidazol (3.0 mmol/l) as catalyst, pH 6.6 with oxalic acid	Peroxyoxalate chemiluminescence detection (4.3–8.3 fmol)	NDMA, N-nitrosopyrrolydine, N-nitrosodiethylamine, N-nitrosopiperidine, N-nitrosodipropylamine, N-nitrosodibutylamine and corresponding secondary amines	64
C_{18} 8.3/4.6/3	Actonitrile/water/ethanol (63.5/35.4/1.0 v/v), containing imidazol (3.0 mmol/l) as catalyst, pH 06.6 with oxalic acid	Peroxyoxalate chemiluminescence detection (bis (2-nitrophenil)oxalato and H_2O_2) Denitrosation with hydrobromic acid-acetic acid and formation of dansyl derivatives (6.5–9.4 fmol)	NDMA, N-nitrosopyrrolidine, N-nitrosodiethylamine, N-nitrosopperidine, N-nitrosodipropylamine, N-nitrosodibutylamine	127
LiChrosorb S:100 25/4.6/5	Acetone/n-hexane (gradient)	TEA (5 μg/kg)	N-Nitrosomethylphenylamine, N-nitrosoethylphenylamine, N-nitrosodicyclohexylamine, N-nitrosodibenzylamine	128
Supelcosil LC-18 25/4.6/5	5% Methanol/water (0.05 M ammonium acetate)	TEA and electron impact MS (5 ng)	N-Nitrosodiethanolamine, N-nitrosomethyl-p-amino-2-ethylhexylbenzoate	129
Waters Xterra C_{18} MS 5/4.6/5	5 mM ammonium acetate/5 mM ammonium acetate in acetonitrile (linear gradient)	Electrospray ionization MS–MS (0.05–1.23 ng/l)	Tobacco-specific nitrosamines	94
Genesis C_{18} 10/3/4	Methanol/10 mM ammonium formate, pH4, in water/methanol (80:20) (gradient)	Electrospray ionization MS–MS (0.005–0.01 μg/g)	Tobacco-specific nitrosamines	130

Column[a]	Mobile phase	Detection (detection limit)	Compounds	Ref.
μ-Bondapak C$_{18}$ 30/3.9/5	Acetonitrile/water (gradient)	Photolysis/MS (2–6 ng)	N-Nitrosodialkylamines	131
LiChrosorb RP-18 (5 μm)	Na$_2$HPO$_4$ (9 g/l)/acetonitrile (95:5)	Photolysis/ED (29 pg)	NDMA	132
μ-Bondapak C$_{18}$ 25/4.1/10	Methanol/acetonitrile/phosphate buffer, pH 3.5 (57:38:5)	ED (0.2 ng)	N-Nitrosopiperidine	133
LiChrosorb-Si60 25/4.6/5	1.5% acetone/n-hexane	TEA (1 ppb)	NDMA, N-nitrosodibutylamine	134
Spherisorb 25/2.1/5	Hexane/isopropanol (96.5:3.5)	CL (0.6–1.2 nmol)	NDMA, N-nitrosopyrrolydine, N-nitrosomorpholine, N-nitrosodiphenylamine	135
Spherisorb 25/4.6/5	7 mM ammonium phosphate, pH 3 1-heptane or 1-octanesulfonate 5 mM	UV	NDMA and its metabolites	136
Shim-pack CLC-ODS 15/4.7/5	0.1 M phosphate buffer, pH 1.5	EC (1×10^{-8} M)	NDMA, N-nitrosodipropylamine, N-nitrosodiphenylamine	137
LiChrospher 100 RP-18e 15/4.0/5	10 mM phosphate buffer/methanol (gradient)	UV (20–250 ng/ml)	4-(Methylnitrosamino)-1-(3-pyridyl)-1-1butanone and metabolites	105

[a] Length/ID/particle size.

appropriate selectivity with low detection limits. The separation of these compounds has been improved by reducing the velocity of the electroosmotic flow.

The selectivity can also be enhanced through modifications of the capillary tube. The experimental set-up is similar to that used for conventional CE except the injection end, a small section of teflon tube which contains stationary phase materials being connected to the analytical column. This design incorporates the advantages of the high efficiency of open-tubular CE as well as the extra interaction provided by the chromatographic stationary phases. Furthermore, the effects of the composition and types of packing materials (C_{18} and silica gel) used in the packed section and the effect of the SDS concentrations in the electrophoretic buffer on the separation and migration times of the nitrosamines have been investigated.[149] Matyska et al.[150] developed a screening method to determine the presence of N-nitrosodiethanolamine in cosmetics by open-tubular capillary electrochromatography. Filho et al.[151-153] investigated the separation of a selected group of nitrosamines in aqueous samples using commercial CE equipment. McCorquodale et al.[154] employed uncoated fused-silica capillaries with a citric acid buffer containing hydroxypropyl-β-cyclodextrin for the resolution of the (R)- and (S)-forms of N-nitrosonicotine.

IV. DETECTION METHODS

Since GC instrumentation is available in most analytical laboratories, it has been the principal method of analysis for volatile N-nitrosamines. Many detectors have been coupled to GC for the detection of N-nitrosamines. The conventional flame ionization detector (FID) was initially used but was found to be limited for N-nitrosamines. Nitrogen-specific detectors such as the Alkali Flame Ionization (AFID), the Coulson Electrolytic Conductivity (CECD) and Hall Electrolytic Conductivity (HECD) are useful for routine screening. Although the HECD is the most selective, it is not specific to N-nitrosamines and an independent confirmation is necessary for each analysis. The efficiency of common GC detectors for the analysis of N-nitrosamines has been compared by several authors.[155]

The high sensitivity and selectivity of mass spectrometer detector (MSD) and TEA make these detectors most popular in the determination of nitrosamines.

A. METHODS BASED ON DENITROSATION

The most sensitive and selective methods for the detection of nitrosamines are based on their denitrosation followed either by a chemiluminescent determination of the nitric oxide (NO) released or by the determination of the resulting nitrite ion or its amino counterpart, with various techniques (spectrophotometry, fluorescence, chemiluminescence, and amperometry).

The denitrosation of N-nitrosamines can be achieved by different procedures including the following:

Pyrolysis. Thermally labile NOCs can be denitrosed by exposure to high temperatures into a flash catalytic pyrolyzer like that designed in the TEA.

Photodecomposition. N-Nitrosamines absorb in the region of 300 to 380 nm and their photodecomposition can be performed by UV irradiation. On photolysis the N-nitroso group is cleaved and the nitrogen oxide is liberated.[125,156] The kinetics of these photolysis reactions are largely dependent on irradiation wavelength, irradiation time, mobile-phase composition, and mobile-phase pH. The generated NO can be detected by chemiluminescence.[125,157] Likewise, the nitrite can be measured as the final product of the photolysis of the NOC solution. This latter photolytic procedure has the advantage of being more specific than pyrolysis, in which a false-positive may result from the pyrolytic decomposition of C- or O-nitroso compounds. Nevertheless, it is less sensitive when it is applied after GC separation.

Chemical Denitrosation. The cleaving of the N−NO bond can also be performed by a chemical denitrosation. Dialkylnitrosamines undergo slow denitrosation in acid conditions catalyzed by nucleophiles such as I⁻, SCN⁻, Br⁻, and Cl⁻.[158]

1.   Thermal Energy Analyzer

The TEA was first described by Fine et al.[159,160] for the analytical determination of nitrosamines. Any gas chromatograph, operating isothermally or with temperature programming, can be interfaced to the TEA. The principle of the operation is shown in Figure 12.2.

The GC–TEA interface operates with gaseous samples, previously separated by GC, and swept through a catalytic pyrolyzer by a carrier gas such as argon or helium. All *N*-nitroso compounds present in the sample are cleaved at the N−NO bond, releasing the nitrosyl radical NO. This radical is separated from organic fragments and other gaseous products by passing through one or more cold traps and/or through a solid-state chemical CTR filter cartridge.

The NO and the carrier gas are introduced into a reduced pressure reaction chamber, where it is oxidized with ozone previously formed by high-voltage electric discharge in pure oxygen. This reaction generates electronically excited nitrogen dioxide (NO_2^*). The NO_2^* instantaneously decays back to the ground state with the emission of a characteristic radiation in the near-infrared region of the spectrum (0.6 to 2.8 μm). The signal is monitored by a photomultiplier tube associated with a red optical filter that eliminates the light of wavelengths shorter than 600 nm. The amplified signal is displayed and integrated, being proportional to the amount of *N*-nitroso compound present in the sample.

TEA is highly selective, although other compounds will also respond to the TEA, such as some organic nitrites, *N*-nitramines, C-nitroso, nitrates, and inorganic nitrite. TEA is very sensitive,

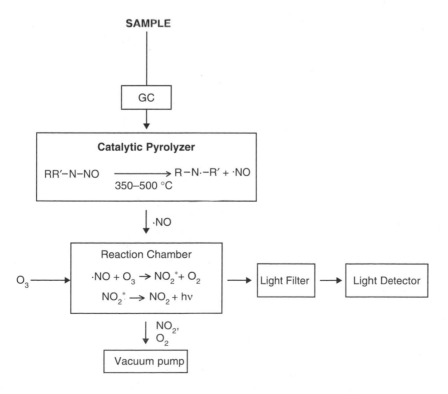

FIGURE 12.2 Scheme of the TEA.

allowing the detection in the range of picomols. A linear response is observed over a wide range of concentrations with two to three orders of magnitude. Because of the high selectivity and sensitivity of TEA, it is possible to analyze samples in the presence of many coeluting compounds that do not interfere with NOC. This advantage leads to a reduction of time in the clean-up procedures.

Various applications of GC–TEA for the determination of volatile N-nitrosamines have been reported in the literature, some of them being included in Table 12.1.

The TEA has also been interfaced to HPLC equipment. The advances in technology for interfacing reversed-phase HPLC with TEA have resulted in the development of two interfaces, a KI/HOAc postcolumn reaction interface[161] and a UV photolysis-based interface.[125] Both interfaces are based on reactions which involve liberation of nitric oxide (NO gas) from the N-nitroso moiety rather than from pyrolysis as performed in the typical GC–TEA mode. The liberated NO (gas) and LC solvent is then swept by a flow of carrier gas into a series of cold traps which remove the LC mobile phase and the residual vapors. The NO (gas) survives the cold traps, enters the TEA detection cell where it is mixed with ozone gas. The resulting chemiluminescence is detected by means of a sensitive photomultiplier tube. These two interfaces allow high sensitivity (1 to 10 ng of total compound injected) and also a high selectivity for a variety of N-nitroso compounds. However, these suffer from several limitations such as the impossibility of using either an aqueous carrier solvent, which might seriously affect the baseline stability, or inorganic buffers, which might result in solid residues accumulating in the pyrolyzer.

Billedeau et al.[129] developed an HPLC–TEA interface utilizing a particle beam type of instrumentation developed initially for interfacing HPLC to MS. The interface incorporates a thermospray (TSP) vaporiser, a desolvatation chamber, a counter flow gas diffusion cell for reducing the LC effluent in a dry aerosol, and a single-stage momentum separator to form a particle beam of the nonvolatile analyte. The high solvent removal efficiency of this interface has made possible HPLC–TEA analysis with reversed-phase solvents without the need for solvent venting[162] or cryogenic trapping techniques[125,161] currently being used as alternatives to HPLC–TEA interfaces.

2. Spectrophotometric Detection

The nitrite ion can be detected spectrophotometrically after the separation of nitrosamines by HPLC followed by photolytic or chemical denitrosation. Postcolumn formation of an azo dye by the reaction of nitrite with a Griess-type reagent allows its spectrophotometric detection at 546 nm.[156,175] The kinetics and mechanisms of the Griess reaction have been extensively studied.[176]

Singer et al.[177] developed a specific method in which a postcolumn reaction detection system is used for HPLC. This system is useful for those compounds which can be hydrolyzed in a dilute acidic solution to give the nitrite ion. This method involves the use of the Griess reagent in the postcolumn reactor for production of chromophores from N-nitrosamines. The theoretical detection limit for this method was reported to be 0.5 nmol. However, owing to the slow reaction kinetics of some nitroso compounds, this technique requires both an air segmentation system and a high-temperature reactor.

Based upon the previous procedure, Bellec et al.[126] described a method for the separation and detection of volatile and nonvolatile N-nitrosamines with colorimetric detection by a Griess reagent of the nitrite generated by the cleave of nitroso compounds with a postcolumn photohydrolysis–UV photoreactor.[156] The yield of the photohydrolysis depends upon pH and time of exposure under UV light. The detection limits reported were 8 pmol for N-dialkylnitrosamines and 20 pmol for N-nitrosamines bearing two phenyl groups.

3. Fluorimetric and Chemiluminescent Detection

These methods are in general based on their photolytic or chemical denitrosation, and the subsequent derivatization of the resulting nitrite ion or the corresponding amine.

Lee and Field[178] reported a selective fluorescence detection method for the determination of some *N*-nitrosamines after a postcolumn reaction. The nitrosamines eluted from the column are first hydrolyzed to produce the nitrite anion, which is then oxidized with Ce^{4+} to give the fluorescent Ce^{3+}. The detection limit for this method is at the ppb level. A more sensitive fluorimetric method has been developed based upon the reaction of nitrite with 2,3-diaminonaphthalene to form the highly fluorescent product 1-(H)-naphthotriazole.[179] About 10 nmol/l of nitrite can be detected by this procedure.

Among the more sensitive methods for the determination of *N*-nitrosamines are the ones based on the denitrosation of *N*-nitroso compounds by hydrobromic-acetic acid,[158,180] and the subsequent detection of the liberated secondary amines via fluorescence derivatization. Precolumn or postcolumn derivatization has been used for the determination of *N*-nitrosamines. The most commonly used fluorescent derivatization reagents are listed in Table 12.4.

Peroxyoxalate chemiluminescence detection has been shown to be a highly sensitive detection method[184–186] for the determination of *N*-nitrosamines and secondary amines[64,127] in combination with reversed-phase HPLC. Fentomole limits of detection can be obtained with conventional instrumentation. The principle of the reaction is illustrated in Figure 12.3.

Fu et al.[64] have reported the determination of six nitrosamines by HPLC combined with a sensitive postcolumn bis(2-nitrophenyl) oxalate–hydrogen peroxide chemiluminescent detection. The sample was first denitrosated with hydrobromic acid–acetic acid to produce the corresponding secondary amines, which were then subjected to reaction with dansyl chloride to form fluorescent dansyl derivatives. The reaction mixtures were separated on a C_{18} column with a mobile phase consisting of acetonitrile–water–ethanol (pH 6.2 with oxalic acid) containing 3.0 mmol/l of imidazole added as a catalyst for the chemiluminescence reaction. The sensitivity of this method was 120 times greater than that of fluorescence detection and four orders of magnitude greater than that of UV–Vis spectrophotometric detection. The detection limits with this procedure at a signal-to-noise ratio of four were between 0.31 and 1.20 pg.

This method was applied simultaneously to the determination of *N*-nitrosamines and the corresponding secondary amines in environmental water samples.[127] The method combines solid-phase extraction using a mini activated carbon column, followed by elution with acetone, concentration of the extracts by denitrosation, and fluorescent derivatization.

4. Electrochemical Detection

The development in the determination of many classes of nonvolatile nitrosamines depends on the development of detectors suitable for the trace analysis of ionizable, ionic, macromolecular, and thermally unstable *N*-nitroso compounds. In particular, the development of detectors compatible with reversed-phase liquid chromatographic conditions is receiving special attention. Among others, the electrochemical detectors are attractive because of their high sensitivity and their ability to operate in different aqueous and mixed aqueous-organic eluents.

Some interesting classes of nitrosamines are decomposable by warm strong acids or by photolytic cleavage in alkaline media so that nitrite or nitrous acid is produced. They can be electrochemically detected as such or after further reaction. The availability of a voltametric detector for flowing solutions equipped with solid electrodes allows the direct anodic detection of the NO_2^- species produced by postcolumn photolysis.[132] The Ce(IV) and iodide in acid medium have been investigated as postcolumn reagents for the oxidation and voltametric detection of the nitrite produced by the warm acid decomposition of nonvolatile nitrosamines.[187–189]

TABLE 12.4
Derivatization Reactions for High Performance Liquid Chromatographic Determination of *N*-Nitrosamines with Fluorescence Detection

Reagent	Compound	Chromatographic Conditions	λ_{em} (λ_{exc}) nm	Detection Limit	Ref.
Acridone-*N*-acetyl chloride	NDMA, *N*-nitrosodiethylamine, *N*-nitrosodipropylamine, *N*-nitrosomethyphenylamine	Spherisorb (200 × 4.6 mm, 5 μm), Binary gradient: (20 m*M* NH$_4$H$_2$PO$_4$ + 9 m*M* triethylamine (pH 6.5)–methanol (95:5))/acetonitrile–water (75:25) 1.0 ml/min	430 (404)	24–128 fmol	78
4-(2-Phthalimidyl)benzoil chloride	NDMA, *N*-nitrosopyrrolidine, *N*-nitrosodiethylamine, *N*-nitrosopiperidine, *N*-nitrosodipropylamine, *N*-nitrosodibuthylamine	Nucleosil C$_{18}$ (125 × 4.6 mm, 5 μm), Acetonitrile–water (48:52) 0.8 ml/min	299 (426)	0.4–1.6 pmol	181
Dansyl chloride	NDMA, *N*-nitrosodiethylamine, *N*-nitrosodipropylamine, *N*-nitrosodibutylamine	Perkin-Elmer C$_{18}$ (125 × 4.6 mm), Acetonitrile–water (73:27) 0.8 ml/min	530 (350)	0.06–0.16 ng	182
—	*N*-nitrosomorpholine, NDMA, *N*-nitrosodiethylamine, *N*-nitrosopyrrolidine, *N*-nitrosopiperidine,	NovaPak C$_{18}$ (150 × 3.9 mm, 5 μm, 60 Å), Acetonitrile–water (55:45) 1.2 ml/min	531 (339)	8–75 pg	92
Lumarin 9	NDMA, *N*-nitrosopyrrolidine, *N*-nitrosoproline, *N*-nitrosodiethanolamine	Nucleosil C$_{18}$ (125 × 4.6 mm), Acetonitrile–water (52:48) 1.0 ml/min	489 (339)	0.4–1.0 pmol	183

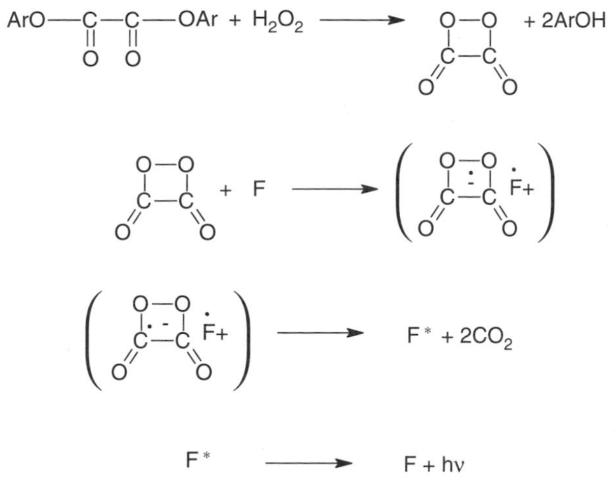

$$\text{ArO}-\underset{\underset{O}{\|}}{C}-\underset{\underset{O}{\|}}{C}-\text{OAr} + H_2O_2 \longrightarrow \underset{\underset{O}{\diagup}}{\overset{O-O}{C-C}}\underset{\diagdown O}{} + 2\text{ArOH}$$

$$\underset{\underset{O}{\diagup}}{\overset{O-O}{C-C}}\underset{\diagdown O}{} + F \longrightarrow \left(\underset{\underset{O}{\diagup}}{\overset{O\cdots O}{C\cdots C}}\underset{\diagdown O}{}\ F+ \right)$$

$$\left(\underset{\underset{O}{\diagup}}{\overset{O-O}{C\cdots C}}\underset{\diagdown O}{}\ F+ \right) \longrightarrow F^* + 2CO_2$$

$$F^* \longrightarrow F + h\nu$$

FIGURE 12.3 Peroxyoxalate chemiluminescence reaction. F, fluorophore; F^*, chemically excited fluorophore.

In the case of the N-nitrosamines, the easy electrochemical reduction of the nitro group has allowed the development of direct electrochemical methods for their determination. The reductive amperometric determination has been described using dropping mercury electrodes or hanging mercury drop electrodes.[133,137,190] Likewise the oxidative amperometric determination using modified glassy carbon electrodes has been reported.[191] In contrast to most of the reported methods for the determination of N-nitrosamines, these electrochemical approaches do not require an initial denitrosation step.

B. MASS SPECTROMETRY

The majority of the analytical methods for the detection of N-nitrosamines have employed GC or liquid chromatography in conjunction with a TEA. The disadvantage of these techniques, however, is that a subsequent confirmation is needed to ensure that the method does not give a false-positive response.

The use of GC–MS or LC–MS–MS instead of GC–TEA provides higher throughput with analyte-specific detection based on both retention time and structurally specific analyte fragmentation information. The identification is usually based on the detection of several dominant ions characteristic of the nitrosamines, such as fragments of $m/z = 75, 74, 59, 42, 41$ for NDMA and $m/z = 103, 102, 71, 57, 56, 42, 41$ for N-nitrosodiethylamine. The ion ratio of the primary ion ($m/z = 74$) to the secondary ion ($m/z = 42$) is generally used to confirm the presence/absence of NDMA in complex samples.[63]

The quantification is based on the intensity of selected ions after calibration with standard nitrosamine solutions. The use of labeled internal standards reduces the variability due to extraction efficiencies or changes in the instrument performance and it also assures the accuracy of quantification. For the analysis of N-nitrosamines several deuterated internal standards, e.g., d_6-N-nitrosodimethylamine, are commercially available.

Most of the N-nitrosamines analyses by GC–MS have been performed using capillary columns. The use of these columns avoids the necessity of intensive clean-up of the extracts prior to the analysis. The clean-up is needed for removing most of the many potentially interfering substances contained in environmental samples. On the other hand, the packed columns present the advantage of their greater sample capacity. The electron impact (EI) ionization mode is the most frequently

TABLE 12.5
Determination of N-nitrosamines in Environmental Samples

Compound	Sample Preparation	Analytical Method	Recovery (%)	LOD	Ref.
Aqueous Samples					
NDMA, Drinking and contaminated groundwaters	Extract with DCM. Concentrated and loaded onto a dual-stage carbon sorbent trap. SPTD	HPLC–CLND	71	2 ng/l	62
Nine nitrosamines on the EPA list, Drinking waters and groundwater	Extract with DCM or solid-phase extraction on two adsorbents (C8 and activated carbon), and elute with ethanol	GC–NPD	78–92 72–97	0.1 ppb	60
NDMA, surface water and wastewater	Extract with DCM. Wastewater (CLLE)	GC–MS–MS	21 ± 10 56 ± 11 (CLLE)	0.09 μg/l	61
NDMA, water	Extract with DCM. Dried with anhydrous sodium sulfate and concentrated in a rotary evaporator	GC–MS-SIM	100 ± 15	0.003 pg/μl	63
Volatile N-nitrosamines, groundwater	Solid-phase extraction onto a mini activated carbon column, elute with acetone. Concentration of the extracts	HPLC–CL	95	4.3–8.3 fmol	64
NDMA, drinking waters	Extract with DCM in the presence of sulfamic acid	GC–TEA GC–MS-SIM	74–105	15 pg/g 1 pg/g	15
N-Nitrosodiphenylamine, waste water	Extract with DCM clean-up: alumine	GC–TEA GC–FID		—	196
N-Nitrosodiphenylamine, water	Extract with DCM	GC–MS HPLC–MS		—	197
NDMA, N-nitrosomorpholine, tap water	Sampling: Ambersorb XE-340, elute with DCM. Concentration of the extracts	GC–TEA GC–MS	58–99	—	65
N-Nitrosodipropylamine, groundwater	Solvent extraction under neutral conditions or at pH 11. Solvent concentrated	GC–MS		10 μg/l	197,198
N-Nitrosodipropylamine, water and wastewater	Solvent extracted, acid washed, column chromatographic clean-up	GC–NPD GC–HECD GC–TEA		0.46 μg/l (GC–NPD)	196

Air Samples

N-Nitrosodiphenylamine, air	Sampling: ThermoSorb/N air sampling cartridge Extract with acetone, methanol, or methanol/dichloromethane	GC–TEA HPLC–TEA	—	—	199,200
Tobacco-specific nitrosamines Mainstream cigarette smoke	Sampling: Cambridge filter pad Extract with dichloromethane, back extracted into aqueous solution. Purified by solid-phase extraction	HPLC–MS–MS	50–66	0.05–1.23 ng/ml	94
NDMA, *N*-nitrosodiethylamine, *N*-nitrosopyrrolidine, *N*-nitrosomorpholine, environmental tobacco smoke	Sampling: ThermoSorb/N cartridge Extract with 33% methanol in chloroform. Purified by alumina-B Sep-Pack Desorbed: 10% chloroform in dichloromethane. Concentrated in a rotary evaporator	GC–TEA	96 ± 5	17,23–23 ng/cigarette	163
NDMA, *N*-nitrosodiethylamine, *N*-nitrosodipropylamine, *N*-nitrosodibutylamine, *N*-nitrosopiperidine, *N*-nitrosopyrrolidine, *N*-nitrosomorpholine	Sampling: ThermoSorb/N flow rates of 1.5 l/min Desorbed: acetone or DCM/methanol 75/25	GC–TEA	—	0.04–0.08 μg/m^3*	169
Volatile and tobacco-specific *N*-nitrosamines, cigarette smoke	Sampling: passing the smoke through 4 consecutive midget impingers, containing 15 ml citrate–phosphate buffer, pH 4.5 and 53 mg ascorbic acid	GC–TEA	89 ± 4 ((NDEA) 92 ± 3 (NNN)	1.1–4.2 ng/cigarette	89
N-nitrosomorpholine, vulcanization fumes	Sampling: trapped in an absorbing solution (Isopropanol 0.001 *M* KOH)	GC–MS	86–130	10 pg/μl	192
NDMA, ambient air	Sampling: Tenax. Thermally desorbed	CG–MS	—	0.3 pg	201
NDMA, ambient air	Sampling in impinger containing KOH. Extracted in solvent and concentrated	GC–MS GC–TEA	—	10 pg (MS) 1 ng/m^3 (TEA)	202,203
N-Nitrosodiphenylamine, air	Sampling: quartz filter followed by an XAD-II trap Extract with DCM and methanol. Solvent concentrated	GC–TEA	—	0.8 pmol	204
Volatile nitrosamines, tire storage air	Sampling: ThermoSorb/N Flow rates of 2 l/min Desorbed: DCM/methanol 75/25	GC–MS	92 ± 5 to 59 ± 13	0.1–0.2 μg/m^3	82

Continued

TABLE 12.5
Continued

Compound	Sample Preparation	Analytical Method	Recovery (%)	LOD	Ref.
Soil and Sediment Samples					
N-Nitrosodiphenylamine, soil	Extract with appropriate solvent. Concentration	GC–TEA HPLC–TEA	—	—	205
NDMA soil	Extract with water, water extracted with solvent and concentrated	GC–TEA	—	—	23
Volatile nitrosamines, soil	Extract with DCM, filtered and concentrated	HPLC/F	—	—	78
N-Nitrosodipropylamine, soil	Homogenized sample mixed paraffin or glycerol, water and NaOH, vacuum distilled at low temperature. Distillate extracted on-column or by shaking with solvent and concentrated	GC–TEA	—	0.05–0.5 μg/g	206
N-Nitrosodipropylamine, soil/sediment	Solvent extracted, column chromatographic clean-up, extract concentrated	GC–MS	—	330 μg/kg	207
N-Nitrosodipropylamine, soil, sludge or solid waste	Solvent extracted by Soxhlet or sonication, extract subjected to column chromatographic clean-up, concentration of extract	GC–MS	—	660 μg/kg	206,207

SPTD: short-path thermal desorption unit; CLND: chemiluminescent nitrogen detector; CL: peroxyoxalate chemiluminescence detection; CLLE: continuous liquid–liquid extraction.

used, nevertheless, the chemical ionization using methanol as the reagent gas has also been reported.[61] Some typical applications of GC–MS are summarized in Table 12.2.

The combination of LC and mass spectrometry (LC–MS) offers significant analytical advantages over the aforementioned techniques. The LC–MS technique can be used to separate and detect volatile and nonvolatile N-nitrosamines without the need of a solvent cold trap. It allows an increase in the use of polar mobile phases when comparing to LC–TEA. Moreover, the LC–MS technique allows the identification of unknown components. This is especially important in the analysis of nonvolatile N-nitrosamines, because contrary to volatile N-nitrosamines, the information regarding their occurrence and concentration in many products is not yet well known.

On the other hand, there are limitations to the LC–MS analysis. The two most common ionization techniques available in LC–MS, TSP and electrospray ionization (ESI), yield primarily molecular weight information; i.e., little fragmentation is observed to confirm the structure of the analyte. Thermally induced decomposition[194] and in-source collision-induced dissociation[195] have been utilized to produce structurally significant ions. However, these techniques are often unreliable and can suffer significant losses in sensitivity. Alternatively, the on-line photolysis can be used to induce photolytic dissociation of different types of compounds. Volmer et al.[131] have reported the simultaneous detection and confirmation of several N-nitrosodialkylamines by on-line coupling of a photolysis reactor with an ESI mass spectrometer.

Liquid chromatography–tandem mass spectrometry has been recently applied to the determination of tobacco-specific N-nitrosamines.[94,130] The resolution power of tandem mass spectrometry has inherent advantages over the current, widely used TEA method for tobacco-specific N-nitrosamines quantitation. These advantages are the ease of sample preparation, the selectivity, and the sensitivity. In addition to unambiguous identification, the use of isotope-labeled analogues as internal standards allows excellent reproducibility and accuracy. Additional applications of LC–MS to the analysis of N-nitrosamines in environmental samples are shown in Table 12.5.

V. APPLICATIONS IN THE ANALYSIS OF THE ENVIRONMENT

Table 12.5 provides a brief description of the extraction and clean-up procedures and the analytical technique applied for the determination of N-nitrosamines in samples of environmental interest.

REFERENCES

1. Magee, P. N. and Barnes, J., *J. Cancer*, 10, 114, 1956.
2. Preussmann, R. and Steward, B. W., In *Chemical Carcinogens, American Chemical Society Monograph No. 182*, 2nd ed., Searle, C. W., Ed., American Chemical Society, Washington, DC, 1984, chap. 12.
3. Tricker, A. R. and Preussmann, R., *Mutat. Res.*, 259, 277, 1991.
4. Preussmann, R. and Eisenbrand, G., In *Chemical Carcinogens, American Chemical Society Monograph No. 182*, 2nd ed., Searle, C. W., Ed., American Chemical Society, Washington, DC, 1984, chap. 13.
5. Hotchkiss, J. H., *Adv. Food Res.*, 31, 54, 1987.
6. Challis, B. C. and Challis, J. A., In *The Chemistry of Amino, Nitroso and Nitro-Compounds and Their Derivatives*, Patai, S., Ed., Wiley, New York, p. 1151, 1982.
7. Mirvish, S. S., *Toxicol. Appl. Pharmacol.*, 31, 325, 1975.
8. Archer, M. C., Yang, H. S., and Okun, J. D., In *Environmental Aspects of N-Nitroso Compounds, IARC Sci. Publ. No. 19*, Walker, E. A., Castegnaro, M., Griciute, L., and Lyle, R. E., Eds., International Agency for research on Cancer, Lyon, France, p. 239, 1978.

9. Mergens, W. J., Kamm, J. J., Newmark, H. L., Fiddler, W., and Pensabene, J., In *Environmental Aspects of N-Nitroso Compound, IARC Sci. Publ. No. 19*, Walker, E. A., Castegnaro, M., Griciute, L. and Lyle, R. E., Eds., International Agency for research on Cancer, Lyon, France, p. 199, 1978.
10. Archer, M. C., Tannenbaum, S. R., Fan, T. Y., and Weisman, M., *J. Intl. Cancer Inst.*, 54, 1203, 1975.
11. Challis, B. C. and Kryptopoulos, S. A., *J. Chem. Soc. Perkin Trans.*, I, 299, 1979.
12. *Scientific and Technical Assessment Report on Nitrosamines* (1977). Office of the Research and Development, U.S. EPA, Washington, DC, EPA-600/6-77-001.
13. Budavani, S. (Ed.), *The Merck Index*, 12th ed., N.J. Rahway, 1996.
14. Fleming, E. C., Pennington, J. C., Wachob, B. G., Howe, R. A., and Hill, D. O., *Hazard J. Mater.*, 51, 151, 1996.
15. Sen, N. P., Baddoo, P. A., Water, D., and Boyle, M., *Inter. J. Environ. Anal. Chem.*, 56, 149, 1994.
16. Jenkins, S. W. D., Koester, C. J., Taguchi, V. Y., Wang, D. T., Palmentier, J. P. F. P., and Hong, K. P., *Environ. Sci. Pollut. Res.*, 2, 207, 1995.
17. Lijinsky, W., *Mutation Res.*, 443, 129, 1999.
18. Levallois, P., Ayotte, P., Louchini, R., Desrosiers, T., Baribeau, H., Phaneuf, D., Gingrass, S., Dumas, P., Zee, J., and Poirier, G., *J. Expo. Anal. Environ. Epidemiol.*, 10, 188, 2000.
19. Levallois, P., Ayotte, P., Van Maanen, J. M. S., Desrosiers, T., Gingras, S., Dallinga, J. W., Vermeer, I. T. M., Zee, J., and Poirier, G., *Food Chem. Toxicol.*, 38, 1013, 2000.
20. Müller, H., Landt, J., Pedersen, E., Jensen, P., Autrup, H., and Jensen, O., *Cancer Res.*, 49, 3117, 1989.
21. Bagardi, I. and Kuzelka, R. D., *Nitrate Contamination: Exposure, Consequence and Control*, Springer-Verlag, Berlin, 1991.
22. Bouchard, D. C., Wiliams, M. K., and Surampalli, R. Y., *J. Am. Water Works Assoc.*, 84, 85, 1992.
23. Mitch, W. A. and Sedlak, D. L., *Environ. Sci. Technol.*, 36, 588, 2002.
24. Choi, J. and Valentine, R. L., *Water Res.*, 36, 817, 2002.
25. Mitch, W. A. and Sedlak, D. L., *Water Sci. Techn. Water Supply*, 2, 191, 2002.
26. Choi, J., Duirk, S. E., and Valentine, R. L., *J. Environ. Monit.*, 4, 249, 2002.
27. Järvholm, B., Zingmark, P. A., and Österdahl, B. G., *Am. J. Ind. Med.*, 19, 237, 1991.
28. Monarca, S., Scassellati-Sforzolini, G., Donato, F., Angeli, G., Spielgelhalder, B., Fatigoni, C., and Pasquini, R., *Environ. Health Perspect.*, 104, 78, 1996.
29. Ayanaba, A., Verstraete, W., and Alexander, M., *Soil Sci. Soc. Am. Proc.*, 37, 565, 1973.
30. Pancholy, S. K., *Soil Biol. Biochem.*, 10, 27, 1978.
31. Tate, R. L. and Alexander, M., *Soil Sci.*, 118, 317, 1974.
32. Grosjean, D., *J. Air Waste Manage. Assoc.*, 41, 306, 1991.
33. Hoffmann, D., Hecht, S. S., Melikian, A. A., Haley, N. L., Brunnemann, K. D., Adams, J. D., and Wynder, C. L., *Biochem. Mol. Epid. Cancer*, 191, 1986.
34. Burton, H. R., Dye, N. K., and Bush, L. P., *J. Agri. Food Chem.*, 42, 2007, 1994.
35. Wiernik, A., Christakopoulos, A., Johansson, L., and Wahlberg, I., In *Recent Advances in Tobacco Science*, Burton, H. R., Ed., The Tobacco Chemists' Research Conference, Lexington, p. 39, 1995.
36. Lewis, R. S., Tannenbaum, S. R., and Deen, W. M., *J. Am. Chem. Soc.*, 117, 3933, 1995.
37. Fischer, S., Spiegelhalder, B., Eisenbarth, J., and Preussman, R., *Carcinogenesis*, 11, 723, 1990.
38. Ruhl, C., Adams, J. D., and Hoffmann, D., *J. Anal. Toxicol.*, 4, 255, 1980.
39. Spiegelhalder, B. and Preussmann, R., *Carcinogenesis*, 4, 1147, 1983.
40. Oury, B., Limasset, J. C., and Protois, J. C., *Int. Arch. Occup. Environ. Health*, 70, 261, 1997.
41. Willoughby, B. G. and Scott, K. W., *Rubber Chem. Technol.*, 71, 310, 1998.
42. Sen, N. P., Seaman, S. W., and Kushwaha, S. C., *J. Assoc. Off. Anal. Chem.*, 70, 434, 1987.
43. Sen, N. P., Kushwaha, S. C., Seaman, S. W., and Clarkson, S. G., *J. Agric. Food Chem.*, 33, 428, 1985.
44. Gray, J. I. and Stachiw, M. A., *J. Assoc. Off. Anal. Chem.*, 70, 64, 1987.
45. Pensabene, J. W., Fiddler, W., and Gates, R. A., *J. Agric. Food. Chem.*, 43, 1919, 1995.
46. Pensabene, J. W. and Fiddler, W., *J. AOAC Int.*, 77, 981, 1994.
47. Fiddler, W., Pensabene, J. W., Gates, R. A., Cluster, C., Yoffe, A., and Phillipo, T., *J. AOAC Int.*, 80, 353, 1997.
48. Loeppky, R. N., Bao, Y. T., Bae, J., Yu, L., and Shevlin, G., *Nitrosamines and Related N-Nitroso Compounds, Chemistry and Biochemistry*, ACS Symposium Series 553, p. 52, 1994.

49. Preston-Martin, S., Pogoda, J. M., Mueller, B. A., Holly, E. A., Lijinsky, W., and Davis, L., *Cancer Epidemiol. Biomarkers Prev.*, 5, 599, 1996.
50. Bos, J. L., *Cancer Res.*, 49, 4682, 1989.
51. Hecht, S. S., *Proc. Soc. Exp. Biol. Med.*, 216, 181, 1997.
52. Walker, R., *Food Addit. Contam.*, 7, 717, 1990.
53. Lijinsky, W., *Chemistry and Biology of N-Nitroso Compounds*, University Press, 1992.
54. Sheweita, S. A. and Mostafa, M. H., *Cancer Lett.*, 99, 29, 1996.
55. Kato, S., Onda, M., Matsukura, N., Tokunaga, A., Tajiri, T., Kim, D. Y., Tsuruta, H., Matsuda, N., Yamashita, K., and Shields, P. G., *Pharmacogenetics*, 5, 141, 1995.
56. Hecht, S. S., *Proc. Soc. Exp. Biol. Med.*, 216, 181, 1997.
57. Loechler, E. L., Green, C. C., and Essigam, J. M., *Proc. Natl. Acad. Sci. USA*, 81, 6271, 1984.
58. Pegg, A. E., Dolan, M. E., and Moschel, R. C., *Prog. Nucleic Acid Res. Mol. Biol.*, 51, 167, 1995.
59. Saffhill, R., *Chem. Biol. Int.*, 53, 121, 1985.
60. Queiroz, M., Wuchner, K., Grob, R., and Mathieu, J., *Analusis*, 20, 12, 1992.
61. Mitch, W. A., Gerecke, A. C., and Sedlak, D. L., *Water Res.*, 37, 3733, 2003.
62. Tomkins, B. A., Griest, W. H., and Higgins, C. E., *Anal. Chem.*, 67, 4387, 1995.
63. Raksit, A. and Johri, S., *J. AOAC Int.*, 84, 1413, 2001.
64. Fu, C. G. and Xu, H. D., *Analyst*, 120, 1147, 1995.
65. Kimoto, W. I., Dooley, C. J., Carré, J., and Fiddler, W., *Water Res.*, 15, 1099, 1981.
66. Fan, T. Y., Ross, R., Fine, D. H., Keith, L. H., and Garrison, A. W., *Environ. Sci. Technol.*, 12, 692, 1978.
67. Nikaido, M. M., Dean-Raymond, D., Francis, A. J., and Alexander, M., *Water Res.*, 11, 1085, 1977.
68. Fine, D. H., Rounbehler, D. P., and Oettinger, P. E., *Anal. Chim. Acta*, 78, 383, 1975.
69. Havery, D. C., Fazio, T., and Howard, J. W., *J. AOAC Int.*, 61, 1374, 1978.
70. Hotchkiss, J. H., Libbey, L. M., and Scanlan, R. A., *J. AOAC Int.*, 63, 74, 1980.
71. Owens, J. L. and Kinast, O. E., *J. Agric. Food Chem.*, 28, 1262, 1980.
72. Greenfield, E. I., Vasco, G. A., and Legette, L., *J. AOAC Int.*, 65, 1316, 1982.
73. Eisenbrand, G., Ellen, G., Preussmann, R., Schuller, P. L., Spiegelhalder, B., Stephany, R. W., and Webb, K. S., In *Environmental Carcinogens: Selected Methods of Analysis*, Egan, H., Ed., IARC Sci. Pub. 45, Lyon, p. 181, 1983.
74. Österdahl, B. G., *Food Addit. Contam.*, 5, 33, 1987.
75. Eisenbrand, G. and Sen, N. P., In *Environmental Carcinogens: Selected Methods of Analysis*, Egan, H., Ed., IARC Sci. Pub. 45, Lyon, p. 149, 1983.
76. Billedeau, S. M., Thompson, H. C. Jr., Hansen, E. B. Jr., and Miller, B. J., *J. AOAC. Int.*, 6, 67, 1984.
77. Sen, N. P., Seaman, S., and Miles, W. F., *J. Agric. Food Chem.*, 27, 1354, 1979.
78. You, J., Fan, X., Lao, W., Ou, Q., and Zhu, Q., *Talanta*, 48, 437, 1999.
79. Ross, R., Morrison, J., and Fine, D. H., *J. Agric. Food Chem.*, 26, 455, 1978.
80. Rounbehler, D. P., Reish, J., and Fine, D. H., *Food Cosmet. Toxicol.*, 18, 147, 1980.
81. Fine, D. H., Rounbehler, D. P., and Goff, U., In *Environmental Carcinogens Methods of Analysis and Exposure Measurement, IARC Science Publication 109*, Seifert, B., Van de Weil, H. J., Dodet, B. and O'Neill, I. K., Eds., International Agency for Research on Cancer, Lyon, France, p. 269, 1993.
82. Marano, R. S., Updegrove, W. S., and Machen, R. C., *Anal. Chem.*, 54, 1947, 1982.
83. Pellizzari, E. D., Bunch, J. E., Berkley, R. E., and McRae, J., *Anal. Lett.*, 9, 45, 1976.
84. Pellizzari, E. D., Bunch, J. E., Bursey, J. T., and Berkley, R. E., *Anal. Lett.*, 9, 579, 1976.
85. Pellizzari, E. D., Bunch, J. E., Berkley, R. E., and Bursey, J. T., *Biomed. Mass Spectrom*, 3, 196, 1976.
86. Rounbehler, D. P., Reisch, J. W., Coombs, J. R., and Fine, D. H., *Anal. Chem.*, 52, 273, 1980.
87. Brunnemann, K. D., Fink, W., and Moser, F., *Oncology*, 37, 217, 1980.
88. Stehlik, G., Richter, O., and Altman, H., *Ecotoxicol. Environ. Saf.*, 6, 495, 1982.
89. Caldwell, W. S. and Conner, J. M., *J. Assoc. Off. Anal. Chem.*, 73, 783, 1990.
90. Chamberlain, W. J. and Arrendale, R. F., *J. Chromatogr.*, 234, 478, 1982.
91. Kataoka, H., Shindoh, S., and Makita, M., *J. Cromatogr. A*, 723, 93, 1996.
92. Cárdenes, L., Ayala, J. H., González, V., and Afonso, A. M., *J. Chromatogr. A*, 946, 133, 2002.
93. Mahanama, K. R. R. and Daisey, J. M., *Environ. Sci. Technol.*, 30, 1477, 1996.
94. Wu, W., Ashley, D. L., and Watson, C. H., *Anal. Chem.*, 75, 4827, 2003.

95. Fine, D. H., Rounbehler, D. P., Belcher, N. M., and Epstein, S. S., *Science*, 192, 1328, 1976.
96. Cucco, J. A. and Brown, P. R., *J. Chromatogr.*, 213, 253, 1981.
97. Belardi, R. P. and Pawliszyn, J. B., *Water Pollut. Res. J. Canada*, 24, 179, 1989.
98. Lauch, D., Motlag, S., and Pawliszyn, J. B., *Anal. Chem.*, 64, 1187, 1992.
99. Zhang, Z. and Pawliszyn, J. B., *Anal. Chem.*, 65, 1843, 1993.
100. Nilsson, T., Pelusio, F., Montanarella, L., Larsen, B., Facchetti, S., and Madsen, J. O., *J. High Resolut. Chromatogr.*, 18, 617, 1995.
101. Cai, Y. and Bayona, J. M., *J. Chromatogr. A*, 696, 113, 1995.
102. Chen, J. and Pawliszyn, J., *Anal. Chem.*, 67, 2530, 1995.
103. Sen, N. P., Seaman, S. W., and Page, B. D., *J. Chromatogr. A*, 788, 131, 1997.
104. Mahapatra, S., Tripathi, R. M., Bhalke, S., and Sadasivan, S., *BARC Newsletter*, 237, 103, 2003.
105. Mullet, W. M., Levsen, K., Borlak, J., Wu, J., and Pawliszyn, J., *Anal. Chem.*, 74, 1695, 2002.
106. US EPA Pollution Prevention Strategy, *Fed. Regist.*, 56, 7849, 1991.
107. McHugh, M. and Barnes, J. M., Supercritical Fluid Extraction, 2nd Ed., Buterworth-Heinemann, Stoneham, MA, 1994.
108. Prokopcyk, B., Hoffmann, D., Cox, J. E., Djordjevic, V., and Brunnemann, K. D., *Chem. Res. Toxicol.*, 5, 336, 1992.
109. Prokopcyk, B., Wu, M., Cox, J. E., Amin, S., Desai, D., Idris, A. M., and Hoffmann, D., *J. Agric. Food Chem.*, 43, 916, 1995.
110. Song, S. and Ashley, D. L., *Anal. Chem.*, 71, 1303, 1999.
111. Reche, F., Garrigós, M. C., Marín, M. L., Cantó, A., and Jiménez, A., *J. Chromatogr. A*, 963, 419, 2002.
112. Reche, F., Garrigós, M. C., Marín, M. L., and Jiménez, A., *J. Chromatogr. A*, 976, 301, 2002.
113. Hoffman, D., Rathkamp, G., and Liu, Y. Y., In *N-Nitroso Compounds in the Environment, IARC Sc. Publ. No. 9*, Bogovsky, P. and Walker, E. A., Eds., International Agency for research on Cancer, Lyon, France, p. 159, 1974.
114. Eisenbrand, G., Von Rappardt, E., Zappe, R., and Preussmann, R., In *Environmental N-Nitroso Compounds Analysis and Formation, IARC Sci. Publ. No. 14*, Walker, E. A., Bogovski, P. and Griciute, L., Eds., International Agency for Research on Cancer, Lyon, France, p. 65, 1976.
115. Alliston, T. G., Cox, B. G., and Kirk, R. S., *Analyst*, 97, 915, 1972.
116. Kimoto, W. I., Silbert, L. S., and Fiddler, W., *J. Assoc. Off. Anal. Chem.*, 67, 751, 1984.
117. Cooper, S. F., Lemoyne, C., and Gauvreau, D., *J. Anal. Toxicol.*, 11, 12, 1987.
118. Janzowski, C., Eisenbrand, G., and Preussmann, R., In *Environmental Carcinogens: Selected Methods of Analysis, Vol. 6, N-Nitroso Compounds*, Egan, H., Ed., IARC Sci. Pub. 45, Lyon, p. 423, 1983.
119. Sen, N. P., Miles, W. F., Seaman, S., and Lawrence, J. F., *J. Chromatogr.*, 128, 169, 1976.
120. Ohshima, H., Matsui, M., and Kawabata, T., *J. Chromatogr.*, 169, 279, 1979.
121. Hildrum, K. I., Scanlan, R. A., and Libbey, L. M., *J. Agric. Food Chem.*, 25, 252, 1977.
122. Lee, J. S., Libbey, L. M., and Scanlan, R. A., In *Environmental Carcinogens: Selected Methods of Analysis, Vol. 6, N-Nitroso Compounds*, Egan, H., Ed., IARC Sci. Pub. 45, Lyon, p. 411, 1983.
123. Massey, R. C., Crews, C., Dennis, M. J., McWeeny, D. J., Startin, J. R., and Knowles, M. E., *Anal. Chim. Acta*, 174, 327, 1985.
124. Castegnaro, M. and Webb, K. S., In *Environmental Carcinogens: Selected Methods of Analysis*, Egan, H., Ed., IARC Sci. Pub. 45, Lyon, p. 439, 1983.
125. Conboy, J. J. and Hotchkiss, J. H., *Analyst*, 114, 155, 1989.
126. Bellec, G., Cauvin, J. M., Salaun, M. C., Le Calvé, K., Dréano, Y., Gouérou, H., Ménez, J. F., and Berthou, F., *J. Chromatogr. A*, 727, 83, 1996.
127. Fu, C., Xu, H., and Wang, Z., *J. Chromatogr.*, 634, 221, 1993.
128. Sen, N. P., Seaman, S. W., and Kushwaha, S. C., *J. Chromatogr.*, 463, 419, 1989.
129. Billedeau, S. M., Heinze, T. M., Wilkes, J. G., and Thompson, H. C., *J. Chromatogr. A*, 688, 55, 1994.
130. Jansson, C., Paccou, A., and Österdahl, B., *J. Chromatogr. A*, 1008, 135, 2003.
131. Volmer, D. A., Lay, J. O., Billedeau, S. M., and Vollmer, D. L., *Anal. Chem.*, 68, 546, 1996.
132. Righezza, M., Murello, M. H., and Siouffi, A. M., *J. Chromatogr.*, 410, 145, 1987.

133. Goicolea, M. A., Gómez de Balugera, Z., Portela, M. J., and Barrio, R. J., *Anal. Chim. Acta*, 305, 310, 1995.

134. Sen, N. P., Kushwaha, S. C., Seaman, S. W., and Clarkson, S. G., *J. Agric. Food Chem.*, 33, 428, 1985.

135. Pinche, C., Billard, J. P., Frasey, A. M., Bargnoux, H., Petit, J., Bergi, J. A., Dang Vu, B., and Yonger, J., *J. Chromatogr.*, 463, 201, 1989.

136. Nims, R. W., Grove, J. F. Jr., Ho, M. Y. K., Streeter, A. J., and Keefer, L. K., *J. Liq. Chromatogr.*, 15, 195, 1992.

137. Samuelsson, R. and Osteryoung, J., *Anal. Chim. Acta*, 123, 97, 1981.

138. Ong, C. P., Ng, C. L., Chong, N. C., Lee, H. K., and Li, S. F. Y., *J. Chromatogr. A*, 516, 263, 1990.

139. Ong, C. P., Lee, H. K., and Li, S. F. Y., *J. Chromatogr. A*, 542, 473, 1991.

140. Balchunas, A. T. and Sepaniak, M. J., *Anal. Chem.*, 59, 1466, 1987.

141. Terabe, S., Ozaki, H., Otsuka, K., and Ando, T., *J. Chromatogr.*, 332, 211, 1985.

142. Terabe, S., Miyashita, Y., Shibata, O., Barnhart, E. R., Alexander, L. R., Patterson, D. G., Karger, B. L., Hosoya, K., and Tanaka, N., *J. Chromatogr.*, 516, 23, 1990.

143. Terabe, S., Shibata, O., and Miyashita, Y., *J. Chromatogr.*, 480, 403, 1989.

144. Guttman, A., Cohen, A. S., Heiger, D. N., and Karger, B. L., *Anal. Chem.*, 62, 137, 1990.

145. Towns, J. K. and Regnier, F. E., *J. Chromatogr.*, 516, 69, 1990.

146. Nashabeh, W. and Rassi, Z. E., *J. Chromatogr.*, 536, 31, 1991.

147. Altria, K. D., *CE Guide Book, Principles, Operation and Applications*, vol. 52, Humana Press Inc, Totowa, NJ, 1996.

148. Dabek-Zlotorzynska, E., *Electroforesis*, 18, 2453, 1997.

149. Ng, C. L., Ong, C. P., Lee, H. K., and Li, S. F. Y., *J. Chromatogr. Sci.*, 32, 121, 1994.

150. Matyska, M. T., Pesek, J. J., and Yang, I., *J. Chromatogr. A*, 887, 497, 2000.

151. Sanches Filho, P. J., Rios, A., Valcárcel, M., and Bastos Caramao, E., *Water Res.*, 37, 3837, 2003.

152. Sanches Filho, P. J., Zanin, K. D., Bastos Caramao, E., Caramao Garcia, R., Rios, A., and Valcárcel, M., *Quim. Nova*, 26, 3837, 2003.

153. Sanches Filho, P. J., Rios, A., Valcárcel, M., Zanin, K. D., and Bastos Caramao, E., *J. Chromatogr. A*, 985, 193, 2003.

154. McCorquodale, E. M., Boutrid, H., and Colyer, C. L., *J. Chromatogr. A*, 496, 177, 2003.

155. Rhoades, J. W., Hosenfeld, J. M., Taylor, J. M., and Johnson, D. E., *N-Nitroso Compounds: Analysis, Formation and Occurrence*, Walker, E. A., Griciute, L., Castegnaro, M. and Börzsönyi, M., Eds., IARC Sci. Pub. 31, Lyon, p. 377, 1980.

156. Shuker, D. E. G. and Tannenbaum, S. R., *Anal. Chem.*, 55, 2152, 1983.

157. Budevska, B. O., Rizov, N. A., and Gheorghiev, G. K., *J. Chromatogr.*, 351, 501, 1986.

158. Downes, M. J., Edwards, M. W., Eisey, T. S., and Walters, C. L., *Analyst*, 101, 742, 1976.

159. Fine, D. H., Rufeh, F., Lieb, D., and Rounbehler, D. P., *Anal. Chem.*, 47, 1188, 1975.

160. Fine, D. H., Lieb, D., and Rufeh, F., *J. Chromatogr.*, 107, 351, 1975.

161. Havery, D. C., *J. Anal. Toxicol.*, 14, 181, 1990.

162. Meyer, T. A. and Powell, J. B., *J. Assoc. Off. Anal. Chem.*, 74, 766, 1991.

163. Mahanama, K. R. R., and Daisey, J. M., *Environ. Sci. Technol.*, 30, 1477, 1996.

164. Scharfe, R. R. and McLenaghan, C. C., *J. Assoc. Off. Anal. Chem.*, 72, 508, 1989.

165. Billedeau, S. M. and Thompson, H. C., *J. Chromatogr.*, 393, 367, 1987.

166. Prokopczyk, B., Wu, M., Cox, J. E., Amin, S., Desai, D., Idris, A. M., and Hoffman, D., *J. Agric. Food Chem.*, 43, 916, 1995.

167. Sen, N. P., Seaman, S. W., and Kushwaha, S. C., *J. Assoc. Off. Anal. Chem.*, 70, 434, 1987.

168. Ducos, P., Gaudin, R., and Francin, J. M., *Intern. J. Environ. Anal. Chem.*, 72, 215, 1999.

169. Oury, B., Limaste, J. C., and Protois, J. C., *Int. Arch. Occup. Environ. Health*, 70, 261, 1997.

170. Rundlöf, T. R., Olsson, E., Wiernnik, A., Back, S., Aune, M., Johansson, L., and Wahlberg, I., *J. Agric. Food Chem.*, 48, 4381, 2000.

171. Oliveira, C. P., Gloria, M. B. A., Barbour, J. F., and Scanlan, R. A., *J. Agric. Food Chem.*, 43, 967, 1995.

172. Fiddler, W., Pensabene, J. W., Gates, R. A., Custer, C., and Yoffe, A., *J. AOAC Int.*, 80, 353, 1997.

173. Pensabene, J. W., Fiddler, W., and Gates, R. A., *J. AOAC Int.*, 75, 438, 1992.

174. Pensabene, J. W., Fiddler, W., Maxwell, R. J., Lightfield, A. R., and Hampson, J. W., *J. AOAC Int.*, 78, 744, 1995.
175. Ohta, T., Goto, N., and Takitani, S., *Analyst*, 113, 1333, 1988.
176. Fox, J. B., *Anal. Chem.*, 51, 1493, 1979.
177. Singer, G. M., Singer, S. S., and Schmidt, D. G., *J. Chromatogr.*, 133, 59, 1977.
178. Lee, S. H. and Field, L. R., *J. Chromatogr.*, 386, 137, 1987.
179. Misko, T. P., Schilling, R. J., Salvemini, D., Moore, W. M., and Currie, M. G., *Anal. Biochem.*, 214, 11, 1993.
180. Drescher, G. S. and Frank, C. W., *Anal. Chem.*, 50, 2118, 1978.
181. Zheng, M., Fu, C., and Xu, H., *Analyst*, 118, 269, 1993.
182. Wang, Z., Xu, H., and Fu, C., *J. Chromatogr.*, 589, 349, 1992.
183. Zhou, Z. Y., Dauphin, C., Prognon, P., and Hamon, M., *Chromatographia*, 39, 185, 1994.
184. Imai, K., Nishitani, A., and Tsukamoto, Y., *Chromatographia*, 24, 77, 1987.
185. Kobayashi, K. and Imai, K., *Anal. Chem.*, 52, 424, 1980.
186. Imai, K. and Wienberger, R., *Trends Anal. Chem.*, 4, 170, 1985.
187. Sacchetto, G. A., Favaro, G., Pastore, P., and Fiorani, M., *Anal. Chim. Acta*, 294, 251, 1994.
188. Favaro, G., Sacchetto, G. A., Pastore, P., and Fiorani, M., *Anal. Chim. Acta*, 273, 457, 1993.
189. Sacchetto, G. A., Pastore, P., Favaro, G., and Fiorani, M., *Anal. Chim. Acta*, 258, 99, 1992.
190. Samuelsson, R., O'Dea, J., and Osteryoung, J., *Anal. Chem.*, 52, 2215, 1980.
191. Gorskl, W. and Cox, J. A., *Anal. Chem.*, 66, 2771, 1994.
192. Aarts, A. J., Benson, G. B., Duchateau, N. L., and Davies, K. M., *Intern. J. Environ. Anal. Chem.*, 38, 85, 1990.
193. Longo, M., Lionetti, C., and Cavallaro, A., *J. Chromatogr. A*, 708, 303, 1995.
194. Tsai, C. P., Sahil, A., McGuire, J. M., Karger, B. L., and Vouros, P., *Anal. Chem.*, 58, 2, 1986.
195. Straub, R. E. and Voyksner, R. D., *J. Am. Soc. Mass Spectrom.*, 4, 578, 1993.
196. Rhoades, J. W., Hosenfeld, J. M., and Taylor, J. M., *IARC Sci. Publ.*, 31, 377, 1980.
197. Eichelberger, J. W., Kerns, E. H., and Olynyk, P., *Anal. Chem.*, 55, 1471, 1983.
198. EPA. *US EPA Contract Laboratory Program. Statement of Work for Organic Analysis*, US EPA, Washington, DC, 1987.
199. Fajen, J. M., Carson, G. A., and Rounbehler, D. P., *Science*, 205, 1262, 1979.
200. NIOSH 1983. *N-Nitroso Compounds in the Factory Environment*. U.S. Department of Health and Human Services, Public Health Sevice, Centers for Disease Control, National Institute for Occupational Safety and Health, Cincinnati, OH. DHHS (NIOSH) publication no 83-114.
201. Webb, K. S., Gough, T. A., and Carrick, A., *Anal. Chem.*, 51, 989, 1979.
202. Fisher, R. L., Reiser, R. W., and Lasoski, B. A., *Anal. Chem.*, 49, 1821, 1977.
203. Fine, D. H., Rounbehler, D. P., and Sawicki, E., *Environ. Sci. Technol.*, 11, 577, 1977.
204. Ding, Y., Lee, M. L., and Eatough, D. J., *Intern. J. Environ. Anal. Chem.*, 69, 243, 1998.
205. NIOSH 1983. *N-Nitroso Compounds in the Factory Environment*. U.S. Department of Health and Human Services, Public Health Sevice, Centers for Disease Control, National Institute for Occupational Safety and Health, Cincinnati, OH. DHHS (NIOSH) publication no. 83-114.
206. Eisenbrand, G., Ellen, G., and Preussmann, R., *IARC Sci. Publ.*, 181, 1983.
207. EPA 1986. *Test Methods for Evaluation Solid Waste. SW-846, 3rd Ed. Method No 8250 and 8270, Vol. IB: Laboratory Manual: Physical/Chemical Methods*, Office of Solid Waste and Emergency Response, U.S. EPA, Washington, DC.

13 Organic Acids

Sigrid Peldszus

CONTENTS

I. GENERAL BACKGROUND

Organic acids are hydrocarbons, which are characterized by their carboxylate function. Their hydrocarbon structure can vary considerably from aliphatic to aromatic, saturated to unsaturated, and straight chain to branched. Other possible variations in structure are chain length, multiple carboxylate groups (e.g., di- or tricarboxylic acids) and substitution with hydroxyl- or ketogroups. Depending on their structure organic acids can differ in their physical and chemical properties. Many short-chain organic acids, including di- and tricarboxylic acids and also substituted acids, play important roles in the metabolism of living organisms. Important sources for mostly short-chain organic acids in the environment are photochemical and biochemical degradation of anthropogenic and natural organic material, anthropogenic emissions (e.g., sewage sludge, landfill leachates, exhaust fumes) and excretion by microorganisms, higher plants, and animals. Short-chain aliphatic organic acids are therefore the most commonly analyzed organic acids in environmental samples. The most prevalent ones are listed together with their pK_a values and structures in Table 13.1. Fatty acids, i.e., longer-chain organic acids, are not included in this chapter since these are predominantly analyzed in biological specimens. Further up-to-date information with respect to fatty acid research and analysis can be found, for example, in a theme issue of *Analytica Chimica Acta* (Vol. 465, 2002).

Organic acids are present in all environmental compartments (e.g., air, water, and soil) and hence, matrices such as, air, rain, groundwater, soil pore water and also wastewater, drinking water, and landfill leachates have been analyzed. Organic acid concentrations can vary considerably ranging from low μg/l to several hundred mg/l. Research topics and applications where organic acids play an important role are numerous. Some of these are listed in Table 13.2. Before discussing

Abbreviations: AEC, anion exchange chromatography; APCI, atmospheric pressure chemical ionization; CI, chemical ionization; conc., concentration; CTAB, cetyltrimethylammonium bromide; CZE, capillary zone electrophoresis; DETA, diethylenetriamin; DOM, dissolved organic matter; DI, deionized water; ECD, electron capture detector; EOF, electroosmotic flow; EI, electron impact ionization; ES, electrospray; FID, flame ionization detector; GC, gas chromatography; HPLC, high performance liquid chromatography; IEC, ion exclusion chromatography; MRL, minimum reporting level; MS, mass spectrometry; MSE, methanesulfonic acid; MTAB, myristyltrimethylammonium bromide; MTAH, myristyltrimethylammonium hydroxide; MtBE, mehtyl-*t*-butylether; MTBSTFA, n-(*tert*-butyldimethyl-silyl)-*N*-methylfluoracetamide; na, not available; OPD, *o*-phenylenediamine dihydrochloride; PDA, 2,6-pyridinedicarboxylic acid; PDAM, 1-pyrenyldiazomethane; PFBHA, *o*-(2,3,4,5,6-pentafluorobenzyl)hydroxylamine; PFBOH, pentafluorobenzyl alcohol; ppbv, parts per billion, volume to volume; pptv, parts per trillion, volume to volume; PZDA, 2,3-pyrazinedicarboxylic acid; RI, refractive index; SPE, solid-phase extraction; SPME, solid-phase microextraction; TBAOH, tetrabutylammonium hydroxide; TCD, thermal conductivity detector; TMA, trimellitic acid = 1,2,4-benzene-tricarboxylic acid; TOPO, tri-*n*-octylphosphine oxide; TRIS, *tris*(hydroxymethyl)aminomethane; TTAB, tetradecyltrimethylammonium bromide; TTAH, tetradecyltrimethylammonium hydroxide; UASB, upflow anaerobic sludge blanket (reactor); UV, ultraviolet; VFA, volatile fatty acids.

TABLE 13.1
Structures and pK_a Values of Organic Acids

	Formula	Structure	pK_{a1}	pK_{a2}	MW
Monocarboxylic Acids					
Formic acid	CH_2O_2	HCOOH	3.75	—	46.03
Acetic acid	$C_2H_4O_2$	CH_3-COOH	4.75	—	60.05
Propionic acid	$C_3H_6O_2$	CH_3-CH_2-COOH	4.87	—	74.08
n-Butyric acid	$C_4H_8O_2$	$CH_3-(CH_2)_2-COOH$	4.81	—	88.12
iso-Butyric acid	$C_4H_8O_2$	$CH_3-CH(CH_3)-COOH$	4.84	—	88.12
n-Valeric acid	$C_5H_{10}O_2$	$CH_3-(CH_2)_3-COOH$	4.82	—	102.13
iso-Valeric acid	$C_5H_{10}O_2$	$CH_3-CH(CH_3)-CH_2-COOH$	4.77	—	102.13
n-Caproic acid	$C_6H_{12}O_2$	$CH_3-(CH_2)_4-COOH$	4.83	—	116.16
n-Heptanoic acid	$C_7H_{14}O_2$	$CH_3-(CH_2)_5-COOH$	4.89	—	130.19
Hydroxyacids					
Glycolic acid[a]	$C_2H_4O_3$	$H_2C(OH)-COOH$	3.83	—	76.05
Lactic acid[b]	$C_3H_6O_3$	$CH_3-HC(OH)-COOH$	3.08	—	90.08
Ketoacids					
Glyoxylic acid[c]	$C_2H_2O_3$	CHO–COOH	3.34	—	74.04
Pyruvic acid[d]	$C_3H_4O_3$	$CH_3-CO-COOH$	2.49	—	88.06
α-Ketobutyric acid	$C_4H_6O_3$	$CH_3-CH_2-CO-COOH$	—	—	102.09
Dicarboxylic Acids					
Oxalic acid[e]	$C_2H_2O_4$	HOOC COOH	1.23	4.19	90.04
Malonic acid[f]	$C_3H_4O_4$	$HOOC-CH_2-COOH$	2.85	5.69	104.06
Succinic acid[g]	$C_4H_6O_4$	$HOOC-(CH_2)_2-COOH$	4.16	5.61	118.09
Glutaric acid[h]	$C_5H_8O_4$	$HOOC-(CH_2)_3-COOH$	4.31	5.41	132.13
Adipic acid[i]	$C_6H_{10}O_4$	$HOOC-(CH_2)_4-COOH$	4.43	5.41	146.14
Pimelic acid[j]	$C_7H_{12}O_4$	$HOOC-(CH_2)_5-COOH$	4.42[k]	5.06[k]	160.17
Other Acids					
Maleic acid	$C_4H_4O_4$	$HOOC-CH{=}CH-COOH$ (cis)	1.83	6.07	116.07
Fumaric acid	$C_4H_4O_4$	$HOOC-CH{=}CH-COOH$ ($trans$)	3.03	4.44	116.07
Malic acid[l]	$C_4H_6O_5$	$HOOC-CHOH-CH_2-COOH$	3.40	5.11	134.09
Tartaric acid[m]	$C_4H_6O_6$	$HOOC-CHOH-CHOH-COOH$	2.98	4.34	150.09
Oxalacetic acid[n]	$C_4H_4O_5$	$HOOC-CO-CH_2$ COOH	2.22	3.89	132.07
Citric acid	$C_6H_8O_7$	$HOOC-CH_2-C(OH)(COOH)-CH_2-COOH$	3.15	4.77	192.14
—	—	—	5.19[o]	—	—

All data except for k from Weast, R. C., Ed., *CRC Handbook of Chemistry and Physics*, 61th ed., CRC Press, Boca Raton, Florida, 1980.

[a] Hydroxyacetic acid.

[b] 2-Hydroxypropanoic acid.

[c] Oxoacetic acid = formylformic acid.

[d] 2-Oxopropanoic acid = α-ketopropionic acid = acetylformic acid = pyroracemic acid.

[e] Ethanedioic acid.

[f] Propanedioic acid = methanedicarboxylic acid.

[g] Butanedioic acid.

[h] Pentanedioic acid.

[i] Hexanedioic acid.

[j] Heptanedioic acid.

[k] Data from Adler, H., Sirén, H., Kulmala, M., and Riekkola, M. L., Capillary electrophoretic separation of dicarboxylic acids in atmospheric aerosol particles, *J. Chromatogr. A*, 990, 133–141, 2003.

[l] Hydroxybutanedioic acid = hydroxysuccinic acid.

[m] L-2,3-Dihydroxybutanedioic acid = D-2,3-dihydroxysuccinic acid.

[n] Oxobutanedioic acid = oxosuccinic acid = ketosuccinic acid.

[o] pK_{a3}.

TABLE 13.2
Examples for Research Topics and Applications Involving Organic Acids in the Environment

Air—Including Atmospheric Precipitation[3-9]	Water[10-18]	Soil[19-23]	Waste[24-32]
Contribution to acidic rain	Formed as ozonation by-product in drinking water treatment from natural organic matter	Sources: Excretion from plant roots and microorganisms. Anaerobic degradation of organic litter	Control and monitoring of anaerobic digestion of organic material in wastewater treatment processes
Role in depletion of ozone layer	Intermediates of anaerobic degradation of hydrocarbons (naturally occurring or at contamination sites) in groundwater	Metal mobilization. Complexation of Al (toxic for plants at certain concentrations)	Monitoring of VFA in landfill leachates as indicators for onset of methane production, metal mobilization, and leaching of other contaminants
Identification of biogenic and anthropogenic sources of organic acids	Metabolic pathways and organic acid turnover rates in deep seawater and sediment	Biological availability of nutrients (e.g., P, Fe)	Intermediates in remediation of organic pollutants with advanced oxidation processes
Photochemical transformation of organic compounds to organic acids	Corrosion control of ultrapure water in industrial processes	Mineralization processes leading to podsolization of soil	Fermentation processes in biological materials

analytical techniques in details, the motivation behind organic acid analysis is briefly summarized by matrix.

A. Air Including Atmospheric Precipitation

Organic acids in rain were first detected in the 1970s.[5] Since then they have been found in tropospheric gas and aqueous phases, and also adsorbed onto tropospheric particulate matter. More specifically, organic acids have been measured in fog water, cloud water, rain, snow, ice, the gas phase, and on particles in varying locations including highly urbanized areas and remote regions. Formic and acetic acid are usually present in higher concentrations than any other organic acids, such as dicarboxylic and/or ketoacids. Organic acids contribute significantly to the total organic carbon content and to the acidity in fog, cloud water, and wet precipitation. It is reported that in North American rain up to one third of the total free acidity is caused by organic acids.[33,34] However, it is postulated that organic acids do not contribute substantially to the long-term acidification of the environment since these are easily biodegradable.[33] Major sources for organic acids in the troposphere are direct biogenic and anthropogenic emissions but also there are indirect contributions through photochemical formation from organic precursors.[3] Biogenic sources include emissions from soil and especially vegetation,[9] whereas anthropogenic emissions can be traced back to biomass combustion (e.g., forest fires, agricultural burnings)[6-8] and incomplete combustion of fuel and fuel additives (e.g., motor exhausts).[35,36] Photochemical processes in the gaseous phase and to some extent in the aqueous phase lead to organic acid formation via radical reactions involving, for example, oxidation of hydrocarbons and aldehydes by free radicals (e.g., $\cdot OH$, $HO_2\cdot$) and other oxidants (e.g., ozone).[3,4,6,34,37,38] The role of particulate matter and therefore of heterogeneous interactions in these processes remains unclear at this time. With the exception of ketoacids which are susceptible to photolysis, main losses of organic acids from the troposphere occur through wet or dry deposition. A more detailed overview of organic acids in the troposphere can be found elsewhere.[3] Current research continues to further catalogue possible emission sources of organic acids, documents their occurrences and residence times, elucidates reaction mechanisms leading to organic acid formation, and investigates their role in the depletion of the ozone layer.[4] Ice-core samples can put current results into a historical context.[6] It should also be noted that organic acids are not regulated as air pollutants.

B. WATER

1. Drinking Water

Short-chain organic acids such as formic, acetic, oxalic, glyoxylic, pyruvic, and ketomalonic acid are formed as by-products in drinking water treatment plants employing ozone.[10,39–44] Reasons for ozone applications include destruction of taste and odor compounds, color removal, and pretreatment.[11] Moreover, ozone is often used as a disinfectant and as such it can at least partially substitute chlorine, thus lowering formation of chlorinated disinfection by-product, which is a major issue for the drinking water industry in North America. During ozonation natural organic matter in raw or partially treated water is oxidized, leading to the formation of a range of by-products dominated by aldehydes and organic acids with the latter formed in higher concentrations.[45] Ozone contactor effluents may have concentrations as high as 250 μg/l for formate or 195 μg/l for acetate.[10] Subsequent treatment steps such as biological filtration can remove aldehydes completely, whereas organic acids may only be partially removed.[10] Smaller concentrations of organic acids can therefore pass into the finished drinking water and serve as nutrients for microorganisms, thus leading to bacterial regrowth in distribution systems which are used to transport the finished drinking water to the end user. Current research continues to assess the formation, removal, and impact of ozonation by-products such as organic acids. Organic acids are not regulated in drinking water.

2. Wastewater

Volatile fatty acids (VFA) are defined as "water soluble fatty acids which can be distilled at atmospheric pressure."[46] These are comprised of aliphatic, short-chain organic acids with chain lengths of up to seven C-atoms and are formed during anaerobic fermentation of organic material. Removal of organic substrate through anaerobic digestion has been applied primarily to waste sludge, but it is also used as pretreatment for high organic waste streams and is even becoming common for the treatment of dilute waste streams.[25] Organic substrate removal is accomplished by fermentation, i.e., break down of larger molecules (e.g., fats, proteins, hydrocarbons) followed by a two-step anaerobic degradation process of these break-down products leading to hydrogen and methane as end products.[24,25] More specifically, acidogenic bacteria convert organic break-down products into VFA and H_2 with acetate being the major product. This is followed by metabolization of VFA by methanogenic bacteria into methane. These two processes have to be in balance to ensure successful treatment. The difficulty lies in the fact that methanogenic bacteria metabolize at a slower rate than the acidogenic bacteria. In addition, methanogenic bacteria are usually more susceptible to sudden changes in their environment than the acidogenic bacteria.[24–26] Hence, it is possible that VFA may not be consumed at the same rate as they are produced with the consequence that VFA concentrations increase. This may disturb the balance between these processes even further. Changes in background concentration and VFA composition are therefore used as indicators in the operation of these anaerobic processes and are monitored for process control purposes.[24–26] VFA concentrations in wastewater and diluted sludge are quite high (medium to high mg/l), which simplifies the analyses of this rather complex matrix (high concentrations of inorganic and organic compounds) requiring only moderate detection limits.[24,25]

3. Groundwater

Organic acids in groundwater can be found close to seeps of naturally occurring hydrocarbons, or in the proximity of sites contaminated with, for example, petroleum hydrocarbons. Biodegradation of hydrocarbons under the anaerobic conditions of the groundwater aquifer leads to a variety of metabolic intermediates, including organic acids.[14,47,48] High concentrations of short-chain aliphatic organic acids are most commonly observed as intermediates, even when aromatic hydrocarbons are the original source. Localized concentration of formate, acetate, and isobutyrate combined may be as high as 9000 μmol/l,[13] but vary considerably depending on the rate of organic acid production and consumption. Hence, factors such as the proximity to the hydrocarbon source, hydrocarbon loading, microbial population, presence of microbial inhibitors, availability of nutrients, and availability of electron acceptors affect the organic acid concentration in groundwater.

High organic acid concentrations in groundwater may cause mineral dissolution. This can lead to changes in soil structure, and complexation of metals such as Fe or Al, which may subsequently be mobilized into the aquifer.[13] Hence, it is important to take the organic acid production and its consequences into account when observing and projecting the transport of hydrocarbon contamination in groundwater,[13] and when estimating changes in the porosity of oil reservoirs due to organic acid formation in oilfield water.[15,49]

4. Seawater

In marine environments such as deep seawater and marine sediment pore water, low-molecular weight (LMW) organic acids are formed as intermediate metabolic break-down products from larger organic molecules under anoxic conditions.[16,17] Depending on environmental conditions, organic acids are often further degraded to CH_4 and CO_2. Research focuses on the elucidation of metabolic pathways and organic acid turnover rates.[17] There are also specialized research interests such as measuring acrylic acid in seawater, algal cultures, and sediment pore water.[50] Acrylic acid may be released from (dimethylsulfonio)propionate, a compound present in many marine algae.

5. Other Water

Organic acids have also been analyzed in various other aqueous matrices. LMW organic acids have been detected as intermediates during remediation of organic pollutants using advanced oxidation processes[30] and upon UV irradiation of aqueous dissolved organic matter.[12,51] Ultrapure water used in certain industries for cooling and production processes (e.g., power generating industry, electronics industry) is monitored for inorganic anions and organic acids, since their presence may lead to corrosion and other disturbances of the production process.[18]

C. Soil

Organic acids play an important role in soils although these typically comprise only 0.5 to 5% of the dissolved organic carbon (DOC) in soil solutions.[21] Acids such as lactic, acetic, oxalic, succinic, fumaric, malic, citric, and aconitic acid have been detected frequently in soil solutions.[19] Main sources are excretion from plant roots (e.g., root exudates), release by microorganisms, and degradation of organic matter (e.g., plant litter). Concentrations of organic acids in soil solutions are in the μmolar range with significant spatial and temporal variations. The highest concentrations are usually found in organic-rich soil layers.[19,21] Concentrations are influenced by a number of very complex processes. These processes are based on proton release by organic acids causing pH change in localized environments, and on the complexing ability of various organic acid anions. Effects associated with these processes include an increase in mineral nutrients (e.g., phosphate, iron) which are biologically available for plants, thus enhancing plant growth; reduction in Al concentrations by complexation which could otherwise be toxic for plants; metal mobilization either by direct interaction (e.g., desorption) or through complexes and subsequent metal transport into deeper soil layers; stimulation of bacterial growth by enhancing the carbon source; and mineral weathering leading to podzolization.[19-21] Research into these very complex processes has expanded over the last decade since more sophisticated analytical techniques became available to measure organic acids in soil matrices.

D. OTHERS

1. Landfill Leachates

The anaerobic degradation of organic waste in landfills passes through several stages with methane and carbon dioxide as end products.[27,28] During the acidogenic phase, organic acids, mainly VFA, are formed in high concentrations and thus contribute significantly to the DOC fraction in landfill leachates, which is the solution collecting at the bottom of a landfill. One concern associated with VFA production is the possible mobilization of heavy metals from the landfill into the underlying aquifer if there is either an incomplete seal of the liner or if there is no liner at all, which is the case for many older landfill sites. Monitoring VFA concentration in landfill leachates allows for the determination of the degradation stage of the organic waste in the landfill and the prediction of the onset of methane production. It also serves as an indicator for the mobilization of organic pollutants and heavy metals.[27,29] When VFA concentrations are rising, precautions can be taken to prevent leakage of pollutants into the groundwater through, for example, collection and treatment of leachates.[28]

2. Biological Material

Matrices covered in this section are very diverse. Examples listed in Table 13.3 include silage juice, fermentation products, hydrolysated biomass residue, cellulose polymer, and also biological specimens.[31,32,52-54] Although these matrices do not strictly classify as environmental samples, some analytical methods dealing with these matrices were included to show their unique approach. Motivations for the analyses are as diverse as the matrices themselves.

II. ANALYSIS

A large number of methods are available to determine organic acids in various environmental matrices. Factors such as type of organic acid, their concentrations, and sample matrix, determine largely which analytical methods are suitable for a certain sample. The overall analysis process comprises a sequence of steps starting with sampling and sampling preservation followed by sample preparation and finally instrumental analysis and data interpretation. All these steps are discussed in the following sections and put in context with environmental matrices. Table 13.3 gives an overview of methods available for organic acid analysis organized by matrix and instrumentation.

A. SAMPLING AND SAMPLE PRESERVATION

Sampling is an important part of the overall analytical process. The ideal sampling technique is adapted to the matrix, prevents contamination, and results in an unaltered sample which reflects the

TABLE 13.3
Organic Acid Method Overview Organized by Matrix

Sample Preparation	Instrumentation	Acids Determined[a]	Detection Limits	Comments	Ref.
		AIR Including Atmospheric Precipitation			
(1) Air					
Annular denuder, aqueous extraction	IEC/cond.	Formic, acetic acid	1–50 mg/l standards	Gas phase, sample preservation with $CHCl_3$	55
Cryogenic trapping	AEC/cond.	Formic, acetic, pyruvic, oxalic acid	b	Gas phase, also detection of Cl^-, NO_2^-, SO_4^{2-}	56
Collection on filters, freezer, aqueous extraction	AEC/cond.	Formic, acetic acid	0.1 ppbv[b]	Gas and particulate phase, filters stored in freezer	57
Collection on filters and denuder, aqueous extraction, $CHCl_3$ preservation	AEC/cond., CZE/ind. UV	Formic, acetic, glycolic, pyruvic, oxalic, β-hydroxybutyric acid	45–102 μg/l (standards)[b]	Gas and particulate phase, confirmation with CZE	58
Scrubber, filters, aqueous extraction, $CHCl_3$ preservation	AEC/cond., IEC/cond.	Formic, acetic, pyruvic, glyoxylic, succinic, malonic, oxalic acid	0.07–0.19 ppbv[b]	Gas and particulate phase organic acids, confirmation with IEC	59
Varied	AEC/cond.	Formic, acetic acid	b	Gas and particulate phase, comparison of different sampling techniques	60
Collection on filters, freezer, ultrasonic extraction	AEC/cond.	C_2–C_5 dicarboxylic, glyoxylic acid	10–50 ng[b]	Particulate phase	61
Collection on filters, ultrasonic aqueous extraction	AEC/cond.	Formic, acetic, oxalic acid	na	Particulate phase, investigated sorochemical processes during extraction	62
Filters, film, aqueous extraction	AEC/cond.	C_2–C_5 dicarboxylic acids	b	Size segregated particulate phase, measured also inorganic anions	63
Aqueous extraction of filters	CZE/ind. UV	C_2–C_{10} dicarboxylic acids	1–9 mg/l	Particulate phase, compared different electrolytes	2
Aqueous extraction of filters, filtration	AEC/cond., CZE/ind. UV	C_1–C_4 moncarboxylic, glycolic, lactic, β-hydroxybutyric acid	0.016–0.082 mg/l, 0.05–0.32 mg/l	Particulate phase, developed and compared AEC and CZE methods, investigated coelution	64

Aqueous extraction of filters, fractionation on SPC18, BF_3/MeOH and BF_3/propanol deriv.	GC/FID, GC/MS	C_{10}–C_{18} monocarboxylic, C_2–C_6 dicarboxylic, glyoxylic, pyruvic acid	0.2–3.0 ng/m³ (for 50 m³ sample vol.)[b]	Aerosols, simultaneous determination on FID and MS by splitting flow after injection	65
Aqueous extraction of filters, BF_3/butanol deriv.	GC/FID	C_2–C_{10} saturated and unsaturated dicarboxylic acids	Measured low ng/m³ conc.	Urban aerosols, confirmation of peak identification by GC/MS	66
Indoor and outdoor smog chamber experiments	GC/MS (EI, CI–CH_4, PFBOH)	C_1–C_6 saturated and unsaturated monocarboxylic, hydroxy, oxocarboxylic, dicarboxylic acids	na	Identification of (a) products of isoprene oxidation, (b) products of toluene/NOx chamber experiments	4
Collection on cartridges, aqueous extraction, $CHCl_3$ preservation	HPLC/UV	Formic, acetic acid	2–2.8 ppbv[b]	Air chamber experiments, focused on carbonyl products in lab study	67

(2) Rain, Fog, Mist, Snow

Direct injection	AEC/cond.	Formic, acetic, pyruvic acid	0.2 μM	Rain, optimized for formic and acetic acid only	68,69
Direct injection	AEC/cond	Formic, acetic, pyruvic acid	0.02–0.1 μM	mist	38
Direct injection	AEC/cond	C_1–C_5 monocarboxylic, 2- and 3-hydroxybutyric, lactic, glycolic, pyruvic, oxalic acid	0.5–1 μM	Fog, lake, sediment pore water, rain, simultaneous detection of F^-, Cl^-, NO_2^-, CO_3^{2-}, PO_4^{3-}, Br^-, SO_3^{2-}, SO_4^{2-}	70
Direct injection	AEC/cond.	Formic, acetic, oxalic acid	20–100 $\mu g/l$	Rain, snow, simultaneous detection of F, Cl^-, NO_2^-, NO_3^-, Br^-, SO_3^{2-}, SO_4^{2-}, PO_4^{3-}	71
Direct injection	AEC/cond.	Formic, acetic, propionic, glycolic, lactic acid	0.2–1.6 $\mu g/l$	Snow, investigates coelution of compounds	72
Addition of HCl to sample, direct injection	IEC	C_1–C_5 monocarboxylic, citric, lactic, glycolic acid	na. but <0.6 mg/l	Rain, describes need for sample preservation with $CHCl_3$	33,73
Concentrated under vacuum at pH 8.5–9, α-p-bromo acetophenone deriv.	GC/FID	C_1–C_9 monocarboxylic acids	~0.1 μM	Rain, fog, confirmation with GC/MS possible	34,74–76

Continued

TABLE 13.3
Continued

Sample Preparation	Instrumentation	Acids Determined[a]	Detection Limits	Comments	Ref.
Concentrated under vacuum at pH 8–9, α,p-bromoacetophenone deriv.	GC/FID or HPLC	C_1–C_5 monocarboxylic acids	na, analyzed low μM conc.	Rain, sewage, soil pore water	77
Concentrated under vacuum at pH 8–9, BF_3/butanol deriv., removal of excess butanol	GC/FID or GC/MS	More than 20 saturated and unsaturated C_2–C_{10} dicarboxylic acids	na, but <0.8 μM	Rain, fog, mist	36,75,78
Concentrated under vacuum at pH 8–9, BF_3/butanol deriv.	GC/FID or GC/MS	Saturated (C_2–C_{11}) and unsaturated (C_4, C_5, C_8) dicarboxylic acids, C_2–C_{10} ω-oxoacids, glyoxylic, pyruvic acid	0.05 μg/l	Rain, snow, aerosols, simultaneous determination of aldehydes, e.g., glyoxal, methylglyoxal	79,80
Derivatization with OPD	HPLC/UV	Oxalic, pyruvic, gyoxylic acid	na, but <2 μM	Rain, fog and mist	81
Direct injection using sample stacking	CZE/ind. UV	Formic, acetic, propionic, oxalic, ten other mostly dicarboxylic acids	10–30 nM	Measured single rain drops, simultaneous detection of Cl^-, NO_3^-, SO_4^{2-}, NO_2^-, Br^-, PO_4^{3-}	82
Direct injection	(a) micro HPLC (b) CZE/ind. UV	Formic, acetic acid	(a) 13–20 μg/l (b) 180–450 μg/l	Measured single rain drops, simultaneous detection of Cl^-, NO_3^-, SO_4^{2-}	83
(3) Ice					
Outer part of core removed, sample melted	AEC/cond.	Formic, acetic, oxalic, glycolic acid	0.2–0.6 ng/g	Ice core, simultaneous detection of F^-, Cl^-, NO_2^-, NO_3^-, SO_4^{2-}	6,84
Anion exchange preconcentrator column	IEC/UV	Formic, acetic, propionic, butyric acid	5.6–9.4 μg/l with 10 ml inj. vol.	Antarctic ice as example, combines AEC concentrator with IEC separation	85
Water					
(1) Drinking Water					
Direct injection	AEC/cond.	Formic, acetic, oxalic acid	2–3 μg/l, MRL 15 μg/l	$HgCl_2$ for preservation[42], Hg^{2+} removal with H^+-cartridge in line[42]	42,44

Sample preparation	Detection	Analytes	Detection limits	Comments	Ref.
Direct injection	AEC/cond.	Formic, acetic, butyric, β-hydroxybutyric, glycolic, pyruvic, α-ketobutyric, oxalic acid	1–9 µg/l	Oxalate in matrices with high sulphate concentrations requires switching technique	43,86,87
Filtration; sulphate, chloride and phosphate removal	AEC/cond.	Formic, acetic, oxalic, glycolic acid	20–40 µg/l. concentrator 0.1–0.5 µg/l	No testing on 'real' drinking water samples described	45
1. Aqueous PFBHA deriv. 2. MtBE extraction 3. CH_2N_2 deriv.	GC/ECD	Glyoxylic, pyruvic, ketomalonic acid	na	Ozonated drinking water	39,40,88
1. Aqueous PFBHA deriv. 2. Diethylether extraction 3. CH_2N_2 deriv.	GC/ECD	Glyoxylic acid	na	Ozonated fulvic acid solutions, simultaneous detection of several aldehydes	89
1. Aqueous PFBHA deriv. 2. RP18 SPE or MtBE extraction 3. CH_2N_2 deriv. 4. MTBSTFA deriv.	GC/MS (EI, CI)	C_2–C_5 ketoacid and others	na	Simultaneous detection of aldehydes and hydroxyl substituted compounds, focused on identification of compounds in ozonated drinking water	41,45
(2) Wastewater					
Centrifugation and filtration, acidification	GC/FID	C_2–C_5 monocarboxylic acids	na	Direct aqueous injection	90–96
Static headspace injection	GC/FID	C_2–C_5 monocarboxylic acids	0.3–3.7 mg/l	2-Ethylbutyric acid as internal standard, weighted least square calibration curve	97
Direct SPME, headspace and gas phase SPME with PDAM derivatization	GC/FID	C_2–C_{10} monocarboxylic acids	0.02–760 µg/l	Only few real samples, detection limits for standards in deionized water	98
Headspace SPME	(a) GC/FID (b) GC/MS	C_2–C_7 monocarboxylic acids	(a) 6–675 µg/l (b) 2–150 µg/l	Few real samples, method development in deionized water	99,100

Continued

TABLE 13.3
Continued

Sample Preparation	Instrumentation	Acids Determined[a]	Detection Limits	Comments	Ref.
(3) Groundwater					
Filtration, direct injection	IEC/cond.	C_2–C_4 monocarboxylic acids	0.5 μM	Microcosm study	101
CH_2Cl_2 extraction at pH < 2	GC/MS	C_2–C_7 monocarboxylic acids	na	Identification of polar degradation products at superfund sites	48
Diethylether extraction of freeze dried samples	GC/FID	C_2–C_7 monocarboxylic acids	na	Groundwater at gas spill site	14,47
(4) Seawater					
Concentrated using static diffusion	GC/FID	C_2–C_5 monocarboxylic, pyruvic, acrylic, benzoic acid	Down to 10 nM	Unusual concentration procedure	16
Extraction, PFBBr derivat.	GC/ECD	Acrylic acid, other acids possible	3 nM	0.5% TOPO in MtBE for extraction	50
Filtration, 2-nitrophenyl hydrazine deriv.	HPLC/VIS	C_1–C_5 monocarboxylic, lactic acid	na, analyzed low nM conc.	Blank problems for acetic and formic acid	17
(5) Other					
None	AEC/cond.	Formic, acetic, oxalic acid	Below 1 μg/l	Power plant water, online preconcentration, simultaneous determination of inorganic anions	18
None	AEC/cond./UV	Formic, acetic, glycolic, oxalic, malonic maleic, fumaric acid	0.001–0.006 mM	By-products of organic pollutant degradation in water	30
None, direct injection	AEC/UV CZE/ind.UV	Formic, acetic, oxalic, malonic, succinic, pyruvic acid	0.2 μM	Photoformation of organic acids from brown water dissolved organic matter; confirmation with CZE	51,102
Addition of octanesulfonate	CZE/ind. UV	Formic, acetic, lactic, oxalic, malonic acid	na, analyzed μg/l conc.	UV irradiated humic surface water	12,103

Sample preparation	Method	Compounds	Concentration	Comments	Ref.
Dilution with trifluoroacetic acid, filtration	IEC/ES-MS	Formic, gyoxylic, oxalic, 2-hydroxy-*iso*-butyric, maleic, succinic, malic, 1-hexanoic, malonic acid	2–8 mg/l	Aqueous standards only, established ES-MS conditions	104
None, direct injection	IEC/cond.	C_1–C_5 monocarboxylic acids	3–79 µg/l	Standards, introduced C_6 aliphatic monocarboxylic acids as eluents	105

Soil

(1) Soil Solutions

Sample preparation	Method	Compounds	Concentration	Comments	Ref.
Online removal of interfering compounds	AEC/cond.	Formic, acetic, lactic, pyruvic, oxalic, malic, citric acid	60–250 n*M*	Sample preparation completely online including extraction with TOPO impregnated membrane	107,108
Several extraction steps, filtration, removal of humic acids	(a) AEC/cond. (b) AEC/UV	(a) 13 aliphatic mono- and dicarboxylic acids (b) 14 aromatic acids	(a) 103–1286 n*M* (b) 37–1729 n*M*	Used one AEC system for aliphatic and the other AEC system for aromatic acids	109
Centrifugation, filtration, cation exchange	IEC/UV	Formic, acetic, propionic, lactic, citric, tartaric, malic, succinic, fumaric, malonic, *trans*-aconitic acid	ne, analyzed 6–6000 µ*M*	Investigated Al complexes in soil, confirmation by CZE	110,111
Centrifugation, filtration, addition of Na₄EDTA	CZE/ind. UV	C_1–C_4 monocarboxylic, lactic, oxalic, malonic, tartaric, malic, succinic, citric acid	0.26–1.77 µl/l	Used different electrolytes for mono- and dicarboxylic acids	112,113
Aqueous extraction, centrifugation, filtration	CZE/ind. UV	Formic, acetic, lactic, tartaric, malic, citric, succinic acid	0.5–6 µ*M*	Also measured root extracts	114

(2) Sediments

Sample preparation	Method	Compounds	Concentration	Comments	Ref.
Centrifugation, concentrated at pH 9, acidification	GC/FID	Acetic, propionic, butyric acid	Analyzed µ*M* to high n*M* conc.	Supernatant from wetland sludge, direct aqueous injection	106
Filtration	IEC/RI	C_1–C_5 monocarboxylic, lactic, oxalic, citric, malic, tartaric, fumaric acids	na. analyzed high µ*M* conc.	Samples stored in freezer	115
Centrifugation, acidification	GC/FID	C_2–C_7 monocarboxylic acids	2–5 µ*M*	Direct aqueous injection	116

Continued

TABLE 13.3
Continued

Other

Sample Preparation	Instrumentation	Acids Determined[a]	Detection Limits	Comments	Ref.
(1) Landfill Leachate					
Centrifugation, filtration, removal of humin-like substances	IEC/UV IEC/cond.	C_1–C_5 monocarboxylic, pyruvic, glyoxylic, glycolic, lactic, glyceric, succinic acid	na, analyzed 50 to 50,000 μM	Developed and compared two IEC methods for organic acids	29
Filtration, dilution, OnGuardP pretreatment, carbonate removal	IEC/UV	C_2–C_6 monocarboxylic acids	5 mg/l	Determined inorganic anions with AEC on split sample	27
Filtration, storage under N_2	IEC/UV	C_1–C_4 monocarboxylic, lactic, pyruvic, glycolic, glyceric, succinic, adipic acid	na, analyzed low mM conc.	Simulation of organics degradation in radio active waste	117
Acidification, distillation	GC/FID	C_2–C_4 monocarboxylic acids	na, analyzed mg/l conc.	Direct aqueous injection	118
Acidification, diethylether extraction	GC/FID	C_2–C_7 monocarboxylic acids	na, analyzed mg/l conc.	Low level radioactive waste leachates	119
Dilution	GC/MS	C_2–C_8 monocarboxylic acids	1–8 mg/l	Landfill monitoring wells, direct aqueous injection, deuterated internal standards	120
Acidification, distillation	HPLC/UV	C_1–C_4 monocarboxylic acids	na, analyzed high mg/l conc.	HPLC/UV less sensitive but formic acid analysis possible, confirmed results with GC/FID	121
(2) Biological Samples					
Filtration, dilution	IEC/UV	C_1–C_4 monocarboxylic, lactic, pyruvic, citric, fumaric, malic, succinic, α-ketoglutaric acid	0.2–150 μM	Fermentation products, used predicted capacity factors to optimize separation of target analytes	32

Sample preparation	Detection	Analytes	Detection limit[b]	Description	Ref.
None	IEC/UV	$C_1–C_4$ monocarboxylic, $C_2–C_5$ dicarboxylic, lactic, glyoxylic, glycolic, tartaric, α-ketoglutaric	0.2–10 μM	Hydrolysated biomass residues	122
Acidification, purging of CO_2	IEC/cond.	Malic, succinic, citric, trans-aconitic acid	na	Root exudates	22,23
None, direct injection	(a) IEC/APCI-MS (b) IEC/cond.	$C_1–C_5$ monocarboxylic acids	(a) 0.7–3.8 μM (b) 0.08–2.3 μM	beverages, developed APCI-MS method, compared MS, conductivity and UV-phodiodearray detection	123
Dilution with oxalic acid (1:10)	GC/FID	$C_2–C_7$ monocarboxylic, lactic acid	na, analyzed mg/l conc.	Silage juice, direct aqueous injection, compared different GC columns and conditions	31
Aqueous extraction of CH_2Cl_2 sol., centrifugation	GC/TCD	$C_1–C_4$ monocarboxylic acids	10 mg/l	Cellulose polymer, only GC method measuring formic acid directly	52
Vacuum distillation, acidification	GC/FID	$C_2–C_7$ monocarboxylic acids	0.05 mM	Biological samples, direct aqueous injection; compared different GC columns and conditions	54
Alcoholic extraction	GC/FID	$C_2–C_6$ monocarboxylic acids	20 nM	Biological specimens, uses precolumn	53
Direct collection on filter paper, acidification, filtration	HPLC/UV	Malic, malonic, lactic, acetic, maleic, citric, cis-, trans-aconitic, succinic, fumaric acid	0.05–24 μM	Root exudates	124

na, not available; APCI, atmospheric pressure chemical ionization; CI, chemical ionization; cond., indirect conductivity detection; CZE, capillary zone electrophoresis; ECD, electron capture detector; EI, electron impact ionization; ES, electrospray; FID, flame ionization detection; GC, gas chromatography; HPLC, high performance liquid chromatography; AEC, anion exchange chromatography; IEC, ion exclusion chromatography; MRL, minimum reporting limit; MS, mass spectrometry; MtBE, methyl-*tert*-butylether; MTBSTFA, *n*-(*tert*-butyldimethylsilyl)-*N*-methylfluoracetamide; OPD, *o*-phenylenediamine dihydrochloride; PDAM, 1-pyrenyldiazomethane; PFBBr, pentafluorobenzyl bromide; PFBHA, *o*-(2,3,4,5,6-pentafluorobenzyl)hydroxylamine; PFBOH, pentafluorobenzyl alcohol; ppbv, parts per billion, volume to volume; RI, refractive index (detection); SPE, solid-phase extraction; SPME, solid-phase microextraction; TCD, thermal conductivity detector; TOPO, tri-*n*-octylphosphine oxide; UV, ultraviolet (detection).

[a] All aliphatic unless mentioned otherwise.
[b] Detection limits vary with sampling time.

average content of the analytes at the time when the sample was taken. Note that contamination can be an issue for matrices with very low analyte concentrations since organic acids are ubiquitous in the environment. Organic acids have been reported to be present in laboratory air,[84] as impurities in chemicals,[106] on glassware,[16,120] and also on human skin.[87,103] Precautions such as wearing gloves, minimal sample exposure to laboratory air, and thorough cleaning of glassware and sampling devices may have to be taken when dealing with these contamination-sensitive matrices. The following subsections discuss aspects of sampling for matrices with special sampling requirements.

1. Air Including Gaseous and Aqueous Phase and Particulate Matter

Air samples are usually taken by pumping ambient air for a defined time interval through a device which retains the organic acids. A filter precedes these devices if particulate phase organic acids are to be analyzed separately from gaseous and aqueous phase organic acids. Devices and processes used for air sampling include scrubbers, denuders, condensation, or cryogenic trapping.[55–57,60,80,125] Solutions used in scrubbers may be water or alkaline.[60,125] Filters may also be impregnated with alkaline solutions[37,57,126] and denuders are coated with either NaOH or KOH.[55,127] All sampling devices are easily contaminated with the ubiquitous organic acids. Precautions include thorough cleaning of the sampling devices accompanied by regular measurements of blanks. Note that formation of artefacts has been reported during long sampling periods in the presence of alkali and/or water. This may lead to difficulties in distinguishing between organic acids trapped from the air and those formed while sampling.[56,60]

Recoveries and detection limits vary considerably between different air sampling techniques. Detection limits can be as low as 10 parts per trillion, volume to volume (pptv) for formic and acetic acid when utilizing the more established scrubber (recoveries 55 to 98%)[56,125] or the newer cryogenic trapping technique (recoveries 100%).[56] When utilizing filters or denuders, detection limits are considerably higher (20 to 450 pptv), although recoveries between 84 and 99% ensure reliable data.[56]

2. Rain

Rain collectors are usually located high above ground, for example on rooftops. They consist of a collection container and a funnel made from inert material such as glass, polyethylene, teflon, or stainless steel. The design of rain collectors varies from a simple wide-neck bottle with a funnel, which is operated manually, to sophisticated automated wet-only collection systems.[5,9,33–35,68,70,71,73,74,78,79,81,128] To avoid dry deposition, rain collectors are kept shut until the rain event starts.[128] Rain collectors need also to be cleaned thoroughly between samplings to avoid contamination and carry over from one sampling event to the next.

Sampling is done either by rain event or, less frequently, in bulk. When sampling by rain event, collectors are exchanged after a single rain event resulting in an average sample of this specific event. When collecting in bulk, rain collectors are replaced after defined extended periods of time. Although this approach makes operation easier, samples represent only the average of all rain events which occurred throughout the collection interval.

It is also possible to collect individual raindrops, classify these by size and analyze for organic acids as well as for inorganic cations and anions.[82,83] This is accomplished with a so-called "Guttalgor." Raindrops entering this collection system during a brief opening period freeze immediately when they come into contact with liquid nitrogen and sink to the bottom of the vessel. The frozen raindrops may then be grouped by size by using sieves with different mesh ranging from 0.1 to 1.0 mm. The analysis of individual raindrops is accomplished by direct introduction into either capillary zone electrophoresis (CZE) or micro HPLC,[82,83] both of which require only small sample volumes for analysis.

3. Ice

Ice is sampled by drilling cores. These cores are divided into segments which are then analyzed individually. By relating each segment to its date of origin, information can be gained about the composition of the atmosphere in the past.[6] Ice-core samples most often originate in remote areas of the world and organic acid concentrations are usually quite low (ng/g), which makes these samples particularly susceptible to contamination. During the sampling process it is unavoidable to contaminate the outer parts of ice cores with organic acids and inorganic ions. These outer layers have therefore to be removed before analyzing the inner parts of the ice core.[84] Moreover, extended exposure of the melted ice to the laboratory atmosphere may result in contamination with organic acids.[84]

4. Drinking Water

Most drinking water treatment plants have special sampling ports at all stages of their treatment process. Before taking samples, the lines leading to these ports are flushed thoroughly until the water at the sampling port has the same composition as the water at this particular stage of the treatment process. In locations which do not have these sampling ports, bailers are used.

5. Wastewater

Obtaining representative samples from waste streams and anaerobic digesters is challenging since these materials are not homogeneous. Automatically collected 24-h composite samples reflect daily average performance, whereas grab samples give an indication of plant performance at the time of sampling.[25]

6. Groundwater

Groundwater samples are taken from wells by using either a bailer or a pump. To obtain a representative sample, water should be pumped until pH and conductivity are stabilized. Only then should the actual sample be taken.[13,14] Contamination can be avoided by working with clean equipment and storing samples in precleaned glass containers.

7. Soil Solutions

Soil sampling may be done by drilling cores or, alternatively, upper soil layers may be removed until the desired depth is reached and samples are taken. Green parts like grass, moss, etc., are removed.[110,111] Soil samples are either stored frozen or processed immediately, usually by centrifugation, to obtain the soil solutions which can then be filtered, preserved, and stored.[189]

8. Preservation

Organic acids are biodegradable and samples should therefore be preserved to ensure stable analyte concentrations during transport and storage. The choice of preservation mode is largely determined by sample matrix and compatibility with the instrumentation used for analysis. Often samples are stored at 4°C, although several authors found this to be insufficient to prevent organic acid losses in rain and drinking water samples.[33,42,73,86,87] In addition to storage at 4°C it is therefore necessary to add preservatives such as chloroform or mercuric chloride.[33] Note that addition of preservative alone is also not sufficient since organic acid degradation has been reported on with chloroform-preserved samples which were stored at room temperature.[103] Chloroform in combination with storage at 4°C is commonly used for samples which will be analyzed by liquid chromatography,[33,38,43,68,69,73,86,87] mainly anion exchange chromatography (AEC) or ion exclusion chromatography (IEC), whereas mercuric chloride is predominantly added to samples which are later analyzed by GC.[34,47,74,75,78,79,81] When using mercuric chloride for samples which will be analyzed by AEC or IEC, the mercury cation has to be removed prior to injection (e.g., by passing the sample through a cation exchanger) in order to prevent poisoning of the AEC or IEC column.[42] Another less common preservative for environmental samples is benzalkonium chloride which was recently employed to stabilize drinking water samples.[44] Freezing samples is another option for sample preservation which was shown to be reliable for long-term storage.[103] Ice cores are usually kept frozen and melted just before their measurement.[6,84] Filters used for sampling air may be stored at $-4°C$ or $-20°C$ until these are extracted with deionized water in preparation for their measurement.[37,57,60] Sometimes aqueous sample solutions (e.g., rain, groundwater, landfill leachates, or wastewater) are also stored at sub-zero temperatures.[14,29,82,83]

B. SAMPLE PREPARATION

Sample preparation procedures are determined by target analytes, their concentrations, the sample matrix, and the analytical instrumentation used for measurement. Procedures can range from a simple filtration to a sequence of steps which may include filtration, centrifugation, extraction, concentration, removal of interfering compounds, and derivatization. Simple uncomplicated sample preparation procedures are preferred since they are less prone to errors, less susceptible to contamination, and less time consuming. Sample preparation can be quite labor-intensive although on-line procedures have also been developed.[42,107,108]

1.   Filtration/Centrifugation

Samples with high solids contents such as, wastewater or landfill leachates are often centrifuged before the supernatant is processed further.[29,90–92] Soil solutions are obtained most commonly by centrifugation of moist soil, usually followed by filtration and sometimes further clean-up steps.[110–113,189] Methods involving direct injections of aqueous samples into AEC, IEC, or GC require the filtration of turbid samples, usually through filters with a standard pore size of 0.45 μm.[70] For some samples such as ultraclean water or drinking water this step may be omitted.[18,42,43,86] Polycarbonate filters and cellulose acetate filters may be used for filtration of samples with low organic acids concentrations since these only leach moderate amounts of organic acids which can be removed by flushing with deionized water (100 ml for polycarbonate, 500 ml for cellulose acetate) prior to sample filtration.[103] Glass fiber filters are not recommended since these release high concentrations of organic acids and colloids.[103]

2.   Extraction/Concentration/Clean up

Extractions are employed either to concentrate organic acids, to remove interfering compounds, or to obtain solutions of organic acids from sampling devices such as air filters.

Air sampling involves trapping of organic acids from the gaseous phase, the aqueous phase, and from airborne particulate matter. Different devices allow for the collection of organic acids either separated out by phase or as a total of all phases combined. If alkaline filters or denuders are used as the main trapping device, sample preparation consists in extraction with either deionized water or eluent followed by preservation.[55,60] Filters which are used to collect particulate organic acids prior to sampling the gaseous and aqueous phases may also be extracted by first wetting them with methanol and then extracting them with either deionized water or eluent.[60] Several applications use ultrasound for enhancing the extraction of organic acids from particulates trapped on filters.[61,63] However, it is reported that extended use of ultrasound may lead to analytical errors and simply soaking filters in water is preferred for organic acids extraction.[58,62]

A concentration step may be required for sample matrices with relatively low organic acid concentrations such as rain, drinking water, or seawater. Different approaches to concentrate samples have been taken depending on sample matrix, the specific organic acids to be analyzed, and analytical instrumentation.

When using AEC most aqueous matrices can be measured directly.[38,42–44,68,70–72,86] In cases where concentration is required, a concentrator column, usually an anion exchange resin, may be used on-line. This approach works well for samples with low inorganic anion concentrations[6,18,84] but problems arise (e.g., breakthrough) when even moderate concentrations of inorganic anions are present (Section II.C.2.a). An alternative for increasing method sensitivity in AEC is large volume injections of up to 1 ml which rely on the so-called "relaunch" effect (Section II.C.2.a),[43,86,87] meaning that in high capacity columns, analytes are collected as a relatively small sample band at the start of the column during injection, resulting in chromatograms without significant peak broadening.[129–131]

When using GC organic acids are typically transferred into an organic solvent followed by derivatization, which makes organic acids suitable for GC measurement. Exceptions are direct aqueous injections of water samples (e.g., wastewater) onto specialty GC columns (Section II.C.1). Transfer into organic solvent may be achieved by liquid–liquid extraction, evaporation at high pH and subsequent solvent addition, solid-phase extraction on anion exchange resins and subsequent elution with solvents, or aqueous derivatization followed by extraction (Section II.B.3.a).

Liquid–liquid extraction of short-chain organic acids, ketoacids, or dicarboxylic acids result in low and often unreproducible extraction yields due to the hydrophilic character of the analytes.[132] However, some authors report reproducible results for short-chain acids at mg/l concentrations after liquid–liquid extraction at pH 2, although extraction yields remain low.[48,119] Note also that organic solvents, namely diethylether, may be contaminated with organic acids.[106] An unusual variation in liquid–liquid extraction is the use of tri-n-octylphosphine oxide (TOPO) in methyl-$tert$-butylether (MtBE) to enhance extraction yields, e.g., of acrylic acid in marine waters[50] and of organic acids in aqueous solutions obtained from air collection chambers.[4] TOPO's very low solubility in water and its high polarity make it suitable for extraction of polar compounds. The extraction yield for acrylic acid was 40% and its detection limit after derivatization with pentafluorobenzyl bromide was estimated to be 3 nM.[50]

Organic acids including saturated and unsaturated mono- and dicarboxylic and also keto- and hydroxy acids may be concentrated by first adjusting the sample pH to approximately 9, thus ensuring dissociation of the acids.[34,74,78,79] Next, samples are concentrated by using a rotary evaporator and then blown off to dryness with nitrogen. The dry extract is redissolved in solvent and derivatized. When determining monocarboxylic acids in rain, good recoveries of 73 to 107% were reported.[74]

Solid-phase extraction on anion exchange resins is very rarely used.[132] Recoveries for the organic acids may show great variability which may be caused by incomplete removal from the resin, especially at low concentrations, or it may be attributed to breakthrough of the carboxylate ions during the extraction process if the matrix contains significant amounts of inorganic anions.

A unique approach has been taken to concentrate nanomolar concentrations of organic acids in seawater by employing static or dynamic diffusion using membranes.[16] Although time consuming, this procedure ensures removal of the majority of salts which might interfere with subsequent analysis if using, for example, AEC. However, in this application the concentrated acids have been measured by GC/FID (flame ionization detector) although AEC could have been employed equally as well.

Complex matrices such as soil solutions or landfill leachates usually require clean up where interfering compounds are removed prior to organic acid analysis. Procedures used include removal of humin-like substance by passing through special cartridges[27,29] or precipitation after acidification,[109] removal of carbonate by purging samples after their acidification,[22,23] and distillation when determining VFA.[54,118,121] Although pretreatment procedures are quite time consuming, only a few attempts have been made to automate these. Removal of Hg^{2+}, which was initially added for preservation purposes, has been automated by putting a cation exchanger in-line between the autosampler and injection loop prior to AEC injection.[42] Another very interesting on-line procedure for soil solutions combines clean up and extraction. Samples are first passed through a cation exchanger to trap metal ions and then through an anion exchanger after acidification to a pH value of 1 to remove interfering anions. This is followed by extraction of the organic acids into a TOPO-impregnated liquid membrane and their subsequent trapping in NaOH solution followed by AEC.[108] Pretreatment and chromatographic analysis take 35 min per sample. This method has been applied successfully in a study investigating low molecular organic acids at μmolar concentrations in soil solutions of beech forest.[107]

3. Derivatizations

Derivatizations are used to make the analytes suitable for a chosen instrumentation and to increase the sensitivity of the overall method. Ideally, derivatizations should be specific to the compounds of interest. They should not result in any by-product formation and achieve reproducible, preferably high yields. The excess reagent should not interfere with the determination of the derivative and the procedure should be quick and easy to execute. Although many derivatizations are available for organic acids,[133–136] only relatively few are used in the analysis of organic acids in environmental matrices.

a. Gas Chromatography (GC)

Many short-chain organic acids are thermostable and sufficiently volatile, thus fulfilling key requirements for GC measurement. However, their high polarity leads to severe peak tailing when employing standard capillary columns. Only when using specialty columns is it possible to analyze a range of these acids directly (Section II.C.1). Organic acids are therefore often derivatized to their less polar corresponding esters prior to their measurement on standard GC columns. Most derivatizations take place in nonaqueous solutions and organic acids have thus to be transferred into suitable solvents by either a concentration step or an extraction procedure (Section II.B.2) prior to their derivatization.

Esterification with alcohols using the Lewis catalyst BF_3 is a well-established procedure.[133] Organic acids are derivatized with BF_3/butanol to their corresponding butylesters and then extracted into hexane. For dicarboxylic acids derivatizations, excess butanol was removed to avoid interference with the analytes.[36,75,78] Butylesters are preferred over methylesters since these are more easily separated from the solvent peak due to their higher boiling points. Dicarboxylic acids as well as α-ketoacids were quantified with this procedure in rainwater samples.[36,75,78–80]

Methylation with diazomethane is rarely applied as the only derivatization step to short-chain organic acids.[132] It is, however, used in combination with o-(2,3,4,5,6-pentafluorobenzyl)-hydroxylamine (PFBHA) for ketoacids (see below). Once diazomethane has been generated, it is added to the extract which can then be injected in the GC, thus making this reagent rather convenient. Heating of the sample mixture is not required since the reaction takes place at low temperatures. The major disadvantages of diazomethane are its hazardous and explosive nature, and the necessity to generate it before it can be applied.[133]

A procedure specifically developed for ketoacids in drinking water utilizes an aqueous derivatization as a first step.[39–41,45,88,89] The ketofunction of the analyte is derivatized with PFBHA to the corresponding oxime which is much less polar than the original ketoacids. The oximes are then extracted at low pH with polar solvents such as MtBE and then dried with sodium sulfate which is followed by methylation of the carboxylate function using either diazomethane or

BF$_3$/methanol. The extracts are measured with GC/ECD (electron capture detector) which detects halogenated derivatives with a high sensitivity.

Derivatizations of organic acids with α,p-bromoacetophenone in the presence of dicyclohexyl-18-crown-16 as a catalyst to their p-bromophenacyl esters have also been employed.[34,74–77] The samples have to be conditioned by a cation exchange procedure prior to the reaction and excess reagent must be removed by passing the sample through a SiO$_2$ column after the reaction. The resulting p-bromophenacyl esters may be measured by GC and/or HPLC.

Studies aiming to identify unknown compounds, including substituted organic acids, often employ a sequence of extraction and derivatization steps where different functional groups are marked with specific reagents and are then measured by GC/MS.[41,45] For example, direct aqueous oximation of carbonyl functions with PFBHA is followed either by liquid–liquid extraction or by solid-phase extraction. Measurements of these extracts are used to identify aldehydes and ketones. Further methylation (e.g., BF$_3$/methanol) of the extracts allows for identification of ketoacids whereas sylilation with n-(tert-butyldimethylsilyl)-N-methylfluoracetamide (MTBSTFA) marks hydroxyl functions. Typical fragments when employing GC/MS with electron impact ionization (EI) are m/z 181 for carbonyl function, m/z 59 for methylesters, and m/z 75 for the sylilated hydroxyl functions. Measurements by GC/MS using chemical ionization (CI) result in dominant M$^+$ ions, which can be used to determine the molecular weight of the unidentified compound. Samples such as ozonated drinking water, ozonated paper pulp, and oxidized isoprene have been investigated with this approach.[41,45]

b. High Performance Liquid Chromatography (HPLC)

HPLC methods are rarely applied to the analysis of short-chain organic acids due to their poor UV-absorbance and their nonfluorescent character. Only high concentrations can be measured directly by HPLC in combination with UV, diode array, or fluorescence detection. To enhance method sensitivity, organic acids may be derivatized in pre- or postcolumn reactions. Although an abundance of derivatization methods for various compounds is available, especially for physiological important acids in biological fluids,[134,135,137,138] only very few have been applied to short-chain organic acids in environmental matrices (Table 13.3).

As mentioned under GC derivatizations, p-bromophenacyl esters of monocarboxylic acids may be measured by HPLC with UV detection.[34,74,77]

o-Phenylenediamine dihydrochloride (OPD) has been used to derivatize α-ketoacids and oxalate to their corresponding quinoxilinols and hydroxy-quinoxilinol.[81] The derivatives absorb UV light and are also fluorescent, allowing for either detection mode. Oxalic, pyruvic, and glyoxylic acid have been measured in rain using UV-detection.[81]

C. SEPARATION TECHNIQUES

Separation of analytes is accomplished by various chromatographic techniques ranging from GC to liquid chromatography (LC) techniques such as AEC, IEC, and HPLC. Also included in this discussion is capillary electrophoresis which has gained in interest.

1. Gas Chromatography

GC is a well-established technique with applications in many fields. Theoretical background and practical applications have been described extensively.[139–144] Organic acids analyzed by GC in environmental samples include VFA (i.e., C$_1$–C$_7$ aliphatic monocarboxylic acids), dicarboxylic acids, and also hydroxy- and ketoacids. An overview of GC applications is given in Table 13.4. Most organic acids are quite stable at high temperatures and some of the shorter-chain acids are even volatile. Thermostability and volatility are some of the key requirements for GC analysis

TABLE 13.4
Overview of GC Methods

AIR Including Atmospheric Precipitation

Matrix	Extraction/ Preparation	Injection[a]	Column[b]	Detector	Gas[c]	Compounds	Detection Limits	Ref.
(1) Air								
Indoor and outdoor smog chamber experiments	Transfer into aqueous phase, TOPO/MtBE extraction, PFBBr deriv.	60–250°C, 1 μl, on-column	RTX-5, 60 m, 0.32 mm i.d., 0.5 μm	Ion trap MS (EI, CI–CH$_4$, PFBOH)	na	C$_1$–C$_6$ saturated and unsaturated mono-carboxylic, hydroxy, oxocarboxylic, dicarboxylic acids	na	4
Aerosols	Extraction of filters, fractionation on SPC18, BF$_3$/ methanol or BF$_3$/propanol deriv.	na	HP Innowax, 30 m, 0.25 mm i.d., 0.25 μm	(a) FID (b) MS	na	C$_{10}$–C$_{18}$ monocarboxylic, C$_2$–C$_6$ dicarboxylic, glyoxalic, pyruvic acid	0.2–3.0 ng/m^3 (for 50 m^3 vol.)	65
Urban aerosols	Filter extraction, BF$_3$/butanol deriv.	na	UP-2, 25 m, 0.3 mm i.d.	FID	na	Saturated and unsaturated C$_2$–C$_{10}$ dicarboxylic acids	Analyzed low ng/m^3	66
(2) Rain, Fog, Mist								
Rain, fog	Concentrated under vacuum at pH 8.5–9, α,p-bromoacetophenone deriv.	200°C, 1 μl	DB5, 30 m, 0.25 mm i.d.	(a) FID, 300°C (b) MS	na	C$_1$–C$_9$ aliphatic acids	~0.1 μM	34,74–76
Rain, snow, aerosols	Concentrated under vacuum at pH 8–9, BF$_3$/butanol deriv.	300°C	(a) HP5, 25 m, 0.32 mm i.d., 0.52 μm (b) DB5, 30 m, 0.25 mm i.d., 0.25 μm	(a) FID (b) MS (EI)	na	C$_2$–C$_{10}$ ω-oxo-acids, pyruvic, saturated (C$_2$–C$_{11}$) and unsaturated (C$_4$, C$_5$, C$_8$) dicarboxylic acids	0.05 μg/l	79,80

Sample	Preparation/Derivatization	Injection	Column	Detector	Carrier gas	Analytes	Concentration	Ref.
Rain, fog, mist	Concentrated under vacuum at pH8–9, 14% BF$_3$/butanol deriv.	250°C, 1 μl, splitless	DB5, 30 m, 0.25 mm i.d.	(a) FID (b) MS	na	More than 20 C$_2$–C$_{10}$ dicarboxylic acids	na, but <0.8 μM	36,75,78
Rain, sewage, soil pore water	Concentrated under vacuum at pH 8–9, α,p-bromoacetophenone deriv.	190°C, 1 μl, on-column	packed glass column, 2 m, 3 mm i.d.	FID, 220°C	N$_2$	C$_1$–C$_5$ monocarboxylic acids	na, analyzed low μM conc.	77
Water								
(1) Drinking Water								
Drinking water	1. Aqueous PFBHA deriv. 2. MTBE extraction 3. CH$_2$N$_2$ deriv.	2 μl	SPB5, 30 m, 0.32 mm i.d., 0.25 μm df	ECD	na	Glyoxylic, pyruvic, ketomalonic acid	na, analyzed μg/l conc.	39,40,88
Ozonated fulvic acid solutions	1. Aqueous PFBHA deriv. 2. Diethylether extraction 3. CH$_2$N$_2$ deriv.	na	(a) CPSil5, 30 m, 0.32 mm i.d. (b) OV170I, 25 m, 0.25 mm i.d.	(a) ECD (b) MS	na	Glyoxylic acid	na	89
Ozonated drinking water	1. Aqueous PFBHA deriv. 2. MtBE extraction 3. CH$_2$N$_2$ deriv.	na	DB5, 30 m, 0.25 μm	MS (EI, CI)	He	C$_2$–C$_5$ ketoacids and others	na	41
Ozonated drinking water	1. Aqueous PFBHA deriv. 2. RP18 SPE 3. CH$_2$N$_2$ deriv. 4. MTBSTFA deriv.	180°C, 1 μl, splitless for 30 s	DB5, 30 m, 0.25 mm i.d., 0.25 μm	MS (EI CI)	He	Ketoacids and others	na	45
(2) Wastewater								
Activated sludge, bench scale exp.	Adjust to pH2 with H$_2$SO$_4$	250°C, direct aqueous injection	Stabilwax-DA, 15 m, 0.53 μm i.d., 0.5 μm	FID, 275°C	H$_2$	C$_2$-C$_6$ monocarboxylic acids	na, analyzed mg/l conc.	93,97
Wastewater, UASB reactor effluent	Filtration 0.45 μm, adjust to pH3 with H$_3$PO$_4$	200°C, direct aqueous injection	HP-FFAP, 10 m, 0.53 mm i.d.	FID, 250°C	He	C$_2$–C$_7$ monocarboxylic acids	na, analyzed mg/l conc.	90

Continued

TABLE 13.4
Continued

Matrix	Extraction/ Preparation	Injection[a]	Column[b]	Detector	Gas[c]	Compounds	Detection Limits	Ref.
Wastewater, bench scale	Centrifugation, supernatant filtered through 0.45 μm filters	250°C, direct aqueous injection	HP-Innowax, 15 m, 0.25 mm i.d., 0.15 μm	FID, 300°C	na	C_2–C_5 mono-carboxylic acids	na, analyzed mg/l conc.	96
Thermophilic aerobic digester	Centrifugation, supernatant acidified with H_3PO_4	150°C, direct aqueous injection	0.3% Carbowax 20 M/0.1% H_3PO_4 on Supelco Carbopack	FID, 200°C	He	C_2–C_4 mono-carboxylic acids	na, analyzed mg/l conc.	92
Wastewater, bench scale	na	Direct aqueous injection	Glass column, 2 m, 3 mm i.d., B-DA/4% Carbowax 20 M on 80/120 mesh Carbopack	FID	He	C_2–C_4 mono-carboxylic acids	na, analyzed mg/l conc.	95
Wastewater, benchscale UASB reactor	Centrifugation, supernatant acidified with 3% formic acid	200°C, direct aqueous injection	Glass column 2 m, 4 mm i.d., 10% Fluorad FC 431 on 100–120 mesh Supelcoport	FID, 280°C	N_2[d]	C_2–C_4 mono-carboxylic acids	na, analyzed mg/l conc.	91
Wastewater, artificial wastewater	$NaHSO_4$ addition, 2-ethylbutyric acid as internal std.	85°C, 30 min, headspace	FFAP, 30 m, 0.25 mm i.d., 0.25 μm	FID, 200°C	na	C_2–C_5 mono-carboxylic acids	0.3–3.7 mg/l	97
Wastewater, deionized water	(a) Direct SPME (b) Headspace SPME with PDAM deriv.	(a) 275°C, 3 min desorpt. (b) 300°C, 4 min desorpt.	SPB-5, 30 m, 0.25 mm i.d., 1 μm	(a) + (b) FID, 300°C	na	C_2–C_{10} mono-carboxylic acids	0.02–760 μg/l (direct SPME in DI standards)	98

Sample	Preparation/extraction	Injection	Column	Detector	Carrier gas	Analytes	Detection limit	Ref.
Wastewater, deionized water	(a) Headspace SPME (b) Headspace SPME with internal standard	(a) 260°C, 3 min desorpt. (b) 300°C, 5 min desorpt.	(a) and (b) FFAP, 30 m, 0.25 mm i.d. 0.25 μm	(a) FID, 260°C (b) MS (PCI–CH$_4$, NCI–NH$_3$)	(a) H$_2$ (b) He	C$_2$–C$_7$ mono-carboxylic acids	(a) 6–675 μg/l (b) 2–150 μg/l	99,100
(3) Other Water								
Groundwater at superfund site	Adjusted to pH < 2, CH$_2$Cl$_2$ extraction	250°C	FFAP, 30 m, 0.53 μm i.d.	MS	He	C$_2$–C$_7$ mono-carboxylic acids	na	48
Groundwater at gas-spill site	Diethylether extraction of freeze dried samples	25–200°C, 0.5 μl,	DB FFAP, 30 m, 0.32 mm id.	FID, 240°C	He	C$_2$–C$_7$ mono-carboxylic acids	na	14,47
Seawater	Concentration by static diffusion, acidification to pH2	Direct aqueous injection, 200°C, 1 μl	FFAP, 30 m, 0.53 μm i.d.	FID, 220°C	He	C$_2$–C$_5$ mono-carboxylic, pyruvic, acrylic, benzoic acid	10 nM, except for acetate 750 nM	16
Seawater, soil pore water	Clean-up with hexane, 0.5% TOPO in MtBE extraction at pH2, PFBBr deriv.	225°C, 0.5 μl		ECD, 325°C	H$_2$	Acrylic acid, other acids possible	3 nM	50
			Soil					
Wetland sediment pore water	Centrifugation, supernatant pH raised to 11, dried at 95°C, redissolved in H$_3$PO$_4$	180°C, 1 μl, aqueous injection	Precolumn: 1 m, 0.53 mm i.d., anal. column: Nukol, 15 m, 0.53 mm i.d.	FID, 180°C	N$_2$	C$_2$–C$_4$ mono-carboxylic acids	na, analyzed μM to high nM	106
Sediment pore water, (a) better than (b)	Centrifugation, storage of frozen supernatant, acidified	(a) 150°C (b) 225°C 0.5–3 μl, splitless, direct aqueous inj.	FFAP-CBwax (HP): (a) 10 m, 0.53 μm i.d., 1 μm (b) 25 m, 0.32 mm i.d., 0.33 μm	FID, (a) 200°C (b) 260°C	He	C$_2$–C$_7$ mono-carboxylic acids	2–5 μM	116

Continued

TABLE 13.4
Continued

Matrix	Extraction/ Preparation	Injection[a]	Column[b]	Detector	Gas[c]	Compounds	Detection Limits	Ref.
			Other					
(1) Landfill Leachate								
Landfill leachates	Acidification, distillation	215°C, aqueous injection	Glass column: 152 mm, 2.0 mm i.d.; 10% SP-1200–1% H_3PO_4 on 80–100 mesh Chromosorb WAW	FID, 250°C	N_2	C_2–C_4 mono-carboxylic acids	na, analyzed mg/l conc.	118
Low level radioactive waste leachates	Acidification, diethylether extraction	250°C, 10 μl	Stainless steel column GP 10% SP-1200/1% H_3PO_4 on 80–100 mesh Chromosorb WAW	FID, 250°C	He	C_2–C_7 mono-carboxylic acids	na, analyzed mg/l conc.	119
Landfill monitoring wells	Dilution	200°C, split injection 1:30	DB-FFAP, 30 m, 0.32 mm i.d., 0.25 μm df	MS	He	C_2–C_8 mono-carboxylic acids	1–8 mg/l	120
(2) Biological Material								
Biological samples, (b) better than (a)	Vacuum distillation, acidification with formic acid	Direct aqueous injections (a) 220°C, 1 μl, split 1:30	(a) HP Supelco-wax 10, 30 m, 0.32 mm i.d., 0.25 μm df	FID (a) 260°C (b) 210°C (c) na	(a) He (b) He (c) He[4]	C_2–C_7 mono-carboxylic acids	0.05 mM	54

Sample	Sample preparation	Injection[a]	Column[b]	Detector	Carrier gas[c]	Analytes	Concentration	Ref.
Biological specimens	Alcoholic extraction, final extract in water/HCl/acetonitrile	1 μl, on-column; (b) 200°C, 1 μl, splitless; (c) 5 μl	Carbowax 20 M, 25 m, 0.32 mm i.d., 0.3 μm; also used precolumn; (b) DB-Wax 15, 15 m, 0.53 mm i.d., 1.0 μm df; (c) glass column packed with Chromosorb 101	FID 300°C	H$_2$	C$_2$–C$_6$ monocarboxylic acids	20 nM	53
Silage juice (b) better than (a)	Filtration, dilution with oxalic acid (1:10)	Direct aqueous injection, 0.5 μl, on-column	(a) DB-Wax, 15 m, 0.53 mm i.d., 1 μm df; (b) FFAP 10 m, 0.53 mm i.d., 1 μm df	FID 200°C	He	C$_2$–C$_7$ monocarboxylic, lactic acid	na, analyzed mg/l conc.	31
Cellulose polymers	Aqueous extraction of CH$_2$Cl$_2$ solution of cellulose, centrifugation	Direct aqueous injection, on-column	Carbowax 20 M, 10 m, 0.53 mm i.d., 1 μm df	Thermal conduct. detector	He	C$_1$–C$_4$ monocarboxylic acids	10 mg/l	52

Information for a certain parameter not available, if not given. na, not available; DI, deionized water; CI, chemical ionization; ECD, electron capture detector; EI, electron impact ionization; FID, flame ionization detector; MS, mass spectrometer; MtBE, methyl *tert*-butylether; MTBSTF, *n*-(*tert*-butyldimethylsilyl)-*N*-methylfluoracetamide; PDAM, 1-pyrenyldiazomethane; PFBBr, pentafluorobenzyl bromide; PFBOH, pentafluorobenzyl alcohol; SPME, solid-phase microextraction; TOPO, tri-*n*-octylphoshine oxide; UASB, upflow anaerobic sludge blanket reactor.

[a] Injector temperature, injection volume, type of injection.
[b] Column brand name, length, inner diameter, film thickness.
[c] Carrier gas.
[d] Saturated with formic acid.

and, hence, the more volatile organic acids may be measured directly. Other acids have to be derivatized in order to increase their volatility, thus making them amenable for GC. Further reasons for employing derivatizations include decrease in polarity to avoid chromatographic complications due to the polar character of these analytes and increase in sensitivity when using halogenated derivatives in conjunction with an ECD. Different derivatization options for GC are discussed in more detail in Section II.B.3.a. Most derivatizations are accompanied by extraction and clean-up steps which are described in Section II.B.1–2.

Derivatized acids include monocarboxylic, dicarboxylic, and keto- and hydroxyacids. These are usually measured under standard GC conditions (Table 13.4) where split, splitless, or on-column injections are used in combination with standard capillary columns. The type of injection employed depends largely on the concentrations to be measured and the solution matrix. Flame ionization detectors (FID) are used as nonspecific detectors,[35,36,65,66,77,80] whereas ECDs are utilized to measure halogenated derivatives, thus improving detection limits.[39,40,45,50] In both cases retention times are used for compound identification. Several applications employ mass spectrometric (MS) detection,[4,34,41,45,65,74,78,79,89,145] especially where identification of new constituents or by-products is required, thus giving additional compound information through mass spectra.

Direct sample introduction without any derivatization is used as an alternative to derivatization and is most commonly applied to VFA. GC introduction techniques include direct aqueous and solvent injections,[31,90,94,96,106,116] headspace,[97] and more recently solid-phase microextraction (SPME).[98–100] Specially designed capillary GC columns with polar phases are typically followed by FID[16,47,90,96,116] and in some cases by MS detection.[48,120]

Aqueous samples are usually acidified prior to direct aqueous injection[16,54,90,106] so that the organic acids are in their protonated form. This increases their volatility, reduces adsorption effects and results in better peak shapes. However, some authors report that acidification leads to a deterioration of the films in capillary columns,[52,97,120] whereas others did not observe any adverse effects.[53] Some sample matrices require clean up and the resulting organic solvent extracts (diethylether or dichloromethane) are injected directly without further derivatization.[14,47,48,53,119]

The separation of underivatized VFA is now commonly accomplished on specialty capillary GC columns. These have replaced packed GC columns which were frequently used in the past.[54,90–92,95,118] Capillary fused silica columns designed for organic acid separation are of high polarity. The chemically bonded film is either comprised of polyethyleneglycol (brand names: DB-Wax, HP Wax, Stabilwax, Supelcowax 10...) or acid-modified polyethyleneglycol (brand names: DB-FFAP, HP-FFAP, Stabilwax-DA, Nukol...). These columns are quite sensitive to heat and oxygen compared to standard columns with less polar films. Their maximum operating temperatures range from $\sim 200°C$ for FFAP-type columns to $\sim 280°C$ for Carbowax-type columns. According to product information supplied by the manufacturer, longer-chain acids can also be separated on these columns. However, attempts to determine dicarboxylic or aromatic acids with these columns resulted in poor chromatography.[14,120]

Separation of VFA on polar capillary GC columns is most often followed by FID.[14,16,31,47,53,54,90,94,96,106,116,119] When employing direct aqueous injections this configuration results in fairly high detection limits (low mg/l). However, it is still adequate for many applications in matrices such as wastewater or diluted sludge which have high organic acid contents (mg/l). Advantages of direct injections on GC/FID are its simplicity, a fast turnaround time and equipment availability. Disadvantages include potential "memory effects," contamination problems, and the fact that formic acid cannot be measured in this configuration.[116] Direct GC measurement of formic acid has only been reported when using a thermal conductivity detector (TCD).[52] Contamination problems, such as deteriorating column performance due to presence of other compounds originating from sample matrix, can be controlled by a regular maintenance program. "Memory effects" may occur when organic acids are adsorbed onto metal or other parts of the GC. A subsequent random release of these acids may lead to irreproducible retention times and unreliable quantification.[53,54,116,120] Measures to control this problem include purging of the adsorbed compounds or contaminants by heating the GC to its maximum temperature at the end of each run,[53] or conditioning of the column by injecting a formic acid solution repeatedly before starting analysis, thus saturating the film and reducing potential irreversible adsorption.[116] Others recommend sample acidification[54] or injections of blanks in between samples[120] whereas elsewhere it is reported that none of this is necessary.[31]

VFA analysis with headspace introduction has been explored as an alternative to direct aqueous injections.[97] The method was developed for routine analysis of wastewater. Carry-over effects were minimized by using three aquatic wash cycles in between sample injections and headspace matrix effects are accounted for by employing 2-ethylbutyric acid as internal standard. However, the authors advise that standard addition should be employed if severe sample matrix effects are observed.[97]

Several papers investigated the use of SPME for VFA analysis in wastewater and in air.[98–100] Briefly, a fiber is exposed to the sample headspace or inserted directly into the sample. Analytes adsorb onto the fiber and are subsequently desorbed at high temperatures in the GC injection port. SPME is a solvent-free technique which introduces less potential contaminants into the GC compared to direct injections. SPME is also rapid since no further sample preparation steps are required. It may be used for routine analysis provided that the specific autosampler required for this method is available and that the optimized method conditions are suitable for autosampler application. Further information on principles and other applications of this technique can be found elsewhere.[146,147] Parameters which have been optimized for VFA analysis are fiber coating, fiber exposure time, sample temperature, sample pH, sample agitation, potential salt addition, and desorption parameters. Surrogate standards employed for VFA analysis were 2-ethylbutyric acids[99,100] for GC/FID or GC/MS and ^{13}C-labeled organic acids[98] for GC/MS. The method was optimized using standards in deionized water and only a few wastewater samples were analyzed as examples.

In general, it is surprising that not even half of the GC methods listed in Table 13.4 make use of internal or surrogate standards. If choosing the right compound, a surrogate standard can account for matrix effects, fluctuating recoveries, or changes in GC conditions. Ideally, it should have properties as similar as possible to the analyte, but it should not be present in the sample matrix itself. Deuterated or carbon-labeled compounds are ideal when employing MS detection.[16,98,120] Fluorinated compounds can be a cost-effective alternative, especially when using an ECD. Other compounds utilized as internal standards are 2-ethylbutyric,[97,99,100] iso-caproic,[14,47,53] 2-methylvaleric,[54] 2-bromodecanoic,[65] and hexanoic acid.[4,14]

A maintenance program for the GC system is crucial for achieving reliable results when using methods with, and especially without, derivatization. Changing of the injection port liner with a clean deactivated one is advisable,[47,120] as is the regular change of injection port septa. A sample filtration or extraction step reduces the amount of particulate matter introduced into the GC system and thus reduces unnecessary maintenance tasks. It is also common practice to cut off a piece at the front end of a capillary column to restore chromatographic performance. Instead, use of a retention gap or precolumn should be considered. A retention gap is a deactivated piece of fused silica which is easily connected to the analytical column with a press-fit connector. By shortening the retention gap, or replacing it altogether, chromatographic performance can be restored while extending the life of the analytical column.[53,141] When using a high polarity solvent such as water, a retention gap with medium to high polarity deactivation should be employed.[141]

2. Liquid Chromatography

LC techniques, i.e., IEC and to a lesser degree HPLC, are widely used for the analysis of organic acids in environmental matrices. In particular, ion exchange chromatography and IEC are well suited for aqueous matrices. These do not exert any thermal stress on the analytes and usually require only simple sample pretreatment compared to GC. In-depth theoretical and practical information with respect to these techniques can be found in various publications.[129–131,148]

a. Anion Exchange Chromatography (AEC)

Separation by ion exchange chromatography is primarily based on the partitioning of the analytes, i.e., ions, between the mobile phase and the ion exchange groups bound to the surface of the stationary phase. Secondary mechanisms for ions with hydrophobic characteristics may involve adsorption processes. Ion exchange chromatography is normally applied to the analysis of inorganic ions. However, organic acids dissociate readily at the high pH of the mobile phase to their corresponding carboxylate anions and thus may be analyzed by AEC. An overview of AEC methods used for organic acid analysis in environmental matrices is given in Table 13.5. Most methods employ direct injection via a sample loop followed by anion exchange separation with suppressed conductivity detection. Organic acids determined by AEC are monocarboxylic, dicarboxylic, and hydroxy- and ketoacids. An additional benefit of using AEC is the fact that, depending on the sample matrix, inorganic anions may be quantified together with organic acids.[56,63,70,71,84]

Most AEC methods use direct sample injections vial a sample loop, thus achieving adequate detection limits.[42,43,69–71,86,87] Aqueous sample matrices usually do not require any sample pretreatment other than possibly filtration prior to injection. For matrices such as airborne particles, soil, or filters from air collection devices, organic acids are commonly transferred into an aqueous solution which can then be injected directly into the ion chromatograph.[57,59,107–109]

Separation of organic acids as carboxylate anions is accomplished on anion exchange columns. The most widely used columns consist of a copolymer core — usually divinylbenzene crosslinked with polystyrene — to which substituted quaternary ammonium groups are bound as anion exchange sites. Columns are characterized by their particle diameter, amount of crosslinking, and the capacity and properties of their exchange sites.[129,130] Column materials are continuously improved and new columns designed for organic acid analysis are regularly introduced. When implementing an organic acid method, these new columns should also be considered beyond the ones mentioned in this text. In order to protect the analytical column, most applications utilize guard columns in-line prior to the analytical column.[42,51,57,58,63,68,69,72,86,109] Guard columns contain the same stationary phase as the analytical column but are much shorter. These can easily be replaced once deterioration in chromatographic performance or excessive pressure build-up is noticed, which is much more cost effective than having to replace the analytical column itself.

Eluents frequently used in conjunction with anion exchange columns are hydroxide, carbonate, bicarbonate, or borate. Eluents have to show an affinity to the sample ion and the stationary phase while being compatible with the detector. Isocratic elution, which has dominated in the past, is adequate when only few anions with the same charge have to be separated.[38,57,68–70] More recent applications often use gradient elution for more complex samples, which makes it possible to separate anions ranging in charge from -1 to -3.[6,44,51,56,61,63,71,86,109] In addition, many applications use eluent purifying columns which are installed in the eluent line prior to the injection loop.[6,43,44,70,86,87,109] These strong anion exchange columns (e.g., ATC-1 or ATC-3, Dionex) will retain impurities such as carbonate, sulfate, and chloride, which are present in even high-grade (e.g., hydroxide) eluents. Without purifying columns these impurities may first concentrate at the head of the guard or analytical column and then with increasing eluent strength elute off the column, resulting in potentially large, interfering peaks.

Virtually all methods (Table 13.5) use suppressed conductivity detection which allows for sensitive detection of anions by suppressing the background conductivity of the mobile phase, usually employing membrane suppressors. In brief, a membrane which is only permeable for protons is placed between the mobile phase and the regenerant which is an acid. Protons pass from the regenerant through the membrane into the eluent, thus neutralizing it just prior to entering the conductivity cell. Different forms of membrane suppressors are available and are described in

TABLE 13.5
Overview of AEC Methods

Matrix	Injection	Column	Eluent	Detector	Compounds	Detection Limits	Comments	Ref
(1) Air								
AIR Including Atmospheric Precipitation								
Gas phase	na	AS11	NaOH gradient	Cond.	Acetic, formic, pyruvic, oxalic acid	a	Also detection of Cl^-, NO_2^-, SO_4^{2-}, sampling with cryogenic trapping	56
Gas and particulate phase	20 μl	AG9, AS9	$Na_2B_4O_7$ 1 M	Cond. A	Formic, acetic acid	0.1 ppbv[a]	Filters stored in freezer	57
Gas and particulate phase	500 μl	AG5-A, AS5-A	na	Cond. A	Formic, acetic, glycolic, pyruvic, oxalic, β-hydroxybutyric acid	45–102 $\mu g/l$ (standards)[a]	Preservation of extracts with $CHCl_3$, confirmation with CZE	58
Gas and particulate phase	100 μl	AG5-A, AS5-A	NaOH 3 mM, 20 mM	Cond. A	Formic, acetic, pyruvic, glyoxylic, succinic, malonic, oxalic acid	0.07–0.19 ppbv[a]	Preservation of extracts with $CHCl_3$, confirmation with IEC	59
Gas and particulate phase	50–500 μl	AG4A, AS4A	$Na_2B_4O_7$ or $Na_2CO_3/NaHCO_3$	Cond. A or B	Formic, acetic acid	a	Comparison of different sampling techniques	60
Particulate phase	na	AS11	NaOH/methanol gradient	na	C_2–C_5 dicarboxylic, glyoxylic acid	10–50 ng[a]	Filters stored frozen	61
Particulate phase	500 μl	AS5A	NaOH gradient	Cond. A	Formic, acetic, oxalic acid	na	Investigated impact of sonication on aqueous extraction of filters	62
Particulate phase	300 or 1000 μl	AG11, AS11	NaOH gradient	na	C_2–C_5 dicarboxylic acids	a	Storage of filters at room temperature, measured also Cl^-, NO_3^-, SO_4^{2-}	63
Particulate phase	na	ATC-5A, AS5	Borate and hydroxide gradients	Cond. A	C_1–C_4 monocarboxylic, glycolic, lactic, β-hydroxybutyric acid	0.016–0.082 mg/l	Developed and compared AEC and CZE methods, investigated coelution	64

Continued

TABLE 13.5
Continued

Matrix	Injection	Column	Eluent	Detector	Compounds	Detection Limits	Comments	Ref
(2) Rain, Fog, Mist								
Rain	na	AG4A, AS4A	$Na_2B_4O_4$ 1.5 mM	Cond. A	Formic, acetice, (pyruvic) acid	0.2 μM	Focus on formic and acetic acid	68
Fog, lake, sediment pore water, rain	10, 25 or 50 μl	ATC-1 (a) AG11, AS11 (b) AG10, AS10	(a) Borate gradient (b) Borate 7 mM	Cond. A, cond. B	C_1–C_5 moncarboxylic, 2- and 3-hydroxy-butyric, lactic, glycolic, pyruvic, oxalic acid	0.5–1 μM	Simultaneous detection of F^-, Cl^-, NO_2^-, CO_3^{2-}, PO_4^{3-}, Br^-, SO_3^{2-}, SO_4^{2-}	70
Rain	10 μl	AG11, AS11	NaOH 0.5 mM	Cond. A	Formic, acetic acid	0.2 μM	Acetic acid not always resolved from hydroflufic acid	69
Mist	na	AS4	$NaHCO_3$ 0.4 mM	Cond. A	Formic, acetic, pyruvic acid	0.02–0.1 μM	Detection limits for 60 min sampling interval: 5–20 pptv	38
Rain, snow	100 μl	AG9, AS9	Na_2CO_3/$NaHCO_3$ gradient	Cond. A	Formic, acetic, oxalic acid	20–100 $\mu g/l$	Simultaneous detection of F^-, Cl^-, NO_2^-, NO_3^-, Br^-, SO_3^{2-}, SO_4^{2-}, PO_4^{3-}	71
Snow	700 μl	(a) AG4A, AS4A (b) AG11, AS11	(a) $NaHCO_3$ 0.5 mM (b) NaOH/methanol gradient	Cond. A	Formic, acetic, propionic, glycolic, lactic acid	0.2–1.6 $\mu g/l$	Investigates compound coelution and its implications	72
(3) Ice								
Ice cores	5 ml	ATC-1; TAC1[b] (a) AG5, AS5 (b) CG10, Pax 500	(a) + (b) NaOH, gradient	Cond. A	Formic, acetic, oxalic, glycolic acid	0.2–0.6 ng/g	Simultaneous detection of F^-, Cl^-, NO_2^-, NO_3^-, SO_4^{2-}	6,84

Water

	Injection	Column	Eluent	Detection	Acids	Concentration/LOD	Comments	Ref.
(1) Drinking Water								
Ozonated drinking water	20 μl	ATC-1, AG11, AS11	NaOH gradient	Cond. C	Formic, acetic, oxalic acid	2–3 μg/l, MRL 15 μg/l	Hg^{2+} for sample preservation,[42] Hg^{2+} removal through H^+-cartridge essential,[42] suppressor cleaned daily with 0.5 N H_2SO_4	42,44
Ozonated drinking water	760 μl	ATC-1, AG10, AS10	NaOH gradient	Cond. C	Formic, acetic, butyric, β-hydroxybutyric, glycolic, pyruvic, α-ketobutyric, oxalic acid	1–9 μg/l	Oxalate determination in matrices with high SO_4^{2-} requires switching technique	43,86
Ozonated model water	25 μl loop, 5–10 ml conc.	AG11[b], AG11, AS11	NaOH gradient	Cond.	Formic, acetic, glycolic, oxalic acid	Loop 20–40 μg/l, conc. 0.1–0.5 μg/l	No testing on 'real' drinking water samples described	45
(2) Other Water								
Power plant water	5 or 10 ml	AC10[b], AS10	NaOH 0.085 mM	Cond. A	Formic, acetic, oxalic acid	Below 1 μg/l	Online preconcentration, simultaneous determination of inorganic anions	18
By-products of organic pollutant degradation in water	50 μl	AS5A	NaOH gradient	Cond. A, UV 200 nm	Formic, acetic, glycolic, oxalic, malonic maleic, fumaric acid	0.001–0.006 mM	Confirmation of identification by using conductivity/UV response ratios	30
Photoformation in brown water DOM	100 μl	AG11, AS11	KOH gradient	Cond. B	Formic, acetic, oxalic, malonic, succinic, pyruvic acid	0.2 μM	Confirmation of identification with CZE	51
Soil								
Soil solutions, root exudates	50 μl	AG11, AS11	NaOH gradient	Cond. B	Formic, acetic, lactic, pyruvic, oxalic, malic, ct.ric acid	50–250 nM	Online sample prep incl. cation and anion exchange columns for removal of interfering compounds, extraction with TOPO impregnated membrane	107,108

Continued

TABLE 13.5
Continued

Matrix	Injection	Column	Eluent	Detector	Compounds	Detection Limits	Comments	Ref
Soil extract	50 μl	ATC-1, AG5A, AS5A	NaOH gradient	Cond. A	Formic, acetic, oxalic, pyruvic, L-malic, succinic, fumaric, tartaric, citric, *trans*-aconitic, gluconic, α-ketoglutaric, glutaric acid	103–1286 nM	Extraction with NaOH, centrifugation, filtration, precipitation of humic acids, reextraction with ethylacetate, transfer into H_2O for injection	109
Soil extract	50 μl	Omnipac Pax 100 guard and analytical column	H_2O/acetonitrile gradient	UV 254 nm	14 different aromatic acids	37–1729 nM	Extraction with NaOH, centrifugation, filtration, precipitation of humic acids, reextraction with ethylacetate, transfer into H_2O for injection	109

na, not available; CZE, capillary zone electrophoresis; DOM, dissolved organic matter; MRL, Minimum reporting level; TOPO, tri-*n*-octylphoshine oxide. Columns manufactured by Dionex if not mentioned otherwise: AG, IonPac A# = guard column, 4 × 50 mm; AS, IonPac AS # = analytical column, 4 × 250 mm; ATC-1, anion trap column used to trap carbonate and other anions from eluent, 9 × 24 mm; TAC1, concentrator, 4 × 35 mm; AC10, concentrator, 4 × 50 mm. Detection: cond., conductivity detection, no further details available; cond. A, chemical suppressed conductivity detector using H_2SO_4 as regenerant; cond. B, autosuppressed conductivity detection in recycle mode; cond. C, autosuppressed conductivity detection in external water mode; UV, ultraviolet.

[a] Detection limits vary with sampling time.
[b] Used as concentrator column.

detail elsewhere.[149,150] Direct and indirect UV detection are rarely used with AEC due to their inferior sensitivity, whereas amperometric detection has recently been gaining some interest.[129–131]

Some matrices such as ice or drinking water contain very low concentrations of organic acids (low μg/l). In order to achieve the necessary detection limits, methods for these matrices make use of either concentrators or large volume injections.

On-line concentrators are elegant tools to increase method sensitivity without using time- and labor-intensive pretreatment steps.[18,45,84] A concentrator column is a short, medium- to high-capacity strong anion exchanger which is installed instead of a sample loop. Volumes of up to 10 ml of aqueous sample have been loaded onto concentrators. Inorganic anions and carboxylate anions are retained on the concentrator during the loading process due to the minimal elution power of water on this resin. Trapped anions are then transferred onto the analytical column by using a hydroxide eluent which also facilitates analyte separation on the analytical column during the chromatographic process. Loading and transfer process are performed countercurrent to each other so that the sample band entering the analytical column remains narrow, thus ensuring acceptable peak shapes in the final chromatogram.[130] Autosampler use is recommended for achieving reproducible injections and for overcoming the increased backpressure caused by the concentrator. Carry-over can be avoided by flushing transfer lines in between injections. Employing a concentrator works well for samples with low organic acid and low inorganic anion concentrations such as ice.[18,84] Detection limits for organic acids can be as low as 1 μg/l with the added benefit that inorganic anions may be quantified at the same time. However, the use of concentrators is limited by the fact that the analytes have to be retained quantitatively. Problems arise when inorganic anions are present even in moderate amounts (e.g., drinking water) since these can cause breakthrough of the more weakly retained carboxylate anions during the loading process, thus leading to irreproducible results. Breakthrough of carboxylate anions should always be considered possible when employing a concentrator — especially in the method development phase and later, if significant changes in the sample matrix are experienced. However, if interfering anions such as chloride and sulfate are removed prior to injection, preconcentration may be applied successfully to the analysis of organic acids.[45]

An alternative for increasing method sensitivity is the injection of larger sample volumes.[43,58,60,62,63,72,86,87] If the capacity of the analytical column is sufficiently high, volumes of up to 1 ml may be injected via a sample loop without significant peak broadening. This "relaunch" effect ensures that the analytes are collected as a relatively small sample band at the start of the column during injection, which is a prerequisite for achieving sharp peaks.[129–131] Method detection limits for this approach are in the low μg/l range for samples such as drinking water and snow. Again, higher concentrations of inorganic anions may interfere with the analysis, this time by coelution and masking of specific organic acids. Careful column selection and optimization of the gradient can alleviate this problem to a certain extent.[86,87] If coelution still occurs, another option is to utilize the "heart-cut" technique which has been successfully used for drinking water analysis of coeluting sulfate and oxalate. In brief, a large volume injection is followed by gradient separation of most carboxylate and inorganic anions. The detector effluent is then redirected to a concentrator for the time window of the coeluting peaks. Sulfate and oxalate are retained on the concentrator and then reinjected onto the same analytical column using sodium hydroxide for the transfer. Separation is achieved by gradient elution optimized for these anions.[43]

Peak identification and potential coelution of organic acids are not always given due consideration, although AEC has much inferior separation efficiencies compared to GC. Ideally, organic acid identification, which is typically done by retention time comparison, should be confirmed with another independent method. Several publications investigating rain and atmospheric precipitation validated their AEC results with either IEC or CZE.[30,51,58,59] Other publications completely ignore the possibility of coelution of different organic acids, although it has been shown to occur.[43,64,72,86,87,116] Separation of coeluting acids may be achieved by taking a structured approach to improving separation by using capacity factors.[64]

Overall, AEC is a well-developed separation technique which is widely applied for organic acid analysis. Some of the current trends include the implementation of eluent generators;[149] development of and increased accessibility to ion chromatograph/MS systems which are now commercially available;[149,151] miniaturization of complete ion chromatographic systems;[152] and the development of monolithic columns which have been used for fast analysis of inorganic anions.[153]

b. Ion Exclusion Chromatography (IEC)

Separation in IEC is a complex process involving Donnan exclusion, size exclusion, adsorption, and polar interactions such as hydrogen bonding. IEC columns used for organic acid analysis are usually comprised of copolymers carrying strong cation exchange groups which are negatively charged. Various inorganic or organic acids are employed as eluents. The charge and the size of the analytes determine if separation can be achieved by IEC. Analytes with low pK_a values such as inorganic acids are predominantly dissociated into their anions at the pH of the eluent and thus repelled by the negative charge of the cation exchange groups (electrostatic or Donnan exclusion). Anions such as chloride or sulfate will therefore elute with the system peak. Larger molecules will also not be able to penetrate the outer layer due to steric hindrance (size exclusion). However, relatively small, neutral molecules such as weak organic acids, which are predominantly in their undissociated form at the pH of the eluent, can pass through the negatively charged outer layer and interact with the stationary phase through hydrophobic, polar, and $\pi-\pi$ interactions, leading to their separation. This specific separation mechanism makes IEC very suitable for the analysis of weak organic acids in complex sample matrices with high ionic strength and it has found widespread use, for example in food analysis.[129,130]

Organic aids monitored by IEC include mono-, di- and tricarboxylic acids, and also hydroxy acids (Table 13.6). The elution order of these acids is quite predictable. The higher the pK_a value and the higher the molecular weight, the longer are the acids in a homologue series retained on the column. Monocarboxylic acids therefore elute before dicarboxylic acids, saturated acids before unsaturated acids, and branched acids before their straight-chain isomers. Because of strong hydrophobic interactions with the resin, aromatic acids always display long retention times unless eluent modifiers are employed.[129,130]

Organic acids with relatively low pK_a values such as oxalic acids elute close to, or even coelute with, the system peak and may therefore not be quantifiable by IEC. Also, weak inorganic acid (e.g., carbonate, borate) may interfere with the analyses of organic acids when using conductivity detection. If present in sufficiently high enough concentrations, these inorganic acids may potentially mask analytes. Some samples such as landfill leachates have carbonate removed prior to their analyses by IEC.[27] Table 13.6 gives an overview of IEC methods utilized for organic acid analysis in environmental matrices. A more detailed overview has been published by Fischer.[155] Many environmental applications involve complex matrices such as landfill leachates or soil solutions. Only occasionally is IEC applied to simpler matrices such as rain or groundwater. IEC systems utilized in these methods comprise injection loop, separation column, usually with an acid as eluent, and either direct UV, nonsuppressed, or suppressed conductivity detection.

Most environmental samples require some pretreatment, usually in the form of filtration and centrifugation, before these are injected via a sample loop. IEC separation of organic acids is achieved on cation exchange columns which are characterized by their particle diameter, substrate crosslinking, ion exchange capacity, type of functional group, and hydrophobicity. The most commonly used columns are comprised of fully sulfonated, crosslinked divinylbenzene/polystyrene copolymers (e.g., HPICE-AS6 or -AS1, Dionex). However, a comparison study found other columns to be equally as effective for organic acid separation.[156] Unmodified silica gel columns

TABLE 13.6
Overview of IEC methods

Matrix	Sample Preparation	Injection	Column	Eluent	Detector	Compounds	Detection Limits	Ref.
AIR Including Atmospheric Precipitation								
Gas and particulate phase	Scrubber, filters, aqueous extraction, $CHCl_3$ preservation	na	HPICE-AS5, Dionex	Heptafluorobutyric acid, 0.7 mM	Cond.	Formic, acetic, pyruvic, glyoxylic, succinic, malonic, oxalic acid	0.07–0.19 ppbv[a]	59
Gas phase only	Annular denuder, aqueous extraction	50 μl	HPICE-AS1, Dionex	Octane sulfonic acid, 1 mM with 2% isopropylalcohol	Cond.	Formic, acetic acid	1–50 mg/l in standards	55
Rain	Acidification	500 μl	Separator: ICE 30580	HCl, 0.002 N	Cond.	C_1–C_5 monocarboxylic, citric, lactic, glycolic, propionic acid	na, ≪0.6 mg/l	73
Antarctic ice	None	10 ml	Anion exch. concentrator, HPX-87H, Biorad	Methanesulfonic acid, 5 mM (a) pH 9 for injection (b) pH 2.7 for separation	UV, 200 nm	C_1–C_4 monocarboxylic acids	5.6–9.4 μg/l	85
Water								
Groundwater	Filtration	na	HPICE-AS1, Dionex	HCl. 1 mM	Cond.	C_2–C_4 monocarboxylic acids	0.5 μM	101
Standards	Dilution with trifluoroacetic acid, filtration	50 μl	HPICE-AS6, Dionex	Trifluoroacetic acid, 0.4 mM	ES-MS	Formic, gyoxylic, oxalic, 2-hydroxy isobutyric, maleic, succinic, malic, 1-hexanoic, malonic acid	2–8 mg/l	104

Continued

TABLE 13.6
Continued

Matrix	Sample Preparation	Injection	Column	Eluent	Detector	Compounds	Detection Limits	Ref.
Standards	None	100 μl	TSKgel SCX, Tosoh	3-methyl-n-valeric, iso-caproic or caproic acid, all 1 mM, 35°C	Cond.	C_1–C_5 mono-carboxylic acids	3–79 μg/l	105
Soil								
Soil solutions	Centrifugation, filtration, cation exchange	50 μl	Supelcogel C610-H, Supelco	H_3PO_4 (85%) 0.2 vol%, 30°C, (a) 30°C (b) 60°C	UV, 210 nm, monitored 200–300 nm	(a) Formic, acetic, propionic, tartaric, malic, succinic, fumaric, malonic, t-aconitic acid (b) Citric, lactic, shikimic acid	0.2–23.9 μM, analyzed 6–6000 μM	110,111
Sediment pore water	Filtration	50 μl	OA-1000, Altech	H_2SO_4, 0.05 M, 35°C	Refractive index	C_1–C_4 mono-carboxylic, lactic, oxalic, citric, malic, tartaric, fumaric	na. analyzed high μM conc.	115
Other								
Landfill leachates	Centrifugation, filtration, removal of humin-like substances	(a) 20 μl (b) 25 μl	(a) Polyspher OA-HY, Merck (b) HPICE-AS6, Dionex	(a) H_2SO_4, 5 mM 45°C, 50 mM 10°C (b) Perfluorobutyric acid, 0.4 mM 60°C, 1.6 mM 10°C	(a) UV, 210 nm (b) Cond.	C_1–C_5 mono-carboxylic, pyruvic, glyoxylic, glycolic, lactic, glyceric, succinic acid	na, analyzed 50 to 50,000 μM	29
Landfill leachate	Filtration, dilution, OnGuardP pretreatment, carbonate removal	na	HPICE-AS1, Dionex	Octanesulphonic acid, 10 mM	Cond.	C_2–C_6 mono-carboxylic acids	5 mg/l	27

Application	Sample preparation	Injection volume	Column	Mobile phase	Detection	Analytes	Detection limit	Ref.
Simulation of organics degradation in radioactive waste	Filtration, storage under N_2	20 μl	Polysher OA-HY, Merck	H_2SO_4, 5.0 mM, 45°C	UV, 210 nm	C_1–C_4 monocarboxylic, lactic, pyruvic, glycolic, glyceric, succinic, adipic acid	na, analyzed low mM conc.	117
Root exudates	None, direct injection	50 μl	HPICE-AS6, Dionex	Fluorobutyric acid, 1 mM	Cond.	Malic, succinic, citric, trans-aconitic acid	na	22,23
Fermentation products	Filtration, dilution	5 μl	Coregel 64H guard and analytical col., Interaction Chromatogr.	H_2SO_4, 7, 10, 13, 16 mN, 40°C, 50°C 60°C	UV, 210 nm	C_1–C_4 monocarboxylic, lactic, pyruvic, citric, fumaric, malic, succinic, α-ketoglutaric acid	0.2–150 μM	32
Hydrolysated biomass residues	None	20 μl	Polysher OA-HY guard and analytical column, Merck	H_2SO_4, 0.005 M, 45°C; H_2SO_4, 0.05 M, 10°C	UV, 210 nm	C_1–C_4 monocarboxylic, C_2–C_5 dicarboxylic, lactic, glyoxylic, glycolic, tartaric, α-ketoglutaric	0.2–10 μM	122,154
Beverages	None, direct injection	100 μl	TSK gel OA pak-A, Tosoh, Japan	Benzoic acid 0.85 mM in 10% aqueous methanol, 40°C	(a) APCI-MS, (b) UV 210–215 nm, cond.	C_1–C_5 monocarboxylic acids	(a) 0.7–3.8 μM (b) 0.08–2.3 μM	123

cond, suppressed conductivity; na, not available; APCI, atmospheric pressure — chemical ionization; ES, electrospray; MS, mass spectrometer; UV, ultraviolet (detection).

a Detection limits vary with sampling time.

have also been investigated recently for their use in IEC with promising results.[157,158] This material is inert towards organic solvents and allows for the addition of high amounts of solvents as eluent modifiers. These are commonly used to reduce retention times of strongly retained acids such as aromatic acids. In the past, the use of solvents as eluent modifiers was accompanied by problems such as swelling when using sulfonated cross-linked polymers. Another approach to reduce swelling in the presence of organic solvents is the use of highly crosslinked copolymers instead of the normally used copolymers with low crosslinkage.[158] However, these newer type columns have not yet found widespread use in routine analysis.

Eluent choice is determined by factors such as acidity, solvation properties, polarity, and especially detection mode. Eluent choice and detection mode are therefore discussed together. Traditionally UV detection predominates, although suppressed conductivity detection using tetrabutylammonium hydroxide as the regenerant and also nonsuppressed conductivity detection are becoming more common. Eluents applied in IEC are usually aqueous solutions of mineral acids or organic acids. The relatively low pH of the eluent leads to a shift in the dissociation equilibrium of weak organic acids towards their nonionized form, thus obtaining narrow peaks. Mineral acids such as sulfuric acid are preferably used as eluents in connection with direct UV detection, whereas strong organic acids, especially aliphatic sulfonic acids (e.g., methanesulfonic or octanesulfonic acid) are used as eluents in combination with suppressed conductivity detection.[130] Nonsuppressed conductivity detection requires eluents with low background conductivity and, hence, weak organic acids such as benzoic acid are preferred. Other weak organic acids more recently suggested include C_6 organic acids,[105] C_7 organic acids,[159] and benzoic acid with β-cyclodextrin.[160] It has also been shown that nonacidic eluents (e.g., polyvinylalcohol/water,[161] butanol,[162] or sucrose/methanol[163]) can be applied successfully to the separation of organic acids in conjunction with nonsuppressed conductivity detection.

Organic solvents such as acetonitrile[164] or various alcohols[165] are used as eluent modifiers to reduce tailing and retention times of more hydrophobic analytes such as aromatic organic acids. The eluent modifiers compete with the analytes, thus reducing interaction between the polymer and the more hydrophobic analytes. Gradient elution is usually not applied in IEC since it has been found that concentration gradients give only very little benefit.[164] However, gradients with increasing modifier amounts achieved better and faster separation of more hydrophobic organic acids than elution under isocratic conditions.[164]

Compound identification in IEC is accomplished by retention time comparison and coelution and/or incomplete separation of organic acids can be expected due to the low separation efficiency of IEC compared to GC, for example. Organic acid pairs known to coelute are fumaric/acetic acid and also succinic/glycolic acid.[64,155] Ideally the identity of a compound should be confirmed with an independent analytical method. It is therefore surprising that only few of the environmental methods listed in Table 13.6 even touch on this issue.[29,59,111]

An interesting approach to lowering detection limits is the use of a concentrator column in place of a sample loop. IEC columns cannot be used as a concentrator since the organic acids would not be retained during the injection process, given that water acts as an eluent. However, an anion exchange concentrator has been combined successfully with an ion exclusion analytical column achieving detection limits of 7 to 10 μg/l.[85] The difficulty in combining these two techniques lies in the choice of eluent which must be able to remove the anions from the concentrator while being suitable for IEC. This is accomplished by using methanesulfonic acid — at pH 9 for the removal of the dissociated analytes from the concentrator and at pH 2.7 for their separation on the IEC column. Unfortunately, the use of anion exchange concentrators is largely restricted to samples with relatively low ionic strength due to potential breakthrough of the organic acids.

AEC has also been coupled with IEC to achieve two-dimensional separations for the quantification of organic acids and inorganic anions.[166] Though AEC/IEC has not found widespread use for routine analysis of environmental samples since this system is complex to operate and prone to contamination problems.[166] Newer AEC methods are capable of determining inorganic anions and organic acids in one run, at least for matrices such as precipitation samples where inorganic anions and organic acids are present in similar concentrations.[18,70,71]

Newer developments in IEC include the introduction of vacancy IEC[167,168] and the development of IEC/MS.[104,123] In vacancy IEC, a new approach was taken to improve detection limits when using nonsuppressed conductivity detection.[167,168] The sample containing weak organic acids is used as eluent and, when injecting water, vacancy peaks appear for each of the analytes. The results are sharp, well-shaped peaks and improved detection limits compared to conventional IEC. Although this is an interesting concept practical applications of this approach still need to be developed.

IEC has been coupled with MS using electrospray,[104] thus achieving detection limits of 2 to 8 ppm for a mixture of ten structurally different organic acids. Care has to be taken to account for matrix effects which can significantly suppress the analyte signal. IEC with atmospheric pressure chemical ionization (APCI) MS was introduced more recently.[123] Detection limits for IEC-APCI-MS were in the same range as for conductimetric and photometric detection. Several wines and vinegars were analyzed as examples demonstrating the usefulness of this instrumentation. However, it is anticipated that IEC/MS may not find widespread use in the near future for routine analysis due to cost factors.

Overall, IEC is well suited for matrices with high ionic strength. IEC is, for example, preferred over AEC in complex matrices such as landfill leachates or samples from biological origins. IEC requires only simple sample pretreatment, does not exert thermal stress on analytes, and is well suited for aqueous samples. No interference from inorganic anions except for very weakly dissociated inorganic acids can be expected. Its drawbacks include only moderate detection limits, coelution of organic acids with low pK_a values with the system peak, and insufficient peak resolution leading to compound identification ambiguities.[155]

c. High Performance Liquid Chromatography (HPLC)

In general, HPLC refers to liquid chromatography employing reversed-phase columns. It is most commonly used for the determination of hydrophobic compounds. After injection, compounds are separated by partitioning between the reversed-phase column surface and the eluent, usually a solvent buffer mixture. Separation is typically followed by UV or fluorescence detection. Numerous publications describe HPLC in theory and practice.[169–171] An abundance of information is also available on the analysis of fatty acids in biological matrices by HPLC,[136,172] but only a few methods use HPLC for short-chain organic acid analysis in environmental matrices (Table 13.7).

All methods listed in Table 13.7 employ sample pretreatment — some without and some with derivatization of the analytes. The different pretreatment procedures for methods without derivatization have in common that they result in fairly clean extracts. Detection limits are adequate for the matrices analyzed and range from mg/l concentrations down to 0.05 μM.[67,121,124] HPLC results for VFA analysis in landfill leachates compared favourably with results obtained by GC,[121] proving that HPLC without a derivatization step can be a suitable method for organic acid analysis in environmental matrices.

Methods employing derivatization aim to change the physical properties of the organic acids, usually with the goal of making them more susceptible to UV or fluorescence detection (Section II.B.3.b). 2-Nitrophenyl hydrazine (NPH),[17] α,p-bromoacetophenone[77] and o-phenylene-diamine dihydrochloride (OPD)[81] have been utilized as derivatizing agents prior to HPLC measurements of organic acids, thus achieving detection limits down to nanomolar concentrations.[17] More details pertaining to these HPLC methods can be found in Table 13.7.

TABLE 13.7
Overview of HPLC Methods

Matrix	Sample Preparation	Inj. (μl)	Column	Eluents	Flow (ml/Min)	Det.	Compounds	Detection Limits	Comments	Ref.
Air chamber exp.	Collection on cartridges, aqueous extraction, $CHCl_3$ preservation	150	PRP-X-100, 25 cm, × 4.1 mm, Hamilton	KH phthalate (1 mM); aceto nitrile, 95:5, pH 4.5	2.0	UV, 280 nm	Formic, acetic acid	2–2.8 ppbv [a]	Focused on carbonyl products of lab study	67
Rain, sewage, soil pore water	Concentrated under vacuum, α,p-bromoaceto phenone deriv.	5	Guard: Perisorb RP-18, 30 μm, 4 cm × 1 mm analytical: RP 18, 10 μm, 25 cm × 3.2 mm, Merck	Methanol:H_2O, 50:50 (v/v)	1.30	UV, 254 nm	C_1–C_5 mono-carboxylic acids	na, analyzed low μM conc.	Confirmation by GC possible [74, 34]	77
Rain, fog mist	OPD deriv.	10–50	C18 column, 5 μm, Alltech	NaH_2PO_4 (0.02 M); aceto nitrile, 97:3 to 25:75 in 37 min	1.0	UV, 320 nm	Oxalic, pyruvic, gyoxylic acid	na, but <2 μM	Fluorimetric detection possible	81
Seawater, sediment pore water	Filtration, 2-Nitrophenyl hydrazine deriv.	500	Guard: C8, 1.5 cm analytical: C8, 22 cm, concentrator: C8; all Brownlee	n-butanol (2.5%), sodium acetate (50 mM), TBAOH (2 mM), TTAB (2 mM), pH 4.5	1.5	VIS, 400 nm	C_1–C_5 mono-carboxylic, lactic acid	na, analyzed low nM conc.	Blank problems for acetic and formic acid	17
Landfill leachates	Acidification and distillation	20	Spherisorb 5 ODS, 250 mm × 4.6 mm, Supelco	Methanol:H_2O (3:97) at pH4	1.0–2.0	UV, 210 nm	C_1–C_5 mono-carboxylic acids	na, analyzed high mg/l conc.	Confirmation by GC/FID	121
Root exudates	Direct collection on filter paper, extraction, acidification, filtration	na	Alltima C18, 5 μm, 25 cm × 4.6 mm, Alltech	KH_2PO_4, 25 mM, pH 2.5	1.0	UV, 210 nm	Malic, malonic, lactic, acetic, maleic, ctric, cis-, trans-aconitic, succinic, fumaric acid	0.05–24 μM	Ran gradient ramping up to 60% methanol after every five inj. to remove hydrophobic compounds	124

na, not available; FID, flame ionization detector; GC, gas chromatography; OPD, *o*-phenylenediamine dihydrochloride; TBAOH, tetrabutylammonium hydroxide; TTAB, tetradecyltrimethylammonium bromide.

[a] Detection limits vary with sampling time.

3. Capillary Zone Electrophoresis (CZE)

Capillary electrophoresis is a comparatively new technique which has generated considerable interest and is therefore included in this chapter. Separation in capillary electrophoresis is based on differences in the electrophoretic mobility of the analytes (e.g., differences in m/z ratios). Migration times are very short and, hence, analyses are completed within minutes, thus allowing for high sample throughput. Other advantages are high separation efficiencies similar to capillary GC, minimum solvent consumption, small sample volume requirements, simultaneous separation of anions and cations, and a simple system configuration, which makes this technique quite economical. The main disadvantages include often fairly high detection limits, compound identification and quantification problems due to matrix effects, and the lack of routine applications for organic acids in environmental matrices. Numerous publications are available describing capillary electrophoresis in more detail.[173-175] This section will focus on CZE, which is well suited for the analysis of smaller ions. Table 13.8 gives an overview of some CZE methods applied to organic acid analysis in environmental matrices.

A CZE system consists of a polyimide-coated fused silica capillary (i.d. 25 to 75 μm) which is filled with an electrolyte, usually a buffer. The ends of the capillary are immersed in reservoirs which contain electrodes and a detector is placed at the cathode end of the capillary. High voltage, up to 30 kV, is applied to the electrolyte and as a consequence the electrolyte is flowing towards the cathode generating an eletroosmotic flow (EOF). Analyte migration times are determined by the apparent or net electrophoretic mobility with which an analyte moves towards the cathode. The net electrophoretic mobility is comprised of overall EOF and the individual eletrophoretic mobility of the analytes. Cations are attracted by the cathode, thus flowing faster than the EOF, whereas anions are moving slower than the EOF since these are attracted to the anode which is opposite to the EOF direction. This makes it possible to determine cations and anions within the same run.[173-175]

Injection volumes are in the nanoliter range to avoid system overloading, since the total volume of the capillary is in the μl range. Direct injection techniques have been developed to ensure efficient and reproducible injection. Techniques employed are electrokinetic injection (i.e., electromigration injection), hydrodynamic injection by pressure or vacuum, and hydrostatic injection by gravity. Organic acids are almost exclusively detected with indirect UV, whereas other analytes have been measured by direct UV[176,177] or conductivity detection.[176,178]

Optimizing separation and selectivity of a CZE system can be accomplished by influencing net electrophoretic mobilities of the analytes through altering parameters such as applied voltage, system pH, type of buffer employed, and addition of electroosmotic modifiers. When primarily analyzing for anions, the EOF is normally reversed towards the anode by employing electroosmotic modifiers, usually cationic surfactants such as tetradecyltrimethylammonium bromide (TTAB). The detector is obviously placed at the anodic side when reversing the EOF.

In comparison to AEC or GC, there are fewer CZE methods for organic acids in environmental matrices (Table 13.8).[176] Organic acids analyzed range from just formic and acetic acid[82] to a whole range of saturated and unsaturated dicarboxylic acids.[83] Interestingly, CZE has been applied to the analysis of individual raindrops.[82,83] Although most often detection limits are quite high when using CZE, in some instances μmolar and even nanomolar concentrations have been reported. Detection limits of 10 to 30 nM are achieved through sample stacking, which allows for concentration of the analytes at the start of the capillary before migration starts. The sensitivity is further enhanced by indirect UV detection with aminobenzoate as the background electrolyte.[82] Matrix effects reported for samples with higher ionic strength include signal suppression[12] and shifts in migration times.[114] To compensate for potential signal suppression it is recommended to prepare standard solutions in matrix water.[12] Most applications also use standard addition for unambiguous compound identification, thus accounting for any changes in the migration time due to matrix effects.[12,114]

TABLE 13.8
Overview of CZE Methods

Matrix	Sample Preparation	Injection	Capillary[a]	Electrolyte	Voltage	Detection	Compounds	Detection Limits	Ref.
Atmospheric particulate matter	Aqueous extraction of filters	50 mbar, 6 s	50 μm, 50 cm	(a) PZDA 4.0 mM, MTAH 0.5 mM, pH 10.6 (b) PDA 4.0 mM, MTAB 0.5 mM, pH 11.0	−21 kV	Ind. UV, (a) 280 nm (b) 266 nm	C_2–C_{10} dicarboxylic acids	(a) 1–9 mg/l (b) 1–7 mg/l	2
Atmospheric particulate matter	Aqueous extraction of filters, filtration	5 in. Hg, 10 s	50 μm	3,5-Dinitrobenzoic acid 10 mM, CTAB 0.1 mM, pH 5–6	−15 kV, 25°C	Ind. UV, 254 nm	C_1–C_4 mono-carboxylic, glycolic, lactic, β-hydroxybutyric acid	0.05–0.32 mg/l	64
Single rain drops	Frozen sample	Sample stacking: 1.5 psi, 30 s	75 μm i.d., 50 cm	p-Aminobenzoic acid 3 mM, Na p-aminobenzoate 4.5 mM, Ba(OH)$_2$ 0.76 mM, TTAH 55 μM, pH 9.6	−30 kV, 25°C	Ind. UV, 264 nm	Formic, acetic, propionic, oxalic, 10 other mostly dicarboxylic acids	10–30 nM	82
Single rain drops	Frozen sample	10 cm, 30 s	75 μm, 55 cm	K$_2$CrO$_4$ 5 mM, TTAB 0.2 mM	−22 kV	Ind. UV, 276 nm	Formic, acetic acid	180, 450 μg/l	83
Humic surface water	Addition of octanesulfonate	−5 kV, 45 s	75 μm, 74 cm	TMA 5 mM, TTAB 0.5 mM, pH 8	−15 kV	Ind. UV, 254 nm	Formic, acetic, lactic, oxalic, malonic acid	na, analyzed μg/l conc.	12,103
(a) Brown water DOM (b) Beer	(a) None (b) Degassing, dilution	50 mbar, 4.0 s, −5 kV, 45 s	75 μm, 72 cm	CTAB 0.5 mM, PDA 5 mM, pH 5.6	−25 kV, 20°C	Ind. UV, 350 nm	Formic, acetic, oxalic, malonic, succinic, pyruvic acid	na, analyzed μM conc.	51,102
Soil solutions	Centrifugation, filtration, cation exchange	(a) −5 kV, 45 s (b) 100 mm, 30 s	75 μm, 74 cm	TMA 5 mM, TTAB 0.5 mM, pH 8	−15 kV	Ind. UV, 254 nm	Formic, acetic, lactic, oxalic, malonic acid	na, analyzed μg/l conc.	110,111

| Soil solutions | Centrifugation, filtration, addition of Na_4EDTA | 0.5 lb/in.² (a) 20 s (b) 10 s | 75 μm (a) 70 cm (b) 57 cm | (a) TRIS 8 mM, TMA 2 mM, TTAB 0.3 mM, pH 7.6 (b) TMA 3 mM, DETA 0.02% (v/v), pH5.8 | −30 kV, 20°C | Ind. UV, 254 nm | (a) Lactic, C_1–C_4 monocarboxylic (b) Oxalic, malonic, tartaric, malic, succinic, citric acid | 0.26–1.77 μM | 112,113 |
| Soil solutions, root extracts | Aqueous extraction, centrifugation, filtration | 20 psig, 5 s | 75 μm, 50 cm | KH phthalate 15 mM, MTAB 5 mM, methanol 5%, pH 5.6 | −20 kV, 25°C^a | Ind. UV, 254 nm | Formic, acetic, lactic, tartaric, malic, citric, succinic acid | 0.5–6 μM | 114 |

ind. UV, indirect ultraviolet (detection); CTAB, cetyltrimethylammonium bromide; DETA, diethylenetriamin; DOM, dissolved organic matter; MTAB, myristyltrimethylammonium bromide; MTAH, myristyltrimethylammonium hydroxide; PDA, 2,6-pyridinedicarboxylic acid; PZDA, 2,3-pyrazinedicarboxylic acid; TMA, trimellitic acid = 1,2,4-benzenetricarboxylic acid; TRIS, tris(hydroxymethyl)aminomethane; TTAB, tetradecyltrimethylammonium bromide; TTAH, tetradecyltrimethylammonium hydroxide.

^{a} Polyimide-coated fused silica capillary, inner diameter, effective length.

Overall, CZE is commonly applied in the analysis of organic acids in matrices such as biological fluids and food,[102,176,179–183] and it is also increasingly used in the analysis of environmental matrices.

III. APPLICATIONS

Numerous methods deal with the determination of organic acids in environmental matrices (Table 13.3). The approaches taken depend to a large extent on matrix characteristics and, hence, applications are discussed in this context.

A. AIR INCLUDING ATMOSPHERIC PRECIPITATION

1. Air

In air, organic acids are present in the gas phase and these are also found associated with atmospheric particles. Some methods combine organic acid measurement for both phases,[57–60] whereas others distinguish between gas or particulate phase organic acids.[55,56,61,62] One application even goes as far as fractionating the atmospheric particles prior to extraction and organic acid quantification.[63] Initially, formic and acetic acid have been the focus of many investigations[55–57,60,62] although later applications included further acids such as hydroxy-, keto-, and dicarboxylic acids.[58,59,61,63,65,66] Concentrations of organic acids in air are very low (ppbv concentrations) — substantially lower than concentrations of trace gases such as NO_x or CO. Sampling and sample preparation usually lead to an enrichment of organic acids in aqueous solutions and, hence, most applications utilize AEC with conductivity detection[56–58,60–64] although other techniques have also been employed (e.g., IEC,[2,55,59] GC/FID,[65,66] GC/MS,[4,65] and CZE[58,64]) (Table 13.3). Using AEC has the additional benefit of being able to measure inorganic anions together with organic acids in the same chromatogram, since inorganic anion concentrations are low enough not to interfere with organic acid analysis by AEC.[56] It is also known that organic acids can coelute, thus potentially leading to misinterpretation.[64] A number of AEC applications confirm, therefore, their organic acids results with a different independent method such as IEC or CZE.[58,59,64]

2. Rain, Fog, Mist, Snow

Rain samples have been analyzed for organic acids since the late 1970s[5] and, as a consequence, numerous methods are available today — mostly using GC and AEC (Table 13.3). Although fog, mist, and snow samples require matrix-specific sample collection and sample preparation steps, the resulting aqueous solutions and therefore the actual organic acid measurements are virtually identical to rain and thus are also covered in this section. At first, organic acid measurement focused on formic and acetic acid,[38,68,69,71] although methods were quickly expanded to include hyroxy-, keto-, and dicarboxylic acids.[33,70,72,73,78] Organic acid content in rain ranges from low to high $\mu g/l$ concentrations depending on location, precipitation event, and type of acid,[5,9,33–35,68,71,73,74,78,79,81,83] whereas the more predominant inorganic anions (e.g., sulfate, sulfite, nitrate, and chloride) have typically somewhat higher concentrations.[56]

AEC is the preferred technique when dealing with large sample numbers[38,68,70–72] since rain samples can be injected directly, perhaps after filtration if particles are present. Organic acids are separated on anion exchange columns and then detected by suppressed conductivity. Detection limits thus achieved are in the low μmolar range, which is adequate for rain samples. Inorganic anion concentrations in rain are not high enough to interfere with organic acid peak separation and, hence, they are even sometimes quantified simultaneously with organic acids, thus further streamlining the analytical process.[70,71] Similarly, methods utilizing CZE are capable of determining inorganic anions and organic acids simultaneously while achieving low detection limits.[82,83] Interestingly, these CZE methods have been employed for measurements of single raindrops.

GC methods tend to be more time consuming due to necessary sample preparation steps (e.g., concentration and derivatization). Some GC methods cover monocarboxylic acids[34,74−77] whereas others focus on dicarboxylic and/or ketoacids.[36,75,78−80] Detection limits are again at low μg/l concentrations. In general, GCs have higher separation efficiencies than ICs and coelution is therefore not as common in GC as in AEC. In addition, extracts may be measured by GC/MS, thus confirming peak identity. With one exception,[72] AEC applications for rain do not discuss a potential coelution of organic acids although other independent methods would be available for confirmation.

3. Ice

Ice cores differ from rain or air samples in that organic acid and also inorganic anion concentrations are very low (i.e., low ng/g concentrations). Direct AEC injection methods are not sensitive enough and, hence, sample introduction onto concentrators has been used for sample enrichment. Anion exchange concentrators are combined successfully with either AEC[6,84] or IEC[85] for compound separation. The newer AEC/AEC combination[6,84] achieves lower detection limits than the AEC/IEC method.[85] Moreover, inorganic anions (F^-, Cl^-, NO_2^-, NO_3^-, SO_4^{2-}) can be determined with the AEC/AEC method within the same chromatogram as the organic acids.[6,84] Overall, concentrator columns are powerful tools for increasing method sensitivity if handled correctly within their limitations (e.g., avoiding breakthrough; Section II.C.2.a).

B. WATER

1. Drinking Water

Over the last two decades organic acids have been measured in finished drinking water as well as in partially treated water. They are formed as by-products during ozonation,[10,39,40,44,88] a well-accepted drinking water treatment technique. Organic acids monitored include short-chain monocarboxylic, keto-, hydroxy-, and also some dicarboxylic acids. Concentrations are usually in the low μg/l range although ozone contactor effluents may reach medium μg/l levels.[10,40,42,44] Different approaches have been taken involving either GC or AEC (Table 13.3). In general, AEC methods are less time consuming than GC methods due to their significantly shorter sample preparation procedures. However, current GC methods quantify ketoacids,[39,40,41,45,88,89] whereas AEC methods focus on formic, acetic, and oxalic acid,[42,44,45] including a few other acids in one of these methods.[43,86,87] When using both techniques, a more complete picture of type and overall quantity of organic acids formed during ozonation may be obtained.

As a prerequisite for GC, organic acids have to be concentrated and derivatized. Extraction is usually accomplished by liquid/liquid extraction or SPE. In all drinking water methods listed in Table 13.3, the extraction step was preceded by aqueous derivatization of a ketofunction, thus limiting these GC methods to ketoacids.[39,40,88,89] Simple, aliphatic mono- and dicarboxylic acids cannot be determined with these methods. However, approaches taken for GC analysis of organic acids in rain could be equally as well applied to drinking water, thus covering additional mono- and dicarboxylic acid.[34,36,74−76,78−80]

AEC methods have also been used for organic acid analysis in drinking water although inorganic anions such as chloride, sulfate, and carbonate are usually present in much higher concentrations (low to medium mg/l) than organic acids (low μg/l range). This can lead to masking or incomplete separation of organic acid peaks and therefore to identification and quantification problems. Hence, it is surprising that IEC, where inorganic anions elute up-front with the system peak, has not been applied to organic acid analysis in drinking water. This may be due to problems anticipated when using IEC, for example interferences from weak inorganic acids such as carbonate, and separation problems between the early eluting system peak and organic acids with low pK_a values such as oxalate or pyruvate.[29,155]

AEC has been predominantly employed for formic, acetic, and oxalic acid,[42–45,86,87] although other acids have also been measured.[43,86,87] Slightly different approaches have been taken, usually keeping sample pretreatment to a minimum. Direct, large volume injections have been combined with the separation on a high capacity column. In this case, sample pretreatment, i.e., filtration, is only necessary if samples are turbid.[43,86,87] However, high sulfate concentrations may interfere with oxalate quantification and therefore this method was further developed utilizing a "heart-cut" technique.[43,87] In another direct injection method, removal of mercuric cations, which were added to the sample for preservation purposes, was necessary to avoid column poisoning. This is accomplished in a time-saving manner by using a cation exchanger in-line between the autosampler and the injection loop.[42,44] A third approach requires the removal of chloride, sulfate, and phosphate prior to injection, which is achieved by pushing the samples through cartridges filled with silver or barium salts,[45] a technique which is often used in bromate analysis.[184–186] All of these relatively fast methods achieve detection limits at very low μg/l concentrations, thus being sensitive enough to detect organic acids in drinking water in the presence of inorganic anions.

2. Wastewater

VFA are aliphatic, short-chain organic acids which are formed during anaerobic digestion of organic material. This process has been applied primarily to waste sludge, but is also used as pretreatment for high organic waste streams and is even applied to dilute waste streams.[25] VFA are frequently monitored for process control purposes and consequently many methods are available, ranging from traditional wet chemistry, e.g., distillation and titration,[46] to GC techniques.[90,91,93,97] Although concentrations of VFA are usually quite high (medium to high mg/l) analytical challenges exist with respect to the complex matrix. Wastewater or sludge contains very high concentrations of organic material and inorganic compounds,[24,25] which can lead to matrix effects potentially influencing extraction efficiencies[119] or compound identification.[98–100] Landfill leachates have a composition very similar to wastewater and methods used for their analyzes may also be applied to wastewater.[28]

The most common technique used for VFA analysis is direct aqueous injection into GC/FID.[26,90–96,106] Sample pretreatment is usually minimal, consisting of dilution when necessary, centrifugation, and filtration followed by acidification. Samples are then injected directly into a GC/FID which is equipped with columns specifically designed for the analysis of acidic compounds. More recently, capillary columns have gained in popularity over packed columns. Usually, aliphatic monocarboxylic acids with carbon chain length from C_2 to C_5 are measured. Formic acid cannot be determined by GC/FID, but alternative methods using GC/TCD[52] or HPLC[121] are available. Method sensitivity is not very high but it is adequate for the determination of the rather high concentrations of VFA in wastewater or diluted sludge. Overall, direct aqueous injection into GC/FID delivers immediate results with adequate accuracy and sensitivity if care is taken to use appropriate GC operating conditions and if the GC system is maintained regularly (Section II.C.1). Alternatives to direct aqueous GC injection include headspace injection[97] and SPME,[98–100] both of which have not found widespread use.

3. Groundwater

Organic acids in groundwater have been measured at sites contaminated with organic compounds (e.g., superfund sites), where these are formed during anaerobic, biological degradation of these organic contaminants. Matrix and organic acid content can vary substantially depending on factors such as type of contamination, proximity to the contamination, and geology. Only short-chain monocarboxylic acids have been monitored, although other types of organic acids are most likely present as well. Organic acid content can fluctuate from low to very high μg/l concentrations.[14,47] Type and concentrations of inorganic anions can also differ widely. Other sample constituents may include organic solvents and/or metals which have to be taken into consideration when developing or implementing analytical methods. Analytical techniques applied to the analysis of organic acids at groundwater sites range from IEC[101] to GC[14,47,48] (Table 13.3).

GC methods employ an extraction step prior to injection on GCs equipped with polar columns which facilitate organic acids separation.[14,48,47] Matrix interferences from inorganic anions, metals, and solvents are mostly eliminated by the extraction step. GC methods used for analysis of rain or air samples should also be suitable for the analysis of groundwater samples.[34,65,74,75,78–80]

IEC has been applied to samples from a groundwater site contaminated with petroleum hydrocarbons.[101] Most inorganic anions elute with the system peak and will therefore not interfere with organic acid analysis. If solvents are present they should be removed prior to injection since they can interfere with the separation process and, if present in large amounts, may cause swelling of conventional IEC columns. IEC methods used in the analysis of air and rain samples may also be applied to groundwater samples.[2,33,73]

4. Seawater

Very high concentrations of inorganic ions and low concentrations of organic acid make the determination of organic acids in this matrix a challenge. Membrane dissociation is one of the more unconventional approaches used to concentrate and separate organic acids from inorganic ions,[16] whereas other methods rely on using GC or HPLC after extraction and derivatization.[17,50]

5. Other Water

Organic acids have occasionally been measured in other matrices such as ultrapure water used in industrial cooling processes,[18] UV-irradiated humic water,[12,51,102,103] or in aqueous samples investigating by-products of organic pollutant degradation.[30] With such different matrices and varying organic acid concentrations it is not surprising that techniques ranging from IEC and AEC to CZE have been used. Interestingly, the investigation into photoformation of organic acids from dissolved organic matter confirmed results obtained by AEC with a second independent method using CZE.[51,102] In the case of ultrapure water, an on-line anion exchange concentrator has been used in conjunction with AEC to obtain the required low detection limits.[18]

C. Soil

Soil solutions, that is water contained within soil samples, have also been screened for their organic acid content. Organic acids measured in these samples include saturated and unsaturated mono- and dicarboxylic acids, keto-, hydroxy-, and also aromatic acids (Table 13.3). Organic acid concentrations are usually in the μM range.

Soil solutions are most often obtained by centrifugation, usually followed by filtration. Other procedures include column displacement techniques and saturation extracts.[189] Depending on soil characteristics, soil solutions can vary in their composition, (e.g., organic acids, inorganic compounds, and complex organics such as humic substances), thus requiring varying degrees of pretreatment. Solutions obtained from sediments are often only filtered prior to their measurement with GC[116] or IEC,[115] whereas more complex soil solutions undergo further pretreatment prior to organic acid measurement employing AEC,[107–109] IEC,[110,111] or CZE.[112,113] In order to streamline operations one application developed an on-line sample pretreatment scheme which includes cation and anion exchange, and also extraction with TOPO- impregnated membranes.[108] This approach has been utilized in a subsequent study, thus demonstrating its usefulness.[107] Because of the similarity in their matrices, methods employed for groundwater may also be applied to soil solutions.[14,47,48,101]

D. Others

Other matrices analyzed for organic acids include landfill leachates and samples derived from biological material such as silage juice, root exudates, or fermentation products.

1. Landfill Leachates

Landfill leachates are similar to wastewater in that these contain very high concentrations of inorganic ions and also organic compounds. The composition of the organic fraction depends on the age of the landfill. Organic acids monitored include VFA but also hydroxy- and ketoacids. Concentrations can be quite high, in the medium to high mg/l range (Table 13.3).

Because of the complex matrix, most of the methods involve a sample preparation step using either distillation[118,121] or extraction,[119] which is then followed by either GC/FID or HPLC measurement of mostly VFAs. GC methods employed for VFA measurement in wastewater should also be applicable to landfill leachates.[90,91,93-96] Most of these wastewater methods do not require an extensive sample preparation step other than filtration, acidification, and perhaps dilution. In fact, one method for landfill leachates took this approach using direct aqueous GC injections after dilution in conjunction with deuterated internal standards.[120]

More recently IEC has also been employed for the analysis of landfill leachates.[27,29,117] Sample pretreatment steps may include centrifugation, filtration, removal of interfering organic compounds, and carbonate removal. An advantage is that these IEC methods analyze for a wider spectrum of organic acids than the GC methods, hence giving a more complete picture of the organic acid composition in landfill leachates.

2. Biological Material

Organic acids have also been measured in samples from various biological origins such as silage juice and other fermentation products, hydrolysated biomass, or root exudates. A few methods covering biological specimens and beverages have also been included in Table 13.3 to demonstrate their unique approaches. It should be noted that there is an abundance of literature available on organic acid analysis in food, beverages, and biological specimens, often covering short-chain organic acids but also longer-chain acids such as fatty acids.[135,138,187,188]

Biological matrices are quite complex, thus usually requiring a sample preparation step for the isolation of the organic acids prior to their measurement. The type of organic acid analyzed is largely dependent on the matrix investigated and the objectives associated with measuring organic acids. Some methods measure just VFA,[53,54] whereas others determine a whole range of mono- and dicarboxylic, keto-, and hydroxy acids.[22,23,32,122,124,154] Depending on the matrix, concentrations vary from nM to mM. Hence, methods employed are very diverse, ranging from aqueous extraction of a dichloromethane solution of a cellulose polymer and GC/TCD measurement[52] to direct injections of beverages into an IEC/MS system.[123]

IV. SUMMARY

Approaches to organic acid analysis in environmental matrices are as varied as the matrices. GC methods often requiring derivatization are in general more time consuming than AEC or IEC methods which frequently allow for direct sample injection. IEC and especially AEC are therefore gaining in popularity over GC applications. An exception is the measurement of high concentrations of VFA by direct aqueous injection into GC, which is well established in the wastewater field. CZE is seldom used on a routine basis although it has been employed as a tool to independently confirm results obtained with another technique. If suitable methods for a certain water matrix do not exist, applications developed for other matrices may give a good starting point on how to approach this problem and, after further development, may result in an appropriate method.

REFERENCES

1. Weast, R. C., Ed., *CRC Handbook of Chemistry and Physics*, 61st ed., CRC Press, Boca Raton, Florida, 1980.
2. Adler, H., Sirén, H., Kulmala, M., and Riekkola, M. L., Capillary electrophoretic separation of dicarboxylic acids in atmospheric aerosol particles, *J. Chromatogr. A*, 990, 133–141, 2003.
3. Chebbi, A. and Carlier, P., Carboxylic acids in the troposphere, occurrence, sources, and sinks: a review, *Atmos. Environ.*, 30, 4233–4249, 1996.
4. Chien, C. J., Charles, M. J., Sexton, K. G., and Jeffries, H. E., Analysis of airborne carboxylic acids and phenols as their pentafluorobenzyl derivatives: gas chromatography/ion trap mass spectrometry with novel chemical ionization reagent, PFBOH, *Environ. Sci. Technol.*, 32, 299–309, 1998.
5. Galloway, J. N., Likens, G. E., and Edgerton, E. S., Acid precipitation in the Northeastern United States: pH and acidity, *Science*, 194, 722–723, 1976.
6. Legrand, M. and De Angelis, M., Origins and variations of light carboxylic acids in polar precipitation, *J. Geophys. Res.*, 100, 1445–1462, 1995.
7. Talbot, R. W., Becher, K. M., Harris, R. C., and Cofer, W. R. III, Atmospheric geochemistry of formic and acetic acids at a mid-latitude temperate site, *J. Geophys. Res.*, 93, 1638–1652, 1988.
8. Helas, G., Bingemer, H., and Andreae, M. O., Organic acids over Equatorial Africa: results from DECAFE 88, *J. Geophys. Res.*, 97, 6187–6193, 1992.
9. Keene, W. C. and Galloway, J. N., Considerations regarding sources for formic and acetic acids in the troposphere, *J. Geophys. Res.*, 91, 14466–14474, 1986.
10. Gagnon, G. A., Booth, S. D. J., Peldszus, S., Mutti, D., Smith, F., and Huck, P. M., Carboxylic acids: formation and removal in full-scale plants, *J. Am. Water Works Assoc.*, 89(8), 88–97, 1997.
11. Montgomery, J. M., *Water Treatment: Principle and Design*, Wiley, New York, 1985.
12. Dahlén, J., Bertilsson, S., and Pettersson, C., Effects of UV-A irradiation on dissolved organic matter in humic surface waters, *Environ. Int.*, 22, 501–506, 1996.
13. McMahon, P. B., Vroblesky, D. A., Bradley, P. M., Chapelle, F. H., and Gullett, C. D., Evidence for enhanced mineral dissolution in organic acid-rich shallow ground water, *Ground Water*, 33, 207–216, 1995.
14. Cozzarelli, I. M., Baedecker, M. J., Eganhouse, R. P., and Goerlitz, D. F., The geochemical evolution of low-molecular-weight organic acids derived from the degradation of petroleum contaminants in groundwater, *Geochim. Cosmochim. Acta*, 58, 863–877, 1994.
15. Stoessel, R. K. and Pittman, E. D., Secondary porosity revisited: the chemistry of feldspar dissolution by carboxylic acids and anions, *Am. Assoc. Petrol. Geol.*, 74, 1795–1805, 1990.
16. Yang, X. H., Lee, C., and Scranton, M. I., Determination of nanomolar concentrations of individual dissolved low molecular weight amines and organic acids in seawater, *Anal. Chem.*, 65, 572–576, 1993.
17. Albert, D. B. and Martens, C. S., Determination of low-molecular-weight organic acid concentrations in seawater ad pore-water samples via HPLC, *Mar. Chem.*, 56, 27–37, 1997.
18. Toofan, M., Stillian, J. R., Pohl, C. A., and Jackson, P. E., Preconcentration determination of inorganic anions and organic acids in power plant waters. Separation optimization through control of column capacity and selectivity, *J. Chromatogr. A*, 761, 163–168, 1997.
19. Jones, D. L., Organic acids in the rhizosphere — a critical review, *Plant Soil*, 205, 25–44, 1998.
20. Lundström, U. S., van Breemen, N., and Bain, D., The podzolization process. A review, *Geoderma*, 94, 91–107, 2000.
21. van Hees, P. A. W., Lundström, U. S., and Giesler, R., Low molecular weight organic acids and their Al-complexes in soil solution — composition, distribution and seasonal variation in three podzolized soils, *Geoderma*, 94, 173–200, 2000.
22. Gaume, A., Mächler, F., and Frossard, E., Aluminum resistance in two cultivars of *Zea mays* L.: root exudation of organic acids and influence of phosphorus nutrition, *Plant Soil*, 234, 73–81, 2001.
23. Gaume, A., Mächler, F., De Leon, C., Narro, L., and Frossard, E., Low P tolerance by maize (*Zea mays* L.) genotypes: significance of root growth, and organic acids and acid phosphatase root exudation, *Plant Soil*, 228, 253–264, 2001.

24. Sawyer, C. N., McCarty, P. L., and Parkin, G. F., *Chemistry for Environmental Engineering*, 4th ed., McGraw-Hill, New York, 1994.
25. Metcalf & Eddy Inc. *Wastewater Engineering — Treatment/Disposal/Reuse*, 3rd ed., McGraw-Hill, New York, 1991.
26. Ahring, B. K., Sandberg, M., and Angelidaki, I., Volatile fatty acids as indicators of process imbalance in anaerobic digestors, *Appl. Microbiol. Biotechnol.*, 43, 559–565, 1995.
27. Manning, D. A. C. and Bewsher, A., Determination of landfill leachates by ion chromatography, *J. Chromatogr. A*, 770, 203–210, 1997.
28. McBean, E. A., Rovers, F. A., and Farquhar, G. F., *Solid Waste Landfill Engineering and Design*, Prentice Hall PTR, NJ, 1995.
29. Fischer, K., Chodura, A., Kotalik, J., Bienik, D., and Kettrup, A., Analysis of aliphatic carboxylic acids and amino acids in effluents of landfills, composting plants and fermentation plants by ion-exclusion and ion-exchange chromatography, *J. Chromatogr. A*, 770, 229–241, 1997.
30. Scheuer, C., Wimmer, B., Bischof, H., Nguyen, L., Maguhn, J., Spitzauer, P., Kettrup, A., and Wabner, D., Oxidative decomposition of organic water pollutants with UV-activated hydrogen peroxide: determination of anionic products by ion-chromatography, *J. Chromatogr. A*, 706, 253–258, 1995.
31. McCalley, D. V., Analysis of volatile fatty acids by capillary gas chromatography using on-column injection of aqueous solutions, *J. High Resolut. Chromatogr.*, 12, 465–467, 1989.
32. Eiteman, M. A. and Chastain, M. J., Optimization of the ion-exchange analysis of organic acids from fermentation, *Anal. Chim. Acta*, 338, 69–75, 1997.
33. Keene, W. C. and Galloway, J. N., Organic acidity in precipitation of North America, *Atmos. Environ.*, 18, 2491–2497, 1984.
34. Sakugawa, H., Kaplan, I. R., and Shepard, L. S., Measurements of H_2O_2, aldehydes and organic acids in Los Angeles rainwater: their sources and deposition rates, *Atmos. Environ.*, 27B, 203–219, 1993.
35. Kawamura, K. and Kaplan, I. R., Biogenic and anthropogenic organic compounds in rain and snow samples collected in Southern California, *Atmos. Environ.*, 20(1), 115–124, 1986.
36. Kawamura, K. and Kaplan, I. R., Motor exhaust emissions as a primary source for dicarboxylic acids in Los Angeles ambient air, *Environ. Sci. Technol.*, 21, 105–110, 1987.
37. Nolte, C. G., Solomon, P. A., Fall, T., Salmon, L. G., and Cass, G. R., Seasonal and spatial characteristics of formic and acetic acids concentration in Southern California atmosphere, *Environ. Sci. Technol.*, 31, 2547–2553, 1997.
38. Talbot, R. W., Mosher, B. W., Heikes, B. G., Jacob, D. J., Munger, J. W., Daube, B. C., Keene, W. C., Maben, J. R., and Artz, R. S., Carboxylic acids in the rural continental atmosphere over the Eastern United States during the Shenandoah cloud and photochemistry experiments, *J. Geophys. Res.*, 100, 9335–9343, 1995.
39. Xie, Y. and Reckhow D. A., A new class of ozonation by-products: the ketoacids. *Proceedings of the American Water Works Association Annual Conference*, pp. 251–265, 1992.
40. Xie, Y. and Reckhow, D. A., Formation of ketoacids in ozonated drinking water, *Ozone Sci. Eng.*, 14, 269–275, 1992.
41. Le Lacheur, R. M., Sonnenberg, L. B., Singer, P. C., Christman, R. F., and Charles, M. J., Identification of carbonyl compounds in environmental samples, *Environ. Sci. Technol.*, 27, 2745–2753, 1993.
42. Kuo, C. Y., Wang, H. C., Krasner, S. W., and Davis, M. K., Ion chromatographic (IC) determination of three short-chain carboxylic acids in ozonated drinking water, *ACS Symp. Ser.*, 649, 350–365, 1996.
43. Peldszus, S., Huck, P. M., and Andrews, S. A., Quantitative determination of oxalate and other organic acids in drinking water at low μg/l concentrations, *J. Chromatogr. A*, 793, 198–203, 1998.
44. Kuo, C. Y., Improved application of ion chromatographic determination of carboxylic acids in ozonated drinking water, *J. Chromatogr. A*, 804, 265–272, 1998.
45. Weinberg, H. S. and Glaze, W. H., A unified approach to the analysis of polar organic by-products of oxidation in aqueous matrices, *Water Res.*, 31, 1555–1572, 1997.

46. *Standard Methods for the Examination of Water and Wastewater. Method 5560 Volatile Organic Acids*, 20th ed., APHA, AWWA, WEF, Washington, DC, 1998.

47. Cozzarelli, I. M., Herman, J. S., and Baedecker, M. J., Fate of microbial metabolites of hydrocarbons in a coastal plain aquifer: the role of electron acceptors, *Environ. Sci. Technol.*, 29, 458–469, 1995.

48. Betowski, L., Kendall, D. S., Pace, S. M., and Donnelly, J. R., Characterisation of groundwater samples from superfund sites by gas chromatography/mass spectrometry and liquid chromatography/mass spectrometry, *Environ. Sci. Technol.*, 30, 3558–3564, 1996.

49. Surdam, R. C. and MacGowan, D. B., Oilfield waters and sandstone diagenesis, *Appl. Geochem.*, 2, 613–619, 1987.

50. Vairavamurthy, A., Andreae, M. O., and Brooks, J. M., Determination of acrylic acid in aqueous samples by electron capture GC after extraction with tri-*n*-octylphosphine oxide and derivatisation with pentafluorobenzyl bromide, *Anal. Chem.*, 58, 2684–2687, 1986.

51. Brinkmann, T., Hörsch, P., Sartorius, D., and Frimmel, F. H., Photoformation of low-molecular-weight organic acids from brown water dissolved organic matter, *Environ. Sci. Technol.*, 37, 4190–4198, 2003.

52. Allen, B. J., Spence, M. H., Lewis, J. S., and Rapid, A., Precise determination of C_1–C_4 normal volatile fatty acids in aqueous solutions by gas chromatography, *J. Chromatogr. Sci.*, 25, 313–314, 1987.

53. Fleming, S. E., Trailer, H., and Koellreuter, B., Analysis of volatile fatty acids in biological specimens by capillary column gas chromatography, *Lipids*, 22, 195–200, 1987.

54. Duisterwinkel, F. J., Wolthers, B. G., van der Slik, W., and Dankert, J., Determination of volatile fatty acids by GC on a capillary and a megabore column, *Clin. Chim. Acta*, 156, 207–214, 1986.

55. Lawrence, J. E. and Koutrakis, P., Measurement of atmospheric formic and acetic acids: methods evaluation and results from field studies, *Environ. Sci. Technol.*, 28, 957–964, 1994.

56. Hofmann, U., Weller, D., Ammann, C., Jork, E., and Kesselmeier, J., Cryogenic trapping of atmospheric organic acids under laboratory and field conditions, *Atmos. Environ.*, 31(9), 1275–1284, 1997.

57. Granby, K., Christensen, C. S., and Lohse, C., Urban and semi-rural observations of carboxylic acids and carbonyls, *Atmos. Environ.*, 31, 1403–1415, 1997.

58. Souza, S. R., Vasconcellos, P. C., and Carvalho, L. R. F., Low molecular weight carboxylic acids in an urban atmosphere: winter measurements in Sao Paulo city, Brazil, *Atmos. Environ.*, 33, 2563–2574, 1999.

59. Khwaja, H. A., Atmospheric concentrations of carboxylic acids and related compounds at a semiurban site, *Atmos. Environ.*, 29, 127–139, 1995.

60. Keene, W. C., Talbot, R. W., Andreae, M. O., Beecher, K., Berresheim, H., Castro, M., Farmer, J. C., Galloway, J. N., Hoffmann, M. R., Li, S. M., Maben, J. R., Munger, J. W., Norton, R. B., Pszenny, A. A. P., Puxbaum, H., Westberg, H., and Winiwarter, W., An intercomparison of measurement systems for vapors and particulate phase concentrations of formic and acetic acids, *J. Geophys. Res.*, 94(D5), 6457–6471, 1989.

61. Röhrl, A. and Lammel, G., Low-molecular weight dicarboxylic acids and glyoxylic acid: seasonal and air mass characteristics, *Environ. Sci. Technol.*, 35, 95–101, 2001.

62. Carvalho, L. R. F., Souza, S. R., Martinis, B. S., and Korn, M., Monitoring of the ultrasonic irradiation effect on the extraction of airborne particulate matter by ion chromatography, *Anal. Chim. Acta*, 317, 171–179, 1995.

63. Kerminen, V. M., Teinilä, K., Hillamo, R., and Mäkelä, T., Size-segregated chemistry of particulate dicarboxylic acids in the Arctic atmosphere, *Atmos. Environ.*, 33, 2089–2100, 1999.

64. Souza, S. R., Tavares, M. F. M., and de Carvalho, L. R. F., Systematic approach to the separation of mono- and hydroxycarboxylic acids in environmental samples by ion chromatography and capillary electrophoresis, *J. Chromatogr. A*, 796, 335–346, 1998.

65. Limbeck, A. and Puxbaum, H., Organic acids in continental background aerosols, *Atmos. Environ.*, 33, 1847–1852, 1999.

66. Kawamura, K. and Ikushima, K., Seasonal changes in the distribution of dicarboxylic acids in the urban atmosphere, *Environ. Sci. Technol.*, 27, 2227–2235, 1993.

67. Grosjean, D., Grosjean, E., and Williams, E. L. II, Atmospheric chemistry of olefins: a product study of the ozone–alkene reaction with cyclohexane added to scavenge OH, *Environ. Sci. Technol.*, 28, 186–196, 1994.

68. Morales, J. A., de Medina, H. L., de Nava, M. G., Velásquez, H., and Santana, M., Determination of organic acids by ion chromatography in rain water in the State of Zulia, Venezuela, *J. Chromatogr. A*, 671, 193–196, 1994.

69. Morales, J. A., de Graterol, L. S., Velásquez, H., de Nava, M. G., and de Borrego, B. S., Determination by ion chromatography of selected organic and inorganic acids in rainwater at Maracaibo, Venezuela, *J. Chromatogr. A*, 804, 289–294, 1998.

70. Amman, A. and Rüttimann, T. B., Simultaneous determination of small organic and inorganic anions in environmental water samples by ion-exchange chromatography, *J. Chromatogr. A*, 706, 259–269, 1995.

71. Hoffmann, P., Karandashev, V. K., Sinner, T., and Ortner, H. M., Chemical analysis of rain and snow samples from Chernogolokov/Russia by IC, TRXF and ICP-MS, *Fresenius J. Anal. Chem.*, 357, 1142–1148, 1997.

72. Jaffrezo, J. L., Calas, N., and Bouchet, M., Carboxylic acids measurements with ionic chromatography, *Atmos. Environ.*, 32, 2705–2708, 1998.

73. Keene, W. C., Galloway, J. N., and Holden, J. D., Measurement of weak organic acidity in precipitation from remote areas of the world, *J. Geophys. Res.*, 88, 5122–5130, 1983.

74. Kawamura, K. and Kaplan, I. R., Capillary G.C. determination of volatile organic acids in rain and fog samples, *Anal. Chem.*, 56, 1616–1620, 1984.

75. Kawamura, K., Steinberg, S., and Kaplan, I. R., Concentration of monocarboxylic and dicarboxylic acids and aldehydes in Southern California wet precipitations: comparison of urban and non-urban samples and compositional changes during scavenging, *Atmos. Environ.*, 30, 1035–1052, 1996.

76. Kawamura, K. and Kaplan, I. R., Stabilities of carboxylic acids and phenols in Los Angeles rainwater during storage, *Water Res.*, 24, 1419–1423, 1990.

77. Barcelona, M. J., Liljestrand, H. M., and Morgan, J. J., Determination of low molecular weight volatile fatty acids in aqueous samples, *Anal. Chem.*, 52, 321–325, 1980.

78. Kawamura, K., Steinberg, S., and Kaplan, I. R., Capillary G.C. determination of short-chain dicarboxylic acids in rain, fog and mist, *Int. J. Environ. Anal. Chem.*, 19, 175–188, 1985.

79. Kawamura, K., Identification of C_2–C_{10} ω-oxocarboxylic acids, pyruvic acid, and C_2–C_3 α-dicarbonyls in wet precipitation and aerosol samples by capillary GC and GC/MS, *Anal. Chem.*, 65, 3505–3511, 1993.

80. Kawamura, K., Kasukabe, H., and Barrie, L., Source and reaction pathways of dicarboxylic acids, ketoacids and dicarbonyls in Arctic aerosols: one year of observations, *Atmos. Environ.*, 30(10/11), 1709–1722, 1996.

81. Steinberg, S., Kawamura, K., and Kaplan, I. R., The determination of α-keto acids and oxalic acid in rain, fog and mist by HPLC, *Int. J. Environ. Anal. Chem.*, 19, 251–260, 1985.

82. Röder, A. and Bächmannn, K., Simultaneous determination of organic and inorganic anions in the sub-μmol/l range in rain water by capillary zone electrophoresis, *J. Chromatogr. A*, 689, 305–311, 1995.

83. Bächmannn, K., Haag, I., Prokop, T., Röder, A., and Wagner, P., Chromatographic methods for the analysis of size-classified and individual drops, *J. Chromatogr. A*, 643, 181–188, 1993.

84. Legrand, M., De Angelis, M., and Maupetit, F., Field investigation of major and minor ions along summit (Central Greenland) ice cores by ion chromatography, *J. Chromatogr. A*, 640, 251–258, 1993.

85. Haddad, P. R. and Jackson, P. E., Studies on sample preconcentration in ion chromatography. VIII. Preconcentration of carboxylic acids prior to ion-exclusion separation, *J. Chromatogr. A*, 447, 155–163, 1988.

86. Peldszus, S., Andrews, S. A., and Huck, P. M., Determination of short-chain aliphatic, oxo-and hydroxy acids in drinking water at low microgram per liter concentrations, *J. Chromatogr. A*, 723, 27–34, 1996.

87. Peldszus S., Huck, P. M., and Andrews S. A., Determination of carboxylic acids in drinking water at low μ/l concentrations: method development and application. Paper P1-h, in: *Proceedings of*

American Water Works Association Water Quality Technology Conference (WQTC), Boston, MA, 1996.

88. Xie Y. and Reckhow, D. A., Identification and quantification of ozonation by-products: ketoacids in drinking water. Paper five, in: *Proceedings of International Ozone Association Pan American Committee Pasadena Conference: Ozonation for Drinking Water Treatment*, 1992.

89. Xiong, F., Croué, J. P., and Legube, B., Long-term ozone consumption by aquatic fulvic acids acting as precursors of radical chain reactions, *Environ. Sci. Technol.*, 26, 1059–1064, 1992.

90. Fang, H. H. P. and Chui, H. K., Maximum COD loading capacity in UASB reactors at 37°C, *J. Environ. Eng.*, 119, 103–119, 1993.

91. Visser, A., Beeksma, I., van der Zee, F., Stams, A. J. M., and Lettinga, G., Anaerobic degradation of volatile fatty acids at different sulfate concentrations, *Appl. Microbiol. Biotechnol.*, 41, 549–556, 1993.

92. Chu, A., Mavinic, D. S., Kelly, H. G., and Ramey, W. D., Volatile fatty acid production in thermophilic aerobic digestion of sludge, *Water Res.*, 28, 1513–1522, 1994.

93. Eilersen, A. M., Henze, M., and Kloft, L., Effect of volatile fatty acids and trimethylamine in activated sludge, *Water Res.*, 28, 1329–1336, 1994.

94. Eilersen, A. M., Henze, M., and Kloft, L., Effect of volatile fatty acids and trimethylamine on denitrification in activated sludge, *Water Res.*, 29, 1259–1266, 1995.

95. Kong, I. C., Hubbard, J. S., and Jones, W. J., Metal-induced inhibition of anaerobic metabolism of volatile fatty acids and hydrogen, *Appl. Microbiol. Biotechnol.*, 42, 396–402, 1994.

96. Yeniguen, O., Kizilguen, K., and Yilmazer, G., Inhibition effects of zinc and copper on volatile fatty acid production during anaerobic digestion, *Environ. Technol.*, 17, 1269–1274, 1996.

97. Cruwys, J. A., Dinsdale, R. M., Hawkes, F. R., and Hawkes, D. L., Development of a static headspace gas chromatographic procedure for the routine analysis of volatile fatty acids in wastewaters, *J. Chromatogr. A*, 945, 195–209, 2002.

98. Pan, L., Adams, M., and Pawliszyn, J., Determination of fatty acids using solid-phase microextraction, *Anal. Chem.*, 67, 4396–4403, 1995.

99. Ábalos, M., Bayona, J. M., and Pawliszyn, J., Development of a headspace solid-phase microextraction procedure for the determination of free volatile fatty acids in waste waters, *J. Chromatogr. A*, 873, 107–115, 2000.

100. Ábalos, M. and Bayona, J. M., Application of GC coupled to chemical ionization mass spectrometry following headspace solid-phase microextraction for the determination of free volatile fatty acids in aqueous samples, *J. Chromatogr. A*, 891, 287–294, 2000.

101. Bradley, P. M., Chapelle, F. H., and Vroblesky, D. A., Does load affect microbial metabolism in aquifer sediments under different terminal electron accepting conditions?, *Geomicrobiol. J.*, 11, 85–94, 1993.

102. Soga, T. and Ross, G. A., Capillary electrophoretic determination of inorganic and organic anions using 2,6-pyridinedicarboxylic acid: effect of electrolyte's complexing ability, *J. Chromatogr. A*, 767, 223–230, 1997.

103. Karlsson, S., Wolrath, H., and Dahlen, J., Influence of filtration, preservation and storing on the analysis of low molecular weight organic acids in natural waters, *Water Res.*, 33, 2569–2578, 1999.

104. Johnson, S. K., Houk, L. L., Feng, J., Johnson, D. C., and Houk, R. S., Determination of small carboxylic acids by ion exclusion chromatography with electrospray mass spectrometry, *Anal. Chim. Acta*, 341, 205–216, 1997.

105. Ohta, K. and Ohashi, M., Separation of C1–C5 aliphatic carboxylic acids on a highly sulfonated styrene-divinylbenzene copolymer resin column with a C6 aliphatic carboxylic acid solution as the mobile phase, *Anal. Chim. Acta*, 481, 15–21, 2003.

106. Westerman, P., The effect of incubation temperature on steady-state concentrations of hydrogen and volatile fatty acids during anaerobic degradation in slurries from wetland sediments, *FEMS Microbiol. Ecol.*, 13, 295–302, 1994.

107. Shen, Y., Ström, L., Jönsson, J. A., and Tyler, G., Low-molecular weight organic acids in the rhizosphere soil solution of beech forest (*Fagus sylvatica l.*) cambisols determined by ion chromatography using supported liquid membrane enrichment technique, *Soil. Biol. Biochem.*, 28, 1163–1169, 1996.

108. Shen, Y., Obuseng, V., Grönberg, L., and Jönsson, J. A., Liquid membrane enrichment for the ion chromatographic determination of carboxylic acids in soil samples, *J. Chromatogr. A*, 725, 189–197, 1996.
109. Baziramakenga, R., Simard, R. R., and Leroux, G. D., Determination of organic acids in soil extracts by ion chromatography, *Soil Biol. Biochem.*, 27, 349–356, 1995.
110. van Hees, P. A. W., Andersson, A. M. T., and Lundström, U. S., Separation of organic low molecular weight aluminum complexes in soil solution by liquid chromatography, *Chemosphere*, 33, 1951–1966, 1996.
111. van Hees, P. A. W., Dahlén, J., Lundström, U. S., Borén, H., and Allard, B., Determination of low molecular weight organic acids in soil solution by HPLC, *Talanta*, 48, 173–179, 1999.
112. Strobel, B. W., Bernhoft, I., and Borggaard, O. K., Low-molecular-weight aliphatic carboxylic acids in soil solutions under different vegetations determined by capillary zone electrophoresis, *Plant Soil*, 212, 115–121, 1999.
113. Westergaard, B., Hansen, H. C. B., and Borggaard, O. K., Determination of anions in soil solutions by capillary electrophoresis, *Talanta*, 123, 721–724, 1998.
114. Li, Y., Huang, B., and Shan, X., Determination of low molecular weight organic acids in soil, plants and water by capillary zone electrophoresis, *Anal Bioanal. Chem.*, 375, 775–780, 2003.
115. Cízková, H., Brix, H., Kopecký, J., and Lukavská, J., Organic acids in the sediments of wetlands dominated by phragmites australis: evidence of phytotoxic concentrations, *Aquat. Bot.*, 64, 303–315, 1999.
116. Hordijk, C. A., Burgers, I., Phylipsen, G. J. A., and Cappenberg, T. E., Trace determination of lower volatile fatty acids in sediments by GC with chemically bonded FFAP columns, *J. Chromatogr.*, 511, 317–323, 1990.
117. Glaus, M. A., van Loon, L. R., Achatz, S., Chodura, A., and Fischer, F., Degradation of cellulosic materials under the alkaline conditions of a cementitious repository for low and intermediate level radioactive waste. Part I: Identification of degradation products, *Anal. Chim. Acta*, 398, 111–122, 1999.
118. Yan, C. T. and Jen, J. F., Determination of volatile fatty acids in landfill leachates by GC with distillation pretreatment, *Anal. Chim. Acta*, 259, 259–264, 1992.
119. Manni, G. and Caron, F., Calibration and determination of volatile fatty acids in waste leachates by gas chromatography, *J. Chromatogr. A*, 690, 237–242, 1995.
120. Beihoffer, J. and Ferguson, C., Determination of selected carboxylic acids and alcohols in groundwater by GC-MS, *J. Chromatogr. Sci*, 32, 102–106, 1994.
121. Jen, J. F., Lin, C. W., Lin, C. J., and Yan, C. T., Determination of volatile fatty acids in landfill leachates by high-performance liquid chromatography, *J. Chromatogr.*, 629, 394–397, 1993.
122. Fischer, K., Bipp, H. P., Bieniek, D., and Kettrup, A., Determination of monomeric sugar and carboxylic acids by ion-exclusion chromatography, *J. Chromatogr. A*, 706, 361–373, 1995.
123. Helaleh, M. I. H., Tanaka, K., Taoda, H., Hu, W., Hasebe, K., and Haddad, P. R., Qualitative analysis of some carboxylic acids by ion-exclusion chromatography with atmospheric pressure chemical ionization MS detection, *J. Chromatogr. A*, 956, 201–208, 2002.
124. Cawthray, G. R., An improved reversed-phase liquid chromatographic method for the analysis of low-molecular mass organic acids in plant root exudates, *J. Chromatogr. A*, 1011, 233–240, 2003.
125. Cofer, W. R. III, Collins, G. C., and Talbot, R. W., Improved aqueous scrubber for collection of soluble atmospheric trace gases, *Environ. Sci. Technol.*, 19, 557–560, 1985.
126. Grosjean, D., Organic acids in southern california air: ambient concentrations, mobile source emissions, in situ formation and removal processes, *Environ. Sci. Technol.*, 23, 1506–1514, 1989.
127. Winiwarter, W., Puxbaum, H., Fuzzi, S., Facchini, M. C., Orsi, G., Beltz, N., Enderle, K., and Jaeschke, W., Organic acid gas and liquid-phase measurements in po-valley fall-winter conditions in the presence of fog, *Tellus*, 40B, 348–357, 1988.
128. Galloway, J. N., Likens, G. E., Keene, W. C., and Miller, J. M., The composition of precipitation in remote areas of the world, *J. Geophys. Res.*, 11, 8771–8786, 1982.
129. Weiss, J., *Ion Chromatography*, 3rd ed., VCH, Weinheim, 2001.

130. Haddad, P. R. and Jackson, P. E., *Ion Chromatography: Principles and Applications*, Elsevier, Amsterdam, 1990.

131. Fritz, J. S. and Gjerde, D. T., *Ion Chromatography*, 3rd ed., VCH, Weinheim, 2000.

132. Richard, J. J., Chriswell, C. D., and Fritz, J. S., Concentration and determination of organic acids in complex aqueous samples, *J. Chromatogr.*, 199, 143–148, 1980.

133. Blau, K. and Halket, J., *Handbook of Derivatives for Chromatography*, 2nd ed., Wiley, Chichester, 1993.

134. Rosenfeld, J. M., Application of analytical derivatisations to the quantitative and qualitative determination of fatty acids, *Anal. Chim. Acta*, 465, 93–100, 2002.

135. Toyo'oka, T., Fluorescence tagging of physiologically important carboxylic acids, including fatty acids, for their detection in liquid chromatography, *Anal. Chim. Acta*, 465, 111–130, 2002.

136. Brondz, I., Development of fatty acid analysis by high-performance liquid chromatgraphy, gas chromatography, and related techniques, *Anal. Chim. Acta*, 465, 1–37, 2002.

137. Nohta, H., Sonoda, J., Yoshida, H., Satozono, H., Ishida, J., and Yamaguchi, M., Liquid chromatographic determination of dicarboxylic acids based on intramolecular excimer-forming fluorescence derivatisation, *J. Chromatogr. A*, 1010, 37–44, 2003.

138. Lima, E. S. and Abdalla, D. S. P., HPLC of fatty acids in biological samples, *Anal. Chim. Acta*, 465, 81–91, 2002.

139. Jennings, W., *Analytical Gas Chromatography*, 2nd ed., Academic Press, San Diego, CA, 1997.

140. Grob, K., *Classical Split and Splitless Injection in Capillary Gas Chromatography*, 2nd ed., Hüthig Verlag, Heidelberg, 1988.

141. Grob, K., *On-Column Injection in Capillary Gas Chromatography. Basic Technique, Retention Gaps, Solvent Effects*, Hüthig Verlag, Heidelberg, 1987.

142. Rood, D., *A Practical Guide to the Care, Maintenance, and Troubleshooting of Capillary Gas Chromatographic Systems*, Hüthig Verlag, Heidelberg, 1991.

143. Hübschmann, H. J., *Handbook of GC/MS: Fundamentals and Applications*, Wiley-VCH, Weinheim, 2001.

144. Grob, R. L., *Modern Practice of Gas Chromatography*, 3rd ed., Wiley, New York, 1995.

145. Wakeham, S. G., Monocarboxylic, dicarboxylic and hydroxy acids released by sequential treatments of suspended particles and sediments of the black sea, *Org. Geochem.*, 30, 1059–1074, 1999.

146. Pawliszyn, J., *Solid Phase Microextraction: Theory and Practice*, Wiley-VCH, New York, 1997.

147. Scheppers-Wercinski, S. A., *Solid Phase Microextraction: A Practical Guide*, Marcel Dekker, New York, 1999.

148. Lucy, C. A., Evolution of ion-exchange: from moses to the manhattan project to modern times, *J. Chromatogr. A*, 1000, 711–724, 2003.

149. Dionex Reference Library, P/N 053891-18, CD-ROM, Dionex Corp., Sunnyvale, CA, March 2003.

150. Buchberger, W. W., Detection techniques in ion analysis: what are our choices?, *J. Chromatogr. A*, 884, 3–22, 2000.

151. Ahrer, W. and Buchberger, W., Analysis of low-molecular-mass inorganic and organic anions by ion-chromatography-atmospheric pressure ionization mass spectrometry, *J. Chromatogr. A*, 854, 275–287, 1999.

152. Boring, C. B., Dasgupta, P. K., and Sjögren, A., Compact field portable capillary ion chromatograph, *J. Chromatogr. A*, 804, 45–54, 1998.

153. Hatsis, P. and Lucy, C. A., Ultra-fast HPLC separation of common anions using a monolithic stationary phase, *Analyst*, 127, 451–454, 2002.

154. Fischer, K., Bipp, H. P., Riemschneider, P., Leidmann, P., Bieniek, D., and Kettrup, A., Utilisation of biomass residues for the remediation of metal-polluted soils, *Environ. Sci. Technol.*, 32, 2154–2216, 1998.

155. Fischer, K., Environmental analysis of aliphatic carboxylic acids by ion-exclusion chromatography, *Anal. Chim. Acta*, 465, 157–173, 2002.

156. Ohta, K., Ohashi, M., Jin, J. Y., Takeuchi, T., Fujimoto, C., Choi, S. H., Ryoo, J. J., and Lee, K. P., Separation of aliphatic carboxylic acids and benzenecarboxylic acids by ion-exclusion chromatography with various cation-exchange resin columns and sulfuric acid eluent, *J. Chromatogr. A*, 997, 117–125, 2003.

157. Ohta, K., Tanaka, K., and Haddad, P. R., Ion-exclusion chromatography of aliphatic carboxylic acids on an unmodified silica gel column, *J. Chromatogr. A*, 739, 359–365, 1996.

158. Klampfl, C. W., Buchberger, W., Rieder, G., and Bonn, G. K., Retention behavior of acids on highly cross-linked poly(styrene-diviniylbenzene)-based and silica-based cation exchangers, *J. Chromatogr. A*, 770, 23–38, 1997.

159. Ohta, K., Towata, A., and Ohashi, M., Ion-exclusion chromatographic separations of C_1–C_6 aliphatic carboxylic acids on a sulfonated styrene-divinylbenzene co-polymer resin column with 5-methylhexanoic acid as eluent, *J. Chromatogr. A*, 997, 107–116, 2003.

160. Tanaka, K., Mori, M., Xu, Q., Helaleh, M. I. H., Ikedo, M., Taoda, H., Hu, W., Hasebe, K., Fritz, J. S., and Haddad, P. R., Ion-exclusion chromatography with conductimetric detection of aliphatic carboxylic acids on a weakly acidic cation-exchange resin by elution with benzoic acid-β-cyclodextrin, *J. Chromatogr. A*, 997, 127–132, 2003.

161. Tanaka, K., Ohta, K., and Fritz, J. S., Ion exclusion chromatography of aliphatic carboxylic acids on a cation-exchange resin by elution with polyvinyl alcohol, *J. Chromatogr. A*, 770, 211–218, 1997.

162. Morris, J. and Fritz, J. S., Eluent modifiers for the liquid chromatographic separation of carboxylic acids using conductivity detection, *Anal. Chem.*, 66, 2390–2396, 1994.

163. Tanaka, K., Ohta, K., Fritz, J. S., Lee, Y. S., and Shim, A. B., Ion-exclusion chromatography with conductimetric detection of aliphatic carboxylic acids on an H^+-form cation-exchange resin column by elution with polyols and sugars, *J. Chromatogr. A*, 706, 385–393, 1995.

164. Widiastuti, R. and Haddad, P. R., Approaches to gradient elution in ion-exclusion chromatography for carboxylic acids, *J. Chromatogr.*, 602, 43–50, 1992.

165. Ohta, K., Towata, A., and Ohashi, M., Ion-exclusion chromatographic behavior of aliphatic carboxylic acids and benzenecarboxylic aids on a sulfonated styrene-divinylbenzene co-polymer resin column with sulfuric acid containing various alcohols as eluent, *J. Chromatogr. A*, 997, 95–106, 2003.

166. Jones, W. R., Jandik, P., and Schwartz, M. T., Automated dual column coupled system for simultaneous determination of carboxylic acids and inorganic anions, *J. Chromatogr.*, 473, 171–188, 1989.

167. Tanaka, K., Ding, M. Y., Takahashi, H., Helaleh, M. I. H., Taoda, H., Hu, W., Hasebe, K., Haddad, P. R., Mori, M., Fritz, J. S., and Sarzanini, C., Vacancy ion-exclusion chromatography of carboxylic acids on a strong acidic cation-exchange resin, *Anal. Chim. Acta*, 474, 31–35, 2002.

168. Tanaka, K., Ding, M. Y., Helaleh, M. I. H., Taoda, H., Takahashi, H., Hu, W., Hasebe, K., Haddad, P. R., Fritz, J. S., and Sarzanini, C., Vacancy ion-exclusion chromatography of carboxylic acids on a weakly acidic cation-exchange resin, *J. Chromatogr. A*, 956, 209–214, 2002.

169. Swadesh, J., *HPLC: Practical and Industrial Applications*, 2nd ed., CRC Press, Boca Raton, FL, 2001.

170. Hanai, T., *HPLC: A Practical Guide*, Royal Society of Chemistry, Cambridge, 1999.

171. Neue, U. D., *HPLC Columns: Theory, Technology and Practice*, Wiley-VCH, New York, 1997.

172. Chen, S. H. and Chuang, Y. J., Analysis of fatty acids by column liquid chromatography. Review, *Anal. Chim. Acta*, 465, 145–155, 2002.

173. Baker, D. R., *Capillary Electrophoresis*, Wiley, New York, 1995.

174. Li, S. F. Y., *Capillary Electrophoresis: Principles, Practice and Applications*, Elsevier, Amsterdam, 1992.

175. Landers, J. P., Ed., *Handbook of Capillary Electrophoresis*, CRC Press, Boca Raton, FL, 1994.

176. Klampfl, C. W. and Buchberger, W., Determination of low-molecular-mass organic acids by capillary zone electrophoresis, *TrAC—Trends in Analytical Chemistry*, 16, 221–229, 1997.

177. Volgger, D., Zemann, A. J., Bonn, G. K., and Antal, M. J., High-speed separation of carboxylic acids by co-electroosmotic capillary electrophoresis with direct and indirect UV detection, *J. Chromatogr. A*, 758, 263–276, 1997.

178. Harrold, M., Stillian, J., Bao, L., Rocklin, R., and Avdalovic, N., Capillary electrophoresis of inorganic anions and organic acids using suppressed conductivity detection. Strategies for selectivity control, *J. Chromatogr. A*, 717, 371–383, 1995.

179. Arellano, M., Andrianary, J., Dedieu, F., Couderc, F., and Puig, P., Method development and validation for the simultaneous determination of organic and inorganic acids by capillary zone electrophoresis, *J. Chromatogr. A*, 765, 321–328, 1997.

180. Buchberger, W., Klampfl, C. W., Eibensteiner, F., and Buchgraber, K., Determination of fermenting acids in silage by capillary zone electrophoresis, *J. Chromatogr. A*, 766, 197–203, 1997.

181. Haddad, P. R., Harakuwe, A. H., and Buchberger, W., Separation of inorganic anionic components of bayer liquor by capillary zone electrophoresis. I. Optimization of resolution with electrolyte-containing surfactant mixtures, *J. Chromatogr. A*, 706, 571–578, 1995.

182. Oeffner, P. J., Surface charge reversed capillary zone electrophoresis of inorganic and organic acids, *Electrophoresis*, 16, 46–56, 1995.

183. Wu, C. H., Lo, Y. S., Lee, Y. H., and Lin, T. I., Capillary electrophoretic determination of organic acids with indirect detection, *J. Chromatogr. A*, 716, 291–301, 1995.

184. Kuo, C. Y., Krasner, S. W., Stalker, G.A., and Weinberg, H.S., Analysis of inorganic disinfection by-products in ozonated drinking water by ion chromatography. *Proceedings of American Water Works Association: Water Quality and Technology Conference (WQTC)*, San Diego, CA, pp. 503–525, 1990.

185. Dionex Application Note 101. Trace Level Determination of Bromate in Ozonated Drinking Water Using Ion Chromatography. Dionex Corp., Sunnyvale, CA, 1995.

186. Slingsby, R. W. and Pohl, C. A., Approaches to sample preparation for ion chromatography, sulfate precipitation on barium-form ion exchangers, *J. Chromatogr. A*, 739, 49–55, 1996.

187. Senorans, F. J. and Ibanez, E., Analysis of fatty acids in foods by supercritical fluid chromatography, *Anal. Chim. Acta*, 465, 131–144, 2002.

188. Ng, L. K., Analysis by gas chromatography/mass spectrometry of fatty acids and esters in alcoholic beverages and tobaccos, *Anal. Chim. Acta*, 465, 309–318, 2002.

189. Sparks, D. L., *Environmental Soil Chemistry*, Academic Press, New York, 1995.

14 BTEX

Nacho Martín García and Leo M.L. Nollet

CONTENTS

I. INTRODUCTION

BTEX is the abbreviation for a group of volatile organic compounds (VOCs) consisting of benzene, toluene, ethylbenzene, and *o*-, *p*-, and *m*-xylenes. The presence of BTEX in the environment is caused mainly by the industrial use as solvents and in the production of organic chemicals such as rubbers, plastics, resins, nylon, lubricants, dyes, detergents, drugs, pesticides, paints, lacquers, and adhesives. BTEX contamination is also related to petroleum due to the high concentration in gasolines and in other derivate products such as diesel fuel, lubricating, and heating oil.

The characteristics and the uses of BTEX can cause them to be present in many environments. This represents a hazard to the environment itself as well as to public health. That is the reason why these are priority pollutants in environmental organizations. In order to limit the impact of BTEX on the environment, it is necessary to know the most accurate, sensitive, and reliable analytical methods to detect their presence.

II. PHYSICOCHEMICAL, TOXICOLOGICAL, AND ECOLOGICAL ASPECTS OF BTEX

Although BTEX compounds have a similar molecular structure they show slightly different physicochemical properties. In normal conditions these are colorless liquids presenting a sweet odor, but the most important characteristics are their high volatility and low solubility in water. These and other physicochemical properties for different BTEX compounds are shown in Table 14.1.[1]

TABLE 14.1
Physicochemical Properties of BTEX Compounds

Property	Benzene	Toluene	Ethylbenzene	*m*-Xylene	*p*-Xylene
Chemical structure					
CAS number	71-43-2	108-88-3	100-41-4	108-38-3	106-42-3
Molecular weight (g/mol)	78.11	92.13	106.16	106.16	106.16
Water solubility (g/l)	0.7	0.5	0.2	0.2	0.2
Vapor pressure (hPa 20°C)	101	29	9.3	8	8.2
Density (20°C g/ml)	0.8786	0.8669	0.8670	0.8642	0.8611
Octanol–water partition coefficient K_{ow} 20°C	135	489	1412	—	1510
Henry's law constant (atm cm^3/mol)	126	340	528	—	831
Melting point (°C)	−11	−95	−95.01	−47.4	13–14
Boiling point (°C)	80.1	110.6	136.2	139.1	138.35

From Valor, I., In *Handbook of Water Analysis*, Nollet, Leo, M. L., Ed., Marcel Dekker, New York, 2000. With permission.

Some aspects of how the different BTEX compounds can affect human health follow.

Benzene. Benzene is recognized as the most toxic compound among BTEX, because it has been proved that breathing very high concentrations of benzene in air can cause death and that long-term exposure to lower levels causes leukemia. The Occupational Safety and Health Administration (OSHA) has set a permissible exposure limit of 1 ppm in the workplace during an 8-h day if 40 h a week are worked.

For water, the U.S. Environmental Protection Agency (EPA) has set the maximum permissible level for benzene in drinking water at 5 $\mu g/l$, while in Europe a maximum concentration of 1 $\mu g/l$ is permitted.

In air, the maximum limit recommended by the European Community is 10 mg/l.

Toluene. Compared to benzene, toluene's toxicity is very low. Whereas very high concentration exposures are needed to cause death, moderate ones may affect the nervous system. There is no evidence for the carcinogenicity of toluene. For air, the OSHA has set a limit of 200 ppm in the workplace and the EPA's drinking water limit is set at 1 mg/l.

Ethylbenzene. Ethylbenzene's toxicity is low and there is no evidence that it causes cancer. EPA drinking water and OSHA occupational exposure limits are set at 0.7 mg/l and 100 ppm (8 h/day per 40 h/week), respectively.

Xylenes. Like toluene, high concentration exposures to xylene can cause death, while moderate ones can affect the brain. No evidence for the carcinogenicity of xylene is found. EPA drinking water and OSHA occupational exposure limits are set at 10 and 100 ppm (8 h/day per 40 h/week), respectively.

III. ANALYSIS OF BTEX IN WATER

BTEX are found in several water matrices; they occur in groundwater due to the leakage from underground storage tanks or from pipeline cracks. Industrial and domestic effluents are also sources of BTEX in surface waters and wastewaters. Very commonly, BTEX in low concentrations have to be monitored when being accompanied by other contaminants in higher concentrations like metyl-*tert*-butyl ether and degradation products in gasoline-contaminated groundwater, and like soluble or insoluble organic matter in surface and wastewaters. In order to detect BTEX in such low levels and separate them from the water matrix, suitable sample preparation methods have to be used. As seen in the next section, most common sample preparation methods for BTEX involve the partial volatilization of samples, taking advantage of their high volatility, to achieve a selective separation from the matrix. Other methods use the low polarity of BTEX to selectively extract them from water.

A. SAMPLE PREPARATION METHODS

Sample preparation methods involve the extraction of volatile compounds from their matrices. Extraction of organic volatile compounds from environmental matrices has been carried out using gases, liquids, or solids. Nowadays, most sample preparation methods for the analysis of VOCs in water use gases or solids as extracting agents because liquid–liquid extraction methods (LLE) present several disadvantages:

1. These methods require several steps, making the cleanup process tedious and difficult to automate.
2. Highly purified solvents, which are expensive to purchase and to dispose of, are needed.
3. Several of these solvents have to be carefully handled because their use is dangerous and exposure to them can affect human health.

In the next sections three solventless methods for analyzing BTEX are presented. Purge and trap (P&T) is the most widely used technique for the analysis of organic volatiles in any kind of water, and it is the official method in many countries. Solid-phase microextraction (SPME) and membrane extraction (ME) are solventless extraction methods being widely used and further investigated.

1. Purge and Trap

In P&T volatile analytes from water samples are stripped with an inert gas, trapped into a solid sorbent, and thermally desorbed into the chromatographic device for separation of the target species. Therefore a P&T method involves three processes carried out in different devices. Figure 14.1[1] shows a schematic drawing of a P&T automated system. In the stripping–adsorption step, after an aliquot of the sample is loaded into the sample chamber, the extracting gas (generally helium or nitrogen) passes through it at a specified flow rate and temperature, and for a specified time. In this way volatile analytes from the aqueous phase are transferred into the gas-phase stream. Then, by making the stripping gas stream flow through a column in which a solid adsorbent is packed (sorbent trap), analytes are separated from the gas phase and retained on a solid phase. Finally, target compounds are desorbed from the trap by a rapid increase in temperature and a small volume of inert gas that carries them into the chromatographic system for further analysis.

a. Factors Affecting the Technique

i. Purge Parameters

(i) *Sample Volume.* For low-solubility, high vapor pressure analytes (relatively small, nonpolar molecules) like BTEX, recoveries from the aqueous phase should not depend on the purging liquid volume.[2] Generally, 5 to 25 ml aliquots of liquid samples are loaded into the purging chamber, choosing a volume to achieve good detection limits for the concentration level of the sample.

(ii) *Extraction Temperature and Stripping Gas Volume.* These are the main parameters affecting the extraction efficiency in the purge process. Better recoveries are achieved by using high temperatures and long purging times (or large stripping gas volumes). High temperatures will transfer excessive moisture to the gas phase leading to problems with the trap, the chromatographic column, and the detector system. On the other hand, very high stripping volumes will cause breakthrough problems. Therefore, a compromise has to be achieved between these two variables. Table 14.2 shows purge conditions applied for the analysis of BTEX and other volatiles in water samples.

ii. Trap Parameters

(i) *Adsorption Temperature.* The adsorption process generally improves with lower temperatures. To minimize breakthrough volumes, the concentration trap should operate at room temperature.

(ii) *Solid Adsorbent.* The choice of the proper adsorbent should be based mainly on the breakthrough volumes for the target analytes, but thermal stability, pure chromatographic blanks, presence of irreversible active sites, and low affinity for water should be taken into account. The most suitable and most often used adsorbent materials for the analysis of BTEX by P&T technique are porous polymers like Tenax, because they eliminate the problems associated with water retention, irreversible adsorption, and thermal decomposition phenomena. In spite of their lower capacity compared to activated charcoal and graphitized sorbents (due to the lower specific surface), porous polymers are excellent adsorbents for nonpolar compounds such as BTEX.

The recommended trap packing by the EPA for the analysis of volatiles in drinking water, when only compounds above 35°C boiling point are to be analyzed, consists of

ADSORPTION STEP

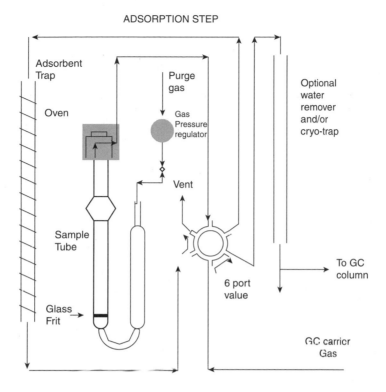

DESORPTION STEP

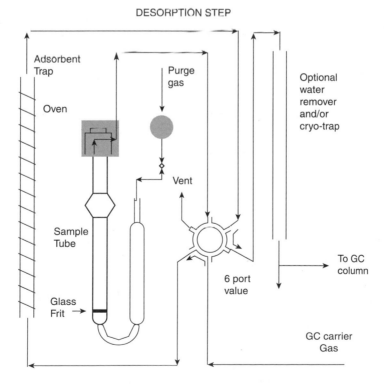

FIGURE 14.1 Schematic drawing of a modern online purge and trap system. (From Valor, I., In *Handbook of Water Analysis*, Nollet, Leo, M. L., Ed., Marcel Dekker, New York, 2000. With permission.)

TABLE 14.2
Purge and Trap Conditions for Water Analysis of BTEX

Water Sample Type	Experimental Conditions	Comments	Reference
Method testing for groundwater analysis	Sample volume: 5 ml Trap: see comments Extraction temperature: 60°C Purge gas: Helium Purge flow: 30 ml/min Purge time: 30 min Desorption temperature: 180°C Desorption time: 4 min	A PTI Sample concentrator (O.I Analytical, TX, USA) equipped with a multibed TENAX/SIGEL/CMS was used	3
Method testing for groundwater analysis	Sample volume: 15 ml Trap: Tenax silica gel Extraction temperature: room Purge gas: helium Purge flow: 30 ml/min Purge time: 13 min Desorption temperature: 225°C Desorption time: 4 min	A commercial Tekmar 3100 purge and trap concentrator coupled to an aquatek 70 liquid autosampler (Tekmar-Dohrmann, USA) was used	4
Heavy loaded aqueous samples	Sample volume: 10 ml Trap: — Extraction temperature: room Purge gas: purified nitrogen Purge flow: 43 ml/min Purge time: 10 min Desorption temperature: 250°C Desorption time: — min	(1) Macrotrap packed with 105 mg Tenax GC and 20 mg of Carbosieve S-III. Desorption time: 10 min (2) Microtrap packed with 15 mg Tenax GC and 20 mg of Carbosieve-SIII. Desorption time: 3 min	5
Method testing	Sample volume: 1 ml Trap: Tenax TA 60–80 mesh Extraction temperature: ambient Purge gas: nitrogen Purge flow: 40 ml/min Purge time: 30 min Desorption temperature: 250°C Desorption time: 10 min	—	6
River water	Sample volume: 4 ml Trap: Tenax Extraction temperature: —°C Purge gas: helium Purge flow: — ml/min Purge time: 11 min Desorption temperature: 180°C Desorption time: 4 min	—	7
River water	Sample volume: 5 ml Trap: VOCARB 3000 trap Extraction temperature: 20°C Purge gas: helium Purge flow: 40 ml/min Purge time: 11 min Desorption temperature: 250°C Desorption time: 2 min	—	8

Continued

TABLE 14.2
Continued

Water Sample Type	Experimental Conditions	Comments	Reference
Groundwater	Sample volume: 13 ml Trap: Tenax silica gel Extraction temperature: ambient Purge gas: helium Purge flow: 35 ml/min Purge time: 11 min Desorption temperature: 225°C Desorption time: 3 min	Helium at 3.5 ml/min was used for desorption	9

Spiked water
  solutions
	Sample volume: 5 ml
	Trap: 0.2–0.45 g of adsorbent
	Extraction temperature: ambient
	Purge gas: helium
	Purge flow: 3 ml/min
	Purge time: 11 min
	Desorption temperature: 190°C
	Desorption time: 5 min
		Helium at 10 ml/min was used
		  for desorption
		Several adsorbents were tested:
		(1) Tenax, Carboxen 569,
		  coconut charcoal,
		  Carbosieve SIII
		(2) Activated carbon from
		  macadamia, hazelnut, and
		  walnut shells
			10

poly(2,6-diphenyl-*p*-phenylene oxide), commercially called Tenax. As seen in Table 14.2, traps can also be packed with multiple solid sorbents in order to recover different types of analytes together with BTEX.[4,5]

iii.   Desorption
The desorption step takes place by a rapid increase in temperature and by a convenient volume of inert gas in order to transport the analytes to the chromatographic system in a short time and through a narrow band. The faster the heating is, the quicker the analytes are desorbed, making possible adsorption in a short time with a small volume of inert gas, which improves the chromatographic separation and analytical signal. Table 14.2 also shows the desorption parameters for the analysis of BTEX in water samples.

2.   Solid-Phase Microextraction

SPME is a solventless sample preparation technique that can be applied to the analysis of BTEX in water, air, and soils. In SPME, analytes from aqueous or gas phases are concentrated by absorption into a solid phase. The sampling device consists of a short, thin rod of fused silica (typically 1 cm length and 0.11 mm diameter), coated with an absorbent polymer (SPME fiber), attached to a metal rod (fiber holder), and surrounded (in the standby position) by a protective sheath. This fiber holder is mounted in a modified gas chromatography (GC) syringe (see Figure 2.12 Chapter 2 of this book).

Two different processes take place in an SPME analysis:

1. Absorption of analytes over the polymeric fiber by putting it in contact with the matrix until equilibrium is reached. It is the most important step influencing the effectiveness of

the analysis, and factors such as fiber selection, sampling method, extraction time, agitation, and temperature should be taken into consideration.

2. Desorption. Once equilibrium is achieved the fiber is withdrawn into the protective sheath. Immediately after, the sheath is inserted into the septum of a GC injector, the plunger is pushed down, and the fiber is forced into the injection insert where the analytes are thermally desorbed and separated in the GC column.

a. Selection of the Solid-Phase Microextraction Fiber

Several characteristics of the polymeric coatings affect the performance of SPME. Using thick coatings requires larger equilibrium and desorption times, and carryover effects may appear. See Chapter 2 of this volume for further details.

Although 100 μm polydimethylsiloxane (PDMS) film seems to be the most suitable and widely used for the analysis of BTEX,[11–13] 75 μm Carboxen–PDMS seems have higher extraction efficiencies for low volatility compounds.[14–18] Figure 14.2 shows higher responses for Carboxen–PDMS than for PDMS fiber, but it is also seen how the response of the former depends too much on the total concentration of analytes, probably because an adsorption or condensation mechanism also takes place in the extraction process. Thus, the repeatability for the 75 μm Carboxen–PDMS is lower than for 100 μm PDMS.

Other SPME fibers have been demonstrated to give better performance than 100 μm PDMS fibers. A porous layer activated charcoal (PLAC) coating showed detection limits between 1.5 and 2 pg/ml for headspace (HS)-SPME analysis of BTEX in water samples, while for PDMS fiber detection limits were in the range of 190 to 700 pg/ml.[19] Polymeric fullerene was also tested for the headspace analysis of BTEX where greater extraction recoveries compared to those obtained for PDMS fiber were observed.[20]

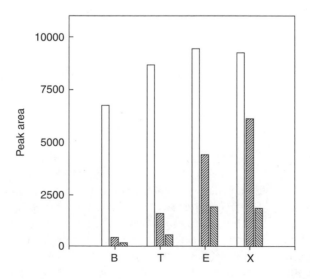

FIGURE 14.2 Effect of fiber coatings and composition on the extracted amount at 0.1 mM. □, 75 μm Carboxen–PDMS fiber, single-component extraction; ▨, 75 μm Carboxen–PDMS fiber, multi component extraction; ▨, 100 μm PDMS fiber, single- and multi component extraction. (Reprinted from Cho, H. J., Baek, K., Lee, H. H., Lee, S. H., and Yang, J. W., *J. Chromatogr. A*, 988(2), 177–184, 2003, Copyright (2003), with permission from Elsevier.)

b. Sampling the Liquid Phase or the Headspace

When SPME is applied to analysis of water samples, the sample is placed in a vial with a cap that contains a septum. In order to extract the analytes from the sample, the sheath is pushed through the septum and the plunger is lowered, forcing the fiber into the vial, where it is immersed into the aqueous sample or its headspace. Then the SPME fiber absorbs organic compounds from the phase that is in contact with it until equilibrium is achieved.

There are many differences between sampling from the liquid phase (direct SPME) and from the headspace (HS-SPME). The factors affecting direct SPME and HS-SPME, and the conditions that lead to the optimum performance of the analytical method, are different due to the nature of each process. In direct SPME the mass transfer rate of analytes is limited by the diffusion in the liquid phase, while in HS-SPME the limiting rate is the transport of analytes from the sample to the headspace. Because diffusion in the liquid phase is much slower than in the headspace and transport of analytes from the liquid to the vapor phase can be accelerated by proper conditions, the time taken to reach equilibrium by HS-SPME is shorter than in direct SPME. A comparative study[11] showed, how for the optimal conditions of each method, the time taken to reach equilibrium in HS-SPME was shorter than for direct SPME (see Table 14.3). Limits of detection were also slightly better for HS-SPME than for direct SPME.

The final selection between direct and HS-SPME depends on the nature of the water matrix and the target compounds being analyzed. HS-SPME is recommended over direct SPME in order to avoid contamination of the fiber when dirty samples are being analyzed. Sampling from the headspace also involves a selective separation of the most volatile compounds like BTEX from the liquid matrix and protects the GC column from high-molecular-mass nonvolatile compounds. Direct SPME is recommended only for clean water samples and for the analysis of high boiling point analytes.

c. Direct Solid-Phase Microextraction

The amount of analyte extracted in direct SPME, N from a volume V_2 of a sample with an initial concentration C_0, can be calculated by the following equation:

$$N = C_0 V_1 V_2 K / (KV_1 + V_2) \tag{14.1}$$

where V_1 is the volume of the fiber and K is the partition coefficient of the analyte between the coating and the aqueous sample. Because the volume of the coating is very small compared to the sample volume ($KV_1 \ll V_2$), the amount of analyte extracted results independent from the volume of the sample.

$$N = C_0 V_1 K \tag{14.2}$$

The parameters of interest in SPME are the ones that increase the diffusion of analytes in the liquid phase without decreasing the concentration of the analytes in it.

i. Agitation
As previously stated, diffusion of the analytes in the liquid phase is the limiting mechanism in the absorption process. Therefore stirring of the samples reduces the time to reach equilibrium.

ii. Temperature
Increase in temperature during the extraction process enhances the diffusion during the extraction process towards the fiber, decreasing the time to reach equilibrium. However, the distribution constants of the analytes decrease with increasing temperature because it also favors evaporation of BTEX towards the headspace, reducing the concentration in the liquid phase. Therefore, as seen in Table 14.3, room temperature is often the most suitable temperature for BTEX analysis.

TABLE 14.3
SPME Conditions for Water Analysis of BTEX

Water Sample Type	Experimental Conditions	Comments	Reference
Method optimization for application in contaminated waters resulting tar residues leaching	Vial volume: 12 ml Sample volume: 10 ml Fiber coating: 100 μm PDMS Extraction temperature: room Extraction time: 10 min Agitation: 900 rpm Desorption temperature: 150°C Desorption time: 3 min	—	11
Method optimization for application in contaminated waters resulting tar residues leaching	Vial volume: 22 ml Sample volume: 11 ml Fiber coating: 100 μm PDMS Extraction temperature: room Extraction time: 4 min Agitation: 900 rpm Desorption temperature: 150°C Desorption time: 3 min	—	11
Method testing	Vial volume: 4 ml Sample volume: 1 ml Fiber coating: 100 μm PDMS Extraction temperature: room Extraction time: 30 min Agitation: Yes Desorption temperature: 150°C Desorption time: 1 min	—	6
Method testing for application of waste water from paintwork industry and same water after passing sewage purification plant	Vial volume: 4 ml Sample volume: 1.3 ml Fiber coating: 100 μm PDMS Extraction temperature: 25°C Extraction time: — min Agitation: Yes Desorption temperature: 180°C Desorption time: 1.5 min	Extraction time was dependent on the BTEX concentration: 2 min for 425 mg/l 5 min for 4250 mg/l	12
Spiked water solutions	Vial volume: 4 ml Sample volume: 2.5 ml Fiber coating: 100 μm PDMS Extraction temperature: 25°C Extraction time: 5 min Agitation: Yes Desorption temperature: 180°C Desorption time: 3 min	Open cap and closed septa vials were tested	13
Spiked water solutions	Vial volume: 8 ml Sample volume: 5 ml Fiber coating: — Extraction temperature: room Extraction time: 60 min Agitation: 1000 rpm Desorption temperature: 250°C Desorption time: 2 min	Many fiber coatings were tested: (1) 7, 30, and 100 mm PDMS (2) 65 mm PDMS–divinylbenzene (3) 85 mm Polyacrylate (4) 65 mm Carbowax–divinylbenzene (5) 75 mm Carboxen–PDMS (6) 65 mm c8 fiber	14

Continued

TABLE 14.3
Continued

Water Sample Type	Experimental Conditions	Comments	Reference
Method testing for application in groundwater	Vial volume: 2 ml Sample volume: 0.6 ml Fiber coating: 75 μm CAR–PDMS Extraction temperature: room Extraction time: 15 min Agitation: yes Desorption temperature: 280°C Desorption time: 15 min	CAR–PDMS stands for carboxen–polydimethylsiloxane 0.15 g of salt was added	15
Spiked water sample with single- and multi-component BTEX	Vial volume: 20 ml Sample volume: 10 ml Fiber coating: — Extraction temperature: 25°C Extraction time: 30 min Agitation: 300 rpm Desorption temperature: — °C Desorption time: 2 min	75 mm CAR–PDMS and 100 mm PDMS fibers were used for comparison. Their desorption temperatures were: 300°C for CAR–PDMS; 250°C for PDMS	16
Groundwater	Vial volume: 22 ml Sample volume: 10 ml Fiber coating: 75 μm CAR–PDMS Extraction temperature: room Extraction time: 30 min Agitation: yes Desorption temperature: — °C Desorption time: — min	—	17
Method testing for application in local tap water, bi-distilled water, deionized water, and parenteral preparation	Vial volume: 70 ml Sample volume: 50 ml Fiber coating: 100 μm PLAC Extraction temperature: 25°C Extraction time: 15 min Agitation: yes Desorption temperature: 280°C Desorption time: 1 min	PLAC stands for porous layer activated charcoal The addition of 15 g of salt improved BTEX recovery	19
Method testing	Capillary length (cm): 5.7, 13.5 and 170 Capillary i.d. (mm): 0.21 Stationary phase: 48 nm PDMS (1% vinyl groups), 0.3% (w/w) crosslinked with dicumyl peroxide Extraction temperature: 20–23°C Extraction time: — Sample volume: 5 ml	Extraction time was about 30 sec for 5.7 and 13.5 capillaries	21

Continued

TABLE 14.3
Continued

Water Sample Type	Experimental Conditions	Comments	Reference

Method testing	Capillary length (cm): 15
	Capillary i.d. (mm): 0.16
	Stationary phase: 0.5 μm PDMS
	  (1% vinyl groups),
	  2% (w/w) crosslinked with
	  dicumyl peroxide
	Extraction temperature:
	  ambient
	Extraction time: 10 sec
	Sample volume: 1 ml
		—	22

Method testing	Capillary length (cm): 70–75
	Capillary i.d. (mm): 0.474
	Stationary phase: PDMS
	  (1% vinyl groups),
	  2% (w/w) crosslinked
	  with dicumyl peroxide
	Extraction temperature:
	  ambient
	Extraction time: —
	Sample volume: 1 ml
		20 extractions of 20 sec
		  each were enough to
		  achieve equilibrium
			23

River water	Capillary length (cm): 2
	Capillary i.d. (mm): 0.32
	Stationary phase: 1.1 μm
	  CP-Sil 5 CB
	  stationary phase
	Extraction temperature: 30°C
	Sample flow rate: 1.5 ml/min
	Sample volume: 24 ml
		—	24

iii.   Salting Out Effect

Addition of salt to aqueous samples is often used to drive polar compounds into the headspace. Therefore addition of salt should not increase the recovery of BTEX compounds when direct SPME is applied.

d.   Headspace Solid-Phase Microextraction

The amount of analyte extracted in HS-SPME, N from the headspace volume V_3 above a volume V_2 of a sample with an initial concentration C_0 can be calculated by the following equation

$$N = C_0 V_1 V_2 K / (K_1 K_2 V_1 + K_2 V_3 + V_2) \qquad (14.3)$$

where K_1 is the partition coefficient between the liquid and the headspace and K_2 the partition coefficient between the headspace and the fiber. As in direct SPME the volume of the fiber is much smaller than the sample and the headspace volumes, and Equation 14.3 can be rewritten as

$$N = C_0 V_1 V_2 K / (K_2 V_3 + V_2) \qquad (14.4)$$

The parameters of interest in HS-SPME are the ones that increase the concentration of analytes to the headspace.

i.   Salting Out Effect
Although adding salt has more effect on polar compounds there are some reports[11,19] that demonstrate greater recoveries for HS-SPME.

ii.   Temperature
Increasing temperature decreases the time needed to reach equilibrium as well as the amount of analyte extracted. Extraction recoveries at a constant temperature increase with exposure time and reach a plateau when equilibrium is established. This can be explained because the rate-limited step, the transport of analytes from the liquid to the headspace, is speeded up. The decrease in the extracted quantities of analytes onto the fiber (specially the less volatile) with increasing temperature is a result of the exothermic process of adsorption.[19]

iii.   Headspace/Sample Volume Ratio
As seen in Equation 14.4, increasing the volume of the liquid sample V_2 and reducing that of the headspace V_3 can increase the amount of analyte extracted onto the fiber. Figure 14.3 shows an increase of the analytical signal for every BTEX with the increase of the liquid sample volume in 4 ml vials.[13]

iv.   Losses of Volatiles
HS-SPME analysis can lead to erroneous results due to noncontrolled analyte losses through cracks in membranes used to seal vials. The cracks result from piercing the membrane with a SPME device needle. Recently open Teflon caps with a narrow bore for introducing the SPME syringe have been used to ensure constant and predictable losses. A comparative study on the performance of opened cap and closed septa vials was carried out by Matisova et al.[12] Good agreement between open cap and closed septa vials confirmed the negligible losses of analytes. It was also observed that repeatability (expressed as the relative standard deviation) was better for open cap than for closed septa vials, probably because holes in the rubber septum produced by the penetration

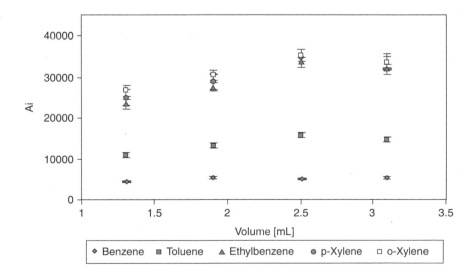

FIGURE 14.3 The dependences of peak area of BTEX determined by headspace SPME-GC with PMDS fiber on the volume of the spiked water sample with concentration level 42 μg l^{-1} in 4 ml vials. (Reprinted from Matisova, E., Medved'ova, M., Vraniakova, J., and Simon, P., *J. Chromatogr. A*, 960(1–2), 159–164, 2002, Copyright (2002), with permission from Elsevier.)

of the fiber needle are different in size and shape, allowing analytes to escape at different rates. In spite of the higher precision achieved by open cap vials, the technique introduces more parameters into the system, especially the time of taking the sample. Therefore, this sampling method is not recommended for the analysis of HS-SPME with classical autosamplers.

Losses of volatiles are more important when SPME onsite sampling is carried out because evaporation to the atmosphere after sampling is unavoidable when GC injection is delayed. An interesting approach to improve volatile recoveries such as BTEX is intube SPME (IT-SPME or capillary extraction),[21–23] in which analytes arc extracted by short open tubular traps (OTTs) consisting of common capillary columns (capillary extractors) similar to those used in GC. During extraction, syringe needles are connected to the extractor through press-fit caps. The syringe connected to the capillary column inlet introduces the sample and the other one collects it at the extractor exit (Figure 14.4). After sampling, the capillary extractor is attached as a GC precolumn.

Capillary extraction is a SPME technique having several advantages[22]:

1. It is very fast, taking just a few seconds.
2. Headspace has no influence in the extraction performance because the capillary is always filled with liquid.
3. Heated injectors are not required because the extractor itself can assume the role of precolumn injector liner.
4. All the extracted analyte is injected when capillary extractors are used, while the injection of the SPME fiber in the gas chromatograph produces a certain amount of analyte losses.
5. Capillary extractors are inert and reusable, because extracting the phase is immobilized and can be cleaned without danger from using high temperature or organic solvents.
6. The technique should not be difficult to automate.

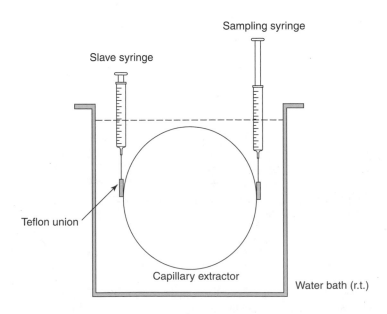

FIGURE 14.4 Set up for isothermal capillary extraction. (Reprinted from Nardi, L., *J. Chromatogr. A*, 985(1–2), 39–45, 2003, Copyright (2003), with permission from Elsevier.)

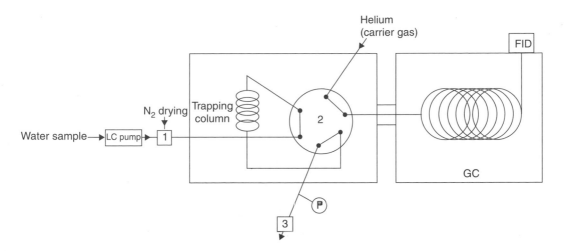

FIGURE 14.5 Schematic drawing of an automatic capillary extractor system. (1) Three-way flow selection valve. (2) Six-port switch selection valve (sampling position). (3) On–off valve and P is a pressure gauge. (Reprinted from Aguilar, C., Janssen, H. G., and Cramers, C. A., *J. Chromatogr. A*, 867(1–2), 207–218, 2000, Copyright (2000), with permission from Elsevier.)

Figure 14.5 shows the schematic diagram of the set up used for the analysis of water samples spiked with BTEX, epichlorohydrin, and dichlorohydrin.[24] The sampling enrichment system consisted of a pump used to make the sample flow through an OTT, a six-port switching valve (2 in Figure 14.5), working in two different positions. The first position is the adsorption position in order to retain target analytes in the trap and to remove the water in the column by purging it with nitrogen. A three-way flow selection valve (1 in Figure 14.5) is used to introduce either the sample or the drying gas into the system. The second is the desorption position in which analytes are desorbed from the trap and carried to the chromatographic column.

3. Membrane Extraction

Membrane extraction (ME) is a solventless extraction method that has gained popularity for the analysis of VOCs in water.[6,25–27] In ME the sample is contacted with a membrane surface. Analytes will migrate according to the polymer affinity from the aqueous phase to the surface of the membrane and selectively permeate through it.

After this isolation step, analytes are usually concentrated in sorbent traps (membrane extraction sorbent interface [MESI]) and thermally desorbed in a cryofocusing device[25] or directly in a gas chromatographic column.[6,26,27] In other cases,[28] no sorbent trap is used, analytes are desorbed directly from the membrane unit cell (thermal membrane desorption [TMD]) and carried by a gas stream onto the front of the chromatographic column of the analytical system.

Figure 14.6 shows the most common ME cells in which polymeric hollow fibers are inserted into capillary tubes. Water samples can be fed into the membrane,[27] in which case the stripping gas will flow through the space between the membrane and the capillary. Otherwise samples can be fed between the membrane and the probe, and the stripping gas passes through the internal membrane space.[25] In any of these configurations aqueous and gas streams flow in countercurrent.

a. Factors Affecting the Technique

The main factors affecting the technique are those with important influence on the mass transfer processes occurring in ME convection and diffusion of analytes through the fluid phases,

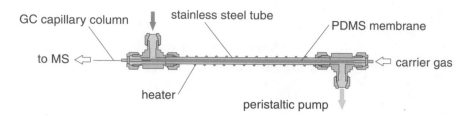

FIGURE 14.6 Membrane probe. (Reprinted from Matz, G., Loogk, M., and Lennemann, F., *J. Chromatogr. A*, 819, 51–60, 1998, Copyright (1998), with permission from Elsevier.)

partitioning of analytes between membrane and fluid phases, and diffusion through the membrane. Diffusion and partition coefficients are temperature dependent. Convection parameters depend on variables such as geometry of the fiber (internal and external diameter, length), velocity of fluid phases through the membrane surface, viscosity and density of fluid phases. The main factors affecting the ME are the fluid phases flow rates, the time of contact between the aqueous sample and the membrane, the temperature, and the composition of the sample.

i.   Water and Stripping Gas Flow Rates

Hauser and Popp[25] have recently found an increase in the analytical signal with increasing water flow rate for an analytical cycle of 10 min. At higher flow rates a larger amount of water sample comes into contact with the membrane surface and a higher mass transfer rate is achieved since the interfacial resistance of the aqueous phase decreases; consequently more analyte can be extracted. It was also observed that there was a trend towards a limit for the analytical signal with increasing water flow as a shorter contact time between water sample and the membrane was achieved. Shoemaker et al.[29] also found no further increase in the extracted amount of volatile analytes out of a flowing water sample when the flow rate of the water reached the same value as the flow rate of the gas on the other side of the membrane. An effect of the water flow rate on the extraction efficiency was also observed in the detection limits, whereas the reproducibility expressed as the relative standard deviation remained constant for each analyte. Sensitivity of the analysis (slope of the calibration lines) also increased with increasing water flow rate.

In order to eliminate the dependence of flow rate or sampling time on the amount of analyte extracted by the membrane, Guo and Mitra[27] developed a pulse introduction membrane extraction (PIME) method where pulse samples were injected into a tubular membrane device. This system does not need to reach steady state because all the analyte is extracted from the sample in the "lag time." This is an important parameter that determines the time of an analytical cycle. The duration of the lag time increased with the sample size and could be reduced by cleaning the membrane with a flow of nitrogen after the sample passed through it. Thus, a large sample with nitrogen purge was used in order to obtain higher sensitivities, avoid sample carryover between two analyses, and break up the boundary layer on the membrane to increase the mass transfer rate for the pulse injection.

ii.   Effects of Temperature and Matrix Composition

The temperature of extraction is limited by the temperature of destruction of the membrane fiber or by the temperature that allows a high solubility of water. In Ref. 28, an increase of the extraction yield with increasing temperature in the range between 10 and 50°C was observed due to the increase of the diffusion coefficients of analytes in the silicone fiber. Higher temperatures had a negative effect because higher amounts of water penetrated the membrane.

These studies also showed that the impact of matrix components such as different pH (2 and 9), varying levels of humic acids (17 mg/l and 130 mg/l), detergent, or a larger amount of organic solvents on the extraction rates of BTEX was relatively small. Particles or ions in the water samples should not affect the analysis as they are not permeable through the membrane.

In Table 14.4 some examples are given for ME conditions of BTEX in water.

B. Separation and Detection Methods

The analytical technique for separating BTEX is GC. Table 14.5 shows information about gas chromatographic columns and conditions, and detection methods used for the analysis of BTEX in water and soil.

A special application of GC, to separate the oxygenates (methyl-*tert*-butyl-ether and ethyl-butyl-ether) and aromatics like BTEX from hydrocarbons coeluents in groundwater due to gasoline spills, is presented by Gaines et al.,[30] where comprehensive two-dimensional GC, in which the eluent of a first column is introduced into a second column with a different separation mechanism, provides additional separation power and precision. A modulator that repeatedly generates sharp concentration pulses from the first column eluent and deposits them onto the second column links the two columns. This modulator consists of a tube with a thick stationary phase to retain and accumulate the first column eluent, and a rotating slotted heater that focuses the analytes and injects them in the second chromatographic column as it passes through the modulator tube.

GC is the most commonly used separation method in the analysis of BTEX from environmental samples. Liquid chromatography (LC) analysis with superheated water[31] or water dimethylsulfoxide (DMSO) mixtures[32] has also been reported. In both cases a reduction in the dielectric constant of the mobile phase for the separation of nonpolar analytes was studied. The results showed how the rise in temperature required a decrease in DMSO in order to achieve the same retention time.

If only water is used as mobile phase, higher temperatures have to be used. The dielectric constant of liquid water decreases with decreasing pressure and increasing temperature. Therefore high temperatures are necessary to decrease the dielectric constant or polarity of liquid water in order to be able to use it as reversed-phase liquid chromatography eluent. If the column is at the same temperature along its length, the pressure drop will cause a decrease in the polarity of superheated water, having the same effect as a decrease in the capacity of the stationary phase of the column. Another problem is the selection of a suitable stationary phase that does not decompose or dissolve in hot water, and whose functional groups are not influenced by temperature or aqueous environment. Zirconia-based stationary phases with elemental carbon and polybutadiene functional groups have good thermal stability and are compatible with water as eluent.

Although UV detectors are most commonly used for LC, FID and MS can also be attached to the chromatographic column.

IV. ANALYSIS OF BTEX IN SOIL

Soils become contaminated with BTEX through spillage of industrial solvents and oils, leakage of petrol from tanks and cracked pipelines, and deposition from contaminated air. Sea sediments can also be contaminated with BTEX due to their deposition after being transported by rivers into the sea. Extraction of analytes form soil samples is more difficult than from water owing to the greater analyte–matrix interaction. Therefore more aggressive sample preparation methods are required. Eliminating this analyte–matrix interaction is very difficult, so analytical methods performance depends very much on soil conditions such as humidity, organic content, or the presence of other pollutants.

TABLE 14.4
Membrane Extraction Conditions for Water Analysis of BTEX

Water Sample Type	Experimental Conditions	Comments	Reference
Method testing	Membrane material: silicone fiber Inner radius: 305 mm Outer radius: 635 mm Length: 4 cm Sample temperature: 70°C Sample volume (inside): 7.5 ml Sample flow or contact time: 5 ml/min Stripping gas (outside): helium Stripping gas flow: 1 ml/min Desorption temperature: — °C Desorption time: — sec	—	6
Method testing for groundwater analysis	Membrane material: silicone fiber Inner radius: 700 mm Outer radius: 800 mm Length: 30 cm Sample temperature: 15–20°C Sample volume: 9.3–280 ml Sample flow or contact time: 560 sec Stripping gas: air Stripping gas flow: 55–60 ml/min Desorption temperature: 350°C Desorption time: 180 sec	—	25
Water spiked with benzene, toluene, ethylbenzene, methanol and isobutyl alcohol	Membrane material: SPME fiber Inner radius: 15 mm Outer radius: 35 mm Length: 4 cm Sample temperature: 25°C Sample volume: 40 ml Sample flow or contact time Stripping gas: nitrogen Stripping gas flow: 2.2 ml/min Desorption temperature: 200°C Desorption time: 60 sec	Membrane was introduced in the sample which was stirred. Estimated speed of sample to the outer surface of the membrane was 55 cm/sec	26
Method testing	Membrane material: composite silicone Inner radius: 240 mm Outer radius: 290 mm Length: 200 cm Sample temperature: 50°C Sample volume: 2 ml Sample flow or contact time: 5 min Stripping gas: nitrogen Stripping gas flow: 5 ml/min Desorption temperature: — °C Desorption time: 1.2 sec	—	27

Continued

TABLE 14.4
Continued

Water Sample Type	Experimental Conditions	Comments	Reference
Aqueous medium in a biogas tower for online monitoring	Membrane material: PDMS Inner radius: 700 mm Outer radius: 900 mm Length: 15 cm Sample temperature: 50°C Sample volume: 20–60 ml Sample flow or contact time: 20 ml/min Stripping gas: air Stripping gas flow: 24 ml/min Desorption temperature: 195°C Desorption time: 60 sec	—	28

A. SAMPLE PREPARATION METHODS

In order to extract BTEX from soil, liquid solvents in combination with sample preparation methods used for water analysis of BTEX are used. In the following sections, P&T and SPME methods applied to soil samples are considered.

1. Purge and Trap

The P&T analysis of soil samples can be done by purging directly from the soil, extracting the target analytes with a solvent, by purging the extract, or just by purging a solvent–soil mixture. Purging directly from the soil requires higher extraction temperatures as well as longer heating and purging times than those for purging the extract or the soil–water mixture. This behavior shows how the use of a solvent helps to extract analytes more easily from the soil due to the decrease of the analyte–matrix interaction.

2. Soild-Phase Microextraction

The extraction of analytes from soil samples takes place in the same way as in water samples. A few grams of the sample are introduced in a vial sealed with a Teflon septa cap (or with an open cap) through which the protective sheath of the SPME syringe is pushed. Then the plunger is lowered, forcing the fiber into the headspace to avoid contamination of the fibers with solid particles.

Although extraction of analytes could be achieved just by sampling the headspace over the soil sample, in several works solvents have been mixed with the soil in order to increase BTEX volatilities by decreasing the analyte–soil interaction.[33,34] Figure 14.7 shows the effect on the response after the addition of different polar solvents to the soil. In all cases the amount of analyte extracted with solvents was greater than by sampling the headspace over the soil. Water was the solvent that achieved the greater recovery of analytes. This behavior can be explained either by a displacement of the analytes from the polar active sites of the soil or by the competitive mechanism between the matrix and the solvent for the analytes. In this report,[33] the effect of the amount of water added to the soil and the headspace volume on the recovery of BTEX was also studied. Whereas the relation between these two variables had an important effect on the amount of analyte extracted when sampling a liquid matrix (Equation 14.4), in the HS-SPME extraction of soil mixed with water no influence was observed. This means that the amount of analyte extracted will not

TABLE 14.5
Gas Chromatographic and Detection Conditions for BTEX Analysis in Water

Injection	Column	Temperature Program	Carrier Gas	Detection Method	Reference
Splitless mode at 260°C	Column: HP-1 capillary column Length (m): 60 I.D. (mm): 0.25 Film thickness (μm): 1	50°C (1 min) to 70°C at 3°C/min; 70°C (5 min) to 210°C at 6°C/min; 210°C (5 min)	Helium at 1.2 ml/min	MS operating, at 150°C and transfer line at 280°C	3
Splitless mode for 2 min	Column: DB-624 fused silica capillary Length (m): 75 I.D. (mm): 0.53 Film thickness (μm): 3	35°C (5 min) to 140°C at 10°C/min; 140°C (5 min) to 180°C at 10°C/min; 210°C (5 min)	Helium at 5 ml/min for 1 min to 35 ml/min at 45 sec until the end	MS operating 200°C and GC interface at 270°C	4
—	Column: HP-5MS capillary column (crosslinked 5% Ph Me silicone) Length (m): 30 I.D. (mm): 0.25 Film thickness (μm): 0.25	40°C (3 min) to 120°C at 4°C/min	Helium	MS	5
Split mode 50:1	Column: DB-624 Length (m): 30 I.D. (mm): 0.32 Film thickness (μm): 1.8	40°C (2 min) to 250°C at 5°C/min	—	MS	6
200°C	Column: DB-1 capillary Length (m): 30 I.D. (mm): 0.32 Film thickness (μm): 0.25	45°C (4 min) to 100°C at 10°C/min; 100°C (2 min)	Helium	FID at 200°C	7
250°C	Column: DB-624 fused silica capillary Length (m): 75 I.D. (mm): 0.53 Film thickness (μm): 3	(−) 20 to 220°C at 8°C/min	Helium at 20 ml/min	MS	8

	Column	Temperature program	Carrier gas	Detector	
Splitless	Column: DB-624 fused silica capillary Length (m): 75 I.D. (mm): 0.53 Film thickness (μm): 3	35°C (5 min) to 70°C at 3°C/min; 70°C (5 min) to 210°C at 6°C/min. 210°C (5 min)	Helium at 3.5 ml/min	MS	9
220°C	Column: DB-624 fused silica capillary Length (m): 75 I.D. (mm): 0.53 Film thickness (μm): 3	For C3–C6: 30°C (6 min) to 150°C at 5°C/min; 150°C (2 min) For C7–C10: 40°C (6 min) to 150°C at 5°C/min. 150°C (4 min)	—	—	10
Splitless mode, purge off for 30 sec at 150°C	Column: DB-WAXETR Length (m): 50 I.D. (mm): 0.32 Film thickness (μm): 1	40°C to 250°C at 10°C/min; 250°C (4 min)	Helium at 1 ml/min	MS	6
150°C	Column: DB-5 fused silica capillary Length (m): 50 I.D. (mm): 0.53 Film thickness (μm): —	45°C (4.5 min) to 80°C at 30°C/min; 80°C (5 min) to 150°C at 60°C/min; 150°C (7 min)	Helium at 2.3 ml/min	FID at 320°C	11
Splitless mode at 180°C	Column: CP SIL 13 CB combined with (1 m × 0.53 I.D.) deactivated empty precolumn Length (m): 25 I.D. (mm): 0.32 Film thickness (μm): 1.2	—	Helium at 27 cm/s linear velocity	FID at 280°C	12
Splitless mode at 180°C	Column: CP SIL 13 CB combined with (1m × 0.53 I.D.) deactivated empty precolumn Length (m): 25 I.D. (mm): 0.32 Film thickness (μm): 1.2	35°C (1.5 min) to 88°C at 35°C/min to 95°C at 2°C/min and to 150°C at 40°C/min	Helium at 29 cm/s linear velocity	FID at 280°C	13

Continued

TABLE 14.5
Continued

Injection	Column	Temperature Program	Carrier Gas	Detection Method	Reference
Splitless mode at 250°C	Column: CP SIL 5 CB Length (m): 25 I.D. (mm): 0.32 Film thickness (μm): 5	30°C (4 min) to 150°C at 10°C/min	Nitrogen at 45 ml/min	FID at 300°C	14
Splitless mode followed by 1:50 split after 0.5 min and 280°C	Column: CP-Select 624 Length (m): 30 I.D. (mm): 0.32 Film thickness (μm): 1.5	35°C (5 min) to 225°C at 10°C/min; 225°C (1 min)	Helium 1.7 ml/min	FID at 300°C. For detector 25 ml/min make up, 30 ml/min of hydrogen 300 ml/min air flow were added	15
300°C (Carboxen–PDMS) and 250°C (PDMS)	Column: HP-1 capillary Length (m): 30 I.D. (mm): 0.25 Film thickness (μm): 1.5	40°C (5 min) to 100°C at 4°C/min100°C (5 min)	Helium 1 ml/min	FID at 300°C	16

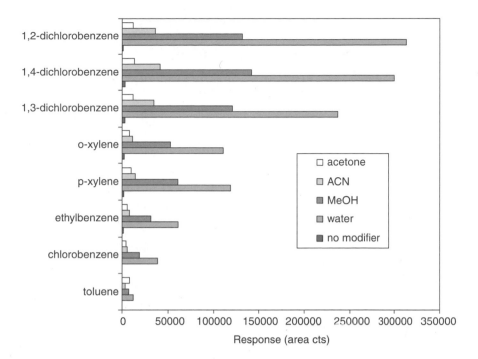

FIGURE 14.7 Effect of the solvent on the extraction efficacy. (Reprinted from Llompart, M., Li, K., and Fingas, M., *Talanta*, 48, 451–459, 1999, Copyright (1999), with permission from Elsevier.)

depend on the moisture content of the soil and that a great headspace volume can be used in order to place the SPME fiber far from the slurry to avoid its contamination. On the other hand, the addition of salt to the sample, as in the HS-SPME analysis of water, had no effect on the extracted amount of nonpolar analytes.

The influence of the organic content of the soil and the temperature on the analyte recovery in the HS-SPME analysis of different soils mixed with water was studied.[34] The increase of organic matter clearly decreased the analyte recovery. Hence it was demonstrated that the organic matter is the principal element responsible for the soil–analyte interaction. To counteract this effect, the soil–water mixtures were heated up to 110 or 120°C for 10 min and cooled rapidly before extraction at 30°C with the SPME fiber. For each type of soil the recoveries increased with the increase in temperature. This effect was clearer for more volatile BTEX than for less volatiles ones.

The experimental conditions for the SPME analysis of BTEX in soil are shown in Table 14.6 together with those of P&T.

B. Separation and Detection Methods

GC is the separation method used for the analysis of BTEX in soils. Table 14.7 shows the chromatographic conditions as well as the detection methods most commonly used.

V. ANALYSIS OF BTEX IN AIR

Air can be polluted by BTEX by combustion of fuels in industrial emissions, vehicle motors, or even by tobacco smoke, and by evaporation from polluted surface water or fuels in gas stations. Analysis of BTEX in air is of special importance because their concentration in urban locations is increasing significantly through human activity.

TABLE 14.6
Purge and Trap and SPME Conditions for Soil Analysis of BTEX

Sample	Sample Preparation Method	Experimental Conditions	Comments	Reference
Soil	Purge and trap	Sample volume: 5 ml Trap: BTEXTRAP Extraction temperature: ambient Purge gas: helium Purge flow: 40 ml/min Purge time: 4.3 min Desorption temperature: 260°C Desorption time: 4 min	100 ml of the methanol extract (for 5 g of soil) and 50 of internal standard were added to 4.85 ml of NaCl solution (4.3 mol/l) to make the 5 ml sample. Before purging sample was heated at 46°C for 7.5 min	35
Soil	Purge and trap	Sample volume: 13 ml Trap: BTEXTRAP Extraction temperature: ambient Purge gas: helium Purge flow: 40 ml/min Purge time: 51.5 min Desorption temperature: 260°C Desorption time: 4 min	Before the soil was directly purged it was heated at 61°C for 42.6 min	35
Soil	Purge and trap	Sample volume: 5 ml Trap: BTEXTRAP Extraction temperature: ambient Purge gas: helium Purge flow: 40 ml/min Purge time: 8.7 min Desorption temperature: 260°C Desorption time: 4 min	5 ml of NaCl solution (0.04 mol/l) water were added to 0.5 g of soil and heated at 77°C for 8.1 min before purging	35
Sediments	Purge and trap	Sample volume: — Trap: VOCARB 4000 Extraction temperature: 70°C Purge gas: helium Purge flow: 20 ml/min Purge time: 30 min Desorption temperature: 250°C Desorption time: —	30 g of sediment and 15 ml of water was the purged mixture	36
Soil	HS-SPME	Vial volume: 22 ml Sample volume: 1 ml Fiber coating: 100 mm PDMS Extraction temperature: 20°C Extraction time: 30 min Agitation: Yes Desorption temperature: 260°C Desorption time: 3 min	1 ml of water was added to 1 g of soil before SPME analysis	33
Soil	HS-SPME	Vial volume: — Sample volume: 5 ml Fiber coating: 100 mm PDMS Extraction temperature: 30°C Extraction time: 8 min Agitation: ultrasonic Desorption temperature: 250°C Desorption time: 4 min	5 ml of water was added to 4 g of soil before SPME analysis Before extraction slurry was heated for 110 or 120°C	34

TABLE 14.7
Gas Chromatographic and Detection Conditions for BTEX Analysis in Soils

Injection	Column	Temperature Program	Carrier Gas	Detection Method	Reference
Splitless injection at 225°C	Column: SPB-1 Length (m): 30 I.D. (mm): 0.53 Film thickness (μm): 1.5	40°C (5 min) to 200°C at 7.5°C/min	7.5 ml/min carrier gas flow	MSD (SIM mode)	33
Splitless injection at 250°C	Column: RTx-5 fused silica capillary of polydimethylsiloxne (5% phenyl groups) Length (m): 31 I.D. (mm): 0.54 Film thickness (μm): 1.6	($-$) 40°C (5 min) to 250°C at 7°C/min	—	Quadrupole MSD	34
Split injection at 250°C	Column: RTx-5 fused silica capillary of polydimethylsiloxne (5% phenyl groups) Length (m): 30 I.D. (mm): 0.32 Film thickness (μm): 1.8	50°C (1 min) to 140°C at 5°C/min 140°C (0.5 min)	Helium at 5.74 ml/min	FID at 280°C	35
200°C	Column: RTx-502.2 Length (m): 60 I.D. (mm): 0.32 Film thickness (μm): 1.8	40°C (2 min) to 200°C at 10°C/min 200°C (5 min)	Helium at 16 psi	MS	36

A. Sampling Techniques and Sample Preparation

Measurements of BTEX and other VOCs in air are performed in order to determine the sources and transport mechanisms of pollution, the compliance with regulated limits, and the health effects of pollutants. In the latter, high time sampling methods like canisters or diffusive sampling are required because peak concentrations are of minor concern and analysis must reflect the cumulative exposure of pollutants. In the first two cases or in process controlling emissions, for example, faster sampling methods such as automated gas chromatographs or manual pumping tube sampling are used in order to measure the fluctuation in the concentration of pollutants. In the following sections these sampling methods and pretreatments prior to chromatographic analysis are described.

1. Canister Sampling

Sampling is performed by introducing air into stainless steel or polyethylene canisters. Canisters of different volumes ranging from 2 to 10 l are commonly used. Air is introduced into the canisters in two ways:

1. *Grab sampling.* An evacuated canister, to which vacuum has been previously applied, is opened until the air fills it up to atmospheric pressure.[37,38]
2. *Pumped sampling.* Air is pumped into the canister over a certain period of time up to a desired pressure. It is preferred when larger volumes of sample have to be used.[39–41]

After field sampling, canisters are brought to the laboratory where 100 to 500 ml volumes are cryogenically concentrated with liquid nitrogen in order to eliminate interferences in the chromatographic separation due to noncondensable compounds (nitrogen, methane, oxygen, etc.). Cryogenic devices usually consist of a stainless steel or glass U-tube filled with 60-80 mesh glass beads partially immersed in the cryogenic fluid. Figure 14.8 shows a cryogenic preconcentrator system for atmospheric hydrocarbon measurements. In the cryoconcentration mode, atmospheric air in the canister is allowed to pass through the U-column. As a result hydrocarbons are condensed and noncondensable compounds pass through the column into the equalizing canister. In the next step, sample hold and heat mode, the valve positions are such that the U is closed in order to evaporate the concentrated sample by heating it without losing analytes. Finally the sample is transferred to the analytical column for analysis. Sometimes sorbent traps are used in the transfer line between the cryogenic device and the chromatographic column to remove carbon dioxide or water vapor from the sample.

2. Sorbent Sampling

In sorbent sampling analytes are sorbed in tubes containing solid adsorbents. Therefore, unlike in canister sampling, sampling and separation are achieved in a single step. Many different techniques exist in which sorbent sampling is applied depending on how each step is performed:

1. The sampling step can be performed by active sampling, in which air is pumped through adsorbent tubes, or by passive sampling, where adsorbent cartridges are exposed to polluted air. Figure 14.9 shows a schematic drawing of a sampling tube used for passive sampling methods.[42]
2. The desorption step to remove trapped vapors for analysis can be performed either by solvent extraction or by thermal desorption.

After desorption target compounds are transferred directly onto the chromatographic system cryofocused or preconcentrated in cryotraps in order to eliminate interferences and to inject a sharp pulse of sample onto the separation column.

The final choice in the analytical method employed for the determination of a volatile compound depends among other things on the required accuracy of the results, the measurement time, and the concentration levels.

In the following sections the principles and the characteristics of sampling and desorption methods employed for the analysis of BTEX in air are discussed.

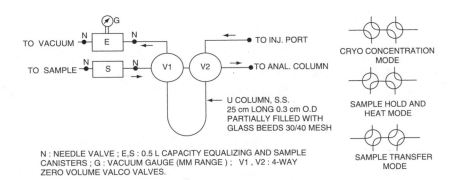

FIGURE 14.8 Schematic drawing of a cryogenic concentrator. (Reprinted from Rao, A. M. M., Pandit, G. G., Sain, P., Sharma, S., Krishnamoorthy, T. M., and Nambi, K. S. V., *Atmos. Environ.*, 31(7), 1077–1085, 1997, Copyright (1997), with permission from Elsevier.)

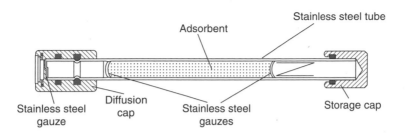

FIGURE 14.9 Schematic drawing of a sampling tube for diffusive sampling. (Reprinted from Ballach, J., Greuter, G., Shultz, E., and Jaeshke, W., *Sci. Total Environ.*, 243/244, 203–217, 1999, Copyright (1999), with permission from Elsevier.)

a. Sorbent Sampling Techniques

Two types of sorbent sample technique are commonly used: active or pumped sampling, and passive or diffusive sampling. Table 14.8 shows the experimental conditions in which BTEX were determined from air using active sorbent sample methods.

i. Active or Pumped Sampling

In active sampling, the most employed sampling method for BTEX analysis, air is forced with a pump to pass for a certain time through adsorbent tubes. The mass flow has to be exactly known and calibrated in order to know the total volume of air from which analytes have been desorbed. Sample volumes ranging from 2 to 10 l are taken in 15 to 60 min. A small amount of sorbent in a small bed (typically less than 2 g of sorbent in tubes of less than 5 mm internal diameter and 15 cm in length) is enough to retain all target analytes due to the fast kinetics of adsorption.

Careful choice of the sampling volume has to be made in order to avoid breakthrough when analyte molecules are detected in the outer stream of the adsorbent tubes, because of the saturation of the bed or displacement by another chemical. In this situation sampling is no longer efficient and as breakthrough progresses the sample will be less representative of the external environment. To avoid breakthrough, larger amounts of adsorbent or even multibed sorbent tubes can be used.

Breakthrough depends further on concentration, temperature, and the presence of other species in the atmosphere, which can reduce the effective capacity of the adsorbent. Breakthrough tests performed in the laboratory by injecting standard gas mixtures in the adsorbent tubes attached to the detection ports[43] or to a second adsorbent tube[44] to measure the amount of analyte are useful to know the volume of air at a given concentration of analytes that can be used for sampling. However, as field sampling will usually provide greater breakthrough[44] due to interfering chemicals (among which water vapor is the most common when hydrophilic adsorbents such as charcoal are used), sampling volumes are set to half or the third part of the breakthrough volumes. Table 14.8 shows the experimental parameters employed in the analysis of BTEX in air by active sampling methods.

ii. Passive or Diffusive Sampling

In diffusive sampling VOCs are extracted from the air by an adsorbent at a rate controlled by Fick's first law of diffusion:

$$dm_i/dt = (D_iA/L)(C_{i0} - C_i) \qquad (14.5)$$

where m_i is the mass of compound i sorbed on the adsorbent, t the time of exposure, C_{i0} the ambient concentration of compound i, C_i the concentration of compound i above the sorbent's surface,

TABLE 14.8
Active Sampling Parameters for the Analysis of BTEX in Air

Air Sample Type	Sampling Parameters	Adsorption Parameters	Desorption Parameters	Reference
VOCs in air at urban site	Air flow: 78 ml/min	Two adsorption tubes in series: 1st: 25 mg of Tenax; 2nd: 10 mg of charcoal	Thermal desorption: 3 min at 160°C (Tenax) and 220°C (charcoal) while no carrier gas flows and injection valve is closed and 4 more minutes with carrier gas and injection valve opened	45
VOCs exposure in urban air	Air flow: 100 ml/min	Commercial Tekmar multisorbent tubes (7 in. × 1/4 in.) packed with Tenax TA and Carbosieve S-III	Thermal desorption: 20 min at 225°C with a 40 ml/min flow of He carrying VOCs to a preconcentrator trap at −40°C. The trap was desorbed at 225°C for 4 min and transferred to the gas chromatograph	46
Benzene in air at urban site	Air flow: 200 ml/min Sampling volume (l): 3.6	Commercial ORBO 402 (Supelco) packed with Tenax TA	Thermal desorption-SPME: a polydimethylsiloxane–divinylbenzene SPME fiber was exposed for 10 min above the headspace of the adsorbent material	47
Benzene emitted from glowing charcoal	Sampled volume (ml): 1 Sampling time (s): 20	Adsorption cartridge containing 0.1 g of Tenax TA 60–80 mesh	Thermal desorption: done in the injector at 220°C with carrier gas	48
BTEX monitoring in ambient air	Air flow: 500 ml/min Sampled volume in rural areas (ml): 151 Sampled volume nearby road (ml): 300	Adsorption cartridge containing 500 mg of Carbopack B and 750 mg of Carbosieve S-III	Thermal desorption: by passing He at 15 cm²/min at 200°C followed by cryofocusing in cold trap with liquid nitrogen	49
Roadside in vehicle concentrations of BTEX	Air flow: 100 ml/min Sampling time (min): 60	1 or 2 (in series) adsorbent cartridges containing 0.16 g of Tenax TA	Thermal desorption: at 250°C for 30 min after which desorbed compounds were cryogenically focused at −30°C. After, rapid heating of cryotrap up to 250°C to volatilize compounds into GC	50
Testing multichannel sampling system	Air flow: 20–30 ml/min Sampling time (min): 60	Stainless steel tubes (9 cm long × 6.3 mm outer diameter) packed with 0.4 g of Carbopack C (60–80 mesh); 0.2 g of Carbopack B (60–80 mesh)	Thermal desorption: at 350°C for 5 min after which desorbed compounds were cryogenically focused at −15°C in the same trap as in adsorption. After, rapid heating of cryotrap up to 300°C for 5 min to volatilize compounds into GC	51
Atmosphere of Athens	Air flow: 100 ml/min Sampling time (min): 30	Stainless tubes (7 in. × 0.25 in.) filled with 0.6 g of Tenax TA (60:80) mesh	Thermal desorption: (1) Desorption at 230°C for 10 min (2) Temperature of external and internal cryotrap: −100°C (3) Final temperature of external cryotrap: 230°C	52

Application	Sampling conditions	Adsorbent/tube	Thermal desorption	Ref.
Urban roadside measurements of aromatic hydrocarbons	Air flow: 100 ml/min	Commercial multibed stainless steel tubes (7 in. × 1/4 in.) packed with Tenax TA and Carbosieve S-III	Thermal desorption: Tekmar 6000 Aerotrap	53
In vehicle and commuting studies of VOCs	Air flow: 10 ml/min; Sampling volume (l): 3 May	Two stainless steel adsorbent l tubes (4 cm i.d. × 10 cm long) packed with 160 mg of Tenax GR (60–80) and 70 mg of Carbosieve S-III (60–80 mesh)	Thermal desorption: cryofocusing	54
Measurements of aromatic VOCs in public transportation	Air flow: 0.15 l/min; Sampling volume (l): 4–9	Stainless steel adsorbent l tubes (0.25 in i.d." × 7 in long) packed with Tenax TA and Carbosieve S-III	Thermal desorption: Tekmar 6000 Aerotrap	55
Aqueous medium in a biogas tower for online monitoring	Air flow: 10–20 ml/min; Sampling time (min): 60	Multiadsorbent tubes (16 cm long × 4 mm i.d.) with: 0.2 g of Carbopack C; 0.2 g of Carbopack B; 0.2 g of Carbotrap S-III	Thermal desorption: (1) Desorption at 230°C for 10 min (2) Desorbed analytes were retrapped at −150°C by a Teflon tube (10 cm long × 1.6 mm i.d.) filled with 2 cm of deactivated glass beds (60-80 mesh) (3) Cold trap was heated at 50°C/sec for 7 min transferring analytes into a moisture control system to remove water	56
Monitoring urban air quality and uptake rates measurements	Air flow: 12 ml/min; Sampling time (h): 12–14; Air flow: 55 ml/min; Sampling time (min): 4–6	Stainless steel tubes (8.9 cm long × 5 mm i.d.) of Chromosorb 106	Two stage desorption with a Perkin-Elmer ATD-400 Thermal desorber: (1) Primary desorption with 25 ml/min of He for 10 min at 230°C (Chromosorb 106) or 280°C (Carbograph) (2) Secondary desorption from the cold trap (25 mg of Tenax at −30°C) was at 300°C	57
Monitoring urban air quality and uptake rates measurements	Sampling volume (l): 5–7; Sampling time (h): 12; Samplers were changed automatically for 1 week of continuous sampling	Perkin-Elmer stainless steel tubes filled with 400 mg of Serdolit AD-4 (60-80) mesh	Two stage desorption with a Perkin-Elmer ATD-400 thermal desorber: At 170°C for 5 min to transfer the desorbed compounds to a cold trap of Tenax at 0°C. Desorption from cold trap at 300°C	42

Continued

TABLE 14.8
Continued

Air Sample Type	Sampling Parameters	Adsorption Parameters	Desorption Parameters	Reference
Vehicular emissions	Sampling volume for car exhaust (l): 5–10 Sampling volume for roadside air (l): 200–300	Charcoal	3 ml of CS_2 added to the chilled charcoal and agitated for 10 min	58
Solvents measured in workplace of solvent industry	Air flow: 1.5 ml/min Sampling time (h): 2–8	Charcoal tubes	1 ml of CS_2	59
Determination of ethyl-benzene, indane, indene and acenaphtene in breathing zone of coke plant	Air flow: 0.5 l/min Sampling time (h): 6	Charcoal	1 ml of CS_2–methanol (60:1, v/v) in an ultrasonic bath for 10 min	60
Monitoring of BTEX in urban and rural sites with two different automatic devices	Siemens: RGC 402 Sampling time (min): 20 Sampling volume (l): 0.7 Air-motec: HC 1010 Sampling time (min): 28.55 Sampling volume (l): 1.5	Double stage preconcentrator with: (1) Poropack Super Q (2) Cryofocusing step (5°C with methanol as refrigerant) Double stage preconcentrator with: (1) Carbosieve S-III (60–80 mesh)/Carbotrap cartridge (2) Cryofocusing step (-20°C with CO_2 as refrigerant) in a Carbopack (60-80 mesh) adsorbent	Thermal desorption at 140°C Thermal desorption at 350°C	61

D_i the diffusion coefficient of compound i in the adsorbent, A the selected area of adsorbent the tube and L the diffusive path of the adsorbent. The term $(D_i A/L)$ is called uptake rate U_i, of the compound i.

Assuming that the adsorbent acts as a perfect sink ($C_i = 0$) if no variation of the ambient concentration occurs, Equation 14.5 can be simplified to

$$C_{i0} = m_i/(tU_i) \qquad (14.6)$$

With this equation, once sampling time is fixed the concentration of an analyte can be calculated by obtaining m_i from the chromatographic analysis if the constant uptake rate is known. The validity of the simplified constant uptake rate model depends on several factors, some of which involve the nature of the diffusive process and others the ambient conditions found in field sampling. For nonideal adsorbents that weakly adsorb analytes, the concentration above the surface will not be zero. Therefore analytes will accumulate in the adsorbent surface until equilibrium between inner and outer diffusion is reached. Their concentration there will increase, making the uptake rate decrease with time as a consequence of the decrease in the concentration gradient ($C_{i0} - C_i$). Thus the applicability of Equation 14.6 for the determination of the concentration of analytes in ambient air is limited to long sampling periods (typically 1 to 4 weeks) in order to reduce the effects of the initial uptake changes.

Figure 14.10 shows the amount of BTEX adsorbed on Carbopack B (60-80 mesh) adsorbent for a period of 28 days, sampling with four tubes for 7 days, seven tubes for 4 days, two tubes for 14 days, and one tube for the whole period. It was observed that the shorter the sampling time is, the higher the extracted amount of BTEX, because uptake rates have not decreased so much. It would be expected that sampling times greater than 28 days will give the same amount of extracted BTEX.

Atmospheric variations in the concentration of target compounds, temperature, humidity, and wind speed can lead to unreliable values of uptake rates:

- When ambient concentration decreases with respect to the concentration above the adsorbent surface, back diffusion takes place, resulting in a loss of analyte from the adsorbent to the atmosphere. Therefore diffusive sampling over a large period of time where concentrations change will give reliable time-weighted results.

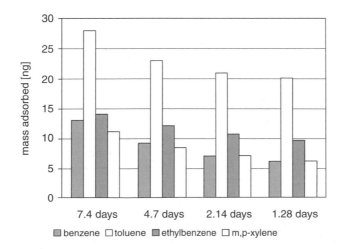

FIGURE 14.10 Adsorbed mass of BTEX in passive samplers for different sampling times. (Reprinted from Tolnai, B., Geleneser, A., and Hlavay, J., *Talanta*, 54, 703–713, 2001, Copyright (2001), with permission from Elsevier.)

- With nonideal sorbents a decrease in the uptake rate with increasing temperature due to a decrease in the sorption coefficient is expected. A 1.1 and 0.6%/K decrease in the uptake rates for benzene and toluene were observed with Serdolit AD-4 (60-80 mesh) as adsorbent.[42]
- Back diffusion and decrease in the uptake rates can also be observed for hydrophilic sorbents at high humidity. A decrease in the uptake rate for benzene from 2.12 to 1 ng/(ppm min) was observed with increasing relative humidity from 50 to 90% when sampling for 2 weeks with Carbopack X.[43]
- The possible effect of wind speed on the uptake rate becomes negligible when sorbent tubes with end caps and wind shields are used.

Two types of passive samplers exist, differing in their geometry:

1. Tube-type samplers, which are also used for active pumping, have a long axial diffusive path and a small cross-sectional area. These are also called axial diffusive samplers because diffusion of the analytes takes place in the axial direction.
2. Badge-type or radial samplers which have a shorter diffusive path length but a high cross-sectional area.

Badge samplers show greater uptake rates than tube-type samplers due to their greater cross-sectional area. This agrees with the given definition of uptake rate, which is directly proportional to the cross-sectional area and inversely proportional to the diffusive path length.

Table 14.9 shows the experimental parameters employed in the analysis of BTEX in air by passive sampling methods.

b. Desorption Methods: Adsorbent Selection

Two types of desorption methods, thermal desorption and solvent desorption, are used with either active or passive sampling.

In the former, sampling tubes are heated in order to release the analytes from the adsorbent and in the latter a solvent extracts the analytes from the adsorbent. The choice for the method of desorption depends on the adsorbent used in the sampling step and on the required accuracy of the analysis.

Two types of solid adsorbents used in sorbent sampling can be distinguished: those with a relatively low and those with a very high specific surface containing micropores. Their adsorption mechanism can be radically different. While in the first one macropores are filled by a molecular coverage (monolayer adsorption), where molecules are held by weak forces, in the latter adsorbed molecules are held by stronger forces. Therefore, solvent extraction of the analytes from the adsorbent has to be applied. The main problem with solvent extraction is the decrease in sensitivity due to the dilution of small amounts of analytes in large amounts of solvent. When a decrease in sensitivity cannot be allowed because very low concentrations of pollutants are to be detected, thermal desorption is the best choice. Solid adsorbents with relatively high temperature resistance used for thermal desorption have lower capacities than adsorbents used for solvent.

i.   Solvent Desorption

Solvent desorption is used with strong adsorbents such as activated charcoal,[58–60, 62–64] which has a high capacity and enhanced catalytic activity. A problem with activated charcoal is that in moderately high relative humidity atmospheres it can absorb large amounts of water which displaces other analytes.

For BTEX analysis in air with either active[58,59] or passive[64] sampling, carbon disulfide is the most used extracting agent due to its high adsorption capacity that displaces other molecules

TABLE 14.9
Passive Sampling Parameters for the Analysis of BTEX in Air

Air Sample Type	Sampling Time	Adsorption Parameters	Desorption Parameters	Reference
Indoor and outdoor air monitoring	4 weeks	ORSA 5 sampler (tube-type diffusive sampler) 400 mg of activated charcoal contained in a 0.5 cm diffusive path and 0.88 cm^2 cross sectional area sampler	Extraction with 2 ml of CS$_2$ by mechanical agitation for 30 min, followed by centrifugation for 5 min at 4000 rpm	62
		OVM 500 sampler (batch-type diffusive sampler) 180 mg of activated charcoal contained in a 1 cm diffusive path and 7.07 cm^2 cross sectional area sampler	Extraction with 1.5 ml of CS$_2$ by mechanical agitation for 30 min	
Residential indoor and outdoor levels of VOCs and personal exposure	24 h	3 M organic vapor monitor #3500 [OVM]	Extraction with 2 ml of CS2 by mechanical agitation for 30–45 min	63
Method optimization for analysis	7 days	Radiello radial diffusive sampler with activated charcoal	Extraction with dichloromethane	64
BTEX in indoor air	30–60 min	3500 OVM diffusive samplers (charcoal)	Extraction with CS$_2$. Xanthation reaction of extract with sodium methanolate. HS-SPME of resulting mixture	65
Indoor and ambient air levels of VOCs	—	4 mm internal diameter cartridge containing 600 mg of Tenax	Thermal desorption: At 150–330°C for 15–30 min with a carrier gas flow of 150 ml/min to transfer the desorbed compounds to a cold trap of Tenax at −30 to 30°C. Desorption from cold trap at 250–300°C	66
Monitoring urban air quality and uptake rates measurements	1–4 weeks	Stainless steel tubes (8.9 cm long × 5 mm i.d.) of either 300 mg of Chromosorb 106 (60-80 mesh) 330 mg of Carbograph TD-1 (20-40 mesh)	Two stage desorption with a Perkin-Elmer ATD-400 Thermal desorber: (1) Primary desorption with 25 ml/min of He for 10 min at 230°C (Chromosorb 106) or 280°C (Carbograph) (2) Secondary desorption from the cold trap (25 mg of Tenax at −30°C) was at 300°C	57
Monitoring urban air quality and uptake rates measurements	7 days	Perkin-Elmer stainless steel tubes filled with 400 mg of Serdolit AD-4 (60-80) mesh	Two stage desorption with a Perkin-Elmer ATD-400 thermal desorber: at 170°C for 5 min to transfer the desorbed compounds to a cold trap of Tenax at 0°C. Desorption from cold trap at 300°C	42

from the activated charcoal achieving better recovery efficiencies. In a desorption efficiency study for the determination of volatiles in the breathing zone of a coke plant, a CS$_2$–methanol mixture (60:1 v/v) extracted 98.6% of ethylbenzene spiked on charcoal tubes.[60] Other advantages of CS$_2$ are that it elutes rapidly from the front of the chromatogram and that it has a very low response on a FID detector.

In spite of these advantages, CS_2 is not suitable for ECD detectors and it is a highly toxic solvent. In trace analysis of BTEX from air samples using low pressure gas chromatography,[64] dichloromethane was preferred as extracting solvent from a Radiello charcoal tube in spite of the coelution with benzene, because the extensive carbon disulfide peak tailing resulted in a poor signal to noise ratio and detection limits.

ii.   Thermal Desorption

Thermal desorption is a high-cost method; it consumes the entire sample in one analysis. Nevertheless it is preferred over solvent extraction in active and passive sampling methods for the analysis of volatiles mainly due to the enhanced.

Some of the adsorbents used for the analysis of BTEX and other volatiles are:

1. *Tenax.* The low adsorption capacity of Tenax makes it useful for collecting samples from low-level concentration atmospheres. The main advantages of Tenax are high temperature stability, hydrophobicity, and rapid desorption kinetics.
2. *Chromosorb 106.* This is a very hydrophobic adsorbent with a greater capacity than Tenax. Therefore it can be used in the analysis of high concentration atmospheres. Drawbacks for Chromosorb 106 are a lower thermal stability than Tenax and the high background levels of aromatics.
3. *Graphitized carbons.* The most common are Carbopack B and Carbopack C. These are usually found in multisorbent tubes, as a pair or also with Carbosieve S-III. Graphitized carbons adsorb some water.
4. *Carbosieve S-III.* This is usually used together with Tenax and graphitized carbons. Carbon molecular sieves such as Carbosieve S-III also collect water.

Thermal desorption time–temperature profiles depend on the adsorption material used, as can be seen in Tables 14.8 and 14.9.

3.   Automatic Instruments

Automatic instruments are used to measure analyte concentrations *in situ*. Sampling, sample preparation, separation, and detection steps are performed onfield. Online gas chromatograph instruments are often used to measure BTEX in ambient air.[44,51,61,67] Table 14.10 shows some commercially available online instruments for BTEX analysis:

Online automatic instruments for measuring BTEX use active sorbent sampling coupled with thermal desorption previous to the separation and the detection steps. The aim of collecting such a low volume of air is to achieve low sampling times with total analysis times of 15 to 30 min. Therefore automatic instruments can be applied to control emissions in order to record the variation in concentration of pollutants during a period.

Good agreement between the results of an online instrument and a pumped-thermal desorption analysis are demonstrated.[44,51]

However, standard deviations and detection limits are better for automatic than for manual analysis; especially if the latter involves a passive sampling–solvent desorption method. This is a consequence of the many sources of errors possible in a manual method which are not present in automatic instruments.

4.   Solid-Phase Microextraction

SPME technique can be applied to the analysis of BTEX and other VOCs in air in different ways:

TABLE 14.10
Characteristics of Some Commercial BTEX Monitors

Firm/Trade Name	Airmotec ag/BTX 1000	AMA Systems/GC 5000 BTX-2	Chrompack/CP 7001	Siemens/U102 BTX	Synthech Spectras/GC855 BTX	Environment SA/Appareil 61 M
Adsorbent	Carbotrap/Carbo-sieve SIII	Multibed of graphitized carbons	Tenax GR	Poropack SuperQ	Tenax GR	Carbotrap B
Sampling	Pump/nozzle	Pump/MFC	Pump/MFC	Pump/MFC	Piston pump	Critical orifice/pump
Sample amount (ml)	500	500	300	390	100	756
Detectors	FID	FID	FID	FID	PID	FID
Focusing system	Cryotrap	Precolumn	Precolumn	Precolumn	Precolumn	Precolumn
GC-column (trade name)	1 μm BGB 2.5 (2.5% phenyl)	DB-624	Ultimetall 1 μm	CP-WAX	1.2 μm AT-5	DB-5

MFC: mass flow controller.

1. Using SPME fibers and SPME sampling chambers in the sampling and preconcentration steps has several advantages over the methods described before:
   (a) Sampling can be done in the field without the use of canisters and air pump.
   (b) No chemical reprocessing or solvent is needed.
   (c) Nonpolar stationary phases used in SPME fibers are not influenced by atmospheric humidity.

In spite of these advantages, the SPME technique has an important drawback. Most SPME fibers show poor storage stability due to the losses of analytes by evaporation.

2. The SPME fiber has also been used as an extracting agent in the desorption step of active and passive sorbent sampling. In the former,[47] the adsorbent material was placed in a vial and sealed with a septum. Different SPME fibers were tested at different temperatures and extracting times. Carboxen–PDMS fiber showed higher recoveries for BTEX, with a maximum between 80 and 140°C. Lower temperatures did not achieve effective desorption from the adsorbent, and higher ones decreased the partition coefficient between the headspace and the fiber. In the latter one,[65] after sampling for different periods of time (2 h, 1 day, and 1 month) with passive monitors, BTEX were extracted with CS_2. Detection limits for this procedure ranged between 0.4 and 2.0 μg/m^3 and between 0.4 and 1.1 μg/m^3 for sampling intervals of 2 h and 1 day, respectively, whereas for the 1 month diffusive sampling with direct injection of extraction solvent onto the gas chromatographic column they were between 0.1 and 0.4 μg/m^3. Therefore SPME provides a useful tool for making diffusive sampling a short-term method for the determination of BTEX.

B. SEPARATION AND DETECTION METHODS

The separation and detection methods most commonly used for the analysis of BTEX in air are GC coupled to mass spectrometric (MS) and flame ionization detectors (FID). Table 14.11 shows

TABLE 14.11
Gas Chromatographic Conditions and Detection Methods for the Analysis of BTEX in Air

Injection	Column	Temperature Program	Carrier Gas	Detection Method	Reference
—	Column: HP-VOC capillary column Length (m): 60 I.D. (mm): 0.32 Film thickness (μm): 1.8	35°C (2 min) to 220°C at 5°C/min; 220°C (10 min)	He	MS	46
Injector in split mode for 3 min at 250°C	Column: MDN-5S capillary column with a poly(5%–diphenyl–95%-dimethylsiloxane) stationary phase Length (m): 30 I.D. (mm): 0.25 Film thickness (μm): 0.25	40°C (1 min) to 70°C at 10°C/min; 70°C (1 min) to 280°C at 35°C/min	He at 1 ml/min	MS	47
220°C	Column: stationary phase cyanopropylsilicone Length (m): 30 I.D. (mm): 0.25 Film thickness (μm): 0.25	−50 to 50°C at 10°C/min; 50 to 250°C at 5°C/min	—	IT-MS	48
—	Column: DB-1 polydimethyl-siloxane Length (m): 30 I.D. (mm): 0.32 Film thickness (μm): 5	—	—	MS	49
Injection at 150°C	Column: Al$_2$O$_3$ PLOT column Length (m): 50 I.D. (mm): 0.53 Film thickness (μm): 1.8	100°C (15 min) to 100–180°C at 5°C/min;100–180°C (20 min) to 180–200°C at 5°C/min; 180°C–200°C (20 min)	He at 14.9 ml/min	FID at 250°C	50

	Column	Temperature program	Carrier gas	Detector	Ref.
—	Column: DB-WAX fused silica capillary Length (m): 60 I.D. (mm): 0.32 Film thickness (μm): 0.5	35°C (2 min) to 250°C at 5°C/min; 250°C (10 min)	Helium at 0.36 bar	MSD	56
250°C	Column: HP-VOC capillary column Length (m): 30 I.D. (mm): 0.2 Film thickness (μm): 0.51	(−) 20°C to 220°C at 8°C/min	He	MSD	53
—	Column: — Length (m): — I.D. (mm): — Film thickness (μm): —	—	—	MS	54
—	Column: HP-VOC capillary column Length (m): 30 I.D. (mm): 0.32 Film thickness (μm): 0.5	35°C (2 min) to 250°C at 8°C/min 250°C (5 min)	He	MS	55
—	Column capillary polyethyleneglycol–TPA modified Length (m): 25 I.D. (mm): 0.2 Film thickness (μm): 0.3	40°C (2 min) to 90°C at 2°C/min	He	MSD	58
Injection at 25°C	Column: OV-1 capillary column Length (m): 25 I.D. (mm): 0.53 Film thickness (μm): 2	35°C (6 min) to 95°C at 10°C/min; 95°C (1 min) to 135°C at 20°C/min	Nitrogen at 0.4 kg/cm²	FID at 25°C. Nitrogen at 40 ml/min and air at 350 ml/min make up	59
1 ml injected at 250°C. Split (1 min), split ratio 1:30	Column: capillary column (cross-linked 5% of phenylmethylsilicone) Length (m): 25 I.D. (mm): 0.32 Film thickness (μm): 0.52	40°C (1 min) to 80°C at 8°C/min; 95 to 220°C at 120°C/min	Helium at 2.5 ml/min	FID at 250°C	60

Continued

TABLE 14.11
Continued

Injection	Column	Temperature Program	Carrier Gas	Detection Method	Reference
—	3 Columns; BP-1 in series with Al$_2$O$_3$/Na$_2$SO$_4$ PLOT. CP-Sil 8 in parallel Length (m): 50, 50, 50 I.D. (mm): 0.32, 0.32, 0.32 Film thickness (μm): 1, 1, 1	—	He	FID	52
Split injection 50:1	Column: dual capillary column: BP-1 and BP-10 Length (m): 50 (BP-1), 50 (BP-10) I.D. (mm): 0.22 (BP-1), 0.22 (BP-10) Film thickness (μm): 1 (BP-1), 10 (BP-10)	50°C (5 min) to 130°C at 5°C/min	—	Dual FID	57

information about gas chromatographic columns, and conditions and detection methods used for the analysis of BTEX in air.

Some special configurations of gas chromatographic columns are reported. The Siemens RGC 402 online gas chromatograph instrument for BTEX measurements reported in Ref. 61 uses a W-COT-CP-WAX CB in series with a WCOT-CP-SIL to achieve in a first step of 4 min the separation of higher volatiles in the second column where these are retained. In the second step, both columns are disconnected from each other in order to continue the separation process of both fractions simultaneously in different columns. For the separation of fractions of compounds of different polarities (BTEX, trichloroetehene, tetrachloroethene, ethylacetate, and nonane), two columns of different polarities, DB-5 (5% phenyl–95% methyl silicone), and DB-1701 (14% cyanopropyl phenyl silicone) were switched in parallel.[62]

A fast gas chromatographic method was tested in order to achieve the full separation of BTEX within 1 min. This reduction in time compared to the conventional chromatographic analysis was achieved with a short (10 m), wide (0.53 mm) capillary (CP-Sil 8 CB with a nonpolar 1 μm stationary phase) between the reduced pressure of the mass spectrometer and the atmospheric pressure in the inlet applying very fast temperature rates (80°C/min).[64]

REFERENCES

1. Valor, I., Analysis of BTEX in water, In *Handbook of Water Analysis*, Leo, M. L. and Nollet, M., Eds., Marcel Dekker, New York, 2000.
2. Leonard, C., Liu, H. F., Brewer, S., and Sacks, R., High-speed gas extraction of volatile and semivolatile organic compounds from aqueous samples, *Anal. Chem.*, 70(16), 3498–3504, 1998.
3. Bianki, F., Careri, M., Marengo, E., and Musci, M., Use of experimental design for the purge-and-trap gas chromatography–mass spectrometry determination of methyl *tert*-butyl ether, *tert*-butyl alcohol and BTEX in groundwater at trace level, *J. Chromatogr. A*, 975(1), 113–121, 2002.
4. Rosell, M., Lacorte, S., Ginebreda, A., and Barcelo, D., Simultaneous determination of methyl *tert*-butyl ether and its degradation products, other gasoline oxygenates and benzene, toluene, ethylbenzene and xylenes in Catalonian groundwater by purge-and-trap–gas chromatography–mass spectrometry, *J. Chromatogr. A*, 995(1–2), 171–184, 2003.
5. Zygmunt, B., Determination of benzene alkyl derivatives in heavily loaded environmental aqueous samples by means of combination of distillation and purge and trap gas chromatography mass spectrometry, *HRC J. High Resolution Chromatogr.*, 20(9), 482–486, 1997.
6. Creaser, C. S., Weston, D. J., Wilkins, J. P. G., Yorke, C. P., Irwin, J., and Smith, B., Determination of benzene in aqueous samples by membrane inlet, solid-phase microextraction and purge and trap extraction with isotope dilution gas chromatography–mass spectrometry, *Anal. Commun.*, 36(11–12), 383–386, 1999.
7. deAndrade, J. B., Pereira, P. A. D., and Oliveira, C. D. L., Determination of volatile organic compounds in groundwater by GC: comparison between headspace and purge and trap, *Energy Sources*, 20(6), 497–504, 1998.
8. Yamamoto, K., Fukushima, M., Kakutani, N., and Kuroda, K., Volatile organic compounds in urban rivers and their estuaries in Osaka, Japan, *Environ. Pollut.*, 95(1), 135–143, 1997.
9. Lacorte, S., Olivella, L., Rosell, M., Figueras, M., Ginebreda, A., and Barcelo, D., Cross validation of methods used for analysis of MTBE and other gasoline components in groundwater, *Chromatogr. A*, 56(11–12), 739–744, 2002.
10. Wartelle, L. H., Marshall, W. E., Toles, C. A., and Johns, M. M., Comparison of nutshell granular activated carbons to commercial adsorbents for purge and trap gas chromatographic analysis of volatile organic compounds, *J. Chromatogr. A*, 879, 169–175, 2000.
11. Menendez, J. C. F., Sanchez, M. L. F., Uria, J. E. S., Martinez, E. F., and Sanz Medel, A., Static headspace, solid phase microextarction and headspace solid-phase microextraction for BTEX determination in aqueous samples, *Anal. Chim. Acta*, 415(1–2), 9–20, 2000.
12. Matisova, E., Sedlakova, J., Simon, P., and Welsch, T., Solid phase microextraction of volatiles from water using open cap vials, *Chromatographia*, 49(9–10), 513–519, 1999.

13. Matisova, E., Medved'ova, M., Vraniakova, J., and Simon, P., Optimization of solid-phase microextraction of volatiles, *J. Chromatogr. A*, 960(1–2), 159–164, 2002.

14. Popp, P. and Paschke, A., Efficiency of direct solid-phase microextraction from water — comparison of different types of fibers including new C8-coating, *Chromatogr. A*, 49(11–12), 686–690, 1999.

15. Pons, B., Fernandez Torroba, M. A., Ortiz, G., and Tena, M. T., Monitoring and evolution of the pollution by VOCs in the groundwater of the Najerilla river basin (Spain), *Int. J. Environ. Anal. Chem.*, 83(6), 495–506, 2003.

16. Cho, H. J., Baek, K., Lee, H. H., Lee, S. H., and Yang, J. W., Competitive extraction of multicomponent contaminants in water by carboxen-polydimethylsiloxane fiber during solid-phase microextraction, *J. Chromatogr. A*, 988(2), 177–184, 2003.

17. Wang, Z. D., Li, K., Fingas, M., Sigouin, M., and Menard, L., Characterization and source identification of hydrocarbons in water samples using multiple analytical techniques, *J. Chromatogr. A*, 971(1–2), 173–184, 2002.

18. Popp, P. and Paschke, A., Solid phase microextraction of volatile organic compounds using carboxenpolydimethylsiloxane fibers, *Chromatographia*, 46(7–8), 419–424, 1997.

19. Djozan, D. and Assadi, Y., A new porous-layer activated-charcoal-coated fused silica fiber: application for determination of BTEX compounds in water samples using headspace solid-phase microextraction and capillary gas chromatography, *Chromatographia*, 45(Suppl.), S183–S189, 1997.

20. Xiao, C. H., Liu, Z. L., Wang, Z. Y., Wu, C. Y., and Han, H. M., Use of polymeric fullerene as a new coating for solid-phase microextraction, *Chromatographia*, 52(11–12), 803–809, 2000.

21. Nardi, L., Capillary extractors for "negligible depletion" sampling of benzene, toluene, ethylbenzene and xylenes by in-tube solid-phase microextraction, *J. Chromatogr. A*, 985(1–2), 85–91, 2003.

22. Nardi, L., In-tube solid-phase microextraction, *J. Chromatogr. A*, 985(1–2), 93–98, 2003.

23. Nardi, L., Determination of siloxane–water partition coefficients by capillary extraction-high-resolution gas chromatography. Study of aromatic solvents, *J. Chromatogr. A*, 985(1–2), 39–45, 2003.

24. Aguilar, C., Janssen, H. G., and Cramers, C. A., Online coupling of equilibrium-sorptive enrichment to gas chromatography to determine low molecular mass pollutants in environmental water samples, *J. Chromatogr. A*, 867(1–2), 207–218, 2000.

25. Hauser, B. and Popp, P., Combining membrane extraction with mobile gas chromatography for the field analysis of volatile organic compounds in contaminated waters, *J. Chromatogr. A*, 909(1), 3–12, 2001.

26. Luo, Y. Z., Adams, M., and Pawliszyn, J., Aqueous sample direct extraction and analysis by membrane extraction with a sorbent interface, *Analyst*, 122(12), 1461–1469, 1997.

27. Guo, X. M. and Mitra, S., Development of pulse introduction membrane extraction for analysis of volatile organic compounds in individual aqueous samples, and for continuous online monitoring, *J. Chromatogr. A*, 826(1), 39–47, 1998.

28. Matz, G., Loogk, M., and Lennemann, F., Online gas chromatography–mass spectrometry for process monitoring using solvent free sample preparation, *J. Chromatogr. A*, 819, 51–60, 1998.

29. Shoemaker, J. A., Bellar, T. A., Fichelberger, J. W., and Budde, W. L., *J. Chromatogr. Sci.*, 31, 279, 1993.

30. Gaines, R. B., Ledford, E. B., and Stuart, J. D., Analysis of water samples for trace levels of oxygenate and aromatic compounds using headspace solid-phase microextraction and comprehensive two dimensional gas chromatography, *J. Microcolumn Separations*, 10(7), 597–604, 1998.

31. Kephart, T. S. and Dagsputa, P. K., Superheated water eluent capillary liquid chromatography, *Talanta*, 56, 977–987, 2002.

32. Kondo, T., Yang, Y., and Lamm, L., Separation of non-polar analytes using dimethyl sulfoxide-modified subcritical water, *Anal. Chim. Acta*, 460, 185–191, 2002.

33. Llompart, M., Li, K., and Fingas, M., Headspace solid phase microextraction (HSSPME) for the determination of volatile and semivolatile pollutants in soil, *Talanta*, 48, 451–459, 1999.

34. Zygmunt, B. and Namiesnik, J., Solid-phase microextraction-gas chromatographic determination of volatile monoaromatic hydrocarbons in soil, *Fresenius J. Anal. Chem.*, 370(8), 1096–1099, 2001.

35. Zuloaga, O., Etxeberria, N., Zubiaur, J., Fernandez, L. A., and Madaraiga, J. M., Multisimplex optimization of purge and trap extraction of volatile organic compounds in soil samples, *Analyst*, 125(3), 477–480, 2000.

36. Roose, P., Dewulf, J., Brinkman, U. A. T., and Van Langenhove, H., Measurements of volatile organic compounds in sediments of the Scheldt Estuary and the Southern North Sea, *Water Res.*, 35(6), 1478–1488, 2001.

37. Lau, W. L. and Chan, L. Y., Commuter exposure to aromatic VOCs in public transportation modes in Hong Kong, *Sci. Total Environ.*, 308(1–3), 143–155, 2003.

38. Bravo, H., Sosa, R., Sanchez, P., Bueno, E., and Gonzalez, L., Concentrations of benzene and toluene in the atmosphere of the Southwestern area at the Mexico City Metropolitan Zone, *Atmos. Environ.*, 36, 3843–3849, 2002.

39. Na, K. and Kim, Y. P., Seasonal characteristics of ambient volatile organic compounds in Seoul, Korea, *Atmos. Environ.*, 35(15), 2603–2614, 2001.

40. Srivastava, P. K., Pandit, G. G., Sharma, S., and Rao, A. M. M., Volatile organic compounds in indoor environments in Mumbai, India, *Sci. Total Environ.*, 255(1–3), 161–168, 2000.

41. Rao, A. M. M., Pandit, G. G., Sain, P., Sharma, S., Krishnamoorthy, T. M., and Nambi, K. S. V., Non-methane hydrocarbons in industrial locations of Bombay, *Atmos. Environ.*, 31(7), 1077–1085, 1997.

42. Ballach, J., Greuter, G., Shultz, E., and Jaeshke, W., Variations of uptake rates in benzene diffusive sampling as a function of ambient conditions, *Sci. Total Environ.*, 243/244, 203–217, 1999.

43. Martin, N. A., Marlow, D. J., Henderson, M. H., Goody, B. A., and Quincey, P. G., Studies using the sorbent Carbopack X for measuring environmental benzene with Perkin–Elmer-type pumped and diffusive samplers, *Atmos. Environ.*, 37(7), 871–874, 2003.

44. Wideqvist, U., Vesely, V., Johansson, C., Potter, A., Brorstrom Lunden, E., Sjoberg, K., and Jonson, T., Comparison of measurement methods for benzene and toluene, *Atmos. Environ.*, 37(14), 1963–1973, 2003.

45. Tran, N. K., Steinberg, S. M., and Johnson, B. J., Volatile aromatic hydrocarbons and dicarboxylic acid concentrations in air at urban site in the Southwestern U.S, *Atmos. Environ.*, 34(11), 1845–1852, 2000.

46. Chan, L. Y., Wang, X., He, Q., Wang, H., Sheng, G., Chang, L. Y., Fu, J., and Blake, D. R., *Atmos. Environ.*, 38, 6177–6184, 2004.

47. Saba, A., Cuzzola, A., Raffaelli, A., Pucci, S., and Salvadori, P., Determination of benzene at trace levels in air by a novel method based on solid-phase microextraction gas chromatography/mass spectrometry, *Rapid Commun. Mass Spectrom.*, 15(24), 2404–2408, 2001.

48. Olsson, M. and Petersson, G., Benzene emitted from glowing charcoal, *Sci. Total Environ.*, 303(3), 215–220, 2003.

49. Keymeulen, R., Görgenyi, M., Heberger, K., Priksane, A., and van Langehove, H., Benzene, toluene, ethylbenzene and xylenes in ambient air and *Pinus sylvestris* L. needles: a comparative study between Belgium, Hungary and Latvia, *Atmos. Environ.*, 35, 6327–6335, 2001.

50. Leung, P.-L. and Harrison, R. M., Roadside and in-vehicle concentrations of monoaromatic hydrocarbons, *Atmos. Environ.*, 33, 191–204, 1999.

51. Kim, K. H., Oh, S. I., and Choi, Y. J., Comparative analysis of bias in the collection of airborne pollutants: tests on major aromatic VOCs using three types of sorbent methods, *Talanta*, 64, 518–527, 2004.

52. Bakeas, E. B. and Siskos, P. A., Volatile hydrocarbons in the atmosphere of Athens, Greece, *Environ. Sci. Pollut. Res.*, 9(4), 234–240, 2002.

53. Wang, X. M., Sheng, G. Y., Fu, J. M., Chan, C. Y., Lee, S. G., Chan, L. Y., and Wang, Z. S., Urban roadside aromatic hydrocarbons in three cities of the Pearl river delta, People's Republic of China, *Atmos. Environ.*, 36(33), 5141–5148, 2002.

54. Batterman, S. A., Peng, C. Y., and Braun, J., Levels and composition of volatile organic compounds on commuting routes in Detroit, Michigan, *Atmos. Environ.*, 36(39–40), 6015–6030, 2002.

55. Chan, L. Y., Lau, W. L., Wang, X. M., and Tang, J. H., Preliminary measurements of aromatic VOCs in public transportation in Guangzhou, China, *Environ. Int.*, 29(4), 429–435, 2003.

56. Wu, C. H., Lin, M. N., Feng, C. T., Yang, K. L., Lo, Y. S., and Lo, J. G., Measurement of toxic volatile organic compounds in indoor air of semiconductor foundries using multisorbent adsorption/thermal desorption coupled with gas chromatography–mass spectrometry, *J. Chromatogr. A*, 996(1–2), 225–231, 2003.

57. Wright, M. D., Plant, N. T., and Brown, R. H., Diffusive sampling of VOCs as an aid to monitoring urban air quality, *Environ. Monit. Assess.*, 52(1–2), 57–64, 1998.

58. Ng, K. C. and Cheng, Z. L., Environmental monitoring of benzene and alkylated benzene from vehicular emissions, *Environ. Monit. Assess.*, 44(1–3), 437–441, 1997.
59. Moon, C. S., Lee, J. T., Chun, J. H., and Ikeda, M., Use of solvents in industries in Korea: experience in Sinpyeong-Jangrim industrial complex, *Int. Arch. Occup. Environ. Health*, 74(2), 148–152, 2001.
60. Bieniek, G., Simultaneous determination of ethylbenzene, indan, indene and acenaphthelne in air by capillary gas chromatography, *J. Chromatogr. A*, 891(2), 361–365, 2000.
61. Rappenglük, B. and Fabian, P., An analysis of simultaneous online GC. Measurements of BTEX aromatics at three selected sites in the greater Munich area, *J. Appl. Meteor.*, 38, 1448–1462, 1998.
62. Begerow, J., Jemann, E., Keles, T., and Dunemann, L., Performance of two different types of passive samplers for the GC/ECD-FID determination of environmental VOCs in air, *Fresenius J. Anal. Chem.*, 636(4), 399–403, 1999.
63. Son, B., Breysse, P., and Yang, W., Volatile organic compounds concentrations in residential indoor and outdoor and its personal exposure in Korea, *Environ. Int.*, 29(1), 79–85, 2003.
64. Joos, P. E., Godoi, A. F. L., De Jong, R., De Zeeuw, J., and Van Grieken, R., Trace analysis of benzene, toluene, ethylbenzene and xylene isomers in environmental samples by low-pressure gas chromatography–ion trap mass spectrometry, *J. Chromatogr. A*, 985(1–2), 191–196, 2003.
65. Elke, K., Jermann, E., Begerow, J., and Dunemann, L., Determination of benzene, toluene, ethylbenzene and xylenes in indoor air at environmental levels using diffusive samplers in combination with headspace solid-phase microextraction and high-resolution gas chromatography-flame ionization detection, *J. Chromatogr. A*, 826(2), 191–200, 1998.
66. Bouhamra, W. S., BuHamra, S. S., and Thomson, M. S., Determination of volatile organic compounds in indoor and ambient air in Kuwait, *Environ. Int.*, 23(1), 197–204, 1997.
67. Czaplicka, M. and Klejnowski, K., Determination of volatile organic compounds in ambient air. Comparison of methods, *J. Chromatogr. A*, 976(1–2), 369–376, 2002.

15 Polycyclic Aromatic Hydrocarbons

Audrey E. McGowin

CONTENTS

I. INTRODUCTION

A. SOURCES

Polycyclic aromatic hydrocarbons (PAHs) are a group of hydrocarbon compounds containing fused aromatic rings with synonyms that include polynuclear aromatic hydrocarbons (PNAs), arenes, or polyarenes. There are 1896 possible structures for PAHs containing two to eight rings.[1] Chemical transformation of PAHs in the environment results in the formation of homocyclic, and heterocyclic derivatives of PAHs containing nitrogen, oxygen, or sulfur atoms. When combined with PAHs, this larger group of aromatic compounds is referred to as polyaromatic compounds (PACs).[2] Figure 15.1 shows the molecular structures of selected PACs. The sixteen PAHs (EPA$_{16}$) marked with an asterisk (*) have been designated by the United States Environmental Protection Agency (USEPA) as priority pollutants.[3] Compounds marked with a dagger (†) have been selected by the European Union (EU) for monitoring,[4] although delisting of fluoranthene has occurred recently.[5] With the exception of benzo[a]pyrene, compound selection was based mostly on the expediency of analysis rather than actual evidence of toxicity or carcinogenicity.

Anthropogenic input of PACs to the environment stems from incomplete combustion of fossil fuels, waste incineration, and industrial operations such as coke oven and aluminum smelter operation.[6] In addition, motor vehicle emissions may contribute up to 35% of PAH input to the environment in industrialized countries. PAHs are also produced when foods, especially meats, are cooked at high temperatures by smoking, roasting, or grilling.[7] Leachate from oil and coal products, including asphalt[8] and creosote,[9] used as a wood preservative, can contain high levels of PAHs. Except for spills and leaching, anthropogenic PAHs enter the environment as air pollutants and are transported over time to water, soil, sediment, and biota.[10] Forest fires, volcanic eruptions, and soil diagenesis (primarily perylene) are the greatest natural sources of PACs.[11]

B. PHYSICAL PROPERTIES

The physical properties of selected PAHs, shown in Table 15.1, vary widely with molecular weight.[12,13] The vapor pressure of PAHs decreases over 11 orders of magnitude as the number of

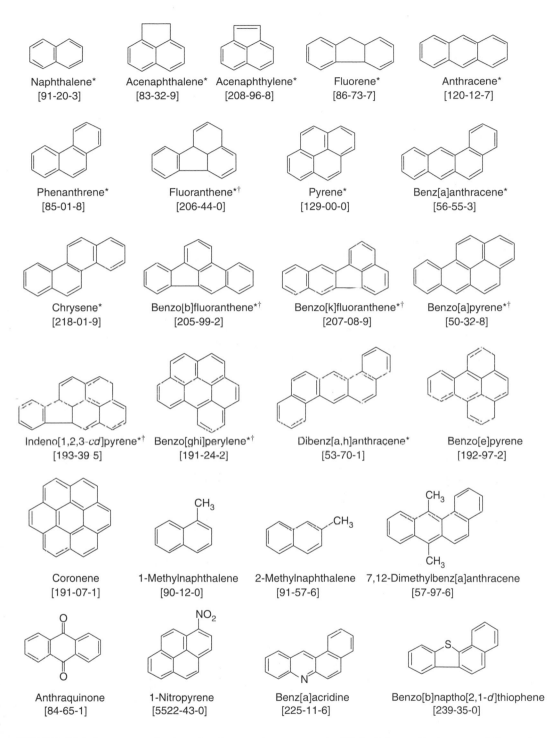

FIGURE 15.1 Structures of selected polyaromatic compounds, *United States Environmental Protection Agency priority pollutants, EPA_{16}, †European Union priority pollutants, EU_6.

TABLE 15.1
Physical Properties of Selected PACs[12,13]

Compound	Abbreviation	MW (g mol^{-1})	MP (°C)	BP (°C)	VP (Pa)	S (g m^{-3})	Log K_{ow}	Log K_{oc}
						At 25°C		
Naphthalene	NAP	128	81	218	10.4	31	3.37	3.11
Acenaphthene	ACE	152	96	278	3×10^{-1}	38	3.92	3.79
Acenaphthylene	ACY	154	92	265	9×10^{-1}	16	4.00	3.83
Fluorene	FLU	166	116	295	9×10^{-2}	1.9	4.18	4.15
Anthracene	ANC	178	216	340	1×10^{-3}	0.045	4.54	4.41
Phenanthrene	PHN	178	101	339	2×10^{-2}	1.1	4.57	4.22
Fluoranthene	FLA	202	111	375	1.2×10^{-3}	0.26	5.22	4.74
Pyrene	PYR	202	156	360	6×10^{-4}	0.13	5.18	4.82
Benz[a]anthracene	B[a]A	228	160	435	2.8×10^{-5}	0.011	5.91	5.66
Chrysene	CHY	228	255	448	5.7×10^{-7}		1.65	5.37
Benzo[b]fluoranthene	B[b]F	252	168	481		0.0015	5.80	5.89
Benzo[k]fluoranthene	B[k]F	252	217	481	5.2×10^{-8}	0.0008	6.00	5.89
Benzo[a]pyrene	B[a]P	252	175	495	7.0×10^{-7}	0.0038	6.04	5.71
Benzo[e]pyrene	B[e]P	252	178		7.4×10^{-7}	0.004		
Indeno[1,2,3-*cd*]pyrene	IND	276	164	536		0.00019	6.58	6.14
Benzo[ghi]perylene	B[ghi]P	276	277			0.00026	6.50	
Dibenz[a,h]anthracene	D[ah]A	278	267	524	3.7×10^{-10}	0.0006	6.75	5.97
Coronene	COR	300	>350	525	2.0×10^{-10}	0.00014	6.75	

MW = molecular weight, MP = melting point, BP = boiling point, VP = vapor pressure of the solid, S = water solubility, K_{ow} = octanol/water partition coefficient, K_{oc} = soil partition coefficient.

fused rings increases from two to seven. Because of naphthalene's high vapor pressure, it tends to partition to a greater degree into the vapor phase in the environment. Larger PAHs (three- and four-ring) will partition between the gaseous and solid phases in the environment. PAHs with five or more aromatic rings are found almost exclusively associated with particulate or solid phases.[14] Water solubility of PAHs with two to six rings decreases over five orders of magnitude with increasing molecular weight. Therefore, two- and three-ring PAHs are more likely to be found in aquatic environments in the dissolved phase than are higher molecular weight PAHs, which tend to be associated with dissolved organic matter (DOM) and solid phases such as soot.[15] The tendency of PAHs to accumulate in soils, sediments, and biota also increases with the size of the molecule. The soil/sediment partition coefficient (K_{oc}) is a measure of the tendency of a compound to partition into natural organic matter and soot in soils and sediments. For PAHs, log K_{oc} values are relatively high, indicating a strong tendency for soils and sediments to become sinks for PAHs even though most PAHs are initially dispersed into the atmosphere.[16]

Since chromatographic separations are primarily based on differences in physical properties, the considerable variability described above for PAHs makes sampling, sample preparation, and analysis especially challenging. When several PAHs are measured as a group, a single sample collection, extraction or analysis method may not be adequate. This is further complicated when the more polar derivatives of PAHs are added to the list of analytes.

C. OCCURRENCE

PAH concentrations in various environmental compartments depend on the proximity of emission sources, meteorological conditions, season, and the physical properties of the compounds themselves.[6] Atmospheric fallout is a major source of PAH input to bodies of water. Water concentrations tend to be extremely low due to the very low water solubility of PAHs. Typical concentrations range from 10 to 50 ng l^{-1} in groundwater to 50 to 250 ng l^{-1} in surface water, although heavily polluted waters often have higher concentrations.[17] Air concentrations range from <1 to 10 ng m^{-3} [18] but polluted air may contain much higher levels. For example, air on a subway train contained 30 ng m^{-3} and 68 ng m^{-3} total PAH measured in winter and summer, respectively.[19] Ambient PAH concentrations for a total of 30 PAHs and alkylated derivatives in Lower Manhattan, NY following the destruction of the World Trade Center were 2207 and 2757 ng m^{-3} for measurements taken two and three weeks following September 11, 2001, respectively.[20] Total PAH concentrations in uncontaminated soils and sediments in rural areas are 0.1 to 100 mg kg^{-1}, yet contaminated soils and sediments may retain levels that are thousands of times higher.[21] However, the strong retention of PAHs in soils and sediments can limit their bioavailability, so total concentrations determined by exhaustive extraction techniques may not reflect the actual risk associated with high concentrations.[22]

D. Environmental Transformations and Toxicity

PAHs are generally unreactive and have low acute toxicities yet degradates and biotransformation products of PAHs can be very potent mutagens and carcinogens.[23] PAHs may induce cancer of the lungs, bladder, and skin. Several PAHs have been classified by the International Agency for Research on Cancer (IARC) as probable human carcinogens. Exposure to high levels of PAHs has been shown to produce immunosuppressive effects.

PAHs require metabolic activation to produce their mutagenic or carcinogenic effects. The primary mechanism of PAH biotransformation in higher organisms is by cytochrome P450-based monooxygenases leading to detoxification and excretion. However, attack by cytochrome P4501A1 can activate certain PAHs such as B[a]P to form a mutagenic diol epoxide capable of forming DNA adducts. The carcinogenesis of nitro-PAHs involves ring oxidation and nitro-reduction to form N-hydroxyamino-PAH intermediates that can bind with DNA. The formation of hydroxy-PAH metabolites allows PAHs to be excreted by higher organisms. PAHs can bioconcentrate or bioaccumulate in aquatic invertebrates such as molluscs that do not posses the ability for their biotransformation while fish can effectively biotransform PAHs, preventing biomagnification up the food chain.[23]

Low molecular weight PAHs are more readily biodegraded than high molecular weight PAHs, which are strongly associated with soil, sediment, and soot particles in the environment.[24] Through the action of bacteria, PAHs can become oxidized by incorporating molecular oxygen across an aromatic ring followed by the formation of hydroxy- or dihydroxy-PAHs. Further oxidation can produce ketone, dione and quinone derivatives, and carboxylic acid derivatives as rings cleave.

Atmospheric transformation of PAHs occurs through reactions with hydroxyl radicals ($\cdot$OH), nitrate radicals ($\cdot NO_3$), and ozone O_3 to produce nitro-PAHs (NPAH), oxy-PAHs (OPAH), and hydroxy-PAHs, although these derivatives can also be formed during combustion processes.[25-28] In addition to nitrated PAHs, oxygenated derivatives include ketones, diones, quinones, and dicarboxylic acid anhydrides. Oxy- and nitro-PAH concentrations in the atmosphere are about an order of magnitude or less than their parent PAHs, yet they can account for a considerable degree of the toxicity and mutagenicity of urban aerosols.[29] Exposure of PAHs to UV radiation in aquatic environments has been shown to produce significant "phototoxicity" to fish and inverte-brates,[23,30,31] and likely results from the formation of PAH photooxidation products.

In general, the degradation products of PAHs in the environment are more polar than their parent PAHs and therefore more water soluble. Unfortunately, the presence of major degradates is not often considered in the course of an environmental analysis. Overestimation of individual PAH concentrations in chromatographic analyses can sometimes be attributed to the coelution of

OPAHs, NPAHs, and alkylated PAH derivatives, as well as other unidentified structural isomers of PAHs.

E. REGULATIONS

There are no specific regulations limiting PAH levels or emissions in the atmosphere, although USEPA[32] and the EU[33] have set limits on the amount of particulate matter (PM) in ambient air. This provides for indirect regulation of PAHs since most are so strongly associated with atmospheric particles. The EU Working Group on Polycyclic Aromatic Hydrocarbons is currently assessing the need for a PAH atmospheric monitoring program.[34] The U.K. Expert Panel of Air Quality Standards (EPAQS) has recommended an annual average of 0.25 ng m^{-3} using B[a]P as a marker.[35–37]

In 1984, the World Health Organization (WHO) set maximum permissible concentration limits in drinking water for the EU$_6$ at 200 ng l^{-1} and for B[a]P at 20 ng l^{-1}.[38] In 1998, the EU set limits for the sum of the EU$_6$, minus FLA which was delisted at 100 ng l^{-1} and for B[a]P at 10 ng l^{-1} to be met by 2003.[39] The USEPA has established a list of drinking water contaminants to be monitored that limits B[a]P concentration to 200 ng l^{-1}.[40] The USEPA Hazardous Constituents List for wastewater and solids contains 19 PAHs and several PACs including 1,4-naphthoquinone, 7,12-dimethylbenz[a]anthracene, and 2-chloronaphthalene.[41] The USEPA Groundwater Monitoring List includes ten of the EPA priority PAHs along with 7,12-dimethyl-benz[a]anthracene,2-methylnaphthalene, and 3-methcholanthrene.[42] Although PACs are monitored in groundwater and wastes, standards have not been set. Limits are determined on a case-by-case basis based on Best Professional Judgment (BPJ) for National Pollutant Discharge Elimination System (NPDES) permitted discharges under the Clean Water Act.[43]

Risk assessment for PAH is complicated by a lack of understanding of the cancer potency of PAH mixtures. Toxicity equivalency factors (TEFs) have been determined for many PAHs relative to B[a]P. The concentration of PAH × TEF for each individual PAH gives a concentration known as the B[a]P equivalent. Since the background level of PAH is generally below 1 mg kg^{-1} for most rural sites, USEPA remediation goals are usually set at that level for B[a]P equivalents, and 10 mg kg^{-1} for industrial sites or well-vegetated areas where human contact with soil is less likely.[44] The EU has set limits for FLA, B[b]F, and B[a]P of 5, 2.5, and 2 mg kg^{-1}, respectively, in sewage sludge to be spread on agricultural land.[45]

F. SAMPLE STORAGE

The photolability of PAHs dictates that samples be protected from light by storage in the dark in glass or PTFE containers.[46,47] Sodium thiosulfate can be added as a preservative to water samples containing residual chlorine. Most water samples require the addition of suitable organic solvents or surfactants to prevent the adsorption of PAHs on the inside of container walls. Aqueous samples can then be frozen for storage but should be analyzed within seven days. Extracted samples should be analyzed within 40 days, and extracts retained on solid sorbents can be stored at 4°C for up to a month.[48] The USEPA recommends a maximum holding time of 14 days for soils and sediments at 4°C, although lower molecular mass PAHs have been shown to biodegrade in some samples under these conditions.[49] Sodium azide poisoning or freezing of samples at −20°C can prevent biodegradation of analytes.

II. SAMPLING OF THE ATMOSPHERE

PAHs are measured directly from emission sources or indirectly as deposition from the atmosphere. Because of the complicated nature of particulate and aerosol analysis, and the sheer number of PAH, OPAH, and NPAH compounds present in atmospheric samples, sample analysis is commonly performed offline.[50] Sampling times range from hours to days depending on PAH concentrations, analytical sensitivity, and atmospheric conditions.

A. Atmospheric Sampling of PAH

PAHs and related compounds will partition in the atmosphere between the gaseous (aerosol) and particulate phases.[51] Naphthalene, a two-ring PAH, is found primarily in the gaseous phase while PAHs with five or more rings are found almost exclusively sorbed to particulate matter (soot). The difficulty comes in the measurement of three- and four-ring PAHs because they tend to partition between the gaseous and particulate phases. A single collection method will be inadequate for the analysis of a variety of PAHs. The method selected to collect atmospheric samples will also influence the results obtained.[50] Bias in sampling caused by mass transfer between the particulate and gaseous phases, disruption of equilibrium from temperature changes during sampling, and relative humidity can have strong influences on the amounts of PAHs determined in each phase. Another confounding factor is the occurrence of ozone oxidation of collected analytes during or after sampling.[52,53] Of course, trapping efficiency of <100% will also affect the results.

The carbon content and amount of particulate matter to which PAHs are sorbed influences their extractability.[54] Recoveries of PAHs from particulates decrease with increasing carbon content and amount of particulate matter. Diesel particulate matter, from which PAHs are extremely difficult to recover, is about 80% carbon.

There are several sampling configurations that are used most frequently. Most sampling trains for atmospheric particulates begin with a cyclone separator to affect a 2-μm particle size cutoff that removes larger particles.[55] Cyclones can collect wet and dry particles. These are inexpensive and simple to operate.[50] A variety of materials have been used as filters for collecting particles containing PAHs such as quartz fiber (QF), glass fiber (GF), PTFE-coated glass fiber, and PTFE. Filters and sorbent materials are scrupulously cleaned prior to use by heating and solvent extraction to avoid contamination. Downstream sorbents are usually polyurethane foam (PUF) or XAD-4 resin. Soxhlet extraction and sonication are most commonly used to extract collected PAHs from filter materials following sample collection. More recently, extraction methods minimizing the use of organic solvents have emerged including supercritical fluid extraction (SFE), accelerated solvent extraction (ASE), and microwave assisted extraction (MAE). These methods are described in greater detail in Section IV.

The most commonly used technique for sampling PAHs in the atmosphere is USEPA Method TO-13A, updated in 1999.[56] Method TO-13A is a filter-adsorbent (FA) method that uses a high-volume sampler connected to a FA sampling train with a quartz fiber filter (QFF), for trapping particulate-bound PAHs, followed by a PUF adsorbent trap to collect the more volatile PAHs. For analysis, the QFF and PUF are extracted together. However, this technique gives no information on partitioning of PAHs between the particulate and gaseous phases. If possible, the QFF and the PUF can be analyzed separately to examine partitioning of PAHs between the two phases, though sampling artifacts are common. On one hand, volatilization of PAH from the QFF "blow off," followed by their sorption onto the PUF can result in overestimation of the amount of PAH that is in the gas phase and underestimation of the amount of particulate-bound PAH. On the other hand, gaseous PAH can adsorb to the filter and be included in the particulate fraction causing the amount of particulate-bound PAH to be overestimated and the concentration of more volatile PAH to be underestimated.[51,55] In addition, NAP, ACE, and ACY have low trapping efficiencies on PUF and significant loss during sample storage. While XAD-2 resin is more efficient at trapping volatile PAHs with higher recoveries, PUF cartridges are easier for field sampling and demonstrate better flow characteristics.

To correct for adsorption artifacts in FA methods, a second filter can be placed in series behind the first. These types of sampling trains have been referred to as filter–filter–adsorbent (FFA) trains. While the first filter collects the particles, both filters should collect equal amounts of gaseous PAHs. The difference gives the amount of PAHs in the particulate phase. If the amounts of gaseous PAHs are not the same in both filters, or PAHs evaporate from the particles on the first filter, bias can be introduced.[50]

Recently, it has been proven that high-volume sampling methods also suffer from a loss of PAHs, especially BaP, by reaction of particulate-sorbed PAHs with ozone and other oxidants (NO_2, NO_3, and $\cdot OH$) in the atmosphere.[52,53] QFFs and PUF were used to measure gaseous and particulate PAHs in Los Angeles, CA, U.S.A. at several locations during a smog episode in 1993.[57] NAP broke through the sampling train and the concentration of PAHs decreased as the number of fused rings increased. Individual three- to five-ring PAH levels of 0.1 to 50 ng m^{-3} were measured by extracting the PUF filter. For four- to six-ring PAHs extracted from the particulate fraction, atmospheric concentration levels from 0.02 to 0.77 ng m^{-3} were measured. OPAHs were detected in the 0.09 to 41 ng m^{-3} range and NPAHs were detected in the 1.62 to 102 ng m^{-3} range. Air samples taken as they were transported downwind showed increasing concentrations of OPAHs and NPAHs with decreasing PAH concentrations. These results indicate that PAHs reacted in the atmosphere as they were transported downwind from their site of formation.

A recent strategy to collect information on partitioning of PAHs between the particulate and gaseous phases employs a glass denuder. Denuders used for PAH analysis have been of the cylindrical or annular types with flow rates of 1 to 20 l min^{-1} and coated with silicone, methylsilicone, Tenax, activated carbon, or polystyrene–divinylbenzene copolymers such as XAD-4 and XAD-2.[58,59] Denuders allow particles to pass through while collecting gases by diffusion.[60] Residence times in typical denuders are about 0.5 sec. Gases will diffuse faster than particles and will sorb onto a coating inside the denuder while particles are trapped on a filter downstream. The denuder will avoid sorption of gases onto the particulate phase thereby reducing artifacts in sampling. In addition, a sorbent phase is placed downstream of the filter to collect evaporated particles from the filter to allow correction for evaporative losses. This type of sampling train has been referred to as a denuder–filter–adsorbent (DFA) type. Sources of bias result from deposition of particles in the denuder and a < 100% trapping efficiency of gaseous PAHs. Denuders are solvent-washed several times to collect the analytes. Decreasing residence time in the denuder will decrease deposition of particles and will also decrease gas sorption.[60]

Another device used to separate particulate and vapor phase PAHs is an electrostatic precipitator, consisting of a conducting surface to which an electric field has been applied.[61,62] A corona is produced, which charges particles and allows them to collect on an oppositely-charged surface. A sorbent is placed downstream of the electrostatic precipitator to collect gaseous PAHs. The electrostatic precipitator is often referred to as an EA method (electrostatic precipitator-adsorbent). EA methods are less susceptible to sorption/desorption. The corona has the potential to destroy PAHs and create other artifacts by reactions with corona-generated ozone.[63]

A selection of recent analysis methods and results for PACs in the atmosphere is presented in Table 15.2.

B. PASSIVE SAMPLING OF ATMOSPHERIC PAH DEPOSITION

Vegetation samples are easier and more economical to collect than air samples. Moss (*Hypnum cupressiforme*) has no root system so pollutant uptake is only from the atmosphere.[74] Pine needles have also been sampled to determine ambient PAH concentrations.[75,76] Although vegetation concentrations do not directly measure atmospheric concentrations, they allow for monitoring PAH deposition over a large area for a long time period.

Semipermeable membrane devices (SPMD) can be used to sample atmospheric PAHs. Vapor-phase PAHs can pass through and accumulate inside a triolein-filled low-density polyethylene layflat tubing bag. SPDMs are extracted by dialysis in organic solvent. PAH levels measured with SPDMs in urban areas were ten times higher than those in rural areas at six sites in and around Bangkok, Thailand.[77] Deposition rates for Σ15 PAHs ranged from 17 to 134 ng d^{-1}, with higher

TABLE 15.2
Measurement of Atmospheric PACs

Sample Type	Sampling Train	Extraction Method	Analysis Method Stationary Phase	PAC Levels (ng m^{-3})	Ref.
Los Angeles, CA, U.S.A. Sept. 8–9, 1993 smog	FA QF PUF	HEX, benzene-2-propanol rinse, DCM	GC–MS DB-1	Σ15PAH 6072 1-NNAP 10.7 2-NNAP 3.81	57
Indoor Laboratory Air Tobacco smoke	DFA XAD-4 PTFE-coated GF XAD-4	Sonication, cyclohexane	HPLC-FLD Vydac 201TP52	Σ13PAH indoor 645 smoke 2863	60
Hazelrigg, U.K. September–October 1998	FFA GF GF PUF DFA XAD-4 GF PUF	Sohxlet, HEX Sohxlet, HEX Sohxlet, HEX Rinse, HEX Sohxlet, HEX Sohxlet, HEX	GC–MS HP-5MS	Σ19PAH FFA 8.5 DFA 41.3	64
Barcelona Spain, winter and summer 2001	High-vol, GF	SFE CO$_2$, PAH toluene: CO$_2$, OPAH, NPAH	GC–MS GC-ECD DB-17	OPAH, winter: ECD 1.02; MS 0.75 OPAH, summer: ECD 0.43; MS 0.38 NPAH, winter: ECD 0.10; MS 0.10 NPAH, summer: ECD 0.02; MS 0.02	65
Car and subway train interiors Germany, summer 1995 and winter 1996	High-vol, GF	Sonication, cyclohexane	HPLC-FLD Nucleosil-5 C18	Car, winter: ΣPAH 28.7; B[a]P 3.2 Car, summer: ΣPAH 10.2; B[a]P 1.0 Subway, winter: ΣPAH, 67.5; B[a]P 4.0 Subway, summer: ΣPAH 30.2; B[a]P 0.7	19

Continued

TABLE 15.2
Continued

Sample Type	Sampling Train	Extraction Method	Analysis Method Stationary Phase	PAC Levels (ng m^{-3})	Ref.
La Porte Airport Houston, TX, U.S.A. August 31, 2000	DDF XAD-4 XAD-4 QF coated with XAD-4	Rinse, HEX:DCM:MeOH Rinse, HEX:DCM:MeOH Sonication, HEX:DCM:MeOH	GC–GM Rtx-5SilMS	Σ17PAH ~360	66
Munich, Germany April 2001 to October 2002	GF and QF	Sonication, DCM:MeOH:tolene	HPLC-FLD Envirosep PP	ΣEPA$_{16}$: Urban 1.9–5.0; Suburban 0.8–2.9	53
Bus depot, truck repair shop, underground tunnel Lausanne city, Switzerland Summer and winter 2001	GF and QF filters	Soxhlet, toluene	GC–MS BPX-50	Σ15PAH, summer: Bus depot 3.85; Truck shop 2.32; Mine tunnel 12.3; Σ15PAH, winter: Bus depot 24.6; Truck shop 26.4	67
Santiago, Chili June 9–August 10, 1997	DFA Activated charcoal PTFE PUF	Rinse, toluene:HEX Reflux, DCM Soxhlet, DCM	GC–ion trap MS HP-5MS	Σ30PAH 0.60–20.17 B[a]P 0.03–0.68	68
University of Helsinki, Helsinki, Finland, 2001	High-vol, QF	SFE in CO_2, DCM modifier	SFE-LC-GC–MS Silica LC HP-5MS	OPAH Verbenone 0.17 9-fluorenone 0.03 anthraquinone 0.21 methylanthraquinone 0.59	69

Source	Sampling media	Extraction	Analysis	Compounds/concentrations	Ref
Mercedes Benz 1980 Model 300SD diesel engine exhaust	DFA, XAD-4, PTFE-coated GF, XAD-4, FFA, PTFE-coated GF, XAD-4, FA Teflon, XAD-4, EA Aluminum foil	Soxhlet, filters DCM rinse, XAD-4	GC–MS	ΣACE, FLU, PHN, FLA, PYR, B[a]A 49,000–53,000	61
Mercedes Benz diesel engine 1980 model 300SD	FFDD, PTFE-coated GF, XAD-4	Soxhlet, HEX:DCM rinse, HEX:DCM	GC–MS DB-5	1-nitronaphthalene gas phase 2245 particle phase 206	70
Garbage truck driver (GTD) air Maintenance worker (MW) air Helsinki, Finland	FA, PTFE, XAD-2	Extraction, cyclohexane sonication, ACN	HPLC-FLD ChromSpher 5	Σ15PAH: GTD air 71–2660; MW air 68–900	71
Paving worker air Finland	FA, PTFE, XAD-2	Sonication, cyclohexane sonication, ACN	HPLC-FLD	Σ15PAH: 870–40,000	72
Tunghai University (TU) Taichung Industrial Park (TIP) Taiwan	FAAA, QF, PUF, XAD-16, PUF	Soxhlet, DCM:HEX	GC–MS DB-5	Σ20PAH: TU 610; TIP 1232	73

QF = quartz fiber, GF = glass fiber, PUF = polyurethane foam, FA = filter–adsorbent, FFA = filter–filter–adsorbent, DFA = denuder–filter–adsorbent, DDF = denuder–denuder–filter, DCM = dichloromethane, ACN = acetonitrile, HEX = n-hexane, GC–MS = gas chromatography–mass spectrometry, GC-ECD = gas chromatography–electron-capture detector, HPLC = high-performance liquid chromatography, FLD = fluorescence detector, SFE = supercritical fluid extraction, EA = electrostatic precipitator.

levels measured in urban areas. High 1-methylphenanthrene/phenanthrene ratios (0.47 to 0.88) indicated that the source was motor vehicle emissions. SPMDs were deployed at nineteen sites in northwest England.[78] Uptake kinetics were controlled by temperature and wind speed. Methyl oleate and oleic acid are impurities in triolein that have been shown to seep out during deployment and cause the exterior of the sampler to become sticky. This allows particulate-sorbed PAHs to be included in the sample. The SPMDs were deployed for 42 to 45 d in five-sided boxes with the open side pointed down to reduce exposure to precipitation, sun, and particles. SPMDs were spiked with deuterated PAHs prior to dialysis for 24 h with hexane (HEX), silica gel/alumina column cleanup, and elution with dichloromethane (DCM)/hexane followed by gel-permeation chromatography (GPC) cleanup. Solvent was exchanged to isooctane before GC–MS analysis. The $\Sigma 15$ PAHs ranged from 2215 to 13,746 ng SPMD^{-1} with PHN as the predominant ($>40\%$) PAH found in the samples. Uptake rates were used to estimate actual air concentrations.

III. SAMPLING AQUEOUS ENVIRONMENTAL SYSTEMS

PAHs can be found in three phases in aquatic systems: dissolved, DOM-associated, and particulate-sorbed. The dissolved fraction is thought to be the more bioavailable and therefore an important factor to measure in toxicity assessment. DOM-associated PAHs tend to be more transportable over greater distances. Particulate phases provide great sinks for PAHs and allow them to be released into the environment for long periods of time while providing exposure to benthic organisms. Although most PAHs are thought to be partitioned to organic matter in sediments, different bonding affinities are found for various sediment types, especially for those containing soot which strongly binds PAHs.[79] Specific PAHs are associated with each phase, and will show different fate and effects in the environment. This affects the toxicity, fate, and transport of PAHs. It is generally believed that only the dissolved PAHs are bioavailable and therefore of the greatest potential harm. The difficulty in measurement of PAHs in water results from three major issues. First, PAHs have very low solubilities in water and range from mg l^{-1} to pg l^{-1}. This is below the detection limit of most analytical instrumentation used for PAH analysis. Second, PAHs tend to exist as aggregates or colloidal particles rather than as dispersed molecules in aqueous solutions. The third important complicating factor is the tendency of PAHs to sorb to glassware walls and other sample containers. Total PAH concentrations in aquatic systems are obtained by exhaustive extraction with a suitable organic solvent. As with all PAH measurements, the sampling approach will determine the ultimate outcome obtained. Most physical means used to separate dissolved and particle or colloid associated fractions will affect the equilibrium, and the results obtained may be biased by the chosen sampling method.

Sampling of precipitation, wet or dry, is complicated by the fact that the concentrations are very low, and there are large spacial and temporal variations including weather and season. There are various forms of precipitation that can be samples such as, rain, fog, snow, and hail, which can move in different directions.[80] Samplers require a large inlet area and capacity and are customarily placed 1.5 m above the surface of the ground in an open area. Runoff can be sampled by burying the sample container in the side of a drainage ditch with the top of the bottle protruding 2 cm from the surface to keep out insects. Samples are stored in the dark at 4°C with the addition of bactericides or frozen at -20°C. Samples should be analyzed as soon as possible or within 48 h if left in solution.

Loss of PAHs from aqueous solutions onto sample container walls is a serious problem. Almost complete loss of PAHs from solution at ng l^{-1} concentration levels has been documented.[81] Polyethylene containers sorbed PAHs from solution within several hours followed by borosilicate glass and PTFE. The addition of acetonitrile at 40% (v/v) prevented sorption losses. Also effective was the addition of surfactants at concentrations above their critical micelle concentrations (CMCs).

A. LIQUID–LIQUID EXTRACTION (LLE)

LLE consists of partitioning of an analyte from water into an immiscible solvent. LLE of organic contaminants from water samples is very much on the decline as a preconcentration technique. This is due to the fact that large volumes of toxic organic solvents, such as methylene chloride, are needed, which creates an expensive waste stream and unnecessarily exposes laboratory workers to hazardous fumes. Furthermore, developments in separation technology have provided several superior methods for use today. Nevertheless, LLE is still being applied to the extraction of PAHs from water samples and an EPA Method still exists for its use.

EPA Methods 610 and 3510C[82,83] are available for LLE of PAHs from water and wastewater samples. Large volumes of sample are collected, acidified to pH < 2, and frozen. For extraction, 1 l of sample is spiked with surrogate standards and extracted three times with 60 ml of DCM. The extract is dried by passing through a column of anhydrous sodium sulfate and rotary evaporated to concentrate the sample prior to GC analysis. If HPLC analysis is desired, the solvent is exchanged with ACN. LLE was used to measure PAHs in surface water and runoff in western Nigeria,[84] where there is no central sewer system and discharges are directed into the rivers during the rainy season. Very low recoveries for the more volatile PAHs were observed, 11% for naphthalene; average corrected recoveries were 89%. Reported measured concentrations were 0.10 to 73.72 mg l^{-1} of predominantly three- and four-ring PAHs. Water samples, from the Izmit Bay in Turkey, on the east side of the Sea of Marmara, were extracted with HEX and analysis was performed by HPLC-FLD.[85] Of 17 PAHs measured, PHN was the most abundant compound followed by CHR both at around 1 ng l^{-1}. Water concentrations measured were 1000 times lower than concentrations measured in mussels. Reported detection limits were 0.55 to 0.004 ng l^{-1}. Almost all PAHs detected were the smaller, more water-soluble ones from NAP to CHY. Water near a pulp and paper factory had the highest levels.

LLE was used to concentrate 14 PAHs from bulk precipitation (wet and dry deposition) and surface water in Northern Greece.[86] Glass funnels were attached to 1 l amber bottles to collect samples. Analysis was done by HPLC-FLD and collected samples were filtered through glass wool, extracted with HEX, and dried with anhydrous sodium sulfate. Rotary evaporation and N$_2$ blowdown reduced the volume to 5 ml, and the residue was redissolved in ACN. The highest values obtained were for NAP, 426 ng l^{-1}, and the lowest for B[k]F, 1.0 ng l^{-1}. The ΣPAH_{14} ranged from 225 to 672 ng l^{-1}. For surface waters, NAP was again the predominant PAH detected at 677 ng l^{-1} and B[k]F was the lowest at 0.2 ng l^{-1}. The ΣPAH_{14} ranged from 184 to 627 ng l^{-1}. Higher concentrations were measured in winter months and concentrations in surface waters were lower than in precipitation. Bulk PAHs were measured in Paris and other sites in rural France.[87] Stainless steel funnels were attached to aluminum collection tanks. LLE with HEX/DCM (85:15, v/v) was carried out after adding sulfides to sequester mercury. Rotary evaporation concentrated the samples and solvent was changed to ACN for HPLC with UV and FLD. The ΣPAH_{14} of the EPA 16 PAHs were quantified and it was also found that winter concentrations were two to three times greater than summer concentrations because of increased fossil fuel use. The six potential carcinogenic PAHs listed by the IARC accounted for 19% of the total, of which 3% was B[a]P. PHN, FLA, PYR, and CHY accounted for 62 to 71% of the total. PAH concentrations decreased proportionally from the main population center, Paris. In winter, ΣPAH_{14} for rural sites ranged from 25 to 30 ng l^{-1} but in Paris the ΣPAH_{14} was 221 ng l^{-1}. In the summer months, ΣPAH_{14} for rural sites averaged 12 to 15 ng l^{-1} and in Paris the ΣPAH_{14} averaged 124 ng l^{-1}.

B. SOLID-PHASE EXTRACTION (SPE)

SPE is currently the most common method in use for the extraction and preconcentration of PAHs from water samples and is the process of selectively removing an analyte from the solution phase through physical or chemical attractive forces into or onto a solid phase. For most nonpolar to slightly polar environmental contaminants, this means partitioning of analytes from water into an immobilized nonpolar phase. The solid phase can be contained in a disposable cartridge or within a membrane. The water passes through the membrane or column containing the sorbent, leaving the analyte, which then can be selectively desorbed with a suitable organic solvent or thermally desorbed directly into an analytical system. The sample is typically drawn through the sorbent bed by vacuum. Various sorbents have been applied to the analysis of PAHs in the aquatic environments including styrene–divinylbenzene (SDB) copolymers, polydimethylsiloxane (PDMS), and octadecylsilane (ODS or C18). Another type of SPE utilizes immunoaffinity to attract analytes.[88] SPE procedures are easily adapted to automation. USEPA Method 3535[89] describes the use of solid phase.

Two important issues must be addressed to obtain consistent results using SPE. First, the stationary phase must be conditioned with a water-miscible organic solvent such as methanol or acetone before the sample is added. Alternatively, 0.5% methanol can be added directly to the aqueous sample. During the conditioning and extraction steps, the solid phase must be kept "wet" with solvent or sample. Second, the flow through the sorbent must be optimized and monitored. After the entire sample has been extracted, the solid phase is either air-dried or dried under a stream of N_2. While in the stationary phase, samples may be stored for longer periods prior to elution and analysis. Typically, a 1 l sample that has been filtered to remove larger particles is extracted for surface and groundwater analysis. After extraction, greater recoveries can be achieved by first rinsing the sorbent with a water-miscible solvent, such as acetone, that will release analytes from pores containing water. The elution solvent is then applied and allowed to soak into the pores of the sorbent for a minute or two. Analytes solutions are usually dried by passage through a column of anhydrous sodium sulfate. Eluted analyte solutions can be concentrated by N_2 blowdown or with a Kuderna–Danish (K–D) concentrator.

Low ng l^{-1} detection limits with recoveries between 70 and 120% have been reported using reverse-phase SPE combined with GC–MS in the single ion mode (SIM).[90] Trace levels of PAHs in surface waters in France were detected in two of three samples at levels in the 0.02 to 0.45 ng l^{-1} range. Surface waters in Lake Balaton, a summer resort in Hungary, were analyzed using SPE coupled with HPLC-FLD. Fifteen sites were sampled, including sediments. Ninety percent of the PAHs measured in the water were two- to three-ring PAHs of which 80% was NAP. PAHs in sediments were predominantly (85 to 90%) the four- to six-ring variety. Levels of PAHs did not vary much in space or time within the lake.

SPE on a C18 cartridge was used to concentrate PAHs from rainwater and runoff collected from the "Threecity" region of Poland.[92] Analytes were eluted with n-pentane:DCM (1:1, v:v) and analysis was performed by GC–MS. Higher concentrations were measured in the winter months. Levels for thirteen PAHs ranged from 0.81 ng l^{-1} for B[ghi]P to 265 ng l^{-1} for NAP in precipitation. Runoff samples had up to 20 times higher PAH concentrations. The highest value was 989 ng l^{-1} for NAP and the lowest was 8.2 ng l^{-1} for D[ah]A. The results were statistically correlated to emissions from coal stoves and vehicles. An electronic continuous wet-only sampling system that concentrated PAHs on a C18 cartridge was described recently.[93] The system collected rain, snow, sleet, and hail with an internal heating system and in situ filtration. Nonpolar organics in precipitation were preconcentrated on a C18 SPE cartridge. By only collecting wet deposition, rain-out and dry deposition were distinguished. Hyamine 1622, a cationic surfactant, was used to treat the system to prevent loss of PAHs on sampler surfaces. Hyamine did not interfere with analysis by HPLC-FLD, but had to be removed by silica gel prior to GC-FID analysis. Good recoveries were obtained for spiked samples.

SPE has been used to estimate the amount of freely dissolved PAHs versus DOM-associated and particulate-associated PAHs. Glass fiber filters used to remove particulate matter from water samples before extraction can also retain some DOM-associated PAHs.[94] It is largely thought that DOM-associated PAHs will pass through reverse-phase SPE cartridges allowing for the measurement of the freely-dissolved fraction, however, some DOM-associated PAHs are retained in SPE leading to an overestimation of dissolved PAHs. Reported levels of freely-dissolved B[k]F of 296 μg l^{-1} and IND of 83 μg l^{-1} are likely overestimations for measurements using SPE in the Jinsha River in southwest China. Very high levels were detected for total PAHs up to 383 μg l^{-1}. PAHs in water samples that are retained on a filter are considered to be associated with the particulate phase. Still, much of the PAHs that pass through a filter can still be associated with colloidal matter or DOM, of the 1-nm to 1-μm size, and not fully dissolved. SPE was used to determine dissolved versus DOM-associated PAHs in urban stormwater runoff in Dunedin, New Zealand.[95] Up to half of the DOM was retained by the SPE cartridge and led to an overestimation of PAHs in the dissolved phase. It has been shown that silanol groups remaining on the surface of silica solid supports can irreversibly bind PAHs and lead to low recoveries. Recoveries can be improved by using thermally-assisted desorption, sonication, and microwave-assisted desorption.[96]

In SPE, a small amount of organic solvent or surfactant is added to collected samples to prevent adsorption to sample containers. To increase recoveries of ng l^{-1} levels of PAHs in SPE, ACN (40%) or surfactants above their CMC can be added to samples prior to preconcentration.[97] Solid supports, chemically modified with copper phthalocyanine trisulfonic acid derivatives for selective sorption of PAHs, have been investigated. The selective interaction is thought to be with the π electrons of the PAHs.[98] Brij-35, a neutral polyoxyethylene lauryl ether surfactant was added above the CMC to water samples to prevent sorption on container walls. Before preconcentration by SPE, samples were diluted to reduce the surfactant concentration to just below the CMC. Recoveries of over 90% for SPE on solids containing copper phthalocyanine trisulfonic derivatives were obtained for spiked water samples at low ng l^{-1} levels, except for NAP, ACE, and FLU. Experiments repeated using a C18 SPE preconcentration sorbent gave >90% recovery for all 16 EPA PAHs, except for ACY. Examples of the use of SDB as an SPE sorbent include the online analysis of seawater from the coast of Catalonia in Spain, where no PAHs above the low ng l^{-1} level were detected,[99] and the analysis of leachate from coal deposits.[100]

PUF has also been used as a solid-phase to sorb PAHs from water samples. PUF sorbents were batch-equilibrated with PAH standard solutions, removed, squeezed with a filter paper, air-dried, and subjected to solid-matrix luminescence (SML). PAH was converted to 3,4 benzopyrene for luminescence analysis.[101] A level of 8 ng ml^{-1} PAH was detected in sewage water from a petroleum refinery.[102] More recently, cotton was applied as a solid-phase sorbent in the online investigation of tap water, sewage treatment plant effluent, and river water in Beijing, China.[103] Water samples, of 100-ml volume and 10% 2-propanol, to prevent sorption on container walls, were preconcentrated using a precolumn attached to an HPLC-FLD system. Recoveries between 92 and 119% were reported at ng l^{-1} spike levels for FLU, PHN, FLA, and B[k]F in real water samples.

Preconcentration of PAHs in rainwater by a factor of 50,000 has been reported by performing a solid-phase extraction on "blue cotton," cellulose fiber containing covalently linked copper (II) phthalocyanine trisulfonate (Cu–PC).[104] Up to 20 l of rainwater was passed twice through a Cu–PC column containing 0.1 g of sorbent. PAHs were eluted with 5 ml THF and 90 μl was added to the eluent. This mixture was then homogenized and, upon addition of 40 ml water, 20 μl of an immiscible phase separated from the mixture. This phase, now containing PAHs originally in the rainwater, was removed via syringe, filtered, and analyzed by HPLC-FLD. B[a]P and B[a]A were detected in rainwater at 0.51 and 0.39 parts per trillion, respectively.

C. Immunosorbents

Most SPE sorbents are nonselective so matrix components can be coextracted and coeluted thus requiring further cleanup steps before analysis. Immunosorbents (IS) allow for class-specific extraction of environmental samples.[105] Separation of PAHs by IS prior to analysis by conventional methods allows individual quantification, whereas immunoassay is used for direct detection without separation. Immunoassay can only give results for all of the molecules in the class that react and crossreact with the enzyme, so quantitation of individual compounds is not possible. Antibodies are often combined to increase retention of more compounds in a class. Initially, polyclonal antibodies were used in method development, however, monoclonal antibodies are easier to reproduce and the use of animals is minimized in largescale production.

USEPA Method 4035 "Soil Screening for Polynuclear Aromatic Hydrocarbons by Immunoassay" was approved in December 1996 for the determination of total PAHs.[106] This method is only to be used for the purposes of screening samples in the field using commercially available ELISA kits. These kits allow for detection of PAHs at the 1 to 10 mg kg^{-1} level in extracted soils. Crossreactivity occurs when antibodies react with other compounds that have similar structures to the target analyte. ELISA kits suffer from high rates of crossreactivity, however this can be a benefit in screening as structurally similar compounds are detected such as alkyl-substituted PAHs. False positive rates are about 10 to 20% but false negative rates are extremely low. Use of ELISA can significantly reduce the number of samples that must be taken back to the laboratory for traditional analysis. ELISA has been applied to the analysis of prefiltered water samples and the results compared directly to those obtained by reverse-phase SPE with GC–MS analysis.[107] Using ELISA, PAH values were greatly overestimated by one order of magnitude or more. Water samples from the Nitra basin in Slovakia contained very low levels of PAHs and other aromatic molecules with crossreactivity producing high results. Unfiltered samples also gave much higher results than filtered samples.

Antifluorene antibodies were immobilized on silica for the selective extraction of the more volatile two- to three-ring PAHs, as well as FLU and FLA.[108] Online extraction and analysis greatly reduced losses of volatile PAHs. Sample sizes were small, 20 ml, while reverse-phase HPLC with diode-array detection (DAD) and FLD allowed quantification to 10 to 20 ng l^{-1}. HPLC-FLD provided good sensitivity while DAD provided identity information when doing online analysis. Instead of adding organic solvent necessary for reverse-phase (RP)-SPE concentration, Brij-35 was added ($3 \times 10^{-4} M$) as a solubilizer to keep PAHs in solution. Recoveries were 15 to 65% for fortified water samples. Antipyrene antibodies immobilized on silica showed a higher affinity for all of the 16 EPA PAHs than antifluorene antibodies.[109] Water samples were first concentrated using RP-SPE and dried, then the residues were redissolved in water with 25% ACN. Recoveries of fortified reagent water samples were 45 to 60% for 11 PAHs not including NAP or PHN. Limits of detection (LODs) were determined to be 1 to 25 ng l^{-1} with HPLC-DAD and FLD. Wastewater in a sewage treatment plant showed no detectable PAHs with recoveries of 13 to 43%. Antipyrene antibodies are better at isolating the more hydrophobic PAHs with four to six rings.[110]

Covalently bonded antibodies show conformational changes with a subsequent loss of activity, while physically sorbed antibodies are prone to leaching. Pyrene-selective IS columns were developed using the sol–gel method for retaining the antibodies.[111] Here, antibodies were encapsulated into the pores of a hydrophilic silica support. Small pore size limited access to the antibodies by larger molecules and bacteria thus increasing selectivity and stability. Columns were regenerated after elution of analytes by rinsing with ACN:water (40:60, v/v) followed by phosphate-buffered saline (PBS) at a pH of 7.6. No leakage of antibodies was observed and there was no change in retention properties. PAHs did sorb nonspecifically to the glass matrix, especially the larger four- to six-ring PAHs. The columns could tolerate 500 ml sample volumes. In general, recoveries for immunoextraction are usually less than 50%.

D. CLOUD-POINT OR MICELLE-MEDIATED EXTRACTION

Cloud-point extraction (CPE) and micelle-mediated extraction are methods that utilize aqueous solutions of nonionic or zwitterionic surfactant concentrations above their critical micelle concentrations. The addition of ions or solvents as well as increasing temperature of the solution can cause a reversible phase separation above the cloud-point temperature.[112] Also, addition of surfactant disrupts DOM-PAH associations so it will affect the distribution of PAHs in the sample. The surfactant will separate from the aqueous phase taking the associated organic contaminants with it. The surfactant is then removed from solution, usually by centrifugation, before analysis by HPLC-FLD. Triton X-114, a nonionic surfactant, was tested for its efficiency in the extraction of the fifteen fluorescence-active EPA PAHs from aqueous samples. Triton X-114 interferes with the fluorescence signal and retention of PAHs by the HPLC column so it had to be removed before injection of the samples. Removal consisted of dissolution of the surfactant in cyclohexane followed by cleanup on a deactivated silica/sodium sulfate column and elution with cyclohexane:DCM (80:20). The extract was evaporated to near dryness and redissolved in ACN. Even with these conditions, NAP coelutes with the surfactant and so it cannot be determined. Recoveries for the other fourteen PAHs ranged from 35 to 103%. Actual river samples were analyzed and levels of PAH ranged from 26.8 mg l^{-1} for FLU to 1.6 ng l^{-1} for B[a]P.

Sodium docecane sulfonic acid (SDSA) is an anionic surfactant that does not interfere with the chromatographic analysis and so does not have to be removed thus eliminating the cleanup step.[113] SDSA separated into the two isotropic phases upon the addition of hydrochloric acid (HCl) at room temperature. To a 30-ml sample of water, 0.05 g SDSA and 15 ml concentrated HCl were added. After a 24-h equilibration time, the solutions were centrifuged to remove the sorbed analytes. ACN was added to the surfactant-rich phase to reduce its viscosity for analysis. FLU, PHN, PYR, B[a]A, and ACE were detected at low ng l^{-1} levels in groundwater in Córdoba, Spain. Seven PAHs were detected in river water at low ng l^{-1} concentrations. Recoveries of spiked samples ranged from 63% for ACE to 106% for B[b]F. Recoveries for volatile PAHs were all 83% or better.

Results using poly(N-isopropylacrylamide) (PNIPAAm) for CPE are more encouraging.[114] PNIPAAm forms a gummy precipitate at a critical micelle temperature of about 32°C and is nontoxic. Recoveries of eight PAHs ranged from 28% for NAP to 97% for perylene. The polymer was dissolved in ACN prior to HPLC-FLD analysis and no spectroscopic and chromatographic interferences were observed. The viability of the method was presented, but no real samples were analyzed. Another nonionic surfactant with encouraging results is Tergitol 15-S-7,[115] which has a cloud-point temperature of about 37°C at 3 weight % and pH 6.8. It is a readily biodegradable secondary ethoxylated alcohol. Cloud-point temperatures were reduced to below room temperature by adding 0.5 M sodium sulfate and allowing for 10-min equilibration times. Centrifugation separated the precipitated analytes and surfactant, and no washing steps were required. Also, there were no spectroscopic interferences and no retention on the analytical column. Polyoxyethylene-10-lauryl ether (POLE) also shows no spectroscopic interferences and low column retention.[116] A 1% (w/v) POLE addition to samples at 95°C for 90 min extracted PAHs from artificial seawater.

E. Semipermeable Membrane Devices (SPMD) for Passive Sampling

Membrane-based passive sampling devices (PSDs), such as the patented SPMD, are deployed in water bodies and left to equilibrate. Layflat polyethylene tubing is heat sealed on the ends and contains triolein into which the PAHs and other pollutants partition. After the sampling time is up, the bag is dialyzed with solvent to recover the contaminants. SPMDs have been successful for screening of environmental contaminants, including PAHs, in water.[117,118] The device is a 86-cm long, 2.54-cm wide piece of lay flat low-density polyethylene (LDPE) tube (50- to 90-μm wall thickness) that contains 1 ml or 0.91 g of triolein. The ends are heat sealed and attached after turning one end 180° to form a Mobius strip giving a surface area to liquid volume ratio of about 450 cm^2 ml^{-1}. Sample cleanup is with GPC and potassium silicate. Measurement of PAH concentrations in fish and other aquatic organisms is complicated by the biotransformation of PAHs with mixed-function oxidase enzymes, so correlation with fish tissues PAH concentration with exposure concentrations is difficult, if not impossible. SPMDs can serve as models for bioconcentration in organisms by concentrating hydrophobic contaminants from water without bioaccumulation. SPMDs made from LDPE approximate the same steric exclusion limits of 10 Angstroms, similar to fish gill membranes. SPMDs are also an effective screening device for a variety of environmental contaminants. Correlation of SPMD contaminant concentrations with actual environmental concentrations can be approximated with partition equations and uptake kinetics. The minimum lower quantifiable limit for the method was reported to be 50 mg per SPMD following washing of the outside of the device to remove attached microscopic and macroscopic organisms. Size exclusion chromatography recovers coextracted lipids, biogenic compounds, tubing contaminants, and additives from the samples. Recoveries from the EPA 16 PAHs ranged from 120% for NAP to 17% for B[ghi]P. Salinity affects analyte water solubility but not SPMD sampling rates. Temperature effects on uptake rates were small. Biofouling caused impedance of uptake, especially at higher temperatures with greater biological activity.[119] Recovery-corrected sampling rates ranged from 1.0 to $8.0 \, l \, day^{-1}$. Sampling rates were independent of water concentration for the 1 to 100 $ng \, l^{-1}$ range and increased with K_{ow} up to PYR but then declined as molecular size began to inhibit their transport across the membrane. SPMDs collected PAHs in proportion to their aqueous concentrations. Log K_{SPMD} values ranged from 3.36 to 5.55 for the EPA 16 PAHs. Analyte concentrations can be estimated by using permeability reference compounds. Also, the more lipophilic PAHs with higher K_{ow} values required longer equilibration times and were more affected by biofouling. There is also the potential for photodegradation of samples. SPMD allowed for measurement of a time-integrated average of dissolved PAHs.[120] PAH levels measured in an urban stream were 20 times greater in concentration at high flow than at base flow conditions. SPMDs were able to measure freely dissolved PAHs where sampling of bulk water leads to overestimation of bioavailability since many PAHs are sorbed to particulate matter and are thus less bioavailable. Each SPMD device was able to extract up to 45 l of water. PAHs with five to six rings were infrequently detected using SPMDs since their water solubilities are much lower than smaller molecular weight PAHs. Residence time for PAHs in water using SPMD were studied[121] for PAHs measured in a canal in the vicinity of a refinery containing elevated aqueous dissolved concentrations. The ratio of a downstream site concentration to an upstream site concentration allowed for lifetime or residence time of chemicals in the dissolved phase to be calculated without directly measuring water concentration at very low concentration levels. Molluscs are commonly used to assess bioavailability of PAHs in aquatic systems because they do not biotransform PAHs. Occurrence and concentration of PAHs in SPMDs and clams deployed in three streams in the Dallas–Fort Worth, Texas (U.S.A.) metropolitan area have been compared.[122] Twenty PAHs were detected at all three sites, but only three at the highest levels were detected in the clams (FLU, PYR, and 2,6-dimethylnaphthaene). Concentrations of B[a]A, B[a]P, and CHR exceeded EPA's health criteria for water. The clams also had a high mortality rate indicating significant distress.

SPMDs were used to assess the bioavailability of PAHs compared to sediment extraction methods.[123] SPMDs were extracted with HEX and analyzed by HPLC-FLD to compare the EPA 16 PAHs accumulation in SPMDs with water, sediment, and fish concentrations. SPMDs gave good sampling precision with <10% variation, and extracts had higher LODs. SPMD estimated bioavailable concentrations of PAHs more accurately than those determined by sediment extraction with organic solvent. SPMD data also gave more realistic estimation of aqueous PAH concentrations than fish data. Permeability reference compounds are useful for correcting for sampling rates, and exchange kinetics. SPMDs are cost-effective tools for time-integrated sampling as well. SPMD was applied to the analysis of PAHs in the secondary sewage treatment process.[124] Much higher levels of PAHs were determined using SPMD than by LLE of water with DCM. Higher levels of organic matter in sediments decreased bioavailability of PAHs. Sampling rates for 28 PAHs and 19 homologues have been reported.[125] PAHs with log K_{ow} > 4.5 demonstrated linear uptake for 30 days when deployed. PAHs with log K_{ow} < 4.5 approached steady state within 15 days. Sampling rates ranged from 2.11 to $6.06 \, l \, d^{-1}$. Quantitative estimates of freely dissolved PAHs and related heterocyclic compounds in laboratory samples has been demonstrated.

F. Solid-Phase Microextraction (SPME)

SPME is performed by a commercially available device consisting of a glass fiber with a sorptive coating for direct sampling of air and water.[126] Since SPME methods do not require exhaustive extraction of analyte, it was determined that SPME can measure analyte levels without significantly affecting the bulk concentration in water samples in equilibrium with sediments and DOM. SPME with external calibration measures freely dissolved PAHs, while SPME with internal calibration measures total concentration of PAHs.[127] Aqueous distribution measurements of freely dissolved vs. reversibly-bound PAHs were performed without doing a physical separation of the solid and liquid phases of aquatic samples, which can affect equilibrium. External standards without DOM are used to calibrate for SPME extractions to determine freely soluble PAH levels. For internal calibration nondeuterated analytes and deuterated standards are assumed to have the same partition coefficients for polydimethylsiloxane (PDMS) coated fibers. The response of nonlabeled to labeled standard multiplied by the labeled standard concentration gives the total concentration of individual PAHs. SPME is limited to low extraction efficiencies because of the very small amount of PDMS (<0.5 μl) that is available for partitioning of analytes. The amount extracted by PDMS depends on the phase ratio of water to PDMS volumes. The K_{ow} of PAHs at equilibrium affects the recovery of PAHs because the sorption process is that of partitioning from the water into the PDMS phase.

PAH concentrations in coal wastewater were determined using internal and external calibration.[128] Distribution coefficients were determined for the EPA$_{16}$ PAHs using 100 μm PDMS and 85 μm PA (polyacrylate) fibers. A linear correlation was found for log K_{ow} vs. log K_{spme} from NAP to CHR. As molecular weight increased, extraction efficiency increased up to PYR and decreased from B[a]A to B[ghi]P on the PA fiber. It was surmised that the highest molecular weight compounds were sorbing on to the surface of the fiber coating rather than absorbing into the polymeric phase allowing for two mechanisms of sorbtion to compete.

SPME with PDMS fibers was employed to measure PAHs and alkylated-PAHs in groundwater samples collected over two years from sites contaminated with coal tar and refinery wastes.[129] NAP was the predominant PAH detected. Heavier PAH concentrations were much lower due to partitioning onto soils and sediments. Source determination by examining isomer ratios was difficult due to the very low concentrations measured. Very low detection limits were achieved, 0.07 ng ml^{-1}.

Certified marine sediment SRM 1941a from NIST was analyzed using micellar microwave-assisted extraction (MAE) with SPME.[130] Polyoxyethane-10-lauryl ether (POLE) was added and microwave energy used to facilitate desorption of PAHs from the sediment prior to extraction by SPME. Desorption of five- to six-ring PAHs was adequate enough to allow for their quantification. For ten of the EPA$_{16}$, recoveries ranged from 59% for PHN to 112% for B[k]F. Equilibration times were over 80 h for higher molecular weight PAHs.

G. Stir Bar Sorptive Extraction (SBSE)

In order to increase the sorption capacity over SPME, a glass-coated stir bar was covered with 1-mm thick silicone (PDMS) tubing to conduct stir bar sorptive extraction or SBSE.[131] The 10- to 40-mm long stir bars were conditioned by heating at 300°C for 4 h. Stirring ensured good contact between the solution and the sorbing surface. After a predetermined equilibration time, the stir bar was removed, dried of water, and transferred to a glass thermal desorption tube. Thermal desorption at 250°C for 5 min into the inlet of a GC–MS allowed for the quantitative and qualitative determination of PAHs. The increased capacity allowed for a 500-fold increase in sensitivity over SPME with low ng l^{-1} detection limits. An extraction time of 2 h gave a recovery of 96% for PAHs up to PYR at a spike level of 30 parts per trillion. Following desorption, stir bars were reused immediately up to 100 times.

SBSE with HPLC-FLD was used to measure PAHs in surface water, groundwater, and precipitation.[132] Surface water and groundwater were analyzed directly and the precipitation was filtered through a glass fiber filter prior to analysis. To condition, stir bars were rinsed in an appropriate solvent, dried in a desiccator, and heated to 280°C for 90 min in an N_2 stream. After extraction, stir bars were dried with a tissue, extracted with 150 μl ACN:H_2O (4:1, v:v), and treated to 10 min of sonication. A 1-h equilibration time for fifteen of the EPA_{16} (ACE is not detectable with FLD) gave estimated limits of detection of 0.2 to 2 mg l^{-1}. Higher recoveries were obtained for the more volatile three- and four-ring PAHs. SBSE was compared to SPE for the analysis of fifteen of the EPA_{16} using HPLC-FLD in the analysis of precipitation from several sites in Halle in Saxony–Anhalt, Germany.[133] For SBSE, the method was nearly solventless with no cleanup needed. By this time, SBSE stir bars with a 0.5-mm PDMS (24 μl) coating thickness had become commercially available and sold by the name of Twister® by Gerstel GmbH, Muhlheim, Germany. Sixty-minute exposures gave estimated LODs between 0.6 and 5.0 ng l^{-1}.

A method was developed for the analysis of 35 priority semivolate pollutants in water with analysis by thermal desorption GC–MS.[134] It was shown that PAHs can be lost from samples by sorption to the glass walls of sample containers, however adding 10% methanol reduced adsorption of five- and six-ring PAHs to glass vial walls by 30 to 100%. Larger PAHs such as IND and B[ghi]P suffer from 6 to 7% carry-over, so stir bars may need to be extracted more than once to achieve adequate recoveries.[135] SBSE-HPLC-FLD applied to the analysis of drinking water samples showed that the method was rapid, free of interferences and carry-over, and precision was good.[136] All values of PAHs detected were below the quantitative limits except for FLU (9 ng l^{-1}) and B[a]P (5 ng l^{-1}) for 15 water samples from various sites in northwestern Spain. SBSE was compared to SPE for the analysis of runoff and river water.[137] C18 SPE cartridges did not quantitatively retain DOM-associated PAHs so their use in sample preparation tends to overestimate free PAHs. This effect tends to be worse for higher molecular weight PAHs and higher levels of DOM. SBSE only extracted free PAHs.

A passive sampling device was constructed using the commercially available Twister® sorbent stir bar by enclosing it inside a dialysis membrane.[138] This device has been called a membrane-enclosed sorptive coating sampler or MESCO. Extraction efficiencies were three orders of magnitude lower than for SPMD due a lower sampling rate, however, the sensitivity was comparable because all of the collected analyte is desorbed into the GC whereas in SPMD, only a small sample in injected for analysis. Twister® stir bars are also much smaller and can be deployed less conspicuously.

Polydimethylsiloxane rods (1 mm × 10 mm) are an effective sorbent for PAHs from aquatic solutions.[139] Each rod is placed in a vial with only 15 ml of sample and shaken for 3 h. The rods are then extracted with 100 ml of ACN:H_2O (4:1, v:v) with 10 min of sonication prior to analysis. When HPLC-FLD is utilized for analysis, recoveries of 62 to 97% were obtained with limits of detection of 0.1 to 1.2 ng l^{-1} when a total of four desorptions were combined. The method was applied to the semiquantitative screening of scrubber dust slurry from copper processing. Further quantitation by pressurized liquid extraction-HPLC-FLD showed that the concentration of B[k]F and PHN were 8.9 mg kg^{-1} and 135.6 mg kg^{-1}, respectively. Since the PDMS rod is thicker than the PDMS coating on typical stir bars, longer equilibration times are required, yet the rods are inexpensive and disposable.

IV. SOIL, SEDIMENT, ATMOSPHERIC PARTICULATE, AND SOLID WASTE EXTRACTION

A. ALKALINE SAPONIFICATION WITH LIQUID–LIQUID EXTRACTION

Alkaline saponification is a sample preparation method used for soils, sediments, and biological samples. Samples are refluxed in 0.5 M KOH in 95% methanol:toluene (2:1, v/v) and cooled to room temperature. The organic layer is separated and extracted with toluene (3 × 50 ml). The combined extract is washed with distilled water, dried over sodium sulfate, filtered, and concentrated. The alkaline digestion is thought to break down humic substances causing them to release contaminants. Sediments, mussels, crabs, and lobsters near a former gasworks site were analyzed with GC-MS for PAHs and alkylated-PAHs after alkaline digestion under reflux, liquid–liquid extraction, and column cleanup.[140] Concentrations of ΣPAH ranged from 4.9 to 6450 μg kg^{-1} for crustacea and mussels, which are known to be limited in their ability to metabolize PAHs. The soil at the former gasworks site had levels of 458,000 μg kg^{-1} dry wt, and mussels at the beach directly below contained 3060 to 6450 μg kg^{-1} wet weight PAHs.

B. SOXHLET EXTRACTION

US EPA Method 3540 is for the Soxhlet extraction of nonvolatile and semivolatile organic pollutants from environmental solids and solid wastes.[141] Solid samples are mixed with anhydrous sodium sulfate and placed in an extraction thimble inside a Soxhlet extraction apparatus. The Soxhlet apparatus allows the solvent to be distilled and recycled through the system thereby introducing fresh solvent with each cycle. Concentration and solvent exchange are usually required before analysis. Surrogate spiking standards are added to indicate recoveries, however, significant losses of more volatile compounds is common. The extraction can take up to 24 h and use up to a liter of organic solvent for each sample. Soxhlet extraction is the gold standard by which all other new extraction techniques are compared. It is a technique that has been in use for over a century[142] and is very efficient at extracting nonpolar PAHs from environmental solids. It is exhaustive but nonselective and produces samples requiring extensive cleanup. Heating required to recirculate and distill the extraction solvent can cause thermal decomposition of analytes, evaporative loss of more volatile analytes, and artifact formation. Samples, however, do not require filtration prior to analysis.

Incineration ashes are used as fill material for roads and in construction.[143] Municipal solid waste, mixed biofuel, and heating plant ashes were analyzed by Sohxlet extraction in toluene, deactivated silica gel cleanup, and GC-MS to show ΣPAH$_{16}$ levels of 140 to 77,000 μg kg^{-1}. The highest levels measured were for ashes from biofuels incineration with NAP and PHN as the predominant PAHs. Since volatile PAHs are lost in the Soxhlet sample preparation process, actual levels of NAP and PHN may have been underestimated. B[a]P ranged from 1 to 1327 μg kg^{-1}. Bottom ashes contained more of the less volatile PAHs as expected. Results for the mixed biofuel ash were in excess of the Swedish EPA soil limits for less sensitive land use of 7 μg kg^{-1} of carcinogens and 40 μg kg^{-1} for noncarcinogenic PAHs.

New developments in Soxhlet extraction have been applied to the analysis of PAHs in soils and sediments.[144] Focused Microwave-Assisted Soxhlet Extraction (FMASE) uses focused microwave irradiation to achieve agitation in the sample. The application of microwaves drastically reduces extraction time (~1 h) and solvent use. Recoveries of PAHs were reported to be 11 to 101% in the extraction step and no filtering was necessary prior to analysis.

A recent report on Soxhlet extraction of PAHs from environmental matrices examined the choice of solvent.[145] Like most nonpolar environmental contaminants, PAHs show time-dependent partitioning into organic matter and micropores, also referred to as "aging." Extracting PAHs from native matrices increases in difficulty with time thus rendering preparation of solid sample the most disconcerting aspect of the analysis. A considerable amount of time, energy, and materials are consumed in the process. The most common extraction solvents used for PAHs are acetone mixed with HEX, cyclohexane, or DCM. Ethyl acetate is more polar than these mixtures and is less toxic. A certified reference soil (CRM 524) was leached with ethyl acetate in an automated Soxhlet extractor without the addition of drying agent, sodium sulfate, and grinding before extraction. Also, silica or alumina column cleanup was eliminated along with the concentration and solvent exchange step necessary when using HPLC-FLD analysis. Recoveries were comparable to those obtained with acetone:HEX (1:1, v/v) in the standard process that included drying, cleanup, and solvent exchange. For each method, the automatic extractor was programmed to begin with the sample immersed in the solvent for 60 min followed by 90 min of normal Soxhlet extraction. A method similar to Soxhlet, fluidized-bed extraction, has been introduced.[146] A two-chamber extractor is separated by a filter. The sample is placed in the upper chamber and the solvent in the lower chamber. The solvent is heated and as it evaporates, the vapor passes up through the sample and condenses on a cooling bar above the sample. As the solvent condenses, it drips on to the sample and collects there. After most of the solvent has evaporated, the heater is turned off and a cooler is turned on causing a pressure drop that pulls the condensed solvent back down to the bottom chamber. This represents one cycle. Recoveries are similar to Soxhlet extraction with soil of the more volatile PAHs but only 60 ml of solvent is used for the extraction in one hour or seven cycles.

C. Ultrasonic Extraction (USE)

Ultrasound-assisted extraction is a solid–liquid leaching technique done in the batch mode. Solvent is added to samples before they are placed in a sonication bath. The procedure is usually repeated two or three times for quantitative extraction. Ultrasonication produces very localized, extremely high effective temperatures that increase extraction efficiencies of solvents, yet it is inexpensive and fast. Combined extracts are centrifuged or filtered, cleaned up, and concentrated by SPE or LLE prior to analysis. US EPA Method 3550B describes a general ultrasonic extraction procedure, though not specific for PAHs.[147] An ultrasonic extraction method was optimized for the EPA_{16} in soils. The method was applied to the analysis of soil from an industrial site in the U.K.[148] Extraction was with acetone for 30 min in a sonication bath. The extract was concentrated by RP-SFE and eluted with acetone:THF (1:1). Fourteen of the EPA_{16} were detected at levels of 0.7 $\mu g\ g^{-1}$ for FLU to 9.7 $\mu g\ g^{-1}$ for NAP. Recoveries were better than when compared to Soxhlet extraction for 8 h.

Forest soils have been shown to have higher PAH concentrations than agricultural soils because leaves and needles concentrate PAHs from the atmosphere.[149] Three soils were analyzed from forests near an aluminum plant in Slovakia using ultrasonic extraction for 1 h at 65°C in a methanolic potassium hydroxide solution followed by LLE with HEX. Samples were analyzed by GC–MS. Coal tar has been used as binder in anodes in aluminum plants thus PAHs are released when the anodes are burned during electrolysis. ΣPAH_{20} ranged from 22,303 $\mu g\ kg^{-1}$ to 192,299 $\mu g\ kg^{-1}$ in soils associated with beech tree roots suggesting that the trees were bioconcentrating PAHs from the atmosphere. Bulk soil ΣPAH_{20} levels ranged from 151 to 93,785 $\mu g\ kg^{-1}$. The organic layer of soils had higher PAH concentrations.

The longer PAHs have to sorb to soils, the harder they are to extract. Potential underestimation of PAH in soils using EPA Method 3550 is possible since no solvent or extraction time is specified.[150] USE was compared to Soxhlet extraction for 8 h of a clay soil with long-time contamination from a tar production facility. DCM:acetone (1:1) was determined to be the best solvent mixture for the extraction of PAHs. Analysis by US EPA Method 8100 (GC-FID) for the EPA_{16} after a 48 h Soxhlet extraction showed mostly four- to six-ring PAHs with the predominant one being B[a]P and a ΣPAH_{16} of 2073 mg kg^{-1}. NAP, ACN, and ACY were not detected in Soxhlet extracted samples. Ultrasonication extraction for 8 h gave ΣPAH_{16} of 2179 mg kg^{-1}. After 4 h of extraction, ultrasonication showed greater recoveries of the more volatile two- and three-ring PAHs that were lost by volatilization during longer extraction times.

NPAHs are emitted by diesel and internal combustion engines, and their presence indicates traffic as a source.[151] In a variation of ultrasonic extraction, a closed, stainless-steel extraction cell contained soil samples while solvent was circulated through the cell as sonication was applied with a titanium probe in a water bath. The extract was then flushed out of the extraction cell into a collection vial external to the water bath. The extract was concentrated by rotary evaporation and the solvent exchanged to ACN before GC–MS–MS analysis. Extraction efficiency was compared to Soxhlet extraction by EPA Method 3540. Results obtained by both methods were similar, however the ultrasonic extraction time was 10 min as compared to Soxhlet extraction, which was 24 h. The amount of solvent used was also ten times less. Levels of NPAHs found in soil were 1-nitronaphthalene, 3-nitrobiphenyl, 2-nitrofluorene, 3-nitrofluoranthene, and 1-nitropyrene at 9, 9, 90, 200, and 5 ng g^{-1}, respectively.

Sediment samples from England and Wales were ultrasonication extracted for 15 min in acetone with anhydrous sodium sulfate.[152] Sediments were frozen when collected and thawed just before extraction. Anthracene-d10 was added as internal standard (IS) and samples were extracted three times. The combined extract was filtered and analyzed by HPLC-FLD with programmable excitation and emission wavelengths. Fifteen EPA$_{16}$ PAHs plus benzo[e]pyrene were measured with maximum concentrations of 611 μg kg^{-1} for D[ah]A to 22,100 μg kg^{-1} for PYR in one sample. Elevated levels of PHN, PYR, B[a]A, and CHR are characteristic of fossil fuel sources. Sediment samples collected from the Shetland and the Orkney Islands in the U.K. were subjected to USE in DCM:methanol and dried over sodium sulfate, then solvent-exchanged into isooctane for GC–MS analysis.[153] Deuterated surrogate standards were added before extraction. GC–MS using SIM was used for separation and detection of PAHs, and LODs ranged from 0.1 ng g^{-1} to 0.3 ng g^{-1}. ΣPAH for parent and branched compounds ranged from not detected to 22,169 ng g^{-1}. Higher concentrations were found near a pier, fish farms, a boat repair yard, and entrances to harbors. Lower concentrations were found in sites with sandy sediments as opposed to sediments higher in organic matter. Orkney sediments showed a higher proportion of lighter alkylated-PAH suggesting input of petrogenic sources. PAH concentrations were classified as low, medium, or high, corresponding to total concentrations of <40, 150 to 500, and >750 ng g^{-1} based on dry weight.

Freeze-dried sewage sludges were extracted by USE (20 ml × 3) in DCM:methanol (2:1) for 20 min.[154] Centrifugation separated the extractant from solids and the combined extract was evaporated to dryness, redissolved in HEX:DCM (19:1), and cleaned up on aluminum oxide/sodium sulfate. The cleanup column was washed with HEX:DCM (19:1) to remove hydrophobic impurities and then eluted with HEX:DCM (1:2). After rotary evaporation and N$_2$ blowdown, the solvent was switched to isooctane for GC-MS analysis. Naphthalene-d8, anthracene-d10, benzo[a]anthracene-d10, and benzo[ghi]perylene-d12 were surrogate standards. Recoveries varied between 60 and 98%. Pyrene-d10 and perylene-d12 served as internal standards. LODs were in the low μg kg^{-1}. The more volatile PAHs were lost in the N$_2$ blowdown step. For six sludges from wastewater treatment plants, ΣPAH$_{16}$ ranged from 100 to 5500 μg kg^{-1}. The EU cutoff limit for 11 PAHs in sewage sludges that are applied to agricultural lands is 6 mg kg^{-1} so these sludges were below this limit.[155] It appeared that PAH levels were higher in sludges treated by aerobic digestion. Extracts of sewage sludges using this same extraction method were subjected to the ToxAlert® 100 bioassay.[156] Total ΣPAH$_{16}$ ranged from 1019 to 5520 μg kg^{-1} for sludges from two wastewater treatment plants. PAH concentrations also correlated with toxicity. FLA/PYR ratios ranged from 0.23 to 0.90 and PHN/ANC ratios ranged from 3.5 to 8.7, which is characteristic for domestic wastewater sources. The EPA$_{16}$ were measured in sewage sludges from fifteen sewage treatment plants in southeastern Poland in the summer.[157] Samples were dried and milled prior to ultrasonic extraction twice with 40 ml DCM. Extracted samples were centrifuged, evaporated to dryness, and then redissolved in (ACN:H$_2$O) (1:1. v/v). Sample preconcentration and cleanup was done using RP-SPE. NAP was not detected in any samples. ΣPAH$_{16}$ ranged from 2,040 to 36,034 μg kg^{-1} with a mean of 11,613 μg kg^{-1}. Higher concentrations were found in sludges from plants that treated industrial and petroleum refining wastewaters. Most of the PAHs detected contained three and four rings with the five- and six-ring PAHs making up about 10 to 20% of the total. ACE, FLA, and B[b]F were the predominant PAHs detected using HPLC-UV. The EU maximum concentrations of FLA, B[a]P, and B[b]F are 5, 2.5, and 2 mg kg^{-1}, respectively. Standards in Austria state that the total FLA, B[b]F, B[k]F, B[ghi]P, and IND cannot exceed 9.6 mg kg^{-1}. These levels were exceeded for several of the sludges tested especially for those with industrial inputs. PHN/ANC ratios > 10 indicate that the PAHs could be of petrogenic origin while PHN/ANC ratios < 10 indicate that the PAHs could by of pyrogenic origin. FLA/PYR ratio values > 1 indicate that PAHs originated from pyrogenic processes, mainly coal burning. Ratios for these sewage sludges indicated pyrogenic origins except for the highest concentrated sample with substantial industrial input.

D. MICELLE-MEDIATED EXTRACTION

PAHs were extracted from certified reference soil (CRM 524), sediment (CRM 535), and sewage sludge (CRM 088) using sodium dodecane sulfate (SDOS) an anionic surfactant.[158] Samples required a prewashing step with hydrochloric acid to remove alkaline and alkaline earth metals, which precipitate anionic surfactants. SDOS was added above the CMC at 2% and the samples were acidified and stirred at 60°C for 1 h. Heating was necessary to increase recoveries of the higher molecular weight compounds. Centrifugation produced three distinct phases: solid, aqueous, and a gel-like surfactant-rich phase that could be physically removed from the samples after cooling to 0°C. The gel was dissolved in 2 ml ACN, warmed to room temperature and analyzed by RP-HPLC-DAD-FLD directly. No further cleanup or concentration steps were required. Cloud point extraction with nonionic surfactant and microwave heating was also investigated but recoveries were slightly lower than for traditional Soxhlet extraction for a spiked sediment sample.[159]

E. CURIE POINT PYROLYSIS

Direct thermal desorption of PAHs onto a GC column was recently reported in a method that uses Curie point pyrolysis.[160] The solid sample is placed in a ferromagnetic coil (pyrofoil) inside a quartz tube connected to the GC column inlet. The pyrofoil is inductively heated by radiofrequency to the Curie point temperature where it loses its magnetic properties. Extreme heating occurs in 0.2 sec, and analytes are vaporized and swept onto the column. Very small (30 mg) samples of CRM 104 were desorbed at 590°C for 10 sec. The method required no solvent, sample concentration or cleanup. A few difficulties were experienced. The column lifetime was shortened by other coextracted compounds and B[k]F was not resolved from B[b]F in the chromatogram. Since the sample is so small, careful attention must be paid to homogenization of the original sample to ensure it is representative of the bulk material. Total recovery of ΣPAH_{16} was 1058 mg kg^{-1} dry weight as compared to results obtained by Soxhlet extraction, ΣPAH_{16} of 867 mg kg^{-1} dry weight. Values were considerably higher for the more volatile two- and three-ring PAHs. Saponification gave more comparable results, ΣPAH_{16} of 1026 mg kg^{-1} dry weight but extensive sample cleanup and long extraction times were required. Higher molecular weight PAHs were not quantifiable at <10 mg kg^{-1} but considerable possibilities exist for improvement.

F. MICROWAVE-ASSISTED EXTRACTION (MAE)

Microwaves are used to generate rapid, localized heating inside the sample matrix. US EPA Method 3546 outlines a procedure for microwave extraction of nonpolar or slightly polar organic compounds from soils, clays, sediments, sludges, and solid wastes.[161] Samples are air-dried and ground or mixed with a drying agent then extracted by microwave heating in acetone:HEX (1:1, v/v) in a sealed vessel for 10 to 20 min. When cooled, samples are filtered and cleanup is performed. Considerable pressure (50 to 150 psi) can build up inside the extraction vessel so extreme caution must be used when handling hot vessels. Seals can also leak causing analyte loss. Results for the MAE of certified reference materials showed recoveries of 75% for 17 PAHs when extracted at 115°C for 10 min. Using commercially available equipment, extractions can be performed simultaneously in several cells.

MAE has been applied to the extraction of PAHs from diesel particulate matter (SRM 1650).[162] The high carbon content of diesel particulate matter results in extremely tight binding of PAH and very low extraction recoveries. Results comparable to certified values were obtained when DCM was used as the solvent at 400 W and triplicate 20 min extractions.

Focused microwave-assisted extraction (FMAE) is an emerging technique that uses open extraction vessels thereby reducing the threat of explosion and leakage.[163] Other advantages include the possibility of extracting much larger samples and greater homogeneity of the electromagnetic field. The device consists of an extraction vessel with a condenser at the top, which is open to the atmosphere. The vessel is submerged in a bath that is subjected to microwave energy. The system operates at atmospheric pressure and at the boiling temperature of the solvent. A method was optimized for PAH extraction and includes 30% water in the sample to take advantage of the high sorption of water of microwaves. Water also helps micropores to swell and increases mass transfer. MAE for 10 min at 30 W of power in DCM achieved recoveries of 85 to 100%, as compared to Soxhlet extraction, of certified reference marine sediment (CRM 1941a) for thirteen PAHs analyzed by GC–MS.[164] Sufficient water in the sample can increase recoveries for the higher molecular weight PAHs.[165]

FMAE was compared with pressurized-liquid, Soxhlet, and ultrasonic extraction for the analysis of sewage sludges.[166] The optimum conditions for FMAE were determined to be 30 W, for 10 min in 30 ml of acetone:HEX (1:1, v/v). Recoveries for certified reference marine sediment, CRM 1941a, were 75, 61, and 56% for FLA, B[b]F, and B[a]P, respectively. Activated copper bars were added to samples to remove sulfur. Mean recovery by FMAE was 70% compared to traditional Soxhlet extraction. FMAE recoveries exceeded those for ultrasonic extraction but better results were obtained by pressurized-liquid extraction (PLE) with the best results obtained by Soxhlet extraction. All extracts required column cleanup, concentration, and solvent exchange for RP-HPLC-FLD analysis.

G. PRESSURIZED-LIQUID EXTRACTION (PLE)

SFE and PLE are continuous extraction techniques that have enhanced mass transfer rates due to an increase of the concentration gradient between the phases, whereas Soxhlet is a batch technique, though it can be considered a continuous technique because freshly distilled solvent is contacted with the solid matrix with each cycle.

PLE has obtained a plethora of names and acronyms including pressurized solvent extraction (PSE), pressurized fluid extraction (PFE), enhanced solvent extraction (ESE), and accelerated solvent extraction (ASE), a brand name introduced by Dionex® although there are several suppliers of extraction systems. This technique requires that the sample and solvent be placed in a high-pressure extraction cell, usually of stainless steel, that is heated to above the boiling point of the solvent for a specified time period. Pressure is applied to the sample or occurs as a result of heating the solvent. The exact pressure is not critical to the procedure as long as there is enough pressure to maintain the solvent in the liquid state except for wet matrices where increased pressure may help to force the solvent deeper into matrix pores. At the usual flow rate of 0.1 to 2 ml min^{-1}, the flow rate in the dynamic mode does not affect recoveries significantly as long as an adequate amount of solvent is used to flush the cell. The solvent is then removed from the cell and cleaned up for analysis. PLE can be performed in the static or dynamic mode or a combination of both. Solvent use can be reduced by up to a factor of ten over Soxhlet extraction with recoveries that are comparable with greatly reduced extraction times, 20 min as opposed to 24 h.[167] At the higher temperatures, the diffusivity of the solvent and solubility of the analytes increase, thereby increasing mass transfer. For PAHs, which are thermally stable and very tightly held by the organic matter in soils and sediments, enhanced recoveries are seen using most PLE techniques. The extraction solvent and the temperature of extraction are the two most important variables to consider in PLE. Temperatures of 250 to 300°C are required for extraction of PAHs from soils and sediments. Also, samples must be homogeneous and carefully packed in the extraction cell to ensure reproducible results.

EPA Method 3545[168] describes the application of PLE to the analysis of environmental solids. The recommended extraction temperature of 100°C is inadequate for the extraction of PAHs. A number of solvents and solvent combinations have been shown to give comparable recoveries for PAHs extracted from contaminated soils and sediments.[169] Acetone, DCM/acetone, DCM, methanol, ACN, and acetone/HEX gave similar recoveries. However, HEX used alone showed poorer recoveries. PAH-contaminated soil from a former gasworks site in Stockholm, Sweden and a certified reference soil, CRM 103-100 (US EPA, RTC, Laramie, Wyoming, U.S.A.) were extracted with various solvents beginning with a dynamic extraction followed by one to three static extraction cycles, and then a rinse of the solid with fresh solvent.[170] The combined extract was fractionated on deactivated silica gel with the first 10 ml of HEX eluent discarded to remove aliphatic hydrocarbons. Smaller samples gave better recoveries and the highest recoveries were obtained using toluene:HEX as the solvent. Low molecular weight PAHs were extracted more efficiently at higher temperatures. The IS added to the top of the extraction cell caused excessive variability in recovery values since it had to pass through the whole sample while the analytes were evenly distributed throughout the sample. Therefore, the IS should be thoroughly mixed with the sample prior to extraction. Adequate solvent must also be used in at least two static cycles and a final rinse of 100% of the extraction cell volume. Larger IS molecules would be retained by the matrix to a greater extent leading to an overestimation of native PAH concentrations. Acetone/HEX makes a good choice because it is not chlorinated and in more amenable to cleanup and analysis procedures.

PLE was applied for the extraction of PAHs from contaminated sediments in the Mankyung River in southeastern Korea.[171] Samples were frozen for storage, then thawed, air-dried, and ground prior to extraction with DCM. The extraction and cleanup were combined into one step by including acid-activated copper, silica gel, and sodium sulfate at the bottom of the cell through which the extracted sediment passed before elution. Mean recovery was 89% compared to standard PLE which was $>90\%$. The addition of tetrasodium ethylenediamine tetraacetic acid (EDTA) as a matrix modifier increased average recoveries by about 4%. PLE was used to measure PAHs in soils from Bayreuth, Germany and the surrounding area.[172] Acetone/HEX (1:2, v/v) at 120°C with two static extraction cycles and a final rinse of 60% of the cell volume was used to extract samples. Combined extracts were cleaned up on silica gel and alumina, and analyzed by GC–MS. The ΣPAH_{20} ranged from 0.2 to 186 mg kg^{-1} and were highest at a roadside grassland, a former landfill, and a former gasworks site. Field-fresh, freeze-dried, and air-dried forest soil samples near a chemical plant site in eastern Slovakia were extracted to determine the effects of sample preparation on PLE efficiencies.[173] PAH concentrations were consistently lower for air- and freeze-dried samples especially for NAP and the more volatile PAHs. Drying also reduced the extractability for all PAHs, although drying increases reproducibility and homogeneity of the samples. Results obtained for field-fresh samples gave ΣPAH_{21} concentrations that ranged from 53 to 6870 μg kg^{-1} with benzofluoranthenes being the predominant compounds. PAH levels also decreased with distance from the factory implicating it as a source. In addition, PAH concentrations were also higher in the organic soil horizons indicating the strong association of PAHs with organic matter in soils and sediments. There was a positive correlation between the water content in the samples and PAH recovery. Mean ΣPAH_{20} for temperate urban soils from Germany and tropical soils from Brazil were 23,000 μg kg^{-1} and 155 μg kg^{-1}, respectively, for soils extracted by PLE.[174] For the temperate soils, PHN and the four- to five-ring PAHs were predominant indicating pyrolytic sources while in the tropical soils, the most abundant PAH was NAP followed by PHN and perylene. Other PAHs were found at extremely low levels when detected.

When water is used as the solvent, PLE is referred to as superheated water extraction, subcritical water extraction (SWE), or pressurized (hot) water extraction (PWE). Hot water is very effective as an extraction solvent for PAHs from soil and sediment.[175] Superheated water is water above the boiling point but below the supercritical point, and under sufficient pressure to maintain the liquid state and prevent boiling. The polarity and viscosity of water decrease with increasing temperature so that its solvent properties resemble those of room temperature methanol, ACN, or even DCM, and it can readily solubilize PAHs. Comparable recoveries to Soxhlet extraction can be obtained very rapidly and with minimal organic solvent use. With SWE at 250°C, PAHs were extracted from soil and urban air particulates with recoveries >97% in 15 min with greater selectivity since *n*-alkanes were not quantitatively extracted concurrently.[176] Degassing the water with nitrogen to remove oxygen can minimize oxidation of analytes.

One disadvantage of using superheated water for extraction is that as the extract solution cools, the extracted PAHs can precipitate or resorb onto the matrix or internal surfaces of extraction equipment.[177] As a result, PAHs must be removed from the water in the cooling step or a small amount of organic solvent must be added to the extract to insure that PAHs remain solubilized. Soil from a former manufactured gas plant (MGP) was static-extracted with water at 250°C and 6 cm × 2 cm pieces of styrene divinylbenzene SPE disc were added to each sample before extraction to provide a matrix for selective sorption of the PAHs upon cooling.[178] After extraction, the SPE material was removed and eluted with acetone/DCM. With internal standards, recoveries of >95% were achieved for urban dust (1649a) and sediment (SRM 1944) reference materials if the extraction cells were agitated during extraction. This procedure did not work well for diesel exhaust particulate matter since the soot was resistant to wetting. A similar method used C18 resin mixed in with the sample before extraction.[177] Following static extraction at 150°C for 20 min, municipal waste compost contained levels of the sum of six PAHs from 7 to 13 μg g^{-1} with the highest value from compost containing sewage sludge.

SWE was compared to extraction with steam and thermal desorption with nitrogen for PAHs on EC-1 sediment.[179] Extracted PAHs were trapped on a Tenax TA solid trap and eluted with 10% ethyl acetate in *n*-heptane. At temperatures >100°C, PEEK and PTFE seals can leak, so care must be taken to use metal seals such as copper. All three methods were comparable at 300°C with steam extraction giving slightly higher recoveries. SWE was coupled with SPME to analyze railroad bed soils in a method that used no organic solvents.[180] Deuterated internal standards were used to compensate for resorption of PAHs to the matrix upon cooling. Recoveries of 60 to 140% were achieved with a static 60 min extraction at 250°C. Recoveries of NAP and ACY were higher than those obtained by Soxhlet extraction as SWE greatly reduced analyte loss through volatilization. Also, alkanes present in urban dust, SRM 1649, were not coextracted. It was reported that anthracene-d$_{10}$ degraded to produce anthraquinone-d$_8$ while nondeuterated anthracene nor any of the other PAHs degraded in the procedure. PAHs were extracted from petroleum waste sludge, crude oil lake bottom dewatered sludge, spent catalyst, and soil from a wood treatment facility (USEPA Lot AQ103) in a dynamic SWE procedure where degassed water was pumped into an extraction cell placed in an oven.[181] A preheating coil, also inside the oven, brought the water to extraction temperature before contacting the sample and a cooling loop outside the oven, followed by a needle valve restrictor allowed samples to be collected in the liquid state. PAHs were also selectively and quantitatively extracted leaving interfering alkanes behind in the matrix at temperatures up to 300°C, while greater temperatures allowed the alkanes to be extracted separately. Contaminated soil required only a 30-min extraction time while spent catalyst required a 60-min extraction time due to tight binding and larger particle size.

Solid-phase trapping of PAHs from subcritical water extracts is effective using a cold silica-bonded ODS sorbent trap allowing for online extraction followed by HPLC analysis.[182] A flow rate of 0.6 ml min^{-1} at temperatures up to 250°C allowed for quantitative extraction in 15 ml of water. By valve switching, PAHs were desorbed from the solid trap with a small amount of mobile phase and swept into the analytical column. SWE coupled with LC–GC was reported to give limits of quantitation below 0.01 μg g^{-1} for sediments.[183] No sample pretreatment was done including no drying. Very small samples, 10 mg, were extracted by SWE at 300°C for 30 min and the effluent was directed onto a Tenax-TA solid trap. The trap was dried with nitrogen, washed with pentane to remove alkanes, and PAHs were eluted with pentane/ethyl acetate into a GC using partially concurrent solvent evaporation (PCSE). The average recovery for fifteen PAHs was 103%. When compared to SFE or Soxhlet, results for the more volatile NAP were much higher.

Increasing the temperature of water extractions also increases the amount of matrix materials that are coextracted, especially for samples high in organic matter. Coextractants can reduce trapping efficiency and reproducibility on solid-phase traps. Using microporous membrane liquid–liquid extraction (MMLLE) in place of a solid trap can more selectively trap PAHs from hot water extracts and minimize or eliminate sample cleanup.[184] The extraction solvent, cyclohexane, is immobilized in the pores of a polypropylene membrane where liquid–liquid mass transfer occurs. Limits of quantitation of about 1 $\mu g \, g^{-1}$ for very small samples (5 to 10 mg) with an average recovery of 41% were demonstrated. Higher results than those obtained using the Soxhlet method for more volatile PAHs were reported also for a 30-min extraction at 1.0 ml min^{-1} and 300°C. The total analysis time was 80-min.

When extracting with superheated water, soil samples do not require predrying.[185] Water is the ideal solvent when performing immunoassay detection because it does not denature the reagents. However, a small amount of methanol can be added to prevent precipitation of the analytes on to container walls.

Water modified with sodium dodecyl sulfate (SDS), an anionic surfactant, can increase extraction of less polar analytes at lower temperatures.[186] Spiked 0.2-g soil samples and certified reference industrial soil (CRM 524) were extracted at 150°C for 15 min static and 10 min dynamic at 3 ml min^{-1}. The 50-ml extract was concentrated on a silical gel column to retain the SDS, then PAHs were eluted with 5 ml HEX and solvent switched to ACN before HPLC analysis. Limits of quantitation were low, 0.026 to 0.094 $\mu g \, ml^{-1}$, for higher molecular weight PAHs. The SDS-modified water allowed for shorter extraction times at lower temperatures and reduced clogging of tubing that can occur upon cooling, especially with samples high in organic matter. The completeness of extraction in the static mode depends on the partition equilibrium of the analytes between the matrix and the aqueous extractant.[187] Using $2.5 \times 10.2 \, M$ SDS in water as the extractant, static, dynamic, and static-dynamic modes of extraction were compared for PAHs in soils, sediments, and fish samples. Dynamic extraction has the advantage of a continuous introduction of fresh solvent into the sample, however, samples are diluted to a greater degree. A short static extraction (5 min) followed by a short dynamic extraction (15 min) gave quantitative extraction in the shortest amount of time with the least dilution of the analytes.

H. Supercritical Fluid Extraction (SFE)

SFE usually refers to the extraction of solid samples with CO_2 above its critical temperature and pressure. SFE greatly reduces extraction time, is inexpensive and nontoxic, and its properties can be tuned by changing temperature and pressure or adding modifiers. USEPA Method 3561 provides a method for the extraction of PAHs from soils, sediments, fly ash, solid-phase media, and other solids.[188] Two- and three-ring PAHs can be extracted in pure CO_2 while the higher molecular mass PAHs require the addition of modifiers. Supercritical CO_2 is similar in solvent properties to HEX and can be modified by the addition of water or organic solvents.

Several types of sewage sludge have been analyzed using supercritical fluid extraction with or without modifiers added to the CO_2.[189] Sludges from more industrial sources contained higher levels of PAHs than those containing primarily domestic waste sludge. The average and highest concentrations were determined to be 6.9 and 22 mg kg^{-1} dry weight, respectively, for the EPA$_{16}$. Copper powder was added to remove interfering sulfur. A short cleanup step on silica and alumina was required when water, methanol, and DCM were added as modifiers. Toluene was used to elute the extracted PAHs from an ODS solid trap. High levels of lipids and variability of sewage sludges causes their analysis to be more challenging.[190] NPAHs and alkyl-PAHs were also extracted. When compared, Soxhlet extraction and SFE of sewage sludges gave similar recoveries. However, Soxhlet extracts had residual solids after solvent evaporation that were difficult to redissolve following nitrogen blowdown. For a native sludge sample, a level of 9.5 mg kg^{-1} was reported for the sum of nine less volatile PAHs when analyzed by HPLC-FLD-DAD.

SFE was compared with Soxhlet, PLE, and SWE for the extraction of PAHs from MGP soil.[191] The extracts obtained using organic solvents were dark colored and had more artifact peaks. Organic solvents and water all extracted soil organic matter. The water extract was much lighter in color and did not contain alkanes. SFE was selective for PAHs with only 8% of the bulk organic matter coextracted yet alkanes were coextracted. So, for PAHs from solid samples, the choice of extraction method will be influenced by the type of matrix and the presence of organic matter and nonpolar co-contaminants. The addition of sodium sulfate as a drying agent did not affect the recoveries by Soxhlet or PLE. Adding organic modifiers will increase recoveries of the higher molecular weight PAHs but will also increase the extraction of matrix components. SWE gives higher yields of the lower molecular weight PAHs, greater selectivity for PAHs, and also removes nitrogen-containing compounds. Samples extracted with pure CO_2 required no sample cleanup. Extraction fractionation can be achieved with SFE by extracting alkanes first at milder conditions with two- and three-ring PAHs, then the four- to six-ring PAHs at higher temperatures and pressures with modifier addition. In addition, SFE of PAHs with pure CO_2 at mild conditions correlated well with the bioavailable fraction of PAHs in soil, based on toxicity to earthworms. Removal of SFE-available molecules eliminated the toxicity to worms.[192] Toxicity was dependent on type of PAH and availability, not on total PAH concentration. Mutagenic five- and six-ring PAHs did not extract under mild SFE conditions, and were also not bioavailable. Mild SFE for 120 min agreed with 120 days of water desorption on the amount of PAH that was bioavailable.

Successful SFE with CO_2 depends on the particle size of the sample, the smaller the better, and on the amount of water in the sample, less is better. Water can coat the surface of particles and hinder CO_2 penetration, therefore samples are usually dried or mixed with a drying agent prior to extraction.[193] Extraction recoveries are matrix dependent so no one method will work best for all matrices. Optimum conditions for SFE of PAHs were reported to be 15 min at 45 MPa and 95°C using CO_2 modified with methanol/DCM 5:1 as recoveries of 70 to 90% for high molecular weight PAHs were achieved. Larger PAHs have an expanded π-electron system and more interaction with surfaces making them more difficult to extract with time. PAHs also migrate into soil micropores from which they must diffuse to be extracted. Comparison of SFE with CO_2 modified with methanol gave as good or better results as sonication or Soxhlet, only faster.[194,195] Decreasing particle size also increased extraction efficiency by SFE. Trapping of analytes is challenging in SFE since they must be collected from the CO_2 as it is converted to the gaseous phase. Liquid trapping is most efficient, especially for more volatile PAHs. Modifiers can also affect solid-phase trapping efficiencies by coating the solid phase and reducing transfer efficiency from the gaseous to the solid phase. NPAH and OPAH can be extracted from urban aerosols with toluene-modified CO_2 in the second step of a sequential extraction beginning with a mild extraction with pure CO_2 to remove alkanes and volatile PAHs.[65] No sample cleanup was needed. A solid C18 trap was used to collect the analytes after a 5 min static, 30 min dynamic extraction at 1 ml min^{-1} flow rate in CO_2 modified with 5% toluene. The glass fiber filters containing the samples were placed directly inside the extraction cell.

Comparison of Soxhlet, ultrasonic, SFE, and PLE for high and low PAH contaminated soils showed that the best results were achieved using PLE with acetone–toluene 1:1 but PLE or ultrasonic extraction with just acetone was adequate when performing subsequent HPLC analysis.[196] Another comparisons of sample preparation of sewage sludge showed no real differences in the recoveries of PAHs by Soxhlet, SFE, USE, PLE, or MAE when each method was optimized.[197]

I. SOLID-PHASE MICROEXTRACTION

For analysis of environmental samples, common goals include decreasing extraction and analysis times, reducing or eliminating organic solvent use, and developing field analysis and screening techniques. Headspace solid-phase microextraction (HS-SPME) can measure concentrations of more volatile contaminants with minimal or no sample pretreatment. Also, more freely available contaminants will be sampled giving information on bioavailability. HS-SPME has been used to sample PAHs from soil at an old gasworks site.[198] PDMS-coated fibers were used to sample for 5 min and analysis had to be within two days. Calibration was done internally by standard addition and externally by spiking sand. Results comparable to those obtained by Soxhlet extraction were obtained for two- and three-ring PAHs. The method was only valuable as a screening tool for PAHs with four or more rings due to their lower vapor pressures. Certified soil samples (CRM-104) were analyzed by direct immersion or headspace SPME following equilibration with water using PDMS and PA fibers.[199] Only FLA and PYR were at high enough concentrations in the headspace to be extracted at 25°C. A temperature of 80°C was necessary to volatilize enough of the five-ring PAHs for successful quantitation. Surfactant addition further enhanced the sensitivity of the method.

V. SAMPLE CLEANUP METHODS

The selectivity of the extraction procedure and analytical technique will determine the sample cleanup needs. As always, the number of additional steps in an analysis must be minimized because concentration and solvent exchange steps can increase the risk for contamination and analyte loss. Particulate filters and waste samples often contain aliphatic hydrocarbons and monoaromatic compounds, while soil and sediment extracts may contain elemental sulfur, lipids, proteins, dissolved soil organic matter, and other high molecular mass substances. Because nonpolar to slightly polar organic solvents are generally chosen for extraction of PACs, these interfering compounds in the matrix may coextract making sample cleanup a necessity.

Silica gel (SiO_2) is the preferred sorbent for the column separation of PAHs from coextracted matrix interferences[52,60,67,68,84,85,200] and separates analytes based on differences in chemical polarity. Silica gel is activated by heating to 150°C to remove sorbed water and can then be partially deactivated by adding up to 10% (w/w) water. Sodium sulfate, a drying agent, is added to the top of the column before the addition of the concentrated extract. PAHs are often eluted with DCM/pentane (2:3) (v/v). DCM alone will coelute PAHs and alkanes from silica gel, while cyclohexane allows alkanes, NPAHs, and OPAHs to be fractionated.[201] Alkanes, branched alkanes, cycloalkanes, monoaromatics, naphthalenes, and three-ring PAHs were fractionated in PLE extracts of petroleum-contaminated sediments on silica gel followed by a silver ion-impregnated silica gel column.[202] The first fraction eluted from the silica gel column in HEX was further fractionated on a silver ion impregnated column into the alkanes (HEX eluate) and the monoaromatics (DCM eluate). The second fraction from the silica gel column, eluted with 1:1 HEX:DCM, was fractionated on another silver ion-silica gel column into two-ring aromatics (9:1 HEX:DCM eluate) and three-ring aromatics (DCM eluate).

Alumina, Al_2O_3, can also be an effective solid-phase for column cleanup.[156,203] EPA Method 3611B, Alumina Column Cleanup and Separation of Petroleum Wastes[204] describes a method for separation of neutral PAHs on a column of alumina covered with a layer of sodium sulfate, a drying agent. Elution of the column with HEX will elute base/neutral aliphatics first followed by DCM to elute base/neutral aromatics. Silica gel and alumina may be used together when necessary.[78,172] Soxhlet extracts of creosote-contaminated soil were fractionated using a strongly basic anion-exchange resin with 10% deactivated silica get to separate PAHs, SPAHs, and OPAHs with HEX:DCM.[205] Neutral and basic NPAHs and neutral metabolites were eluted with DCM:methanol, and acidic metabolites were eluted with HCl in methanol. Neutral and basic NPAHs and neutral metabolites were further fractionated on a strongly acidic cation-exchange resin where the neutral analytes were not retained but the basic NPAHs were eluted with ammonia in methanol.

SPE on ODS[69,152] or Tenax®[183] can be used for sample cleanup in online extraction-analysis systems. Cleanup of extracts in polar solvents is generally performed using C_{18},[177,186,187]

Tenax®,[179,183,206] or styrene divinylbenzene.[178] The elution solvent is then chosen to be compatible with the analytical technique.

For the removal of lipids, proteins, dissolved soil organic matter, and other high molecular mass substances in soil and sediment extracts, GPC can separate compounds in extracts by size-exclusion on hydrophobic polymer gels of crosslinked styrene divinylbenzene porous copolymers.[207,208] PAHs can be eluted as a separate band although they may still coelute with chlorinated aromatics, pesticides, and nitroaromatics, thus necessitating another cleanup step prior to analysis. Another strategy is to separate the organic acids and phenols from soil/sediment extracts by performing a liquid–liquid extraction with 10 N NaOH.[209] The solvent extract is shaken in a separatory funnel with three aliquots of concentrated NaOH. The combined extracts containing the acids and phenols is discarded while the extract, in DCM, is dried and concentrated for GC analysis or solvent exchanged for another analysis method.

Elemental sulfur, in soil/sediment and waste extracts, can be removed by reaction with clean copper metal.[166] USEPA Method 3660B calls for the extract to be concentrated to 1.0 ml and shaken with 2 grams of copper powder to remove elemental sulfur.[210] The sample could then be fractionated to remove alkanes or other interferences.

VI. ANALYSIS METHODS

The analysis of PACs in the environment is challenging because of the complexity of environmental samples, the lack of availability of suitable reference standards and the presence of numerous isomers of alkylated and high molecular mass compounds that are difficult to differentiate. Separation and detection of hundreds of possible PACs with widely varying concentrations and properties in environmental samples requires some compromise if quantitation of the maximum number of PACs is desired. PAHs with 24 or fewer carbons atoms (seven or fewer rings) are sufficiently volatile for analysis by gas chromatography (GC).[211] Even though the efficiency of GC is considerably higher, high-performance liquid chromatography (HPLC) and supercritical fluid chromatography (SFC) have advantages over GC such as simpler sample preparation, greater selectivity for isomers, and the ability to separate less volatile and high molecular mass compounds.

A. Gas Chromatography (GC)

Cross-linked fused-silica capillary columns containing nonpolar or slightly polar stationary phases, such as 5% phenyl methylpolysiloxane or 100% methylpolysiloxane, are the most widely utilized in separating PAHs.[58,64–66,68,73,90,153,170,172,189,212] Five percent phenyl methylpoly-siloxane capillary columns are available from a number of suppliers as HP-5 (Agilent), DB-5 (J&W Scientific), or PTE-5 QTM (Supelco). In general, elution is by increasing boiling point or molecular weight. Cold oncolumn injection can improve resolution of low molecular weight PAHs.[213] Mass spectrometry (MS) is the most commonly reported detector for capillary GC with helium, nitrogen, or hydrogen as the carrier gas. One of the most referenced methods for PAH analysis is EPA Method 8270C for semivolatiles in water, wastewater, environmental samples, and solid wastes.[214] This method recommends using a 30 m × 0.25 or 0.32 mm i.d. 5% phenyl methylpolysiloxane fused-silica capillary column with 1-μm film thickness, a flow rate of 30 cm sec^{-1}, and a column temperature set at 40°C for 4 min, then programmed at 10°C min^{-1} up to 270°C and held until B[ghi]P elutes. The injection volume is 1 to 2 μl in the splitless mode. The MS scan rate is set at 1 sec scan^{-1} and the mass range is 35 to 500 amu. While the mass spectrometric detector is sensitive and selective, resolution of isomers is still a challenge when electron impact (EI) ionization (70 V) is used in detection. This method is also approved for the analysis of PAH derivatives such as 2-methylnaphthylene and 1,4-naphthoquinone. Using deuterated internal standards, detection limits are in the low ppb level. When used in the single-ion mode (SIM) an order of magnitude improvement in the LOD can be achieved.[64,90,153] USEPA Method 525 is a standard method for the determination of organic compounds in drinking water by liquid–solid extraction and GC–MS.[215] Closely eluting pairs, PHN/ANC, B[a]A/CHR, B[b]F/B[k]F, and IND/D[ah]A are usually not completely resolved using capillary gas chromatography on nonpolar and slightly polar stationary phases. The use of 50-m columns can increase resolution but results in much longer analysis times.[216] EPA Method 8275A is recommended for the analysis of PAHs in environmental solids and solid wastes.[217] Similar operating conditions are described in EPA Method 8275A as for EPA Method 8270C for GC–MS with the inclusion of operating parameters for a thermal extraction sample introduction unit. The estimated quantitation limit is listed as 1.0 mg kg^{-1} (dry weight) for this method. Wet samples can be analyzed when naphthalenes are the target analytes but generally, samples are dried in a fume hood at room temperature and ground before thermal extraction. Organic matter in the sample can complicate the analysis process and cause the capillary column to be overloaded.

USEPA Methods 610[218] and 8100[219] describe standard methods using a 1.8 m × 2 mm i.d. glass column packed with 3% OV-17 on Chromosorb W-AW-DCMS or a 30-m fused-silica capillary GC with flame ionization detection (GC-FID). When used to analyze the EPA$_{16}$ the closely eluting pairs, PHN/ANC, B[a]A/CHR, B[b]F/B[k]F, and IND/ D[ah]A are not completely resolved. Internal standards are added to correct for variability in instrument performance and injection volume.

Analysis times can be almost an hour per injection by capillary GC. Fast-GC methods for the analysis of PAHs include the use of more selective stationary phases, two-dimensional GC (GC × GC), short narrow-bore capillary columns, selective detection, advanced MS detection techniques, and faster temperature programming. Faster column heating at 100°C min^{-1} from 50 to 320°C has been achieved with resistive heating of a 5 m × 0.25 mm i.d. 5% phenyl methylpolysiloxane capillary column inside a metal tube.[220] This apparatus was used to separate a mixture of the EPA$_{16}$ in under 4 min. Although geometric isomers were not resolved, the method could be an effective screening tool. With FID detection, LODs of 5 pg μl^{-1} were demonstrated for PAHs.

Liquid crystalline stationary phases have a layered structure that resembles the ODS structure of stationary phases used to separate PAHs by reverse-phase HPLC. Isomer separation is achieved when the isomer of larger length-to-breadth ratio is more strongly retained. Although liquid crystalline stationary phases area able to separate PAH isomers more effectively than traditional nonpolar phases, their use has been limited due to the instability of the column over time that results in irreproducible results, low separation efficiencies, and a limited temperature range. A 12 m × 0.2 mm i.d., 0.15-μm film thickness smectic liquid crystalline capillary column was compared to a typical nonpolar 50-m DB-5 fused-silica column for the separation of several sets of PAH isomers.[221] PAHs were much more strongly retained on the smectic phase than on the nonpolar phase and all of the isomers in the EPA$_{16}$ PAH priority list were baseline separated, although internal standards were used to compensate for changing retention times resulting from column bleed. Side-chain liquid crystalline monomers grafted on to polysiloxane polymer (SCLCP) backbones as stationary phases in capillary chromatography are able to better separate geometric isomers of PAHs.[222] PAHs in coal tar were separated on a capillary column coated with poly (4-[(2-(2-(2-methylethoxy) ethoxy) ethoxy) carbonyl] phenyl 4-[4-(allyloxy)-phenyl] benzoate) (PBPBE3). Better resolution than on a nonpolar phase was obtained for PHN/ANC, B[a]A/CHY, and B[b]F/B[k]F. Good stability was reported for side-chain liquid crystalline polysiloxane polymer phases grafted with nonpolar, nonreactive hydrocarbon side chains.[223] Better separation of isomeric PAHs was achieved, but lower molecular weight PAHs were poorly separated. When used in conjunction with conventional nonpolar stationary phases, adequate separation of a mix of 21 PAHs may be possible. In addition, good column stability was reported for up to 500 injections. Packed column GC liquid crystalline stationary phases are commercially available, although two- and three-ring PAHs are not adequately resolved while larger PAHs are too strongly retained.[224]

B. Supercritical Fluid Chromatography (SFE)

SFC of PAHs has been extensively reviewed.[225] SFC combines the speed and efficiency of GC with the stationary-phase interactions possible in HPLC. PYR, B[a]A, 1-nitropyrene, and B[a]P in an explosives residue were separated on a 6 m × 50 μm i.d. open tubular column coated with 0.1 μm p,p-cyanobiphenyl polysiloxane with supercritical CO_2 as the mobile phase.[226] Solid-phase injection was required due to the effects of solvent on selectivity in SFC. Seven PAHs were identified in a mixture of explosives on a 25 cm × 250 μm microcolumn packed with poly(octylhydrosiloxane) polymer stationary phase.[227] No modifier was added to pure supercritical CO_2 at 100°C and a density program of 0.14 to 0.64 g ml^{-1}. Separation was achieved in approximately 60 min. Five 20 cm × 4.6 mm Hypersil silica columns connected in series served as the stationary phase for a separation using CO_2 at a flow rate of 1.5 ml min^{-1} and 40°C modified with methanol from 2 to 10%, and an initial pressure of 100 bar increasing to 150 bar.[228] PAHs in water have been measured using SFC with diode array detection (DAD) following online concentration by solid-phase extraction.[229] A 15 cm × 4.6 mm Spherisorb 5-μm ODS2 column and a 125 mm × 4.6 mm Envirosep-PP 5-μm column were connected in series. They were operated at 40°C and 200 bar. A flow rate of 2.5 ml min^{-1} of CO_2 modified with a methanol gradient allowed for resolution of the EPA$_{16}$ in 14 min with LODs of 0.1 to 0.8 mg l^{-1}. Preconcentration on C18 extraction disks showed recoveries of >70% with a concentration factor of about 1000.

C. High-Performance Liquid Chromatography (HPLC)

HPLC has adequate selectivity to separate PAH isomers and high molecular weight PAHs with more than 24 carbon atoms that cannot be separated using gas chromatography. USEPA Methods 550 and 8310 are HPLC methods for the determination of PAHs in drinking water,[230] groundwater and wastes,[231] respectively, with ultraviolet (UV) and fluorescence (FLD) detection. Most samples are extracted or preconcentrated first, then cleaned up on a silica gel column, if necessary. The recommended analytical column is a reverse-phase ODS of 5 μm particle size, 25 cm × 2.6 mm i.d. Decafluorobiphenyl is a commonly selected surrogate since it is not found naturally and does not interfere with the analysis. With 40 to 100% ACN in water as the gradient mobile phase in reverse-phase liquid chromatography (RPLC), all sixteen of the EPA priority PAHs can be resolved with LODs in the low and sub μg l^{-1} range in 30 to 40 min.[19,71,85–87,91,95,98,148,152,157,177,182,185,186]

Selectivity varies continuously with temperature.[232] Increasing the temperature can significantly reduce analysis times for HPLC, however operation at subambient temperatures (10 to 15°C) increases separation of some closely eluting compounds, including geometric isomers. Using a steeper gradient for ACN: water can increase separation of D[ah]A and B[ghi]P. Separation of the EPA$_{16}$ was performed on a Zorbax C$_{18}$ column, with FLU and ACE coeluting, and a mixture of six methylchrysene isomers was separated on a Bakerbond widebore C$_{18}$ column at −8°C.[233] There appears to be an enhancement of shape selectivity on C$_{18}$ phases at subambient temperatures. The use of smaller particle size can increase the efficiency of a column and reduce analysis times. Separations such as this are often referred to as "fast" or "rapid" LC. Use of a 3-μm, C$_{18}$ stationary phase in a 5 cm × 4.6 mm i.d. column can shorten the analysis time to 4 min for the EPA$_{16}$ at a flow rate of 3 ml min^{-1}.[234]

1. Novel HPLC Stationary Phases

Crosslinked chitosan stationary phases were shown to effectively separate PAH mixtures.[235] Chitosan was prepared by deacetylation of the acetylamino groups on chitin extracted from the cell walls of crustaceans. Chitosan, 95% crosslinked with EOAD and packed into a fused-silica capillary column (150 mm × 0.53 mm i.d.), successfully separated a number of PAHs by micro-column chromatography with an elution rate of methanol:water (80:20) of 2 μl min^{-1}. The behavior of PAHs on the chitosan phase was very similar to the behavior of PAHs on C$_{18}$ columns by conventional techniques.

Zirconia stationary phases are gaining in popularity due to their thermal stability. Separation of 16 out of 20 PAHs in 15 min was accomplished on a 15 cm × 4.6 mm ZirChrom-PS, a polystyrene-coated zirconia, at 80°C with ACN in water as the mobile phase.[236] Neither 1-methylnaphthalene and 2-methylnaphthalene were resolved, nor were B[b]F and B[k]F, or B[a]A and CHY.

A separation of the EPA$_{16}$ was achieved using a 10 cm × 4.6 mm i.d. polybutadiene-coated zirconia (2.5-μm particle size) reversed-phase column.[237] Resolution and retention times were compared for this column at 30°C and at 100°C. The separation at 30°C and 1 ml min^{-1} required 70 min while at 100°C and 3 ml min^{-1} it was achieved in less than 4 min. Two isomeric pairs (B[a]A/CHY and B[b]F/B[k]F) were not resolved at either temperature, but peak symmetry was superior at the higher temperature and flow rate. ACE and FLU were also not resolved under the "fast" conditions.

2. Novel HPLC Mobile Phases

Hot pressurized liquid water (HPLW), also called subcritical water, is hot water under enough pressure to maintain the liquid state even when heated to above the boiling point. Higher temperature reduces the viscosity of the mobile-phase water and increases column efficiency. Further, a reduction in the dielectric constant of water at higher temperatures decreases its polarity such that the amount of organic modifier can be reduced, therefore HPLW has the potential to replace solvent programming in reversed-phase liquid chromatography. The column is placed inside an oven that can be temperature programmed and a restrictor or back-pressure regulator is installed after the column to keep the water in the liquid state. Unfortunately, currently available ODS stationary phases do not have sufficient thermal stability to withstand the temperatures necessary (>200°C) for water to be used as the sole mobile phase for extended time periods.

In an effort to reduce the use of toxic mobile phases such as methanol and ACN, dimethyl sulfoxide (DMSO)-modified HPLW was used as the mobile phase for the reversed-phase separation of a four-PAH mixture.[238] A typical 25-cm ODS column was employed as the stationary phase and temperature was increased from ambient to 125°C as DMSO content in the mobile phase was decreased from 80 to 66%. PAHs have been separated on a 25-cm Zorbax ODS column using HPLW modified with methanol as the mobile phase.[239] NAP, PHN, PYR, CHY, and B[a]P were separated in 20 min at 140°C in 62% methanol with reasonable column stability reported. An increase of 4 to 5°C was equivalent to a 1% increase in the methanol content of the mobile phase. The quality of the chromatograms also improved with heating.

The addition of surfactants to water above the CMC can influence retention and selectivity in reverse-phase HPLC in a method termed micellar liquid chromatography (MCL).[240] Above the CMC, surfactants form aggregates called micelles into which analytes partition thus increasing their "apparent" solubility in the mobile phase and allowing for a reduction in the use of organic solvent which must be kept below 20% in order to prevent micelle disruption. Surfactants selected must have a small molar absorptivity in the ultraviolet-visible range. Micellar stabilization of analytes can also enhance fluorescence signals. Separation of the EPA$_{16}$ was conducted on a 15 cm × 4.6 mm, 5-μm Hypersil PAH column at 65°C and 0.6 ml min^{-1} using 0.10 M cetyltrimethylammonium chloride (CTAC, a cationic surfactant) +9% 1-pentanol +3% 1-propanol in water mobile phase.[241] Following optimization, the mixture was resolved into ten separate peaks. Lower CTAC concentrations, above the CMC, gave better resolution and longer retention times.

D. CAPILLARY ELECTROPHORESIS (CE)

Separation in capillary electrophoresis (CE) depends on the velocity of charged solutes in an electric field. When a voltage is applied to opposite ends of a short length of fused-silica capillary, a surface charge forms on the inside wall of the capillary that results in flow of buffer solution through the system or electroosmosis. Charged species contained in the buffer will then migrate at different rates allowing for their separation displayed as an electropherogram. A recent review of CE in environmental analysis is available.[242-244] The advantages of CE include high efficiencies, small sample size, low reagent consumption, and the ability to utilize chemical selectivity to aid separation. Although numerous methods can be found in the literature for the CE separation of PAHs, practical applications have been few. PAHs are neutral molecules that must be associated with a charged species in order the separate them by CE. Nonionic molecules will migrate at the same rate in CE. Five amino-PAHs, were separated as protonated amines by CE at pH 2.3 in phosphate buffer:30% methanol. Complete separation was achieved in less than 6 min with LODs in the mg l^{-1} range using UV detection at 220 nm.[245] The effective length of the capillary was 56 cm and the separation was performed at 30 kV and 25°C.

Large ionic substances that associate with PAHs can be added to the running buffer causing them to undergo electrophoretic migration. Micellar electrokinetic chromatography (MEKC) is a CE technique in which PAHs are associated with charged surfactant above its CMC.[243] The degree of partitioning of individual PAHs between the aqueous phase and the micellar phase, a pseudo-stationary phase, determines their migration rate in the running buffer. Analytes that do not interact with the micellar phase migrate at the electroosmotic mobility. The elution window, a limiting factor for analytes, is between these two points. The addition of organic modifiers or mixed micelles can increase the size of the window. Organic solvents can also increase sensitivity by increasing analyte solubility in the running buffer thereby increasing sample capacity and efficiency.[244] Ion-pair reagent, 100 mM tetrahexylammonium cation (THA$^+$) in 50 mM ammonium acetate was added to complex with 13 PAHs (two to seven rings) in methanol that had been preconcentrated by SPME from water. The effective length, injection point to detector, was 40 cm and detection was by UV at 254 nm. ANC/PHN and perylene/benzo[e]pyrene/B[a]P were not resolved. THA$^+$ in ammonium acetate was used for the solvochromatic association of six PAHs and two methylated PAHs in ACN/water in a 100 cm × 25 μm i.d. fused-silica capillary at 20 kV.[246] The larger, more hydrophobic solutes migrated faster due to stronger associations with THA$^+$. Laser-induced fluorescence (LIF) with excitation at 275 and 325 nm from a He/Cd laser was employed for detection. Although limits of detection were estimated to be in the sub ppb range, a 12-min separation using this MEKC-modifier method could not resolve structural isomers.

Negatively-charged additives show less interaction with capillary walls than positively-charged additives. Excellent separation of 23 nonionic aromatic compounds, including 17 PAHs, was achieved in 22 min by adding 50 mM sodium dioctyl sulfosuccinate (DOSS) and 8 mM sodium borate to 40% (v/v) ACN in water at pH 9 with 20 kV applied voltage.[247] Detection was by UV at 250 nm. Larger PAHs associated more strongly with the DOSS phase than smaller PAHs and therefore had longer migration times. Another solvochromatic association method involving planar organic cations, tropylium and 2,4,6-triphenylpyrylium ions used in combination increased resolution compared to their use separately, yet B[a]P/CHY and ACE/PHN were not resolved at 30 kV in a capillary of 90 cm effective length with detection at 254 nm. In general, larger PAHs migrated faster due to their increased interaction with the cations.

MS detection is difficult in MEKC because the ion source is contaminated by the surfactants. The addition of charged, water-soluble cyclodextrin (CD) as a modifier in MEKC was introduced to improve the resolution of PAHs by allowing for their differential partitioning between the micellar and the aqueous/CD phase.[248] The chiral character of CD increased selectivity based on size and shape. Cyclodextrins, produced by the enzymatic digestion of starches, have a toroidal shape with a hydrophobic cavity and a hydrophilic exterior. Nonpolar solutes can enter the cavity and form inclusion complexes. CD-PAH interactions reduce retention times by inhibiting solute-micelle association. CDs with eight glycopyranose units (γ-CD) were added to 0.02 mM SDS in 5% 2-propanol/water to aid in the separation of B[a]P and six methylated benzo[a]pyrenes. In a phosphate/borate running buffer at 20 kV with an effective capillary length of 60 cm, the separation was achieved in 13 min.

PHN, ACE, PYR, CHY, B[a]P, and benzo[e]pyrene were separated in a 50 mM borate buffer (pH 9) containing a mixture of 20 mM neutral methyl-β-cyclodextrin (MβCD) and 25 mM anionic sulfobutylether-β-cyclodextrin (SBβCD) at 30 kV and 30°C.[249] B[a]P and benzo[e]pyrene were successfully resolved with the other compounds in under 11 min in a 50-cm effective length of capillary without micelles in the mobile phase. The system was also less sensitive to temperature and separation potential. LIF detection with excitation at 325 nm at 2.5 mW from a He/Cd laser coupled to an optical fiber allowed for detection limits in the sub ppb range. The method described above was applied to the analysis of contaminated soil that had been extracted by supercritical CO_2 for 20 min at 120°C and collected in methanol/DCM.[250] Of the 16 EPA PAH mixtures, eleven compounds were detectable in the low ppb range. Ten of the eleven detectable compounds were measured in the soil extract. When compared to RP-HPLC, CE values were slightly lower but only six compounds were detected by HPLC-FLD. No direct relationship between PAH molecular size, polarity, or volatility with migration order was observed and B[b]F/B[k]F isomers were readily separated.

SPME was coupled with cyclodextrin-modified CE in the development of a method for the EPA$_{16}$.[251] A PDMS-coated SPME fiber was contacted with a low ppb level aqueous solution of PAHs and then placed directly in the inlet of the separation capillary. The running buffer at pH 9 contained 35 mM SBβCD, 10 mM MβCD, and 4 mM MαCD. At 30 kV and a 60 cm × 50 μm i.d., 350 μm o.d. fused-silica capillary, ACE, NAP, and FLU coeluted. With UV detection, sensitivity was only slightly less than with LIF detection.

Increased selectivity was achieved by combining β- and γ-CDs with SDS and urea to the borate running buffer to increase CD and PAH solubility in the aqueous phase.[252] At 30°C and 15 kV in a capillary of 50.5 cm effective length, 20 PAHs were resolved into 18 peaks with perylene/ACE and B[a]P/B[a]A coeluting in a 54 min separation. Baseline resolution of all 16 EPA PAHs was finally achieved using 0.50% (w/v) poly(sodium undecylenic sulfate) or poly-SUS as the micellar phase.[253] The running buffer was 12.5 mM sodium phosphate/borate at pH 9.2 with 40% ACN in water. Elution order was generally by increasing length-to-breadth ratio for the 30 min run with UV detection.

Capillary electrochromatography (CEC) on a 100 μm i.d. × 20 cm effective length fused-silica capillary column packed with 1.5-μm nonporous ODS particles also separated the 16 EPA priority PAHs, but in under 10 min.[254] The mobile phase consisted of 65% ACN with 2 mM tris solution at 29 kV. B[b]F and B[k]F were resolved by 2 min and LIF excitation at 257 nm and detection at 280 and 600 nm provided adequate sensitivity for low mg l^{-1} LODs. Separation of B[a]P and the 12 possible methylated B[a]P isomers into seven peaks has been reported.[255] The stationary phase was 3 μm ODS particles slurry pressure packed into a 25 cm × 75 μm i.d. × 363 μm o.d. fused-silica capillary. The mobile phase was 75% (v/v) ACN and 25% 12.5 mM tris at pH 8.0. Electrophoresis was conducted at 30 kV and 25°C.

VII. DETECTION

A. GASEOUS EFFLUENTS

1. Flame Ionization Detection (FID)

FID is commonly utilized for detection of gas chromatography effluents.[57,68,128,150,169,256,257] USEPA Methods 610[218] and 8100[219] apply FID to the analysis of gas chromatography effluents. Sensitivity for PAHs is good with LODs in the low parts-per-billion levels.

2. Fourier Transform Infrared (FTIR) Detection

FTIR detection of PAHs upon elution from capillary GC separation is described in USEPA Method 8410.[258] The GC conditions recommended are quite similar to those for EPA Method 8270C however, compound class assignments for group absorption frequencies are made using FTIR. Eleven of the less volatile EPA$_{16}$ can be quantified by this method along with 2-methylnaphthylene and 2-chloronaphthalene. This method can be applied to wastewater, soil, sediment, and solid wastes. LODs are in the low ppb range. The FTIR spectrometer must be capable of collecting one scan set per second at 8 cm^{-1} resolution.

3. Gas Chromatography–Mass Spectrometry (GC–MS)

The detection method most applied to capillary gas chromatography of PAHs in environmental samples is MS using electron-impact (EI) ionization.[214,215] Fragmentation of PAHs is minimal under the 70 eV conditions generally used for EI therefore single-ion monitoring (SIM) of the molecular ion provides maximum sensitivity although identification of individual isomers is virtually impossible.[77,90,94,138,143,153,154,169,170,172] LODs in the low to sub parts-per-billion level are routinely achieved.

Collision-induced dissociation (CID) can enhance the degree of ionization and increase the sensitivity for MS detection to the ng l^{-1} level by increasing the abundance of confirmatory ions.[259] Time-of-flight-mass spectrometry (TOF-MS) coupled with fast GC can achieve separation and detection of PAHs in 3 to 5 min at the parts-per-billion level.[260] While the sensitivity is slightly better than for single-quadrupole MS selection, coelution of isomers is still observed as B[b]F and B[k]F are not resolved. NPAHs have been detected in soil extracts by MS with CID and MS–MS to achieve low picogram detection limits.[151] Positive-ion chemical ionization has been shown to differentiate all five groups of isomers in the 16 EPA list.[261] Ion-molecule reactions of PAHs with ionized dimethyl ether formed fragment-molecule adducts that were used to distinguish between isomers at an ionization energy of 90 eV. A separation time of 53 min was necessary on a 30 m capillary column. NPAHs were also detected using negative-ion chemical ionization-mass spectrometry (NCI-MS) with methane as the reagent gas.[262] Most of the 2-nitrofluoranthene extracted from PM collected in winter in Saitama City, Japan, was found in the 0.5 μm fraction, ~34 μg g^{-1} PM.

Ion Trap MS (GC–ITMS) can lower detection limits for GC–MS even further. GC–ITMS, when combined with SPE, can achieve upper parts-per-quadrillion (ppq) level detection limits.[259] Method detection limits (MDLs) of ~1 ng l^{-1} were demonstrated for seventeen priority PAHs when CID was used with low resolution MS.

4. Electron-Capture Detection (ECD)

Detection of nitrated PAHs (NPAHs) is possible using ECD and can be made more sensitive by preparing the fluorinated derivatives of NPAHs with heptafluorobutyric anhydride (HFBA).[263] The fine fraction (<0.5 μm) of urban air particulate matter in Saitama, Japan contained about ten times the amount of NPAHs as the course fraction with levels as high as 111 pg m^{-3} for 1-nitropyrene and 18 pg m^{-3} for 1-nitronaphthalene.

B. DETECTION OF PAHs IN LIQUID MATRICES

1. Ultraviolet

UV detection of PAHs at 254 nm is simple and sensitive (nanogram level LODs) but is not sufficiently selective for compound identification.[148,157,177,182] Variable wavelength diode array detection (DAD) is preferable since the UV spectrum of a PAH can serve as a fingerprint indicating its identity.[105,185] To improve on the selectivity of DAD, second-order bilinear calibration was applied to the analysis of coeluted peaks from an HPLC separation of PAHs and Sulfonated PAHs (SPAHs).[264] The Generalized Rank Annihilation Method (GRAM) requires two data matrices, calibration data from spiked samples to minimize matrix effects and the unknown data. The method was applied to spiked marine sediments with reasonable success, although no analytes were detected in unspiked samples.

2. Fluorescence Techniques

Most PAHs are highly fluorescent following excitation at their characteristic wavelengths allowing for their selective detection in environmental samples. Fluorescence detection (FLD) methods are at least an order or more in magnitude more sensitive than UV detection (picogram level LODs) and have also shown utility in the direct detection of PAHs in complex mixtures.[265] Typical excitation and emission wavelengths used for fixed wavelength FLD are 280 to 300 nm and 400 nm, respectively.[95,186] Variable wavelength fluorescence detectors can improve on sensitivity and selectivity even further for individual PAHs. Table 15.3 is a list of excitation and emission wavelengths that have been applied to the detection of PAHs in environmental samples. Some variability is expected since FLD is sensitive to the analyte's microenvironment including the presence of quenchers and other molecules that fluoresce such as humic substances. NAP, ACE, FLU, and especially ACY, give relatively weak fluorescence signals, hence coupling DAD for these compounds with FLD can provide a powerful combination for achievement of optimal sensitivity and selectivity.[95,105,108,265,267]

Fiberoptic sensors can be used to measure PAHs *in situ*.[268] A xenon lamp is a common excitation source when connected to optical fiber to excite a sample remotely and induce fluorescence. The fluorescence light emitted is collected by another optical fiber and transmitted to a photomultiplier detector. Sensors are chemically modified to enhance selectivity and sensitivity. β-cyclodextrins and immunochemical reagents on optical fiber probes can selectively determine individual PAHs but with some crossreactivity. Fluorescence probes such as perylene can be imbedded in polymers affixed to the end of the probe.

TABLE 15.3
Fluorescence Wavelengths for the EPA$_{16}$ PAHs

PAH	Excitation λ, Emission λ(nm)							
	Ref. 266	Ref. 98	Ref. 103	Ref. 108	Ref. 132	Ref. 137	Ref. 152	Ref. 166
NAP	280, 340	218, 357		220, 340	221, 337		220, 330	
ACY	289, 321	218, 357		220, 340			220, 330	
ACE	289, 321	226, 359		240, 340	227, 315		220, 330	
FLU	289, 321	226, 359	215, 320	240, 340	227, 315		250, 410	
PHN	249, 362	250, 350	250, 360	240, 340	252, 372		250, 410	
ANT	250, 400	250, 425		240, 440	252, 372		250, 410	
FLA	285, 450	234, 440	234, 445	240, 440	237, 440	284, 464	250, 410	230, 410
PYR	333, 390	234, 440		240, 440	237, 440		250, 410	
B[a]A	285, 385	286, 405		280, 398	277, 393	274, 414	275, 375	
CHY	260, 381	265, 405		280, 398	277, 393		275, 375	
B[b]F	295, 420	250, 420		238, 416	258, 442	300, 446	280, 420	250, 420
B[k]F	296, 405	238, 460	250, 425	238, 416	266, 415	300, 446	280, 420	250, 420
B[a]P	296, 405	238, 460		238, 416	266, 415	296, 406	280, 420	
D[ah]A	296, 405	298, 420		296, 420	399, 425	300, 470	280, 420	
B[ghi]P	380, 405	298, 420		296, 420	295, 425	300, 470	280, 420	
IND	300, 500	246, 490			251, 510	300, 470	280, 420	

Laser sources that enhance sensitivity by providing more intense excitation and fiberoptics for remote sensing are two recent developments that have greatly improved the fluorescence detection of PAHs and related compounds. LIF takes advantage of the intensity of a laser to cause increased excitation of the target analyte and therefore increase sensitivity. PAHs extracted from creosote-contaminated soil were separated by CEC on a polyimide-coated fused-silica capillary packed with 3 μm ODS porous particles and detected at 10 kV in a 27-cm column with 75% ACN and sodium tetraborate (pH 9).[269] Ten of sixteen PAHs were identified in the soil extract in a 60 min separation. Laser-induced dispersed fluorescence detection using a liquid-nitrogen cooled charge-coupled device allowed for deconvolution of coeluting peaks at an excitation wavelength of 257 nm from an argon ion laser. LIF spectroscopy with a fiber optic probe was applied to the analysis of PAHs in sediments in Milwaukee Harbor *in situ*.[270] The optical fiber was contained in a watertight steel probe with a mirror at the end to turn the excitation beam 90° before it exited the probe through a sapphire window which illuminated the sediment directly on the outside of the window. Fluorescence emission scattered back to the return fiber. Pulsed XeCl eximer laser at 50 Hz emitted 308 nm light as the excitation source. Measurements were compared with PLE-GC–MS analysis of sediment cores indicating a relative error for the probe techniques of about 30%. Measured levels of PAHs ranged from10 to 65 μg g^{-1}.

3. Selective Fluorescence Quenching (SFQ)

Selective quenching agents can suppress fluorescence of coeluting peaks from HPLC or CEC during fluorescence detection. There are two classes of PAHs: alternant and nonalternant. Alternant PAHs have completely conjugated aromatic systems while nonalternant PAHs have interrupted aromaticity as a result of five-member rings in their structure. Selective quenching agents can suppress the fluorescence signal of one PAH class without affecting the fluorescence signal of the other class of coeluting analytes. Diisopropylamine can be used as a selective quencher for nonalternant PAHs in ACN.[271] It does absorb in the excitation region so sensitivity is reduced. Micellar cetylpyridinium chloride (CPC) is a selective fluorescence quenching agent for alternant PAHs.[272] Pyridinium chloride (PC) is another selective quenching agent for alternant PAHs which is more sensitive than nitromethane and can be used to suppress alternant PAHs in HPLC effluents.[273]

4. Room Temperature Phosphorescence (RTP)

Most PAHs are phosphorescent, although this is not detectable at room temperature in liquid solvents. However, enhanced phosphorescence can be achieved for PAHs sorbed to solids with the addition of phosphorescence enhancers. Solid-phase extraction has been coupled with solid-surface room-temperature phosphorescence (SPE-SS-RTP) for PAH screening in water.[274] This method demonstrates the potential for SS-RTP to detect pg ml^{-1} levels of PAHs directly on the surface of membranes through which water samples have been filtered to concentrate PAHs. A right-angle excitation-emission configuration was used to collect data following addition of TlNO$_3$ and SDS as phosphorescence enhancers.

5. Mass Spectrometry of Liquid Effluents

High-molecular-mass PAHs (302 amu or greater) are not normally detected by GC–MS owing to their low volatility. These groups of PAH isomers are present at very low levels yet they have been shown to be much more potent carcinogens. MS detection of liquid effluents allows for the detection of very high molecular mass PAHs such as coronene that are not sufficiently volatile for GC separation. The interface between the HPLC or CE column and the mass spectrometer represents a challenge for the analysis of neutral and nonpolar molecules. Also, the use of water as a mobile phase in reversed-phase LC is problematic due to its low volatility. Heated pneumatic nebulizers can take LC effluents from microbore columns and introduce them directly into an ionization source, however moving belt and particle beam interfaces perform poorly for PAC analysis.[275] Ionization and fragmentation of PAHs does not occur to any great extent although PAH transformation products (OPAHs, NPAHs, and SPAHs) are more polar and form ions more readily.

a. Electrospray Ionization (ESI)

HPLC with ESI produces mostly molecular ions and few to no molecular fragments. This limits the use of LC–MS in the differentiation of isomers. Two atmospheric pressure ionization (API) interfaces allow for the formation of molecular ion, $[PAH]^+$, yet in general, derivatization and additives are required to induce fragmentation. ESI is an interface that transfers ions from the mobile phase into the gaseous phase for introduction into the mass spectrometer so that atmospheric pressure chemical ionization (APCI) can cause ionization of chemical species in the gaseous phase. Both techniques can be operated in the positive- or negative-ion mode. Reports exist for the ionization of PAHs by both techniques, although there are few actual applications to the analysis of real environmental samples.

Electrospray ionization in the presence of tropylium cation can induce ion formation in PAHs and allow for the detection of larger PAHs.[276] Coronene, PYR, B[a]P, and 1-nitropyrene were detected in HPLC effluent at low to sub ng levels with tropylium tetrafluoroborate added as a postcolumn reagent. Using MS–MS (tandem MS), stable molecular ions can be selected for subsequent fragmentation.[277] Tropylium ion, $[TR]^+$, complexes with PAHs to form $[PAH\text{-}TR]^+$ that undergoes charge transfer to form $[PAH]^+$ in the electrospray interface. ANC, PYR, and 1,2-benzoanthracene were detected in reverse-phase HPLC effluent using SIM at 178, 202, and 228 m/z, respectively. Although the molecular ion peaks were not completely resolved, adequate fragmentation was achieved in the collision cell between the mass spectrometers at 100 eV with argon. The unusual stability of PAHs to ionization was circumvented by the addition of silver nitrate solution before ESI.[278] A bench-top quadrupole MS could distinguish between four groups of PAH isomers, thirteen total PAHs, by forming silver ion adducts with PAHs preceding ESI. Relative intensity ratios of the $[PAH\text{-}Ag]^+$, $[2PAH\text{-}Ag]^+$, and $[PAH]^{+\cdot}$ ions peaks were different for the various ions.

b. Atmospheric-Pressure Chemical Ionization (APCI)

In APCI, the eluent from the HPLC is vaporized and carried through the APCI source by heated nitrogen gas. A corona discharge is used to produce reagent ions that in turn ionize the analytes for introduction into the mass spectrometer. APCI-MS has been applied to the analysis of seawater and sediments.[279] HPLC separation was used with 100% methanol as the mobile phase and water was added postcolumn prior to ionization in the positive mode. SIM of $[PAH]^+$ was employed for detection. No PACs were detected in seawater, but B[b]F, B[k]F, and B[a]P were detected in sediments with complete resolution. No signal was observed for lower molecular mass PACs. FLA, B[b]F, B[k]F, B[a]P, B[ghi]P, and IND were analyzed in sewage sludge extracts by HPLC-APCI-MS with < mg kg^{-1} LODs.[280] The results compared well with concurrent analysis by GC–MS and showed PAH levels of 0.2 to 0.7 mg kg^{-1}. Air PM was analyzed for B[a]P and its oxidized diones by HPLC-APCI-MS.[281] Samples collected in Munich, Germany had levels of B[a]P diones of 8 to 605 pg m^{-3} with higher levels measured in daytime samples indicating photochemical formation.

HPLC-APCI-MS was applied to the analysis of PAHs of molecular masses 326, 350, and 374 amu in air particulate, zebra mussels, and coal tar from Hamilton, Ontario, Canada.[282] Better instrument response was obtained with 100% ACN in the mobile phase although 100% DCM was required to elute the highest molecular mass PAHs. LODs for B[a]P were 200 pg. PAHs with masses up to 450 amu were analyzed in air, water, and soil samples in an area surrounding a chemical plant that makes pitch black pellets in the Czech Republic.[283] Analysis by LDI-TOF-MS and LC-APCI-MS were compared. SPMDs were used to sample air and water while PLE with DCM was used to extract soils. High levels of PAHs were measured all around the plant with the major components identified as benzo[a]coronene, naphtho[1,2,3,4-ghi]perylene, and dibenzo[a,e]-pyrene. Still, the insufficient resolving power of microbore LC and the lack of available standards limited the identification and quantitation of most other components.

APCI processes are not well understood for PACs. Sensitivity of APCI for detection of HPLC and SFE effluents is poor for low molecular mass PACs and it depends strongly on the solvents selected and the source gas.[284] Charge exchange and proton transfer are competing mechanisms in the ionization of PAHs. The addition of carbon dioxide to the nebulizer gas with nitrogen for the other gas streams increased sensitivity dramatically. Eliminating water from the mobile phase also increased sensitivity. PACs separated by normal-phase liquid chromatography using HEX or isooctane may give the optimum sensitivity with the inclusion of carbon dioxide in the eluent streams.

SFC-APCI-MS using methanol–water-modified carbon dioxide as the mobile phase has been applied to the analysis of PACs.[285] A direct fluid introduction interface with flow splitting removed two thirds of the SFC effluent transferred to the APCI-MS system. PACs were separated on a C_{18} packed column at 45°C and LODs were similar to those obtained with HPLC-UV. SFC-APCI-MS–MS using microbore packed columns applied to the analysis of PACs in coal tar was able to provide structural information for isomer differentiation.[286] LODs in the low ng range were obtained by SIM.

LC–MS–MS using a high-pressure quadrupole collision cell, following APCI-MS, produced collision-induced dissociation with adequate fragmentation so that isomers could be identified by peak-area ratios.[287] This technique was applied to the analysis of PAHs in coal tar extract SRM 1597 with standard addition. Twelve PAHs were detected at levels that were close to or within certified limits.

c. Atmospheric-Pressure Photoionization (APPI)

Photoionization has been used as a detection technique in GC for decades. It has only recently been utilized for the production of ions for mass spectral analysis. APPI of PAHs has been reported and in the future may prove to be useful in the separation of PAH isomers. In APPI, the corona discharge used in APCI is replaced by a photon-emitting gas discharge lamp. A vacuum-ultraviolet (VUV) photoionization lamp can supply 10 eV photons to analytes from LC effluents once the solvent is evaporated. The ionization source is operated at atmospheric pressure, allowing a high collision rate. Toluene, as a dopant, produces photoions that can be transferred to PAHs efficiently allowing for sensitivity increases of almost an order of magnitude compared to corona discharge-APCI. APPI has been applied to the analysis of 12 PAHs in sediments with detection limits of 0.4 to 4.6 ng ml^{-1} by SIM.[288] Although sensitivity was increased, resolution of isomers has not been demonstrated.

d. Immunoassay

The use of immunoassay techniques for the determination of PAHs has been reviewed.[289] Immunoassay is based on the coupling of a specific biological antibody in the detection device with the analyte either directly in water or extracted from solid samples and diluted in buffer solution. Enzyme-linked immunosorbent assay (ELISA) is the most common immunoassay technique employed in commercially available test kits. Water samples or soil extracts are added with an enzyme conjugate reagent to immobilized antibodies where the conjugate competes with PAHs for binding to the antibodies. ELISA test kit sensitivity and crossreactivity depends on the PAH used to raise the antibody. Antiphenanthrene or antifluoranthene antibodies raised in host animals are the most commonly employed. Test kits will be most sensitive to the PAH from which the antibody was derived and others with closely related structures (crossreactivity). Most commercially available test kits are more sensitive to three-, four-, and five-ring PAHs, so response to NAP, D[ah]A, IND, and B[ghi]P will be minimal. Combining antibodies can achieve more universal detection for PAHs. The crossreactivity allows for the detection of additional PAHs, PAH degradates and derivatives that are not selected for measurement using traditional chromatographic methods. As a result, total PAH concentrations are usually higher when measured by ELISA compared with chromatographic methods. Very small samples are required and many samples can be analyzed concurrently in this field-portable technique making immunoassay an excellent technique for rapid screening on site. Reproducibility is not as good as when using GC, HPLC, or CEC and since no single antibody can detect all or only one PAH, a second confirmatory analysis method must be performed on representative samples.

USEPA Method 4035 describes the use of ELISA test kits for the screening of soil for PAHs.[290] The method recommends that positive test results be verified using chromatographic techniques. Inaccurate standards, supplied with test kits, has resulted in gross overestimation of PAHs in environmental samples.[291] Soils were extracted with methanol prior to analysis with results reported as concentration ranges. Results for spiked samples were within the ranges detected by the test kits for semiquantitative analysis. Sensitivity was about 100 parts per billion. River water samples from the Nitra basin in Slovakia were analyzed with a commercial test kit and the results compared to analysis of the same samples by GC–MS.[292] ELISA results were higher by a factor of two or more, probably due to the crossreactivity of other structurally similar molecules. HPLW extraction is well suited for the preparation of solid samples for immunoassay and was applied to the analysis of industrial soils and certified marine sediment, HS-3.[293] The extract was collected in methanol to maintain the solubility of the analytes. Immunoassay results were slightly higher than results determined by HPLC for the same samples. Total PAH measured in the soil by HPLC was 27.4 mg kg^{-1} while immunoassay results were 47.7 mg kg^{-1} indicating some crossreactivity to the antiphenanthrene antibodies of PAH analogs. Predrying of samples was not necessary and the more volatile PAHs were retained in the sample. Water samples around the island of Oahu in Hawaii, U.S.A., screened by ELISA and compared with results obtained by GC–MS, gave results that were considerably higher using ELISA.[294] A positive result by ELISA indicated PAH concentrations >1.4 ng ml^{-1}. There were no false negatives but positive samples analyzed by GC–MS gave values <0.1 ng ml^{-1} for the ΣEPA$_{16}$. Sediment samples were extracted by SFE with CO$_2$ also showed higher results when ELISA was used for detection (259 to 531 ng g^{-1} by ELISA versus 166 to 356 ng g^{-1} by GC–MS as B[a]P equivalents).

Groundwater, at a former manufactured gas plant, soils, and landfill leachates were analyzed by ELISA using polyclonal antipyrene antibodies and by HPLC.[295] The PYR degradate, 1-hydroxypyrene, has a crossreactivity of 180% compared to PYR causing a positive bias in ELISA results following biodegration. ELISA underestimated PAH concentrations in landfill leachates because the samples contained more two- and three-ring PAHs which have substantially lower crossreactivities to the antipyrene antibody. Site-specific calibration can reduce false negative results. Structure and crossreactivity were evaluated for an ELISA test kit using antiphenanthrene antibodies.[296] Comparing the structure of various PAHs and PACs with the structure of PHN and crossreactivities showed that the kit was more sensitive to molecules with structural features resembling PHN. For example, three-ring PAHs, with segments of a third fused ring (9-methylanthracene and 1-methylanthracene) with the same configuration as PHN, had crossreactivities of >100%. The kit was much less sensitive for two-, five-, and six-ring PAHs. An amperometric immunosensor using antiphenanthrene antibodies coated on screen-printed carbon electrodes demonstrated a detection limit of 5 ng ml^{-1} for PHN in spiked river- and tap-water samples.[297] Amperometric detection at +300 mV versus Ag/AgCl with enzyme alkaline phosphatase and p-aminophenyl phosphate as substrate was applied. Anthracene and chrysene were highly crossreactive although ACY, B[ghi]P, and D[ah]A showed no crossreactivity.

VIII. FUTURE TRENDS

A. MULTIDIMENSIONAL SEPARATION TECHNIQUES

In a technique referred to as GC × GC or two-dimensional GC, direct coupling of two columns with different selectivities allows for a continuous separation of closely eluting or coeluting peaks; therefore, it readily separates PAH isomers and many other species that coelute under ordinary capillary column conditions. A GC × GC–MS system has been described where the first separation was on a 20 m × 0.25 mm i.d. 5% phenyl dimethylpolysiloxane column with 0.25 μm film thickness. Eluate from the first column passed through a cryogenic modulator onto a 1.2 m × 0.1 mm i.d. polar 14% cyanopropyldiphenylmethylpolysiloxane column with 0.1 μm film thickness.[298] The modulation time was 5 sec with He carrier gas at a constant flow for both systems. Samples of urban aerosols were collected on glass fiber filters with a high-volume sampler in Helsinki, Finland. After sonication in n-HEX:acetone (1:1), filter extracts were cleaned up and fractionated on a silver-impregnated silica column. Reported levels of PAH and OPAH ranged from 0.5 to 5.5 mg m^{-3} during winter. The quadrupole MS scan rate was adequate for accurate identification of 23 unknowns including methylfluorene, trimethylnaphthalene, and methylpyrene. Quantitation was done by integration of peak volumes. PYR levels were high, >5 ng m^{-3}, indicating that combustion sources were contributing substantially to the PAH loading. Another report of GC × GC-FID analysis of PAHs extracted from soils and sediments by PLE used almost identical conditions as were described in the previous reference.[299] The first column was of low polarity and the second was liquid crystalline. Two-dimensional plots of retention times on column one versus retention times on column two showed particular zones of groupings of PAHs based on number of aromatic rings. In-cell cleanup during PLE was successful by adding silica gel to the outlet end of the extraction cell. When compared to GC–MS, GC × GC gave lower values for PYR and larger PAHs than GC–MS which may have resulted from the use of a greater number of internal standards in GC–MS. The liquid crystalline column also suffered from column bleed, lack of resolution of the benzofluoroanthene isomers, and a narrow operating temperature range. Coupling of a nonpolar polydimethylsiloxane phase with a polar14% cyanopropylphenyl column or a chiral γ-cyclodextrin column was shown to resolve hundreds of individual compounds in a previously unresolved chromatographic "hump" from aged petroleum-contaminated sediment extracts.[202] PAHs, PAH derivatives and degradates, alkanes, cycloalkanes, and monoaromatics in the extract were fractionated on silica gel and silver-impregnated silica gel prior to GC × GC–MS analysis.

A two-dimensional HPLC (HPLC × HPLC) system has been described that was able to separate alkanes, alkylbenzenes, PAHs, and NPAHs in gasoline exhaust particulate matter extracts.[300] Of the 12 different stationary phases evaluated, a 15 cm × 4.6 mm i.d. pentabromobenzyl (PBB) column was coupled with two 5 cm × 4.6 mm i.d. C$_{18}$ columns connected in parallel. A switching valve directed effluent from the first column alternately every 0.2 min to either of the second columns. UV absorption at 254 nm was monitored for 0.2 min alternately from each of the C$_{18}$ columns to generate two-dimensional chromatograms. One hundred and fifty 2-D chromatograms were collected for every 30 min run on the first column. B[a]P was not completely resolved on this system, although good resolution was achieved for PAHs in general from the alkanes thus eliminating sample cleanup steps. Another two-dimensional HPLC system has been described where a 25 cm × 4.6 mm i.d. C$_{18}$ column was coupled with a 15 cm × 4.6 mm i.d. PBB column. This system was used to separate 1-nitropyrene in diesel and gasoline exhaust with fluorescence detection.[301] No sample cleanup was required and limits of detection were in the range of 0.01 to 0.3 ng ml^{-1}. Emission levels were determined to be 3.0 and 0.02 μg km^{-1} for diesel and gasoline engines, respectively.

Online LC–GC-ITD-MS was used to identify alkyl-PAH, OPAH, and NPAH in supercritical fluid extracts of particulate matter on quartz filters, collected in DCM, and solvent exchanged into

HEX for injection onto a normal-phase 10 mm × 2 mm i.d. silica HPLC column.[302] The LC mobile phase consisted of n-pentane/DCM. A UV detector monitored absorption of the eluent at 245 nm to allow for selection of the aromatic fraction onto a deactivated capillary precolumn where LC mobile phase was evaporated. The GC separation was performed on a 25 m × 0.32 mm i.d. fused-silica capillary column coated with dimethylsiloxane. Ion-trap mass spectrometry served as the detector. Compounds appeared to be separated, based on polarity, since some of the alkylated PAHs coeluted with the unsubstituted PAHs and two-ring NPAHs coeluted with large PAHs such as coronene. Totally automatic analysis of PAHs in particulate matter was reported using SFE-LC–GC–MS.[303] Sections of filter containing particulates were extracted with CO_2 and the extracts were collected on an ODS solid trap. When extraction was complete, the trap was rinsed onto a 10 cm × 2.1 mm i.d. normal-phase cyano HPLC column with n-pentane/ethyl acetate as the mobile phase. The fraction containing PAHs was directed on to a 10 m × 0.53 mm i.d. diphenyltetramethyldisilazane (DPTMDS) deactivated precolumn and then on to a 20 m × 0.25 mm 5% dimethylpolysiloxane fused-silica capillary column with 0.25 μm film thickness. Complete separation of isomers was not attained on this system, yet LODs were in the sub ng m^{-3} range. Total measured PAH concentrations for 17 PAHs ranged from 0.81 to 5.68 ng m^{-3}.

B. DIRECT ANALYSIS

Direct laser desorption ionization of aerosols (<2.5 μm diameter) collected on Teflon® filters and time-of-flight (TOF) mass spectrometry proved to be a useful screening tool for PAHs and NPAHs in particulate matter collected in the summer at a bus terminal.[304] PAHs were detected at 10 ng m^{-3} of air using positive-ion TOF-MS and NPAHs were detected at levels of <100 pg m^{-3} by negative-ion TOF-MS. Two-step laser mass spectrometry (L2MS) desorbs neutral molecules from a surface with a pulsed infrared laser and then ionizes the desorbed molecules with a pulse from a tunable UV laser for resonance-enhanced multiphoton ionization (REMPI).[305] This soft ionization prevents fragmentation of the analytes. The delay of a few microseconds between both lasers is optimized for maximum signal intensities. When combined with TOF-MS, PACs and other organic compounds can be characterized directly in particulate samples without sample preparation. L2MS was extended to the analysis of ng l^{-1} levels of PAHs in wastewater samples extracted by adsorption onto a PVC membrane.[306] Analysis was by mass only because isomers were not differentiated. PHN and alkylated phenanthrenes were the most abundant PAHs detected. PAH levels increased by an order of magnitude in the primary sedimentation basin directly after a rainstorm indicating their presence in runoff from streets. L2MS has been combined with a rotating drum impactor (RDI), which allowed for the sampling and analysis of PAHs in ambient aerosols every 20 min over a three-day period.[307] Samples were collected on strips of aluminum foil inside the RDI. Real-time chemical characterization of PAHs in diesel engine exhaust particles as a function of particle size using tandem MS[308] and direct analysis of PAHs on soil surfaces using real-time laser desorption/ionization MS analysis have been demonstrated.[309] Both methods demonstrate the potential of doing direct analysis of PAHs without sample preparation and could prove to be powerful screening tools.

Photoionization aerosol mass spectrometry (PIAMS) is a recently described technique that can analyze aliphatic and aromatic compounds in ambient aerosol particulates at the low picogram range.[310] Particles were collected on a probe near the source region of the mass spectrometer. A pulsed infrared laser irradiates the probe to vaporize the sample which is then ionized by a vacuum UV laser for soft ionization. The viability of the method was tested by sampling emissions from a diesel-powered bus, a gasoline engine, a wood stove, cooking fires, and cigarettes. The sensitivity of the method must be improved, however, for the detection of PAHs at normal, ambient levels.

C. Multicomponent Analysis

Multicomponent analysis methods are continually being developed and refined, although they are still only recommended as potential screening methods. Multiparameter measurement allows for the direct analysis of PAHs in complex matrices. Parameters including the excitation wavelength, the emission wavelength, and the fluorescence lifetime can be measured and evaluated with chemometrics to deconvolute complicated fluorescence spectra of mixtures, although very few examples of their application to real environmental samples have been reported to date. Fluorescence lifetimes for PAHs are characteristic, so they can provide information on the identity on particular PAHs, but they must be determined in each particular case due to their dependence on the surrounding environment.

Synchronous fluorescence spectroscopy (SFS) is performed by synchronously scanning both the excitation and emission wavelengths with a constant wavelength offset. The offset chosen is normally the difference between the λ_{max} in the absorption and emission, or a mean value for a mixture of PAHs. Two monochromators scanning at the same rate will produce a constant wavelength interval, $\Delta\lambda$. Emission spectra are greatly simplified with narrower spectral bands with the intensity directly proportional to concentration. SFS has been used for the direct analysis of spiked water samples containing NAP, PHN, ACE, and PYR.[311] Sorption was on to an 8-mm thick OV1 polydimethylsiloxane block immersed in the solution for 50 min. The offset, $\Delta\lambda$, was 70 nm with estimated detection limits of 0.2 parts per billion. An SFS method was developed to detect B[a]P, CHY, and FLU in a mixture using 1-cm quartz cuvettes. LODs were estimated to be in the ng l^{-1} range.[312] A screening method for the analysis of the EU_6 PAHs, in waters from Galicia, Spain called for a Xenon discharge lamp, quartz cuvettes, and a $\Delta\lambda$ of 120 nm.[313] One liter of water was extracted with HEX and concentrated. PAHs were detected in only one-third of the 404 samples that were analyzed and at levels that were well below the 200 mg l^{-1} allowable limit. The LOD was 6 ng l^{-1} and the average recovery for spiked samples was 94%. Chemometric techniques are necessary for resolving the spectra of multicomponent fluorescence methods. Synchronous fluorescence spectra of a mixture of ten PAHs in spiked water samples used partial least-squares regression (PLSR).[314] An aqueous micellar medium of 40 CMC Brij-35, polyoxyethylenelaurlether was added to increase the fluorescence signal and a $\Delta\lambda$ of 50 and 100 nm were measured. Recoveries for most PAHs were 80 to 120% but FLU and NAP were quite low, 7 to 81%. The second derivative of the synchronous fluorescence spectrum can yield lower LODs and increase precision.[315] To optimize, each component was scanned using its optimum $\Delta\lambda$ and LODs were in the $\mu g\,l^{-1}$ range. Second-order constant energy synchronous luminescence (SDCESL) and constant wavelength synchronous luminescence (CWSL) with $\Delta\lambda$ at 140 nm were applied to the analysis of riverwater samples in Spain.[316] Samples were extracted with HEX and a xenon discharge lamp was used as the excitation source. The second derivatives at 260 and 264 nm were analyzed. No B[a]P was detected in any of the samples analyzed.

Time-resolved laser-induced fluorescence spectroscopy (TR-LIF) has been used to measure PAHs in Boston Harbor sediments at ng l^{-1} levels.[317] Water column PAH concentrations may be a better indicator of bioavailability of PAHs in aquatic environments. The fluorescence of humic substances in aquatic environments can interfere with PAH fluorescence since humic substances are present in much greater amounts. Time-resolved measurements can distinguish between PAHs and humic substances because PAHs have much longer fluorescence lifetimes and greater fluorescence quantum efficiencies. Also PAHs have characteristic emission spectra in the 360 to 420 nm range while those of humic substances are less well defined. A pulsed nitrogen laser was used as the excitation source with fiber optic probes to obtain a 8 ng l^{-1} LOD for PYR. SPE of water samples on ODS membranes concentrated samples and provided a surface for the enhancement of the fluorescence signal of PAHs.[318] Excitation by pulsed Nd:YAG pumped tunable dye laser provided high-excitation energy with narrow bandwidth. Direct determination of B[a]P in a complex mixture with high organic matter content and 13 other PAHs was accomplished with LODs estimated at the parts-per-trillion level. The resolution was not as good as with the Shpol'skii technique described below,[319] but no cryogenic equipment or organic reagents were required.

EEMF is also known as total fluorescence spectroscopy and is a surface plot of emission intensities at all excitation and emission wavelengths. The excitation wavelength is held constant while the emission wavelength is scanned. A whole series of scans at increasing excitation wavelengths will generate the necessary three-dimensional data. A characteristic contour plot is created for individual PAHs. Combined with multidimensional calibration, resolution of ten PAHs was achieved successfully in a method with potential for screening of water samples.[320] Algorithms were necessary for the resolution of the data for spiked tap and mineral waters but spectra could be obtained in 2 min. Sub parts-per-billion detection levels for the EPA$_{16}$, except ACY, were measured in water samples extracted with an ODS membrane, eluted with HEX, and analyzed by laser excited time-resolved Shpol'skii spectrometry.[319] Analysis time was 5 min per sample with one end of a bifurcated optical fiber probe frozen into the sample matrix by liquid nitrogen. For detection of individual PAHs, the length of the solvent molecule is selected to match the effective length of the PAH to permit the greatest interaction between the guest and host molecules. Solvents included n-pentane, HEX, n-heptane, and n-octane. Direct analysis of B[a]P was done on petroleum refinery wastewater using a cryogenic fiber optic probe and EEMF of the octane extract layer.[321] Pulsed tunable dye laser excitation produced spectra of excitation wavelength, emission wavelength, and fluorescence lifetime in 8 min per sample.

In addition to increased isomer and derivative separation, real-time measurement, and direct measurement, future developments will continue to focus on source apportionment, bioavailability, and the analysis of the more toxic PAH degradates and derivatives.

REFERENCES

1. Zander, M., Physical and chemical properties of polycyclic aromatic hydrocarbons, In *Handbook of Polycyclic Aromatic Hydrocarbons*, Bjorseth, A., Ed., Marcel Dekker, New York, pp. 1–25, 1983.
2. Vo-Dinh, T., Monitoring and characterization of polyaromatic compounds in the environment, *Talanta*, 47, 943–969, 1998.
3. United States Environmental Protection Agency, Toxics criteria for those states not complying with the Clean Water Act section 303(c)(2)(B), 40 CFR 131.36, pp. 531–538, 1995.
4. Council of European Communities, Directives 75/440/EEC, 79/869/EEC, and 80/778/EEC.
5. *Off. J. Eur. Commun.* L330/32, p. 11, 1998.
6. Harvey, R. G., *Polycyclic Aromatic Hydrocarbons*, Wiley-VCH, New York, pp. 8–11, 1997.
7. Janoszka, B., Warzecha, I., Blaszczyk, U., and Bodzek, D., *Acta Chromatogr.*, 14, 115–128, 2004.
8. Norin, M. and Stromvall, A. M., Leaching of organic contaminants from storage of reclaimed asphalt pavement, *Environ. Technol.*, 25, 323–340, 2004.
9. Becker, L., Matuschek, G., Lenoir, D., and Kettrup, A., Leaching behavior of wood treated with creosote, *Chemosphere*, 42, 301–308, 2001.
10. Pozzoli, L., Gilardoni, S., Perrone, M. G., De Gennaro, G., De Rienzo, M., and Vione, D., Polycyclic aromatic hydrocarbons in the atmosphere: monitoring, sources, sinks and fate. I: monitoring and sources, *Annali di Chimica*, 94, 17–32, 2004.
11. Jiang, C. Q., Alexander, R., Kagi, R. I., and Murray, A. P., Origin of perylene in ancient sediments and its geological significance, *Org. Geochem.*, 31, 1545–1559, 2000.
12. MacKay, D., Shiu, W. Y., and Ma, K. C., *Chemicals, Illustrated Handbook of Physical–Chemical Properties and Environmental Fate for Organic*, Vol. II, Lewis, Boca Raton, pp. 246–252, 1992.
13. Sverdrup, L. E., Nielsen, T., and Krogh, P. H., Soil ecotoxicity of polycyclic aromatic hydrocarbons in relation to soil sorption, lipophilicity, and water solubility, *Environ. Sci. Technol.*, 36, 2429–2435, 2002.
14. Gundel, L. A., Lee, V. C., Mahanama, K. R. R., Stevens, R. K., and Daisey, J. M., Direct determination of the phase distributions of semi-volatile polycyclic aromatic hydrocarbons using annular denuders, *Atmos. Environ.*, 29, 1719–1733, 1995.

15. Chin, Y.-P., Aiken, G. R., and Danielsen, K. M., Binding of pyrene to aquatic and commercial humic substances: the role of molecular weight and aromaticity, *Environ. Sci. Technol.*, 31, 1630–1635, 1997.

16. Krauss, M. and Wicke, W., Predicting soil-water partitioning of polycyclic aromatic hydrocarbons and polychlorinated biphenyls by desorption with methanol–water mixtures at different temperatures, *Environ. Sci. Technol.*, 35, 2319–2325, 2001.

17. World Health Organization. *Guidelines for drinking-water quality, Vol. 1. Recommendations*: World Health Organization, Geneva, pp. 47–102, 1984.

18. Fernández, P., Carrera, G., Grimalt, J. O., Ventura, M., Camarero, L., Catalan, J., Nickus, U., Thies, H., and Psenner, R., Factors governing the atmospheric deposition of polycyclic aromatic hydrocarbons to remote areas, *Environ. Sci. Technol.*, 37, 3261–3267, 2003.

19. Fromme, H., Oddoy, A., Piloty, M., Krause, M., and Lahrz, T., Polycyclic aromatic hydrocarbons (PAH) and diesel engine emission (elemental carbon) inside a car and a subway train, *Sci. Total Environ.*, 217, 165–173, 1998.

20. Swartz, E., Stockburger, L., and Vallero, D. A., Polycyclic aromatic hydrocarbons and other semivolatile organic hydrocarbons collected in New York city in response to the events of 9/11, *Environ. Sci. Technol.*, 37, 3537–3546, 2003.

21. Wilcke, W., Krauss, M., and Amelung, W., Carbon isotope signature of polycyclic aromatic hydrocarbons (PAHs): evidence for different sources in tropical and temperate environments? *Environ. Sci. Technol.*, 36, 3530–3535, 2002.

22. Hawthorne, S. B., Poppendieck, D. G., Grabanski, G. B., and Loehr, R. C., PAH release during water desorption, supercritical carbon dioxide extraction, and field bioremediation, *Environ. Sci. Technol.*, 35, 4577–4583, 2001.

23. Yu, H., Environmental carcinogenic polycyclic hydrocarbons: photochemistry and phototoxicity, *J. Environ. Sci. Health Part C — Environ. Carcinog. Ecotoxicol. Rev.*, C20(2), 149–183, 2002.

24. Mrozik, A., Piotrowska-Seget, Z., and Labuzek, S., Bacterial degradation and bioremediation of polycyclic aromatic hydrocarbons, *Pol. J. Environ. Stud.*, 12, 15–25, 2003.

25. Esteve, W., Budzinski, H., and Villenave, E., Heterogeneous reactivity of OH radicals with phenenthrene, *Polycyclic Aromat. Compd.*, 23, 441–456, 2003.

26. Sasaki, J., Arey, J., and Harger, W. P., Formation of mutagens from the photooxidations of 2-4-ring PAH, *Environ. Sci. Technol.*, 29, 1324–1335, 1996.

27. Reisen, F. and Arey, J., Reactions of hydroxyl radicals and ozone with acenaphthene and acenaphthylene, *Environ. Sci. Technol.*, 36, 4302–4311, 2002.

28. Allen, J. O., Dookeran, N. M., Lafleur, A. L., Smith, K. A., and Sarofim, A. F., Measurement of oxygenated polycyclic aromatic hydrocarbons associated with a size-segregated urban aerosol, *Environ. Sci. Technol.*, 31, 2064–2070, 1997.

29. Durant, J. L., Busby, W. F. Jr., Lafleur, A. L., Penman, B. W., and Crespi, C. L., Human cell mutagenicity of oxygenated, nitrated and unsubstituted polycyclic aromatic hydrocarbons associated with urban aerosols, *Mutat. Res.*, 371, 123–157, 1996.

30. Hatch, A. C. and Burton, G. A., Phototoxicity of fluoranthene to two freshwater crustaceans, *Hyalella azteca* and *Daphnia magna*: measured of feeding inhibition as a toxicological endpoint, *Hydrobiologia*, 400, 243–248, 1999.

31. Diamond, S. A., Milroy, N. J., Mattson, V. R., Heinis, L. J., and Mount, D. R., *Environ. Toxicol. Chem.*, 22, 2752–2760, 2003.

32. United States Clean Air Act. U.S. CFR 42 (85)(I)(C)(i)(7473)(b)(2), (2003).

33. Council Directive 1999/30/EC, Relating to limit values for sulphur dioxide, oxides of nitrogen, particulate matter and lead in ambient air, *Off. J. Eur. Comm.* L163/41 of 29.6.1999.

34. Parsons, B. and Salter, L. F., Air quality effects of traffic in a canyon-like street (Falmouth, U.K.), *Environ. Monit. Assess.*, 82, 63–73, 2003.

35. European Union Working Group on Polycyclic Aromatic Hydrocarbons, Ambient Air Pollution by Polycyclic Aromatic Hydrocarbons (PAH), Position Paper, July 27, 2001.

36. WHO. *Air Quality Guidelines for Europe 2001*, WHO Regional Publication, World Health Organization, Geneva, 2001.

37. WHO. *International Programme on Chemical Safety*, Environmental Health Criteria 202. Selected non-heterocyclic polycyclic aromatic hydrocarbons, World Health Organization, Geneva, 2001.

38. World Health Organization, *Guidelines for drinking water quality — recommendations*, Vol. 1, World Health Organization, Geneva, p. 66, 1984.

39. *Off. J. Eur. Commun.* L330/32, p. 1, 1998.

40. USEPA, National primary drinking water regulations, 40 CFR Pt 141.61, 2002.

41. USEPA, Hazardous Constituents, 40 CFR Pt 261, App VIII, 2000.

42. USEPA, Ground-water Monitoring List, 40 CFR Pt 264, App IX, 2000.

43. USEPA. Clean Water Act section 301, 306, OR 402(A)(1), 40 CFR Pt 131, 2000.

44. LaGoy, P. K. and Quirk, T. C., Establishing generic remediation goals for the polycyclic aromatic hydrocarbons: critical issues, *Environ. Health Perspect.*, 102, 348–352, 1994.

45. Miége, C., Bouzige, M., Nicol, S., Dugay, J., Pichon, V., and Hennion, M. C., Selective immunoclean-up followed by liquid or gas chromatography for the monitoring of polycyclic aromatic hydrocarbons in urban waste water and sewage sludges used for soil amendment, *J. Chromatogr. A*, 859, 29–39, 1999.

46. United States Environmental Protection Agency, Methods for the determination of organic compounds in drinking water — Supplement 1, EPA-600/4-90/020. USEPA Office of Research and Development, Cincinnati, OH, pp. 143–168, 1990.

47. Greenberg, A. E., Clesceri, L. S., and Eaton, A. D., Standard methods for the analysis of water and waste water, In *American Water Works Association, Water Environment Federation, Method 6440 Polynuclear aromatic hydrocarbons*, 18th ed., American Public Health Association, Washington, pp. 6/96–6/101, 1992.

48. Sliwka-Kaszynska, M., Kot-Wasik, A., and Namiesnik, J., Preservation and storage of water samples, *Crit. Rev. Environ. Sci. Technol.*, 33, 31–44, 2003.

49. Rost, H., Loibner, A. P., Hasinger, M., Braun, R., and Szolar, O. H. J., *Chemisphere*, 49, 1239–1246, 2002.

50. McMurry, P. H., A review of atmospheric aerosol measurements, *Atmos. Environ.*, 34, 1959–1999, 2000.

51. Volckens, J. and Leith, D., Effects of sampling bias on gas-particle partitioning of semi-volatile compounds, *Atmos. Environ.*, 37, 3385–3393, 2003.

52. Tsapakis, M. and Stephanou, E. G., Collection of gas and particle semi-volatile organic compounds: use of an oxidant denuder to minimize polycyclic aromatic hydrocarbons degradation during high-volume air sampling, *Atmos. Environ.*, 37, 4935–4944, 2003.

53. Schauer, C., Niessner, R., and Pöschl, U., Polycyclic aromatic hydrocarbons in urban air particulate matter: decadal and seasonal trends, chemical degradation, and sampling artifacts, *Environ. Sci. Technol.*, 37, 2861–2868, 2003.

54. Piñeiro-Iglesias, M., López-Mahía, P., Vázquez-Blanco, E., Muniategui-Lorenzo, S., and Prada-Rodríguez, D., Problems in the extraction of polycyclic aromatic hydrocarbons from diesel particulate matter, *Polycyclic Aromat. Compd.*, 22, 129–146, 2002.

55. Turpin, B. J., Saxena, P., and Andrews, E., Measuring and simulating particulate organics in the atmosphere: problems and prospects, *Atmos. Environ.*, 34, 2983–3013, 2000.

56. United States Environmental Protection Agency Method TO-13A. 625/R-96/010b, Determination of polycyclic aromatic hydrocarbons (PAHs) in ambient air using gas chromatography/mass spectrometry (GC/MS), USEPA Office of Research and Development, Cincinnati, OH, pp. 1–78, 1999.

57. Fraser, M. P., Cass, G. R., Simoneit, B. R. T., and Rasmussen, R. A., Air quality model evaluation data for organics. 5. C-6-C-22 nonpolar and semipolar aromatic compounds, *Environ. Sci. Technol.*, 32, 1760–1770, 1998.

58. Krieger, M. S. and Hites, R. A., Measurement of polychlorinated biphenyls and polycyclic aromatic hydrocarbons in air with a diffusion denuder, *Environ. Sci. Technol.*, 28, 1129–1133, 1994.

59. Kloskowski, A., Pilarczyk, M., and Namiesnik, J., Denudation — A convenient method of isolation and enrichment of analytes, *Crit. Rev. Anal. Chem.*, 32, 301–335, 2002.

60. Gundel, L. A., Lee, V. C., Mahanama, K. R. R., Stevens, R. K., and Daisey, J. M., Direct determination of the phase distributions of semi-volatile polycyclic aromatic hydrocarbons using annular denuders, *Atmos. Environ.*, 29, 1719–1733, 1995.

61. Volckens, J. and Leith, D., Comparison of methods for measuring gas-particle partitioning of semivolatile compounds, *Atmos. Environ.*, 37, 3177–3188, 2003.

62. Volckens, J. and Leith, D., Electrostatic sampler for semivolatile aerosols: chemical artifacts, *Environ. Sci. Technol.*, 36, 4608–4612, 2002.
63. Volckens, J. and Leith, D., Filter and electrostatic samplers for semivolatile aerosols: physical artifacts, *Environ. Sci. Technol.*, 36, 4613–4617, 2002.
64. Peters, A. J., Lane, D. A., Gundel, L. A., Northcott, G. L., and Jones, K. C., A comparison of high volume and diffusion denuder samplers for measuring semivolatile organic compounds in the atmosphere, *Environ. Sci. Technol.*, 34, 5001–5006, 2000.
65. Castells, P., Santos, F. J., and Galceran, M. T., Development of a sequential supercritical fluid extraction method for the analysis of nitrated and oxygenated derivatives of polycyclic aromatic hydrocarbons in urban aerosols, *J. Chromatogr. A*, 1010, 141–151, 2003.
66. Swartz, E. and Stockburger, L., Recovery of semivolatile organic compounds during sample preparation: implications for characterization of airborne particulate matter, *Environ. Sci. Technol.*, 37, 597–605, 2003.
67. Sauvain, J.-J., Vu Duc, T., and Guillemin, M., Exposure to carcinogenic polycyclic aromatic compounds and health risk assessment for diesel-exhaust exposed workers, *Int. Arch. Occup. Environ. Health*, 76, 443–455, 2003.
68. Kavouras, I. G., Lawrence, J., Koutrakis, P., Stephanou, E. G., and Oyola, P., Measurement of particulate aliphatic and polynuclear aromatic hydrocarbons in Santiago do Chile: source reconciliation and evaluation of sampling artifacts, *Atmos. Environ.*, 33, 4977–4986, 1999.
69. Shimmo, M., Anttila, P., Hartonen, K., Hyötyläinen, T., Paatero, J., Kulmala, M., and Riekkola, M.-L., Identification of organic compounds in atmospheric aerosol particles by on-line supercritical fluid extraction-liquid chromatography–gas chromatography-mass spectrometry, *J. Chromatogr. A*, 1022, 151–159, 2004.
70. Feilberg, A., Kamens, R. M., Strommen, M. R., and Nielsen, T., Modeling the formation, decay, and partitioning of semivolatile nitro-polycyclic aromatic hydrocarbons (nitronaphthalenes) in the atmosphere, *Atmos. Environ.*, 33, 1231–1243, 1999.
71. Kuusimäki, L., Peltonen, Y., Kyyrö, E., Mutanen, P., Peltonen, K., and Savela, K., Exposure of garbage truck drivers and maintenance personnel at a waste handling center to polycyclic aromatic hydrocarbons derived from diesel exhaust, *J. Environ. Monit.*, 4, 722–727, 2002.
72. Väänänen, V., Hämeilä, M., Kontsas, H., Peltonen, K., and Heikkilä, P., Air concentrations and urinary metabolites of polycyclic aromatic hydrocarbons among paving and remixing workers, *J. Environ. Monit.*, 5, 739–746, 2003.
73. Fang, G.-C., Wu, Y.-S., Fu, P., Yang, I.-L., and Chen, M.-H., Polycyclic aromatic hydrocarbons in the ambient air of suburban and industrial regions of central Taiwan, *Chemosphere*, 54, 443–452, 2004.
74. Ötvös, E., Kozák, I. O., Fekete, J., Sharma, V. K., and Tuba, Z., Atmospheric deposition of polycyclic aromatic hydrocarbons (PAHs) in mosses (*Hypnum cupressiforme*) in Hungary, *Sci. Total Environ.*, 330, 89–99, 2004.
75. Tremolada, P., Burnett, V., Calamari, D., and Jones, K. C., Spatial distribution of PAHs in the U.K. atmosphere using pine needles, *Environ. Sci. Technol.*, 30, 3570–3577, 1996.
76. Lehndorff, E. and Schwark, L., Biomonitoring of air quality in the cologne conturbation using pine needles as a passive sampler–Part II: polycyclic aromatic hydrocarbons (PAH), *Atmos. Environ.*, 38, 3793–3808, 2004.
77. Soderstrom, H. S. and Bergqvist, P. A., Polycyclic aromatic hydrocarbons in semiaquatic plant and semipermeable membrane devices exposed to air in Thailand, *Environ. Sci. Technol.*, 37, 47–52, 2003.
78. Lohmann, R., Corrigan, B. P., Howsam, M., Jones, K. C., and Ockenden, W. A., Further developments in the use of semipermeable membrane devices (SPDMs) as passive air samplers for persistent organic pollutants: field application in a spatial survey of PCDD/Fs and PAHs, *Environ. Sci. Technol.*, 35, 2576–2582, 2001.
79. Naes, K., Axelman, J., Näf, C., and Broman, D., Role of soot carbon and other carbon matrices in the distribution of PAHs among particles, DOC, and the dissolved phase in the effluent and recipient waters of an aluminum reduction plant, *Environ. Sci. Technol.*, 32, 1786–1792, 1998.
80. Grynkiewicz, M., Polkowska, Z., Zygmunt, B., and Namiesnik, J., Atmospheric precipitation for analysis, *Pol. J. Environ. Stud.*, 12, 133–140, 2003.

81. López García, A., Blanco Gonzalez, E., García Alonso, J. I., and Sanz-Medel, A., Potential of micelle-mediated procedures in the sample preparation steps for the determination of polynuclear aromatic hydrocarbons in waters, *Anal. Chim. Acta*, 264, 241–248, 1992.
82. U.S. Environmental Protection Agency, Methods for organic chemical analysis of municipal and industrial wastewater, 40 CFR Appendix A to Pt. 136, pp. 1–24.
83. U.S. Environmental Protection Agency Test Method 3510C: Separatory Funnel Liquid–Liquid Extraction, Third revision, U.S. Government Printing Office, Washington, pp. 1–8, 1996.
84. Ogunffowokan, A. O., Asubiojo, O. I., and Fatoki, O. S., Isolation and determination of polycyclic aromatic hydrocarbons in surface runoff and sediments, *Water, Air, Soil Pollut.*, 147, 245–261, 2003.
85. Telli-Karakoc, F., Tolun, L., Henkelmann, B., Klimm, C., Okay, O., and Schramm, K.-W., Polycyclic aromatic hydrocarbons (PAHs) and polychlorinated buphenyls (PCBs) distributions in the Bay of Marmara sea: Izmit Bay, *Environ. Pollut.*, 119, 383–397, 2002.
86. Manoli, E., Samara, C., Konstantinou, I., and Albanis, T., Polycyclic aromatic hydrocarbons in the bulk precipitation and surface waters of Northern Greece, *Chemosphere*, 41, 1845–1855, 2000.
87. Garban, B., Blanchoud, H., Motelay-Massei, A., Chevreuil, M., and Ollivon, D., Atmospheric bulk deposition of PAHs into France: trends from urban to remote sites, *Atmos. Environ.*, 36, 5395–5403, 2002.
88. Smith, R., Before the injection — modern methods of sample preparation for separation techniques, *J. Chromatogr. A.*, 1000, 3–27, 2003.
89. U.S. Environmental Protection Agency Test Method 3535: Solid-Phase Extraction (SPE), U.S. Government Printing Office, Washington, 1–13, 1996.
90. Lacorte, S., Guiffard, I., Fraisse, D., and Barceló, D., Broad spectrum analysis of 109 priority compounds listed in the 76/464/CEE, council directive using solid-phase extraction and GC/EI/MS, *Anal. Chem.*, 72, 1430–1440, 2000.
91. Kiss, G., Varga-Puchony, Z., Gelencsér, A., Krivácsy, Z., Molnár, Á., and Hlavay, J., Survey of concentration of polycyclic aromatic hydrocarbons in Lake Balaton by HPLC with fluorescence detection, *Chromatographia*, 48, 149–153, 1998.
92. Grynkiewicz, M., Polkowska, Z., and Namiesnik, J., Determination of polycyclic aromatic hydrocarbons in bulk precipitation and runoff waters in an urban region (Poland), *Atmos. Environ.*, 36, 361–369, 2002.
93. Cereceda-Balic, F., Kleist, E., Prast, H., Schlimper, H., Engel, H., and Günther, K., Description and evaluation of a sampling system for long-time monitoring of PAHs wet deposition, *Chemosphere*, 49, 331–340, 2002.
94. Huang, J., Zhang, Z., and Yu, G., Occurrence of dissolved PAHs in the Jinsha River (Panzhihua) — upper reaches of the Yangtze River, southwest China, *J. Environ. Monit.*, 5, 604–609, 2003.
95. Brown, J. N. and Peake, B. M., Determination of colloidally-associated polycyclic aromatic hydrocarbons (PAHs) in fresh water using C_{18} solid phase extraction disks, *Anal. Chim. Acta*, 486, 159–169, 2003.
96. Hagestuen, E. D. and Campiglia, A. D., Near approach for screening polycyclic aromatic hydrocarbons in water samples, *Talanta*, 49, 547–560, 1999.
97. López García, A., Blanco Gonzalez, E., García Alonso, J. I., and Sanz-Medel, A., Potential of micelle-mediated procedures in the sample preparation steps for the determination of polynuclear aromatic hydrocarbons in waters, *Anal. Chim. Acta*, 264, 241–248, 1992.
98. Brouwer, E. R., Hermans, A. N. J., Lingeman, H., and Brinkman, U. A. Th., Determination of polycyclic aromatic hydrocarbons in surface water by column liquid chromatography with fluorescence detection, using on-line micelle-mediated sample preparation, *J. Chromatogr. A*, 669, 45–57, 1994.
99. Gimeno, R. A., Altelaar, A. F. M., Marcé, R. M., and Borrull, F., Determination of polycyclic aromatic hydrocarbons and polycyclic aromatic sulfur heterocycles by high-performance liquid chromatography with fluorescence and atmospheric pressure chemical ionization mass spectrometry detection in seawater and sediment samples, *J. Chromatogr. A*, 958, 141–148, 2002.
100. McElmurry, S. P. and Voice, T. C., Screening methodology for coal-derived organic contaminants in water, *Int. J. Environ. Anal. Chem.*, 84, 227–287, 2004.

101. Dimitrienko, S. G., Ya Gurariy, E., Nosov, R. E., and Zolotov, Yu A., Solid-phase extraction of polycyclic aromatic hydrocarbons from aqueous samples using polyurethane foams in connection with solid-matrix spectrofluorimetry, *Anal. Lett.*, 34, 425–438, 2001.

102. Dimitrienko, S. G., Shapovalova, E. N., Ya Gurarii, E., Kochetiva, M. V., Shpigun, O. A., and Zolotov, Yu A., Preconcentration of polycyclic aromatic hydrocarbons on polyurethane foams and their determination in waters with the use of luminescence and high-performance liquid chromatography, *J. Anal. Chem.*, 57, 1009–1016, 2002.

103. Liu, J. F., Chi, Y. G., Jiang, G. B., Tai, C., and Hu, J. T., Use of cotton as a sorbent for on-line column enrichment of polycyclic aromatic hydrocarbons in waters prior to liquid chromatography determination, *Microchem. J.*, 77, 19–22, 2004.

104. Akiyama, R., Takagai, Y., and Igarashi, S., Determination of lower sub ppt levels of environmental analytes using high-powered concentration system and high-performance liquid chromatography with fluorescence detection, *Analyst*, 129, 396–397, 2004.

105. Pichon, V., Bouzige, M., and Hennion, M.-C., New trends in environmental trace-analysis of organic pollutants: class-selective immunoextraction and clean-up in one step using immunosorbents, *Anal. Chim. Acta*, 376, 21–35, 1998.

106. U.S. Environmental Protection Agency Test Method 4035: Soil screening for polynuclear aromatic hydrocarbons by immunoassay, U.S. Government Printing Office, Washington, pp. 1–10, 1996.

107. Barceló, D., Oubina, A., Salau, J. S., and Perez, S., Determination of PAHs in river water samples by ELISA, *Anal. Chim. Acta*, 376, 49–53, 1998.

108. Bouzige, M., Machtalère, G., Legeay, P., Pichon, V., and Hennion, M.-C., Online coupling of immunosorbent and liquid chromatography analysis for the selective extraction and determination of polycyclic aromatic hydrocarbons in water samples at the ng l^{-1} level, *J. Chromatogr. A*, 823, 197–210, 1998.

109. Miege, C., Bouzige, M., Nicol, S., Dugay, J., Pichon, V., and Hennion, M. C., Selective immunoclean-up followed by liquid or gas chromatography for the monitoring of polycyclic aromatic hydrocarbons in urban waste water and sewage sludges used for soil amendment, *J. Chromatogr. A*, 859, 29–39, 1999.

110. Perez, S. and Barceló, D., Evaluation of anti-pyrene and anti-fluorenc immunosorbent clean-up for PAHs from sludges and sediment reference materials followed by liquid chromatograph and diode array detection, *Analyst*, 125, 1273–1279, 2000.

111. Cichna, M., Knopp, D., and Niessner, R., Immunoafinity chromatography of polycyclic aromatic hydrocarbons in columns prepared by the sol–gel method, *Anal. Chim. Acta*, 339, 241–250, 1997.

112. Ferrer, R., Beltrán, J. L., and Guiteras, J., Use of cloud point extraction methodology for the determination of PAHs priority pollutants in water samples by high-performance liquid chromatography with fluorescence detection and wavelength programming, *Anal. Chim. Acta*, 330, 199–206, 1996.

113. Sicilia, D., Rubio, S., Pérez-Bendito, D., Maniasso, N., and Zagatto, E. A. G., Anionic surfactants in acid media: a new cloud point extraction approach for the determination of polycyclic aromatic hydrocarbons in environmental samples, *Anal. Chim. Acta*, 392, 29–38, 1999.

114. Saitoh, T., Yoshida, Y., Matsudo, T., Fujiwara, S., Dobashi, A., Iwaki, K., Suzuki, Y., and Matsubara, C., Concentration of hydrophobic organic compounds by polymer-mediated extraction, *Anal. Chem.*, 71, 4506–4512, 1999.

115. Bai, D., Li, J., Chen, S. B., and Chen, B.-H., A novel cloup-point extraction process for preconcentrating selected polycyclic aromatic hydrocarbons in aqueous solution, *Environ. Sci. Technol.*, 35, 3936–3940, 2001.

116. Pino, V., Ayala, J. H., Afonso, A. M., and González, V., Determination of polycyclic aromatic hydrocarbons in seawater by high-performance liquid chromatography with fluorescence detection following micelle-mediated preconcentration, *J. Chromatogr. A*, 949, 291–299, 2002.

117. Huckins, J. N., Tubergen, M. W., and Manuweera, G. K., Semipermeable membrane devices containing model lipid: a new approach to monitoring the bioavailability of lipophilic contaminants and estimating their bioconcentration potential, *Chemosphere*, 20, 533–552, 1990.

118. Lebo, J. A., Zajicek, J. L., Huckins, J. N., Petty, J. D., and Peterman, P. H., Use of semipermeable membrane devices for *in situ* monitoring of polycyclic aromatic hydrocarbons in aquatic environments, *Chemosphere*, 25, 691–718, 1992.

119. Huckins, J. N., Petty, J. D., Orazio, C. E., Lebo, J. A., Clark, R. C., Gibson, V. L., Gala, W. R., and Echols, K. R., Determination of uptake kinetics (sampling rates) by lipid-containing semipermeable membrane devices (SPMDs) for polycyclic aromatic hydrocarbons (PAHs) in water, *Environ. Sci. Technol.*, 33, 3918–3923, 1999.

120. Crunkilton, R. L. and DeVita, W. M., Determination of aqueous concentrations of polycyclic aromatic hydrocarbons (PAHs) in an urban stream, *Chemosphere*, 35, 1447–1463, 1997.

121. Prest, H. F. and Jacobson, L. A., Passive water sampling for polynuclear aromatic hydrocarbons using lipid-containing semipermeable membrane devices (SPMDs): application to contaminant residence times, *Chemosphere*, 35, 3047–3063, 1997.

122. Moring, J. B. and Rose, D. R., Occurrence and concentrations of polycyclic aromatic hydrocarbons in semipermeable membrane devices and clams in three urban streams of the Dallas–Fort Worth metropolitan area, *Texas. Chemosphere*, 34, 551–566, 1997.

123. Verweij, F., Booij, K., Satumalay, K., van der Molen, N., and van der Oost, R., Assessment of bioavailable PAH, PCB, and OCP concentrations in water, using semipermeable membrane devices (SPMDs), sediments and caged carp, *Chemosphere*, 54, 1675–1689, 2004.

124. Wang, Y., Wang, Z., Ma, M., Wang, C., and Mo, Z., Monitoring priority pollutants in a sewage treatment process by dichloromethane extraction and triolein-semipermeable membrane device (SPMD), *Chemosphere*, 43, 339–346, 2001.

125. Luellen, D. R. and Shea, D., Calibration and field verification of semipermeable membrane devices for measuring polycyclic aromatic hydrocarbons in water, *Environ. Sci. Technol.*, 36, 1791–1797, 2002.

126. Pawliszyn, J. and Potter, D. W., Rapid-determination of polyaromatic hydrocarbons and polychlorinated-biphenyls in water using solid-phase microextraction and GCMS, *Environ. Sci. Technol.*, 28, 298–305, 1994.

127. Pörschmann, J., Kopinke, F.-D., and Pawliszyn, J., Solid-phase microextraction for determining the binding state of organic pollutants in contaminated water rich in humic organic matter, *J. Chromatogr. A*, 816, 159–167, 1998.

128. Doong, R. and Chang, S., Determination of distribution coefficients of priority polycyclic aromatic hydrocarbons using solid-phase microextraction, *Anal. Chem.*, 72, 3647–3652, 2000.

129. Havenga, W. J. and Rohwer, E. R., The determination of trace-level PAHs and diagnostic ratios for source identification in water samples using solid-phase microextraction and GC/MS, *Polycyclic Aromat. Compd.*, 22, 327–338, 2002.

130. Pino, V., Ayala, J. H., Afonso, A. M., and González, V., Micellar microwave-assisted extraction combined with solid-phase microextraction for the determination of polycyclic aromatic hydrocarbons in a certified marine sediment, *Anal. Chim. Acta*, 477, 81–91, 2003.

131. Baltussen, E., Sandra, P., David, F., and Cramers, C., Stir bar sorption extraction (SBSE), a novel extraction technique for aqueous samples: theory and principles, *J. Microcolumn Sep.*, 11, 737–747, 1999.

132. Popp, P., Bauer, C., and Wennrich, L., Application of stir bar sorptive extraction in combination with column liquid chromatography for the determination of polycyclic aromatic hydrocarbons in water samples, *Anal. Chim. Acta*, 436, 1–9, 2001.

133. Niehus, B., Popp, P., Bauer, C., Peklo, G., and Zwanziger, H., Comparison of stir bar sorptive extraction and solid phase extraction as enrichment techniques in combination with column liquid chromatography for the determination of polycyclic aromatic hydrocarbons in water samples, *Int. J. Anal. Chem.*, 82, 669–676, 2002.

134. León, V. M., Álvarez, B., Cobollo, M. A., Muñoz, S., and Valor, I., Analysis of 35 priority semivolatile compounds in water by stir bar sorptive extraction — thermal desorption — gas chromatography — mass spectrometry, *J. Chromatogr. A*, 999, 91–101, 2003.

135. Kolahgar, B., Hoffmann, A., and Heiden, A. C., Application of stir bar sorptive extraction to the determination of polycyclic aromatic hydrocarbons in aqueous samples, *J. Chromatogr. A*, 963, 225–230, 2002.

136. García-Falcón, M. S., Cancho-Grande, B., and Simal-Gándara, J., Stirring bar sorptive extraction in the determination of PAHs in drinking waters, *Water Res.*, 38, 1670–1684, 2004.

137. García-Falcón, M. S., Pérez-Lamela, C., and Simal-Gándara, J., Strategies for the extraction of free and bound polycyclic aromatic hydrocarbons in run-off waters rich in organic matter, *Anal. Chim. Acta*, 508, 177–183, 2004.

138. Vrana, B., Popp, P., Paschke, A., and Schüürmann, G., Membrane-enclosed sorptive coating, an integrative passive sampler for monitoring organic contaminants in water, *Anal. Chem.*, 73, 5191–5200, 2001.

139. Popp, P., Bauer, C., Paschke, A., and Montero, L., Application of a polysiloxane-based extraction method combined with column liquid chromatography to determine polycyclic aromatic hydrocarbons in environmental samples, *Anal. Chim. Acta*, 504, 307–312, 2004.

140. Law, R. J., Kelly, C. A., Baker, K. L., Langford, K. H., and Bartlett, T., Polycyclic aromatic hydrocarbons in sediments, mussels and crustacea around a former gasworks site in Shoreham-by-Sea, U.K, *Mar. Pollut. Bull.*, 44, 903–911, 2002.

141. U.S. Environmental Protection Agency Test Method 3540C: Soxhlet Extraction, Third revision, U.S. Government Printing Office, Washington, pp. 1–8, 1996.

142. Luque de Castro, M. D. and García-Ayuso, L. E., Soxhlet extraction of solid materials: an outdated technique with a promising innovative future, *Anal. Chim. Acta*, 369, 1–10, 1998.

143. Johansson, I. and van Bavel, B., Levels and patterns of polycyclic aromatic hydrocarbons in incineration ashes, *Sci. Total Environ.*, 311, 221–231, 2003.

144. García-Ayuso, L. E., Luque-García, J. L., and Luque de Castro, M. D., Approach for independent-matrix removal of polycyclic aromatic hydrocarbons from solid samples based on microwave-assisted Soxhlet extraction with on-line fluorescence monitoring, *Anal. Chem.*, 72, 3627–3634, 2000.

145. Szolar, O. H., Rost, H., Braun, R., and Loibner, A. P., Analysis of polycyclic aromatic hydrocarbons in soil: minimizing sample pretreatment using automated Soxhlet with ethyl acetate as extraction solvent, *Anal. Chem.*, 74, 2379–2385, 2002.

146. Gfrerer, M., Serschen, M., and Lankmayr, E., Optimized extraction of polycyclic aromatic hydrocarbons from contaminated soil samples, *J. Biochem. Biophys. Methods*, 53, 203–216, 2002.

147. U.S. Environmental Protection Agency Test Method 3550B: Ultrasonic Extraction. U.S. Government Printing Office, Washington, pp. 1–14, 1996.

148. Sun, F., Littlejohn, D., and Gibson, M. D., Ultrasonication extraction and solid phase extraction clean-up for determination of U.S. EPA 16 priority pollutant polycyclic aromatic hydrocarbons in solids by reversed-phase liquid chromatography with ultraviolet detection, *Anal. Chim. Acta*, 364, 1–11, 1998.

149. Wilcke, W., Zech, W., and Kobza, J., PAH-pools in soils along a PAH-deposition gradient, *Environ. Pollut.*, 92, 307–313, 1996.

150. Guerin, T. F., The extraction of aged polycyclic aromatic hydrocarbon (PAH) residues from a clay soil using sonication and a Soxhlet procedure: a comparative study, *J. Environ. Monit.*, 1, 63–67, 1999.

151. Priego-Lopez, E. and Luque de Castro, M. D., Ultrasound-assisted extraction of nitropolycyclic aromatic hydrocarbons from soil prior to gas chromatography-mass detection, *J. Chromatogr. A*, 1018, 1–6, 2003.

152. Woodhead, R. J., Law, R. J., and Matthiessen, P., Polycyclic aromatic hydrocarbons in surface sediments around England and Wales, and their possible biological significance, *Mar. Pollut. Bull.*, 38, 773–790, 1999.

153. Webster, L., Fryer, R. J., Dalgarno, E. J., Megginson, C., and Moffat, C. F., The polycyclic aromatic hydrocarbon and geochemical biomarker composition of sediments from voes and coastal areas in the Shetland and Orkney Islands, *J. Environ. Monit.*, 3, 591–601, 2001.

154. Pérez, S., Guillamón, M., and Barceló, D., Quantitative analysis of polycyclic aromatic hydrocarbons in sewage sludge from wastewater treatment plants, *J. Chromatogr. A*, 938, 57–65, 2001.

155. European Union Draft Directive on Sewage Sludge. Brussels, 27/04/2000, pp. 1–20.

156. Pérez, S., la Farré, M., García, M. J., and Barceló, D., Occurrence of polycyclic aromatic hydrocarbons in sewage sludge and their contribution to its toxicity in the ToxAlert® 100 bioassay, *Chemosphere*, 45, 705–712, 2001.

157. Baran, S. and Oleszczuk, O., The concentration of polycyclic aromatic hydrocarbons in sewage sludge in relation to the amount and origin of purified sewage, *Pol. J. Environ. Stud.*, 12, 523–529, 2003.

158. Merino, F., Rubio, S., and Pérez-Benito, D., Acid-induced cloud point extraction and preconcentration of polycyclic aromatic hydrocarbons from environmental solid samples, *J. Chromatogr. A*, 962, 1–8, 2002.

159. Gulmini, M., Bianco Prevot, A., Pramauro, E., and Zelano, V., Surfactant micellar solutions as alternative solvents for microwave-assisted extraction of polycyclic aromatic hydrocarbons from a spiked river sediment, *Polycyclic Aromat. Compd.*, 22, 55–70, 2002.
160. Buco, S., Moragues, M., Doumenq, P., Noor, A., and Mille, G., Analysis of polycyclic aromatic hydrocarbons in contaminated soil by Curie point pyrolysis coupled to gas chromatography-mass spectrometry, an alternative to conventional methods, *J. Chromatogr. A*, 1026, 223–229, 2004.
161. U.S. Environmental Protection Agency Test Method 3546: Microwave extraction, U.S. Government Printing Office, Washington, pp. 1–13, 2000.
162. Pinero-Iglesias, M., Lopez-Mahia, P., Vazquez-Blanco, E., Muniategui-Lorenzo, S., and Prada-Rodriguez, D., Problems in the extraction of polycyclic aromatic hydrocarbons from diesel particulate matter, *Polycyclic Aromat. Compd.*, 22, 129–146, 2002.
163. Budzinski, H., Baumard, P., Papineau, A., Wise, S., and Garrigues, P., Focused microwave assisted extraction of polycyclic aromatic compounds from standard reference materials, sediments and biological tissues, *Polycyclic Aromat. Compd.*, 9, 225–232, 1996.
164. Budzinski, H., Letellier, M., Garrigues, P., and Le Menach, K., Optimization of the microwave-assisted extraction in open cell of polycyclic aromatic hydrocarbons from solid and sediments: study of moisture effects, *J. Chromatogr. A*, 837, 187–200, 1999.
165. Shu, Y. Y., Lai, T. L., Lin, H., Yang, T. C., and Chang, C., Study of factors affecting on the extraction efficiency of polycyclic aromatic hydrocarbons from soils using open-vessel focused microwave-assisted extraction, *Chemosphere*, 52, 1667–1676, 2003.
166. Flotron, V., Houessou, J., Bosio, A., Delteil, C., Bermond, A., and Camel, V., Rapid determination of polycyclic aromatic hydrocarbons in sewage sludges using microwave-assisted solvent extraction, comparison with other extraction methods, *J. Chromatogr. A*, 999, 175–184, 2003.
167. Ramos, L., Kristenson, E. M., and Brinkman, U. A. Th., Current use of pressurized liquid extraction and subcritical water extraction in environmental analysis, *J. Chromatogr. A*, 975, 3–29, 2002.
168. U.S. Environmental Protection Agency Test Method 3545: Pressurized Fluid Extraction (PFE), U.S. Government Printing Office, Washington, pp. 1–9, 1996.
169. Saim, N., Dean, J. R., Abdullah, Md. P., and Zakaria, Z., An experimental design approach for the determination of polycyclic aromatic hydrocarbons from highly contaminated soil using accelerated solvent extraction, *Anal. Chem.*, 70, 420–424, 1998.
170. Lundstedt, S., van Bavel, B., Haglund, P., Tysklind, M., and Öberg, L., Pressurized liquid extraction of polycyclic aromatic hydrocarbons from contaminated soils, *J. Chromatogr. A*, 883, 151–162, 2000.
171. Kim, J. H., Moon, K., Li, Q. X., and Cho, J. Y., One-step pressurized liquid extraction method for the analysis of polycyclic aromatic hydrocarbons, *Anal. Chim. Acta*, 498, 55–60, 2003.
172. Krauss, M., Wilcke, W., and Zech, W., Availability of polycyclic aromatic hydrocarbons (PAHs) and polychlorinated biphenyls (PCBs) to earthworms in urban soils, *Environ. Sci. Technol.*, 34, 4335–4340, 2000.
173. Wilcke, W., Krauss, M., and Barancíková, G., Persistent organic pollutant concentrations in air- and freeze-dried compared to field-fresh extracted soil samples of an eastern Slovak deposition gradient, *J. Plant Nutr. Soil Sci.*, 166, 93–101, 2003.
174. Wilcke, W., Krauss, M., and Amelung, W., Carbon isotope signature of polycyclic aromatic hydrocarbons (PAHs): evidence for different sources in tropical and temperate environments? *Environ. Sci. Technol.*, 36, 3530–3535, 2002.
175. Smith, R. M., Extractions with superheated water, *J. Chromatogr. A*, 975, 31–46, 2002.
176. Hawthorne, S. B., Yang, Y., and Miller, D. J., Extraction of organic pollutants from environmental solids with sub- and supercritical water, *Anal. Chem.*, 66, 2912–2920, 1994.
177. McGowin, A. E., Adom, K. K., and Obubuafo, A. K., Screening of compost for PAHs and pesticides using static subcritical water extraction, *Chemosphere*, 45, 857–864, 2000.
178. Hawthorne, S. B., Trembley, S., Moniot, C. L., Grabanski, C. B., and Miller, D. J., Static subcritical water extraction with simultaneous solid-phase extraction for determining polycyclic aromatic hydrocarbons on environmental solids, *J. Chromatogr. A*, 886, 237–244, 2000.
179. Andersson, T., Hartonen, K., Hyötyläinen, T., and Riekkola, M.-L., Pressurized hot water extraction and thermal desorption of polycyclic aromatic hydrocarbons from sediment with use of a novel extraction vessel, *Anal. Chim. Acta*, 466, 93–100, 2002.

180. Hageman, K. J., Mazeas, L., Grabanski, C. B., Miller, D. J., and Hawthorne, S. B., Coupled subcritical water extraction with solid-phase microextraction for determining semivolatile organics in environmental solids, *Anal. Chem.*, 68, 3892–3898, 1996.

181. Yang, Y., Hawthorne, S. B., and Miller, D. J., Class-selective extraction of polar, moderately polar, and nonpolar organics from hydrocarbon wastes using subcritical water, *Environ. Sci. Technol.*, 31, 430–437, 1997.

182. Yang, Y. and Li, B., Subcritical water extraction coupled to high-performance liquid chromatography, *Anal. Chem.*, 71, 1491–1495, 1999.

183. Hyötyläinen, T., Andersson, T., Hartonen, K., Kuosmanen, K., and Riekkola, M.-L., Pressurized hot water extraction coupled on-line with LC–GC : determination of polyaromatic hydrocarbons in sediment, *Anal. Chem.*, 72, 3070–3076, 2000.

184. Kuosmanen, K., Hyötyläinen, T., Hartonen, K., and Riekkola, M.-L., Analysis of polycyclic aromatic hydrocarbons in soil and sediment with on-line coupled pressurized hot water extraction, hollow fiber microporous membranes liquid–liquid extraction and gas chromatography, *Analyst*, 128, 434–439, 2003.

185. Kipp, S., Peyrer, H., and Kleiböhmer, W., Coupling superheated water extraction with enzyme immunoassay for an efficient and fast PAH. Screening in soil, *Talanta*, 46, 385–393, 1998.

186. Fernández-Pérez, V. and Luque de Castro, M. D., Micelle formation for improvement of continuous subcritical water extraction of polycyclic aromatic hydrocarbons in soil prior to high-performance liquid chromatography-fluorescence detection, *J. Chromatogr. A*, 902, 357–367, 2000.

187. Morales-Muñoz, S., Luque-García, J. L., and Luque de Castro, M. D., Pressurized hot water extraction with on-line fluorescence monitoring: a comparison of the static, dynamic, and static–dynamic modes for the removal of polycyclic aromatic hydrocarbons from environmental solid samples, *Anal. Chem.*, 74, 4213–4219, 2002.

188. U.S. Environmental Protection Agency Test Method 3561: Supercritical fluid extraction of poly-nuclear aromatic hydrocarbons, U.S. Government Printing Office, Washington, pp. 1–14, 2000.

189. Berset, J. D. and Holzer, R., Quantitative determination of polycyclic aromatic hydrocarbons, polychlorinated biphenyls and organochlorine pesticides in sewage sludges using supercritical fluid extraction and mass spectrometric detection, *J. Chromatogr. A*, 852, 545–558, 1999.

190. Miège, C., Dugay, J., and Hennion, M.-C., Optimization and validation of solvent and supercritical-fluid extractions for the trace-determination of polycyclic aromatic hydrocarbons in sewage sludges by liquid chromatography coupled to diode-array and fluorescence detection, *J. Chromatogr. A*, 823, 219–230, 1998.

191. Hawthorne, S. B., Grabanski, C. B., Martin, E., and Miller, D. J., Comparisons of Soxhlet extraction, pressurized liquid extraction, supercritical fluid extraction and subcritical water extraction for environmental solids: recovery, selectivity and effects on sample matrix, *J. Chromatogr. A*, 892, 421–433, 2000.

192. Hawthorne, S. B., Poppendieck, D. G., Grabanski, C. B., and Loehr, R. C., PAH release during water desorption, supercritical carbon dioxide extraction, and field bioremediation, *Environ. Sci. Tech.*, 35, 4577–4583, 2001.

193. Goncalves, C., de Rezende Pinto, M., and Alpendurada, M. F., Benefits of a binary modifier with balanced polarity for an efficient supercritical fluid extraction of PAHs from solid samples followed by HPLC, *J. Liq. Chrom. Rel. Technol.*, 24, 2943–2959, 2001.

194. Hartonen, K., Bøwadt, S., Dybdahl, H. P., Nylund, K., Sporring, S., Lund, H., and Oreld, F., Nordic laboratory intercomparison of supercritical fluid extraction for the determination of total petroleum hydrocarbons, polychlorinated byphenyls and polycyclic aromatic hydrocarbons in soil, *J. Chromatogr. A*, 958, 239–248, 2002.

195. Librando, V., Hutzinger, O., Tringali, G., and Aresta, M., Supercritical fluid extraction of polycyclic aromatic hydrocarbons from marine sediments and soil samples, *Chemosphere*, 54, 1189–1197, 2004.

196. Hollender, J., Koch, B., Lutermann, C., and Dott, W., Efficiency of different methods and solvents for the extraction of polycyclic aromatic hydrocarbons from soils, *Intern J. Environ. Anal. Chem.*, 83, 21–32, 2003.

197. Miège, C., Dugay, J., and Hennion, M. C., Optimization, validation and comparison of various extraction techniques for the trace determination of polycyclic aromatic hydrocarbons in sewage

sludges by liquid chromatography coupled to diode-array and fluorescence detection, *J. Chromatogr. A*, 995, 87–97, 2003.

198. Eriksson, M., Fäldt, J., Dalhammar, G., and Borg-Karlson, A.-K., Determination of hydrocarbons in old creosote contaminated soil using headspace solid phase microextraction and GC-MS, *Chemosphere*, 44, 1641–1648, 2001.

199. Doong, R., Chang, S., and Sun, Y., Solid-phase microextraction and headspace solid-phase microextraction for the determination of high molecular-weight polycyclic aromatic hydrocarbons in water and soil samples, *J. Chromatogr. Sci.*, 38, 528–534, 2000.

200. U.S. Environmental Protection Agency Test Method 3630C, Silica Gel Cleanup, U.S. Government Printing Office, Washington, pp. 1–15, 1996.

201. Prycek, J., Ciganek, M., and Simek, Z., Development of an analytical method for polycyclic aromatic hydrocarbons and their derivatives, *J. Chromatogr. A*, 1030, 103–107, 2004.

202. Frysinger, G. S., Gaines, R. B., Xu, L., and Reddy, C. M., Resolving the unresolved complex mixture in petroleum-contaminated sediments, *Environ. Sci. Technol.*, 37, 1653–1662, 2003.

203. Hubert, A., Wenzel, K.-D., Manz, M., Weissflog, L., Engewald, W., and Schüürmann, G., High extraction efficiency for POPs in real contaminated soil samples using accelerated solvent extraction, *Anal. Chem.*, 72, 1294–1300, 2000.

204. U.S. Environmental Protection Agency Test Method 3611B, Alumina Column Cleanup and Separation of Petroleum Wastes, U.S. Government Printing Office, Washington, pp. 1–7, 1996.

205. Meyer, S., Cartellieri, S., and Steinhart, H., Simultaneous determination of PAHs, hetero-PAHs (N, S, O), and their degradation products in creosote-contaminated soils, method development, validation, and application to hazardous waste sites, *Anal. Chem.*, 71, 4023–4029, 1999.

206. Lüthje, K., Hyötyläinen, T., and Reikkola, M.-L., Comparison of different trapping methods for pressurized hot water extraction, *J. Chromatogr. A*, 1025, 41–49, 2004.

207. U.S. Environmental Protection Agency Test Method 3640A, Gel-Permeation Cleanup, U.S. Government Printing Office, Washington, pp. 1–24, 1994.

208. Berset, J. D., Kuehne, P., and Shotyk, W., Concentrations and distributions of some polychlorinated biphenyls (PCBs) and polycyclic aromatic hydrocarbons (PAHs) in an ombrotrophic peat bog profile of Switzerland, *Sci. Total Environ.*, 267, 67–85, 2001.

209. U.S. Environmental Protection Agency Test Method 3650B, Acid-Base Partition Cleanup, U.S. Government Printing Office, Washington, pp. 1–7, 1996.

210. U.S. Environmental Protection Agency Test Method 3660B, Sulfur Cleanup, U.S. Government Printing Office, Washington, pp. 1–6, 1996.

211. Vo-Dinh, T., Fetzer, J., and Campiglia, A. D., Monitoring and characterization of polyaromatic compounds in the environment, *Talanta*, 47, 943–969, 1998.

212. Casellas, M., Fernandez, P., Bayone, J. M., and Solanas, A. M., Bioassay-directed chemical analysis of genotoxic compounds in urban airborne particulate matter from Barcelona, *Chemosphere*, 30, 725–740, 1995, Spain.

213. de Boer, J. and Law, R. L., Developments in the use of chromatographic techniques in marine laboratories for the determination of halogenated contaminants and polycyclic aromatic hydrocarbons, *J. Chromatogr. A*, 1000, 223–251, 2003.

214. U.S. Environmental Protection Agency Test Method 8270C: Semivolatile organic compounds by gas chromatography/mass spectrometry (GC/MS), U.S. Government Printing Office, Washington, DC., pp. 1–54, 1996.

215. U.S. Environmental Protection Agency Method 525: Determination of organic compounds in drinking water, Supplement III, USEPA Cincinnati, OH, EPA-600/R-95/131, 1995.

216. Castello, G. and Gerbino, T. C., Analysis of polycyclic aromatic hydrocarbons with ion-trap mass detector and comparison with other gas-chromatographic and high-performance liquid-chromatographic techniques, *J. Chromatogr. A*, 642, 351–357, 1993.

217. U.S. Environmental Protection Agency Test Method 8275A: Semivolatile organic compounds (PAHs and PCBs) in soils/sludges and solid wastes using thermal extraction/gas chromatography/ mass spectrometry (TE/GC/MS), U.S. Government Printing Office, Washington, pp. 1–23, 1996.

218. U.S. Environmental Protection Agency Method 610: Methods for organic chemical analysis of municipal and industrial wastewater, U.S. Environmental Protection Agency, Cincinnati, OH, U.S.A., pp. 441–454, 1992.

219. U.S. Environmental Protection Agency Test Method 8100: Polynuclear Aromatic Hydrocarbons, U.S. Government Printing Office, Washington, pp. 1–10, 1986.

220. Dallüge, J., Ou-Aissa, R., Vreuls, J. J., Brinkman, U. A. Th., and Veraart, J. R., Fast temperature programming in gas chromatography using resistive heating, *J. High Resolut. Chromatogr.*, 22, 459–464, 1999.

221. Berset, J. D., Holzer, R., and Häni, H., Solving coelution problems on a smectic-liquid crystalline polysiloxane capillary column for the determination of priority polycyclic aromatic hydrocarbons in environmental samples, *J. Chromatogr. A*, 823, 179–187, 1998.

222. Chang-Chien, G., Terminally carboxyl oligo(ethylene oxide) monomethyl ether-substituted side chain liquid crystalline polysiloxane polymer as stationary phase in capillary gas chromatography for the separation of polynuclear aromatic hydrocarbons, *J. Chromatogr. A*, 808, 201–209, 1998.

223. Lee, W. and Chang-Chien, G., All-hydrocarbon liquid crystalline polysiloxane polymer as stationary phase in gas chromatography capillary column for separation of isomeric compounds of polynuclear aromatic hydrocarbons, *Anal. Chem.*, 70, 4094–4099, 1998.

224. Supelco, Sigma-Aldrich Co., Bellefonte, PA U.S.A.

225. Shariff, S. M., Robson, M. M., and Bartle, K. D., Supercritical fluid chromatography of polycyclic aromatic compounds: a review, *Polycyclic Aromat. Compd.*, 12, 147–185, 1997.

226. Koski, I. J., Lee, E. D., Ostrovsky, I., and Lee, M. L., Solid-phase injector for open tubular column supercritical fluid chromatography, *Anal. Chem.*, 65, 1125–1129, 1993.

227. Payne, K. M., Tarbet, B. J., Bradshaw, J. S., Maekides, K. E., and Lee, M. L., Simultaneous deactivation and coating of porous silica particles for microcolumn supercritical fluid chromatography, *Anal. Chem.*, 62, 1379–1384, 1990.

228. Toribio, L., del Nozal, M. J., Bernal, J. L., Jiménez, J. J., and Serna, M. L., Packed-column fluid chromatography coupled with solid-phase extraction for the determination of organic micro-contaminants in water, *J. Chromatogr. A*, 823, 163–170, 1998.

229. Bernal, J. L., del Nozal, M. J., Toribio, L., Serna, M. L., Borrull, F., Marcé, R. M., and Pocurull, E., Determination of polycyclic aromatic hydrocarbons in waters by use of supercritical fluid chromatography coupled on-line to solid-phase extraction with disks, *J. Chromatogr. A*, 778, 321–328, 1997.

230. U.S. Environmental Protection Agency Test Method 550: Methods for the determination of organic compounds in drinking water — Supplement I. EPA-600/4-90/020, USEPA, Office of Research and Development, Cincinnati, OH, pp 121–168, 1990.

231. U.S. Environmental Protection Agency Test Method 8310: Polynuclear Aromatic Hydrocarbons, U.S. Government Printing Office, Washington, pp. 1–13, 1986.

232. Hesselink, W., Schiffer, R. H. N. A., and Kootstra, P. R., Separation of polycyclic aromatic hydrocarbons on a wide-pore polymeric C_{18} bonded phase, *J. Chromatogr. A*, 697, 165–174, 1995.

233. Sander, L. C. and Wise, S. A., Subambient temperature modification of selectivity in reversed-phase liquid chromatography, *Anal. Chem.*, 61, 1749–1754, 1989.

234. Ascah, T. and Kraft, E., Rapid HPLC analysis of PAH compounds, using porous 3 μm particles, *The Supelco Rep.*, 16(4), 5, 1997.

235. Saito, Y., Nojiri, M., Shimizu, Y., and Jinno, K., Liquid chromatographic retention behavior of polycyclic aromatic hydrocarbons on newly-synthesized chitosan stationary phases crosslinked with long aliphatic chains, *J. Liq. Chromatogr. Relat. Technol.*, 25, 2767–2779, 2002.

236. Stoll, D. and Johnson, K. S., *PAH on ZirChrom-PS and ZirChrom-PDB*, ZirChrom Separations, Inc., Anoka, MN, U.S.A., 2002, September 19.

237. Li, J., Hu, Y., and Carr, P. W., Fast separations at elevated temperatures on polybutadiene-coated zirconia reverse-phase material, *Anal. Chem.*, 69, 3884–3888, 1997.

238. Kondo, T., Yang, Y., and Lamm, L., Separation of polar and non-polar analytes using dimethyl sulfoxide-modified subcritical water, *Anal. Chim. Acta*, 460, 185–191, 2002.

239. Jones, A. and Yang, Y., Separation of nonpolar analytes using methanol–water mixtures at elevated temperatures, *Anal. Chim. Acta*, 485, 51–55, 2003.

240. Khaledi, M. G., Micelles as separation media in high-performance liquid chromatography and high-performance capillary electrophoresis: overview and perspective, *J. Chromatogr. A*, 780, 3–40, 1997.

241. Mao, C., McGill, K. E., and Tucker, S. A., Optimization of micellar liquid chromatographic separation of polycyclic aromatic hydrocarbons with the addition of second organic additive, *J. Sep. Sci.*, 27, 991–996, 2004.

242. Martínez, D., Cugat, M. J., Borrull, F., and Calull, M., Solid-phase extraction coupling to capillary electrophoresis with emphasis on environmental analysis, *J. Chromatogr. A*, 902, 65–89, 2000.

243. Song, L., Xu, Z., Kang, J., and Cheng, J., Analysis of environmental pollutants by capillary electrophoresis with emphasis on micellar electrokinetic chromatography, *J. Chromatogr. A*, 780, 297–328, 1997.

244. Koch, J. T., Beam, B., Phillips, K. S., and Wheeler, J. F., Hydrophobic interaction electrokinetic chromatography for the separation of polycyclic aromatic hydrocarbons using non-aqueous matrices, *J. Chromatogr. A*, 914, 223–231, 2001.

245. Pumera, M., Muzikár, J., Barek, J., and Jelínek, I., Determination of amino derivatives of polycyclic aromatic hydrocarbons using capillary electrophoresis, *Anal. Lett.*, 34, 1369–1375, 2001.

246. Nie, S., Dadoo, R., and Zare, R. N., Ultrasensitive fluorescence detection of polycyclic aromatic hydrocarbons in capillary electrophoresis, *Anal. Chem.*, 65, 3571–3575, 1993.

247. Shi, Y. and Fritz, J. S., HPCZE of nonionic compounds using a novel anionic surfactant additive, *Anal. Chem.*, 67, 3023–3027, 1995.

248. Copper, C. L. and Sepaniak, M. J., Cyclodextrin-modified micellar electrokinetic capillary chromatography separations of benzopyrene isomers: correlation with computationally derived host-guest energies, *Anal. Chem.*, 66, 147–154, 1994.

249. Szolar, O. H. J., Brown, R. S., and Luong, J. H. T., Separation of PAHs by capillary electrophoresis with laser-induced fluorescence (LIF) detection using mixtures of neutral and anionic β-cyclodextrins, *Anal. Chem.*, 67, 3004–3010, 1995.

250. Brown, R. S., Luong, J. H. T., Szolar, O. H. J., Halasz, A., and Hawari, J., Cyclodextrin-modified capillary electrophoresis: determination of polycyclic aromatic hydrocarbons in contaminated soils, *Anal. Chem.*, 68, 287–292, 1996.

251. Nguyen, A.-L. and Luong, J. H. T., Separation and determination of polycyclic aromatic hydrocarbons by solid phase microextraction/cyclodextrin-modified capillary electrophoresis, *Anal. Chem.*, 69, 1726–1731, 1997.

252. Jiménez, B., Patterson, D. G., Grainger, J., Liu, Z., González, M. J., and Marina, L. M., Enhancement of the separation selectivity of a group of polycyclic aromatic hydrocarbons using mixed cyclodextrin-modified micellar electrokinetic chromatography, *J. Chromatogr. A*, 792, 411–418, 1997.

253. Shamsi, S. A., Akbay, C., and Warner, I. M., Polymeric anionic surfactant for electrokinetic chromatography: separation of 16 priority polycyclic aromatic hydrocarbon pollutants, *Anal. Chem.*, 70, 3078–3083, 1998.

254. Dadoo, R., Zar, R. N., Yan, C., and Anex, D. S., Advances in capillary electrochromatography: rapid and high-efficiency separations of PAHs, *Anal. Chem.*, 70, 4787–4792, 1998.

255. Norton, D., Zheng, J., and Shamisi, S. A., Capillary electrochromatography of methylated benzo[a]pyrene isomers, *J. Chromatogr. A*, 1008, 205–215, 2003.

256. Sharma, V. K., Hicks, S. D., Rivera, W., and Vasquez, F. G., Characterization and degradation of petroleum hydrocarbons following an ion spill into a coastal environment of south Texas, U.S.A, *Water, Air Soil Pollut.*, 134, 111–127, 2002.

257. Grote, C., Levsen, K., and Wünsch, G., An automatic analyzer for organic compounds in water based on solid-phase microextraction coupled to gas chromatography, *Anal. Chem.*, 71, 4513–4518, 1999.

258. U.S. Environmental Protection Agency Test Method 8410: Gas chromatographic/Fourier transform infrared (GC/FT-IR) spectrometry for semivolatile organics: capillary column, U.S. Government Printing Office, Washington, pp. 1–18, 1994.

259. Crozier, P. W., Plomley, J. B., and Matchuk, L., Trace level analysis of polycyclic aromatic hydrocarbons in surface waters by solid phase extraction (SPE) and gas chromatography-ion trap mass spectrometry (GC-ITMS), *Analyst*, 126, 1974–1979, 2001.

260. Vreuls, R. J. J., Dallüge, J., and Brinkman, U. A. Th., Gas chromatography-time-of-flight mass spectrometry for sensitive determination of organic microcontaminants, *J. Microcolumn Sep.*, 11, 663–675, 1999.

261. Riahi, K. and Sellier, N., Separation of isomeric polycyclic aromatic hydrocarbons by GC-MS: differentiation between isomers by positive chemical ionization with ammonia and dimethyl ether as reagent gases, *Chromatographia*, 47, 309–312, 1998.

262. Jinhui, X. and Lee, F. S. C., Analysis of nitrated polynuclear aromatic hydrocarbons, *Chemosphere*, 42, 245–250, 2001.

263. Kawanaka, Y., Matsumoto, E., Sakamoto, K., Wang, N., and Yun, S.-J., Size distribution of mutagenic compounds and mutagenicity in atmospheric particulate matter collected with a low-pressure cascade impactor, *Atmos. Environ.*, 38, 2125–2132, 2004.

264. Gimeno, R. A., Comas, E., Marcé, R. M., Ferré, J., Rius, F. X., and Borrull, F., Second-order linear calibration for determining polycyclic aromatic compounds in marine sediments by solvent extraction and liquid chromatography with diode-array detection, *Anal. Chim. Acta*, 498, 47–53, 2003.

265. Patra, D., Applications and new developments in fluorescence spectroscopic techniques for the analysis of polycyclic aromatic hydrocarbons, *Appl. Spectrosc. Rev.*, 38, 155–185, 2003.

266. Wise, S. A., Sander, L. C., and May, W. E., Determination of polycyclic aromatic hydrocarbons by liquid chromatography, *J. Chromatogr.*, 642, 329–349, 1993.

267. Bouzige, M., Machtalère, G., Legeay, P., Pichon, V., and Hennion, M.-C., New methodology for a selective on-line monitoring of some polar priority industrial chemicals in waste water, *Waste Manage.*, 19, 171–180, 1999.

268. Patra, D. and Mishra, A. K., A novel disposable sensor head for a fiber optic spectrofluorimeter, *Instrum. Sci. Technol.*, 30, 31–41, 2002.

269. Garguillo, M. G., Thomas, D. H., Anex, D. S., and Rakestraw, D. J., Laser-induced dispersed fluorescence detection of polycyclic aromatic compounds in soil extracts separated by capillary electrochromatography, *J. Chromatogr. A*, 883, 231–248, 2000.

270. Gundl, T. J., Aldstadt, J. H., Harb, J. G., Germain, R. W. St., and Schweitzer, R. C., Demonstration of a method for the direct determination of polycyclic aromatic hydrocarbons in submerged sediments, *Environ. Sci. Technol.*, 37, 1189–1197, 2003.

271. Mao, C., Larson, C. L., and Tucker, S. A., Spectrochemical evaluation of diisopropylamine as a selective fluorescence quenching agent of polycyclic aromatic hydrocarbons in acetonitrile, *Polycyclic Aromat. Compd.*, 22, 99–110, 2002.

272. Pandey, S., Acree, W. A., and Fetzer, J. C., Cetylpyridinium chloride micelles as a selective fluorescence quenching solvent media for discriminating between alternant versus nonalternant polycyclic aromatic hydrocarbons, *Talanta*, 45, 39–45, 1997.

273. Mao, C. and Tucker, S. A., High-performance chromatographic separation of polycyclic aromatic hydrocarbons using pyridinium chloride as a selective fluorescence quencher to aid detection, *J. Chromatogr. A*, 966, 53–61, 2002.

274. Hagestuen, E. D. and Campiglia, A. D., New approaches for screening polycyclic aromatic hydrocarbons in water samples, *Talanta*, 49, 547–560, 1999.

275. Anacleto, J. F., Ramaley, L., Benoit, F. M., Boyd, R. K., and Quilliam, M. A., Comparison of liquid chromatography/mass spectrometry interfaces for the analysis of polycyclic aromatic compounds, *Anal. Chem.*, 67, 4145–4154, 1995.

276. Moriwaki, H., Electrospray mass spectrometric determination of 1-nitropyrene and non-substituted polycyclic aromatic hydrocarbons using tropylium cation as a post-column HPKC reagent, *Analyst*, 125, 417–420, 2000.

277. Airiau, C. Y., Brereton, R. G., and Crosby, J., High-performance liquid chromatography/electrospray tandem mass spectrometry of polycyclic aromatic hydrocarbons, *Rapid Commun. Mass Spectrom.*, 15, 135–140, 2001.

278. Ng, K. M., Ma, N. L., and Tsang, C. W., Differentiation of isomeric polyaromatic hydrocarbons by electrospray Ag(I) cationization mass spectrometry, *Rapid Commun. Mass Spectrom.*, 17, 2082–2088, 2003.

279. Gimeno, R. A., Altelaar, A. F. M., Marce, R. M., and Borrull, F., Determination of polycyclic aromatic hydrocarbons and polycyclic aromatic sulfur heterocycles by high-performance liquid chromatography with fluorescence and APCI mass spectrometry detection in seawater and sediment samples, *J. Chromatogr. A*, 958, 141–148, 2002.

280. Perez, S. and Barcelo, D., Determination of polycyclic aromatic hydrocarbons in sewage reference sludge by liquid chromatography-atmospheric-pressure chemical-ionization mass spectrometry, *Chromatographia*, 53, 475–480, 2001.

281. Koeber, R., Bayona, J. M., and Niesser, R., Determination of benzo[a]pyrene diones in air particulate matter with liquid chromatography mass spectrometry, *Environ. Sci. Technol.*, 33, 1552–1558, 1999.

282. Marvin, C. H., Smith, R. W., Bryant, D. W., and McCarry, B. E., Analysis of high-molecular-mass polycyclic aromatic hydrocarbons in environmental samples using liquid chromatography-atmospheric pressure chemical ionization mass spectrometry, *J. Chromatogr. A*, 863, 13–24, 1999.

283. Cáslavsky, J. and Kotlaríková, P., High-molecular-weight polycyclic aromatic hydrocarbons in the area and vicinity of the Deza chemical plant, Czech Republic, *Polycyclic Aromat. Compd.*, 23, 327–352, 2003.

284. Kolalowski, B. M., Grossert, J. S., and Ramaley, L., Studies on the positive-ion mass spectra from atmospheric pressure chemical ionization of gases and solvents used in liquid chromatography and direct liquid injection, *J. Am. Soc. Mass Spectrom.*, 15, 311–324, 2004.

285. Moyano, E., McCullagh, M., Galceran, M. T., and Games, D. E., Supercritical fluid chromatography-atmospheric pressure chemical ionization mass spectrometry for the analysis of hydroxy polycyclic aromatic hydrocarbons, *J. Chromatogr. A*, 777, 167–176, 1997.

286. Anacleto, J. F., Ramaley, L., Boyd, R. K., Ploeasance, S., Quilliam, M. A., Sim, P. G., and Benoit, F. M., Analysis of polycyclic aromatic compounds by supercritical fluid chromatography mass spectrometry using atmospheric-pressure chemical ionization, *Rapid Commun. Mass Spectrom.*, 5, 149–155, 1991.

287. Mansoori, B. A., Isomeric identification and quantification of polycyclic aromatic hydrocarbons in environmental samples by liquid chromatography tandem mass spectrometry using a high pressure quadrupole collision cell, *Rapid Commun. Mass Spectrom.*, 12, 712–728, 1998.

288. Moriwaki, H., Ishitake, M., Yoshikawa, S., Miyakoda, H., and Alary, J. F., Determination of polycyclic aromatic hydrocarbons in sediment by liquid chromatography-atmospheric pressure photoionization-mass spectrometry, *Anal. Sci.*, 20, 375–377, 2004.

289. Fähnrich, K. A., Pravda, M., and Guilbault, G. G., Immunochemical detection of polycyclic aromatic hydrocarbons (PAHs), *Anal. Lett.*, 35, 1269–1300, 2002.

290. U.S. Environmental Protection Agency Test Method 4035: Soil screening for polynuclear aromatic hydrocarbons by immunoassay, U.S. Government Printing Office, Washington, pp. 1–10, 1996.

291. Waters, L. C., Palausky, A., Counts, R. W., and Jenkins, R. A., Experimental evaluation of two field test kits for the detection of PAHs by immunoassay, *Field Anal. Chem. Technol.*, 1, 227–238, 1997.

292. Barceló, D., Qubiña, A., Salau, J. S., and Perez, S., Determination of PAHs in river water samples by ELISA, *Anal. Chim. Acta*, 376, 49–53, 1998.

293. Kipp, S., Peyrer, H., and Kleiböhmer, W., Coupling superheated water extraction with enzyme immunoassay for an efficient and fast PAH screening in soil, *Talanta*, 46, 385–393, 1998.

294. Li, K., Woodward, L. A., Karu, A. E., and Li, Q. X., Immunochemical detection of polycyclic aromatic hydrocarbons and 1-hydroxypyrene in water and sediment samples, *Anal. Chim. Acta*, 419, 1–8, 2000.

295. Knopp, D., Seifert, M., Vaananen, V., and Niessner, R., Determination of polycyclic aromatic hydrocarbons in contaminated water and soil samples by immunological and chromatographic methods, *Environ. Sci. Technol.*, 34, 2035–2041, 2000.

296. Nording, M. and Haglund, P., Evaluation of the structure/crossreactivity relationship of polycyclic aromatic compounds using an enzyme-linked immunosorbent assay kit, *Anal. Chim. Acta*, 487, 43–50, 2003.

297. Fähnrich, K. A., Pravda, M., and Guilbault, G. G., Disposable amperometric immunosensor for the detection of polycyclic aromatic hydrocarbons (PAHs) using screen-printed electrodes, *Biosens. Bioelectron.*, 18, 73–82, 2003.

298. Kallio, M., Hyötyläinen, T., Lehtonen, M., Jussila, M., Kartonen, K., Shimmo, M., and Riekkola, M. L., Comprehensive two-dimensional gas chromatography in the analysis of urban aerosols, *J. Chromatogr. A*, 1019, 251–260, 2003.

299. Ong, R., Lundstedt, S., Haglund, P., and Marriott, P., Pressurized liquid extraction-comprehensive two-dimensional gas chromatography for fast-screening of polycyclic aromatic hydrocarbons in soil, *J. Chromatogr. A*, 1019, 221–232, 2003.

300. Murahashi, T., Comprehensive two-dimensional high-performance liquid chromatography for the separation of polycyclic aromatic hydrocarbons, *Analyst*, 128, 611–615, 2003.

301. Murahashi, T., Tsuruga, F., and Sasaki, S., An automatic method for the determination of carcinogenic 1-nitropyrene in extracts from automobile exhaust particulate matter, *Analyst*, 128, 1346–1351, 2003.

302. Lewis, A. C., Robinson, R. E., Bartle, K. D., and Pilling, M. J., On-line coupled LC–GC-ITD/MS for the identification of alkylated, oxygenated, and nitrated polycyclic aromatic compounds in urban air particulate extracts, *Environ. Sci. Technol.*, 29, 1977–1981, 1995.

303. Shimmo, M., Adler, H., Hyotylainen, T., Hartonen, K., Kulmala, M., and Riekkola, M. L., Analysis of particulate polycyclic aromatic hydrocarbons by on-line coupled supercritical fluid extraction-liquid chromatography-gas chromatography-mass spectrometry, *Atmos. Environ.*, 36, 2985–2995, 2002.

304. Bezabeh, D. Z., Jones, A. D., Ashbaugh, L. L., and Kelly, P. B., Screening of aerosol filter samples for PAHs and nitro-PAHs by laser desorption ionization TOF mass spectrometry, *Aerosol. Sci. Technol.*, 30, 288–299, 1999.

305. Haefliger, O. P., Bucheli, T. D., and Zenobi, R., Laser mass spectrometric analysis of organic atmospheric aerosols. 1. Characterization of emission sources, *Environ. Sci. Technol.*, 34, 2178–2183, 2000.

306. Emmenegger, C., Kalberer, M., Morrical, B., and Zenobi, R., Quantitative analysis of polycyclic aromatic hydrocarbons in water in the low-nanogram per liter range with two-step laser mass spectrometry, *Anal. Chem.*, 75, 4508–4513, 2003.

307. Emmenegger, E., Kalberer, M., Samburova, V., and Zenobi, R., Analysis of size-segregated aerosol-bound polycyclic aromatic hydrocarbons with high time resolution using two-step laser mass spectrometry, *Analyst*, 129, 416–420, 2004.

308. Reilly, P. T. A., Gieray, R. A., Whitten, W. B., and Ramsey, J. M., Real-time characterization of the organic composition and size of individual diesel engine smoke particles, *Environ. Sci. Technol.*, 32, 2672–2679, 1998.

309. Rodgers, R. P., Lazar, A. C., Reilly, P. T. A., Whitten, W. B., and Ramsey, J. M., Direct determination of soil surface-bound polycyclic aromatic hydrocarbons in petroleum-contaminated soils by real-time aerosol mass spectrometry, *Anal. Chem.*, 72, 5040–5046, 2000.

310. Oktem, B., Tolocka, M. P., and Johnston, M. V., On-line analysis of organic components in fine and ultrafine particles by photoionization aerosol mass spectrometry, *Anal. Chem.*, 76, 253–261, 2004.

311. Algarra, M., Radin, C., Fornier de Violet, Ph., Lamotte, M., Garrigues, Ph., Hardy, M., and Gillard, R., Direct fluorometric analysis of PAHs in water and in urine following liquid solid extraction, *J. Fluoresc.*, 10, 355–359, 2000.

312. Miller, J. S., Determination of polycyclic aromatic hydrocarbons by spectrofluorimetry, *Analytical Chimica Acta*, 388, 27–34, 1999.

313. López de Alda Villaizán, M. J., Simal Lozano, J., and Lage Yusty, M. A., Determination of polycyclic aromatic hydrocarbons in drinking water and surface waters from Galicia (N.W. Spain) by constant-wavelength synchronous spectrofluorimetry, *Talanta*, 42, 967–970, 1995.

314. Ferrer, R., Bertrán, J. L., and Guiteras, J., Multivariate calibration applied to synchronous fluorescence spectrometry. Simultaneous determination of polycyclic aromatic hydrocarbons in water samples, *Talanta*, 45, 1073–1080, 1998.

315. López de Alda Villaizán, M. J., García Falcón, M. S., González Amigo, S., Simal Lozano, J., and Lage Yusty, M. A., Second-derivative constant-wavelength synchronous scan spectrofluorimetry for determination of benzo[b]fluoranthene, benzo[a]pyrene and indeno[1,2,3-cd]pyrene in drinking water, *Talanta*, 43, 1405–1412, 1996.

316. Andrade Eiroa, A., Vazquez, E., Lopez Mahia, P., Muniategui Lorenzo, S., and Prada Rodriguez, D., Resolution of benzo[a]pyrene in complex mixtures of other polycyclic aromatic hydrocarbons. Comparison of two spectrofluorimetric methods applied to water samples, *Analyst*, 125, 1321–1326, 2000.

317. Andrade Eiroa, A., Vazquez, E., Lopez Mahia, P., Muniategui Lorenzo, S., and Prada Rodriguez, D., Resolution of benzo[a]pyrene in complex mixtures of other polycyclic aromatic hydrocarbons. Comparison of two spectrofluorimetric methods applied to water samples, *Analyst*, 125, 1321–1326, 2000.

318. Whitcomb, J. L., Bystol, A. J., and Campiglia, A. D., Time-resolved laser-induced fluorimetry for screening polycyclic aromatic hydrocarbons on solid-phase extraction membranes, *Anal. Chim. Acta*, 464, 261–272, 2002.

319. Bystol, A. J., Whitcomb, J. L., and Campiglia, A. D., Solid–liquid extraction laser excited time-resolved Shpol'skii spectrometry: a facile method for the direct detection of 15 priority pollutants in water samples, *Environ. Sci. Technol.*, 35, 2566–2571, 2001.

320. Beltrán, L., Ferrer, R., and Guiteras, J., Multivariate calibration of polycyclic aromatic hydrocarbon mixtures from excitation-emission fluorescence spectra, *Anal. Chim. Acta*, 373, 311–319, 1998.

321. Bystol, A. J., Thorstenson, T. and Campiglia, A. D., Laser-induced multidimensional fluorescence spectroscopy in Shpol'skii matrices in liquid helium (4.2 K) for the analysis of polycyclic aromatic hydrocarbons in HPLC fractions and complex environmental matrices, *Environ. Sci. Technol.*, 36, 4424–4429, 2002.

16 Volatile Organic Compounds in the Atmosphere*

Paul V. Doskey

CONTENTS

I. INTRODUCTION

Volatile organic compounds (VOCs), which are sometimes referred to as nonmethane organic compounds (NMOCs), exist entirely in the gas phase in the atmosphere. This condition requires the nonpolar species to have vapor pressures greater than about 10 Pa. The classes of organic substances categorized as VOCs include chemicals containing only carbon and hydrogen (hydrocarbons) and oxygen-, nitrogen-, sulfur-, and halogen-substituted hydrocarbons. This chapter discusses three classes of VOCs: nonmethane hydrocarbons (NMHCs), oxygenated hydrocarbons (OxHCs), and halogenated hydrocarbons (HaHCs).

Observed levels of selected chemicals of the three classes of VOCs are presented in Table 16.1 to Table 16.3.

The VOCs originate from anthropogenic and biogenic sources; many species are important reactants in the formation of photochemical smog. The reactive species are readily oxidized by hydroxyl radical (OH), forming a complex mixture of peroxy radicals which oxidize NO to NO_2 without consuming O_3 and thus allowing O_3 to increase in the daytime atmospheric boundary layer (see, e.g., Refs. 1,2). The compositions, concentrations, and reactivities of the VOCs which

---

*The submitted manuscript has been created by the University of Chicago as operator of Argonne National Laboratory under Contract No. W-31-109-ENG-38 with the U.S. Department of Energy. The U.S. government retains for itself, and others acting on its behalf, a paid-up, nonexclusive, irrevocable worldwide license in said article to reproduce, prepare derivative works, distribute copies to the public, and perform publicly and display publicly, by or on behalf of the government.

TABLE 16.1
Observed Levels of Selected NMHCs at an Urban Industrial Site in Houston, Texas[166] and a Rural Site near Kinterbish, Alabama[10]

Molecule	Mixing Ratio (ppbv)	
	Urban	Rural
Saturated		
Ethane	1.97–38	—
Propane	1.19–12	1.2
2-Methylpropane	0.48–4.3	0.20
n-Butane	0.64–4.2	0.30
2-Methylbutane	1.16–7.1	0.25
n-Pentane	0.39–2.8	0.09
2-Methylpentane	0.33–2.1	0.05
3-Methylpentane	0.19–1.2	0.02
n-Hexane	0.17–1.4	0.02
2,2,4-Trimethylpentane	0.22–1.2	
Unsaturated		
Acetylene	1.5–2.9	—
Ethene	0.64–5.5	—
Propene	0.35–1.1	—
2-Methylpropene/1-butene	0.33–0.76	—
1,3-Butadiene	0.09–0.21	—
trans-2-Butene	0.04–0.28	—
cis-2-Butene	0.03–0.27	—
3-Methyl-1-butene	0.02–0.13	—
2-Methyl-1-butene	0.05–0.44	—
Isoprene	0.07–0.23	6.3
trans-2-Pentene	0.06–0.74	—
cis-2-Pentene	0.05–0.36	—
2-Methyl-2-butene	0.04–0.78	—
Aromatic		
Benzene	0.25–1.4	0.10
Toluene	0.69–3.9	0.08
Ethylbenzene	0.09–0.63	0.01
p-Xylene/*m*-xylene	0.24–1.9	0.02
o-Xylene	0.09–0.72	0.01

compose biogenic and anthropogenic emissions vary greatly (see, e.g., Ref. 3). The NMHCs are emitted during the production, refining, and use of petroleum fuels. In urban areas, mobile sources contribute the greatest amount of reactive NMHC emissions.

Mixing ratios of the individual NMHCs vary greatly, from several parts per trillion by volume (pptv) to several parts per billion by volume (ppbv). Variations in the reactivities of NMHCs are also substantial; for example, isoprene (2-methyl butadiene), which is emitted by deciduous vegetation, has an atmospheric lifetime with respect to oxidation by OH of about 20 min in polluted air ($[OII] = 10^7$ radicals cm^{-3}). Monoterpenes ($C_{10}H_{16}$), which are emitted from coniferous trees, react extremely rapidly with OH and O_3 and have lifetimes of minutes. The atmospheric lifetimes of 2-methylpropene, 2-methylbutane, and the xylenes, which are found in vehicle emissions, are approximately 30 min, 7 h, and 1.5 h, respectively. Oxidation of the terpenes

TABLE 16.2
Observed Levels of Selected OxHCs at Urban and Rural Locations

Molecule	Mixing Ratio (ppbv)	
	Urban	Rural
Alcohols and Ethers		
Methanol	5.6–31[a]	11.0[b]
Ethanol	<1.0–22[a]	1.2[b]
2-Propanol	<1.0–7.9[a]	—
Methyl *tert*-butyl ether	<0.2–2.8[a]	—
Carbonyls		
Formaldehyde	3.2–16.8[c]	0.5–4.41[d]
Acetaldehyde	1.4–7.3[c]	0.09–1.92[d]
Propionaldehyde	<0.5–0.7[c]	0.004–0.067[d]
Acetone	12.4–94[a]	0.61–4.30[d]
Methacrolein	<0.7[c]	0.09–5.79[e]
Methyl vinyl ketone	—	0.27–3.77[e]
Glyoxal	—	0.01–0.29[f]
Methyl Glyoxal	—	0–0.92[f]
Glycolaldehyde	—	0–2.82[f]
Methyl ethyl ketone	0.47–58[a]	0.49[b]
Acids		
Formic	4.1–8.6[c]	0.1–9.8[e]
Acetic	0.5–6.7[c]	0.1–6.5[e]
Propionic	0.6[c]	—
Pyruvic	0.4–2.1[c]	0.01–0.40[e]

[a] Urban industrial location in Houston, Texas.[74]
[b] Forested site near Kinterbish, Alabama.[10]
[c] Albuquerque, New Mexico.[51]
[d] Agricultural and forested sites in Ontario, Canada.[27]
[e] Forested site near State College, Pennsylvania.[25]
[f] Aircraft platform <1900 m above Nashville, Tennessee.[54]

and monoaromatic hydrocarbons produce hygroscopic aerosols which influence radiative forcing of Earth's climate.[4–7]

Methanol, ethanol, and methyl tertiary butyl ether (MTBE) are OxHCs added to fuels to decrease tailpipe emissions of NMHCs and CO.[8,9] Some alcohols, aldehydes, and ketones are emitted from biogenic sources; others are also produced in the atmosphere through photochemical oxidation. The levels of some of the light OxHCs (e.g., acetaldehyde, methanol, ethanol, and acetone) are substantial. In the rural atmosphere, these can dominate the VOCs distribution (see, e.g., Ref. 10). Although the reactivity of acetaldehyde is very high, the other OxHCs have lifetimes of several days or more, facilitating global-scale transport and enabling the species to contribute to oxidant formation in remote areas.[11] Organic acids like formic and acetic acids are generated by anthropogenic and biogenic sources, and these are very soluble in water, and represent the major acidic species in rainwater of remote areas.[12]

The HaHCs include the chlorofluorocarbons (CFCs) and their replacements — the hydrochlorofluorocarbons (HCFCs) and the hydrofluorocarbons (HFCs). Also included in the HaHC category are bromo- and iodo-substituted organic compounds and various chlorinated hydrocarbons (CHCs) such as chloroform, 1,1,1-trichloroethane, carbon tetrachloride,

TABLE 16.3
Observed Levels of Selected HaHCs at Surface Locations

Molecule	Formula	Mixing Ratio (pptv)
CFCs, HFCs, HCFCs		
CFC-11	CCl_3F	$270-280^a$
CFC-12	CCl_2F_2	$520-530^a$
CFC-113	CCl_2FCClF_2	$83-86^a$
HFC-134a	CH_2FCF_3	$1-3^b$
HFC-141b	CH_3CCl_2F	$0.5-2^c$
HFC-142b	CH_3CF_2Cl	$3.8-6.8^c$
HCFC-22	$CHClF_2$	$110-135^a$
Bromo-[d]		
Methyl bromide	CH_3Br	$9.8-11$
Halon-1211	$CBrClF_2$	2.5
Halon-1301	$CBrF_3$	2.0
Halon-2402	$CBrF_2CBrF_2$	0.47
Dibromomethane	CH_2Br_2	$1.4-3.7$
Bromochloromethane	CH_2BrCl	$0.5-2.9$
Bromoform	$CHBr_3$	$0-10$
Dibromochloromethane	$CHBr_2Cl$	<1
Bromodichloromethane	$CHBrCl_2$	$0.6-3.8$
1,2-Dibromoethane	CH_2BrCH_2Br	$0-17.7$
Iodo-[e]		
Methyl iodide	CH_3I	$<0.004-5.0$
Ethyl iodide	C_2H_5I	$<0.03-0.31$
Chloroiodomethane	CH_2ICl	$<0.004-0.21$
Bromoiodomethane	CH_2IBr	$<0.02-0.32$
Diiodomethane	CH_2I_2	$<0.02-1.02$
CHCs[f]		
Chloroform	$CHCl_3$	$<3-1800$
1,1,1-Trichloroethane	CH_3CCl_3	$200-4600$
Carbon tetrachloride	CCl_4	$100-500$
Tetrachloroethylene	C_2Cl_4	$<0.4-200$

[a,b,c] Remote surface locations in Northern and Southern Hemisphere.[36,48,167]

[d] Various remote locations reported in Ref. 14.

[e] Various remote marine locations reported in Ref. 16.

[f] Urban location in Richmond, Virginia.[22]

trichloroethylene, and perchloroethylene which are used as vapor degreasers and for dry cleaning of garments.[13] The CFCs have been used as propellants, refrigerants, and blowing agents for producing polyurethane foam. These are chemically inert in the troposphere and thus have very long atmospheric lifetimes; however, their very rapid decomposition in the stratosphere leads to stratospheric ozone depletion (see, e.g., Ref. 2). CFCs are also considered as potent greenhouse gases. The HCFCs and HFCs were manufactured as CFC substitutes. Addition of an extractable H to the molecule makes them more reactive and thus more easily removed from the atmosphere by oxidation. Methyl bromide (CH_3Br) is used as a fumigant for soils and shipments of fruits and vegetables. The fumigant is used domestically to control termites. Methyl bromide is the most abundant organobromine compound in the atmosphere but dibromethane and chlorobromomethane are also found in significant concentrations.[14] Halons are used as fire suppressants.

Iodine-containing organics such as CH_3I, $ClCH_2I$, and CH_2IBr are believed to be generated by biological processes in the ocean and are transferred to the marine atmosphere through atmosphere–water exchange.[15,16] Once in the atmosphere, these are readily photolyzed and generate aerosol particles that affect Earth's climate.[17,18]

Selection of the appropriate VOC-sampling method depends upon the physicochemical properties of the target analytes. Samples are collected for VOCs analysis in containers, on solid sorbents, or in a scrubbing solution. The VOC analytes which are amenable to processing by thermal methods are injected directly, preconcentrated from whole-air samples, or desorbed thermally from sorbent cartridges and analyzed by gas chromatography (GC). Polar VOC analytes that are derivatized on solid sorbents during sample collection or collected in a solvent containing a derivatizing agent are eluted or directly injected, respectively, and analyzed by high-pressure liquid chromatography (HPLC). The organic acids are collected in aqueous solution and injected directly into an ion chromatograph (IC). Gas chromatographic analysis can be by one-dimensional or two-dimensional approaches. The choice of a detection system is based on selectivity and sensitivity to a particular class of VOCs. Detectors based on principles of operation such as flame ionization, electron capture, reduction of a gas by the analyte, photoionization, and helium ionization are used to quantify VOCs by GC. Ultraviolet–visible and conductivity detectors are used in HPLC and IC applications. Mass spectrometers are interfaced to GCs and HPLCs for positive confirmation of molecular composition and also for quantification. Sampling and analytic instrumentation which use the various methodologies have been deployed on surface, balloon, and aircraft platforms to measure the temporal and spatial variability of atmospheric VOCs.[11,16,19–66,167]

II. SAMPLING TECHNIQUES

Samples for whole-air analysis of VOCs are typically collected in a polymeric bag, a glass bulb, or a canister, made of treated stainless steel or glass-lined stainless steel. Direct-sampling instruments inject a measured volume of air directly into the analytic instrument or preconcentrate the VOCs prior to injection by cryogenic or solid-sorbent approaches. Cartridges containing multiple solid sorbents have also been used to sample a wide range of VOCs. Many of the OxHCs require derivatization during sample collection. The derivatizing agents can be coated on a solid sorbent or dissolved in an organic solution.

A. WHOLE-SAMPLE COLLECTION METHODS

Whole samples for VOCs analysis can be collected in containers, preconcentrated in sampling loops, or measured in a sampling loop of known volume and injected directly into the analytic instrument. The materials of the containers and sampling loops are chosen for their compatibilities with the target analytes. Teflon and Tedlar polymeric bags have been used for selected NMHCs, OxHCs, and HaHCs with success to some extent. However, the permeabilities of the polymers vary, and analytes can be exchanged across the surface of the bag and equilibrate with the atmosphere outside the bag over extended storage periods. In addition, the analyte can also dissolve in the polymeric material. For example, Andino and Butler[67] found that methanol was stable in Tedlar bags up to 6 h, however, 10 to 15% of the methanol was lost after 25 h. Wang et al.[68] found losses of 5, 5, and 10% for 1,1-dichloroethane, trichloroethylene, and toluene in Tedlar bags, respectively. Tests of several different types of hardware for the sampling valves revealed that the VOCs were sorbed to the hardware. Carbon tetrachloride has been found to be stable in Tedlar bags for a period of 57 days.[69] Martin et al.[70] observed that monoterpenes were stable for periods up to 1 week when these were stored in Tedlar bags at room temperature in the dark. Doskey et al.[71] observed that, low aqueous solubility decreased the efficiency of transfer of VOCs in open tubular materials composed of various polymers of Teflon. For example, losses of tetrachloroethylene, benzene, toluene, and 1, 2-dimethylbenzene were 20, 5, 11, and 20%, respectively, for samples having a residence time of 2 min in Teflon PFA (perfluoroalkoxy). The 1,2-dimethylbenzene and tetrachloroethylene have similar aqueous solubility, exhibited the greatest losses, and were the least water soluble of the analytes.

Analyte stability in treated stainless steel canisters is related to the polarity of the molecule and the number of water molecules in the matrix air. At a relative humidity of about 10%, sufficient water molecules are available to form a monolayer on the inner surface of a 6 l spherical canister. Water molecules are thought to coat the active sites and minimize sorption of surface-active VOCs. However, if samples of air at high relative humidity are pressurized, water can condense in the container, and polar molecules can partition into the liquid water. As the air sample is withdrawn for analysis, the water will evaporate, liberating polar molecules dissolved in the water and increasing analyte concentrations in subsequent samplings of the container.

Methanol, ethanol, acetaldehyde, and acetone are stable in treated stainless steel canisters for at least 4 days when the matrix air has a relative humidity of at least 10%.[72-76] Batterman et al.[77] observed losses of C_4-C_9 n-aldehydes, benzaldehyde, and monoterpenes in humidified air stored in electropolished canisters and computed half-lives of 18 days for the target analytes. Approximately 19% of the losses occurred in the first 60 min of storage. The VOCs which are susceptible to oxidation by O_3 may also be affected by the composition of the sampling system. Greenberg et al.[78] found that O_3 in dry or humidified air in stainless steel canisters decomposes to an undetectable level within 90 min of entering the container. Fukui and Doskey[79] used the data of Ref. 78 for the exponential decay of O_3 concentrations in a stainless steel canister for a theoretical evaluation of the stability of a biogenic VOC mixture in a treated stainless steel canister containing O_3 at 50 ppbv. The modeling experiment revealed that, trans-3-hexen-1-ol, trans-2-hexenylacetate, and the acyclic and monocyclic monoterpenoids were the least stable analytes in the mixture, most of the losses occurred in the first 60 min of sample storage. Some reactive olefins are also oxidized rapidly by O_3, and comparisons of direct sampling approaches with sampling into treated stainless steel canisters showed lower levels of C_5-C_6 olefins for samples collected in the canisters.[80] Product literature for glass-lined stainless steel canisters indicates VOC stability as superior to that in treated stainless steel containers. However, O_3 is stable in glass containers, though it may be destroyed rapidly on metal surfaces in the sampling system before reaching the glass-lined canister. This necessitates a thorough evaluation of the stability of VOCs in entire sampling systems. Decomposition of VOCs by O_3 is more critical in direct sampling approaches which cryogenically preconcentrate the VOCs at $-185°C$ in a loop during sample collection.[39] Ozone is also condensed and must be removed before preconcentration to prevent oxidation of reactive species to the analytic instrumentation during thermal extraction and transfer of the VOCs through open tubular materials.

Interpretation of VOC data can be confounded by improper techniques which are used to clean the polymeric bag, glass bulb, or stainless steel canister; however, cleaning methods are rarely published. To certify the containers as "clean" depends upon the level of VOCs in the air to be sampled and thus is more stringent for sampling in remote areas than for sampling in urban locations. Containers are cleaned in several pressurization/evacuation cycles using clean humidified air. For some treated stainless steel canisters, because residues of the solution used to passivate the interior surface remain in the canister, most exhaustive cleaning is needed when the canister is obtained from the manufacturer. In our laboratory, canisters received from the manufacturer are opened to the atmosphere and heated at 105°C for 7 days. The canisters are then attached to an automated system, evacuated to 130 Pa at room temperature, and filled with humidified, ultrazero air. The cycle is repeated a total of 100 times and then the canisters are evacuated to 1.3 Pa for sampling. This initial cleaning process has proven to be effective for canisters to be used in the cleanest environments. After sampling, canisters are cleaned at room temperature in a total of three pressurization/evacuation cycles. Water can be injected into the canister after the final evacuation which is necessary to prepare for sampling in a very dry environment. Since organic aerosols can accumulate in canisters over time and release organic vapors into samples, use of an in-line particle filter in canister sampling systems is also essential. Elevated levels of VOCs can result from vaporization of organic substances from particles in "clean" canisters which are heated as part of the cleaning process.

Samples for VOC analysis can be collected in a container passively or actively. Passive collection systems are very simple and minimize the surfaces to which the analytes are exposed during sampling. The evacuated-chamber technique is used to fill polymeric bags passively.[81] Placing the bag in an airtight chamber and evacuating the chamber at a controlled rate allows the bag to fill at the rate of chamber evacuation. Canisters can be filled passively at a controlled rate by using a metering valve or an open-tube capillary. Rossner et al.[82] used deactivated fused-silica capillaries of various lengths and diameters as a variation of a sharp-edge orifice flow controller to fill canisters passively. When a critical orifice is used as a flow controller, the sampling rate varies with the difference in pressure between the sampled air and the container. Thus, the sampling objective and the expected variations in VOC concentrations are crucial in the design of sampling strategies which use this technique. Differential pressure flow controllers are available to maintain constant flow during the passive filling of canisters,[83] but the surfaces of the flow controller may be more active than those of a deactivated fused-silica capillary.

Active collection systems use pumps with diaphragms and valves to pressurize samples in canisters. For many VOC applications, the diaphragm and valve are of solid Viton- or Teflon-coated neoprene and solid Teflon, respectively. The automated sampler of Doskey and Bialk[83] consists of a Viton diaphragm pump; a metal bellows metering valve; a two-position, three-port valve and a multiposition valve; and a digital valve sequence programmer. Valcon M, a material which is very impermeable to light gases at room temperature, was chosen for the valve rotors to eliminate leakage of air into the preevacuated canisters while those are open to the multiposition valve before sample collection. An in-line, glass-treated 5-μm filter is used to remove large particles from the airstream. The pressure in the canister is monitored with a vacuum/pressure gauge. All tubing and fittings are treated with glass. A treatment process which deposits the thickest layer of glass should be used for stainless steel fittings because interior surfaces of fittings are rougher than the surfaces of stainless steel tubing. The valves are electrically actuated, and the sequence is controlled with a four-interval digital valve sequence programmer. During automated operation, the two-position, three-port valve directs the sample flow either to vent the pump or to the multiposition valve to collect a sample in the preevacuated canister. After the sample is collected, the multiposition valve advances to the next position in preparation for subsequent sample collections. This sequence of operation requires use of three of the four timing intervals on the digital sequence programmer. Each of the intervals can be set in ranges of 0 to 99 sec, 0 to 9.9 min, or 0 to 99 min.

The position and type of flow controller used in a VOC sampling system affect the sampler's ability to maintain constant flow over the collection period and to prevent water from condensing in the system. Diaphragm pumps should be mounted with the sampling ports facing downward to minimize condensation of water in the pump head, which causes deterioration of the diaphragm and valves. Locating a differential-pressure flow controller downstream of the pump also leads to condensation of water in the system when sampling occurs under very humid conditions. To prevent water condensation, a stainless steel tee and a micrometering valve can be inserted into the sampling system at the lowest position of the connecting tubing to vent water from the sample stream. Alternatively, a metal bellows metering valve can be located as a critical orifice upstream of the pump to control the flow, as is done in a passive collection system. The flow rate will vary with the difference in pressure between the sampled air and the canister. Thus, data interpretation may be confounded if VOC levels are highly variable during the sampling period. By sampling ambient air at a height of 3 m over five 1-hour intervals,[83] comparisons could be made between passive and active collection systems using differential pressure flow controllers, the active collection system using a critical orifice. The flow was controlled at 200 ml min^{-1} and 100 ml min^{-1} for the active and passive collection systems, respectively, filling 6 l canisters to pressures of 203 kPa and 88 kPa over the

1-hour sampling period. The sampling precisions of the three approaches were very similar for a mixture of acetylene, alkanes, alkenes, aromatics, and OxHCs in the sampled air.

B. SOLID-SORBENT COLLECTION METHODS

Tenax (2,6-diphylphenyleneoxide polymer), graphitized carbon solid and molecular sieve, and octadecyl- C_{18} bonded silica coated with a derivatizing agent are routinely used to sample VOCs. Tenax, the first sorbent used for VOC collection continues to be used to collect some classes of VOCs.[84,85,165] Simultaneously collected water is a common interference in VOC analytic techniques, and Tenax has low water retention capacity.[86] Tenax has been used for collection of HaHCs and $C_4 - C_{12}$ VOCs.[15,87-91,109] The dimensions of the sorbent bed, sample flow rate, and total volume of air to be sampled are determined by the volatility and concentration of the target analyte or the mixture of target analytes. Sorbent beds of Tenax typically contain about 0.5 g of material, permitting flow rates of about 500 ml min^{-1}. The sampled volumes are restricted to approximately 1 to 2 l, to prevent breakthrough of VOC analytes.

Graphitized-carbon solid sorbents and molecular sieves, and multiple beds of carbon-based materials, are now widely used for VOC collection and have replaced Tenax in many applications.[92] The materials have been used to sample a wide variety of VOCs, including $C_4 - C_{14}$ NMOCs and CHCs.[52,53,63,66,69,93] The graphitized carbon sorbents are prepared for sampling by purging the material in a stainless steel or glass sampling cartridge at 315 to 350°C with an inert gas such as N_2 or He free from O_2.[69,94] The cartridges are placed in leak-tight glass storage vessels (e.g., vials with septum-lined caps) to prevent contamination. Multiple beds of the carbon-based materials are designed so that the most volatile analytes pass through to the last bed in the sampled airstream. This is typically a molecular sieve which is the most adsorbent of the materials. Flow rates of 250 to 500 ml min^{-1} through multiple beds containing 600 mg of carbon-based sorbents have been used with 2.5 l sample volumes to sample $C_4 - C_{14}$ NMOCs in forest air.[53,93] Helmig and Greenberg[92] developed a direct sampling system using a multisorbent to collect the VOCs and a mass spectrometer for analysis. An automated multisorbent-tube sampling system has been used to measure vertical profiles of VOCs from a tethered balloon.[63] A multiposition valve with a mass flow controller on the downstream, controlled by a laptop computer with a serial valve interface are used to select and maintain the flow and control the sampling cycle.

Many carbonyl compounds can be collected effectively with C_{18}-bonded silica coated with 2,4-dinitrophenylhydrazine (2,4-DNPH).[27,95-97,151] The carbonyls react with the derivatizing agent as these are sampled, forming the corresponding 2,4-dinitrophenylhydrazones. In preparation for sampling, the C_{18}-bonded silica is cleaned and wetted with HPLC grade methanol and acetonitrile. The 2,4-DNPH is recrystalized twice from methanol and dissolved in HPLC grade acetonitrile, containing concentrated phosphoric acid, the solution which is used to impregnate the cartridges with derivatizing agent. The cartridges are dried overnight under vacuum in a desiccator which contains passive samplers of 2,4-DNPH-coated filter paper to prevent contamination. The cartridges are placed in leak-tight containers to prevent contamination before sampling. The sampling system includes a pump and mass flow controller to generate and maintain flow of $0.5-2$ l min^{-1}. Typical volumes of sampled air are approximately 60 l.

Annular diffusion denuders coated with KOH and NaOH have been used to sample gaseous formic and acetic acid.[98,99] The samples are collected at a rate of about 9 l min^{-1} over a period of 3 to 4 h. Kawamura et al.[100] used filters, impregnated with KOH, for the same purpose. A standard filter was placed in front of the KOH impregnated filter, to prevent collection of particle-phase formic and acetic acid. Quantitative collections were accomplished for sample volumes of 2.4 to 14 m^3 at sampling rates of 10 l min^{-1}. Andreae et al.[101] used a fast-flow cellulose filter, impregnated with K_2CO_3 in glycerol, to trap formic, acetic, and pyruvic acid. The oxidation of some aldehydes and peroxyacetylnitrate to form formic and acetic acid[102] is a suspected artifact associated with supports coated with a chemical base. In an intercomparison study of organic acid collection techniques, acetic acid was formed on a K_2CO_3-impregnated filter when formaldehyde was injected into a sampled airstream; however, the KOH-coated denuder technique exhibited no artifact formation.[102]

Many techniques have been used to remove O_3 from the sampled airstream and prevent decomposition of reactive VOCs during collection (see, e.g., Ref. 103). Conflicting data have been reported on the effects of O_3 on VOCs when sampled with solid sorbents. Several investigators reported losses of monoterpenes sampled on Tenax and graphitized carbon without an O_3 scrubber,[104–107] but others reported no losses.[88,90,108,163] Pellizzari et al.[110] reported losses of styrene and cyclohexene by reactions with O_3 during sampling. Roberts et al.[109] reported formation of monoaromatics, n-alkanes, and aldehydes when Tenax was exposed to O_3. Unsaturated compounds adsorbed on Tenax have been observed to react with nitrogen oxides, Cl_2, and O_3 to produce a mixture of OxHCs.[110,111] Decomposition of the Tenax by O_3 can also occur to produce benzaldehyde, phenol, and acetophenone; however, graphitized carbon sorbents appear more robust.[94] Interferences of O_3 with carbonyl sampling by 2,4-DNPH-coated cartridges include depletion of the 2,4-DNPH, degradation and formation of carbonyl hydrazones, and formation of unknown 2,4-DNPH products.

C. Scrubbing Solution and Condensed-Water Collection Methods

Bubblers and impingers, which absorb the analyte into a scrubbing solution, were the first sampling methods used for volatile carbonyl compounds.[112,113] The nebulization-reflux concentrator[114] and the scrubber coil[115,116] used to collect various OxHCs are more recent and more efficient versions of these techniques. The nebulization-reflux concentrator samples formic, acetic, and pyruvic acid by drawing air through a nebulizing nozzle at about $8 \, l \, min^{-1}$. The process aspirates an aqueous solution from a reservoir into the nozzle, where the solution is atomized into small droplets by impaction, forming a mist. The large surface of the mist efficiently extracts the organic acids from the sampled air. The scrubbing solution is aspirated from the reservoir at about $2 \, ml \, min^{-1}$. The mist impinges on a hydrophobic membrane and forms droplets which fall back into the solution as the sampled air exits the concentrator through the membrane. The technique has been applied to measure formic and acetic acid from surface and aircraft platforms.[51,70,117–122]

Scrubber coils containing solutions of 2,4-DNPH in an aqueous acetonitrile mixture, acidified to pH 2.50 have been designed to sample carbonyls exhibiting a Henry's law solubility similar to or greater than that of formaldehyde.[115,116]. For example, Lee et al.[54] used the scrubber-coil technique to measure formaldehyde, glycolaldehyde, glyoxal, and methylglyoxal from aircraft. The sampler has three parts: the inlet, an all-pyrex ten-turn coil (0.2 cm i.d., 2 cm helix diameter), and a glass reservoir downstream of the coil for gas–liquid separation. The small inlet is fabricated from a short piece of Teflon tubing to minimize sorption of analytes. Air enters the sampler through a pinhole in the tubing. A peristaltic pump is used to generate a flow of scrubbing solution ($0.3 \, ml \, min^{-1}$) through Teflon tubing (0.05 cm i.d.) into the sampling inlet. The scrubbing solution is drawn out from the base of the separator by another peristaltic pump. A mass-flow controller and metal bellows pump are connected to the top of the separator to maintain and generate a flow of air ($2.3 \, l \, min^{-1}$) through the glass coil sampler. Sampling periods are normally about 20 min.

Condensed-water collectors have been used to sample formaldehyde and formic and acetic acid.[123] The condensation collectors are constructed from chromium-plated copper. A refrigeration unit controls the temperature of the collection surface by circulating ethylene glycol through the plates. The ambient air temperature and dew point are monitored continuously and the plate temperature is maintained below the dew point so that water from the atmosphere condenses. To facilitate uniform wetting of the plate, the sampler is cleaned by first immersing the plates in boiling water and then by washing them with soap followed by rinsing with deionized water. The plates are tilted in the collector to collect the condensate in a graduated cylinder. Condensed water is collected at approximately 10 to 15 ml h^{-1} which is equivalent to sampling air at a rate >10 l min^{-1}. Episodic artifacts in the measurement of formic acid were observed when the condensed-water sampler was compared with the nebulization-reflux concentrator.[102]

III. SAMPLE PREPARATION

Sample preparation schemes are designed to be compatible with the collection method and the analytic technique. Whole-air analysis of samples, containing low levels of VOCs, requires extraction of a measured volume of air from a canister and preconcentration of the VOCs by condensing these on an inert, solid substrate or by collection on a sorbent material. For samples containing high concentrations of VOCs no preparation is necessary and the air is sampled directly into a loop of ml volume. Many of the VOCs collected on solid sorbents can be processed by thermal extraction with subsequent cryogenic preconcentration before injection into the analytic instrument. Some of the OxHCs collected on solid sorbents are eluted and injected directly into the analytic instrument. Analytes collected in scrubbing solutions do not require a processing step.

A. WHOLE-SAMPLE ANALYSIS

Whole-sample analysis of air collected in canisters requires extraction of a measured volume of air from the container. A sample canister, a 2.8 l reservoir, a cryogenic trap, and a GC connected by a two-position, six-port valve constitutes the simplest sample-processing system available (Figure 16.1).[83,124,125] The valve and all connecting glass-treated tubes are contained in a temperature-controlled enclosure maintained at 60°C. To measure the sample aliquot, a 2.8 l reservoir connected to the vent port of the valve is evacuated with a Teflon diaphragm pump to a pressure of 20 mm of Hg. The valve on the sample canister is opened, and the sample flows through the inlet and cryogenic trap at 40 to 80 ml min^{-1}. A micrometering valve positioned on the vacuum side of the preconcentration system controls the flow rate and ensures that the sample makes contact with only the treated surfaces. However, if canister pressures are greater than 2×10^5 Pa and the

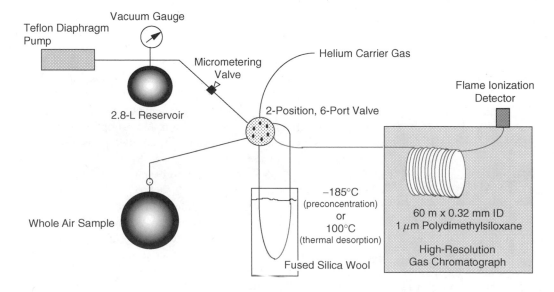

FIGURE 16.1 Cryogenic preconcentration/high-resolution gas chromatographic system for the analysis of whole-air samples collected in canisters.

micrometering valve is on the downstream of the cryogenic trap, O_2 may condense and introduce a measurement error in the sample volume. The pressure in the reservoir is monitored with a high-precision absolute-pressure gauge. In direct-sampling instruments, a mass-flow controller is used to maintain flow through the cryogenic trap, and a timing circuit controls the valve switching, so that the sample volume can be calculated.

The VOCs are cryogenically preconcentrated in a glass-treated tube (16 cm × 0.216 cm i.d.) packed with 9 cm of fused-silica wool and immersed in liquid Ar (at − 185°C). The analytes are thermally desorbed from the trap for 2 min by replacing the liquid Ar with boiling water. The preconcentration unit is interfaced to a high-resolution GC with an uncoated, deactivated fused-silica transfer line (0.53 mm × 1.2 m). The transfer line is contained within a stainless steel tube heated to 60°C with heat tape. A glass union is used to connect the transfer line to the analytic column. The temperature and composition of the transfer line are critical for quantitative transfer of the VOCs to the GC. A temperature of 60°C prevents excessive amounts of H_2O from being transferred to the GC. Deactivated fused silica is used in our laboratory because it is flexible and relatively inert for most VOC analytes. Results from a comparison of several types of transfer-line materials suggest that higher temperatures are required to transfer VOCs through glass-treated tubings because of the increased surface area produced by the treatment process.[126]

Sample preparation schemes which include a cryogenic preconcentration step sometimes require H_2O and CO_2 management steps. If direct sampling is used, removal of O_3 from the sampled air is required, however, some VOC analytes may also be removed by these treatments. Water can plug cryogenic traps, extinguish the H_2 flame in a flame ionization detector (FID), and interfere with the response of an electron capture detector (ECD). The resolution of VOCs on some porous-layer-open-tubular (PLOT) columns can be affected by CO_2. In direct-sampling instruments, O_3 is cryogenically preconcentrated and will decompose reactive VOCs when these are thermally extracted from the trap.[39]

Permeable membrane dryers composed of Nafion (a copolymer of tetrafluoroethylene and fluorosulfonyl monomer), coaxially mounted in stainless steel tubing, have been used to remove H_2O.[127,128] The sample is directed through the interior of the Nafion tubing, and H_2O permeates the membrane and enters a stream of dry He flowing through the annular space. The dryer is conditioned between analyses by directing dry He through the Nafion while it is heated to 100°C. However, Nafion may not be suitable for some VOC analytic schemes. The dryer is known to cause rearrangement of monoterpenes and to remove C_5-C_6 OxHCs.[129] Tenax and graphitic carbon have been used to preconcentrate VOCs because the sorbents are hydrophobic and water passes through the sorbent bed as the analytes are adsorbed.[86] Water can also be removed by passing the sample through tubing cooled to − 50°C, a technique which removes sufficient H_2O for cryogenic preconcentration of VOCs and analyzation by GC-FID.[66] Traps of NaOH have been used to remove CO_2. The most frequently used and commonly investigated O_3 scrubbers are Na_2CO_3, Na_2SO_3, $Na_2S_2O_3$, and KI.[103] Efficient O_3 removal and VOC transfer requires the scrubbers to be heated to approximately 50°C.[31,39] Obviously, for whole-sample analytic schemes which use high resolution GC-FID when sensitivity is not an issue, it is best to leave the H_2O and CO_2 in the sample and reduce the sample size to prevent plugging by ice and extinguishing the H_2 flame.

B. THERMAL DESORPTION, SOLVENT EXTRACTION, AND SUPERCRITICAL-FLUID EXTRACTION

Thermal desorption, solvent elution, and solvent extraction are used in VOC preparation schemes for samples collected on solid sorbents. Thermal desorption methods require determination of the sensitivity of the target analytes to the desorption temperature. It is also critical to remove all traces of O_2 from the gas used to purge the sorbent and transfer the analytes to the preconcentration trap. Quantitative recovery of monoterpenes from Tenax is accomplished at thermal desorption temperatures of 250°C.[90] Multibed sorbent tubes consisting of graphitic carbon solids and molecular sieve have been used to sample C_3-C_{15} VOCs. After sampling, H_2O is removed from the multisorbent by heating the bed to 45 to 50°C and purging with dry He in the direction used during sampling. The multisorbent is then heated to 250°C at 50°C min^{-1} and backflushed with He at 25 ml min^{-1} to transfer the VOCs to the preconcentration trap.[94]

Carbonyl compounds collected on 2,4-DNPH cartridges are eluted with 2 ml of HPLC grade acetonitrile in preparation for analysis.[151] Mixtures of C_1-C_{10} organic acids are extracted from inorganic-base impregnated filters and derivatized in preparation for analysis.[100,130] The filters are extracted by ultrasonication in purified water, and the extracts are centrifuged and passed through a cation exchange column. The eluent pH is adjusted to between 8.0 and 8.5 with HCl, and the solution is dried in a rotary evaporator. The carboxylates are dissolved in acetonitrile and esterified at 80°C using α, p-dibromoacetophenone reagent and a dicyclohexyl-18-crown-6 catalyst. The p-bromophenacyl esters are purified on a silica gel column in preparation for analysis. The organic acid preparation scheme of Ref. 130 converts the carboxylates to their pentafluorobenzyl derivatives. The organic acids are extracted with a 0.5% mixture of tri-n-octylphosphine oxide in MTBE. The extract is acidified with two drops of concentrated H_2SO_4, dried with anhydrous Na_2SO_4, and evaporated to dryness with a stream of N_2, then the residue is dissolved in acetone. The carboxylates are derivatized with α-bromo-2,3,4,5,6-pentafluorotoluene (PFBBr; pentafluorobenzyl bromide) reagent in a vial containing powdered K_2CO_3, heated to 40 to 60°C for 2 to 3h, and then filtered.

Supercritical fluid extraction (SFE) is useful for the extraction of high molecular weight organic compounds from a variety of matrices (see, e.g., Ref. 131). Few applications for VOCs have been reported, but the technique may be proved to be useful for VOCs which are thermally labile or are inefficiently transferred by thermal methods in sample preparation systems.[132] High pressures and moderate temperatures are required in SFE. Static extractions with supercritical fluid CO_2 are effective for most of the nonpolar organic compounds. Methanol can be added to the supercritical fluid as a polar modifier to facilitate the extraction of polar substances.[131] A temperature of 50°C and a pressure of 9.60×10^6 Pa are required to produce a supercritical fluid mixture of methanol and CO_2. To prepare the sample for analysis by GC, derivatizing agents can be dissolved in the polar modifier and introduced on-line to the supercritical fluid by using a microdelivery pump.[133]

IV. CHROMATOGRAPHY

High-resolution analysis of the entire suite of C_2-C_{12} NMHCs by conventional one-dimensional gas chromatography is typically performed with various siloxane-coated fused-silica capillaries or PLOT columns. Designing a single GC analysis which will adequately separate the entire complex mixture is very difficult to achieve, because ethene, acetylene, and ethane are difficult to separate, even with a 100 m siloxane-coated column. Furthermore, separations with a column of this length take approximately 70 min. The PLOT columns can completely resolve the C_2-C_3 NMHCs, but resolution of the C_6-C_{12} NMHCs is poor. If the sample is analyzed separately in a 60 m siloxane column, mounted in one GC and a PLOT column mounted in another, the analyses can be completed in approximately 45 min.[83] The preconcentrated sample is transferred to a 60 m × 0.32 mm i.d. fused-silica capillary column coated with a 1 μm-thick film of poly-dimethylsiloxane (PDMS) and a 30 m × 0.53 mm i.d., PLOT column coated with alumina to resolve the C_4-C_{12} and C_2-C_3 NMHCs, respectively. The oven temperature for the PDMS column is held at −50°C for 2 min, then increased to 210°C at 8°C min^{-1} and thereafter to 250°C at 20°C min^{-1}. It is then held at 250°C for 5 min.[134] The oven temperature for the alumina column is held at −50°C for 2 min and then increased to 200°C at 10°C min^{-1} and held at 200°C for 5 min. A chromatogram of air from Mexico City showing the resolution of approximately 250 peaks with a PDMS column is presented in Figure 16.2. Identifications presented in Table 16.4 were made by comparing retention times of standards, prepared in static dilution bottles with samples injected into

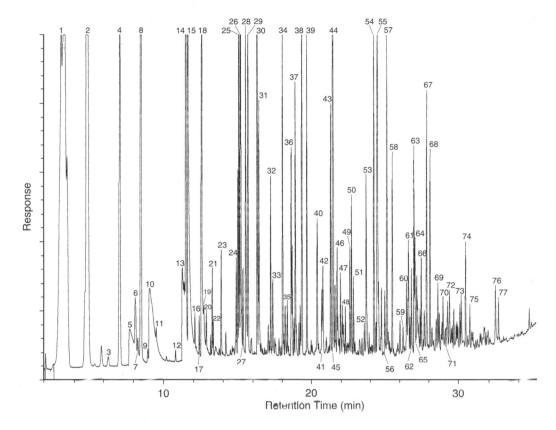

FIGURE 16.2 High-resolution gas chromatogram of Mexico City air.[166]

a GC-FID, under identical chromatographic conditions in our laboratory and by mass spectral analysis of samples from Mexico City by Lonneman.[135] Water was not removed during the sample pretreatment. Thus, before H_2O was vaporized in the PDMS column, it acted as a stationary phase, and the OxHCs partitioned into the liquid H_2O and eluted from the column as tailing peaks. Many of the peaks in the chromatogram of Figure 16.2 are only partially resolved, which complicates the integration of peaks. Techniques are now available in data processing software which can integrate the most complex chromatograms. If peaks are only partially resolved, imprecision and inaccuracies in quantitation are observed.[83] For example, analytic precisions for methanol and ethanol are about $\pm 20\%$, while precisions for well-resolved NMHCs are approximately $\pm 5\%$.

Conventional and comprehensive two-dimensional gas chromatographic (2DGC) techniques have been developed to increase the speed and resolution of chromatographic separations. Conventional 2DGC techniques, commonly known as heart-cutting, couple two chromatographic columns of different polarities in series to improve resolution of complex mixtures. Pierotti[136] reported use of a polar precolumn and dual analytic columns to analyze light nonpolar NMHCs, OxHCs, and monoaromatics. A switching valve was used to direct flows from the precolumn to the analytic columns. The C_2–C_5 NMHCs elute rapidly from a polar precolumn and can be cryo-focused and resolved on a PDMS column.[137,138] A short section of a PLOT column cooled to $-120°C$ can effectively focus the C_2 NMHCs.[139] Light OxHCs like acetaldehyde, acetone, methacrolein, methyl vinyl ketone, methyl ethyl ketone, and C_6–C_{12} NMHCs elute after the C_2–C_5 NMHCs; these are cryo-focused in a section of uncoated, deactivated fused-silica capillary tubing cooled to $-150°C$ and are thermally transferred to the second PDMS column. Deactivated fused

TABLE 16.4
Identifications for the VOCs in the High-Resolution Gas Chromatogram of Figure 16.2

1.	Acetylene, Ethene, Ethane	40.	2,2-Dimethylhexane
2.	Propene, Propane	41.	2,5-Dimethylhexane
3.	Methyl Chloride	42.	2,4-Dimethylhexane
4.	2-Methylpropane	43.	2,3,4-Trimethylpentane
5.	Acetaldehyde	44.	Toluene
6.	2-Methylpropene/1-Butene	45.	2,3-Dimethylhexane
7.	1,3-Butadiene	46.	2-Methylheptane
8.	*n*-Butane	47.	3-Methylheptane
9.	*trans*-2-Butene	48.	1-Octene
10.	Methanol	49.	2,2,5-Trimethylhexane
11.	*cis*-2-Butene	50.	*n*-Octane
12.	3-Methyl-1-butene	51.	Tetrachloroethylene
13.	Ethanol	52.	2,4-Dimethylheptane
14.	2-Methylbutane	53.	2,5-Dimethylheptane
15.	Acetone	54.	Ethylbenzene
16.	1-Pentene	55.	*p*-Xylene/*m*-xylene
17.	2-Methyl-1-butene	56.	Styrene/Heptanal
18.	*n*-Pentane	57.	*o*-Xylene
19.	Isoprene	58.	*n*-Nonane
20.	*trans*-2-Pentene	59.	Isopropylbenzene
21.	Dichloromethane	60.	2,6-Dimethyloctane
22.	2-Methyl-2-butene	61.	α-Pinene
23.	2,2-Dimethylbutane	62.	*n*-Propylbenzene
24.	2,3-Dimethylbutane	63.	*m*-Ethyltoluene
25.	Methyl-*t*-butyl ether	64.	*p*-Ethyltoluene
26.	2-Methylpentane	65.	1,3,5-Trimethylbenzene
27.	Butanal	66.	β-Pinene
28.	2-Butanone	67.	*t*-Butylbenzene/1,2,4-trimethylbenzene
29.	3-Methylpentane	68.	*n*-Decane
30.	*n*-Hexane	69.	1,2,3-Trimethylbenzene
31.	*trans*-2-Hexene	70.	Limonene
32.	Methylcyclopentane	71.	1,3-Diethylbenzene
33.	2,4-Dimethylpentane	72.	*n*-Butylbenzene
34.	Benzene	73.	1-Undecene
35.	Cyclohexane	74.	Undecane
36.	2-Pentanone/cyclohexene	75.	1,2,4,5-Tetramethylbenzene
37.	2-Methylhexane/3-methylhexane	76.	1-Dodecene
38.	2,2,4-Trimethylpentane	77.	*n*-Dodecane
39.	*n*-Heptane		

silica is chosen for the cryofocuser to facilitate thermal extraction of the polar analytes. However, the monoaromatics may not be completely separated from some of the C_6–C_{12} OxHCs in the PDMS column. Effective removal of H_2O by the polar precolumn improves the resolution of the OxHCs in the PDMS column, which is backflushed in preparation for the next sample injection.

High-speed chromatography combines column selectivity and special selectivity adjustment techniques with conventional 2DGC to improve the speed of analysis in what is essentially a rapid, heart-cutting technique.[140] Rapid chromatography requires cryofocusing of analytes in the narrowest volume possible, prior to injection and rapid heating of the cryofocuser to inject the sample onto the analytic column.[139,141] This requires the H_2O and CO_2 to be removed from the sample. To be effective, the temperature of the cryofocuser must be at least 100°C below the boiling point of the analyte.[142] Focusing the lightest NMHCs in open tubes under these conditions is not possible. Instead, sections of PLOT columns are used to focus the analytes.[60] However, thermal extraction of some polar analytes from cryofocusers fabricated from PLOT columns might not be possible. High-speed chromatographs use serially linked analytic columns with stationary phases of different polarities to tune the selectivity of a separation. The selectivity is also adjusted by changing the relative lengths of the columns or the pressures at the junction points between the columns. The technique uses rapid, partial separation of the analyte mixture on a nonpolar capillary column. The partially separated components are transferred to either a polar or a nonpolar capillary column to complete the separation and to separate detectors to monitor the effluents. High-speed chromatography can separate 20 to 30 analytes in less than 1 min.[140] Column stationary phases include PDMS, phenyl methyl polysiloxane (PMP), trifluoropropoyl methyl polysiloxane (TFP), bis cyanopropyl polysiloxane, and polyethylene glycol (PEG). The inner diameters of the columns are normally 0.25 or 0.32 mm. A wide tuning range for VOCs is obtained through a serial connection of a DMP and PEG columns. For example, a mixture of 21 components, ranging in volatility from methylene chloride to *o*-xylene, can be separated in approximately 40 sec.

Comprehensive GC × GC requires the use of primary and secondary columns.[143,144] An on-column, thermal modulator collects sample from the primary column at a frequency of about 1 Hz, which is more rapid than high-speed chromatographic techniques, and transfers the plug to a secondary column. Plotting the variation of retention time in the secondary column as a function of the retention time in the primary column produces an orthogonal or 2D chromatogram. Fast carrier gas flows and short columns (typically 5 m) result in extremely rapid chromatography. Thermal and differential-flow modulators have been developed to sample the primary column. Thermal modulators allow the entire sample to pass through to the secondary column.[143] The most effective thermal modulators spray liquid N_2 or CO_2 onto a section of the primary column to freeze the sample.[145,146] A slotted heater then sweeps a short section of the capillary to inject the sample rapidly into the secondary column. Lewis et al.[147] were able to separate over 500 VOCs in an urban air sample by using the comprehensive GC × GC technique with thermal modulation. The primary and secondary columns were a nonpolar 50 m × 0.53 mm i.d. column with a 5-μm film thickness of PDMS and a moderately polar 2.2 m × 0.15 mm i.d. column with a 0.2-μm film thickness of 50% phenyl-polysilphenylene siloxane. These workers were able to resolve over 100 multisubstituted monoaromatic hydrocarbons and OxHCs. Superior resolution is obtained with the column combination; however, analysis times are very long with a 50-m primary column.

A comprehensive GC × 2GC technique with differential flow modulation has been developed for rapid analysis of VOCs.[148,164] In comprehensive GC × 2GC, two secondary columns of different polarities are used, requiring a GC with two detectors to monitor the column effluents. The flow-switching device is a modification of the unit developed by Sacks and Akard[140] for high-speed chromatography. The flow of carrier gas to the columns is controlled by a three-port solenoid valve placed outside the GC oven and not in the sample flowpath. Parallel and serially positioned tee unions control the flows of carrier and auxiliary gases and interface the primary column effluent to the secondary columns. Bueno and Seeley[148] used a 5.0 m × 0.25 mm i.d. capillary column with a 1.4-μm film thickness of 6% cyanopropylphenyl and 94% dimethylpolysiloxane for separation of complex mixtures of VOCs. Two 5.2 m × 0.25 mm i.d. capillary columns, one with a 0.10-μm film thickness of PEG and the other with a 0.25 μm film thickness of TFP, were used as the secondary columns. A VOCs mixture of 41 NMHCs and OxHCs was completely separated in 5 min using this combination of analytic columns. A chromatogram of forest air is presented in Figure 16.3.

Derivatized carbonyls are analyzed by HPLC with isocratic and gradient elution methods. Grosjean et al.[96] used a 4.6-mm × 125-mm C_{18} column and isocratically eluted the 2,4-dinitrophenylhydrazone derivatives with an acetonitrile/water eluent (55/45% by volume) at

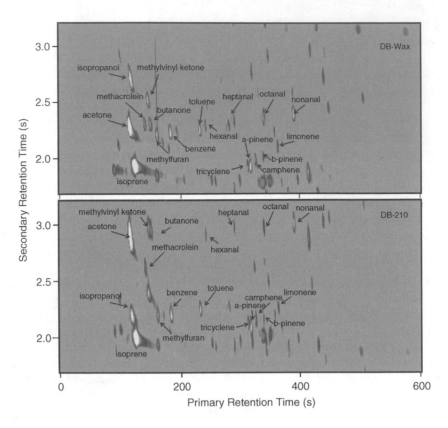

FIGURE 16.3 A two-dimensional gas chromatogram of forest air developed with a GC × 2GC technique.[148]

a flow rate of 1 ml min^{-1}. A 30/70 (v/v) water/methanol mixture at a flow rate of 1.5 ml min^{-1} has also proven effective at resolving the hydrazones.[149] Kuwata et al.[150] used a 4.0-mm × 150-mm C$_{18}$ column and isocratic elution at 1.0 ml min^{-1} with a mobile phase of 65/35 (v/v) acetonitrile/water. An example of the resolution of a complex mixture of carbonyls by this elution scheme is presented in Figure 16.4. Identifications of the peaks in the chromatogram are presented in Table 16.5. Shepson et al.[27] used two 4.6 mm × 250 mm C$_{18}$ columns connected in series and a gradient elution scheme with a mobile phase mixture of methanol/water that was solvent programmed from 60 to

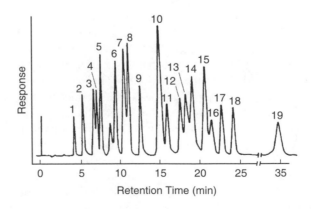

FIGURE 16.4 High-pressure liquid chromatogram of a mixture of the 2,4-dinitrophenylhydrazone derivatives of a standard carbonyl mixture.[150]

TABLE 16.5
Identifications for the Carbonyls in the High-Pressure Liquid Chromatogram of Figure 16.4

1.	Formaldehyde	11.	*n*-Valeraldehyde
2.	Acetaldehyde	12.	*o*-Tolualdehyde
3.	Acrolein	13.	*m*-Tolualdehyde
4.	Acetone	14.	*p*-Tolualdehyde
5.	Propionaldehyde	15.	Methyl-*i*-butyl ketone/Methyl-*sec*-butyl ketone
6.	Crotonaldehyde	16.	Methyl-*t*-butyl ketone
7.	Methyl ethyl ketone	17.	Methyl-*n*-butyl ketone
8.	*i*-*n*-Butraldehyde	18.	*n*-Caproaldehyde
9.	Benzaldehyde	19.	Methyl-*n*-amyl ketone
10.	*i*-Valeraldehyde		

90% methanol. Zielenska et al.[151] used a 3.9 mm × 150 mm C_{18} column and a gradient elution scheme. The gradient started with a 100% mixture of water/acetonitrile/tetrahydrofuran, 60/30/10 (v/v/v), for 2 min. The gradient was increased linearly to a 100% mixture of acetonitrile/water, 60/40 (v/v) over 10 min, then was held at the 100% mixture of acetonitrile/water, 60/40 (v/v), for 8 min. Lee and Zhou[115] used a 250 mm C_{18} column and a ternary gradient of water, acetonitrile, and methanol to analyze samples collected by the glass coil scrubbing technique. Formic, acetic, and pyruvic acid are routinely analyzed by ion exchange chromatography with a weak bicarbonate (0.45 to 0.75 mM) or borate (0.125 to 4.000 mM) eluent (see, e.g., Ref. 102). The organic acids can also be resolved by ion exclusion chromatography with 1.0 to 2.0 mM HCl. An aliquot of the aqueous scrubbing solution is injected directly into the instrument.

V. DETECTION MODES

Detection systems for GC are chosen for their sensitivity and selectivity for a particular class of VOCs. Detectors for GC include FID, the ECD, the photoionization detector (PID), the pulsed discharge detector (PDD), and the reduction gas detector (RGD). A variety of mass spectrometers can also be interfaced with a GC for confirmation of molecular structure and quantitation. Single-wavelength ultraviolet-visible detectors (190 to 600 nm) and diode array detectors are used to detect carbonyls as their 2,4-dinitrophenylhydrazone derivatives. The absorption maxima for aliphatic carbonyls, aromatic carbonyls, and dicarbonyls are near 360 nm, 385 to 390 nm, and 415 to 430 nm, respectively. Formic, acetic, and pyruvic acid are detected by ion conductivity.

The most common GC detector for trace analysis is the FID. The FID is a mass-sensitive detector that responds to nearly all organic compounds with roughly the same sensitivity. The linear dynamic range of the FID is 10^6. It has moderately high sensitivity of 4 pg of carbon (C). The FID operates by mixing the effluent of the analytic column with H_2 and burning the compound at the tip of a jet in an excess of air. The FID is a destructive detector. Organic matter entering the detector is combusted in the H_2 flame, forming ions that are collected at an electrode above the jet. The current is amplified and recorded. The detector response is proportional to the mass of effective C atoms (basically the number of $-CH_2-$ groups in the molecule). Thus, organic compounds having double and triple bonds and heteroatoms exhibit diminished response. However, empirical methods have been used to determine the effective carbon numbers for many different molecules.[152,153] If the sensitivity of the FID is sufficient for the target analytes and their effective carbon numbers have been determined, the detector can be calibrated with a single saturated hydrocarbon.

The ECD is a selective detector that responds primarily to compounds having functional groups with high electron affinities. The ECD is most sensitive to HaHCs but also responds to N-substituted hydrocarbons, highly substituted OxHCs, and some aromatic compounds. The ECD is extremely sensitive to many HaHCs and has a detection limit of about 0.08 pptv and 3.3 pptv for carbon tetrachloride and trichloromethane, respectively.[69,81] Thus, many of the HaHCs in ambient air can be sampled directly into a milliliter-sized loop and injected into the analytic column of a GC-ECD without preconcentration.[44,154] The ECD must be calibrated for each substance, because the response is compound specific and may vary by orders of magnitude, even for compounds having similar structures. The dynamic linear range is reported to be about 10^4; however, it may extend only to about 10 ng for many HaHCs.[69,81] The ECD operates by introducing the effluent of the analytic column into a cell containing a β-radiation source, typically radioactive ^{63}Ni. A detector make-up gas (N_2 or CH_4-Ar) or the column carrier gas (N_2) is ionized in the detector, producing a standing current of electrons. When a compound with high electron affinity enters the detector, the standing current is decreased. The linearity of the detector has been improved by adding to the effluent of the analytic column a temperature-controlled make-up gas that maintains a constant standing current and minimizes baseline drift. The detector is sensitive to the concentration of the analyte in the cell. Therefore, the flow rate of the make-up gas through the detector is critical to sensitivity and recent designs of the detector incorporate a smaller cell volume.

In the PID, the organic compound is ionized by UV radiation. The PID is more sensitive than the FID for compounds having photoionizable moities, such as conjugated double bonds and some heteroatoms.[155,156] Thus, the PID can be used to obtain some information on molecular structure. The PID is typically used to measure aromatic and some olefinic VOCs. The selectivity of the detector can be adjusted by using UV lamps of different energies (e.g., 11.8, 10.6, 9.5, and 8.4 eV). The PID's sensitivity to compounds having conjugated double bonds is in general about 1 ppb.

The PDD is a multiple-mode detector that operates in the electron capture, He ionization, and photoionization modes. The PDD is starting to replace the ECD, because it can operate in the electron capture mode without a sealed radioactive source and is safe for commercial carriers to transport. The detector uses a stable, low-powered, pulsed-DC direct current in He as the ionization source. Noble gases are added to the He and excited to higher electronic-energy states. Organic molecules entering the detector that have ionization potentials less than or equal to that of the excited noble gas are ionized, producing an electric current that can be monitored. In the ECD mode, the PDD is sensitive to HaHCs with minimum detectable quantities (MDQs) of 10^{-15} to 10^{-12} g. In the He ionization mode, the PDD is a universal, nondestructive detector. The standing current is increased, with MDQs in the low picograms or less. When the He discharge gas is doped with a noble gas (e.g., Ar, Kr, Xe), the PDD acts as a specific PID for determination of aliphatics and aromatics.

The RGD has been used to measure selected OxHCs like acetaldehyde and acetone[157] and also olefinic compounds like isoprene.[158] Analytes in the He effluent of the analytic column reduce HgO solid to Hg vapor; emitting UV radiation that is detected. The temperature of the detector must be kept to 200°C. The detector is concentration sensitive, with MDQs for acetaldehyde and acetone in the sub-p*mole* range.

Mass spectral analysis is used for confirmation of molecular structure and also for quantitation. However, the response of mass spectrometric detectors is compound specific and it must be calibrated for each target analyte. Mass spectrometers based on the linear quadrupole, quadrupole ion trap, time-of-flight (TOF), and proton transfer have been developed for VOCs analysis. The quadrupole mass spectrometers typically use electron impact to generate fragment ions of the VOC analyte, although chemical ionization is used in some applications. The quadrupole mass spectrometers are path-stability mass spectrometers or mass filters. An ion beam is injected into a dynamic arrangement of electromagnetic fields. The linear quadrupole consists of four electrically conducting, parallel rods of hyperbolic cross section. A radio frequency (RF) field is formed by application of DC and RF voltages. Ideally only one mass will take a stable path and pass through the rods; all others will take an unstable path and collide with the rods. The DC and RF voltages are varied while their ratio is kept constant so that ions of increasing mass will take the stable path. Gas chromatographs with linear quadrupoles have been developed for direct sampling of VOCs from surface and aircraft platforms.[62,66,92] The quadrupole ion trap is essentially a three-dimensional quadrupole consisting of a ring electrode and two end-cap electrodes. The ion trap is essentially a linear quadrupole rotated around an axis perpendicular to a line through the centers of a pair of opposing equipotential hyperbolic rods. The ion trap is about 10 to 100 times more sensitive than the linear quadrupole. However, ion or molecular reactions with air and water in the sample can complicate mass spectra. Fukui and Doskey[79] used an ion trap to identify a mixture of C_6 OxHCs and monoterpenoids in emissions from herbaceous plants. Kuster et al.[66] report good agreement for VOC quantitation by direct sampling GCs with an ion trap mass spectrometer, a linear quadrupole mass spectrometer, and an FID. Chemical ionization techniques can be used in quadrupole mass spectrometers to improve the identification and quantitation of analytes. Chien et al.[130] derivatized carboxylic acids in the free gas phase in the cell of an ion trap mass spectrometer. Pentafluorobenzyl alcohol was added to the transfer line going into the cell, and CH_4 was introduced through the chemical ionization inlet as a stabilization reagent gas to convert the carboxylic acids to their pentafluorobenzyl derivatives.

A TOF-mass spectrometer (TOF–MS) is equipped with a modified electron impact source and a long, straight flight tube. The fragment ions are detected by their unique arrival times at the detector at the end of the drift tube. The advantage of TOF/MS is its rapid scanning ability, which has proven useful for 2DGC applications.[7] For example, the TOF/MS has been used to collect spectra of analytes in cigarette smoke at a rate of 500 Hz.[159]

A mass-spectral technique that has recently found wide application for its direct sampling and real-time analytic capabilities is proton transfer-chemical ionization mass spectrometry (PT-CIMS). In PT-CIMS, organic trace gases are ionized by a proton transfer reaction in an ion drift tube. Protonated water is produced in a sidearm of the drift tube. Water vapor is diluted in a stream of He, and electrons from a resistively heated filament ionize the gas mixture. The electrons accelerate toward a grid a few millimeters downstream of the filament. A voltage difference of 20 V is used to maximize the ion count rates. A feedback circuit operating at 10 to 15 μA minimizes ion-molecule reactions and leads to the production of mainly H_3O^+ ions. Ion clustering is prevented, and the ion yield is maximized in the sidearm by reducing the residence times of the ions and producing an electric field in the sidearm. The H_3O^+ ions flow from the sidearm and react with the analytes in the stream of air in the main drift tube. The drift tube increases the ion kinetic energy and inhibits formation of cluster ions. At pressures of 1 to 2 Torr and a homogeneous electric field of approximately 35 to 70 V cm^{-1} in the drift tube, ion clusters are not stable and fragment to yield protons. The primary and product ions are extracted from the drift tube and analyzed by quadrupole mass spectrometry. The method can detect VOCs with proton affinities greater than that of water. The technique has been validated for a mixture of OxHCs and NMHCs including methanol, acetonitrile, acetaldehyde, acetone, benzene, toluene, and a suite of substituted aromatic compounds.[160] The PT-CIMS has been used in fast-response mode to measure biogenic emissions of methanol, acetaldehyde, acetone, and butanone from the cutting and drying of agricultural crops.[161,162]

VI. CONCLUSIONS

Molecular characterization of the complex mixture of atmospheric VOCs is essential for studies of atmospheric chemistry and air quality, as well as to evaluate the effects of VOCs on climate. Because of the myriad VOCs present in the atmosphere, some form of chromatographic separation is needed to quantify the molecular composition of VOCs even when a mass spectrometer is used as a detector. Chromatographic analysis of atmospheric VOCs is evolving toward direct-sampling instruments that use rapid, high-resolution gas chromatographic approaches like comprehensive 2DGC. However, the wide range of volatilities and polarities represented by VOCs makes it difficult to (1) sample the atmosphere by a single approach and (2) perform rapid molecular characterization of the entire suite of VOCs in a single chromatographic analysis.

Direct sampling into an open tube-packed with fused-silica wool or beads, or glass beads and cooled to $-185°C$, is the most efficient, cleanest approach to preconcentrating VOCs from the atmosphere. However, direct cryogenic preconcentration requires removal of O_3 to prevent decomposition of reactive VOCs, and this may also remove some target analytes. In addition, desorbing the VOCs from the cryogenic trap by thermal methods and transferring the analytes to the GC is inefficient for some polar VOCs and can cause catalytic decomposition and molecular rearrangement of thermally labile VOCs exposed to untreated metal surfaces. Rapid, comprehensive 2DGC of a cryogenically preconcentrated sample requires some type of H_2O and CO_2 management, which may also remove some VOCs. Multiple beds of sorbents can be a suitable alternative to direct cryogenic preconcentration; however, catalytic decomposition and molecular rearrangement of some reactive VOCs is possible during the thermal desorption. Supercritical-fluid extraction is an alternative preparative technique to thermal desorption. Combining SFE with an on-line derivatization method can enable the NMHCs and volatile carbonyl species to be analyzed by GC, making separate collection of the carbonyls and analysis by HPLC unnecessary. However, a more thorough evaluation of the technique is required for quantitative extraction of thermally sensitive and oxygenated species from sorbent materials.

Of currently available GC separation methods, comprehensive 2DGC approaches can perform the most rapid molecular characterization of the entire suite of VOCs in a single chromatographic analysis. A GC × GC technique produces adequate resolution when a mass spectrometer is used as a detector, because co-eluting species can be discerned by their mass spectra. However, GC × 2GC approaches produce superior resolution and are preferred for use with detectors other than a mass spectrometer. The GC × 2GC technique requires monitoring of the secondary-column effluents by two detectors operating on the same principle. Dual FID and dual PDD systems would be useful for analyses of (1) the NMHCs and OxHCs and (2) the HaHCs, respectively. However, the very different sensitivities and magnitudes of the linear ranges of the PDD and FID complicate quantitation of species of the three categories of VOCs in a single analysis. Accomplishing this would require accurate splitting of a single injection to an arrangement of two primary columns and four secondary columns in a dual FID–dual PDD chromatographic system.

The challenge of quantifying the complex mixture of atmospheric VOCs, which are highly variable in their physicochemical properties and ambient levels, has produced numerous advancements in sampling techniques and preparation schemes, chromatographic analysis, and, detection systems. In addition, accurate and precise measurement of the temporal and spatial variability of atmospheric VOCs has led to the development of rapid sampling and high-throughput analytic schemes that are specific to various classes of VOCs. New advances in the chromatographic analysis of atmospheric VOCs will result from the continued development of field-deployable instrumentation that can rapidly sample and analyze several classes of VOCs at the same time from surface and aircraft platforms.

ACKNOWLEDGMENTS

Special thanks to William A. Lonneman for his guidance and many helpful discussions on VOC analysis. Karen Haugen's assistance with editing is greatly appreciated. This work was supported by the U.S. Department of Energy under contract W-31-109-Eng-38, as part of the Atmospheric Science Program (ASP) of the Office of Science, Office of Biological and Environmental Research, Climate Change Research Division.

REFERENCES

1. Seinfeld, J. H. and Pandis, S. N., *Atmospheric Chemistry and Physics*, Wiley, New York, 1998.
2. Finlayson-Pitts, B. J. and Pitts, J. N. Jr., *Chemistry of the Upper and Lower Atmosphere*, Academic Press, San Diego, CA, 2000.
3. Hewitt, C. N., *Reactive Hydrocarbons in the Atmosphere*, Academic Press, San Diego, CA, 1999.
4. Kavouras, I. G., Mihalopoulos, N., and Stephanou, E. G., Formation of atmospheric particles from organic acids produced by forests, *Nature*, 395, 683–686, 1998.
5. Yu, J., Griffin, R. J., Cocker, D. R. III, Flagan, R. C., Seinfeld, J. H., and Blanchard, P., Observations of gaseous and particulate products of monoterpene oxidation in forest atmospheres, *Geophys. Res. Lett.*, 26, 1145–1148, 1999.
6. Claeys, M., Graham, B., Vas, G., Wang, W., Vermeylen, R., Pashynska, V., Cafmeyer, J., Guyon, P., Andreae, M. O., Artaxo, P., and Maenhaut, W., Formation of secondary organic aerosols through photooxidation of isoprene, *Science*, 303, 1173–1176, 2004.
7. Hamilton, J., Webb, P., Lewis, A., Hopkins, J., Smith, S., and Davy, P., Partially oxidized organic components in urban aerosol using GC × GC-TOF/MS, *Atmos. Chem. Phys. Discuss.*, 4, 1393–1423, 2004.
8. Chang, T. Y., Hammerle, R. H., Japar, S. M., and Salmeen, I. T., Alternative transportation fuels and air quality, *Environ. Sci. Technol.*, 25, 1190–1197, 1991.
9. Hoekman, S. K., Speciated measurements and calculated reactivities of vehicle exhaust emissions from conventional and reformulated gasolines, *Environ. Sci. Technol.*, 26, 1206–1216, 1992.
10. Fehsenfeld, F., Calvert, J., Fall, R., Goldan, P., Guenther, A. B., Hewitt, C. N., Lamb, B., Liu, S., Trainer, M., Westberg, H., and Zimmerman, P., Emissions of volatile organic compounds from vegetation and the implications for atmospheric chemistry, *Global Biogeochem. Cycles*, 6, 389–430, 1992.
11. Singh, H., Chen, Y., Staudt, A., Jacob, D., Blake, D., Heikes, B., and Snow, J., Evidence from the Pacific troposphere for large global sources of oxygenated organic compounds, *Nature*, 410, 1078–1081, 2001.
12. Keene, W. C. and Galloway, J. N., Considerations regarding sources for formic and acetic acids in the troposphere, *J. Geophys. Res.*, 91, 14466–14474, 1986.
13. Aronian, P. F., Scheff, P. A., and Wadden, R. A., Wintertime source-reconciliation of ambient organics, *Atmos. Environ.*, 23, 911–920, 1989.
14. Wamsley, P. R., Elkins, J. W., Fahey, D. W., Dutton, G. S., Volk, C. M., Myers, R. C., Montzka, S. A., Butler, J. H., Clarke, A. D., Fraser, P. J., Steele, L. P., Lucarelli, M. P., Atlas, E. L., Schauffler, S. M., Blake, D. R., Rowland, F. S., Sturges, W. T., Lee, J. M., Penkett, S. A., Engel, A., Stimpfle, R. M., Chan, K. R., Weisenstein, D. K., Ko, M. K. W., and Salawitch, R. J., Distribution of halon-1211 in the upper troposphere and lower stratosphere and the 1994 total bromine budget, *J. Geophys. Res.*, 103, 1513–1526, 1998.
15. Yokouchi, Y., Mukai, H., Yamamoto, H., Otsuki, A., Saitoh, C., and Nojiri, Y., Distribution of methyl iodide, ethyl iodide, bromoform, and dibromomethane over the ocean (east and southeast Asian seas and the western Pacific), *J. Geophys. Res.*, 102, 8805–8809, 1997.
16. Carpenter, L. J., Sturges, W. T., Penkett, S. A., Liss, P. S., Alicke, B., Hebestreit, K., and Platt, U., Short-lived alkyl iodides and bromides at Mace Head, Ireland: links to biogenic sources and halogen oxide production, *J. Geophys. Res.*, 104, 1679–1689, 1999.
17. Kolb, C. E., Iodine's air of importance, *Nature*, 417, 597–598, 2002.
18. O'Dowd, C. D., Jimenez, J. L., Bahreinl, R., Flagan, R. C., Seinfeld, J. H., Hamerl, K., Pirjola, L., Kulmala, M., Jennings, S. G., and Hoffmann, T., Marine aerosol formation from biogenic iodine emissions, *Nature*, 417, 632–636, 2002.
19. Rasmussen, R. A. and Khalil, M. A. K., Isoprene over the Amazon Basin, *J. Geophys. Res.*, 93, 1417–1421, 1988.
20. Hartmann, W. R., Andreae, M. O., and Helas, G., Measurements of organic acids over Central Germany, *Atmos. Environ.*, 23, 1531–1533, 1989.
21. Kanakidou, M., Bonsang, B., and Lambert, G., Light hydrocarbon vertical profiles and fluxes in a French rural area, *Atmos. Environ.*, 23, 921–927, 1989.

22. McClenny, W. A., Oliver, K. D., and Pleil, J. D., A field strategy for sorting volatile organics into source-related groups, *Environ. Sci. Technol.*, 23, 1373–1379, 1989.

23. Greenberg, J. P., Zimmerman, P. R., and Haagenson, P., Tropospheric hydrocarbon and CO profiles over the U.S. West Coast and Alaska, *J. Geophys. Res.*, 95, 14015–14026, 1990.

24. Bonsang, B., Martin, D., Lambert, G., Kanakidou, M., Le Roulley, J. C., and Sennequier, G., Vertical distribution of non methane hydrocarbons in the remote marine boundary layer, *J. Geophys. Res.*, 96, 7313–7324, 1991.

25. Martin, R. S., Westberg, H., Allwine, E., Ashman, L., Farmer, J. C., and Lamb, B., Measurement of isoprene and its atmospheric oxidation products in a central Pennsylvania deciduous forest, *J. Atmos. Chem.*, 13, 1–32, 1991.

26. Mowrer, J. and Lindskog, A., Automatic unattended sampling and analysis of background levels of C_2-C_5 hydrocarbons, *Atmos. Environ.*, 25A, 1971–1979, 1991.

27. Shepson, P. B., Hastie, D. R., Schiff, H. I., Polizzi, M., Bottenheim, J. W., Anlauf, K., Mackay, G. I., and Karecki, D. R., Atmospheric concentrations and temporal variations of C_1-C_3 carbonyl compounds at two rural sites in central Ontario, *Atmos. Environ.*, 25A, 2001–2015, 1991.

28. Blake, D. R., Hurst, D. F., Smith, T. W. Jr., Whipple, W. J., Chien, T.-Y., Blake, N. J., and Rowland, F. S., Summertime measurements of selected NMHCs in the Arctic and Subarctic during the 1988 Arctic Boundary Layer Expedition (ABLE 3A), *J. Geophys. Res.*, 97, 16559–16588, 1992.

29. Doskey, P. V. and Gaffney, J. S., Nonmethane hydrocarbons in the arctic atmosphere at Barrow, Alaska, *Geophys. Res. Lett.*, 19, 381–384, 1992.

30. Khalil, M. A. K. and Rasmussen, R. A., Artic haze: patterns and relationships to regional signatures of trace gases, *Global Biogeochem. Cycles*, 7, 27–36, 1993.

31. Montzka, S. A., Trainer, M., Goldan, P. D., Kuster, W. C., and Fehsenfeld, F. C., Isoprene and its oxidation products, methyl vinyl ketone and methacrolein, in the rural troposphere, *J. Geophys. Res.*, 98, 1101–1111, 1993.

32. Penkett, S. A., Blake, N. J., Lightman, P., Marsh, A. R. W., Anwyl, P., and Butcher, G., The seasonal variation of NMHCs in the free troposphere over the north Atlantic Ocean: possible evidence for extensive reaction of hydrocarbons with the nitrate radical, *J. Geophys. Res.*, 98, 2865–2885, 1993.

33. Andronache, C., Chameides, W. L., Rodgers, M. O., Martinez, J., Zimmerman, P., and Greenberg, J., Vertical distribution of isoprene in the lower boundary layer of the rural and urban southern United States, *J. Geophys. Res.*, 99, 16989–16999, 1994.

34. Davis, K. J., Lenschow, D. H., and Zimmerman, P. R., Biogenic nonmethane hydrocarbon emissions estimated from tethered balloon observations, *J. Geophys. Res.*, 99(25587), 25598, 1994.

35. Farmer, C. T., Milne, P. J., Riemer, D. D., and Zika, R. G., Continuous hourly analysis of C_2-C_{10} nonmethane hydrocarbon compounds in urban air by GC-FID, *Environ. Sci. Technol.*, 28, 238–245, 1994.

36. Montzka, S. A., Myers, R. C., Butler, J. H., and Elkins, J. W., Early trends in the global tropospheric abundance of hydrochlorofluorocarbon-141b and 142b, *Geophys. Res. Lett.*, 21, 2483–2486, 1994.

37. Riemer, D. D., Milne, P. J., Farmer, C. T., and Zika, R. G., Determination of terpene and realted compounds in semi-urban air by GC-MSD, *Chemosphere*, 28, 837–850, 1994.

38. Yokouchi, Y., Seasonal and diurnal variation of isoprene and its reaction products in a semi-rural area, *Atmos. Environ.*, 28, 2651–2658, 1994.

39. Goldan, P. D., Kuster, W. C., Fehsenfeld, F. C., and Montzka, S. A., Hydrocarbon measurements in the southeastern United States: The Rural Oxidants in the Southern Environment (ROSE) Program 1990, *J. Geophys. Res.*, 100, 25945–25963, 1995.

40. Goldstein, A. H., Daube, B. C., Munger, J. W., and Wofsy, S. C., Automated *in situ* monitoring of atmospheric nonmethane hydrocarbon concentrations and gradients, *J. Atmos. Chem.*, 21, 43–59, 1995.

41. Lawrimore, J. H., Das, M., and Aneja, V. P., Vertical sampling and analysis of NMHCs for ozone control in urban North Carolina, *J. Geophys. Res.*, 100, 22785–22793, 1995.

42. Singh, H. B., Kanakidou, M., Crutzen, P. J., and Jacob, D. J., High concentrations and photochemical fate of oxygenated hydrocarbons in the global troposphere, *Nature*, 378, 50–54, 1995.

43. Blake, D. R., Chen, T. Y., Smith, T. W. Jr., Wang, C. J.-L., Wingenter, O. W., Blake, N. J., Rowland, F. S., and Mayer, E. W., Three-dimensional distribution of NMHCs and halocarbons over the

northwestern Pacific during the 1991 Pacific Exploratory Mission (PEM-West A), *J. Geophys. Res.*, 101, 1763–1778, 1996.

44. Elkins, J. W., Fahey, D. W., Gilligan, J. M., Dutton, G. S., Baring, T. J., Volk, C. M., Dunn, R. E., Myers, R. C., Montzka, S. A., Wamsley, P. R., Hayden, A. H., Butler, J. H., Thompson, T. M., Swanson, T. H., Dlugokencky, E. J., Novelli, P. C., Hurst, D. F., Hurst, J. M., Lobert, J. M., Ciciora, S. J., McLaughlin, R. J., Thompson, T. L., Winkler, R. H., Fraser, P. J., Steele, L. P., and Lucarelli, M. P., Airborne gas chromatography for *in situ* measurements of long-lived species in the upper troposphere and lower stratosphere, *Geophys. Res. Lett.*, 23, 347–350, 1996.

45. Grosjean, E., Grosjean, D., Fraser, M. P., and Cass, G. R., Air quality model evaluation data for organics. 2. C_1–C_{14} carbonyls in Los Angeles air, *Environ. Sci. Technol.*, 30, 2687–2703, 1996.

46. Guenther, A., Baugh, W., Davis, K., Hampton, G., Harley, P., Klinger, L., Vierling, L., Zimmerman, P., Allwine, E., Dilts, S., Lamb, B., Westberg, H., Baldocchi, D., Geron, C., and Pierce, T., Isoprene fluxes measured by enclosure, relaxed eddy accumulation, surface layer gradient, mixed layer gradient, and mixed layer mass balance techniques, *J. Geophys. Res.*, 101, 18555–18567, 1996.

47. Laurila, T. and Hakola, H., Seasonal cycle of C_2–C_5 hydrocarbons over the Baltic Sea and northern Finland, *Atmos. Environ.*, 30, 1597–1607, 1996.

48. Montzka, S. A., Butler, J. H., Myers, R. C., Thompson, T. M., Swanson, T. H., Clarke, A. D., Lock, L. T., and Elkins, J. W., Decline in the tropospheric abundance of halogen from halocarbons: implications for stratospheric ozone depletion, *Science*, 272, 1318–1322, 1996.

49. Biesenthal, T. A., Wu, Q., Shepson, P. B., Wiebe, H. A., Anlauf, K. G., and Mackay, G. I., A study of relationships between isoprene, its oxidation products, and ozone, in the lower Fraser Valley, BC, *Atmos. Environ.*, 31, 2049–2058, 1997.

50. Blake, N. J., Blake, D. R., Chen, T.-Y., Collins, J. E. Jr., Sachse, G. W., Anderson, B. E., and Rowland, F. S., Distribution and seasonality of selected hydrocarbons and halocarbons over the western Pacific basin during PEM-West A and PEM-West B, *J. Geophys. Res.*, 102, 28315–28331, 1997.

51. Gaffney, J. S., Marley, N. A., Martin, R. S., Dixon, R. W., Reyes, L. G., and Popp, C. J., Potential air quality effects of using ethanol–gasoline fuel blends: a field study in Albuquerque, New Mexico, *Environ. Sci. Technol.*, 31, 3053–3061, 1997.

52. Helmig, D., Balsley, B., Davis, K., Kuck, L. R., Jensen, M., Bognar, J., Smith, T. Jr., Arrieta, R. V., Rodriguez, R., and Birks, J. W., Vertical profiling and determination of landscape fluxes of biogenic NMHCs within the planetary boundary layer in the Peruvian Amazon, *J. Geophys. Res.*, 103, 25519–25532, 1998.

53. Helmig, D., Greenberg, J., Guenther, A., Zimmerman, P., and Geron, C., VOCs and isoprene oxidation products at a temperate deciduous forest site, *J. Geophys. Res.*, 103, 22397–22414, 1998.

54. Lee, Y.-N., Zhou, X., Kleinman, L. I., Nunnermacker, L. J., Springston, S. R., Daum, P. H., Newman, L., Keigley, W. G., Holdren, M. W., Spicer, C. W., Young, V., Fu, B., Parrish, D. D., Holloway, J., Williams, J., Roberts, J. M., Ryerson, T. B., and Fehsenfeld, F. C., Atmospheric chemistry and distribution of formaldehyde and several multioxygenated carbonyl compounds during the 1995 Nashville/Middle Tennessee Ozone Study, *J. Geophys. Res.*, 103, 22449–22462, 1998.

55. Apel, E. C., Calvert, J. G., Riemer, D., Pos, W., Zika, R., Kleindienst, T. E., Lonneman, W. A., Fung, K., Fujita, E., Shepson, P. B., Starn, T. K., and Roberts, P. T., Measurements comparison of oxygenated volatile organic compounds at a rural site during the 1995 SOS Nashville Intensive, *J. Geophys. Res.*, 103, 22295–22316, 1998.

56. Blake, N. J., Blake, D. R., Wingenter, O. W., Sive, B. C., Kang, G. H., Thornton, D. C., Bandy, A. R., Atlas, E., Flocke, F., Harris, J. M., and Rowland, F. S., Aircraft measurements of the latitudinal, vertical, and seasonal variations of NMHCs, methyl nitrate, methyl halides, and DMS during the First Aerosol Characterization Experiment (ACE 1), *J. Geophys. Res.*, 104, 21803–21817, 1999.

57. Doskey, P. V. and Gao, W., Vertical mixing and chemistry of isoprene in the atmospheric boundary layer: aircraft-based measurements and numerical modeling, *J. Geophys. Res.*, 104, 21263–21274, 1999.

58. Lamanna, M. S. and Goldstein, A. H., *In situ* measurements of C_2–C_{10} volatile organic compounds above a Sierra Nevada ponderosa pine plantation, *J. Geophys. Res.*, 104, 21247–21262, 1999.

59. Rappengluck, B. and Fabian, P., An analysis of simultaneous online GC measurements of BTEX aromatics at three selected sites in the greater Munnich area, *J. Appl. Meteor.*, 38, 1448–1462, 1999.

60. Goldan, P. D., Parrish, D. D., Kuster, W. C., Trainer, M., McKeen, S. A., Holloway, J., Jobson, B. T., Sueper, D. T., and Fehsenfeld, F. C., Airborne measurements of isoprene, CO, and anthropogenic hydrocarbons and their implications, *J. Geophys. Res.*, 105, 9091–9105, 2000.

61. Rappengluck, B., Reitmayer, H., and Fabian, P., On the use of manned hydrogen-gas ballooning in boundary layer studies, *Environ. Sci. Pollut. Res.*, 7, 211–218, 2000.

62. Apel, A. C., Hills, A. J., Lueb, R., Zindel, S., and Eisele, S., A fast-GC/MS system to measure C_2 to C_4 carbonyls and methanol aboard aircraft, *J. Geophys. Res.*, 108(D20), 8794, 2003, doi:10.1029/2002JD003199.

63. Karbiwnyk, C. M., Mills, C. S., Helmig, D., and Birks, J. W., Use of chlorofluorocarbons as internal standards for the measurement of atmospheric nonmethane volatile organic compounds sampled onto solid adsorbent cartridges, *Environ. Sci. Technol.*, 37, 1002–1007, 2003.

64. Kawamura, K., Umemoto, N., Mochida, M., Bertram, T., Howell, S., and Huebert, B. J., Water-soluble dicarboxylic acids in the tropospheric aerosols collected over east Asia and western North Pacific by ACE-Asia C-130 aircraft, *J. Geophys. Res.*, 108(D23), 8639, 2003, doi:10.1029/2002JD003256.

65. Price, H. U., Jaffe, D. A., Doskey, P. V., McKendry, I., and Anderson, T. L., Vertical profiles of O_3, aerosols, CO and NMHCs in the northeast Pacific during the TRACE-P and ACE-ASIA Experiments, *J. Geophys. Res.*, 108, 8799, 2003, doi:10.1029/2002JD002930.

66. Kuster, W. C., Jobson, B. T., Karl, T., Riemer, D., Apel, E., Goldan, P. D., and Fehsenfeld, F. C., Intercomparison of volatile organic carbon measurement techniques and data at La Porte during the TexAQS2000 Air Quality Study, *Environ. Sci. Technol.*, 38, 221–228, 2004.

67. Andino, J. M. and Butler, J. W., A study of the stability of methanol-fueled vehicle emissions in Tedlar bags, *Environ. Sci. Technol.*, 25, 1644–1646, 1991.

68. Wang, Y., Raihala, T. S., Jackman, A. P., and John, R. St, Use of Tedlar bags in VOC testing and storage: evidence of significant VOC losses, *Environ. Sci. Technol.*, 30, 3115–3117, 1996.

69. Doskey, P. V., Costanza, M. S., Hansen, M. C., and Kickels, W. T., Solid sorbent method for the collection and analysis of volatile halogenated organic compounds in soil gas, *J. Chromatogr. A*, 738, 73–81, 1996.

70. Martin, R. S., Villanueva, I., Zhang, J., and Popp, C. J., Nonmethane hydrocarbon, monocarboxylic acid, and low molecular weight aldehyde and ketone emissions from vegetation in central New Mexico, *Environ. Sci. Technol.*, 33, 2186–2192, 1999.

71. Doskey, P. V., Aldstadt, J. H., Kuo, J. M., and Costanza, M. S., Evaluation of an *in situ*, on-line purging system for the cone penetrometer, *J. Air Waste Manag. Assoc.*, 46, 1081–1085, 1996.

72. Gholson, A. R., Storm, J. F., Jayanty, R. K. M., Fuerst, R. G., Logan, T. J., and Midgett, M. R., Evaluation of canisters for measuring emissions of volatile organic air pollutants from hazardous waste incineration, *J. Air Pollut. Control Assoc.*, 39, 1210–1217, 1989.

73. Pate, B., Jayanty, R. K. M., Peterson, M. R., and Evans, G. F., Temporal stability of polar organic compounds in stainless steel canisters, *J. Air Waste Manag. Assoc.*, 42, 460–462, 1992.

74. Kelly, T. J., Callahan, P. J., Pleil, J., and Evans, G. F., Method development and field measurements for polar volatile organic compounds in ambient air, *Environ. Sci. Technol.*, 27, 1146–1153, 1993.

75. Brymer, D. A., Ogle, L. D., Hones, C. J., and Lewis, D. L., Viability of using SUMMA polished canisters for the collection and storage of parts per billion by volume level volatile organics, *Environ. Sci. Technol.*, 30, 188–195, 1996.

76. Fukui, Y. and Doskey, P. V., An enclosure technique for measuring nonmethane organic compound emissions from grasslands, *J. Environ. Qual.*, 25, 601–610, 1996.

77. Batterman, S. A., Zhang, G.-Z., and Baumann, M., Analysis and stability of aldehydes and terpenes in electropolished canisters, *Atmos. Environ.*, 32, 1647–1655, 1998.

78. Greenberg, J. P., Zimmerman, P. R., Pollock, W. F., Lueb, R. A., and Heidt, L. E., Diurnal variability of atmospheric methane, nonmethane hydrocarbons, and carbon monoxide at Mauna Loa, *J. Geophys. Res.*, 97, 10395–10413, 1992.

79. Fukui, Y. and Doskey, P. V., Identification of nonmethane organic compound emissions from grassland vegetation, *Atmos. Environ.*, 34, 2947–2956, 2000.

80. Apel, E. C., Calvert, J. G., Gilpin, T., and Fehsenfeld, F. C., The nonmethane hydrocarbon intercomparison experiment (NOMHICE): Task 4, ambient air, *J. Geophys. Res.*, 108, 4300, 2003.

81. Fukui, Y. and Doskey, P. V., A measurement technique for organic nitrates and halocarbons in ambient air, *J. High Resolut. Chromatogr.*, 21, 201–208, 1998.

82. Rossner, A., Farant, J. P., Simon, P., and Wick, D. P., Development of a flow controller for long-term sampling of gases and vapors using evacuated canisters, *Environ. Sci. Technol.*, 36, 4912–4920, 2002.

83. Doskey, P. V. and Bialk, H. M., Automated sampler for the measurement of nonmethane organic compounds, *Environ. Sci. Technol.*, 35, 591–594, 2001.

84. Bertsch, W., Chang, R. C., and Zlatkis, A., The determination of organic volatiles in air pollution studies: characterization of profiles, *J. Chromatogr. Sci.*, 12, 175–182, 1974.

85. Pellizzari, E. D., Bunch, J. E., Carpenter, B. H., and Sawicki, E., Collection and analysis of trace organic vapor pollutants in ambient atmospheres: technique for evaluating concentration of vapors by sorbent media, *Environ. Sci. Technol.*, 9, 552–555, 1975.

86. Helmig, D. and Vierling, L., Water adsorption capacity of the solid adsorbents Tenax TA, Tenax GR, Carbotrap, Carbotrap C, Carbosieve SIII, and Carboxen 569 and water management techniques for the atmospheric sampling of volatile organic trace gases, *Anal. Chem.*, 67, 4380–4386, 1995.

87. Holdren, M. W., Westberg, H. H., and Zimmerman, P. R., Analysis of monoterpene hydrocarbons in rural atmospheres, *J. Geophys. Res.*, 84, 5083–5088, 1979.

88. Roberts, J. M., Fehsenfeld, F. C., Albritton, D. L., and Sievers, R. E., Measurement of monoterpene hydrocarbons at Niwot Ridge, Colorado, *J. Geophys. Res.*, 88, 10667–10678, 1983.

89. Shepson, P. B., Kleindienst, T. E., and McElhoe, H. B., A cryogenic trap/porous polymer sampling technique for the quantitative determination of ambient volatile organic compound concentrations, *Atmos. Environ.*, 21, 579–587, 1987.

90. Kesselmeier, J., Schafer, L., Ciccioli, P., Brancaleoni, E., Cecinato, A., Frattoni, M., Foster, P., Jacob, V., Denis, J., Fugit, J. L., Dutaur, L., and Torres, L., Emission of monoterpenes and isoprene from a mediterranean oak species *Quercus ilex* L. measured within the BEMA (Biogenic Emissions in the Mediterranean Area) Project, *Atmos. Environ.*, 30, 1841–1850, 1996.

91. Starn, T. K., Shepson, P. B., Bertman, S. B., White, J. S., Splawn, B. G., Riemei, D. D., Zika, R. G., and Olszyna, K., Observations of isoprene chemistry and its role in ozone production at a semirural site during the 1995 Southern Oxidants Study, *J. Geophys. Res.*, 103, 22425–22435, 1998.

92. Helmig, D. and Greenberg, J. P., Automated *in situ* gas chromatographic–mass spectrometric analysis of ppt level volatile organic trace gases using multistage solid-adsorbent trapping, *J. Chromatogr. A*, 677, 123–132, 1994.

93. Seeley, J. V., Kramp, F. J., Sharpe, K. S., and Seeley, S. K., Characterization of gaseous mixtures of organic compounds with dual-secondary column comprehensive two-dimensional gas chromatography, *J. Sep. Sci.*, 25, 53–59, 2002.

94. Helmig, D., Artifact-free preparation, storage and analysis of solid adsorbent sampling cartridges used in the analysis of volatile organic compounds in air, *J. Chromatogr. A*, 732, 414–417, 1996.

95. Druzik, C. M., Grosjean, D., van Neste, A., and Parmer, S. S., Sampling of atmospheric carbonyls with small DNPH-coated C18 cartridges and liquid chromatography analysis with diode array detection, *Int. J. Environ. Anal. Chem.*, 38, 495–512, 1990.

96. Grosjean, D., Miguel, A. H., and Tavares, T. M., Urban air pollution in Brazil: acetaldehyde and other carbonyls, *Atmos. Environ.*, 24B, 101–106, 1990.

97. Zhou, X. and Mopper, K., Measurement of sub-parts-per-billion levels of carbonyl compounds in marine air by a simple cartridge trapping procedure followed by liquid chromatography, *Environ. Sci. Technol.*, 24, 1482–1485, 1990.

98. Rosenberg, C., Winiwarter, W., Gregori, M., Pech, G., Casensky, V., and Puxbaum, H., Determination of inorganic and organic volatile acids, NH_3, particulate, SO_4^{-2}, NO_3^- and Cl^- in ambient air with an annular diffusion denuder system, *Fresenius Z. Anal. Chem.*, 331, 1–7, 1988.

99. Puxbaum, H., Rosenberg, C., Gregori, M., Lanzerstorfer, C., Ober, E., and Winiwarter, W., Atmospheric concentrations of formic and acetic acid and related compounds in eastern and northern Austria, *Atmos. Environ.*, 22, 2841–2850, 1988.

100. Kawamura, K., Ng, L.-L., and Kaplan, I. R., Determination of organic acids (C_1–C_{10}) in the atmosphere, motor exhausts, and engine oils, *Environ. Sci. Technol.*, 19, 1082–1086, 1985.

101. Andreae, M. O., Talbot, R. W., and Li, S.-M., Atmospheric measurements of pyruvic and formic acid, *J. Geophys. Res.*, 92, 6635–6641, 1987.

102. Keene, W. C., Talbot, R. W., Andreae, M. O., Beecher, K., Berresheim, H., Castro, M., Farmer, J. C., Galloway, J. N., Hoffman, M. R., Li, S.-M., Maben, J. R., Munger, J. W., Norton, R. B., Pszenny, A. A. P., Puxbaum, H., Westberg, H., and Winiwarter, W., An intercomparison of measurement systems for vapor and particulate phase concentrations of formic and acetic acids, *J. Geophys. Res.*, 94, 6457–6471, 1989.

103. Helmig, D., Ozone removal techniques in the sampling of atmospheric volatile organic trace gases, *Atmos. Environ.*, 31, 3635–3651, 1997.

104. Juttner, F., A cryo-trap technique for the quantitation of monoterpenes in humid and ozone-rich forest air, *J. Chromatogr.*, 442, 157–163, 1988.

105. Janson, R. W., Monoterpene emissions from Scots pine and Norwegian spruce, *J. Geophys. Res.*, 98, 2839–2850, 1993.

106. Peters, R. J. B., Duivenbode, J., Duyzer, J. H., and Verhagen, H. L. M., The determination of terpenes in forest air, *Atmos. Environ.*, 28, 2413–2419, 1994.

107. Hoffmann, T., Adsorptive preconcentration technique including oxidant scavenging for the measurement of reactive natural hydrocarbons in ambient air, *Fresenius J. Anal. Chem.*, 351, 41–47, 1995.

108. Steinbrecher, R., Eichstadter, G., Schurmann, W., Torres, L., Clement, B., Simon, V., Kotzias, D., Daiber, R., and van Eijk, J., Monoterpenes in air samples: European intercomparison experiments, *Int. J. Environ. Anal. Chem.*, 54, 283–297, 1994.

109. Roberts, J. M., Fehsenfeld, F. C., Albritton, D. L., and Sievers, R. E., Sampling and analysis of monoterpene hydrocarbons in the atmosphere with Tenax gas chromatographic porous polymer, *Identification and Analysis of Organic Pollutants in Air*, Butterworth, Boston, pp. 371–387, 1984.

110. Pellizzari, E. D., Demian, B., and Krost, K., Sampling of organic compounds in the presence of reactive inorganic gases with Tenax GC, *Anal. Chem.*, 56, 793–798, 1984.

111. Pellizzari, E. D. and Krost, K. J., Chemical transformations during ambient air sampling for organic vapors, *Anal. Chem.*, 56, 1813–1819, 1984.

112. Kuwata, K., Uebori, M., and Yamaski, Y., Determination of aliphatic and aromatic aldehydes in polluted airs as their 2,4-dinitrophenylhydrazones by high performance liquid chromatography, *J. Chromatogr. Sci.*, 17, 264–268, 1979.

113. Grosjean, D. and Fung, K., Collection efficiencies of cartridges and microimpingers for sampling of aldehydes in air as 2,4-dinitrophenylhydrazones, *Anal. Chem.*, 54, 1221–1224, 1982.

114. Cofer, W. R. III, Collins, V. G., and Talbot, R. W., Improved aqueous scrubber for collection of soluble atmospheric trace gases, *Environ. Sci. Technol.*, 19, 557–560, 1985.

115. Lee, Y.-N. and Zhou, X., Method for the determination of some soluble atmospheric carbonyl compounds, *Environ. Sci. Technol.*, 27, 749–756, 1993.

116. Lee, Y.-N., Zhou, X., Leaitch, W. R., and Banic, C. M., An aircraft measurement technique for formaldehyde and soluble carbonyl compounds, *J. Geophys. Res.*, 101, 29075–29080, 1996.

117. Andreae, M. O., Talbot, R. W., Andreae, T. W., and Harris, R. C., Formic and acetic acids over the central Amazon region, Brazil, *J. Geophys. Res.*, 93, 1616–1624, 1988.

118. Talbot, R. W., Beecher, K. M., Harris, R. C., and Cofer, W. R. III, Atmospheric geochemistry of formic and acetic acids at a mid-latitude temperate site, *J. Geophys. Res.*, 93, 1638–1652, 1988.

119. Andreae, M. O., Talbot, R. W., Berresheim, H., and Beecher, K. M., Precipitation chemistry in central Amazonia, *J. Geophys. Res.*, 95, 16987–16999, 1990.

120. Talbot, R. W., Andreae, M. O., Berresheim, H., Jacob, D. J., and Beecher, K. M., Sources and sinks of formic, acetic, and pyruvic acids over Central Amazonia: 2. Wet Season, *J. Geophys. Res.*, 95, 16799–16811, 1990.

121. Sanhueza, E., Figueroa, L., and Santana, M., Atmospheric formic and acetic acids in Venezuela, *Atmos. Environ.*, 30, 1861–1873, 1996.

122. Talbot, R. W., Dibb, J. E., Lefer, B. L., Scheuer, E., Bradshaw, J. D., Sandholm, S. T., Smyth, S., Blake, D. R., Blake, N. J., Sachse, G. W., Collins, J. E. Jr., and Gregory, G. L., Large scale distributions of tropospheric nitric, formic, and acetic acids over the western Pacific basin during wintertime, *J. Geophys. Res.*, 102, 28303–28313, 1997.

123. Farmer, J. C. and Dawson, G. A., Condensation sampling of soluble atmospheric trace gases, *J. Geophys. Res.*, 87, 8931–8942, 1982.

124. Lonneman, W. A., Kopczynski, S. L., Darley, P. E., and Sutterfield, F. D., Hydrocarbon composition of urban air pollution, *Environ. Sci. Technol.*, 8, 229–236, 1974.

125. Seila, R. L. and Rickman, E. E. Jr., *Research Protocol Method For Analysis of C_2-C_{12} Hydrocarbons in Ambient Air by GC with Cryogenic Concentration, ASRL–ACPD–RPM 002*, U.S. Environmental Protection Agency, Research Triangle Park, NC, 1986.

126. Helmig, D., Revermann, T., Pollmann, J., Kaltschmidt, O., Jiménez Hernández, A., Bocquet, F., and David, D., Calibration system and analytic considerations for quantitative sesquiterpene measurements in air, *J. Chromatogr. A*, 1002, 193–211, 2003.

127. Foulger, B. E. and Simmonds, P. G., Drier for field use in the determination of trace atmospheric gases, *Anal. Chem.*, 51, 1089–1090, 1979.

128. Pleil, J. D., Oliver, K. D., and McClenny, W. A., Enhanced performance of Nafion dryers in removing water from air samples prior to gas chromatographic analysis, *J. Air Pollut. Control Assoc.*, 37, 244–248, 1987.

129. Burns, W. F., Tingey, D. T., Evans, R. C., and Bates, E. H., Problems with a Nafion membrane dryer for drying chromatographic samples, *J. Chromatogr.*, 269, 1–9, 1983.

130. Chien, C.-J., Charles, M. J., Sexton, K. G., and Jeffries, H. E., Analysis of airborne carboxylic acids and phenols as their pentafluorobenzyl derivatives: gas chromatography/ion trap mass spectrometry with a novel chemical ionization reagent, PFBOH, *Environ. Sci. Technol.*, 32, 299–309, 1998.

131. Janda, V., Bartle, K. D., and Clifford, A. A., SFE in environmental analysis, *J. Chromatogr.*, 642, 283–299, 1993.

132. Hong, J. G., Maguhn, J., Freitag, D., and Kettrup, A., Detection of volatile organic peroxides in indoor air, *Fresenius J. Anal. Chem.*, 371, 961–965, 2001.

133. Hills, J. W., Hill, H. H. Jr., and Maeda, T., Simultaneous supercritical fluid derivatization and extraction, *Anal. Chem.*, 63, 2152–2155, 1991.

134. Fukui, Y. and Doskey, P. V., Air–surface exchange of NMOCs at a grassland site: seasonal variations and stressed emissions, *J. Geophys. Res.*, 103, 13153–13168, 1998.

135. Lonneman, W. A., *Personal communication*, Research Triangle Park, NC, 2004.

136. Pierotti, D., Analysis of trace oxygenated hydrocarbons in the environment, *J. Atmos. Chem.*, 10, 373–382, 1990.

137. Jonsson, A. and Berg, S., Determination of low-molecular-weight oxygenated hydrocarbons in ambient air by cryo-gradient sampling and two-dimensional gas chromatography, *J. Chromatogr.*, 279, 307–322, 1983.

138. Berg, S. and Jonsson, A., Two-dimensional GC for determination of volatile compounds in ambient air, *J. High Resolut. Chromatogr. & Chromatogr. Commun.*, 7, 687–695, 1984.

139. Klemp, M., Peters, A., and Sacks, R., High-speed GC analysis of VOCs: sample collection and inlet systems, *Environ. Sci. Technol.*, 28, 369A–376A, 1994.

140. Sacks, R. and Akard, M., High-speed GC analysis of VOCs: tunable selectivity and column selection, *Environ. Sci. Technol.*, 28, 428A–433A, 1994.

141. Klemp, M. A., Akard, M. L., and Sacks, R. D., Cryofocusing inlet with reverse flow sample collection for gas chromatography, *Anal. Chem.*, 65, 2516–2521, 1993.

142. Jennings, W., *GC with Glass Capillary Columns*, Academic Press, New York, 1980.

143. Liu, Z. and Phillips, J. B., Comprehensive two-dimensional GC using an on-column thermal modulator interface, *J. Chromatogr. Sci.*, 29, 227–231, 1991.

144. Phillips, J. B. and Xu, J., Comprehensive multi-dimensional gas chromatography, *J. Chromatogr. A*, 703, 327–334, 1995.

145. Kinghorn, R. M. and Marriott, P. J., Comprehensive GC using a modulating cryogenic trap, *J. High Resolut. Chromatogr.*, 21, 620–622, 1998.

146. Kinghorn, R. M. and Marriott, P. J., Modulation and manipulation of chromatographic bands using thermal means, *Anal. Sci.*, 14, 651–660, 1998.

147. Lewis, A. C., Carslaw, N., Marriot, P. J., Kinghorn, R. M., Morrison, P., Lee, A. L., Bartle, K. D., and Pilling, M. J., Larger pool of ozone-forming carbon compounds in urban atmospheres, *Nature*, 405, 778–781, 2000.

148. Bueno, P. A. Jr. and Seeley, J. V., Flow-switching device for comprehensive two-dimensional gas chromatography, *J. Chromatogr. A*, 1027, 3–10, 2004.

149. Fung, K. and Grosjean, D., Determination of nanogram amounts of carbonyls as 2,4-dinitrophenylhydrazones by high-performance liquid chromatography, *Anal. Chem.*, 53, 168–171, 1981.

150. Kuwata, K., Uebori, M., Yamasaki, H., and Kuge, Y., Determination of aliphatic aldehydes in air by liquid chromatography, *Anal. Chem.*, 55, 2013–2016, 1983.

151. Zielenska, B., Sagebiel, J. C., Harshfield, G., Gertler, A. W., and Pierson, W. R., VOCs up to C_{20} emitted from motor vehicles: measurement methods, *Atmos. Environ.*, 30, 2269–2286, 1996.

152. Scanlon, J. T. and Willis, D. E., Calculation of FID relative response factors using the effective carbon number concept, *J. Chromatogr. Sci.*, 23, 333–340, 1985.

153. Jorgensen, A. D., Picel, K. C., and Stamoudis, V. C., Prediction of GC FID response factors from molecular structures, *Anal. Chem.*, 62, 683–689, 1990.

154. Romashkin, P. A., Hurst, D. F., Elkins, J. W., Dutton, G. S., Fahey, D. W., Dunn, R. E., Moore, F. L., Myers, R. C., and Hall, B. D., *In situ* measurements of long-lived trace gases in the lower stratosphere by gas chromatography, *J. Atmos. Oceanic Technol.*, 18, 1195–1204, 2001.

155. Cox, R. D. and Earp, R. F., Determination of trace level organics in ambient air by high-resolution GC with simultaneous photoionization and flame ionization detection, *Anal. Chem.*, 54, 2265–2270, 1982.

156. Nutmagul, W. and Cronn, D. R., Determination of selected atmospheric aromatic hydrocarbons at remote continental and oceanic locations using photoionization/flame — ionization detection, *J. Atmos. Chem.*, 2, 415–433, 1985.

157. O'Hara, D. and Singh, H. B., Sensitive gas chromatographic detection of acetaldehyde and acetone using a reduction gas detector, *Atmos. Environ.*, 22, 2613–2615, 1988.

158. Greenberg, J. P., Zimmerman, P. R., Taylor, B. E., Silver, G. M., and Fall, R., Sub-parts per billion detection of isoprene using a RGD with a portable gas chromatograph, *Atmos. Environ.*, 27A, 2689–2692, 1993.

159. Dalluge, J., van Stee, L. L. P., Beens, J., and Tijssen, R., Unraveling the composition of very complex samples by comprehensive GC coupled to time of flight mass spectrometry: cigarette smoke, *J. Chromatogr. A*, 974, 169–184, 1997.

160. Warneke, C., de Gouw, J. A., Kuster, W. A., Goldan, P. D., and Fall, R., Validation of atmospheric VOC measurements by proton-transfer-reaction mass spectrometry using a gas-chromatographic preseparation method, *Environ. Sci. Technol.*, 37, 2494–2501, 2003.

161. de Gouw, J. A., Howard, C. J., Custer, T. G., Baker, B. M., and Fall, R., Proton-transfer chemical-ionization mass spectrometry allows real-time analysis of volatile organic compounds released from cutting and drying of crops, *Environ. Sci. Technol.*, 34, 2640–2648, 2000.

162. Warneke, C., Luxembourg, S. L., de Gouw, J. A., Rinne, H. J. I., Guenther, A. B., and Fall, R., Disjunct eddy covariance measurements of oxygenated volatile organic compounds fluxes from an alfalfa field before and after cutting, *J. Geophys. Res.*, 107(D8)2002, 10.1029/2001JD000594.

163. Caligirou, A., Larson, B. R., Brussol, C., Duane, M., and Kotzias, D., Decomposition of terpenes by ozone during sampling on Tenax, *Anal. Chem.*, 68, 1499–1506, 1996.

164. LaClair, R. W., Bueno, P. A. Jr., and Seeley, J. V., A systematic analysis of a flow-switching modulator for comprehensive two-dimensional gas chromatography, *J. Sep. Sci.*, 27, 389–396, 2004.

165. Pellizzari, E. D., Bunch, J. E., Carpenter, B. H., and Sawicki, E., Collection and analysis of trace organic vapor pollutants in ambient atmospheres: thermal desorption of organic vapors from sorbent media, *Environ. Sci. Technol.*, 9, 556–560, 1975.

166. Doskey, P.V., Unpublished results, Argonne National Laboratory, Argonne, IL, 2004.

167. Montzka, S. A., Myers, R. C., Butler, J. H., Elkins, J. W., Lock, L. T., Clarke, A. D., and Goldstein, A. H., Observations of HFC-134a in the remote troposphere, *Geophys. Res. Lett.*, 23, 169–172, 1996.

17 Halogenated VOCs

Filip D'hondt, Haytham Chahin, and Mohammad Ghafar

CONTENTS

I. PHYSICAL, CHEMICAL AND ENVIRONMENTAL PROPERTIES AND SAFETY ASPECTS OF HALOGENATED VOLATILE ORGANIC COMPOUNDS (VOCs)

Halogenated VOCs are used in industry, although various consumer products may also contain them. Some of these compounds, such as chloroform, may originate from natural processes, others exist only as a result of man-made chemical production. For example, natural sources of carbon tetrachloride have not been reported in the literature. Likewise, methylene chloride, 1,2-dichloroethane, and trichloroethylene do not occur naturally in the environment, but are released from man-made sources and have become air, water and soil pollutants.[1] The persistence of these pollutants in the environment is described by the term, half-life, which gives the period in which half of the amount of a particular pollutant has decomposed through various chemical reactions. The half-lives vary from a few hours to years, even hundreds of years for some fully halogenated chlorofluorocarbons (freons) (CFC). The Montreal Protocol (1989) is an international agreement to regulate the disposal of substances which may deplete the ozone layer. In this agreement the governments promised to reduce or ban the use and release into the environment of specified halogenated hydrocarbons. Carbon tetrachloride and 1,1,1-trichloroethane are among those substances. Most of the chlorinated hydrocarbons have narcotic effects and may affect heart and damage the liver. Some of these may penetrate the skin sufficiently to cause health hazards. It should be kept in mind that toxic effect may even result from repeated exposures to amounts which are too low to produce acute symptoms and give a warning of danger. Also there is a wide range of human responses to these chemicals — some workers may be seriously affected by exposure to concentrations which seem to have no effect on others. Among halogenated compounds ethylene dibromide (mainly used as soil fumigant), vinyl bromide (used in the plastic industry), and epichlorohydrin (used in epoxy resin production) are classified by IARC (International Agency for Research on Cancer) as probably carcinogenic to human. The following compounds are classified as possibly carcinogenic: bromodichloromethane, carbon tetrachloride, chloroform, 1,2-dichloroethane, 1,3-dichloropropene, methylene chloride, and tetrachloroethylene.

In animal tests, some of the compounds of this group have also demonstrated, in addition to carcinogenic effects, changes in the reproductive process. More research and follow-up of the long-term effects of human exposure are necessary. Halogenated hydrocarbon compounds decompose producing toxic and corrosive gases when brought into contact with open flames or hot surfaces. Some of the halo-solvents are also highly flammable. In Europe about one million people (in the European Union countries) risk exposure to these compounds at work. In the U.S.A., it has been

Abbreviations: ACGIH, American Conference of Governmental Industrial Hygienists; ADI, average daily intake; BCEE, bis(2-chloroethyl) ether; BEI, biological exposure indices; BOD, biochemical oxygen demand; CMS, carbon molecular sieve; CFC, chlorofluorocarbon; DME, dynamic membrane extraction; ECD, electron capture detection; FID, flame ionization detection; fr., flow rate; GC, gas chromatography; HS, static headspace; IARC, international agency for research on cancer; IT-MS, ion trap mass spectrometer; MS, mass spectrometer; na, not available; PAT, purge and trap; PCBs, polychlorinated biphenyls; PDMS, polydimethylsiloxane; PTFE, polytetrafluoroethylene (Teflon); SPME, solid-phase microextraction; SPE, solid-phase extraction; UV, ultraviolet; VOCs, volatile organic compounds; TLV, threshold limit values.

estimated that five million workers are exposed to four of the most commonly used solvents: methylene chloride, trichloroethane, trichloroethylene, and tetrachloroethylene.

An undetermined amount of secondary exposure of occupational origin is caused to the general public. Occupational exposure to these compounds can be expected in the following industrial activities: the agriculture, rubber, and plastics industry; the chemical industry; manufacturing of pharmaceuticals and hygiene products for consumers; metalwork; and the assembly of instruments and equipment; and dry-cleaning.

A. CARBON TETRACHLORIDE

CAS 56-23-5

Synonyms and trade names: methane tetrachloride, tetra chloromethane, Fasciolin, Freon 10, Necatorine, Tetrafinol, Tetraform, Tetrasol.

Molecular formula: $C-Cl_4$
Molecular weight: 153.82
Boiling point: 76.8°C
Melting point: −23°C

Carbon tetrachloride is a volatile, clear, colorless, noncombustible liquid with an ethereal odor. When the odor is detected, the recommended occupational exposure limit value has already been exceeded. Carbon tetrachloride is insoluble in water, miscible with most aliphatic solvents and has a good dissolving ability. For example, it can dissolve asphalt, benzyl chloride polymers, chlorinated rubber, ethyl cellulose, resins, waxes, and fats. Stabilizers are added to commercially used grades to prevent slow decomposition due to light. It is used as a solvent in the manufacture of cables and semiconductors. Carbon tetrachloride is an intermediate in the production of CFC refrigerants for air conditioning and cooling equipment. It is used in the production of paints and plastics and in formulation of petrol additives. Previously it was used in fire extinguishers, but this is not recommended, because heat may decompose it and the resulting products are highly toxic. Formerly it was used in dry-cleaning, degreasing, and as a grain fumigant. These uses, as well as its use in consumer products, are now restricted in many countries. Carbon tetrachloride was first produced commercially in 1907. Production and extensive use have declined since reaching a peak in 1974 because of its adverse effects on human health and the environment.[2,12]

1. Health Effects

There are numerous reports of injury and death following acute and chronic exposure to carbon tetrachloride. Intoxications result from inhalation of vapor, but it is also readily absorbed through the gastrointestinal tract and is able to penetrate through healthy skin. Volunteers exhaled easily measurable amounts of carbon tetrachloride after an experiment involving a 30 min thumb immersion in this substance. Carbon tetrachloride is an irritant to the skin, eyes, nose, and throat. Exposure to this solvent and its vapor causes nausea, vomiting, and abdominal pain, which some individuals experience even at low vapor concentrations. Carbon tetrachloride has anesthetic properties which affect the central nervous system causing dizziness, vertigo, headache, depression, mental confusion, and poor coordination. Higher concentration levels lead to unconsciousness and cardiac problems. Health effects may be immediate or delayed. Chronic exposure damages the kidneys, liver, and bone marrow, causes visual disturbances, and bronchitis. Carbon tetrachloride produces toxic effects on offspring in animal tests and it may have adverse effects on human reproduction. The ingestion of fat or alcohol together with carbon tetrachloride enhances the toxic health effects. There is inadequate evidence of carcinogenicity in humans caused by carbon tetrachloride. There is sufficient evidence in experimental animals for carcinogenicity caused by carbon tetrachloride. IARC has classified carbon tetrachloride as possibly carcinogenic to humans.[12]

Threshold limit values: 8 h TWA 5 ppm

2. Effects on the Environment

Carbon tetrachloride production and use as a powerful solvent for asphalt, benzyl resin, gums, and rosin; a cleaning agent for machinery and electrical equipment; and in the synthesis of nylon and other chlorination processes may result in its release into the environment through various waste streams. Vapor-phase carbon tetrachloride will be degraded in the atmosphere by reaction with photochemically produced hydroxyl radicals; the half-life for this reaction in air is estimated to be 366 years. Direct carbon tetrachloride photolysis is not important in the troposphere, but irradiation at higher energies (195 to 254 nm) such as found in the stratosphere, results in degradation and leads to ozone depletion. It is mentioned in the Montreal Protocol among chemicals which deplete the ozone layer and are subject to efforts of reducing their use by substituting with less hazardous solvents. If released into soil, carbon tetrachloride has a high mobility. It may volatilize from dry soil surfaces based upon its vapor pressure. Aerobic degradation of carbon tetrachloride is more then 87% in 7 days. If released into water, carbon tetrachloride is not adsorbed by suspended solids and sediment in water and it can be found at trace concentrations. However, it is not formed during the purification of drinking-water using chlorine. It rapidly evaporates into the atmosphere from industrial effluents. Carbon tetrachloride residing in water is resistant to biological and chemical degradation. Hence, carbon tetrachloride is a marine pollutant.[12]

B. CHLOROBENZENES

Monochlorobenzene   *1,2-Dichlorobenzene*

Molecular formula: C_6H_5Cl Molecular form: $C_6H_4Cl_2$
Molecular weight: 112.6   Molecular weight: 147.01
Boiling point: 132°C   Boiling point: 181°C
Melting point: −45.6°C   Melting point: −17°C

Monochlorobenzene and 1,2-dichlorobenzene are both colorless, flammable liquids with a sweet, penetrating odor. These are soluble in benzene and alcohols, but are insoluble in water. The vapors are heavier than air and may travel a considerable distance from the source of evaporation. Ignition and flashback to the source may lead to a fire, releasing in toxic and corrosive gases. Monochlorobenzene is a process solvent for methylene diisocyanate, the latter being used as a solvent in adhesives, polishes, waxes, pharmaceutical products, and natural rubber. 1,2-dichlorobenzene is used as a solvent for a wide range of organic materials and nonferrous metal oxides. It is also used as a solvent carrier in the production of toluene diisocyanate, as a fumigant and insecticide. Chlorobenzenes are chemically active and used as intermediates in many industrial processes, such as the production of pesticides, silicones, and dyes. Other applications include degreasing of hides and wool, in metal polishing products, industrial odor control, and in cleaning agents for drains.[14]

1. Health Effects

Chlorobenzenes are classified as very toxic and there are numerous reports of injury caused by exposure to them. Inhalation of chlorobenzene vapor causes headache, dizziness, drowsiness, and stomach irritation. It also affects the central nervous system. It irritates skin and eyes. Chlorobenzene intoxication causes pain and numbness of fingers, and muscle spasms. High doses damage the kidneys and liver, and lead to unconsciousness. Chlorobenzenes may affect the blood causing anemia. Acute leukemia has been attributed to 1,2-dichlorobenzene exposure by absorption through the skin (case studies). There are no human data and inadequate animal data of carcinogenicity.[13]

Threshold limit values: 8 h TWA 5 ppm

2. Effects to the Environment

Chlorobenzenes enter the environment from manufacture and dispersion in different formulations, but natural sources have not been identified. Chlorobenzene decomposes mainly in biological processes, such as microbial activity. In the air, photo degradation takes place. In the soil, only volatilization from the top layer plays an important role in the removal of chlorobenzenes. Chlorobenzenes are adsorbed strongly by soil. Monochlorobenzene moves through soil containing 1% organic matter ten times slower than through water, and so probably does not constitute a contamination hazard to deep groundwater. Aerobic biodegradation decreases and anaerobic biodegradation increases with increasing number of chlorine substituents. The overall half-life is a function of the number of chlorine atoms in the molecule. Chlorobenzene in water will be redistributed to air and sediment. The amount of chlorobenzene in sediment may be 1000 times higher than that of the contaminated water. As in soil, aerobic biodegradation decreases and anaerobic biodegradation increases with the increasing number of chlorine substituents. In the latter situation, the resulting lower chlorinated benzenes can be transported to deeper soil layers with percolating water. These are moderately toxic to aquatic life and marine pollutants, therefore, their discharge into water should be prohibited. Chlorobenzenes usually occur in the atmosphere in a gaseous state. Four processes are important in removal of chlorobenzenes from atmosphere, namely: dry deposition, wet deposition, reactions with OH radicals, and photolysis.[3]

C. CHLOROFORM

CAS 67-66-3

Synonyms and trade names: methane trichloride, methyl trichloride, trichloromethane, Freon 20, and R20.

Molecular formula: $C-H-Cl_3$
Molecular weight: 119.38
Boiling point: 61.2°C
Melting point: −63.2°C

Chloroform is a very volatile, highly refractive, colorless, nonflammable liquid with a burning sweet taste. It has an etheric, nonirritating odor, but when it is detected, the recommended exposure limit value (10 ppm) has already been exceeded. Chloroform is miscible with most organic solvents (alcohol, ether, benzene, carbon tetrachloride, fixed and volatile oils) but it is only slightly soluble in water. Pure chloroform decomposes slowly in sunlight producing highly toxic gases (phosgene, chlorine, and hydrogen chloride). To prevent this decomposition, commercial chloroform contains a stabilizing agent, usually 0.5 to 1% ethanol. It is used as a solvent and an insecticidal fumigant. Further it is used in formulation of pesticides, drugs, and flavors and as an intermediate in the production of fluorocarbon refrigerant, such as, HCFC22. Chloroform is a source material for polymer production, such as, Teflon (PTFE). It was once used as an inhalation anesthetic but has now been replaced by safer substances.[15]

1. Health Effects

Chloroform is well absorbed into animals and humans by ingestion and inhalation. It penetrates skin, especially when it is wet or humid. Chloroform is distributed throughout the whole body, the highest concentrations being found in fat, blood, liver, kidneys, lungs, and the nervous system. Liquid chloroform is irritating to eyes and skin and may cause chemical burns, although the vapor has not been reported to have irritant effects. It is more irritating to eyes than most of the commonly used solvents, and splashes may cause inflammation. The acute toxicity of chloroform shows considerable species-, strain-, and sex-dependence and this appears to be due to differences in tissue distribution and metabolism. Acute effects depend on the concentration and duration of exposure, and include headache, drowsiness, dizziness, nausea, and a feeling of drunkenness. Higher concentrations lead to unconsciousness, respiratory depression, and heart failure. Liver and kidney damage arise from continued exposure. Damaging effects on liver and kidneys were tested in rats and several mice strains. Toxicity varies depending upon the strain, sex, and vehicle. Damage to the liver is the most universally observed toxic effect of chloroform. In animal tests, it has been shown to have harmful effects on the reproductive process and fetus. Pregnant and nursing mothers should not come into contact with this solvent. IARC has assessed chloroform as *possibly carcinogenic* to humans. The average daily intake (ADI) has been estimated at 0.04 mg/day.[4,5]

Threshold limit values: 8 h TWA 10 ppm

2. Effects on the Environment

Chloroform is released into the environment from industrial production processes, transport, and use. It is present in water as a result of chlorination of drinking water and from industrial sources. Due to its high volatility it is rapidly transferred from surface water and surface soils into the air. Chloroform has a residence time of several months in the atmosphere and is removed from the atmosphere through chemical transformation. It is resistant to biodegradation by aerobic microbial populations of soils. Biodegradation may occur under anaerobic conditions. It may also be a product of atmospheric reactions involving other solvent vapors, such as 1,1,1-trichloroethylene vapors. The concentrations in tap water can considerably contribute to the quality of indoor air and to the general daily intake. In soil, chloroform is highly mobile and may reach ground water. Chloroform is also a marine pollutant. The discharge of ballast water or tank washing with residues or mixtures containing chloroform into sea should be prohibited.[5]

D. 1,2-DICHLOROETHANE

CAS 107-06-2

Synonyms and trade names: Ethane dichloride, ethylene dichloride, glycol dichloride, EDC, Gaz olefiant, Freon 150, Dutch liquid, and Dutch oil.

Molecular formula: $C_2-H_4-Cl_2$
Molecular weight: 98.96
Boiling point: 83.5°C
Melting point: −35.3°C

1,2-Dichloroethane is a volatile, clear, colorless, oily. and highly flammable liquid. It has a sweet taste and an odor similar to chloroform. It is miscible with other chlorinated hydrocarbons and soluble in most commonly used solvents, but it is only slightly soluble in water. The vapor is heavier than air and may move along the ground to a distant ignition point with possible flashback to the container. 1,2-Dichloroethane may explode when its vapor comes into contact and mixes with powdered metals, such as, dusts of magnesium or aluminum. It reacts violently with ammonia. Mixtures of 1,2-dichloroethane and nitric acid may detonate due to heat, impact, or friction. It corrodes aluminum, iron, zinc, and some plastics and these materials are not therefore suitable to be used as storage containers for 1,2-dichloroethane. The commercial end-products are stabilized to prevent the slow decomposition of 1,2-dichloroethane by air, moisture, and light usually by adding small amounts of alkylamines. If not stabilized, it may contain toxic and corrosive gases (chlorine and hydrogen chloride). The production volume of 1,2-dichloroethane is one of the largest of all globally produced chemicals. It is available for industrial use in various volumes; in tank wagons, drums, cans, and bottles. The main use of 1,2-dichloroethane is in the production of other chemicals, such as vinyl chloride, 1,1,1-trichloroethane, ethylamines, and tri- or tetra-chloroethylenes. Gasoline may contain this solvent. It is used in textiles, varnishes, and paint formulations as well as in the processing of pesticides and resins. As a solvent, it is used in extraction of fats and oils and in degreasing operations. When used as a fumigant it may contain carbon tetrachloride (up to 25%) to reduce the risk of a fire hazard.[16]

1. Health Effects

Indoor and outdoor air are the predominant sources of exposure by the general population to 1,2-dichloroethane and only minor amounts are contributed by drinking water. Intake of 1,2-dichloroethane from food is probably negligible. 1,2-Dichloroethane is readily absorbed into the body through the skin, by inhalation of the vapor, by ingestion or dermal exposure and is rapidly and widely distributed throughout the body. The first symptoms of acute intoxication are headache, dizziness, weakness, muscular spasms and vomiting, irritation of mucous membranes of the eyes and respiratory tract. Exposure can also lead to changes in blood and heart rhythm (cardiovascular insufficiency) which may be fatal. Poisonings by inhalation or skin exposure have frequently been reported from workplaces where 1,2-dichloroethane is used as a solvent or fumigant. Accidental ingestion of 20 to 50 g of 1,2-dichloroethane has been identified as a cause of death, with a delay of 6 h to 6 days. Symptoms may also be delayed. The results of short-term and subchronic studies in several species of experimental animals indicate that the liver and kidneys are the target organs. Morphological changes in liver were observed in several species following subchronic exposure to airborne concentrations. Increases in the relative liver weight have been observed in rats following subchronic oral administration. 1,2-Dichloroethane has produced cancer in animal tests and IARC has classified it as *possibly carcinogenic* to humans.[16]

Occupational exposure standards: TLV: 8 h TWA 10 ppm

2. Effects on the Environment

1,2-Dichloroethane is a man-made chemical. It does not occur naturally in nature, and hence it can only be found as a result of releases from industrial production processes. Large amounts may be released through disposal of waste from vinyl chloride industries, since it is the main component of the waste products (EDC-tar). Being volatile it evaporates from water into the atmosphere, where it decomposes fairly quickly in sunlight to oxides and acidic gases of hydrogen chloride. The decomposition is sufficiently rapid to prevent accumulation of the compound in the atmosphere. 1,2-Dichloroethane is slightly toxic to the aquatic environment. It is also considered to be a marine pollutant. It poses a real hazard only in the case of an accident or inappropriate disposal. In some countries authorization for its disposal is needed.[16]

E. Bis(2-chloroethyl)ether

CAS 111-44-4
  Synonyms: dichloroethyl ether or 2,2-dichlorodiethyl ether.

  Molecular formula: $C_4-H_8-Cl_2-O$
  Molecular weight: 143.01
  Boiling point: 178.5°C
  Melting point: −51.9°C

Bis(2-chloroethyl)ether is a clear, colorless, combustible liquid with a pungent odor. It is soluble in oxygenated and aromatic solvents and reacts in contact with water or steam producing toxic and corrosive fumes. It is chemically active. In contact with metal powders, or strong acids it produces heat inducing a fire hazard. Inhibitors have been added to commercial formulations to prevent formation of peroxides and polymerization. Bis(2-chloroethyl)ether may decompose forming peroxides in sunlight and during storage, particularly if the container has been opened. Peroxides are explosive and in liquid form may be detonated by friction, impact or heating. Bis(2-chlorocthyl)ether is used as solvent for special lacquers, resins, and oils. Because of its strong dissolving power it is also used as penetrant in spot removing and dewaxing agents.[17]

1. Health Effects

In the atmosphere, bis(2-chloroethyl)ether is expected to exist primarily in the vapor phase. Bis(2-chloroethyl)ether is poisonous by any type of entry route into the body. Even a diluted solution, in sufficienty large amounts may penetrate skin to cause toxic effects without noticeable irritation. The respiratory reaction following inhalation of the vapor may be severe and symptoms delayed. Direct contact with the liquid or vapor irritates the skin, eyes, and mucous membranes, causing coughing, nausea, and vomiting. Bis(2-chloroethyl)ether is a *probable* human carcinogen, classified as weight-of-evidence Group B2 under the EPA Guidelines for Carcinogen Risk Assessment. Quantitative information on the kinetics and metabolism of bis(2-chloroethyl)ether in humans is not available. Limited data show that radioactive BCEE, administered to rats by inhalation or gavage is rapidly absorbed.[17]

Occupational exposure standards: TLV: 8 h TWA 10 ppm

2. Effects on the Environment

Bis(2-chloroethyl)ether's former production and use in the textile industry and as solvent in natural and synthetic resins may result in its release to the environment through various waste streams. If released to air, a vapor pressure of 1.55 mm Hg at 25°C indicates that bis(2-chloroethyl)ether will exist solely as vapor in the ambient atmosphere. Vapor-phase bis(2-chloroethyl)ether will be degraded in the atmosphere by reaction with photochemically produced hydroxyl radicals; the half-life for this reaction in air is estimated to be 5 days. If released into soil, bis(2-chloroethyl) ether has a high mobility. Many ethers are known to be resistant to biodegradation. Volatilization from moist soil surfaces is an important fate process. If released into water, bis(2-chloroethyl)ether is not adsorbed by suspended solids and sediment in water. Volatilization from water surfaces is an important fate process. The volatilization half-life from a model river and a model lake is estimated as approximately 40 h and 16 days. Bis(2-chloroethyl) ether is a marine pollutant and its release to the sea is prohibited by the International Convention since 1973.[17]

F. 1,2-DICHLOROPROPANE

CAS 78-87-5
   Synonyms and trade names: propylene dichloride or ENT 15,406.

Molecular formula: $C_3-H_6-Cl_2$
Molecular weight: 112.99
Boiling point: 96.4°C
Melting point: -100.4°C

1,2-Dichloropropane is a colorless, highly flammable liquid. It is soluble in water to a certain extent (2.7 g/l), as well as in alcohol, ether, benzene, and chloroform. The vapor is nearly four times heavier than air and may travel along the ground and ignite from a distant source. In the presence of moisture it forms highly corrosive hydrochloric acid. 1,2-Dichloropropane should be kept separate from aluminum, o-dichlorobenzene and 1,2-dichloroethane. Strong acids induce the decomposition of 1,2-dichloropropane. Contact with strong oxidizing agents, such as nitric acid and chlorates may lead to fire and explosion. 1,2-Dichloropropane is used as a component in spot and paint removers, dry-cleaning, and furniture finishing products. It is a solvent for metal degreasing, oil, resin, and gum processing. Other applications of this substance are in rubber compounding and vulcanizing operations. It is also a source material and intermediate of chemical industry processes, such as the manufacture of tetrachloroethylene and propylene oxide. It is used in the extraction processes of fats, oils, lactic acid, and petroleum waxes. 1,2-Dichloropropane mixed with 1,3-dichloropropane in soil fumigants is used to control nematodes in vegetables, potatoes, and tobacco. It has been used in agriculture as a fumigant for grain, fruit and nut crops, peach trees and for insect control. 1,2-Dichloropropane is also an additive for anti-knock fluids in fuels for motor vehicles. This solvent should not be used without proper precaution, ventilation, protective clothing, and methods of disposal, although it is not among the most toxic of the halogenated solvents.[18]

1. Health Effects

Exposure of the general population to 1,2-dichloropropane via air and water is unlikely, except in areas where there is extensive use. The vapor and the liquid irritate the eyes. Several cases of dermatitis and skin reactions have been reported from repetitive exposure of workers using 1,2-dichloropropane and mixtures of solvents containing it. Inhalation of low concentration causes irritation of the respiratory tract, coughing, and sneezing. High concentrations are narcotic and, depending on the dose, these lead to other symptoms, such as weight loss, drowsiness, and vertigo. Headache may be persistent and delayed from the initial exposure. Several cases of acute poisoning have been reported due to accidental or intentional (suicide) overexposure to 1,2-dichloropropane. It is not classifiable as a human carcinogen.[18]

Threshold limit values: 8 h TWA 75 ppm

2. Effects on the Environment

1,2-Dichloropropane's production and use as a solvent may result in its release to the environment through various waste streams. If released to air, a vapor pressure of 53.3 mm of Hg at 25°C indicates that 1,2-dichloropropane will exist solely as a vapor in the ambient atmosphere. Vapor-phase 1,2-dichloropropane will be degraded in the atmosphere by reaction with photochemically produced hydroxyl radicals; the half-life for this reaction in air is estimated to be 36 days. Photolysis is not expected to be an important environmental fate process, as vapor-phase photolysis under simulated sunlight did not occur after prolonged exposure. If released to soil, 1,2-dichloropropane has a very high mobility and is persistent. In an experiment, 98% of the applied 1,2-dichloropropane was recovered 12 to 20 weeks after soil treatment. Volatilization from moist soil surfaces is an important fate process and it may volatilize from dry soil surfaces based upon its vapor pressure. Biodegradation will not be an important environmental fate process in soil or water given a 0% theoretical BOD using the MITI test. The solvent may leak from the soil to contaminate ground water. If released into water, 1,2-dichloropropane is not adsorbed by suspended solids and sediment. Volatilization from water surfaces will be an important fate process based upon this compound. 1,2-Dichloropropane may be taken up by plants and food crops in small amounts. It is a marine pollutant. Trace amounts of 1,2-dichloropropane have been measured in the atmosphere.[18]

G. Methylene Chloride

CAS 75-09-2

Synonyms and trade names: dichloromethane, methylene dichloride, methane dichloride, Aerothene MM, Freon 30, and Solmethine.

Molecular formula: $C-H_2-Cl_2$
Molecular weight: 84.93
Boiling point: 39.75°C
Melting point: −95°C

Methylene chloride is a volatile, colorless, nonflammable liquid. It is slightly soluble in water and miscible with many other solvents, such as acetone, chloroform, carbon tetrachloride, and alcohol. Under specific conditions it may burn. Its commercial formulations for paint stripping are particularly flammable. Methylene chloride is a widely used solvent where quick drying (i.e., high volatility) is required. Such application areas include adhesives, cellulose acetate fiber production, blowing of polyurethane foams, and metal and textile treatment. It dissolves oils, fats, waxes, many plastics, bitumen, and rubber. This property is used in paint stripper formulations. It is used as an aerosol solvent, and for extraction operations in the pharmaceutical industry. It was previously used in fire-extinguishing products.[19]

1. Health Effects

The acute toxicity of methylene chloride is one of the lowest in the family of halogenated hydrocarbon solvents. Workers and their associates are mainly exposed through the inhalation of vapors. Its narcotic effects and drunkenness-related symptoms depend on the degree of exposure. In the body, methylene chloride is transformed to release carbon monoxide, which replaces oxygen in the blood. This has resulted in deaths from cardiac failure, which have been reported in spray painting. Methylene chloride is irritating to the eyes. It may penetrate healthy skin, and is irritating on repeated contact. It is moderately toxic if ingested. Liver and kidney damage is less likely at low exposure levels. However, methylene chloride is used mainly in mixtures in which other components may increase adverse effects. Methylene chloride passes through the placenta and has been found to accumulate in the fetal tissue and breast milk. Increased rates of spontaneous abortions were reported among female pharmaceutical workers, indicating a hazardous effect in pregnancy. Repeated exposure to methylene chloride has been shown to be carcinogenic in animal tests, and IARC has classified this compound as possibly carcinogenic to humans.[19]

Threshold limit values: 8 h TWA 50 ppm

2. Effects on the Environment

Dichloromethane's production and use as solvent, chemical intermediate, grain fumigant, paint stripper and remover, metal degreaser, and refrigerant may result in its release to the environment through various waste streams. If released to air, a vapor pressure of 435 mm of Hg at 25°C indicates methylene chloride will exist solely as a vapor in the ambient atmosphere. Vapor-phase methylene chloride will be degraded in the atmosphere by reaction with photochemically produced hydroxyl radicals; the half-life for this reaction in air is estimated to be 119 days. It will not be subject to direct photolysis. If released into soil, methylene chloride has a very high mobility. It may volatilize from dry soil surfaces. Biodegradation in soil may occur based on activated sludge studies. If released into water, methylene chloride is not adsorbed by suspended solids and sediment in water. Its degradation in surface waters is slow and it is removed from water by evaporation. Its presence in drinking water results from chlorination of water or is due to contamination.[19]

H. 1,1,2,2-TETRACHLOROETHANE

CAS 79-34-5

Synonyms and trade names: Tetrachloroethane, sym-Tetrachloroethane, acetylene tetrachloride, Bonoform, Cellon.

Molecular formula: $C_2-H_2-Cl_4$
Molecular weight: 167.85
Boiling point: 46.5°C
Melting point: −43.8°C

1,1,2,2-Tetrachloroethane is volatile, colorless (pure) to pale-yellow (technical), heavy, mobile, nonflammable liquid. Its suffocating odor does not provide any warning as it is smelled only at concentrations exceeding the occupational exposure limit. It is slightly soluble in water, but dissolves in a wide range of organic solvents, such as, alcohol, acetone, chloroform, and ether. In the presence of moisture tetrachloroethane decomposes slowly and produces corrosive gases. It reacts with alkali metals and their alloys, many metal powders, sodium or potassium hydroxide, bromoform and nitrogen tetroxide, producing explosive compounds. 1,1,2,2-Tetrachloroethane is used in the production of tetrachloroethylene and trichloroethylene. It has a relatively high boiling point and is a solvent for paints and lacquers. It is used in metal degreasing and finishing (e.g., in jewelry production) and also in extraction of fats and oils, and for production of insecticides and herbicides. It dissolves sulfur, gums and resins, bitumen, pitch, and tarry materials and is used in fumigation products for greenhouses and grain. At present it has been replaced in most of its applications by less toxic solvents and is recommended to be used only in closed processes.[20]

1. Health Effects

1,1,2,2-Tetrachloroethane is considered to be one of the most poisonous of the common chlorinated hydrocarbon solvents. Exposure may arise from inhalation of vapor or on skin contact. The slow removal processes of the human body contribute to the high toxicity. It is a powerful poison which has a narcotic effect on the central nervous system, kidneys, and liver. In animal tests it has shown to be two to three times more effective than chloroform in causing narcosis. Cases of chronic intoxication from artificial silk and leather manufacturing have shown two main types of effects: those originating from the central nervous system, such as tremor, vertigo, and headache; those originating from the gastrointestinal tract and liver, such as nausea, vomiting, stomach pain, and jaundice from liver damage. Tetrachloroethane causes changes in the blood, the adverse effects of which may be delayed. In animal tests it adversely affects the fetus. In humans there is inadequate evidence for the carcinogenicity of 1,1,2,2-tetrachloroethane. However, there is limited evidence in experimental animals for the carcinogenicity of 1,1,2,2-tetrachloroethane. Overall evaluation: 1,1,2,2-tetrachloroethane is not classifiable as having carcinogenicity to humans.[20]

Threshold limit values: 8 h TWA 1 ppm

2. Effects on the Environment

Most of the released 1,1,2,2-tetrachloroethane enters the atmosphere where it is extremely stable (half-life > 2 years). Some of the chemicals will eventually diffuse into the stratosphere where it will rapidly photodegrade. There is evidence that 1,1,2,2-tetrachloroethane slowly biodegrades. A product of biodegradation under anaerobic conditions is 1,1,2-trichloroethane, a chemical which is resistant to further biodegradation. Under alkaline conditions, 1,1,2,2-tetrachloroethane may be expected to hydrolyze. When disposed of on soil, part of the 1,1,2,2-tetrachloroethane may leach into groundwater. 1,1,2,2-Tetrachloroethane which is released into water will primarily be lost by volatilization in a matter of days to weeks. 1,1,2,2-Tetrachloroethane is not expected to partition from the water column to organic matter contained in sediments and suspended solids.[20]

I. TETRACHLOROETHYLENE

CAS 127-18-4

Synonyms and trade names: perchloroethylene, ethylene tetrachloride, tetrachloroethane, 1,1,2,2-tetrachloroethylene, Antisal 1, Dee-solv, Didakene, Dow-per, ENT1860, NeMa, Perchlor, Persec, Tetlen, Tetravec, and Tetropil.

Molecular formula: C_2-Cl_4
Molecular weight: 165.83
Boiling point: 121.3°C
Melting point: -22.3°C

Tetrachloroethylene is colorless, nonflammable liquid. The odor is pleasantly ethereal, resembling that of chloroform. It saturates the sense of smell and does not provide a warning and a person can suffer from overexposure without smelling it. It is miscible with many organic solvents (alcohol, ether, chloroform, benzene, hexane), but insoluble in water. Commercial grades contain stabilizers, as pure tetrachloroethylene is decomposed slowly by light and in contact with moisture. Stabilized products are stable up to 140°C. Tetrachloroethylene is an important solvent in dry-cleaning fluids; its share is more than 90% of solvents used in this application sector. As a solvent, it is also applied in metal degreasing, rubber and resin production and can dissolve silicones. It has been used to replace hazardous heat transfer liquids, PCBs, in transformers. Under proper conditions tetrachloroethylene reacts violently with some metals, strong alkalis and nitrogen tetroxide gas. Like all other halogenated solvents, tetrachloroethylene decomposes in an open flame or on a very hot surface producing corrosive and toxic gases. Tetrachloroethylene which is collected as waste can be recycled.[21]

1. Health Effects

The most common route of occupational exposure is via inhalation of vapor which is readily absorbed from the lungs into the blood circulation. The symptoms of exposure to low concentrations are related to effects on the central nervous system, where it causes dizziness, confusion, headache, and nausea. Inhalation of high concentrations may lead to unconsciousness. The acute toxicity is relatively low when compared to many other commonly used chlorinated solvents. The adverse health effects which are experienced depend on increased concentration inhaled and the length of the exposure. Repeated exposure to (concentrations over 100 ppm) tetrachloroethylene over months or years may lead to adverse effects on the nervous system, respiratory tract, liver and kidneys, and heart functions (cardiac arrhythmias). Tetrachloroethylene degreases the skin causing irritation by repeated contact. It passes into breast milk and through placental barrier to the fetus. In animal tests, it has shown to have effects on reproduction. ACGIH has established biological exposure indices (BEI) for tetrachloroethylene. It can be measured from exhaled air (BEI 5 ppm) or blood samples (BEI 0.5 ppm) prior to the last shift of the working week, or from urine at the end of the working week (BEI as trichloroacetic acid 3.5 mg/l). Tetrachloroethylene has caused cancer in animal tests and is classified as possibly carcinogenic to humans.[21]

Threshold limit values: 8 h TWA 25 ppm

2. Effects on the Environment

Tetrachloroethylene's production and its use as a dry-cleaning agent, degreasing agent and as a chemical intermediate in production of fluorocarbons will result in its release into the environment through various waste streams. Tetrachloroethylene has no natural sources. If released to air, a vapor pressure of 18.5 mm of Hg at 25°C indicates that tetrachloroethylene will exist solely as vapor in the ambient atmosphere. Vapor-phase tetrachloroethylene will be degraded in the atmosphere by reaction with photochemically produced hydroxyl radicals; the half-life for this reaction in air is estimated to be 96 days. Direct photolysis is not expected to be an important environmental fate process since this compound only absorbs light weakly in the environmental UV spectrum. If released to soil, tetrachloroethylene has a moderate mobility and is often detected in groundwater. Volatilization from moist soil surfaces is expected to be an important fate process. Tetrachloroethylene may volatilize from dry soil surfaces based upon its vapor pressure. Volatilization half-lives in the range of 1.2 to 5.4 h were measured for tetrachloroethylene from a sandy loam soil surface and volatilization half-lives of 1.9 to 5.2 h were measured from an organic top-soil. Biodegradation is expected to occur slowly in soils under aerobic and anaerobic conditions. If released into water, tetrachloroethylene is adsorbed by suspended solids and sediments in water. It has been also found in foods, such as dairy products, meat, oils and fat, fruits and vegetables, and fresh bread.[21]

J. 1,1,1-TRICHLOROETHANE

CAS 71-55-6

Synonyms and trade names: methyl chloroform, methyl trichloromethane, chloroethane, chlorothene, Aerothene TT, Alpha-T, Genklene, and Inhibisol.

Molecular formula: $C_2-H_3-Cl_3$
Molecular weight: 133.42
Boiling point: 74.0°C
Melting point: −30.4°C

1,1,1-Trichloroethane is a volatile, colorless liquid and its vapor is heavier than air. It has a sweet, ethereal odor. It is poorly soluble in water but dissolves in other solvents, such as acetone, benzene, ethanol, and carbon tetrachloride. The vapors of 1,1,1-trichloroethane are heavier than air and may travel a considerable distance from the evaporation source. In contact with water or humidity, trichloroethane decomposes slowly yielding corrosive acids. In normal working conditions 1,1,1-trichloroethane presents no risk of flammability. However, when heated, and particularly when in contact with some metal salts, it decomposes producing very toxic and corrosive gases, as may be the case when welding in confined places containing trichloroethane vapor. Commercial products are stabilized to avoid corrosion of storage containers. The stabilizers (3 to 8%) may have toxic health effects. When in contact with a strong alkali, such as calcium hydroxide, 1,1,1-trichloroethane undergoes a hazardous reaction producing extremely flammable, volatile, and hazardous compounds. It has properties similar to those of the more hazardous solvents, viz., carbon tetrachloride and trichloroethylene. It has replaced the use of the latter and is extensively used as a solvent for rubber, bitumen, mineral and vegetable oils, stearic acid, lanolin, polystyrene, polyvinyl acetate, acrylic resins, and as a coolant in metal cutting oils and in lubricants. It finds usages in dry-cleaning formulations, in cleaning of metal and plastic surfaces, as a solvent in printing inks and consumer products, such as adhesives and correction fluids like Tipp-Ex. It can be found in aerosols and as an additive raises the flash point of many flammable solvents. It was previously used as a solvent in insecticides, and in the treatment of citrus fruits and strawberries.[22]

1. Health Effects

Among the chlorinated hydrocarbon solvents, 1,1,1-trichloroethane is generally accepted to be the least toxic to humans. It is mainly absorbed through inhalation, although it may also enter the body through the skin and as a result of ingestion. It has fairly low acute and chronic toxicity, and is relatively safe when used as a solvent. Concentration levels under 1350 mg/m^3 (250 ppm) are estimated to cause no adverse effects. The odor is smelled before the occupational exposure limit is exceeded and may serve as a warning. The potential adverse health effects in short-term and repeated exposure vary from changes in behavior to symptoms originating from the central nervous system: depression, unconsciousness and heart failure, kidney, liver and lung damage, at high concentrations (above 350 to 500 ppm). Inhalation of the vapor causes headache, dizziness, and drowsiness. The seriousness of the effects depends on individual susceptibility and the dose. Alcohol drinking combined with exposure to 1,1,1-trichloroethane may increase the toxic effects. Skin and eyes are irritated if in direct contact with liquid or vapor. Highly volatile 1,1,1-trichloroethane's vapor is nearly five times heavier than air. Poorly ventilated or confined spaces, such as tanks and vaults, may contain a very high concentration of solvent vapor. This has led to serious accidents.[22]

Threshold limit values: 8 h TWA 350 ppm

2. Effects on the Environment

Spillage of 1,1,1-trichloroethane evaporates (over 99% of the lost solvent) to the atmosphere and slowly decomposes. It resides in the atmosphere for an estimated time of about 6 years. It also reaches the upper parts of the atmosphere in significant amounts, where decomposition releases reactive chlorine, resulting in depletion of the protective ozone layer causing global warming, but with lesser potential than chlorofluorohydrocarbons (freons). The large-scale releases of this solvent into the atmosphere raise concern for global atmospheric effects. The Montreal Protocol on substances which deplete the ozone layer includes 1,1,1-trichloroethane and its use is going to be reduced. Finally it is to be substituted by other solvents. It is heavier than water and is not water soluble. It is not absorbed on to soil particles and thus leaks readily into groundwater where it remains as persistent contamination. Traces of 1,1,1-trichloroethane have been detected in all water sources; groundwater, drinking rainwater, seawater, and sewage. It has been found in seawater organisms, and fresh- and seawater birds and their eggs. Although being a widely distributed contaminant in nature, 1,1,1-trichloroethane is considered not to bioaccumulate[22] but it has been classified as a marine pollutant.

K. 1,1,2-TRICHLOROETHANE

CAS 79-00-5

Synonyms and trade names: ethane trichloride or beta-T.

Molecular formula: $C_2-H_3-Cl_3$
Molecular weight: 133.42
Boiling point: 113.8°C
Melting point: −36.6°C

1,1,2-Trichloroethane is a clear, colorless, and nonflammable liquid, with a pleasant odor. It is practically insoluble in water but miscible with ethanol, chloroform, ether and chlorinated solvents. It is mainly used in the manufacturing of vinylidene chloride but may act as a specialty solvent for adhesives, fats, and resins.

1. Health Effects

1,1,2-Trichloroethane is generally accepted as more toxic than 1,1,1-trichloroethane. Humans are exposed to 1,1,2-Trichloroethane from ambient air, particularly near sources of emission and from contaminated drinking-water supplies. Exposure is possible for workers in blast furnaces, steel mills, and engineering and scientific instrument manufacturing. Due to hazardous liver toxicity and suspicion of carcinogenicity, its use has been replaced, often by the less toxic 1,1,1-trichloroethane.[23]

Threshold limit values: 8 h TWA 10 ppm

2. Effects on the Environment

1,1,2-Trichloroethane will enter the atmosphere from its use in the manufacture of vinylidene chloride and its use as a solvent. Once in the atmosphere, 1,1,2-trichloroethane will photodegrade slowly by reaction with hydroxyl radicals (half-life 24 to 50 days in unpolluted atmospheres and within a few days in polluted atmospheres). The soil partition coefficient of 1,2-trichloroethane is low and it will readily leach in the case of eventual, very slow biodegradation. Bioconcentration is not a significant process. It will also be discharged in wastewater associated with these uses and in leachates and volatile emissions from landfills. Releases to water will primarily be lost through evaporation.[23]

L. TRICHLOROETHYLENE

CAS 79-01-6

Synonyms and trade names: ethylene trichloride, acetylene trichloride, trichloroethene, TCE, Tre, Algylen, Benzinol, Blancosolv, Chlorilen, Circosolv, Dow-tri, Fleck-Flip, Lanadin, Perm-A-Chlor, Petzinol, Trethylene, Triasol, Trichloran, Tri-Clene, Trilene, TRI-plus, Vestrol, Vitran.

Molecular formula: C_2-H-Cl_3
Molecular weight: 131.39
Boiling point: 87.2°C
Melting point: −84.7°C

Trichloroethylene is a volatile, clear, colorless liquid, sometimes dyed blue. It is nonflammable in normal working conditions and combustible under specific conditions. The odor is slightly sweetish and it is smelled at concentrations low enough to provide a warning of exposure. Trichloroethylene is poorly soluble in water but dissolves in ethanol, diethyl ether, and other chlorinated hydrocarbons. Accidents have been recorded when toxic and corrosive decomposed gases have been inhaled. These gases may be formed in contact with hot surfaces and open welding flames. Stabilized commercial products are prevented by stabilizers (below 1%) against undergoing any chemical changes in contact with air, humidity or light. However, some of the stabilizers used are very toxic. Contact with strong alkaline substances, such as soda lime, particularly if heated above 70°C, or in presence of epoxides, causes trichloroethylene to react, producing a highly reactive and explosive gas (dichloroacetylene), which is acutely damaging and causes permanent nerve injury to humans and animals. Trichloroethylene is a good solvent and is used in the degreasing of metals, textiles, leather, and wool. Commercial grades are available and some of these are for specified use, such as the cleaning of ferrous metals. Aluminium and magnesium metals should not be cleaned with the same grade but with a neutral, stabilized grade designed for nonferrous metals. Trichloroethylene is used in extraction processes and as a solvent for various materials, such as waxes, oils, resins, rubber, lacquers and paints, printing inks, adhesives, fluid silicones, cellulose esters and ethers, and sulfur. Another use of trichloroethylene has been dry-cleaning, in insecticides and fungicides, where it is a vehicle for dental use. Less toxic solvents have replaced it in several places of usage. Trichloroethylene from industrial processes can be recovered by distillation and may be recycled.[24]

1. Health Effects

Exposure to trichloroethylene occurs mainly through inhalation and skin contact. Overexposure affects the central nervous system causing headache, drowsiness, depression, vertigo, and lack of coordination. It also causes alcohol intolerance. Sudden deaths have been caused by heart arrhythmia. Repeated exposure may cause memory impairment and changes in behavior, intellectual functions, and reaction speed. Termination of exposure usually results in rapid recovery of normal behavior. There are reports of damage to the liver and kidneys. Trichloroethylene is an irritant to the skin, mucous membranes, and the eyes. ACGIH has given BEI values to trichloroethylene metabolite products measured in urine (trichloroacetic acid and trichloroethanol, 300 mg/g creatinine) and blood (free trichloroethanol, 4 mg/l), analyzed from samples taken at the end of the work-shift. Trichloroethylene has induced cancer in animal tests. Until now, there has been no evidence of the carcinogenic effects in humans. However, more research is being carried out in this field.[24]

Threshold limit values: 8 h TWA 50 ppm

2. Effects on the Environment

Trichloroethylene has no natural sources, but may be detected in the environment as a result of human activities and an estimated 60% of the annual work production is lost to the environment (millions of tons evaporate into the atmosphere yearly). If released to air, a vapor pressure of 69 mm of Hg at 25°C indicates that trichloroethylene will exist solely as a vapor in the ambient atmosphere. Vapor-phase trichloroethylene will be degraded in the atmosphere by reaction with photochemically produced hydroxyl radicals; the half-life for this reaction in air is estimated to be 7 h. If released to soil, trichloroethylene is expected to have high mobility. Volatilization from moist soil surfaces is an important fate process. Trichloroethylene is expected to volatilize from dry soil surfaces based upon its vapor pressure. Cometabolic biodegradation of trichloroethylene has been reported under aerobic conditions where additional nutrients have been added. Under anaerobic conditions, as might be seen in flooded soils, sediments, or aquifer environments, trichloroethylene is slowly biodegraded via reductive dechlorination; the extent and rate of such degradation is dependent upon the strength of the reducing environment. Trichloroethylene half-lives in the field for aquifer studies range from 35 days to over 6 years. If released into water, trichloroethylene does not to adsorb to suspended solids and sediments. Volatilization from water surfaces will be an important fate process. It has been detected in traces in surface water, soil, and sediment. Consumer products, such as dairy products, meat, oils and fat, beverages, fruit, and vegetables have also been found to contain trace amounts but there is no direct evidence of its bioaccumulation in the human food chain. Trichloroethylene is a marine pollutant.[24]

M. 1,2,3-TRICHLOROPROPANE

CAS 96-18-4

Synonyms: allyl trichloride or glycerol trichlorohydrin.

Molecular formula: $C_3-H_5-Cl_3$
Molecular weight: 147.43

Boiling point: 156.85°C
Melting point: −14.7°C

A combustible liquid, 1,2,3-trichloropropane is colorless to straw-colored. It has a strong acid odor similar to that of trichloroethylene or chloroform. It is soluble in water, alcohol, ether, and slightly soluble in chloroform. The vapor is heavier than air and may travel along the ground to a distant ignition point and may form explosive mixtures with air. It is a solvent for oils, fats, waxes, chlorinated rubber, and resins. In addition, 1,2,3-trichloropropane is used in paint and varnish removers in degrading agents.[25]

1. Health Effects

Inhalation and penetration of 1,2,3-trichloropropane through the skin are common exposure routes at work. It is irritating to the skin and the respiratory tract and acts as a severe irritant to the eyes. It has a narcotic effect, headache being the symptom of exposure to low concentrations. High concentrations lead to unconsciousness. Repeated exposure may cause liver damage. There is inadequate evidence in humans for the carcinogenicity of 1,2,3-trichloropropane. There is sufficient evidence in experimental animals for the carcinogenicity of 1,2,3-trichloropropane. The overall evaluation is that 1,2,3-trichloropropane is probably carcinogenic to humans.[25]

Threshold limit values: 8 h TWA 10 ppm

2. Effects on the Environment

As a result of its manufacture, transport, storage, and use as a solvent for oils, fats, waxes, chlorinated rubber, and resins and in the synthesis of some elastomers, 1,2,3-trichloropropane may be released to the environment in emissions and wastewater. In the atmosphere, 1,2,3-trichloropropane will react with photochemically produced hydroxyl radicals. If released to soil, 1,2,3-trichloropropane would be expected to leach and volatilize from dry and moist soil. In one study, its volatilization half-life from soil was 2.7 days. Based upon results of a study, 1,2,3-trichloropropane volatilizes readily from water. Its estimated half-life in a model river and model lake is 6.7 h and 5.7 days, respectively. Due to its low adsorptivity, it would not adsorb to sediment or particulate matter in the water column. It is resistant to biodegradation and hydrolysis and therefore these fate processes should not be important in soil or the aquatic environments. It would not be expected to bioconcentrate in aquatic organisms. Due to its high water solubility, 1,2,3-trichloropropane will be subject to wash-out by rain. The combination of volatilization, wash-out, and resistance to degradation should result in a recycling of 1,2,3-trichloropropane among the environmental compartments. It is a marine pollutant.[25]

II. ANALYSIS OF HALOGENATED VOCs

All analysis methods for halogenated VOCs found in literature, are based on gas chromatography (GC) in combination with different detectors. In Table 17.1 a selection of GC analysis procedures for halogenated VOCs are tabulated.

For a description of the various sample preparation techniques, such as, solid-phase extraction (SPE), solid-phase microextraction (SPME), headspace and purge and trap (P&T), dynamic membrane extraction (DMA) and the different detection methods, the reader is directed to the other chapters of this book and especially the chapter on sample preparation methods.

TABLE 17.1
Overview GC Methods, Including Sample Preparation and Detailed Conditions

Compound(s)	Source	Extraction Technique	Injection Temperature	Carrier Gas	Column	Temperature Program	Detection Technique	LOD	RSD%	Refs.
Chlorobenzenes	Soil	Headspace SPME 100 μm PDMS fiber	280°C	He	Varian 3400 CX fused silica capillary column	50°C → 90°C (20°C/min) → 150°C (3°C/min) → 180°C (25°C/min)	IT–MS	0.055–1 ng/g	2–8%	7
Carbon tetrachloride	Soil	P&T (dry samples)	na	He (1 ml/min)	HP-6890	35°C → 200°C (10°C/min)	MS	na	na	8
	Soil	Headspace (dry samples)	na	He (12 ml/min)	HP 5890 Series II fused silica capillary column	90°C	ECD		na	8
Chloroform, 1,1,1-trichloroethane, 1,1,2-trichloroethane, trichloroethylene and 1,2-dichlorobenzene	Water	P&T (methanol extract)	na	He (1 ml/min)	HP-6890	35°C → 200°C (10°C/min)	MS	na	na	8
		P&T	200°C	He (3.8 ml/min)	Perkin Elmer	35°C → 100°C (4°C/min)	MS	0.36 μg/l, 0.30 μg/l, 0.64 μg/l, 0.36 μg/l, 0.80 μg/l	na	11
Chloroform, 1,1,1-trichloroethane, 1,1,2-trichloroethane, tetrachloroethylene, trichloroethylene, carbon tetrachloride and 1,1,2,2-tetrachloroethane	Water	P&T	200°C	He	HP 5890 Series II fused silica capillary column	40°C → 100°C (3°C/min), 100°C → 180°C (5°C/min)	MS	0.025 μg/l, 0.02 μg/l, 0.05 μg/l, 0.02 μg/l, 0.02 μg/l, 0.025 μg/l, 0.025 μg/l	na	6
Chlorobenzenes	Water	DME (silicone hollow fiber)	350°C	na	AirmoBTX HC 1000	40°C → 140°C (20°C/min)	FID	0.5 μg/l (fr.:30 ml/min)	na	11
Chloroform and 1,1,1-trichloroethane	Marine biota	P&T	200°C	He	na	40°C → 200°C (10°C/min)	MS	0.2 ng/g, 0.19 ng/g	50–200	10
Chloroform, 1,1,1-trichloroethane, carbon tetrachloride	Landfill leachates	HS-SPME	250°C	He (1 ml/min)	HP fused silica capillary column	45°C → 150°C (15°C/min)	MS	0.10 ng/ml, 0.05 ng/ml, 0.10 ng/ml	16%, 11%, 12%	9

Chloroform, 1,1,1-trichloroethane, 1,1,2-trichloroethane, carbon tetrachloride, methylene chloride, tetrachloroethylene, 1,1,2,2-tetrachloroethane, 1,2-dichlorobenzene, and 1,2-dichloroethane	Air	Draw a sample of ambient air through a sampling train comprised of components that regulate the rate and duration of sampling into a pre-evacuated SUMMA passivated canister	na	He (2 ml/min)	General OV-1 cross-linked methylsilicone	−50°C, 2 min → 150°C, 15 min (8°C/min)	MS	>1 Ppb	90–110	26,27
Chloroform, 1,1,1-trichloroethane, 1,1,2-trichloroethane, carbon tetrachloride, methylene chloride, tetrachloroethylene, 1,2-dichlorobenzene and 1,1,2,2-tetra-chloroethane	Air	P&T	150°C	He	HP methyl silicone cross-linked fused silica	na	ECD/FID	1–5 ng/l (FID)	100	27,28
Chloroform, carbon tetrachloride, 1,2-dichloroethane, 1,1,1-trichloroethane, monochlorobenzene, trichloroethylene and tetrachloroethylene	Air	Draw ambient air through a cartridge containing approximately 1–2 g of Tenax	na	He (1–2 ml/min)	SE-30 or alternative coating, glass capillary or fused silica	Depends on the specific compound of interest General: 30°C → 200°C (8°C/min)	MS	General: 20 ng/l or less	na	27,29
Methylene chloride, 1,2-dichloroethane, 1,1,1-trichloroethane, carbon tetrachloride	Air	Draw ambient air through a cartridge containing approximately 0.4 g of a CMS	na	He (2–3 ml/min)	SE-30 or alternative coating, glass capillary or fused silica	Depends on the specific compound of interest. General: −70°C → 150°C (8°C/min)	MS	na	85	27,30

Continued

TABLE 17.1
Continued

Compound(s)	Source	Extraction Technique	Injection Temperature	Carrier Gas	Column	Temperature Program	Detection Technique	LOD	RSD%	Refs.
Carbon tetrachloride, monochlorobenzene, chloroform, tetrachloroethylene, 1,1,1-trichloroethane, trichloroethylene	Air	Place the front and back sorbent sections of the sampler tube in separate vials. Discard the glass wool and foam plugs. Add 1 ml carbon disulfide to each vial.	200°C	He (2,6 ml/min)	Capillary, fused silica	35°C, 3 min → 150°C, (8°C/min)	FID	4.0 $\mu g/.$, 0.6 $\mu g/l$, 0.8 $\mu g/l$, 2.0 $\mu g/l$, 1.0 $\mu g/l$, 0.6 $\mu g/l$	na	27,28
1,2-dichlorobenzene, 1,1,2-trichloroethane, 1,2,3-trichlororopane	Air	Place the front and back sorbent sections of the sampler tube in separate vials. Discard the glass wool and foam plugs. Add 1 ml carbon disulfide to each vial	225°C	He (4,7 ml/min)	Capillary, fused silica	35°C 3 min → 190°C (8°C/min)	FID	0.8 $\mu g/l$, 1.0 $\mu g/l$, 1.0 $\mu g/l$	na	27,28

REFERENCES

1. Chemical safety, *International Programme on Chemical Safety (IPCS)*, Supplement 1/1998.
2. Anonymous, *Environmental Health Criteria*, 208, 165, 1999.
3. Bremmer, H.J., Hesse, J.M., Matthijsen, A.J., and Slooff, W., National Institute for Public Health and Environmental Protection (RIVM), The Netherlands, Vol:Report No 710401015, p. 135, 1991.
4. Standring, P., Cartlidge, G.D., and Meldrum, M., Carbon tetrachloride, chloroform, *HSE Toxicity Rev.*, 23, 36,1991.
5. WHO Working Group, *Environmental Health Criteria*, 163, 174, 1994.
6. Kostopoulou, M. N., Golfinopoulos, S. K., Nikolaou, A. D., Xilourgidis, N. K., and Lekkas, T. D., Volatile organic compounds in the surface waters of Northern Greece, *Chemosphere*, 40, 527–532, 2000.
7. Sarrión, M. N., Santos, F. J., and Galceran, M. T., Strategies for the analysis of chlorobenzenes in soils using solid phase microextraction coupled with gas chromatography-ion trap mass spectrometrie, *J. Chromatogr. A*, 819, 197–209, 1998.
8. Alvarado J.S., Spokas K., and Taylor J.D. Analytical methods for the determination of carbon tetrachloride in soils, Argonne National Lab, *The Second International Symposium on Integrated Technical Approaches to Site Characterization*, Chicago, IL, June 7–9, 1999.
9. Florés Menéndes, J. C., Fernández Sánchez, M. L., Fernández Martinez, E., Sánchez Uria, J. E., and Sanz-Medel, A., Static Headspace versus head space SPME(HS SPME) for the determination of volatile organochlorine compounds in landfill leachates by gaschromatography, *Talanta*, 63, 809–814, 2004.
10. Roose, P. and Brinkman, U. A. Th., Determination of volatile organic compounds in marine biota, *J. Chromatogr. A*, 799, 233–248, 1998.
11. Kuo, H. W., Chiang, T. F., Lai, J. S., Chan, C. C., and Wang, J. D., VOC concentration in Taiwan's household water, *Sci. Total Environ.*, 208, 41–47, 1997.
12. http://toxnet.nlm.nih.gov/cgi-bin/sis/search/f?./temp/~SgjY30:1 (5 September 2004).
13. http://toxnet.nlm.nih.gov/cgi-bin/sis/search/f?./temp/~m992Eo:1 (4 September 2004).
14. http://toxnet.nlm.nih.gov/cgi-bin/sis/search/f?./temp/~NqNcRC:1 (2 September 2004).
15. http://toxnet.nlm.nih.gov/cgi-bin/sis/search/f?./temp/~IY3cUj:7 (3 September 2004).
16. http://toxnet.nlm.nih.gov/cgi-bin/sis/search/f?./temp/~Al5hFb:1 (4 September 2004).
17. http://toxnet.nlm.nih.gov/cgi-bin/sis/search/f?./temp/~Vg3UYt:1 (6 September 2004).
18. http://toxnet.nlm.nih.gov/cgi-bin/sis/search/f?./temp/~fehud4:1 (8 September 2004).
19. http://toxnet.nlm.nih.gov/cgi-bin/sis/search/f?./temp/~ImghDV:1 (8 September 2004).
20. http://toxnet.nlm.nih.gov/cgi-bin/sis/search/f?./temp/~QAk8Gl:1 (9 September 2004).
21. http://toxnet.nlm.nih.gov/cgi-bin/sis/search/f?./temp/~cmpAHP:1 (9 September 2004).
22. http://toxnet.nlm.nih.gov/cgi-bin/sis/search/f?./temp/~kEhxST:1 (10 September 2004).
23. http://toxnet.nlm.nih.gov/cgi-bin/sis/search/f?./temp/~WyGYib:1 (10 September 2004).
24. http://toxnet.nlm.nih.gov/cgi-bin/sis/search/f?./temp/~EanifA:1 (22 September 2004).
25. http://toxnet.nlm.nih.gov/cgi-bin/sis/search/f?./temp/~QDmBwI:1 (23 September 2004).
26. http://www.epa.gov/ttn/amtic/files/ambient/airtox/to-14ar.pdf (22 July 2004).
27. http://www.epa.gov/ttn/amtic/files/ambient/airtox/to-3.pdf (22 July 2004).
28. http://www.cdc.gov/niosh/nmam/pdfs/1003.pdf (23 July 2004).
29. http://www.epa.gov/ttn/amtic/files/ambient/airtox/to-1.pdf (23 July 2004).
30. http://www.epa.gov/ttn/amtic/files/ambient/airtox/to-2.pdf (26 July 2004).

18 Polychlorobiphenyls

Alessio Ceccarini and Stefania Giannarelli

CONTENTS

I. INTRODUCTION

Polychlorobiphenyls (PCBs) are a group of 209 related compounds, known as congeners, which differ only in terms of number of chlorine atoms attached to the parent biphenyl molecule (Table 18.1). Rather than using the complete IUPAC name, these are commonly referred to with a number from PCB 1 to PCB 209.

They were synthesized in the late 19th century and were produced at an industrial level from around 1930 and marketed under various trade names, e.g., Acelor, Aroclor, Clophen, Delor, Fenclor, Kanechlor, Montar, PCBs, Phenoclor, Sovol, Turbinol, etc.,[1] but they have since been banned from use because of the potential health problems associated with them. The reaction and separation conditions for production of each commercial mixture favor the synthesis of certain congeners, giving each a unique signature or pattern based on its congener composition. Each is a complex mixture of isomers, which include four to five congener classes, and has a chlorine content ranging from 21 to 68%. For instance, no Aroclor contains all 209 congeners; in fact, 110 to 120 congeners typically account for over 95% of the total mass in each Aroclor.[2] Table 18.2 shows, as an example, the typical percentage composition of a few commercial PCB mixtures.

These compounds are a class of nonpolar, nonflammable, industrial fluids with good thermal and chemical stability, and electrical insulating properties which meant that these could be used as dielectric fluids in transformers and capacitors, as heat-transfer and hydraulic fluids, as plasticizers in paints, adhesives, sealants and plastics, and in the formulation of lubricating and cutting oils.[3]

TABLE 18.1
Congener Classes, Number of Isomers and IUPAC Numbering

Congener Class	Number of Isomers	IUPAC Nos.
Monochloro	3	PCB1–PCB3
Dichloro	12	PCB4–PCB15
Trichloro	24	PCB16–PCB39
Tetrachloro	42	PCB40–PCB81
Pentachloro	46	PCB82–PCB127
Hexachloro	42	PCB128–PCB169
Heptachloro	24	PCB170–PCB193
Octachloro	12	PCB194–PCB205
Nonachloro	3	PCB206–PCB208
Decachloro	1	PCB209
Total	209	

The estimated world production in the period 1930 to 1974 is about 1.2×10^6 tons,[4] of this about one third has been released into the environment without any precautions regarding toxic effects on biota and any care to prevent environmental pollution. This has led to the widespread occurrence of PCBs all over the world, even in remote areas.[5,6] The U.S. Environmental Protection Agency (EPA), under the provisions of the Toxic Substances Control Act, specifically banned most of the uses of PCBs in 1997. Current releases of PCBs are mainly as a result of the cycle of these persistent contaminants from soil to air and back to soil again. Other possible sources of contamination, such as leaching, occurs. Moreover, PCBs can be unintentionally produced as by-products in a wide variety of chemical processes which contain chlorine and hydrocarbon sources, during water chlorination, and by thermal degradation of other chlorinated organics.[7]

TABLE 18.2
Typical Percentage Composition of Some Commercial PCB Mixtures

Congener Class	Aroclor					Clophen		Kanechlor		
	1016	1242	1248	1254	1260	A30	A60	300	400	500
Mono-CBs	2	1	—	—	—	—	—	—	—	—
Di-CBs	19	13	1	—	—	20	—	17	3	—
Tri-CBs	57	45	21	1	—	52	—	60	33	5
Tetra-CBs	22	31	49	15	—	22	1	23	44	26
Penta-CBs	—	10	27	53	12	3	16	1	16	55
Hexa-CBs	—	—	2	26	42	1	51	—	5	13
Hepta-CBs	—	—	—	4	38	—	28	—	—	—
Octa-CBs	—	—	—	—	7	—	—	—	—	—
Nona-CBs	—	—	—	—	1	—	—	—	—	—
Deca-CB	—	—	—	—	—	—	—	—	—	—

"—" means less than 1%.

PCBs are chemically stable and their half-life time in the environment is connected to the number of chlorine atoms in the biphenyl structure. The half-life time of PCBs ranges from a few days to about 10 years for mono- to pentachlorobiphenyls, and it can be as high as 20 years for higher substituted congeners, making PCBs one of the most persistent widespread class of environmental pollutants typically associated with organic matter. Traditionally, PCBs have been considered resistant to biodegradation but they do in fact biodegrade in the environment, although at a very low rate, depending on chlorine contents and position.[8] The vapor pressures of PCBs are quite low, and hence the evaporation of neat PCBs tends to be minimal, especially for high-molecular-weight (highly chlorinated) species. Their affinity for fine particulate matter strongly favors partitioning into sediment phase rather than the aqueous phase. Unfortunately, like many other types of hydrocarbons, PCBs are hydrophobic and lipophilic. These tend to accumulate in the lipid-rich tissue and organs of biota,[9] and may act as cancer initiators. There is also evidence that PCBs may cause reproductive failure in animals.

The development of sensitive and specific analytical methods for their detection led to the growing awareness of their increasing presence in the ecosystem. In addition, because PCBs are a mixture of up to 209 distinct congeners, laboratory analysis and risk assessment of PCBs are particularly challenging.

II. PHYSICAL AND CHEMICAL PROPERTIES OF POLYCHLOROBIPHENYLS

PCBs are noninflammable and water-insoluble compounds. Their water solubility decreases from about 6 to 0.08 mg l^{-1} for mono- and dichloro congener classes, respectively, and it ranges from 0.175 to 0.007 mg l^{-1} for all other classes. These are chemically inert under acid and alkali conditions and very stable to oxidation (the thermal decomposition rise over 1000°C). These have high boiling points and low electrical conductivity. The boiling point, vapor pressure, and octanol–water partition coefficient (K_{OW}) of PCBs vary with the degree of chlorination and with the position of chlorine atoms in the biphenyl structure. Both vapour pressure and octanol–water partition coefficients (K_{OW}) are largely used in diffusion model of PCBs in the environment. The mean value of log K_{OW} varies quite linearly with the number of chlorine atoms from 4.1 to 9. Table 18.3 presents the physical properties of PCBs that are important in understanding their chemical properties. Density and viscosity also increase with the degree of chlorination. Congeners with one to four chlorine atoms are oily fluids, pentachlorobiphenyls are honey-like oils, and the higher chlorinated PCBs are greases and waxy substances. Table 18.4 shows the apparent color, distillation range, molecular weight average, density, and viscosity for various Aroclor mixtures. The commercial products existed as mobile oils (A1221 to A1248), viscous liquids or sticky resins (A1250 to A1262), or solids (A1268). Their vapor pressures were such that volatilization was possible. Likewise, their water solubility, although low, was sufficient to allow movement in water. All PCB products had good solubility in organic solvents, oils, and fats.

III. TOXICOLOGICAL EFFECTS

Human are exposed to PCBs through various pathways, e.g., air, water, sediment, soil, and food. The PCB level may vary over several orders of magnitude, often depending on proximity to a source of release into the environment.[10,11] PCBs are soluble in fatty and lipid-rich tissues and organs of biota where they are accumulated and may cause a variety of adverse health effects. Both metabolic and elimination processes are slow and strongly affected by the chlorination level. In fact, bioaccumulation through the food chain is more efficient for congeners of higher chlorine content, producing residues which have a very different congener distribution with respect to that of commercial mixtures.[12] Since a few more toxic congeners are preferentially retained, PCB

TABLE 18.3
Physical and Chemical Properties of Some PCB Congeners

IUPAC No.	Compound	Boiling Point (°C)	Vapor Pressure (mm Hg, 25°C)	Log K_{OW}	Solubility mg l^{-1}
	Biphenyl	255	9.5×10^{-3}	4.10	7.2
	Monochlorobiphenyls				
1	2	274	8.4×10^{-3}	4.56	5.9
2	3	284–5	1.5×10^{-3}	4.72	3.5
3	4	291	4.6×10^{-3}	4.69	1.19
	Dichlorobiphenyls				
4	2,2'	—	1×10^{-3}	5.02	1.50
5	2,3	172	—	—	—
7	2,4	—	1.8×10^{-3}	5.15	1.40
8	2,4'	—		<5.32	1.88
9	2,5	171	1.4×10^{-3}	5.18	0.59
11	3,3'	322–4	6.8×10^{-4}	5.34	—
12	3,4	195–200	—	—	—
14	3,5	166	—	—	—
15	4,4'	315–9	1.9×10^{-5}	5.28	0.08
	Trichlorobiphenyls				
18	2,2',5	—	9×10^{-5}	5.65	0.14
33	2',3,4	—	7.7×10^{-5}	6.1	0.078
28	2,4,4'	—	—	5.74	0.085
29	2,4,5	—	3.3×10^{-4}	5.77	0.092
30	2,4,6	—	8.8×10^{-4}	—	—
31	2,4',5	—	3.0×10^{-4}	5.77	—
33	2',3,4	—	—	—	0.078
37	3,4,4'	—	—	5.90	0.015
	Tetrachlorobiphenyls				
40	2,2',3,3'	—	7.3×10^{-5}	6.67	0.034
44	2,2',3,5'	—		6.67	0.170
47	2,2',4,4'	—	8.6×10^{-5}	6.44	0.068
52	2,2',5,5'	—	3.7×10^{-5}	6.26	0.046
53	2,2',5,6	—	2.1×10^{-4}	—	—
54	2,2',6,6'	—	—	5.94	—
60	2,3,4,4'	—	—	—	0.058
61	2,3,4,5	—	—	6.39	0.019
66	2,3',4,4'	—	4.6×10^{-5}	6.67	0.058
70	2,3',4',5	—	4.4×10^{-6}	6.39	0.041
77	3,3',4,4'	—	2.3×10^{-6}	6.52	0.175
80	3,3',5,5'	—	—	6.58	—
	Pentachlorobiphenyls				
86	2,2',3,4,5	—	5.8×10^{-7}	6.38	0.0098
87	2,2',3,4,5'	—	1.6×10^{-5}	6.58	0.022
88	2,2',3,4,6	—	—	7.51	0.012
99	2,2',4,4',5	—	2.1×10^{-5}	—	—
101	2,2',4,5,5'	—	9.0×10^{-6}	6.85	0.031
105	2,3,3',4,4'	—	6.8×10^{-6}	—	—
116	2,3,4,5,6	—	—	6.85	0.0068
118	2,3',4,4',5	195–220	9.0×10^{-6}	—	—

Continued

TABLE 18.3
Continued

IUPAC No.	Compound	Boiling Point (°C)	Vapor Pressure (mm Hg, 25°C)	Log K_{OW}	Solubility mg l^{-1}
	Hexachlorobiphenyls				
128	2,2',3,3',4,4'	—	2.6×10^{-6}	7.44	0.00044
129	2,2',3,3',4,5	—	—	8.26	—
134	2,2',3,3',5,6	—	—	8.18	0.00091
138	2,2',3,4,4',5'	—	4.0×10^{-6}	—	—
149	2,2',3,4',5',6	—	1.1×10^{-5}	—	—
153	2,2',4,4',5,5'	—	5.2×10^{-6}	7.44	0.0088
					0.0013
155	2,2',4,4',6,6'	—	1.3×10^{-5}	7.12	0.00091
156	2,3,3',4,4',5	—	1.6×10^{-6}	—	—
	Heptachlorobiphenyls				
170	2,2',3,3',4,4',5	—	6.3×10^{-7}	—	—
171	2,2',3,3',4,4',6	—	1.8×10^{-6}	—	—
180	2,2',3,4,4',5,5'	240–280	9.7×10^{-7}	—	—
185	2,2',3,4,5,5',6	—	—	7.93	0.00048
187	2,2',3,4',5,5',6	—	2.3×10^{-6}	—	—
	Octachlorobiphenyls				
194	2,2',3,3',4,4',5,5'	—	—	8.68	0.0070
					0.0014
202	2,2',3,3',5,5',6,6'	—	—	8.42	0,00018
	Nonachlorobiphenyls				
206	2,2',3,3',4,4',5,5'6	—	—	9.14	0.00011
209	Decachlorobiphenyl	—	—	9.60	0.015
					0.00049

bioaccumulation seems to amplify the toxicity calculated according to the composition of commercial PCB mixtures.[13]

Congener analysis provides a wealth of data about the composition of PCBs in environmental samples, but risk assessors currently cannot assess the health risk for humans, from exposure to each

TABLE 18.4
Characteristic of Aroclor Mixtures

	Apparent Color	Distillation Range (°C)	Molecular Weight Average	Density (g ml^{-1} at 20 °C)	Viscosity (Saybolt Universal sec.) at 98.9°C
Aroclor 1260	Light yellow, soft, sticky resin	385–420	366–372	1.62	72–78
Aroclor 1254	Light yellow, viscous liquid	365–390	326.4–327	1.54	44–58
Aroclor 1248	Colorless mobile oil	340–375	291.9–288	1.44	36–37
Aroclor 1242	Colorless mobile oil	325–366	257.5–261	1.38	34–35
Aroclor 1016	Colorless mobile oil	323–356	—	1.37	—

individual congener because toxicity data are not available for most congeners. Historically, PCB risk assessments at Comprehensive Environmental Response, Compensation, and Liability Act (CERCLA; "Superfund") sites have focused on risks from exposure to total PCBs using the results of Aroclor analysis. More recently, however, the focus has been on a subset of 12 PCB congeners (WHO-12 PCB: PCB 77, 81, 105, 114, 118, 123, 126, 156, 157, 167, 169, 189) which have demonstrated a toxicity in mammals, including humans, similar to that of 2,3,7,8-tetrachloro-dibenzo-p-dioxin (TCDD, "dioxin"), and other chlorinated dibenzo-p-dioxins and dibenzofurans (CDDs/CDFs).[9,14] The most toxic congener is PCB 126 with a lowest observed adverse effect level (LOAEL) of 0.74 μg kg^{-1} day^{-1}, which was approximately 1/50 of the LOAEL of 39 μg kg^{-1} day^{-1} for PCB 105 (the next most toxic congener) and 1/500 of the LOAEL of 425 μg kg^{-1} day^{-1} for PCB 128 (the least toxic congener). Considering dose–response and severity of liver effects, the order of toxicity was PCB 126 > PCB 105 > PCB 118, PCB 77 > PCB 153, PCB 28 > PCB 128. Risk assessors evaluating sites with PCB contamination have begun to evaluate risks from these "dioxin-like" congeners separately from the risks associated with total PCBs, using toxicity equivalency factors (TEFs) to estimate a dioxin toxicity equivalence for the dioxin-like congeners.[15] Thus, a congener analysis performed with appropriate specificity and accuracy is useful for improved estimates of total PCBs and for the determination of the 12 dioxin-like congeners. When the toxicity of a complex mixture such as PCBs has to be evaluated, the TEQ concept is introduced. Tetrachlorodibenzodioxin equivalent quantity (TEQ) is defined as the quantity of tetrachloro-dibenzodioxin (TCDD) which gives the same toxic effect as the mixture considered. TEQ is calculated by adding up values obtained by multiplying the concentration of each PCB toxic congener in the sample by its appropriate TEF to TCDD.[16]

The publication of TEFs[17] has facilitated assessment of dioxin-like risks. Table 18.5 shows the TEF values for a few non-*ortho* and mono-*ortho* substituted PCBs as well as their relative toxicity factor (RTF). RTFs are evaluated against 2,3,7,8-tetrachlorodibenzo-p-dioxin, whose value is assumed to be 100. If we consider that the concentration ratio between PCBs and dioxins in biota is

---

TABLE 18.5
WHO-TEF and Relative Toxicity Factors to 2,3,7,8-Tetrachloro-dibenzo-p-dioxin Values for a Few Non-*ortho* and Mono-*ortho* Substituted PCBs

IUPAC No.	Compound	WHO-TEF[a]	RTF[b]
—	2,3,7,8-Tetrachlorodibenzo-p-dioxin	1	100
15	4,4′-Dichlorobiphenyl	—	0.1
37	3,4,4′-Trichlorobiphenyl	—	0.1
77	3,3′,4,4′-Tetrachlorobiphenyl	0.0001	1.0
81	3,4,4′,5-Tetrachlorobiphenyl	0.0001	0.1
105	2,3,3′,4,4′-Pentachlorobiphenyl	0.0001	—
114	2,3,4,4′,5-Pentachlorobiphenyl	0.0005	—
118	2,3′,4,4′,5-Pentachlorobiphenyl	0.0001	—
123	2′,3,4,4′,5-Pentachlorobiphenyl	0.0001	—
126	3,3′,4,4′,5-Pentachlorobiphenyl	0.1	10
156	2,3,3′,4,4′,5-Hexachlorobiphenyl	0.0005	—
157	2,3,3′,4,4′,5′-Hexachlorobiphenyl	0.0005	—
167	2,3′,4,4′,5,5′-Hexachlorobiphenyl	0.00001	—
169	3,3′,4,4′,5,5′-Hexachlorobiphenyl	0.01	5.0
189	2,3,3′,4,4′,5,5′-Heptachlorobiphenyl	0.0001	—

[a] World Health Organization Toxicity Equivalent Factor.
[b] Relative Toxicity Factor.

higher than 100, and can reach values as high as 10,000, it follows that the TEF of PCBs may be much higher than that of dioxins, even for those congeners whose RTF is 0.1. All these factors support the need for reliable analytical procedures to determine the contents of planar and nonplanar congeners.

The congener composition of Aroclors has been studied extensively by Frame et al.[2,18–21] These and other studies[22] established the composition of Aroclor to approximately the 0.01% [100 parts per million (ppm)] level, but this level of scrutiny is insufficient for detecting and quantifying concentrations of certain dioxin-like congeners that are present at ppm and sub-ppm concentrations in pure Aroclor.

There have been reports[23–25] of PCBs being involved in a plethora of short-term and long-term toxicological effects, including skin rashes (e.g., chloracne,[26] itching and burning, eye irritation, skin and fingernail pigmentation changes, disturbances in liver function), as endocrine disrupters and environmental estrogens,[27] and as inducers of cancer, neurobehavioral changes, cognitive dysfunction, reproductive and developmental defects, and immunological abnormalities.[28]

The effect of PCB congeners is not the result of direct DNA reactivity, but involves epigenetic mechanisms based on the induction of the Ah receptor.[23] A relationship between PCB structure and its ability to stimulate oncogene expression reduces the gap-junction protein level in rat livers and induces mini-satellite mutations in the germ-line of male mice have been reported.[29–31] The toxicity of some PCB congeners is correlated with induction of mixed-function oxidases; some congeners are phenobarbital-type inducers, some are 3-methylcholanthrene-type inducers, and some have mixed inducing propertics. The latter two groups most resemble 2,3,7,8-tetrachlorodibenzo-*p*-dioxin in structure and toxicity. In fact, the binding to a specific receptor, namely the Ah receptor, seems to be a common pathway in biological systems for dioxins and PCBs. It is thought that endocrine disrupter compounds (EDCs) may be responsible for some reproductive failures in both women and men as well as for the increases in the frequency of certain types of cancer. EDCs have also been linked to developmental deficiencies and learning disabilities in children. Because hormone receptor systems are similar in human and animals, effects observed in wildlife species raise concerns of potential effects on human health. Animals may be exposed to relatively high concentrations of EDCs because they persist in environment and when ingested, these may be concentrated in fat tissue and released when the fat is mobilized during pregnancy or lactation, thus exposing embryos and neonates to relatively high concentrations. These stages of development are particularly susceptible to EDC effects. During fetal development and early childhood, low-dose exposure to EDCs may have profound effects which are not observed in adults, such as reduced mental capacity and genital malformations. There are extensive human data that show a strong association of low birth weights and shortened gestation with PCB exposure in human.[32,33] In addition, extensive neurological testing of children who experienced exposure to PCBs prior to birth revealed impaired motor function and learning disorders.[34]

The EPA's recent assessment of the risks of PCBs used the toxic equivalent method, which involved multiplying concentration of each PCB by a weighting factor (i.e., the TEQ factor, TEF) to give the TEQ of each congener.[23,35]

A very exhaustive report on the toxicological profile for PCBs has recently been published by the Syracuse Research Corporation.[36]

A. POLYCHLOROBIPHENYLS REGULATION

There are very few regulations and laws dealing with the presence of PCBs in water either at an international or national level. In these few cases the total PCB concentration is always indicated, but no mention is made about specific congeners. In particular, the U.S. Environmental Protection Agency (EPA) has fixed the maximum concentration level of PCBs in drinking water at $0.5 \ \mu g \ l^{-1}$.[37]

Moreover, a recent Italian regulation fixed the PCB water quality criteria to be reached in the Venetian Lagoon ecosystem at 0.04 ng l^{-1} (law approved in April 1998). The FDA established a 2.0 ppm tolerance for Great Lake fish. National regulatory limits for PCBs in fish and shellfish range from 500 μg kg^{-1} to 5000 μg kg^{-1}, with various countries setting their limits depending on fishing grounds and species. Regulatory limits for PCBs in dairy milk in various countries vary from 20 μg kg^{-1} to 60 μg kg^{-1}.[23] The Ministry of Agriculture, Fisheries and Food of the U.K. (MAFF) reported that there was a large decline in the estimated average U.K. dietary intake of PCBs from 1.0 mg/person/day in 1982 to 0.34 mg/person/day in 1992.[38] A tolerable daily intake level of 1 μg kg^{-1} body weight per day of total PCBs is also currently under review in Canada.[39]

An European Community Directive[40] on the disposal of PCBs, with its requirements for the preparation of inventories, labeling of all significant PCB holdings, and the tighter regulation of PCB treatment facilities, is committed to phasing out identifiable PCBs by 2010.

Since PCBs produce numerous adverse biological and ecological effects, state and federal governments have a significant interest in the safety issues associated with current landfilling practices of PCB-contaminated wastes. Concern over the toxicology of these compounds has led to international efforts to control their use and disposal, and to understand their global distribution, fate, and behavior. In 1998, the United Nations Economic Commission for Europe (UNECE) protocol (UNECE, 1998) on the long-range trans-boundary air pollution of persistent organic pollutants (POPs) was signed by 36 countries, although it is still to be ratified. Risk criteria were used to identify 16 substances for inclusion on the UNECE POP list. The protocol aims to eliminate their use and or discharges and emissions. International measures have since moved to a larger arena. The United Nations Environment Programme (UNEP) is developing a similar protocol, aimed at the global elimination of certain POPs.

The UNEP Governing Council included PCBs among the 12 POPs identified for remedial international action.[41]

Council Regulation (EEC) 315/93 of 8th February 1993, which lays down Community procedures for contaminants in food, stipulates that food containing a contaminant in an amount which is unacceptable from the public health viewpoint, and in particular at a toxicological level, shall not be placed on the market; contaminant levels shall be kept as low as can reasonably be achieved by following good practices; and maximum levels must be set for certain contaminants in order to protect the public health.[42-47]

In June 1997, an expert meeting was organized by the World Health Organization (WHO) to reach a consensus about TEFs for human and wildlife of polychlorinated dibenzo-p-dioxins and -furans (PCDD/Fs) and PCBs.[14,48-51] Twenty-nine congeners of these compound classes were reevaluated for TEFs. The criteria for including a compound in the TEF scheme and adding it to the list of dioxin-like compounds (DLCs) were:

1. A compound which shares certain structural relationships to PCDD/Fs
2. A compound must bind to the Ah receptor
3. A compound must elicit Ah receptor-mediated biochemical and toxic responses
4. A compound must be persistent and accumulate in the food chain[14,48]

In May 2001, the Scientific Committee on Food (SCF) of the EU expressed strong concern about the risk assessment of dioxins and dioxin-like PCBs. The Committee established a tolerable weekly intake (TWI) of 14 pg WHO-TEQ/kg bodyweight/week, which is in line with the provisional tolerable monthly intake (TMI) of 70 pg kg^{-1} bodyweight/month established in 2001 by the Joint FAO/WHO Expert Committee on Food Additives (JECFA) and concurs with the lower end of the ranged TDI of 1 to 4 pg WHO-TEQ/kg bodyweight/day established by the WHO Consultation in 1998. The SCF (May 2001) established a maximum intake of 2 pg kg^{-1} bodyweight/day, JECFA (June 2001) 2.3 pg kg^{-1} bodyweight/day and the WHO Consultation

(May 1998) a range of 1 to 4 pg kg^{-1} bodyweight/day. Although the SCF has concluded that a considerable part of the European population is exceeding the tolerable intake, the Committee also states that this does not necessarily mean that there is an appreciable risk to the health of individuals, because the TWI includes a safety factor. It is important to state that the dietary intake of individuals varies widely among Europeans because of different eating habits and different food sources. For example, in a diet consisting mainly of fish from highly contaminated areas such as the Baltic, the risk is much higher than that associated with a varied diet in southern Europe.[52]

1. The Presence of Polychlorobiphenyls in the Terrestrial Ecosystem

After being deposited in the environment, PCBs may remain resident in a given area for a period of time or be subject to redistribution and alteration by biogeochemical processes.

There are two primary pathways through which PCBs can be biodegraded: aerobic breakdown and anaerobic dechlorination. During aerobic breakdown, individual congeners are transformed into chlorobenzoic acid (CBA) in a multistep process.[53,54] Anaerobic dechlorination is the primary mechanism for breakdown of PCBs. During this process, highly chlorinated congeners are transformed into lower chlorinated analogs[55–59] and also lower chlorinated congeners that can undergo aerobic breakdown. Verification of the PCB sources and further insights into the exact nature of the congener distributions have usually been obtained through the relative abundances of congeners and by matching these concentration relationships to analogous relationships from possible Aroclor sources.[60,61] Such methods of source apportionment have yielded some useful insights, but confusion with regard to the actual PCB sources can occur when more than one possible input to an area is present. In addition, general knowledge of PCB use in an area does not necessarily correlate well with contamination in that area. Biodegradation may alter congener concentration relationships between source and sample, further confounding correlations.[12,62] Furthermore, although different Aroclor mixtures have different overall ranges of chlorination, many of the mixtures contain the same compounds.[19] In recent years, the field of stable isotope geochemistry has advanced from investigating bulk sample properties to the analysis of the stable isotopic composition of individual compounds within a mixture. Compound specific isotope analysis (CSIA) has the potential to identify individual compounds from source material through environmental and trophic transport. By measuring the isotopic composition of congeners from potential source material and contaminants extracted from the environment, confirmation or rejection of the linkages could be achieved. Contaminant–source linkages could even be achieved when local processes are altering some of the contaminants because unaltered compounds would provide the linkage the information. Stable isotope analysis could also provide information about the occurrence and type of alteration active in an area by comparing individual congeners.[62,63] CSIA has been used to investigate the isotopic signatures of individual PCB congeners from a variety of PCB manufacturers.[63,64]

Of the 209 PCB congeners, 78 display axial chirality in their nonplanar conformations. Kaiser[65] predicted that of these atropisomers (as these conformational isomers are known), the 19 congeners with three or four *ortho* chlorine atoms exist as pairs of stable enantiomers at ambient temperatures as a result of restricted rotation about the C–C biphenyl bond.[66] The enantiomeric composition may give information about the enantioselective biodegradation of organochlorine compounds.[67]

2. Analytical Methods

For resolution of an analytical problem the choice of a suitable procedure is vital in terms of accuracy and reproducibility required.

A typical analytical procedure for the quantification of organic analytes in environmental matrices can be summarized with the following steps:

- Sampling and storage
- Extraction of PCBs and their preconcentration
- Cleanup
- Instrumental analysis
- Data evaluation

a. Sampling and Storage

The soundness of analytical data directly depends on a correct sampling procedure in order to minimize the variation of analytical information when the sample is isolated from its environment. Moreover, sampling is often the only step which is not possible to repeat if, at the end of the analytical procedure, doubtful data are obtained.

Whatever chemical species have to be monitored in a given system, the sampling procedure depends on:

- the physical state of the sample and nature of the matrix analyzed
- the size of the sample, which depends on the homogeneity of the system studied and on the analyte concentration with respect to the sensitivity of the instrument being used
- the minimal number of samples that allow the required information to be obtained in order to solve the analytical problem

Before planning the sampling program, all the information available on the studied area should also be collected, and different chemical, physical, and biological parameters which may affect the concentration level of the analytes in the sample should be monitored. Based on this preliminary study, the minimum number of sampling stations, their spatial position, and their time frequency can be defined.

Generally, only the sampling operations are performed in the field and other operations are performed in the laboratory. However, when the PCB concentrations are so low that large amounts of sample have to be collected, the extraction and preconcentration of analytes can be performed in the field in order to facilitate the storage and transport in the laboratory.

b. Extraction

PCBs are nonpolar compounds and consequently can be extracted from samples with nonpolar solvents. The efficiency of extraction procedure depends on the nature of the sample, which in turn determines the availability of native analyte toward the extraction process. For example, in sediment or soil samples, analytes tend to be very tightly bound to the matrix and the yield of the extraction may be lower than the same analyte added as spike, so it should be noted that spiking the samples tends to give higher recovery values. In such cases, the availability of a reference material may be crucial in order to evaluate the correct recovery of the extraction process.

Solvents used in the extraction can be a source of contamination. Generally, in a sample preparation procedure hundreds of milliliters of solvents are reduced in volume to hundreds of microliters with an enrichment factor of about 1000. To avoid contamination, all solvents used in sample treatment processes must be pesticide residue grade and blanks must be checked frequently.

Liquid extraction uses large volumes of solvent compared with the injection volume. Large solvent volumes can be concentrated with a rotary evaporator or with a Kuderna-Danisch concentrator. Volumes lower than 10 ml can be concentrated up to almost dry under a gentle stream of nitrogen.

c. Cleanup of Extract and Preparation of Sample Solution

GC analysis is the only one that guarantees separation and an accurate determination of a large class of compounds such as PCBs, but the extraction methods used are not sufficiently selective to isolate exclusively PCBs from the sample. Because of the complex matrices that often characterize environmental samples and only trace levels concentration of analytes, extract solutions are generally incompatible with the chromatographic system so that they do not allow direct injection. In spite of the high separation power of the GC techniques, after extraction of samples with a matrix content, a cleanup of the organic solution is necessary for the purification and fractionation of the analyte in order to isolate PCBs as much as possible prior to instrumental analysis.[68] Sample preparation thus plays an important role in the whole analysis. It consists of a very complex procedure. This means that reliable data can only be obtained if a suitable program for analytical quality control and quality assurance is run in the laboratory. In particular, it is important to reduce any possible interference from other organochlorine compounds which may be co-extracted with PCBs.

The cleanup is a critical step since not only does it increase the total time of analysis but it can also change the analytical information contained in the sample. For example, in the procedures for concentration of organic solvents even relatively nonvolatile compounds may be partially or totally lost.[68]

Cleanup is performed by column chromatography. Silica/allumina or Florisil (synthetic magnesium silicate salt), deactivated or suitably activated (for instance 130 °C for 12 h[69]) are the most frequently used stationary phases. Their performances are checked by standard solution in order to find out the best solvent or mixture of solvents and the optimum volume to be used for selectively eluting PCBs and leaving interferents in the column.[70,71] N-Hexane and dichloromethane are the most widely used solvents. Better precision on cleanup results has been observed when Florisil cartridges are used instead of silica gel ones.[72] In some cases, in order to minimize interference from other non-PCB organic compounds, additional fractionations are performed.

HPLC not only provides a very good cleanup of the sample performing separation of PCBs from other similar organic compounds such as polycyclic aromatic hydrocarbons (PAHs),[73] but it also allows the separation of coplanar non-*ortho* substituted PCBs.[74–78] Coupling of the extraction system with an online LC–GC system allows the whole analysis to be done in a closed system, minimizing the risk of sample contamination or loss of analytes.[68] Moreover, automating the cleanup procedure may reduce the time of analysis and increase overall reproducibility.

For each category of environmental samples an overview of the most representative procedures for sample preparation is given in Table 18.6.

i. Air

The determination of PCB concentrations in the atmosphere generally entails sampling volumes of air ranging from 50 m^3 up to 13,000 m^3 with a rate of uptake of up to 800 l min^{-1}.[72,79–82]

Because of the high volumes of sample taken up, the organic substances in the ambient air are preconcentrated on suitable adsorption materials directly on the field. A wide range of solid sorbents are commonly used for air sampling: polyurethane foam,[82,83] silica gel, Florisil, Amberlite XAD-2, or functionalized styrene–divinylbenzene.[84]

The adsorption material is preceded by a filtering system in order to differentiate the content associated with the particulate matter from that present in the gas phase. The filtering material generally used is glass fiber,[82,83,85] quartz fiber filter,[84] or Teflon with 0.2 to 0.45 μm pore diameter.

TABLE 18.6
Analytical Methods for PCBs Determination in Environmental Matrices

Congeners	Matrix	Sample Extraction and Cleanup	Reference
		Air	
8, 18, 28/31, 44, 52, 66, 70, 77, 97, 101, 105, 118, 138, 151, 153, 170, 180	Urban air	250 m³ of air were sampled by using a high-volume sampler loaded with a glass fiber filter and a polyurethane foam plug. The foam was Soxhlet extracted with n-hexane. Extracts were cleaned through mini-columns with 1.5 g of 6% (w/w) deactivated Florisil; PCBs were eluted with 6 ml of hexane and 4 ml of dichloromethane in hexane (10%)	83
Total	Air	The sample was collected using a high-volume air sampler, in which 13,000 m³ of air was sampled over a 7-day period. Air was aspirated through a glass fiber filter and two polyurethane foam plugs. Samples were subjected to Soxhlet extraction	82
17, 18, 28, 16 + 32, 31, 53, 52, 49, 47 + 48 + 75, 44, 54, 70, 74, 66, 155, 90, 95, 101 + 90, 99, 110, 116, 118, 123, 118, 149, 153, 132, 105, 138 + 164 + 163, 158 + 160, 180, 185, 199, 194	Atmospheric aerosol	Up to 1700 m³ of air were filtered on a glass fiber filter and then drowned through a polyurethane foam plug. The foam and the glass filter were Soxhlet extracted with n-hexane for 24 h. The extract was concentrated to 4 ml by rotary evaporation and then to 1 ml under a gentle nitrogen stream at ambient temperature. The sample was then loaded onto an activated silica gel column. PCBs were eluted with n-hexane. The eluted fraction was reduced in volume and treated with concentrated sulphuric acid. Subsequently, the n-hexane phase was further cleaned on a multi-layer column packed with anhydrous sodium sulphate, silica gel impregnated with sulphuric acid, silica gel impregnated with KOH, and deactivated silica. PCBs were eluted with n-hexane. The fraction was reduced in volume for GC analysis	85
Noncoplanar polychlorinated biphenyls (PCBs)	Indoor air	Samples are filtered through a quartz filter and drowned through a SPE cartridge packed with functionalized styrene–divinylbenzene. PCBs are eluted with n-hexane. Airborne particulate matter is microwave extracted in using 15 ml hexane–acetone (1:1) mixture	84
	Urban air	Air was filtered on a quartz fiber filter and drowned through a polyurethane foams at a flow rate of 100 l min⁻¹. The foam was extracted with toluene. The sample was then cleaned by chromatography on silica gel impregnated with sulphuric acid	196
		Water	
28, 52, 44, 70, 101, 118, 128, 138, 153, 170, 180, 187	Seawater	50 to 100 l of seawater was passed through a column filled with XAD 2 resin at a flow rate of 400 ml min⁻¹. PCBs were eluted from the resin with methanol followed by dichloromethane. The methanolic fraction was concentrated and extracted with n-hexane. The hexane extract was combined with the dichloromethane fraction and dried with anhydrous sodium sulphate. The extract was then fractionated by column chromatography on activated alumina. PCBs were eluted with n-hexane. Elemental sulphur was removed by shaking the eluate with activated copper	97,98

101, 118, 136, 138, 151, 153	Seawater	Water was passed through a column filled with XAD 2. PCBs were extracted with methanol followed by dichloromethane and hexane. The extracts were then transferred to hexane, dried with anhydrous sodium sulphate, treated with activated Cu powder and cleaned up by column chromatography on alumina and silica	89
8, 18, 22, 26, 28, 31, 44, 49, 52, 70, 77, 101, 105, 110, 118, 126, 128, 138, 149, 153, 156, 157, 167, 169, 170, 180, 183, 187, 189, 194, 199	Seawater	Water was filtered and passed through a column filled with Amberlite XAD 2 resin at a flow rate of 30 l h^{-1}. The analytes adsorbed were Soxhlet extracted with acetonitrile containing 15% water for 6 h. Extract was concentrated and extracted with n-hexane. The hexane, dried on anhydrous sodium sulphate, was cleaned by HPLC (column: Nucleosil 100-5; eluent: 20% dichloromethane 80% pentane). The fraction containing PCBs was concentrated at 20 to 50 μl by a gentle nitrogen stream at room temperature	92
52, 101, 118, 153, 138, 180	Seawater	Filtered water was extracted with pentane. The organic extract was reduced in volume and treated with 0.01N NaOH. PCBs in the organic phase were separated from pesticides by alumina column chromatography using 1% Ethanol in hexane as eluent and then on silica column using 3% water in n-hexane	90
18, 31 + 28, 52, 44, 101, 149, 118, 153, 138, 180, 194	Seawater, sediment	Subsurfaces were filtered and extracted using a SPE system. The cartridges were eluted with ethyl acetate. Extract was treated with anhydrous sodium sulphate and purified in a silica gel column. After a first elution with n-hexane, and a second elution with benzene to obtain PAHs, PCBs were eluted with dichloromethane. Suspended particulate matter and sediment samples were extracted in an ultrasonic bath, with n-hexane for 30 min. The extracts were purified as extracts from water samples	197
8, 18, 52, 44, 66, 101, 77, 118, 153, 105, 138, 126, 187, 128, 201, 180, 170, 195, 206, 209	Ocean water, wetland water	PCBs were extracted using solid-phase SPME with a 100 μm poly(di-methylsiloxane) (PDMS) fiber	100
28, 52, 101, 138, 153, 180	River water	A 20 ml headspace vial was filled with 15 ml of the aqueous sample. A membrane bag (4 cm long with a wall thickness of 0.03 mm and an internal diameter of 6 mm) which is attached to a steel funnel and fixed with a PTFE ring is placed into the vial and filled with 800 μl of cyclohexane. The extraction takes place inside a stirrer. After extraction, the organic phase was withdrawn with a syringe from the membrane bag and injected into the inlet of the gas chromatograph	198
18, 28, 33, 44, 52, 70, 101, 105, 118, 128, 138, 153, 170, 180, 187, 194, 195, 199, 206 and 209	River water	Microwave-assisted headspace solid-phase microextraction (MA-HS-SPME). A 100 μm poly(dimethylsiloxane) SPME fiber was exposed to headspace over the water sample. After extraction the fiber was transferred inside the GC injector port and desorbed at 270°C for 3 min	199

Continued

TABLE 18.6
Continued

Congeners	Matrix	Sample Extraction and Cleanup	Reference
28, 52, 101, 138, 153	River water	Filtered water was passed throughout a SPE cartridge. PCBs were eluted from the cartridges with dichloromethane and in a second step with dichloromethane/n-hexane (1:1) mixture. The collected fractions were dried and cleaned on a column containing anhydrous sodium sulphate and silica gel. PCBs were eluted with 5% 2-propanol in n-hexane. Then the solution was concentrated under a gentle stream of nitrogen for GC analysis	87
1, 8, 18, 28, 29, 44, 50, 52, 66, 77, 87, 101, 105, 118, 126, 128, 138, 153, 154, 170, 180, 187, 188, 195, 200, 206, 209	Rain water	Rainwater was added with 20% of NaCl and extracted with isooctane in a liquid–liquid microextraction apparatus. The organic solution was injected without further purification	93
28, 52, 101, 105, 118, 128, 138, 149, 153, 156, 170, 180	Stormwater	The sample was extracted with a hexane/dichloromethane (85:15, v/v) mixture in a separating funnel. The organic phases were reduced and treated with sulphuric acid. The extract was reduced to a final volume of approximately 50 μl with a gentle stream of nitrogen	96
Sediment and soil			
8, 18, 28, 29, 44, 50, 52, 66, 77, 87, 101, 104, 105, 118, 126, 128, 138, 153, 154, 170, 180, 187, 188, 195, 201, 206, 209	Storm discharged particulate	The sample was thawed, homogenized and centrifuged to remove pore water. Samples were mixed with anhydrous sodium sulphate, and solvent extracted with dichloromethane. Sulphur was removed with activated copper, and extracts were cleaned on an alumina:silica (2:1) column. PCBs were eluted with 1:6 hexane/dichloromethane mixture and concentrated to 1 ml prior to instrumental analysis	102
153/132, 138/160, 136, 180, 118, 107, 110, 170, 190, 187, 149, 123, 77, 81, 194, 126, 169, 105	Sediment	Freeze-dried sediment was Soxhlet-extracted with dichloromethane. Concentrated extracts were fractionated by chromatography on alumina/silica gel column. After elution of the aliphatic fraction with pentane, PCBs were eluted with 1:1 pentane–dichloromethane mixture. The fractions were then concentrated to 1 ml using Kuderna-Danish tubes heated in a water bath at 60°C	107
Total	River sediment	Lyophilised sediments were Soxhlet-extracted for 8 h with n-hexane. Concentrated extracts were purified on a Florisil column with a layer of activated Cu powder on top. PCBs were eluted with n-hexane and the eluate was concentrated by rotary evaporation for GC analysis	106

Continued

18, 28, 31, 44, 52, 66, 70, 74, 87, 99, 101, 110, 105, 118, 123, 128, 138, 149, 153, 170, 180, 187	Sediment	The sample was thawed at room temperature, partially air-dried in an oven at 40°C and mixed with anhydrous sodium sulphate. The samples were Soxhlet-extracted with toluene and treated with Cu and Hg. The extracts were cleaned on a multi-layer column packed with acidic, neutral and basic silica	125
8, 18, 28/31, 44, 52, 66, 70, 77, 97, 101, 105, 118, 138, 151, 153, 170, 180	Surface soil	Samples were Soxhlet-extracted with n-hexane. Extracts were cleaned through mini-columns with 1.5 g of 6% (w/w) deactivated Florisil. PCBs were eluted with 6 ml of hexane and 4 ml of dichloromethane in hexane (10%)	83
18, 26, 52, 49, 44, 101, 151, 149, 118, 153, 105, 138, 187, 183, 128, 180, 170, 194	Estuary sediment	PCBs were Soxhlet-extracted with a hexane/acetone mixture for 24 h. The extracts were dried over sodium sulphate, followed by cleanup with hexane. The extracts were subjected to a further cleanup with sulphuric acid. Sulphur was eliminated with copper	110
31, 28, 52, 77, 101, 105, 110, 118, 126, 128, 138, 149, 156, 180, 153, 169, 170, 187, 194	Sediment	Freeze-dried samples were submitted to a two-stage Soxhlet extraction with n-hexane/acetone (80:20, v/v) mixture for 3 h. Extracts were purified on an alumina/silica column by elution with n-pentane. Coplanar PCBs were isolated from the other PCB by a high performance liquid chromatography (HPLC) fractionation on a PYE [2-(1-$pyrenyl$)ethyldimethylsilylated silica gel] column; n-hexane was used as eluent	77
5 + 8, 15 + 18, 17, 16 + 32, 26, 31, 28, 20 + 33 + 53, 45, 52, 49, 47, 48, 44, 37, 41, 96, 74, 70, 66 + 88, 101, 77 + 110	Contaminated soil	The samples were Soxhlet-extracted and analyzed by gas chromatography	8
16, 24, 28, 31, 32, 44, 52, 74, 87, 99, 101, 110, 118, 138, 149, 163, 174, 180, 182, 187, 194, 195, 201, 206	Lake sediment	Frozen sediment samples were thawed, thoroughly mixed, air-dried and extracted with dichloromethane in an accelerated solvent extractor. Solvent extracts were reduced in volume under gentle stream of nitrogen and redissolved in iso-octane. Extracts were cleaned on deactivated Florisil. Two fraction for PCBs analysises were obtained by elution with a 50% pentane/dichloromethane mixture followed by elution with only dichloromethane. The first fraction was treated with mercury to remove sulphur	104
28, 52, 101, 118, 153, 138, 180, $\Sigma7$, Total	Marine sediments	Freeze-dried sediments were sieved through a 2 mm brass sieve and Soxhlet-extracted using a 2:1 hexane:acetone mixture. Sulphur-containing compounds were removed by reaction with elemental copper during the extraction. The extract was reduced in volume and cleaned up by adsorption chromatography on sulphuric acid alumina and silver nitrate alumina columns	103
ΣPCB	Surface sediment	The sample was Soxhlet-extracted with toluene. The organic phase was cleaned on a 10% deactivated silica column followed by HPLC separation of PCBs from PAHs (hexane was used as eluent). The organic fraction containing PCBs was further cleaned on an acid/basic silica column	73

TABLE 18.6
Continued

Congeners	Matrix	Sample Extraction and Cleanup	Reference
45, 84, 88, 91, 95, 131, 132, 135, 136, 139, 144, 149, 171, 174, 175, 176, 183, 196, 197. Chiral PCBs	Soil	Sediments were added with anhydrous sodium sulphate and extracted on an accelerated solvent extractor with dichloromethane. Extracts were mixed with mercury to remove sulphur. They were then transferred into hexane and cleaned by chromatography on activated silica column. PCBs were eluted with hexane. Extracts were then carefully reduced in volume and taken up in isooctane	66,109
Total	Soil, sediment	The sample was submitted to MAE using acetone/n-hexane (1:1, v/v) mixture. Extract solution was centrifuged and evaporated just to dryness. Residues were dissolved in 1 ml of n-hexane. Cleanup was performed on a SiO_2/Al_2O_3. PCBs were eluted with n-hexane	112
8, 28, 20, 52, 35, 101, 118, 153, 138, 180	Certified soil (CRM 481)	*Microwave assisted extraction*: the sample was mixed with anhydrous sodium sulphate and extracted in a microwave assisted extraction system using an acetone/n-hexane (74:26, v/v) mixture *Soxhlet extraction*: the sample was mixed with anhydrous sodium sulphate and extracted with an acetone/hexane (75:25, v:v) mixture For both the extraction procedures the organic solution was filtered through glass wool and concentrated by using a gentle stream of nitrogen. The extract was cleaned on a Florisil cartridge. PCBs were eluted with 10 ml of n-hexane. The eluate was concentrated until almost dry and dissolved in isooctane for GC–MS analysis	111
Biota			
Total	Human serum	Serum was extracted with hexane in a rotor running. The extraction solution was evaporated in a hot water bath at 40°C and transferred on top of a Florisil/acid silica column. PCBs were eluted from the column with hexane. The hexane solution was concentrated for GC analysis	133
		The sample was denatured with formic acid and extracted with cyclohexane. The organic extracts were purified on a Power Prep automated sample cleanup system with acidic silica gel, acid/base/neutral silica, and a carbon column. PCBs were eluted using dichloromethane and cyclohexane. The eluates were concentrated using a nitrogen evaporator for analysis by GC-HRMS	130
1, 5, 29, 47, 98, 154, 171, 200	Blood plasma	The sample was extracted using LPME in conjunction with a hollow fiber membrane (HFM). An eight PCB congener mixture was spiked into 2.5 ml of blood plasma, and the solution was then adjusted to pH 10.5 with a salinity of 20% (w/v) prior to making the total volume to 5 ml with ultrapure water. The porous HFM, filled with 3 μl of organic solvent, was then immersed into the solution, which was continuously agitated at 700 rpm for 30 min. Extract was injected into a GC–MS without further pretreatment	200
38 PCB	Fish	Liver and ovary were Soxhlet-digested with 1 N KOH/ethanol. Extract was transferred to n-hexane and cleaned on a silica gel, column. PCBs were eluted with n-hexane, and the eluates were concentrated for GC analysis	136

Congeners	Sample	Method	Ref.
50 congeners	Krill, silverfish	The sample was homogenized with sodium sulphate and Soxhlet-extracted with dichloromethane/hexane (3:1) mixture. The extract was concentrated on a rotavapor at 40°C, and cleaned on a multi-layer silica gel column (silica, 40% acidic silica, silica and a thin layer of sodium sulphate at the top). PCBs were eluted with n-hexane	188
61 congeners	Chicken and pork fat, chicken feed	Fat was melted (50°C for chicken fat and 80°C for pork fat, respectively) and solubilized in n-hexane. The hexane solution was cleaned over sulphuric acid/silica (1:1,w/w) topped with 0.5 g anhydrous sodium sulphate and eluted with n-hexane. The eluate was concentrated until almost dry and solubilised in isooctane for GC analysis	147
34 congeners	Human adipose tissues	The sample was mixed with anhydrous sodium sulphate and Soxhlet extracted with hexane/acetone/dichloromethane (3:1:1, v/v) mixture. The solution was cleaned by SPE on cartridges containing acid silica and acid silica:neutral silica:deactivated basic alumina (from top to bottom). After a first elution with hexane, PCBs were eluted with hexane:dichloromethane (1:1, v/v) mixture. Extract was then concentrated under a gentle stream of nitrogen for GC analysis	128, 201, 202
	Hair	Hair samples were overnight incubated at 40°C in 3 N HCl. The solution was extracted with hexane:dichloromethane (4:1) mixture. The organic solvents were purified on a column filled with alumina deactivated with 10% of water, acidified silica and anhydrous sodium sulphate. PCBs were eluted with n-hexane. The eluate was concentrated under a gentle stream of nitrogen for GS analysis	121
70, 74, 87, 99, 101, 77, 105, 118, 126, 128, 138, 151, 153, 156, 169, 170, 180, 183, 187, 191, 194, 205, 206, 208, 209	Human milk	Solid-phase extraction. Three columns were used sequentially, and those were a Bond Elut C18, a Sep-Pak Plus NH$_2$ and a Bond Elut PCB cartridge	203
40 PCB	Fish	Tissue was ground in a mortar with anhydrous sodium sulphate and extracted with dichloromethane: hexane (50:50 v/v). Extracts were cleaned by gel permeation chromatography followed by chromatography on activated Florisil	124
	Benthic species and sediments	Samples were mixed and homogenized with anhydrous sodium sulphate and directly introduced to a multi layer column filled with anhydrous sodium sulphate, activated silica, 40% H$_2$SO$_4$ silica, 33% KOH silica and a top layer of anhydrous sodium sulphate. The sample was extracted with a mixture of cyclohexane/dichloromethane (80:20 v/v). The extract was purified on a column filled with activated Florisil using 1% of dichloromethane in n-hexane as eluent	143
Total	Mussels (fortified samples and standard reference material)	Freeze-dried samples were Soxhlet extracted with hexane/dichloromethane (1:1) mixture. The extracts were then concentrated and eluted on Florisil with a mixture of 120 ml acetonitrile and 30 ml hexane-washed water. The eluents were collected in a separatory funnel containing 100 ml of hexane and 600 ml of hexane-washed water. After shaking and phase separation, the hexane layer was concentrated and treated with concentrated sulphuric acid and cleaned-up on activated Florisil. PCBs were eluted with n-hexane	69

Continued

TABLE 18.6
Continued

Congeners	Matrix	Sample Extraction and Cleanup	Reference
18, 28, 31, 44, 52, 66, 70, 74, 87, 99, 101, 110, 105, 118, 123, 128, 138, 149, 153, 170, 180, 187	Fish tissues	Samples were dried with sodium sulphate and extracted with dichloromethane–hexane (50:50 v/v). Lipids were removed from the extract by gel permeation chromatography (GPC). The fraction containing PCBs was cleaned on a multi-layer column packed with acidic, neutral and basic silica	125
18, 28, 31, 44, 52, 66, 70, 74, 87, 99, 101, 110, 105, 118, 123, 128, 138, 149, 153, 170, 180, 187	Fish tissue	Tissue samples were ground with anhydrous sodium sulphate and quartz sand and Soxhlet extracted with n-hexane/dichloromethane (3:1) mixture. Lipids were removed by size exclusion chromatography (GPC) on Bio-Beads S-X8 with n-hexane/acetone (3:1, v:v) or Bio-Beads S-X3 (Bio-Rad Laboratories, Hercules, CA, USA) using cyclohexane/acetone (3:1, v:v) as eluent. The eluted solution was further fractionated by micro preparative NP-HPLC on 3-chloropropylsiloxane or 3-aminopropylsiloxane. PCBs were eluted with n-hexane. The NP-HPLC fractions were spiked with the internal quantification standards PCB 103 or TCN, and concentrated prior to HRGC analysis	126
18, 31, 28, 52, 49, 47, 44, 66, 101, 99, 87, 110, 118, 105, 149, 151, 153, 156, 138, 180, 170, 199, 195, 194	Aquatic organism	Tissue samples were ground in a mortar with anhydrous sodium sulphate and Soxhlet extracted with a 50:50 of hexane and dichloromethane mixture for 8 h. Extracts were concentrated and lipids were removed by gel permeation chromatography (GPC) in Bio Beads S-X3 and extracts were subfractionated by silica gel chromatography	127
All 209 congeners	Seal blubber	A blubber sample was macerated with anhydrous sodium sulphate and it was extracted with acetone/n-hexane (5:2, v/v) followed by n-hexane/diethyl ether mixture (9:1, v/v). The lipids were removed using a multi-layer column containing, from the bottom, basic silica (KOH treated), activated silica, 40% sulphuric acid impregnated silica (w/w), 20% sulphuric acid impregnated acidic silica (w/w) and a thin layer of anhydrous sodium sulphate. The PCBs were eluted with n-hexane. Isooctane was added and the extract was concentrated to 1 ml for GC × GC analysis	152,153
91, 95, 135, 136, 149, 174, 176, 183	Wolverines	Liver sample was homogenized with anhydrous sodium sulphate and Soxhlet-extracted for 8 h with dichloromethane. Lipids were removed using gel permeation chromatography. Analytes from each sample were separated on a 100%-activated silica gel (8 g) column into two fractions: n-hexane (65 ml; Fraction 1) and n-hexane:dichloromethane (95 ml; 50:50 by volume; Fraction 2). Samples were transferred to isooctane and concentrated to 100 ml	138
31, 28, 52, 77, 101, 105, 110, 118, 126, 128, 138, 149, 156, 180, 153, 169, 170, 187, 194	Fish tissue	Freeze-dried samples were submitted to a two-stage extraction in a Soxtec System HT6 Tecator (France) with n-hexane–acetone (80:20, v/v) mixture for 3 h. Extracts were purified on an alumina/silica column by elution with n-pentane. Coplanar PCBs were isolated from the other PCBs by a HPLC fractionation on a PYE [2-(1-pyrenyl)ethyldimethylsilylated silica gel] column; n-hexane was used as eluent	77

8, 20, 28, 52, 101, 118, 138, 180, total	Cormorant eggs	Whole eggs were homogenized with anhydrous sodium sulphate and extracted with hexane/petroleum ether 1:1 mixture. Lipids were removed from extracts by treatment with concentrated sulphuric acid. A further cleanup was performed in a glass column packed with Florisil and sodium sulphate. The purified sample was concentrated for CG analysis	134
PCBs 8, 18, 28, 33, 44, 47, 52, 66, 74, 77, 101, 105, 118, 126, 128, 138, 153, 169, 170, 180, 187, 195, 206, 209	Serum	2 ml of serum sample was added with 2 ml of IS in methanol and 6 ml of diethyl ether–hexane (1:1, v/v). The sample was mixed for 30 min and after separation of the phases, the organic layer was mixed with 2.5 ml sulphuric acid to remove fat and polar materials. The organic layer (2 ml) was dried on anhydrous sodium sulphate and then was cleaned on a silica column. The PCBs were eluted with 2 ml of hexane. The sample was concentrated to 250 μl and transferred to autosampler bottles for gas chromatography	135
18, 28, 44, 66, 101, 118, 126, 128, 156, 180, 169	Mussels	Supercritical fluid extraction combined with matrix solid-phase dispersion with Florisil as sorbent	115
	Human hair	Hair samples were decomposed with 2N KOH and extracted with toluene under reflux. After the addition of n-decane as keeper solvent, the extract was concentrated and residue was concentrated until almost dry and dissolved with n-hexane. The solution was purified on a multi-layer column containing anhydrous sodium sulphate, 10% (w/w) AgNO3–silica mixture, silica, 22%(w/w) H2SO4–silica, 44%(w/w)H2SO4–silica, silica and 2%(w/w)KOH–silica. PCEs were eluted with n-hexane. PCBs were fractionated in coplanar and noncoplanar by chromatography on an activated alumina column by successive elution with 2% methylene chloride in n-hexane, 50% methylene chloride in n-hexane	78
Total 28, 52, 101, 105, 118, 138, 153, 180	Fish	Homogenization of the sample in n-hexane:acetone (2 5) and extracted with n-hexane/MTBE (9:1). The samples were treated with concentrated sulphuric acid. Cleanup on silica gel impregnated with sulphuric acid (2:1, w:w) and elution with n-hexane	145,146, 148
Chiral PCBs		The sample was homogenized with anhydrous sodium sulphate and Soxhlet extracted with dichloromethane for 16 h. Lipids were removed by gel permeation chromatography followed by chromatography on 100% activated silica. PCBs were eluted with n-hexane. Extracts were exchanged into isooctane, and concentrated under a gentle stream of nitrogen for GC analysis	108,109
Total	Mussels	The sample was submitted to MAE using acetone–n-hexane (1:1, v/v) mixture for sediment samples and methanol or methanolic 1 M KOH for biological samples. Extract solution was centrifuged and evaporated just to dryness. Residues were dissolved in 1 ml of n-hexane. Cleanup was performed on a SiO2/Al2O3. PCBs were eluted with n-hexane	112
209 congeners	Duck muscle, liver and egg, grass carp	The sample was ground with anhydrous sodium sulphate and Soxhlet extracted with n-hexane/dichloromethane (1:1,v/v) mixture for 8 h.The extract was concentrated and treated with concentrated sulphuric acid (96%)	119

Continued

TABLE 18.6
Continued

Congeners	Matrix	Sample Extraction and Cleanup	Reference
8,18, 44, 49, 50, 52, 66, 87, 101, 105, 110, 118, 128, 138, 149, 151, 153, 157, 160, 169, 170, 173, 180, 194, 195, 206, 209	Dolphin blubber	A sample of blubber was ground with anhydrous sodium sulphate and Soxhlet extracted for 8 h using n-hexane/dichloromethane (1:1, v/v) mixture. The extract was concentrated and treated with concentrated sulphuric acid. The lipid-free extract was directly analyzed in a gas chromatograph	122
Total	Fish tissue	Samples were mixed with anhydrous sodium sulphate and extracted with methylenechloride. Extracts were concentrated and cleaned on a multi-layer column containing 30% (w/w) sulphuric acid/silica gel and a second layer of potassium hydroxide-impregnated silica gel and eluted with methylene chloride. After concentration the eluates were further purified on a second column containing a first layer of 40% (w/w) sulphuric acid/silica gel a second layer of silica gel. The PCBs were eluted from these absorbents with 0.5% benzene/99.5% hexane (v/v) mixture. Clean sample extracts were solvent exchanged to isooctane for GC analysis	120

To avoid loss of the volatile fraction of PCBs, filter and adsorbing materials are generally stored at temperatures below 0°C in stainless steel or glass containers.

PCBs are quantitatively recovered from the absorbent material by liquid extraction in a Soxhlet apparatus. Also, particulate matter is generally Soxhlet-extracted or extracted in an ultrasonic bath.[84] The most used solvents for the extraction are n-hexane, petroleum ether, benzene, ethyl ether, and acetone.[83,84,86]

ii.   Water, Snow, and Ice

A few liters of surface water can be sampled directly with cleaned glass bottles that can also be used for sample storage.[87] For sampling at different depths, Go-flo or Niskin bottles are more suitable, and allow sampling volumes of up to 50 l.[88] If it is not possible to extract the samples immediately after sampling, these are stored at temperatures below 0°C in stainless steel or glass containers, rinsed beforehand with pesticide grade acetone followed by n-hexane.

For large-volume sampling, water samples can be also collected by a Teflon or stainless steel pumping system without any lubricant or oil in order to avoid contamination. The sample volume may vary from a few to a thousand liters.[89–91]

Ice cores of up to about 20 m in depth are collected using a metal, hand-operated, ice-coring auger with a diameter of about 8 cm. Usually, the auger is painted on the outside with a PCB-free epoxy paint, and before use, it is rinsed thoroughly with hexane and dichloromethane. For deeper samplings an engine moving system is used. The cores, 10 to 80 cm in length, are wrapped in precleaned aluminum foil and returned to the laboratory, where they are unwrapped, the surface layer scraped with a clean metal scraper, sectioned, and analyzed.

Large quantities of snow are collected at each sampling site into Teflon bags of 2 mm thickness placed inside protective containers. A stainless steel shovel, prerinsed with pesticide-grade acetone, is used to place the snow in the containers, while the operator must use a complete clean-room dressing to avoid any sample contamination. Potential contamination from surface snow can be minimized by removing the top 2 to 3 cm of snow and by sampling upwind of the landing site. The samples are generally stored at -20°C.

The occurrence of PCBs in contaminated waters is usually in parts per trillion (ppt, ng kg^{-1}) levels so an appropriate sample treatment is usually required to extract and concentrate the analytes prior to the chromatographic determination. The following techniques are used for this purpose:

- Liquid–liquid extraction (LLE)
- Solid-phase extraction (SPE)
- Solid-phase microextraction (SPME)
- Direct immersion SPME (DI-SPME)
- Headspace SPME (HS-SPME)
- Microwave-assisted extraction (MAE)
- Coupling MAE and SPME (the SPME fiber is directly immersed into the aqueous media or exposed to the headspace of the sample followed by MAE)

When large volumes of water are sampled, the extraction of analytes is directly performed in the field: water is passed through a cartridge containing a suitable stationary phase, for instance XAD-2 resin.[89,92] Generally, before the preconcentration cartridge, a filtering system, with a pore size lower than 1 μm, is positioned,[87] and the particulate matter recovered is stored and analyzed separately.

The most widely used solvents in water extraction are: n-hexane,[70] isooctane,[93] dichloromethane,[94,95] or pentane,[90] or a mixture of them.[96] SPE and elution with different mixtures of solvents are also very common.[87] XAD-2,[89,92,97,98] and C18-bonded silica[97] are the most widely used adsorbing resins. The adsorbing material is generally supported inside a column or fixed on a membrane disk. SPE has several advantages, i.e., use in field applications, easy automation, low solvent consumption, and a less critical cleanup of the eluate. SPE has some drawbacks which might limit its application to water samples, such as low capacity for samples which have a high content of organic matter, and the need for critical calibration procedures for quantitative determinations.[99]

SPME is a modified SPE procedure based on the use of a coated fiber made of fused silica. After the extraction the fiber is directly introduced into the injector of the GC instrument to allow the direct transfer of the analytes into the chromatographic column, thus avoiding the use of organic solvents.[100,101] Chromatographic stationary phases, such as poly(methylsiloxane), are used as chemically bonded coatings of the fiber. SPME is an inexpensive and easily automated technique, but its most important drawbacks are the poor detection limits compared to SPE and the time required for sorption on the fiber.

Ice and snow samples are allowed to melt in a clean laboratory, and extraction is undertaken as soon as they have melted, following the same procedures as those applied to water samples.

iii.   Sediments/Soil

Surface sediments are collected by a stainless steel grab[102] or Craib corer,[103] while a box-corer system is used if a depth profile is required.[104] A stainless steel shovel can be used to collect soil samples. Potential contamination by exhaust on surface soil can be minimized by removing the top 2 to 3 cm of soil and by sampling upwind of the landing site. Sediment samples can be collected in stainless steel containers, glass jars,[104,105] or in polypropylene boxes[73] and frozen immediately below 0°C for transport and storage in the laboratory. All the sampling devices and storage containers must be prerinsed beforehand with pesticide-grade solvents, such as acetone, followed by n-hexane. To reduce possible contamination only the part of the sample which is not in contact with the wall of the sample device can be stored.[102]

Before the extraction, sediment samples are dried by homogenization with anhydrous sodium sulphate[68] or by air-drying.[104]

There are several solvents that are used for the extraction of PCBs from sediment and soil. The most common are: n-hexane,[83,106] dichloromethane,[102,104,107–109] cyclohexane, and mixtures of them.[68] Ultrasonic treatment can improve the interaction between the solid and the liquid phase. Extraction is generally performed in a Soxhlet apparatus,[83,106,107] with an extraction time which can vary from 2 h to 24 h.[110]

More efficient devices are commercially available. One of these is the Accelerated Solvent Extractor sold by DIONEX,[104,108,109] where the extraction takes place in a closed vessel in which temperatures of 100°C and pressures up to 2000 psi are reached even with low boiling solvents.

Microwave-assisted techniques (MATs) can also be used in the extraction process. In MAE, microwaves improve efficiency with respect to Soxhlet extraction with a lower extraction time.[111] Microwave-assisted decomposition (MAD) has been applied in PCB determination in soil[111,112] and sediment.[112]

An alternative to liquid–solid extraction is supercritical fluid extraction (SFE) which allows the extraction of analytes from solid samples, i.e., marine sediments, to be performed faster and more efficiently since these have a lower viscosity and higher diffusivity than liquid solvents.[113] CO_2 is the most widely used supercritical fluid with or without a modifier, e.g., methanol and toluene. SFE can be combined with solid-phase trapping.[114–116] Compared with Soxhlet extraction, SFE gave similar yields, but the extracts were much cleaner and it was not necessary to clean the extracts before GC analysis.[114]

Extracts from sediment, and sometimes also from soils, generally contain amounts of elemental sulphur, which, in addition to compromising the chromatographic separation, may damage the chromatographic column.[68] The typical methods for sulphur removal are treatment with copper powder[103] and/or mercury,[70,104,108,109,117] which can be admixed directly to the sample during extraction or added to the extraction solution in a separate step. Copper powder is activated beforehand with HCl (18%) and washed with acetone and finally with n-hexane.

With SFE, copper powder can be directly admixed in the extraction cell.[114] Alternatively the extract can be treated with tetrabutylammonium that converted sulphur in thiosulphate insoluble in the organic phase.[77,118]

iv. Biological Samples

This is a very broad matrix category, and sampling procedures accordingly vary greatly. In general, the analysis is performed on selected tissues and organs, where PCBs are accumulated due to their lipophilic characteristic, or on the whole sample for very small biota. Immediately after sampling, tissues, organs, or the whole sample should be frozen, homogenized, freeze-dried, and finally stored at $-20°C$ in stainless steel or glass containers.

PCBs can be extracted from biological samples with a Soxhlet or by performing homogenization in the presence of the organic solvent. After the extraction, the sample is homogenized with anhydrous sodium sulphate. Dichloromethane is the most commonly used solvent,[102,108,109,119,120] by itself or in a mixture with *n*-hexane.[69,96,121–127] Other solvent mixtures can also be used: hexane: acetone:dichloromethane (3:1:1, v/v, respectively,) mixture,[128] *n*-hexane–acetone,[77] acetonitrile,[129] cyclohexane,[130] or cyclohexane/acetone,[131] and pentane/dichloromethane or hexane,[132,133,147] hexane/petroleum ether mixture,[134] diethyl ether–hexane,[135] and 1 *N* KOH/ethanol mixture.[136]

For biological samples, MATs not only improve the extraction efficiency but can also decompose the matrix to increase the availability of the analyte. Microwave-assisted saponification (MAS) and MAD have been applied in the PCB determination in mussels.[112] SFE can be used combined with matrix solid-phase dispersion with Florisil as sorbent.[115]

Extracts from biota samples can be treated with sulphuric acid,[122,134,137] can be cleaned up by gel permeation chromatography[108,109,117,124–127,138–140] or by polyethylene film dialyses[141,142] for elimination of the lipid content. Lipids can also be eliminated by alkaline alcohol digestion of the sample.[78,112,136] After lipid elimination the extract can be further fractionated by column chromatography on Florisil,[69,124,134,143,144] alumina/silica,[77,136,138,145–147] or silica gel impregnated with sulphuric acid.[145,146,148]

3. Instrumental Analysis

GC continues to play an important role in the identification and quantification of ubiquitous pollutants in the environment. GC coupled with an electron capture detector (ECD) or a mass spectrometric detector (MSD) has been widely applied for the quantification of PCBs.

GC on a fused silica capillary column should be used, whenever possible, with a MSD. In fact, it allows the extremely high resolution of GC to be combined with the very high sensitivity and identification power of mass spectrometry (MS), which makes it possible to determine an analyte at trace level (low ng kg^{-1}) in the sample solution.

The sample solution may contain over 100 chemically similar compounds, so an accurate control of the experimental parameter is crucial for a good separation. In addition to the selection of the most suitable stationary phase and column dimension and optimization of the oven temperature program, other experimental parameters are decisive for the band broadening such as a proper injection procedure and a proper carrier gas flow rate. Instead of helium, hydrogen can be used as a carrier gas, thus allowing an increase in the gas flow rate without a loss in column efficiency.[69,77,108,109,120,149,150]

A split/splitless injector is more frequently used than on-column injection. When this injection system is used in splitless mode, the initial column temperature should be at least 20°C below the boiling point of the solvent and the injection temperature should be over 200°C. To increase reproducibility, the sample is injected slowly by setting a high carrier gas flow rate during the injection time.[151] On-column injection is also often used, in which case the initial temperature is 10°C to 15°C lower than the boiling point of the solvent.

For low polarity stationary phases the retention time depends on the boiling point of the analyte, and thus the retention time of PCBs increases with increasing chlorine content. For stationary phases with a higher polarity, several low chlorinated PCBs are retained in the column stronger than high chlorinated ones. This is more evident for compounds with none or one *ortho*-chlorine substituted.

A 30 to 50 m fused silica capillary column with a 5% phenyl–95% methyl-polysiloxane chemically bonded stationary phase (DB-5, CP Sil-8, HP Ultra 2, PTE-5) is very often used, while several oven temperature programs have been applied for PCB analysis. Table 18.7 shows a selection of combinations of column lengths, stationary phases, oven temperature programs, and detectors. Table 18.8 shows the stationary phase composition of the capillary columns listed in Table 18.7. Cochran et al.[149] have reviewed the most recent developments for the capillary GC of PCBs with detailed lists of PCB retention times on common capillary columns.

Although a large variety of stationary phases has been used in the literature, only multidimensional GC, a powerful two-column technique, allows complete separation of all 209 PCB congeners.[149,152–154] In this technique two capillary columns are arranged in series, and congeners, coeluted from the first column, are transferred to a second column with a different selectivity. Pressure or valve control at the midpoint of a tandem-column system can be switched to cause effluent from the precolumn to travel to a monitor detector or to the analytical column and its detector.

Complete PCB analysis has been proposed which entails the simultaneous injection of the sample into two chromatographic columns with different polarities.[155]

a. Electron Capture Detector

ECD, together with MS, is the most commonly used detection method for trace PCB analysis due to its low cost and high sensitivity and selectivity towards polyhalogenated compounds.[70] ECD, as used for PCB analysis, has two major problems: nonlinear response behavior across a relatively narrow range of amounts,[151,156] and wide variation in response within a homologous group of PCBs.[149]

When the detector shows a narrow linear dynamic range, several standards covering the concentration range of interest have been analyzed and sample solutions have been analyzed at several dilution levels since congeners exist in a wide concentration range in the same sample. As the various PCBs could be present in samples at different concentrations, this method could sometimes result in a substantial amount of work because several dilutions and injections had to be made. Later, it appeared to be more effective and more precise to work with multi-level calibration.[157] Five or six different dilutions were randomly spread over a series of samples and injected. Several options exist to fit the calibration curve (quadratic, exponential, point-to-point).[156] The differences resulting from the use of various curve-fitting methods were small.[158] This method of multi-level calibration is now accepted, and has also been used with MS detection, although the linearity of MS detection is greater than that of ECD.

Recently, a new micro-ECD cell size of 150 μl (1/10 the size of the previous model) with improved linearity response was developed.[159,160] A sampling rate up to 50 Hz makes it suitable for high-speed GC.

Because of the variability associated with ECD relative response factors (RRFs), attempts at using one RRF for a homologous group, or using published RRF data instead of measuring RRFs, can lead to serious errors in quantitation.[149]

Moreover, one disadvantage of ECD due to the absence of substantial qualitative data is the identification of PCBs congeners solely on retention time. ECD is unable to differentiate between PCBs and coeluting interferences such as 4,4′-DDE.[161] More seriously, it cannot resolve PCB congener pairs such as congeners 77 and 110.[162]

TABLE 18.7
HRGC Experimental Conditions for PCB Analysis

Stationary-Phase Composition	Column Length, Internal Diameter, Film Thickness	Temperature Program	Detector	Reference
AT-5	10 m, 0.10 mm, 0.10 μm	90°C (1 min), 50°C min^{-1} to 200°C (0.5 min), 25°C min^{-1} to 250°C (0.2 min), 75°C min^{-1} to 280°C (2 min)	MS	128
AT-5	10 m, 0.10 mm, 0.10 μm	90°C (1 min), 50°C min^{-1} to 200°C (0.5 min), 25°C min^{-1} to 250°C (0.2 min), 75°C min^{-1} to 280°C (2 min)	MS	201
AT-5	25 m, 0.32 mm, 0.52 μm	100°C (1 min), 5°C min^{-1} to 140°C (1 min), 1.5°C min^{-1} to 250°C (1 min), 10°C min^{-1} to 300°C (10 min)	ECD	122
BPX-5	60 m, 0.25 mm, 0.25 μm	140°C (1 min), 20°C min^{-1} to 200°C (1 min), 3°C min^{-1} to 300°C (10 min)	μECD	204
BPX-5	60 m, 0.25 mm, 0.25 μm	—	ECD	127
Chirasil-Dex	25 m, 0.25 mm, 0.25 μm	60°C (2 min), 10°C min^{-1} to 150°C, 1°C min^{-1} to column maximum temperature, 180–250°C (20 min)	ECD	66
Cyclosil-B	30 m, 0.32 mm, 0.25 μm			
B-PA Chiraldex Astec	30 m, 0.32 mm, 0.25 μm			
B-DM Chiraldex Astec	20 m, 0.25 mm, 0.25 μm			
G-TA Chiraldex Astec	30 m, 0.32 mm, 0.125 μm			
B-PH Chiraldex Astec	30 m, 0.32 mm, 0.125 μm			
G-PT Chiraldex Astec	12 m, 0.25 mm, 0.125 μm			
CP-Sil 5 with 10% octadecyl (C18) chains incorporated in the methyl siloxane	50 m, 0.25 mm, 0.25 μm	—	ECD	121
CP-SIL19 CB. Hydrogen was used as carrier gas	60 m, 0.25 mm, 0.15 μm	75°C (2 min), 30°C min^{-1} to 180°C, 2.5°C min^{-1} to 280°C (2 min), 10°C min^{-1} to 300°C	ECD	77

Continued

TABLE 18.7
Continued

Stationary-Phase Composition	Column Length, Internal Diameter, Film Thickness	Temperature Program	Detector	Reference
CP-SIL5C18 CB. Hydrogen was used as carrier gas	100 m, 0.25 mm, 0.10 μm	75°C (1 min), 45°C min⁻¹ to 180°C, 2.5°C min⁻¹ to 280°C, 3°C min⁻¹ to 300°C (2 min)	ECD	77
CP-Sil-8 CB	30 m, 0.32 mm, 2.0 μm	60°C, 20°C min⁻¹ to 180°C, 1.5°C min⁻¹ to 270°C	ECD	106
CP-Sil-8	50 m, 0.22 mm, 0.2 μm	80°C (1 min), 3°C min⁻¹ to 270°C	ECD	90
Cyclosil-B	30 m, 0.25 mm, 0.25 μm	—	MS	108,109
DB-1	25 m, 0.25 mm, 0.25 μm	90°C (1 min) then 15°C min⁻¹ to 275°C (10 min)	MS	121
DB-1. DB-17 for nonorto	30 m, 0.25 mm, 0.25 μm	70°C (1 min), 15°C min⁻¹ to 200°C, 4°C min⁻¹ to 270°C (15 min)	HRMS	205
DB-5	30 m, 0.32 mm, 0.25 μm	100°C (1 min), 5°C min⁻¹ to 150°C (1 min), 10°C min⁻¹ to 300°C (5 min)	ECD	107
DB-5	50 m, 0.2 mm, 0.33 μm	50°C (2 min), 10°C min⁻¹ to 300°C (3 min)	MS	200
DB-5	30 m, 0.25 mm, 0.25 μm	—	ECD	188
DB-5	60 m, 0.25 mm, 0.25 μm	—	ECD	124
DB-5	30 m, 0.25 mm, 0.25 μm	—	—	125
DB-5	30 m, 0.25 mm, 0.25 μm	90°C, 20°C min⁻¹ to 170°C, 4°C min⁻¹ to 280°C (8 min)	MS (SIM)	83
DB-5	60 m, 0.25 mm, 0.25 μm	60°C, 15°C min⁻¹ to 100°C, 6°C min⁻¹ to 300°C (10 min)	FID MS	97,98
DB-5	30 m, 0.32 mm, 0.25 μm	GC–ECD: 150°C (2 min), 5°C min⁻¹ to 200°C (45 min), 10°C min⁻¹ to 270°C (3 min) GC–MS: 55°C (2 min), 5°C min⁻¹ to 210°C (20 min), 20°C min⁻¹ to 270°C (4 min)	ECD MS	134
DB-5	60 m, 0.25 mm, 0.25 μm	90°C (2 min), 20°C min⁻¹ to 200°C (25 min), 4°C min⁻¹ to 290°C (10 min)	ECD	115

Column	Dimensions	Temperature program	Detector	Ref.
DB-5	30 m, 0.32 mm, 0.25 μm	140°C (1 min), 20°C min^{-1} to 220°C, 4°C min^{-1} to 250°C, 20°C min^{-1} to 310°C (2 min)	HRMS	78
DB-5	60 m, 0.25 mm, 0.25 μm	—	ECD	102
DB-5	60 m, 0.25 mm, 0.25 μm	100°C (3 min), 3°C min^{-1} to 180°C (1 min), 3°C min^{-1} to 230°C (1 min), 20°C min^{-1} to 285°C (50 min)	HRMS	119
DB-5	60 m, 0.32 mm, 0.25 μm	100°C (15 min), 3°C min^{-1} to 180°C (5 min), 1.5°C min^{-1} to 230°C (20 min), 20°C min^{-1} to 285°C (35 min)	MS	119
DB-5	60 m 0.25 mm, 0.25 μm	60°C (1 min), 20°C min^{-1} to 210°C (8 min), 2°C min^{-1} to 250°C (17 min), 4°C min^{-1} to 260°C (15 min)	—	110
DB-5	30 m, 0.25 mm, 0.25 μm	80°C (2 min), 6°C min^{-1} to 300°C (10 min)	ECD	206
DB-5	25 m, 0.32 mm, 0.52 μm	130°C (1 min), 5°C min^{-1} to 140°C (1 min), 1.5°C min^{-1} to 250°C (1 min), 10°C min^{-1} to 300°C (10 min)	ECD	122
DB-5	60 m, 0.25 mm, 0.25 μm	60°C, 15°C min^{-1} to 100°C, 6°C min^{-1} to 300°C (10 min)	—	97, 98
DB-5 hydrogen as carrier gas	57 m, 0.25 mm, 0.25 μm	—	—	120,150
DB-5 hydrogen was used has carrier gas	60 m, 0.25 mm, 0.25 μm	80°C (2 min), 10°C min^{-1} to 150°C, 2°C min^{-1} to 280°C	ECD	108,109
DB-5 MS	30 m, 0.25 mm, 0.25 μm	70°C (2 min), 20°C min^{-1} to 180°C (2 min), 10°C min^{-1} to 300°C (4 min)	MS	115
DB-5 nitrogen was used as carrier gas	60 m, 0.25 mm, 0.25 μm	80°C (2 min), 30°C min^{-1} to 185°C (3 min), 1.58°C min^{-1} to 230°C (15 min), 5°C min^{-1} to 270°C (15 min)	μECD	204
DB-5 nitrogen was used as carrier gas	60 m, 0.25 mm, 0.25 μm	—	ECD	127
DB-5MS	30 m, 0.25 mm, 0.25 μm	90°C (2 min), 25°C min^{-1} to 250°C (2.4 min), 5°C min^{-1} to 200°C (16.7 min), 8°C min^{-1} to 280°C (10 min), 20°C min^{-1} to 300°C (1 min)	MS	203

Continued

TABLE 18.7
Continued

Stationary-Phase Composition	Column Length, Internal Diameter, Film Thickness	Temperature Program	Detector	Reference
DB-5MS	30 m, 0.25 mm, 0.25 μm	100°C (0.60 min), 25°C min^{-1} to 200°C (5 min), 4°C min^{-1} to 250°C (5 min), 35°C min^{-1} to 320°C (3 min)	HRMS	130
DB5-MS	30 m, 0.25 mm, 0.25 μm	100°C (1 min), 2.5°C min^{-1} to 285, 10°C min^{-1} to 310°C (3 min)	LRMS	144
First-dimension column: Chirasil-Dex CB	10 m, 0.10 mm, 0.1 μm	*First oven:* 80°C (2 min), 30°C min^{-1} to 110°C, 0.5°C min^{-1} to 155°C when ECD detector was used or 180°C when TOF-MS was used (0 min), 10°C min^{-1} to 250°C (5 min)	MECD TOF-MS	153
Second-dimension column: LC-50 or VF-23MS	1.4 m, 0.15 mm, 0.1 μm 1.5 m, 0.1 mm, 0.1 μm	*Second GC oven* was ramped in a similar way, but with a 50°C and 70°C offset for the VF-23MS and LC-50 columns, respectively. The final temperature of the second GC oven was limited to 275°C for the LC-50 and 280°C for the VF-23MS		
First-dimension column: HP-1	26.7 m, 0.32 mm, 0.25 μm	—	—	154
Second-dimension column: BPX-50 HP 1 nitrogen was used as carrier and makeup gas	46 cm, 0.1 mm, 0.1 μm 25 m, 0.32 mm, 0.17 μm	90°C (2 min), 20°C min^{-1} to 170°C (7.5 min), 3°C min^{-1} to 280°C (5 min)	ECD MS	145,146,148
HP PAS-1701	30 m, 0.25 mm, 0.25 μm	*GC–MS:* 80°C, 30°C min^{-1} to 180°C, 4°C min^{-1} to 188°C (9 min), 5°C min^{-1} to 230°C (25 min) *GC–ECD:* 80°C, 30°C min^{-1} to 180°C, 4°C min^{-1} to 190°C (9 min), 5°C min^{-1} to 230°C (50 min)	MS ECD	135
HP Ultra-2	50 m, 0.2 mm, 0.33 μm	80°C (1 min), 10°C min^{-1} to 280°C	ECD	207

Column	Dimensions	Temperature program	Detector	Ref.
HP-5	30 m, 0.25 mm, 0.25 μm	60°C, 25°C min⁻¹ to 170°C, 4°C min⁻¹ to 190°C, 10°C min⁻¹ to 230°C, 2°C min⁻¹ to 240°C, 10°C min⁻¹ to 270°C (5 min)	ECD	87
HP-5	30 m, 0.25 mm, 0.25 μm	70°C (1 min), 10°C min⁻¹ to 200°C (0.5 min), 3°C min⁻¹ to 230°C, 20°C min⁻¹ to 280°C (2 min)	ECD	112
HP-5	25 m, 0.32 mm, 0.17 μm	60°C, 25°C min⁻¹ 130°C, 8°C min⁻¹ to 320°C	ECD	100
HP-5MS hydrogen was used as carrier gas	60 m, 0.25 mm, 0.25 μm	40°C, 20°C min⁻¹ to 150°C (5 min), 2°C min⁻¹ to 260°C, 10°C min⁻¹ to 290°C (10 min)[a]	ECD	69
HP-5MS	30 m, 0.25 mm, 0.25 μm	90°C (0.5 min), 30°C min⁻¹ to 225°C, 5°C min⁻¹ to 300°C, (13 min)	ECD	199
HT-5	25 m, 0.25 mm, 0.1 μm	90°C (2 min), 20°C min⁻¹ to 200°C (7.5 min), 3°C min⁻¹ to 280°C (20 min)	ECD MS	208
HT-8	30 m, 0.22 mm, 0.25 μm	GC–ECD: 90°C (1 min), 15°C min⁻¹ to 180°C (1 min), 3°C min⁻¹ to 250°C, 15°C min⁻¹ to 290°C (6 min); GC–MS: 90°C (1 min), 15°C min⁻¹ to 275°C (10 min)	ECD MS	147
HT-8	50 m, 0.22 mm, 0.25 μm	90°C (1 min), 15°C min⁻¹ to 170°C (2 min), 4°C min⁻¹ to 290°C (14 min)	MS	128
HT-8	25 m, 0.22 mm, 0.25 μm	90°C (1 min) then 15°C min⁻¹ to 180°C (1 min) then 3°C min⁻¹ to 250°C, then 25°C min⁻¹ to 20°C (6 min)	μECD	121
HT-8	50 m, 0.22 mm, 0.25 μm	90°C (1 min), 15°C min⁻¹ to 170°C (2 min), 14°C min⁻¹ to 290°C (4 min)		201
OV210	25 m, 0.32 mm, 0.25 μm; 30 m, 0.32 mm, 0.25 μm	140°C, 4°C min⁻¹ to 250°C 160°C (20 min), 4°C min⁻¹ to 230°C	FID	22,155
PAS 5	30 m, 0.32 mm	160°C, (1 min), 5°C min⁻¹ to 260°C (20 min)	ECD	133
PTE 5	30 m, 0.25 mm, 0.25 μm	80°C, 50°C min⁻¹ to 195°C (10 min), 2°C min⁻¹ to 225°C, 12°C min⁻¹ to 300°C	MS	93
PTE-5	30 m, 0.25 mm, 0.25 μm	—	MS	73

Continued

TABLE 18.7
Continued

Stationary-Phase Composition	Column Length, Internal Diameter, Film Thickness	Temperature Program	Detector	Reference
Restek RTX-1701	30 m, 0.32 mm	50°C (2 min), 25°C min^{-1} to 150°C, 4°C min^{-1} to 270°C (20 min)	ECD	132
Restek RTX-5	30 m, 0.32 mm			209
RTX-1	15 m 0.53 mm	140°C, 2°C min^{-1} to 200°C, 4°C min^{-1} to 240°C	ECD	
Rtx-CL pesticides	30 m, 0.25 mm, 0.25 μm	80°C (10 min), 2 to 250°C min^{-1} (20 min)	MS	210
SBP-5	30 m, 0.25 mm, 0.25 μm	—	ECD	136
SE-54	60 m, 0.25 mm, 0.25 μm	—		96
Sil-88	50 m, 0.25 mm, 0.20 μm	90°C (2 min), 20°C min^{-1} to 150°C (7.5 min), 3°C min^{-1} to 240°C	ECD MS	208
SPB 5	30 m, 0.25 mm, 0.25 μm	50°C, 15°C min^{-1} to 200°C (1 min), 8°C min^{-1} to 300°C (3 min)	MS	198
TRB-1701	30 m, 0.25 mm, 0.25 μm	80°C (1 min), 25°C min^{-1} to 170°C, 3°C min^{-1} to 260°C (2 min)	—	111
Ultra 2	50 m, 0.2 mm, 0.33 μm	—	MS	103
β-TBDM	—	60°C (2 min), 10°C min^{-1} to 150°C (180 min), 0.5°C min^{-1} to 170°C, 10°C min^{-1} to 220°C (40 min)	ECD	67
CP-Sil 8CB	50 m, 0.25 mm, 0.25 μm	60°C (2 min), 15°C min^{-1} to 180°C (6 min), 4°C min^{-1} to 220°C (2 min), 5°C min^{-1} to 280°C (25 min)	MS	70
SE-54	60 m, 0.25 mm, 0.15 μm	110°C (2 min), 10°C min^{-1} to 180°C (8 min), 4°C min^{-1} to 220°C (5 min), 4°C min^{-1} to 270°C	ECD	89
SE-54	25 m, 0.32 mm, 0.25 μm	140°C, 4°C min^{-1} to 250°C	FID	92

TABLE 18.8
Stationary Phase Composition of Some Capillary Columns

Column	Stationary Phase (nomenclature)
AT-5	5% Phenyl polydimethyl siloxane
B-DM Chiraldex Astec	2,3-Di-*O*-methyl β-cyclodextrin
B-PA Chiraldex Astec	Permethyl β-cyclodextrin
B-PH Chiraldex Astec	(*S*)-2-Hydroxypropyl methyl ester β-cyclodextrin
Chirasil-Dex	Polysiloxane derivatized with 2,3,6-tri-*O*-methyl β-cyclodextrin
Cyclosil-B	30% 2,3-Di-*O*-methyl-6-*O*-*tert*butyl dimethylsilyl β-cyclodextrin in DB-1701
CP-Sil 8CB	95% Dimethyl–5% phenyl polysiloxane
DB-1	Methyl polysiloxane
DB-5/DB-5 MS	(5%-Phenyl)-methylpolysiloxane
G-PT Chiraldex Astec	Hydroxypropyl-permethyltrifluroacetyl γ-cyclodextrin
G-TA Chiraldex Astec	2,6-Di-*O*-pentyl-3-trifluoroacetyl γ-cyclodextrin
HP 1	100% Dimethyl polysiloxane
HP PAS-1701	Cyanopropylphenyl–dimethyl (14:86) polysiloxane
HP Ultra 2	5% Phenyl–95% methylpolysiloxane (which corresponds to DB-5 or a CP Sil-8)
HT-5	1,2-Dicarba-closo-dodecarborane dimethyl polysiloxane
HT-8	1,7-Dicarba-closo-dodecarborane 8% phenylmethyl siloxane
LC-50	Poly (50% liquid crystalline/50% dimethyl) siloxane
OV210	50% Cyanopropyl–methil 50% phenylmethyl polysiloxane
PTE-5	95% Dimethyl–5% phenyl polysiloxane
Restck RTX-1701	14%-Cyanopropyl-phenyl)-methylpolysiloxane
Restek RTX-5	(5%-Phenyl)-methylpolysiloxane
SE-54	95% Dimethyl–5% phenyl polysiloxane
Sil-88	*Bis*cyanopropyl phenyl polysiloxane
VF-23MS	>75% *bis*cyanopropyl polysiloxane
β-TBDM	35% Heptakis(6-*O*-*tert*-butyldimethylsilyl-2,3-di*O*-methyl)-b-cyclodestrine in OV-1701
BP-5	95% Dimethyl–5% phenyl polysiloxane
CP-Sil 19CB	14% Cyanopropyl 1% vinyl 85% methyl polysiloxane

ECD is also subject to other types of interference which do not give specific signals, such as elemental sulphur.[7]

A detection limit of 0.05 to 0.5 pg of each congener injected in the GC can be obtained.

While ECD is selective towards PCBs and other planar halogenated hydrocarbons (PHHs), samples with complex matrixes can confound measurements of PCBs, so the atomic emission detector (AED) was investigated. AED is an element-sensitive technique which employs plasma to excite atoms of GC-eluted compounds, which are subsequently detected by diodes. There are two advantages of AED. First, its operation in chlorine mode makes it selective for PCBs and other chlorinated compounds. Second, element detection may allow quantitative identification of PCBs and their quantitative determination with only a few standards being required. The biggest obstacle of element-selective detection for PCB analyses has been the poor response for chlorine. This problem has been overcome in the second generation of the HP AED on-column plasma system, which has a PCB detection limit of ~1 pg.[163] AED was selectively used to detect methylsulfonyl PCBs, an important class of bioaccumulatable PHHs, in gray-seal tissue using the sulfur channel of the AED.[164]

b. Mass Spectrometric Detector

The MSD is very useful for detecting PCBs since they generally have a very intense molecular ion, along with a typical chorine cluster associated with the two naturally occurring chlorine isotopes (^{35}Cl and ^{37}Cl). It does not have many of the drawbacks of ECD, and can be very easily interfaced to a gas chromotograph. At present, due to their relatively low cost, robustness, and ease of operation, quadrupole- and ion trap-based instruments are the most popular and are included in the basic instrumentation of most public and private analytical laboratories.

GC–MS is a reliable technique for planar PCBs quantitation because of its improved selectivity, particularly given the availability of ^{13}C-labeled PCB standards.[7] Using isotope dilution, each individual sample (i.e., unknown samples, calibration standards, quality controls, and blanks) is enriched with stable isotope-labeled analogs of analytes of interest, usually ^{13}C-labeled for PCBs and pesticides. Chemically, the analytes and labeled analogues behave identically; however, they can be distinguished based on their mass differences, thus allowing a complete and automatic recovery correction for each analyte in individual sample. These analyses are typically more accurate, selective, and sensitive than GC–ECD analyses; to obtain the sensitivity needed, the mass spectrometers must be operated in the selected ion monitoring (SIM) mode.

Mass spectrometry may be divided into the following categories, according to the ionization process and the polarity of the ions detected:

- Electron impact ionization (positive ion detection) (EIMS)
- Chemical ionization (positive ion detection) (PICI-MS)
- Chemical ionization (negative ion detection) (NICI-MS)

Since PCBs produce an abundance of molecular ions by electron impact, PICI-MS does not give any further improvement, and thus only EIMS and NICI-MS will be discussed.

The use of MS detection in electron ionization (EI) mode increases selectivity with respect to ECD and enhances analyte identification potential conjointly with the GC retention time information.[165,166] It can be used either in the full scan mode (observation of the full mass range) from which specific ions can be extracted, or in the SIM mode (observation of a few selected ions).

The EI mass spectra of PCBs are characterized by the cluster of the chlorine isotopic distribution (i.e., 75.8% ^{35}Cl and 24.2% ^{37}Cl), which is very useful for the identification of chlorinated species. The most abundant fragments are obtained by chlorine elimination; the odd-electron species are favored (i.e., $[M]^{+}$, $[M - Cl_2]^{+}$, $[M - Cl_4]^{+}$. Asymmetrically substituted *ortho*-chloro PCBs only exhibit the $[M - 35]^{+}$ fragment ion. For less chlorinated isomers, the loss of HCl is also observed.

Table 18.9 and Table 18.10 show the mean relative response of each congener class and the abundance of $[M - 35]^{+}$, and $[M - 70]^{+}$ ions, relative to $[M]^{+}$ ion, for many PCB congeners, and the relative intensity of $[M + 2]^{+}$, and $[M + 4]^{+}$ ions, relative to $[M]^{+}$ ion, for each PCB congener class, respectively.

When SIM is performed on a magnetic sector mass spectrometer, each group of masses holds the field on the magnet at a constant setting; changing the acceleration voltage allows different masses to be monitored. The lowest mass in each group uses the highest acceleration voltage, and this voltage is decreased to analyze higher masses. The ratio of the highest to the lowest mass in each group should typically be under 1.5 to prevent loss of sensitivity and stability at larger mass ratios. PCBs tend to elute from the GC column in the order of increasing mass. They also tend to yield strong molecular ions $[M]^{+}$ when analyzed by ionization with low electron energy. Conversely, the organochlorine pesticides (OCPs) in the same extracts do not always elute with increasing mass and only produce lower molecular mass fragment ions with little apparent molecular ion.

TABLE 18.9

Mean Relative Response of PCB Homologue Classes and Relative Abundance of Molecule and Fragment Ions of Some PCB Congeners in EI-SIM Mass Spectra

IUPAC No.	Compound	Relative Abundance % (referred to molecular ion)		Mean Relative Response
		M-35	M-70	
	Monochlorobiphenyls	—	—	3.331
1	2-	12	—	—
2	3-	10	—	—
3	4-	12	—	—
	Dichlorobiphenyls	—	—	2.027
4	2,2'-	2.8	79	—
7	2,4-	1.5	39	—
10	2,6-	2.4	35	—
11	3,3'-	6.1	38	—
12	3,4-	6.7	35	—
15	4,4'-	1.4	38	—
	Trichlorobiphenyls	—	—	1.573
28	2,4,4'-	1	35	—
30	2,4,6-	1	36	—
	Tetrachlorobiphenyls	—	—	0.951
47	2,2',4,4'-	40	56	—
52	2,2',5,5'-	13	71	—
54	2,2',6,6'-	2.0	75	—
61	2,3,4,5-	1.0	38	—
65	2,3,5,6-	1.5	44	—
66	2,3',4,4'-	0.5	35	—
77	3,3',4,4'-	0.5	30	—
80	3,3',5,5'-	1.0	33	—
	Pentachlorobiphenyls	—	—	0.720
	Hexachlorobiphenyls	—	—	0.514
133	2,2',3,3',5,5'-	2	54	—
138	2,2',3,4,4',5'-	5	36	—
153	2,2',4,4',5,5'-	2	52	—
155	2,2',4,4',6,6'-	1	56	—
169	3,3',4,4',5,5'-	1	31	—
	Heptachlorobiphenyls	—	—	0.361
	Octachlorobiphenyls	—	—	0.253
194	2,2',3,3',4,4',5,5'-	6	53	—
197	2,2',3,3',4,4',6,6'-	3	38	—
202	2,2',3,3',5,5',6,6'-	8	45	—
	Nonachlorobiphenyls	—	—	0.230
209	Decachlorobiphenyls	1	65	0.213

The detection limit of EIMS in the SIM mode is in the range of a few pg of each congener injected in the GC.

The determination of PCBs by conventional EI-MS, even in the selected ion-monitoring (SIM) mode, exhibits higher detection limits than ECD.[7]

A study of EI response for all 209 PCBs reported that molecular ion response decreased with increasing chlorine number.[149]

TABLE 18.10
Relative Intensity of Molecular Ions of PCB Homolog Classes According to the ^{35}Cl and ^{37}Cl Isotope Abundance in the Mass Spectra Obtained by Electron Impact in the Selected Ion Monitoring Mode (EI-MS-SIM)

Homolog	[M]$^+$	[M + 2]$^+$	Relative Intensity	[M + 4]$^+$	Relative Intensity
Mono	188	190	33	—	—
Di	222	224	66	226	11
Tri	256	258	99	260	33
Tetra	292	290	76	294	49
Penta	326	328	66	324	61
Hexa	360	362	82	364	36
Hepta	394	396	98	398	54
Octa	430	432	66	428	87
Nona	464	466	76	462	76
Deca	498	500	87	496	68

The main alternative MS techniques currently used for analyzing PCBs at trace level are negative chemical ionization MS (NCI-MS)[167–169] and high-resolution MS (HRMS).[170,171] Usually, both techniques provide much higher sensitivity than EI-MS but the complexity of use and the high cost of purchase and maintenance restrict their use in routine analyses of PCBs. The recent introduction of improved bench-top MS for electron capture nagative ionization (ECNI) may increase the use of this technique.[149]

Negative chemical ionization is one of the soft ionization techniques which produces the fewest fragments, and favors the molecular ion. NCI generates relatively simple mass spectra which may be affected by the physical and geometrical parameters of the ion source,[172–175] including temperature and reagent pressure of the ion source. The presence of water and oxygen might also affect the ionization process.[176,177] NCI mass spectra of mono-, di-, and trichlorobiphenyls are dominated by m/z 35 and 37; whereas, the molecular ion M$^-$ is the most abundant one in those PCBs with more than four chlorine atoms. This is attributed to the stabilization effect of a negative charge due to the higher number of chlorine atoms.[178] The detection limit of NCI-MS in the SIM is in the range 0.05 to 0.1 pg of each congener with more than four chlorines injected into the GC.

Negative ions produced when PCBs capture low-energy electrons generated from a buffer gas (e.g., methane, hydrogen, or argon) were studied by employing ECNI.[35]

Electron impact (EI) and ECNI have been used as ionization methods in low-resolution GC–MS for PCB detection. EI is the most selective method. ECNI is more sensitive, but only for molecules which contain five or more chlorine atoms.[179] Occasionally, and depending on substitution pattern, molecules with four chlorine atoms can also be detected by ECNI-MS. High resolution EI-MS has also been applied for PCB determination, but on a relatively small scale as the costs of using this technique are higher than for using ECD and low-resolution MS (LRMS).

Recently, tandem MS (MS–MS) analysis by ion trap MS (ITMS) systems has become a competitive technique for the determination of PCDD/Fs and PCBs.[180]

Ion-trap MS (IT-MS) is an ion manufacture/storage, three-dimensional quadrupole, sensitive, mass-analysis device which, when coupled with MS–MS (for improved selectivity), has been used to analyze the more toxic coplanar and mono-*ortho* PCB congeners in mussels and fish, with lower detection limits than MS alone.[180] IT-MS has been used for determination of hydroxylated PCBs.[149,181] IT-MS–MS and ECNI-LRMS have given comparable results for the enantiomeric analysis of methylsulfonyl PCBs extracted from arctic ringed seal and polar bear adipose tissue.[182]

EPA draft method 1668 has been expanded to include the analysis of all 209 PCB congeners by HRMS.[183] The technique permits quantitation of lower PCBs in coeluting pairs, where the PCBs differ by two chlorines because the high resolving power allows unbiased measurements of ions.[149] Recently, a combination of laser-induced resonance enhanced multi-photon ionization (REMPI) and time-of-flight-MS (TOF-MS), in conjunction with a postcolumn hydrodechlorination reactor, was developed to measure PCBs, with higher selectivity against interference from other chlorinated compounds. High-speed GC-TOF-MS needs to be developed for analysis of closely eluting PCB congeners as it permits the recording of hundreds of mass spectra per second.[149] Often, more than one method of determination is needed. For example, HRGC-ECD PCB congener results are confirmed by HRGC-LRMS in the SIM mode.[184,185] HPLC has also been used for the semiquantitative detection of PCBs, because it has been demonstrated that these compounds absorb UV well enough at short wavelengths.[186]

4. Data Evaluation and Analytical Quality Control

The main difficulty in comparing analytical data from different bibliographic sources is related above all to the method of PCB quantification since there is not yet a widely accepted standard procedure.

The individual concentration of only a limited number of selected congeners is often given. The most widely measured congeners are PCB 28, PCB 52, PCB 101, PCB 118, PCB 138, PCB 153, and PCB 180. On the basis of the concentration of the most representative congeners, the total PCB concentration can also be expressed as the equivalent quantity of an Aroclor, Clophen J or the commercial PCB mixture that shows a distribution pattern near to that observed in the sample. The value is calculated on the assumption that the concentration pattern in the sample does not significantly change from that of a pure commercial mixture.

In this respect, it is worth mentioning that the experimental total PCB concentration, as obtained by measuring a suitable number of the most abundant congeners (about 30 to 60, depending on the concentration level), has been in quite good agreement with the value obtained by calculating the sum of the seven selected congeners (Σ 7) and multiplying it by a factor of 4 for air and water samples and a factor of 3.5 for both biological and sediment samples.[6] Since mass spectrometers have become commonly used as GC detectors, recent literature tends to give individual concentrations of all the congeners identified; in this case, the total PCB content and the congener class distribution can be easily obtained.

Only the quantification of a large number of PCB congeners will give useful information about the source of contamination and about their environmental fate, which is related to the chemical and physical properties of each individual congener.

In the latter case, PCB congeners are first identified by GC/MSD, by the analysis of a standard solution of, for example, several Aroclors (e.g., 1221, 1232, 1248, and 1260). The relative retention time (RRT) for each identified congener is then calculated by using one or more internal standards (ISs). Finally, RRTs are applied for chromatographic peak assignment of real samples, which can be analyzed either by GC/MSD or GC/ECD. Experimental response factors (RFs) are generally obtained for a limited number of selected PCB congeners and for the internal standard (IS) in a suitably selected concentration range. Comparing these RRTs with data reported in the literature[187] it is possible to extrapolate the RRTs for all other congeners. If a MS detector in SIM mode is used, at least three ions should be selected: one as target and two as qualifiers. Using this method, it is much easier to evaluate the content of each congener class, from trichloro- to nonachloro-biphenyl, and to get a more accurate determination of the total PCB content of the sample. Final extracts of real samples are analyzed after adding a known amount of an IS.

IS can be added directly to the sample or after a critical step in analytical procedure such as reduction of solvent volume, changing solvent, and cleanup. An IS is a substance not present in the original sample but added to it in order to control loss of sample during a single step or during the whole analytical procedure. An ideal IS should have the same behavior as the analyte during the analytical procedure; if added to the sample (spike) it should be dispersed in the matrix like the analyte. The IS should be chromatographically resolved from the other components injected. The wide range of PCBs congeners usually found in environmental samples make it difficult to meet all these requirements.

PCBs which are not observed in real samples can be added as IS: PCB 29,[69] PCB 30[83,108,109,188] PCB 46, 43,[128] PCBs 54, 155,[85] PCB 103,[107,126] PCB 116,[126] PCB 130,[138] PCB 166,[108,109,138] PCB 198,[107] PCB 204,[83,138] PCB 209.[87,126] Even though these are more expensive, when a mass spectrometer is used, isotopically labeled PCBs should be preferred[130]: $^{13}C_{12}$-PCB.[125]

Other organic compounds which can be used as IS are: octachloronaphthalene,[108,109] dibromooctafluorobiphenyl (DBOFB),[107] 1,2,3,4-tetrachloronaphthalene[126] hexachlorocyclohexane,[126] 1,3,5-tribromobenzene,[124] Endrin ketone[138] 1,3-dibromobenzene,[138] pentachlorotoluene,[87] and dibutyl chlorendate and 2,4,5,6-tetra-chloro-m-xylene.[111]

There are some specific data-evaluation methods according to the nature of the sample. For instance, in the case of sediment samples, it can be assumed that every particle is adsorbed.

Since PCBs are adsorbed on the particle surface, normally coated with a thin layer of organic matter such as humic acid, the concentration in sediment and soil samples is much more likely to be related to the particle surface area per volume unit than to the mass unit.[70] For this reason, the concentration of each sample, expressed in pg g^{-1} dry weight, is normalized by dividing it by the relevant calculated specific surface area, expressed in square meters of surface per cubic centimeter of dry sample (m^2 cm^{-3}), as obtained by particle size analysis.[70] Comparisons among concentration values of organic pollutants relevant to samples with different particle size distribution may lead to erroneous conclusions if these are expressed in a conventional way.[70]

For biological samples the PCB concentration is expressed as g g^{-1} of wet or dry weight of sample, and as g g^{-1} of extractable lipids since, as already stated, PCBs are lipophilic in nature and accumulate in lipid-rich tissues and organs. The latter allows better data comparison.

Although representing complex mixtures of neutral, polar, free, and bound lipid forms, the total lipid content of a sample is usually measured by simple gravimetric techniques whereby residues are extracted with organic solvents and the solvent-free extracts determined by weighing.[124]

Several studies have shown that different mixtures of solvents used to extract the sample give different total lipid contents.[189-191] Different results are also obtained if different extraction techniques are applied to the same sample.[192] The chloroform/methanol mixture that quantitatively extracted all lipid classes has produced the highest lipid yields.[190,192]

As already mentioned, the quantification of PCBs in environmental matrices is particularly difficult owing to the complexity of their pattern of peaks (209 possible congeners), the low detection limits often required (pg g^{-1} to pg kg^{-1}) and the time-consuming sample preparation (a large number of interfering compounds are present). Analytical quality control procedures allow data to be obtained within assigned values of accuracy and precision.

The analysis of certified reference materials is considered the best way to ensure the accuracy and precision of analytical methods.[193,194] Spiked samples might be an alternative when reference materials are not available,[193,194] although it must be remembered that spiked analytes generally behave differently from native ones.[195] For a correct use of certified reference materials, their analysis should be scheduled within the same time sequence used for the analysis of real samples, and the results should be reported, for example, on a working analytical control chart.

Participation in intercomparison exercises is also a valid opportunity for a laboratory to assess the quality of its analytical capability. Intercomparison exercises allow the interlaboratory coefficients of variation to be estimated for that specific analysis.

Calibration solutions are very useful for optimizing and routinely testing an analytical procedure, but their preparation and storage are still one of the main sources of error in these analyses and certified analytes in neat form (purity higher than 99%) are preferable for preparing calibration solutions following suitable procedures.[193]

REFERENCES

1. Polychlorinated biphenyls. CASRN: 1336-36-3. Hazardous Substances Database (HSDB), 2000, http://toxnet.nlm.nih.gov.
2. Frame, G. M. and Cochran, J. W., and Bowadt, S. S., *J. High Resolution Chromatogr.*, 19, 657–668, 1996.
3. Moffat, C. F., Whittle, K. J., Eds., *Environmental Contaminants in Food*, Sheffield Academic Press, England, p. 500, 1999, chap. 13.
4. IPCS. *Environmental Health Criteria 140, Polychlorinated Biphenyls and Terphenyls*, WHO, Geneva, 1993.
5. IARC. *IARC Monographs on the Evaluation of Carcinogenic Risks to Humans, Polychlorinated Dibenzo-para-dioxins and Polychlorinated Dibenzofurans*, Vol. 69, International Agency for Research on Cancer, Lyon, 1997.
6. Fuoco, R. and Ceccarini, A., Polychlorobiphenyls (PCBs) in Antarctica, In *Environmental Contamination in Antarctica A Challenge to Analytical Chemistry*, Caroli, S., Cescon, P., and Walton, D. W. II., Eds., Elsevier Science, Amsterdam, pp. 237–273, 2001.
7. Erickson, M. D., *Analytical Chemistry of PCBs*, 2nd ed., CRC Lewis Publishers, Boca Raton, FL, 1997.
8. Katánek, F., Demnerová, K., Pazlarová, J., Burkhard, J., and Maléterová, Y., *Int. Biodeterioration Biodegradation*, 44(1), 39–47, 1999.
9. Safe, S., *Environ. Health Perspect.*, 100, 259–268, 1992.
10. ATSDR (Agency for Toxic Substances and Disease Registry). *Toxicological Profile for Polychlorinated Biphenyls*, ATSDR, Atlanta, 1993, TP-92/16, update.
11. WHO (World Health Organization). Polychlorinated biphenyls and terphenyls, *Environmental Health Criteria 140*, 2nd ed., WHO, Geneva, 1993.
12. Lake, J. L., McKinney, R., Lake, C. A., Osterman, F. A., and Heltshe, J., *Arch. Contam. Toxicol.*, 29, 207–220, 1995.
13. Aulerich, J. R., Ringer, R. K., and Safronoff, J., *Arch. Environ. Contam. Toxicol.*, 15, 393–399, 1986.
14. Van den Berg, M., Birnbaum, L., Bosveld, A. T. C., Brunstrom, B., Cook, P., Feeley, M., Giesy, J. P., Hanberg, A., Hasegawa, R., Kennedy, S. W., Kubiak, T., Larsen, J. C., Rolaf van Leeuwen, F. X., Djien Liem, A. K., Nott, C., Peterson, R. E., Poellinger, L., Safe, S., Tillit, D., Tysklind, M., Younes, M., Waern, F., and Zacharewski, T., *Environ. Health Perspect.*, 106, 775–792, 1998.
15. PCBs: Cancer Dose-Response Assessment and Application to Environmental Mixtures, EPA/600/P-96/001F, available from the National Technical Information Service, PB-104616.
16. Ramos, L., Hernandez, L. M., and Gonzalez, M. J., *Int. J. Environ. Anal. Chem.*, 67, 1, 1998.
17. Ahlborg, U. G., Becking, G. C., Birnbaum, S., Brouwer, A., Derks, H. J. G. M., Feeley, M., Golor, G., Hanberg, A., Larsen, J. C., Liem, A. K. D., Safe, S. H., Schlatter, C., Waern, F., Younes, M., and Yrjänheikki, E., *Chemosphere*, 28, 1048–1067, 1994.
18. Frame, G. M., *Fresenius J. Anal. Chem.*, 357, 701–713, 1997.
19. Frame, G. M., *Fresenius J. Anal. Chem.*, 357, 714–722, 1997.
20. Frame, G. M., *Anal. Chem.*, 69, 468A–475A, 1999.
21. Frame, G. M., *J. High Resolution Chromatogr.*, 22, 533–540, 1999.
22. Duinker, J. C., Schulz-Bull, D. E., and Petrick, G., *Anal. Chem.*, 60, 478, 1988.

23. Ahmed, F. E., In *Environmental Contaminants in Food*, Moffat, C. F. and Whittle, K. J., Eds., Sheffield Academic Press, England, p. 500, 1999, chap. 13.
24. Ahmed, F. E., *TrAC Trends Anal. Chem.*, 22(3), 170–185, 2003.
25. Survey of Currently Available NonIncineration of PCB Destruction Technologies, United Nations Environment Programme, August 2000.
26. Guo, Y. L., Yu, M.-L., and Hsu, C.-C., *Environ. Health Perspect.*, 107(9), 715–719, 1999.
27. National Research Council. *Hormonally Active Agents in the Environment*, National Academy Press, Washington, DC, p. 15, 1999.
28. Brunner, M. J., Sullivan, T. M., Singer, A. W., Ryan, M. J., Toft, J. D. II, Menton, R. S., Graves, S. W., and Peters, A. C., An assessment of the chronic toxicity and oncogenicity of Aroclor-1016, Aroclor-1242, Aroclor-1254, and Aroclor-1260 administered in diet to rats, Study No. SC920192, Chronic toxicity and oncogenicity report, Battelle, Columbus, OH, 1996.
29. Gribaldo, L., Sacco, M. G., Cusati, S., Zucchi, I., Dosanjh, M. K., Catalani, P., and Marafante, E., *J. Toxicol. Environ. Health*, 55, 121, 1998.
30. Bager, Y., Kato, Y., Kenne, K., and Warngard, L., *Chem. Biol. Interact.*, 103, 199, 1997.
31. Hedenskong, M., Sjogren, M., Cederberg, H., and Rannug, U., *Environ. Mol. Mutagen.*, 30, 254, 1997.
32. Patandin, S. C., Koopman-Esseboom, M. A., de Ridder, N., Weisglas-Kuperus, G., and Sauer, P. J., *Pediatr. Res.*, 44, 538–545, 1998.
33. Moline, J. M., Golden, A. L., and Bar-Chama, N., *Environ. Health Perspect.*, 108, 1–20, 2000.
34. Jacobson, J. L., Jacobson, S. W., and Humphrey, H. E., *N. Engl. J. Med.*, 335, 783–789, 1996.
35. Wells, D. E. and de Boer, J., In *Environmental Contaminants in Food*, Moffat, C. F. and Whittle, K. J., Eds., Sheffield Academic Press, England, p. 305, 1999, chap. 9.
36. Syracuse Research Corporation. *Toxicological Profile for Polychlorinated Biphenyls (PCBs)*, Under Contract No. 205-1999-00024 Prepared for: US Department of Health and Human Services, Public Health Service Agency for Toxic Substances and Disease Registry, November, 2000.
37. EPA. *National Primary and Secondary Drinking Water Regulations, 40 CFR part 141 — 2: M.D. Erickson, Analytical Chemistry of PCBs*, 2nd ed., CRC Lewis Publishers, Boca Raton, FL, 1997.
38. Ministry of Agriculture Fisheries and Food, *Polychlorinated Biphenyls in Food — U.K. Dietary Intakes*, Food Surveillance Information Sheet No. 89, MAFF, London, May 1996.
39. Feeley, M. M. and Grant, D. L., *Reg. Toxicol. Pharmacol.*, 18, 428, 1993.
40. Council Directive 96/59/EC of 16 September 1996, *Official Journal L*, 243, 24/09/1996, 31–35, http://europa.eu.int/eur-lex/en/lif/reg/en_register_15103030.html.
41. Survey of Currently Available NonIncineration of PCB Destruction Technologies, United Nations Environment Programme, August 2000.
42. European Committee for Standardization. *CEN Guideline EN 1528-1, Fatty Food — Determination of Pesticides and Polychlorinated Biphenyls (PCBs), General*, Beuth, Berlin, 1997.
43. European Committee for Standardization. *CEN Guideline EN 1528-2, Fatty Food — Extraction of Fat, Pesticides and PCBs and Determination of Fat Content, General*, Beuth, Berlin, 1997.
44. European Committee for Standardization. *CEN Guideline EN 1528-3, Fatty Food — Determination of Pesticides and Polychlorinated Biphenyls (PCBs). Cleanup Methods, General*, Beuth, Berlin, 1997.
45. European Committee for Standardization. *CEN Guideline EN 1528-4, Fatty Food — Determination, Confirmatory, Tests, Miscellaneous, General*, Beuth, Berlin, 1997.
46. European Commission, *Offic. J. Eur. Commission L*, 310, 62, 1999.
47. Bernard, A., Hermans, C., Broeckaert, F., De Poorter, G., De Cock, A., and Housins, G., *Nature*, 401, 231, 1999.
48. Van den Berg, M., Peterson, R. E., and Schrenk, D., *Food Addit. Contam.*, 17, 347–358, 2000.
49. World Health Organization (WHO). *Dioxins and their Effects on Human Health*, Fact sheet no. 225, June 1999, www.who.int/inf-fs/en/fact225.html.
50. Larsen, J. C., Farland, W., and Winters, D., *Food Addit. Contam.*, 17(4), 359–369, 2000.
51. Environmental Protection Agency (EPA) and Bushman, D. R., Guidance documents for dioxin and dioxin-like compounds and other persistent bioaccumulative toxic (PBT) chemicals. Community right-to-know toxic chemical release reporting, *Fed. Reg.*, 65(116), 37548, June 2000, www.epa.gov/tri.

52. Alcock, R. E., Behnisch, P. A., Jones, K. C., and Hagenmaier, H., *Chemosphere*, 37(8), 1457–1472, 1998.
53. Flanagan, W. P. and May, R. J., *Environ. Sci. Technol.*, 27, 2207–2212, 1993.
54. Williams, W. A. and May, R. J., *Environ. Sci. Technol.*, 31, 3491–3496, 1997.
55. Berkaw, M., Sowers, K. R., and May, H. D., *Appl. Environ. Microbiol.*, 62, 2534–2539, 1996.
56. Brown, J. F. Jr., Bedard, D. L., Brennan, M. J., Carnahan, J. C., Feng, H., and Wagner, R. E., *Science*, 236, 709–712, 1987.
57. Quensen, J. F. III, Tiedje, J. M., and Boyd, S. A., *Science*, 242, 752–754, 1988.
58. Bedard, D. L., Van Dort, H. M., May, R. J., and Smullen, L. A., *Environ. Sci. Technol.*, 31, 3308–3313, 1997.
59. Pulliam-Holoman, T. R., Elberson, M. A., Cutter, L. A., May, H. D., and Sowers, K. R., *Appl. Environ. Microbiol.*, 64, 3359–3367, 1998.
60. Sawhney, B. L., Frink, C. R., and Glowa, W., *J. Environ. Quality*, 10, 444–448, 1981.
61. Kannan, K., Maruya, K. A., and Tanabe, S., *Environ. Sci. Technol.*, 31, 1483–1488, 1997.
62. Drenzek, N. J., Eglinton, T. I., Wirsen, C. O., May, H. D., Wu, Q., Sowers, K. R., and Reddy, C. M., *Environ. Sci. Technol.*, 35, 3310–3313, 2001.
63. Jarman, W. M., Hilkert, A., Bacon, C. E., Collister, J. W., Ballschmiter, K., and Risebrough, R. W., *Environ. Sci. Technol.*, 32, 833–836, 1998.
64. Drenzek, N. J., Tarr, C. H., Eglinton, T. I., Heraty, L. J., Sturchio, N. C., Shiner, V. J., and Reddy, C. M., *Org. Geochem.*, 33, 437–444, 2002.
65. Kaiser, K. L. E., *Environ. Pollut.*, 7, 93–101, 1974.
66. Wong, C. S. and Garrison, A. W., *J. Chromatogr. A*, 866(2), 213–220, 2000.
67. Klobes, U., Vetter, W., Luckas, B., Skírnisson, K., and Plötz, J., *Chemosphere*, 37(9–12), 2501–2512, 1998.
68. Hyötyläinen, T. and Hartonen, K., *TrAC Trends Anal. Chem.*, 21(1), 13–30, 2002.
69. Fillmann, G., Galloway, T. S., Sanger, R. C., Depledge, M. H., and Readman, J. W., *Anal. Chim. Acta*, 461(1), 75–84, 2002.
70. Fuoco, R. and Colombini, M. P., *Microchem. J.*, 51, 106–121, 1995.
71. Peltonen, K. and Kuljukka, T., *J. Chromatogr.*, 710, 93, 1995.
72. García-Alonso, S., Pérez-Pastor, R. M., and Quejido-Cabezas, A. J., *Anal. Chim. Acta*, 440(2), 223–230, 2001.
73. Pettersen, H., Axelman, J., and Broman, D., *Chemosphere*, 38(5), 1025–1034, 1999.
74. Fuoco, R., Colombini, M. P., and Samcova, E., *Chromatographia*, 36, 65, 1993.
75. Choi, J. W., Miyabara, Y., Hashimoto, S., and Morita, M., *Chemosphere*, 47, 591–597, 2002.
76. Abad, E., Llerena, J. J., Sauló, J., Caixach, J., and Rivera, J., *Organohalogen Compounds*, 46, 439–442, 2000.
77. Jaouen-Madoulet, A., Abarnou, A., Le Guellec, A.-M., Loizeau, V., and Leboulenger, F., *J. Chromatogr. A*, 886(1–2), 153–173, 2000.
78. Nakao, T., Aozasa, O., Ohta, S., and Miyata, H., *Chemosphere*, 48(8), 885–896, 2002.
79. Bidleman, T. F., Patton, G. W., Hinckley, D. A., Walla, M. D., Cotham, W. E., and Hargrave, B. T., In *Long-range Transport of Pesticides*, Kurtz, D. A., Ed., Lewis Pub. Inc., Chelsea, MI, p. 347, 1990, chap. 23.
80. Weber, R. R. and Montone, R. C., Distribution of organochlorines in the Atmosphere of the South Atlantic and Antarctic oceans, In *Long-range Transport of Pesticides*, Kurtz, D. A., Ed., Lewis Pub. Inc., Chelsea, MI, p. 462, 1990.
81. Screitmüller, J. and Ballshmiter, K., *Fresenius J. Anal. Chem.*, 348, 226–239, 1994.
82. Hung, H., Halsall, C. J., Blanchard, P., Li, H. H., Fellin, P., Stern, G., and Rosemberg, B., *Environ. Sci. Technol.*, 35, 1303–1311, 2001.
83. García-Alonso, S., Pérez-Pastor, R. M., and Quejido-Cabezas, A. J., *Talanta*, 57(4), 773–783, 2002.
84. Criado, R. M., Pereiro, R. I., and Torrijos, C. R., *J. Chromatogr. A*, 963(1–2), 65–71, 2002.
85. Mandalakis, M., Tsapakis, M., and Stephanou, E. G., *J. Chromatogr. A*, 925, 183–196, 2001.
86. Stanley, J., Haile, C., Small, A., and Olson, E., Sampling and analysis procedures for assessing organic emissions from stationary combustion sources for exposure evaluation division studies. *Methods Manual*, US Environmental Protection Agency, 1982, EPA-560/5-82-014.

87. Sun, C., Dong, Y., Xu, S., Yao, S., Dai, J., Han, S., and Wang, L., *Environ. Pollut.*, 117(1), 9–14, 2002.
88. Fuoco, R., Colombini, M. P., and Abete, C., *Int. J. Environ. Anal. Chem.*, 55, 15–25, 1994.
89. Gupta, S. R., Sarkar, A., and Kureishey, T. W., *Deep-Sea Res. II*, 43, 119–126, 1996.
90. Kelly, A. G., Cruz, I., and Wells, D. E., *Anal. Chim. Acta.*, 276, 3, 1993.
91. Ewen, A., Lafontaine, H. J., Hidesheim, K.-T., and Stuurmen, W. H., *Int. Chromatogr. Lab.*, 14, 4, 1993.
92. Schulz-Bull, D. E., Petrick, G., Kannan, N., and Duinker, J. C., *Marine Chem.*, 48, 245, 1995.
93. Guidotti, M., Giovinazzo, R., Cedrone, O., and Vitali, M., *Environ. Int.*, 26, 23–28, 2000.
94. EPA National Primary and Secondary Drinking water regulations, 40 CFR appendix to Part 136, method 608.
95. Pearson, R. F., Hornbuckle, K. C., Eisenreich, S. J., and Swackhamer, D. L., *Environ. Sci. Technol.*, 30, 1429, 1996.
96. Rossi, L., de Alencastro, L., Kupper, T., and Tarradellas, J., *Sci. Total Environ*, 322, 179–189, 2004.
97. Dachs, J. and Bayona, J. M., *Chemosphere*, 35, 1669–1679, 1997.
98. Dachs, J., Bayona, J. M., and Albaigés, J., *Marine Chem.*, 57, 313, 1997.
99. Zhang, Z., Yang, M. J., and Pawliszyn, J., *Anal. Chem.*, 66, 844A, 1994.
100. Yang, Y., Miller, D. J., and Hawthorne, S. B., *J. Chromatogr. A*, 800, 257–266, 1998.
101. Poole, G., Thibert, B., Lemaire, H., Sheridan, B., and Chiu, C., *Organohalogen Compounds*, 24, 47, 1995.
102. Schiff, K. and Bay, S., *Marine Environ. Res.*, 56, 225–243, 2003.
103. Miller, B. S., Pirie, D. J., and Redshaw, C. J., *Marine Pollut. Bull.*, 40(1), 22–35, 2000.
104. Marvin, C. H., Murray, S. P., Charlton, N., Fox, M. E., and Thiessen, P. A. L., *Chemosphere*, 54, 33–40, 2004.
105. Frignani, M., Bellucci, L. G., Carraro, C., and Raccanelli, S., *Chemosphere*, 43(4–7), 567–575, 2001.
106. Camusso, M., Galassi, S., and Vignati, D., *Water Res.*, 36(10), 2491–2504, 2002.
107. Barakat, A. O., Kim, M., Qian, Y., and Wade, T. L., *Baseline/Marine Pollut. Bull.*, 44, 1421–1434, 2002.
108. Wong, C. S., Hoekstra, P. F., Karlsson, H., Backus, S. M., Mabury, S. A., and Muir, D. C. G., *Chemosphere*, 49(10), 1339–1347, 2002.
109. Rose, N., Heidikarlsson, S., and Muir, D. G., *Environ. Sci. Technol.*, 35, 1312–1319, 2001.
110. Gil, O. and Vale, C., *Marine Pollut. Bull.*, 42(6), 452–460, 2001.
111. Zuloaga, O., Etxebarria, N., Fernández, L. A., and Madariaga, J. M., *Talanta*, 50(2), 345–357, 1999.
112. Xiong, G., He, X., and Zhang, Z., *Anal. Chim. Acta*, 413(1–2), 49–56, 2000.
113. de Castro, L. M. D., Valcàrcel, M., and Tena, M. T., *Analytical supercritical fluid extraction*, Springer, Berlin, 1994.
114. Fuoco, R., Ceccarini, A., Onor, M., and Lottici, S., *Anal. Chim. Acta*, 346, 81–86, 1997.
115. Ling, Y.-C. and Teng, H.-C., *J. Chromatography A*, 790, 153–160, 1997.
116. Anitescu, G. and Tavlarides, L. L., *J. Supercrit. Fluids*, 14(3), 197–211, 1999.
117. Kennicutt, M. C., McDonald, S. J., Sericano, L. J., Boothe, P., Oliver, J., Safe, S., Presley, B. J., Liu, H., Douglas, W., Wade, T. L., Crockett, A., and Bockus, D., *Environ. Sci. Technol.*, 29, 1279–1287, 1995.
118. Hartonen, K., Bowadt, S., Hawthorne, S. B., and Riekkola, M.-L., *J. Chromatogr. A*, 774, 229, 1997.
119. Yanik, P. J., O'Donnell, T. H., Macko, S. A., Qian, Y., and Kennicutt, M. C., *Org. Geochem.*, 34, 239–251, 2003.
120. Zajicek, J. L., Tillitt, D. E., Schwartz, T. R., Schmitt, C. J., and Harrison, R. O., *Chemosphere*, 40(5), 539–548, 2000.
121. Covaci, A., Tutudaki, M., Tsatsakis, A. M., and Schepens, P., *Chemosphere*, 46(3), 413–418, 2002.
122. Yogui, G. T., de Oliveira Santos, M. C., and Montone, R. C., *Sci. Total Environ.*, 312, 67–78, 2003.
123. Herman, D. P., Effler, J. I., Boyd, D. T., and Krahn, M. M., *Marine Environ. Res.*, 52(2), 127–150, 2001.
124. Drouillard, K. G., Hagen, H., and Haffner, G. D., *Chemosphere*, 55, 395–400, 2004.

125. Fournier, M., Pellerin, J., Lebeuf, M., Brousseau, P., Morin, Y., and Cyr, D., *Aquatic Toxicol.*, 59, 83–92, 2002.

126. Froescheis, O., Looser, R., Cailliet, G. M., Jarman, W. M., and Ballschmiter, K., *Chemosphere*, 40, 651–660, 2000.

127. González Sagrario, M., de los, A., Miglioranza, K. S. B., Aizpún de Moreno, J. E., Moreno, V. J., and Escalante, A. H., *Chemosphere*, 48(10), 1113–1122, 2002.

128. Covaci, A., de Boer, J., Ryan, J. J., Voorspoels, S., and Schepens, P., *Environ. Res.*, 88(3), 210–218, 2002.

129. O'Neal, J. M., Benson, W. H., and Allgood, J. C., Analysis of organochlorine pesticides and PCBs in fish and shrimp tissues, In *Techniques in Aquatic Toxicology*, Ostrander, G. K., Ed., Lewis Pub., Boca Raton, FL, pp. 561–565, 1996.

130. Barr, J. R., Maggio, V. L., Barr, D. B., Turner, W. E., Sjödin, A., Sandau, C. D., Pirkle, J. L., Needham, L. L., and Patterson, D. G. Jr., *J. Chromatogr. B*, 794(1), 137–148, 2003.

131. Kleivane, L., Espeland, O., Fagerheim, K. A., Hylland, K., Polder, A., and Skaare, J. U., *Marine Environ. Res.*, 43, 117–130, 1997.

132. Licata, P., Di Bella, G., Dugo, G., and Naccari, F., *Chemosphere*, 52, 231–238, 2003.

133. Ahmed, M. T., Loutfy, N., and El Shiekh, E., *J. Hazard. Mater.*, 89(1), 41–48, 2002.

134. Konstantinou, I. K., Goutner, V., and Albanis, T. A., *Sci. Total Environ.*, 257, 61–79, 2000.

135. Kontsas, H. and Pekari, K., *J. Chromatogr. B*, 791, 117–125, 2003.

136. Corsi, I., Mariottini, M., Sensini, C., Lancini, L., and Focardi, S., *Oceanol. Acta*, 26, 129–138, 2003.

137. Harding, G. C., LeBlanc, R. J., Vass, W. P., Addison, R. F., Hargrave, B. T., Pearre, S. Jr., Dupuis, A., and Brodie, P. F., *Marine Chem.*, 56(3–4), 145–179, 1997.

138. Hoekstra, P. F., Braune, B. M., Wong, C. S., Williamson, M., Elkin, B., and Muir, D. C. G., *Chemosphere*, 53, 551–560, 2003.

139. Yang, Y.-H., Chang, Y.-S., Kim, B.-H., Shin, D.-C., and Ikonomou, M. G., *Chemosphere*, 47(10), 1087–1095, 2002.

140. Rantalainen, A. L., Ikonomou, M. G., and Rogers, I. H., *Chemosphere*, 37, 1119–1138, 1998.

141. Huckins, J. N., Tubergen, M. W., Lebo, J. A., Gale, R. W., and Schwartz, T. R., *J. Assoc. Offic. Anal. Chem.*, 73, 290, 1990.

142. Bavel, B. N. C., Bergqvist, P. A., Broman, D., Lundgren, K., Papakosta, O., Rolff, C., Strandberg, B., Zebòhr, Y., Zook, D., and Rappe, C., *Marine Pollut. Bull.*, 32, 210–218, 1995.

143. Fernández, M. A., Hernández, L. M., González, M. J., Eljarrat, E., Caixach, J., and Rivera, J., *Chemosphere*, 36(14), 2941–2948, 1998.

144. Aries, E., Anderson, D. R., Ordsmith, N., Hall, K., and Fisher, R., *Chemosphere*, 54(1), 23–31, 2004.

145. Olsson, A., Vitinsh, M., and Plikshs, M., Bergman Å, *Sci. Total Environ.*, 239(1–3), 19–30, 1999.

146. Olsson, A., Valters, K., and Burreau, S., *Environ. Sci. Technol.*, 34, 4878–4886, 2000.

147. Covaci, A., Ryan, J. J., and Schepens, P., *Chemosphere*, 47, 207–217, 2002.

148. Jensen, S., Reutergårdh, L., and Jansson, B., *FAO Fish. Tech. Paper*, 212, 21–33, 1983.

149. Cochran, J. W. and Frame, G. M., *J. Chromatogr. A*, 843(1–2), 323–368, 1999.

150. Steingraeber, M. T., Schwartz, T. R., Wiener, J. G., and Lebo, J. A., *Environ. Sci. Technol.*, 28(4), 707–714, 1994.

151. Poole, C. F. and Schuette, S. A., *Contemporary Practice of Chromatography*, Elsevier Science, Amsterdam, 1984.

152. Harju, M., Danielsson, C., and Haglund, P., *J. Chromatogr. A*, 1019, 111–126, 2003.

153. Harju, M., Bergman, A., Olsson, M., Roos, A., and Haglund, P., *J. Chromatogr. A*, 1019(1–2), 127–142, 2003.

154. Kristenson, E. M., Korytár, P., Danielsson, C., Kallio, M., Brandt, M., Mäkelä, J., Vreuls, R. J. J., Beens, J., and Brinkman, U. A. Th., *J. Chromatogr. A*, 1019, 65–77, 2003.

155. Storr-Hansen, E., In *Environmental Analytical Chemistry of PCBS/Current Topics in Environmental and Toxicological Chemistry*, Albaigés, J., Ed., Gordon and Breach Science Pub., Amsterdam, p. 25, 1993.

156. Storr-Hansen, E., *J. Chromatogr.*, 558, 375, 1991.

157. Smedes, F. and de Boer, J., *Trends Anal. Chem.*, 16, 503, 1997.

158. Anon. Report QUASIMEME Workshop on PCB Analysis, Roskilde, Denmark, June, FRS Marine Laboratory, Aberdeen, UK, 1995.
159. Korytar, P., Leonards, P. E. G., de Boer, J., and Brinkman, U. A. Th., *J. Chromatogr. A*, 958, 203, 2002.
160. Chanel, I. and Chang, I., Analysis of Organochlorine Pesticides and PCB Congeners with HP 6890 Micro-ECD, Hewlett-Packard Application Note 228–384, June 1997.
161. Jones, K. C., Burnett, V., Duarte-Davidson, R., and Waterhouse, K. S., *Chem. Br.*, 27, 435, 1991.
162. Harrad, S. J., Stewart, A. S., Boumphrey, R., Duarte-Davidson, R., and Jones, K. C., *Chemosphere*, 24, 1147, 1992.
163. Asp, T. N., Pedersen-Bjergaard, S., and Greibrokk, T., *J. High Resolution Chromatogr.*, 20, 201, 1997.
164. Janak, K., Becker, G., Colmsjo, A., Ootman, C., Athanassiadory, M., Valters, K., and Bergman, A., *Environ. Toxicol. Chem.*, 17, 1046, 1998.
165. Safe, S., Overview of analytical identification and spectroscopic properties, In *National Conference on Polychlorinated Biphenyls, November 19–21, 1975, Chicago, IL*, Ayer, F. A., Ed., EPA-560/6-75-004; NTIS No. PB 253–248, March 1976.
166. Stan, H.-J., In *Pesticide Analysis*, Das, K. G., Ed., Marcel Dekker, Inc., New York, 1981, chap. 9.
167. Bagheri, H., Leonards, P. E. G., Ghijsen, R. T., and Brinkman, U. A. Th., In *Environmental Analytical Chemistry of PCBS/Current Topics in Environmental and Toxicological Chemistry*, Albaigés, J., Ed., Gordon and Breach Science Pub., Amsterdam, p. 221, 1993.
168. Chaler, R., Vilanova, R., Santiago-Silva, M., Fernandez, P., and Grimalt, J. O., *J. Chromatogr. A*, 823, 73, 1998.
169. Rothweiler, B. and Berset, J. D., *Chemosphere*, 38, 1517, 1999.
170. Ferrario, J., Byrne, C., and Dupuy, A. E., *Chemosphere*, 34, 2451, 1997.
171. Gardinali, P. R., Wade, T. L., Chambers, L., and Brooks, J. M., *Chemosphere*, 32, 1, 1996.
172. Stemmler, E. A., Hites, R. A., Arbogast, B., Budde, W. L., Deinzer, M. L., Dougherty, R. C., Eichelberger, J. W., Foltz, R. L., Grimm, C., Grimsrud, E. P., Sakashita, C., and Sears, L. J., *Anal. Chem.*, 60, 781, 1988.
173. Stemmler, E. A. and Hites, R. A., *Biomed. Environ. Mass Spectrom.*, 15, 659, 1988.
174. Stemmler, E. A. and Hites, R. A., *Biomed. Environ. Mass Spectrom.*, 17, 311, 1988.
175. Oehme, M., Stocki, D., and Knoppei, H., *Anal. Chem.*, 58, 554, 1986.
176. Stemmler, E. A. and Hites, R. A., *Anal. Chem.*, 57, 684, 1985.
177. Budzikiewicz, H., *Mass Spectrom. Rev.*, 5, 354, 1986.
178. Bagheri, H., Leonards, P. E. G., Ghijsen, R. T., and Brinkman, U. A. Th., *Int. J. Environ. Anal. Chem.*, 50, 257, 1993.
179. Wester, P. G., de Boer, J., and Brinkman, U. A. Th., *Environ. Sci. Technol.*, 30, 473, 1996.
180. Leonards, P. E. G., Brinkman, U. A. Th., and Cofino, W. P., *Chemosphere*, 32, 2381, 1996.
181. Abraham, V. M. and Lynn, B. C. Jr., *J. Chromatogr. A*, 790, 131, 1997.
182. Wiberg, K., Letcher, R., Sandau, C., Duffe, J., Norston, R., Haglund, P., and Bidleman, T., *Anal. Chem.*, 70, 3845, 1998.
183. Rahman, M. S., Bowadt, S., and Larsen, B., *J. Chromatogr. A*, 655, 275, 1993.
184. Ramos, L., Torre, M., La borda, F., and Marina, M. L., *J. Chromatogr. A*, 823, 365–372, 1998.
185. Ramos, L., Torre, M., and Marina, M. L., *J. Chromatogr. A*, 815, 272–277, 1998.
186. Hayteus, D. L. and Duffield, D. A., *J. Chromatogr. B*, 705, 362, 1998.
187. Mullin, M. D., Pochini, C. M., McCrindle, S., Romkes, M., Safe, S. H., and Safe, L. M., *Environ. Sci. Technol.*, 18, 468–476, 1984.
188. Corsolini, S., Romeo, T., Ademollo, N., Greco, S., and Focardi, S., *Microchem. J.*, 73, 187–193, 2002.
189. de Boer, J., *Chemosphere*, 17, 1803–1810, 1988.
190. Randall, R. C., Young, D. R., Lee, I. H., and Echols, S. F., *Environ. Toxicol. Chem.*, 17, 788–791, 1998.
191. Manirakiza, P., Covaci, A., and Schepens, P., *J. Food Compos. Anal.*, 14, 93–100, 2001.
192. Smeds, A. and Saukko, P., *Chemosphere*, 44(6), 1463–1471, 2001.
193. Wells, D. E., Maier, E. A., and Griepink, B., *Int. J. Environ. Anal. Chem.*, 46, 255, 1992.

194. Jacob, J., In *Quality Assurance for Environmental Analysis*, Vol. 17, Quevauviller, P., Maier, E. A. and Griepink, B., Eds., Elsevier Science, Amsterdam, p. 649, 1995.

195. Langenfeld, J. J., Hawthorne, S. B., and Miller, D. J., *Anal. Chem.*, 67, 1727–1739, 1995.

196. Toshihiko, O., Kazuhiro, H., Taku, O., and Yukio, N., *Hiroshima-ken Hoken Kankyo Senta Kenkyu Hokoku*, 9, 91–99, 2001.

197. Zhou, J. L., Maskaoui, K., Qiu, Y. W., Hong, H. S., and Wang, Z. D., *Environ. Pollut.*, 113(3), 373–384, 2001.

198. Schellin, M. and Popp, P., *J. Chromatogr. A*, 1020(2), 153–160, 2003.

199. Shu, Y. Y., Wang, S. S., Tardif, M., and Huang, Y., *J. Chromatogr. A*, 1008, 1–12, 2003.

200. Basheer, C., Lee, H. K., and Obbard, J. P., *J. Chromatogr. A*, 1022, 161–169, 2004.

201. Covaci, A., de Boer, J., Ryan, J. J., Voorspoels, S., and Schepens, P., *Environ. Res. Sec. A*, 88, 210–218, 2002.

202. Covaci, A., de Boer, J., Ryan, J. J., Voorspoels, S., and Schepens, P., *Anal. Chem.*, 74, 790–798, 2002.

203. Dmitrovic, J. and Chan, S. C., *J. Chromatogr. B*, 778(1–2), 147–155, 2002.

204. Gómara, B., Ramos, L., and González, M. J., *J. Chromatogr. B: Anal. Technol. Biomed. Life Sci.*, 766(2), 279–287, 2002.

205. Taniyasu, S., Kannan, K., Holoubek, I., Ansorgova, A., Horii, Y., Hanari, N., Yamashita, N., and Aldous, K. M., *Environ. Pollut.*, 126, 169–178, 2003.

206. Sala, M., Ribas-Fitó, N., Cardo, E., de Muga, M. E., Marco, E., Mazón, C., Verdú, A., Grimalt, J. O., and Sunyer, J., *Chemosphere*, 43(4–7), 895–901, 2001.

207. Çok, I., Hakan, M., and Satiroglu, S., *Environ. Int.*, 30, 7–10, 2004.

208. Larsen, B., Bøwadt, S., and Tilio, R., In *Environmental Analytical Chemistry of PCBS/Current Topics in Environmental and Toxicological Chemistry*, Albaigés, J., Ed., Gordon and Breach Science Pub., Amsterdam, p. 3, 1993.

209. Royal, C. L., Preston, D. R., Sekelsky, A. M., and Shreve, G. S., *Int. Biodeterioration Biodegradation*, 51, 61–66, 2003.

210. Liu, S. and Pleil, J. D., *J. Chromatogr. B*, 769, 155–167, 2002.

19 Peroxyacyl Nitrates, Organic Nitrates, and Organic Peroxides (AIR)*

Jeffrey S. Gaffney and Nancy A. Marley

CONTENTS

*The submitted manuscript has been created by the University of Chicago as operator of Argonne National Laboratory under Contract No. W-31-109-ENG-38 with the U.S. Department of Energy. The U.S. government retains for itself, and others acting on its behalf, a paid-up, nonexclusive, irrevocable worldwide license in said article to reproduce, prepare derivative works, distribute copies to the public, and perform publicly and display publicly, by or on behalf of the government.

I. THE PEROXYACYL NITRATES

A. INTRODUCTION: PEROXYACETYL NITRATE, AN HISTORICAL PERSPECTIVE

But if any man undertake to write a history that has to be collected from materials gathered by observation and the reading of works not easy to be got in all places, nor written always in his own language, but many of them foreign and dispersed in the other hands, for him, undoubtedly, it is in the first place and above all things most necessary to reside in some city of good note, addicted to liberal arts, and populous; where he may have plenty of all sorts of books, and upon inquiry may hear and inform himself of such particulars as, having escaped the pens of writers, are more faithfully preserved in the memories of men, lest his work be deficient in many things, even those which it can least dispense with. But for me, I live in a little town, where I am willing to continue, lest it should grow less.

Plutarch (AD 46 – 120)

Coincidentally, the place Plutarch suggested as best for writing a history, a city, or urban area, is where photochemical smog and associated air pollutants were first observed and peroxyacetyl nitrate (PAN) was first identified as a potentially important air pollutant. The discovery of PAN has an interesting history. It is a story based on collaboration and discussion between biologists and chemists, leading to identification of a unique family of molecules, the peroxyacyl nitrates (PANs), which are now just beginning to be appreciated as important compounds for measurement in the free troposphere, as well as in urban air.

This chapter is a brief overview of the history of the discovery of the PANs and their connection to plant damage observed in southern California, their chemical and physical properties, and the chromatographic measurement techniques developed for their measurement in the atmosphere, with a special focus on the simplest analog in the group (PAN). The phytotoxicity of PAN is briefly discussed, in terms of its discovery and its relationship with other atmospheric trace species formed in air pollution chemistry (smog chemistry), namely the organic peracids, peroxides, and nitrates. Work performed by pioneers in this research area is highlighted, particularly the first uses of long-path infrared spectroscopy and gas chromatography (GC) with electron capture detection. The development of these techniques has enabled further exploration of PAN chemistry since the 1950s. The more recent use of other chromatographic detectors, as described, has improved understanding of the key roles which PAN and its analogs play in urban, regional, and global tropospheric chemistry. The advances addressed include synthetic procedures for the establishment of calibration standards, laboratory determination of properties of PANs (rates of reaction, ultraviolet photolysis, aqueous solubilities), current measurement techniques, and some examples of recent field measurements. The measurement methods employed for the associated organic nitrates, peroxides, and peracids are also described briefly.

1. First Observation in the Environment

The south coast air basin in southern California was known for its "photochemical smog" during the early 1940s.[1] Eye irritation was reported during the smog episodes, along with formation of a characteristic brown haze, leading to visibility reduction in the area. In the late 1940s and early 1950s, farmers began to report serious damage to some crops in the San Gabriel valley, especially to romaine lettuce and parsley. Rubber products were found to deteriorate at an accelerated rate. John Middleton, a professor of plant pathology at the University of California, and his research group first eliminated sulfur dioxide (SO_2) and hydrogen fluoride (HF) as the cause of the damage, determining that oxidation, not reduction, was at hand.[2,3] Various atmospheric oxidants were suspected to be the cause. As Edgar Stephens pointed out in his paper on smog studies of the 1950s,[1] "… the state of the art of analytical chemistry in 1945 was inadequate to deal with the problem." The problem was to identify the key oxidants causing the damage by using the analytical tools available at that time.

Attempting with little success to determine the cause of plant leaf bronzing, Ellis Darley of the Citrus Experiment Station at the University of California, Riverside (UCR), began to expose plants to a wide variety of chemicals by using a fumigation system. Arie J. Haagen-Smit, a perfumery chemist at the California Institute of Technology in Pasadena, noticed that the smog smelled like his terpene laboratory. In 1950, Haagen-Smit fumigated a test plant with the products from the ozonolysis of gasoline vapors and was able to reproduce the plant damage caused by leaf bronzing.[4] His theory stated that partially ozonized hydrocarbons were the cause of the observed damage. Haagen-Smit followed these studies with photochemical investigations using nitrogen dioxide (NO_2) and hydrocarbons known to be emitted from motor vehicles. Ozone (O_3) was produced, and the observed damage to plants and rubber was replicated.[5,6] The demonstration that olefins are more reactive than other hydrocarbons in producing smog symptoms led to the pursuit of the sources of these emissions. The two likely candidate sources were automobiles and oil refineries. Haagen-Smit's studies raised questions about the products of the photochemical reactions, specifically the products responsible for triggering smog production and whether O_3 caused the plant damage and eye irritation observed, or there were other reactive gases responsible. The answers to these questions led to the discovery of a new class of air pollutants, the PANs.

2. The Discovery of Peroxyacetyl Nitrate

Edgar Stephens, Philip Hanst, Robert Doerr, and William Scott of the Franklin Institute Laboratories in Philadelphia used the first long-path infrared gas cell, constructed in 1954, to identify products of the photooxidation of some simple organics in the presence of NO and NO_2, collectively known as NO_x. This instrumentation made use of a long-path optical cell and a conventional scanning dispersive infrared spectrometer and allowed the infrared spectrum of the photolyzed olefin–NO_x mixtures to be obtained.[7,8] A very interesting, unexpected set of infrared bands appeared in photolyzed mixtures containing 3-methylpentane. These bands were very strong when biacetyl ($CH_3CO)_2$ was studied.[8] The strong bands at 1740 cm^{-1} and 1841 cm^{-1} were quite unique, and the molecule responsible was referred to as "Compound X" by the Franklin Institute research team. These workers attempted to isolate compound X from the biacetyl reaction, but a two-drop sample trap exploded violently before these could obtain a mass spectrum. Five structures were suggested for compound X, all of which proved to be incorrect. The correct structure was not determined until the leading candidate, acetyl nitrate, was synthesized and characterized by Edward Schuck and George Doyle. Their results, presented in 1959,[9] showed similarities with compound X but also proved that acetyl nitrate was not identical to the unknown nitrate. In 1961, Philip Leighton's classic textbook, *Photochemistry of Air Pollution*,[10] did a remarkable job of laying out the series of events which led Edgar Stephens to propose the correct structure for compound X: $CH_3CO–OO–NO_2$. This structure was accepted as correct and confirmed in 1974 by infrared studies of isotopically labeled PAN.[11]

Further studies initiated in 1958 by O. Clifton Taylor at the UCR Horticulture Department identified PAN as the key compound causing the bronzing of leaves of petunias and other susceptible plants.[12] These studies led to the gas chromatographic separation of PAN from the photochemical reaction mixture. Subsequent characterization of the product identified PAN as a potent eye irritant. The collaboration between chemists and plant pathologists at the UCR Statewide Air Pollution Research Center, led to the determination that PAN and some of its analogs are 10 to 50 times more toxic than O_3 to plants. This important collaboration was one of the first to apply inter-disciplinary work to air quality problems and to demonstrate the value of and need for this approach.

B. PHYSICAL AND CHEMICAL PROPERTIES

1. Formation in the Atmosphere

Researchers recognized in the 1950s that the complex interactions of organics and NO_x in the presence of sunlight led to a unique chemistry which resulted in enhanced O_3 production. The photolysis of O_3 was found to form singlet oxygen [$O(^1D)$]. This species reacts rapidly with water vapor to form two hydroxyl radicals (OH), which can initiate chain reactions with organic radicals in the atmosphere. The peroxy radicals (e.g., HO_2, RO_2) can then react with NO from vehicle emissions to form hydroxyl radical, alkoxy radical, NO_2, and subsequently O_3:

$$NO_2 + h\nu \rightarrow NO + O(^3P) \tag{19.1}$$

$$O(^3P) + O_2 \rightarrow O_3 \tag{19.2}$$

The atmospheric formation of the PANs is a direct consequence of this peroxy radical (RO_2) chemistry.[13]

The PANs are formed in the troposphere during the photochemical oxidation of organic molecules which contain more than one carbon atom. This photooxidation produces a variety of free radicals. The immediate precursors of the PANs are the peroxyacyl radicals (RCO_3). For the simplest member of the family, PAN, this is the peroxyacetyl radical (CH_3CO_3). The peroxyacetyl radical (CH_3CO_3) can be formed by the direct oxidation of reactive organics or by the reaction of acetaldehyde (CH_3CHO) with hydroxyl radical via abstraction of the aldehydic hydrogen followed by addition of molecular oxygen. Once formed, the peroxyacetyl radical can react with NO_2 to form PAN by the following reaction:

$$CH_3CO_3 + NO_2 \rightarrow CH_3COO_2NO_2 \tag{19.3}$$

The PANs are trapped radical species which are in thermal equilibrium with peroxy radicals. This thermal equilibrium chemistry is important in determining the atmospheric lifetimes of the PANs.

The current convention is that, the products of the reactions of atmospheric NO_x with organic radicals, hydroxyl radical, and hydroperoxy radical (HO_2) are known collectively as NO_y. Two of the major NO_y species are nitric acid (HNO_3) and the PANs ($RCO-OO-NO_2$). Some of the more common PANs are listed in Table 19.1, and the most stable structures for the first three analogs are

TABLE 19.1
Chemical Formulas and Names of Some of the More Common Peroxyacyl Nitrates (PANs)

Name	Chemical Formula	Abbreviation
Peroxyacetyl nitrate	$CH_3C=OO_2NO_2$	PAN
Peroxypropionyl nitrate	$CH_3CH_2C=OO_2NO_2$	PPN
Peroxybutyryl nitrate	$CH_3CH_2CH_2C=OO_2NO_2$	PBN
Peroxyisobutyryl nitrate	$(CH_3)_2CHC=OO_2NO_2$	PiBN
Peroxybenzoyl nitrate	$C_6H_5C=OO_2NO_2$	PBzN
Peroxymethacroyl nitrate	$CH_2C(CH_3)C=OO_2NO_2$	MPAN
Trifluoroperoxyacetyl nitrate	$CF_3C=OO_2NO_2$	FPAN

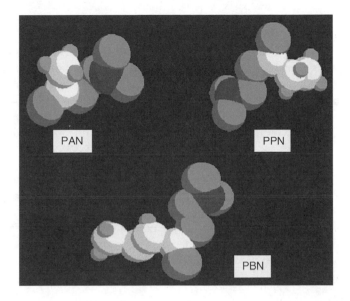

FIGURE 19.1 Structures for the first three analogs of the PANs: C, ▢; H, ▨; N, ▉; O, ▩.

shown in Figure 19.1. PAN, where R is a methyl group, is the simplest and commonest member of the PANs. Peroxypropionyl nitrate (PPN), where R is an ethyl group, is the next most abundant member in the troposphere, occurring at concentrations usually about 10% of the observed PAN concentrations. Higher analogs of the PANs, such as peroxybutyryl nitrate (PBN), peroxybenzoyl nitrate (PBzN), and peroxymethacroyl nitrate (MPAN), have been identified and measured in the atmosphere, but at much lower levels than the two major components of the PANs.

Originally, PAN, H_2O_2, and NO_2 were all identified as important atmospheric pollutants in the first air pollution criteria documents, but these were later omitted from the Clean Air Act of 1970 because of the difficulties at that time in measuring them and synthesizing standards. Currently, PAN is not considered a criteria pollutant, but it is monitored as a key indicator species of atmospheric RO_2 chemistry and as a significant component of the photochemical products from NO emissions.

2. Structure and Properties

The PANs were initially thought to be rather soluble in water, because PAN was observed to undergo rapid base hydrolysis.[13] The reaction of PAN in aqueous solution with base leads to the formation of acetate, nitrite, and $O(^1D)$.[14] This reaction has been used as a means of determining PAN concentrations in standards by bubbling the air containing PAN through a pH 12 solution, with subsequent analysis of the solution for nitrite and acetate products to determine the original PAN concentrations. Further studies have found that PAN is not very soluble in water and is in fact highly soluble in nonpolar organic solvents. At more normal atmospheric pH levels the PANs do not undergo rapid hydrolysis. The aqueous-phase solubility of PAN has been determined to be 3 to 5 $M\,atm^{-1}$, more soluble than NO or NO_2 but less soluble, by orders of magnitude, than HNO_3.[15,16] Thus, aqueous loss of PAN in the troposphere is not an important process.

The very simple ultraviolet absorption spectrum of PAN, first observed by Stephens for samples obtained from ethyl nitrate photolysis with gas chromatographic separation, was confirmed in the early 1980s by using high purity samples.[15,16] Interestingly, PAN has no strong structural features in its ultraviolet spectrum and does not absorb above 290 nm.[15] The PAN molecule is not readily photolyzed in the atmosphere at altitudes below 5 to 7 km.[14] Photolytic lifetimes for PAN are calculated to be on the order of $20 \times 10^{-8}\ sec^{-1}$.[15] The reaction of PAN with hydroxyl radical is also quite slow ($<3 \times 10^{-14}$ cm^3 molecule^{-1} sec^{-1}) at room temperature and is of little importance as a loss mechanism in the troposphere.[17]

In the laboratory, PAN has been observed to decompose thermally to form the major products methyl nitrate (CH_3ONO_2) and carbon dioxide (CO_2). This decomposition reaction is slow and is not likely to be of major importance for PAN in the troposphere. A cyclic intermediate for PAN has been proposed to explain this unimolecular reaction. The chemical and physical properties of PAN suggest that it likely exists in two conformational forms: a cyclic conformation and a more linear structure (Figure 19.2). The cyclic structure is stabilized by the strongly electron-withdrawing oxygen atoms in the molecule which cause an induced-potential charge interaction between the methyl group and the oxygen on the nearby nitrogen. The methyl group is depleted of electron density, and the oxygen is likely to be slightly enriched in electron density. Such a cyclic structure yields a less polar molecule than the linear structure; this may explain the high solubility of PAN in nonpolar solvents and its low solubility in water. It also may explain the featureless ultraviolet spectrum, as well as the C=O stretch observed for PAN at 1841 cm^{-1}, which is remarkably similar to the C=O stretch observed for acid fluorides which also contain strongly electronegative and electron-withdrawing groups.

A thermal decomposition reaction of much greater atmospheric importance is that of PAN, in thermal equilibrium with NO_2 and the peroxyacetyl radical[14,15]:

$$CH_3COO_2NO_2 \rightarrow CH_3CO_3 + NO_2 \qquad (19.4)$$

This is, of course, the reverse of Reaction 19.3 The overall atmospheric lifetime of PAN depends on the ratio of NO to NO_2 and the abundance of peroxyacetyl radical, because the reverse reaction to form PAN is also important. The forward rate for the unimolecular decomposition reaction (Reaction 19.4) is 3.3×10^{-4} sec^{-1} at 298 K.[14] The temperature dependence of the thermal equilibrium is quite strong, with an activation energy of approximately 25 kcal. At the cold temperatures found at higher altitudes and in winter time, PAN is quite stable in the atmosphere, while at lower altitudes in the summer PAN has a fairly short lifetime (<1 h). These observations have implications for sampling and chromatographic analysis of PAN in warm temperatures.

The peroxyacetyl radical can be removed from the atmosphere by NO and hydroperoxy radical via the following reactions:

$$NO + CH_3CO_3 \rightarrow NO_2 + CH_3COO \qquad (19.5)$$

$$HO_2 + CH_3CO_3 \rightarrow CH_3COO_2H + O_2 \qquad (19.6)$$

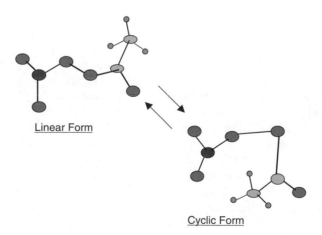

Linear Form

Cyclic Form

FIGURE 19.2 Linear and cyclic conformational forms of PAN: C, ▨; H, ▨; N, ▨; O, ▨.

The peroxyacetyl radical decomposes in the presence of oxygen to form the methylperoxy radical (CH_3OO) and CO_2. The methylperoxy radical can react further with NO to form methoxy radical (CH_3O), which in turn can react with O_2 to yield formaldehyde (CH_2O), a known air toxic, and hydroperoxy radical. The hydroperoxy radical reacts with NO to form NO_2 and hydroxyl radical. The thermal decomposition of the PANs during night time can result in the production of hydroxyl radical if sufficient NO is available, as happens in urban or suburban environments. This process can lead to significant night-time conversion of NO to NO_2.

Because PAN is in thermal equilibrium with NO_2 and the peroxyacetyl radical, it can act as a means of transporting these more reactive species over long distances.[18] The NO_2 released by thermal decomposition of PAN is photolyzed rapidly in the troposphere to form O_3 by Reaction 19.1 and Reaction 19.2. Ozone is a criteria air pollutant and is a major health concern. Thus, the PANs play important roles as a chemical means of transporting key species such as NO_2 and formaldehyde to remote locations. As such, PANs are globally important atmospheric molecules, as well as urban air pollutants. Since the original observation of PANs in Los Angeles photochemical smog, PANs have been measured in every corner of the world.[15,16]

C. SYNTHESIS AND ISOLATION

Any analytical measurement technique used routinely for monitoring PANs in the atmosphere first requires a procedure for synthesis and isolation of the compounds to accomplish calibration. The original procedures for synthesis of the PANs were gas-phase photolysis reactions which mimicked the natural atmospheric formation processes. The first routine synthetic method was developed by Edgar Stephens and colleagues at UCR.[19] Ethyl nitrite (CH_3CH_2ONO) was photolyzed in air to produce PAN by the following reactions:

$$CH_3CH_2ONO + h\nu \rightarrow CH_3CH_2O + NO \tag{19.7}$$

$$CH_3CH_2O + O_2 \rightarrow HO_2 + CH_3CHO \tag{19.8}$$

$$HO_2 + NO \rightarrow NO_2 + OH \tag{19.9}$$

$$OH + CH_3CHO \rightarrow CH_3CO + H_2O \tag{19.10}$$

$$CH_3CO + O_2 + NO_2 \rightarrow CH_3CO-OO-NO_2 \tag{19.11}$$

The PAN was isolated from the numerous side products by gas chromatography on preparatory-sized columns and collected by cryogenic trapping. The PAN was then placed in large air canisters, diluted with zero air, and stored in a cold room for future use. Safety precautions are required with this method, because explosive accidents have been reported.[15] The cause of the explosions is believed to be condensation of PANs in vacuum or pressure gauge systems. Like all nitrates, the peroxy nitrate PAN has explosive potential, and care must be taken when handling PAN on metal surfaces. The Stephens' synthetic approach illustrated by Reaction 19.7 to Reaction 19.11 was quite successful, and a number of publications on the toxicity of PAN and its chemical and physical properties resulted from the use of the scheme.[15,16]

Other gas-phase photochemical production methods which have been used for the synthesis of PAN follow a general reaction scheme in which halogens remove the aldehydic proton from the corresponding aldehyde (X = Cl, Br). Reaction 19.12 and Reaction 19.13 are the key processes leading to the formation of PAN in the troposphere when X is replaced by hydroxyl radical.

$$X_2 + h\nu \rightarrow 2X \tag{19.12}$$

$$X + CH_3CHO \rightarrow CH_3CO + HX \tag{19.13}$$

$$CH_3CO + O_2 + NO_2 \rightarrow CH_3CO-OO-NO_2 \tag{19.14}$$

Similarly, nitrate radical (NO_3) has been used to produce PAN through abstraction of the aldehydic proton, with formation of HNO_3. The NO_3 is usually generated by reaction of O_3 and NO_2 in the dark and is stored as nitrogen pentoxide, which is in equilibrium with NO_3 and NO_2.[15]

Other possible photochemical approaches use biacetyl ($CH_3CO-COCH_3$) or acetone (CH_3COCH_3) photolysis in the presence of NO_2 and oxygen.[20]

$$CH_3COCH_3 + h\nu \rightarrow CH_3COO_2 + CH_3 \qquad (19.15)$$

$$CH_3CO + O_2 \rightarrow CH_3COO_2 \qquad (19.16)$$

$$CH_3COO_2 + NO_2 \rightarrow CH_3CO-OO-NO_2 \qquad (19.17)$$

This method has been used to generate a continuous flow of PAN for instrument calibration during aircraft sampling. All of the gas-phase synthetic methods require a photochemical apparatus and chromatographic or distillation equipment to isolate the PAN for instrument calibration or laboratory studies.

The PANs can be viewed as the mixed anhydrides of the peracids or peroxyacids and HNO_3.[16] This fact led Nielson and coworkers[21] to use strong acid nitration of peracetic acid in aqueous solution to synthesize PAN, with the peracetic acid being formed by the reaction of H_2O_2 with acetic anhydride. The PAN formed was extracted from the aqueous solution into a normal alkane solvent, typically n-hexane.[15] The high-volatility alkane solvents used in these syntheses present safety hazards because of the combination of volatile hydrocarbons and active oxidants. The substitution of n-tridecane or other heavy lipid solvents in these procedures has been demonstrated to overcome these difficulties, producing high purity PAN samples with minimal contamination from the solvent.[22] The storage of PAN is then easily accomplished by freezing the solution of PAN in n-tridecane. When a PAN sample is required for calibration of a PAN monitoring device, the solution is thawed.

Avoiding potential safety problems, this approach also minimizes solvent contamination of the sample. Because PAN has a vapor pressure of approximately 30 torr at room temperature and n-tridecane has a vapor pressure of a few millitorr, PAN will distill out of the solution as a highly pure standard with minimal contamination from the solvent. For easy instrument calibrations, the liquid solutions of PAN in n-tridecane are placed in an open diffusion tube which is dropped vertically into a glass U tube capped with Teflon screw caps. The U tube is then place in an ice bath to slow the diffusion of PAN from the solution. Carrier gas, typically zero air or nitrogen, passing through the U tube at different rates enables dilution of the PAN leaving the solution to the desired concentration. A diagram of such a diffusion system is in Figure 19.3.

This method has been used by numerous researchers and has become the simplest and most cost-effective method for the synthesis of PAN, PPN, and PBN calibration standards. For the synthesis of other important PANs this method has some problems. Low aqueous solubilities of some anhydrides, as for PBzN, make the generation of the corresponding peracids problematic. For the unsaturated analogs (e.g., MPAN), polymerization reactions in the acidic solutions used for the synthesis compete with peracid formation. The photochemical procedures remain more successful for the production of these higher analogs.

D. ATMOSPHERIC SAMPLING

The PANs are quite surface active and can easily be lost on to unconditioned reactive surfaces. Figure 19.4 demonstrates the loss of PAN signal for continuous flows of PAN standard over different sampling materials. Teflon is the best material for use in sampling lines. Other materials, like Tygon, will cause loss of the PANs in the sample lines. Some materials, such as stainless steel and aluminum, are more reactive with PANs than Teflon but can be conditioned for use by exposure to high PAN levels for a few hours. For this reason, sampling lines are usually made from

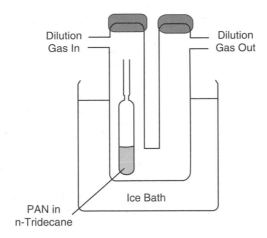

FIGURE 19.3 Apparatus for generation of PAN gas standards for instrument calibration. The system is constructed of glass with Teflon-lined screw caps for ease in inserting the diffusion tube with the PAN solution into the U tube gas dilution system.

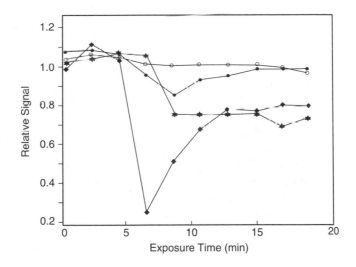

FIGURE 19.4 Loss of PAN on to different surfaces after exposure to a constant flow of standard: ●, aluminum; ○, Teflon; ◆, Tygon; ✳, copper.

high-density Teflon tubing. Glass or Teflon is also recommended for sample manifolds and inlets to minimize sampling loss of the PANs. Any heated surface will cause PAN loss due to thermal decomposition (Reaction 19.4), so sampling lines and inlets should be kept at temperatures below 25°C if possible.

E. ANALYSIS OF THE PANs

1. Detection Methods

As noted earlier, PAN will decompose rapidly in basic conditions. In the early years, instrument calibration was usually accomplished by using basic hydrolysis in a pH 12 solution to yield acetate and nitrite, followed by quantitation of the nitrite by colorimetric analysis to determine the original PAN concentration. The analysis of nitrite and acetate by basic hydrolysis can now be accomplished with ion chromatography. Wet chemical methods are not routinely used for the detection of PANs, principally because of a lack of selectivity and sensitivity. Nevertheless, these methods have been used in the laboratory as a quick way to verify standard concentrations.

The first laboratory measurements of PAN were made by using long-path infrared absorption techniques, which worked reasonably well at the higher concentrations found in Los Angeles during strong smog episodes.[23,24] The infrared instruments of the early 1960s clearly were not capable of routine monitoring of PAN, because the detection limits for PAN in the infrared required a kilometer path length to obtain sufficient sensitivity for ambient PAN levels. The general application of long-path infrared spectroscopy was limited until the development of Fourier transform techniques in the early 1970s. The two infrared bands used most commonly for the quantitation of the PANs are the NO scissors band at 793.9 cm^{-1} and the C–O stretch at 1165 cm^{-1}.[15] These bands were chosen because of the minimal interference from water and CO_2 in this region. A long-path optical cell, consisting of an eight-mirror gold-plated multiple-reflection system, can now achieve path lengths of up to 2 km,[15] resulting in detection limits as low as 3 ppbv. The long-path cells can be used either in a closed configuration for instrument calibration or laboratory studies or in an open path for ambient *in-situ* monitoring. Infrared techniques are considered to be the primary standard for calibration of PAN instruments and evaluation of the purity of standards. These techniques are not suitable for chromatography detectors because of the long paths required and the difficulties in measuring levels lower than parts per billion.

Mass spectrometry of PAN was first reported in electron impact mode in 1969 and in chemical ionization mode in 1976.[15,16] Initial work was done in positive-ion mode, and consequently, the molecular ions observed were at very low intensities. Recently, negative-ion chemical ionization mass spectrometry (NICI/MS) has been used for the ambient detection of PAN[25] and as a gas chromatographic detector for PAN, PPN, and MPAN.[26,27] Because of its very large electron capture cross section, the NICI/MS sensitivity for PAN is quite high. The fragment ions: CH_3COO^-, with a mass-to-charge ratio (m/z) of 59, NO_3^-, with $m/z = 62$, $CH_3CO_3^-$, with $m/z = 75$, and NO_2^-, with $m/z = 46$ have been identified. Among the four fragment ions, NO_3^- gives the best signal-to-noise ratio, with a detection limit for PAN of 15 pptv and an accuracy of approximately 20%. However, the high costs of these instruments and the need for skilled operators have prevented their wide use as chromatographic detectors.

By far the most widely used detector for the analysis of the PANs is the electron capture detector (ECD) invented by James Lovelock, Ellis Darley, and coworkers in 1963.[28] The ECD uses a nickel-63 source (usually in the form of a foil) to produce a standing current of electrons. Nitrogen (or another carrier gas) flowing into the detector slows the electrons down to give a standing current. The electron current is monitored by applying a pulsed voltage across two electrodes in the detector. When electrophilic compounds enter the detector with the carrier gas, these can "capture" electrons and affect the current. The ECD standing currents are affected by any strong electron capture agent in the gas stream. Oxygen, a strong electron capture agent, is a major interference with PAN analysis by ECD and is the major source of background caused by tailing of its very large peak. The ECD was invented to detect oxygen in planetary atmospheres. Other common atmospheric interferences include freons, nitrous oxide, carbon dioxide, and water vapor.

One of the drawbacks of the classical ECD for routine analyses, is the radioactive source it contains. A number of conditions must be met if the instrument, with its source, is to be shipped for field study or used in aircraft or mobile-source applications. Because of these problems and the oxygen and freon interferences with the ECD method, other detection schemes have been developed for PAN analysis to improve response time and avoid potentially hazardous materials.

Another method for the detection of the PANs involves thermal decomposition to NO_2 and direct measurement with an NO_x chemiluminescence monitor.[16] This instrument relies on the reaction of NO with O_3 to produce excited NO_2. The NO_2 emission is a broadband chemiluminescence starting at about 600 nm and peaking at $\lambda = 1.27\ \mu m$ (1270 nm). A red-sensitive photomultiplier is required to monitor the emission. To monitor NO_2, the sample is first passed over a hot catalyst which converts the NO_2 to NO. A number of atmospheric nitrates, including HNO_3 and the PANs, are also decomposed by the hot catalyst to NO. The main drawback of this method for PAN detection is its sensitivity, which is limited to 1 to 2 ppbv because of the lack of sensitivity of commercial phototubes at the emission wavelength. This method is useful for calibration of other detectors with high purity standards under controlled laboratory conditions.

NO_2 and the PANs have been shown to react with luminol (5-amino-2,3-dihydro-1,4-phthalazinedione) in a gas–liquid reaction to produce intense chemiluminescence emission with a maximum at 425 nm. Air is passed over a glass fiber wick which is kept wet with a continuously flowing solution of luminol. The chemiluminescent reaction, which takes place on the wet wick surface, is monitored by using a standard photomultiplier detector. This method has an inherent advantage for detection of the PANs in that the emission at 425 nm is easily monitored with high sensitivity by using commercially available photomultiplier tubes. Detection limits with this system are on the order of 5 to 10 pptv. This application has been used as a chromatography detector,[29–31] and lack of interferences greatly shortens analysis times vs. ECD detection. Because of its small size and rapid response, this detector has advantages for applications such as aircraft measurements.

Because the luminol detection system is also sensitive to NO_2, chemical amplification methods have been attempted to further decrease detection limits for PANs below the pptv range for trace-level measurements.[32] With this approach, the PANs are thermally decomposed to NO_2 in the presence of large amounts of NO (6 ppm) and CO (8%). Thermal decomposition of the PANs yields peroxy radicals which initiate a free-radical chain oxidation of NO to NO_2, producing several NO_2 molecules (approximately 180 (20) for each PAN decomposed. This technique has been used as a gas chromatography detector to achieve ultratrace detection limits without sample preconcentration. The detector exhibits a slightly nonlinear response relative to conventional ECD, attributed to the nonlinear response of the luminol reaction in the presence of NO at 6 ppm.

2. Chromatographic Separation

Gas chromatography with electron capture detection (GC/ECD) is the method used most often for detection of PAN and its analogs. A nonpolar column is preferred, because the PANs are not very water soluble but are readily soluble in nonpolar media. As noted in Section I.B, this property can be very useful for isolating and storing calibration standards. For analytical purposes, a nonpolar Carbowax 400-packed column or, more recently, a DB1 capillary column is quite adequate for separation of the PANs from interfering species. The column and injector should be kept at room temperature, because the PANs are quite unstable thermally and rapidly equilibrate to the peroxyacyl radicals and NO_2 (Reaction 19.4) at temperatures significantly above room temperature. The detector temperature can be elevated slightly but should not exceed 100°C, or loss of PAN will occur.[33] Thermal decomposition and loss of some PANs on the column may occur even at room temperature during the time required for analysis with conventional GC/ECD methods. In general, the longer the RT for the PANs, the greater will be their destruction on the column and the lower the sensitivity which can be achieved.

Ultra high purity nitrogen is most often used as the carrier gas for GC/ECD analyses. Other appropriate ECD gases, such as helium with a methane-argon make-up gas, can also be used.[33] Because the ECD is very sensitive to oxygen impurities, all trace levels of oxygen in the carrier gas should be removed, or background levels will be extremely high. An oxygen scrubber or oxy-trap system placed in the carrier gas stream at the supply tank will achieve this. Commercial traps which use pyrophoric agents in a contained system are most suitable and effective. The oxygen removal traps are usually good for about two or three tanks of gas, depending on the oxygen impurity levels; the traps should be changed when these are exhausted.

A simple PAN analysis system, or "PANalyzer," then uses a small, portable GC with ECD and an integrator set up for continuous analysis. An automated sampling valve is used to inject the air sample, contained in a continuously filled sample loop, on to the column. For most applications a sample loop with volumes of 1 to 5 cm^3 is typical, with larger volumes used in environments where low levels of PAN are anticipated. The smaller volumes are used downwind of polluted urban environments where levels are expected to be higher. For trace analysis in remote areas, cryogenic preconcentration of the samples can be used in combination with larger sampling loops, up to 280 cm^3 in size, to improve detection limits for the PANs. With cryogenic sample concentration and larger sample loops, detection sensitivities of 0.02 to 10 pptv have been demonstrated for the PANs and alkyl nitrates.[34,35]

The typical instrumental conditions used for GC/ECD analysis of the PANs are given in Table 19.2. With a packed column, a carrier gas flow rate of 30 cm^3 min^{-1} is recommended, and a 0.25 in. × 6-ft (6.3 mm × 2-m) column made of Teflon or glass is sufficient to separate the PANs from the interfering oxygen and freons in the air. Typically, RT = 3 to 4 min for PAN under these conditions. A 33-ft (10-m) capillary column with a carrier gas flow rate of 30 cm^3 min^{-1} typically yields RT = 4 to 6 min for PAN.

Figure 19.5 shows a typical chromatogram for PAN analysis with a packed column and ECD. This chromatogram, obtained in Mexico City, shows the high levels of PANs observed in the air during a photochemical episode recorded in February 1997.[36] The large peak at RT = 0.454 min is due to oxygen, and the small peak at RT = 1.350 min is one of the freons. The water peak is usually observed as a negative response at around 15 min. Peaks corresponding to methyl nitrate (CH$_3$ONO$_2$), PAN, PPN, and PBN are superimposed on the background generated by the tailing of the large oxygen peak. It takes at least 15 min for the background levels from the oxygen peak to decrease sufficiently to allow the next sample injection. The presence of this large interfering peak limits the achievable detection limits and the analysis times required for detection of the PANs by GC/ECD. Shorter analysis times have been achieved by using a combination of two short megabore capillary columns with different polarities. Separating interfering peaks, including oxygen, on a precolumn (Rtx-1) and venting prior to separation of the PANs on the main column (Rtx-1701) can achieve analysis times of 4 to 5 min.[37]

TABLE 19.2
Typical Conditions Used for Gas Chromatographic Analysis of PANs with ECD

Column Type and Size	Carrier Gas	Flow Rate (cc min^{-1})	Injector Temperature (°C)	Column Temperature (°C)	Detector Temperature (°C)
Packed Column					
Carbowax 400, 6.3 mm × 2 m Teflon/glass	Nitrogen, UHP[a]	25–60	25–30	25–30	45–100
Capillary Column					
DB-1, 0.53 mm × 10 m, 2.65 μm film	Nitrogen, UHP[a], or 10% CH$_4$/Ar/He[a]	5–35	25–35	35	45–100

[a] Ultra high purity gases are used with oxygen traps to remove trace impurities of oxygen.

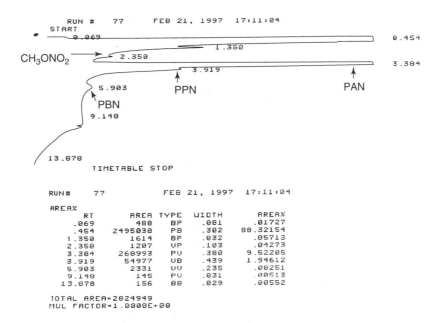

FIGURE 19.5 Typical gas chromatogram for analysis of the PANs obtained in Mexico City in February 1997 with a packed column and ECD. Peaks at RT = 0.454 min and 1.350 min are for oxygen and a freon, respectively. Other peaks shown correspond to methyl nitrate (RT = 2.350 min), PAN (RT = 3.384 min), PPN (RT = 3.919 min), and PBN (RT = 5.903 min).

Because the ECD is sensitive to many species containing electronegative atoms, peaks which coelute with the PANs are sometimes reported when the oven temperature is maintained at the standard 30°C. Reducing the oven temperature to 10°C with liquid nitrogen cryogenic cooling improves separation of the various freons, organic nitrates, and PANs, achieving baseline resolution of otherwise coeluting peaks with a standard DB1 capillary column.[33] As anticipated, the sensitivity for PAN detection is also increased by an average of 55% because of the compound's enhanced stability at the lower temperature and reduced loss on the column during the shorter analysis time. Lower oven temperatures (e.g., −50°C) achieve extremely high sensitivities for PAN, but the ECD response is nonlinear.[33]

Changes in ambient relative humidity can cause RT for PAN to change and can also cause peak tailing. This is especially noticeable with packed columns. These effects are likely due to water–column interactions. As the ambient water increases, it can coat the column and make it less active toward PAN. Because the PANs have extremely low water solubilities, any wetting of the nonpolar liquid support will change the PAN-column adsorption properties. PAN retention times become shorter and peaks become sharper as the relative humidity rises. Use of an integrator for PAN determinations and use of peak areas instead of peak heights for quantitation during routine atmospheric analysis circumvents this problem. Air dryers have been used to avoid this effect, but loss of PAN on the dryer can occur.

Because luminol detection for the PANs does not suffer from the oxygen interference which affects ECD, separation of NO_2 and the PANs can be accomplished in a much shorter analysis time with luminol detection and the same nonpolar column materials. Figure 19.6 shows three replicate analyses of room air, every 30 sec, with fast gas chromatography and luminol chemiluminescence detection (GC/LCD), a capillary column, and a 5-cm^3 sample loop. The column used in Figure 19.6 is a 10 m × 0.53 mm i.d. DB1 capillary column maintained at room temperature (25°C). A carrier gas mixture of 5% O_2 in helium was used at a flow of 60 cm^3 min^{-1}. The arrows in Figure 19.6 indicate the NO_2 peak (first elution) at a RT of approximately 6 sec and

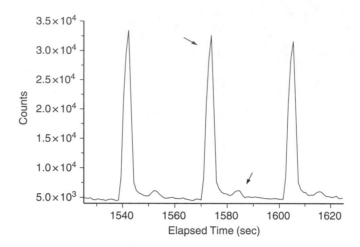

FIGURE 19.6 Replicate room air analyses obtained every 30 sec with fast gas chromatography with luminol detection. Arrows point to NO_2 (RT = 6 sec) and PAN (RT = 16 sec) peaks.

the PAN peak (second elution) at a RT of approximately 16 sec. This response corresponds to concentrations of about 0.3 ppb for NO_2 and about 40 ppt for PAN.

Comparison studies between luminol detection and ECD for PAN analysis have shown that either method can yield accurate, reliable, sensitive measurements of ambient PAN concentrations, with typical sensitivities on the order of 10 ppt.[31,38] The advantages of the luminol detector over GC/ECD lie in the faster analysis times achievable. This makes luminol an attractive alternative for aircraft measurements, where time resolution translates into spatial resolution.[39] The luminol detector also enables the simultaneous measurement of PANs and NO_2 to monitor decomposition and formation processes in the atmosphere.[40,45]

F. ATMOSPHERIC CONCENTRATIONS

In recent years, PAN levels have dropped significantly in urban U.S. regions like the Los Angeles air basin and surrounding air shed. Table 19.3 lists some typical PAN concentrations determined in U.S. cities during recent field studies. Maximum observed concentrations are rarely above 5 ppb, a tenfold decrease compared to the 1960s and 1970s. In other areas of the world, PANs can still be produced in large quantities if NO_x levels and the reactive hydrocarbon species leading to PAN and O_3 formation are not controlled. In Mexico City in the spring of 1997, total PANs were found to exceed 40 ppb.[36]

Figure 19.7 shows results of simultaneous 1-min analyses of PAN and NO_2 by GC/LCD in Centerton, New Jersey, as part of the Northeast Oxidant and Particulate Study (NEOPS) during

TABLE 19.3
Typical PAN Levels (ppbv) Observed in Various U.S. Cities during Some Recent Field Studies

City	Date	Maximum	Median	Reference
Centerton, NJ	7/30–8/11, 1999	2.23	0.675	41
Houston, TX	8/8–9/15, 2000	5.38	0.299	42
Phoenix, AZ	5/16–6/11, 1998	1.09	0.094	43
Phoenix, AZ	6/15–6/28, 2001	1.91	0.075	44
Salt Lake City, UT	9/30–10/28, 2000	4.35	0.613	45

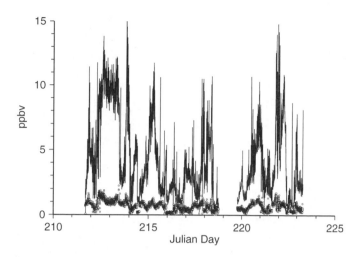

FIGURE 19.7 PAN (∘) and NO₂ (—) concentrations (ppbv) determined by GC/LCD at Centerton, New Jersey, on July 30–August 11, 1999.

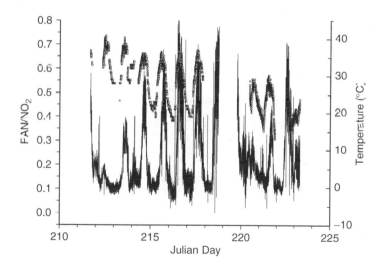

FIGURE 19.8 The ratio of PAN/NO₂ concentrations (—) and ambient temperature (∘) at Centerton, New Jersey, on July 30–August 11, 1999.

the summer of 1999.[41] These simultaneous analyses enabled the determination of PAN/NO₂ ratios shown in Figure 19.8. This ratio is an indication of PAN formation and thermal decomposition rates according to Reaction 19.4. The ratio shows a strong diurnal variation, with higher values during the day, when PAN is formed by the photooxidation of atmospheric organic species, and lower values at night, when the thermal loss of PAN leads to NO₂ formation.

II. THE ORGANIC NITRATES

A. PHYSICAL AND CHEMICAL PROPERTIES

The large group of atmospheric species known as organic nitrates, of which the PANs are members, also contains alkyl nitrates, aryl nitrates, alkyl dinitrates, peroxy nitrates, multifunctional organic nitrates (hydroxynitrates, ketonitrates, etc.), and the sister species — the organic nitrites. All of these species are formed primarily as secondary air pollutants in photochemical smog.[14,16] Though they are trace components in the atmosphere, they are also considered as NO_y species. The formation and destruction of these components can influence tropospheric O_3 formation.

1. Alkyl Nitrates

Alkyl nitrates have the chemical formula $R-ONO_2$. The key reactions leading to their formation involve RO_2 and NO. Alkyl nitrates are formed as minor products in these reactions, though the larger alkyl peroxy radicals yield larger amounts of alkyl nitrates. The yield of formation increases continually with increasing alkyl chain length. As the R group increases in size, the probability of organic nitrate formation increases. Unlike the PANs, the alkyl nitrates are thermally stable, the major atmospheric loss processes being photolysis and reaction with hydroxyl radical.[46] The alkyl nitratres are fairly slow to react with hydroxyl radical, as the electron-withdrawing nitrate group decreases the reactivity for H atom abstraction reactions. Alkyl nitrates can act as local sinks for NO_x, with typical lifetimes of days to weeks.

The atmospheric removal processes also increase with increasing alkyl chain length.[47] The simplest of the organic nitrates, methyl nitrate (CH_3ONO_2), is the slowest of the group to react with hydroxyl radical. It is also formed at a slower rate from the reaction of NO with methylperoxy radical, CH_3O_2.[14] The next analog, ethyl nitrate ($CH_3CH_2ONO_2$), is more reactive than methyl nitrate with hydroxyl radical, and its formation, by the reaction of NO with ethylperoxy radical ($CH_3CH_2O_2$), is faster than the formation of methyl nitrate. As the R group increases in size, the loss rate, as determined by reactivity with hydroxyl radical, and the formation rate from the $RO_2 + NO$ reactions tend to balance each other, leading to the stable concentrations of these species observed in urban and rural air.[47–50] The alkyl nitrates are typically seen at levels in the low ppt in remote atmospheres and in the high ppt in urban atmospheres.[14]

2. Alkyl Nitrites, Peroxy Nitrates, and Other Organic Nitrates

The alkyl nitrites have the chemical formula $R-ONO$. They are associated with the production of the alkyl nitrates and the PANs. However, the alkyl nitrites absorb light quite strongly in the actinic region and are very rapidly photolyzed to form RO and NO. These are not generally seen in the atmosphere at significant levels.

The peroxy nitrates are associated with the PANs, as these are formed by the addition reaction of RO_2 and NO_2 under high concentrations of NO_2:

$$RO_2 + NO_2 \leftrightarrow RO_2NO_2 \tag{19.18}$$

Because of their short lifetimes at room temperature, the peroxy nitrates have been assumed not to act as key storage modes for peroxy radicals and NO_2 in the lower atmosphere.[51] At middle latitudes in the wintertime these may have lifetimes that approach days. Further, like the PANs these might be reformed to actively transport NO_2 and peroxy radicals over long distances, depending upon the NO, hydroperoxy radical, and NO_2 concentrations. With the possible exception of very cold air masses, these compounds are typically not present in significant concentrations in the troposphere because of rapid thermal decomposition to form NO_2 and RO_2. At room temperature they would be lost in sampling lines or during analysis.

Because of these problems, neither the alkyl nitrites nor the peroxy nitrates will be dealt with here. We will focus here on the alkyl nitrates and their analysis. Other multifunctional organic nitrates (e.g., hydroxynitrates, ketonitrates) will be included with the alkyl nitrates, because these are determined with the same analytical methods.

B. Synthesis and Isolation

A few of the simple alkyl nitrates (e.g., isopropyl nitrate, *n*-propyl nitrate, isobutyl nitrate) are commercially available. Most of the other alkyl nitrates require simple synthesis. Alkyl nitrates are normally synthesized in gram or larger quantities by the reaction of an alcohol with concentrated HNO_3 in the presence of urea.[52] Large amounts of the alkyl nitrates must be handled with care. Like other nitro compounds, some can be explosive. As an alternative to the large-scale synthetic methods, several microscale techniques have been used to avoid the safety problems associated with handling large amounts of product. Reference solutions in the parts per million can be easily obtained by the two-phase esterification of a dilute solution of the corresponding alcohol in dichloromethane (4 to 10 mmol l^{-1}) with a mixture (1:1) of concentrated HNO_3 and H_2SO_4.[52,53] The alkyl dinitrates can be prepared in this manner by using the dialcohols, and chiral alkyl nitrates can be synthesized by using chiral alcohols.[54] This procedure is not applicable for the C_1–C_3 alkyl nitrates when ECD is used, because the dichloromethane solvent coelutes with the alkyl nitrates.

Alkyl nitrates are also formed by the reaction of $AgNO_3$ with alkyl bromides in an acetonitrile solvent.[52] The drawback of this exchange reaction is that in most cases the yield is only about 60 to 80%. Dichloromethane extraction with solvent removal under reduced pressure has been employed to isolate the alkyl nitrates as pure reference compounds, as well as to synthesize the hydroxyalkyl nitrates from the hydroxyhalides.[55] Direct nitration of the alkanes is achieved with 12% HNO_3 in the presence of solid copper.[53] This method is especially useful in preparing mixtures of alkyl nitrates from mixtures of the starting alkanes.

C. Atmospheric Sampling

Few studies have completely determined alkyl nitrates in the field, mainly because of the complexity of the analysis and the time required for sample pretreatment. Typical reported atmospheric concentrations for some alkyl nitrates are in Table 19.4. The values for the higher alkyl nitrates ($C \geq 4$) are the sums for all isomers. As the very low values indicate, sample preconcentration is necessary to achieve low detection limits with ECD or mass spectrometry. Sample concentration most commonly employs solid sorbent materials to trap the alkyl nitrates. Pumping of large volumes of air (100 to 1000 m^3) through charcoal or silica gel is followed by

---

TABLE 19.4
Typical Concentration Levels (pptv) Observed for Some Alkyl Nitrates in Two U.S. Cities, Two German Cities, and One Remote Marine Site

Species	Las Vegas (56)	Salt Lake City (56)	Juelich (52)	Ulm (52)	Hawaii (52)
Methyl nitrate	—[a]	—	109	—	—
Ethyl nitrate	—	—	47.7	—	—
2-Propyl nitrate	—	—	27.4	13.1	3.2
1-Propyl nitrate	—	—	8.5	3.5	1.1
Butyl nitrates	—	—	27.3	13.3	2.7
Pentyl nitrates	—	—	42.0	9.3	1.1
Hexyl nitrates	0.25	0.91	—	0.39	—
C_7 nitrates	0.53	3.03	—	—	—
C_8 nitrates	0.27	1.91	—	—	—
C_9 nitrates	0.49	2.09	—	—	—
C_{10} nitrates	1.36	3.07	—	—	—

[a] No value given.

solvent extraction to remove the alkyl nitrates from the solid material.[50-59] If Tenax is used as a sorbent, the alkyl nitrates can be released by thermal desorption, eliminating the need for solvent extraction.[58] This technique is faster and less labor intensive than with other solid adsorbents. Alternately, the sample can be concentrated by cryogenic methods. One cryogenic method makes use of a 2-mm diameter quartz tube (150 mm long) filled with 0.25-mm glass beads and immersed in liquid nitrogen.[50] After the sampling period, the length of which depends on the ambient concentrations of alkyl nitrates to be determined, the tube is heated to 70°C for 1 min, and the sample is injected directly onto the gas chromatograph. This method has been used successfully for the determination of C_1–C_8 alkyl nitrates.

D. Analysis of the Alkyl Nitrates

The major difficulty with the analysis of alkyl nitrates is the great complexity of analytes because of the large number of possible analogs. As an example of this complexity, Table 19.5 gives the structures of and numbering conventions for the C_1–C_6 alkyl nitrate isomers.[52] With increasing numbers of carbon atoms in the R group, the number of possible isomers increases exponentially. Furthermore, nitrates with more than one functional group (e.g., dinitrates, hydroxynitrates, ketonitrates) are detected by the same methods. If these multifunctional analogs are included, 27 analogs are possible for the C_4 organic nitrate alone.[55] Methods of analysis for the alkyl nitrates must be able to handle a very large number of distinct species with very similar chemical properties. Because their concentrations in the atmosphere are in the low parts per trillion by volume (ppt), the samples typically must be preconcentrated before analysis, either by adsorption on solid sorbents (e.g., charcoal, Tenax) or by cryogenic trapping.[52-60]

1. Detection Methods

The alkyl nitrates are quite readily detected by the methods used for the PANs, as both groups have high electron capture cross sections. The most common detection method for the alkyl nitrates is ECD, followed by mass spectrometry. Intercomparison between these two methods showed them to yield similar results.[61] The major disadvantage of ECD for the alkyl nitrates is the interference from coeluting halocarbons.[58] An ECD is more sensitive to chlorinated compounds than to alkyl nitrates, therefore any coeluting halocarbon will swamp the alkyl nitrate signal. This is particularly a problem with the methyl and ethyl nitrates.[62] This interference is eliminated by using mass spectrometry detection tuned to the mass fragment with $m/z = 46$, which arises from the NO_2 fragment and is very specific to the alkyl nitrates.[48] With mass spectrometry as a detector, the presence of other fragments in the spectrum can help to identify the structures of the eluting species in the complex mixture of analytes.

The alkyl nitrates also can be detected by chemiluminecent reactions similar to those for the PANs. The alkyl nitrates are converted to NO by hot catalysts, and the NO is detected by O_3 chemiluminescence as described in Section I.E.1. The detector can be calibrated with an NO standard, which is more readily available, more reliable, and more easily transported than an alkyl nitrate standard.[50,63] However, O_3 chemiluminescence is less sensitive, by one or two orders of magnitude, than ECD.[58] The alkyl nitrates can also be converted to NO_2 by thermal decomposition with a postcolumn pyrolyzer consisting of a heated quartz tube. The NO_2 can then be determined by luminol chemiluminescence.[64] The sensitivity of this type of detector is comparable to that of the ECD. Both types of chemiluminescence detectors respond only to oxidized nitrogen species, therefore the halocarbon interference is eliminated.

2. Chromatographic Separation

The chromatographic separation of the alkyl nitrates, like that of the PANs, is accomplished with gas chromatograph capillary columns with a nonpolar stationary phase. Because the alkyl nitrates

TABLE 19.5
Structure and Numbering Conventions of C_1-C_6 Alkyl Nitrate Isomers[52]

Structure	Name
CH_3ONO_2	Methyl nitrate
$CH_3CH_2ONO_2$	Ethyl nitrate
$CH_3CH_2CH_2ONO_2$	1-Propyl nitrate
$CH_3CH(ONO_2)CH_3$	2-Propyl nitrate
$CH_3CH_2CH_2CH_2ONO_2$	1-Butyl nitrate
$CH_3CH_2CH(ONO_2)CH_3$	2-Butyl nitrate (secondary butyl nitrate)
$CH_3CH(CH_3)CH_2ONO_2$	2-Methyl-1-propyl nitrate (isobutyl nitrate)
$CH_3C(CH_3)_2ONO_2$	1,1-Dimethyl-1-ethyl nitrate (tertiary butyl nitrate)
$CH_3CH_2CH_2CH_2CH_2ONO_2$	1-Pentyl nitrate
$CH_3CH_2CH_2CH(ONO_2)CH_3$	2-Pentyl nitrate
$CH_3CH_2CH(ONO_2)CH_2CH_3$	3-Pentyl nitrate
$CH_3CH_2CH(CH_3)CH_2ONO_2$	2-Methyl-1-butyl nitrate
$CH_3CH_2C(CH_3)(ONO_2)CH_3$	2-Methyl 2-butyl nitrate
$CH_3CH(CH_3)CH(ONO_2)CH_3$	2-Methyl-3-butyl nitrate
$CH_3CH(CH_3)CH_2CH_2ONO_2$	2-Methyl-4-butyl nitrate
$CH_3C(CH_3)_2CH_2ONO_2$	2,2-Dimethyl-1-propyl nitrate
$CH_3CH_2CH_2CH_2CH_2CH_2ONO_2$	1-Hexyl nitrate
$CH_3CH_2CH_2CH_2CH(ONO_2)CH_3$	2-Hexyl nitrate
$CH_3CH_2CH_2CH(ONO_2)CH_2CH_3$	3-Hexyl nitrate
$CH_3CH_2CH_2CH(CH_3)CH_2ONO_2$	2-Methyl-1-pentyl nitrate
$CH_3CH_2CH_2C(CH_3)(ONO_2)CH_3$	2-Methyl-2-pentyl nitrate
$CH_3CH(CH_3)CH(ONO_2)CH_2CH_3$	2-Methyl-3-pentyl nitrate
$CH_3CH(CH_3)CH_2CH(ONO_2)CH_3$	2-Methyl-4-pentyl nitrate
$CH_3CH(CH_3)CH_2CH_2CH_2ONO_2$	2-Methyl-5-pentyl nitrate
$CH_3CH_2CH(CH_3)CH_2CH_2ONO_2$	3-Methyl-1-pentyl nitrate
$CH_3CH_2CH(CH_3)CH(ONO_2)CH_3$	3-Methyl-2-pentyl nitrate
$CH_3CH_2C(CH_3)(ONO_2)CH_2CH_3$	3-Methyl-3-pentyl nitrate
$CH_3CH_2C(CH_3)_2CH_2ONO_2$	2,2-Dimethyl-1-butyl nitrate
$CH_3C(CH_3)_2CH(ONO_2)CH_3$	2,2-Dimethyl-3-butyl nitrate
$CH_3C(CH_3)_2CH_2CH_2ONO_2$	2,2-Dimethyl-4-butyl nitrate
$CH_3CH(CH_3)CH(CH_3)CH_2ONO_2$	2,3-Dimethyl-1-butyl nitrate
$CH_3CH(CH_3)C(CH_3)(ONO_2)CH_3$	2,3-Dimethyl-2-butyl nitrate
$CH_3CH_2CH(CH_3CH_2)CH_2ONO_2$	2-Ethyl-1-butyl nitrate

are not as thermally unstable as the PANs, the separation is typically accomplished by using temperature programming techniques. A number of columns have been reported as successful in separating the alkyl and bifunctional nitrates. The columns and the temperature programming conditions reported for each group are listed in Table 19.6 and Table 19.7. Injection port temperatures of up to 190°C can be used, with optimum values of 150 to 190°C, while detector temperatures for the ECD are typically 220 to 260°C.[52] Several carrier gasses have been used. Hydrogen has been reported to improve separation for the alkyl nitrates.[55]

A summary of the retention indices of more than 80 alkyl nitrates has been reported.[66] In general, the separation of the alkyl nitrates follows the order of boiling points in Table 19.8. The 1-*n*-alkyl nitrate is always the last member of each group to elute. Thus, it can be used as a window marker or retention index marker compound.[52] The most abundant alkyl nitrates found in atmospheric samples are typically the secondary nitrates. For this reason, the 2-C_5 and 2-C_{10} alkyl nitrates have also been suggested as marker compounds.[66]

TABLE 19.6
Columns Reported for Gas chromatographic Analysis of Alkyl Nitrates

Type	Column	Manufacturer	Length (m)	I.D. (mm)	Film Thickness (μm)	Reference
C_1–C_5	HP-1	Hewlett-Packard	25	0.32	1.00	65
C_1–C_{14}	SIL5	Chrompack	105	0.32	0.25	52
C_6–C_{17}	HP-5MS	Hewlett-Packard	5	0.25	0.25	53
C_6–C_{17}	CP-SIL5	Chrompack	10	0.25	0.25	53
Alkyl nitrates	CP-SIL2	Chrompack	50	0.32	0.25	55
	DB-1	J&W	60	0.25	0.25	
Aryl nitrates	SIL13	Chrompack	50	0.34	1.20	52
Dinitrates	DB-5	J&W	60	0.32	0.25	55
Hydroxynitrates	MN1701	Macherey-Nagel	50	0.32	0.25	55
	CP-SIL88	Chrompack	50	0.22	0.20	

TABLE 19.7
Column Temperature Programming Schemes Used for Gas Chromatographic Analysis of Alkyl Nitrates with Columns Listed in Table 19.6

Type	Starting Temperature (°C)	Time (min)	Heating Rate (°C min^{-1})	Final Temperature (°C)	Time at Final Temperature (min)	Reference
C_1–C_5	30	35	70	100	10	65
C_1–C_{14}	40	3	4	180	3	52
C_6–C_{17}	40	1	20, 2[a]	140, 240[b]	5	53
C_6–C_{17}	40	1	6	240	5	53
Alkyl nitrates	40	3	5	200	10	55
Aryl nitrates	40	10	4	260	3	52
Dinitrates	40	3	3	200	15	55
Hydroxynitrates	40	5	3	200	15	55

[a] Rate to intermediate temperature, followed by rate to final temperature.
[b] Intermediate temperature, followed by final temperature.

Because the conditions for the separation of the alkyl nitrates are very similar to those used for the PANs, both groups can be determined in a single analysis. In a typical temperature program for analysis of the C_1–C_5 alkyl nitrates and the PANs on a nonpolar HP1 capillary column, one would hold the column at 30°C for 20 to 35 min to allow the PANs to elute first, then heat the column to 100°C at a rate of 70°C min^{-1} to allow the alkyl nitrates to be removed and to recondition the column.[58] The ECD must be held at 45°C to prevent thermal loss of the PANs. The detector temperature can be increased periodically to 150°C for reconditioning.

Because of the vast complexity of the alkyl nitrate species to be determined, a group separation by normal-phase liquid chromatography has been used to reduce this complexity and help remove interferences due to coeluting peaks in the final analysis by GC/ECD or gas chromatography with mass spectrometry detection.[57,67] Fractions from adsorbent cartridges are extracted with pentane–acetone and concentrated to a volume of 200 μl for group separation on gravity-run silica gel columns. The alkyl nitrates and dinitrates elute from the column with hexane as solvent, while more polar hydroxynitrates and ketonitrates elute with dichloromethane, and the aryl nitrates elute with a hexane–dichloromethane (1:1) mixture. Adapting this group separation technique

TABLE 19.8
Boiling Points of Some Alkyl Nitrates[52]

Name	Boiling Point (°C)
Methyl nitrate	64.6
Ethyl nitrate	87.2
2-Propyl nitrate	101.7
1-Propyl nitrate	110
2-Methyl-1-propyl nitrate	123
2-Butyl nitrate	124
2-Pentyl nitrate	144
3-Methyl-1-butyl nitrate	147
1-Pentyl nitrate	157
1-Hexyl nitrate	171
1-Heptyl nitrate	215
1-Octyl nitrate	240
Cyclohexyl nitrate	181
Benzyl nitrate	101
p-Methyl-benzyl nitrate	115
2-Nitroxy ethanol	75
3-Nitroxy ethanol	103

to normal-phase high-performance liquid chromatography (HPLC) required columns with stationary phases having properties similar to those of the alkyl nitrates. Because stationary phases with nitro-oxy groups are not commercially available, surface-modified silica columns have been developed for this application. By using hexane followed by hexane–dichloromethane as eluents, a nitrated silica ester stationary phase synthesized by reacting a commercial polyol phase (Serva, Polyol Si 100, 5 m) with a mixture of H_2SO_4 and HNO_3 enabled complete separation of alkyl mononitrates, dinitrates, and hydroxynitrates.[67] However, the dinitrates and ketonitrates coeluted. A β-cyclodextrin silica stationary phase which has also been synthesized improves separation between the alkyl mononitrates, dinitrates, and aryl nitrates, with pentane as eluent.[68]

A number of the alkyl nitrates can exist as diastereomeric or enantiomeric forms. The ability to separate the diastereomers of the alkyl nitrates has potential for identifying biogenic or anthropogenic sources of the hydrocarbon precursors, as well as for investigating biogenic decomposition processes. High-resolution gas chromatography has been used to achieve the enantioselective separation of a synthetic mixture of alkyl nitrates with a LIPODEX-D capillary column (heptakis[3-O-acetyl,- 2-6-di-O-pentyl]-β-cyclodextrin) which was developed as a chiral selector for γ-lactones.[66] The complexity of mixtures in atmospheric samples makes initial separation of interfering species necessary. Some alkyl nitrates found in air samples partly coeluted on the LIPODEX-D column. The application of two-dimensional gas chromatography, in which a polar achiral stationary phase like polyalkylene glycol was coupled with a LIPODEX-D column, was required to achieve chiral separation of alkyl nitrates in atmospheric samples.[66]

III. THE ORGANIC PEROXIDES AND PERACIDS

A. PHYSICAL AND CHEMICAL PROPERTIES

1. Formation in the Atmosphere

Organic peroxides have the chemical formula R–OOH. When R is the acyl group (RCO–), the resulting family of compounds is the organic peracids, which have the formula RCO–OOH.

The organic peracids and peroxides are produced by the same chemistry which forms the PANs and the alkyl nitrates; the alkyl nitrates and PANs are formed from peroxyacyl radicals (RCO_3) and peroxy radicals (RO_2) under high-NO_x conditions, whereas the peracids and peroxides are formed under low-NO conditions through reactions with hydroperoxy radical:

$$RCO-OO + HO_2 \rightarrow RCO-OOH + O_2 \tag{19.19}$$

$$RO_2 + HO_2 \rightarrow RO_2H + O_2 \tag{19.20}$$

Note that H_2O_2 is formed by Reaction 19.20 when R is H.

As the available NO is converted to NO_2 by reaction with peroxy radicals (RO_2 and HO_2) and the resulting NO_2 is tied up either as PANs through reaction with peroxyacyl radical (RCO_3) or HNO_3 through reaction with hydroxyl radical, the atmospheric NO levels drop significantly, and Reaction 19.19 and Reaction 19.20 become more important. The result is the formation of the organic peroxides and peracids.[14] Organic peroxides and peracids are known to be produced along with H_2O_2 in rural air masses, where low NO levels allow the buildup of hydroperoxy radical.

2. Structure and Properties

The structures of some of the peracids are shown in Table 19.9. As with the PANs, the most common peracid is peracetic acid, with the structure $CH_3CO-OOH$. This compound is used in many commercial applications as a strong, yet selective, oxidizing agent for organic syntheses, particularly as an epoxidizing agent. Although formation of atmospheric peracids according to Reaction 19.19 has been demonstrated in the laboratory, their relative atmospheric abundances have not been measured.[69] The structures of some atmospherically relevant organic peroxides are given in Table 19.10. The most abundant organic peroxides in the atmosphere are methyl hydroperoxide, hydroxymethyl hydroperoxide, and bis-hydroxymethyl peroxide. Their atmospheric concentrations, which are on the same order as that of H_2O_2, can reach levels of parts per billion.[70–72]

With a Henry's law constant of $1 \times 10^5\ M^{-1}\ atm^{-1}$, H_2O_2 is quite water soluble, as compared to the relatively insoluble PANs ($H_{PAN} = 5\ M^{-1}\ atm^{-1}$).[14] Organic peracids and peroxides have intermediate water solubilities of 3×10^2 to $8 \times 10^2\ M^{-1}\ atm^{-1}$.[14] The peracids and peroxides can be present both in the aqueous phase, such as wet aerosols, cloud droplets, and precipitation and in the gas phase.

The organic peracids and peroxides are important oxidants in the atmosphere. These are responsible for the formation of H_2SO_4 in the aqueous phase and are thought to have toxic effects on plants.[70] They function as reservoirs for the peroxy radicals and reflect the radical levels of the atmosphere. The organic peracids and peroxides should be given serious attention, especially as we begin to control NO_x emission levels in an attempt to reduce urban and regional O_3 levels.

TABLE 19.9
Chemical Formulas and Names of Some of the More Common Peracids

Name	Chemical Formula	Abbreviation
Peroxyacetic acid	$CH_3C=OO_2H$	PAA
Peroxypropionic acid	$CH_3CH_2C=OO_2H$	PPA
Peroxybutyric acid	$CH_3CH_2CH_2C=OO_2H$	PBA
Peroxyisobutyric acid	$(CH_3)_2CHC=OO_2H$	PiBA
Peroxybenzoic acid	$C_6H_5C=OO_2H$	PBzA
Peroxymethacrylic acid	$CH_2C(CH_3)C=OO_2H$	PMA
Peroxyformic acid	$CH(O)OOH$	PFA

TABLE 19.10
Chemical Formulas and Names of Some of the More Common Organic Peroxides[70,71]

Name	Chemical Formula	Abbreviation
Methyl hydroperoxide	CH_3OOH	MHP
Hydroxymethyl hydroperoxide	$CH_2(OH)OOH$	HMHP
Ethyl hydroperoxide	CH_3CH_2OOH	EHP
Hydroxyethyl hydroperoxide	$CH_3CH(OH)OOH$	HEHP
Bis-hydroxymethyl peroxide	$HOCH_2OOCH_2OH$	BHMP
Hydroxymethylmethyl peroxide	$CH_2(OH)OOCH_3$	HMMP
1-Hydroxyethyl hydroperoxide	$CH_3CH(OH)OOH$	HEHP
2-Hydroxyethyl hydroperoxide	$HOCH_2CH_2OOH$	2HEHP
Bis-1-hydroxyethyl peroxide	$CH_3CH(OH)OOCH(OH)CH_3$	BHEP
1-Hydroxypropyl hydroperoxide	$CH_3CH_2CH_2(HO)OOH$	1HPHP
2-Hydroxypropyl hydroperoxide	$CH_3CH(OH)CH_2OOH$	2HPHP
3-Hydroxypropyl hydroperoxide	$HOCH_2CH_2CH_2OOH$	3HPHP

B. SYNTHESIS AND ISOLATION

The organic peracids and peroxides are all strong oxidizing agents. In addition, the organic peroxides can be explosive in their pure form. Their concentrated solutions should not be mixed with reducing agents or organic substances. For this reason, only dilute solutions are usually used. Peroxyacetic acid is available commercially. The simple peracids (peracetic acid, peroxypropionic acid, peroxybutyric acid) can be synthesized by the reaction of H_2O_2 with the corresponding anhydride in aqueous solution, as described in Section I.C for the synthesis of the PANs.[21] Because only dilute H_2O_2 is available, the product of this reaction is at best 50% organic peracid and 50% organic acid (if 50% H_2O_2 is used). Purification of the peracid from this solution is difficult. The peracids can also be synthesized by reaction of the carboxylic acid with H_2O_2 in concentrated H_2SO_4.[71,73] The solution must be cooled to $-5°C$ to $-10°C$, and the reaction mixture is diluted with acetonitrile–water (80:20) to a concentration of 10 to 100 mM.

The organic peroxides can be synthesized by the reaction of H_2O_2 with an organic precursor under basic conditions. Because organic peroxides are known to be explosive, the synthesis is usually carried out in dilute solution. Methyl hydroperoxide is synthesized from dimethylsulfate in the presence of 40% KOH.[70] The product can be separated from the side product CH_3OOCH_3 by stripping the solution with argon gas and trapping the vapor in a collection trap at 0°C.[74] In a similar procedure, ethyl hydroperoxide is formed by the reaction of H_2O_2 with diethyl sulfate. Hydroxymethyl hydroperoxide can be synthesized by the reaction of H_2O_2 and *para*formaldehyde at pH 10, while hydroxyethyl hydroperoxide can be synthesized from H_2O_2 and acetaldehyde.[70]

A major difficulty arises in the purification of the organic peracids and peroxides from these reactions, as some H_2O_2 is always present in the final mixture. The H_2O_2 can be removed by using bovine catalase enzyme immobilized on a syringe filter.[71] The H_2O_2 can be removed from a working solution of calibration standards by passing the solution through the filter at a rate of 5 ml min^{-1}. This procedure can also be used to remove H_2O_2 from samples. Excessive use of the catalyst can destroy the organic peroxides.

C. ATMOSPHERIC SAMPLING

Nearly all techniques used for the determination of the organic peracids and peroxides employ sampling devices which collect the sample in aqueous solution prior to analysis. Such sampling techniques include impingers, the mist chamber, the membrane diffusion scrubber, the scrubbing glass coil, and the cryogenic trap.[70] The continuous scrubbing glass coil is the most often used sampling system. In this system, sample air and the collection solution flow together through a glass coil. Collection efficiency for the organic peroxides for all aqueous trapping methods will vary according to the respective Henry's law constants. The collection efficiency for hydroxymethyl hydroperoxide with the continuous scrubbing glass coil is 100% at 10°C, the same is 78% for methyl hydroperoxide, 81% for ethyl hydroperoxide, and 91% for peracetic acid, respectively.[70] Corrections should be made for these differences in collection efficiencies. Because the collection efficiency is temperature dependent, the solution temperature should be recorded during sampling.[75]

The cryogenic trapping techniques have a higher concentration factor than the aqueous trapping methods, such as the continuous scrubbing coil, and therefore have been applied to improve detection limits when ambient concentrations are low. The aqueous coil collection method typically has a preconcentration factor of about 5,000 l air to 1 l water, whereas the cryogenic trap typically has a preconcentration factor of about 15,000.[76] This increase is advantageous in HPLC techniques, as the separation dilutes the sample before it is quantified. The increase in sample concentration compensates for the eluent dilution factor and results in detection limits for HPLC separation which are comparable to those for methods of measuring total peroxides. For cryogenic trapping of the organic peroxides, interferences have been reported which result from surface reactions of O_3 with hydrocarbons in the sample.[70] Caution should be used in interpreting organic peroxide data obtained by cryotrapping methods.

All of the sample collection methods suffer from inlet losses, as the reactive organic peracids and peroxides tend to decompose on unconditioned surfaces. Aerosols deposited in sampling lines can also react with the peroxides, with subsequent loss of sample. For this reason, sample lines must be kept clean and as short as possible. A surfaceless intake scrubber has been designed to avoid surface loss.[77] This system minimizes surface contact of the sample by eliminating all surfaces at the intake.

The most noticeable negative interference with aqueous collection methods in urban environments is due to the reaction with SO_2 in the aqueous phase, producing H_2SO_4. Because collection efficiencies for organic peroxides are enhanced at high pH and the solubility of SO_2 decreases at high pH, collection solutions should be maintained at pH 3 in high SO_2 environments.[75]

D. ANALYSIS OF THE PEROXIDES AND PERACIDS

1. Detection Methods

As the organic peroxides and peracids are formed by reaction with hydroperoxy radical (Reaction 19.19), these typically occur in air masses with appreciable amounts of H_2O_2 generated by reaction of hydroperoxy radical with itself (Reaction 19.20 with $R = H$). Methods have been developed for measuring the peracids in the presence of significant amounts of H_2O_2. Most methods for the determination of the peracids are based on their redox properties. For example, potentiometric detection with the glassy carbon electrode has been applied to the determination of peracetic acid, with sensitivities in the micromolar concentration range.[78] The high sensitivity to the peracid is based on the faster rate of reaction with iodide for peracetic acid vs. H_2O_2.

Because direct photometric detection of the organic peracids is not possible, several indirect photometric methods use reagents which are selectively oxidized by the peracids to form colored products. One method is based on the iodide-catalyzed selective oxidation of 2,2'-azino-bis(3-ethylbenzthiazoline)-6-sulfonate (ABTS) by the peracid to a green radical cation.[73] The product is detected by absorption at 405 to 815 nm, with detection limits in the range of 1×10^{-6} mol l^{-1} (76 μg l^{-1} for peracetic acid). A similar method is based on the selective oxidation of p-tolyl sulfide (MTS) to the corresponding sulfoxide (MTSO) by the peracids.[79] The product MTSO is detected by ultraviolet absorption.[73] More recently, an azo dye functionalized sulfide reagent, 2-([3-{2-(4-amino-2-[methylsulfanyl]phenyl)-1-diazenyl}phenyl]sulfonyl)-1-ethanol (ADS) provided lower detection limits and higher selectivity, with a similar reaction yielding the sulfoxide product, ADSO.[80]

Chemical ionization mass spectrometric detection has been explored for the detection of methyl hydroperoxide.[74] However, fluorometry has dominated the current detection schemes for the organic peroxides. Typically, a nonfluorescent substrate is oxidized by the peroxide to generate a fluorescent product. These methods are sufficiently sensitive for accurate measurement of the peroxides in the low ppt by volume. For example, the peroxidase-catalyzed dimerization of p-hydroxyphenylacetic acid (POPHA) occurs in the presence of a peroxy group at elevated pH. The formation of the fluorescent dimer, detected by excitation at 310 nm and emission at 405 nm, is proportional to the concentration of the peroxide.[71] The most common peroxidase catalyst used for this reaction is horseradish peroxidase (HRP). Cost and stability issues with the use of HRP led to the use of other catalysts, such as metalloporphyrins or phthalocyanine complexes.[81] Another fluorescent reaction scheme involves the oxidation of the nonfluorescent thiamine (vitamin B_1) to the fluorescent thiochrome by the peroxide group. This reaction is catalyzed by bovine hematin.[81] This reaction is 25-fold more sensitive for H_2O_2 than for the organic peroxides.

Luminol chemiluminescence detection, which can be very sensitive for the peracids and peroxides and for H_2O_2, has been used for the continuous measurement of H_2O_2 at a detection limit of about 0.5 ppb.[82] The peroxidase-catalyzed oxidation of luminol with m-chloroperoxybenzoic acid has been reported to generate enhanced chemiluminescence vs. H_2O_2.[83] This approach has not been pursued, because many atmospheric oxidants generate luminol chemiluminescence, depending on the solution conditions. This detection system has promise for future development as a chromatographic detector for the organic peracids and peroxides.

2. Chromatographic Separation

Gas chromatographic techniques have been used for the determination of high-molecular-weight organic peroxides which are used as oxidants in industrial processes. These techniques are not suitable for the direct determination of the low-molecular-weight organic peroxides expected in atmospheric samples. The high temperatures required to volatilize the organic peroxides for gas chromatography, lead to decomposition of the atmospherically relevant species in the metal injection systems common to most commercial gas chromatographs. Because these are relatively strong oxidants, these compounds often react with the column materials as well. They are also much more polar than PANs and tail strongly on the columns, making separation more problematic. Separation methods other than gas chromatography have been used for these compounds.

One approach to separation of the peroxides involves the precolumn reaction with MTS described in Section III.D.1, with subsequent separation of the products by HPLC and standard ultraviolet detection at 225 nm.[84] Separation of the H_2O_2 in the sample is accomplished by precolumn reaction with triphenylphosphine (TPP) and postcolumn detection of the oxidation product, triphenylphosphine oxide (TPPO). A nucleosil C8 reversed-phase column (70 × 3 mm, 5 μm particle size) was used with acetonitrile–water gradient elution (40 to 75 to 40%) at a flow rate of 1 ml min^{-1} to separate the products. Because the reaction with MTS precedes separation, this method gives a measurement of total organic peroxides.

More recently, the use of HPLC with postcolumn reaction with ABTS has successfully determined the peracids up to C_{12}.[73] A Merck LiChroSorb RP18 reversed-phase column (125 × 4 mm, 5 μm particle size) was used with acetonitrile–water gradient elution (25 to 100 to 25%) at a flow rate of 1.4 ml min^{-1}. To optimize peak shape, 2% acetic acid and 1% tetrahydrofuran were added to the water eluent. A turbomixing chamber was used to mix the eluent with the ABTS reagent for postcolumn reaction. The resulting oxidation product (green radical cation) was detected by conventional ultraviolet-visible spectroscopy. Detection limits were in the low micromolar range.[73]

Mixtures of H_2O_2 with ten low-molecular-weight $(C_1–C_3)$ organic peroxides and peracids have been separated successfully by HPLC with POPHA postcolumn reaction and fluorescence detection[71] with an Inertsil ODS C18 column $(250 \times 4.6$ mm, 5 μm particle size). The eluent used was 10^{-3} M H_2SO_4 with 1×10^{-4} M ethylenediaminetetraacetic acid tetrasodium salt (EDTA) at a flow rate of 0.60 ml min^{-1}. Calibrations for the individual peroxides were referenced to H_2O_2, greatly simplifying calibration of the analysis. The detection limit for this technique is 9×10^{-8} M for H_2O_2 and 2×10^{-8} M for the organic peroxides.

IV. FUTURE APPLICATIONS AND NEEDS FOR MEASUREMENT IN THE ENVIRONMENT

Recently, considerable interest has been focused on the use of biofuels in motor vehicles.[85] If ethanol usage as a biofuel or in fuel blends with gasoline is increased, the primary emissions of acetaldehyde might enhance levels of PANs in urban environments, because the production of PANs is a direct process when aldehydes react with hydroxyl radical to abstract the aldehydic hydrogen (Reaction 19.10 and Reaction 19.11). The use of oxygenated fuels which increase emissions of aldehydes or aldehyde precursors such as ethanol or methyl-t-butyl ether (MTBE) should continue to be accompanied by assessment for PANs and other oxidants.[85,86] The potential for long-range transport of PANs on continental scales and for PANs to function as a source of regional O_3 should be kept in mind.[15,35,36]

The PANs are known to be quite sensitive to walls in laboratory studies, and therefore are likely to react on aerosol surfaces. The PANs are very soluble in nonpolar organics. PANs can undergo important oxidation reactions on soot surfaces, leading to the formation of oxidized and nitrated polynuclear aromatic hydrocarbons which can be highly mutagenic.[14] The measurement of the PANs, as well as more usual oxidants such as O_3, nitrate radical, and hydroxyl radical, is an important part of the characterization of potentially hazardous air pollutants.

With the reduction of reactive hydrocarbons and the lowering of NO levels, concentrations of urban PANs are continually being decreased. PANs are still an important part of overall NO_x transport and deposition processes on regional and global scales. With the anticipated decrease in NO levels due to improved control technologies and mitigation strategies for O_3, the PANs and associated organonitrates are expected to decrease in industrialized nations. Developing nations are likely to see higher levels of PANs as industrialization continues and fosters increases in fossil fuel usage and vehicular traffic, especially in absence of control technologies to limit reactive hydrocarbon and NO emissions. Higher levels of PANs have been observed in recent years in a number of urban centers, most notably Mexico City, Mexico, and Santiago, Chile, where levels of PANs exceeding 20 ppb have been reported.

Lowered NO levels will lead to the formation of organic peroxides and peracids, and better instrumentation will be needed for their speciated measurement. These compounds have high biological activities and are used as disinfectants. Their increased wet deposition in certain environments could decrease microbial populations. Sensitive, selective detection methods for organic peroxides and peracids will be needed in the future. The application of chromatography towards detection of PANs, alkyl nitrates, organic peroxides, and peracids in the past has been productive in improving our basic understanding of the roles of these compounds in atmospheric chemistry. New highly sensitive, highly selective spectroscopic methods for postcolumn detection (e.g., quantum cascade lasers, mass spectrometry, chemiluminescence) are expected to spur continued analytical development in the future. We have known about the PANs for over 40 years, but our fundamental understanding of these fascinating molecules and our ability to produce standards and measure them is just now expanding. Interestingly, many physical chemists and biochemists are beginning to explore the unusual properties of these highly energetic molecules, along with the associated alkyl nitrates and organic peroxides and peracids. The future of PAN chemistry, opened to us by Edgar Stephens and his colleagues, looks bright, and its further exploration should lead us to a much better understanding of the key oxidation reactions occurring in the troposphere.

ACKNOWLEDGMENTS

The authors' work is supported by the U.S. Department of Energy, Office of Science, Office of Biological and Environmental Research, Atmospheric Science Program.The authors wish to thank Peter Lunn and Rick Petty of the U.S. Department of Energy for their continuing encouragement. This paper is dedicated to the late Dr. Edgar Stephens and to Dr. James Lovelock, whose pioneering work on the PANs continues to inspire workers in the field who are developing new, faster methods for analysis of this important class of pollutants. In this brief chapter, we might not have been able to give credit to all who are deserving and we apologize to any pioneers in the measurement and study of PANs, organic nitrates, and organic peracids whom we have not mentioned in this work.

REFERENCES

1. Stephens, E. R., Smog studies of the 1950s, *EOS*, 68, 91–93, 1987.
2. Middleton, J. T., Kendrick, J. B., and Schwalm, H. W., Smog in the south coastal area of California, *Agriculture*, 4, 7, 1950.
3. Middleton, J. T., Kendrick, J. B., and Schwalm, H. W., Injury to herbaceous plants by smog or air pollutants, *Plant Dis. Rep.*, 34, 245–252, 1950.
4. Haagen-Smit, A. J., The air pollution problem in Los Angeles, *Eng. Sci.*, 14, 1, 1950.
5. Haagen-Smit, A. J., Bradley, C., and Fox, M. M., Formation of ozone in Los Angeles smog, *Proceedings of the Second National Air Pollution Symposium*, 1952, pp. 54–56.
6. Haagen-Smit, A. J., Darley, E. F., Zaitlin, M., Hull, H., and Noble, W., Investigation on injury to plants from air pollution in the Los Angeles basin, *Plant Physiol.*, 27, 18–34, 1952.
7. Stephens, E. R., Hanst, P. L., Doerr, R. C., and Scott, W. E., Reactions of nitrogen dioxide and organic compounds in air, *Eng. Chem.*, 48, 1498, 1956.
8. Stephens, E. R., Scott, W. E., Hanst, P. L., and Doerr, R. C., Auto exhaust: composition and photolysis products, *J. Air Pollut. Control Assoc.*, 6, 159, 1956.
9. Schuck, E.A. and Doyle, G.J., *Photooxidation of Hydrocarbons in Mixtures Containing Oxides of Nitrogen and Sulfur Dioxide: Report 29*, Air Pollution Foundation, San Marino, CA, 1959.
10. Leighton, P. A., *Photochemistry of Air Pollution*, Academic Press, New York, 1961.
11. Varetti, E. L. and Pimentel, G. C., The infrared spectrum of ^{15}N-labeled PAN in an oxygen matrix, *Spectrochim Acta Part A*, 30, 1069, 1974.
12. Talyor, O. C., Stephens, E. R., Darley, E. F., and Cardiff, E. A., Effect of air-borne oxidants on leaves of pinto bean and petunia, *Proc. Am. Soc. Hort. Sci.*, 75, 435, 1960.
13. Nicksic, S. W., Harkins, J., and Mueller, P. K., Some analyses for PAN and studies of its structure, *Atmos. Environ.*, 1, 11–18, 1966.
14. Finlayson-Pitts, B. J. and Pitts, J. N. Jr., *Chemistry of the Upper and Lower Atmosphere*, Academic Press, San Diego, CA, 2000.
15. Gaffney, J. S., Marley, N. A. and Prestbo, E. W., *Peroxyacyl Nitrates (PANs): Their Physical and Chemical Properties*, The Handbook of Environmental Chemistry, Part B (Air Pollution), Vol. 4, Springer, Berlin, pp. 1–38, 1989.
16. Roberts, J. M., The atmospheric chemistry of organic nitrates: review article, *Atmos. Environ.*, 24, 243–287, 1990.
17. Talkudar, R. K., Herndon, S. C., Burkholder, J. B., Roberts, J. M., and Ravishankara, A. R., Investigation of the loss processes for PAN in the atmosphere: UV photolysis and reaction with OH, *J. Geophys. Res.*, 100, 14163–14173, 1995.

18. Singh, H. B., Reactive nitrogen in the troposphere: chemistry and transport of NO_x and PAN, *Environ. Sci. Technol.*, 21, 320–327, 1987.

19. Stephens, E. R. and Price, M. A., Analysis of an important air pollutant: peroxyacetyl nitrate, *J. Chem. Ed.*, 50, 351–354, 1973.

20. Warneck, P. and Zerbach, T., Synthesis of PAN in air by acetone photolysis, *Environ. Sci. Technol.*, 26, 74–79, 1992.

21. Nielson, T., Hansen, A. M., and Thompson, E. L., A convenient method for preparation of pure standards of PAN for atmospheric analysis, *Atmos. Environ.*, 16, 2447–2450, 1982.

22. Gaffney, J. S., Fajer, R., and Senum, G. I., An improved procedure for high purity gaseous PAN production: use of heavy lipid solvents, *Atmos. Environ.*, 18, 215–218, 1984.

23. Tuazon, E. C., Graham, R. A., Winer, A. M., Easton, R. R., Pitts, J. N. Jr., and Hanst, P. L., A kilometer path length Fourier-transform infrared system for the study of trace pollutants in ambient and synthetic atmospheres, *Atmos. Environ.*, 12, 865–875, 1978.

24. Hanst, P. L., Wong, N. W., and Bragin, J., A long-path infrared study of Los Angeles smog, *Atmos. Environ.*, 16, 969–981, 1982.

25. Hansel, A. and Wisthaler, A., A method for real-time detection of PAN, PPN and MPAN in ambient air, *Geophys. Res. Lett.*, 27, 895–898, 2000.

26. Tanimoto, H., Hirokawa, J., Kajii, Y., and Akimoto, H., A new measurement technique of PAN at parts per trillion by volume levels: gas chromatography/negative ion chemical ionization mass spectrometry, *J. Geophys. Res.*, 104, 21343–21354, 1999.

27. Tanimoto, H., Hirokawa, J., Kajii, Y., and Akimoto, H., Characterization of gas chromatography/ negative ion chemical ionization mass spectrometry for ambient measurements of PAN: potential interferences and long-term sensitivity drift, *Geophys. Res. Lett.*, 27, 2089–2092, 2000.

28. Darley, E. F., Kettner, K. A., and Stephens, E. R., Analysis of peroxyacyl nitrates with electron capture detection, *Anal. Chem.*, 35, 589–591, 1963.

29. Burkhardt, M. R., Maniga, N. I., Stedman, D. H., and Paur, R. J., Gas chromatographic method for measuring nitrogen dioxide and PAN in air without compressed gas cylinders, *Anal. Chem.*, 60, 816–819, 1988.

30. Gaffney, J. S., Marley, N. A., and Drayton, P. J., Fast gas chromatography with luminol detection for measurement of nitrogen dioxide and PANs, *Proceedings Sixth U.S./German Workshop on Ozone/ Fine Particle Science*, EPA/600/R-00/076, Riverside, CA, pp. 110–117, 1999.

31. Gaffney, J. S., Bornick, R. M., Chen, Y.-H., and Marley, N. A., Capillary gas chromatographic analysis of nitrogen dioxide and PANs with luminol chemiluminescent detection, *Atmos. Environ.*, 32, 1445–1454, 1998.

32. Blanchard, P., Shepson, P. B., Schiff, H. I., and Drummond, J. W., Development of a gas chromatograph for trace level measurement of PAN using chemical amplification, *Anal. Chem.*, 65, 2472–2477, 1993.

33. Danalatos, D. and Glavas, S., Improvement in the use of capillary columns for ambient air PAN monitoring, *J. Chromatogr. A*, 786, 361–365, 1997.

34. Moschonas, N. and Glavas, S., Simple cryoconcentration technique for the determination of peroxyacyl and alkyl nitrates in the atmosphere, *J. Chromatogr. A*, 902, 405–411, 2000.

35. Singh, H. B., O'Hara, D. O., Herlth, D., Bradshaw, J. D., Sandholm, S. T., Gregory, G. L., Sachse, G. W., Blake, D. R., Crutzen, P. J., and Kanakidou, M. A., Atmospheric measurements of PAN and other organic nitrates at high latitudes: possible sources and sinks, *J. Geophys. Res.*, 97, 16511–16522, 1992.

36. Gaffney, J. S., Marley, N. A., Cunningham, M. M., and Doskey, P. V., Measurements of peroxyacyl nitrates (PANs) in Mexico City: implications for mega city air quality impacts on regional scales, *Atmos. Environ.*, 33, 5003–5012, 1999.

37. Schrimpf, W., Muller, K. P., Johnen, F. J., Lienaerts, K., and Rudolph, J., An optimized method for airborne PAN (PAN) measurements, *J. Atmos. Chem.*, 22, 303–317, 1995.

38. Blanchard, P., Shepson, P. B., So, K. W., Schiff, H. I., Bottenheim, J. W., Gallant, A. J., Drummond, J. W., and Wong, P., A comparison of calibration and measurement techniques for gas chromatographic determination of atmospheric PAN (PAN), *Atmos. Environ.*, 24A, 2839–2846, 1990.

39. Gaffney, J. S., Marley, N. A., and Steele, H. D., Aircraft measurements of nitrogen dioxide and peroxyacyl nitrates using luminol chemiluminescence with fast gas chromatography, *Environ. Sci. Technol.*, 33, 3285–3289, 1999.

40. Gaffney, J. S., Bornick, R. M., Chen, Y. H., and Marley, N. A., Capillary gas chromatographic analysis of nitrogen dioxide and PANs with luminol chemiluminescent detection, *Atmos. Environ.*, 32, 1445–1454, 1998.

41. Marley, N. A., Gaffney, J. S., Drayton, P. J., and Ravelo, R. M., Northeast oxidant and particulate study: preliminary results from the Centerton, New Jersey field site, *Millennium Symposium on Atmospheric Chemistry: Past, Present, and Future*, 81st American Meteorological Society National Meeting, Albuquerque, NM, pp. 29–33, 2001.

42. Marley, N. A. and Gaffney, J. S., Measurements of NO_2, PANs and ozone at Deer Park, Texas, *Nobel Symposium Atmospheric Chemistry Abstract*, The 37th Western Regional Meeting of the American Chemical Society, WERM: "An Earth Odyssey", Santa Barbara, CA, pp. 99–100, 2001.

43. Gaffney, J. S., Marley, N. A., Drayton, P. J., Kotamarthi, V. R., Cunningham, M. M., Baird, J. C., Dintamin, J., and Hart, H. L., Field observations of regional and urban impacts on NO_2, ozone, UVB, and nitrate radical production rates in the Phoenix air basin, *Atmos. Environ.*, 36, 825–833, 2002.

44. Marley, N. A. and Gaffney, J. S., Air quality measurements in Phoenix, AZ, *Millennium Symposium on Atmospheric Chemistry: Past, Present, and Future*, 83rd American Meteorological Society National Meeting, Long Beach, CA, preprint P1.10, 2003.

45. Marley, N. A., Gaffney, J. S., White, R. V., Rodriguez-Cuadra, L., Herndon, S. E., Dunlea, E., Volkamer, R. M., Molina, L. T., and Molina, M. J., Fast gas chromatography with luminol chemiluminescence detection for the simultaneous determination of nitrogen dioxide and peroxyacetyl nitrate in the atmosphere, *Review of Scientific Instruments*, 75, 4595–4605, 2004.

46. Talukdar, R. K., Burkholder, J. B., Hunter, M., Gilles, M. K., Roberts, J. M., and Ravishankara, A. R., Atmospheric fate of several alkyl nitrates, *J. Chem. Soc. Faraday Trans.*, 93, 2797–2805, 1997.

47. Schneider, M., Luxenhoffer, O., Deissler, A., and Ballschmiter, K., C_1–C_{15} alkyl nitrates, benzyl nitrate, and bifunctional nitrates: measurements in California and South Atlantic air and global comparison using C_2C_{14} and $CHBr_3$ as marker molecules, *Environ. Sci. Technol.*, 32, 3055–3062, 1998.

48. Atlas, E., Evidence for $> C_3$ alkyl nitrates in rural and remote atmospheres, *Nature*, 331, 426–428, 1988.

49. Fisher, R. G., Kastler, J., and Ballschmitter, K., Levels and patterns of alkyl nitrates, multifunctional alkyl nitrates, and halocarbons in the air over the Atlantic Ocean, *J. Geophys. Res.*, 105, 14473–14494, 2000.

50. Flock, F., Voltz-Thomas, A., Buers, H. J., Patz, W., Garthe, J. J., and Kley, D., Long-term measurements of alkyl nitrates in southern Germany. 1. General behavior and seasonal and diurnal variation, *J. Geophys. Res.*, 103, 5729–5746, 1998.

51. Edney, E. O., Spence, J. W. and Hanst, P. L., Synthesis and thermal stability of peroxy alkyl nitrates, *J. Air Pollut. Control Assoc.*, 29, 741–743, 1979.

52. Luxenhofer, O., Schneider, E., and Ballschmiter, K., Separation, detection and occurrence of (C2–C8)-alkyl and phenyl–alkyl nitrates as trace compounds in clean and polluted air, *Fresenius J. Anal. Chem.*, 350, 384–394, 1994.

53. Luxenhofer, O., Schneider, M., Dambach, M., and Ballschmiter, K., Semivolatile long chain C6–C17 alkyl nitrates as trace compounds in air, *Chemosphere*, 33, 393–404, 1996.

54. Schneider, M. and Ballschmiter, K., Alkyl nitrates as achiral and chiral solute probes in gas chromatography. Novel properties of a β-cyclodextrin derivative and characterization of its enantioselective forces, *J. Chromatogr. A*, 852, 525–534, 1999.

55. Kastler, J. and Ballschmitter, K., Bifunctional alkyl nitrates — Trace constituents of the atmosphere, *Fresenius J. Anal. Chem.*, 360, 812–816, 1998.

56. Kastler, J., Jarman, W., and Ballschmiter, K., Multifunctional organic nitrates as constituents in European and U.S. urban photo-smog, *Fresenius J. Anal. Chem.*, 368, 244–249, 2000.

57. Wodich, S., Froescheis, O., Luxenhofer, O., and Ballschmiter, K., EI- and NCI-mass spectrometry of arylalkyl nitrates and their occurrence in urban air, *Fresenius J. Anal. Chem.*, 364, 91–99, 1999.

58. Ostling, K., Kelly, B., Bird, S., Bertman, S., and Pippin, M., Fast-turnaround alkyl nitrate measurements during the PROPHET 1998 summer intensive, *J. Geophys. Res.*, 106, 24439–24449, 2001.

59. Atlas, E. and Schauffler, S., Analysis of alkyl nitrates and selected halocarbons in the ambient atmosphere using a charcoal preconcentration technique, *Environ. Sci. Technol.*, 25, 61–67, 1991.

60. Bertman, S. B., Buhr, M. P., and Roberts, J. M., Automated cryogenic trapping technique for capillary GC analysis of atmospheric trace compounds requiring no expendable cryogens: application to the measurement of organic nitrates, *Anal. Chem.*, 65, 2944–2946, 1993.

61. Stroud, C. A., Roberts, J. M., Williams, J., Gouldan, P. D., Kuster, W. C., Ryerson, T. B., Sueper, D., Parrish, D. D., Trainer, M., Fehsenfeld, F. C., Flocke, F., Schauffler, S. M., Stroud, V. R. F., and Atlas, E., Alkyl nitrate measurements during STERAO 1996 and NARE 1997: intercomparison and survey of results, *J. Geophys. Res.*, 106, 23043–23053, 2001.

62. Schneider, M. and Ballschmiter, K., C1–C13-alkyl nitrate in remote South Atlantic air, *Chemosphere*, 38, 233–244, 1999.

63. Weiner, A. M., Peters, J. W., Smith, J. P., and Pitts, J. N. Jr., Response of commercial chemiluminescent NO–NO$_2$ analyzers to other nitrogen-containing compounds, *Environ. Sci. Technol.*, 8, 1118–1121, 1974.

64. Hao, C., Shepson, P. B., Drummond, J. W., and Muthuramu, K., Gas chromatographic detector for selective and sensitive detection of atmospheric organic nitrates, *Anal. Chem.*, 66, 3737–3743, 1994.

65. Glavas, S., Analysis of C2–C4 peroxyacyl nitrates and C1–C5 alkyl nitrates with a non-polar capillary column, *J. Chromatogr. A*, 915, 271–274, 2001.

66. Schneider, M. and Ballschmeter, K., Separation of diastereomeric and enantiomeric alkyl nitrates — Systematic approach to chiral discrimination on cyclodextrin LIPODEX-D, *Chem. Eur. J.*, 2, 539–544, 1996.

67. Kastler, J., Dubourg, V., Deisenhofer, R., and Ballschmiter, K., Group separation of organic nitrates on a new nitric acid ester NP–LC stationary phase, *Chromatographia*, 47, 157–161, 1998.

68. Wößner, M. and Ballschmiter, K., New stationary phase based on β-cyclodextrin for normal-phase HPLC group-separation of organic nitrates, *Fresenius J. Anal. Chem.*, 366, 346–350, 2000.

69. Orlando, J. J. and Tyndall, G. S., Gas phase UV absorption spectra for peracetic acid, and for acetic acid monomers and dimers, *J. Photochem. Photobiol.*, 157, 161–166, 2003.

70. Lee, M., Heikes, B. G., and O'Sullivan, D. W., Hydrogen peroxide and organic hydroperoxide in the troposphere: a review, *Atmos. Environ.*, 34, 3475–3494, 2000.

71. Kok, G. L., McLaren, S. E., and Staffelbach, T. A., HPLC determination of atmospheric organic hydroperoxides, *J. Atmos. Ocean Technol.*, 12, 282–289, 1995.

72. Jacob, P., Wehling, B., Hill, W., and Klockow, D., Feasibility study of Raman spectroscopy as a tool to investigate the liquid-phase chemistry of aliphatic organic peroxides, *Appl. Spectrosc.*, 51, 74–80, 1997.

73. Effkemann, S., Pinkernell, U., Neumuller, R., Schwan, F., Engelhardt, H., and Karst, U., Liquid chromatographic simultaneous determination of peroxycarboxylic acids using postcolumn derivatization, *Anal. Chem.*, 70, 3857–3862, 1998.

74. Messer, B. M., Stiestra, D. E., Cappa, C. D., Scholtens, K. W., and Elrod, M. J., Computational and experimental studies of chemical ionization mass spectrometric detection techniques for atmospherically relevant peroxides, *Int. J. Mass Spectrom.*, 197, 219–235, 2000.

75. Lee, M., Noone, B. C., O'Sullivan, D., and Heikes, B. G., Method for the collection and HPLC analysis of hydrogen peroxide and C$_1$ and C$_2$ hydroperoxides in the atmosphere, *J. Atmos. Ocean Technol.*, 12, 1060–1070, 1995.

76. Campos, T. L. and Kok, G. L., Evaluation of Horibe traps for cryogenic collection of hydrogen peroxide and methyl hydroperoxide, *Atmos. Environ.*, 30, 2575–2582, 1996.

77. Lee, J. H., Leahy, D. F., Tang, I. N., and Newman, L., Measurement and speciation of gas phase peroxides in the atmosphere, *J. Geophys. Res.*, 98, 2911–2915, 1993.

78. Awad, M. I. and Ohsaka, T., Potentiometric analysis of peroxyacetic acid in the presence of a large excess of hydrogen peroxide, *J. Electroanal. Chem.*, 544, 35–40, 2003.

79. Minning, S., Weiss, A., Bornscheuer, U. T., and Schmid, R. D., Determination of peracid and putative enzymatic peracid formation by an easy colorimetric assay, *Anal. Chim. Acta*, 378, 293–298, 1999.

80. Effkemann, S., Brodsgaard, S., Mortensen, P., Linde, S. A., and Karst, U., Determination of gas phase peroxyacetic acid using precolumn derivatization with organic sulfide reagents and liquid chromatography, *J. Chromatogr. A*, 855, 551–561, 1999.

81. Li, J. and Dasgupta, P. K., Measurement of atmospheric hydrogen peroxide and hydroxymethyl hydroperoxide with a diffusion scrubber and light emitting diode-liquid core waveguide-based fluorometry, *Anal. Chem.*, 72, 5338–5347, 2000.

82. Kok, G. L., Holler, T. P., Lopez, M. B., Nachtrieb, H. A., and Yuan, M., Chemiluminescence method for determination of hydrogen peroxide in the ambient atmosphere, *Environ. Sci. Technol.*, 12, 1072–1076, 1978.

83. Yeh, H.-C. and Lin, W.-Y., Enhanced chemiluminescence for the oxidation of luminol with *m*-chloroperoxybenzoic acid catalyzed by microperoxidase 8, *Anal. Bioanal. Chem.*, 372, 525–531, 2002.

84. Pinkernell, U., Effkemann, S., and Karst, U., Simultaneous HPLC determination of peroxyacetic acid and hydrogen peroxide, *Anal. Chem.*, 69, 3623–3627, 1997.

85. Gaffney, J. S. and Marley, N. A., Alternative fuels, In *Air Pollution Reviews, The Urban Air Atmosphere and Its Effects*, Vol. 1., Brimblecombe, P. and Maynard, R., Eds., Imperial College Press, London, pp. 195–246, 2000.

86. Gaffney, J. S., Marley, N. A., Martin, R. S., Dixon, R. W., Reyes, L. G., and Popp, C. J., Potential air quality effects of using ethanol–gasoline blends: a field study in Albuquerque, New Mexico, *Environ. Sci. Technol.*, 31, 3053–3061, 1997.

20 Speciation in Environmental Samples

Willy Baeyens, Marjan De Gieter, Martine Leermakers, and Isabelle Windal

CONTENTS

I. INTRODUCTION

The complexity of analyses of organometallic compounds in environmental samples depends on a number of factors such as: (1) the amount of sample and sample concentrations. The lower the concentrations of compounds and smaller the amount of sample (this is for example, often the case in blood monitoring studies), the more difficult the analysis becomes. (2) The matrix of the compartment — a homogeneous matrix (e.g., fish tissue) is easier to handle than a heterogeneous

Abbreviations: AB, Arsenobetaine; AC, Arsenocholine; AES, Atomic Emission Spectrometry; CGC, Capillary Gas Chromatography; CRM, Certified Reference Materials; CV-AFS, Cold Vapor-Atomic Fluorescence Spectrometry; DBT, Dibutyltin; DDT, 1,1,1-trichloro-2,2-di(*p*-chlorophenyl)-ethane; DDTC, Diethyldithiocarbamate; DMA, Dimethyl Arsenic; DMHg, Dimethylmercury; DMT, Dimethyltin; DOC, Dissolved Organic Carbon; DOT, Dioctyltin; DSMA, Disodium methanearsonate; ECD, Electron Capture Detection; EDTA, Ethylene diamine tetraacetatic Acid; F/GF/CV/QF/ET-AAS, Flame/Graphite Furnace/Cold Vapor/Quartz Furnace/Electrothermal-Atomic Absorption Spectrometry; FAO, Food and Agricultural Organization; FAPES, Furnace Atomization Plasma Emission Spectrometry; FEP, Fluorocarbon Polymer; HG, Hydride Generation; HPLC, High Pressure Liquid Chromatography; IARC, International Agency for Research on Cancer; IC, Ion Chromatography; ICP, Inductively Coupled Plasma; IUPAC, International Union of Pure and Applied Chemistry; LOD, Limit of Detection; MIP, Microwave Induced Plasma; MMA, Monomethyl Arsenic; MMHg, Monomethylmercury; MS, Mass Spectrometry; MSMA, Monosodium methanearsonate; MBT, Monobutyltin; MMT, Monomethyltin; MOT, Monooctyltin; PDMS, Polydimethylsiloxane; PFPD, Pulsed Flame Photometric Detector; PPM, Parts per million; PTFE, Polytetrafluoroethylene; QC/QA, Quality Control/Quality Assurance; RSD, Relative Standard Deviation; SFE, Supercritical Fluid Extraction; SIDMS, Speciated Isotope Dilution Mass Spectrometry; SPME-GC, Solid Phase Micro Extraction-Gas Chromatography; TBT, Tributyltin; TcHT, Tricyclohexyltin; TEMA, Tetramethylarsonium ion; TePhT, Tetraphenyltin; TPhT, Triphenyltin; THF, Tetrahydrofuran; TMAH, Tetramethylammoniumhydroxide; TMAO, Trimethylarsine Oxide; TMDTC, Tetramethylenedithiocarbamate; TOF-MS, Time of Flight-Mass Spectrometry; USN, Ultrasonic Nebulization; WHO, World Health Organization.

one (e.g., a sediment sample). (3) The accuracy and precision aimed at (qualitative, semiquantitative, or quantitative). Screening techniques can be used when only a qualitative assessment of the compounds is required, making the analysis much easier.

Although in recent years more and more attempts are made for determining various organometals in one run,[1] most of the time too many problems have to be solved with natural samples making these approaches cumbersome. Existing procedures often require large amounts of sample and tedious separation–preconcentration steps. No standard QC/QA protocols exist either for the assessment of extraction and clean-up recoveries of analytes or for that of derivatization efficiencies, both depending on the matrices involved. Ultra trace analyses of organometals in remote environments have to face the problems of low blank values. The sampling conditions in the field, the quality and purity of chemical reagents including gasses and filters and the cleaning of small laboratory material all contribute to increased blank values. Uncertainties on low natural concentrations of organometals can be very large but are often considered not straightforward. Too often, the variability on instrumental replicates is reported as the procedural uncertainty, but is often many times lower than the real one.

In this overview, we have limited our discussion to the speciation of four important organometals, each presenting specific analytical features, in environmental samples: organomercury, organoarsenical, organotin, and organolead compounds. Their uses, presence in the environment, toxicity and threat to human health are also briefly discussed.

II. EXTRACTION RECOVERIES

The extraction of organometallic compounds from environmental samples is a very complex matter in which two conflicting issues need to be addressed; obtaining an adequate recovery, and preventing losses, especially destruction of the compound(s).[2] The extraction should be performed in such a way that the analyte is separated from the interfering matrix without loss, contamination, or change of the speciation, and with a minimum of interferences. Extractions of organometals from an aqueous solution or a biological tissue are much easier to realize than from sediments. The former compartments are much more homogeneous than sediments and metal-bonds in a given sample and are quite similar. In sediments, metals are distributed over the various fractions and metal-bonds to these fractions can be very different. A lot of effort has been dedicated to assess and to better understand those distributions, because this information is of utmost importance for the design of organometals extraction procedures.

Since the early 1980s, single and sequential extraction schemes have been designed for the speciation of total metal amounts in sediments.[3] The speciation in this case aims at understanding the distribution of a metal over various sedimentary substrates such as carbonates, iron- and manganese-oxyhydroxides, organic matter, sulfides, silicates, etc. Under particular conditions, some of these substrates will dissolve or release adsorbed metals. For example, the oxyhydroxides under reducing conditions and the carbonates under acidic conditions will dissolve and when the electrolytic strength of solution is increased it will release adsorbed metals. It is possible by carefully selecting the composition of extraction solutions, also called extractants, to destroy selectively specific soil or sediment substrates such as, for example, reduced or oxidized forms.

When testing a five-step extraction scheme on the Mn and Fe release, from a natural sediment, it appeared that the first three extraction steps are of similar aggressiveness and by far less aggressive than the last two.[4] The first extraction step is carried out at an almost neutral pH and without oxidant/reductant, the second step at the same pH but with a reductant and the third step at a pH 2 but without oxidant/reductant. The fourth step is an oxidation in acidic environment, and when it is performed in first place, the Mn and Fe contents in the last two steps (four and five) amount to more than 95% of the total burden of those metals. This implies that steps one to three of the standard scheme should be performed before the acid-oxidation step. The spike recoveries of oxidized Mn and Fe compounds in the second (reducing) step were also verified. Spike recoveries were between $100 \pm 10\%$.

It is clear that, extraction of organometals from sedimentary phases should be carried out with extractants which liberate organometals as much as possible without destroying them. Mineralization techniques, dissolving all or a large part of the sediment matrix should be avoided because the metal–carbon bonds of the organometal species will also be broken. The ideal extraction scenario is where:

(1) Organometal species remain unaltered.
(2) Sediment substrates (e.g., carbonates, Fe- and Mn-oxyhydroxides, amorphous sulfides, etc.) are as much as possible solubilized.
(3) The adsorption–desorption equilibrium of organometal compounds is shifted towards the dissolved phase, for example, by adding appropriate complexating ligands to the extraction solution.

All these efforts, however, do not guarantee that there is no loss (e.g., degradation) or incomplete recovery of a given compound from the matrix (aqueous solution, biological tissue or sediment). It is current practice to apply compensation for these losses by correcting the results with a recovery factor in order to achieve a better approximation to the true value in a material. These correction factors are established after undertaking recovery studies, which are an essential component of the validation of extraction-based techniques. As described below, this practice is not without problems, and the most critical aspect is the lack of common strategies for assessment of recovery and the way in which corrections have to be applied. The following definitions are adapted from a IUPAC[5] document.

Recoveries: Recovery is the proportion of the amount of analyte, present or added to the test material, which is extracted and presented for measurement and can be determined by the analysis of Certified Reference Materials (CRMs). The recovery being the ratio of found analyte content to the certified values. A number of drawbacks exist:

(1) The range of CRMs available for organometallic determinations is rather limited.
(2) The matrix of the CRMs may not match with that of the sample.
(3) The form in which the CRMs exist (generally a finely grained dry powder) can differ from that of the sample (e.g., a fresh biological tissue).

The use of CRMs to calculate "correction factors", which can then be applied to unknown samples, is still a controversial matter.

Recovery evaluation using surrogates or spikes implies the assumption that the extraction of spike is equivalent of the native analyte. In practice, it is often difficult to demonstrate that equivalence, and the only solution left is to accept the above assumption (extraction of spike is equivalent to that of native compound). A special form of this method is the standard addition method, where spiking at different levels is performed. Depending on the number of levels (i.e., two, three, or more) and/or the concentration jump chosen for the spiking experiment, a different recovery evaluation can be obtained.

With regard to the concentration levels (concentration jump), it is acceptable to double the concentration of the analyte(s). In many other cases reported in the literature, spike concentrations ten times higher than the concentrations of the native compounds are generally applied. In this case, the risk of overestimation of a recovery is quite high.

If native compounds and spikes are extracted with different efficiencies (with different percentages) the real extraction efficiency can not be assessed.

A third method is the use of the same organometal compound but containing a strongly enriched metal isotope, different from the major natural one. Recoveries can be assessed as long as the native analyte and the spike come into equilibrium. The latter is impossible to verify and this approach, although state-of-the-art, may still yield a biased recovery estimate. This procedure is limited by the availability and cost of isotopically enriched organometallic compounds and the instrumentation for their determination. Some specific recommendations have been formulated by[6]:

- A long equilibration time in case of solid samples (e.g., soils, sediments, biota) is usually necessary to simulate as well as to make possible a "natural" adsorption of the spike.
- An equilibration time of 24 h, or at least "overnight", is recommended. However, the equilibration time should be decided on case to case basis, taking into consideration the nature of the species and the matrix to be spiked.
- Recovery tests are to be carried out for each single species, since the extraction efficiency changes from one species to another.
- Real matrices should be used which are similar to the unknown sample. First, the level of the incurred (native) compound(s) must be accurately evaluated before spiking, and the percentage recovery is usually referred to as the sum of original content plus the spike or directly to the spike amount after subtraction of the original content.

Uncorrected results would likely to yield an unacceptable bias which would hamper mutual recognition and comparability of data. Methods of correction may be applied widely to speciation analysis, as soon as these are based on commonly agreed protocols applied with techniques under control. The correction factors have to be calculated for each matrix since these may vary from matrix to matrix and for different levels of contents. The wide uncertainty of correction factors may yield a high relative uncertainty in the final results compared to uncorrected results (for which the uncertainty is merely related to the analysis alone).

A. Properties of As, Hg, Sn, and Pb

In Table 20.1 physical and chemical properties of the four elements are summarized.

TABLE 20.1
Properties of As, Hg, Sn, and Pb

	As	Hg	Sn	Pb
Element Characteristics				
Group	Va	IIb	IVa	IVa
Atomic number	33	80	50	82
Mass number	74.92	200.59	118.69	207.2
Natural isotopes	1 (74.92)	7 (between 196 and 204)	9 (between 112 and 124)	4 (between 204 and 208)
Oxidation state	$+5, +3, 0, -3$	$+2, +1, 0$	$+4, +2, 0$	$+4, +2, 0$
Allotropic forms	Steel grey, semimetallic solid	Silver white metallic liquid	Grey cubic or white tetragonal metal	Bluish-white, soft, and malleable metal
Physicochemical Properties				
Density (g cm^{-3})	5.78	13.55	7.28	11.3
Melting temperature (°C)	—	-38.87	231.97	327.5
Boiling temperature (°C)	613 (sublimation)	356.58	2270	1740
Electronic Configuration	[Ar] $3d^{10} 4s^2 4p^3$	[Xe] $4f^{14} 5d^{10} 6s^2$	[Kr] $4d^{10} 5s^2 5p^2$	[Xe] $5d^{10} 6s^2 6p^2$

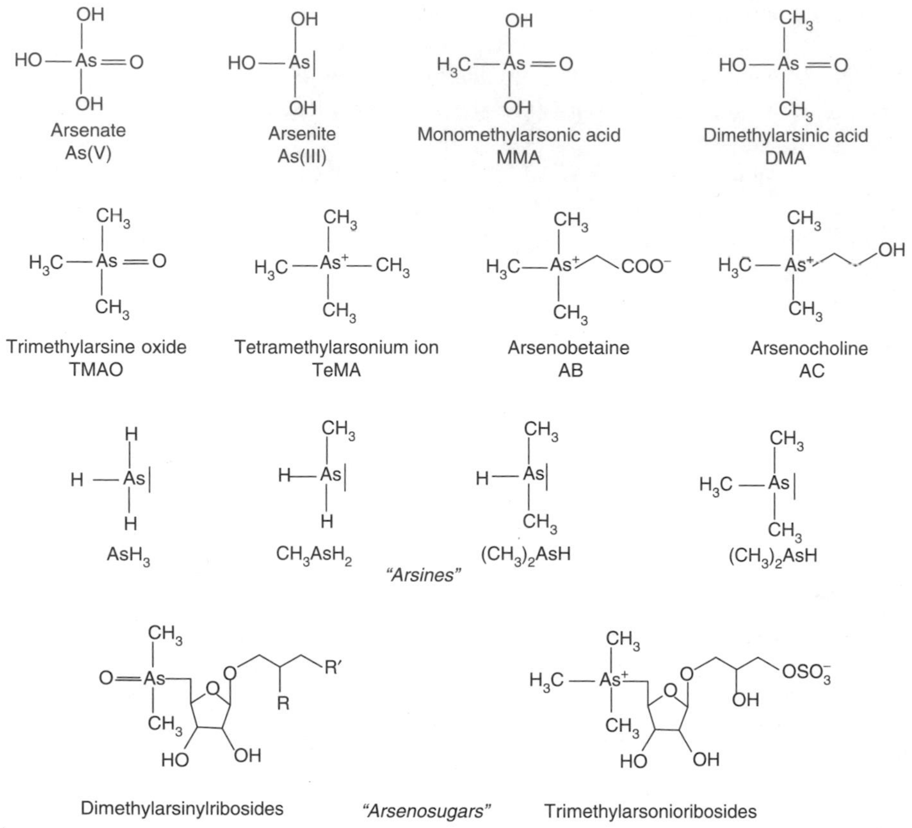

FIGURE 20.1 Structures of the most common arsenicals in the environment.

1. Organometallic Forms

The As species which are most frequently found in the environment are presented in Figure 20.1.

Organomercury compounds which may be encountered in the environment:

Monomethylmercury (MMHg) as well as dimethylmercury (DMHg) are the classic compounds which are encountered in aquatic systems. Benzoic mercury, mersalytic acid, ethyl- and phenylmercury, ethoxyethylmercury, nitromersol, and thimerosal are mercury derivatives, applied in different fields.

Organotin compounds which may be encountered in the environment:
Monosubstituted: Monomethyltin (MMT), Monobutyltin (MBT), Monooctyltin (MOT)
Disubstituted: Dimethyltin (DMT), Dibutyltin (DBT), Dioctyltin (DOT)
Trisubstituted: Tributyltin (TBT), Tricyclohexyltin (TcHT), Triphenyltin (TPhT)
Tetrasubstituted: Tetraphenyltin (TePhT)

Alkyllead compounds which may be encountered in the environment are:
Monosubstituted (RPb^{3+}): MPb^{3+}, $EtPb^{3+}$
Disubstituted (R_2Pb^{2+}): M_2Pb^{2+}, $MEtPb^{2+}$, Et_2Pb^{2+}
Trisubstituted (R_3Pb^+): M_3Pb^+, M_2EtPb^+, MEt_2Pb^+, Et_3Pb^+
Tetrasubstituted (R_4Pb): M_4Pb, M_3EtPb, M_2Et_2Pb, MEt_3Pb, Et_4Pb

B. Natural Occurrences

Arsenic is a major compound in 20 species of minerals, of which one elemental As, 23 alloys and arsenides, 49 sulfides and sulfosalts, 119 arsenates, seven arsenites, three silicates, and five other oxygen containing compounds.[7] Especially high concentrations of As are found in minerals containing sulfides and in the mixed sulfides of the M(II)AsS type, where M(II) stands for two-valent metals such as iron, silver, aluminum, copper, manganese, magnesium, nickel, and lead. Arsenopyrite, realgar, and orpiment are by far the three most abundant minerals containing As as a major element. These minerals are still less abundant than "arsenian" pyrite (FeS_2), in which, due to its chemical resemblance to sulfur, As apparently substitutes for S in the crystal structure of the mineral. This "arsenian" pyrite is probably the most important source of As in ore zones. Arsenic can also substitute for Si^{4+}, Al^{3+}, Fe^{3+}, Ti^{4+} in mineral structures and is present in many other rock-forming minerals, albeit at much lower concentrations.[8]

Mercury has an average crustal abundance of 0.05 to 0.10 mg kg^{-1}, the majority of which occurs as the mineral cinnabar.[9] Mercury is quite different from other metals in several respects:

(1) It is the only metal which is liquid at room temperature.
(2) It is the only metal which boils at temperatures below 650°C.
(3) It exists in oxidation states zero (Hg°) and one (Hg_2^{+2}) and two (Hg^{2+}).
(4) It is chemically quite inert, having a higher ionization potential than other electropositive elements. Natural sources include ocean emission, degassing of the earth's crust, weathering, emission from volcanoes, geothermal zones, and Hg mineralized areas.

Tin is a relatively scarce element with an average abundance in the Earth's crust of about two parts per million (ppm) compared with 94 ppm for zinc, 63 ppm for copper, and 12 ppm for lead.[10] Tin is produced from lode (hard-rock) deposits and placer deposits derived from the lodes. The tin mineral cassiterite (SnO_2) is the source of most tin production. A notable exception is the complex tin sulfide minerals in the subvolcanic or tin–silver lode deposits in Bolivia.[11] Cassiterite has a high specific gravity (6.8 to 7.1), a Moh's scale hardness of 6 to 7, and is usually a dark brown or black color with an adamantine luster.

Lead has been known since ancient times. It is sometimes found free in nature, but is usually obtained from the ores galena (PbS), anglesite ($PbSO_4$), cerussite ($PbCO_3$), and minum (Pb_3O_4). Although lead makes up only about 0.0013% of the earth's crust, it is not considered to be a rare element, since it is easily mined and refined. Most lead is obtained by roasting galena in hot air, although nearly one third of the lead used in the United States is obtained through recycling efforts. Environmental methylation of inorganic lead[12] may be a possible natural source of organic lead.

C. Anthropogenic Uses

Most As is produced as a by-product during mining of other metals. Arsenic trioxide was first produced from smelting of gold and silver; nowadays it is recovered from the dust and sludge associated to the smelting of copper and lead. Arsenic trioxide formed during the smelting process, volatilizes and concentrates primarily in the flue dust. Arsenic trioxide of commercial purity is then obtained by stepwise distillation.

Over the years, there have been large shifts in the anthropogenic uses of As. According to Loebenstein,[13] the largest U.S. demand for As in the 1920s stemmed from the manufacturing of insecticides, particularly calcium arsenate, which was used to fight boll weevil in cotton plants. Nowadays, the world As market is dominated by its use as a wood preservative. From the 1970s onwards, the As demand from agriculture has steeply decreased, due to the stricter environmental regulations on the use of inorganic arsenicals, and due to the rise of synthetic organic pesticide DDT in 1967. The use of inorganic arsenic as pesticide was restricted in 1991. The only remaining major agricultural use for As is, in the herbicides monosodium methanearsonate (MSMA) and disodium methanearsonate (DSMA). These organic pesticides have the advantage that they require application in much lower concentrations (2 to 4 kg ha^{-1}) than their inorganic equivalents (10 to 1000 kg ha^{-1}).[14] As also finds uses in the manufacture of glass, metal alloys, semiconductors, fodder additions, and veterinary chemicals.

Anthropogenic inputs of Hg to the environment largely exceed natural inputs. Hg emissions to air are comparable to direct inputs to aquatic environment and about half of the direct releases to terrestrial environment.[15] Fuel combustion, waste incineration, industrial processes (chlor-alkali plants), metal ore roasting, refining and processing are the largest anthropogenic point source categories on a global scale. Diffuse sources include fluorescence lamps, dry cell battery disposal, historical industrial discharges to waterways and waste deposits, landfill gas emissions, Hg from discarded manometer, passive emissions from latex paint (in which phenylmercury served as a biocide), motor vehicle emissions (fuel combustion and lubrication oil), gold mining waste, agricultural and lumber fungicides. Because of the wide use and volatility of some species, mercury is now a global pollutant which has been measured in the deep oceans, atmosphere, Antarctica, and the Arctic.[9] The main exposure pathway of Hg to humans is through the consumption of marine fishery products (fish, shellfish, crustaceans). In most foodstuffs, Hg is predominantly in the inorganic form and in low concentrations (<20 ng g^{-1}).[16] In fish and fish products, however, mercury occurs primarily in the methylated form and levels greater than 1200 ng g^{-1} have been found in edible portions of shark, swordfish, and tuna. Similar levels have also been found in fish of affected fresh-water systems, which have led to the introduction of fish consumption advisory limits in countries such as Canada, Sweden, and the U.S.A.

Tin and inorganic tin compounds are used in a variety of products and processes. Tin–niobium wires have interesting properties in the field of superconductive magnets which generate enormous field strengths but use almost no power.

Some 25 to 40 years ago, the introduction of organotin compounds into a large variety of industrial products contributed to the development of a number of important sectors within the chemical industry. The high toxicity of the trisubstituted species to a variety of organisms was quickly established, and the materials were immediately used in a large variety of biocidal applications. The most successful of these was tributyltin, in self-polishing, antifouling paint formulations, and these materials were employed on major sea-going vessels. Leisure boats also adopted this most efficient paint, and have unwittingly contributed to contamination of coastlines throughout the world.

By the start of the 1980s, 140,000 tons of TBT-based antifouling paints were consumed each year in boating and the merchant marine in the United States. During the same period, it was estimated that such paints covered 70% of the global fleet. This success was due to their high efficiency and long lifetime (4 to 7 years).[17] Between the 1980s and 1990s, because of their toxicity, the sale and use were severely restricted in Europe, the U.S.A., and New Zealand.

The use of TBT (mainly as an antifouling agent on boats) has been drastically restricted, whereas TPhT is still frequently used in agriculture. The industrial use of organotins is very different, and is directly related to chemical formulation of the compounds. Organotins are mainly used as stabilizers, catalysts, biocides, or pesticides.

Lead is a soft, malleable, and corrosion resistant material. Lead's high density makes it useful as a shield against x-ray and gamma-ray radiation and is used in x-ray machines and nuclear reactors. Lead is also used as a covering on some wires and cables to protect these from corrosion, as a material to absorb vibration and sound, and in the manufacture of ammunition. Today most of the lead is used in production of lead-acid storage batteries, such as the batteries in automobiles.

Several lead alloys are widely used, such as solder (used to join metallic items), type metal (used in printing presses and plates) and babbit metal (used to reduce friction in bearings).

Lead forms many useful compounds used in paints such as lead sulfate ($PbSO_4$), also known as anglesite, and is used as sublimed white lead and lead chromate ($PbCrO_4$) also known as crocoite, and used as chrome yellow paint. Lead dioxide (PbO_2) is a brown material which is used in lead-acid storage batteries. Trilead tetraoxide (Pb_3O_4), also known as red lead, is used to make a reddish-brown paint which prevents rust on outdoor steel structures. Lead nitrate ($Pb[NO_3]_2$) is used to make fireworks and other pyrotechnics. Lead monoxide (PbO) and lead silicate ($PbSiO_3$) are used to make some types of glass and in the production of rubber and paints.

Environmental pollution by organic lead is almost entirely due to the manufacture and use of tetra-alkyllead (R_4Pb) compounds as petrol additives a few years ago, and the toxicity of these species is well documented. Varying proportions of the five R_4Pb compounds (tetramethylated, tetraethyllead, and their mixed derivaties) are added to petrol to increase the octane rating of fuels for high-compression internal combustion engines. In the environment, R_4Pb compounds decompose to inorganic lead with trialkyllead (R_3Pb^+) and dialkyllead (R_2Pb^{2+}) compounds as fairly persistent intermediates.[18]

D. TOXICITY

As is one of the few chemical elements which is universally associated with the word "poison". This reputation is not undeserved, for over the centuries, many deaths can be attributed to the administration of arsenic trioxide as "inheritance powder". The metalloid has a dual reputation, for its use in medicine in earlier times is equally well documented. As was widely prescribed to treat skin diseases, fevers, malarial disorders, syphilis, lumbago, epilepsy, anemia, ulcers, etc.[19]

Nowadays, much more is known about the toxicity effects of As. Although some authors suggested that, at low concentrations, As is an essential element for organisms.[20] The toxicity arising from ingestion of inorganic As is believed to manifest into a series of effects, most importantly in systemic effects involving skin, cardiovascular and neurological systems.[21] Additionally, the International Agency for Research on Cancer (IARC) concluded that there is enough evidence to associate exposure of inorganic As to skin, lung and bladder cancers and classified As as a so called "group A" carcinogen to humans.[22] Dimethylarsinic (DMA) is equally shown to induce organ-specific lesions in the lungs.[23] These authors also mentioned dose-dependent increases in urinary bladder tumors upon lifetime exposure to DMA from diet or drinking water. DMA is believed to act either as a cancer promoter or as a complete carcinogen in different animals. As is the case for many environmental pollutants, extensive toxicity studies of As, have shown that different forms exhibit different toxicities. In contrast to mercury and lead, inorganic As species are more toxic than organic compounds. With the exception of tetramethylarsonium ion, a general toxicity order can be set as decreasing with the increasing degree of methylation (Table 20.2). From this point of view, the need for adequate, sensitive and accurate speciation methods for As is self-evident.

The toxic effects of Hg are dependent on the chemical form. $Hg°$ is readily adsorbed from the respiratory tract, but only slightly from the gastro-intestinal tract. After adsorption, Hg vapor is dissolved in blood and may pass the blood–brain barrier into the nervous system; the major part of body burden is however in the kidney. Hg^{2+} is only partially adsorbed from the gastro-intestinal tract and is mainly accumulated in the kidneys; it passes the blood–brain barrier to a limited extent. MMHg compounds are considerably more toxic than elemental Hg and its inorganic salts. MMHg is efficiently adsorbed from the gastro-intestinal tract, it passes the blood–brain and placenta barriers. MMHg primarily affects the central nervous system. In severe cases, specific anatomical areas of the brain are affected, causing irreversible damage. The first symptom at the lowest doses is paresthesias, an abnormal sensation or loss of sensation in the extremities of hands and feet. The intake associated with a 5% risk of mild neurological symptoms (the lowest effect intake) is approximately 4 to 8 $\mu g\,kg^{-1}\,day^{-1}$.[24] Based on this value, a joint FAO/WHO expert group, estimated a tolerable intake of 5 $\mu g\,kg^{-1}$ per week, with no more than 3.3 $\mu g\,kg^{-1}$ per week as MMHg for the adult population.[25] Prenatal life is more susceptible to brain damage from MMHg, as compared to adults. MMHg is believed to inhibit

TABLE 20.2
Experimental LD$_{50}$ Values from Oral Administration of Arsenic Species to Mice[189]

Arsenic Species	LD$_{50}$ (mg kg^{-1})
Arsenite: sodium arsenite	4.5
Arsenite: arsenic trioxide	34.5
Arsenate: sodium arsenate	14 to 18
Tetramethylarsonium ion	890
Dimethylarsinic acid	1200
Monomethylarsonic acid	1800
Arsenocholine	6500
Trimethylarsine oxide	10,600
Arsenobetaine	>10,000

cellular processes basic to cell division and neuronal migration.[26] Fetal MMHg exposure at low concentrations may cause metal retardation.[27,28] These effects may appear at intakes of five to ten times lower than that observed with adult exposure. Diet also influences the tolerance of humans for MMHg, and such variability must be taken into account when evaluating the toxic levels of organic mercurials. Some epidemiological studies suggest that high levels of Se may reduce the toxic effects of MMHg or produce a modification in the methylated activity with less conversion of inorganic Hg to MMHg. This antagonistic behavior is well known, mainly by formation of mercury selenide.

Well known outbreaks of Hg poisoning are the contamination of Minamata Bay by an acetaldehyde plant from 1948 to 1960[29] and the poisoning of bread in Iraq in 1972 after grain seeds had been treated with organomercury fungicides.[27] As a result of these outbreaks, the use of organomercurials in agriculture has been banned in most countries and its use in pharmaceuticals has dropped significantly. Measures taken in several industrial sectors (Chlor-alkali industry, phosphate industry, paper and pulp industry) such as the replacement of Hg-containing products in electrical components and batteries, the improvement of gas flue cleaning technologies, and the decreased use of Hg in dentistry and pharmaceuticals have led to a significant reduction in anthropogenic Hg inputs to the environment.[28,29]

During the last decade improvements in analytical techniques, speciation and reaction-oriented environmental Hg research has considerably improved knowledge of Hg-biogeochemical cycling. The main transformation pathways between the various Hg species in different environmental compartments have been identified (Figure 20.2), although the reaction mechanisms and biological species involved in the interconversion of Hg species in the ocean remain uncertain. The *in situ* (bacterial) conversion of inorganic Hg species to MMHg is an important feature of the Hg cycle in aquatic systems, as it is the first step in the bioaccumulation process. Methylation occurs in the water column as in the sediments (its origin in the atmosphere is still unknown) and has been shown to be predominant due to sulfate reducing bacteria in freshwater and estuarine systems. In the ocean DMHg is the main methylated compound in contrast to freshwater systems where DMHg is not found, suggesting that probably other species are responsible for the formation of MMHg in the oceans.[30] The *in situ* production and air–water exchange of Hg in surface waters[15,31] exerts a major influence on the fate of Hg in the environment. Volatilization of Hg competes with MMHg formation for the available Hg(II) substrate for reduction and methylation.

Although all forms of Hg are poisonous, the effects of mercury on ecology and human health are related to environmental transformations of inorganic mercury to the toxic and biomagnification-prone compound, monomethylmercury. Many Hg speciation methods are only focused on the

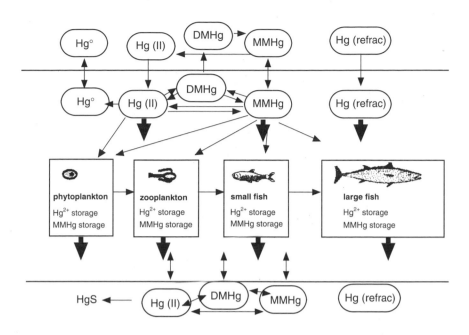

FIGURE 20.2 Hg transformation in the aquatic environment (adapted from Fitzgerald, W. F. and Mason, R., In *Global and Regional Mercury Cycles: Sources, Fluxes and Mass Balances*, Baeyens, W., Ebinghaus, R., and Vasiliev, O., Eds., Kluwer Academic Publishers, Dordrecht, The Netherlands, p. 85, 1996).

determination of MMHg.[32–37] Although DMHg has been found in fish, water and sediments,[33,34,38] most analytical protocols used would not provide reliable results for DMHg. Only very few studies deal with other Hg species (monoethylmercury, monophenylmercury, methoxyethylmercury, thimerosal,…).

The most dramatic noticeable toxic effect of TBT on the ecosystem (biota) is imposex, and although this was first recorded on European coastlines, it has now been reported throughout the world. This phenomenon appears when only 2 ng l^{-1} of TBT (as Sn) is present in the water. This very low concentration is sufficient to induce changes in the sexual characteristics of marine snails (dog-whelks), leading to sterility and a decline in the population. As in many environmental crises, biological responses provide the earliest evidence of a major disorder. Tremendous efforts were made by the international community on various issues to fully address this problem. Progress was made by paint manufacturers, resulting in considerably reduced releases of these compounds into the environment.

Tin has no known biochemical function. Tin toxicity includes growth depression and anaemia, and can modify the activity of several enzymes by interfering with the metabolism of Zn, Cu, and Ca. Neil et al.[39] investigated the capacity of Sn (II) and Sn (IV) ions (as $SnCl_2$ and $SnCl_4$) to activate heme oxygenase in cardiac tissue. The results from this investigation were that Sn (II) ions were more potent activators than Sn (IV). When compared with inorganic tin, organotin compounds are highly toxic and attack the central nervous system.[40] These studies have employed rats, but epidemiological studies suggest a similar behavior in humans. There exists, however, a specific need for elemental tin speciation in human studies.

Lead is considered to be a toxic element as a consequence of a variety of biochemical effects. Among these are included neurological problems, haematological effects, renal dysfunction, hypertension and cancer (the IARC has classified Pb as a carcinogen), for which there is evidence in animals but not yet in humans.[41] The latter study, also involving humans, however suggested that the increases in lead uptake have occurred when dietary Fe was low.

The harmful effects of organolead compounds are considered to be much greater than those of inorganic lead. The toxicity of alkyllead species diminishes in the sequence $R_4Pb \rightarrow R_3Pb^+ \rightarrow R_2Pb^{2+} \rightarrow Pb^{2+}$ (where R is a methyl or ethyl group). About 150 fatal cases of human intoxication with Et_4Pb have been reported in the literature. These were related to accidental exposures, but long term environmental exposure to low levels has been associated with a wide range of metabolic disorders and neurophysical deficits, especially in children.[42]

In mammals, inhalation or adsorption of R_4Pb compounds results in the formation of trialkyllead in tissues and body fluids. With the exception of methylleads found in the blood of petrol workers, these are usually below the detection limits in blood and urine samples.[43]

III. ARSENIC SPECIATION

A. SAMPLE PREPARATION

1. Storage

While for arsenic, contamination problems will only rarely occur, as long as trace element standard procedures are complied with, it is rather the preservation of samples which will be one of the troublesome steps in As speciation analysis. Events like changes in oxidation state, changes induced by microbial activity, or losses by volatilization or adsorption have to be kept from occurring. Total As aqueous samples are not subject to losses during storage, when kept in acid-washed glass, polytetrafluoroethylene (PTFE) or polyethylene containers.[44] Regarding storage for As speciation experiments on the other hand, there is at present no excess of studies which provide information on appropriate storage conditions for As. An overview of the influence of critical factors for species stability (pH, temperature, light and container material) and of procedures for preservation of the integrity of species is given in Ariza et al.[45] Recommended procedures are freezing, cooling, acidification, sterilization, deaeration, addition of ascorbic acid and/or storage in the dark. There is however, no general agreement on these procedures and reports are sometimes even conflicting. This is especially true for complex solid matrices such as soils, sediments and biological tissues. Nonetheless, for samples where bacteria will exist naturally, storage at low temperatures is required to prevent biological activity which might otherwise modify the sample's nature. For aqueous samples, time and temperature studies provided information that at higher concentrations (20 $\mu g\,l^{-1}$), immediate storage of filtered (0.45 μm) natural waters, at about 5°C, will preserve the As(III) and As(V) concentrations for about 30 days.[46] Samples with lower As concentrations are advised to be kept in the dark, at 4°C.[47]

2. Extraction

Most of the present-day techniques allow for assessment of As species in aqueous samples, without the requirement of any preconcentration or pretreatment. This is particularly true when coupled to hydride generation, which has shown to significantly increase the sensitivity of the analytical method, due to high selectivity of hydride formation and separation of the analytical compound from the matrix. This method is based on the very first report of As species separation, based on the ability of As species to form volatile hydrides and on their differences in boiling points (Figure 20.3).[48]

Sediment samples on the other hand may be treated in different ways, depending on the information required. Water contained in sediments can be removed and analyzed by the same methods as for aqueous samples, but speciation information on the solid fraction is more difficult to acquire. Series of sequential extractions are already employed to acquire the information needed to understand the cycling of As in sediments, for example, on water-soluble, phosphate-exchangeable, organically-bound and residual phases in such media.[49,50] Literature on extraction procedures for detection, hyphenated by chromatographic techniques, is much more recent; extractions of

$$(CH_3)_nAsO(OH)_{3-n} + H^+ + BH_4^- \rightarrow (CH_3)_nAs(OH)_{3-n} + H_2O + BH_3$$

$$(CH_3)_nAs(OH)_{3-n} + (3-n)H^+ + (3-n)BH_4^- \rightarrow (CH)_nAsH_{3-n} + (3-n)H_2O + (3-n)BH_3$$

As compound	Reaction product	Boiling point
Arsenite H_3AsO_4	AsH_3	$-55°C$
Monomethylarsonic acid $CH_3AsO(OH)_2$	CH_3AsH_2	$2°C$
Dimethylarsinic acid $(CH_3)_2AsOOH$	$(CH_3)_2AsH$	$36°C$
Trimethylarsine oxide $(CH_3)_3AsO$	$(CH3)_3As$	$70°C$

FIGURE 20.3 Arsine generation and corresponding boiling points.

sediments with phosphoric acid and hydroxylamine hydrochloride are reported to allow high pressure liquid chromatography (HPLC) measurement of labile As, since they preserve the two redox states of As in this fraction[51]; Garcia-Manyes et al.[52] reported a similar procedure with phosphoric acid and ascorbic acid; Montperrus et al.[53] efficiently extracted As from sediment and sludge using orthophosphoric acid, but reported a higher efficiency from old formation soils, with oxalate. Other extraction chemicals employed are methanol/hydrochloric acid/water,[54] acetone and hydrochloric acid[55] and 1,3-propanedithiol or 1,2-ethanedithiol.[56]

For biological materials, several extraction methods for As speciation, focusing on lower solvent volumes and reduced extraction times, have been developed. Biological tissues with high fat content may need to be defatted, prior to extraction of As. Solvents such as acetone or ether have been used for this purpose.[57,58] Extraction of solid samples is almost always aided by techniques such as shaking, heating or sonification. The latter is most popular, as the dispersion of tissue is thought to be maximized. Chemical mixtures for extraction of biological material are most often mixtures of methanol/water or methanom/chloroform,[59] but due to the difficulty in handling chloroform, it is the water/methanol mixture which is widely applied. An alternative possibility is enzymatic digestion with trypsin,[60] but changes in activity of the enzyme have to be checked, as differences in activity can lead to poor reproducibility. The extract may need further treatment prior to separation, seeing that next to the species of interest, matrix components are often also extracted. Cleanup of the extracts on a silica column was proven to be a possible source of loss of compounds; filtration with C_{18} or Florisil cartridges was observed to be efficient.[47] Solid particles that could damage chromatographic columns should also be removed.

B. CHROMATOGRAPHIC SEPARATION

Since most environmental As species are present in soluble forms, liquid chromatography (LC) is the most popular technique for the separation of As species. Also, HPLC allows easy coupling with element-specific detectors; only a simple interface is required. The chromatographic separation is pH dependent. At neutral pH, arsenate ($pK_{a_1} = 2.3$), monomethylarsonic acid (MMA, $pK_a = 3.6$) and DMA acid ($pK_a = 6.2$) are present as anions; arsenocholine (AC), trimethylarsine oxide (TMAO, $pK_a = 3.6$) and tetramethylarsonium ion (TeMA) as cations, arsenobetaine (AB, $pK_a = 2.18$) as zwitterion; arsenite (As(III), $pK_{a_1} = 9.3$) as an uncharged species. As a result, the anion exchange for separation of As(V), As(III), MMA and DMA and cation exchange for separation of AB, AC, TMAO, and TeMA, are commonly used. The use of reversed-phase ion pair HPLC, with appropriate counterions in the mobile phase (e.g., tetramethylammonium cation or heptanesulfonate) is also documented; Le et al.[61] reported the separation of seven As species on a reversed-phase C18 column with hexanesulfonate containing mobile phase. An overview of the different types of columns and mobile phases, applied in ion exchange and ion pair chromatography, is given in Gong et al.[62] Nakazato et al.[63] obtained separation of As(V), As(III) and MMA on an ion exclusion column packed with sulfonate polystyrene resin, with dilute nitric acid as mobile phase. The authors state that this technique enables robust and efficient separation of As in highly saline matrices such as seawater or human urine. Indeed, other LC techniques have been widely used for separation of As species in all kinds of environmental samples, but reports on successful separation of As in seawater appear to be rare because of deterioration of the separation performance by large amounts of chloride ions.

Based on the original method of Braman and Foreback,[48] although not as common as HPLC, some authors apply hydride formation in conformation with purge and trap-gas chromatography. Volatile hydrides which are formed from reaction with sodium borohydride and hydrochloric acid, are swept from solution and separated gas chromatographically. Slejkovec et al.[64] reported such a method of trapping the gaseous arsines in a liquid cooled U-tube packed with Chromosorb W. Upon comparison with LC, the authors describe a significant increase in detection limit due to the much larger sample volumes, up to 100 ml, whereas typically only 100 μl of sample is injected in LC. Gas chromatographic separation of As compounds in gaseous emissions is a much more common application of GC; the compounds do not require additional derivatization prior to separation.[65]

C. DETECTION METHODS

Over the years, in trace and major element analysis in general, three detectors which are especially suitable for element-specific detection have been developed: inductively coupled plasma mass spectrometry (ICP-MS), inductively coupled plasma atomic emission spectrometry (ICP-AES) and graphite furnace atomic absorption spectrometry (GFAAS).

1. Atomic Absorption Spectrometry

Until the 1980s, flame atomic absorption spectrometry (FAAS) was extensively used for As detection, but because FAAS suffers from low sensitivity (detection limit for As: 1 mg l^{-1}) and high background noise from flame, graphite furnace AAS (GFAAS) was introduced. An improvement of a factor of 10 to 100 in analytical sensitivity was obtained by avoiding dilution of atoms of the element in flame; drying, ashing, and atomizing the sample was done in a small heated graphite tube. Most often, to exclude any interferences, GFAAS was combined with HG for detection of total As. Additional research was conducted on the use of this technique, in combination with HPLC. However, because of the long drying–ashing–atomizing cycle of GFAAS, a direct coupling to HPLC was proved to be difficult. Procedures of collecting chromatographic fractions, followed by batch analysis of each fraction, and online methods in which effluent fractions are collected and periodically analyzed, have been developed.[66] Nevertheless, the method requires large chromatographic peaks, because 30 to 60 sec are needed for each determination. Another interface, a flow-injection system involving the use of HG for postcolumn hydride generation, has resulted in real-time signals and low μg l^{-1} detection limits for As.[67] Most probably due to these difficulties in coupling and the multielement possibilities of techniques such as ICP, atomic absorption methods do not seem to gain interest in the field of As detection and especially not in the scope of speciation.

2. Inductively Coupled Plasma

Among the various methods of excitation of elements for emission spectrometry, ICP is one of the most efficient. Also looking at the compilation of records on analytical techniques for As, it has found a widespread use. In spite of MS being clearly the predominant detection, coupled to the plasma method, some reports on the use of AES have also been made. Chasseau et al.[68] came to the conclusion that HPLC-ICP-AES may be a reliable technique for As speciation, when low detection limits are not required; they obtained detection limits ranging from 7 μg l^{-1} for As(III) to 18 μg l^{-1}

for As(V). Such results could already be expected, since quantification in ICP-AES is performed by monitoring a specific spectral line emitted by an atom. The most sensitive line for As lies in the UV region, at 193.7 nm, and large amounts of organic matrix may interfere with As analysis because of the emission by carbon at 193.1 nm. Arsenic's emission line at 228.8 nm, equally receives interference; cadmium also has a strong emission line at this wavelength. The sensitivity of ICP-MS, for many elements, exceeds that of ICP-AES by more than two orders of magnitude; detection limits for As are in the range of $\mu g \, l^{-1}$,[69] but major spectroscopic and nonspectroscopic interferences are encountered. The most obvious interference in the case of As is that caused by the formation of $^{40}Ar^{35}Cl^+$ in the plasma. Moreover, this interference is proportional to the concentration of Cl^- ions in the sample. As arsenic is monoisotopic, it is impossible to avoid this isobaric overlap with conventional quadrupole mass analyzers. Research in the field nevertheless revealed that the formation of $ArCl^+$ can be suppressed by the addition of nitrogen into the plasma.[70] By making an additional coupling with hydride generation between the chromatographic column and the detector, only arsines are carried to the detector, while the Cl^- ions remain in solution. Yet, this latter technique has the disadvantage as As species which do not form volatile arsines cannot be detected. In recent years, a second generation of ICP-MS, with high resolution mode $(M/\Delta M = 7800)$, has been developed and widely accepted due to their outstanding performance characteristics. This high mass resolution ICP-MS technique allows the spectral separation of $ArCl^+$ interference and the accompanying reduction in sensitivity results in detection limits from routine As analyses of 0.7 $\mu g \, l^{-1}$. Klaue and Blum[71] nevertheless compared ICP-MS in high resolution mode to HG-ICP-MS in low resolution mode. While the first method allows the precise determination of As above 1 $\mu g \, l^{-1}$, the HG-ICP-MS method resulted in an over 2000-fold increase in relative sensitivity; it offers detection limits in the 0.2 ng l^{-1} range and enables sub $\mu g \, l^{-1}$ measurements of As in drinking water.

Despite the drawbacks of interferences, ICP-MS is extensively used as detector in As speciation research. The consumption rate of liquid sample by ICP is often similar to that of HPLC and the combination of instrumentations can easily be achieved, simply by connecting the effluent line of the column to the input line of the nebulizer. Nevertheless, HPLC column conditions are selected primarily to get optimum separation, but of course, here too, the limiting factors of ICP-MS must be taken into account; for instance the use of phosphate buffer as mobile phase is not suitable for ICP-MS.

3. Atomic Fluorescence Spectrometry

Research for improvement of detection limits in As detection and speciation has resulted in the optimization of atomic fluorescence spectrometry (AFS). The advantages of AFS over AAS have already been described theoretically and experimentally.[72,73] However, it was not until recently that commercial AFS instruments have been developed. Subsequently, AFS became a promising technique for analysis of some environmentally important elements, amongst which is arsenic. The most widely used AFS systems are almost exclusively couplings to HG; as for AAS, hydride generation eliminates light scattering and background interferences from the matrix and increases the sensitivity of ASF significantly. One method which is not based on HG has been described in literature: the research group of Mester and Fodor applied ultrasonic nebulization (USN) as the interface between HPLC and AFS, and achieved accurate separation and determination of As(III), As(V), MMA and DMA[74] and of AB and AC,[75] down to absolute detection limits between 8.9 and 50 ng (250 μl injection volume). Several couplings of HPLC to HG-AFS are equally applied; Gomez Ariza et al.[76] described an anion exchange HPLC–HG-AFS system for the speciation of As(III), As(V), DMA and MMA and described detection limits of respectively 0.17, 0.38, 0.45, and 0.30 $\mu g \, l^{-1}$. Arsenobetaine, a nonhydride forming As species, was also detected by introducing online UV photo-oxidation prior to hydride generation. Other researchers, used cation exchange HPLC coupled to UV-HG-AFS.[77] Le et al.[61] described online microwave derivatization coupled to

HPLC–HG-AFS; they studied the separation of 11 As compounds by using ion pair chromatography. This speciation technique was successfully applied in the study of arsenosugars, which is often the common As constituents in seaweed products. All of these couplings are reported to be easily achieved. Owing to this diversity and due to its high selectivity and sensitivity, AFS together with ICP-MS, has become an important and promising technique for the speciation of As. Additionally, techniques nowadays include an extensive range of reports on different couplings and methods and on accurate determinations of As species in different environmental matrices.

IV. MERCURY SPECIATION

A. SAMPLE PREPARATION

Sample preparation is an extremely critical step in the course of a speciation analysis, as the original sample has to be "transformed" into a form which can be subjected to analysis, while the original distribution of an element over its various chemical forms may not be altered. In general, determination of Hg species involves the following steps: (1) sample collection/pretreatment/ preservation/storage; (2) extraction of Hg from the matrix/cleanup/preconcentration; (3) separation of Hg species of interest; (4) detection.

The appropriate analytical methods depend on the nature of the sample and concentration level.

1. Water

Rigorous cleaning procedures must be used for all equipment and laboratory ware that comes into contact with samples, especially for the speciation of Hg at low concentrations, such as in water samples. The best materials for sample storage and processing are Pyrex and Teflon (PTFE or FEP). Several cleaning procedures can be used (aqua regia, chromic acid, nitic acid, BrCl,...). A final soaking of Teflon in hot 70°C 1%HCl removes all traces of oxidizing compounds (e.g., chlorine) which can destroy MMHg in solution.[36]

The most volatile forms present in water are Hg° and DMHg. These should be removed from the samples immediately after collection by aeration with collection on gold (for total gaseous Hg) and Carbotrap or Tenax (for DMHg). After filtration, samples should be preserved prior to storage. For total Hg, samples can be acidified with HCl or HNO_3 or with the addition of an oxidant (BrCl), whereas for MMHg the samples can be acidified with HCl or stored unpreserved deep frozen.[36,78]

2. Air

Although the analysis of total gaseous Hg and particulate phase Hg in air can be conducted with high accuracy and precision,[79] there are still many problems related to the separation of specific Hg compounds in air. In general two approaches can be used: (1) selective adsorption methods, in which separation is operationally defined and (2) chromatographic methods which allow identification of the organomercury compounds. Selective adsorption methods allow the operational separation of Hg°, Hg (II), MMHg and DMHg and have been reviewed extensively in the literature.[80] Gas chromatographic techniques are limited to the determination of Hg°, MMHg and DMHg. MMHg and DMHg can be trapped using Carbotrap or Tenax. Hg° is retained by gold amalgamation.

3. Biota

Sample preparation must be carried out under clean mercury-free conditions to avoid contamination by inorganic Hg. Significant external contamination with MMHg is unlikely to occur. Relatively, little is known of the effect of storage on the stability of methylmercury in biological samples. Fresh samples are usually stored deep frozen or lyophilized in darkness or are sometimes sterilized. For some organisms methylmercury may decompose with repeated freezing and unfreezing (particularly in bivalves).

4. Sediments

In sediments and soils, the percentage of MMHg is usually very low, resulting from equilibrium between methylation and demethylation reactions. Samples are usually analyzed fresh, or if long-term storage is required, samples should be kept in the dark at low temperatures or lyophilized. There is still much debate on the effect of sample pretreatment on the MMHg levels obtained. In some cases no differences were found between fresh and dried (lyophilized) sediments,[81,82] whereas in other cases much higher results were found in dried compared to wet sediments.[82] Preliminary tests show that the presence of oxygen and porewater during sample preparation may also play a role.[82] Further investigation in this field is required.

B. Extraction Procedures

The most commonly used procedures for the extraction of organomercury species from environmental samples are acid extraction (mostly combined with solvent extraction), distillation and alkaline extractions.

Acid digestion combined with solvent extraction was first proposed by Westöö[83] for the extraction of MMHg in foodstuffs. The method involved leaching the mercury compounds from the sample using concentrated hydrochloric acid, followed by extraction of the metal chloride into benzene. The mercury species were then taken into an aqueous phase by conversion to hydroxide using ammonium hydroxide, saturated with sodium sulfate. Cysteine, thiosulfate, or other thiol-containing reagents are now more commonly used to facilitate phase transfer.[33] Because GC was used to separate the species, the aqueous phase was acidified with concentrated hydrochloric acid and back-extracted with benzene prior to injection. Later, many modifications of Westöö's methods were proposed for selective extraction of methylmercury from a mineral acidic medium containing NaCl,[84,85] KBr,[86,87] and iodoacetic acid,[88,89] using successive extractions with organic solvents, e.g., benzene, toluene, chloroform or dichloromethane. Many studies were afterwards carried out to improve the efficiency of MMHg extraction in foodstuffs (e.g., Refs. 90–92).

For sediments, several acids have been proposed. Bloom et al.[93] used $5\%H_2SO_4$ in combination with $CuSO_4$ and KBr. $4 M HNO_3$ and $4 M HCl$ have been used by Tseng et al.[92] and Leermakers et al.[94] Either room-temperature procedures[93] or procedures at elevated temperatures using either conventional heating or microwave assisted heating have been used.[92,94] The microwave assisted acid extraction[92,94] and microwave assisted organic solvent extraction[95] have been used for the extraction of MMHg from sediments.

In water samples, MMHg complexed to organic ligands may be extracted by $HCl/KCl/CH_2Cl_2$ followed by back extraction in water.[96] Recently an alternative method was proposed for the simultaneous extraction of Hg^{2+} and MMHg in natural waters at pg l^{-1} levels. Hg^{2+} and MMHg are extracted into toluene as dithizonates after acidification of the water sample, followed by back extraction into an aqueous solution of Na_2S, removal of H_2S by purging with N_2.[97]

Vapor distillation, in a stream of air or nitrogen at 150°C, of a homogenate of the solid sample in diluted H_2SO_4 or HCl with excess of NaCl was fist proposed by Nagase et al.[98] and Horvat et al.[99] for the nonchromatographic separation of inorganic Hg and MMHg. In combination with the ethylation technique, Carbotrap or Tenax preconcentration, GC separation and AFS detection[90,96] this became the method of choice for the extraction of MMHg in sediments because of its high efficiency (MMHg recoveries practically 100%), elimination of inorganic Hg in the extract and the formation of clean aqueous extracts which eliminate interferences in the ethylation step. However, investigations in the mid-1990s showed that the distillation procedure used to separate methylmercury from water and sediment samples generates artificially MeHg aided by the presence of natural organic substances (see further).

Alkaline digestion and extraction: extractions in KOH methanol[38] and tetramethylammonium-hydroxide (TMAH)[100] have been proposed to release MMHg from biological samples and sediments while maintaining original Hg–C bonds. This is the most efficient method for extraction of MMHg from biological samples, but for sediments serious problems are encountered in subsequent steps (preconcentration, separation or detection) due to high levels of organic matter, sulfides or ferric ions coextracted with the sought methylmercury species using this sample treatment.[90]

Supercritical fluid extraction (SFE) has also been used for the extraction of MMHg from sediments.[91,101] Lorenzo et al.[101] compared manual, microwave assisted techniques and SFE for the extraction of Hg from aquatic sediments. Higher recoveries were obtained with microwave extraction techniques compared to manual extraction techniques and SFE.

Not all the available methods extract the mercury species from solid samples (soil, sediment or biological material) with acceptable efficiency. The procedure giving the best recovery for methyl mercury from soil (95 ± 4%) is the distillation method,[90] and from fish tissue it is alkaline digestion using tetramethylammonium hydroxide with focused microwave power (95 to 105%).[92] There is no standardized method to assess the extraction efficiency of a particular method, but several options are described in the subchapter on extraction efficiencies.

C. METHYLATION

Significant artificial methylmercury production during analysis was first highlighted by different groups at the Fourth Conference on Mercury as a Global Pollutant in Hamburg in 1996.[93,102] The production of artificial MMHg during the analytical procedure is a problem reported especially for sediments and may result in a significant bias of measurements. Natural sediments often contain very low amounts of MMHg, representing only 0.1 to 1.5% of total mercury. Even if artificial mercury methylation occurs in the small proportion of 0.02 to 0.03% of inorganic mercury only, this can result in 30 to 80% overestimation of MMHg concentrations in sediment.

In the aftermath of these early investigations, critical comments concerning the certified MMHg values in reference materials were made. The controversy was serious enough for the European Commission to finance a workshop, the conclusions of which were summarized in a special issue of *Chemosphere,* published in 1999.[103] The causes and factors involved in methylmercury formation during analysis were systematically evaluated. A series of different techniques commonly used to extract MMHg from various matrixes were screened and tested to evaluate their potential to accidentally generate MMHg from inorganic Hg^{2+} during sample preparation. The results highlighted the assumption that mercury species transformations were occurring during the sample pretreatment step, specially with distillation-based methods. The magnitude of artificial methylation using the distillation procedure increased linearly with the total Hg content, with increasing DOC content (in water samples); it was highest in the presence of carboxylic acids, humic materials, degraded terrestrial leaves or particles with large surface area, whereas it was not observed in fresh plant material or in biological tissues and showed a time–temperature dependence. The observation was not limited to sediment distillates. Methylation artifacts were also observed during hot alkaline digestion and SFE. Acid leaching with $H_2SO_4/KBr/CuSO_4$ at room temperature or with diluted HNO_3 (short microwave extraction procedure) followed by CH_2Cl_2 extraction and back-extraction in water did not give rise to methylation artifacts.[92–94,104] Later experiments showed that the methylation artifact was linked to the amount of reactive Hg in leachate or distillate.[105] Using speciated isotope dilute ICPMS coupled to capillary GC, Rodriguez et al.[106] conclude that the amount of inorganic Hg present in the final derivatization and extraction step is the determining factor for methylation artifacts and that transalkylation reactions in final organic phase are the most plausible mechanisms. In that work the derivatized compounds were extracted in an organic phase for injection in CGC in contrast to the other studies which use either Tenax collection, thermosorption, GC, pyrolysis and cold vapor-atomic fluorescence spectrometry (CVAFS)[93,105]; or ICPMS[102]; or headspace injection of the derivatized compounds[94,104] or cryogenic-trapping GC-AAS.[92]

Methylation artifacts have also been shown to occur during derivatization due to the presence of small impurities of methyl groups in derivatization reagents[90] and during separation due to the silanizing agent (demethyldisilzane) used to prepare the GC column.[92]

D. CHROMATOGRAPHIC SEPARATION

1. Gas Chromatography

Apart from the problems associated with the extraction of organomercurials, problems also exist with the chromatography of the organomercury halides. The different packed and capillary columns used have been reviewed by Baeyens.[32] To prevent ion-exchange and adsorption processes on the column (which cause undesirable effects such as tailing, changing of the retention time and decrease of peak areas/heights) a passivation of packing material is needed with Hg (II)-chloride in benzene (or toluene). Moreover the more common GC detectors, e.g., ECD, may be lacking the required selectivity to be used in the speciation of Hg in environmental samples. Its unselective response required laborious cleanup processes of the extract in the organic phase.

To overcome these problems, precolumn derivatization of Hg species was applied to transform these to nonpolar derivatives. These can then be separated on nonpolar packed[88,96,107] or capillary columns.[85] Iodation with acetic acid,[88,89] hydration with $NaBH_4$,[92,108-110] aqueous phase ethylation with $NaBEt_4$[90,96] and derivatization with a Grignard reagent (ethylation, butylation, propylation,...)[85] are the most commonly used methods.

Aqueous phase ethylation, room temperature precollection, separation by GC with CVAFS detection has become the most frequently used in laboratories involved in studies of the biogeochemical cycle of mercury. The ethylated species are volatile including elemental Hg and dimethylmercury and these can be purged from solution at room temperature and collected on sorbents such as Carbotrap or Tenax. After thermal release the mercury compounds are transferred to a (packed) gas chromatographic column (OV3 on Chromosorb W). Individual mercury compounds are separated either by cryogenic,[96] isothermal[111] or temperature programmed GC.[93] Instead of collection on Carbotrap or Tenax, the ethylated compounds may be injected directly on the GC column by headspace injection[94,104] or cryotrapped on a fused-silica column and desorbed by flash heating.[100,112] As the Hg species are eluted from the column these are thermally decomposed in a pyrolytic column (900°C) before it is measured by a Hg specific detector (CVAFS, CVAAS, QFAAS [quartz-furnace AAS], MIP-AES, ICP-MS,...). Very low detection limits can be achieved, particularly if methylmercury is preseparated by distillation (6 pg l^{-1} for water and 1 pg g^{-1} for biota and sediment samples).[90] The critical part of this procedure is the sample preparation prior to ethylation. Methylmercury compounds must be removed from their binding sites to facilitate the ethylation reaction and interfering compounds such as chlorides and sulfides must also be removed.[90,96] Inorganic Hg and MMHg can be determined simultaneously. It should be mentioned that ethylation cannot be used for determination of ethylmercury compounds because these can not be distinguished from Hg^{2+} after derivatization. The usefulness of other derivatization agents such as sodiumtetrapropylborate (e.g., Refs. 113,114) and sodiumteraphen-hylborate (e.g., Refs. 114,115) has been investigated. Sodium borohydride may also be used to form volatile methylmercury hydride which is then quantified by gas chromatography in line with a Fourier-transform infrared spectrophotometer.[109]

Especially when using Grignard derivatization, sample preparation may be laborious and time consuming and extraction of the organometallic compounds from the concominant matrix, derivatization and further cleanup is required.

Several techniques have been used to overcome the problem of low column loadings on capillary columns. Capillary columns have been used after preconcentration of the alkylderivatives on a wide-bore fused-silica column[112] or by solid-phase microextraction (SPME).[114] Large volume injection techniques have been applied on capillary columns coated with 0.25 μm DB-5.[116] Multicapillary GC (MCGC) (919 capillaries, 1 m $\times$ 40 μm i.d. coated with 0.2 μm SE 30 stationary phase (Alltech)) coupled to ICPMS[112,117]: allows column loadings and carrier gas flow rates to approach those of packed columns. Basic and unique features are the high speed of separation at large sample injection volumes with the exceptional high range of volumetric velocities of the carrier gas at which the column retains its high efficiency. This makes plasma source detection ideally suited for MCGC, leading to a coupled technique with a tremendous potential for separation analysis. Several applications involve the coupling of MCGC with MIP-AES[118] or with ICPMS.[117]

SPME capillary gas chromatography (SPME-GC) can be used for the extraction of organometal compounds after these have been derivatized to a sufficiently volatile form (see also organotin speciation). A silica fiber coated with polydimethylsiloxane (PDMS) is brought into the (headspace) of the sample. After exposure, the fiber is inserted into the GC injection port and the compounds are thermally desorbed for subsequent analysis. This method has higher sensitivity compared to the injection of solvent on a capillary column (usually 1 μl) but requires the use of standard addition as a calibration method. After GC separation, analysis can be performed by furnace atomization plasma emission spectrometry (FAPES).[114]

2. Liquid Chromatography

Until recently the main disadvantage of this technique was the poor sensitivity of the detectors. Development of more sensitive detectors such as a reductive amperometric electrochemical detection, ultraviolet detection, ICP-AES, ICPMS, AFS, and AAS has resulted in wider applications in environmental studies. The main advantage of liquid chromatography is the possibility to separate a great variability of organomercury compounds. Applications of HPLC for Hg speciation studies have been reviewed by Harrington.[37]

Practically all HPLC methods for Hg speciation reported in the literature were based on reversed-phase separations, involving use of a silica-bonded phase column and a mobile phase containing an organic modifier, a chelating or ion-pair reagent and in some cases, a pH buffer. The interface to couple HPLC columns with the atomizer can be very simple, with the direct connection of the exit of the column with the nebulizer of the AAS or plasma detector. Unfortunately, the efficiency of the nebulizer is very low (1 to 3%) limiting the sensitivity, especially for flame AAS. A general way out of this lack of sensitivity is post column derivatization to form cold vapor of Hg. However, generation of cold vapor from organomercury species requires an extra step for conversion to Hg (II), otherwise the response will depend on the species present. This conversion is usually online, involving oxidation with potassium dichromate, with UV light and with acidic potassium persulfate, sometimes in the presence of additional reagents (e.g., Refs. 119–122). In an effort to analyze low levels of mercury species, some workers have developed on- and offline sample preconcentration methods.[119,120,123]

Besides reversed-phased HPLC, ion chromatography (IC) has also been used to separate Hg species.[124,125] IC provides the possibility to separate more polar and ionic species directly, so that sample pretreatment can be simplified. The coupling of IC with CV-ICPMS allows very low detection limits to be obtained.[125]

E. DETECTION METHODS

The analytical sensitivity and selectivity requirements for reliable Hg speciation analysis are achieved with the use of hyphenated techniques. Most chromatographic detectors incorporated in commercial instruments are either universal or selective, but lack the necessary specificity for Hg. The first work on Hg speciation was performed using GC with ECD detection. The nonspecific character of the detector favoured the use of GC-MIP-AES because of its high element specificity towards Hg.[89,91,110,126–127] The availability of a commercial instrument and its higher sensitivity compared to direct nebulization in an ICP-AES has made this a very popular instrument. FAPES[91, 114,128] and quartz furnace atomic absorption spectrometry[100] have also been used. The development of a commercial, relatively inexpensive, extremely sensitive and selective CV-AFS instrumentation in the late 1980s and 1990s[129,130] made this the most popular detector for laboratories working on the biogeochemical cycling of Hg. In recent years, the use of ICPMS in speciation analysis has increased tremendously (e.g., see Ref. 131 for a review). Besides its high sensitivity and selectivity, ICPMS offers the opportunity to perform speciated isotope dilution mass spectrometry (SIDMS).[102,106,132,133] Not only is this technique highly accurate and precise, but also the isotopically enriched isotopes can be used as tracers to check for species transformations and extraction recoveries. Based on SIDMS, Gelaude[134] recently reported the separation and quantification of inorganic Hg and MMHg in solid samples after a thermal liberation of the compounds with an electrothermal graphite furnace.

The detection systems used with HPLC can be broadly divided into three approaches: photometry, plasma techniques (ICPAES, ICPMS), and cold vapour atomic absorption and fluorescence spectroscopy (CV-AAS, CV-AFS). The method with the lowest limits of detection (LOD) with sample introduction via a direct injection nebulizer used ICP-MS.[135] An HPLC system coupled to atmospheric pressure chemical ionization MS was used to identify methyl mercury spiked into a fish tissue CRM (DORM-1, NRCC).[136] This type of system has a significant advantage over elemental detection methods because identification of the species present is based on their structure, rather than matching the analyte's retention time to that of a standard.

The use of cold vapor generation coupled to ICP-MS lowers the detection limit by about a factor of 15, facilitating the detection of mercury species in ocean water samples,[137] which is not possible with conventional nebulization. The use of cold vapor AAS allows for the detection of mercury compounds down to between 0.1 and 1 ng for Hg (II), methyl mercury, and ethyl mercury (depending on the system).

In comparison with HPLC–ICPMS, CGC–ICPMS offers a higher resolving power and 100% introduction efficiency, it allows more stable plasma and gives origin to fewer spectral interferences as the result of the plasma being dry and finally leads to less sampling cone and skimmer wear.[131] The coupling is somewhat more complicated. Usually a heated transfer line is used to avoid condensation of the species[138] for multielement speciation. For Hg this is not mandatory. Simultaneous speciation analysis of mercury and tin in biological samples using CGC–ICPMS has been performed by Montperrus et al.[139]

Time-of-flight mass spectrometry (TOFMS) is an alternative to scanning-based mass analyzers. Coupled to ICP it can produce a complete mass spectrum in less than 50 μsec. CGC combined with ICP-TOFMS has been developed for the speciation of Hg[140] and later improved by MCGC-ICP-TOFS[141] allowing complete chromatographic separation within a chromatographic runtime of less than 1 min.

V. TIN SPECIATION

A. SAMPLE PREPARATION

Although a selection of commonly used sample preparation methods (extraction and derivatization) are presented and their specific application for the determination of butyltin in sediments, it is not our intention to compare and discuss all their advantages and disadvantages. Instead, we will focus on one preparation method which seems to us relatively simple, reliable and sensitive. The instrumentation needed is a GC coupled to either a MS (Ion-Trap or Quadrupole), a Pulsed

Flame Photometric Detector (PFPD) or an ICP-MS. The various sample treatment steps, liberation of the compounds, their derivatization and preconcentration via headspace on to a SPME phase, all occur in the same vial, limiting contamination and loss risks. SPME is a solvent-free sample preparation method in which a fused-silica fiber coated with a polymeric organic stationary phase is used to extract organic compounds directly from aqueous or gaseous samples.[142,143] Further GC separation of the compounds and MS, FPD, or ICP-MS detection allows very sensitive determinations. This method will be further referred to as the SPME method.

1. Extraction and Preconcentration

Owing to the low volatility of the organotins in environment, a derivatization reaction is required before a gas chromatography-based technique is to be applied. Furthermore, derivatization also permits reduction in the occurrence of possible interferences during subsequent analytical steps, and particularly at the detection stage. It is important to know which kind of derivatization (hydride generation with $NaBH_4$, ethylation with $NaBEt_4$, and alkylation with Grignard reagents) one is willing to carry out on the organotin compounds. Hydridization (in basic solution) and ethylation (buffered at a pH around four to five) can be directly performed on the aqueous phase, while alkylation with a Grignard reagent should take place in an anhydrous environment. Often a one-step simultaneous aqueous phase extraction and hydridization or ethylation and a (back-) extraction of the derivatized organotin compounds into an organic solvent is applied.

Sediments and biota. Several sample digestion procedures (mostly between 0.1 and 1 g of freeze-dried sample) can be applied such as: (1) enzymatic digestion with lipase and protease; (2) digestion with TMAH or KOH−EtOH or NaOH in methanol−water with or without microwave oven assistance; (3) digestion with HCl and NaCl aqueous solution, with addition of ethyl acetate and methanol; (4) extraction by supercritical CO_2; (5) extraction by acetic acid. Often these digestions and extractions are accompanied by simultaneous extraction into an organic (or different organic) solvent such as hexane, dichloromethane, acetic acid-tropolone in hexane, and tropolone in dichloromethane, methanol or diethyl ether.

In the SPME method, sample digestion was based on the work of Nagase and Hasebe.[144] A 0.5-g amount of freeze-dried sample was placed in a 40-ml vial with a Teflon stirring bar and 5 ml of TMAH solution or KOH−EtOH solution were added. The mixture was then heated (60°C) for 1 h and afterwards cooled to room temperature before derivatization of the organotin compounds.

2. Derivatization

The derivatization reactions applied most commonly for organotin analysis are hydride generation with $NaBH_4$, ethylation with $NaBEt_4$, and alkylation with Grignard reagents.[145] It is worth stressing that derivatization and extraction must be considered one of the most critical steps in organotin analysis. Low yields in derivatization and degradation phenomena (especially for phenyltins) can heavily affect quality of the results. A validation, or at least a careful study, of the procedure as used in the laboratory for the particular matrix, along with its particular interferences, is necessary.[146] This validation is always hindered by a lack of commercially available derivatized standards.

Recently, the synthesis of derivatized standards for ethylation and Grignard derivatization has been carried out at the Free University of Amsterdam[147] within the framework of an EC-funded certification project (CRM 477) allowing the establishment of optimization and validation studies. The purpose was to prepare highly purified butyltin and phenyltin compounds (in the form of salts) and their ethylated and pentylated derivatives for use as calibration and recovery tests.

The study on derivatization demonstrated that the yields obtained by Grignard reactions were acceptable and that no systematic error could be suspended.[145]

Organotin compounds in aqueous samples can be directly derivatized by adding an appropriate amount of $NaBH_4$ or $NaBEt_4$ to a buffered solution (see Table 20.3). Alkylation with

TABLE 20.3
Examples of Experimental Conditions for Derivatization Reactions Used in Butyltin Determinations in Sediment

Reaction	Experimental Conditions
Hydride generation (with $NaBH_4$)	10% $NaBH_4$ in 1% NaOH in milli-Q water (after acetic acid extraction)
Grignard derivatization	4% $NaBH_4$ in seawater (after acetic acid extraction)
	Ethylation with EtMgCI (2 mol^{-1}) in tetrahydrofuran
	Pentylation with PeMgBr (1 mol^{-1}) in diethyl ether
	Pentylation with PeMgBr (2 mol^{-1}) in diethyl ether
Ethylation	Ethylation with 2% $NaBE_4$

Adapted from Quevauviller.[149]

Grignard reagents should take place in an organic, waterfree solvent such as tetrahydrofuran, diethyl ether, hexane, dichloromethane, etc.

In the SPME method, after cooling the sample to room temperature, 30 ml of ammonium buffer (pH 8) were added. To neutralize the excess of TMAH or KOH, an appropriate amount of hydrochloric acid (12 $mol\,l^{-1}$) was slowly added to the mixture until the pH was restored to eight to nine. Buffer was added before the acid to avoid very low pH values and high temperatures on a local scale in the solution. In the next step, 500 μl of NaBEt4 solution (1%) were added and the vial was placed in a thermostatically controlled bath at 85°C for 15 min. Subsequently, during 15 min the compounds were sampled from the headspace by means of headspace SPME.

B. CHROMATOGRAPHIC SEPARATION

1. Liquid Chromatography

LC, in particular HPLC, offers the possibility of avoiding the time-consuming step of derivatization, minimizing the number of processes involved in the determination, which makes the procedure less prone to contamination or loss of analyte, except for the SPME Headspace-GC method. Several methods have been published following this approach (see Ref. 148).

The liquid chromatographic methods can be classified mainly into three categories: (1) Cation-exchange chromatography; (2) reversed-phase chromatography, and (3) normal-phase chromatography.

The detectors employed ranges between nonelement specific ones (UV absorption detectors, differential-pulse voltammetric detection systems or reverse-pulse amperometric detectors) and element-specific ones, such as atomic absorption (graphite furnace or flame), laser enhanced ionization, laser excited atomic fluorescence, fluorescence, inductively coupled plasma-atomic emission or mass spectrometry. Different approaches have been employed for the coupling of the chromatographic systems to the detector to improve the sensitivity of the overall system (hydride generation (HG)) or to make parts compatible (bleeding oxygen into the nebulizer gas in ICP-MS to avoid deposition of carbon on the surface of the cones). Frequently-used columns include Spherisorb ODS and Partisil-10SCX with methanol + either acetate or acetic acid or citrate as the mobile phase. The elution is normally isocratic, but a pH gradient from six to three or a methanol gradient can improve the elution of MBT or DBT.

While liquid chromatographic approaches obviate the need for a derivatization step, LC has a number of limitations. Many methods were developed on standards, but few on real environmental samples. Resolution is unfortunately poorer than for GC, creating an interest in developing supercritical fluid chromatography and capillary electrophoresis as alternatives to traditional LC approaches. Also the detection limits are far from as good as with GC coupling. Increasingly inductively coupled plasma mass spectrometry is seen as the detector of choice as its excellent sensitivity enables measurements at the $ng\,l^{-1}$ and $ng\,g^{-1}$ level, often observed in real environmental samples.

2. Gas Chromatography

Gas chromatographic separation of derivatized organotin compounds is the most popular and common method. Differences between the various GC procedures are located on the derivatization level rather than the type of capillary column. A selection of typical techniques which have been tested for sediment analysis at the occasion of three interlaboratory studies (TBT-spiked sediment, harbor sediment and coastal sediment) is given below.[149]

Since calibration is of paramount importance, participants were provided with pure calibrants of tributyltin chloride (synthesized by TNO, the Netherlands) for the verification of their own calibrants.

3. Hydride Generation/CGC-FPD

Approximately 2 g sediment were wetted with 2 ml water, extracted with 8 ml of 0.1% NaOH in methanol and back-extracted in 2 ml hexane. Derivatization was by hydride generation, using $NaBH_4$, followed by back-extraction into hexane. Extraction recoveries were assessed using TBTCl, $DBTCl_2$ and tripropyltin as internal standard; these ranged from $(59 \pm 3)\%$ for TPrT, $(87 \pm 3)\%$ for DBT and $(98 \pm 5)\%$ for TBT. Separation was by CGC (column of 25 m length, 0.32 mm internal diameter, 5% phenylmethylsilicone as stationary phase, 0.52 μm film thickness; H_2 as carrier gas at 2 ml min^{-1}; N_2 as make-up gas at 30 ml min^{-1}; injector temperature of 40°C; column temperature ranging from 40 to 250°C) and detection was by FPD (detector temperature of 250°C).

4. Ethylation/CGC-FPD

Approximately 1 g sediment was extracted with 5 ml HCl and 10 ml toluene, mechanically shaking for 15 h. The recovery was verified by spiking a marine sediment and was $(98.3 \pm 1.3)\%$ for TBT. Ethylation was performed with 150 μl of 2% $NaBEt_4$. Separation was by CGC (column of 25 m length, 0.32 mm internal diameter, CPSIL-5 as stationary phase, 0.14 μm film thickness; H_2 as carrier gas at 10 ml min^{-1} and make-up gas at 40 ml min^{-1}; injector temperature at 200°C; column temperature ranging from 70 to 200°C). Detection was by FPD (detector temperature at 250°C). Calibration was by calibration graph, using TBTAc, $DBTCl_2$ and $MBTCl_3$ as calibrants.

5. Pentylation/CGC–MS

First method. Approximately 1 g sediment was extracted ultrasonically with 20 ml diethyl ether/HCl in tropolone (the recovery ranged from 97 to 108% for the three butyltin compounds as assessed by spiking the CRM with the respective compounds). Derivatization was performed by addition of 2 *mol* l^{-1} PeMgBr in diethyl ether. The final extract was cleaned up with silica gel. Separation was by CGC (column of 25 m length, 0.32 mm internal diameter, methylsilicone as stationary phase, 0.8-μm film thickness; He as carrier gas at 1 ml min^{-1}; injector temperature at 280°C; column temperature ranging from 80 to 280°C; detector (transfer line) temperature at 280°C). Detection was by mass spectrometry. Calibration was by standard additions, using TBTCl, $DBTCl_2$, and $MBTCl_3$ as calibrants.

Second method. Approximately 0.5 g sediment was extracted with methanol/tropolone after addition of HCl. Tripropyltin was added as internal standard. Derivatization was performed by addition of pentylmagnesium bromide ($2 \, mol \, l^{-1}$) followed by cleanup with silica gel. Separation was by CGC (column of 25 m length, 0.2 mm internal diameter, methylphenylsilicon as stationary phase, 0.11 μm film thickness; He as carrier gas at 130 ml min^{-1}; injector temperature at 260°C; column temperature ranging from 80 to 280°C; detector (transfer line) temperature at 280°C). Detection was by mass spectrometry. Calibration was by calibration graph, using butyltin chloride compounds as calibrants.

The SPME method in combination with the headspace technique is a recently developed method which offers new possibilities. Headspace means mass transfer from the liquid to the gaseous phase and this mass transfer can be very slow for some compounds at room temperature. Therefore, to increase the efficiency of scavenging of the OT compounds on the solid phase higher temperatures are applied. Other important factors are (1) the fiber diameter (determining the fiber capacity); (2) the fiber type (determining the affinity for the compounds) and (3) the time of extraction.

C. GC-Detection Methods

1. GC-MS

Quite recently, mass spectrometry is emerging for analysis of organotin compounds in various matrices: sediments, water, biological samples and sewage sludge (Table 20.4).

For most of the proposed methods, the mass spectrometer is operating in positive electron impact ionization (EI) and single ion recording mode. Two compounds specific ions are followed for the correct identification of organotin compounds. The quadrupole and ion trap mass spectrometer have been used with LOD quite similar to those reported for other techniques and good linear range. Thanks to the selectivity of MS, perdeuterated organotin compounds, which perfectly mimics the behavior of native organotin compounds can be used as internal standard.[150,151] The high relative standard deviation associated with analysis by GC–MS which has been reported previously[152] has significantly decreased and is, to date, usually less than 10% RSD.

The use of GC–MS/MS instead of GC–MS, as proposed by Tsunoi et al.[151] allows a significant decrease of the noise, and of the LOD, (instrumental LOD < 0.5 pg injected). Large volume injection[153] is an interesting alternative: for the same solution, the LOD is decreased since a higher volume of solution is injected (50 μl compared to 1 μl). High resolution GC–MS (resolution > 10,000) has been investigated by Ikonomou et al.[154] but the advantage of higher sensibility is moderated by the cost of investment. Compared to electron impact ionization, chemical ionization (CI)[155] offers few improvements in term of LOD for methyl or hydride derivative.

Two original approaches have been proposed for more automation of the analysis. First, the combination of *in situ* aqueous ethylation followed by headspace SPME/GC–MS (see also above) provides a simple, cheap and rapid technique for the analysis of organotin compounds in sediments.[156] Second, Eiden et al.[157] proposed the *in situ* ethylation of organotin compounds in water, followed by purge online with helium and cryofocusing in the modified split/splitless injector of a GC–MS. Once the derivatizing agent is added, the method is entirely automated (LOD = 1 to 2 ng l^{-1} [as Sn]).

When the SPME method is combined with a GC–MS (quad) system,[156] instrumental conditions as specified in Table 20.5 can be used.

The repeatability (RSD) of five successive SPME extractions of blank buffer and NaBEt$_4$ are: 5% for MBT, 7% for DBT and 11% for TBT. This results in the following detection limits (pg g^{-1} as Sn dry weight), calculated as three times the baseline noise ($S/N = 3$): MBT (730), DBT (969) and TBT (806). The linearity of the method was investigated for a range of standard solutions between 30 to1000 ng l^{-1} and resulted in correlation coefficients from 0.9918 to 0.9957.

TABLE 20.4
Analysis of Organotin Compounds by GC–MS

Matrix	Analytes	Derivatization	Determination	Method LOD (as Sn) and RSD of the Method	Comments	References
River and sea water (500 ml)	MBT, DBT, TBT, MPT, DPT, TPT	Alkylation (Grignard)	GC–MS–MS (ion trap)	0.26 to 0.84 ng l^{-1} RSD = 6.5 to 11%	—	150
Water (100 ml) Tissue (2 g), Sediments (2 g)	MBT, DBT, TBT, TeBT, MPT, DPT, TPT, TCyT, DCyT	Ethylation	GC–MS (quad) GC–HRMS (10,000 resolution)	For HRMS: LOD water = 7 to 29 ppt; LOD (tissue/sediment) = 0.35 to 1.45 ng g^{-1}	Sensitivity 4 to 10 times greater for HRMS than MS	153
Sediments (1 g)	MBT, DBT, TBT	Ethylation	Headspace SPME-GCMS (quad)	730 to 806 pg g^{-1}	—	153
Mussel (0.2 g dry tissue)	MBT, DBT, TBT, TPhT	Ethylation	GC–MS (quad)	Inst LOD: 2 to 6 pg	—	186
Biol samples (500 mg)	MBT, DBT, TBT, MPT, DPT, TPT	Ethylation	GC–MS (quad)	4 to 78 ng g^{-1} wet weight RSD = 0.3 to 2.3%	—	149
Sea water (200 ml)	TBT, TPhT	Extraction of the chloride form	GC–MS (NCI or EI) quad	LOD <1 ng l^{-1} Inst LOD (NCI) = 25fg	Doping of the GC with HBr-methanolic solution	187
Water (800 ml)	Methyl and butyl tin	Ethylation	GC–MS ion trap	0.6 to 2.2 ng l^{-1} RSD = 10 to 16%	Cold-trap-thermal desorption system	156
Water (60 ml), sediments (1 to 5 g), sewage sludge	MBT, DBT, TBT, MPT, DPT, TPT, tricyclohexyltin	Ethylation	Large volume injection (50μl) GC–MS (quad)	Water: 0.5 to 1.5 ng l^{-1} sediments: 0.2 to 2 ng g^{-1} precision <7%	Cleaner extract with SPE, but absolute recovery are lower than LLE	152
Water (200 ml)	MBT, DBT, TBT	Methylation or hydridation	GC–MS ion trap	Inst LOD (EI): 0.9 to 24 pg inst. LOD (CI): 1.3 to 32 pg	—	154
Biological tissue (2 g)	MBT, DBT, TBT, MPT, DPT, TPT, fenbutatin, DCT, TCT	Methylation by Grignard reagent	GC–MS quad and ion trap	Inst. LOD: 1 to 10 pg RSD = 15 to 30% for real samples		188

EI, Electron Impact; CI, Chemical Ionization; SPE, Solid Phase Extraction; SPME, Solid Phase MicroExtraction; LLE, Liquid–Liquid Extraction.

TABLE 20.5
Instrumental Parameters for GC–MS

Gas Chromatograph

Column	PTE-5 fused silica, 30 m × 0.32 mm i.d.,
	0.25 μm film thickness
Injection system	Splitless, 250°C
Oven temperature program	50°C (1 min), 10°C min^{-1} to 250°C (10 min),
	20°C min^{-1} to 290°C (4 min)

MS

Ionization energy	70 eV
Transfer line temperature	200°C
Quadrupole temperature	100°C
Monitorized m/z values	233 and 235 (for MBT); 261 and 263
	(for DBT and TBT)

The reliability was tested by analyzing a certified marine sediment reference material CRM 462. The TBT and DBT concentrations found[156] were in good agreement with the certified values. The MBT concentration was not certified in that sediment.

2. SPME-ICP-MS

Specific conditions of the GC system, coupled to an ICP-MS[158] are summarized in Table 20.6.

A repeatability of 8% using ten standard solutions of 150 ng l^{-1} TPhT and TCT was found, which is comparable to other values reported for SPME.[1,159–161] The limit of detection for TPhT was calculated in two ways[158]: (1) a procedural LOD was calculated by analyzing ten blanks using the 3s criterion, a value of 125 pg l^{-1} was obtained; (2) an instrumental LOD was calculated using the standard deviation of ten "blank areas" within the chromatogram of the standard. A value of 2 pg l^{-1} was obtained based on the 3s criterion. This type of LOD should be regarded as an indication of the instrumental possibilities of this technique.

TABLE 20.6
Instrumental Parameters for GC–ICP-MS

Gas Chromatograph

Column	FSOT, DB-1 (polydimethylsiloxane),
	30 m × 0.25 mm i.d., 0.50 μm film thickness
d_f Injection system	SPME liner, 270°C
Oven temperature program	80°C (1 min) → 220°C (2 min) → 290°C
	(2 min) temperature ramp = 45°C min^{-1}
Carrier gas/inlet pressure	H_2 (99%) + Xe (1%)/30 psi
Transfer line temperature	270°C

ICP-MS

Rfpower	1000 W
Carrier gas flow rate	0.8 l min^{-1}, Ar
Auxiliary gas flow rate	1.2 l min^{-1}, Ar
Plasma gas flow rate	15 l min^{-1}, Ar
Dwell times	^{120}Sn: 60 ms ^{126}Xe: 10 ms

TABLE 20.7
Instrumental Parameters for GC-PFDP

Gas chromatograph	Varian 3800
Column	Capillary column (30 m × 0.25 mm i.d.,) coated with methylsilicone (0.25 μm film thickness)
Injection system	Varian 1079 split/splitless, 2 μl injection volume
Oven temperature program	80°C (1 min) → 180°C temperature ramp = 30°C min^{-1} → 270°C temperature ramp = 10°C min^{-1}
Carrier gas	Nitrogen
PFPD temperature	270°C
PFPD gas flow rates (Sn–C emission)	22 ml min^{-1}, Air1; 30 ml min^{-1}, Air2; 25 ml min^{-1}, H$_2$

3. SPME-PFPD

The SPME method combined with a GC-PFPD system is still in full development. A similar method without the SPME preconcentration step is described by Bancon-Montigny et al.[162] The instrumental conditions are specified in Table 20.7.

In the flame organotins give rise to Sn–C bonds which emit in the blue at 390 nm and Sn–H bonds which emit in the red at 610 nm. Since Sn–C emission is more intense than Sn–H emission we only mention the absolute LOD reported for the Sn–C bond emission.[162] These LODs vary between 0.07 and 0.10 pg for MBT, DBT and TBT. Repeatability (%) for $n = 6$, varies between 3 and 7% for the former compounds. At 390 nm, however, sulfur species can interfere with the Sn–C emission. SPME would not only lower the LODs by preconcentration, but also avoid interference from sulfur.

VI. LEAD SPECIATION

A. SAMPLE PREPARATION

1. Extraction and Preconcentration

a. Air

The most widely used technique for trapping of tetraalkyllead compounds from an air stream is their collection on GC-packing in a cooled U-tube[163,164] or on an appropriate solid sorbent. By heating the U-tube or the solid sorbent the species are desorbed into a GC column or directly into the detector system for analysis.

b. Water

Tetraalkyllead compounds are either removed from water phase by the purge and trap method (similar to its trapping from air) or by solvent extraction. In the latter method, the species are quantitatively extracted from water saturated with NaCl into a smaller amount of hexane.[165]

The characteristics of di- and trialkyllead species (salts) do not allow these to be directly extracted by any organic solvent, but chelating agent assisted extraction gives better results. Dithizone, diethyldithiocarbamate (DDTC) and tetramethylenedithiocarbamate (TMDTC) are most often used as reagents while pentane, hexane or benzene are used as extraction solvents.

Dithizone is apparently the least convenient chelating agent of the three, due to its nonquantitative extraction efficiency, low stability and the need of addition of polar solvents.[166–169] The unique advantage of the dithizone method is, however, the possibility of reextraction of ionic organolead into the aqueous (dilute HNO_3) phase. Extraction of the complexes of ionic organolead species with DDTC at pH six to nine into benzene[170] or hexane[165,171,172] was found to be more efficient as only one extraction was necessary to obtain quantitative recovery. Pentane may also be used as the extraction solvent, but the extraction has to be carried out at pH nine.[169] Dithiocarbamates are not as sensitive to light as dithizonates which makes the handling easier and the procedure more reliable. Controversy exists about possible interference of the coextracted inorganic lead with subsequent derivatization and determination.[173] The high selectivity of the hexane-TMDTC extraction system for ionic alkylleads over Pb^{2+} greatly facilitates the determination of these analytes in matrices contaminated with high levels of Pb^{2+}. Inorganic interferences with subsequent derivatization may be effectively masked with EDTA.

c. Sediments and Biota

After homogenization of a dried sample, the organolead compounds are extracted in the same way as these are removed from the water phase: tetraalkyllead species are quantitatively extracted from water saturated with NaCl into a smaller amount of organic solvent, while for ionic alkyllead species chelating agents strongly improve the extraction efficiency. For biological materials the extraction is often preceded by a hydrolysis step with either TMAH or enzymes.

2. Derivatization

The nonpolar character, volatility and thermal stability of tetraalkyllead species make these suitable for gas chromatographic separation. The ionic organolead compounds must be converted into forms accessible to gas chromatography by means of derivatization. The derivative chosen must preserve the structure of the lead–carbon bonds to ensure that the integrity of the species remains unaltered. Attempts to employ hydride generation failed apparently due to insufficient reproducibility, abundant interferences and instability of the organolead hydrides.[168,174] The most common derivatization procedures involve ethylation with sodium tetraethylborate (NaBEt$_4$)[107] and propylation,[165,171,172] butylation,[165,169] and phenylation[173] using the Grignard reaction.

Several inorganic lead species form the same end-species after derivatization which makes ethylation useless for these compounds. An interesting feature of ethylation is that it may be performed directly in the aqueous phase using the relatively water stable NaBEt$_4$.[107] To distinguish the various organolead species, propylation and butylation using the Grignard reaction are apparently the best choice at present. Smaller molecular weight and larger volatility of propylated species make their gas chromatographic separation faster than that of butylated species with less column carryover problems associated with derivatized inorganic lead. From a resolution point of view, butylation is preferable to propylation. In the latter case several pairs of products, including mostly mixed methylethyllead species, are poorly resolved whereas the only resolution problem after butylation seems to exist between M_2Bu_2Pb and Et_3BuPb. The use of high performance capillary columns alleviates the resolution problems. The use of a phenylation procedure seems to be less convenient owing to increased formation of artifacts which is probably due to the relatively large stability of the phenyl radical which promotes redistribution reactions.[173] The unreacted Grignard reagent must be destroyed prior to injection on to a column of the derivatized extract which is obtained by shaking the organic phase with dilute sulfuric acid. The organic phase is finally dried over anhydrous Na_2SO_4 and injected on to a GC column. The injected extract must not contain substances such as salts which remain in the injector after analyzing a series of samples.

Recoveries of Grignard derivatization (propylation and butylation) of ionic alkyllead species are reported by Radojevic.[175]

High recoveries of R_3Pb^+ species by butylation and propylation method and of R_2Pb^{2+} by the propylation method have been found by all workers, though there is disagreement regarding the recovery of R_2Pb^{2+} compounds after butylation. Chau et al.[176] and Chakraborti et al.[169] found almost quantitative recoveries of R_2Pb^{2+} species, but in three other studies low recoveries were found for similar procedures.[165,177] Chau et al.[176] extracted the ionic alkyllead species into benzene, whereas Chakraborti et al.[169] employed pentane as the solvent, and evaporated it prior to butylation.

B. Chromatographic Separation

1. Gas Chromatography

a. Injection

Conventional packed column injection ports are used for the sample introduction on packed and megabore columns. In case of capillary columns, unless a special injection technique is used, the low maximum allowable sample volume which may be introduced on the column negatively affects the experimental detection limits as only a tiny fraction of the derivatized extract is finally processed in the hyphenated system. Online preconcentration and injection of derivatized organolead species can solve that problem.[171] It consists of three consecutive processes taking place in the injection liner: sample injection, solvent venting and release of the analytes on to the column. Up to 25 μl can be processed at a time and larger amounts can be handled by successive injections of 20 to 25 μl volumes at 1 min intervals to remove the solvent.[172]

b. Columns

Nonpolar phases have been recommended in the literature for separation of derivatized organolead species.[175] Loadings of 3 to 10% OV-101 on Chromosorb W have been most frequently used.[165,169,178] There is a strong trend to replace these by open-tube megabore or capillary columns with polymethoxysilane coatings (DB-1, HP-1, RSL-150). Capillary columns have been used mostly in combination with MIP-AES and recently ICP-MS,[1,178-182] but very seldom with GF-AAS[183] or QF-AAS.[184] Packed columns do not allow for effective resolution between M_2Pb^{2+} and Et_3Pb^+ when butylation is used as the derivatization technique. In the case of propylation, mixed methylethyl species may interfere in the determination of Et_3Pb^+ due to the same number of carbon atoms but this problem has never been investigated.

2. Liquid Chromatography

This technique has not been widely used for organolead analysis because the detection limits obtained with HPLC/atomic spectrometry techniques are not sufficiently sensitive for general environmental analysis and they have only been successfully applied to the determination of R_4Pb compounds in petrol. HPLC–ICP-MS can solve the sensitivity problem. Trones et al.[185] solved the HPLC–ICP-MS coupling problem by using a home-made microconcentric nebulizer as interface. The LODs for tetraethyllead and tetramethyllead were found to be 5 pg as Pb ($S/N = 3$). Also Acon et al.[186] using a micronebulizer and direct nebulization into the base of the argon plama of the ICP-MS, obtained absolute detection limits for trietyllead and triphenyllead in the low to subpicogram range.

C. GC-Detection Methods

Detectors usually employed for hyphenated organolead determinations are based on atomic spectrometry: electrothermal atomization-atomic absorption spectrometry (ETA-AAS) and

ICP-AES. These detectors are element specific, avoiding most chemical and spectral interferences, and also sufficiently sensitive to quantify the organolead compounds in most environmental samples. The LODs obtained on trimethyllead after ethylation with CGC-MIP-AES[187] or with Headspace-SPME-ICP-MS[1] are comparable (0.20 ng l^{-1}). However, ICP-MS becomes increasingly popular for all element determinations, because it allows applying isotope dilution. In this way extraction and derivatization efficiencies can be estimated, especially if organolead species, including only one particular Pb isotope, will be available. In addition, the procedure can be incorporated into an isotope dilution calibration strategy which has the advantage of increased accuracy and precision. Typically for lead the single measurement of m/z 206 or 208 may produce precision (RSD %) of 0.5 to 1.0%; the isotope ratio 206/208 would produce precision of 0.1 to 0.2%.[150]

1. Application of AAS, MS, and ICP-MS Detection in an Intercalibration Study for Trimethyllead in Artificial Rainwater[150]

Several laboratories, each using their own analytical methods, participated in an intercalibration study of EC (SM&T) for trimethyllead in artificial rainwater.[150] The results produced by the participating laboratories were of high quality. It appeared that the conditions of the Grignard reaction in terms of temperature, concentration, and length of the alkyl chain were key factors which required careful control. The risk for degradation of the compound and for GC peak broadening increases with the length of the alkyl-chain of the Grignard reagent. An example of an ethylation, propylation, and butylation derivatization procedure is given.

2. Ethylation/CGC-QF AAS

A 100-ml intake was used for analysis. Extraction was performed with 50 ml NaOH and 2 ml sodium acetate/acetic acid ($2 \ mol \ l^{-1}$) in 10 ml hexane. Cleanup was carried out using silica gel, followed by preconcentration over a N_2 stream to a volume of 1 ml. Derivatization was performed with 10% NaBE$_4$ in acetic acid at pH 4. Separation was by CGC (column of 30 m length, 0.32 mm internal diameter, DB-5 as stationary phase, 0.25-μm film thickness; He as carrier gas; air/H$_2$ as make-up gases; injector temperature at 80°C). Detection was by QFAAS at 283.3 nm (detector temperature at 750°C). Calibration was by standard additions, using M$_3$PbCl provided by SM&T.

3. Propylation/CGC–ICP-MS

Approximately, 20 ml was used for the analysis to which were added 2 ml EDT N ammonia/citrate buffer solution to pH 8. Five hundred microliters of hexane were added, containing Bu$_4$Pb as internal standard. The mixture was mechanically shaken for 10 min, decanted for 5 min (phase separation), and 400 μl of the hexane phase was derivatized by addition of 40 ml of $2 \ mol \ l^{-1}$ propylmagnesium bromide in diethyl ether. The excess Grignard reagent was destroyed with 2 ml of 0.1 $mol \ l^{-1}$ H$_2$SO$_4$. Separation was by CGC (column of 30 m length, 0.25 mm internal diameter, RSL-150 as stationary phase, 0.50-μm film thickness; He was used as carrier gas; the column temperature ranged from 60 to 230°C). Detection was by ICP-MS. Calibration was by standard additions using ^{208}Pb isotope. The recovery (assessed by spiking) was 93%.

4. Butylation/CGC–MS

Approximately 75 ml solution was used for the analysis. Buffering was performed with ammonium citrate at pH 9, followed by addition of 0.5 ml of 0.25 $mol \ l^{-1}$ diethyldithiocarbamate. Extraction was carried out with 5 ml pentane. The extract was dried with Na$_2$SO$_4$, evaporated to 0.5 ml and redissolved into hexane. Derivatization involved the addition of 0.5 ml of $2 \ mol \ l^{-1}$ BuMgCl in THF, followed by addition of H$_2$SO$_4$.Separation was by CGC (fused-silica column of 60 m length,

0.25 mm internal diameter, DB-1 as stationary phase, 0.25-μm film thickness; He was used as carrier gas at 110 ml min^{-1}; injector temperature at 250°C; column temperature ranging from 50 to 260°C). Detection was by MS (detector temperature at 280°C), monitoring the ions 208, 223, and 253 (M_3BuPb), 208, 237, and 295 (E_4Pb), and 208 and 379 (Bu_4Pb). Calibration was by calibration graph, using M_3PbCl as calibrant and the addition of E_4Pb as internal standard.

REFERENCES

1. Moens, L., De Smaele, T., Dams, R., Van Den Broeck, P., and Sandra, P., *Anal. Chem.*, 69, 1604, 1997.
2. Quevauviller, Ph., Maier, E. A., and Griepink, B., *Element Speciation in Bioinorganic Chemistry*, Vol. 135, Caroli, S., Ed., Wiley, New York, p. 195, 1996.
3. Tessier, A., Campbell, P. G. C., and Bisson, M., *Anal. Chem.*, 51, 844, 1979.
4. Baeyens, W., Monteny, F., Leermakers, M., and Bouillon, S., *J. Anal. Bioanal. Chem.*, 376, 890, 2003.
5. IUPAC, Harmonised guidelines for the use of recovery information in analytical measurement, 1996 (in Quevauviller and Morabito, 2000).
6. Morabito, R., *Microchem. J.*, 51, 198, 1995.
7. Stunz, H., *Mineralogische Tabellen*, 5th ed., Akademische Verslaggesellschaft, Leipzig, 1973.
8. Medley, S. P. L. and Kinniburgh, D. G., *Appl. Geochem.*, 17, 517, 2002.
9. Fergusson, J. E., *The Heavy Elements: Chemistry, Environmental Impact and Health Effects*, Pergamon Press, Oxford, 1990.
10. Lee, T. and Yao, C. L., *Int. Geol. Rev.*, 12, 778, 1970.
11. Carlin, J. F. Jr., Geological Survey Circular 1196-K, U.S.G.S., Reston, VA, 2003, Version 1.0 Published online in the Eastern Region, Reston, VA.
12. Harrison, R. M. and Laxen, D. P. H., *Nature*, 275, 738, 1978.
13. Loebenstein, J. R., The material flow of arsenic in the United States, Bureau of Mines Information Circular 9382, United States Department of the Interior, 1994.
14. Ashby, J. R. and Craig, P. J., Organometallic compounds in the environment, In *Pollution: Causes, Effects and Control*, Harrison, R. M., Ed., Royal Society of Chemistry, London, pp. 309–342, 1990.
15. Fitzgerald, W. F. and Mason, R., The global mercury cycle: oceanic and anthropogenic aspects, In *Global and Regional Mercury Cycles: Sources, Fluxes and Mass Balances*, Baeyens, W., Ebinghaus, R., and Vasiliev, O., Eds., Kluwer Academic Publishers, Dordrecht, The Netherlands, p. 85, 1996.
16. WHO. Environmental Health Criteria 101, Methylmercury, World Health Organization, Geneva, 1990, p. 144.
17. Bushong, S. J., Ziegenfuss, M. C., Unger, M. A., and Hall, L. W., *Environ. Toxicol. Chem.*, 9, 359, 1990.
18. Harrison, R.M., Hewitt, C.N., and Radojevic, M., International Conference on Chemicals in the Environment, Lisbon, Portugal, Selper, London, p. 110, 1986.
19. Przygoda, G., Feldmann, J., and Cullen, W. R., *Appl. Organomet. Chem.*, 15, 457, 2001.
20. Uthus, E. O., Arsenic essentiality and factors affecting its importance, In *Arsenic Exposure and Health*, Chappell, W. R., Abernathy, C. O., and Cothern, C. R., Eds., Science and Technology Letters, Northwood, pp. 199–208, 1994.
21. Mandal, B. K. and Suzuki, K. T., *Talanta*, 58, 201, 2002.
22. International Agency for Research on Cancer, Monograph "Arsenic and its Compounds", Vol. 23, IARC, Lyons, 1980, pp. 39–141.
23. Kenyon, E. M. and Hughes, M. F., *Toxicology*, 160, 227, 2001.
24. Nriagu, J. O. and Pacyna, J. M., *Nature*, 333, 134, 1988.
25. Mahafey, R. K., Rice, G. E., and Schoeny, R., Mercury study report to congress volume IV: Characterization of human health and wildlife risk from mercury exposure in the United States, EPA-452/R-97-0091997, Washington, DC, December.
26. Clarkson, T. W., Environmental Health Perspectives, 75, 59, 1987.

27. Marsh, D., Clarkson, T., Cox, C., Myers, G., Amin-Zaki, L., and Al Tikriti, S., *Arch. Neurol.*, 44, 1017, 1987.
28. McKeown-Eyssen, G., Ruedy, J., and Neims, A., *Am. J. Epidemiol.*, 18, 470, 1983.
29. EPA Mercury Study Report to the Congress, EPA 452/R-97-003, EPA, U.S.A., December 1997.
30. Mason, R., Rolfhus, K., and Fitzgerald, W., *Water Air Soil Pollut.*, 80, 775, 1995.
31. Baeyens, W. and Leermakers, M., *Mar. Chem.*, 60, 257, 1998.
32. Baeyens, W., *TrAC*, 11, 245, 1992.
33. Puk, R. and Weber, J. H., *Appl. Organomet. Chem.*, 8, 293, 1994.
34. Quevauviller, Ph., Donard, O. F. X., Maier, E. A., and Griepink, B., *Mikrochim. Acta*, 109, 169, 1992.
35. Sanchez Uria, J. E. and Sanz-Medel, A., *Talanta*, 47, 509, 1998.
36. Horvat, M., Mercury analysis and speciation in environmental samples, In *Global and Regional Mercury Cycles: Sources, Fluxes and Mass Balances*, Baeyens, W., Ebinghaus, R., and Vasiliev, O., Eds., Kluwer Academic Publishers, Dordrecht, The Netherlands, p. 1, 1996.
37. Harrington, *TrAC*, 19, 167, 2000.
38. Bloom, N. S., *Can. J. Fish Aquat. Sci.*, 49, 1010, 1992.
39. Neil, K., Abraham, N. G., Levere, R. D., and Kappas, A., *J. Cell. Biochem.*, 57, 409, 1995.
40. Fritsch, P., De Saint Blanquat, G., and Derache, R., *Toxicology*, 8, 165, 1977.
41. World Health Organization. *Trace Elements in Human Nutrition and Health*, WHO, Geneva, 1996.
42. Lobinski, R., Dirkx, W. M. R., Spuznar-Lobinska, J., and Adams, F. C., *Quality Assurance for Environmental Analysis*, Quevauviller, Ph., Maier, E., and Griepink, B., Eds., Elsevier, Amsterdam, pp. 319–356, 1995.
43. Gercken, B. and Barnes, R. M., *Anal. Chem.*, 63, 283, 1991.
44. Chaem, V. and Agemina, H., *Analyst*, 105, 737, 1980.
45. Ariza, J. L. G., Morales, E., Sanchez-Rodas, D., and Giraldez, I., *TrAC*, 19, 200, 2000.
46. Hall, G. E. M., Pelchat, J. C., and Gauthier, J. A. A. S., *JAAS*, 14, 205, 1999.
47. Lagarde, F., Amran, M. B., Leroy, M. J. F., Demesmay, C., Ollé, M., Lamotte, A., Muntau, H., Michel, P., Thomas, P., Caroli, S., Larsen, E., Bonner, P., Rauret, G., Foulkes, M., Howard, A., Griepink, B., and Maier, E. A., *Fresenius J. Anal. Chem.*, 363, 5, 1999.
48. Braman, R. S. and Foreback, C. C., *Science*, 182, 1247, 1973.
49. González, J. C., Lavilla, I., and Bendicho, C., *Talanta*, 59, 525, 2003.
50. Száková, J., Tlustos, P., Balík, J., Pavlíková, D., and Vanek, V., *Fresenius J. Anal. Chem.*, 363, 594, 1999.
51. Elwood, M. J. and Maher, W. A., *Anal. Chim. Acta*, 477, 279, 2003.
52. Garcia-Manyes, S., Jiminez, G., Padro, A., Rubio, R., and Rauret, G., *Talanta*, 58, 97, 2002.
53. Montperrus, M., Rodriguez Martin-Doimeadios, R., Scancar, J., Amouroux, D., and Donard, O. F. X., *Anal. Chem.*, 75, 4095, 2003.
54. Yehl, P. M. and Tyson, J. F., *Anal. Commun.*, 34, 49, 1997.
55. Yehl, P. M., Gurleyuk, H., Tyson, J. F., and Uden, P. C., *Analyst*, 126, 1511, 2001.
56. Szostek, B. and Aldstadt, J. H., *J. Chromatogr. A*, 807, 253, 1998.
57. McKiernan, J. W., Creed, J. T., Brockhoff, C. A., Caruso, J. A., and Lorenzana, R. M., *JAAS*, 14, 607, 1999.
58. Beauchemin, D., Bednas, M. E., Berman, S. S., McLaren, J. W., Siu, K. W., and Sturgeon, R. E., *Anal. Chem.*, 60, 2209, 1989.
59. Alberti, J., Rubio, R., and Rauret, G., *Fresenius J. Anal. Chem.*, 351, 420, 1995.
60. Branch, S., Ebdon, L., and O'Neill, P., *JAAS*, 9, 33, 1994.
61. Le, X. C., Mingsheng, M. A., and Wong, N. A., *Anal. Chem.*, 68, 4501, 1996.
62. Gong, Z., Lu, X., Ma, M., Watt, C., and Le, X. C., *Talanta*, 58, 77, 2002.
63. Nakazato, T., Tao, H., Taniguchi, T., and Isshiki, K., *Talanta*, 58, 121, 2002.
64. Slejkovec, Z., van Elteren, J. T., and Byrne, A. R., *Anal. Chim. Acta*, 358, 51, 1998.
65. Prohaska, T., Pfeffer, M., Tulipan, M., Stingeder, G., Mentler, A., and Wenzel, W. W., *Fresenius J. Anal. Chem.*, 364, 467, 1999.
66. Brinckman, F. E., Blair, W. R., and Iverson, W. P., *J. Chromatogr. Sci.*, 15, 493, 1977.
67. Zhang, X., Cornelis, R., de Kimpe, J., and Mees, L., *JAAS*, 11, 1075, 1996.
68. Chasseau, M., Roussel, C., Gilon, N., and Mermet, J. M., *Fresenius J. Anal. Chem.*, 366, 476, 2000.
69. Vanhoe, H., Goosens, J., Moens, L., and Dams, R., *JAAS*, 9, 177, 1994.

70. Ebdon, L., Fisher, A., Roberts, N., and Yoqoob, M., *Appl. Organomet. Chem.*, 13, 183, 1999.
71. Klaue, B. and Blum, J. D., *Anal. Chem.*, 71, 1408, 1999.
72. West, C. D., *Anal. Chem.*, 46, 797, 1974.
73. Thompson, K. C. and Godden, R. G., *Analyst*, 100, 544, 1975.
74. Woller, A., Mester, Z., and Fodor, P., *JAAS*, 10, 609, 1995.
75. Mester, Z. and Fodor, P., *JAAS*, 12, 363, 1997.
76. Gomez-Ariza, J. L., Sanchez-Rodas, D., Beltran, R., Corns, W., and Stockwell, P., *Appl. Organomet. Chem.*, 12, 1, 1998.
77. Van Elteren, J. T. and Slejkovec, Z., *J. Chromatogr. A*, 789, 339, 1997.
78. Leermakers, M., Lansens, P., and Baeyens, W., *Fresenius J. Anal. Chem.*, 336, 655, 1990.
79. Fitzgerald, W. F. and Gill, G. A., *Anal. Chem.*, 51, 1714, 1979.
80. Schroeder, W. H., *Environ. Sci. Technol.*, 16, 362A, 1982.
81. Muhaya, B., Leermakers, M., and Baeyens, W., *Water Air Soil Pollut.*, 107, 277, 1998.
82. Foucher, D., Ph.D. thesis. Universite des Sciences et Technologies de Lille (Fr.), p. 258, 2002.
83. Westöö, G., *Acta Chem. Scand.*, 20, 2131, 1966.
84. Padberg, S., Burrow, M., and Frech, W., *Anal. Chim. Acta*, 249, 686, 1991.
85. Bulska, E., Daxter, D. C., and Frech, W., *Anal. Chim. Acta*, 249, 545, 1991.
86. Rezende, M., Campos, R., and Curtius, A., *JAAS*, 8, 247, 1993.
87. Alli, A., Jaffe, R., and Jones, R., *J. High Resolut. Chromatogr.*, 17, 745, 1994.
88. Decadt, G., Baeyens, W., Bradley, D., and Goeyens, L., *Anal. Chem.*, 57, 2788, 1985.
89. Lansens, P. and Baeyens, W., *Anal. Chim. Acta*, 228, 93, 1990.
90. Horvat, M., Bloom, N. S., and Liang, L., *Anal. Chim. Acta*, 281, 135, 1993.
91. Emteborg, H., Björklund, E., Odman, F., Karlsson, L., Mathiasson, L., Frech, W., and Baxter, D., *Analyst*, 121, 19, 1996.
92. Tseng, C. M., de Diego, A., Martin, F. M., and Amouroux, D., *JAAS*, 12, 743, 1997.
93. Bloom, N., Colman, J., and Barber, L., *Fresenius J. Anal. Chem.*, 358, 371, 1997.
94. Leermakers, M., Nguyen, H. L., Kurunczi, S., Vanneste, B., Galletti, S., and Baeyens, W., *J. Anal. Bioanal. Chem.*, 377, 327, 2003.
95. Vazquez, M. J., Carro, A. M., Lorenzo, R. A., and Cela, R., *Anal. Chem.*, 69, 221, 1997.
96. Bloom, N., *Can. J. Fish. Aquat. Sci.*, 46, 1131, 1989.
97. Logar, M., Horvat, M., Akagi, H., and Pihlar, B., *J. Anal. Bioanal. Chem.*, 374, 1015, 2002.
98. Nagase, H., Ose, Y., Sabo, T., and Ishikawa, T., *Int. J. Environ. Anal. Chem.*, 7, 261, 1980.
99. Horvat, M., May, K., Stoeppler, M., and Byrne, A. R., *Appl. Organomet. Chem.*, 2, 515, 1988.
100. Tseng, C. M., de Diego, A., Pinaly, H., Amouroux, D., and Donard, O. X. F., *JAAS*, 13, 755, 1998.
101. Lorenzo, R., Vazquez, M., Carro, A., and Cela, R., *TrAC*, 18, 410, 1999.
102. Hintelmann, H., Falter, R., Ilgen, G., and Evans, D., *Fresenius J. Anal. Chem.*, 358(3), 363, 1997.
103. Falter, R., *Chemosphere*, 39(7), 1037, 1999.
104. Baeyens, W., Leermakers, M., Molina, R., Holsbeek, L., and Joiris, C., *Chemosphere*, 7, 1107, 1999.
105. Hammerschmidt, C. and Fitzgerald, W. F., *Anal. Chem.*, 73, 5930, 2001.
106. Rodriguez Martin-Doimeadios, R. C., Monperrus, M., Krupp, E., Amouroux, D., and Donard, O. F. X., *Anal. Chem.*, 75, 3202, 2003.
107. Rapsomanikis, S., Donard, O. F. X., and Weber, J. H., *Anal. Chem.*, 58, 35, 1986.
108. Craig, P. J., Mennie, D., Needham, M. I., Donard, O. F. X., and Martin, F., *J. Organomet. Chem.*, 447, 5, 1993.
109. Filipelli, M., Baldi, F., and Weber, J. H., *Environ. Sci. Technol.*, 26, 1457, 1992.
110. Dietz, C., Madrid, Y., Camara, C., and Quevauviller, P., *JAAS*, 14, 1349, 1999.
111. Liang, L., Horvat, M., and Bloom, N., *Talanta*, 41, 371, 1994.
112. Lobinski, R., Rodriguez, P., Chassaigne, H., Wasik, A., and Szpunar, J., *JAAS*, 13, 859, 1998.
113. De Smaele, T., Moens, L., Dams, R., Sandra, P., Van de Eycken, J., and Vandyck, J., *J. Chromatogr. A*, 793, 99, 1998.
114. Grinberg, P., Campos, R., Mester, Z., and Sturgeon, R., *Spectrochim. Acta Part B*, 58, 427, 2003.
115. Cai, Y., Monsalud, S., and Furton, K., *Chromatographia*, 52, 82, 2000.
116. Hänström, S., Briche, C., Emteborg, H., and Baxter, D., *Analyst*, 121, 1657, 1996.
117. Slaets, S., Adams, F., Rodriguez Pereiro, I., and Lobinski, R., *JAAS*, 14, 851, 1999.
118. Rodriguez, I., Wasik, A., and Lobinski, R., *J. Chromatogr.*, 795, 4799, 1998.

119. Munaf, E., Haraguchi, H., Ishii, D., Takeuchi, T., and Goto, M., *Anal. Chim. Acta*, 235, 399, 1990.
120. Wu, G., *Spectrosc. Lett.*, 24, 681, 1991.
121. Falter, R. and Scholer, H., *J. Chromatogr. A*, 675, 253, 1994.
122. Falter, R. and Scholer, H., *Fresenius J. Anal. Chem.*, 348, 253, 1994.
123. Falter, R. and Scholer, H., *Fresenius J. Anal. Chem.*, 353, 34, 1995.
124. Schlegel, D., Mattusch, J., and Dittrich, K., *J. Chromatogr. A*, 683, 261, 1994.
125. Qiang, T., Johnson, W., and Buckley, B., *JAAS*, 18, 696, 2003.
126. Gebersmann, C., Heisterkamp, M., Adams, F., and Broekaert, J., *Anal. Chim. Acta*, 350, 273, 1997.
127. Aguerre, S., Lepes, G., Desauziers, V., and Potin Gautier, M., *JAAS*, 16(3), 263, 2001.
128. Frech, W., Snell, J., and Sturgeon, R., *JAAS*, 13, 1347, 1998.
129. Bloom, N. S. and Fitzgerald, W. F., *Anal. Chim. Acta*, 208, 151, 1988.
130. Stockwell, P. B., Thompson, K. C., Henson, A., Temmerman, E., and Vandecasteele, C., *Int. Labmate*, 14, 45, 1989.
131. Vanhaeke, F. and Moens, L., *Fresenius J. Anal. Chem.*, 364, 440, 1999.
132. Heumann, K., Gallus, S., Rädlinger, G., and Vogl, J., *Spectrochim. Acta Part B*, 53, 273, 1998.
133. Clough, R., Belt, S., Evans, E., Fairman, B., and Catterick, T., *JAAS*, 18, 1033, 2003.
134. Gelaude, I., Ph.D. thesis. University of Ghent (Belgium), p. 254, 2003.
135. Shum, S., Pang, H., and Houk, R., *Anal. Chem.*, 64, 2444, 1992.
136. Harrington, C., Romeril, J., and Catterik, T., *Rapid Commun. Mass Spectrom.*, 12, 911, 1998.
137. Wan, C. C., Chen, C. S., and Jiang, S., *JAAS*, 12, 683, 1997.
138. De Smaele, T., Verrept, P., Moens, L., and Dams, R., *Spectrochim. Acta B*, 50, 1409, 1995.
139. Montperrus, M., Bohari, Y., Bueno, M., Astruc, A., and Astruc, M., *Appl. Organomet. Chem.*, 16, 347, 2002.
140. Leenaerts, J., Van Mol, W., Goegana Infante, H., and Adams, F., *JAAS*, 17, 1492, 2002.
141. Jitaru, P., Goenaga Infante, H., and Adams, F., *Anal. Chim. Acta*, 489, 45, 2003.
142. Zhang, Z. and Pawliszyn, J., *Anal. Chem.*, 65, 1843–1852, 1993.
143. Boyd-Boland, A. A., Magdic, S., and Pawliszyn, J., *Analyst*, 121, 929–938, 1996.
144. Nagase, M. and Hasebe, K., *Anal. Sci.*, 9, 517, 1993.
145. de la Calle-Guntinàs, M. B., Scerbo, R., Chiavarini, S., and Quevauviller, Ph., *Appl. Organomet. Chem.*, 11, 693, 1997.
146. Ritsema, R., Martin, F. M., and Quevauviller, Ph., *Quality Assurance for Environmental Analysis*, Quevauviller, Ph., Maier, E. A., and Griepink, B., Eds., Elsevier, Amsterdam, 1995, chap. 19.
147. Ariese, F., Cofino, W., Gòmez-Ariza, J.-L., Kramer, G., and Quevauviller, Ph., *J. Environ. Monit.*, 1, 191, 1999.
148. Ebdon, L., Hill, S. J., and Rivas, C., *TrAC*, 17, 277, 1998.
149. Quevauviller, Ph., *Method Performance Studies for Speciation Analysis*, The Royal Society of Chemistry, Cambridge, UK, p. 271, 1998.
150. Looser, P. W., Berg, M., Fent, K., Mühlemann, J., and Schwarzenbach, R. P., *Anal. Chem.*, 72, 5136, 2000.
151. Tsunoi, S., Matoba, T., Shioji, H., Giang, L. T. H., Harino, H., and Tanaka, M., *J. Chromatogr. A*, 962, 197, 2002.
152. Abalos, M., Bayona, J.-M., Compañó, R., Granados, M., Leal, C., and Prat, M.-D., *J. Chromatogr. A*, 788, 1, 1997.
153. Arnold, C. G., Berg, M., Müller, S. R., Dommann, U., and Schwarzenbach, R. P., *Anal. Chem.*, 70, 3094, 1998.
154. Ikonomou, M. G., Fernandez, M. P., He, T., and Cullon, D., *J. Chromatogr. A*, 975, 319, 2002.
155. Plzák, Z., Polanská, M., and Suchánek, M., *J. Chromatogr. A*, 699, 241, 1995.
156. Cardellicchio, N., Giandomenico, S., Decataldo, A., and Di Leo, A., *Fresenius J. Anal. Chem.*, 369, 510, 2001.
157. Eiden, R., Scholer, H. F., and Gastner, M., *J. Chromatogr. A*, 809(1–2), 151, 1998.
158. Vercauteren, J., De Meester, A., De Smaele, T., Vanhaecke, F., Moens, L., Dams, R., and Sandra, P., *JAAS*, 15, 651–656, 2000.
159. Zhang, Z., Yang, M. J., and Pawliszyn, J., *Anal. Chem.*, 66, 844A, 1994.
160. Wittkamp, B. L., Hawthorne, S. B., and Tilotta, D. C., *Anal. Chem.*, 69, 1197, 1997.
161. Martos, P. A. and Pawliszyn, J., *Anal. Chem.*, 69, 206, 1997.

162. Bancon-Montigny, Ch., Lespes, G., and Potin-Gautier, M., *J. Chromatogr. A*, 896, 149, 2000.
163. Allen, A. G., Radojevic, M., and Harrison, R. M., *Environ. Sci. Technol.*, 22, 517, 1988.
164. Hewitt, C. N. and Harrison, R. M., *Environ. Sci. Technol.*, 21, 260, 1987.
165. Radojevic, M., Allen, A., Rapsomanikis, S., and Harrison, M., *Anal. Chem.*, 58, 658, 1986.
166. Forsyth, D. S. and Marshall, W. D., *Environ. Sci. Technol.*, 20, 1033, 1986.
167. Uthe, J. F. and Chou, C. L., *Sci. Total Environ.*, 71, 67, 1988.
168. Blais, J. S. and Marshall, W. D., *JAAS*, 4, 271, 1989.
169. Chakraborti, D., De Jonghe, W. R. A., Van Mol, W. E., Van Cleuvenbergen, R. J. A., and Adams, F. C., *Anal. Chem.*, 55, 2692, 1984.
170. Chau, Y. K., Wong, P. T. S., Bengert, G. A., and Dunn, J. L., *Anal. Chem.*, 56, 271, 1984.
171. Lobinski, R. and Adams, F. C., *JAAS*, 7, 987, 1992.
172. Lobinski, R., Boutron, C., Candelone, J. P., Hong, S., Szpunar-Lobinska, J., and Adams, F., *Anal. Chem.*, 65, 2510, 1993.
173. Blais, J. S. and Marshall, W. D., *J. Environ. Qual.*, 15, 255, 1986.
174. Yamauchi, H., Arai, F., and Yamamura, Y., *Ind. Health*, 19, 115, 1981.
175. Radojevic, M., *Environmental Analysis Using Chromatography Interfaced with Atomic Spectroscopy*, Harrison, R. M. and Rapsomanikis, S., Eds., Horwood, Chichester, p. 237, 1989.
176. Chau, Y. K., Wong, P. T. S., and Kramar, O., *Anal. Chim. Acta*, 146, 211, 1983.
177. Harrison, R. M. and Radojevic, M., *Environ. Technol. Lett.*, 6, 129, 1985.
178. Brunetto, M. R., Burguera, J. L., Burguera, M., and Chakraborti, D., *At. Spectrosc.*, 13, 123, 1992.
179. Greenway, G. M. and Barnett, N. W., *JAAS*, 4, 783, 1989.
180. Scott, A. E., Uden, P. C., and Barnes, R. M., *Anal. Chem.*, 53, 1336, 1981.
181. Uden, P., *Anal. Proc.*, 18, 189, 1981.
182. Estes, S. A., Uden, P. C., and Barnes, R. M., *Anal. Chem.*, 54, 2402, 1982.
183. Nygren, O., *JAAS*, 2, 801, 1987.
184. Forsyth, D. S., *Anal. Chem.*, 59, 1742, 1987.
185. Trones, R., Tangen, A., Lund, W., and Greibrokk, T., *J. Chromatogr. A*, 835, 105, 1999.
186. Acon, B. W., McLean, J. A., and Montaser, A., *JAAS*, 16, 852, 2001.
187. Ceulemans, M. and Adams, F. C., *JAAS*, 11, 201, 1996.
188. Gallina, A., Magno, F., Tallandini, L., Passaler, T., Caravello, G. U., and Pastore, P., *Rapid Commun. Mass Spectrom.*, 14, 373, 2000.
189. Shiomi, K., *Arsenic in the environment. Part II: Human Health and Ecosystem Effects*, Vol. 27, Nriagu, J. O., Ed., Wiley, New York, 1994.
190. Stäb, J. A., Brinkman, U. A. Th., and Cofino, W. P., *Appl. Organomet. Chem.*, 8, 577, 1994.

21 Isocyanates

Paal Molander

CONTENTS

I. BACKGROUND

A. CHEMICAL PROPERTIES AND USE

The discovery of the procedure for the production of polyurethane (PUR), by the reaction of a diisocyanate with a polyfunctional alcohol in 1937, has positively influenced the modern way of living by introducing a polymeric material with unique properties and widespread use. In our daily life we are surrounded by products made from PURs such as foams, paints, lacquers, inks, adhesives, insulating materials, sealants, varnishes, rubber modifiers, and bonding and vulcanizing agents.[1] In 1999, the global production of PURs was eight million tonnes, with a computed annual increase of 5%.[2] Various diisocyanates or polyisocyanates serve as raw materials in the production of different PURs and have, over the years, become major industrial chemicals. Isocyanates contain the highly unsaturated N=C=O group, and PUR is formed by the reaction of a difunctional isocyanate and a polyfunctional alcohol. A wide range of PURs can be tailored by reacting different diisocyanates and polyols or simply by varying the physical conditions controlling the polymerization process.[3] Both aliphatic and aromatic isocyanates are used for the production of PUR; hexamethylene diisocyanate (HDI), toluene diisocyanate (TDI), and, methylene diisocyanate (MDI) are the most frequently used diisocyanates for this purpose and account for more than 90% of the total world consumption.[1] Their chemical structures are shown in Figure 21.1. Monofunctional isocyanates are mainly used as intermediates in the pharmaceutical and agricultural industries.[4] The chemical bond between an

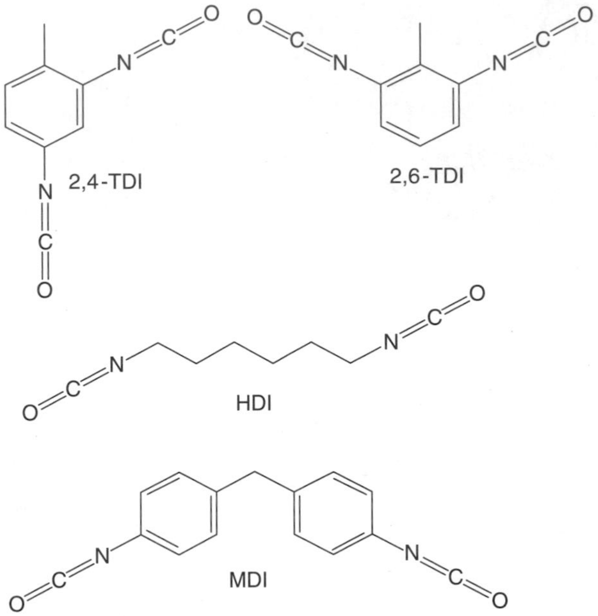

FIGURE 21.1 Chemical structures of 2,4- and 2,6 TDI, HDI, and MDI.

isocyanate and a polyol in PUR products is not thermally stable, and can potentially be broken by treatment at elevated temperatures to release compounds containing isocyanate or amino groups,[5] such as the diisocyanate building bricks of the polymer and the aliphatic monoisocyanate methyl isocyanate (MIC) and isocyanic acid (ICA).

B. HEALTH EFFECTS AND EXPOSURE

Diisocyanates are highly toxic substances which act as respiratory irritants and skin and respiratory sensitizers, with the possibility of causing diseases like bronchitis, pulmonary emphysema, and asthma[6-8] in addition to allergic reactions.[7] Furthermore, diisocyanates have a mutagenic potential through their ready reaction with DNA to form adducts.[9] Short-term occupational exposures to MIC primarily cause eye and mucous membrane irritations. However, no scientific studies have documented a connection between occupational exposure for MIC and chronic health effects; the knowledge on toxicological effects of MIC is primarily documented in follow-up studies concerning the Bhopal accident, where the population was exposed to extreme concentrations of MIC in combination with phosgene, methylamine, and hydrogen cyanide. This led to a unfortunate lethal outcome for a large number of people.[10] Health risks in relation to ICA exposure are not documented.

Health hazards are not related to the common use of PUR products by consumers or private households, but rather to the production or processing of PURs in different industries, such as foaming or spray-painting processes. Monitoring of isocyanates in workroom air is important to industrial hygiene. In addition, exposure is likely during working procedures causing thermal degradation of PUR at temperatures above 150°C, such as the production of PUR coated wires and processing of PUR-coated metal sheets. So far as occupational exposure limits (OELs) are concerned, isocyanates are among the substances with the lowest OELs. In most European countries, the OELs for isocyanates in air is 5 ppb for an 8 h average, and similar OELs are recommended for several of the isocyanate monomers in the U.S.A. by the National Institute for Occupational Safety and Health (NIOSH).

C. MONITORING OF OCCUPATIONAL EXPOSURE

As distinct from MDI, the monomers of HDI and the TDI isomers, as well as MIC, are volatile, but may still be present in workroom air as nonvolatile dimers, trimers or prepolymers.[4,11] Prepolymers of the volatile diisocyanates, such as biuret, allofanat, and isocyanurate adducts, exhibit substantially lower vapor pressures than the diisocyanates, reducing the gaseous phase exposure levels. Highly reactive isocyanates in workroom air may exist as vapors, aerosols, or in mixed phases, rendering sampling of isocyanates in workroom air a complicated task.

Isocyanate monitoring has been performed by air sampling strategies, by direct reading instruments, or by measuring biomarkers in biological fluids of exposed personnel, but the air sampling methods are by far the most used technique for this purpose. Due to high reactivity of the isocyanates, the most commonly used sampling method for isocyanates in workroom air includes a derivatization step, by pumping a volume of air through an amine reagent coated filter or an impinger solution containing an amine reagent with a filter attached up-stream.[12] A number of different amine reagents have been explored for the sampling of isocyanates, while numerous liquid chromatographic (LC) methods with ultraviolet (UV), mass spectrometric (MS), fluorescence (F), or electrochemical (EC) detection have been presented for determination of the isocyanate derivatives.[4,11] Gas chromatographic (GC) and LC methods have been employed in biomonitoring methods.[4]

D. SCOPE

The aim of this chapter is to account for employment of chromatographic techniques for determination of airborne isocyanates in working environments.

II. DETERMINATION OF ISOCYANATES IN AIR

A. SAMPLING DEVICES

The ideal sampling strategy for isocyanates in workroom air should possess capabilities for measurement of short-term (peak exposure) and long-term isocyanate exposure of gas or vapor phase, aerosol phase, and mixed phases simultaneously. Personal sampling equipment located close to the breathing zone is preferred over stationary equipment, in order to obtain a best possible estimate of the chemical compound inhaled by the worker. An 8-h long-term sampling interval is an adequate international standard, alternatively a 2-h sampling interval (to be multiplied by four) if the working atmosphere is presupposed to be homogenous over cumulative 8 h elapsed time may be considered. However, regarding isocyanates, shorter than 2-h sampling intervals can beneficially be considered so that the impinger solutions do not evaporate and the problems related to local depletion of aerosol particles on the filters are reduced. With respect to exposure from thermally released isocyanates, the sampling time should be at least 5 min in order to sample a required total air volume of not less than 10 l. Finally, the limit of quantification (LOQ) of the analysis method should be at least 1/10 of the OEL. Despite the reactive and unstable characters of isocyanates, the low OELs for isocyanates and the complex mixture of numerous isocyanates in different phases which potentially can be present in workroom atmospheres make the requirement for sampling devices with high sampling efficiencies and subsequent analysis methods with capabilities for selective trace determinations unambiguous.

After the introduction of the first method for determination of isocyanates in air by Marcali in 1957, based on colorimetric analysis, a number of different collection devices and amine derivatization agents have been tailored for measurement of isocyanates in air.[13] The major effort in developing sampling procedures has been on "wet" vs. "dry" sampling techniques, where "wet" impinger flasks, containing an amine reagent solution, and "dry" amine reagent impregnated filter methods are the most common sampling techniques. A schematic diagram of impinger (a) and

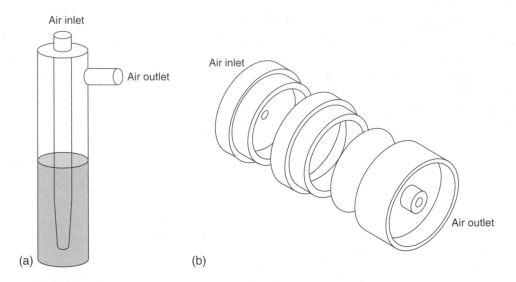

FIGURE 21.2 Schematic of (a) an impinger sampler and (b) filter sampler.

filter (b) sampling devices are presented in Figure 21.2, where a measured volume of air is drawn through the devices by use of pumps with flow rates typically in the order of 1 to 2 l/min. Regarding the filter methods, glass fiber filters of diameters of 25 or 37 mm are in general use, while glass containers with 10- ml reagent solution volume are traditionally used as impingers. Both techniques have certain advantages and drawbacks, and neither of these provides fully suitable performance characteristics for representative simultaneous sampling of a wide range of isocyanates present in different phases.[4,11,14] Nevertheless, impinger methods and filter methods have very high collection efficiencies of isocyanates present in gas or vapor phase.[4,11,15] For the most volatile isocyanates, e.g., MIC, a double filter with increased reagent film thickness is required[16] if "dry" methods are employed, as compared to gaseous diisocyanate sampling only.

Sampling of isocyanates present in aerosols is a more complicated task. Impregnated filters are highly suited for sampling of particles of all sizes. However, the amount of reagent available for active isocyanates is limited as compared to impinger solutions. This is especially true for larger particles and/or particles containing a reacting mixture, e.g., a curing paint, possibly leading to losses of isocyanates due to continued curing in aerosol droplets trapped on the filter. Furthermore, only the isocyanates which are directly in contact with the filter are available for reaction with the impregnated amine reagent, possibly leading to underestimation of isocyanates located inside large particles or on the side of the particle surface which is not in contact with the filter.[17,18] This effect has been reported to be more pronounced when the particle concentration is increased.[19] Such effects can, to some extent, be reduced by the use of filters with thicker reagent films and depth filters at shorter sampling intervals with increased air sampling flow rates, in addition to immediate transfer of the filters into reagent solutions after completed air sampling.[4,11,15,19] In general, filter methods are more user-friendly than impinger methods, making them especially suitable for personal sampling.

Particle size selective OEL have already been established to address the problems associated with specific health effects, especially within fields where solid aerosol particles are to be measured. A number of filter cassette designs have been tailored to sample specific size ranges of the total aerosol fraction, in order not to overestimate the exposure causing health effects by inhalation.[20] Since only particles less than approximately 100 μm can be inhaled and the inhaled aerosol fraction is most certainly the main contributor to specific health effects among workers exposed to isocyanates, it seems reasonable to use sampling devices designed for sampling of the inhalable fraction of aerosol when applied for gaseous, aerosol, and mixed phase sampling of isocyanates with filter methods. To date, a standard filter cassette with capabilities of sampling the total aerosol fraction is in general use.[12] Several inhalable air sampling devices are commercially available, but some of these may under-sample the inhalable aerosol fraction due to inner-wall losses of droplets or particles inside the sampler before they reach the filter. Since the well-known IOM cassette is designed to include the inner-wall deposited aerosol mass in the measurements of the inhalable fraction, a quantitative recovery of this deposited mass may be very difficult, as the isocyanate aerosol matrix is likely to be polymerized. If such inner-wall deposition of a sampling device is considered to be an important issue, a sampler design where only the aerosol collected on the filter is used to assess the inhalable fraction should be preferred. Examples of such devices are the German GSP[21] and the Dutch PAS-6.[22] The Nordic Network on Isocyanates recently extended a general invitation to the researchers in this field for evaluation of such samplers for isocyanate monitoring in workroom atmospheres.[14] Such samplers should, in theory, be very suitable for nondiscriminative sampling of relevant aerosol fractions, in addition to being well suited for sampling of isocyanates in gaseous phase, especially if equipped with double filters as used in a standard cassette in a Finnish study.[16]

Impingers were not originally constructed for aerosol sampling, but in theory are also suited for sampling of size-limited fractions of particles. Impingers are constructed to have an ideal collecting effect when using a flow rate of 2 l/min.[23] However, toluene is the most used impinger solvent, putting limits on the maximum allowed air sampling flow rate to avoid evaporation of toluene. Several other higher boiling solvents have been evaluated with various amine reagents including dimethyl sulfoxide, butyl acetate, and octane,[4] to avoid this problem, but toluene has remained by far the most commonly used for this purpose. An air sampling flow rate of 1 l/min is used with toluene or other volatile solvents as the impinger solvent, leading to breakthrough and insufficient collection of submicron particles which often are present in, e.g., thermal degradation processes.[24] This effect can, however, be reduced by attaching a filter downstream to the impinger solution.[24,25] In cases where volatile amine reagents are used and preimpregnated filters are not available, as with the case when using dibutylamine (DBA) as derivatizing agent, the filter will then be continuously coated with amine reagent in toluene, evaporating from the impinger solution. Unfortunately, impinger flasks are fragile, cumbersome for the workers to wear them, and their use as personal samplers is not compatible with several working procedures. The use of toluene can create additional exposure or fire hazards, because it evaporates during sampling, additionally limiting the maximum sampling time available. Finally, impingers require manipulation of solvents by industrial hygienists in the field and subsequent transportation of solvents. Nevertheless, the use of impingers for isocyanate sampling is widespread.[16,17,23,24,26–32] The establishment of impinger samplers has especially been pronounced after the successful introduction of DBA as a derivatizing agent by Skarping and coworkers,[24] which until now has only been used with impingers due to the reagent's high volatility. As a general rule, new amine derivatizing agents are initially evaluated in impingers, and in the case of successful tests these are evaluated on solid supports to develop sampling strategies more suitable for personal monitoring close to the breathing zone.

Other procedures than filter- and impinger methods have also been developed for the collection of airborne isocyanates, however to a lesser extent, and these are all based on amine reagent derivatization. Among these, a number of solid supports and adsorbents impregnated with the amine reagents and packed in tubes with inner diameter of 4–10 mm and lengths of 20–100 mm, for pumped sampling have been evaluated, including glass beads, glass wool, silica gels, and synthetic materials such as XAD resins (styrene–divinylbenzene).[2,4] Denuders coated with chemisorptive stationary phases and amine derivatizing agents have also been presented for collection of TDI isomers in gaseous phase,[33,34] possibly providing improved mass-exchange conditions. An interesting approach for sampling of mixed phases was introduced by Rando and Poovey, who designed a dichotomous vapor/aerosol sampler for measurement of HDI derived total reactive isocyanate groups.[35] The sampler consisted of an impactor or a cyclone inlet, followed by an annular denuder, and a glass fiber filter backup. Vapor-phase HDI was completely collected by the diffusional denuder, and when a mixture of HDI-biuret and HDI was nebulized and collected with the dichotomous sampler, approximately 78% of the HDI was in the vapor phase, whereas about 22% was associated with the aerosol fraction.

All the above-mentioned sampling procedures have been based on active pumping of volumes of air through the samplers. More recently, however, diffusive samplers have been used for sampling of isocyanates in gaseous phase. One of these approaches has been the use of solid phase micro extraction (SPME) fibers, where the polydimethylsiloxanc-divinylbenzene fiber was coated with the amine reagent, producing sampling efficiencies comparable to other methods.[36,37] Furthermore, Levin and coworkers used a commercially available diffusive sampling device equipped with a glass fiber filter impregnated with amine reagents to collect MIC, and obtained comparable results to pumped-reference methods.[38,39] The fact that diffusive samplers do not require pumping make them especially user-friendly and attractive for personal monitoring. That these types of samplers are only suitable for sampling of isocyanates in gaseous phase is an obvious drawback and attention must be directed towards establishment of individual diffusion coefficients for quantitative measurements.

B. DERIVATIZATION AGENTS

Once the isocyanate species have been collected, they must be efficiently derivatized to accomplish stabilization of the highly reactive isocyanates and to improve their detectability and specificity in the subsequent laboratory analyses. Excessive amounts of derivatizing agents in solution are typically dissolved in the impinger solution or coated on the glass fiber filters. A number of derivatization agents have been evaluated for fast and efficient conversion of isocyanates into more stable derivatives. The most important features of a successful derivatization agent are the inherent reactivity of the agent with the isocyanate groups and the ability of the collection medium (e.g., impinger solution or impregnated filter) to dissolve or disperse collected particles or droplets, in addition to making the agent accessible to the isocyanate groups. The derivatization agents in use today are most commonly primary or secondary amines, creating stable and non-volatile urea derivatives. Figure 21.3 shows an example of the reaction between a diisocyanate

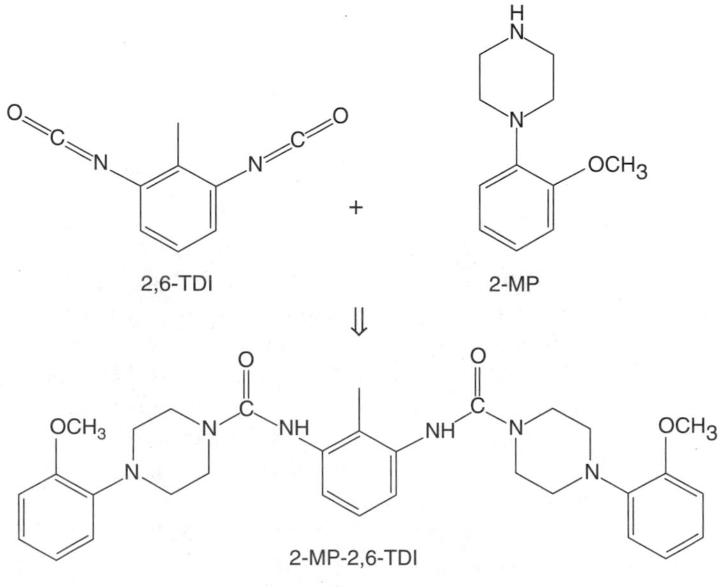

FIGURE 21.3 Derivatization reaction between 2,6 TDI and 2 MP.

and an amine reagent. Lately, the amine reagents N-([4-nitrophenyl]methyl)propylamine (NITRO),[35] 9-(N-methylaminomethyl)antracene (MAMA),[26] 1-(2-methoxyphenyl)piperazine (2 MP),[12,14,16,23,40,41] 1-(2-pyridyl)piperazine (2PP),[42] 1-(9-antraceneylmethyl)piperazine (MAP),[26] tryptamin (3-[2-aminoethyl]indol) (TRYP),[43] di-n-butylamine (DBA),[2,14,16,23,27–31,37] 9-antracenylmethyl-1-piperazinecarboxylate (PAC),[44] 4-nitro-7-piperazino-benzo-2-oxa-1,3-diazol (NBDPZ),[39,45,46] and 4-methoxy-6-(4-methoxy-1-naphthyl)-1,3,5-triazine-2-(1-piperazine) (MMNTP)[47] have been reported in the literature for isocyanate derivatization. Their chemical structures are shown in Figure 21.4. Their inherent reactivities with isocyanates typically differ by a factor of five or less.[11,16,23,26,48] Tremblay et al. recently pointed out, in a comparative study of reaction rates of various derivatizing agents in solutions, that the relative difference in reactivity is a function of the isocyanate and the solvent used, but the reaction rates in general are in the order DBA > MAP > 2MP > MAMA. Furthermore, they observed that hindered aromatic diisocyanates (TDI and MDI) show a greater difference in reactivity with the derivatization agents.[26] Streicher et al., however, correctly states that the differences in reaction rates most probably is unimportant, as the efficiency of the mixing of the collected particles and derivatizing agents is the

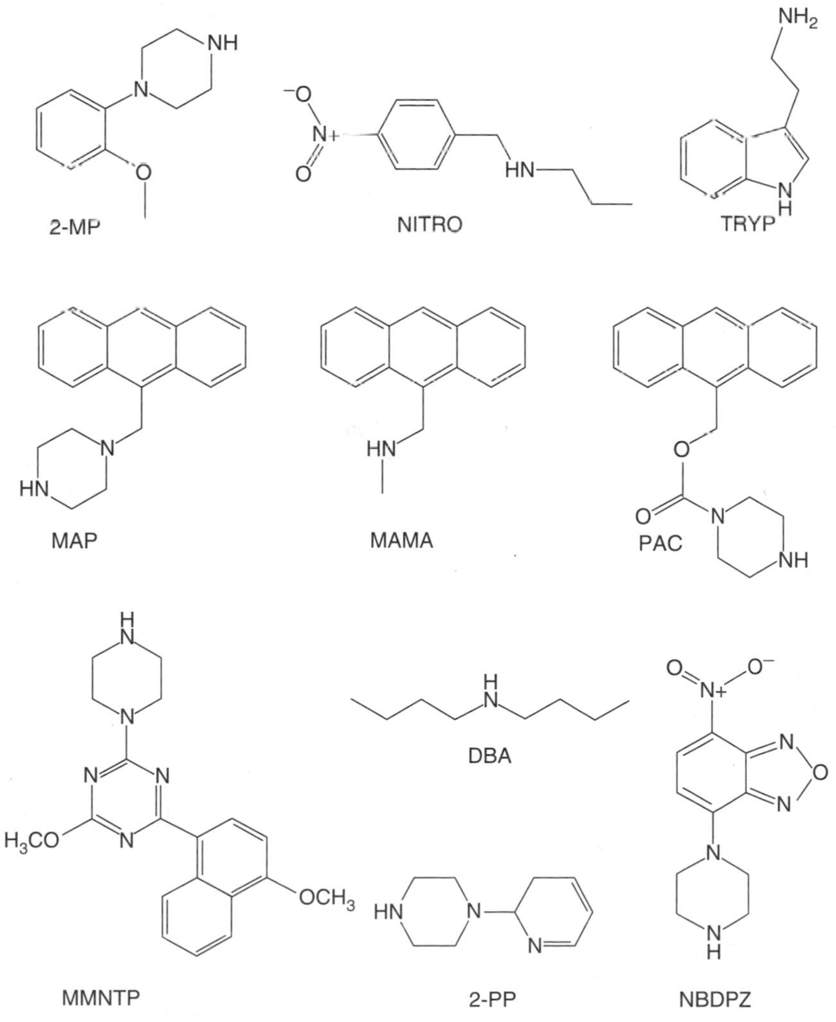

FIGURE 21.4 Chemical structures of derivatizing agents for reaction with isocyanates.

limiting factor in this process.[11] Furthermore, differences in reaction rates are often overruled by employment of excessive amounts of derivatizing agents. These statements are supported by the fact that often only small differences in sampling efficiency among various samplers with different amine reagents are observed in comparative field and laboratory studies.[17,23,26] However, special attention must be paid to the choice of sampling strategy when sampling complex isocyanate atmospheres with mixed phases and variance in aerosol particle diameters. In such cases, the BDA method introduced in 1996 by Skarping and coworkers, where an impinger with an upstream filter, continuously coated with DBA from impinger solution, has showed promising performance in several studies.[24,27-31] Unfortunately, impinger methods are not attractive for personal sampling, and are often inhibit the worker from performing standard working procedures during sampling.

Some of the amine reagents have been reported to be of limited stability in certain environments. For instance, the MAMA reagent is known to be light sensitive,[49] while the NITRO reagent should avoid being used in oxidative or reducing atmospheres.[2] Furthermore, the 2PP reagent has been reported to show substantial loss from the impregnated filters during storage.[11] However, adequate storage of amine-impregnated filters in a freezer prior to sampling minimizes such problems.[11]

C. Sample Treatment

In order to avoid problems related to slow reaction kinetics and sample storage stability, precautions are often taken in the field to minimize such effects. Furthermore, sample preparation is often required to transfer the collected isocyanate derivatives to solvents, which are compatible with the analysis method of choice, or to preconcentrate the sample to improve the sensitivity of the method.

During the last decade, reversed phase LC has evolved to be the primary method of choice for determination of collected isocyanate derivatives. Regarding impinger sampling, this implies, especially when using toluene as impinger solvent, that the sample must be dried by evaporation prior to redissolution in a solution that is miscible with the mobile phases used with this technique. Such solutions may be mixtures of water and acetonitrile or methanol, preferably in a composition which is equal to the initial composition of the mobile phase gradient, if employed. However, some of the isocyanate derivatives are not easily dissolved in aqueous solutions and dissolution in neat acetonitrile or methanol may be required. This procedure might, however, limit the maximum allowed injection volume of the LC method, as the sample is then introduced in a solvent with elution strength substantially higher than that of the mobile phase. If the evaporated sample is redissolved in a volume which is substantially lower than the initial volume of the impinger solution, a preconcentrating effect is obtained resulting in lowered limit of detection (LOD) of the method. This advantage is further exploited if the redissolution volume is so small that the total volume is allowed to be injected into the chromatographic system without overloading the column. In any case, the redissolution step is very critical with regard to losses of samples and consequent reductions in component recoveries. As the resolution of various derivatives might be different, precautions must be taken in order to secure that all the components are fully dissolved in the final sample solution. This precaution usually favors the use of pure acetonitrile as final sample solvent.[12] If such undesired effects occur, ultrasonication of the reconstituted sample solution facilitates the solvatization process.[11]

All filter methods require extraction of the filter prior to the analysis. Immediately after finalized sampling, filters are favorably transferred into solutions containing excessive amounts of the derivatizing agent, in order to improve the reaction between the derivatizing agent and aerosol particles which are not easily accessible.[11] If this derivatizing agent-containing solution is not compatible with the analysis method of choice, evaporation to dryness and redissolution in a compatible solvent or solvent mixture are necessary. This might again be favorable if sample concentration is desired.

Removal of excessive amounts of derivatizing agent from the impinger solution or the filter extract prior to any evaporation might also be favorable, and is usually performed by acetylation of the reagent with acetic anhydride, or by the use of solid-phase extraction (SPE) cartridges.[11] Excessive amounts of reagent typically preclude the quantitative chromatographic determinations, as the large amounts of amines typically elute subsequent to the void volume as a large tailing peak. Furthermore, repeated injections of large amounts of unreacted amines might degrade the analytical column. In order to extend the column lifetime, all final sample solutions are preferentially filtered prior to injection.

D. DETERMINATION OF ISOCYANATE DERIVATIVES

1. Reversed-Phase Liquid Chromatography

Currently, determination of isocyanate derivatives is to a great extent performed by the use of reversed-phase LC,[4,11,12,14,16,23–25,29,40,45,47] where the stationary phase is hydrophobic and the percolating mobile phase is hydrophilic. Separation of the compounds to be determined is obtained according the solvophobic theory, resulting in increased retention with increased hydrophobicity. The isocyanates are converted to nonvolatile relatively high molecular weight compounds of hydrophobic nature upon their derivatization during sampling. Such capabilities make them especially suited for reversed phase LC separations. The stationary phases used in reversed phase LC, are predominantly of the silica backbone type with chemically bonded alkyl ligands of various lengths, usually C_8 or C_{18}, where the retaining interactions with the solutes occur. Over the years, numerous silica-based reversed-phase materials have been launched, with special emphasis on the use of higher purities and lower metal content of the silica backbone, smaller particle sizes and extended end-capping or shielding of residual silanol groups on the silica particles, to prevent undesired secondary interactions or to extend the applicable pH range of the stationary phase materials. Modern generation silica-based chemically-bonded reversed-phase stationary materials with particle sizes of $3-5$ μm, in combination with appropriate mobile phases, have in numerous studies proven to provide the efficiency and selectivity for time-efficient separation of isocyanate derivatives.[4,11,12,14,16,23–25,29,40,45,47] Given the fact that most modern generation reversed-phase LC stationary phases provide fair separation of the plentiful isocyanate derivatives of interest with often only slight differences in selectivity or efficiency among them, no detailed presentation of the numerous stationary phases which have been employed for this purpose will be presented here. As a general rule it is important to realize that different stationary phases of the same type, e.g., C_{18}, potentially can result in different retention and selectivity. Such consideration should be made while separating isocyanate derivatives. For instance, if strongly retained oligomeric isocyanate derivatives are to be determined, a stationary phase with a corresponding mobile phase providing low retention certainly is to be preferred. On the other hand, if low-retained monomeric isocyanate derivatives are to be determined, e.g., MIC derivatives or derivatives of TDI or HDI, a stationary phase providing high retention of these species is to be preferred, in order to separate the compounds of interest from the matrix front. The review paper by Guglya effectively covers different separation methods for isocyanate determinations up to year 1998.[4]

The mobile phases in reversed-phase LC are usually mixtures of water and acetonitrile or methanol. The various isocyanate derivatives are often ionizable, requiring buffered mobile phases for pH control and separation optimization. Buffers used for this purpose are mostly formate or acetate buffers, which are volatile and especially suited for MS detection. Other buffers, such as phosphate buffers which cover a wide pH range, can also be employed, if another detection technique than MS is used. Separation of complex mixtures of isocyanate derivatives frequently requires gradient action to resolve all the solutes. Gradient action is typically obtained by the use of mobile phase gradients, where the mobile phase composition is gradually changed during the chromatographic run towards higher elution strengths by a LC pump with gradient capabilities.

This effect is usually obtained by gradually increasing the content of the organic modifier in the mobile phase, although pH gradients have also been explored for separation of isocyanate MAP derivatives.[11] Furthermore, Molander et al. explored temperature gradients, more conveniently used in GC, in miniaturized reversed-phase LC systems with isocratic mobile phases for separation of 2 MP derivatives.[40,41] Gradient action typically reduces analysis times and improves peak shapes, often resulting in improved LODs. Unfortunately, mobile phase gradients are not always compatible with the detectors in use. For instance, it is well known that mobile phase gradients are troublesome in combination with EC detectors, and may give rise to evolving base lines with UV detectors, precluding integration of small peaks. In such cases, temperature programming is an interesting approach.[50]

2. Detection of Isocyanate Derivatives in Liquid Chromatography

In typical chromatographic analyses of environmental contaminants, analytical standards exist for the analyte of interest. The analyte is identified in a real sample if its chromatographic retention time matches that of the analytical standard. However, for isocyanate species, pure analytical standards exist only for derivatized monomers. Yet, in many environments, monomers contributes very little to the total isocyanate (NCO) groups present. The analysis of a derivatized bulk of prepolymeric isocyanate product can be useful in identifying nonmonomeric isocyanate species in real samples collected during use of that product, but there are limitations of using such products as analytical standards for identification and quantification. Not all isocyanate species to which a worker may be exposed are present in the product. When isocyanate products are used, the components are typically undergoing curing reactions with polyols, so new species containing isocyanate groups are generated. Isocyanate-containing species are also generated during thermal breakdown of PUR. Chromatographic retention times are not available to identify these new species. To identify all isocyanate species (monomers or oligomers, those present in the bulk product, and those newly generated) a means of identification other than chromatographic retention time is necessary. Correct identification of unknown isocyanate species requires that the detection scheme be selective or provide some qualitative information about the species in question. Isocyanate methods generally use derivatizing reagents that are responsible for the detectability of the reagent/isocyanate derivative. Total isocyanate methods generally seek to identify all compounds labeled with the derivatizing reagent. Knowledge of the work environment is required to discount any nonisocyanate species which may react with the derivatizing reagent and give a signal in the sample chromatogram. Once these compounds are accounted for, it is assumed that all other compounds in the chromatogram which contain the reagent label are derivatized isocyanates.[11]

Two different detection principles in series operating within their linear range are required when determination of isocyanate derivatives without available standards need to be performed. Such approaches will provide a constant detector response ratio for any given compound, allowing compounds to be identified by that ratio.[11] Instead of using two one-dimensional detectors in series, some methods use multidimensional detectors for identification. It has been found that photodiode array (PDA) detection, which provides an entire UV spectrum of a chromatographic peak, is useful in identifying 2 MP derivatized isocyanates.[51] Furthermore, several researchers have investigated MS as a detector for derivatized isocyanates with great success.[16,27–29,30,31,46,52,53]

Once a chromatographic peak has been correctly identified as isocyanate-derived, it must be quantified. For methods used to determine monomeric isocyanates only, identification is generally based on retention time, and analytical standards exist which enable direct construction of a calibration curve for quantification. LOD is the major factor in choosing a derivatizing reagent/detector combination. LODs for reagent/detector combinations are presented in a review paper by Streicher et al.[11] To quantify compounds for which analytical standards are not available, the detector response factor for the unknown derivatized isocyanate species must be the same as that of the derivatized monomer. This is achieved by choosing a derivatizing reagent/detector combination such that nearly all the detector response is attributable to the derivatization reagent label and that response does not change. Illustratively, the UV detector is not very suitable for quantification of aromatic TDI oligomer derivatives based on the TDI monomer derivatives, as the increased number of aromatic rings in the oligomers will contribute substantially to the UV signal as compared to the monomeric species.[54]

In addition to providing fast reaction rates with the isocyanate functional groups, several of the derivatizing agents are especially designed to provide high intensities with different LC detection principles. In general, many of the LC detection principles are applicable for several of the different derivatives, but with different signal intensities. When keeping in mind the very low OELs for isocyanates and the often short sampling times in order to reveal peak exposures, it is obvious that detection principles providing high sensitivity are required.

Methods employing the MAMA- and MAP-reagents usually use UV and F detection, and the very small compound-to-compound UV-response variability of MAMA- and MAP-derivatized isocyanates, in addition to the fairly good UV sensitivity for these compounds at 256 nm, makes this detector attractive for quantification. However, the F sensitivity for the two reagents is superior for any other derivative/detector combinations, but the compound-to-compound F response is unfortunately too low for quantitative measurements.[11] 2 MP isocyanate derivatives are routinely determined using UV and EC detection,[12] while methods based on TRYP derivatization often are recommended with F and EC detection in series, providing improved low compound-to-compound variabilities and improved sensitivity and selectivity as compared to UV detection.[11] F or UV detection is often recommended for determination of 2PP derivatives, while UV detection usually has been employed for NITRO derivatives, unfortunately with relative poor sensitivity.[11] The PAC reagent includes an anthracene-group, which makes it especially suited for F detection. Regarding the newly introduced MMNTP reagent, the urea derivatives of phenylisocyanate (PI), HDI, TDI, and MDI monomers showed good spectroscopic properties with small compound-to-compound variabilities with UV and F detection.[47] For another newly launched derivatizing agent for isocyanates, NBDPZ, diode array, and F detection provided high sensitivity of the derivatives.[45] In contrast to established derivatizing agents for the analysis of isocyanates, NBDPZ provides increased selectivity due to the favorable detection of wavelengths in the visible range. In addition, the high molar absorptivities of this reagent and the derivatives provide excellent sensitivity which is superior to most of the published methods. This derivatizing agent has also been used with MS detection with atmospheric pressure chemical ionization (APCI) in the negative mode.[46]

The most pronounced direction in LC detection the last few years has been the establishment of ESI and APCI MS techniques as a sensitive and selective detection principle, where the ionization takes place at atmospheric pressure. LC–MS has evolved to be a robust and routinely used technique in many laboratories during the past few years, making available structure elucidation and qualitatively screening. The use of tandem quadropole–quadropole or iontrap mass analyzers makes further specificity available through fragmentation and extraction of compound-selective daughter ions, also leading to increased sensitivity. This general trend has appeared with regard to isocyanate measurements, especially after the introduction of the DBA derivatizing agent, for which MS detection is by far the best detection principle.[24] As a direct consequence of this introduction, MIC and ICA were for the first time determined in working atmospheres.[30,31] The trend of using LC–MS for determination of isocyanate derivatives will most probably continue to evolve for years to come, especially in hyphenation with the nondestructive UV and F detectors, and has already successfully been explored with the DBA, 2 MP and the NBDPZ derivatives.[16,23,24,27–31,46,47,52,53] Typical selected ion reaction (SIR) mass chromatogram of monomeric and polymeric HDI DBA derivatives from a 6-liter air sample taken in a car-painting workshop during a spray-painting process is shown in Figure 21.5, illustrating the excellent selectivity obtained by LC–MS.[29]

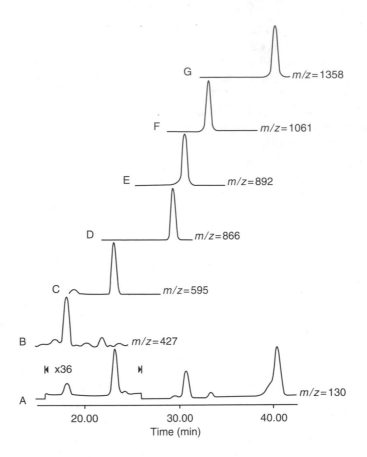

FIGURE 21.5 LC–ESI mass chromatograms of airborne aliphatic isocyanates collected with an impinger containing DBA in toluene. The 6-1 (1 l/min) sample was taken in a car-painting workshop during spray-painting. The presence of six different isocyanates was observed. Trace A indicates the aliphatic isocyanate [DBA + H]$^+$ ions. The extracted SIR [M + H]$^+$ ions of HDI–DBA, HDI-dimer (uretidone)–DBA, HDI-biuret–DBA, HDI-isocyanurate–DBA, HDI-isocyanurate-uretidone–DBA and HDI-diisocyanurate–DBA are showed in traces A, B, C, D, E, F, and G, respectively. The HDI–DBA peak reflects an air concentration of about 2 μg/m^3. (The figure is reprinted with permission from Karlsson, D., Spanne, M., Dalene, M., and Skarping, G., *Analyst*, 123, 117–123, 1998.)

3. Miniaturized Liquid Chromatography

There is an increasing demand for analytical techniques which can measure components at low concentrations in often limited amounts of environmental and biological samples, e.g., air monitoring of isocyanates in occupational settings. Miniaturization is the keyword when it comes to development of such new techniques, including the recent work on nanoparticle technology,[55] chemistry-on-a-chip,[56] in addition to the general trend of utilization of microscaled separation systems and detectors. Among these miniaturized techniques, packed-capillary LC, with column inner diameters ranging from 500 to 50 μm, has shown considerable progress and advances of practical value.[57] The use of miniaturized columns in LC offers enhanced mass sensitivity due to reduced dilution of the chromatographic band as compared to the use of conventional columns.[58] Further improvements in concentration sensitivity are possible if focusing techniques are employed on sample introduction, making available total sample exploitation. On-column focusing is traditionally performed by dissolving the sample in solvent compositions of noneluting properties.[40,41,57] By use of focusing techniques, sample volumes up to 1 ml have efficiently been loaded onto packed capillaries.[59] However, loading of such large injection volumes is a time consuming process considering the low flow rates used in packed-capillary LC, typically in the range 1 to 10 μl/min. Hence, capillary scale precolumn switching systems have been explored, where large sample volumes are loaded on to shorter precolumns generating low back-pressures, allowing efficient sample loading at increased flow rates prior to column-switching back-flushed solute elution at linear velocities close to optimal.[41,59] Furthermore, column-switching systems have potential for allowing elution from the analytical column simultaneous to precolumn loading of the next sample, in addition to the capability of online sample cleanup implementation. The low flow rates used with miniaturized LC are especially attractive with regard to ESI–MS coupling,[60] and this characteristic in combination with the requirement for high sensitivity has made this instrumental combination also attractive for isocyanate measurements.[14,40,41,61] The general trend in LC towards the use of narrower columns has escalated lately due to the high sensitivity required for separations within fields related to proteomics, and this trend will probably also be more prominent in the future for isocyanate analysis, along with the development of more user-friendly and robust miniaturized LC instrumentation.

4. Other Separation Principles

Although reversed-phase LC with various detectors is the dominant technique at the present stage for isocyanate derivative separations, other principles such as normal phase LC, thin layer chromatography (TLC) and GC, usually with derivatization to improve the volatility of the derivatives, have been explored for this purpose over the years. A detailed review of such methods for this purpose can be found elsewhere.[4] For the most volatile isocyanates, however, GC separations have appeared in recent literature, and MIC and ICA have been determined by GC–MS as DBA and 2 MP derivatives.[16] Furthermore, capillary electrophoresis (CE) has been explored for separation of isocyanate 2 MP derivatives.[62,63] CE is a highly miniaturized technique, and provides the same attractive improved mass sensitivity as miniaturized LC. However, in contrast to miniaturized LC, the small injection volumes allowed with this technique make the concentration sensitivity somewhat limited.

5. Direct Reading Instruments

In order to get instantaneous results, continuous filter tape instruments have been developed based on modification of the colorimetric Marcali method.[13] The instruments have been used for continuous air monitoring of production sites and for personal monitoring of isocyanates. A limitation with the filter tape instruments is that no compound-specific information is given in means of retention times or detector specific structural information. Since the instruments have to be calibrated for the different isocyanates, quantitative estimation of mixed isocyanate exposures is troublesome. Problems associated with influence of interfering compounds, humidity, and collection of particles have been reported.[15,64–66]

III. BIOMONITORING METHODS

Biomonitoring involves the analysis of human tissues, blood, and excreta for evidence of exposure to chemical substances and may involve the direct measurement of a chemical or a metabolite in a biological matrix, or an indirect measurement of a biochemical or physiological change which occurs in response to the exposure. To obtain a measure of the total exposure to an individual of a chemical substance from all routes including inhalation, dermal, and oral pathways, biomonitoring methods are preferred over the more commonly used air monitoring methods. Furthermore, effects of exposure can differ greatly between individuals due to differences in toxicokinetcis and toxicodynamics. Individual dose monitoring will probably play an increasingly important role in modern occupational toxicology.

Metabolites of isocyanates in biological fluids and their protein adducts have been used as biomarkers of exposure.[67-87] However, biomarkers are only available for a limited number of isocyanates at present. Several studies have suggested that metabolites of isocyanates in hydrolyzed urine may be used as an indicator of exposure over a work-shift, and some useful relationships between levels in air and urine have been presented for TDI, HDI, and MDI (as monomeric diisocyanates).[74-77,79,81,84,88-92] Furthermore, metabolites of diisocyanates in hydrolyzed plasma and or as hemoglobin (Hb) adducts can be used as an indicator of accumulated dose over about a month.[9,73-75,77,79-82,90,93]

While biomarkers can be used for inter- and intraindividual comparisons and for studies of time-trends and the effects of personal protection equipment, biomarkers are at present not suited for assessment of outcome, in terms of health effects, due to lack of knowledge. If peak exposures are significant for the outcome of exposure to isocyanates, biomarkers may also have a poor relationship to outcome. Nevertheless, biomarkers are determined in samples taken after exposure, which makes it possible to analyze samples taken by the initiative of the individual worker, physician, or industrial hygienist. In addition, biomonitoring is an asset in the establishment of human dose–response models, and further research in this field is likely to progress in the future.

The analytical techniques employed for biomonitoring of isocyanates have primarily been GC and LC, often in combination with MS detection. These are techniques which are widely adapted in the field of bioanalysis. The corresponding diamine metabolites of the respective monomeric diisocyanates have in most studies been determined in urine or plasma, or as blood protein metabolites, after sample preparation often based on hydrolysis, extraction, and subsequent derivatization of polar functional groups, in order to increase the volatility in cases where GC was employed.[79-82]

IV. OCCUPATIONAL EXPOSURE

Although there has been a great interest in development of analytical methodology for air sampling of isocyanates in workroom atmospheres and biomonitoring of isocyanates, only a limited number of studies appear in the scientific literature, where these methods have been used for exposure assessment at workplaces on a larger scale. It appears that occupational hygienists working within this field have summarized their findings to some extent, but not in the form of reports which are easily accessible or understandable. These reports are often written in native languages and have not been critically evaluated through peer-review procedures. This observation might be corroborated by the fact that analytical chemists, to a higher extent, are working within a scientific tradition, while occupational hygienists often have more practical working methods. Furthermore, such exposure measurements are performed at industrial sites often financed by the industries, and it is not always in the interests of the companies to publicly broadcast the findings. Such monitoring studies are likely to appear to a greater extent in the years to come, due to the recent successful establishment of analytical procedures for such measurements and the increased interest in isocyanates as workplace hazards. Nevertheless, some scientifically evaluated studies in journals with peer-review procedures of exposure measurements of isocyanates in workroom atmospheres and biomonitoring studies of occupationally exposed personnel, are still present, and will be presented within the context of this chapter.

The study of Myer et al. in 1993 summarized the results of industrial hygiene surveys performed between 1979 and 1987 in paint manufacturing and applications using PUR coatings containing HDI and HDI-based polyisocyanates.[18] A total of 466 HDI-based polyisocyanate samples and 457 HDI samples were collected from 47 operations, most of which were in application. The application surveys covered manufacture and refinishing of transportation vehicles, painting of large military and civilian equipment, industrial finishing operations, and maintenance and construction operations. The primary objective of the surveys was to assess the potential exposure to HDI and HDI-based polyisocyanate. In more than 60% of the surveys concentrations of airborne organic solvents also were monitored. Isocyanates were sampled using toluene/NITRO in impingers, while solvents were collected using charcoal tubes. These were analyzed using LC or GC, respectively. The authors pointed out that the data from these workplace situations showed some potential for isocyanate overexposure of unprotected workers which is greater in spray than in nonspray operations.

In 1996, Maitre et al. evaluated two air sampling methods for monitoring the level and the variability of a painter's exposure to HDI polyisocyanates while spraying Tolonate-based paints in typical auto body repair shop paint booths.[17] Personal air samples were collected by impingers containing 2 MP absorber solution and 2 MP-impregnated filters. HDI-biuret and isocyanurate, the principal polyisocyanates in Tolanate paint systems, were analyzed concurrently by two different laboratories. Potential exposure to HDI polyisocyanates measured by impinger devices ranged from 0.25 to 3 mg/m^3, while impregnated filters significantly underestimated the atmospheric concentrations of HDI polyisocyanates in the painter's breathing zone. In addition, the use of an appropriate half-face mask with 90% efficiency was evaluated, measuring the air levels beneath the mask, resulting in significant lowering of the exposure level.

Comparative air measurements of TDI were performed by Tinnerberg et al. in 1997 in a 5.6-m^3 standard atmosphere and at a TDI flexible foam plant.[77] Air samples were collected in impinger flasks containing MAMA in toluene and on 13-mm glass fiber filters impregnated with MAMA and glycerol. The samples were analyzed by LC–UV and with filter-tape instruments. In the laboratory study the average amounts of the TDI–MAMA derivatives determined were higher for filters compared to impingers, when tested at concentrations between 16 and 150 μg/m^3 ($n = 29$). At the TDI foaming plant the amount of TDI–MAMA collected on the filters compared with impingers showed higher TDI values at low concentrations and lower values at higher concentrations. Similar behavior was also observed for the filter-tape measurements, but for two samples at very low concentrations the response was much lower. The average air concentration was 29.8 μg/m^3 (12.5 to 79.9; $n = 12$). The highest exposure peak measured was approximately 3 mg TDI/m^3. Furthermore, the TDI metabolites 2,4- and 2,6-toluene diamine (TDA) in urine and in plasma from four exposed workers and one volunteer were determined after strong acid hydrolysis as their pentafluoro-propionic anhydride (PFPA) derivatives using GC–MS. The plasma TDA among the workers varied between 1 and 38 μg/l and between 7 and 24 μg/l for 2,4- and 2,6-TDA, respectively. The individual plasma levels among the workers over the three-day periods varied between 7 and 73%. For the volunteer, plasma TDA reached a maximum about 24 h after the last exposure, while the half-time was about 10 days. The urine TDA levels varied greatly with time and exposure, and high peaks were found during or shortly after the exposure.

Rudzinski et al. explored several methods for the sampling and analysis of airborne HDI and polyisocyanates during spray-painting operations in 1995.[94] An impinger filled with 2 MP in toluene for collection and derivatization followed by LC–UV–EC determination was compared directly to a glass fiber filter coated with 2PP subsequent to LC–UV determination. The results for HDI monomer demonstrated that the 2 MP impinger LC–UV–EC appeared to give higher results than those obtained using 2 PP filter LC–UV, especially when the total particulate concentration was high. Furthermore, field studies showed that polyisocyanate concentrations during spray-paint operations might exceed a concentration of 1 mg/m^3, which is believed to be hazardous. Preliminary results indicate that the true concentration of polyisocyanate in air, in all cases but one, was within two times the theoretical concentration based on the total particulate mass.

Karlsson et al. have presented a methodology for workplace air monitoring of aromatic and aliphatic, mono- and polyisocyanates by derivatization with DBA, using impinger flasks containing 10 ml of 0.01 M DBA in toluene and a glass fiber filter in series after the impinger flask, thereby providing the possibility of collecting isocyanates in the gaseous and particle phases.[95]

Quantification was made by LC–MS, monitoring the molecular ions $[M + H]^+$. Air samples taken with this method in car repair shops showed that many different isocyanates are formed during thermal decomposition of PUR coatings. In addition to isocyanates such as HDI, isophorone (IPDI), TDI and MDI, monoisocyanates such as MIC, ethyl isocyanate (EIC), propyl isocyanate (PIC), butyl isocyanate (BIC), and PI were also reported. In many air samples the aliphatic monoisocyanates dominated, and the highest levels of isocyanates were observed during cutting and welding operations. In a single air sample from a welding operation in a car repair shop, the highest concentrations were 290, 60, 20, 9, 27, 105, 39, 4, and 140 μg/m^3 with respect to MIC, EIC, PIC, BIC, phenyl isocyanate (PhI), HDI, IPDI, MDI, and TDI, respectively. Monitoring of the particle size distribution and concentration during grinding, welding, and cutting operations showed that, ultra fine particles ($<$0.1 μm) were formed at high concentrations. Isocyanates with low volatility were mainly found in the particle phase, but isocyanates with a relatively high volatility, such as TDI, were found in the particle and gaseous phases. Furthermore, the same authors have reported that when mineral wool with a phenol–formaldehyde–urea (PFU) resin was thermally degraded, 0.1% m/m of MIC was released.[30] In air samples taken on top of a new electric oven insulated with mineral wool, MIC was found in the range 0.13 μg/m^3. No MIC in air was found from a preheated oven. In 2001, the same innovative research group presented a method for the determination of ICA in air samples based on BDA impinger derivatization and LC–MS determinations.[31] ICA was emitted during thermal degradation of PFU resins and PUR lacquers from car metal sheets. ICA was the most dominant isocyanate and in PUR coating up to 8% of the total weight was emitted as ICA, and for PFU resins up to 14% was emitted as ICA. When air samples were collected in an iron foundry during casting in sand moulds with furan resins, concentrations of ICA in the range 50–700 μg/m^3 were found in the working atmosphere. Another study[96] reported the occurrence of thermal degradation products of PUR in high concentrations during welding in district heating pipes and PUR-coated metal sheets. Three amines, five aminoisocyanates and 11 isocyanates were identified. The concentrations of isocyanates, aminoisocyanates, and amines in samples collected in the smoke close to the welding spot were in the ranges 150–650, 4–290, and 1–70 ppb, respectively. In samples collected in the breathing zone, isocyanates and aminoisocyanates were observed in the ranges 9–120 and 4–19 ppb, respectively. The compounds were present in gaseous and particle phases. Volatile compounds dominated in the gaseous phase, whereas less volatile compounds dominated in the particle phase.

Henriks-Eckerman et al. also investigated the thermal degradation products of PURs and exposure to isocyanates by stationary and personal measurements in five different occupational environments.[97] Isocyanates were collected on glass fiber filters impregnated with 2 MP and in impingers containing DBA in toluene, connected to a glass fiber post filter. The derivatives formed were analyzed by liquid chromatography (2 MP derivatives with UV–EC detection and DBA derivatives with MS detection). The release of aldehydes and other volatile organic compounds into the air was also studied. In a comparison of the two sampling methods, the 2 MP method yielded about 20% lower concentrations for MDI than the DBA method. In car repair shops, the median concentration of diisocyanates (given as NCO groups) in the breathing zone was 1.1 μg/m^3 NCO during grinding and 0.3 μg/m^3 NCO during welding, with highest concentrations of 1.7 and 16 μg/m^3 NCO, respectively. High concentrations of MDI, up to 25 and 19 μg/m^3 NCO, respectively, were also measured in the breathing zone during welding of district heating pipes and turning of a PUR coated metal cylinder. During installation of PUR-coated floor covering, small amounts of aliphatic diisocyanates were detected in the air. MIC and ICA were detected only during welding and turning operations. The diisocyanate concentrations were in general higher near the emission source than in the workers breathing zone, illustrating the importance of using samplers suitable for personal monitoring in order to obtain representative measurements.

Kaaria et al. measured occupational exposure to MDI during molding of rigid PUR foam.[92] Airborne MDI was sampled on 2 MP impregnated glass fiber filters and determined by LC–UV–EC. Workers ($n = 57$) from three different factories participated in the study.

The MDI concentrations were below the LOD in most (64%) of the air samples collected in the workers' breathing zone. Furthermore, urine samples from the workers were collected, and the metabolite 4,4'-methylenedianiline (MDA) was measured by GC–MS. Detectable amounts of MDA were found in 97% of the urine samples. Monitoring of urinary MDA appears to be an appropriate method of assessing MDI exposure in work environments with low or undetectable MDI concentrations in the workplace air. The same authors have also reported measurement of occupational exposure to the TDI isomers during the production of flexible foam, by means of air measurements of TDIs and biomonitoring of the urinary TDA metabolites.[84] Again, airborne TDI was sampled on 2 MP impregnated glass fiber filters and determined by LC–UV–EC, based on personal sampling of 17 workers. There was a trend for linear correlation between urinary TDA concentration and the product of airborne TDI concentration vs. sampling time. Measurement of TDA in urine was proposed as a practical method for assessing personal exposure in workers exposed intermittently to TDI.

Blomqvist et al. quantified isocyanates, aminoisocyanates, and amines from the combustion of 24 different materials or products typically found in buildings.[98] Small-scale combustion experiments were conducted in a cone calorimeter, where generally well-ventilated combustion conditions are attained. Measurements were further made in two different full-scale experiments. Isocyanates and amino-compounds were sampled using an impinger-filter sampling system with a reagent solution of DBA in toluene prior to determination by LC–MS. Isocyanates were produced from the majority of the materials tested, and the highest concentration was found for glass wool insulation and PUR products. The distribution of isocyanates between the particulate and fluid phases varied for the different materials and a tendency toward enrichment of particles was seen for some of the materials. When comparing the potential health hazard between isocyanates and other major fire gases it was reported that, isocyanates in several cases represented the greatest hazard.

From the end of the 1980s, Skarping and coworkers introduced biological monitoring of diisocyanates as their respective diamine metabolites by GC–MS in single ion monitoring (SIM) mode after derivatization in a series of papers.[68–83] These methods were subsequently applied for exposure assessment on occupationally exposed personnel. In one study, exposure to TDI in a factory producing flexible PUR foam was studied for 48 h and biological samples were collected from five PUR workers, two white-collar workers and two volunteers.[99] The concentrations of TDI in air were determined by MAMA derivatization on a filter, followed by LC–UV determination, while the TDA isomers in urine and plasma samples were determined after hydrolysis as PFPA derivatives by GC–MS in the negative chemical ionization mode. The concentration of TDI in air was 0.4 to 4 $\mu g/m^3$. The five male PUR workers showed the highest average urinary elimination rate of TDA. Two PUR workers and the two white collar workers had an elimination rate of 20 to 70 ng on average for the sum of 2,6-TDA per hour and 2,4-TDA per hour, and three PUR workers had an average of 100 to 300 ng TDA per hour. The elimination rate curves for all the studied subjects had a linear relation with exposure to TDI. The concentrations of 2,4- and 2,6-TDA in plasma for the PUR factory employees were virtually stable. No relation between the elimination rates of TDA in urine and plasma concentrations of TDA was found. The five PUR workers showed plasma concentrations of the sum of 2,4- and 2,6-TDA in the range 1 to 8 ng/ml, while the TDA level in the plasma of two white collar workers, present only occasionly in the factory, was 0.2 to 1 ng/ml. The two volunteers showed an increasing concentration of TDA in plasma with time. In another study, they used an LC–MS approach for the determination of MDA in hydrolyzed urine as a biomarker for exposure to MDI.[68] The concentration of MDA was 4 $\mu g/l$ in pooled urine samples from ten workers exposed to thermal degradation products of a MDI based PUR. Several MDA isomers and oligomers were observed in the samples. In another study a GC–MS approach was used with derivatization for the same purpose with quite similar results.[100] In 2000, they reported a relation among exposure to PUR glue, biomarkers of exposure and effect, and work-related symptoms which occurred at least once a week in a cross-sectional study where 152 workers in a factory were exposed to sprayed and heated PUR glue. Furthermore, plasma and urine samples were examined with GC–MS with respect to determination of HDA and MDA in relation to symptoms of the eyes, airways, and lung function.[83] Plasma MDA was detected in 65% of the workers, while HDA in none. The authors established relations among exposures to sprayed and heated PUR glue based on MDI and HDI, concentrations of MDA in plasma and urine, and work related symptoms.

Schutze et al. also monitored MDA as Hb adducts and urine metabolites from 27 workers exposed to MDI.[93] The samples were analyzed by GC–MS after hydrolysis, extraction, and derivatization with heptafluoro-butyric anhydride (HFBA). Exposure levels, as monitored using personal air direct reading samplers, were below the detection limit of 3 $\mu g/m^3$, with the exception of three individuals. In ten of the MDI workers, hydrolysable Hb adducts of MDA were found. Except for four subjects, the presence of MDA (0.007–0.14 nM) was detected in all urine samples after base treatment. They also investigated biological samples from a group of 20 workers exposed to MDI vapor during the manufacture of PUR products.[90] The blood and urine samples were analyzed for the presence of adducts and metabolites using GC–MS methods. Urinary base-extractable metabolites were found above control levels in 15 of the 20 workers and ranged from 0.035 to 0.83 nM MDA. MDA was detected as Hb adduct in all of the 20 subjects, and the level ranged from 70 to 710 fmol/g Hb. The plasma MDA levels ranged from 0.25 to 5.4 nM, and up to 120 fmol/mg were covalently bound to albumin.

To assess the exposure of sprayers employed in motor vehicle repair shops, Williams et al. also explored biomonitoring methods to measure urinary HDA by GC–MS.[91] Samples were collected among sprayers wearing personal protective equipment and spraying in booths or with local exhaust ventilation, from bystanders, and from unexposed subjects. HDA was detected in four sprayers and one bystander out of 22 workers, while it was not detected in samples from unexposed persons. They concluded that exposure to isocyanates still occurred despite the use of personal protective equipment and the use of a booth or extracted space, and that health surveillance is likely to be required to provide feedback on the adequacy of controls, even if such precautions are used, and to identify cases of early asthma.

In a recent study by Rosenberg et al.[101] exposure to diisocyanates was assessed by biological monitoring among workers, exposed to thermal degradation products of PUR in five PUR-processing environments. The processes included: grinding and welding in car repair shops; milling and turning of PUR-coated metal cylinders; injection molding of thermoplastic PUR; welding and cutting of PUR-insulated district heating pipes during installation and joint welding; heat-flexing of PUR floor covering.

Isocyanate-derived amines in acid-hydrolyzed urine samples were analyzed as perfluoro-acylated derivatives by GC–MS in negative chemical ionization mode. TDA and MDA were detected in urine samples from workers in car repair shops, and MDA in samples from workers welding district heating pipes. The 2,4-TDA isomer accounted for about 80% of the total TDA detected. No 2,6-TDA was found in the urine of nonexposed workers. The highest measured urinary TDA and MDA concentrations were 0.79 n*mol*/m*mol* creatinine and 3.1 n*mol*/m*mol* creatinine, respectively. The concentrations found among nonexposed workers were 0.08 n*mol*/m*mol* creatinine for TDA and 0.05 n*mol*/m*mol* creatinine for MDA (arithmetic means). They concluded that, exposure to diisocyanates originating from the thermal degradation of PURs are often intermittent and of short duration. Nevertheless, monitoring diisocyanate-derived amines in acid-hydrolyzed urine samples can identify exposure to aromatic diisocyanates.

After the establishment of LC–MS as a widely adapted analytical technique, an APCI–LC–MS in positive mode approach has also been reported for biomonitoring of urinary TDA for workers exposed to TDI, after acid hydrolysis and extraction with dichloromethane.[102] TDA isomers in the urine of exposed workers as determined by LC–MS correlated well with those obtained by GC–MS. 2,6- and 2,4-TDA were not detected in nonexposed subjects, whereas exposed workers showed urinary levels up to 250 and 63 $\mu g/l$, respectively. LC–MS will probably be used in the years to come to an increasing extent within the field of biomonitoring of isocyanates, avoiding time and labor consuming derivatization.

REFERENCES

1. LeSage, J., Goyer, N., Desjardins, N., Vincent, J. Y., and Perrault, G., Workers exposure to isocyanates, *Am. Ind. Hyg. Assoc. J.*, 53, 46–153, 1992.
2. Karlsson, D., *Airborne Isocyanates, Aminoisocyanates and Amines*, Doctoral dissertation, ISBN 91-7874-163-7, Lund University, Lund Sweden, 2001.
3. *Kirk-Othmer Encyclopedia of Chemical Technology*, 4th ed., Vol. 14, Wiley, New York, pp. 902–934, 1995.
4. Guglya, E. B., Determination of isocyanates in air, *J. Anal. Chem.*, 55, 508–529, 2000.
5. *Kirk-Othemer Encyclopedia of Chemical Technology*, 4th ed., Vol. 24, Wiley, New York, pp. 695–727, 1995.
6. Musk, A. W., Peters, J. M., and Wegman, D. H., Isocyanates and respiratory disease – Current status, *Am. J. Ind. Med.*, 13, 331–349, 1988.
7. Baur, X., Marek, W., Ammon, J., Czuppon, A. B., Marczynski, B., Raulf-Heimsoth, M., Roemmelt, H., and Fruhmann, G., Respiratory and other hazards of isocyanates, *Int. Arch. Environ. Health*, 66, 141–152, 1994.
8. Redlich, C. A., Karol, M. H., Graham, C., Homer, R. J., Holm, C. T., Wirth, J. A., and Cullen, M. R., Airway isocyanate-adducts in asthma induced by exposure to hexamethylene diisocyanate, *Scand. J. Work Environ. Health*, 23, 227–231, 1997.
9. Sabbioni, G., Hartley, R., Henschler, D., Höllrigl-Rosta, A., Kober, R., and Schneider, S., Isocyanate-specific hemoglobin adduct in rats exposed to 4,4'-methylenediphenyl diisocyanate, *Chem. Res. Toxicol.*, 13, 82–89, 2000.
10. Metha, P., Metha, A., Metha, S., and Makhijani, A., Bhopal tragedys helath effects – A review of methyl isocyanate toxicity, *J. Am. Med. Assoc.*, 264, 2781–2787, 1990.
11. Streicher, R. P., Reh, C. M., Key-Schwartz, R. J., Schlecht, P. C., Cassinelli, M. E., and O'Connor, P. F., Determination of airborne isocyanate exposure: considerations in method selection, *Am. Ind. Hyg. Assoc. J.*, 61, 544–556, 2000.
12. Health and Safety Executive Health and Safety Laboratory, MDHS 25/3. *Methods For the Determination of Hazardous Substances; Organic Isocyanates in Air*, Health and Safety Executive/ Occupational Safety and Hygiene Laboratory, Sheffield, 1999.
13. Marcali, K., Microdetermination of toluenediisocyanates in atmosphere, *Anal. Chem.*, 29, 552–558, 1957.
14. Molander, P., Levin, J. O., Ostin, A., Rosenberg, C., Henriks-Eckerman, M. L., Brodsgaard, S., Hetland, S., Thorud, S., Fladseth, G., and Thomassen, Y., Harmonized nordic strategies for isocyanate monitoring in workroom atmospheres, *J. Environ. Monit.*, 4, 685–687, 2002.
15. Levine, S. P., Hillig, K. J. D., Dharmarajan, V., Spence, M. W., and Baker, M. D., Critical-review of methods of sampling, analysis, and monitoring for TDI and MDI, *Am. Ind. Hyg. Assoc. J.*, 56, 581–589, 1995.
16. Henriks-Eckerman, M.-L., Välimaa, J., and Rosenberg, C., Determination of airborne methyl isocyanate as dibutylamine or 1-(2-methoxyphenyl)piperazine derivatives by liquid and gas chromatography, *Analyst*, 125, 1949–1954, 2000.
17. Maître, A., Leplay, A., Perdrix, A., Ohl, G., Boinay, P., Romazini, S., and Aubrun, J. C., Comparison between solid sampler and impinger for evaluation of occupational exposure to 1,6-hexamethylene diisocyanate polyisocyanates during spray painting, *Am. Ind. Hyg. Assoc. J.*, 57, 153–160, 1996.
18. Myer, H. E., O'Block, S. T., and Dharmarajan, V., A survey of airborne HDI, HDI-based polyisocyanate and solvent concentrations in the manufacture and application of polyurethane coatings, *Am. Ind. Hyg. Assoc. J.*, 54, 663–670, 1993.
19. Rudzinski, W. E., Dahlqvist, B., Svejda, S. A., Richardson, A., and Thomas, T., Sampling and analysis in spraypainting operations, *Am. Ind. Hyg. Assoc. J.*, 56, 284–289, 1995.
20. Vincens, J. H., *Aerosol Sampling*, Wiley, Chichester, U.K., 1989.
21. Breuer, D., Measurement of vapour-aerosol mixtures?, *J. Environ. Monit.*, 1, 299–305, 1999.
22. Kenny, L. C., Aitken, R., Chalmers, C., Fabriés, J. F., Gonzalez-Fernandez, E., Kromhout, H., Lidén, G., Mark, D., Riediger, G., and Prodi, V., A collaborative European study of personal inhalable aerosol sampler performance, *Ann. Occup. Hyg.*, 41, 135–153, 1997.

23. Ekman, J., Levin, J. O., Lindahl, R., Sundgren, M., and Ostin, A., Comparison of sampling methods for 1,6-hexamethylene diisocyanate, (HDI) in a commercial spray box, *Analyst*, 127, 169–173, 2002.

24. Spanne, M., Tinnerberg, H., Dalene, M., and Skarping, G., Determination of complex mixtures of airborne isocyanates and amines - Part 1: Liquid chromatography with ultraviolet detection of monomeric and polymeric isocyanates as their dibutylamine derivatives, *Analyst*, 121, 1095–1099, 1996.

25. Streicher, R. P., Kennedy, E. R., and Lorbeau, C. D., Strategies for the simulatneous collection of vapors and aerosols with emphasis on isocyanate sampling, *Analyst*, 119, 89–97, 1994.

26. Tremblay, P., Lesage, J., Ostiguy, C., and Van Tra, H., Investigation of the competitive rate of derivatization of several secondary amines with phenylisocyanate (PHI), hexamethylene-1,6-diisocyanate (HDI), 4,4′-methylenebis(phenyl isocyanate) (MDI) and toluene diisocyanate (TDI) in liquid medium, *Analyst*, 128, 142–149, 2003.

27. Tinnerberg, H., Spanne, M., Dalene, M., and Skarping, G., Determination of complex mixtures of airborne isocyanates and amines - Part 3: Methylenediphenyl diisocyanate, methylenediphenylamino isocyanate and methylenediphenylidiamine and structural analogues after thermal degradation of polyurethane, *Analyst*, 122, 275–278, 1997.

28. Tinnerberg, H., Karlsson, D., Dalene, M., and Skarping, G., Determination of toluene diisocyanate in air using di-n-butylamine and 9-n-methylaminomethyl-anthracene as derivatization reagents, *J. Liq. Chromatogr. Relat. Technol.*, 20, 2207–2219, 1997.

29. Karlsson, D., Spanne, M., Dalene, M., and Skarping, G., Determination of complex mixtures of airborne isocyanates and amines - Part 4: Determination of aliphatic isocyanates as dibutylamine derivatives using liquid chromatography and mass spectrometry, *Analyst*, 123, 117–123, 1998.

30. Karlsson, D., Dalene, M., and Skarping, G., Determination of complex mixtures of airborne isocyanates and amines - Part 5: Determination of low molecular weight aliphatic isocyanates as dibutylamine derivatives, *Analyst*, 123, 1507–1512, 1998.

31. Karlsson, D., Dalene, M., Skarping, G., and Marand, A., Determination of isocyanic acid in air, *J. Environ. Monit.*, 3, 432–436, 2001.

32. Bello, D., Streicher, R. P., and Woskie, S. R., Evaluation of the NIOSH draft method 5525 for determination of the total reactive isocyanate group (TRIG) for aliphatic isocyanates in autobody repair shops, *J. Environ. Monit.*, 4, 351–360, 2002.

33. Nordqvist, Y., Nilsson, U., and Colmsjo, A., Evaluation of denuder sampling for a mixture of three common gaseous diisocyanates, *Anal. Bioanal. Chem.*, 375, 786–791, 2003.

34. Nordqvist, Y., Melin, J., Nilsson, U., Johansson, R., and Colmsjo, A., Comparison of denuder and impinger sampling for determination of gaseous toluene dilsocyanate (TDI), *Fresenius J. Anal. Chem.*, 371, 39–43, 2001.

35. Rando, R. J., and Poovey, H. G., Development and application of a dichotomous vapor/aerosol sampler for HDI-derived total reactive isocyanate group, *Am. Ind. Hyg. Assoc. J.*, 60, 737–746, 1999.

36. Battle, R., Colmsjo, A., and Nilsson, U., Development of a personal isocyanate sampler based on DBA derivatization on solid-phase microextraction fibers, *Fresenius J. Anal. Chem.*, 371, 514–518, 2001.

37. Battle, R., Colmsjo, A., and Nilsson, U., Determination of gaseous toluene diisocyanate by use of solid-phase micoextraction with on-fibre derivatization, *Fresenius J. Anal. Chem.*, 369, 524–529, 2001.

38. von Zweigbergk, P., Lindahl, R., Ostin, A., Ekman, J., and Levin, J. O., Development of a diffusive sampling method for determination of methyl isocyanate in air, *J. Environ. Monit.*, 4, 663–666, 2002.

39. Henneken, H., Lindahl, R., Ostin, A., Vogel, M., Levin, J. O., and Karst, U., Diffusive sampling of methyl isocyanate using 4-nitro-7-piperazinobenzo-2-oxa-1,3-diazole (NBDPZ) as derivatizing agent, *J. Environ. Monit.*, 5, 100–105, 2003.

40. Molander, P., Haugland, K., Fladseth, G., Lundanes, E., Thorud, S., Thomassen, Y., and Greibrokk, T., Determination of 1-(2-methoxyphenyl)piperazine derivatives of isocyanates at low concentrations by temperature-programmed miniaturized liquid chromatography, *J. Chromatogr. A*, 892, 67–74, 2000.

41. Molander, P., Karlsen, A., Fladseth, G., Lundanes, E., Thorud, S., Thomassen, Y., and Greibrokk, T., Determination of 1-(2-methoxyphenyl)-piperazine derivatives of airborne diisocyanates by packed capillary liquid chromatography with pre-column large-volume enrichment, *J. Sep. Sci.*, 24, 947–954, 2001.

42. Wang, L., Air sampling methods for diisocyanates: dynamic evaluation of SUPELCO ORBO (TM)-80 coated filters, *Am. Ind. Hyg. Assoc. J.*, 59, 490–494, 1998.

43. Wu, W. S. and Gaind, V. S., Application of tryptamine as a derivatizing agent for the determination of airborne isocyanates - 6: confirmation of the concept of isolation of a selected PI-system of a derivative for specific high-performance liquid-chromatographic detection through analyzing bulk materials for total isocyanate content, *Analyst*, 119, 1043–1045, 1994.

44. Roh, Y. M., Streicher, R. P., and Ernst, M. K., Development of a new approach for total isocyanate determination using the reagent 9-anthracenylmethyl 1-piperazinecarboxylate, *Analyst*, 125, 1691–1696, 2000.

45. Vogel, M. and Karst, U., 4-Nitro-7-piperazino-2,1,3-benzoxadiazole as a reagent for monitoring of airborne isocyanates by liquid chromatography, *Anal. Chem.*, 74, 6418–6426, 2002.

46. Hayen, H., Jachmann, N., Vogel, M., and Karst, U., LC-electron capture-APCI(-)-MS determination of nitrobenzoxadiazole derivatives, *Analyst*, 128, 1365–1372, 2003.

47. Werlich, S., Stockhorst, H., Witting, U., and Binding, N., MMNTP - a new tailor-made modular derivatization agent for the selective determination of isocyanates and diisocyanates, *Analyst*, 129, 364–370, 2004.

48. Kuck, M., Balle, G., and Slawyk, W., Sampling of diisocyanates (HDI, TDI) in air by derivatisation with secondary amines as reagents - Part 1: Partial rate factors (PRF) of reagents, *Analyst*, 124, 933–939, 1999.

49. Sango, C. and Zimerson, E., A new reagent for determination of isocyanates in working atmospheres by HPLC using UV or fluorescence detection, *J. Liq. Chromatogr.*, 3, 971–990, 1980.

50. Nordstrom, O., Molander, P., Greibrokk, T., Blomhoff, R., and Lundanes, E., Evaluation of temperature programming for gradient elution in packed capillary liquid chromatography coupled to electrochemical detection, *J. Microcolumn Sep.*, 13, 179–185, 2001.

51. Key-Schwartz, R. J., Analytical problems encountered with NIOSH method-5521 for total isocyanates, *Am. Ind. Hyg. Assoc. J.*, 56, 474–479, 1995.

52. Vangronsveld, E. and Mandel, F., Workplace monitoring of isocyanates using ion trap liquid chromatography/tandem mass spectrometry, *Rapid Commun. Mass Spectrom.*, 17, 1685–1690, 2003.

53. Gagne, S., Lesage, J., Ostiguy, C., and Tra, H. V., Determination of unreacted 2,4-toluene diisocyanate (2,4TDI) and 2,6-toluene diisocyanate (2,6TDI) in foams at ultratrace level by using HPLC-CIS-MS-MS, *Analyst*, 128, 1447–1451, 2003.

54. Streicher, R. P., Arnold, J. E., Cooper, C. V., and Fischbach, T. J., Investigation of the ability of MDHS method-25 to determine urethane-bound isocyanate groups, *Am. Ind. Hyg. Assoc. J.*, 56, 437–442, 1995.

55. Lu, Y. F., Fan, H. Y., Stump, A., Ward, A., Rieker, T. L., and Brinker, C. J., Aerosol-assisted self-assembly of mesostructured spherical nanoparticles, *Nature*, 398, 223–226, 1999.

56. Bousse, L., Cohen, C., Nikiforov, T., Chow, A., Kopfsill, A. R., Dubrow, R., and Parce, J. W., Electrokinetically controlled microfluidic analysis systems, *Annu. Rev. Biophys. Biomol.*, 29, 155–181, 2000.

57. Vissers, J. P. C., Recent developments in microcolumn liquid chromatography, *J. Chromatogr. A*, 856, 117–143, 1999.

58. Ishii, D., Ed., *Introduction to Microscale High Performance Liquid Chromatography*, p. 7, 1998.

59. Holm, A., Molander, P., Lundanes, E., and Greibrokk, T., Determination of rotenone in river water utilizing packed capillary column switching liquid chromatography with UV and time-of-flight mass spectrometric detection, *J. Chromatogr. A*, 983, 43–50, 2003.

60. Abian, J., Oosterkamp, A. J., and Gelpi, E., Comparison of conventional, narrow-bore and capillary liquid chromatography mass spectrometry for electrospray ionization mass spectrometry: practical considerations, *J. Mass Spectrom.*, 34, 244–254, 1999.

61. Brunmark, P., Dalene, M., Sango, C., Skarping, G., Erlandsson, P., and Dewaele, C., Determination of toluene diisocyanate in air using micro liquid-chromatography with UV-detection, *J. Microcolumn Sep.*, 3, 371–375, 1991.

62. Rudzinski, W. E., Pin, L., Sutcliffe, A., and Thomas, T., Determination of hexamethylene diisocyanate in spary-painting operations using capillary zone electrophoresis, *Anal. Chem.*, 66, 1664–1666, 1994.

63. Rudzinski, W. E., Yin, E., England, E., and Charlton, G., Determination of hexamethylene diisocyanate-based isocyanates in spray-painting operations - Part 2: Comparison of high performance liquid chromatography with capillary zone electrophoresis, *Analyst*, 124, 119–123, 1999.

64. Dharmarajan, V. and Rando, R. J., Critival-evaluation of continous monitors for toluene diisocyanate, *Am. Ind. Hyg. Assoc. J.*, 41, 869–878, 1980.

65. Mazur, G., Baur, X., and Rommelt, H., Determination of toluene diisocyanate in air by HPLC and band-tape monitors, *Int. Arch. Occup. Environ. Health*, 58, 269–276, 1986.

66. Dharmarajan, V., Evaluation of personal continuous paper-tape monitors for toluenediisocyanate, *Am. Ind. Hyg. Assoc. J.*, 57, 68–71, 1996.

67. Brunmark, P., Persson, P., and Skarping, G., Determination of 4,4′-methylenedianiline in hydrolyzed human urine by micro liquid-chromatography with ultraviolet detection, *J. Chromatogr. B*, 579, 350–354, 1992.

68. Skarping, G., Dalene, M., and Brunmark, P., Liquid-chromatography and mass-spectrometry determination of aromatic-amines in hydrolyzed urine from workers exposed to thermal degradation products of polyurethane, *Chromatographia*, 39, 619–623, 1994.

69. Skarping, G., Dalene, M., and Tinnerberg, H., Biological monitoring of hexamethylene-diisocyanate and isophorone-diisocyanate by the determination of hexamethylene-diamine and isophorone-diamine in hydrolyzed urine using liquid-chromatography and mass-spectrometry, *Analyst*, 119, 2051–2055, 1994.

70. Skarping, G., Dalene, M., Svensson, B. G., Littorin, M., Akesson, B., Welinder, H., and Skerfving, S., Biomarkers of exposure, antibodies, and respiratory symptoms in workers heating polyurethane glue, *Occup. Environ. Med.*, 53, 180–187, 1996.

71. Dalene, M., Skarping, G., and Brunmark, P., Assessment of occupational exposure to 4,4′-methylenedianiline by the analysis of urine and blood-samples, *Int. Arch. Occup. Environ. Health*, 67, 67–72, 1995.

72. Brunmark, P., Bruze, M., Skerfving, S., and Skarping, G., Biomonitoring of 4,4′-methylene dianaline by measurement in hydrolyzed urine and plasma after epicutneous exposure in humans, *Int. Arch. Occup. Environ. Health*, 67, 95–100, 1995.

73. Dalene, M., Jakobsson, K., Rannug, A., Skarping, G., and Hagmar, L., MDA in plasma as a biomarker of exposure to pyrolysed MDI-based polyurethane: correlations with estimated cumulative dose and genotype for N-acetylation, *Int. Arch. Occup. Environ. Health*, 68, 165–169, 1996.

74. Dalene, M., Skarping, G., and Lind, P., Workers exposed to thermal degradation products of TDI- and MDI-based polyurethane: biomonitoring of 2,4-TDA, 2,6-TDA, and 4,4′-MDA in hydrolyzed urine and plasma, *Am. Ind. Hyg. Assoc. J.*, 58, 587–591, 1997.

75. Brorson, T., Skarping, G., and Sangö, D., Biological monitoring of isocyanates and related amines - Part 4: 2,4-Toluenediamine and 2,6-toluenediamine in hydrolyzed plasma and urine after test-chamber exposure of humans to 2,4-toluenediisocyanate and 2,6- toluenediisocyanate, *Int. Arch. Occup. Environ. Health*, 63, 253–259, 1991.

76. Brorson, T., Skarping, G., and Sangö, D., Biological monitoring of isocyanates and related amines - Part 2: Test chamber exposure of humans to 1,6′hexamethylene diisocyanate (HDI), *Int. Arch. Occup. Environ. Health*, 62, 385–389, 1990.

77. Tinnerberg, H., Dalene, M., and Skarping, G., Air and biological monitoring of toluene diisocyanate in a flexible foam plant, *Am. Ind. Hyg. Assoc. J.*, 58, 229–235, 1997.

78. Tinnerberg, H., Skarping, G., Dalene, M., and Hagmar, L., Test chamber exposure of humans to 1, 6′hexamethylene diisocyanate and isophorone diisocyanate, *Int. Arch. Occup. Environ. Health*, 67, 367–374, 1995.

79. Lind, P., Dalene, M., Lindstrom, V., Grubb, A., and Skarping, G., Albumin adducts in plasma from workers exposed to toluene diisocyanate, *Analyst*, 122, 151–154, 1997.

80. Lind, P., Dalene, M., Skarping, G., and Hagmar, L., Toxicokinetics of 2,4- and 2,6-toluenediamine in hydrolysed urine and plasma after occupational exposure to 2,4- and 2,6- toluene diisocyanate, *Occup. Environ. Med.*, 53, 94–99, 1996.

81. Lind, P., Dalene, M., Tinnerberg, H., and Skarping, G., Biomarkers in hydrolysed urine, plasma and erythrocytes among workers exposed to thermal degradation products from toluene diisocyanate foam, *Analyst*, 122, 51–56, 1997.

82. Lind, P., Skarping, G., and Dalene, M., Biomarkers of toluene diisocyanate and thermal degradation products of polyurethane, with special reference to the sample preparation, *Anal. Chim. Acta*, 333, 277–283, 1996.

83. Littorin, M., Rylander, L., Skarping, G., Dalene, M., Welinder, H., Strömberg, U., and Skerfving, S., Exposure biomarkers and risk from gluing and heating of polyurethane: a cross sectional study of respiratory symptoms, *Occup. Environ. Med.*, 57, 396–405, 2000.

84. Kaaria, K., Hirvonen, A., Norppa, H., Piirila, P., Vainio, H., and Rosenberg, C., Exposure to 2,4-and 2, 6-toluene diisocyanate (TDI) during production of flexible foam: determination of airborne TDI and urinary 2,4-and 2,6-toluenediamine (TDA), *Analyst*, 126, 1025–1031, 2001.

85. Sennbro, C. J., Lindh, C. H., Tinnerberg, H., Gustavsson, C., Littorin, M., Welinde, H., and Jonsson, B. A. G., Development, validation and characterization of an analytical method for the quantification of hydrolysable urinary metabolites and plasma protein adducts of 2,4-and 2,6-toluene diisocyanate, 1, 5-naphthalene diisocyanate and 4,4′-methylenediphenyl diisocyanate, *Biomarkers*, 8, 204–217, 2003.

86. Rosenberg, C., Nikkila, K., Henriks-Eckerman, M. L., Peltonen, K., and Engstrom, K., Biological monitoring of aromatic diisocyanates in workers exposed to thermal degradation products of polyurethanes, *J. Environ. Monit.*, 4, 711–716, 2002.

87. Pauluhn, J. and Lewalter, J., Analysis of markers of exposure to polymeric methylene-diphenyl diisocyanate (pMDI) in rats: a comparison of dermal and inhalation routes of exposure, *Exp. Toxicol. Pathol.*, 54, 135–146, 2002.

88. Rosenberg, C. and Savolainen, H., Determination of occupational exposure to toluene diisocyanate by biological monitoring, *J. Chromatogr.*, 367, 385–392, 1986.

89. Maitre, A., Berode, M., Perdrix, A., Stoklov, M., Mallion, J. M., and Savolainen, H., Urinary hexane diamine as an indicator of occupational exposure to hexamethylene diisocyanate, *Int. Arch. Occup. Environ. Health*, 69, 65–68, 1996.

90. Sepai, O., Henschler, D., and Sabbioni, H., Albumin adducts, hemoglobin adducts and urinary metabolites in workers exposed to 4,4′methylenediphenyl diisocyanate, *Carcinogenesis*, 16, 2583–2587, 1995.

91. Williams, N. R., Jones, K., and Cocker, J., Biological monitoring to assess exposure from use of isocyanates in motor vehicle repair, *Occup. Environ. Med.*, 56, 598–601, 1999.

92. Kaaria, K., Hirvonen, A., Norppa, H., Piirila, P., Vainio, H., and Rosenberg, C., Exposure to 4,4′-methylenediphenyl diisocyanate (MDI) during moulding of rigid polyurethane foam: Determination of airborne MDI and urinary 4,4′-methylenedianiline (MDA), *Analyst*, 126, 476–479, 2001.

93. Schutze, D., Sepai, O., Lewalter, J., Miksche, L., Henschler, D., and Sabbioni, G., Biomonitoring of workers exposed to 4,4′methylenedianiline or 4,4′methylenediphenyl diisocyanate, *Carcinogenesis*, 16, 573–582, 1995.

94. Rudzinski, W. E., Sutcliffe, R., Dahlquist, B., and Key-Schwartz, R., Evaluation of tryptamine in an impinger and on XAD-2 for the determination of hexamethylene-based isocyanates in spray-painting operations. *Analyst*, 122, 605–608, 1997.

95. Karlsson, D., Spanne, M., Dalene, M., and Skarping, G., Airborne thermal degradation products of polyurethane coatings in car repair shops, *J. Environ. Monit.*, 2, 462–469, 2000.

96. Karlsson, D., Dahlin, J., Dalene, M., and Skarping, G., Determination of isocyanates, aminoisocyanates and amines in air formed during the thermal degradation of polyurethane, *J. Environ. Monit.*, 4, 216–222, 2002.

97. Henriks-Eckerman, M. L., Valimaa, J., Rosenberg, C., Peltonen, K., and Engstrom, K., Exposure to airborne isocyanates and other thermal degradation products at polyurethane-processing workplaces, *J. Environ. Monit.*, 4, 717–721, 2002.

98. Blomqvist, P., Hertzberg, T., Dalene, M., and Skarping, G., Isocyanates, aminoisocyanates and amines from fires – a screening of common materials found in buildings, *Fire Mater.*, 27, 275–294, 2003.

99. Persson, P., Dalene, M., Skarping, G., Adamsson, M., and Hagmar, L., Biological monitoring of occupational exposure to toluene diisocyanate – Measurement of tluenediamine in hydrolyzed urine and plasma by gas chromatography-mass spectrometry, *Br. J. Ind. Med.*, 50, 1111–1118, 1993.

100. Brunmark, P., Dalene, M., and Skarping, G., Gas-chromatography negative-ion chemical-ionization mass-spectrometry of hydrolyzed human urine and blood-plasma for the biomonitoring of occupational exposure to 4,4′-methylenebisaniline, *Analyst*, 120, 41–45, 1995.

101. Rosenberg, C., Nikkila, K., Henriks-Eckerman, M. L., Peltonen, K., and Engstrom, K., Biological monitoring of aromatic diisocyanates in workers exposed to thermal degradation products of polyurethanes, *J. Environ. Monit.*, 4, 711–716, 2002.
102. Sakai, T., Morita, Y., Kim, Y., and Tao, Y. X., LC-MS determination of urinary toluenediamine in workers exposed to toluenediisocyanate, *Toxicol. Lett.*, 134, 259–264, 2002.

22 Chromatographic Analysis of Insecticides Chlorinated Compounds in Water and Soil

Michela Maione and Filippo Mangani

CONTENTS

I. INTRODUCTION

Several kinds of insecticides or pesticides have been used over the decades in the attempt to defeat the huge number of crop-eating insects, approximately 700 species worldwide, that caused infective and parasitic diseases to humans and loss to harvest.

The use of pesticides has contributed to the drastic reduction of the diseases transmitted by insects, most of these are life-threatening while also protecting crops during their growth and storage.

Before World War II the selection of insecticides was more or less the same as available a thousand and more years before. It was in the 1940s and in the 1950s, a new concept of pest control that emerged, initiating a new era of synthetic, highly effective compounds.

The extensive use of synthetic pesticides were greeted at first with lot of enthusiasm, but in a few years it appeared clear that these pesticides and their residues contaminate the soil and ground- as well as surface water as a consequence of their great persistence and stability.

Their use has gradually been phased out in industrial nations; however, some of these agents continue to be used in developing countries (for example in tropical regions). Interest in these compounds is still considerable, their occurrence in environmental elements such as soil and water, and also in previously pristine environments such as the Arctic and Antarctic, being just one aspect of the problem. In fact, as a consequence of bioaccumulation processes, in several organisms organochlorines are found in a higher concentration than in the environment they live in. That is particularly true in predators and other species at the top of the food chain.

In most cases these substances are organochlorinated compound insecticides, also known as chlorinated hydrocarbons, chlorinated organics, chlorinated insecticides, chlorinated synthetics, and organochlorine pesticides (OCPs) (Table 22.1).

A. PHYSICAL AND CHEMICAL PROPERTIES

The popularity of chlorinated pesticides was based on some important properties, in fact most of these are extremely stable, show very low solubility in water, high solubility in organic media, and high toxicity to insects but low toxicity to humans. However, some of these properties are at the base of their hazardousness. Hazards associated with these pollutants are persistence in the environment, bioaccumulation potential in the tissues of animals and humans through the food chain, and the toxic properties for humans and wildlife.

Indeed, OCPs, once released into the environment, are distributed into various environmental compartments (e.g., water, soil, and biota) as a result of complex physical, chemical, and biological processes. In order to perform appropriate exposure and risk assessment analyses, multimedia models of pollutant partitioning in the environment have been developed.[1] Properties which are at the base of such a partitioning are water solubility (WS), octanol–water partition coefficient (K_{ow}), soil adsorption (K_d), and bioconcentration factors (BCFs) in aquatic organisms, following these four equilibriums:

1. WS   pure chemical ↔ aqueous solution
2. K_{ow} chemical in organic solvent ↔ aqueous solution
3. BCF chemical in organism ↔ aqueous solution
4. K_d chemical adsorbed on soil ↔ aqueous solution

Water solubility can be regarded as the partition of a chemical between itself and water and bioconcentration factors as a partition between the water, lipid, and protein phases in an organism. Adsorption of nonionic chemicals by water from soil can also be regarded as an organic phase– water partition, the close correlation between K_d and the organic matter content of the soil having been recognized. Thererfore, the four properties can be considered to be the expressions of essentially the same process, i.e., partitioning between an aqueous and an organic phase. For that reason, measurement or calculation of one of these four properties allows prediction of the other three to within an order of magnitude.[2]

K_{ow} of a number of pesticides, including OCPs, are reported in a paper by Finizio et al.,[3] who determined K_{ow} values by means of three different estimation methods, including calculation from water solubility, then compared the obtained results with experimental values measured with Slow Stirring or Shake Flask methods.

It should be remembered that the amount of pesticide coming in direct contact with target pests was estimated to be lower than 0.3% of the amount applied[4] and this is important so far as toxicity is concerned. Therefore the use of pesticides inevitably leads to exposure of nontarget organisms

TABLE 22.1
List of Selected Organochlorinated Pesticides

Pesticide	CASRN	Formula	MW	Structural Formula
Aldrin	309-00-2	$C_{12}H_8Cl_6$	364.9	
Chlordane	57-74-9	$C_{10}H_6Cl_8$	409.8	
Chlordecone	143-50-0	$C_{10}Cl_{10}O$	490.6	
Dieldrin	60-57-1	$C_{12}H_8Cl_6O$	377.9	
o,p'-DDE	3424-82-6	$C_{14}H_8Cl_4$	315.9	
p,p'-DDE	72-55-9	$C_{14}H_8Cl_4$	315.9	
o,p'-DDD	72-54-8	$C_{14}H_{10}Cl_4$	318.0	
p,p'-DDD	72-54-8	$C_{14}H_{10}Cl_4$	318.0	
o,p'-DDT	789-02-6	$C_{14}H_9Cl_5$	354.5	
p,p'-DDT	50-29-2	$C_{14}H_9Cl_5$	354.5	

Continued

TABLE 22.1
Continued

Pesticide	CASRN	Formula	MW	Structural Formula
Endosulfan	115-29-7	$C_9H_6Cl_6O_3S$	406.9	
Endosulfan sulfate	1031-07-8	$C_9H_6Cl_6O_4S$	422.9	
Endrin	72-20-8	$C_{12}H_8Cl_6O$	380.9	
Heptachlor	76-44-8	$C_{10}H_5Cl_7$	373.3	
Heptachlor epoxide	1024-57-3	$C_{10}H_5Cl_7O$	385.8	
Hexachloro benzene (HCB)	118-74-1	C_6Cl_6	284.8	
Lindane	58-89-9	$C_6H_6Cl_6$	290.8	
Methoxychlor	72-43-5	$C_{16}H_{15}Cl_3O_2$	345.7	
Mirex	2385-85-5	$C_{10}Cl_{12}$	545.6	

(including humans). These may take up pesticides through ingestion of food and water, respiration, and through contact with the skin.[5]

The toxicity of a chemical compound is expressed in terms of *lethal dose* (LD) and, in particular, as LD_{50}, which represents the amount of substance necessary to kill half of the laboratory animals treated with that particular chemical. The LD_{50} values, expressed in milligrams of substance per kilogram of weight of the organism, are available for most pesticides. However, data on chronic toxicity in humans leading to phenomena such as carcinogenesis, immunodysfunction, mutagenesis, neurotoxicity, and teratogenesis are insufficient.[6]

Physicochemical characteristics of selected OCPs are reported in Table 22.2 together with toxicity data.

Characteristics and toxicity of the main OCPs are reported in the following.

Dichlorodiphenyltrichloroethane (*DDT*).   This is by far the most notorious pesticide ever used, its production starting in 1943. DDT was used extensively as an agricultural and vector control pesticide. In the United States its peak production occurred in 1963. Although it has been banned since the 1970s, it is still used in many developing countries. DDT is a very toxic compound. Exposure to high doses can affect the Central Nervous System (CNS). In moderately severe poisoning cases, cardiac and respiratory failure can occur. DDT breaks down to dichlorodiphenyldichloroethylene (DDE) and dichlorodiphenyldichloroethane (DDD), with the parent/metabolite ratio decreasing with time. DDD itself was also used for controlling a number of insects. DDE is a metabolite of DDT as well as an impurity, so its presence in the environment is strictly correlated to the use of DDT.

Methoxychlor.   Its chemical structure and properties are similar to those of DDT, but it biodegrades more easily. Aquatic organisms metabolize it and transform it into other less toxic substances and therefore it does not lead to significant bioaccumulation phenomena.

Lindane.   This is the γ isomer of hexachlorocyclohexane (HCH) and it is commonly used externally to prevent animals from infestation by lice and ticks and internally to discourage the propagation of parasites. Photodegradation is not a major environmental fate process. Bioconcentration is low, but present. Short-term exposure interferes with transmission of nerve impulses, while long-term exposure leads to liver and kidney damage.

Cyclodienes (chlordane, aldrin, dieldrin, heptachlor, endrin, mirex, endosulfan, chlordecone). This new class of pesticides appeared on the scene following World War II. Most of them are very stable to sunlight and persistent in soil and they were used to control termites and other insects. Their effectiveness leads to insect resistance and bioaccumulation in the food chain and for these reasons their use was banned between 1984 and 1988. These compounds affect the CNS in the same way, causing tremors, convulsions, and prostration to the maximum extent, depending on the rate and time of exposure.

Hexachlorobenzene.   This was used to protect crops against fungi until 1965. Because of its high toxicity, it has been withdrawn or severely restricted in most countries. Hexachlorobenzene is a very stable compound and, consequently, very persistent in the environment. It enters the food chain and bioaccumulates in plants and animals. It may affect the CNS to different degrees and it is suspected to be a carcinogen.

B. DISTRIBUTION IN THE ENVIRONMENT

As a consequence of the above cited properties, OCPs are widely distributed among the different environmental compartments and represent a global contamination problem.

Pesticides which reach the target soil or plant begin to disappear by degradation or dispersion. However, a significant part may volatilize into the air, runoff or leach into surface water and groundwater, be taken up by plants or soil organisms, or remain in the soil. Marine sediments act as an ultimate sink for persistent pollutants brought into the aquatic environment from direct discharges, surface run-off, and atmospheric fall out.[7] OCPs can also be transported for long

TABLE 22.2
Some Physical and Chemical Properties of the Selected Organochlorinated Pesticides

Pesticide	Log K_{ow} (log P)	Solubility in Water at 25°C (Ppm)	Melting Point (°C)	Vapor Pressure (mPa) at 25°C	LD$_{50}$ mg/kg
Aldrin	6.5	0.01–0.2	104	—	—
Chlordane	6.16	0.1	104–107	61	Rats 457–590
Chlordecone	5.41	7.6	350	—	—
Dieldrin	5.4	0.186 at 20°C	176	0.39	Rats 46
o,p'-DDE	6.51	—	—	—	—
p,p'-DDE	6.51	—	88–90	2.09	—
o,p'-DDD	6.02	0.1	—	—	—
p,p'-DDD	6.02	0.05	88–90, 109–112	0.62	—
o,p'-DDT	—	—	—	—	—
p,p'-DDT	6.91	0.0077 at 20°C	108.5	0.025 at 20°C	Rats 115
Endrin	5.2	0.23	200	0.026	—
Endosulfan	(I) 3.83	(I) 0.32 at 22°C (II) 0.33 at 22°C	(I) 70–100 (II) 213.3	0.0012 at 80°C	Rats 80–110 TC
Endosulfan sulfate	3.66	0.117	181	—	—
Heptachlor	6.1	0.18	46–74	40	Rats 147–220
Heptachlor epoxide	4.98	—	—	—	—
Hexachlorobenzene (HCB)	5.73	0.006	226	1.45 at 20°C	Rats 10,000
Hexachlorocyclohexane (HCB)	4.26	—	112	5.6 at 20°C	—
Lindane	3.72	7	112	5.6 at 20°C	Rats 88–270, Mice 59–246
Methoxychlor	5.08	0.12	89	—	Rats 6000 TC
Mirex	—	7×10^{-5}	485	0.1	—

distances by air, rivers, and ocean currents, and contaminate regions remote from their sources, as evidenced by their occurrence in Arctic snow.[8] Studies on the levels of OCPs in the global environment show that emission sources of a number of them in the last 20 years have shifted from industrialized countries of the Northern Hemisphere to developing countries in tropical and subtropical regions, including India and China, owing to the late production ban or still being used both legally or illegally in agriculture and for the control of diseases like malaria, typhus, and cholera.[9] However, tropical regions seem to act as a global source and sink for OCPs, since removal processes may be faster compared to temperate and Arctic regions.[10] As a matter of fact, in the global environment, regions with different ambient temperatures are linked by large-scale atmospheric and oceanic movements which might lead to a process of global distillation in which chemicals volatilize in a warm region and condense and accumulate in a cold one.[11,12]

C. RULES ON THE USE OF OCPs

1. International Protocols

The environmental concerns caused by widespread use of potentially toxic substances such as OCPs gave rise to restrictions in their production and use, first in developed countries, and more recently in developing nations as well.

Therefore, OCPs are controlled and governed by numerous international legal instruments:

(1) The Prior Informed Consents (PIC) Convention objective is to promote shared responsibility and cooperative efforts among parties in the international trade of hazardous chemicals in order to protect human health and the environment from potential harm and to contribute to their environmentally sound use. In 1998, governments decided to strengthen the procedure by adopting the Rotterdam Convention, which makes PIC legally binding. The Rotterdam Convention entered into force on February 24th, 2004. It was signed and ratified by several countries on a worldwide basis.[13]

(2) The Protocol on Persistent Organic Pollutants (POPs) is in the framework of the 1979 Geneva Convention on Long-range Transboundary Air Pollution (LRTAP). The Convention on LRTAP entered into force in 1983. It has been extended by eight specific protocols, one of which is the 1998 Aarhus Protocol on POPs, which entered into force on October 23rd, 2003. It focuses on 16 substances, most of these being chlorinated pesticides, which have been singled out according to agreed risk criteria. The ultimate objective is to eliminate any discharges, emissions, and losses of POPs. The Protocol bans the production and usage of some products outright, while others are scheduled for elimination at a later stage. Currently, the Protocol has been signed by 36 countries and ratified by 19.[14]

(3) The United Nations Environmental Programme, Mediterranean Action Plan (UNEP MAP) Barcelona Convention for the Protection of the Marine Environment and the Coastal Region of the Mediterranean, is a 1995 revision of the Convention for the Protection of the Marine Environment and the Coastal Region of the Mediterranean came in force since 1978 (this revision is still under ratification). This Programme, involving 21 countries bordering the Mediterranean Sea and the European Union, is aimed at meeting the challenges of environmental degradation in the sea, coastal areas, and inland and linking sustainable resource management with development in order to protect the Mediterranean region. The Land-Based Sources (LBS) protocol entered into force in June 1983 and requires the contracting parties to take all appropriate measures to prevent, abate, combat, and to the fullest possible extent eliminate pollution of the Mediterranean Sea area and to draw up and implement plans for the reduction and phasing out of substances which are toxic, persistent, and liable to bioaccumulate arising from LBS.[15]

(4) The Helsinki Convention on the Protection of Baltic Sea (European Union Council Decisions 94/156/EC, and 94/157/EC 21 February 1994) is aimed at reducing pollution in the Baltic Sea area. The Parties to the Convention undertake to ban the use of a series of hazardous substances,

among which are DDT, DDE, and DDD, in the Baltic Sea area. Parties must also take all appropriate measures and work together to control and minimize pollution of a number of substances, including pesticides, from LBS. Large quantities of the substances referred to may not be introduced without a prior special permit issued by the appropriate national authority.[16]

(5) The Great Lakes Water Quality Agreement of the International Joint Commission (IJC), Canada and United States, first signed in 1972 and renewed in 1978, expresses the commitment of each country to restore and maintain the chemical, physical, and biological integrity of the Great Lakes Basin Ecosystem and includes a number of objectives and guidelines to achieve these goals. It reaffirms the rights and obligation of Canada and the United States under the Boundary Waters Treaty (1909). In 1987, a Protocol was signed amending the 1978 Agreement where some specific objectives are identified, i.e., the concentration or quantity of a substance or level of effect which the Parties agree, after investigation, to recognize as a maximum or minimum desired limit for a defined body of water or portion thereof, taking into account the beneficial usages or level of environmental quality which the Parties desire to secure and protect. Within such objectives, OCPs are recognized as critical pollutants.[17]

(6) The OSPAR Convention for the Protection of the Marine Environment of the North-East Atlantic opened for signature at the ministerial meeting of the Oslo and Paris Commissions in September 1992 and entered into force on March 25th 1998. The aim of the Convention is to take all possible steps to prevent and eliminate pollution and take the necessary measures to protect the maritime area against the adverse effects of human activities so as to safeguard human health, to conserve marine ecosystems and restore marine areas which have been adversely affected. The objective of the Commission with regard to hazardous substances, including OCPs, is to prevent pollution of the maritime area by continuously reducing discharges, emissions, and losses of hazardous substances with the ultimate aim of achieving concentrations in the marine environment near background values for naturally occurring substances and close to zero for manmade synthetic substances.[18]

(7) The UNEP POPs Convention, known as the Stockholm Convention on Persistent Organic Pollutants (POPs), was signed on May 23rd 2001 by 120 countries. The Convention entered into force on May 17th 2004, i.e., 90 days after France became the 50th state to ratify the agreement. The aim of the Convention is to seek continuous minimization, and wherever feasible, ultimate elimination of POPs. In its first phase, the agreement targets only twelve POP candidates, known as the "dirty dozen." The dirty dozen includes, besides PCBs, dioxins and furans, the main chlorinated pesticides, i.e., aldrin, chlordane, DDT, dieldrin, endrin, heptachlor, hexachlorobenzene, mirex, and toxaphene. The Convention will require all member states to stop producing the pesticides aldrin, dieldrin, and heptachlor and require those wishing to use remaining supplies to register publicly for exemptions. Countries with exemptions will have to restrict their use of these chemicals for narrowly allowed purposes for limited time periods. Production and use of chlordane, hexachlorobenzene, and mirex will be limited to narrowly prescribed purposes and to countries who have registered for exemptions. So far as DDT is concerned, its production and use will be limited to controlling disease vectors such as malarial mosquitoes; DDT may also be allowed to be used as an intermediate in the production of the pesticide dicofol in countries who have registered for this exemption.[19]

Even if the agricultural use of DDT is almost totally banned worldwide, the urgent and immediate need to maintain the reliance on DDT for indoor residual spraying to control insect vectors, particularly of malaria, is still recognized due to the current lack of effective and affordable alternatives. An action plan for the reduction of reliance on DDT in disease vector control was devised in 2001 by the World Health Organization (WHO)[20] in the spirit of the Stockholm Convention. It stems from the need to accelerate the research and development of safe and effective alternatives to DDT with a view to improving Stockholm Convention Member States' vector control programs in the medium term through the adoption and use of such alternatives. Furthermore, it recognizes the need to work towards a longer-term goal of reducing reliance of vector control programmes on pesticides in general and DDT in particular to safeguard the ecosystem and human health from the insidious effects of POPs pesticides. Guidelines for reduction and elimination of the use of POPs, foreseeing alternative strategies for sustainable pest and vector management, were also published by the Inter Organization Programme for the Sound Management of Chemicals (IOMC), i.e., a cooperative agreement among UNEP, International Labor Organization (ILO), Food and Agriculture Organization (FAO), WHO, United Nations Industrial Development Organization (UNIDO), United Nations Institute for Training and Research (UNITAR), and Organization for Economic Cooperation and Development (OECD).[21]

Notwithstanding the existence of all the above cited Protocols and Conventions, unfortunately, DDT is still diverted illegally from government health programmes to agricultural use on a regular basis. This is known or suspected to have happened in Bangladesh, Belize, Ecuador, India, Kenya, Madagascar, Mexico, and Tanzania.[22]

2. United States Legislation

In the United States (U.S.), the first OCP to be banned in 1972 was DDT. The use of other compounds, like aldrin, dieldrin, and chlordane, was initially limited to termite control and subsequently extended to all uses. Lindane has not been produced in the U.S. since 1977 but it is still imported to and formulated in the U.S.

The U.S. Department of Health and Human Services (DHHS) regards DDT as being "reasonably anticipated to be a human carcinogen."[23]

Many federal governmental agencies set recommended limits for the amounts of OCPs to be found in different matrices (drinking water, workplace air, raw food) to protect human health. Table 22.3 reports limits as set by different agencies for selected OCPs, as well as the year in which their utilization was stopped in the U.S.

II. CHROMATOGRAPHIC ANALYSIS

The environmental relevance of OCPs yielded a number of research works aimed at their determination in the different environmental matrices down to trace levels, using gas chromatography (GC) as the chosen separation technique. In the following, the latest analytical methodologies to be used for the determination of OCPs in two such important environmental matrices as water and soil will be described.

The overall analytical procedure to be used in the determination of OCPs present in water and soils, requires extraction from the matrix, clean-up, and enrichment preanalytical procedures prior to the analysis by GC–ECD or GC–MS.

As previously stated, OCPs represent a toxic and ubiquitous class of compounds. Therefore, they are included among those chemical species which need to be analyzed, even though present in very low concentrations such as parts per billion (ppb). Furthermore, when the molecular weight increases, the chromatographic separations of the compounds of interest from interfering compounds become more and more difficult, particularly when the chemical structure of the interfering compounds is similar to that of the analytes.

For these reasons and for the need to push the analytical procedure to the minimum detection limits technologically available, the analysis of such compounds requires highly sophisticated instrumentation and, more important, careful clean-up and preconcentration procedures.

The final aim of preconcentration procedures is to obtain the sample to be analyzed in a suitable solvent, within a concentration range compatible with the sensitivity and detection limits of the instrumentation used. One of the major problems related to preconcentration methods is the possibility of severe and/or nonreproducible losses of the analytes during sample manipulation. Thus, the procedure adopted should be carefully evaluated for sample losses and reproducibility. In the following sections the various preconcentration techniques formerly and currently applied to the analysis of OCPs will be examined.

TABLE 22.3
Recommended Limits for OCPs as Set by U.S. Federal Governmental Agencies

Pesticide	Year of Phase-Out	US-EPA Limit in Drinking Water ($\mu g\ l^{-1}$)	US-EPA Spills into the Environment to be Reported (g)	OSHA Limit in Workplace Air in an Eight-Hour Shift or 40 h Week (mg m^{-3})	FDA Residues Limit in Raw Food ($\mu g\ g^{-1}$)
Aldrin	1987	1	—	0.25	0–0.1
Chlordane	1988	2	≥453.6	0.5	0.1–0.3
Chlordecone	1978	10^{-3}	≥453.6	—	4×10^{-5}
DDT	1972	—	—	0.1	—
Dieldrin	1987	2	—	0.25	0–0.1
Endosulfan	Still used	74 (surface water)	—	—	0.1–24
Endrin	1986	2×10^{-1}	—	0.1	—
Heptachlor	1988	2.78×10^{-3}	≥453.6	0.5	0–10×10^{-3}
Hexachlorobenzene	—	1	≥453.6	—	—
Lindane	Still used	2×10^{-1}	≥453.6	0.5	—
Methoxychlor	Still used	40	—	15	4×10^{-2}
Mirex	1978	10^{-3} (surface water)	—	—	10^{-5}

U.S.-EPA: United States Environmental Protection Agency; OSHA: Occupational Safety and Health Administration; FDA: Food and Drug Administration.
Source: Agency for Toxic Substances and Disease Registry (ATSDR).

A. EXTRACTION FROM WATER

1. Liquid–Liquid Extraction (LLE)

LLE is based on the low value of the partition coefficient for the organic compounds between water and organic solvents, and is particularly advantageous in trace analysis when the compounds of interest have a very high solubility in the organic solvent and a very low solubility in water, as in the case of OCPs. Extraction methods of OCPs from water grab samples, which made use of separative funnels to perform a LLE of the organic compounds from the aqueous matrix, were described in the early 1960s.[24] EPA method 508[25] still involves LLE, with the use of methylene chloride as the organic solvent to be added, in a separative funnel, to a 1 l sample of water. Threefold replicate extractions must be performed, adding 60 ml of solvent each time. The extract is then dried and exchanged to hexane during concentration to a volume of *ca.* 10 ml.

More recently, a micro-LLE method was described. Extraction is performed on 400 ml water samples extracted once with 500 μl toluene. Extracts are then analyzed directly, without any further treatment by GC–ECD.[26]

In order to carry out large-sample preconcentration, the Goulden large-sample extractor (GLSE), a continuous-flow liquid–liquid extractor, was designed by Goulden et al.[27] to effectively preconcentrate trace organic substances, including many pesticides, from large volumes of water providing very low detection levels. GLSE performance was compared in terms of concentration factor enhancements relative to conventional 1 l continuous liquid–liquid extractor and of reproducibility of the GLSE method in pesticide isolation from replicate 35 l water samples.[28] Results confirmed that such a device is very effective in lowering estimated detection levels. On the other hand, the precision for replicate extractions of the same sample seems to be inversely proportional to sample volume. Such an extraction method was also applied by Foreman and Gates[29] in a study aimed at assessing the matrix-enhanced degradation of *p,p'*-DDT during GC analysis. In that case filtered water samples of 4 to 112 l contained in stainless steel cans were extracted with dichloromethane using the GLSE. The undried GLSE extracts were stored up to 5 months at 4°C. The extracts were prepared for analysis by removal of residual water, followed by solvent exchange to toluene and reduction to 500 μl. Extracts were analyzed for determination of 68 pesticides of various chemical classes by GC/electron impact mass spectrometry (EIMS) without additional clean-up steps.

As a matter of fact, the use of LLE as an extraction technique shows several drawbacks, i.e., the long time necessary to perform the extraction, the use of relatively large volumes of expensive and potentially toxic solvents, formation of emulsions resulting in analyte losses, extensive use of glassware which can contaminate the sample, magnification of solvent impurities, need of sample preconcentration prior to analysis, evaporative losses of analytes, not always satisfactory repeatability, and loss of sensitivity as a consequence of the injection of only fractions of the extracted compounds.

However, an interesting evolution of LLE as an extraction technique was developed by Cramers and coworkers.[30] The method implies extraction of the analytes of interest from an aqueous matrix by passing the water sample through a sorption cartridge containing particles consisting of a polymeric liquid phase. As the extraction phase, polydimethylsiloxane (PDMS) was used, which appears to be a solid but has sorptive characteristics similar to those of a liquid phase. Retention of analytes is not based on adsorption of the solutes onto the surface of the PDMS material; rather, the solutes are dissolved (partitioned) into the bulk of this high-viscosity liquid phase. The extraction cartridges are then thermally desorbed to ensure full transfer of all analytes onto the GC column. In this way the consumption of organic solvents is eliminated and maximum sensitivity is attained since all solutes trapped from the sample are actually introduced into the GC column. In this respect, such a technique showed itself to be more advantageous than solid-phase extraction (SPE) and solid-phase micro-extraction (SPME) described in the following sections. In SPE, based on adsorption of analytes onto an active surface, only a fraction of the desorption liquid is injected, thus affecting the sensitivity of the method. In SPME, the active part is a sorbent material very similar to that used in this "new" LLE. The two methods differ dramatically since SPME is an equilibrium method, while in LLE the extraction is complete, thus enhancing the overall sensitivity. Such a novel method was shown to be particularly suitable for the analysis of OCPs in water.[31] For these relatively apolar compounds, quantitative extraction can be obtained for sample volumes up to at least 100 ml, obtaining detection limits in the low ppt range. The method can therefore be considered as an attractive alternative for certain LLE, SPE, and SPME applications.

2. Solid-Phase Extraction

SPE is based on the principle of liquid–solid chromatography. A polypropylene cartridge (Figure 22.1) is filled with an adsorbent in a fine mesh range (150 to 400). Water passed through such a cartridge should not be retained and it should not yield significant modification in the physicochemical properties of the adsorbent which, instead, should completely retain the organics dissolved in the water. Moreover, water should not behave as an eluent for the organics under analysis. In this way, large amounts of polluted water can be passed through the cartridge, even in liter quantities, leaving the organic compounds fully adsorbed. Once the desired amount of water is passed through the cartridge, the latter is dried by means of a nitrogen flow and is eluted with an organic solvent. In this situation the retention volume of the organics should be nearly zero, so that the pollutants can be eluted rapidly with a retention volume only a few times higher than the dead volume of the cartridge. Advantages of such a technique are:

It is fast, simple, and low-cost.
Sample handling is reduced.
Extraction and preconcentration are performed in a single step.
Formation of emulsions does not take place.
A very small volume of solvents is used.
The solvent removal step is eliminated.
Extraction can be performed online.
Sample storage space is considerably reduced.
At least a 1000-fold enrichment is achieved in a single step.
The recovery of the organics from the adsorbent is complete and reproducible.

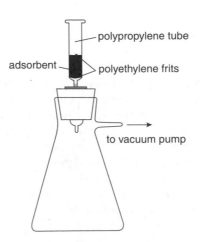

FIGURE 22.1 Solid-phase extraction apparatus.

In the early times, when SPE was introduced, porous polymers such as Tenax or polystyrenedivinylbenzene (XAD-2) were used as adsorbents.[10,32] With the very important and rapid progress made in the synthesis and availability of new materials for reverse-phase HPLC, the C8–C18 bonded silica, these materials have become very popular for SPE.[33,34] These materials have a smooth surface which is made nonpolar by the attached carbon chain. Absence of small pores is another advantage. Furthermore, unlike porous polymers and resins, once cleaned by passing water and some organic solvent in the SPE bed, they do not give rise to contamination of the water and solvents. A slightly different sorbent, named tC18, is characterized by trifunctional bonding chemistry, smaller particle size, but with similar pore size. This sorbent can be submitted to prolonged exposures of acidic solutions without the risk of releasing the C18 functional group; this might be particularly interesting when there is a need to process large volumes of water for measuring ppt-levels of contaminants and/or when the extracts are due to be kept for a certain time before eluting them. Extraction efficiencies on this sorbent were compared by Fernandez and coworkers[35] to those of a conventional LLE method. The results did not differ significantly from those obtained with the LLE method but with all the SPE advantages, like the possibility of automation and the use of low amounts of solvents.

Another class of adsorbents, graphitized carbon blacks (GCB), exhibit the same positive characteristics as the bonded silicas, with the difference of being more retentive, absolutely nonpolar and nonporous. With GCBs, quite good results have been obtained for classic chlorinated pesticides.[36] Recoveries of selected OCPs from different adsorbents are reported in Table 22.4.

In 1996, a new hydrophilic–lipophophilic balance (HLB) adsorbent was introduced, Oasis® HLB (Waters Corporation, Millford, MA).[37] This macroporous copolymer (poly[divinylbenzene-co-N-vinylpyrrolidone]) exhibits hydrophilic and lipophilic retention characteristics, its major features being the abilities to remain wetted with water and to retain a wide spectrum of polar and nonpolar compounds, among which are OCPs.[38,39]

Thanks to the greater attention given to choice of the adsorbents and solvents, the trend is now to use a small amount of water and a small amount of solvent. In 1988, Junk and Richard[40] were able to determine many pesticides present in concentrations of about 0.1 ng/ml, using 100 ml water extracted by means of a cartridge containing only 100 mg silica C18 with 0.1 ml ethyl acetate. The advantages of this procedure are in the combination of small water volumes, fast flow rates, small columns, and small eluate volumes. Of course, the use of this method is only possible

TABLE 22.4
Recoveries of Some Chlorinated Pesticides from Different Adsorbents

	Recovery %			
Pesticide	GCB	Tenax	Porapack P	C$_{18}$
α-BHC	93	81	55	95
β-BHC	100	81	60	93
γ-BHC	100	77	51	93
Heptachlor	97	94	70	96
δ-BHC	97	94	50	96
Aldrin	96	88	71	88
Heptachlor epoxide	95	100	63	99
4,4′-DDE	100	87	77	93
Dieldrin	100	95	80	95
Endrin	99	89	75	94
4,4′-DDD	100	83	63	92
4,4′-DDT	100	86	55	95

when the determination is limited to within the ppb range. More recently, Barcelò and coworkers[38] optimized an automated offline SPE method to trap 109 compounds, among which were several OCPs. Samples were extracted with the automated system ASPEC XL (Gilson, Villiers-le-Bel, France). For each sample, 200 ml was extracted at neutral pH, and 200 ml was acidified with $2N$ HCl to pH 2. Oasis® polymeric sorbent 60 mg cartridges (3 ml syringe volume) were conditioned with 6 ml dichloromethane, 6 ml acetonitrile, and 6 ml HPLC water. Samples were percolated through the cartridge at a flow rate of 6 ml min^{-1}. Immediately after, cartridges were rinsed with 1 ml HPLC water at a flow rate of 30 ml min^{-1}. Elution was carried out with 2.5 ml acetonitrile–dichloromethane (1:1) and 3.2 ml dichloromethane, pushing at each step the residual solvent by applying air at a flow rate of 3 ml min^{-1}. After elution, the extract was transferred to vials and the excess solvent was evaporated under nitrogen stream to a weight of 500 μg.

Online application of SPE methodology in combination with GC originated from coupled LC–GC, where small volumes from the LC eluent were transferred through a modified GC auto-sampler. Benefits are in the automation of SPE procedures. In online SPE trace-enrichment techniques the extraction cartridges are called precolumns. In this approach, small columns (1 to 4.6 mm i.d., 2 to 10 mm length) replace the conventional injection loop of a six- or ten-port valve. Using a second pump, a large volume sample (10 to 100 ml) is loaded onto the precolumn at a flow rate of 1 to 10 ml min^{-1}. Before desorption to the GC column can take place, the water remaining in the precolumn must be removed, because the introduction of even a small amount of water destroys the deactivation layer of the retention gap. Online coupling of SPE to GC applied to chlorinated compounds was first achieved by Noroozian et al.[41] They used a four-valve system and a 4 mm × 1 mm i.d. micro-precolumn which was built in a six-port valve. After sample loading and drying by means of a nitrogen purge, desorption took place with n-hexane, which was a convenient solvent for the GC introduction. Recoveries of more than 95% were observed for the majority of the analytes and detection limits were in the order of 1 ng l^{-1}. Noy et al. obtained similar results with the same class of compounds.[42]

SPE can be performed also using membrane disks instead of cartridges. Disks are made of a network of PTFE fibers in which C18-bonded silica are enmeshed to form a strong porous membrane.[43–47] These are advantageous with respect to the SPE cartridge, whose narrow i.d. limits the flow rate, and whose small cross-sectional area is easily clogged by suspended solids prolonging the extraction of a large volume sample. Using membrane disks, which have smaller particles, large diameter, and short length, the extraction process is faster while maintaining its effectiveness. It has also been stated that the performance of membrane disks is enhanced by placing two or three of them in series. These can be used online (Figure 22.2). A semiautomated SPE system with reverse-phase disks was evaluated for the extraction and

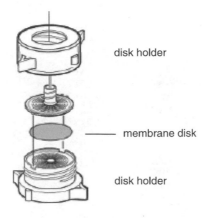

disk holder

—— membrane disk

disk holder

FIGURE 22.2 Inline extraction disk holders.

concentration of different OCPs from freshwater by Ridal and coworkers.[48] A semiautomated device was used which pumps up to six aqueous samples simultaneously through SPE disks at preselected constant flow rates of up to 60 ml/min. The disk pretreatment, sample loading, and various rinse, pause, dry, and elute conditions are chosen by the analyst and loaded into the instrument microprocessor from a method diskette created on a nondedicated PC. Another advantage of this system is the use of positive pressure for disk loading and elution processes. Many other disks use suction to draw fluids through the adsorbent. However, suction systems are prone to air voids, dry surfaces, and variable flow rates, all of which can lead to variable recoveries of analytes. Satisfactory recoveries (64 to 91%) of those OCPS with log K_{ow} 3.8 to 6.4 when dissolved in Milli-Q water were obtained. As expected, lower recoveries (42 to 72%) were obtained for compounds dissolved in lake water, while more analytes (58 to 98%) were recovered from lake water samples by a shake-flask LLE method; a four-fold time saving with the automated SPE method over LLE was estimated. The presence of dissolved organic compounds resulted in increased breakthrough for samples prepared with lake water compared with Milli-Q water.

SPE based on cartridges or disks was accepted as the EPA method for the determination of OCPs and other organic compounds in drinking water.[49]

One of the major problems connected with the analysis of organics in water using SPE as preanalytical procedure is that the breakthrough volume (BTV) of the compounds of interest may become much lower than forecast by spiking experiments due to the presence of naturally occurring substances, such as fulvic and humic acids, humins, and dissolved organic matter (DOM). In fact, binding interactions with such substances influence the solubility and particle adsorption of hydrophobic compounds, thus affecting their extraction efficiencies from water. Extraction efficiency can be further worsened under acidic conditions. In that case, humic acids are protonated and increase their hydrophobic character, which leads to higher interactions with the more apolar compounds, and lower recoveries are obtained.[38] Johnson et al.[50] has effectively shown that the SPE of water containing a commercially available humic acid failed when silica C18 cartridges containing 500 mg of the adsorbent were used, thus yielding inferior recoveries with respect to LLE, and that the recovery changed according to the different pesticides tested. However, such difficulty can be overcome either by using a larger amount of adsorbent or by eliminating the interfering compounds prior to extraction by chemical oxidation. When particulate matter is present in water, which is quite normal when analyzing lowlands river or lake waters, dwell water, or marine water, most organics are adsorbed on the particulate itself which passes undisturbed through the adsorbing bed. However, particulate can be easily eliminated by placing a fiber glass filter before the SPE cartridge or high-density glass beads to be used on top of SPE extraction disks. Conditions affecting SPE were investigated together with the chemical stability of some pesticides extracted from seawater on C18 disks.[51] Unfiltered seawater samples exhibited a lower recovery compared to filtered seawater, with recoveries ranging from 49.8 to 72.3% and from 64.8 to 110.1% for unfiltered and filtered water, respectively. Therefore, filtered seawater showed similar recoveries when compared to spiked clean water samples. Filtration appears to be necessary when the level of humic acids or particulate matter is higher than 10%. For samples heavily contaminated with particulates, these should be analyzed separately from the water for adsorbed OCPs in order to determine the pollution load for the whole sample. It was observed that recoveries can vary between low and high K_{ow} compounds. Therefore, it is recommended to use surrogate or matrix spikes of compounds similar to the target analytes to determine possible differences in extraction efficiencies between samples from natural waters. The use of appropriate internal standards following extraction is also recommended to control for losses during sample concentration steps.[48]

In conclusion, it can be stated that SPE is an effective extraction method for OCPs from water, but problems linked to the presence of DOM and/or particulate must be taken in account.

3. Solid-Phase Micro-Extraction

SPME was introduced as an extraction technique for organic trace pollutants from aqueous matrices at the end of the 1980s by Pawliszyn and coworkers.[52–54] SPME involves exposing a fused silica fiber which has been coated with a nonvolatile polymeric coating to a sample or its headspace (Figure 22.3). The absorbed analytes are then thermally desorbed in the injector of a gas chromatograph for separation and quantitation, not requiring solvent extraction. The fiber is mounted in a syringe-like holder, which protects the fiber during storage and penetration of septa in the sample vial and in the GC injector. This device is operated like an ordinary GC syringe for sampling and injection. The extraction principle can be described as an equilibrium process in which the analyte partitions between the fiber and the aqueous phase. Since SPME is a process dependent on equilibrium more than total extraction; the amount of analyte extracted at a given time is dependent upon the mass transfer of an analyte through the aqueous phase. Therefore, a shorter equilibrium time can be attained by simply agitating the solution by means of a magnetic stirrer. Main advantages of SPME over other preanalytical methodologies can be summarized as follows: it is fast, simple, inexpensive, and solventless, cannot be plugged, can be easily automated, is portable and therefore amenable to field use, compatible with both GC and LC, has large linear dynamic range while retaining excellent detection limits, can be used as a fast screening technique and in quantitative analyses.

Since its introduction, SPME has found numerous applications in the analysis of different classes of compounds present in various matrices. Several analytical methods[55–57] for the determination of OCPs in water samples which make use of SPME as extracting and preconcentrating technique have been described. Magdic and Pawliszyn[55] analyzed environmental water samples for the determination of OCPs using a PDMS-coated fiber (film thickness 100 μm). PDMS was preferred to other commercially available coating, i.e., polyacrylate, the latter being

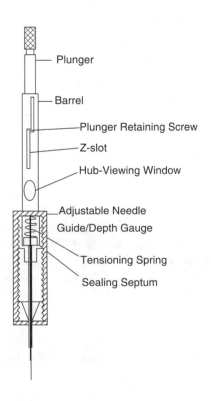

FIGURE 22.3 Solid-phase micro-extraction device.

more polar. In fact, because of the relatively high octanol–water coefficient of OCPs, these analytes are expected to partition more readily into a more nonpolar fiber than a polar one. Optimization of extraction conditions by means of matrix modification was investigated as well. In fact, the more soluble is the analyte in water, the lower is the affinity of the analyte towards the fiber. The amount of analyte extracted can be increased by decreasing its solubility in water. This can be achieved by altering the ionic strength by addition of salt to the matrix or by adjusting the pH of the water. Eighteen OCPs were detected by using such a SPME method coupled either with ECD or MS, obtaining appreciable results. A slight modification of the method was recently proposed in which the classical agitation with a stirring bar was replaced by immersion of the whole PDMS-coated fiber sample system in an ultrasonic bath at 50°C, which was able to accelerate the extraction of the analytes.[58] The equilibrium times obtained for chlorinated pesticides vary between 15 and 20 min. Therefore, a time of 20 min was adopted which was considered sufficient to reach a thermodynamic equilibrium for the partitioning of the majority of the studied analytes. Other advantages are avoiding the risk of breakage of the fiber in the classical agitation with a stirring bar as well as the imperfection of this type of agitation.

In order to improve the SPME technique, porous stationary phases for larger surface area and higher absorption were studied. In particular, porous multifibers were developed in order to achieve the largest contact surface and absorption amount without diffusion limit in the stationary phase.[59] Porous multifibers were prepared as follows. Glass fibers were coated with a porous layer by applying a thin film of epoxy glue over the fibers and pressing onto a C18-bonded silica particle bed. The coated fibers were dried at ambient temperature and then heated to 100°C under nitrogen stream. After cooling to room temperature, 15 pieces of the coated fibers were attached to a stainless steel tube by gluing the uncoated ends of the fibers into the tubing with high-temperature epoxy. The fibers were then cut to a total length of 2.5 cm with a coated length of 2 cm (see Figure 22.4.). The porous multifibers showed larger absorption capacity, higher absorption rate,

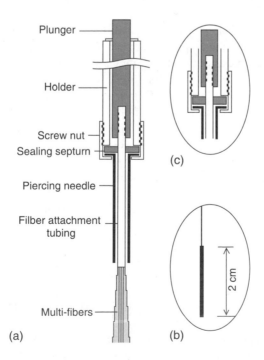

FIGURE 22.4 Porous multifiber solid-phase micro-extraction syringe. (From Ridal, J. J., Fox, M. E., Sullivan, C. A., Maguire, R. J., Mazumder, A., and Lean, D. R. S., *Anal. Chem.*, 69, 711–717, 1997. With permission.)

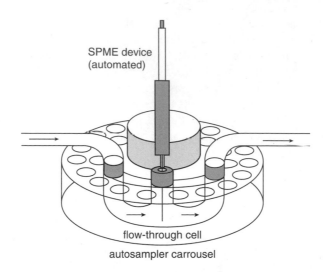

SPME device
(automated)

flow-through cell

autosampler carrousel

FIGURE 22.5 Automated solid-phase micro-extraction apparatus. (From Veningerova, M., Prachar, V., Kovacicova, J., and Uhnak, J., *J. Chromatogr. A*, 774, 333–347, 1997. With permission.)

and stronger analyte interactions compared to the polymer-coated single fibers. The absorption rate was ten times higher than that of the 100 μm PDMS-coated fibers. The desorption temperature (280°C) indicated that the analyte interaction with the C18-bonded silica was stronger than that with the PDMS polymer (injection temperature 250°C). However, multifibers directly exposed to a water solution tended to stick together because of the strong hydrophobic property of C18-bonded silica. This greatly reduced the surfaces exposed to the water phase. For that reason, multifibers were wetted in acetone for 1 min and then immediately placed into the water sample for analyte extraction.

A fully automated analytical method based on inline coupling of SPME to GC for a continuous analysis of OCPs and other organic contaminants present in surface and sewage water was described.[56] The water sample is pumped continuously through a flow-through cell mounted on a commercial GC autosampler and the fiber is dipped at regular intervals into the flowing sample, thus allowing a continuous monitoring of OCPs in aqueous systems (Figure 22.5).

Since equilibrium times quoted for OCPs fall in the range of 30 to 180 min, nonequilibrium SPME can be used for a fast screening of such compounds.[57] If a highly sensitive detection system (such as ECD) is available, a reduction in extraction time is possible; in fact, linear responses having good precision are possible even by using extraction times shorter than equilibrium times. In this paper, an extraction time of 2 min was used, thus attaining a further reduction of the sample preparation time.

SPME is a partition process and interferences due to dissolved and non-DOM are not likely to occur.

4. Stir Bar Sorptive Extraction (SBSE)

SBSE is a novel sample preparation method introduced by Baltussen et al.[60] based on the same mechanisms as SPME. In SBSE, a magnetic stirring bar coated with PDMS is added to water samples of 10 to 200 ml to promote the transport of analytes into the coating polymer. After a predetermined extraction period, the analytes are thermally desorbed in the GC injector or solvent extracted for HPLC analysis. The main advantage of SBSE is that 25 to 100 μl PDMS polymer is used instead of 0.5 μl as in SPME. The applications developed with SBSE have shown low detection limits (sub-ng^{-1} to ng l^{-1} levels) and good repeatability, confirming to the great potential of this technique. An application has been presented by León and coworkers for the analysis of 35 priority semivolatile pollutants, among which are several OCPs.[61]

The optimized conditions include extraction of 100 ml water samples fortified with 20% NaCl using 20 mm × 30.5 mm commercial stir bars (Gerstel, Mülheim a/d Ruhr, Germany) agitated at 900 rpm for 14 H at ambient temperature. After this, the stir bars are thermally desorbed using a commercial thermal desorption unit TDS-2 (Gerstel) in the splitless mode at 280°C for 6 min, and the analytes are transferred with a helium flow rate of 75 ml min^{-1} to the PTV injector which remains at 20°C. Finally, the PTV is ramped to 280°C to transfer the analytes to the GC–MS column. Matrix modifications by salting out are performed in order to modify extraction efficiency, which is affected by the K_{ow} of the analyte.[62] For the 35 compounds investigated, K_{ow} ranged from 10^2 for the polar compounds to 10^7 for the apolar ones. Extraction efficiency of polar compounds is lower than those of the more apolar ones. To enhance the extraction efficiency particularly for polar compounds, the effect of NaCl addiction at concentrations from 0 to 30% was studied. The increase of ionic strength favors the recovery of polar compounds but reduces the extraction of apolar compounds such as OCPs. Since such a method was developed for simultaneous determination of 35 priority pollutants, a 20% NaCl addition was finally selected as a compromise. Under these conditions, the good repeatability, high analyte recoveries, robustness, simplicity, and automation make SBSE a powerful tool for the routine quality control analysis of organic pollutants, including OCPs, in water samples.

5. Passive Sampling

The passive sampling approach was based on the observation that dialysis membranes filled with hexane accumulate persistent lipophilic pollutants in a way similar to that of aquatic organisms. The uptake of low-molecular-weight lipophilic compounds seems to be a passive process governed by partitioning mechanisms. A technique based on the principle that the partitioning of lipophilic substances in a two-phase system, consisting of water and an organic solvent, is displaced toward the solvent was developed by Södergren in 1987,[63] using a dialysis membrane filled with about 3 ml n-hexane and exposed for about 1 week to different organochlorine pollutants, including p,p'-DDE, p,p'-DDT, and hexachlorobenzene, in static and continuous-flow systems. A similar approach was used later on by Huckins and coworkers,[64] which used a semipermeable membrane device (SPMD) for passive *in situ* monitoring of organic contaminants in water. The device consists of a thin film of neutral lipid (molecular mass generally ≥ 600 Da), such as triolein, enclosed in thin-walled lay-flat tubing made of low-density polyethylene or another nonporous polymer (Figure 22.6). However, the authors stated that such an approach is useful for the estimation of average concentrations of dissolved organic contaminants in water and for the prediction of contaminant uptake by aquatic organisms, but it is not able to function as quantitative monitoring. Nevertheless, in a study concerning the temporal monitoring of OCPs following a flooding episode in Western Europe, such a device was successfully used and several OCPs were subsequently determined in two sites in the Lysekil archipelago.[65]

B. Extraction from Soil

1. Soxhlet Extraction

This method, developed at the end of the 19th century, is still the most widely used when organic compounds have to be extracted from solid materials, like dusts, sand, soil, and marine sediments. It is particularly suitable when the organic material is strongly adsorbed on a porous solid matrix. Such a simple method presents several advantages[66]: the sample is repeatedly brought into contact with fresh portions of the solvent and no filtration is required after the leaching step, simultaneous extraction in parallel can be performed since the basic equipment is inexpensive, and finally it has the possibility to extract more sample mass than most of the latest methods [microwave extraction,

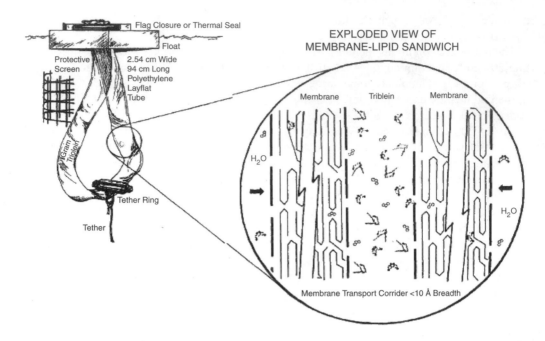

FIGURE 22.6 One of many possible configurations of lipid-containing SPMDs. The exploded view of the membrane–lipid–membrane sandwich illustrates the torturous membrane transport corridors that prevent significant losses of triolein to the environment yet allow permeation of smaller analyte molecules. (From Arthur, C. L., Potter, D., Buchholz, K., Motlagh, S., and Pawliszyn, J., *LC-GC*, 10, 656–661, 1992. With permission.)

supercritical fluids (SFs), etc.]. For these reasons, there is a wide variety of official methods involving a sample preparation step based on Soxhlet extraction.[67–69] Disadvantages are the prolonged time requirement, the large amount of solvent wasted, and the adsorption of the analytes on the glass container walls. Therefore, many authors have tried to improve the conventional Soxhlet technique and some Soxhlet methods used quite recently are described as follows.

A Soxhlet technique was used for determination of soil-bound DDT residues which may be released gradually from the soil increasing the insecticide load of the soil. The bound residues so released may be available for uptake by the biota.[70,71] In this application, 50 g samples of air-dried soil, in triplicate, were extracted by three volumes of methanol in Soxhlet apparatus for 24 h (72 cycles). Sulphuric acid was used for clean-up of the extract since it did not affect DDT and its metabolites. The released residues were taken up in HPLC-grade methanol for analyses by HPLC.

Khim and coworkers utilized Soxhlet extraction for the characterization of trace organic pollutants, among them OCPs, in marine sediments.[72] The described method implied that sediment samples (20 g + 100 g of Na_2SO_4) were Soxhlet extracted for 20 h using 400 ml of high-purity dichloromethane. Extracts were treated with acid-activated copper granules to remove sulphur and concentrated to 1 ml. Extracts were then passed through 10 g of activated Florisil (60 to 100 mesh size; Sigma, St. Louis, MO) packed in a glass column (10 mm i.d.) for clean-up and fractionation. The first fraction eluted with 100 ml high-purity hexane contained PCBs, HCB, and *p,p′*-DDE. Remaining OCPs and PAHs were eluted in the second fraction using 100 ml of 20% dichloromethane in hexane.

A substantially similar procedure was used by Barra et al.[73] Wet samples were homogenized with anhydrous Na_2SO_4 and extracted for 18 h using *n*-hexane. Extracts were then concentrated in a rotary evaporator and subjected to sulphuric acid clean-up. The organic residue was then concentrated to about 2 ml and passed through a Florisil (1.5 g) column (8 mm i.d.). OCPs were eluted with 60 ml of *n*-hexane and eluates were evaporated in a rotary evaporator and finally under nitrogen stream to *ca.* 0.5 ml. Aliquots (2 μl) were injected in a GC–ECD system. Results obtained analyzing a lake sediment core demonstrated the relatively new occurrence of *pp'*-DDT in the watershed, even if OCPs have been banned in Chile since 1985.

A faster extraction time was used by Nhan et al.[74] in research concerning the occurrence of OCPs in sediments from the Hanoi region. Sediments were wet sieved through a metallic sieve and only the fraction of particle size less than 0.1 mm was recuperated. The sediment samples obtained were dried by mixing with anhydrous Na_2SO_4 (sediment:salt ratio 1:3, w/w). The dry sediments were then put in *n*-hexane washed and dried glass bottles and stored deep frozen ($-20°C$) until analysis. Trichorobiphenyl was added to each sample as an internal standard. Samples were Soxhlet extracted with *n*-hexane for 8 h. Extracts obtained were subject to concentration in a rotary evaporator down to about 10 ml and then further evaporated to exactly 1 ml under nitrogen stream. Thus extract was purified to remove organosulphur compounds by treatment with freshly prepared copper powder and then cleaned-up and separated into two fractions by using a silica gel packed column. The first fraction, eluted with *n*-hexane, contained almost all the OCPs. Results obtained showed that, even if the number of OCPs is frequently below the detection limit, the DDTs were detected in relatively high concentrations in all samples.

To extract DDT from soil and sediment from an endemic leishmaniasis area in Rio de Janero city, the following procedure was used[75]: 2 g of wet sample were mixed with 4 g of silica gel 60 (70 to 230 mesh). This mixture underwent a continuous Soxhlet extraction with 20 ml of hexane:cyclohexane in 3:1 proportion and 2 ml of isooctane as solvents in a water bath for 2 H.[76] Clean-up was performed on a chromatographic glass column filled with an Al_2O_3, NaOH, and Na_2SO_3 mixture (7 g), with 20 ml of hexane as solvent. As internal standard, octachloronaphtalene was added. Extracts were analyzed by GC–ECD. Results showed that in the superficial soils a high percentage of DDE was observed as expected, since aerobic environments promote a significant DDT degradation to DDE instead of DDD, which is more likely to occur under reducing conditions. In fact, in sediments collected in the streams nearby, a higher percentage of DDD was observed. The not too high DDT concentrations found in this area give an indication that DDT contamination is diminishing, as confirmed by the decrease of the T/E ratio (i.e., the *p,p'*-DDT/*p,p'*-DDE ratio) in the sampling campaign carried out in 1999 with respect to that carried out in 1997. In fact, such a ratio, decreasing with time, may provide an estimate of DDT input over time.

In order to measure air–soil exchange of OCPs in agricultural soils in southern Ontario,[77] soil samples were extracted as follows. Soils were mixed with Na_2SO_4 and ground using a mortar and pestle until a granular consistency was achieved. They were fortified with a recovery spike containing α-HCH-d_6 and *p,p'*-DDT-d_8 and extracted in a Soxhlet apparatus for 20 h using dichloromethane. The extracts were reduced in a rotary evaporator under vacuum to 0.5 ml, transferred to a vial and reduced under a gentle nitrogen flow to 0.5 to 1 ml, and then cleaned on a 1 g alumina column. The column was washed first with two volumes (10 ml) of dichloromethane 5% in petroleum ether and, after application of the sample, was eluted with 10 ml of 5% dichloromethane in petroleum ether under a minimal nitrogen flow. The sample was then reduced again to 1 ml and transferred to a 2 ml GC vial using isooctane. The sample was further reduced to 0.5 ml, and mirex was added as the internal standard for volume correction. The soil extracts were analyzed by GC–ECD.

2. Microwave-Assisted Extraction (MAE)

MAE is, like Soxhlet extraction, classified among the leaching methods. Such a method was initially applied to sample mineralization.[78] Nevertheless, in recent years numerous applications have reported the use of microwaves for assisting the extraction of organic from solid matrices. MAE consists in heating the extractant (mostly liquid organic solvents) in contact with the sample with microwave energy. The partitioning of the analytes of interest from the sample matrix to the extractant depends on the temperature and the nature of the extractant. Unlike classical heating, microwaves heat all the sample simultaneously without heating the vessel. Therefore, the solution reaches its boiling point very rapidly, leading to very short extraction times. Microwaves for sample extraction were used for the first time by Ganzler and Salgo.[79] They used a domestic microwave oven in order to demonstrate a higher extraction of polar compounds by this leaching technique as compared to the Soxhlet method. In order to avoid degradation of the target analytes the authors carried out their process in several short heating and cooling cycles. Afterwards, a number of researchers have used microwaves; as an example, Onuska and Terry[80] used microwave energy to extract OCPs from sediment samples using a cyclic mode. They reported quantitative recoveries and no compound breakdown due to sample exposure to microwaves. The procedure consisted in weighing 1.0 g of sample in a 5 ml vial. Prior to extraction, samples were saturated with distilled water and 2 ml of iso-octane. Sample was extracted in a conventional home oven for 30 sec using maximum power, not allowing the vial content to boil. After 30 sec the vial was immersed into an ice bath for 2 to 5 min. The extraction step and cooling were repeated up to five times. Clean-up was performed as follows. Extracted sediments and solvent were centrifuged for 10 min and solvent was filtered through filter paper on a Buchner funnel under partial vacuum. The filter paper and the sediment cake were rinsed with the solvent. GC–ECD analysis was then performed.

However, the use of domestic microwave ovens can pose serious hazards in the application of microwave energy to flammable organic compounds (such as solvents). For that reason, *ad hoc* commercial devices were developed using only the 2450 MHz frequency. Noticeable contributions to expand the use of microwaves as an alternative to conventional methods have been reported by López-Avilla.[81,82] The authors used a MES-1000 microwave sample extraction system (CEM Corp., Matthews, NC) as shown in Figure 22.7. It uses the technology with closed vessels under

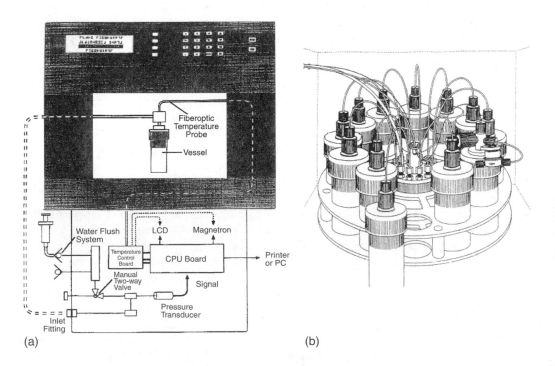

FIGURE 22.7 (a) Schematic diagram of the temperature/pressure control system for the MDS-2000 microwave system; (b) view of the 12-lined digestion vessels, containment vessel, and temperature and pressure probes. (From Singh, D. K. and Agarwal, H. C., *J. Am. Food Chem.*, 40, 1713–1716, 1992. With permission.)

controlled pressure, commonly called pressurized MAE (PMAE), where the solvent can be heated above its boiling point at atmospheric pressure enhancing extraction speed and efficiency. Such system permits temperature control of the extraction process. In addition, sample throughput is increased if several vessels are used, thus allowing simultaneous extractions. As the electric field is nonhomogeneous in the cavity, the vessels are placed on a turntable, as in domestic microwave ovens.

This system is equipped with an inboard pressure and Fluoroptic temperature control system for regulating sample extraction conditions via magnetron power output control. Temperature and pressure control set points could be programmed in five separate heating stages. The instrument controls either pressure or temperature, depending on which parameter reached its control set point first. Twelve-lined digestion vessels (110 ml volume) were used for extractions. The following extraction procedure was used. A 5 g portion of each matrix was accurately weighed and quantitatively transferred to the Teflon-lined extraction vessel. To prepare the wet samples, the calculated volume of water was added and allowed to equilibrate with the matrix for 10 min. Then 30 ml of hexane/acetone (1:1) were added and the extraction vessel was closed. Extractions were performed at 115°C for 10 min at 100% power. After extraction, the vessels were allowed to cool to room temperature for 20 min before these were opened. The supernatant was filtered through glass wool prewashed with hexane–acetone and was then combined with the 2 to 3 ml hexane–acetone rinse of the extracted sample. The extract was concentrated to 5 ml under nitrogen flow and was centrifuged twice for 10 min at 2300 rpm to separate the fine particulates. The extract was concentrated to 1 ml for GC/ECD analysis. Recoveries for topsoil samples spiked with selected OCPs at 50 pg/kg were at least 7% higher for MAE than for either the Soxhlet or the sonication extraction technique, except for 4,4′-DDT, where the recoveries were above 100% for all three techniques, and for dieldrin, where the Soxhlet recoveries were 12% higher than MAE and sonication. Such results are summarized in Figure 22.8. MAE technique was also applied successfully to the elution of several OCPs from C18 membrane disks.[83]

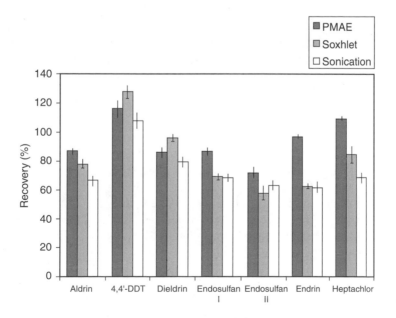

FIGURE 22.8 Comparison of MAE and traditional extraction techniques for the determination of several pollutants in soils and sediments. Spiked OCPs from a topsoil PMAE conditions: 5 g matrix, 30 ml hexane–acetone (1:1), 1000 W, 115°C, 10 min. (From Khim, J. S., Kannan, K., Villeneuve, D. L., Koh, C. H., and Giesy, J. P., *Environ. Sci. Technol.*, 33, 4199–4205, 1999. With permission.)

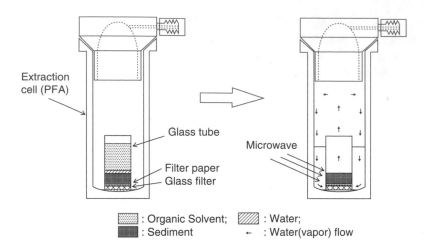

FIGURE 22.9 Principle of the microwave-assisted steam distillation. (From Japenga, J., Wagenaar, W. J., Smedes, F., and Salomons, W., *Environ. Technol. Lett.*, 8, 9–29, 1987. With permission.)

Advantages of microwave-assisted leaching vs. conventional Soxhlet can be ascribed to the performance of this heating source based on dielectric loss. The heat appears in the bulk of the irradiated material, thus giving rise to an inverse temperature gradient; that is, volume rather than surface heating. Thus, both the extraction time and the volume of solvent required are dramatically reduced.[84,85] However, the efficiency of microwaves can be very poor when either the target analytes or the solvents are nonpolar or of low polarity, when these are volatile, and when the solvents used have low dielectric constants. In these cases, Soxhlet extraction is superior to MAE.

A very recent development of MAE is microwave-assisted steam distillation (MASD), a combination of MAE and steam distillation methods.[86] Wet sediment samples were heated by microwave irradiation, and the analytes (PCBs and OCPs) were desorbed from the matrix (sediments) with water vapor. The condensed analytes were trapped in a small amount of nonpolar organic solvent. The apparatus used is reported in Figure 22.9. A sintered glass filter attached on the bottom of a glass tube (50 mm, 20 mm i.d.) was covered with a piece of filter paper. The sediment was weighted in the glass tube, which was put into a Teflon PFA extraction cell (140 mm, 32 mm i.d.) containing 3.0 ml of water. After the sample soaked up the water, the surrogates solution prepared from solutions of each isotope-labeled compounds (135 mg 2,2,4-trimethyl-pentane solution) and 10 ml nonpolar solvent was added to the glass tube. The cells were covered by a pressure-resistant holder and were heated using a microwave (MARSX, CEM Co, NC) at 110 to 170°C for 10 to 90 min. Upon termination of the microwave irradiation, the cells were air-cooled then opened, and the glass tubes placed inside were removed. To dry the organic layer, 1.0 g anhydrous sodium sulfate was added to the cell, then the organic solvent layer was recovered with a pipette, and the inner wall of the cell was rinsed with a small amount of hexane to recover the whole extract. The combined extract was analyzed with GC/MS without any clean-up procedure.

3. Accelerated Solvent Extraction (ASE) or Pressurized Liquid Extraction (PLE)

ASE (also known as PLE) constitutes a leaching technique which is based on principles similar to those of MAE, but microwave energy is replaced in ASE by conventional heating in an oven. Like PMAE, ASE uses organic solvents at high pressures and temperatures above the boiling point. With ASE, a solid sample is enclosed in a sample cartridge which is filled with an extraction fluid

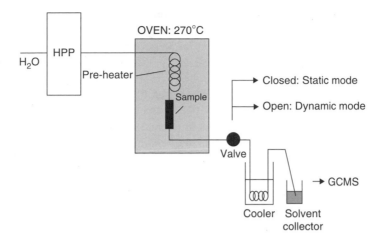

FIGURE 22.10 Subcritical water-extraction manifold. HPP, high-pressure pump. (From Meijer, S. N., Shoeib, M., Jantunen, L. M. M., Jones, K. C., and Harner, T., *Environ. Sci. Technol.*, 37, 1292–1299, 2003. With permission.)

and used to statically extract the sample under elevated temperature (50 to 200°C) and pressure (500 to 3000 psi) conditions for short time periods (5 to 10 min). Compressed gas is used to purge the sample extract from the cell into a collection vessel. A schematic diagram of the subcritical water-extraction manifold is reported in Figure 22.10. Richter and coworkers[87] reported details of ASE together with the effect of experimental parameters on recovery. Furthermore, they studied thermal degradation during extraction. They concluded that ASE allowed an enhanced extraction of analytes from solid samples to be achieved, giving recoveries comparable to those obtained with Soxhlet and other techniques in use while spending only a fraction of the time and solvents needed for those techniques. Some applications of ASE for the analysis of OCPs are reported here. Popp et al.[88] used ASE for extracting OCPs and other organic contaminants from solid wastes. The extractions were carried out using a Dionex (Dionex Co., Sunnyvale, CA) ASE 200 Accelerated Solvent Extractor with 11, 22 or 33 ml stainless steel extraction cells. Solvents were acetone–hexane (1:1, v/v). Static extraction of 2 × 5 min performed at an oven temperature of 100°C was sufficient. However, in the case of extremely contaminated soils, toluene as solvent is more suitable, with an oven heat-up time of 8 min and an oven temperature of 175–200°C. For the determination of OCPs, ASE extracts were concentrated to 1 ml and directly injected into the GC–ECD.

Efficiency of ASE was compared to that of Soxhlet extraction in a study carried out in 1997 and aimed at determining whether such a method could replace established EPAV (EPA Victoria, Australia) methods for soil screening.[89] Instrumental conditions implied the use of a Dionex ASE 200 Instrument and the extraction of 10 g of soil for 10 min with 1:1 dichloromethane/acetone at 100°C and 10 MPa pressure in an 11 ml cell. Extracts were analyzed by GC–MS injecting 1 μl samples with an autosampler. Results confirmed that ASE performance for the general screening of contaminated soils for semivolatile organic pollutants is at least equivalent to Soxhlet extraction. For the spiked uncontaminated soil, ASE recoveries were mostly higher than those from Soxhlet extraction, especially for OCP-contaminated soils.

Other authors reported that, using three sequential static phases, ASE (here defined as PLE) removed an equivalent quantity of DDT and its metabolites as Soxhlet extraction in less time and with less solvent and that recovery was almost quantitative when the sample was appropriately worked-up and manipulated.[90] Hubert and coworkers investigated the dependence of the extraction efficiency of POPs, including OCPs, on the ASE operating variables solvent and temperature. Mixed soil samples from two locations with considerable differences in soil properties and contamination were used. The objective was to optimize ASE for the extraction of POPs from real soil samples and to improve on the results achieved with Soxhlet extraction. The authors found that, after optimization, ASE represents an exceptionally effective extraction technique compared to other extraction techniques,[91] especially for highly contaminated sample material and a complicated soil matrix. Toluene was demonstrated to be the best extraction solvent. The extractions were performed in two temperature steps at 80 and 140°C for an extraction time of 3 to 10 min. The extraction temperatures of 80 and 140°C represent the key temperatures for the extraction of POPs from the soil matrix at which the highest extraction efficiency can be achieved depending on the soil parameters, the degree of multiple contamination of the sample, and the physicochemical properties of the substances to be analyzed.

The advantages of ASE are summarized as low solvent consumption, easy optimization, the main parameters to be considered being the extraction solvent and the extraction temperature. As a general strategy, the solvent can also be used at a temperature above its boiling point and a pressure high enough to keep it as a liquid at that temperature. The nature of the matrix and its physical characteristics should also be considered.

Proper solvent selection can help to eliminate the influence of the matrix properties on the ASE recoveries. The selectivity of the ASE can also be improved by adding an appropriate sorbent to the extraction cell for simultaneous clean-up. Therefore ASE combines good recoveries and adequate precision with rapid and rather selective extraction, while the sample handling is less time consuming than with classical procedures. Disadvantages are the high initial investment costs, some practical problems associated with the homogeneous and reproducible packing of heterogeneous samples in the smaller-size PLE extraction cells, and the limited possibility of carrying out selective extractions of organic compounds from complex samples.[92]

4. Supercritical Fluid Extraction (SFE)

SFE is one of the most successful contributions to leaching techniques. A SF is defined as a fluid which is above its critical pressure and temperature, and possesses unique properties depending on the pressure, temperature, and composition. In particular, density may be adjusted by correctly choosing both the pressure and temperature. The SFE system consists of a high-pressure pump which delivers the fluids and an extraction cell containing the sample that is maintained at correct temperature and pressure. The SFE may be carried out in either a static or a dynamic mode. Carbon dioxide is the most common extraction fluid due to its unique properties: first, its critical parameters (72.8 bar and 31.1°C) are easily accessible, it is not toxic, nonflammable, nonreactive, gaseous at atmospheric pressure, and is easily removable from extracts. The main drawback is its nonpolar character, which precludes efficient extraction of polar compounds, unlike the case of OCPs. Therefore, OCPs are sufficiently nonpolar to be easily extracted using carbon dioxide as extraction fluid, with no addition of any modifier, as shown in a study by Lohleit et al.[93] However, this is true when an inert spiked matrix is extracted, while recoveries from spiked soil are dramatically decreased because the interactions of the soil solid matrix and OCPs are also dependent on pesticides polarity, identified as K_{ow}. Analytes with higher K_{ow}, i.e., nonpolar, partition more readily into the soil,[94] especially those with higher organic content.[95] For that reason several authors used organic modifiers to enhance SFE.[96–98]

A paper by Snyder and coworkers[97] describes the use of SFE to extract organochlorine from a variety of soils, including sand, top soil, clay, and river sediment.

SFE conditions were the following: the soils were loaded into 2 or 10 ml stainless steel extraction vessels depending on the sample size and extracted at 350 atm pressure and 50°C. Carbon dioxide premixed with 3% methanol was used as the extraction fluid. Samples were extracted statically at 350 atm for 10 min. After 10 min, the system was extracted dynamically at 350 atm. A flow of about

1 ml min^{-1} of CO_2 in the supercritical state was obtained. The pesticide analytes were collected by bubbling the vented CO_2 through 5 ml of methyl *tert*-butyl ether (MTBE). No concentration of the extracts was required, and each MTBE extract was diluted to an exact final volume of 5 ml. The overall average recovery for all four soil types and the pesticides was 94%. The overall precision for all pesticides and soils was 5.1%. Poorer recoveries were obtained from soil aged by storage at 4°C for 8 months. The authors concluded that SFE method was faster and easier than other extraction techniques like sonication or Soxhlet. Further, less solvent was consumed by SFE, and as a consequence of the small amount of solvent used, no solvent concentration step was necessary. That is particularly important since, during such a step, the chance for analyte loss due to evaporation, breakdown, or reaction of the compound is greatly increased.

Since the presence of sulphur in soil matrices can deteriorate the separation capability of GC capillary columns and the sensitivity of ECD, it is necessary to minimize sulphur interferences in SFE extracts before GC. Therefore, a study was carried out in order to investigate the influences of spiking method, soil composition, moisture, and grain size on SFE extraction of 16 organochlorine pesticides from sulfur-containing soils,[99] finding that spiking method significantly influences SFE extraction efficiency.

The above cited studies are limited to the analysis of spiked samples. However, the importance of differences between spiked and real samples must be taken into account. Van der Velde and coworkers[98] found that the optimal conditions achieved for spiked samples did not yield maximum concentrations for real samples. In the latter case, stronger extraction conditions were needed to overcome interactions between matrix and analytes. Longer dynamic extraction times and the use of modifiers were necessary, while an increase in extraction pressure did not have any influence. After evaluation of different experimental conditions, the following final SFE conditions were considered the most appropriate: soil was extracted six times at 20 MPa, 50°C, 30 min dynamic extractions using 20 μl methanol as modifier of the CO_2 extraction fluid with flow 160 to 180 μl min^{-1}. The influence of the amount of methanol modifier on extraction is reported in Table 22.5. Real soil samples from a contaminated soil were extracted for the determination hexachlorocyclohexane isomers using SFE followed by solvent trapping using toluene as collection solvent.[96] In this case, toluene was also used as modifier, adding 200 μl directly into the cell. Later on, the same group presented an upgrading of the above method based on a fractional factorial approach to optimization and modifier selection for extracting chlorinated benzenes and hexachlorocyclohexanes from soil.[100]

TABLE 22.5
Influence of the Amount of Methanol Modifier on Orchard Soil

Component	5 μl, MeOH, 2.4%[a], $n=3$		10 μl, MeOH, 4.8%[a], $n=3$		15 μl, MeOH, 7.1%[a], $n=3$		20 μl, MeOH, 9.5%[a], $n=3$		25 μl, MeOH, 11.9%[a], $n=3$		30 μl, MeOH, 14.3%[a], $n=3$	
	ng g^{-1}	SD	ng g^{-1}	SD	ng g^{-1}	SD	ng g^{-1}	SD	ng g^{-1}	SD	ng g^{-1}	SD
p,p'-DDE	53.9	1.1	55.9	2.3	63.3	4.7	72.0	2.9	61.3	5.2	65.3	1.0
TDE	17.1	0.5	17.1	0.4	16.6	0.5	20.6	0.8	16.3	0.6	17.3	0.2
o,p'-DDT	42.9	1.0	44.5	1.9	52.3	3.6	61.8	1.5	51.8	5.0	56.6	0.8
p,p'-DDT	250.0	3.0	261.0	9.0	311.0	22.0	355.0	8.0	315.0	23.0	344.0	1.0

SFE conditions: 20 MPa; 50°C; CO_2; 30 min dynamic; flow *ca.* 160 to 180 μl min^{-1}. SD is expressed in ng g^{-1}.
[a] % (v/v) of the modifier with respect to CO_2, on basis of free cell volume.

Source: From Richter, B. E., Jones, B. A., Ezzell, J. L., Porter, N. L., Avdalovic, N., and Pohl, C., *Anal. Chem.*, 68, 1033–1039, 1996. With permission.

Advantages and drawbacks of SFE as an extraction technique follow. The main advantages over the Soxhlet extraction are shorter extraction time, lower solvent waste, the possibility of easily altering extraction conditions, and therefore less necessity for clean-up due to the high selectivity achieved by manipulating pressure and temperature. Among the three trapping systems usually employed for analyte collection after extraction, namely, liquid collection, cryogenic trapping, and solid-phase trapping, the last is considered by some authors[101] to be the least simple but most effective as it allows simultaneous collection, clean-up, and concentration prior to either individual GC separation or direct detection. Main drawbacks are as follow: the big discrepancies in efficiency between spiked and natural samples, the need of modifier to overcome strong interactions which occur in environmental matrices, and the high investment costs.[84,102]

C. GAS CHROMATOGRAPHIC COLUMNS

The chromatographic separation of chlorinated pesticides has been performed with packed columns until the late 1970s to early 1980s, the liquid phases most commonly used being siliconic phases. The operating isothermal temperature was in most cases around 200°C. Nowadays for the GC analysis of chlorinated pesticides fused silica capillary columns are exclusively used. The nonpolar siliconic stationary phases (methylphenylsilicon polymer) are the most frequently used.

Table 22.6 shows summarized stationary phases and columns more recently used for the separation of OCPs in water and soil.

High-resolution capillary GC with electron capture detection analysis is one of the most popular technique for OCPs analysis due to the outstanding sensitivity of ECD to halogenated compounds (see next section). ECD is not able to positively identify analytes. Because of the compound complexity in some sample matrices, the possibility of false-positive identifications from interfering unknown compounds becomes significant. To reduce the incidence of false-positive identifications and improve the quality of the data, a dual-column, dual-detector GC system was developed in 1990 to simultaneously identify and confirm OCPs in environmental sample extracts.[113] A sample extract was split between two capillary columns of different polarities (a nonpolar column and a more polar one), which were connected to two detectors of slightly different specificity to chlorinated compounds, i.e., ECD and electrolytic conductivity detector (ELCD), the latter being a halogen-compound-specific detector which does not response to nonhalogenated compounds which the ECD recognizes. Thus, many of the interfering compounds typically detected by GC–ECD in organically complex samples are not detected by GC–ELCD. With this method, identification and confirmation of analytes were obtained in one GC run. The dual-column, dual-detector methodology was applied in a study for the characterization of 13 PCBs and 103 pesticides (including OCPs).[114] In this case, the second detector was a nitrogen phosphorous detector (NPD), useful for detection of the organophosphorous and the organonitrogen compounds, which were determined together with OCPs. The paper provides retention times for two different GC columns of different polarity for a great number of pesticides. Such a dual-column approach was widely applied to the analysis of OCPs in various matrices also with ECD detection only[10,48,59,100,115–117] and it is included in the U.S. EPA method 608.[118] In this case, the ECD signal from the nonpolar column is used for quantitative analysis, while the ECD signal from the more polar one is used as confirmation.

D. DETECTORS

The most common method for detecting OCPs is ECD. Its high sensitivity and excellent selectivity have made it the detector of choice in pesticide analysis. However, as stated above, GC–ECD analysis is often unlikely to provide unambiguous pesticide identification in the presence of a wide range of chlorinated organics. In this case the use of MS as a confirmation method is required.

TABLE 22.6
Capillary Columns Commonly Used for the Separation of OCPs

Column	Injection Volume	Detector	Reference	Year
25 m, 0.22 mm i.d., 0.25 μm df 5% phenyl polysilphenylene-siloxane	1.0 μl	MS	89	1997
30 m, 0.25 mm i.d., 0.25 μm df 20% permethylated, cyclodextrin in SPB-35	2.0 μl	MS	103	1998
30 m, 0.25 mm i.d., 0.25 μm df 20% tert-butyldimethylsilylated, -cyclodextrin	2.0 μl	MS	103	1998
60 m, 0.32 mm i.d., 0.25 μm df bonded poly(5% diphenyl/95% dimethylsiloxane)	—	MS	65	1998
30 m, 0.25 mm i.d., 0.25 μm df bonded 5% phenyl/95% dimethylpolysiloxane	PDMS capillary	MS	31	1998
	1.0 μl	MS	104, 105	1998, 2000
	0.5 μl	MS	90	2000
	0.6 μl	ECD	47	2000
	2.0 μl	ECD	73	2001
	SPME	MS	106	2001
	NA	ECD	107	2002
	SBSE	MS	61, 108	2003
	2.0 μl	MS	109	2003
30 m, 0.25 mm i.d., 0.25 μm df bonded poly(5% diphenyl/95% dimethylsiloxane)	SPME	ECD	58	2002
60 m, 0.25 mm i.d., 0.25 μm df bonded 5% phenyl/95% dimethylpolysiloxane	NA	ECD	77	2003
30 m, 0.32 mm i.d., 0.25 μm df 5% dimethyl polysiloxane	3.0 μl	AED	110	2003
	1.0 μl	ECD	111	1999
50 m, 0.22 mm i.d., 0.25 μm df 8% Phenyl polycarborane-siloxane	1.0 μl	MS	86	2003
30 m, 0.25 mm i.d., 0.25 μm df (5%-phenyl)-methylpolysiloxane		MS	112	2003

NA = not available.

1. Electron Capture Detector (ECD)

Since its introduction in the early 1960s, the ECD has played a major role as a gas chromatographic detection technique in the analysis of OCPs and other halogenated organic pollutants. First applications of ECD to environmental problems appeared in 1961 when two papers were simultaneously published which showed the ubiquitous distribution of chlorinated pesticides.[119,120] These works, possible only because of the availability of the ECD, had an enormous impact on the scientific world and on public opinion, giving rise to great interest in the fate of such compounds of environmental concern.

Since the introduction of ECD, the ^{63}Ni β-ray radioactive electron source has remained unchanged, and considerable efforts have been made to develop nonradioactive alternatives. A new version of such a detector — the pulsed discharge electron capture detector (PDECD) — employs a pulsed discharge in helium as the primary source of electron generation. A modified version of PDECD which makes use of methane as the dopant gas and of a sapphire and quartz insulation was used for detecting OCPs.[121,123] The relatively low ionization potential of methane allows reduction of the interference from extraneous ionization peaks enhancing sensitivity, while the highly inert sapphire and quartz insulation allows operating at a temperature up to 400°C. Minimum detectable quantities (MDQs) of OCPs obtained with PDECD are in the midfemtogram range (Table 22.6), and a linear dynamic range of over three to four orders of magnitude is attained.

In conclusion, it can be stated that chlorinated pesticides analysis must be considered one of the most important applications of the ECD, since no other detectors have competitive sensitivity and selectivity. For this reason, ECD is still widely and successfully used for routine analysis of OCPs[47,57,58,73,75,77,107,116,117] (these references mention just some among the latest applications of ECD). To overcome problems related to its inability in providing positive identification of the different compounds, especially in complex environmental matrices, this dual-column approach can be used. However, because sometimes an extremely high degree of specificity is required for the analysis of certain compounds, MS is the accepted instrumentation for many controversies.

B. Mass Spectrometric Detection (MSD)

MS has always been seen as one of the most conclusive techniques for positive identification of organic compounds. The availability, since the beginning of the 1980s, of benchtop GC–MS systems based on quadrupole mass analyzers (GC-Q-MS) made such an analytical tool extremely popular also for routine applications. However, when GC-Q-MS is operated in the full scan mode, limits of detection (LODs) are too high, especially in trace analysis, and its use is seldom restricted to a confirmation technique.[123] When the selected ion monitoring mode (SIM) is employed, the sensitivity is dramatically enhanced. On the other hand, SIM implies the detection of specific analytes with the consequent loss of all other information.

Nevertheless, the capability of SIM is confirmed by a study conducted by Barceló and coworkers, who achieved LODs at the ng l^{-1} range for 109 priority pollutants (including OCPs),[38] with easy identification of the compounds through library search. Such a result was obtained using SPE followed by GC/MS, with electron impact (EI) as ionization technique. GC/EI/MS generated the molecular [M] + ion as base peak for most families of compounds or other diagnostic ions formed by the loss of group-specific components, depending on the molecule type. OCPs exhibited losses of two to five chlorine atoms, giving characteristic fragments. In most cases the molecular ion was the base peak and it was the ion chosen for the SIM program. Even though the U.S. EPA recommends the use of three ions for confirmation and quantification of target analytes, in this application a single ion per compound was selected to avoid a decrease in sensitivity due to the large number of ions scanned. In this case, by selecting the appropriate ion of each compound and by retention time comparison, automatic identification of the analytes was achieved, and quantification was accurately performed from the SIM chromatogram. The corresponding scan chromatogram permitted the confirmation of specific analytes by the different fragment ions formed and by comparison with an authentic standard. Table 22.7 reports instrumental detection limits, molecular weight, and the three most important fragment ions for 21 OCPs.

An ionization technique alternative to EI is positive chemical ionization (CI), where the amount of energy involved in the ionization process is lower, so that the molecular ion formed is less likely to fragment as compared with that formed using EI. Therefore, CI is a very useful for obtaining the molecular ion, and consequently, the molecular weight of compounds. However, compounds with high electron affinity show higher sensitivity and selectivity when electron capture negative chemical ionization (ECNCI) is used. Advantages of ECNCI over EI and CI include selective ionization in the presence of complex matrices and greater sensitivity. Ideally, the negative molecular ion is stable enough or undergoes only minor decomposition to high mass fragments, but unfortunately sometimes the negative molecular ions decompose to low-mass fragments which are representative of only the electrophilic moiety of the original molecule. For this reason, the use of a derivative which under ECNI conditions would produce a stable molecular ion or fragment ion consisting of a major portion of the molecule of interest is recommended. As an example, the

TABLE 22.7
Instrument Detection Limits (IDL, pg) Obtained with GC/EI/MS, Molecular Weight (M_w), and the Three Most Abundant Fragment Ions of Selected OCPs

Compounds	IDL	M_w	m/z		
Hexachloro benzene	0.57	282	142 (34) [M-4Cl]$^+$	249 (22) [M-Cl]$^+$	284 (100) [M]$^+$
α-Hexachloro-cyclohexane	1.49	291	111 (48) [M-5Cl]$^+$	181 (95) [M-3Cl]$^+$	219 (100) [M-2Cl]$^+$
β-Hexachloro cyclohexane	1.02	291	109 (80) [M-5Cl]$^+$	181 (98) [M-3Cl]$^+$	219 (100) [M-2Cl]$^+$
Lindane	1.03	291	111 (55) [M-5Cl]$^+$	181 (100) [M-3Cl]$^+$	219 (79) [M-2Cl]$^+$
δ-Hexachloro-cyclohexane	1.13	291	109 (100) [M-5Cl]$^+$	181 (64) [M-3Cl]$^+$	219 (42) [M-2Cl]$^+$
Heptachlor	1.82	371	100 (100) ni	237 (50) [M-4Cl]$^+$	272 (92) [M-3Cl]$^+$
Aldrin	1.39	362	66 (100) [C$_5$H$_6$]$^+$	91 (53) ni	263 (30) [M-C$_5$H$_6$-Cl]$^+$
Isodrine	0.90	362	66 (75) [C$_5$H$_6$]$^+$	193 (100) [M-C$_5$H$_6$-3Cl]$^+$	263 (50) [M-C$_5$H$_6$-Cl]$^+$
o,o'-DDE	0.60	316	176 (30) [M-4Cl]$^+$	246 (100) [M-2Cl]$^+$	318 (43) [M]$^+$
α-Chlordane	nq	410	237 (30) [M-5Cl]$^+$	274 (26) [M-4Cl]$^+$	373 (100) [M-Cl]$^+$
α-Endosulfan	3.93	407	207 (100) ni	239 (84) ni	277 (58) ni
o,p'-DDE	1.02	316	176 (26) [M-4Cl]$^+$	246 (100) [M-2Cl]$^+$	318 (34) [M]$^+$
γ-Chlordane	nq	410	237 (26) [274-Cl]$^+$	272 (25) [M-4Cl]$^+$	373 (100) [M-Cl]$^+$
Dieldrin	3.77	378	79 (100) [C$_2$H$_4$ClO]$^+$	263 (33) ni	279 (23) [M-C$_5$H$_6$-Cl]$^+$
p,p'-DDE	0.37	316	176 (35) [M-4Cl]$^+$	246 (100) [M-2Cl]$^+$	318 (80) [M]$^+$
o,p'-DDD	0.90	318	165 (30) [235-2Cl]$^+$	199 (12) [235-HCl]$^+$	235 (100) [M-CHCl$_2$]$^+$
Endrin	16.8	378	197 (73) ni	237 (100) [M-HCl-3Cl]$^+$	265 (53) [M-CHCl$_2$CH$_2$O]$^+$
β-Endosulfan	4.49	407	195 (100) ni	207 (78) ni	237 (79) ni
p,p'-DDD	1.39	318	165 (44) [235-2Cl]$^+$	199 (17) [235-HCl]$^+$	235 (100) [M-CHCl$_2$]$^+$
o,p'-DDT	1.65	352	165 (40) [235-2Cl]$^+$	199 (16) [235-HCl]$^+$	235 (100) [M-CCl$_3$]$^+$
p,p'-DDT	1.23	352	165 (33) [235-2Cl]$^+$	199 (10) [235-HCl]$^+$	235 (100) [M-CCl$_3$]$^+$

ni = not identified; nq = not quantified.

From Goulden, P. and Anthony, D. H. J., Design of a Large Sample Extractor for the Determination of Organics in Water, Report No. 85-121, Environment Canada, National Water Research Institute, Burlington, Ontario, 1985. With permission.

oxygen-induced dechlorination of DDT and its metabolites (p,p'-DDE and p,p'-DDD) can be mentioned.[124]

In the early 1990s the introduction of ion-trap detectors (ITD) coupled to GC showed itself to be an extremely useful tool also for routine analysis of OCPs in various matrices.[106,125–129] An important feature of the ion-trap detector is that there is no loss in sensitivity when going from full scan data acquisition to single ion monitoring data. Further, in ion-trap technology, switching from full scan electron EI to CI can be achieved in a very easy way, rapidly providing enough information for the identification and quantitation of pesticides and metabolites.[131] Moreover, enhanced selectivity and confirmation can be obtained in the MS/MS mode.[132–136] Ion-trap mass spectrometric detection was included in several U.S. EPA methods for pesticide determination.[137]

A drawback of the instrument is that sensitivity is dependent on the amount of ions present in the trap and additional requirements for either calibration procedures (matrix-modified) or clean-up are required.

In recent years both ITD and benchtop quadrupole instruments have been improved in their detector design and operation and acquisition software leading to the widespread use of bench-top mass spectrometers in routine laboratories, and it seems that nowadays both types of instruments yield comparable performance. As compared to ECD, LODs were reported to be comparable, but with much more qualitative reliability.[138]

Even if GC is the separation technique of choice for OCPs, a MULTIANALYSIS system for the automated analysis of environmental water samples has been described.[130] In such a system, SPE-GC analyses were performed simultaneously with SPE-LC analyses employing a single mass spectrometric detector operated in total ion current (TIC) mode, using a 4 to 400 atomic mass unit (amu) scan range for positive-ion EI, and 65 to 400 amu for negative chemical ionization (NCI) detection. For target analysis SIM mode was used. Prior to entering the MS, the LC eluent was allowed to pass through the flow cell of an UV diode-array detector (DAD). However, the UV spectra not being of high information content, DAD was mainly used for a first screening to find suspect samples or peaks. Furthermore, LC–DAD data provided additional means for quantitation and yield complementary spectral information. The goal of this study was to integrate the three different techniques (GC–MS, LC–MS, and LC–DAD) in one system. The relative standard deviations (RSDs) of retention times were lower than 0.2% in all systems, while RSDs of peak areas ranged from 5 to 15%. Detection limits (DLs) in total ion current mode were below 0.1 $\mu g/l$ for GC–MS (10 ml samples). For LC–MS, 0.5– and 0.05–1 $\mu g/l$ DL values were obtained in TIC and SIM mode, respectively. Negative chemical ionization with methane as reagent gas improved the sensitivity of halogenated compounds 3- to 30-fold and provided relevant information for structural elucidation of unknown compounds.

REFERENCES

1. Cohen, Y., Tsal, W., Chetty, S. L., and Mayer, G. J., Dynamic partitioning of organic chemicals in regional environments: a multimedia screening-level modeling approach, *Environ. Sci. Technol.*, 24, 1549–1558, 1990.
2. Briggs, G. G., Theoretical and experimental relationships between soil adsorption, octanol-water partition coefficients, water solubilities, bioconcentration factors, and the parachor, *J. Agric. Food Chem.*, 29, 1050–1059, 1981.
3. Finizio, A., Vighi, M., and Sandroni, D., Determination of *n*-octanol/water partition coefficient (Kow) of pesticide, critical review and comparison of methods, *Chemosphere*, 34, 131–161, 1997.
4. Pimentel, D., Amounts of pesticides reaching target pests: environmental impacts and ethics, *J. Agric. Environ. Ethics*, 8, 17–29, 1995.
5. Spear, R., Recognized and possible exposure to pesticides, In *Handbook of Pesticide Toxicology*, Hayes, W. J. and Laws, E. R., Eds., Academic Press, San Diego, CA, pp. 245–274, 1991.
6. van der Werf, H. M. G., Assessing the impact of pesticides on the environment, *Agric. Ecosyst. Environ.*, 60, 81–96, 1996.
7. Gibbs, R. J., Mechanisms of trace metal transport in rivers, *Science*, 180, 71–72, 1973.
8. Gregor, D. J. and Gummer, W. D., Evidence of atmospheric transport and deposition of organochlorine pesticides and polychlorinated biphenyls in Canadian arctic snow, *Environ. Sci. Technol.*, 23, 561–565, 1989.
9. Iwata, H., Tanabe, S., Sakai, N., Nichimura, A., and Tatsukawa, R., Geographical distribution of persistent organochlorines in air, water and sediments from Asia and Oceania and their implications for global redistribution from lower latitudes, *Environ. Pollut.*, 85, 15–33, 1994.
10. Karlsson, H., Muir, D. C. G., Teixiera, C. F., Burniston, D. A., Strachan, W. M. J., Hecky, R. E., Mwita, J., Bootsma, H. A., Grift, N. P., Kidd, K. A., and Rosenberg, B., Persistent chlorinated pesticides in air, water, and precipitation from the lake Malawi area, Southern Africa, *Environ. Sci. Technol.*, 34, 4490–4495, 2000.
11. Mackay, D. and Wania, F., Transport of contaminants to the Arctic: partitioning, processes and models, *Sci. Total Environ.*, 160/161, 25–38, 1995.
12. Wania, F. and Mackay, D., A global distribution model for persistent organic chemicals, *Sci. Total Environ.*, 160/161, 211–232, 1995.
13. PIC web site, http://www.pic.int/.

14. LRTAP-POPs web site, http://www.unece.org/env/lrtap/pops_h1.htm.

15. Barcelona Convention web site, http://www.unep.ch/seas/main/med/medconvii.html.

16. Helsinki Convention web site, http://europa.eu.int/scadplus/leg/en/lvb/l28089.htm.

17. IJC web site, http://www.ijc.org/en/home/main_accueil.htm.

18. OSPAR Convention web site, http://www.ospar.org/eng/html/welcome.html.

19. Stockholm Convention web site, http://www.pops.int/.

20. World Health Organization, Sustainable Development and Healthy Environment, Report WHO/SDE/ WHS/01.5 in http://www.who.int/docstore/water_sanitation_health/vector/ddt2.htm.

21. UNEP web site, http://www.chem.unep.ch/pops/pdf/redelipops/redelipops.pdf.

22. WWF, Resolving the DDT Dilemma, WWF U.S. and WWF Canada, Washington and Toronto, p. 52, 1998.

23. Eighth Report on Carcinogens, 1998 Summary, U.S. Department of Health and Human Services, DHHS, p. 252, 1998.

24. Teasley, J. I. and Cox, W. S., Determination of pesticides in water by micro colorimetric gas chromatography after liquid–liquid extraction, *J. Am. Water Works Assoc.*, 55, 1093–1098, 1963.

25. Determination of Chlorinated Pesticides in Water by Gas Chromatography with an Electron Capture Detector, Method 508, rev 3.1, National Exposure Research Laboratory, Office of Research and Development, U.S. Environmental Protection Agency, Cincinnati, Ohio 45268, EPA-600/R-95/131, August 1995.

26. Zapf, A., Heyer, R., and Stan, H. J., Rapid micro LLE for trace analysis of organic contaminants in drinking water, *J. Chromatogr. A*, 694(2), 453–461, 1995.

27. Goulden, P. and Anthony, D. H. J., Design of a Large Sample Extractor for the Determination of Organics in Water, Report No. 85-121, Environment Canada, National Water Research Institute, Burlington, Ontario, 1985.

28. Foster, G. D., Gates, P. M., Foremany's, W. T., McKenzie, S. W., and Rinella, F. A., Determination of dissolved-phase pesticides in surface water from the Yakima river basin, Washington, using the Goulden large-sample extractor and gas chromatography/mass spectrometry, *Environ. Sci. Technol.*, 27, 1911–1917, 1993.

29. Foreman, W. T. and Gates, P. M., Matrix-enhanced degradation of *p,p'*-DDT during gas chromatographic analysis: a consideration, *Environ. Sci. Technol.*, 31, 905–910, 1997.

30. Baltussen, E., Janssen, H. G., Sandra, P., and Cramers, C. A., A novel type of liquid/liquid extraction for the preconcentration of organic micropollutants from aqueous samples: application to the analysis of PAHs and OCPs in water, *J. High Resolut. Chromatogr.*, 20, 395–399, 1997.

31. Baltussen, E., David, F., Sandra, P., Janssen, H.-G., and Cramers, C. A., Retention model for sorptive extraction-thermal desorption of aqueous samples: application to the automated analysis of pesticides and polyaromatic hydrocarbons in water samples, *J. Chromatogr. A*, 805, 237–247, 1998.

32. Leoni, V., Puccetti, G., and Grella, A., Preliminary results on the use of Tenax for the extraction of pesticides and polynuclear aromatic hydrocarbons from surface and drinking waters for analytical purposes, *J. Chromatogr.*, 106(1), 119–124, 1975.

33. Moltò, J. C., Albeda, C., Font, G., and Mañes, J., SPE of organochlorine pesticides from water samples, *J. Environ. Anal. Chem.*, 41, 21–26, 1990.

34. Sun, C., Dong, Y., Xu, S., Yao, S., Dai, J., Han, S., and Wang, L., Trace analysis of dissolved polychlorinated organic compounds in the water of the Yangtse river (Nanjing, China), *Environ. Pollut.*, 117, 9–14, 2002.

35. Fernandez, M. J., Garcia, C., Garcia-Villanova, R. J., and Gomez, J. A., Evaluation of liquid–solid extraction with a new sorbent and LLE for multiresidue pesticides. Determination in raw and finished drinking waters, *J. Agric. Food Chem.*, 44, 1790–1795, 1996.

36. Bruner, F., Crescentini, G., Mangani, F., and Petty, R., Comments on sorption capacities of graphitized carbon black in determination of chlorinated pesticides traces in water, *Anal. Chem.*, 55, 793–795, 1983.

37. Patent No. 5,882,521, 1996; Patent No. 5,976,376, 1998; Patent No. 6,106,721, 1999; Patent No. 6, 254,780, 2001; Patent No. 6,322,695, 2001.

38. Lacorte, S., Guiffard, I., Fraisse, D., and Barceló, D., Broad spectrum analysis of 109 priority compounds listed in the 76/464/CEE council directive using solid-phase extraction and GC/EI/MS, *Anal. Chem.*, 72, 1430–1440, 2000.

39. Wong, J. W., Webster, M. G., Halverson, C. A., Hengel, M. J., Ngim, K. K., and Ebeler, S. E., Multiresidue pesticide analysis in wines by solid-phase extraction and capillary gas chromatography-mass spectrometric detection with selective ion monitoring, *J. Agric. Food Chem.*, 5, 1148–1161, 2003.

40. Junk, G. A. and Richard, J. J., Organics in water: SPE on a small scale, *Anal. Chem.*, 60, 451–454, 1988.

41. Noroozian, E., Maris, F. A., Nielen, M. W. F., Frei, R. W., deJong, G. J., and Brinkman, U. A. Th., Liquid chromatographic trace enrichment with online capillary gas chromatography for the determination of organic pollutants in aqueous samples, *J. High Resolut. Chromatogr. Chromatogr. Commun.*, 10, 17–24, 1987.

42. Noy, T. H. M., Weiss, E., Herps, T., van Cruchten, H., and Rijks, J., Online combination of liquid chromatography and capillary gas chromatography. Preconcentration and determination of organic compounds in aqueous samples, *J. High Resolut. Chromatogr. Chromatogr. Commun.*, 11(2), 181–1866, 1988.

43. Hagen, D. F., Markell, C. G., Schmitt, G. A., and Blevins, D. D., Membrane approach to solid phase extraction, *Anal. Chim. Acta*, 236, 157–164, 1990.

44. Tomkins, B. A., Merriwaether, R., and Jenkins, R. A., Determination of eight organochlorine pesticides at low ng/l concentrations in groundwater using filter disk extraction and gas chromatography, *J. AOAC Int.*, 75, 1091–1099, 1992.

45. Veningerova, M., Prachar, V., Kovacicova, J., and Uhnak, J., Analytical methods for the determination of organochlorine compounds. Application to environmental samples in the Slovak Republic, *J. Chromatogr. A*, 774, 333–347, 1997.

46. Fernandez, M., Alonso, C., Gonzales, M. J., and Hernandez, L. M., Occurrence of organochlorine insecticides, PCBs and PCB congeners in waters and sediments of the Ebro River (Spain), *Chemosphere*, 38, 33–43, 1999.

47. Fernandez, M., Cuesta, S., Jimenez, O., Garcia, M. A., Hernandez, L. M., Marina, M. L., and Gonzales, M. J., Organochlorine and heavy metal residues in the water/sediment system of the southeast Regional Park in Madrid, Spain, *Chemosphere*, 41, 801–812, 2000.

48. Ridal, J. J., Fox, M. E., Sullivan, C. A., Maguire, R. J., Mazumder, A., and Lean, D. R. S., Evaluation of automated extraction of organochlorine contaminants from freshwater, *Anal. Chem.*, 69, 711–717, 1997.

49. Determination of Organic Compounds in Drinking Water by Liquid Solid Extraction and Capillary Column Gas Chromatography/Mass Spectrometry, Method 525.2, rev 2.0, National Exposure Research Laboratory, Office of Research and Development, U.S. Environmental Protection Agency, Cincinnati, Ohio 45268, EPA-600/R-95/131, August 1995.

50. Johnson, W. E., Fendinger, N. J., and Plimmer, J. R., SPE of pesticides from water: possible interferences from dissolved organic material, *Anal. Chem.*, 63, 1510–1513, 1991.

51. Chee, K. K., Wong, M. K., and Lee, H. K., Determination of organochlorine pesticides in water by membranous solid-phase extraction, and in sediment by microwave-assisted solvent extraction with gas chromatography and electron capture detector and mass spectrometric detection, *J. Chromatogr. A*, 736, 211–218, 1996.

52. Arthur, C. L. and Pawliszyn, J., Solid phase microextraction with thermal desorption using fused silica optical fibers, *Anal. Chem.*, 62, 2145–2148, 1990.

53. Arthur, C. L., Potter, D., Buchholz, K., Motlagh, S., and Pawliszyn, J., Solid phase microextraction for the direct analysis of water: theory and practice, *LC-GC*, 10, 656–661, 1992.

54. Zhang, Z., Yang, M., and Pawliszyn, J., Solid phase microextraction: a new solvent-free alternative for sample preparation, *Anal. Chem.*, 66, 844A–853A, 1994.

55. Magdic, S. and Pawliszyn, J., Analysis of organochlorine pesticides by solid phase microextraction, *J. Chromatogr. A*, 723, 111–122, 1996.

56. Eisert, R. and Levsen, K., Development of a prototype system for quasi-continuous analysis of organic contaminants in surface or sewage water based on inline coupling of solid-phase microextraction to gas chromatography, *J. Chromatogr. A*, 737, 59–65, 1996.

57. Jackson, G. P. and Andrews, A. R. J., A new fast screening method for organochlorine pesticides in water by using solid-phase microextraction with fast gas chromatographic and a pulsed-discharge electron capture detector, *The Analyst*, 123, 1085–1090, 1998.

58. Boussahel, R., Bouland, S., Moussaoui, K. M., Bauduc, M., and Montiel, A., Determination of chlorinated pesticides in water by SPME/GC, *Water Res.*, 36, 1909–1911, 2002.

59. Xia, X.-R. and Leidy, R. B., Preparation and characterization of porous silica-coated multifibers for solid-phase microextraction, *Anal. Chem.*, 73, 2041–2047, 2001.

60. Baltussen, E., Sandra, P., David, F., and Cramers, C., Stir bar sorptive extraction (SBSE), a novel extraction technique for aqueous samples: theory and principles, *J. Microcolumn Sep.*, 11(10), 737–747, 1999.

61. Leon, V. M., Alvarez, B., Cobollo, M. A., Munoz, S., and Valor, I., Analysis of 35 priority semivolatile compounds in water by stir bar sorptive extraction-thermal desorption-gas chromatography-mass spectrometry. I. Method optimization, *J. Chromatogr. A*, 999, 91–101, 2003.

62. Baltussen, E., Sandra, P., David, F., Janssen, H.-G., and Cramers, C., Study into the equilibrium mechanism between water and poly(dimethylsiloxane) for very apolar solutes: adsorption or sorption, *Anal. Chem.*, 71, 5213–5216, 1999.

63. Södergren, A., Solvent-filled dialysis membranes simulate uptake of pollutants by aquatic organisms, *Environ. Sci. Technol.*, 21/9, 855–859, 1987.

64. Huckins, J. N., Manuweera, G. K., Petty, J. D., Mackay, D., and Lebo, J. A., Lipid-containing semipermeable membrane devices for monitoring organic contaminants in water, *Environ. Sci. Technol.*, 27, 2489–2496, 1993.

65. Bergquist, P.-A., Strandberg, B., Ekelund, R., Rappe, C., and Granmo, A., Temporal monitoring of organochlorine compounds in seawater by semipermeable membranes following a flooding episode in western Europe, *Environ. Sci. Technol.*, 32, 3887–3892, 1998.

66. Luque de Castro, M. D. and García-Ayuso, L. E., Soxhlet extraction of solid materials: an outdated technique with a promising innovative future, *Anal. Chim. Acta*, 369, 1–10, 1998.

67. U.S. EPA Method 8100, U.S. Government Printing Office, Washington, 1986.

68. U.S. EPA Method 3540, U.S. Government Printing Office, Washington, 1995.

69. AOAC Method 963.15, Association of Official Analytical Chemist, U.S.A, 1990.

70. Singh, D. K. and Agarwal, H. C., Chemical release and nature of soil-bound DDT residues, *J. Am. Food Chem.*, 40, 1713–1716, 1992.

71. Singh, D. K. and Agarwal, H. C., Persistence of DDT and Nature of Bound Residues in Soil at Higher Altitude, *Environ. Sci. Technol.*, 29, 2301–2304, 1995.

72. Khim, J. S., Kannan, K., Villeneuve, D. L., Koh, C. H., and Giesy, J. P., Characterization and distribution of trace organic contaminants in sediment from Masan Bay, Korea. 1. Instrumental analysis, *Environ. Sci. Technol.*, 33, 4199–4205, 1999.

73. Barra, R., Cisternas, M., Urrutia, R., Pozo, K., Pacheco, P., Parra, O., and Focardi, S., First report on chlorinated pesticide deposition in a sediment core from a small lake in central Chile, *Chemosphere*, 45, 749–757, 2001.

74. Nhan, D. D., Carvalho, F. P., Am, N. M., Tuan, N. Q., Yen, N. T. H., Villeneuve, J.-P., and Cattini, C., Chlorinated pesticides and PCBs in sediments and molluscs from freshwater canals in the Hanoi region, *Environ. Pollut.*, 112, 311–320, 2001.

75. Vieira, E. D. R., Torres, J. P. M., and Malm, O., DDT environmental persistence from its use in a vector control program: a case study, *Environ. Res., Sect. A*, 86, 174–182, 2001.

76. Japenga, J., Wagenaar, W. J., Smedes, F., and Salomons, W., A new, rapid clean-up procedure for the simultaneous determination of different groups of organic micropollutants in sediments: application in two European estuarine sediment studies, *Environ. Technol. Lett.*, 8, 9–29, 1987.

77. Meijer, S. N., Shoeib, M., Jantunen, L. M. M., Jones, K. C., and Harner, T., Air–soil exchange of organochlorine pesticides in agricultural soils. 1. Field measurements using a novel in situ sampling device, *Environ. Sci. Technol.*, 37, 1292–1299, 2003.

78. Abu-Samra, A., Morris, J. S., and Koirtyohann, S. R., Wet ashing of some biological samples in a microwave oven, *Anal. Chem.*, 47, 1475–1477, 1975.

79. Ganzler, K., Salgo, A., and Valko, K., Microwave extraction. A novel sample preparation method for chromatography, *J. Chromatogr.*, 371, 299–306, 1986.

80. Onuska, F. I. and Terry, K. A., Extraction of pesticides from sediments using a microwave technique, *Chromatographia*, 36, 191–194, 1993.

81. Lopez-Avila, V., Young, R., and Beckerl, W. F., Microwave-assisted extraction of organic compounds from standard reference soils and sediments, *Anal. Chem.*, 66, 1097–1106, 1994.

82. Lopez-Avila, V., Young, R., Benedicto, J., Ho, P., Kim, R., and Beckert, W. F., Extraction of organic pollutants from solid samples using microwave energy, *Anal. Chem.*, 67, 2096–2102, 1995.

83. Chee, K. K., Wong, M. K., and Lee, H. K., Microwave assisted solvent elution technique for the extraction of organic pollutants in water, *Anal. Chim. Acta*, 330, 217–227, 1996.

84. Luque de Castro, M. D. and García-Ayuso, L. E., Soxhlet extraction of solid materials: an outdated technique with a promising innovative future, *Anal. Chim. Acta*, 369, 1–10, 1998.

85. Pastor, A., Vazquez, E., Ciscar, R., and De La Guardia, M., Efficiency of the microwave assisted extraction of hydrocarbons and pesticides from sediments, *Anal. Chim. Acta*, 344, 241–249, 1997.

86. Numata, M., Yarita, T., Aoyagi, Y., and Takatsu, A., Microwave-assisted steam distillation for simple determination of polychlorinated biphenyls and organochlorine pesticides in sediments, *Anal. Chem.*, 75, 1450–1457, 2003.

87. Richter, B. E., Jones, B. A., Ezzell, J. L., Porter, N. L., Avdalovic, N., and Pohl, C., Accelerated solvent extraction: a technique for sample preparation, *Anal. Chem.*, 68, 1033–1039, 1996.

88. Popp, P., Keil, P., Möder, M., Paschke, A., and Thuss, U., Application of accelerated solvent extraction followed by gas chromatography, high-performance liquid chromatography and gas chromatography-MS for the determination of polycyclic aromatic hydrocarbons, chlorinated pesticides and polychlorinated dibenzo-*p*-dioxins and dibenzofurans in solid wastes, *J. Chromatogr. A*, 774, 203–211, 1997.

89. Fisher, J. A., Scarlett, M. J., and Stott, A. D., Accelerated solvent extraction: an evaluation for screening of soils for selected U.S. EPA semivolatile organic priority pollutants, *Environ. Sci. Technol.*, 31, 1120–1127, 1997.

90. Fitzpatrick, L. J., Dean, J. R., Comber, M. H. I., Harradine, K., and Evans, K. P., Extraction of DDT [1,1,1,-trichloro-2,2-*bis*(*p*-chlorophenyl)ethane] and its metabolites DDE [1,1-dichloro-2,2-*bis* (*p*-chlorophenyl)-ethylene] and DDD [1,1-dichloro-2,2-*bis* (*p*-chlorophenyl)-ethane]) from aged contaminated soil, *J. Chromatogr. A*, 874, 257–264, 2000.

91. Hubert, A., Wenzel, K.-D., Manz, M., Weissflog, L., Engewald, W., and Schüürmann, G., High extraction efficiency for POPs in real contaminated soil samples using accelerated solvent extraction, *Anal. Chem.*, 72, 1294–1300, 2000.

92. Ramos, L., Kristenson, E. M., and Brinkman, U. A. Th., Current use of PLE and subcritical water extraction in environmental analysis, *J. Chromatogr. A*, 975, 3–29, 2002.

93. Lohleit, M., Hillmann, R., and Baechmann, K., The use of supercritical-fluid extraction in environmental analysis, *Fresenius' J. Anal. Chem.*, 339, 470–474, 1991.

94. Liu, M. H., Kapila, S., Yanders, A., Clevenger, T. E., and Elseewi, A. A., Role of entrainers in supercritical fluid extraction of chlorinated aromatics from soils, *Chemosphere*, 23(8–10), 1085–1095, 1991.

95. Dean, J. R., Barnabas, I. J., and Owen, S. P., Influence of pesticide-soil interactions on the recovery of pesticides using supercritical fluid extraction, *Analyst*, 121(4), 465–4688, 1996.

96. Wenclawiak, B. W., Maio, G., Holst, Ch. v., and Darskus, R., Solvent trapping of some chlorinated hydrocarbons after supercritical fluid extraction from soil, *Anal. Chem.*, 66, 3581–3586, 1994.

97. Snyder, J. L., Grob, R. L., McNally, M. E., and Oostdykt, T. S., Comparison of supercritical fluid extraction with classical sonication and soxhlet extractions for selected pesticides, *Anal. Chem.*, 64, 1940–1946, 1992.

98. van der Velde, E. G., Dietvorst, M., Swart, C. P., Ramlal, M. R., and Kootstra, P. R., Optimization of supercritical fluid extraction of organochlorine pesticides from real soil samples, *J. Chromatogr. A*, 683, 167–174, 1994.

99. Ling, Y.-C. and Liao, J.-H., Matrix effect on supercritical fluid extraction of organochlorine pesticides from sulfur-containing soils, *J. Chromatogr. A*, 754, 285–294, 1996.

100. Maio, G., von Holst, C., Wenclawiak, B. W., and Darskus, R., Supercritical fluid extraction of some chlorinated benzenes and cyclohexanes from soil: optimization with fractional factorial design and simplex, *Anal. Chem.*, 69, 601–606, 1997.

101. Bøwadt, S. and Johansson, B., Analysis of PCBs in sulfur-containing sediments by offline supercritical fluid extraction and HRGC-ECD, *Anal. Chem.*, 66, 667–673, 1994.

102. Camel, V., The determination of pesticide residues and metabolites using supercritical fluid extraction, *Trends Anal. Chem.*, 16(6), 351–369, 1997.

103. Aigner, E. J., Leone, A. D., and Falconer, R. L., Concentrations and enantiomeric ratios of organochlorine pesticides in soils from the U.S. corn belt, *Environ. Sci. Technol.*, 32, 1162–1168, 1998.

104. Pyle, S. M., Marcus, A. B., and Robertson, G. L., ECD dual-column pesticide method verification by ion trap GC/MS and GC/MS/MS, *Environ. Sci. Technol.*, 32, 3213–3217, 1998.

105. Fitzpatrick, L. J. and Dean, J. R., Extraction solvent selection in environmental analysis, *Anal. Chem.*, 74, 74–79, 2002.

106. Kim, D.-G., Paeng, K.-J., Cheong, C., and Hong, J., Systematic approach to determination of pesticides in water with solid phase microextraction combined GC/Ion trap MS, *Anal. Sci.*, 17, a53–a56, 2001.

107. Barakat, A. O., Kim, M., Qian, Y., and Wade, T. L., Organochlorine pesticides and PCB residues in sediments of Alexandria Harbor, Egypt, *Mar. Pollut. Bull.*, 44, 1421–1434, 2002.

108. Sandra, P., Tienpont, B., and David, F., Multi-residue screening of pesticides in vegetables, fruits and baby food by stir bar sorptive extraction-thermal desorption-capillary gas chromatography-mass spectrometry, *J. Chromatogr. A*, 1000, 299–309, 2003.

109. Wong, J. W., Webster, M. G., Halverson, C. A., Hengel, M. J., Ngim, K. K., and Ebeler, S. E., Multiresidue pesticide analysis in wines by solid-phase extraction and capillary gas chromatography-mass spectrometric detection with selective ion monitoring, *J. Agric. Food Chem.*, 51, 1148–1161, 2003.

110. Viñas, P., Campillo, N., Lóperz-García, I., Aguinaga, N., and Hernández-Córdoba, M., Capillary gas chromatography with atomic emission detection for pesticide analysis in soil samples, *J. Agric. Food Chem.*, 51, 3704–3708, 2003.

111. Mottaleb, M. A. and Abedin, M. Z., Determination of chlorinated pesticides in soil by solid phase extraction-gas chromatography, *Anal. Sci.*, 15, 283–288, 1997.

112. Richter, P., Sepulveda, B., Oliva, R., Calderon, K., and Seguel, R., Screening and determination of pesticides in soil using continuous subcritical water extraction and gas chromatography-mass spectrometry, *J. Chromatogr. A*, 994, 169–177, 2003.

113. Durell, G. S. and Sauer, T. C., Simultaneous dual-column, dual detector gas chromatographic determination of chlorinated pesticides and polychlorinated biphenyls in environmental samples, *Anal. Chem.*, 62, 1867–1871, 1990.

114. Bernal, J. L., Del Nozal, M. J., Atienza, J., and Jimenez, J. J., Multidetermination of PCBs and pesticides by use of a dual GC column-dual detector system, *Chromatographia*, 33, 67–76, 1992.

115. Rovedatti, M. G., Castañé, P. M., Topalián, M. L., and Salibián, A., Monitoring of organochlorine and organophosphorus pesticides in the water of the Reconquista river, Buenos Aires, Argentina, *Water Res.*, 35, 3457–3461, 2001.

116. Pedersen, J. A., Yeager, M. A., and Suffet, I. H., Xenobiotic organic compounds in runoff from fields irrigated with treated wastewater, *J. Agric. Food Chem.*, 51, 1360–1372, 2003.

117. Tomkins, B. A. and Barnard, A. R., Determination of organochlorine pesticides in ground water using solid-phase microextraction followed by dual-column gas chromatography with electron-capture detection, *J. Chromatogr. A*, 964, 21–33, 2002.

118. U.S. EPA Method 608, U.S. Government Printing Office, Washington, 1997.

119. Goodwin, E. S., Goulden, R., and Reynolds, J. G., Gas Chromatography with Electron Capture Ionization Detector for Rapid Identification of Pesticide Residues in Crops, *Proceedings of the 18th International Congress on Pure and Applied Chemistry*, Montreal, August, 1961.

120. Watts, J. D. and Klein, A. K., Determination of chlorinated pesticides residues by electron capture gas chromatography, *Proceedings of the 75th Annual Meeting of the Association of Official Agricultural Chemists*, Washington, DC, October, 1961.

121. Cai, H., Wenthworth, W. E., and Stearns, S. D., Characterization of the pulsed discharge electron capture detector, *Anal. Chem.*, 68, 1233–1244, 1996.

122. Jackson, G. P. and Andrews, A. R. J., A new fast screening method for organochlorine pesticides in water by using solid-phase microextraction with fast gas chromatographic and a pulsed-discharge electron capture detector, *The Analyst*, 123, 1085–1090, 1998.

123. Senseman, S. A., Lavy, T. L., and Daniel, T. C., Monitoring groundwater for pesticides at selected mixing/loading sites in Arkansas, *Environ. Sci. Technol.*, 31, 283–288, 1997.

124. Lépine, F. L., Milot, S., and Mamer, O. A., Regioselectivity of oxygen addition induced dechlorination of PCBs and DDT metabolites in electron capture mass spectrometry, *J. Am. Soc. Mass Spectrom.*, 7, 66–72, 1996.

125. Kodakami, K., Sato, K., Hanada, Y., Shinohara, R., Koga, M., Minoru, S., and Shiraishi, H., Simultaneous determination of 266 chemicals in water at ppt levels by GC-ion trap MS, *Anal. Sci.*, 11, 375–384, 1995.

126. Bao, M. L., Pantani, F., Barbieri, K., Burrini, D., and Griffini, O., Multi-residue pesticide analysis in soil by solid-phase disk extraction and gas chromatography/ion-trap mass spectrometry, *Int. J. Environ. Anal. Chem.*, 64, 223–245, 1996.

127. Patsias, J. and Papadopoulou-Mourkidou, E., Rapid method for the analysis of a variety of chemical classes of pesticides in surface and ground waters by offline solid-phase extraction and gas chromatography-ion trap mass spectrometry, *J. Chromatogr. A*, 740, 83–98, 1996.

128. Papadopoulou-Mourkidou, E., Patsias, J., and Kotopoulou, A., Determination of pesticides in soils by gas chromatography-ion trap mass spectrometry, *J. AOAC Int.*, 80, 447–454, 1997.

129. Bergamaschi, B. A., Baston, D. S., Crepeau, K. L., and Kuivila, K. M., Determination of pesticides associated with suspended sediments in the San Joaquin River, California, U.S.A, using gas chromatography-ion trap mass spectrometry, *Toxicol. Environ. Chem.*, 69, 305–319, 1999.

130. Slobodník, J., Hogenboom, A. C., Louter, A. J. H., and Brinkman, U. A. Th., Integrated system for online gas and liquid chromatography with a single mass spectrometric detector for the automated analysis of environmental samples, *J. Chromatogr. A*, 730, 353–371, 1996.

131. Hernando, M. D., Aguera, A., Fernandez-Alba, A. R., Piedra, L., and Contreras, M., Gas chromatographic determination of pesticides in vegetable samples by sequential positive and negative chemical ionization and tandem mass spectrometric fragmentation using an ion trap analyzer, *Analyst*, 26, 46–51, 2001.

132. Louter, A. J. H., Van Doornmalen, J., Vreuls, J. J., and Brinkmann, U. A. Th., Online solid phase extraction-thermal desorption-gas chromatography with ion trap detection tandem MS for the analysis of microcontaminants in water, *J. High Res. Chromatogr.*, 19, 679–685, 1996.

133. Pyle, S. M. and Marcus, A. B., Rapid and sensitive determination of pesticides in environmental samples by accelerated solvent extraction and tandem mass spectrometry, *J. Mass Spectrom.*, 32, 897–898, 1997.

134. Fernandez-Alba, A. R., Agüera, A., Contreras, M., Peñuela, G., Ferrer, I., and Barceló, D., Comparison of various sample handling and analytical procedures for the monitoring of pesticides and metabolites in ground waters, *J. Chromatogr. A*, 823, 35–47, 1998.

135. Geppert, H. and Kern, H., Ion trap GC-MS/MS determination of chlorinated pesticides in drinking water. An additional dimension with more reliability, *GIT Labor Fachzeitschrift*, 44(8), 896–897, 2000.

136. Steen, R. J. C. A., Freriks, I. L., Cofino, W. P., and Brinkman, U. A. Th., Large volume injection in gas chromatography-ion trap tandem MS for the determination of pesticides in the marine environment at the low ng/l level, *Anal. Chim. Acta*, 353, 153–163, 1997.

137. U.S. EPA Method 525, 507, 508 and 515.1, U.S. Government Printing Office, Washington.

138. van der Hoff, G. R. and van Zoonen, P., Trace analysis of pesticides by gas chromatography, *J. Chromatogr. A*, 843, 301–322, 1999.

23 Organophosphorus Compounds in Water, Soils, Waste, and Air

Roger Jeannot and Thierry Dagnac

CONTENTS

I. INTRODUCTION

The large-scale use of several hundreds of active substances in pesticides has led to their appearance in the environment: in water, soil, air, and waste (by which we mean tank residues, packaging, unused chemicals, etc.). Their dissemination in all of these situations has been the object of numerous studies. During spreading, pesticides are emitted in the form of droplets that can be adsorbed on aerosols or carried away by the wind to deposition zones. After spreading, they continue to be emitted from soil or plants, mainly by volatilization. The replacement of persistent biocumulative organochlorinated pesticides, now outlawed, by other less toxic families of products with shorter lifetimes, including notably organo-phosphorus pesticides, is the current trend.

The regulatory thresholds set by the European Economic Community (EEC), the World Health Organization (WHO) and the Environmental Protection Agency (EPA) for drinking water and water to be treated for drinking requires sensitive, reproducible, and robust analytical methods.

Organophosphorus pesticides (OPPs) have been used extensively for agricultural purposes for more than 40 years. There are some 200 different compounds on the market, accounting for 45% of the registered pesticides in the U.S.A. alone. The use of this class of pesticides is favored because of their ability to degrade more easily in the environment. OPPs have been found in groundwater, surface water, and drinking water under various conditions and there is now an increasing environmental concern with regard to these compounds.

In the European Union, because of the general view that there should be no pesticides in drinking water, a precautionary principle is applied, and standards are set as low as is reasonably achievable. EEC directive 98/83/CE for drinking water therefore set limit values at 0.1 μg/l for each individual pesticide and 0.5 μg/l for total pesticides. For surface water used to produce drinking water, these values are 2 μg/l for each individual substance and 5 μg/l for total pesticides.

In the U.S., the U.S. EPA Office of Ground Water and Drinking Water (OGWDW) has also established drinking water regulation and health advisory levels for individual pesticides. For organophosphorus compounds, the health advisory level is 3 μg/l for diazinon, 2 μg/l for parathion-methyl, 1 μg/l for disulfoton, and 2 μ/or fenamiphos.

In Australia, the limit for drinking water is 2 μg/l for azinphos-methyl, 1 μg/l for disulfoton, 1 μg/l for ethoprophos, 0.3 μg/l for parathion-ethyl, 0.5 μg/l for terbufos, and 2 μg/l for tetrachlorvinphos.

In Canada, the maximum acceptable concentrations for drinking water are 90 μg/l for chlorpyriphos, 190 μg/l for malathion, and 50 μg/l for parathion.

In addition, EPA has set ambient water quality criteria for the protection of aquatic organisms: 0.041 μg/l for chlopyriphos, 0.1 μg/l for demeton, 0.1 μg/l for malathion, and 0.013 μg/l for parathion.

OPPs have been detected more often and at higher concentrations in surface water than in groundwater. In the U.S., the United States Geological Survey (USGS) reports that 66 million pounds of OPPs are used each year and that in surface-water samples, the compounds most often detected were diazinon, chlorpyriphos, and malathion. The maximum concentrations for chlorpyriphos were 0.3 μg/l.[1] In France, Institut Français de l'Environnement (IFEN) reports that chlorfenvinphos and chlorpyriphos ethyl, both of which are on the European priority list, have rarely been detected in surface and ground-water.[2]

Of the EEC's 33 priority substances on the list drawn up by the European Parliament and the Council on November 20, 2001, which modified Directive 2000/60/CE, two are organophosphate pesticides: chlorfenvinphos and chlorpyriphos.

On the former black list of Directive 74/464/CE, there were 132 substances, including the following OPPs: azinphos ethyl, azinphos methyl, coumaphos, demeton, dichlorvos, disulfoton, fenitrothion, fenthion, malathion, methamidophos, mevinphos, omethoate, parathion, dimethoate, oxydemeton methyl, phoxim, trichlorfon, and triazophos.

The 22 following OPPs and metabolites are on the U.S. National Pesticide Survey list: diazinon, dichlorvos, dicrotophos, dimethoate, diphenamiphos sulfone, disulfoton, disulfoton sulfone, disulfoton sulfoxide, fenamiphos, fenamiphos sulfone, fenamiphos sulfoxide, fenitrothion, methyl paraoxon, mevinphos, monocrotophos, omethoate, parathion ethyl, phosphamidon, stirofos, terbufos, tetrachlorvinphos, and merphos.

II. PHYSICOCHEMICAL PROPERTIES OF ORGANOPHOSPHORUS PESTICIDES

A. CLASSIFICATION

Organophosphate pesticides are synthetic and are usually esters, amides, or thiol derivatives of phosphoric, phosphonic, phosphorothioic, or phosphonothioic acids. Over 100 organo phosphorus compounds, representing a variety of chemical, physical, and biological properties, are currently on the market (Table 23.1).

OPPs have the following general structure:

$$R_3 - O \diagdown \quad \diagup R1 \atop R_3 - O \diagup P \diagdown R_2 R$$

Chemical structures of selected OPPs of the main classes of organophosphorus compounds are showed in Figure 23.1.

The main classes of OPPs are[3]:

- *Phosphates* [*R1, R2*] $= O$: chlorfenvinphos, dichlorvos, heptenophos
- *Thiophosphates* (*phosphorothioates*) [*R2* $= N,S$, or O]: chlorpyriphos-methyl, diazinon, fenitrothion, fenthion, omethoate, oxydemeton-methyl, parathion, vamidothion
- *Dithiophosphates* (*phosphorodithioates*) [*R1, R2* $= S$]: azinphos-methyl, ethion, dimethoate, disulfoton, formothion, malathion, phorate, phosalone, methidathion, terbufos
- *Phosphonates* [*P−C bond*]: trichlorfon
- *Phosphoramides* [*containing NH$_2$ as R group and O as R1, R2*]: fenamiphos, methamidophos, isofenphos

B. TOXICITY

These products, in spite of their various chemical structures, all come from phosphorus chemistry. The first OPP, parathion, appeared on the market in 1944. They are phosphates, thiophosphates, dithiophosphates, phosphonates, etc., halogenated or not, alkylated, heterocyclic, etc. They affect the nervous system of pest insects by acting on cholinergic synapses to inhibit cholinesterase.

TABLE 23.1
Physicochemical Properties of OPPs

Pesticide	Solubility at 20 to 25°C (mg/l)	Vapor Pressure (Pa) at 20 to 25°C	Log K_{ow}	$t^{1/2}$ (days)	K_{oc} (cm^3/g)	GUS
Azinphos ethyl	44	3.2×10^{-4}	3.4	52	1465	1.43
Azinphos-methyl	28	1.8×10^{-4}	2.7	52	1465	1.43
Bromophos-ethyl	2	6.1×10^{-3}	5.7	—	—	—
Chlorfenvinphos	145	1.0×10^{-3}	3.8	—	—	—
Chlormephos	60	7.6	ND	—	—	—
Chlorpyriphos	1.4	2.7×10^{-3}	4.96	94	4981	2.57
Chlorpyriphos-methyl	4	5.6×10^{-3}	4.3	—	—	—
Coumaphos	1.5	1.3×10^{-5}	4.13	—	—	—
Cyanophos	46	0.1	2.65	—	—	—
Demeton-S-methyl	22,000	0.04	1.3	—	—	—
Diazinon	60	1.2×10^{-2}	3.3	23	272	2.13
Dichlorvos	18,000	2.1	1.9	—	—	—
Dimethoate	23	1.1×10^{-3}	0.7	7	20	—
Disulfoton	12	7.2×10^{-3}	3.95	30	600	—
Ethion	2	2.0×10^{-4}	5.07	150	10,000	—
Ethoprophos	700	0.046	3.6	31	101	2.99
Fenamiphos	700	0.12×10^{-3}	3.3	16	267	1.89
Fenthion	4.2	7.4×10^{-4}	4.84	34	1500	—
Fonofos	13	2.8×10^{-2}	3.94	40	870	—
Formothion	2600	1.13×10^{-4}	ND	—	—	—
Malathion	145	5.3×10^{-3}	2.75	1	1800	—
Methamidophos	>200 g/l	2.3×10^{-3}	0.8	2.6	1.7	1.56
Methidathion	200	2.5×10^{-4}	2.2	4.5	163	1.17
Mevinphos	Miscible	1.7×10^{-2}	0.13	3	44	—
Monocrotophos	Miscible	2.9×10^{-4}	−0.22	30	1	—
Omethoate	Soluble	3.3×10^{-3}	−1.1	—	—	—
Oxydimeton-methyl	Miscible	3.8×10^{-3}	−0.7	3.4	75	1.13
Parathion	11	8.9×10^{-4}	3.83	14	5000	—
Parathion-methyl	55	0.2×10^{-3}	3.0	18.5	236	2.06
Phorate	50	8.5×10^{-2}	3.9	60	1000	—
Phosalone	17	$<6.7 \times 10^{-5}$	3.3	21	18,000	—
Phosmet	25	6.5×10^{-5}	2.95	—	—	—
Phosphamidon	Miscible	2.2×10^{-3}	0.79	—	—	—
Phoxim	1.5	2.1×10^{-3}	3.4	—	—	—
Pirimiphos-ethyl	2.3	0.68×10^{-3}	5.0	—	—	—
Pirimiphos-methyl	9.9	2.0×10^{-3}	4.2	10	1000	—
Pyrazaphos	4.2	0.22×10^{-3}	3.8	—	—	—
Temephos	0.03	ND	4.9	—	—	—
Terbuphos	4.5	3.46×10^{-2}	4.5	5	500	—
Tetrachlorvinphos	11	5.6×10^{-6}	ND	—	—	—
Thiometon	200	2.3×10^{-2}	3.5	—	—	—
Triazophos	30	0.39×10^{-3}	3.3	—	—	—
Trichlorfon	120 g/l	2.1×10^{-4}	0.43	29	29	3.71
Vamidothion	Very soluble (4 kg/l)	Negligible	1.08	—	—	—

From Barcelo, D. and Hennion, M. C., *Determination of Pesticides and Their Degradation Products in Water, Techniques and Instrumentation in Analytical Chemistry*, Vol. 19, Elsevier, Amsterdam, 1997.

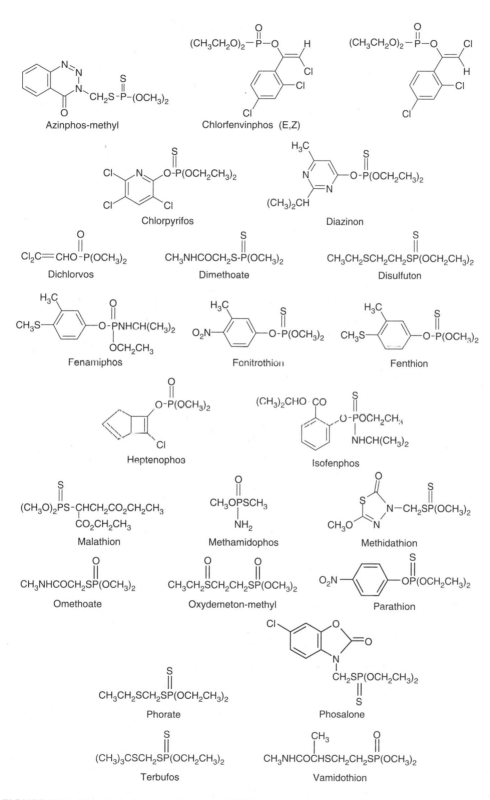

FIGURE 23.1 Chemical structure of selected OPPs of the main classes of organophosphorus compounds.

Acute poisoning is mainly a result of binding to cholinesterase and build-up of acetylcholine. Delayed-onset peripheral neuropathy has also been associated with exposure to these compounds. They have a certain liposolubility.[4]

The so called "external insecticides" coat the surface of the plant and have more or less penetration power (e.g., azinphos-methyl and -ethyl, dichlorvos, diazinon, malathion, parathion-methyl, and -ethyl). Others, called "systemic insecticides," penetrate the plant tissue, are transported by the sap, and diffuse within the plant (e.g., dimethoate, mevinphos, omethoate, vamidothion). Mites have developed a resistance to OPPs, most of which are no longer marketed as acaricides. OPPs are now found in products developed to treat plants, soil, buildings, and foods.

C. PHYSICOCHEMICAL PARAMETERS: DEFINITIONS AND DATA[3-6]

Definitions

K_{ow}	Octanol–water partition coefficient. This parameter is usually reported as log K_{ow} or log P_{ow}. It is defined as the ratio of the equilibrium concentrations of a compound in two phases — water and n-octanol.
S_w	Water solubility is the maximum amount of a substance, which can be dissolved in water at equilibrium at a given temperature and pressure.
K_h	Henry's law constant. The concentration ratio of a substance in water and in the air directly above (i.e., at equilibrium).
P_v	Vapor pressure: this is the partial pressure of a chemical in the gas phase, in equilibrium with the pure chemical. The pressure (usually expressed in millimeters of mercury) is a characteristic at any given temperature of a vapor in equilibrium with its liquid or solid form.
K_{oc}	Soil sorption (adsorption/mobility): this parameter provides information on the potential mobility of a chemical in soil.
$T_{0.5}$	Field half-life: the time required for the chemical to degrade half of its initial concentration.
Groundwater Ubiquity Score (GUS)	Mobility index that relates pesticide persistence (half-life) and sorption (K_{oc}) in soil. It is used to rate pesticides for their potential to move towards groundwater.

GUS = log $T_{0.5}$(4-log K_{oc})

Most OPPs are slightly water-soluble, have a high K_{ow}, and a low vapor pressure. Some are very miscible with water (trichlorfon, vamidothion, mevinphos, monocrotophos, omethoate, oxydemeton-methyl, phosphamidon). Most, with the exception of dichlorvos, are of comparatively low volatility, and are all degraded by hydrolysis, yielding water-soluble products. Parathion, for example, is freely soluble in alcohol, esters, ethers, ketones, and aromatic hydrocarbons, but is almost insoluble in water (20 ppm), petroleum ether, kerosene, and spray oils.

Some are transient leachers (chlorpyriphos-methyl, diazinon, parathion-methyl). Others are potential leachers (trichlorfon).

A pesticide can potentially reach groundwater if the following conditions are met[6,22]:

- Water solubility S_w > 30 ppm
- Henry's law constant K_h < 10^3 Pa
- Hydrolysis half-life > 25 weeks
- Photolysis half-life > 3 days
- Soil sorption K_{oc} < 300
- Solid–water distribution coefficient K_d < 5
- Field dissipation half-life > 2 to 3 weeks

D. DEGRADATION

Degradation includes both chemical biodegradation and transformation. It can lead to a diversified pollution that is not always easily perceptible. Indeed, there are relatively little available toxicological and ecotoxicological data on metabolites.

The principal degradation pathways for OPPs are: oxidation, demethylation, hydrolysis, and isomerization.

For some organophosphorus compounds the metabolites are:

- *Parathion*: paraoxon, aminoparathion, 4-nitrophenol
- *Diazinon*: diazoxon
- *Azinphos-methyl*: desethyl azinphos-ethyl
- *Chlorpyriphos*: 3,5,6-trichloropyridin-2-ol
- *Fenamiphos*: fenamiphos sulfone, fenamiphos sulfoxide
- *Malathion*: malaoxon
- *Fenitrothion*: fenitrooxon, 3-methyl-4-nitrophenol
- In environmental and biological systems, organophosphorus compounds are readily metabolized into alkylphosphates (O,O-diethyldithiophosphate, diethyl phosphate, O,O-dimethyl phosphorothioate, O,O-diethyl phosphorothioate).

III. SAMPLE HANDLING TECHNIQUES

A. WATER SAMPLES

The procedure for determining OPPs in water involves several different steps, each of which has a determining incidence on the signification and interpretability of results.

These steps are:

- Sampling
- Water sample storage and transport
- Substance extraction
- Concentration of extracts and transfer of solvent prior to analysis
- Purification of concentrated extracts, if necessary
- Analysis of extracts by separative methods coupled with identification or detection techniques
- Identification and quantification of detected substances

Pesticides and their metabolites must be determined at very low concentrations, the threshold value set by European Directive 98/83 for each individual substance being 0.1 μg/l. To be detected with the analytical techniques commonly used in most laboratories, pesticides in very diluted environments must, therefore, be isolated and concentrated at least 1000-fold beforehand.

1. Sampling

Because of the multiplicity of phenomena that can affect the result as soon as the sample is bottled — adsorption, hydrolysis, photolysis, volatilization, biodegradation, as well as the sampling equipment, itself — particular attention must be paid to the nature and cleanliness of the containers used, to pretreatment operations, and to transport times and conditions.

The cleanliness of bottle walls must be ensured by rigorous procedures (washing with tensio-active solutions, rinsing with dionized water, rinsing with acetone and hexane). The stopper materials that will come into contact with the extraction solvent must be chosen with care because some materials can contaminate the extracts, possibly causing significant analytical errors. Glass bottles, preferably inactinic to avoid photolysis, are generally used.

2. Storage

Various chemical reactions can occur in samples during storage, leading to the loss or transformation of substances. These include:

- Oxidation reactions
- Hydrolysis
- Photolysis
- Volatilization of substances with low boiling points
- Adsorption of substances on bottle walls
- Biodegradation if there is a large and acclimated bacterial load.

For OPPs, the extraction solvent should be added when the sample is collected. Under these conditions, the maximum conservation time under refrigeration is 14 days before and 28 days after extraction.[6]

The following OPPs were taken off the U.S. National Pesticide Survey list due to significant losses when stored at 4°C: parathion-ethyl, parathion-methyl, azinphos-methyl, fenitrothion, demeton, fenthion, diazinon, fonofos, malathion, disulfoton, terbufos, phosmet, and ethion.[6]

Preconcentrating on solid-phase extraction (SPE) cartridges to conserve substances could be an interesting alternative. One study[6] has shown that preconcentration on precolumns of silica grafted by octadecyl groups, used with a Prospekt online system coupled with liquid chromatography–diode array detection (LC–DAD) followed by storage at −20°C, enabled a good stabilization of the following pesticides for 8 months: fensulfothion, azinphos-methyl, fenamiphos, pyridafenthion, parathion-methyl, malathion, fenitrothion, azinphos-ethyl, chlorfenvinphos, fenthion, parathion-ethyl, coumaphos, fonofos, EPN, chlorpyriphos, temephos. Dichlorvos, phosmet, and mevinphos were shown to be unstable.

Another study[26] showed that nine OPPs (fonofos, diazinon, parathion-methyl, malathion, ethion, phosmet, EPN, azinphos-methyl, and azinphos-ethyl) extracted from 4 l of water on large particle-size graphitized carbon black (GCB) cartridges (Carbopack B 60–80 mesh) were stable after a storage time of 2 months at −20°C.

3. Extraction

The principal extraction methods enabling standard 1000-fold concentration of pesticides in a final reduced volume of solvent (generally 1 ml) are described below.

a. Liquid–Liquid Extraction

Liquid–liquid extraction (LLE) is based on the principle of a distribution of pesticides between an aqueous phase and a nonwater-miscible organic solvent.[7] This partition is modeled by the Nernst equation. The distribution of a substance between two nonmiscible phases (an organic and an aqueous phase) can be defined by the following equation:

$$K_p = C_o/C_w$$

where K_p is the equilibrium constant called a partition coefficient, C_o is the concentration of the compound in the organic phase, and C_w is the concentration of the compound in the aqueous phase.

The solvent is chosen from a wide range of products with different polarities and densities. Extraction efficiency will depend, in addition to K_p, on pH, ionic strength, the solvent/aqueous phase (v:v) ratio, and the number of successive extractions.[5] This efficiency is illustrated by the following equation:

$$E = 1 - [1/(1 + K_p V)]^n$$

where E is the fraction of the compound extracted from the aqueous phase, $V = V_o/V_w$ is the ratio between the organic phase and the aqueous phase, and n is the number of extractions.

The criteria for selecting a solvent or solvent mixture are:

- The distribution coefficient of the substances between the aqueous phase and the extraction solvent
- The selectivity of the solvent or its wide spectrum of efficiency

Extraction can be done either in discontinuous mode in separatory funnels with manual or mechanical shaking and small volumes of water (100 ml to 1 l), or in continuous mode using pieces of apparatus specially designed for large volumes of water (up to 20 l) (Table 23.2).[7,26] When the distribution coefficient is low and the sample very dilute, a large volume of water is needed and extraction should be done by the continuous liquid–liquid method.

A satisfactory experimental protocol for a given matrix/substance pair is not necessarily suitable for all matrixes, for all products in the same family, or for metabolites or transformation products.

i   Liquid–Liquid Extraction Steps

1. *Extraction* by mixing the solvent with the water sample/addition of a surrogate spiking solution
2. *Drying* the extract using a drying column containing 5 g of anhydrous sodium sulphate or by deep-freezing; addition of a keeper solvent and a surrogate
3. *Concentration* of the solvent extract to 1 ml with a rotary evaporator, Kuderna-Danish, or under nitrogen flow
4. *Clean-up* of the extract using Florisil or another clean-up adsorbent
5. *Concentration* of the different fractions and addition of an internal standard

The principal solvents used for the extraction of OPPs are dichloromethane, ethyl acetate, or a mixture of dichloromethane and hexane.

A Standard Committee of Analysis in the United Kingdom (SCA) method using a dichloromethane–hexane mixture enables a better recuperation of the least polar OPPs: chlorpyriphos, fenitrothion, carbophenothion, and pirimifos-methyl.

Microextraction methods (EPA method 505), which use small quantities of sample (30 ml) and a limited volume of solvent (2 ml), exist for organochlorinated pesticides and can be used to extract OPPs.[11]

A study done on water by microextraction of 2 to 5 ml sample by 1 ml methyl tertiary–buthylether (MTBE) followed by injection of 200 μl extract into a GC–FPD analytical system enabled the determination of 16 OPPs, with detection limits of between 5 and 100 ng, by eliminating the evaporation–concentration step.[11]

The concentration of the substances extracted in the final solvent by evaporation in a rotary evaporator, in a Kuderna-Danish system, or in an automatic evaporation system under nitrogen stream is a crucial step that can cause losses of the most volatile compounds. It must be done slowly and care must be taken to avoid contamination. Five hundred microliters of a slightly volatile solvent (isooctane) should be added before evaporation in order to avoid evaporation to dryness, which might entail losses by volatilization.[32] The quantification limits obtained with LLE vary,

TABLE 23.2
Analytical Methods Using Liquid–Liquid Extraction to Determine OPPs in Water

Method	Pesticides of Interest	Extraction Solvent	Analysis Conditions	Estimated Quantification Limits ($\mu g/l$)
EPA 507: Determination of 46 nitrogen- and phosphorus-containing pesticides in water	Organophosphorus: diazinon, dichlorvos, disulfoton, disulfoton sulfone, disulfoton sulfoxide, ethoprop, fenamiphos, merphos, methyl paraoxon, mevinphos, terbufos	1 l water/dichloromethane and transfer in MTBE	GC–NPD	0.1 to 4.5
EPA 8141 B: Organophosphorus by gas chromatography	Organophosphorus: azinphos-methyl, azinphos-ethyl, carbophenothion, chlorvinphos, chlorpyriphos, chlorpyriphos methyl, coumaphos, demeton-O, demeton-S, diazinon, dichlorofenthion, dichlorvos, dicrotophos, dimethoate, disulfoton, EPN, ethion, ethoprop, fenitrothion, fensulfothion, fenthion, fonophos, malathion, merphos, mevinphos, monocrotophos, naled, parathion ethyl, parathion, phorate, phosmet, phosphamidon, tetrach lorvinphos, terbuphos, trichlorfon, trichloronate	1 l water/dichloromethane transfer in hexane	GC–FPD or NPD	0.1 to 0.8
ISO 10695: Water quality — determination of selected nitrogen- and phosphorus-containing organic compounds	Organophosphorus: parathion ethyl, parathion, dimethoate	1 l water/dichloromethane or ethyl acetate transfer in acetone	GC–NPD	0.02 to 0.1

EN 12918: Water quality — determination of parathion, parathion-methyl and some other organophosphorus compounds in water by dichloromethane extraction and gas chromatographic analysis	Organophosphorus: parathion ethyl, parathion, azinphos-ethyl, azinphos-methyl, chlorvirphos, diazinon, dichlorvos, fenitrothion, fenthion, malathion, propetamphos	1 l water/dichloromethane transfer in 2,2,4-trimethyl-pentane	GC–NPD	0.02 to 0.1
Phorate, dimethoate, fonofos, chlorpyriphos-methyl, parathion-methyl, fenitrothion, malathion, chlorpyriphos, fenthion, chlorfenvinphos, methidathion, fenamiphos, ethion, phosalone, azinphos-methyl[a]	Phorate, dimethoate, fonofos, chlorpyriphos-methyl, parathion-methyl, fenitrothion, malathion, chlorpyriphos, fenthion, chlorfenvinphos, methidathion, fenamiphos, ethion, phosalone, azinphos-methyl	2 to 5 ml water sample/1 ml MTBE	GC–FPD large volume injection (LVI): 200 μl	0.005 to 0.1

[a] From Lopez, F. J., Beltran, J., Forcada, M., and Hernandez, F., *J. Chromatogr. A*, 823, 25–33, 1998.

TABLE 23.3
Recoveries (%) of OPPs from Water Samples Using Liquid–Liquid Extraction (EPA method 8141)

Compounds	Solvent	Reference	Recovery (%)	CV
Azinphos-methyl	Dichloromethane	EPA 8141	109.7	7
Azinphos-ethyl	—	EPA 8141	92.8	8.1
Chlorfenvinphos	—	EPA 8141	90.1	6
Chlorpyriphos	—	EPA 8141	77.5	4.2
Chlorpyriphos	Dichloromethane	6	86	3
Chlorpyriphos methyl	—	EPA 8141	59.4	7.5
Coumaphos	—	EPA 8141	100.8	13.5
Demeton	—	EPA 8141	73.8	5.1
Diazinon	—	EPA 8141	70.0	5.0
Dichlorvos	—	EPA 8141	90.1	7.9
Dimethoate	—	EPA 8141	76.7	9.5
Disulfoton	—	EPA 8141	79.5	6.1
Ethion	—	EPA 8141	79.2	6.5
Fenamiphos	—	6	87	9
Fenamiphos sulfone	—	6	98	10
Fenamiphos sulfoxide	—	6	97	6
Fenitrothion	—	EPA 8141	85.0	5.0
Fenthion	—	EPA 8141	79.5	4.3
Fonophos	—	EPA 8141	81.6	3.6
Malathion	—	EPA 8141	78.0	8.7
Mevinphos	—	EPA 8141	96.8	6.7
Parathion ethyl	—	EPA 8141	69.6	8.1
Parathion methyl	—	EPA 8141	83.6	4.7
Phosmet	—	EPA 8141	90.3	10.7
Terbuphos	—	EPA 8141	78.0	3.7
Tetrachlorvinphos	—	6	92	4
Trichlorfon	—	EPA 8141	45.6	6.9

depending on the compound, between 0.02 and 4.5 μg/l for EPA method 507.[12] The extraction yields are between 60 and 100% with variation coefficients of 3 to 10%. Of the 27 molecules tested, only trichlorfon has a low extraction yield (45%) (Table 23.3). The addition of deuterated internal standards, right from the extraction phase, in accordance with the liquid–liquid method or SPE, should enable researchers to improve the quantitative analyses of OPPs by GC–MS by correcting the variability resulting from nonrepeatable losses occurring during extraction. A feasibility study of the use of deuterated trichlorfon, dichlorvos, demeton-O-methyl, and demeton-O as internal standards for the analysis of these pesticides in water by GC–MS after LLE and SPE showed that they improved the reproducibility of the quantitative analyses (CV with and without internal standard) by a factor of three to four.[13]

The principal disadvantages of LLE are:

- The presence of impurities in the extraction solvent
- The risk of losses or contamination during evaporation and transfer
- The large volumes of solvent required
- Low productivity
- Poor efficiency for the extraction of the most polar substances, notably metabolites

b.   Liquid–Solid or Solid-Phase Extraction

Because of the disadvantages of LLE, alternative liquid–solid extraction (LSE) methods, also called SPE, which almost totally eliminate the use of solvents, have been developed in recent years.[31] Three main types of methods are now operational:

- Liquid–solid cartridge extraction (cartridges packed with a solid phase with adsorbing properties)
- Liquid–solid disk extraction
- Solid-phase microextraction (SPME) on fiber impregnated with a coating of sorbent

i.   Cartridge Extraction

There are two SPE methodologies: offline and online. In offline methodologies, samples are percolated through a sorbent packed in a cartridge or column.

The syringe is usually made out of polypropylene, filled with 100 to 1000 mg of sorbent having a particle size of 30 to 120 μm, and equipped with frits to contain the packing.

The principal phases of an SPE analytical sequence, similar in principle to a chromatographic process where the adsorbent plays a role of the stationary phase and water the role of the mobile phase, are:

- Conditioning: activation by solvent (methanol or acetonitrile–water)
- Sample application
- Clean-up: washing with a water–organic solvent mixture
- Drying
- Desorption with solvent (2 to 5 ml)

When simultaneously extracting compounds with very different polarities such as OPPs, the difficulty resides in the three types of behaviors observed during their extraction. The most polar compound, monocrotophos, is quantitatively extracted by small volumes of eluent (low break-through volume), moderately polar paraoxon is quantitatively extracted by volumes ten times greater, and the extraction efficiency of apolar bromophos-ethyl increases with the volume of eluent.[28]

The main sorbents on the market are C18- and C8-bonded silica, copolymers (PLRP-S, specific surface area: 500 m^2/g, Envi Chrom P, specific surface area: 900 m^2/g), and graphitized carbon black (GCB). Highly cross-linked styrene divinylbenzene (Enviro-Chrom P) and chemically modified polymeric resins with a functional group (OASIS-HLB, specific surface area: 800 m^2/g), which have breakthrough volumes higher than those obtained with their unmodified analogues have been developed. Other sorbents, such as LiChrolut EN, Isolute ENV, and HYsphere-1, have a higher degree of cross-linking and high porosity materials, which increases their specific area (1060 m^2/g for Isolute, $>$1000 m^2/g for Hysphere-1 and 1200 m^2/g for Lichrolut) and allows greater π–π interactions between analytes and sorbent. N-Alkyl silica sorbents are usually suitable for substances whose log $K_{ow} > 3$ (the majority of organophosphorus compounds), whose breakthrough volumes are high. For substances with a log $K_{ow} < 3$ (e.g., dichlorvos, trichlorfon, mevinphos, oxydemeton methyl) whose breakthrough volumes are low, the polymeric phases allow higher percolation volumes.

The recovery of OPPs by SPE depends on the physicochemical properties of the analytes, the nature of the water (pH, organic matter load, ionic strength, etc.), the volume of sample percolated, the percolation rate, the type and mass of sorbent used, and the desorption method (eluent, contact time, flow rate).[21,28]

In offline techniques, the desorption solvent most often used for SPE with C18 sorbent is ethyl acetate,[34] sometimes associated with isooctane.[14] MTBE is also used.[78] In the case of polymeric phases, methanol, acetonitrile, and acetone are used.[15]

Table 23.4 shows recoveries of organophosphorus compounds using C18 octadecyl silica and different polymeric sorbents. Recoveries of 63 to 120% on C18,[14,34,87] 68 to 72% on XAD-2,[17] and 80 to 109% on Lichrolut[15] have been reported. Some metabolites have been determined with recoveries of 80 to 148%.[34,87] The chromatographic methods used with SPE were liquid chromatography–electrospray ionization–mass spectrometry (LC–ESP–MS),[15,87] liquid chromatography–ultraviolet/diode array detection (LC–UV/DAD)[15,24] gas chromatography–flame thermoionic detection (GC–NPD),[14,18] large volume injection–gas chromatography–flame photometry detection/flame thermoionic detection (LVI–GC–FPD/NPD),[78] and gas chromatography–mass spectrometry (GC–MS).[17,34] Using SPE detection limits range from 3 to 200 ng/l and in some cases up to 1000 ng/l depending of the sample volumes, the nature of the sorbent used for the extraction, and the analytical methods (Table 23.4).

EPA method 525 recommends using SPE to extract, from groundwater and drinking water, several families of organic compounds including certain OPPs (chlorpyriphos, dichlorvos, diazinon, disulfoton, disulfoton sulfone, disulfoton sulfoxide, fenamiphos, mevinphos, methyl paraoxon, stirofos, terbufos) on cartridges packed with octadecyl silica or disks coated with octadecyl silica followed by desorption of the compounds by ethyl acetate and dichloromethane, and analysis by GC–MS. Recoveries for OPPs in the Table 23.4 range from 80 to 148%. The detection limits of the EPA 525 method using SPE on C18 cartridges range from 0.05 to 1.6 μg/l.

Likewise, EPA method 3535A describes the SPE analytical procedure using disks for eight families of analytes in groundwater, wastewater, and landfill leachate. For the particular case of OPPs, the following protocol is described:

- No pH adjustment
- Use of styrene divinylbenzene reversed-phase sulfonated disks (SDB-RPS)
- First washing step with 5 ml of acetone
- Second washing step with 5 ml of methanol
- Conditioning with 5 ml of methanol and 20 ml of water
- Filtering of the sample
- Elution with 0.6 ml of acetone and 2 × 5 ml of MTBE

Laboratory Services Branch (LSB) method E3389 of the Canadian Ministry of the Environment's Laboratory Services Branch is used for testing drinking water. The principle of the method is LC–UV after SPE followed by desorption of the OPPs with ethyl acetate. The sample extracts are evaporated to dryness and reconstituted in 1 ml acetonitrile. Limits of quantification range from 0.5 μg/l for phorate to 28 μg/l for temephos. The other pesticides analyzed with this method are: azinphos-methyl, chlorpyriphos, diazinon, dimethoate, parathion, malathion, terbufos.

LSB method E 3347 was used for the quantification of the same pesticides by the LC–MS/MS technique.

All of these solid phases have been tested with satisfactory recoveries (%) for each of the OPPs on various types of water (groundwater, surface water) at various concentrations between 0.1 and 10 μg/l.

In an online SPE configuration, extraction is done in a solid phase packed in a precolumn placed at the loop of a six-port switching valve. This method is easily automated, and preconcentration and chromatographic analysis can be done in the same run. It allows high throughput analysis, timesaving, improved precision, and accuracy. Its disadvantages are systematic errors and recovery error. It can be used in combination with GC or LC.[6]

Online SPE–LC–UV/DAD with 40 μm octadecyl silica precolumns was developed for the determination of 11 organophosphorus compounds in ground- and wastewater samples with average error varying from 4 to 65%. The main problems encountered were the determination of dichlorvos, mevinphos, and malathion due to their low UV absorbance, the early breakthroughs of dichlorvos and mevinphos, and interferences in wastewater samples. SPE–LC–APCI/MS with 10 μm

TABLE 23.4
Recoveries (%) and Quantification Limits (LQ) of Organophosphorus Pesticides Extracted from Different Types of Water Sample Using SPE Cartridges with Different Sorbents

Nature of the sorbents	C18 SPE cartridge 1000 mg Octadecylsilica from JT Baker	Styrene–divinylbenzene 200 mg Lichrolut EN from Gilson	XAD-2 6 ml from Fluka		C 18 SPE cartridge 200 mg from Varian		Sep Pak C18 Cartridge 360 mg from Waters	Lichrolut-EN 200 mg From Merck (pH = 4)		C 18 SPE Cartridge 1000 mg		Oasis HLB 200 mg from Waters
Sample volume	1000 ml	100 ml	1000 ml		5 ml		1000 ml	200 ml		1000 ml		500 ml
Desorption solvent	Ethyl acetate + isooctane	:Acetonitrile	:Acetone		MTBE		Acetone, acetone:hexane (1:1) and hexane	Methanol		Ethyl acetate		Methanol + MTBE (10:90)
Analytical method	GC–NPD	LC–UV/ DAD/MS	GC–MS (SIM)		GC–NPD		GC–NPD	LC–ESP/MS		GC–MS		GC–NPD
Concentrations	0.1 µg/l	40 µg/l	0.1 µg/l		0.1 µg/l		0.12 to 1 mg/l	0.2 µg/l		5 µg/l		250 µg/l
Types of water	Nanopure water	Groundwater	Groundwater		Surface water		Milli-Q water	Groundwater		Reagent water		Drinking water
References	14	15	17		78		18	87		34		33
	Recoveries (%)	Recoveries (%)	Recoveries (%)	LQ (µg/l)	Recoveries (%)	LQ (ng/l)	Recoveries (%)	Recoveries (%)	LQ (µg/l)	Recoveries (%)	LQ (µg/l)	Recoveries (%)
Compounds	(%)	(%)	(%)	(µg/l)	(%)	(ng/l)	(%)	(%)	(µg/l)	(%)	(µg/l)	(%)
Dichlorvos	91 ± 4		72 ± 5	0.04				67	0.06	110	0.09	86.7 ± 7.9
Mevinphos	71 ± 3							79	0.03	114	0.08	99.6 ± 7.8
Ethoprophos	90 ± 2		68 ± 9	0.03						123	0.05	110 ± 8.9
Dimethoate	63 ± 18											
Diazinon	88 ± 2						130 ± 15	89	0.03	83	0.11	
Parathion-methyl	92 ± 2	90 ± 13						65	0.01			113 ± 11
Parathion-ethyl		109 ± 9					87 ± 3					112 ± 9.8
Malathion	93 ± 1				106 ± 4	4	86 ± 4					
Chlorpyriphos	88 ± 3				107 ± 2	4	85 ± 3			110	0.12	120 ± 11
Methidathion	95 ± 3				100 ± 3	7	83 ± 4					
Phosmet	72 ± 11						91 ± 7					
Azinphos-methyl	95 ± 1						93 ± 7					113 ± 9.3
Azinphos-ethyl		80 ± 10										
Fenitrothion					113 ± 3	4	87; ± 3					
Fenamiphos	71 ± 5		71 ± 5	0.13	116 ± 4	:5		71	0.01	103	1.60	
Monocrotophos	68 ± 7		68 ± 7	0.08								

Continued

TABLE 23.4
Continued

Compounds	Recoveries (%)	Recoveries (%)	Recoveries LQ (%)	Recoveries LQ (μg/l)	Recoveries (%)	Recoveries LQ (%)	Recoveries LQ (ng/l)	Recoveries (%)	Recoveries (%)	Recoveries LQ (%)	Recoveries LQ (μg/l)	Recoveries (%)	Recoveries LQ (%)	Recoveries LQ (μg/l)	Recoveries (%)
Isophenphos		69 ± 8		0.05											
Phorate					127 ± 1		4								88.2 ± 14
Fonofos					123 ± 1		3	86 ± 2							
Chlorfenvinphos					105 ± 2		7								
Ethion					108 ± 2		4								
Phosalone					97 ± 2		10	90 ±							
Oxydemeton-methyl									96		0.01				
Triclorfon									65		0.10				
Fenthion								82 ± 3	57		0.20				112 ± 8.5
Demeton-S-methyl									70		0.02				
Fenamiphos sulfoxide									95		0.10				
Fenamiphos sulfone									87		0.10				
Fenitrooxon									90		0.20				
Disulfoton										96	1.30				115 ± 12
Disulfoton sulfone										80	0.07				
Disulfoton sulfoxide										148	0.18				
Terbufos								113 ± 10		123	0.08				
Formothion															113 ± 9
Coumaphos															110 ± 9.9
Demeton															
Phosphamidon															118 ± 7.9
Trichloronate															107 ± 9.9
Merphos															

Lichrospher Si 100 RP-18 precolumns was used to determine 11 organophosphorus compounds in groundwater[25] with recoveries ranging from 94 to120% and LOQ ranging from 5 to 37 ng/l.

Another study[19] on 16 OPPs in water using SPE on 200 mg C18 and 30 mg Oasis HLB cartridges with final desorption by 2 ml of MTBE, combined with large volume injection (LVI) (200 μl) in a GC–FPD system showed recoveries ranging from 80 to 129% for C18 cartridges (except for fenamiphos) and 67 to 110% for Oasis HLB cartridges. Detection limits ranged from 1 to 6 ng/l using 50 ml of water sample extracted in a 2 ml final volume of MTBE and final injection volumes of 200 μl (Table 23.4 *bis*). Figure 23.2 shows a chromatogram obtained in these conditions.

ii.   Solid-Phase Extraction on Disks
A variation of the extraction cartridge is a disk in which the sorbent is embedded in a web of polytetrafluoroethylen (PTFE) or glass fiber. Glass fiber disks are thicker and more rigid than PTFE membranes, enabling higher flow rates. The sorbent particles embedded in the disks are smaller than those in the cartridges (8 μm in diameter rather than 40 μm). The most frequently used disk size is 47 mm, suitable for handling 0.5 to 1 l sample volumes (Table 23.5).

The membrane is placed in a filtration apparatus, the disk is conditioned with 10 ml of methanol and 10 ml of water, and the water sample is filtered through it. The retained compounds are eluted with two 5 ml aliquots of an organic solvent (methanol, ethyl acetate, etc.).

Table 23.5 shows recoveries of organophosphorus compounds using C18 octadecyl silica and different polymeric sorbent disks. Recoveries of 53 to 164% on C18 except for dimethoate (0%)[6,14,34] and 64 to 99% on polymers[24,36] have been reported. Some metabolites have been determined with recoveries of 97 to 100%.[34] The chromatographic methods used in combination with SPE were LC–UV/DAD[24] GC–NPD,[34] and GC–MS.[33,34,36] Detection limits using SPE with disks range from 0.02 to 4.3 μg/l depending of the sample volumes used for the extraction and of the analytical methods.

Online SPE–GC/NPD with 4.2 mm diameter and 0.5 mm thick Empore extraction disks consisting of C18 or XAD enmeshed in a PTFE matrix was used to determine 15 organophosphorus

---

TABLE 23.4 *BIS*
Mean Recoveries (%) Obtained after SPE with 200 mg C18 and 30 mg Oasis HLB Cartridges of 50 ml of Groundwater Sample Spiked with 1 ng of Each Pesticide (Analysis by Large Volume Injection 200 μl Combined with GC-FPD)[19]

Compounds	C18 SPE Cartridge 200 mg from Varian Elution: MTBE	Oasis HLB 30 mg from Waters Elution: MTBE	Detection Limits (ng/l)
Phorate	89 ± 8	81 ± 9	1
Dimethoate	83 ± 10	98 ± 4	1
Fonofos	110 ± 5	96 ± 6	1
Chlorpyriphos-methyl	108 ± 5	110 ± 3	1
Parathion-methyl	98 ± 5	86 ± 3	1
Fenitrothion	108 ± 5	89 ± 4	1
Malathion	110 ± 8	94 ± 4	1
Chlorpyriphos	39 ± 8	33 ± 9	1
Fenthion	67 ± 8	75 ± 9	2
Chlorfenvinphos	88 ± 4	80 ± 4	1
Methidathion	106 ± 8	89 ± 6	1
Fenamiphos	Interferences	67 ± 9	5
Ethion	80 ± 12	52 ± 9	1
Phosmet	129 ± 8	82 ± 9	4
Phosalone	116 ± 10	91 ± 3	4
Azinphos-methyl	100 ± 10	67 ± 3	6

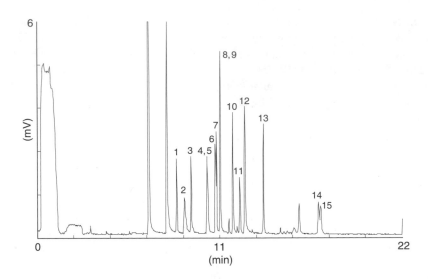

FIGURE 23.2 Chromatogram obtained after manual LLME with 1 ml of MTBE of 5 ml of groundwater fortified at 0.1 μg/l level and injection of 200 μl of sample extract in GC–FPD. (1) Phorate, (2) dimethoate, (3) fonofos, (4) chlorpyriphos-methyl, (5) parathion-methyl, (6) fenitrothion, (7) malathion, (8) chlorpyriphos, (9) fenthion, (10) chlorfenvinphos, (11) methidathion, (12) fenamiphos, (13) ethion, (14) phosalone, (15) azinphos-methyl. From Ref. [19].

compounds in aqueous samples.[16] Ethyl acetate was used for desorption. With a sample volume of 2.5 ml, detection limits of 10 to 30 ng/l were achieved in tap water and 50 to 100 ng/l in river water with recoveries better than 95%, except for methamidophos and dichlorvos.

iii. Other Sorbents
Some carbon-based sorbents are now available for SPE in water. The most well known is GCB, characterized by a low surface area of around 100 m^2/g. The high efficiency of these nonporous sorbents has been demonstrated for polar pesticides. The mechanisms involved are based on hydrophobic and electronic interactions.

Desorption of the trapped compounds requires using a mixture of methylene chloride–methanol (80:20, v/v). GCB has been used to determine multiresidues including some organophosphorus compounds with recoveries better than 80% (omethoate, monocrotophos, dicrotophos, vamidothion, dimethoate, mevinphos, phosphamidon, dichlorvos).[6] Nine OPPs were extracted from 1 l with GCB Carbopack B (120–400 mesh) with recoveries ranging from 57 to 107% using LVI–GC/MS.[27] Phosmet was not recovered in the spiked surface water.[6] With the LVI technique combined with GC–MS after SPE using GCB Carbopack B, detection limits ranged from 1 to 2 ng/l for azinphos-methyl, azinphos-ethyl, diazinon, fonofos, malathion, and parathion-methyl.

Omethoate and monocrotophos were extracted using GCB with recoveries of 83 and 98%, compared to 58 and 68% using LLE.[6]

Porous graphitized carbon (PGC) is also available in SPE cartridges. PGC has a two-dimensional graphite structure made of layers of hexagonally arranged carbon atoms. It has an average specific area of 120 m^2/g. This is due to the large layers of carbons containing delocalized π electrons and the high polarizability.[29]

A small number of OPPs with similar properties (water solubility, K_{ow}) is missing in the methods described in literature or is incidentally taken into consideration, and includes: acephate,methamidophos, omethoate, vamidothion, oxydemeton-methyl, and monocrotophos. These compounds are thermally labile or very polar. Moreover, acephate, methamidophos, monocrotophos, and omethoate are extremely water-soluble and not extractable using the more

TABLE 23.5
Recoveries (%) and Quantification Limits (μg/l) Obtained from Different Types of Water Samples Using SPE on Disks with Different Sorbents

Nature of the sorbent	C18 Silica Empore disks 47 mm diameter	Styrene-divinylbenzene Empore disks 47 mm diameter 500 mg S-DVB	C 18 Silica Disks 47 mm diameter	SDB-RPS Disks 47 mm diameter	C18 Silica Empore disks 4.2 mm diameter
Sample volume	1000 ml	500 ml	1000 ml	250 ml	2.5 ml
Desorption solvent	Ethyl acetate + isooctane	Acetonitrile	Ethyl acetate + dichloromethane	Acetone + MTBE	Ethyl acetate
Analytical method	GC–NPD	HPLC–UV/DAD	GC–MS	GC–NPD	GC–NPD on line
Concentrations	0.1 μg/l	4 μg/l	5 μg/l	10 to 250 μg/l	0.1 μg/l
Types of water	Nanopure water	River water	Reagent water	Groundwater	Tap water
References	14	24	34	10, 36	16

Compounds	C18 Silica — Recoveries (%)	Styrene-divinylbenzene — Recoveries (%)	Styrene-divinylbenzene — LOD (μg/l)	C 18 Silica Disks — Recoveries (%)	C 18 Silica Disks — LOD (μg/l)	SDB-RPS — Recoveries (%)	SDB-RPS — LOD (μg/l)	C18 Silica 4.2 mm — Recoveries (%)	C18 Silica 4.2 mm — LOD (μg/l)	C18 Silica 4.2 mm — LOD (μg/l)
Dichlorvos	53 ± 1					108	0.09	88.1 ± 11.5	2.2	0.10
Mevinphos	68 ± 12					143	0.08	57.9 ± 6.9	2.0	0.05
Ethoprophos	70 ± 2					138	0.05	95.6 ± 4.1	0.9	
Dimethoate	0							99.3 ± 1.8	2.3	
Diazinon	62 ± 10					109	0.11	91.7 ± 4.7	1.1	0.02
Parathion-methyl	77 ± 4							93.9 ± 5.8	1.2	0.02
Parathion-ethyl		71 ± 10	1.0					76.7 ± 9.6	1.8	
Malathion	83 ± 7							79.5 ± 6.9	2.1	
Chlorpyriphos	76 ± 9	67 ± 9	0.5					98.8 ± 5.7	1.0	
Methidathion	70 ± 3									
Phosmet	82 ± 14							66.1 ± 17.7	3.0	0.02
Azinphos-methyl	69 ± 14							83.0 ± 13.4	2.4	0.02
Fenitrothion		64 ± 6	0.5					91.2 ± 8.8	1.3	
Fenchlorphos		88 ± 7	1.0							
Disulfoton						96	1.30			0.025
Disulfoton sulfone						164	0.07			
Disulfoton sulfoxide						136	0.18			

Continued

TABLE 23.5
Continued

Compounds	Recoveries (%)	LOD (µg/l)	Recoveries (%)	LOD (µg/l)	Recoveries (%)	LOD (µg/l)	LOD (µg/l)
Terbufos			123	0.08	87.1 ± 10.5	0.9	0.9
Methyl paraoxon			122	0.25			
Azinphos-ethyl					88.3 ± 10.8	2.4	
Phorate					92.3 ± 7.1	1.5	
Fonofos					91.0 ± 8.0	0.9	
Chlorfenvinphos					87.8 ± 10.2	1.7	
Ethion					85.5 ± 10.6	1.6	
Trichlorfon					72.7 ± 13.5	1.0	
Fenthion	0.025				91.2 ± 5.4	1.7	
Disulfoton					93.2 ± 7.6	1.5	
Coumaphos	0.01				84.3 ± 8.7	4.3	
Demeton (S) (O)	0.04				93.6 ± 4.5	1.2	
Phosphamidon					86.2 ± 11.2	1.4	
Trichloronate					95.3 ± 4.5	0.9	
Triazophos	0.015						
Vadimothion	0.2						

common LLE or SPE procedure. LC–MS/MS has demonstrated its applicability for analyzing these compounds by direct injection of large volume (1 ml) of water samples onto a RP18 HPLC column with a polar endcapping.[89] The detection limits were in the range of 0.01 to 0.03 μg/l (Table 23.15 and Figure 23.5).

c. Solid-Phase Micro-Extraction (SPME)

The principle of a new technique, SPME, developed in 1990 by Pawlyszin et al.,[52] is based on the adsorption of organic compounds present in a small volume of water (2 to 5 ml) on a fused silica fiber coated with a micro-layer of solid phase (7 to 100 μm thick). The extraction can be described as an equilibrium process in which analytes partition between the aqueous phase and the solid phase immobilized on the fiber, governed by Nernst's law:

$$n = K_d V_f C_w V_w / K_d V_f + V_w$$

where

- V_f is the volume of the fiber-impregnated phase
- V_w is the volume of the liquid phase
- C_w is the concentration of the analyte in the aqueous phase
- K_d is the distribution coefficient of the analyte between the immobile phase and the aqueous phase
- n is the number of molecules of analyte adsorbed on the fiber-impregnated phase

Knowing that K_d is very high, $K_d V_f$ will outweigh V_w.
The equation can be simplified to:

$$n = C_d V_w$$

The number of molecules adsorbed on the coating of the stationary phase after shaking reaches an equilibrium value proportional to the concentration of the analyte in solution.
The coating on the fused silica fiber includes various stationary phases:

- Polydimethylsiloxane (PDMS) used as 7, 30, and 100 μm thick coatings
- The copolymer vinyl-divinylbenzene (DVB)–polydimethylsiloxane (PDMS): PDMS–DVB, a 65 μm thick coating
- Polyacrylate polymer (PA), a 85 μm thick coating
- Carbowax–divinylbenzene mixture (CW–DVB), a 65 μm thick coating
- Mixture of Carboxen molecular sieve and PDMS: CX–PDMS, a 75 μm thick coating

The choice depends on the nature of the analytes. The less-polar PDMS and PDMS–DVB coatings are used to extract nonpolar OPPs, whereas the more polar PA coating is more suitable for the more polar pesticides. The fibers containing divinylbenzene, CW–DVB, and PDMS–DVB have the highest affinity for the more polar pesticides.
The SPME steps are:

- Exposure of the fiber to a water sample until equilibrium is reached
- Desorption by liquid or gas chromatography

Gas chromatography (GC) desorption conditions are governed by temperature and time.
SPME enables better quantification limits than other methods, as shown by results obtained using different fibers (Table 23.6). These limits range from 1 to 400 ng/l (except for disulfoton sulfoxide, mevinphos, disulfoton sulfone whose LQ reach 1000 ng/l), depending on the compound.

TABLE 23.6
Quantification Limits (ng/l) Obtained from Water Samples Analysis Using SPME with Different Fibers

	PDMS	PA	PDMS	PDMS	PDMS–DVB
Nature of the fiber coating	PDMS	PA	PDMS	PDMS	PDMS–DVB
Film thickness (µm)	100	85	100	100	60
Time of extraction (min)	20	45	60 (15% NaCl)	45 (Head–space)	60
Desorption temperature	220	250	270	240	250
Desorption time (min)	5	4	4	10	5
Analytical method	GC–MS	GC–MS	GC–NPD	GC–FTD	GC–NPD
Types of water sample	Drinking water	Surface water	Groundwater	Natural waters	Ground and drinking waters
References	44	75	50	49	48
Dichlorvos	80	6			10
Disulfoton sulfoxide	8100				
Mevinphos	4300				
Ethoprophos	10	100			
Terbufos	10				
Diazinon	10	2		10	
Disulfoton	10	3			3
Disulfoton sulfone	1100				
Stirophos	20				
Merphos	20				20
Azinphos-methyl		3			
Chlorpyriphos		2	30		

Compound				
Dimethoate	73			400
Ethyl-parathion	5		20	4
Fenchlorvos	2			
Fenitrothion	4	30	25	10
Methyl parathion	11			1
Phorate	2	20		
Fenamiphos		50		6
Phosalone		40		
Fenthion		30	20	
Fonofos		20		1
Malathion		40		5
Chlorfenvinphos		40		5
Methidathion		500		
Ethion		40	15	
Bromophos methyl			20	
Bromophos ethyl			20	
Tetrachlorvinphos				3

The stationary phases PDMS, PA, and PDMS–DVB are efficient using the immersion extraction mode (the headspace mode being used in Ref. [49]. Addition of NaCl enabled better partitioning.[50] Optimization of the parameters shows an extraction time ranging from 20 to 60 min, a desorption temperature varying from 220 to 250°C, and a desorption time of 4 to 5 min.

SPME can be interfaced with LC. The setup consists of a custom-made desorption chamber and a six-port injection valve. The desorption chamber is placed on the injection valve where the injection loop normally is. When the injection valve is in the "load" position, the fiber enters the desorption chamber under ambient pressure. The valve is then switched to "inject" and the desorbed analytes are transferred to the column.[52]

In one study,[51] OPPs were extracted with SPME (85 μm of polyacrylate coating) by the immersion technique at 75°C for 60 min. Desorption was done in a desorption device by supercritical fluid carbon dioxide (temperature 50°C; pressure 306 atm) prior to online introduction into LC. The detection limits were 300 μg/l for diazinon, 40 μg/l for EPN, and 60 μg/l for chlorpyrifos, with recoveries ranging from 62 to 64%.

d.   Stir Bar Sorptive Extraction (SBSE)

The classical techniques for sample pretreatment — LLE and SPE — are time consuming and hard to automate. The need for new, rapid, and easily automated techniques for the analysis of semivolatile compounds in water is evident. Sorptive sample enrichment techniques, SPME, and the novel SBSE are, without a doubt, attractive alternatives.

In addition to the fact that it is a solvent-free sampling technique, SPME has the advantage of being simple, very sensitive, and requiring only a small sample size (500 ml for SPE vs. 5 ml for SPME).

SBSE is a novel technique recently developed by Sandra and coworkers,[38,42] which is based on the same mechanisms as SPME.

The main advantage of SBSE is that 25 to 300 μl of PDMS polymer is used instead of 0.5 μl in the case of SPME, which increases the sensitivity.

Ref. [42] describes a multiresidue method for the optimization and validation of 35 semivolatile compounds including OPPs using SBSE–GC–MS.

Ref. [41] describes a procedure developed for the determination of eight organophosphorus insecticides in natural waters using SBSE combined with thermal desorption–GC–atomic emission detection (AED). Optimization of the extraction and thermal desorption conditions showed that an extraction time of 50 min and a desorption time of 6 min were sufficient. Addition of salt and adjustment of the pH were not necessary. Recoveries of seven of the compounds studied between 62 and 88%. For fenamiphos, which is highly water-soluble, recovery was only 15%. The very low detection limits, between 0.8 ng/l (ethion) and 15.4 ng/l (fenamiphos), indicate that the SBSE–GC–AED procedure is suitable for sensitive detection of OPPs in natural waters.

4.   Other Methods: Membrane Extraction, Immunoextraction

New methods emerge to selectively extract organophosphorus compounds from water samples without interferences. One example of this is the immunoextraction technique, at present used only for some triazines and phenylureas.[54] These sorbents have been obtained by covalent bonding of polyclonal antibodies to a silica matrix. The antibodies have a strong affinity for the triazines and the phenylureas.

Another new technique is membrane extraction, developed in combination with gas chromatography–pulsed photometric flame detection (GC–PFPD).[53] It uses a surface-modified acetic cellulose membrane. Like SPME and SPE, it greatly simplifies the extraction process and uses significantly smaller amounts of organic solvent. Acetic cellulose membranes, 47 mm in diameter with an average pore size of 0.45 μm, were used to prepare different surface-modified membranes with plant hydrolases, glutaraldehyde, polyethylene glycol, polyvinyl alcohol, sodium alinate, and chitose acetate.

Water samples (800 ml) were filtered through the membranes and were extracted by methanol after addition of sodium sulphate. Dimethoate, parathion, and parathion-methyl were tested. The method enabled the determination of these compounds with a detection limit of 0.05 $\mu g/l$ and recoveries ranging from 66 to 94%.

Other selective extraction methods applied to some pesticides, such as molecularly imprinted polymer extraction,[55] could be extended to organophosphorus compounds.

The requirements for an optimal analytical performance are:

- Samples should be preserved and stored at 4°C and for no more than 14 days. Some unstable compounds should be extracted immediately.
- For each analyte and surrogate, the mean accuracy, expressed as a percentage of the true value, should be 70 to 130% and the RSD should be <30%.
- The internal standard should be >70%.
- With each batch of samples processed, a laboratory reagent blank should be analyzed to determine the background system contamination.
- With each batch of samples processed, one laboratory fortified blank should be analyzed.
- For LC or GC, two columns with different polarities should be used in case of processing analysis with specific detectors (FPD, NPD, UV, etc.) to confirm the identity of the detected compounds.
- A system of control charts should be developed and maintained to plot the precision and accuracy of analyte and surrogate measurements as a function of time.

B. Soil Analysis

Detecting and quantitatively determining OPPs in soil involves the following steps:

- Sample preparation involving (depending on the nature of the sample) air drying, drying at moderate temperature (<40°C), or freeze-drying, followed by sieving, grinding, homogenization, and quartering
- Extraction of pesticides by various procedures using various solvents: acetone, methanol, hexane–acetone mixture, dichloromethane–acetone mixture, or supercritical fluid
- Evaporation of the extracts obtained
- Clean-up in order to eliminate coextracted substances that might interfere with analyses
- Transfer into a solvent compatible with the analytical technique (GC or LC).

1. Sample Preparation

The extraction of soil samples aims, first of all, to dissolve all of the pollutants in a solvent prior to analysis. Samples must be pretreated in order to ensure the representativity of the final test portions. Pretreatment involves:

- Lyophilization, air drying, or drying at a temperature below 40°C
- Breaking up clumps and sieving at 2 mm (ISO) or 1 mm (EPA)
- Grinding, if needed, of the oversize to 2 or 1 mm
- Division or quartering to obtain a final representative sample for analysis

An alternative is cryogenic crushing of samples as described in ISO method 14507.

"Recovery" of organic compounds can be defined as the efficiency of extraction of analytes from a solid matrix or, more simply, as the quantity of pollutant extracted from the soil sample with a given technique. It is usually measured with a certified reference material. Note that recovery depends not only on the extraction technique used, but also on the soil's physicochemical composition. For example, we know that pollutants in a soil having high clay content will be more difficult to extract than those in a sandy soil.

The different extraction methods are:

- Soxhlet and a variant Soxtec
- Shaking
- Ultrasonic extraction
- Accelerated solvent extraction (ASE), also called pressurized fluid extraction (PFE)
- Microwave-assisted extraction (MAE)
- Supercritical fluid extraction (SFE)
- SPE associated with methanol extraction
- SPME in headspace mode

2. Extraction

a. Shaking Method (Project ISO 11264)

The principle of the classical cold shaking method is based on LSE. This sample method consists in shaking the sample in a solvent at atmospheric pressure and room temperature and is of interest because it gives results similar those of the Soxhlet method. It requires large volumes of organic solvent (200 ml of acetone) and long extraction times (several hours) and includes a distribution step with water and dichloromethane.

ISO 11246 recommends using this method for the extraction of a wide variety of herbicides.

A study carried out on soils using the shaking method with acetate ethyl obtained recoveries of 85% for diazinon and 88% for chlorpyriphos[73] (Table 23.7).

b. Soxhlet Extraction with Solvent (EPA 3540)

The Soxhlet method is the reference method for organic molecules because it is the oldest and most commonly used method in laboratories. Soxhlet extraction is described in various EPA methods and in some ISO methods. The principle is based on LSE by continuous recirculation through the soil sample of a condensed solvent.

The sample is placed in a cellulose acetate cartridge repeatedly filled with the condensing liquid of a solvent boiled in a distilling flask. As it condenses, the solvent descends and soaks the sample. When the liquid in the cartridge reaches a certain level, it is siphoned off into a recovery vessel. This ensures a progressive extraction of the pollutants and a continuous regeneration of the solvent. It requires large volumes of solvent (200 to 300 ml per sample) and at least 16 h of extraction. This method was used to extract 18 OPPs in soils and recoveries of over 60% were obtained for 12 compounds with quantification limits of 2 to 40 μg/kg (Table 23.7).

c. Automated Soxhlet Extraction: Soxtec (EPA 3541)

A faster variant of the Soxhlet method using less solvent (40 ml instead of 200 to 300 ml) was developed under the generic name Soxtec. The extraction time with this technique drops from 16 h (for Soxhlet extraction) to 3 h.

d. Microwave-Assisted Extraction: MAE (EPA 3546)

Microwaves are nonionizing electromagnetic rays. The frequency used in MAE is 2450 MHz. The principle is based on wave–matter interaction, which transforms electromagnetic energy into heat,

TABLE 23.7
The Main Extraction Methods Used for Determining Organophosphorus Pesticides in Soils and Some Data

Compounds	Extraction Methods	Operating Conditions	Results	References
Diazinon, chlorpyriphos	Shaking	5 g of soil shaken with ascorbic acid at pH 2.15. Extraction of liquid phase and solid residue by 10 ml of ethyl acetate Evaporation of ethyl acetate and transfer to acetone Analysis: GC–AED	Recoveries Diazinon: 85.4% ± 10.2% Chlorpyriphos: 88.1% ± 3.2%	73
		10 g of sediment with 30 ml acetone for 10 min Acetone combined with 200 ml 5% NaCl solution and liquid–liquid extraction by 50 ml dichloromethane	Diazinon: 97% ± 6.2%	
Diazinon, malathion, EPN, methidathion, salithion, phosalone, phosmet	Stirring, shaking and ultrasonication	Evaporation of organic phase, cleanup on silica column with acetone:hexane (10:90) and transfer to dichloromethane Analysis: GC–MS	Malathion: 93% ± 14.8% EPN: 125 ± 11.7% Methidathion: 133% ± 20.4% Salithion: 102% ± 17.7% Phosalone: 166% ± 15.5% Phosmet: 113% ± 6.4%	72
	Microwave (MAE)	Temperature: 100 to 115°C Pressure: 50 to 150 psi Extraction time at recommended temperature: 10 to 20 min Cooling: room temperature Filtration/rinsing: with the extraction solvent Extraction solvent: hexane/acetone (1:1 v/v) Sample intake: 2–20 g	250 to 2500 µg/kg	EPA 3546
	Soxhlet	Dichloromethane/acetone (1:1 v/v) Transfer in hexane		EPA 3540

Continued

TABLE 23.7
Continued

Compounds	Extraction Methods	Operating Conditions	Results	References
	PFE	Sample intake: 10 g Oven temperature: 100°C Pressure: 14 MPa (2000 psi) Oven preheating: 5 min Static time: 5 min Flush volume: 60% of extraction cell volume Nitrogen purge: 1 MPa (150 psi) for 60 sec Solvent: dichloromethane/acetone (1:1 v/v)	250–2500 μg/kg	EPA 3545
Dichlorvos, diazinon, ronnel, parathion-ethyl, methidathion, tetrachlorvinphos	SFE	Sample intake: 10 g Supercritical carbon dioxide with 3% methanol 350 atm at 50°C	Recoveries Dichlorvos: 61% ± 4% Diazinon: 84% ± 3% Ronnel: 98% ± 5% Parathion-ethyl: 94% ± 1.2% Methidathion: 106% ± 3.3% Tetrachlorvinphos: 109% ± 6.3%	74
Dichlorvos, ethoprop, methyl parathion, malathion, ethion	SPE	Shaking 10 g of soil with 5 ml water for 1 h Two sonications for 15 min with 15 ml methanol Filtration and dilution to 1 ml with water Acidification at pH < 3 SPE on 500 mg C18 Empore disks, 47 mm diameter Desorption by 2 × 5 ml ethyl acetate Analysis: GC–MS	Recoveries ranging from 42% for dichlorvos to 76% for ethion	71

Parathion-methyl, chlorpyriphos, methidathion, carbophenothion	SPME	0.5 g of soil in 5 ml methanol, dilution to 10% methanol with NaCl 10% w/v Fiber: PDMS 100 μm Immersion time: 30 min Desorption : 250°C for 5 min Analysis: GC–ECD and GC–MS	Recoveries at 33 μg/kg: 72 to 90%	70
Phorate, diazinon, disulfoton, malathion, parathion	Headspace SPME	3.5 g of soil with 3 5 ml water Temperature: 80°C Time: 60 min Fiber: 85 μm polyacrylate used in headspace mode Desorption: 250°C for 2 min Analysis: GC–FID and GC–MS.	Detection limits: 14.3 to 28.6 μg/kg	65

freeing the pollutants adsorbed on the soil. This heating method is more homogeneous than classical conduction heating, which creates a decreasing temperature gradient between the sample surface and its center. Furthermore, the fact that these extractions are done in a sealed flask enables researchers to reach high pressures and raise the temperature of the extraction solvent beyond its boiling point at atmospheric pressure.

Extraction conditions (temperature, time, and microwave power) must be controlled.

Extraction is done in hermetically sealed, heated, pressurized PTFE flasks, with a controlled and regulated solvent temperature.

The extraction conditions recommended by EPA 3546 are given in Table 23.7.

e. Pressurized Fluid Extraction: PFE (EPA 3545A)

The principle of PFE is based on liquid–solid extraction using solvents at high temperature (100°C) and under high pressure (2000 psi). Temperature modifies the properties of the solvent. An increase in temperature increases its solvation capability and its diffusion capability in the solid matrixes, yielding better recoveries, and considerably reducing extraction times and the amount of solvent needed. The high pressure also maintains the solvent in the liquid form at high temperatures. Recovery varies from 95 to 100% with the PFE method, using a dichloromethane/acetone (1:1, v/v) mixture with quantification limits of 2 to 40 μg/kg, except for monocrotophos (Table 23.8).

f. Ultrasonic Extraction (EPA 3550)

Ultrasounds are used to increase the speed of pollutant extraction from soil samples. The frequency and wattage of the ultrasounds and the application time must be defined and monitored in such a way as to ensure good repeatability. At least 300 W are required.

TABLE 23.8
Organophosphorus Recoveries (%) in Soils Using Soxhlet Extraction and PFF

Compounds	Recoveries by Soxhlet (%)	Recoveries by PFE (%)	Quantification Limits (μg/kg) with FPD Detector	References
Azinphosl-methyl	110 ± 6	100	5	EPA 8141
Chlorpyriphos	66 ± 17	98.3	5	EPA 8141
Coumaphos	89 ± 11	100	10	EPA 8141
Demeton	64 ± 6	103.7	6	EPA 8141
Diazinon	96 ± 3	97.6	10	EPA 8141
Dichlorvos	39 ± 21	100	40	EPA 8141
Dimethoate	48 ± 7	92.5	13	EPA 8141
Disulfoton	78 ± 6	121.8	3.5	EPA 8141
EPN	93 ± 8	85.8	2	EPA 8141
Fenthion	43 ± 7	95.9	5	EPA 8141
Malathion	81 ± 8	ND	5.5	EPA 8141
Mevinphos	71	100.6	25	EPA 8141
Monocrotophos	Not recovered	100	—	EPA 8141
Parathion ethyl	80 ± 8	95.4	3	EPA 8141
Parathion methyl	41 ± 3	94.9	6	EPA 8141
Phorate	77 ± 6	98.8	2	EPA 8141
Tetrachlorvinphos	81 ± 7	93.8	40	EPA 8141
Trichloronate	53	Not recovered	40	EPA 8141

However, the use of this extraction method is not recommended by EPA for OPPs due to the risk of decomposition of certain compounds such as parathion.

g. Other Methods: SFE/SPE/SPME

Other methods for extracting organophosphorus compounds from soil, waste, and sediment samples have been described in recently published articles. They include supercritical fluid (such as carbon dioxide) extraction (SFE) with cosolvents such as methanol,[66,68,69,74] SPE after extraction by methanol,[67,71] and SPME in headspace or immersion mode after extraction by methanol.[70,75] The recoveries obtained with these techniques for OPP extraction in soil samples are given in Table 23.7.

h. Clean-Up Procedures

To avoid or minimize analytical interferences often encountered in wastewater, contaminated surface water, polluted soils, and wastes, a clean-up step is sometimes necessary to produce a clean extract that can be injected directly into the chromatograph.

Several methods are suitable, depending on the analytes to be analyzed, their concentration, and the nature of the matrix. These methods use different mechanisms:

- Adsorption chromatography with polar sorbents such as alumina, silica gel, or florisil. The extract dissolved in a nonpolar solvent such as hexane is percolated through the sorbent and the different compounds are eluted, depending on their polarity, with solvents of increasing polarity.
- Gel-permeation. This clean-up procedure is based on the separation by molecular size.
- Acid–base partition. This method is used to separate basic and acid compounds.
- Chemical methods such as sulfur clean-up.

The clean-up procedure recommended by EPA 3620 for organophosphorus compounds uses Florisil. It involves reducing the sample extract volume in hexane, transferring this concentrated extract onto the Florisil column and eluting the organophosphorus compounds in diethyl ether/hexane (30:70, v/v).

C. WASTE

Waste can be a very complex matrix, which produces not only liquid and solid samples, but also multiphase samples. The critical step when dealing with waste is preparing multiphase samples for extraction. Extraction, clean-up, and analytical methods are, otherwise, similar to those used for soils, except that pollutant concentrations can be very high.

A specific method of preparation by dilution is described in EPA method 3580 A.

A clean-up step using florisil is often required.

D. AIR ANALYSIS

To determine OPPs in gaseous matrixes, they must be trapped on polymeric resins or polyurethane foam (PUF), then desorbed. The extracts are analyzed by GC–FPD, GC–NPD, or CG–MS.

1. The Sorbents

Sampling systems used to collect pesticides in the atmosphere are usually made up of a pump, a glass cartridge, a particle filter, and sorbents that trap the compounds in the gaseous phase. The sorbents can be polymeric resins or PUF. Silica gel has also been used. Of the resins, studies have been done with Tenax, Florisil, Chromosorb, XAD-2, and XAD-4. PUF and XAD-2 resins are the most suitable for collecting pesticides in the air. Tenax resins might also be used together with a thermal desorption unit coupled with a GC–MS system.

Two types of sorbent cartridges are used for sampling: cartridges containing PUF cylinders and quartz microfiber prefilters, and cartridges containing PUF/XAD-2 polymer "sandwiches" and quartz microfiber prefilters.

2. Sampling

EPA method TO 10 describes low-volume sampling and analysis of pesticides in the air. The pumps used should provide constant airflow with a flow rate of 1 to 5 l/min. The tubes or cartridges used to trap the pesticides have a 32 mm diameter quartz microfiber filter cartridge attached to a glass tube containing either a 22 mm diameter, 7.6 cm long PUF cylinder (plug), or a PUF (22 mm OD × 30 mm long)/XAD-2 resin (1.5 g)/(22 mm OD × 30 mm long) "sandwich." NIOSH standards 5600 and 5602 recommend the use of tubes equipped with 11 mm diameter quartz filters and XAD-2 resin at flow rates between 0.2 and 1 l/min.

3. Desorption of Pesticides Trapped on the Sorbent Phase

The compounds trapped on the sorbent are extracted either by Soxhlet with a hexane/diethylether (1:1, v/v) mixture,[90] or by ultrasonic extraction with an acetone/toluene (10:90, v/v) mixture for 30 min.[94] The extraction method using PFE can also be used, according to EPA method 3545A.

The extract is dried, filtered, and evaporated in order to obtain a final volume of 1 ml.

Desorption efficiencies using acetone/toluene (10:90) of the analytes from the XAD-2 sorbent and the quartz microfiber prefilters are equivalent to >90%, except for methamidophos (68%) and dichrotophos (86%).[94]

The compounds trapped on the sorbent phases have been shown to be stable after 30 days of storage under refrigeration.[94]

4. Analysis

Analyses are done by GC–FPD using a DB-1 capillary column (30 m × 0.32 mm × 0.25 μm thick), as described in NIOSH 5600 for OPPs. This method enables determination of 19 OPPs with quantification limits in the liquid extract of between 0.07 and 1.0 mg/l: methamidophos, mevinphos, ethoprop, dicrotophos, monocrotophos, phorate, fonofos, terbufos, disulfoton, diazinon, parathion-methyl, ronnel, malathion, parathion-ethyl, chlorpyriphos, fenamiphos, ethion, sulprofos, and azinphos-methyl.

The detection limits for diazinon and diazoxon are 400 ng/m^3 for 120 l of air sample.

IV. ANALYTICAL METHODS FOR THE DETERMINATION OF ORGANOPHOSPHORUS PESTICIDES

Chromatographic techniques, mainly GC, LC, and HPTLC, are the most suitable for determining OPPs in the extracts obtained with the methods described above from water, soil, waste, and air samples. The hydrophobic character of most OPPs, their volatility and their thermal stability enable easy determination by GC. For thermolabile and polar pesticides, LC is the most appropriate method. High performance thin layer chromatography (HPTLC), although marginal, might be used for complex mixtures. Immunoassays and biosensors complete the range of analytical techniques available. These are used for specific applications limited to one compound and include onsite, rapid response analyses, and continuous analysis.

A. GAS CHROMATOGRAPHY

GC is the most commonly used technique. It has, thanks to capillary columns, a very good resolution and enables, when coupled with other specific detectors such as the electron capture detector (ECD), nitrogen phosphorus detector (NPD), flame photometric detector (FPD), pulsed flame photometer (PFPD) and AED separation, identification, and quantification of OPPs containing halogenated groups, or phosphorus or sulfur atoms.

Specifications of the GC detectors used for OPP determination (from Ref. [6]) are found in Table 23.9.

The results of the GC method coupled with its specific detectors must be confirmed on a second column with a different polarity. A flow divider can be used to automatically direct equal portions of the injected volume to two columns of different polarity (Table 23.10).

This need to confirm test results for quality assurance has led to the development and dissemination of reliable identification techniques, coupling gas chromatography with mass spectrometry (GC–MS), which do not require repeated analysis. The most commonly used ionization interface is the electron impact ion source and pesticides are identified by obtaining a mass spectrum. An example of a mass spectrum of malathion obtained in GC–EI/MS is presented in Figure 23.3. Chemical ionization, in either positive (PCI) or negative (NCI) mode, can also be used. Chemical ionization is a "soft" ionization technique, which uses methane, isobutane, or ammonia as a reactant gas in a low-pressure source (0.1 to 0.2 mm Hg). The advantage of using GC–MS with NCI to analyze OPPs having a nitro or chloroaromatic group or thiophenolate anions (parathion, fenitrothion, tetrachlorvinphos, etc.) is that these groups capture electrons. Identification is even more precise when GC–MS/MS is used with a triple quadrupole or an ITD. Table 23.11 lists the quantitation ions that can be used in SIM mode and the qualifier ions that can be used to identify substances.

The stationary phases used to separate OPPs are either polar or moderately polar. Columns are 30 to 60 m long, generally 0.25 to 0.32 mm in diameter, and have a 0.25 μm thick coating.

Detection limits are around 0.1 μg/l using offline extraction from 1000 ml of water combined with GC–NPD, GC–FPD, GC–PFPD, or GC–MS. GC methods are not suitable for thermolabile and highly polar compounds such as dichlorvos, temephos, trichlorfon, oxydemeton-methyl, or mevinphos. These detection limits can be improved using LVI, which enables the injection of 20 to 200 μl, compared to 1 to 2 μl in split–splitless mode.

TABLE 23.9
Specification of the GC Detectors Used for Organophosphorus Pesticides Determination[6]

Detector	Functional Group	Sensitivity (pg/s)	Linearity
Flame photometric (FPD)	P	0.5 to 0.9	5×10^4
	S	5 to 20	5×10^2
Electron capture (ECD)	Cl, Br, compounds capturing electrons	0.1 (lindane)	10^3
Nitrogen-phosphorus (NPD)	N	0.2 to 0.4	10^4
	P	0.1 to 0.2	10^4
Atomic emission (AED)	S	1.7	10^4
	P	1.5	
	Cl	39	
	C	0.5	
	H	2.2	
	Br	79	

TABLE 23.10
Specification of the GC Detectors Used for Organophosphorus Pesticides in Environmental Samples

Columns	Detectors	Injection Systems	Chromatographic Conditions
100% Dimethylpolysiloxane	NPD	Split–splitless (1 to 2 μl)	Column temperature: 40/110 to 270/300°C (5°C/min)
Nonpolar	FPD	Large volume injection (20 to 200 μl)	Injection temperature: 250°C
DB-1; HP-1, Ultra-1, SPB-1, CP-Sil 5 CB, RS/L-150, RS/L-160, Rtx-1, ZB-1, CB-1, OV-1, PE-1, CP Sil 5 CB MS, SE-30, SP2 100; BP-1	AED	Oncolumn (1 μl) PTV (1 to 2 μl)	Detector temperature: 250°C
5% Phenyl 95% dimethylpolysiloxane (35% phenyl)-methylpolysiloxane mid-polarity	EI–MS CI–MS		
DB-35, DB-35ms, MDN-35, Rtx-35, SPB-35, AT-35, PE-35, HP-35 (50% trifluoropropyl–50% methyl) methylpolysiloxane DB 210 (14%-cyanopropyl-phenyl)-methylpolysiloxane DB-1701, Rtx-1701, SPB-7, HP-1701, CP-Sil 19CB			

The criteria for the identification of target compounds in various matrices by GC/MS are[79]:

- The retention time measured in the sample does not differ by more than $\pm 0.2\%$ from the retention time determined in the last measured external standard solution.
- The relative intensities of all of the selected diagnostic ions measured in the sample do not deviate by more than $\pm (0.1 \times I_{std} + 10)\%$ from the relative intensities determined in the external standard solution (I_{std} being the relative intensity of the diagnostic ion in the external standard solution).

B. LIQUID CHROMATOGRAPHY

LC separates and detects some OPPs using methanol–water, methanol–acetonitrile–water, or acetonitrile–water gradient mobile phases. LC is suitable for determining some thermolabile and polar compounds such as dichlorvos, temephos, trichlorfon, and oxydemeton-methyl (Table 23.12).

This technique can be used for OPP determination with an ultraviolet–diode array detector (UV/DAD) and ESI–MS, APCI–MS, ESI–MS/MS, and APCI–MS/MS. The separation mechanisms used for the separation of organophosphorus compounds are reverse-phase chromatography with alkyl-bonded silicas or apolar copolymers.

1. LC–UV/DAD

The OPPs with an aromatic ring (e.g., azinphos-methyl, azinphos-ethyl, coumaphos, diazinon, fenitrothion, fenthion, parathion, paraoxon, tetrachlorvinphos, trichloronate) can be determined with UV/DAD because they contain good chromophores. Problems are encountered with pesticides such as, dimethoate, disulfoton, malathion, or trichlorfon, which have no chromophores. The wavelength data for analyzing OPPs using LC–UV/DAD are given in Table 23.13.

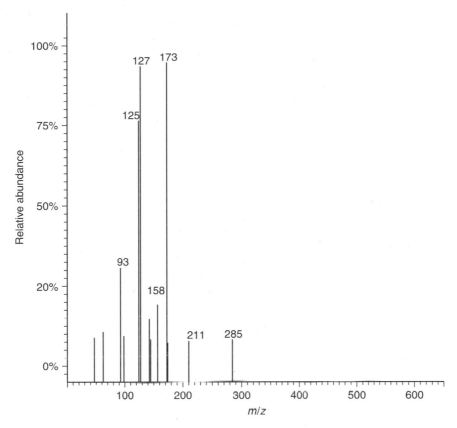

FIGURE 23.3 Example of a mass spectrum of malathion obtained in GC–EI/MS (m/z: 93 + 125 + 127 + 173).

2. LC–MS

Thermolabile and polar OPPs can now be determined using LC–MS thanks to recently developed ESI and APCI interfaces, which enable the transfer of analytes from the liquid phase to a high-vacuum gaseous phase by desolvation and ionization. It is now possible to determine the following pesticides with low quantifications limits (0.01 to 0.2 μg/l): demeton-S-methyl, diazinon, dichlorvos, dimethoate, fenamiphos, fenitrothion, fenthion, mevinphos, oxydemeton-methyl, and trichlorfon.[6,81,83,85–87] In APCI–MS configuration, heat and pneumatic nebulization are applied with a high voltage corona discharge. Figure 23.4 shows an example of a mass spectrum of oxydemethon-methyl obtained in LC–APCI–MS using the positive ionization mode. In ESI–MS configuration, ions are generated during the evaporation of ions from charged droplets.

The mass spectra resulting from the use of these ion sources are generally composed of $[M + H]^+$, $[M - H^+]^-$, $[M + Na]^+$ coming from the protonation or deprotonation of the analytes. Some analyte fragmentation can be induced with APCI–MS and ESI–MS by collision-induced dissociation (CID) on octapole, hexapole, or cone devices at the input of the mass spectrometer. The newly developed direct electron ionization interface (DEI)[82] involves the direct introduction of a nano-LC system working with a mobile phase flow rate of between 0.3 and 1.5 μl/min into a mass spectrometer equipped with an electron ionization interface. It has been used to determine and identify several OPPs in water samples. Electron ionization generates spectra that can be interpreted using commercially available documentation (Wiley or NIST).

Four out of ten selected pesticides characterized by low polarity and low thermal stability (dimethoate, paraoxon, azinphos-methyl, azinphos-ethyl, parathion-methyl, parathion-ethyl,

TABLE 23.11
Mass Spectra for Organophosphorus Pesticides Obtained in EI and NCI Modes Using GC–MS[5,6,10]

Compound	Quantitation m/z in EI–MS	Other m/z Fragments in EI–MS	NCI
Azinphos ethyl	132	105, 129, 160	185
Azinphos methyl	160	77, 132, 93, 104, 105	—
Chlorfenvinphos	267	81, 109, 269, 323	—
Chlorpyriphos ethyl	197	97, 199, 314	169, 212, 313
Chlorpyriphos methyl	286	125, 197, 201, 290	—
Coumaphos	362	226, 210, 364, 97, 109	169, 362
Demeton-o	88	89, 60, 61, 115, 171	—
Demeton-s	88	60, 81, 89, 114, 115	—
Diazinon	137	179, 199, 276, 304	169
Dichlorvos	109	145, 185, 79	125, 134, 170
Dicrotophos	127	67, 72, 109, 193, 237	—
Dimethoate	87	93,125, 143, 229	—
Disulfoton	88	89, 97, 142, 186	185
Ethion	231	97, 121, 125, 153, 384	—
EPN	157	169, 185, 141, 323	—
Fenitrothion	125	79, 109, 260, 277	169, 293
Fenthion	278	125, 109, 169, 153	—
Fonofos	109	137, 174, 246	109, 169
Malathion	173	125, 127, 93, 158	157
Mevinphos	127	109, 67, 192, 164	—
Monocrotophos	127	67, 97, 192, 109	—
Parathion ethyl	291	97, 109, 139, 155	154
Parathion methyl	109	125, 263, 79, 93	247
Phosalone	182	121, 97, 184, 154, 367	—
Phosmet	160	77, 93, 317, 76	—
Propetamphos	138	194, 222, 236	—
Pirimiphos ethyl	168	318, 152, 304, 180	—
Pirimiphos methyl	290	276, 125, 305, 233	—
Terbufos	231	57, 97, 153, 103	—
Tetrachlorvinphos	331	109, 329, 79, 333	—
Triazophos	161	257, 285, 313	—
Vamidothion	87	109, 145, 169	141

From Barcelo, D. and Hennion, M. C., *Determination of Pesticides and Their Degradation Products in Water, Techniques and Instrumentation in Analytical Chemistry*, Vol. 19, Elsevier, Amsterdam, 1997.

malathion, diazinon, phorate, phoxim) were detected at concentrations of approximately 3 ng/l after preconcentration by a factor of 2000 on an extraction cartridge filled with Carbograph 1 from Alltech.

Table 23.14 gives m/z values used for quantification in LC–MS using APCI or ESI interfaces in positive or negative ionization mode depending on the pesticides.[15,81,83]

LC–MS/MS is increasingly used to determine pesticides in water. Using this technique in multiple reaction modeling (MRM) mode results in better sensitivity, better quantification limits and improves the identification performance. Generally, in the MS/MS configuration, a triple quadrupole or an ion trap is used.

For each OPP, the $[M + H^+]^+$ or $[M - H^+]^-$ ion is chosen as the precursor ion, a collision energy is applied producing daughter fragments from the parent ion and a product ion is selected.

TABLE 23.12
LC Techniques Used to Determine Organophosphorus Pesticides in Environmental Samples

Compounds	Column	Detector	Chromatographic Conditions: Mobile Phase	References
Trichlorfon, dichlorvos, dimethoate, oxydemeton-methyl, mevinphos, demeton-S-methyl, fenamiphos, fenitrothion, fenthion, diazinon	15 cm × 2.1 mm i.d., 5 μm Zorbax coated with a cyanopropyl phase	ISP–MS	Methanol–water	87
Dimethoate, fenitrothion, fenthion, diazinon, dichlorvos, malathion, parathion-methyl, parathion, methidathion, EPN, disulfoton, ethion, chlorpyriphos-methyl, chlorpyriphos, propaphos, pyridaphenthion, edifenphos, dimethylvinphos, isoxathion, phenthoate	30 cm × 3.9 mm i.d. with 10 μm μBondapack C18	APCI–MS	Ammonium acetate 10 mM methanol	83
Methamidophos, acephate, omethoate, monocrotophos, oxydemeton-methyl, vamidothion	15 cm × 4.6 mm i.d. with 5 μm C18	APCI–MS	Water–methanol/ 0.1% acetic acid	88, 89
Mevinphos, dichlorvos, azinphos-methyl, azinphos-ethyl, parathion-methyl, parathion-ethyl, malathion, fenitrothion, chlorvinphos, fenthion, diazinon	25 cm × 4.6 mm i.d. with 5 μm C8	APCI–MS	Water–methanol/ 1% acetic acid	81
Paraoxon, methyl-parathion, ethyl-parathion, fenitrothion	25 cm × 4.6 mm i.d. with sphere-5 RP-18	UV/DAD	Water–methanol + $2.5 \times 10^{-2} M$ acetic acid–acetate buffer	84
Azinphos-methyl, dichlorvos, fenitrothion, malathion, mevinphos, chlorfenvinphos, diazinon, azinphos-ethyl, fenthion, parathion-ethyl, parathion-methyl	25 cm × 4.6 mm i.d. with SupelCosil 5 μm C18	UV/DAD	Acetonitrile–methanol–water	25
Chlorpyriphos-methyl, fenitrothion, fenchlorphos, parathion-ethyl	25 cm × 4.6 mm i.d. with 5 μm Spherisorb ODS-2	UV/DAD	Buffer pH 7 + acetonitrile	24
Parathion-methyl, fenitrothion, parathion-ethyl, paraoxonmethyl, fenitrooxon, paraoxon methyl	15 cm × 4.6 mm i.d. with 5 μm Hypersil green C18	UV/DAD and APCI–MS	Water–acetonitrile	15
Azinphos-methyl, phosmet, parathion-methyl, azinphos-ethyl, fenitrothion, parathion, diazinon	15 cm × 4.6 mm i.d. with 5 μm Nucleosil C18	UV/DAD	Acetonitrile–water–acetic acid 0.5%	20
Azinphos-methyl, dimethoate, malathion	15 cm × 4.6 mm i.d. with 5 μm Spherisorb ODS-2	UV/DAD	Acetonitrile–water	23

TABLE 23.13
Maximum Absorption Wavelength (nm) and Laboratory Measurement Wavelength (nm) Used in LC–UV/DAD to Determine Organophosphorus Pesticides in Environmental Samples

Compounds	Wavelength (nm)	Detection Limits (ng/l) (20, 24, 25)
Azinphos-ethyl	300, 230	50[a]
Azinphos-methyl	300, 229	59[a]
Chlorfenvinphos	205, 247	90[a]
Chlorpyriphos	289, 230	—
Chlorpyriphos-methyl	289	500[b]
Coumaphos	280	—
EPN	220	—
Fenthion	254	—
Fenitrothion	254	—
Fensulfothion	254	—
Fonofos	254	—
Malathion	220	—
Mevinphos-*cis/trans*	220	—
Parathion	280	48[a]
Parathion-methyl	280	48[a]
Phosmet	220	—
Diazinon	248, 288	60[a]
Dichlorvos	<200	100[a]
Dimethoate	<200	—
Fenamiphos	248	—
Pirimiphos-ethyl	240, 300	—
Temephos	<200, 250	—

[a] Online.
[b] Offline.

From Ingelse, B. A., Van Dam, R. C. J., Vreeken, R. J., Mol, H. G. J., and Steijger, O. M., *J. Chromatogr. A*, 918, 67–78, 2001.

Table 23.13 gives the precursor ion and the product ion monitored for each substance determined using LC–APCI–MS/MS.[88,89] Vamidothion, monocrotophos, oxydemeton-methyl, omethoate, acephate, methamidphos were detected on a tandem mass spectrometer operated in multiple reaction monitoring mode and detection limits range of 0.01 to 0.03 μg/l by injecting directly 1 ml of water samples on an RP 18 HPLC column with a polar endcapping (Figure 23.5, Table 23.15).[89]

C. Immunoassays

There is a growing demand for more rapid and economical methods for determining pesticide residues. Immunoassays are proving to be a suitable complement to traditional methods for monitoring a small number of compounds with a rapid response.

Immunoassay techniques are based on the antigen–antibody interaction. These techniques involve a competitive reaction between antigen molecules of the target molecules and labeled antigen molecules for a limited number of antibodies. Enzyme-linked immunosorbent assays (ELISA) in which antibodies are immobilized on a solid phase are the most popular for pesticide detection. As pesticides are small molecules, in order to synthesize antibodies, pesticide derivatives (haptens) must be synthesized and coupled to carrier proteins.[6]

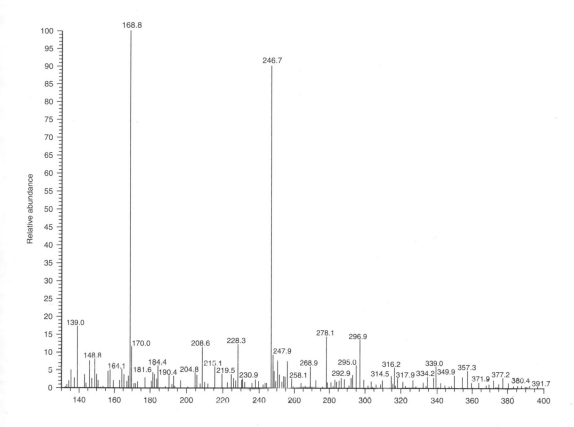

FIGURE 23.4 Example of a mass spectrum of oxydemethon-methyl obtained using LC–APCI$^+$–MS.

ELISA kits are available for the detection of the following organophosphorus compounds in water samples with quantification limits around 0.1 μg/l[6,99,100] (except for bromophos [7 μg/l]):

- Bromophos: the main cross-reactants are chlorpyriphos and fenitrothion[98]
- Chlorpyrifos: the main cross-reactants are fenchlorphos, diazinon and chlorpyriphos-methyl
- Diazinon: the main cross-reactant is diazoxon
- Parathion-methyl: the main cross-reactant is parathion-ethyl
- Tolclofos-methyl

There is also a global screening method for carbamate/OPPs using a test based on acetylcholinesterase inhibition.

D. SENSORS

Biosensors include immunosensors, enzyme biosensors, and microbial sensors.

For pesticide analysis, the potential of enzyme biosensors has been tested. In this field, biosensors based on the inhibition of acetylcholinesterases, acylcholinesterases, or butylrylcholinesterases by organophosphorus compounds are widely used. Their specific activity can be monitored by electrochemical methods such as the ion-selective electrode[6] and the ion-selective field effect transistor (ISFET).

A recent study has demonstrated that new biosensors for the detection of organophosphorus compounds should enable continuous monitoring of water quality. These biosensors, enzymatic field

TABLE 23.14

LC–MS Conditions for Organophosphorus Pesticides Determination: Ionization Modes, m/z Used for Quantification and Detection Limits[15,81–83,87]

Compounds	m/z	Interface	Ionization Mode	Detection Limits (ng/l)
Azinphos-ethyl	160 (+); 185 (−)	APCI	+ and −	30
Azinphos-methyl	160 (+); 157 (−)	APCI	+ and −	22
Chlorfenvinphos	155	APCI	+	5
Chlorpyriphos	330	APCI	−	50
Chlorpyriphos methyl	302	APCI	−	50
Demeton-s-methyl	253[a]	ESI	+	20
Diazinon	305	APCI	+	10
Diazinon	327[a]	ESI	+	10
Dichlorvos	221	APCI	+	2
Dichlorvos	243[a]	ESI	+	60
Dimethoate	230	APCI	+	2
Dimethoate	252[a]	ESI	+	30
Disulfoton	245	APCI	−	20
Ethion	355	APCI	−	10
Fenamiphos	326[a]	ESI	+	10
Fenamiphos sulfone	358[a]	ESI	+	100
Fenamiphos sulfoxide	342[a]	ESI	+	100
Fenitrothion	248 (+); 168 (−); 262 (−); 152 (−)	APCI	+ and −	2/9
Fenitrooxon	284[a]	ESI	+	200
Fenitrooxon	246; 157	APCI	−	
Fenthion	279[a]	APCI	+	20
Malathion	127 (+); 331 (+)157 (−)	APCI	+ and −	5/10
Medidathion	287	APCI	−	10
Mevinphos	193	APCI	+	10
Mevinphos	247[a]	ESI	+	30
Oxydemeton-methyl	269[a]	ESI	+	10
Paraoxon-methyl	141; 185	APCI	−	—
Parathion-methyl	234 (+); 154 (−); 248 (−); 262 (−)	APCI	+ and −	22
Paraoxon-ethyl	152; 246; 274	APCI	−	—
Parathion-ethyl	262; 169; 154	APCI	−	37
Trichlorfon	279[a]	ESI	+	100

[a] [M + Na]+

effect transistors (ENFET), are based on the inhibition of acylcholinesterases (acetylcholinesterase and butyrylchlinesterase) by organophosphorus compounds, the inhibition of enzyme phosphatases, or the direct detection of organophosphorus compounds by organophosphorus hydrolase.[96]

E. HPTLC

Automated multiple development (AMD) is a new technique for determining pesticides suitable for several organic plant protection agents and some of their main metabolites in drinking water and groundwater with a quantification limit ranging from 0.05 to 0.1 μg/l.

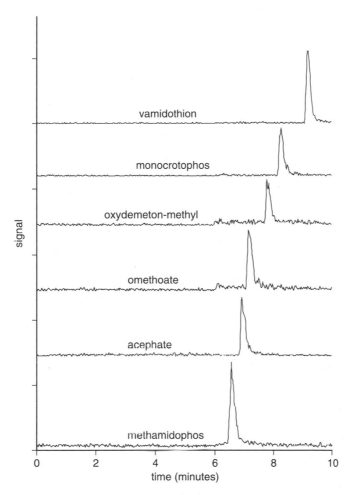

FIGURE 23.5 Example of MS–MS chromatograms of a surface water sample fortified with 0.05 μg/l of each of the OPPs. Acquisition by monitoring in MRM mode of the following pairs: vadimothion (m/z 288/146); monocrotophos (m/z 224/127); oxydemeton-methyl (m/z 247/168); omethoate (m/z 214/125); acephate (m/z 184/143); methamidophos (m/z 142/94). (From Ingelse, B. A., Van Dam, R. C. J., Vreeken, R. J., Mol, H. G. J., and Steijger, O. M., *J. Chromatogr. A*, 918, 67–78, 2001.)

It is described in ISO method 11370 entitled "Determination of selected organic plant protection agents—automated multiple development (AMD) technique".

It consists of SPE extraction on C18 sorbent followed by a desorption step with various solvents: methanol, acetonitrile, ethyl acetate, and hexane.

The solvent is concentrated to dryness and the residue is dissolved in 200 μl of methanol.

This extract is analyzed by thin-layer high-performance chromatography with UV detection at several wavelengths: 190, 220, 240, 280, and 300 nm.

Some OPPs are included in the list analyzed by this method: azinphos-methyl, coumaphos, parathion-methyl, and parathion-ethyl.

V. CONCLUSIONS

OPPs are found on various lists of priority substances and included in numerous environmental quality monitoring programs. They belong to the lists of substances to be monitored in various

TABLE 23.15
Precursor Ion m/z and Product Ion *m/z* Monitored in MRM Mode for Some Organophosphorus Pesticides Determined by LC–MS/MS[89]

Compounds	Precursor/Product Ion Pairs	Interface	Collision Energy (V)	Ionization
Acephate	184/143	APCI	11	+
Methamidophos	142/94	APCI	19	+
Monocrotophos	224/127	APCI	21	+
Omethoate	214/125	APCI	29	+
Oxydemeton-methyl	247/168	APCI	19	+
Vamidothion	288/146	APCI	17	+

From Ingelse, B. A., Van Dam, R. C. J., Vreeken, R. J., Mol, H. G. J. and Steijger, O. M., *J. Chromatogr. A*, 918, 67–78, 2001.

countries including the EEC member states, the U.S., and Canada for water quality. They must be determined at very low concentrations (20 to 100 ng/l for water samples, several μg/kg for soil, and several ng/m^3 for gaseous samples). Preconcentration and extraction methods (liquid/liquid, SPE on disks and cartridges containing silica grafted by octadecyl groups, copolymeric sorbent or GCB, SPME, SBSE, etc.) coupled with gas and liquid chromatography methods and mass spectrometry (GC–MS and LC–MS), enable reliable identification and quantification of a wide range of OPPs. Some of their degradation products at the concentrations stipulated in the various regulations for risk assessment and prevention.

NOMENCLATURE

RP	reverse phase
LLE	liquid/liquid extraction
MTBE	methyl tertiary-butyl ether
EPA	Environmental Protection Agency
AFNOR	Agence Française pour la Normalisation
SCA	Standard Committee of Analysts in the United Kingdom
ISO	International Organization for Standardization
IFEN	Institut Français de l'Environnement
USGS	U.S. Geological Survey
HAL	health advisory levels
K_d	solid–water distribution ratio
K_h	Henry's law constant
K_{oc}	soil organic carbon sorption coefficient
K_{ow}	octanol–water partition coefficient
K_p	partition coefficient
S_w	water solubility
LSE	liquid–solid extraction
EPN	*O*-ethyl *O*-(4-nitrophenyl) phenylphosphonothionate
OGWDW	Office of Ground Water and Drinking Water
OPPs	organophosphorus pesticides
LLME	liquid–liquid micro extraction
LPME	liquid phase micro extraction

SBSE	stir bar sorptive extraction
SPE	solid-phase extraction
SPME	solid-phase micro-extraction
GC–MS	gas chromatography–mass spectrometry
GC–MS/MS	gas chromatography–tandem mass spectrometry
LC–APCI–MS	liquid chromatography–atmospheric pressure chemical ionization–mass spectrometry
LC–ESP–MS	liquid chromatography–electrospray ionization–mass spectrometry
LC–MS	liquid chromatography–mass spectrometry
LC–MS/MS	liquid chromatography–tandem mass spectrometry
LC-UV/DAD	liquid chromatography–ultraviolet/diode array detection
LSB	Legal Services Branch
LVI	large volume injection
HPLC	high performance liquid chromatography
DAD	diode array detection
ITD	ion trap detector
NPD	nitrogen phosphorus detector
ECD	electron capture detector
FPD	flame photometry detector
FTD	flame thermoionic detector
PTV	programmed temperature vaporizer
EI	electronic impact
CI	chemical ionization
HPTLC	high performance thin layer chromatography
LOD	limit of detection
LOQ	limit of quantification
ASE	accelerated solvent extraction
SFE	supercritical fluid extraction
MAE	microwave assisted extraction
PFE	pressurized fluid extraction
GCB	graphitized carbon black
MIP	molecular imprinted polymer
SBSE	stir bar sorptive extraction
PS–DVB	polystyrene–divinylbenzene
PDMS	polydimethylsiloxane
PDMS–DVB	polydimethylsiloxane–divinylbenzene
CW–DVB	carbowax–divinylbenzene
CX–PDMS	carboxen–polydimethylsiloxane
PA	polyacrylate
HRMS	high resolution mass spectrometry
ESP	electrospray ionization
API	atmospheric pressure ionization
MRM	multiple reaction monitoring
SIM	single reaction monitoring
PELMO	pesticide leaching model

REFERENCES

1. USGS, *Organophosphorus Pesticide Occurrence and Distribution in Surface and Ground Water of the United States, 1992–1997*, p. 4, 2000.
2. IFEN, Pesticides in Water, *Annual report*, p. 23, 2002.
3. Tomlin, C. D. S., Ed., *The Pesticide Manual*, 12th ed., British Crop Protection Council, 2000.
4. Index phytosanitaire ACTA, 39ème édition, *Association de coordination technique agricole 149*, rue de Bercy, 75 595 Paris Cedex 12, 2004.
5. Culea, M. and Gocan, S., Organophosphates, In *Handbook of water Analysis*, Nollet, Leo M. L., Ed., Marcel Dekker, New York, pp. 571–608, 1997.
6. Barcelo, D. and Hennion, M. C., *Determination of Pesticides and Their Degradation Products in Water, Techniques and Instrumentation in Analytical Chemistry*, Vol. 19, Elsevier, Amsterdam, 1997.
7. Barcelo, D., *Techniques and Instrumentation in Analytical Chemistry, Techniques, Applications and Quality Assurance*, Vol. 13, Elsevier, Amsterdam, 1993.
8. EN 12918, *Water Quality — Determination of Parathion, Parathion-Methyl and Some Other Organophosphorous Compounds in Water by Dichloromethane Extraction and Gas Chromatographic Analysis*, October 1999.
9. ISO 10695, *Water Quality — Determination of Selected Organonitrogen and Organophosphorus Compounds — Gas Chromatographic Methods*, 2000.
10. EPA Method 8141B, *Organophosphorus Compounds by Gas Chromatography*, 1998.
11. EPA Method 505, *Determination of Organohalide Pesticides in Water*.
12. EPA 507, *Determination of Nitrogen- and Phosphorus-Containing Pesticides in Water By GC–NPD*.
13. Grasso, P., Benfenati, E., Terreni, M., Pregnolato, M., Natangelo, M., and Pagani, G., Deuterated internal standards for gas chromatographic–mass spectrometric analysis of polar organophosphorus pesticides in water samples, *J. Chromatogr. A*, 822, 91–99, 1998.
14. Baez, M. E., Rodriguez, M., Lastra, O., and Contreras, P., Solid-phase extraction of organophosphorus, triazine, and triazole-derived pesticides from water samples. A critical study, *J. High Resolut. Chromatogr.*, 20, 591–596, 1997.
15. Castillo, M., Domingues, R., Alpendurada, M. F., and Barcelo, D., Persistence of selected pesticides and their phenolic transformation products in natural waters using offline liquid solid extraction followed by liquid chromatography techniques, *Anal. Chim. Acta*, 353, 133–142, 1997.
16. Kwakman, P. J. M., Vreuls, J. J., and Brinkman, U. A. T., Determination of organophosphorus pesticides in aqueous samples by online membrane disk extraction and capillary gas chromatography, *Chromatographia*, 34, 41–47, 1992.
17. Psathaki, M., Manoussaridou, E., and Stephanou, G. E., Determination of organophosphorus and triazine pesticides in ground- and drinking water by solid-phase extraction and gas chromatography with nitrogen-phosphorus detection, *J. Chromatogr. A*, 667, 241–248, 1994.
18. de la Colina, C., Pena Heras, A., Dios Cancela, G., and Sanchez Rasero, F., Determination of organophosphorus and nitrogen-containing pesticides in water by solid-phase extraction with gas chromatography with nitrogen–phosphorus detection, *J. Chromatogr. A*, 655, 127–132, 1993.
19. Lopez, F. J., Beltran, J., Forcada, M., and Hernandez, F., Comparison of simplified methods for pesticide residue analysis Use of large-volume injection in capillary gas chromatography, *J. Chromatogr. A*, 823, 25–33, 1998.
20. Driss, M., Hennion, M.-C., and Bouguerra, M. L., Determination of carbaryl and some organophosphorus pesticides in drinking water using online liquid chromatographic preconcentration techniques, *J. Chromatogr. A*, 639, 352–358, 1993.
21. Pichon, V., Multiresidue solid-phase for trace analysis of pesticides and their metabolites in environmental water, *Analusis*, 26, M91–M98, 1998.
22. Sabik, H., Jeannot, R., and Rondeau, B., Multiresidue methods using solid-phase extraction techniques for monitoring priority pesticides, including triazines and degradation products, in ground and surface waters, *J. Chromatogr. A*, 885, 217–236, 2000.
23. Jimenez, J. J., Bernal, J. L., del Nozal, M. J., and Rivera, J. M., Determination of pesticide residues in waters from small loughs by solid-phase extraction and combined use of gas chromatography with electron-capture and nitrogen–phosphorus detection and high-performance liquid chromatography with diode array detection, *J. Chromatogr. A*, 778, 289–300, 1997.

24. Aguilar, C., Borrul, F., and Marcé, R. M., Online and offline solid-phase extraction with styrene-divinylbenzene membrane extraction disks for determining pesticides in water by reversed-phase liquid chromatography-diode array detection, *J. Chromatogr. A*, 754, 77–84, 1996.

25. Lacorte, S. and Barcelo, D., Improvements in the determination of organophosphorus pesticides in ground- and wastewater samples from interlaboratory studies by automated online liquid–solid extraction followed by liquid chromatography-diode array detection, *J. Chromatogr. A*, 725, 85–92, 1996.

26. Sabik, H. and Jeannot, R., Stability of organophosphorus insecticides on graphitized carbon black extraction cartridges used for large volumes of surface water, *J. Chromatogr. A*, 879, 73–82, 2000.

27. Jeannot, R., Sabik, H., Amalric, L., Sauvard, E., Proulx, S., and Rondeau, B., Ultra-trace analysis of pesticides by solid-phase extraction of surface water with carbopack B cartridges, combined with large-volume injection in gas chromatography, *Chromatographia*, 54, 236–240, 2001.

28. Pichon, V., Solid-phase extraction of multiresidue analysis of organic contaminants in water, *J. Chromatogr. A*, 885, 195–225, 2000.

29. Hennion, M. C., Graphitized carbon for solid-phase extraction, *J. Chromatogr. A*, 885, 73–95, 2000.

30. Huck, C. W. and Bonn, G. K., Recent developments in polymer-based sorbents for solid-phase extraction, *J. Chromatogr. A*, 885, 51–72, 2000.

31. Thurman, E. M. and Mills, M. S., *Solid-Phase Extraction. Principles and Practice*, Wiley, 1998.

32. Bruchet, A., Patty, L., and Jaskulké, E., *Optimization and Evaluation of Multiresidue Methods for Priority Pesticides in Drinking and Related Waters (European Project SMT4-CT96-2142)*, 1998.

33. Waters Corporation. *Organophosphorus Pesticides in Drinking Water and Fruit. Environmental and Agrochemical Applications Notebook*, p. 9, 2002.

34. EPA 525, *Determination of Organic Compounds in Drinking-Water by Liquid–Solid Extraction and GC–MS*.

35. Chang, H. Y., Wu, H. C., lin, W. C., Wu, P. L., and Kuei, C. H., Determination of organophosphorus pesticides metabolites in surface water by use of a strong anion-exchange disk and in-vial derivatization, *Chromatographia*, 51, 630–633, 2000.

36. EPA 3535 A, *Solid-Phase Extraction (SPE)*, Revision 1, 1–19, 1998.

37. Arthur, C. L. and Pawliszyn, J., Solid-phase Microextraction with thermal desorption using fused silica optical fibers, *Anal. Chem.*, 62, 2145–2148, 1990.

38. Balthussen, B., Sandra, P., David, F., and Cramers, C., Stir Bar Sorptive Extraction (SBSE), a novel extraction technique for aqueous samples. Theory and principles, *J. Microcolumn Sep.*, 11, 737–747, 1999.

39. Valor, I., Molto, J. C., Apralz, D., Apraiz, D., and Font, G., Matrix effects on Solid-phase Microextraction (SPME) of organophosphorus pesticides from water, *J. Chromatogr. A*, 767, 195–203, 1997.

40. Souza, D. A. and Lanças, F. M., Solventless sample preparation for pesticides analysis in environmental samples by SPME-HRGC/MS, *J. Environ. Sci. Health B-Pesticides, Food contaminants*, 38, 417–428, 2003.

41. Mothes, S., Popp, P., and Wennrich, R., Analysis of organophosphorus insecticides in natural waters by use of Stir-Bar-Sorptive extraction then gas chromatography with atomic emission detection, *Chromatographia*, 57(Suppl.), S/249–S/252, 2003.

42. David, F., Tienpont, B., and Sandra, P., Stir-Bar Sorptive Extraction of trace organic compounds from aqueous matrices, *LC–GC*, 2003, July 2–7, 2003.

43. Lopez-Avila, V., Young, R., and Beckert, W. F., Online determination of organophosphorus pesticides in water by solid-phase microextraction and gas chromatography with thermoionic-selective detection, *J. High Resolut. Chromatogr.*, 20, 487–492, 1997.

44. Miege, C. and Dugay, J., Solid-phase microextraction and gas chromatography for rapid analysis of pesticides, *Analusis*, 26, M137–M143, 1998.

45. Sng, M. T., Lee, F. K., and Lakso, H. A., Solid-phase microextraction of organophosphorus pesticides from water, *J. Chromatogr. A*, 759, 225–230, 1997.

46. Tomkins, B. A. and Ilgner, R. H., Determination of atrazine and four organophosphorus pesticides in ground water using solid-phase microextraction (SPME) followed by gas chromatography with selected-ion monitoring, *J. Chromatogr. A*, 972, 183–194, 2002.

47. Beltran, J., Lopez, F. J., Cepria, O., and Hernandez, F., Solid-phase microextraction of organopho-sphorus pesticides in environmental water samples, *J. Chromatogr. A*, 808, 257–263, 1998.

48. Gonçalves, C. and Alpendurada, M. F., Multiresidue method for the simultaneous determination of four groups of pesticides in ground and drinking waters, using soild-phase microextraction — gas chromatography with electron-capture and thermoionic specific detection, *J. Chromatogr. A*, 968, 177–190, 2002.

49. Lambropoulou, D. A. and Albanis, T. A., Optimization of headspace solid-phase microextraction conditions for the determination of organophosphorus pesticides in natural waters, *J. Chromatogr. A*, 922, 243–255, 2001.

50. Beltran, J. P., Lopez, F. J., and Hernandez, F., SPME in pesticide residues analysis, *J. Chromatogr. A*, 885, 389–404, 2000.

51. Shamsul Hairi, S., Yoshihiro, S., Yoshiaki, K., and Kiyokatsu, J., Solventless sample preparation procedure for organophosphorus pesticides analysis using solid-phase microextraction and online supercritical fluid extraction/high performance liquid chromatography technique, *Anal. Chim. Acta*, 433, 207–215, 2001.

52. Pawliszyn, J., *Solid-Phase Microextraction. Theory and Practice*, Wiley-VCH, 1997.

53. Pengxiang, X., Dongxin, Y., Shuming, Z., and Quingmei, L., Determination of organophosphorus pesticides in water samples by membrane extraction and gas chromatography, *Environ. Monit. Assess.*, 87, 155–168, 2003.

54. Bouzigue, M. and Pichon, V., Immunoextraction of pesticides at the trace level in environmental samples, *Analusis*, 26, M112–M117, 1998.

55. Ensing, K., Beggren, C., and Majors, R. E., Selective sorbents for solid-phase extraction based on molecularly imprinted polymers, *LC–GC*, 2002, January 2–8, 2002.

56. ISO 14507, *Pretreatment of Samples for Determination of Organic Contaminants*, pp. 1–15, 2003.

57. Dionex, *Extraction of Organophosphorus Pesticides Using Accelerated Solvent Extraction (ASE)*, Application note 319, 1–3, 1996.

58. EPA 3540 C, *Soxhlet Extraction*, Revision 3, 1–8, 1996.

59. EPA 3620 B, *Florisil Clean-Up*, Revision 2, 1–25, 1996.

60. EPA 3580 A, *Waste Dilution*, Revision 1, 1–4, 1992.

61. EPA 3546, *Microwave Extraction*, Revision 0, 1–13, 2000.

62. EPA 3550 B, *Ultrasonic Extraction*, Revision 2, 1–14, 1996.

63. EPA 3545, *Pressurized Fluid Extraction (PFE)*, Revision 0, 1–9, 1996.

64. EPA 3541, *Automated Soxhlet*, Revision 0, 1–10, 1994.

65. Wei Fang, Ng., Mui Jun Karen, T., and Hans-Ake, L., Determination of organophosphorus pesticides in soil by headspace solid-phase microextraction, *Fresenius J. Anal. Chem*, 363, 673–679, 1999.

66. Yu-Tzu, H., Jui-Hung, Y., and Yei-Shung, W., Study on the extraction of organophosphorus insecticides from soil by supercritical fluid method, *Anal. Sci.*, 13(Suppl.), 137–140, 1997.

67. Bao, M. L., Pantani, F., Barbieri, K., Burrini, D., and Griffini, O., Multi-residue pesticide analysis in soil by solid-phase disk extraction and gas chromatography/ion-trap mass spectrometry, *Int. J. Environ. Anal. Chem.*, 64, 23–245, 1996.

68. Camel, V., Supercritical fluid extraction as a useful method for pesticide determination, *Analusis*, 26, M99–M111, 1998.

69. Ho Kim, D., Suk Heo, G., and Woon Lee, D., Determination of organophosphorus pesticides in wheat flour by supercritical fluid extraction and gas chromatography with nitrogen–phosphorus detection, *J. Chromatogr. A*, 824, 63–70, 1998.

70. Bouaid, A., Ramos, L., Gonzalez, M. J., Fernandez, P., and Camara, C., Solid-phase microextraction method for the determination of atrazine and four organophosphorus pesticides in soil samples by gas chromatography, *J. Chromatogr. A*, 939, 13–21, 2001.

71. Bao, M. L. and Pantani, F., Multi-residue pesticide analysis in soil by solid-phase disk extraction and gas chromatography/ion-trap mass spectrometry, *Int. J. Environ. Anal. Chem.*, 64, 233–245, 1996.

72. Okumura, T. and Nishikawa, Y., Determination of organophosphorus pesticides in environmental samples by capillary gas chromatography–mass spectrometry, *J. Chromatogr. A*, 709, 319–331, 1995.

73. Vinas, P., Campillo, N., Lopez-Garcia, I., Aguinaga, N., and Hernandez-Cordoba, M., Capillary gas chromatography with atomic emission detection for pesticide analysis in soil samples, *J. Agric. Food Chem.*, 51, 3704–3708, 2003.

74. Snyder, J. L., Grob, R. L., McNally, M. E., and Oostdyk, T. S., Comparison of supercritical fluid extraction with classical sonication and soxhlet extractions for selected pesticides, *Anal. Chem.*, 64, 1940–1946, 1992.

75. Magdic, S., Boyd-Boland, A., Jinno, K., and Pawliszyn, J. B., Analysis of organophosphorus insecticides from environmental samples using solid-phase microextraction, *J. Chromatogr. A*, 736, 219–228, 1996.

76. Castro, J., Sanchez-Brunete, C., and Tadeo, J. L., Multiresidue analysis of insecticides in soil by gas chromatography with electron-capture detection and confirmation by gas chromatography–mass spectrometry, *J. Chromatogr. A*, 918, 371–380, 2001.

77. Skopec, Z. V., Clark, R., Harvey, P. M. A., and Wells, R. J., Analysis of organophosphorus pesticides in rice by supercritical fluid extraction and quantification using an atomic emission detector, *J. Chromatogr. Sci.*, 31, 445–449, 1993.

78. Forcada, M., Beltran, J., Lopez, F. J., and Hernandez, F., Multiresidue procedures for determination of triazine and organophosphorus pesticides in water by use of large-volume PTV injection in gas chromatography, *Chromatographia*, 51, 362–368, 2000.

79. ISO/CD 22892, *Soil Quality — Guideline for the GC/MS identification of target compounds*, p. 16, 2002.

80. Mallet Victorin, N., Duguay, M., Bernier, M., and Trottier, N., An evaluation of high performance liquid chromatography-UV for the multi-residue analysis of organophosphorous pesticides in environmental water, *Int. J. Environ. Anal. Chem.*, 39, 271–279, 1990.

81. Lacorte, S. and Barcelo, D., Determination of parts per trillion levels of organophosphorus pesticides in groundwater by automated online liquid-solid extraction followed by liquid chromatography/ atmospheric pressure chemical ionization mass spectrometry using positive and negative ion modes of operation, *Anal. Chem.*, 68, 2464–2470, 1996.

82. Cappiello, A., Famiglini, G., Palma, P., and Mangani, F., Trace level determination of organophosphorus pesticides in water by the new direct-electron ionization LC/MS interface, *Anal. Chem.*, 74, 3547–3554, 2002.

83. Kawasaki, S., Ueda, H., Itoh, H., and Tadano, J., Screening of organophosphorus pesticides using liquid chromatography–atmospheric pressure chemical ionization mass spectrometry, *J. Chromatogr. A*, 595, 193–202, 1992.

84. Carabias Martinez, R., Rodriguez Gonzalo, E., Amigo Moran, M. J., and Hernandez Mendez, J., Sensitive method for the determination of organophosphorus pesticides in fruits and surface waters by high-performance liquid chromatography with ultraviolet detection, *J. Chromatogr. A*, 607, 37–45, 1992.

85. Ferrer, I. and Barcelo, D., LC–MS methods for the determination of pesticides in environmental samples, *Analusis*, 26, M118–M122, 1998.

86. Aguera, A. and Fernandez-Alba, A., GC–MS and LC–MS evaluation of pesticide degradation products generated through advanced oxidation processes: an overview, *Analusis*, 26, M123–M130, 1998.

87. Molina, C., Grasso, P., Benfenati, E., and Barcelo, D., Automated sample preparation with extraction columns followed by liquid-chromatography–ionspray mass spectrometry interferences, determination and degradation of polar organophosphorus pesticides in water samples, *J. Chromatogr. A*, 737, 47–58, 1996.

88. Mol, H. G. J., Van Dam, R. C. J., and Steijger, O. M., Determination of polar organophosphorus pesticides in vegetables and fruits using liquid chromatography with tandem mass spectrometry: selection of extraction solvent, *J. Chromatogr. A*, 1015, 119–127, 2003.

89. Ingelse, B. A., Van Dam, R. C. J., Vreeken, R. J., Mol, H. G. J., and Steijger, O. M., Determination of polar organophosphorus pesticides in aqueous samples by direct injection using liquid chromatography with tandem mass spectrometry, *J. Chromatogr. A*, 918, 67–78, 2001.

90. LIG'AIR, Les pesticides en milieu atmosphérique: étude en région Centre, Réseau de surveillance de la qualité de l'air en région Centre, Orléans, France, p. 57, 200–2000.

91. Method EPA TO 4, *Determination of Pesticides and Polychlorinated Biphenyls In Ambient Air Using High Volume Polyurethane Foam (PUF) Sampling Followed By Gas Chromatography/Multidetector*, US EPA.

92. Method EPA TO 10, *Determination Of Pesticides and Polychlorinated Biphenyls In Ambient Air Using Low Volume Polyurethane Foam (PUF) Sampling Followed By Gas Chromatography/ Multidetector*, US EPA.

93. NIOSH 5600, *Organophosphorus Pesticides*, 4th ed., 1994.

94. Kennedy, E. R., Abell, M. T., Reynolds, J. and Wickman, D., A sampling and analytical method for the simultaneous determination of multiple organophosphorus pesticides in air, *Am. Ind. Hyg. Assoc. J.*, 55, 1172–1177, 1994.

95. ISO 11 370, *Water Quality — Determination Of Selected Organic Plant Protection Agents — Automated Multiple Development (AMD) Technique*, 1999.

96. Jaffrezic-Renault, N., New trends in biosensors for organophosphorus pesticides, *Sensors*, 1, 60–74, 2001.

97. Marty, J. L., Leca, B. and Noguer, T., Biosensors for the detection of pesticides, *Analusis*, 26, M144–M149, 1998.

98. Park, W. C., Cho, Y. A., Kim, Y. J., Hammock, B. D., Lee, Y. T. and Sung-Lee, H., Development of an enzyme-linked immunosorbent assay for the organophosphorus insecticide bromophos, *Bull. Korean Chem. Soc.*, 23, 1399–1403, 2002.

99. Hennion, M. C., Applications and validation of immunoassays for pesticide analysis, *Analusis*, 26, M149–M155, 1998.

100. Wong, J. M., Li, Q. X., Hammock, B. D. and Seiber, J. N., Method for the analysis of 4-Nitro-phenol and parathion in soil using supercritical fluid extraction and immunoassay, *J. Agric. Food Chem.*, 39, 1802–1807, 1991.

24 Chromatographic Determination of Carbamate Pesticides in Environmental Samples

Evaristo Ballesteros Tribaldo

CONTENTS

I. INTRODUCTION

A number of pesticides are used in vast amounts in agriculture and horticulture each year. As a result, waters, soils, and plants are frequently contaminated with these substances, which therefore constitute one of the major sources of potential environmental hazards to man and animals through their presence and concentration in the food chain.[1] Pesticides are classified according to their chemical structure into organochlorines, organophosphates, carbamates, triazine herbicides, inorganic, etc. Organochlorines, organophosphates, and carbamates exhibit high persistence in the environment, which causes various health and safety problems.[2]

Analytical work on pesticide residues in the diet began in the middle of the last century in response to the alarming findings on environmental hazards. Before 1960, most analyses were focused on individual pesticides and involved relatively nonspecific methods including spectrometry, total halogen methods (for chlorinated pesticides), and biochemical methods based on the inhibition of thin-layer plates of the enzyme cholinesterase (for carbamate and organophosphate pesticides).[3] The field of pesticide residue analysis was revolutionized in the late 1960s with the introduction of gas chromatography (GC). The inception in the late 1970s of sensitive, selective GC detectors facilitated the widespread application of this technique to pesticide residue analysis. The use of liquid chromatography (LC) for pesticide analysis has also grown fast since its introduction in the late 1970s. LC is especially suitable for pesticides that cannot be determined directly by GC.[4]

This chapter deals with the properties of carbamate pesticides, many of which influence the design and choice of analytical methods for their determination. It reviews the chromatographic techniques used to quantify carbamate pesticides in environmental samples, with special emphasis on GC and LC. Other chromatographic techniques, viz., capillary electrophoresis (CE), thin-layer chromatography (TLC), supercritical fluid chromatography (SFC), and sample preparation procedures, are also discussed.

II. PHYSICO-CHEMICAL PROPERTIES

Carbamate pesticides have the general structure depicted in Figure 24.1, where R_1 and R_2 denote aromatic and aliphatic moieties, respectively. Table 24.1 shows the most important carbamates, which can be classified into nine groups, namely: N-methylcarbamates, aminophenyl N-methylcarbamates, oxime N-methylcarbamates, N,N-dimethylcarbamates, N-phenylcarbamates, benzimidazole carbamates, thiocarbamates, dithiocarbamates, and ethylenebisdithiocarbamates. The table summarizes the physical and chemical properties of the pesticides. Pure carbamate pesticides are generally white, crystalline, almost odorless solids exhibiting good shelf stability by virtue of their high melting point and low vapor pressure. They are usually low sparsely soluble in water, but readily dissolved in polar organic solvents such as methanol, ethanol, or acetone. Most carbamates are moderately soluble in solvents of medium polarity such as benzene, toluene, xylene, chloroform, dichloromethane, or 1,2-dichloromethane, and poorly soluble in nonpolar organic solvents such as petroleum ether or n-hexane.[3-6]

A. ENVIRONMENTAL FATE OF CARBAMATE PESTICIDES

The environmental fate of pesticides can be predicted from the following physical parameters[7]:

Octanol-water partition coefficient (K_{ow}), which describes the partitioning of a pesticide between octanol and water. High K_{ow} values are typical of persistent compounds that are largely (bio)accumulated in the fat portion of organisms.

Dissociation constant (K_a). The degree of ionization of pesticides affects such processes as photolysis, solubilization, evaporation from water, soil sorption, etc.

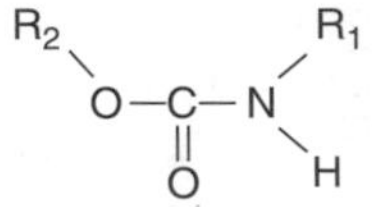

FIGURE 24.1 General structure of carbamate pesticides.

TABLE 24.1
Names and Properties of Selected Carbamate Pesticides

No.	Common Name / Other Names / CAS Registry Number[a]	Chemical Name / Molecular Formula	Physical Form / Melting Point (°C) / Vapor Pressure (25°C) / Water Solubility (25°C) / K_{ow}[b]/K_{oc}[c]	Use[d]	Toxicity to Mammals, LD_{50} (acute) (mg/kg) Oral (rat)	Dermal (rat)
	N-methylcarbamates					
1	Bendiocarb Bendiocarbe, Ficam 22781-23-3	2,3-Isopropylidenedioxyphenyl *N*-methylcarbamate $C_{11}H_{13}NO_4$	Colorless crystals 124.6 to 128.7 4.6 mPa 0.28 g/l (20°C) 1.72 (pH 6.55)/28 to 40	I	40 to 156	566 to 600
2	Benfuracarb 82560-54-1	Ethyl *N*-[2,3-dihydro-2,2-dimethylbenzofuran-7-yloxycarbonyl(methyl aminothio)]-*N*-isopropyl-β-alaninate $C_{20}H_{30}N_2O_5S$	Viscous reddish brown liquid — 26.5 μPa (20°C) 8.0 mg/l (20°C) 4.3 (20 to 22°C, pH 7)/—	I	138	>2,000
3	Bufencarb Bunfencarbe, Bux, Metalkamate 8065-36-9	3:1 Mixture of 3-(1-methylbutyl) phenyl and 3-(1-ethylpropyl)phenyl *N*-methylcarbamates $C_{13}H_{19}NO_2$	Yellow amber solid 26 to 39 4.0 mPa (at 30°C) <0.005% —/—	I	87	680 (rabbit)
4	Carbanolate Banol, Chlorxylam 671-04-5	2-Chloro-4,5-dimetylphenyl *N*-methylcarbamate $C_{10}H_{12}ClNO_2$	White crystals 130 to 133 — — 2.3/—	I	30 to 55	—
5	Carbaryl Sevin[R], Dicarbam, Carbicide 63-25-2	1-Naphthyl *N*-methylcarbamate $C_{12}H_{11}NO_2$	Colorless crystals 142 41 μPa (23.5°C) 120 mg/l (20°C) 1.59/—	I	500 to 850	>4,000 > 2,000 (rabbit)

Continued

TABLE 24.1
Continued

No.	Common Name Other Names CAS Registry Number[a]	Chemical Name Molecular Formula	Physical Form Melting Point (°C) Vapor Pressure (25°C) Water Solubility (25°C) K_{ow}[b]/K_{oc}[c]	Use[d]	Toxicity to Mammals, LD_{50} (acute) (mg/kg) Oral (rat)	Dermal (rat)
6	Carbofuran Furadan[R], NIA-10242 1563-66-2	2,3-Dihydro-2,2-dimethylbenzofuran-7-yl N-methylcarbamate $C_{12}H_{15}NO_3$	Colorless crystals 153 to 154 0.031 mPa 320 mg/l (20°C) 1.52 (20°C)/22	I N	8 to 14	>3,000
7	Landrin[R]	4:1 Mixture of 3,4,5-trimethylphenyl and 2,3,5-trimethylphenyl N-methylcarbamate $C_{11}H_{15}NO_2$	Buff crystals 105 to 114 5×10^{-5} mm Hg (23°C) 60 mg/l (23°C) —/—	I	208	2,500 (rabbit)
8	Methiocarb Mesurol[R], Metmercapturon, Mercaptodimethur 2032-65-7	4-Methyl-3,5-xylyl N-methylcarbamate $C_{11}H_{15}NO_2S$	Colorless crystals 119 0.015 mPa 27 mg/l (20°C) 3.34/—	I M A BR	100	350 to 400
9	Mobam MCA-600 —	4-Benzothienyl N-methylcarbamate $C_{10}H_9NO_2S$	White crystals 128 1×10^{-8} mm Hg (25°C) <0.1% —/—	I	20 to 125	—
10	Propoxur Baygon[R], Arprocarb, Blattanex, Unden, Sendran 114-26-1	2-Isopropoxyphenyl N-methylcarbamate $C_{11}H_{15}NO_3$	Colorless crystals 90 2.8 mPa 1.9 g/l (20°C) 1.56/—	I	128	>5,000

Aminophenyl N-methylcarbamates

No.	Common name, synonyms, CAS	Chemical name, Formula	Physical properties						Class		
11	Aminocarb, Matacil[R], Aminocarbe, 2032-59-9	4-Dimethylamino-m-tolyl N-methylcarbamate, $C_{11}H_{16}N_2O_2$	Tan crystals	93 to 94	Nonvolatile	Slight	1.73/—		I	30	275
12	Mexacarbate, Zectran[R], 315-18-4	4-Dimethylamino-3,5-xylyl N-methylcarbamate, $C_{12}H_{18}N_2O_2$	White crystals	85	<0.1 mm Hg (139°C)	—	—/—		I	24	>500

Oxime N-methylcarbamates

No.	Common name, synonyms, CAS	Chemical name, Formula	Physical properties						Class		
13	Aldicarb, Temik[R], UC-21149, 116-06-3	2-Methyl-2-(methylthio)propionaldehyde o-methylcarbamoyloxime, $C_7H_{14}N_2O_2S$	Colorless crystals	98 to 100	13 mPa (20°C)	4.9 g/l	1.08/—		I A N	0.9	20 (rabbit)
14	Methomyl, Lannate, Methavin, 16752-77-5	1-(Methylthio)acetaldehyde o-methylcarbamoyloxime, $C_5H_{10}N_2O_2S$	Colorless crystals	78 to 79	6.65 mPa	57.9 g/l	0.09972		I A	17 to 24	>5,000 (rabbit)
15	Oxamyl, Vydate, DPX-1410, 23135-22-0	2-Dimethylamino-1-(methylthio)glyoxal o-methylcarbamoyloxime, $C_7H_{13}N_3O_3S$	Colorless crystals	100 to 102	31 mPa	280 g/l	−0.44 (pH 5)/25		I A N	5.4	>2,000 (rabbit)
16	Thiodicarb, Larvin, 59669-26-0	3,7,9,13-Tetramethyl-5,11-dioxa-2,8,14-trithia-4,7,9,12-tetra-azapentadeca-3,12-diene-6,1-dione, $C_{10}H_{18}N_4O_4S_3$	Colorless crystals	173 to 174	5.7 mPa (20°C)	35 mg/l	—/—		I M	66	>2,000 (rabbit)

Continued

TABLE 24.1
Continued

No.	Common Name Other Names CAS Registry Number[a]	Chemical Name Molecular Formula	Physical Form Melting Point (°C) Vapor Pressure (25°C) Water Solubility (25°C) K_{ow}[b]/K_{oc}[c]	Use[d]	Toxicity to Mammals, LD$_{50}$ (acute) (mg/kg) Oral (rat)	Dermal (rat)
17	Thiofanox Dacamox, Thiofanocarb 39196-18-4	3,3-Dimethyl-1-(methylthio)-2-butanone o-methylcarbamoyloxime $C_9H_{18}N_2O_2S$	Colorless crystals 56.5 to 57.5 22.6 mPa 5.2 g/l (22°) —/—	I A	8.5	39 (rabbit)
		N-N-Dimethylcarbamates				
18	Dimetilan Snip 644-64-4	1-Dimethylcarbamoyl-5-methylpyrazoyl-3-yl *N,N*-dimethylcarbamate $C_{10}H_{16}N_4O_3$	Colorless crystals 68 to 71 1×10^{-4} mm Hg (20°C) 24% —/—	I	<50	>2,000
19	Pirimicarb Pirimicarbe, Pirimor, Aphox, Fernos 23103-98-2	2-Dimethylamino-5,6-dimethylpirimidin-4-yl *N,N*-dimethylcarbamate $C_{11}H_{18}N_4O_2$	Colorless crystals 90.5 0.97 mPa 3.0 g/l (20°C) 1.7/—	I	147	>500
		N-Phenylcarbamates				
20	Chlorpropham Chlorpropame, CIPC, Chloro-IPC 101-21-3	Isopropyl 3-chlorocarbanilate, (isopropyl *m*-chlorocarbanilate) $C_{10}H_{12}ClNO_2$	Colorless solid 38.5 to 40 10^{-5} mm Hg (25°C) 89 mg/l —/—	H	5,000	2,000 (dog)

No.	Name, synonyms, CAS, formula	Chemical name, formula	Physical data	Type		
21	Propham Prophame, IPC, Banhoe, Tuberit 122-42-9	1-Methyl phenylcarbamate (isopropyl carbanilate) $C_{10}H_{13}NO_2$	Colorless crystals 87 to 87.6 — 250 mg/l (20°C) 2.6/—	H	5,000	6,800 (rabbit)
22	Swep 1918-18-9	Methyl 3,4-dichlorophenyl carbamate (methyl 3,4-dichlorocarbanilate) $C_8H_7Cl_2NO_2$	White solid 112 to 114 — — 2.8/—	H	552	—

Benzimidazole carbamates

No.	Name, synonyms, CAS, formula	Chemical name, formula	Physical data	Type		
23	Benomyl Arylate, Benlate, Tersan 17804-35-2	Methyl 1-(butylcarbamoyl) benzimidazol-2-yl-carbamate $C_{14}H_{18}N_4O_3$	Colorless crystals 140 $< 4.9\ \mu Pa$ 4 mg/kg (pH 3 to 10) 2.12/1,900	F	>10,000	>10,000 (rabbit)
24	Carbendazim Carbendazime, Derosal, carbendazol, Bavistin, MBC, BMC 10605-21-7	Methyl benzimidazol-2-yl-carbamate $C_9H_9N_3O_2$	Crystalline powder 302 to 307 0.09 mPa (20°C) 29 mg/l (pH 4) 1.38 (pH 5)/200 to 250	F	>15,000	>2,000

Thiocarbamates

No.	Name, synonyms, CAS, formula	Chemical name, formula	Physical data	Type		
25	Butylate Sutan 2008-41-5	S-Ethyl di-isobutylthiocarbamate $C_{11}H_{23}NOS$	Colorless liquid — 1.73 Pa 36 mg/l (20°C) 1.146/—	H	5,366	>5,000 (rabbit)
26	Cycloate Ro-Neet, Hexylthiocarbam 1134-23-2	S-Ethyl ciclohexyl(ethyl) thiocarbamate $C_{11}H_{21}NOS$	Colorless liquid 11.5 2.13 mPa 75 ng/l (20°C) 3.88/—	H	3,160	>5,000 (rabbit)

Continued

TABLE 24.1
Continued

No.	Common Name / Other Names / CAS Registry Number[a]	Chemical Name / Molecular Formula	Physical Form / Melting Point (°C) / Vapor Pressure (25°C) / Water Solubility (25°C) / K_{ow}^b/K_{oc}^c	Use[d]	Toxicity to Mammals, LD$_{50}$ (acute) (mg/kg) Oral (rat)	Dermal (rat)
27	Diallate Avadex 2303-16-4	S-2,3-Dichloroallyl di-isopropyl (thiocarbamate) $C_{10}H_{17}Cl_2NOS$	Yellowish oily liquid 1.5×10^{-4} mm Hg 14 mg/l —/—	H	395	>2,000 (rabbit)
28	EPTC Eptam 759-94-4	S-Ethyl dipropyl thiocarbamate $C_9H_{19}NOS$	Colorless liquid − 30 0.01 mPa 375 mg/l 3.2/—	H	1,367	>2,000
29	Molinate Ordram 2212-67-1	S-Ethyl N,N-hexamethylelenethiocarbamate $C_9H_{17}NOS$	Clear liquid — 746 mPa 88 mg/l (20°C) 2.88/0.74 to 2.04	H	369 to 450	>4,640 (rabbit)
30	Pebulate Tillan 1114-71-2	S-Propyl butylethylthiocarbamate $C_{10}H_{21}NOS$	Colorless or yellow liquid — 9 Pa (30°C) 60 mg/l (20°C) 3.83/—	H	1,120	4,640 (rabbit)
31	Thiobencarb Thiobencarbe, Benthiocarb 28249-77-6	S − 4-chlorobenzyl diethylthiocarbamate $C_{12}H_{16}ClNOS$	Pale yellow liquid 3.3 2.2 Pa (23°C) 30 mg/l (20°C) 3.42/—	H	1,300	>2,000

No.	Name / CAS	Formula	Description / properties	Use		
32	Tiocarbazil 36756-79-3 *S*-Benzyl di-*sec*-butylthiocarbamate	$C_{16}H_{25}NOS$	Colorless liquid — 93 mPa (50°C) 2.5 mg/l (30°C) 4.40/1711	H	>10,000	>1,200
33	Triallate Tri-allate, Avadex BW 2303-17-5 *S*-2,3,3-Trichloroallyl di-isopropylthiocarbamate	$C_{10}H_{16}Cl_3NOS$	Oily, amber liquid 29 to 30 16 mPa 4 mg/l —/2,400	H	1,100	8,200 (rabbit)
34	Vernolate Vernam 1929-77-7 *S*-Propyl dipropylthiocarbamate	$C_{10}H_{21}NOS$	Clear liquid — 1.39 Pa 90 mg/l (20°C) 3.84 (20°C)/—	H	1,500 to 1,550	>5,000 (rabbit)

Dithiocarbamates

No.	Name / CAS	Formula	Description / properties	Use		
35	Ferbam Ferbame 14484-64-1 Ferric dimethyldithiocarbamate	$C_9H_{18}FeN_3S_6$	Black powder Decompose > 180°C Negligible (20°C) 130 mg/l 0.80/—	F	>4,000	—
36	Thiram Thirame, Thiuram, TMTD 137-26-8 Tetramethylthiuram disulfide	$C_6H_{12}N_2S_4$	Colorless crystals 155 to 156 2.3 mPa 18 mg/l 1.73/—	F	1,800	>1,000
37	Ziram Zirame, Milbam, Zerlate 137-30-4 Zinc dimethyldithiocarbamate	$C_6H_{12}N_2S_4Zn$	White powder 246 < 1 µPa 0.05 mg/l (20°C) 1.03/—	F R	320	>6,000

Continued

TABLE 24.1
Continued

No.	Common Name Other Names CAS Registry Number[a]	Chemical Name Molecular Formula	Physical Form Melting Point (°C) Vapor Pressure (25°C) Water Solubility (25°C) K_{ow}[b]/K_{oc}[c]	Use[d]	Toxicity to Mammals, LD$_{50}$ (acute) (mg/kg) Oral (rat)	Dermal (rat)
		Ethylenebisdithiocarbamates				
38	Mancozeb Dithane M-45, Manzeb 8018-01-7	Manganese ethylenebisdithio carbamate (polymeric) complex with zinc salt —	Greyish-yellow powder Decomposes without melting Negligible 6 to 20 mg/l —/>2,000	F	>5,000	>10,000 (rabbit)
39	Maneb Dithane M-22, Manzate, MEB 12427-38-2	Manganese ethylenebisdithiocarbamate (polymeric) $C_4H_6MnN_2S_4$	Yellow crystalline solid Decomposes without melting Negligible Slightly soluble —/—	F	6,750	>5,000
40	Nabam Dithane D-14, Parzate, nabame 142-59-6	Disodium ethylenebisdithiocarbamate $C_4H_6N_2Na_2S_4$	Colorless crystals Decomposes without melting Negligible 200 g/l	F Al	395	—
41	Zineb Dithane Z-78, Zinèbe 12122-67-7	Zinc ethylenebisdithiocarbamate (polymeric) $C_4H_6N_2S_4Zn$	Light-colored powder Decomposes without melting <0.01 mPa (20°C) 10 mg/l <1.30 (20°C)/—	F	>5,200	>6,000

[a] CAS registry number = Chemical Abstract Service registry number

[b] K_{ow}: Partition coefficient between n-octanol and water (reported as log P_{ow})

[c] K_{oc}: Soil organic carbon sorption coefficient

[d] Use: I = insecticide; N = nematicide; M = molluscicide; A = acaricide; BR = bird repellent; H = herbicide; F = fungicide; R = repellent; Al = algicide

Soil sorption coefficient (K_{oc}), which is calculated by measuring the ratio of the distribution constant of the sorbed to soluble pesticide fractions upon equilibration in a water–soil slurry and dividing it into the weight fraction of organic carbon in soil. Pesticides that are strongly sorbed by soil particles are likely to the more persistent because binding protects them from degradation and volatilization.

Vapor pressure, which describes the contribution to the pressure of the pesticide in the gas phase at a given temperature. It influences the ease of volatilization of the pesticide from water.

Bioconcentration factor (*BCF*), which describes the affinity of the pesticide for aquatic organisms and measures the accumulation of toxins in fish relative to the water in which they swim.

Water solubility. Solubility values provide valuable insight into the fate of pesticides in the environment and losses during processing of residue-containing crops.

Table 24.1 gives the values of selected physical parameters for carbamate pesticides. A pesticide can reach ground water if its water solubility is higher than 20 mg/l, its soil sorption coefficient lower than 300 to 500 cm^3/g, its soil half-life longer than about 2 to 3 weeks, its hydrolysis half-life (Table 24.2) longer than approximately 6 months, and its photolysis half-life longer than 3 days.[8] The Commission of the European Community[9] has published a report on a comprehensive study based on the GUX index (Ground Water Ubiquity Score). The index is a simple mathematic model that measures the likelihood of leaching of a given pesticide: GUS = $\log t_{1/2}x - (4 - \log K_{oc})$. Thus, pesticides are classified as probable leachers (GUS > 2.8), transient leachers (1.8 ≤ GUS ≤ 2.8), and improbable leachers (GUS < 1.8). Only eight carbamate pesticides (viz., aldicarb, carbaryl, carbendazim, EPTC, maneb, methiocarb, propham, and ziram) have a GUS index higher than 1.8 and are thus potential contaminants of ground water.[10]

Alternatively, carbamate pesticides can enter the atmosphere in different ways (viz., through application drift during spraying operations, wind erosion of soil, or volatilization), and it appears that pesticide concentration in the atmosphere is an important problem for human health and forest ecosystems.

B. DEGRADATION OF CARBAMATE PESTICIDES IN THE ENVIRONMENT

Carbamate pesticides are quite labile in the environment as compared to the persistent organochlorines. However, they are somewhat more persistent than organophosphate pesticides.

The factors effecting pesticide degradation can be chemical, physical, or biological. Light and heat are the main physical agents affecting pesticide degradation. The photolysis of residues on plants, on soil surfaces, and in water contributes significantly to pesticide dissipation.[11] Carbamates are metabolized chemically or biochemically (via enzyme-catalyzed reactions) through hydrolysis, oxidation, and conjugation. Table 24.2 lists the principal factors influencing stability of carbamate pesticides. As can be seen, most pesticides are broken down in aqueous solutions, especially in conjunction with extreme pH values. Oxidation products are often generated by reaction with oxygen and its reactive forms (e.g., ozone, superoxides, peroxides). Microorganisms such as bacteria, fungi, and actinomycetes constitute the most important group of pesticide degraders in soil and water.[12] Table 24.2 also lists the half-lives for selected carbamates in water and soil. Some studies on the persistence of carbamates in water have shown that they have short half-lives in water, but may be prone to total destruction, and small amounts may persist over long periods.[3–6,13]

Understanding the transport and fate of pesticides in the environment is of great importance with a view to their efficient use and regulation. The interaction between dissolved organic matter

TABLE 24.2
Stability of Carbamate Pesticides in the Environment

Carbamate Pesticide	Stability[a]
Aldicarb	Stable in neutral, acid, and weakly alkaline media. Hydrolyzed by concentrated alkalis. Decomposes above 100°C. Rapidly converted by oxidizing agents to the sulfoxide, which is more slowly converted into the sulfone. Aldicarb has been shown to decompose more slowly in soils than in plants, and to have a $t_{1/2}$ of 7 to 12 days depending on soil type
Aminocarb	$t_{1/2}$ (stream water: environmental conditions) 8.7 days (pH 7.1) and (pond water: environmental conditions) 4.4 days (pH 5.5)
Bendiocarb	Hydrolyzed rapidly in alkaline media, and more slowly in neutral and acid media. Stable to light and heat. Rapidly degraded in soil, via hydrolysis of the methylcarbamate and heterocyclic rings. $t_{1/2}$ (water: 25°C) 4 days (pH 7). $t_{1/2}$ (soil) ranges from 0.5 to 10 days depending upon soil type, moisture and temperature
Benfuracarb	Stable in neutral and weakly alkaline media, but unstable in acid and strongly alkaline media. Decomposes at 225°C. Degraded on glass plates by sunlight. The principal hydrolytic products are carbofuran phenol and 3-hydroxy- and 3-ketophenol. $t_{1/2}$ (water) 3 h; (soil) 4 to 28 h
Benomyl	Decomposed by strong acids and strong alkalis. Decomposes slowly in the presence of moisture. Decomposes on storage in contact with water and under moist conditions in soil. Benomyl is rapidly converted into carbendazim in the environment, with a $t_{1/2}$ of 2 and 19 h in water and in soil, respectively
Bufencarb	Unstable in highly alkaline media
Butylate	Hydrolyzed by strong acids and alkalis, and in aqueous solutions in sunlight. Thermally stable up to 200°C. In soil, microbial degradation involves hydrolysis to ethylmercaptan, carbon dioxide and di-isobutylamine. $t_{1/2}$ (soil) 1.5 to 10 weeks
Carbanolate	Unstable in highly alkaline media
Carbaryl	Stable under neutral and weakly acid conditions. Hydrolyzed in alkaline media to 1-naphthol. Stable to light and heat. $t_{1/2}$ (sea water: 20°C) 4 days (pH 8.0), (river water: environmental conditions) 4.6 days (pH 7.5). Under aerobic conditions, carbaryl (1 mg/l) degrades with $t_{1/2}$ 7 to 14 days in sandy loam and 14 to 28 days in clay loam
Carbendazim	Slowly decomposed in alkaline solutions (22°C). Stable in acids, forming water-soluble salts. Decomposes at melting point. Stable for at least 2 years below 50°C. $t_{1/2}$ (water) >350 days (pH 5 and pH 7), 124 days (pH 9). $t_{1/2}$ (soil) 8 to 32 days under outdoor conditions
Carbofuran	Unstable in alkaline media. Stable in acid and neutral media. Decomposes above 150°C. Most important metabolite is CO_2, formed by microbiological degradation of the phenol compounds. $t_{1/2}$ (river water: environmental conditions) 13.5 days (pH 7.5), and (pond water: 26 to 30°C) 2.3 days (pH 7.8 to 8.5), and (deionized water: 27 ± 2°C) 36 days (pH 7), and (deionized water: 27 ± 2°C) 1.2 h (pH 10). $t_{1/2}$ (soil) 30 to 60 days
Chlorpropham	Hydrolyzed slowly in acid and alkaline media. Stable to UV light. Decomposes above 150°C. In soil, microbial degradation yields 3-chloroaniline via an enzymatic hydrolysis reaction with release of CO_2. $t_{1/2}$ (distilled water) 4 weeks. $t_{1/2}$ (soil) 65 days (15°C), 30 days (29°C)

Continued

TABLE 24.2
Continued

Carbamate Pesticide	Stability[a]
Cycloate	Hydrolyzed by strong acids and alkalis. Thermally stable (120°C). Microbial degradation is largely responsible for the disappearance of cycloate from soil. $t_{1/2}$ (soil) 4 to 8 weeks
Diallate	$t_{1/2}$ (soil: heavy clay) 5 to 6 weeks, (soil: loam) 4 weeks
Dimetilan	Hydrolyzed by boiling with strong acids and alkalis
EPTC	Hydrolyzed by strong acids on heating. Stable up to 200°C. In soil, it rapidly undergoes microbial degradation to a mercaptan residue, an amino residue, and CO_2. $t_{1/2}$ (soil: heavy clay) 4 to 5 weeks, (soil: loam) 4 weeks
Ferbam	Stable to storage in closed containers. Tends to decompose on exposure to moisture and heat, and on prolonged storage
Landrin	$t_{1/2}$ (water: 38°C) 42 h (pH 8)
Mancozeb	Stable under normal, dry storage conditions. Slowly decomposed by heat and moisture. Rapidly degrades in the environment by hydrolysis, oxidation, photolysis, and metabolism. $t_{1/2}$ (water: 25°C) 20 days (pH 5) and 34 h (pH 9). $t_{1/2}$ (soil) 6 to 15 days
Maneb	Stable to light. Decomposes on prolonged exposure to air or moisture. Rapidly degraded in the environment by hydrolysis, oxidation, photolysis and metabolism. $t_{1/2} < 24$ h (pH 5, 7 or 9). $t_{1/2}$ (soil) 25 days (loamy sand in dark, aerobic conditions)
Methiocarb	Unstable in highly alkaline media. Photodegradation contributes to the overall elimination of methiocarb from the environment. Major metabolites are methylsulfinylphenol and methylsulfonylphenol. $t_{1/2}$ (water: 22°C) > 1 year (pH 4), <35 days (pH 7), 6 h (pH 9). Degradation in soil is rapid
Methomyl	At room temperature, aqueous solutions undergo slow decomposition. The rate of decomposition increases at higher temperatures, in the presence of sunlight, on exposure to air, and in alkaline media. Rapidly degrades in soil. $t_{1/2}$ (ground water) <5 h
Mexacarbate	$t_{1/2}$ (nonsterile river water: 20°C) 9.1 days (pH 8.2), (sterile river water: 20°C) 6.2 days (pH 8.2 to 8.4), (buffered water: 12 to 13°C) 2 weeks (pH 7.4), (buffered water: 20°C) 25.7 days (pH 7.0), (buffered water: 12 to 13°C) 2 days (pH 9.5)
Mobam	Unstable in alkaline media
Molinate	Relatively stable to hydrolysis by acids and alkalis (pH 5 to 9) at 40°C. Stable for at least 2 years at room temperature and at least 2 months at 120°C. Unstable to light. In soil, microbial degradation involves hydrolysis to ethyl mercaptan, CO_2 and dialkylamine. $t_{1/2}$ (aerobic soil: pH 5 to 6) 8 to 25 days, (flooded soil) 40 to 160 days
Nabam	Stable as an aqueous solution. Decomposed by light, moisture and heat. On aeration, aqueous solutions deposit yellow mixtures of which the main fungicidal components are sulfur and etem
Oxamyl	Solid and formulations are stable; aqueous solutions decompose slowly. Decomposition is accelerated by aeration and sunlight. $t_{1/2}$ (water) >31 days (pH 5), 8 days (pH 7), 3 h (pH 9). $t_{1/2}$ (soil) 7 days
Pebulate	Stable up to 200°C. In soil, it disappears mainly through microbial degradation to mercaptan, ethylbutylamine and CO_2. $t_{1/2}$ (water: 40°C) 11 days (pH 4 and pH 10), 12 days (pH 7). $t_{1/2}$ (soil: heavy clay) 2 to 3 weeks, (soil: loam) 2 to 3 weeks

Continued

TABLE 24.2

Continued

Carbamate Pesticide	Stability[a]
Pirimicarb	Stable for more than 2 years under normal storage conditions. Hydrolyzed by boiling with strong acids and alkalis. Aqueous solutions are unstable to UV light. $t_{1/2}$ (water: 20°C) <1 day (pH 5, 7 or 9). $t_{1/2}$ (soil) 7 to 234 days, depending on soil type
Propham	Stable up to 100°C. Hydrolyzed slowly in acid and alkaline media. Not sensitive to light. In soil, it is degraded by microorganisms, with enzymatic hydrolysis of the ester bond and degradation of the unstable N-phenylcarbamic acid to aniline and CO_2. $t_{1/2}$ (distilled water) 8.5 weeks. $t_{1/2}$ (soil) 15 days (16°C), 5 days (29°C)
Propoxur	Hydrolyzed by strong alkalis. $t_{1/2}$ (river water: environmental conditions) 16.1 days (pH 7.5), (buffered water: 20°C) 16 days (pH 8.0), (buffered water: 20°C) 1.6 days (pH 9.0), (buffered water: 20°C) 4.2 h (pH 10.0). Direct photodegradation is not a major contributor to the overall elimination of propoxur from the environment ($t_{1/2}$ 5 to 10 days); indirect photodecomposition (addition of humic acid) is more rapid ($t_{1/2}$ 88 h). The compound rapidly degrades in different soils
Swep	Hydrolyzed slowly in acid and alkaline media
Thiobencarb	Stable in water at pH 5 to 9 for 30 days at 21°C. Degradation is primarily via microbial breakdown, with little loss from volatilization or photodegradation. $t_{1/2}$ (soil) 2 to 3 weeks (aerobic conditions) or 6 to 8 months (anaerobic conditions)
Thiodicarb	Stable at pH 6, rapidly hydrolyzed at pH 9 and slowly at pH 3 ($t_{1/2}$ 9 days). Stable up to 60°C. Aqueous suspensions are decomposed by sunlight. Rapidly degraded in soils under both aerobic and anaerobic conditions, by hydrolysis and by photolysis. $t_{1/2}$ (soil) 3 to 8 days
Thiofanox	Stable under normal storage conditions. Relatively stable to hydrolysis at pH 5 to 9 (under 30°C). Decomposed by strong acids and alkalis. In soil, the methylthio group is rapidly oxidized to the sulfoxide and, further, to the sulfone
Thiram	Decomposed in acid media. Some deterioration on prolonged exposure to heat, air or moisture. $t_{1/2}$ (water: 22°C) 128 days (pH 4), 18 days (pH 7), 9 h (pH 9). $t_{1/2}$ (sandy soil: pH 6.7) 12 h
Tiocarbazil	Stable to hydrolysis at pH 5.6 to 8.4. Slightly decomposed after 30 days at 40°C in aqueous ethanol at pH 1.5. Stable to storage for 60 days at 40°C, and for 100 h in aqueous solution exposed to sunlight. Strongly adsorbed in soil, undergoes rapid degradation upon attack by soil microorganisms. $t_{1/2}$ (soil/water of a rice field) 8 to 15 days
Triallate	Stable under normal storage conditions. Hydrolyzed by strong acids and alkalis. Stable to light. Decomposition temperature >200°C. Main loss from soil via microbial degradation. $t_{1/2}$ (soil: heavy clay) 10 to 12 weeks, (soil: loam) 8 to 10 weeks
Vernolate	Stable in neutral media, and relatively stable in acid and alkaline media. Stable up to 200°C. Decomposed by sunlight. In soil, it undergoes microbial degradation to mercaptan, amine, CO_2 and isopropanol. $t_{1/2}$ (water: 40°C) 13 days (pH 7). $t_{1/2}$ (soil) 8 to 16 days (27°C), >2 months (4°C)
Zineb	Unstable to light, moisture and heat on prolonged storage (decomposition is reduced by stabilizers). When precipitated from a concentrated solution, a polymer is formed which is less fungicidal
Ziram	Decomposed in acid media and by UV irradiation

[a] $t_{1/2}$: half-life of the carbamate

and carbamate pesticides not only changes the solubility and mobility of the pesticides in the environment, but also affects their photodegradation and hydrolysis rate.[14]

Composting, which involves vigorous biological activity, can be expected to accelerate the natural degradation of pesticides in soil. Therefore, composting can be used to reduce the toxicity and hazards of pesticide-containing materials. However, the effects of composting are not always favorable as some pesticides cannot be readily degraded, and can in fact be concentrated through composting as organic matter decomposes and the composting substrate decreases in dry mass and volume.[15]

C. Toxicity of Carbamate Pesticides

Although designed to control pests, pesticides can also be toxic to nontarget organisms, including humans, since a number of species ranging from insects to man share the same basic enzyme, hormone, and other biochemical systems. The toxicity of carbamate pesticides, like that of organophosphorus compounds, is due to the inhibition of the enzyme acetylcholinesterase. With carbamates, however, the inhibition is reversible, so this pesticide class is less toxic to mammals.[16]

The acute toxicity of carbamates ranges from high to low or even zero. The World Health Organization (WHO) has divided pesticides by hazard into five classes on the basis of their LD_{50} values (LD_{50} being the amount of a pesticide needed to cause the death of 50% of the laboratory animals — usually rats — in a test batch), namely: (Ia) extremely hazardous (<5 mg/kg), (Ib) highly hazardous (5 to 50 mg/kg), (II) moderately hazardous (50 to 500 mg/kg), (III) slightly hazardous (>500 mg/kg), and (III +) unlikely to present hazard in normal use (>2.000 mg/kg).[17] Thus, some carbamates such as aldicarb, oxamyl, and carbofuran, with an LD_{50} of 0.9, 5.4, and 8 to 14 mg/kg, respectively (i.e., Ia and Ib class pesticides), are highly toxic, whereas others such as benomyl and carbendazim, with a LD_{50} of 10.000 and 15.000 mg/kg body weight, respectively, (III + class) are virtually nontoxic (Table 24.1). The acute dermal toxicity of carbamates is generally low to moderate, one exception is aldicarb with LD_{50} 20 mg/kg for the rabbit.[3–6]

D. Regulations and Legislation

The rapid increase in the use of pesticides in agriculture after the Second World War led many governments to enforce regulations on their sale and use in order to protect users of pesticides, consumers of treated foodstuffs, domestic animals, and, at a later stage, the environment. Thus, the presence of pesticides in the environment has compelled official international institutions to establish maximum allowable concentration levels of pesticides in drinking water and foods.[18]

The U.S.A., through the National Pesticide Survey,[19,20] which was organized by the Environmental Protection Agency (US-EPA), established a list of compounds based on the amount used (>7000 Tons), water solubility (>30 mg/l), and hydrolysis half-life (>25 weeks). The list includes some carbamate pesticides (e.g., aldicarb, propoxur, carbaryl, carbofuran, methiocarb, methomyl, oxamyl, cycloate, butylate, propham, and swep) and various derivatives (e.g., aldicarb sulphone, aldicarb sulphoxide, and 3-hidroxycarbofuran). In Europe, a list of priority pollutants including pesticides was established in order to protect the environment from the adverse ecological impact of these compounds.[21]

The enforcement of this legislation has led to an increasing need from analysts to develop reliable, effective methods for qualitative and quantitative pesticide residue analysis in environmental and food matrices. EEC Directive 80/778, which is concerned with the quality of water designated for human consumption, has established the maximum admissible concentration of each individual pesticide at 0.1 μg/l and the total amount of pesticides at 0.5 μg/l.[9,22] Other countries (e.g., U.S.A. and Australia) have established concentration limits based on the values recommended by WHO.[23] Such values are based on the acceptable daily intake (ADI), which is calculated as the 20% ADI for a person of 70 kg drinking 2 l of water per day.[10]

III. SAMPLE PREPARATION

The final goals of pesticide analyses are to obtain the cleanest possible samples, to determine the minimum possible concentration with the lowest limits of detection, and to avoid pesticide degradation during transfer to the laboratory. All this means that the accuracy and precision of a method for pesticide analysis will be directly dependent on the sample preparation procedure used.[24] This operation is the most time-consuming and labor-intensive task in the analytical scheme. In response to the need for effective, robust, reliable sample preparation, a number of procedures have developed for fast, simple, and, if possible, solvent-free or solvent-minimized operation. Most such procedures, both conventional and new, are used for the analysis of pollutants in air, water, soils, sediments, and biota.[25]

Sample pretreatment can be implemented in three steps:

(a) Extracting traces of the target pesticides from the environmental samples
(b) Removing coextracted and coconcentrated components from the matrix to avoid potential interferences with analyses (i.e., cleanup)
(c) Derivatizing pesticides to aid separation and/or detection.

The extraction, cleanup and derivatization of carbamate pesticides are reviewed in this section. Their automation and online coupling with chromatographic instruments is also dealt with here.

A. EXTRACTION

The procedure to be used to extract carbamate pesticides from environmental samples depends on their polarity and on the type of sample matrix involved. Various choices exist for the extraction of pesticides ranging from conventional procedures (e.g., Soxhlet extraction, liquid–liquid extraction (LLE), evaporation, steam distillation) to new methodologies including solid-phase extraction (SPE), solid-phase microextraction (SPME), supercritical fluid extraction (SFE), matrix solid-phase dispersion (MSPD), accelerated solvent extraction (ASE) and microwave-assisted extraction.[24]

1. Liquid–Liquid Extraction

LLE is a widely used technique among the official US-EPA methods for the preconcentration of pesticides in liquid samples. Nonpolar solvents for the LLE of pesticides include *n*-hexane, benzene, and ethyl acetate. Water-miscible solvents for this purpose include dichloromethane, methanol, acetonitrile, acetone, and water, which have been employed for the extraction of residues from high-moisture commodities. Mixed solvents have often been used to finely adjust the solvent strength. Thus, various carbamate pesticides were extracted from aqueous environmental samples with chloroform and determined by HPLC with a mean recovery of 71%.[26] Also, a method based on the extraction by sonication of solid samples placed in small columns with a low volume of ethyl acetate was developed for the extraction of thiocarbamates and other herbicides from soil with recoveries between 89 and 109%.[27]

Although LLE is a simple, easy pesticide preconcentration technique, it has a number of drawbacks including the formation of emulsions, the risk of losses and contamination through solvent evaporation, the need to use toxic or flammable solvents, and its difficult automation, which make it a labor-intensive, time-consuming, expensive technique for this purpose.

2. Solid-Phase Extraction

SPE was introduced to avoid or to minimize the shortcomings of LLE. SPE can be used directly as an extraction technique for liquid matrices or as a cleanup method for solvent extracts. A typical SPE sequence includes the activation of the sorbent bed (wetting), removal of excess activating solvent (conditioning), application of the sample, removal of interferences (cleanup) and water, and elution of the sorbed analytes with a small volume of an appropriate solvent.[28] SPE is performed on three different types of supports, namely: cartridges, columns, and disks. The solid phases generally employed in SPE are similar to those used in column LC and include activated charcoal, alumina, silica gel, magnesium silicate (Florisil), chemically bonded silica phases, and polymers (e.g., styrene divinylbenzene copolymers such XAD-2 and PRP-1).[29] The most popular sorbents used with carbamate pesticides are octadecyl- and octyl-silica, styrene divinylbenzene copolymers, and activated carbon black.[30] New sorbents have recently been employed as alternatives to conventional SPE sorbents with the aim of achieving more selective preconcentration. Immunosorbents, which rely on reversibility and highly selective antigen–antibody interactions, and synthetic antibody material such as molecularly imprinted polymers (MIP), are excellent candidates. Thus, an MIP sorbent has been used for the preconcentration of pirimicarb from water samples with recoveries ranging from 76 to 102%.[31]

3. Solid-Phase Microextraction

SPME was first used by Pawliszyn et al. in 1990.[32] It is a two-step process conductive to the simultaneous extraction and preconcentration of analytes form sample matrices. In the first step, a fused-silica fiber coated with a polymeric stationary phase is exposed to the sample matrix where the analyte partitions between the matrix and the polymeric stationary phase. In the second step, the fiber/analyte is transferred to the analytical instrument for desorption, separation, and quantification.[33] SPME has a number of advantages over traditional extraction techniques for pesticides. In fact, it is fast, simple, solvent-free, and easily automated for both GC and HPLC instruments. It exhibits good linearity and sensitivity.[34] Thus, carbamate and organophosphorus pesticides in golf course samples were successfully extracted by SPME and analyzed by HPLC by Jinno et al.[35]

4. Supercritical Fluid Extraction

SFE is an alternative to traditional liquid extraction for pesticide residue analysis. Compared to conventional Soxhlet, LLE, and SPE procedures, SFE has several major advantages, namely:

(a) Less solvent is more expeditious.
(b) It exhibits a higher selectivity that can be controlled by changing the fluid pressure and temperature, or by adding small amounts of modifiers.
(c) It is amenable to automation (equipment is now available allowing multiple, simultaneous, or sequential sample extractions).
(d) It uses minimal amounts of solvent to collect the extracted material.[36,37] For example, N-methylcarbamates have been extracted from soils and cereals by SFE with recoveries ranging from 39.6 to 91.7% for soils, and from 30 to 75% for cereals.[38]

5. Other Extraction Techniques

Other techniques such as, MSPD, ASE, subcritical water extraction (SWE) and microwave-assisted Soxhlet extraction have recently been used for the extraction of carbamate pesticides from environmental samples.

MSPD involves blending a solid sample with solid particles, which simultaneously disrupt and disperse the sample. The underlying mechanisms of MSPD include sample homogenization, cellular disruption, exhaustive extraction, fractionation, and purification in a simple process.[39] One carbamate (pirimicarb) and 12 organophosphorus pesticides were extracted by MSPD from honey, the sample was mixed with Florisil and anhydrous sodium sulfate in small glass columns, and subsequently extracted with a volume of n-hexane-ethyl acetate.[40]

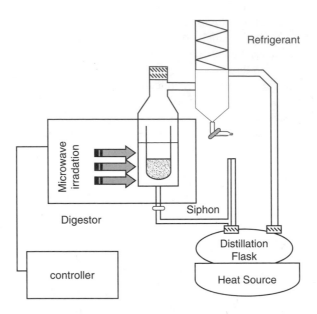

FIGURE 24.2 Scheme of the prototype operation. (Reprinted from Prados-Rosales, R. C., Herrera, M. C., Luque-García, J. L., and Luque de Castro, M. D. *J. Chromatogr. A*, 953, 133–140, 2002. copyright 2002, with permission of Elsevier Science.)

ASE has been used for about 8 years. Extractions can be carried out at temperatures ranging from room level (very gentle conditions) to 200°C in order to accelerate extraction, and at pressures over the range 5 to 200 atm in order to maintain the extraction solvent in a liquid state.[41] Residual *N*-methylcarbamate pesticides in food samples were determined by ASE and LC. The pesticides were extracted with acetonitrile at 100°C at a 2000 psi pressure.[42]

SWE is based on a principle similar to that of ASE, but uses water as solvent as a high temperature and pressure strongly reduces its dielectric constant, viscosity, and surface tension. A SWE device was recently used for the extraction of carbofuran and other pesticides from soils with recoveries of 97.3 and 106.7%, depending on the extraction time.[43]

In 1998, Luque et al.[44] developed a new device called the "focused microwave-assisted Soxhlet extractor" (FMASE) with the aim of overcoming the main drawbacks of conventional Soxhlet extraction (viz., long extraction times and a high organic solvent consumption) while maintaining its advantages (viz., fresh sample-solvent contact throughout the extraction step, no filtration required after extraction, and easy manipulation). Recently, an FMASE device for the extraction of *N*-methylcarbamates from soil was reported.[45] Figure 24.2 illustrates the operation of the overall extraction assembly, which includes a conventional Soxhlet extractor modified to accommodate the sample cartridge compartment in the irradiation zone of a microdigest device, and a microprocessor programmer to control the microwave unit.

B. CLEANUP

Most environmental sample extracts require cleanup prior to their chromatographic analysis. This step is usually intended to remove coextracted compounds that might interfere with the chromatographic determination or damage the analytical instrumentation. Cleanup requirements depend strongly on the selectivity and sensitivity of the detection technique subsequently used to determine the pesticide residues.[46]

Liquid chromatographic cleanup has traditionally been used with normal phase (Florisil, alumina, silica) or reversed-phase (C$_{18}$) columns. In most cases, the aim is to remove the bulk of coextracted materials prior to a more refined cleanup preceding to the final determination.[47] However, SPE (see Section III.A.2) and gel permeation chromatography (GPC) are becoming increasingly popular for the cleanup of environmental sample extracts.

GPC using cross-linked dextran gels has been widely used to separate molecules in aqueous of buffered solvents on the basis of molecular size, partition, and adsorption. The GPC column retains molecules that are small enough to enter the pores of polymer beads. Thus, the molecular mass of most synthetic pesticides is between 200 and 400, whereas that of most lipids ranges from 600 to 1500.[48] Therefore, this cleanup technique can be used to remove lipids from pesticide samples because lipid molecules are too large to enter polymer pores, so they are not retained.[49] Thus, plant samples containing carbamates and other pesticides have been cleaned-up by using Bio Beads S-X3 (the most frequent choice used in GPC) as gel and ethyl acetate as eluent.[50] Cleanup by GPC is effective for GC and LC in most cases. However, some overlap of the large lipid chromatographic band with that for the pesticide fraction typically occurs and additional cleanup by adsorption chromatography on a mini-column is necessary in some cases.[46]

C. Derivatization

Analytical derivatization converts the analyte into a product with greater stability or better chromatographic properties, or one that can be detected with higher sensitivity. It is also a subset of functional group analysis. As such, it boosts the selectivity of quantitative determinations by labeling only those compounds reacting with the derivatizing reagent. In some instances, derivatization is essential in order to isolate the analytes from the sample matrix.[51,52] However, the derivatization procedure involves several manipulations that are potential sources of error. Therefore, the analytical chemist must ensure, for instance, that no impurities are present in the solvents and the reagents. Also, a large number of derivatives are unstable (e.g., they undergo hydrolysis under the influence of moisture or some other degradation process).

The reasons for derivatizing carbamates, apart from their thermolability during the gas chromatographic analysis, include increased detector sensitivity (particularly with electron capture detection), increased volatility, better chromatography separation, applicability to multiresidue and confirmatory analysis, and enhanced compound stability.[3] There are two general approaches to the analysis of N-methylcarbamates by derivatization, namely: derivatization of the intact pesticides and derivatization of a hydrolysis product (one of which will always be the volatile methylamine). The reactions typically used to obtain derivatives of both intact and hydrolysis products of N-methylcarbamates are methylation, silylation, halogenation, acylation, and esterification.[3,51,52] Some of derivatizing reagents (e.g., heptafluorobutyric anhydride and pyridine) are also used as SFE modifiers for the simultaneous extraction and derivatization of carbamates from the sample matrix. Finally, derivatized carbamates can be determined by GC with electron capture or mass spectrometric detection.[53]

The principal aim of derivatization in LC is to improve the response of an analyte to a specific detector. Less frequently, the aim is to improve the stability of the analyte against a specific separation system used in the chromatographic separation of a mixture yielding overlapping peaks. There are two methods of derivatization in LC, viz., precolumn labeling of substances prior to separation on the column and postcolumn derivatization of substances in eluates from the column.[54–56] Most derivatization procedures introduce chromophores or fluorescent groups into functionalized molecules of the analytes.[57] Thus, carbaryl, pirimicarb, and aldicarb were extracted from soils by SFE, and determined by HPLC with fluorescent detection, using postcolumn derivatization with o-phthalaldehyde and 2-mercaptoethanol.[58] This reaction was also used for the derivatization of N-methylcarbamates following micellar electrokinetic chromatography separation and thermal decomposition, the resulting derivatives being determined fluorimetrically with detection limits better than 0.5 ppm.[59]

D. Online Combinations of Sample Preparation and Chromatographic Techniques

Demands such as (i) improved analyte detectability and separation power (or sensitivity and selectivity) next to (ii) real-time confirmation of analyte identity and quantification, and (iii) the generally recognized need to increase sample throughput, have triggered the development of online (preferably automated) chromatography-based systems. The benefits of integrated analytical procedures, where the emphasis is on the inclusion of the sample preparation step, have frequently been discussed in the literature.[60,61]

The online coupling of SPE and LC is particularly easy in the laboratory, and has been extensively described in general reviews dealing with the online preconcentration of organic compounds in environmental or biological samples.[61-63] Automated assemblies combining SPE-based sample preparation and chromatographic separation online have been introduced by some companies. Such is the case with the Prospekt from Spark Holland, OSP-2 from Merck, ASPEC from Gilson, and HP1090 from Hewlett Packard. Some authors have used these commercial systems for the determination of carbamate pesticides[64,65] and for the onsite monitoring of pesticides in surface waters as an early warning alarm system.[66] Figure 24.3 depicts the online SPE-liquid chromatography-particle beam mass spectrometry (MS) system for the determination of carbamate pesticides.[65] The HP1090 system is used as a central control unit. The remote control port of the HP1090 starts up the MS at a preprogrammed time, LC pumps and switching valve V_1 are controlled by electronic event contact signals (on and off positions). Finally, valve V_2 is controlled directly via the HP1090 instrument software.

A continuous SPE system coupled to a gas chromatograph was used for the preconcentration and determination of N-methylcarbamates in water samples.[13] The continuous system (Figure 24.4) comprised a peristaltic pump furnished with pumping tubes, an injection valve, and a laboratory-made adsorption column placed in the loop of the injection valve. Also, a LLE derivatization module coupled online to a gas chromatograph was used for the determination of carbamate pesticides in aqueous samples.[67] The manifold consisted of a peristaltic pump, pumping

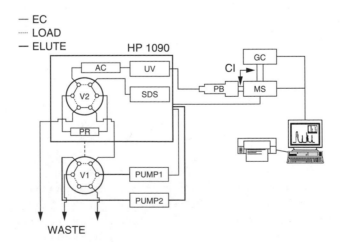

FIGURE 24.3 Setup for automated online SPE–LC–PB–MS analysis of carbamate pesticides. HP1090 liquid chromatograph. V_1, V_2: automatic six-port switching valves; Load/Elute, positions of V_1 and V_2; EC: electronic connections; Pump 1, Pump 2: preparative LC pumps; PR: precolumn; AC: analytical column; SDS: solvent-delivery system of HP1090 liquid chromatograph; PB: particle beam interface; GC: gas chromatograph; CI: chemical ionization reagent gas inlets; MS: mass spectrometer. (Reprinted from Slobodník, J., Hoekstra-Oussoren, S. J. F., Jager, M. E., Honing, M., van Baar, B. L. M., and Brinkman, U. A. Th. *Analyst*, 121, 1327–1334, 1996, copyright 1996, with permission of The Royal Society of Chemistry.)

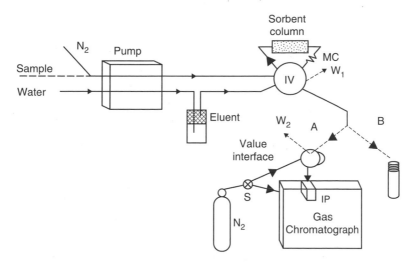

FIGURE 24.4 Flow manifold for the preconcentration of *N*-methylcarbamates and their phenols in water samples. (A) online and (B) offline mode. IV: injection valve; MC: mixing coil; S: tube stopcock; IP: injection port; W: waste. Sample, nitrogen, and eluent flow rate: 3.5, 3.5, and 0.10 ml/min, respectively. (Reprinted from Ballesteros, E., Gallego, M. and Valcárcel, M. *Environ. Sci. Technol.*, 30, 2071–2077, 1996, copyright 1996, with permission of the American Chemical Society.)

tubes, and a custom-made phase separator furnished with a fluoropore membrane. In both methods,[13,67] the interface unit between the continuous system and the gas chromatograph was an injection valve. Recently, Vreuls et al.[68] reviewed different methods for the online combination of sample preparation and GC including LLE–GC, SPE–GC, SPE–thermal desorption GC, and SPM GC.

Combinations of continuous-flow systems with CE equipment can be characterized in terms of the degree of integration between the two units, which can be coupled offline, atline, online, or inline. Dialysis, evaporation, SFE, and SPE, among other techniques, have been coupled to CE.[69] Thus, Hinsmann et al.[70] used an automatic online SPE capillary electrophoresis system for the preconcentration and determination of pesticides in fortified water samples. The manifold (Figure 24.5) consisted of three peristaltic pumps, an automatic ten-port switching valve, and a programmable arm for sample injection. The whole system was automatically controlled via an electronic interface. The mechanical interface used to couple the continuous system to the CE instrument was a laboratory-made programmable arm.[71]

Most SPME applications involve GC. Following extraction, the analytes are thermally desorbed in the chromatograph injector. More recently, the scope of application has been extended to nonvolatile and thermally unstable compounds by coupling SPME to LC. Desorption is performed at an appropriate interface consisting of a standard six-way HPLC injector with a special fiber-desorption chamber used instead of the sample loop.[72] A different approach to SPME–LC called "in-tube SPME" has also been developed, it uses an open tubular fused-silica capillary column instead of the typical SPME fiber.[73] This latter method has been used for the determination of carbamate pesticides in water with relative standard deviations of 1.5 to 4.6%.[74]

IV. ANALYTICAL METHODS

Accurate, sensitive analytical methods are required for reliable environmental control analyses. GC is currently the most flexible and sensitive method for residue analysis. One alternative technique in growing use for the determination of carbamates is LC (or high-performance liquid

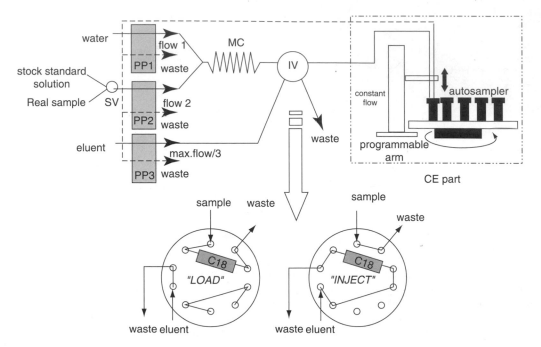

FIGURE 24.5 Continuous-flow manifold for the preconcentration of the pesticides. MC: mixing coil; IV: injection valve; SV: switching valve. (Reprinted from Hinsmann, P., Arce, L., Ríos, A., and Valcárcel, M. *J. Chromatogr. A*, 866, 137–146, 2000, copyright 2000, with permission of Elsevier Science.)

chromatography [HPLC]). Other chromatographic techniques, such as supercritical fluid and thin-layer chromatographies, and CE, have also been used for the determination of carbamate pesticides in environmental samples. This section describes various methodologies for the determination of carbamate pesticides based on chromatographic techniques, with emphasis on GC and LC.

A. GAS CHROMATOGRAPHY

GC was one of the earliest chromatographic separation techniques and continues to be a prominent choice. The popularity of GC relies on a favorable combination of very high selectivity and resolution, good accuracy and precision, a wide dynamic concentration range, and high sensitivity.[75] Unsurprisingly, this technique has traditionally been the most commonly employed in the determination of pesticides. However, direct GC analyses of carbamate pesticides often cause their breakdown at the injection port or on the column during analysis. There are two possible solutions to this problem, namely: (a) using more stable derivatives (see Section III.C), and (b) using lower temperatures and shorter analysis times. Cold oncolumn injection[76] or the use of electronic pressure programming at the GC inlet[77] reduces thermal degradation of pesticides at the injection port. Santos Delgado et al.[78] studied the stability of carbamate pesticides using two different detectors, viz., a nitrogen–phosphorus detector (NPD) and a flame ionization detector (FID), and concluded that the degradation of carbamate pesticides could be minimized by using a temperature program with a gradual gradient, syringes of 70 mm (i.e., longer than the conventional 51 mm), and columns whose stationary phase had been less modified by use.

Carbamate pesticides in environmental samples have long been determined by GC, using packed columns. Methyl silicones (SE-30, OV-101),[79,80] methyl phenyl silicones (OV-17),[81,82] and fluoropropylsilicones (QF-1)[82] are among the stationary phases most frequently used for this purpose (Table 24.3). In the past decade, fused capillary columns of variable polarity were widely used in carbamate residue analyses. These capillary columns are open tubular columns (10 to 60 m long, 0.2 to 0.53 mm i.d.) packed with cross-linked stationary phases such as:

(a) The nonpolar HP-1,[13,27,83,84] DB-5,[53,85-91] HP-5,[43,92] Ultra-2,[93] and BP-5[94]
(b) The medium polar HP-17,[67] DB-17,[95] DB-1701,[27,85,96] and BP-10[97]
(c) The polar Supelcowax-10.[98]

The increased sensitivity and resolution achieved with capillary columns has led them to supersede packed columns.

Splitless injection is generally the preferred choice for the analysis of pesticides by virtue of its robustness. However, oncolumn and programmed temperature vaporizer (PTV) injection have also been used for this purpose. One important, interesting approach here is the direct injection of large sample volumes using a PTV injector or an Autoloop. These interfaces have been used for the determination of pesticides in water samples.[99]

GC coupled to MS is the most common choice for the analysis of pollutants in environmental samples.[100] The vast number of applications developed to date is the result of the efficiency of GC separation and of the qualitative information and high sensitivity provided by MS.[27,43,53,83,84,87,88,90-93] The NPD is very well suited to these as it is selective for phosphorus- and nitrogen-containing compounds.[27,79,80,82,85,86,89,91,94-98] Other detection systems such as the electron capture detector (ECD),[81,85,89,94-96] FID,[13,67,85] and flame photometric detector (FPD)[101] have also been used in this context. A comparative study of the determination of organonitrogen and organophosphorus pesticides by GC using different detectors (viz., NPD, ECD, MS, and FPD) was conducted by Lartiges and Garrigues.[102]

Recently, Kochman et al.[103] developed a new instrument termed "supersonic GC–MS" that affords the fast, sensitive confirmatory and quantitative analysis of a broad range of pesticides in complex agricultural matrices. Figure 24.6 illustrates the determination of 13 typical pesticides. As can be seen, the analysis time is 6 min and the resolution very good, even with a chromatographic column as short as 6 m.

A comprehensive review about two-dimensional gas chromatography (GC × GC) has been published,[104] which includes an exhaustive study of the principles, advantages, and characteristics of GC × GC. This new analytical tool, in the GC × GC–TOF MS mode (TOF MS denotes time-of-flight mass spectrometry) has been used for the determination of pesticides with a linear range from 0.1 to 3 ng, and detection limits between 5 and 23 pg.[105]

B. LIQUID CHROMATOGRAPHY

LC is being extensively used in pesticide chemistry and related areas where the chemicals of interest are frequently of low volatility, thermally unstable, or (very) polar for GC separation. The earliest LC determinations of carbamate pesticides were reported 30 years ago.[106,107] Since then, methods for the LC determination of this type of pesticides have been the subject of excellent reviews.[30,108-110]

Table 24.4 summarizes the features of selected methods for the LC determination of carbamates in environmental samples. As can be seen, the most used reversed-phase chromatography is with C_{18} or C_8 columns and aqueous mobile phases. Other reversed-phase chromatography such as with phenyl[111] has also been used. Some methods use normal-phase silica[107] or diol and nitrile.[112] Sparcino and Hines[113] studied the retention and resolution of 30 N-methylcarbamates in the normal (e.g., silica, cyanopropyl, and propylamine) and reversed-phase mode (C_{18}). The authors concluded that the reversed-phase columns generally provided better results and acetonitrile–water gave the best overall results on a C_{18} column.

TABLE 24.3
Analytical Methods for the Pretreatment of Environmental Samples and Gas Chromatographic Determination of Carbamate Pesticides

Carbamate Pesticides[a]	Environmental Samples	Extraction Technique[b]	Cleanup Technique[c]	Derivatization Reagent/Other Pretreatment[d]	Detection[e]	Chromatographic Column	Analytical Figures of Merit[f]	Ref.
6	Water	—		—	NPD	OV-101 (1.2 m)	Lin: 0.34 to 43.03 mg/l	79
6	Aqueous solution	LLE (CH$_2$Cl$_2$)	Sep-Pak C$_{18}$ cartridge	—	NPD	OV-101 (1.2 m)	DL: 33 ug/l; Rec: >90%; RSD: 3.1%	80
5, 1-naphthol	Distilled, river, well water	LLE (CH$_3$Cl)	XAD-8	Heptafluorobutyric anhydride/pyridine	ECD	OV-17, XF-1105, diethylene glycol succinate (2 m)	DL: 2.5 to 10 ppb; Rec: 82 to 102%	81
6	Water, soil	LLE (CHCl$_3$)	—	Hydrolysis (Na$_2$CO$_3$)-derivatization (1-fluoro-2,4-dinitrofluorobenzene)	NPD	OV-17-QF-1 (1:1) (1.8 m)	DL: 0.08 to 4 Ppb	82
5,6,10, 1-naphthol, 3-hydroxycarbofuran, 2-isopropoxyphenol	River, pond, waste water	SPE (XAD-2, C$_{18}$) online, eluted with ethyl acetate	—		FID	HP-1 (15 m)	DL: 0.7 to 1 μg/l; Rec: 94 to 103.5%; RSD: 1.9 to 3.9%	13
28,29,31,33, other	Soil	LLE (ethyl acetate) Ultrasonic water bath	—		NPD, MS	HP-1, HP-1701 (30 m)	DL: 3 to 7 ng/g; Rec: 95 to 106%; RSD: 3 to 8%	27
5, other	Water	LLE (n-hexane)	—	Concentration with nitrogen	MS	HP-1 (12 m)	DL: 0.11 μg/l; Rec: 97%	83
35,36,37	Occupational hygiene sampling devices	LLE (Isooctane)	—	ATD	MS	HP-1 (60 m)	DL: 0.05 to 3 μg; Rec: 70 to 110%; RSD: <15%	84
5,6,8,13, 3-hydroxycarbofuran	Solution	SFE (CO$_2$)	—	Heptafluorobutyric anhydride/pyridine	MS	DB-5 (30 m)	QL: 0.48 to 22 ng/ml; Rec: 98.3 to 98.7%	53
5,6, other	Water	SPE (C$_{18}$, XAD-2) eluted with ethyl acetate	—	—	ECD, NPD, FID	DB-5 OV-1701 (30 m)	Rec: 83 to 92%	85
5, other	Pond, well water	SPE (C$_8$) eluted with acetonitrile-n-hexane-CH$_2$Cl$_2$	—	—	NPD	DB-5 (30 m)	Rec: 84 to 106%	86

5,6,25, other	Surface water	SPE (XAD-2, XAD-7) eluted with CH_2Cl_2	—		MS	DB-5 (15 m)	DL: 0.005 to 1 μg/l; Rec: 80.4 to 101.1%	87
5,6,8,10,13,14,15, 3-hydroxycarbofuran	Aqueous solutions	—	—		MS	DB-5 (10 m)	RSD: 12 to 20.5%	88
5,19,29,31, other	Water	SPE (C_8, C_{18} columns and disks eluted with ethyl acetate–n-hexane	—		ECD, NPD	DB-5 (30 m)	Rec: 51 to 92%; RSD: 4 to 11%	89
6,29, other	Soil	LLE acetonitrile–water	LLE petroleum ether–diethyl ether		MS	DB-5 (29 m)	DL: 5 ppb; Rec: 81 to 131%	90
6,28, other	River water	SPM	—		NPD, MS	DB-5 (30 m)	DL: 0.01 to 0.03 μg/l; Rec: 87 to 96%; RSD: 7 to 9%	91
6, other	Soil	SWE	—		MS	HP-5 (30 m)	DL: 30.9 μg/kg; Rec: 97.3 to 106.7%; RSD: 2.3%	43
10, other	Indoor air	Adsorption on glass glass–fiber filter or PUF plug	—	Extraction with ethyl acetate (ultrasonic extraction)	MS	HP-5 (60 m)	DL: 0.1 to 5 ng/m^3; Rec: 73 to 107%; RSD: 3.2 to 9.5%	92
5,8,10,24, isoprocarb	Water, sediment	LLE (CH_2Cl_2) Ultrasonic extraction	SPE (Florisil)	Trifluoroacetic anhydride or diazomethane	MS	Ultra-2 (25 m)	DL: 0.014 to 0.18; 18 ppb; DL: 1.6 to 4.6 ng/g; Rec: 55 to 129%; RSD: 2.6 to 22.6%	93
5,6,19,29, vegadex, other	River, lake, irrigation channel water	SPE (C_{18}) eluted with ethyl acetate–n-hexane	—		NPD, ECD	BP-5 (25 m)	DL: 0.006 to 1.049 μg/l	94
1,5,6,8,10,11	Water	LLE (ethyl acetate) online	—	Acetic anhydride online	FID	HP-17 (10 m)	DL: 0.2 to 0.4 mg/l; RSD: 2.1 to 3.9%	67
6,19,29,31, other	Soil	Ultrasonic extraction (water)	SPE (C_8) eluted with ethyl acetate		NPD, ECD	DB-17 (30 m)	DL: 8.5 to 24.2 ng/g; Rec: 72.7 to 100.1%; RSD: 4.2 to 12.8%	95
5,29,31, other	River water, agricultural drains	SPE (XAD-4)	—	Eluted with CH_2Cl_2-acetone-methanol by Soxhlet extraction	NPD, ECD	DB-1701 DB-1 (30 m)	DL: 25 to 50 ng/l; Rec: 47.3 to 90.6%	96

Continued

TABLE 24.3
Continued

Carbamate Pesticides[a]	Environmental Samples	Extraction Technique[b]	Cleanup Technique[c]	Derivatization Reagent/Other Pretreatment[d]	Detection[e]	Chromatographic Column	Analytical Figures of Merit[f]	Ref.
5	Ground water	SPE (C_{18}) Eluted with CH_2Cl_2	—	—	NPD	BP-10 (12 m)	DL: 1 ppb; Rec: 106%	97
6, other	Soil	LLE (CH_2Cl_2)	SPE (Silica gel)	—	NPD	Supelcowax-10 (30 m)	DL: 20 ng/g	98
14	Soil, sediment, water	LLE (CH_2Cl_2)	SPE (Florisil)	—	FPD	—	—	101

[a] Number of carbamate pesticides (see Table 24.1); other: other noncarbamate pesticides

[b] LLE: liquid–liquid extraction; SPE: solid-phase extraction; SPM: solid-phase microextraction; SWE: subcritical water extraction; MSPD: matrix solid-phase dispersion; SFE: supercritical fluid extraction; PUF: polyurethane foam

[c] SPE: solid-phase extraction; LLE: liquid–liquid extraction; GPC: gel permeation chromatography

[d] ATD: automatic thermal desorption

[e] ECD: electron capture detector; NPD: nitrogen–phosphorus detector; FID: flame ionization detector; MS: mass spectrometry; FPD: flame photometric detector

[f] Lin: linear range; DL: detection limit; QL: quantification limit; Rec: recovery; RSD: relative standard deviation; PC: percent conversion

[g] DL: detection limits in soil

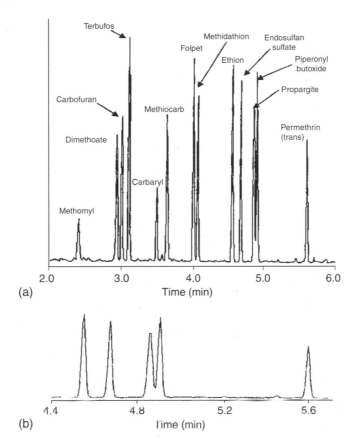

(a)

(b)

FIGURE 24.6 Fast GC–MS analysis of the indicated 13 pesticides obtained with Supersonic GC–MS. (b) This is a zoom of the upper trace (a) in order to demonstrate the symmetric tailing-free peak shapes. A 6 m × 0.2-mm I.D., capillary column with a 0.33-μm DB-5 MS film was used at a 10-ml/min He flow rate. A 1-μl sample volume was injected with an initial concentration of 7 ppm. Methomyl and carbaryl slowly degraded in the methanol solution and their concentrations was assumed to be 3 ppm. The optic injector initial temperature was 100°C and raised to 260°C at a rate of 4°C/sec. The GC oven was started at 80°C for 1 min, followed by a temperature ramp of 35°C/min to 310°C. (Reprinted from Kochman, M., Gordin, A., Goldshlag, P., Lehotay, S. J., and Amirav, A. *J. Chromatogr. A*, 974, 185–212, 2002, copyright 2002, with permission of Elsevier Science.)

The wide scope of application, long-term stability, ease of use, low cost, and improved selectivity (diode array) of UV have turned it into the most widely used detection mode in the determination of carbamate pesticides.[74,80,112,114–125] However, confirmation of pesticides of the same class is made difficult by the high degree of similarity between their UV spectra. Therefore, LC–UV methods are most effective for fast screening of samples, but usually require additional confirmatory analyses for positive samples.

Fluorescence detection is not nearly as widely used as UV detection since most carbamates possess no native fluorescence. However, the structure of these pesticides contains a *N*-methyl substituted urethane with variations in the ester moiety. The common methylamine functionality allows the detection of compounds via a two-stage postcolumn reaction. Carbamates in the column effluent are first hydrolyzed with NaOH at a high temperature to form methylamine, which is converted into a fluorophore compound by addition of *o*-phthaldehyde (OPA) and 2-mercaptoethanol, thiofluor or thiolactic acid.[38,45,58,64,107,108,124–129] The excitation and emission wavelengths vary among compounds (see Table 24.4).

TABLE 24.4
Analytical Methods for the Pretreatment of Environmental Samples and Liquid Chromatographic Determination of Carbamate Pesticides

Carbamate Pesticides[a]	Environmental Samples	Extraction Technique[b]	Cleanup Technique[c]	Derivatization Reagent/Other Pretreatment[d]	Detection[e]	Chromatographic Column/Mobile Phase	Analytical Figures of Merit[f]	Ref.
2,4,5,8,10,13,14,15, butacarb, promecarb	Water, soil	LLE (CH$_2$Cl$_2$)	SPE (silica)	OPA	FL	Silica	—	107
24, other	Drinking, river and estuarine water	SPE disk (C$_{18}$)	—	—	DAD, MS (TSP)	Phenyl	DL: 5 to 2000 ng/l; Rec: 40 to 120%	111
5,8,20,21, barban, promecarb	Surface water	In-tube SPM	—	Online	UV (220 nm)	C$_{18}$ (15 cm) Water–acetonitrile	DL: 0.02 to 0.26 µg/l; RSD: 1.5 to 4.6	74
6	Aqueous solution	LLE (CH$_2$Cl$_2$)	Sep-Pak C$_{18}$ cartridge	—	UV (280 nm)	C$_{18}$ (30 cm) Water–acetonitrile	DL: 37 µg/l; Rec: >90%; RSD: 2.2%	80
5,6,10,20,21,25, captan, barban	Drinking water	SPE (C$_{18}$, C$_8$, TP-201)	—	Online	UV (220 nm)	C$_8$ (25 cm), C$_{18}$ (15 cm) Water–acetonitrile	DL: 10 to 460 pg/ml	114
5,6,14,15,20,21, other	Drinking water	SPE (graphitized carbon black cartridge)	—	—	UV (225 nm)	C$_{18}$ (25 cm) Water–acetonitrile	DL: 0.003 to 0.04 µg/l; Rec: 86 to 101%	115
5,6,14,15,24,31, other	Lake water	SPE (C$_8$)	—	—	UV (210 nm)	C$_{18}$ (25 cm) Water–acetonitrile	LD: 0.07 to 0.21 µg/l; Rec: 52 to 95%; RSD: 3 to 9%	116
11,14,15, other	River, ground water	SPE (C$_{18}$, PLRP to S)	—	Online	UV (220 nm)	C$_{18}$ (10 cm) Acetonitrile–phosphate buffer	DL: 0.1 µg/l; Rec: 23 to 100%	117
5,6 other	Atmospheric gas, aerosol	Collector (XAD-2, glass fiber filters)	Soxhlet (n-hexane-CH$_2$Cl$_2$)	—	UV (254 nm)	C$_{18}$ (30 cm) Water–methanol–acetonitrile	DL: 70 to 440 pg/m^3; RSD: <10%	118
6, other	Water	SPE (PolyF + LiChrolut EN)	—	Online TAD	UV/DAD (223 nm)	C$_{18}$ (12.5 cm) Water–acetonitrile	DL: <100 ng/l	119
8,10,20,21, thiuram	Soil	MAE	—	—	UV	C$_{18}$	Rec: 70 to 99%	120
6,13, metabolites, other	River, sea water	SPE (C$_{18}$) disk	—	—	DAD	C$_{18}$ (12.5 cm)	DL: 0.01 to 3 µg/l; Rec: 74 to 125%; RSD: 5 to 10%	121
6,8,13,14, other	Ground, drinking, sea water	SPE (C$_{18}$)	—	—	DAD 212 nm	C$_{18}$ (15 cm) Water–methanol–acetonitrile	DL: 0.1 to 0.3 µg/l; Rec: 14 to 110%; RSD: 3.3 to 22.4%	122

Analytes	Matrix	Extraction		Derivatization	Detection	Column / Mobile phase	Performance	Ref.
6,33, other	Water	SPE (C$_{18}$, PSDB)	—	—	DAD (220 nm)	C$_{18}$ (25 cm) Water–acetonitrile	DL: 144 to 1057 µg/l; Rec: 35 to 107%; RSD: 4.1 to 14.7%	123
5, 1-naphthol	Water	SPE (C$_{18}$)	—	Online	DAD/FL	C$_{18}$	DL: 10 to 50 ng/l; Rec: 93 to 102%; RSD: 1 to 4%	124
1,5,6,8,10,13,14,15,16,21, other carbamates	Surface, ground water	SPE (C$_{18}$)	—	OPA—thiofluor	DAD/FL	C$_{18}$ (15 cm) Water–methanol	Rec: 63 to 100%; RSD: 1.9 to 8.1%	125
5,8,10,11	Soil	SFE (CO$_2$)	—	OPA	FL	C$_{18}$ (25 cm) Water–methanol	Rec: 39.6 to 91.7%	38
5,6,15, dioxacarb, metolcarb	Soil	FMASE	—	OPA/Thiolactic acic Postcolumn (Fl)	FL (340 and 445 nm)	C$_{18}$ (25 cm) Water–methanol	Rec: 75 to 80%; RSD: 2.34 to 7.53%	45
5,13,19	Soil	SFE (CO$_2$) with dimethyl sulfoxide	—	OPA/mercaptoethanol Post column	FL (330 and 450 nm)	C$_{18}$ (15 cm) Water–methanol	Rec: 91.5 to 107.8%	58
1,3,5,8,10,13,14,15, 17, other	River, lake water	SPE (PLRP-S, C$_{18}$) Online (Prospect)	—	OPA	FL	C$_8$ (25 cm)	DL: 30 to 50 ng/l; Rec: 60 to 108.4%; RSD: 2 to 10%	64
5	Water	SPE (C$_{18}$) Online	—	Catalytic hydrolysis, OPA Postcolumn	FL	C$_{18}$	DL: 0.4 to 2 ng; Rec:104 to 106%; RSD: 2%	126
1,3,4,5,7,8,10,13,14, 15,17, other carbamates	River, lake water	SPE (C$_{18}$, C$_{18}$-OH)	—	Hydrolysis, OPA/ mercaptoethanol Postcolumn	FL (340 and 445 nm)	C$_8$ (25 cm) Water–methanol–acetonitrile	DL: 20 to 30 ng/l; Rec: 76 to 106%; RSD: 0.5 to 0.8%	127
5,6,13, other	Drinking water	SPE (C$_{18}$) Online	—	OPA Postcolumn	FL (330 and 465 nm)	C$_8$ (25 cm) Water–acetonitrile	DL: 0.01 to 0.03 µg/l	128
11,12, metabolites	Natural water, soil	—	—	OPA/mercapto ethanol Postcolumn	FL (230 and 418 nm)	C$_8$ (20 cm) Water–acetonitrile	DL: 0.1 to 0.4 ng; Rec: 72 to 98.4%; RSD: 5 to 10.7%	129
1,3,4,5,6,8,10,11,12,13,14, 15,17,19,20,21,33, other carbamates	Surface water	SPE (C$_{18}$/OH)	—	Online	MS (PB) GC/MS	C$_{18}$ (25 cm) Water–methanol	DL: 0.1 to 8 µg/l; RSD: 0.05 to 0.2%	65
5,6,13,14,15,20,21, other	River water	SPE (graphitized carbon black cartridges)	—	—	MS (PB)	C$_{18}$ (25 cm) Water–acetonitrile	DL: 0.2 to 30 ppb; Rec: 85 to 101%	130

Continued

TABLE 24.4
Continued

Carbamate Pesticides[a]	Environmental Samples	Extraction Technique[b]	Cleanup Technique[c]	Derivatization Reagent/Other Pretreatment[d]	Detection[e]	Chromatographic Column/Mobile Phase	Analytical Figures of Merit[f]	Ref.
20,22, other	River, ground, tap water	SPE (C$_{18}$ cartridge)	—	—	MS (FAB with FRIT interface)	C$_8$ (15 cm) Water–methanol	DL: 0.01 to 0.15 µg/l; Rec: 92.9 to 98.7%	131
5,6,8,10,19,20,21, other	Surface, drinking water	SPE (PLRP-S, PRP-1, C$_8$) Online	—	—	MS (TSP)	C$_{18}$ (12.5 cm) Water–methanol	DL: 1 to 6 ng/l; Rec: 59 to 81%; RSD: 1 to 15%	132
5,6,8,10,12,13,14,15, other	Lake, well, cistern, pond, tap water	SPE (C$_{18}$) Online	—	—	MS (TSP)	C$_{18}$ (15 cm) Water–acetonitrile–Acetic acid	DL: 41 to 210 pg/ml; Rec: 75 to 124%; RSD: 11 to 16%	133
5,8,20,21, barban, promecarb, other	Tap, surface, well water	In-tube SPM	—	Online	MS (ESI) UV	C$_{18}$ (5 cm) Water–acetonitrile	DL: 0.01 to 1.2 ng/ml; RSD: 2.7 to 6.3%	134
5,6,13,14,15,19	Water	SPE (C$_{18}$, PSDB, NVPDB)	—	—	MS (ESI)	C$_{18}$ (10 cm) Water–methanol	DL: 0.1 to 0.5 µg/l; Rec: 73.7 to 92.6%; RSD: 11.1 to 14.7%	135
5,11,20,24, other carbamates	Water	—	—	—	EL (CV)	C$_{18}$ (25 cm) Phosphate buffer–acetonitrile	DL: 40 to 150 pg; RSD: 1 to 2%	136
5,11,20,24, other carbamates	River water	SPE (C$_{18}$)	—	—	EL (MAECFD)	C$_{18}$ (25 cm) Phosphate buffer–methanol	DL: 2.6 to 22 ppb; Rec: next to 100%; RSD: 3.4 to 17%	137

[a] Number of carbamate pesticides (see Table 24.1); other: other noncarbamate pesticides

[b] LLE: liquid–liquid extraction; SPE: solid-phase extraction; SPM: solid-phase microextraction; MAE: microwave-assisted extraction; SFE: supercritical fluid extraction; FMASE: focused microwave-assisted Soxhlet extraction; PSDB: polystyrene-divinylbenzene; NVPDB: *N*-vinylpyrrolidane–divinylbenzene

[c] SPE: solid-phase extraction; LLE: liquid–liquid extraction; GPC: gel permeation chromatography

[d] OPA: *o*-phthalaldehyde reagent; TAD: thermally assisted desorption; FI: flow injection

[e] FL: fluorescence; DAD: diode array detector; UV: UV detector; MS: mass spectrometry; PB: particle beam interface; TSP: thermospray interface; ESI: electrospray ionization; EC: electrochemical detection; CV: cyclic voltammetry; MAECFD: micro-array electrochemical flow detector

[f] DL: detection limit; Rec: recovery; RSD: relative standard deviation

MS is becoming the detection system of choice for LC by virtue of its flexibility and high selectivity for individual solutes.[138,139] However, LC–MS is always less sensitive than GC–MS as a result of the need to transfer the analytes from the liquid phase into a high-vacuum gas phase. Other limitations of LC–MS combination include the inability to use nonvolatile buffers, the narrow optimum range for eluent flow rate influence of the proportion of organic modifier on the sensitivity, and the narrow choice of ionization methods.[140] Nevertheless, LC–MS has been widely accepted as an advantageous choice for the determination of carbamate pesticides in water matrices, which is more robust and flexible in the absence of derivatization.[65,111,130–135] Thermospray and particle-beam interfaces are probably most commonly used for offline and online determination of carbamates in water.[65,111,130,132,133] Atmospheric pressure sources such as electrospray ionization (ESI) have some advantages as samples are ionized directly in the liquid phase at a *quasi*-ambient temperature, thereby minimizing the degradation of thermally labile compounds.[134,135] Figure 24.7a shows the selective ion mass (SIM) chromatograms for four different types of water spiked with six carbamates determined by automated in-tube SPME coupled with LC/MS (in-tube SPME–HPLC–MS). This combination is more selective and sensitive than in-tube SPME–HPLC–UV (see Figure 24.7b). Heated nebulizers in the atmospheric chemical ionization mode (APCI) have also been used in the determination of carbamates.[141]

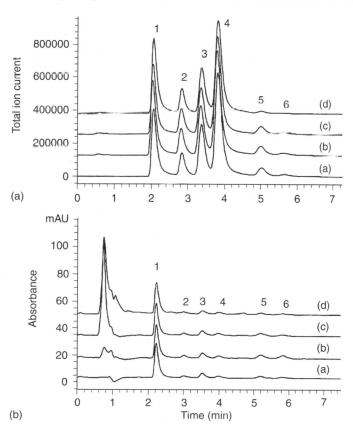

FIGURE 24.7 Chromatograms for six carbamates spiked into the different water samples obtained by (a) in-tube SPME–HPLC–ESI–MS under the total selected ion monitoring (SIM) mode and (b) by in-tube SPME–HPLC–UV. Twenty nanograms of each pesticide were spiked to each of the 1-ml water samples. Water samples: (a) pure water, (b) well water, (c) tap water, (d) surface (lake) water. Peak identification: (1) carbaryl, (2) propham, (3) methiocarb, (4) promecarb, (5) chlorpropham, (6) barban. (Reprinted from Wu, J., Tragas, C., Lord, H., and Pawliszyn, J. *J. Chromatogr. A*, 976, 357–367, 2002, copyright 2002, with permission of Elsevier Science.)

Electrochemical detection is possible for analytes amenable to oxidation or reduction at moderate electrode potentials. The selectivity can to some extent be adjusted through electrode potential. However, this methodology has scarcely been used for the detection of carbamate pesticides. Thus, Anderson et al. determined various carbamates in water samples by LC using a cyclic voltammetry detector[136] or a micro-array electrochemical flow detector.[137]

Multidimensional chromatographic techniques such as online coupled liquid chromatography–liquid chromatography (LC–LC), also called "LC with column switching", and liquid chromatography–gas chromatography (LC–GC), are excellent tools for the determination of pesticides in environmental and biological samples. The advantages and disadvantages of these combinations are discussed in two reviews.[142,143]

C. THIN-LAYER CHROMATOGRAPHY

In the past, TLC and spectroscopy were the firsts analytical techniques used for the determination of carbamate pesticides, with their inherent difficulties (particularly their limited sensitivity). With the inception of GC and LC, TLC applications were restricted to sample cleanup, metabolic studies, and, perhaps most important, confirmatory analyses in clinical and medical-legal cases requiring positive, unambiguous identification, and quantitative determination.[144,145] At present, TLC continues to offer attractive features, such as, parallel sample processing for a high sample throughput, accessibility of the sample for postchromatographic evaluation free of time constraints, detection in the presence of the stationary phase, which is somewhat independent of mobile phase and normally used only once. Table 24.5 shows selected examples of the use of TLC for the determination of carbamates.[146–151]

High performance thin-layer chromatography (HPTLC) has also been used for the determination of carbamate pesticides.[152] Thus, TLC methods provide increased selectivity through silica derivatization, as well as higher analytical precision and sensitivity with high-performance plates. Butz and Stan[153] reported an HPTLC system with automated multiple development (AMD–HPTLC) to screen water samples for pesticides. The method was applied to the determination of 265 pesticides in drinking water spiked with 100 ng/l of each analyte.

D. SUPERCRITICAL FLUID CHROMATOGRAPHY

SFC has received increasing attention in recent years, especially in the environmental, pharmaceutical, petroleum, and polymer fields. Because supercritical fluids are like gas in some aspects and liquid in others (e.g., they are typically 10 to 100 times less viscous than liquids), they can be used as mobile phases, thus providing supplementary aids for HPLC and GC.[36] The most salient advantages of SFC include:

(i) Shorter retention times in the analysis of moderately polar and thermally labile pesticides;
(ii) Flexibility in the separation conditions (modifier, stationary phase);
(iii) Compatibility with most LC and GC (UV, FID, NPD, MS) detectors.[154]

Many analytes that are not amenable to GC (e.g., thermolabile compounds) can be separated by SFC. Separations via SFC are often more efficient and faster than traditional LC analyses. Wider coverage of supercritical fluid methods can be found in the literature on SFC.[36,154–156] This technique has also been employed for the determination of carbamate pesticides in environmental samples (see Table 24.5).[157–160]

Packed-column supercritical fluid chromatography (PSFC) is currently competitive with LC and GC as it combines the speed and efficiency of GC with the extensive selectivity adjustment capabilities of LC, thereby facilitating the determination of polar and thermolabile compounds.[161] Thus, carbamate pesticides have been determined by PSFC in river water[162] and soil.[163]

TABLE 24.5
Other Chromatographic Methods for the Determination of Carbamate Pesticides in Environmental Samples

Carbamate Pesticides[a]	Environmental Samples	Extraction/Cleanup Techniques[b]	Derivatization Reagent/Other Pretreatment[c]	Detection[d]	Stationary Phase-/Column/Mode[e]	Mobile Phase[f]	Analytical Figures of Merit[g]	Ref.
Thin-Layer Chromatography								
5,10	Water, soil	LLE (CH$_2$Cl$_2$)	—	—	—	—	—	146
5,8,10,11,12	River water	LLE (CHCl$_3$)/H$^+$	—	—	—	—	—	147
5,9	Water	LLE (ethyl ether), pH 2	—	—	Alumina	—	—	148
6, metabolites	Water, soil, plants	LLE (CH$_2$Cl$_2$)/Florisil	NBDF, DBBQC	—	Silica Gel G	CH$_2$Cl$_2$–acetonitrile, n-hexane–ethyl acetate	DL: 0.1 to 0.3 ppm	149
5,6,10,24, other	Water	—	—	—	Silica gel G, alumina	Acetone, benzene, CHCl$_3$, CCl$_4$, ethanol	—	150
5,6,10	Forensic samples	—	4-aminoantipyrine	—	—	—	—	151
5,6,8,10	Water	SPE (C$_{18}$)	NBDF	Densitometric scanning	HP-plate, Silica gel, (HPTLC)	—	Rec: 82.5 to 112%; RSD: 7.5%	152
1,5,6,8,10,11,13, 14,15,17,19, 20,21,22,23, 25,26,27,28, 33, other	Drinking water	SPE (C$_{18}$)	—	UV, 190–298 nm	HP-plate, 60 F 254 Silica gel, (AMD–HPTLC)	Acetonitrile–CH$_2$Cl$_2$	DL: 5 to 250 ng; RSD: 1 to 2%	153
Supercritical Fluid Chromatography								
5,10,20, phenmedipham	Water	—	—	FID	SE-54 (0.9 m)	CO$_2$	—	157
1,5, other	Water	—	—	DAD	PMS (12 m)	CO$_2$	Lin: 3.8 to 150 ng; RSD: 1.2%	158
1,5,8,13, other	Water	—	—	MS (EI)	Capillary	CO$_2$	RSD: 6.4	159

Continued

TABLE 24.5
Continued

Carbamate Pesticides[a]	Environmental Samples	Extraction/Cleanup Techniques[b]	Derivatization Reagent/Other Pretreatment[c]	Detection[d]	Stationary Phase/-Column/Mode[e]	Mobile Phase[f]	Analytical Figures of Merit[g]	Ref.
5	Water	SPE, Online	—	DAD (210 nm)	—	—	DL: 5 ppb	160
6,20,21, aldicarb sulphone, other	River water	SPE (C$_{18}$, PLRP-S, LiChrolut EN), Online	—	DAD, (210–220 nm)	Hypersil silica, (20 cm) packed	CO_2–methanol	DL: 1.3 to 2.5 µg/l; Rec: 99.3 to 108.2%; RSD: 5.56 to 14.02%	162
6, other	Soil	LLE, (methanol–CH_2Cl_2), sonication	—	MS (APCI)	Cyanopropylsilica (25 cm) packed	CO_2–methanol	DL: 285 pg; RSD: 8.6%	163
Capillary Electrophoresis								
5, other	River water	SPE (C$_{18}$) Online	—	DAD, (226 nm)	Fused-silica, Lt: 47 cm, MEKC	Phosphate–SDS–acetonitrile, pH 9.5/25 kV	DL: 0.02 µg/ml; QL: 0.05 µg/ml; Rec: 90.3 to 113.5%	70
6,10,21, other	Aqueous solution	—	—	UV (200–300 nm)	Fused-silica, Lef:63 cm, MEKC	Borate-SDS buffer pH 8/30 kV	DL: 0.08 to 0.13 mg/l; RSD: 1.7 to 5.1%	167
5,6,10,11,13, 14,15, other carbamates	Drinking water	SPE (C$_{18}$)	Oncolumn staking	UV	Fused-silica, Lt: 90 cm, MEKC	Borate-SDS buffer pH 7/18 kV	DL: 0.1 1 ppb	168
5,6,8,10,11, metabolites	Well, river, pond water	SPE (LiChrolut EN)	Oncolumn staking	DAD (202, 214 nm)	Fused-silica, Lef: 50 cm, MEKC	Borate-phosphate-SDS buffer, pH 89/15 kV	DL: 22 to 85 ng/l; Rec: 82.2 to 108.2%; RSD: 2.6 to 7.4%	169
1,5,6,8,10,11, 13,14,15	Tap water	—	—	DAD, (266 nm)	C$_{18}$, Lef: 25 cm, CEC	Phosphate–ammonium acetate buffer–acetonitrile, pH 6/ 20 kV	QL: 10^7 to 10^8 M; RSD: 1.5 to 2.9%	170

5,6,8, other	Tap, lake water	—	DAD, (208, 282 nm), z-cell	C_{18}, C_8, Lef: 25 cm, CEC	Ammonium acetate buffer–acetonitrile, pH 6/20 kV	DL: 3.8 to 8.9 $\times 10^8\ M$	171
5,6,10,11,13, 14,15,19	Water	—	MS (ESI)	Fused-silica (uncoated), AMPS-coated, Lt: 88 cm, MEKC	SDS- ammonium acetate buffer, pH 5, 8.5 or 9/25 kV	DL: 0.04 to 2 µg/l	172
6,10,11,13, 14,15, other carbamates	Water	OPA 2-mercapto-ethanol, Oncolumn TD	FL (340 ard 450 nm)	Fused-silica, Lt: 90 cm, MEKC	Borate-CTAB buffer, pH 9/20 kV	DL: < 0.5 ppm	59

[a] Number of carbamate pesticides (see Table 24.1); other: other noncarbamate pesticides

[b] LLE: liquid–liquid extraction; SPE: solid-phase extraction

[c] NBDF: 4-nitrobenzenediazonium fluoroborate; DBBQC: 3,5-dibromo-p-benzoquinonechlorimine; OPA: o-phthalaldehyde; Orcolumn-TD: oncolumn thermal decomposition

[d] DAD: diode array detector; UV: UV detector MS: mass spectrometry; EI: electron ionization; ESI: electrospray ionization; FID: flame ionization detector; APCI: atmospheric pressure chemical ionization; z-cell: z-cell configuration; FL: fluorescence detection

[e] HP-plate: high performance preadsorbed plates; AMD–HPTLC: automated multiple development–high performance thin-layer chromatography; PMS: 5% phenylmethylsilicone; Lef: effective length; Lt: total length; MEKC: micellar electrokinetic capillary chromatography; CEC: capillary electrochromatography; AMPS-coated: fused silica capillary coated with poly(sodium 2-acrylamide-2-methylpropanesulfonate)

[f] SDS: sodium dodecyl sulfate; CTAB: cetyltrimethyl–ammonium bromide

[g] DL: detection limit; QL: quantification limit; Rec: recovery; RSD: relative standard deviation; Lin: linear range

E. CAPILLARY ELECTROPHORESIS

CE has become a major choice for the analytical laboratory. Initially introduced as a technique for the separation of biological macromolecules, CE has since attracted much interest in other areas including the determination of pesticides.[164] The advantages of CE over conventional chromatographic techniques include:

(i)   The need for no organic solvents to prepare the running buffer;
(ii) The use of small volumes;
(iii) The low cost of capillaries relative to LC or GC columns.

In some cases, organic solvents can be used as modifiers. However, they never account for more than 5 to 30% of the total amount of solvent. The use of small sample volumes (ca. 1 to 10 nl) in the determination of pesticides can result in inadequate detection sensitivity. The sensitivity of CE can be enhanced by using a more sensitive detector or by inserting a sample-enrichment step (e.g., SPE, LLE, or SPME) before separation.[165]

Most CE work so far has been done using the capillary zone electrophoresis (CZE) mode, where analytes are separated on the basis of differences in electrophoretic mobility, which is related to charge density. The separation is carried out in a capillary filled with a continuous background electrolyte (buffer). Micellar electrokinetic capillary chromatography (MEKC or MECC) is one other CE method based on differences in the interaction of the analytes with micelles present in the separation buffer, which can easily separate both charged and neutral solutes with either hydrophobic or hydrophilic properties. An alternative to MEKC is capillary

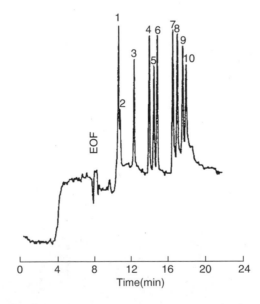

FIGURE 24.8 Typical MEKC separation of ten N-methylcarbamates (5 ppm each) with oncolumn thermal decomposition and fluorescence detection. Running electrolyte: 10 mM borate 40 mM CTAB (pH 9.0) containing 28 ppm OPA and 200 ppm 2-mercaptoethanol. Capillary: i.d. 50 μm, total length 90 cm. The heating site and fluorescence detection window were located at 65 and 77 cm, respectively, from the capillary inlet. Applied voltage, 20 kV. Temperature of the heating circulator, 120°C. Fluorescence detection: $\lambda_{ex} = 340$ nm, $\lambda_{em} = 450$ nm. Peak identification: (1) oxamyl, (2) methomyl, (3) aldicarb, (4) propoxur, (5) carbofuran, (6) aminocarb, (7) isoprocarb, (8) trimethacarb, (9) fenobucarb, (10) promecarb. (Reprinted from Wu, Y. S., Lee, H. K. and Li, S. F. Y. *Anal. Chem.*, 72, 1441–1447, 2000, copyright 2000, with permission of the American Chemical Society.)

electrochromatography (CEC), which uses a stationary phase (C_{18}) rather than a micellar *pseudo*-stationary phase. CEC is a hybrid technique that combines the selectivity of LC and the separation efficiency of CE.[166]

Table 24.5 summarizes the features of selected CE methods for the determination of carbamate pesticides in environmental samples. Most use UV or DAD for detection.[70,167–171] MS[172] and fluorescence detectors[59] have also been used for this purpose. A typical electropherogram for the fluorescence detection of ten *N*-methylcarbamates at a concentration of 5 ppm using MEKC with oncolumn thermal decomposition is shown in Figure 24.8.[59]

V. APPLICATIONS IN THE ANALYSIS OF ENVIRONMENTAL SAMPLES

As can be seen from Tables 24.3–24.5, most of the matrices where carbamate pesticides are determined are waters (e.g., drinking, tap, well, river, stream, lake, pond, ground, sea, and waste). Also, most of the chromatographic methods reviewed in this chapter use some pretreatment technique such as SPE, LLE, or SPME to preconcentrate and eliminate matrix effects. Most have precision and sensitivity values compliant with existing regulations (see Section II.D). Their accuracy has been frequently assessed in recovery studies involving the use of spiked samples or comparing the results with those provided by another independent method. Some interlaboratory exercises have been made to determine pesticides in water. Thus, the BCR (Community Bureau of Reference) of the European Union (currently the Measurement and Testing Program) launched a project with a view to preparing a reference material for pesticides (e.g., carbaryl, atrazine, simazine, propanil, and linuron) in freeze-dried water samples.[173,174]

Analyses of soil require stronger sample pretreatments, such as, SFE, MASE, and ultrasonic LLE, owing to the increased complexity of the matrix. These operations are prone to error, so they detract somehow from precision and accuracy.

While air and atmospheric analyses have been the subjects of many research articles, chromatographic determinations of pesticides in these types of sample are scant. Various methods for sampling pesticides in air have been reported, which use silica gel, Tenax, Chromosorb, XAD-2, polyurethane foam plug, glass fiber filters, and activated carbon or carbon fiber filters as sampling media.[92,118,175] Pesticides are subsequently desorbed by LLE, LLE in an ultrasonic bath,[92] or Soxhlet extraction.[118]

VI. CONCLUSIONS

This chapter reviews the wide variety of methods available for the determination of carbamate pesticides in environmental samples using chromatographic techniques. This is a consolidated field of study, where much room remains for further work. Particularly useful are bound to be additional developments in sensitive and selective detection systems minimizing the need to pretreat samples (a source of major errors). Also, new procedures using low solvent volumes and nonpollutant reagents are desirable in order to efficiently protect the environment.

Coupled chromatographic techniques are gaining significance for the determination of pesticides in environmental and biological samples. The main advantages of multidimensional chromatographic techniques are enhanced sensitivity, improved sensitivity (desired from the use of large-volume injections combined with peak compression), and automatability (online systems). On the other hand, the use of pretreatment sampling systems coupled with chromatographic instruments reduces sample and reagent consumption, minimizes or even avoids human participation in preliminary analytical operations, increases sample throughput, and provides results of a higher quality.

Finally, the development of methods for the quantitation of a relatively large number of pesticides with widely different chemical properties in one run is essential as multiresidue methods often involve a compromise between the number of analytes that can be determined and the selectivity that can be achieved.

ACKNOWLEDGMENTS

The author thanks the University of Jaén (Spain) for financial support (Plan de Ayuda para el Fomento de la Investigación, UJA 2004).

REFERENCES

1. Landis, W. G. and Yu, M. H., *Introduction to Environmental Toxicology*, 2nd ed., Lewis Publishers, Boca Raton, 1999, chap. 3.
2. van der Hoff, G. R. and van Zoonen, P., Trace analysis of pesticides by gas chromatography, *J. Chromatogr. A*, 843, 301–322, 1999.
3. Ripley, B. D. and Chau, A. S. Y., Carbamate pesticides, In *Analysis of Pesticides in Water, Nitrogen-Containing Pesticides*, Vol. III, Robinson, J. W., Chau, A. S. Y. and Afghan, B. K., Eds., CRC Press, Boca Raton, FL, pp. 1–182, 1982.
4. Ballesteros Tribaldo, E., Residue analysis of carbamate pesticides in water, In *Handbook of Water Analysis*, Nollet, L. M. L., Ed., Marcel Dekker, New York, 2000, chap. 27.
5. WHO. *Environmental Health Criteria 64, Carbamate Pesticides: A General Introduction*, World Health Organization, Geneva, 1986.
6. Tomlin, C., *The Pesticide Manual*, 10th ed., British Crops Protection Council, Surrey, 1994.
7. Hartley, G. S., Graham-Bryce, I. J., *Physical Principles of Pesticides Behavior*, Vol. 2, Academic Press, New York, 1980.
8. Barceló, D., Occurrence, handling, and chromatographic determination of pesticides in the aquatic environment, *Analyst*, 116, 681–689, 1991.
9. Fielding, M., Barceló, D., Helwey, A., Galassi, S., Tortenson, L., van Zoonen, P., Wolter, R., and Angetti, G., *Pesticides in Ground and Drinking Water, Water Pollutions Research Report, No. 27*, Commission of the European Community, Brussels, 1992, pp. 1–136.
10. Martín-Esteban, A., Fernández, P., Fernández-Alba, A., and Cámara, C., Analysis of polar pesticides in environmental waters: a review, *Quim. Anal.*, 17, 51–66, 1998.
11. Zepp, R. G., Photochemical fate of agrochemicals in natural waters, In *Pesticide Chemistry: Advances in International Research, Development, and Legislation*, Frehse, H., Ed., VCH, Weinheim, pp. 329–346, 1991.
12. Racke, K. D. and Coats, J. R., Eds., *Enhanced Biodegradation of Pesticides in the Environment, ACS Symposium Series 426*, American Chemical Society, Washington, DC, 1990.
13. Ballesteros, E., Gallego, M., and Valcárcel, M., Online preconcentration and gas chromatographic determination of *N*-methylcarbamates and their degradation products in aqueous samples, *Environ. Sci. Technol.*, 30, 2071–2077, 1996.
14. Fang, F., Kanan, S., Patterson, H. H., and Cronan, C. S., A spectrofluorimetric study of the binding of carbofuran, carbaryl, and aldicarb with dissolved organic matter, *Anal. Chim. Acta*, 373, 139–151, 1998.
15. Büyüksönmez, F., Rynk, R., Hess, T. F., and Bechinski, E., Occurrence, degradation, and fate of pesticides during composting. Part I: composting, pesticides, and pesticide degradation, *Compost Sci. Utiliz.*, 7, 66–82, 1999.
16. Corbett, J. R., Wright, K., and Baillie, A. C., *The Biochemical Mode of Action of Pesticides*, 2nd ed., Academic Press, London, 1984, chap. 3.
17. FAO/WHO. *Pesticide residues in food. Report of the 1983 Joint Meeting of the FAO Panel of Experts on Pesticide Residues in Food and the Environment and the WHO, Expert Group on Pesticide Residues*, Food and Agriculture Organization of the United Nations, Rome, 1984, FAO Plant Production and Protection Paper 56.
18. Hajslová, J., Pesticides, In *Environmental Contaminants in Food*, Moffat, C. F. and Whittle, K. J., Eds., CRC Press, Boca Raton, FL, 1999, chap. 7.

19. U.S. Environment Protection Agency. *National Pesticide Survey, Phase I. Report PB-91-125765*, National Technical Information Service, Springfield, VA, 1990.

20. U.S. Environment Protection Agency. *National Pesticide Survey, Phase II Report EPA 570/9-91-020*, National Technical Information Service, Springfield, VA, 1992.

21. Economic European Communities. *Pollution caused by certain dangerous substances discharged into the aquatic environment of the Community (Black list), Off. J. Eur. Commun. No. L129/7. Directive 76/464/ECC*, ECC, Brussels, 1976.

22. Economic European Communities. *Drinking Water Directive, Off. J. Eur. Commun. No. 299/11. Directive 80/778/ECC*, ECC, Brussels, 1980.

23. World Health Organization. *Drinking Water Quality: Guidelines for Selected Herbicides*, WHO, Copenhagen, 1987.

24. Majors, R. E., Trends in sample preparation, *LC–GC Europe*, 16(2), 71–81, 2003.

25. Santos, F. J. and Galceran, M. T., The application of gas chromatography to environmental analysis, *Trends Anal. Chem.*, 21, 672–685, 2002.

26. Bellar, T. A. and Budde, W. L., Determination of nonvolatile organic compounds in aqueous environmental samples using liquid chromatography/mass spectrometry, *Anal. Chem.*, 60, 2076–2083, 1988.

27. Sánchez-Brunete, C., Pérez, R. A., Miguel, E., and Tadeo, J. L., Multiresidue herbicide analysis in soil samples by means of extraction in small columns and gas chromatography with nitrogen–phosphorus and mass spectrometry detection, *J. Chromatogr. A*, 823, 17–24, 1998.

28. Berrueta, L. A., Gallo, B., and Vicente, F., A review of solid phase extraction: basic principles and new developments, *Chromatographia*, 40, 474–483, 1995.

29. Thurman, E. M. and Mills, M. S., Eds., *Solid-Phase Extraction, Principles and Practice*, Wiley, New York, 1998.

30. Soriano, J. M., Jiménez, B., Font, G., and Moltó, J. C., Analysis of carbamate pesticides and their metabolites in water by solid phase extraction and liquid chromatography: a review, *Crit. Rev. Anal. Chem.*, 31, 19–52, 2001.

31. Mena, M. L., Martínez-Ruiz, P., Reviejo, A. J., and Pingarrón, J. M., Molecularly imprinted polymers for online preconcentration by solid phase extraction of pirimicarb in water samples, *Anal. Chim. Acta*, 451, 297–304, 2002.

32. Arthur, C. L. and Pawliszyn, J., Solid phase microextraction with thermal desorption using fused silica optical fibers, *Anal. Chem.*, 62, 2145–2148, 1990.

33. Pawliszyn, J., *Solid Phase Microextraction: Theory and Practice*, Wiley, New York, 1997.

34. Krutz, L. J., Senseman, S. A., and Sciumbato, A. S., Solid-phase microextraction for herbicide determination in environmental samples, *J. Chromatogr. A*, 999, 103–121, 2003.

35. Jinno, K., Muramatsu, T., Saito, Y., Kiso, Y., Magdic, S., and Pawliszyn, J., Analysis of pesticides in environmental water samples by solid-phase micro-extraction–high-performance liquid chromatography, *J. Chromatogr. A*, 754, 137–144, 1996.

36. Smith, R. M., Ed., *Supercritical Fluid Chromatography*, The Royal Society of Chemistry, London, 1988.

37. Luque de Castro, M. D., Valcárcel, M., and Tena, M. T., *Analytical Supercritical Fluid Extraction*, Springer, Berlin, 1994.

38. Izquierdo, A., Tena, M. T., Luque de Castro, M. D., and Valcárcel, M., Supercritical fluid extraction of carbamate pesticides from soils and cereals, *Chromatographia*, 42, 206–212, 1996.

39. Ahmed, F. E., Analyses of pesticides and their metabolites in foods and drinks, *Trends Anal. Chem.*, 20, 649–661, 2001.

40. Sánchez-Brunete, C., Albero, B., Miguel, E., and Tadeo, J. L., Determination of insecticides in honey by matrix solid-phase dispersion and gas chromatography with nitrogen–phosphorus detection and mass spectrometry confirmation, *J. AOAC Int.*, 85, 128–133, 2002.

41. Wenzel, K. D., Hubert, A., Manz, M., Weissflog, L., Engewald, W., and Schüürman, G., Accelerated solvent extraction of semivolatile organic compounds from biomonitoring samples of pine needles and mosses, *Anal. Chem.*, 70, 4827–4835, 1998.

42. Okihashi, M., Obana, H., and Hori, S., Determination of N-methylcarbamate pesticides in foods using an accelerated solvent extraction with a mini-column cleanup, *Analyst*, 123, 711–714, 1998.

43. Richter, P., Sepúlveda, B., Oliva, R., Calderón, K., and Seguel, R., Screening and determination of pesticides in soil using continuous subcritical water extraction and gas chromatography–mass spectrometry, *J. Chromatogr. A*, 994, 169–177, 2003.

44. García-Ayuso, L. E., Sánchez, M., Fernández de Alba, A., and Luque de Castro, M. D., Focused microwave-assisted Soxhlet: an advantageous tool for sample extraction, *Anal. Chem.*, 70, 2426–2431, 1998.

45. Prados-Rosales, R. C., Herrera, M. C., Luque-García, J. L., and Luque de Castro, M. D., Study of the feasibility of focused microwave-assisted Soxhlet extraction of *N*-methylcarbamates from soil, *J. Chromatogr. A*, 953, 133–140, 2002.

46. Tekel, J. and Hatrík, S., Pesticide residue analyses in plant material by chromatographic methods: clean-up procedures and selective detectors, *J. Chromatogr. A*, 754, 397–410, 1996.

47. Wells, D. E., Extraction, clean-up and recoveries of persistent trace organic contaminants from sediment and biota samples, In *Environmental Analysis: Techniques, Applications, and Quality Assurance*, Barceló, D., Ed., Elsevier, Amsterdam, 1996, chap. 3.

48. Steinwandter, H., Contributions to the application of gel chromatography in residue analysis. I. A new gel chromatographic system using acetone for the separation of pesticide residues and industrial chemicals, *Fresenius J. Anal. Chem.*, 331, 499–502, 1988.

49. Balinova, A., Multiresidue determination of pesticides in plants by high-performance liquid chromatography following gel permeation chromatographic clean-up, *J. Chromatogr. A*, 823, 11–16, 1998.

50. Specht, W., Pelz, S., and Gilsbach, W., Gas-chromatographic determination of pesticide residues after clean-up by gel-permeation chromatography and mini-silica gel-column chromatography, *Fresenius J. Anal. Chem.*, 353, 183–190, 1995.

51. Blau, K. and Halket, J. M., *Handbook of Derivatives for Chromatography*, 2nd ed., Wiley, Chichester, 1993.

52. Toyo'oka, T., *Modern Derivatization Methods for Separation Sciences*, Wiley, Chichester, 1999.

53. King, J. W. and Zhang, Z., Derivatization reactions of carbamate pesticides in supercritical carbon dioxide, *Anal. Bioanal. Chem.*, 374, 88–92, 2002.

54. Krull, I. S., *Reaction Detection in Liquid Chromatography*, Marcel Dekker, New York, 1986.

55. Iwai, K. and Toyo'oka, T., Design and choice of suitable labelling reagents for liquid chromatography, In *Selective Sample Handling and Detection in High-Performance Liquid Chromatography*, Frei, R. W. and Zech, K., Eds., Elsevier, Amsterdam, 1988, chap. 4.

56. Lingeman, H. and Underberg, W. J. M., *Detection-Oriented Derivatization Techniques in Liquid Chromatography*, Marcel Dekker, New York, 1990.

57. Lun, G. and Helwig, L. C., *Handbook of Derivatization Reactions for HPLC*, Wiley, New York, 1998.

58. Stuart, I. A., Ansell, R. O., Maclachlan, J., Bather, P. A., and Gardiner, W. P., Five-way ANOVA interactions analysis of the selective extraction of carbaryl, pirimicarb, and aldicarb from soils by supercritical fluid extraction, *Analyst*, 122, 303–308, 1997.

59. Wu, Y. S., Lee, H. K., and Li, S. F. Y., A fluorescence detection scheme for capillary electrophoresis of *N*-methylcarbamates with oncolumn thermal decomposition and derivatization, *Anal. Chem.*, 72, 1441–1447, 2000.

60. Barceló, D. and Hennion, M. C., Online handling strategies for the trace-level determination of pesticides and their degradation products in environmental waters, *Anal. Chim. Acta*, 318, 1–41, 1995.

61. Hyötyläinen, T. and Riekkola, M. L., Online combinations of pressurized hot-water extraction and microporous membrane liquid–liquid extraction with chromatography, *LC–GC Europe*, 15(5), 298–306, 2002.

62. Nielen, M. W. F., Frei, R. W., and Brinkman, U. A. Th., Online handling and trace enrichment in liquid chromatography. The determination of organic compounds in water samples, In *Selective Sample Handling and Detection in High-Performance Liquid Chromatography*, Frei, R. W. and Zech, K., Eds., Elsevier, Amsterdam, 1988, chap. 1.

63. Hennion, M. C. and Scribe, P., Sample handling strategies for the analysis of organic compounds from environmental water samples, In *Environmental Analysis: Techniques, Applications, and Quality Assurance*, Barceló, D., Ed., Elsevier, Amsterdam, 1996, chap. 2.

64. Hiemstra, M. and de Kok, A., Determination of *N*-methylcarbamate pesticides in environmental water samples using automated online trace enrichment with exchangeable cartridges and high-performance liquid chromatography, *J. Chromatogr.*, 667, 155–166, 1994.

65. Slobodník, J., Hoekstra-Oussoren, S. J. F., Jager, M. E., Honing, M., van Baar, B. L. M., and Brinkman, U. A. Th., Online solid-phase extraction–liquid chromatography–particle beam mass spectrometry and gas chromatography–mass spectrometry of carbamate pesticides, *Analyst*, 121, 1327–1334, 1996.

66. Slobodník, J., Groenewegen, M. G. M., Brouwer, E. R., Lingeman, H., and Brinkman, U. A. Th., Fully automated multiresidue method for trace level monitoring of polar pesticides by liquid chromatography, *J. Chromatogr.*, 642, 359–370, 1993.

67. Ballesteros, E., Gallego, M., and Valcárcel, M., Automatic determination of *N*-methylcarbamate pesticides by using a liquid–liquid extractor derivatization module coupled online to a gas chromatograph equipped with a flame ionization detector, *J. Chromatogr.*, 633, 169–176, 1993.

68. Vreuls, J. J., Louter, A. J. H., and Brinkman, U. A. Th., Online combination of aqueous-sample preparation and capillary gas chromatography, *J. Chromatogr. A*, 856, 279–314, 1999.

69. Valcárcel, M., Arce, L., and Ríos, A., Coupling continuous separation techniques to capillary electrophoresis, *J. Chromatogr. A*, 924, 3–30, 2001.

70. Hinsmann, P., Arce, L., Ríos, A., and Valcárcel, M., Determination of pesticides in waters by automatic online solid-phase extraction capillary electrophoresis, *J. Chromatogr. A*, 866, 137–146, 2000.

71. Arce, L., Ríos, A., and Valcárcel, M., Flow injection capillary electrophoresis coupling to automate online sample treatment for the determination of inorganic ions in waters, *J. Chromatogr. A*, 791, 279–287, 1997.

72. Zambonin, C. G., Coupling solid-phase microextraction to liquid chromatography. A review, *Anal. Bioanal. Chem.*, 375, 73–80, 2003.

73. Eisert, R. and Pawliszyn, J., Automated in-tube solid-phase microextraction coupled to high-performance liquid chromatography, *Anal. Chem.*, 69, 3140–3147, 1997.

74. Gou, Y. and Pawliszyn, J., In-tube solid-phase microextraction coupled to capillary LC for carbamate analysis in water samples, *Anal. Chem.*, 72, 2774–2779, 2000.

75. Guiochon, G. and Guillemin, C. L., *Quantitative Gas Chromatography for Laboratory Analyses and Online Process Control*, Elsevier, Amsterdam, 1988.

76. Mueller, H. M. and Stan, H. J., Thermal degradation observed with different injection techniques: quantitative estimation by the use of thermolabile carbamate pesticides, *J. High Resolut. Chromatogr.*, 13, 759–763, 1990.

77. Wylie, P. L., Kein, K. J., Thompson, M. Q., and Hemann, B. W., Using electronic pressure programming to reduce the decomposition of labile compounds during splitless injection, *J. High Resolut. Chromatogr.*, 15, 763–768, 1992.

78. Santos Delgado, M. J., Rubio Barroso, S., Toledano Fernández-Tostado, G., and Polo-Díez, L. M., Stability studies of carbamate pesticides and analysis by gas chromatography with flame ionization and nitrogen–phosphorus detection, *J. Chromatogr. A*, 921, 287–296, 2001.

79. Ling, C. F., Pérez-Melian, G., Jímenez-Conde, F., and Revilla, E., Comparative study of some columns for direct determination of carbofuran by gas chromatography with nitrogen-specific detection, *J. Chromatogr.*, 519, 359–362, 1990.

80. Ling, C. F., Pérez-Melian, G., Jiménez-Conde, F., and Revilla, E., Determination of carbofuran in a nutrient solution by GLC–NPD (nitrogen–phosphorus detection) and HPLC, *Chromatographia*, 30, 421–423, 1990.

81. Nagasawa, K., Uchiyama, H., Ogamo, A., and Shinozuka, T., Gas chromatographic determination of microamounts of carbaryl and 1-napthol in natural water as sources of water supplies, *J. Chromatogr.*, 144, 77–84, 1977.

82. Bilikova, A. and Kuthan, A., Determination of carbofuran in water, *Vodni Hospod. B*, 33, 215–219, 1983.

83. Muino, M. A. F., Gandara, J. S., and Lozano, J. S., Simultaneous determination of pentachlorophenol and carbaryl in water, *Chromatographia*, 32, 238–240, 1991.

84. Matthew, R., Coldwell, M. R., Pengelly, I., and Rimmer, D. A., Determination of dithiocarbamate pesticides in occupational hygiene sampling devices using the isooctane method and comparison with an automatic thermal desorption (ATD) method, *J. Chromatogr. A*, 984, 81–88, 2003.

85. Junk, G. A. and Richard, J. J., Organics in water: solid phase extraction on a small scale, *Anal. Chem.*, 60, 451–454, 1988.

86. Beyers, D. W., Carlson, C. A., and Tessari, J. D., Solid-phase extraction of carbaryl and malathion from pond and well water, *Environ. Toxicol. Chem.*, 10, 1425–1429, 1991.

87. Mattern, C. G., Louis, J. B., and Rosen, J. D., Multipesticide determination in surface water by gas chromatography–chemical ionization mass spectrometry–ion-trap detection, *J. Assoc. Off. Anal. Chem.*, 74, 982–986, 1991.

88. Wigfield, Y. Y., Grant, R., and Snider, N., Gas chromatographic and mass spectrometric investigation of seven carbamate insecticides and one metabolite, *J. Chromatogr.*, 657, 219–222, 1993.

89. Viana, E., Redondo, M. J., Font, G., and Moltó, J. C., Disks versus columns in the solid-phase extraction of pesticides from water, *J. Chromatogr. A*, 733, 267–274, 1996.

90. Papadopoulou-Mourkidou, E., Patsias, J., and Kotopoulou, A., Determination of pesticides in soils by gas chromatography–ion trap mass spectrometry, *J. AOAC Int.*, 80, 447–454, 1997.

91. Lambropoulou, D. A., Sakkas, V. A., Hela, D. G., and Albanis, T. A., Application of solid-phase microextraction in the monitoring of priority pesticides in the Kalamas River (NW Greece), *J. Chromatogr. A*, 963, 107–116, 2002.

92. Elflei, L., Berger-Preiss, E., Lewsen, K., and Wünsch, G., Development of a gas chromatography–mass spectrometry method for the determination of household insecticides in indoor air, *J. Chromatogr. A*, 985, 147–157, 2003.

93. Okumura, T., Imamura, K., and Nishikawa, Y., Determination of carbamate pesticides in environmental samples as their trifluoroacetyl or methyl derivatives by using gas chromatography–mass spectrometry, *Analyst*, 120, 2675–2681, 1995.

94. Picó, Y., Moltó, J. C., Redondo, M. J., Viana, E., Mañes, J., and Font, G., Monitoring of the pesticide levels in natural waters of the Valencia Community (Spain), *Bull. Environ. Contam. Toxicol.*, 53, 230–237, 1994.

95. Redondo, M. J., Ruíz, M. J., Boluda, R., and Font, G., Optimization of a solid-phase extraction technique for the extraction of pesticides from soil samples, *J. Chromatogr. A*, 719, 69–76, 1996.

96. Woodrow, J. E., Majewski, M. S., and Seiber, J. N., Accumulative sampling of trace pesticides and other organics in surface water using XAD-4 resin, *J. Environ. Sci. Health, Part B*, 21, 143–164, 1986.

97. Brooks, M. W., Tessier, D., Soderstrom, D., Jenkins, J., and Clark, J. M., Rapid method for the simultaneous analysis of chlorpyrifos, isofenphos, carbaryl, iprodione, and triadimefon in groundwater by solid-phase extraction, *J. Chromatogr. Sci.*, 28, 487–489, 1990.

98. Getzin, L. W., Cogger, C. G., and Bristow, P. R., Simultaneous gas-chromatographic determination of carbofuran, metalaxyl, and simazine in soils, *J. Assoc. Off. Anal. Chem.*, 72, 361–364, 1989.

99. Mol, H. G. J., Janssen, H. G. M., Cramers, C. A., Vreuls, J. J., and Brinkman, U. A. Th., Trace level analysis of micropollutants in aqueous samples using gas chromatography with online enrichment and large volume injection, *J. Chromatogr. A*, 703, 277–307, 1995.

100. Santos, F. J. and Galceran, M. T., Modern developments in gas chromatography–mass spectrometry-based environmental analysis, *J. Chromatogr. A*, 1000, 125–151, 2003.

101. Reeves, R. G. and Woodham, D. W., Gas chromatographic analysis of methomyl residues in soil, sediment, water, and tobacco utilizing the flame photometric detector, *J. Agric. Food Chem.*, 22, 76–78, 1974.

102. Lartiges, S. B. and Garrigues, P., Gas chromatographic analysis of organophosphorus and organonitrogen pesticides with different detectors, *Analusis*, 23, 418–421, 1995.

103. Kochman, M., Gordin, A., Goldshlag, P., Lehotay, S. J., and Amirav, A., Fast, high-sensitivity, multipesticide analysis of complex mixtures with supersonic gas chromatography–mass spectrometry, *J. Chromatogr. A*, 974, 185–212, 2002.

104. Dallüge, J., Beens, J., and Brinkman, U. A. Th., Comprehensive two-dimensional gas chromatography: a powerful and versatile analytical tool, *J. Chromatogr. A*, 1000, 69–108, 2003.

105. Dallüge, J., Vreuls, R. J. J., Beens, J., and Brinkman, U. A. Th., Optimization and characterization of comprehensive two-dimensional gas chromatography with time-of-flight mass spectrometric detection (GC × GC–TOF MS), *J. Sep. Sci.*, 25, 201–214, 2002.

106. Thurston, A. D., *Liquid Chromatography of Carbamate Pesticides, EPA-R2-72-079*, National Environmental Research Center, U.S. Environmental Protection Agency, Corvallis, 1972.

107. Frei, R. W., Lawrence, J. F., Hope, J., and Cassidy, R. M., Analysis of carbamate insecticides by fluorigenic labelling and high-speed liquid chromatography, *J. Chromatogr. Sci.*, 12, 40–45, 1974.

108. McGarvey, B. D., High-performance liquid chromatographic methods for the determination of *N*-methylcarbamate pesticides in water, soil, plants, and air, *J. Chromatogr.*, 642, 89–105, 1993.

109. Yang, S. S., Goldsmith, A. I., and Smetena, I., Recent advances in the residue analysis of *N*-methylcarbamate pesticides, *J. Chromatogr. A*, 754, 3–16, 1996.

110. Hogendoorn, E. and van Zoonen, P., Recent and future developments of liquid chromatography in pesticide trace analysis, *J. Chromatogr. A*, 892, 435–453, 2000.

111. Salau, J. S., Alonso, R., Batllo, G., and Barceló, D., Application of solid-phase disk extraction followed by gas and liquid-chromatography for the simultaneous determination of the fungicides — captan, captafol, carbendazim, chlorothalonil, ethirimol, folpet, metalaxyl, and vinclozolin in environmental waters, *Anal. Chim. Acta*, 293, 109–117, 1994.

112. Fogy, I., Schmid, E. R., and Huber, J. F. K., Determination of carbamate pesticides in fruits and vegetables by multidimensional high-pressure liquid chromatography, *Z. Lebensm. Unters. Forsch.*, 170, 194–199, 1980.

113. Sparcino, C. M. and Hines, J. W., High-performance liquid chromatography of carbamate pesticides, *J. Chromatogr. Sci.*, 14, 549–565, 1976.

114. Marvin, C. H., Brindle, I. D., Hall, C. D., and Chiba, M., Development of an automated high-performance liquid chromatographic method for the online preconcentration and determination of trace concentrations of pesticides in drinking water, *J. Chromatogr.*, 503, 167–176, 1990.

115. di Corcia, A. and Marchetti, M., Multiresidue method for pesticides in drinking water using a graphitized carbon black cartridge extraction and liquid chromatographic analysis, *Anal. Chem.*, 63, 580–585, 1991.

116. Jiménez, B., Moltó, J. C., and Font, G., Influence of dissolved humic material and ionic strength on C8 extraction of pesticides from water, *Chromatographia*, 41, 318–324, 1995.

117. Guenu, S. and Hennion, M. C., Online trace-enrichment of polar pesticides or degradation products in aqueous samples: analytical columns, *Anal. Methods Instrum.*, 2, 247–253, 1995.

118. Sanusi, A., Millet, M., Wortham, H., and Mirabel, P., A multiresidue method for determination of trace levels of pesticides in atmosphere, *Analusis*, 25, 302–308, 1997.

119. Renner, T., Baumgarten, D., and Unger, K. K., Analysis of organic pollutants in water at trace levels using fully automated solid-phase extraction coupled to high-performance liquid chromatography, *Chromatographia*, 45, 199–205, 1997.

120. Sun, L. and Lee, H. K., Microwave-assisted extraction behavior of nonpolar and polar pollutants in soil with analysis by high-performance liquid chromatography, *J. Sep. Sci.*, 25, 67–76, 2002.

121. Durand, G., Chiron, S., Bouvot, V., and Barceló, D., Use of extraction disks for trace enrichment of various pesticides from river and sea-water samples, *Int. J. Environ. Anal. Chem.*, 49, 31–42, 1992.

122. Parrilla, P., Martínez Vidal, J. L., Martínez Galera, M., and Frenich, A. G., Simple and rapid screening procedure for pesticides in water using SPE and HPLC/DAD detection, *Fresenius J. Anal. Chem.*, 350, 633–637, 1994.

123. Schülein, J., Martens, D., Spitzauer, P., and Kettrup, A., Comparison of different solid phase extraction materials and techniques by application of multiresidue methods for the determination of pesticides in water by high-performance liquid chromatography (HPLC), *Fresenius J. Anal. Chem.*, 352, 565–571, 1995.

124. Hidalgo, C., Sancho, J. V., Roig-Navarro, A., and Hernández, F., Rapid-determination of carbaryl and 1-naphthol at ppt levels in environmental water samples by automated online SPE–LC–DAD–FD, *Chromatographia*, 47, 596–600, 1998.

125. García de Llasera, M. P. and Bernal-González, M., Presence of carbamate pesticides in environmental waters from the Northwest of Mexico: determination by liquid chromatography, *Water Res.*, 35, 1933–1940, 2001.

126. She, L. K., Brinkman, U. A. Th., and Frei, R. W., Liquid-chromatographic residue analysis of carbaryl based on a postcolumn catalytic reactor principle and fluorigenic labelling, *Anal. Lett.*, 17, 915–931, 1984.

127. de Kok, A., Hiemstra, M., and Brinkman, U. A. Th., Low ng/l level determination of twenty *N*-methylcarbamate pesticides and twelve of their polar metabolites in surface water via offline

solid-phase extraction and high-performance liquid chromatography, *J. Chromatogr.*, 623, 265–276, 1992.

128. Chiron, S. and Barceló, D., Determination of pesticides in drinking water by online solid-phase disk extraction followed by various liquid-chromatographic systems, *J. Chromatogr.*, 645, 125–134, 1993.

129. Sundaram, K. M. S. and Curry, J., High-performance liquid-chromatographic methods for the analysis of aminocarb, mexacarbate, and some of their *N*-methylcarbamate metabolites by postcolumn derivatization with fluorescence detection, *J. Chromatogr. A*, 672, 117–124, 1994.

130. Cappiello, A., Famiglini, G., and Bruner, F., Determination of acidic and basic/neutral pesticides in water with a new microliter flow rate LC/MS particle beam interface, *Anal. Chem.*, 66, 1416–1423, 1994.

131. Okkura, T., Takechi, T., Deguehi, S., Ishimaru, T., Maki, T., and Inouye, H., Simultaneous determination of herbicides in water by FRIT–FAB LC–MS, *J. Toxicol. Environ. Health*, 40, 266–273, 1994.

132. Sennert, S., Volmer, D., Levsen, K., and Wünsch, G., Multiresidue analysis of polar pesticides in surface and drinking water by online enrichment and thermospray LC–MS, *Fresenius J. Anal. Chem.*, 351, 642–649, 1995.

133. Wang, N. and Budde, W. L., Determination of carbamate, urea, and thiourea pesticides and herbicides in water, *Anal. Chem.*, 73, 997–1006, 2001.

134. Wu, J., Tragas, C., Lord, H., and Pawliszyn, J., Analysis of polar pesticides in water and wine samples by automated in-tube solid-phase microextraction coupled with high-performance liquid chromatography–mass spectrometry, *J. Chromatogr. A*, 976, 357–367, 2002.

135. Nogueira, J. M. F., Sandra, T., and Sandra, P., Considerations on ultra trace analysis of carbamates in water samples, *J. Chromatogr. A*, 996, 133–140, 2003.

136. Anderson, J. L. and Chesney, D. J., Liquid chromatographic determination of selected carbamate pesticides in water with electrochemical detection, *Anal. Chem.*, 52, 2156–2161, 1980.

137. Anderson, J. L., Whiten, K. K., Brewster, J. D., Ou, T. Y., and Nonidez, W. K., Micro-array electrochemical flow detectors at high applied potentials and liquid chromatography with electrochemical detection of carbamate pesticides in river water, *Anal. Chem.*, 57, 1366–1373, 1985.

138. Barceló, D., *Applications of LC–MS in Environmental Chemistry*, Elsevier, Amsterdam, 1996.

139. Niessen, W. M. A., *Liquid Chromatography–Mass Spectrometry*, Marcel Dekker, New York, 1999.

140. Honing, M., Barceló, D., van Baar, B. L. M., and Brinkman, U. A. Th., Limitations and perspectives in the determination of carbofuran with various liquid chromatography–mass spectrometry interfacing systems, *Trends Anal. Chem.*, 14, 496–504, 1995.

141. Nunes, G. S., Alonso, R. M., Ribeiro, M. L., and Barceló, D., Determination of aldicarb, aldicarb sulfoxide, and aldicarb sulfone in some fruits and vegetables using high-performance liquid chromatography–atmospheric pressure chemical ionization mass spectrometry, *J. Chromatogr. A*, 888, 113–120, 2000.

142. van Zoonen, P., Hogendoorn, E. A., van der Hoff, G. R., and Baumann, R. A., Selectivity and sensitivity in coupled chromatographic techniques as applied in pesticide residue analysis, *Trends Anal. Chem.*, 11, 11–17, 1992.

143. Hyötyläinen, T. and Riekkola, M. L., Online coupled liquid chromatography–gas chromatography, *J. Chromatogr. A*, 1000, 357–384, 2003.

144. Rathore, H. S., Chromatographic and related spot tests for the detection of water pollutants, *J. Chromatogr. A*, 733, 5–17, 1996.

145. Ameno, K., Lee, S. K., In, S. W., Yang, J. Y., Yoo, Y. C., Ameno, S., Kubota, T., Kinoshita, H., and Ijiri, I., Blood carbofuran concentrations in suicidal ingestion cases, *Forensic Sci. Int.*, 116, 59–61, 2001.

146. Abbott, D. C., Blake, K. W., Tarrant, K. R., and Thomson, J., Thin-layer chromatographic separation, identification, and estimation of residues of some carbamates and allied pesticides in soil and water, *J. Chromatogr.*, 30, 136–141, 1967.

147. Eichelberger, J. W. and Lichtenberg, J. J., Persistence of pesticides in river water, *Environ. Sci. Technol.*, 5, 541–545, 1971.

148. MacNeil, J. D., Frei, R. W., and Frei-Häusler, M., Electron donor–acceptor reagents in the analysis of pesticides, *Int. J. Environ. Anal. Chem.*, 2, 323–326, 1973.

149. Abdel-Kader, M. H. K., Stiles, D. A., and Ragab, M. T. H., Thin-layer chromatographic separation and identification of carbofuran and two metabolites and their dinitrophenyl esters, *Int J. Environ. Anal. Chem.*, 18, 281–286, 1984.

150. Rathore, H. S. and Begum, T., Thin-layer chromatographic behavior of carbamate pesticides and related compounds, *J. Chromatogr.*, 643, 321–329, 1993.

151. Sevalkar, M. T., Patil, V. B., and Garad, M. V., Thin-layer chromatographic method for detection and identification of carbaryl, propoxur, and carbofuran by use of 4-aminoantipyrine, *J. Planar. Chromatogr.*, 13, 235–237, 2000.

152. McGinnis, S. C. and Sherma, J., Determination of carbamate insecticides in water by C_{18} solid-phase extraction and quantitative HPTLC, *J. Liq. Chromatogr.*, 17, 151–156, 1994.

153. Butz, S. and Stan, H. J., Screening of 265 pesticides in water by thin-layer chromatography with automated multiple development, *Anal. Chem.*, 67, 620–630, 1995.

154. Chester, T. L., Pinkston, J. D., and Raynie, D. E., Supercritical fluid chromatography and extraction, *Anal. Chem.*, 70, 301R–319R, 1998.

155. Hutchinson, K. W., *Innovations in Supercritical Fluids: Science and Technology*, ACS Symposium Series, No. 608, American Chemical Society, Washington, 1996.

156. Smith, R. M., Supercritical fluids in separation science — the dreams, the reality and the future, *J. Chromatogr. A*, 856, 83–115, 1999.

157. Wright, B. W. and Smith, R. D., Rapid capillary supercritical-fluid chromatographic analysis of carbamate pesticides, *J. High Resolut. Chromatogr. Commun.*, 8, 8–11, 1985.

158. France, J. F. and Voorhees, K. J., Capillary supercritical-fluid chromatography with ultra-violet multichannel detection of some pesticides and herbicides, *J. High Resolut. Chromatogr. Commun.*, 11, 692–696, 1988.

159. Murugaverl, B., Voorhees, K. J., and Deluca, S. J., Utilization of a benchtop mass-spectrometer with capillary supercritical fluid chromatography, *J. Chromatogr.*, 633, 195–205, 1993.

160. Medvedovici, A., David, V., and Sandra, P., Fast analysis of carbaryl by online SPE–SFC–DAD, *Rev. Roum. Chim.*, 45, 827–833, 2000.

161. Combs, M. T., Ashraf-Khorassani, M., and Taylor, L. T., Packed column supercritical fluid chromatography–mass spectroscopy: a review, *J. Chromatogr. A*, 785, 85–100, 1997.

162. Toribio, L., del Norzal, M. J., Bernal, J. L., Jiménez, J. J., and Serna, M. L., Packed-column supercritical fluid chromatography coupled with solid-phase extraction for the determination of organic microcontaminants in water, *J. Chromatogr. A*, 823, 163–170, 1998.

163. Dost, K., Jones, D. C., Auerbach, R., and Davidson, G., Determination of pesticides in soil samples by supercritical fluid chromatography–atmospheric pressure chemical ionization mass spectrometric detection, *Analyst*, 125, 1751–1755, 2000.

164. Shintani, H. and Polonski, J., Eds., *Handbook of Capillary Electrophoresis Applications*, Chapman & Hall, London, 1997.

165. Picó, Y., Rodríguez, R., and Mañes, J., Capillary electrophoresis for the determination of pesticide residues, *Trends Anal. Chem.*, 22, 133–151, 2003.

166. Landers, J. P., *Handbook of Capillary Electrophoresis. Part I*, CRC Press, Boca Raton, FL, 1994.

167. Süsse, H. and Müller, H., Application of micellar electrokinetic capillary chromatography to the analysis of pesticides, *Fresenius J. Anal. Chem.*, 352, 470–473, 1995.

168. Wu, Y. S., Lee, H. K., and Li, S. F. Y., Separation and determination of pesticides by capillary electrophoresis. II. Determination of N-methylcarbamates in drinking water by micellar electrokinetic chromatography with SPE and oncolumn enrichment, *J. Microcolumn Sep.*, 10, 529–535, 1998.

169. Molina, M., Pérez-Bendito, D., and Silva, M., Multiresidue analysis of N-methylcarbamate pesticides and their hydrolytic metabolites in environmental waters by use of solid-phase extraction and micellar electrokinetic chromatography, *Electrophoresis*, 20, 3439–3449, 1999.

170. Tegeler, T. and El Rassi, Z., Oncolumn trace enrichment by sequential frontal and elution electrochromatography. I. Application to carbamate insecticides, *Anal. Chem.*, 73, 3365–3372, 2001.

171. Tegeler, T. and El Rassi, Z., Oncolumn trace enrichment by sequential frontal and elution electrochromatography. II. Enhancement of sensitivity by segmented capillaries with z-cell configuration — application to detection of dilute samples of moderately polar and nonpolar pesticides, *J. Chromatogr. A*, 945, 267–279, 2002.

172. Molina, M., Wiedmer, S. K., Jussila, M., Silva, M., and Riekkola, M. L., Use of a partial filling technique and reversed migrating micelles in the study of *N*-methylcarbamate pesticides by micellar electrokinetic chromatography–electrospray ionization mass spectrometry, *J. Chromatogr. A.*, 927, 191, 2002.

173. Barceló, D., House, W. A., Maier, E. A., and Griepink, B., Preparation, homogeneity, and stability studies of freeze-dried water containing pesticides, *Int. J. Environ. Anal. Chem.*, 57, 237–254, 1994.

174. Martín-Esteban, A., Fernández, P., Cámara, C., Kramer, G. N., and Maier, E. A., The preparation of a Certified Reference Material of polar pesticides in freeze-dried water (CRM-606), *Fresenius J. Anal. Chem.*, 363, 632–640, 1999.

175. Winegar, E. D. and Keith, L. H., Eds., *Sampling and Analysis of Airborne Pollutants*, Lewis Publishers, Boca Raton, FL, 1993.

25 Analysis of Urea Derivative Herbicides in Water and Soil

Sara Bogialli, Roberta Curini, Antonio Di Corcia, and Manuela Nazzari

CONTENTS

I. PHYSICOCHEMICAL PROPERTIES

Herbicides derived from urea form a large group of chemical compounds widely used in agriculture to control weeds in cereal, vegetable, and fruit tree crops. On the basis of their chemical natures, use, and mode of action, substituted urea herbicides can be divided into two main groups — phenylureas and sulfonylureas.

Although phenylurea herbicides (PUHs) were introduced more than 40 years ago, they are still widely used. Phenylureas can be divided into three subgroups:

1. *N*-phenyl-*N'* *N'*-dialkylureas, such as, chlorotoluron, isoproturon, and diuron
2. *N*-phenyl-*N'*-alkyl-*N'*-methoxyureas, such as, linuron, monolinuron, and metobromuron
3. Phenylureas containing a heterocyclic group, the major exponent being methabenz-thiazuron.

The chemical structures of the main representatives of the three phenylurea subgroups are depicted in Figure 25.1. The common names, water solubility, half-life in soil, and leaching potential through the soil (when available) of the most widely used phenylureas are presented in Table 25.1.

PUHs are generally absorbed through the roots of plants and transported via the transpiration system. The mode of action of PUHs seems to be due to the combined effects of the inhibition of photosynthesis and the irreversible injury of the plant photosynthesis system via inhibition of $NADPH_2$.[1]

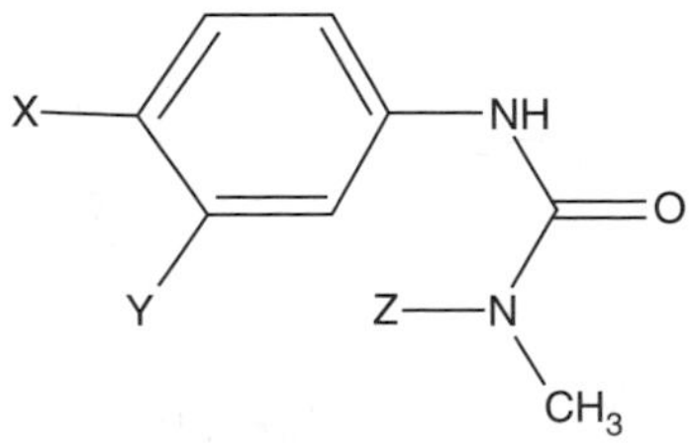

	X	Y	Z
Metoxuron	OCH_3	Cl	CH_3
Chlortoluron	Cl	Cl	CH_3
Isoproturon	*i*-C_3H_7	H	CH_3
Diuron	Cl	Cl	CH_3
Neburon	Cl	Cl	C_4H_9
Linuron	Cl	Cl	OCH_3
Metobromuron	Br	H	OCH_3
Monolinuron	Cl	H	OCH_3

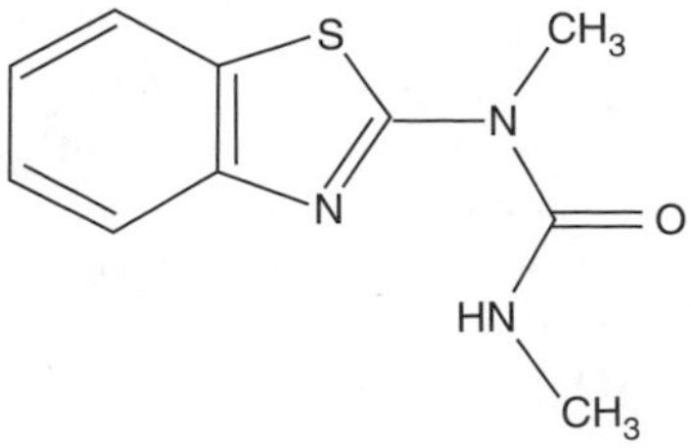

Methabenzthiazuron

FIGURE 25.1 Structures of some selected phenylureas.

TABLE 25.1
Physicochemical Properties of Phenylurea Herbicides

Common Name	IUPAC Name, M.F.[a]	Molecular Weight	Water Solubility (mg/l)	DT$_{50}$ (days)	K_{oc}[a] (ml/g)
Chlorbromuron	3-(4-Bromo-chlorophenyl)-1-methoxy-1-methylurea C$_9$H$_{10}$BrClN$_2$O	293.5	35 (20°C)	56 to 196	908
Chlorotoluron	3-(3-Chloro-p-tolyl)-1,1-dimethylurea C$_{10}$H$_{13}$ClN$_2$O	212.7	74 (25°C)	30 to 40	n.f.
Diuron	3-(3,4 Dichlorophenyl)-1,1-dimethylurea C$_9$H$_{10}$Cl$_2$N$_2$O	233.1	36.4 (25°C)	90 to 180	400
Fluometuron	1,1-Dimethyl-3-(α,α,α-trifluoro-m-tolyl)urea C$_{10}$H$_{11}$F$_3$N$_2$O	232.2	110 (20°C)	10 to 100	31 to 117
Isoproturon	3-(4-Isopropylphenyl)-1,1-dimethylurea C$_{12}$H$_{18}$N$_2$O	206.3	65 (22°C)	6 to 28	n.f.
Linuron	3-(3,4-Dichlorophenyl)-1-methoxy-1-methylurea C$_9$H$_{10}$Cl$_2$N$_2$O$_2$	249.1	64 (25°C)	82 to 150	75 to 250
Methabenzia-thiazuron	1-(1,3-Benzothiazol-2-yl)-1,3-dimethylurea C$_{10}$H$_{11}$N$_3$OS	221.3	59 (20°C)	n.f.	n.f.
Metobromuron	3-(4-Bromophenyl)-1-methox-1-methylurea C$_9$H$_{11}$BrN$_2$O$_2$	259.1	330 (20°C)	30	184
Monolinuron	3-(4-Chlorophenyl)-1-methoxy-1-methylurea C$_9$H$_{11}$ClN$_2$O$_2$	214.6	735 (25°C)	45 to 60	250 to 500
Neburon	1-Butyl 3 (3,4-dichlorophenyl)-1-methylurea C$_{12}$H$_6$Cl$_2$N$_2$O	275.2	5 (25°C)	n.f.	n.f.

M.F., molecular formula; DT$_{50}$, time for 50% loss; n.f., not found.

[a]K_{oc} = distribution coefficient (K_d) between soil and water adjusted for the proportion of organic carbon in water. It is a measure of the relative affinities of the pesticide for water and soil surface. As such, it indicates the tendency of a certain pesticide to leach through the soil and reach ground waters. Roughly, K_{oc} values higher than 100 indicate a low potential leaching.

Source: From Tomlin, C., In *The Pesticide Manual*, British Crop Protection Council, Farnham, Surrey, 1994.

Phenylureas reaching the environment are gradually decomposed over a short or a longer period, the steps and the rate of decomposition depending on the stability of the molecule and on the medium. The active substance on the soil surface or reaching aquifers is chemically decomposed by UV radiation or components of the soil. PUHs absorbed by the plants or in the soil are biodegraded by stepwise demethylation or demethoxylation of the urea moiety followed by generation of aromatic amines. These species are the endproducts of microbial activity.[1] Some of the phenylureas and their related endproducts are suspected to induce cancer.[2,3]

Sulfonylurea herbicides (SUHs) are relatively new herbicides, introduced in the 1980s. Chlorsulfuron was the first sulfonylurea marketed in the United States, in 1982. World-wide, 19 sulfonylureas had been commercialized by 1994, and five more are being developed. This rapid increase is due to their very high and specific herbicidal activity, which results in extremely low application rates of 10 to 40 g/ha. Furthermore, as compared to other herbicides, sulfonylureas are less toxic and degrade more rapidly. Chemical structures of some representative sulfonylureas are presented in Figure 25.2. From a chemical point of view, these herbicides are labile and weakly acidic compounds. The common names, chemical formulas, water solubility, pKa, half-life in soil, and leaching potential through the soil (when available) of the most representative sulfonylureas are reported in Table 25.2.

THIFENSULFURON METHYL
(M.W. = 387 Da, pK_a 4.0)

METSULFURON METHYL
(M.W. = 381 Da, pK_a 3.3)

TRIASULFURON
(M.W. = 402 Da, pK_a 4.6)

CHLORSULFURON
(M.W. = 358 Da, pK_a 3.6)

RIMSULFURON
(M.W. = 431 Da, pK_a 4.0)

BENSULFURON METHYL
(M.W. = 410 Da, pK_a 5.2)

TRIBENURON METHYL
(M.W. = 395 Da, pK_a 5.0)

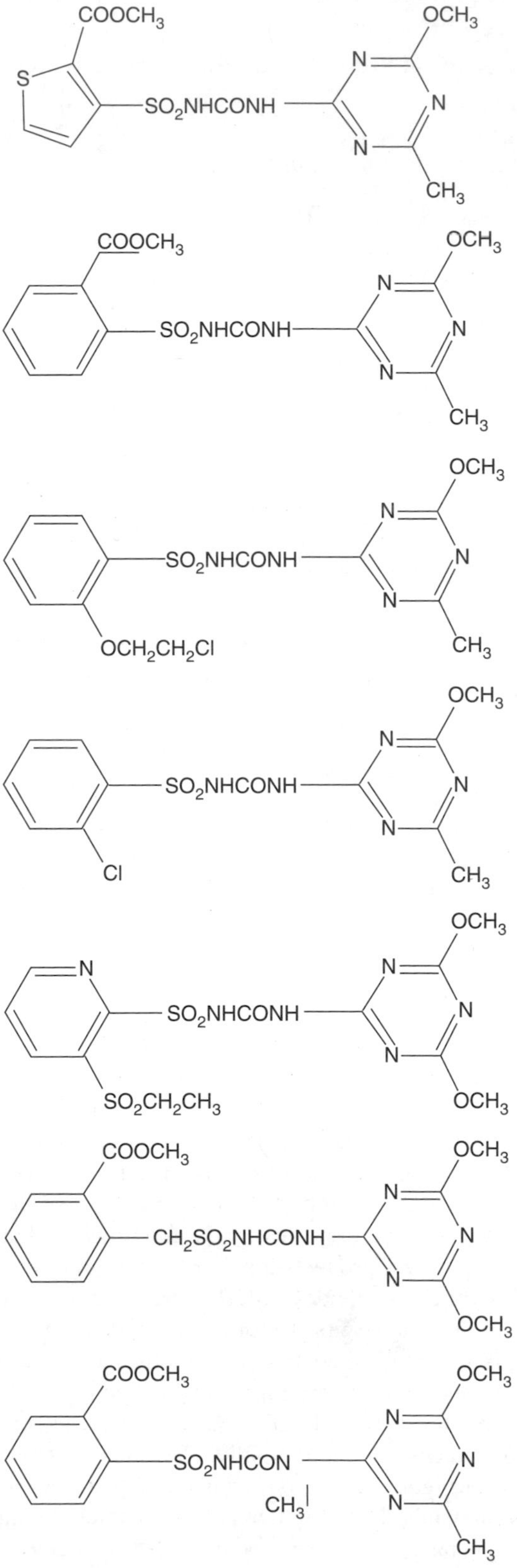

FIGURE 25.2 Structures of some selected sulfonylureas.

TABLE 25.2
Physicochemical Properties of Sulfonylurea Herbicides

Common Name	IUPAC Name, M.F.	Molecular Weight	pK_a	Water Solubility (g/l; 25°C, pH 7)	DT_{50} (days)	K_{oc} (ml/g)
Bensulfuron-methyl	α(4,6-Dimethoxypyrimidin-2-ylcarbamoylsulfamoyl)-o-toluic acid $C_{16}H_{18}N_4O_7S$	410.4	5.2	0.12	28 to 140	n.f.
Chlorsulfuron	1-(2-Chlorophenylsulfonyl)-3-(4-methoxy-6-methyl-1,3,5-triazin-2-yl)urea $C_{12}H_{12}ClN_5O_4S$	357.8	3.6	3.1	28 to 42	40
Metsulfuron-methyl	2-(4-Methoxy-6-methyl-1,3,5-triazin-2-ylcarbamoylsulfamoyl)benzoic acid $C_{14}H_{15}N_5O_6S$	381.4	3.3	2.8	7 to 35	35
Nicosulfuron	2-(4,6-Dimethoxypyridin-2-ylcarbamoylsulfamoyl)-N,N-dimethylnicotinamide $C_{15}H_{18}N_6O_6S$	410.4	4.6	12	26 to 67	n.f.
Primisulfuron-methyl	2-[4,6-bis(difluoromethoxy)pyrimidin-2-ylcarbomoylsulfamoyl]benzoic acid $C_{15}H_{12}F_4N_4O_7S$	468.3	3.5	0.24	4 to 29	n.f.
Rimsulfuron	1-(4,6-Dimethoxypyrimidin-2-yl)-3-(3-ethylsulfonyl-2-pyridylsulfonyl)urea $C_{14}H_{17}N_5O_7$	431.4	4.0	7.3	10 to 20	n.f.
Sulfometuron-methyl	2-(4,6-Dimethylpyrimidin-2-yl)benzoic acid $C_{15}H_{16}N_4O_5S$	364.4	5.2	0.24	ca. 28	85
Thifensulfuron-methyl	3-(4-Methoxy-6-methyl-1,3,5-triazin-2-ylcarbamoylsulfamoyl)thiophen-2-carboxylic acid $C_{12}H_{13}N_5O_6S_2$	387.4	4.0	6.3	6 to 12	n.f.
Triasulfuron	1-[2-(2-Chloroethoxy)phenylsulfonyl]-3-(4-methoxy-6-methyl-1,3 5-2-yl)urea $C_{14}H_{16}ClN_5O_5S$	401.8	4.6	0.82	19	n.f.
Tribenuron-methyl	2-[4-Methoxy-6-methyl-1,3,5-triazin-2-yl(methyl)carbamoylsulfamoyl]benzoic acid $C_{15}H_{17}N_5O_6S$	395.4	5.0	2.0	1 to 7	n.f.

Source: From Tomlin, C., In *The Pesticide Manual*, British Crop Protection Council, Farnham, Surrey, 1994; for explanation of acronyms, see Table 25.1.

Sulfonylureas are systemic herbicides absorbed by the foliage and roots. They act by inhibiting acetolactate synthase, a key enzyme in the biosynthesis of branched chain aminoacids.[4] This results in stopping cell division and plant growth. The most important degradation pathways of sulfonylureas are chemical hydrolysis and microbial degradation.

II. REGULATIONS

Published information shows that the consumption of pesticides for agricultural and industrial purposes is increasing. According to a report published by the United States Environmental Protection Agency (USEPA), a total of 5×10^8 kg of pesticides was used and dispersed into the environment in 1985. By various transport mechanisms, pesticides can reach and contaminate surface waters, groundwaters, and, ultimately, drinking waters. In the United States, 101 pesticides and 25 related degradation products (DPs) are included in the list of priority pollutants, which are to be monitored in water destined for human consumption.[5,6] Four PUHs, namely, diuron, fluometuron, linuron, and neburon are present in this list. So far, no SUH has been included in this list. The selection of the different pesticides was based on the use of at least 7×10^6 kg in 1982, a water solubility larger than 30 mg/l, and hydrolysis half-life longer than 25 weeks. Pesticides and pesticide DPs previously detected in groundwater, as well as pesticides regulated under the Safe Drinking Water Act, were automatically included in the list of priority analytes.

In the 15 European countries making up the European Community, several priority lists of contaminants, which include many pesticides, have been published to protect the quality of drinking and surface waters. Diuron and isoproturon are included in the list of priority substances of the Decision 2455/2001/EC on pollution caused by certain dangerous substances discharged into the aquatic environment of the Community.[7] In order to prevent the contamination of groundwater and drinking water in Western Europe, a priority list was published, which considers pesticides, the use of which exceeds 50,000 kg per year, and their capacity for probable or transient leaching.[8] Chlorotoluron, diuron, isoproturon, and methabenzthiazuron are present in this list.

The 98/83/EC Directive on the Quality of Water Intended for Human Consumption states a maximum admissible concentration of 0.1 μg/l for individual pesticides and 0.5 μg/l for total pesticides, regardless of their toxicity.[9] Table 25.3 lists the acute oral toxicity of phenylurea and SUHs for rats.

III. ANALYTICAL METHODS

A. SAMPLING

1. Water Sampling

The volume of water sample to be collected depends on the amount of water needed to perform the analysis at the required limit of quantification and, eventually, to duplicate the analysis.

As sampling containers, the best choice is amber glass bottles. When analyzing polar and medium polar compounds, such as PUHs and SUHs, nonfragile and lighter containers such as those made of plastic, can be a desirable alternative. However, it should be remembered that the latter type of containers, except for Teflon, can leach analytical interferences. Whatever the type of aqueous matrix and sample containers, it is good practice to triple-rinse the containers three times with the sample and then to collect it. Another good practice is to sample by using new containers to avoid memory effects.

TABLE 25.3
Toxicity Data and Tolerances in Drinking Water of Selected Phenylurea and Sulfonylurea Herbicides

Compound	Toxicity[a]		MAC (ng/l) in Drinking Water	
	Rats, LD_{50} (mg/kg)	Rainbow Trout, LC_{50} (96 h; mg/l)	EU	U.S.A.
Phenylureas				
Chlorbromuron	>5000	5.0	100	n.c.
Chlorotoluron	>10000	35	—	n.c.
Diuron	3400	5.6	—	To be set
Fluometuron	>6000	47	—	To be set
Isoproturon	2420	37	—	n.c.
Linuron	4000	3.2	—	To be set
Methabenziathiazuron	>2500	16	—	n.c.
Metobromuron	2623	36	—	n.c.
Metoxuron	3200	19	—	n.c.
Monolinuron	2215	56 to 75	—	n.c.
Neburon	>11000	0.6 to 0.9	—	To be set
Sulfonylureas				
Bensulfuron-methyl	>5000	>150	—	n.c.
Chlorsulfuron	>5000	>250	—	n.c.
Metsulfuron-methyl	>5000	>150	—	n.c.
Nicosulfuron	>5000	>1000	—	n.c.
Primisulfuron-methyl	>5000	70	—	n.c.
Rimsulfuron	>5000	>390	—	n.c.
Sulfometuron-methyl	>5000	12.5	—	n.c.
Thifensulfuron-methyl	>5000	>100	—	n.c.
Triasulfuron	>5000	>100	—	n.c.
Tribenuron-methyl	>5000	>1000	—	n.c.

MAC, maximum admissible concentration; LD_{50}, dose required to kill 50% of the test organism; LC_{50}, concentration required to kill 50% of the test organism; EU, European Union; n.c., not considered as a priority pollutant.

[a] *Source*: From Tomlin, C., In *The Pesticide Manual*, British Crop Protection Council, Farnham, Surrey, 1994.

a. Well Water Sampling

When conducting a pesticide monitoring campaign, water in shallow wells located within an area of heavy pesticide use should not be sampled. The collection of a well water sample is usually performed after eliminating stagnant water. The collection of a homogeneous sample can be accomplished by measuring the stability of some parameters of interest (pH, conductivity). The stability of such parameters implies sample homogeneity. Usually, a representative sample is obtained after purging three to ten well volumes. Other details of well water sampling can be found elsewhere.[10]

b. Potable Water Sampling

Sampling from potable water is usually simplified by collecting water from an existing tap. Before sampling, water is flushed for about 10 min to eliminate sediments and gas pockets in the pipes.

If any water treatment device exists, representative water sampling should be made before the treatment unit. These devices contain ion exchangers and active carbons able to strongly adsorb organic compounds.

c. Surface Water Sampling

The composition of stream water is both flow- and depth-dependent. Analyte concentration gradients are not present in shallow lakes, because of the action of wind, as well as in rapidly flowing shallow streams. When sampling water from deep-water bodies and a single intake point is used, it should be located at about 60% of the stream depth, where complete mixing occurs. Samples from surface waters can be collected by automatic sample devices. Depending on the device, samples can be collected at individual specified times or a composite sample can be accumulated over a specified time period (24 h, usually). In some studies, manual collection could be made, making sure that the sampler entering the water approaches from downstream of the sample point. When representative depth-integrated water samples are to be collected, the methods of Nordin[11] and Meade[12] should be followed.

2. Soil and Sludge Sampling

Often, inherent sample heterogeneity of soils and sludge, related to their chemical composition and climatic events, causes problems of representativeness, which largely exceeds those associated to collection. Therefore, the best choice is to collect a sample as large as practical for sample preparation. An extract will be more homogeneous and provide more reproducible aliquots than a smaller portion of the sample.[13]

B. SAMPLE STORAGE

1. Aqueous Samples

Extensive environmental surveys require the analysis of a large number of samples. Once samples are collected, containers are shipped to a laboratory, where the rest of the analytical procedure is carried out. In order to avoid possible chemical and biochemical analyte alterations, field samples should be analyzed immediately after collection. Since it is impossible to do this for many environmental laboratories, serious problems of sample stability arise. A traditional way of preserving samples is to place them immediately after collection in insulated bags filled with ice, "blue" ice, or dry ice until arrival at the laboratory, and then to store bottles in a refrigerator at 4°C. Hypochlorite in drinking water samples can continue to degrade pesticides by oxidation or chlorination reactions. Many PUHs were demonstrated to degrade rapidly in the presence of the residual chlorine disinfectant in drinking water.[14] The addition of a tris buffer[14] or sodium thiosulfate[15] to water eliminated this problem. Transfer of a groundwater sample from an anaerobic ambient to an aerobic one may initiate biodegradation of some pesticides, which continues during transportation and storage. In this case, addition of biological inhibitors can prevent analyte biodegradation. During the National Pesticide Survey conducted by the EPA over a two-year period, 1349 well water samples spiked with phenylureas were preserved by the addition of 10 mg/l $HgCl_2$.[16] Stability study results showed that no significant loss of fluometuron, diuron, linuron, and neburon occurred by storing $HgCl_2$-containing sample bottles at 4°C for 14 days. On the other hand, $HgCl_2$ can increase degradation rates of pesticides, other than PUHs, subject to acid-catalyzed hydrolysis. During the initial 14 day storage stability study, 26 of 147 target analytes had 100% loss in recovery, presumably due to preservation with $HgCl_2$.[16]

Other negative aspects of mercury compounds are high human toxicity and the expense of hazardous waste disposal. During method development of EPA Method 532 (for PUHs), copper sulfate has been successfully used as an antimicrobial agent.[14,17] Later, another effective strategy for analyte preservation will be illustrated.

2. Solid Samples

Whatever the nature of the sample, the best choice for retarding analyte degradation is that of freezing the samples after placing them in tightly sealed containers with the precaution of minimizing headspace or wrapping them with aluminum foils.

C. EXTRACTION

Before accomplishing aqueous sample extraction, one or more surrogates should be added to the sample. A surrogate analyte is defined by the EPA as "a pure analyte, which is extremely unlikely to be found in any sample aliquot in known amount before extraction and is measured with the same procedure used to measure other sample components. The purpose of a surrogate analyte is to monitor method performance with each sample." Surrogates have an important role in assessing the effectiveness of a sample preparation procedure. For drinking water analysis of PUHs by Liquid Chromatography (LC), surrogates suggested by the EPA are monuron (an obsolete PUH) and carbazole.[14]

1. Water Samples

Methods for the extraction of pesticides from water exploit the partitioning of analytes between the aqueous phase and a water-immiscible solvent (Liquid–Liquid Extraction [LLE]) or an adsorbent material (Solid-Phase Extraction [SPE]). The SPE technique has been shown to offer several advantages over LLE and it is included in most of recent analytical protocols devoted to analyzing contaminants in water samples. Nevertheless, the LLE technique is still used in many environmental laboratories.

a. Liquid–Liquid Extraction

Among the solvents used for the extraction of phenylureas, dichloromethane is the most preferred, owing to its effectiveness in extracting compounds having a broad range of polarity. Solvent extraction is usually carried out in a separatory funnel, which is vigorously shaken to increase the contact area between the two liquids. This operation enhances the extraction rate and yield. The LLE technique has been proposed for isolating PUHs[18,19] SUHs,[20–22] PUHs and SUHs[23] from water samples. In all these methods, dichloromethane has been used as the extracting solvent. For efficiently extracting sulfonylureas that are weakly acidic in nature, the pH of the water sample was adjusted in advance to three. Table 25.4 lists selected LLE-based extraction procedures for urea herbicides.

The drawbacks of this technique are that it is labor intensive and time consuming. When performing trace analysis of pesticides, the extensive use of glassware may result in cumulative loss by adsorption on glass of hydrophobic pesticides. This technique requires the use of relatively large amounts of pesticide-grade solvents, which are expensive as well as flammable and toxic. Even by using pesticide-grade solvents, the concentration step by a factor of 1000 or more can introduce analyte interferences by residual solvent impurities. Vigorous shaking of solvent and water, especially surface water, may create serious problems of emulsions, owing to the presence in the sample of natural or synthetic surfactants. Emulsions can be eliminated only by additional time-consuming operations.

TABLE 25.4
Selected Liquid–Liquid Extraction Procedures for Extracting Urea Herbicides from Water Samples

Compound	Water Volume (l)	Solvent (ml)	Solvent Exchange	Quantification Technique	Recovery (%)	Ref.
Phenylureas	0.5 (pH 7)	3 × 50 CH_2Cl_2	CH_2Cl_2	GC–MS	84 to 100	18
Phenylureas	1 (pH 7)	3 × 60 CH_2Cl_2	Acetonitrile	LC-UV–MS	88 to 90	19
Chlorsulfuron	1 (pH 3)	3 × 70 CH_2Cl_2	Toluene	GC-ECD	101 to 105	20
Sulfonylureas	0.5 (pH 3)	100 CH_2Cl_2	Ethyl acetate	GC-ECD	80 to 92	22
Phenylureas	1 (pH 4)	3 × 100 CH_2Cl_2	CH_2Cl_2	LC–MS	96 to 107	23

ECD = electron capture detector.

b. Solid-Phase Extraction

Since the 1970s, as an alternative to LLE, the method of combined extraction and preconcentration of organic compounds in water by passing the sample through a short column of an adsorbing medium followed by desorption with a small quantity of an organic solvent has attracted the attention of many researchers. In the past 15 years, the availability of small-size particle (ca. 40 μm) adsorbents in inexpensive cartridges has largely contributed to the dramatic expansion of the SPE technique (Figure 25.3). This technique appears especially appealing to researchers and analysts, and it is rapidly replacing LLE in official methods.[24–26] Besides solving many problems associated with LLE, the SPE technique is particularly attractive because it lends itself to coupling with chromatographic systems for online applications.

Sample stability and storage space are problems that many environmental laboratories must address when collecting, storing, and analyzing water samples. With the large numbers of samples typical of environmental studies, the use of bulky glass bottles for sampling, transport, and storage also becomes a hindrance. One of the most impressive features of the SPE technique is that small adsorbent traps can be deployed in the field by using newly available submersible instrumentation. In this way, combined sampling, extraction, and preconcentration are done at the sampling site, thus eliminating most contamination and handling problems associated with sample collection. The small-volume trap could be sealed and shipped to the laboratory for elution and chromatographic analysis, or it could be frozen in a small storage place, until analysis.[27,28] PUHs extracted from a river water sample by means of an extraction cartridge filled with a sample of graphitized carbon black (GCB), namely Carbograph 1, cartridge showed them to be stable on this adsorbent for over 15 days of storage, even at ambient temperature.[29]

i. Adsorbing Materials for SPE

Typical adsorbents for SPE are silica, chemically modified with a C_{18} alkyl chain, commonly referred to as C-18, highly cross-linked styrene-divinylbenzene copolymers (PS–DVB), commonly referred to as PRP-I, Envichrom P, Lichrolut, RP-102, and GCBs, commonly referred to as Carbopack, Envicarb, Carbograph 1, and Carbograph 4. All these materials are commercially available in medical-grade polypropylene housing and polyethylene frits. In spite of some limitations in extracting polar compounds from large water volumes, C-18 is still the most commonly used material, and it has been considered for introduction into official methods by European and American environmental agencies.

With the view of selectively extracting target compounds from environmental waters, there has been a certain interest in developing and employing selective adsorbents based on analyte–antibody interactions achieved by immunosorbents. In the immunosorbent, the antibody

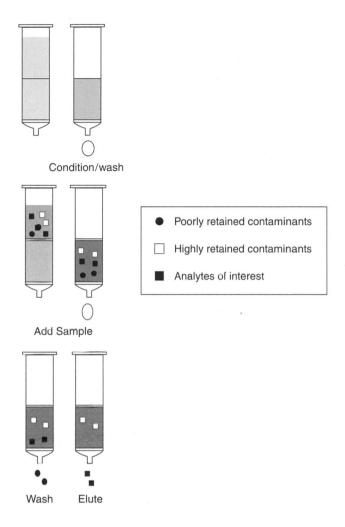

Condition/wash

- ● Poorly retained contaminants
- □ Highly retained contaminants
- ■ Analytes of interest

Add Sample

Wash   Elute

FIGURE 25.3 Schematic of the offline SPE process with cartridges.

is immobilized on to a silica support and used as an affinity ligand to extract selectively the target analyte from complex matrices. Taking advantage of the cross-reactivity of polyclonal antibodies, selective extraction can be achieved for a group of analytes having similar structures such as a class of pesticides. SPE by immunosorbents containing either antiisoproturon or antichlorotoluron antibodies succeeded in efficiently extracting many, but not all, of the phenylureas from spiked samples of river waters.[30–32] The production of antibodies is, however, laborious, time consuming, and expensive. Furthermore, antisera obtained by different researchers may have different affinity and selectivity, making it difficult to standardize and implement such methods at present.

Baker's yeast cells (*Saccharomices cerevisae*) were successfully immobilized on to silica gel and used in selective online trace enrichment of selected pesticides, including linuron, in various types of natural waters.[33] This technique relies upon the fact that microorganisms are able to absorb pesticides from water in the environment. Cell membranes contain many classes of lipids and lipoproteins, which contribute to pesticide absorption. Since the diffusion rate across membranes is inversely proportional to molecular size, low-molecular weight compounds, such as pesticides, can be extracted from water and isolated by naturally occurring high-molecular weight substances, such as humic acids, which are abundantly present in environmental waters.

A rather new rapidly growing trend in SPE technique is the design and use of synthetic antibody mimics, such as molecularly imprinted polymers (MIPs). Molecular imprinting is an increasingly applied technique allowing the formation of selective recognition sites in a stable polymer matrix. In this technique, polymerizable functional monomers are prearranged around a template molecule by noncovalent or covalent interactions prior to initiation of polymerization. A rigid, highly cross-linked macroporous polymer is formed, which contains sites that are complementary to the template molecule both in shape and in the arrangement of functional groups. After removal of the template molecule by extraction, the MIPs may then be used as an artificial receptor to selectively rebind the template from a mixture of chemical species. The advantages of the MIPs are their high selectivity, high affinity constants, and stability. The latter feature was positively tested by consecutive percolation of water sample, and it was shown that the performance of the MIP did not vary even after 200 enrichment and desorption cycles.[34] Because of their compatibility with organic solvents, MIPs have attracted considerable attention as SPE sorbents for the cleanup of target compounds. A tailor made MIP has been successfully involved in the development of a method for monitoring five SUHs in environmental waters.[34]

ii.   OffLine SPE with Cartridges

Offline SPE of analytes from water samples with cartridges is commonly accomplished by attaching the cartridge to the outlet of a separatory funnel containing the sample. More simply, water can be transferred from the sample bottle to the cartridge by attaching it to Teflon tubing put into the water sample. For routine analysis, devices allowing simultaneous extraction of several aqueous samples are supplied by several companies selling chromatographic supplies.

In any case, water is forced to pass through the cartridge by vacuum created by a water pump. Before pumping water, the cartridge is first washed with the eluent phase, to eliminate possible contaminants, and then with distilled water. After the water sample is passed through, the cartridge is washed with a little distilled water. Following this passage, water contained in the cartridge is in part eliminated by decreasing the pressure in the extraction apparatus. Before analyte reextraction, when possible (that is, when analytes are strongly retained by the sorbent material by nonspecific or specific interactions), cartridge washing by a suitable solvent mixture is useful to eliminate compounds that can interfere with the analysis.[35] Analyte desorption is accomplished by flowing slowly 4 to 8 ml of a suitable solvent or a solvent mixture through the adsorbent bed, and collecting the eluate in a vial. This is placed in a water bath at 25 to 50°C (depending on the analyte volatility), and, via a gentle stream of nitrogen, the extract is concentrated down to dryness, or to 50 to 100 μl if analytes are rather volatile.

Depending upon the type of adsorbent and the final destination of the extract, various solvents or solvent mixtures are used to reextract pesticides from adsorbent cartridges. With both C-18 and PS–DVB materials, methanol or acetonitrile is the eluent of choice, when analyzing via LC instrumentation. With C-18 cartridges and Gas Chromatography (GC) instrumentation, ethyl acetate is preferred, although methylene chloride was chosen in EPA Method 525 for eluting nonpolar and medium polar compounds.[24] With GCB cartridges, a CH_2Cl_2/CH_3OH (80:20, v/v) mixture offers quantitative desorption of neutral pesticides having a broad range of polarity. For eluting acidic analytes not desorbed by the foregoing solution, CH_2Cl_2/CH_3OH (80:20, v/v) acidified with trifluoroacetic acid (TFA), 10 mmol/l, can be used. When analyzing weakly acidic pesticides, such as sulfonylureas, in humic acid-rich aqueous environmental samples, more selective elution can be obtained by replacing TFA with a weakly acidic agent, such as acetic acid.[35] By doing so, a large fraction of humic acids coextracted with weakly acidic pesticides are not reextracted from the GCB cartridge and do not interfere with the analysis (Figure 25.4).

SPE cartridges filled with C-18 material analysis have been adopted for the offline extraction of some phenylurea[36–41] and sulfonylureas herbicides from drinking water.[42–47]

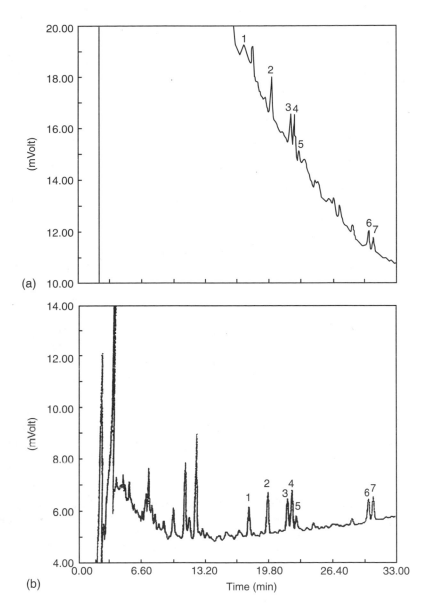

FIGURE 25.4 LC-UV chromatograms obtained by extracting seven sulfonylureas added at the level of 2 μg/l to 0.2 l of a simulated surface water sample and reextracting with a CH_2Cl_2/CH_3OH mixture (80:20, v/v) acidified with (a) 10 mmol/l trifluoroacetic acid and (b) 10 mmol/l acetic acid. Column: Alltima 250 × 4.6 mm i.d. containing 5 μm C-18. Elution: CH_3CN/H_2O (both containing 3 mmol/l trifluoroacetic acid) linear gradient elution from 32:68 to 62:38 (v/v) in 40 min. Peak numbering: 1, thifensulfuron methyl; 2, metsulfuron methyl; 3, triasulfuron methyl; 4, chlorsulfuron; 5, rimsulfuron; 6, bensulfuron methyl; 7, tribenuron methyl. (From Klaffenbach, P. and Holland, P. T., *J. Agric. Food Chem.*, 41, 396, 1993. With permission.)

Carbograph 1 (100 m²/g, surface area) cartridges proved to be a valuable material for quantitatively extracting phenylureas from large volumes of environmental waters.[29,39,48–51] Carbograph 4, having a surface area two times larger than that of Carbograph 1, has been involved in a method for analyzing PUHs,[15,52] PUHs and their metabolites in water.[53] With this material, quantitative extraction of PUHs for 4 l drinking water, 2 l groundwater, and 0.5 l river water has been achieved.[15] With respect to a C-18 cartridge, Carbographs had far better extraction efficiency for polar phenylureas.[29]

By means of a Carbograph 4 cartridge and within a single step, seven commonly used sulfonylureas were extracted from large volumes of various types of water, and isolated from coextracted nonacidic compounds and humic acids.[35] Before extraction, no adjustment of the water pH was needed.

A single multiresidue method was developed to determine 109 priority compounds, including PUHs, listed in the 76/464/EEC Council Directive on Pollution of the European Union.[54] For trapping analytes, automated offline SPE, using polymeric sorbent Oasis 60 mg cartridges, was optimized. SPE with a PS–DVB material (Lichrolut EN) has been combined offline with Liquid Chromatography–tandem Mass Spectrometry (LC–MS–MS) for ultra-trace determination of phenylureas and their transformation products in various natural waters.[55] A proposed method for analyzing PUHs and SUHs involves analyte preconcentration with 300 mg combined polystyrene–divinylbenzene and methacrylate macroporous resins.[56] SPE have been used for offline trapping some sulfonylureas from 250 ml of water samples from various sources.[57] Recovery of three analytes out of four was better than 85%, while recovery of tribenuron-methyl was about 75%. In both cases, the pH of the water samples was adjusted to three before analyte extraction. A multianalyte method for the confirmation and quantification of 16 selected sulfonylurea, imidazolinone, and sulfonamide herbicides in surface water has been proposed.[58] This method is based on analyte extraction with a polymeric material (RP-102) and extract cleanup with a strong anion exchanger (SAX) cartridge stacked on top of an alumina cartridge. Analysis of the final extract was performed by LC–MS. An analytical protocol was developed for determining sulfonylureas in surface and ground water in the midwestern U.S.A.[59,60] Extraction of the analytes was performed by two stacked in-series SPE cartridges, the upper one containing a SAX material and the lower one filled with RP-102 PS–DVB resin. The first cartridge served to block dissolved humic acids that can interfere with the analysis. Acidic herbicides, including sulfonylureas, were extracted from water by using 0.5 g of a synthetic adsorbent, that is, Porapak Rdx.[61]

iii. Offline SPE with Adsorbents Imbedded in Membranes
In recent years, commercially available filter disks containing both C-18 and PS–DVB materials with particle sizes finer than those used with cartridges (8 μm against 40 μm) imbedded in a Teflon matrix have been used for both offline and online SPE of pesticides. The specific advantages claimed for disk design over cartridge design are as follows:

- Shorter sampling flow rate due to faster mass transfer and lack of channeling effects;
- Decreased plugging by particulate matter due to the large cross-sectional area;
- Cleaner background interferences.

The last advantage derives from the fact that, unlike SPE with cartridges, the extraction apparatus with disks consists of glass. The use of an extraction disk is rather simple. The membrane is placed in a filtration apparatus connected to a vacuum source by a water pump. After the disk has been washed/conditioned with 10 ml of methanol and 10 ml of distilled water, the aqueous sample is passed through the disk (Figure 25.5). After eliminating part of the water by vacuum, the assembly supporting the disk is transferred to a second vacuum flask containing a vial. Then, 5 to 10 ml of the eluent phase, usually methanol or acetonitrile, is slowly drawn onto the membrane by moderate vacuum. The vacuum is interrupted for 2 to 4 min to allow the liquid to soak the membrane. Thereafter, analytes are eluted and collected into the vial. This operation is repeated by applying another 5 ml aliquot of the eluent phase to the top of the disk.

C-18 Empore extraction disks were used for the isolation and trace enrichment of linuron in 4 l of river water.[62] At the 0.25 pg/l level, recovery of this phenylurea was 94% with 6%

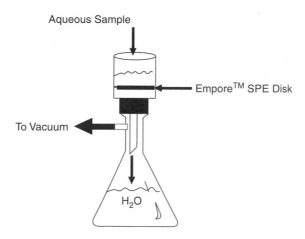

Aqueous Sample

Empore™ SPE Disk

To Vacuum

H₂O

FIGURE 25.5 Schematic of an offline SPE device with membranes. (From Howard, A. L. and Taylor, L. T., *J. Chromatogr. Sci.*, 30, 374, 1992. With permission.)

coefficient of variation. A procedure based on C-18 extraction disks and GC after derivatization has been elaborated for analyzing traces of chlorsulfuron and metsulfuron methyl in water.[63] A combination of the SPE disk technology for extracting analytes in water and supercritical fluid extraction technology for reextracting analytes from disks has been applied to the analysis of sulfuron methyl and chlorsulfuron.[64] A method for analyzing PUHs in water by extracting them by an automated offline SPE with 47 mm C-18 Empore extraction disks has been proposed.[65,66] Phenylureas were isolated from water and soil extracts by SPE using a layered system of two extraction disks.[67] The first disk consisted of a SAX material imbedded in Teflon, and the second disk was made of C-18 particles also imbedded in Teflon. The purpose of using the first disk was that of blocking humic and fulvic acids, thus resulting in greater sensitivity for diode array detection (DAD).

iv.   Online SPE

With the offline SPE technique, a certain skill and care is required of the analyst. Moreover, rapid screening procedures of many samples for monitoring pesticides require analysis automation. In recent years, fully automated analysis of contaminants in water by the online coupling of SPE to LC or GC instrumentation has received increasing attention. Besides allowing rapid analysis, additional positive features of online SPE are that analyte loss due to evaporation does not occur and that the entire sample is introduced into the chromatographic instrumentation, instead of a fraction as with offline procedures. In this way, the sampled volume can be drastically reduced, thus lowering the costs of cooled sample transportation and storage. Automatic devices that couple online the sample pretreatment by SPE-LC in one analytical run are nowadays commercially available. With online SPE, the water sample is pumped through a short precolumn (typically 10 mm length × 2 mm ID) filled with small particles (15 to 25 μm) of either C-18 or PS–DVB adsorbing media. Solutes are trapped, while water is wasted. Eventually, the precolumn can be washed with a small volume of a water/methanol mixture. By a system of switching valves, the solutes are then removed from the precolumn by the LC mobile phase itself and transported into the LC column (Figure 25.6). When using a precolumn packed with an adsorbent having a larger affinity for analytes than that filling the analytical column, broad peaks for the last-eluted analytes are obtained. In this case, analyte backward-elution from the precolumn with the LC mobile phase can eliminate peak broadening.

A fully automated sample handling system has been developed for LC combining the advantages of a disposable cartridge system and precolumn technology with its high automation

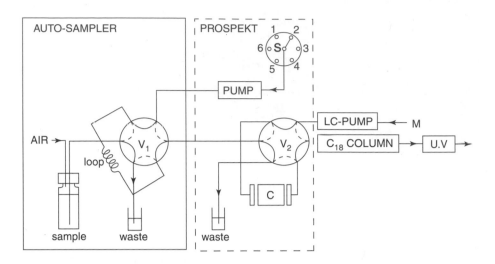

FIGURE 25.6 Design of the automated sample handling (PROSPEKT) system. V_1, injection valve; V_2, high-pressure switching valve; S, six-port solvent selection valve; M, mobile phase; C, 10×2 mm i.d. cartridge packed with 40 μm C_8 silica. (From Nielen, M. W. P., Valk, A. J., Frei, R. W., Brinkman, D. A. Th., Mussche, Ph., De Nijs, R., Ooms, B., and Smink, W., *J. Chromatogr.*, 393, 69, 1987. With permission.)

potential. As an example, this device was applied to the analysis of some phenylureas in river water.[68] A precolumn packed with a PS–DVB copolymer or a stack of eight 4.6 mm diameter C-18-loaded membrane extraction disks have been coupled to LC–MS I for trace analysis of 15 phenylureas in surface water and drinking water.[69] By sampling only 50 ml water, limits of detection (LODs) ranged between 5 and 20 ng/l. For analyzing a test mixture of 17 pesticides, which included three phenylureas, in tap water and river water, the analytes were preconcentrated on a short analytical LC column and then gradient-eluted into a LC–MS apparatus.[70,71] Automated online trace enrichment, LC analysis with DAD was investigated for the determination of widely used pesticides (19 phenylureas were included in this list) in environmental waters.[72] Detection limits of 0.1 μg/l were obtained using 150 ml of river waters. Mills proposed a similar device for routine determination of PUHs in drinking water.[73] For the purpose of developing a single-class Multiple Reaction Monitoring (MRM) involving online SPE in combination with LC–MS for sulfonylureas in aqueous samples, eight sulfonylureas were selected.[74] Trace enrichment of herbicides in river water, including several phenylureas, on a precolumn packed with Polygosil C-18 material was combined online with LC–Fourier-transform infrared spectrometry.[75] A system consisting of sorptive enrichment in columns packed with 100% polydimethylsiloxane particles online coupled with LC analysis was evaluated for the purpose of analyzing the most polar phenylureas in various types of water.[76] A device consisting of 1 mm i.d. $\times$ 10 mm Zorbax 3.5 μm SB-C-18 guard column (extraction cartridge) online coupled to a microbore LC-Electrospray (ESI)/ MS has been applied to the determination of 16 pesticides in natural waters. In this list of pesticides widely used in the U.S.A., diuron, linuron, fluometuron, and siduron were included.[77] A methanol gradient does not effectively desorb analytes from a polymeric precolumn onto a C-18 analytical column. This problem was resolved by desorbing analytes from the polymeric material with 0.3 ml of acetonitrile and mixing it with the mobile phase by inserting a T-piece inbetween the trap and the analytical column.[78] Use of short turbulent-flow chromatography (TFC) columns as extraction cartridges enabling fast online SPE at high sampling flow-rate (5 ml/min) was proposed.[79] Polymeric and carbon based TFC columns (Oasis HLB, Cyclone, Hypercarb) allowed complete extraction of priority pesticides (including some phenylureas) from 50 ml of water. Online coupling to LC was performed with remixing of the organic TFC eluate with water in front of the analytical column to ensure efficient band focusing.

v.   Online Extraction with Liquid Membranes

Sample preparation by means of liquid membrane extraction is a technique that in essence contains two LLE extractions in one step. The setup is easily automated, and sample preparation is performed in a closed system, thus minimizing the risk for contamination and losses during the process. Because the extraction is made from an aqueous phase (donor) to a second, also aqueous phase (acceptor), further enrichment on a precolumn is possible before injection into the LC apparatus. Liquid membranes were used for enrichment of metsulfuron-methyl and chlorsulfuron from clean aqueous samples[80] and natural waters.[81] A similar device was developed by a Chinese research group for monitoring sulfonylureas in river water.[82,83]

vi.   Solid-Phase Microextraction (SPME)

SPME was introduced at the end of the 1980s by Pawliszyn and coworkers as a technique for extracting organic micropollutants from aqueous matrices. A 0.5 to 1 mm i.d. uncoated fiber or coated with suitable immobilized liquid phase (in the second case this technique should be more correctly called liquid-phase microextraction) is immersed in a continuously stirred water sample. After equilibrium is reached (a good exposure time takes 15 to 25 min), the fiber is introduced into the injection port of a gas chromatograph, where analytes are thermally desorbed and analyzed. Positive features of this technique are that it is rapid and very simple, and does not use any solvent. In addition, like online SPE, this technique requires small sample volumes (2 to 5 ml) because of all the sample extract is injected into the analytical column. SPME is often used in combination with GC for analyzing pesticides in water. Because many pesticides and their polar metabolites are not amenable to GC, separations are often performed by LC. Therefore, a special interface was constructed, which allows offline coupling of SPME with fibers to LC. However, the efficiency of this analytical method relies on manual operation. Another approach is the use of open-tubular capillaries coated with a stationary phase for the extraction of analytes from aqueous samples. The main advantage is the ease of coupling capillaries online to an LC system (Figure 25.7). Additionally, the surface area of a capillary and the amount of coating are greater than those of a fiber of the same length, which can be varied for capillaries. An automated in-tube SPME has been proposed for analyzing six phenylureas in aqueous samples.[84] The authors used an ordinary GC capillary coated with Omegawax (0.25-μm film thickness) for automated in-tube SPME with UV detection. The LODs for PHUs ranged between 2.7 to 4.1 μg/l. The sorption rate was rather low, varying between 2 and 5%. This was mainly due to the fact that Omegawax is a stationary phase for GC separation of nonpolar compounds. To increase the extraction efficiency of in-tube SPME for PUHs, the capillary has been coated with a special polyacrylate.[85] This material has a larger affinity for polar compounds, such as several PUHs, than Omegawax. By this device, LODs for phenylureas ranged between 10 to 90 ng/l, this fulfilling the requirements of the European regulations for drinking water (100 ng/l tolerance level). For analyzing five PUHs and chlorsulfuron, the extraction efficiencies of two polydimethylsiloxanes and a polyacrylate fiber were compared.[86] The extraction time, addition of NaCl to the water and the influence of humic acids on the extraction efficiency were evaluated in order to obtain a sensitive method.

Table 25.5 summarizes the use of various extraction procedures for assaying phenylurea and SUHs in environmental waters.

c.   Soil Samples

i.   Liquid–Solid Extraction (LSE)

Current methodology frequently applies methods for the analysis of PUHs in soil involving classic LSE with organic solvents at room temperature. Henze et al.[87] performed the extraction of linuron and its metabolites from soil samples by LSE with acetone followed by SPE cleanup. Performing LSE, Liegeois et al.[88] proposed a quantification procedure for isoproturon in soil samples using an enzyme-linked immunosorbent assay technique. Perez et al.[89] isolated chlorotoluron, isoproturon,

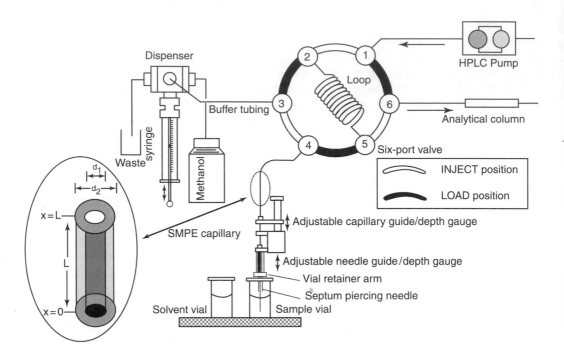

FIGURE 25.7 Instrumental setup of the online SPME-LC interface based on an in-tube SPME capillary technique. A piece of GC column (in-tube SPME) hosts in the position of the former needle capillary. The aqueous sample is frequently aspirated from the sample vial through the GC column and dispensed back to the vial (INJECT position) by movement of the syringe. After the extraction step, the six-port valve is switched to the LOAD position for the desorption of the analytes from the in-tube SPME by flushing 100% methanol from another vial through the SPME capillary. The volume is transferred to the loop. After switching the Valco valve to the INJECT position, an isocratic separation using a mixture of 60:40 acetonitrile/water was performed. A detailed view of the in-tube SPME capillary is included at the left side of the figure. (From Eisert, R. and Pawliszyn, J., *Anal. Chem.*, 69, 3140, 1997. With permission.)

and metoxuron in soil by solvent extraction. A method based on LSE has been developed for the simultaneous determination of ten PUHs in soil.[90] Eight commonly used SUHs were isolated from soil using 90:10 acetonitrile/water.[91] LSE with CH_2Cl_2 and CH_3OH separately and assisted by sonication for 30 min has been adopted for isolating SUHs from soil.[92] Taking advantage of the acidic nature of SUHs, several searchers have proposed extraction of these herbicides from soil using water buffered at pH 7 to pH 8 alone[93–98] or mixed with 20% acetone.[99] The main advantage of using buffered water as extractant is that the extract can be readily purified by SPE after extract acidification.

ii. Microwave-Assisted Solvent Extraction (MASE)

A number of disadvantages have been noticed with conventional LSE methods. They are laborious, time-consuming, and expensive. They are also subject to problems arising from the formation of emulsions, the evaporation of large solvent volumes, and the disposal of toxic or inflammable solvents. The drawback of using water buffered at pH 7 to pH 8 for extracting SUHs is that many acidic soil interferences, e.g., humic acids, are coextracted. The need for more efficient and more economical methods for the extraction of organic pollutants from soil has generated (semi) automated techniques, such as MASE. This technique permits a reduction of solvent consumption and extraction time. With MASE, the typical extraction solvent mixture is CH_2Cl_2/CH_3OH (90:10, v/v). The combination of MASE and LC

TABLE 25.5
Solid-Phase Extraction of Phenylurea and Sulfonylurea Herbicides from Water Samples

Compound	Sample	Mode	Sorbent	Eluent Phase	Ref.
14 Phenylureas	2 l DW, 2 l GW, 1 l RW	Offline	Carbograph 1, 0.25 g	6 ml CH_2Cl_2/CH_3OH (95:5)	29
16 Phenylureas	20 ml DW, 20 ml RW	—	Anti-chortoluron antibody on silica support	CH_3CN/0.5% acetic acid	31
16 Phenylureas	1 l DW, 1 l RW	—	Molecularly imprinted polymers	2 ml CH_2Cl_2/CH_3OH (90:10)	34
7 Sulfonylureas	4 l DW, 2 l GW, 0.2 l RW	—	Carbograph 4, 0.5 g	8 ml CH_2Cl_2/ CH_3OH (80:20) acidified with 10 mM acetic acid	35
Linuron, monolinuron in MRM	0.25 l PW	—	C-18, 0.5 g	2 ml Ethyl acetate	38
12 Sulfonylureas	0.5 l MW (pH 3)	—	PS–DVB, 0.5 g	10 ml CH_3OH	46
Fenuron, metoxuron in MRM	2 l DW	—	Carbograph 1, 1 g	6 ml CH_2Cl_2/CH_3OH (80:20)	50
9 Phenylureas and related chloroanilines	4 l DW, 2 l GW, 0.5 l RW	—	Carbograph 4, 0.5 g	1.5 ml CH_3OH, then 6 ml CH_2Cl_2/CH_3OH (80:20) acidified with 10 mM HCl	53
Phenylureas in MRM	1 l DW	—	Copolymer, 60 mg	2.5 ml acetonitrile/CH_2Cl_2 (50:50), then 3.2 ml CH_2Cl_2	54
Phenylureas	1 l RW	—	GCB disk, 0.5 g; PS–DVB, 0.2 g	2 × 5 ml CH_2Cl_2/CH_3OH (80:20); 3 × 3 ml CH_3OH	55
16 Sulfonylureas	250 ml RW, DW, MW	—	PS–DVB, then SAX	10 ml CH_3OH, then 17 ml CH_2Cl_2	58
Linuron in MRM	4 l RW, 4 l simulated SW	—	C-18 disks	—	62
Chlorsulfuron, sulfometuron methyl	1 l PW	—	C-18 disks	2% CH_3OH-modified CO_2	64
5 Phenylureas	10 ml RW	Online	C-18	CH_3OH /20 mM phosphate buffer (45:55)	68
15 Phenylureas	50 ml RW	—	PLRP-S or several C-18 disks	CH_3OH/0.1 M $AcNH_4$ (gradient elution in the backflush mode)	69
Monuron, diuron, neburon in MRM	100 ml DW, 100 ml RW	—	PLRP-S	CH_3CN/H_2O, gradient elution	70
16 Phenylureas in MRM	150 ml RW	—	PLRP-S or C-18	CH_3CN/phosphate buffer, gradient elution	72
5 Phenylureas	10 ml PW, DW, RW,	—	PDMS	CH_3CN/H_2O, gradient elution	76
3 Phenylureas in MRM	50 ml DW, RW	—	Copolymer HLB, Cyclone or Hypercarb	CH_3OH	79
Metsulfuron, ethametsulfuron	20 ml DW, SW	—	Liquid membrane (PTFE)	CH_2Cl_2/phosphate buffer	82
6 Phenylureas	25 μl PW	—	SPME with a glass capillary coated with Carbowax	38 μl CH_3OH	84

DW, drinking water; GW, ground water; RW, river water; MRM, multiresidue method; PW, pure water; MW, marsh water; SW, sea water.

with UV detection has been investigated for the efficient determination of sulfonylureas[100] and phenylureas[101] in soil.

iii. Hot Water Extraction

Hawthorne et al.[102] reported that subcritical water efficiently extracted chlorophenols, alkylbenzenes, polycyclic aromatic hydrocarbons, and *n*-alkanes from soil. Class-selective extraction of these organics was simply achieved by adjusting the water temperature (50 to 300°C). This finding was explained by considering that the polarity of water steadily decreases as the temperature is increased (note that at 190°C, the polarity of water equals that of the methanol), thus making water more and more capable of competing with nonpolar organics for adsorption on soil particles or soil organic matter. Like CO_2 used in supercritical fluid extraction, water is an environmentally acceptable solvent, it is cost effective, and hot water conditions are easily achieved with commercial laboratory equipment. Another advantage of using hot water as extractant is that the extract can be easily coupled online to a sorbent cartridge for analyte trapping and sample purification. Alternatively, large aliquots of the aqueous extract can be directly injected in a reversed-phase LC column.

The feasibility of extracting selectively and rapidly herbicide residues, including some phenylureas, in soil by water heated at 90°C and collecting analytes with a Carbograph 4 SPE cartridge set online with the extraction cell (Figure 25.8) was evaluated.[103] Recovery of neutral and acidic herbicides ranged between 81 and 93%, except for two acidic herbicides (63%). For the analytes considered, comparison of methods showed hot water extraction was overall more efficient than Soxhlet and sonication extraction techniques. Rapid trace analysis of polar and

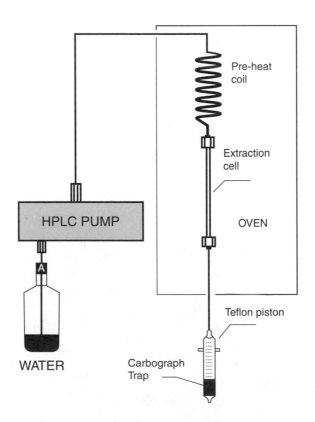

FIGURE 25.8 Schematic view of the laboratory-made extraction device. (From Crescenzi, C., D'Ascenzo, G., Di Corcia, A., Nazzari, M., Marchese, S., and Samperi, R., *Anal. Chem.*, 71, 2157, 1999. With permission.)

medium polar contaminants, including several PUHs, in soil was achieved by coupling online a hot phosphate buffered water extraction apparatus to a LC–MS system.[104] Coupling was accomplished by using a small C-18 sorbent trap for collecting analytes and two six-port valves. The efficiency of this device was evaluated on extracting 13 selected pesticides from 200 mg of laboratory-aged soils by varying the extraction temperature, the extractant volume, and the flow rate at which the extractant passed through the extraction cell and the sorbent trap. In terms of extraction efficiency, robustness of the method, and extraction time, the best compromise was that of using 8 ml of extractant at 90°C and 0.5 ml/min flow rate. Under this condition, recoveries of 11 out of 13 analytes ranged between 82 and 103%, while those of the least hydrophilic pesticides, i.e., neburon and prochloraz, were 73 and 63%, respectively. By increasing the extractant volume to 60 ml, additional amounts of the two latter compounds could be recovered. Under this condition, however, the most hydrophilic analytes were in part no longer retained by the C-18 sorbent trap. From a naturally 1.5 years-aged soil, hot phosphate buffered water removed larger amounts of three herbicides and hydroxyterbuthylazine (a terbuthylazine DP) than pure water and Soxhlet extraction. This result seems to confirm that hot phosphate buffer is able to remove from soil also those fractions of contaminants which, on aging, are sequestered into the humic acid framework.

Table 25.6 summarizes the use of various extraction procedures for assaying phenylurea and SUHs in soil samples.

D. Separation and Detection Methods

1. Gas Chromatography

a. Phenylurea Herbicides

For determining phenylureas in water, the GC technique with the use of selective detectors, such as the electron capture (ECD), nitrogen phosphorous (NPD), and, chiefly, MS detectors, has attracted the attention of many searchers. Direct GC analysis of PUHs has been reported in the literature.[54,105] However, it is well recognized that a significant number of these compounds as well as sulfonylureas cannot be analyzed in this way owing to their thermal instability. They partly decompose into isocyanates and amines, the main contributory factor

---

TABLE 25.6
Extraction of Phenylurea and Sulfonylurea Herbicides from Soil Samples

Compound	Extracting Technique	Extracting Phase	Cleanup	Ref.
8 Sulfonylureas	LSE	4 ml CH_3OH	10 ml CH_3CN/H_2O (90:10) + 3 ml CH_2Cl_2–	91
3 Sulfonylureas	—	10 ml CH_3OH or CH_2Cl_2	—	92
Primisulfuron, triasulfuron	—	100 ml CH_3OH 0.1 M p.b. (50:50)	3 × 30 ml + 2 × 5 ml CH_2Cl_2	95
5 Sulfonylureas	MASE	20 ml CH_2Cl_2/CH_3OH (90:10)	—	100
4 Phenylureas	HWE	25 ml H_2O at 90°C	SPE, carbograph 4	103
5 Phenylureas	—	p.b. 0.5 mM 8 ml at 90°C + 8 ml 130°C	—	104

LSE, liquid solid extraction; p.b., phosphate buffer; MASE, microwave assisted solvent extraction; HWE, hot water extraction; SPE, solid phase extraction.

being the NH moiety. Some methods have relied on quantification of the DP formed into the injection port.[86,106,107]

To make phenylureas amenable to GC analysis, various derivatization procedures have been elaborated. It has to be pointed out that all these reactions consist in substitution of the free hydrogen attached to the nitrogen atom close to the aromatic moiety by different groups. These procedures can be grouped into three reaction classes:

i. Direct Acylation

The reaction most frequently carried out is perfluoroacylation, by reacting the analyte with trifluoroacetic anhydride (TFAA) or heptafluorobutyric anhydride (HFBA).[18,108,109] These derivatization agents are chosen for exploiting sensitive ECD (Figure 25.9).

ii. Indirect Acylation

Here, phenylureas are first converted to their anilines, and then the later are reacted with the aforementioned derivatization agents.[110] Aniline derivatives are more readily formed than the corresponding phenylureas, but derivative preparation is time consuming. Figure 25.10 shows a chromatogram obtained by following this derivatization procedure.

iii. Alkylation

The reagents most frequently used are trimethylaniline hydroxide (TMAH)[111] and alkyl iodide.[36,51,112] The reaction with TMAH can be carried out on column by direct injection of a mixture of the phenylureas and the reagent in methanol. Figure 25.11 shows a typical GC chromatogram of some phenylureas after their conversion to alkylated species.

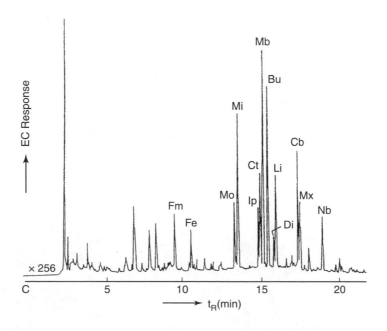

FIGURE 25.9 Capillary GC with fused-silica column coated with CP-Sil 5 (analogous to SE 30 and OV-101) of HFB derivatives of 13 phenylureas obtained after extraction of a Bosbaan river water sample spiked at 1 μg/l level, and direct derivatization with HFBA. Injected amount corresponds to 0.1 ng of each herbicide. Symbol explanation: Fm, fluometuron; Fe, fenuron; Mo, monuron; Ml, monolinuron; Ip, isoproturon; Ct, chlorotoluron; Mb, metobromuron; Bu, buturon; Di, diuron; Li, linuron; Cb, chlorbromuron; Mx, metoxuron; Nb, neburon. (From Brinkman, U. A. Th., de Kok, A., and Geerdink, R. E., *J. Chromatogr.*, 283, 113, 1984. With permisssion.)

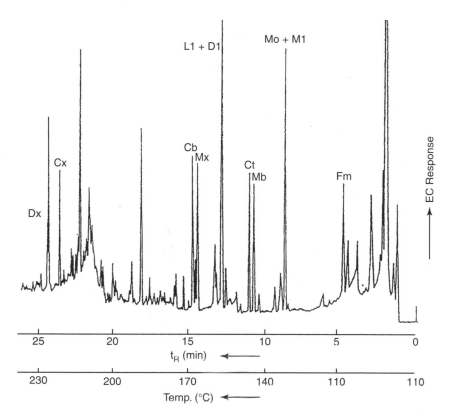

FIGURE 25.10 Capillary GC with fused-silica column coated with CP-Sil 5 (analogous to SE 30 and OV-101) of HFB-anilines obtained after extraction of a pure water sample spiked with 1 μg/l parent herbicides, and subsequent hydrolysis and derivatization. Stationary phase: CP-Sil 5. Symbol explanation: Fm, fluometuron; Mo, monuron; Ml, monolinuron; Mb, metobromuron; Ct, chlorotoluron; Di, diuron; Li, linuron; Mx, metoxuron; Cb, chlorbromuron;. Cx, chloroxuron; Dx, difenoxuron. (From de Kok, A., Roorda, I. M., Frei, R. W., and Brinkman, U. A. Th., *Chromatographia*, 14, 579, 1981. With permission.)

b. Sulfonylureas

Two GC methods have been elaborated for analyzing chlorsulfuron and metsulfuron methyl in water.[22,63] Sulfonylureas are even less volatile and more thermally labile than phenylureas, so they need to be converted to more volatile compounds before GC analysis. Diazomethane has been used to convert the two sulfonylureas to their stable N,N'-dimethyl derivatives (Figure 25.12).

Table 25.7 summarizes selected methods for analyzing urea herbicides in water by the GC technique.

2. Supercritical Fluid Chromatography (SFC)

The SFC technique is closely related to LC, but it is three to five times faster, allows more rapid generation of high efficiency, and can be used with both typical GC and LC detectors, simultaneously. Higher efficiency coupled with multiple detection simplifies the task of resolving complex mixtures, such as in screening for the presence or absence of a large number of target compounds without the expense of a mass spectrometer.

SFC coupled to MS has been proposed for analyzing benzsulfuron methyl, chlorsulfuron, and metsulfuron methyl in soil.[92] A method based on SFC coupled online to SPE has been

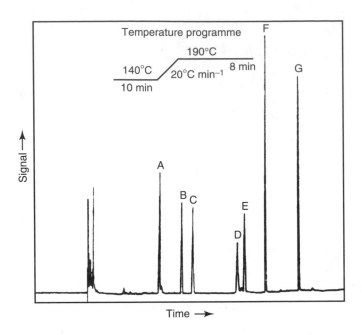

FIGURE 25.11 GC with a fused-silica capillary column and a NPD detector of some urea herbicides after conversion to their methylated forms. Symbol explanation: A, monuron; B, isoproturon; C, chlorotoluron; D, linuron; E, diuron; F, methabenzthiazuron; G, tebuthiuron. (From Scott, S., *Analyst*, 118, 1117, 1993. With permission.)

proposed for monitoring four sulfonylureas in natural waters (Figure 25.13) with detection limits as low as 50 ng/l.[113]

3. Liquid Chromatography

Liquid chromatographic systems for environmental pesticide analysis have been extensively reviewed in a previous paper.[114] Nowadays, LC is the technique of choice for analyzing those pesticides which, being thermolabile, are not amenable to direct GC analysis, such as phenylurea and SUHs. LC methods of analysis also have the important advantage over GC methods in that online pre- and postcolumn reaction systems are compatible with LC instrumentation. Furthermore, the LC apparatus can easily be coupled online with the enrichment step using SPE on precolumns, thereby making the analysis fully automated.

In many LC methods involving the use of UV,[21,29,35,40,45,48–50,64,68,72,73,80,81,84,90,99–101] diode array,[30,31,33,34,39,57,65,67,82,83,115,116] photoconductivity,[44] and Fourier-transform infrared spectrometry,[75] after postcolumn photochemical reaction fluorescence[41] detectors have been developed for analyzing phenylurea and SUHs in water samples. As examples, Figure 25.14 to Figure 25.16 show LC chromatograms obtained by injecting extracts of real water samples spiked with trace amounts of urea herbicides.

In Table 25.8, selected LC methods for assaying urea pesticides in water and soil are listed.

4. Capillary Electrophoresis (CE)

Electrophoresis is a process in which charged species are separated according to differences in their electrophoretic mobilities, and these are related to their charge densities. In the mid 1980s, instruments able to fractionate charged analytes into a capillary column were introduced.

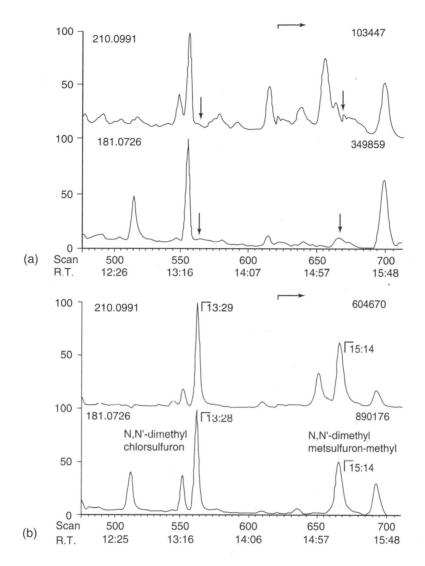

FIGURE 25.12 GC–MS SIM chromatograms with a HP-5 capillary column of pure water sample extracts, (a) blank and (b) spiked with 0.1 μg/l of chlorsulfuron and metsulfuron methyl. (From Klaffenbach, P., and Holland, P. T., *J. Agric. Food Chem.*, 41, 396, 1993. With permission.)

This technique is called CE or capillary zone electrophoresis (CZE). The electrophoresis process enables also simultaneous separation of both uncharged and charged species. This technique is called micellar electrokinetic chromatography (MECK). In MECK, surfactants are added to the electrolyte in concentrations high enough to form micelles. Neutral as well as ionic solutes are separated on the basis of their different distribution between a fast moving aqueous phase, migrating with the electroosmotic flow velocity, and a micellar pseudo-stationary phase with a slower migration velocity. Versatility is due to various surfactants and modifiers, which can be selected in order to optimize separation.

With both CZE and MECK, analyte fractionation is usually carried out in a short fused-silica capillary filled with a buffer solution. Typically, capillary columns are 25 to 100 cm in length with i.d. ranging between 25 and 100 μm. Electrodes are usually platinum and are connected to a power supply able to provide constant voltages up to 30 kV and currents up to 100 μA. A particular

TABLE 25.7
Selected Capillary Column Gas Chromatographic Methods for Determining Urea Herbicides in Water Samples

Compound	Derivatizing Agent	Column Characteristics	Injection Device	Detector	LOD	Ref.
Monolinuron, linuron	Direct analysis	HP-5MS 30 m × 0.25 mm 0.25 μm film thickness	Splitless	MS	<5 pg	54
8 Phenylureas, chlorsulfuron	Analysis of degradation products	BP10 on 30 m × 0.25 mm 0.25 μm	On column	NPD, ECD	<5 ng	107
15 Phenylureas	HFBA	CP-Sil 5 on 25 m × 0.22 mm	Splitless	ECD, MS	1 pg	108,110
Isoproturon, chlorotoluron, linuron, diuron	Iodomethane	BP1 on 25 m × 0.22 mm, 0.25 μm film thickness	Split	NPD, MS	<0.1 μg/l	37
Metsulfuron-methyl, chlorsulfuron	Hydrolysis	DB-17 on Megabore column	On column	ECD	0.01 μg/l	22
Metsulfuron-methyl, chlorsulfuron	Diazomethane	HP-5 on 25 m × 0.22 mm, 0.33 μm film thickness	Splitless	ECD, NPD	<0.1 μg/l	63

ECD, electron capture detector; HFBA, heptafluorobutyric anhydride; MS, mass spectrometry; NPD, nitrogen–phosphorous detector.

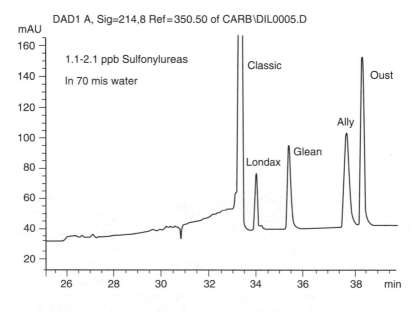

FIGURE 25.13 SFC-UV chromatogram of online SPE extract of 70 ml of 1.1 to 2.1 μg/l of sulfonylureas in water. Column: eight standard LC column, each 200 × 4.6 mm i.d. with 5 μm silica particles. The first column was C-18, while the rest were Hypersil silica. The mobile phase was 1% CH$_3$OH in CO$_2$. After a 4 min hold, CH$_3$OH was programmed to 16% at 0.5%/min, then held. Temperature was 60°C. (From Berger, T. A., *Chromatographia*, 41, 133, 1995. With permission.)

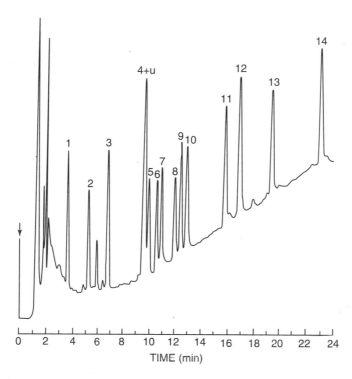

FIGURE 25.14 LC–UV chromatogram obtained on analyzing 2 l of a municipal water (Rome) specimen spiked with 30 ng/l of each phenylurea. Column: 250×4.6 mm i.d. packed with 5 μm C-18. Elution: CH_3OH-CH_3CN (85:15)/H_2O linear gradient elution from 47% organic modifier to 70% in 20 min. Peak numbering: 1, fenuron; 2, metoxuron; 3, monuron; 4, monolinuron; 5, fluometuron; 6, chlorotoluron; 7, metobromuron; 8, difenoxuron; 9, isoproturon; 10, diuron; 11, linuron; 12, chlorbromuron; 13, chloroxuron; 14, neburon; u, unknown compound. (From Di Corcia, A. and Marchetti, M., *J. Chromatogr.*, 541, 365, 1991. With permission.)

characteristic of the electro-osmotic flow is that the profile of the liquid front is practically flat, instead of being parabolic, because it occurs when a liquid is forced to pass through a tube by hydrodynamic pressure. This effect, coupled to the absence of any resistance to the mass transfer, enables CE to separate compounds in 10 min with an efficiency of more than 200,000 plates. Extremely sharp peaks for the analytes also reflect that CE instruments equipped with UV detectors are able to detect analyte quantities as low as 0.2 pg. On the other hand, only a few nanoliters of a sample volume can be injected into the capillary without affecting the electrophoretic process. This results in method detection limits of several hundreds of ppb, which are too high for practical environmental applications. Several techniques have been reported for on column concentration to enhance detection in CZE. Among these, the field-amplified technique seems to offer the best possibilities, in terms of sensitivity. By this expedient, a ten fold analyte concentration can be reached, provided the sample volume occupies only a small section of the capillary.

Sulfonylureas are ionogenic compounds, and thus they lend themselves to analysis by CE. Some CE procedures involving UV detection have been developed for analyzing sulfonylureas in natural waters[46,56,93,117,118] and soil.[94,95,96] A typical electropherogram of sulfonylureas is shown in Figure 25.17. One study has described the use of CE coupled with MS for the rapid online separation and characterization of sulfonylureas as synthetic mixtures.[119] The feasibility of using the MEKC technique for the multiresidue monitoring of herbicides, including some PUHs and SUHs, has been evaluated.[56]

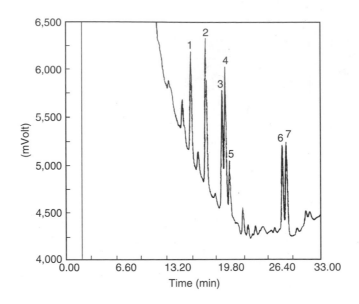

FIGURE 25.15 LC-UV chromatogram obtained by analyzing 4 l of drinking water spiked with seven sulfonylureas at the individual level of 10 ng/l. Column: Alltima 250×4.6 mm i.d. containing 5 μm C-18. Elution: CH_3CN/H_2O (both containing 3 mmol/l trifluoroacetic acid) linear gradient elution from 32:68 to 62:38 (v/v) in 40 min. Peak numbering: 1, thifensulfuron methyl; 2, metsulfuron methyl; 3, triasulfuron methyl; 4, chlorsulfuron; 5, rimsulfuron; 6, bensulfuron methyl; 7, tribenuron methyl. (From Di Corcia, A., Crescenzi, C., Samperi, R., and Scappaticcio, L., *Anal. Chem.*, 69, 2819, 1997. With permission.)

5. Liquid Chromatography–Mass Spectrometry

A serious weakness of methods based on LC with conventional detectors, such as UV, DAD, and other ones, is that they lack sufficient specificity for showing without doubt the presence of traces of target compounds in complex aqueous matrices. If a photodiode array is employed as the LC detector, UV spectra can be used to confirm peak identification. However, the UV spectra of many pesticides belonging to the same compound class are very similar, and differences between compound classes are frequently small. This limits the use of these spectra for peak confirmation. Furthermore, peak overlapping precludes quantification of target compounds, even by the use of diode array detectors. Public Environmental Agencies in many countries relies on detection by MS for unambiguous confirmation of contaminants in the environment. The Commission Decision 93/256/EEC states that "Methods based only on chromatographic analysis without the use of molecular spectrometric detection are not suitable for use as confirmatory methods." GC–MS is still the technique of choice, as it has been routinely used in the last 35 years for analyzing an enormous number of compounds in a variety of matrices. However, many pollutants are very polar and thermally unstable compounds, thus complicating or precluding their analysis by GC. Research in new methodologies in MS, notably LC–MS, has greatly benefited from the international need of protecting food quality, and now can serve to fulfill the goals initially sought by such a technique, which is monitoring nonvolatile and polar target compounds with the specificity and sensitivity similar to GC–MS. In the past 20 years, a large variety of interfaces have been developed to make the high vacuum of the mass analyzer compatible with the large amounts of liquids coming out from the LC column. LC–MS has been extensively reviewed in the past years. Several books[120–123] and review papers[124–129] devoted to illustrate principles, instrumentations, and applications of LC–MS were published.

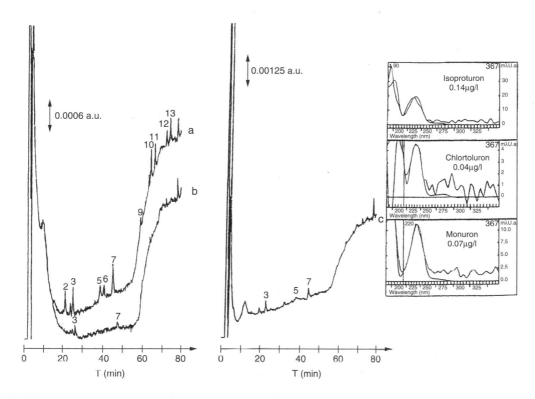

FIGURE 25.16 LC-DAD analysis after online preconcentration of 50 ml of river Seine water spiked with 0.1 µg/l with (a) phenylureas and (b) nonspiked on a precolumn containing 0.22 g of silica bonded with antichlorotoluron antibodies, and (c) identification of three compounds in the nonspiked river water. Detection was performed at 244 nm. Column: 250 × 4.6 mm i.d. packed with Baker Narrow Pore C-18. Peak numbering: 1, fenuron; 2, metoxuron; 3, monuron; 4, methabenzthiazuron; 5, chlortoluron; 6, fluometuron; 7, isoproturon; 8, difenoxuron; 9, buturon; 10, linuron; 11, chlorbromuron; 12, diflubenzuron; 13, neburon. (From Pichon, V., Chen, L., Durand, N., Le Goffic, P., and Hennion, M.-C., *J. Chromatogr.*, 725, 107, 1996. With permission.)

Among the various interfaces developed in the past for coupling LC to MS, only the ESI and Atmospheric Pressure Chemical Ionization (APCI) sources are nowadays commercially available.

The ESI interface is the youngest device introduced for LC–MS coupling. ESI has opened new and exciting perspectives to the LC–MS technique. It is sufficient to say that the ESI interface enables LC–MS analysis of compounds having molecular weights up to 4,000,000 Da, as the ESI process is able to form multiply charged ions, depending on the acid/base chemistry and hydration energy of the molecules. The ability to increase charge (z) permits the analysis of large molecular masses on a conventional quadrupole limited to m/z of 2000 Da for singly charged ions. The versatility of this interface has made it extremely popular among both analytical chemists and biochemists.

The ESI source apparently suffers from the limitation that it cannot accept more than 40 to 50 µl/min of the LC mobile-phase. These flow rates are compatible with 1 mm i.d. LC columns. Alternatively, the effluent from a conventional 4.6 mm i.d. LC column can be partially diverted by a split device to the ESI source. As the ESI-MS arrangement is a concentration-sensitive detector, diverting only a fraction of the LC mobile-phase does not affect sensitivity. Another way of overcoming the problem of coupling LC with 4.6 mm i.d. conventional columns is that of inducing

TABLE 25.8
Liquid Chromatographic Methods for Determining Urea Herbicides in Water and Soil Samples

Compound	Matrix	Column	Mobile Phase	Detector	LOD (ppt)	Ref.
14 Phenylureas	Water	C-18 (5 μm) in 25 cm × 4.6 mm	H_2O/CH_3OH–CH_3CN (85:15), gradient elution	UV-250 nm	1	29
5 Phenylureas	—	C-18 (5 μm) in 25 cm × 4.6 mm	H_2O/CH_3CN, gradient elution	DAD/UV detection at 244 nm	100	30
7 Sulfonylureas	—	C-18 (5 μm) in 25 cm × 4.6 mm	H_2O/CH_3CN, both containing 3 mM TFA, gradient elution	UV-230 nm	0.6 to 2 in drinking water	35
4 Phenylureas	—	C-18 in 15 cm × 3.9 mm	CH_3CN/H_2O + 0.01M p.b. (pH 7)/(40:60), isocratic elution	Fluorimetric. detector	1000	41
12 Phenylureas	—	C-18 (5 μm) in 25 cm × 4.6 mm	1 mM p.b. (pH 7)/CH_3CN, gradient elution	UV-220 nm	3 to 15	49
3 Phenylureas	—	C-18 (5 μm) in 25 cm × 4.6 mm	50 mM p.b. (pH 7)/CH_3CN, gradient elution	DAD/UV at 220 nm	*ca.* 100	72
Phenylureas	—	C-18 (5 μm) in 20 cm × 2.1 mm	CH_3CN/H_2O + 0.01M p.b. (pH 7)/ (40:60), isocratic elution	FT-IR	1000	75
4 Phenylureas	—	C-18 (3 μm) in 15 cm × 2.1 mm	H_2O/CH_3CN (70.30) + ACH (pH 3), isocratic elution	UV-240 nm	50 to 100	81
5 Sulfonylureas	Water, soil	C-18 (3 μm) in 25 cm × 4.6 mm	H_2O/CH_3CN, 3 mM TFA gradient elution	DAD	2 to 14 water, 5000 to 12000	34
8 Phenylureas	Water, soil	C-18 (3 μm) in 25 cm × 4.6 mm	H_2O/CH_3OH/H_2O: CH_3OH:CH_3CN (60:35:5) gradient elution	DAD	50 to 100	67
5 Sulfonylureas	Soil	C-18 (3 μm) in 10 cm × 4.6 mm	H_2O 0.01% phosphoric acid/CH_3OH gradient elution	UV-226 nm	1000	100
5 Phenylureas	Soil	C-18 (5 μm) in 15 cm × 4.6 mm	H_2O/CH_3CN (55.45) isocratic elution	UV-244 nm	10000	101

DAD, diode array detector; p.b., phosphate buffer; FT-IR, Fourier transform infrared spectrometry.

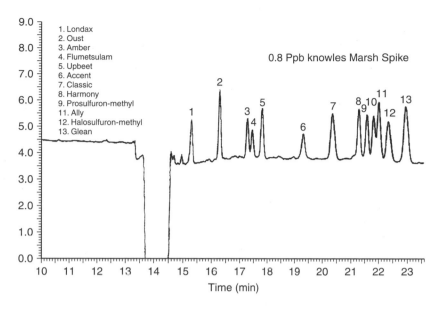

FIGURE 25.17 Electropherogram of 0.8 μg/l Knowles Marsh water spike. CE conditions: 161 nl injection; 240 nm UV detection; mobile phase: 50 mmol/l ammonium acetate at pH 4.75, with 12% acetonitrile added to inlet buffer vial; 30 kV, 30 μA, 30°C. Capillary: length 122 cm (100 cm effective length) × 75 μm i.d. bare fused silica high sensitivity optical cell. (From Krynitsky, A., *J. Assoc. Off. Anal. Chem.*, 80, 108, 1997. With permission.)

analyte ionization by gas-phase ion-molecule reactions, under APCI conditions. Reactant ion formation is achieved by the introduction of electrons from a corona discharge located in the chamber at atmospheric pressure. In this way, reversed-phase LC effluents as high as 2 ml/min can be handled.

The most serious drawback of the ESI-MS system is that it cannot accommodate LC effluents containing relatively high salt concentrations. With such solutions, signal instability and plugging of the small orifice of the sample cone occur. Recently, negative effects provoked by the presence of nonvolatile additive in the LC mobile phase have been eliminated by flowing the electro-sprayed solution orthogonal to the sample cone and washing the orifice continuously with a small flow of water.

ESI is a soft ionization technique generating $[M + H]^+$ in the positive-ion (PI) mode or $[M - H]^-$ in negative-ion (NI) mode, even for the most thermally labile and nonvolatile compounds. In some cases, spectra from nonbasic nonionic analytes display intense signals for Na^+, K^+, NH_4^+ adduct ions, in addition to that of the protonated molecule. These cations are always present as impurities in organic solvents used as organic modifiers of the LC mobile-phase. We noted that the relative abundance of cationized molecules depends mainly on the particular design of the ESI interface.

A very interesting option offered by the ESI-MS system is that, by raising the electrical field in the intermediate region of the mass analyzer, protonated molecules can be accelerated to such a point that multiple collisions with residual molecules from the drying gas generate characteristic fragment ions (Figure 25.18). The rate of fragmentation is strictly dependent on the potential difference between the sample cone and the skimmer lens. Provided the target compound is not coeluted with nontarget compounds, the resulting "in-source" collision-induced dissociation (CID) spectra closely resemble those obtained by the more costly tandem MS technique. The appearance of fragment ions in spectra from analytes is of paramount

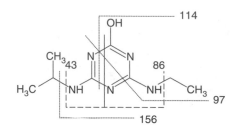

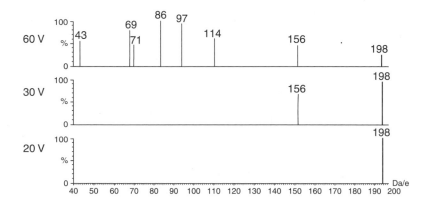

FIGURE 25.18 In-source CID spectra of hydroxyatrazine (MW = 197) taken at various sample cone voltages.

importance, considering that legal criteria for testing the presence of contaminants in real matrices usually accept, among other conditions, spectra displaying the molecular ion plus, at least, two fragment ions.

Finally, ESI-MS is a rugged technique. Since ionization occurs at atmospheric pressure, there is no worry about vacuum failure. On a daily basis, the ionization chamber and the counter electrode can be easily checked and cleaned in a matter of minutes, while vacuum is maintained in the transport and mass analyzer regions. Pump oil changes are about as frequent as with GC–MS (about three to four months), far less than in thermospray and particle beam LC–MS.

APCI is another very soft ionization technique and has many similarities to ESI. Ionization takes place at atmospheric pressure and the ions are extracted into the mass detector in the same way as in ESI. Similarly, $[M + H]^+$ and $[M - H]^-$ ions are usually formed to give molecular weight information, and, when using a single quadrupole, fragmentation of the precursor ions can be induced in the source by increasing the cone voltage. Yet, the APCI process differs from the ESI one mainly in that:

(i)   The high voltage is applied to a corona pin, not to the probe insert capillary.
(ii) The solvent evaporation and ion formation processes are separated.
(iii) The APCI process does not yield multiply charged ions for high mass molecules.

Using APCI, the liquid flow from the LC column is nebulized and rapidly evaporated by a coaxial nitrogen flow (nebulizing gas) and heating the nebulizer to high temperatures (350 to 500°C). Although these temperatures may degrade the analytes, the high flow rates of the LC mobile phase and coaxial nitrogen flow prevent breakdown of the molecules. Preformed ions

TABLE 25.9
Selected LC–MS Methods for Analyzing Urea Pesticides in Water and Soil

Compound	Matrix	Interface	Acquisition Mode	LOD (ppt)	Ref.
Chlortoluron, isoproturon, diuron, linuron, diflubenzuron	Water	ESI	Full-scan	0.6 to 8	131
Monuron, diuron, neburon	—	ESI	Full-scan one-ion SIM	7 to 3000, 0.1 to 200	71
Chlortoluron, isoproturon, diuron,	—	APCI	SIM	7	19
7 Sulfonylureas	—	ESI	Three-ion SIM	0.5 to 3	36
5 Phenylureas	—	APCI	SRM one transition	30	78
4 Sulfonylureas	—	ESI	SRM one transition	2 to 5	47
Monuron, diuron, isoproturon	—	ESI	Ion Trap-MS	80 to 200	133
8 Sulfonylureas	Soil	ESI	SRM two–four transitions	n.r.	91
4 Sulfonylureas	Soil	ESI	Three-ion SIM	100	97
6 Sulfonylureas	Soil	ESI	SRM one transition	130	98

ESI, electrospray; SIM, selected ion monitoring; APCI, atmospheric pressure chemical ionization; SRM, selected reaction monitoring.

can be carried into the gas phase, while ionization of analyte molecules is achieved using a corona discharge (3 to 6 kV) in the spray. The corona discharge produced by this high voltage causes solvent molecules entering the source to be ionized. In the atmospheric pressure region around the corona pin, a series of reactions occur, which give rise to stable solvent reagent ions. Any analyte molecules eluting from the column and passing through this region of solvent ions can be ionized by the transfer of a proton to form $[M + H]^+$ and $[M - H]^-$ ions. This is a form of chemical ionization, hence the name of the technique, APCI. Compared to traditional chemical ionization, the APCI process is more efficient, since it occurs at a higher pressure, this resulting in a higher collision frequency.

Another major difference between APCI and ESI can be found in LC flow rates used. APCI is a technique with optimal performance at high flow rates (1 ml/min and higher). Lower flow rates can also be used. However, when flow rates are too low, the stability of the corona discharge may become problematic.

For analyzing PUHs and SUHs in water and soil, many LC–MS methods based on various detection and quantitation systems, summarized in Table 25.9, have been published.[15,19,30,35,38,39,46,47,53,55,58–61,65,66,70,71,76–79,85,91,97,98,116,130–133] Figure 25.19 to Figure 25.22 show LC-single quadrupole MS chromatograms resulting from analysis of PUHs,[15] phenylureas and their metabolites,[53] SUHs,[35] and SUHs and PUHs[52] in real water samples. Finally, Figure 25.23 shows two LC–MS–MS chromatograms achieved by injecting 1 ppb levels of SUHs in soil, using one tuning period or four different tuning periods.[91]

IV. CONCLUSIONS

Liquid–liquid extraction with traditional solvents is still used for the isolation of pesticides from water samples. It tends to consume large volumes of high-purity solvents, which may have significant health hazards and disposal costs associated with their use. Furthermore, it is frequently plagued by problems, such as emulsion formation. The SPE technique with various adsorbing materials packed in cartridges or imbedded in membranes and used in the

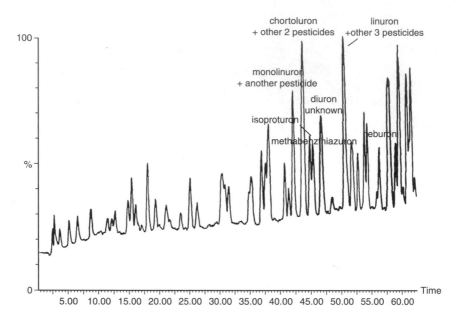

FIGURE 25.19 Full-scan LC–MS chromatogram obtained by analyzing 4 l of a drinking water sample spiked with 45 pesticides (including 7 PUHs) at the individual level of 50 ng/l. (From Crescenzi, C., Di Corcia, A., Guerriero, E., and Samperi, R., *Environ. Sci. Technol.*, 31, 479, 1997. With permission.)

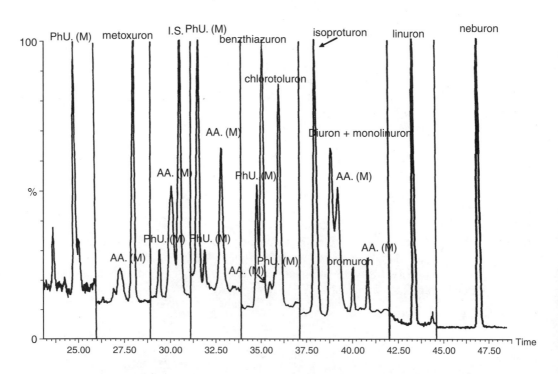

FIGURE 25.20 SIM LC-ES/MS of an extract of 4 l drinking water spiked with PUHs and their metabolites. Individual spike level: 2 ng/l. MS data acquisition was performed by using 8 retention windows. Acronyms: AA, aromatic amine; M, metabolite; PhU, partial or total dealkylated phenylurea herbicid; IS, internal standard (monuron). (From Di Corcia, A., Costantino, A., Crescenzi, C., and Samperi, R., *J. Chromatogr.*, 852, 465, 1999.)

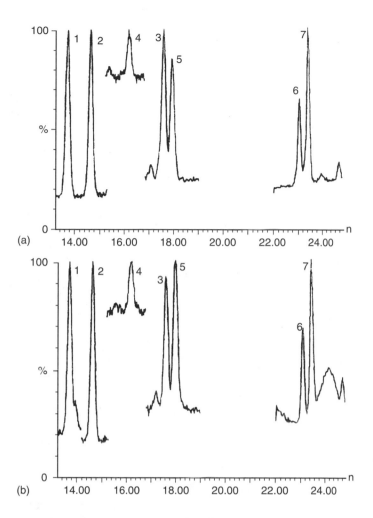

FIGURE 25.21 Time-scheduled three-ion SIM LC–MS chromatograms obtained by analyzing (a) 4 l of drinking water spiked with seven sulfonylureas at the individual level of 3 ng/l and (b) 0.2 l Tiber river water sample spiked with the analytes at the individual level of 60 ng/l. Column: Alltima 250 × 4.6 mm i.d. containing 5 μm C-18. Elution: CH_3CN/H_2O (both containing 3 mmol/l trifluoroacetic acid), linear gradient elution from 32:68 to 62:38 (v/v) in 40 min. Peak numbering: 1, thifensulfuron methyl; 2, metsulfuron methyl; 3, triasulfuron methyl; 4, chlorsulfuron; 5, rimsulfuron; 6, bensulfuron methyl; 7, tribenuron methyl. (From Di Corcia, A., Crescenzi, C., Samperi, R., and Scappaticcio, L., *Anal. Chem.*, 69, 2819, 1997. With permission.)

offline or online mode, as shown by this review, is now definitely preferred to LLE. As to PUHs and SUHs in soil, the new emerging extraction techniques using microwave, accelerated solvent extraction, and hot water have shown to be excellent alternatives to Soxhlet extraction.

The broad spectrum of well-established GC methods with selective detectors available today allows the identification and determination of hundreds of contaminants in environmental waters and soil. However, several classes of pesticides, among these those ones considered in this review, are not amenable to GC without time-consuming derivatization procedures. For such compounds, the LC technique seems to be the most appropriate separation method. In recent years, with the exception of few methods proposing GC or capillary electrophoresis, many LC applications have been used for analyzing phenylurea and SUHs in water

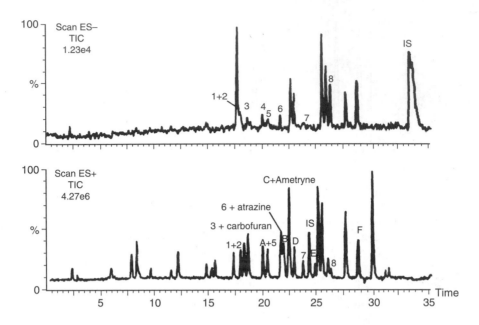

FIGURE 25.22 Typical mass chromatogram obtained by simultaneous acquisition of positive and negative ions and relative to analysis of 4 l of drinking water spiked with 39 pesticides (including selected phenylurea and sulfonylurea herbicides) at an individual level of 50 ng/l. LC eluent: water-methanol (gradient elution) both acidified with 1 mmol/l formic acid. A, monolinuron; B, methabenzthiazuron; C, isoproturon; D, diuron; E, linuron; 1, thifensulfuron; 2, triasulfuron; 3, metsulfuron; 4, chlorsulfuron; 5, rimsulfuron; 6, tribenuron; 7, bensulfuron; 8, primisulfuron. (From Di Corcia, A., Nazzari, M., Rao, R., Samperi, R., and Sebastiani, E., *J. Chromatogr.*, 878, 87, 2000.)

and soil extracts. Several of these applications rely on the use of conventional UV detectors. A serious weakness of these methods is that UV detection does not provide qualitative information sufficient to recognize μg/l or sub-μg/l levels of target compounds in complex mixtures with a low probability of false positives. In such cases, the use of a diode array detector can give some relief to this problem. In terms of qualitative and quantitative analysis, this discussion has shown that monitoring of urea herbicides in water and soil can greatly benefit from the use of LC–MS. In the last 20 years, many sensitive and selective LC–MS methods making use of different interfaces have been proposed. Today, only the electrospray ion source is definitely considered to have a very promising future. It is expected that the very recent introduction of less expensive, easy-to-use benchtop LC-ESI–MS/MS instrumentation will further stimulate practical applications of the recently developed analytical methodologies, enabling sensitive and reliable monitoring of the aforementioned compounds in environmental matrices. The governments of some European countries, namely, Denmark and Sweden, are considering decreasing the maximum admissible concentration of an individual pesticide in drinking water from 100 to 10 ng/l and including in the list of undesired compounds those pesticide DPs which are toxic in nature. It is possible in the near future that other European countries will follow this strategy. This will urge European analytical chemists to develop new analytical LC-ESI/MS/MS methodologies which, in addition to being more sensitive than most of the actual ones, will also be capable of simultaneously analyzing pesticides and their health-hazardous DPs, such as the case of phenylureas and their related chloroanilines. The latter compounds are more toxic than the parent compounds, and some of them are suspected of inducing cancer.

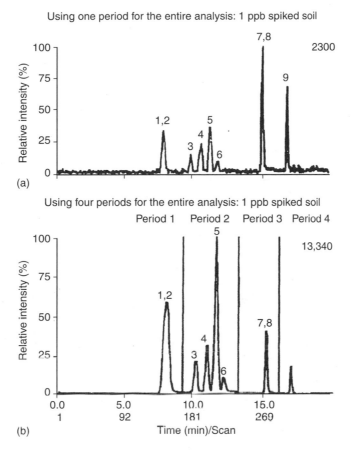

FIGURE 25.23 Total selected ion current chromatogram (TIC) of 1 ppb levels of SUHs in soil. The upper chromatogram (a) was obtained using one tuning period, and the lower chromatogram (b) was obtained using four different tuning periods for the entire MRM LC/MS analysis. Peak numbering: 1, nicosulfuron; 2, D6-nicosulfuron; 3, thifensulfuron methyl; 4, metsulfuron methyl; 5, sulfometuron methyl; 6, chlorsulfuron; 7, bensulfuron methyl; 8, tribenuron methyl; 9, chlorimuron ethyl. (From Li, L. Y. T., Campbell, D. A., Bennett, P. K. and Henion, J., *Anal. Chem.*, 68, 3397, 1996. With permission.)

REFERENCES

1. Matolcsy, Gy., Nadasy, M., and Andriska, V., *Pesticide Chemistry*, Elsevier, Amsterdam, 1988, pp. 682–684.
2. Gosselin, R. E., Smith, R. P., and Hodge, H. C., *Clinical Toxicology of Commercial Products*, 5th ed., Williams and Wilkins, Baltimore, p. 11, 1984.
3. Scott, T. S., *Carcinogenic and Toxic Health of Aromatic Amines*, Elsevier, New York, 1962.
4. Tomlin, C., Ed., *The Pesticide Manual*, British Crop Protection Council, Farnham, Surrey, 1994.
5. Munch, D. J., Graves, R. L., Maxey, R. A., and Engel, T. M., *Environ. Sci. Technol.*, 24, 1446, 1990.
6. U.S. Environmental Protection Agency, National Survey of Pesticides in Drinking Water Wells, Phase II Report, EPA 570/9-91-020, National Technical Information Service, Springfield, VA (1992).
7. Decision No. 2455/2001/EC of the European Parliament and of the Council of 20 November 2001 establishing the list of priority substances in the field of water policy and amending Directive 2000/60/EC.

8. Fielding, M., Barceló, D., Helweg, A., Galassi, S., Thorstensson, L., Van Zoonen, P., Wolter, R., and Angeletti, G., Pesticides in Ground and Drinking Water (Water Pollution Research Report, 27), Commission of the European Communities, Brussels, pp. 1–136, 1992.

9. Council Directive 98/83/EC of 3 November 1998 on the quality of water intended for human consumption.

10. Smith, J. S., *Principles of Environmental Sampling*, Keith, L. H., Ed., American Chemical Society, Washington, DC, p. 225, 1988.

11. Nordin, C. P., *Proceedings of International Symposium on River Sedimentation, 2nd, Beijing*, Water Resources and Electric Power Press, Nanjing, China, p. 1145, 1983.

12. Meade, R. H., Suspended Sediments in the Amazon River and its Tributaries in Brazil During 1982–1984, Open File Report, U.S. Geological Survey No. 85–492, 1985.

13. Bone, L. T., In *Principles of Environmental Sampling*, Keith, L. H., Ed., American Chemical Society, Washington, DC, p. 409, 1988.

14. Basset, M. V., Wedelken, S. C., Dattilio, T. A., Pepich, B. V., and Munch, D. J., *Environ. Sci. Technol.*, 36, 1809, 2002.

15. Crescenzi, C., Di Corcia, A., Guerriero, E., and Samperi, R., *Environ. Sci. Technol.*, 31, 479, 1997.

16. Munch, D. J. and Frebis, C. P., *Environ. Sci. Technol.*, 26, 921, 1992.

17. Winslow, S. D., Pepich, B. V., Basset, M. V., Wedelken, S. C., Munch, D. J., and Sinclair, J. L., *Environ. Sci. Technol.*, 35, 4103, 2001.

18. Charrêteur, C., Colin, R., Morin, D., and Péron, J. J., *Analusis*, 26, 8, 1998.

19. Jeannot, R. and Sauvard, E., *Analusis*, 27, 271, 1999.

20. Ahmad, I., *J. Assoc. Off. Anal. Chem.*, 70, 745, 1987.

21. Leoni, V., Cremisini, C., Casuccio, A., and Gullotti, A., *Pestic. Sci.*, 31, 209, 1991.

22. Thompson, D. G. and McDonald, L. M., *J. Assoc. Off. Anal. Chem.*, 75, 1084, 1992.

23. Spliid, N. H. and Køppen, B., *Chemosphere*, 37, 1307, 1998.

24. Method 525.1, Determination of Organic Compounds in Drinking Water by Liquid–Solid Extraction and Capillary Column Gas Chromatography/Mass Spectrometry (Revision 2.2), Environmental Monitoring Systems Laboratory, Office of Research and Development, U.S. Environmental Protection Agency, Cincinnati, OH.

25. Basset, M. V., Wedelken, S. C., Dattilio, T. A., Pepich, B. V., and Munch, D. J., EPA Method 532, National Exposure Research Laboratory, Office of Research and Developments, U.S. EPA, Cincinnati, OH, 2000, http://www.epa.gov./ogwdw000/methods/532.pdf.

26. Winslow, S. D., Prakash, B., Domino, M. M., Pepich, B. V., Munch, D. J., EPA Method 526, National Exposure Research Laboratory, Office of Research and Developments, U.S. EPA, Cincinnati, OH, 2000, http://www.epa.gov./ogwdw000/methods/526.pdf.

27. Crescenzi, C., Di Corcia, A., Madbouly, M. D., and Samperi, R., *Environ. Sci. Technol.*, 29, 2185, 1995.

28. Senseman, S. A., Lavy, T. L., Mattice, J. D., Myers, B. M., and Skulman, B. W., *Environ. Sci. Technol.*, 27, 516, 1993.

29. Di Corcia, A. and Marchetti, M., *J. Chromatogr.*, 541, 365, 1991.

30. Pichon, V., Chen, L., Hennion, M.-C., Daniel, R., Martel, A., Le Goffic, P., Abian, J., and Barceló, D., *Anal. Chem.*, 67, 2541, 1995.

31. Pichon, V., Chen, L., Durand, N., Le Goffic, P., and Hennion, M.-C., *J. Chromatogr.*, 725, 107, 1996.

32. Ferrer, I., Hennion, M.-C., and Barceló, D., *Anal. Chem.*, 69, 4508, 1997.

33. Martin-Esteban, A., Fernandez, P., and Camera, C., *Anal. Chem.*, 69, 3267, 1997.

34. Zhu, Q. Z., Degelmann, P., Niessner, R., and Knopp, D., *Environ. Sci. Technol.*, 36, 5411, 2002.

35. Di Corcia, A., Crescenzi, C., Samperi, R., and Scappaticcio, L., *Anal. Chem.*, 69, 2819, 1997.

36. Scott, S., *Analyst*, 118, 1117, 1993.

37. Incorvia Mattina, M. J., *J. Chromatogr.*, 549, 237, 1991.

38. Giraud, D., Ventura, A., Camel, V., Bermond, A., and Arpino, P., *J. Chromatogr.*, 777, 115, 1997.

39. Jeannot, R., Sabik, H., Sauvard, E., and Genin, E., *J. Chromatogr.*, 879, 51, 2000.

40. van der Heeft, E., Dijkman, E., Baumann, R. A., and Hogendoorn, E. A., *J. Chromatogr.*, 879, 39, 2000.

41. Muñoz de la Peña, A., Mahedero, M. C., and Bautista Sánchez, A., *Talanta*, 60, 279, 2003.

42. Zhanow, E. W., *J. Agric. Food Chem.*, 33, 479, 1985.
43. Wells, M. J. M. and Michael, J. L., *J. Chromatogr. Sci.*, 25, 345, 1987.
44. Cotterill, E. G., *Pestic. Sci.*, 34, 291, 1992.
45. Galletti, G. C., Bonetti, A., and Dinelli, G., *J. Chromatogr.*, 692, 27, 1995.
46. Krynitsky, A., *J. Assoc. Off. Anal. Chem.*, 80, 1084, 1997.
47. Bossi, R., Køppen, B., Spliid, N. H., and Streibig, J. C., *J. AOAC Int.*, 81, 775, 1998.
48. Di Corcia, A. and Marchetti, M., *Anal. Chem.*, 63, 580, 1991.
49. Di Corcia, A. and Marchetti, M., *Environ. Sci. Technol.*, 26, 66, 1992.
50. Di Corcia, A., Marcomini, A., Samperi, R., and Stelluto, S., *Anal. Chem.*, 65, 907, 1993.
51. Gerecke, A. C., Tixier, C., Bartels, T., Schwarzenbach, R. P., and Muller, S. R., *J. Chromatogr.*, 930, 9, 2001.
52. Di Corcia, A., Nazzari, M., Rao, R., Samperi, R., and Sebastiani, E., *J. Chromatogr.*, 878, 87, 2000.
53. Di Corcia, A., Costantino, A., Crescenzi, C., and Samperi, R., *J. Chromatogr.*, 852, 465, 1999.
54. Lacorte, S., Guiffard, I., Fraisse, D., and Barceló, D., *Anal. Chem.*, 72, 1430, 2000.
55. Steen, R. J. C. A., Hogenboom, A. C., Leonards, P. E. G., Peerboom, R. A. L., Cofino, W. P., and Brinkman, U. A. Th., *J. Chromatogr.*, 857, 157, 1999.
56. Stutz, H. and Malissa, H. Jr., *J. AOAC Int.*, 82, 1510, 1999.
57. Young, M. S., *J. AOAC Int.*, 81, 99, 1998.
58. Rodriguez, M. and Orescan, D. B., *Anal. Chem.*, 70, 2710, 1998.
59. Furlong, E. T., Burkhardt, M. R., Gates, P. M., Werner, S. L., and Battaglin, W. A., *Sci. Total Environ.*, 248, 135, 2000.
60. Battaglin, W. A., Furlong, E. T., Burkhardt, M. R., and Peter, C. J., *Sci. Total Environ.*, 248, 123, 2000.
61. Køppen, B. and Spliid, N. H., *J. Chromatogr.*, 803, 157, 1998.
62. Barceló, D., Durand, G., Bouvot, V., and Nielen, M., *Environ. Sci. Technol.*, 27, 271, 1993.
63. Klaffenbach, P. and Holland, P. T., *J. Agric. Food Chem.*, 41, 396, 1993.
64. Howard, A. L. and Taylor, L. T., *J. Chromatogr. Sci.*, 30, 374, 1992.
65. Ruberu, S. R., Draper, W. M. and Perera, S. K., *J. Agric. Food Chem.*, 48, 4109, 2000.
66. Draper, W. M., *J. Agric. Food Chem.*, 49, 2746, 2001.
67. Ferrer, I., Barceló, D., and Thurman, E. M., *Anal. Chem.*, 71, 1009, 1999.
68. Nielen, M. W. P., Valk, A. J., Frei, R. W., Brinkman, D. A. Th., Mussche, Ph., De Nijs, R., Ooms, B., and Smink, W., *J. Chromatogr.*, 393, 69, 1987.
69. Bagheri, H., Brouwer, E. R., Ghijsen, R. T., and Brinkman, V. A. Th., *Analysis*, 20, 475, 1992.
70. Slobodnik, J., Hogenboom, A. C., Vreuls, J. J., Rontree, J. A., van Baar, B. L. M., Nissen, W. M. A., and Brinkman, D. A. Th., *J. Chromatogr.*, 741, 59, 1996.
71. Hogenboom, A. C., Niessen, W. M. A. and Brinkman, U. A. Th., *J. Chromatogr.*, 794, 201, 1998.
72. Pichon, V. and Hennion, M. C., *J. Chromatogr.*, 665, 269, 1994.
73. Mills, G. R., *J. Chromatogr.*, 813, 63, 1998.
74. Volmer, D., Wilkes, J. G. and Levsen, K., *Rapid Commun. Mass Spectrom.*, 9, 767, 1995.
75. Somsen, G. W., Jagt, I., Gooijer, C., Velthorst, N. H., Brinkman, U. A. Th., and Visser, T., *J. Chromatogr.*, 756, 145, 1996.
76. Baltussen, E., Snijders, H., Janssen, H. G., Sandra, P., and Cramers, C. A., *J. Chromatogr.*, 802, 285, 1998.
77. Wang, N. and Budde, W. L., *Anal. Chem.*, 73, 997, 2001.
78. Geerdink, R. B., Kooistra-Sijpersma, A., Tiesnitsch, J., Kienhuis, P. G. M., and Brinkman, U. A. Th., *J. Chromatogr.*, 863, 147, 1999.
79. Asperger, A., Efer, J., Koal, T., and Engewald, W., *J. Chromatogr.*, 960, 109, 2002.
80. Nilvé, G. and Stebbins, R., *Chromatographia*, 32, 269, 1991.
81. Nilvé, G., Knutsson, M. and Jonsson, J. A., *J. Chromatogr.*, 688, 75, 1994.
82. Chao, J., Liu, J., Wen, M., Liu, J., Cai, Y., and Jiang, G., *J. Chromatogr.*, 955, 183, 2002.
83. Liu, J., Chao, J., Jang, G., Cai, Y., and Liu, J., *J. Chromatogr.*, 995, 21, 2003.
84. Eisert, R. and Pawliszyn, J., *Anal. Chem.*, 69, 3140, 1997.
85. Hartmann, H., Burhenne, J., Müller, K., Frede, H. G., and Spiteller, M., *J. AOAC Int.*, 83, 762, 2000.

86. Berrada, H., Font, G., and Moltó, J. C., *J. Chromatogr.*, 890, 303, 2000.
87. Henze, G., Meyer, A. and Hausen, J., *Fresenius J. Anal. Chem.*, 346, 761, 1993.
88. Liegeois, E., Dehon, Y., de Brabant, B., Perry, P., Portetelle, D., and Copin, A., *Sci. Total Environ.*, 123/124, 17, 1992.
89. Perez, S., Garcia-Baudin, J. M. and Tadeo, J. L., *Fresenius J. Anal. Chem.*, 339, 413, 1991.
90. Berger, B., *J. Chromatogr.*, 769, 338, 1997.
91. Li, L. Y. T., Campbell, D. A., Bennett, P. K., and Henion, J., *Anal. Chem.*, 68, 3397, 1996.
92. Dost, K., Jones, D. C., Auerbach, R., and Davidson, G., *Analyst*, 125, 1751, 2000.
93. Dinelli, G., Vicari, A., and Catizone, P., *J. Agric. Food Chem.*, 41, 742, 1993.
94. Dinelli, G., Vicari, A., and Brandolini, V., *J. Chromatogr.*, 700, 201, 1995.
95. Penmetsa, K. V., Leidy, R. B., and Shea, D., *J. Chromatogr. A*, 766, 225, 1997.
96. Chen, Z. L., Kookana, R. S., and Naidu, R., *Chromatographia*, 52, 142, 2000.
97. Marek, L. J. and Koskinen, W. C., *J. Agric. Food Chem.*, 44, 3878, 1996.
98. Laganà, A., Fago, G., Marino, A., and Penazzi, V. M., *Anal. Chem. Acta*, 415, 41, 2000.
99. Powley, C. R. and de Bernard, P. A., *J. Agric. Food Chem.*, 46, 514, 1998.
100. Font, N., Hernandez, F., Hogendoorn, E. A., Baumann, R. A., and van Zoonen, P., *J. Chromatogr.*, 798, 179, 1998.
101. Molins, C., Hogendoorn, E. A., Dijkman, E., Heusinkveld, H. A. G., and Baumann, R. A., *J. Chromatogr.*, 869, 487, 2000.
102. Hawthorne, S. B., Yang, Y. and Miller, D. J., *Anal. Chem.*, 66, 2912, 1994.
103. Crescenzi, C., D'Ascenzo, G., Di Corcia, A., Nazzari, M., Marchese, S., and Samperi, R., *Anal. Chem.*, 71, 2157, 1999.
104. Crescenzi, C., Di Corcia, A., Nazzari, M., and Samperi, R., *Anal. Chem.*, 72, 3050, 2000.
105. Grob, K. Jr., *J. Chromatogr.*, 208, 217, 1981.
106. Tadeo, J. L., García-Baudin, J. M., Matienzo, T., Pérez, S., and Sixto, H., *Chemosphere*, 18, 1673, 1989.
107. Berrada, H., Moltó, J. C., and Font, G., *Chromatographia*, 54, 360, 2001.
108. Brinkman, U. A. Th., de Kok, A., and Geerdink, R. E., *J. Chromatogr.*, 283, 113, 1984.
109. Perez, S., Matienzo, M. T., and Tadeo, J. L., *Chromatographia*, 36, 195, 1993.
110. de Kok, A., Roorda, I. M., Frei, R. W., and Brinkman, U. A. Th., *Chromatographia*, 14, 579, 1981.
111. Ogierman, L., *Fresenius Z. Anal. Chem.*, 320, 365, 1985.
112. Pérez, S., García-Baudín, J. M. and Tadeo, J. L., *Fresenius Z. Anal. Chem.*, 339, 413, 1991.
113. Berger, T. A., *Chromatographia*, 41, 133, 1995.
114. Barceló, D., *Analyst*, 116, 681, 1991.
115. Jimenez, J. J., Bernal, J. L., del Nozal, M. J., and Rivera, J. M., *J. Chromatogr.*, 778, 289, 1997.
116. Scribner, E. A., Thurman, E. M. and Zimmerman, L. R., *Sci. Total Environ.*, 248, 157, 2000.
117. Dinelli, G., Boetti, A., Catizone, P., and Galletti, G., *J. Chromatogr. B*, 656, 275, 1994.
118. Hickes, H. and Watrous, M., *J. AOAC Int.*, 82, 1523, 1999.
119. Garcia, F. and Henion, J., *J. Chromatogr.*, 606, 237, 1992.
120. Niessen, W. M. A. and Van der Greef, J., *Liquid-Chromatography–Mass Spectrometry*, *Chromatographic Science Series*, Vol. 58, Marcel Dekker, New York, 1992.
121. Lamoree, M. H., Ghijsen, R. T., and Brinkman, U. A. Th., In *Environmental Analysis: Techniques, Applications and Quality Assurance*, Barcelo, D., Ed., Elsevier, Amsterdam, p. 521, 1993.
122. Cole, R. B., Ed., *Electrospray Ionization Mass Spectrometry Fundamentals, Instrumentation & Applications*, Wiley, New York, 1997.
123. Niessen, W. M. A., *Liquid Chromatography–Mass Spectrometry, Chromatographic Science Series*, 2nd edn., Vol. 79, In Niessen, W. M. A., Ed., Marcel Dekker, Inc., New York, 1999.
124. Voyksner, R. D. and Keever, J., In *Analysis of Pesticides in Ground and Surface Water*, Stan, H. J., Ed., Springer-verlag, Berlin Heidelberg, p. 110, 1995.
125. Volmer, D. and Levsen, K., *Analysis of Pesticides in Ground and Surface Water*, Stan, H. J., Ed., Springer-Verlag, Berlin Heidelberg, p. 133, 1995.
126. Barcelo, D., Ed., Application of LC–MS in Environmental Chemistry, *J. Chrom. Library*, 59, Elsevier, Amsterdam, 1996.
127. Alomirah, H. F., Alli, I. and Konishi, Y., *J. Chromatogr. A*, 893, 1, 2000.

128. Picó, Y., Font, G., Moltò, J. C., and Manes, J., *J. Chromatogr.*, 882, 153, 2000.
129. Hogendoorn, E. and Van Zoonen, P., *J. Chromatogr.*, 892, 435, 2000.
130. Molina, C., Durand, G. and Barceló, D., *J. Chromatogr.*, 712, 113, 1995.
131. Aguilar, C., Ferrer, I., Borrul, F., Marcé, R. M., and Barceló, D., *J. Chromatogr.*, 794, 147, 1998.
132. Hogenboom, A. C., Speksnijder, P., Vreken, R. J., Niessen, W. M. A., and Brinkman, U. A. Th., *J. Chromatogr.*, 777, 81, 1997.
133. Baglio, D., Kotzias, D., and Larsen, B. R., *J. Chromatogr. A*, 854, 207, 1999.

26 Herbicide Residues in the Environment

Thierry Dagnac and Roger Jeannot

CONTENTS

I. INTRODUCTION

More than 500 referenced substances are classified in 39 herbicide families. Thus, it is unrealistic to claim covering the broad range of applications described for all these families. This chapter will mainly focus on triazines, chloroacetanilides and chlorophenoxy acids, some of the most used herbicides, with some scarce views on phenylureas and carbamates when multiresidue methods are mentioned.

Herbicides are the most used pesticides in the world for more than 40 years, with 45% of the total market value in 1993.[1] Among the top ten herbicides used in the world, atrazine and glyphosate are used worldwide, and interesting differences appear between the U.S.A and Europe regarding this "top ten" list. Urea herbicides belong to this list in Europe and contaminate many water sources whereas in the U.S.A these are not used at all. More than 80% of the herbicide use is concentrated in three agriculture areas: North America, western Europe, and east Asia. 22% of the total herbicides are also found for nonagricultural uses with a lot of triazines and ureas in Europe.[1]

Phenoxyacid herbicides are in wide use because of their relative cheapness and effectiveness in controlling broad-leaved weeds and other vegetations in crops. These herbicides are very potent even at low concentrations. Because of their high water solubility and toxicological risk, monitoring of groundwater and surface water is required.

Chloroacetanilide herbicides (e.g., alachlor, metolachlor, and acetochlor) are of an important class of herbicides used to control grass weeds in various crops. Alachlor and metolachlor have been widely used (both in the U.S.A and in Europe) for more than 20 years.[2] Acetochlor, a herbicide used for maize, has been on the United States market since 1994, following approval by the U.S. Environmental Protection Agency. This approval will be renewed, however, only if the total quantity of other herbicides used on this crop, including atrazine, decreases. Acetochlor was approved in France in 2000 and is now used in substitution programs.

A. PHYSICAL AND CHEMICAL PROPERTIES

Two main physico-chemical parameters for herbicides are the water solubility and the water-octanol partition coefficient (K_{ow}). Water solubility (expressed in mg/l) is used to assess herbicide removal from soil and reachment of surface water. K_{ow} is the characteristic of liphophility of the molecule and indicates that it may accumulate in membranes of living organisms.[3] Herbicides with log K_{ow} higher than three can exhibit accumulation. In terms of polarity, logs K_{ow} above 4 to 5 are specific of nonpolar compounds, whereas logs K_{ow} below 1 to 1.5 correspond to polar compounds. Together with water solubility, log K_{ow} allows assessment of herbicide behavior and fate in the

TABLE 26.1
Properties of Herbicides, Indicating Their High Groundwater Contamination Potential

Parameter	Value
Water solubility	>30 mg/l
K_d	<5, usually <1
K_{oc}	<300
Henry's law constant	$<10^{-2}$ atm m^3/mol
Speciation	Negatively, fully or partially charged at ambient pH
Hydrolysis half-life	>25 weeks
Photolysis half-life	>1 week
Field dissipation half-life	>3 weeks

environment. Other important factors are the vapor pressure of the herbicides and the Henry law constant, which allow prediction of the herbicide volatilization.

The soil sorption coefficient (K_{oc}) represents the herbicide partition between the solid and the liquid phases in the soil. This coefficient is normalized as a function of the organic matter content that plays a very important role for the nonionized herbicides at natural pH on soils. The higher is K_{oc} the more sorbed is the molecule. However, the exact composition of the organic matter is another relevant criterion to assess sorption mechanisms, as shown by Dousset for humic substances.[4,5] One relevant parameter linked to chemical properties is the Gustafon Ubiquity Score (GUS) factor, which defines mobility index taking into account the half-life time and K_{oc}. The molecules with a GUS above 2.8 are very likely to reach groundwater and those with a GUS below 1.8 are considered nonleacher to groundwater.

Compounds having high water solubility and slight soil adsorption (K_d) will move easily to the groundwater.

Table 26.1 summarizes the properties of herbicides and their high groundwater contamination potential.[6]

B. OCCURRENCE IN THE ENVIRONMENT

1. Water and Soil Compartments

The three main compartments in which the fate of herbicides is investigated are the "root zone," the unsaturated zone, and the saturated zone. In the first zone, volatilization, biodegradation, and sorption processes take place. In the second and third zones, other degradation and sorption reactions occur, but with lower kinetics and dissociation constants.

Herbicides are subjected to several biotic and abiotic degradation processes, such as, photolysis, hydrolysis, oxidation, or dealkylation. Data on the fate of herbicides and metabolites in the environment are supplied by information on the rates of degradation in the soil; the nature and the persistence of metabolites; and the distribution of herbicides via leaching (contamination of aquatic compartment), runoff, and volatilization. Runoff occurring within a few days of a pesticide application typically removes about 1% of the amount present in soil.[7] Volatilization is strongly dependent on the vapor pressure of each compound and those with a high Henry constant are the most volatile. An interesting study showed that different formulations of atrazine and alachlor strongly influence their volatilization rate.[8]

The fifth national report of IFEN[9] showed that 159 different pesticides were found in surface water and 144 in groundwater of France. Triazines systematically occur on a massive scale, along with their degradation products, with deethylatrazine and atrazine as main compounds in surface water and groundwater.[10] 2-Hydroxy atrazine quantified in 50% of the surface water samples. Many herbicides such as atrazine, cyanazine, alachlor, and some of their degradation products have also been frequently detected in surveys performed by USGS.[1,11] In streams of the U.S. Midwest, seven of nine herbicides (mainly triazines and acetanilides) detected in more than 50% of the samples were metabolites and their total concentration was significantly greater than the total concentration of parent compounds.[12] In soils, hydroxy atrazine metabolites were the most frequently found and chemical hydrolysis is considered the predominant degradation pathway for atrazine into the environment, HA being the major abiotic degradation product.[13] Herbicides were detected in about 50% of wells sampled in Iowa in 1996, with herbicide degradation products being detected in some 75%.[14] The frequency of detection for cyanazine or some of its degradates alone was more than 12 fold over that of cyanazine alone.[15] Deethylcyanazine acid (DCAC) and cyanazine acid (CAC) has similar detection frequencies, however DCAC was generally present in higher concentrations.[15]

A seven year study on the groundwater in the Paris region of France, revealed DEA was present at a concentration above that of its parent compound.[16] The atrazine degradation pathway and the higher solubility of DEA in water may explain this finding. Recent work showed that photolysis of triazines and acetanilides followed pseudo first order kinetics, and the photodegradation in soils was accelerated as the content of organic matter increased.[17] Another study showed that humic substances enhanced terbutylazine photolysis.[18]

The use of atrazine is, therefore, strictly controlled in some countries (Denmark) and completely banned in others (e.g., Germany, Italy, Austria, Sweden, and Norway). In France, because of national pressure, atrazine was restricted to agriculture uses in February 1997, with dose limitations of 1000 g/ha. Despite this restriction, atrazine continues to be detected in groundwater. Consequently, authorities in some regions have decided to ban all use of atrazine, and have set up substitution programs. Sales of atrazine will be forbidden in France after June 30, 2003.

Although the sorption, leaching, and degradation of herbicides in soils have been studied extensively over the last decade, few data are available for chloroacetanilides in general. Some laboratory studies have been carried out, but field studies are rare despite the fact that this is the only way to take into account all the processes which do or could control the fate of agrochemicals in the environment. However, using percolation experiments on soil columns, some authors[19] showed that acetochlor, like the other chloroacetanilides (metolachlor and alachlor), is a potential groundwater contaminant. Zheng and Ye[20] also concluded, based on adsorption and thin layer chromatography experiments, that acetochlor presents a risk of groundwater contamination, particularly in sandy soil or if aquifers are shallow. Barbash et al.[10] sampled groundwater in 20 major hydrologic basins in the United States and detected acetochlor in some wells tapping shallow aquifers just one year after the first applications, thus confirming the potential mobility of the molecule. The parent chloroacetanilide agrochemicals may degrade in soils and water to form oxanilic and sulfonic acid metabolites.[2] In the U.S.A, these compounds have been detected in groundwater more often and at higher concentrations than their parent compounds.[12,14,21] Information on how and why these molecules reach groundwater is still very scarce, partly because of a lack of data concerning their presence and fate in the soil and unsaturated zone overlying aquifers.[22,23] Indeed, a great number of papers have been published on the determination of the chloroacetanilide metabolites in water,[24–28] but very few studies have been carried out on their determination in soils and solids.[29]

Some recent surveys in the Mississippi River basin pointed out the importance of chloro-acetanilide metabolites in surface water. Thus, sulfonic acid and oxanilic acid (OXA) averaged 70% of the total herbicide concentration in samples from the upper Mississippi River, whereas this proportion was much less in the Missouri and Ohio Rivers, 24% and 41%, respectively.[30]

Biodegradation is much dependent on the pH, moisture, temperature, and type of soil. Jurado-Exposito[31] has shown that alachlor degradation increases with temperature and soil moisture. A laboratory study on the biodegradation of butachlor and acetochlor in soils showed that reaction kinetics followed first order rates. The main detected metabolites were hydroxybutachlor, hydroxy acetochlor, 2,6-diethylaniline, and 2-methyl-6-ethylaniline.[32]

Due to their polar nature and water solubility, phenoxy acid herbicides and their degradates are broadly dispersed in the environment, in particular via leaching and runoff processes.[33,34] In soils, their persistence is estimated to six to eight weeks with a result of both microbial action and photodegradation.[35,36] Some of their metabolites, in particular 2,4-dichlorophenol and 4-chloro-2-methylphenol, can be more toxic and persistent than parent compounds.[35]

From the available literature, it can be underlined that the toxicity of major herbicide metabolites is not yet well known and analytical standards are not always available. The reduced molecular weight and high polarity of metabolites make them more difficult to analyze, and their removal from drinking water plants is not easy to achieve. However, some of them can be analyzed in the frame of "normalized" methods, in particular those compounds with physico chemical properties very similar to these of their parent compounds.

The main investigated herbicide metabolites are presented in Table 26.2.

2. Dispersion and Contamination of Herbicides in the Atmosphere

Herbicide occurrence in air is mainly related to volatilization, soil erosion, and emissions after spreading.[37–39] Volatilization from plant surfaces is one of the main pathways of pesticide emission to the environment, and may lead to contamination by long range transport and deposition at locations remote from their application.[40,41] The main factors affecting volatilization of pesticides from crops are their physicochemical properties, their persistence on the plant surface, and the environmental conditions. The persistence on the leaf surface depends on various dissipation processes, such as photodegradation, washoff from the leaves by rainfall or irrigation, and penetration into the plant leaves.[42] It should be stressed that there is a lack of data regarding the photolytic degradation of herbicides into the environment by photolysis, radical degradation, or ozone reaction.

During spreading, spontaneous emissions of herbicides to the atmosphere can reach 30% of the applied dose, and they depend on several factors such as meteorological conditions and spread droplet size. A study pointed out the importance of not using droplets with a size below 100 μm to avoid their dispersion from the spreading point (730). Taylor and Spencer[43,44] showed that losses of

TABLE 26.2
Main Investigated Metabolites of Herbicides

Herbicide Family	Some Main Parent Compounds	Some Main Metabolites
Triazines	Atrazine, simazine, cyanazine, etc.	Deethylatrazine, deisopropylatrazine or deethylsimazine, deethyldeisopropylatrazine, hydroxyatrazine, hydroxydeethylatrazine, hydroxydeisopropylatrazine, hydroxy-deethyldeisopropylatrazine, cyanazine acid, deethylcyanazine acid, cyanazine 2-methylpropionic
	Terbutylazine	Deethylterbutylazine, hydroxyterbutylazine
Phenylureas	Isoproturon	Monodemethylisoproturon, didemethylisoproturon, 4-isopropylaniline
	Diuron	3,4-Dichloroaniline
	Chlortoluron	3-Chloro-4-methylaniline
Chloroacetanilides	Alachlor, metolachlor, acetochlor, etc.	2,6-Diethylaniline, hydroxy forms, sulfonic and oxanilic acid forms
Phenoxyalcanoic acids	(Mecoprop, MCPA, 2,4-D, 2,4,5-T, etc.)	Chlorophenols (4-chloro-2-methylphenol; 2,4-dichlorophenol)

herbicides after spreading can reach 80 to 90%, and this phenomenon is very intense in the 4 h following the spreading or just few hours after precipitations. Of course, the incorporation of herbicides to the soil dramatically decreases the volatilization process, but this is very dependent on the weather conditions. Once into the atmosphere, herbicides can be transported by air masses to large distances depending on their stability and configuration of the atmospheric layers. Heptachlor was detected in air samples from Arctic and in precipitations in Antarctic.[45] Emission and diffusion of herbicides into atmosphere can induce concentrations close to several ng/m³, in urban as well as in agricultural environments.[38,39,46–49]

One of the main objectives of the EC project APECOP[50] was to develop process descriptions for pesticide volatilization from plants and to include them in the current PEC models (predicting environmental concentrations of pesticides), PEARL, PELMO, and MACRO.[51] As a screening-level approach for estimating the initial volatilization rate after plant application, a correlation between physicochemical pesticide properties and measured volatilization fluxes was used.[52] For the prediction of cumulative losses from plant surfaces, a similar estimation method was developed by Smit.[53] Despite intense research in recent years, including the development of numerous laboratory and field methods to measure volatilization rates,[43,54,55] knowledge of rate-determining processes is currently not sufficient for developing a reliable, physically-based model approach to predict fluxes of pesticide volatilization from plant surfaces.

II. REGULATIONS

The European Union (EU) directive (98/83) states that the pesticide level must not exceed 0.1 μg/l, for individual compounds and some of their metabolites (0.5 μg/l for all compounds), in water intended for human consumption, including groundwater.[56] In Canada and the U.S.A., neither health advisory levels (HALs) nor maximum contamination levels (MCLs) have yet been set for triazine degradation products. The possibility of summing parent and degradation products to meet the health advisory limit is being considered. The U.S. EPA estimates that a drinking water exposure to 200 ppb of atrazine poses a one-in-a-million lifetime cancer risk. At the proposed HAL of 3 ppb, consumption of atrazine in drinking water poses a risk of about one-in-one-hundred-thousand. In the U.S.A, maximum authorized concentrations have been set for a list of herbicides, including, atrazine 3 μg/l, simazine 4 μg/l, and alachlor 2 μg/l.

Guideline values proposed by the World Health Organization based on a toxicological approach for each substance are 2 μg/l for simazine and atrazine, and 20 μg/l for alachlor. Directive 91/414/CE "harmonizes" at the European scale the principle of authorization delivery of active substances. For example, a herbicide can be introduced in the positive list of the directive Annex 1 if only the scenario takes into account toxicity, ecotoxicity, and study on presence and fate in the environment, in order to exclude all risks of groundwater contamination at levels higher than 0.1 μg/l. The laboratory performance is set by these directives, in terms of limits of quantification LQ (25% of the parametric value) and of measurement accuracy (25% at the LQ level).

As regards the atmospheric compartment, there is no regulation setting the maximum level of herbicides in air, rainwater, or fogs.

A. NORMALIZED METHODS FOR PESTICIDE ANALYSIS

1. French and European Guidelines: AFNOR/ISO/CEN

The main following methods are applied in an international framework of regulation for several herbicide families:

NF EN ISO 10695 (AFNOR T 90-121): Water Quality — Determination of selected organonitrogen and organophosphorus compounds — Gas chromatographic methods.

NF EN ISO 11369 (AFNOR T 90-123): Water Quality — Determination of selected plant treatment agents — Method using high performance liquid chromatography with UV detection after solid-phase extraction.

ISO 15913: Water quality — Determination of selected phenoxyalcanoic herbicides, including bentazones and hydroxybenzonitriles, using gas chromatography/mass spectrometry after solid-phase extraction and derivatization.

2. US EPA Guidelines

EPA Method 507: Determination of nitrogen- and phosphorus-containing pesticides in water by GC-NPD.

EPA Method 508: Determination of chlorinated pesticides in groundwater by GC-ECD.

EPA Method 515.1: Determination of chlorinated acids in groundwater by GC-ECD;

EPA Method 548: Determination of endothal in drinking water by aqueous derivatization, liquid–solid extraction, and GC-ECD.

EPA Method 525: Determination of organic compounds in drinking water by liquid–solid extraction and GC-MS.

3. Guidelines from Quebec for Surface and Ground Water

Méthode M.403-PEST 4.0: Eaux — Détermination des pesticides de types organophosphorés, triazines, carbamates, urées substituées, phtalimides et pyréthrinoïdes. Extraction in situ avec dichlorométhane; dosage par chromatographie en phase gazeuse couplée à un spectromètre de masse.

Méthode M.403-PEST 3.0: Eaux — Détermination des pesticides de types organophosphorés, triazines, carbamates, urées substituées, phtalimides et pyréthrinoïdes. Extraction avec C18; dosage par chromatographie en phase gazeuse couplée à un spectromètre de masse.

B. European Project "Optimization and Evaluation of Multiresidue Methods for Priority Pesticides in Drinking and Related Water"

This European project aimed at developing extraction methods for pesticides and metabolites (without using halogened solvents), and qualitative and quantitative analyses whose performance complied with the requirements of the European directive 98/83. The following three methods have been validated:

PL 95-3327 SMT4-CT96-2142: Determination of priority herbicides and insecticides by gas chromatography with mass spectrometric detection after solid–liquid extraction.

PL 95-3327 SMT4-CT96-2142: Determination of priority herbicides, insecticides, and fungicides by high performance liquid chromatography with UV detection after solid–liquid extraction.

PL 95-3327 SMT4-CT96-2142: Determination of priority herbicides, insecticides, and fungicides by high performance liquid chromatography with mass spectrometric detection after solid–liquid extraction.

III. SAMPLE PREPARATION

The analytical process for the determination of herbicides in water contains several steps with a significant incidence of each one on the result interpretation:

– Sampling
– Storage and shipment of water samples

- Extraction of the substances from the water
- Extract concentration before analysis
- Extract cleanup
- Extract analyses by separation and detection methods
- Identification and quantification of the detected substances.

Herbicide analysis consists of detecting and quantifying traces of hundred of substances with very different physico-chemical properties at very low concentration levels (ng/l to μg/l). Herbicides must be isolated and concentrated, in particular with solid-phase extraction methods, then separated and detected with sensitive, selective, and robust mass spectrometric or spectrophotometric methods.

A. SAMPLING AND SAMPLE STORAGE FOR WATER

1. Water Sampling

Water is a complex medium where numerous exchanges occur. So strict rules for sampling, conditioning, storage, and transport must be followed from the sampling site to the analytical laboratory. Sampling operations can explain 80% of the analytical errors. The selection of the sampling sites, the frequency, and the sampling periods are mandatory requirements prior to the implementation of a survey (monitoring) strategy. The main errors come from the following factors:

- Sampling material
- Sampling mode
- Sample pretreatment (filtration, input of solvent, or inhibitor of micro-organisms)

Concerning the herbicide analyses in surface water, results can be very different depending on the sampling:

- of surface films (herbicide contents can be 10,000 higher in this film),
- in deep layers to study interfaces between water and sediments,
- taking into account the water flow,
- random mode,
- automatic mode with selection of the sampling frequencies and of the sampling volumes.

2. Sampling Flasks

Cautions must be taken when choosing sampling flasks to avoid adsorption, hydrolysis, photolysis, volatilization, and biodegradation processes. Generally speaking, amber glass containers or flasks can be used after appropriate cleaning, excepted for some herbicides such as diquat, paraquat, glyphosate, AMPA, glufosinate, and aminotriazole, for which either polyvinylchloride (PVC) or deactivated glass (by silanization) are strictly required. These restrictions for polar and water-soluble herbicides are due to irreversible adsorption of these hydroxylated compounds on the glass silanol groups.

3. Sample Storage

Herbicides such as, chloroacetanilides, triazines, and chlorophenoxyalcanoic acids can be stored in amber glass flasks, at 4°C during 48 h, before analysis. A study on the storage of 147 herbicides carried out by EPA,[57] showed that all these molecules were stable if they were stored immediately at 4°C after being transferred into an extraction organic solvent. The same behavior was observed for phenoxyalcanoic and amide herbicides. Mouvet et al. at Bureau de Recherche Géologiques et Minières (BRGM), France,[58] studied some herbicide stability in three types of water. They showed that terbutylazine and isoproturon were stable after 30 days at 4°C, and that alachlor was quite unstable with 50% of degradation after only 14 days at 4°C.

B. Soil Samples

Typically, sampling takes place in a small hydrogeological basin by selecting very different soils according to FAO classification. This sampling scheme followed guidelines implemented in the frame of the PEGASE European project.[59] Soil cores were collected, each plot was divided into four equal area subplots, and four samples were taken from each subplot during each campaign, thus giving a total of 16 cores per plot per campaign.[23] A 10 cm diameter percussion corer was used for sampling and the maximum sampled depth was 1.0 m. The cores were sent to the laboratory and cut, after the outer layer had been removed, into segments corresponding to depth intervals of 0 to 5, 5 to 10, 10 to 20, 20 to 30 cm, etc. Each sample was dried at 40°C for 3 days then ground to 2 mm. For each layer, mixing equal weights of the 16 individual samples made a composite sample.[23]

C. Air Sampling Methodologies

Devices used for herbicide sampling from air are made with sorbents and glass or quartz filters. Polyurethane foams (PUF), silica gels, and resins such as Tenax, Florisil, Chromosorb, XAD-2, or XAD-R have been used as sorbents. PUF and XAD-2 gave the best results for herbicide sampling,[60–62] but it should be noted that Tenax can be used in a thermal desorption injection system coupled with GC-MS (730). In order to determine the amount of herbicides picked up from air, sampling flow and time are two main parameters. The EPA TO-4 method describes a high flow method at around 15 m^3/h with PUF (6 cm × 7.6 cm) and quartz filters of 102 mm in diameter, for 24 h.[63] Another EPA method (TO-10) is devoted to low flow samplings with only several liters per minute on small quartz filters (32 mm) and PUF (0.22 cm × 7.6 cm).[64] Two other American guidelines (NIOSH 5600 and 5602) suggest quartz filters of 11 mm in diameter and flow samplings included between 0.2 and 1 l per minute, by using a sandwich cartridge of XAD-2/PUF.[65,66] Herbicides sampled from air can be quantitatively extracted by using ultrasonic or mechanical techniques with methylene chloride or a mixture of hexane/methylene chloride.[38,47,67] EPA methods suggest Soxhlet extraction for 24 h with mixtures of hexane/ether diethyl oxyde or hexane/ methylene chloride. However, for liquid samples, Soxhlet extractions are time and solvent consuming, so the PFE method was tested by Foreman on 47 herbicides and recoveries were quantitative in particular for alachlor and atrazine.[68]

In order to ensure the detection of both parent triazines and their degradation products, XAD-2 resins were used for two days to sample the atmospheric phase. Analytical measurements were performed by GC-MS/MS and detection limits ranged from 0.8 to 15 pg/m^3. Only atrazine was quantified in the gas phase at 180 pg/m^3, whereas atrazine, DEA, and terbutylazine were quantified in the particulate phase between 180 and 870 pg/m^3.[69]

As an extension of the relevant lysimeter concept[70] within the framework of the APECOP project,[50] a glass wind tunnel[71,72] was set up above a lysimeter with a soil surface area of 0.5 m^2 to measure the gaseous emissions of the applied pesticide. Realistic conditions are simulated inside this wind tunnel by a continuous, automatic adjustment of the air temperature to the outdoor situation. Due to the glass design, sufficient light intensity is ensured, so that experiments after application on plant surfaces can also be performed. ^{14}C-labeled pesticide in the exhaust air was sampled with a High Volume Sampler (HVS) equipped with an adhesive-free glass fiber filter to trap particulate matter followed by three polyurethane foam plugs. The maximum sampling rate was 50 m^3/h, corresponding to 3 to 10% of the total airflow through the wind tunnel. The sampling period of the HVS ranged between 1 and 24 h. ^{14}C-labelled carbon dioxide, formed from the mineralization of ^{14}C-labeled pesticide, was collected with a Medium Volume Sampler (MVS) at a sampling rate of 1.0 to 3.5 l/Min over a maximum sampling interval of 48 h, (sampling rate of 10 m^3 in 48 h). In order to ensure sampling of [14]CO_2 only, volatile organic compounds were trapped with two cartridges filled with XAD-4 resin. Then the air sample was dried intensively by silica gel and phosphorus pentoxide. [14]CO_2 was subsequently absorbed by 2-methoxy-propylamine (Carbosorb E+) using a cooled intensive-wash bottle. Losses of highly volatile 2-methoxy-propylamine were minimized by intensive cooling at a reflux temperature of $-40°C$. Soil and plant samples, glass fiber filters, and XAD cartridges were extracted with methanol in a Soxhlet apparatus for 16 h. The active ingredients of the samples were characterized by radio-HPLC and radio-TLC in combination with a Bio-Imaging Analyzer.[50]

IV. EXTRACTION

A. LIQUID–LIQUID EXTRACTION

1. Classical Methods

Most officially sanctioned methods for the analysis of herbicides, including triazines, in water still use LLE techniques based on the distribution of herbicides between the aqueous phase and an immiscible organic solvent. Conventional methods take samples of < 1 l (up to 1 l and pH adjusted to 7), which are shaken with an immiscible organic solvent (more or less selective) such as, methylene chloride, hexane, ether diethyl oxyde, chloroform, or octanol. For pg/l or ng/l levels, larger sample volumes (up to 120 l) have been extracted using the Goulden large sample extractor.[73,74] There are disadvantages to LLE techniques: they cannot extract polar herbicides like degradation products; they are laborious, time consuming, expensive, and subject to problems arising from the formation of emulsion. Moreover, the evaporation of large solvent volumes and the disposal of toxic or inflammable solvent are needed.

2. Liquid Phase Microextraction (LPME)

Recent research trends involve miniaturization of the traditional liquid–liquid extraction principle by reducing the volumetric ratio of the acceptor-to-donor phase. One emerging technique is the LPME, in which a hollow fiber impregnated with an organic solvent is used to accommodate microvolumes of acceptor solution. This is an extremely simple, low cost, and solvent free sample preparation technique with a high degree of selectivity and enrichment, eliminating the possibility of carry over between run. The method has been mainly applied to drug substances, however triazines were successfully extracted with 3 μl of toluene and the limits of detection in GC/MS were included between 0,04 and 0.18 μg/l.[75,76] Very similar detection limits (0.03 μg/l) were obtained in another study on seven alkylthio-s-triazines in river water samples.[77]

3. Supercritical Fluid Extraction (SFE) in Water

Some triazines, including atrazine and simazine, were extracted from water by SFE after preconcentration on solid-phase extraction disks.[78,79] The freeze-dried residue or SPE disk was then introduced into the extraction cell and eluted with either pure CO_2 or methanol or acetone-modified CO_2. SFE has been applied in combination with online solid-phase extraction for pesticides, including triazines.[80] However, this technique is still used mainly on solid matrixes. The main limitation with aqueous matrixes remains the miscibility of water with supercritical carbon dioxide.[81] Recoveries of nonpolar pesticides, extracted by SFE technique using octacedyl-bonded silica, are generally effective with CO_2 elution alone. This is not the case for semipolar and polar compounds, such as triazines and their degradation products. For example, the addition of

10% methanol to CO_2 is necessary to reach acceptable recoveries for atrazine and some degradation products like HA and DAHT. Atrazine, simazine, DEA, and DIA were extracted from cartridges filled with granular activated carbon (GAC) using the SFE technique.[82] Pure CO_2 was insufficient to elute these chemicals because of the interactions between GAC sites and the compounds. The addition of 50% of acetone was necessary to obtain acceptable recoveries.

B. SOLID/LIQUID EXTRACTION

Due to the numerous drawbacks of L–L extraction procedures, Liquid–Solid (L–S) methods using very low volumes of solvent were introduced since the beginning of the 1980s. These include SPME and SPE, and currently the emphasis is on automation of the whole SPE procedure. The ideal methodology for sample preparation is fast, accurate, precise, and consumes little solvent. Furthermore, it is easily adapted for fieldwork, and requires less costly materials. The SPE method may be the isolation technique capable of meeting all these expectations.[83]

1. Solid-Phase Extraction

SPE cartridges and disks are available from many suppliers and represent a variety of chemical matrixes. In conventional SPE, a liquid is passed over a sorbent packed in a glass or polypropylene cartridge, or embedded in a disk. Because of the strong attraction between them, the analytes are retained on the sorbent, which is later washed with a small volume of solvent to disrupt the bonds between analytes and itself. The selection of an SPE method depends upon the herbicide under evaluation, expected concentrations, and the water volume being processed. Disk extraction has been reported to use 90% less solvent than LLE and up to 20% less solvent than cartridges, and it eliminates the problem of channeling associated with cartridges.

Sabik and Jeannot recently published an exhaustive review on SPE and multiresidue methods for the monitoring of priority herbicides in water.[6] The same year, Thurman and Snavely proposed some views and advances related to disk extraction methods for environmental applications, including herbicides.[26]

Methanol is usually utilized to prewet the C_{18} Bond-Elut columns and opens the hydrophobic chains to increase the effective surface area. Water samples are also fortified with at least 1% methanol to continuously wet the stationary phase. This can improve recovery rates for a large number of herbicides, including triazines. By contrast, degradation products, which are often more polar than parent compounds, may not be retained as effectively in the presence of a modifier. Ground and surface water must always be filtered prior to the extraction of pesticides with the SPE technique. Prefiltering will not affect the determination of herbicides and their degradation products, since these compounds exhibit a log K_{oc} near 2 and consequently they are largely (99.5%) distributed in water in the dissolved phase.[84]

a. Sorbents

Sorbent–analyte interactions fall into three categories: nonpolar, polar, and ionic. Nonpolar sorbents are generally selected for extracting triazines from water. By contrast, degradation products containing polar functional groups such as, hydroxyl, carbonyls, amines, and sulfhydryls, need polar sorbents.

Different types of sorbents have been employed in SPE techniques to extract triazines, phenylureas, some methylcarbamates, triazoles, chlorophenoxy acid compounds, and their degradation products from water. The most widely used are C_8 and C_{18} chemically-bonded to silica,[85,86] carbon black,[87,88] and polymeric resins (such as PLRP-S).[89] The most polar compounds, like DIA, DEA, HA, and metribuzin, have low breakthrough volumes with these sorbents, except for carbon material[90,91] and some highly cross-linked styrene-divinylbenzenes (Envi-chrom P).[92]

In recent years, chemically modified polymeric resins with a polar functional group have been developed and used in the SPE of these compounds, and the breakthrough volumes were higher than those obtained with their unmodified analogues.[91,93,94] New cross-linked styrene-divinylbenzene packing materials, such as LIChrolut EN,[93,95,96] Styrosob, and Macronet Hypersol,[97] are now available. These sorbents have a higher degree of cross-linking and, thereby, an open structure (high-porosity materials) that increases their specific area (400 to 1000 m^2/g) and allows greater $\Pi-\Pi$ interactions between analytes and sorbent. This means that the breakthrough volumes will be higher than those obtained when the cross-linked sorbents are used. The three sorbents allowed the same percent recoveries for atrazine and simazine (80 to 86%) in water.[98]

A multiresidue method based on offline SPE mode with GC/MS, LC/UV/DAD. and LC/MS was developed in Europe for monitoring pesticides on the priority list. Various sorbents were tested: Isolut C$_{18}$, Lichrolut, Envi 18, SDB, OASIS, Envi-chrom. and Envi-carb. The SDB, OASIS, Envi-chrom. and Envi-carb appear to be the most promising for extracting polar compounds, including triazines and their degradation products.[99]

b. Cartridges

SPE cartridges are available in a wide range of sizes, with volumes ranging from less than 1 ml to over 50 ml. When selecting the optimum cartridge size for a particular application, factors to be considered are: ability to retain all analytes in a sample, volume of original sample, and final volume of the purified sample after elution. In general, the mass of the analytes and interfering compounds retained by the sorbent should be less than 5% of the mass of the sorbent. A good rule of thumb is that the elution volume should be 2 to 5 times the bed volume of the cartridge. This volume may be higher depending on the properties of the selected pesticides, the nature of the adsorbent, the type of eluent. and the analytical technique used.[90]

The extraction and quantification of the OXA and ethane sulfonic acid (ESA) metabolites of the acetochlor and metolachlor herbicides were described in water samples.[92] Extractions were performed at pH 3 by using PS-DVB (Chrom-P ENVI) extraction cartridges and by loading 500 ml of water samples. Each of the different water matrixes was spiked in triplicate with standards of each degradation compound at concentrations of 0.2, 0.8. and 4 μg/l. The average recoveries range from 80 to 120% for both metabolites, with relative standard deviations lower than 15%. Twelve acetanilide degradates were extracted by SPE from 100 ml of water samples using carbon cartridges with mean of recoveries above 90% and relative standard deviations lower than 16%.[100]

A multiresidue method was developed to determine 22 pesticides in drinking water of the area of Barcelona, including triazines and chloroacetanilides. A relevant feature of this interlaboratory study was the estimation of the expanded uncertainties, ranging from 10 to 20%, with the SPE procedure as main source of uncertainty.[101]

A new selective enrichment technique was investigated for the sample preparation for GC-MS analysis of 16 acidic herbicides in water.[102] By using a dynamic ion-exchange solid-phase extraction (DIE-SPE) combined with reverse-phase SPE, interference by humic substances could be reduced and most of the acidic herbicides were extracted with recoveries above 70%.[102]

c. Disks

A variation on the extraction cartridge is the disk in which the sorbent (on a polymer or silica substrate) is embedded in a web of PTFE or glass fiber. Glass fiber disks are thicker and more rigid, providing higher flow-rates than with PTFE membranes as illustrated by the high throughputs used with laminar extraction disks. The sorbent particles embedded in the disks are smaller than those found in the cartridges (8 μm diameter rather than 40 μm). The short sample path and small particle size allow efficient trapping of analytes with a relatively high flow rate through the sorbent, as compared to the cartridges. The disks are primarily used to reduce analysis time when handling large volumes of aqueous environmental samples.[26] The most frequently used disk size is the 47 mm, suitable for standard methods (0.5 to 1 l water sample volumes). D. Barceló et al.[103] showed that the recoveries on C_{18} Empore disks (1 to 4 l) were very high for a large number of pesticides, including atrazine, simazine. and cyanazine (80 to 125% for the parent compounds). However, recoveries only ranged from 3 to 17% for degradation products (DEA and DIA). The extraction disks allowed relatively high flow rates, compared to cartridges using this same material because of the absence of channeling and the faster mass transfer provided by the smaller particle sizes.[103] Viana et al.[104] have demonstrated that C_8 disks allow better recoveries for atrazine, prometryn. and propazine (87 to 93%) than do C_{18} disks (66 to 67%). Pichon et al.[105] employed a multiresidue method using a new laminar extraction disk in combination with LC, a Baker Speedisk DVB for polar compounds, and a C_{18} silica disk for nonpolar compounds. They achieved rapid handling of 1-l sample volumes, with DLs ranging from 0.01 to 0.05 μg/l.

Table 26.3 presents numerous works involving SPE techniques, using either cartridges or disks, for the determination of herbicides in ground and surface water.

d. Online and Offline SPE Procedures

Online SPE/GC (equipped with thermo-ionic, electron capture, or mass spectrometer detectors) and SPE/LC (equipped with PDA, fluorescence, or mass spectrometer detectors) are the methods of choice for the trace-level determination of herbicides. In general, the combination of SPE and LC is an important improvement over GC applications, since it is not necessary to remove all residual water from sorbents, because elution solvents (c.g., methanol and acetonitrile) are compatible with the final separation method.[103] The development of a large volume injection system in GC (10 to 250 μl) has partly closed this gap, however. There have been a number of reports in the literature of methods employing online and offline procedures for determining priority pesticides, including triazines and degradation products, in water.[131–134] Some of the online methods are summarized in Table 26.3. Several studies, using precolumn (10 to 20 mm length × 1 to 4.6 mm i.d., 5 to 10 μm packing gradually replaced by 15 to 40 μm packing with C_8, C_{18}. and silica-divinylbenzene [S-DVB]) and membrane disks (diameter, 3 to 4.6 mm packing with C_{18} and S-DVB), have compared different sorbent materials for online SPE/LC. SPE methods can now be easily converted into fully automated online systems coupled to LC or GC techniques. With these methods, small sample volumes (0.001 to 0.010 l) are sufficient to obtain 0.01 to 0.1 μg/l for a large variety of compounds, including triazines and their degradation products.[131,135] This is an enormous advantage over offline procedures. Regardless, online procedures benefit from the absence of contamination or loss of analytes during solvent evaporation, while offline procedures are favorable for their applicability to onsite sampling and the opportunity to inject the same extract twice.

The reproducibility, sensitivity. and robustness of a fully online SPE and LC/PAD (SAMOS) system have been demonstrated for monitoring pesticides in surface water. The robustness of the SAMOS system was illustrated by the fact that no major problem was encountered in the course of over 1000 determinations.[105]

e. Onsite Extraction and Stability of Herbicides on SPE Materials

Very few studies have reported on the stability of herbicides, including triazines and degradation products, on SPE materials.[87,136,137] In addition to time and space savings, the stabilization of pesticides on these materials makes it possible to use SPE techniques for onsite extraction. Sabik et al.[87] demonstrated the stability of 20 urea and triazine herbicides, including four degradation products, on GCB material over a two month period. Liška and Bilikova[137] studied the stability of 16 polar pesticides including triazines, carbamates, and phenylureas sorbed on to a polymer sorbent. They found that most of these remained stable over a seven-week period.[137] Crescenzi et al.[136] studied the stability of 34 pesticides, including atrazine, metamitron. and metribuzin, on C_{18}

TABLE 26.3
SPE Applications for the Determination of Herbicides and Metabolites in Water

Herbicide Family	Sample Type, Volume, and Pretreatment	Sorbent	Elution	Extract Treatment	LOD	Recovery (%)	RSD (%)	Detection Technique	Year	Ref.
Triazines	River and reservoir, 1 l	C18 (360 mg)	2 × 2 ml ethyl acetate	+0.05 g Na_2SO_4 Concentration to 30 μl	0.75 to 12 ng/l	83 to 94	3.2 to 16.1	GC-MS/MS	2004	106
Triazines	Groundwater, 250 ml	C18	5 ml acetone and 5 ml methanol	Evaporation to dryness + 0.1 ml methanol	0.05 to 0.09 μg/l	>87 (except DEA, 52%)	—	MEKC-DAD	2004	107
Carbamates, phenylureas, Triazines	Drinking and surface water, 50 ml	C18 (250 mg)	3 × 1 ml methanol/ acetonitrile (1:1)	Evaporation to dryness + 0.2 ml water	0.5 to 13 ng/l	67.7 to 105.2	<12.6	LC-ESI-MS (SIM)	2004	108
Phenoxy acid and phenylureas	Natural water, 200 ml	GCB (0.25 g)	2 ml methylene chloride/ methanol (6:4) and KOH (0.016 M)	Evaporation to dryness + 0.1 ml sodium dodecyl sulphate	—	92 to 98	—	MEKC-UV	2004	109
Acidic herbicides	Surface and agricultural water, 250 ml, pH 3	C18 (200 mg)	2 × 0.5 ml Methanol	Evaporation to dryness + 0.5 ml mobile phase	LOQ, 0.1 to 0.5 μg/l	85.7 to 110	1.8 to 13.4	LC-UV	2003	110
Triazines and metabolites and ureas	Natural water, 250 ml	PS-DVB (200 mg)	5 ml methanol and 5 ml ethyl acetate	Evaporation to dryness + 0.5 ml methanol	0.13 to 2.7 μg/l	35 to 115	6 to 17	MEKC-DAD	2003	111
Chlorotriazines, methylthiotriazines and methoxytriazine	Drinking water, 2 l	GCB (1 g)	1 ml methanol + 9 ml methylene chloride/ methanol (8:2)	Evaporation to dryness + 1 ml acetonitrile	1.9 to 8.4 ng/l	80 to 97	0.6 to 13	LC-UV	2003	112

Analyte	Matrix	Sorbent	Elution	Evaporation/reconstitution	LOD	Recovery (%)	RSD	Detection	Year	Ref.
Chlorotriazines, methylthiotriazines and methoxytriazine	Drinking water, 2 l	PS-DVB (1 g)	2 × 4 ml methanol/ethyl acetate (2:1)	Evaporation to dryness + 1 ml acetonitrile	—	79 to 100	5.4 to 21	LC-UV	2003	112
Triazines	Reservoir and river water (1 l)	C18 (360 mg)	2 ml ethyl acetate	Dryness with Na_2SO_4 + Evaporation to 30 µl water/methanol (9:1)	1.7 ng/l	90.5	3.2	GC-MS (SIM)	2003	113
Triazines, ureas and metabolites	Surface and groundwater, 250 ml	PS-DVB (200 mg)	5 ml methanol and 5 ml ethyl acetate	Evaporation to dryness +0.5 ml mobile phase	0.01 to 0.1 µg/l	68 to 109	7.6 to 17.7	LC-DAD	2002	114
Phenoxyacid herbicides and metabolites	Surface and drinking water	GCB (0.5 g)	1 ml methanol and 8 ml methylene chloride/methanol	Evaporation to dryness +0.2 ml water/methanol (1:1)	0.1 ng/l (5 to 10 ng/l for metabolites)	>85	2 to 10	LC-ESI-MS/MS	2002	115
Dimethenamid, flufenacet and oxanilic and sulfonic degradates	Natural water (123 ml)	C18 (500 mg)	3.2 ml ethyl acetate	Evaporation to 70 µl ethyl acetate	0.02 to 0.04 µg/l	103 to 107	—	GC-MS	2002	116
			3.5 ml methanol	Evaporation to dryness and reconstitution: 125 µl (0.3/24/35.7/40): acetic acid/methanol/water/acetonitrile	0.01 to 0.07 µg/l	76 to 98		LC-MS		
Neutral and acidic herbicides	Rain water, 500 ml	Oasis HLB (200 mg)	10 ml methanol	Evaporation to dryness + 1 ml water/methanol (9:1)	5 to 23 ng/l 3 to 59 ng/l	Neutral: 50 to 109 Acidic: 19 to 96	6 to 33 9 to 32	LC-MS/MS	2002	117

Continued

TABLE 26.3
Continued

Herbicide Family	Sample Type, Volume, and Pretreatment	Sorbent	Elution	Extract Treatment	LOD	Recovery (%)	RSD (%)	Detection Technique	Year	Ref.
Chlorophenoxy acids and metabolites	River and drinking water (1 or 2 l)	Carbograph (120/400 mesh)	8 ml methylene chloride/methanol (80:20) + formic acid 50/mmol	Evaporation to 100 μl + 100 μl acetonitrile/water (50:50) + formic acid 1/mmol	2 to 75 ng/l	84 to 95	6 to 10	LC-MS/MS	2002	118
Twelve chloro-acetabilide degradates	Tap water (200 ml)	GCB (0.25 g)	5 ml, 10 mM ammonium acetate in methanol	Evaporation to dryness + 1 ml ammonium acetate in water	<0.1 $\mu g/l$	73 to 109	0.8 to 24	LC-ESI/MS/MS	2002	100
Phenoxyacetic acids and metabolites	Natural water, 50 ml, pH 3	C18 (100 mg)	0.2 ml methanol and 0.6 ml water	—	LOQ 5 ng/l	63 to 109	2 to 17	LC-ESI-MS/MS	2001	119
Chloroacetanilide metabolites	Groundwater (500 ml), pH 3	PS-DVB (500 mg)	2 × 5 ml methanol	Evaporation to 500 μl methanol	10 to 40 ng/l	80 to 120	<15	LC-ESI/MS	2001	92
Triazines, phenylureas, and other priority herbicides	Ground and surface water, 1.33 ml	Online C18 (back flush elution)	—	—	0.5 to 60 ng/l	55 to 116	2 to 21	LC-ESI-MS/MS	2001	120
22 pesticides including triazines and chloroacetanilides	Drinking water	PS-DVB (200 mg)	2 × 2.5 ml ethyl acetate	Evaporation and reconstitution with 500 μl of isooctane	0.025 $\mu g/l$ (except metribuzin, 0.035 $\mu g/l$)	73 to 131	<12.5	GC-MS	2001	101
Choroacetanilides and metabolites	Surface and groundwater, 1 l	GCB (0.5 g)	Sequential elution for parent compounds and metabolites	Evaporation to dryness + 100 μl of solvent	1 to 8 ng/l (parent compounds)	76 to 100 (parent compounds)	<12%	GC-MS and LC-UV	2000	24

Analyte	Sample	SPE	Elution	Derivatization/cleanup	LOD/LOQ	Recovery (%)	RSD	Detection	Year	Ref
Triazines	Surface water (10 ml)	Online Immuno-affinity SPE	—	—	1 to 90 ng/l (metabolites)	41 to 96 (metabolites)	—	GC-NPD	1999	121
Triazines	Water + 20 ppm humic acids (200 ml)	Online MIP/C18	—	—	1.5 ng/l	64 to 88	—	LC-UV	1999	122
Triazines (chloro and hydroxy)	Tap and river water (200 ml)	PS-DVB (200 mg)	4 ml methanol/acetone (3:2)	Hydroxytriazines: Evaporation to dryness + 100 μl water; Chlorotriazines: Evaporation to dryness + 20 μl toluene	0.1 to 0.25 μg/l	Hydroxytriazines: 43.8 to 93.4; Chlorotriazines: 96.3 to 124.8	Hydroxy triazines: 5.7 to 14.8; Chlorotriazines: 4.6 to 11.8	GC-MS and CE-UV	1999	123
Triazines, phenylureas	Surface water (500 ml)	Double disk (500 mg SAX +C18)	—	—	—	85 to 110 (DIA 25)	—	LC-UV	1999	124
Choroacetanilide metabolites	Groundwater, 100 ml	PS-DVB (1 g)	5 ml methanol/water (7:3)	—	LOQ 0.1 μg/l	>89	<10	LC-MS/MS	1998	25
Acidic herbicides	Drinking water, pH 2, 11	C18 disks (47 mm)	2 $\times$ 25 ml ethyl acetate	Evaporation to 2 or 1 ml Derivatization with diazomethane	2 to 9 ng/l	51 to 140	~20	GC-MS	1998	125
Triazines, phenylureas	Surface water (4 ml)	Online C18	—	—	0.1 μg/l	—	—	LC-MS/MS	1998	126
Fifteen herbicides including triazines	Groundwater (200 ml)	Online C18	—	—	0.8 ng/l	—	—	LC-MS	1998	127
Triazines	Surface water (18 l)	GCB (0.5 g)	—	—	3 to 52 ng/l	51 to 84 (metribuzin 5%)	—	GC-NPD and LC-MS	1998	239

Continued

TABLE 26.3
Continued

Herbicide Family	Sample Type, Volume, and Pretreatment	Sorbent	Elution	Extract Treatment	LOD	Recovery (%)	RSD (%)	Detection Technique	Year	Ref.
Choroacetanilides and metabolites	Surface and groundwater, 100 ml	C18 (360 mg)	3 ml ethylacetate (parent compounds) + methanol (metabolites)	Evaporation to dryness + 75 μl mobile phase	LOQ 0.01 μg/l	98	—	LC-MS	1997	27
Triazines	Surface (1 l) and groundwater (4 l)	GCB (0.5 g)	—	—	—	80 to 101	—	LC-MS	1997	128
Triazines	Surface water (20 ml)	Online Immuno-affinity SPE	—	—	—	86 to 103 (DIA 0%)	—	LC-MS	1997	129
Twelve herbicides including triazines	Surface water (100 to 500 ml)	Online and offline PS-DVB disks	—	—	Offline 0.05 to 0.1 μg/l Online 0.03 μg/l	74 to 92	—	LC-UV	1996	130

and GCB materials, reporting that selected triazines and triazinone remained stable on both materials during the test period (three weeks). Other onsite preconcentration techniques have been reported, including those involving cartridges and disks.[138] An automated online SPE/LC/PDA method for onsite pesticide monitoring in surface water has also been used in the basin of the Rhine River.[139]

2. Immuno-Extraction

The immuno-extraction technique consists of using SPE cartridges filled with antibody materials bonded onto silica-based sorbents. These materials, called immuno-affinity sorbents, have been used to extract triazines from water samples.[140] However, as they were specific to the target compounds, DIA was not recovered with antiatrazine immunosorbents, while HA and prometon were not recovered with antisimazine immunosorbents.[140] The comparison of an antiatrazine immunosorbent and a PLRP-S sorbent for the extraction of triazine from the Seine River (50 ml), using the SPE technique, has demonstrated the high selectivity and efficiency of the immunosorbent.[140] More recently, Dallüge et al.[121] reported on the use of an online coupling of immunoaffinity-based solid-phase extraction and gas chromatography for the determination of s-triazines in aqueous samples. These sorbents are expected to undergo further refinement for other classes of herbicides. This would allow the extraction of some high polar compounds from water because the antigen-antibody interaction is not based on the hydrophobic process.

Enzyme-Linked Immunoassays (ELISA) were used to determine chloroacetanilide levels in water, with detection limits of 0.06, 0.3, and 0.4 μg/l for metolachlor, alachlor, and acetochlor, respectively.[141]

3. Molecularly-Imprinted Polymer (MIP)

The concept of this technique was inspired by Pauling's antibody formation theory. An antigen is used as a template to aid in the rearrangement of antibody polypeptide chains, so that the antibody having a three-dimensional configuration complements the antigen molecule.[142] Successful imprints on synthetic organic polymers were achieved in the 1990 s,[143,144,145] and the MIP technique has become increasingly popular in recent years. It has already been used in different applications as a drug-retaining matrix, in the enantioseparation of drugs, and as a solid-phase extraction material for hydroxycoumarin extraction, showing its considerable potential for selective extraction. It is expected to be beneficial for the extraction and cleanup of various polar herbicides from complex matrixes. Certain applications have already been performed, mainly on triazines, using an offline system.[146–148] By coupling a MIP-SPE column online with a C_{18} column, Bjarnason et al.[122] distinguished triazines from humic acid, reaching an enrichment factor of up to 100 with satisfactory recoveries of 74 to 77%. Lanza and Sellergren[149] tested six functional monomers of MIPs — methacrylic acid (MAA), methyl methacrylate (MMA), hydroxyethyl methacrylate (HEMA), N-vinyl-a-pyrrolidone (NVP), (trifluoromethyl) acrylic acid (TFM), and 4-vinylpyridine — and found that MAA was more suitable for the extraction of chlorotriazines. Further optimization of MIPs may lead to more efficient matrix discrimination and allow for the extraction of some high polar compounds from water, as polymer-molecule interaction is not based on the hydrophobic process. By choosing 2,4,5-trichlorophenoxyacetic acid as a template, MIP was used to concentrate chlorophenoxy acid herbicides from river water samples,[150] at concentration levels of ng/ml. Quantitative recoveries were achieved, close to those obtained with C_{18} sorbents on these substances (worst recovery of 81% for Fenoprop).

4. Stir Bar Sorptive Extraction

Stir Bar Sorptive Extraction (SBSE) was described by Baltussen in 1999 as a novel extraction approach.[151] This is a similar technique to SPME where the fiber is substituted by a stir bar also coated with polydimethylsiloxane, but 50 to 300 μl of PDMS polymer can be used instead of 0.5 μl in the case of SPME, hereby increasing the sensitivity. SBSE coated with 50 μl of PDMS was applied to the extraction of triazines in aliquots of 20 ml of water for 4 h. Analyses performed by thermal desorption with a cryofocusing step and gas chromatography/mass spectrometry allowed detecting triazines below 0.05 μg/l.[152] A SBSE on PDMS polymer was recently applied to the extraction of 35 priority semivolatile compounds, including eight triazine herbicides. The optimized conditions consisted of a 100-ml water sample with 20% of NaCl extracted with 20-mm-length film thickness stir bars, at 900 rpm for 14 h. Analysis was performed by desorption at 280°C for 6 min on a PTV-GC/MS system in full scan mode.[153]

C. Solid-Phase Micro-Extraction

Solid-phase micro-extraction (SPME) first became available to analytical researchers in 1989.[154] The technique consists of two steps: first, a fused-silica fiber coated with a polymeric stationary phase is exposed to the sample matrix where the analyte partitions between the matrix, and the polymeric phase. In the second step, there is thermal desorption of analytes from the fiber into the carrier gas stream of a heated GC injector, then separation and detection. Headspace (HS) and direct insertion (DI) SPME are the two fiber extraction modes, whereas the GC capillary column mode is referred to as in-tube SPME.[155] The thermal desorption in the GC injector facilitates the use of the SPME technology for thermally stable compounds. Otherwise, the thermally labile analytes can be determined by SPME/LC or SPME/GC (e.g., if an *in situ* derivatization step in the aqueous medium is performed prior to extraction). Different types of commercially-available fibers are now being used for the more selective determination of different classes of compounds: 100 μm polydimethylsiloxane (PDMS), 30 μm PDMS, 7 μm PDMS, 65 μm carbowax-divinylbenzene (CW-DVB), 85 μm polyacylate (PA), 65 μm PDMS-DVB, and 75 μm carboxen-polydimethyl-siloxane (CX-PDMS).[156,157] PDMS, which is relatively nonpolar, is used most frequently. Since SPME is an equilibrium extraction rather than an exhaustive extraction technique, it is not possible to obtain 100% recoveries of analytes in samples, nor can it be assessed against total extraction. Method validation may thus include a comparison of the results with those obtained using a reference extraction technique on the same analytes in a similar matrix.

Boyd-Boland and Pawliszn reported the first application of SPME to the analysis of herbicide residues in 1995, for the simultaneous determination of nitrogen-containing herbicides in soil, water, and wine samples.[158] Herbicides have been extracted following the three extraction modes (DI, HS, and in-tube), but direct insertion mode was the most used for these compounds. Krutz et al. have recently published an exhaustive review dealing with SPME for herbicide determination in environmental samples.[159]

For example, 22 compounds including triazines and chloroacetanilides were simultaneously quantified with this mode and the limits of detection were between ng and sub-ng/l.[160] Later, the in-tube SPME method, first developed for phenylureas, was used to identify phenoxy acid herbicides in water samples.[161] In water, 81 compounds from 14 herbicide families have been quantified by SPME, with simultaneous determination of triazines, phenylureas, phenoxy acids, and carbamates in some cases. In evaluating the performance of a 65-μm CW-DVB fiber combined with SPME/GC/NPD, it was shown that this fiber is most sensitive to 12 herbicides, including atrazine, prometon, and terbutryn.[156] PDMS-DVB, CW-DVB, and PA are more appropriate for polar, nitrogen-containing herbicides. An interlaboratory trial involving the analysis of triazines and their degradation products demonstrated the validity of SPME, using CW-DVB fiber in association with added NaCl and in combination with a GC system.[162] The results obtained with these methods demonstrated that SPME is a robust, reproducible, and sensitive method for the analysis of triazines and two metabolites, DEA and DIA. SPME alachlor extraction from water was successfully optimized, with the extraction time and sample volume as the only statistically significant factors.[163]

Photo-SPME was applied to investigate the photochemical degradation of various priority pesticides, including atrazine and alachlor.[164] In addition, aqueous photo-degradation was performed and compared with "on-fiber photo-degradation" (photo-SPME) to confirm the potential of this new technique. Photoreaction kinetics of herbicides were monitored by studying the influence of the irradiation time on the extent of photodegradation. The analytes were first extracted and then the PDMS fiber was exposed to 254 nm radiations for the designed time. Immediately, GC-MS analysis was carried out. Atrazine was quickly photodegraded in aqueous photodegradation experiments whereas Photo-SPME degradation was slower (less than 25% of atrazine remained in 120 min). The only photoproduct identified after atrazine photodegradation was generated through substitution of chlorine by OH radical. Alachlor was photodegradated through aqueous photodegradation and photo-SPME (254 nm), and photodegradation kinetics in both experiments were similar. The photoproduct generation mechanisms for alachlor involved successive reduction, cyclation, and dechlorination reactions.[164]

Chlorophenol metabolites of chlorophenoxy acid herbicides were successfully extracted from water and soils by SPME.[165,166] The optimization of the derivatization-SPME in water for these metabolites (among 30 phenolic compounds), showed that CW-PDMS (85 μm) was the most suitable fiber, with quantification limits ranging from 1 to 15 ng/l.[167]

Since 1995, 21 compounds from five herbicide families have been quantified in soils by SPME. Researchers originally used soil/water suspension samples, either by DI- or HS- SPME.158,160, 168. K_{oc} for six triazines in soils and sediments have been determined by SPME. Recoveries were satisfactory in the organic carbon range 0.2 to 2.4%, with detection limits included between 50 and 500 ng/g.[169] Recent works suggest that the most reliable method consists of performing a DI-SPME of a diluted organic extract obtained after conventional solid–liquid extraction.[170]

Advantages of SPME to traditional extraction methods should promote advances in the field of herbicide chemistry. However, SPME has some limitations, such as analyte carryover, fiber damage at extreme pH, salt-related problems, and low sensitivity in some complex soil samples. Advancements are being made in the refinement of the SPME technique. The HPLC/SPME interface has then been improved, and new mixed phases based on solid/liquid sorption (e.g., CW-DVB and PDMS-DVB) have been developed in recent years for the analysis of compounds by HPLC. A modified accessory to the HPLC system, called in-tube SPME, was developed. This device aspirates and dispenses samples from vials with the syringe in the inject position and then desorbs with aspirated solvent in the load position. Returning the valve to the inject mode will transfer analytes to the analytical column.[161]

Some examples of herbicide determination in water by SPME are presented in Table 26.4.

D. SOLID SAMPLES

Extraction of herbicides from solid matrixes has frequently been done by Soxhlet extraction, which required large volumes of solvent and was a time consuming process. Therefore, new extraction techniques have been developed and applied for the past ten years. Herbicides and their main metabolites can be extracted from solid samples by these new methods such as, SFE,[82] subcritical water extraction (SWE), microwave-assisted extraction (MAE), or pressurized fluid extraction (PFE).[185] From a general point of view, Camel evaluated potentials and pitfalls of SFE, MAE, and PFE.[186]

The limited availability of certified reference materials for herbicides in soils is detrimental to the development of robust extraction methods. In many works, soils are then spiked with known quantities of herbicides and recoveries are calculated to check the applicability of the extraction method. However, although recoveries are often quantitative, they must be interpreted with caution because herbicides have no time to interact properly with the soil matrix constituents. Bearing in mind these limitations, several techniques to extract herbicides from solid matrixes will be described in this chapter.

TABLE 26.4
SPME Applications for the Determination of Herbicides and Metabolites in Water

Herbicide Family	Sample Type	Fiber	LOD	RSD (%)	Detection Technique	Year	Ref.
Several families, including triazines, and chloroacetanilides	Groundwater 3 ml	PDMS-DVB	0.2 to 40 ng/l	2.9 to 27	GC-MS-MS	2004	171
Triazines, chloroacetanilides, phenylureas	River and well water, 5 ml + NaCl	PA	0.02 to 0.11 μg/l	<26%	GC-MS	2003	172
Priority herbicides including triazines, and chloroacetanilides	River water, 5 ml	PDMS	0.01 to 0.09 μg/l	7 to 14	GC-MS	2003	173
Four groups of pesticides including triazines	Ground and drinking water; 1.5 ml	PDMS-DVB	1 to 50 ng/l	6.9 to 39	GC-ECD	2002	174
Phenoxy acid herbicides	Surface water	In tube DB-WAX	5 to 30 ng/l	2.5 to 4.1	LC-MS	2001	175
Phenoxy acid herbicides	Water	Derivatization + PDMS	0.6 to 2.3 μg/l	23.6 to 53.3	GC-MS	2001	176
Triazines and chloroacetanilides	Groundwater	CW-DVB	5 to 20 ng/l	0.4 to 2 (except fluroxypyr, 30)	GC-ECD	2001	177
Chloroacetanilides	Surface water	CW-DVB	0.3 μg/l	—	LC-UV	2000	163
Thiocarbamates and triazines	Surface and groundwater	CW-DVB	10 to 60 ng/l	<10	GC-MS	2000	170
Triazines	Soil leachates	PA	<1 μg/l	—	GC-MS	2000	178
Herbicides including triazines	River water	PDMS and PA	0.1 to 3 μg/l (NPD) 0.002 to 0.48 μg/l 4(ECD)	5 to 16 3 to 25	GC-NPD and GC-ECD	2000	179
Amides and chloroacetanilides	Deionized water	CW-DVB	2 to 15 ng/l	4 to 12	GC-MS/MS	1999	180
Triazines and carbamates	Soil leachates	CW-DVB	0.5 to 10 μg/l	<10	LC-MS	1999	168

Analyte	Matrix	Phase	Concentration		Technique	Year	Ref.
Phenoxy acid herbicides and metabolites	Drinking water	PDMS and PA + derivatization with diazomethane	10 to 30 ng/l	<12	GC-MS	1998	181
Phenoxy acid herbicides	Deionized water	Derivatization with benzyl bromide + PDMS	0.1 to 1 μg/l	14 to 42	GC-MS	1998	182
Herbicides including triazines	Tap and river water	PA	2 to 20 ng/l	10 to 24	GC-MS (SIM)	1998	183
Triazines	Well and stream water	PDMS	<U.S.A EPA 507 detection limits	—	GC-MS/MS and GC-NPD	1996	184
60 pesticides including chloroacetanilides	Groundwater	PDMS and PA	~ng/l	—	GC-MS	1996	160

In some cases SFE, MAE, or PFE cannot be used because of the weakness of the targeted compounds. An analytical method based on a "soft" extraction procedure was applied to the extraction and quantification of the OXA and ESA acid metabolites of the acetochlor and metolachlor herbicides in soil samples.[29] The extractions were performed by using 50 or 100 g of soil and a solvent extraction method with a mixture of acetonitrile/water (60/40) in acid medium. Each of the four different soil matrixes was spiked in triplicate with standards of each degradation compound, at three concentration levels between 2 and 80 μg/kg. The average recoveries range from 90 to 120% for both metabolites, with relative standard deviations lower than 15%. The limits of detection were about 0.5 and 2 μg/kg for the ESA and the OXA metabolites, respectively.[29]

Sample preparation and chromatographic analysis of acidic herbicides in soils and sediments were recently reviewed by Macutkiewicz et al.[187]

Some relevant applications of SPME, SFE, PFE, and MAE for the extraction of herbicides in solids are presented in Table 26.5.

1. Supercritical Fluid Extraction

Temperature and pressure thresholds at which gases become fluids are named the critical parameters. Physical properties of these supercritical fluids give them many advantages over liquids like a lower viscosity that aids the penetration of the fluid into the solid matrix, enhancing extraction efficiency.[203] Several gases can be used in their supercritical state to extract analytes from solid matrixes. These include N_2O, pentane, CO_2, and NH_3. Organic solvents are often added to these gases to increase their polarity. Some workers studied the SFE of herbicides in soils by using CO_2 modified with methanol at temperature close to 50°C (201).

In case of phenoxyacetic acids, an ion-pairing or derivatization reagent may be added to enable their extraction. With these herbicides, stronger modifications of the fluid should be used, all owing complex formation or *in situ* derivatization prior to the extraction.[204] Another study reported methanol or the mixture acetone–water–triethylamine (90/10/1.5 v/v/v) to enhance extraction of 2,4-D from soils.[205–207] Subcritical water extraction in continuous mode at a flow rate of 1 ml/min and 85°C was used for the determination of acidic herbicides in soils. Leaching, filtration, preconcentration steps, and chromatographic separation were coupled. Recoveries of targeted compounds ranged between 94 and 113%.[192]

Regarding triazines, methanol modified CO_2 enabled the extraction of atrazine, deethylatrazine, and deisopropylatrazine from spiked sediment samples,[208] while methanol containing 2% (v/v) water was efficient for atrazine and 2-hydroxyatrazine in a spiked soil (4% organic matter).[209] However, more stringent conditions were required for bound residues. For example, 30% methanol was needed to efficiently extract bound atrazine from a mineral soil, along with high pressure (350 bar) and temperature (125°C).[207]

Recently, a multivariate optimization scheme has been applied to the SFE of residues of atrazine, diuron, and bensulfuron-methyl from soils, using a quadratic model and a central composite design, considering two groups of independent variables (soil environmental variables and SFE parameters).[210] The analyte residence time in the soil was the most significant environmental factor. For aged samples (12 months), the soil organic matter and clay minerals content had a negative effect on the recoveries due to stronger analyte-matrix interactions (especially for bensulfuron-methyl). Considering the SFE parameters, solubility of the pesticides in the fluid was crucial with freshly spiked soils. On the opposite, the diffusion processes were the limiting factor for aged soils. In that case, the extraction was favored upon elevation of the temperature or addition of a modifier. In particular, a surfactant (Triton X-100) was more efficient than acetonitrile or methanol as a modifier, possibly because of a better swelling of the matrix or the formation of nonionic reverse-micelle.[211]

TABLE 26.5
Extraction Techniques for the Determination of Herbicides and Metabolites in Solid Matrixes

Herbicide Family	Matrix	Extraction Technique	Operating Conditions	Extract Treatment	Recovery (%)	RSD (%)	LOD	Detection Technique	Year	Ref.
Triazines, ureas, and metabolites and chloroacetanilides	Soil (15 g)	PFE	30 ml acetone, 100 atm, 60°C, 3 × 5 min	Evaporation to dryness and reconstitution to 1 ml, ethyl acetate	Triazines and chloro-acetanilides: 83 to 120; Ureas: 65 to 120	<15	Triazines and chloroaceta-nilides: 0.3 to 1.6 ng/g; Ureas: 2 to 10 ng/g	GC-MS/MS and LC-MS	2004	188
Chlorophenoxy acid and metabolites	Soil	Ultrasonic extraction	Methylene chloride, 60 min	Filtration, evaporation to dryness + 1 ml methanol	85 to 111	—	0.03 µg/g	LC-UV	2003	189
Triazines	Soil (1 to 10 g)	MAE	25 ml water (1% methanol), 105°C, 3 min	SPME (CW-DVB)	76.1 to 87.2	2.1 to 6.7	2 to 4 ng/g	GC-MS	2003	190
Chloroacetanilide metabolites	Soil (50 g)	Mechanical Shaking	200 ml acetonitrile/water (6:4); 30 min	Centrifugation, evaporation and reconstitution to ~3 ml methanol	90 to 120	<15	0.3 to 0.7 ng/g	LC-MS	2002	29
Chloroacetanilides and triazines	Soil (10 g)	MAE	20 ml acetonitrile, 5 min, 80°C	Centrifugation, evaporation to dryness + 0.2 ml ethyl acetate	>80	<20	1 to 5 ng/g	GC-MS and GC-NPD	2002	191
Phenoxy acids	Soil (5 g)	SWE	Continuous mode, 1 ml/min, 85°C	Preconcentration, C18	94 to 113	0.6 to 6.9	—	LC-UV	2002	192
Phenoxy acids	Soil (10 g)	MAE	50 ml, Water/methanol (5:5), 80°C, 10 min	Online SPE, C18	~80	1 to 9	LOQ 20 to 50 ng/g	LC-UV	2002	193
Chlorophenoxy acids and their esters	Soil (1 g)	Ultrasonic extraction	5 ml acetonitrile, 3 × 5 min	Centrifugation, and dilution to 100 ml deionized water	72 to 97	1 to 4	0.3 to 0.5 µg/g	LC-UV	2002	194

Continued

TABLE 26.5
Continued

Herbicide Family	Matrix	Extraction Technique	Operating Conditions	Extract Treatment	Recovery (%)	RSD (%)	LOD	Detection Technique	Year	Ref.
Metribuzin and metabolites	Soil (40 g)	PFE	35 ml methanol/water (75:25), 60°C	Filtration	50 to 75	—	1.2 to 12.5 ng/g	LC-MS/MS	2002	195
Chlorophenoxy acid and metabolites	Soil (1.25 g)	Mechanical Shaking	25 ml KOH (0.5 M), 60 min	pH 2.3, Centrifugation, cleanup, C18	75 to 91	4 to 10	~0.3 ng/g	LC-MS/MS	2001	119
Chloroacetanilides and nitrogen heterocyclic herbicides	Soil (5 g)	PFE	Pretreatment with 37.6% of water. 32 ml acetone, 1500 psi, 100°C	Dried over Na_2SO_4, filtered, evaporated to 1 ml, and adjusted to 5 ml with MTBE	93 to 103	1 to 7	—	GC-MS	2000	196
Thiocarbamates and Triazines	Soil (5 g)	SPME (CW-DVB) and MAE	20% of power, 5 ml methanol, 1.5 min	Centrifugation and 2 ml of extract diluted to 20 ml water before SPME	77 to 86	<10	1 to 10 ng/g	GC-MS	2000	170
Triazines	Soil (1 to 4 g)	MAE	600 W, 30 ml water or organic solvents	Centrifugation, evaporation to dryness + 0.5 ml hexane	89 to 101	2.5 to 7.5	—	GC-NPD	1999	197
Chloroacetanilides and triazines	Soil (2 × 20 g)	PFE	30 ml methanol, 10—Mpa, 125°C, 15 min	Evaporation to dryness and reconstitution to 0.25 ml, ethyl acetate	47 to 99	2.7 to 39.4	0.1 to 0.5 ng/g	GC-NPD	1999	198
Chlorophenoxy acid	Soil (2 g)	SFE	CO_2 + 10% TMPA in methanol, 400 atm, 80°C, 15 min	5 ml methanol	69 to 89	—	—	GC-MS and GC-ECD	1998	199

Triazines and metabolites	Sediment	SFE	CO_2 + 10% methanol containing 2% water, 300 atm, −65°C	—	>	—	LC-MS	1996	200
Triazines and ureas	Soil	SFE	CO_2 + 10% methanol, 50°C, 250 kg/cm^2, 45 min	3 to 4 ml methylene chloride	60 to 73	—	LC-UV	1996	201
Triazines, chloroacetanilides, and 2,4-D	Soil	SFE	CO_2 + 5% methanol, 80°C, 400 atm, 30 min	2 ml methanol	—	—	Enzyme immunoassay	1994	202

2. Microwave Assisted Extraction (MAE)

By using MAE, the organic solvent and the sample are subjected to radiation from a magnetron in either a sealed vessel (pressurized) or an open vessel under atmospheric pressure.[212] Pressurized MAE allows a large number of samples to be extracted simultaneously, whereas atmospheric pressure MAE is limited to six simultaneous vessels. The main drawback of this technique is that the solvent needs to be removed from the sample matrix upon extraction completion prior to analysis.

MAE has been applied to the extraction of triazines and their metabolites in soils in closed-vessel systems.[197,213] Xiong and coworkers compared water, methanol, acetone–hexane (1:1), and methylene chloride for the extraction of triazines in soils. They showed that water was as efficient as an organic solvent.[197] Quantitative triazine recoveries were achieved (97 to 103%) for freshly spiked soils as well as for aged soils (300 days) by using methylene chloride/methanol (90:10).[214] In addition, the results were similar to those from classical extractions, showing that the analytes were not degraded under microwave energy. Besides, MAE extracts could be analyzed directly by gas chromatography without any cleanup step. However, for soils containing more than 5% organic matter, performance of a cleanup step was advisable for better chromatographic performance.[215] Methylene chloride allowed the extraction of triazines from spiked soil samples using a focused microwave-assisted Soxhlet system.[216] The use of aqueous solvents alone, instead of the common mixtures water/methanol or water/acetonitrile, seemed to yield cleaner extracts. Another MAE procedure was applied to the extraction of triazines and chloroacetanilides in soils containing 1.5% and 3% of organic matter. Acetonitrile was used for 5 min at 80°C and mean recoveries were above 80% with detection limits of 5 ng/g.[191] MAE applied to triazine extraction from soils was also investigated by using water containing 1% of methanol as extractant at 105°C for 3 min. Recoveries were included between 76 and 87%, with low detection limits close to 3 ng/g.[190]

MAE parameters have been optimized for the extraction of phenoxyalcanoic acid herbicides in soils, with two levels of organic matter (1.5 and 3.5%). Herbicide recoveries were around 80% in both cases with detection limits close to 30 ng/g.[193]

Imidazolinone herbicides (especially imazethapyr) have been extracted from soils. Water alone gave low recoveries from soils and among the extractants tested. The mixture water/triethylamine (TEA) gave excellent extractability, even though matrix materials (mainly humic acids) were also extracted.[217] However, to achieve good extraction along with high selectivity, the use of ammonium acetate buffer (pH 10) was recommended.

3. Pressurized Fluid Extraction

PFE uses organic solvent under pressure and relatively high temperatures to sequentially extract organic pollutants from solid matrixes. There are many advantages related to PFE which requires little solvent remaining in liquid state at high temperatures. The viscosity decrease gives rise to a best diffusion into the matrix and the desorption kinetics of the compounds are increased. Typically, the sample cell vessel is filled with the chosen solvent, the cell is heated to a preset temperature and pressure, and held between 5 and 10 min. Static valves are released and clean solvent is passed through the cells to recover the extracted substances.

There are fewer applications of PFE to the extraction of herbicides in soils,[185,218] being mainly used until now for PAH and PCB extractions. Guzella and Pozzoni performed one of the first successful studies with PFE, by comparing it with Soxhlet procedures on triazine and chloroacetanilide extractions from agricultural soils.[198] A PFE method was developed to extract metribuzin and three metabolites (deaminometribuzin, deaminodiketometribuzin, and diketometribuzin) with methanol–water (75:25) at 60°C. Recoveries were about 75% with detection limits of 1 ng/g, except for diketometribuzin with only 50% of recovery and detection limit higher than 10 ng/g.[195]

Zhu et al. have shown that water is the most effective modifier of PFE for quantitative recoveries of alachlor, metribuzin, and hexazinone in four Hawaiian clayey soils.[196]

In the frame of the PEGASE project,[59] PFE was optimized (Doehlert design) to perform the extraction of chloroacetanilides, ureas, triazines, and their metabolites from soils and solids recovered from the deeper unsaturated zone (between 5 and 100 cm in depth). Extractions were performed by mixing 15 g of dried soil with 30 ml of acetone and by taking into account the overall PFE procedure a set of 24 samples could be processed in 6 h.[188] Each of the five representative soil matrixes used as blanks (especially selected as a function of the depth) was spiked in triplicate with standards of each parent and degradation compound at three concentration levels (5, 30, and 90 μg/kg). For each experiment, isoproturon-D6, atrazine-D5, and pretilachlor were used as surrogates for the three pesticide families of concern. The excellent average recovery for the surrogate atrazine D5 (87%) after 57 extraction experiments argues for the method reliability.

V. ANALYTICAL METHODS

As mentioned before, analytical methods required for herbicide determination must be very sensitive, selective, and robust. Normalized methods generally use liquid and gas chromatography techniques with detectors more or less specific. Sample pretreatment such as derivatization steps or cleanup of the extracts are sometimes mandatory prior to analytical measurement.

A. PRETREATMENT

1. Cleanup of the Extracts

Cleanup steps sometimes follow extraction procedures (e.g., water with a high content of organic matter), in order to remove coextracted substances that can interfere with the measurement of herbicides of interest. Among the most used cleanup methods of the extracts, we can emphasize:

- Extract percolation on solid phases such as silica gel, alumina, alumina silicate, which retain interfering substances by adsorption mechanism;
- Liquid–liquid partition based on the affinity differences for extracted substances in several solvents;
- Removal of coextracted molecules separated by their sizes, with gel permeation.

2. Derivatization

In the case where liquid chromatography is not available, acidic herbicides need to be derivatized because they can dissociate in water and are not usually volatile to be analyzed by gas chromatography. The basic methods used for chlorophenoxy acid herbicides are esterification, silylation, and alkylation, as described in a recent exhaustive review.[111] The derivatization step is performed after preconcentration and cleanup. The step consists of the formation of esters and ethers from the carboxyl and phenol groups of the acidic herbicides.[111] A lot of reagents and chemical mechanisms can be used to perform derivatization reactions. The most employed derivatization reagents are diazomethane, methyliodide, trimethylsulfonium (or anilinium) hydroxide, bis (trimethylsilyl) trifluoroacetamide (BSTFA), pentafluorobenzyl bromide, and anhydride acetate. It should be noted that explosive and hazardous diazomethane was replaced by safer agents. Authors also underline that surface water generally contains humic substances, which can interfere with the derivatization reaction.[111]

Another way of performing derivatization is to undertake an *in situ* esterification followed by an in-vial liquid–liquid extraction, using dimetylsulfate (DMS) for methylation and tetrabutylammonium salts as ion pairing agents. The miniaturization of both methylation and extraction steps could be implemented because of the use of large volume oncolumn injection and mass spectrometric detection.[219]

B. LIQUID CHROMATOGRAPHY–UV SPECTROMETRY

High performance liquid chromatography coupled with UV and fluorometric detections were the most used methods until the middle of the 1990s. These techniques are well suited to polar and thermolabile, but not to volatilizable substances, such as chlorophenoxy acid herbicides or phenylureas.[110,194,220] 2,4-D and its metabolites (i.e., 2,4-DCP) were determined in soil extract samples by LC-UV, and detection limits were about 20 ng/g.[189] However, in order to validate the results, an additional analysis on another column of different polarity is needed. Therefore, to comply with the quality control criteria, new sensitive analytical techniques had to be implemented to identify these herbicides, without the need to duplicate the analysis. Thus, high performance liquid chromatography coupled with diode array detection (DAD) became one of the methods of choice in routine laboratories. As in the case of the previous cited methods, this spectrometric method only provides UV spectra of each eluted compound, but the only required identification criterion consists of checking the purity of each substance. In the case of the monitoring of herbicides in surface water, good correlations were observed between the results obtained in LC-DAD and LC-MS for 60 water samples.[221] Very satisfying detection limits, ranging from 65 to 280 ng/l, have been reached for the determination of chlorotriazines, methylthiotriazines, and methoxytriazines in water with SPE-LC-DAD.[112]

C. THIN LAYER CHROMATOGRAPHY

Thin layer chromatography (TLC) and high-performance thin layer chromatography (HPTLC) complement other methods used for herbicide residue determinations[222] because of their specific advantages:

- High sample throughput and low operating costs
- Selective and sensitive detection and identification with numerous chromogenic, fluorogenic, and biological reagents
- High resolution and accurate quantification achieved on HPTLC plates.

Thin layer radiochromatography (TLRC) is used for the study of metabolism and breakdown of pesticides in the environment. For the characterization and separation of metabolites, TLC is usually combined with UV, IR, and MS. A quantitative HPTLC method was validated for assessment of photodegradation of bensulfuron-methyl on silica gel.[222] This method uses silica gel 60-F with irradiation with a sunlight-stimulating xenon arc lamp, a mobile phase of methylene chloride–acetone–methanol 9 M aqueous ammonia (45/15/10/1), and scanning of fluorescence-quenched zones of samples and standards at 240 nm. The overall method, faster and less expensive than LC, is suitable for the estimation of photodegradability of some herbicides in the adsorbed state in the environment. The binding of amino acids to 2,4-dichlorophenoxyacetic acid (2,4-D) was studied by charge transfer TLC performed on diatomaceous layers covered with different amounts of 2,4-D and salt solutions as mobile phases. Principal component analysis proved that salt effect was negligible and that concentration of 2,4-D had the highest impact on the interaction. This result suggests that amino acid residues account for the binding of 2,4-D to proteins and can play an important role in detoxification processes by forming conjugates with 2,4-D.[223]

D. CAPILLARY ELECTROPHORESIS

Capillary Electrophoresis (CE) is a fast and powerful technique for a variety of compounds.[107] It was much more used in biochemistry than in environmental chemistry. Although CE is not a very sensitive technique, it has been applied for the separation and quantification of herbicides at legal levels after preconcentration procedures.[224-226] This low-cost technique gives very good separation in a short time, in particular for compounds with enantiomeric isomers. Moreover, CE can be more easily coupled to mass spectrometry due to the improvement of electrospray ionization interfaces. In 1984, micellar electrokinetic capillary chromatography (MEKC) was introduced by Terabe[227] to separate neutral molecules. In this mode, a surfactant is added to the buffer at a concentration above its critical micellar concentration. Separation is based on the differential partition between the micelle and the surrounding aqueous phase, diode array being the detection method. Concentrations and types of organic solvents and organic modifiers are very influential on the optimization of the efficiency and separation selectivity. Several authors[123,228] have proposed the separation of some triazines and metabolites by CE and MEKC. The latter combined with SPE procedure for the analysis of seven triazines was validated with a "water certified reference material" (CRM 606).[229] More recently, 11 triazine compounds, 13 triazine and urea herbicides as well as phenoxy acid herbicides were separated and quantified by MEKC DAD after solid-phase extraction.[107,109,230] Capillary zone electrophoresis (CZE) in conjunction with MEKC was applied to the determination of hydrophilic degradation products of hydroxytriazines in water, after SPE on a styrene divinyl benzene sorbent.[123] The separation and determination of chloro- and methylthiotriazines in natural water samples by both MEKC and nonaqueous CZE were compared.[231] After SPE on PSD-DVB sorbents, electropherograms obtained with nonaqueous CZE contained less interference than those acquired with MEKC.

E. MASS SPECTROMETRIC METHODS

1. Gas Chromatography/Mass Spectrometry

Mass spectrometry coupled with gas chromatography (GC-MS) is a very powerful tool to identify and quantify a broad variety of thermally stable herbicides in complex environmental matrixes, as shown by the numerous literature published in the recent years.[24,101,113,116,123,125,153,160,170,172,173,176,178,181,182,190,191,199,219,232]

The most widely used ionization technique in GC-MS is the electronic impact (EI) mode where a specific fragmentation pattern is obtained for each targeted molecule. This mass spectrum allows identification of each compound of the complex mixture, in particular by comparing these spectra with library spectra acquired in full scan mode. The identity of each molecule must be firmly established before starting quantitation procedures in order to avoid the so-called "false positive" effect. The chemical ionization (CI) mode can also be used to identify herbicides with halogen substituents by providing better sensitivity than electronic impact ionization. When using the EI mode, quantification of herbicides is performed by summing the main fragments of mass/charge ratio m/z, typically up to 3 m/z for each compound, relatively to an internal standard carefully chosen. During the last few years, a tendency towards the use of GC-MS in selected ion monitoring mode (SIM) has been observed for quantitative purposes, without the need of confirmatory techniques usually required after GC-NPD (nitrogen phosphorus detection) or GC-ECD (electron capture detection) modes. However, SIM has to be used with caution and should always be associated with monitoring in full scan mode.

Tandem mass spectrometry (MS/MS), offering a high degree of sensitivity and selectivity, enables the herbicide analysis at trace levels even if interfering compounds, with possibly the same parent mass, are coeluted.[69,106,171,180,184,233] In the MS/MS mode, identification and quantification of compounds are performed based on specific transitions between parent ions and products ions, which confer high selectivity and sensitivity (reduction of the background) and a high degree of certainty for the identification of the compounds of interest. MS/MS analyses can be performed with two kinds of instruments: ion trap and triple quadrupole systems. Ion trap has existed for more than 20 years, but one of the most relevant technological improvements has been the conception of ion traps with external ion sources, very useful and suited to the analysis of extract samples from complex matrixes. Best analytical performance can be obtained with triple quadrupole instruments, the use of which is much extended in the case of coupling with liquid chromatography. Triple quadrupole-based technique is able to perform multiple reaction monitoring and its large dynamic range is favorable for accurate quantitative analyses. Ion trap detectors are the cheapest and the most used systems, but triple quadrupole instruments are now more frequently encountered in analytical labs, in particular because of their dramatic cost decrease. The time of analysis of 72 pesticides, including herbicides, was significantly reduced with low-pressure gas chromatography linked to ion-trap tandem mass spectrometry; 32 min instead of 72 min with a conventional GC-MS/MS system.[233]

2. Liquid Chromatography/Mass Spectrometry

In conjunction with the development of numerous solid-phase extraction procedures (see Section IV.B) for the main herbicide families, the major instrumental improvement arose from the implementation of robust atmospheric ionization interfaces, such as particle-beam (PB), thermospray (TS), electrospray ionization (ESI), atmospheric pressure chemical ionization (APCI), and atmospheric pressure photo ionization (APPI).[234] Most atmospheric pressure ionization (API) mass spectrometers now offer both ESI and APCI interfaces, in negative and positive ionization modes. ESI is more suited to polar and ionic compounds, whereas APCI is used for moderately nonpolar compounds. For nonionic and nonpolar molecules, a third ionization mode (APPI) introduced by Robb[235] became recently commercially available, but very few papers have been devoted to this promising technique.[236,237] During the past five years, there has been an increase in the scientific publications dealing with the coupling of liquid chromatography with mass spectrometry (LC-MS) for the determination of herbicides in the environment.[27–29,92,100,108,116, 127–129,168,175,238,239] Reverse-phase liquid chromatography is the most used technique in LC-MS (or tandem LC-MS/MS), in which acidic herbicides, such as chloroacetanilide metabolites and phenoxy acidic compounds, are commonly analyzed by ESI-MS by adding acetic or formic acid or ammonium salts in the mobile phase. On the contrary, triazines, ureas, and their main metabolites are analyzed by using an APCI-MS system, often without any modificator in the mobile phase, consisting of gradient of methanol/acetonitrile/water.

Figure 26.1 shows a chromatogram corresponding to the separation of 29 herbicides in spiked water, in positive APCI-MS full scan mode.

As for gas chromatography coupling systems, tandem mass spectrometry is now the method of choice for identification and quantitation purposes of herbicides, with ion trap or triple-quadrupole based systems.[25,115,117–120,126,195,221,240–243] Although triple quadrupole MS/MS is more sensitive and robust, ion trap detection can be the method of choice to many identification and screening purposes. The ion trap instrument can provide full scan mass spectra in MS^n mode, which cannot be achieved with a triple quadrupole system. The second-order spectra acquired by the fragmentation of specific parent ions allow building up a database as for GC-MS applications.

Two reviews have been recently devoted to the determination of herbicides in water by LC-MS (/MS),[244,245] highlighting the growing interest of the researchers in this technique.

Analysis of acetochlor metabolites (ESA and OXA) in groundwater was carried out by reverse-phase liquid chromatography and detection by electrospray ionization/mass spectrometry (ESI/MS) in single ion monitoring and negative modes.[92]

Figure 26.2 shows chromatograms for the separation of four chloroacetanilide metabolites in an extract sample from a spiked groundwater, in negative ESI-MS mode.

The LC separation of these herbicides exhibits two peaks for each acetochlor metabolite, corresponding to the diastereomers only partially separated at room temperature. By working at 60°C, these diastereomers could be eluted in one single peak for each metabolite.

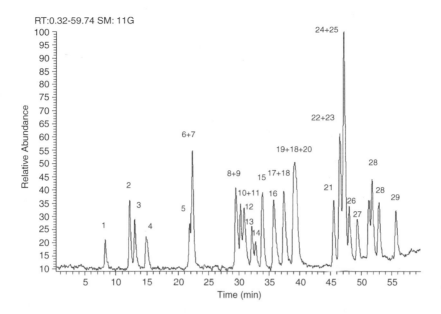

FIGURE 26.1 Full scan LC-APCI-MS of 1 l of water spiked with 5 µg/l of each herbicide. LC conditions: LC column Chromspher Pesticides, 250 mm × 4.6 mm × 5 µm. Gradient profile (water 85%/acetonitrile 15%): 15 to 60% in 45 min and then to 100% in 15 min. Flow rate: 1 ml/min 20 µl injected. Full scan MS chromatogram was obtained by scanning the single quadrupole (SSQ 7000 Thermo Finnigan) between m/z 120 and 650 in 1 s. Peaks: 1. DIA; 2. metamitron; 3. DEA; 4. carbendazim; 5. cyanazine; 6. simazine + 7. metribuzin; 8. atrazine + 9. chlortoluron; 10. secbumeton; 11. methabenzthiazuron; 12. desmetryn; 13. isoproturon; 14. diuron; 15. metazachlor; 16. terbumeton; 17. ametryn; 18. triadimenol (1); 19. terbutylazine; 18. triadimenol (2); 20. linuron; 21. napropamide; 22. metolachlor + 23. penconazole; 24. tebuconazole; 25. flusilazole; 26. neburon; 27. hexaconazole; 28. propiconazole; 29. prochloraz.

In order to determine ureas and metabolites of triazines and ureas in agricultural soils, reverse-phase liquid chromatography was used with atmospheric pressure chemical ionization/mass spectrometry (APCI/MS), in positive mode.[188] Gas chromatography/ion trap mass spectrometry was used in MS/MS mode to analyze the parent compounds of the triazines and chloroacetanilides. Method performance was much better in GC-MS/MS than in LC-MS, and acetochlor LOQ was dramatically improved by using GC-MS/MS, this herbicide being very poorly ionized into the APCI interface.

The use of mass spectrometry/negative chemical ionization in SIM mode, coupled with liquid chromatography by means of a particle beam interface, allowed sensitive determination of polar herbicide residues in soils at the µg/kg level.[246] Thurman et al. made a complete comparative study on 75 pesticides to rationalize the selection of either ESI or APCI.[247] They pointed out that the suppression of the dissociation of ionic molecules for enhanced chromatographic retention might reduce ESI-MS sensitivity. However, since the pH at the surface of an electrospray droplet may differ significantly from the pH in the eluent solution, it is difficult to predict the molecule behavior in the ESI interface. Twelve chloroacetanilide degradates were quantitatively extracted from tap water and a novel liquid chromatographic separation of sulfonic and OXA metabolites of acetochlor and alachlor was developed. Authors used a gradient of ammonium acetate–methanol combined with heating of the analytical column at 70°C, before ESI-MS/MS detection by means of a triple quadrupole instrument.[100] Five polar herbicides were separated using high-speed analytical countercurrent chromatography with a standard isocratic biphasic solvent system of hexane/ethyl acetate/methanol/water

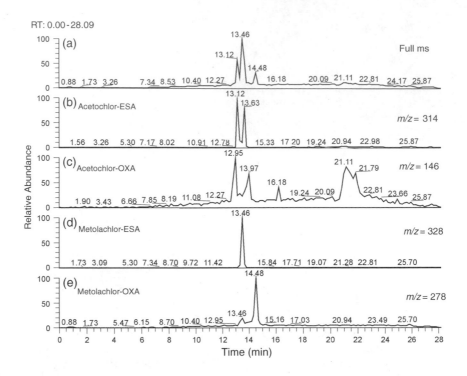

FIGURE 26.2 (a) Chromatogram from a LC-ESI full scan MS analysis of a groundwater sample spiked with 0.8 μg/l of ethane sulfonic acids and oxanilic acids of acetochlor and metolachlor. (b) Reconstructed ion chromatogram for a LC-ESI full scan MS analysis (at m/z 314) of a groundwater sample spiked with 0.8 μg/l of acetochlor ethane sulfonic acid. (c) Reconstructed ion chromatogram for a LC-ESI full scan MS analysis (at m/z 146) of a groundwater sample spiked with 0.8 μg/l of acetochlor oxanilic acid. (d) Reconstructed ion chromatogram for a LC-ESI full scan MS analysis (at m/z 328) of a groundwater sample spiked with 0.8 μg/l of metolachlor ethane sulfonic acid. (e) Reconstructed ion chromatogram for a LC-ESI full scan MS analysis (at m/z 278) of a groundwater sample spiked with 0.8 μg/l of metolachlor oxanilic acid.

in reverse phase, in conjunction with online negative ESI-MS with a triple quadrupole instrument.[248]

A comparison between APCI-LC/MS and PB-LC/MS was conducted for the determination of a priority group of herbicides in water, including phenylureas, triazines, and chlorophenoxy acids. The potential of both ionization techniques for the monitoring of environmental matrixes was demonstrated. However, APCI was approximately ten-fold more sensitive than PB, both in negative and positive ionization mode.[242]

In a SPE-LC-APCI-MS/MS study of 37 polar herbicides in water, (mainly triazines, phenylureas, and phenoxy acids), a 5- to 50-fold improved detectability was observed for the combined acetonitrile desorption–methanol gradient procedure, compared to the traditional "acetonitrile only" approach.[241]

A recent study investigated the effects of matrix interferences on the analytical performance of a LC-MS/MS triple quadrupole system for the determination of acidic herbicides in water. Salinity provided a dramatic decrease in response for early eluting compounds, but LC–LC coupled column configuration efficiently eliminated the matrix effect. This approach of LC–LC-ESI-MS/MS using a Supelco ABZ + (amide modified) as a second column was the most favorable as regards matrix effects, with reliable quantification of herbicides at the level 0.4 μg/l.[243]

3. High Resolution Mass Spectrometry

Higher resolution for identification purposes is provided by the orthogonal accelerating time-of-flight (TOF) mass spectrometers and their resolving power is sufficient to give a molecular formula confirming or denying a suggested structure. In terms of qualitative analysis, the TOF instrument is also very convenient because it can provide full scan spectra with high sensitivity. The Q-TOF MS (MS) combines the simplicity of a quadrupole instrument with the ultra high efficiency of a TOF mass analyzer. The TOF side of Q-TOF MS/MS achieves simultaneous detection of ions across the full mass range at all times. In contrast, tandem quadrupoles must scan one mass at a time and for this reason Q-TOF MS/MS are more sensitive than the third quadrupole of the triple quadrupole MS/MS. The TOF side of the Q TOF instrument has the same sensitivity in scan mode as in SIM (single reaction monitoring) mode. This is not true for the triple quadrupole which needs to work in MRM (multiple reaction monitoring) mode to increase sensitivity. For screening purposes, it has been shown that TOF instruments working at their higher resolutions allowed the unambiguous detection of some pesticides in river water, even accompanied by isobaric compounds.[249] Only few references exist in the literature regarding the application of LC-TOF-MS to the analysis of herbicides and metabolites in the environment.[250–252] Analyses of ESA degradation products of alachlor and acetochlor in groundwater were reported by Thurman, using TOF/MS in both negative and positive ionization modes.[253] The same authors address the discovery of new secondary amide metabolites (ethane sulfonic acid degradates) of alachlor and acetochlor in water, by using a LC-TOF ion trap MS/MS system.[254] Photodegradation products of diuron and exact mass measurements of phenylureas and carbamates were investigated by LC-TOF-MS instruments.[250,255]

Hogenboom and his coworkers showed the applicability of accurate mass measurements by LC-TOF-MS, to the analysis of herbicides in surface water.[249] They also investigated the photodegradation products of alachlor in water by this powerful technique.[256]

The possibility of screening surface water and identifying unknown compounds by Q-TOF/MS was explored.[257] The authors developed a model based on known herbicides, including atrazine and metribuzin. They successfully applied their model for the structural elucidation of three unknown compounds in native surface water.

Finally, Q-TOF/MS was very recently applied to study the transformation products of terbutylazine, simazine, terbutryn, and terbumeton in water after UV exposure.[258] MS/MS helped to elucidate structures of degradation products (even those with only 2% of the total peak area) and to differentiate isomeric metabolites.

4. Perspectives on Mass Spectrometry for Herbicide Analysis

Some trends can be drawn according to the exhaustive literature and recent reviews[234,244,245] cited in this chapter, about the determination of herbicides in the environment by mass spectrometric methods. Among the overrepresented studies on triazines, phenylureas, and phenoxy acids, there is still a lack of reliability and robustness of the published methods for the phenoxy herbicides.

Even with the new sophisticated MS/MS techniques, sample volumes still have to be included at least between 5 and 10 ml. Therefore, a trace-enrichment step is usually performed and online procedures can be recommended. GC procedure still offers better separation efficiency and could be the method of choice if a derivatization step is not required.

As regards identification purposes, the use of exhaustive libraries in GC/MS became a routine operation in most laboratories. In contrast, these libraries are still scarce in LC-MS applications, mainly because ESI-MS and APCI-MS spectra are strongly influenced by the instrument settings, the LC conditions, and the sample type. Therefore, much more effort should be paid to define identification and confirmation criteria for herbicide analysis by these ionization techniques. In this way, a confirmation criterion was successfully evaluated on triazines by Flow Injection Analysis

(FIA)-APCI-MS/MS. This procedure consisted of comparing four diagnostic transitions for each analyte in a real sample, with the relative ion abundance of the same transitions in a standard solution.[259]

Speed and sensitivity of recent TOF-MS/MS methods have been highlighted earlier in this chapter. Nevertheless, analytical chemists should always maintain a critical view to interpret the structural information obtained with these powerful instruments.

Finally, the combined influence of the LC conditions on the herbicide response, and the ion suppression by matrix effect into the API interface, make the quantification procedure quite tricky. The former needs much more theoretical and experimental data to be clearly understood, whereas the use of "in-matrix" calibrations (standard additions) often helps to take into account matrix effects.

NOMENCLATURE

2,4-D	2,4-dichlorophenoxyacetic acid
2,4-DCP	2,4-dichlorophenol
AFNOR	agence française pour la normalisation
AMPA	aminomethylphosphonic acid
APCI	atmospheric pressure chemical ionization
API	atmospheric pressure ionization
APPI	atmospheric pressure photo ionization
BRGM	bureau de recherche géologiques et minières
BSTFA	bis (trimethylsilyl) trifluoroacetamide
CE	capillary electrophoresis
CI	chemical ionization
CRM	certified reference material
CW-DVB	carbowax-divinylbenzene
CX-PDMS	carboxen-polydimethylsiloxane
CZE	capillary zone electrophoresis
DAD	diode array detection
DAHT	diaminohydroxy-s-triazine
DEA	deethylatrazine
DIA	disopropylatrazine
ECD	electron capture detector
EI	electronic impact
EPA	environmental protection agency
ESA	ethane sulfonic acid
ESI	electrospray ionization
FAO	united nations food and agricultural organization
GAC	granular activated carbon
GCB	graphitized carbon black
GC–MS/MS	gas chromatography–tandem mass spectrometry
GC–MS	gas chromatography–mass spectrometry
HA	hydroxyatrazine
HAL	health advisory levels
HPLC	high performance liquid chromatography
HPTLC	high performance thin layer chromatography
HRMS	high resolution mass spectrometry
HVS	high volume sampler
IR	infrared
K_d	solid–water distribution ratio

K_{oc}	soil organic carbon sorption coefficient
K_{ow}	octanol–water partition coefficient
LC–MS/MS	liquid chromatography–tandem mass spectrometry
LC–MS	liquid chromatography–mass spectrometry
LLE	liquid–liquid extraction
LOD	limit of detection
LOQ	limit of quantification
LPME	liquid phase microextraction
MAE	microwave assisted extraction
MCL	maximum contamination level
MEKC	micellar electrokinetic capillary chromatography
MIP	molecular imprinted polymer
MRM	multiple reaction monitoring
MTBE	methyl tertiary-butyl ether
MVS	medium volume sampler
NPD	nitrogen phosphorus detector
OXA	oxanilic acid
PA	polyacrylate
PB	particle beam
PDMS	polydimethylsiloxane
PEARL	pesticide emission assessment at regional and local scale
PELMO	pesticide leaching model
PFE	pressurized fluid extraction
PS–DVB	polystyrene–divinylbenzene
PTV	programmed temperature vaporizer
Q-TOF	orthogonal accelerating time-of-flight
RP	reverse phase
SBSE	stir bar sorptive extraction
SFE	supercritical fluid extraction
SIM	single reaction monitoring
SPE	solid-phase extraction
SPME	solid-phase micro extraction
SWE	subcritical water extraction
TLC	thin layer chromatography
TLRC	thin layer radiochromatography

REFERENCES

1. Barceló, D. and Hennion, M. C., *Introduction, Trace Determination of Pesticides and their Degradation Products in Water*, Elsevier, Amsterdam, NL, 1997.
2. Kalkhoff, S. J., Kolpin, D. W., Thurman, E. M., Ferrer, I., and Barcelo, D., Degradation of chloroacetanilide herbicides: the prevalence of sulfonic and oxanilic acid metabolites in Iowa groundwater and surface water, *Environ. Sci. Technol.*, 32, 1738–1740, 1998.
3. Finizio, A., Vighi, M., and Sandromi, D., Determination of *n*-octanol/water partition co-efficient (Kow) of pesticide critical review and comparison of methods, *Chemosphere*, 34, 131–161, 1997.
4. Dousset, S., Mouvet, C., and Schiavon, M., Sorption of terbuthylazine and atrazine in relation to the physico-chemical properties of the three soils, *Chemosphere*, 28, 467–476, 1994.
5. Dousset, S., Mouvet, C., and Schiavon, M., Leaching of atrazine and some of its metabolites in undisturbed field lysimeters of three soil types, *Chemosphere*, 30, 511–524, 1995.

6. Sabik, H., Jeannot, R., and Rondeau, B., Multiresidue methods using solid-phase extraction techniques for monitoring priority pesticides, including triazines and degradation products, in ground and surface water, *J. Chromatogr. A*, 885, 217–236, 2000.

7. Wauchope, R.D., Pesticides in runoff: measurement, modeling, and mitigation, *J. Environ. Sci. Health* 31, Part B: Pesticides, Food Contaminants, and Agricultural Wastes, 31, pp. 337–344, 1996.

8. Wienhold, B. J., Sadeghi, A. M., and Gish, T. J., Effect of starch encapsulation and temperature on volatilization of atrazine and alachlor, *J. Environ. Qual.*, 22, 162–166, 1993.

9. Pesticides in Water, Fifth Annual Report, 2001 Data, IFEN, July 2003, ISBN: 2-911089 58-8.

10. Barbash, J. E., Thelin, G. P., Kolpin, D. W., and Gilliom Major, R. J., Herbicides in Ground Water: Results from the National Water-Quality Assessment, *J. Environ. Qual.*, 30, 831–845, 2001.

11. Thurman, E. M. and Meyer, M. T., Introduction and overview, Ch 1, In *Herbicide Metabolites in Surface Water and Groundwater*, Meyer, M. T. and Thurmann, E. M. Eds., American Chemical Society Symposium Series, p. 318, 1996.

12. Kalkhoff, S. J., Lee, K. E., Porter, S. D., Terrio, P. J., and Michael Thurman, E., Herbicides and herbicide degradation products in upper midwest agricultural streams during August base-flow conditions, *J. Environ. Qual.*, 32, 1025–1035, 2003.

13. Lerch, R. N., Donald, W. W., Li, Y.-X., and Alberts, E. E., Hydroxylated atrazine degradation products in a small Missouri stream, Ch 19, In *Herbicide Metabolites in Surface Water and Groundwater*, Meyer, M. T. and Thurmann, E. M. Eds., American Chemical Society Symposium Series, p. 318, 1996.

14. Kolpin, D. W., Thurman, E. M., and Linhart, S. M., The environmental occurrence of herbicides: the importance of degrades in ground water, *Arch. Environ. Contam. Toxicol.*, 35, 385–390, 1998.

15. Kolpin, D. W., Thurman, E. M., and Linhart, S. M., Occurrence of cyanazine compounds in groundwater: degrades more prevalent than the parent compound, *Environ. Sci. Technol.*, 35, 1217–1222, 2001.

16. Dupas, S., Guenu, S., Pichon, V., Montiel, A., Welte, B., and Hennion, M. C., Long term monitoring of pesticides and polar transformation products in ground water using automated online trace-enrichment and liquid chromatography with diode array detection, *Int. J. Environ. Anal. Chem.*, 65, 53–68, 1996.

17. Konstantinou, I. K., Zarkadis, A. K., and Albanis, T. A., Photodegradation of selected herbicides in various natural water and soils under environmental conditions, *J. Environ. Qual.*, 30, 121–130, 2001.

18. Sanlaville, Y., Guittonneau, S., Mansour, M., Feicht, E. A., Meallier, P., and Kettrup, A., Photosensitized degradation of terbuthylazine in water, *Chemosphere*, 33, 353–362, 1996.

19. Balinova, A. M., Acetochlor: a comparative study on parameters governing the potential for water pollution, *J. Environ. Sci. Health*, 32, 645–658, 1997.

20. Zheng, H. and Ye, C., Adsorption and mobility of acetochlor and butachlor on soil, *Bull. Environ. Contam. Toxicol.*, 68, 509–516, 2002.

21. Potter, T. L. and Carpenter, T. L., Occurrence of alachlor environmental degradation products in groundwater, *Environ. Sci. Technol.*, 29, 1557–1563, 1995.

22. Hernandez, F., Beltran, J., Forcada, M., Lopez, F., and Morell, I., Experimental approach for pesticide mobility studies in the unsaturated zone, *Int. J. Environ. Anal. Chem.*, 71, 87–103, 1998.

23. Baran, N., Mouvet, C., Dagnac, T., and Jeannot, R., Infiltration of acetochlor and two of its metabolites in two contrasting soils, *J. Environ. Qual.*, 33, 241–249, 2004.

24. Heberle, S. A., Aga, D. S., Hany, R. H., and Müller, S. R., Simultaneous quantification of acetanilide herbicides and their oxanilic and sulfonic acid metabolites in natural water, *Anal. Chem.*, 72, 840–845, 2000.

25. Vargo, J. D., Determination of sulfonic acid degrades of chloroacetanilides and chloroacetamide herbicides in groundwater by LC/MS/MS, *Anal. Chem.*, 70, 2699–2703, 1998.

26. Thurman, E. M. and Snavely, K., Advances in solid-phase extraction disks for environmental chemistry, *Trends Anal. Chem.*, 19, 18–26, 2000.

27. Ferrer, I., Thurman, E. M., and Barceló, D., Identification of ionic chloroacetanilide herbicide metabolites in surface water and groundwater by HPLC/MS using negative ion spray, *Anal. Chem.*, 69, 4547–4553, 1997.

28. Hostetler, K. A. and Thurman, E. M., Determination of chloroacetanilide herbicide metabolites in water using high performance liquid chromatography diode array detection and high performance liquid chromatography mass spectrometry, *Sci. Total Environ.*, 248, 147–156, 2000.

29. Dagnac, T., Jeannot, R., Mouvet, C., and Baran, N., Determination of oxanilic and sulfonic acid metabolites of acetochlor in soils by liquid chromatography-electrospray ionization mass spectrometry, *J. Chromatogr. A*, 957, 69–77, 2002.

30. Rebich, R. A., Coupe, R. H., and Thurman, E. M., Herbicide concentrations in the Mississippi River Basin—the importance of chloroacetanilide herbicide degradates, *Sci. Total Environ.*, 321, 189–199, 2004.

31. Jurado-Exposito, M. and Walker, A., Degradation of isoproturon, propyzamide and alachlor in soil with constant and variable incubation conditions, *Weed Res.*, 38, 309–318, 1998.

32. Chang-ming, Y. E., Xing-jun, W., and He-hui, Z., Biodegradation of acetanilide herbicides acetochlor and butachlor in soil, *J. Environ. Sci.*, 14, 524–529, 2002.

33. Felding, G., Brande Sørensen, J., Bügel Mogensen, B., and Christian Hansen, A., Phenoxyalkanoic acid herbicides in runoff, *Sci. Total Environ.*, 175, 207–218, 1995.

34. Klöppel, H., Kördel, W., and Stein, B., Herbicide transport by surface runoff and herbicide retention in a filter strip — rainfall and runoff simulation studies, *Chemosphere*, 35, 129–141, 1997.

35. Crespín, M. A., Gallego, M., Valcárcel, M., and Luis González, J., Study of the degradation of the herbicides 2,4-D and MCPA at different depths in contaminated agricultural soil, *Environ. Sci. Technol.*, 35, 4265–4270, 2001.

36. Romero, E., Dios, G., Mingorance, M. D., Matallo, M. B., Peña, A., and Sánchez-Rasero, F., Photodegradation of mecoprop and dichlorprop on dry, moist and amended soil surfaces exposed to sunlight, *Chemosphere*, 37, 577–589, 1998.

37. Chérif, S. and Wortham, H., A new laboratory protocol to monitor the volatilization of pesticides from soils, *Int. J. Anal. Chem.*, 68, 199–212, 1997.

38. Haraguchi, K., Kitamura, E., Yamashita, T., and Kido, A., Simultaneous determination of trace pesticides in urban air, *Atmos. Environ.*, 28, 1319–1325, 1994.

39. Trevisan, M., Montepiaini, C., Ragossa, L., Bartoletti, C., Ioannilli, E., and Del Re, A. M., Pesticides in rainfall and air in Italy, *Environ. Pollut.*, 80, 31–39, 1993.

40. Eisenreich, S. J., Looney, B. B., and Thorton, J. D., Airborne organic contaminants in the Great Lakes ecosystem, *Environ. Sci. Technol.*, 15, 30–38, 1981.

41. Van den Berg, F., Kubiak, R., Benjey, W. G., Majewski, M. S., Yates, S. R., Reeves, G. L., Smelt, J. H., and Van der Linden, A. M. A., Emission of pesticides into the air, *Water Air Soil Pollut.*, 115, 195–210, 1999.

42. Bedos, C., Cellier, P., Calvet, R., Barriuso, E., and Gabrielle, B., Mass transfer of pesticides into the atmosphere by volatilization from soils and plants: overview, *Agronomie*, 22, 21–33, 2002.

43. Taylor, A. W. and Spencer, W. F., Volatilization and vapor transport processes, In *Pesticides in the Soil Environment: Processes, Impacts, and Modeling*, Cheng, H. H., Ed., Soil Society of America, Madison, WI, pp. 214–269, 1990, chap. 7.

44. Spencer, W. F. and Cliath, M. M., Movement of pesticides from soil to the atmosphere, In *Long Range Transport of Pesticides*, Kurtz, D. A., Ed., Lewis Publishers, Chelsea, MI, pp. 1–16, 1990.

45. Bidleman, T., Walla, M. D., Roura, R., Carr, E., and Schmidt, S., Organochlorine pesticides in the atmosphere of the Southern Ocean and Antarctica, *Mar. Pollut. Bull.*, 26, 258–262, 1993.

46. Millet, M., Etude de la composition chimique des brouillards et analyse de pesticides dans les phases liquide, gazeuse et particulaire de l'atmosphère. Ph.D. thesis, University of Strasbourg I, 1994.

47. Chevreuil, M., Gamouna, M., Teil, M. J., and Chesterikoff, A., Occurrence of organochlorines (PCBs, pesticides, and herbicides (triazines, phenyluras)) in the atmosphere in the fallout form urban and rural stations of the Paris area, *Sci. Total Environ.*, 182, 25–37, 1996.

48. Sanusi, A., Millet, M., Wortham, H., and Mirabel, P., Atmospheric contamination by pesticides: determination in the liquid, gaseous, and particule phases, *Environ. Sci. Pollut. Res.*, 4, 172–180, 1997.

49. Kuang, Z., McConnell, L. L., Torrents, A., Meritt, D., and Tobash, S., Atmospheric deposition of pesticides to an agricultural waterhed of the Chesapeake Bay, *J. Environ. Qual.*, 32, 1611–1622, 2003.

50. Vanclooster, M., Armstrong, A., Baouroui, F., Bidoglio, G., Boesten, J. J. T. I., Burauel, P., Capri, E., De Nie, D., Fernandez, E., Jarvis, N., Jones, A., Klein, M., Leistra, M., Linnemann, V., Pineros-Garcet, J. D., Smelt, J. H., Tiktak, A., Trevisan, M., Van den Berg, F., Van der Linden, A. M. A., Vereecken, H., and Wolters, A., Effective approaches for predicting environmental concentrations of pesticides: the APECOP project, 2003.

51. Dubus, I. G. and Brown, C. D., Sensitivity and first-step uncertainty analyses for the preferential flow model MACRO, *J. Environ. Qual.*, 31, 227–240, 2002.

52. Woodrow, J. E. and Seiber, J. N., Correlation techniques for estimating pesticide volatilization flux and downwind concentrations, *Environ. Sci. Technol.*, 31, 523–529, 1997.

53. Smit, A. A. M. F. R., Leistra, M., and Van den Berg, F., *Estimation method for the volatilization of pesticides from plants, Environmental Planning Bureau series 4*, DLO Winand Staring Center, Wageningen, The Netherlands, 1998.

54. Stork, A., Witte, R., and Führ, F., A wind tunnel for measuring the gaseous losses of environmental chemicals from the soil/plant system under field-like conditions, *Environ. Sci. Pollut. Res.*, 1, 234–245, 1994.

55. Wolters, A., Kromer, T., Linnemann, V., Schäffe, A., and Vereecken, H., A new tool for laboratory studies on volatilization: extension of applicability of the photovolatility chamber, *Environ. Toxicol. Chem.*, 22, 791–797, 2003.

56. EEC Drinking Water Guideline 80/779/EEC, EEC No. L229/11-29, Brussels, 1980.

57. Munch, D. J. and Frebis, C. P., Analyte stability studies conducted during the national pesticide survey, *Environ. Sci. Technol.*, 67, 3064–3068, 1992.

58. Moreau, C. and Mouvet, C., Stability of isoproturon, bentazone, terbuthylazine, and alachlor in natural groundwater, surface water, and soil water samples stored under laboratory conditions, *J. Environ. Qual.*, 26, 416–424, 1997.

59. PEGASE project, contract EVK1-CT1999-00028, financed by the EU through its 5th PCRDT. (Pesticides in European Groundwater: detailed study of aquifers and simulation of possible evolution scenarios).

60. Millet, M., Wortham, H., Sanusi, A., and Mirabel, P., A multiresidue method for the determination of trace levels of pesticides in air and water, *Arch. Environ. Contam. Toxicol.*, 31, 543–556, 1996.

61. Palm, W. U. and Zetzsch, C., Investigation of the photochemistry and quantum yields of triazines using polychromatic irradiation and UV-spectroscopy as analytical tool, *Int. J. Environ. Anal. Chem.*, 65, 313–329, 1996.

62. Sanusi, A., Millet, M., Wortham, H., and Mirabel, P., A multiresidue method for determination of trace levels of pesticides in atmosphere, *Analusis*, 25, 302–308, 1997.

63. Method EPA TO 4, Determination of pesticides and polychlorinated biphenyls in ambient air using high volume polyurethane foam (PUF) sampling followed by Gas Chromatographic/Multidetector U.S. Environmental Protection Agency.

64. Method EPA TO 10, Determination of Pesticides and Polychlorinated Biphenyls in Ambient Air Using Low Volume Polyurethane Foam (PUF) sampling Followed By Gas Chromatographic/ MultiDetector U.S. Environmental Protection Agency.

65. Niosh 5600, Organophosphorus Pesticides, Fourth Edition 8/15/94.

66. Niosh 5602, Chlorinated and Organonitrogen herbicides (Air Sampling), Fourth Edition.

67. Haraguchi, K., Kitamura, E., Yamashita, T., and Kido, A., Simultaneous determination of trace pesticides in urban precipitation, *Atmosph. Environ.*, 29, 247–253, 1995.

68. Foreman, B., An accelerated solvent extraction GC/MS method for the determination of multiple classes of pesticide degradation products in ambient air. Backround Information. USGS. 2000.

69. Sauret, N., Millet, M., Herckes, P., Mirabel, P., and Wortham, H., Analytical method using gas chromatography and ion trap tandem mass spectrometry for the determination of S-triazines and their metabolites in the atmosphere, *Environ. Pollut.*, 110, 243–252, 2000.

70. Führ, F., Burauel, P., Dust, M., Mittelstaedt, W., Pütz, T., Reinken, G., and Stork, A., Comprehensive tracer studies on the environmental behavior of pesticides: The Lysimeter Concept, In *The lysimeter concept, ACS Symp. Ser. 699*, Führ, F., Hance, R. J., Plimmer, J. R. and Nelson, J. O., Eds., Am. Chem. Soc., Washington, pp. 1–20, 1998.

71. Linnemann, V., Transport of volatile hydrocarbons through an undisturbed soil core into the atmosphere after contamination of the groundwater with the fuel additive methyl-tertbutyl ether (MTBE). Ph.D. thesis, 2002, University of Bonn, Germany.

72. Stork, A., Wind tunnel for the measurement of gaseous losses of environmental chemicals from the soil/plant system under practice-like conditions with direct air analytics using 14C labeled chemicals. Ph.D. thesis, 1995, University of Bonn, Germany.

73. Sabik, H., Fouquet, A., and Proulx, S., Ultratrace determination of organophosphorus and organonitrogen pesticides in surface water, *Analusis*, 25, 267–273, 1997.

74. Headley, J. V., Dickson, L. C., Swyngedouw, C., Crosley, B., and Whitley, G., Evaluation of the Goulden Large Sample Extractor for acidic compounds in natural water, *Environ. Toxicol. Chem.*, 15, 1937–1944, 1996.

75. Psillakis, E. and Kalogerakis, N., Developments in liquid-phase microextraction, *Trends Anal. Chem.*, 22, 565–574, 2003.

76. Shen, G. and Lee, H. K., Hollow fiber-protected liquid-phase microextraction of triazine herbicides, *Anal. Chem.*, 74, 648–654, 2002.

77. Megersa, N. and Joensson, J. A., Trace enrichment and sample preparation of alkylthio-s-triazine herbicides in environmental water using a supported liquid membrane technique in combination with high performance liquid chromatography, *Analyst*, 123, 225–231, 1998.

78. Alzaga, R., Durand, G., Barceló, D., and Bayona, J. M., Comparison of supercritical fluid extraction and liquid–liquid extraction for isolation of selected pesticides stored in freeze-dried water samples, *Chromatographia*, 38, 502–508, 1994.

79. Barnabas, I. J., Dean, J. R., Hitchen, S. M., and Owen, S. P., Selective supercritical fluid extraction of organochlorine pesticides and herbicides from aqueous samples, *J. Chromatogr. Sci.*, 32, 547–551, 1994.

80. Ho, J. S., Tang, P. H., Eichelberger, J. W., and Buddle, W. L., Liquid–solid disk extraction followed by SFE and GC-ion-trap MS for the determination of trace organic pollutants in water, *J. Chromatogr. Sci.*, 33, 1–8, 1995.

81. Bernal, J. L., Jiménez, J. J., Rivera, J. M., Toribio, L., and del Nozal, M. J., Online solid-phase extraction coupled to supercritical fluid chromatography with diode array detection for the determination of pesticides in water, *J. Chromatogr. A*, 754, 145–157, 1996.

82. Camel, V., Supercritical fluid extraction as a useful method for pesticides determination, *Analusis*, 26, M99, 1998.

83. Concetta Bruzzoniti, M., Sarzanini, C., and Mentasti, E., Preconcentration of contaminants in water analysis, *J. Chromatogr. A*, 902, 289–309, 2000.

84. Pham, T. T., Rondeau, B., Sabik, H., Proulx, S., and Cossa, D., Lake Ontario: the predominant source of triazine herbicides in the St. Lawrence River, *Can. J. Fish. Aquat. Sci.*, 57, 78–85, 2000.

85. Chiron, S., Fernandez Alba, A., and Barcelo, D., Comparison of online solid-phase disk extraction to liquid-liquid extraction for monitoring selected pesticides in environmental water, *Environ. Sci. Technol.*, 27, 2352–2359, 1993.

86. Somsen, G. W., Jagt, I., Gooijer, C., Velthorst, N. H., Brinkman, U. A. Th., and Visser, T., Identification of herbicides in river water using online trace enrichment combined with column liquid chromatography-Fourier-transform infrared spectrometry, *J. Chromatogr. A*, 756, 145–157, 1996.

87. Sabik, H., Jeannot, R., and Sauvard, E., Stability of herbicides and their degradation products on graphitized carbon black extraction cartridges used for large volumes of surface water, *Analysis*, 28, 835–842, 2000.

88. Crescenzi, C., Di Corcia, A., Passariello, G., Samperi, R., and Turnes Carou, M. I., Evaluation of two new examples of graphitized carbon blacks for use in solid-phase extraction cartridges, *J. Chromatogr. A*, 733, 41–55, 1996.

89. Marcé, R. M., Prosen, H., Crespo, C., Calull, M., Borrull, F., and Brinkman, U. A. Th., Online trace enrichment of polar pesticides in environmental water by reversed-phase liquid chromatography-diode array detection-particle beam mass spectrometry, *J. Chromatogr. A*, 696, 63–74, 1995.

90. Sabik, H., Graphitized carbon black cartridges for monitoring polar pesticides in large volumes of surface water, *Int. J. Environ. Anal. Chem.*, 72, 113–128, 1998.

91. Slobodník, J., Öztezkizan, Ö., Lingeman, H., and Brinkman, U. A. Th., Solid-phase extraction of polar pesticides from environmental water samples on graphitized carbon and Empore-activated carbon disks and online coupling to octadecyl-bonded silica analytical columns, *J. Chromatogr. A*, 750, 227–238, 1996.

92. Dagnac, T., Jeannot, R., Mouvet, C., and Baran, N., Determination of chloroacetanilide metabolites in water and soils by LC/ESI-MS. p. 103 in 2nd MGPR international Symposium of pesticides in food and the environment, Valencia (Spain), 9–12 may 2001.

93. Masqué, N., Galia, M., Marcé, R. M., and Borrull, F., Solid-phase extraction of phenols and pesticides in water with a modified polymeric resin, *Analyst*, 122, 425–428, 1997.

94. Masqué, N., Galià, M., Marcé, R. M., and Borrull, F., New chemically modified polymeric resin for solid-phase extraction of pesticides and phenolic compounds from water, *J. Chromatogr. A*, 803, 147–155, 1998.

95. Fiehn, O. and Jekel, M., Comparison of sorbents using semipolar to highly hydrophilic compounds for a sequential solid-phase extraction procedure of industrial wastewater, *Anal. Chem.*, 68, 3083–3089, 1996.

96. Junker-Buchheit, A. and Witzenbacher, M., Pesticide monitoring of drinking water with the help of solid-phase extraction and high performance liquid chromatography, *J. Chromatogr. A*, 737, 67–74, 1996.

97. Tsyurupa, M. P., Ilyin, M. M., Andreeva, A. I., and Davankov, V. A., Use of the hyper-crosslinked polystyrene sorbents "Styrosorb" for solid-phase extraction of phenols from water, *Fresenius J. Anal. Chem.*, 352, 672–675, 1995.

98. Masqué, N., Marcé, R. M., and Borrull, F., Comparison of different sorbents for online solid-phase extraction of pesticides and phenolic compounds from natural water followed by liquid chromatography, *J. Chromatogr. A*, 793, 257–263, 1998.

99. Pichon, V., Multiresidue solid-phase extraction for trace-analysis of pesticides and their metabolites in environmental water, *Analusis*, 26, M91, 1998.

100. Shoemaker, J. A., Novel chromatographic separation and carbon solid-phase extraction of acetanilide herbicide degradation products, *J. AOAC Int.*, 85, 1331–1337, 2002.

101. Quintana, J., Martí, I., and Ventura, F., Monitoring of pesticides in drinking and related water in NE Spain with a multiresidue SPE-GC-MS method including an estimation of the uncertainty of the analytical results, *J. Chromatogr. A*, 938, 3–13, 2001.

102. Li, N. and Lee, H. K., Sample preparation based on dynamic ion-exchange solid-phase extraction for GC/MS analysis of acidic herbicides in environmental water, *Anal. Chem.*, 72, 3077–3084, 2000.

103. Barceló, D., Durand, G., Bouvot, V., and Neilen, M., Use of extraction disks for trace enrichment of various pesticides from river water and simulated seawater samples followed by liquid chromatography-rapid-scanning UV-visible and thermospray-mass spectrometry detection, *Environ. Sci. Technol.*, 27, 271–277, 1993.

104. Viana, E., Redond, M. J., Font, G., and Moltó, J. C., Disks versus columns in the solid-phase extraction of pesticides from water, *J. Chromatogr. A*, 733, 267–274, 1996.

105. Pichon, V., Charpak, M., and Hennion, M. C., Multiresidue analysis of pesticides using new laminar extraction disks and liquid chromatography and application to the French priority list, *J. Chromatogr. A*, 795, 83–92, 1998.

106. Cai, Z., Wang, D., and Ma, W. T., Gas chromatography/ion trap mass spectrometry applied for the analysis of triazine herbicides in environmental water by an isotope dilution technique, *Anal. Chim. Acta*, 503, 263–270, 2004.

107. Frías, S., Sánchez, M. J., and Rodríguez, M. A., Determination of triazine compounds in ground water samples by micellar electrokinetic capillary chromatography, *Anal. Chim. Acta*, 503, 271–278, 2004.

108. Nogueira, M. F., Sandra, T., and Sandra, P., Multiresidue screening of neutral pesticides in water samples by high performance liquid chromatography-electrospray mass spectrometry, *Anal. Chim. Acta*, 505, 209–215, 2004.

109. Farran, A. and Ruiz, S., Application of solid-phase extraction and micellar electrokinetic capillary chromatography to the study of hydrolytic and photolytic degradation of phenoxy acid and phenylurea herbicides, *J. Chromatogr. A*, 1024, 267–274, 2004.

110. Zanella, R., Primel, E. G., Goncalves, F. F., Kurz, M. H. S., and Mistura, C. M., Development and validation of a high performance liquid chromatographic procedure for the determination of herbicide residues in surface and agriculture water, *J. Sep. Sci.*, 26, 935–938, 2003.

111. Martyna, R., Ewa, K., and Bogdan, Z., Dramatization in gas chromatographic determination of acidic herbicides in aqueous environmental samples, *Anal. BioAnal. Chem.*, 377, 590–599, 2003.

112. Dopico G, M. S., González R, M. V., Castro R, J. M., González S, R. E., Pérez I, S. J., Rodríguez T, I. M., Calleja, A., and Vilariño, J. M. L., Determination of chlorotriazines, methyl-thiotriazines and one methoxytriazine by SPE-LC-UV in water samples, *Talanta*, 59, 561–569, 2003.

113. Ma, W. T., Fu, K. K., Cai, Z., and Jiang, G. B., Gas chromatography/mass spectrometry applied for the analysis of triazine herbicides in environmental water, *Chemosphere*, 52, 1627–1632, 2003.

114. Rita, C.-M., Encarnacion, R.-G., Eliseo, H.-H., Roman Francisco Javier, S.-S., and Prado Flores, M., Determination of herbicides and metabolites by solid-phase extraction and liquid chromatography evaluation of pollution due to herbicides in surface and groundwater, *J. Chromatogr. A*, 950, 157–166, 2002.

115. Stefano, M., Daniela, P., Alessandra, G., Guiseppe, D., and Angelo, F., Determination of phenoxyacid herbicides and their phenolic metabolites in surface and drinking water, *Rapid Commun. Mass Spectrom.*, 16, 134–141, 2001.

116. Zimmerman, L. R., Schneider, R. J., and Thurman, E. M., Analysis and detection of the herbicides dimethenamid and flufenacet and their sulfonic and oxanilic acid degradates in natural water, *J. Agric. Food Chem.*, 50, 1045–1052, 2002.

117. Bossi, R., Vejrup, K. V., Mogensen, B. B., and Asman, W. A. H., Analysis of polar pesticides in rainwater in Denmark by liquid chromatography-tandem mass spectrometry, *J. Chromatogr. A*, 957, 27–36, 2002.

118. Aldo, L., Alessandro, B., Leva Ilaria, De., Angelo, F., Giovanna, F., and Alessandra, M., Occurrence and determination of herbicides and their major transformation products in environmental water, *Anal. Chim. Acta*, 462, 187–198, 2002.

119. Pozo, O., Pitarch, E., Sancho, J. V., and Hernández, F., Determination of the herbicide 4-chloro-2-methylphenoxyacetic acid and its main metabolite, 4-chloro-2-methylphenol in water and soil by liquid chromatography-electrospray tandem mass spectrometry, *J. Chromatogr. A*, 923, 75–85, 2001.

120. Hernández, F., Sancho, J. V., Pozo, O., Lara, A., and Pitarch, E., Rapid direct determination of pesticides and metabolites in environmental water samples at sub-g/l level by online solid-phase extraction-liquid chromatography-electrospray tandem mass spectrometry, *J. Chromatogr. A*, 939, 1–11, 2001.

121. Dallüge, J., Hankemeier, T., Vreuls, R. J. J., and Brinkman, U. A. Th., Online coupling of immunoaffinity-based solid-phase extraction and gas chromatography for the determination of s-triazines in aqueous samples, *J. Chromatogr. A*, 830, 377–386, 1999.

122. Bjarnason, B., Chimuka, L., and Ramström, O., On line solid-phase extraction of triazine herbicides using a molecularly imprinted polymer for selective sample enrichment, *Anal. Chem.*, 71, 2152–2156, 1999.

123. Loos, R. and Niessner, R., Analysis of atrazine, terbutylazine and their N-dealkylated chloro and hydroxy metabolites by solid-phase extraction and gas chromatography-mass spectrometry and capillary electrophoresis-ultraviolet detection, *J. Chromatogr. A*, 835, 217–229, 1999.

124. Ferrer, I., Barceló, D., and Thurman, E. M., Double-disk solid-phase extraction: simultaneous cleanup and trace enrichment of herbicides and metabolites from environmental samples, *Anal. Chem.*, 71, 1009–1015, 1999.

125. Thompson, T. S. and Miller, B. D., Use of solid-phase extraction disks for the GC-MS analysis of acidic and neutral herbicides in drinking water, *Chemosphere*, 36, 2867–2878, 1998.

126. Hogenboom, A. C., Niessen, W. M. A., and Brinkman, U. A. Th., Rapid target analysis of microcontaminants in water by online single short-column liquid chromatography combined with atmospheric pressure chemical ionization ion-trap mass spectrometry, *J. Chromatogr. A*, 794, 201–210, 1998.

127. Albanis, T. A., Hela, D. G., Sakellarides, T. M., and Konstantinou, I. K., Monitoring of pesticide residues and their metabolites in surface and underground water of Imathia (N. Greece) by means of solid-phase extraction disks and gas chromatography, *J. Chromatogr. A*, 823, 59–71, 1998.

128. Di Corcia, A., Crescenzi, C., Guerriero, E., and Samperi, R., Ultratrace determination of atrazine and its six major degradation products in water by solid-phase extraction and liquid chromatography-electrospray/mass spectrometry, *Environ. Sci. Technol.*, 31, 1658–1663, 1997.

129. Ferrer, I., Hennion, M.-C., and Barceló, D., Immunosorbents coupled online with liquid chromatography/atmospheric pressure chemical ionization/mass spectrometry for the part per trillion level determination of pesticides in sediments and natural water using low preconcentration volumes, *Anal. Chem.*, 69, 4508–4514, 1997.

130. Aguilar, C., Borrull, F., and Marcé, R. M., Online and offline solid-phase extraction with styrene-divinylbenzene-membrane extraction disks for determining pesticides in water by reversed-phase liquid chromatography-diode-array detection, *J. Chromatogr. A.*, 754, 77–84, 1996.

131. Barceló, D. and Hennion, M. C., *Sample Handling Techniques (extraction and cleanup of samples)*, *Trace Determination of Pesticides and their Degradation Products in Water*, Elsevier, Amsterdam, NL, pp. 249–349, 1997.

132. Pichon, V., Chen, L., Guenu, S., and Hennion, M. C., Comparison of sorbents for the solid-phase extraction of the highly polar degradation products of atrazine (including ammeline, ammelide and cyanuric acid), *J. Chromatogr. A*, 711, 257–267, 1995.

133. Crescenzi, C., Di Corcia, A., Guerriero, E., and Samperi, R., Development of a multiresidue method for analyzing pesticide traces in water based on solid-phase extraction and electrospray liquid chromatography mass spectrometry, *Environ. Sci. Technol.*, 31(479), 479–488, 1997.

134. Barceló, D. and Hennion, M. C., *Online Strategies, Trace Determination of Pesticides and their Degradation Products in Water*, Elsevier, Amsterdam, NL, pp. 357–422, 1997.

135. Pichon, V., Cau dit Coumes, C., Chen, L., and Hennion, M. C., Solid phase extraction, cleanup, and liquid chromatography for routine multiresidue analysis of neutral and acidic pesticides in natural water in one run, *Int. J. Environ. Anal. Chem.*, 65, 11–25, 1996.

136. Crescenzi, C., Di Corcia, A., Mabdouly, M. D., and Samperi, R., Pesticide stability studies upon storage in a graphitized carbon black extraction cartridge, *Environ. Sci. Technol.*, 29, 2185–2190, 1995.

137. Liška, I. and Bilikova, K., Stability of polar pesticides on disposable solid-phase extraction precolumns, *J. Chromatogr. A*, 795, 61–69, 1998.

138. Alcaraz, A., Hulsey, S. S., Haas, J. S., Riley, M. O., and Andresen, B. D., Development of solid-phase extraction methods for CW onsite sample preparation in support of the Cooperative OnSite Analysis Exercise (COSAX) project, Govt-Reports-Announcements-&-Index-(GRA&I),-Issue-22, 1995.

139. Brinkman, U. A. Th., Slobodník, J., and Vreuls, J. J., Trace-level detection and identification of polar pesticides in surface water: the SAMOS approach, *Trends Anal. Chem.*, 13, 373–381, 1994.

140. Bouzige, M. and Pichon, V., Immunoextraction of pesticides at the trace level in environmental matrices, *Analusis*, 26, M1121998.

141. Casino, P., Morais, S., Puchades, R., and Maquieira, A., Evaluation of enzyme-linked immunoassays for the determination of chloroacetanilides in water and soils, *Environ. Sci. Technol.*, 35, 4111–4119, 2001.

142. Pauling, L., Theory of the structure and process of formation of antibodies, *J. Am. Chem. Soc.*, 62, 2643–2657, 1940.

143. Steinke, J., Sherrington, D. C., and Dunkin, I. R., Imprinting of synthetic polymers using molecular templates, *Adv. Polym. Sci.*, 123, 81–125, 1995.

144. Mayes, A. G. and Mosbach, K., Molecularly imprinted polymers: useful materials for analytical chemistry? *Trends Anal. Chem.*, 16, 321–332, 1997.

145. Sellergren, B., Noncovalent molecular imprinting: antibody-like molecular recognition in polymeric network materials, *Trends Anal. Chem.*, 16, 310–320, 1997.

146. Siemann, M., Andersson, L. I., and Mosbach, K., Selective recognition of the herbicide atrazine by noncovalent molecularly imprinted polymers, *J. Agric. Food Chem.*, 44, 141–145, 1996.

147. Muldoon, M. and Stanker, L., Polymer synthesis and characterization of a molecularly imprinted sorbent assay for atrazine, *J. Agric. Food Chem.*, 43, 1424–1427, 1995.

148. Matsui, J., Doblhoff-Dier, O., and Takeuchi, T., Atrazine-selective polymer prepared by molecular imprinting technique, *Chem. Lett.*, 24, 489–496, 1995.

149. Lanza, F. and Sellergren, B., Method for synthesis and screening of large groups of molecularly imprinted polymers, *Anal. Chem.*, 71, 2092–2096, 1999.

150. Baggiani, C., Giovannoli, C., Anfossi, L., and Tozzi, C., Molecularly imprinted solid-phase extraction sorbent for the cleanup of chlorinated phenoxyacids from aqueous samples, *J. Chromatogr. A*, 938, 35–44, 2001.

151. Baltussen, E., Sandra, P., David, F., and Cramers, C., Stir bar sorptive extraction (SBSE), a novel extraction technique for aqueous samples: theory and principles, *J. Microcolumn Sep.*, 11, 737–747, 1999.

152. Valor Herencia, I., Stir bar sorptive extraction coupled to gas chromatography mass spectrometry for multiresidue analysis in water samples at ng/l levels. p. 174 in Extech 2001, Advances in extraction technologies, Barcelona (Spain), 17–19 September 2001.

153. León, V. M., Álvarez, B., Cobollo, M. A., Muñoz, S., and Valor, I., Analysis of 35 priority semivolatile compounds in water by stir bar sorptive extraction-thermal desorption-gas chromatography-mass spectrometry: I. method optimization, *J. Chromatogr. A*, 999, 91–101, 2003.

154. Belardi, R. P. and Pawliszyn, J., The application of chemically modified fused silica fibers in the extraction of organics from water matrix samples and their rapid transfer to capillary columns, *Water Poll. Res. J. Can.*, 24, 179, 1989.

155. Pawliszyn, J., *Solid Phase Microextraction, Theory and Practice*, Wiley-VCH, New York, 1997.

156. Dugay, J., Miège, C., and Hennion, M. C., Effect of the various parameters governing solid-phase microextraction for the trace-determination of pesticides in water, *J. Chromatogr. A*, 795, 27–42, 1998.

157. Miège, C. and Dugay, J., Solid-phase microextraction and gas chromatography for rapid analysis of pesticides, *Analusis*, 26, M137, 1998.

158. Boyd-Boland, A. A. and Pawliszyn, J. B., Solid-phase microextraction of nitrogen-containing herbicides, *J. Chromatogr. A*, 704, 63–172, 1995.

159. Krutz, L. J., Senseman, S. A., and Sciumbato, A. S., Solid-phase microextraction for herbicide determination in environmental samples, *J. Chromatogr. A*, 999, 103–121, 2003.

160. Boyd-Boland, A. A., Magdic, S., and Pawliszyn, J. B., Simultaneous determination of 60 pesticides in water using solid-phase microextraction and gas chromatography-mass spectrometry, *Analyst*, 121, 929–937, 1996.

161. Eisert, R. and Pawliszyn, J., Automated in-tube solid-phase microextraction coupled to high performance liquid chromatography, *Anal. Chem.*, 69, 3140–3147, 1997.

162. Ferrari, R., Nilsson, T., Arena, R., Arlati, P., Bartolucci, G., Basla, R., Cioni, F., Del Carlo, G., Dellavedova, P., Fattore, E., Fungi, M., Grote, C., Guidotti, M., Morgillo, S., Muller, L., and Volante, M., Interlaboratory validation of solid-phase microextraction for the determination of triazine herbicides and their degradation products at ng/l level in water samples, *J. Chromatogr. A*, 795, 371–376, 1998.

163. González-Barreiro, C., Lores, M., Casais, M. C., and Cela, R., Optimization of alachlor solid-phase microextraction from water samples using experimental design, *J. Chromatogr. A*, 896, 373–379, 2000.

164. Sánchez-Prado, L., Lores, M., García-Jares, C., Cela, R., and Llompart, M., Photo-SPME, a fast and simple procedure to study the photochemical behavior of pesticides. p. 62 in 3rd MGPR International Symposium of pesticides in food and the environment, Aix-en Provence (France), 20–24 May 2003.

165. Llompart, M., Blanco, B., and Cela, R., Determination of phenols in soils by *in situ* acetylation headspace solid-phase microextraction, *J. Microcolumn Sep.*, 12, 25–32, 2000.

166. Rodriguez, I., Llompart, M., and Cela, R., Solid-phase extraction of phenols, *J. Chromatogr. A*, 885, 291–304, 2000.

167. Llompart, M., Lourido, M., Landín, P., García-Jares, C., and Cela, R., Optimization of a derivatization-solid phase microextraction method for the analysis of 30 phenolic pollutants in water samples, *J. Chromatogr. A*, 963, 137–148, 2002.

168. Möder, M., Popp, P., Eisert, R., and Pawliszyn, J., Determination of polar pesticides in soil by solid phase microextraction coupled to high performance liquid chromatography-electrospray/mass spectrometry, *Anal. Bioanal. Chem.*, 363, 680–685, 1999.

169. Zambonin, C. G., Catucci, F., and Palmisano, F., Solid phase microextraction coupled to gas chromatography-mass spectrometry for the determination of the adsorption coefficients of triazines in soil, *Analyst*, 123, 2825–2828, 1998.

170. Hernandez, F., Beltran, J., Lopez, F. J., and Gaspar, J. V., Use of solid-phase microextraction for the quantitative determination of herbicides in soil and water samples, *Anal. Chem.*, 72, 2313–2322, 2000.

171. Gonçalves, C. and Alpendurada, M. F., Solid-phase micro-extraction-gas chromatography-(tandem) mass spectrometry as a tool for pesticide residue analysis in water samples at high sensitivity and selectivity with confirmation capabilities, *J. Chromatogr. A*, 1026, 239–250, 2004.

172. Carabias-Martinez, R., Garcia-Hermida, C., Rodriguez-Gonzalo, E., Soriano-Bravo, F., and Hernandez-Mendez, E., Determination of herbicides, including thermally labile phenylureas, by solid-phase microextraction and gas chromatography-mass spectrometry, *J. Chromatogr. A*, 1002, 1–12, 2003.

173. Lambropoulou, D. A., Sakkas, V. A., Hela, D. G., and Albanis, T. A., Application of solid-phase microextraction in the monitoring of priority pesticides in the Kalamas River (N.W. Greece), *J. Chromatogr. A*, 963, 107–116, 2002.

174. Gonçalves, C. and Alpendurada, M. F., Multiresidue method for the simultaneous determination of four groups of pesticides in ground and drinking water, using solid-phase microextraction-gas chromatography with electron-capture and thermionic specific detection, *J. Chromatogr. A*, 968, 177–190, 2002.

175. Takino, M., Daishima, S., and Nakahara, T., Automated online in-tube solid-phase microextraction followed by liquid chromatography/electrospray ionization-mass spectrometry for the determination of chlorinated phenoxy acid herbicides in environmental water, *Analyst*, 126, 602–608, 2001.

176. Henriksen, T., Svensmark, B., Lindhardt, B., and Juhler, R. K., Analysis of acidic pesticides using in situ derivatization with alkylchloroformate and solid-phase microextraction (SPME) for GC–MS, *Chemosphere*, 44, 1531–1539, 2001.

177. Ramesh, A. and Elumalai Ravi, P., Applications of solid-phase microextraction (SPME) in the determination of residues of certain herbicides at trace levels in environmental samples, *J. Environ. Monit.*, 3, 505–508, 2001.

178. Carlo, P., Zambonin, G., and Palmisano, F., Determination of triazines in soil leachates by solid-phase microextraction coupled to gas chromatography-mass spectrometry, *J. Chromatogr. A*, 874, 247–255, 2000.

179. Sampedro, M. C., Martín, O., López de Armentia, C., Goicolea, M. A., Rodríguez, E., Gómez de Balugera, Z., Costa-Moreira, J., and Barrio, R. J., Solid-phase microextraction for the determination of systemic and nonvolatile pesticides in river water using gas chromatography with nitrogen-phosphorous and electron-capture detection, *J. Chromatogr. A*, 893, 347–358, 2000.

180. Natangelo, M., Tavazzi, S., Fanelli, R., and Benfenati, E., Analysis of some pesticides in water samples using solid-phase microextraction-gas chromatography with different mass spectrometric techniques, *J. Chromatogr. A*, 859, 193–201, 1999.

181. Lee, M.-R., Lee, R.-J., Lin, Y.-W., Chen, C.-M., and Hwang, B.-H., Gas-phase postderivatization following solid-phase microextraction for determining acidic herbicides in water, *Anal. Chem.*, 70, 1963–1968, 1998.

182. Nilsson, T., Baglio, D., Galdo-Miguez, I., Øgaard Madsen, J., and Facchetti, S., Dramatization/solid-phase microextraction followed by gas chromatography-mass spectrometry for the analysis of phenoxy acid herbicides in aqueous samples, *J. Chromatogr. A*, 826, 211–216, 1998.

183. Aguilar, C., Peñalver, S., Pocurull, E., Borrull, F., and Marcé, R. M., Solid-phase microextraction and gas chromatography with mass spectrometric detection for the determination of pesticides in aqueous samples, *J. Chromatogr. A*, 795, 105–115, 1998.

184. Choudhury, T. K., Gerhardt, K. O., and Mawhinney, T. P., Solid-phase microextraction of nitrogen- and phosphorus-containing pesticides from water and gas chromatographic analysis, *Environ. Sci. Technol.*, 30, 3259–3265, 1996.

185. Dean, J. R. and Fitzpatrick, L. J., *Pesticides Defined by Matrix, Environmental Analysis, Handbook of Analytical Separations*, Vol. 3, Elsevier Science, New York, pp. 123-173, 2001.

186. Camel, V., Recent extraction techniques for solid matrices supercritical fluid extraction, pressurized fluid extraction and microwave-assisted extraction: their potential and pitfalls, *Analyst*, 126, 1182–1193, 2001.

187. Macutkiewicz, E., Rompa, M., and Zygmunt, B., Sample preparation and chromatographic analysis of acidic herbicides in soils and sediments, *Crit. Rev. Anal. Chem.*, 33, 1–17, 2003.

188. Dagnac, T., Jeannot, R., Bristeau, S., Mouvet, C., and Baran, N., Determination of chloro-acetanilides, triazines and ureas and some of their metabolites in soils by pressurized fluid extraction, GC/MS/MS AND LC-APCI/MS. *J. Chromatogr. A*, 1067, 225–233, 2005.

189. de Amarante, O. P. Jr., Brito, N. M., dos Santos, T. C. R., Nunes, G. S., and Ribeiro, M. L., Determination of 2,4-dichlorophenoxyacetic acid and its major transformation product in soil samples by liquid chromatographic analysis, *Talanta*, 60, 115–121, 2003.

190. Sgen, G. and Lee, H. K., Determination of triazines in soil by microwave-assisted extraction followed by solid-phase microextraction and gas chromatography-mass spectrometry, *J. Chromatogr. A*, 985, 167–174, 2003.

191. Zisis, V. and Papadopoulou-Mourkidou, E., Determination of triazine and chloroacetanilide herbicides in soils by microwave-assisted extraction (MAE) coupled to gas chromatographic analysis with either GC-NPD or GC-MS, *J. Agric. Food Chem.*, 50, 5026–5033, 2002.

192. Luque-Garcia, J. L. and Luque de Castro, M. D., Coupling continuous subcritical water extraction, filtration, preconcentration, chromatographic separation and UV detection for the determination of chlorophenoxy acid herbicides in soils, *J. Chromatogr. A*, 959, 25–35, 2002.

193. Patsias, J., Papadakis, E. N., and Papadopoulou-Mourkidou, E., Analysis of phenoxyalkanoic acid herbicides and their phenolic conversion products in soil by microwave assisted solvent extraction and subsequent analysis of extracts by online solid-phase extraction-liquid chromatography, *J. Chromatogr. A*, 959, 153–161, 2002.

194. Rosales-Conrado, N., Leon-Gonzalez, M. E., Perez-Arribas, L. V., and Polo-Diez, L. M., Determination of chlorophenoxy acid herbicides and their esters in soil by capillary high performance liquid chromatography with ultraviolet detection, using large volume injection and temperature gradient, *Anal. Chim. Acta*, 470, 147–154, 2002.

195. Henriksen, T., Svensmark, B., and Juhler, R. K., Analysis of metribuzin and transformation products in soil by pressurized liquid extraction and liquid chromatography-tandem mass spectrometry, *J. Chromatogr. A*, 957, 79–87, 2002.

196. Zhu, Y., Yanagihara, K., Guo, F., and Li, Q. X., Pressurized fluid extraction for quantitative recovery of chloroacetanilide and nitrogen heterocyclic herbicides in soil, *J. Agric. Food Chem.*, 48, 4097–4102, 2000.

197. Xiong, G., Tang, B., He, X., Zhao, M., Zhang, Z., and Zhang, Z., Comparison of microwave-assisted extraction of triazines from soils using water and organic solvents as the extractants, *Talanta*, 48, 333–339, 1999.

198. Guzella, L. and Pozzoni, F., Accelerated solvent extraction of herbicides in agricultural soil samples, *Int. J. Environ. Anal. Chem.*, 74, 123–133, 1999.

199. Cserháti, T. and Forgács, E., Phenoxyacetic acids: separation and quantitative determination, *J. Chromatogr. B: Biomedical Sciences and Applications*, 717, 157–178, 1998.

200. Papilloud, S., Haerdi, W., Chiron, S., and Barcelo, D., Supercritical fluid extraction of atrazine and polar metabolites from sediments followed by confirmation with LC-MS, *Environ. Sci. Technol.*, 30, 1822–1826, 1996.

201. Dean, J. R., Barnabas, I. J., and Owen, S. P., Influence of pesticide-soil interactions on the recovery of pesticides using supercritical fluid extraction, *Analyst*, 121, 465–468, 1996.

202. Lopez Avila, V. and Charan, C., Using supercritical fluid extraction and enzyme immunoassays to determine pesticides in soils, *Trends Anal. Chem.*, 13, 118–126, 1994.

203. Camel, V., The determination of pesticide residues and metabolites using supercritical fluid extraction, *Trends Anal. Chem.*, 16, 351–369, 1997.

204. Luque de Castro, M. D. and Tena, M. T., Strategies for supercritical fluid extraction of polar and ionic compounds, *Trends Anal. Chem.*, 15, 32–37, 1996.

205. Stearman, G. K., Wells, M. J., Adkisson, S., and Ridgill, T., Supercritical fluid extraction coupled with enzyme immunoassay analysis of soil herbicides, *Analyst*, 120, 2617–2621, 1995.

206. Lopez-Avila, V., Charan, C., and Beckert, W. F., Using supercritical fluid extraction and enzyme immunoassays to determine pesticides in soils, *Trends Anal. Chem.*, 13, 118–125, 1994.

207. Khan, S. U., Supercritical fluid extraction of bound pesticide residues from soil and food commodities, *J. Agric. Food Chem.*, 43, 1718–1723, 1995.

208. Cassada, D. A., Spalding, R. F., Cai, Z., and Gross, M. L., Determination of atrazine, deethylatrazine and deisopropylatrazine in water and sediment by isotope dilution gas chromatography-mass spectrometry, *Anal. Chim. Acta*, 287, 7–15, 1994.

209. Papilloud, S. and Haerdi, W., Supercritical fluid extraction of triazine herbicides from solid matrixes, *Chromatographia*, 38, 514–519, 1994.

210. Zhou, M., Trubey, R. K., Keil, Z. O., and Sparks, D. L., Study of the effects of environmental variables and supercritical fluid extraction parameters on the extractability of pesticide residues from soils using a multivariate optimization scheme, *Environ. Sci. Technol.*, 31, 1934–1939, 1997.

211. Jimenez-Carmona, M. M. and Luque de Castro, M. D., Reverse-micelle formation: a strategy for enhancing CO2-supercritical fluid extraction of polar analytes, *Anal. Chim. Acta*, 358, 1–4, 1998.

212. Camel, V., Microwave-assisted solvent extraction of environmental samples, *Trends Anal. Chem.*, 19, 229–248, 2000.

213. Xiong, G., Liang, J., Zou, S., and Zhang, Z., Microwave-assisted extraction of atrazine from soil followed by rapid detection using commercial ELISA kit, *Anal. Chim. Acta*, 371, 97–103, 1998.

214. Molins, C., Hogendoorn, E. A., Heusinkveld, H. A. G., Van Harten, D. C., Van Zoonen, P., and Baumann, R. A., Microwave assisted solvent extraction (MASE) for the efficient determination of triazines in soil samples with aged residues, *Chromatographia*, 43, 527–532, 1996.

215. Molins, C., Hogendoorn, E. A., Heusinkveld, H. A. G., Van Beuzekom, A. C., van Zoonen, P., and Baumann, R. A., Effect of organic matter content in the trace analysis of triazines in various types of soils with GC-NPD, *Chromatographia*, 48, 450–456, 1998.

216. Garcia-Ayuso, L. E., Sanchez, M., Fernandez de Alba, A., and Luque de Castro, M. D., Focused microwave-assisted Soxhlet: an advantageous tool for sample extraction, *Anal. Chem.*, 70, 2426–2431, 1998.

217. Stout, S. J., Dacunha, A. R., and Allardice, D. G., Microwave-assisted extraction coupled with gas chromatography/electron capture negative chemical ionization mass spectrometry for the simplified determination of imidazolinone herbicides in soil at the ppb level, *Anal. Chem.*, 68, 653–658, 1996.

218. Dean, J. R. and Xiong, G., Extraction of organic pollutants from environmental matrices: selection of extraction technique, *Trends Anal. Chem.*, 19, 553–564, 2000.

219. Catalina, M. I., Dalluge, J., Vreuls, R. J. J., and Brinkman, U. A. T., Determination of chlorophenoxy acid herbicides in water by in situ esterification followed by in-vial liquid-liquid extraction combined with large-volume oncolumn injection and gas chromatography-mass spectrometry, *J. Chromatogr. A*, 877, 153–166, 2000.

220. Juhler, R. K., Sorensen, S. R., and Larsen, L., Analyzing transformation products of herbicide residues in environmental samples, *Water Res.*, 35, 1371–1378, 2001.

221. Jeannot, R., Sabik, H., Sauvard, E., and Génin, E., Application of liquid chromatography with mass spectrometry combined with photodiode array detection and tandem mass spectrometry for monitoring pesticides in surface water, *J. Chromatogr. A*, 818, 51–71, 2000.

222. Sherma, J., Recent advances in the thin-layer chromatography of pesticides: a review, *J. AOAC Int.*, 86, 603–611, 2003.

223. Forgács, E., Cserháti, T., and Barta, I., The binding of amino acids to the herbicide 2,4-dichlorophenoxy acetic acid, *Amino acids*, 18, 69–79, 2000.

224. Chicharro, M., Zapardiel, A., Bermejo, E., and Moreno, M., Determination of 3-amino-1,2,4-triazole (amitrole) in environmental water by capillary electrophoresis, *Talanta*, 59, 37–45, 2003.

225. Chicharro, M., Zapardiel, A., Bermejo, E., and Sánchez, A., Simultaneous UV and electrochemical determination of the herbicide asulam in tap water samples by micellar electrokinetic capillary chromatography, *Anal. Chim. Acta*, 469, 243–252, 2002.

226. Núñez, O., Moyano, E., and Galceran, M. T., Solid-phase extraction and sample stacking-capillary electrophoresis for the determination of quaternary ammonium herbicides in drinking water, *J. Chromatogr. A*, 946, 275–282, 2002.

227. Terabe, S., Otsuka, K., Ichikawa, K., Tsuchiya, A., and Ando, T., Electrokinetic separations with micellar solutions and open-tubular capillaries, *Anal. Chem.*, 56, 111–113, 1984.

228. Dinelli, G., Bonetti, A., Catizone, P., and Galletti, G. C., Separation and detection of herbicides in water by micellar electrokinetic capillary chromatography, *J. Chromatogr. B*, 656, 275–280, 1994.

229. Turiel, E., Fernández, P., Pérez-Conde, C., and Cámara, C., Online concentration in micellar electrokinetic chromatography for triazine determination in water samples: evaluation of three different stacking modes, *Analyst*, 125, 1725–1731, 2000.

230. Carabias-Martinez, R., Rodriguez-Gonzalo, E., Revilla-Ruiz, P., and Dominguez-Alvarez, R., Solid-phase extraction and sample stacking-micellar electrokinetic capillary chromatography for the determination of multiresidues of herbicides and metabolites, *J. Chromatogr. A*, 990, 291–302, 2003.

231. Carabias-Martínez, R., Rodríguez-Gonzalo, E., Domínguez-Alvárez, J., and Hernández-Méndez, J., Comparative study of separation and determination of triazines by micellar electrokinetic capillary chromatography and nonaqueous capillary electrophoresis: Application to residue analysis in natural water, *Electrophoresis*, 23, 494–501, 2002.

232. Hong, S. and Lemley, A. T., Gas chromatographic mass spectrometric determination of alachlor and its degradation products by direct aqueous injection, *J. Chromatogr. A*, 822, 253–261, 1998.

233. González-Rodríguez, J., Garrido-Frenich, A., Arrebola, F. J., and Martínez-Vidal, J. L., Evaluation of low pressure gas chromatography linked to ion-trap tandem mass spectrometry for the fast trace analysis of multiclass pesticide residues, *Rapid Commun. Mass Spectrom.*, 16, 1216–1224, 2002.

234. Reemtsma, T., Liquid chromatography-mass spectrometry and strategies for trace-level analysis of polar organic pollutants, *J. Chromatogr. A*, 1000, 477–501, 2003.

235. Robb, D. B., Covey, T. R., and Bruins, A. P., Atmospheric pressure photoionization: an ionization method for liquid chromatography-mass spectrometry, *Anal. Chem.*, 72, 3653–3659, 2000.

236. Rauha, J.-P., Vuorela, H., and Kostiainen, R., Effect of eluent on the ionization efficiency of flavonoids by ion spray, atmospheric pressure chemical ionization, and atmospheric pressure photoionization mass spectrometry, *J. Mass Spectrom.*, 36, 1269–1280, 2001.

237. Kauppila, T. J., Kuuranne, T., Meurer, E. C., Eberlin, M. N., Kotiaho, T., and Kostiainen, R., Atmospheric pressure photoionization mass spectrometry. Ionization mechanism and the effect of solvent on the ionization of naphthalenes, *Anal. Chem.*, 74, 5470–5477, 2002.

238. Jeannot, R. and Sauvard, E., High performance liquid chromatography coupled with mass spectrometry applied to analyses of pesticides in water, *Analusis*, 27, 271–280, 1999.

239. Sabik, H. and Jeannot, R., Determination of organonitrogen pesticides in large volumes of surface water by liquid-liquid and solid-phase extraction using gas chromatography with nitrogen-phosphorus detection and liquid chromatography with atmospheric pressure chemical ionization mass spectrometry, *J. Chromatogr. A*, 818, 197–207, 1998.

240. Hogenboom, A. C., Niessen, W. M. A., and Brinkman, U. A. Th., Online solid-phase extraction short-column liquid chromatography combined with various tandem mass spectrometric scanning strategies for the rapid study of transformation of pesticides in surface water, *J. Chromatogr. A*, 841, 33–44, 1999.

241. Geerdink, R. B., Kooistra-Sijpersma, A., Tiesnitsch, J., Kienhuis, P. G. M., and Brinkman, U. A. Th., Determination of polar pesticides with atmospheric pressure chemical ionization mass spectrometry-mass spectrometry using methanol and/or acetonitrile for solid-phase desorption and gradient liquid chromatography, *J. Chromatogr. A*, 863, 147–155, 1999.

242. Aguilar, C., Ferrer, I., Borrull, F., Marcé, R. M., and Barceló, D., Comparison of automated online solid-phase extraction followed by liquid chromatography-mass spectrometry with atmospheric pressure chemical ionization and particle beam mass spectrometry for the determination of a priority group of pesticides in environmental water, *J. Chromatogr. A*, 794, 147–163, 1998.

243. Dijkman, E., Mooibroek, D., Hoogerbrugge, R., Hogendoorn, E., Sancho, J. V., Pozo, O., and Hernandez, F., Study of matrix effects on the direct trace analysis of acidic pesticides in water using various liquid chromatographic modes coupled to tandem mass spectrometric detection, *J. Chromatogr. A*, 926, 113–125, 2001.

244. Geerdink, B., Niessen, W. M. A., and Brinkman, U. A. Th., Trace-level determination of pesticides in water by means of liquid and gas chromatography, *J. Chromatogr. A*, 970, 65–93, 2002.

245. Hogenboom, A. C., Niessen, W. M. A., and Brinkman, U. A. Th., The role of liquid chromatography-mass spectrometry in environmental trace level analysis. Determination and identification of pesticides in water, *J. Sep. Sci.*, 24, 331–354, 2001.

246. Jimenez, J. J., Bernal, J. L., Del Nozal, M. J., and Martin, M. T., Use of a particle beam interface combined with mass spectrometry/negative chemical ionization to determine polar herbicide residues in soil by liquid chromatography, *J. AOAC Int.*, 83, 756–761, 2000.

247. Thurman, E. M., Ferrer, I., and Barceló, D., Choosing between atmospheric pressure chemical ionization and electrospray ionization interfaces for the HPLC/MS analysis of pesticides, *Anal. Chem.*, 73, 5441–5449, 2001.

248. Kidwell, H., Jones, J. J., and Games, D. E., Separation and characterization of five polar herbicides using countercurrent chromatography withy detection by negative ion electrospray ionization mass spectrometry, *Rapid Commun. Mass Spectrom.*, 15, 1181–1186, 2001.

249. Hogenboom, A. C., Niessen, W. M. A., Little, D., and Brinkman, U. A. Th., Accurate mass determinations for the confirmation and identification of organic microcontaminants in surface water using online solid-phase extraction liquid chromatography electrospray orthogonal-acceleration time-of-flight mass spectrometry, *Rapid Commun. Mass Spectrom.*, 13, 125–133, 1999.

250. Maizels, M. and Budde, W. L., Exact mass measurements for confirmation of pesticides and herbicides determined by liquid chromatography/time-of-flight mass spectrometry, *Anal. Chem.*, 73, 5436–5440, 2001.

251. Ferrer, I. and Thurman, E.M., Liquid chromatography/time-of-flight/mass spectrometry (LC/TOF/MS) for the analysis of emerging contaminants, *Trends Anal. Chem.*, 22, 750–756, 2003.

252. Ferrer, I., and Thurman, E. M., Eds., *Liquid Chromatography Mass Spectrometry/Mass Spectrometry, MS/MS and Time of Flight MS: Analysis of Emerging Contaminants*, American Chemical Society Symposium Series 80, Oxford University press, New York, 2003.

253. Thurman, E. M., Ferrer, I., and Parry, R., Accurate mass analysis of ethanesulfonic acid degradates of acetochlor and alachlor using high performance liquid chromatography and time-of-flight mass spectrometry, *J. Chromatogr. A*, 957, 3–9, 2002.

254. Thurman, E. M., Ferrer, I., and Furlong, E. T., In *Liquid Chromatography Mass Spectrometry/Mass Spectrometry, MS/MS and Time of Flight MS: Analysis of Emerging Contaminants*, Ferrer, I. and Thurman, E. M., Eds., American Chemical Society Symposium Series 80, Oxford University press, New York, 2003, chap. 8.

255. Malato, S., Albanis, T., Piedra, L., Aguera, A., Hernando, D., and Fernandez-Alba, A., In *Liquid Chromatography Mass Spectrometry/Mass Spectrometry, MS/MS and Time of Flight MS: Analysis of Emerging Contaminants*, Ferrer, I. and Thurman, E. M., Eds., American Chemical Society Symposium Series 80, Oxford University press, New York, 2003, chap. 5.

256. Hogenboom, C., Niessen, W. M. A., and Brinkman, U. A. Th., Characterization of photodegradation products of alachlor in water by online solid-phase extraction liquid chromatography combined with tandem mass spectrometry and orthogonal-acceleration time-of-flight mass spectrometry, *Rapid Commun. Mass Spectrom.*, 14, 1914–1924, 2000.

257. Bobeldijk, I., Vissers, J. P. C., Kearney, G., Majorand, H., and van Leerdam, J. A., Screening and identification of unknown contaminants in water with liquid chromatography and quadrupole-orthogonal acceleration-time-of-flight tandem mass spectrometry, *J. Chromatogr. A*, 929, 63–74, 2001.

258. Ibanez, M., Sancho, J. V., Pozo, O. J., and Hernandez, F., Use of quadrupole time-of-flight mass spectrometry in environmental analysis: elucidation of transformation products of triazine herbicides in water after UV exposure, *Anal. Chem.*, 76, 1328–1335, 2004.

259. Geerdink, R. B., Wilfried Niessen, M. A., and Udo, A. Th., Brinkman, Mass spectrometric confirmation criterion for product-ion spectra generated in flow injection analysis. Environmental application, *J. Chromatogr. A*, 910, 291–300, 2001.

27 Oil and Petroleum Product Fingerprinting Analysis by Gas Chromatographic Techniques

Zhendi Wang and Merv Fingas

CONTENTS

I. INTRODUCTION

The word *petroleum* means "rock oil" (from the Greek words *petros* [rock] and *elaion* [oil]). Petroleum or crude oil is a complex mixture of thousands of different organic compounds. Petroleum is generally believed to be derived from a variety of organic materials that are chemically converted under differing geological and thermal conditions over long periods of time (hundreds of million years). Crude oils contain primarily carbon and hydrogen (which form a wide range of hydrocarbons from light gases to heavy residues), but also contain smaller amounts of sulfur, oxygen and nitrogen as well as metals such as nickel, vanadium and iron. As the number of carbon atoms in, for example, the paraffin series increases, the complexity of petroleum mixtures also rapidly increases. The infinitely variable nature of these factors results in distinct chemical differences between oils. Refined petroleum products are fractions usually derived by distillation of crude oil. Because of dissimilarities in characteristics of crude oil feed stocks and variations in refinery processes, refined products differ in their chemical compositions. Thus, all crude oils and petroleum products, to some extent, have chemical compositions that differ from each other. This variability in chemical compositions results in unique chemical "fingerprints" for each oil and provides a basis for identifying the source(s) of the spilled oil.

Liquid petroleum (crude oil and the products refined from it) plays a pervasive role in our modern society. For example, about 286,000 tonnes of oil and petroleum products are used in Canada every day. The United States uses about 10 times this amount and, worldwide, about 11 million tonnes are used per day. Extraction, transportation, and widespread use of petroleum inevitably result in intentional and accidental releases to the environment. In addition, natural seepage of crude oil from geologic formations below the seafloor to the sea surface also contributes to pollution of the marine environment. Based on analysis of data from a wide variety of sources, each year on average about 260,000 tonnes of petroleum spills into the waters off North America. Annual worldwide estimates of petroleum input into the sea exceed 1,300,000 tonnes.[1] In Canada, about 12 spills of more than 4000 l are reported each day, of which only about one spill is into navigable waters and most spills take place on land. In the U.S.A., about 25 such spills occur each day into navigable waters and about 75 occur on land.[2]

The most recent examples of large scale marine spills are the "*Erika*" and "*Prestige*" spills. On December 12, 1999, the Maltese tanker Erika broke into two during a fierce storm about 110 km south of Brest, France. An estimated 10,000 tonnes (2.8 million gallons) of heavy fuel oil spilled, and an equal amount remained aboard the sunken stern. A deadly storm after the spill hurled the sticky, heavily emulsified oil from the Erika ashore, churning tar into sandy beaches, splattering cliffs, roads, and car parks. This incident became France's most damaging oil spill in 20 years, causing an environment of devastation on the French Coast. The Prestige spill occurred on November 2002. After a six-day effort to salve the vessel and its cargo, the aging single-hulled tanker Prestige, carrying 77,000 tonnes (22.6 million gallons) of heavy fuel oil (more than twice the oil that the *Exxon Valdez* spilled in Alaska in 1989), broke in two under stormy conditions and sank in roughly 3000 m of water about 240 km off the northwest coast of Spain on November 19, 2002. Thousands of square kilometers of sea area was covered by the oil slick. An estimated 5000 to 25,000 tonnes of thick gooey oil was washing up on the shoreline, tarring birds and other wildlife, and threatening fishing and shellfish industries. The incident became the worst oil spill in more than a decade.

It should be realized, however, that despite the large spills in the marine environment are from tankers, these spills only make up about 5% of all oil pollution entering the sea; most oil pollution in the oceans comes from the runoff of oil and fuel from land-based sources (such as broken or leaking pipelines that cross the land, and release of hydrocarbons from the thousands of underground tanks and giant above-ground storage containers) rather than from accidental spills.[2]

Oil spills cause extensive damage to marine life, terrestrial life, human health, and natural resources, and have resulted in legal battles amounting annually to billions of dollars in casualty and punitive payments. Therefore, to unambiguously characterize spilled oils and to link them to the known sources is extremely important for environmental damage assessment, understanding the fate and behavior and predicting the potential long term impact of spilled oils on the environment, selecting appropriate spill response and taking effective cleanup measures. In addition, successful forensic investigation and analysis of oil and refined product hydrocarbons in contaminated sites and receptors yield a wealth of chemical "fingerprinting" data. These data, in combination with historic, geological, environmental, and any other related information on the contaminated site can, in many cases, help to settle legal liability and to support litigation against the spillers.

II. ADVANCES IN OIL HYDROCARBON FINGERPRINTING TECHNIQUES

A. OIL CHEMISTRY

Crude oil is extremely complex mixtures of hydrocarbons. Oil hydrocarbons range from small, volatile compounds to very large, nonvolatile compounds. For example, over 300 compounds have been identified in the Alberta Sweet Mixed Blend using GC–MS and by comparison of GC retention data with authentic standards and calculation of retention index values.[3] In general, the oil hydrocarbons are characterized and classified by their structures, including saturates, olefins, aromatics, polar compounds (wide variety of compounds containing sulfur, oxygen, and nitrogen), and asphaltenes.

Saturates are a group of hydrocarbons composed of only carbon and hydrogen with no double carbon–carbon bond. They are the predominant hydrocarbon classes that comprise crude oil. Saturates include straight chain, branched chain, and cyclo alkanes (paraffins).

(1) Normal alkanes (normal paraffins) ranging from C_5 to C_{40} are often the most abundant constituents in many oils. Large *n* alkanes ($>C_{18}$) are often referred to as waxes.

(2) Isoalkanes are hydrocarbons containing branched carbon chains. They are also a major group of constituents of oil. Five most abundant and important oil isoprenoid compounds are farnesane (i-C_{15}: 2, 6, 10-trimethyl-dodecane), trimethyl-tridecane (i-C_{16}), norpristane (i-C_{18}: 2, 6, 10-trimethyl-pentadecane), pristane (i-C_{19}: 2, 6, 10, 14-tetramethyl-pentadecane) and phytane (i-C_{20}: 2, 6, 10, 14-tetramethyl-hexadecane).

(3) Cycloalkanes consist of rings of carbon atoms joined by single atomic bond. The most abundant cycloalkanes (also called naphthenes) are the single ring cyclopentane (C_5H_{10})

and cyclohexane (C_6H_{12}), and their alkylated (from C_0 to C_{14}) homologues (alkyl-cyclopentanes and alkyl-cyclohexanes).

(4) Terpanes and steranes are branched cycloalkanes consisting of multiple condensed five or six carbon rings. They have been increasingly used in recent years as marker compounds for source identification and differentiation of oils, and monitoring the weathering and degradation process of oil hydrocarbons under a wide variety of conditions.

Alkenes, commonly referred to as olefins, are partially unsaturated straight-chain hydrocarbons characterized by one or two double carbon-to-carbon bonds in their molecules. Concentrations of olefins are generally very low in crude oils. Significant amounts of olefins are found only in some refined products.

Aromatic hydrocarbons are cyclic, planar compounds that are stabilized by a delocalized π electron system. Aromatics include the mono aromatic hydrocarbons such as BTEX (benzene, toluene, ethylbenzene, and *o*-, *m*-, and *p*-xylenes) and other alkyl-substituted benzene compounds (C_n-benzenes), and polycyclic aromatic hydrocarbons (including oil-characteristic alkylated PAH homologues and the other U.S. EPA priority PAHs). Benzene is the simplest one ring aromatic compound. The commonly analyzed PAH compounds range from two ring PAHs (such as naphthalene) up through six ring PAHs (benzo (*g*, *h*, *i*) perylene). BTEX and PAHs are of concern because of their toxic properties in the environment.

Polar compounds are those with distinct regions of positive and negative charge, because of bonding with atoms such as oxygen, sulfur, or nitrogen. The "polarity" or charge that the molecules carry result in behavior that, under some circumstances, is different from that of nonpolar compounds. In the petroleum industry, the smaller polar compounds are called "resins," which are largely responsible for oil adhesion. The large polar compounds are called "asphaltenes" because they often make up the largest percentage of the asphalt commonly used for road construction.

Resin compounds include heterocyclic hydrocarbons (such as sulfur, oxygen, and nitrogen containing PAHs), phenols, acids, alcohols, and monoaromatic steroids. Because of their polarity, these compounds are more soluble in polar solvents. Sulfur is typically the most abundant element in petroleum and may be present in several forms, including elemental sulfur, hydrogen sulfide, mercaptans, thiophenes (thiophene and its alkylated homologues), and dibenzothiophenes (dibenzothiophene and its alkylated homologues). The sulfur content in most crude oils varies from about 0.1% to 3% and 5% to 6% for some heavy oils and bitumen. Most organic nitrogen hydrocarbons in crude oils are present as alkylated aromatic heterocycles with a predominance of neutral pyrrole and carbazole structures over basic pyridine and quinoline forms. They are chiefly associated with high boiling fractions, and much of the nitrogen in petroleum is in the asphaltenes. Oxygen reacts with hydrocarbons to form various oxygen-containing hydrocarbons, such as phenols, cresols, and benzofurans. Compared to PAHs, the concentrations of these nitrogen and oxygen-containing compounds are generally very low.

Asphaltenes are a class of very large and complex compounds, precipitated from oils in laboratory by addition of excess *n* pentane or *n* hexane. Despite a considerable volume of relevant analytical data, very little is known about molecular configuration of asphaltenes. From x-ray diffraction patterns of solid asphaltenes, it has been inferred that crystallographic organization can be represented by an asphaltene "macromolecule," in which clusters of partly ordered aromatic matter and carrying aliphatic chains of varying length are associated in micelles or particles. If abundant in oil, they have a significant effect on oil behavior.

Table 27.1 summarizes the typical composition of some oils and petroleum products.

B. Physical Properties of Oil

Physical properties of the almost limitless variety of crude oils are generally correlated with aspects of chemical composition. The properties briefly discussed here are viscosity, density, specific

TABLE 27.1
Typical Composition of Some Oils and Petroleum Products (%)

Group	Compound Class	Gasoline	Diesel	Light Crude	Heavy Crude	IFO	Bunker C
Saturates		50 to 60	65 to 95	55 to 90	25 to 80	25 to 45	20 to 40
	Alkanes	45 to 55	35 to 45				
	Cyclo-alkanes	~5	30 to 50				
	Waxes		0 to 1	0 to 20	0 to 10	2 to 10	5 to 15
Olefins		5 to 10	0 to 10				
Aromatics		25 to 40	5 to 25	10 to 35	15 to 40	40 to 60	30 to 50
	BTEX	15 to 35	0.5 to 2	0.1 to 2.5	0.01 to 2	0.05 to 1	0 to 1
	PAHs		0.5 to 5	0.5 to 3	1 to 4	1 to 5	1 to 5
Polar Compounds			0 to 2	1 to 15	5 to 40	15 to 25	10 to 30
	Resins		0 to 2	0 to 10	2 to 25	10 to 15	10 to 20
	Asphaltenes			0 to 10	0 to 20	5 to 10	5 to 20
Sulphur		<0.05	0.05 to 0.5	0 to 2	0 to 5	0.5 to 2	2 to 4
Metals (Ppm)				30 to 250	100 to 500	100 to 1000	100 to 2000

gravity, flash point, distillation, interfacial tension, and vapor pressure. These properties for the oils are listed in Table 27.2.

Solubility in water is the measure of how much of oil will dissolve in the water column on a molecular basis at a known temperature and pressure. The more polar the compound, the more soluble in water. BTEX compounds are so frequently encountered in ground water in part due to their high water solubility. BTEX solubility in water is dependent on the nature of the multi component mixture, such as gasoline, diesel, or crude oil. The solubility of a constituent within a multicomponent mixture may be orders of magnitude lower than the aqueous solubility of the pure constituent in water. Oil is a complex mixture of many compounds each of which partitions uniquely between oil and water, therefore different oils have different water solubility. The solubility of oil in water is very low, generally less than 100 parts per million. However, solubility is important because the dissolved oil components are often toxic to aquatic life, especially at higher concentrations.

Viscosity is the resistance to flow in a liquid. The lower the viscosity, the more readily the liquid flows. The viscosity of oil is a function of its composition; therefore, crude oil has a wide range of viscosities. For example, the viscosity of Federated oil from Alberta is five mPa, while a Sockeye oil from California is 45 mPa at 15°C. In general, the greater the fraction of saturates and aromatics and the lower the amount of asphaltenes and resins, the lower the viscosity. As oil weathers, the evaporation of the lighter components leads to increased viscosity.

As with other physical properties, viscosity is affected by temperature, with a lower temperature giving a higher viscosity. For most oils, the viscosity varies as the logarithm of the temperature, which is a very significant variation. Oils that flow readily at high temperature can become a slow moving, viscous mass at low temperature. In terms of oil spill cleanup, viscous oils do not spread rapidly, do not penetrate soils rapidly, and affect the ability of pumps and skimmers to handle the oil. The dynamic viscosity of oil (in mPa s) is conveniently measured by a viscometer using a variety of cup-and-spindle sensors at very strictly controlled temperatures.

Density is the mass of a given volume of oil and is typically expressed in grams per cubic centimeter (g/cm^3). It is the property used by the petroleum industry to define light or heavy crude oils. Density is also important because it indicates whether particular oil will float or sink in water. As the density of water is 1.0 g/cm^3 at 15°C and the density of most oils ranges from 0.7 to 0.99 g/cm^3, most oils will float on water. As the density of seawater is 1.03 g/cm^3, even heavier oils

TABLE 27.2
Typical oil properties

Property	Units	Gasoline	Diesel	Light Crude	Heavy Crude	Intermediate Fuel Oil	Bunker C	Crude Oil Emulsion
Viscosity	mPa s at 15°C	0.5	2	5 to 50	50 to 50,000	1,000 to 15,000	10,000 to 50,000	20,000 to 100,000
Density	g/mL at 15°C	0.72	0.84	0.78 to 0.88	0.88 to 1.00	0.94 to 0.99	0.96 to 1.04	0.95 to 1.0
Flash Point	°C	−35	45	−30 to 30	−30 to 60	80 to 100	>100	>80
Solubility in Water	Ppm	200	40	10 to 50	5 to 30	10 to 30	1 to 5	—
Pour Point	°C	not relevant	−35 to −1	−40 to 30	−40 to 30	−10 to 10	5 to 20	>50
API Gravity		65	35	30 to 50	10 to 30	10 to 20	5 to 15	10 to 15
Interfacial Tension	mN/m at 15°C	27	27	10 to 30	15 to 30	25 to 30	25 to 35	not relevant
Distillation Fraction	% distilled at							
	100°C	70	1	2 to 15	1 to 10	—	—	not relevant
	200°C	100	30	15 to 40	2 to 25	2 to 5	2 to 5	
	300°C		85	30 to 60	15 to 45	15 to 25	5 to 15	
	400°C		100	45 to 85	25 to 75	30 to 40	15 to 25	
	Residual			15 to 55	25 to 75	60 to 70	75 to 85	

will usually float on it. Only certain Bitumen and very heavy residual oils such as Bunker C have densities greater than water and may submerge in water. The density of oil increases with time, as the light fractions evaporate.

Another measure of density is specific gravity, which is an oil's relative density compared with that of water at 15°C. The American Petroleum Institute (API) uses the API gravity as a measure of density for petroleum:

$$\text{API gravity} = (141.5 \div (\text{density at } 15.6°C)) - 131.5$$

Pure water has an API gravity of 10°. Oils with progressively lower specific gravities have higher API gravities. The scale is commercially important for ranking oil quality. Heavy inexpensive oils are <25° API; medium oils are 25 to 35° API; and light commercially valuable oils are 35 to 45° API. API gravities vary inversely with viscosity, asphaltic matter content (which increase from 4% to 8% at 40° to ~50% at 10° to 15° API), and N-content (which rises from 0.08% to 0.20% to ~1% over the same interval).

The density of oil is determined using an acoustic cell density meter following the American Society for Testing and Materials (ASTM) method D5002.

The *flash point* of oil is the temperature at which the vapor over the liquid will ignite upon exposure to an ignition source. A liquid is considered to be flammable if its flash point is less than 60°C. Flash point is a very important factor in relation to the safety of spill cleanup operations. Gasoline and other light fuels can ignite under most ambient conditions and therefore are a serious hazard when spilled. Many freshly spilled crude oils also have low *flash points* until the lighter components have evaporated or dispersed. On the other hand, Bunker C and heavy crude oils generally are not flammable when spilled. The flash point of oil with lower viscosity is determined following the ASTM method D1310, while that of heavier oils is determined following the ASTM method D93.

The *pour point* of oil is the temperature at which no flow of the oil is visible over a period of five seconds from a standard measuring vessel. The pour point of crude oils generally varies from 6 to 30°C. Lighter oils with low viscosities generally have lower pour points. As oils are made up of hundreds of compounds, some of which may still be liquid at the pour point, the pour point is not the temperature at which oil will no longer pour. The pour point represents a consistent temperature at which oil will pour very slowly and therefore has limited use as an indicator of the state of the oil. For example, waxy oils can have very low pour point, but may continue to spread slowly at that temperature and can evaporate to a significant degree. The pour point of oil is determined by following ASTM method D97.

Distillation fractions of an oil represent the fraction (generally measured by volume) of an oil that is boiled off at a given temperature. This data is obtained on most crude oils so that oil companies can adjust parameters in their refineries to handle the oil. This data also provides environmentalists with useful insights into the chemical composition of oils. For example, while 70% of gasoline will boil off at 100°C, only 5% of a crude oil will boil off at that temperature and an even smaller amount of a typical Bunker C. The distillation fractions correlate strongly to the composition as well as to other physical properties of the oil.

The *oil/water interfacial tension*, sometimes called surface tension, is the force of attraction or repulsion between the surface molecules of oil and water. The SI units for interfacial tension are milliNewtons per meter (mN/m). Together with viscosity, surface tension is an indication of how rapidly and to what extent oil will spread on water. The lower interfacial tensions with water, the greater the extent of spreading of oil. In actual practice, the interfacial tension must be considered along with the viscosity because it has been found that interfacial tension alone does not account for spreading behavior. Surface tensions change in smaller degree from one oil to another oil, but larger changes can accompany changes in temperature. Interfacial tensions can be measured by following closely ASTM method D971: using a Krüss K-10 Tensionmeter by the de Noüy ring method.

Odor of oil is a "quality" parameter, not a quantitative parameter. Oils that contain significant amount of certain types of unsaturated nitrogenous compounds, and sulfur containing compounds such as mercaptans tend to possess a pervasive H_2S like odor. In contrast, oil mainly composed of light hydrocarbons, containing high proportions of aromatics, or composed of mix of paraffins and naphthenes possess a sweet gasoline-like odor.

C. NONSPECIFIC METHODS FOR OIL ANALYSIS

In the last two decades, a wide variety of instrumental and noninstrumental techniques have been developed and used in the analysis of oil hydrocarbons. They include gas chromatography (GC), gas chromatography mass spectrometry (GC–MS), high performance liquid chromatography (HPLC), size exclusion HPLC, infrared spectroscopy (IR), supercritical fluid chromatography (SFC), thin layer chromatography (TLC), ultraviolet (UV) and fluorescence spectroscopy, stable isotope ratio mass spectrometry, and gravimetric methods. Of all these techniques, GC techniques are the most widely used technique to measure a wide range of oil hydrocarbons from volatile to high molecular weigh organic compounds. Compared to the molecular measurements two decades ago, GC methods have now been enhanced by more sophisticated analytical techniques, such as capillary GC-mass spectrometry (GC–MS), which is capable of analyzing the oil specific biomarker compounds and PAH hydrocarbons. The accuracy and precision of analytical data has been improved and optimized by a series of quality assurance/quality control measures, and the laboratory data handling capability has been greatly increased through advances in computer technology.

Depending on chemical/physical information needs, the point of application and the level of analytical detail, the methods used for oil spill study can be, in general, divided into two categories: nonspecific methods and specific methods for detailed chemical component analysis.

The conventional nonspecific methods include field screening gas chromatography with flame ionization detector (GC–FID) and photo ionization detectors (GC-PID); gravimetric and IR determinations (such as the U.S. Environmental Protection Agency (EPA) Method 418.1 and Method 9071, and American Society for Testing and Materials (ASTM) Method 3414 and 3921); ultraviolet fluorescence spectroscopy, thin layer chromatography used for the oil component class (saturated, aromatic, resin and asphaltene fraction) characterization; HPLC; size exclusion chromatography with fluorescence detection; and supercritical fluid chromatography (SFC). Compared to the specific methods, these nonspecific methods require shorter preparation and analytical time and are less expensive to use. These techniques have been used to screen sediments for petroleum saturate and aromatic compounds, to measure total petroleum hydrocarbons (TPH), to assess site contamination and remediation, to determine the presence and type of petroleum products that may exist in soil or water, and to qualitatively examine and compare oil weathering/degradation.

The major shortcoming associated with the nonspecific methods is that the data generated from these methods generally lack detailed individual component and petroleum source specific information, and therefore these methods are of limited value in many environmental forensic cases, for spilled oil characterization and source identification.

D. SPECIFIC METHODS FOR DETAILED OIL COMPONENT ANALYSIS

In response to the oil spill identification need and specific site investigation needs, attention has focused on the development of flexible, tiered analytical approaches, which facilitate the detailed compositional analysis by GC–MS, GC–FID, and other analytical techniques that quantitatively determine a broad range of individual petroleum hydrocarbons. A variety of diagnostic ratios, especially ratios of PAH and biomarker compounds, for interpreting chemical data from oil spills have been proposed.

1. Selected U.S. EPA Methods and Their Limitations for Oil Analysis

Table 27.3 summarizes selected EPA methods, major applications, and limitations of these methods for oil analysis. These EPA methods have been used as routine procedures for determination of volatile and semivolatile aromatic hydrocarbons presented in spilled oil and petroleum product samples. However, these methods were originally designed for measuring a wide variety of discrete industrial chemicals in wastewater and industrial waste. The fundamental shortcoming with these methods is that none of these standard EPA methods can provide information on detailed chemical

TABLE 27.3
Major Applications and Limitations of Standard EPA Methods for Oil Analysis

EPA Standard Method	Target Compounds and Application	Limitation for Oil Work
EPA 418.1	TPH by IR spectroscopy	Inherent accuracy of the method (positive or negative biases)
		Subject to various interferences
		Lack of effective reference standards
EPA 1664	n-Hexane extractable materials and silica gel treated n-hexane extractable material by extraction and gravimetry	Only measures total extractable materials in aqueous matrices
		Heavy interference
		Low molecular weight hydrocarbons could be lost during distillation
EPA 600 series (method standards for waste water)		
602	Purgeable aromatics, by GC/FID	600 and 8000 series were originally designed for waste water and industrial waste
610	16 polycyclic aromatics, by HPLC/GC	Can not provide detailed composition information of spilled oil
624	Purgeable volatiles, by GC/MS	Only BTEX measured, do not measure over 100 important oil hydrocarbons
625	Semi volatiles and pesticides, by GC/MS	
EPA 8000 series (method standards for solid waste, SW 846)		
8015	Non halogenated volatiles by GC/FID	Do not measure dominated alkylated PAH homologues, aliphatics and biomarkers in oils
8020	Aromatic volatiles by GC	Provide little diagnostic source information
8100	Volatiles by Capillary GC/MS	
8260	24 PAHs by GC/FID	
8270	Semi volatiles by capillary GC/MS	

TPH, Total petroleum hydrocarbons; PAH, polycyclic aromatic hydrocarbons.

components, which comprise the complex spill oil, or petroleum derived samples. The data generated from these methods are generally insufficient to answer the fundamental questions (such as type and source, weathering status of spilled oil, potential spillers, and so on) raised in an oil spill liability investigation. Of the more than 160 EPA priority pollutant organic compounds determined by these methods, only 20 are petroleum related hydrocarbons. Further, only half of these 20 compounds are found in significant quantities in oils and petroleum products. Also, the PAH compounds in oils are dominated almost exclusively by the C_1 to C_4 alkylated homologues of the parent PAH, in particular, naphthalene, phenanthrene, dibenzothiophene, fluorene and chrysene, none of which are measured by the standard EPA methods. Other important classes of petroleum hydrocarbons (e.g., aliphatics and biomarkers) are not measured by these methods at all.

Another example is the use of the EPA 418.1 method to determine TPH content. The EPA 418.1 method, based on measuring the absorption of C–H bond in the 3200 to 2700 wave number range, was originally intended for use only with liquid waste but had been one of the most widely used methods for the determination of TPH in soils before its demise because of the use of a chloro-fluoro carbon extracting. For some site assessments, Method 418.1 was the sole criterion for verification of site cleanup. However, there were some problems associated with this method such as inherent inaccuracy in the method (i.e., positive or negative biases caused by various factors) and the lack of effective reference standards when working with an unknown.

In recent several years, many EPA and ASTM methods have been modified (such as the modified EPA method 8015, 8260 and 8270; and the modified ASTM methods 3328-90, 5037-90 and 5739-95) to improve specificity and sensitivity for measuring spilled oil and petroleum products in soils and waters by environmental chemists. For example, EPA Method 8270 has been modified to increase analytical sensitivity and to expand the analyte list to include petroleum specific compounds such as the alkylated PAHs, sulfur, and nitrogen containing PAHs, and biomarker triterpane and sterane compounds. The principal modification to EPA Method 8270 is the use of the high resolution GC–MS selected ion mode (SIM) analysis that offers increased sensitivity relative to the full scan mode. Many environmental laboratories have used the modified EPA Method 8270, combined with column cleanup and rigorous QA measures, to identify and quantify low levels of hydrocarbons.

2. Selection of Source Specific Target Analytes

In addition, in the determination of groups or fractions of oil hydrocarbons,[4] oil spill identification requires further elaboration of oil target analytes to include identification of the individual specific target compounds and isomeric groups. The selection of appropriate target oil analytes is dependent mainly on the type of oil spilled, the particular environmental compartments being assessed, and on expected needs for current and future data comparison. In general, the major petroleum specific target analytes that may be needed to be chemically characterized for oil source identification and environmental assessment include the following:

(1) Individual saturated hydrocarbons including n-alkanes (n-C_8 through n-C_{40}) and selected isoprenoids pristane (2,6,10,14-tetramethyl-pentadecane) and phytane (2,6,10,14-tetra-methyl-hexadecane). In some cases, another three highly abundant isoprenoid compounds: farnesane (2,6,10-trimethyl-C_{12}), 2,6,10-trimethyl-C_{13}, and norpristane (2,6,10-trimethyl-C_{15}) are also included;

(2) Alkyl (C_1–C_{14}) cyclo-hexane homologous compound series. These homologous compounds exhibit a characteristic distribution patter in m/z 83 mass chromatograms for different types of fuels, providing another useful fingerprint for characterizing petroleum derivatives;

(3) The volatile hydrocarbons including BTEX (benzene, toluene, ethylbenzene, and 3 xylene isomers) and alkylated benzenes (C_3- to C_5-benzenes), naphthenes, and volatile paraffins and isoparaffins;

(4) The EPA priority parent PAHs and, in particular, the petroleum specific alkylated (C_1 to C_4) homologues of selected PAHs (that is, alkylated naphthalene, phenanthrene, dibenzothiophene, fluorene, and chrysene series). These alkylated PAH homologues (Table 27.4) are the backbone of chemical characterization and identification of oil spill assessments;

(5) Biomarker terpane and sterane compounds (Table 27.5). Analysis of selected ion peaks produced by these characteristic, environmentally persistent compounds generates information of great importance in determining source(s), weathered state and potential treatability;

(6) Measurements of bulk hydrocarbon groups including TPH, the unresolved complex mixtures (UCM), and the total saturates and total aromatics, contents of asphaltenes and resins,

(7) Additives to petroleum products. They include alkyl lead additives (tetramethyl lead and trimethylethyl lead at m/z 253 and 223, dimethyldiethyl lead at m/z 267 and 223, methyltriethyl lead at m/z 281 and 223, tetraethyl lead at 295 and 237); oxygenates including substances such as ethanol, methanol, methyl tertiary butyl ether (MTBE), ethyl tertiary butyl ether (ETBE), and tertiary amyl methyl ether (TAME); fuel dyes used for differentiation among fuel grades; and antioxidant compounds added to fuels to retard auto oxidation;

(8) Measurement of stable carbon isotope ratio ($\delta^{13}C$) is also included in many oil spill studies.

Another potentially valuable hydrocarbon group for oil spill identification is nitrogen and oxygen heterocyclic hydrocarbons. These heterocyclic hydrocarbons are generally only present in oils at quite relatively low concentrations compared to PAHs. However, they become enhanced with weathering because they are biorefractory and persistent in the environment. Most organic nitrogen hydrocarbons in crude oils are present as alkylated aromatic heterocycles with a predominance of neutral pyrrolic structures over basic pyridine forms. They are chiefly associated with high boiling fractions, much of the nitrogen in petroleum being in asphaltenes. Individual and alkyl homologues of carbazole, quinoline, and pyridine have been identified in many crude oils. These compounds may provide important clues for potential sources of hydrocarbons in the environment and for tracing petroleum molecules back to their biological precursors. Compared to the PAHs and biomarkers, the application of nitrogen and oxygen containing heterocyclic hydrocarbons in source identification is still in its infancy, and more research is clearly needed.

3. Oil Spill Identification Protocol

The oil spill identification system currently used is largely based on GC–FID and GC–MS techniques. Data produced from these two methods are used to compare spill samples with samples taken from suspected sources. Very recently, SINTEF Applied Chemistry of Norway and Battelle of the U.S.A. published the "Improved and standardized methodology for oil spill fingerprinting,"[5] which include four "levels" of analyses and data treatment. The recommended methodology approach is a result of documented and analytical improvements and a quantitative treatment of analytical data from GC–FID and GC–MS and the operational experiences over past few years among the participating forensic laboratories. Figure 27.1 presents the modified "Protocol/decision chart for the oil spill identification methodology." The final assessment is concluded by the four operational and technical defensible identification terms: *positive match, probable match,*

TABLE 27.4
Source-Specific Target PAHs and Alkylated Homologous PAHs for Oil Spill Studies

Oil-characteristic alkylated PAHs

Compound	Code	Ring Numbers	Target Ions
Naphthalenes			
C$_0$-naphthalene	C$_0$N	2	128
C$_1$-naphthalenes	C$_1$N	2	142
C$_2$-naphthalenes	C$_2$N	2	156
C$_3$-naphthalenes	C$_3$N	2	170
C$_4$-naphthalenes	C$_4$N	2	184
Phenanthrenes			
C$_0$-phenanthrene	C$_0$P	3	178
C$_1$-phenanthrenes	C$_1$P	3	192
C$_2$-phenanthrenes	C$_2$P	3	206
C$_3$-phenanthrenes	C$_3$P	3	220
C$_4$-phenanthrenes	C$_4$P	3	234
Dibenzothiophenes			
C$_0$-dibenzothiophene	C$_0$D	3	184
C$_1$-dibenzothiophenes	C$_1$D	3	198
C$_2$-dibenzothiophenes	C$_2$D	3	212
C$_3$-dibenzothiophenes	C$_3$D	3	226
Fluorenes			
C$_0$-fluorene	C$_0$F	3	166
C$_1$-fluorenes	C$_1$F	3	180
C$_2$-fluorenes	C$_2$F	3	194
C$_3$-fluorenes	C$_3$F	3	208
Chrysenes			
C$_0$-chrysene	C$_0$C	4	228
C$_1$-chrysenes	C$_1$C	4	242
C$_2$-chrysenes	C$_2$C	4	256
C$_3$-chrysenes	C$_3$C	4	270

Other EPA priority PAH pollutants

Compound	Code	Ring Numbers	Target Ions
Biphenyl	Bph	2	154
Acenaphthylene	Acl	3	152
Acenaphthene	Ace	3	153
Anthracene	An	3	178
Fluoranthene	Fl	4	202
Pyrene	Py	4	202
Benzo[a]anthracene	BaA	4	228
Benzo[b]fluoranthene	BbF	5	252
Benzo[k]fluoranthene	BkF	5	252
Benzo[e]pyrene	BeP	5	252
Benzo[a]pyrene	BaP	5	252
Perylene	Pe	5	252
Indeno[1,2,3-cd]pyrene	IP	6	276
Dibenz[a,h]anthracene	DA	5	278
Benzo[ghi]perylene	BP	6	276

Surrogates and Internal Standard

Compound	Code	Ring Numbers	Target Ions
[^{2}H$_{10}$]Acenaphthene			164
[^{2}H$_{10}$]Phenanthrene			188
[^{2}H$_{12}$]Benz[a]anthracene			240
[^{2}H$_{12}$]Perylene			264
[^{2}H$_{14}$]Terphenyl			244

TABLE 27.5
Source-Specific Target Biomarker Terpane and Sterane Compounds for Oil Spill Studies

Peak	Compound	Empirical Formula	Target Ions
	Sesquiterpanes		123
	Diamondoids		
	Adamantanes		135, 136, 149, 163, 177
	Diamantanes		187, 188, 201, 205, 229
	Terpanes		
1	C_{19} tricyclic terpane	$C_{19}H_{34}$	191
2	C_{20} tricyclic terpane	$C_{20}H_{36}$	191
3	C_{21} tricyclic terpane	$C_{21}H_{38}$	191
4	C_{22} tricyclic terpane	$C_{22}H_{40}$	191
5	C_{23} tricyclic terpane	$C_{23}H_{42}$	191
6	C_{24} tricyclic terpane	$C_{24}H_{44}$	191
7	C_{25} tricyclic terpane	$C_{25}H_{46}$	191
8	C_{24} tetracyclic terpane $+C_{26}$ (S + R) tricyclic terpanes (triplet)	$C_{24}H_{42} + C_{26}H_{48}$	191
9	C_{28} tricyclic terpane 1	$C_{28}H_{52}$	191
10	C_{28} tricyclic terpane 2	$C_{28}H_{52}$	191
11	C_{29} tricyclic terpane 1	$C_{29}H_{54}$	191
12	C_{29} tricyclic terpane 2	$C_{29}H_{54}$	191
13	Ts: $18\alpha(H),21\beta(H)$-22,29,30-trisnorhopane	$C_{27}H_{46}$	191
14	$17\alpha(H),18\alpha(II),21\beta(H)$-25,28,30-trisnorhopane	$C_{27}H_{46}$	191, 177
15	Tm: $17\alpha(H),21\beta(H)$-22,29,30-trisnorhopane	$C_{27}H_{46}$	191
16	$17\alpha(II),21\beta(H)$-25,30-bisnorhopane	$C_{28}H_{48}$	191, 177
17	$17\alpha(H),18\alpha(H),21\beta(H)$-28,30-bisnorhopane	$C_{28}H_{48}$	191
18	$17\alpha(H),21\beta(H)$-30-norhopane	$C_{29}H_{50}$	191
19	$18\alpha(H),21\beta(H)$-30-norneohopane (C_{29}Ts)	$C_{29}H_{50}$	191
20	$18\alpha(H)$ and $18\beta(H)$-oleanane	$C_{30}H_{52}$	191
21	$17\alpha(H),21\beta(H)$-hopane	$C_{30}H_{52}$	191
22	$17\beta(H),21\alpha(H)$-hopane (moretane)	$C_{30}H_{52}$	191
23	$22S - 17\alpha(H),21\beta(H)$-30-homohopane	$C_{31}H_{54}$	191
24	$22R$-$17\alpha(H),21\beta(H)$-30-homohopane	$C_{31}H_{54}$	191
25	Gammacerane	$C_{30}H_{52}$	191
26	$17\beta(H),21\beta(H)$-hopane	(Internal standard)	191
27	$22S$-$17\alpha(H),21\beta(H)$-30,31-bishomohopane	$C_{32}H_{56}$	191
28	$22R$-$17\alpha(H),21\beta(H)$-30,31-bishomohopane	$C_{32}H_{56}$	191
29	$22S$-$17\alpha(H),21\beta(H)$-30,31,32-trishomohopane	$C_{33}H_{58}$	191
30	$22R$-$17\alpha(H),21\beta(H)$-30,31,32-trishomohopane	$C_{33}H_{58}$	191
31	$22S$-$17\alpha(H),21\beta(H)$-30,31,32,33-tetrakishomohopane	$C_{34}H_{60}$	191
32	$22R$-$17\alpha(H),21\beta(H)$-30,31,32,33-tetrakishomohopane	$C_{34}H_{60}$	191
33	$22S$-$17\alpha(H),21\beta(H)$-30,31,32,33,34-pentakishomohopane	$C_{35}H_{62}$	191
34	$22R$-$17\alpha(H),21\beta(H)$-30,31,32,33,34-pentakishomohopane	$C_{35}H_{62}$	191
	Steranes		
35	C_{20} $5\alpha(H),14\alpha(H),17\alpha(H)$-sterane	$C_{20}H_{34}$	217 & 218
36	C_{21} $5\alpha(H),14\beta(H),17\beta(H)$-sterane	$C_{21}H_{36}$	217 & 218
37	C_{22} $5\alpha(H),14\beta(H),17\beta(H)$-sterane	$C_{22}H_{38}$	217 & 218
38	C_{27} $20S$-$13\beta(H),17\alpha(H)$-diasterane	$C_{27}H_{48}$	217 & 218
39	C_{27} $20R$-$13\beta(H),17\alpha(H)$-diasterane	$C_{27}H_{48}$	217 & 218
40	C_{27} $20S$-$13\alpha(H),17\beta(H)$-diasterane	$C_{27}H_{48}$	217 & 218
41	C_{27} $20R$-$13\alpha(H),17\beta(H)$-diasterane	$C_{27}H_{48}$	217 & 218
42	C_{28} $20S$-$13\beta(H),17\alpha(H)$-diasterane	$C_{28}H_{50}$	217 & 218

Continued

TABLE 27.5

Continued

Peak	Compound	Empirical Formula	Target Ions
43	C_{29} 20S-13β(H),17α(H)-diasterane	$C_{29}H_{52}$	217 & 218
44	C_{29} 20R-13α(H),17β(H)-diasterane	$C_{29}H_{52}$	217 & 218
45	C_{27} 20S-5α(H),14α(H),17α(H)-cholestane	$C_{27}H_{48}$	217 & 218
46	C_{27} 20R-5α(H),14β(H),17β(H)-cholestane	$C_{27}H_{48}$	217 & 218
47	C_{27} 20S-5α(H),14β(H),17β(H)-cholestane	$C_{27}H_{48}$	217 & 218
48	C_{27} 20R-5α(H),14α(H),17α(H)-cholestane	$C_{27}H_{48}$	217 & 218
49	C_{28} 20S-5α(H),14α(H),17α(H)-ergostane	$C_{28}H_{50}$	217 & 218
50	C_{28} 20R-5α(H),14β(H),17β(H)-ergostane	$C_{28}H_{50}$	217 & 218
51	C_{28} 20S-5α(H),14β(H),17β(H)-ergostane	$C_{28}H_{50}$	217 & 218
52	C_{28} 20R-5α(H),14α(H),17α(H)-ergostane	$C_{28}H_{50}$	217 & 218
53	C_{29} 20S-5α(H),14α(H),17α(H)-stigmastane	$C_{29}H_{52}$	217 & 218
54	C_{29} 20R-5α(H),14β(H),17β(H)-stigmastane	$C_{29}H_{52}$	217 & 218
55	C_{29} 20S − 5α(H),14β(H),17β(H)-stigmastane	$C_{29}H_{52}$	217 & 218
56	C_{29} 20R-5α(H),14α(H),17α(H)-stigmastane	$C_{29}H_{52}$	217 & 218
	Monoaromatic steranes		253
	Triaromatic steranes		231

inconclusive or *mismatch*. These categories represent degrees of differences between the analyses of two oils[5] according to the present criteria (e.g., ASTM Method D3328):

(1) *Positive match*: The chromatographic patterns of the samples submitted for comparison are virtually identical and the observed differences between the spill sample and suspected source are caused and can be explained by the acceptable analytical variance and weathering effects.

(2) *Probable match*: The chromatographic patterns of the spill sample is similar to that of the samples submitted for comparison, except: (a) for obvious changes which could be attribute to weathering, or (b) differences attributable to specific contamination.

(3) *Inconclusive*: The chromatographic patterns of the spill sample is somewhat similar to that of the sample submitted for comparison, except for certain differences that are of such magnitude that it is impossible to ascertain whether the unknown is the same oil heavily weathered, or a totally different oil.

(4) *Mismatch*: Unlike the samples submitted for comparison.

4. Tiered Analytical Approach

Tiered analytical approaches have been increasingly applied for oil spill identification in recent years. Depending on the needs of spilled oil characterization, support for biological studies, monitoring weathering effects on chemical composition changes, or source differentiation, the tiered analytical approaches may vary. The tiered approach used by the Environment Canada Oil Spill Research Program[6] includes the following:

- Tier 1, determination of hydrocarbon groups in oil residues;
- Tier 2, product screening and determination of *n*-alkanes and TPH;
- Tier 3, distribution pattern recognition of target PAHs and biomarker components (sometimes the volatile hydrocarbons are monitored);

- Tier 4, determination and comparison of diagnostic ratios of the "source specific marker" compounds with the potential source oil and with the corresponding data from database; and
- Tier 5, determination of weathered percentages of the residual oil.

In this tiered analytical approach, the high resolution capillary GC–FID analysis is applied to evaluate hydrocarbon groups (including TPH, UCM, the total saturates and total aromatics), to determine concentrations of n-alkanes and major isoprenoid compounds, and to characterize the product types in fresh to highly weathered oil samples. The GC–MS analysis provides data on the "source specific" marker compounds including the target alkylated PAH homologues and other

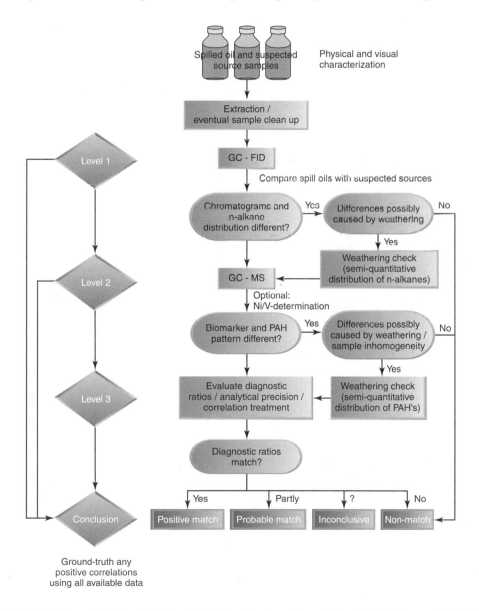

FIGURE 27.1 Protocol/decision chart for oil spill identification. (From Daling, P. S., Faksness, L. G., Hansen, A. B., and Stout, S. A., In *Proceedings of the 25th Arctic and Marine Oil Spill Program (AMOP) Technical Seminar*, Environment Canada, Ottawa, pp. 429–454, 2002. With permission.)

EPA priority PAHs, and biomarker terpane and sterane compounds to support the GC–FID results and to provide additional information for low concentration hydrocarbon contaminated samples.

Stout et al.[7] applied a similar tiered approach to hydrocarbon forensic analysis. The progression of analytical techniques used at each step of their tiered approach focuses on the detailed analysis of particular hydrocarbon boiling point ranges (e.g., volatile range hydrocarbons that comprise light distillates, or semivolatile range hydrocarbons that comprise middle and residual distillates of petroleum or coal liquids) and particular classes of chemical compounds that comprise those fractions. This tiered approach gives environmental forensic investigator the flexibility to gather information necessary to address site or incident specific questions about the nature and extent, and ultimately source(s), of hydrocarbon contamination.

5. Spilled Oil Sample Collection and Preparation

a. Sampling

For most purposes, a comparison is made between spilled samples and suspected source candidates. In the case of a ship being the suspected source, it is essential that reference samples be taken of all the oils carried on board the vessel, which might include cargo oils, fuel oil, lubricating oils and waste oils. In many cases attempts may be made to imply that not all the oil pollution came from one source, careful and detailed examination of contaminated sites such as beaches should be made to determine the uniformity of the spilled oil deposit. Any apparent variation in the type of oil should be sampled, the extent noted and supported by photographs.

Collecting a sample of spilled oil and then transporting it to a laboratory for subsequent analysis is common practice. While there are many procedures for collecting oil samples, it is always important to ensure that the oil is not tainted from contact with other materials and that the sample bottles are precleaned with solvents, such as hexane, and are suitable for the oil.

To collect oil-contaminated soil samples, common tools such as shovels, trowels, scoops, hand-operated auger coring devices are suitable for the top 30 cm. From 30 to 100 cm, one can manually remove the top layer of soil and then use the common tools as described above. For oil deposited on solid surfaces such as wood, rock, and concrete, it can be scrapped off the solid surfaces and placed directly into a sample container. On prolonged weathering at sea, oil tends to form blackish, semi-solid tar balls (in diameter of 1 mm to 300 mm). They can be collected by hand and placed into sample containers without difficulty. If freshly spilled oils or refined products have been absorbed and penetrated into sand or soil, representative oil-contaminated sand or sediment samples from various sites and varying depth should be collected.

When oil spreads to a thin film on water surface, it is often difficult to obtain a representative sample. In the absence of specialized equipment, oil-absorbing materials and wide-necked glass jars are often used to skim samples from the water surface. Highly viscous oils and emulsions tend to be much more concentrated on the sea surface and can be usually be scooped up easily.

b. Sample Preparation

The various oil samples are prepared following the ETC Method 5.3/1.3/M.[3,8]

The oil samples are directly dissolved in hexane at a concentration of 50 to 100 mg/ml.

The water or soil samples are spiked with appropriate amounts of deuterated surrogate standard compounds (*ortho*-terphenyl, a mixture of d_{10}-acenaphthene, d_{10}-phenanthrene, d_{12}-benz (a) anthracene, and d_{12}-perylene) prior to extraction. Water samples are extracted according to the EPA Method 3510: Separator Funnel Liquid-Liquid Extraction.

Soil samples are dried with anhydrous sodium sulphate and extracted with methylene chloride (DCM) using Soxhlet extraction, or serially extracted 3 times with 1:1 hexane/DCM, DCM, and DCM using sonication. If there is color in the third extraction, an additional extraction is performed. The extracts are combined, further dried with sodium sulphate, and filtered with a glass fiber filter.

The extracts are concentrated to appropriate volumes and solvent-exchanged with hexane by rotary evaporation and nitrogen blow down. An aliquot of the concentrated extract is then evaporated to dryness and weighed to obtain the total solvent extractable materials (TSEM) or the total oil weight.

A microcolumn packed with 3 g of activated silica gel is used for sample cleanup and fractionation. An aliquot of the extract (containing about 20 mg oil or TSEM) is then transferred to the silica gel cleanup column to remove polar components and other interferences. The column is eluted first with hexane, which recovers saturated hydrocarbons as Fraction 1 (F1). The mixture of hexane/DCM or hexane/benzene (1:1, v:v) is used to elute the aromatic compounds as Fraction 2 (F2). Half of F1 and half of F2 are combined into the Fraction 3 (F3). These three fractions are concentrated to appropriate volume (0.5 to 1.0 ml) by nitrogen purge. The quantitation internal standards are 5-α-androstane, d_{14}-terphyl, and C_{30} 17β(H), 21β(H)-hopane. The Fraction 3 is analyzed for quantitation of the TPH and the UCM by GC–FID. The Fraction 1 is analyzed for determination of n-alkanes by GC–FID and biomarker terpanes and steranes by GC–MS. The Fraction 2 is analyzed for the determination of alkylated PAH homologues and other EPA priority unsubstantiated PAHs by GC–MS.

For analysis of BTEX and other alkyl benzenes, all oil samples are directly weighed and dissolved in n-pentanc to an approximate concentration of 2 mg/ml. Prior to analysis, the tightly capped oil solutions are put in a refrigerator for 30 min to precipitate the asphaltenes to the bottom of the vials in order to avoid performance deterioration of the column.

c. GC Analysis

The GC–FID analysis is conducted by injection of 1 to 2 μl of F1 or F3 into a gas chromatograph equipped with a high resolution capillary column (operated in splitless injection mode). The injector and detector temperatures are set at 290 and 300°C, respectively. The GC temperature program[3,8] is selected to achieve near-baseline separation of all of the saturated hydrocarbons. Quantitation of the individual components is performed by the internal standard method. The relative response factor (RRF) for each component is calculated relative to the internal standard. The TPH is also quantified by the internal standard method using the baseline corrected total area of the chromatogram and the average hydrocarbon response factor determined over the entire analytical range.[3,8]

The GC–MS analysis is conducted by injection of 1 μL of F1 or F2 into a gas chromatograph-mass spectrometer. The MS detector is operated in the scan mode to obtain spectral data for identification of components and in the selected ion mode (SIM) for quantitation of target compounds. An appropriate temperature program is selected to achieve near-baseline separation of all of the target components. Quantitation of the alkalized PAH homologues, other EPA priority PAHs and biomarker compounds are performed by the internal standard method with the RRFs for each compound determined during the instrument calibration. The ions monitored for alkylated PAH and biomarker analyses are listed in Table 27.4 and Table 27.5, respectively.

The quantitation report limits of 0.2 μg/l in water and 0.1 μg/g in sediment (dry weight) for n-alkanes, 0.01 μg/l in water and 0.0005 μg/g in sediment for PAHs and biomarkers, 50 μg/l in water and 10 μg/g in sediment for TPH may be achieved.

The analysis of BTEX and alkyl benzene compounds is performed on a GC–MS. The MSD is operated in the SIM mode. A fused-silica column with dimension of 30 m × 0.25 mm (i.e., 0.25-μm film) is used. The temperature program used is the following: 35°C for 2 min, ramp at 10°C/min to 300°C, and hold for 10 min. A standard solution composed of five BTEX compounds, six C_3-benzene compounds, two C_4-, one C_5-, and one C_6-benzene is used to determine the response factor relative to the internal standard d_{10}-ethylbenzene for each target compound.

6. Quality Assurance and Quality Control

The reliability of analytical methods is dependent on the quality control procedures followed. In order to support spilled oil forensic investigations, well-defined quality management (including a quality assurance and quality control system, updated standard operational procedures, personnel training program and record, up-to-date methodology, equipment management, sample management, and data management) must be a fundamental element of any analytical lab program. The chemical measurements must be conducted within the framework of highly stringent, defensible, and reliable QA and QC programs.

Prior to individual components or TPH analysis, a five-point response calibration curve is established to demonstrate the linear-range of the analysis. The calibration solution for GC–FID analysis is composed of C_8 through C_{40} n-alkanes, plus pristane and phytane. The calibration solutions for GC–MS analysis are composed of C_{30} 17β(H), 21α(H)-hopane, C_{30} 17β(H), 21β(H)-hopane and C_{29} 20R ααα sterane for F1; and the National Institute of Standards and Technology ((NIST) certified standard reference materials SRM 1491 (including priority PAH components plus dibenzothiophene) for F2, respectively. Check standards at the midpoint of the established calibration curves are analyzed before and after each analytical batch of samples (seven to ten samples) to validate the integrity of the initial calibration. If the response of the check standards varies from the historical average response by more than 25%, the test should be repeated using a fresh calibration standard. The RRF stability is a key factor in maintaining the quality of the analysis. A control chart for RRF values should be prepared and monitored. The RRFs for n-C_8 to n-C_{34}, and pristane and phytane should be 0.95 ± 0.1 relative to the internal standard 5-α-androstane. Mass discrimination for high molecular weight n-alkanes in the injection port must be carefully monitored. If there is a problem with mass discrimination, it can be minimized by trimming the column and by replacing the liner. All samples and quality control samples (procedural blank, matrix spike samples, duplicate, and reference oil sample) are spiked with appropriate surrogates to measure individual sample matrix effects associated with sample preparation and analysis. Surrogate and matrix spike recoveries should be within 60% to 120%, and duplicate relative reference values should be less than 25%. Method detection limits (MDLs) studies of target compounds are performed according to the procedure described in the EPA protocol titled "Definition and Procedure for the Determination of the Method Detection Limit" (Code of Federal Regulations 40CFR Part 136).

Besides these routine quality control measures required by standard EPA and ASTM methods, some refinements may further implemented for the purposes of unambiguous spill source identification and for environmental samples with low concentrations of hydrocarbons. The key refinements may include the following:

- Applying more rigorous calibration check standards of ±15 percentage
- Quantifying the alkylated PAH homologues using RRFs obtained directly from authentic alkylated PAH standards, rather than the standard parent PAH compounds
- Quantifying alkylated isomer PAH series at various alkylation levels by manually setting the integration baselines
- Increasing sample size and to reduce the sample extract preinjection volume for those sediment samples with very low concentrations of hydrocarbons.

III. OIL AND PETROLEUM PRODUCT TYPE SCREENING AND DIFFERENTIATION

Generally, the type and identity of fresh to mildly weathered oils and petroleum products can be readily revealed from their GC–FID traces, especially where the spilled oil or petroleum product is heavy and background hydrocarbon levels are low in an impacted environment. In addition to measuring TPH and other hydrocarbon groups in samples, GC–FID chromatograms provide a distribution pattern of petroleum hydrocarbons (e.g., carbon range and profile of UCM), fingerprints of the major oil components (e.g., individual resolved n-alkanes and major isoprenoids), and information on the weathering extent of the spilled oil. Comparing biodegradation indicators (such as n-C_{17}/pristane and n-C_{18}/phytane) for the spilled oil with the source oil can be also used to monitor the effect of microbial degradation on the loss of hydrocarbons at the spill site for short time periods.

A. GENERAL CHEMICAL COMPOSITION OF CRUDE OIL

Crude oil compositions vary widely. Depending on the sources of carbon from which the oils are generated and the geologic environment in which they migrated and from which reservoir, they can have: (1) dramatically varied compositions in the C_5 to C_{42} carbon range such as relative amounts of paraffinic, aromatic and asphaltenic compounds; (2) large differences in the n-alkane and cyclic-alkanes (such as alkyl cyclo-hexanes) concentrations and distribution patterns and UCM profiles; (3) significantly different relative ratios of isoprenoids to normal alkanes; and (4) large differences in distribution patterns and concentrations of oil-characteristic long-side-chain n-alkyl benzenes (the carbon number in the alkyl side chain can be up to C_{27} for some oils), alkylated PAH homologues (many of four- to six-ring unsubstantiated PAHs are only minor components in oils), and biomarkers.

In general, most crude oils exhibit an n-alkane distribution profile of decreasing abundances with increasing carbon number. The carbon preference index (CPI) values of most oils are ~ 1. Oils with CPI values greater than 1 are often derived from source rock strata that contained relatively abundant land plant organic components including leaf waxes. CPI is defined as the total of n-alkanes with odd carbon numbers divided by the total of n-alkanes with even carbon numbers in the carbon range of C_8 to C_{40}. The CPI values can be used as an indicator for distinguishing input of petrogenic or biogenic (CPI > 2) hydrocarbons in contaminated samples.

$$CPI = (\text{the sum of odd } n\text{-alkanes})/(\text{the sum of even } n\text{-alkanes})$$

or in the simplified formula:

$$CPI = (C_{23} + C_{25} + C_{27} + C_{29} + C_{31} + C_{33})/(C_{24} + C_{26} + C_{28} + C_{30} + C_{32} + C_{34})$$

Figure 27.2 shows GC–FID chromatograms (by high temperature program) for six different oils. Table 27.6 summarizes the hydrocarbon group analysis results for these six oils. Clearly, these six oils are very different, as not only are there large differences in the n-alkane distributions and UCM profiles, but also differences in hydrocarbon group composition and in relative ratios of isoprenoids to normal alkanes. Note that the Orinoco oil (a Bitumen oil from Venezuela) has nearly no n-alkanes in its GC–FID chromatogram.

B. GENERAL CHEMICAL COMPOSITION FEATURES OF REFINED PRODUCTS

Refined petroleum products are obtained from crude oil through a variety of refining processes including[9]:

Distillation: separation of oil fractions from light to heavy based on boiling point and volatility of oil components under atmospheric and vacuum pressure;

Cracking: the process of cracking large hydrocarbons into smaller ones;

Catalytic reforming: the process to rearrange the molecular structure of oil hydrocarbons using either heat or a catalyst, such as enrichment of monoaromatics from alkane and cycloalkane compounds by catalytic reforming processes;

Isomerization: reversible conversion leading ultimately to thermodynamic equilibrium mixture of isomers or of compounds of the same molecular formula but with different

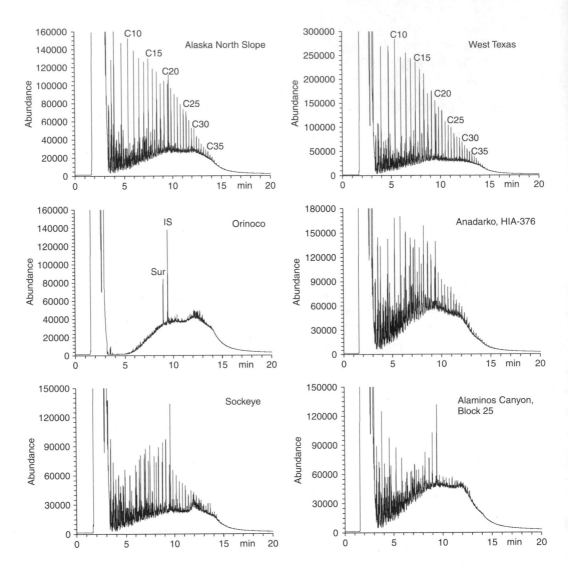

FIGURE 27.2 GC–FID chromatograms of six oils. These six oils are different, as not only are there large differences in the *n*-alkane distributions and UCMs, but also in relative ratios of isoprenoids to normal alkanes (see Table 27.6). Note that the Orinoco sample has nearly no *n*-alkanes on its GC–FID chromatogram.

    arrangements (for example, isomerization of straight-chain alkanes to branched ones for octane-number enhancement of gasoline);

Alkylation: the reaction of alkanes and olefins with different alkylating agents. It includes alkylation of alkanes with alkenes, alkylation of alkenes with alkenes, and alkylation of aromatics. Acid-catalyzed alkylation, particularly alkene–alkene reactions, are of great practical importance in upgrading motor fuels; and

Blending.

    Depending on the chemical composition of their "parent" crude oil feedstock, varying refining approach and conditions, wide range of applications, regulatory requirements, and economic requirements, refined products can have wide variety in chemical compositions. However, they can be still categorized into the following broad classes based on their general chemical composition features.[7,10]

TABLE 27.6
Hydrocarbon Group Analysis Results for Six Different Crude Oils

Hydrocarbon Groups	Alaska North Slope	West Texas	Orinoco	Anadarko (HIA-376)	Sockeye	Alaminos Canyon (Block 25)
Saturates (%)	75.0	78.5	44.6	88.6	49.2	79.0
Aromatics (%)	15.0	14.8	27.3	8.1	17.2	13.2
Resins (%)	6.1	6.0	13.3	3.3	15.1	7.1
Asphaltenes (%)	4.0	0.7	14.8	0.0	18.5	0.7
Resolved peaks/ GC-TPH (%)	19.8	24.2	3.0	14.8	20.0	9.6
GC-UCM/GC-TPH (%)	80.2	75.8	97.0	85.2	80.0	90.4
Total n-alkanes (mg/g oil)	63.4	94.7	—	54.1	26.0	22.4
n-C17/pristane	1.62	2.40	—	0.82	0.78	0.27
n-C18/phytane	1.90	1.83	—	0.74	0.66	0.40
Pistane/phytane	1.35	1.07	—	1.14	0.94	1.16
Total BTEX and C3-benzenes[a] (μg/g oil)	21920	33560	250	15700	14040	7950
Total PAHs						
Five alkylated PAH homologues (μg/g oil)	10493	7841	3672	9771	5149	6116
Other EPA priority PAHs (μg/g oil)	204	106	55	204	84	79

[a] BTEX: benzene, toluene, ethylbenzene, and xylenes; C3-benzenes include eight isomers.

1. Light Distillates

Light distillates are typically products in the C_3 to C_{12} carbon range. They include aviation gas (gasoline-type jet fuel which has a wider boiling range than kerosene-type jet fuel and includes some gasoline fractions), naphtha (a liquid petroleum product that boils from about 30°C to approximately 200°C, the term petroleum solvent is often used synonymously with naphtha), and automotive gasoline. The GC trace of fresh light distillates is featured with dominance of light-end, resolved hydrocarbons and a minimal UCM. As an example, Figure 27.3 presents the GC–FID chromatogram of a gasoline.

Gasoline is the generic term used to describe volatile, inflammable petroleum fuels used primarily for internal combustion engines. It is a complex mixture of hundreds of different hydrocarbons predominately in C_4 to C_{13} range, with the nominal boiling point range of 40 to 180°C or, at most, below 200°C. The composition of gasoline is best expressed in five major hydrocarbon classes: paraffins, isoparaffins (branched alkanes), naphthenes (cyclo-alkanes), aromatics, and olefins (PIANO). The bulk PIANO composition provides a useful cumulative parameter for fuel type (such as gasoline, aviation gasoline, or jet fuel) differentiation. Gasoline contains considerable BTEX and alkylated benzene compounds.

The properties of gasoline are quite diverse, and the principal properties affecting the performance of gasoline are volatility and combustion characteristics. The knock rating of a gasoline, one of the combustion characteristics, is expressed as "octane number" and it is the percentage by volume of iso-octane (2,2,4-trimethyl-pentane, octane number 100, by definition) in admixture with normal heptane (octane number 0, by definition) that has the same knock characteristics as the gasoline being assessed. In general, the higher the relative content of iso-octane and aromatics, particularly toluene, the higher the octane rating of the fuel. In the northern regions of Canada, gasoline used in winter (where the temperature can be as low as −40°C) can be

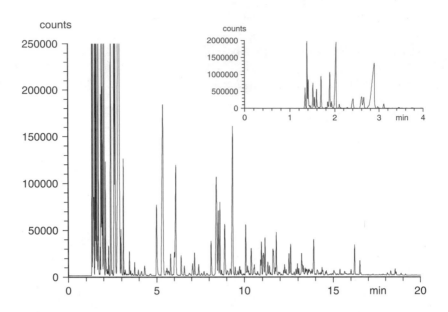

FIGURE 27.3 GC–FID chromatogram of a pure gasoline. The chromatogram was obtained by using the DB-5HT column (30 m × 0.32 mm id). The temperature program used is the following: 5-min hold at 30°C, ramp to 250°C at 7.5°C/min, and hold for 1 min.

quite different from that used in the summer (where the temperature can be as high as 30°C). In order to improve some specific properties such as the engine efficiency and antiknock properties, certain chemical compounds, *additives*, are often added to gasoline or other petroleum products. They may include octane-boosting additives (such as methyl tertiary butyl ether, MTBE), oxidation inhibitors (such as aromatic amines and hindered phenols), corrosion inhibitors (such as carboxylic acids and carboxylates), anti-icing additives (such alcohols, glycols, and surfactants), antiknocking lead alkyls, and dyes (oil-soluble solid and liquid dyes: *red*, alkyl derivatives of azobenzene-4-azo-2-naphthol; *orange*, benzene-azo-naphthol; *yellow*, *para*-diethylaminoazobenzene, and *blue*, 1,4-diisopropyl-aminoanthraquinone) for identification of different gasoline.

Since the 1970s, the lead level in refined products in Canada and the U.S.A. has decreased substantially. Use of leaded gasoline in cars was completely banned in Canada and the U.S.A. in 1990 and 1996, respectively.

2. Mid-Range Distillates

Mid-range distillates are typically products in a relatively broader carbon range (C_6 to C_{26}) and include kerosene (a flammable pale yellow or colorless oily liquid with a characteristic odor intermediate in volatility between gasoline and diesel oil that distills between 125 and 260°C), aviation jet (turbine) fuels, and lighter diesel products.

Jet fuel is kerosene-based aviation fuel. It is medium distillate used for aviation turbine power units and usually has the same distillation characteristics and flash point as kerosene. Jet fuels are manufactured predominately from straight-run kerosene or kerosene-naphtha blends in the case of wide cut fuels that are produced from the atmospheric distillation of crude oil. Jet fuels are similar in gross composition, with many of the differences in them attributable to additives designed to control some fuel parameters such as freeze and pour point characteristics. For example, the chromatogram (Figure 27.4) of a commercial jet fuel (Jet A) is dominated by GC-resolved *n*-alkanes in a narrow range of *n*-C_7 to *n*-C_{18} with maximum being around *n*-C_{11}. The UCM is well defined.

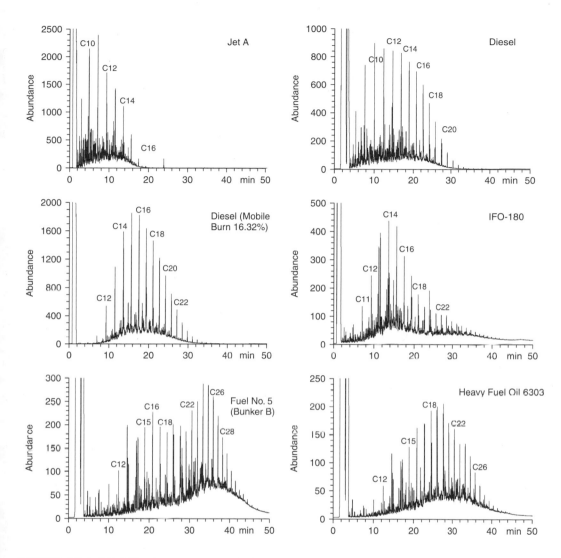

FIGURE 27.4 GC–FID chromatograms of six petroleum products (Jet fuel, Diesel, weathered Diesel, IFO-180, Fuel No. 5 (Bunker B), and Heavy Fuel Oil), illustrating differences of these products in the chromatographic profiles, carbon range, and UCM distribution patterns.

In chemical composition, military-grade jet fuels are slightly more widely varying than the commercial Jet A. The military jet fuels are generally fall into three categories: (1) wide cut-type fuels (including JP-4 and JP-6); (2) kerosene type fuels (including JP-1, JP-2, JP-3, JP-5, JP-7, and JP-8); and (3) selected component fuels (including JP-9 and JP-10).

Diesel fuels originally were straight-run products obtained from the distillation of crude oil. Currently, diesel fuel may also contain varying amounts of selected cracked distillates to increase the volume available. The boiling range of diesel fuel is approximately 125 to 380°C. One of the most widely used specifications (ASTM D-975) covers three grades of diesel fuel oils: diesel fuel #1, diesel fuel #2, and diesel fuel #4. Grades #1 and #2 diesels are distillate fuels; they are most commonly used in high-speed engines of the mobile type, in medium speed stationary engines, and in railroad engines. Grade #4 diesel covers the class of more viscous distillates and, at times, blends of these distillates with residual fuel oils. The marine fuel specifications have four categories of distillate fuels and fifteen categories of fuels containing residual components (ASTM D-2069 Method).

Diesel consists of hydrocarbons in a carbon range of C_8 to C_{28} and has significantly high concentrations of n-alkanes, alkyl-cyclohexane, and PAHs (Figure 27.4). The properties of a given diesel are largely a function of the crude oil feedstock. The GC chromatogram of the diesel fuel #2 is generally dominated by a nearly normal distribution of n-alkanes with maxima being around n-C_{11} to n-C_{14}. In addition, a central UCM "hump" is obvious.

3. Classic Heavy Residual Fuels

Classic heavy fuel types include fuel No. 5 and No. 6 (also known as Bunker C) fuel. The heavy residual fuels are largely used in marine diesel and industrial power generation. The residual fuel oil is fuel oil that is manufactured from the distillation residuum, and the various grades of heavy fuel oils are produced to meet rigid specifications to ensure suitability for their intended purpose.

For many years, the term "Bunker C fuel oil" has been widely used to designate the most viscous residual fuels for general land and marine use. Recognizing the need for a wide and more accurate classification for residual fuels, particularly when used for marine applications, in 1977 the major oil companies in cooperation with other interested parties introduced an expanded list of grading for residual oils. The intermediate fuel oil (IFO) grades are based upon the kinetic viscosity in centistokes (cSt) at 50°C.

The chemical composition of Bunker C (or IFO 380) can vary widely and remarkably, depending on production oilfields, production years, and processes it has undergone. Currently, many Bunker type fuels are produced by blending residual oils with diesel fuels or other lighter fuels in various ratios to produce residual fuel oil of acceptable viscosity for marine or power plant use.

For comparison, the chromatograms of an IFO 180, a lighter residual fuel No. 5 (also called Bunker B) and a Heavy Fuel Oil 6303 (also called Bunker C or Land Bunker, from Imperial Oil Ltd., Nova Scotia, Canada) are also shown in Figure 27.4. Figure 27.5 depicts graphically the quantitative distribution of n-alkanes for these products. The differences in the chromatographic profiles, carbon range, the shapes of UCM, distribution of n-alkanes and major isoprenoids, and diagnostic ratios of target alkanes (such as n-C17/pristane and n-C18/phytane) among these products are obviously considerable.

4. Lubricating Oil

Lubricating oil is used to reduce friction and wear between bearing metallic surfaces that are moving with respect to each other by separating the metallic surface with a thin film of the oil. Petroleum derived lubricating oil is a mixture produced by distillation of selected paraffinic and naphthenic crude oils, after which chemical changes may be required to produce the desired properties in the product. The production of lubricating oils is well established and consists of four basic procedures[10]:

1. Distillation and deasphalting to remove the lighter constituents of the feedstock
2. Solvent refining and hydrogen treatment to remove the nonhydrocarbon constituents and to improve the feedstock quality
3. Dewaxing to remove the wax constituents and to improve the low-temperature properties
4. Clay treatment or hydrogen treatment to prevent instability of the product.

Lubricating oils may be divided into many categories according to the type of services and applications, such as motor oil, transmission oil, crankcase oil, hydraulic fluid, cutting oil, turbine oil, heat-transfer oil, electrical oil, and many others. However, there are two main groups: (1) oils used in intermittent service, such as motor and aviation oils, and (2) oils designed for continuous

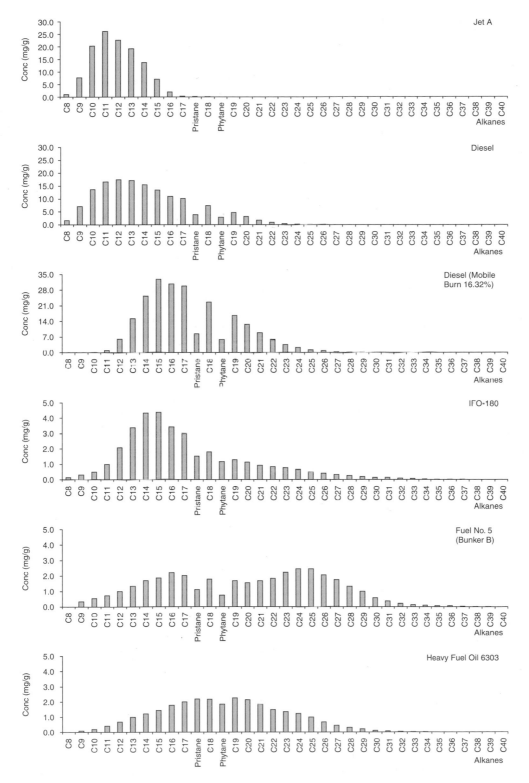

FIGURE 27.5 Quantitative distribution of *n*-alkanes of the six petroleum products mentioned in Figure 27.4, illustrating distinguishing features of *n*-alkane distribution patterns between these products.

service, such as turbine oils. Chemical additives are often added to base oil to enhance the properties and to improve such characteristics as oxidation resistance and corrosion resistance of lubricating oil. Small scale lubricating oil spills and contaminations are quite common due to their wide application.

Figure 27.6 shows the high temperature (from 40 to 325°C) GC–FID chromatograms for six different lubricating oils. In general, lubricating oils have broad GC profiles in the carbon range of C_{18} to C_{40} with boiling points greater than 340°C. Lubricating oil does not contain lower boiling portion of petroleum hydrocarbons. It is largely composed of saturated hydrocarbons, and its GC trace is often dominated by the UCM of hydrocarbons with very small amount of resolved peaks being present. In lubricating oil such as hydraulic fluid, for example, the PAH concentrations can be very low, while the concentration of multi condensed-ring biomarker compounds could be very high, in comparison with most other refined products. Therefore, determination of these source specific marker compounds often allow for successful identification and correlation between refined products from different sources.

5. GC–MS Analysis for *n*-Alkanes and Alkyl Cyclo-Hexanes

Figure 27.7 presents the GC–MS (at m/z 85 for *n*-alkane distribution) chromatograms of Arabian Light crude oil, Jet A, Diesel Fuel No. 2, and Heavy Fuel Oil 6303. For comparison, the corresponding GC–MS chromatograms (at m/z 83) showing the characteristic distribution of alkyl (C_0- to C_{15}-) cyclo-hexane homologous series are also presented in Figure 27.7. Clearly, the m/z 83 chromatograms provide another useful fingerprint for characterizing petroleum products.[11] Due to the increased resolution and sensitivity of the MS detector, the distribution of low-abundance *n*-alkanes and alkyl cyclo-hexanes are more clearly distinguished. In contrast to the corresponding GC–FID chromatograms, the GC–MS (SIM) chromatograms at m/z 85 and 83 have much simpler and clear traces for the target saturated hydrocarbons. The unresolved complex mixture "envelope" seen in the GC–FID chromatograms is greatly reduced. Differentiation between samples is considerably simplified by comparing the chromatogram profiles and distribution patterns of selected ions. The Jet A and Diesel No.2 can be readily distinguished from each other by the *n*-alkane and cyclo-hexane distribution ranges (C_8–C_{17} for Jet A Fuel and C_8–C_{25} for Diesel No.2, respectively) and patterns and relative amounts of pristane and phytane. The HFO 6303 eluted late in the 10 to 50 minutes range, and its m/z 83 chromatogram is dominated by a broad hump of unresolved saturated hydrocarbons and can be readily distinguished from the Jet A and Diesel No.2. The resolved *n*-alkanes and alkyl cyclo-hexanes in HFO 6303 distributed in a wide range with the maximum being around in C_{18} to C_{20}. The chromatograms of the Arabian Light oil are significantly different from the refined products. This oil is characterized by the *n*-alkane and alkyl cyclo-hexane distribution in a much wider carbon range from C_8 to C_{40} for *n*-alkanes and from alkyl C_1- to C_{16}- for cyclo-hexane homologous series, respectively.

IV. EFFECTS OF WEATHERING ON OIL CHEMICAL COMPOSITION AND PHYSICAL PROPERTY CHANGES

A. OIL WEATHERING

When crude oil or petroleum products are accidentally released to the environment, whether on water or land, they are immediately subject to a wide variety of changes in physical and chemical properties that in combination are termed "weathering". The weathering processes include (1) evaporation, (2) emulsification, (3) natural dispersion, (4) dissolution, (5) microbial degradation, (6) photo oxidation, (7) other processes such as sedimentation, adhesion on to surface of suspended particulate materials, and oil-fine interaction. The rate of weathering of oil can be very different, depending on the type of oil spilled and the local environmental conditions during and after spillage.

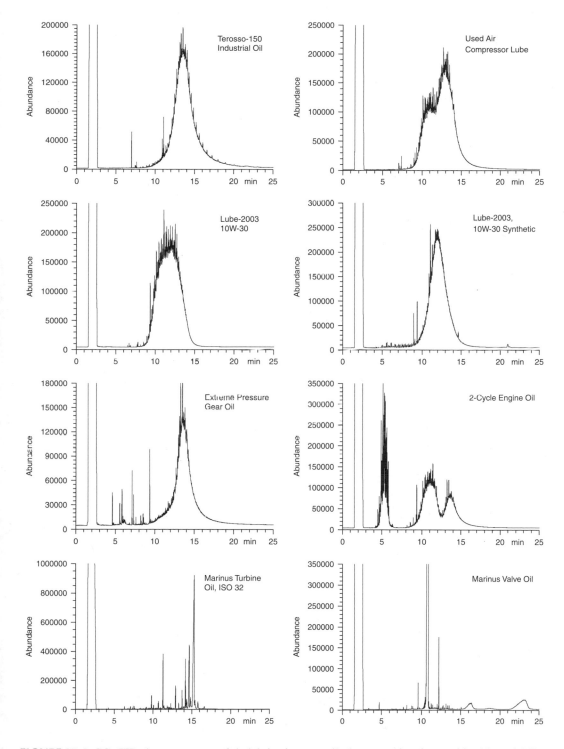

FIGURE 27.6 GC–FID chromatograms of six lubricating type oils demonstrating the considerable variability among this group of petroleum products.

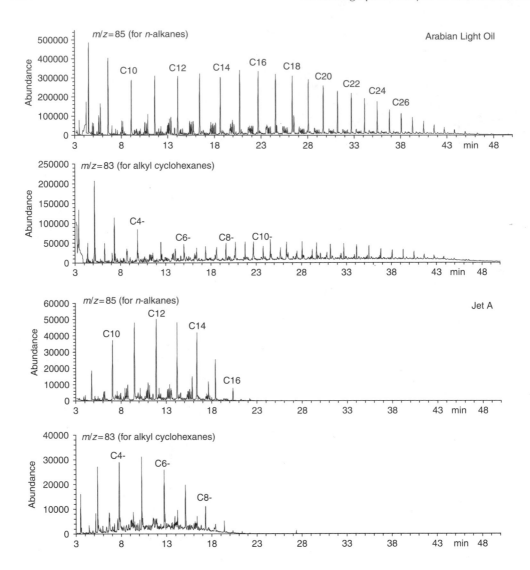

FIGURE 27.7 GC–MS (m/z 85 and 83) chromatograms of Arabian Light Oil, Jet A fuel, Diesel No.2, and HFO6303, illustrating distinguishing features of n-alkane and alkyl-cyclohexane distribution patterns between these oil and oil products.

Evaporation: In the short term after a spill, evaporation is the single most important and dominant weathering process, in particular for the light petroleum products. In the first few days following a spill, the loss can be up to 70% and 40% of the volume of light crude and petroleum products, and gasoline can evaporate completely above zero degrees. For heavy or residual oils such as Bunker C oil, the losses are only about several percentage of volume. The rate at which oil evaporates depends primarily on the oil composition. The more volatile components an oil or fuel contains, the greater the extent and rate of its evaporation.

The rate of evaporation is very rapid immediately after a spill and then slows considerably. The properties of oil can change significantly with the extent of evaporation. If about 40% (by weight) of oil evaporates, its viscosity could increase by as much as a thousand-fold, its density could rise by as much as 10% and its flash point by as much as 400%. The extent of evaporation can be the

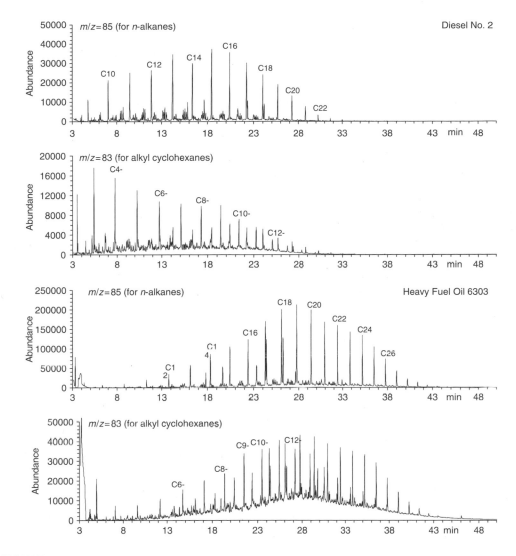

FIGURE 27.7 Continued.

most important factor in determining properties of oil at a given time after the spill and in changing the behavior of the oil.

Emulsification: Emulsification is considered the second most important weathering process after a marine spill by which water is dispersed into oil in the form of small droplets. The mechanism of water-in-oil emulsion formation is not yet fully understood, but it probably starts with sea energy forcing the entry of small water droplets, about 10 to 25 μm in size, into the oil. Emulsions of many types contain about 70% water. In general, water-in-oil emulsion can be categorized into four types: (1) unstable: oil simply does not hold water; (2) entrained: water droplets are simply held in the oil by viscosity to form an unstable emulsion, and it breaks down into water and oil within minutes or a few hours at most; (3) semistable or meso-stable: the small droplets of water are stabilized to a certain extent by a combination of the viscosity of the oil and interfacial action of asphaltenes and resins. For this to happen, the asphaltenes or resin content of the oil must be at least 3% by weight. The viscosity of meso-stable emulsions is 20 to 80 times higher than that of the starting oil. These emulsions generally break down into oil and water or sometimes into water, oil, and stable emulsion within a few days; (4) stable emulsions: they form in a way similar to meso-stable emulsions except that the oil must contain at least 8% asphaltenes. The viscosity of stable emulsions is 500 to 1200 times higher than that of the starting oil and the emulsion will remain stable for weeks and even months after formation. Stable emulsions are reddish-brown in color and appear to be nearly solid. These emulsions do not spread and tend to remain in lumps or mats on the sea or shore.

Emulsion formation changes the fate of the oil. It has been noted that when oil forms stable or meso-stable emulsions, evaporation slows considerably. Biodegradation also appears to slow down.

Dissolution: Solubility is defined as the amount of a substance (solute) that dissolves in a given amount of another substance (solvent). Through the process of dissolution, some of the soluble components of the oil are lost to the water column. The amount of the oil hydrocarbons dissolving in the water phase from oil slicks largely depend on the molecular structure and polarity of a given oil component, and the relative solubility of the oil component in water phase versus its solubility in the oil phase. In general, (1) the aromatic hydrocarbons are more soluble than aliphatic hydrocarbons, (2) solubility increases as the alkylation degrees of alkylated benzene or PAHs decrease, (3) the lower molecular weight hydrocarbons are more soluble than the high molecular weight hydrocarbons in that class, (4) the polar S, N, and O-containing compounds are more soluble than hydrocarbons. Therefore, it can be readily understood why the BTEX and lighter alkyl-benzene compounds and some smaller PAH compounds such as naphthalene are particularly susceptible to dissolution or "water washing." The significance of dissolution is that the soluble aromatic compounds are toxic to fish and other aquatic life.

Biodegradation: Biodegradation of hydrocarbons by natural populations of microorganisms (such as many species of bacteria, fungi, and yeasts) represents one of the primary mechanisms by which petroleum and other hydrocarbon pollutants are eliminated from the environment.[12,13] The biodegradation of petroleum and other hydrocarbons in the environment is a long-term complex process, whose quantitative and qualitative aspects depend on the type, nature and amount of the oil or hydrocarbon present, the ambient and seasonal environmental conditions (such as temperature, oxygen, nutrients, water activity, salinity, and pH), and the composition of the autochthonous microbial community.

Hydrocarbons differ in their susceptibility to microbial attack. In general, the degradation of hydrocarbons is ranked in the following order of decreasing susceptibility: *n*-alkanes > branched alkanes > low-molecular-weight aromatics > high-molecular-weight aromatics and cyclic alkanes. Hydrocarbons metabolized by microorganisms are generally converted to an oxidized compound, which may be further degraded, may be soluble, or may accumulate in the remaining oil. The aquatic toxicity of the biodegradation products is sometimes greater than that of the parent compounds.

Biodegradation is generally considered a slow process. Under less optimal conditions, it may take many years to biodegrade to certain degrees for some oils.

Photooxidation: Photooxidation is considered another factor involved in the transformation of crude oil or its products released into the marine environment. The photochemical degradation can yield a variety of oxidized compounds that are highly soluble in water. However, it should be noted that for most oils, photooxidation is probably a minor process in terms of changing their fate or mass balance after a spill.

Sedimentation and oil-fines interaction: Sedimentation is the process by which oil is deposited on the bottom of the sea or other water body. Once oil is on the bottom, it is usually covered by other sediments and degraded very slowly. Following the Exxon Valdez oil spill in Prince William Sound, Alaska in March 1989, the process of oil-mineral fine aggregates was discovered as a mechanism affecting the rate of natural cleansing of oil residues from shorelines.[14] Oil-mineral aggregates were found to result from interactions among the oil residues, fine mineral particle, and seawater. Particles of mineral with oil attached may be heavier than water and sink to the bottom as sediment or the oil may detach and refloat. Oil-fines interaction does not generally play a significant role in the fate of most oil spills in their early stages, but can have impact on the rejuvenation of an oiled shoreline over the long-term.

B. Chemical Composition Changes due to Weathering

Weathering causes considerable changes in the chemical and physical properties of spilled oils. The degree (lightly, moderately, or severely weathered) and rate of weathering is different for each spill and is controlled by a number of spill conditions and natural processes such as:

- type of the spilled oil and the chemical composition and concentrations of the spill oil hydrocarbons,
- spill site and environmental conditions (such as temperature, pH, water level, salinity, soil type, air, and nutrients), and
- natural population of indigenous microbial and microbiological activities.

Major chemical compositional changes due to weathering can be summarized as the following:

1. For lightly weathered oils and refined products (for example, <15% weathered), the abundances of low end n-alkanes are significantly reduced. However, the ratios of n-C_{17}/pristane and n-C_{18}/phytane are virtually unaltered. The losses of BTEX (benzene, toluene, ethylbenzene, and xylenes) and C_3-benzene compounds (8 isomers) are obvious, and the most abundant 2-ring alkylated naphthalene series appear slightly enriched.
2. For moderately weathered oils and refined products (for example, ~15 to 30% weathered), significant losses occur in n-alkanes and relatively low-molecular-weight isoprenoid compounds. Rapid loss of volatile aromatic compounds is clear. When oils are weathered to a certain degree (approximately in the range of 20 to 25% weathering for most oils), the BTEX and C_3-benzenes could be completely lost,[15] and the loss of C_0 and C_1- naphthalenes can be significant. The ratio of GC-resolved peaks to UCM can be considerably decreased due to the preferential loss of resolved hydrocarbons over the unresolved complex hydrocarbons. The biomarker compounds are enriched.
3. For severely weathered oils and refined products:
   - not only n-alkanes but branched and cyclo-alkanes are heavily or completely lost, and the UCM becomes extremely pronounced, resulting in a significant increase in relative ratios of UCM/GC-TPH and in a substantial decrease in relative ratios of resolved peaks to GC-TPH
   - the BTEX and alkyl benzene compounds are completely lost
   - pronounced decrease in the alkylated naphthalene series relative to other alkylated PAH homologous series
   - development of a profile in each alkylated PAH family showing the distribution of C_0- < C_1- < C_2- < C_3-
   - enhancement of the alkylated chrysene series relative to other PAH series and significant decrease in the relative ratios of the sum of alkylated naphthalenes, phenanthrenes, dibenzothiophenes, and fluorenes to chrysene series
   - significant concentration in the relative abundances of biomarker compounds (terpanes and steranes) because of their refractory nature and high resistance to biodegradation, while the relative ratios of paired terpane compounds including Ts/Tm (see Table 27.5 for definition of Ts and Tm), C_{23}/C_{24}, C_{29}/C_{30}, C_{31} 22S/(22S + 22R), C_{32} 22S/(22S + 22R), and C_{33} 22S/(22S + 22R) are not altered in most cases.

As an example, Figure 27.8 shows the GC–FID chromatograms of the saturated fractions of the weathered source oil, the moderately weathered sample #1, and the severely weathered sample #2, and Figure 27.9 shows the GC–MS chromatograms (in SIM mode) of the aromatic fractions of the weathered source oil, the moderately weathered sample #1, and the severely weathered sample #6 from the Baffin Island Oil Spill project conducted at the northern end of Baffin Island in the Canadian Arctic from 1980 to 1983. Figure 27.8 and Figure 27.9 clearly illustrated the effects of over 15 years field weathering on chemical composition changes and the weathering trend of the spilled oil.[16] For the severely weathered spill samples #2 and #6, the loss of n-alkanes and isoprenoids, lighter BTEX and alkylbenzene compounds, and even alkylated PAH (naphthalene, phenanthrene, dibenzothiophene, and fluorene) homologues is very pronounced in comparison with the lightly weathered source oil and moderately weathered sample #1.

C. ABIOTIC AND BIOTIC WEATHERING

The "weathering," as discussed above, is the term in combination of a wide variety of physical and chemical processes of spilled oil in the environment. The weathering process can be generally categorized into two types: abiotic (physical) weathering and biotic (microbial) weathering. Too often, the term "weathering" is misunderstood by some to mean processes that are biological or entirely physical by others.

The abiotic weathering is more predictable, especially for the alkanes and PAHs. The study[17] of the effects of physical weathering (evaporation) on chemical composition changes of the Alberta Sweet Mixed Blend (ASMB) oil at various weathering degrees (0 to 45% loss by mass) reveals: (1) ratios of n-C_{17}/pristane, n-C_{18}/phytane, and pristane/phytane were virtually unaltered; (2) the sum of n-alkanes from C_8 to C_{40} showed little change (in the range of 70 to 76 mg/g oil) as the physical weathering percentages increased from 0% to 45%; (3) isomeric distributions within C_1-phenanthrenes (4 isomers), C_1-dibenzothiophenes (3 isomers), C_1-fluorenes (3 isomers), and C_1-naphthalenes (2-methyl- and 1-methyl-naphthalene) exhibited great consistency in their relative ratios as the weathering percentages increased from 0% to 45%; (4) biomarker compounds were concentrated in proportion with the increase of the weathering percentages; (5) the weathering degree can be readily checked by integrating of n-alkanes in the GC–FID chromatograms, and determination and comparison of the weathering index (WI), defined as the sum of n-C_8, n-C_{10}, n-C_{12}, and n-C_{14} concentrations divided by the sum of n-C_{22}, n-C_{24}, n-C_{26}, and n-C_{28} concentrations:

$$WI = (n\text{-}C_8 + n\text{-}C_{10} + n\text{-}C_{12} + n\text{-}C_{14})/(n\text{-}C_{22} + n\text{-}C_{24} + n\text{-}C_{26} + n\text{-}C_{28})$$

Figure 27.10 and Figure 27.11 show n-alkane distribution and target PAH distribution of the ASMB oil at weathered percentages of 0% and 45%, respectively, to illustrate the effects of physical weathering on the oil chemical compositions.

Biodegradation, or biotic weathering, of hydrocarbons by natural microbial represents one of the primary mechanisms by which oil and oil-related hydrocarbons are eliminated from a contaminated environment. The biodegradation is generally a long-term weathering process, affects straight-chain n-alkanes more than branched alkanes, alkanes more than other hydrocarbon classes, GC-resolved compounds more than GC-unresolved complex hydrocarbons, small aromatics more than large aromatic compounds. Another important feature of oil biodegradation is that it is usually more isomer specific. The general susceptibility of oil hydrocarbon classes to biodegradation can be summarized as the following:

n-alkanes > BTEX and other monoaromatic compounds > branched and cyclo-alkanes > PAHs (lighter PAHs are more susceptible than larger PAHs, and increase in alkylation level in the same PAH series decreases susceptibility to microbial attack) > biomarker terpanes and steranes

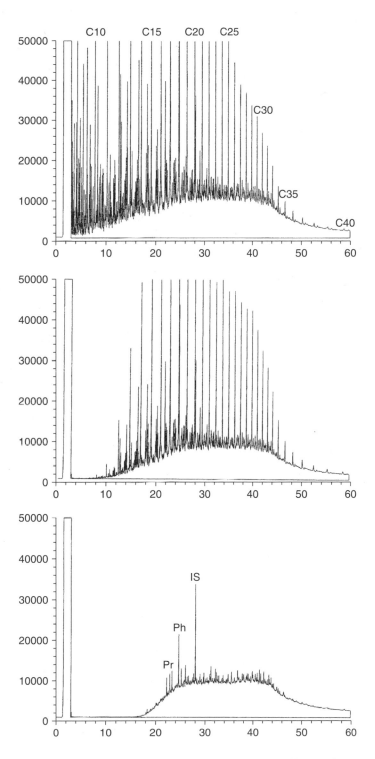

FIGURE 27.8 GC traces of the saturated fractions of the weathered source oil (top), the moderately weathered sample #1 (middle), and the severely weathered sample #2 (bottom) from the Baffin Island Oil Spill project conducted at Baffin Island in the Canadian Arctic from 1980 to 1983, illustrating the effects of over 15 years field weathering on chemical composition changes and the weathering trend of the spilled oil. IS, pr, and ph represent the internal standard, pristane, and phytane, respectively.

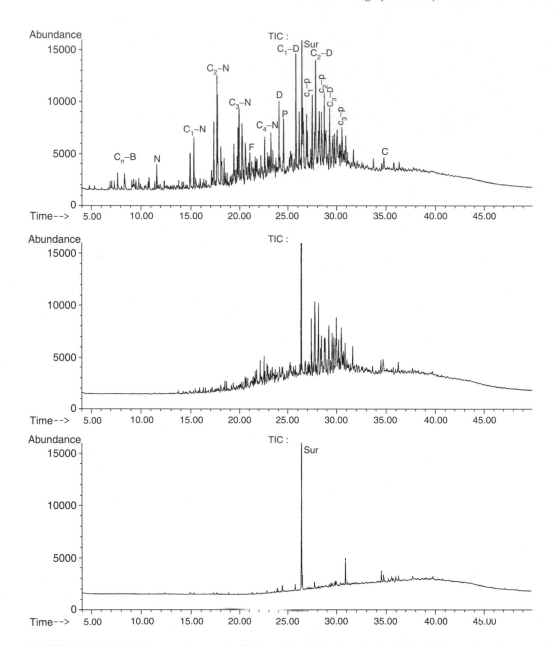

FIGURE 27.9 GC–MS chromatograms (in SIM mode) of the aromatic fractions of the weathered source oil (top), the moderately weathered sample #1 (middle), and the severely weathered sample #6 (bottom). Cn-B represents alkyl-benzene compounds. N, P, D, F, and C represent naphthalene, phenanthrene, dibenzothiophene, fluorene, and chrysene, respectively; 0–4 represent carbon numbers of alkyl groups in alkylated PAH homologues.

It should be noted, however, this sequence of increasing susceptibility only represents a general biodegradation trend, it does not mean that the more resistant class of hydrocarbons start to be biodegraded only after the less resistant class be completely degraded. Instead, there is always some overlap among different classes of hydrocarbons during the biodegradation process.

As an example, Figure 27.12 and Figure 27.13 show GC–MS chromatograms for a Prudhoe Bay oil biodegradation series incubated under fresh water using the standard inoculums.[18]

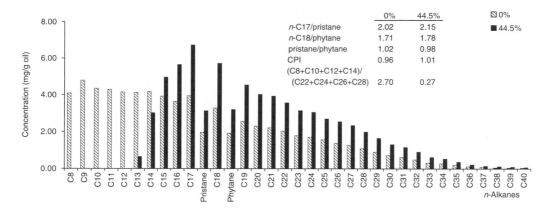

FIGURE 27.10 Comparison of n-alkane distribution for the ASMB oil at weathered percentages of 0% and 45%, to illustrate the effects of physical evaporative weathering on the oil chemical composition changes.

The circled peaks in Figure 27.13 indicated preferential biodegradation of these isomers over those noncircled isomers within the same isomeric group. This study clearly reveals a pattern of distinct oil composition changes due to biodegradation, which is significantly different from the pattern due to physical or short-term weathering. It is important to be able to distinguish between these two forms of loss, so that the loss due to physical weathering is not interpreted as loss due to biodegradation.

The alterations in chemical composition of naturally weathered spilled oils are generally resulted from the combined effects of abiotic and biotic weathering, as Figure 27.8 and Figure 27.9 shows. The transformations of oil hydrocarbons by biodegradation are likely to occur stepwise, producing alcohols, phenols, aldehydes, and carboxylic acids in sequence.

D. ESTIMATION OF WEATHERING DEGREE AND WEATHERING RATE

As discussed above, oil weathering is a very complex process. The weathering degree and the weathering rate are determined by many factors. In the early biodegradation studies, the ratios of biodegradable to less biodegradable compounds such as $n\text{-}C_{17}$/pristane and $n\text{-}C_{18}$/phytane were largely used for estimating biodegradation degree and for comparing the weathering state of

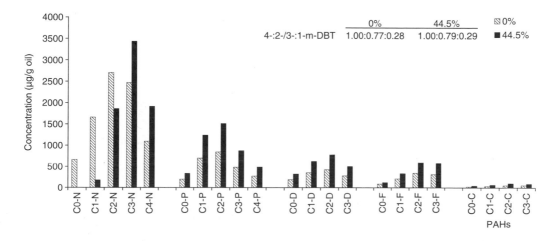

FIGURE 27.11 Comparison of target PAH distribution for the ASMB oil at weathered percentages of 0% and 45%, to illustrate the effects of physical evaporative weathering on the oil chemical composition changes.

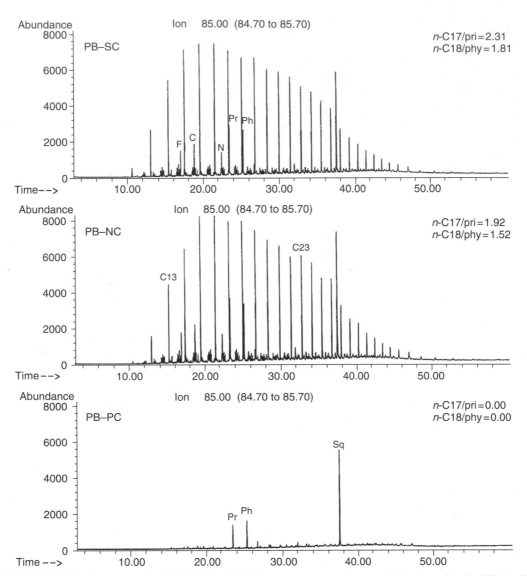

FIGURE 27.12 Representative GC–MS (m/z 85) chromatograms of saturated fractions for the SC (sterile control, top), NC (negative control, middle) and PC (positive control, bottom) of a Prudhoe Bay (PB) oil biodegradation series under the standard inoculum conditions. Peaks F, C, N, Pr, and Ph represent the most abundant 5 isoprenoids (farnasane, trimethyl-C_{13}, norpristane, pristane, and phytane) in the oil. Sq represents the surrogate squalane.

spilled oil. For lightly and some moderately weathered oils, these ratios provide a useful tool for estimation of weathering degree and for oil source identification and differentiation. In severely weathered oils, however, the n-alkanes and even the isoprenoids (including pristane and phytane) may be partially or completely lost. Under such circumstances, the use of the traditional measure of n-C_{17}/pristane and n-C_{18}/phytane might substantially underestimate the extent of biodegradation and weathering degree because isoprenoids also degrade to a significant degree. Later, highly degradation-resistant components such as C_{30} 17α(H), 21β(H)-hopane are selected to serve as conserved "internal standards" for determining rate and extent of weathering for the spilled residual oil[19–21]:

$$P\,(\%) = (1 - C_s/C_w) \times 100$$

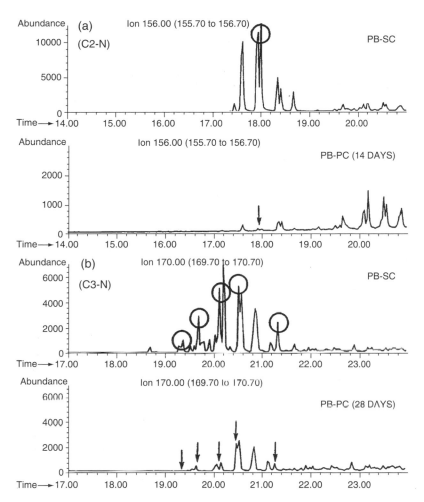

FIGURE 27.13 Extracted ion chromatograms for C_2-naphthalenes (ion 156, at 14 days, a), C_3-naphthalenes (ion 170, at 28 days, b), C_1-fluorenes (ion 180, at 14 and 28 days, c), C_1-phenanthrenes (ion 192, at 14 and 28 days, d), and C_1-dibenzothiophenes (ion 198, at 14 and 28 days, e) in the source sterile control PB oil and the corresponding positive controls under standard fresh water biodegradation conditions. The circled peaks indicate the characteristic changes in distribution within isomers that were preferentially altered by biodegradation and/or co-metabolic oxidation.

where P is the weathered percentages of the weathered samples, C_s and C_w are the concentrations of C_{30} $\alpha\beta$-hopane in the source oil and weathered samples, respectively. A number of studies have demonstrated that this method can provide a more accurate representation of the degree of biodegradation than do the traditional alkane/isoprenoid hydrocarbon ratios.

For refined products, such as diesel and jet fuel samples, which may not contain significant quantities of biomarker compounds and chrysenes, less "conservative" PAHs with a high degree of alkylation such as C_4 or C_3-phenanthrenes can be selected and used as alternative internal standards to evaluate the weathering degrees.

Short and Heintz[22] have developed a first-order loss-rate (FOLR) kinetic model of PAH weathering based on molecular size to evaluate environmental samples collected for the Exxon Valdez oil spill for the presence of spilled oil. They found that the predictability of the model is sufficiently robust that the initial PAH composition of oil can be inferred from analysis of a

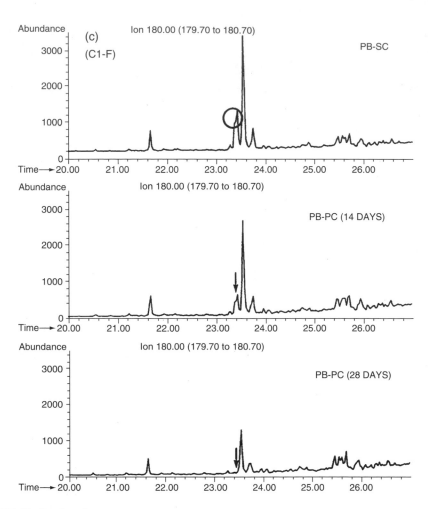

FIGURE 27.13 Continued.

weathered sample, thereby considerably increasing the number of analytes that may be used to evaluate a suspected source. This approach has been recently illustrated by application to four independent case studies.[23]

E. Spill Age Dating

Age dating of a spill is a very complex issue and difficult task, and it should be dealt within case by case way in most situations and with caution. For example, we found during the 25-year-old Nipisi spill study that the weathering degrees of the 25-year-old residual oil samples collected from same sampling spots but with different depths were dramatically different in chemical composition from sample to sample[24]: surface samples (0 to 2 cm) were severely weathered with all n-alkanes and isoprenoids being completely lost, subsurface samples from 10 to 15 cm were moderately weathered, while subsurface samples from 30 to 40 cm were only lightly weathered, indicated by the existence of large quantities of BTEX and alkylbenzene compounds, and the n-alkane distribution being almost not changed in comparison with the reference oil (Figure 27.14).

Christensen and Larsen[25] assembled data from 12 diesel spill sites with known spill time for each spill in northern Europe. The data from these sites ultimately yielded a linear correlation

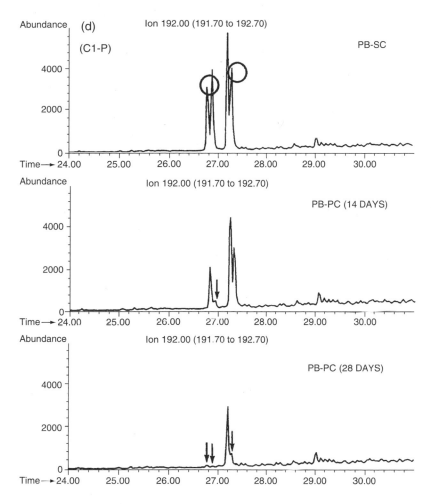

FIGURE 27.13 Continued.

between the "time since the release in years" and "average n-C_{17}/pristane ratio". Kaplan et al.[26] extracted the plotted data points from Christensen and Larsen and published a linear equation used for weathering age determination of diesel. However, it should be noted that their calibration chart was developed for particular spill sites under a unique set of environmental conditions. Bacterial activity could be greatly different from site to site, resulting in very different rate of biodegradation. Other environmental factors such as temperature, water level, salinity, and many others will also play key roles in determining the rate of biodegradation. Stout et al.[7] have discussed in detail the numerous caveats that must be considered in this age-dating method; otherwise, it might lead to over simplication of the very complex issue of age dating diesel or other oil product contamination.

V. CHARACTERIZATION AND IDENTIFICATION OF SPILLED OILS

A. DISTINGUISHING BIOGENIC HYDROCARBONS FROM PETROGENIC HYDROCARBONS

Characterization and differentiation of hydrocarbons from different sources is an essential part of any objective oil spill study. After oil spills, oil hydrocarbons often mix with other background

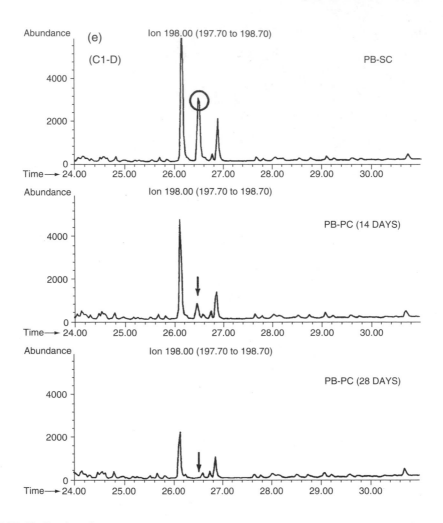

FIGURE 27.13 Continued.

hydrocarbon sources in the impacted area. One of the potential sources of hydrocarbons contributing to the background is biogenic hydrocarbons. Hydrocarbons from anthropogenic and natural sources including biogenic source are very common in the marine and inland environments.

Biogenic hydrocarbons are generated either by biological processes or in the early stages of diagenesis in recent marine sediments. Most soils and sediments contain some fraction of organic matter derived from biological sources including land plants, phytoplankton, animals, bacteria, macroalgae, and microalgae.

It has been recognized[27–31] that the biogenic hydrocarbons have the following chemical composition characteristics:

FIGURE 27.14 GC–FID (Figure 27.14A, left panel) and GC-SIM-MS (Figure 27.14B, right panel) chromatograms of saturated and aromatic fractions for the Nipisi reference source oil PL-B, and samples N2-1A (0–2 cm), N2-1B (12–16 cm), and N2-1C (30–40 cm), illustrating the sample depths on chemical composition changes of aliphatics, and alkylbenzenes and alkylated PAHs, respectively. Sur. and IS represent surrogate and internal standard. B and N represent benzene and naphthalene, respectively; n, 0, 1, 2, and 3 represent carbon numbers of alkyl groups in alkylbenzenes and alkylated naphthalene homologues.

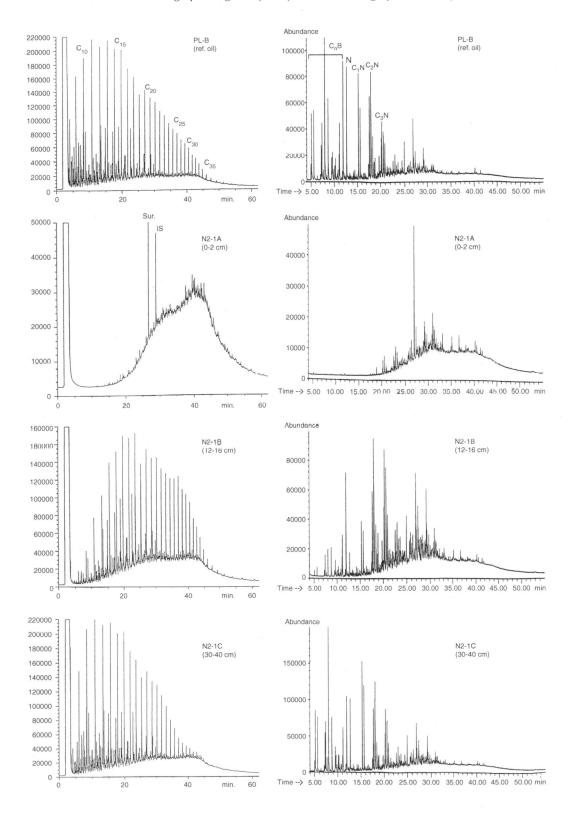

(1) n-alkanes show a distribution pattern of odd carbon-numbered alkanes being much more abundant than even carbon-numbered alkanes in the range of n-C_{21} to n-C_{33}, resulting in unusually high carbon preference index (CPI) values, which is defined as the sum of the odd carbon-numbered alkanes to the sum of the even carbon-numbered alkanes (petroleum oils characteristically have CPI values around 1.0)

(2) notable absence of the "unresolved complex mixture (UCM)" hump in the chromatograms

(3) pristane is often more abundant than phytane, suggesting a phytoplankton input and resulting in abnormally high pristane/phytane ratio values

(4) presence of a "biogenic cluster" (identified as olefinic hydrocarbons of biogenic origin) in the gas chromatograms of the aromatic fractions

(5) wide distribution of the biogenic PAH perylene, an unsubstituted PAH produced in subtidal sediments by a process known as early diagenesis

(6) presence of plant terpenoid biomarker compounds on occasion.

In some environmental forensic investigation, the CPI values were used in recognizing the contribution of modern hydrocarbons derived from modern plant leaf debris in soil and sediments. The presence of modern plant leaf waxes can impart a strong odd-carbon dominance (CPI > 2) that is unrelated to the petroleum contamination.[32] In the study of hydrocarbon biogeochemical setting of the Baffin island oil spill (BIOS) experimental site, Cretney et al.[27] found that the BIOS subtidal samples showed very high pristane/phytane ratios (5 to 15) and CPI values (3 to 11). High concentrations of pristane relative to phytane in most of beach and subtidal sediments indicate biological hydrocarbon input from a marine biological source. In addition, the GC chromatograms of the aromatic fractions were typified by the olefinic hydrocarbon clusters. This cluster is a common feature of coastal marine subtidal sediments and is believed to be of marine biological (planktonic or bacterial) origin. The possibility of *in situ* genesis of PAHs is indicated by the presence of perylene as a major PAH in almost all the beach and subtidal sediments. However, it should be noted that it cannot be used alone as a definitive source identification criterion because perylene is also produced in combustion processes.

During the years 1970 to 1972, the Nipisi, Rainbow, and Old Peace River pipeline spills occurred in the Lesser Slave Lake area of northern Alberta. The Nipisi spill is one of the largest land spills in Canadian history. The most recent field survey was conducted in 1995 in order to determine which cleanup methods were most successful, and to provide up-to-date information about any changes in residual oil and vegetative recovery 25 years after the spills. The comprehensive chemical data[24] from analysis of the Nipisi samples indicate the following:

(1) The Nipisi samples can be categorized into three groups plus the background group, according to the contamination level and degradation degree of the samples.

(2) The background samples showed typical biogenic n-alkane distribution in the range of C_{21} to C_{33} with abundances of odd-carbon-number n-alkanes being much higher than that of even-carbon-number n-alkanes. The biogenic cluster was also obvious and no UCM was observed (Figure 27.15). No petrogenic hydrocarbons, in particular no alkylated PAH homologues and petroleum-characteristic biomarker compounds such as pentacyclic hopanes and C_{27} to C_{29} steranes were detected.

(3) In addition, three plant terpenoid biomarker compounds with remarkable abundances were detected and they were identified as 12-oleanene ($C_{30}H_{50}$, MW $= 410.7$, RT $= 42.27$ min), 12-ursene ($C_{30}H_{50}$, MW $= 410.7$, RT $= 42.74$ min), and 3-friedelene ($C_{30}H_{50}$, MW $= 410.7$, RT $= 44.26$ min). Formation of a six-member ring E from the baccharane precursor leads to the oleanane group. Oleananes and their derivatives form the largest group of triterpenoids and occur in the plant kingdom, specifically from higher plants.[33]

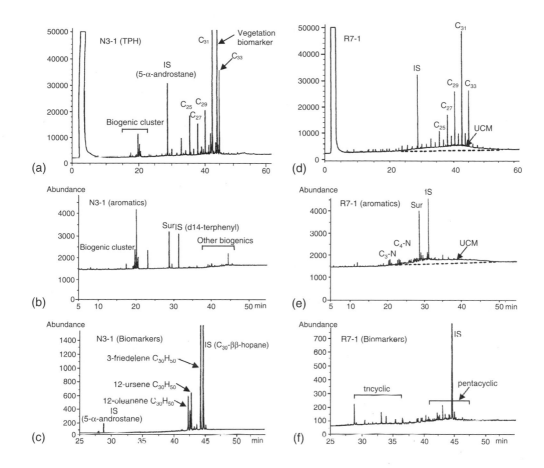

FIGURE 27.15 GC chromatograms of two representative Nipisi samples: background sample N3-1 (left, biologic hydrocarbons) and Group 3 sample R7-1 (right, mixture of vegetation hydrocarbons and lightly contaminated oil hydrocarbons), illustrating differences between petrogenic and biogenic hydrocarbon distributions. Top panel (a and d): GC–FID chromatograms for TPH analysis. Sample R7-1 showed distribution of mixed petrogenic and biogenic n-alkanes, while Sample N3-1 showed only typical biogenic n-alkane distribution in the range C21 to C33. The biogenic cluster was also obvious and no UCM was observed. Middle panel (b and e): total ion GC–MS chromatograms of aromatic fractions. Bottom panel (c and f): GC–MS chromatograms (m/z 191) of saturated fractions. Petrogenic alkylated PAH homologues and biomarkers were detected in Sample R7-1. In contrast, No petrogenic PAHs and biomarkers were detected in Sample N3-1. However, three vegetation biomarkers (C30H50: 12-oleanene, 12-uresene, and 3-friedelene) with remarkable abundances were detected.

(4) Hydrocarbons in the Group 3 subsurface samples taken from a depth of 40 to 100 cm were identified to be mixtures of vegetation hydrocarbons and lightly contaminated oil hydrocarbons.

B. Distinguishing Pyrogenic Hydrocarbons from Petrogenic Hydrocarbons

PAH, distributions are the most useful tool in distinguishing pyrogenic hydrocarbons from petrogenic hydrocarbons. The differences in PAH distribution between petrogenic and pyrogenic PAH sources were first recognized in modern sediment studies, and then expanded to the environmental forensic interpretation of petrogenic, pyrogenic and biogenic PAHs.

As discussed above, petrogenic materials (crude oils and refined products), in general, exhibit alkylated PAH distribution patters where the C_{1^-}, C_{2^-}, and C_3-PAHs are more abundant than the parent (C_{0^-}) and C_4-PAHs. This kind of characteristic PAH distribution profile has been termed as "*bell shaped*." By weathering or degradation, the "bell shaped" distribution can be readily modified to the distribution profile of $C_{0^-} < C_{1^-} < C_{2^-} < C_{3^-}$ (*inverse-sloped*) in most alkylated PAH homologous families. In contrast, pyrogenic materials generally exhibit alkylated PAH homologue distribution patterns in which the parent PAH is often the most abundant. The composition features of pyrogenic PAHs can be summarized as follows:

(1) The dominance of the unsubstituted compounds over their corresponding alkylated homologues, and this kind of PAH distribution profile of $C_0 \gg C_1 > C_2 > C_3 > C_4$ has been generically termed as *skewed* or *sloped*,[34]
(2) The dominance of the high molecular mass 4- to 6-ring PAHs over the low molecular mass 2- to 3-ring PAHs, and
(3) On the gross level PAH can comprise a much higher mass percentage in most pyrogenic source materials than in most petrogenic source materials.[35-38]

As an example, Figure 27.16 compares PAH fingerprints for the 1994 Mobile Burn starting oil, burn residue, and soot sample, illustrating the distinguishing features of pyrogenic PAH distribution from the petrogenic PAH distribution. Figure 27.17 compares extracted ion chromatograms at m/z 178, 228, 252, and 276 for the 1994 Mobile diesel, residue sample MB-16 and soot sample TSP-B3. The changes in relative distribution patterns of selected PAHs clearly demonstrate the formation of pyrogenic PAH from 3-ring anthracene to 6-ring indeno (1,2,3-cd) pyrene and benzo (ghi) perylene due to combustion. The concentrations of PAHs with five or more rings were many times greater in the smoke than in oil

Numerous quantitative diagnostic ratios have been defined to differentiate pyrogenic PAHs from other hydrocarbon sources,[35-38] including phenanthrene/anthracene (Ph/An), phenanthrene/methyl-phenanthrene (Ph/m-Ph), fluoranthene/pyrene (Fl/Py), benz(a)anthracene/chrysene (BaA/Ch), Ph/(Ph + An), and BeP/(BeP + BaP). Recently, a new "Pyrogenic Index" was proposed as a quantitative indicator for identification of pyrogenic PAHs and for differentiation of pyrogenic and petrogenic PAHs.[39] The Pyrogenic Index (PI) is defined as the ratios of the total of the other EPA priority unsubstituted 3- to 6-ring PAHs to the five targets alkylated PAH homologues:

$$PI = \Sigma(\text{other 3- to 6-ring EPA PAHs})/\Sigma(\text{5 alkylated PAHs})$$

Table 27.7 summarizes ranges of the PAH quantitation results and the "Pyrogenic Index" values for various oil-related samples. For comparison purposes, the ratios of phenanthrene/anthracene are also listed in Table 27.7. Compared to other diagnostic ratios obtained from individual compounds, this index ratio has its own distinct advantages:

(1) as discussed above, petrogenic and pyrogenic PAHs are characterized by dominance of 5 alkylated PAH homologous series and by dominance of unsubstituted high-molecular-weight PAHs, respectively, therefore, determination of the changes in this ratio more truly reflects the difference in the PAH distribution between these two sets of hydrocarbons;

FIGURE 27.16 PAH fingerprints and distinguishing features of distribution patterns between petrogenic and pyrogenic PAHs for the starting oil, burn residue, and soot samples from 1994 Mobile burn study. The abbreviations from Acl to BgP represent the other EPA priority unsubstituted PAHs (please refer to Table 27.4 for the full names of these PAHs). For comparison, the fingerprints of the other 3- to 6-ring PAHs have been enlarged and shown in the left insets. Note that for clarity, different *y*-axis scales are used for soot samples.

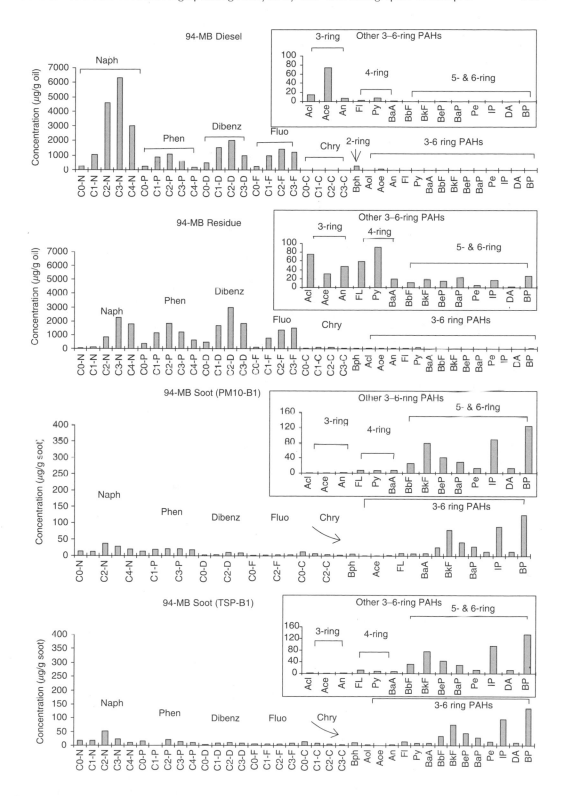

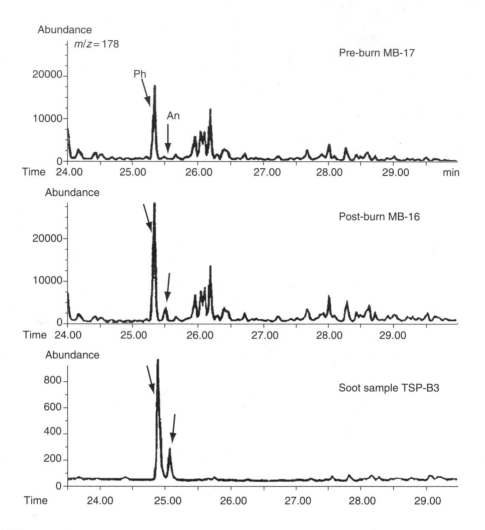

FIGURE 27.17 Comparison of representative extracted ion chromatograms at m/z 178, 228, 252, and 276 for the starting oil, burn residue MB-16, and soot sample TSP-B3, illustrating changes in the relative distribution of unsubstituted PAH isomers and demonstrating the formation of pyrogenic PAHs from 3-ring anthracene (An) to 6-ring indeno(1,2,3-cd)pyrene (IP) and benzo(ghi)perylene (BP) due to combustion.

(2) determination of these two sets of PAHs has become conventional measurement for many environmental labs, and this ratio can offer better accuracy with less uncertainty than those relative ratios determined from individual PAH compounds; and

(3) this ratio shows great consistency from sample to sample and is subject to little interference from the concentration fluctuation of individual components within the PAH series. In addition, long-term natural weathering and biodegradation (such as the biodegraded Alberta Sweet Mixed Blend (ASMB) oil series and nine biodegraded Alaska oil series) only slightly alter the values of this ratio (Table 27.7), but the ratio will be dramatically altered by combustion.

Therefore, this index ratio can be used as a general and effective criterion to unambiguously differentiate pyrogenic PAHs and petrogenic PAHs. The usefulness of the Pyrogenic Index in environmental forensic investigations for input of pyrogenic PAHs and spill source identification has been clearly demonstrated in several recent spill case studies.[40–42]

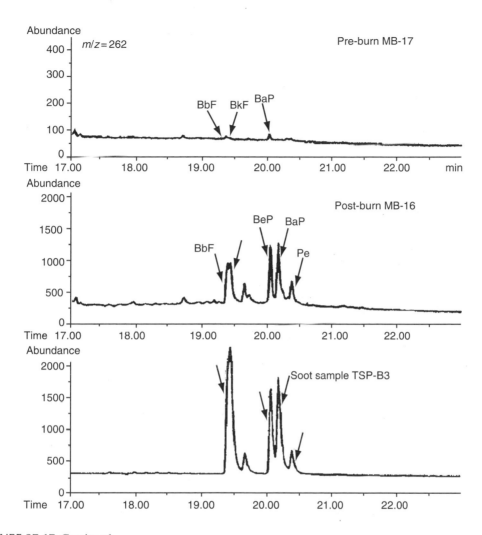

FIGURE 27.17 Continued.

Figure 27.18 depicts the "Pyrogenic Index" versus the relative ratios of phenanthrene to anthracene for over 60 oils and refined products (including jet fuel, diesel, lube oil, Bunker C, and heavy fuel) analyzed. As Figure 27.18 shows, the "Pyrogenic Index" exclusively falls in the range of 0.01 to 0.05 for oils and refined products, while it dramatically increased to a range of 0.8 to 2.0 for the six 1994 Mobile burn soot samples. The difference in the magnitude of the data is very significant.

It is also seen from Figure 27.18 that the jet fuel, diesel, and most crude show the "Pyrogenic Index" ratios smaller than 0.01 with ratios of phenanthrene/anthracene being very scattered. But heavy oils (such as Cold Lake Bitumen and Orimulsion) and heavy fuels (such as IFO-180, A-02, IF-30, and Bunker C type fuels) show significantly higher ratios (falling in the range of 0.01 to 0.05, clusters 1 and 2), indicating that this index ratio can be also used as a screening tool to distinguish heavy oils and heavy fuels from most crude oils and light petroleum products. For example, as Figure 27.18 shows, the ratios for the unknown tarball samples collected from the coasts of British Columbia (BC) and California (CA) in 1996 and Newfoundland in 1997 fell in the ratio range for Bunker C type fuels, implying that these tarballs might be from a source of heavy fuels. A comprehensive study using GC–MS and isotopic techniques has revealed that the tarball samples from BC and CA were chemically similar and both originated from bunker type fuels.[43]

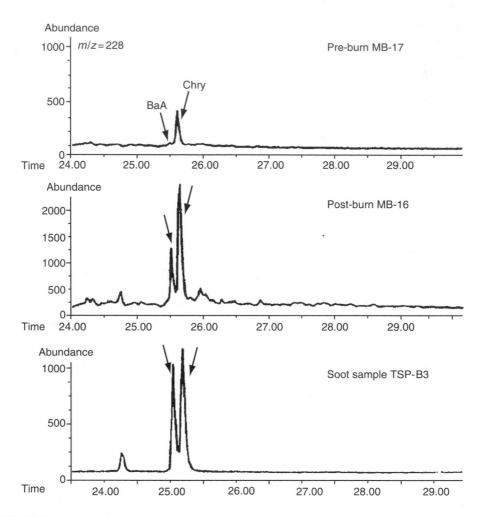

FIGURE 27.17 Continued.

C. PAH FINGERPRINTING ANALYSIS FOR OIL CORRELATION AND SOURCE IDENTIFICATION

In many instances where the chemical similarity/difference between spilled oil and the suspected source(s) is not obvious, or a large number of candidate sources are involved, or spilled oil has undergone severe weathering and significant alteration in its chemical compositions, the qualitative and semiquantitative approach (such as visual comparison of GC chromatograms, estimation of weathering degrees, and *n*-alkane distributions) would be difficult to defend, and therefore the quantitative fingerprinting analysis of degradation-resistant PAH and biomarker compounds becomes not only useful, but necessary.

1. PAH Distribution Pattern Recognition

PAH compounds are probably the most studied hydrocarbon group in crude oils and refined products. Crude oils and refined products from different sources can have very different PAH distributions. In addition, many PAH compounds are more resistant to weathering than their saturated hydrocarbon counterparts (*n*-alkanes and isoprenoids) and volatile alkylbenzene compounds, thus making PAHs one of the most valuable fingerprinting classes of hydrocarbons

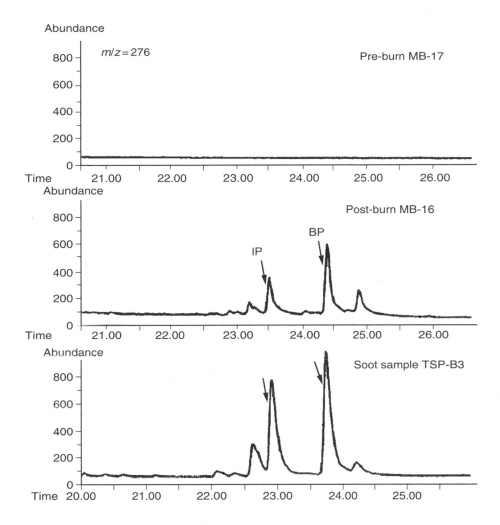

FIGURE 27.17 Continued.

for oil identification. Even differences between the same types of products are discernible through examination of the PAH distribution. Examples of PAH distribution of some oils and petroleum products are illustrated in Figure 27.19. The oil products differ significantly in the PAH concentrations and distribution patterns from the crude oils and from each other. Generally, in unweathered crude oils, the alkylated naphthalenes and alkylated chrysenes are the most and least abundant PAHs among the five targets alkylated PAH homologues, while many of 4- to 6-ring unsubstituted PAHs, are only minor components or even absent in oils. As discussed above, the PAHs in crude oils and refined products typically exhibit a characteristic "*bell shaped*" profile within each alkylated homologous series.

Jet B fuel has extremely high content of the naphthalene series (99%) among the five target alkylated PAH homologues, with the other four alkylated PAH series being only 1% in total. In addition, no 4- to 6-ring PAHs were detected of the other 15 EPA priority PAHs. Diesel No. 2 has high naphthalene content (86%), low phenanthrene content (5%), and no chrysenes. In the No. 5 fuel and HFO 6303, the unusually high contents of the alkylated naphthalene and chrysene series are very pronounced. In the Orimulsion 400, the concentrations of the alkyl phenanthrenes and dibenzothiophenes are very high, accounting for approximately 38% and

TABLE 27.7
PAH Quantitation Results and "Pyrogenic Index" (PI) Values for over 60 Oils and Refined Products, Artificially Weathered Oil Series, Severely Weathered Spill Samples, Biodegraded Oils, and Oil Burn Products

Oil Type	Sum of Five Alkylated PAH Series (μg/g)	Sum of Other Three to Six Ring PAHs (μg/g)	Phen/Antha[a]	Σ(other 3-6 ring PAHs) / Σ(5 alkylated PAH series)[b]
Oil and refined products				
Oils (40 oils)	4,000–45,000	25–160	13–350	<0.010 (0.002–0.010)
Lube oil	352	0		0.0000
Heavy fuels (9 fuels)	2,000–33,000	30–700	7–20	0.014–0.051
Bunker C (4 Bunker C)	12,000–32,000	550–860	11–25	0.022–0.047
Oil contaminated birds (1995)	9,700–15,500	300–540	11–19	0.031–0.0410
Tarballs (BC and CA, 1996)	3,800–6,300	80–130	13–18	0.020–0.022
Tarballs (NF, 1997)	12,500–14,700	380–430	7–10	0.026–0.030
Artificially weathered oils				
ASMB oil series (0–45%)	13,400–18,800	65–82	c	0.004–0.005
California oil series (0–15%)	4,900–6,100	40–50	c	0.008–0.010
25-year-old Nipisi spill samples	800–12,000	7–45	c	0.004–0.012
Biodegraded oils				
ASMB oil series	7,000–17,000	45–110	1–46	0.005–0.009
3 North slope oil series	5,700–18,000	40–82	(anth: under detection limit)	0.004–0.007
3 Cook inlet oil series	4,600–15,500	60–130	0.4–60	0.007–0.014
Jet fuel B series	4,200–28,100	21–53	2–50	0.002–0.009
Diesel No. 2 series	7,000–25,700	15–120	3–34	0.002–0.005
Bunker C/diesel mixture series	7,000–12,500	350–590	2.7–25	0.047–0.060
1993 NOBE burn samples				
Starting oil and preburn oils	11,300–11,900	97–117	c	0.008–0.010
Residues	2,900–4,300	220–650	c	0.067–0.223
1994 Mobile burn samples				
Starting diesel and preburn diesels	27,000–29,000	107–110	30–37	0.0040
Residues	20,400–24,300	220–430	8–17	0.009–0.019
Soot samples	120–300	450–750	4–8	0.81–1.94

[a] phen/anth: phenanthrene/anthracene; [b] See Table 27.4 for definition for other 3-6 ring EPA PAHs; [c] Not determined.

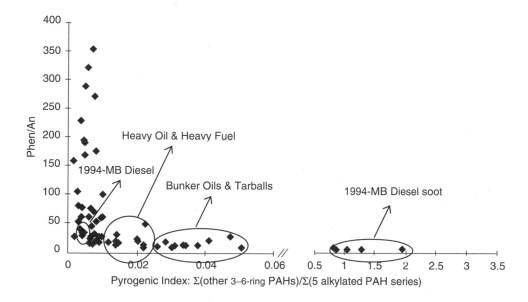

FIGURE 27.18 Plot of the relative ratios of Σ(other 3- to 6-ring PAHs)/Σ(5 alkylated PAHs) over the relative ratios of Phenanthrene/Anthracene for over 60 oils and petroleum products. Lighter petroleum products and most crude oils show the ratios of Σ(other 3- to 6-ring PAHs)/Σ(5-alkylated PAHs) falling into a range of 0 to 0.01, while heavy oils and heavy fuels show significantly higher ratios in the range of 0.01 to 0.05. The soot samples show the most striking increase in the ratio, indicated by the right circle.

22%, respectively, of the total PAHs. In addition, a profile in each alkylated PAH family showing the distribution of $C_0 < C_1 < C_2 < C_3$ is very apparent, similar to the severely weathered oil.

2. Diagnostic Ratios of PAH Compounds

A number of diagnostic ratios of target alkylated PAH species have been successfully used for source identification and differentiation, establishing statistical models for source allocation, distinguishing inputs of pyrogenic hydrocarbons from petrogenic hydrocarbons, markers of biodegradation, and weathering indicator. These are briefly summarized in Table 27.8. A benefit of comparing diagnostic ratios of spilled oil and suspected source oils is that any concentration effects are minimized. In addition, the use of diagnostic ratios to correlate and differentiate oils tends to induce a self-normalizing effect on the data since variations due to instrument operating conditions, operators, or matrix effects are minimized.

A method using the double ratio plots of alkylated PAH homologues, C_2D/C_2P versus C_3D/C_3P (the ratios of alkylated dibenzothiophenes to alkylated phenanthrenes), for identification and differentiation of petroleum product sources has been developed. Due to that these ratios remain relatively stable over a wide range of degree of weathering (that is, these PAH groups tend to weather at comparable rate), they were extensively used in the studies of the 1989 Exxon Valdez oil spill to distinguish Alaska North Slope (ANS) crude, its weathering products, and diesel refined from ANS feed stock from other petrogenic hydrocarbons including the sulfur-depleted tertiary oil seeps in the region.[30,31,44-47] Table 27.9 lists the double ratio values for some representative crude oils and petroleum products from light jet fuel to heavy Bunker C fuel.[48] Table 27.9 clearly shows how different these ratios between oils and refined products are. Furthermore, Douglas et al.[48] defined the C_3D/C_3P and C_3D/C_3C (the ratios of alkylated

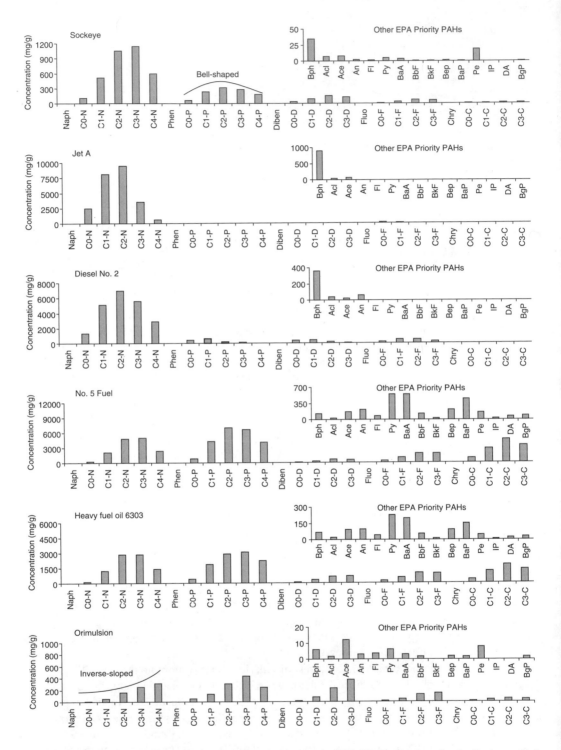

FIGURE 27.19 Alkylated homologous PAH and other EPA priority PAH distributions for the Sockeye oil, Jet A, Diesel No. 2, No. 5 Fuel, HFO 6303, and Orimulsion, illustrating differences in PAH distribution features between different oil and oil products. Note that for clarity, different scales are used for y-axis.

TABLE 27.8
Diagnostic Parameters of PAHs Used in Oil Spill Fingerprinting

Diagnostic Ratios	Ion Monitored
Double ratios	
C2D/C2P vs. C3D/C3P	212, 206, 226, 220
C3D/C3P vs. C3D/C3P	226, 220, 270
Pyrogenic index	Ions for target PAHs
C0C:C1C:C2C:C3C	228, 242, 256, 270
Reten/C4-phen (Reten: 7-isopropyl-1-methyl-phen)	270
Ratios between alkylated PAH series	
Σphens/Σdibenz, phen/Σphens	128, 142, 156, 170, 184
Σnaphs/Σchrys, Σphens/Σchrys	178, 192, 206, 220, 234
Σdibenzs/Σchrys, Σfluos/Σchrys	184, 198, 212, 226
	166, 180, 194, 208
	228, 242, 256, 270
Ratios of isomeric PAHs	
Methyl-dibenzothiophenes (4-:2-/3-:1-m-DBT)	198
Methyl-phenanthrenes (3- + 2-m-P)/(4-/9-m- +1-m-P)	192
2-m-N/1-m-N	142
An/Phen and An/(An + Phen)	178
Fluoranthene/Pyrene (Fl/Py)	202
BaA/Chry	228
BeP/BaP and BeP/(BeP + BaP)	252
Isomers in C3-naphs and C4-naphs	156, 170
Isomers in C2-phens and C4-phens	206, 234
Isomers in C1-fluorenes	180

2-m-N and 1-m-N represent 2-methyl-naphthalene and 1-methyl-naphthalene, respectively.
See Table 27.4 for definitions of all other PAH abbreviations shown in this Table.

dibenzothiophenes to alkylated chrysenes) as "source ratios" (the ratios that be almost constant because the compounds degraded at the same rate) and "weathering ratios" (the ratios that change substantially with weathering and biodegradation), respectively. They were applied to describe oil depletion and to identify sources in subtidal sediment data from the Exxon Valdez spill and a North Sea oil spill. Hostettler et al.[49] reported a method using the PAH refractory index, the ratio of two of the most refractory constituents of most oils (triaromatic steranes and methylchrysenes) as a source discriminant of hydrocarbon input for differentiation of three different oils (Exxon Valdez oil, Katalla oil, and PWS sediment hydrocarbons).

In the studies of characterization of spilled oil residues and identification of unknown spill samples, Wang et al.[41,43,50,51] utilized a number of diagnostic ratios of selected source-specific alkylated PAHs in combination with determination of ratios of selected paired biomarkers for source identification and differentiation, determination of weathering extent and degree of surface and subsurface samples, and distinguishing between composition changes due to physical weathering and biodegradation.

3. PAH Isomer Analysis

The use of the sum of the alkylated PAHs as multicomponent analytes in deriving diagnostic ratios for oil characterization and spill assessment have made considerable advances, as

TABLE 27.9
C_2D/C_2P and C_3D/C_3P Ratios of Representative Crude Oils and Petroleum Products

Oil Type	C_2D/C_2P	C_3D/C_3P
JP 4 fuel	0.14	0.26
Jet A fuel	1.09	0.00
No. 1 Arctic diesel	0.98	1.02
No. 2 fuel oil	0.54	0.74
No. 2 EPA fuel	0.32	0.58
Union 76 diesel	0.85	1.39
Alaska diesel	0.61	0.62
Diesel fuel marine	0.41	0.64
No. 4 fuel oil (1% sulphur)	0.22	0.28
EPA Bunker C residual oil	1.05	1.03
No. 6 fuel oil	0.29	0.20
Lube oil	0.36	0.41
Coal tar	0.09	0.15
Texas intermediate crude	0.61	0.54
Argo merchant cargo oil	0.74	1.08
API Ref Arabian light crude oil	3.68	3.99
Merban crude oil	3.77	4.59
Karachaganak condensate	6.72	11.47
Alaska North slope oil	0.87	1.08
Cook inlet oil	0.11	0.12

Adapted from Douglas, G. S., Bence, A. E., Prince, R. C., McMillen, S. J., and Butler, E. L., *Environ. Sci. Technol.*, 30, 2332–2339, 1996. With permission.

described above. In recent several years, research has been further expanded to use individual source-specific isomers within the same alkylation level and to determine the relative isomer-to-isomer distribution for oil spill source identification.

As the alkylation levels increase, more isomers are detected (for example, the C_3-dibenzothiophenes, as a group, contain more than 20 individual isomers with different relative abundances). The differences between the isomer distributions reflect the differences of the depositional environment during oil formation. Compared to PAH homologous groups at different alkylation levels, higher analytical accuracy and precision may be achieved due to the close match of physical/chemical properties of the isomers. In addition, the relative distribution of isomers is subject to little interference from weathering in short-term or lightly weathered oils. Hence, this approach can be positively used for oil spill identification. On the other hand, it has been demonstrated that the position of alkylation on the PAHs can influence the biodegradation rate of the isomers within an isomer group. This information can be used to sort environmental factors such as the impact of biodegradation on the PAH distribution and to differentiate oil compositional changes due to physical weathering from those due to biodegradation. For example, the ratios among methyl dibenzothiophenes, methyl-phenanthrenes, and methyl and dimethyl naphthalenes have been thoroughly studied and used for environmental forensic investigations.

(1) Methyl phenanthrenes. All oils contain four methyl-phenanthrenes (3-, 2-, 4-/9-, and 1-m-P). Ratios among four methyl-phenanthrene isomers have been demonstrated to be related to the thermal history of crude oils and its source strata, and numerous methyl-phenanthrene indices have

been defined for monitoring the thermal maturities of oils[52]:

$$MPI\ 1 = 1.5(2\text{-m-P} + 3\text{-m-P})/(P + 1\text{-m-P} + 9\text{-m-P})$$

$$MPI\ 2 = 3(2\text{-m-P})/(P + 1\text{-m-P} + 9\text{-m-P})$$

$$MPR = 2\text{-m-P}/1\text{-m-P}$$

Another relative ratio of methyl-phenanthrene isomers, (3- + 2-m-P)/(4-/9- + 1-m-P), has been increasingly used for spill oil source correlation and differentiation, and monitoring oil biodegradation.[6,43,50,51]

(2) Methyl-dibenzothiophenes. Chromatographically well-resolved C_1-dibenzothiophene isomers[53,54] are present in all oils at relatively high concentrations. Their relative abundance distributions vary significantly from different sources:

$$C_1\text{-dibenzothiophene distribution index} = (4\text{-}:2\text{-}/3\text{-}:1\text{-m-DBT})$$

A database of the relative ratios of the C_1-DBT isomers for several hundred crude, weathered and biodegraded oils, and petroleum products has been established.[54] Figure 27.20 plots 2-/3-methyldibenzothiophene versus 1-methyldibenzothiophene (both isomers are normalized relative to 4-methyldibenzothiophene) for some oils and oil products. Figure 27.20 shows how scattered the data points representing the various oils are. Another pronounced feature observed from Figure 27.20 is that related oils produce tight clusters on the plot. The use of these ratios complements existing methods of oil characterization, but has its own distinct advantages for discrimination of different oils.

(3) Other relative ratios of PAH isomers. In addition of relative ratios of (3- + 2-m-P)/(4-/9- + 1-m-P) and C_1-dibenzothiophene distribution index (4-:2-/3-:1-m-DBT), the selected PAH isomers[41-43,50,51] for oil fingerprinting studies include 3 isomers each within C_3-naphthalenes

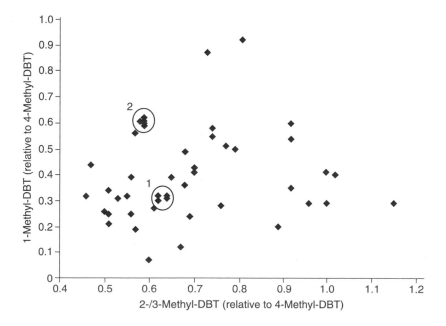

FIGURE 27.20 Plot of the relative ratios of 2-/3-methyldibenzothiophene to 4-methyldibenzothiophene versus the relative ratios of 1-methyldibenzothiophene to 4-methyldibenzothiophene for 49 different oils and oil products. The circles 1 and 2 indicate related samples of North Slope and Terra Nova, respectively.

(m/z 156) and C_4-naphthalenes (m/z 170), 4 isomers within C_2-phenanthrenes (m/z 206), 2 isomers within C_4-phenanthrenes (m/z 234), 3 isomers within C_1-fluorenes (m/z 180), 2-m-naphthalene/ 1-m-naphthalene (m/z 128), anthracene/phenanthrene (m/z 178), BaA/Chrysene (m/z 228), BeP/ BaP (m/z 252), and indeno(1,2,3-cd)pyrene/benzo(ghi)perylene (m/z 276).

D. BIOMARKER FINGERPRINTING ANALYSIS FOR OIL CORRELATION AND SOURCE IDENTIFICATION

Biological markers or biomarkers are complex molecules derived from formerly living organisms. As an example, Figure 27.21 shows molecular structures of representative cyclic biomarker compounds. Biomarkers are found in rocks and sediments and show little or no changes in structures from their parent organic molecules in living organisms. Biomarkers are useful because they retain all or most of original carbon skeleton of the original natural product and this structural similarity reveals more information about their origins than other compounds.

Biomarker fingerprinting has historically been used by petroleum geochemists[55] in: (1) characterization of oils in terms of the type(s) of precursor organic matter in the source rock (such as bacteria, algae, marine algae, or higher plants, because each type of organism may have different biomarkers); (2) correlation of oils with each other and oils with their source rock, (3) determination of depositional environmental conditions (such as marine, deltaic, or hypersaline environments); (4) assessment of thermal maturity and thermal history of oil during burial; and (5) evaluation of migration and the degree of biodegradation.

The conversion of a vast number of the precursor biochemical compounds from living organisms into biomarkers creates a vast suite of compounds in crude oils that have distinct structures. Further, due to the wide variety of geological conditions and ages under which oil has formed, every crude oil exhibits an essentially unique biomarker "fingerprint." Therefore, chemical

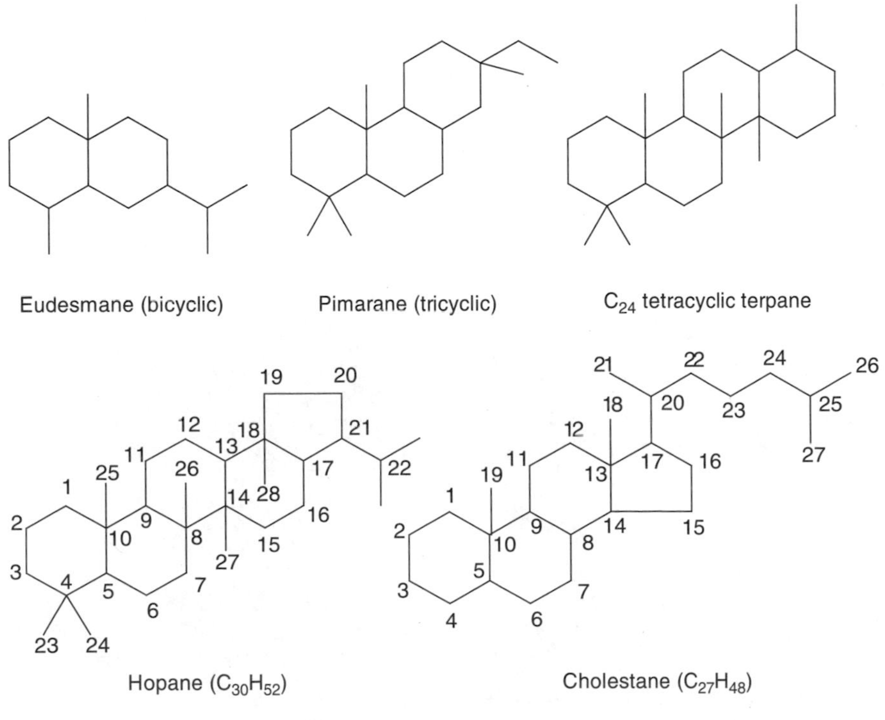

Eudesmane (bicyclic)   Pimarane (tricyclic) C_{24} tetracyclic terpane

Hopane ($C_{30}H_{52}$)   Cholestane ($C_{27}H_{48}$)

FIGURE 27.21 Molecular structures of representative cyclic biomarker compounds.

analysis of source-characteristic and environmentally persistent biomarkers generates information of great importance in determining the source of spilled oil, differentiating oils, monitoring the degradation process, and weathering state of oils under a wide variety of conditions. In the past decade, use of biomarker fingerprinting techniques to study spilled oils has greatly increased, and biomarker parameters have been playing a prominent role in almost all oil-spill work.

1. Biomarker Distributions

Generally, oil biomarkers of interest to environmental forensic investigation can be categorized into two classes: acyclic or aliphatic biomarkers, and cyclic biomarkers.

(1) Acyclic biomarkers. Isoprenoids including pristane, phytane, botryococcane (C_{30}), and bis-phytane (C_{40}) are one important group of commonly studied acyclic biomarker compounds.

(2) Cyclic biomarkers. Terpanes and steranes are most studied and most useful cyclic biomarker compounds (Figure 27.21).

The terpanes in petroleum include sesqui- (C_{15}), di- (C_{20}), sester- (C_{25}), and triterpanes (C_{30}). Many of the terpanes in petroleum originate from bacterial (prokaryotic) membrane lipids. Because bacteria are ubiquitous in sediments, terpanes are found in nearly all oils. These bacterial terpanes include several homologous series, including bicyclic, tricyclic, tetracyclic, and pentacyclic (e.g., hopanes) compounds. The hopanes (with 30 carbon atoms or less) are composed of three stereoisomeric series, namely 17 $\alpha\beta$-, 17 $\beta\beta$-, and 17 $\beta\alpha$-hopanes. Compounds in the $\beta\alpha$ series are called moretanes. Hopanes with the 17 $\alpha\beta$-configuration in the range of C_{27} to C_{35} are characteristic of petroleum because of their greater thermodynamic stability compared to other epimeric ($\beta\beta$ and $\beta\alpha$) series. The $\beta\beta$ series is not found in petroleum because it is thermally very unstable even during early catagenesis stage.

The steranes are a class of biomarkers containing 21 to 30 carbons that are derived from sterols, and they include regular steranes, rearranged diasteranes, and mono-aromatic and tri-aromatic steranes. Among them, the regular $C_{27}-C_{28}-C_{29}$ homologous sterane series are the most common and useful steranes because they are highly specific for correlation. The concentrations of C_{29} steranes (24-ethylcholestanes) compared to the C_{27} and C_{28}-steranes may indicate a land plant source.

The chemistry and formation of terpanes and steranes has been thoroughly studied.[55] These cyclic biomarkers are generally stable and relatively resistant to degradation. As discussed in the previous sections, characterization of these compounds are achieved by using GC–MS in the selected ion monitoring mode: m/z 191 for tricyclic, tetracyclic and pentacyclic terpanes, m/z 123 for bicyclic sesquiterpanes, m/z 217 and 218 for steranes, m/z 217/259 for diasteranes, m/z 253 for mono aromatic steranes, and m/z 231 for tri-aromatic steranes. For many oils, the GC–MS chromatograms of terpanes (m/z 191) are characterized by the terpane distribution in a wide range from C_{20} to C_{30} often with C_{23} and C_{24} tricyclic terpanes and C_{29} $\alpha\beta$- and C_{30} $\alpha\beta$- pentacyclic hopanes being prominent. As for steranes (at m/z 217 and 218), the dominance of C_{27}, C_{28}, and C_{29} 20S/20R homologues, particularly the epimers of $\alpha\beta\beta$-steranes, among the C_{20} to C_{30} steranes is often apparent.

The distribution patterns of biomarkers are, in general, different from oil to oil and from oil to refined products. Figure 27.22 and Figure 27.23 show GC–MS chromatograms at m/z 191 and 218 for Sockeye (California), Orimulsion-400 (Venezuela), HFO 6303, and Diesel No.2 (Ontario). The differences in the relative distribution of terpanes and steranes between Sockeye oil and Orimulsion are very apparent. For Sockeye, C_{28}-bisnorhopane, C_{29} and C_{30} $\alpha\beta$ hopane are the most abundant with the concentration of C_{28} being even higher than C_{29} and C_{30} hopane. For Orimulsion, C_{23} terpane is the most abundant and the concentration of C_{29} is lower than C_{30} hopane. For HFO 6303,

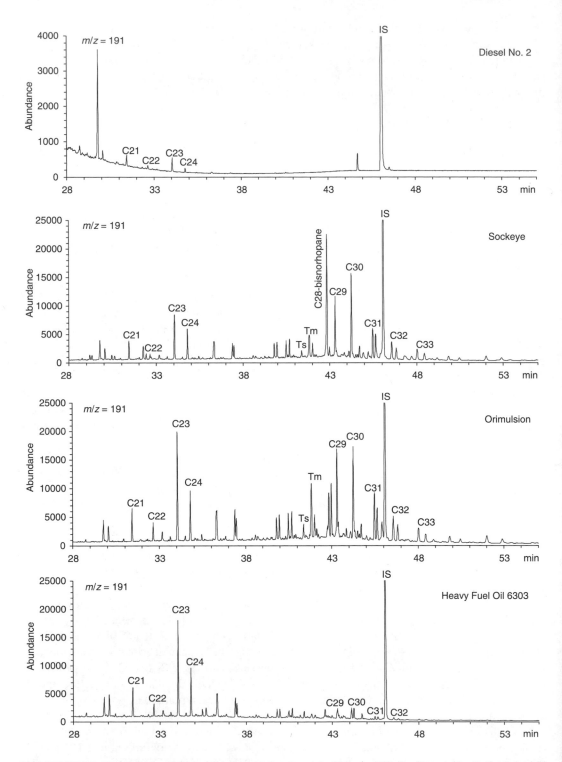

FIGURE 27.22 Distribution of biomarker terpane compounds (at m/z 191) for Diesel No. 2, Sockeye oil, Orimulsion, and HFO 6303, illustrating the differences in the relative distribution of terpanes between oils and oil products.

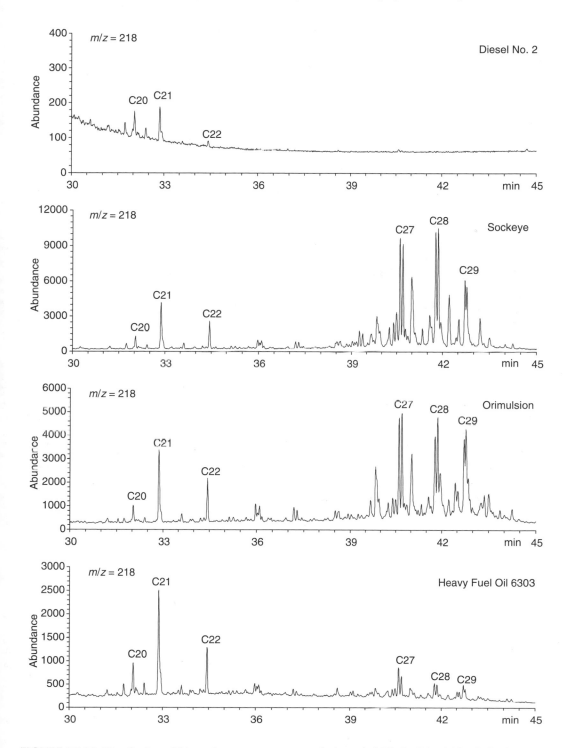

FIGURE 27.23 Distribution of biomarker sterane compounds (at m/z 218) for Diesel No. 2, Sockeye oil, Orimulsion, and HFO 6303, illustrating the differences in the relative distribution of steranes between oils and oil products.

C_{23} terpane is the most abundant, but nearly no homohopanes of C_{31} to C_{35} were detected. Different from most Bunker C type oils, the concentrations of terpanes and steranes are quite low in HFO 6303. As for refined products, only traces of C_{20} to C_{24} terpanes and C_{20} to C_{22} steranes were detected in Diesel No. 2. In contrast, most lube oils contain very high quantity of biomarkers. Obviously, refining processes have removed or concentrated high molecular mass biomarkers from the corresponding crude oil feed stocks.

a. Biomarker Distribution in Weathered Oil

Biomarkers are source characteristic and highly degradation-resistant. Therefore, for severely weathered oils, they may exhibit completely different GC–FID chromatograms and n-alkane profiles or isoprenoid distribution from their source oil, but their biomarker distribution patterns may be the same. Thus the fingerprinting of terpane and sterane biomarkers provides us with a powerful tool for tracking the source of the long term weathered oil. Characterization of many long-term spilled oils, as described above, demonstrated[6,24,50,66] that the chromatograms for n-alkane and isoprenoid distribution of weathered oil samples can be completely different from their corresponding source oils, that is, the source oils had significant distribution of n-alkanes and isoprenoids in the carbon range of $C_8–C_{42}$, while n-alkanes and isoprenoids in severely weathered samples were completely lost because of the effects of many years' field weathering and degradation. However, the profiles of their GC–MS fingerprints at m/z 191 and 217/218 were nearly identical. Furthermore, the computed diagnostic ratios of a series of target pairs of biomarker compounds were also nearly identical, clearly indicating that these samples were from the same source.

b. Biomarker Distribution in Petroleum Products with Similar Chromatographic Profiles

On some other cases where two oils may exhibit similar or even nearly identical n-alkane and isoprenoid distributions, however, their biomarker distribution may be markedly different. Thus, the successful forensic investigation will require performing detailed analysis and comparison of not only the concentrations but also the diagnostic ratios of biomarkers among similar product types. Figure 27.24 shows the GC–FID chromatograms of three unknown oil samples received from Montreal on March of 2001 for product identification and differentiation. Three samples show nearly identical GC chromatographic profiles and distribution patters. The relative ratios of low-abundant hydrocarbons n-C_{17}/n-C_{18}, n-C_{17}/pristane and n-C_{18}/phytane are also very similar. All the GC trace features suggest that the oils are hydraulic fluid-type products. The questions which must be answered at this stage are: do these three oil samples really come from the same source or not? The GC–MS analysis for biomarker characterization (Figure 27.25) indicates that samples #1 and #2 are nearly identical in distribution patterns and profiles of terpanes and steranes, but sample #3 shows markedly different distribution pattern of biomarkers, in particular the terpanes, from sample #1 and #2. The concentrations of C_{23} and C_{24}, and the sum of C_{31} through C_{35} homohopanes in sample #3 are significantly lower and higher than the corresponding compounds in samples #1 and #2, respectively. The biomarker analysis results, in combination with other GC analysis results, clearly demonstrated that sample #1 and #2 are identical and from the same source, and the sample #3 is different from sample #1 and #2 and does not come from the same source as samples #1 and #2 do, but it has bulk group hydrocarbon composition very similar to samples #1 and #2.

2. Sesquiterpane and Diamondoid Biomarkers in Oils and Lighter Petroleum Products

As described above, refining processes have removed most high molecular weight biomarkers from the corresponding crude oil feed stocks. Therefore the high boiling point pentacyclic triterpanes and steranes are generally absent in lighter petroleum products, jet fuels, and most diesels. However, several bicyclic sesquiterpanes (Figure 27.26) including eudesmane and

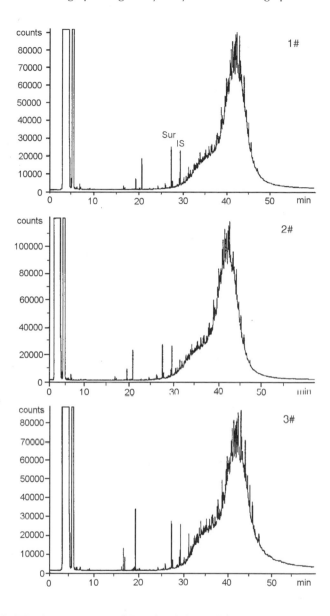

FIGURE 27.24 GC–FID chromatograms of Fraction 3 for *n*-alkane and TPH analysis of three unknown oil samples. These three samples show very similar GC chromatographic profiles and distribution patters, featured by dominance of unresolved complex mixture (UCM) of hydrocarbons with very small amount of resolved peaks.

drimane (C_{15}), which boil in diesel range and are ubiquitous in sediments and crude oils, are still present in diesels and certain jet fuels with significant abundances after refining processes. Examination of GC–MS chromatograms of these bicyclic biomarkers at m/z 123, 179, 193, and 207 can provide a comparable and highly diagnostic means of comparison for diesel type products.[55,56]

Another promising group of low-boiling cyclic biomarkers of interest for environmental forensic investigators are "diamondoid" hydrocarbons (the collective term of adamantane (C_{10}), diamantane (C_{14}), and their alkyl-homologous series). Diamondoid hydrocarbons are rigid, three-dimensionally fused cyclohexane-ring alkanes that have a diamond-like (cage-like) structure. The diamondoids found in petroleum are thought to be formed from rearrangements of suitable organic

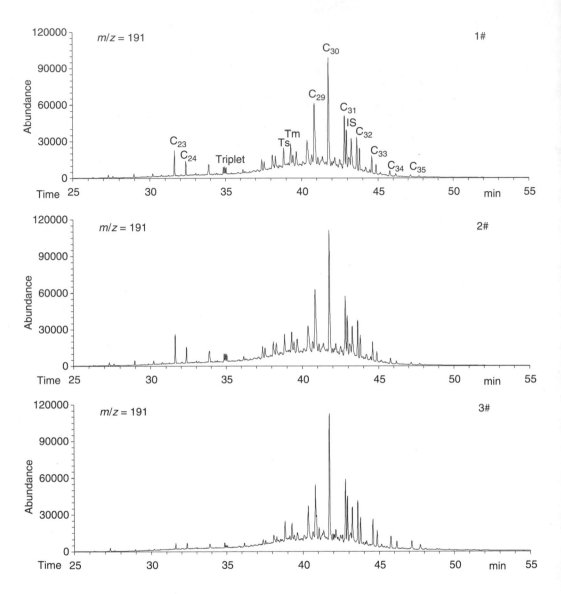

FIGURE 27.25 Distribution of biomarker terpane compounds (m/z 191) for the three unknown oil samples. Samples #1 and #2 are nearly identical in distribution patterns and profiles of terpanes, but sample #3 shows markedly different distribution pattern of biomarkers.

precursors (such as multi-ring terpene hydrocarbons) with strong Lewis acids acting as catalysts during oil generation.[57,58] The diamond structure endows these molecules with an unusually high thermal stability and high resistance to biodegradation. Adamantanes and diamantanes can be examined using m/z 135, 136, 149, 163, 177, and 191 mass chromatograms (Figure 27.27) and m/z 187, 188, 201, 215 and 229 mass chromatograms (Figure 27.28), respectively. Two diamondoid hydrocarbon ratios have been developed and used as novel high-maturity indices to evaluate the maturation and evolution of crude oils and condensates in several Chinese basins.[58] The lab thermal-cracking experiments[57] showed that the increase in methyldiamantane (C_{15}) concentration is directly proportional to the extent of cracking, indicating that under the conditions of the experiments, diamondoids are neither destroyed nor created. Instead, they are conserved and concentrated, and hence can be considered a naturally occurring

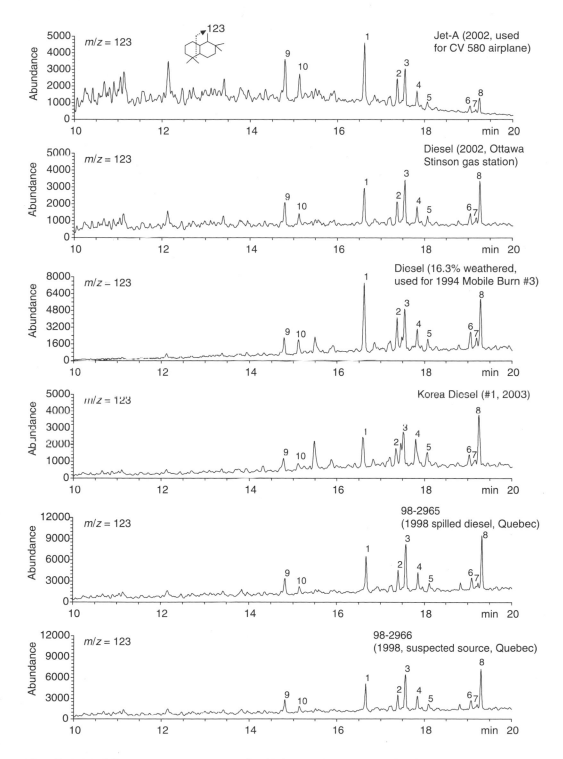

FIGURE 27.26 GC–MS chromatograms at m/z 123 for sesquiterpane analysis of Jet A (2002), Diesel No. 2 (2002, from Ottawa Stinson Gas Station), Diesel No. 2 (for 94 Mobile Burn, 16.3% weathered), a Korean diesel (#1, 2003, from Korea), 1998-spilled-diesel (from Quebec) and 1998-suspected-source diesel (from Quebec). The different distributions of the sesquiterpanes demonstrate the differences between diesels and between diesel and jet fuel.

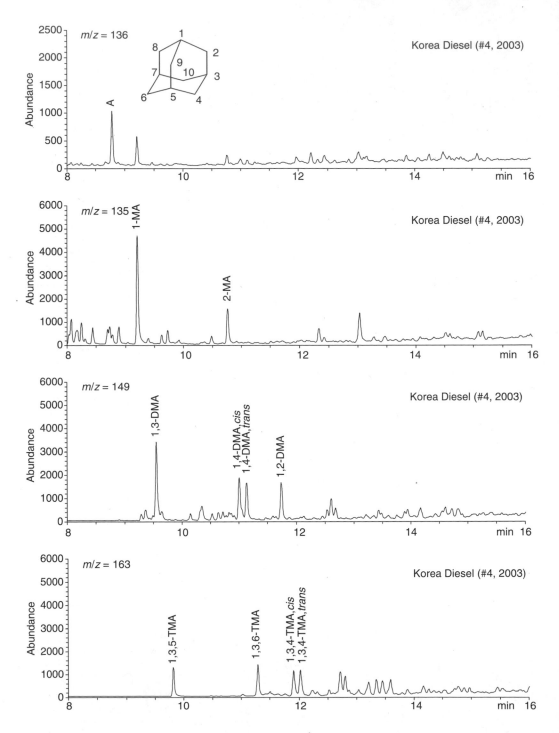

FIGURE 27.27 Distribution of biomarker adamantanes in Korean diesel (#4, 2003). The identified adamantanes include adamantane (*m/z* 136), methyl- (*m/z* 135), dimethyl- (*m/z* 149), and trimethyl-adamantanes (*m/z* 163).

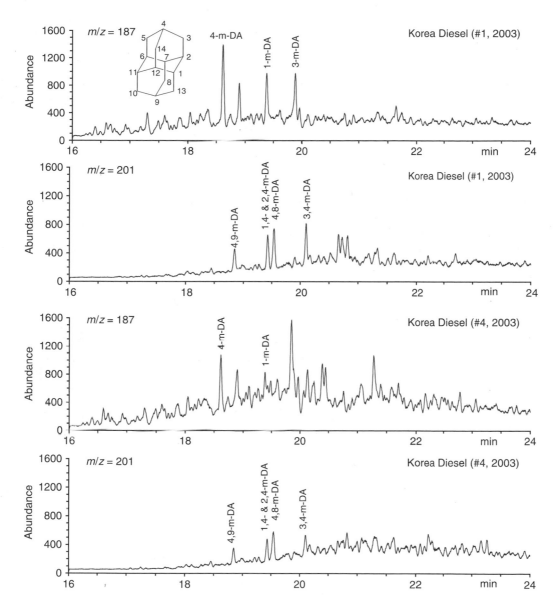

FIGURE 27.28 Comparison of distribution of biomarker diamantanes (at m/z 187 and 201) in Korean diesels (#1 and #4).

"internal standard" by which the extent of oil destruction can be determined. The extent of cracking is equal to $(1 - (C_0/C_C)) \times 100$, where C_0 is the concentration of methyldiamantanes in the uncracked samples and C_C is the methyldiamantane concentration of any cracked samples derived from the same starting oil. This principle should be also applied to determine the weathering percentages of diesels and similar light refined products, but C_0 and C_C in the equation should represent the concentrations of selected methyldiamantane in the fresh and weathered oil, respectively.

The distributions and relative ratios of sesquiterpanes and diamondoids may also have potential applications in investigation of oil and refined product spills. Stout et al.[7] compared the chromatographic distributions of bicyclic sesquiterpanes[59] of two weathered diesel fuel samples from two adjacent petroleum terminal properties. The samples have been highly weathered with n-alkanes being completely lost. However, GC–MS (m/z 123) analysis results showed very different distribution profiles of sesquiterpanes between two samples, strongly indicating that two sources of diesel existed in the study area.

Figure 27.26 shows the GC–MS chromatograms at m/z 123 for sesquiterpane analysis of a jet fuel, Diesel No.2 (2002, from an Ottawa Stinson Gas Station), Diesel No.2 (for 94 Mobile Burn, 16.3% weathered), Diescl No.2 (2003, from Korea), 1998-spilled-diesel (from Quebec) and 1998-suspected source diesel (from Quebec). The different distributions of the sesquiterpanes demonstrate the differences between diesels and between diesel and jet fuel. Figure 27.26 also clearly shows the 1998-spilled-diesel (from Quebec) had nearly identical GC chromatogram (m/z 123) with the 1998-suspected source diesel (from Quebec). These similarities, in combined with other GC analysis results (such as hydrocarbon groups, n-alkanes and PAHs), argued strongly that the spilled diesel was from the suspected source.

The smaller sesquiterpane and diamondoid biomarkers with lower boiling points (Figures 27 and 28) are useful and promising for source correlation and differentiation of refined products, in particular for the weathered refined products, because of their stability and resistance to bio-degradation. It can be anticipated that more work will be published in this fertile area of research.

3. Unique Biomarker Compounds

Biomarker terpanes and steranes are common constituents of crude oils. However, a few "specific" biomarker compounds including several geologically rare acyclic alkanes are found to exist only in certain oils and, therefore, can be used as unique markers to provide an interpretational advantage in fingerprinting sources of spilled oils and to provide additional diagnostic information on the types of organic matter that give rise to the crude oil. For example, the geologically rare acyclic alkane botryococcane ($C_{34}H_{70}$) was used to identify a new class of Australian nonmarine crude oils.[60] The presence of botryococcane indicates that the source rock contains remains of the algae *Botryococcus braunii*. The biomarkers 18α(H)-oleanane and 17α(H),18α(H),21β(H)-28, 30-bisnorhopane have been of special interest. The presence of biomarker 18α(H)-oleanane in benthic sediments in PWS, coupled with its absence in Alaska North Slope crude and specifically in Exxon Valdez oil and its residues, confirmed another petrogenic source.[44]

Other "specific" pentacyclic terpanes include C_{30} 17α (H)-diahopane (it may be related to bacterial hopanoid precursors that have undergone oxidation and rearrangement by clay mediated acidic catalysis), β-carotane ($C_{40}H_{78}$, highly specific for lacustrine deposition, highly abundant in Green River Shale), gammacerane ($C_{30}H_{52}$, it has been tentatively suggested as a marker for hyper saline episodes of source rock deposition), lupanes and bisnorlupanes (they are believed to indicate terrestrial organic matter input), and bicadinanes (highly specific for resinous input from higher plants).

4. Diagnostic Ratios of Biomarkers

Biomarker diagnostic parameters have been long established and are widely used by geochemists[55] for oil correlation (oil source rock correlation and oil–oil correlation), determination of organic input and depositional environment; for assessment of thermal maturity; and for evaluation of oil biodegradation. Many of biomarker diagnostic parameters currently used in oil spill studies are originated from geochemistry parameters. Table 27.10 lists some of the primary diagnostic ratios of biomarkers frequently used by the environmental chemists for spilled oil identification, correlation, and differentiation.

During the Arrow oil spill work, the ratio of the most abundant C_{29} to C_{30} hopane was defined and used as a reliable source indicator.[50] Zakaria et al.[61] studied oil pollution in the Straits of Malacca. Various samples including Malaysia oil, Middle East crude oils, South East Asian crude

TABLE 27.10
Diagnostic Parameters of Biomarkers Most Used in Oil Spill Fingerprinting

Diagnostic Ratios	Ion Monitored
Terpanes	
Tricyclic C23/C24	191
C29 $\alpha\beta$/C30 $\alpha\beta$ hopane	191
C23/C30 and C24/C30$\alpha\beta$ hopane	191
C30 $\beta\alpha$/C30 $\alpha\beta$ hopane	191
Oleanane/C30 $\alpha\beta$ hopane	191
Gamacerane/C30 $\alpha\beta$ hopane	191
Homohopane (C31 to C35) distribution	191
C30 $\alpha\beta$ hopane/homohopanes (C31 to C35)	191
Homohopanes 22S/(22S + 22R)	191
Ts/Tm & Ts/(Ts + Tm)	191
Triplet ratio:	
C24 tertracyclic terpane/C26 tricyclic (S)/C26 tricyclic (R) sterane	191
C25 norhopane/C30 hopane	191, 177
Sterane	
Regular sterane distribution (C27–C28–C29) $\alpha\alpha\alpha$ and $\alpha\beta\beta$ steranes (20S + 20R)	227, 228
Regular steranes/C30 $\alpha\beta$ hopane	227, 228
Diasteranes/regular steranes	227, 228
C27 $\alpha\beta\beta$ steranes/C29 $\alpha\beta\beta$ steranes	227, 228
C27–C28–C29: $\alpha\beta\beta$/($\alpha\alpha\alpha$ + $\alpha\beta\beta$) and 20S/(20S + 20R)	227, 228
Aromatic steranes	
C26–C27–C28 triaromatic steranes (TA)	231
Triaromatic steranes: 20R/20S and 20R/(20R + 20S)	231
C27–C28–C29 monoaromatic steranes (MA)	253
TA/(TA + MA)	231, 253

oils, tarballs, sediments, and mussels were collected and analyzed. The analytical results in this study demonstrated the utility of C_{29}/C_{30} ratio, $\Sigma(C_{31} - C_{35})/C_{30}$, and homohopane index as molecular tools to distinguish the source of petroleum in the Straits of Malacca. Barakat et al.[62] studied biomarker properties of five crude oils from the Gulf of Suez, Egypt. The results reveal significant difference in biomarker distribution and diagnostic ratios within the oils that suggest two oil types and one mixed type. The triplet ratio, in general, varies in oils from different sources and is dependent upon sources, depositional environment, and maturity. The triplet ratio was first used in a chemistry study of North Slope crude by Kvenvolden et al.,[63] in which the ratio is ~2. Exxon Valdez oil (an Alaska North Slope crude) and its residues also have triplet ratios of ~2; in contrast, many tarballs collected from the shorelines of the Sound have triplet ratios of ~5.

Table 27.11 summarizes quantitation results and diagnostic ratios of target biomarkers for three unknown oil samples having very similar bulk chemical compositions and nearly identical GC–FID chromatographic profiles, as discussed in Section 6.1.2. All the biomarker fingerprinting evidences, in combination with other GC analysis results, unambiguously point toward to the conclusion that samples #1 and #2 are identical and from the same source. However, the sample #3 is indeed different and is not from the same source as samples #1 and #2.

During January and February 1996, a significant number of tarball incidents[43] occurred along the coasts of Vancouver Island in British Columbia (BC), Washington (WA), Oregon (OR), and California (CA). The diagnostic values of "source specific" biomarker and PAH isomer compounds

TABLE 27.11
Quantitation Results and Diagnostic Ratios of Major Biomarkers

	#1	#2	#3
Quantitation results (μg/g oil)			
C_{23}	145.3	151.7	38.6
C_{24}	82.5	87.0	35.9
C_{29}	549.9	554.4	523.9
C_{30}	1049.5	1054.8	1164.9
$C_{31(S)}$	415.8	427.2	504.3
$C_{31(R)}$	259.4	262.8	333.8
$C_{32(S)}$	231.4	224.9	330.7
$C_{32(R)}$	149.0	148.0	196.4
$C_{33(S)}$	126.2	128.7	219.6
$C_{33(R)}$	73.7	75.0	136.7
$C_{34(S)}$	45.1	48.3	130.6
$C_{34(R)}$	22.0	22.1	63.8
$C_{35(S)}$	19.7	22.3	87.2
$C_{35(R)}$	15.2	16.8	68.4
Ts	133.5	136.9	137.2
Tm	148.0	148.3	128.9
C_{27}aßß-sterane	529.2	534.0	596.5
C_{29}aßß-sterane	705.4	715.7	766.9
Sum of C_{31} to C_{35} homohopanes	1358	1376	2072
Total of target biomarkers	4701	4759	5464
Diagnostic ratios			
C_{23}/C_{24}	1.76	1.74	1.08
C_{23}/C_{30}	0.14	0.14	0.03
C_{24}/C_{30}	0.08	0.08	0.03
C_{29}/C_{30}	0.52	0.53	0.45
Triplet (RT = ~35 min)	1.14:1.08:1.00	1.14:1.12:1.00	2.22:1.09:1.00
$C_{31}(S)/C_{31}(R)$	1.60	1.63	1.51
$C_{32}(S)/C_{32}(R)$	1.55	1.52	1.68
$C_{33}(S)/C_{33}(R)$	1.71	1.72	1.61
$C_{34}(S)/C_{34}(R)$	2.05	2.19	2.05
$C_{35}(S)/C_{35}(R)$	1.30	1.33	1.28
$C_{30}/(C_{31} + C_{32} + C_{33} + C_{34} + C_{35})$	0.77	0.77	0.56
Ts/Tm	0.90	0.92	1.06
C_{27}aßß-steranes/C_{29}aßß-steranes	0.75	0.75	0.78

of representative tarball samples and the suspected source Alaska North Slope (ANS) oil were calculated, tabulated and compared (Table 27.12). The results clearly reveals the following: (1) almost all diagnostic ratios for the ANS oil differ significantly from those of the tarball samples, indicating the ANS oil was not the source oil of tarball samples; (2) all the relative ratios are almost identical for sample BC-1 and BC-2, indicating they were from the same source; and (3) the tarball sample from CA was very similar in concentrations and diagnostic ratios of target biomarkers with the samples BC-1 and BC-2, but it had markedly different PAH isomeric ratios from samples BC-1 and BC-2, indicating CA tarballs may have another source different than the BC samples.

The biomarker fingerprinting results described above strongly suggest a *"basic rule"* in the environmental forensic investigations: a negative correlation of biomarkers is strong evidence for

TABLE 27.12
Comparison of Diagnostic Ratios of Biomarkers and PAH Isomers within Various Alkylation Levels Between BC and CA Tarball Samples, and the Suspected Source Alaska North Slope Oil

	BC-1	BC-2	CA-1	ANS
	Biomarkers			
Terpane C23/C24	2.15	2.19	2.07	1.69
Hopane C29/C30	0.84	0.84	0.84	0.61
Ts/Tm	0.31	0.33	0.29	0.50
C32(S)/C32(R)	1.46	1.49	1.52	1.46
C33(S)/C33(R)	1.56	1.56	1.52	1.44
C27$\alpha\beta\beta$/C29$\alpha\beta\beta$ steranes	1.11	1.09	1.14	0.84
	PAH isomers			
Alkylated naphthalenes				
C3N: isomer 1/isomer 2/isomer 3	4.37:3.42:1.0	4.46:3.50:1.0	3.17:2.35:1.0	4.58:3.16:1.0
C4N: isomer 1/isomer 2/isomer 3	1.0:0.68:0.37	1.0:0.72:0.37	1.0:0.50:0.44	1.0:1.05:0.58
Alkylated phenanthrenes				
C1P: (3 + 2-methyl-P)/(4/9 + 1-methyl-P)	1.37	1.42	0.89	0.74
C2P: isomer 2/isomer 1	3.11	3.00	4.85	4.10
isomer 3/isomer 1	1.74	1.70	2.50	2.09
C4P: isomer 1/isomer 2	1.43	1.48	0.95	0.61
Alkylated fluorenes				
C1F: isomer 1/isomer 2/isomer 3	1.0:1.02:0.43	1.0:1.00:0.42	1.0:1.42:0.53	1.0:1.94:0.42
Alkylated dibenzothiophenes				
C1D: 4-: 2/3 : 1-methyl-DBT	1.0:0.92:0.55	1.0:0.93:0.53	1.0:0.92:0.60	1.0:0.64:0.32

lack of relationship between samples, and a positive correlation of biomarkers is not necessarily "proof" that samples are related because some oils from different sources can show similar characteristics of biomarkers. In order to reliably and defensively correlate or differentiate samples, the "multi parameter approach" must be initiated, that is, analyses of more than one suite of analytes must be performed.

5. Biodegradation of Biomarkers

It has well recognized that terpane and sterane compounds are very resistant to biodegradation. In laboratory studies of biodegradation of nine Alaska oils and oil products[64] and eight Canadian oils[18] by a defined bacterial consortium incubated under freshwater and cold/marine conditions, it is found that the fingerprint patterns of triterpanes and steranes showed no changes after incubation, despite extensive saturate and aromatic losses, and the ratios of selected paired biomarkers also remained constant. Therefore, biomarkers can, and in many cases, have been used as conserved internal references for estimation of oil weathering percentages.

Cyclic biomarkers are highly resistant to biodegradation, but it does not mean they cannot be biodegraded. Actually, they are still biodegradable in severe weathering conditions. Based on several geochemical studies, Peters and Moldowan[55] have created a "quasi-stepwise" sequence for assessing the extent to which oil has been biodegraded as the follows:

acyclic isoprenoids > hopane (25-norhopanes present) ≥ steranes > hopanes(no 25-norhopanes)

~ diasteranes > aromatic steroids > porphyrins (least susceptible)

Munoz et al.[65] found that isoprenoids were severely degraded and biomarkers were more or less altered eight years after an oil spill in a peaty mangrove in a tropical ecosystem. They also found that norhopanes were the most biodegradation resistant among the studied terpane and sterane groups and the C_{30} $\alpha\beta$-hopane appeared more sensitive to weathering than its higher homologues.

In a very recent study on long-term fate and persistence of the spilled Metula oil in a marine marsh environment, Wang et al.[66] found that in highly degraded asphalt pavement samples, even the most refractory biomarker compounds showed some degree of biodegradation. The degree of biodegradation of biomarkers was not only molecular mass and size dependent, but also stereoisomer dependent. The biomarkers were generally degraded in the declining order of importance as diasterane $> C_{27}$ steranes $>$ tricyclic terpanes $>$ pentacyclic terpanes $>$ norhopanes $\sim C_{29}$ $\alpha\beta\beta$-steranes. The degradation of steranes was in the order of $C_{27} > C_{28} > C_{29}$ with the stereochemical degradation sequence 20R $\alpha\alpha\alpha$ steranes $>$ (20R + 20S) $\alpha\beta\beta$ steranes $>$ 20S $\alpha\alpha\alpha$ steranes. For the pentacyclic homohopanes, degradation of $C_{35} > C_{34} > C_{33} > C_{32} > C_{31}$ was apparent with significantly preferential degradation of the 22R epimers over 22S epimers. C30 $\alpha\beta$ hopane appeared more degradable than the 22S epimers of C_{31} and C_{32} homohopanes, but had roughly the same biodegradation rate as the 22R epimers of C_{31} and C_{32} homohopanes. C29-18α(H), 21β(H)-30 norneohopane and C29 $\alpha\beta\beta$ 20R and 20S stigmastanes appeared to be the most biodegradation resistant terpane and sterane compounds, respectively, among the studied target biomarkers.

E. CHARACTERIZATION OF ADDITIVES FOR SOURCE IDENTIFICATION OF REFINED PRODUCTS

As described in Section 3.2.1, certain chemical compounds, *additives*, are often added to petroleum products in order to improve some specific properties such as the engine efficiency and antiknock properties.

The additives added to gasoline (such as lead alkyls and dyes) are, in general, less volatile than the volatile gasoline components, therefore, characterization and determination of the distribution patterns of additive lead alkyls in gasoline and gasoline contaminated samples may provide beneficial information in certain environmental forensic investigation, such as source(s), history and age.[7] For example, analysis of dye additives using thin layer chromatography in dispersed gasoline and free products may allow differentiation between gasoline grades and manufacturers and establish a source relationship.[67] The lead alkyls can be analyzed using GC–MS at m/z 253 and 223 for tetramethyl lead and trimethylethyl lead, at m/z 267 and 223 for dimethyldiethyl lead, at m/z 281 and 223 for methyltriethyl lead, and at m/z 295 and 237 for tetraethyl lead.

Recently, Wang et al.[68] identified major unknown compounds with remarkable abundances in three oil samples (they were identified to be hydraulic fluids) with very similar bulk chemical compositions and nearly identical GC–FID chromatograms. Three major unknown compounds were identified as 2,6-bis(1,1-dimethylethyl)-phenol, butylated hydroxytoluene or 2,6-di-*tert*-butyl-4-methylphenol, and N-phenyl-1-naphthalenamine. Samples #1 and #2 contained these three compounds with nearly equal concentrations. However, butylated hydroxytoluene (BHT) was not detected in the sample #3. The sample #3 only contained the first and the last compound with the abundance of 2,6-bis(1,1-dimethylethyl)-phenol being markedly higher than N-phenyl-1-naphthalenamine. All these three compounds identified are antioxidant compounds, also called inhibitors. They are added to oxidizable organic materials (such as lubricants and gasoline) to retard auto-oxidation and, in general, to prolong the useful life of the substrates. The identification and differentiation of these additives clearly supports the general conclusion obtained from TPH, PAH and biomarker characterization, that is, the sample #3 is different from samples #1 and #2 and does not come from the same source as do samples #1 and #2.

F. Two-Dimensional GC for Oil Spill Studies

Comprehensive two-dimensional gas chromatography (GC × GC) is another hyphenated technique where two different chromatographic separation mechanisms act in concert to greatly improve component separation and identification. To date, GC × GC has been used to analyze light and middle distillate petroleum products. GC × GC has successfully separated and quantitated oxygenates, BTEX, and heavier aromatics in gasoline. GC × GC has been used to study the composition of kerosene, gas oil, cycle oil, and for forensic fingerprinting of marine diesel fuel spills.

The basic GC × GC system consists of a gas chromatograph containing two chromatography columns with different stationary phases and selectivity, a thermal modulator assembly, and a flame ionization detector (FID). The thermal modulator is the key part of the GC × GC system, and it is used to transfer analyte between the two chromatographic dimensions in GC × GC. The main elements of the thermal modulator are a small section of capillary column called the "modulator tube" and a rotating slotted heater. The modulator tube is a short, 8-cm section of capillary column that provides significant analyte retention, typically with a capacity factor greater than 40. The slotted heater, with the temperature 100°C greater than the modulator tube temperature, periodically rotates over the modulator tube to desorb the analyte, focus it, and inject it to the second column. The bandwidth of the injected band is less than 100 ms. The modulator repeats the injection every 5 seconds.

Gains et al.[69] described application of two-dimensional GC–GC for a spill-source identification. In this study, each analyte in oil is subject to two different separations achieved using two GC columns connected serially by the thermal modulator. Compounds of similar chemical structure were grouped together in ordered two dimensional chromatograms. The GC–GC analysis resulted in a match between the spill samples and one of the source samples. This result was consistent with the results obtained by GC–MS. Recently, the same group[70] applied the same GC × GC technique to investigate the chemical composition of the unresolved complex mixture of hydrocarbons (UCM) in petroleum contaminated marine sediments. Prior to GC × GC analysis, the UCM hydrocarbons were extracted and separated with silica- and silver-impregnated silica gel chromatography to yield four fractions. GC × GC separations used a nonpolar poly-dimethylsiloxane stationary phase for volatility based selectivity on the first dimension and a 14% cyanopropylphenyl polysiloxane phase for polarity based selectivity on the second dimension to fully resolve all chemical groups of the UCM including the alkanes, branched alkanes, one-, two-, and multi-ring cycloalkanes; and one-, two-, and multi-ring aromatics.

VI. CONCLUSIONS

The advances in petroleum hydrocarbon fingerprinting and data interpretation methods and approaches in the last two decades have now allowed for detailed qualitative and quantitative characterization of spilled oils. Chemical fingerprinting is a powerful tool for hydrocarbon source identification and differentiation when it is applied properly. However, in many cases, particularly for complex hydrocarbon mixtures or extensively weathered and degraded oil residues, there is no single fingerprinting analysis which can meet the objectives of forensic investigation and quantitatively allocate hydrocarbons to their respective sources. Under such circumstances, integrated "multiple parameter" approaches are always needed and used, more than one suite of analytes must be performed, and other independent techniques such as isotope analysis may be applied to support correlations. If large number of spill and source candidate samples are involved, statistical and numerical analysis techniques (such as principal component analysis) for data analysis are always performed.

Development in hydrocarbon fingerprinting techniques will continue as analytical and statistical techniques evolve. It can be anticipated that these developments will further enhance the utility and defensibility of oil hydrocarbon fingerprinting.

ACKNOWLEDGMENTS

We thank Dr. Chun Yang and Mr. Mike Landriault, of Emergencies Science and Technology Division, for performing some laboratory work and working on some graphics.

REFERENCES

1. National Research Council. *Oil in the Sea III: Inputs, Fates, and Effects*, The National Academies Press, Washington, 2002.
2. Fingas, M., *The Basics of Oil Spill Cleanup*, 2nd ed., Lewis Publishers, New York, 2001.
3. Wang, Z. D., Fingas, M., and Li, K., Fractionation of ASMB oil, identification and quantitation of aliphatic, aromatic and biomarker compounds by GC/FID and GC/MSD, *J. Chromatogr. Sci.*, 32, 361–366 (Part I) and 367–382 (Part II), 1994.
4. Jokuty, P., Whiticar, S., Wang, Z. D., Fingas, M., Fieldhouse, B., Lambert, P., and Mullin, J., *Properties of Crude Oils and Oil Products*, *Report Series No. EE-165*, Environment Canada, Ottawa, ON, 1999, Also available in Environmental Technology Center (ETC) web site: http://www.etc-cte.ec.gc.ca/databases/spills_e.html (2003).
5. Daling, P. S., Faksness, L. G., Hansen, A. B., and Stout, S. A., Improved and standardized methodology for oil spill fingerprinting. *Environmental Forensics*, 3, pp. 263–278, 2002.
6. Wang, Z. D., Fingas, M., and Page, D., Oil spill identification, *J. Chromatogr.*, 843, 369–411, 1999.
7. Stout, S. A., Uhler, A. D., McCarthy, K. J., and Emsbo-Mattingly, S., In *Introduction to Environmental Forensics*, Murphy, B. L. and Morrison, R. D., Eds., Academic Press, London, 2002, Chap. 6.
8. Wang, Z. D., *Analytical Methods for Determination of Oil Components*, ETC Method No. 5.3/1.3/M, 2002, Environmental Technology Centre, Environment Canada, Ottawa, Ont., 2002, updated version.
9. Olah, G. A. and Molnar, A., *Hydrocarbon Chemistry*, Wiley-Interscience, New York, 1995.
10. Speight, J. G., *Handbook of Petroleum Product Analysis*, Wiley-Interscience, Hoboken, NJ, 2002.
11. Kaplan, I. R., Galperin, Y., Alimi, H., Lee, R., and Lu, S., Pattern of chemical changes during environmental alteration of hydrocarbon fuels, *Ground Water Monit. Rem.*, 113–124, 1996, Fall.
12. Leahy, J. G. and Colwell, R. R., Microbial degradation of hydrocarbons in the environment, *Microbial. Rev.*, 54, 305–315, 1990.
13. Prince, R. C., Petroleum spill bioremediation in marine environment, *Crit. Rev. Microbial.*, 19, 217–242, 1993.
14. Bragg, J. R. and Owens, E. H., Clay-oil flocculation as a natural cleansing process following oil spill: Part 1 – studies of shoreline sediments and residues from past spills, in: *Proceedings of the 17th Arctic and Marine Oil Spill Program (AMOP) Technical Seminar*, Environment Canada, Ottawa, Ont., pp. 1–25, 1994.
15. Wang, Z. D., Fingas, M., Landriault, M., Sigouin, L., and Xu, N., Identification of alkylbenzenes and direct determination of BTEX and (BTEX + C_3-Benzenes) in oils by GC/MS, *Anal. Chem.*, 67, 3491–3500, 1995.
16. Wang, Z. D., Fingas, M., and Sergy, G., Chemical characterization of crude oil residues from an Arctic beach by GC/MS and GC/FID, *Environ. Sci. Technol.*, 29, 2622–2631, 1995.
17. Wang, Z. D. and Fingas, M., Study of the effects of weathering on the chemical composition of a light crude oil using GC/MS and GC/FID, *J., Microcolumn Sep.*, 7, 617–639, 1995.
18. Wang, Z. D., Fingas, M., Blenkinsopp, S., Sergy, G., Landriault, M., Sigouin, L., Foght, J., Semple, K., and Westlake, D. W. S., Comparison of oil composition changes due to biodegradataion and physical weathering in different oils, *J. Chromatogr.*, 809, 89–107, 1998.

19. Butler, E. L., Douglas, G. S., Steinhauter, W. S., Prince, R. C., Axcel, T., Tsu, C. S., Bronson, M. T., Clark, J. R., and Lindstrom, J. E., In *On-site Reclamation*, Hinchee, R. E. and Olfenbuttel, R. F., Eds., Butterworth-Heinemann, Boston, MA, pp. 515–521, 1991.

20. Douglas, G. S., Prince, R. C., Butler, E. L., and Steinhauer, W. G., *In-situ and On-site Bioremediation Conference Proceedings*, San Diego, CA, April 5–8, 1993.

21. Prince, R. C., Elmendorf, D. L., Lute, J. R., Hsu, C. S., Haith, C. E., Senius, J. D., Dechert, G. J., Douglas, G. S., and Butler, E. L., $17\alpha(H)$, $21\beta(H)$-Hopane as a conserved internal marker for estimating the biodegradation of crude oil, *Environ. Sci. Technol.*, 28, 142–145, 1994.

22. Short, J. W. and Heintz, R. A., Identification of Exxon Valdez oil in sediments and tissues from Prince William Sound and the Northwestern Gulf of Alaska based on a PAH weathering model, *Environ. Sci. Technol.*, 31, 2375–2384, 1997.

23. Short, J. W., Oil identification based on a goodness-of-fit metric applied to hydrocarbon analysis results, *Environmental Forensics*, 3, 349–356, 2002.

24. Wang, Z. D., Fingas, M., Blenkinsopp, S., Sergy, G., Landriault, M., and Sigouin, L., Study of the 25-year-old Nipisi oil spill: persistence of oil residues and comparisons between surface and subsurface sediments, *Environ. Sci. Technol.*, 32, 2222–2232, 1998.

25. Christensen, L. B. and Larsen, T. H., Method for determining the age of diesel oil spills in the soil, *Ground Water Monit. Rem.*, 142–149, 1993, Fall.

26. Kaplan, I. R., Alimi, M., Galperin, Y., Lee, R. P., and Lu, S., SPE Paper No. 29754, Presented at the 1995 SPE/EPA Exploration and Production Environmental Conference, Houston, TX, March 27–29, 1995.

27. Cretney, W. J., Green, D. R., Fowler, B. R., Humphrey, B., Fiest, D. L., and Boehm, P. D., Hydrocarbon biogeochemical setting of the Baffin Island oil spill experimental site, 1, Sediments, *Arctic*, 40, 51–55, 1987.

28. Venkatesan, M. I., Occurrence and possible sources of perylene in marine sediments - a review, *Mar. Chem.*, 25, 1–27, 1988.

29. Kolattukudy, P. E., *Chemistry and Biochemistry of Natural Waxes*, Elsevier, New York, 1976.

30. Page, D. S., Boehm, P. D., Douglas, G. S., and Bence, A. E., In *Exxon Valdez Oil Spill: Fate and Effects in Alaska Waters, ASTM STP 1219*, Wells, P. G., Butler, J. N. and Hughes, J. S., Eds., ASTM, Philadelphia, PA, pp. 41–83, 1995.

31. Bence, A. E. and Burns, W. A., In *Exxon Valdez Oil Spill: Fate and Effects in Alaska Waters, ASTM STP 1219*, Wells, P. G., Butler, J. N. and Hughes, J. S., Eds., ASTM, Philadelphia, PA, pp. 84–140, 1995.

32. Stout, S. A., Uhler, A. D. and McCarthy, K. J., Recognizing the confounding influences of "background" contamination in "fingerprinting" investigation, *Soil Sediment Groundwater*, 35–38(Febraury/March), 2000.

33. Connolly, J. D. and Hill, R. A., *Dictionary of Terpenoids*, Vol. 1 and 2, Eds., Chapman and Hall, New York, 1991.

34. Sauer, T. C. and Uhler, A. D., Pollutant source identification and allocation: advances in hydrocarbon fingerprinting, *Remediation*, 25–50, Winter Issue, 1994–1995.

35. Blumer, M. and Youngblood, W. W., Polycyclic aromatic hydrocarbons in soils and recent sediments, *Science*, 188, 53–55, 1975.

36. Bjøeseth, A., In *Handbook of Polycyclic Aromatic Hydrocarbons*, Bjøeseth, A. and Ramdahl, T., Eds., Marcel Dekker, New York, pp. 1–20, 1985.

37. Benlahcen, K. T., Chaoui, A., Budzinski, H., Bellocq, J., and Garrigues, P., Distribution and sources of polycyclic aromatic hydrocarbons in some Mediterranean coastal sediments, *Mar. Pollut. Bull.*, 34, 298–305, 1997.

38. Sicre, M. A., Marty, J. C., Salion, A., Aparicio, X., Grimalt, J., and Albaiges, J., Aliphatic and aromatic compounds in aerosols, *Atmos. Environ.*, 21, 2247–2259, 1987.

39. Wang, Z. D., Fingas, M., Shu, Y. Y., Sigouin, L., Landriault, M., and Lambert, P., Quantitative characterization of PAHs in burn residue and soot samples and differentiation of pyrogenic PAHs from petrogenic PAHs - the 1994 Mobile burn study, *Environ. Sci. Technol.*, 33, 3100–3109, 1999.

40. Meniconi, M. G., Gabardo, I. T., Carneiro, M. E., Barbanti, S. M., Silva, G. C., and Massone, C. G., Brazilian oil spills chemical characterization - case studies, *Environ. Forensics*, 3, 303–322, 2002.

41. Wang, Z. D., Fingas, M., and Sigouin, L., Characterization and Identification of a "mystery" Oil Spill from Quebec (1999), *J. Chromatogr.*, 909, 155–169, 2001.

42. Wang, Z. D., Fingas, M., and Lambert, P., *Proceedings of the 26th Arctic and Marine Oil Spill Program (AMOP) Technical Seminar*, Environment Canada, Ottawa, pp. 169—192, 2003.

43. Wang, Z. D., Fingas, M., Landriault, M., Sigouin, L., Castle, B., Hostetter, D., Zhang, D., and Spencer, B., Identification and linkage of tarballs from the coasts of Vancouver Island and Northern California using GC/MS and isotopic techniques, *J. High Resolut. Chromatogr.*, 21, 383–395, 1998.

44. Bence, A. E., Kvenvolden, K. A., and Kennicutt, M. C. II, Organic geochemistry applied to environmental assessments of Prince William Sound, Alaska, after the Exxon Valdez oil spill - a review, *Organic Geochemistry*, 24, 7–42, 1996.

45. Boehm, P. D., Page, D. S., Gilfillan, E. S., Bence, A. E., Burns, W. A., and Mankiewicz, P. J., Study of the fate and effects of the Exxon Valdez oil spill on benthic sediments in two bays in Prince William Sound, Alaska. 1. Study design, chemistry and source fingerprinting. *Environ. Sci. Technol.*, 32, 567–576, 1998.

46. Boehm, P. D., Page, D. S., Burns, W. A., Bence, A. E., Mankiewicz, P. J., and Brown, J. S., Resolving the origin of the petrogenic hydrocarbon background in Prince William Sound, Alaska, *Environ. Sci. Technol.*, 35, 471–479, 2001.

47. Page, D.S., Bence, A.E., Burns, W.A., Boehm, P.D., and Douglas, G.S., A holistic approach to hydrocarbon source allocation in the subtidal sediments of Prince, William Sound, Alaska, Embayments. *Environmental Forensics*, 3, 331–340, 2002.

48. Douglas, G. S., Bence, A. E., Prince, R. C., McMillen, S. J., and Butler, E. L., Environmental stability of selected petroleum hydrocarbon source and weathering ratios, *Environ. Sci. Technol.*, 30, 2332–2339, 1996.

49. Hostettler, F. D., Rosenbauer, R. J., and Kvenvolden, K. A., PAH refractory index as a source discriminant of hydrocarbon input from crude oil and coal in Prince William Sound, Alaska, *Org. Geochem.*, 30, 873–879, 1999.

50. Wang, Z. D., Fingas, M., and Sergy, G., Study of 22-year-old Arrow oil samples using biomarker compounds by GC/MS, *Environ. Sci. Technol.*, 28, 1733–1746, 1994.

51. Wang, Z. D., Fingas, M., Landriault, M., Sigouin, L., Feng, Y., and Mullin, J., Using systematic and comparative analytical data to identify the source of unknown oil on contaminated birds, *J. Chromatogr.*, 775, 251–265, 1997.

52. Radke, M. D., Welte, H., and Willsch, H., Maturity parameters based on aromatic hydrocarbons: influence of the organic matter type, *Organic Geochem.*, 10, 51–63, 1986.

53. Fayad, N. M. and Overton, E., A unique biodegradation pattern of the oil spilled during the 1991 Gulf war, *Mar. Pollut. Bull.*, 30, 239–246, 1995.

54. Wang, Z. D. and Fingas, M., Use of methyldibenzothiophenes as markers for differentiation and source identification of crude and weathered oils, *Environ. Sci. Technol.*, 29, 2841–2849, 1995.

55. Peters, K. E. and Moldowan, J. W., *The Biomarker Guide: Interpreting Molecular Fossils in Petroleum and Ancient Sediments*, Prentice Hall, New Jersey, 1993.

56. Stout, S. A., McCarthy, K. J., Seavey, J. A., and Uhler, A. D., *Proceedings of the Nineth Annual West Coast Conference on Contaminated Soils and Water*, 1999 March, Abstract.

57. Dahl, J. E., Moldowan, J. M., Peters, K. E., Claypool, G. E., Rooney, M. A., Michael, G. E., Mello, M. R., and Kohnen, M. L., Diamondoid hydrocarbons as indicators of natural oil cracking, *Nature*, 399, 54–57, 1999.

58. Chen, J., Fu, J., Sheng, G., Liu, D., and Zhang, J., Diamondoid hydrocarbon ratios: novel maturity indices from highly mature crude oils, *Org. Geochem.*, 25, 179–190, 1996.

59. Noble, R. A., Alexander, R., Kagi, R. I., and Knox, J., Identification of some diterpenoids hydrocarbons in petroleum, *Org. Geochem.*, 10, 825–829, 1986.

60. McKirdy, D. M., Cox, R. E., Volkman, J. K., and Howell, V. J., Botryococcane in a new class of Australian non-marine crude oils, *Nature*, 320, 57–59, 1986.

61. Zakaria, M. P., Horinouchi, A., Tsutsumi, S., Takada, H., Tanabe, S., and Ismail, A., Oil pollution in the Straits of Malacca, Malaysia: Application of molecular markers for source identification, *Environ. Sci. Technol.*, 34, 1189–1196, 2000.

62. Barakat, A. O., Mostafa, A., El-Gayar, M. S., and Rullkotter, J., Source-dependent biomarker properties of five crude oils from the Gulf of Suez, Egypt, *Org. Geochem.*, 26, 441–450, 1997.

63. Kvenvolden, K. A., Rapp, J. B. and Bourell, J. H., In *Alaska North Slope Oil/Rock Correlation Study*, Magoon, L. B. and Claypool, G. E., Eds., *American Association of Petroleum Geologists Studies in Geology, No. 20*, pp. 593–617, 1985.

64. Wang, Z. D., Blenkinsopp, S., Fingas, M., Sergy, G., Landriault, M., Sigouin, L., Foght, J., Semple, K., and Westlake, D. W. S., Preprints of Symposia, Chemical composition changes and biodegradation potentials of nine Alaska oil under freshwater incubation conditions, *American Chemical Society*, 43, 828–835, 1997.

65. Munoz, D., Guiliano, M., Doumenq, P., Jacquot, F., Scherrer, P., and Mille, G., Long term evolution of petroleum biomarkers in mangrove soil, *Mar. Pollut. Bull.*, 34, 868–874, 1997.

66. Wang, Z. D., Fingas, M., Owens, E. H., Sigouin, L., and Brown, C. E., Long-term fate and persistence of the spilled Metula oil in a marine salt marsh environment: degradation of petroleum biomarkers, *J. Chromatogr.*, 926, 290–275, 2001.

67. Kaplan, I. R., Galperin, Y., Lu, S., and Lee, R. P., Forensic environmental geochemistry differentiation of fuel-types, their sources, and release time, *Org. Geochem.*, 27, 289–317, 1997.

68. Wang, Z. D., Fingas, M., and Sigouin, L., Using multiple criteria for fingerprinting unknown oil samples having very similar chemical composition, *Environ. Forensics*, 3, 251–262, 2002.

69. Gains, R. B., Frysinger, G. S., Hendrick-Smith, M. S., and Stuart, J. D., Oil spill source identification by comprehensive two-dimensional gas chromatography, *Environ. Sci. Technol.*, 33, 2106–2112, 1999.

70. Frysinger, G., Gains, R. B., Xu, L., and Reddy, C. M., Resoling the unresolved complex mixture in petroleum-contaminated sediments, *Environ. Sci. Technol.*, 37, 1653–1662, 2003.

28 Phthalate Esters

Maria Llompart, Carmen García-Jares, and Pedro Landín

CONTENTS

I. INTRODUCTION

Phthalate is the term commonly employed to refer to the dialkyl or alkyl aryl esters of 1,2-benzenedicarboxylic acid (phthalic acid). These are primary synthesized using esterification of phthalic anhydride and the corresponding oxo alcohol in the presence of an acid catalyst such as sulphuric or *p*-toluene sulfonic acid. Although there are a high number of different phthalates, only about 60 have industrial applications. Among then, only a small number, those with alkyl chains from 1 to 13 carbons, are produced in large scale, di(2-ethylhexyl) phthalate being the most widely produced phthalate. Chemical names, abbreviations, Chemical Abstract Registry (CAS) and European Inventory of Existing Commercial Substance (EINECS) numbers, chemical formulations, and molecular weights for important phthalate esters are summarized in Table 28.1. Chemical structures for twenty of these chemicals are shown in Figure 28.1.

In the "OXO" industry, the term "iso" denotes a mixture of isomers and does not refer to the IUPAC definition. Therefore, the abbreviations included in Table 28.2 indicate when a phthalate ester is a mixture of branched or linear isomers (i.e., DNP for linear di-*n*-nonylphthalate, and DINP for branched diisononylphthalate). With the exception of di-(2-ethylhexyl) phthalate, higher molecular weight phthalate esters (alkyl chains >6 carbons) are mixtures based on the alcohols used for its production. For example, DINP is a mixture of di-C8-C10 branched alkylesters, containing principally isomers with nine carbon alkyl chains. In the same way, the term DIDP refers to a mixture of di-C9-11 branched alkyl esters (C10-rich) of phthalic acid.

A. USES

Because of their high solubility in polymeric materials, its inertness, fluidity, low water solubilities and low volatilities, high molecular weight phthalate esters are widely employed in the manufacture of plastics, as nonreactive plasticizers, to make polymers softer, more flexible and workable. When used as plasticizers, phthalate esters can represent 5% to 60% of the total weight of the plastics and resins. Its main plasticizer application is the production of polyvinylchloride (PVC), although they are employed in the manufacture of other polymeric material, such as epoxy and polystyrene resins, chlorinated, natural and synthetic rubbers, polysulfide, nitrocellullulose, ethylcellullose, and polyurethane. Plastics and resins which contain phthalates, have a broad spectrum of applications including toys, rainwear, shower curtains, films for food packaging, carpets, wall coverings, shoes, cable and medical tubing, automobile, and furniture upholstery... Besides their main applications as plasticizers, phthalates are used as industrial solvents, as additives in the textile industries, as components of dielectric fluids, lubricants, fragrances, hairsprays, nail polish, deodorants, paints, glues, pesticide formulations ...etc.[1–4]

The annual global production of phthalates in the 1990s was approximately four million tonnes,[5] and about one million tonnes are produced each year in Western Europe, of which about 850,000 tonnes of phthalates are used in the plasticization of PVC. In the European Union (E.U.),

Abbreviations: APCI, atmospheric pressure chemical ionization; ASE, accelerated solvent extraction; CE, capillary electrophoresis; CGC, capillary gas chromatography; CI, chemical ionization; DCM, dichloromethane; ECD, electron capture detector; EI, electron ionization; ESI, electrospray ionization; EtAc, ethyl acetate; FID, flame ionization detector; FTIR, Fourier-transform infrared; gal, gallons; GC, gas chromatography; GC-MS, gas chromatography coupled to mass spectrometry; GC-MS-MS, gas chromatography coupled to in tandem mass spectrometry; GPC, gel permeation chromatography; HRMS, high resolution mass spectrometry; IS, internal standard; IT, ion trap; KD, kuderna–danish; LC, liquid chromatography; LC-UV/VIS, liquid chromatography with ultraviolet/visible detector; LLE, liquid–liquid extraction; LOD, limits of detection; MASE, solvent extraction assisted by microwaves; MeOH, methanol; MSD, mass spectrometry detector; PA, phthalic acid; PGC, packed gas chromatography; PGC, packed gas chromatography; PID, photoionization detector; PMEs, phthalic acid monoesters; PS-DVB, polyestyrene–divynylbenzene; PUF, polyurethane foam; RP, reverse phase; RSD, relative standard deviation; SPE, solid-phase extraction; SPME, solid-phase microextraction.

TABLE 28.1
Chemical Names, Abbreviations, Chemical Abstract Registry (CAS) and European Inventory of Existing Commercial Substance (EINECS) Numbers, Chemical Formulations, and Molecular Weights for Phthalate Esters

Name	Abbreviation	CAS No.	EINECS No.	Formulation	Molecular Mass
Dimethyl phthalate	DMP	131-11-3	205-011-6	$C_{10}H_{10}O_4$	194.2
Diethyl phthalate	DEP	84-66-2	201-550-6	$C_{12}H_{14}O_4$	222.2
Diallyl phthalate	DAP	131-17-9	205-016-3	$C_{14}H_{14}O_4$	246.3
Di-n-propyl phthalate	DPP	131-16-8	205-015-8	$C_{14}H_{18}O_4$	250.3
Diisopropyl pthalate	DIPP	605-45-8	210-086-3	$C_{14}H_{18}O_4$	250.3
Di-n-Butyl phthalate	DBP	84-74-2	201-557-4	$C_{16}H_{22}O_4$	278.3
Diisobutyl phthalate	DIBP	84-69-5	201-553-2	$C_{16}H_{22}O_4$	278.3
Di(2-methoxyethyl) phthalate	DMEP	117-82-8	204-212-6	$C_{14}H_{18}O_6$	282.3
Dipentyl phthalate	DAMP	131-18-0	205-017-9	$C_{18}H_{26}O_4$	306.4
Diisopentyl phthalate	DIAMP	605-50-5	210-088-4	$C_{18}H_{26}O_4$	306.4
Di(2-ethoxyethyl) phthalate	DEEP	605-54-9	210-090-5	$C_{16}H_{22}O_6$	310.3
Butylbenzyl phthalate	BBP	85-68-7	201-622-7	$C_{19}H_{20}O_4$	312.4
Dicyclohexyl phthalate	DCHP	84-61-7	201-545-9	$C_{20}H_{26}O_4$	330.4
Butyl 2-ethylhexyl phthalate	BOP	85-69-8	201-623-2	$C_{20}H_{30}O_4$	334.5
Di-n-hexyl phthalate	DHP	84-75-3	201-559-5	$C_{20}H_{30}O_4$	334.5
Diisohexyl phthalate	DIHP	71850-09-4, 68515-50-4	276-090-2, 271-093-5	$C_{20}H_{30}O_4$	334.5
Di-n-heptyl phthalate	DHpP	3648-21-3	222-885-4	$C_{22}H_{34}O_4$	362.5
Diisoheptyl phthalate	DIHpP	7188-89-6, 6815-44-6	276-15-8, 271-086-7	$C_{22}H_{34}O_4$	362.5
Hexyl 2-ethylhexyl phthalate	HEHP	75673-16-4	—	$C_{22}H_{34}O_4$	362.5
Di(2-butoxyethyl) phthalate	DBEP	117-83-9	204-213-1	$C_{20}H_{30}O_6$	366.5
Di-n-octyl phthalate	DOP	117-84-0	204-214-7	$C_{24}H_{38}O_4$	390.6
Diisooctyl phthalate	DIOP	27554-26-3	248-523-5	$C_{24}H_{38}O_4$	390.6
Di(2-ethylhexyl) phthalate	DEHP	117-81-7	204-211-0	$C_{24}H_{38}O_4$	390.6
Di(n-hexyl, n-octyl, n-decyl) phthalate	D610P	25724-58-7, 68515-51-5	247-210-0, 271-094-0	$C_{25}H_{40}O_4$	404.6
Di(n-heptyl, n-nonyl, n-undecyl) phthalate	D711P	3648-20-2, 111381-89-6 68515-44-6, 111381-90-9 68515-45-7, 111381-91-0	222-884-9 271-086-7 271-087-2	$C_{26}H_{42}O_4$	418.6

Continued

TABLE 28.1
Continued

Name	Abbreviation	CAS No.	EINECS No.	Formulation	Molecular Mass
Di-*n*-nonyl phthalate	DNP	84-76-4	201-560-0	$C_{26}H_{42}O_4$	418.6
Diisononyl phthalate	DINP	28553-12-0, 68515-48-0	249-079-5, 271-090-9	$C_{26}H_{42}O_4$	418.6
		68515-45-7	271-087-2		
Di-*n*-decyl phthalate	DDP	84-77-5	201-561-6	$C_{28}H_{46}O_4$	446.7
Diisodecyl phthalate	DIDP	26761-40-0, 68515-49-1	247-977-1, 271-091-4	$C_{28}H_{46}O_4$	446.7
Di-*n*-undecyl phthalate	DUP	3648-20-2	222-884-9	$C_{30}H_{50}O_4$	474.7
Diisoundecyl phthalate	DIUP	96507-86-7, 85507-79-5	306-165-8, 287-401-6	$C_{30}H_{50}O_4$	474.7
Di-*n*-tridecyl phthalate	DTDP	119-06-2	204-294-3	$C_{34}H_{58}O_4$	530.8
Diisotridecyl phthalate	DITDP	27253-26-5, 68515-47-9	248-368-3, 271-089- 3	$C_{34}H_{58}O_4$	530.8

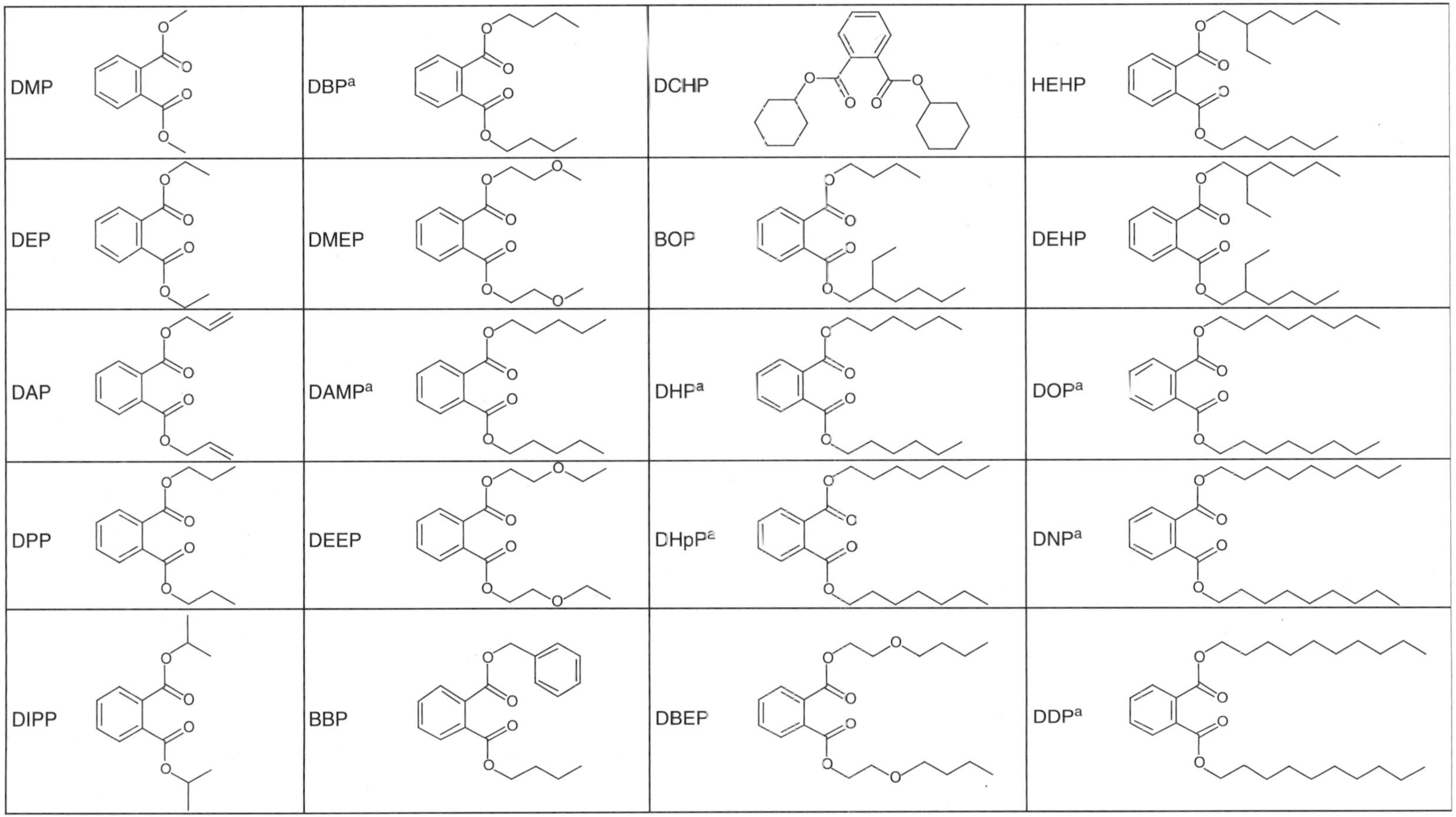

FIGURE 28.1 Phthalate esters chemical structure ([a]the side chains may be branched and therefore, several isomeric forms of the phthalate ester exist). See Table 28.1 for compound name abbreviations.

TABLE 28.2
Physical–Chemical Data Summary for Phthalate Esters. Solubility, Vapour Pressure, Partition Coefficient (Log K_{OW}) and Henry's Law Constant (H) [*]

Name	Abbreviation	Specific Gravity (20°C)	Melting Point (°C)	Solubility (mg/l)	Vapor Pressure (Pa)	Log K_{OW}	H (Pa m^3/mol)
Dimethyl phthalate	DMP	1.189	5.5	5220	0.263	1.61	9.78×10^{-3}
Diethyl phthalate	DEP	1.118	-40	591	6.48×10^{-2}	2.54	2.44×10^{-2}
Diallyl phthalate	DAP	1.121	-70	156	2.71×10^{-2}	3.11	4.28×10^{-2}
Di-n-propil phthalate	DPP	1.077	—	1.3	1.28×10^{-3}	5.12	0.302
Diisobutyl phthalate	DIBP	1.039	-64	9.9	4.73×10^{-3}	4.27	0.103
Butylbenzyl phthalate	BBP	1.114	-35	3.8	2.49×10^{-3}	4.70	0.205
Butyl 2-Ethylhexyl phthalate	BOP	0.993	-37	0.385	5.37×10^{-4}	5.64	0.466
Di-n-Hexyl phthalate	DHP	1.011	-27.4	0.159	3.45×10^{-4}	6.00	0.726
Diisoheptyl phthalate	DIHpP	1.00	-45	2.00×10^{-2}	9.33×10^{-5}	6.87	1.69
Diisooctyl phthalate	DIOP	0.986	-45	2.49×10^{-3}	2.52×10^{-5}	7.73	3.95
Di(2-ethylhexyl) phthalate	DEHP	0.986	-47	2.49×10^{-3}	2.52×10^{-5}	7.73	3.95
Di(n-hexyl, n-octyl, n-decyl) phthalate	D610P	0.97	-4	8.76×10^{-4}	1.31×10^{-5}	8.17	6.05
Di(n-heptyl, n- nonyl, n-undecyl) phthalate	D711P	0.97	< -50	3.08×10^{-4}	6.81×10^{-6}	8.60	9.26
Diisononyl phthalate	DINP	0.975	-48	3.08×10^{-4}	6.81×10^{-6}	8.60	9.26
Diisodecyl phthalate	DIDP	0.967	-48	3.81×10^{-5}	1.84×10^{-6}	9.46	21.6
Di-n-undecyl Phthalate	DUP	0.953	-9	4.41×10^{-6}	4.97×10^{-7}	10.33	50.5
Di-n-tridecyl Phthalate	DTDP	0.951	-37	7.00×10^{-8}	3.63×10^{-8}	12.06	275

[*] Values taken from Cousin, I. T. and Mackay, D., *Chemosphere*, 41, 1389–1399, 2000.

Council Regulation 793/93[6] divides existing chemicals into two categories: High Production Volume Chemicals (HPVC), produced or imported in quantities exceeding 1000 tonnes per year and Low Production Volume Chemicals (LPVC), produced or imported in quantities between 10 and 1000 tonnes per year. The list of HPVCs includes 22 phthalic acid esters, both single compounds and technical mixtures with different isomers. A further 11 phthalates are compiled in the list of LPVC. The most commonly used phthalate esters are DEHP (which accounts for around half of consumption in Western Europe, DINP and DIDP, which represents the 52.2% of phthalate consumption in the United States[7] and more than 85% of phthalate esters production in Western Europe.[8]

B. PROPERTIES

Phthalate esters have a wide range of physical chemical properties, summarized in Table 28.2. These chemicals are from colorless to faint yellow, oily liquids at room temperature with a slight aromatic odor, and molecular weights ranging from 194 (DMP) to 530 g/mol (DTDP). Phthalate esters have boiling points varying from approximately 230°C to 486°C.[9] According to their water solubility, phthalates may be classified from moderately soluble (5.2 g/l for the DMP) to practically insoluble (0.1 ng/l for DTDP); being less soluble in saltwater than in freshwater.[10,11] In natural water, complexation of these chemicals with humic substances, such as fulvic acid, can increase their solubilization.[1] In addition, the linear isomers are less soluble than the branched chain analogues.[12] The Log K_{OW} values (ranging from 1.61 for DMP to 12.06 for DTCP)[12] indicate that phthalate esters are very hydrophobic, especially those with a higher number of carbons of alcohol moiety. The lower molecular weight phthalate esters have high vapor pressures, so in pure states these volatilize rapidly. However, these have low Henry's law constants (H), thus evaporation from water is a slow process. Based on their H values, higher molecular weight phthalate esters will potentially volatilize more rapidly from water, however, because of their hydrophobycity, these phthalates, predominantly bond to suspended particulate matter, so they are not available for migration from natural water to the atmosphere.[13] In general, as molecular weight increases, volatility and water solubility of phthalate esters decrease, meanwhile hydrophobicity increases.

C. ENVIRONMENTAL FATE

Release of phthalate esters to the environment can occur during their production and incorporation into plastic materials. Since phthalate esters are not chemically bound to the polymeric matrix, they are fluid within the material to which they are added, so they can diffuse from these media into the environment. Therefore, the main portion of phthalate esters released to the environment is due to evaporation from consumption products during their use, despite their relatively low vapor pressures. Their mobility to the surrounding media combined with their high volume production and large spectrum of applications, made phthalates ubiquitous in today's environment.

As stated above, diffusion into the air is the major route by which phthalate esters enter the environment. Phthalates have relatively low vapor pressure and Henry's law constants, as well as relatively high K_{OW} and K_{OC}, so these compounds are found to only a limited extent in air. Nonetheless, phthalate esters are present in air, in both the vapor phase and associated with particulates, at concentrations generally at the low ng/m^3 level. Indoor concentrations may be several orders of magnitudes higher.[1] The discovery of some phthalate esters in Antarctic surface snow and in pack ice suggest that they can be transported for long distance.[14] Photodegradation by radical oxidation is expected to be the dominant degradation pathway in the atmosphere, with estimated half lives of less than one day for most of the phthalate esters.[9] Atmospheric fallout, via wet and dry deposition, is correlated with temperature. Since a higher portion of atmospheric phthalates are in the vapor phases in summer than in winter, lower amounts are subjected to fallout in the warmer season.[1]

Because of their hydrophobycity, when entering the water compartment phthalate esters have the tendency to bind suspended and particulate matter in such a way that concentrations in suspended particulate matter and sediments are several orders of magnitude higher than in the dissolved form.[15] Biodegradation under aerobic conditions can be an important fate process for phthalate esters in water, anaerobic and chemical degradation being much slower. It is generally accepted that biodegradation of the dissolved phthalates in water is relatively fast. Furtman[16] found that more than 90% of phthalates in river water was degraded within five days, although DEHP was slower to break down, particularly at low concentrations. Therefore, concentrations of these chemicals in natural waters are in the range of low μg/l to undetectable. In soil, phthalate esters are expected to have limited mobility, based on their K_{OC} values. Nonetheless, percolation of phthalates through the soil to groundwater may be enhanced by the presence of organic solvents, such as alcohols and ketones, usually found in hazardous waste sites. Aerobic biodegradation half lives in soils, as in natural waters, tends to increase with increasing alkyl chain length. Staples et al.[9] have estimated aerobic biodegradation half lives in natural waters from <1 day to about two weeks, and half lives in soils from 4 to 250 days for these chemicals. They also suggest, based on limited data available, that primary biodegradation of phthalate esters in sediments is slower than in soils and of the order of several months. Bioconcentration of phthalate esters has been documented in biota, including plants, fish, rats, cows, and humans. However, phthalates are rapidly metabolized to its corresponding alcohol monoesters and further oxidized and conjugated metabolites, which are rapidly excreted through the urine.[1,9] Therefore, accumulation of this class of compounds will be minimized by rapid metabolism in higher organisms. In this way, instead of biomagnification, biodilution is expected to occur as the phthalates are transferred through the food chain.[17]

D. TOXICOLOGY

Toxicological studies have linked some phthalate esters to liver and kidney damage, and to possible testicular or reproductive birth defect problems, characterizing them as endocrine disruptors. In this way, up to 12 phthalate esters, such as DBP, BBP, DEHP, DIDP, and DINP are within the list of the proposed substances suspected to produce endocrine alterations published by the EU.[18] The endocrine disruption potential of pthalate esters was recently reviewed by Harris and Sumpter.[19] The U.S. Agency for Toxic Substances and Disease Registry (ASTDR),[1-4] the World Health Organization (WHO),[20-23] the U.S. Department of Health and Human Services (DHHS),[24-30] and the EU[31-33] have undergone comprehensive risk assessments regarding human health aspects for some phthalates. Based on rat and mice studies, the U.S. Environmental Protection Agency (EPA), concluded that DEHP and BBP are *probable human carcinogens* (group B2) and a *possible human carcinogen* (group C), respectively. EPA classifies other phthalate esters, such as DBP and DEP, into group D (inadequate or no human and animal evidence of carcinogenicity). On the other hand, although initial evaluation stated DEHP as *possibly carcinogenic to humans*, a more recent International Agency for Research on Cancer (IARC) evaluation, in 2000,[34] classified DEHP in group 3 (*not classifiable as to its carcinogenicity to humans*) together with BBP.[35]

E. REGULATIONS

The adverse health effects of plastics containing phthalates has prompted the EU to ban DBP, BBP, DEHP, DOP, DINP, and DIDP in baby toys that may be introduced into the mouths of children under three years.[36] In addition, several proposals for the prohibition of DEHP use in medical materials have been made. In 1998, the Convention for the Protection of the Marine Environment of the Northeast Atlantic established the objective of eliminating the emissions and release of all hazardous substances into the environment by 2020, including DBP and DEHP in the list of chemicals for priority action.[37] The EU has also included DEHP in the list of 33 substances of priority or possible priority substances in the field of water policy.[38] According to section 307 of the U. S. Clean Water Act, DEP, DMP, DEHP, BBP, DBP, and DOP should be considered Priority Toxic Pollutants.[39] The WHO has established a guideline value of 8 μg/l for DEHP for fresh and drinking water,[40] which is similar to the Maximum Contaminant Level (MCL) for DEHP set by the EPA (6 μg/l).[41] In addition, the EPA has established a Maximum Concentration Level Goal of zero for DEHP.[42] The Netherlands National Institute of Public Health and Environment has derived environmental risk levels (ERLs) in water of 10 and 0.19 μg/l for DBP and DEHP, respectively. For sediment and soils with 10% of organic matter, the ERLs are 0.7 and 1 mg/Kg dry weight (dw), respectively.[43] The Danish EPA has issued groundwater quality criteria of 10 μg/l for phthalates (not including DEHP) and 1 μg/l for DEHP, and quality criteria for soils of 250 and 25 mg/Kg dw, respectively.[44] In the U.S.A., three phthalate esters (DMP, DBP, and DEHP) are included in the list of hazardous air pollutants issued in the Clean Air Act.[45] The U.S. Occupational Safety and Health Administration (OSHA), establish Time Weight Average limits (TWA) of 5 mg/m^3 for these phthalates.[46] The same TWA value is established by the U.S. National Institute for Occupational Safety and Health (NIOSH) and the American Conference of Government of Industrial Hygienists (ACGIH) for these and other phthalate esters.

II. STANDARD ANALYTICAL PROCEDURES

A. U.S. EPA PROCEDURES

The EPA has published numerous analytical procedures dealing with the determination of phthalate esters: methods 506[47] and 525[48] for the determination of phthalate esters[47] and organic compounds[48] in drinking water; methods 606,[49] 625[50] and 1625[51] for the determination of phthalate esters,[49] base or neutral acids,[50] and semivolatile compounds[51] in municipal and industrial wastewaters; and methods 8061,[52] 8270,[53] and 8410 for the quantification of phthalate esters[52] and semivolatile organics[53,54] in aqueous and solid matrices including groundwater, leachate, soil, sludge, and sediment. Method 8270 can also be employed for quantification of semivolatile species in air. Table 28.3 recompiles basic information of these analytical procedures.

1. Sampling, Handling, and Extraction Procedures

a. Water

In accord with these methods, when sampling matrices such as drinking water, groundwater, leachates, and municipal and industrial wastewater, usually 1 l of sample is collected in amber glass bottles[47–54] with screw caps with Teflon lined septa. Under ideal conditions, the containers must not be prerinsed with sample before collection and the sample should be collected by completely filling the container. Automatic sampling is also permitted, but devices should be as free as possible of potential contamination sources such as plastic tubing. For dechlorination of the samples, sodium thiosulfate[47,50–54] or sulfite[48] is added. Hydrochloric acid may be added to retard microbiological degradation.[48] Samples can be iced or refrigerated at +4°C until extraction. Any adjustment of the sample pH should take place after the surrogates and matrix spiking compounds are added, so that these are affected by the pH in the same manner as the target analytes. Method 8061 employs NaOH or H$_2$SO$_4$ to adjust the pH of aqueous samples to five to seven prior to extraction, since phthalate esters tend to hydrolyze below pH 5 and above pH 7.[52] However, method 8410 does not recommended adjusting the pH of the sample[54]; and in methods 8270 and 625, extraction is accomplished after adjusting samples to the basic pH.[50,53] Furthermore, according to method 525, samples should be acidified at pH < 2. Method 506 includes the addition of NaCl to the sample prior to the extraction to produce a salting-out effect.[47] In solid phase extraction (SPE) procedures, methanol should usually be added to the sample prior to extraction.[47,48,52,53]

TABLE 28.3A
Procedures for Sample Extraction Included in Official Methods Used in the Analysis of Phthalate Esters

Organization Method #	Analytes	Matrix	Sampling/Preservation/Pretreatment	Extraction
NIOSH 5020	DBP, DEHP	Air	6–200 l with 0.8 μm cellulose ester membrane at 1–3 ml/min	Elution with 2 ml of CS_2 (30 min in ultrasonic bath)
OSHA PV2076	DIHP, DHP	Air	240 l collected on OVS-TENAX sampling tubes at 1 ml/min	Elution with 4 ml of toluene (30 min with occasional shaking)
OSHA 104	DMP, DEP, DBP, DEHP, DOP	Air	240 l collected on OVS-TENAX sampling tubes at 1 ml/min	Elution with 4 ml of toluene (30 min in a mechanical shaker)
EPA 506 1.1	DMP, DEP, DBP, BBP, DEHP, DOP	Drinking Water	1 l in amber glass containers, for dechlorination: 80 mg/l of $Na_2S_2O_3$; storage at $\leq 4°C$ free from light	Separatory funnel LLE: Add 50 g of NaCl; and extract with 3×60 ml with DCM followed by 40 ml with hexane SPE on C_{18} disk, Add 5 ml of MeOH to the sample, and extract on C_{18} disk. After extraction elute with 5 ml of acetonitrile followed by 10 ml of DCM SPE on C_{18} cartridge, after extraction elute with 10 ml of DCM
EPA 525.2	DMP, DEP, DBP, BBP, DEHP	Drinking Water	1 l in amber glass containers, or with automatic sampling equipment; for dechlorination add 40–50 mg/l of Na_2SO_3; adjust to pH < 2 with HCl; storage at $\leq 4°C$ free from light; add 5 ml/l of MeOH to the sample	SPE on C_{18} cartridge, after extraction elute with 5 ml EtAc followed by 5 ml of DCM SPE on C18 disk, after extraction elute with 5 ml of EtAc; then 5 ml DCM and finally with 3 ml of EtAc:DCM (1:1)
APHA 6410B	DMP, DEP, DBP, BBP, DEHP, DOP	Municipal and industrial discharges	1–2 l in amber glass container or > 250 ml with automatic sampling equipment; for dechlorination add 80 mg/l of $Na_2S_2O_3$; storage at $\leq 4°C$ free from light; adjust to pH > 11 with NaOH	Separatory funnel LLE with 3×60 ml DCM Continuous LLE: with 200–500 ml of DCM (24 h)

Method	Analytes	Matrix	Sample container and preservation	Extraction
EPA 606	DMP, DEP, DBP, BBP, DEHP, DOP	Municipal and industrial wastewater	1 l in amber glass containers or >250 ml with automatic sampling equipment; storage at ≤4°C	Separatory funnel LLE with 3 × 60 ml with DCM
EPA 625	DMP, DEP, DBP, BBP, DEHP, DOP	Municipal and industrial wastewater	1–2 l in amber glass containers or >250 ml with a automatic sampling equipment; for dechlorination add 80 mg/l of $Na_2S_2O_3$; storage at ≤4°C; adjust pH > 11 with NaOH	Separatory funnel LLE with 3 × 60 ml with DCM (for 1 l of sample) Continuous LLE with 200–500 ml of DCM (24 h)
EPA 1625.2	DMP, DEP, DBP, BBP, DEHP, DOP	Municipal and industrial discharges	1 l in amber glass containers; or automatic sampling equipment for dechlorination add 80 mg/l of $Na_2S_2O_3$; storage at 0–4°C. adjust pH > 12–13 with NaOH	Continuous LLE with 200–300 ml of DCM (18–24 h)
DOE OM100R	DMP, DEP, DBP, BBP, DEHP, DOP	Ground water, solid waste, soils	No specified	No specified
EPA 8061A	DMP, DEP, DBP, BBP, DEHP, DOP, DBEP, DEEP, DMEP, DAMP, DCHP, DHP, DIHP, DIBP, DNP, HEHP	Groundwater, wastewater, leachates Sediments, soils, sludges, solid waste	1 gal, 2 × 0.5 gal, or 4 × 1 l amber glass container with Teflon-lined lid; for dechlorination add 3-ml/gal 10% sodium thiosulfate; storage at ≤4°C; adjust pH to 5–7 with H_2SO_4 or with NaOH Collect samples in 250-ml widemouth glass container with Teflon-lined lid; storage at ≤4°C; samples should be air-dried and ground to a fine powder prior to extraction. Alternatively, samples may be mixed with Na_2SO_4.	Separatory funnel LLE (EPA 3510C) with 3 × 60 ml of DCM SPE on C_{18} disk (EPA 3535A). Add MeOH to the samples and, after extraction, elute with 5 ml acetone followed by 15 ml acetonitrile Soxhlet Extraction (EPA 3540C): 20 g of dried and ground material is extracted with 300 ml of acetone/hexane (1:1) or DCM/acetone (1:1). (16–24 h) ASE (3545A): 10–30 g of dried and ground material is extracted, with acetone/hexane (1:1) or DCM/acetone (1:1) at 100°C and 1500–2000 psi for 5 min Ultrasonic Extraction (EPA 3550.B): 30 g of sample, extract for 3 min with 3 × 100 ml of acetone/hexane (1:1) or DCM/acetone (1:1)

Continued

TABLE 28.3A
Continued

Organization Method #	Analytes	Matrix	Sampling/Preservation/Pretreatment	Extraction
EPA 8270D	DMP, DEP, DBP, BBP, DEHP, DOP	Air	Trap semivolatile analytes on 20 g XAD-2 resin mounted on a multicomponent sampling train (EPA Method 0010)	Soxhlet extraction (EPA 3542)
		Groundwater, wastewater, leachates	1 gal., 2 × 0.5-gal., or 4 × 11 amber glass container with Teflon-lined lid; for dechlorination add 3-ml/gal 10% sodium thiosulfate; storage at ≤4°C; after extracting acids at pH < 2, adjust to pH > 11 with NaOH	Separatory funnel LLE (EPA 3510C) with 3 × 60 ml of DCM Continuous LLE: (EPA Method 3520C) with 300–500 ml of DCM (18–24 h) SPE on C_{18} disk (EPA 3535A). Add MeOH to the samples and, after extraction, elute with 5 ml acetone followed by 15 ml acetonitrile
			1 gal., 2 × 0.5-gal., or 4 × 11 amber glass container with Teflon-lined lid; for dechlorination add 3-ml/gal 10% sodium thiosulfate; storage at ≤4°C; adjust pH to 5–7 with H_2SO_4 sulfuric acid or with NaOH	
		Sediments, soils, sludges, solid waste	Collect samples in 250-ml widemouth glass container with Teflon-lined lid; storage at ≤4°C; samples should be air-dried and ground to a fine powder prior to extraction. Alternatively, samples may be mixed with Na_2SO_4	Soxhlet Extraction (EPA 3540C): 20 g of dried and ground material is extracted with 300 ml of acetone/hexane (1:1) or DCM/acetone (1:1). (16–24 h) Automated Soxhlet Extraction (EPA 3541): 10 g of dried and ground material is extracted with 50 ml of acetone/hexane (1:1) or DCM/acetone (1:1). (120 min) ASE (EPA 3545A): 10–30 g of dried and ground material is extracted, with acetone/hexane (1:1) or DCM/acetone (1:1) at 100°C and 1500–2000 psi for 5 min Ultrasonic Extraction (EPA 3550.B): 30 g of sample, extract for 3 min with 3 × 100 ml of acetone/hexane (1:1) or DCM/acetone (1:1)

	Solid waste	Collect samples in 250-ml widemouth glass container with Teflon-lined lid; storage at ≤4°C; samples should be air-dried and ground to a fine powder prior to extraction. Alternatively, samples may be mixed with Na₂SO₄	Waste dilution (EPA 3580A): 1 g of sample is diluted with 10 ml of DCM and shaked in the presence of Na₂SO₄ (2 min)
EPA 8410	DMP, DEP, DPP, DBP, BBP, DEHP, DOP	Groundwater, wastewater, leachates	Collect samples in 250-ml widemouth glass container with Teflon-lined lid; storage at ≤4°C; samples should be air-dried and ground to a fire powder prior to extraction. Alternatively, samples may be mixed with Na₂SO₄	Separatory funnel LLE (EPA 3510C) with 3 × 60 ml of DCM Continuous LLE: (EPA Method 3520c) with 300–500 ml of DCM (18–24 h)
		Sediments, soils, sludges, solid waste	1-gal. 2 × 0.5-gal, or 4 × 1-l amber glass container with Teflon-lined lid; for dechlorination add 3-ml 10% sodium thiosulfate per gallon; storage at ≤4°C. No adjust pH	Soxhlet Extraction (EPA 3540C): 20 g of dried and ground material is extracted with 300 ml of acetone/hexane (1:1) or DCM/acetone (1:1). (16–24 h) Automated Soxhlet Extraction (EPA 3541): 10 g of dried and ground material, extract with 50 ml of acetone/hexane (1:1) or DCM/acetone (1:1). (120 min) Ultrasonic Extraction (EPA 3550.B): 30 g of sample, extract for 3 min with 3 × 100 ml of acetone/hexane (1:1) or DCM/acetone (1:1)
			Collect samples in 250-ml widemouth glass container with Teflon-lined lid; storage at ≤4°C; samples should be air-dried and ground to a fine powder prior to extraction. Alternatively, samples may be mixed with Na₂SO₄	
		Solid waste	Collect samples in 250-ml widemouth glass container with Teflon-lined lid; storage at ≤4°C; samples should be air-dried and ground to a fine powder prior to extraction. Alternatively, samples may be mixed with Na₂SO₄	Waste dilution (EPA 3580A): 1 g of sample is diluted with 10 ml of DCM and shaken in the presence of Na₂SO₄ (2 min)

TABLE 28.3B
Procedures for Cleanup and Analysis Included in Official Methods Used in the Analysis of Phthalate Esters

Organization Method #	Clean-Up	Separation	Stationary Phase	Detector
NIOSH 5020	NO	PGC	2 m × 3 mm OD stainless steel column, packed with 5% OV-101 on Chromosorb W-HP (100/120 mesh)	FID
OSHA PV2076	NO	GC	DB-5, 60-m × 0.32-mm i.d., 1.5 μm film thickness	FID
OSHA 104	NO	GC	HP-1, 5 m × 0.53 mm i.d., 2.65 μm film thickness	FID
APHA 6410B	NO	PGC	1.8 m × 2 mm i.d. glass column, packed with 3% SP-2250 on Supelcoport (100/200 mesh)	MSD
EPA 506 1.1	Florisil and/or alumina	GC	DB-1 or DB-5, 30 mm × 0.32 mm i.d., 0.25 μm film thickness	PID
EPA 525.2	NO	GC	DB-5MS, 30 mm × 0.25 mm i.d., 0.25 μm film thickness	MSD
EPA 606	Florisil and/or Alumina	PGC	1.8 m × 4 mm i.d. glass column, packed with 3% OV-1 or 1.5% SP-2250/1.95% SP-2401 on Supelcoport (100/200 mesh)	ECD
EPA 625	NO	PGC	1.8 m × 2 mm i.d. glass column, packed with 3% SP-2250 on Supelcoport (100/200 mesh)	MSD
EPA 1625.2	NO	GC	DB-5, 30 ± 5 m × 0.25 ± 0.02 mm i.d.	MSD
EPA 8061A	Alumina(EPA 3610); florisil (EPA 3620); gel permeation clean-up (EPA 3640); and/or sulfur removal (EPA 3660)	GC	DB-5, 30 m × 0.53 mm i.d., 1.5 μm film thickness or DB-1701 30 m × 0.53 mm i.d. × 1 μm film thickness	ECD
DOE OM100R	Not specified		DB-5MS 30 m long × 0.25 mm i.d., 0.25-μm film thickness	MSD
EPA 8270D	Alumina(EPA 3610); florisil (EPA method 3620); and/or gel permeation clean-up (EPA 3640)	GC	DB-5, 30 m long × 0.25 mm i.d. (or 0.32 mm i.d.), 1.0 μm film thickness	MSD
EPA 8410	Gel permeation clean-up (EPA 3640)	GC	DB-5, 30 m long × 0.32 mm i.d., 1.0 μm film thickness	FTIR

Phthalate esters can be extracted by separatory funnel liquid–liquid extraction (LLE) with methylene chloride,[47,49,50,53,54] following EPA method 3501.[55] To prevent the formation of emulsions continuous LLE extraction can be used,[50,51,53,54] as described in standard method 3520.[56] However, according to method 8061, continuous LLE should be avoided since phthalate esters with longer chains tend to absorb into the glassware and consequently extraction recoveries are less than 40%.[52] In addition to LLE techniques, EPA procedures include SPE with C_{18} disk[47,48,52,53] and cartridge.[47,48] To elute trapped analytes, method 506[47] employs acetonitrile followed by methylene chloride (disk procedure) or methylene chloride alone (cartridge procedure), meanwhile method 525[48] employs ethyl acetate followed by methylene chloride. According to method 3535B[57] (included in methods 8061, and 8270), the use of two water-miscible solvents, such as acetone and acetonitrile, improves the recovery of analytes trapped in water-filled pores of the sorbent.

b. Air

Method 8270 may be used for the determination of semivolatile compounds such as phthalate esters in air. These chemicals, that should be sampled following method 0010,[58] which uses a multicomponent sampling train where semivolatile analytes are trapped on XAD-2 resin. After sampling, the resin is extracted in a Soxhlet extractor according to EPA methods 3542[59] and 3540.[60]

c. Solid Samples

Solid samples, such as sediments, soils, waste samples and dry waste samples amenable to grinding, should be collected in 250-ml widemouth glass containers with Teflon lined lids. Samples should be air dried and ground to a fine powder prior to extraction. Alternatively, samples can be mixed with anhydrous sodium sulfate. Gummy, fibrous, or oily materials not amenable to grinding should be broken up to allow mixing and maximum exposure of the sample surfaces for extraction. The addition of anhydrous sodium sulfate to the sample 1:1 can make the mixture amenable for grinding. Usually acetone or hexane (1:1) or acetone or methylene chloride (1:1) are the solvents of choice for the extraction of solid matrices.[52–54] Methylene chloride or acetone solvent mixture has generally been found to be more effective extracting the analytes of interest from solid matrices, and hexane or acetone solvent system can be appropriate where specific interferences are expected.[52] Conventional extraction procedures such as Soxhlet extraction (method 3540[60]) or automatic Soxhlet extraction (method 3541[61]) may be used. Method 3541 is recommended when shorter extraction time is required, since this procedure permits reducing the time from 18 to 24 h to a couple of hours and the solvent volumes from 300 to 50 ml. Additional methodologies such as accelerated solvent extraction (ASE) at 100°C and 1500 to 2000 psi (method 3545A[62]); and ultrasonic extraction (method 3550B[63]) are also included as extraction procedures. Methods 8270 and 8410 also incorporate a waste dilution procedure (method 3580A[64]) for solid waste matrices that may contain organic chemicals at a concentration greater than 20 g/Kg and that are soluble in the dilution solvent. In this case, 1 g of sample is diluted with 10 ml of methylene chloride in the presence of anhydrous sodium sulphate and shaken for two minutes. After filtering, the extract is ready for cleanup or analysis.

2. Concentration, Change of Solvent and Cleanup of the Extracts

Usually, extracts obtained by the described extraction procedures need to be dried and concentrated. Anhydrous sodium sulphate is the drying agent of choice in all procedures. For sample concentration, a Kuderna–Danish (K–D) concentrator is generally used, while it is also possible to achieve further concentration under a stream of N_2. In addition, the exchange of solvents, for cleanup procedures or analysis, if required, is also performed in the K–D concentrator. Cleanup, can be performed with alumina,[47,49,52,53] florisil,[47,49,52,53] by gel permeation chromatography (GPC),[52-54] and by sulfur removal procedures,[52] as described in methods 3610,[65] 3620,[66] 3640 [67] and 3660,[68] respectively. If organochlorine pesticides are known to be present in the extract, florisil cartridges are recommended instead of alumina cartridges,[66] As indicated in method 8061, alumina and florisil should be employed cautiously, since these materials can be contaminated with phthalate esters. The heating at 320°C for florisil and 210°C for alumina before its use is recommended. Phthalate esters were detected in florisil cartridge methods blanks at concentrations ranging from 10 to 460 ng. According to this EPA method, complete removal of phthalate esters from florisil cartridges does not seem possible, it being desirable to keep the steps to a minimum.[52] Sulfur removal can be an appropriate procedure of eliminating interferences when a electron capture detector (ECD) is to be employed (e.g., method 8061). In addition, GPC may be accomplished when analyzing samples that contain high amounts of lipids and waxes.

3. Separation and Detection

Gas chromatography (GC), using capillary[47,48,51-54] or packed columns,[49,50] is the separation technique of choice in all discussed EPA methods. Recommended stationary phases are: 100% dimethylpolysiloxane,[47] (5%-phenyl)-methylpolysiloxane,[47,48,51-54] 3% OV-1,[49] 1.5% SP-2250/ 1.95% SP-2401[49] and 3% SP-2250.[50] Detectors used by EPA standards procedures, include photoionization (PID),[47] electron capture (ECD),[49,52] Fourier transform infrared spectrometry (FTIR),[54] and mass spectrometry detectors (MSD).[48,50,51,53] Method 8061 employs an ECD, so identification of the phthalate esters should be supported by al least one additional qualitative technique. This method also describes the use of an additional column (14% cyanopropyl phenyl polysiloxane) and dual ECD analysis, which fulfills the above mentioned requirement. Among MSDs, most of the procedures employ electron impact (EI) ionization,[48,50,51,53] but chemical ionization (CI)[50] is also employed. In all MSD methods, except 1625, quantitative analysis is performed using internal standard techniques with a single characteristic *m/z*. Method 1625 is an isotope dilution procedure. The use of a FTIR detector (method 8410) allows the identification of specific isomers that are not differentiated using GC-MSD.

B. OTHER STANDARD PROCEDURES

In addition to EPA procedures, other agencies and organisms such as the U.S. Department of Energy (DOE), the American Public Health Association (APHA), NIOSH and OSHA have issued standard procedures for dealing with phthalate ester determination.

In accordance to U.S. 40 Code Federal Regulation part 136,[69] standard method 6410B[70] of the APHA, the American Water Works Association (AWWA), and the Water Environment Federation (WEF), as well as EPA methods 606, 625, and 1625, can be employed for the analysis of phthalates in municipal and industrial discharges. As equivalent to EPA 625, method 6410 procedure employs LLE with methylene chloride for the isolation of phthalates and other base or neutral analytes at pH > 11. Alternatively, water samples may be subjected to continuous LLE for 24 h. Phthalates, together with other extracted analytes are separated on a chromatographic column packed with 3% SP-2250 or equivalent, and quantified on a MS analyzer.

Similar to EPA method 8270, the U.S. DOE method OM100R[71] describes the determination of semivolatile organic compounds, including phthalate esters, in extracts from all types of solid waste matrices, soils, and ground water. Method OM100R incorporates the use of *an ion trap mass detector* in place of the quadrupole, typically used in earlier versions of the EPA method.

To monitorize the exposure of workers OSHA and NIOSH have developed some analytical procedures for the determination of phthalate esters in workers' environment. In NIOSH method 5020.2,[72] DBP and DEHP are sampled from air with a 0.8-μm cellulose ester membrane and trapped analytes are extracted from the filter with carbon sulfide in an ultrasonic bath for 30 min. Separation of the analytes is accomplished on a column packed with 5% OV-101 or, alternatively, a DB-1 capillary column. Quantification with an internal standard is carried out with a flame ionization detector (FID). In OSHA methods 104[73] and PV2076,[74] phthalate esters are collected from air in OVS-Tenax tubes containing a glass fiber filter in front of two resin beds of Tenax. Target analytes are extracted in toluene for 30 min, and the extracted analytes, separated in a HP-1[73] or DB-5[74] column, are quantified with a FID.

III. SAMPLE HANDLING AND SAMPLE PREPARATION

A. CONTAMINATION OF THE SAMPLES

Phthalates are widely produced and used and due to their ubiquity, can be found everywhere, including common laboratory equipment and reagents. In consequence, the main problem in phthalate analysis is external contamination coming from the sampling and sample preparation procedure and even the chromatographic analysis.

The analysis of blanks is of great importance, as are all the precautions in the treatment of the material and reagents used in any step of the analytical process. Several recommendations should be followed to minimize contamination[75-77]:

- The use of plastic materials should be avoided.
- The sample preparation procedure should be as simple as possible with minimal extraction steps, minimal glassware use, and minimal extract concentration.
- Glassware should be properly cleaned by solvent rinsing and thermal treatment at 400°C. Prior to use, the glassware should be rinsed with blank tested organic solvent (cyclohexane or isooctane) to deactivate the surface.
- Organic solvents and laboratory grade water usually contain traces of phthalates, even the ones commonly available for trace analysis, and these must be checked to establish background levels. Also, reagents need to be checked.
- Additional contamination of material, water solvents, and reagents can occur due to the lab air. The material should be stored in a closed container or wrapped in aluminium foil to avoid adsorption of phthalates from the air. The bottles of solvents and reagents should be kept closed until use.
- Precautions should be taken with the cleaning products used in the lab and with personal hygiene products, because these often contain phthalates.
- Phthalates can be present in the chromatographic system. The most important contamination is located in the inlet and gas supply system. Split or splitless inlets may contain septa, liners and o-rings that are contaminated with phthalates. Another critical factor is the quality of caps for autosampler vials. These caps can also contain phthalates. As general precaution, only one injection should be made from each vial.

B. SAMPLE PREPARATION FOR WATER ANALYSIS

1. Water Blanks

One of the most important problems in the analysis of phthalates from water samples is the detection of these compounds in the samples used as blanks.[78-86] Phthalates have been detected in purified water commonly used in laboratories, including water distilled in a glass distillation apparatus, Milli-Q water, and commercially available water specially for VOC (*Volatile organic compounds*) determination.[86] Therefore, special caution should be taken with the experimental use of water in laboratories. Some authors have reported the levels of phthalate esters found in the purified water employed in its studies (see Table 28.4). The concentrations found are frequently

TABLE 28.4
Phthalate Ester Concentrations Found in Purified Waters

Water Type	Method	Concentration, $\mu g/l$					Ref.
		DMP	DEP	DBP	BBP	DEHP	
Deionized	SPME-GC-ECD	nd	0.04	0.15	0.005	0.49	84
Distilled	HPLC-GC-MS	—	—	0.005	—	0.002	82
Distilled	LLE-GC-FID	0.10	0.06	1.24	—	4.21	86
Milli-Q	Online SPE-GC-MS	—	—	0.5	0.02	0.5	83
Purified	LLE-GC-MS	<0.01	0.29	1.58	0.02	1.06	85
Redistilled	LLE-GC-FID	nd	0.14	3.28	—	0.93	86
Commercial for VOC	LLE-GC-FID	nd	nd	10.58	—	nd	86
Treated for TOC	LLE-GC-FID	nd	nd	4.16	—	nd	86
Retreated for TOC	LLE-GC-FID	nd	nd	1.8	—	nd	86

high considering the levels of concentration at which these compounds must be controlled in water samples. These contamination levels forced the limits of detection achieved, mainly for DBP and DEHP, the most ubiquitous phthalate esters.

2. Sample Pretreatment

Storage of the samples at neutral or acid pH can be done at 4°C. Since phthalates undergo biodegradation, the storage of aqueous samples at this temperature should be not for longer than four days. Chemical preservation may be performed by addition of 500 mg sodium azide per liter of sample.[75,87] Storage at pH 9, even at 4°C, should be avoided because most target compounds show more than 50% decreases in concentration after seven days of storage.[90]

Before extraction, samples are often filtered. Filtration is usually required for waters containing high levels of suspended solids, especially when extraction is carried out by SPE.

3. Extraction

The procedures most often used for isolation and preconcentration of phthalate esters from water are LLE and SPE. Both techniques are included in official methods to perform the extraction of phthalates (see Table 28.3A). In the last years, solid phase microextraction (SPME) have acquired an increased importance in the analysis of semivolatile compounds, such as phthalates, in water. After extraction, the final analysis is usually carried out using a chromatographic technique.

a. LLE

In most of the applications of LLE to phthalate analysis, extraction was performed in discontinue mode using separatory funnels. Table 28.5 summarizes some applications of this technique to phthalate analysis.[85–92] In most cases, the volume of sample extracted is large (from 0.1 to 5 l). Frequently, samples are acidified and NaCl is added to favor the transfer of the analytes into the organic solvent. The solvents most frequently used are dichloromethane and hexane. In most cases, the extracts are dried with anhydrous sodium sulphate and concentrated to achieve high sensitivity. Nevertheless, the concentration factor is limited due to the presence of trace levels of phthalates in commercially available solvents, even in solvents for trace analysis. In consequence, accurate determinations below 0.1 $\mu g/l$ are questionable with this extraction technique.[75] The extracts obtained are usually analyzed without a cleanup step.

TABLE 28.5
Liquid–Liquid Extraction Methods for the Analysis of Phthalate Esters in Water Samples

Year	Analyte	Sample	Sample Volume and Pretreatment	Extraction	Concentration and Treatment	Determination	Recovery %	LOD	RSD%	Ref.
2003	DMP, DEP, DBP, BBP, DEHP, DDP, PME, PA	Leachates from landfills	pH < 1	Addition of NaCl and extraction with ether	Concentration (N_2) + silylation	GC-MS	—	LOQ ∼ 1 $\mu g/l$ (DEHP = 20 ug/l)	<20	85
2002	DMP, DEP, DIBP, DBP, DEHP, PME	River, tap, well, mineral, distilled water	5000 ml	Addition of 50 g NaCl and extraction with hexane-ethyl acetate 8:2	Drying with anhydrous sodium sulphate (20 g), vacuum concentration and addition of hexane	GC-FID	DBP = 97, DEHP = 95	0.03 $\mu g/l$	—	86
2002	DMP, DEP, DIBP, DBP, DEHP, PME	River, tap, well, mineral, distilled water	950 ml	Addition of 150 g NaCl and extraction with 25 ml hexane-ethyl acetate 8:2 by stirring (4 h)	Drying with anhydrous sodium sulphate (20 g), vacuum concentration and addition of hexane	GC-FID	DBP = 90, DEHP = 47	0.05 $\mu g/l$	—	86
1995	DEHP	Surface water	100 ml	6 × 10 ml hexane	Drying with anhydrous magnesium sulphate, evaporation (vacuum rotary evaporator), addition of acetonitrile, filtration (0.45 μm). Final volume = 25 ml	LC-UV	—	—	—	88
1994	DMP, DEP, DBP, DMEP, DEEP, DCP, DEHP, DOP, BBP, DOP	—	250 ml	12 + 6 + 6 ml DCM (EPA SW-846 method 3510)	Addition of IS and concentration to 0.1–1 ml (N_2)	GC-MS-MS	57–124	—	2.5–7.1	89

Continued

TABLE 28.5
Continued

Year	Analyte	Sample	Sample Volume and Pretreatment	Extraction	Concentration and Treatment	Determination	Recovery %	LOD	RSD%	Ref.
1991	DMP, DEP, DIBP, DBP, DMPP, DMEP, DAP, DEEP, HEHP, DHP, BBP, DBEP, DEHP, DCP, DOP, DNP	Estuarine leachate, groundwater (spiked)	1 l + surrogates	3 × 60 ml DCM (EPA method 3510) or DCM in continuous (EPA method 3520)	Drying with anhydrous sodium sulphate (20 g), concentration to <10 ml, solvent exchange to 50 ml hexane, concentration to 2 ml (N₂)	GC-ECD	60–117 (EPA method 3510)	26–320 ng/l	<28 (EPA method 3510)	90
1991	DMP, DEP, DIBP, DBP, DEHP	Estuary water	2.7 l + surrogates	DCM	Drying and concentration to 300 μl	GC-MS	—	—	—	91
1990	DEP, DBP, DEHP, DIOP	River water, sewage effluents	1–2 l pH 2, stored at 5°C	Addition of 150–300 g NaCl and extraction with 3 × 60 ml chloroform	Washing with 3 × 10 ml 0.1 M Na₂CO₃), solvent evaporation, addition of 1 ml DCM, clean-up with silica gel, elution with benzene-ethyl acetate, and concentration to 1 ml	GC-MS	—	—	—	92

PME, Phthalic acid monoesters; PA, Phthalic acid.

b. SPE

Nowadays, SPE is the most frequently used extraction technique for the extraction and concentration of trace organic analytes from water samples. This technique has been applied to the determination of phthalates in water samples[80–83,87,90,93–99] as well as in biological fluids.[100,101]

The advantage of SPE over LLE is that large concentration factors can be obtained with little or no solvent concentration, and thus there is no concentration of phthalate traces from the organic solvent. The major limitation of SPE is in the extraction of water samples containing solids or heavily contaminated samples. In the first case, samples have to be filtered and the phthalates measured separately in the aqueous and the solid phases. In the second case, low recoveries could be achieved as a result of the incomplete enrichment of the sorbent material.[75]

Table 28.6 describes applications of this technique to water analysis. The combination of SPE with GC-MS is one of the most common alternatives for the analysis of phthalates in water samples. Some official methods are based on this combination (see Table 28.3). SPE has also been combined with LC using UV and MS detectors.

The sorbent materials most commonly used for the extraction of phthalate esters from water are C-18 and polymeric phases based on PS-DVB. Recently, a new polymeric phase, Oasis, has been applied for the concentration of phthalates, but the recoveries achieved with these polymers were lower than those obtained with C18.[80] Nevertheless, this extracting phase was suitable for the extraction of phthalate metabolites from biological fluids.[102–104]

The SPE technique can be used offline as well as online.

i. Offline SPE

In offline applications, large volumes of water (250 to 1000 ml) are passed through disposable cartridges usually packed with 200 to 500 mg of sorbent material. To improve extraction efficiency, a certain amount of water-miscible organic solvents such as methanol or acetonitrile, can be added to the sample before enrichment. Most cartridges are made from polyethylene or polypropylene barrels and this may cause relatively high background levels of phthalates. Alternatively, glass barrels that can be packed with different materials are available. Before elution, cartridges are usually dried under nitrogen.

Elution of the sorbent is carried out using organic solvents such as ethylacetate, methanol, acetonitrile, or mixtures of solvents. Frequently, the eluent is concentrated to achieve higher sensitivity and the final extract is analysed without any additional treatment. Nevertheless, when analysing heavily contaminated samples, some authors have included a extract cleanup step using activated alumina.[87,90] The recoveries obtained were satisfactory in several applications, as can be seen in Table 28.6.

An alternative to the SPE cartridges is the use of membrane disks.[90,98] Advantages of using membranes are that sampling flow is higher and large samples can be processed faster. One of the first applications of membrane disks to phthalate analysis was performed by Lopez-Avila et al.[90] Concentration of phthalates in aqueous samples on C18-membrane disks followed by extraction with acetonitrile yielded good recoveries and repeatability, and was, therefore, incorporated as an option in the revised EPA method 8061 (see Table 28.3A).

ii. Online SPE

In recent years, fully automated analysis of contaminants in water by the online coupling of SPE to LC or GC instrumentation has received increasing attention. Besides allowing rapid analysis, in online SPE there is reduced handing of samples and the consumption of organic solvents is also reduced. In addition, the total amount of analytes extracted is introduced into the chromatograph and, consequently, sample volume can be drastically reduced.

SPE has been coupled online with GC-MS for the determination of phthalates in water analysis. With online SPE, the water sample is pumped through a short precolumn filled with small particles (15 to 25 μm) of either C-18[82] or PS-DVB[81,83] adsorbing media. Different interfaces have been

TABLE 28.6
SPE Methods for the Analysis of Phthalate Esters in Water Samples

Year	Analyte	Sample	Sample Volume and Pretreatment	Sorbent Conditioning	Sorbent Material	Elution	Extract Treatment	Determination	Recovery %	Linear Range	LOD	RSD %	Ref.
2003	DMP, DEP, DBP, BBP, DEHP	Bottled and distribution water	250 ml + IS	15 ml methanol + 15 ml deionised water	Oasis 200 mg or C18 500 mg	5 ml DCM-hexane 4:1 + 5 ml methanol-DCM 9:1	Evaporation to dryness (N_2), redissolution with 0.3 ml ethyl acetate + IS	GC-MS	Oasis = 21–88, C18 = 42–92	0.002–4 µg/ml	0.002–0.005 µg/ml	Oasis = 4.1–8.5 C18 = 0.2–7.4	80
2003	Total phthalates	Drinking, well and river water	—	2 ml methanol + 2 ml distilled water	C18	1.5 ml methanol	Formation of a complex with terbium (III)	Luminescence	91.5–102	—	—	—	93
2003	DBP, BBP, DEHP	River, sea, and tap water, irrigation stream and sewage water	15 ml + 50% methanol	3 ml methanol + 3 ml water	(Online SPE) PS-DVB (PLRP-S 100 Å)	300 µl ethyl acetate	—	GC-MS	72–93	0.003–10 µg/l	1–36 ng/l	Repeat. = 1–8 (DEHP = 20) Reprod. = 3–25	81
2003	DMP, DEP, BBP, DBP, DEHP, DOP	Coastal water and waste water	100 ml + 20% acetonitrile, filtered through 0.45 µm	10 ml acetonitrile + 10 ml water	PS-DVB (Isolute ENV +) 200 mg	10 ml acetonitrile	Evaporation to dryness and redissolution in 0.5 ml acetonitrile	LC-MS	32–95	0.1,4–100 µg/l	0.01–1 ug/l	Repeat. = 12–21 Reprod. = 15–25	94
2003	DBP, DCHP, DOP, DDC, DNP	Tap and river water	500 ml	5 ml acetonitrile + 10 ml water	800 mg PTFE turnings	10 ml acetonitrile	—	LC-UV	92.1–127.5	10–200 µg/l	3.1–5.8 µg/l	—	95
2002	DMP, DEP, DBP, BBP, DEHP, PMEs, PA	Landfill lea chates	pH 0.9 with HCl to prevent ionization of PMEs and PA	5 ml ethyl acetate + 5 ml methanol + 5 ml acidified water	2 PS-DVB cartridges (Isolute 101and Isolute ENV +)	1 + 1 ml of different solvents	Concentration to 200 µl (N_2) + IS, sylylation	GC-MS	61–89	—	—	<7 (real samples)	96

Year	Compounds	Sample	Volume	SPE	Elution solvent	Treatment	Detection	Recovery				Ref.
2002	DBP, BBP, DEHP	Tap, river and coastal water	15 ml + 50% methanol	3 ml methanol + 3 ml water	(Online SPE) PS-DVB (PLRP-S 100 Å)	100 µl ethyl acetate	—	GC-MS	50–65	0.05–5 µg/l	0.1–7 ng/l	Repeat. = 15–22 Reprod. = 14–16	83
2002	DMP, DEP, DBP, DEHP	River and marine water	11	Deionized water + 2 ml methanol	Envi. C18 1 g	2 ml methanol-DCM 50:50	Evaporation to dryness, redissolution in 1 ml methanol + IS	GC-FID	83–96	—	27.2–60.6 ng/l	—	97
2001	—	Potable and mineral water	11 + surrogates	5 ml methanol	C18 disks	2 × 5 ml DCM	Filtered through anhydrous sodium sulphate, evaporation to dryness and redissolution in 1 ml acetone	GC-ECD	89.3–98.7	—	0.5–5.0 ng/l	—	98
2000	BBP, DEHP	Aquarium water	11 + IS, filtrated (1 µm glass fiber or 100 µm cellulose)	6 ml DCM + 6 ml methanol + 6–7 ml water	PS-DVB 500 mg (Isolute ENV +)	2 × 2.5 ml methanol + DCM 1:1	Concentration to 500 µl	LC-UV	99–104	1–200 µg/l BBP, 0.1–3 µg/l DEHP	0.05 µg/l BBP, 0.1 µg/l DEHP	BBP P = 0 .4–4.8, DEHP = 1.0–8.7	99
1997	DEP, DCHP, DBP, DEHP, DINP	Ground water and surface water (river)	10 ml + IS		RP C18 LC column	Water–methanol 15:85 (100 µl/min)	—	GC-MS	—	—	5 ng/l DEHP, 10 ng/l DBP	DEHP = 7.5, DBP = 12	82
1994	DMP, DEP, DBP, BBP, DPP, DEHP, DMPP, DCHP, DOP	River water, landfill leachates and waste water	250 g	5 ml ethyl acetate, dryness (N₂) + methanol	C-18 250 mg	2 ml ethyl acetate	+IS. For leachates and waste water clean-up with alumina	GC-MS	91–108	20 ng/l–20 µg/l	0.01–0.02 µg/l	3–8	87
1991	DMP, DEP, DIBP, DBP, DMPP, DMEP, DAP, DEEP, HEHP, DHP, BBP, DBEP, DEHP, DCP, DOP, DNP	Estuarine leachate and ground water	500 ml + 2.5 ml methanol + surrogates	10 ml methanol + 10 ml LC grade water	C8 and C18 disks	10 ml acetonitrile	Concentration to 1 ml (N₂) and clean-up with florisil or alumina	GC-ECD	59.5–82	—	26–320 ng/l	<28	90

PME, Phthalic acid monoesters; PA, Phthalic acid.

developed and used for this purpose. Hyötyläinen et al.[82] have used a vaporizer chamber or precolumn solvent split or gas discharge interface to perform the online analysis. The method was applied to the analysis of phthalates in drinking and surface water. The volume of sample was only 10 ml and detection limits were 5 to 10 ng/l.

Brossa et al.[81] developed an automated SPE-GC-MS method for the determination of endocrine disrupting compounds including six phthalate esters. The interface device was a programmed temperature vaporizer (PTV), whose liner was packed with Tenax. The samples were spiked with 50% of methanol and 15 ml of this mixture were preconcentrated. Before elution, the precolumn was dried with nitrogen. The analytes were desorbed in the backflush mode with three ethyl acetate fractions of 100 μl and online transferred to the GC system. The performance of the method was tested with several environmental water samples. The recoveries achieved were satisfactory and the detection limits were between 1 to 36 ng/l.

SPE has also been coupled online to LC for the analysis of phthalates and its metabolites in biological fluids.[100,101,105,106]

c. SPME and Related Techniques

SPME is a solvent-free extraction technique that allows the performance of sampling, extraction and concentration in one step.[107] Applications of this technique to phthalate analysis[78,79,84,108–112] are summarized in Table 28.7. This technique is an interesting alternative for the determination of phthalates in aqueous samples, because the risk of contamination during sample handling, which is the major problem in phthalate analysis, is significantly reduced. In addition, the elimination of organic solvents in the sample preparation process could reduce the phthalate background levels. Nevertheless, the main problems for applying SPME to phthalate analysis are the levels of phthalates found in blanks of purified water and even commercial water (especially for DBP and DEHP). Most authors have taken blanks into account in calibration curves and estimating LODs.

One of the first applications of SPME to phthalate analysis was the development of a method for the extraction of DEP from water.[112] The final analysis was done by LC-UV. Different parameters were optimized including four types of fibers. Carbowax-template resin (CW-TRP) and polydimethylsiloxane–divinylbenzene (PDMS-DVB) were found suitable to perform phthalate extraction. The other two fibers, polyacrylate (PA) and polydimethylsiloxane (PDMS), were discarded due to low response and broad peaks, respectively. Samples were extracted at room temperature by direct exposition of the fiber to the sample, previously enriched with 25% of NaCl. The linearity achieved was good from 5 to 50 μg/l. Higher concentrations show a lost of linearity that could suggest the saturation of the fiber coating. Detection limit was 1 ng/ml.

SPME in combination with LC was also used for the extraction of DPP, DIBP and DCHP using CW or TRP and dynamic desorption mode.[108] Good recoveries were obtained for water spiked at 100 μg/l (88.5 to 106.8%). Detection limits were between 4 to 9.5 μg/l.

When SPME is coupled to GC-MS, detection limits in the ng/l level have been reported. Peñalver et al.[110,111] studied the SPME of six phthalate esters in water. In a first study, the authors developed a method based on PA fibers, while in a second study, different extraction coatings were tested: PDMS, PDMS-DVB, PA and carbowax-divinylbenzene (CW-DVB). The final extraction conditions were as follows: PDMS-DVB fibers, 80°C extraction temperature, 30 min extraction time, and desorption at 250°C (3 to 5 min.). Some of the analytes were found in the blanks (Milli-Q water) and, in consequence, the responses obtained for blanks were considered for establishing LODs. Luks-Betlej et al.[78] also studied the extraction of phthalate esters using different commercial fibers: 100 and 7 μm PDMS, PDMS-DVB (StableFlex), PA, CW-DVB (StableFlex), divinyl-benzene–carboxen–polydimethylsiloxane (DVB-Carboxen-PDMS) (StableFlex). The highest extraction efficiency was achieved with the fibers containing a DVB phase. Taking into account the repeatability obtained, CW-DVB fibers were the ones recommended by the authors (4.4% to 20.3% RSD). The proposed extraction conditions were: 25°C extraction temperature, 60 min

TABLE 28.7
SPME Methods for the Analysis of Phthalate Esters in Water Samples

Year	Analyte	Sample	Sample Volume and Pretreatment	Fiber Type	Determination	Linear Range	LOD	RSD%	Ref.
2003	DPP, DIBP, DCHP	Spiked wastewater, tap and river water	10 ml, filtered (0.45 μm Millipore cellulose membranes)	CW-TPR	LC-UV	20–200 μg/l	4–9.5 μg/l	1–9	108
2002	DEP, DBP, BBP, DEHP, DOP	Estuaries	30 ml	PDMS	GC-MS	—	0.07–3.15 μg/l	2–23	79
2002	DHP, DEHP, DOP	River water and domestic waste water	0.2–0.8 ml, filtered (2 glass fiber filters, 1 and 0.3 μm, respectively)	Polymeric-coated synthetic fibers	LC-UV	—	LOQ 0.15–0.20 μg/l	<3	109
2002	DMP, DEP, DBP, BBP, DEHP, DOP	Potable tap water, deionized water from purification systems, and spring water	3 ml	PA	GC-ECD	0.001–10 μg/l	LOQ 0.001–0.050 μg/l	4–10	84
2001	DMP, DEP, DBP, BBP, DEHP, DOP	River and sea water (industrial and fishing port)	3.5 ml, filtered (0.45 μm nylon membrane filters)	PDMS-DVB	GC-IR	0.1–10 μg/l	2–27 ng/l	13–18	110
2001	DMP, DEP, DBP, BBP, DEHP, DOP, DNP	Drinking water	5 ml	PDMS-DVB	GC-MS	0.02–10 μg/l	0.005–0.04 μg/l	4.4–28.3	78
2000	DMP, DEP, DBP, BBP, DEHP, DOP	Tap, commercial mineral, river and sea water (industrial and fishing port)	3.5 ml filtered (0.45 μm nylon membrane filters)	PA	GC-MS	0.02–10 μg/l	0.006–0.17 μg/l	Repeat. = 10–19 Reprod. = 10–21	111
1999	DEP	Mineral water	10 ml	PDMS-DVB (CW-TRP)	LC-UV	5–75 μg/l	1 μg/l	0.11–5.22	112

extraction time, and 270°C desorption temperature (5 min). The levels of phthalates found in the blanks were also considered in the calibration curves. Polo et al.[113] have optimized the factors affecting the SPME of phthalates using a factorial design. One of the factors studied by these authors was the sampling mode: direct SPME and headspace SPME. Headspace mode was more favorable for DEHP but it gave low responses for other target compounds such as BBP. Figure 28.2 shows a chromatogram obtained for a water sample spiked with 0.5 μg/l of phthalates. The extraction conditions were the following: PDMS-DVB fiber, 100°C extraction temperature, direct sampling, and 20 min extraction time.

A screening method for the analysis of 16 PAHs, 6 PCBs and 6 phthalate esters have been recently developed[79] using a multisimplex strategy to optimize experimental parameters affecting PDMS SPME. Due to blank problems, detection limits for some phthalates were quite high, especially for DEHP (3.15 μg/l).

Phthalates have also been analysed by SPME-GC using an ECD detector. Prokupkova et al.[84] studied different SPME parameters including two fibers: PDMS and PA. PA extraction at room temperature for 20 min was the experimental condition selected. Samples were stirred during extraction. Desorption was performed at 250°C for 50 min to minimize carryover. Quantification was based on calibration curves for distilled water after subtraction of blanks.

Recently, Psillakis et al.[114] have developed a liquid phase microextraction (LPME) technique using a hollow fiber membrane in conjunction with GC-MS for the extraction and analysis of phthalates. The resulting method was validated and compared with SPME. Both techniques showed comparable performance and were considered suitable for trace analysis of phthalates in water.

A miniaturized sample preparation technique based on polymeric coated synthetic fibers made up of several hundreds of fine fibrous materials has been developed.[109] The extraction capillary was installed in a liquid chromatograph as a sample loop of the injection valve to perform online analysis. The online coupling of the microscale sample preparation step with a micro-LC made it possible not only to significantly reduce solvent consumption, but also to improve the quantification

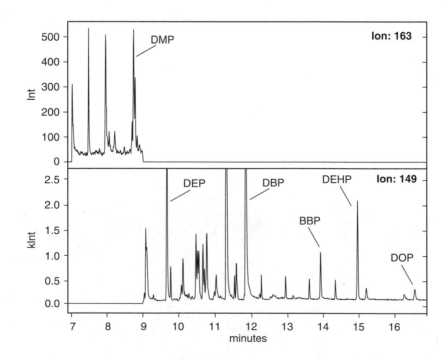

FIGURE 28.2 SPME of a water sample spiked with 0.5 μg/l of phthalate esters (see text for more details and Table 28.1 for abbreviations).

limits based on its higher extraction efficiency. The method was validated for phthalate analysis and quantification limits under 1 μg/l could be achieved.

4. Extract Cleanup

In most cases, the extracts obtained after LLE or SPE were analyzed without including a cleanup step. Laborious cleanup steps are necessary when analysis is performed by GC-FID or GC-ECD, but nowadays the most common determination technique is GC-MSD and, due to its selectivity, there is no need for a cleanup step for surface water samples. So, the cleanup step is not part of the routine methods for phthalate water analysis. It is only necessary if high background levels or high amounts of polar substances are found in the samples, as could be the case in extracts from landfill leachates of wastewaters.[75,87] Cleanup is generally performed using a polar sorbent, such as activated alumina or florisil. Lopez-Avila et al.[90] have compared florisil and alumina columns to perform the cleanup step. Alumina was preferred over florisil mainly because it allows recovery of all target compounds in the elution step, while three of the 16 phthalates included in the study could not be recovered with florisil clean up.

C. Sample Preparation for Solid Samples Analysis

Phthalate esters were analyzed in several kinds of environmental solid matrices such as soils, sediment of different origins (mainly marine sediments), and sewage sludge. Some applications of phthalate analysis of solid matrices are summarized in Table 27.8.[77,88–90,94,115–124]

Sampling is highly dependent on the type of solid. To study the evolution of pollution from phthalate esters, terrestrial soil core samples were collected by means of stainless steel drills, divided into core sections representing the soil profile, and transferred to cleaned glass bottles provided with PTFE-lined screw caps.[118] Soil samples can also be taken by shallow excavations as well as drilling.[88] To collect sediment samples, a Ponar grab or a petit Ponar can be used.[115,119] To analyze phthalate esters, sewage sludge is currently collected in clean dark glass-stoppered bottles. Once taken, sludge samples are immediately transported to the laboratory and stored at 4°C for not more than four days before analysis.[123]

The amount of solid sampled for the analysis is quite variable (see Table 28.8), depending on factors such as sensitivity of the analytical technique.

Prior to analysis, solid samples are usually homogenized, dried or freeze-dried and ground with a mill to obtain particles with a diameter of less than 1 mm or 0.2 mm. Sometimes, surrogates are added before the extraction process.

1. Extraction

Extraction of phthalate esters from solid samples is mostly performed by Soxhlet extraction or sonication. In addition, microwave assisted solvent extraction, supercritical fluid extraction, and ASE are used to extract phthalate esters from environmental solid matrices (see Table 28.8).

Soxhlet extraction is one of the most conventional techniques for extraction of organic compounds from solid matrices. It was applied early to the analysis of DEP and other phthalate esters in sediments and soils, using different extraction solvents such as hexane, dichloromethane or even acetonitrile, and is still considered as the reference method for the extraction of semivolatile compounds from solid environmental samples. The major disadvantages to Soxhlet extraction are length of time and consumption of solvent.

A typical Soxhlet procedure was employed by Ruminski et al.[88] to extract DEHP from very polluted soil using hexane as the extraction solvent and changing to methanol for LC analysis. Soxhlet was also used to extract phthalate esters from dust and from airborne fallout.[116,124] In the latter case, the material was trapped in nylon nets impregnated with silicone oil (SE-30) and further cleanup processes were needed afterwards to separate the oil from the phthalate esters extracted.

TABLE 28.8
Extraction Methods for the Analysis of Phthalate Esters in Environmental Solid Samples

Year	Analyte	Sample	Extraction	Extract Treatment	Determination	Recoveries %	Linear Range	LOD	RSD%	Ref.
2003	DMP, DEP, BBP, DBP, DEHP, DOP	Marine sediment (1 g dried sample)	Sonication with 2 × 5 ml acetonitrile (10 min/step) at room temperature	Centrifugation and concentration to dryness (rotary evaporator), addition of 0.5 ml acetonitrile, filtration through 0.2 μm nylon syringe	LC-MS	>70	0.1–20 mg/Kg (std addition)	—	Repeat. = 7–10, Reprod. = 10–19	94
2003	DEP, DBP, DEHP, DMP, DOP, BBP	Sediments (2 g + surrogate + 15–20 g sodium sulphate)	Sonication with 50 ml DCM-hexane 1:1 (10 min) and shaking (10 min). Repeated 3 times and the extracts were combined	Concentration to 5 ml (N$_2$). Treatment with 15 g deactivated alumina, anhydrous sodium sulphate, concentration to dryness, addition of 2 ml methanol containing IS	GC-MS, LC-MS	71–106	—	GC-MS: 0.3–3.3 ng/g, LC-MS: 2.5–4.2 ng/g	<15	115
2003	DEP, DBP, BBP, DEHP, DCHP, DPP, DIBP	Indoor dust (4 g)	Soxhlet extraction with 6% ether in hexane (16 h)	Concentration to 10 ml, 1 ml cleaned-up with florisil and concentrated to 2 ml with 10% diethyl ether in hexane	GC-MS	40–220	—	—	<20	116

Year	Compounds	Sample	Extraction	Cleanup/concentration	Detection	Recovery (%)	Concentration	Range	Reference
2002	DMP, DEP, DBP, DEHP, DOP	Greenhouse soil	Sonication with 3 × 20% acetone in petroleum ether	Filtration, rinsing with water, concentration to 1 ml, clear-up with silicagel, elution with 6 + 10 ml petroleum ether/diethyl ether (10:3). Concentration to dryness + 1 ml methanol	LC-UV	86.4–97.6	—	—	117
2002	DBP, DPP, DEHP, DOP, DNP, DINP, BBP	Soils (50 g + deuterated surrogates)	Shaking with 100 ml DCM (2 h)	Concentration to dryness (N_2), addition of hexane and IS	GC-MS	—	—	—	118
2001	DMP, DEP, DIBP, DBP, DMEP, DMPP, DEEP, DAMP, DHP, BBP, HEHP, DBEP, DEHP, DCHP, DOP, DNP	Sewage sludge (2 g freeze-dried, grinded to <1 mm)	Addition of 7 ml ethyl acetate and 1 ml of IS (D_4-DBP, D_4-DEHP), shaking at room temperature (1 h)	Centrifugation	GC-MS	—	10.1–632 μg/Kg (dry matter)	4–18	77
2001	DEP, BBP, DBP, DEHP	—	SFE with CO_2 at 80°C (dynamic mode, 2 ml/min, max extraction volume = 25 ml). Final hexane volume = 12 ml	—	GC-MS	—	—	—	119
2001	DEP, BBP, DBP, DEHP, DHP, DAMP, DPP, DCHP, DIBP	Dust (1.4–12.1 g)	After collected in a cellulose thimble, dust is Soxhlet extracted with 6% ether in hexane (16 h)	Addition of anhydrous sodium sulphate, concentration to 2.5 ml, cleanup with florisil, concentration to 2 ml in 10% diethyl ether in hexane and silylation	GC-MS	110–378	—	12–175	120

Continued

TABLE 28.8
Continued

Year	Analyte	Sample	Extraction	Extract Treatment	Determination	Recoveries %	Linear Range	LOD	RSD%	Ref.
2000	DEP, DBP, DEHP	Sewage sludge (2 g lyophilized)	Sonication with 20 ml methanol-DMC 7:3, and centrifugation. Repeated 3 times and the extracts were combined	Concentration to 1 ml (rotary vacuum evaporator) and redissolution in 200 ml HPLC water. Cleanup and fractionation in 500 mg C18 cartridges. Evaporation to dryness + 1 ml methanol	LC-MS	78–91	—	15–50 ng/g	—	121
1996	DEP, DBP, DEHP, DMP, DAP, BBP	Marine sediments and soil (5 g)	Soxhlet with: 300 ml DCM (16 h)	Preconcentration and solvent change to hexane, concentration to 1 ml	GC-MS, GC-ECD	65.5–89.5	—	—	—	122
1996	DEP, DBP, DEHP, DMP, DAP, BBP	Marine sediments and soil (5 g)	2 × 50 ml DCM and sonication (15 min)	Filtration, preconcentration and solvent change to hexane, concentration to 1 ml	GC-MS, GC-ECD	64.6–88.6	—	—	5.7–13.4	122
1996	DEP, DBP, DEHP, DMP, DAP, BBP	Marine sediments and soil (5 g)	MASE with 30 ml acetone-hexane 1:1 (115°C, 10 min)	Filtration, preconcentration and solvent change to hexane, oncentration to 1 ml	GC-MS, GC-ECD	0.1–91	—	—	3–8.9	122
1995	DEHP	Soil (20 g)	Soxhlet with 50 ml hexane (6 h)	Concentration to 1 ml, addition of methanol, filtration (0.45 μm), addition of methanol up to 25 ml	LC-UV	—	—	—	—	88

Year	Compounds	Matrix	Method	Clean-up	Detection				Ref
1994	DMP, DEP, DBP, DMEP, DEEP, DCP, DEHP, DOP, BBP, BOP, DAMP	Soil (10 g)	Method EPA SW-846 method 3550: addition of 30 ml DCM-acetone 1:1, sonication, filtration, drying (N$_2$), and addition of 1 ml hexane	Clean-up with florisil for soils heavily contaminated + IS, and concentration to 0.1–1 ml (N$_2$)	GC-MS-MS	22–68, excepting for DBP and DEHP (found in blanks)	—	15–25	89
1991	DMP, DEP, DIBP, DBP, DMPP, DMEP, DAP, DEEP, HEHP, DHP, BBP, DBEP, DEHP, DCP, DOP, DNP	Sediments, sludge, sandy loam soil (10–30 g + surrogates)	EPA method 3550: Sonication with 3 × 100 ml DCM-acetone 1:1 (3 min/step) or EPA method 3540: Soxhlet with 350 ml hexane–acetone 1:1 (18 h)	Filtration through filter paper, drying through anhydrous sodium sulphate, concentration to <10 ml, solvent exchange to 50 ml hexane, concentration to 2 ml (N$_2$). Clean-up with florisil or alumina (10 g)	GC-ECD	20–155 (Sonication) 53.5–135 (Soxhlet)	6–60 μg/Kg (clean samples)	4.2–24.6 (Sonication) 19.8–46.9 (Soxhlet)	90
1990	DMP, DEP, DBP, BBP, DEHP	Sewage sludge (1 g)	Sonication with 40 ml DCM (30 min) or Soxhlet with 30 ml DCM (48 h)	Centrifugation, filtration (glass fibre), concentration to 1–2 ml, clean-up with deactivated alumina and florisil, evaporation to dryness + IS	GC-ECD	DEHP = 92.5 (Soxhlet), 109.4 (Sonication)	—	DEHP = 6.4 (Soxhlet), 8.1 (Sonication)	123
1990	DBP, DEHP	Airborne fallout	Particles are trapped in a nylon net impregnated of silicone, and device is Soxhlet extracted with hexane	Clean-up with fuming concentrated sulphuric acid, and separation of phthalates from the oil	GC-MS GC-FID	—	—	—	124

Solvent extraction using ultrasound is widely applied. Zurmühl[123] determined phthalate esters in sewage sludge after extraction of freeze-dried samples with *dichloromethane (DCM)* in an ultrasonic water bath for 30 min. The extract was centrifuged, filtered through a glass fiber filter, and concentrated. As the analysis was performed by GC-ECD, further cleanup and fractionation of the extracts with alumina and florisil was needed. Reported recoveries were quantitative for the five phthalate esters considered in the study (see Table 28.8). Brumley et al.[89] extracted soil samples using a modified EPA method (SW-846 Method 3550). After extraction with 1:1 DCM or acetone and cleanup of the extracts with florisil, phthalate esters were analyzed by GC-tandem MS. Recoveries obtained were between 22 and 68%, excepting for phthalates found in blanks (DBP and DEHP).

Other methods based on ultrasonic extraction use different solvent mixtures as well as diverse cleanup processes, leading to good recoveries of the analytes from solid samples, such as soils and sediments.[115,117]

Petrovic and Barceló[121] developed a method for the simultaneous determination of anionic and nonionic surfactants, the corresponding degradation products, and endocrine disrupting compounds in sewage sludge. Phthalate esters are among the contaminants which were determined. Sludge samples were sonicated with a mixture of methanol-DCM and the extract was separated by centrifugation. The overall procedure was repeated three times and the extracts were combined, concentrated to 1 ml and dissolved in 200 ml HPLC water. Extracts were fractionated using C18 cartridges, eluting the different fractions with solvent mixtures of different polarities. Final extracts were evaporated to dryness and reconstituted with methanol for LC-MS analysis. Phthalate esters and other less polar compounds were determined using an APCI interface working in the positive ionization mode.

A comparison of sonication and Soxhlet extraction methods was made by Lopez-Avila et al.[90] EPA methods 3550 and 3540 were used to extract some phthalate esters from solid samples, such as sandy loam soil and municipal sludge. The mean recoveries for the method 3540 (Soxhlet extraction) were, in general, similar to those obtained for method 3550 (ultrasonic extraction); however, its RSD was much higher.

A microwave-assisted solvent extraction (MASE) method was optimized by Chee et al.[122] to extract phthalate esters from marine sediments, soils, and results were compared with those obtained by the same authors using conventional Soxhlet and sonication techniques. The analysis was performed by GC-ECD or GC-MS. The overall optimal conditions for the extraction of phthalate esters by MASE included the use of 1:1 acetone or hexane at 115°C for 10 min. Recoveries for six individual phthalate esters (DMP, DEP, DAP, DBP, BBP, DEHP) ranged from 71% to 91%, and were better than those obtained with Soxhlet (66% to 90%) or sonication (65% to 89%). The authors stated that advantages of MASE extraction over sonication or Soxhlet are larger sample throughput, lower usage of hazardous solvents, and less laborious cleanup steps.

A simple procedure to determine phthalate esters in dry sewage sludge samples, based on solvent extraction by mechanical shaking, was developed by Berset and Etter-Holzer.[77] Samples were shaked with ethyl acetate at room temperature for one hour. The solution was centrifuged and an aliquot of the supernatant was immediately analyzed by GC or MS without further cleanup. Recoveries were quantitative with precision ranging between 4% and 18% depending on the analyte. LODs based on real sample were 10.1 to 632 $\mu g/Kg$ expressed in a dry weight matter basis.

According to Vikelsoe et al.[118] cleanup is not necessary when high resolution mass spectrometry (HRMS) is used as detector due to the high selectivity of the technique. Therefore, soil samples were spiked with the deuterated surrogates and extracted by shaking in DCM for 2 h. The extract was concentrated under nitrogen, redissolved in hexane and analyzed by GC or HRMS. Blank responses were subtracted from sample responses.

McDowell and Metcalfe[119] proposed the use of supercritical fluid extraction (SFE) with ultra high purity CO_2 as supercritical fluid for the extraction of phthalate esters from sediment samples. Extracts were collected by bubbling the vented gas through 12 ml hexane, and solvent volume was reduced to 0.5 ml by evaporation in a heated water bath. Due to the presence of coextracted compounds that interfered with the analysis, a silica gel cleanup step was necessary. Phthalate esters were eluted from the silica column with DCM and the extracts were concentrated before GC-MS. Analysis of blanks indicated that there was background contamination in the samples mainly due to DEHP and DBP. At the optimal extraction conditions, extraction efficiencies ranged between 70% and 90%. Method detection limits were high for individual phthalates (0.81 μg/g for DEHP, 0.18 μg/g for DEP). The advantages of SFE in the analysis of phthalate esters in solid samples include reduced use of glassware and organic solvents, rapidity in comparison with other techniques such as Soxhlet, and the use of only small amounts of sample (<1 g).

D. SAMPLE PREPARATION FOR AIR ANALYSIS

Air samples are usually collected by pumping air through a collecting device for a period of time (hours, days or even months), which depends on the purpose of the study, the pollution levels, and the detection limits of the analytical method. In Table 28.9, details on the analysis of phthalate esters in air samples are illustrated.[116,117,120,124-126] For indoor air samples, between 1 and 6 m^3 of air are required,[120,125] whereas the analysis of atmospheric samples require 300 to 400 m^3 air.[124] To sample industrial emissions, where phthalates are found at much higher concentrations, a few liters of air are usually enough.[126] The devices currently employed to retain the target compounds are cartridges filled with sorbent material retained by glass wool. Such material can be polyurethane foam (PUF), octadecylsylane modified silica, charcoal, GDX-102 resin, or combinations of various sorbents like PUF and XAD resin. To prevent possible contamination, sorbent materials are usually preextracted by Soxhlet using different solvents or solvent mixtures.[120]

Breakthrough air volumes for each analyte need to be previously determined to select the maximum sample volume that can be concentrated.

Sometimes, phthalates in particulate matter are also the object of analysis. Collection of solid particles can be accomplished by placing a particle filter in front of the sorbent,[124] or using special devices, such as cellulose extraction thimbles.[116,120] Sample preparation of solid samples collected in this way has been discussed in the section devoted to environmental solid samples.

1. Extraction

Desorption of phthalate esters from cartridges can be performed by extraction with organic solvents or by thermal desorption. Solvent extraction can be made using direct elution, Soxhlet extraction or extraction assisted by ultrasounds.

Fischer et al.[126] proposed a simple method for recovering analytes based on backflushing of the sampling cartridge (filled with C18) with 20 ml methanol or n-hexane. In their pioneer study on the presence of phthalate esters in the Swedish atmosphere, Thurén and Larsson used polyurethane filters connected in series.[124] Compounds adsorbed to the PUF filters were extracted with acetone hexane in an ultrasonic bath. More recently, Otake et al.[125] extracted the esters adsorbed in a charcoal tube by sonication with 1 ml of toluene for 10 min. These authors proved that longer sonication times did not improve the efficiency of the extraction (97.5 to 115%). Soxhlet can also be used to extract the analytes from the sorbent cartridges. Rudel et al.[116,120] performed an extraction for 16 h with a quartz filter + PUF + XAD sampling cartridges using 200 ml of 6% ether in n-hexane. Prior to the extraction, p-terpenyl-d$_{14}$ was added as a surrogate. DCM was the solvent selected by Wang et al.[117] to Soxhlet extract the phthalate esters retained on GDX-102 resin.

After extraction of the target compounds, extracts are usually concentrated to achieve sufficient overall method sensitivity or for solvent exchanging for further analysis. Before concentration, the addition of anhydrous sodium sulphate avoids the presence of residual water traces in the organic extracts. Either a gentle stream of nitrogen or Kuderna–Danish can be used for the

TABLE 28.9
Extraction Methods for the Analysis of Phthalate Esters in Air Samples

Year	Analyte	Samplers and Sampling Rate	Desorption	Extract Treatment	Determination	Recovery %	Linear Range	Limits of Detection	RSD%	Ref.
2003	DEP, DBP, BBP, DEHP, DCHP, DPP, DIBP	Indoor air (10–14 m³)	URG personal pesticide sampling cartridges (impactor inlet followed by a cartridge fitted with quartz fiber filter, XAD-2 resin and PUF plugs) (8–9 l/min)	Soxhlet with 150 ml of 6% ether in hexane (16 h)	Addition of sodium sulphate and concentration to 2 ml 10% ether in hexane	GC-MS	40–220	—	2–75 ng/m³	15–25	116
2002	DMP, DEP, DBP, DEHP, DOP	Plastic film greenhouse air	GDX-102 resin	Soxhlet with DCM, 6 h	Filtration, concentration (50°C). Cleanup on silicagel, dryness (N_2) + 1 ml methanol	LC-UV	—	—	—	—	117
2001	DEP, BBP, DBP, DEHP, DHP, DAMP, DPP, DCHP, DIBP	Indoor air (0.29–5.9 m³)	Cartridge filled with quartz fiber, XAD-2 resin and PUF (3.8 l/min)	Soxhlet with 200 ml of 6% ether in hexane (16 h)	Addition of sodium sulphate and concentration to 1 ml 10% ether in hexane. Silylation	GC-MS	95–129 (DEP)	—	0.0045–1.64 µg/ extract (BBP, present in the blanks)	0–8	120
2001	DEP, DBP, BBP, DEHP	Indoor air (4.3 m³)	Cartridge filled with charcoal granules in 2 layers, one with 100 mg for sampling and other with 50 mg for breakthrough (1 l/min, during 3 days)	Sonication with 1 ml toluene	Centrifugation	GC-MS, GC-FPD	97.5–115	0.6×10^{-3} –23 µg/m³	0.0256–0.1186 µg/m³	~10	125

Year	Compound	Sample	Trapping	Extraction	Clean-up	Determination	Recovery (%)	Range	Detection limit		Ref.
1993	DBP, DEHP, DIDP	Industrial emissions (4–20 l)	Silica-cart™ cartridge (1 ml) packed with C18 (0.5 l/min)	20 ml methanol or hexane	Drying with N₂ to almost dryness + 1 ml hexane (for NP-LC chromatography) or + 1 ml 75% 2-propanol in water (for RP-LC)	LC-UV	92.8–98.7	0–500 µg/ml	0.1–0.3 µg	—	126
1990	DBP, DEHP	Air (300–400 m³)	Particle filter and 1–2 PUF filters connected in series (4.5 m³/day, 3 months)	Sonication with acetone–hexane	Treatment with fuming concentrated sulphuric acid	GC-MS GC-FID	—	—	—	—	124

concentration of the extracts.[117,126] Different cleanup procedures can be performed. Some of them include the use of fuming concentrated sulphuric acid, or silicagel columns.[117,124]

Thermal desorption of the sampling cartridges presents some advantages over the solvent-based extraction methods. As all the retained compounds are thermally desorbed into the GC, higher sensitivity can be achieved. Nevertheless, some limitations deal with the high temperatures needed for quantitative desorption of less volatile analytes from typical sorbents, such as Tenax or carbon materials. An alternative to these sorbents could be the use of silicones as sorptive material. A procedure based on the use of this material for enrichment, thermal desorption-GC-MS was described.[75] The LODs achieved sampling 15 l air ranged between 1 and 10 ng/m^3.

IV. SEPARATION TECHNIQUES

GC and liquid chromatography (LC) are the usual techniques for the determination of phthalate esters in environmental and other types of samples. For quantification, the addition of internal standards is highly recommended to account for variability during the sample introduction process. The use of surrogate standards allow correction of the variations produced throughout the analytical process. The best internal and surrogate standards are isotopically labeled phthalates, which are especially suitable with mass spectrometry detection.

A. GC

Phthalate diesters are sufficiently volatile and thermally stable to be analyzed by GC.

Several types of GC detectors, such as infrared (IR),[110] flame photometric (FPD),[125] and electron capture (ECD)[84,90,98,122,123] have been applied to the GC analysis of phthalates in environmental samples. Although FID was used for environmental applications [86,97,123], nowadays, the use of this detector is basically limited to analyzing other kind of samples (toys, plastics), in which phthalate esters are found in higher concentrations.[127–130] EPA methods 606 and 8060 consider the use of ECD for the determination of phthalates (Table 28.3B). Nevertheless, this detector shows a more sensitive response for halogenated compounds, which are currently present in polluted samples. Most of the recently proposed methods for phthalate esters analysis in environmental samples involve the use of MSD working in the electron ionization mode (EI).[77–81,83,85,87,91,92,96,111,114,118–120,122,124,125] Phthalates fragmentize with characteristic ions, such as $m/z = 149$ (DEP, DBP, BBP, DEHP, DIBP), $m/z = 163$ (DMP) and $m/z = 293$ (DINP, DIDP), allowing a very sensitive and selective detection, particularly when operating in the selected ion monitoring (SIM) mode. Tandem mass spectrometry can also be used,[89] as well as CI.[89,131] Mass analyzers are usually low resolution spectrometers and both quadrupole and ion trap configurations have been used as GC detectors.

Separation columns are usually 25 to 30 m × 0.25 to 0.32 mm i.d. coated with phenyl methylpolysiloxane or dimethylpolysiloxane stationary phases, which allow program separations in a wide range of temperatures (typically, from about 50 to 300°C at 10°C/min) with low bleeding. These general analytical conditions are suitable to obtain good resolution in a short analytical time. Nevertheless, chromatographic separation of all isomers is not possible and thus, for analysis involving complex separations, selective mass detection should be employed.

The analysis of phthalate metabolites has recently become of environmental concern. Primary metabolites of phthalate diesters are monoesters, in which one ester function is hydrolyzed. These monoesters have a carboxylic function that is currently derivatized to avoid adsorption during GC determination. Several derivatization agents have been used to block the free acid group. Methyl-esterification was used by Hashizume et al.[86] and Suzuki et al.[132] Silylation was preferred by Jonsson and Borén,[96] and Jonsson et al.[85] to simultaneously determine the diesters, monoesters and phthalic acid in the same analysis. Wahl et al.[133] performed the analysis of three acid metabolites of

DEHP previously converted in its *tert.*-bytyldimethylsilyl derivatives. For the analysis of the intact conjugates of DEHP metabolites by GC, Egestad et al.[134] prepared the methyl ester trimethylsylil ether derivatives of the major glucuronides. On the other hand, Pietrogrande et al.[135] performed direct GC analysis of the enzymatic hydrolysis products of phthalate esters conjugates without any derivatization step.

B. LIQUID CHROMATOGRAPHY

Phthalates can also be analyzed by liquid chromatography. With this technique, ultraviolet absorption can be used for the detection of phthalates in environmental samples[86,88,99,108,109,112,117,126] as well as in other matrices, such as biological samples and plastics.[136–139] Recently, mass spectrometric detection was applied to the analysis of phthalate esters, operating with a single spectrometer,[94,100,101,104,115] or using mass spectrometry in tandem.[101–103,106,121,140] Compared to GC-MS analysis, lower sensitivity is generally obtained, although the LC-MS approach presents some advantages, such as higher selectivity, with molecular weight information for the isomeric mixtures, more reliable quantification of the phthalate esters isomeric mixtures, and simpler cleanup procedures and shorter analysis times.[75] With mass detection, isotope dilution is particularly suitable for the quantification of phthalate esters and its metabolites in complex samples.

Separation of phthalates by LC is usually performed in the reverse phase mode using C18 columns, although the use of other stationary phases can also be found in literature.[141] Both isocratic and gradient elution modes were described.

Determination of primary and secondary metabolites of phthalates has been proposed to establish the exposure of general population to phthalates avoiding external contamination problems.[101,142] Monoesters can be analyzed by LC without derivatization. In some studies regarding the problem of phthalate ester migration from plastic material, the monitoring of DEHP and its metabolite MEHP was performed using a UV detector.[143–145] Nevertheless, the most recent studies focusing on the analysis of metabolites are carried out using mass spectrometric detectors. Inoue et al.[100] present a method based on LC-electrospray (ESI)-MS working in the positive and the negative ionization modes to determine DEHP and its primary metabolite in blood samples. Anderson et al.[146] determine isotopically labeled monoesters of phthalic acid by LC-MS using atmospheric pressure chemical ionization (APCI) in the negative mode. In addition, tandem mass spectrometry was used to enhance selectivity.[100,102,103–105,140,146–148]

C. MISCELLANEOUS TECHNIQUES

Some studies on the application of electrophoretic techniques to the separation of phthalate esters can be found in the literature.[149,150] In these cases, micellar electrokinetic chromatography (MEKC) allowed the study of the migration of phthalate esters in different electrophoretic media.

Supercritical fluid chromatography was coupled online to Proton High Field Nuclear Magnetic Resonance Spectroscopy by an specially designed pressure-proof continuous-flow probe head. Separation of phthalate esters was carried out under supercritical conditions using carbon dioxide as eluent.[151]

The potential of ion mobility spectrometry combined with solid phase extraction for field screening of organic pollutants, such as phthalate esters, in water was also studied.[152]

A method based on direct time resolved fluoroimmunoassy (TR-FIA) was developed to selectively recognized phthalate esters, obtaining a sensitivity of 0.5 pmol/ml.[153]

Luminiscence is other technique that has been proposed for the determination of total phthalate esters in water.[93] After hydrolysis of the ester function, a chelate with terbium (III) was formed, and luminescence measurements were obtained in the time resolved mode. The method is useful for screening purposes in water samples.

V. PRESENCE OF PHTHALATE ESTERS IN THE ENVIRONMENT

As previously pointed out, phthalate esters have a wide variety of industrial uses, and are produced in large quantities all over the world. These compounds are primarily used as plasticizers and are by far the largest class of plastic additives. Because of several anthropogenic inputs, phthalates have been detected everywhere in the world, contaminating aquatic systems, air, wildlife, plants, sediments and soils. In consequence, there is significant concern for their ubiquitous presence in the environment. Scientists, clinicians, and regulatory agencies currently debate the potential for adverse health effects on humans. In addition, phthalate esters are chemicals with potential endocrine-disrupting properties. In the Netherlands, van Wezel et al.[43] derived the environmental risk limits (ERLs) for these compounds in water as 10 μg/l for DBP and 0.19 for DEHP μg/l. The authors used these ERLs as the estimated ecosystem no-effect concentration, concluding that these limits provide sufficient protection against endocrine-disrupting effects. It is clear that the environmental analysis of this group of compounds is of major concern[85] and phthalate esters are commonly among the list of typical pollutants in the study of specific environmental problems.

Table 28.10 reports some examples of the levels of phthalates found in different environmental matrices, such as river, coastal, waste and drinking water, soils, sediments, sludges, and air. The phthalates presented in this table are the six congeners included in the U.S. EPA Priority Pollutant list.

The presence of phthalates in coastal and river water samples is due to anthropogenic inputs from various resources, which include sewage treatment plants, industries that use phthalate esters, and leaching from disposed plastic wastes. The levels found in surface waters for DBP are in general below the ERL, but for DEHP the levels reported are often considerably higher than their corresponding ERL (see Table 28.10A).

Phthalates were measured in some 400 water samples from the Rhine river and its main affluents in North Rhine-Westfalia.[87] DEP, DMPP, DBP, and DEHP were found in almost all samples and their levels were quite stable throughout the year. The mean values found were at the sub μg/l. Fatoki and Noma[97] studied the levels of phthalate esters in river and marine water samples from the Eastern Cape Province (South Africa), and found that samples were grossly polluted with several phthalate esters. The relatively high levels in the rivers were not unexpected because many of them receive effluents from industries and municipal sewage works with partial treatment or no treatment at all. The high levels of phthalate esters recorded in this study raise some concern.

DEHP is the most widely used phthalate ester and the U.S. EPA has established its MAC for drinking water at 6 μg/l. In addition, the U.S. EPA suggests that DEHP concentrations in potable water above 0.6 μg/l should be closely monitored.[154] Determination of DEHP in potable waters has often reached this limit (see Table 27.10A). The difficulty in controlling this level in water (0.6 μg/l), as well as the ERL indicated above (0.2 μg/l), is that background and detection limits in many studies frequently reach or surpass these limits.

The presence of phthalate esters in distribution water comes from a variety of sources, such as existing phthalate ester concentrations in groundwater or leaching from reservoirs and pipes containing plastic, epoxy resins or paints.[80] These compounds are also found in bottled water. It should be taken into consideration that the sources of organic pollutants in this type of water are mainly attributed to (i) compounds which are directly present in the aquifer as contaminants; (ii) contamination from the bottling plant and (iii) migration from containers, especially during storage. In general, storage of mineral water in deficient conditions, e.g., close to high temperatures, increases migration of plastic components.[80] It should be mentioned that caps are often responsible for phthalate contamination of bottled mineral water. Even commercially available mineral waters

TABLE 28.10A
Phthalate Ester Concentrations Found in Environmental Water Samples

Sample	Year	N	Method	Concentration (μg/l)						Ref.
				DMP	DEP	DBP	BBP	DEHP	DOP	
Coastal Water	2003	1	Online SPE-GC-MS	—	—	—	0.15	8.10	—	81
Coastal water	2002	1	Online SPE-GC-MS	—	—	0.48	0.08	0.12	—	83
Coastal water	2002	4	SPME-GC-MS	—	—	0.8–1.9	nq	2.5–10	—	79
Coastal water	2002	28	SPE-GC-FID	0.03–351	0.03–398	1.0–1028	—	0.06–2307	—	97
Coastal water	2001	2	SPME-GC-MS	1.6–2.1	1.4–1.8	1.3–1.9	0.5–1.1	2.1–3.2	0.8–1.5	110
Coastal water	2000	2	SPME-GC-MS	nd	0.39–0.62	0.12–0.16	nd	1.62–2.12	nd	111
Ground water	1997	2	LC-GC-MS	—	—	0.01–0.24	—	0.1	—	82
Lake water	1995	8	LLE-LC-UV	—	—	—	—	124–645	—	88
River water	2003	1	Online SPE-GC-MS	—	—	—	nq	2.10	—	81
River water	2002	1	Online SPE-GC-MS	—	—	0.08	nq	nq	—	83
River water	2002	2	LC-UV	—	—	—	—	0.17–1.63	0.24	109
River water	2002	13	SPE-GC-FID	0.03–19.4	0.03–35.6	0.04–75.6	—	4.6–90.5	—	97
River water	2002	9	LLE-GC-FID (LC-UV)	nd	nd-0.03	0.29–3.90	—	nd-0.83	—	86
River water	2001	1	SPME-GC-MS	—	0.6	0.4	—	1.1	—	110
River water	2000	1	SPME-GC-MS	nd	0.26	nq	nd	0.70	nd	111
River water	1997	2	LC-GC-MS	—	—	0.015–0.27	—	0.04–0.095	—	82
River water	1994	400	SPE-GC-MS	<0.02–0.61	<0.02–1.8	<0.03–1.3	<0.04–49.0	0.11–10.3	<0.03–0.9	87
River water	1990	4	LLE-GC-MS	—	0.4–0.6	12.1–33.5	—	nd-1.6	—	92
Spring water	2002	1	SPME-GC-MS	0.1	0.04	0.02	0.002	2.88	nd	84
Spring water	2002	1	LLE-GC-MS	—	—	nd	nd	0.14	nd	84
Tap water	2003	1	Online SPE-GC-MS	—	—	—	nq	4.26	—	81
Tap water	2003	1	LPME-GC-MS	—	0.30	1.04	—	0.93	—	114
Tap water	2003	1	SPME-GC-MS	—	0.11	0.44	—	0.87	—	114
Tap water	2003	7	SPE-GC-MS	nd-0.004	nd-0.09	nd-0.032	nd-0.017	nd-0.331	—	80
Tap water	2002	1	SPME-GC-MS	0.08	0.07	0.05	0.002	0.66	nd	84
Tap water	2002	1	LLE-GC-MS	—	0.02	0.05	nd	0.24	nd	84
Tap water	2002	1	Online SPE-GC-MS	—	—	0.3	—	0.1	—	83

Continued

TABLE 28.10A
Continued

Sample	Year	N	Method	Concentration (µg/l)						Ref.
				DMP	DEP	DBP	BBP	DEHP	DOP	
Tap water	2002	7	LLE-GC-FID (LC-UV)	nd-0.08	nd-0.02	0.57–9.26	—	nd-5.22	—	86
Tap water	2001	2	SPME-GC-MS	—	0.16–0.20	0.38–0.64	0.02–0.05	0.05–0.06	—	78
Bottled water	2003	9	SPE-GC-MS	nd	nd-0.139	nd-0.072	nd	nd-0.188	—	80
Bottled water	2002	2	LLE-GC-FID (LC-UV)	nd	nd	0.19–0.52	—	nd-0.42	—	86
Mineral water	2003	2	LPME-GC-MS	—	0.05–0.13	0.32–0.51	—	0.57–0.65	—	114
Mineral water	2003	1	SPME-GC-MS	—	0.07–0.12	0.08–0.14	—	0.36–0.46	—	114
Mineral water	2002	1	SPME-GC-MS	nd	nd	0.18	nd	9.78	nd	84
Mineral water	2002	1	LLE-GC-MS	—	—	0.37	nd	9.93	nd	84
Landfill leachates	2003	17	LLE-GC-MS	nd	2–33	1–23	2–7	3–460	—	85
Waste water	2003	1	Online SPE-GC-MS	—	—	—	nq	3.97	—	81
Waste water	2003	1	SPE-LC-MS	—	nq	2.2	0.2	3.8	3.4	94
Waste water	2002	1	LC-UV	—	—	—	—	1.38	—	109
Waste water	1990	1	LLE-GC-MS	—	0.4	6.0	—	1.9	—	92

N, Number of samples; nd, Not detected; nq, Not quantified.

TABLE 28.10B
Phthalate Ester Concentrations Found in Environmental Solid and Air Samples

Sample	Year	N	Method	Concentration (solid samples, mg/Kg; air, $\mu g/m^3$)						Ref.
				DMP	DEP	DBP	BBP	DEHP	DOP	
Marine sediment	2003	1	Sonication-LC-MS	0.35	0.41	0.42	0.13	9.4	—	94
Marine sediment	2001	5*	SFE-GC-MS	—	nd	nd	nd	6.5–29.7	—	119
Marine sediment	1996	3	MASE-GC-MS (ECD)	—	—	0.74–1.60	—	0.94–2.79	—	122
Marine sediment	1996	3	Sonication-GC-MS (ECD)		—	0.69–1.35	—	0.89–2.60	—	122
Soil	2002	8	SE-GC-MS			0.0003–0.453	0.00001–0.032	0.012–1.9	0.00061–0.067	118
Soil	2002	8	Sonication-LC-UV			0.9–3.6	—	0.8–2.9	—	117
Soil	1996	3	MASE-GC-MS (ECD)		—	0.68–0.98	—	0.16–1.06	—	122
Soil	1996	3	Sonication-GC-MS (ECD)		—	0.60–0.80	—	0.15–0.88	—	122
Soil	1995	38	SE-LC-UV		—	—	—	83–45720	—	88
Sewage sludge	2001	12	SE-GC-MS	nd-0.027	nd-0.145	0.193–1025	nd	21.6–113.9	nd-0.629	77
Sewage sludge	1990	9*	Sonication-GC-ECD	nd	nd	2.3–256.0	0.2–0.7	65.8–480.6	—	123
Dust	2001	6	GC-MS		1.01–3.58	11.1–59.4	12.1–524	69.4–524.0	—	120
Greenhouse air	2002	2*	SPE-Soxhlet-LC-UV	nd-56	nd-32	0.224–1.910	—	0.056–0.550	—	117
Workplace atmosphere	1993	—	LC-UV			400–8200	—	4200–58400	—	126
Workplace and residential air	2001	7	GC-MS		0.236–1.29	0.101–0.431	0.01–0.172	0.02–0.114	—	120
Indoor air	2001	6*	SPE-Sonication-GC-MS (FPD)		0.05–C.19	0.11–0.60	<0.0012–0.10	0.04–0.23	—	125

N, Number of samples or *number of sampling sites; SE, Solvent extraction; nd, Not detected.

distributed in glass bottles are suspected of such contamination. It has been reported that metal caps may be sealed with PVC inserts contributing to a high level of DEHP contamination.[84] Analysis of bottled water indicated that the type of packing material could affect the phthalate concentrations. Peñalver et al.[111] analyzed a variety of commercial water samples stored in containers made from different materials, such as polyethylene terephthalate (PET), PVC, glass and tetra-brick, and the influence of the material on the concentration of phthalates was also evaluated. DEP, DBP and DEHP were found in all samples. DEHP was the phthalate ester with the highest concentrations (about 1 to 2 μg/l). On the basis of its results, the authors conclude that the container material affect the concentrations of phthalates in water. Glass and tetra-brick bottled water showed lower concentrations of some phthalates.

Sediment samples are used to study the historical phthalate contamination and for the determination of local contamination and biodegradation. Concentration levels of phthalates largely depend on the sampling site. The average concentrations are of the same order of magnitude as the soil samples[75] (see Table 28.10B).

Local episodes of high pollution by phthalate esters in solid environmental samples are reported in the literature. Ruminski et al.[88] found high DEHP concentrations in the soil around a big factory of synthetic polymers in Poland that may be the result of many years of pollution. Concentrations varied from 0.07 to 45.7 g/Kg. Analysis of water samples taken from a lake nearby also showed the presence of DEHP in the range 0.12 to 0.65 ppm. Concentrations in bottom mud from this lake were found to be about two magnitudes higher than the levels in water.

Phthalates have also been analyzed in marine sediments. In a recent study, marine surface sediments and biota samples from various locations were collected to assess the sources and distribution profiles of phthalate esters in a specific ecosystem (False Creek Harbour, Vancouver, BC, Canada). Concentrations of all phthalate ester congeners combined ranged from 2.0 to 3.6 ppm on a dry weight basis.[115] The phthalate ester composition found was in some way similar to the North American per capita phthalate ester consumption levels reported.[155]

Sludge samples from wastewater treatment are important samples to monitor input sources and to study biodegradation of phthalates and other pollutants. Phthalates tend to be concentrated in sewage sludge and are listed in the group of compounds found at the highest concentrations in these samples. Typical phthalate concentrations in treated sewage sludge are between 10 and 100 mg/Kg dry weight.[75] Petrovic and Barceló[121] analyzed several phthalate esters in sludge produced by several municipal sewage treatment plants. The concentrations found for DBP and DEHP were from 0.25 to 9.7 and from 8 to 27 mg/Kg dry weight, respectively. Berset et al.[77] analyzed different sludge samples and these found that the levels of various phthalate esters reflect its production volumes quite well. Clearly, DEHP was the dominant phthalate in all the sewage sludges (21 to114 mg/Kg), accounting for more than 90% of the total.

The use of sewage sludge in agriculture could lead to human exposure, either directly or through the introduction of these compounds into the food chain.[118] The European Union has presented an initiative for the purpose of improving the quality of sludge for recycling. This includes the establishment of concentration limits for organic compounds, including DEHP, which should not be exceeded for sewage sludge used in agricultural land.[156] In an interesting study, Vikelsoe et al. measured the concentration of phthalates in depth profiles of eight differently treated soils.[118] A significant correlation was found between the concentration of phthalates in the soil profiles, and the treatment method. Heavy sludge amendment leads to significant concentrations of phthalates in soil, which persists even eight years after the amendment had ceased. The maximum concentrations of DBP and DEHP found in this study exceeded the recommended Danish soil quality criteria.[44]

Many building materials contain important amounts of phthalates and inhalation exposure to these compounds can constitute one of the main intake sources for humans. In addition, people working in industrial plants producing plasticizers or living near such plants may be exposed to levels exceeding the TWA limits.[46] In consequence, the levels of phthalates in air, especially in indoor environments, must be controlled. In various studies reported in the literature, the phthalate concentration found in indoor air[125,126] demonstrated that the exposure to phthalate esters via indoor air inhalation could constitute a significant contribution to the total daily intake.

Methods for analysing organic pollutants in environmental samples have mostly focused on phthalate esters, but degradation products have rarely been considered. Nevertheless, some ecotoxicological studies have shown that some of the degradation products, i.e., monoesters, may be toxic for mammals. Recently, Jonsson et al.[85] studied the levels of phthalates in 17 leachates from landfills in Europe. The concentrations found were from 1 to 460 $\mu g/l$. Some degradation products, including monoesters and phthalic acid were also found in concentrations of 1 to 20 $\mu g/l$ and 2 to 880 $\mu g/l$, respectively. Hashizume et al.[86] investigated the level of phthalate ester pollution in various water samples and also the biodegradation products and pathways of phthalate esters in river water.

In a recent study, Clark et al.[157] have compiled and analyzed measured concentration data of six phthalate esters in seven environmental media including water, sediment, soil, air, dust, food, wastewater, sewage sludge, and rainwater. The data are predominantly from Europe, the United States, Canada, and Japan. The complete database, with references, was presented in a report to the American Chemistry Council. The reported concentrations vary widely; as an example, the overall mean concentration of DMP in surface water in Canada (1.40 $\mu g/l$) is three orders of magnitude higher than that found in the U.S. (0.0017 $\mu g/l$). The authors consider that this wide distribution is due to several factors including analytical error, sample contamination, and proximity to a variety of past and present phthalate sources.

REFERENCES

1. Agency for Toxic Substances and Disease Registry (ATSDR), Toxicological Profile for di-(2-ethylhexyl) phthalate, U.S. Department of Health and Human Services, Public Health Service, Atlanta, 2002.
2. Agency for Toxic Substances and Disease Registry (ATSDR), Toxicological Profile for diethyl phthalate, U.S. Department of Health and Human Services, Public Health Service, Atlanta, 1995.
3. Agency for Toxic Substances and Disease Registry (ATSDR), Toxicological Profile for di-*n*-butyl phthalate, U.S. Department of Health and Human Services, Public Health Service, Atlanta, 2001.
4. Agency for Toxic Substances and Disease Registry (ATSDR), Toxicological Profile for di-*n*-octyl phthalate, U.S. Department of Health and Human Services, Public Health Service, Atlanta, 1997.
5. Lin, Z. P., Ikonomou, M. G., Jing, H., Mackintosh, C., and Gobas, F. A. P. C., Determination of phthalate ester congeners and mixtures by LC/ESI-MS in sediments and biota of an urbanized marine inlet, *Environ. Sci. Technol.*, 37, 2100–2108, 2003.
6. The Council of the European Communities, Council Regulation (EEC) 793/93 of 23 March 1993 on the Evaluation and Control of the Risks of Existing Substances, *Off. J. L.*, 084, 0001–0075, 1993.
7. Stanley, M. K., Robillard, K. A., and Staples, C. A., Introduction, *The Handbook of Environmental Chemistry*. Part Q. Phthalate Esters, Vol. 3, Staples, C. A., Ed., Springer-Verlag, Berlin, pp. 1–7, 2003.
8. PricewaterhouseCoopers, Eco-profile of High Volume Commodity Phthalate esters (DEHP/DINP/DIDP), European Council for Plasticizers and Intermediates (ECPI), January 2001.
9. Staples, C. A., Peterson, D. R., Parkerton, T. F., and Adams, W. J., The environmental fate of phthalate esters: a literature review, *Chemosphere*, 35, 667–749, 1997.
10. Syracuse Research Corporation, Measurement of the Water Solubility of Phthalate esters, Chemical Manufacturers Association, Washington, DC. U.S.A, 1983.
11. Turner, A. and Rawling, M. C., The behaviour of di(2-ethylhexyl) phthalate in estuaries, *Mar. Chem.*, 68, 203–217, 2000.
12. Cousin, I. T. and Mackay, D., Correlating the physical–chemical properties of phthalate esters using the three solubility' approach, *Chemosphere*, 41, 1389–1399, 2000.

13. Cousin, I. T., Mackay, D., and Parkerton, T. F., Physical–chemical properties and evaluative fate modelling of phthalate esters, In *The Handbook of Environmental Chemistry,* Part Q. Phthalate Esters, Vol. 3, Staples, C. A., Ed., Springer-Verlag, Berlin, pp. 57–84, 2003.

14. Desideri, P., Lepri, L., Checchini, L., and Santianni, D., Organic compounds in surface and deep Antarctic snow, *Int. J. Environ. Anal. Chem.*, 55, 33–46, 1994.

15. Long, J. L. A., House, W. A., Parker, A., and Rae, J. E., Micro-organic compounds associated with sediments in the Humber river, *Sci. Tot. Environ.*, 210/211, 229–253, 1998.

16. Furtmann, R. N. K., Phthalates in the Aquatic Environment, Report no. 6/93, European Chemical Industry Council, European Council for Plasticisers and Intermediates (ECPI), Brussels, 1996.

17. Cousins, I. and Mackay, D., Review of EUSES modelling for di-2-ethylhexyl phthalate (DEHP), Final Report, CEMC Report No. 200101, European Chemical Industry Council (CEFIC), 2001.

18. Comisión de las Comunidades Europeas. Comunicación de la Comisión al Consejo y al Parlamento Europeo: Aplicación de la Estrategia Comunitaria en Materia de Alteradores Endocrinos-Sustancias de las que se Sospecha Interfieren en los Sistemas Hormonales de Seres Humanos y Animales, COM (1999) 706. COM 262 final. Bruselas, 2001.

19. Harris, C. A. and Sumpter, J. P., The endocrine disrupting potential of phthalates, In *The Handbook of Environmental Chemistry, Part L, Endocrine Disruptors, Part I*, Vol. 3, Metzler, M., Ed., Springer-Verlag, Berlin, pp. 169–201, 2001.

20. International Programme on Chemical Safety (IPCS), Environmental Health Criteria 189: di-n-butyl phthalate, World Health Organization, Geneva, 1997.

21. International Programme on Chemical Safety (IPCS), Concise International Chemical Assessment Document 52: diethyl phthalate, World Health Organization, Geneva, 2003.

22. International Programme on Chemical Safety (IPCS), Environmental Health Criteria 131: diethylhexyl phthalate, World Health Organization, Geneva, 1992.

23. International Programme on Chemical Safety (IPCS), Concise International Chemical Assessment Document 17: butyl benzyl phthalate, World Health Organization, Geneva, 1997.

24. National Toxicology Program-Center for the Evaluation of Risks to Human Reproduction (NTP-CERHR), NTP-CERHR Monograph on the Potential Human Reproductive and Developmental Effects of di-*n*-butyl phthalate (DBP), U. S. Department of Health and Human Service, 2003.

25. National Toxicology Program-Center for the Evaluation of Risks to Human Reproduction (NTP-CERHR), NTP-CERHR Monograph on the Potential Human Reproductive and Developmental Effects of butyl benzyl phthalate (BBP), U.S. Department of Health and Human Service, 2003.

26. National Toxicology Program-Center for the Evaluation of Risks to Human Reproduction (NTP-CERHR), NTP-CERHR Monograph on the Potential Human Reproductive and Developmental Effects of di-*n*-hexyl phthalate (DnHP), U.S. Department of Health and Human Service, 2003.

27. National Toxicology Program-Center for the Evaluation of Risks to Human Reproduction (NTP-CERHR), Expert Panel Report on di(2-ethylhexyl)phthalate, U.S. Department of Health and Human Service, 2000.

28. National Toxicology Program-Center for the Evaluation of Risks to Human Reproduction (NTP-CERHR), NTP-CERHR Monograph on the Potential Human Reproductive and Developmental Effects of di-*n*-octyl phthalate (DNOP), U.S. Department of Health and Human Service, 2003.

29. National Toxicology Program-Center for the Evaluation of Risks to Human Reproduction (NTP-CERHR), NTP-CERHR Monograph on the Potential Human Reproductive and Developmental Effects of di-isononylphthalate (DINP), U.S. Department of Health and Human Service, 2003.

30. National Toxicology Program-Center for the Evaluation of Risks to Human Reproduction (NTP-CERHR), NTP-CERHR Monograph on the Potential Human Reproductive and Developmental Effects of di-isodecylphthalate (DIDP), U.S. Department of Health and Human Service, 2003.

31. European Commission, European Union Risk Assessment Report, 1st Priority List, Vol. 29, Dibutyl phthalate, Joint Research Centre Institute for Health and Consumer Protection, European Chemicals Bureau (ECB), Office for Official Publications of the European Communities, 2003.

32. European Commission, European Union risk assessment report, 2nd Priority List, Vol. 35, 1,2-benzenedicarboxylic acid, di-C8-10-branched alkyl esters, C9-rich and di-"isononyl" phthalate (DINP), Joint Research Centre Institute for Health and Consumer Protection, European Chemicals Bureau (ECB), Office for Official Publications of the European Communities, 2003.

33. European Commission, European Union risk assessment report, 2nd Priority List, Vol. 36, 1,2-benzenedicarboxylic acid, di-C9-11-branched alkyl esters, C10-rich and di-"isodecyl" phthalate (DIDP), Joint Research Centre Institute for Health and Consumer Protection European Chemicals Bureau (ECB), Office for Official Publications of the European Communities, 2003.

34. International Agency for Research on Cancer (IARC), IARC Monographs Programme on the Evaluation of Carcinogenic Risks to Humans, Some industrial chemicals, Vol. 77, World Health Organization, p. 41, 2000, http://193.51.164.11/htdocs/monographs/vol77/77-01.html.

35. International Agency for Research on Cancer (IARC), IARC Monographs Programme on the Evaluation of Carcinogenic Risks to Humans, Some chemicals that cause tumors of the kidney or urinary bladder in rodents and some other substances, Vol. 73, World Health Organization, 1999, p. 115, http://193.51.164.11/htdocs/monographs/vol73/73-04.html.

36. European Union Commission, Decision 2003/113/EC amending Decision 1999/815/EC concerning measures prohibiting the placing on the market of toys and childcare articles intended to be placed in the mouth by children under three years of age made of soft PVC containing certain phthalates, *Off. J. Eur. Commun. L.*, 46, 27–28, 2003.

37. OSPAR Commission, OSPAR list of chemicals for priority action, OSPAR convention for the protection of the marine environment of the North–East Atlantic, Meeting of the OSPAR Commission, Summary Record OSPAR 2002, OSPAR 02/21/1-E, Annex 5, Amsterdam, 24–28 June 2002.

38. European Union Commission, Decision No 2455/2001/EC of the European Parliament and of the Council of 20 November 2001 establishing the list of priority substances in the field of water policy and amending Directive 2000/60/EC, *Off. J. Eur. Commun.*, L., 331, 1–4, 2001.

39. U.S. Environmental Protection Agency, Introduction to water policy standards, Office of Water, 1999.

40. World Health Organization (WHO), Guidelines for drinking-water quality, 3rd edition (Draft), Chapter 8: Chemical Aspects, World Health Organization, pp. 254, 2003, http://www.who.int/docstore/water_sanitation_health/GDWQ/Updating/draftguidel/2003gdwq8.pdf.

41. U.S. Environmental Protection Agency (EPA), National primary drinking water regulations: maximum contaminant levels for organic contaminants, U.S. Environmental Protection Agency, Code of Federal Regulations 40 CFR 141.61, pp. 426–428, 2002.

42. U.S. Environmental Protection Agency (EPA), National primary drinking water regulations: maximum contaminant level goals for organic contaminants, Code of Federal Regulations 40 CFR 141.50, pp. 424–425, 2002.

43. van Wezel, A. P., van Vlaardingen, P., Posthumus, R., Crommentuijn, G. H., and Sijm, D.T.H.M., Environmental risk limits for two phthalates, with special emphasis on endocrine disruptive properties, *Ecotoxicol. Environ. Saf.*, 46, 305–321, 2000.

44. Danish Environmental Protection Agency, Guidelines on remediation of contaminated sites, Environmental Guidelines no. 7, 2002, http://www.mst.dk/udgiv/publications/2002/87-7972-280-6/pdf/87-7972-281-4.pdf.

45. U.S. Code, Clean Air Act, Section 112, Hazardous air pollutants, 42 USC 7412, 1990.

46. U.S. Occupational Safety and Health Administration (OSHA), OSHA general industry standards, Toxic and hazardous substances, Air contaminants, Code of Federal Regulations 29 CFR 1910.1000, http://www.setonresourcecenter.com/29CFR/1910/19101000.htm.

47. Method 506: determination of phthalate and adipate esters in drinking water by liquid–liquid extraction or liquid–solid extraction and gas chromatography with photoionization detection, In *Revision 1.1, National Exposure Research Laboratory, Office of Research and Development*, Munch, J. W., Ed., Environmental Protection Agency (EPA), Cincinnati, 1995.

48. Method 525.2: determination of organic compounds in drinking water by liquid–solid extraction and capillary column gas chromatography/mass spectrometry, In *Revision 2.0, National Exposure Research Laboratory, Office of Research and Development*, Munch, J. W., Ed., U.S. Environmental Protection Agency (EPA), Cincinnati, 1995.

49. U.S. Environmental Protection Agency (EPA), Methods for organic chemical analysis of municipal and industrial wastewater, Method 606-Phthalate ester, Code of Federal Regulations 40 CFR 141.136, Appendix A, pp. 89–99, 2001.

50. U.S. Environmental Protection Agency (EPA), Methods for organic chemical analysis of municipal and industrial wastewater, Method 625: base/neutrals and acids, Code of Federal Regulations 40 CFR 141.136, Appendix A, pp. 200–229, 2001.

51. U.S. Environmental Protection Agency (EPA), Method 1625 revision B, Semivolatile organic compounds by isotope dilution GC/MS, Code of Federal Regulations 40 CFR 141.136, Appendix A, pp. 287–310, 2002.

52. U.S. Environmental Protection Agency (EPA), Method 8061A. Phthalate esters by gas chromatography with electron capture detection (GC/ECD), Revision 1, In *Online Test Methods for Evaluating Solid Waste Physical/Chemical Methods (SW-846)*, Office of Solid Waste, 1996, http://www.epa.gov/epaoswer/hazwaste/test/pdfs/8061a.

53. U.S. Environmental Protection Agency (EPA), Method 8270D, Semivolatile organic compounds by gas chromatography/mass spectrometry (GC/MS), Revision 4, In *Online Test Methods for Evaluating Solid Waste Physical/Chemical Methods (SW-846)*, Office of Solid Waste, 1998, http://www.epa.gov/epaoswer/hazwaste/test/pdfs/8270d.

54. U.S. Environmental Protection Agency (EPA), Method 8410, Gas chromatography/fourier transform infrared (GC/FT-IR) spectrometry for semivolatile organics: capillary column, Revision 0, In *Online Test Methods for Evaluating Solid Waste, Physical/Chemical Methods (SW-846)*, Office of Solid Waste, 1994, http://www.epa.gov/epaoswer/hazwaste/test/pdfs/8410.

55. U.S. Environmental Protection Agency (EPA), Method 3510C, Separatory funnel liquid–liquid extraction, Revision 3, In *Online Test Methods for Evaluating Solid Waste Physical/Chemical Methods (SW-846)*, Office of Solid Waste, 1996, http://www.epa.gov/epaoswer/hazwaste/test/pdfs/3510c.

56. U.S. Environmental Protection Agency (EPA), Method 3520C, Continuous liquid–liquid extraction, Revision 3, In *Online Test Methods for Evaluating Solid Waste Physical/Chemical Methods, (SW-846)*, Office of Solid Waste, 1996, http://www.epa.gov/epaoswer/hazwaste/test/pdfs/3520c.

57. U.S. Environmental Protection Agency (EPA), Method 3535A, Solid-phase extraction (SPE), Revision 1, In *Online Test Methods for Evaluating Solid Waste Physical/Chemical Methods (SW-846)*, Office of Solid Waste, 2000, http://www.epa.gov/epaoswer/hazwaste/test/pdfs/3535a.

58. U.S. Environmental Protection Agency (EPA), Method 0010: Modified Method 5 Sampling Train, Revision 0, In *Online Test Methods for Evaluating Solid Waste Physical/Chemical Methods (SW-8469)*, Office of Solid Waste, 1986, http://www.epa.gov/epaoswer/hazwaste/test/pdfs/0010.

59. U.S. Environmental Protection Agency (EPA), Method 3542: Extraction of Semivolatile Analytes Collected Using Method 0010 (Modified Method 5 Sampling Train), Revision 0, In *Online Test Methods for Evaluating Solid Waste Physical/Chemical Methods (SW-846)*, Office of Solid Waste, 1996, http://www.epa.gov/epaoswer/hazwaste/test/pdfs/3542.

60. U.S. Environmental Protection Agency (EPA), Method 3540C, Soxhlet Extraction, Revision 3, In *Online Test Methods for Evaluating Solid Waste Physical/Chemical Methods (SW-846)*, Office of Solid Waste, 1996, http://www.epa.gov/epaoswer/hazwaste/test/pdfs/3540c.

61. U.S. Environmental Protection Agency (EPA), Method 3541: Automated Soxhlet Extraction, Revision 0, In *Online Test Methods for Evaluating Solid Waste Physical/Chemical Methods (SW-846)*, Office of Solid Waste, 1994, http://www.epa.gov/epaoswer/hazwaste/test/pdfs/3541.

62. U.S. Environmental Protection Agency (EPA), Method 3545A: Pressurized Fluid Extraction (PFE), Revision 1, In *Online Test Methods for Evaluating Solid Waste Physical/Chemical Methods (SW-846)*, Office of Solid Waste, 2000, http://www.epa.gov/epaoswer/hazwaste/test/pdfs/3545.

63. U.S. Environmental Protection Agency (EPA), Method 3550B: Ultrasonic Extraction, Revision 2, In *Online Test Methods for Evaluating Solid Waste Physical/Chemical Methods*, SW-846, Office of Solid Waste, 1996, http://www.epa.gov/epaoswer/hazwaste/test/pdfs/3550b.

64. U.S. Environmental Protection Agency (EPA), Method 3580A, Waste Dilution, Revision 1, In *Online Test Methods for Evaluating Solid Waste Physical/Chemical Methods (SW-846)*, Office of Solid Waste, 1992, http://www.epa.gov/epaoswer/hazwaste/test/pdfs/3550a.

65. U.S. Environmental Protection Agency (EPA), Method 3610B: Alumina Cleanup, Revision 2, In *Online Test Methods for Evaluating Solid Waste Physical/Chemical Methods*, SW-846, Office of Solid Waste, 1996, http://www.epa.gov/epaoswer/hazwaste/test/pdfs/3610b.

66. U.S. Environmental Protection Agency (EPA), Method 3620B: Florisil Cleanup, Revision 2, In *Online Test Methods for Evaluating Solid Waste Physical/Chemical Methods (SW-846)*, Office of Solid Waste, 1996, http://www.epa.gov/epaoswer/hazwaste/test/pdfs/3620b.

67. U.S. Environmental Protection Agency (EPA), Method 3640A: Gel-Permeation Cleanup, Revision 1, In *Online Test Methods for Evaluating Solid Waste Physical/Chemical Methods (SW-846)*, Office of Solid Waste, 1994, http://www.epa.gov/epaoswer/hazwaste/test/pdfs/3640a.

68. U.S. Environmental Protection Agency (EPA), Method 3660B, Sulfur Cleanup, Revision 2, In *Online Test Methods for Evaluating Solid Waste Physical/Chemical Methods (SW-846)*, Office of Solid Waste, 1996, http://www.epa.gov/epaoswer/hazwaste/test/pdfs/3660b.

69. U.S. Environmental Protection Agency (EPA), 40 CFR parts 136, 141, and 143, Guidelines establishing test procedures for the analysis of pollutants under the Clean Water Act; National Primary Drinking Water Regulations; and National Secondary Drinking Water Regulations; Methods Update; Final Rule, Federal Registry, Vol. 67, No. 203, October, 23, pp. 65220–65253, 2002.

70. American Public Health Association (APHA), American Water Works Association (AWWA), Water Environment Federation (WEF), Method 6410B: Extractable Base/Neutrals and acids, In *Standard Methods for the Examination of Water and Wastewater*, 18th ed., Greenberg, A. E., Clesceri, L. S., and Eaton, A. D., Eds., American Public Health Association, Washington, 1992.

71. U.S. Department of Energy (DOE), DOE methods for evaluating environmental and waste management samples, Method OM100R — analysis of semivolatile organic compounds using capillary gas chromatography with ion trap mass spectrometric detection (Draft), Pacific Northwest National Laboratory, http://infotrek.er.usgs.gov/intermedia/nemi_port_read/mediaget/nemi_get_blob/42.

72. U.S. Institute for Occupational Safety and Health (NIOSH). Dibutyl Phthalate and Di(2-Ethylhexyl) Phthalate: Method 5020.2, In *Dibutyl Phthalate and Di(2-Ethylhexyl) Phthalate: Method 5020.2, NIOSH manual of analytical methods (NMAM)*, 4th ed., Cassinelli, M. and O'Connor, P. F., Eds., 1994.

73. U.S. Occupational Safety and Health Administration (OSHA), Method 104: Dimethyl phthalate (DMP), diethyl phthalate (DEP), dibutyl phthalate (DBP), di-2-ethylhexyl phthalate (DEHP), di-n-octyl phthalate (DNOP), http://www.osha-slc.gov/dts/sltc/methods/organic/org104/org104.html.

74. U.S. Occupational Safety and Health Administration (OSHA), Method PV2076: Dihexyl Phthalate (branched and linear isomers), di-n-hexyl phthalate, http://www.osha-slc.gov/dts/sltc/methods/partial/pv2076/pv2076.html.

75. David, F., Sandra, P., Tienpont, B., and Vanwalleghem, F., Analytical methods review, In *The Handbook of Environmental Chemistry,* Part Q, Phthalate Esters, Vol. 3, Staples, C. A., Ed., Springer-Verlag, Berlin, 2003.

76. Gómez-Hens, A. and Aguilar-Caballos, M. P., Social and economic interest in the control of phthalic acid esters, *Trends Anal. Chem. (TrAC)*, 22, 847–857, 2003.

77. Berset, J. D. and Etter-Holzer, R., Determination of phthalates in crude extracts of sewage sludges by high-resolution capillary gas chromatography with mass spectrometric detection, *J. AOAC Int.*, 84, 383–391, 2001.

78. Luks-Betlej, K., Popp, P., Janoszka, B., and Paschke, H., Solid-phase microextraction of phthalates from water, *J. Chromatogr. A.*, 938, 93–101, 2001.

79. Cortazar, E., Zuloaga, O., Sanz, J., Raposo, J. C., Etxebarria, N., and Fernández, L. A., MultiSimplex optimization of the solid-phase microextraction-gas chromatographic-mass spectrometric determination of polycyclic aromatic hydrocarbons, polychlorinated biphenyls and phthalates from water samples, *J. Chromatogr A.*, 978, 165–175, 2002.

80. Casajuana, N. and Lacorte, S., Presence and release of phthalic esters and other endocrine disrupting compounds in drinking water, *Chromatographia*, 57, 649–655, 2003.

81. Brossa, L., Marcé, R. M., Borrull, F., and Pocurull, E., Determination of endocrine-disrupting compounds in water samples by Online solid-phase extraction-programmed-temperature vaporisation-gas chromatography-mass spectrometry, *J. Chromatogr A.*, 998, 41–50, 2003.

82. Hyötyläinen, T., Grob, K., Biedermann, M., and Riekkola, M. L., Reversed phase HPLC coupled online to GC by the vaporizer/precolumn solvent split/gas discharge interface; analysis of phthalates in water, *J. High Resolut. Chromatogr.*, 20, 410–416, 1997.

83. Brossa, L., Marcé, R. M., Borrull, F., and Pocurull, E., Application of online solid-phase extraction-gas chromatography-mass spectrometry to the determination of endocrine disruptors in water samples, *J. Chromatogr. A.*, 963, 287–294, 2002.

84. Prokupková, G., Holadová, K., Poustka, J., and Hajslová, J., Development of a solid-phase microextraction method for the determination of phthalic acid esters in water, *Anal. Chim. Acta*, 457, 211–223, 2002.

85. Jonsson, S., Ejlersson, J., Ledin, A., Mersiowsky, I., and Svensson, B. H., Mono- and diesters from *O*-phthalic acid in leachates from different European landfills, *Water Res.*, 37, 609–617, 2003.

86. Hashizume, K., Nanya, J., Toda, C., Yasui, T., Nagano, H., and Kojima, N., Phthalate esters detected in various water samples and biodegradation of phthalates by microbes isolated from river water, *Biol. Pharm. Bull.*, 25, 209–214, 2002.

87. Furtmann, K., Phthalates in surface water — a method for routine trace level analysis, *Fresenius J. Anal. Chem.*, 348, 291–296, 1994.

88. Ruminski, J. K., Dejewska, B., and Wojtanis, J., Environmental Research, Part I, Investigation on dioctyl phthalate (DEHP) pollution in soil and surface water near Wabrzezno (Torun district), Poland, *J. Environ. Studies*, 4, 65–69, 1995.

89. Brumley, W. C., Shafter, E. M., and Tillander, P. E., Determination of phthalates in water and soil by tandem mass spectrometry under chemical ionisation conditions with isobutene as reagent gas, *J. AOAC Int.*, 77, 1230–1236, 1994.

90. López-Avila, V. and Milanes, J., Single-laboratory evaluation of method 8060 for the determination of phthalates in environmental samples, *J. AOAC*, 74, 793–808, 1991.

91. Law, R. J., Fileman, T. W., and Matthiessen, P., Phthalate esters and other industrial organic chemicals in the north and irish seas, *Water. Sci. Tech.*, 24, 127–134, 1991.

92. Fatoki, O. S. and Vernon, F., Phthalate esters in rivers of the greater Manchester area, *U.K. Sci. Total Environ.*, 95, 227–232, 1990.

93. Casas-Hernández, A. M., Aguilar-Caballos, M. P., and Gómez-Hens, A., Application of time-resolved luminescence methodology to the determination of phthalate esters, *Anal. Lett.*, 36, 1017–1027, 2003.

94. Gimeno, R. A., Marcé, R. M., and Borrull, F., Determination of plasticizers by high-performance liquid chromatography and atmospheric pressure chemical ionization mass spectrometry in water and sediment samples, *Chromatographia*, 58, 37–41, 2003.

95. Cai, Y., Jiang, G., and Liu, J., Solid-phase extraction of several phthalate esters from environmental water samples on a column packed with polytetrafluoroethylene turnings, *Anal. Sci.*, 19, 1491–1494, 2003.

96. Jonsson, S. and Borén, H., Analysis of mono- and diesters of *O*-phthalic acid by solid-phase extractions with polystyrene–divinylbenzene-based polymers, *J. Chromatogr. A.*, 963, 393–400, 2002.

97. Fatoki, O. S. and Noma, A., Solid phase extraction method for selective determination of phthalate esters in the aquatic environment, *Water, Air Soil Pollut.*, 140, 85–98, 2002.

98. Lourciro, I., de Andrade Brüning, I. M. R., and Moreira, I., Phthalate contamination in potable waters of Rio de Janeiro City, *Progress in Water Resources, 3(Water Pollution VI)*, pp. 347–355, 2001.

99. Jara, S., Lysebo, C., Greibrokk, T., and Lundanes, E., Determination of phthalates in water samples using polystyrene solid-phase extraction and liquid chromatography quantification, *Anal. Chim. Acta*, 407, 165–171, 2000.

100. Inoue, K., Kawaguchi, M., Okada, F., Yoshimura, Y., and Nakazawa, H., Column-switching high-performance liquid chromatography electrospray mass spectrometry coupled with Online of extraction for the determination of mono- and di-(2-ethylhexyl) phthalate in blood samples, *Anal. Bioanal. Chem.*, 375, 527–533, 2003.

101. Koch, H. M., Rossbach, B., Drexler, H., and Angerer, J., Internal exposure of the general population to DEHP and other phthalates-determination of secondary and primary phthalate monoester metabolites in urine, *Environ. Res.*, 93, 177–185, 2003.

102. Blound, B. C., Milgram, K. E., Silva, M. J., Malek, N. A., Reidy, J. A., Needham, L. L., and Brock, J. W., Quantitative detection of eight phthalate metabolites in human urine using HPLC-APCI-MS/MS, *Anal. Chem.*, 72, 4127–4134, 2000.

103. Kato, K., Shoda, S., Takahashi, M., Doi, N., Yoshimura, Y., and Nakazawa, H., Determination of three phthalate metabolites in human urine using Online solid-phase extraction-liquid chromatography-tandem mass spectrometry, *J. Chromatogr. B.*, 788, 407–411, 2003.

104. Blount, B. C., Silva, M. J., Caudill, S. P., Needham, L. L., Pirkle, J. L., Sampson, E. J., Lucier, G. W., Jackson, R. J., and Brock, J. W., Levels of seven urinary phthalate metabolites in human reference population, *Environ. Health Perspect.*, 108, 979–982, 2000.

105. Koch, H. M., González-Reche, L. M., and Angerer, J., Online clean-up by multidimensional liquid chromatography-electrospray ionization tandem mass spectrometry for high throughput quantification of primary and secondary phthalate metabolites in human urine, *J. Chromatogr. B*, 748, 169–182, 2003.

106. Kato, K., Yamauchi, T., Higashiyama, K., and Nakazawa, H., High throughput analysis of di-(2-ethylhexyl)phthalate metabolites in urine for exposure assessment, *J. Liq. Chromatogr. Related Technol.*, 26, 2167–2176, 2003.

107. Pawliszyn, J., *Solid Phase Microextraction: Theory and Practice*, Wiley-VCH, New York, 1997.

108. Cai, Y., Jiang, G., and Liu, J., Solid-phase microextraction coupled with high performance liquid chromatography-UV detection for the determination of di-n-propyl-phthalate, di-iso-butyl-phthalate, and di-cyclohexyl-phthalate in environmental water samples, *Anal Lett*, 36, 389–404, 2003.

109. Saito, Y., Nojiri, M., Imaizumi, M., and Nakao, Y., Polymer-coated synthetic fibers designed for miniaturized sample preparation process, *J. Chromatogr. A*, 975, 105–112, 2002.

110. Peñalver, A., Pocurull, E., Borrull, F., and Marcé, R. M., Comparison of different fibers for the solid-phase microextraction of phthalate esters from water, *J. Chromatogr. A*, 922, 377–384, 2001.

111. Peñalver, A., Pocurull, E., Borrull, F., and Marcé, R. M., Determination of phthalate esters in water samples by solid-phase microextraction and gas chromatography with mass spectrometric detection, *J. Chromatogr. A*, 872, 191–201, 2000.

112. Kelly, M. T. and Larroque, M., Trace determination of diethylphthalate in aqueous media by solid-phase microextraction-liquid chromatography, *J. Chromatogr. A.*, 841, 177–185, 1999.

113. Polo, M., Macías, S., Salgado, C., Llompart, M., García-Jares C., and Cela, R., Analysis by solid-phase microextraction of phthalate esters in water samples, Extech 2002, *Advances in Extraction Technologies*, Paris, 2002.

114. Psillakis, E. and Kalogeranis, N., Hollow-fibre liquid-phase microextraction of phthalate esters from water, *J. Chromatogr. A*, 999, 145–153, 2003.

115. Lin, Z. P., Ikonomou, M. G., Jing, H., Mackintosh, C., and Gobas, F. A. P. C., Determination of phthalate ester congeners and mixtures by LC/ESI-MS in sediments and biota of an urbanized marine inlet, *Environ. Sci. Technol.*, 37, 2100–2108, 2003.

116. Rudel, R. A., Camann, D. E., Spengler, J. D., Korn, L. R., and Brody, J. G., Phthalates, alkylphenols, pesticides, polybrominated diphenyl ethers, and other endocrine-disrupting compounds in indoor air and dust, *Environ. Sci. Technol.*, 37, 4543–4553, 2003.

117. Wang, X. K., Guo, W. L., Meng, P. R., and Gan, J. A., Analysis of phthalate esters in air, soil and plants in plastic film greenhouse, *Chin. Chem. Lett.*, 13, 557–560, 2002.

118. Vikelsoe, J., Thomsen, M., and Carlsen, L., Phthalates and nonylphenols in profiles of differently dressed soils, *Sci. Total Environ.*, 296, 105–116, 2002.

119. McDowell, D. C. and Metcalfe, C. D., Phthalate esters in sediments near a sewage treatment plant outflow in Hamilton harbour, Ontario: SFE extraction and environmental distribution, *J. Great Lakes Res.*, 27, 3–9, 2001.

120. Rudel, R. A., Brody, J. G., Spengler, J. D., Vallarino, J., Geno, P. W., Sun, G., and Yau, A., Identification of selected hormonally active agents and animal mammary carcinogens in commercial and residential air and dust samples, *J. Air Waste Manage. Assoc.*, 51, 499–513, 2001.

121. Petrovic, M. and Barceló, D., Determination of anionic and non-ionic surfactants, its degradation products, and endocrine-disrupting compounds in sewage sludge by liquid chromatography/mass spectrometry, *Anal. Chem.*, 72, 4560–4567, 2000.

122. Chee, K. K., Wong, M. K., and Lee, H. K., Microwave extraction of phthalate esters from marine sediment and soil, *Chromatographia*, 42, 378–384, 1996.

123. Zurmühl, T., Development of a method for the determination of phthalate esters in sewage sludge including chromatographic separation from polychlorinated biphenyls, pesticides and polyaromatic hydrocarbons, *Analyst*, 115, 1171–1175, 1990.
124. Thuren, A. and Larsson, P., Phthalate esters in the Swedish atmosphere, *Environ. Sci. Technol.*, 24, 554–559, 1990.
125. Otake, T., Yoshinaga, J., and Yanigisawa, Y., Analysis of organic esters of plasticizer in indoor air by GC-MS and GC-FPD, *Environ. Sci. Technol.*, 35, 3099–3102, 2001.
126. Fischer, J., Ventura, K., Prokes, B., and Jandera, P., Method for determination of plasticizers in industrial emissions, *Chromatographia*, 37, 47–50, 1993.
127. Triantafillou, V. I., Akrida-Demertzi, K., and Demertzis, P. G., Migration studies from recycled paper packaging materials: development of an analytical method for rapid testing, *Anal. Chim. Acta*, 467, 253–260, 2002.
128. Rastogi, S. C., Gas chromatographic analysis of phthalate esters in plastic toys, *Chromatographia*, 47, 724–726, 1998.
129. Page, B. D. and Lacroix, G. M., The occurrence of phthalate ester and di-2-ethylhexyl adipate plasticizers in Canadian packaging and food sampled in 1985–1989: a survey, *Food Addit Contam.*, 12, 129–151, 1995.
130. Marín, M. L., López, J., Sánchez, A., Vilaplana, J., and Jiménez, A., Análisis of potentially toxic phthalate plasticizers used in toy manufacturing, *Bull. Environ. Contam. Toxicol.*, 60, 68–73, 1998.
131. George, C., Prest, H., A new approach to the analysis of phthalate esters by GC/MS, Application note, Agilent Technologies, March 2001.
132. Suzuki, T., Yaguchi, K., Suzuki, S., and Suga, T., Monitoring of phthalic acid monoesters in river water by solid-phase extraction GC-MS determination, *Environ. Sci. Technol.*, 35, 3757–3763, 2001.
133. Wahl, H. G., Hong, Q., Stübe, D., Maier, M. E., Häring, H. U., and Liebich, H. M., Simultaneous analysis of the di(2-ethylhexyl)phthalate metabolites 2-ethylhexanoic acid, 2-ethyl-3-hydroxyhexanoic acid and 2-ethyl-3-oxohexanoic acid in urine by gas chromatography-mass spectrometry, *J. Chromatogr. B*, 758, 213–219, 2001.
134. Egestad, B., Green, G., Sjöberg, P., Klasson-Wehler, E., and Gustafsson, J., Chromatographic fractionation and analysis by mass spectrometry of conjugated metabolites of bis(2-ethylhexyl)phthalate in urine, *J. Chromatogr.*, 677, 99–109, 1996.
135. Pietrogrande, M. C., Rossi, D., and Paganetto, G., Gas chromatographic-mass spectrometric analysis of di(2-ethylhexyl) phthalate and its metabolites in hepatic microsomal incubations, *Anal. Chim. Acta.*, 480, 1–10, 2003.
136. Kambia, K., Dine, T., Gressier, B., Germe, A. F., Luyckx, M., Brunet, C., Michaud, L., and Gottrand, F., High-performance liquid chromatography method for the determination of di(2-ethylhexyl) phthalate in total parenteral nutrition and in plasma, *J. Chromatogr. B.*, 755, 297–303, 2001.
137. Dine, T., Luyckx, M., Cazin, M., Brunet, C. I., Cazin, J. C., and Goudaliez, F., Rapid determination by high performance liquid chromatography pf di-2-ethylhexyl phthalate in plasma stored in plastic bags, *Biomed. Chromatogr.*, 5, 94–97, 1991.
138. Aignasse, M. F., Prognon, P., Stachowicz, M., Gheyouche, R., and Pradeau, D., A new simple and rapid HPLC method for determination of DEHP in PVC packing and releasing studies, *Int. J. Pharm*, 113, 241–246, 1995.
139. Fung, Y. S. and Tang, A. S. K., Liquid chromatographic determination of the plasticizers di(2-ethylhexyl) phthalate (DEHP) in PVC plastics, *Fresenius J. Anal. Chem.*, 350, 721–723, 1994.
140. Silva, M. J., Malek, N. A., Hodge, C. C., Reidy, J. A., Kato, K., Barr, D. B., Needham, L. L., and Brock, J. W., Improved quantitative detection of 11 urinary phthalate metabolites in humans using liquid chromatography-atmospheric pressure chemical ionization tandem mass spectrometry, *J. Chromatogr. B*, 789, 393–402, 2003.
141. Penner, N. A., Nesterenko, P. N., Ilying, M. M., Tsyurupa, M. P., and Davankov, V. A., Investigation of the properties of hypercrosslinked polyestirene as a stationary phase for high-performance liquid chromatography, *Chromatographia*, 50, 611–620, 1999.
142. Koch, H. M., Bolt, H. M., and Angerer, J., Di-(2ethylhexyl)phthalate (DEHP) metabolites in human urine and serum after a single oral dose of deuterium-labelled DEHP, *Arch. Toxicol.*, 78, 123–130, 2004.

143. Snell, R. P., Solid-phase extraction and liquid chromatographic determination of monophthalates and phthalide extracted from solution administration sets, *J. AOAC Int.*, 76, 531–534, 1993.

144. Shintani, H., Pretreatment and chromatographic analysis of phthalate esters, and its biochemical behaviour in blood products, *Chromatographia*, 52, 721–726, 2000.

145. Paris, I., Ruggieri, F., Mazzeo, P., and Carlucci, G., Simultaneous determination of di(2-ethylhexyl)phthalate and mono(2-ethylhexyl)phthalate in human plasma by high-performance liquid chromatography, *Anal. Lett.*, 36, 2649–2658, 2003.

146. Anderson, W. A. C., Barnes, K. A., Castle, L., Damant, A. P., and Scotter, M. J., Determination of isotopically labelled monoester phthalates in urine by high performance liquid chromatography-mass spectrometry, *Analyst*, 127, 1193–1197, 2002.

147. Kato, K., Silva, M. J., Brock, J. W., Reidy, J. A., Malek, N. A., Hodge, C. C., Nakazawa, H., Needham, L. L., and Barr, D. B., Quantitative detection of nine phthalate metabolites in human serum using reversed-phase high-performance liquid chromatography-electrospray ionization-tandem mass spectrometry, *J. Anal. Toxicol.*, 27, 284–289, 2003.

148. Hoppin, J. A., Brock, J. W., Davis, B. J., and Baird, D. D., Reproducibility of urinary phthalate metabolites in first morning urine samples, *Environ. Health Perspect.*, 110, 515–518, 2002.

149. Ong, C. P., Lee, H. K., and Li, S. F. Y., Separation of phthalates by micellar electrokinetic chromatography, *J. Chromatogr.*, 542, 473–481, 1991.

150. Takeda, S., Wakira, S., Yamade, M., Kawahara, A., and Higashi, K., Migration behaviour of phthalate esters in micellar electrokinetic chromatography with or without added methanol, *Anal. Chem.*, 65, 2489–2492, 1993.

151. Albert, K., Braumann, U., Tseng, L. H., Nicholson, G., Bayer, E., Spraul, M., Hofmann, M., Dowle, C., and Chippendale, M., Online coupling of supercritical fluid chromatography and proton high-field nuclear magnetic resonance spectroscopy, *Anal. Chem.*, 66, 3042–3046, 1994.

152. Pozlomek, E. J. and Eiceman, G. A., Solid-phase enrichment, thermal desorption, and ion mobility spectrometry for field screening of organic pollutants in water, *Environ. Sci. Technol.*, 26, 1313–1318, 1992.

153. Ius, A., Bacigalupo, M. A., Meroni, G., Pistillo, A., and Roda, A., Development of a time resolved fluoroinmmunoassay for phthalate esters in water, *Fresenius J. Anal. Chem.*, 345, 589–591, 1993.

154. National Primary Drinking Water Regulations; Federal Regulations Part 12, 40 CFR Part 141, U.S. Environmental Protection Agency, Washington, p. 395, July 1991.

155. Parkerton, T. F. and Konkel, W. J., Application of quantitative structure-activity relationships for assessing the aquatic toxicity of phthalate esters, *Ecotoxicol. Environ. Saf.*, 45, 61–78, 2000.

156. Feuille, A.M., Association Francaise de Normalization, Sewage Sludge Management, Revision of Directive 86/278/CE, pp. 1–14, 1999.

157. Clark, K., Cousins, I. T., and Mackay, D., Assesment of critical exposure pathways, In *The Handbook of Environmental Chemistry*, Part Q, Phthalate Esters, Vol. 3, Staples, C. A., Ed., Springer-Verlag, Berlin, 2003.

29 Humic Substance

Fengchang Wu and Congqiang Liu

CONTENTS

Humic substance (HS) is the most commonly naturally occurring organic material in the environment, and is considered to result from the chemical and biological degradation of plant and animal residues.[1,2] Being ubiquitous, heterogeneous, mixture, high molecular weight, yellow to black in color and refractory are its most apparent characteristic features. Traditionally, HS can be separated into three fractions: humin, humic acid (HA), and fulvic acid (FA).

HS represents the majority of dissolved organic matter in waters and of organic material in soils and sediments.[2,3] Its interactions with metals and organic compounds may modify their bioavailability, speciation, and toxicity, thus affecting their ultimate fate in the environment. Because of its biogeochemical significance, it has been of great interest to study the nature of HS. However, HS is in fact an extremely polydisperse mixture of macromolecules with ill-defined structures, variable molecular size, and subsequently uncertain and variable chemical properties, its formation is highly dependent on the composition of original plant material and origin of biological activities, and pathways of decomposition. It, to some extent, is still like mysterious material, and it is difficult to explore its structures, properties and environmental processes without thermally degrading, separating or fractionating it into smaller fractions based on a chemical or physical property. As a result, a variety of analytical methods has been developed to fractionate or degrade HS, including ultra filtration,[4] flow field-flow fractionation,[5] reversed-phase high-performance liquid chromatography (RP-HPLC),[6] immobilized metal ion affinity chromatography (IMAC),[7–9] gel electrophoresis,[3] size exclusion chromatography (SEC),[10,11] and pyrolysis gas chromatography (Py–GC).[2,12] The advantages, limitations, and applications of these methods have been currently reviewed.[3] The principles and applications of some of these methods are summarized in Table 29.1.

Among these methods, liquid chromatographic separation and analysis have become the major chromatographic method for HS. While among the liquid chromatographic methods, those based on

TABLE 29.1
Comparison of Main Chromatographic Methods (RP-HPLC, IMAC, SEC, and GC–MS):
Principle and Main Applications

	Theory or Principle	Main Applications	References
RP-HPLC	Solvophobic interactions between the solute and the stationary phase	Fractionation and characterization of HS	3,6,23–31
IMAC	Affinity to immobilized metal ions or compounds	Fractionation of HS, its characterization, and its complexation with trace metals	7–9
SEC	Size exclusion of nonfitting molecules in stationary phase	Calculation of molecular size and its distribution of HS, and their characterization	3,10,11,33–39
GC–MS	Separated by GC and identified by MS	Identification of chemical structure of HS	3,49,50

size exclusion effect play the most important role. They have been not only used for the calculation of molecular size, but also widely coupled with other offline or online advanced analytical methods, e.g., UV–Vis absorbance spectroscopy, MS, ICP-MS, fluorescence, and dissolved organic carbon (DOC) detectors, whose coupling provides new analytical windows to investigate the nature of HS and its binding with trace metals.

In this chapter, we first discuss HS isolation procedure, and then focus on its four major chromatographic methods: RP-HPLC, IMAC, SEC, and Py–GC. Their advances and applications since 1990, in terms of the understanding of physical and chemical properties and structures of HS, are mainly stressed.

I. HUMIC SUBSTANCE ISOLATION PROCEDURE

Although original water samples can apply for some techniques, the isolation procedure is essential for better HS characterization with most analytical techniques. The traditional HS isolation method probably is XAD adsorption, which has been widely used for decades in the HS isolation from waters, soils, and organisms.[1] This method can not only isolate HS, but also further separate HS into two fractions (humic acid and fulvic acid). Detailed isolation procedures have been reviewed in previous articles.[1,3] XAD™ resins are styrene–divinylbenzene or methyl methacrylate polymer with various hydrophobicities and cross linkages. The resins adsorb dissolved organic matter mainly by hydrophobic binding or weak interactions such as Van Der Waals force, but the exact mechanism of adsorption is still unknown.

The adsorption capacity and efficiency of XAD resins depend on the surface area and pore size of the resins used. Among various XAD resins, XAD-2 and XAD-8 seem to be often used in the isolation of seawater and aquatic HS.[1] XAD-2 has been widely used in HS isolation of seawater, but the recovery of organic fractions was very low.[13] By using XAD-8 and XAD-4 in tandem or mixed resins, the efficiency can be improved.[14] It was reported that XAD-16 and XAD-2010 resins had better adsorption properties than XAD-4 and XAD-2.[13] Better fractionation of HS can be achieved using a gradient elution. Despite the frequent use of the XAD method, its disadvantages should be always kept in mind. First, in the isolation procedure, water samples have to be adjusted to pH 2 and 12 using HCl and NaOH, and these extreme chemical conditions may potentially alter the nature of HS.[15] Second, usually only less than 40 to 60% of DOM can be recovered as HS using this method. Third, the resin-based system requires extensive precleaning with solvent to reduce the bleeding of organics, and resins are conditioned with strong acids and bases. These processes are not only time consuming, but also create a considerable amount of hazardous waste.[16]

In addition to the standard XAD resins, other materials such as diethylaminoethyl (DEAE)-cellulose, a weak anion exchanger, have also been developed for HS isolation. The DEAE-cellulose is particularly convenient for HS isolation from large volumes of samples.[17,18] Studies have demonstrated that DEAE isolates and the main XAD fractions consist of similar organic compounds.[17-19] DEAE-cellulose has several advantages over the macroporous XAD resins in that it allows a higher flow rate, it does not require preacidification of the water samples, and the absorption efficiency is relatively high. About 76% of the adsorbed HS can be recovered by elution with 0.1 M NaOH.[18] However, if pure HS samples are required, further purified procedures such as reprecipitation and HCl/HF treatment for removal of inorganic impurities have to be carried out with both methods.[20]

II. REVERSED-PHASE HIGH-PERFORMANCE LIQUID CHROMATOGRAPHY

This is a chromatographic technique where retention is governed by solvophobic interactions between the solute and the stationary phase.[21,22] Its first application for HS was reported in 1984.[23] Since then, numerous applications and review articles have been prepared by many authors.[3,6,24-32]

Table 29.2 summarizes the experimental conditions (solvent and mobile phases, columns and detectors, etc) used in RP-HPLC for studies of HS. Most conditions are originally used for the analyses of organic acids with low molecular weight. For instance, the mobile phases are usually prepared from water, methanol, acetonitrile, or tetrahydrofuran, and pH is mostly buffered at pH6 to 8 by phosphate or milli-Q water. The packing materials used in RP columns have pore diameters of 4 to 10 nm, the estimated exclusion limit for typical bare silica having 10-nm averages pore diameter which is close to a relative molecular size of 5000 Da. It has been reported that RP-HPLC may suffer some limitations when these typical conditions are applied to the investigation of HS.[3,30] First, HS consists of polyfunctional macromolecules; the retention behavior of large and small molecules may be different. Second, macromolecules may undergo conformational and structural changes or aggregation within the changing mobile-phase environment and after interaction with the stationary phase. The molecular size of aquatic HS may change in hydrophobic interaction chromatography. Third, size exclusion effects may exist in the separation process, since HS has a high molecular weight range. Preuße et al.[28] reported that recovery, chromatography, and the abundance of different fractions of HS depend on injected sample amounts in RP-HPLC, and they suggested that a RP-HPLC standard method must include injected sample volume and sample concentration consistent with the usual parameters, e.g., pH, ionic strength, elution gradient, and flow rate. Linear gradient, commonly employed in HPLC systems does not seem effective for HS fractionation; this is probably because HS consists of a great variety of components with continuous polar characteristics. Thus, a stepwise gradient is usually used to unfold macromolecules into several constituents in application studies for HS.

To eliminate the effect of interactions between molecules and stationary phase, pH suppression of columns was used. For example, Saleh et al.[6] demonstrated that better resolution can be obtained when using acidic pH 4 and a Novapak column compared to pH 7. However, in ion-suppression chromatography, HS may be highly charged and could be excluded from analyses due to the size exclusion effect. Instead, Smith and Warwick[25] used ion-pair chromatography together with a wide-pore polymeric RP column to separate fulvic acids. Tetra butyl ammonium hydroxide was added in the buffer as an ion-pair reagent, which reacts with solvated acid and stationary phase to produce an ion pair. The ion pair has a stronger hydrophobic character than that of the original acid, thus limiting the effect of pH changes on the stationary phase and solvated molecule. Hutta and Gora[30] demonstrated that dimethyl formamide (DMF) can be used as a solvent to improve the surface interactions of the analytic due to its

TABLE 29.2
Experimental Conditions in the RP-HPLC Application for HS Studies

Analyte	Column	Mobile Phase	pH	Flow Rate	Detectors	Aim of the Study	References
Soil fulvic acid tC18 Sep-Pak processed	5 μm octadecyl-bonded silica	Acetonitrile and 0.05 M NaAc or 0.004 M KCl buffer	7.0	0.2 to 0.5 ml min^{-1}	UV 254 nm	Characterization of HSs	26,27
Sedimentary fulvic acid	μBondapack C18	2-Propanol and water	8.0	1.0 to 1.5 ml min^{-1}	UV 254 nm	Characterization of HSs	23
Humic acid in environmental samples	ODS column	Mixture of methanol, tetrahydrofuran and acetonitrile, and 0.003% ammonia solution	8.0	1.0 ml min^{-1}	Fluorescence (Ex/Em 340/418 nm)	Determination of HSs	24
Suwannee River fulvic acid	Novapak C18–5 μm, Hypersil ODS C18	Acetonitrile and water, or pH 7 phosphate or acetic acid	7.0	0.3 to 0.5 ml min^{-1}	Photodiode array detection	Fractionation of FA	6
Commercial Na-humic acid	Nucleosil C18 with particle size of 7 μm and pore size of 100 nm	Acetonitrile and water	—	0.6 to 1.0 ml min^{-1}	UV 254 nm	Retention behavior of HSs	28
Groundwater fulvic acid	PLRP-S (300 Å) column	Water–acetonitrile mixture and tetrabutyl ammonium hydroxide in phosphate buffer	7.0	—	Photodiode array detection	Characterization of HSs	25
Aquatic fulvic acid	Nova-Pak C18, 4 μm	Acetonitrile and phosphate (pH = 6.8) buffer	6.8	2 ml min^{-1}	UV 254 nm and fluorescence detection	Characterization of HS fractions	29
Commercial humic acids	LiChroCART column filled by wide pore octadecylsilica LiChrospher WP 300 RP-18, 5 μm spherical particles	Dimethylformamide (DMF) and phosphate buffer (pH-3.0–0.05 mM) containing 1% DMF	3.0	0.5 to 1.0	UV absorbance and fluorescence detection	Separation of HSs	30

excellent solvating and disaggregating properties for HS, and the use of a wide-pore RP sorbent can greatly eliminate the influence of size exclusion effects. They also reported that distinct features of HS can be obtained using stepwise gradients of DMF even at trace concentration level with excellent reproducibility within the detection limit of 3.3 μg ml^{-1}. However, the effect of pH changes on stationary phase and solvated molecules still needs further investigation.

Recent studies have shown that RP-HPLC is a useful tool, although not ideal, for the separation and characterization of HS of various origins when coupled or combined with other analytical methods, e.g., photodiode array detection (PDA) and three-dimensional excitation emission fluorescence matrix (3DEEM) detection. RP-HPLC with fluorescence detection was first used to determine the concentration of humic acid in environmental samples, e.g., coral skeletal matter, sea-water, river water, soils, and plant matter. The detection limit reached 15 ng with a relative standard deviation of 1.9%.[24] Hayase and Tsubota[23] fractionated HS by ultrafiltration, the fractions were characterized by gradient RP-HPLC on a μBondapack C18 column, and they found that fulvic acids extracted from marine sediments exhibited an increasing hydrophilic character with increasing molecular size. Further, Wu et al.[29] characterized fulvic acids, first fractionated by IMAC, by gradient RP-HPLC on a C18 column and by SEC, the results also showed that the fractions with higher affinity had higher molecular size and exhibited stronger hydrophobic character.

Saleh et al.[6] studied Suwannee River fulvic acids (SRFA) using stepwise RP-HPLC and PDA detection. They reported that the hydrophilic constituents represented about 40% fulvic acid and can be resolved into at least six peaks, while the hydrophobic constituents represented about 30% and can be resolved into 12 peaks. The retention and UV–Vis spectra of the resolved peaks were characteristic of aliphatic organic acids in the hydrophilic fraction and of conjugated aliphatic ketones and phenols in the hydrophobic fraction.

RP-HPLC and online 3DEEM detection were recently combined to study fluorescence properties of HS fractions as a function of polarity.[31] SRFA and Aldrich humic acid (AHA) can be separated by stepwise RP-HPLC into several fractions, a red-shifted fluorescence maximum pattern and an increase in molecular size were found when the fraction changing from hydrophilic to hydrophobic. In a similar study, Lombardi and Jardim[26] separated a soil and marine fulvic acid into several classes of fractions with different fluorophores, which were related to their distinct origins.

III. GEL PERMEATION CHROMATOGRAPHY/SIZE EXCLUSION CHROMATOGRAPHY

Gel permeation chromatography/size exclusion chromatography (GPC/SEC) is probably the most widely used method for characterization of HS. Its principle merits and limits have been well reviewed by many authors.[3,10,33] Here, we first briefly summarize these principle, merits, limits, and most importantly, solutions; then we focus on its applications for HS separation and characterization when coupled with other analytical methods, and discuss some recent results based on these new applications.

A. The Principle and Merits of GPC/SEC Methods

In GPC/SEC methods, the determination of the molecular weight or molecular weight distribution is based on the key presumption that the differential permeation of molecules or molecular sieving effect is solely responsible for the separation.[3] Variable path-length HS is created through the column packing material according to differences in molecular size. Large molecules cannot penetrate into the stationary phase pores and they are eluted in a shorter time. Smaller molecules, on the other hand, penetrate more deeply into the pores and thus their pathway is longer and this results in a longer retention time. However, the molecules larger than the average gel pore diameter are excluded from the column, and are beyond the method. It should be noted that the term "molecular weight" (MW, expressed in $Da = g\, mol^{-1}$) is commonly used interchangeably with the term molecular "size" ($g\, mol^{-1}\, m^{-3}$). However, molecular separation by GPC/SEC is based on hydro-dynamic volume rather than pure MW.

GPC/SEC was first developed with a soft gel, e.g., Sephadex.[34] Recently, this method has advanced dramatically with the development of HPLC instrumentation, column packing materials, and detectors. Several components comprise a standard HPLC system including solvent pumps, a degasser for the solvent, the column, and a detector. The most important component of the HPLC system is the packing material within the column, referred to as the stationary phase. Rigid inorganic material, e.g., modified silica and nonrigid organic material, e.g., polymer, is commonly used for the tiny gel beads to make up a high-performance SEC stationary phase. The gel bead diameters typically range from 3 to 10 μm with a smaller gel bead diameter having a greater resolution. The most crucial parameter determining the MW separation range and resolution is the average gel bead pore size and the distribution of its size. Most column manufacturers provide columns with average gel bead pore sizes of 60 to 300 Å, and the distribution of pore sizes depends upon individual manufacturers.[35] With the advance of the high performance mode of GPC/SEC, the separation efficiency has been enhanced significantly with more stability, reproducibility, and versatility.

B. LIMITATIONS AND SOLUTIONS

Although GPC/SEC is a powerful method for HS separation, limits are evident, and most of them are related to the complicated nature of HS itself. HS is an extreme polydisperse mixture with varying molecular size, substructures, and functional groups, its molecular size strongly depends on its surrounding conditions, e.g., pH, ionic strength, buffer, degree of complexation with metal ions, and degree of hydration. For example, functional groups can be dissociated/protonated when pH changes, dissociated functional groups carry negative charges, electrostatic repulsion between close negatively-charged sites causes the stretching of the HS molecules.[3] As a result, and the same HS may have different molecular size when conditions change. Therefore, the molecular size determined by GPC/SEC method is always condition dependent, and is generally called "apparent molecular size".[3,36]

GPC/SEC is also complicated by the no-size exclusion effects. In order to achieve true size exclusion, it is essential to eliminate those effects due to the chromatographic interactions. There are two main types of chromatographic interactions concerned: (a) ionic exclusion; and (b) specific adsorption. Ionic exclusion is the result of negative or positive charges on the stationary phase and the solute particle, creating repulsion between the solute particle and the partially charged stationary phase, resulting in a smaller retention time. This effect can be overcome by using a low molecular weight electrolyte such as NaCl or $NaClO_4$. Ions such as Na^+ and Cl^- or ClO_4^- can efficiently reduce the existing ionic repulsion between the stationary phase and solute particle. While specific adsorption is the result of interactions between the analyte and the stationary phase. The silica- or polymer-based support materials commonly used in the SEC columns usually carry negative charges and have hydrophobic sites, which can bind macromolecules of HS. Borate buffer is always chosen as an eluent since it eliminates the aromatic interactions between HS and the column materials.[29] Some researchers use urea or a 10% methanol eluent to improve conditions for materials with strong adsorption.[35] Thus, the choice of an appropriate eluent is important to ensure those no size exclusion effects are minimal.

The accurate calculation of molecular size is affected by the lack of molecular weight standards. For compounds, e.g., proteins, sodium polystyrene sulfonates (PSS) or dextrans, a relationship between molecular size and weight is easily established. However, due to the lack of molecular conformity among molecules of HS, it is difficult to derive suitable molecular weight standards. Hence, the molecular size of HS must be determined with standards that may not be exactly representative of the structure of HS. As a result, the molecular weight standards used in analysis simply act as an arbitrary basis of comparison in order to estimate representative values of molecular size. Despite the level of inaccuracy associated with using standards with different compositions such as proteins or PSS, relative changes and trends between macromolecules can be established. PSS standards are believed to best represent the structure of HS, and are recommended as calibration standards for the analyses of HS and other naturally dissolved organic matter.[10,33]

The accurate calculation of molecular size is also affected by the detection method chosen. There are three commonly used detectors in HPSEC, e.g., ultraviolet absorbance (UVA), fluorescence and DOC detectors. These detectors measure different properties of HS molecules, for example, UVA absorbance detector detects absorbing compounds (mostly π bonded molecules), and fluorescence detects fluorescing chromophores, while DOC detects all of the organic carbon. Although these properties are closely related to one another, molar absorptivity and fluorescence efficiency of different molecular size fractions in HS are not equal. Therefore, the calculation of molecular size using different detectors is quite different. It has been reported that fluorescence detection usually generates much smaller values than UVA detection as fluorescence efficiency increases with increasing molecular size of HS.[29,36] Chin et al.[10] reported that molecular size measured by HPSEC with the most commonly used UVA detector generated higher values than those measured with other methods, e.g., vapor pressure osmometry and field-flow fractionation. This is because larger molecular size fractions have a greater molar absorptivity than lower molecular size fractions, larger molecular size fractions appear to be more abundant than they actually are, and thus lower molecular size fractions are lower in concentration. The wavelength of the UV detector chosen also affects the determination of molecular size. O'Loughlin and Chin[37] reported that the calculated molecular size values increased with increasing wavelength chosen. It seems that UV absorbance and fluorescence detectors are inherently inaccurate to generate "true" molecular size values. UV absorbance detection tends to be biased towards the larger molecular size fractions, while fluorescence detection towards the smaller molecular size fractions. Her et al.[38,39] demonstrated the estimation with online DOC detection may better represent the molecules of HS since the DOC detection analyzes virtually all of the organic compounds in HS, and showed that HPSEC with DOC detection, for Suwannee River humic acid and fulvic acid, displayed higher averaged MW lower number-averaged MW as well as higher polydispersivity than HPSEC with UVA detection.

C. APPLICATIONS OF GPC/SEC METHODS

Although GPC/SEC has some limitations in the accurate calculation of molecular size or its distribution of HS, it has found wide applications (Table 29.3) in the research of HS, advancing our knowledge of the nature of HS, e.g., structural and compositional properties, when coupled with other detection methods, e.g., ICP-MS, electrospray ionization-mass spectrometry (ESI-MS), DOC analysis, and fluorescence.[31,38-45] This coupling offers several advantages for the study of HS over individual methods: (a) The fractionation of HS according to their hydrodynamic volume adds an additional dimension to ICP-MS, ESI-MS, DOC and fluorescence analyses, and provides new insight into structural and compositional information versus molecular weight. (b) The GPC/SEC separation provides the opportunity for the large molecular size fraction of HS to be analyzed using fluorescence and ESI-MS. In bulk HS, these fractions are difficult to ionize and analyze by the infuse approach in ESI-MS, and are less sensitive to measurement using fluorescence due to the low efficiency. (c) A separation of inorganic sample constituents from the organic compounds enables to study the complexation between trace metals and HS.

ICP-MS is a powerful analytical method for trace elements due to its unique combination of high selectivity and sensitivity, wide linear dynamic range, nearly interference free operation and multi element capabilities. Its coupling of SEC with other detection methods allows simultaneous

Chromatographic Analysis of the Environment

TABLE 29.3
Separation and Characterization of HS with the Aid of GPC/SEC Coupled with Other Online Detections

Analyzed Material	Column	Mobile Phase	Online Detectors	Aim of the Study	References
Aquatic HS	TSK G2000 SW column, and TSK HW 40(S)	2.5 mM phosphate buffer (pH 7.0)	UV–Vis, DAD, fluorescence, and DOC detectors	Compare HS with different isolated methods and origin	40
Aquatic HS	HEMA SEC BIO 300 column	Milli-Q water	UV absorbance at 254 nm, and ICP-MS with isotope dilution technique	Trace metal complexation with HS	41,42
Aquatic HS	Sephadex G-25 and G-75	0.01 M Tris–HCl buffer	ICP-MS with hydride generation technique	Methy-Hg binding to HS	43
Aquatic HS	TSK-50S column	Phosphate buffer (pH = 6.8, 0.025 M Na$_2$SO$_4$)	UVA and DOC detectors	Calculation of molecular size using DOC detection	38,39
HS from compost	Superdex Peptide column	0.01 M sodium pyrophosphate decahydrate or CAPS buffer (pH 10.3–10.4)	DAD and ICP-MS detectors	Elemental binding to HS	46
Aquatic HS	PL Aquagel-OH 30 SEC column	80/20 (v/v) Water/methanol mixture with 10 mM NH$_4$HCO$_3$	DAD and ESI-MS detectors	Structure of HS	45
HS from compost	Superdex HR 10/30 peptide column	0.01 M Tris–HCl buffer with 0.01 M NaCl (pH 8.0)	DAD and ICP-MS detectors	The effect of metal ions upon the molecular size distribution of HS	44
Aquatic HS	TSK-gel G2500PWXL column	Phosphate buffer (0.2 M NaCl, pH = 6.8)	DAD and 3DEEM fluorescence detectors	Separation and characterization of HS	31

element specific determination in addition to HS separation based on molecular size, thus allowing to study the speciation of trace elements, and their complexation with HS fractions.[41,43,46] For example, Sadi et al.[46] used HPSEC coupled with online UV–Vis absorbance and ICP-MS to study the complexation of metals with HS in compost extract, and they found the following affinity order for HS: Cu > Ni > Co > Pb > Cd > (Cr, U, Th) ≫ (As, Mn, Mo, Zn). Vogl and Heumann[41] used HPSEC coupled with online UV–Vis absorbance and isotope dilution mass spectrometric method (ICP-IDMS) to simultaneously determine the concentrations of trace elements and their binding with HS. O'Driscoll and Evans[43] reported the binding of freshwater humic and fulvic acids to methyl mercury using GPC with hydride generation ICP-MS. It was demonstrated that the binding of fulvic acids to MeHg increased at higher molecular size, but the binding of humic acids was not related to molecular size, and the reason they argued is that the largest molecules of the fulvic fraction shared some common characteristics with the humic fraction as the binding capacities became similar with increasing molecular size. Wrobel et al.[44] studied the effect of metal ions on the molecular size distribution of HS from municipal compost using HPSEC with online UVA absorbance and ICPMS. They demonstrated that molecular size of dissolved HS was reduced when two complexing agents (citrates and EDTA) were added; their results indicated that bridging between small molecules and complexation/chelation by individual molecules was involved in the binding of metal ions with HS.

A sensitive and fast method for direct determination of DOC using ICP-IDMS was developed.[42] This coupling with HPSEC offers some unique advantages for the study of complexation between trace metals and HS. First, ICP-IDMS used for online DOC detection is better than conventional DOC methods, since ICP-IDMS is matrix free, independent of the type of the dissolved organic compounds, and uses internal calibration methods, while the conventional DOC methods must be calibrated by external standard solutions. Second, the coupling enables quantitative measurement of metal complexes with HS. Despite this, only limited applications have been reported[41,42] probably due to the difficult setup of instrumental optimization for the C^{12} and C^{13} detection, the interference of inorganic carbon and the high background signal in the HPSEC and ICP-IDMS coupling system.

GPC/SEC coupled with fluorescence detection was often used to study fluorescence properties of HS as a function of molecular size. Artinger et al.[36] used GPC with UV–Vis absorbance and fluorescence detection to characterize groundwater humic and fulvic acids of different origin. However, the fluorescence detection at specific excitation/emission wavelength provides limited information and obtains poor separation resolution as compared to UVA detection. Variability in fluorescence properties among different MS fractions using offline fluorescence spectroscopy has been reported.[47,48] Wu et al.[31] used HPSEC with advanced online 3DEEM fluorescence detection to study the fluorescence properties of HS fractions as a function of molecular size. With the online 3DEEM detector, more fluorescence information, e.g., fluorescence intensity, fluorescence maximum pattern and water Raman Scattering, can be obtained in addition to the molecular size. It was reported that there existed subtle differences and trends in fluorescence maxima across all size classes of HS, and smaller molecular weight fractions tended to have a shorter wavelength fluorescence maxima. The results suggested that HPSEC might be better for characterizing major fulvic like fluorescence and smaller MS fractions, but not for those having humic- and protein-like fluorescence materials due to their strong hydrophobic nature.

HPSEC coupled with ESI-MS was also developed to study the structure of HS.[45] It was demonstrated that the MS scan analysis and fragmentation of molecular anions in the MS/MS model can provide novel insight into the structural information on the building blocks of HS, which helps to elucidate the structure of HS. The results suggested that high molecular weight fractions of HS consisted of several subunits that originated from the low molecular weight fractions, and a class of well-defined polycarboxylated molecules, which occurs in all fulvic acid fractions was observed.[45]

IV. IMMOBILIZED METAL ION OR CHELATE AFFINITY CHROMATOGRAPHY

IMAC is a separation technique that utilizes the differential affinity of macromolecules for immobilized or coated metal ions or metal oxides to effect their separation. This differential affinity is derived from the coordination of metal ions or oxides chelated by appropriately prepared adsorbent and electron donor groups on the surface of macromolecules. Since the interaction between the immobilized or coated metal ions and the electron donor functional groups has a readily reversible character, it can be utilized for adsorption and then be disrupted by changing conditions, e.g., pH or ion strength, or using competing ligands. Since its introduction in 1975, IMAC has been well known for its various applications in biochemical fields, particularly in the separation and purification of biotic macromolecules, proteins, and peptides. However, the application of IMAC to HS fractionation just begins.[7–9,29] At present, there are two types of IMAC commonly used in HS separation, one is the conventionally used metal ions chelating absorbents, e.g., iminodiacetate (IDA) gel, and the other is the iron (III)-loaded ion exchanger.

In the conventionally used metal ions chelating IDA-Sepharose gel, histidine, cysteine, or tryptophan are the usual functional groups responsible for coordination with chelated metal ions, and Cu(II), Hg(II), Zn(II), Co(II), Ni(II), and Cd(II) ions are often used for chelated metal ions because of their strong affinity with HS, and acidic eluent (pH = 2) and glycine are used to elute the retained HS.[9] In a previous study for IMAC,[9] four steps were involved in the IMAC preparation: (1) conditioning of the IMAC column; (2) immobilized copper affinity adsorption of HS solution; (3) elution of HS from the IMAC column; and (4) regeneration of the column. Briefly, 20 ml of IDA-Sepharose gel (Pharmacia, Uppsala, Sweden) was used in a column as the solid matrix for IMAC. The column was equilibrated with Milli-Q water, and then borate buffer. 10 ml of 0.2 M solution of the relevant metal ion was loaded, and was subsequently rinsed with the borate buffer. HS solution was loaded at a flow rate of 20 ml cm^{-2} h^{-1}. After that, the column was first eluted with the borate buffer, and was then eluted with acidic eluent. After elution, the column was regenerated by washing with six column volumes of a solution containing 0.05 M ethylenediaminetetraacetic acid (EDTA) and 0.5 M NaCl to remove any metal ions from the column.

pH and ionic strength of the buffer solution have major effects on the retention of HS in IMAC for copper. It was reported that abundance and retention of retained HS increased with increasing pH and ionic strength,[9] indicating that salts can promote the complexation of HS in IMAC. Due to the extreme heterogeneous properties of HS, the salt promotion may be considered to result from the combined effect arising from the increased accessibility of hydrophobic residues of HS and charge shielding effects at higher salt concentration. In comparison with various metal ions for IMAC, abundance of retained HS and retention would increase when metal ion was changed from Cd^{2+} to Co^{2+} to Ni^{2+} to Cu^{2+}, consistent with the Irving-Williams series for the binding strength of bivalent metal ion complexes of the first transition series with a given ligand, and with proteins. Figure 29.1 shows the HS fractionation using stepwise decreasing pH gradient on IMAC for copper (II). It was reported that stronger affinity HS fractions eluted with lower pH-value eluent were associated with a higher molecular size, stronger hydrophobic character, and a higher proportion of absorbing and fluorescing materials at longer wavelength.[9,29]

In the iron (III)-loaded ion exchangers used for IMAC, iron hydroxides are often coated on the native cellulose fibres (Cell-Fe [III]). In a previous study for the fractionation of HS, Cell-Fe (III) was carefully prepared as described by Kuckuk and Burba.[7] Briefly, 20 g of highly purified cellulose powder stabilized by cross linking with formaldehyde was suspended in 300 ml Milli-Q water. Under magnetic stirring, 300 mg Fe (III) dissolved as FeCl$_3$ in 30 ml of 22 M HCl was added. The suspended cellulose was homogeneously coated with Fe (III)-hydroxide slowly precipitated by stepwise addition of 5 M NaOH solution up to the pH range 6 to 8. The combined Cell-Fe (III) was filled into a preparative HPLC column. In terms of binding mechanism, the strong hydrogen bridges between the hydroxo groups of the iron hydroxide and the phenolic and carboxylic functionalities of the HS molecules may play a key role, their binding strength depends on pH, and HS retained on

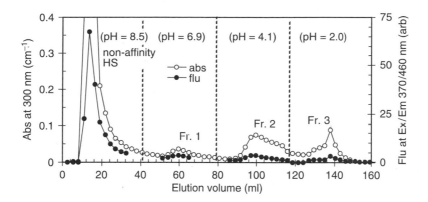

FIGURE 29.1 Chromatograms of HS macromolecules in Cu-IMAC experiments. Retained HS was eluted at decreasing pH-value eluent. Cited from Wu, F. C., Evans, R. D., and Dillon, P. J., *Anal. Chim. Acta*, 452, 85–93, 2002.

the column can be eluted using increasing pH.[7] Thus, fractions eluted include not only HS but also their metal complexes. HS was fractionated using stepwise increased pH values (pH 8 to 12, borate buffer), and the HS fractions were found to be different in their complexation capacity, absorbance ratio and Fourier transform infrared spectra.[7] Burba et al.[8] further developed a more advanced IMAC procedure coupled with ICP-MS for the study of fractionation and characterization of HS, and its complexation with metals.

Although IMAC is a powerful tool for HS fractionation, there are several potential problems needing further investigation: (1) decreasing or increasing pH eluent, for example, using pH 2 or 12 eluent, may cause structural or conformational changes of HS macromolecules; (2) competing ligand eluent may contaminate the fractionated HS itself and hamper further characterization; (3) IMAC was originally developed for the separation of small molecules, e.g., proteins or peptides, thus size-exclusion effects may exist when used for HS separation; (4) metal leaching still remains the common problem of IMAC and limits its further characterization though it can be solved using a post-column trap for leaching metal ions consisting of a column packed with a strong chelating adsorbent. Furthermore, IMAC is time consuming, particularly in the preparation and regeneration processes. Nevertheless, recent preliminary results[7–9,29] suggest that IMAC would find much broader use for prefractionation of complex HS prior to analyses, e.g., MS, CE, HPLC, and ICP-MS, and can be certainly used to study the interactions between metals and HS, and provide some insight on the information of HS fractions which have affinity for metals.

V. GAS CHROMATOGRAPHY–MASS SPECTROMETRY

HS is generally recalcitrant to any analytical approach, and its chemical structure can only be analyzed after it is broken into low molecular weight compounds by some kinds of degradation. Among the various methods, pyrolysis–gas chromatography–mass spectrometry (Py–GC–MS) is currently most commonly used, in which HS is thermally degraded by pyrolysis, the pyrolysate is separated by a gas chromatogram column, and identified by mass spectrometry.

However, Py–GC–MS is not perfect from the beginning and suffers limitations. Those pitfalls and limitations have been stressed in previous review articles.[3,49,50] Briefly, those limitations are not from the GC separation or MS detection methods themselves, but mainly from the analytical pyrolysis process; and pyrolysate only partially reflects the structure of the original building blocks. It has been well proven that the naturally occurring units can be altered before or after their breakdown from macromolecules due to the thermal reactions and configuration of the pyrolysis units in chromatographic systems.[49,50] Despite its limitations, Py–GC–MS has been greatly improved in the last decades and some solutions have been found, and demonstrated as an important tool to provide clues for understanding the chemical structure of HS. In this chapter, the limitations, recent advances, in terms of the possible solutions to the problems, and new information obtained from these advances, are stressed.

VI. LIMITATIONS AND SOLUTIONS

It is well known that the original building blocks can be modified or altered mostly due to thermal degradation and secondary reactions in pyrolysis. For example, the structures of pyrolysis products from polysaccharides, proteins, lignins, and fatty acids are significantly different from those of parent units.[50] Due to thermal degradation, carboxylic acids decarboxylate upon pyrolysis and produce alkanes and alkenes instead. Thus, it makes it difficult to analyze polar functional groups (–COOH, –OH) and fatty acids in pyrolysis, resulting in the loss of essential information about the structural composition of the original macromolecules.[49–51] This was somewhat frustrating before 1990, in that carboxyl-containing aromatic structures are obvious in HS from the studies of NMR and fluorescence spectroscopy,[52,53] while those groups cannot be detected by Py–GC–MS. It should also be mentioned that volatilization tends to decrease with increasing polarity because of the intermolecular forces. Therefore, polar degradation products are poorly represented in pyrolyzate and badly separated from the GC column.

To overcome some of these problems, a derivatization technique involving simultaneous pyrolysis/derivatization with tetramethylammonium hydroxide (TMAH) was proposed.[51] It appears that this method resulted in hydrolysis and methylation of the polar components, forming methyl esters of the polybasic acid, long-chain fatty acids and polyhydric alcohols. These methyl esters can be easily detected by GC–MS. Saiz-Jimenez et al.[54] first applied TMAH pyrolysis to the chemical structural investigation of FA, and they identified additional furancarboxylic acids, benzene carboxylic acids, and aliphatic dicarboxylic acids as their respective methyl esters, which were not previously identified using conventional pyrolysis, and other independent nondestructive methods confirmed their results. More interestingly, they identified the chemical differences between podzol and meadow fulvic acids: phenolic and benzene carboxylic acids constituted a significant part of the aromatic pyrolysis products in podzol fulvic acid; while lignin phenols constituted a significant part of the aromatic pyrolysis in meadow fulvic acid. Lehtonen et al.[17] used TMAH-Py–GC–MS to characterize the structural composition of lake HS and differences between different HS. It was demonstrated that the main degradation products were, in addition to varying proportions of different nitrogen and sulfur compounds, methyl derivatives of different phenols, alkylphenols, phenolic acids, and aliphatic mono- and dicarboxylic acids, indicating the occurrence of the original building blocks; conventional XAD isolate (HA and FA) was different from the other two XAD fractions (hydrophobic neutrals and after extended handling of the XAD effluent with Amberlite IRA-67 hydrophilic acid), which seem to belong to different structural categories yielding more alkylbenzenes/styrenes, alkanes, and aliphatic monocarboxylic acids. The DEAE and XAD isolate consisted of similar organic matter dominated by different phenols and aliphatic acids.

The second limitation of Py–GC–MS is that the complex pyrolysate was not just pyrolysis products; it consisted of evaporation and combustion products of HS.[50] It was reported that free compounds, e.g., alkanes, and fatty acids in HS macromolecules evaporated quickly under pyrolysis, and structural units split off through burning in the presence of oxygen and can be further incorporated into HS. For example, lipids, e.g., alkanes, fatty acids, dicarboxylic acids, and ketones were often found as free or solvent-extractable compounds in soils and soil HS. These compounds can be synthesized by microorganisms and plants,[55] and can occur upon combustion of fossil fuels and biomasses.[50] Alkylfurans and methoxylated phenols were considered pyrolysis products of

HS and combustion products of biomasses,[50] and alkylbenzenes and thiophenes can be both evaporation/pyrolysis products from humic substances.[56] At present, solvent extraction and low temperature desorption followed by TMAH Py–GC–MS have been proposed to distinguish combustion, evaporation and pyrolysis products in the pyrolysis.[55,57]

The third limitation is that pyrolysis products are highly dependent on pyrolysis temperature in that the pyrolysis efficiency is different at different temperatures, and pyrolysis behavior of HS may depend on its origin and structure. Experiments have shown that HS subjected to different temperatures (358, 510, 610, and 770°C) produced very distinct classes of evaporation and pyrolysis products.[50,52,58] HS with higher molecular weight generally has a structure that is more complex and needs higher pyrolysis temperature. For example, pyrolysis at 500 and 610°C was often used for fulvic and humic acid, respectively.[50] Five hundred to 600°C was the commonly used temperature range for pyrolysis investigation, but sometimes, higher temperatures, e.g., at 770°C were applied for more resistant moiety, e.g., aliphatic biopolymer and residual humic acid.[50] Lehtonen et al.[17,59] found that a temperature of 600°C yielded greater amounts of aromatic structural constituents compared with lower temperatures, and suggested that the use of high temperature pyrolysis is essential, since TMAH pyrolysis at lower temperature is basically not sufficient to split off the most strictly-bound aromatic subunits of the heterogeneous HS. They further proposed two types of cores in HS structural models: an ester/ether bond link core, and a more resistant alkyl aromatic network.

The fourth limitation is that only a small part of the original building blocks is available for Py–GC–MS analysis, the main pyrolysate is a carbonaceous residue. This resulted from the splitting off functional groups accompanied by cross linking reactions, and low molecular weight waste, e.g., water and CO_2, which have no value for the structural information.[50] These unwanted reactions and byproducts seem inevitable in current pyrolysis. Therefore, the understanding of a chemical structure of the whole HS is limited, and confirmation from different analytical methods is necessary in order to avoid misinterpretations of structural identification.

VII. RECENT PROGRESS AND NEW INFORMATION OBTAINED

In spite of the limitations, rapid progress for the Py–GC–MS method has been made recently to better characterize the chemical structure of HS, including new preparation methods, instrumentation, and multi-method combination and online coupling approaches. This development has greatly extended our knowledge of HS structure.

Pyrolysis with *in situ* methylation in the presence of TMAH is now commonly applied for the structural investigation of HS. It has been reported, however, that TMAH not only methylates polar pyrolysate but also assists in bond cleavage.[17] For example, TMAH was found as effective at 300°C as at 700°C for the production of some volatile products from HS, indicating that pyrolysis occurs with equal effectiveness at subpyrolysis temperature of 300°C.[60,61] It is believed that TMAH pyrolysis is actually a thermally assisted chemolysis rather than pure pyrolysis and it can cause hydrolytic ester and ether bond cleavage even at lower temperature, resulting in some unwanted side reactions, e.g., artificial formation of carboxylic groups from aldehydes.[62,63] Therefore, TMAH thermochemolysis at low temperature, e.g., 300°C has been proposed.[61,64,65] This technique offers several advantages over classical flash pyrolysis or preparative pyrolysis apparatus[61,64,65]: (a) It can induce ether/ester cleavage, and eliminate reactions with either concomitant or subsequent methylation of oxygen functionalities; (b) Due to the low temperature used, this technique can be performed offline in a sealed tube, so specialized pyrolysis equipment is not essential. Pyrolysis experiments can be implemented in a batch model in which internal standards can be used for quantification; (c) This enables to obtain a high quantity of thermolysis products in one experiment for chromatographic separation and quantitative determination at the same time. It was stated that the thermochemolysis procedure is presented follows[64]: The samples were placed in a ceramic boat after overnight moistening with 2 ml of a 50% (w/w) methanol solution of TMAH. For the treatment with tetraethyl ammonium acetate (TEAAc), the samples were moistened overnight with 3 ml of a 25% (w/w) ethanol solution of TEAAc. Each sample was then transferred into a 60 × 3 cm i.d. Pyrex tube and heated at 400°C. Thermochemolysis products were swept by helium to two successive traps containing chloroform cooled at −200°C. After evaporation of the solvent, trapped pyrolysate was combined and further separated on a SiO_2 column.[64] The new preparative thermochemolysis technique was also used to investigate ester and ether groups in humic acids and humin, and the results indicated that lignin and lipid biopolymers contributed highly to the formation of complex organic matter in soil, and ester and ether groups were noticeably involved in the structure of humin and humic acids.[64]

In order to discriminate between free and esterified mono- and dicarboxylic acids present in the structure of HS, a combination of TMAH thermochemolysis and TEAAc was used as TMAAc was proven a selective reagent for the methylation of free acids and their salts in the presence of fatty acid esters, while all fatty acid esters were hydrolyzed and methylated by TMAH.[66,67]

In order to obtain information on functional groups and structures linked to the core structure of HS, acid-catalyzed transesterification/esterification (TE/E) followed by GC–MS and GC–FID, involving transesterification of ester and amide bound structures, and esterification of free carboxylic groups using acid catalyzed methanolysis followed by GC–MS and GC–FID analysis, was proposed.[68] It was demonstrated that TE/E reaction was typically performed in an acid methanolic solution, hereby, free carboxylic groups were esterified, existing ester and amide bonds interchanged to methyl esters, and saccharides were transformed into methyl glycosides. Therefore, by adjusting the reaction condition it is possible to distinguish between ester and amide interchange. The results suggested the presence of multifunctional hydroxyl substituted benzoic acids, hexoses, and long-chain fatty acids in HS of different origin, and showed that the hydrophobic long-chain fatty acids constituted about one-third of the low molecular weight moieties.[68]

Other recent advance mainly focuses on the combination or coupling of Py–GC–MS with other advanced techniques online or offline. GC-separated individual pyrolysis products can be identified online by various instruments, e.g., mass spectrometry, Fourier Transform Infrared spectrometry, and atomic emission detection.[68–70]

At present, Py–MS, Py–GC–MS in the presence of TMAH and solid state ^{13}C NMR are considered the most fundamental techniques for structural studies of HS. Their combination, in terms of the limitations of individual technique, is of great advantage as they complement and confirm each other in the interpretations of chemical structure. For instance, NMR spectroscopy is nondestructive and provides the overall structure of the whole HS, while Py–MS and Py–GC–MS can only characterize the pyrolyzable compounds or moieties. Among the various GC–MS instruments, pyrolysis and field ionization mass spectrometry (Py–FIMS) is limited by the low volatility of highly polar constituents and cross-linked portions in the HS structure, while curie-point Py–GC–MS with rapid transfer of flash pyrolysis to the analytical specimen produces preferentially smaller products compared to Py–FIMS, and result in large amount of residuals without analysis. The combination of Py–FIMS and Py–GC–MS would allow a more reliable identification of the pyrolysis products.[70,71] The usefulness of the combination of ^{13}C-NMR and Py–GC–MS has been demonstrated by many authors.[72,73] It was reported that relative contribution of paraffinic structures determined by cross polarization magic angle spinning ^{13}C-NMR was in good agreement with the relative abundance of unbranched aliphatic hydrocarbons released by Py–GC–MS, and both techniques confirmed the importance of polymethylene structures in humin and humic acid. However, the content of aromatic carbons observed in NMR spectra is not correlated with the presence of aromatic hydrocarbons released by pyrolysis. This is because aromatic hydrocarbons are stable thermal degradation products of many types of materials including aliphatic structures.[72] Gonzalez et al.[73] reported that although the data obtained from NMR and pyrolysis, in terms of the quantitative distribution of carbon atoms pertaining to alkyl and aromatic structures, were not in complete agreement and pyrolysis seemed to be biased towards an "enrichment" of aliphatic moieties. The two techniques gave similar information on the overall structural composition of HS, and agreed that FA contained less aromatics and more aliphatic than HA, whereas the content in *O*-alkyl structures was similar in the two fractions.

Curie-point-gas chromatography–combustion-isotope ratio mass spectrometry (GC–C-IRMS) in combination with Py–FIMS and Py–GC–MS has been recently developed to investigate the structure of HS and track the pathways and origin of pyrolysis products.[70,71] It was demonstrated that the $\delta^{13}C$-values of the pyrolysis products were in agreement with generally accepted data for carbohydrates, lignins, and benzenes from biological sources, implying that the pyrolysis step had no isotopic effect. While some of the thermal products, e.g., 5-methyl-2-furancarboxaldehyde and benzene methyl had stronger depletions in ^{13}C in comparison to the mean $\sigma^{13}C$-values of furans and benzenes, respectively. The results contradicted the enrichment of ^{13}C due to trophic effects and therefore suggested the incorporation of high carbon sources, such as respiratory CO_2, methane, or anthropogenic pollution by fossil fuels, in the humification processes.[70,71]

The combination of analytical pyrolysis, molecular modeling, and computational chemistry has also been stressed in investigating the structure of HS.[71] It was reported that computational chemistry which allows to draw, construct and optimize in 3D space biomacromolecules, e.g., aquatic and terrestrial humic substances, with precise bond distances, bond angles, torsion angles, nonbonded distances, hydrogen bonds, charges, and chirality is a powerful tool, and molecular visualization and simulation can also be used to further understand the structure and dynamics of humic and dissolved organic matter.

Overall, current GC–MS technique together with other complementary analytical approaches has the capability to provide new structural information for building blocks, and to create more sophisticated structural models of HS. Contemporary structural models suggested the presence of the aromatic and heterocyclic cores in HS; the core units were linked to each other via a net of alkyl chains of different length HS, which were highly substituted with functional groups, e.g., carboxylic, aliphatic, and aromatic. Carbohydrates and residues of amino acids are the important structural component of HS.[70,71] However, due to the extreme complexity of the ill-defined structural and chemical composition of HS of different origins, pyrolysate is still like a mysterious box, there is a lot to be known and explored in terms of the thermal degradation, secondary reactions, derivatization, and chromatographic and detection systems. Thus, the current experiments and data, in terms of the complete chemical structure of HS, should be always interpreted with great caution, and interpretations could be revisited in the near future.

REFERENCES

1. Aiken, G. R., Isolation and concentration techniques for aquatic humic substances, In *Humic Substances in Soil, Sediment, and Water*, Aiken, G. R., McKnight, D. M., Wershaw, R. L. and MacCarthy, P., Eds., Wiley, New York, pp. 363–385, 1985.
2. Hayes, M. H. B., MacCarthy, P., Malcolm, R. L., and Switt, R. S., Eds., *Search of Structure, Humic Substances*, Vol. II, Wiley, New York, 1989.
3. Janoš, P., Separation methods in the chemistry of humic substances, *J. Chromatogr. A*, 983, 1–18, 2003.
4. Cai, Y., Size distribution measurements of dissolved organic carbon in natural waters using ultrafiltration technique, *Water Res.*, 33, 3056–3060, 1999.
5. Amarasiriwardena, D., Siripinyanond, A., and Barnes, R. M., Flow field-flow fractionation–inductively coupled plasma-mass spectrometry (FFFF–ICP-MS): a versatile approach for characterization of trace metals complexed to soil-derived humic acids, In *Humic Substances: Versatile Components of Plants, Soil and Water*, Ghabbour, E. A. and Davies, G., Eds., Royal Society of Chemistry, Cambridge, UK, pp. 215–226, 2000.
6. Saleh, F. Y., Ong, W. A., and Chang, D. Y., Structural features of aquatic fulvic acids: analytical and preparative reversed phase high performance liquid chromatography separation with photodiode array detection, *Anal. Chem.*, 61, 2792–2800, 1989.

7. Kuckuk, R. and Burba, P., Analytical fractionation of aquatic humic substances by metal affinity chromatography on iron (III)-coated cellulose, *Fresenius J. Anal. Chem.*, 366, 95–101, 2000.

8. Burba, P., Jakubowski, B., Kuckuk, R., Küllmer, K., and Heumann, K. G., Characterization of aquatic humic substances and their metal complexes by immobilized metal-chelate affinity chromatography on iron (III)-loaded ion exchangers, *Fresenius J. Anal. Chem.*, 368, 689–696, 2000.

9. Wu, F. C., Evans, R. D., and Dillon, P. J., Fractionation and characterization of fulvic acid by immobilized metal ion affinity chromatography, *Anal. Chim. Acta*, 452, 85–93, 2002.

10. Chin, Y. P., Alken, G., and O'Loughlin, E., Molecular weight, polydispersity, and spectroscopic properties of aquatic humic substances, *Environ. Sci. Technol.*, 28, 1853–1856, 1994.

11. Peuravuori, J. and Pihlaja, K., Molecular size distribution and spectroscopic properties of aquatic humic substances, *Anal. Chim. Acta*, 337, 133–149, 1997.

12. Saiz-Jimenez, C. and de Leeuw, J. W., Chemical characterization of soil organic matter by analytical pyrolysis–gas chromatography–mass spectroscopy, *J. Anal. Appl. Pyrolysis*, 9, 99–119, 1986.

13. Lepane, V., Comparison of XAD resins for the isolation of humic substances from seawater, *J. Chromatogr. A*, 845, 329–335, 1999.

14. Watt, B. E., Malcolm, R. L., Hayes, M. H. B., Clark, N. W. E., and Chipman, J. K., Chemistry and potential mutagenicity of humic substances in waters from different watersheds in Britain and Ireland, *Water Res.*, 30, 1502–1516, 1996.

15. Aiken, G. R., A critical evaluation of the use of macroporous resins for the isolation of aquatic humic substances, In *Humic Substances and their Role in the Environment*, Frimmel, F. H. and Christman, R. F., Eds., Wiley, New York, pp. 15–28, 1988.

16. Cho, J. W., Amy, G., and Pellegrino, J., Membrane filtration of natural organic matter: comparison of flux decline, NOM rejection, and foulants during filtration with three UF membranes, *Desalination*, 127, 283–298, 2000.

17. Lehtonen, T., Peuravuori, J., and Pihlaja, K., Characterization of lake-aquatic humic matter isolated with two different sorbing solid techniques: tetramethylammonium hydroxide treatment and pyrolysis–gas chromatography/mass spectrometry, *Anal. Chim. Acta*, 424, 91–103, 2000.

18. Hejzlar, J., Szpakowska, B., and Wershaw, R. L., Comparison of humic substances isolated from peatbog water by sorption on DEAE-cellulose and amberlite XAD-2, *Water Res.*, 28, 1961–1970, 1994.

19. Peuravuori, J. and Pihlaja, K., Multi method characterization of lake aquatic humic matter isolated with two different sorbing solids, *Anal. Chim. Acta*, 363, 235–247, 1998.

20. Velthorst, E., Nakken-Brameijer, N., and Mulder, J., Fraction of soil organic matter, *Int. J. Environ. Anal. Chem.*, 73, 237–251, 1999.

21. Horvath, C. S. and Melander, W., *J. Chromatogr. Sci.*, 15, 393–402, 1977.

22. Colin, H. and Guiochon, G., *J. Chromatogr.*, 141, 289–385, 1977.

23. Hayase, K. and Tsubota, H., Reversed-phase liquid chromatography of molecular weight-fractionated sedimentary fulvic acid, *J. Chromatogr.*, 295, 530–532, 1984.

24. Susic, M. and Boto, K. G., High performance liquid chromatographic determination of humic acid in environmental samples at the nanogram level using fluorescence detection, *J. Chromatogr.*, 482, 175–187, 1989.

25. Smith, B. and Warwick, P., Analysis of fulvic acids by ion-pair chromatography, *J. Chromatogr. A*, 547, 203–210, 1991.

26. Lombardi, A. T. and Jardim, W. F., Fluorescence spectroscopy of high performance liquid chromatography fractionated marine and terrestrial organic materials, *Water Res.*, 33, 512–520, 1999.

27. Lombardi, A. T., Morelli, E., Balestreri, E., and Seritti, A., Reverse phase high performance liquid chromatographic fractionation of marine organic matter, *Environ. Technol.*, 13, 1013–1021, 1992.

28. Preuße, G., Friedrich, S., and Salzer, R., Retention behaviour of humic substances in reversed phase HPLC, *Fresenius J. Anal. Chem.*, 368, 268–273, 2000.

29. Wu, F. C., Evans, R. D., and Dillon, P. J., High performance liquid chromatographic fractionation and characterization of fulvic acid, *Anal. Chim. Acta*, 464, 47–55, 2002.

30. Hutta, M. and Gora, R., Novel stepwise gradient reversed-phase liquid chromatography separation of humic substances, air particulate humic-like substances and lignins, *J. Chromatogr. A*, 1012, 67–79, 2003.

31. Wu, F. C., Evans, R. D., and Dillon, P. J., Separation and characterization of NOM by high performance liquid chromatography and online three dimensional excitation emission matrix fluorescence detection, *Environ. Sci. Technol.*, 37, 3687–3693, 2003.
32. Sontheimer, H., *Vom Wasser*, 75, 183–200, 1990.
33. Hongve, D. J., Baann, J., Becher, G., and Lomo, S., Characterization of humic substances by means of highperformance size exclusion chromatography, *Environ. Int.*, 22, 489–494, 1996.
34. Gjessing, E. T., Use of "Sephadex" gel for the estimation of molecular weight of humic substances in natural water, *Nature*, 208, 1091–1092, 1965.
35. Wu, C. S., *Handbook of Size Exclusion Chromatography*, Marcel Dekker Inc., New York, 1995.
36. Artinger, R., Buckau, G., Kim, J. I., and Geyer, S., Characterization of groundwater humic and fulvic acids of different origin by GPC with UV/Vis and fluorescence detection, *Fresenius J. Anal. Chem.*, 364, 737–745, 1999.
37. O'Loughlin, E. and Chin, Y. P., Effect of detector wavelength on the determination of the molecular weight of humic substances by high pressure size exclusion chromatography, *Water Res.*, 35, 333–338, 2000.
38. Her, N., Amy, G., Foss, D., Cho, J., Yoon, Y., and Kosenka, P., Optimization of method for detecting and characterizing NOM by HPLC-size exclusion chromatography with UV and online detection, *Environ. Sci. Technol.*, 36, 1069–1076, 2002.
39. Her, N., Amy, G., Foss, D., and Cho, J., Variations of molecular weight estimation by HP-size exclusion chromatography with UVA versus online DOC detection, *Environ. Sci. Technol.*, 36, 3393–3399, 2002.
40. Frimmel, F. H., Gremm, T., and Huber, S., Liquid chromatographic characterization of refractory organic acids, *Sci. Total Environ.*, 117, 197–206, 1992.
41. Vogl, J. and Heumann, K. G., Determination of heavy metal complexes with humic substances by HPLC/ICP-MS coupling using online isotope dilution technique, *Fresenius J. Anal. Chem.*, 359, 438–441, 1997.
42. Vogl, J. and Heumann, K. G., Development of an ICP-IDMS method for dissolved organic carbon determinations and its application to chromatographic fractions of heavy metal complexes with humic substances, *Anal. Chem.*, 70, 2038–2043, 1998.
43. O'Driscoll, N. and Evans, R. D., Analysis of methyl mercury binding to freshwater humic and fulvic acids by gel permeation chromatography/hydride generation ICP-MS, *Environ. Sci. Technol.*, 34, 4039–4043, 2000.
44. Wrobel, K., Sadi, B. B. M., Wrobel, K., Castillo, J. R., and Caruso, J. A., Effect of metal ions on the molecular weight distribution of humic substances derived from municipal compost: ultrafiltration and size exclusion chromatography with spectrophotometric and inductively coupled plasma-MS detection, *Anal. Chem.*, 75, 761–767, 2003.
45. Reemtsma, T. and These, A., Online coupling of size exclusion chromatography with electrospray ionization-tandem mass spectrometry for the analysis of aquatic fulvic and humic acids, *Anal. Chem.*, 75, 1500–1507, 2003.
46. Sadi, B. B. M., Wrobel, K., Wrobel, K., and Kannamkumarath, S. S., SEC–ICP-MS studies for elements binding to different molecular weight fractions of humic substances in compost extract obtained from urban solid waste, *J. Environ. Monit.*, 4, 1010–1016, 2002.
47. Alberts, J. J., Takacs, M., and Egeberg, P. K., Total luminescence spectral characteristics of natural organic matter (NOM) size fractions as defined by ultrafiltration and high performance size exclusion chromatography (HPSEC), *Org. Geochem.*, 33, 817–828, 2002.
48. Alberts, J. J., Takacs, M., McElvaine, M., and Judge, K., Apparent size distribution and spectral properties of natural organic matter isolated from six rivers in southeastern Georgia, USA, In *Humic Substances, Structures, Models and Functions*, Ghabbour, E. and Davies, G., Eds., The Royal Society of Chemistry, Cambridge, UK, pp. 179–190, 2001.
49. Saiz-Jimenez, C., Pyrolysis/methylation of soil fulvic acids: benzenecarboxylic acids revisited, *Environ. Sci. Technol.*, 28, 197–200, 1994.
50. Saiz-Jimenez, C., Analytical pyrolysis of humic substances: pitfalls, limitations, and possible solutions, *Environ. Sci. Technol.*, 28, 1773–1780, 1994.
51. Challinor, J. M., A pyrolysis-derivation–gas chromatography technique for the structural elucidation of some synthetic polymers, *J. Anal. Appl. Pyrolysis*, 16, 323–333, 1989.

52. Saiz-Jimmenez, C., Boon, J. J., Hedges, J. J., Hessels, J. K. C., and de Leeuw, J. W., Chemical characterization of recent and buried woods by analytical pyrolysis. Comparison of pyrolysis data with ^{13}C NMR and wet chemical data, *J. Anal. Appl. Pyrolysis*, 11, 437–450, 1987.

53. Senesi, N., Molecular and quantitative aspects of the chemistry of fulvic acid and its interactions with metal ions and organic chemicals: part II. The fluorescence spectroscopy approach, *Anal. Chem. Acta*, 232, 77–106, 1990.

54. Saiz-Jimenez, C., Hermosin, B., and Ortega-Calvo, J. J., Pyrolyis/methylation: a method for structural elucidation of the chemical nature of aquatic humic substances, *Water Res.*, 27, 1693–1696, 1993.

55. Grimalt, J. O., Hermosin, B., Yruela, I., and Saiz-Jimenez, C., *Sci. Total Environ.*, 81, 421–428, 1989.

56. Saiz-Jimenez, C., The origin of alkylbenzenes and thiophenes in pyrolysis of geochemical samples, *Org. Geochem.*, 23, 81–85, 1995.

57. Jandl, G., Schulten, H. R., and Leinweber, P., Quantification of long-chain fatty acids in dissolved organic matter and soils, *J. Plant Nutr. Soil Sci.*, 165, 133–139, 2002.

58. Saiz-Jimenez, C. and de Leeuw, J. W., Chemical characterization of soil organic matter by analytical pyrolysis–gas chromatography–mass spectroscopy, *J. Anal. Appl. Pyrolysis*, 9, 99–119, 1986.

59. Lehtonen, T., Peuravuori, J., and Pihlaja, K., Degradation of TMAH treated aquatic humic matter at different temperatures, *J. Anal. Appl. Pyrolysis*, 55, 151–160, 2000.

60. Hatcher, P. G. and Clifford, D. J., Flash pyrolysis and *in situ* methylation of humic acids from soil, *Org. Geochem.*, 21, 1081–1902, 1994.

61. McKinney, D. E., Carson, D. M., Clifford, D. J., Minard, R. D., and Hatcher, P. G., Offline thermochemolysis versus flash pyrolysis for the in situ methylation of lignin: is pyrolysis necessary?, *J. Anal. Appl. Pyrolysis*, 34, 41–46, 1995.

62. Tanczos, I., Schoflinger, M., Schmidt, H., and Balla, J., Cannizzaro reaction of aldehydes in TMAH thermochemolysis, *J. Anal. Appl. Pyrolysis*, 42, 21–31, 1997.

63. Tanczos, I., Rendl, K., and Schmidt, H., The behavior of aldehydes-produced as primary pryrolysis products-in the thermochemolysis with tetramethylammonium hydroxide, *J. Anal. Appl. Pyrolysis*, 49, 319–327, 1999.

64. Grasset, L. and Ambles, A., Structural study of soil humic acids and humin using a new preparative thermochemolysis technique, *J. Anal. Appl. Pyrolysis*, 47, 1–12, 1998.

65. Filley, T. R., Minard, R. D., and Hatcher, P. G., Tetramethylammonium hydroxide (TMAH) thermochemolysis: proposed mechanisms based upon the application of 13C-labeled TMAH to a synthetic model lignin dimer, *Org. Geochem.*, 30, 607–621, 1997.

66. Hardell, H. L. and Nilvebrant, N. O., A rapid method to discriminate between free and esterified fatty acids by pyrolytic methylation using tetramethylammonium acetate and hydroxide, *J. Anal. Appl. Pyrolysis*, 52, 1–14, 1999.

67. Grasset, L., Guignard, C., and Ambles, A., Free and esterified aliphatic carboxylic acids in humin and humic acids from a peat sample as revealed by pyrolysis with tetramethylammonium hydroxide or tetraethylammonium acetate, *Org. Geochem.*, 33, 181–188, 2002.

68. Rozenbaha, I., Odham, G., Jarnberg, U., Alsberg, T., and Klavins, M., Characterization of humic substances by acid catalysed transesterification, *Anal. Chim. Acta*, 452, 105–114, 2002.

69. Pörschmann, J., Kopinke, F. D., Remmler, M., Mackenzie, K., Geyer, W., and Mothes, S., Hyphenated techniques for characterizing coal wastewaters and associated sediments, *J. Chromatogr. A*, 750, 287–301, 1996.

70. Schulten, H. R. and Gleixner, G., Analytical pyrolysis of humic substances and dissolved organic matter in aquatic systems: structure and origin, *Water Res.*, 33, 2489–2498, 1999.

71. Schulten, H. R., Analytical pyrolysis and computational chemistry of aquatic humic substances and dissolved organic matter, *J. Anal. Appl. Pyrolysis*, 49, 385–415, 1999.

72. Fabbri, D., Mongardi, M., Montanari, L., Galletti, G. C., Chiavari, G., and Scotti, R., Comparison between CP/MAS 13C-NMR and pyrolysis-GC/MS in the structural characterization of humans and humic acids of soil and sediments, *Fresenius J. Anal. Chem.*, 362, 299–306, 1998.

73. Gonzalez-Via, F. J., Lankes, U., and Ludemann, H. D., Comparison of the information gained by pyrolytic techniques and NMR spectroscopy on the structural features of aquatic humic substances, *J. Anal. Appl. Pyrolysis*, 58/59, 349–359, 2001.

30 Surfactants

Bjoern Thiele

CONTENTS

I. INTRODUCTION

Surfactants have amphiphilic structures consisting of a hydrophilic and a hydrophobic part. These special structures cause their surface-active properties like concentration at surfaces, reduction of the surface tension, and formation of micelles in bulk solution. Therefore, they are widely used in formulations for washing, wetting, emulsifying, and dispersing. Laundry detergents, cleaning agents, and personal care products are by far the largest class of surfactant containing products for domestic use. After use, they are mainly discharged into municipal wastewaters which enter sewage treatment plants. The different ingredients of a detergent formulation are eliminated there by biodegradation or adsorption. In the case of insufficient biological degradability, however, they are potential sources of environmental pollution. Tetrapropylenebenzene sulfonate (TPS) is a typical example of a persistent anionic surfactant which was used in detergents between 1946 and 1965. As a consequence of rising TPS concentrations in German rivers during dry years of 1959/1960, visible foam formed on the water surface.

As a reaction, strict standards were applied to surfactants with regard to their biodegradability. In a directive of the European Community (73/404/EEC),[1] an average biodegradation rate of at least 90% for all surfactants (referring to a certain residence time in a municipal sewage treatment plant) is required. Consequently, TPS was replaced by readily biodegradable linear alkylbenzene-sulfonates (LAS) in the 1960s. The dramatic increase in the production of detergents during the second part of the last century still has an enormous impact on the environment. In order to evaluate the ecological risks of the different components of detergent formulations their levels in the different environmental compartments have to be determined. The analytical methods for the determination of surfactants as the main risk factors in environmental matrices have been continuously improved with regard to reproducibility, selectivity, and sensitivity over last few years. This chapter describes the broad spectrum of different analytical methods for these analytes beginning with correct sampling, followed by matrix-specific enrichment procedures, and finally the determination by colorimetric, spectroscopic, electrochemical, or chromatographic methods.

A. GENERAL REMARKS

Depending on the nature of the hydrophilic groups of surfactants, they can be divided into anionic, nonionic, cationic, and amphoteric surfactants. The last-mentioned class only plays a minor role with respect to domestic and industrial applications and practically no methods for the environmental analysis of amphoteric surfactants have been published so far.

1. Anionic Surfactants

The hydrophilic groups of anionic surfactants consist in most cases of sulfonate, sulfate, or carboxyl groups (Table 30.1). Amongst them, LAS are produced in the largest quantities worldwide. These are mainly used in powdery and liquid laundry detergents and household cleaners.

Abbreviations: AEO, alcohol ethoxylates; AES, alcohol ethoxy sulfates; AP, alkylphenols; APCI, atmospheric pressure chemical ionization; APEC, alkylphenoxy carboxylates; APEO, alkylphenol ethoxylates; APG, alkyl polyglucosides; AS, alcohol sulfates; BGE, background electrolyte; BiAS, bismuth active substance; CAD, collisionally activated decomposition; CI, chemical ionization; DBAS, disulphine blue active substances; DEEDMAC, diethylester dimethylammonium chloride; DEQ, diesterquaternary; DSDMAC, distearyldimethylammonium chloride; DTDMAC, ditallowdimethylammonium chloride; ECD, electron capture detector; EI, electron impact ionization; ESI, electrospray ionization; FAB, fast atom bombardment; FD, field desorption; FID, flame ionization detector; GC, gas chromatography; GCB, graphitized carbon black; HPLC, high performance liquid chromatography; IR, infrared; LAB, linear alkylbenzenes; LAS, linear alkylbenzene sulfonates; LC, liquid chromatography; MBAS, methylene blue active substances; MS, mass spectrometry; NCI, negative chemical ionization; NMR, nuclear magnetic resonance spectroscopy; NP, nonylphenols; NPEC, nonylphenoxy carboxylates; NPEO, nonylphenol ethoxylates; SAS, secondary alkane sulfonate; SFC, supercritical fluid chromatography; SFE, supercritical fluid extraction; SIM, selected ion monitoring; SPC, sulphophenyl carboxylates; SPE, solid-phase extraction; SPME, solid-phase micro-extraction; TPS, tetrapropylenebenzene sulfonate; UV, ultraviolet.

TABLE 30.1
Classification of Anionic Surfactants

Type	Formula	
Linear alkylbenzene sulfonates (LAS)		$R = C_{10}-C_{13}$
Alkylsulfonates	NaO_3S-R	$R = C_{11}-C_{17}$
α-Olefine sulfonates	$NaO_3S-(CH_2)_mHC=CH(CH_2)_nCH_3$	$m+n = 9-15$
Alkylsulfates	$NaO_3S-O \overset{R}{\diagup}$	$R = C_{11}-C_{17}$
Fatty alcohol ether sulfates	$NaO_3S-O \overbrace{(CH_2CH_2O)}_{n} R$	$R = C_{12}-C_{14}; n = 1-4$
α-Sulfo fatty acid methyl esters	$NaO_3S-\overset{COOCH_3}{\underset{R}{\diagup}}$	$R = C_{14}-C_{16}$
Sulfo succinate esters	$\overset{NaO_3S}{\underset{NaOOC}{\diagdown}}\diagup\diagdown COOR$	$R-C_{12}$
Soaps	$NaOOC-R$	$R = C_{10}-C_{16}$

2. Nonionic Surfactants

The hydrophilic behavior of nonionic surfactants is caused by polymerized glycol ether or glucose units (Table 30.2). They are almost exclusively synthesized by addition of ethylene oxide or propylene oxide to alkylphenols (AP), fatty alcohols, fatty acids, or fatty acid amides. Nonionic surfactants found major applications as detergents, emulsifiers, wetting agents, and dispersing agents. They are used in many sectors, including household, industrial and institutional cleaning products, textile processing, pulp and paper processing, emulsion polymerization, paints, coatings, and agrochemicals.

3. Cationic Surfactants

Cationic surfactants contain quaternary ammonium ions as their hydrophilic parts (Table 30.3). This class of surfactants has gained importance because of its bacteriostatic properties. Therefore, cationic surfactants are applied as disinfectants and antiseptic components in personal care products and medicine. Because of their high adsorptivity to a wide variety of surfaces, they are used as antistatic agents, textile softeners, corrosion inhibitors, and flotation agents.

II. SAMPLING

Correct sampling and storage of environmental samples are indispensable in environmental analysis. On the one hand, the samples must be representative of the environmental compartment from which they were taken and, on the other hand, it must be guaranteed that the chemical composition of the samples does not change during storage. The main problem in the analysis of surfactants is that they tend to concentrate at all interfaces due to their amphiphilic nature. Consequently, losses from aqueous solutions occur because of adsorption of the surfactants to

TABLE 30.2
Classification of Nonionic Surfactants

Type	Formula	
Alkylphenolethoxylates (APEO)		$R = C_8-C_{12}; n = 3-40$
Alcoholethoxylates (AEO)	$R-O\left(CH_2CH_2O\right)_n H$	$R = C_9-C_{18}; n = 1-40$
Fatty acid ethoxylates		$R = C_{12}-C_{18}; n = 4$
Fatty acid alkanolamide ethoxylates		$R = C_{11}-C_{17}; m = 0, 1;$ $n = 1, 2$
Fatty alcohol polyglycol ethers	$R-O\left(CH_2CH_2O\right)_m O\left(CH_2CHO\right)_n H$ with CH_3	$R = C_8-C_{18}; m = 3-6;$ $n = 3-6$
Alkylpolyglucosides (APG)		$R = C_8-C_{16}; x = 1-4$

laboratory apparatus or suspended particles. Especially for matrices like sewage sludges, sediments, and biological samples, the quantitative recovery of the analytes becomes a major problem. For this reason, internal standards are added to the samples in order to correct for nonquantitative recovery. This approach, however, is restricted to chromatographic determination methods because less selective methods such as the determination of summary parameters cannot discriminate surfactant initially present from added internal standards. Table 30.4 contains a selection of internal standards used in surfactant analysis.

Irrespective of the surfactants to be determined, water samples are immediately preserved upon collection by the addition of formaldehyde up to a concentration of 1% and stored at 4°C in the dark.[2–4] In order to prevent adsorption of LAS to laboratory apparatus, sodium dodecylsulphate is added to water samples.[5]

Sewage sludges are either preserved like water samples by the addition of formaldehyde up to 1% and storage at 4°C in the dark[6] or immediately filtrated and air-dried.[3]

Fertilization of agricultural land with sewage sludge has resulted in the need to monitor surfactant concentrations in sludge-amended soils. Soil samples are collected from the upper 5 cm with a stainless steel corer, dried at 60°C, pulverized, and stored at 4°C in the dark.[7]

III. ISOLATION AND ENRICHMENT

The concentrations of surfactants in environmental samples are usually below the limit of the analytical method. Therefore, preconcentration is necessary before analysis. Interfering substances

TABLE 30.3
Classification of Cationic Surfactants

Type		Formula
Tetraalkylammonium salts		$R^1, R^2 = C_1, C_{16}-C_{18}$ $R^1, R^2 = C_{16}-C_{18}$ $R^1 = C_8-C_{18}, R^2 = CH_2C_6H_5$
Alkylpyridinium salts		$R = C_16-C_{18}$
Imidazoliumquaternary- ammonium salts		$R = C_{16}-C_{18}$

TABLE 30.4
Selected Internal Standards Used in Determination Procedures for Surfactants in Different Environmental Matrices

Surfactant	Matrix	Determination Method	Internal Standard	Reference
LAS	Water	HPLC	C_9-, C_{15}-LAS or 1-C_8-LAS, 3-C_{15}-LAS	6,57
LAS	Water	GC–MS	CF_3CH_2-LAS	78
AEO	Sewage sludge, water	GC	1-Octanol and 1-eicosanol	67
AEO, APEO	Water	LC–MS	Hexylphenol5EO and ethylphenol5EO	33
APEO, AP	Sewage sludge, water	GC	n-Nonylbenzene or tribromophenol	31,43
APEO, AP	Water	HPLC	2,4,6-Trimethylphenol	2
NPEO, NP	Water, sediments	LC–MS	4-n-NP3EO, 4-n-NP	32

from the matrix have to be removed in an additional prepurification step prior to quantitative determination of the surfactants.

A. SOLID-PHASE EXTRACTION

Solid-phase extraction (SPE) has gained importance for the extraction and isolation of surfactants from aqueous samples over the last few years. It has advantages of very low solvent consumption, little time consumption, easy handling, and a broad spectrum of different exchange resins with regard to polarities and functionalities. SPE works on the principle that organic substances adsorb from aqueous solutions to exchange resin. The adsorbed substances are then eluted with small amounts of organic solvents.

1. Anionic Surfactants

Anionic surfactants are efficiently concentrated at reversed-phase (RP) materials consisting of silica gel modified with alkyl groups of different chain lengths or graphitized carbon black (GCB). LAS have been extracted by C2-,[8] C8-,[3,9] or C18-silica gels,[10-13] as well as by GCB stationary phases.[14]

The RP cartridges are usually rinsed with methanol/water before the adsorbed LAS is eluted with methanol. For further purification, these extracts are passed through an anionic exchange resin.[12,15] After passing water samples through GCB cartridges coextracted matrix substances are washed out by a formic acid-acidified solvent mixture. LAS are then eluted by CH_2Cl_2:methanol (9:1) containing 10 mM tetramethylammoniumhydroxide.[16] C_2 resins have been applied for the enrichment of alcohol ethoxy sulfates (AES) and alcohol sulfates (AS) from water. Afterwards the analytes have been eluted with methanol/2-propanol (8:2).[17] Marcomini et al. have developed a method for the simultaneous determination of LAS and nonylphenol ethoxylates (NPEO) as well as their metabolites sulphophenyl carboxylates (SPC) and nonylphenoxy carboxylates (NPEC), respectively. Wastewater or river water samples are adjusted to pH 2 with HCl and passed through C18 cartridges. The adsorbed analytes are eluted with methanol.[18] Solid-phase micro-extraction (SPME) has been proved an alternative technique for extraction of LAS. Desorption of the extracted LAS from a Carbowax/Templated Resin-coated fiber in a specially designed SPME–LC interface enable the analysis with HPLC and ESI–MS.[19]

2. Nonionic Surfactants

Nonionic surfactants like alkylphenol ethoxylates (APEO) and their biodegradation products alkylphenol diethoxylate (AP2EO), alkylphenol monoethoxylate (AP1EO), and AP are isolated from aqueous solutions with a number of different stationary phases. Kubeck et al.[20] used C18 cartridges to adsorb NPEO, but first the water samples were passed through a mixed-bed ion exchange resin to remove all ionic species. For SPE of alcohol ethoxylates (AEO) C8 cartridges have been successfully applied from which the surfactants were eluted with methanol followed by 2-propanol.[21] Alkyl polyglucosides (APG) are becoming more and more interesting because of their production from renewable raw materials (fatty alcohol and glucose or starch) and their good toxicological, dermatological, and ecological properties. Of the few analytical methods presently available for APG, C18 cartridges are employed to enrich APG from water. Desorption from the cartridges is carried out with methanol.[22] Amberlite XAD-2 and XAD-4 have been proved to extract APEO and AP from water samples with high selectivity. These resins are based upon a styrene structure cross-linked with divinylbenzene. Water samples saturated with NaCl are passed through a XAD-2 column, and the analytes are eluted with acetone/water (9:1) with a recovery of 91 to 94%.[23] Isolute ENV is a hyper-cross-linked hydroxylated poly(styrene–divinylbenzene) copolymer, which allows the extraction of APEO/AP from large sample volumes with similar recoveries compared to C18 cartridges.[24] GCB is a nonporous material with positively charged active centers on the surface. Therefore, it is employed for separation of NPEO/nonylphenol (NP) from acidic NPEC as well as LAS and SPC. The procedure involves the stepwise desorption of the adsorbed analytes from the GCB cartridges with different solvent systems.[25,26] SPME coupled to GC–MS was developed for analysis of NP in water. Optimal conditions were found with an 85 μm polyacrylate fiber, 1 g NaCl per 9.5 ml water sample, pH 2 and an extraction time of 1 h at 30°C.[27]

B. Liquid–Liquid Extraction

The attempt to extract surfactants directly from aqueous solutions into organic solvents without auxiliary measures is usually futile. The tendency of surfactants to concentrate at phase boundaries leads to the formation of emulsions and phase separation becomes very difficult.

Formation of lipophilic ion pairs between ionic surfactants and suitable counterions, however, avoids these problems. Hon-Nami et al.[28] developed a method of extracting LAS as these ion pairs with methylene blue using chloroform from river water. This method is also often applied to purify LAS extracts. Afterwards the ion pair is cleaved on a cationic exchange resin.[29]

Analogously to anionic surfactants, cationic surfactants are also extracted, e.g., into methylene chloride by the formation of ion pairs with LAS.[4,30]

Because of the formation of emulsions, the liquid–liquid extraction (LLE) of nonionic surfactants, e.g., APEO, is restricted to these less surface-active metabolites, i.e., APEO with one to three ethoxy units, APEC, and AP. Noncontinuous LLE of water samples with methylene chloride using a separatory funnel has been applied for NP and NPEO (one to three ethoxy units).[31,32] In addition, an ultrasonic bath has been shown to be suitable for the LLE of APEOs and AEOs form water samples.[33] Continuous LLE (percolation) has been successfully used for concentration of short-chained APEO and AP too.[31] Steam distillation/solvent extraction using an apparatus designed by Veith and Kiwus[34] is a sophisticated method of concentrating steam-distillable AP and APEO (one to three ethoxy units) from water samples.[2,35] AEOs have been efficiently extracted by combination of reflux hydrolysis with sulfuric acid and steam distillation with a "Karlsruhe Apparatus."[36]

C. SOLVENT SUBLATION

Solvent sublation is a technique capable of selectively concentrating surfactants free from nonsurface-active materials. In the original procedure by Wickbold,[37] the water sample is placed into a sublation apparatus and overlaid by ethyl acetate. Then ethyl acetate-saturated nitrogen is purged through the liquids whereupon surfactants are enriched at the gas–liquid phase boundary and carried by gas stream into the organic layer. This method has often been applied for the enrichment of nonionic surfactants and has now been standardized.[38] Waters et al.[39] optimized the Wickbold procedure and additionally purified the sublation extracts by passing them through a cation/anion exchanger.

Kupfer[40] applied the same sublation procedure for isolation of cationic surfactants. For separation of anionic and nonionic surfactants, the sublation extract is passed through a cation exchanger. Afterwards, the adsorbed cationic surfactants are eluted with methanolic HCl.

D. SOLID–LIQUID EXTRACTION

The method of choice for the extraction of surfactants from sewage sludges or sediments is solid–liquid extraction (SLE). In most cases, however, further purification of the extracts is necessary prior to quantitative determination. LAS are desorbed from sewage sludge either in a noncontinuous procedure by extraction into chloroform as ion pairs with methylene blue[41] or in a continuous procedure by the application of a Soxhlet apparatus and addition of solid NaOH to the dried sludge in order to increase extraction efficiency.[6] Heating of sludge or sediment samples in methanol under reflux for 2 h is also sufficient to extract LAS with recoveries of 85%.[3]

Extraction of APEO from solid matrices is performed in the same way as for LAS, i.e., Soxhlet extraction with methanol in combination with NaOH.[6] In addition to methanol, methanol:ethylene chloride (1:2)[23] and hexane[42] are used as extraction solvents. Steam distillation–solvent extraction is especially suitable for extraction of the APEO metabolites AP and APEO (one to three ethoxy units) from solid matrices.[2,43]

Quite drastic conditions are required to desorb cationic surfactants from solids. Extraction with methanolic HCl resulted in optimum recovery.[44,45] However, the extract has to be purified by extraction into chloroform in the presence of disulphine blue[44] or LAS.[45] Finally, cleavage of the ion pairs is done on ion exchangers. Hellmann[46] used an Al_2O_3 column to purify sewage sludge extracts. In this way, he was not only able to separate impurities but also to elute cationic and anionic surfactants stepwise with different solvent systems.

Supercritical fluid extraction (SFE) turns out to be very effective in the isolation of all three surfactant classes from solid matrices. While supercritical CO_2 alone did not affect significant recovery of surfactants, the addition either of modifiers or of reactants resulted in nearly quantitative recoveries. Thus, LAS and secondary alkane sulphonates (SAS) are extracted from sewage sludges in the form of tetrabutylammonium ion pairs.[47] Lee et al. extracted NP from sewage sludge spiked with acetic anhydride and a base with supercritical CO_2. In this way NP is, *in situ*, converted into its acetyl derivative.[48] Ditallowdimethylammonium chloride (DTDMAC) is quantitatively extracted from digested sludges and marine sediments using supercritical CO_2 modified with 30% methanol.[49]

IV. DETERMINATION PROCEDURES

A. COLORIMETRY/TITRIMETRY

Nonspecific analytical methods, such as colorimetry and titrimetry, for determination of summary parameters were the earliest attempts to analyze surfactants in the environment. The main disadvantage of these methods is that, apart from surfactants, other interfering organic compounds from the environmental matrices are recorded too, resulting in systematic errors. Nevertheless, colorimetric and titrimetric methods are still widely used for determination of anionic, nonionic, and cationic surfactants because of their easy handling and the need for relatively simple apparatus.

1. Anionic Surfactants

Anionic surfactants are determined with methylene blue. The procedure is based on the formation of ion pairs between the cationic dye methylene blue and anionic surfactants, which are extractable into chloroform. The concentrations of anionic surfactants are determined colorimetrically at 650 nm after separation of the organic phase.[38] Other anionic organic compounds also form extractable complexes with methylene blue resulting in high values for methylene blue active substances (MBAS). On the other hand, cationic substances lead to low values because of formation of ion pairs with anionic surfactants. Osburn, therefore, eliminated interfering compounds by several clean-up steps. Concentration of all organic compounds on an XAD-2 resin eliminates inorganic salts; the following anion exchange step separates all interfering cationic surfactants.[50]

2. Nonionic Surfactants

The bismuth active substances (BiAS) method for the determination of nonionic surfactants with barium tetraiodobismuthate ($BaBiI_4$, modified Dragendorff reagent) is used in the standardized (DIN-Norm) procedure in Germany,[38] as well as in other countries. Ba^{2+} as a hard Lewis acid forms cationic coordination complexes with the polyethoxylate chain of the nonionic surfactants, which are precipitated by $[BiI_4]^{2-}$ in the presence of acetic acid. The orange precipitate is then dissolved with ammonium tartrate solution, and the released bismuth ions are determined by potentiometric titration with pyrrolidinedithiocarbamate solution.[38,51] Waters et al.[39] optimized the BiAS procedure by introduction of a cation/anion exchange clean-up of the sublation extracts. The BiAS procedure fails to determine ethoxylates with less than five ethoxy units because these compounds are not precipitated by barium tetraiodobismuthate. Thus, this procedure is not suitable for determination of APEO metabolites, i.e., the shorter APEO and AP.[31]

3. Cationic Surfactants

Cationic surfactants form ion pairs with suitable anionic dyes that are extractable into organic solvents. The anionic dye most widely used is disulphine blue. After extraction of the ion pair into chloroform the extinction is determined at 628 nm. The presence of anionic surfactants results in serious interferences, and therefore they have to be separated by anion exchange before the addition of disulphine blue.[52,53] The determination of cationic surfactants is hampered by some problems not encountered with MBAS. In particular, cationic surfactants are strongly adsorbed to almost any surface, so that all apparatus has to be specially pretreated.

B. HIGH-PERFORMANCE LIQUID CHROMATOGRAPHY (HPLC)

The ultimate goal in environmental analysis is the quantification of individual compounds separated from all their isomers and/or homologues. Chromatographic methods like HPLC, GC, or SFC are amongst the most powerful analytical instruments with regard to separation efficiency and sensitivity. Because of the low volatility of surfactants, HPLC is used far more often than GC. Since the launch of atmospheric pressure ionization (API) interfaces, LC–MS coupling is increasingly used for determination of surfactants (Table 30.5).

1. Anionic Surfactants

The majority of HPLC applications in determination of anionic surfactants are only concerned with the analysis of LAS, which are surfactants in the largest quantities in present detergent formulations. Individual homologues of LAS are typically separated on reversed-phase columns with a $NaClO_4$-modified mobile phase using UV or fluorescence detection. Application of C18 columns with gradient elution results in the separation not only of the LAS homologues but also of their isomers (Figure 30.1).[3,6,54,55] While information on individual isomers could be valuable for studies on the biological degradation of LAS this is a hindrance in routine trace analysis because of the high number of peaks resulting in higher detection limits. By the use of short-chain alkyl bonded reversed phases like C8[6,11,56] and C1 columns[57] or long-chain C18 phases with isocratic elution,[58,59] however, the isomers of every single LAS homologue are eluted as one peak. Thus, the interpretation of the chromatograms becomes easier because of a greatly reduced number of peaks. Fluorescence detection is more selective and more sensitive than UV detection resulting in lower detection limits. Detection limits of 2 $\mu g/l$ for water using fluorescence detection[57] compared to 10 $\mu g/l$ for water using UV detection[3] have been reported for determination of LAS by HPLC.

For the analysis of aliphatic anionic surfactants by HPLC other detection systems than UV or fluorescence detection have to be used because of the lack of chromophoric groups. Refractive index detection and conductivity detection provide a solution for this type of anionic surfactants but their detection limits are rather high and gradient elution is not usually possible. Another possibility is the application of indirect photometric detection which is based on the formation of ion pairs between UV-active cationic compounds, such as N-methylpyridinium chloride, used as mobile-phase additives and the anionic surfactants followed by UV detection.[60] Gradient elution with indirect photometric detection is possible in principle but the detection limits increase considerably.[61] A selective and sensitive method for the determination of aliphatic anionic surfactants is reversed-phase HPLC combined with postcolumn derivatization and fluorescence detection.[62] After HPLC separation of the surfactants on a C1 column an UV-active cationic dye is added to the eluate in order to form fluorescent ion pairs. Then $CHCl_3$ is added to the eluent stream as the extraction solvent for the ion pairs. The two phases are conducted through a sandwich-type phase separator where the major part of the organic phase is separated. Finally, the amount of ion pairs extracted into $CHCl_3$ is determined by a fluorescence detector.

Simultaneous determination of LAS and their main metabolites SPC was enabled by LC–MS with an electrospray ionization (ESI) interface. Problems with high salt loads of the mobile phase due to the ion pair reagent have been overcome by incorporation of a suppressor between the LC column and the mass spectrometer.[63] A LC–MS method for the determination of AES and AS was introduced by Popenoe et al.[17] After separation on a C8 column the analytes are determined by ion spray LC–MS. The mass chromatograms obtained give information about both the distribution of the alkyl homologues and distribution of the oligomeric ethoxylates as well.

2. Nonionic Surfactants

The main nonionic surfactants are AEO, APEO, and recently APG. The hydrophobic part of AEO consists of n-alkanols with chain lengths between 8 and 20, typical AP are branched-chain octyl- or

TABLE 30.5
HPLC Methods for the Analysis of Surfactants

Compound	Matrix	Column	Mobile Phase	Derivatization Detector	LOD [µg/l]	Ref.
			Anionic Surfactants			
LAS	Sewage sludge	C-18 (Spherisorb S3 ODS II, 3 µm), 250 × 4 mm; C-8 (LiChrosorb RP8, 10 µm), 100 × 4 mm	A: iPrOH; B: H_2O; C: CH_3CN:H_2O (45:55) + 0.02 M $NaClO_4$	UV (225 nm) or Fluorescence (230/295 nm)[a]	80 ng	6
LAS	River water	C-18 (µ-Bondapak, 10 µm), 300 × 3.9 mm	A: H_2O + 0.15 M $NaClO_4$; B: CH_3CN:H_2O (70:30) + 0.15 M $NaClO_4$	UV (230 nm)	10	3
LAS	Sea water	C-18 (Spherisorb S3 ODS II, 3 µm), 250 × 4 mm	A: CH_3CN; B: CH_3CN:H_2O (25:75) + 10 g/l $NaClO_4$	Fluorescence (225/295 nm)[a]	-	54
LAS	River water	C-18 (Chromasil), 250 × 3.1 mm	A: CH_3CN:H_2O (50:50); B: CH_3CN:H_2O (70:30) Both containing 0.1 M $NaClO_4$	UV (225 nm)	100 (C-11 LAS)	55
LAS, SPC	Sea water	C-8 (LiChrosorb RP-8, 10 µm), 250 × 4.6 mm, gradient elution	A: MeOH:H_2O (80:20) + 1.25 mM TEAHS[b]; B: H_2O	Fluorescence (225/295 nm)[a]	0.2	11
LAS	River water, waste water	C-8 (C$_8$-DB, 5 µm), 250 × 4.6 mm	H_2O:MeOH (20:80) + 0.1 M $NaClO_4$ Isocratic elution	Fluorescence (225/290 nm)[a]	0.8	56
LAS	River water, waste water	C-1 (Spherisorb, 5 µm), 250 × 4 mm	THF:H_2O (45:55) + 0.1 M $NaClO_4$ Isocratic elution	Fluorescence (225/290 nm)[a]	2.0	57
LAS, SPC	Waste water	C-8 (Eclipse XDB, 3.5 µm), 150 × 3 mm	MeOH:0.01 M CH_3COONH_4 (75:25) Isocratic elution	Fluorescence (220/290 nm)[a]	5.0	19

Analyte	Matrix	Column	Mobile phase	Detection	LOD	Ref.
LAS, SPC	River water	C-18 (LiChrospher 100 RP-18, 5 μm), 250 × 4 mm	A: CH_3CN; B: 0.008 M KH_2PO_4 + H_3PO_4 (pH 2.2)	UV (215 nm)	20 (SPC)	13
LAS, SPC	River water	C-18 (Hypersil ODS, 5 μm), 250 × 2.1 mm	A: H_2O + 5 mM TEAAc; B: CH_3CN:H_2O (80:20) + 5 mM TEAAc	ESI-MS (suppressor before MS), full-scan m/z 170–400		63
SAS, AS	Water	C-1 (Spherisorb S5-C1), 40 × 4 mm	A: 0.01 M trisodium citrate − 5 μM HCl; B: CH_3CN:H_2O (50:50) + 0.01 M trisodium citrate + 5 μM HCl	*Post-column derivatization with CTBI[c]* Fluorescence (285/485 nm)[a]	3–30 ng	62
AS, AES	Waste water	C-8 (Baker, 5 μm), 250 × 4.6 mm	A: CH_3CN:H_2O (20:80) + 0.3 mM CH_3COONH_4; B: CH_3CN:H_2O (80:20) + 0.3 mM CH_3COONH_4	Ion spray-MS	-	17
			Nonionic Surfactants			
AP, APEO	Waste water	C-8 (LiChrosorb RP8, 10 μm), 250 × 3 mm	MeOH:H_2O (8:2) Isocratic elution	UV (277 nm)	0.5	43, 2
AP, APEO	Waste water	NH_2 (LiChrosorb- NH_2, 10 μm), 250 × 4.6 mm	A: hexane; B: hexane:iPrOH (1:1)	UV (277 nm)	0.5	2
APEO	Waste water	NH_2 (Hypersil APS, 3 μm), 100 × 4 mm	A: hexane:iPrOH (98:2); B: iPrOH:H_2O (98:2)	UV (277 nm)	10	64
APEO	Waste water	NH_2 (Zorbax NH_2), 250 × 4.6 mm	A: MTBE[d] + 0.1% acetic acid; B: CH_3CN:MeOH (95.5) + 0.1% acetic acid	Fluorescence (230/302 nm)[a]	0.2 ng	65

Continued

TABLE 30.5
Continued

Compound	Matrix	Column	Mobile Phase	Derivatization Detector	LOD [μg/l]	Ref.
APEO, LAS	Waste water	C-18 (LiChrosopher RP-18, 5 μm), 250 × 4 mm	A: MeOH B: H_2O + 0.14 g/l trifluoroacetic acid C: H_2O + 14 g/l $NaClO_4$ D: H_2O	Fluorescence-detection[a]	-	18, 66
NPEO, AEO	Waste water	C-18 (Phenomex Luna, 5 μm), 250 × 2 mm	A: H_2O B: MeOH Both containing 5 mM CH_3COONH_4 and 0.5 mM trichloroacetic acid	ESI-MS: m/z 300–1400	1–10	33
NPEO, NP	Waste water	Poly(vinylalcohol) (Shodex MSpak GF-310 4D), 150 × 4.6 mm	A: H_2O B: MeOH Both containing 5 μM CH_3COONH_4	[$^{13}C_6$]NP and [$^{13}C_6$]NPEO as surrogate standards ESI-MS	1–55 pg	32
OPEO, NPEO, OP, NP	River water	Poly(vinylalcohol) (Shodex MSpak GF-310 4D), 150 × 4.6 mm	A: H_2O:MeOH (50:50) + 10 mM CH_3COONH_4 B: MeOH	ESI-MS/MS: e.g. m/z 219 → 133 (NP), m/z 205 → 133 (OP) in the electrospray negative mode	0.1–9 pg	24
NPEO, AEO, LAS	Waste water	C-18 (Nucleosil C_{18}, 5 μm), 250 × 4.6 mm	A: H_2O:CH_3CN (20:80) + 0.5 mM CH_3COONH_4 B: H_2O:CH_3CN (80:20) + 0.5 mM CH_3COONH_4	APCI-MS and -MS/MS	-	71, 72
AEO	Waste water	C-18 (μBondapak C_{18}), 300 × 3.9 mm	A: H_2O B: MeOH	*Derivatization with phenylisocyanate* UV (235 nm)	100	67, 68, 69
APG	Technical APGs	C-18 (LiChrospher RP-18)	MeOH:H_2O (80:20) Isocratic elution	Refractive index	-	70

Cationic Surfactants

DTDMAC, DSDMAC	River water	NH$_2$/CN (Partisil PAC, 5 or 10 μm), 250 mm	CHCl$_3$:MeOH (80:20)	Conductivity	3–16	4, 73, 74
DTDMAC, DSDMAC	River water	NH$_2$/CN (Partisil PAC, 10 μm), 250 × 4.6 mm	Isocratic elution A: CHCl$_3$ B: MeOH C: CH$_3$CN	Post-column ion pair formation with methyl orange or DAS[e] Fluorescence (383/452 nm)	0.01 by use of DAS[e]	45, 49, 75
DTDMAC, DEEDMAC, DEQ	River water	NH$_2$/CN (Partisil PAC, 5 μm), 150 × 1 mm	A: CHCl$_3$ + 4% CH$_3$CN	ESI-MS	-	30
	Waste water		B: MeOH + 2% CH$_3$CN			

[a] Fluorescence ($\lambda_{ex}/\lambda_{em}$).
[b] TEAHS: Tetraethylammonium hydrogensulfate.
[c] CTBI: 1-Cyano-[2-(2-trimethylammonio)ethyl]benz(f)isoindole.
[d] MTBE: Methyl *tert*-butyl ether.
[e] DAS: 9,10-Dimethoxyanthracene-2-sulfonate.

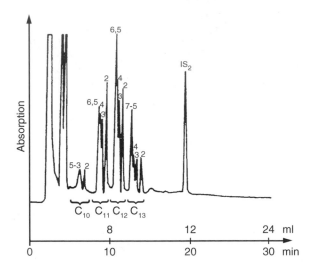

FIGURE 30.1 Reversed-phase high performance liquid chromatogram of LAS from digested sludge. C_{10}, C_{11}, C_{12}, and C_{13}: LAS homologues; the numbers above the LAS peaks indicate the position of the phenyl group on the alkyl chain; IS_2: 3-pentadecylbenzenesulphonate (3-C_{15}-LAS). (From Marcomini, A. and Giger, W:, *Anal. Chem.*, 59, 1709–1715, 1987.)

nonylphenol, and APG typically have alkyl groups with chain lengths in the range of 8 to 18. The degrees of polymerization of the polyethoxylate chains of AEO and APEO vary from 3 to 40 ethoxy units, while the average polymerization degree of APG is in the range of 1.3 to 1.7 moles glucose per mole of fatty alcohol. Consequently, HPLC separation of these surfactants into individual compounds is a two-dimensional problem best solved by the use of different HPLC stationary phases. Reversed-phase columns separate these compounds by their interaction with the hydrophobic alkyl chains, only eluting the hydrophilic oligomers as a single peak, while normal phase columns separate them by interaction with the hydrophilic polyethoxylate and polyglucoside chains without resolving the hydrophobes. Giger et al.[2,43] described a reversed-phase HPLC method for the determination of APEO on a C8 column with isocratic water/methanol elution and UV detection at 277 nm. Under these conditions, the homologous compounds octylphenol ethoxylates (OPEO) and NPEO are separated into two peaks. Normal phase HPLC is mostly applied to obtain information about the ethoxylate chain distribution of APEO. Aminosilica columns with gradient elution and UV detection are well suited to determine the individual oligomers of APEO.[2,6,64] An increase in sensitivity and selectivity for APEO is attained using a fluorescence detector. Thus, each single oligomer of APEO is determined by normal phase HPLC and fluorescence detection with a minimum detection of 0.2 ng.[65] Fluorescence detection is also used for the simultaneous determination of LAS and APEO as well as these corresponding metabolites SPC and NPEC, respectively, by reversed-phase HPLC and gradient elution.[18,66]

AEO can be sensitively determined in the form of these corresponding UV-active phenylisocyanate derivatives by UV detection. In this case, the residue of the extraction of a water sample or a solid matrix is dissolved in dichloromethane or dichloroethane. This solution is mixed with phenylisocyanate as well as 1-octanol and/or 1-eicosanol as internal standards and heated to 55 to 60°C for 45 to 120 min. Then the AEO derivatives are separated either by reversed-phase HPLC with regard to different alkyl chain lengths[67–69] or by normal phase HPLC with regard to different ethoxylate oligomers.[67,69] The addition of the internal standard is imperative for quantitative determination because derivatization is not completed even after 2 h.[69]

HPLC analysis of APG is carried out with C8[22] or C18 columns[70] by use of a refractive index detector[70] or a conductivity detector after the addition of 0.3 mol/l NaOH to the eluate in a postcolumn reactor.[22]

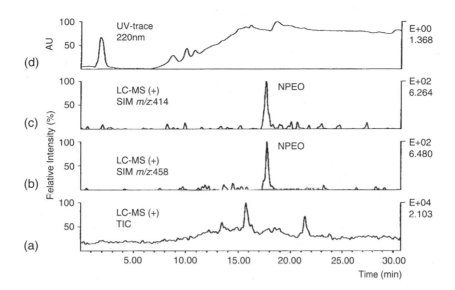

FIGURE 30.2 LC–APCI–MS total ion current chromatogram of wastewater (a); LC–MS mass trace m/z 458 (b); LC–MS mass trace m/z 414 (c); UV trace (220 nm) (d). (From Li, H. Q., Jiku, F., and Schröder, H. F., *J. Chromatogr. A*, 889, 155–176, 2000.)

Several LC–MS methods using an ESI interface have been published for the analysis of APEO and AEO. The formation of crown ether-type complexes between the ethoxylate chain and cations like NH_4^+ or Na^+ leads to efficient ion formation of the APEO and AEO surfactants during the electrospray process.[24,32,33] By use of a C-18 HPLC column NPEO and AEO are separated according to these aliphatic chain lengths. In the subsequent MS analysis, coeluting ethoxylate homologues are individually detected because of their mass differences of 44 mass units (CH_2CH_2O, m/z 44).[33] The comprehensive analysis of APEO and AP by LC–ESI–MS is enabled in a single chromatographic run by mixed-mode HPLC, using a Shodex MSpak GF-310 4D gel filtration column. This column operates with size-exclusion and reversed-phase mechanisms.[24,32] Complex water samples have been analyzed by LC–APCI–MS–MS in order to characterize the different surfactant classes (APEO, AEO, LAS) with the help of parent-ion and neutral-loss scans (Figure 30.2).[71,72]

3. Cationic Surfactants

DTDMAC and distearyldimethylammonium chloride (DSDMAC), which have long been amongst the most important cationic surfactants, are traditionally analyzed by normal phase HPLC with conductivity detection.[4,73,74] However, with conductivity detection an isocratic elution mode is mandatory, resulting in a steady broadening of the peaks with increasing retention time thus leading to higher detection limits. An alternative method for the quantitative analysis of cationic surfactants is the combination of HPLC separation with postcolumn ion pair formation and fluorescence detection.[45,49,75] Analogous to the method described for anionic surfactants (see above), an UV-active anionic dye is added to the HPLC eluate. The ion pairs formed are extracted online into a nonpolar organic phase in a phase separator and detected by a fluorescence detector. The application of LC–ESI–MS has enabled the homologue-specific analysis of esterquats and DTDMAC in environmental samples.[30]

C. Gas Chromatography (GC)

As a separation technique GC is inherently more powerful than HPLC; however, it is limited by the volatility of the compounds to be analyzed. For this reason, only nonionic surfactants with low degrees of ethoxylation are amenable to direct determination using GC. High-molecular nonionic surfactants as well as ionic surfactants must be derivatized prior to GC analysis in order to transform them into more volatile compounds. Apart from the flame ionization detector (FID), MS is increasingly becoming the dominant determination method for surfactants in environmental matrices. MS is not only a very sensitive and selective detection method but also provides valuable information on the molecular weight and structure of separated compounds (Table 30.6).

1. Anionic Surfactants

GC analysis of LAS is only possible after derivatization into volatile derivatives. Desulfonation of LAS in the presence of strong acids like phosphoric acid leads to linear alkylbenzenes (LAB). The identification of every single LAB isomer by GC–FID is achieved with detection limits lower than 1 μg/l.[76] In an alternative derivatization method, LAS are converted into their alkylbenzene sulfonyl chlorides by PCl_5, which can be directly analyzed by GC–FID.[41] Derivatization reactions for aliphatic anionic surfactants have mainly been described for product analysis. Among the very few methods for environmental analysis, the derivatization of alkyl sulfates to their corresponding trimethylsilylesters followed by determination with GC–FID is mentioned here.[77]

Several GC–MS methods are described for LAS in the literature. McEvoy et al. accomplished GC analysis by formation of the corresponding sulfonyl chlorides and subsequent mass spectrometric detection employing electron impact ionization (EI) and chemical ionization (CI) modes. The mass chromatograms obtained are complementary with regard to their qualitative and quantitative information. In the EI modus the mass spectra are characterized by fragment ions, which allow conclusions to be drawn on the distribution of LAS isomers, whereas CI-induced mass spectra give very reliable information on homologous distributions due to the presence of protonated molecular ions $(M + 1)^+$.[41] In other GC–MS methods LAS are converted in a two-step derivatization procedure to the corresponding trifluoroethyl sulfonate derivatives which are analyzed by GC–MS with EI and low-pressure CI modes[78,79] or with negative chemical ionization (NCI) mode in order to enhance sensitivity and selectivity due to the high electron affinity of the CF_3 group.[9] Direct derivatization in the hot injection port is carried out with LAS–tetraalkylammonium ion pairs to form the corresponding alkyl esters, which are subsequently determined by GC–MS.[14,47] Suter et al. developed a GC–MS–MS method to differentiate LAS and branched alkylbenzenesulfonates (ABS). Despite partial overlapping of LAS and ABS homologues, tandem mass spectrometric detection enabled the homologue-specific determination of these compounds due to their different fragmentation behaviors (Figure 30.3).[79]

2. Nonionic Surfactants

APEO analysis by GC without derivatization has been mainly used on the more volatile biodegradation products like NPEO (one to four ethoxy units) and NP. Using capillary columns a complex pattern is obvious for every ethoxylate oligomer, indicating that each single alkyl chain isomer is separated.[31,80] Quantification is performed by the addition of internal standards with a detection limit of 10 μg/l.[31] Derivatization of APEO not only increases their volatility but also, by an intelligent choice of derivatization reagent, more specific or sensitive detectors can be used. Thus, using perfluoroacid chlorides to derivatize NPEO the resulting perfluoroesters can be detected with the very sensitive electron capture detector (ECD) achieving detection limits lower than 1 μg/l.[81]

Because of the low volatility of APG, high-temperature GC with temperature programs up to 400°C in combination with silylation prior to GC analysis is required for these determination. The GC system allows detection of the separated oligomeric glucosides up to five units. While monoglucosides are well separated into these individual isomers, glucosides with higher degrees of polymerization are not resolved.[22]

TABLE 30.6
GC Methods for the Analysis of Surfactants

Analyte	Matrix	Injector	Column	Oven Program	Derivatization Detector	LOD [µg/L]	Ref.
				Anionic Surfactants			
LAS	River water	Splitless (1 µl), 275°C	OV-101 (30 m × 0.5 mm)	140°C, 3°C/−240°C (4′)	Desulfonation with phosphoric acid FID	1.0	76
LAS	Sewage sludge	Splitless (0.5–1 µl), 275°C	Fused silica coated with PS 255 (19 m × 0.31 mm)	50°C, 4°C/−300°C	Formation of sulfonyl chlorides with PCl$_5$ FID	-	41
LAS	River water, waste water	230°C	DB-5 (15 m × 0.25 mm, 0.25 µm film)	125°C (1′), 5°C/−230°C (5′)	Two-step PCl$_5$ - trifluoroethanol derivatization MS (NCI): SIM m/z 380, 394, 408, 422, 436	1.0	9
LAS, SAS	Sewage sludge	Split (1:7)	HP-5 (20 m × 0.2 mm, 0.33 µm film)	110°C, 10°C/−220°C 6°C/−300°C	Injection port derivatization with tetraalkylammonium salts MS (EI): Full scan m/z 50−400	-	47
LAS	River water	Large-volume (10–20 µl) direct sample introduction	DB-5MS (30 m × 0.25 mm, 0.25 µm film)	100°C (3′), 7°C/−300°C (7′)	Injection port derivatization with tetraalkylammonium salts MS (EI): Full scan m/z 50−550	0.1	14
LAS, SPC	River water	Splitless (1 µl) 250°C	DB-5MS (30 m × 0.25 mm, 0.25 µm film)	60°C (2′), 8°C/−180°C, 3°C/−230°C, 10°C/−250°C (10′)	Two-step thionyl chloride - trifluoroethanol derivatization MS (EI): Full scan m/z 50−500 MS (CI): Methane as reagent gas	0.01	78
LAS, ABS	standards	On-column (1 µl)	Fused silica coated with PS089 (15 m × 0.25 mm)	60°C, 8°C/−180°C, 3°C/−230°C	Two-step thionyl chloride - trifluoroethanol derivatization MS (CI): Isobutane as reagent gas, full scan m/z 80–500 MS/MS (CI): Argon as collision gas, m/z 295 → 167 (LAS), m/z 295 → 181 (ABS)	-	79
AS	Waste water	Splitless (1 µl) 200°C	Rt$_x$ − 1 (60 m × 0.25 mm, 0.25 µm film)	50°C (1′), 10°C/−215°C (20′)	Sylylation with BSTFA[a] + 1% TMCS[b] FID	1 ng	77

Continued

TABLE 30.6
Continued

Analyte	Matrix	Injector	Column	Oven Program	Derivatization Detector	LOD [µg/L]	Ref.
Nonionic Surfactants							
NPEO, NP	Waste water	Splitless (1–2 µl), 250°C	Glass capillary coated with OV-73 (15 m × 0.3 mm)	50°C, 2°C/–280°C	FID	10	31,80
NPEO, NP	Waste water	Splitless (1–2 µl) 280°C	Glass capillary coated with OV-1 (20 m × 0.3 mm)	50°C, 3°C/–270°C	MS (EI): Full scan m/z 45–480	–	35, 2
NPEO, NP	Waste water	Splitless (2 µl) 250°C	SGE BP-1 (25 m × 0.22 mm, 0.25 µm film)	80°C (1'), 30°C/– 210°C, 10°C/–300°C (15')	Derivatization with pentafluorobenzoyl chloride MS (EI) MS (CI): methane as reagent gas	0.1 (NP) 0.2–1 (NPEO)	81
OPEO, OP, AEO	Waste water	-	DB-5 (30 m × 0.25 mm, 0.25 µm film)	70°C (1'), 3°C/–300°C (10')	MS (EI): Full scan m/z 45–500 MS (CI): methane as reagent gas	–	82, 83
NP	Effluent water	Splitless 250°C	HP-5-MS (30 m × 0.25 mm, 0.25 µm film)	70°C (1'), 30°C/–160°C, 5°C/–240°C	In situ derivatization with acetic anhydride during extraction MS (EI): SIM m/z 107, 121, 135, 1613, 191, 262	0.1	48
NP	Biological samples	Splitless (2 µl), PTV[c]: 50°C (0.6'), 12°C/s–285°C	DB-5-MS (60 m × 0.25 mm, 0.25 µm film)	50°C (0.8'), 20°C/–110°C (1'), 4°C/–230°C, 20°C/–285°C (20')	MS (EI): SIM m/z 121, 135, 149, 163, 177, 191 (NP); m/z 107, 220 (4-n-NP)	15 ng (NP)	84
NPEO, NP, NPEC	River water, sewage effluent	Large-volume (10 µl) direct sample introduction	DB-5MS (30 m × 0.25 mm, 0.25 µm film)	100°C (5'), 8.5°C/–280°C (15')	Derivatization of NPEC to the propyl esters MS (EI): Full scan m/z 50–500 MS (CI): Methane or acetone as reagent gases	0.01	85
NP	Waste water	SPME, desorption at 280°C for 3 min	HP-5 MS (30 m × 0.25 mm, 0.25 µm film)	50°C (4'), 20°C/–140°C (1'), 10°C/–280°C (8')	MS (EI): SIM m/z 107, 135 (NP); m/z 107, 220 (4-n-NP)	0.2–0.8	27

[a] BSTFA: Bis(trimethylsilyl)trifluoroacetamide.
[b] TMCS: Trichloromethylsilane.
[c] PTV: programmed temperature vaporization.

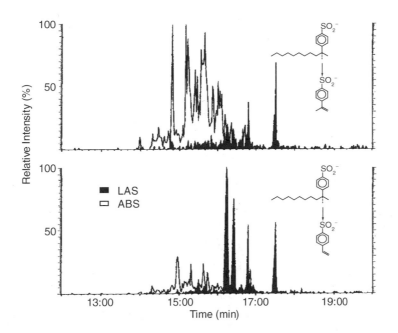

FIGURE 30.3 Superimposed reconstructed GC–MS–MS chromatograms of LAS obtained in the negative CI mode (parent ion m/z 295). The top trace corresponds to m/z 295 → 181 and the bottom trace to m/z 295 → 167, both recorded for C_{11}-LAS (solid peaks) and C_{11}-ABS (open peaks). (From Suter, M. J. F., Reiser, R., and Giger, W., *J. Mass Spectrom.*, 31, 357–362, 1996.)

GC–MS in the EI mode is well established for the identification and sensitive quantification of APEO and AP in environmental matrices.[31,35] Moreover, the fragmentation patterns in the mass spectra allow the structural characterization of the nonyl side-chain isomers; however, valuable information on the distribution of the oligomeric ethoxylates is lost due to very weak intensities of the molecular ions. The distribution of the ethoxylates is determined by CI–MS as a complementary method to EI–MS because of the presence of intensive adduct ions like, e.g., $(MH)^+$.[82,83] Lee et al. developed an *in situ* derivatization procedure in which NP is simultaneously extracted and converted into the corresponding acetyl derivatives. Quantification of NP from effluent water and sewage sludge is carried out by GC–EI–MS in the selected ion monitoring (SIM) mode with detection limits of 0.1 μg/l and 0.1 μg/g.[48] Günther et al. used an off-line coupling of normal phase HPLC and GC–EI–MS in the SIM mode to determine the individual isomers of NP in biological matrices. The HPLC step serves as clean-up of the extracts by collection of the NP containing eluate after passing the HPLC column.[84] Simultaneous determination of NPEO and their degradation products, NP and NPEC, is accomplished by GC–MS with EI, CI, and CI–MS–MS modes. Prior to the GC analysis NPEC is derivatized with propanol/acetyl chloride. Sensitivity has been increased by use of a large-volume direct sample introduction device.[85]

3. Cationic Surfactants

GC analysis is not of practical relevance for the determination of cationic surfactants in environmental matrices.

D. Supercritical Fluid Chromatography (SFC)

SFC combines the advantages of HPLC and GC into one method. Gases above their critical temperatures and conditions are used as mobile phases in order to separate analytes with a

TABLE 30.7
CE Methods for the Analysis of Surfactants

Analyte	Matrix	Injection	Column	BGE	Detection	LOD (μg/L)	Reference
				Anionic Surfactants			
LAS	Detergents	Large volume sample stacking: Sample injection (4 psi/90 sec) followed by injection of a buffer plug, stacking voltage of 15 kV at reversed polarity, voltage of 20 kV at normal mode	Fused silica (60 cm × 50 μm i.d., 50 cm eff.)	20 mM sodium tetraborate + 30% acetonitrile, pH 9.0	UV (200 nm)	2 to 10	90
LAS, SPC	Wastewater	Pressure (0.5 psi/20 sec)	Fused silica (80 to 100 cm × 75 μm i.d.)	10 mM ammonium acetate + 16% CH$_3$CN, pH 9.8	ESI–MS: iPrOH:H$_2$O (80:20) + 0.1% ammonia as makeup solvent	4 to 23	91
LAS	Wastewater	Pressure (5 sec)	Fused silica (57 cm × 75 μm i.d., 50 cm eff.)	250 mM borate + 30% CH$_3$CN, pH 8.0	UV (200 nm)	1000	12
LAS, aliphatic anionic surfactants	Detergents	Pressure (50 mbar/4 sec)	Fused silica (48.5 cm × 75 μm i.d., 40 cm eff.)	NACE[a]: 15 mM naphthalene sulfonic acid, 15 mM triethylamine in CH$_3$CN: MeOH (75:25)	Indirect UV (280 nm)	—	94
LAS	Detergents, river water	Pressure (5 sec)	Fused silica (57 cm × 25, 50 or 75 μm i.d., 50 cm eff.)	A: 50 mM borate, pH 8.2 B: 100 mM phosphate + 30% CH$_3$CN, pH 6.8 C: 100 mM phosphate + 30% CH$_3$CN 20 mM α-CD[b], pH 6.8	UV (200 nm)	5900 (C11-LAS)	55,88
AS	Detergents	Pressure (5 or 10 sec)	Fused silica (57 cm × 75 μm i.d., 50 cm eff.)	20 mM salicylate + 30% CH$_3$CN, pH 6	Indirect UV (214 nm)		88

SAS, tetraalkylammonium halides	Surfactants	Pressure (25 mbar/12 sec)	Fused silica (60 cm × 50 μm i.d., 60 cm eff.)	20 mM NaF, 1 mM triethanolamine + 10% CH$_3$CN	Indirect conductivity	6000	89
Nonionic Surfactants							
NPEO	Surfactants	Pressure (5 to 10 sec)	Fused silica (57 cm × 75 μm i.d., 50 cm eff.)	10 mM phosphate, 70 mM SDS[c] +35% CH$_3$CN, pH 6.8	UV (200 nm)	—	88
Cationic Surfactants							
Alkylbenzylammonium salts, alkyl pyridinium salts	Detergents	Pressure (5 to 10 sec)	Fused silica (57 cm × 75 μm i.d., 50 cm eff.)	50 mM phosphate + 58% THF, pH 6.8	UV (200 nm)	—	92
Alkyltrimethylammonium salts	Detergents	Pressure (5 sec)	Fused silica (57 cm × 75 μm i.d., 50 cm eff.)	20 mM phosphate, 5 mM C$_{12}$-benzyl-DMA[d] +50% THF, pH 4.4	UV (214 nm)	—	93

[a] NACE: Nonaqueous CE.
[b] α-CD: α-Cyclodextrin.
[c] SDS: Sodium dodecyl sulfate.
[d] C$_{12}$-benzyl-DMA: Dodecylbenzyldimethylammonium salt.

conventional HPLC column. Under these conditions the supercritical fluids have densities of liquids while retaining the diffusion coefficients of typical gases. The universal and sensitive FID detector can be applied to SFC. Consequently, no derivatization of analytes is required, either to increase volatility or to increase detectability.

Until now applications of SFC have been limited to product analysis of, e.g., nonionic surfactants but here with great success.[86,87] No reports on the determination of surfactants in environmental matrices using SFC is known to the authors.

E. CAPILLARY ELECTROPHORESIS (CE)

CE is a separation technique which uses empty capillaries to effect separation by the electrophoretic movement of charged compounds. Therefore, CE is not a chromatographic method in the strict sense. Recently CE has been applied for the separation and determination of all three surfactant classes (Table 30.7).

1. Anionic Surfactants

LAS are analyzed in river water by CE using UV detection. The efficiency of separating LAS homologues and isomers significantly depends on the addition of organic modifiers to the buffers. In phosphate and borate buffers without an organic modifier only one peak is obtained in the electropherogram for all LAS isomers and homologues.[55,88] The addition of 20 to 30% acetonitrile to the buffer leads to a separation of homologues and with buffers containing α-cyclodextrin (α-CD) even a complete separation of isomers is possible (Figure 30.4).[55,88] Aliphatic anionic surfactants can be determined by CE with indirect UV detection using salicylate as chromophore in the buffer[88] or indirect conductivity detection.[89] CE of LAS with large-volume sample stacking technique has been shown to improve the peak shapes, the efficiency, and the sensitivity.[90] CE–ESI–MS has been used for the simultaneous determination of LAS and their metabolites, SPC. Limits of detection of 4.4 to 23 $\mu g/l$ could be reached for the quantification of LAS homologues.[91]

2. Nonionic Surfactants

Nonionic surfactants of the ethoxylate type are not so efficiently separated compared to ionic surfactants.[88] The complexity of the surfactant mixtures and the lack of charge leads to insufficient peak resolution and high detection limits.

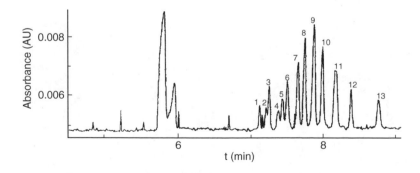

FIGURE 30.4 CE electropherogram of a LAS detergent (Marlon A-390), buffer: 100 mM phosphate, pH 6.8, 15 mM α-CD, 20% (v/v) acetonitrile. Numbered peaks correspond to LAS isomers (From Heinig, K., Vogt, C., and Werner, G., *J. Chromatogr. A*, 745, 281–292, 1996): (1) 2-C_{13}, (2) 3-C_{13}, (3) 2-C_{12}, (4) 4-C_{13}, (5) 3-C_{12}, 5-C_{13}^*, (6) 2-C_{11}, 5-C_{13}^*, 4-C_{12}^*, (7) 5-C_{13}^*, 4-C_{12}^*, (8) 3-C_{11}, 6-C_{13}, 4-C_{12}^*, (9) 4-C_{11}, 2-C_{10}, 5-C_{12}, 7-C_{13}^*, (10) 3-C_{10}, 6-C_{12}^*, 7-C_{13}^*, (11) 5-C_{11}, 4-C_{10}, 6-C_{12}^*, 7-C_{13}^*, (12) 6-C_{11}, (13) 5-C_{10} (*denotes supposed).

3. Cationic Surfactants

Cationic surfactants are separated using direct UV detection[92] or indirect UV detection with a chromophore as electrolyte additive.[93] The addition of organic solvents as modifiers to the electrolytes is essential to obtain efficient separations because of the ability of cationic surfactants to adsorb onto the capillary surface.

F. MASS SPECTROMETRY (MS)

MS is a reliable method for the determination of molecular weight distributions of homologous and/or oligomeric surfactants as well as for the determination of molecular structures, e.g., the position of side chains or the degree of branching. Soft ionization methods like fast atom bombardment (FAB) or field desorption (FD) are well suited for the formation of molecular ions of high molecular surfactants. For this reason, they are not only used in product analysis for the determination of molecular weight distributions but also in biodegradation studies of surfactants.

1. Anionic Surfactants

FAB–MS was successfully employed for the identification of LAS in groundwater. The mass spectra obtained from the samples, which were slurred in glycerol as matrix show molecular ions $(M)^+$ separated by 14 mass units corresponding to the different LAS homologues.[8] Triethanolamine or thioglycerol in combination with NaCl is alternatively used as matrix but then quasimolecular ions $(M + H)^+$ and $(M + Na)^+$, respectively, are formed.[95] Moreover, FAB spectra exhibit fragment ions, which are in part structure specific.[96] FD-MS spectra obtained in the positive or negative mode only contain quasimolecular ions while fragment ions are missing.[96] Therefore, FD spectra are well suited for determining the molecular weight distribution of surfactants but less suited for structure elucidation.

2. Nonionic Surfactants

FAB–MS spectra of APEO and AEO are preferentially obtained by thioglycerol saturated with NaCl as matrix due to the formation of strong $(M + Na)^+$ ions.[95,97,98] The characteristic appearance of these spectra is a series of $(M + Na)^+$ ions separated by 44 units corresponding to different degrees of ethoxylation. Cleavage of the alkyl constituents and the ethoxylate chains lead to fragmentation patterns in the lower mass range, which make it possible to elucidate the structures of nonionic surfactants. The clarity of FD–MS spectra due to the dominance of quasimolecular ions $(M + H)^+$ and missing fragment ions caused Levsen et al.[99] to monitor the biodegradation of NPEO in surface water. FD–MS is also used for the identification of APEO in water samples after separation by reversed-phase HPLC and collection of the APEO-containing eluate.[100,101]

3. Cationic Surfactants

Conventional ionization techniques like EI or CI are less well suited for the characterization of quaternary amines, which are the most common cationic surfactants. Because of their thermal instability and low volatility their corresponding mass spectra only show decomposition products and fragment ions which make it impossible to analyze environmental samples of unknown composition. By the use of FAB–MS and FD–MS, however, ionization of quaternary amines can be achieved without decomposition. FAB spectra are characterized by strong quasimolecular ions as well as structure specific ions.[95,102] FAB in combination with collisionally activated decomposition (CAD) in a tandem mass spectrometer enables a clear differentiation between quasimolecular and fragment ions, which is often difficult using FAB alone.[102] FD spectra of quaternary amines are dominated by quasimolecular ions as already described for other surfactant types.[102] By combining FD and CAD in a tandem MS it is even possible to obtain fragment ions for the structure elucidation of individual cationic surfactants in environmental samples.[103]

Quantitative determinations of surfactants by FAB or FD–MS are rather difficult because of the need for isotopically labeled internal standards.

G. Infrared Spectroscopy (IR)

IR spectroscopy is used for the qualitative identification of surfactants and for differentiating between them and nonsurfactant compounds. Prior to IR spectroscopy, however, separation of the organic compound complex into different fractions, performed by, e.g., the use of thin-layer chromatography, is required to obtain meaningful spectra.[104,105] By comparing the IR spectra of the isolated fractions with IR spectra of standard compounds with regard to characteristic bands, the qualitative determination of surfactants in environmental samples is possible. The method is equally applicable to anionic,[105] nonionic,[104] and cationic surfactants.[106] The prerequisite for a clear identification of surfactants, however, is the availability of suitable standards. Moreover, considerable experience and knowledge are needed to interpret IR spectra of environmental samples.

H. Nuclear Magnetic Resonance Spectroscopy (NMR)

NMR spectra regularly contain far more information on the molecular structure of the particular compound investigated than IR spectra. However, the complex compound mixture in environmental samples has to be thoroughly separated in order to obtain meaningful NMR spectra. Furthermore, the amount of analyte needed for NMR is relatively high; therefore, NMR spectroscopy is exclusively used in product analysis for the characterization of pure compounds and is of no importance in environmental analysis.

REFERENCES

1. EEC (European Economic Community) (73/404/EEC), Off. J.E.C. No. L 347/51, 1973.
2. Ahel, M. and Giger, W., *Anal. Chem.*, 57, 1577–1583, 1985.
3. Matthijs, E. and De Henau, H., *Tenside Surf. Det.*, 24, 193–199, 1987.
4. Wee, V. T. and Kennedy, J. M., *Anal. Chem.*, 54, 1631–1633, 1982.
5. Marcomini, A., Capri, S., and Giger, W., *J. Chromatogr.*, 403, 243–252, 1987.
6. Marcomini, A. and Giger, W., *Anal. Chem.*, 59, 1709–1715, 1987.
7. Marcomini, A., Capel, P. D., Liechtensteiger, T., Brunner, P. H., and Giger, W., *J. Environ. Qual.*, 18, 523–528, 1989.
8. Field, J. A., Barber, L. B. II, Thurman, E. M., Moore, B. L., Lawrence, D. L., and Peake, D. A., *Environ. Sci. Technol.*, 26, 1140–1148, 1992.
9. Trehy, M. L., Gledhill, W. E., and Orth, R. G., *Anal. Chem.*, 62, 2581–2586, 1990.
10. Kikuchi, M., Tokai, A., and Yoshida, T., *Water Res.*, 20, 643–650, 1986.
11. Leon, V. M., Gonzalez-Mazo, E., and Gomez-Parra, A., *J. Chromatogr. A*, 889, 211–219, 2000.
12. Heinig, K., Vogt, C., and Werner, G., *Analyst*, 123, 349–353, 1998.
13. Sarrazin, L., Arnoux, A., and Rebouillon, P., *J. Chromatogr. A*, 760, 285–291, 1997.
14. Ding, W. H. and Chen, C. T., *J. Chromatogr. A*, 857, 359–364, 1999.
15. Schöberl, P., Klotz, H., Spilker, R., and Nitschke, L., *Tenside Surf. Det.*, 31, 243–252, 1994.
16. Crescenzi, C., DiCorcia, A., Marchiori, E., Samperi, R., and Marcomini, A., *Water Res.*, 30, 722–730, 1996.
17. Popenoe, D. D., Morris, S. J. III, Horn, P. S., and Norwood, K. T., *Anal. Chem.*, 66, 1620–1629, 1994.
18. Marcomini, A., Di Corcia, A., Samperi, R., and Capri, S., *J. Chromatogr.*, 644, 59–71, 1993.
19. Ceglarek, U., Efer, J., Schreiber, A., Zwanziger, E., and Engewald, W., *Fresenius J. Anal. Chem.*, 365, 674–681, 1999.
20. Kubeck, E. and Naylor, C. G., *J. Am. Oil Chem. Soc.*, 67, 400–405, 1990.

21. Evans, K. A., Dubey, S. T., Kravetz, L., Dzidic, I., Gumulka, J., Mueller, R., and Stork, J. R., *Anal. Chem.*, 66, 699–705, 1994.
22. Steber, J., Guhl, W., Stelter, N., and Schröder, F. R., *Tenside Surf. Det.*, 32, 515–521, 1995.
23. Valls, M., Bayona, J. M., and Albaiges, J., *J. Environ. Anal. Chem.*, 39, 329–348, 1990.
24. Loyo-Rosales, J. E., Schmitz-Afonso, I., Rice, C. P., and Torrents, A., *Anal. Chem.*, 75, 4811–4817, 2003.
25. Di Corcia, A., Samperi, R., and Marcomini, A., *Environ. Sci. Technol.*, 28, 850–858, 1994.
26. Crescenzi, C., Di Corcia, A., Samperi, R., and Marcomini, A., *Anal. Chem.*, 67, 1797–1804, 1995.
27. Braun, P., Moeder, M., Schrader, S., Popp, P., Kuschk, R., and Engewald, W., *J. Chromatogr. A*, 988, 41–51, 2003.
28. Hon-Nami, H. and Hanya, T., *J. Chromatogr.*, 161, 205–212, 1978.
29. Takada, H. and Ishiwatari, R., *Environ. Sci. Technol.*, 24, 86–91, 1990.
30. Radke, M., Behrends, T., Förster, J., and Herrmann, R., *Anal. Chem.*, 71, 5362–5366, 1999.
31. Stephanou, E. and Giger, W., *Environ. Sci. Technol.*, 16, 800–805, 1982.
32. Ferguson, P. L., Iden, C. R., and Brownawell, B. J., *J. Chromatogr. A*, 938, 79–91, 2001.
33. Cohen, A., Klint, K., Bowadt, S., Persson, P., and Jönsson, J. A., *J. Chromatogr. A*, 927, 103–110, 2001.
34. Veith, G. D. and Kiwus, L. M., *Bull. Environ. Contam. Toxicol.*, 17, 631–636, 1977.
35. Giger, W., Stephanou, E., and Schaffner, C., *Chemosphere*, 10, 1253–1263, 1981.
36. Meissner, C. and Engelhardt, H., *Chromatographia*, 49, 12–16, 1999.
37. Wickbold, R., *Tenside Surf. Det.*, 8, 61–63, 1971.
38. DIN 38409, Teil 23, 1980.
39. Waters, J., Garrigan, J. T., and Paulson, A. M., *Water Res.*, 20, 247–253, 1986.
40. Kupfer, W., *Tenside Surf. Det.*, 19, 158–161, 1982.
41. McEvoy, J. and Giger, W., *Environ. Sci. Technol.*, 20, 376–383, 1986.
42. Marcomini, A., Pavoni, B., Sfriso, A., and Orio, A. A., *Mar. Chem.*, 29, 307–323, 1990.
43. Giger, W., Brunner, P. H., and Schaffner, C., *Science*, 225, 623–625, 1984.
44. Oshurn, Q. W., *J. Am. Oil Chem. Soc.*, 59, 453–457, 1982.
45. De Ruiter, C., Hefkens, J. C. H. F., Brinkman, U. A. Th., Frei, R. W., Evers, M., Matthijs, E., and Meijer, J. A., *Int. J. Environ. Anal. Chem.*, 31, 325–339, 1987.
46. Hellmann, H., *Z. Wasser Abwasser Forsch.*, 22, 4–12, 1989.
47. Field, J. A., Miller, D. J., Field, T. M., Hawthorne, S. B., and Giger, W., *Anal. Chem.*, 64, 3161–3167, 1992.
48. Lee, H. B. and Peart, T. E., *Anal. Chem.*, 67, 1976–1980, 1995.
49. Fernandez, P., Alder, A. C., Suter, M. J. F., and Giger, W., *Anal. Chem.*, 68, 921–929, 1996.
50. Osburn, Q. W., *J. Am. Oil Chem. Soc.*, 63, 257–263, 1986.
51. Wickbold, R., *Tenside Surf. Det.*, 9, 173–177, 1972.
52. Waters, J. and Kupfer, W., *Anal. Chim. Acta*, 85, 241–251, 1976.
53. DIN 38409, Teil 20, 1989.
54. Marcomini, A., Stelluto, S., and Pavoni, B., *Int. J. Environ. Anal. Chem.*, 35, 207–218, 1989.
55. Vogt, C., Heinig, K., Langer, B., Mattusch, J., and Werner, G., *Fresenius J. Anal. Chem.*, 352, 508–514, 1995.
56. Di Corcia, A., Marchetti, M., Samperi, R., and Marcomini, A., *Anal. Chem.*, 63, 1179–1182, 1991.
57. Castles, M. A., Moore, B. L., and Ward, S. R., *Anal. Chem.*, 61, 2534–2540, 1989.
58. Nakae, A., Tsuji, K., and Yamanaka, M., *Anal. Chem.*, 52, 2275–2277, 1980.
59. Holt, M. S., Matthijs, E., and Waters, J., *Water Res.*, 23, 749–759, 1989.
60. Liebscher, G., Eppert, G., Oberender, H., Berthold, H., and Hauthal, H. G., *Tenside Surf. Det.*, 26, 195–197, 1989.
61. Pietrzyk, D. J., Rigas, P. G., and Yuan, D., *J. Chromatogr. Sci.*, 27, 485–490, 1989.
62. Schoester, M. and Kloster, G., *Fresenius J. Anal. Chem.*, 345, 767–772, 1993.
63. Knepper, T. P. and Kruse, M., *Tenside Surf. Det.*, 37, 41–47, 2000.
64. Ahel, M. and Giger, W., *Anal. Chem.*, 57, 2584–2590, 1985.
65. Holt, M. S., McKerrel, E. H., Perry, J., and Watkinson, R. J., *J. Chromatogr.*, 362, 419–424, 1986.
66. Marcomini, A., Tortato, C., Capri, S., and Liberatori, A., *Ann. Chim.*, 83, 461–484, 1993.

67. Schmitt, T. M., Allen, M. C., Brain, D. K., Guin, K. F., Lemmel, D. E., and Osburn, Q. W., *J. Am. Oil Chem. Soc.*, 67, 103–109, 1990.
68. Kiewiet, A. T., van der Steen, J. M. D., and Parsons, J. R., *Anal. Chem.*, 67, 4409–4415, 1995.
69. Nitschke, L. and Huber, L., *Fresenius J. Anal. Chem.*, 345, 585–588, 1993.
70. Spilker, R., Menzebach, B., Schneider, U., and Venn, I., *Tenside Surf Det.*, 33, 21–25, 1996.
71. Schröder, H. F., *J. Chromatogr. A*, 777, 127–139, 1997.
72. Li, H. Q., Jiku, F., and Schröder, H. F., *J. Chromatogr. A*, 889, 155–176, 2000.
73. Wee, W. T., *Water Res.*, 18, 223–225, 1984.
74. Nitschke, L., Müller, R., Metzner, G., and Huber, L., *Fresenius J. Anal. Chem.*, 342, 711–713, 1992.
75. Schoester, M. and Kloster, G., *Vom Wasser*, 77, 13–20, 1991.
76. Waters, J. and Garrigan, J. T., *Water Res.*, 17, 1549–1562, 1983.
77. Fendinger, N. J., Begley, W. M., McAvoy, D. C., and Eckhoff, W. S., *Environ. Sci. Technol.*, 26, 2493–2498, 1992.
78. Ding, W. H., Lo, J. H., and Tzing, S. H., *J. Chromatogr. A*, 818, 270–279, 1998.
79. Suter, M. J. F., Reiser, R., and Giger, W., *J. Mass Spectrom.*, 31, 357–362, 1996.
80. Ahel, M., Conrad, T., and Giger, W., *Environ. Sci. Technol.*, 21, 697–703, 1987.
81. Wahlberg, C., Renberg, L., and Wideqvist, U., *Chemosphere*, 20, 179–195, 1990.
82. Stephanou, E., *Chemosphere*, 13, 43–51, 1984.
83. Stephanou, E., *Org. Mass Spectrom.*, 19, 510–513, 1984.
84. Günther, K., Dürbeck, H. W., Kleist, E., Thiele, B., Prast, H., and Schwuger, M. J., *Fresenius J. Anal. Chem.*, 371, 782–786, 2001.
85. Ding, W. H. and Tzing, S. H., *J. Chromatogr. A*, 824, 79–90, 1998.
86. Brossard, S., Lafosse, M., and Dreux, M., *J. Chromatogr.*, 591, 149–157, 1992.
87. Silver, A. H. and Kalinoski, H. T., *J. Am. Oil Chem. Soc.*, 69, 599–608, 1992.
88. Heinig, K., Vogt, C., and Werner, G., *J. Chromatogr. A*, 745, 281–292, 1996.
89. Gallagher, P. A. and Danielson, N. D., *J. Chromatogr. A*, 781, 5331997.
90. Ding, W. H. and Liu, C. H., *J. Chromatogr. A*, 929, 143–150, 2001.
91. Riu, J. and Barcelo, D., *Analyst*, 126, 825–828, 2001.
92. Heinig, K., Vogt, C., and Werner, G., *Fresenius J. Anal. Chem.*, 358, 500–505, 1997.
93. Heinig, K., Vogt, C., and Werner, G., *J. Chromatogr. A*, 781, 17–22, 1997.
94. Grob, M. and Steiner, F., *Electrophoresis*, 23, 1921–1927, 2002.
95. Ventura, F., Caixach, J., Figueras, A., Espalder, I., Fraisse, D., and Rivera, J., *Water Res.*, 23, 1191–1203, 1989.
96. Schneider, E., Levsen, K., Dähling, P., and Röllgen, F. W., *Fresenius J. Anal. Chem.*, 316, 488–492, 1983.
97. Ventura, F., Figueras, A., Caixach, J., Espadaler, I., Romero, J., Guardiola, J., and Rivera, J., *Water Res.*, 22, 1211–1217, 1988.
98. Ventura, F., Fraisse, D., Caixach, J., and Rivera, J., *Anal. Chem.*, 63, 2095–2099, 1991.
99. Schneider, E. and Levsen, K., In *Identification of Surfactants and Study of These Degradation in Surface Water By MS*, Bjoerseth, A. and Angeletti, G., Eds., D. Reidel Publ., Dordrecht, p. 14, 1986.
100. Otsuki, A. and Shiraishi, H., *Anal. Chem.*, 51, 2329–2332, 1979.
101. Shiraishi, H., Otsuki, A., and Fuwa, K., *Bull. Chem. Soc. Jpn.*, 55, 1410–1415, 1982.
102. Schneider, E., Levsen, K., Dähling, P., and Röllgen, F. W., *Fresenius J. Anal. Chem.*, 316, 277–285, 1983.
103. Weber, R., Levsen, K., Louter, G. J., Henk Boerboom, A. J., and Haverkamp, J., *Anal. Chem.*, 54, 1458–1466, 1982.
104. Hellmann, H., *Fresenius Z. Anal. Chem.*, 321, 159–162, 1985.
105. Hellmann, H., *Tenside Surf. Det.*, 28, 111–117, 1991.
106. Hellmann, H., *Z. Wasser Abwasser Forsch.*, 16, 174–179, 1983.

31 Determination of Flame Retardants in Environmental Samples

Tuulia Hyötyläinen

CONTENTS

I. PHYSICAL AND CHEMICAL PROPERTIES OF FLAME RETARDANTS (FRs)

FRs are chemicals that can inhibit ignition and/or reduce the burning rate of a product. These are often used by manufacturers to help them meet fire safety standards set down by law for products such as electrical equipment, furnishings, and vehicles. These compounds are added to polymers, paint, textiles, and other materials to improve their fireproof properties. The main applications are in plastic housings of electronic products such as TV sets and computers, car parts, circuit boards, electric components, and cables. There are currently over 100 different substances used as FRs and these can be classified in four categories (Figure 31.1):

1. Inorganic compounds
2. Organophosphorus (mainly phosphate esters)
3. Halogenated organic compounds (mainly brominated or chlorinated)
4. Nitrogen based compounds

Further, the flame retardants can be divided into reactive and additive flame retardants according to their use. The reactive chemicals are covalently bonded to the polymer and therefore

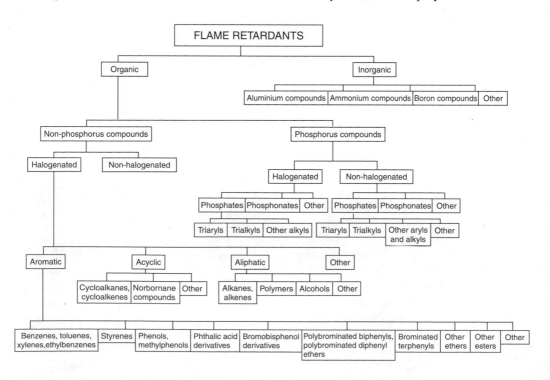

FIGURE 31.1 The classification of FRs (Modified from Kemi, *The Flame Retardants Project*, Swedish National Chemicals Inspectorate, (KemI) Report 5/96).

less likely to leach out to the environment until the product is decomposed or burnt. The additive compounds, on the other hand, are only mixed with or dissolved in the material and can more easily migrate out of the product.

The scale of production and use of flame retardants has grown dramatically along with growth in the use of synthetic polymers and the introduction of more rigorous fire safety requirements. This growth is reflected in their increasing levels in the environment. The FRs enter the environment directly from point sources during manufacture or disposal or are released from products over their lifetimes. Many flame retardants are also inherently stable as they are intended to exist in the treated article for its whole lifetime. In addition, burning products containing halogenated FRs can also release toxic by-products. Particular attention has been drawn to the production of polybrominated dibenzofurans (PBDF) and polybrominated dibenzo-*p*-dioxins (PBDD), which can be formed during high temperature processing (e.g., as part of production or recycling) as well as during combustion.

A. Types of FRs

The most important organic FRs are halogenated (brominated or chlorinated) compounds and phosphate esters.[2] The major inorganic products are aluminum hydroxide (alumina trihydrate, ATH), antimony oxides and borates. Other inorganic compounds used as FRs include molybdenum compounds, magnesium hydroxide, ammonium polyphosphate, and red phosphorus. Some of these inorganic compounds function as synergists rather than directly as FRs, enhancing the effectiveness. Synergists are not usually used alone unless the chemical nature of the polymer provides some innate flame retardancy.

1. Inorganic FRs

Inorganic FRs include metal compounds, boron compounds, and others. Of the metal compounds, antimony compounds and metal hydroxides have the highest rate of consumption.[2] Boric acid and sodium borate are frequently used in cellulosics, especially cotton and paper, where the presence of hydroxyl groups contributes to the effectiveness of these FRs. Other inorganic FRs include phosphorus-containing compounds, which are used in cellulosic textiles. Phosphoric acid itself has been used to treat cellulosics. Ammonium polyphosphates of varying chain lengths are used in many applications, particularly in coatings, and paints. Other FRs are frequently used in conjunction with them.

2. Organophosphorus Compounds

Organophosphorus compounds (OPs) are utilized on a large scale as flame retarding agents and plasticizers in a variety of products, such as plastic materials, rubbers, varnishes, lubricants, hydraulic fluids, and other industrial applications. This family of chemicals consists of alkylated and arylated phosphate or phosphonate esters and related compounds such as phosphites, phosphines, and related dimeric forms as well as ionic forms (Figure 31.2).[2–4] The low volatility of phosphoric acid and derivatives makes it the preferred choice of the phosphorus based FRs. These FRs are most effective in polymers that char readily. Also halogenated phosphate esters, such as tris(1-chloroisopropyl) phosphate (TCPP), and tris(2-chloroethyl) phosphate (TCEP), are widely used. These combine the properties of both the halogen and the phosphorus compounds.

3. Organohalogens

The organohalogen FRs can be classified into three groups: aromatic, aliphatic, and cycloaliphatic compounds. The halogen is either chlorine or bromine. Fluorinated compounds are expensive and generally not effective; iodinated compounds are effective but unstable and are therefore not used.

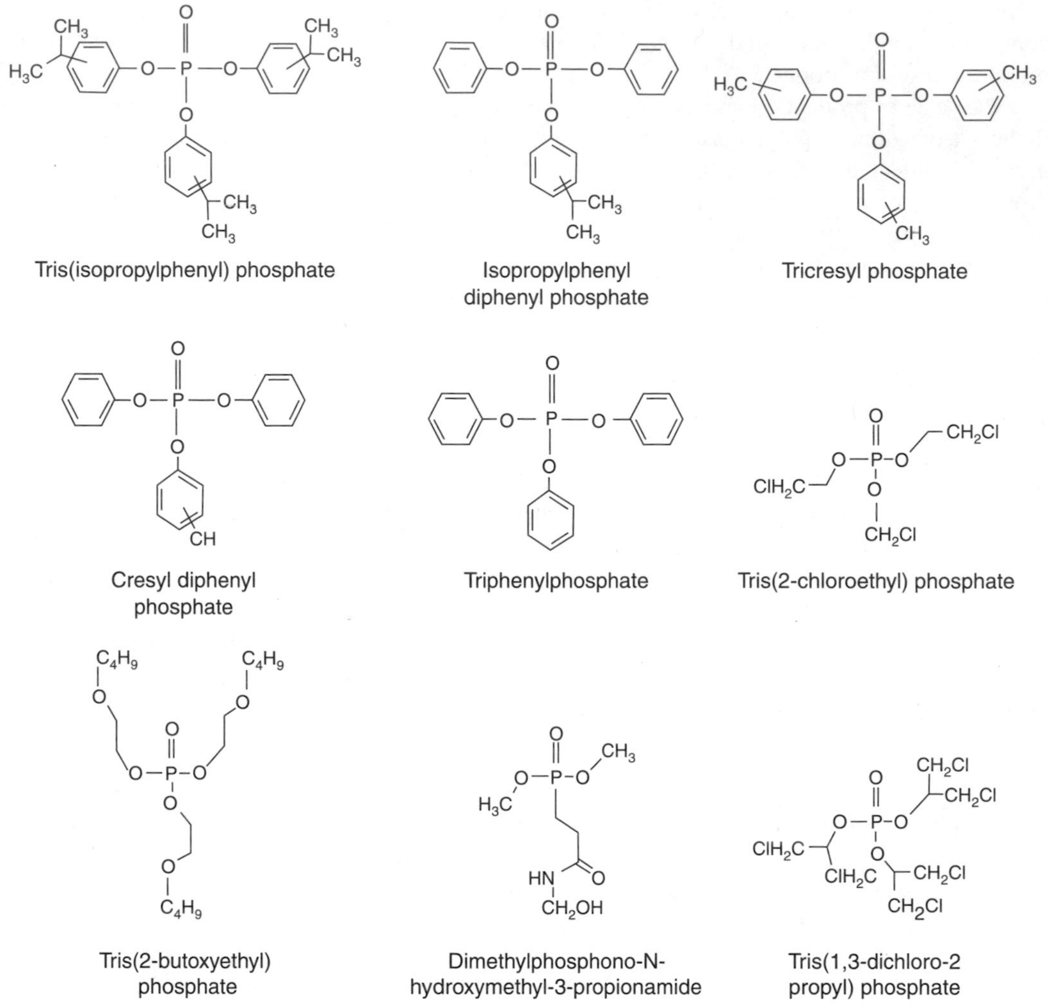

Tris(isopropylphenyl) phosphate

Isopropylphenyl diphenyl phosphate

Tricresyl phosphate

Cresyl diphenyl phosphate

Triphenylphosphate

Tris(2-chloroethyl) phosphate

Tris(2-butoxyethyl) phosphate

Dimethylphosphono-N-hydroxymethyl-3-propionamide

Tris(1,3-dichloro-2 propyl) phosphate

FIGURE 31.2 Structures of selected organophosphorus FRs.

A wide variety of organohalogens are used as additives or reactive FRs, and a few can be used as either, depending upon the application.[2,3]

a. Brominated FRs

Brominated flame retardants (BFRs) are a structurally diverse group of compounds including aromatics, cyclic aliphatics, phenolic derivatives, aliphatics, and phthalic anhydride derivatives (Figure 31.3).[5–7] The most common BFRs are tetrabromobisphenol A (TBBPA), polybrominated diphenyl ethers (PBDE), hexabromocyclododecane (HBCD), and polybrominated biphenyls (PBB). The primary use of TBBPA is as reactive additive in epoxy resin circuit boards, while decabromodiphenyloxide (DBDO) is primarily used in high impact polystyrene for electronic enclosures. PBDEs are typically used as the additive type of flame retardant in high impact polystyrene, acrylonitrile butadiene styrene, flexible polyurethane foam, textile coatings, wire and cable insulation and electrical connectors.

Structurally, the PBBs and PBDEs both comprise 209 congeners, in a similar manner to PCBs, and the same numbering system is used for the individual PPB and PBDE congeners as for

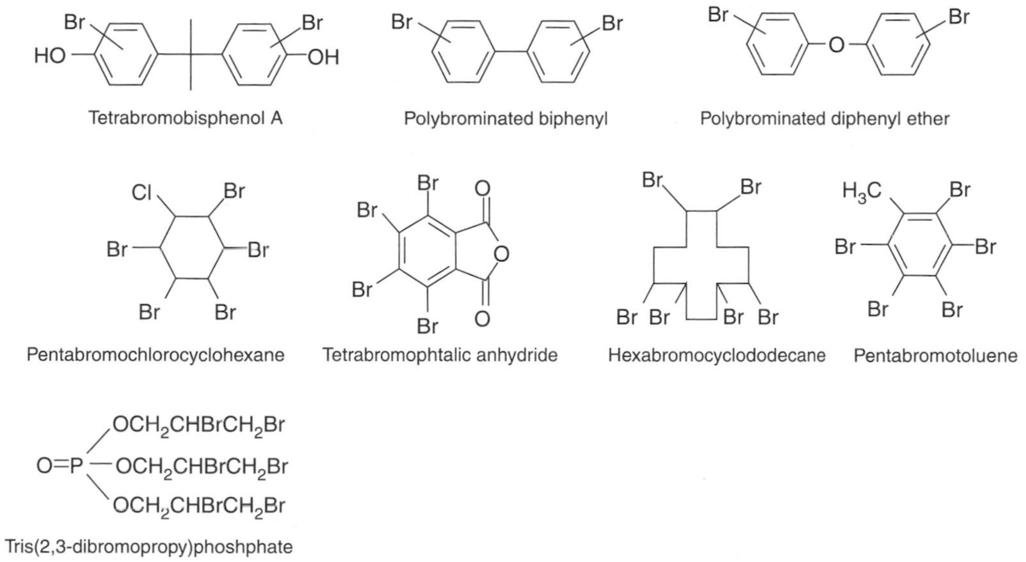

Tetrabromobisphenol A   Polybrominated biphenyl   Polybrominated diphenyl ether

Pentabromochlorocyclohexane   Tetrabromophtalic anhydride   Hexabromocyclododecane   Pentabromotoluene

Tris(2,3-dibromopropy)phoshphate

FIGURE 31.3 General structures of the most common PBDEs.

the PCBs. Present commercially available PBDE products primarily consist of highly brominated compounds, such as penta-, octa-, or decabromodiphenyl ethers. For example, the commercial technical PBDE mixtures generally contain less than ten congeners, while commercial technical PCBs are mixtures of perhaps 80 congeners. PBBs and PBDEs share many of the properties of PCBs and PCDDs, which make them long-lived, bioaccumulating, environmental pollutants.

b. Chlorinated Organics

Chlorinated flame retardants are much less effective than the equivalent brominated compounds, and are declining in use.[2] The level of chlorination necessary for sufficient flame retardancy often has a detrimental effect on the desired properties of the polymer. Chlorinated paraffins (aliphatics) are used in plastics, textiles, and coatings. Chlorinated aromatics are not used as FRs.

4. Nitrogen-Based FRs

Nitrogen-based flame retardants are used mainly in polymers in which nitrogen is present (e.g., polyurethanes, polyamides), and also polyolefins.[2] The most important inorganic nitrogen–phosphorus compound used as a flame retardant is ammoniumpolyphosphate which is applied in intumescent coatings and in rigid polyurethane foams. The most important organic nitrogen compound used as a flame retardant is melamine: melamine or derivatives are added to intumescent varnishes or paints. At present, its main applications are melamine for polyurethane flexible foams, melamine cyanurate in nylons, melamine phosphates in polyolefins, melamine and melamine phosphates, or dicyandiamide in intumescent paints, guanidine phosphates for textiles and guanidine sulphamate for wallpapers. Ammonium sulphate and sulphamate and the ammonium halides are also used as FRs in various cellulosic products (textiles, paper, and wood), and in fighting forest fires.

B. Occurrence in Environment

The environmental fate of the many of the FRs is not well documented. The water solubilities and vapor pressures of many of the FRs are very low (Table 31.1), so that, when released to the environment, these compounds are likely to quickly adsorb onto solid particles of sediment

TABLE 31.1
Chemical Properties of Most Common FRs Divided to the Following Groups: B, brominated; C, chlorinated; CP, chlorophosphates; N, nitrogen containing; P, organophosphates; and IO, inorganic FRs

Type	Name and Abbreviation	Water Solubility (mg/l)	Log K_{ow}[a]
B	Tetrabromobisphenol-A (TBBPA)	0.08	4.54
B	Hexabromobenzene	1.60×10^{-4}	6.07
B	2,4,6-Tribromophenol	50	4.13
B	Tris(2,3-dibromopropyl)phosphate	8	3.71
B	2,4-Dibromophenol	1.90×10^{3}	3.22
B	Dibromoneopentylglycol	2.00×10^{4}	1.06
B	HBCD	3.40×10^{-3}	5.63
B	Octabromodiphenylether	5.00×10^{-4}	6.29
B	Pentabromophenol	—	—
B	Pentabromotoluene	Insoluble	—
B	Polybrominated di-Ph ethers (PBDEs)	20 to 30 μg/l (deca)	5.24 (deca)
		9×10^{-7} mg/l (penta)	5.5 (octa)
		4.8 mg/l (mono)	6.86 to 7.92 (hexa)
			6.64 to 6.97 (penta)
			5.87 to 6.16 (tetra)
			5.47 to 5.58 (tri)
			5.03 (di)
			4.28 (mono)
B	PBBs	0.65 mg/l (mono)	4.59 to 4.96 (mono)
		0.006 mg/l (di)	5.72 to 5.78 (di)
		0.016 mg/l (tri)	6.03 to 6.42 (tri)
		0.004 mg/l (tetra)	6.5 to 7.42 (tetra)
		0.0004 mg/l (penta)	7.1 (penta)
		0.00056 mg/l (hexa)	7.2 (hexa)
			8.58 (deca)
B	Tetrabromopthalic anhydride	Insoluble (<0.01)	—
B	Pentabromochlorocyclohexane	—	—
C	Tetrachlorobisphenol-A (TCBPA)	—	—
C	Hexachlorocyclopentadiene	1.80	5.04
C	Dodecachlorooctahydrometheno-1H-cyclobutapentalene	0.085	5.28
CP	Tris(2-chloroethyl)phosphate(TCEP)	7.82×10^{3}	1.78
CP	Tris(2-chloroisopropyl)phosphate(TCPP)	1.60×10^{3}	2.59
CP	Tris(3-Bromo-2,2(Bromomethyl)Propyl) Phosphate	0.9	3.7
N	Melamine	3.24×10^{3}	-1.37
P	Triethylphosphate	5.00×10^{5}	0.8
P	Tris(2-ethylhexyl)phosphate	0.6	4.22
P	Tri-butoxyethylphosphate(TBEP)	1.20×10^{3}	3.65
P	Triphenylphosphate	1.9	4.59
P	Tri-n-butylphosphate	280	4
P	Trimethylphosphate	5.00×10^{5}	-0.65
P	2-Ethylhexyldiphenyl phosphate	1.90	5.73
P	Tricresylphosphate	0.36	5.11
P	Cresyldiphenyl phosphate	0.24	4.51
P	Isodecyldiphenylphosphate	0.75	5.44

Continued

TABLE 31.1
Continued

Type	Name and Abbreviation	Water Solubility (mg/l)	Log K_{ow}[a]
P	Tris(isopropylphenyl)phosphate	1.0	5
IO	Aluminium trihydroxide	Insoluble	—
IO	Magnesium hydroxide	Insoluble	—
IO	Phosphorus	Insoluble	—
IO	Zinc borate	Insoluble	—

[a] Approximately, slightly different values are given in literature.

Data from WHO/ICPS. *Environmental Health Criteria 192: Flame Retardants — General Introduction*, World Health Organization, Geneva, 1997; WHO/ICPS. *Environmental Health Criteria 209: Flame Retardants: Tris [chloropropyl]phosphate and Tris [chloroethyl]phosphate*, World Health Organization, Geneva, 1998; WHO/ICPS. *Environmental Health Criteria 218: Flameretardants tris[2 butoxyethyl] phosphate,tris[2 ethylhexyl]phosphate and tetrakis [hydroxymethyl] phosphonium salts*, World Health Organization, Geneva, 2000; WHO/ICPS. *Environmental Health Criteria 162: Brominated Diphenyl Ethers*, World Health Organization, Geneva, 1994.

and soil. In addition, several FRs, such as PBBs and PBDEs are highly lipophilic and resistant to degradative processes, so these compounds are also expected to bioaccumulate easily. The main emphasis in environmental analysis of FRs has been on the determination of organohalogen based FRs, particularly brominated FRs, and organophosphate FRs, as these compounds are considered to possess the most serious environmental hazard. The main reason for this is that many of these compounds are persistent, can be toxic, and as they have relatively high soil adsorption coefficients, they can be accumulated into environment. Some of the FRs that are analyzed in the sample are not in use at present (e.g., PBBs), but as the lifetimes of many products are long, and the analytes are persistent, they still can pose a risk for the environment.

The increasing scale of production and use of FRs is reflected in its increasing level in the environment. FRs enter the environment from products over their entire lifetimes, not just from point sources during manufacture. In addition, heating of, e.g., PBBs and PBDEs may lead to the formation of brominated dioxins and furans. FRs have been found throughout the world in air, sediment, sewage sludge, fish tissue, bird eggs, whale, dolphin and seal fat, mussels and sediment, in human serum, milk and tissue.[8-64] Congener patterns in the environmental samples do not always match those of technical products, indicating an environmental alteration, possibly by photochemical reactions. As a result of bioaccumulation the concentrations in higher levels of the food chain are sometimes substantial. For example, human milk samples collected in Sweden between 1972 and 1997 showed continuously increasing levels of brominated diphenyl ethers, in contrast to the decline seen for other organohalogenated compounds such as DDTs, PCBs, and PCNs.[45]

Inorganic FRs are incorporated into the plastics as fillers and these are usually considered immobile in the plastics, in contrast to the organic additives. Emissions during use can be considered negligible. In addition, most inorganic FRs have presumably insignificant environmental effects. Aluminum trihydroxide, for example, is not released during use, it has minimum human and environmental toxicity, it suppresses formation of hazardous fumes and decomposes into nonhazardous substances.[2] Also nitrogen-based compounds can be considered relatively environmentally friendly because they are of low toxicity, are usually in a solid state and, in case of fire, do not produce dioxin and halogen acids or large amounts of smoke.[2] Therefore, these types of FRs are not analyzed widely in environmental samples.

Brominated flame retardents (BFRs) can accumulate to the environment and have been found in water, biota, soil and sediment. Several BFRs undergo photochemical reactions by UV radiation, and therefore, various reaction products of the BFRs can also be found in the environment.[64]

For example, in sunlight deca-BDE easily degrades to lower brominated congeners, which themselves readily bioaccumulate.[7] It is unclear at present what proportion of the tetra- to hexa-BDEs found in the environment are breakdown products of deca-BDE congeners and what proportion comes from commercial penta-BDE mixtures. The impact on health and the environmental characteristics of BFRs are generally not well known. The acute toxicity of most of the BFRs has shown to be fairly low, but some BFRs have shown similar toxic effects to PCBs and polychlorinated dibenzo-p-dioxins and furans.[7,65] The available data suggest, for example, that the lower PBDE congeners (tetra to hexa) are likely to be carcinogens, endocrine disrupters, and neurodevelopment toxicants.[64,65] Deca-BDE, which is the major commercial product, is presumed to be a less active congener than the lower BDEs because of its lower bioavailability and poor gastrointestinal adsorption.[30,31,66] Some studies have shown that PBDEs can be metabolized to hydroxylated compounds, and as such, these polybrominated phenoxyphenols may compete with thyroxin for the binding of the thyroxin transporting protein transthyretin.[67] PBDEs and hydroxylated PBDEs are also reported to possess estrogenic activity. In addition, HBCD has been shown to induce intragenic recombination in mammalian cells, indicating that it is carcinogenic.[67]

C. REGULATIONS

Only part of the FRs have been evaluated in detail so far (PBB, PBDE, and chlorinated paraffins), and have been found to be harmful for the environment. Some of these have not been recommended for use. Several countries have developed regulations affecting the production, use and disposal of FRs.

Several countries have been given restrictions on the use of compounds because of potential toxic effects in humans. In the European Community, the use of tris(2,3-dibromopropyl) phosphate (EC Directive 76/769/EEC) and tris(1-aziridinyl)phosphine oxide (EC Directive 83/264/EEC) in textiles has been banned. In 1977, the U.S. Consumer Product Safety Commission banned the use of tris(2,3-dibromopropyl) phosphate in children's clothing (ICPS, 1995). The European Community has also banned the use of PBBs in textiles (EC Directive 83/264/EEC). Several countries have either taken or proposed regulatory actions on PBBs. In addition, controls on the emissions of dioxins and furans from municipal solid waste incinerators have been implemented in the United Kingdom under the Environmental Protection Act (1990). Germany has developed rules for the maximum content of selected 2,3,7, and 8 substituted polychlorinated dibenzo-$para$-dioxins and dibenzofurans in products. Recently, the European Commission has issued a proposal to ban the production and use of PentaBDE. In U.S.A., on the other hand, there are currently no regulations on PBDE production or use. PBBs have not been used widely in Europe and also in U.S.A. the production of the main mixture, hexabromobiphenyl (Firemaster BP-6), ceased in 1974, after the Michigan disaster.[5]

II. SAMPLE PREPARATION

In the analytical scheme, besides the sampling and the final analysis, the sample preparation and cleanup are also crucial. Sample preparation plays an important role in the analysis of FRs in environmental samples because of the complex matrices and only trace levels of analytes. Solid and semisolid samples are usually first dried and homogenized. Then the FRs are extracted from the sample (solid or liquid), and the extract is usually purified, fractionated, and concentrated before the final analysis, which is typically performed with gas or liquid chromatography. The extraction procedure is dependent on the sample matrix; different methods are used for sediment, tissue, and liquid samples. After extraction, it will usually be necessary to purify and fractionate the extract, because most extraction methods are insufficiently selective and the separation power of the analytical technique not sufficient. Extracts typically contain several analytes similar to the FRs, which may be present in much higher quantities. The fractionation procedures are similar for the different types of extracts. Typical analytical procedures are given in Tables 31.2 to 31.6.

TABLE 31.2
General Sample Pretreatment Procedures for Determination of BFRs and OPS in Environmental Samples

Sample Type	Sediment, Soil, Sludge	Air Particles (collected into filter or adsorbent)	Biological Tissue (fish, mussels, eggs, blubber, liver, adipose tissue)	Biological Fluid (serum, urine, milk)	Water
Pretreatment	Grinding, homogenization, drying	Not required	Homogenization	Not required	Not required
Extraction	Soxhlet, LSE, PLE, SAE, MAE, SFE	Soxhlet, LSE, PLE, SAE, MAE	Scxhlet, LSE, PLE, SAE, MAE, SFE	LLE, SPE	LLE, SPE
Halogenated FRs, solvent	Hex, acetone, tol, hex: acetone and DCM: cyclohexane	Hex, acetone, tol, hex: acetone DCM	Hex or acetone, hex:ACN, hex:DEE, hex:DCM, hex, DCM	Hex or acetone, hex: MTBE	Hex:acetone hexane:MTBE
OPs, solvent	Acetone and ACN:DCM	Hex, acetone, toluene, hex:acetone DCM	Hex:acetone, hex:ACN, hex:diethyl ether, hex: DCM, hex, DCM	DCM, tol, DCM: ACN or DCM: CCl_4	DCM, tol, DCM: ACN, DCM: CCl_4
Sulphur removal	Treatment with conc. H_2SO_4, Cu powder: TBA:sulphite	Not required	Not required	Not required	Not required
Lipid removal	Not required	Not required	Treatment with conc. H_2SO_4:GPC	Treatment with conc. H_2SO_4:GPC	Not required
Liquid partitioning (not always required)	Not required	KOH or EtOH partitioning (neutral compounds)	KOH or EtOH partitioning (neutral compounds)	KOH or EtOH partitioning (neutral compounds) (SPE)	Not required

Solvents: ACN, acetonitrile; DCM, dichloromethane; DEE, diethyl ether; EtOH, ethanol; hex, hexane; MTBE, methyl-*tert*-butyl ether; tol, toluene.
Techniques: LLE, liquid–liquid extraction; LSE, liquid–solid extraction; MAE, microwave assisted extraction; PLE, pressurized liquid extraction; SAE, sonication assisted extraction; SFE, supercritical fluid extraction; SPE, solid phase extraction; TBA, tetrabutylammonium hydroxide.

TABLE 31.3
Approaches for the Extraction, Cleanup Procedures and Analysis for Soil, Sediment and Sewage Sludge

Sample	Analytes	Pretreatment	Extraction	Cleanup	Analysis	References
Sewage sludge	PBDEs, PBDFs, PBDDs	Drying, powdering	Soxhlet extraction with toluene, 18 h	(1) 30 × 2 cm silica column: 2 g of silica gel + 15 g silicagel/44% conc. H_2SO_4 2 g of silica gel, elution with 100 ml C_6H_{14} or DCM, 8/2 (2) 30 × 2 cm column packed with 30 g alumina, impregnated with 10% aq. solution of 5% $AgNO_3$, elution with 100 ml hexane or acetone, 96/4 (3) 30 × 2.5 cm GPC, BioBeads S-X3, elution with cyclohexane:EtAc, 1:1, fraction 100 to 180 ml (4) 25 × 0.8 cm HPLC, Nucleosil-5 NO_2, elution with hexane, fraction 10 to 42 ml (5) 11 × 0.07 mm column packed with 2 g basic Alumina B-Super I, elution with C_6H_6 (PCBs, CBs) and hexane: DCM, 98/2 (PCBs, CBs) and hexane: CH_2Cl_2, 1:1 PBDEs	GC–EI–MS	69
Sediment, sewage sludge	PBDEs, HBCD, TBBPA, DDT, DDE, PCBs, PCNs	Centrifugation	(1) 60 min with 40 ml acetone (2) 50 ml 0.2 M NaCl in 0.1 M Na_2HPO_4 (3) 30 min with 40 ml acetone:n-hexane, 1/3 (4) 10 ml n-hexane:DEE: undecane, 90/10/2	(1) Concentration to 2 ml (2) Mixing with 4 ml of 2-propanol: TBA-sulphite (1/1), washing with water, centrifugation (3) Treatment with H_2SO_4	GC–NCI–MS (methane)	54,55

| Sediment | DPDE, PBBs, PCBs, CBs | Drying, homogenization, mixing with Cu-powder | SFE: CO$_2$, 20 min static and 40 min dynamic extraction at 120 EC and 374 bar, trapping in C$_{18}$ column, elution with heptane:EtAc, 98:2 | Not required | GC–MS | 70 |
| Sediment | PBCCH, HpBB, PBT, TBBPA, TDBPP | Drying, homogenization | PHWE: 40 min, water at 325°C, 120 bar, trapping with Tenax TA, elution with pentane-EtAc | Not required | GC–EI–MS | 71 |

Analytes: CB, chlorobenzenes; DDE, 1,1-dichloro-2,3-bis(4-chlorophenyl)ethane; DDT, 1,1-dichloro-2,3-bis(4-chlorophenyl)ethylene; DBF, decabromobiphenyls; HBCD, Hexabromocyclododecane; HpBB, heptabromobiphenyl; PBB, polybrominated biphenyl; PBCCH, pentabromochlorccyclohexane; PBDE, polybrominated diphenyl ether; PBT, Polybutylene terephthalate; PCB, Polychlorinated biphenyl; PBDD, polybrominated dibenzo-*p*-dioxins; PBDF, polybrominated dibenzofuranes; PCN, polychlorinated naphthalenes; PCP, polychlorinated phenols; PBB, polybrominated biphenyl; PeBDE, pentabromodiphenyl ether; PET, Polyethylene terephthalate; PXDDs, polyhalogenated dibenzo-*p*-dioxins; PXDFs, polyhalogenated dibenzofurans; TBBPA, tetrabromobisphenol A; TBPA, tetrabromophthalic anhydride; TCBPA, tetrachlorobisphenol A; TDBPP, tris(2,3-dibromopropyl)phosphate.

Solvents: ACN, acetonitrile; DCM, dichloromethane; DEE, diethyl ether; EtAc, ethyl acetate; EtOH, ethanol; MTBE, methyl-*tert*-butyl ether; tol, toluene.

Other: PHWE, pressurized hot water extraction; SFE, supercritical fluid extraction; TBA, tetrabutylammonium hydroxide.

TABLE 31.4
Extraction, Cleanup Procedures and Analysis for Air Particle Samples

Sample, Analytes	Analytes	Extraction	Cleanup	Detector	References
Indoor air, Ioniser, aluminium collector cup	PBDEs, TBBPA, PBBs, PCP	Particles were wet wiped with glass wool dipped in CH_2Cl_2, dissolving to hexane	(1) Partitioning with KOH:EtOH (2) Concentration to 200 μl, derivatization with diazomethane (3) Treatment with conc. H_2SO_4	GC–NPD, GC–NCI–MS	74
Outdoor air, quartz fibre filter, XAD-2 adsorbent	PBDEs, pesticides, DDT, DDT, PCBs, HCB	Soxhlet extraction (24 h, acetone:hexane, 1/1) of filters and adsorbents	(1) Concentration, solvent exchange to hexane (2) Silica gel clean up, elution with hexane (PCBs, HCB, DDE) and hexane:CH_2Cl_2, 1/1 (PBDEs, pesticides)	GC–ECD and GC–EI–MS	58
Indoor air, quartz filter	OPEs	Dynamic SAE (3 min, hexane:MTBE) at 70°C, flow rate 0.2 ml/min, 120 kW, 35 kHz	—	GC–NPD	78
Indoor air, quartz filter, PUF	OPEs	SAE 2 × 20 min, 2 × 5 ml, DCM, at 50 W and a frequency of 48 kHz	Concentration	GC–NPD, GC–MS	15
Indoor air, adsorbed in charcoal tube	OPEs	SAE, 10 min, 1 ml toluene at 24 kHz	Centrifugation	GC–FPD	77
Indoor air, adsorbed in PUF	OPEs	SAE, 2 × 10 min, acetone: cyclohexane	Concentration	GC–MS	36
Indoor air, adsorbed in PUF	OPEs	Cold extraction, 12 h, water: acetone	Concentration	GC–MS	36
Indoor air, adsorbed in PUF	OPEs	Soxhlet, 8 h, hexane:acetone	Concentration	GC–MS	36

Analytes: DDE, 1,1-dichloro-2,3-bis(4-chlorophenyl)ethane; DDT, 1,1-dichloro-2,3-bis(4-chlorophenyl)ethylene; HCB, hexachlorinated benzenes; PBB, polybrominated biphenyl; OPE, organophosphorus esters; PBDE, polybrominated diphenyl ether; PCP, polychlorinated phenols.
Other: DCM, dichloromethane; SAE, sonication-assisted extraction.

TABLE 31.5
Extraction, Cleanup Procedures and Analysis for Biota Samples

Sample, Analytes	Pretreatment	Extraction	Cleanup	Analysis	References
Fish muscle, PBDEs, PCBs, FCDBFs, DBFs	Homogenization	Soxhlet extraction with DCM, 20 h	(1) 3 × treatment with conc. H_2SO_4 (2) 30 × 0.2 cm column packed with 2 g 10% $AgNO_3$–silica, 3 g 44% H_2SO_4–silica and 2 g KOH–silica; elution with 200 ml hexane: CH_2Cl_2, 5/95 (3) 30 × 0.1 cm column, 3 g alumina, elution with 30 ml hexane:CH_2Cl_2, 4/96 (PBDEs) and 20 ml hexane:DCM, 1/1 (PBDDs, PBDFs, PCDDs, PCDFs)	GC–EI–MS	82
Fish tissue, PBDEs	Homogenization with anhyd. Na_2SO_4	Soxhlet extraction with n-hexane:acetone, 1:1, 4 h	(1) Evaporation to 1 ml (2) Passing through 3 g alumina column, elution with first 4 ml of hexane (3) Concentration and change of solvent to iso-octane: 1 ml	GC–ECD, conformation with GC–EI–MS	12
Fish, PBDEs, PCBs	Homogenization with dry ice	Extraction with DCM in a column	(1) GPC (2) Silica column, 5 g, elution with 50 ml hexane	GC–ITMS	43
Fish tissue, PBDEs, HBCD	Homogenization	Extraction with hexane: acetone, then with hexane:diethyl ether	(1) Washing with NaCl:$Na_2H_2PO_4$ (2) Treatment with conc. H_2SO_4	GC–NCI–MS	55
Whale, PBDEs	Homogenization with anhyd. Na_2SO_4	Extraction with hexane: DCM (1/1)	(1) Multilayer silica column (H_2SO_4–silica:neutral activated silica:KOH–silica), elution with hexane	GC–EI–MS	41
Seal blubber, toxaphene, chlordanes, PBDEs	Homogenization	Extraction with 35 ml hexane:acetone (5/2) and then with 25 ml of hexane	(1) Extraction with 10 ml 0.1 M H_3PO_4 with 1% NaCl (2) Silica gel column, washing with hexane, elution with hexane:diethyl ether (3/1) (3) Alumina column, elution with hexane	GC–ECD	80

Continued

TABLE 31.5
Continued

Sample, Analytes	Pretreatment	Extraction	Cleanup	Analysis	References
Whale blubber, PBDEs	Homogenization with Na_2SO_4, mixing with AlO_x	SFE with CO_2, at 40°C and 281 bar, flow rate 2 ml/min, trapping into C_{18} column, elution with 2 ml of hexane and 2 ml of DCM	—	GC–EI–MS	83
Human breast and adipose tissue, PBDEs	Homogenization	Hexane:DCM	(1) GPC (2) Florisil column	GC–NCI–MS	81
Human adipose tissue, PBDEs, PCBs, pesticides	Homogenization with Na_2SO_4, mixing with AlO_x	SFE with CO_2, at 40 DBF, and 300 atm, flow rate 2 ml/min, trapping into PX–21/C_{18} column, elution with 10 ml hexane:DCM	—	GC–EI–MS, GC–TOFMS	109
Human adipose tissue, PBDEs	Homogenization with Na_2SO_4	Hot Soxhlet, 2 h with 75 ml of hexane:acetone: dichlormethane	(1) Evaporation (2) Acid silica:neutral silica:deactivated basic alumina, elution with hex and hexane:DCM (3) Concentration	LVI–GC–MS	16

Analytes: DBF, decabromobiphenyl; HBCD, Hexabromocyclododecane; PBCDE, polybrominated or chlorinated diphenyl ether; PBDE, polybrominated diphenyl ether; PCB, polychlorinated biphenyl; PBDD, polybrominated dibenzo-*p*-dioxins; PBB, polybrominated biphenyl; PeBDE, pentabromodiphenyl ether.

Other: DCM, dichloromethane; SFE, supercritical fluid extraction; LVI, large volume injection.

TABLE 31.6
Extraction, Cleanup Procedures and Analysis for Liquid Samples

Sample, Analytes	Pretreatment	Extraction	Cleanup	Analysis	References
Human milk, PBDEs	Mixing 10 ml milk with 10 ml formic acid and 5 g Lipidex 5000	Packing the mixture to a glass column, washing with 40 ml MeOH:H$_2$O, 30/70, 40 ml MeOH:H$_2$C, 1/1 and 60 ml MeOH:CHCl$_3$:hexane, 1/1/1, elution with 90 ml ACN	(1) 5 g aluminium oxide, elution with hexane, fraction 20 to 30 ml (2) Silica column, 0.6 g, elution with 5 ml hexane or DCM, 75/25 after washing with 4 ml hexane and 5 ml hexane or DCM, 75/25 (3) GPC, 9 g Bio beads S-X3, elution with hexane or DCM, fraction 28 to 38 ml	GC–EI–MS	45
Human milk, PBDEs	—	Extraction with 2 × hexane: acetone, 1/1	Treatment with conc. H$_2$SO$_4$	Dual capillary– GC–ECD	18
Serum, PBPs, TCBPA, TBBPA	Formic acid:2-propanol (4/1) treatment in ultrasonic bath, dilution with H$_2$O: propanol, 19/1	SPE, polystyrene-vinylbenzene (1) Conditioning (MeOH, CH$_2$Cl$_2$, MeOH or CH$_2$Cl$_2$, MeOH. H$_2$O) (2) Loading 15 ml sample, drying (3) Washing (H$_2$O or propanol), drying (4) Lipid decomposition with conc. H$_2$SO$_4$ (5) Washing (H$_2$O, NaAc, H$_2$O, MeOH or H$_2$O), drying (6) Elution with 6 ml MeOH or DCM, 1/1	(1) Concentration to 30 μl (2) Derivatization with diazomethane, evaporation of excess regent	GC–EI–MS	103
Serum, PBDEs	HCl:2-propanol treatment	Extraction with 2 × MTBE or acetone, 1/1	(1) Washing with 0.5 M KOH or EtOH (2) Treatment with conc. H$_2$SO$_4$ (3) Elution with hexane from silica gel–H$_2$SO$_4$ column (0.5 g, 2:1)	GC–NCI–MS	57

Continued

TABLE 31.6
Continued

Sample, Analytes	Pretreatment	Extraction	Cleanup	Analysis	References
Water, fire retardants, pesticides, plasticizers	—	100 l water pumped at 1.5 l/min through XAD-2 resin, Soxhlet extraction first with acetone and then with hexane	(1) Evaporation to 1 ml (2) Florisil column cleanup	GC–MS	88
Water, OPEs		SPE	—	GC–MS	24
OPE metabolites in urine	Acidification by 4 M hydrochloric acid	Membrane solvent: 6-undecanone acceptor phase: 1 mM borate buffer of pH 9.2, extraction time 12 min, extracted volume 1 ml	—	LC–MS	85

Analytes: PBDE, polybrominated diphenyl ether; PBB, polybrominated biphenyl; PBP, pentabromophenyl; TBBPA, tetrabromobisphenol A; TCBPA, tetrachlorobisphenol A; TDBPP, tris(2,3-dibromopropyl)phosphate.

Others: ACN, acetonitrile; DCM, dichloromethane; DEE, diethyl ether; GPC, gel permeation chromatography; MTBE, methyl-*tert*-butyl ether; SPE, solid phase extraction.

A. Drying and Homogenization of Solid Samples

Drying of solid samples, such as soil, sediment, and sewage sludge is usually the first step in the analysis. Dry samples are more effectively homogenized, allowing accurate subsampling for parallel analyses for other determinants. In addition, the absence of water in the samples makes the sample matrix more accessible to organic solvents. Because some of the FRs are relatively volatile, both losses and uptake of compounds from air can occur if the drying is done at room temperature or in a heated oven ($<40°C$). Freeze-drying (water evaporation below $0°C$ under vacuum conditions) is a more gentle option. Also chemical drying of samples can be performed by grinding with anhydrous Na_2SO_4. Intensive grinding and the addition of sufficient quantity of drying salt to obtain a free-flowing powder are of vital importance for a complete extraction. Drying with water-adsorbing materials (alumina, silica, etc.) may also be an alternative, but in this case water is not bound irreversibly and can easily be released only when polar solvents are used for extraction. The use of a mixture of less polar solvents (e.g., hexane, dichloromethane) may help to avoid these problems.

B. Extraction

During the extraction step, the contaminants are isolated from the matrix and transferred to a suitable organic solvent. Different types of extraction procedures are used for different types of sample matrixes and analytes.

1. Soil, Sediment, and Sewage Sludge Samples

Soil, sediment, and sewage sludge samples require highly efficient methods of extraction, because the analytes tend to be very tightly bound to the sample matrix.

Traditional liquid–solid extraction continues to be used in the sample pretreatment of environmental samples. For the extraction of BFRs, hexane, acetone, hexane:acetone, and dichloromethane:cyclohexane has been used as the extraction solvents.[54,55] Similar solvents have been used also for chlorinated FRs. Liquid–solid extraction with acetone and acetonitrile: dichloromethane has been used in the extraction of TCEP in sediments.[3]

Typical solvents in Soxhlet extraction of PBBs, PBDEs, PBDFs, and PBDDs from soil, sediment, and sewage sludge samples have been hexane, toluene, and hexane:acetone mixtures.[11, 17,68,69] The extraction time has varied from 4 to 24 h. Before the extraction, the sample usually is mixed with anhydrous sodium sulphate or other drying agent.

Supercritical fluid extraction (SFE) with solid-phase trapping has been used for the extraction of DBDE and PBBs together with PCBs and chlorinated benzenes from sediment samples, with CO_2 as the supercritical fluid.[70] Before the extraction, the sediment sample can be mixed with copper powder and sodium sulphate for the removal of moisture and sulphur. Usually, the extraction combines static and dynamic extraction. The time required for the extraction ranges from 40 to 60 min, the extraction temperature is around 120°C and the pressure 374 bar. Compared with Soxhlet extraction, SFE gives similar yields, but the extracts are generally much cleaner and it might not be necessary to clean the extracts before GC analysis.

Pressurized hot water extraction (PHWE) has also been used for the analysis of several brominated analytes in sediment.[71,72] The extracted analytes were trapped into a solid-phase trap (Tenax TA), from which these were eluted with pentane:ethyl acetate mixture after drying the trap with nitrogen. No further cleanup of the extract was required. Best results were obtained at 325°C, using a pressure of 118 bar and an extraction time of 40 min. Compared with Soxhlet extraction, extraction yields were clearly better, and the extract was much cleaner.

2. Air Particles

Relatively few methods have been developed for air samples. Typically, the particles, which have been collected into a filter, an adsorbent, or a similar system, are extracted with suitable solvent, such as hexane (for PBDEs, TBBPA, organophosphate esters), hexane:acetone mixture (PBDEs), or dichloromethane (polybrominated and polybromochlorinated dibenzo-*p*-dioxins and dibenzo-*p*-dibenzofurans).[73–75] Often, BFRs have been extracted using Soxhlet extraction, but in most recent studies, sonication assisted extraction (SAE) has also been used.[56,58] For organophosphates Soxhlet extraction, microwave assisted extraction (MAE), and SAE has been applied.[36,76–79] In both methods, similar solvents, such as dichloromethane, acetone or toluene have been used as solvent. Soxhlet extraction requires 15 to 24 h while pressurized liquid extraction (PLE), SAE, and MAE can be accomplished in less than 30 min.

3.   Tissue Samples

FRs have been studied in various tissue samples, including fish, bird eggs, dolphin, seal, and whale fat. Usually the samples have first been homogenized, for example, with anhydrous Na_2SO_4.

Liquid extraction has been applied to homogenized tissue samples from fish and bird eggs. Typically hexane:acetone, hexane:acetonitrile, hexane:diethyl ether, hexane:dichloromethane mixtures, hexane, and dichloromethane have been used for the extraction.[14,41,43,55,80,81] The extract can be purified with further extraction, e.g., with the use of NaCl or NaH_2PO_4 buffer. Lipids are typically broken with concentrated sulphuric acid. Soxhlet extraction has been applied to fish and human tissue samples.[10,12,38,40,68,73,82] Usually, the solvents have been the same as for sediment samples, i.e., toluene, hexane, and hexane:acetone mixture. The extraction time has varied from 4 to 24 h. SFE is also a good choice for the extraction of FRs from solid biological samples, although it has not yet been utilized very widely.[83] The benefit of SFE is that the cleanup can be combined with the extraction. For example, the sample may be mixed with copper powder to remove the sulphur from the sample (sediments) or aluminum oxide may be used to retain and separate the lipids (biological samples).[70,83] In addition, the extract can be further cleaned and fractionated with the use of selective solid phase trapping.

4.   Liquid Samples

Liquid samples typically analyzed for FRs include water, wastewater, plasma, urine, and milk.

Slightly different solvents have been used for the LLE of different types of FRs. For the extraction of organophosphorus FRs, dichloromethane, toluene, mixtures of dichloromethane and acetonitrile, or dichloromethane:CCl_4 have been used. For brominated flame retardants (PBDEs), mixtures of hexane:acetone and hexane:MTBE have been applied. Pressurized solvent extraction has been used in the extraction of PBDEs in human milk.[84] In addition, membrane extraction utilizing a hollow fiber extractor has been developed for the extraction of OP metabolites in urine.[85]

Several SP materials have been used for the extraction of FRs from aqueous samples, plasma and milk (Table 31.7). Similar materials have been used for all FRs. Typical SP materials include C_{18} and C_8 bonded to porous silica, highly cross-linked poly(styrene divinylbenzene) (PS-DVB), and graphitized carbon black (GCB).[61,62,86,87] It is also possible to use XAD-2 resin for extraction of various FRs, pesticides, and plastic additives from large volumes of water (100 l). The analytes can then be either eluted from the resin by acetone:hexane mixture, or Soxhlet extracted with acetone and hexane.[88] For a specific determination of diphenyl phosphate in water and urine, molecularly imprinted polymers have been used in the solid phase extraction.[86] The imprinted polymer was prepared using 2-vinylpyridine as the functional monomer, ethylene glycol dimethacrylate as the cross linker, and a structural analog of the analyte as the template molecule. Elution was done with methanol triethylamine as solvent. Also solid phase microextraction (SPME) has been applied in the analysis of PBDEs in water samples. The extraction has been done from a headspace of a heated water sample (100°C) using polydimethylsiloxane (PDMS) or polyacryl (PA) as the fiber material.[89]

TABLE 31.7
SPE Material Used in the Extraction of FRs in Liquid Samples

Sorbent	Elution	Sample	Analytes	Recovery	References
C$_{18}$ LiChrolut RP 18	MeOH	Urine	Diphenyl phosphate	—	86
MIP column	—	Urine	Diphenyl phosphate	83%	86
SAX Extract-Clean column	1% TFA in methanol	Urine	Diphenyl phosphate	—	86
NH$_2$ Isolute SPE column	1% TEA in methanol	Urine	Diphenyl phosphate	—	86
Oasis MAX SPE	Methanol–TFA (98:2)	Urine	Diphenyl phosphate	102%	86
PS–DVB (Isolute ENV)	DCM:MeOH(1:1, v:v)	Plasma (5 g + formic acid–iPr, 9:1, v:v)	PBBs, TCBPA, TBBPA	54 to 92%	62,90
Divinylbenzene-N-vinylpyrrolidone copolymer (Oasis HLB)	DCM:MeOH (1:1, v:v)	Milk (formic acid–iPr, 9:1, v:v)	PBDEs, TBBPA	49 to 83%	90
Envi Carb (graphitized carbon black)	DCM	Sea water, 1 l	Pesticides, organophosphates	>80%	87
PS–DVB (LiChrolut ENV)	Ethyl acetate	Sea water, 1 l	Pesticides, organophosphates	>78%	87
PDMS fibre (SPME)	Thermal desorption	Water	PBBs, PBDEs	—	89

Solvents: ACN, acetonitrile; DCM, dichloromethane; iPR, isopropanol; MeOH, methanol; TEA, triethylamine; TFA, trifluoroacetate.
Other abbreviations: MAX, mixed anion exchange; MIP, molecularly imprinted polymer; PBB, polybrominated biphenyl; PBDE, polybrominated diphenyl ether; PBP, polybrominated phenols; PDMS, polymethyldisiloxane; PS-DVB, polystyrene-divinyl benzene; SAX, stron anion exchange; SPE, solid-phase extraction; TBBPA, tetrabromobisphenol A; TCBPA, tetrachlorobisphenol A.

It is also possible to combine SPE and lipid decomposition in the same procedure, if polymeric SP materials, which can tolerate low pH, are used for the extraction. Typically, concentrated sulphuric acid which is added directly to the SP cartridge after application of the sample is used for the decomposition.

C. Extract Cleanup

Particularly after Soxhlet or other enhanced liquid extraction, the extracts may be too dirty to allow direct analysis of analytes of interest. The extract requires a cleanup as many other compounds such as humic acids and lipids, are typically coextracted with the analytes. If selective extraction, such as SFE, is used, separate cleanup procedures are often not necessary.

1. Lipid Removal from Biological Extracts

The extracts of biological samples usually contain high concentration of lipids which must be removed before the analysis. Particularly if GC is used in the analysis, efficient removal of lipids is crucial. As the concentrations of many liphophilic FRs are related to the amount of lipids, the lipid content is often measured gravimetrically prior to the cleanup, or determined separately by a total lipid determination. Lipids can be removed by destructive or nondestructive methods. For serum or plasma samples, the lipid determination can be conveniently done on separate aliquots by enzymatic tests.

Although treatment with concentrated sulphuric acid is frequently used for the removal of lipids, it may destroy some of the compounds. Alumina columns offer less harsh treatment for lipid removal, and these are also often used for further cleanup of sediment extracts. Gel permeation chromatography (GPC) offers another approach for the removal of lipids from the biological extracts. For more selective removal of the lipids it can also be used in combination with florisil columns.

2. Sulphur Removal from Sediment Extracts

Sediment extracts, and sometimes also soils and sewage sludges, often contain relatively large amounts of elemental sulphur, which would disturb the GC analysis and must be removed. The typical methods for sulphur removal are treatment with concentrated sulphuric acid, copper powder, and tetrabutylammonium hydroxide/sulphite.

3. Fractionation

Florisil and silica columns are used, often in combination with alumina columns, to fractionate the extract into different classes of compound. Both pure silica and acid-treated silica are used for fractionation. High performance liquid chromatography (HPLC) and GPC have also been used for cleanup and fractionation of the extract.

Relatively simple cleanup steps have been used in some analytical methods for BFRs. For example, in determination of PBDEs in human milk, the dominating PCB congeners in the LLE extract were removed by passage through a silica column.[18] A similar method was used for PBDEs, toxaphene, and chlordane compounds in seal blubber extracts.[80] The liquid extracts after treatment with sulphuric acids in hexane were purified twice by silica gel column, and after elution of PCBs with hexane the analytes were eluted with a mixture of hexane and diethyl ether. PBBs and DeBDE have also been eluted from silica gel with isooctane, and other PBDEs with diethyl ether:isooctane, as was done with Soxhlet extracts of various marine mammals after treatment of the extract with sulphuric acid.[40] Alumina columns have been used in a similar manner for Soxhlet extracts of sediment samples in the determination of PBDEs. Concentrated acetone:hexane extract was passed through an alumina column and BFRs were eluted with hexane.[12]

Sometimes very complicated fractionation procedures are required. For example, in the determination of polyhalogenated dibenzo-*p*-dioxins, dibenzofurans, and diphenylethers, five different columns were used for the fractionation. The Soxhlet extract of sewage sludge was purified first with a use of a multilayer silica column. The analyte fraction was eluted with hexane:dichloromethane and transferred to a macro alumina column where it was eluted with hexane:acetone and it was further purified by GPC with use of cyclohexane:ethyl acetate for elution of the analytes. The sample was then fractionated by HPLC using a NO$_2$ column and rechromatographed with a micro alumina column with benzene and two mixtures of hexane and dichloromethane as eluents. The benzene and the first hexane and dichloromethane fractions were used for the determination of PCBs and chlorobenzenes, and the last hexane and dichloromethane fraction was used for the determination of PBDEs. After an additional cleanup step, polyhalogenated dibenzo-*p*-dioxins (PXDDs), polyhalogenated dibenzofurans (PXDFs), and octachlorothiantrene (OcCTA) were determined in the third fraction.[69]

A three-step purification was used to fractionate the PBDEs in a milk extracted with acetonitrile. First the extract was purified on an aluminum oxide column, and the eluate in the second hexane fraction was then purified with silica gel, from which the analytes were eluted with dichloromethane:hexane mixture after cleanup with hexane. The final fractionation was done with a GPC column from which the analytes were eluted with dichloromethane:hexane.[45]

Particularly in the purification of biological extracts, liquid–liquid partitioning is often used before column chromatography. For example, the LLE extract of serum samples was purified by partitioning the extract in hexane with ethanolic KOH solution. The neutral fraction was treated with concentrated H$_2$SO$_4$ and passed through a silica gel:sulphuric acid column with hexane.[57] SPE extract of human milk (an *N*-vinylpyrrolidone-divinylbenzene copolymer, on column lipid destruction, elution with hexane) was treated in a similar manner. Phenolic compounds were first partitioned into ethanolic potassium hydroxide, the organic phase containing neutral substances was removed, and the aqueous phase was thereafter extracted with hexane. The pooled organic phases were designated as the neutral fraction. After acidification of the aqueous phase with hydrochloric acid, phenolic compounds were extracted into hexane:MTBE and derivatized diazomethane. The phenolic derivatives were then subjected to further cleanup on a silica gel:sulphuric acid column, employing DCM as the mobile phase. Neutral compounds were cleaned up on a silica gel:sulphuric acid column, using hexane as the mobile phase.[92]

D. Derivatization

Most FRs can be analyzed with GC without derivatization. For some of the analytes, however, derivatization can improve its GC analysis. For brominated and chlorinated derivatives of bisphenol A, for example, a silylation step using *N,O*-bis(trimethylsilyl)trifluoroacetamide as reagent can be carried out after clean up. Phenolic compounds, such as TCBPA, TBBPA, and brominated phenols, can also be methylated with, e.g., diazomethane.[103]

III. ANALYTICAL METHODS

Gas chromatography is the most popular analytical technique for the determination of FRs in environmental samples. Also liquid chromatography has been used for separation, but to a lesser extent due to its relatively low separation efficiency in comparison with GC.

A. Gas Chromatography

Gas chromatographic methods used for the analysis of FRs in environmental samples are similar to those developed for other organic pollutants.[8,93,94] The methods are summarized in Table 31.8.

TABLE 31.8
GC Methods for the Analysis of BFRs and OPs in Environmental Samples

Sample, Compounds	Injection, v (μl), T, (°C)	Column, Dimensions, Film Thickness	Detector	References
Sediment, fish (1–2)Tetra- to octa-BDEs (3) BDE 209	(1 to 2) Splitless, 1 μl (270°C) (3) Splitless (110°C)	(1) HP-5, 50 m × 0.2 mm, 0.25 μm (2) HP-1701, 60 m × 0.25 mm, 0.25 μm (3) HP-5, 25 m × 0.2 mm, 0.33 μm	Dual ECD	11
Sediment, mussels (1) Tetra to hexa-BDEs (2) BDE 209	Splitless Cold splitless (110°C)	(1) DB-5, 60 m × 0.25 mm, 0.25 μm (2) DB-1, 15 m × 0.25 mm, 0.10 μm	ECNI–MS	98
Mussels; Tri- to deca-BDEs	Splitless (270°C)	CP Sil-8 25 m × 0.25 mm, 0.25 μm	ECNI–MS	14
Fish; Di- to deca-BDEs	Splitless, 2 μl (275°C)	DB-1 15 m × 0.25 mm, 0.25 μm	ECNI–MS	9
Fish (1) Tetra- to hepta-BDEs (2) BDE 209	Splitless	(1) DB-5MS, 30 m × 0.25 mm, 0.25 μm (2) DB-5MS, 15 m × 0.25 mm, 0.25 μm	(1) EI–LRMS (2) ECNI–MS	58,110
Fish; tetra- to hexa-BDEs	Splitless	DB-5, 60 m × 0.25 mm × 0.32 μm	GC–ECD, EI–LRMS	27
Fish, human milk, vegetables; Tri- to hexa-BDEs	Splitless, 2 μl (260°C)	SPB-5, 30 m × 0.32 mm, 0.25 μm	EI–LRMS	49
Seal, fish, crab, porpoise (1) Mono to nona BDEs (2) BDE 209	Splitless, 1 μl (300°C)	(1) DB-5, 30 m × 0.25 mm, 0.25 μm (2) DB-5-HT, 15 m × 0.25 mm, 0.10 μm	EI–HRMS	34
Milk; Tri to hexa-BDEs	Splitless	Quadrex 007–525 m × 0.32 mm, 0.25 μm	EI–HRMS	45
Milk; serum; Tri to hepta-BDEs	Pulsed splitless, 1.5 μl (250°C)	CP Sil-5CB 30 m × 0.25 mm, 0.25 μm	ECNI–MS	63,103
Fish; mono to hepta BDEs	Oncolumn, 1 μl (100°C)	RTX-5 60 m × 0.25 mm, 0.25 μm DB-1701 60 m × 0.25 mm, 0.25 μm	EI–HRMS	73,42
Air; tetra to deca BDE, TBBPA	Oncolumn (60°C)	DB5-HT, 15 m × 0.25 mm, 0.10 μm	ECNI–MS	56
Chicken fat; Mono to deca BDEs	Oncolumn	DB-5MS 30 m	EI–HRMS	99
Sediment; DPDEs, PBBs, PCBs, CBs	Oncolumn, 1 μl (80°C)	2.5 m DPTDMS ret. gap + DB-5 and DB-17, 2 × 60 m × 0.25 mm i.d., 0.17 μm	GC–ECD	70
Sediment; PBCCH, HxBB, PBT, TBBPA, TDBPP	Oncolumn, 3 μl (80°C)	3 m DPTDMS ret. Gap + BGP-5, 20 m × 0.25 mm i.d., 0.25 μm	EI–MS	71

Porpoise, cormorant				
(1) Tetra to octa BDEs	(1) PTV (50°C)	(1) DB-5, 50 m × 0.2 mm, 0.25 μm	(1) ECNI–MS	40,68
(2) BDE 209	(2) PTV (70°C)	(2) HP-1, 15 m × 0.25 mm, 0.10 μm	(2) ECD	16
Human adipose; Tri to hepta BDEs	PTV, large volume, 20 μl (70°C)	AT-5 10 m × 0.10 mm, 0.10 μm	EI–MS	
Air, OPs	PTV, 600 μl (70°C)	DB-5MS, 30 m × 0.32 mm, 0.1 μm	GC–FPD	79
Air, OPs	Splitless, 1 μl, 280°C	DB-5, 30 m × 0.25 mm, 0.1 μm	GC–NPD, GC–MS (PICI), GC–AED	15
Air, OPs	Pulsed splitless, 2 μl, 250°C	HP-5MS, 30 m × 0.32 mm, 0.25 m	GC–FPD	77
Indoor air, OPs	Splitless, 2 μl	DB-5.625 column, 30 m × 0.25 mm, 0.5 μm	GC–MS (EI)	36
Indoor air, OPs,	Splitless, 1 μl	DB-1, 30 m × 0.25 mm i.d., 0.25 μm	GC–MS (EI)	105

Analytes: CB, chlorobenzenes; BDE, brominated diphenyl ether; HBB, hexabromobiphenyl; OP, Organophosphorus FRs; PBB, polybrominated biphenyl; PBCCH, pentabromochlorocyclohexane; PBDE, polybrominated diphenyl ether; PBT, Polybutylene terephthalate; PCB, polychlorinated biphenyls; TBBPA, tetrabromobisphenol A; TDBPP, tris(2,3-dibromopropyl)phosphate.

Others: EI, electron impact ionization; ECNI, electron capture negative ionization; HR, high resolution; LR, low resolution; LVI, large volume injection; PICI, positive ion chemical ionization.

1. Injection Techniques

The most common injection methods in the determination of FRs are splitless injection and on column injection. On column injection is suitable especially to thermally labile analytes, and it suits very well to quantitative analysis. However, the sample extract should be clean from nonvolatile matrix components in on column injection. Split injection is not recommended because of its low sensitivity and strong discrimination effects which can occur during the injection. Large volume injection techniques have also been applied in the analysis of FRs.[16,72,78,79]

a. Splitless Injection

In splitless injection, the injected volume is typically around 1 μl and the injector temperature ranges from 250 to 300°C.[16,81,95] In many GC apparatus it is possible to use a pressure pulse during splitless injection for improving the transfer of sample vapors to the column. The use of an autosampler is a prerequisite for obtaining an acceptable reproducibility. Too high injector temperature can cause problems. For example, in the injection of octa- to deca-BDEs, some degradation may take place in the hot injector if the residence time in the liner is too long.[16,95] Also some organophosphorus FRs can degradate during splitless injection, when hydrogen has been used as carrier gas, probably due to hydrolysis of the phosphate esters to phosphoric acid in the injector.[15]

b. On-Column Injection

In on-column injection, the injected volume is in conventional injection typically 1 to 3 μl. The oven temperature during injection varies, depending on the volatility of solvent and analytes. If the analytes of interest are volatile, also the solvent should be volatile and the oven temperature during injection should be below the (pressure corrected) boiling point of the solvent. Then solvent trapping takes place during the injection concentrating the bands of volatile solutes. If the analytes are not volatile, solvent trapping is not required and higher temperatures can be used during injection, as has been done, e.g., in the injection of several BFRs (solvent:pentane:ethylacetate, and temperature 80°C).[71]

c. Large Volume Injection

In large volume injection, the programmable temperature vaporizer (PTV) has been the most popular technique, although also on-column injection can be applied, if the sample extract is sufficiently clean.

In PTV injection, split injection to a cold liner is usually applied. During injection, the split exit is open, and the injector is kept at a low temperature to minimize losses of volatile analytes. The optimum temperature during the injection is dependent on the volatility of the sample solvent and the analytes, and on the split flow rate. Typically, high split flows (50 to 200 ml/min) are used, as high flow rates decrease the dew point of the solvent, and lower injector temperatures can be used. The solvent elimination time used should be sufficiently long to allow all the solvent to evaporate but not too long to avoid losses of volatile analytes. For example, using isooctane as solvent, the injector temperature 70°C which is some 30°C below its boiling point has been suitable in the injection of relatively nonvolatile PBDEs.[68] In the injection of organophosphorus esters in methyl-*tert*-butyl ether (MTBE), the injection temperature was 60°C, while it was 70°C when the solvent was a mixture of hexane and MTBE.[78,79] If empty liner is used, only some 5 μl can be injected as once, due to limited capacity of the liner. In the case of an empty liner, automated multiple injection can be used for increasing the injected volume. A better option is to use a packed liner (e.g., glass wool) and speed-controlled injection. In this way, injection volumes can be up to milliliters.

On-column injection can also be applied for large volume sample introduction. However, especially when the sample volume is large, special attention has to be paid on the cleanup of the sample. On-column injection has also been used as a transfer technique in an online combination of extraction, liquid chromatographic cleanup and gas chromatography.[72]

2. Choice of the Column

In most cases, single capillary column GC is sufficient for the separation of FRs, if sufficiently long columns (30 to 60 m) and small diameters (i.d. 0.25 mm) are used. Good resolution may also be obtained utilizing fast GC, i.e., short, narrow-bore columns (i.d. = 0.1 mm, length > 10 m) and very fast temperature programming.[16]

Typically, FRs are determined on nonpolar or semipolar columns such as 100% methyl polysiloxane type (DB-1) and 5% phenyl dimethyl polysiloxane type (DB-5, CP-Sil 8, and AT-5). Also semipolar phases, such as 8% phenyl-polycarborane-siloxane HT-8, 14% cyanopropylphenyl 86% dimethyl polysiloxane (CP-Sil 19, HP-1701, and DB-1701) have been utilized.

Some typical problems with resolution have been noticed. For example, coelution of CB-194 and BDE-100 can occur on a 60-m DB-5 column.[29] In addition, as the BFRs are often analyzed together with other organic pollutants, coelution of, e.g., BDE congeners can occur with organochlorine pesticides or PCBs, with a particular concern addressed to CB 180 and BDE 47.[8] Coelution is usually not a problem if mass spectrometric detection is used, but it can cause problems with ECD detection.

GC analysis of less volatile FRs require special attention. Particularly, analysis of BDE-209 can be critical because it can degradate at higher temperatures.[95] The GC column should be relatively short, preferably 10 to 15 m, the film thickness should be small and the carrier gas flow rate should be relatively high in order to reduce the residence time of the less volatile analytes in the column. This means that the analysis of less volatile FRs must be done separately from the analysis of the other FRs.

Another problematic brominated flame retardant in GC analysis is HBCD, because the mutual ratio of the HBCD diastereomers can change at temperatures over 160°C and this feature makes it very difficult to analyze the three diastereomers.[96] The maximum oven temperature should be 300°C. It should be applied only for a short time at the end of the oven temperature program.[8] Therefore, HPLC–MS is a better alternative for the HBCD determination, although the sensitivity is significantly lower than with GC–MS.

3. Detection

With gas chromatography, the most widely used detectors for the analysis of flame retardants are mass spectrometer, electron capture detector for halogenated flame retardants, flame photometric detector, and nitrogen phosphorus detector for organophosphorus FRs.

a. Electron Capture Detector (ECD)

ECD is very sensitive for organohalogenated compounds and it has been used in the detection of brominated and chlorinated FRs and organophosphorus FRs.[11,68,70,97] The main benefits of ECD are its relative low purchase and maintenance cost, combined with a relatively good sensitivity for compounds with four or more halogen atoms. It should be noted, however, that the sensitivity is not influenced only by the halogen load, but also by the substitution pattern of the compound. This results in unequal responses for the different congeners. Furthermore, ECD is linear only over a limited concentration range. Another drawback is the lack of selectivity. Any halogen containing molecules, such as PCBs, will produce a signal and therefore disturb the analysis of halogenated FRs, especially when PCBs are present at high concentrations. This limits the use of the ECD to samples where it has been verified, e.g., by random screening by MS, that co eluting compounds are not present. Many of these coelutions can be avoided by selective cleanup and appropriate selection of the column and oven temperature. Also, dual columns with different stationary phases and two ECD systems have been used.

b. Flame Photometric Detector

The flame photometric detector (FPD) is selective to sulphur- or phosphorus-containing compounds and it has been used in the determination of organophosphorus FRs.[77,79] In FPD, the emitter for phosphorus compounds in the flame is excited HPO (λ_{max} = doublet 510 to 526 nm) and detection requires a suitable interference filter for isolation of the emission band. For organophosphorus FRs, a phosphorus filter at 526 nm has been used in the detection. The detection limits for the organophosphates have been on the ng level.

c. Nitrogen Phosphorus Detector

The nitrogen phosphorus detector (NPD) is selective towards nitrogen and phosphorus containing FRs, and it has been used for the detection of organophosphate FRs.[15] For OPs, limits of detection (LOD) have been in the range of a few nanograms. A comparison of NPD and MS detection in the analysis of organophosphates in indoor air samples is shown in Figure 31.4. In the GC–MS chromatogram, all of the dominating peaks were identified as phthalates and other non phosphorus-containing organic compounds and only the most abundant of the organophosphorus esters could be identified as being a phosphate.[15]

d. Mass Spectrometry

Two types of mass spectrometric systems can be applied, namely low and high resolution MS. The low resolution MS (LRMS) is cheaper and is easier to use, while high resolution MS (HRMS) gives more sensitive and selective results, but requires more experienced users. In ionization, electron impact ionization (EI), chemical ionization (CI), and electron capture negative ionization (ECNI) can be used. Electron capture ionization is a "soft" ionization technique that takes advantage of the interactions between thermal energy electrons and electrophilic molecules, such as PBDEs. In ECNI, the low energy electrons (thermal electrons) generated by interactions between a high energy electron beam and a moderating gas, react with the analytes to form negative ions. The electron energy should be very low to facilitate electron capture, and the specific energy required for electron capture depends on the molecular structure of the analyte.

i.   MS Detection of Brominated FRs

The use of mass spectrometric detection for BFRs in combination with GC has utilized mainly electron impact (EI) and electron capture (EC) mass spectrometry (MS).[9,14,34,42,49,62,63,73,91,98–101] Both LRMS and HRMS has been used. In the detection of BFRs, HRMS has a number of advantages over LRMS, such as increased sensitivity and selectivity, but it is almost exclusively operated in EI mode. For LRMS, ECNI, in addition to EI, can be applied to obtain an increased sensitivity for higher brominated compounds. Each technique has its own advantages and disadvantages. Electron capture negative ionization (ECNI), although generally more sensitive and less costly than other ionization methods for PBDE analysis, does not provide information on the molecular ion cluster (as required for qualitative identification), is more subject to brominated interferences, and does not allow the use of ^{13}C-labeled standards for quantitation. Conversely, EI methods suffer from fragmentation of the molecular ions, creating difficulties in both identification and quantitation of congeners in full scan and single ion monitoring (SIM) modes, respectively. For example, loss of Br atoms from PBDE congeners during EI ionization may lead to incorrect identification of the parent ion as a lower brominated congener. In addition, the relatively unpredictable fragmentation during EI or EC

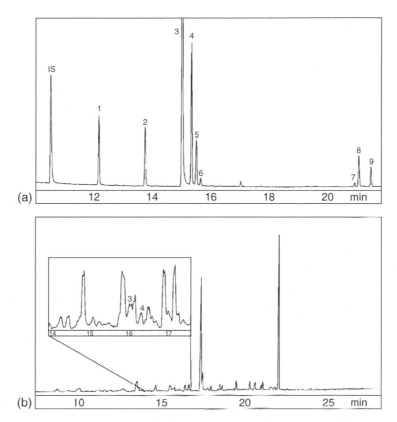

FIGURE 31.4 Comparison of selectivity of MS and NPD detection. (a) GC–NPD chromatogram of an air sample, collected in an office building. IS, tripropyl phosphate (internal standard); one, a tributyl phosphate isomer; two, tri(*n*-butyl) phosphate; three, tri(2-chloroethyl) phosphate; four to six, isomers of tri(chloropropyl) phosphate; seven, triphenyl phosphate; eight, tri(2-butoxyethyl) phosphate; nine, tri(2-ethylhexyl) phosphate. (b) GC–MS–EI chromatogram of the same sample as in (a). The peaks marked three and four are tri(2-chloroethyl) phosphate and isomers of tri(chloropropyl) phosphate, respectively. (From Carlsson, H., Nilsson, U., Becker, G., and Östman, C., *Environ. Sci. Technol.*, 31, 2931–2936, 1997.)

restricts the utility of applying relative response factors of one congener for which an analytical standard is available (e.g., BDE47) to other members of its homolog group (e.g., tetra-BDEs).

In EI–MS, the major ions formed from PBDEs are the M^+ and the $[M - 2Br]^+$, which can be used for its identification and quantitation. This ionization technique facilitates the analysis of BDE congeners in the presence of possible coeluted compounds (such as PCBs). EI–LRMS is not routinely used for the PBDE analysis, because of its relatively low sensitivity, especially for the analysis of higher brominated BDE congeners (hepta- to deca-DBE). However, this ionization mode allows the acquisition of full scan spectra, thus offering a multiple choice in ion selection than ECNI mode. Yet, ECNI is more selective towards aromatic brominated compounds.

ECNI has proven to be highly sensitive, especially for compounds with more than four bromine atoms. The sensitivity for these compounds is approximately tenfold that obtained with ECD.[93] Methane and ammonia have been used as reaction gases. GC or ECNI–MS spectra of most brominated compounds are dominated by the intense fragment ions of the bromine isotopes $[^{79}Br]^-$ and $[^{81}Br]^-$ found in the typical isotopic distribution of 0.505 to 0.495. Molecular ions or fragment ions at high mass are either scarcely found or found at very low intensity relative to the bromide ion. The drawback is that exclusive monitoring of m/z 79 and 81 allows no further identification of the hydrocarbon backbone of a brominated compound. However, some brominated compounds also form $[HBr_2]^-$ (161 type) or $[Br_2]^-$ (160 type) fragment ions, while other brominated compounds do not (79 type). These additional fragment ions can be used in distinguishing between different classes of brominated compounds. For example, when methane has been used as the moderating gas in ECNI, intense $[HBr_2]^-$ fragment ions are obtained for PBDE congeners, while using ammonia as the moderating gas results in intense $[Br_2]^-$ fragment ions.[102] In a comparative study of GC–MS for the detection of PBDEs with EI or ECNI mode, the two modes gave almost equally good results in respect of response, detection limits and quantification of standard solutions. However, the EI mode offered higher selectivity because PBDEs are detected as molecular or higher mass fragmented ions, leading to a higher certainty of identification. Also, it is easier to find suitable internal standards for GC–EI–MS.[92,103] In Figure 31.5, the difference of EI and ECNI ionization is visualized for the tetrabrominated metabolite of PBDE.

Another promising analytical tool used for identification of PBDEs is metastable atom bombardment (MAB) in MS, which has been shown to offer a high degree of ionization and fragmentation selectivity for a variety of analytes, including halogenated aromatics. Such selectivity results from the variation and quantization of the energy transferred upon ionization, allowing a range of 8 to 20 eV to be transferred, depending on the gas (He, Ne, Ar, Kr, Xe, or

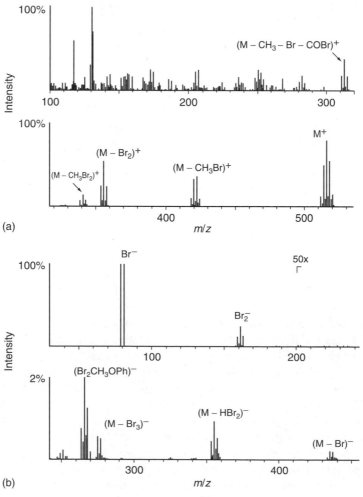

FIGURE 31.5 (a) EI and (b) ECNI mass spectra of a tetrabromo compound detected in ringed seal blubber. (From Haglund, P., Zook, D., Buser, H.R., and Hu, J., *Environ. Sci. Technol.*, 31, 3281–3287, 1997.)

N_2) used to generate the metastable beam. For PBDEs, best results with MAB have been obtained using N_2 which results the molecular ion as the base peak, with little fragmentation taking place. Even though the MAB–MS detection gives lower sensitivity than EI–MS, MAB-N_2 had a lower limit of detection for tetra- and penta-BDEs than EI because of reduced background noise to the detector. In Figure 31.6, the difference in ionization of BDE25 and BDE35 using EI and MAB-N_2 is shown.[35]

In recent years, MS–MS detection utilizing, e.g., ion trap MS has also been applied for the BRFs. For PBDEs and PBBs, the main reaction pattern is the loss of two bromine atoms $[M - 2Br]^+$ (for PBB 15, BDE 47, BDE 100, BDE 99, BDE 153, and BDE 154), as previously observed for PCBs. However, for PBB 49 (a tetrabromobiphenyl), the main path is the loss of a single bromine atom $[M - Br]^+$, and for BDE-3 (a monobromodiphenyl ether), together with the $[M - Br]^+$ ion, two intense ions corresponding to $[M - CO]^+$ and $[M - COBr]^+$ are observed, in a way that resembles the dissociation pattern of PCDDs or PCDFs.[89]

ii. MS Detection of Organophosporus FRs

Both electron impact mode (EI) and positive ion chemical ionization mode (PICI) has been used for the detection of organophosphorus FRs.[15,36,104,105] In the case of PICI, methane has been utilized as reagent gas.[15]

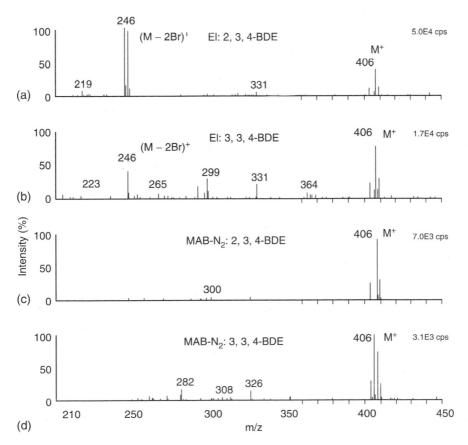

FIGURE 31.6 EI and MBI ionization of BDE 25 and BDE 35. (From Ikonomou, M., *Anal. Chem.*, 74, 5263–5272, 2002.)

By use of SIM, the selectivity and thereby the sensitivity of the MS for organophosphates can be enhanced. For alkylated phosphates, no or very weak molecular ions are observed, while a m/z 99 fragment corresponding to protonated phosphoric acid can be considered to be characteristic. Due to the extensive fragmentation of alkylated phosphates using GC–MS–EI, it is often necessary to apply the GC–MS–PICI technique for identification. Arylated phosphates, on the other hand, exhibit an intense molecular ion and do not show the fragment of m/z 99 in its MS–EI spectra.[15]

e. Other Detectors

Other detectors, such as the flame ionization detector, atomic emission detector, Fourier transform infrared spectrometer and nuclear magnetic resonance detectors, have been used in GC for the detection of various flame retardants.[15,72,94,106] However, these have not been very widely utilized.

B. LIQUID CHROMATOGRAPHY

Liquid chromatographic techniques have not been applied widely in the analysis of FRs in environmental samples. The main reason for this is that the separation efficiency of LC is often not sufficient for multicomponent analysis. HPLC techniques have been applied mainly in the analysis of selected analytes, which would be difficult to analyze with GC or in other special cases, where the concentrations of analytes are relatively high. LC analyses of FRs in environmental samples have been performed with RPLC or, less commonly, with ion chromatography.

As was pointed out earlier, the GC analysis of HBCD is problematic. RPLC–MS has been used in the analysis of HBCD and tetrabromobisphenol-A (TBBPA). The benefit, in comparison with GC–MS is that liquid chromatographic determination circumvents problems of thermally induced reactions and isomeric rearrangements. The drawback is the poorer sensitivity.[107]

RPLC analysis utilizing a C_{18} column has been used for the analyzing of radiolabeled ^{14}C tris(1,3-dichloro-2-propyl)phosphate (TDCP) from skin samples from mouse exposed to TDCP and DBDO. In the analysis of radiolabeled DBDPO, a C_8 column was used. In the previous analysis, a gradient elution with water and acetonitrile was applied, while in the latter, isocratic conditions were used. In both cases, a flow scintillation analyzer was used for detection.[66]

In the analysis of ionic organophosphate diesters, RPLC using a Hypercarb column and MS detection has been applied. The retention behavior and selectivity of this Hypercarb column differ substantially from those of silica- and polymer-based sorbents and it has several advantages, including stability over the entire pH range and strong retention of polar and ionic compounds. Although the organophosphate diesters are anions at neutral pH, these can be separated without ion pairing agents or derivatization steps. To elute the strongly retained ionic compounds from the column, NH_3 can be added to the mobile phase, which typically is a water organic solvent (e.g., THF) mixture. Alkaline conditions also ensure full ionization of the acidic compounds.[85] A similar system has also been used in the determination of diphosphate esters.[86] Mass spectrometry was performed with an ion trap instrument and LC separation was accomplished with a porous C_{18} column. The mobile phase was a methanol–water gradient containing ion pair additive in order to enable retention of the highly acidic analytes on RPLC stationary phase. Of the two additives tested, namely triethylamine (TEA) and ammonium acetate, best signal-to-noise ratios were obtained for TEA. However, for both TEA and ammonium acetate, the intensities of the signal were shown to decrease with increasing concentrations of the ion pairing agent due to the large amounts of anions competing for the MS detection. Ions were formed in negative electrospray ionization (ESI) mode and selected ion monitoring (SIM) was performed at m/z 249.2 for diphenyl phthalate and m/z 277.2 for ditolyl phosphate.[86]

1. Detection

With liquid chromatography, mainly mass spectrometry with electrospray ionization has been used for the detection of FRs in environmental samples.[66,85,86] As the LC–MS analyses have been used in the determination of organophosphates, alkaline conditions during the ES ionization have typically been applied. Specific detection utilizing a flow scintillation analyzer and radioactively labeled analytes has also been used. UV or VIS detection is generally not sensitive enough for the detection of FRs in environmental samples, but the sensitivity can be enhanced by using specific post-column reactions, as has been done in the analysis of THPC and THPS.

C. FUTURE DEVELOPMENTS

Several options are available for the analysis of flame retardants in environmental samples. Good results can be obtained with most of the methods, as has been shown in several interlaboratory studies.[19,36,108] For example, in an interlaboratory study of PBDEs in sediments and biological samples, various methods were used both for sample pretreatment and final analysis. In the extraction, Soxhlet, PLE, sonication assisted extraction, and SFE were used, utilizing different solvents. The cleanup procedures varied largely as well. In the final analysis, GC–MS with either HRMS or LRMS was used, with varying column dimensions and stationary phases. The results agreed well with all the methods used, although some problems were noticed in the analysis of BDE-209.[108]

Although conventional extraction methods, such as Soxhlet, do provide accurate results, it can be assumed that the faster and less laborious extraction techniques such as PLE, sonication assisted extraction, and SFE will gain popularity. The relevant figures of merit of the techniques utilized in the extraction of FRs in environmental samples are summarized in Table 31.9. Gas chromatography is the obvious choice for the final separation of the analytes. Selective and sensitive detection, i.e., MS or ECD, is required, as the levels of the FRs tend to be very low and other compounds are typically present in much higher concentrations.

It can also be expected that comprehensive two dimensional gas chromatography (GC × GC) will be applied to the analyses of FRs in near future. The GC × GC method has been successfully applied to the determination of other complex samples, such as toxaphene, PCBs and dioxins, and furans. Thus, it should be well suited also for the determination of halogenated FRs.

IV. APPLICATIONS IN THE ANALYSIS OF THE ENVIRONMENT

Presentative examples of analytical procedures for various PBDEs are given in Table 31.2 to Table 31.6.

Several approaches are available for the analysis of brominated FRs in environmental samples. For extraction, Soxhlet extraction and liquid solid extraction are the most typical procedures. For example, in the analysis of PBDEs in fish tissue, the tissue sample was first homogenized with a food processor. The homogenized tissue (300 g) was then mixed with diatomaceous earth (300 g) and extracted three times with diethyl ether:hexane (1:3) mixture. The extract was filtrated and evaporated to dryness. The residue was redissolved in hexane and washed with 5% NaCl in water. The organic phase was dried and evaporated to dryness again and the lipid content was then determined gravimetrically. The residue was dissolved into 7 ml of acetone:cyclohexane (3:7). Five ml of the extract was then injected into GPC column (CLN pak EV-2000, 300 × 20 mm i.d.) using acetone:cyclohexane (3:7) as eluent. The fraction between 15 to 28 min was collected (65 ml), evaporated to dryness and redissolved into 1 ml of hexane. The solution was injected into a minicolumn packed with H_2SO_4 treated silica gel sandwiched between two layers of pure silica gel (0.5 g, 2 × 0.25 G). The fraction containing the PBDEs was eluted with 10 ml of hexane and

TABLE 31.9
Relevant Figures of Merit of Extraction Methods

	Soxhlet	LSE or LLE	PLE	SAE	MAE	SFE	PHWE	SPE	SPME
Sample size (g)	10	1 to 10	5 to 50	1 to 30	1 to 10	1 to 10	0.5 to 10	0.1 to 1000	1 to 1000
Sample type	Solid	Solid or liquid	Solid	Solid or liquid	Solid	Solid	Solid	Liquid, gaseous	Liquid, gaseous
Extraction time	6 to 24 h	5 min to 12 h	5 to 40 min	3 to 40 min	5 to 40 min	20 to 60 min	20 to 60 min	A few minutes	30 to 80 min
Solvent	Organic, >100 ml	Organic, >50 ml	Organic, <50 ml	Organic, <50 ml	Organic, <50 ml	0 to 10 ml	1 to 5 ml	1 to 5 ml	No solvent
Selectivity	Low	Low	Low	Low	Low	High	High	High	High
Instrumentation cost	Low	Low	High	Moderate	Moderate	High	High	Low	Low
Level of automation	Low	Low	High	Moderate	Moderate	High	High	High	Moderate
Operator skill	Low	Low	Moderate	Moderate	Moderate	High	High	Moderate	Low

LLE, liquid–liquid extraction; LSE, liquid–solid extraction; MAE, microwave-assisted extraction; PHWE, presurized hot water extraction; PLE, pressurized liquid extraction; SAE, sonication-assisted extraction; SFE, supercritical fluid extraction; SPE, solid-phase extraction; SPME, solid-phase micro extraction.

evaporated to dryness and redissolved into 0.2 ml of nonane. The analysis was carried out by GC–NCI–MS, using 15 m × 0.25 mm i.d. DB-1 column. Splitless injection was used for sample introduction. NCI was used for ionization, using isobutane as reaction gas, and ions at m/z 79 and 81 were monitored. Fifteen PBDE congeners could be identified and the recoveries ranged from 88 to 128% and the LODs were 0.01 to 0.2 ng/g lipid weight. The most abundant BDE congener found in the fish samples, collected in Japan, was BDE-47, which was found in all samples with concentration ranging from 1.2 to 2100 ng/g lipid weight.[9]

A similar method utilizing MAE in extraction has been developed for the PBDEs in marine mammals. Tissue samples (1.5 g) were extracted using MAE and ethyl acetate:cyclohexane as solvent (8 ml, 1:1, v:v). The extraction cycle took 38.5 min. The extract was then filtered through Na_2SO_4 and cleaned up using GPC (lipid removal). In GPC, 60 × 2.5 cm i.d. column filled with bio beads S-X3 was used and the eluent was cyclohexane:ethyl acetate. The GPC fraction was concentrated to 1 ml and the extract was then fractionated by silica gel adsorption chromatography for the separation of PCBs and similar compounds from the more polar aliphatic or alicyclic chloropesticides and brominated compounds. Column chromatography was performed in a column packed with 3 g deactivated silica gel and the column was eluted with 60 ml n-hexane. Extracts were reduced at first by rotary evaporation and finally carefully blown down with nitrogen. The recoveries were quantitative. The analysis was carried out by GC–ECNI–MS using a splitless injection. Two types of columns were applied, namely a 30 m, 0.25 mm i.d. column coated with a chiral stationary phase consisting of 25% randomly *tert*-butyldimethylsilylated-cyclodextrin diluted in PS086 and a HP-5 column (30 m, 0.25 mm i.d). Methane was used as the CI moderating gas. In the total ion current mode, m/z 50 to 650 was scanned. In the selected ion monitoring mode, the following m/z values (corresponding compositions are given in parentheses) were recorded in parallel: m/z 79 ($[^{79}Br]^-$) and 81 ($[^{81}Br]^-$), m/z 158 ($[^{79}Br^{79}Br]^-$), m/z 160 ($[^{79}Br^{81}Br]^-$), m/z 159 ($[^1H^{79}Br^{79}Br]^-$), m/z 161 ($[^1H^{79}Br^{81}Br]^-$), m/z 114 ($[^{79}Br^{35}Cl]^-$), m/z 116 ($[^{79}Br^{37}Cl]^-$, and ($[^{81}Br^{35}Cl]^-$), as well as m/z 115 $[^1H^{79}Br^{35}Cl]^-$, m/z 117 ($[^1H^{79}Br^{37}Cl]^-$), and ($[^1H^{81}Br^{35}Cl]^-$). PBDE 47 and 99 were the major brominated contaminants found in both the adipose tissue of a polar bear collected in pack ice close to Iceland and gray seals from the German coast of the Baltic Sea, concentrations ranging from 5 to 37 g/kg.[102]

SFE with solid-phase trapping has been used for the extraction of DBDE and PBBs together with PCBs and chlorinated benzenes from sediment samples, with CO_2 as the supercritical fluid.[70] Before the extraction, the sediment sample was mixed with copper powder and sodium sulphate. Total extraction time was 60 min (20 min static, 40 min dynamic). The extraction temperature was 120°C and the pressure 374 bar. Use of CO_2 with modifiers (diethylamine, methanol, and acetone) gave only slightly better extraction yields than CO_2 alone. Compared with Soxhlet extraction, SFE gave similar yields, but the extracts were much cleaner and it was not necessary to clean the extracts before GC–ECD analysis. In GC a two-channel system was used for verification of the identification. The on-column injector was connected to a retention gap, which was split with a t-piece to two columns (60 m × 0.25 mm i.d.) with different stationary phases (DB 5 and DB 17). The method was proven to be quantitative and sensitive, with LODs ranging from 0.01 to 0.84 ng/g. The SFE extraction was compared with Soxhlet extraction (20 h, acetone:hexane, silica column purification, sulphur removal by TBA sulphite method). Several of the analytes could not be identified and quantified from the Soxhlet extract by the dual GC–ECD because of coeluting matrix compounds while with SFE, most analytes could be determined from the sediment samples.[70]

In the analysis of PBDEs in biological tissues, hot Soxhlet extraction was used. Tissue samples were first mixed with anhydrous Na_2SO_4 and then extracted by automated hot Soxhlet for 2 h with 75 ml of hexane:acetone:dichloromethane. The extract was evaporated to dryness, and the extracted lipids were determined gravimetrically. Two successive SPE cartridges containing acid silica:neutral silica:deactivated basic alumina (from top to bottom), respectively, were used for cleanup. PBDEs, together with PCBs and DDTs, were eluted from the second cartridge using 15 ml of hexane and 20 ml of hexane:dichloromethane (1:1, v/v). The eluate was concentrated to near dryness and redissolved in isooctane. The analysis was done with GC using large volume injection and LRMS with EI ionization. PTV injection was used in sample introduction and the total sample volume was 20 μl. In GC separation a narrow bore capillary was used (10 m $\times$ 0.1 mm i.d., AT-5, 0.1 μm) and fast temperature programming (40 and 25°C/min). As can be seen in Figure 31.7, extremely narrow peaks with a peak width of <1 sec were obtained with fast GC, and mass sensitivity was accordingly increased. Total analysis time was less than 10 min. In a interlaboratory study, the results obtained this system correlated well with results obtained using other detectors, such as HRMS or ECNI–LRMS.[16]

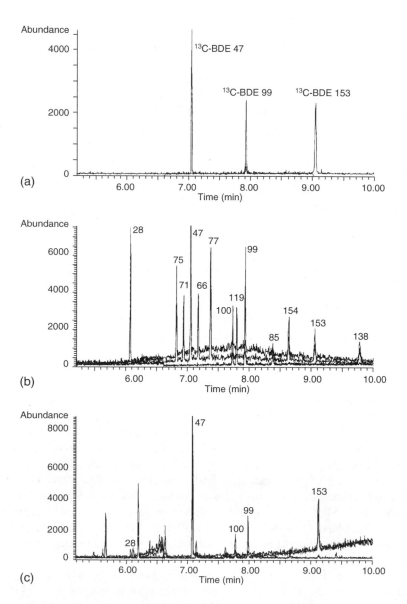

FIGURE 31.7 Selected ion chromatograms (EI–MS) of a standard mixture (^{13}C-labeled BDEs (a) and target BDEs (b)) and human adipose tissue extract (c) analyzed on a 10 m $\times$ 0.10 mm i.d. AT-5 capillary column. (From Covaci, A., de Boer, J., Ryan, J. J., Voorspoels, S., and Schepens, P., *Anal. Chem.*, 74, 790–798, 2002.)

Organophosphate esters have been analyzed mainly in the indoor air samples. SAE has been relatively popular in the extraction of organophosphates from air samples, which have been collected either on filters or adsorbents. Both static and dynamic extraction can be used. An example of dynamic SAE (DSAE) is the extraction of OPEs from quartz filters by hexane MTBE (7:3). The flow rate was 0.2 ml/min, and the total extraction time was only 3 min, at a temperature of 70°C. The recoveries were compared with static SAE (2 × 20 min) and PLE and the recoveries obtained with DSAE (>95%) were at the same level or better than those obtained with other methods. No further purification or concentration was needed before GC–NPD analysis of organophosphates. The GC column was a 30 m × 0.32 mm i.d. DB% column with a phase thickness of 0.1 μm. Splitless injection was applied in the sample introduction. The LODs were better than 0.4 ng/m^3. The system was developed further, and the DSAE was connected online with GC using PTV and large volume injection during the transfer. The most abundant compound found in the air was tri(n-butyl) phosphate.[78,76]

NOMENCLATURE

ABS	acrylonitrile butadiene styrene
ASE	accelerated solvent extraction
BFR	brominated flame retardant
CB	chlorobenzenes
CEN	European Committee for Standardization
DBDO	decabromodiphenyloxide
DDE	1,1-dichloro-2,3-bis(4-chlorophenyl)ethane
DDT	1,1-dichloro-2,3-bis(4-chlorophenyl)ethylene
DeBB	decabromobiphenyl
DeBDE	decabromodiphenyl ether
Deca-BB	commercial decabrominated biphenyl
Deca-BDE	commercial decabrominated diphenyl ether
DiBB	dibromobiphenyl
DiBDE	dibromodiphenyl ether
DMAE	dynamic microwave-assisted extraction
DSAE	dynamic sonication-assisted extraction
ECD	electron capture detector
ECNI	electron capture negative ionization
EHC	environmental health criteria
EI	electron impact ionization
FIA	flow injection analysis
FID	flame ionization detection
FR	flame retardant
GC	gas chromatography
GPC	gel permeation chromatography
HBCD	hexabromocyclododecane
HCB	hexachlorinated benzenes
HGAAS	hydride generation atomic absorption spectroscopy
HpBB	heptabromobiphenyl
(HP)LC	(high performance) liquid chromatography
HpBDE	heptabromodiphenyl ether
HRMS	high resolution mass spectrometry
HxBB	hexabromobiphenyl

IC	ion chromatography
IEC	international Electrotechnical Commission
IPLC	ion pair liquid chromatography
ISO	International Organization for Standardisation
ITMS	ion trap mass spectrometry
KemI	Chemicals Inspectorate (in Sweden)
LC	column liquid chromatography
LLE	liquid–liquid extraction
LRMS	low resolution mass spectrometry
LVI	large volume injection
MAE	microwave assisted extraction
MeO-PBDE	methoxylated polybrominated diphenyl ether
MS	mass spectrometry
NCI	negative chemical ionization
NP	normal phase
NPD	nitrogen phosphorus detector
NPLC	normal phase liquid chromatography
OcBB	octabromobiphenyl
OcBDE	octabromodiphenyl ether
OcCTA	octachlorothiantrene
Octa-BDE	commercial octabrominated diphenyl ether
OP	organophosphorus
OPE	organophosphorus esters
PAH	polycyclic aromatic hydrocarbon
PBB	polybrominated biphenyl
PBCCH	pentabromochlorocyclohexane
PBDE	polybrominated diphenyl ether
PBT	polybutylene terephthalate
PBT	pentabromotoluene
PCB	polychlorinated biphenyl
PCDD	polychlorinated dibenzo-p-dioxins
PCN	polychlorinated naphthalenes
PCP	polychlorinated phenols
PDMS	polymethyldisiloxane
PeBB	pentabromobiphenyl
PeBDE	pentabromodiphenyl ether
PET	polyethylene terephthalate
PHWE	pressurized hot water extraction
PLE	pressurized liquid extraction
POP	persistent organic pollutant
PS-DVB	polystyrenedivinylbenzene
PTV	programmed temperature vaporizer
PUR	polyurethane
PXDDs	polyhalogenated dibenzo-p-dioxins
PXDFs	polyhalogenated dibenzofurans
RP	reversed phase
RSD	relative standard deviation
SAE	sonication assisted extraction
SFE	supercritical fluid extraction
SIM	selective ion monitoring
SPME	solid phase micro extraction

SPE	solid phase extraction
TBBPA	tetrabromobisphenol A
TBPA	tetrabromophthalic anhydride
TCBPA	tetrachlorobisphenol A
TD	thermal desorption
TDBPP	tris(2,3-dibromopropyl)phosphate
TeBB	tetrabromobiphenyl
TeBDE	tetrabromodiphenyl ether
TOFMS	time-of-flight mass spectrometry

REFERENCES

1. Kemi, *The Flame Retardants Project*, Swedish National Chemicals Inspectorate, (KemI) Report 5/96.
2. WHO/ICPS. *Environmental Health Criteria 192: Flame Retardants — General Introduction*, World Health Organization, Geneva, 1997.
3. WHO/ICPS. *Environmental Health Criteria 209: Flame Retardants: Tris[chloropropyl]phosphate and Tris[chloroethyl]phosphate*, World Health Organization, Geneva, 1998.
4. WHO/ICPS. *Environmental Health Criteria 218: Flameretardants tris[2 butoxyethyl] phosphate, tris[2 ethylhexyl]phosphate and tetrakis [hydroxymethyl] phosphonium salts*, World Health Organization, Geneva, 2000.
5. WHO/ICPS. *Environmental Health Criteria 162: Brominated Diphenyl Ethers*, World Health Organization, Geneva, 1994.
6. BSEF, Bromine science and environmental forum, *An introduction to Brominated Flame Retardants*, P.4, www.bsef.com
7. Danish Environmental Protection Agency, Project No. 494, *Brominated Flame Retardants*, 1999.
8. Covaci, A., Voorspoels, S., and de Boer, J., Determination of brominated flame retardants, with emphasis on poly brominated diphenyl ethers in environmental and human samples, *Environ. Int.*, 29, 735–756, 2003.
9. Akutsu, K., Obana, H., Okihashi, M., Kitagawa, M., Nakazawa, H., and Matsuki, Y., GC/MS analysis of polybrominated diphenyl ethers in fish collected from the inland Sea of Seto, Japan, *Chemosphere*, 44, 1325–1333, 2001.
10. Christensen, J. H., Glasius, M., Pecseli, M., Platz, J., and Pritzl, G., Polybrominated diphenyl ethers (PBDEs) in marine fish and blue mussels from southern Greenland, *Chemosphere*, 47, 631–638, 2002.
11. Allchin, C. and de Boer, J., Results of a comprehensive survey for PBDEs in the river Tees, U.K, *Organohalogen Compd.*, 52, 30–34, 2001.
12. Allchin, C. R., Law, R. J., and Morris, S., Polybrominated diphenyl ethers sediments and biota downstream of potential sources in the U.K, *Environ. Pollut.*, 105, 197–207, 1999.
13. Asplund, L., Athanasiadou, M., Sjödin, A., Bergman, Å., and Börjeson, H., Organohalogen substances in muscle, egg, and blood from healthy Baltic Salmon(*Salmo salar*) and Baltic Salmon that produced offspring with the M74 syndrome, *Ambio*, 28, 67–76, 1999.
14. Booij, K., Zegers, B. N., and Boon, J. P., Levels of some polybrominated diphenyl ether flame retardants along the Dutch Coast as derived from its accumulation in SPMDs and blue mussels (*Mytilus edulis*), *Chemosphere*, 46, 683–688, 2002.
15. Carlsson, H., Nilsson, U., Becker, G., and Östman, C., Organophosphate ester flame retardants and plasticizers in the indoor environment: analytical methodology and occurrence, *Environ. Sci. Technol.*, 31, 2931–2936, 1997.
16. Covaci, A., de Boer, J., Ryan, J. J., Voorspoels, S., and Schepens, P., Determination of polybrominated diphenyl ethers and polychlorinated biphenyls in human adipose tissue by large volume injection narrow bore capillary gas chromatography/electron impact low resolution mass spectrometry, *Anal. Chem.*, 74, 790–798, 2002.
17. Covaci, A., de Boer, J., Ryan, J. J., Voorspoels, S., and Schepens, P., Distribution of organobrominated and organochlorinated contaminants in Belgian human adipose tissue, *Environ. Res.*, 88, 210–218, 2002.

18. Darnerud, P. O., Atuma, S., Aune, M., Cnattingius, S., Wernroth, M. L., and Wicklund Glynn, A., *Organohalogen Compd.*, 35, 411–414, 1998.

19. de Boer, J., van der Meer, J., and Brinkman, U. A. Th., Determination of chlorobiphenyls in seal blubber, marine sediment, and fish: interlaboratory study, *J. AOAC Int.*, 79, 83–96, 1996.

20. de Boer, J., Wester, P., Klamer, J., Lewis, W. E., and Boon, J. P., Do flame retardants threaten ocean life? *Nature*, 394, 28–29, 1998.

21. de Boer, J., Wester, P. G., van der Horst, A., and Leonards, P. E. G., Polybrominated diphenylethers in suspended particulate matter, sediments, sewage treatment plant influents and effluents and biota from The Netherlands, *Environ. Pollut.*, 122, 63–74, 2003.

22. de Wit, C., An overview of PBDEs in the environment, *Chemosphere*, 46, 583–624, 2002.

23. Dodder, N., Strandberg, B., and Hites, R. A., Concentrations and spatial variations of polybrominated diphenyl ethers in fish and air from the North Eastern United States, *Environ. Sci. Technol.*, 36, 146–151, 2002.

24. Fries, E. and Puettmann, W., Monitoring of the three organophosphate esters TBP, TCEP and TBEP in river water and ground water (Oder, Germany), *J. Environ. Monit.*, 5, 346–352, 2003.

25. Fukazawa, H., Hoshino, K., Shiozawa, T., Matsushita, H., and Terao, Y., Identification and quantification of chlorinated bisphenol A in wastewater from wastepaper recycling plants, *Chemosphere*, 44, 973–979, 2001.

26. Haglund, P., Zook, D., Buser, H. R., and Hu, J., Identification and quantification of polybrominated diphenyl ethers and methoxy polybrominated diphenyl ethers in Baltic biota, *Environ. Sci. Technol.*, 31, 3281–3287, 1997.

27. Hamm, S., Strikkeling, M., Ranken, P. F., and Rothenbacher, K., Determination of polybrominated diphenyl ethers and PBDD/Fs during the recycling of high impact polystyrene containing decabromodiphenyl ether and antimony oxide, *Chemosphere*, 44, 1353–1360, 2001.

28. Hagmar, L., Jakobsson, K., Thuresson, K., Rylander, L., Sjödin, A., and Bergman, Å., Computer technicians are occupationally exposed to polybrominated diphenyl ethers and tetrabromobisphenol A, *Organohalogen Compd.*, 47, 202–205, 2000.

29. Hale, R. C., La Guardia, M. J., Harvey, E. P., Matteson Mainor, T., Duff, W. H., and Gaylor, W. O., Polybrominated diphenyl ether flame retardants in Virginia freshwater fishes (U.S.A.), *Environ. Sci. Technol.*, 35, 4585–4591, 2001.

30. Hardy, M. L., The toxicology of the commercial polybrominated diphenyl oxide flame retardants: DBDPO, OBDPO, PeBDPO, *Organohalogen Compd.*, 47, 233–236, 2000.

31. Hooper, K. and McDonald, A., The PBDEs: an emerging environmental challenge and another reason for breast milk monitoring programs, *Environ. Health Perspect.*, 108, 387–392, 2000.

32. Hovander, L., Athanasiadou, M., Asplund, L., Jensen, S., and Klasson Wehler, E., Extraction and cleanup methods for analysis of phenolic and neutral organohalogens in plasma, *J. Anal. Toxicol.*, 24, 696–703, 2002.

33. Hovander, L., Malmberg, T., Athanasiadou, I., Rahm, S., Bergman, Å., and Klasson Wehler, E., Identification of hydroxylated PCB metabolites and other phenolic halogenated pollutants in human blood plasma, *Arch. Environ. Contam. Toxicol.*, 42, 105–117, 2002.

34. Ikonomou, M. G., Rayne, S., and Addison, R. F., Exponential increase of the brominated flame retardants, polybrominated diphenyl ethers, in the Canadian Arctic from 1981 to 2000, *Environ. Sci. Technol.*, 36, 1886–1892, 2002.

35. Ikonomou, M., Chromatographic and ionization properties of polybrominated diphenyl ethers using GC/high resolution MS with metastable atom bombardment and electron impact ionization, *Anal. Chem.*, 74, 5263–5272, 2002.

36. Ingerowski, G., Friedle, A., and Thumulla, J., Chlorinated ethyl and isopropyl phosphoric acid triesters in the indoor environment — an inter laboratory exposure study, *Indoor Air*, 11, 145–149, 2001.

37. Jakobsson, K., Thuresson, K., Rylander, L., Sjödin, A., Hagmar, L., and Bergman, Å., Exposure to polybrominated diphenyl ethers and tetrabromobisphenol A among computer technicians, *Chemosphere*, 46, 709–716, 2002.

38. Johnson, A. and Olson, N., Analysis and occurrence of polybrominated diphenyl ethers in Washington state freshwater fish, *Arch. Environ. Contam. Toxicol.*, 41, 339–344, 2001.

39. Kierkegaard, A., Sellström, U., Bignert, A., Olsson, M., Asplund, L., Jansson, B., and de Witt, C., Temporal trends of a polybrominated diphenyl ether (PBDE), a methoxylated PBDE, and hexabromocyclododecane in *Swedish biota*, *Organohalogen Compd.*, 40, 367–370, 1999.

40. Law, R. J., Alchin, C. R., Bennet, M. E., Morris, S., and Rogan, E., Polybrominated diphenyl ethers in two species of marine top predators from England and Wales, *Chemosphere*, 46, 673–681, 2002.

41. Lindström, G., Wingfors, H., Dam, M., and van Bavel, B., Identification of 19 polybrominated diphenyl ethers (PBDEs) in long finned pilot whale (*Globicephala melas*) from the Atlantic, *Arch. Environ. Contam. Toxicol.*, 36, 355–363, 1999.

42. Luross, J. M., Alaee, M., Sergeant, D. B., Cannon, C. M., Whittle, D. M., Solomon, K. R., and Muir, D. C. G., Spatial distribution of polybrominated diphenyl ethers and polybrominated biphenyls in lake trout from the Laurentian Great Lakes, *Chemosphere*, 45, 665–672, 2002.

43. Manchester-Neesvig, J. B., Valters, K., and Sonzogni, W. C., Comparison of polybrominated diphenyl ethers (PBDEs) and polychlorinated biphenyls (PCBs) in Lake Michigan salmonids, *Environ. Sci. Technol.*, 35, 1072–1077, 2001.

44. Meironyte, D., Bergman, Å., and Noren, K., Polybrominated diphenyl ethers in Swedish human liver and adipose tissue, *Arch. Environ. Contam. Toxicol.*, 40, 564–570, 2001.

45. Meironyte, D., Noren, K., and Bergman, Å., Analysis of polybrominated diphenyl ethers in Swedish human milk, *J. Toxicol. Environ. Health A*, 58, 101–113, 1999.

46. Meneses, M., Wingfors, H., Schuhmacher, M., Domingo, J. L., Lindström, G., and van Bavel, B., Polybrominated diphenyl ethers in human adipose tissue from Spain, *Chemosphere*, 39, 2271–2278, 1999.

47. Nylund, K., Asplund, L., Janssen, B., Jonsson, P., Litzen, K., and Sellström, U., Analysis of some polyhalogenated organic pollutants in sediment and sewage sludge, *Chemosphere*, 24, 1721–1730, 1992.

48. Öberg, K., Warman, K., and Öberg, T., Distribution and levels of brominated flame retardants in sewage sludge, *Chemosphere*, 48, 805–809, 2002.

49. Ohta, S., Ishizuka, D., Nishimura, H., Nakao, T., Aozosa, O., and Shimidzu, Y., Comparison of polybrominated diphenyl ethers in fish, vegetables, and meat and levels in human milk of nursing women in Japan, *Chemosphere*, 46, 689–696, 2002.

50. Petreas, M., She, J., Brown, F. R., Winkler, J., Visita, P., Li, C., Chand, D., Dhaliwal, J., Rogers, E., Zhao, G., and Charles, M. J., High PBDE concentrations in California human and wildlife populations, *Organohalogen Compd.*, 58, 177–180, 2002.

51. Pijnenburg, A. M. C. M., Everts, J. W., de Boer, J., and Boon, J. P., Polybrominated biphenyl and diphenylether flame retardants: analysis, toxicity and environmental occurrence, *Rev. Environ. Contam. Toxicol.*, 141, 1–26, 1995.

52. Poliakova, O. V., Lebedev, A. T., and Hänninen, O., Organic pollutants in snow of urban and rural Russia and Finland, *Toxicol. Environ. Chem.*, 75, 181–194, 2000.

53. Sellström, U., Jansson, B., Kierkegaard, A., and de Wit, C., Polybrominated diphenyl ethers in biological samples from the Swedish environment, *Chemosphere*, 26, 1703–1718, 1993.

54. Sellström, U. and Jansson, B., Analysis of tetrabromobisphenol in a product and environmental samples, *Chemosphere*, 31, 3085–3092, 1995.

55. Sellström, U., Kierkegaard, A., de Wit, C., and Jansson, B., Polybrominated diphenyl ethers and hexabromo cyclododecane in sediment and fish from a Swedish river, *Environ. Toxicol. Chem.*, 17, 1065–1072, 1998.

56. Sjödin, A., Carlsson, H., Thuresson, K., Sjölin, S., and Bergman, Å., Flame retardants in indoor air at an electronics recycling plant and at other work environments, *Environ. Sci. Technol.*, 35, 448–454, 2001.

57. Sjödin, A., Hagmar, L., Klasson Wehler, E., Kronholm Diab, K., Jakobsson, E., and Bergman, Å., Flame retardant exposure: polybrominated diphenyl ethers in blood from Swedish workers, *Environ. Health Perspect.*, 107, 643–648, 1999.

58. Strandberg, B., Dodder, N. G., Basu, I., and Hites, R. A., Concentrations and spatial variations of polybrominated diphenyl ethers and other organohalogen compounds in Great Lakes air, *Environ. Sci. Technol.*, 35, 1078–1083, 2001.

59. Strandman, T., Koistinen, J., Kiviranta, H., Vuorinen, P. J., Tuomisto, J., Tuomisto, J., and Vartiainen, T., Levels of some polybrominated diphenyl ethers in fish and human adipose tissue in Finland, *Organohalogen Compd.*, 40, 355–358, 1999.
60. Thomsen, C., Leknes, H., Lundanes, E., and Becher, G., Brominated flame retardants in laboratory air, *J. Chromatogr. A*, 923, 299–304, 2001.
61. Thomsen, C., Lundanes, E., and Becher, G., A simplified method for determination of tetrabromobisphenol A and polybrominated diphenyl ethers in human plasma and serum, *J. Sep. Sci.*, 24, 282–290, 2001.
62. Thomsen, C., Lundanes, E., and Becher, G., Brominated flame retardants in plasma samples from three different occupational groups in Norway, *J. Environ. Monit.*, 3, 366–370, 2001.
63. Simonsen, F. A., Moller, L. M., Madesn, T., and Stavnsbjerg, M., *Brominated flame retardants: toxicity and ecotoxicity*, Project No. 568, Danish Environmental Protection Agency, 2000.
64. Vetter, W., Scholz, E., Gaus, C., Muller, J. F., and Haynes, D., Anthropogenic and natural organohalogen compounds in blubber of dolphins and dugongs (*Dugong dugon*) from Northeastern Australia, *Arch. Environ. Contam.*, 41, 221–231, 2001.
65. Burreu, S., Broman, D., and Örn, U., Tissue distribution of 2,2′,4,4′tetrabromo[14C]diphenyl ether ([14C]-PBDE 47) in pike (*Esox lucius*) after dietary exposure — a time series study using whole body autoradiography, *Chemosphere*, 40, 977–985, 2000.
66. Hughes, M. F., Edwards, B. C., Mitchell, C. T., and Bhooshan, B., In vitro dermal absorption of flame retardant chemicals, *Food Chem. Toxicol.*, 39, 1263–1270, 2001.
67. Bergman, Å., Brominated flame retardants — a burning issue, *Organohalogen Compd.*, 47, 36, 2000.
68. de Boer, J., Allchin, C., Law, R., Zegers, B., and Booij, J. P., Method for the analysis of polybrominated diphenyl ethers in sediments and biota, *Trends Anal. Chem.*, 20, 591–599, 2001.
69. Hagenmeier, H., She, J., Benz, T., Dawidowsky, N., Düsterhöft, L., and Lindig, C., Analysis of sewage sludge for polyhalogenated dibenzo-*p*-dioxins, dibenzofurans, and diphenylethers, *Chemosphere*, 25, 1457–1467, 1992.
70. Hartonen, K., Bowadt, S., Hawthorne, S. B., and Riekkola, M. L., Supercritical fluid extractions with solid phase trapping of chlorinated and brominated pollutants from sediment samples, *J. Chromatogr. A*, 774, 229–242, 1997.
71. Hyötyläinen, T., Hartonen, K., Säynäjoki, S., and Riekkola, M. L., Pressurised hotwater extraction of brominated flame retardants in sediment samples, *Chromatographia*, 53, 301–305, 2001.
72. Kuosmanen, K., Hyötyläinen, T., Hartonen, K., and Riekkola, M. L., Pressurised hot water extraction coupled on line with liquid chromatography–gas chromatography for the determination of brominated flame retardants in Sediment Samples, *J. Chromatogr. A*, 943, 113–122, 2002.
73. Alaee, M., Sergeant, D. B., Ikonomou, M. G., and Luross, J. M., A gas chromatography or high resolution mass spectrometry (GC or HRMS) method for determination of polybrominated diphenyl ethers in fish, *Chemosphere*, 44, 1489–1495, 2001.
74. Soni, B. G., Philp, A. R., Foster, R. G., and Knox, B. E., Novel retinal photoreceptors, *Nature*, 94, 27–28, 1998.
75. Bergman, Å., Ostman, C., Nybom, R., Sjödin, A., Carlsson, H., and Nilsson, U., *Organohalogen Compd.*, 33, 414–417, 1997.
76. Ericsson, M. and Colmsjö, A., Dynamic microwave assisted extraction coupled online with solid phase extraction and large volume injection gas chromatography: determination of organophosphorus esters in air samples, *Anal. Chem.*, 75, 1713–1719, 2001.
77. Otake, T., Yoshinaga, J. and Yanagisawa, Y., Analysis of organic esters of plasticizer in indoor air by GC–MS and GC–FPD, *Environ. Sci. Technol.*, 35, 3099–3102, 2001.
78. Sanchez, C., Ericsson, M., Carlsson, H., Colmsjo, A., and Dyremark, E., Dynamic sonication assisted solvent extraction of organophosphate esters in air samples, *J. Chromatogr. A*, 957, 227–234, 2002.
79. Sanchez, C., Ericsson, M., Carlsson, H., and Colmsjo, A., Determination of organophosphate esters in air samples by dynamic sonication assisted solvent extraction coupled online with large volume injection gas chromatography utilizing a programmed temperature vaporizer, *J. Chromatogr. A*, 993, 103–110, 2003.
80. Andersson, Ö. and Wartanian, A., Levels of polychlorinated camphenes (toxaphene), chlordane compounds and PBDE in seals from Swedish waters, *Ambio*, 21, 550–552, 1992.

81. Björklund, J., Tollbäck, P., Hiärne, C., Dyremark, E., and Östman, C., Influence of injection technique and column system on gas chromatographic determination of polybrominated diphenyl ethers (PBDE), *J. Chromatogr. A*, 1041, 201–210, 2004.

82. Loganathan, B. G., Kannan, K., Watanabe, I., Kawano, M., Irvine, K., Kumar, S., and Sikka, H. C., *Environ. Sci. Technol.*, 29, 1832–1838, 1995.

83. Van Bavel, B., Sundelin, E., Lillbäck, J., Dam, M., and Lindströrm, G., Supercritical fluid extraction of polybrominated diphenyl ethers from long finned pilot whale (*Globicephala melas*) from the Atlantic, *Organohalogen Compd.*, 40, 359–362, 1999.

84. Guillamon, M., Martinez, E., Eljarrat, E., and Lacorte, S., Development of an analytical procedure based in accelerated solvent extraction and GC–EI–MS for the analysis of 40 PBDEs in human milk, *Organohalogen Compd.*, 55, 199–202, 2002.

85. Jonsson, O. B., Nordloef, U., and Nilsson, U. L., The XT tube extractor: a hollow fiber based supported liquid membrane extractor for bioanalytical sample preparation, *Anal. Chem.*, 75, 3506–3511, 2003.

86. Möller, K., Crescenzi, C., and Nilsson, U., Determination of a flame retardant hydrolysis product in human urine by SPE and LC–MS. Comparison of molecularly imprinted solid phase extraction with a mixed mode anion exchanger, *Anal. Bioanal. Chem.*, 378, 197–204, 2004.

87. Tolosa, I., Douy, B., and Carvalho, F. P., Comparison of the performance of graphitized carbon black and poly(styrene divinylbenzene) cartridges for the determination of pesticides and industrial phosphates in environmental waters, *J. Chromatogr. A*, 864, 121–136, 1999.

88. Oros, D. R., Jarman, W. M., Lowe, T., David, N., Lowe, S., and Davis, J. A., Surveillance for previously unmonitored organic contaminants in the San Francisco Estuary, *Mar. Pollut. Bull.*, 46, 1102–1110, 2003.

89. Polo, M., Gomez Noya, G., Quintana, J. B., Llompart, M., Garcia Jares, C., and Cela, R., Development of a solid phase microextraction gas chromatography or tandem mass spectrometry method for polybrominated diphenyl ethers and polybrominated biphenyls in water samples, *Anal. Chem.*, 76, 1054–1062, 2004.

90. Thomsen, C., Leknes, H., Lundanes, E., and Becher, G., A new method for determination of halogenated flame retardants in human milk using solid phase extraction, *J. Anal. Toxicol.*, 126, 129–138, 2002.

91. Thomsen, C., Haug, L. S., Leknes, H., Lundanes, E., Becher, G., and Lindström, G., Comparing electron ionization high resolution and electron capture low resolution mass spectrometric determination of polybrominated diphenyl ethers in plasma, serum and milk, *Chemosphere*, 46, 641–648, 2002.

92. Thomsen, C., Leknes, H. M., Lundanes, E., and Becher, G., A new method for determination of halogenated flame retardants in human milk using solid phase extraction, *J. Anal. Toxicol.*, 26, 129–137, 2002.

93. de Boer, J., Capillary gas chromatography for the determination of halogenated micro contaminants, *J. Chromatogr. A*, 843, 179–198, 1999.

94. Hyötyläinen, T. and Hartonen, K., Determination of brominated flame retardants in environmental samples, *Trends Anal. Chem.*, 21, 13–29, 2002.

95. Peled, M., Scharia, R., and Sondack, D., *Thermal rearrangement of hexabromocyclododecane*, *Advances in Organic Chemistry*, Vol. II, Elsevier, New York, p. 92, 1995.

96. Barontini, F., Cozzani, V., and Petarca, L., Thermal stability and decomposition products of hexabromo cyclododecane, *Ind. Eng. Chem. Res.*, 40, 3270–3280, 2001.

97. Alaee, M., Backus, S., and Cannon, C., Potential interference of PBDEs in the determination of PCBs and other organochlorine contaminants using electron capture detection, *J. Sep. Sci.*, 24, 465–469, 2001.

98. Christensen, J. H. and Platz, J., Screening of polybrominated diphenyl ethers in blue mussels, marine and freshwater sediments in Denmark, *J. Environ. Monit.*, 3, 543–547, 2001.

99. Huwe, J. K., Lorentzsen, M., Thuresson, K., and Bergman, Å., Analysis of mono to deca brominated diphenyl ethers in chickens at the part per billion level, *Chemosphere*, 46, 635–640, 2002.

100. Sjödin, A., Jakobsson, E., Kierkegaard, A., Marsh, G., and Sellström, U., Gas chromatographic identification and quantification of polybrominated diphenyl ethers in a commercial product, Bromkal 70 to 5DE, *J. Chromatogr. A*, 822, 83–89, 1998.

101. Zafra, A., del Olmo, M., Suarez, B., Hontoria, E., Navalon, A. O., and Vilchez, J. L., Gas chromatographic mass spectrometric method for the determination of bisphenol A and its chlorinated derivatives in urban wastewater, *Water Res.*, 37, 735–742, 2003.
102. Vetter, W. A., GC or CNI–MS method for the identification of lipophilic anthropogenic and natural brominated compounds in marine samples, *Anal. Chem.*, 73, 4951–4957, 2001.
103. Thomsen, C., Janak, K., Lundanes, E., and Becher, G., Determination of phenolic flame retardants in human plasma using solid phase extraction and gas chromatography–electron capture mass spectrometry, *J. Chromatogr. B*, 750, 1–11, 2001.
104. Yasuhara, Akio Shiraishi, Hiroaki Nishikawa, Masataka Takashi Yamamoto Uehiro, Takashi Nakasugi, Osami Okumura, Tameo Katashi Kenmotsu Fukui, Hiroshi Nagase, Makoto Ono, Yusaku Yasunori Kawagoshi Baba, Kenzo Noma, and Yukio, Determination of organic components in leachates from hazardous waste disposal sites in Japan by gas chromatography–mass spectrometry, *J. Chromatogr. A*, 774, 321–332, 1997.
105. Yoshida, T., Matsunaga, I., and Oda, H., Simultaneous determination of semivolatile organic compounds in indoor air by gas chromatography–mass spectrometry after solid phase extraction, *J. Chromatogr. A*, 1023, 255–269, 2004.
106. Thruston, A. D. Jr., Richardson, S. D., McGuire, J. M., Collette, T. W., and Trusty, C. D., Multispectral identification of alkyl and chloroalkyl phosphates from an industrial effluent, *J. Am. Soc. Mass Spectrom.*, 2, 419–426, 1991.
107. Morris, S., Allchin, R. C., Bersuder, P., Zegers, N. B., Hafta, J. H. J., Boon, P. J., Brandsma, H. S., Kruijt, W. A., van der Veen, I., van Hesselingen, J., and de Boer, J., A new LC–MS method for the detection and quantification of hexabromocyclododecane diastereoisomers and tetrabromobisphenol flame retardants in environmental samples, *Organohalogen Compd.*, 60, 436–439, 2003.
108. de Boer, J. and Cofino, W. P., First worldwide interlaboratory study on PBDEs, *Chemosphere*, 46, 625–633, 2002.
109. Jansson, B., Andersson, R., Asplund, L., Bergman, Å., Litzen, K., Nylund, K., Reutergårdh, L., Sellström, U., Uvemo, U.-B., Wahlberg, C., and Wideqvist, U., Multiresidue method for the gas chromatographic analysis of some PCB and PBB pollutants in biological samples, *Fresenius J. Anal. Chem.*, 340, 439–445, 1991.
110. Dodder, N., Strandberg, B., and Hites, R. A., Concentrations and spatial variations of PBDE in fish and air from the North Eastern United States, *Environ. Sci. Technol.*, 36, 146–151, 2002.

32 Chromatographic Analysis of Endocrine Disrupting Chemicals in the Environment

Guang-Guo Ying

CONTENTS

I. INTRODUCTION

Since the publication in 1996 of the book *Our Stolen Future*,[1] there has been increasing concern in the general public and scientific community that some natural and synthetic chemicals can interfere with the normal functioning of endocrine systems, thus affecting reproduction and development in wildlife and human beings. The chemicals causing such effects are generally referred to as endocrine disruptors or endocrine disrupting chemicals (EDCs). The chemicals identified or suspected of being EDCs in the literature are summarized in Table 32.1 and include a wide variety of compounds such as pesticides (e.g., DDT, endosulfan, and atrazine), pharmaceuticals (e.g., ethinyl estradiol and mestranol), and industrial chemicals (e.g., bisphenol-A, polychlorinated biphenyls (PCBs) and nonylphenols). Many of these compounds have little in common structurally or in terms of their chemical properties, but evoke agonistic or antagonistic responses, possibly through comparable mechanisms of action.[3]

The endocrine system consists of glands and the hormones they produce that guide the development, growth, reproduction, and behavior of humans and animals. Hormones are biochemicals produced by endocrine glands in one part of the body that travel through the bloodstream and cause responses in other parts of the body. They act as chemical messengers and interact with specific receptors in cells to trigger responses and prompt normal biological functions such as growth, reproduction, and development. There are several ways that chemicals can interfere with the endocrine system.[3] They can mimic or block natural hormones, alter hormonal levels, and affect the functions that these hormones control.

There is compelling evidence on the effects of exposure to EDCs on wildlife. These include imposex of molluscs by organotin compounds,[4–6] developmental abnormalities, demasculization, and feminization of alligators in Florida by organochlorines,[7,8] feminization of fish by wastewater effluent from sewage treatment plants, paper mills,[9,10] and hermaphrodism in frogs from pesticides such as atrazine.[11]

There are also reports that human testicular and breast cancer rates have been increasing during the last four decades, especially in developed countries.[12–18] However, except in a few cases (e.g., diethylstilbestrol), a causal relation between exposure to chemicals and adverse health effect in humans has not been firmly established.

These chemicals with endocrine disrupting properties are released from a wide variety of sources such as domestic sewage, intensive agriculture, animal waste, industrial wastes, mining activity, and landfills. Suspected or known EDCs can be found in every compartment of our environment (air, water, soil, sediment, and biota), in industrial products and household items, and even in the food we eat.[2,19,20] Measurement and identification of EDCs in the different environmental compartments are critical to assessment of the potential risk to humans and wildlife. This chapter will give an introduction to the compounds with known/potential endocrine disrupting properties, and then discuss the analytical techniques used for various classes of those EDCs in the environment, including sample preparation and instrumental analysis.

II. ENDOCRINE DISRUPTING CHEMICALS

A. ALKYLPHENOLS AND ALKYLPHENOL ETHOXYLATES

Alkylphenol polyethoxylates (APEs) are among the most commonly used nonionic surfactants with a wide variety of commercial and domestic applications, such as in the manufacturing of pulp and

TABLE 32.1
List of Suspected or Known Endocrine Disrupting Chemicals (EDCs)

Classification	Endocrine Disrupting Chemicals	
Pesticides	2,4-D	Kepone (Chlordecone)
	Atrazine	Lindane
	Benomyl	Malathion
	Carbaryl	Mancozeb
	Cypermethrin	Methomyl
	Chlordane (γ-HCH)	Methoxychlor
	DDT and its metabolites	Mirex
	Dicofol	Parathion
	Dieldrin or Aldrin	Pentachlorophenol
	Endosulfan	Permethrin
	Endrin	Toxaphene
	Heptachlor	Trifluralin
	Hexachlorobenzene (HCB)	Vinclozolin
	Iprodione	
Organohalogens	Dioxins and furans	Polychlorinated biphenyls
APs	Nonylphenol (NP)	Nonylphenol ethoxylate (NPE)
	Octylphenol (OP)	Octylphenol ethoxylate (OPE)
Heavy metals	Cadmium	Mercury
	Lead	Arsenic
Organotins	Tributyltin (TBT)	Triphenyltin (TPT)
Phthalates	Bis(2-ethylhexyl)phthalate (DEHP)	Di-hexyl phthalate
		Di-propyl phthalate
	Butyl benzyl phthalate (BBP)	Dicyclohexyl phthalate
	Di-n-butyl phthalate (DBP)	Diethyl phthalate (DEP)
	Dimethyl phthalate (DMP)	Di-n-octyl phthalate (DOP)
Natural hormones	17β-Estradiol (E2)	Estriol (E3)
	Estrone (E1)	Testosterone
Pharmaceuticals	Ethinyl estradiol (EE2)	Tamoxifen
	Mestranol	DES
Phytoestrogens	Isoflavonoids	Zearalenone
	Coumestans	β-Sitosterol
	Lignans	
Phenols	Bisphenol-A (BPA)	Bisphenol-F (BPF)
Aromatic hydrocarbons	Benzo[a]pyrene	Anthracene
	Benz[a]anthracene	Pyrene
	Benzo[b/h]fluoranthene	Phenanthrene
	6-Hydroxy-chrysene	n-Butyl benzene

Source: Ying, G. G. and Kookana, R. S., *AWA Water J.*, 29(9), 42–45, 2002.

paper, textiles, paints, adhesives, leather products, rubber, plastics, pesticides, and cosmetics. Annual global production of APEs is over 500,000 tonnes, consisting of approximately 80% nonylphenol ethoxylates (NPEs), 15% octylphenol ethoxylates (OPEs), and the remainder as dodecylphenol and dinonylphenol ethoxylates.[21] These chemicals are mainly introduced into the environment by industrial and domestic effluents as well as sewage sludges discharged to surface waters and land.

Concern has increased recently about the wide usage of APEs because of these relatively stable biodegradation products, including 4-nonylphenol (4-NP), 4-*tert*-octylphenol (4-*t*-OP), nonylphenol monoethoxylate (NPE1) and nonylphenol diethoxylate (NPE2). These compounds have been widely found in sewage effluents and sludges, surface water and groundwater, and aquatic sediments in many countries.[19] Alkylphenols (APs) and APEs have been known not only to be toxic to marine and freshwater species, but also to induce estrogenic responses in fish.[22,23]

B. BISPHENOL-A AND BISPHENOL-F

Bisphenol-A (BPA) and bisphenol-F (BPF) are manufactured in high quantities, with 90% or more being used as a monomer for the production of polycarbonate and epoxy resins, unsaturated polyester–styrene resins and flame retardants. The final products are used as coatings on cans, as powder paints, as additives in thermal paper, in dental fillings and as antioxidants in plastics.[24] Their release into the environment is possible during manufacturing processes and by leaching from final products. Being widely used in households and industry, they can be expected to be present in raw sewage, wastewater effluents, and concentrated in sewage sludge.[25] Because of their similar chemical properties (log K_{ow} 3.06; water solubility 360 mg/l), the distribution and fate of BPF should be comparable to that of BPA.

BPA and BPF have shown weak estrogenic activity at concentrations below acute toxic levels.[26–28] Although in 1938 Dodds et al.[29] noted estrogenic activity of BPA, it is only in the last few years this compound has received attention. The relative potency of BPA ranges from approximately 1×10^{-6} to 5×10^{-7} times less than 17β-estradiol.[30] Based on in vitro receptor-interaction studies, the activity of BPA was estimated to be 2×10^{-3}-fold lower than that of 17β-estradiol.[31]

C. ORGANOTIN COMPOUNDS

Organotin compounds, especially butyl and phenyl species, are used in many human activities. The annual production of organotins in the world was estimated to reach more than 50,000 tonnes.[32] The plastics industry is responsible for the largest single usage of organotins, amounting to approximately two thirds of the total consumption. Most of the remainder is accounted for by biocides for antifouling paints, crop protection, and wood preservation. The wide use of organotins by these industries leads to the release of these compounds into the environment, especially the aquatic environment.[33]

Of particular importance to the environment is the high toxicity of tributyltin (TBT) and triphenyltins (TPT) and these degradation derivatives: mono- and di-butyltins (MBT and DBT), and mono- and di-phenyltins (MPT and DPT). They are highly toxic to fish species and other aquatic animals. A widespread deleterious effect induced by organotin contamination is imposex, a superimposition of male sex organs on some marine organisms such as female sea snails.[33] It is well known that TBT and TPT are EDCs and induce imposex in marine organisms even at a concentration as low as 1 ng/l.[34]

D. PESTICIDES

Pesticides are widely used around the world for agricultural and nonagricultural purposes. In the United States alone, over 800 pesticide active ingredients are formulated in about 21,000 different commercial products.[35] The use of pesticides provides unquestionable benefits in increasing agricultural production and controlling various diseases. Despite the obvious benefits of pesticides, their potential impact on the environment and public health is substantial because of exposure of humans and wildlife to pesticide residues in the environment and food. According to European Community (EC) directives, a pesticide residue must not be present at a concentration greater than 0.1 μg/l in drinking water and the requirements for surface water are 1 to 3 μg/l.[36,37] Some pesticides, including organochlorines, carbamates, triazines, 2,4-D, vinclozolin, malathion, and parathion, have been found or suspected to possess endocrine disrupting properties.[2,38,39]

E. Phthalates

Phthalic acid diesters, commonly known as phthalates, are produced all over the world in large quantities, and these are widely used in different industrial activities. Phthalates are used primarily as plasticisers in plastics, mainly in polyvinylchloride (PVC) products.

Because of its widespread use, relatively large amounts of these compounds are released into the environment and some of them enter the food chain. They have become ubiquitous in the environment.[40–43] The most commonly used phthalates have been included in the list of priority pollutants in several countries. These are: dimethyl- (DMP), diethyl- (DEP), di-*n*-butyl- (DnBP), butylbenzyl- (BBP), bis(2-ethylhexyl)- (DEHP), and di-*n*-octyl (DnOP) phthalates. The United States Environmental Protection Agency (US EPA) has established a maximum admissible concentration (MAC) in water of 6 μg/l for DEHP.[44] In recent years, considerable attention has been paid to human exposure to phthalates because of the suspicion of its carcinogenic and estrogenic properties.[30,45,46]

F. Phytoestrogens

Phytoestrogens are members of classes of polyphenolic compounds synthesized by plants. These include isoflavones and other flavoids, lignans, coumestanes, stilbenes, and zearalenones.[47] Phytoestrogens are found in plants and in many food products as glycosidic conjugates. The common isoflavones include genistein and daidzein, and their 4-methyl ethers biochanin A and formononetin, respectively. Equol and *O*-desmethylangolensin are common metabolites of daidzein and formononetin. Lignans are polyphenolic compounds linked by a four-carbon bridge. Flaxseed is particularly enriched in the lignans matairesinol and secoisolariciresinol and these are converted by bacteria in the mammalian gastrointestinal tract to enterolactone and enterodiol, respectively. Other members of the bioflavonoids that have estrogen-like properties include kaempferol, quercetin, apigenin, and 8-prenylnaringenin. Coumestanes, of which coumestrol is the most common, are present in plants such as alfalfa. Trans-resveratrol is a stilbene present in red wine. Zearalenone is found in fungi on plants.

Phytoestrogens have estrogenic activities with potencies between 10^{-1} and 10^{-4} of the activity of 17β-estradiol, and are thus more potent than man made chemicals.[48,49] Exposure to these compounds may affect humans and wildlife.

G. Polychlorinated Compounds (PCBs and Dioxins)

Polychlorinated biphenyls (PCBs) and dioxins are two groups of toxic organic contaminants that are widespread throughout the ecosystem as a consequence of its persistence and potential for bioaccumulation in the environment. These occur in water, air, soil, sediment, and biota in different areas around the world.[50–57] In general, water birds and marine mammals have accumulated the dioxins and dioxin-like PCBs with much higher concentrations than humans, implying higher risk from exposure to wildlife.[58]

Dioxin is a generic term given to polychlorinated dibenzo-*p*-dioxins (PCDDs) and polychlorinated dibenzofurans (PCDFs). Dioxins are the unwanted byproducts of the manufacture of certain industrial chemicals or are produced during various combustion and incineration processes. The US EPA has estimated that 70% of all quantifiable environmental emissions were contributed by air emissions from just three source categories: municipal waste incinerators; backyard burning; and medical waste incinerators.[59] 2,3,7,8-Tetrachlorodibenzo-*p*-dioxin (TCDD) and 16 other dioxins which contain chlorine at positions 2, 3, 7, and 8 of the molecule are toxic, although the other 16 have been found to be less toxic than TCDD.[54] The concentrations of the 17 dioxins are each multiplied by a weighting factor based on its relative toxicities to give total dioxin content. The sums of these values have been termed "Toxic Equivalents" (TEQs).[54] Two schemes are commonly used. One is the International Toxic Equivalency Factor (TEF) scheme denoted as I-TEF.

The other is the WHO scheme, which has different toxicity factors for humans, mammals, fish, and birds.[54] Most dioxins do not contain chlorine at the toxic combination of positions 2, 3, 7, and 8 of the molecule, and so are believed to present no significant biological activity or safety risk.

PCBs have been mainly used for industrial purposes such as dielectric fluids in transformers and capacitors. They are a group of 209 related industrial compounds which differ only in the number and pattern of chlorine atoms attached to the biphenyl molecule. These compounds are termed congeners of PCBs. The patterns of chlorine substitution ultimately modify and dictate each congener's environmental fate and toxicity. Of the 209 congeners of PCBs, some have been reported as having toxicological effects.[54] As with the dioxins, the concentrations of the dioxin-like congeners of concern are assigned a weighting factor based on its relative toxicities, and the sum of the weighted concentrations forms the TEQ, which seems to be extremely useful for risk assessment. The limit concentrations of PCBs used for regulatory purposes are based either on the "total PCB" level or, more recently, on "standard" individual congeners[28,52,101,138,153,180] chosen in order to cover a wide range of chlorination (from three to seven chlorine atoms) and taking into consideration their relatively high levels in samples.[60]

The wide dispersal of the dioxin-like chemicals throughout the environment is primarily the result of atmospheric transport and deposition. Eventually the dioxin-like chemicals become adsorbed to dust particles and surfaces and are deposited in sediments. The two primary pathways for dioxin-like chemicals to enter the food chain are from the air-to-plant-to-animal and from water- and sediment-to-fish.[59] A third route for dioxin-like chemicals to enter the food chain is through the accidental contamination incidents resulting from inappropriate handling and processing of feed and food substances. It has been estimated that more than 90% of human exposure to dioxin and dioxin-like chemicals is through the ingestion of contaminated food substances.[59]

PCDDs, PCDFs, and dioxin-like PCBs caused adverse effects on organisms through the action of the aryl hydrocarbon receptor (AhR), a cytosolic protein that binds these compounds with high affinity.[61] Dioxins and dioxin-like PCBs have been recognized as carcinogens and teratogens and more recently as endocrine disruptors.[62] TCDD is a potent inhibitor of estrogen-mediated activity that has been shown to be AhR dependent, whereas certain PCBs or mixtures of PCBs exhibit no antiestrogenic activity but are actually estrogenic.[63,64] The adverse effects of PCBs and dioxins on reproduction in mammals have been observed in the Great Lakes polluted with PCBs and other compounds.[62,65,66]

H. Polycyclic Aromatic Hydrocarbons (PAHs)

PAHs are widespread environmental contaminants resulting from incomplete combustion or high-temperature pyrolytic processes involving organic materials, and are thus generated whenever fossil fuels or vegetation are burned. It has been estimated that 230,000 metric tons of PAHs enter the global environment annually from spills and seeps of petroleum, direct discharges from industrial/domestic sources, aerial transport, and biosynthesis.[67] Since these compounds are long lasting, poorly degradable pollutants, they accumulate in soil and sediments, surface water, and the atmosphere as well as organisms. PAHs are potentially carcinogenic, mutagenic,[68,69] and some PAHs (e.g., dibenz[a,h]anthracene, benzo[a]pyrene, and benzo[ghi]perylene) even showed endocrine disrupting effects.[70,71] Sixteen PAHs have been selected by US EPA as priority pollutants for regulatory purposes.[72] These are: naphthalene, acenaphthylene, acenaphthene, fluorene, phenanthrene, anthracene, fluoranthrene, pyrene, benz[a]anthracene, chrysene, benzo[b]fluoranthene, benzo[k]fluoranthene, benzo[a]pyrene, indeno[1,2,3-cd]pyrene, dibenz[a,h]anthracene, benzo[ghi]perylene, and benzo[ghi]perylene.

I. Steroid Hormones

Natural and synthetic steroids have become a major subject of worldwide growing concern because these compounds may interfere with the normal reproduction of human, livestock, and wildlife.

One of the groups of compounds under investigation is the natural estrogens, primarily synthesized in the female body and essential for female characteristics and reproduction, and closely related synthetic hormones.[73] Many estrogenic effects observed in the aquatic environment, for instance the feminization of male fish as indicated by vitellogenin production by sewage effluents, have been identified in rivers worldwide.[74,75] To date, estrogenic effects on aquatic wildlife have not been conclusively linked to only one particular compound, but some chemicals are primarily responsible for causing endocrine disruption. Among them, the natural estrogens estrone (E1), 17β-estradiol (E2), estriol (E3), and the exogenous 17α-ethinylestradiol (EE2), the active ingredient in oral contraceptive pills, possess the highest estrogenicity.

Synthetic steroids are also widely used in humans as therapeutic drugs (e.g., estrogens and progestogens), and in livestock as growth promoters (e.g., E2, progesterone, testosterone, zeranol, trenbolone actate, melengestrol acetate, and their metabolites). There is little information in the literature on the fate and effects of those drugs in the environment.

III. SAMPLE PREPARATION

In this section, we will introduce in general the extraction and clean-up techniques used for EDCs in aqueous, solid, and biological samples. Only those commonly used sample preparation methods will be discussed in the following.

A. AQUEOUS SAMPLES

Various extraction techniques are applied to isolate EDCs in aqueous samples, such as liquid–liquid extraction (LLE), solid-phase extraction (SPE), and solid-phase microextraction (SPME). LLE is frequently used in the extraction of EDCs with water immiscible organic solvents, most commonly with hexane or dichloromethane.[43,53,67,76–78] However, LLE produces emulsions and different extraction efficiencies for various compounds; it also requires large amounts of solvent and is slow, laborious, and difficult to automate.

Because of the formation of emulsions at phase boundaries for APE surfactants, LLE is limited to the degradation products APs, alkylphenol monoethoxylate to triethoxylate (APE(1–3)) and alkylphenol ethoxy carboxylate (APEC).[79] Dichloromethane and hexane are the solvents commonly used in the extraction of APs and APE(1–3) from liquid samples.[80–82] For phenolic compounds including BPA, OP, and NP, water samples are often acidified to pH < 4 with hydrochloric acid. Acidification of water samples suppresses the dissociation of phenols and prevents the ionization of the analytes, which increased the efficiency of the extraction.[83] Del Olmo et al.[84] studied the effect of pH on extraction of BPA using sodium hydroxide and hydrochloric acid for adjustment. The result obtained showed that the extraction efficiency remains constant for pH values lower than 6.5, decreasing sharply for higher values. This behavior agrees with the weak acid nature of BPA.

Ionic strength can also affect the extraction efficiency.[83–85] The extraction efficiency increased with the NaCl concentration, remaining constant at sodium chloride concentrations higher than 0.5 M.[84] Helaleh et al.[83] tested the effect of salt and found that NaCl, Na_2CO_3, and Na_2SO_4 gave poorer recoveries while NaBr and KI gave the highest recovery for all phenols. However, NaCl and NaBr gave comparable recoveries for BPA. NaBr was chosen for the extraction of all phenols.[83]

The extraction of organotin compounds in aqueous samples is different from the other EDCs. Organotin compounds in aqueous samples can be extracted and derivatized at the same time.[86–90] The organic solvent used in the extraction can be dichloromethane, hexane, or isooctane. A general procedure of extraction and derivatization is as follows: 1 l water is first buffered to pH 4.8 with 10 ml acetate buffer and spiked with internal standard tripropyltin chloride (TPrT) (about 100 ng). Samples are simultaneously extracted with 25 ml dichloromethane (or 25 ml hexane or isooctane) and derivatized with 1 to 3 ml of sodium tetraethylborate (NaBEt$_4$) by shaking manually for 5 min (or longer 30 min) in 2 l separatory funnels (or other extraction vessels). After phase separation, the organic phase is collected. The extraction is repeated with an additional 25 ml dichloromethane. The extracts are reduced to 1 ml by rotary evaporation at 30°C.

Among the extraction techniques for aqueous samples, SPE is attracting increasing attention and constitutes an alternative to LLE.[91] Desorption of retained organic compounds can be carried out by elution with a suitable solvent. SPE is widely used for the trace enrichment of very dilute solutions such as natural waters, where large sample volumes may have to be processed, to yield concentrations of analytes sufficient for detection. The extracts can be eluted from cartridges or disks using various solvents such as ethyl acetate, dichloromethane, methanol, and acetone. Among various solid sorbents, C$_{18}$ in cartridges or disks is the most widely used sorbent in the extraction of pesticides,[36,37,91,92] phenolic compounds,[93-95] phthalates,[42] steroid hormones,[96] phytoestrogens,[97,98] PAHs,[99] and chlorinated compounds[100] from aqueous samples. More organic solvent is required to elute chemicals from membrane extraction disks than from cartridges.[92,101,102]

Other extraction techniques are also applied to EDCs in aqueous samples. Steam distillation and solvent sublation were commonly used in the extraction of APs and APEs from water samples.[79,103] SPME coupled to gas chromatography (GC) and GC–MS was successfully used in the analysis of phthalates in water samples.[44,104,105]

B. SOLID SAMPLES

The extraction techniques that can be used to extract EDCs in solid samples (soil, sediment, or sludge) include: Soxhlet,[24,81,106-110] sonication,[43,93,96,111,112] accelerated solvent extraction (ASE),[113,114,115] microwave-assisted solvent extraction (MASE),[56] and supercritical fluid extraction (SFE).[81,116,117] Various solvents such as methanol, dichloromethane, hexane:acetone (1:1), ethyl acetate, and ACN may be used as extraction solvents[118,119]; the choice of solvent depends on the physiochemical properties of the target compounds.

A clean-up step is often necessary to remove interfering compounds from the extracts of sediment and sludge samples. The extracts are usually cleaned on a silica, alumina, or Florisil column,[50,81,106,108,109,120] SPE cartridges (C$_{18}$, NH$_2$, CN, and Carbograph),[93,107,110,111,116,121] or by liquid chromatography (LC).[111]

For organotin compounds in sediment or soil samples, two extraction methods can be used: methanol/hydrochloric acid digestion, or ethanoic acid digestion.[89,122] A prewetting step is needed for the HCl–methanol method (for 1 h). The experiments showed that significant degradation was observed over 0.1 mol/l HCl concentrations, but the use of 0.1 mol/l HCl avoided any degradation and seemed convenient for extraction with recoveries of 108 and 92% for TBT and TPT, respectively.[122] Ethanoic acid digestion also gave high recoveries (105 and 103% for TBT and TPT, respectively).[122] The extraction mixtures (solid to solvent, 1:20 w/v) were sonicated for 1 h and then shaken at 420 rpm for 4 to 12 h. The suspension was centrifuged at 3000 rpm for 15 min. The extracts (1 to 3 ml) were directly introduced into derivatization vessels. Organotin compounds were ethylated in 100 ml acetate buffer (pH 4.8) with 0.1 ml NaBEt$_4$ solution in the presence of 0.4 ml isooctane. The mixture was shaken at 420 rpm for 30 min. Isooctane extracts (2 to 5 μl) were directly injected into a GC. HCl-based extractions (HCl methanol and HCl ethyl acetate) were also used for biological samples.[123]

C. BIOLOGICAL SAMPLES

The extraction of EDCs from biological samples is often more complex than from aqueous and solid samples because of matrix interferences. Various techniques like acidic digestion, saponification, liquid–liquid partition, matrix solid-phase dispersion (MSPD), as well as Soxhlet extraction, sonication, microwave extraction, and SFE are employed in the extraction of biological samples.[35,93,118,124-129]

Clean-up is the most important step for biological samples because they are rich in fat and lipids, etc. Liquid–solid chromatography and gel permeation chromatography (GPC) are widely used for the clean-up of extracts.[51,117,125,128-133] GPC is very useful in removing fats from the extracts of biological samples. The most widely used gel column is SX-3 BioBeads (200-400 mesh) in a range of column sizes and solvents. The eluents used in GPC are mostly mixtures such as cyclohexane–ethyl acetate, cyclohexane–dichloromethane, toluene–ethyl acetate and 2-propanol–heptane.[60,134]

Chemical treatment using acid or alkali can be applied to eliminate interfering substances. Acidic or alkaline digestion is often used during extraction.[117,135,136] Chemical treatment with sulfuric acid or KOH can also be used in combination with other clean-up steps such as silica gel.[54,55,137,138]

IV. ANALYSIS

EDCs in the environment are often analyzed using GC or LC based instrumental techniques. GC coupled with an electron capture detector (ECD), a nitrogen–phosphorus detector (NPD), or mass spectrometry (MS) has been the preferred method due to its excellent sensitivity and separation capability on a capillary column. High performance liquid chromatography (HPLC) with various detectors such as ultraviolet detection (UV), fluorescence detection (FLD), MS, and more recently tandem MS (MS/MS) has also been used for analysis of some EDCs, especially for the polar compounds. Analytical techniques for each class of EDCs will be discussed in the following section.

A. ALKYLPHENOLS AND ALKYLPHENOL ETHOXYLATES

1. GC–MS

GC coupled with mass spectrometry (GC–MS) has been applied to the analysis of free and derivatized APs and APEs with shorter ethoxy units (<4). On a capillary column, NP and NPEs are separated into a cluster of peaks with only one peak for OP since it is a single isomer (4-t-OP) (Figure 32.1). The molecular peaks for APs (M$^+$ 220 for NP and 206 for OP) and APEs are weak under electron ionization (EI). The characteristic peaks for NP and OP are as follows: m/z 135, 107,

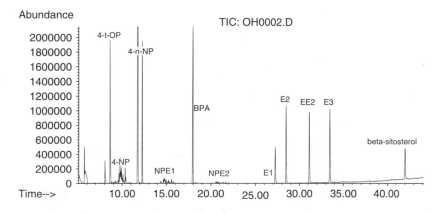

FIGURE 32.1 Total ion chromatogram (TIC) of TMS derivatives of some EDCs. GC–MS conditions: column HP-5MS; injector temperature 280°C; 120 to 190°C at 10°C/min, to 300°C (5 min) at 3°C/min; carrier gas He flow rate 1.1 ml/min; MS source temperature 230°C.

121, and 149, which are generated by the loss of the alkyl moiety. The EI mass spectrum of APE is dominated by typical features such as the occurrence of the $[CH_2CH_2OH]^+$ ion (*m/z* 45) from the ethoxylate chain, and the $[M - 85]^+$ and $[M - 71]^+$ ions corresponding to the loss of C_6H_{13} or C_5H_{11} fragments from the alkyl moiety.[106] The recurring mass increment of 44 is due to the ethoxylate (C_2H_4O) unit difference between oligomers present in mixtures.

APE analysis by GC without derivatization has been mainly limited to the more volatile biodegradation products like OP, NP, NPE1, and NPE2.[139] Acetylation or silylation of APs and APEs increases its volatility, sensitivity, and separation on capillary GC–MS.[80,81,108] The derivatized compounds all have retention times longer than the corresponding free compounds. Because of the presence of APEs with longer ethoxylate chains in environmental samples, GC–MS is not suitable for the analysis of total APEs.

2. HPLC and LC–MS

HPLC coupled with various detectors (UV, FLD, or MS) is a very versatile technique in the analysis of APs and APEs. The major advantage of HPLC is its ability to separate and quantify the various homologues and oligomers by the length of the alkyl and ethoxylate chains. Reversed-phase HPLC resolves the various alkyl homologues, whereas normal-phase HPLC provides information on the ethoxylate oligomer distribution. For environmental samples, FLD or MS is commonly employed to detect these compounds due to its sensitivity and selectivity.[81,107,110,140–142] APs and APEs can be detected using a fluorescence detector with an excitation wavelength of 230 nm and an emission wavelength of 290 nm (Figure 32.2). NPEs or OPEs are eluted as a single peak on a reversed-phase column like C_{18} column, which is very convenient for quantification of total APE concentration in environmental samples.[143] APE oligomers can be separated on a normal-phase column such as NH_2 or CN columns by using mixtures of hexane, isopropanol, and water as the mobile phase.[80,140,144,145] APE oligomers were successfully separated on an NH_2-Hypersil column (100×4.6, 3 μm) by using a mobile phase gradient of hexane–isopropanol–water 93.1:6.8:0.1 (v/v/v) to 44.1:49.9:6.0 in 20 min.[146]

Although different ionization techniques such as particle beam (PB), thermospray (TS), and atmospheric pressure chemical ionization (APCI) have been attempted, electrospray

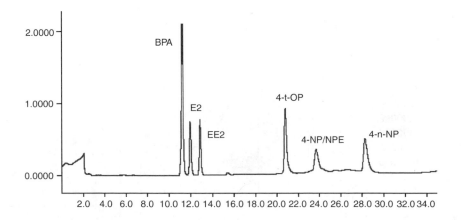

FIGURE 32.2 Liquid chromatogram of some EDCs. HPLC conditions: column Adsorbosphere C_{18} (250×4 mm, 5μ); fluorescence detector, excitation 230 nm and emission 290 nm; mobile phase Milli-Q water and ACN delivered at a constant flow rate of 1 ml/min. The gradient program of the mobile phase was as follows: 30% ACN and 70% water at 0 min, 40% ACN and 60% water at 5 min, 60% ACN and 40% water at 10 min, 80% ACN and 20% water at 20 min and isocratic purge until 30 min, and increasing to 100% ACN and 0% water at 35 min.

ionization (ESI) has recently become very popular for the analysis of surfactants as it produces minimal fragmentation of the molecular ion. HPLC–ESI–MS offers excellent sensitivity for APEs as well as APs in environmental samples.[110,111] However, it is necessary to perform two separate analyses for each sample in order to quantify APs and APEs. APs and alkylphenoxy ethoxy acetic acid (APE2C) and alkylphenoxy acetic acid (APE1C) as well as halogenated APs were detected in negative mode as $[M-H]^-$ because of its extremely low ionization efficiency under positive mode.[110,111,141] APEs were analyzed in positive ion mode as sodium adducts $[M + Na]^+$ because polyethoxylated compounds have a high affinity for alkali metal ions.[110,111] Even in the absence of added electrolyte, APEs can be detected as Na^+ adducts, presumably due to the ubiquity of this metal in the solvents and surface employed.[111] However, it may be necessary to fortify samples with 10 μM sodium acetate prior to injection because of possible reduction in APE ionization, especially for minor APEs. Normal- or reversed-phase HPLC can be used in combination with ESI–MS. Although all APE oligomers are coeluted as a single chromatographic peak on a reversed-phase column, the oligomeric distribution can be easily obtained by extracting chromatograms of selected ions from the total ion chromatogram.[103] Reversed-phase HPLC–ESI–MS is most useful for analyzing APE metabolites in environmental samples, while normal-phase HPLC–ESI–MS is more appropriate for analyzing less degraded APE mixtures in the environment.[111]

B. BISPHENOL-A AND BISPHENOL-B

1. GC–MS

Although the volatility and thermal stability presented by BPA and BPF make them suitable for detection and quantification by GC–MS, a derivatization procedure can improve the selectivity, sensitivity, and performance of the chromatographic properties. Trimethylsilylation of organic compounds containing labile hydrogen atoms is extensively used in analytical chemistry. Trimethylsilylimidazole (TMSI) and bis(trimethylsilyl)trifluoroacetamide (BSTFA) are commonly selected as silylation reagents.[83,85] The silylate peaks are several times greater than the peaks without silylation and the phenol silylate peaks have significantly better peak shape than free phenols and are separated more efficiently.

Without derivatization, the molecular ion of BPA appears at m/z 228, while the base peak corresponding to lose a methyl group appears at m/z 213.[84] With derivatization, the selected ions for selective ion monitoring (SIM) mode operations are: m/z 344 $[M^+]$ and 345 for silylated BPF; m/z 357 $[M - CH_3]^+$ and 372 $[M^+]$ for silylated BPA; m/z 181, 197, and 312 for the three isomers of BPF diglycidyl ether (BPF-DGE) and m/z 325 and 340 for BPA diglycidyl ether (BPA-DGE).[85]

Zafra et al.[147] developed simultaneous determinations of trace amounts of endocrine disruptors such as BPA and its monochloro, dichloro, trichloro, and tetrachloro derivatives in wastewater using GC–MS. Silylated BPA shows the base peak at m/z 372 corresponding to the molecular ion and it was used as the target ion and the peak at m/z 357 $[M - 15]$ was the qualifier ion. The base peak of silylated ClBPA, Cl_2BPA, Cl_3BPA, and Cl_4BPA are at m/z 391, 425, 459, and 493, respectively, which correspond to the loss of a benzylic methyl group $[M - 15]$ and are used as target ions. The molecular peaks of the above silylated chlorinated compounds appear at m/z 406, 440, 476, and 508, respectively. These are selected as the qualifier ion.

2. HPLC and LC–MS

Liquid chromatography coupled with mass spectrometry (LC–MS) has been applied in the determination of BPA, OP, and NP in the environment. The advantage of this analytical method is the capability of directly determining nonvolatile or polar compounds by using ESI or atmospheric pressure chemical ionization (APCI) techniques as an interface between LC and MS.[148,149]

Pedersen and Lindholst[149] applied LC–APCI–MS for the determination of 4-t-OP and BPA in water. Samples were analyzed at a fragmentation voltage of 80V and 100V for 4-t-OP and BPA, respectively. Quantitative analysis was carried out using selected ion monitoring (SIM) in negative mode for the ions m/z 205 [M−H]$^-$ (4-t-OP), 241 (BPA-d$_{16}$) and 227 [M−H]$^-$ (BPA). The limit of quantification for 4-t-OP and BPA was approximately 0.1 μg/l (based on 100 ml sample size). The application of MS considerably enhanced the sensitivity by up to 40 times compared to the HPLC coupled to either UV or FLD.[149]

A rapid and sensitive analytical method based on column-switching semimicrocolumn HPLC–ESI–MS was developed by Motoyama et al.[148] for determining trace levels of BPA and NP in river water. An aliquot of sample solution was directly injected into the precolumn packed with Capcellpak MF-Ph for sample cleanup and enrichment. The compounds of interest were then transferred to a C$_{18}$ analytical column for main separation through a change in flow path by a programmed switching valve. BPA, NP, and interfering substances were satisfactorily separated with a simple gradient elution complete within 35 min. Detection of these deprotonated molecules (m/z 227 for BPA and 219 for NP) was conducted in negative ion mode. However, this method gave detection limits of 0.5 ng/ml for BPA and 10 ng/ml for NP, which is not sensitive enough to directly monitor these compounds in the environment. Therefore, preconcentration using LLE or SPE is necessary in order to detect these compounds in environmental samples.

Deprotonated molecule peaks for BPA (and NP) were predominant in the ESI spectra, while APCI spectra indicated slight thermal fragmentation.[148,149] Signal intensities and signal-to-noise (S/N) values, based on mass chromatograms for [M−H]$^-$ of each analyte, were 50 to 100 times larger in the ESI mode than those obtained in the APCI mode. This indicates that ESI is preferred to APCI for accurate quantification and sensitive detection of the target compounds.[148]

Detectability in ESI–MS is also affected by mobile-phase composition including pH and buffers. The highest response for BPA (and NP) was obtained when a carrier containing NH$_3$ was used, but the reproducibility of the signals was decreased.[148] Therefore, a simple mobile phase (water/acetonitrile) with no additive is preferred in the experiments.

C. Organotin Compounds

The recognition of organotin toxicity at low concentration levels has stimulated the development of accurate and sensitive analytical methods for organotin determinations. Many efficient analytical procedures combine chromatographic separation and various detection techniques such as atomic absorption spectrometry (AAS), MS, atomic emission spectrometry (AED), or flame photometric detection (FPD). Among them, GC–MS has been generally used due to its low detection limit, high selectivity, and wide availability. A GC–MS technique with SIM mode offers the distinct advantage that all compounds of interest can be selectively determined in a single analysis. The technique proved to be sensitive with aqueous detection limits of the individual compounds typically below 2 ng/l (as cation), which is comparable with GC–FPD.[86]

D. Pesticides

1. GC and GC–MS

Capillary GC coupled with various detectors (ECD, NPD, FPD, AED, and MSD) has been widely used for the analysis of pesticides because of its high-resolution power.[125,150] There are many classes of pesticides with different chemical structures and properties requiring different detection techniques (Table 32.2). ECD has been the most widely used detector in pesticide residue analysis. It presents very high sensitivity to organochlorines and other chlorinated pesticides.[36,37,132,153,155] NPD is selective to nitrogen and phosphorus containing compounds while FPD in phosphorus mode is very sensitive to organophosphorus pesticides.[37,125,153,155–157] A newly introduced detector AED is increasingly used in the pesticide multiresidue screening due to its selective detection

TABLE 32.2
Analytical Methods for Pesticides

Pesticides	Analytical Techniques for Environmental Samples
Organochlorines (e.g., DDT, endosulfan, heptachlor, lindane and methoxychlor)	GC–ECD, GC–MS
Triazines (e.g., atrazine and simazine)	GC–NPD, GC–MS, GC–ECD
Organophosphorus (e.g., malathion methyl parathion and parathion)	GC–NPD, GC–MS
Organonitrogens (e.g., molinate, trifluralin and pendimethalin)	GC–NPD, GC–MS
Chlorinated pesticides (e.g., trifluralin, alachlor, vinclozolin and pentachlorophenol)	GC–ECD, GC–MS
Phenoxy acid herbicides (e.g., 2,4-D and 2,4,5-T)	LC–MS, GC–MS or GC–ECD after derivatization
Carbamates (e.g., carbaryl, mancozeb and methomyl)	LC–MS, LC–PFD (postcolumn derivatization and FLD)

Sources: Albanis, T. A. and Hela, D. G., *J. Chromatogr. A*, 707, 283–292, 1995; Barcelo, D., *J. Chromatogr.*, 64, 117–143, 1993; Crespo, C., Marce, R. M., and Borrull, F., *J. Chromatogr. A*, 670, 135–144, 1994; Van Rhijn, J. A., Traag, W. A., Kulik, W., and Tuinstra, L. G. M. Th., *J. Chromatogr.*, 595, 289–299, 1992; Di Corcia, A., Nazzari, M., Rao, R., Samperi, R., and Sebastiani, E., *J. Chromatogr. A*, 878, 87–98, 2000; Hong, J., Eo, Y., Rhee, J., and Kim, T., *J. Chromatogr.*, 639, 261–271, 1993; Lacassie, E., Marquet, P., Gaulier, J.-M., Dreyfuss, M. F., and Lachatre, G., *Forensic Sci. Int.*, 121, 116–125, 2001.

of the elements fluorine, chlorine, bromine, iodine, phosphorus, sulfur, and nitrogen,[125] but it requires multiple runs to cover all these elements.

Confirmation as well as quantification of pesticide residues in environmental samples is commonly performed by GC coupled with MS or MSD due to its high sensitivity and specificity.[101,132] MS can be operated in the full scan or in the SIM mode. The full scan acquisition mode is widely used because it reveals structural information about the different compounds through the spectra but it is of limited sensitivity and so SIM acquisition is mostly used for target compounds analysis.[132,158] GC–MS with the electron impact (EI) mode is the most routine confirmatory method, but GC–MS with positive ion (PCI) and negative ion chemical ionization (NCI) can increases its already high identification power.[101,130,133] For organophosphorus pesticides, it has been extensively reported that GC–MS in the NCI mode is generally more sensitive than either the PCI or EI mode.[101,119,133]

2. HPLC and LC–MS

The use of HPLC is important for pesticide analysis because it is suitable for determining thermally labile and polar pesticides which require prior derivatization if these are to be determined by GC. HPLC methods for the determination of pesticides in environmental samples could employ reversed-phase chromatography with C_{18} or C_8 columns and an aqueous mobile phase, followed by UV, fluorescence, or mass spectrometric detection.[125] In addition to GC–MS or GC–ECD methods after derivatization for phenoxy acid herbicides (e.g., 2,4-D),[37,151,159] LC–ESI–MS with negative ion mode has been developed to analyze these acidic herbicides.[152] HPLC–FLD is often applied for the determination of the carbamate pesticides by postcolumn derivatization [*o*-phthalaldehyde (OPA) derivatives].[37,151] LC–MS with PCI mode has also successfully been employed for the analysis of thermolabile carbamates.[152,157]

E. PHTHALATES

1. GC and GC–MS

Phthalates can be analyzed by GC with an ECD[105,160] or a flame ionization detector.[42,161] Phthalates in the extracts are identified by their respective GC retention time as well as by coelution of standard phthalate esters which are added to sample extracts prior to GC analysis. However, in doubtful cases, GC–MS analysis may be required for the identification of phthalate esters.

GC–MS has been commonly chosen for determination of phthalates because of its high specificity, sensitivity, and its wide availability. It was used to determine phthalates in air, water, effluent, sludge, and sediments as well as childcare articles.[24,40,43,104,162,163] Quantitative analysis is often conducted in SIM mode after recording full-scan spectra from m/z 50 to 500 for qualitative analysis. The monitoring target ions are as follows: DMP: m/z 163, 149; DEP: m/z 149, 177; DBP: m/z 149, 223; BBP: m/z 149, 206, 91; DEHP: m/z 167, 149, 279; DOP: m/z 149, 279; DNP: m/z 149, 167.

In the EI mass spectra of all phthalates with the exception of DMP, the base peak is indicated at m/z 149, which is characteristic of phthalate fragmentation. Other characteristic peaks can be explained by double hydrogen transfer (DHT).[163] The transfer of single hydrogen gives rise to even mass ions that are of low intensity and may not be observed in the mass spectra. In the fragmentation pattern of DEHP, two hydrogen atoms are transferred from the parent ion to the fragment ion of m/z 279. This ion is two mass units heavier than the cleaved fragment would be if unmodified. The molecular ion peak for phthalate compounds with long chain alkyl groups is usually weak and not always present in the mass spectra, but the $[M - R]^+$ and $[M - OR]^+$ (R = alkyl group) fragments can be a secondary form of identification. In the mass spectrum of BBP, the presence of a highly abundant peak at m/z 91 is characteristic of the resonance stabilized benzyl cation.

2. HPLC and LC–MS

In addition to GC and GC–MS, LC with UV or MS detection can be used to analyze phthalates.[164–168] A TS LC–MS method was developed to detect mono- and di-alkyl phthalates using PCI mode.[167] It was found that all of the phthalates gave strong $[M + H]^+$ pseudomolecular ions, which are the base peak of the spectra in most cases. Most of the compounds also exhibited a moderately strong $[M + NH_4]^+$ because of the ammonium acetate used in the TS ionization process. In the analysis of phthalates using LC–ESI–MS/MS in the negative mode, $[M-H]^-$ was chosen as the parent ion for the MS–MS fragmentation of all analytes.[168]

F. PHYTOESTROGENS

1. GC–MS

For GC–MS, the extract is dried under a stream of N_2 at 40°C, and phytoestrogens are converted into these trimethylsilyl (TMS) derivatives using a derivatizing reagent [N-methyl-N-(TMS) trifluoroacetamide or DTE/BSTFA]. TMS derivatives are analyzed using a nonpolar capillary column and a linear temperature gradient. GC–MS with the EI ionization mode has been used in phytoestrogen analysis.[48,169] EI mass spectra can be used to determine the molecular structures of phytoestrogen metabolites.

2. HPLC and LC–MS

HPLC with UV and FLD as well as electrochemical detection (ED) has been extensively used to analyse phytoestrogens, especially in foods.[170–172] HPLC methods with gradient elution or isocratic conditions have been developed for the determination of phytoestrogens.[170–172] The phytoestrogens (isoflavones) can be monitored with a diode array detector (DAD) at 260 nm.

Fluorescence and ED can provide better sensitivity compared to UV detection, but only some phytoestrogens like daidzein, formononetin, and coumestrol have fluorescence response while using ED at an operating potential above 1.2 V creates baseline instability.[47] The weaknesses of these detection methods are their low sensitivity and nonspecificity leading to the possibility of sample matrix interference.[47,173]

However, LC–MS and LC–MS/MS offer a better way to detect and identify phytoestrogens at low concentrations. An isoflavonoid genistein was detected by LC–ESI–MS and positively identified by LC–ESI–MS–MS in bleached Kraft mill effluent.[98] Genistein was quantified at a concentration of 30 μg/kg in air-dried wood pulp and concentrations of 13.1 and 10.5 μg/l in untreated and treated (final) effluent, respectively. These concentrations could contribute to the alterations in sex steroid levels and reduced reproductive capacity observed in fish capture near the discharges of pulp mills.

G. POLYCHLORINATED COMPOUNDS (PCBs AND DIOXINS)

GC coupled with ECD and/or MS is the main method in the determination of PCBs and dioxins.[57,137,174] The most frequently used detector for PCBs is ECD because of its sensitivity and selectivity for chlorinated compounds. Despite the selectivity, many nonhalogenated compounds such as phthalates as well as halogenated compounds such as DDT may substantially interfere with the determination.[60] The mass spectrometer is a very useful technique when the matrices such as sewage sludge and fly ash contain large amounts of chlorinated organic compounds.[117,135]

Although GC–ECD is still used in many laboratories, high resolution gas chromatography–high resolution mass spectrometry (HRGC–HRMS) is increasingly used to analyze polychlorinated compounds such as PCDDs, PCDFs, and dioxin-like PCBs, as well as polybrominated biphenyls (PBBs) and polybrominated diphenyl ethers (PBDEs) due to its high specificity and sensitivity.[175,176] It has a very high capacity to remove the contribution of matrix interfering compounds in the determination of the analytes. HRGC–HRMS with EI (electron energy 38 eV) at a resolving power of 10,000 is the reference method for the analysis of PCDD or Fs and nonortho-PCBs in EPA Methods 1613 and 1668, and the European Standard EN1948-1/2/3.[177,178,179] Quantification is performed by SIM and isotope dilution using isotope-labeled ^{13}C$_{12}$ analogues of PCDD or Fs and dioxin-like PCBs.

H. POLYCYCLIC AROMATIC HYDROCARBONS

There are a wide variety of analytical techniques available for the characterization and quantification of PAHs,[67,72,180] but the most commonly used ones are GC–MS and HPLC-UV/FLD. GC–MS has the advantage of providing comprehensive information that allows qualitative identification and quantification of the analytes of interest. It also provides a wide range of PAHs in a single run.

1. GC–MS

GC–MS is widely used for the analysis of PAHs because of its sensitivity and selectivity.[68,69,99,120,181–186] SIM mode is often carried out in the quantification of PAHs in environmental samples. Under EI spectra, PAHs can be characterized by their strong molecular mass ions. Solvent choice and temperature program used for the PAHs analysis may affect the separation and quantification of the 16 priority PAHs.[68] For the eight late-eluting PAHs, higher boiling solvents such as toluene and xylene give more enhanced signals than those in solvents such as hexane, isooctane, DCM, and benzene. The highest response for the 16 PAHs was obtained under the optimum conditions of an injector temperature of 260°C and an initial column temperature of 120°C.[68]

2. HPLC-UV/FLD

Reversed-phase liquid chromatography coupled ultraviolet and FLD is another analytical technique widely used for the measurement of PAHs in environmental samples.[67,72,77,180,187–189] All 16 PAHs can be separated by HPLC on a polymeric C_{18} phase.[72] However, on the monomeric C_{18} phase, the four-ring isomers chrysene and benz[a]anthracene are unresolved, and the six-ring isomers benzo[ghi]perylene and indeno[1,2,3-cd]pyrene, the five-ring isomers benzo[k]fluoranthene and benzo[b]fluoranthene, and fluorine and acenaphthene are only partially resolved.[72] PAHs can be detected by a UV detector at 254 nm or by a programmable fluorescence detector. The excitation and emission wavelengths are changed during the chromatographic run to achieve optimal sensitivity and selectivity for individual PAHs.[67,72,188,189] Among the 16 PAH pollutants, acenaphthylene has low fluorescent properties and therefore it is excluded from the analysis by HPLC–FLD.[67] A UV detector provides nearly universal detection of all PAHs; however, for complex environmental samples, the fluorescence detector offers more sensitivity and selectivity than UV and, therefore, is more suitable for determining PAHs in those complex mixtures.[180,188] The use of HPLC-UV is not recommended without the use of multichannel wavelength detection to check peak purity. Although separation of the 16 priority PAH pollutants is easily achieved by RP-LC on appropriate C_{18} column, PAHs in environmental samples are extremely complex due to numerous isomeric structures including alkyl-substituted isomers.[67] Thus, the analysis of those PAH mixtures requires the use of more selective detection like MS.

I. STEROID HORMONES

1. GC–MS and GC–MS/MS

GC–MS or GC–MS/MS has been widely used to analyze steroid hormones in the environmental samples.[73,190–195] Because of its polarity, steroids are derivatized by various derivatization agents such as BSTFA, N-(tert-butyldimethylsilyl)-N-methyltrifluoroacetamide (MTBSTFA), or a mixture N-methyl-N-(TMS)trifluoroacetamide (MSTFA):TMSI:dithioerythritol (DTE) (1000:2:2, v/v/w) prior to GC separation. The selected ion masses for quantification in each case vary depending on the derivatization reaction performed. Table 32.3 lists the selected ion masses for TMS derivatives of steroids.

2. HPLC–FLD, LC–MS, and LC–MS/MS

Reversed-phase HPLC coupled with FLD has been used to determine estrogens (e.g., E1/E2/E3 and EE2) in water and sediments.[112,198] The wavelength employed is as follows: 230 nm excitation and 290 nm emission. The limit of quantification is around 10 ng/ml for estrogens. In order to have higher specificity and sensitivity, LC–MS or LC–MS/MS is often used to analyze steroid hormones in environmental samples.[95,96,192,199,200]

Unlike GC–MS, HPLC enables the determination of steroids without derivatization. Two common LC–MS techniques for steroids are APCI and ESI.[199,201] Ma and Kim[201] have compared the two techniques for the analysis of steroids. The steroids were classified into three major groups based on the spectra and the sensitivity observed: (I) those containing a 3-one, 4-one functional group (e.g., testosterone, progesterone), (II) those containing at least one ketone group without conjugation (e.g., E1), and (III) those containing hydroxy groups only (e.g., estradiol). In the PCI mode, the APCI spectra were characterized by $[M + H]^+$, $[MH - H_2O]^+$, $[MH - 2H_2O]^+$, etc., with the degree of H_2O loss being compound dependent: group I steroids produced stable $[M + H]^+$ and group III steroids showed extensive water loss. ESI spectra are characterized by the abundance of $[M + Na]^+$ for the three groups: the group I steroids provided the best sensitivity, followed by group II, and then group III steroids.

TABLE 32.3
Selected Masses for Steroid TMS-Derivatives

Compound	GC–MS	GC–MS
Estrone E1	414, 399; 218, 258, 342[a]	242, 257
Estradiol-17β (E2)	416, 326; 285, 416, 285, 416	285, 326
Estriol E3	504, 414; 147, 311	—
Ethinyl estradiol (EE2)	268, 368; 285, 425	193, 231, 303
Mestranol	227, 367	349, 193
Testosterone	432, 417; 227, 258, 360[a]	—
Progesterone	458, 443	—
Zeranol	538, 433	—
Trenbolone-17β	442, 380	—
Melengestrol	570, 555	—

[a] Derivatized with BSTFA; others derivatized with MSTFA–TMSI–DTE (1000:2:2, v/v/w).

Sources: Belfroid, A. C., van der Horst, A., Vethaak, A. D., Schafer, A. J., Rijs, G. B. J., Wegener, J., and Cofino, W. P., *Sci. Total Environ.*, 225, 101–108, 1999; Jeannot, R., Sabik, H., Sauvard, E., Dagnac, T., and Dohrendorf, K., *J. Chromatogr. A*, 974, 143–159, 2002; Ternes, T. A., Stumpf, M., Mueller, J., Haberer, K., Wilken, R. D., and Servos, M., *Total Environ.*, 225, 81–90, 1999; Lai, K. M., Johnson, K. L., Scrimshaw, M. D., and Lester, J. N., *Environ. Sci. Technol.*, 34, 3890–3894, 2000; Hartmann, S. and Steinhart, H., *J. Chromatogr. A*, 704, 105–117, 1997; Marchand, P., le Bizec, B., Gade, C., Monteau, F., and Andre, F., *J. Chromatogr. A*, 867, 219–233, 2000.

For the determination of estrogens, ESI operating in the negative ion (NI) mode has been the most widely used interface because of its better sensitivity compared to the APCI interface.[199,202] The base peak of ESI spectra is $[M-H]^-$ for estrogens under the NI mode: E1, m/z 269; E2, m/z 271; E3, m/z 287; EE2, m/z 295; diethylstilbestrol (DES), m/z 267.

The LC–MS/MS methods for estrogens show the greatest sensitivity: LC–MS/MS > GC–MS > LC–MS.[192] In LC–MS/MS analysis, the $[M-H]^-$ species used as the parent ions for estrogens gave the following characteristic product ions in the collision-induced dissociation (CID) spectra: m/z 145 and 143 for E1; m/z 183, 145 and 143 for E2; m/z 171, 145, and 143 for E3; m/z 199, 183, 159, 145, and 143 for EE2.[192] Selected reaction monitoring (SRM) mode was chosen for quantitation with the following SRM pairs: E1, 269/145, and 269/143; E2, 271/183, and 271/145; E3, 287/171, and 287/145; EE2, 295/159, and 295/145.[200] The estrogens were separated on an Alltima C_{18} column (250 × 4.6 mm i.d., 5 μm) with a mobile phase of acetonitrile (ACN) and water, which was programmed from 30% ACN to 70% ACN after 24 min at a flow rate of 1 ml/min.[200] A methanolic ammonia solution (40 mmol/l) was added postcolumn to the LC column effluent at a flow rate of 0.11 ml/min to promote deprotonation of the very weakly acidic estrogens, this resulting in a drastic increase in the response of the ESI–MS system.[200] Because of the low concentrations of steroids in environmental samples, LC–MS/MS is the preferred technique to analyze these compounds.

ACKNOWLEDGMENTS

The author would like to thank Dr. Annette Nolan and Dr. Steve Rogers for their critical review and Dr. Rai Kookana for his useful comments.

REFERENCES

1. Colborn, T., Dumanoski, D., and Myers, J. P., *Our Stolen Future*, Plume/Penguin Book, New York, 1996.
2. Ying, G. G. and Kookana, R. S., Endocrine disruption: an Australian perspective, *AWA Water J.*, 29(9), 42–45, 2002.
3. Sonnenschein, C. and Soto, A. M., An updated review of environmental estrogen and androgen mimics and antagonists, *J. Steroid Biochem. Mol. Biol.*, 65(1–6), 143–150, 1998.
4. Alzieu, C., Impact of tributyltin on marine invertebrates, *Ecotoxicology*, 9, 71–76, 2000.
5. Gibbs, P. E., Bryan, G. W., Pascoe, P. L., and Burt, G. R., Reproductive abnormalities in female *Ocenebra erinacea* (Gastropoda) resulting from tributyltin-induced imposex, *J. Marine Biol. Assoc. UK*, 70, 639–656, 1990.
6. Horiguchi, T., Shiraishi, H., Shimizu, M., and Morita, M., Imposex and organotin compounds in *Thais claviger* and *T. bronni* in Japan, *J. Marine Biol. Assoc. UK*, 74, 651–669, 1994.
7. Guillette, L. J., Gross, T. S., Masson, G. R., Matter, J. M., Percival, H. F., and Woodward, A. R., Developmental abnormalities of gonad and abnormal sex hormone concentrations in juvenile alligators from contaminated and control lakes in Florida, *Environ. Health Perspect.*, 102, 680–688, 1994.
8. Guillette, L. J., Crain, D. A., Gunderson, M. P., Kools, S. A., Milnes, M. R., Orlando, E. F., Rooney, A. A., and Woodward, A. R., Alligators and endocrine disrupting contaminants: a current perspective, *Am. Zool.*, 40, 438–452, 2000.
9. Jobling, S., Nolan, M., Tyler, C. R., Brighty, G., and Sumpter, J. P., Widespread sexual disruption in wild fish, *Environ. Sci. Technol.*, 32, 2498–2506, 1998.
10. Bortone, S. A., Davis, W. P., and Bundrick, C. M., Morphological, and behavioural characters in mosquito fish as potential bioindicators of exposure to kraft mill effluent, *Bull. Environ. Contam. Toxicol.*, 43, 370–377, 1989.
11. Hayes, T. B., Collins, A., Lee, M., Mendoza, M., Noriega, N., Stuart, A. A., and Vonk, A., Hermaphroditic, demasculinised frogs after exposure to the herbicide atrazine at low ecologically relevant doses, *PANS*, 99, 5476–5480, 2002.
12. Brown, L. M., Pottern, L. M., Hoover, R. N., Devesa, S. S., Aselton, P., and Flannery, T., Testicular cancer in the US trends in incidence and mortality, *Int. J. Epidemiol.*, 15, 164–170, 1986.
13. Hakulinen, T., Andersen, A. A., Malker, B., Rikkala, E., Shou, G., and Tulinius, H., Trends in cancer incidence in the Nordic countries. A collaborative study of the five Nordic Cancer Registries, *Acta Pathol. Microbiol. Immunol. Scand. (Sect. A)*, 94(Suppl. 288), 1–151, 1986.
14. Adami, H. O., Bergstron, R., Mohner, M., Zatonski, W., Storm, H., Ekbom, A., Tretli, S., Teppo, L., Ziegler, H., Rahu, M., Gurevicius, R., and Stengrevics, A., Testicular cancer in nine northern European countries, *Int. J. Cancer*, 59, 33–38, 1994.
15. Feuer, E. J., State bite: incidence of testicular cancer in US men, *J. Natl Cancer Inst.*, 87, 405, 1995.
16. Moller, H., Clues to the aetiology of testicular germ cell tumours from descriptive epidemiology, *Eur. Urol.*, 23, 8–15, 1993.
17. Ries, L. A. G., Hankey, B. F., and Miller, B. A., *Cancer Statistics Review 1973–88, DHEW (NIH) Publ. No. 91-2789*, US Government Printing Office, Washington, DC, 1991.
18. Wolff, M. S., Toniolo, P. G., Lee, E. W., Rivera, M., and Dubin, N., Blood levels of organochlorine residues and risk of breast cancer, *J. Natl Cancer Inst.*, 85, 648–662, 1993.
19. Ying, G. G., Williams, B., and Kookana, R., Environmental fate of alkylphenols and alkylphenol ethoxylates — a review, *Environ. Int.*, 28(3), 215–266, 2002.
20. Ying, G. G., Kookana, R. S., and Ru, Y. J., Occurrence and fate of hormone steroids in the environment, *Environ. Int.*, 28(6), 545–551, 2002.
21. Hawrelak, M., Bennett, E., and Metcalfe, C., The environmental fate of the primary degradation products of alkylphenol ethoxylate surfactants in recycled paper sludge, *Chemosphere*, 39(5), 745–752, 1999.
22. Comber, M. H. I., Williams, T. D., and Stewart, K. M., The effects of nonylphenol on *Daphnia magna*, *Water Res.*, 27, 273–276, 1993.
23. Jobling, S. J. and Sumpter, J. P., Detergent components in sewage effluent are weakly oestrogenic to fish: an in vitro study using rainbow trout hepatocytes, *Aquat. Toxicol.*, 27, 361–372, 1993.

24. Fromme, H., Kuchler, T., Otto, T., Pilz, K., Muller, J., and Wenzel, A., Occurrence of phthalates and bisphenol A and F in the environment, *Water Res.*, 36, 1429–1438, 2002.
25. Furhacker, M., Scharf, S., and Weber, H., Bisphenol A: emissions from point sources, *Chemosphere*, 41, 751–756, 2000.
26. Perez, P., Pular, R., Olea-Serrano, F., Villalobos, M., Rivas, A., Metzuerenfred, M., Pedraza, V., and Olea, N., The estrogenicity of bisphenol A-related diphenylalkanes with various substituents at the central carbon and the hydroxy group, *Environ. Health Perspect.*, 106, 167–174, 1998.
27. Behnisch, P. A., Fujii, K., Shiozaki, K., Kawakami, I., and Sakai, S. I., Estrogenic and dioxin-like potency in each step of a controlled landfill leachate treatment plant in Japan, *Chemosphere*, 43, 977–984, 2001.
28. Rehmann, K., Schramm, K. W., and Kettrup, A. A., Applicability of a yeast oestrogen screen for the detection of oestrogen-like activities in environmental samples, *Chemosphere*, 38, 3303–3312, 1999.
29. Dodds, E. C., Goldberg, L., Lawson, W., and Robinson, R., Oestrogenic activity of certain synthetic compounds, *Nature*, 141, 247–248, 1938.
30. Harris, C. A., Henttu, P., Parker, M. G., and Sumpter, J. P., The estrogenic activity of phthalate esters in vitro, *Environ. Health Perspect.*, 105, 802–811, 1997.
31. BUA. In *Bisphenol A BUA — Stoffbericht 203*, Chemiker, B. f. u. A. B. d. G. D., Ed., Hirzel, Stuttgart, 1997.
32. Fent, K., Organotin compounds in municipal wastewater and sewage sludge: contamination, fate in treatment process and ecotoxicological consequences, *Sci. Total Environ.*, 185, 151–159, 1996.
33. Hoch, M., Organotin compounds in the environment — an overview, *Appl. Geochem.*, 16, 719–743, 2001.
34. Horiguchi, T., Shiraishi, H., Shimizu, M., and Masatoshi, M., Imposex in sea snails, caused by organotin (tributyltin and triphenyltin) pollution in Japan: a survey, *Appl. Organomet. Chem.*, 11, 451–455, 1997.
35. Barr, D. B. and Needham, L. L., Analytical methods for biological monitoring of exposure to pesticides: a review, *J. Chromatogr. B*, 778, 5–29, 2002.
36. Aguilar, C., Borrull, F., and Marce, R. M., Determination of pesticides in environmental waters by solid phase extraction and gas chromatography with electron-capture and mass spectrometry detection, *J. Chromatogr. A*, 771, 221–231, 1997.
37. Vassilakis, I., Tsipi, D., and Scoullos, M., Determination of a variety of chemical classes of pesticides in surface and ground waters by off-line solid phase extraction, gas chromatography with electron-capture and nitrogen–phosphorus detection, and high-performance liquid chromatography with post-column derivatization and fluorescence detection, *J. Chromatogr. A*, 823, 49–58, 1998.
38. Bisson, M. and Hontela, A., Cytotoxic and endocrine-disrupting potential of atrazine, diazinon, endosulfan, and mancozeb in adrenocortical steroidogenic cells of rainbow trout exposed in vitro, *Toxicol. Appl. Pharmacol.*, 180, 110–117, 2002.
39. Kelce, W. R., Monosson, E., Gamcsik, M. P., Laws, S. C. Jr., and Gray, L. E., Environmental hormone disruptors: evidence that vinclozolin developmental toxicity is mediated by antiandrogenic metabolites, *Toxicol. Appl. Pharmacol.*, 126, 276–285, 1994.
40. Otake, T., Yoshinaga, J., and Yanagisawa, Y., Analysis of organic esters of plasticizer in indoor air by GC–MS and GC–FPD, *Environ. Sci. Technol.*, 35, 3099–3102, 2001.
41. Blount, B. C., Silva, M. J., Caudill, S. P., Needham, L. L., Pirkle, J. L., Simpson, E. J., Lucier, G. W., Jackson, R. J., and Brock, J. W., Levels of seven urinary phthalate metabolites in a human reference population, *Environ. Health Perspect.*, 108, 979–982, 2000.
42. Fatoki, O. S. and Noma, A., Solid-phase extraction method for selective determination of phthalate esters in the aquatic environment, *Water Air Soil Pollut.*, 140, 85–98, 2002.
43. Vitali, M., Guidotti, M., Macilenti, G., and Cremisini, C., Phthalate esters in freshwaters as markers of contamination sources — a site study in Italy, *Environ. Int.*, 23(3), 337–347, 1997.
44. Penalver, A., Pocurull, E., Borrull, F., and Marce, R. M., Determination of phthalate esters in water samples by solid-phase microextraction and gas chromatography with mass spectrometric detection, *J. Chromatogr. A*, 872, 191–201, 2000.
45. Jobling, S., Reynolds, T., White, R., Parker, M. G., and Sumpter, J. P., A variety of environmentally persistent chemicals, including some phthalate plasticisers, are weakly estrogenic, *Environ. Health Perspect.*, 103(6), 582–587, 1995.

46. Moore, N. P., The oestrogenic potential of the phthalate esters, *Reprod. Toxicol.*, 14, 183–192, 2000.
47. Wang, C. C., Prasain, J. K., and Barnes, S., Review of the methods used in the determination of phytoestrogens, *J. Chromatogr. B*, 777, 3–28, 2002.
48. Foster, W. G., Chan, S., Platt, L. Jr., and Hughes, C. L., Detection of phytoestrogens in samples of second trimester human amniotic fluid, *Toxicol. Lett.*, 129, 199–205, 2002.
49. Mazur, W. and Adlercreutz, H., Overview of naturally occurring endocrine-active substances in the human diet in relation to human health, *Nutrition*, 16, 654–687, 2000.
50. Alcock, R. E., Bacon, J., Bardget, R. D., Beck, A. J., Haygarth, P. M., Lee, R. G. M., Parker, C. A., and Jones, K. C., Persistence and fate of polychlorinated biphenyls (PCBs) in sewage sludge-amended agricultural soils, *Environ. Pollut.*, 93, 83–92, 1996.
51. Bavel, B. V., Naf, C., Bergqvist, P.-A., Broman, D., Lundgren, K., Papakosta, O., Rolff, C., Strandberg, B., Zebuhr, Y., Zook, D., and Rappe, C., Levels of PCBs in the aquatic environment of the gulf of Bothnia: benthic species and sediments, *Marine Pollut. Bull.*, 32(2), 210–218, 1995.
52. Chang, Y. S., Kong, S. B., and Ikonomou, M. G., PCBs contributions to the total TEQ released from Korean municipal and industrial waste incinerators, *Chemosphere*, 39(15), 2629–2640, 1999.
53. Hope, B., Scatolini, S., Titus, E., and Cotter, J., Distribution patterns of polychlorinated biphenyl congeners in water, sediment, and biota from Midway Atoll (North Pacific Ocean), *Marine Pollut. Bull.*, 34(7), 548–563, 1997.
54. Dyke, P. H. and Stratford, J., Changes to the TEF schemes can have significant impacts on regulation and management of PCDD/F and PCB, *Chemosphere*, 47, 103–116, 2002.
55. McLachlan, M. S., Sewart, A. P., Bacon, J. R., and Jones, K. C., Persistence of PCDD/Fs in a sludge-amended soil, *Environ. Sci. Technol.*, 30, 2567–2571, 1996.
56. Thompson, S., Budzinski, H., Garrigues, P., and Narbonne, J. F., Comparison of PCB and DDT distribution between water-column and sediment-dwelling bivalves in Arcachon Bay, France, *Marine Pollut. Bull.*, 38(8), 655–662, 1999.
57. Vetter, W., Natzeck, C., Luckas, B., Heidemann, G., Kiabi, B., and Karami, M., Chlorinated hydrocarbons in the blubber of a seal (*Phoca caspica*) from the Caspian Sea, *Chemosphere*, 30, 1685–1696, 1995.
58. Tanabe, S., Contamination and toxic effects of persistent endocrine disrupters in marine mammals and birds, *Marine Pollut. Bull.*, 45, 69–77, 2002.
59. Van Overmeire, I., Clark, G. C., Brown, D. J., Chu, M. D., Cooke, W. M., Denison, M. S., Baeyens, W., Srebrnik, S., and Goeyens, L., Trace contamination with dioxin-like chemicals: evaluation of bioassay-based TEQ determination for hazard assessment and regulatory responses, *Environ. Sci. Policy*, 4, 345–357, 2001.
60. Lang, V., Polychlorinated biphenyls in the environment, *J. Chromatogr.*, 595, 1–43, 1992.
61. Safe, S., Polychlorinated biphenyls (PCBs), dibenzo-*p*-dioxins (PCDDs), dibenzofurans (PCDFs), and related compounds: environmental and mechanistic considerations which support the development of toxic equivalency factors (TEFs), *Crit. Rev. Toxicol.*, 21, 51–88, 1990.
62. Aoki, Y., Polychlorinated biphenyls, polychlorinated dibenzo-*p*-dioxins, and polychlorinated dibenzofurans as endocrine disrupters — what we have learned from Yusho disease, *Environ. Res. Sect. A*, 86, 2–11, 2001.
63. Safe, S., Polychlorinated biphenyls (PCBs): environmental impact, biochemical and toxic response, and implications for risk assessment, *Crit. Rev. Toxicol.*, 24, 87–149, 1994.
64. Gierthy, J. F., Arcaro, K. F. and Floyd, M., Assessment of PCB estrogenicity in a human breast cancer cell line, *Chemosphere*, 34, 1495–1505, 1997.
65. Hartsough, G. R., Great Lakes fish now suspect as mink food, *Am. Fur Breeder*, 38, 25–27, 1965.
66. Hochstein, J. R., Aulerich, R. J., and Bursian, S. J., Acute toxicity of 2,3,7,8-tetrachlorodibenzo-*p*-dioxin to mink, *Arch. Environ. Contam. Toxicol.*, 17, 33–37, 1988.
67. Williamson, K. S., Petty, J. D., Huckins, J. N., Lebo, J. A., and Kaiser, E. M., HPLC–PFD determination of priority pollutant PAHs in water, sediment, and semipermeable membrane devices, *Chemosphere*, 49, 703–715, 2002.
68. Brindle, I. and Li, X. F., Investigation into the factors affecting performance in the determination of polycyclic aromatic hydrocarbons using capillary gas chromatography–mass spectrometry with splitless injection, *J. Chromatogr.*, 498, 11–24, 1990.

69. Thompson, D., Jolley, D., and Maher, W., Determination of polycyclic aromatic hydrocarbons in oyster tissues by high-performance liquid chromatography with ultraviolet and fluorescence detection, *Microchem. J.*, 47, 351–362, 1993.

70. Tran, D. Q., Ide, F., McLachlan, J. A., and Arnold, S. F., The anti-estrogenic activity of selected polynuclear aromatic hydrocarbons in yeast expressing human estrogen receptor, *Biochem. Biophys. Res. Commun.*, 224(1), 102–108, 1996.

71. Chaluopka, K., Krishnan, V., and Safe, S., Polynuclear aromatic hydrocarbons as antiestrogens in MCF-7 human cancer cell: role of the Ah receptor, *Carcinogenesis*, 13(12), 2233–2239, 1992.

72. Wise, S. A., Sander, L. C., and May, W. E., Determination of polycyclic aromatic hydrocarbons by liquid chromatography, *J. Chromatogr.*, 642, 329–349, 1993.

73. Belfroid, A. C., van der Horst, A., Vethaak, A. D., Schafer, A. J., Rijs, G. B. J., Wegener, J., and Cofino, W. P., Analysis and occurrence of estrogenic hormones and its glucuronides in surface water and waste water in the Netherlands., *Sci. Total Environ.*, 225, 101–108, 1999.

74. Purdom, C. E., Hardiman, P. A., Byre, V. J., Eno, N. C., Tyler, C. R., and Sumpter, J. P., Estrogenic effects of effluents from sewage treatment works, *Chem. Ecol.*, 8, 275–285, 1994.

75. Sumpter, J. P. and Jobling, S., Vitellogenesis as a biomarker for estrogenic contamination of the aquatic environment, *Environ. Health Perspect.*, 103, 173–178, 1995.

76. Fukazawa, H., Hoshino, K., Shiozawa, T., Matsushita, H., and Terao, Y., Identification and quantification of chlorinated bisphenol A in wastewater from wastepaper recycling plants, *Chemosphere*, 44, 973–979, 2001.

77. Law, R. J., Dawes, V. J., Woodhead, R. J., and Matthiessen, P., Polycyclic aromatic hydrocarbons (PAH) in seawater around England and Wales, *Marine Pollut. Bull.*, 34(5), 306–322, 1997.

78. Mori, S., Identification and determination of phthalate esters in river water by high-performance liquid chromatography, *J. Chromatogr.*, 129, 53–60, 1976.

79. Thiele, B., Gunther, K., and Schwuger, M. J., Trace analysis of surfactants in environmental matrices, *Tenside Surf. Det.*, 36, 8–18, 1999.

80. Rudel, R. A., Melly, S. J., Geno, P. W., Sun, G., and Brody, J. G., Identification of alkylphenols and other estrogenic phenolic compounds in wastewater, septage, groundwater on Cape Cod, Massachusetts, *Environ. Sci. Technol.*, 32, 861–869, 1998.

81. Bennie, D. T., Sullivan, C. A., Lee, H. B., Peart, T. E., and Maguire, R. J., Occurrence of alkylphenols and alkylphenol mono- and diethoxylates in natural waters of the Laurentian Great Lakes basin and the upper St Lawrence River, *Sci. Total Environ.*, 193, 263–275, 1997.

82. Espejo, R., Valter, K., Simona, M., Janin, Y., and Arrizabalaga, P., Determination of nineteen 4-alkylphenol endocrine disrupters in Geneva municipal sewage wastewater, *J. Chromatogr. A*, 976, 335–343, 2002.

83. Helaleh, M. I. H., Takabayashi, Y., Fujii, S., and Korenaga, T., Gas chromatographic–mass spectrometric method for separation and detection of endocrine disruptors from environmental water samples, *Anal. Chim. Acta*, 428, 227–234, 2001.

84. Del Olmo, M., Gonzalez-Casado, A., Navas, N. A., and Vilchez, J. L., Determination of bisphenol A (BPA) in water by gas chromatography–mass spectrometry, *Anal. Chim. Acta*, 346, 87–92, 1997.

85. Vilchez, J. L., Zafra, A., Gonzalez-Casado, A., Hontoria, E., and del Olmo, M., Determination of trace amounts of bisphenol F, bisphenol A and their diglycidyl ethers in wastewater by gas chromatography–mass spectrometry, *Anal. Chim. Acta*, 431, 31–40, 2001.

86. Tolosa, I. and Readman, J. W., Simultaneous analysis of the antifouling agents: tributyltin, triphenyltin and IRGAROL 1051 in marine water samples, *Anal. Chim. Acta*, 335, 267–274, 1996.

87. Schubert, P., Rosenberg, E., and Grasserbauer, M., Comparison of sodium tetraethylborate and sodium tetra(*n*-propyl)borate as derivatization reagent for the speciation of organotin and organolead compounds in water samples, *Fresenius J. Anal. Chem.*, 366, 356–360, 2000.

88. Thomaidis, N. S., Adams, F. C., and Lekkas, T. D., A simple method for the speciation of organotin compounds in water samples using ethylation and GC-QFAAS, *Mikrochim. Acta*, 136, 137–141, 2001.

89. Bancon-Montigny, C. H., Lespes, G., and Potin-Gautier, M., Improved routine speciation of organotin compounds in environmental samples by pulsed flame photometric detection, *J. Chromatogr. A*, 896, 149–158, 2000.

90. Bancon-Montigny, C., Lespes, G., and Potin-Gautier, M., Optimisation of the storage of natural freshwaters before organotin speciation, *Water Res.*, 35, 224–232, 2001.

91. Font, G., Manes, J., Molto, J. C., and Pico, Y., Solid-phase extraction in multi-residue pesticide analysis of water, *J. Chromatogr.*, 642, 135–161, 1993.

92. Albanis, T. A. and Hela, D. G., Multi-residue pesticide analysis in environmental water samples using solid-phase extraction discs and gas chromatography with flame thermionic and mass-selective detection, *J. Chromatogr. A*, 707, 283–292, 1995.

93. Blackburn, M. A., Kirby, S. J., and Waldock, M. J., Concentrations of alkylphenololyethoxylates entering UK estuaries, *Marine Pollut. Bull.*, 38(2), 109–118, 1999.

94. Pedersen, S. N. and Lindholst, C., Quantification of the xenoestrogens 4-*tert*-octylphenol and bisphenol A in water and in fish tissue based on microwave assisted extraction, solid-phase extraction and liquid chromatography–mass spectrometry, *J. Chromatogr. A*, 864, 17–24, 1999.

95. Sole, M., Lopez de Alda, M. J., Castillo, M., Porte, C., Ladegaard-Pedersen, K., and Barcelo, D., Estrogenicity determination in sewage treatment plants and surface waters from the Catalonian area (NE Spain), *Environ. Sci. Technol.*, 34, 5076–5083, 2000.

96. Lopez de Alda, M. and Barcelo, D., Use of solid phase extraction in various of its modalities for sample preparation in the determination of estrogens and progestogens in sediment and water, *J. Chromatogr. A*, 938, 145–153, 2001.

97. Franke, A. A. and Custer, L. J., High-performance liquid chromatographic assay of isoflavonoids and coumestrol from human urine, *J. Chromatogr. B*, 662, 47–60, 1994.

98. Kiparissis, Y., Hughes, R., and Metcalfe, C., Identification of the isoflavonoid genistein in bleached Kraft mill effluent, *Environ. Sci. Technol.*, 35, 2423–2427, 2001.

99. Dobosiewicz, K., Luks-Betlej, K., and Bodzek, D., Concentration of polycyclic aromatic hydrocarbons and base cations in the air and drinking water of the Silesian Rehabilitation Centre, Poland, *Water Air Soil Pollut.*, 118, 101–113, 2000.

100. Font, G., Manes, J., Molto, J. C., and Pico, Y., Current developments in the analysis of water pollution by polychlorinated biphenyls, *J. Chromatogr. A*, 733, 449–471, 1996.

101. Crespo, C., Marce, R. M., and Borrull, F., Determination of various pesticides using membrane extraction discs and gas chromatography–mass spectrometry, *J. Chromatogr. A*, 670, 135–144, 1994.

102. van der Hoff, G. R. and van Zoonen, P., Trace analysis of pesticides by gas chromatography, *J. Chromatogr. A*, 843, 301–322, 1999.

103. Lee, H. B., Review of analytical methods for the determination of nonylphenol and related compounds in environmental samples, *Water Qual. Res. J. Can.*, 34(1), 3–35, 1999.

104. Luks-Betlej, K., Popp, P., Janoszka, B., and Paschke, H., Solid-phase microextraction of phthalates from water, *J. Chromatogr. A*, 938, 93–101, 2001.

105. Prokupkova, G., Holadova, K., Poustka, J., and Hajslova, J., Development of a solid-phase microextraction method for the determination of phthalic acid esters in water, *Anal. Chim. Acta*, 457, 211–223, 2002.

106. de Voogt, P., de Beer, K., and van der Wielen, F., Determination of alkylphenol ethoxyaltes in industrial and environmental samples, *Trends Anal. Chem.*, 16(10), 584–595, 1997.

107. Marcomini, A., Pojana, G., Sfriso, A., and Quiroga Alonso, J. M., Behaviour of anionic and non-ionic surfactants and its persistent metabolites in the Venice lagoon, Italy, *Environ. Toxicol. Chem.*, 19(8), 2000–2007, 2000.

108. Hawrelak, M., Bennett, E., and Metcalfe, C., The environmental fate of the primary degradation products of alkylphenol ethoxylate surfactants in recycled paper sludge, *Chemosphere*, 39, 745–752, 1999.

109. Khim, J. S., Villeneuve, D. L., Kannan, K., Lee, K. T., Snyder, S. A., Koh, C. H., and Giesy, J. P., Alkyphenols, polycyclic aromatic hydrocarbons, and organochlorines in sediment from Lake Shihwa, Korea: instrumental and bioanalytical characterisation, *Environ. Toxicol. Chem.*, 18, 2424–2432, 1999.

110. Shang, D. Y., Ikonomou, M. G., and Macdonald, R. W., Quantitative determination of nonylphenol polyethoxylate surfactants in marine sediment using normal-phase liquid chromatography–electrospray mass spectrometry, *J. Chromatogr. A*, 849, 467–482, 1999.

111. Ferguson, P. L., Iden, C. R., and Brownawell, B. J., Analysis of alkylphenol ethoxylate metabolites in the aquatic environment using liquid chromatography–electrospray mass spectrometry, *Anal. Chem.*, 72, 4322–4330, 2000.

112. Ying, G. G. and Kookana, R. S., Degradation of selected five endocrine disrupting chemicals in seawater and marine sediment, *Environ. Sci. Technol.*, 37, 1256–1260, 2003.

113. Björklund, E., Nilsson, T., and Bøwadt, S., Pressurised liquid extraction of persistent organic pollutants in environmental analysis, *Trends Anal. Chem.*, 19(7), 434–445, 2000.

114. Dean, J. R. and Xiong, G., Extraction of organic pollutants from environmental matrices: selection of extraction technique, *Trends Anal. Chem.*, 19(9), 553–564, 2000.

115. Saim, N., Dean, J. R., Abdullah, M. P., and Zakaria, Z., Extraction of polycyclic aromatic hydrocarbons from contaminated soil using Soxhlet extraction, pressurized and atmosphereic microwave-assisted extraction, supercritical fluid extraction and accelerated solvent extraction, *J. Chromatogr. A*, 791, 361–366, 1997.

116. Bruno, F., Curini, R., Di Corcia, A., Fochi, I., Nazzari, M., and Samperi, R., Determination of surfactants and some of their metabolites in untreated and anaerobically digested sewage sludge by subcritical water extraction followed by liquid chromatography–mass spectrometry, *Environ. Sci. Technol.*, 36, 4156–4161, 2002.

117. Berset, J. D. and Holzer, R., Quantitative determination of polycyclic aromatic hydrocarbons, polychlorinated biphenyls, and organochlorine pesticides in sewage sludges using supercritical fluid extraction and mass spectrometric detection, *J. Chromatogr. A*, 852, 545–558, 1999.

118. Motohashi, N., Nagashima, H., Parkanyi, C., and Subrahmanyam, B., Official multiresidue methods of pesticide analysis in vegetables, fruits and soil, *J. Chromatogr. A*, 754, 333–346, 1996.

119. Durand, G. and Barcelo, D., Confirmation of chlorotriazine pesticides, their degradation products and organophosphorus pesticides in soil samples using gas chromatography–mass spectrometry with electron impact and positive- and negative-ion chemical ionization, *Anal. Chim. Acta*, 24, 259–271, 1991.

120. Bodzek, D., Janoszka, B., Dobosz, C., Warzecha, L., and Bodzek, M., Determination of polycyclic aromatic compounds and heavy metals in sludges from biological sewage treatment plants, *J. Chromatogr. A*, 774, 177–192, 1997.

121. Jeannot, R., Sabik, H., Sauvard, E., Dagnac, T., and Dohrendorf, K., Determination of endocrine-disrupting compounds in environmental samples using gas and liquid chromatography with mass spectrometry, *J. Chromatogr. A*, 974, 143–159, 2002.

122. Carlier-Pinasseau, C., Lespes, G., and Astruc, M., Determination of butyl- and phenyltin compounds in sediments by GC-FPD after NaBEt$_4$ ethylation, *Talanta*, 44, 1163–1171, 1997.

123. Simon, S., Bueno, M., Lespes, G., Mench, M., and Potin-Gautier, M., Extraction procedure for organotin analysis in plant matrices: optimization and application, *Talanta*, 57, 31–43, 2002.

124. Sabik, H., Gagne, F., Blaise, C., Marcogliese, D. J., and Jeannot, R., Occurrence of alkylphenol polyethoxylates in the St Lawrence River and their bioconcentration by mussels (*Elliptio complanata*), *Chemosphere*, 51, 349–356, 2003.

125. Torres, C. M., Pico, Y., and Manes, J., Determination of pesticide residues in fruit and vegetables, *J. Chromatogr. A*, 754, 301–331, 1996.

126. Rodriguez, P. and Permanyer, J., Confirmation method for the identification and determination of some organophosphorus and organochlorine pesticides in cocoa beans by gas chromatography–mass spectrometry, *J. Chromatogr.*, 562, 547–553, 1991.

127. Isobe, T., Nishiyama, H., Nakashima, A., and Takada, H., Distribution and behaviour of nonylphenol, octylphenol, and nonylphenol monoethoxylate in Tokyo metropolitan area: their association with aquatic particles and sedimentary distributions, *Environ. Sci. Technol.*, 35, 1041–1049, 2001.

128. Snyder, S. A., Keith, T. L., Pierens, S. L., Snyder, E. M., and Giesy, J. P., Bioconcentration of nonylphenol in fathead minnows (*Pimephales promelas*), *Chemosphere*, 44, 1697–1702, 2001.

129. Bennett, E. R. and Metcalfe, C. D., Distribution of degradation products of alkylphenol ethoxylates near sewage treatment plants in the lower Great Lakes, North America, *Environ. Toxicol. Chem.*, 19, 784–792, 2000.

130. Jover, E. and Bayona, J. M., Trace level determination of organochlorine, organophosphorus and pyrethroid pesticides in lanolin using gel permeation chromatography followed by dual gas chromatography and gas chromatography-negative chemical ionization mass spectrometric confirmation, *J. Chromatogr. A*, 950, 213–220, 2002.

131. Pang, G.-F., Can, Y.-Z., Fan, C.-L., Zhang, J.-J., Li, X.-M., Mu, J., Wang, D.-N., Liu, S.-M., Song, W. B., Li, H.-P., Wong, S.-S., Kubinec, R., Tekel, J., and Tahotna, S., Interlaboratory study of identification and quantitation of multiresidue pyrethroids in agricultural products by gas chromatography–mass spectrometry, *J. Chromatogr. A*, 882, 231–238, 2000.

132. Gelsomino, A., Petrovicova, B., Tiburtini, S., Magnani, E., and Felici, M., Multiresidue analysis of pesticides in fruits and vegetables by gel permeation chromatography followed by gas chromatography with electron-capture and mass spectrometric detection, *J. Chromatogr. A*, 782, 105–122, 1997.

133. Lacorte, S., Molina, C., and Barcelo, D., Screening of organophosphorus pesticides in environmental matrices by various gas chromatographic techniques, *Anal. Chim. Acta*, 281, 71–84, 1993.

134. Van Rhijn, J. A., Traag, W. A., Kulik, W., and Tuinstra, L. G. M. Th., Automated clean-up procedure for the gas chromatographic–high-resolution mass spectrometric determination of polychlorinated dibenzo-*p*-dioxins and dibenzofurans in milk, *J. Chromatogr.*, 595, 289–299, 1992.

135. Beard, A., Naikwadi, K., and Karasek, F., Comparison of extraction methods for polychlorinated dibenzo-*p*-dioxins and dibenzofurans in fly ash using gas chromatography–mass spectrometry, *J. Chromatogr.*, 589, 265–270, 1992.

136. Abad, E., Saulo, J., Caixach, J., and Rivera, J., Evaluation of a new automated cleanup system for the analysis of polychlorinated dibenzo-*p*-dioxins and dibenzofurans in environmental samples, *J. Chromatogr. A*, 893, 383–391, 2000.

137. Jasinski, J. S., Multiresidue procedures for the determination of chlorinated dibenzodioxins and dibenzofurans in a variety of foods using capillary gas chromatography–electron-capture detection, *J. Chromatogr.*, 478, 349–367, 1989.

138. Fytianos, K. and Fr Schröder, H., Determination of polychlorinated dibenzodioxins and dibenzofurans in fly ash, *Chromatographia*, 46, 280–284, 1997.

139. Blackburn, M. A. and Waldock, M. J., Concentrations of alkylphenols in rivers and estuaries in England and Wales, *Water Res.*, 29, 1623–1629, 1995.

140. Ahel, M. and Giger, W., Determination of alkylphenols, and alkylphenol mono, and diethoxylates in environmental samples by high performance liquid chromatography, *Anal. Chem.*, 57, 1577–1583, 1985.

141. Maruyama, K., Yuan, M., and Otsuki, A., Seasonal changes in ethylene oxide chain length of poly(oxyethylene)alkylphenyl ether non-ionic surfactants in three main rivers in Tokyo, *Environ. Sci. Technol.*, 34, 343–348, 2000.

142. Ferguson, P. L., Iden, C. R., and Brownawell, B. J., Distribution and fate of neutral alkylphenol ethoxylate metabolites in a sewage-impacted urban estuary, *Environ. Sci. Technol.*, 35, 2428–2435, 2001.

143. Ying, G. G., Kookana, R., and Chen, Z. L., On-line solid-phase extraction and fluorescence detection of selected endocrine disrupting chemicals in water by high-performance liquid chromatography, *J. Environ. Sci. Health B*, 37(3), 225–234, 2002.

144. Lee, H. B., Peart, T. E., Bennie, D. T., and Maguire, R. J., Determination of nonylphenol polyethoxylates and their carboxylic acid metabolites in sewage treatment plant sludge by supercritical carbon dioxide extraction, *J. Chromatogr. A*, 785, 385–394, 1997.

145. Scarlett, M., Fisher, J. A., Zhang, H., and Ronan, M., Determination of dissolved nonylphenol ethoxylate surfactants in waste waters by gas stripping and isocratic high performance liquid chromatography, *Water Res.*, 28, 2109–2116, 1994.

146. Kiewiet, A. T. and de Voogt, P., Chromatographic tools for analysing and tracking non-ionic surfactants in the aquatic environment, *J. Chromatogr. A*, 733, 185–192, 1996.

147. Zafra, A., del Olmo, M., Suarez, B., Hontoria, E., Navalon, A., and Vilchez, J. L., Gas chromatographic–mass spectrometric method for the determination of bisphenol A and its chlorinated derivatives in urban wastewater, *Water Res.*, 37, 735–742, 2003.

148. Motoyama, A., Suzuki, A., Shirota, O., and Namba, R.. Direct determination of bisphenol A and nonylphenol in river water by column-switching semi-microcolumn liquid chromatography or electrospray mass spectrometry, *Rapid Commun. Mass Spectrom.*, 13, 2204–2208, 1999.

149. Pedersen, S. N. and Lindholst, C., Quantification of the xenoestrogens 4-*tert*-octylphenol and bisphenol A in water and in fish tissue based on microwave assisted extraction, solid-phase extraction and liquid chromatography–mass spectrometry, *J. Chromatogr. A*, 864, 17–24, 1999.

150. Jimenez, J. J., Bernal, J. L., del Nozal, M. J., and Rivera, J. M., Determination of pesticide residues in waters from small loughs by solid-phase extraction and combined use of gas chromatography with electron-capture and nitrogen–phosphorus detection and high-performance liquid chromatography with diode array detection, *J. Chromatogr. A*, 778, 289–300, 1997.

151. Barcelo, D., Environmental Protection Agency and other methods for the determination of priority pesticides and their transformation products in water, *J. Chromatogr.*, 64, 117–143, 1993.

152. Di Corcia, A., Nazzari, M., Rao, R., Samperi, R., and Sebastiani, E., Simultaneous determination of acidic and non-acidic pesticides in natural waters by liquid chromatography–mass spectrometry, *J. Chromatogr. A*, 878, 87–98, 2000.

153. Hong, J., Eo, Y., Rhee, J., and Kim, T., Simultaneous analysis of 25 pesticides in crops using gas chromatography and their identification by gas chromatography–mass spectrometry, *J. Chromatogr.*, 639, 261–271, 1993.

154. Lacassie, E., Marquet, P., Gaulier, J.-M., Dreyfuss, M. F., and Lachatre, G., Sensitive and specific multiresidue methods for the determination of pesticides of various classes in clinical and forensic toxicology, *Forensic Sci. Int.*, 121, 116–125, 2001.

155. Navarro, S., Barba, A., Navarro, G., Vela, N., and Oliva, J., Multiresidue method for the rapid determination — in grapes, must and wine — of fungicides frequently used on vineyards, *J. Chromatogr. A*, 882, 221–229, 2000.

156. Sabik, H., Fouquet, A., and Proulx, S., Ultratrace determination of organophosphorus and organonitrogen pesticides in surface water, *Analusis*, 25, 267–273, 1997.

157. Psathaki, M., Manoussaridou, E., and Stephanou, E. G., Determination of organophosphorus and triazine pesticides in ground- and drinking water by solid-phase extraction and gas chromatography with nitrogen–phosphorus or mass spectrometric detection, *J. Chromatogr. A*, 667, 241–248, 1994.

158. Liu, S. and Pleil, J. D., Human blood and environmental media screening method for pesticides and polychlorinated biphenyl compounds using liquid extraction and gas chromatography–mass spectrometry analysis, *J. Chromatogr. B*, 769, 155–167, 2002.

159. Johnson, P. D., Rimmer, D. A., and Brown, R. H., Adaptation and application of a multi-residue method for the determination of a range of pesticides, including phenoxy acid herbicides in vegetation, based on high-resolution gel permeation chromatographic clean-up and gas chromatographic analysis with mass-selective detection, *J. Chromatogr. A*, 765, 3–11, 1997.

160. Ritsena, R., Cofino, W. P., Frintop, P. G. M., and Brinkman, U. A. Th., Trace level analysis of phthalates in surface waters and suspended particulate matter by means of capillary gas chromatography with electron capture detection, *Chemosphere*, 18, 2161–2175, 1989.

161. Rastogi, S. C., Gas chromatographic analysis of phthalate esters in plastic toys, *Chromatographia*, 47, 724–726, 1998.

162. Lega, R., Ladwig, G., Meresz, O., Clement, R. E., Crawford, G., Salemi, R., and Jones, Y., Quantitative determination of organic priority pollutants in sewage sludge by GC/MS, *Chemosphere*, 34, 1705–1712, 1997.

163. Earls, A. O., Axford, I. P., and Braybrook, J. H., Gas chromatography–mass spectrometry determination of the migration of phthalate plasticisers from polyvinyl chloride toys and childcare articles, *J. Chromatogr. A*, 983, 237–246, 2003.

164. Otsuki, A., Reversed-phase adsorption of phthalate esters from aqueous solutions and their gradient elution using a high-performance liquid chromatograph, *J. Chromatogr.*, 133, 402–407, 1977.

165. Persiani, C. and Cukor, P., Liquid chromatographic method for the determination of phthalate esters, *J. Chromatogr.*, 109, 413–417, 1975.

166. Mori, S., Identification and determination of phthalate esters in river water by high-performance liquid chromatography, *J. Chromatogr.*, 129, 53–60, 1976.

167. Baker, J. K., Characterization of phthalate plasticizers by HPLC- or thermospray mass spectrometry, *J. Pharm. Biomed. Anal.*, 15, 145–148, 1996.

168. Koch, H. M., Gonzalez-Reche, L. M., and Angerer, J., On-line clean-up by multidimensional liquid chromatography electrospray ionization tandem mass spectrometry for high throughput quantification of primary and secondary phthalate metabolites in human urine, *J. Chromatogr. B*, 784, 169–182, 2003.

169. Schaefer, W. R., Hermann, T., Meinhold-Heerlein, I., Deppert, W. R., and Zahradnik, H., Exposure of human endometrium to environmental estrogens, antiandrogens, and organochlorine compounds, *Fertility Sterility*, 74(3), 558–563, 2000.

170. Franke, A. A. and Custer, L. J., High-performance liquid chromatographic assay of isoflavonoids and coumestrol from human urine, *J. Chromatogr. B*, 662, 47–60, 1994.

171. Setchell, K. D. R. and Welsh, M. B., High-performance liquid chromatographic analysis of phytoestrogens in soy protein preparations with ultraviolet, electrochemical and TS mass spectrometric detection, *J. Chromatogr.*, 386, 315–323, 1987.

172. Hutabarat, L. S., Mulholland, M., and Greenfield, H., Development and validation of an isocratic high-performance liquid chromatographic method for quantitative determination of phytoestrogens in soya bean, *J. Chromatogr. A*, 795, 377–382, 1998.

173. Wilkinson, A. P., Wahala, K., and Williamson, G., Identification and quantification of polyphenol phytoestrogens in foods and human biological fluids, *J. Chromatogr. B*, 777, 93–109, 2002.

174. Juan, C.-Y., Thomas, G. O., Semple, K. T., and Jones, K. C., Methods for the analysis of PCBs in human food, faeces, and serum, *Chemosphere*, 39, 1467–1476, 1999.

175. Van Rhijn, J. A., Traag, W. A., van de Spreng, P. F., and Tuinstra, L. G. M. Th., Simultaneous determination of planar chlorobiphenyls and polychlorinated dibenzo-*p*-dioxins and -furans in Dutch milk using isotope dilution and gas chromatography–high-resolution mass spectrometry, *J. Chromatogr.*, 630, 297–306, 1993.

176. Santos, F. J. and Galceran, M. T., Modern developments in gas chromatography–mass spectrometry-based environmental analysis, *J. Chromatogr. A*, 1000, 125–151, 2003.

177. USEPA. USEPA Method 1613, Tetra- through octa-chlorinated dioxin and furan by isotope dilution HRGC or HRMS. Revision A, United States Environmental Protection Agency, Washington, DC, 1990.

178. USEPA. USEPA Method 1668, Toxic polychlorinated biphenyls by isotope dilution high gas chromatography or high resolution mass spectrometry, United States Environmental Protection Agency, Washington, DC, 1997.

179. EU European Standard EN1948-1/2/3. Stationary source emissions-determination of the mass concentration of PCDD/Fs, European Committee for Standardization, Brussels, 1996.

180. Lee, H. K., Recent applications of gas and high-performance liquid chromatographic techniques to the analysis of polycyclic aromatic hydrocarbons in airborne particulates, *J. Chromatogr. A*, 710, 79–92, 1995.

181. Gfrerer, M., Serschen, M., and Lankmayr, E., Optimized extraction of polycyclic aromatic hydrocarbons from contaminated soil samples, *J. Biochem. Biophys. Methods*, 53, 203–216, 2002.

182. Nicol, S., Dugay, J., and Hennion, M. C., Simultaneous determination of polycyclic aromatic hydrocarbons and their nitrated derivatives in airborne particulate matter using gas chromatography–tandem mass spectrometry, *J. Sep. Sci.*, 24, 451–458, 2001.

183. Tan, Y. L., Rapid simple sample preparation technique for analyzing polynuclear aromatic hydrocarbons in sediments by gas chromatography–mass spectrometry, *J. Chromatogr.*, 176, 319–327, 1979.

184. Perez, S., Guillamon, M., and Barcelo, D., Quantitative analysis of polycyclic aromatic hydrocarbons in sewage sludge from wastewater treatment plants, *J. Chromatogr. A*, 938, 57–65, 2001.

185. Dugay, A., Herrenknecht, C., Czok, M., Guyon, F., and Pages, N., New procedure for selective extraction of polycyclic aromatic hydrocarbons in plants for gas chromatographic–mass spectrometric analysis, *J. Chromatogr. A*, 958, 1–7, 2002.

186. Moreda, J. M., Arranz, A., de betono, S. F., Cid, A., and Arranz, J. F., Chromatographic determination of aliphatic hydrocarbons and polyaromatic hydrocarbons (PAHs) in a sewage sludge, *Sci. Total Environ.*, 220, 33–43, 1998.

187. Benlahcen, K. T., Chaoui, A., Budzinski, H., Bellocq, J., and Garrigues, P. H., Distribution and sources of polycyclic aromatic hydrocarbons in some Mediterranean coastal sediments, *Marine Pollut. Bull.*, 34(5), 298–305, 1997.

188. Codina, G., Vaquero, M. T., Comellas, L., and Broto-Puig, F., Comparison of various extraction and clean-up methods for the determination of polycyclic aromatic hydrocarbons in sewage sludge-amended soils, *J. Chromatogr. A*, 673, 21–29, 1994.

189. Woodhead, R. J., Law, R. J., and Matthiessen, P., Polycyclic aromatic hydrocarbons in surface sediments around England and Wales, and their possible biological significance, *Marine Pollut. Bull.*, 38(9), 773–790, 1999.

190. Ternes, T. A., Stumpf, M., Mueller, J., Haberer, K., Wilken, R. D., and Servos, M., Behaviour and occurrence of estrogens in municipal sewage treatment plants — I. Investigations in Germany, Canada, and Brazil, *Sci. Total Environ.*, 225, 81–90, 1999.

191. Lai, K. M., Johnson, K. T., Scrimshaw, M. D., and Lester, J. N., Binding of waterborne steroid estrogens to solid phases in river and estuarine systems, *Environ. Sci. Technol.*, 34, 3890–3894, 2000.

192. Croley, T. R., Hughes, R. J., Koenig, B. G., Metcalfe, C. D., and March, R. E., Mass spectrometry applied to the analysis of estrogens in the environment, *Rapid Commun. Mass Spectrom.*, 14, 1087–1093, 2000.

193. Kuch, H. M. and Ballschmiter, K., Determination of endocrine-disrupting phenolic compounds and estrogens in surface and drinking water by HRGC–(NCI)-MS in the picogram per liter range, *Environ. Sci. Technol.*, 35, 3201–3206, 2001.

194. Mol, H. G. J., Sunarto, S., and Steijger, O. M., Determination of endocrine disruptors in water after derivatization with *N*-methyl-*N*-(*tert*-butyldimethyltrifluoroacetamide) using gas chromatography with mass spectrometric detection, *J. Chromatogr. A*, 879, 97–112, 2000.

195. Kelly, C., Analysis of steroids in environmental water samples using solid-phase extraction and ion-trap gas chromatography–mass spectrometry and gas chromatography–tandem mass spectrometry, *J. Chromatogr. A*, 872, 309–314, 2000.

196. Hartmann, S. and Steinhart, H., Simultaneous determination of anabolic and catabolic steroid hormones in meat by gas chromatography–mass spectrometry, *J. Chromatogr. A*, 704, 105–117, 1997.

197. Marchand, P., le Bizec, B., Gade, C., Monteau, F., and Andre, F., Ultra trace detection of a wide range of anabolic steroids in meat by gas chromatography coupled to mass spectrometry, *J. Chromatogr. A*, 867, 219–233, 2000.

198. Snyder, S. A., Keith, T. L., Verbrugge, D. A., Snyder, E. M., Gross, T. S., Kannan, K., and Giesy, J. P., Analytical methods for detection of selected estrogenic compounds in aqueous mixtures, *Environ. Sci. Technol.*, 33, 2814–2820, 1999.

199. Lopez de Alda, M. and Barcelo, D., Determination of steroid sex hormones and related synthetic compounds considered as endocrine disrupters in water by liquid chromatography–diode array detection–mass spectrometry, *J. Chromatogr. A*, 892, 391–406, 2000.

200. Baronti, C., Curini, R., D'Ascenzo, G., Di Corcia, A., Gentili, A., and Samperi, R., Monitoring natural and synthetic estrogens at activated sludge sewage treatment plants and in a receiving river water, *Environ. Sci. Technol.*, 34, 5059–5066, 2000.

201. Ma, Y. C. and Kim, H. Y., Determination of steroids by liquid chromatography or mass spectrometry, *J. Am. Soc. Mass Spectrom.*, 8, 1010–1020, 1997.

202. Petrovic, M., Eljarrat, E., Lopez de Alda, M., and Barcelo, D., Recent advances in the mass spectrometric analysis related to endocrine disrupting compounds in aquatic environmental samples, *J. Chromatogr. A*, 974, 23–51, 2002.

Index

A